COMPREHENSIVE MEDICINAL CHEMISTRY II

COMPREHENSIVE MEDICINAL CHEMISTRY II

Editors-in-Chief

Dr John B Taylor

Former Senior Vice-President for Drug Discovery, Rhône-Poulenc Rorer, Worldwide, UK

Professor David J Triggle

State University of New York, Buffalo, NY, USA

Volume 3

DRUG DISCOVERY TECHNOLOGIES

Volume Editor

Professor Hugo Kubinyi

University of Heidelberg, Heidelberg, Germany

ELSEVIER

AMSTERDAM BOSTON HEIDELBERG LONDON NEW YORK OXFORD
PARIS SAN DIEGO SAN FRANCISCO SINGAPORE SYDNEY TOKYO

Elsevier Ltd.
The Boulevard, Langford Lane, Kidlington, Oxford OX5 1GB, UK

First edition 2007

British Library Cataloguing in Publication Data
A catalogue record for this book is available from the British Library

Library of Congress Catalog Number: 2006936669

ISBN-13: 978-0-08-044513-7
ISBN-10: 0-08-044513-6

Printed and bound in Spain

06 07 08 09 10 10 9 8 7 6 5 4 3 2 1

Disclaimers

Contents

Contents of all Volumes xi

Preface xix

Preface to Volume 3 xxi

Editors-in-Chief xxiii

Editor of Volume 3 xxiv

Contributors to Volume 3 xxv

Target Search

3.01 Genomics 1
J S CALDWELL, S K CHANDA, J IRELAN, and R KOENIG, *Genomics Institute of the Novartis Research Foundation, San Diego, CA, USA*

3.02 Proteomics 27
H VOSHOL, S HOVING, and J VAN OOSTRUM, *Novartis Institutes for BioMedical Research, Genome and Proteome Sciences, Basel, Switzerland*

3.03 Pharmacogenomics 51
K LINDPAINTNER, *F. Hoffman-La Roche, Basel, Switzerland*

3.04 Biomarkers 69
S E DEPRIMO, *Pfizer Global Research and Development, San Diego, CA, USA*

3.05 Microarrays 87
D AMARATUNGA, *Johnson & Johnson Pharmaceutical Research & Development LLC, Raritan, NJ, USA,* and H GÖHLMANN and P J PEETERS, *Johnson & Johnson Pharmaceutical Research & Development, Beerse, Belgium*

3.06 Recombinant Deoxyribonucleic Acid and Protein Expression 107
F BERNHARD, C KLAMMT, and H RÜTERJANS, *University of Frankfurt/Main, Germany*

3.07 Chemical Biology 129
R FLAUMENHAFT, *Harvard Medical School, Boston, MA, USA*

Target Validation

3.08 Genetically Engineered Animals 151
B BOLON, *GEMpath Inc., Cedar City, UT, USA*

3.09 Small Interfering Ribonucleic Acids 171
L GIANELLINI and J MOLL, *Nerviano Medical Sciences S.r.l, Nerviano, Italy*

3.10 Signaling Chains 189
M-H TEITEN, R BLASIUS, F MORCEAU, and M DIEDERICH, *Hôpital Kirchberg, Luxembourg City, Luxembourg,* and M DICATO, *Centre Hospitalier de Luxembourg, Luxembourg City, Luxembourg*

3.11 Orthogonal Ligand–Receptor Pairs 215
F C ACHER, *Université René Descartes – Paris V, Paris, France*

Informatics and Databases

3.12 Chemoinformatics 235
V J GILLET, *University of Sheffield, Sheffield, UK* and A R LEACH, *GlaxoSmithKline, Stevenage, UK*

3.13 Chemical Information Systems and Databases 265
T ENGEL, *Chemical Computing Group AG, Köln, Germany* and E ZASS, *Informationszentrum Chemie Biologie Pharmazie, Zürich, Switzerland*

3.14 Bioactivity Databases 293
M OLAH and T I OPREA, *University of New Mexico School of Medicine, Albuquerque, NM, USA*

3.15 Bioinformatics 315
T LENGAUER and C HARTMANN, *Max Planck Institute for Informatics, Saarbrücken, Germany*

3.16 Gene and Protein Sequence Databases 349
M-J MARTIN, T KULIKOVA, M PRUESS, and R APWEILER, *EMBL Outstation European Bioinformatics Institute, Hinxton, Cambridge, UK*

3.17 The Research Collaboratory for Structural Bioinformatics Protein Data Bank 373
P E BOURNE, W F BLUHM, N DESHPANDE, and Q ZHANG, *University of California, La Jolla, CA, USA,* and H M BERMAN and J L FLIPPEN-ANDERSON, *Rutgers – The State University of New Jersey, Piscataway, NJ, USA*

3.18 The Cambridge Crystallographic Database 389
F H ALLEN, G M BATTLE, and S ROBERTSON, *Cambridge Crystallographic Data Centre, Cambridge, UK*

Structural Biology

3.19 Protein Production for Three-Dimensional Structural Analysis 411
M M T BAUER and G SCHNAPP, *Boehringer Ingelheim Pharma, Biberach, Germany*

3.20 Protein Crystallization 433
G E SCHULZ, *Albert-Ludwigs-Universität, Freiburg im Breisgau, Germany*

3.21 Protein Crystallography 449
M T STUBBS II, *Martin Luther University, Halle, Germany*

3.22 Bio-Nuclear Magnetic Resonance 473
T CARLOMAGNO, M BALDUS, and C GRIESINGER, *Max Planck Institute for Biophysical Chemistry, Göttingen, Germany*

3.23 Protein Three-Dimensional Structure Validation 507
R P JOOSTEN, *Centre for Molecular and Biomolecular Informatics, Nijmegen, The Netherlands*, and G CHINEA, *Center for Genetic Engineering and Biotechnology, Havana, Cuba*, and G J KLEYWEGT, *University of Uppsala, Uppsala, Sweden*, and G VRIEND, *Centre for Molecular and Biomolecular Informatics, Nijmegen, The Netherlands*

3.24 Problems of Protein Three-Dimensional Structures 531
R A LASKOWSKI and G J SWAMINATHAN, *European Bioinformatics Institute, Wellcome Trust Genome Campus, Hinxton, Cambridge, UK*

3.25 Structural Genomics 551
W SHI and M R CHANCE, *Case Western Reserve University, Cleveland, OH, USA*

Screening

3.26 Compound Storage and Management 561
W W KEIGHLEY, T P WOOD, and T J WINCHESTER, *Pfizer Global Research and Development, Sandwich, UK*

3.27 Optical Assays in Drug Discovery 577
B SCHNURR, T AHRENS, and U REGENASS, *Discovery Partners International AG, Allschwil, Switzerland*

3.28 Fluorescence Screening Assays 599
D ULLMANN, *Evotec AG, Hamburg, Germany*

3.29 Cell-Based Screening Assays 617
A WEISSMAN, J KEEFER, A MIAGKOV, M SATHYAMOORTHY, S PERSCHKE, and F L WANG, *NovaScreen Biosciences Corporation, Hanover, MD, USA*

3.30 Small Animal Test Systems for Screening 647
M MUDA, S McKENNA, and B G HEALEY, *Serono Research Institute, Rockland, MA, USA*

3.31 Imaging 659
P M SMITH-JONES, *Memorial Sloan Kettering Cancer Center, New York, NY, USA*

3.32 High-Throughput and High-Content Screening 679
M CIK, *Johnson & Johnson Pharmaceutical Research & Development, Beerse, Belgium* and M R JURZAK, *Merck KGaA, Darmstadt, Germany*

Chemical Technologies

3.33 Combinatorial Chemistry 697
P SENECI, *Desenzano del Garda, Italy*

3.34 Solution Phase Parallel Chemistry 761
M ASHTON and B MOLONEY, *Evotec (UK) Ltd, Abingdon, UK*

3.35 Polymer-Supported Reagents and Scavengers in Synthesis 791
S V LEY, I R BAXENDALE, and R M MYERS, *University of Cambridge, Cambridge, UK*

3.36 Microwave-Assisted Chemistry 837
C O KAPPE, *Karl-Franzens-University Graz, Graz, Austria*

3.37 High-Throughput Purification 861
D B KASSEL, *Takeda San Diego, Inc., San Diego, CA, USA*

Lead Search and Optimization

3.38 Protein Crystallography in Drug Discovery 875
T HOGG and R HILGENFELD, *University of Lübeck, Lübeck, Germany*

3.39 Nuclear Magnetic Resonance in Drug Discovery 901
J KLAGES and H KESSLER, *TU München, Garching, Germany*

3.40 Chemogenomics 921
H KUBINYI, *University of Heidelberg, Heidelberg, Germany*

3.41 Fragment-Based Approaches 939
W JAHNKE, *Novartis Pharma AG, Basel, Switzerland*

3.42 Dynamic Ligand Assembly 959
O RAMSTRÖM, *Royal Institute of Technology, Stockholm, Sweden* and
J-M LEHN, *ISIS Université Louis Pasteur, Strasbourg, France*

Subject Index 977

Contents of all Volumes

Volume 1 Global Perspective

Historical Perspective and Outlook
1.01 Reflections of a Medicinal Chemist: Formative Years through Thirty-Seven Years Service in the Pharmaceutical Industry
1.02 Drug Discovery: Historical Perspective, Current Status, and Outlook
1.03 Major Drug Introductions
The Impact of New Genomic Technologies
1.04 Epigenetics
1.05 Personalized Medicine
1.06 Gene Therapy
Sources of New Drugs
1.07 Overview of Sources of New Drugs
1.08 Natural Product Sources of Drugs: Plants, Microbes, Marine Organisms, and Animals
1.09 Traditional Medicines
1.10 Biological Macromolecules
1.11 Nutraceuticals
Animal Experimentation
1.12 Alternatives to Animal Testing
The Role of Industry
1.13 The Role of Small- or Medium-Sized Enterprises in Drug Discovery
1.14 The Role of the Pharmaceutical Industry
1.15 The Role of the Venture Capitalist
1.16 Industry–Academic Relationships
Drug Discovery: Revolution, Decline?
1.17 Is the Biotechnology Revolution a Myth?
1.18 The Apparent Declining Efficiency of Drug Discovery
Healthcare in the Social Context
1.19 How Much is Enough or Too Little: Assessing Healthcare Demand in Developed Countries
1.20 Health Demands in Developing Countries
1.21 Orphan Drugs and Generics
Ethical Issues
1.22 Bioethical Issues in Medicinal Chemistry and Drug Treatment
1.23 Ethical Issues and Challenges Facing the Pharmaceutical Industry
Funding and Regulation of Research
1.24 The Role of Government in Health Research
1.25 Postmarketing Surveillance

Intellectual Property
1.26 Intellectual Property Rights and Patents
Subject Index

Volume 2 Strategy and Drug Research

Introduction
2.01 The Intersection of Strategy and Drug Research
2.02 An Academic Perspective
2.03 An Industry Perspective
Organizational Aspects and Strategies for Drug Discovery and Development
2.04 Project Management
2.05 The Role of the Chemical Development, Quality, and Regulatory Affairs Teams in Turning a Potent Agent into a Registered Product
2.06 Drug Development
2.07 In-House or Out-Source
2.08 Pharma versus Biotech: Contracts, Collaborations, and Licensing
2.09 Managing Scientists, Leadership Strategies in Science
2.10 Innovation (Fighting against the Current)
2.11 Enabling Technologies in Drug Discovery: The Technical and Cultural Integration of the New with the Old
2.12 How and Why to Apply the Latest Technology
2.13 How and When to Apply Absorption, Distribution, Metabolism, Excretion, and Toxicity
2.14 Peptide and Protein Drugs: Issues and Solutions
2.15 Peptidomimetic and Nonpeptide Drug Discovery: Receptor, Protease, and Signal Transduction Therapeutic Targets
2.16 Bioisosterism
2.17 Chiral Drug Discovery and Development – From Concept Stage to Market Launch
2.18 Promiscuous Ligands
Targets
2.19 Diversity versus Focus in Choosing Targets and Therapeutic Areas
2.20 G Protein-Coupled Receptors
2.21 Ion Channels – Voltage Gated
2.22 Ion Channels – Ligand Gated
2.23 Phosphodiesterases
2.24 Protein Kinases and Protein Phosphatases in Signal Transduction Pathways
2.25 Nuclear Hormone Receptors
2.26 Nucleic Acids (Deoxyribonucleic Acid and Ribonucleic Acid)
2.27 Redox Enzymes
List of Abbreviations
List of Symbols
Subject Index

Volume 3 Drug Discovery Technologies

Target Search
3.01 Genomics
3.02 Proteomics
3.03 Pharmacogenomics
3.04 Biomarkers
3.05 Microarrays
3.06 Recombinant Deoxyribonucleic Acid and Protein Expression
3.07 Chemical Biology

Target Validation
3.08 Genetically Engineered Animals
3.09 Small Interfering Ribonucleic Acids
3.10 Signaling Chains
3.11 Orthogonal Ligand–Receptor Pairs
Informatics and Databases
3.12 Chemoinformatics
3.13 Chemical Information Systems and Databases
3.14 Bioactivity Databases
3.15 Bioinformatics
3.16 Gene and Protein Sequence Databases
3.17 The Research Collaboratory for Structural Bioinformatics Protein Data Bank
3.18 The Cambridge Crystallographic Database
Structural Biology
3.19 Protein Production for Three-Dimensional Structural Analysis
3.20 Protein Crystallization
3.21 Protein Crystallography
3.22 Bio-Nuclear Magnetic Resonance
3.23 Protein Three-Dimensional Structure Validation
3.24 Problems of Protein Three-Dimensional Structures
3.25 Structural Genomics
Screening
3.26 Compound Storage and Management
3.27 Optical Assays in Drug Discovery
3.28 Fluorescence Screening Assays
3.29 Cell-Based Screening Assays
3.30 Small Animal Test Systems for Screening
3.31 Imaging
3.32 High-Throughput and High-Content Screening
Chemical Technologies
3.33 Combinatorial Chemistry
3.34 Solution Phase Parallel Chemistry
3.35 Polymer-Supported Reagents and Scavengers in Synthesis
3.36 Microwave-Assisted Chemistry
3.37 High-Throughput Purification
Lead Search and Optimization
3.38 Protein Crystallography in Drug Discovery
3.39 Nuclear Magnetic Resonance in Drug Discovery
3.40 Chemogenomics
3.41 Fragment-Based Approaches
3.42 Dynamic Ligand Assembly
Subject Index

Volume 4 Computer-Assisted Drug Design

Introduction to Computer-Assisted Drug Design
4.01 Introduction to the Volume and Overview of Computer-Assisted Drug Design in the Drug Discovery Process
4.02 Introduction to Computer-Assisted Drug Design – Overview and Perspective for the Future
4.03 Quantitative Structure–Activity Relationship – A Historical Perspective and the Future
4.04 Structure-Based Drug Design – A Historical Perspective and the Future

Core Concepts and Methods – Ligand-Based
4.05 Ligand-Based Approaches: Core Molecular Modeling
4.06 Pharmacophore Modeling: 1 – Methods
4.07 Predictive Quantitative Structure–Activity Relationship Modeling
4.08 Compound Selection Using Measures of Similarity and Dissimilarity
Core Concepts and Methods – Target Structure-Based
4.09 Structural, Energetic, and Dynamic Aspects of Ligand–Receptor Interactions
4.10 Comparative Modeling of Drug Target Proteins
4.11 Characterization of Protein-Binding Sites and Ligands Using Molecular Interaction Fields
4.12 Docking and Scoring
4.13 De Novo Design
Core Methods and Applications – Ligand and Structure-Based
4.14 Library Design: Ligand and Structure-Based Principles for Parallel and Combinatorial Libraries
4.15 Library Design: Reactant and Product-Based Approaches
4.16 Quantum Mechanical Calculations in Medicinal Chemistry: Relevant Method or a Quantum Leap Too Far?
Applications to Drug Discovery – Lead Discovery
4.17 Chemogenomics in Drug Discovery – The Druggable Genome and Target Class Properties
4.18 Lead Discovery and the Concepts of Complexity and Lead-Likeness in the Evolution of Drug Candidates
4.19 Virtual Screening
4.20 Screening Library Selection and High-Throughput Screening Analysis/Triage
Applications to Drug Discovery – Ligand-Based Lead Optimization
4.21 Pharmacophore Modeling: 2 – Applications
4.22 Topological Quantitative Structure–Activity Relationship Applications: Structure Information Representation in Drug Discovery
4.23 Three-Dimensional Quantitative Structure–Activity Relationship: The State of the Art
Applications to Drug Discovery – Target Structure-Based
4.24 Structure-Based Drug Design – The Use of Protein Structure in Drug Discovery
4.25 Applications of Molecular Dynamics Simulations in Drug Design
4.26 Seven Transmembrane G Protein-Coupled Receptors: Insights for Drug Design from Structure and Modeling
4.27 Ion Channels: Insights for Drug Design from Structure and Modeling
4.28 Nuclear Hormone Receptors: Insights for Drug Design from Structure and Modeling
4.29 Enzymes: Insights for Drug Design from Structure
New Directions
4.30 Multiobjective/Multicriteria Optimization and Decision Support in Drug Discovery
4.31 New Applications for Structure-Based Drug Design
4.32 Biological Fingerprints
Subject Index

Volume 5 ADME-Tox Approaches

Introduction
5.01 The Why and How of Absorption, Distribution, Metabolism, Excretion, and Toxicity Research
Biological and In Vivo Aspects of Absorption, Distribution, Metabolism, Excretion, and Toxicity
5.02 Clinical Pharmacokinetic Criteria for Drug Research
5.03 In Vivo Absorption, Distribution, Metabolism, and Excretion Studies in Discovery and Development
5.04 The Biology and Function of Transporters
5.05 Principles of Drug Metabolism 1: Redox Reactions

5.06 Principles of Drug Metabolism 2: Hydrolysis and Conjugation Reactions
5.07 Principles of Drug Metabolism 3: Enzymes and Tissues
5.08 Mechanisms of Toxification and Detoxification which Challenge Drug Candidates and Drugs
5.09 Immunotoxicology
Biological In Vitro Tools in Absorption, Distribution, Metabolism, Excretion, and Toxicity
5.10 In Vitro Studies of Drug Metabolism
5.11 Passive Permeability and Active Transport Models for the Prediction of Oral Absorption
5.12 Biological In Vitro Models for Absorption by Nonoral Routes
5.13 In Vitro Models for Examining and Predicting Brain Uptake of Drugs
5.14 In Vitro Models for Plasma Binding and Tissue Storage
5.15 Progress in Bioanalytics and Automation Robotics for Absorption, Distribution, Metabolism, and Excretion Screening
Physicochemical tools in Absorption, Distribution, Metabolism, Excretion, and Toxicity
5.16 Ionization Constants and Ionization Profiles
5.17 Dissolution and Solubility
5.18 Lipophilicity, Polarity, and Hydrophobicity
5.19 Artificial Membrane Technologies to Assess Transfer and Permeation of Drugs in Drug Discovery
5.20 Chemical Stability
5.21 Solid-State Physicochemistry
In Silico Tools in Absorption, Distribution, Metabolism, Excretion, and Toxicity
5.22 Use of Molecular Descriptors for Absorption, Distribution, Metabolism, and Excretion Predictions
5.23 Electrotopological State Indices to Assess Molecular and Absorption, Distribution, Metabolism, Excretion, and Toxicity Properties
5.24 Molecular Fields to Assess Recognition Forces and Property Spaces
5.25 In Silico Prediction of Ionization
5.26 In Silico Predictions of Solubility
5.27 Rule-Based Systems to Predict Lipophilicity
5.28 In Silico Models to Predict Oral Absorption
5.29 In Silico Prediction of Oral Bioavailability
5.30 In Silico Models to Predict Passage through the Skin and Other Barriers
5.31 In Silico Models to Predict Brain Uptake
5.32 In Silico Models for Interactions with Transporters
5.33 Comprehensive Expert Systems to Predict Drug Metabolism
5.34 Molecular Modeling and Quantitative Structure–Activity Relationship of Substrates and Inhibitors of Drug Metabolism Enzymes
5.35 Modeling and Simulation of Pharmacokinetic Aspects of Cytochrome P450-Based Metabolic Drug–Drug Interactions
5.36 In Silico Prediction of Plasma and Tissue Protein Binding
5.37 Physiologically-Based Models to Predict Human Pharmacokinetic Parameters
5.38 Mechanism-Based Pharmacokinetic–Pharmacodynamic Modeling for the Prediction of In Vivo Drug Concentration–Effect Relationships – Application in Drug Candidate Selection and Lead Optimization
5.39 Computational Models to Predict Toxicity
5.40 In Silico Models to Predict QT Prolongation
5.41 The Adaptive In Combo Strategy
Enabling Absorption, Distribution, Metabolism, Excretion, and Toxicity Strategies and Technologies in Early Development
5.42 The Biopharmaceutics Classification System
5.43 Metabonomics

5.44 Prodrug Objectives and Design
5.45 Drug–Polymer Conjugates
List of Abbreviations
List of Symbols
Subject Index

Volume 6 Therapeutic Areas I: Central Nervous System, Pain, Metabolic Syndrome, Urology, Gastrointestinal and Cardiovascular

Central Nervous System
6.01 Central Nervous System Drugs Overview
6.02 Schizophrenia
6.03 Affective Disorders: Depression and Bipolar Disorders
6.04 Anxiety
6.05 Attention Deficit Hyperactivity Disorder
6.06 Sleep
6.07 Addiction
6.08 Neurodegeneration
6.09 Neuromuscular/Autoimmune Disorders
6.10 Stroke/Traumatic Brain and Spinal Cord Injuries
6.11 Epilepsy
6.12 Ophthalmic Agents
Pain
6.13 Pain Overview
6.14 Acute and Neuropathic Pain
6.15 Local and Adjunct Anesthesia
6.16 Migraine
Obesity/Metabolic Disorders/Syndrome X
6.17 Obesity/Metabolic Syndrome Overview
6.18 Obesity/Disorders of Energy
6.19 Diabetes/Syndrome X
6.20 Atherosclerosis/Lipoprotein/Cholesterol Metabolism
6.21 Bone, Mineral, Connective Tissue Metabolism
6.22 Hormone Replacement
Urogenital
6.23 Urogenital Diseases/Disorders, Sexual Dysfunction and Reproductive Medicine: Overview
6.24 Incontinence (Benign Prostatic Hyperplasia/Prostate Dysfunction)
6.25 Renal Dysfunction in Hypertension and Obesity
Gastrointestinal
6.26 Gastrointestinal Overview
6.27 Gastric and Mucosal Ulceration
6.28 Inflammatory Bowel Disease
6.29 Irritable Bowel Syndrome
6.30 Emesis/Prokinetic Agents
Cardiovascular
6.31 Cardiovascular Overview
6.32 Hypertension
6.33 Antiarrhythmics
6.34 Thrombolytics
Subject Index

Volume 7 Therapeutic Areas II: Cancer, Infectious Diseases, Inflammation & Immunology and Dermatology

Anti Cancer
7.01 Cancer Biology
7.02 Principles of Chemotherapy and Pharmacology
7.03 Antimetabolites
7.04 Microtubule Targeting Agents
7.05 Deoxyribonucleic Acid Topoisomerase Inhibitors
7.06 Alkylating and Platinum Antitumor Compounds
7.07 Endocrine Modulating Agents
7.08 Kinase Inhibitors for Cancer
7.09 Recent Development in Novel Anticancer Therapies
Anti Viral
7.10 Viruses and Viral Diseases
7.11 Deoxyribonucleic Acid Viruses: Antivirals for Herpesviruses and Hepatitis B Virus
7.12 Ribonucleic Acid Viruses: Antivirals for Human Immunodeficiency Virus
7.13 Ribonucleic Acid Viruses: Antivirals for Influenza A and B, Hepatitis C Virus, and Respiratory Syncytial Virus
Anti Fungal
7.14 Fungi and Fungal Disease
7.15 Major Antifungal Drugs
Anti Bacterials
7.16 Bacteriology, Major Pathogens, and Diseases
7.17 β-Lactam Antibiotics
7.18 Macrolide Antibiotics
7.19 Quinolone Antibacterial Agents
7.20 The Antibiotic and Nonantibiotic Tetracyclines
7.21 Aminoglycosides Antibiotics
7.22 Anti-Gram Positive Agents of Natural Product Origins
7.23 Oxazolidinone Antibiotics
7.24 Antimycobacterium Agents
7.25 Impact of Genomics-Emerging Targets for Antibacterial Therapy
Drugs for Parasitic Infections
7.26 Overview of Parasitic Infections
7.27 Advances in the Discovery of New Antimalarials
7.28 Antiprotozoal Agents (African Trypanosomiasis, Chagas Disease, and Leishmaniasis)
I and I Diseases
7.29 Recent Advances in Inflammatory and Immunological Diseases: Focus on Arthritis Therapy
7.30 Asthma and Chronic Obstructive Pulmonary Disease
7.31 Treatment of Transplantation Rejection and Multiple Sclerosis
Dermatology
7.32 Overview of Dermatological Diseases
7.33 Advances in the Discovery of Acne and Rosacea Treatments
7.34 New Treatments for Psoriasis and Atopic Dermatitis
Subject Index

Volume 8 Case Histories and Cumulative Subject Index

Personal Essays
8.01 Introduction
8.02 Reflections on Medicinal Chemistry Since the 1950s

8.03 Medicinal Chemistry as a Scientific Discipline in Industry and Academia: Some Personal Reflections
8.04 Some Aspects of Medicinal Chemistry at the Schering-Plough Research Institute
Case Histories
8.05 Viread
8.06 Hepsera
8.07 Ezetimibe
8.08 Tamoxifen
8.09 Raloxifene
8.10 Duloxetine
8.11 Carvedilol
8.12 Modafinil, A Unique Wake-Promoting Drug: A Serendipitous Discovery in Search of a Mechanism of Action
8.13 Zyvox
8.14 Copaxone
8.15 Ritonavir and Lopinavir/Ritonavir
8.16 Fosamax
8.17 Omeprazole
8.18 Calcium Channel α_2–δ Ligands: Gabapentin and Pregabalin
Cumulative Subject Index

Preface

The first edition of *Comprehensive Medicinal Chemistry* was published in 1990 and was intended to present an integrated and comprehensive overview of the then rapidly developing science of medicinal chemistry from its origins in organic chemistry. In the last two decades, the field has grown to embrace not only all the sophisticated synthetic and technological advances in organic chemistry but also major advances in the biological sciences. The mapping of the human genome has resulted in the provision of a multitude of new biological targets for the medicinal chemist with the prospect of more rational drug design (CADD). In addition, the development of sophisticated in silico technologies for structure–property relationships (ADMET) enables a much better understanding of the fate of potential new drugs in the body with the subsequent development of better new medicines.

It was our ambitious aim for this second edition, published 16 years after the first edition, to provide both scientists and research managers in all relevant fields with a comprehensive treatise covering all aspects of current medicinal chemistry, a science that has been transformed in the twenty-first century. The second edition is a complete reference source, published in eight volumes, encompassing all aspects of modern drug discovery from its mechanistic basis, through the underlying general principles and exemplified with comprehensive therapeutic applications. The broad scope and coverage of *Comprehensive Medicinal Chemistry II* would not have been possible without our panel of authoritative Volume Editors whose international recognition in their respective fields has been of paramount importance in the enlistment of the world-class scientists who have provided their individual 'state of the science' contributions. Their collective contributions have been invaluable.

Volume 1 (edited by Peter D Kennewell) overviews the general socioeconomic and political factors influencing modern R&D in both the developed and developing worlds. Volume 2 (edited by Walter H Moos) addresses the various strategic and organizational aspects of modern R&D. Volume 3 (edited by Hugo Kubinyi) critically reviews the multitude of modern technologies that underpin current discovery and development activities. Volume 4 (edited by Jonathan S Mason) highlights the historical progress, current status, and future potential in the field of computer-assisted drug design (CADD). Volume 5 (edited by Bernard Testa and Han van de Waterbeemd) reviews the fate of drugs in the body (ADMET), including the most recent progress in the application of 'in silico' tools. Volume 6 (edited by Michael Williams) and Volume 7 (edited by Jacob J Plattner and Manoj C Desai) cover the pivotal roles undertaken by the medicinal chemist and pharmacologist in integrating all the preceding scientific input into the design and synthesis of viable new medicines. Volume 8 (edited by John B Taylor and David J Triggle) illustrates the evolution of modern medicinal chemistry with a selection of personal accounts by eminent scientists describing their lifetime experiences in the field, together with some illustrative case histories of successful drug discovery and development.

We believe that this major work will serve as the single most authoritative reference source for all aspects of medicinal chemistry for the next decade and it is intended to maintain its ongoing value by systematic electronic upgrades. We hope that the material provided here will serve to fulfill the words of Antoine de Saint-Exupery (1900–44) and allow future generations of medicinal chemists to discover the future.

'As for the future, your task is not to foresee it but to enable it'
Citadelle (1948)

John B Taylor and David J Triggle

Preface to Volume 3

Drug discovery is a scientific and intellectual challenge that needs contributions from physics, chemistry, biochemistry, molecular biology, pharmacology, toxicology, and last but not least chemo- and bioinformatics. Whereas a car or even a moon rocket can be constructed, based on solid theoretical and experimental knowledge, the complexity of drug research is without any analogy. Correspondingly, besides the intellectual challenge, it is a huge financial endeavor. These are the reasons why drug research is almost exclusively performed by the pharmaceutical industry. In addition to all scientific problems, there is a high risk of failure, in early stages as well as in clinical trials, even after market introduction. Whereas most drug properties, especially specificity, bioavailability, pharmacokinetics, metabolism, drug–drug interactions, and toxic side effects can be investigated prior to market introduction, some extremely rare but fatal toxic effects in humans may show up only after market introduction, that is after broad therapeutic application.

Volume 3 of the new *Comprehensive Medicinal Chemistry II* is dedicated to Drug Discovery Technologies. The organization of this volume reflects the steps that are followed in drug discovery; target search, target validation, informatics and databases, structural biology, screening, chemical technologies, and lead search and optimization are the titles of the sections. The term 'technologies' may be misleading. Most approaches that are described and discussed in this volume are highly sophisticated research, and the underlying science is often still under active development.

Target search uses information from genomics, proteomics, and eventually pharmacogenomics to discover new genes and proteins that are dysregulated, genetically altered by mutation, or otherwise responsible for a certain disease. Biomarkers and microarrays aid in the identification of new targets and their proper characterization. Once a certain target is identified, recombinant DNA and protein expression enable the production of protein for screening and three-dimensional (3D) structure determination. Chemical biology serves for target identification as well as for the discovery of new modulators of such targets.

Target validation is a most important step in drug discovery. Although a final proof is often delayed till clinical phase II or III, genetically engineered animals and the new techniques of RNA interference enable the investigation of the phenotypic effects of target modulation before any inhibitor or antagonist of a target is identified. Very often, proteins are implemented in signaling chains, where understanding of the interacting macromolecules helps to select or discard a certain target. Orthogonal ligand–receptor pairs are another approach to study the effect of protein modulation without the need of designing a highly specific ligand.

Informatics and databases are of utmost importance in performing drug research. Bioinformatics aids in the interpretation of the vast amount of information from genomics and proteomics; chemoinformatics deals with small molecules. Both disciplines transform data into information, and information into knowledge, to support target and lead identification and validation, as well as lead optimization. The chapters on chemical information systems and databases and on gene and protein sequence databases describe sources and retrieval methods for all relevant information on targets and on small molecules. The 3D structural Protein Data Bank (PDB) and the Cambridge crystallographic database are the sources of experimental 3D structures of targets and organic molecules, respectively, at atomic resolution.

The section on structural biology contains chapters on protein production for 3D structural analysis, protein crystallization, protein crystallography, and bio-NMR, as important technologies for protein 3D structure determination. Much room is reserved for a discussion on protein 3D structure validation and problems of protein 3D structures, to make the reader aware of potential pitfalls in working with protein 3D structures from the data bank. Worldwide structural genomics initiatives aim to generate as many diverse protein 3D structures as possible.

The screening section starts with a chapter on compound storage and management, followed by the most important screening technologies: optical screening assays, fluorescence screening assays, partially overlapping, and cell-based screening assays. In addition to simple in vitro systems, small animal test systems as well as imaging techniques become more and more important in drug discovery. This section is concluded with a chapter on high-throughput and high-content screening.

There is no drug research without chemistry. As chemistry and medicinal chemistry are well-established disciplines, only newer developments are included in this volume: combinatorial chemistry on solid phase, solution phase parallel chemistry, and polymer-assisted chemical syntheses. Another exciting development in organic synthesis is microwave-assisted chemistry. Applications in drug research paved the way for all these methods into classical organic synthesis. In contrast to early combinatorial chemistry, where (undefined) mixtures of compounds were produced and screened, high-throughput purification methods are nowadays used to guarantee a high degree of purity of single compounds.

Lead search and optimization are the domain of medicinal chemistry. Whereas many computer-aided techniques, such as pharmacophore generation and searches, docking and scoring, and virtual screening, are included in Volume 4, experimental techniques, such as protein crystallography and NMR applications in drug discovery, are discussed here. Chemogenomics is an approach to discover selective modulators of evolutionary related targets (target families) in a more systematic manner than so far. Two other approaches, which are not yet fully exploited, are fragment-based drug discovery and dynamic ligand assembly.

It is hoped that this volume contributes to a comprehensive overview of the most important technologies in drug discovery. I would like to thank all authors and co-authors for their engaged work, for good cooperation, and for contributing such excellent chapters. My thanks go also to Gerlinde Ranzinger, Elsevier, and to the editors David Triggle and John Taylor, for having invited me to participate, as well as to Mireille Yanow, Andrew Lowe, and many other members of the Elsevier team for their ongoing support. Of course, I have to thank all former colleagues and many friends for having educated me in medicinal chemistry and drug discovery.

Hugo Kubinyi

Editors-in-Chief

John B Taylor, DSc, was formerly Senior Vice President for Drug Discovery at Rhône-Poulenc Rorer. He obtained his BSc in chemistry from the University of Nottingham in 1956 and his PhD in organic chemistry at the Imperial College of Science and Technology with Nobel Laureate Professor Sir Derek Barton in 1962. He subsequently undertook postdoctoral research fellowships at the Research Institute for Medicine and Chemistry in Cambridge (US) with Sir Derek and at the University of Liverpool (UK), before entering the pharmaceutical industry.

During his career in the pharmaceutical industry Dr Taylor spent more than 30 years covering all aspects of research and development in an international environment. From 1970 to 1985 he held a number of positions in the Hoechst Roussel organization, ultimately as research director for Roussel Uclaf (France). In 1985 he joined Rhône-Poulenc Rorer holding various management positions in the research groups worldwide before becoming Senior Vice President for Drug Discovery in Rhône-Poulenc Rorer.

Dr Taylor is the co-author of two books on medicinal chemistry and has more than 50 publications and patents in medicinal chemistry. He was joint executive editor for the first edition of Comprehensive Medicinal Chemistry, a visiting professor for medicinal chemistry at the City University (London) from 1974 to 1984 and was awarded a DSc in medicinal chemistry from the University of London in 1991.

David J Triggle, PhD, is the University Professor and a Distinguished Professor in the School of Pharmacy and Pharmaceutical Sciences at the State University of New York at Buffalo. Professor Triggle received his education in the UK with a BSc degree in chemistry at the University of Southampton and a PhD degree in chemistry at the University of Hull working with Professor Norman Chapman. Following postdoctoral fellowships at the University of Ottawa (Canada) with Bernard Belleau and the University of London (UK) with Peter de la Mare he assumed a position in the School of Pharmacy at the University at Buffalo. He served as Chairman of the Department of Biochemical Pharmacology from 1971 to 1985 and as Dean of the School of Pharmacy from 1985 to 1995. From 1996 to 2001 he served as Dean of the Graduate School and from 1999 to 2001 was also the University Provost. He is currently the University Professor, in which capacity he teaches bioethics and science policy, and is President of the Center for Inquiry Institute, a secular think tank located in Amherst, New York.

Professor Triggle is the author of three books dealing with the autonomic nervous system and drug–receptor interactions, the editor of a further dozen books, some 280 papers, some 150 chapters and reviews, and has presented over 1000 invited lectures worldwide. The Institute for Scientific Information lists him as one of the 100 most highly cited scientists in the field of pharmacology. His principal research interests have been in the areas of drug–receptor interactions, the chemical pharmacology of drugs active at ion channels, and issues of graduate education and scientific research policy.

Editor of Volume 3

Hugo Kubinyi is a medicinal chemist with 35 years of industrial experience, at KNOLL AG and BASF AG, Ludwigshafen, Germany. Since 1987, until his retirement in summer 2001, he was responsible for the Molecular Modelling, X-ray Crystallography, and Drug Design group of BASF, since early 1998 also for Combinatorial Chemistry in the Life Sciences. He is Professor of Pharmaceutical Chemistry at the University of Heidelberg, former Chair of The QSAR and Modelling Society (1995–2000; from 2000–2010 Advisor to the Chair), and IUPAC Fellow. In 2006 he got the Herman Skolnik Award (CINF, ACS).

From his scientific work resulted more than 100 publications and seven books on QSAR, 3D-QSAR, Drug Design, Chemogenomics, and Drug Discovery Technologies (Volume 3 of Comprehensive Medicinal Chemistry II). He is a member of several Scientific Advisory Boards, coeditor of the Wiley-VCH book series "Methods and Principles in Medicinal Chemistry," and member of the Editorial Boards of several scientific journals.

Contributors to Volume 3

F C Acher
Université René Descartes – Paris V, Paris, France

T Ahrens
Discovery Partners International AG, Allschwil, Switzerland

F H Allen
Cambridge Crystallographic Data Centre, Cambridge, UK

D Amaratunga
Johnson & Johnson Pharmaceutical Research & Development LLC, Raritan, NJ, USA

R Apweiler
EMBL Outstation European Bioinformatics Institute, Hinxton, Cambridge, UK

M Ashton
Evotec (UK) Ltd, Abingdon, UK

M Baldus
Max Planck Institute for Biophysical Chemistry, Göttingen, Germany

G M Battle
Cambridge Crystallographic Data Centre, Cambridge, UK

M M T Bauer
Boehringer Ingelheim Pharma, Biberach, Germany

I R Baxendale
University of Cambridge, Cambridge, UK

H M Berman
Ruters – The State University of New Jersey, Piscataway, NJ, USA

F Bernhard
University of Frankfurt/Main, Germany

R Blasius
Hôpital Kirchberg, Luxembourg City, Luxembourg

W F Bluhm
University of California, La Jolla, CA, USA

B Bolon
GEMpath Inc., Cedar City, UT, USA

P E Bourne
University of California, La Jolla, CA, USA

J S Caldwell
Genomics Institute of the Novartis Research Foundation, San Diego, CA, USA

T Carlomagno
Max Planck Institute for Biophysical Chemistry, Göttingen, Germany

M R Chance
Case Western Reserve University, Cleveland, OH, USA

S K Chanda
Genomics Institute of the Novartis Research Foundation, San Diego, CA, USA

G Chinea
Center for Genetic Engineering and Biotechnology, Havana, Cuba

M Cik
Johnson & Johnson Pharmaceutical Research & Development, Beerse, Belgium

S E DePrimo
Pfizer Global Research and Development, San Diego, CA, USA

N Deshpande
University of California, La Jolla, CA, USA

M Dicato
Centre Hospitalier de Luxembourg, Luxembourg City, Luxembourg

M Diederich
Hôpital Kirchberg, Luxembourg City, Luxembourg

T Engel
Chemical Computing Group AG, Köln, Germany

R Flaumenhaft
Harvard Medical School, Boston, MA, USA

J L Flippen-Anderson
Rutgers – The State University of New Jersey, Piscataway, NJ, USA

L Gianellini
Nerviano Medical Sciences S.r.l, Nerviano, Italy

V J Gillet
University of Sheffield, Sheffield, UK

H Göhlmann
Johnson & Johnson Pharmaceutical Research & Development LLC, Raritan, NJ, USA

C Griesinger
Max Planck Institute for Biophysical Chemistry, Göttingen, Germany

C Hartmann
Max Planck Institute for Informatics, Saarbrücken, Germany

B G Healey
Serono Research Institute, Rockland, MA, USA

R Hilgenfeld
University of Lübeck, Lübeck, Germany

T Hogg
University of Lübeck, Lübeck, Germany

S Hoving
Novartis Institutes for BioMedical Research, Genome and Proteome Sciences, Basel, Switzerland

J Irelan
Genomics Institute of the Novartis Research Foundation, San Diego, CA, USA

W Jahnke
Novartis Pharma AG, Basel, Switzerland

R P Joosten
Centre for Molecular and Biomolecular Informatics, Nijmegen, The Netherlands

M R Jurzak
Merck KGaA, Darmstadt, Germany

C O Kappe
Karl-Franzens-University Graz, Graz, Austria

D B Kassel
Takeda San Diego, Inc., San Diego, CA, USA

J Keefer
NovaScreen Biosciences Corporation, Hanover, MD, USA

W W Keighley
Pfizer Global Research and Development, Sandwich, UK

H Kessler
TU München, Garching, Germany

J Klages
TU München, Garching, Germany

C Klammt
University of Frankfurt/Main, Germany

G J Kleywegt
University of Uppsala, Uppsala, Sweden

R Koenig
Genomics Institute of the Novartis Research Foundation, San Diego, CA, USA

H Kubinyi
University of Heidelberg, Heidelberg, Germany

T Kulikova
EMBL Outstation European Bioinformatics Institute, Hinxton, Cambridge, UK

R A Laskowski
European Bioinformatics Institute, Wellcome Trust Genome Campus, Hinxton, Cambridge, UK

A R Leach
GlaxoSmithKline, Stevenage, UK

J-M Lehn
ISIS Université Louis Pasteur, Strasbourg, France

T Lengauer
Max Planck Institute for Informatics, Saarbrücken, Germany

S V Ley
University of Cambridge, Cambridge, UK

K Lindpaintner
F. Hoffman-La Roche, Basel, Switzerland

M-J Martin
EMBL Outstation European Bioinformatics Institute, Hinxton, Cambridge, UK

S McKenna
Serono Research Institute, Rockland, MA, USA

A Miagkov
NovaScreen Biosciences Corporation, Hanover, MD, USA

J Moll
Nerviano Medical Sciences S.r.l, Nerviano, Italy

B Moloney
Evotec (UK) Ltd, Abingdon, UK

F Morceau
Hôpital Kirchberg, Luxembourg City, Luxembourg

M Muda
Serono Research Institute, Rockland, MA, USA

R M Myers
University of Cambridge, Cambridge, UK

M Olah
University of New Mexico School of Medicine, Albuquerque, NM, USA

T I Oprea
University of New Mexico School of Medicine, Albuquerque, NM, USA

P J Peeters
Johnson & Johnson Pharmaceutical Research & Development, Beerse, Belgium

S Perschke
NovaScreen Biosciences Corporation, Hanover, MD, USA

M Pruess
EMBL Outstation European Bioinformatics Institute, Hinxton, Cambridge, UK

O Ramström
Royal Institute of Technology, Stockholm, Sweden

U Regenass
Discovery Partners International AG, Allschwil, Switzerland

S Robertson
Cambridge Crystallographic Data Centre, Cambridge, UK

H Rüterjans
University of Frankfurt/Main, Germany

M Sathyamoorthy
NovaScreen Biosciences Corporation, Hanover, MD, USA

G Schnapp
Boehringer Ingelheim Pharma, Biberach, Germany

B Schnurr
Discovery Partners International AG, Allschwil, Switzerland

G E Schulz
Albert-Ludwigs-Universität, Freiburg im Breisgau, Germany

P Seneci
Desenzano del Garda, Italy

W Shi
Case Western Reserve University, Cleveland, OH, USA

P M Smith-Jones
Memorial Sloan Kettering Cancer Center, New York, NY, USA

M T Stubbs II
Martin Luther University, Halle, Germany

G J Swaminathan
European Bioinformatics Institute, Wellcome Trust Genome Campus, Hinxton, Cambridge, UK

M-H Teiten
Hôpital Kirchberg, Luxembourg City, Luxembourg

D Ullmann
Evotec AG, Hamburg, Germany

J van Oostrum
Novartis Institutes for BioMedical Research, Genome and Proteome Sciences, Basel, Switzerland

H Voshol
Novartis Institutes for BioMedical Research, Genome and Proteome Sciences, Basel, Switzerland

G Vriend
Centre for Molecular and Biomolecular Informatics, Nijmegen, The Netherlands

F L Wang
NovaScreen Biosciences Corporation, Hanover, MD, USA

A Weissman
NovaScreen Biosciences Corporation, Hanover, MD, USA

T J Winchester
Pfizer Global Research and Development, Sandwich, UK

T P Wood
Pfizer Global Research and Development, Sandwich, UK

E Zass
Informationszentrum Chemie Biologie Pharmazie, Zürich, Switzerland

Q Zhang
University of California, La Jolla, CA, USA

3.01 Genomics

J S Caldwell, S K Chanda, J Irelan, and R Koenig, Genomics Institute of the Novartis Research Foundation, San Diego, CA, USA

3.01.1	**Introduction**	**1**
3.01.1.1	What Is a Good Drug Target?	2
3.01.1.2	Druggability	2
3.01.1.3	Validation	3
3.01.1.4	Discovering Drug Targets in the Post Genome Era	5
3.01.2	**Models**	**5**
3.01.2.1	The Use of Functional Genomics in Model Organisms for Identification of Drug Targets	5
3.01.2.2	*Saccharomyces cerevisiae*	5
3.01.2.3	*Caenorhabditis elegans*	8
3.01.2.4	*Drosophila melanogaster*	9
3.01.2.5	*Danio rerio*	10
3.01.2.6	*Mus musculus*	10
3.01.2.6.1	Gain-of-function analysis: the transgenic mouse model	11
3.01.2.6.2	Loss-of-function analysis: the knockout/reverse genetics	11
3.01.2.6.3	Forward genetics: chemical mutagenesis in the mouse	13
3.01.2.7	Understanding the Genetic Basis of Complex Traits: Quantitative Trait Loci (QTL) Analysis	14
3.01.3	**Exploiting the Human Cell for Drug Targets**	**14**
3.01.3.1	Functional Screening of Complementary Deoxyribonucleic Acid	14
3.01.3.2	Ribonucleic Acid Interference and Functional Screens	17
3.01.3.3	Large-Scale Virus Vector Based Libraries	18
3.01.4	**Conclusion**	**19**
References		**19**

3.01.1 Introduction

The realization of the human genome sequence has electrified the life sciences community, providing a fundamental shift in experimental design toward more global analyses of human physiology, while simultaneously providing a litany of putative novel therapeutic drug targets. For the first time, access to a list of nearly all human genes is available for scrutiny by basic researchers to applied scientists well versed in the practices and pursuit of drug discovery. Some estimate that as many as 5000–10 000 new drug targets[1] may have come to light based on the realization of the full genome sequence, while more conservative estimates place the number in the thousands.[2] The magnitude of the change in number of possible targets has lead to the general and not necessarily unfounded feeling within pharmaceutical companies that there are now more targets than can be dealt with effectively. While on the one hand small-molecule screening groups with industrial high-throughput screening (HTS) robots are prepared for the onslaught of targets, and may welcome the challenge, many drug discovery practitioners are threatened by the real possibility that the expansive number of targets might clog the pipeline, inflate R&D costs, and generally make drug discovery less efficient. Given the highly competitive quest for the next blockbuster compound, pharmaceutical companies must weigh the opportunity cost of delaying a new program in order to do more validation studies versus starting prematurely on an undervalidated target. The fact that the industry as a whole only delivers two to three novel target-based drugs to market per year[3] may indicate an increased difficulty/decreased probability of success and/or the aggregate decision to avoid such targets.

The question and the opportunity then posed by the genome is how to optimally and rapidly translate the human genomic information into validated targets and eventually drugs without sacrificing efficiency. The human genome project has done its job to highlight in an abstract sense that there are additional drug targets 'out there,' and in doing so has effectively shifted the bottleneck from gene identification to target validation, posing a new challenge to the drug discovery industry. The types of questions that require addressing fall into two main categories: first, druggability (a primary concern of the chemist), or what is the likelihood of achieving a small-molecule binder that preferentially modulates the target? And second, validation (of primary concern to the biologist), or what is the likelihood that pharmacological modulation of the target will result in a tangible alleviation of disease or symptoms? In other words, how is the function of the putative target linked to the disease in question? Is the gene expressed in the relevant tissue type? What happens when you genetically delete the gene in physiologically relevant cellular or animal models? Is there any human genetic information correlating gene function with a disease? The target windfall begs for a framework toward understanding which 'targets' represent both 'druggable' and 'validated' targets, an important question given the associated risk involved in launching drug discovery campaigns on poorly validated targets.

3.01.1.1 What Is a Good Drug Target?

Before addressing this question an understanding of which criteria define a 'good' drug target and how such targets have been discovered historically is instructive. While biologics, such as peptide and antibody-based drugs, address targets as well, the scope of this chapter is confined to the discussion of drug targets addressed by small molecules. A cynic's definition of a good drug target would be a target which is already successfully addressed by a marketed small molecule such as 3-hydroxy-3-methylglutaryl coenzyme A (HMG-CoA) reductase with Lipitor,[4] or *BCR-abl* with Gleevec.[5] However apt this definition may be, it is too subjective and relies heavily on circular reasoning. Instead, a pragmatist would define a molecular target as one with a proven crucial role in causation or symptoms of a human disease, which when targeted in the clinic with a small molecule leads to reversal or diminishment of the disease and/or symptoms. Again, this definition invokes the requirement of a small molecule a priori to define what makes a good drug target, and with good reason. Many of the first and best drug targets were discovered with a proven small molecule in hand. For instance, the well-known immunosuppressant drug cyclosporine A (CsA) was used as a probe to purify its target cyclophilin which modulates FK506 binding protein (FKBP) activity for the treatment of transplant rejection.[6] Metformin (glucophage) which modulates adenosine monophosphate-activated protein kinase (AMPK) and acetyl-CoA carboxylase activity in the treatment of diabetes[7] has led to programs targeting each with small-molecule modulators. Many successful antibiotics such as the beta-lactams and even penicillin were used to discover their targets which led to subsequent target-based lead finding programs.[8] So, while it is often the case that drugs themselves provide direct rationale for their targeted gene products as drug targets, these define a limited set of validated targets, totaling approximately 120 based on all marketed drugs.[2] In these instances, the question of druggability and target validity are answered simultaneously, but in order to evaluate novel putative targets, one needs to separate out these two requirements of a 'good' drug target and systematically mine and analyze the genome for optimal target opportunities.

3.01.1.2 Druggability

The list of known drugs provide a means to a reverse definition of a druggable target, but in light of the genome project, and a now extended list of gene products with variable chances of being targeted with a small molecule, how do we determine a 'forward' definition of a druggable target, and triage this to a workable list? In other words, can we come up with a reasonable set of criteria which is predictive of druggability and relevance to disease? By way of an example, an analogy in the chemical world was provided by Chris Lipinski who established a set of criteria based on physicochemical and pharmacokinetic characteristics of known drugs to define the 'rule-of-five' for oral bioavailability.[9] This study showed that poor absorption or permeability of a drug is more likely when there are more than five hydrogen-bond donors, the lipophilicity is high (ClogP >5), the molecular weight is greater than 500 Da, and the sum of all nitrogen and oxygen atoms is greater than 10. Now, drug discovery groups use the 'rule-of-five' as a guideline for prioritizing drug leads to go into animal efficacy studies or even into humans. By analogy, then, in the biological world, are there guidelines to be gleaned from the list of known druggable targets, based on gene family, structure, expression, tissue distribution, function, biological pathway connectivity, etc. to help narrow the list of genes in the genome for drug targeting exercises?

An attempt to assess the number of putative targets that represent an opportunity for therapeutic intervention was made by Hopkins and Groom. Applying the constraint that a druggable target must be physically capable of binding compounds, preferably to those with druglike features as set forth by the Lipinski 'rule-of-five,' this group determined that only 3051 of the predicted ~30 000 or so genes code for a protein with some precedent (or likelihood, by analogy to family members), of binding a druglike small molecule.[2] Almost half of these druggable genes fall into only six of 130 protein families, including the usual suspects G protein-coupled receptors (GPCRs), tyrosine and serine/threonine kinases, zinc metallopeptidases, serine proteases, nuclear hormones, and phosphodiesterases. Of course, these are purely estimates based on a small number of proteins which are documented to bind small molecules, whereas the true number of druggable entities may be much larger. Furthermore, this analysis does not take into account those targets for which chemical modulation has not been profoundly demonstrated nor those which are composed of different material (DNA, RNA, lipid, etc.).

Another issue tangentially related to druggability, but germane to the discussion of what constitutes a 'good' drug target, is the potential selectivity of a small molecule inhibitor against a chosen target. Although historically GPCRs and enzymes such as kinases, proteases, phosphodiesterases, etc. have been the preferred targets of the drug industry, many of the compounds developed for such targets have considerable cross-reactivity with other family members, often leading to untoward and significant side effects. Conventional wisdom suggests that the increased specificity of a compound against its target increases the likelihood of efficacy, a position supported by the considerable cross-reactivity of some of the kinase inhibitors which have efficacy but dose-limiting toxicity ascribed to their lack of selectivity.[10] However, some marketed drugs may have positive attributes based on their lack of selectivity. Even Gleevec, the first molecularly-targeted anticancer drug designed to inhibit *BCR-abl* in chronic myelogenous leukemia, has cross-reactivity with platelet-derived growth factor receptor (PDGFR) and v-kit Hardy–Zuckerman 4 feline sarcoma viral oncogene homolog (c-kit) which may contribute to possible future success in treating other types of cancer, and potentially asthma[11] and diabetes.[12,13] Of course, this type of serendipity is not unprecedented as many drugs are efficacious for off-label indications, likely based on cross-reactivity and off-target effects. The extent to which this strategy will be leveraged in target prioritization is yet to be determined. Drug discovery companies now routinely profile compounds in vitro against panels of related targets, such as kinases, GPCRs, and proteases to find either potential sources of toxicity or new opportunities for a compound of interest. What is clear is that the genome project has significantly expanded and ostensibly defined the number of possible cross-reactive targets within a family or enzyme class, and thus enabled a more complete assessment of a compounds selectivity-based advantages and disadvantages.

3.01.1.3 Validation

While it is possible to create a rough sketch of what a 'good' target should look like by analogy to other proven drug targets, this triage method satisfies only one part of the equation, omitting functional definition of the target within the context of physiology and disease. Admittedly, fewer than half of all proteins encoded by the genome have any documented functional annotation.[14] How then does one cull the extensive list of putative targets further within the framework of function? Is there a high-throughput methodology to ascribe functional definition to a gene to the degree necessary to launch a drug discovery effort? Although the methods that embody the field of genetics which provide means to study physiological effects of single-gene alteration (gene knockout, transgenesis, and organismal mutagenesis) have a long history of success in finding putative targets based on function, the success may be offset by the associated burden of low throughput and evolutionary distance from humans (e.g., yeast, worms, flies, fish, and mice). As a compromise, several groups have focused target identification and validation efforts on human cells which may represent to varying extents models of human disease. These groups have attempted to determine gene function on a genome-wide level by arraying thousands of distinct human genes one by one in mammalian expression vectors in microtiter plates and transfecting massively in parallel into cellular reporter assays and testing for functional effects of overexpression. The process coined genome functionalization through arrayed cDNA transduction (GFAcT)[15] or gene-by-gene screening[16] has been used to functionally annotate hundreds of genes in assays for transcription factor activation, cell proliferation, cell cycle arrest, and other phenotypes that can be easily read out in cells. In an analogous process, comprehensive libraries of short interfering RNAs (siRNAs) targeting individual genes of the human genome for knockdown or reduction of messenger RNA (mRNA) levels have been arrayed and screened for function in cell-based assays as well.[17] These methods have tremendous potential for reducing the genome down to smaller more manageable subpopulations based on activity in functional assays relevant to pathways, physiology, and disease. However, outside the context of the whole organism, there are pitfalls, so although genome- and proteome-wide analysis of genetic and physical interactions, often across multiple biological systems, have begun to reveal functional roles for specific genes and gene products, they fall somewhat short of squarely placing a gene within the global organization of physiology that describes how biological states give rise to phenotypes such as disease.

The importance of defining gene function and testing validation hypotheses in whole organisms is highlighted by the theory of biological connectivity and the concept of biological nodes.[18] Just as is observed with any other network in the real world, such as the internet or telecommunications systems, biological networks responsible for carrying out measurable phenotypes are comprised of at least two distinct types of interactions: a low number of highly linked nodes that interface with a high number of lowly linked nodes, and a high number of lowly linked nodes that interface with a small number of highly linked nodes (**Figure 1a**). An example of a highly linked node is nuclear factor kappa B (NFκB), which can be modulated by dozens of stimuli (such as interleukin-1 (IL1), tumor necrosis factor (TNF), lipopolysaccharide (LPS), sorbitol, hydrogen peroxide, ultraviolet radiation (UV), etc.), perturbations and physiological states by virtue of interaction with a large number of lowly linked nodes, or proteins specific to a given signal transmission pathway (**Figure 1b**).[19] Because of its high connectivity, perturbing the activity of transcription factor NFκB can result in a wide variety of phenotypic effects, of which some are desirable (i.e., disease alleviation) and some undesirable (i.e., toxic). Other examples of highly linked nodes include p53, activating protein-1 (AP-1), and ras, which represent gathering points for many signaling pathways involved in cell growth, proliferation, cytokine production, and adhesion, in a variety of systems regulating oncologic, inflammatory, metabolic, neuronal, and other important physiological areas.[20–22] Studies routinely suggest that perturbation of these highly linked nodes or 'hubs' may produce a desired biological effect but also a considerable number of undesirable effects outside the desired target biological system. The importance of the node theory is highlighted by the likely mechanism-based cardiac side effects reported with inhibitors of the cyclooxygenases, which control both inflammatory and proliferation processes.[23] In contrast, perturbation of a lowly linked node may have fewer untoward side effects while still affecting a desired biological outcome. For instance, perturbation of the IL1 receptor activated kinase (IRAK) would likely affect NFκB only in the context of IL1 signaling as opposed to perturbing NFκB function through alternative upstream inputs. Therefore, identification of lowly linked nodes in phenotypic cellular and organismal assays could simultaneously achieve functional definition of a gene or gene product within a physiological and/or disease context and allude to 'connectivity' in order to help prioritize targets for drug discovery.

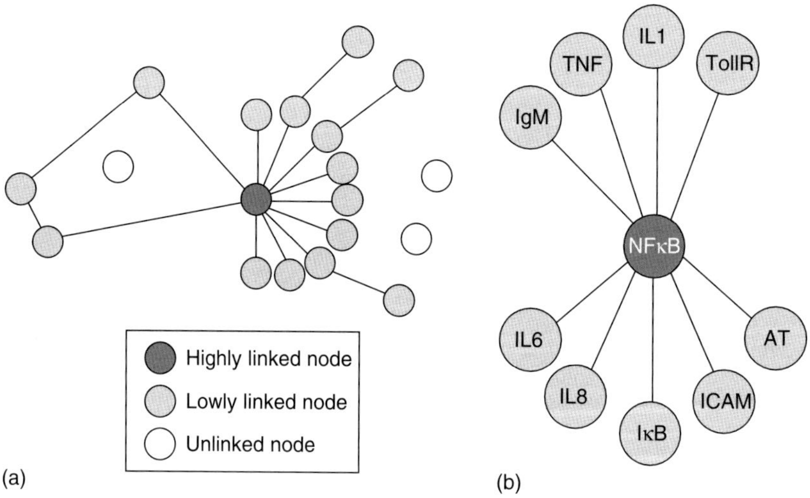

Figure 1 Node network theory schematic. (a) Biological networks can be represented as an array of nodes including highly-linked nodes (dark gray circles), lowly linked nodes (light gray circles), and unlinked nodes (white circles). Theoretically, highly linked nodes would control multiple processes and thus connect with many other nodes. Perturbation of such highly linked nodes would likely have pleiotropic effects because of their influence on many processes. Perturbation of lowly linked nodes would likely have fewer pleiotropic effects because of their isolation (as would be the case of perturbing the disconnected or unconnected nodes). (b) Examples of a highly connected node would be the transcription factor complex NFκB which is involved in regulating many processes including proliferation, inflammation, and retroviral infection. Disruption of a lowly linked node upstream of NFκB activation such as interleukin-1 (IL1) inhibition would not impede NFκB activation through other effectors such as tumor necrosis factor (TNF) and toll-like receptor (TollR) and thus have fewer side effects than direct inhibition of NFκB.

3.01.1.4 Discovering Drug Targets in the Post Genome Era

The endeavor to discover 'good' targets for future efficient drug discovery programs will rely heavily on the application of new methodologies which put genes into the context of both druggability and functional relevance to human disease. The previous section has outlined several of the measurements used for druggability (precedence, binding pockets, target class, etc.) as well as a conceptual framework for assessing the overall role of a putative target in physiology and disease (function, protein family, connectivity), but has only briefly touched upon the vessels, including organisms and cells which are used every day to identify protein function and thus putative target function. Since the completion of the human genome sequence, a multitude of so-called 'post genome' technologies have been built either as a result or in anticipation of exploiting this tremendous resource. These new technologies may pose viable strategies for target discovery, validation, and triage. In fact, advances in HTS, computation chemical libraries, proteomics, gene expression microarrays, chemical–genomic profiling, cell biology, and genetic libraries in a range of organisms and mammalian cells may provide the tool-basis for such an enterprise. But the basis for many of these tools is largely an extension of foundational methods and principles exacted in the past 100 years of genetics. Many of these model organism's genomes had been sequenced prior to that of the human, and a framework for how to exploit complete genomic information has been to some extent established already. Therefore the next sections focus first on the organism-based genetic models, which have a long history in the drug discovery industry, that have served as the basis for and form the foundation of the field of target identification and validation, and then on how these same principles are now being applied to human cellular systems. Careful attention to these methods and their superimposition in the coming age of human genetic biology will likely highlight the new line of scientific inquiry broadly characterized as target discovery and the forthcoming conceptual framework, approaches, and technological advances that are shaping this area, allowing drug discoverers to make rational decisions as to which targets to pursue in lengthy, expensive drug discovery campaigns.

3.01.2 Models

At the time of publication of the finished human genome sequence there were approximately 22 000 identified genes; since these accounted for more than 99% of the previously identified complementary DNAs (cDNAs) we can consider the list to be essentially complete, although refinements in gene prediction will continue.[24] One benefit of having this list is that we can now identify all potential targets for a given class of drugs; for example, the entire set of suspected human kinases numbers 518,[25] allowing for comprehensive target discovery efforts focused on the human kinome.[26] Another potential benefit is the ability to monitor the expression of the entire set of genes at the RNA level or protein level in both healthy and diseased states, in a wide variety of tissues, in the hopes of discovering the critical changes that cause the disease state. This approach has been used extensively in oncology, where samples from tumors from a large numbers of patients can be compared to each other and to normal tissues. One success story using this approach is in the identification and validation of the receptor tyrosine kinase FLT3 as a target for treating MLL-rearranged acute lymphoblastic leukemias.[27,28] Unfortunately, the more common result of these types of experiments is a set of expression differences that is too large to facilitate identification of single drug targets, although insight into druggable pathways may be obtained.[29] Since these types of studies are essentially descriptive, providing information on where and when genes are expressed but not on what is causing those expression changes, they can only establish circumstantial evidence for the importance of a potential drug target by inferring its function in disease. Thus the key to unlocking the potential of the genome sequence lies in directly assessing gene function.

3.01.2.1 The Use of Functional Genomics in Model Organisms for Identification of Drug Targets

With the completion of the human genome one might have expected that the model organisms traditionally used to identify and characterize basic biological processes and pathways might fade into obsolescence. On the contrary, researchers using the budding yeast, nematode, fruit fly, and zebrafish model systems have pioneered the development of techniques that enable systematic, whole-genome assessment of gene function, and these techniques and model systems are being adopted for more focused target identification efforts by the pharmaceutical industry.

3.01.2.2 *Saccharomyces cerevisiae*

Due to its ease of manipulation, the budding yeast *Saccharomyces cerevisiae* has long been the eukaryotic model system of choice for conducting systematic genetic studies on the most fundamental, and thus highly conserved, biological processes and signaling pathways. One indication of the utility of yeast for modeling human diseases is the observation that approximately one-third of a sample of 170 known human disease genes have a close match in the yeast genome.[30] Classical studies in yeast involved a 'forward genetics' approach, screening for random mutations in the genome to produce a phenotype of interest, then laboriously tracking down the cognate gene; perhaps the best-known work in this area, on cell cycle control, was honored with the 2001 Nobel Prize in Medicine.[31] In the area of drug discovery, yeast has been used to develop antifungal drugs as well as to identify and characterize targets of therapeutic interest in humans. The target of the immunosuppressant rapamycin was discovered in yeast,[32] and yeast has been used to study conserved pathways such as mitogen activated protein kinase (MAPK) signaling, to identify cell-cycle inhibitors,[33] and to identify many of the client partners of the oncologically interesting target hsp90.[34]

The budding yeast genome was the first eukaryotic genome sequence completed,[35] and the research community quickly took on the task of characterizing the functions of every gene, developing and improving many genomics technologies in the process. A large number of transcriptional profiling studies have been used to demonstrate putative drug mechanism of action.[36] Another example of whole genome characterization is embodied by the yeast two-hybrid assay, whereby separate fusions of 'bait' protein fragments and a library of 'prey' protein fragments to a pair of protein domains that generate an assayable signal only when bait and prey can bind to each other, was used to generate large-scale protein interaction maps.[37,38] Construction of fusions of nearly every yeast protein to a fluorescent protein allowed assignment of subcellular localization.[39] Many of these data sets and resources are maintained in publicly available form and are continuously updated (**Table 1**).

While such genome-wide descriptive studies are informative, the real advances in the yeast field have been in pioneering the technologies for directly assessing gene function. Within a few years of completion of the genome sequence, a panel of strains deleted for each of the approximately 6000 genes was developed. This allows for systematic, whole-genome 'reverse genetic' screening, wherein the identity of the gene mutated is known and the resulting phenotypes are assessed. By flanking each deletion with a molecular bar code consisting of a unique oligonucleotide sequence, the presence or absence of a particular deletion in a pool of strains can be detected by isolation of DNA and hybridization to microarrays consisting of oligonucleotides complementary to each bar code (**Figure 2**).[40,41] When subjected to various environmental stresses in liquid culture, such as addition of a drug, individual yeast deletion strains within the pool exhibit differential fitness. During growth, strains deleted for genes encoding proteins involved in defense against that stressor are depleted from the pool, resulting in a decreased hybridization signal on the oligonucleotide array. Deletions that nonspecifically affect uptake or metabolism of a class of compounds can be rapidly identified by comparison of hybridization profiles for different drug treatments and thus eliminated from consideration as drug targets. This valuable tool has been used to discover a host of novel players in previously well-defined processes, including DNA repair,[42–44] mitochondrial function,[45] polyglutamine disease,[46] and viral replication.[47] Interestingly, the results from some of these genome-wide functional screens did not correlate well with the results from genome-wide transcriptional profiling using the same condition.[48,49] For example, only 7% of genes that were upregulated under glucose deprivation were actually necessary for growth in limiting glucose, and more importantly, only a minority of genes that were required in the functional assay showed significant gene expression changes, providing solid evidence of the advantages of the functional approach to target discovery.[48]

Another advance in functional assessment of the yeast genome has come from the development of a system for systematically testing pairwise combinations of gene deletions.[50] The concept of 'synthetic lethality,' where two nonlethal mutations become lethal when combined in the same strain, was used in classical genetics to group genes by function and to discover novel players in known pathways. It is of particular interest in oncology target discovery, where inactivation of one such nonessential gene product by a drug in combination with an otherwise nonlethal tumor-specific mutation could lead to tumor-specific lethality.[51] In the yeast system, a haploid single-deletion strain of interest is mated to an array of all other single-deletion haploid strains on agar plates, and when the resulting diploid undergoes

Table 1 Yeast genomics resources

Resource	*Site*	*URL*
Genome sequence	*Saccharomyces* Genome Database	http://www.yeastgenome.org
Transcription profiling	Stanford Microarray Database	http://genome-www5.stanford.edu
Protein interactions	Database of Interacting Proteins	http://dip.doe-mbi.ucla.edu/
Protein localization	Yeast GFP Fusion Localization Database	http://yeastgfp.ucsf.edu
Deletion panel	*Saccharomyces* Genome Deletion Project	http://www-sequence.stanford.edu/group/yeast_deletion_project/

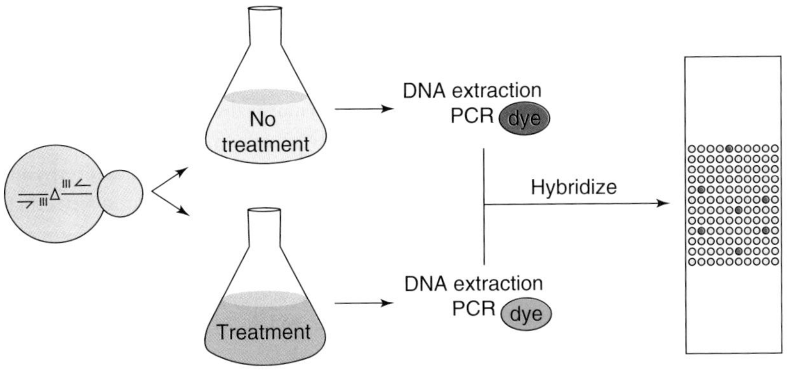

Figure 2 Use of the bar-coded yeast deletion collection in competitive growth experiments. For each deletion strain, the gene deletion (△) is flanked by a unique oligonucleotide sequence (vertical bars) which in turn is flanked by common polyamerase chain reaction (PCR) primer sequences (arrows). A pool comprising of the entire set of possible gene deletions is grown for a limited number of generations under treatment or control conditions. DNA is then extracted and PCR-amplified with dye-labeled primers (red or green) that are specific to the treated or untreated sample. Quantitation of the relative amount of each deletion strain in each condition is achieved by hybridization to a microarray consisting of spotted oligonucleotides complementary to each unique bar code. The rare gene deletion that leads to selective growth inhibition under treatment shows up as a single dye (red) spot, in contrast to most spots which hybridize to DNA labeled with both dyes (yellow).

meiosis a selection is applied that allows only the double-deletion strain to grow (synthetic lethality is indicated by the absence of growth). This approach has allowed for large-scale testing of genes of interest for synthetic lethal partners,[52] and with the advent of a method for conducting such studies in the pooled format,[53] it is likely that eventually most pairwise interactions will be tested and a complete synthetic lethal interaction map will be produced. This data set should prove useful for target identification: by comparing the set of haploid deletion strains inhibited by a drug to the compendium of known synthetic lethal interactions, one should be able to identify the processes or pathways targeted by the drug.[54]

In a sense, these marked deletions mimic a perfectly specific and efficacious drug, in that only one protein is affected, and its function is completely eliminated; however, only genes completely nonessential to yeast, can be assayed. In the real world, partial inhibition of essential functions, or inhibition of proteins that have redundancy in higher eukaryotes but not in yeast, might be therapeutically effective. Thus, a panel of diploid, heterozygous deletions has also been constructed. This collection has proven especially useful for probing the target(s) of therapeutically interesting compounds. This approach, referred to as induced haploinsufficiency or chemical genomics, is based on the concept that reduction of the copy number of a gene encoding a drug target from two to one in the heterozygous deletion strain should render that strain hypersensitive to the drug; in essence, the concentration of the drug target has been halved, therefore the effective drug concentration is reduced.[51,55] There are some obvious limitations to such an approach; the drug target must be a single protein present in yeast cells, and the compound must be able to cross the yeast cell wall. Nevertheless, the potential utility of chemical genomic profiling in yeast was validated by testing panels of commercially available drugs used to treat a wide array of human diseases, as well as antifungal agents, against the collection of heterozygous deletion strains.[56,57] Compounds were tested at concentrations that partially inhibit growth of wild-type yeast against the panel of heterozygous deletion strains in the pooled format, resulting in most cases in selective inhibition of growth of a distinct subset of deletion strains. In this way, the known targets or pathways for a majority of drugs tested were correctly identified, although a large minority of compounds did not selectively inhibit any of the deletion strains. Perhaps more importantly, these studies also identified potentially novel targets for known drugs, indicating usefulness in testing interesting classes of compounds for novel targets, as well as identifying potentially desirable or undesirable off-target effects. For example, both studies found that a common oncology chemotherapeutic, the thymidylate synthetase inhibitor 5-fluoruracil, inhibited growth of strains heterozygous for genes involved in ribosome biogenesis and ribosomal RNA processing, but not strains heterozygous for thymidylate synthetase, suggesting an alternative hypothesis for the mechanism of action of this drug. Also of interest was the observation that three different compounds, with very different therapeutic targets but a similar core structure, exhibited similar chemical genomic profiles, suggesting a potential application for chemical genetic profiling in structure–activity relationship studies.[56] Subsequent work has used automation and miniaturization to increase throughput and decrease expense, leading to identification of sphingolipid biosynthesis as the target of a tumor cell invasion inhibitor, dihydromotuporamine C,[58] and to further elucidate the pathways targeted by rapamycin.[59]

Yeast has also proven valuable as an experimental platform for genome-scale functional studies of human genes. The endogenous signaling pathway used for mating in yeast can be engineered such that the activation of heterologously expressed human GPCRs leads to an assayable phenotype.[60] Several sophisticated systems have been used to study various aspects of GPCR signaling, including identification of previously unknown ligands, or deorphaning,[61,62] and isolation of mammalian cDNAs that modulate GPCR signaling from genomic libraries.[63,64] Similarly, heterologous expression of the proapoptotic protein Bax in yeast results in an apoptosis-like lethal phenotype, allowing for efficient screening for repressors of Bax-induced apoptosis.[65] An extension of the yeast two-hybrid system that links a compound to a fusion DNA binding protein allows for target identification by screening a library of human cDNAs fused to a transcriptional activator; only those fusions producing a protein that binds the compound can activate transcription. This 'three-hybrid' system has been validated by identifying murine dihydrofolate reductase cDNAs in a screen for methotrexate targets,[66] and has been used to identify novel targets of known cyclin-dependent kinase inhibitors.[67]

3.01.2.3 *Caenorhabditis elegans*

The yeast model system has a number of limitations in its uses for human drug target discovery, including its evolutionary distance from humans and the fact that it is a single-celled organism. Thus the nematode worm, a simple metazoan with several complex organ systems, has become useful as a model for discovering types of drug targets that are not likely to be present in yeast. *Caenorhabditis elegans* was pioneered as a model system by Sydney Brenner for the purpose of studying development in a genetically tractable metazoan; this work was honored with the Nobel Prize in Medicine in 2002.[68] In addition to a small size (<1 mm), short generation time (3 days), and general ease of handling, nematodes have the ability to reproduce as hermaphrodites, facilitating genetic analyses, especially the production of homozygous mutants. Furthermore, the small number of somatic cells (959), optically clear body, and invariant cell lineage pattern allows for mapping of the entire set of cell divisions that go into making the adult animal, which in turn allows for easy interpretation of the cause of developmental defects in mutant strains. Worms have proven especially useful in basic research on cell–cell signaling, apoptosis, neurogenesis, and aging. In terms of target discovery, worms have been used to identify the target of a nematicide used to treat river blindness,[69] to identify targets of volatile general anesthetics,[70] to identify genetic modifiers of behavioral responses to alcohol,[71] and to screen for small molecules that suppress muscle degeneration in a muscular dystrophy model.[72]

The *C. elegans* genome was the first completed sequence of a multicellular organism; approximately one-third of its 20 000 genes have close homologs in humans.[73] Many of the genomics resources available in the yeast system are also available in worms, publicly accessible from a central database website called wormbase.[74] Resources unique to the worm system include complete anatomies of the entire set of somatic cells, as well as a complete connectivity map of the neural cells. Some of the pioneering work in developing large protein interaction maps was conducted in worms,[75,76] and a consortium is in the processing of knocking out every gene.[77]

Perhaps the most important tool for conducting functional genomics studies in multicellular eukaryotes is RNA interference (RNAi) technology. A key advance in development of RNAi came with the observation that injection of double-stranded RNA (dsRNA) efficiently and specifically knocked down cognate mRNA levels in worms.[78] Although similar gene silencing phenomena had been previously described in plants and filamentous fungi, this was the first indication that the mechanism might involve enzymatic processes and that a generalized gene knockdown technology based in RNA interference could be developed for metazoans. Subsequent studies revealed a common mechanism for RNAi in eukaryotes: long dsRNA is processed into 21–23 bp siRNA by the RNase III-like enzyme Dicer. These siRNAs are recognized by the RNA-induced silencing complex (RISC), which interacts with the homologous mRNA resulting in its degradation and thus downregulation of the respective protein.[79]

Technological advances in the study of worms improved delivery of the interfering RNA; worms can either be soaked in a solution of dsRNA or simply fed bacteria overexpressing target dsRNA to cause efficient mRNA knockdown.[80,81] The so-called feeding libraries are a very cost-effective and reproducible means of standardizing the RNAi platform, as the bacterial clones can be frozen and expanded at will. This type of library was used for some of the first large-scale RNAi screens in any organism, wherein basic growth, reproduction, morphology, and behavior phenotypes were assessed after application of dsRNAs covering approximately 90% of the genome.[82,83] A compendium of the phenotype data from these and other functional screens is maintained at the wormbase website.[74] Since worms have been studied extensively by classical genetics, this system allows for a rigorous assessment of the false negative rate in RNAi studies; unfortunately, for about 30% of known essential genes and about 60% of genes with known postembryonic developmental phenotypes, the dsRNA feeding approach gave no observable phenotype.[83] This problem should eventually be alleviated by enhancements of RNAi efficacy, such as using a mutant strain that is hypersensitive to RNAi-mediated knockdown.[84]

One example of the power of whole-genome functional screens for target identification using feeding libraries comes in the area of obesity research; by using a live stain for fat droplets, Nile red, genes involved in storage and mobilization of fat reserves could be identified by RNAi.[85] From a whole-genome feeding library, dsRNAs targeting approximately 300 genes exhibited a significant and specific fat deposition phenotype in this assay. Further testing of these dsRNA clones on mutants defective in known players of fat metabolism, such as the insulin and serotonin signaling pathways, allowed for pathway mapping of the targets and resulted in identification of potentially novel targets that affect fat regulation independently of the known pathways. This type of mixed approach, using genomics technologies to identify novel potential targets and using mutants derived from classical genetics studies to validate and characterize them, is extremely rapid and cost-effective in nematodes.

3.01.2.4 *Drosophila melanogaster*

The fruit fly *Drosophila melanogaster* was chosen by Thomas Hunt Morgan nearly 100 years ago as a model system for working out the basic mechanisms of inheritance, for which he was awarded the Nobel Prize in 1933. Relative to nematodes, there is a more comprehensive set of classical genetic tools available due to the long history of genetic research in flies, and sequence conservation between flies and humans is somewhat higher.[86] Its small body size, short generation time (2 weeks), large generation size (~ 50 eggs per day), and the comprehensive set of genetic tools such as balancer chromosomes and a dense genetic and physical map have proven invaluable in conducting genetic analyses, especially forward genetic screens. The instigation of saturating screens for fly embryogenesis mutants in the early 1980s marked a milestone in genetic analysis in metazoans, providing much of the basis for our current understanding of cell signaling pathways.[87]

As with other model organisms, forward genetic screens have been used to hunt for potential drug targets that play a role in fly models of human diseases. Of particular usefulness in this regard is the fact that the signaling pathways involved in normal eye development are well studied, and can be manipulated to produce an easily scored phenotype. Such engineered eye screens have been used to identify suppressors of polyglutamine-induced neurodegeneration in fly models for Huntington's disease[88] and spinocerebellar ataxia 1.[89] The genetic analysis of Ras signaling in flies has been used as a model for cancer; receptor tyrosine kinase activation of Ras, leading to activation of a MAPK pathway, plays a role in normal eye development, and the fly equivalent of oncogenic, constitutively active Ras causes overproliferation of certain cells in the eye.[90] Screens for suppressors or enhancers of this cancerlike phenotype have identified a number of key players in Ras signaling, including novel potential oncology targets.[91,92]

The fly genome sequence was completed in 2000; at approximately 14 000 genes, it is somewhat smaller than the nematode genome.[93] Matching a sample of 929 human disease genes to the fly genome revealed that 77% of the human genes had at least one fly homolog, indicating the utility of fly genomics to target discovery.[94] Many of the typical genomics tools and datasets are available at flybase[95] and the Berkeley Drosophila Genome Project.[96] A project is underway to systematically disrupt every gene via transposon insertions.[97] An ingenious early application of genomics technologies in combination with classical genetic tools for studying fly development used automated embryo sorting and gene expression profiling to study the effects of perturbations of the twist transcription factor on mesoderm development.[98] As in nematodes, RNAi can be triggered by dsRNA, but in flies most genome-scale screening is done with cultured cells arrayed in microwell plates rather than in the whole organism. Some fly cell lines can naturally take up dsRNA, others require addition of a transfection reagent. Large-scale genome screens have been conducted to study hedgehog signaling pathways,[99] cell proliferation and viability,[99] cell morphology,[100] cardiac development,[101] and Wnt signaling.[102] In the Wnt study, 238 potential regulators were identified, including both known pathway components and some genes without previously assigned function. Several regulators were validated by testing for effects on Wnt signaling in *Drosophila* embryos as well as in zebrafish embryos and cultured mammalian cells, indicating that these genes have evolutionarily conserved functions in Wnt signaling.

The potential power of whole-genome screening for novel drugs and simultaneously identifying the targeted pathway was demonstrated in a parallel screen for small-molecule and dsRNA inhibitors of cytokinesis.[103] A dsRNA library covering more than 90% of the fly genome and a small-molecule library consisting of 51 000 compounds was screened in a cell line for the binucleate phenotype indicative of a cytokinesis block by automated fluorescent microscopy. Fifty compounds and 214 dsRNAs gave a detectable cytokinesis block. The 25 most potent compounds and 40 best dsRNA inhibitors were analyzed further by testing for the immunolocalization pattern of a set of 15 proteins involved in various aspects of cytokinesis. By comparing the patterns of localization phenotypes among the dsRNA and small-molecule cytokinesis inhibitors, a match was made between one compound and the Aurora B kinase, suggesting that this compound targets either Aurora B or an upstream member of the Aurora pathway. This type of pattern matching approach, utilizing a combination of small molecule and RNAi perturbations and assessing multiple phenotypic changes, should prove to be a powerful method for target identification in the future.

3.01.2.5 *Danio rerio*

Although flies and worms allow for efficient functional genomics studies of disease-relevant genes and pathways in the context of a multicellular organism, the zebrafish *Danio rerio* is becoming the system of choice for identifying novel targets in the context of a whole vertebrate animal. The zebrafish was originally developed by George Streisinger as model system for doing forward genetic screens in vertebrates and has proven to be a powerful tool for understanding early development.[104] Its relatively short generation time (3–4 months), large generation size (up to 200 eggs per clutch), optically clear embryos, and small body size (4 cm) make it uniquely suited among vertebrates for large-scale genetic analysis, especially phenotypic screening.[105,106] The genome sequence is nearly complete,[107] and many of the standard genetics and genomics tools are available, including transcription profiling with DNA microarrays,[108] a cDNA collection,[109] and high-resolution physical maps to facilitate positional cloning.[110,111] Antisense phosphorodiamidate morpholino oligonucleotides (morpholinos) can be used for efficient gene knockdown,[112] possibly allowing for genome-wide, reverse-genetic functional screening in the future, although the current mode of delivery (microinjection) is rate-limiting. A reverse-genetics strategy allowing for efficient recovery of randomly generated mutations in specific genes, called targeting induced local lesions in genomes (TILLING)[113] has been used successfully in zebrafish.[114] High-throughput, whole-organism compound screens can be conducted as well by simply arraying embryos and compounds together in multiwell plates.[115]

The zebrafish system has proven useful for the identification of novel drug targets by the application of classical forward genetics (or knockdown by morpholino) to identify mutations that mimic human disease phenotypes. The obvious advantage of zebrafish over other model systems is that organ-level and vertebrate-specific diseases can be modeled. The first demonstration of the utility of zebrafish for identification of human disease genes was in the area of anemia; the iron transporter ferroportin 1 was originally discovered in zebrafish[116] and subsequently found to be mutated in the hereditary iron overload disease hemochromatosis.[117,118] Models for polycystic kidney disease,[119–121] melanoma,[122] and various leukemias[123–125] have been developed. Complex nervous system and sensory defects have been modeled in zebrafish, including neurodegeneration caused by heterologous expression of mutant tau[126] and various forms of blindness and deafness.[127,128]

The zebrafish system has been particularly useful in the area of heart disease. Mutations in TBX5 result in congenital defects in cardiac development in both humans and fish,[129,130] and mutations in the muscle filament titin, which cause familial cardiomyopathies in humans,[131] exhibit a similar phenotype in zebrafish.[132] Mutations in the ether-a-go-go related gene hERG cause arrhythmias in both humans and zebrafish, and drugs targeting the potassium channel encoded by hERG can induce arrhythmia in both organisms.[133,134] Since many commonly used drugs have the side effect of causing arrhythmia via this mechanism, the zebrafish system may prove useful in screening out cardiotoxic drug candidates.

A powerful strategy for identification of novel drugs in zebrafish uses a disease model to identify small molecules that suppress the disease phenotype. In this scenario, the target must still be identified by traditional means, so the advantage of using fish lies as much in the presumptive quality of leads as in target identification, since issues of toxicity and metabolism of the drug are reduced by whole organism screening. The utility of this approach has been demonstrated in a heart disease model.[135] The hypomorphic *gridlock* allele of the zebrafish *hey2* (*hairy/enhancer of split-related*) transcription factor produces congenital aortic defects that are phenotypically similar to aortic coarctation in humans: constriction of the aorta upstream of the aortic branch leading to the trunk results in reduced circulation. In the zebrafish screen, mutant embryos arrayed in 96-well plates were treated with a set of 5000 compounds, and a pyridinyl-thioether compound was found that suppresses the effects of the *hey2* mutation, leading to restoration of proper aortic development. In a first step toward addressing the mechanism of action, candidate genes were tested for expression levels, and vascular endothelial growth factor (VEGF) was found to be upregulated by treatment with the drug in both zebrafish and cultured human cells, leading to increased angiogenic potential in both systems. Although the target of this compound remains to be identified, it should prove useful for investigating angiogenic mechanisms. This disease-suppression approach should at least prove valuable for identifying novel pathways to target, and perhaps in the future the application of focused morpholino libraries or panels of candidate mutants produced by TILLING combined with a chemical genomics approach will speed target identification in zebrafish.

3.01.2.6 *Mus musculus*

Although worms, flies, and now fish accommodate efficient functional genomic and disease-relevant target identification studies in the context of a multicellular entity, their resemblance to humans and thus their relevance to human disease biology may be limited, especially as pertains to neuronal function, immunological status, and other highly evolved physiological systems. And although human cells are components of higher-order systems, such as organs, and integrate environmental signals from numerous other cell types, it is still argued that cellular assays, which normally interrogate a single cell type, may not accurately recapitulate function in the context of an organism (in vivo). To perform genetic analysis in whole animal systems, biologists frequently utilize *Mus musculus*, the common house mouse, as a useful experimental model for understanding human gene function because it is anatomically and physiologically similar to *Homo sapiens*. The similarity of human and mouse genomes is approximately 85%, and more than 100 mouse genetic models mirror human diseases attributed to mutations in the same gene.[136] Experimental advantages to using mice include short generation (breeding) time, relative ease of genetic manipulation, and the availability of comprehensive sequence data and genetically defined inbred strains. Three primary approaches are utilized to study gene function in rodents: targeted manipulation of the genome, random mutagenesis, and examination of naturally occurring genetic variations which result in phenotypic disparities.

3.01.2.6.1 Gain-of-function analysis: the transgenic mouse model

As in cellular systems, alterations in gene dosage levels can provide important insights into the function of a gene in the context of a whole organism. The approach is particularly useful in studying diseases such as cancer, where mutations result in the activation of genes that contribute to disease pathology. Transgenic models are established by injecting a DNA construct (usually a cDNA) into the pronuclei of fertilized mouse eggs at the single cell stage (**Figure 3a**). The introduced DNA randomly integrates into the mouse genome, and the eggs are transferred into a pseudopregnant female mouse for the gestational development. Approximately 10–20% of the pups born to this female will have the foreign DNA inserted into their genome, and each can be used to establish an independent transgenic strain.

Each cell of these transgenic animals will contain the introduced DNA; however, the expression of the encoded protein is dependent upon the promoter used to express the gene of interest. Constitutive ectopic expression, or expression in every cell of the organism, is achieved mainly though use of promoters derived from viruses such as simian virus 40 (SV40), herpes simplex virus (HSV), and mouse mammary tumor virus (MMTV).[137] Direct expression of transgenes to specific tissues has been performed by incorporating promoter/enhancer elements derived from genes whose expression is known to be restricted to those compartments. For example, the CD19 protein is expressed almost exclusively in B cells, and regulatory elements from the CD19 promoter have been utilized to restrict transgenic expression of several heterologous genes strictly to B cells.[138] Alternatively, transgenic constructs have been placed under the control of 'inducible' promoters, i.e., promoters that are activated in response to exogenous substances. These include the metallothionein MMTV-LTR, and OAS promoters, which respond to heavy metals, glucocorticoids, and interferon stimulation, respectively.[139–141] Since the DNA construct used for transgenesis integrates randomly into the mouse genome, it is subject to regulation by neighboring chromosomal elements. This issue can make it difficult to achieve complete temporally and/or spatially restricted expression. In general, using these techniques, mouse geneticists can direct expression of a gene in a whole animal, or in a selected tissue, or at select times. Transgenic models are useful tools for biological discovery and recapitulating human disease in an experimental in vivo system. For example, the *BCR-abl* gene of the Philadelphia chromosome was first definitively shown to cause chronic myelogenous leukemia using a transgenic model.[142] However, since this approach entails expressing genes in nonphysiological contexts and with synthetic promoters, a biologist must examine critically whether each model faithfully represents the biological function of the transgene.

3.01.2.6.2 Loss-of-function analysis: the knockout/reverse genetics

In the late 1980s, scientists developed techniques to selectively remove a gene from the genome of an embryonic stem cell (ES), a process known as gene targeting. This modified stem cell can then be inserted into a blastocyst, which develops into a mouse that is haploid for the inactivated gene. Inbreeding of these offspring results in a fraction of the pups containing a homozygous deletion in the locus of interest, allowing the scientist to study the effects of the disruption of this gene in a whole animal. In contrast to transgenesis, this procedure utilizes genomic fragments homologous to the gene of interest which is modified to contain a selectable marker and translational termination sequences. This DNA fragment is electroporated into ES cells, and through a process known as homologous recombination, the modified DNA replaces the endogenous sequence (**Figure 3b**). Thus, the chromosomal and promoter context of the gene is maintained, but no protein product can be produced. Variations to this approach include the incorporation of subtle mutations, as opposed to the complete ablation of the gene product. This approach can be used to assess the contribution of particular amino acids or domains toward the in vivo function of a protein. For example, to model Huntington's disease, which is associated with the extended repeats of glutamines in the relevant gene, Detloff and colleagues inserted a stretch of CAG repeats into the mouse Huntington's disease gene using this technique.[143] Furthermore, a 'knock-in' approach, which employs similar strategies, can be used to place a cDNA under

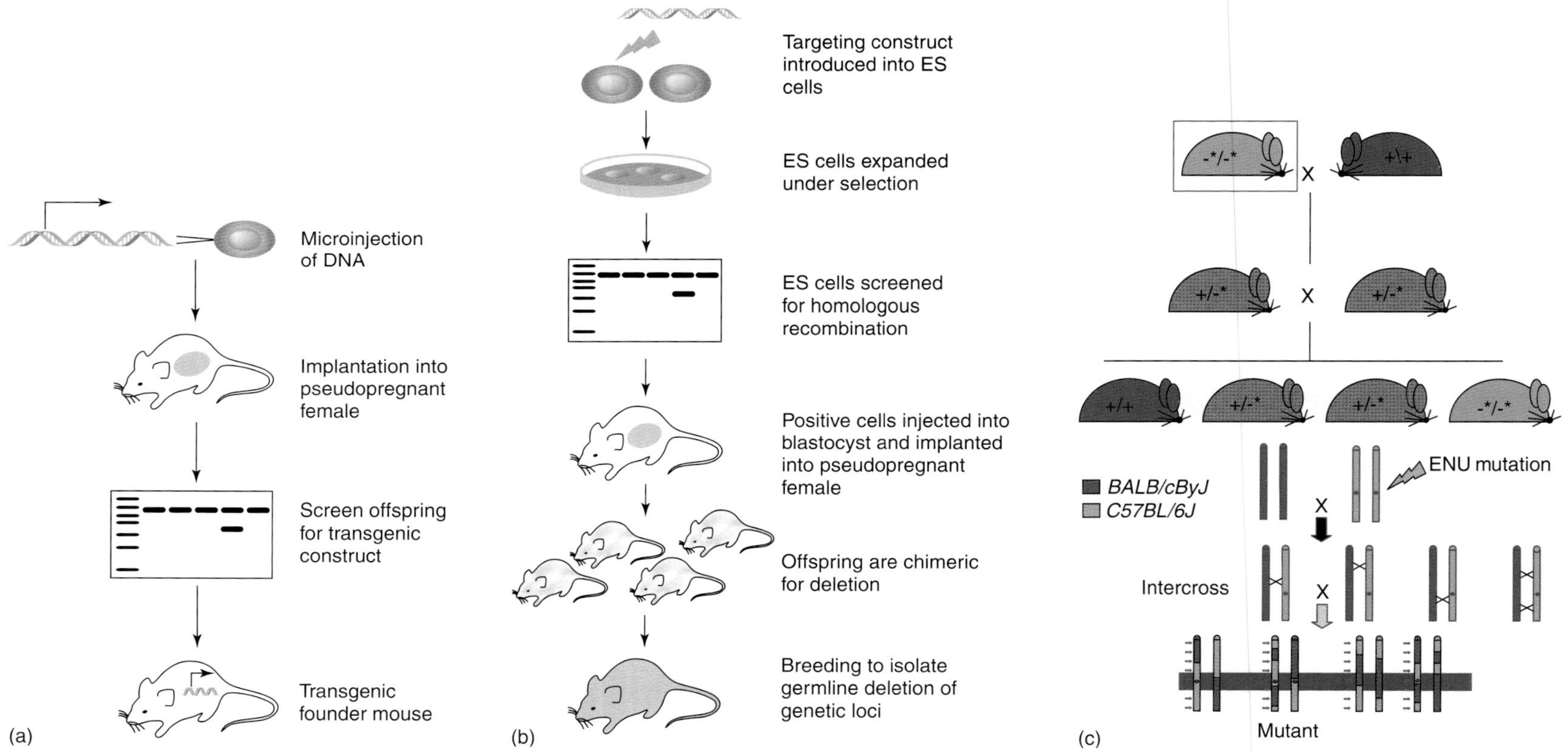

Figure 3 Mouse genetic systems. (a) Transgenic: typically, the DNA construct of interest is microinjected into the pronuclei of fertilized eggs. In a percentage of cases, the foreign DNA is randomly integrated in the genome of the egg, which is transplanted into a pseudopregnant female. The offspring determined to have received the transgene through genetic analysis (e.g., PCR or Southern blot), and positive animals are utilized to establish founder colonies for the transgenic lines. (b) Knockout: to generate a genetically null mutant, first a targeting vector containing regions homologous to the gene of interest, an incorporated mutation (e.g., stop codon) and a positive selectable marker must be constructed. This is then introduced into embryonic stem (ES) cells, which are subsequently propagated in media containing the corresponding chemical selection. ES cells in which the targeting vector has replaced the endogenous genetic loci through homologous recombination are identified through either Southern blot or PCR analysis, and subsequently injected into a mouse blastocyst. These are then transplanted into a pseudopregnant female, and chimeric offspring of the targeting vector are bred to isolate germline deletions of a genetic loci. (c) ENU mutagenesis. Top panel: a homozygous (ENU) mutant is crossed to a wild-type mouse from another genetic background (e.g., C57/BL6/J mutant crossed to BALB/CByJ wild-type), producing mice that are heterozygous for the mutation in a mixed background (F_1 progeny). These offspring are inbred and the mutation is mapped through segregation analysis of single nucleotide polymorphisms (SNPs) and the mutant phenotype. Bottom panel: a depiction of the same process tracking a possible chromosomal mutation.

the control of an endogenous promoter. This strategy, although more time-consuming and labor-intensive than transgenesis, can enable robust temporal and spatial expression of a transgene.

In some cases, the disruption of a gene results in embryonic lethality, making it impossible to study the function of this gene in the adult organism. To circumvent this event, geneticists employ a technique known as conditional knockouts. In this approach, two recombination sites are incorporated into a homologous targeting construct, specifically engineered not to effect the expression of the encoded protein until the expression of a recombinase protein. The mouse that is homozygous for this insert construct is crossed to a mouse that transgenically expresses a recombinase protein, such as Cre or Flp, specifically in the adult or in particular tissues.[144,145] Thus, the gene of interest remains intact during embryogenesis; however, the temporally or spatially restricted expression of a recombinase enzyme directs the excision of the sequences between the engineered recombination sites. This results in a loss of function of the gene in specified cells or at appropriate times, enabling the bypass of embryonic lethality.

These knockout techniques have led to assignment of function to hundreds of previously unannotated genes, and in some cases have lead to supporting evidence for target validation and subsequent drug finding activities. Zambrowicz *et al.* performed a retrospective study to determine how accurately the biology of established drug target knockout strains predicts the phenotype of the small-molecule drug.[3] This study revealed that of the 100 best-selling drugs only 43 distinct targets are represented, and of those 43 only 34 have been knocked out by homologous recombination. Surprisingly, 29 of the 34 have been informative in terms of describing the putative function and therapeutic potential of the target, of which many showed a direct correlation. Specifically, knockout of the hydrogen/potassium ATPase, the target for drugs such as Prilosec and Prevacid for the treatment of gastroesophageal reflux disease (GERD), exhibited the expected increase in gastric pH levels (pH = 6.9 in the knockout versus 3.19 in the wild-type). Similarly, knockout of estrogen receptors alpha and beta, targets of the highly marketed menopause and osteoporosis agonist drugs Evista and Premarin, have the expected sterility and absence of breast tissue development. In contrast, peroxisome proliferator-activated receptor gamma (PPAR-γ) agonists used to increase insulin sensitivity in the treatment of type II diabetes have the opposite effect of the observation that PPAR-γ heterozygous mice have an insulin-sensitive phenotype. And although finding a correlative phenotype between a mouse harboring a gene deletion and a small molecule that antagonizes the respective gene product is sometimes akin to finding a needle in a haystack (the correlation is not always obvious), companies such as Lexicon Genetics are taking it upon themselves to knock out the druggable genome one at a time to harvest more validated targets.

Although gene knockout and its derivative technologies present biologists with powerful tools to study gene function and model disease, these approaches also have certain limitations. These include phenomena referred to as redundancy and compensation. In certain instances, when a gene is disrupted, the organism will either already possess or will activate another gene to fulfill the function of the lost protein, thus making it complicated to decipher the true function of the protein of interest. Also, engineering these models requires significant investment in time, infrastructure, and other costs, thus limiting the ability to conduct these types of studies on a genome scale. Other technologies, including the use of siRNA transgenesis,[146] which mimics the effects of a gene knockout at tremendous time and cost savings, gene trap mutagenesis,[147] and proposed large-scale efforts,[148] may enable a genome-saturating analysis of gene inactivation.

3.01.2.6.3 Forward genetics: chemical mutagenesis in the mouse

Identification of the genetic basis of heritable mutant phenotypes in model organisms is a powerful methodology used to determine the function of genes involved in the specification of biological characteristics. For example, Mendelian inheritance of coat color in mice indicates that this trait is controlled by one or more genes. In most mouse strains, black or brown pigmentation predominates throughout most of the animal's coat with the exception of a thatch of yellow banded hairs found on the belly. Certain mouse mutant strains, known as 'lethal yellow' and 'viable yellow,' maintain an exclusively yellow fur. The genetic locus responsible for this phenotype was identified as Agouti, and its encoded protein was subsequently shown to cause hair follicles to synthesize a yellow pigment.[149]

In order to take advantage of this forward genetic approach, methods have been established for efficient mutagenesis and screening of mutated mice strains for unique phenotypes correlative with disease or suppressors of a studied disease state (**Figure 3c**). For instance, to accelerate the process of creating heritable mutations, geneticists typically use chemical agents such as ethylmethane sulfonate (EMS) or *N*-ethyl-*N*-nitrosourea (ENU) to induce genetic lesions in progenitor cells, such as those of the male germline. EMS and ENU induce point mutations in DNA, which can lead to several alterations of a gene's function. These include gain-of-function (hypermorphic), loss-of-function, reduction of function (hypomorphic), or alteration of function (neomorphic) effects. Male mice treated with these mutagenic reagents are used to inseminate female to produce colonies of pups carrying heritable mutations in their genomes. These animals are then screened for phenotypes of interest (e.g., high cholesterol, high blood pressure, or obesity). Once a mutant has been selected, a series of interbreedings are done to identify the chromosomal locus of the mutated gene.[150] For example, Wen and co-workers employed a genome-wide recessive screening strategy to identify the inositol (1,4,5) trisphosphate 3 kinase B (Itpkb) gene as an important player in T-cell activation,[151] which can be potentially exploited as a therapeutic target for immunosuppression. While a number of new technologies are facilitating the identification of genomics sequences mutated in these screens, identification of the causative mutation of the observed phenotype remains a challenging endeavor.

3.01.2.7 Understanding the Genetic Basis of Complex Traits: Quantitative Trait Loci (QTL) Analysis

While most of the approaches discussed thus far are extremely useful in identifying and characterizing the role of genes in vivo, they are of limited use in determining the genetic basis of phenotypes that are determined by the interaction of a number of genes (epistasis). Quantitative trait loci (QTL) mapping is a technique used to identify chromosomal regions that contribute to such multigenic phenotypes. This approach takes advantage of naturally occurring genetic variations (e.g., in strains of mice). Once a phenotype of interest is selected, these mice are then crossed to map the association of the chosen trait(s) to chromosomal markers throughout the genome. For example, Plum and co-workers performed QTL mapping on an inbred obese mouse strain (New Zealand) with a high incidence of diabetes.[152] They performed iterative matings with these mice and a lean inbred strain (SJL), and monitored levels of blood sugar in successive generations. Using statistical analysis, they were able to identify a region in mouse chromosome 4 as a major susceptibility locus for the development of hyperglycemia. QTLs provide information on genomic loci that contribute to particular phenotypes, but often these regions contain 500 or more genes. While the mouse genome sequencing project and single nucleotide polymorphism (SNP) data have facilitated efforts, understanding the molecular mechanisms underlying individual QTL remains a significant challenge.

3.01.3 Exploiting the Human Cell for Drug Targets

The past section has provided a brief overview of common techniques used to understand the function of genes in the context of a whole-model organism and how these techniques have been used to identify drug targets when applicable. While a number of other methodologies also exist, almost all can be classified into two paradigms: forward genetics, which starts with an observed phenotype and endeavors to elucidate contributing genes, and reverse genetics, which studies the phenotypic effect of gene disruptions. Each of these approaches has associated advantages and disadvantages; it is likely that a combination of techniques and emerging technologies will facilitate the understanding of gene function in vivo at the level of the genome. The impact of these technologies on human biology is immense; however, the level to which they can be superimposed on the study of the human organism as a whole is currently limited for obvious ethical and technical reasons. Fortunately, however, the tools pioneered in the aforementioned section on organismal genetics can be used in the context of the human cell, where they provide significant opportunities for target identification. Postgenomic tools have been developed to begin to annotate functionally the human genome, including functional cDNA, siRNA, focused small-molecule library screens, gene expression profiling, proteomic applications, etc. In parallel, advances in HTS, automation, cell biology, imaging, combinatorial libraries, and computation have allowed researchers to ask once again fundamental biological questions in the context of the human cell in new and exciting ways, while simultaneously identifying druggable targets for future generation therapeutics.

3.01.3.1 Functional Screening of Complementary Deoxyribonucleic Acid

Genetic dissection of signaling pathways relevant to human disease represents the next challenge to drug target hunters. The relatively recent ability to generate libraries of cDNAs from human tissue sources and to clone such entities into mammalian expression vectors has enabled this area. In addition, the ability to use viruses to introduce these libraries wholesale into cells, coupled with advances in human tissue culture systems and phenotypic readouts representative of cell proliferation, metabolism, pathogenic infection, inflammation, etc, have brought on the age of human cell genetics. In the late 1980s Brian Seed pioneered this expression screening approach in mammalian cells and set the foundation for numerous studies identifying genes relevant to most areas of signaling biology.[153] Examples include the cloning of *Toso*, a T-cell surface receptor that blocks FAS-induced apoptosis,[154] isolation of *JAK* and *STAT* genes involved in interferon signaling,[155] and *NFκB* pathway genes such as *IKK-gamma*.[156]

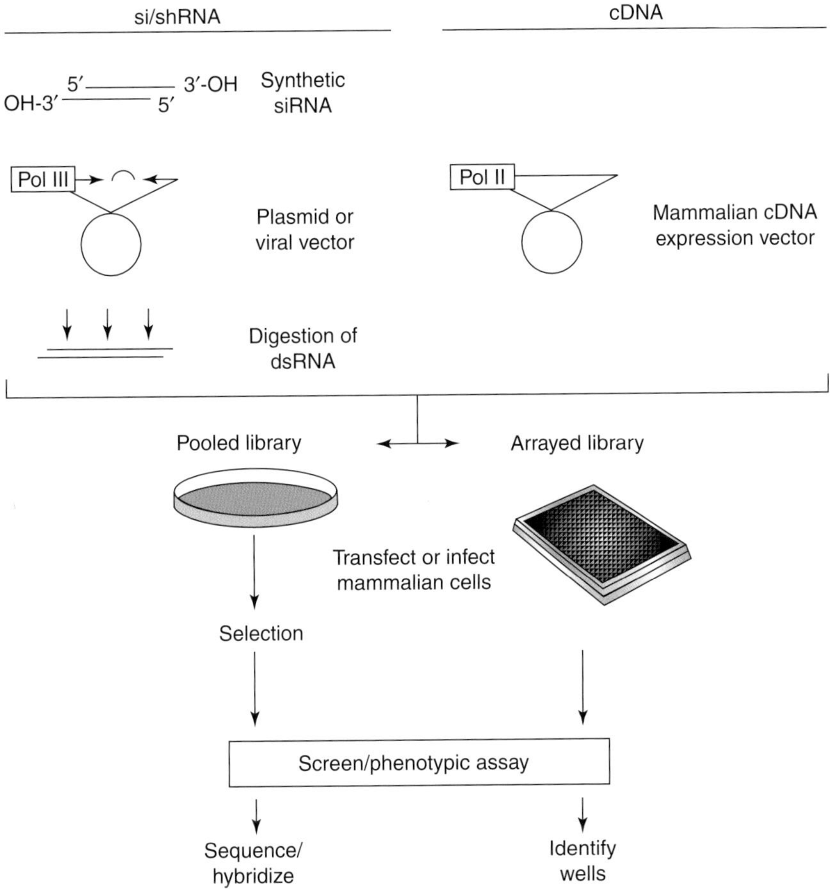

Figure 4 Schematic summary of large-scale screening platforms for mammalian cells. Gain-of-function screens utilizing expressible cDNAs (top right) or loss-of-function si/shRNA screens (top left) can by employed. For the latter, libraries may consist of chemically synthesized short interfering RNAs (synthetic siRNAs) or DNA with an RNA polymerase III promoter driving expression of shRNAs (plasmid or viral vector) or RNA fragments derived from a longer dsRNA (digestion of dsRNA). The libraries may be deployed either as defined pools or individually arrayed in a multiwell format (middle). After introduction into mammalian cells, a selection step is required for the pooled approach and hits are identified by sequencing or by amplification and hybridization to a bar code array as in **Figure 2**. In case of the arrayed approach, identification of the hits is based on well position within the plate.

These cell-based functional screens were performed by overexpression of pools of ectopic cDNAs, and identifying interesting phenotypes conferred by a cDNA that is missing or limited in a specific cell type of interest (gain-of-function screen) (**Figure 4**). This powerful functional genomics technology has been widely used to define the role of many genes by virtue of their overexpression phenotype in cells; however, this approach suffers certain drawbacks in terms of comprehensive scans for gene function. These limitations include: (1) the gene of interest may be represented once in a background of millions of genes in the pool, a situation akin to that of a needle in a haystack, thus putting an undue burden on the optimization of signal-to-noise ratio for the cellular reporter assay, which is an insurmountable task for most readouts; (2) library members are not normalized, thus favoring selection of the more abundant genes and potentially missing important but rare ones; (3) cDNA libraries derived from tissues contain only a fraction of genes in the genome; (4) many of the genes are not full length and thus may not act in their physiological manner; and (5) there is a requirement for laborious isolation of phenotypically altered clonal cell lines followed by subsequent retrieval and sequencing of the causal cDNA.

Fortunately, another approach has been developed to avoid the deficits of pooled library screens. The second approach addresses gain-of-function phenotypes in an array-based format. Full-length cDNAs, encoded by mammalian expression vectors, corresponding to known and predicted genes are arrayed in individual wells (for instance 384-well microtiter plates), and systematically introduced in parallel into reporter cells which are subsequently interrogated for

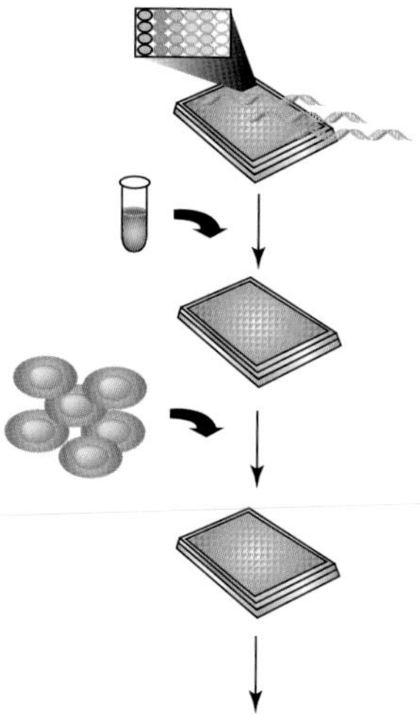

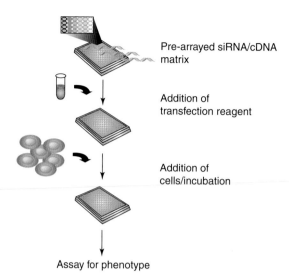

Figure 5 For high-throughput cellular genetic analysis, discrete cDNAs or siRNAs are arrayed in multiwell format. Transfection reagent is then added to the nucleic acids, and the transfection is completed by the addition of cells (retrotransfection). Effects of gain-of-function (cDNAs) and loss-of-function (siRNAs) of genes across the genome on a particular cellular phenotype are then monitored.

phenotypic effects of overexpression (**Figure 5**). The advantages of arrayed-based formats over pooled-library approaches are: (1) the reporter assays do not require a very high signal-to-noise ratio, since each gene is tested individually for function, and the background of cells incorporating 'other' genes is reduced to zero; (2) the cDNAs are normalized such that each well contains the same amount of cDNA; and (3) since the genes are arrayed, the identity of each gene in each well is known, so hit deconvolution can be done in silico in real time. Importantly, as a result of the completed genome sequencing effort, and efforts to clone all human genes, a number of large collections of expression-ready full-length cDNAs for mouse and human genes are available to greatly facilitate genome-wide functional analysis in mammalian cells. These include the Mammalian Genome Collection with more than 12 000 human and 11 000 mouse genes[157,158] and the Origene collection with 24 000 partially sequenced, unique human cDNAs in expression vectors.[159] These cDNAs are provided in mammalian expression vectors for high-level expression via transient transfection (the pooled approach heavily favors stable expression for the oft-lengthy selection process), or can be transferred to virus-based vectors, which offer the advantages of high gene transfer efficiency into diverse, difficult-to-transfect cell types, including primary cells.

These tools allow for functional profiling strategies[160] to analyze gene function on a genome-wide scale by cDNA overexpression in cellular models of human disease. The first study of systematic functional characterization of cDNAs identified the cellular localization of more than 100 fluorescent-tagged proteins of unknown function through transient transfection of mammalian cells.[161] Although this set of experiments did not shed light on the function of these genes per se, it did highlight the utility of the arrayed screening approach, as one could quickly determine gene localization patterns without subsequent deconvolution of gene indentity. Using an arrayed adenoviral expression library (cloned from a primary placenta cDNA library) containing approximately 13 000 cDNAs, Michiels *et al.* identified novel regulators of osteogenesis, metastasis, and angiogenesis. There are several reports on the interrogation of various signal transduction pathways on a genome-scale using an unbiased arrayed cDNA platform of approximately 20 000 characterized individual genes.[15,16,162] Chanda *et al.* reported a strategy to characterize previously unreported effectors of a AP-1 mediated growth and mitogenic response pathways by using a reporter vector containing AP-1 DNA binding sites upstream of a gene encoding luciferase.[15] Iourgenko *et al.* used a similar methodology to identify genes which activation expression of interleukin-8, which is associated with the development of diseases such as asthma, arthritis, and cancer. A protein with previously unknown function, TORC1, was identified as the founding member of a conserved family of coactivators of the cAMP response element binding protein (CREB).[16] Other reports utilize various reporter constructs in genome-wide functional analysis to analyze NFκB signaling pathways,[163] to identify p53 regulators,[164] or to identify new components of the Wnt signaling pathway, which is important for control of embryonic development.[165] In addition, several disease-relevant biological assays related to T-cell activation or diabetes were run by analyzing supernatants from cells that were transfected with approximately 8000 cDNAs representing secreted proteins.[166,167] Using this approach, Chen *et al.* discovered a new role for bone morphogenetic protein-9 (BMP-9) in glucose homeostasis that may be exploited for the treatment of diabetes,[166] and Fiscella *et al.* discovered a T-cell factor, TIP, that has a protective effect in a mouse acute graft-versus-host disease model.[167] The salient advantage of the arrayed approach is that one typically identifies far more signaling pathway members than from the pooled approach, thus increasing the chances of finding one that satisfies 'good' drug target criterion.

3.01.3.2 Ribonucleic Acid Interference and Functional Screens

Although arrayed cDNA overexpression screens have utility in functionally characterizing the genome, screening of arrayed RNAi libraries represents a highly complementary platform technology. Since RNAi inhibition of gene expression in the loose sense is analogous to inhibition of the same gene product by a small-molecule antagonist, the utility of RNAi for drug target identification and validation has caught the interest of the drug discovery community. As previously described, the phenomenon of RNAi, first discovered in the nematode *C. elegans*,[78] has become one of preferred methods for studying loss-of-function phenotypes in a high-throughput and unbiased manner. Even though genome-wide RNAi screens in model organisms have provided insight into the function of homologous human genes, the direct use of RNAi in mammalian tissue culture cells has now become a feasible approach for functional genomics studies which dissect complex biological processes. Advances in efficient design and delivery of siRNA to mammalian cells, as well as the availability of the human genome sequence, have unlocked the potential of applying RNAi to genome-wide loss-of-function screens in mammalian cells.

The availability of the human genome sequence enables design of specific inhibitory RNA sequences against all of the predicted human genes, and advances in screening methodologies and automation allow one to interrogate the effects of knocking down every gene, one by one, in parallel in cellular assays just as described for cDNAs. Administration of RNA interference in mammalian cells comes in a variety of forms and flavors. In mammalian cells, gene silencing can be achieved transiently by transfection of synthetic short double-stranded siRNA,[168,169] in vitro synthesized siRNA,[170] and plasmid-based short hairpin RNAs (shRNA).[171–174] The short hairpins are then processed by the ribonuclease III activity of the Dicer enzyme to generate effective 21–22 nt siRNAs.[175] The most widely used promoters to drive shRNA expression are RNA polymerase III dependent promoters such as U6 or H1, since they are active in all cells and efficiently direct the synthesis of small transcripts with well-defined ends. Viral vector-mediated expression of shRNAs facilitates delivery of RNAi molecules into difficult to transfect cell types and generates a sustained silencing effect beyond the 5-day transient effect of siRNAs.[176] Besides retroviral delivery of shRNAs[171,172] alternative viral delivery system such as adenoviral RNAi vectors[177] or lentiviral vectors[178] can provide additional benefits. Lentiviruses, in contrast to retroviruses, have the capacity to integrate into the host genome of nondividing cells[179,180] and enable stable expression of the delivered shRNAs. Adenoviral vectors are also used because of their ability to achieve long-term expression and high gene-transfer efficiency. By the same token, both types of RNAi reagents, including chemically synthesized siRNAs and vector-encoded shRNAs, have their own limitations that can impact the success of the screen. For instance, it has been shown that some siRNAs, and more often shRNAs, are prone to induce a nonspecific interferon-mediated response,[181,182] which can obviously complicate readouts, especially those related to viral infection and immunoresponses. In addition, excessive concentrations of RNAi reagents can lead to significant off-target effects.[183] The ability to control the concentration of exogenous siRNA in contrast to the noncontrollable effective dose of shRNA transcribed inside the cell minimizes this risk of nonspecific effects by using siRNAs.

Initial proof of concept for an arrayed synthetic siRNA library and the feasibility of large-scale RNAi screens in mammalian cells was first demonstrated by Aza-Blanc *et al.*[168] siRNA oligos targeting 510 human kinases were tested for their ability to modulate the induction of apoptosis by TRAIL (TNF-related apoptosis inducing ligand, a 'biologic' drug in clinical development for oncology applications) in a cell viability assay. The goal of the screen was to identify genes which when selectively knocked down could enhance sensitivity of the cells to TRAIL-mediated apoptosis, or alternatively suppress apoptosis. A variety of known and previously uncharacterized genes, including *DOBI* (downstream of bid) and *MIRSA*, were identified. In addition, a functional linkage between *MYC*, *WNT*, *JNK*, and *BMK1/ERK5* genes to the TRAIL-mediated response pathways was suggested. Another study described an HTS for genes involved in endocytosis, using high-content image analysis to detect inhibitory events caused by any member of an arrayed siRNA library targeting the entire set of kinases ($\sim$510). The identification of kinases involved in endocytosis was monitored in relation to the entry of two different viruses into host cells. Vesicular stomatitis virus (VSV) enters cells via clathrin-mediated endocytosis into early and late endosomes whereas SV40 uses the caveolae/raft-mediated pathway. An unexpectedly high number of kinases were shown to be involved in the process of endocytosis, and specific and coordinate regulation of the two pathways was demonstrated.[184] A genome-wide library comprising 48 746 synthetic siRNAs spotted in an arrayed format (two siRNAs per gene) has also been designed.[17]

An artificial intelligence algorithm which computationally predicts 21-nt siRNA sequences that have an optimal knockdown effect for a given gene was used, implementing a neuronal network based on the Stuttgart Neural Net Simulator. This collection was used to interrogate the response to hypoxia by induction/suppression of hypoxia inducible factor (HIF-1α) levels measured by the use of a luciferase reporter gene containing the hypoxia-response element. HIF-1α and ARNT (HIF-1β), which heterodimerizes with HIF-1α upon hypoxic conditions, scored amongst the top hits, thus validating this approach. These examples demonstrate the powerful nature of siRNA design when coupled with genome-wide screening.

Investigators have demonstrated that there are a substantial number of variations to the siRNA design and delivery theme which lend themselves to different applications. A methodology for the synthesis of RNAi reagents based upon endonuclease-derived siRNAs (esiRNAs) from cDNAs has been described.[170] Controlled RNase III digestion of long dsRNA molecules produced by in vitro transcription results in a pool of heterogenous sequences targeting the same transcript. The potency of the siRNAs is therefore increased while off-target effects of each individual molecule are diluted.[185] The generation of a genome-scale arrayed library of esiRNAs from a sequence-verified complementary DNA collection representing 15 497 human genes was reported by Kittler *et al.*[186] Genes required for cell division in HeLa cells were screened by using a high-throughput cell viability assay followed by a high-content videomicroscopy assay to quantify the frequency of cells in mitosis for arrest phenotypes. Thirty-seven genes were identified, including several splicing factors whose silencing generated mitotic spindle defects. Other groups reported alternative methods to generate targeted RNAi libraries derived directly from source cDNA. In this approach, digestion of a cDNA library with DNase I[187] or with restriction enzymes[188,189] to generate small DNA fragments that can be further modified and incorporated into a hairpin vector to produce shRNA. These libraries demonstrate biases toward highly expressed genes that are best represented in the cDNA library, and results of large-scale screens using these approach have yet to be reported. Finally, an example of a screen using arrayed DNA-based siRNA libraries was reported by Zheng *et al.*[173] A dual promoter system was developed, wherein sense and antisense strands were transcribed by two opposing promoters. HTS of this library identified novel components of the NFκB signaling pathways as well as previously recognized genes. Another example of a vector library directed against the family of deubiquitinating enzymes revealed *CYLD*, the familial cylindromatosis tumour suppressor gene, as a suppressor of NFκB, thus establishing a direct link between a tumour suppressor gene and NFκB signaling.[190] It was also shown that inhibition of *CYLD* enhanced protection from apoptosis and can be reversed by aspirin derivates, such as salicylate, which are established NFκB inhibitory molecules. This result led to the suggestion of a therapeutic strategy for treating cylindromatosis with existing drugs and provides an insight as to how genetic screening approaches can reveal potential drug targets and intervention strategies.

3.01.3.3 Large-Scale Virus Vector Based Libraries

One major criticism of cell-based screening in general is that the cell lines normally used to measure gene function are too far removed from physiologically relevant cell types such as those derived directly from primary tissue. This is partly due to the difficulty of obtaining a primary cell type with the longevity in culture required for engineering, if necessary, or for the screen itself, and due to the difficulty of delivering genetic material to these cell types. Most applications involve transient delivery of genetic material using lipid-based transfection reagents to which many primary cell types are resistant. To overcome this obstacle for more physiologically relevant screens, vector-based RNA interference libraries have been constructed in recombinant retroviral vectors.[171,172] Retroviruses afford the ability to introduce transgenes via stable infection to a broad range of primary tissue types (and cell lines) with high efficiency. For example, Berns *et al.* synthesized shRNA encoding retroviruses against 7914 human genes (NKi library) involved in major cellular pathways, including cell cycle, transcription regulation, stress signaling, signal transduction, and other important biological processes and human diseases.[171] Validation of the library was demonstrated by screening for modulators of p53-dependent proliferation arrest, resulting in successful identification of five new genes involved in the p53 tumor suppressor pathway and p53 itself. Concurrently, Paddison *et al.* reported an shRNA library covering more than 9600 human and 5500 mouse genes.[172] The library was designed with a retroviral vector pSHAG MAGIC that contains the hairpin sequence under the control of the U6 RNA polymerase III promoter and homologous recombination sites that allow easy shuttling between different vector backbones by a bacterial mating system. Paddison *et al.* have provided validation for the library by identifying components of the proteasome complex using a modified green fluorescent protein that is degraded by the proteasome, confirming 15 known proteasome subunits. Recently, a second-generation library comprising 62 000 constructs covering over 28 000 human genes, based on the library used by Paddison *et al.*, developed in collaboration with Hannon Elledge was made commercially available from Open Biosystems. Using this resource, Westbrook *et al.* were able to identify genes that suppress oncogenic transformation in a human mammary epithelial cells (HMECs).[191] Immortalized HMECs expressing naturally high levels of Myc exhibit a proliferation arrest in absence of extracellular matrix proteins. Knockdown of genes leading to anchorage-independent growth were assessed. Two established tumor suppressors, TGFBR2 and PTEN, were identified as was REST/NRSF, a transcriptional repressor of neuronal gene expression, a potential new candidate tumor suppressor gene. Kolfschoten *et al.* utilized the library of Bernards and colleagues in a screen for novel tumor-suppressor genes which, when silenced, substituted for the activity of oncogenic RAS.[192] This screen led to the identification of PITX1 as a promoter of anchorage-independent growth in fibroblasts.

As a result of the power of retroviral-based genetic screens, several groups have combined their efforts to expand this resource. The RNAi consortium (TRC), a collaborative effort among six research institutions and five international life sciences organizations, released a lentiviral shRNA library consisting of approximately 35 000 shRNA constructs targeting 5300 human and 2200 mouse genes. The first-generation library named MISSION TRC-Hs 1.0 (Sigma-Aldrich) or Expression Arrest TRC (Open Biosystems) is based on the self-inactivating lentiviral vector pLKO.1, wherein expression of the hairpin sequence is driven by the U6 promoter.[193] The effectiveness of this newly released lentiviral library has not been reported; however, the results of the aforementioned screens indicate the potential utility of this resource.

3.01.4 Conclusion

The completion of the human genomic sequence has revealed that the number of possible drug targets is represented by a finite yet significant number of genes. The sheer magnitude of possible starting points requires validation of potential drug targets via a multidisciplinary science involving genetics, molecular biology, chemistry, computation, and automation. The trick is somehow to winnow down the genome into consumable fractions that can be addressed efficiently. A framework for culling the list has been established, based on methods learned in genetic studies of lower eukaryotic organisms from yeast to mouse, and methods developed in the post genome era, including cell-based genetic screens. Ultimately, genome-wide analysis using functional genomics technologies will provide a comprehensive understanding of gene function which will, in turn, facilitate the discovery of novel targets for therapeutic intervention. Already, entire disease-relevant pathways have been mapped using a combination of these methodologies. The direct application of functional genomics to drug discovery has been demonstrated in a parallel RNA interference and small molecule screening in *Drosophila* cells that identified a small molecule inhibitor of the Aurora B kinase.[103] This approach provides a framework in which small molecules can be linked to cognate targets in mammalian cells, and further demonstrates the value and feasibility of utilizing both chemical biology and functional genomics toward the identification of a lead compound. In conclusion, by first placing a gene into the context of function within an organism or complex cellular process, then assigning some probability of identifying a useful modulatory small molecule tool or drug candidate, not only does the list become more manageable, but the probability of finding good drugs against good targets will rise. And instead of clogging the drug discovery pipeline with poor targets, or possibly worse, avoiding novel targets and thus missing opportunities, perhaps the chemical and functional genomic tools of the post genome era will manifest as an increased incidence of new blockbuster drugs for important therapeutic indications.

References

1. Bailey, D.; Zanders, E.; Dean, P. *Nat. Biotechnol.* **2001**, *19*, 207–209.
2. Hopkins, A. L.; Groom, C. R. *Nat. Rev. Drug Disc.* **2002**, *1*, 727–730.
3. Zambrowicz, B. P.; Sands, A. T. *Nat. Rev. Drug Disc.* **2003**, *2*, 38–51.
4. Krause, B. R.; Newton, R. S. *Atherosclerosis* **1995**, *117*, 237–244.
5. Druker, B. J.; Tamura, S.; Buchdunger, E.; Ohno, S.; Segal, G. M.; Fanning, S.; Zimmermann, J.; Lydon, N. B. *Nat. Med.* **1996**, *2*, 561–566.
6. Handschumacher, R. E.; Harding, M. W.; Rice, J.; Drugge, R. J.; Speicher, D. W. *Science* **1984**, *226*, 544–547.
7. Zhou, G.; Myers, R.; Li, Y.; Chen, Y.; Shen, X.; Fenyk-Melody, J.; Wu, M.; Ventre, J.; Doebber, T.; Fujii, N.; Musi, N. et al. *J. Clin. Invest.* **2001**, *108*, 1167–1174.
8. Ghuysen, J. M. *J. Gen. Microbiol.* **1977**, *101*, 13–33.
9. Lipinski, C. A.; Lombardo, F.; Dominy, B. W.; Feeney, P. J. *Adv. Drug Deliv. Rev.* **2001**, *46*, 3–26.
10. Fischer, P. M. *Curr. Med. Chem.* **2004**, *11*, 1563–1583.
11. Berlin, A. A.; Lukacs, N. W. *Am. J. Respir. Crit. Care Med.* **2005**, *171*, 35–39.
12. Veneri, D.; Franchini, M.; Bonora, E. *N. Engl. J. Med.* **2005**, *352*, 1049–1050.
13. Couzin, J. *Science* **2005**, *307*, 1711.
14. Su, A. I.; Wiltshire, T.; Batalov, S.; Lapp, H.; Ching, K. A.; Block, D.; Zhang, J.; Soden, R.; Hayakawa, M.; Kreiman, G. et al. *Proc. Natl. Acad. Sci. USA* **2004**, *101*, 6062–6067.
15. Chanda, S. K.; White, S.; Orth, A. P.; Reisdorph, R.; Miraglia, L.; Thomas, R. S.; DeJesus, P.; Mason, D. E.; Huang, Q.; Vega, R. et al. *Proc. Natl. Acad. Sci. USA* **2003**, *100*, 12153–12158.

16. Iourgenko, V.; Zhang, W.; Mickanin, C.; Daly, I.; Jiang, C.; Hexham, J. M.; Orth, A. P.; Miraglia, L.; Meltzer, J.; Garza, D. et al. *Proc. Natl. Acad. Sci. USA* **2003**, *100*, 12147–12152.
17. Huesken, D.; Lange, J.; Mickanin, C.; Weiler, J.; Asselbergs, F.; Warner, J.; Meloon, B.; Engel, S.; Rosenberg, A.; Cohen, D. et al. *Nat. Biotechnol.* **2005**, *8*, 995–1001.
18. Clemons, P. A. *Curr. Opin. Chem. Biol.* **2004**, *8*, 334–338.
19. Ghosh, S.; Karin, M. *Cell* **2002**, *109* (Suppl.), S81–S96.
20. Gomez-Lazaro, M.; Fernandez-Gomez, F. J.; Jordan, J. *J. Physiol. Biochem.* **2004**, *60*, 287–307.
21. Shaulian, E.; Karin, M. *Nat. Cell Biol.* **2002**, *4*, E131–E136.
22. Olson, M. F.; Marais, R. *Semin. Immunol.* **2000**, *12*, 63–73.
23. Edwards, I. R. *Drug Saf.* **2005**, *28*, 651–658.
24. Consortium, I. H. G. S. *Nature* **2004**, *431*, 931–945.
25. Manning, G.; Whyte, D. B.; Martinez, R.; Hunter, T.; Sudarsanam, S. *Science* **2002**, *298*, 1912–1934.
26. Johnson, S. A.; Hunter, T. *Nat. Methods* **2005**, *2*, 17–25.
27. Armstrong, S. A.; Staunton, J. E.; Silverman, L. B.; Pieters, R.; den Boer, M. L.; Minden, M. D.; Sallan, S. E.; Lander, E. S.; Golub, T. R.; Korsmeyer, S. J. *Nat. Genet.* **2002**, *30*, 41–47.
28. Armstrong, S. A.; Kung, A. L.; Mabon, M. E.; Silverman, L. B.; Stam, R. W.; Den Boer, M. L.; Pieters, R.; Kersey, J. H.; Sallan, S. E.; Fletcher, J. A. et al. *Cancer Cell* **2003**, *3*, 173–183.
29. Segal, E.; Friedman, N.; Kaminski, N.; Regev, A.; Koller, D. *Nat. Genet.* **2005**, *37* (Suppl.), S38–S45.
30. Foury, F. *Gene* **1997**, *195*, 1–10.
31. Hartwell, L. H. *Biosci. Rep.* **2002**, *22*, 373–394.
32. Crespo, J. L.; Hall, M. N. *Microbiol. Mol. Biol. Rev.* **2002**, *66*, 579–591.
33. Moorthamer, M.; Panchal, M.; Greenhalf, W.; Chaudhuri, B. *Biochem. Biophys. Res. Commun.* **1998**, *250*, 791–797.
34. Zhao, R.; Davey, M.; Hsu, Y. C.; Kaplanek, P.; Tong, A.; Parsons, A. B.; Krogan, N.; Cagney, G.; Mai, D.; Greenblatt, J.; Boone, C. et al. *Cell* **2005**, *120*, 715–727.
35. Goffeau, A.; Barrell, B. G.; Bussey, H.; Davis, R. W.; Dujon, B.; Feldmann, H.; Galibert, F.; Hoheisel, J. D.; Jacq, C.; Johnston, M. et al. *Science* **1996**, *274*, 546, 563–567.
36. Hughes, T. R.; Marton, M. J.; Jones, A. R.; Roberts, C. J.; Stoughton, R.; Armour, C. D.; Bennett, H. A.; Coffey, E.; Dai, H.; He, Y. D. et al. *Cell* **2000**, *102*, 109–126.
37. Ito, T.; Tashiro, K.; Muta, S.; Ozawa, R.; Chiba, T.; Nishizawa, M.; Yamamoto, K.; Kuhara, S.; Sakaki, Y. *Proc. Natl. Acad. Sci. USA* **2000**, *97*, 1143–1147.
38. Uetz, P.; Giot, L.; Cagney, G.; Mansfield, T. A.; Judson, R. S.; Knight, J. R.; Lockshon, D.; Narayan, V.; Srinivasan, M.; Pochart, P. et al. *Nature* **2000**, *403*, 623–627.
39. Huh, W. K.; Falvo, J. V.; Gerke, L. C.; Carroll, A. S.; Howson, R. W.; Weissman, J. S.; O'Shea, E. K. *Nature* **2003**, *425*, 686–691.
40. Shoemaker, D. D.; Lashkari, D. A.; Morris, D.; Mittmann, M.; Davis, R. W. *Nat. Genet.* **1996**, *14*, 450–456.
41. Winzeler, E. A.; Shoemaker, D. D.; Astromoff, A.; Liang, H.; Anderson, K.; Andre, B.; Bangham, R.; Benito, R.; Boeke, J. D.; Bussey, H. et al. *Science* **1999**, *285*, 901–906.
42. Birrell, G. W.; Giaever, G.; Chu, A. M.; Davis, R. W.; Brown, J. M. *Proc. Natl. Acad. Sci. USA* **2001**, *98*, 12608–12613.
43. Chang, M.; Bellaoui, M.; Boone, C.; Brown, G. W. *Proc. Natl. Acad. Sci. USA* **2002**, *99*, 16934–16939.
44. Huang, M. E.; Rio, A. G.; Nicolas, A.; Kolodner, R. D. *Proc. Natl. Acad. Sci. USA* **2003**, *100*, 11529–11534.
45. Steinmetz, L. M.; Scharfe, C.; Deutschbauer, A. M.; Mokranjac, D.; Herman, Z. S.; Jones, T.; Chu, A. M.; Giaever, G.; Prokisch, H.; Oefner, P. J. et al. *Nat. Genet.* **2002**, *31*, 400–404.
46. Willingham, S.; Outeiro, T. F.; DeVit, M. J.; Lindquist, S. L.; Muchowski, P. J. *Science* **2003**, *302*, 1769–1772.
47. Panavas, T.; Serviene, E.; Brasher, J.; Nagy, P. D. *Proc. Natl. Acad. Sci. USA* **2005**, *102*, 7326–7331.
48. Giaever, G.; Chu, A. M.; Ni, L.; Connelly, C.; Riles, L.; Veronneau, S.; Dow, S.; Lucau-Danila, A.; Anderson, K.; Andre, B. et al. *Nature* **2002**, *418*, 387–391.
49. Birrell, G. W.; Brown, J. A.; Wu, H. I.; Giaever, G.; Chu, A. M.; Davis, R. W.; Brown, J. M. *Proc. Natl. Acad. Sci. USA* **2002**, *99*, 8778–8783.
50. Tong, A. H.; Evangelista, M.; Parsons, A. B.; Xu, H.; Bader, G. D.; Page, N.; Robinson, M.; Raghibizadeh, S.; Hogue, C. W.; Bussey, H. et al. *Science* **2001**, *294*, 2364–2368.
51. Hartwell, L. H.; Szankasi, P.; Roberts, C. J.; Murray, A. W.; Friend, S. H. *Science* **1997**, *278*, 1064–1068.
52. Tong, A. H.; Lesage, G.; Bader, G. D.; Ding, H.; Xu, H.; Xin, X.; Young, J.; Berriz, G. F.; Brost, R. L.; Chang, M. et al. *Science* **2004**, *303*, 808–813.
53. Ooi, S. L.; Shoemaker, D. D.; Boeke, J. D. *Nat. Genet.* **2003**, *35*, 277–286.
54. Parsons, A. B.; Brost, R. L.; Ding, H.; Li, Z.; Zhang, C.; Sheikh, B.; Brown, G. W.; Kane, P. M.; Hughes, T. R.; Boone, C. *Nat. Biotechnol.* **2004**, *22*, 62–69.
55. Giaever, G.; Shoemaker, D. D.; Jones, T. W.; Liang, H.; Winzeler, E. A.; Astromoff, A.; Davis, R. W. *Nat. Genet.* **1999**, *21*, 278–283.
56. Giaever, G.; Flaherty, P.; Kumm, J.; Proctor, M.; Nislow, C.; Jaramillo, D. F.; Chu, A. M.; Jordan, M. I.; Arkin, A. P.; Davis, R. W. *Proc. Natl. Acad. Sci. USA* **2004**, *101*, 793–798.
57. Lum, P. Y.; Armour, C. D.; Stepaniants, S. B.; Cavet, G.; Wolf, M. K.; Butler, J. S.; Hinshaw, J. C.; Garnier, P.; Prestwich, G. D.; Leonardson, A. et al. *Cell* **2004**, *116*, 121–137.
58. Baetz, K.; McHardy, L.; Gable, K.; Tarling, T.; Reberioux, D.; Bryan, J.; Andersen, R. J.; Dunn, T.; Hieter, P.; Roberge, M. *Proc. Natl. Acad. Sci. USA* **2004**, *101*, 4525–4530.
59. Xie, M. W.; Jin, F.; Hwang, H.; Hwang, S.; Anand, V.; Duncan, M. C.; Huang, J. *Proc. Natl. Acad. Sci. USA* **2005**, *102*, 7215–7220.
60. King, K.; Dohlman, H. G.; Thorner, J.; Caron, M. G.; Lefkowitz, R. J. *Science* **1990**, *250*, 121–123.
61. Chambers, J. K.; Macdonald, L. E.; Sarau, H. M.; Ames, R. S.; Freeman, K.; Foley, J. J.; Zhu, Y.; McLaughlin, M. M.; Murdock, P.; McMillan, L. et al. *J. Biol. Chem.* **2000**, *275*, 10767–10771.
62. Brown, A. J.; Goldsworthy, S. M.; Barnes, A. A.; Eilert, M. M.; Tcheang, L.; Daniels, D.; Muir, A. I.; Wigglesworth, M. J.; Kinghorn, I.; Fraser, N. J. et al. *J. Biol. Chem.* **2003**, *278*, 11312–11319.
63. Cismowski, M. J.; Takesono, A.; Ma, C.; Lizano, J. S.; Xie, X.; Fuernkranz, H.; Lanier, S. M.; Duzic, E. *Nat. Biotechnol.* **1999**, *17*, 878–883.
64. Takesono, A.; Cismowski, M. J.; Ribas, C.; Bernard, M.; Chung, P.; Hazard, S., III,; Duzic, E.; Lanier, S. M. *J. Biol. Chem.* **1999**, *274*, 33202–33205.

65. Xu, Q.; Reed, J. C. *Mol. Cells* **1998**, *1*, 337–346.
66. Henthorn, D. C.; Jaxa-Chamiec, A. A.; Meldrum, E. *Biochem. Pharmacol.* **2002**, *63*, 1619–1628.
67. Becker, F.; Murthi, K.; Smith, C.; Come, J.; Costa-Roldan, N.; Kaufmann, C.; Hanke, U.; Degenhart, C.; Baumann, S.; Wallner, W. et al. *Chem. Biol.* **2004**, *11*, 211–223.
68. Brenner, S. *Biosci. Rep.* **2003**, *23*, 225–237.
69. Dent, J. A.; Davis, M. W.; Avery, L. *EMBO J.* **1997**, *16*, 5867–5879.
70. Hawasli, A. H.; Saifee, O.; Liu, C.; Nonet, M. L.; Crowder, C. M. *Genetics* **2004**, *168*, 831–843.
71. Davies, A. G.; Pierce-Shimomura, J. T.; Kim, H.; VanHoven, M. K.; Thiele, T. R.; Bonci, A.; Bargmann, C. I.; McIntire, S. L. *Cell* **2003**, *115*, 655–666.
72. Gaud, A.; Simon, J. M.; Witzel, T.; Carre-Pierrat, M.; Wermuth, C. G.; Segalat, L. *Neuromusc. Disord.* **2004**, *14*, 365–370.
73. Consortium C. e.g. *Science* **1998**, *282*, 2012–2018.
74. WormBase Home Page. http://www.wormbase.org/ (accessed April 2006).
75. Walhout, A. J.; Sordella, R.; Lu, X.; Hartley, J. L.; Temple, G. F.; Brasch, M. A.; Thierry-Mieg, N.; Vidal, M. *Science* **2000**, *287*, 116–122.
76. Davy, A.; Bello, P.; Thierry-Mieg, N.; Vaglio, P.; Hitti, J.; Doucette-Stamm, L.; Thierry-Mieg, D.; Reboul, J.; Boulton, S.; Walhout, A. J. et al. *EMBO Rep.* **2001**, *2*, 821–828.
77. The *C. elegans* Gene Knockout Consortium Home Page. http://celeganskoconsortium.omrf.org/ (accessed April 2006).
78. Fire, A.; Xu, S.; Montgomery, M. K.; Kostas, S. A.; Driver, S. E.; Mello, C. C. *Nature* **1998**, *391*, 806–811.
79. Hannon, G. J. *Nature* **2002**, *418*, 244–251.
80. Timmons, L.; Court, D. L.; Fire, A. *Gene* **2001**, *263*, 103–112.
81. Tabara, H.; Grishok, A.; Mello, C. C. *Science* **1998**, *282*, 430–431.
82. Fraser, A. G.; Kamath, R. S.; Zipperlen, P.; Martinez-Campos, M.; Sohrmann, M.; Ahringer, J. *Nature* **2000**, *408*, 325–330.
83. Kamath, R. S.; Fraser, A. G.; Dong, Y.; Poulin, G.; Durbin, R.; Gotta, M.; Kanapin, A.; Le Bot, N.; Moreno, S.; Sohrmann, M. et al. *Nature* **2003**, *421*, 231–237.
84. Simmer, F.; Moorman, C.; van der Linden, A. M.; Kuijk, E.; van den Berghe, P. V.; Kamath, R. S.; Fraser, A. G.; Ahringer, J.; Plasterk, R. H. *PLoS Biol.* **2003**, *1*, E12.
85. Ashrafi, K.; Chang, F. Y.; Watts, J. L.; Fraser, A. G.; Kamath, R. S.; Ahringer, J.; Ruvkun, G. *Nature* **2003**, *421*, 268–272.
86. Rubin, G. M.; Yandell, M. D.; Wortman, J. R.; Gabor Miklos, G. L.; Nelson, C. R.; Hariharan, I. K.; Fortini, M. E.; Li, P. W.; Apweiler, R.; Fleischmann, W. et al. *Science* **2000**, *287*, 2204–2215.
87. Nusslein-Volhard, C.; Wieschaus, E. *Nature* **1980**, *287*, 795–801.
88. Kazemi-Esfarjani, P.; Benzer, S. *Science* **2000**, *287*, 1837–1840.
89. Fernandez-Funez, P.; Nino-Rosales, M. L.; de Gouyon, B.; She, W. C.; Luchak, J. M.; Martinez, P.; Turiegano, E.; Benito, J.; Capovilla, M.; Skinner, P. J. et al. *Nature* **2000**, *408*, 101–106.
90. Fortini, M. E.; Simon, M. A.; Rubin, G. M. *Nature* **1992**, *355*, 559–561.
91. Karim, F. D.; Chang, H. C.; Therrien, M.; Wassarman, D. A.; Laverty, T.; Rubin, G. M. *Genetics* **1996**, *143*, 315–329.
92. Therrien, M.; Chang, H. C.; Solomon, N. M.; Karim, F. D.; Wassarman, D. A.; Rubin, G. M. *Cell* **1995**, *83*, 879–888.
93. Adams, M. D.; Celniker, S. E.; Holt, R. A.; Evans, C. A.; Gocayne, J. D.; Amanatides, P. G.; Scherer, S. E.; Li, P. W.; Hoskins, R. A.; Galle, R. F. et al. *Science* **2000**, *287*, 2185–2195.
94. Reiter, L. T.; Potocki, L.; Chien, S.; Gribskov, M.; Bier, E. *Genome Res.* **2001**, *11*, 1114–1125.
95. FlyBase. http://flybase.bio.indiana.edu/ (accessed April 2006).
96. Berkeley Drosophila Genome Project. http://www.fruitfly.org/ (accessed April 2006).
97. Spradling, A. C.; Stern, D.; Beaton, A.; Rhem, E. J.; Laverty, T.; Mozden, N.; Misra, S.; Rubin, G. M. *Genetics* **1999**, *153*, 135–177.
98. Furlong, E. E.; Andersen, E. C.; Null, B.; White, K. P.; Scott, M. P. *Science* **2001**, *293*, 1629–1633.
99. Boutros, M.; Kiger, A. A.; Armknecht, S.; Kerr, K.; Hild, M.; Koch, B.; Haas, S. A.; Consortium, H. F.; Paro, R.; Perrimon, N. *Science* **2004**, *303*, 832–835.
100. Kiger, A.A.; Baum, B.; Jones, S.; Jones, M.R.; Coulson, A.; Echeverri, C.; Perrimon, N. *J. Biol.* **2003**, *2*, 27.1–15.
101. Kim, Y. O.; Park, S. J.; Balaban, R. S.; Nirenberg, M.; Kim, Y. *Proc. Natl. Acad. Sci. USA* **2004**, *101*, 159–164.
102. DasGupta, R.; Kaykas, A.; Moon, R. T.; Perrimon, N. *Science* **2005**, *308*, 826–833.
103. Eggert, U. S.; Kiger, A. A.; Richter, C.; Perlman, Z. E.; Perrimon, N.; Mitchison, T. J.; Field, C. M. *PLoS Biol.* **2004**, *2*, E379.
104. Grunwald, D. J.; Eisen, J. S. *Nat. Rev. Genet.* **2002**, *3*, 717–724.
105. Haffter, P.; Granato, M.; Brand, M.; Mullins, M. C.; Hammerschmidt, M.; Kane, D. A.; Odenthal, J.; van Eeden, F. J.; Jiang, Y. J.; Heisenberg, C. et al. *Development* **1996**, *123*, 1–36.
106. Driever, W.; Solnica-Krezel, L.; Schier, A. F.; Neuhauss, S. C.; Malicki, J.; Stemple, D. L.; Stainier, D. Y.; Zwartkruis, F.; Abdelilah, S.; Rangini, Z. et al. *Development* **1996**, *123*, 37–46.
107. Welcome Trust Sanger Institute Home Page. http://www.sanger.ac.uk (accessed April 2006).
108. Ton, C.; Stamatiou, D.; Dzau, V. J.; Liew, C. C. *Biochem. Biophys. Res. Commun.* **2002**, *296*, 1134–1142.
109. Zebrafish Gene Collection. http://zgc.nci.nih.gov/ (accessed April 2006).
110. Geisler, R.; Rauch, G. J.; Baier, H.; van Bebber, F.; Bross, L.; Dekens, M. P.; Finger, K.; Fricke, C.; Gates, M. A.; Geiger, H. et al. *Nat. Genet.* **1999**, *23*, 86–89.
111. Hukriede, N. A.; Joly, L.; Tsang, M.; Miles, J.; Tellis, P.; Epstein, J. A.; Barbazuk, W. B.; Li, F. N.; Paw, B.; Postlethwait, J. H. et al. *Proc. Natl. Acad. Sci. USA* **1999**, *96*, 9745–9750.
112. Nasevicius, A.; Ekker, S. C. *Nat. Genet.* **2000**, *26*, 216–220.
113. Henikoff, S.; Till, B. J.; Comai, L. *Plant Physiol.* **2004**, *135*, 630–636.
114. Wienholds, E.; van Eeden, F.; Kosters, M.; Mudde, J.; Plasterk, R. H.; Cuppen, E. *Genome Res.* **2003**, *13*, 2700–2707.
115. Peterson, R. T.; Link, B. A.; Dowling, J. E.; Schreiber, S. L. *Proc. Natl. Acad. Sci. USA* **2000**, *97*, 12965–12969.
116. Donovan, A.; Brownlie, A.; Zhou, Y.; Shepard, J.; Pratt, S. J.; Moynihan, J.; Paw, B. H.; Drejer, A.; Barut, B.; Zapata, A. *Nature* **2000**, *403*, 776–781.
117. Montosi, G.; Donovan, A.; Totaro, A.; Garuti, C.; Pignatti, E.; Cassanelli, S.; Trenor, C. C.; Gasparini, P.; Andrews, N. C.; Pietrangelo, A. *J. Clin. Invest.* **2001**, *108*, 619–623.
118. Njajou, O. T.; Vaessen, N.; Joosse, M.; Berghuis, B.; van Dongen, J. W.; Breuning, M. H.; Snijders, P. J.; Rutten, W. P.; Sandkuijl, L. A.; Oostra, B. A. et al. *Nat. Genet.* **2001**, *28*, 213–214.

119. Liu, S.; Lu, W.; Obara, T.; Kuida, S.; Lehoczky, J.; Dewar, K.; Drummond, I. A.; Beier, D. R. *Development* **2002**, *129*, 5839–5846.
120. Hostetter, C. L.; Sullivan-Brown, J. L.; Burdine, R. D. *Dev. Dynam.* **2003**, *228*, 514–522.
121. Sun, Z.; Amsterdam, A.; Pazour, G. J.; Cole, D. G.; Miller, M. S.; Hopkins, N. *Development* **2004**, *131*, 4085–4093.
122. Patton, E. E.; Widlund, H. R.; Kutok, J. L.; Kopani, K. R.; Amatruda, J. F.; Murphey, R. D.; Berghmans, S.; Mayhall, E. A.; Traver, D.; Fletcher, C. D. et al. *Curr. Biol.* **2005**, *15*, 249–254.
123. Kalev-Zylinska, M. L.; Horsfield, J. A.; Flores, M. V.; Postlethwait, J. H.; Vitas, M. R.; Baas, A. M.; Crosier, P. S.; Crosier, K. E. *Development* **2002**, *129*, 2015–2030.
124. Onnebo, S. M.; Condron, M. M.; McPhee, D. O.; Lieschke, G. J.; Ward, A. C. *Exp. Hematol.* **2005**, *33*, 182–188.
125. Langenau, D. M.; Traver, D.; Ferrando, A. A.; Kutok, J. L.; Aster, J. C.; Kanki, J. P.; Lin, S.; Prochownik, E.; Trede, N. S.; Zon, L. I. et al. *Science* **2003**, *299*, 887–890.
126. Tomasiewicz, H. G.; Flaherty, D. B.; Soria, J. P.; Wood, J. G. *J. Neurosci. Res.* **2002**, *70*, 734–745.
127. Goldsmith, P.; Harris, W. A. *Semin. Cell Dev. Biol.* **2003**, *14*, 11–18.
128. Whitfield, T. T. *J. Neurobiol.* **2002**, *53*, 157–171.
129. Basson, C. T.; Bachinsky, D. R.; Lin, R. C.; Levi, T.; Elkins, J. A.; Soults, J.; Grayzel, D.; Kroumpouzou, E.; Traill, T. A.; Leblanc-Straceski, J. et al. *Nat. Genet.* **1997**, *15*, 30–35.
130. Garrity, D. M.; Childs, S.; Fishman, M. C. *Development* **2002**, *129*, 4635–4645.
131. Gerull, B.; Gramlich, M.; Atherton, J.; McNabb, M.; Trombitas, K.; Sasse-Klaassen, S.; Seidman, J. G.; Seidman, C.; Granzier, H.; Labeit, S. et al. *Nat. Genet.* **2002**, *30*, 201–204.
132. Xu, X.; Meiler, S. E.; Zhong, T. P.; Mohideen, M.; Crossley, D. A.; Burggren, W. W.; Fishman, M. C. *Nat. Genet.* **2002**, *30*, 205–209.
133. Curran, M. E.; Splawski, I.; Timothy, K. W.; Vincent, G. M.; Green, E. D.; Keating, M. T. *Cell* **1995**, *80*, 795–803.
134. Langheinrich, U.; Vacun, G.; Wagner, T. *Toxicol. Appl. Pharmacol.* **2003**, *193*, 370–382.
135. Peterson, R. T.; Shaw, S. Y.; Peterson, T. A.; Milan, D. J.; Zhong, T. P.; Schreiber, S. L.; MacRae, C. A.; Fishman, M. C. *Nat. Biotechnol.* **2004**, *22*, 595–599.
136. Bedell, M. A.; Largaespada, D. A.; Jenkins, N. A.; Copeland, N. G. *Genes Dev.* **1997**, *11*, 11–43.
137. Ristevski, S. *Mol. Biotechnol.* **2005**, *29*, 153–163.
138. Maas, A.; Dingjan, G. M.; Grosveld, F.; Hendriks, R. W. *J. Immunol.* **1999**, *162*, 6526–6533.
139. Peters, M.; Odenthal, M.; Schirmacher, P.; Blessing, M.; Fattori, E.; Ciliberto, G.; Meyer zum Buschenfelde, K. H.; Rose-John, S. *J. Immunol.* **1997**, *159*, 1474–1481.
140. Lee, F.; Hall, C. V.; Ringold, G. M.; Dobson, D. E.; Luh, J.; Jacob, P. E. *Nucleic Acids Res.* **1984**, *12*, 4191–4206.
141. Whyatt, L. M.; Duwel, A.; Smith, A. G.; Rathjen, P. D. *Mol. Cell Biol.* **1993**, *13*, 7971–7976.
142. Daley, G. Q.; Van Etten, R. A.; Baltimore, D. *Science* **1990**, *247*, 824–830.
143. Lin, C. H.; Tallaksen-Greene, S.; Chien, W. M.; Cearley, J. A.; Jackson, W. S.; Crouse, A. B.; Ren, S.; Li, X. J.; Albin, R. L.; Detloff, P. J. *Hum. Mol. Genet.* **2001**, *10*, 137–144.
144. Farley, F. W.; Soriano, P.; Steffen, L. S.; Dymecki, S. M. *Genesis* **2000**, *28*, 106–110.
145. Dymecki, S. M. Site-Specific Recombination in Cells and Mice. In *Gene Targeting: A Practical Approach*; Al, J., Ed.; Oxford University Press: New York, 2000, pp 37–100.
146. Tiscornia, G.; Singer, O.; Ikawa, M.; Verma, I. M. *Proc. Natl. Acad. Sci. USA* **2003**, *100*, 1844–1848.
147. Weber, F.; de Villiers, J.; Schaffner, W. *Cell* **1984**, *36*, 983–992.
148. Austin, C. P.; Battey, J. F.; Bradley, A.; Bucan, M.; Capecchi, M.; Collins, F. S.; Dove, W. F.; Duyk, G.; Dymecki, S.; Eppig, J. T. et al. *Nat. Genet.* **2004**, *36*, 921–924.
149. Barsh, G. S. *Trends Genet.* **1996**, *12*, 299–305.
150. O'Brien, T. P.; Frankel, W. N. *J. Physiol.* **2004**, *554*, 13–21.
151. Wen, B. G.; Pletcher, M. T.; Warashina, M.; Choe, S. H.; Ziaee, N.; Wiltshire, T.; Sauer, K.; Cooke, M. P. *Proc. Natl. Acad. Sci. USA* **2004**, *101*, 5604–5609.
152. Plum, L.; Giesen, K.; Kluge, R.; Junger, E.; Linnartz, K.; Schurmann, A.; Becker, W.; Joost, H. G. *Diabetologia* **2002**, *45*, 823–830.
153. Allen, J. M.; Seed, B. *Science* **1989**, *243*, 378–381.
154. Hitoshi, Y.; Lorens, J.; Kitada, S. I.; Fisher, J.; LaBarge, M.; Ring, H. Z.; Francke, U.; Reed, J. C.; Kinoshita, S.; Nolan, G. P. *Immunity* **1998**, *8*, 461–471.
155. Darnell, J. E., Jr.; Kerr, I. M.; Stark, G. R. *Science* **1994**, *264*, 1415–1421.
156. Yamaoka, S.; Courtois, G.; Bessia, C.; Whiteside, S. T.; Weil, R.; Agou, F.; Kirk, H. E.; Kay, R. J.; Israel, A. *Cell* **1998**, *93*, 1231–1240.
157. Strausberg, R. L.; Feingold, E. A.; Grouse, L. H.; Derge, J. G.; Klausner, R. D.; Collins, F. S.; Wagner, L.; Shenmen, C. M.; Schuler, G. D.; Altschul, S. F. et al. *Proc. Natl. Acad. Sci. USA* **2002**, *99*, 16899–16903.
158. Gerhard, D. S.; Wagner, L.; Feingold, E. A.; Shenmen, C. M.; Grouse, L. H.; Schuler, G.; Klein, S. L.; Old, S.; Rasooly, R.; Good, P. et al. *Genome Res.* **2004**, *14*, 2121–2127.
159. Origene. http://www.origene.com/cdna/ (accessed April 2006).
160. Orth, A. P.; Batalov, S.; Perrone, M.; Chanda, S. K. *Expert Opin. Ther. Targets* **2004**, *8*, 587–596.
161. Simpson, J. C.; Wellenreuther, R.; Poustka, A.; Pepperkok, R.; Wiemann, S. *EMBO Rep.* **2000**, *1*, 287–292.
162. Conkright, M. D.; Canettieri, G.; Screaton, R.; Guzman, E.; Miraglia, L.; Hogenesch, J. B.; Montminy, M. *Mol. Cells* **2003**, *12*, 413–423.
163. Matsuda, A.; Suzuki, Y.; Honda, G.; Muramatsu, S.; Matsuzaki, O.; Nagano, Y.; Doi, T.; Shimotohno, K.; Harada, T.; Nishida, E. et al. *Oncogene* **2003**, *22*, 3307–3318.
164. Huang, Q.; Raya, A.; DeJesus, P.; Chao, S. H.; Quon, K. C.; Caldwell, J. S.; Chanda, S. K.; Izpisua-Belmonte, J. C.; Schultz, P. G. *Proc. Natl. Acad. Sci. USA* **2004**, *101*, 3456–3461.
165. Liu, J.; Bang, A. G.; Kintner, C.; Orth, A. P.; Chanda, S. K.; Ding, S.; Schultz, P. G. *Proc. Natl. Acad. Sci. USA* **2005**, *102*, 1927–1932.
166. Chen, C.; Grzegorzewski, K. J.; Barash, S.; Zhao, Q.; Schneider, H.; Wang, Q.; Singh, M.; Pukac, L.; Bell, A. C.; Duan, R. et al. *Nat. Biotechnol.* **2003**, *21*, 294–301.
167. Fiscella, M.; Perry, J. W.; Teng, B.; Bloom, M.; Zhang, C.; Leung, K.; Pukac, L.; Florence, K.; Concepcion, A.; Liu, B. et al. *Nat. Biotechnol.* **2003**, *21*, 302–307.
168. Aza-Blanc, P.; Cooper, C. L.; Wagner, K.; Batalov, S.; Deveraux, Q. L.; Cooke, M. P. *Mol. Cells* **2003**, *12*, 627–637.
169. Hsieh, A. C.; Bo, R.; Manola, J.; Vazquez, F.; Bare, O.; Khvorova, A.; Scaringe, S.; Sellers, W. R. *Nucleic Acids Res.* **2004**, *32*, 893–901.

170. Yang, D.; Buchholz, F.; Huang, Z.; Goga, A.; Chen, C. Y.; Brodsky, F. M.; Bishop, J. M. *Proc. Natl. Acad. Sci. USA* **2002**, *99*, 9942–9947.
171. Berns, K.; Hijmans, E. M.; Mullenders, J.; Brummelkamp, T. R.; Velds, A.; Heimerikx, M.; Kerkhoven, R. M.; Madiredjo, M.; Nijkamp, W.; Weigelt, B. et al. *Nature* **2004**, *428*, 431–437.
172. Paddison, P. J.; Silva, J. M.; Conklin, D. S.; Schlabach, M.; Li, M.; Aruleba, S.; Balija, V.; O'Shaughnessy, A.; Gnoj, L.; Scobie, K. et al. *Nature* **2004**, *428*, 427–431.
173. Zheng, L.; Liu, J.; Batalov, S.; Zhou, D.; Orth, A.; Ding, S.; Schultz, P. G. *Proc. Natl. Acad. Sci. USA* **2004**, *101*, 135–140.
174. Brummelkamp, T. R.; Bernards, R.; Agami, R. *Science* **2002**, *296*, 550–553.
175. Bernstein, E.; Caudy, A. A.; Hammond, S. M.; Hannon, G. J. *Nature* **2001**, *409*, 363–366.
176. Holen, T.; Amarzguioui, M.; Wiiger, M. T.; Babaie, E.; Prydz, H. *Nucleic Acids Res.* **2002**, *30*, 1757–1766.
177. Arts, G. J.; Langemeijer, E.; Tissingh, R.; Ma, L.; Pavliska, H.; Dokic, K.; Dooijes, R.; Mesic, E.; Clasen, R.; Michiels, F. et al. *Genome Res.* **2003**, *13*, 2325–2332.
178. Rubinson, D. A.; Dillon, C. P.; Kwiatkowski, A. V.; Sievers, C.; Yang, L.; Kopinja, J.; Rooney, D. L.; Ihrig, M. M.; McManus, M. T.; Gertler, F. B. et al. *Nat. Genet.* **2003**, *33*, 401–406.
179. Naldini, L.; Blomer, U.; Gage, F. H.; Trono, D.; Verma, I. M. *Proc. Natl. Acad. Sci. USA* **1996**, *93*, 11382–11388.
180. Naldini, L.; Blomer, U.; Gallay, P.; Ory, D.; Mulligan, R.; Gage, F. H.; Verma, I. M.; Trono, D. *Science* **1996**, *272*, 263–267.
181. Bridge, A. J.; Pebernard, S.; Ducraux, A.; Nicoulaz, A. L.; Iggo, R. *Nat. Genet.* **2003**, *34*, 263–264.
182. Sledz, C. A.; Holko, M.; de Veer, M. J.; Silverman, R. H.; Williams, B. R. *Nat. Cell Biol.* **2003**, *5*, 834–839.
183. Persengiev, S. P.; Zhu, X.; Green, M. R. *RNA* **2004**, *10*, 12–18.
184. Pelkmans, L.; Fava, E.; Grabner, H.; Hannus, M.; Habermann, B.; Krausz, E.; Zerial, M. *Nature* **2005**, *436*, 78–86.
185. Jackson, A. L.; Bartz, S. R.; Schelter, J.; Kobayashi, S. V.; Burchard, J.; Mao, M.; Li, B.; Cavet, G.; Linsley, P. S. *Nat. Biotechnol.* **2003**, *21*, 635–637.
186. Kittler, R.; Putz, G.; Pelletier, L.; Poser, I.; Heninger, A. K.; Drechsel, D.; Fischer, S.; Konstantinova, I.; Habermann, B.; Grabner, H. et al. *Nature* **2004**, *432*, 1036–1040.
187. Shirane, D.; Sugao, K.; Namiki, S.; Tanabe, M.; Iino, M.; Hirose, K. *Nat. Genet.* **2004**, *36*, 190–196.
188. Sen, G.; Wehrman, T. S.; Myers, J. W.; Blau, H. M. *Nat. Genet.* **2004**, *36*, 183–189.
189. Luo, B.; Heard, A. D.; Lodish, H. F. *Proc. Natl. Acad. Sci. USA* **2004**, *101*, 5494–5499.
190. Brummelkamp, T. R.; Nijman, S. M.; Dirac, A. M.; Bernards, R. *Nature* **2003**, *424*, 797–801.
191. Westbrook, T. F.; Martin, E. S.; Schlabach, M. R.; Leng, Y.; Liang, A. C.; Feng, B.; Zhao, J. J.; Roberts, T. M.; Mandel, G.; Hannon, G. J. et al. *Cell* **2005**, *121*, 837–848.
192. Kolfschoten, I. G.; van Leeuwen, B.; Berns, K.; Mullenders, J.; Beijersbergen, R. L.; Bernards, R.; Voorhoeve, P. M.; Agami, R. *Cell* **2005**, *121*, 849–858.
193. Stewart, S. A.; Dykxhoorn, D. M.; Palliser, D.; Mizuno, H.; Yu, E. Y.; An, D. S.; Sabatini, D. M.; Chen, I. S.; Hahn, W. C.; Sharp, P. A. et al. *RNA* **2003**, *9*, 493–501.

Biographies

Jeremy S Caldwell, PhD, currently heads Lead and Target Discovery at GNF, overseeing multidisciplinary projects encompassing a range of therapeutic areas including metabolic and cardiovascular disease, infectious disease, arthritis/immune diseases, and oncology. He also directs technology efforts in high-throughtput small molecule and functional genomic screening and profiling. Prior to GNF, Jeremy was a staff scientist at Rigel Inc., where he developed high-throughtput FACS-based intracellular peptide phenotypic screening strategies. Jeremy received his PhD from Stanford University in Molecular Pharmacology in 1997 establishing the theoretical and technical framework for high-efficiency retroviral cDNA expression library screening. Prior to his doctoral work, Jeremy obtained his BS in Molecular and Cellular Biology from the University of California – Berkeley in 1992.

Sumit K Chanda, PhD, is a Group Leader in the Division of Genomics at the Genomics Institute for the Novartis Research Foundation in San Diego, CA. Chanda has trained at the Salk Institute in the field of Gene Therapy and Retrovirology, and received his PhD from Stanford University, where he performed early experimental collaborations to help establish the use of microarrays as a robust functional genomics tool in mammalian studies. At GNF, he has pioneered several technologies around genome-wide siRNA and cDNA functional analysis, including high-content imaging and high-throughput viral production. His group has established a fully automated functional genomics screening facility, and has run over 100 genome-wide functional analysis assays.

Jeff Irelan, PhD, is currently a Postdoctoral Associate in Oncology at the Genomics Institute of the Novartis Research Foundation, specializing in the application of functional genomics approaches to study oncogenic and tumor suppressor mechanisms in mammalian cell systems. Jeff gained expertise in the use of fungal model systems to study epigenetics and chromatin biology as a Damon Runyon–Walter Winchell Cancer Research Fund Postdoctoral Fellow at the University of California, Santa Barbara, and in receiving his BS from the University of Iowa and PhD from the University of Oregon.

Renate Koenig, PhD, is focusing on the application of functional genomics technologies to discover novel viral host–cell factors as potential drug targets as a Postdoctoral Associate at the Genomics Institute of the Novartis Research Foundation. As an award recipient of the Elizabeth Glaser Pediatric AIDS Foundation, Renate gained proficiency in studying HIV host–cell interactions at the Salk Institute for Biological Studies and in investigating pathogenic mechanism of SIV in monkey animal models at the Paul-Ehrlich-Institute, the Federal Agency for Sera and Vaccines in Germany. She received her PhD in Biochemistry at the University of Frankfurt.

Comprehensive Medicinal Chemistry II
ISBN (set): 0-08-044513-6

ISBN (Volume 3) 0-08-044516-0; pp. 1–25

3.02 Proteomics

H Voshol, S Hoving, and J van Oostrum, Novartis Institutes for BioMedical Research, Genome and Proteome Sciences, Basel, Switzerland

3.02.1	**Introduction: Proteins and Proteomics**	**28**
3.02.1.1	Proteins and Proteomics	28
3.02.1.2	Proteome Complexity	29
3.02.2	**Proteomics in Practice – Technology Overview**	**29**
3.02.2.1	Protein Extraction and Prefractionation	30
3.02.2.1.1	Cell and tissue extraction	30
3.02.2.1.2	Subcellular fractionation	31
3.02.2.1.3	Depletion and other chromatographic methods	31
3.02.2.2	Mass Spectrometric Identification of Proteins	32
3.02.2.2.1	Overview of the instrumentation	32
3.02.2.2.2	Peptide mass fingerprinting	33
3.02.2.2.3	Tandem mass spectrometry sequencing	33
3.02.2.3	The Protein-Based Approach – Two-Dimensional-Polyacrylamide Gel Electrophoresis	34
3.02.2.3.1	Practical considerations	34
3.02.2.3.2	Protein quantitation	35
3.02.2.3.3	Identification of proteins separated by two-dimensional-polyacrylamide gel electrophoresis	36
3.02.2.3.4	Pros and cons of two-dimensional-polyacrylamide gel electrophoresis in expression proteomics	36
3.02.2.4	The Peptide-Based Approach – Combined Liquid Chromatography/Mass Spectrometry Methods	37
3.02.2.4.1	Shotgun proteomics	37
3.02.2.4.2	Reducing peptide complexity	37
3.02.2.4.3	Quantitation and differential analysis based on peptides	38
3.02.2.5	Protein- versus Peptide-Based Approach?	39
3.02.2.6	Other Global Profiling Approaches	39
3.02.2.6.1	Imaging mass spectrometry	39
3.02.2.6.2	Surface-enhanced laser desorption ionization mass spectrometry (SELDI-MS)	40
3.02.2.6.3	Protein arrays	40
3.02.2.7	Beyond Protein Identification	41
3.02.2.7.1	Protein characterization	41
3.02.2.7.2	Protein–protein interactions – toward validation	41
3.02.3	**Applications of Proteomics in Drug Discovery and Development**	**42**
3.02.3.1	Target Discovery and Validation	42
3.02.3.2	Chemical Proteomics and Compound Profiling	44
3.02.3.3	Biomarkers and Molecular Diagnostics	45
3.02.4	**Perspective – From Proteomics to Systems Biology**	**46**
3.02.4.1	Discovery – Mapping the Proteome	46
3.02.4.2	Description – Scoring the Proteome	46
3.02.4.3	Integration and Prediction	46
References		**47**

3.02.1 Introduction: Proteins and Proteomics

3.02.1.1 Proteins and Proteomics

All living organisms, from the simplest prokaryotes to primates, share the same essential mechanisms and building blocks to store genetic information, transfer that to the next generation, and, most importantly, translate it into functional proteins. Within this universal framework nucleic acids, DNA and RNA, are primarily the carriers of information, whereas proteins are the functional entities in virtually all biological processes. The term 'protein' is a reflection of that key role in biological systems. It was the Dutch chemist Gerardus Johannes Mulder (1802–1880) who in 1838, on the suggestion of Jons Jakob Berzelius, introduced the term 'protein,' derived from the late Greek word *proteios*, meaning literally 'of the first rank.[1] In comparison to that, the word 'proteome' is very young, with its first appearance in the literature (PubMed) in 1995, e.g., in a paper about the first proteome, which was ever cataloged to a significant degree, that of *Mycoplasma genitalium*.[2] Co-author on this paper is Marc Wilkins, who is generally considered as the first person to use the term proteome in the 1994 Siena meeting.[3] At that meeting the proteome was defined as the PROTEin complement of the genOME, the set of proteins that are present in a cell, tissue, or organism at a certain timepoint. Consequently, proteomics is the analysis of the proteome. Patterson and Aebersold[3] provide a more pragmatic, operational definition of proteomics, saying that "proteomics is the systematic study of the many and diverse properties of proteins in a parallel manner with the aim of providing detailed descriptions of the structure, function and control of biological systems in health and disease." Truly a very ambitious outlook, which will require the effort of several generations of proteomicists. While studying protein expression, or expression proteomics,[4] is still a central aspect, gradually additional flavors have been added to the field. Large-scale efforts to determine protein structures are now referred to as structural proteomics or structural genomics, while interaction proteomics aims to elucidate protein function by uncovering networks of interacting proteins.[5,6] Another recently added, but already frequently used, family member is functional proteomics,[7] which encompasses the transition of the current mainly descriptive phase to a more applied stage, where expression data are placed in their biological context. In practice the term proteomics is often used to describe what used to be known as 'protein biochemistry.' Patterson and Aebersold[3] elegantly describe where they see the (fuzzy) borderlines between the two disciplines, the key feature being that proteomic approaches always address multiple proteins in a parallel fashion, while protein biochemistry addresses them one by one. In any case, the scientific literature has quickly embraced the term proteome and its relatives, resulting in an exponential rise in citations since its inception. Nevertheless, there is still a long way to go to reach the level of genomics, if one uses the numbers of references as a measure (**Figure 1**).

For the purpose of this chapter, the discussion about proteomics will be restricted to protein expression and biological function without addressing structural aspects in any detail. These will be discussed in other chapters in this volume (*see* 3.21 Protein Crystallography; 3.22 Bio-Nuclear Magnetic Resonance; 3.25 Structural Genomics). And, consistent with the scientific literature in the proteomics field, the term 'expression' will be used liberally to describe protein abundance, although protein translation would be a more correct designation as opposed to gene expression.

The fact that this chapter considers proteomics more or less in isolation is slightly at odds with the current research philosophy where increasingly the so-called postgenomic disciplines[8] are joining forces. In pharmaceutical research this concept of an integrated 'functional genomics' research group is already widespread.[9] A key role in this integration process is reserved for bioinformatics, the 'glue' that keeps the experimental disciplines together. Within the proteomics field itself, bioinformatics is omnipresent as well (*see* 3.15 Bioinformatics; 3.16 Gene and Protein Sequence

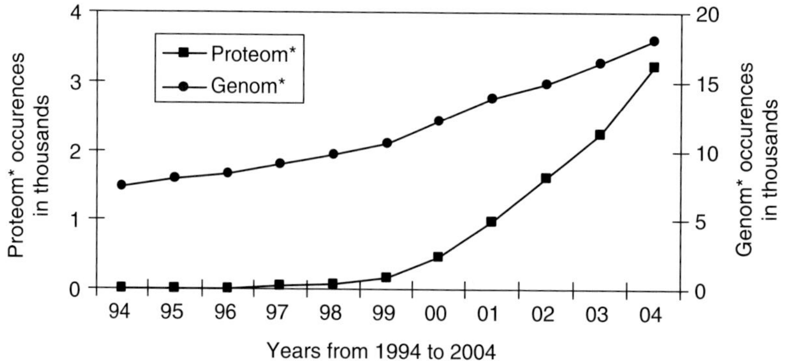

Figure 1 Numbers of PubMed citations using the search terms 'Genom*' or 'Proteom.*'

Databases). Collecting and analyzing protein sequences actually predates nucleic acid sequence analysis by almost two decades[10] and has culminated in the highly curated UniProt database.[124] Today every protein identification by mass spectrometry (*see* Section 3.02.2.2) depends on advanced bioinformatic tools to query the sequence databases with peptide masses and sequences.[3,7] In terms of bioinformatic aspects, an important difference between proteomics and the nucleic acid-based sciences is the enormous heterogeneity of protein data, which ranges from primary sequences through posttranslational modifications, three-dimensional structures to protein–protein interactions in functional networks.[5,6] The ultimate goal should be to capture all these diverse data sources and translate them into relevant biological and medical information.[10]

3.02.1.2 Proteome Complexity

Now that genome sequencing projects of many species, including those of human, mouse, and rat, have been completed or are in advanced stages, increasingly refined estimates of the numbers of encoded genes are becoming available. While those set the bottom line for the number of different proteins that can potentially be expressed, several layers of complexity are added before arriving at the level of the functional protein. The final number of potential protein species is closely correlated with the complexity of the organism itself (**Table 1**). In mammals, just the number of different polypeptide sequences that can be derived from one gene can easily be more than 10 by alternative splicing and single nucleotide polymorphisms that code for an amino acid change.[11,12] Adding to this the posttranslational modifications that can decorate the protein, such as phosphorylation and glycosylation, the potential extent of proteome complexity becomes apparent. For protein expression analysis, however, theoretical proteomes have limited practical significance, because only a fraction of those proteins will be present at any point in time in a specific cell, tissue, or organism. The largest 'experimental proteomes,' i.e., sets of identified proteins, are currently in the range of 2000–2500 unique polypeptides.[13,14] It should be noted that these large data sets are not proteomes in the strict sense, since they combine multiple cell types[13] or even lifecycle stages.[14] Because there is no reliable estimate of the actual numbers of protein species expressed in cells or tissues of higher organisms, currently it is difficult to assess the true depth of such analyses or their shortcomings in terms of coverage of the proteome. Expression analysis of mRNA (transcriptomics) is one possible proxy for proteome size, because a very significant part of the transcriptome can already be probed with microarrays (*see* 3.05 Microarrays). Extrapolation of recent transcriptome analyses[15,16] suggests that even a single cell contains several tens of thousands of different protein species.

On top of this complexity, the dynamic range of protein abundances within a proteome presents a huge challenge to proteomics technology. The concentrations of the most and the least abundant proteins are estimated to extend over 6–10 orders of magnitude in mammalian cells or plasma, respectively.[3,17,18] Consequently, even the most extensive proteomes identified to date show a clear bias toward abundant proteins.[19] As will become clear in the next sections in this chapter, the two factors, complexity and dynamic range, are the central aspects that have to be taken into account in practical applications of proteomics technologies.

3.02.2 Proteomics in Practice – Technology Overview

Albeit somewhat of a simplification, one could say that there are two main, quite different approaches to expression proteomics that, interestingly, employ some of the same key tools (**Figure 2**). On one hand there is the approach that builds on separation and quantitation at the protein level, converting proteins to peptides at the identification stage. Traditionally this approach is based on two-dimensional polyacrylamide gel electrophoresis (2D-PAGE) as the separation method. 2D-PAGE was developed 30 years ago,[20] but has only gained wider acceptance in the last decade,

Table 1 Illustration of the expected proteome complexity in different organisms, showing that it is determined by the number of variants for each protein rather than the number of proteins

Organism	*Genome size (Mbp)*	*Estimated number of genes*	*Estimated number of proteins per gene*
Escherichia coli	4.6	4 300	1–2
Saccharomyces cerevisiae	12	6 300	2–3
Drosophila melanogaster	100	14 000	5–10
Homo sapiens	3000	24 000	>10

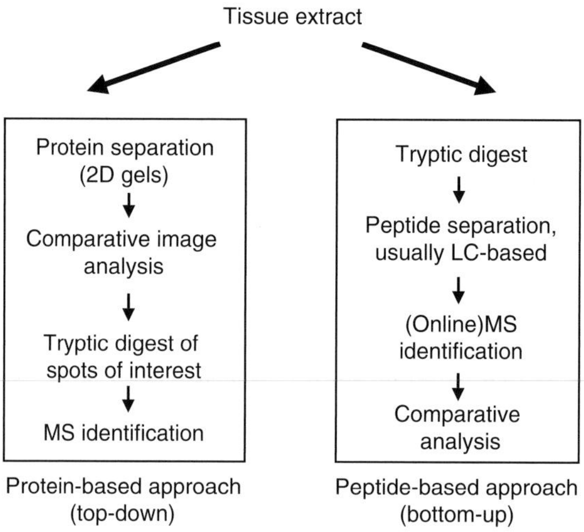

Figure 2 Schematic representation of the main workflows in expression proteomics. For further details, see text.

once its unrivaled protein separation power could be combined with efficient protein identification by mass spectrometry (MS).[3] The alternative approach, which has been even more closely linked to the recent spectacular advances in MS of proteins and peptides,[21,22] is based on – mostly proteolytic – conversion of proteomes to peptides as a first step. In-depth fractionation and identification take place at the peptide level. Quantitation, if performed, is the last step in the process (**Figure 2**). Recently, a more fashionable terminology has gained acceptance, namely top-down (for protein-based) and bottom-up (for peptide-based) approaches.

Before describing these two fundamentally different approaches in more detail, the elements that both workflows have in common will be discussed: sample preparation and prefractionation on one hand and protein identification by MS on the other.

3.02.2.1 Protein Extraction and Prefractionation

Although there is a certain level of arbitrariness in that definition, prefractionation can be distinguished from the core proteomic profiling in two ways. Firstly, in most cases it usually takes place at the level of the intact cellular substructures or at least of the native, possibly even still functional, proteins. Secondly, in an operational sense it yields protein mixtures which are not suitable for direct quantitation or identification by MS, but require further fractionation, generally using one of the two primary profiling methods (**Figure 2**). Some aspects of protein extraction/ sample preparation are discussed here at the same time since the method by which a cell or tissue extract is prepared imposes direct limitations on the subsequent prefractionation method and vice versa. At this point it is important to keep the almost contrary goals of the two steps in mind. By extracting as many proteins as possible, in a sense one aims to achieve maximum sample complexity. Subsequent fractionation steps should result not only in a reduction of that complexity, but also in a higher concentration of the remaining proteins to enable an increased analysis depth. A complicating factor is that each additional fractionation step will not only reduce the complexity but at the same time add to the variability of the sample, which might be a severe drawback for a comparative quantitative analysis of different samples. Ultimately, this crucial balance between complexity and reproducibility is different for every experiment, depending on the availability and the source of the sample, the proteomics technique applied, and the question to be answered.

3.02.2.1.1 Cell and tissue extraction

Rather than providing an extensive review of the decades of literature on protein extraction, this section will focus on the aspects that are most pertinent for a successful, reliable proteomics experiment. For more information on general methods for tissue homogenization and extraction, the reader is referred to the many excellent textbooks on this subject, e.g., in the *Methods in Enzymology* series.[23] A priori an important consideration is that proteomics, like any expression-profiling method, is most powerful when it is performed on a homogeneous sample. Only in that ideal case there will be no danger that significant changes in one cell type or subpopulation are 'diluted' by mixing with other, unaffected cells and disappear under the threshold of significance. Unfortunately, not even a sample consisting of single (cultured) cells is truly homogeneous, unless the cells are perfectly synchronized. Protein expression shows clear variations with the cell cycle, especially when it comes to regulatory modifications such as phosphorylation.[24] At the same time it is clear that synchronization is not an ideal solution to this problem, because it induces an array of proteomic alterations by itself, being achieved by a strong environmental signal such as starvation or compound treatment.[25] Still, in practice a sample consisting of single cells is the closest to the ideal situation one can get. Compared to that, proteomic profiling of complex tissues is a major challenge. Laser capture microdissection (LCM) provides a potential solution to this problem, by enabling the selective dissection of cells of a given type or morphology.[26] Early studies showed that protein extraction and fractionation, e.g., by 2D-PAGE, is feasible from fixed, stained, and laser-dissected material, although the dissection time needed to obtain enough tissue for a single 2D gel ranged from several hours to 2 days.[27] Thus, initially LCM appeared to have the most potential in combination with highly sensitive antibody-based proteomic methods,[28] as well as in gene expression profiling,[29,30] where amplification of the minute amounts of starting material can be used. For comprehensive protein profiling of most cell types, it did not appear to be a realistic option, especially if cells were to be selected singly as opposed to 'cutting' along tissue margins. The extent to which proteomic analysis of LCM-derived material has progressed in a short time is witnessed by recent experiments using more sophisticated detection methods, such as 2D-PAGE with fluorescently labeled samples[31] or liquid chromatography methods directly coupled with MS (LC/MS) analysis of isotope-labeled tryptic peptides.[32] In both cases differential analyses were carried out using 5000–10 000 laser-dissected cells, obtained in a timeframe of 30–60 min.Thus, LCM might play an important role in sample preparation for proteomics analyses in the near future, and turn protein profiling into a tool for molecular pathology of clinical samples.[33]

After dissection, irrespective of whether a native or denaturing extraction is performed, classical tools such as a glass-Teflon homogenizer, a French press, or a sonifier are used to disrupt cells or tissue. Some further points that have to be kept in mind, especially in the case of cultured cells, is the danger of contamination with noncellular proteins, e.g., from the culture medium. Finally, every care should be taken to avoid modifications in the proteome after 'freezing' the biological state by dissecting or harvesting the material. Here a completely denaturing sample preparation such as is done for 2D gel electrophoresis is convenient, because under those conditions virtually all enzymatic activity is stopped instantaneously. Additional aspects of sample preparation that are of particular relevance for each specific fractionation or separation method are discussed in the sections below.

3.02.2.1.2 Subcellular fractionation

Subcellular fractionation is one of the most attractive strategies for prefractionation, because it provides potentially valuable information on protein localization on top of reduced sample complexity. Most of the methodology for organelle proteomics[34] is textbook material, based on decades of experience in the enrichment of subcellular fractions for the isolation and characterization of proteins, often enzymes.[23] At least in theory, subcellular fractionation is the perfect prefractionation strategy, because it combines reduction of complexity and often dynamic range with enrichment of the proteins of interest. Unfortunately, the practice is not quite as ideal. Since no method can provide a complete separation between subcellular fractions, there will always be a more or less significant overlap between them, resulting in the distribution of many proteins across different fractions. A nice illustration of this aspect is the in-depth analysis of nuclear and cytoplasmic fractions of several human cell lines, which showed almost 60% overlap in protein identities between the two fractions.[35] Obviously, a significant part of this overlap reflects the biological situation, where proteins do move from one compartment to the other. However, a quantitative analysis of protein expression would require a large number of replicate analyses to discriminate between biological and artificial distribution. The more sophisticated the fractionation and hence the reduction in complexity, the more successful it is in terms of increasing proteome coverage, but the more difficult it will be to reproduce quantitatively. At this point in time there are not sufficient data available to decide where the optimal balance is.

3.02.2.1.3 Depletion and other chromatographic methods

Chromatography is as old as protein biochemistry and hence it is not surprising that there are almost as many chromatographic applications in proteomics as there are proteins. Since recent reviews[34,36,37] describe quite a few of those, the discussion will be limited here to chromatographic depletion and affinity chromatography, possibly the two most useful and widely used approaches in proteomics.

While some form of enrichment is definitely the most straightforward approach to delving deeper into the proteome, there are cases where the opposite, the depletion of highly abundant proteins, can be a great help. The main application for depletion strategies is the analysis of body fluids, particularly serum and plasma. As mentioned before, the plasma (and serum) proteome is probably the most extreme example in terms of dynamic range of protein concentrations, with two components, albumin and immunoglobulin G (IgG) comprising close to 75% of the total protein content.[38,39] Further 20 proteins add up to 99%, implying that all the other proteins, of which over 1000 have already been described,[40] occupy the remaining 1%. With such a relatively small number of proteins making up a vast majority of the total protein content, depletion of some of them is almost inevitable. While many depletion strategies were restricted to the two most abundant species, serum albumin and/or antibodies,[39] the latest development is the depletion of six of the most abundant plasma proteins (albumin, IgG, IgA, transferrin, haptoglobin, and antitrypsin) with one multiaffinity column.[41] The efficiency and reproducibility of this or even more elaborate depletion approaches will open new avenues in the analysis of the blood proteome.

Owing to their exquisite specificity, antibodies have been traditionally used as tools for the purification and detection of individual proteins. In proteomics, antibody-based affinity chromatography has been extended to include the purification of noncovalent protein complexes, by using antibodies to one or more of the complex components. Thus a highly enriched fraction is obtained, containing only a relatively small number of different proteins, which can usually be resolved efficiently by sodium dodecylsulfate PAGE (SDS-PAGE).

3.02.2.2 Mass Spectrometric Identification of Proteins

3.02.2.2.1 Overview of the instrumentation

In the decade since its definition, progress in proteomics has been tightly linked to the technological advances in MS-based identification of proteins and peptides. While these developments show no signs of slowing down, already at the current level protein identification rarely constitutes a bottleneck in proteomics. In this section, we will briefly review the MS instrumentation most commonly encountered in proteomics. For readers who require more detailed information, several excellent and recent reviews on mass spectrometric applications in proteomics are available.[3,21,22,42]

Despite the fact that we always refer to it as protein identification by MS, in reality the identities are derived from the analysis of peptide fragments, in most cases generated by digestion with trypsin. There are many reasons why it is more practical to implement protein identification at the peptide level, the most important of those being that with current mass spectrometers it is simply not (yet) possible to get the necessary mass and sequence information from intact proteins. Furthermore, the mass of an intact protein, even if it could be determined with the necessary accuracy, is not very informative, since it will hardly ever correspond to the predicted mass of the polypeptide chain, because of co- and posttranslational modifications that are essentially unpredictable.

Key to the application of MS in the analysis of proteins and peptides was the development of soft ionization methods, such as electrospray[43] and matrix-assisted laser desorption ionization (MALDI).[44] In this context soft ionization means that the (majority of) the molecules remain intact upon conversion into the ionized form and are thus amenable to subsequent analysis in the mass spectrometer. For a MALDI-MS analysis, the analyte – in proteomics the mixture of tryptic peptides from a gel piece or a high-performance liquid chromatography (HPLC) fraction – is mixed and co-crystallized with a large excess of a light-absorbing matrix, e.g., α-cyano-4-hydroxycinnamic acid, usually on to a metal plate (the target). In the mass spectrometer, pulses of laser light are fired at the co-crystals of analyte and matrix, resulting in the desorption and ionization of the analyte, after which the mass measurement can take place. The matrix assists in this process by absorbing laser energy and transferring it to the analyte.

In electrospray, ionization is accomplished by injecting and electrostatically dispersing the acidified peptide mixture into the mass spectrometer. The solvent evaporates from the fine droplets that are formed – the spray – leading to desorption of the analyte ions.[42] One difference between the two ionization methods is that MALDI generates mainly singly-charged ions, while electrospray produces multiply-charged species. Consequently, the mass-over-charge (m/z) values in a MALDI mass spectrum correspond to the actual molecular mass plus a proton, while in electrospray the charge state of the ion has to be determined before a molecular mass assignment can be made. Importantly, different subpopulations of peptides respond optimally to the different ionization methods, which means that with identical samples the results of a MALDI- or electrospray-MS experiment will be different and some peptides will be exclusively detected with only one of the two types of instruments.

Apart from their ionization mode, mass spectrometers are distinguished by the mass detectors they employ. Different mass detectors can be combined with the two ionization methods and different types can even be assembled in one instrument. One popular type is the time-of-flight (ToF) detector, which basically determines the m/z ratio of an analyte molecule by measuring the time it needs to ‘fly’ from one end of a vacuum tube to the other. Improvements to this basic technique to keep the ions focused during their flight through the detector have enabled accurate and highly resolving mass measurements using ToF instruments. The second common type is the quadrupole, which can filter ions of a certain m/z ratio by applying a specific electric potential. By varying that potential, the whole m/z range can be scanned and a mass spectrum of all ions generated.

Summarizing, it is important to stress that the choice of ionization method and mass detector depends strongly on the question, the coupling to the protein/peptide separation method, and the requirements in terms of throughput, sensitivity, resolution, and accuracy. Typically MALDI-based instruments with ToF detectors are the instruments of choice for the identification of proteins separated on 2D gels in a so-called 'offline' fashion: protein/peptide separation is decoupled from the mass spectrometric analysis. In contrast, electrospray-based instruments are often used online with HPLC separations: the effluent of the column flows directly into the mass spectrometer, where the analysis is carried out in real time. Two typical instrumental setups that represent the predominant protein identification methods in proteomics will be discussed in more detail in the two sections below.

3.02.2.2.2 **Peptide mass fingerprinting**

Depending on the instrumental setup, protein identification by MS employs mass or sequence information or a combination of the two. A classical application of MALDI-ToF instruments is the so-called peptide mass fingerprinting (PMF). In PMF the masses of the tryptic peptides derived from a protein are measured, and the list of peptides masses is submitted to a database, in which proteins are stored as a collection of peptides according to the cleavage specificity of trypsin. It is the combination of a large enough number of accurately measured peptide masses that allows unambiguous identification of proteins based only on this 'fingerprint,' without having any sequence information. The importance of mass accuracy was nicely illustrated by Clauser *et al.*[45] using a peptide mass fingerprint set consisting of 23 peptide masses derived from a 2D gel spot of bovine apolipoprotein A I (Apo A I). When interrogating the database (GenBank translated into proteins), assuming a mass accuracy of 2 Da, more than 150 000 different entries had at least one matching peptide mass and no fewer than 25 000 entries had five matches to the measured fingerprint. At a roughly 100-fold better mass accuracy of 10 p.p.m., at the limit of what is routinely achievable in high-throughput MALDI-ToF instruments, a query for five matching peptides retrieved only the three entries corresponding to the correct protein, bovine Apo A I. In other words, the 100-fold increase in accuracy decreased the number of hits by a factor of 10 000. Because the number of peptides is the second important factor in reducing the 'hit rate,' in practice PMF is not optimal for small proteins, simply because fewer peptides are available. The intact mass and isoelectric point (pI) of the parent protein can be used as additional criteria, e.g., the observed mass to decide whether it is the full-length protein or a fragment, or the observed pI to judge whether it is posttranslationally modified. Although most analysis software allows subtraction of already assigned peaks, the interpretation of peptide mass fingerprints will become increasingly difficult when dealing with mixtures of more than a few components. Therefore, PMF is a very suitable tool in combination with a highly resolving protein separation method such as a 2D gel, where it is also beneficial that MALDI is quite tolerant to contamination of the protein sample with salts or detergents.

3.02.2.2.3 **Tandem mass spectrometry sequencing**

While PMF is a powerful and readily accessible method, ultimately sequence information is still the most effective way to identify a protein unambiguously, because in principle the sequence of a single specific peptide can be sufficient. For peptide sequencing the configuration of the mass spectrometer has to be extended compared to the setup for mass measurement. An additional mass analyzer is added, hence the name tandem MS (or MS/MS) and a collision cell, in which the peptide ions are broken apart by colliding them with inert gas molecules. In the tandem MS configuration, the first mass analyzer is used to select a peptide at a certain *m/z* ratio for sequencing. This peptide is then fragmented in the collision cell and the fragments analyzed in the second stage of the instrument. How peptides fragment depends on the distribution of charges across the molecule,[42] but generally breakage at the amide bonds is one of the major cleavage patterns. In that case the resulting read-out consists of a series of fragments that differ in one amino acid, from which the parent sequence can be deduced. Because electrospray generates multiply charged ions, fragmentation is more extensive and hence more informative, but also more difficult to interpret than with MALDI methods. Database searches now combine sequence and mass information of parent peptide and fragments to identify the protein.[42] The typical, though certainly not exclusive, setup for peptide sequencing is a so-called LC/MS/MS experiment, in which peptides are separated on a reversed phase HPLC column, directly coupled to an electrospray tandem-MS instrument. In this setup, large amounts of sequence data can be generated in an essentially automated fashion. As an extension of PMF, the MALDI instrument can be equipped with two ToF detectors (MALDI-ToF-ToF[46]), allowing for tandem MS analysis of a number of selected peptides.

Finally, an additional option for peptide sequencing using MALDI is to take advantage of postsource decay (PSD). As the name suggests, PSD is the phenomenon that in MALDI a certain degree of fragmentation of peptide ions occurs after the initial ionization, due to a combination of factors that are not fully understood.[47] Since this fragmentation again occurs preferably at the amide bonds, sequence information can be derived from the fragments, without the need for a more complex instrumental setup with a collision cell. However, PSD spectra are generally complex because of many side reactions; hence true tandem MS, as described above, is the more widespread approach to sequencing.

3.02.2.3 The Protein-Based Approach – Two-Dimensional-Polyacrylamide Gel Electrophoresis

Thirty years after the establishment[20] of the basic method as it is used today, 2D-PAGE is still one of the core methodologies in proteomics. The high resolution of 2D-PAGE is a result of the orthogonal separation (**Figure 3**) along two largely independent dimensions: first by charge using isoelectric focusing (IEF) and subsequently by size with SDS-PAGE. In the first dimension, proteins are placed in a pH gradient in which they will migrate upon the application of an electrical field until they reach a stationary position where their net charge is zero: the isoelectric point or pI. To get an electrophoretic separation, which correlates well with size (molecular weight) in the second dimension, the effects of charge and shape of the protein molecules must be canceled out. This is accomplished by the addition of the anionic detergent SDS that binds tightly to proteins, thus masking their native charges and resulting in the formation of anionic micelles with a constant net charge per mass unit. In addition, variations in the shape of the protein molecules, which would lead to differential migration in the polyacrylamide gel, are reduced to a minimum because the interaction with SDS and the reduction of disulfide bonds cause the unfolding of the proteins. Recent reviews[48,49] provide a detailed overview of current and past 2D-PAGE methodology. Here the discussion will be limited to some of the key factors in successful application of 2D-PAGE for proteomic studies.

3.02.2.3.1 Practical considerations

One inherent advantage of 2D-PAGE is that it is performed under denaturing conditions and hence the risk of proteomic changes during sample preparation, e.g., by remaining enzymatic activity in the sample, is very limited. A typical sample buffer for 2D-PAGE contains chaotropes, such as (thio)urea and nonionic or zwitterionic detergents like 3-[(3-cholamidopropyl)dimethylammonio]-1-propanesulfonate (CHAPS) to improve the solubilization of proteins and the transfer of proteins to the second-dimension gel.[50]

In the most basic application of 2D-PAGE all proteins are separated in a single wide-range gradient of pH 3–10. Using that approach the resolution is limited, resulting in the visualization of around 1500–2500 protein species on a 20×25 cm gel. There are two ways of increasing the resolution and thereby the depth of analysis. One is to use a customized setup with larger gels, in which up to 9000 proteins have been reported to be detectable.[51] This variant of 2D-PAGE has been restricted to a few labs, because the required equipment is not accessible through a major vendor. A more straightforward alternative is to use standard-sized gels, but in combination with commercially available

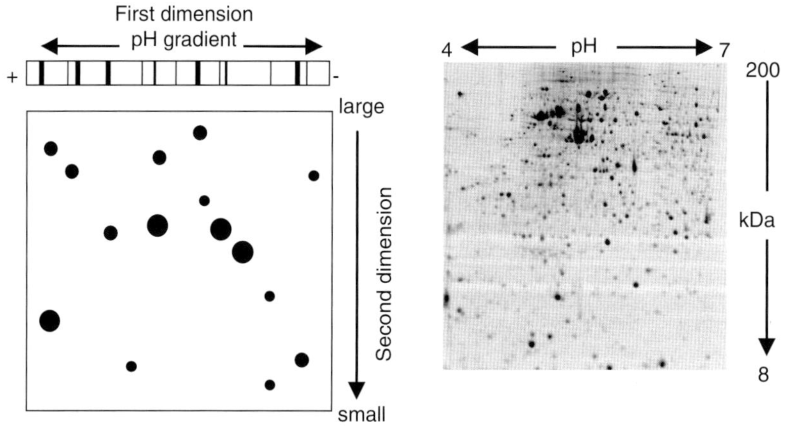

Figure 3 Two-dimensional electrophoresis. On the left the schematic representation of first and second dimension demonstrates the two orthogonal separation steps. In the first dimension proteins are applied to a pH gradient and migrate to their isoelectric point. Subsequently proteins are separated by size because large proteins move slowly through the second-dimension gel and smaller ones much faster. On the right, a real-life 2D gel of a cell lysate, in the pH range from 4 to 7.

narrow-range pH gradients.[52,53] In the most elaborate version, this allows a resolution of over 10 000 protein species from a single sample.[54] Moreover, these narrow-range gradients have a high loading capacity, which permits detection of proteins down to low ng levels corresponding to a few hundred copies per cell, even with postseparation stains.[55]

3.02.2.3.2 Protein quantitation

One of the attractive features of 2D-PAGE is that proteins on the gel can be immediately visualized and quantitated, provided that a staining (or, more accurately, a detection method) with a reasonable linearity and reproducibility is used. Other requirements to gel stains include high sensitivity, a wide dynamic range, and compatibility with downstream mass spectrometric identification. Postseparation staining with silver, Coomassie blue, or a fluorescent dye such as Sypro Ruby continues to be the most common way of detecting protein spots on gel. Silver staining is the most sensitive of those (**Table 2**), with a detection limit in the range of 1 ng protein. Unfortunately, it is essentially incompatible with a quantitative analysis because it is not an endpoint staining. That is a serious drawback, because it implies that incubation times and temperatures of the critical staining steps have been tightly controlled to reproduce the staining quantitatively. Thus for a comparative analysis of different samples the other staining methods are more suitable, especially if a significant throughput is required. Fluorescent dyes are probably the state of the art, because they combine acceptable sensitivity and reproducibility with good linearity and a wide dynamic range.[56] Coomassie blue, on the other hand, appears to have the least interference with tryptic digestion and MS.[57,58] Clearly, the choice of a postseparation stain is still a compromise, because none of them meets the desired specifications fully. The obvious alternative for poststaining is labeling of the proteins before separation. Radioactive metabolic labeling prior to separation allows even lower detection limits,[54] but is not feasible for many types of samples. Chemical labeling on the other hand is not without pitfalls, because it can affect many basic properties of the protein, including solubility, molecular weight, and charge. Moreover, the chemistry has to be robust to avoid adding complexity and to maintain predictable peptide masses for identification. The most sophisticated of the current detection methods is undoubtedly the difference gel electrophoresis (DIGE) approach. This method, originally proposed by Unlu *et al.*,[59] uses a family of (currently three) carefully matched fluorescent cyanine dyes, with different spectral properties. Hence, three samples labeled with different colors can be loaded simultaneously on one gel, thus alleviating the problem of gel-to-gel variation, one of the potential sources of error in 2D-PAGE.[60,61] In the most popular form of DIGE, minimal labeling of amino groups is used to label only a few percent of each protein, while the majority remains unaltered and available for mass spectrometric analysis.[62] Unfortunately, the labeled and unlabeled forms of a protein do not migrate to the exact same position on the gel, because the dye molecule adds approximately 0.5 kDa to the protein. Therefore, accurate excision of spots of interest still requires some form of postseparation staining, which can be further facilitated by spiking additional unlabeled material into the sample. In spite of the limited labeling, this DIGE application still offers detection levels in the range of silver staining (**Table 2**). Meanwhile a saturation labeling variant of DIGE based on derivatization of cysteine residues is available, which yields exceptional sensitivity, albeit at the cost of suboptimal resolution.[63]

In most cases protein spots of interest are selected by comparison of the 2D gel images of the different states under investigation (**Figures 3** and **4**). The specialized software that is employed for the image analysis of 2D gels has to perform two key operations: spot detection and matching of the spots across the individual gels. Owing to the fact that

Table 2 Protein detection on 2D gels

Method	*Detection limit (ng)*	*Dynamic range*	*Compatibility with MS*	*References*
Colloidal Coomassie blue	15	+/−	++	119,120
Sypro Ruby	5	++	+/−	56
Silver nitrate	1	−	+/−	121
Ammoniacal silver	0.5	−	−	122
DIGE, minimal labeling	2	++	NA	62
DIGE, saturation labeling	<1	++	−	63,123

Most common detection methods for 2D gels in proteome analysis. The detection limit is the approximate amount of a 50 kDa protein required to yield a protein spot which is detectable for image analysis software. −, poor; +/−, acceptable; +, good; ++, very good. NA, not applicable. For further details see text.

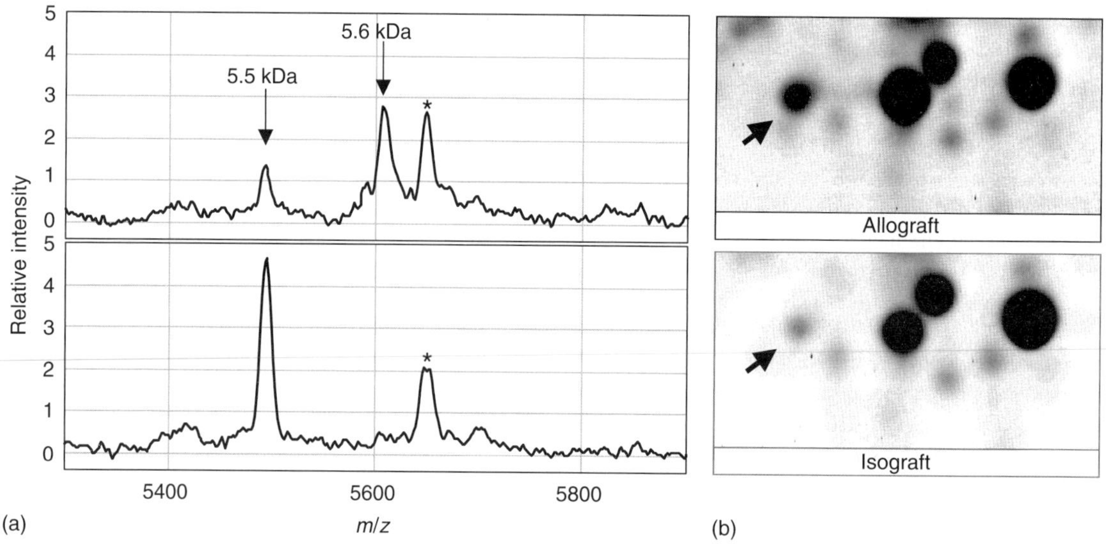

Figure 4 Illustration of differential proteome analysis based on (A) SELDI-MS data or (B) 2D-PAGE. Panel A shows SELDI (*see* Section 3.02.2.6.2) traces of serum of kidney-transplanted rats in an allogeneic setting (top) compared to the syngeneic control (bottom). The differentially regulated proteins at 5.5 and 5.6 kDa (arrows) could serve as indicators of the acute rejection process that is ongoing in the allotransplanted animals. The same samples are compared by 2D-PAGE in panel B. One of the spots visible in this segment of the gel is strongly upregulated in the allotransplanted animals (arrow). This study is described in detail by Voshol *et al.*[114]

preparation of 2D gels can not yet be automated, there is a certain degree of variability, with respect to both intensity of the spots and their precise location. Therefore, even in the most sophisticated analysis programs, a fair degree of user intervention is still required. A consequence of the variability of 2D gels is that small differences can not be reliably detected unless a sufficient number of technical and biological replicates is used. How high that number should be depends on the type of sample and the detection method used. In our hands, for the most homogeneous samples (e.g., cell lines), five replicates per sample allow reliable detection of twofold changes in spot intensity. With DIGE labeling, that number can be as low as 2–3, because of the possibility of including an internal standard in one of the three colors.[60,61] As discussed before, human tissue samples represent the other end of the variability spectrum, because of the many sources of individual variation. In those cases technical variability is usually not a limiting factor and thus reliable comparisons require a sufficiently large group size rather than a large number of replicates.

3.02.2.3.3 Identification of proteins separated by two-dimensional-polyacrylamide gel electrophoresis

After selection by image analysis, protein spots of interest are excised from the gel, digested at specific cleavage sites, usually with trypsin, and analyzed by MS. Spots from 2D gels are ideal samples for MALDI analysis, since they usually contain only one or, at most, a few detectable proteins owing to the high resolution of the separation. With modern MALDI instruments operating at 200 or more laser shots per second, the typical number of 1000 spectra required for a peptide mass fingerprint can be collected in a couple of seconds. In the authors' laboratory a MALDI-ToF-ToF instrument[46] is operated, which automatically obtains sequence information from 4 to 5 peaks of the fingerprint by subjecting them to MS/MS. This increases the success rate and the reliability of identification significantly, especially for low-molecular-weight proteins, which yield a limited number of peptide peaks for the fingerprint. The whole process from spot excision, through digestion, to MS can be automated and run at a throughput of hundreds to thousands of samples per day. Some laboratories even refrain from analysis of gel images altogether, instead using the MS data of every spot on the gel to match the protein patterns of the different samples.

3.02.2.3.4 Pros and cons of two-dimensional-polyacrylamide gel electrophoresis in expression proteomics

The better accessibility of the methodology and the increased need for a highly resolving protein separation method as a 'front end' in proteomic research have led to a much more widespread use of 2D-PAGE in the last years. Paradoxically, in spite of extensive use, the acceptance of 2D electrophoresis as a routine procedure has hardly improved. Instead, it seems to polarize proteomic researchers into two camps: the ones that consider it the ultimate method and the ones that are desperately looking for an alternative.[54]

Two-dimensional electrophoresis is not equally suitable for all classes of proteins. The main limitation of the method is the lack of coverage of molecules with extreme physicochemical properties, such as membrane proteins or highly charged polypeptides. These limitations are fundamental to separation at the protein level, because basic characteristics of proteins like charge, size, and solubility are so variable that a 'universal' protein separation technique does not exist. Further, although in our experience 2D-PAGE is not a particularly difficult method if one sticks to standardized protocols, practice shows that it is not at all trivial for a novice to establish. Clearly IEF is the trickiest part because of its sensitivity to sample composition, sample application, and even the equipment used.[64] While DIGE has alleviated some of the reproducibility issues, it will require improved equipment and further standardization of methods to enable a much wider audience to access the full potential of 2D-PAGE as a core proteomics method.

3.02.2.4 The Peptide-Based Approach – Combined Liquid Chromatography/Mass Spectrometry Methods

3.02.2.4.1 Shotgun proteomics

No matter how 'unfavorable' the properties of a protein in terms of hydrophobicity, size, or charge, there will almost always be a number of (tryptic) peptides that are amenable to HPLC separation and analysis by MS. In that sense LC/MS methods do not suffer from a bias toward 'soluble' polypeptides. For simplicity, we will mostly refer to the coupling of an HPLC and an MS instrument as LC/MS, although usually a tandem MS instrument is used. On top of the reduced bias, instrumental setups for LC/MS can be fully automated and run 24 h a day with relatively little need for operator intervention. Therefore, concomitant with the rapid technical developments, the peptide-based approach (**Figure 2**), in other words the tryptic conversion of whole proteomes to peptides and subsequent analysis at the peptide level, has evolved to a widely used strategy in proteomics. However, because trypsin will generate at least 30 peptides for a typical protein, sample complexity is even more of an issue for bottom-up approaches than for protein-based methods. So a 'whole proteome' can easily consist of several hundreds of thousands of peptides. To obtain the highest possible resolution the separation usually entails a 2D HPLC separation linked to an electrospray-based MS/MS instrument. Peptides are first separated into ion-exchange fractions, which are then injected one by one on a reversed-phase HPLC column coupled to the MS instrument. This can be done in an 'online' fashion where the two chromatographic phases are combined in one column or they can be physically separated (offline). A classical, online implementation of this approach is the so-called multidimensional protein identification technology (MudPIT), first applied to identify close to 1500 proteins of the yeast proteome in the laboratory of Yates.[19] Analogous to the genome sequencing, MudPIT type analyses are also referred to as 'shotgun proteomics,' because a more or less random sample of peptides is identified and mapped back to the parent proteins. A crucial difference is that shotgun genome sequencing is carried out to a level of 10x theoretical coverage to get enough data for the assembly,[65] whereas the proteome coverage of a MudPIT experiment is in fact negligible. Notwithstanding the impressive number of up to 2500 identified proteins,[11] it is important to realize that the vast majority of these are represented by one, or at most a few, peptides. In the MudPIT analysis of a low-complexity sample such as yeast, even making use of prefractionation, the average number of peptides per protein was only 4.[19] That would put the proteome coverage at less than 15% for the 1500 proteins identified and significantly below 5% for the theoretical yeast proteome, which counts over 6000 predicted proteins.

3.02.2.4.2 Reducing peptide complexity

So, in contrast to the genomics variant, with current technology shotgun proteomics is not capable of providing a map of complex proteomes, but a glimpse at best. The bottleneck is the time the mass spectrometer needs to record MS and tandem MS data. Even after the two chromatographic dimensions, the flow of peptides into the instrument is so rich that during each measurement a significant number of peptides pass by without being identified. Prefractionation can help to limit the number of peptides that ultimately end up in the MS instrument simultaneously. That can be done at the stage of tissue extraction or at the peptide level, e.g., by selecting a subset of the tryptic peptides before it is subjected to LC/MS. Certainly the most renowned example of the latter is the isotope-coded affinity tag (ICAT) method, originally designed by Gygi and coworkers.[66] ICAT elegantly combines affinity-based enrichment with an isotope label that can be used for quantitation. The ICAT reagent consists of three functional entities: (1) a thiol-reactive moiety that reacts with reduced Cys residues of proteins; (2) a biotin group that is used for the selection; and (3) a linker group that can be substituted with ^{12}C or ^{13}C, resulting in a mass difference of 9 Da between the 'heavy' and 'light' versions. The initial version of the reagent contained an 8 Da deuterium label,[66] but this was abandoned because the ^{2}H and ^{1}H peptide variants did not co-elute exactly on the HPLC. In addition it turned out that a mass difference of 8 Da was not always easy to interpret because of confusion with other modifications. Despite the fairly drastic, roughly 10-fold, selection of peptides by the ICAT method,[66] it has not emerged as a breakthrough in global proteomic-profiling methods yet. When dealing with complex samples, overlaps between runs are still only around 50%, even using comprehensive 2D LC/MS/MS, and the total number of proteins with detected peptides hits the ceiling at around 1000.[67] Furthermore, the anticipated 'unbiased' coverage of different protein classes, compared to 2D gels, appears to be wishful thinking rather than practical reality.[68]

The ultimate form of peptide prefractionation is represented by one of the combined fractional diagonal chromatography (COFRADIC) approaches developed by Gevaert *et al.*[69] The idea behind this method is that minimal complexity is combined with maximal information on presence or absence of proteins, when every protein is represented by exactly one peptide. To achieve this, N-terminal peptides are selected by a combination of chemical derivatization and chromatographic steps, followed by a (reverse-phase) LC/MS analysis. While in theory a compelling concept, with a theoretical 15–20-fold reduction in the number of peptides, in practice the analysis depth turned out to be rather modest when applied to subcellular fractions of human thrombocytes.[69] The same authors have published methods for selection of Met- and Cys-containing peptides,[70,71] but in none of these variants did the total numbers of identified proteins reach the level of the shotgun studies. Apparently the selected peptides can not be spread out enough to compensate for the loss of an LC dimension. Possibly the combination of one of these selection approaches with a multidimensional LC separation system would further increase the effectiveness of the COFRADIC technology.

As mentioned several times, one of the limitations of these peptide-based approaches is that the mass spectrometer has difficulties to deal with the flood of peptides that elute from the HPLC column. With the increased speed of the latest generation of MALDI mass spectrometers, a new solution to that problem has emerged. By mixing the eluate from the HPLC with matrix and spotting it on to MALDI targets, one can decouple the HPLC separation from the MS measurement and thus 'store' the HPLC run for later analysis. That has the advantage that one can first perform a survey MS scan of the entire fractionated sample to determine how the peptides are distributed over the spots to decide which of them to select for sequencing and from which spot to sequence them. The whole process can still not match the speed of LC/MS setups, but it is just a matter of time before lasers will get fast enough to do so. In any case both approaches have enough complementarity to co-exist, since the different ionization methods will cover different subsets of peptides.

3.02.2.4.3 Quantitation and differential analysis based on peptides

As already illustrated by ICAT, which combines a tag for affinity selection with an isotope label (*see* Section 3.02.2.4.2), peptide-based proteomics requires the introduction of some form of labeling with stable isotopes to enable quantification. MS itself is not quantitative, because of a number of factors, including suppression, varying detector response, and differential ionization yields for different substances. Therefore, no matter which ionization method is used, the intensity of the peptide peaks in the mass spectra are not reliable enough to allow direct quantification. Observed peak ratios for isotopic analogs, however, are highly accurate, because there are no chemical differences between the species, and they can be mixed and analyzed in the same experiment, much like the DIGE electrophoresis method (*see* Section 3.02.2.3.2).

Isotope labels can be introduced in a number of ways, the most obvious of them being metabolic labeling. However, in contrast to classical pulse labeling with radioactive isotopes, the substitution with the nonnatural isotope has to be near quantitative to be useful in mass spectrometric analysis. Otherwise a more or less significant part of the peptides would have the same mass as in the nonlabeled sample and the quantitation would be impossible. The earliest examples of this approach employed ^{15}N-enriched culture media to label and quantify bacterial proteins.[72] For mammalian cells the so-called stable isotope labeling by amino acids in cell culture (SILAC) approach has proven to be the most practical option.[73] In SILAC, ^{13}C-labeled essential amino acids are introduced in the medium, resulting in a complete replacement of unlabeled amino acids after five doubling times of the cells.

Chemical labeling, e.g., using the ICAT method, is less cumbersome than SILAC, but is limited to peptides with specific reactive groups, which can be both an up- and a downside. The most interesting extension of the repertoire of chemical isotope labels is the iTRAQ[74] reagent. This family of reagents consists of four different isobaric tags, which all release a different signature ion in MS/MS, thus allowing the multiplexing of up to four different samples.

A further, relatively simple way of introducing heavy isotopes is to use trypsin-mediated incorporation of ^{18}O atoms at the C-terminus of the newly formed peptides.[75] Originally, this method was used to distinguish the C-terminal peptide from the others, because there is no hydrolysis and hence no incorporation at the C-terminus. Under certain conditions, prolonged incubation of the peptides with trypsin in $H_2^{18}O$ can lead to the exchange of both oxygen atoms of the carboxyl-group, adding 4 Da to the peptide mass, compared to digestion in normal water.[75] This method can be extended to a differential mode where one sample is processed in $H_2^{16}O$ and one in $H_2^{18}O$, after which the heavy and light samples are mixed and analyzed together. Relative quantitation can be based on the intensities of the corresponding peptides which are 4 Da apart in the mass spectrum.

3.02.2.5 Protein- versus Peptide-Based Approach?

After this discussion of the two main approaches to performing proteomic profiling in an open manner, the question arises: which is the best method? An interesting observation in that respect is that authors describing LC/MS-based methods often seem to feel the need to emphasize that their results are superior to those obtained by gel-based approaches,[69–71,76] while the opposite appears to be quite rare. An indication perhaps that the gel-based approach is the de facto gold standard? The truth is that both approaches are so fundamentally different that it is futile to compare them. It is the application, the biological question, that defines what the best selection from the proteomic toolbox is. Generating a global catalog of protein identifications is certainly an area where peptide-based methods are superior. The miniaturization and large degree of automation of LC/MS are certainly not within the reach of 2D gels, where a certain level of manual handling and the associated variability will remain necessary for the foreseeable future. At the same time, the analysis of a large number of samples by LC/MS will require sequential runs, while a 2D gel experiment can easily accommodate 20–50 samples in parallel. Unfortunately, 2D gels have a clear bias when it comes to protein properties. Proteins with extreme isoelectric points or those with transmembrane domains are poorly accessible. The power of 2D gels is the inherent quantitation by labeling or staining of the separated proteins and the availability of information at the intact protein level, which allows the differentiation of isoforms resulting from co- or posttranslational modifications. In the most extreme case, a protein can be 'identified' by a single peptide using LC/MS without any guarantee that anything else than this single peptide was ever present. Information on protein isoforms differing in such crucial modification such as phosphorylation is lost at the peptide level, unless the modified peptides happen to show up in the analysis. Therefore the choice of the optimal profiling method can only be made with the application in mind, with respect to the starting material (numbers of samples, available amounts), the specific proteins of interest (subcellular localization, modifications), and the questions to be answered (catalog, differential analysis).

3.02.2.6 Other Global Profiling Approaches

2D-PAGE and LC/MS clearly constitute the mainstream approaches to expression proteomics, covering probably over 90% of the field between them. Nevertheless, several alternative global profiling methods are worth mentioning. The most interesting one is a hybrid using a regular one-dimensional (1D) SDS-PAGE gel as the first protein separation step, followed by excision of the separated protein bands, tryptic digestion, and LC/MS/MS analysis of the resulting peptide mixtures. Because a lane of a 1D gel is typically gridded into 30–60 bands, requiring the same number of LC runs for a full analysis, this approach is certainly not a 'shortcut.' In terms of numbers of identified proteins, it is a close second behind the shotgun approaches, with around 1800 reported IDs from a HEK293 cell line, albeit using prefractionation into cytoplasmic and nuclear fractions.[35] Choosing SDS-PAGE as the first dimension before performing tryptic digestion has the advantage that at least the information on the molecular weight of the parent protein is preserved and the peptides from one protein end up in the same LC run. Further, because of the superior solubilizing properties of SDS, the physicochemical properties of the proteins almost become irrelevant, a feature that predestines this approach for the analysis of membrane proteins.[77] Obviously, staining of the gel cannot be used for quantitation, since a single gel band can contain up to 50 different proteins. Therefore, similar labeling procedures are required, as discussed for the shotgun-type methods. Alternatively, extensive prefractionation can be applied before the SDS-PAGE separation, and comparative analysis carried out based on the 1D gel patterns.[78] This last approach is gaining interest, because of the ease with which a large number of 1D runs can be performed in parallel.

3.02.2.6.1 Imaging mass spectrometry

In the ideal situation a proteomics analysis of tissue samples would have single-cell resolution to monitor very specific changes in protein expression. To achieve that goal without having to disrupt the tissue entirely, methods have been developed to 'image' tissue with a (MALDI) mass spectrometer.[79,80] Frozen tissue sections with a thickness of 10–20 μm are immobilized on to a MALDI target and covered with matrix solution. The MALDI instrument now scans the whole section with the laser, recording mass spectra at every point. Thus it is possible to make true molecular images of most tissues. The resolution of the image, currently around 50 μm, is determined by the size of laser beam and the matrix crystals. Improvements in laser technology and matrix spotting, e.g., with acoustic spotters,[80] are likely to improve that down to the low μm range, enough for single cell imaging.[79] While the method with the currently used detectors is mainly suitable for the imaging of peptides and small proteins, larger proteins can also be visualized as long as they are relatively abundant.[81,82] The potential of MALDI imaging is that in theory with each laser shot a comprehensive picture is obtained of all molecules present at that position. For example, in brain tissue one could perform spatially resolved measurements of both neurotransmitters and their receptors. Until that is really possible, there are still important obstacles in terms of mass and image resolution, sensitivity, and background that need to be solved, but it is a safe assumption that mass spectrometric imaging is going to play a major role in the analysis of complex tissues.

3.02.2.6.2 Surface-enhanced laser desorption ionization mass spectrometry (SELDI-MS)

While the aim of proteomic studies generally is to obtain profiling data on as many protein species as possible, there are examples of successful molecular characterization of a disease state with a limited analysis depth. Such 'proteomic signatures' would primarily have diagnostic applications, because they highlight only a small subset of the expressed proteins. The proteomic technology that sparked interest in this concept of protein-based molecular diagnosis is surface-enhanced laser desorption ionization MS (SELDI-MS) on ProteinChips.[83,84] In this method, protein extracts, often derived from biological fluids, are fractionated on chromatographic surfaces and analyzed by MALDI-ToF MS. Unfortunately, the initial reported success in the SELDI-MS-based diagnosis of ovarian cancer[83] was later criticized extensively for flaws in the experimental design and consequently the method was all but discounted by many experts in the field.[85,86] However, if one is aware of the limitations of the method, especially the limited analysis depth, SELDI-MS can be a useful tool in proteomics. It is a rapid method, which has enough sensitivity to extract information even from laser capture microdissected samples.[87] Due to the detection principle used in MALDI, the method is most suited for monitoring peptides and small proteins up to 20 kDa, a similar range as MALDI imaging (*see* Section 3.02.2.6.1). Ultimately the most promising application of SELDI-MS might be the screening of large sets of clinical blood sample to search for patterns that distinguish clinically relevant subgroups of patients, which can then be subjected to focused in-depth proteomic analysis.

3.02.2.6.3 Protein arrays

Often referred to as 'protein chips,' different types of protein arrays are starting to emerge as another tool for which molecular diagnosis is the main application, which in this context can mean anything from determining cytokine levels in blood to monitoring phosphorylation states in signaling pathways.[88] Protein array is actually a description of a group of quite diverse tools, which have in common that the proteins they carry are immobilized on a solid phase, sometimes covalently, but more often by noncovalent forces. The function of the carrier is to allow rapid separation of bound from nonbound proteins, either by washing or by some form of selective detection at the surface of the solid phase. While the specific format of a protein array can vary from a classical planar 'chip' to a bead-type carrier, all are geared toward simultaneously assaying many analytes, a large number of samples, or a combination of the two. Functionally there are three main types: (1) the antibody array;[88] (2) the reverse array;[89,90] and (3) the protein array.[91] The antibody array is nothing else than a miniaturized sandwich-enzyme-linked immunosorbent assay (ELISA) with the inherent disadvantage that it requires two different antibodies against each analyte for optimal performance, one each for capture and detection. Because of the inherent selectivity of the capture step, the antibody array is an ideal tool for the analysis of biofluids, where low-level analytes must be detected among high levels of irrelevant proteins. Besides the antibodies themselves, the only limitation is that the samples have to be contained in a nondenaturing medium to allow binding to the antibodies, thereby excluding the detection of epitopes that are not exposed in native proteins.

Reverse (or reverse-phase) arrays are proteome rather than protein arrays, because they are based on a spotted cell or tissue lysate. The most basic form is the dot blot, in which drops of extract are applied to a membrane, e.g., of the type that is commonly used in Western blotting, or a coated glass slide.[89] A highly sophisticated form of reverse array has recently been introduced, combining an optimized chip surface with the exquisite sensitivity of the planar waveguide.[90] Reverse arrays are the most flexible tools for expression profiling, since even fully denatured extracts can be used and only a single antibody is required for each analyte. Among the most interesting applications is the mapping of protein phosphorylation sites to elucidate signaling pathways.[89]

Protein arrays, finally, are not expression-profiling tools in the strict sense, because they contain a large array of functional proteins, which can be applied for proteome-wide interaction studies.[91]

A fundamental difference with the other types of proteomic expression profiling is that all types of antibody-based arrays function as a closed system, probing for known molecular entities, much like DNA microarrays. Concomitant with the improved availability of suitable antibodies, the application of protein arrays is quickly gaining ground, because they combine excellent sensitivity[90] with the potential for a very high throughput, currently unattainable by any other proteomics approach.

3.02.2.7 Beyond Protein Identification

3.02.2.7.1 Protein characterization

Owing to the impressive technological developments in MS, it is safe to say that protein identification is no longer a bottleneck in proteomics. However, ideally a proteomics experiment should go beyond mere protein identification toward full characterization of differentially expressed proteins, in particular with respect to functionally important modifications such as phosphorylation. In this context it is not always obvious, especially to nonspecialists, that there is still a large gap between identification and protein characterization. In contrast to the situation in genomics, the goal of (high-throughput) protein identification is not to obtain a complete sequence, but just enough information to put a name on the molecule of interest. Moreover, when experimentally obtained peptide masses are searched in a protein database, a peptide with a (posttranslational) modification will not be retrieved, because the database only contains the primary amino acid sequences. A further technical issue is that negatively charged modifications such as phosphorylation are heavily underrepresented (suppressed) in mass spectrometric analysis, because that is geared toward positively charged ions, such as tryptic peptides which end in Lys or Arg.[92] The bottom line is that in routine operation only a limited part of the amino acid sequence of a protein is characterized and very few, if any, posttranslational modifications are detected. That is a major issue, certainly in mammalian systems where signal transduction in regulatory networks is often based on modification of the existing pool of proteins instead of altering the abundance. Affinity-based enrichment of modified peptides is a prerequisite to characterize them efficiently from a complex mixture. For example, for phosphorylation several such methods exist, including antibody-based purification and metal affinity chromatography.[92]

Here protein-based approaches, such as 2D-PAGE, have the advantage that they provide an inherent resolution of protein isoforms, as long as these are accompanied by a change in the charge or the molecular weight of the parent protein. Often one observes that in a 2D gel experiment protein spots containing different isoforms of the same protein change in opposite directions. An illustration is the characterization of stathmin isoforms, derived from a differential proteomics study in cells of paclitaxel (Taxol) treatment.[93] In human B-lymphoma cells paclitaxel induced the upregulation of a number of spots, which were all shown to contain the same protein, stathmin/OP-18. A combination of MS methods and different proteolytic digestions finally revealed that the upregulated spots comprised a range of different phosphorylated isoforms. This example shows that, regardless of the technological advances, characterization is still a slow process, in which usually individual proteins have to be tackled according to a tailor-made strategy. In this specific example, the fact that most isoforms of stathmin were already resolved by 2D-PAGE was of great value. Without that, one could have analyzed the phosphopeptides, but it would have been impossible to reconstruct the pattern of phosphoisoforms at the protein level. A useful method to generate isoform patterns of proteins was recently developed in our laboratories.[94] It consists of a blotting method preceded by a high-resolution IEF step. Even though the mass resolution is lost, this method provides a good alternative to 2D gels for the analysis of differentially charged protein isoforms, particularly when a higher throughput is required.

Much of the method development regarding posttranslational modifications has focused on phosphorylation,[92] glycosylation,[95] and, more recently, histone modifications.[96] Nevertheless, even for this limited subset of the over 100 posttranslational modifications, high-throughput strategies are just beginning to be developed.[96] Undoubtedly, the continuous improvements in resolution and accuracy in MS will also accelerate this area over the next years, but novel biochemical approaches for sample preparation and prefractionation are going to be equally crucial.

3.02.2.7.2 Protein–protein interactions – toward validation

So far, in this chapter we have extensively discussed the available technologies for generating protein expression data. However, the 10 years or so of proteomics have already shown that the availability of these protein profiles usually raises more questions than it solves. The reason is that the 'high-throughput biology' that is required to turn this data into meaningful information on disease mechanisms and biological pathways is almost completely lacking. One route to address that is to identify interacting partners for relevant proteins in order to establish their role within the network of cellular pathways. The method of choice to explore interacting partners of a protein is (co-)immunoprecipitation, often referred to as pull-down: the protein of interest is captured from a cell or tissue lysate with an antibody, with the expectation that at least some of the native interacting partners will remain bound during the process. Subsequently the immune complex can be dissociated and separated by electrophoresis and the proteins identified. To avoid interference between capturing antibody and interaction partners, the 'bait' protein is often 'tagged' with a synthetic antigen, which can be used for the pull-down. In yeast such 'fishing' experiments with bait proteins have been carried out on a proteome scale, resulting in interaction maps with thousands of members.[5,6] However, subsequent comparison of these data and other interaction data derived from two-hybrid experiments showed that there was very little overlap between the reported interaction maps. Apparently, these high-throughput interaction proteomics methods have a large potential for 'false positives,'[97] illustrating that it is only with extensive parallel validation that the generation of such large-scale interaction networks makes sense. Recent data from some of the same groups suggest that with a more directed approach, combining known entry points with cross-validation, it is possible to extend pathways with new members using this pull-down strategy.[98] Nevertheless it is important to note that only relatively stable protein complexes can be analyzed with this approach, while many biologically relevant interactions are more transient in nature. To capture those, chemical[99] or enzymatic crosslinking[100] might become an important complementary tool to classical pull-downs.

3.02.3 Applications of Proteomics in Drug Discovery and Development

From the original concept of proteomics as a description of protein expression as a function of perturbance in the biological system (**Figure 4**), the field has expanded to encompass many aspects of protein science. In the context of this chapter, the discussion of proteomics applications will be restricted to the key question in relation to medicinal chemistry: how can proteomics technologies contribute to the development of a successful drug? As a starting point, let's consider the different stages in drug development in more detail and define for which specific tasks proteomics experiments could be useful. Potentially the most important role could be at the very beginning of the drug development pipeline (**Figure 5**), at the stage that can be summarized as target discovery and validation. After a target for a therapeutic approach has been validated and one or several leads identified, compounds have to be profiled and optimized. There is a clear tendency in the pharmaceutical industry to complement the traditional tools, such as pharmacology and pathology, with expression-profiling data already early in compound development. That becomes even more pronounced at the final stages of compound development: preclinical and clinical testing (**Figure 5**), where the aim is to describe patient populations in as much detail as possible, e.g., to diagnose more effectively, and to predict and monitor therapeutic responses. These descriptions range from mapping of genetic polymorphisms (usually SNPs), also referred to as pharmaco-genetics/-genomics,[101] to profiling of mRNA and protein expression, activities often summarized as biomarker discovery.[102] For proteomics the focus on preclinical and clinical biomarkers is in fact a revival, since preclinical compound profiling and toxicology were among the first applications where protein profiling was systematically applied.[103] More recently the search for protein biomarkers, especially in cancer, has become one of the most debated issues in the field.[104]

In the next sections these areas will be illustrated with some examples from the authors' laboratory and the recent literature. It will become evident that the borderlines between those main applications are not rigid, because target and biomarker discovery are often different outcomes of the same process. Generally, in the target discovery phase the desired analysis depth is high and the number of samples small, while that balance shifts more toward higher throughput and inherently less comprehensive characterization during the biomarker phase.

3.02.3.1 Target Discovery and Validation

In pharmaceutical proteomics the questions related to drug targets can be distinguished in two main types: compound mode of action on the one hand and disease mechanisms on the other. Sometimes an interesting substance is discovered in a phenotypic screening process, as it is used in the search for cytotoxic or cytostatic drugs, e.g., in oncology or anti-infectives research. The output of such a screening is a compound or a small family of compounds with the desired phenotype, i.e., it kills tumor cells or bacteria, but the mode of action, in other words the relevant cellular target, is unknown. A typical case could be a natural product, either purified or even as component in an extract from traditional medicine. Without knowledge about the target, medicinal chemistry efforts to improve the properties of the compound are not very fruitful. There are several ways of addressing this question. One is the chemical proteomics

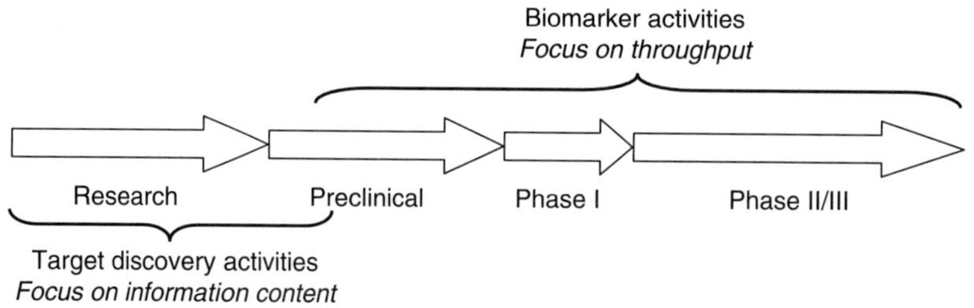

Figure 5 Illustration of the different proteomics activities in relation to the drug discovery process.

approach: the compound itself is used as an affinity bait to capture the target protein from an extract of the appropriate cell or tissue. This strategy will be discussed in more detail in the next section. More often, expression profiling is employed in the hope that the observed changes in protein expression will lead to the target protein. An illustration of how this can work is the identification of methionine aminopeptidase (MetAp) as the target of bengamides, a study carried out in the authors' laboratory.[94] Bengamides are natural products originally isolated from marine sponges with potent anti-tumor activity in vitro and in vivo. Their pattern of inhibition in a panel of cancer cell lines turned out to be quite distinct from other compound classes, suggesting that the activity of bengamides was due to the interaction with a novel target. A protein expression analysis of the H1299 small lung cell carcinoma cell line treated with the lead bengamide derivative LAF389 revealed alterations in a subset of about 20 proteins by 2D electrophoresis, which were readily identified by PMF. In-depth biochemical and mass spectrometric analysis[94] showed that LAF389 blocked the removal of the N-terminal methionine on one of the identified proteins, 14-3-3γ. This pointed to a direct or indirect inhibition of the involved enzyme, MetAp. Follow-up experiments showed that the N-terminal modification was indeed specific for this compound class and the ultimate proof was provided by the observation that the compound inhibited and cocrystallized with the purified MetAp enzyme. The discovery of MetAp as the target for bengamides was a major step forward for the medicinal chemists, who could now further optimize the compound, starting from the crystal structure of the target instead of a cell proliferation assay. At the same time, it is clear that much more effort would be required to unravel the actual mechanism of cytotoxicity by the bengamides. Parallel proteomic approaches could be employed, e.g., whether the different forms of 14-3-3γ interact with different proteins, or whether there is a larger set of proteins with altered N-termini, among which the true effector of cell cycle arrest could be found.

Probably the foremost expectation that accompanied the advent of large-scale expression-profiling methods was that completely novel biological insights would be generated at a much higher speed, including a better understanding of disease mechanisms, especially in complex diseases, where current therapies are often unspecific or only symptomatic. A good example of such a disease is schizophrenia.[105] Schizophrenia has a notable genetic disease component and hence a significant number of chromosomal 'hot spots' have been identified in families with a distinct prevalence of the disease. However, because of the complexity and heterogeneity of the disease, none of these hot spots has yielded a clear connection to a disease mechanism. A plausible reason for these inconsistencies is that those linkage studies are limited to the genetic predisposition for the disease, not taking the important environmental aspect into account. And of course schizophrenia is just one, albeit highly complex, example out of a long list of diseases that result from the interaction of predisposition and environment, including cancer, cardiovascular disease, and depression. Therefore it is not surprising that for many investigators expression profiling represented a paradigm shift, the first 'top-down approach' to complex disease: the investigation of the disease phenotype(s) at the molecular level, more or less decoupled from all the factors that led to that disease state.

Consequently, many proteomic studies have been undertaken to characterize psychiatric disease,[105] cardiovascular disease,[106] and, most of all, cancer,[107] just to name the areas of human disease, where arguably the highest need exists for novel therapeutic approaches. If all these studies have anything in common, it is probably the fact that virtually none of them has gone beyond the stage of describing differences. Hence the biggest lack is the validation of the long lists of potential biomarkers or even targets to prove that they are consistently (dys-)regulated and thus might be part of a disease mechanism or at least a useful diagnostic. There are several reasons why it is so difficult to turn protein expression data into relevant biological information. The choice of samples is one of them. Tissue samples from patients are clearly the most relevant but unfortunately also the most difficult to deal with, because of the inherent variability. And, as discussed before, expression profiling is most powerful with homogeneous samples. At the protein level that is even more pronounced than at the mRNA level, because the proteome is at least an order of magnitude larger, while at the same time the number of variables (i.e., proteins) that can actually be measured is much smaller. Moreover, if the number of proteins is not enough of a problem in its own right, there are many levels of additional heterogeneity, both with respect to the patients (age, diagnosis, comorbidities) as well as to the samples themselves (dissection, postmortem effects, mixture of different cell types). In other words, at the current level of coverage and throughput, a proteomics analysis of a disease phenotype in patient tissue is still a major challenge. In animal models many of the problems of heterogeneity are significantly reduced. However, the caveat is that if one is to extrapolate the molecular phenotype of an animal model to human disease, there has to be a certain level of overlap between the pathways involved. And that is hard to achieve if there is little knowledge about the disease mechanism in the first place. As more knowledge on biological pathways becomes available,[98] it will become easier to derive such animal models and more rewarding to study them with expression-profiling methods.

In conclusion, target discovery is probably the one area where proteomics has not yet fulfilled the sometimes unrealistic expectations. With current technology, a comprehensive analysis of protein expression is feasible in simple homogeneous systems, such as prokaryotes, and a substantial coverage can still be obtained in cell culture models derived from higher organisms. However, there is still some ground to cover before we will be able to perform an in-depth investigation of patient material to discover novel therapeutic targets. Moreover, analysis of protein expression levels is only the starting point for proteomics, because most of the functionally relevant events occur posttranslationally. Therefore, expression proteomics experiments should be combined with approaches that aim to elucidate relevant biological pathways in order to deliver a meaningful contribution to target discovery.

3.02.3.2 Chemical Proteomics and Compound Profiling

In theory the most direct way to gain more insight into the mode of action of a compound is to use it as an affinity bait to capture its interaction partners from a cell or tissue extract, analogous to the protein pull-downs discussed before (*see* Section 3.02.2.7.2). This approach is often referred to as chemical proteomics (or chemoproteomics). It is important to realize that the read-out of this type of experiment is quite different from a differential expression analysis after compound treatment, even though both types of experiment share some common goals. Chemical proteomics will provide an affinity profile for a compound, which will generally include the primary target and other proteins, some of which might even be indicative of potential unwanted biological (side) effects. However, from the profile alone it is not straightforward to deduce the relative importance of these different interactions. Moreover, binding of the compound to a protein has to be strong and fairly stable to be able to perform the pull-down. Relevant interactions that are more transient will not be captured. An expression proteomics experiment might yield only indirect evidence for what the primary target actually is, but it will give a better idea of the impact of all these interactions and the pathways they might affect. A significant practical disadvantage of a chemical proteomics experiment is that it will always require some manipulation of the compound to allow its immobilization. A prerequisite for that is sufficient insight in the structure–activity relationship (SAR), in order not to lose the relevant activity upon modification. Expression profiling, however, can be done with the compound itself without the need for any prior knowledge of the SAR.

Notwithstanding the limitations discussed above, integration of chemical proteomics approaches in the compound profiling process could yield valuable information for compound selection. Currently, selectivity of compounds for a specific target is generally evaluated by testing against closely related proteins, e.g., members of the same or a related protein family as the target. Usually this is done in a closed fashion, i.e., by measuring the affinity toward defined individual proteins. Proteomic methods allow us to complement that information by an affinity signature against a complex protein mixture such as a cell or tissue extract, either modeling the target tissue or a potential site of unwanted effects. The added value is that a biologically relevant signature is obtained, which takes into account the relative abundances of the primary target(s) and the, potentially unintended, secondary interactors. Furthermore, compounds are tested in an open system, thus maximizing the chances of finding unexpected events. While technically feasible, the obstacle to a systematic implementation of affinity signatures in compound profiling is their interpretation, in other words the discrimination of relevant signals from the background noise of unspecific interactions. As in any signature-type approach, a significant investment has to be made upfront to generate enough data for a knowledge base to support data analysis. And that in turn requires a large chemistry effort, which traditionally in pharmaceutical industry is primarily dedicated to generating novel compounds, rather than profiling the existing ones.

Among the most comprehensive chemical proteomics studies are those performed with kinase inhibitors.[108,109] These investigations are an excellent illustration of both the power as well as the inherent limitations of affinity profiles of compounds. In the case of the p38 inhibitor SB203580, even with low-resolution protein separation and identification methods, close to 20 interacting proteins were identified in a single cell line, several of them potential kinase targets not described before. It is certainly not surprising that kinase inhibitors show a significant degree of cross-reactivity because of the structural similarities within the kinase family. Nevertheless this interaction profile could be a valuable starting point to study potential off-target effects. The limitation of the approach is evident from the fact that the list of binders per se contributes little to understanding the biological relevance of all these interactions, including those with other proteins than kinases. For that purpose, it would have to be combined with an expression-profiling experiment, ideally including one that is focused on phosphorylation events. Probably, chemical proteomics is most powerful in a comparative mode, putting interaction profiles of different related compounds side by side to find the most selective one, or to explain specific biological effects of one of them. That will also allow for the filtering of unspecific binders, which will be largely the same for a set of related compounds. A second type of comparative experiment would be to 'fish' with one compound in multiple tissue or cell types, to derive tissue-specific targets or cross-reactions.

Particularly in the case of enzymes, protein abundance is not always correlated with the actual biological activity, because that can depend on posttranslational modifications, such as processing or phosphorylation. To capture that information, chemical proteomics strategies have been developed to perform activity-based protein profiling (ABPP). In this approach, active-site directed chemical probes are used that target several members of an enzyme class in their native environment, as part of a cell or tissue extract.[110] The ABPP probe binds covalently to the active site and contains a tag that can be visualized or used for the subsequent purification of the targets. Because the probe is covalently linked to the enzyme, the stability of the interaction becomes almost irrelevant and transient interactions can be captured efficiently. Once suitable probes are available for a larger number of enzyme classes, the ABPP chemical proteomics approach will undoubtedly contribute significantly to the profiling of enzyme inhibitors. The downside is that compounds have to be modified even more extensively than for immobilization. Moreover, contrary to affinity or expression profiling, ABPP is limited to studying enzyme inhibition.

3.02.3.3 Biomarkers and Molecular Diagnostics

In contrast to targets, where a direct link to the disease model or the mechanism of action of the compound needs to be validated, the criteria for accepting a protein as a biomarker are less strict. In a sense, initially every difference in the proteome that is reproducibly detected in a given situation could be considered a marker for that situation. However, only a degree of specificity for a defined biological state turns a biomarker into a useful diagnostic tool. In contrast to markers, targets are not necessarily differentially expressed, as illustrated by the bengamide study discussed in the section on target discovery and validation, above. MetAp, the target of the compound, is inhibited, resulting in the differential expression of protein species such as the novel 14-3-3γ isoform, which becomes a typical biomarker for bengamides: it appears reproducibly after compound treatment, although there is no direct link to the phenotype (in this case cell death). Because they are primary diagnostic tools, biomarkers are often sought in circulating or excreted biofluids, whereas targets are inherently restricted to the affected tissue itself. The proteomic analysis of biofluids, especially blood, has attracted a great deal of attention lately, because of the notion that the human plasma proteome is likely to contain most, if not all, human proteins, and that almost any disease state causes some specific protein expression changes in the blood.[111] To a lesser degree this is true as well for other body fluids such as urine or cerebrospinal fluid. Since a peripheral marker is almost a prerequisite for clinical application, proteomic-profiling methods for and protein catalogs of human body fluids have been established.[111–113] Still it is questionable whether biofluids are the right compartment to start the search for markers. Most likely, the more specific changes, whether related to disease or treatment, will be found in the affected tissue itself. From there a certain number will diffuse into the circulation, together with markers of tissue damage and inflammation, which might obscure the specific signals.

Along with the conceptual differences between targets and markers go a number of practical nuances. Because of the strict requirements, targets are rare, typically one or a few in each case. Biomarkers, on the other hand, can be quite abundant, as exemplified by a process like allograft rejection, where depending on the stage 50–100 biomarkers can be readily detected.[114] Therefore, although the technical approaches for target and biomarker discovery are the same, the required analysis depth is inversely proportional to their abundance. In other words, a priori successful target discovery requires a comprehensive proteome analysis, whereas in many cases valid biomarkers can be found without digging that deep. That has led to the development of several biomarker platforms, which focus on a subset of the proteome that is readily accessible for mass spectrometric identification, such as peptides and small proteins. Current variants include LC/MS,[115] CE (capillary electrophoresis)-MS[116] and the already discussed SELDI-MS.

Besides sampling only a subset of the proteome, these methods have in common the notion that a diagnosis can be based on a consistent pattern of protein expression changes, a proteomic signature. For diagnostic purposes the underlying identities of the proteins contributing to the signature are not absolutely required, and are in fact, particularly in the case of SELDI, not always easy to obtain.[85] The idea of using proteomic signatures has been around for a while. Spot patterns on 2D gels were considered a potential diagnostic tool soon after the first reports on 2D-PAGE,[117] but were never truly brought to bear for a long time, because of the variability associated with 2D gels. Only the advent of more robust profiling methods based on MS has revived the interest in proteomic signatures. Especially in the oncology field, several reports claim better performance of such multiparameter biomarkers, compared to the traditional individual proteins, such as prostate-specific antigen for prostate cancer or CA125 for ovarian tumors.[85] The ultimate proof of concept, validation in a sample set independent from the one in which they were identified, is still lacking. The portability from one sample set or one lab to another is the main issue of this proteomic signature concept. Ultimately, that will almost always require identification of the contributing proteins, to allow transfer of the signature to more robust, clinically validated platforms like immunoassays, which are still the most suitable for the desired large-population studies. Already now, the implementation of immunoassays in a chip-based format makes it possible to get a simultaneous read-out on the levels of 20–50 protein species. Upon the availability of more and highly specific antibodies, high-throughput diagnostic assays can be developed to analyze signatures consisting of hundreds of different proteins. That should open a whole new avenue of diagnostic possibilities for complex diseases, which by definition depend on multiple parameters. Notwithstanding the current transition phase in which robust discovery and validation tools are still being developed, protein biomarkers are undoubtedly going to be of crucial importance for the development of specific, personalized therapies. For a more detailed discussion on 'biomarkers,' readers are referred to Chapter 3.04.

3.02.4 Perspective – From Proteomics to Systems Biology

After pointing out repeatedly that proteomics is still only beginning to mature to a stage where it will impact the understanding of biological systems, it seems almost inappropriate to speculate about what comes next. However, the one thing that the sequential waves of genomics and proteomics have convincingly shown is that the time has come for integration. Only integration of experimental data at the level of genes, proteins, and metabolites with insights from bioinformatics can advance us significantly toward a comprehensive description of biological systems and ultimately toward understanding them. One could separate the path from proteomics, genomics, or metabonomics to systems biology in four phases: (1) discovery; (2) description; (3) integration; and the last phase that distinguishes systems biology from the '-omics' disciplines in a fundamental way (4) prediction. Below we will briefly discuss these different phases for proteomics.

3.02.4.1 Discovery – Mapping the Proteome

One aspect that we have not discussed extensively is representative for the discovery phase, in which proteomics currently is: proteome mapping, generating a comprehensive list of all expressed proteins in a cell, tissue, or organism. Unfortunately, at the protein level there is no real analog of the human genome in sight, an endpoint where we can claim to have discovered the entire human proteome. Nevertheless efforts are being made to consolidate the collective protein identification efforts of the proteome community into a single database called PeptideAtlas,[118] which currently already represents the products of around 6000 genes. Obviously, merging protein identifications from different platforms and labs is a major challenge, because of the variable confidence levels associated with different experimental methods and database search tools. A key question is whether it is not too early to focus so much effort on the human proteome, while we have not yet succeeded in completely discovering any of the simplest prokaryotic proteomes. Since all the next phases build on a reliable and exhaustive discovery phase, a relevant model system will also help to define the size of the required effort in more complex organisms. Importantly, the discovery phase is not limited to protein expression. At the same time, regulatory modifications and protein–protein interactions have to be charted to delineate biological pathways, particularly with respect to their rate-limiting components.

3.02.4.2 Description – Scoring the Proteome

One of the leading proteome scientists of today, Ruedi Aebersold, has elegantly described why at some point it becomes inevitable to leave the discovery phase behind and move to the description of proteomes under different conditions, e.g., the changes that are induced by specific perturbations. He also suggests a technological implementation for such a scoring platform based on one or more representative isotope-labeled reference peptides for each protein, the so-called proteotypic peptides.[118] These peptides would be selected based on the experimental data compiled in the PeptideAtlas database, and hence be readily detectable by a mass spectrometer. By spiking sets of proteotypic peptides into samples, quantitation of expressed proteins could be performed in high throughput. While an interesting concept, the success of this approach depends completely on the quality of the information generated in the discovery phase. The main issue is to generate specific reference peptides not only for each protein, but also for all of its relevant isoforms. Again, a model system with limited numbers of protein isoforms (**Table 1**) might be a good starting point for a proof of concept of proteome description approaches.

Whether with proteotypic peptides or with another concept, it is clear that current proteome analysis techniques are not fast enough to make the transition from the discovery to the description phase. And while high-throughput solutions at the (protein identification) back end are starting to appear at the horizon, reproducible sample preparation and fractionation procedures to achieve the required analysis depth are not yet in sight and should be the focus of technology development in the near future.

3.02.4.3 Integration and Prediction

Ultimately, a prerequisite for a valid model of a biological state is the integration of information at the genomic, proteomic, and metabonomic levels. The key question is how much detail we need at each level to build a valid model.

Probably the first models should center on metabolic pathways, because these are mostly restricted to a single compartment in a cell. In contrast, signaling pathways often have the additional complexity of involving different locations in a cell, a model of which would require a much better insight in the mechanisms of transport or transmission of signals. Summarizing, it is evident that the moment where the different levels of expression profiling can converge into systems biology is still a while in the future. Proteomics can deliver an important contribution, but to achieve that it will require the use of parallel approaches as opposed to a focus on a single 'best' technology.

References

1. Mulder, G. J. *J. Prakt. Chemie* **1839**, *16*, 129–152.
2. Wasinger, V. C.; Cordwell, S. J.; Cerpa-Poljak, A.; Yan, J. X.; Gooley, A. A.; Wilkins, M. R.; Duncan, M. W.; Harris, R.; Williams, K. L.; Humphery-Smith, I. *Electrophoresis* **1995**, *16*, 1090–1094.
3. Patterson, S. D.; Aebersold, R. H. *Nat. Genet.* **2003**, *33*, 311–323.
4. Blackstock, W. P.; Weir, M. P. *Trends Biotechnol.* **1999**, *17*, 121–127.
5. Gavin, A. C.; Bosche, M.; Krause, R.; Grandi, P.; Marzioch, M.; Bauer, A.; Schultz, J.; Rick, J. M.; Michon, A. M.; Cruciat, C. M. et al. *Nature* **2002**, *415*, 141–147.
6. Ho, Y.; Gruhler, A.; Heilbut, A.; Bader, G. D.; Moore, L.; Adams, S. L.; Millar, A.; Taylor, P.; Bennett, K.; Boutilier, K. et al. *Nature* **2002**, *415*, 180–183.
7. Pandey, A.; Mann, M. *Nature* **2000**, *405*, 837–846.
8. Collins, F. S.; Green, E. D.; Guttmacher, A. E.; Guyer, M. S. *Nature* **2003**, *422*, 835–847.
9. Dyer, M. R.; Cohen, D.; Herrling, P. L. *Drug Disc. Today* **1999**, *4*, 109–114.
10. Boguski, M. S.; McIntosh, M. W. *Nature* **2003**, *422*, 233–237.
11. Karchin, R.; Diekhans, M.; Kelly, L.; Thomas, D. J.; Pieper, U.; Eswar, N.; Haussler, D.; Sali, A. *Bioinformatics* **2005**, *21*, 2814–2820.
12. Nakao, M.; Barrero, R. A.; Mukai, Y.; Motono, C.; Suwa, M.; Nakai, K. *Nucleic Acids Res.* **2005**, *33*, 2355–2363.
13. Koller, A.; Washburn, M. P.; Lange, B. M.; Andon, N. L.; Deciu, C.; Haynes, P. A.; Hays, L.; Schieltz, D.; Ulaszek, R.; Wei, J. et al. *Proc. Natl. Acad. Sci. USA* **2002**, *99*, 11969–11974.
14. Florens, L.; Washburn, M. P.; Raine, J. D.; Anthony, R. M.; Grainger, M.; Haynes, J. D.; Moch, J. K.; Muster, N.; Sacci, J. B.; Tabb, D. L. et al. *Nature* **2002**, *419*, 520–526.
15. Lee, M. S.; Hanspers, K.; Barker, C. S.; Korn, A. P.; McCune, J. M. *Int. Immunol.* **2004**, *16*, 1109–1124.
16. Moran, L. B.; Duke, D. C.; Deprez, M.; Dexter, D. T.; Pearce, R. K.; Graeber, M. B. *Neurogenetics* **2006**, 7, 1–11.
17. Anderson, N. L; Anderson, N. G. *Electrophoresis* **1998**, *19*, 1853–1861.
18. Corthals, G. L.; Wasinger, V. C.; Hochstrasser, D. F.; Sanchez, J. C. *Electrophoresis* **2000**, *21*, 1104–1115.
19. Washburn, M. P.; Wolters, D.; Yates, J. R., III *Nat. Biotechnol.* **2001**, *19*, 242–247.
20. O'Farrell, P. H. *J. Biol. Chem.* **1975**, *250*, 4007–4021.
21. Aebersold, R.; Mann, M. *Nature* **2003**, *422*, 198–207.
22. Wu, C. C.; Yates, J. R., III *Nat. Biotechnol.* **2003**, *21*, 262–267.
23. Deutscher, M. P., Ed. *Guide to Protein Purification. Methods in Enzymology*; Academic Press: San Diego, CA, 1990; Vol. 182.
24. Flory, M. R.; Aebersold, R. *Prog. Cell Cycle Res.* **2003**, *5*, 167–171.
25. Boonstra, J. *J. Cell. Biochem.* **2003**, *90*, 244–252.
26. Emmert-Buck, M. R.; Bonner, R. F.; Smith, P. D.; Chuaqui, R. F.; Zhuang, Z.; Goldstein, S. R.; Weiss, R. A.; Liotta, L. A. *Science* **1996**, *274*, 998–1001.
27. Craven, R. A.; Totty, N.; Harnden, P.; Selby, P. J.; Banks, R. E. *Am. J. Pathol.* **2002**, *160*, 815–822.
28. Knezevic, V.; Leethanakul, C.; Bichsel, V. E.; Worth, J. M.; Prabhu, V. V.; Gutkind, J. S.; Liotta, L. A.; Munson, P. J.; Petricoin, E. F., III.; Krizman, D. B. *Proteomics* **2001**, *1*, 1271–1278.
29. Bonaventure, P.; Guo, H.; Tian, B.; Liu, X.; Bittner, A.; Roland, B.; Salunga, R.; Ma, X. J.; Kamme, F.; Meurers, B. et al. *Brain Res.* **2002**, *943*, 38–47.
30. Ma, X. J.; Salunga, R.; Tuggle, J. T.; Gaudet, J.; Enright, E.; McQuary, P.; Payette, T.; Pistone, M.; Stecker, K.; Zhang, B. M. et al. *Proc. Natl. Acad. Sci. USA* **2003**, *100*, 5974–5979.
31. Greengauz-Roberts, O.; Stoppler, H.; Nomura, S.; Yamaguchi, H.; Goldenring, J. R.; Podolsky, R. H.; Lee, J. R.; Dynan, W. S. *Proteomics* **2005**, *5*, 1746–1757.
32. Zang, L.; Palmer, T. D.; Hancock, W. S.; Sgroi, D. C.; Karger, B. L. *J. Proteome Res.* **2004**, *3*, 604–612.
33. Espina, V.; Geho, D.; Mehta, A. I.; Petricoin, E. F., III; Liotta, L. A.; Rosenblatt, K. P. *Cancer Invest.* **2005**, *23*, 36–46.
34. Stasyk, T.; Huber, L. A. *Proteomics* **2004**, *4*, 3704–3716.
35. Schirle, M.; Heurtier, M. A.; Kuster, B. *Mol. Cell. Proteomics* **2003**, *2*, 1297–1305.
36. Lescuyer, P.; Hochstrasser, D. F.; Sanchez, J. C. *Electrophoresis* **2004**, *25*, 1125–1135.
37. Righetti, P. G.; Castagna, A.; Herbert, B.; Reymond, F.; Rossier, J. S. *Proteomics* **2003**, *3*, 1397–1407.
38. Anderson, N. L.; Anderson, N. G. *Mol. Cell. Proteomics* **2002**, *1*, 845–867.
39. Baussant, T.; Bougueleret, L.; Johnson, A.; Rogers, J.; Menin, L.; Hall, M.; Aberg, P. M.; Rose, K. *Proteomics* **2005**, *5*, 973–977.
40. Anderson, N. L.; Polanski, M.; Pieper, R.; Gatlin, T.; Tirumalai, R. S.; Conrads, T. P.; Veenstra, T. D.; Adkins, J. N.; Pounds, J. G.; Fagan, R. et al. *Mol. Cell. Proteomics* **2004**, *3*, 311–326.
41. Bjorhall, K.; Miliotis, T.; Davidsson, P. *Proteomics* **2005**, *5*, 307–317.
42. Steen, H.; Mann, M. *Nat. Rev. Mol. Cell Biol.* **2004**, *5*, 699–711.
43. Fenn, J. B.; Mann, M.; Meng, C. K.; Wong, S. F.; Whitehouse, C. M. *Science* **1989**, *246*, 64–71.
44. Karas, M.; Hillenkamp, F. *Anal. Chem.* **1988**, *60*, 2299–2301.
45. Clauser, K. R.; Baker, P.; Burlingame, A. L. *Anal. Chem.* **1999**, *71*, 2871–2882.
46. Medzihradszky, K. F.; Campbell, J. M.; Baldwin, M. A.; Falick, A. M.; Juhasz, P.; Vestal, M. L.; Burlingame, A. L. *Anal. Chem.* **2000**, *72*, 552–558.
47. Gevaert, K.; Vandekerckhove, J. *Electrophoresis* **2000**, *21*, 1145–1154.
48. Görg, A.; Obermaier, C.; Boguth, G.; Harder, A.; Scheibe, B.; Wildgruber, R.; Weiss, W. *Electrophoresis* **2000**, *21*, 1037–1053.

49. Görg, A.; Weiss, W.; Dunn, M. J. *Proteomics* **2004**, *4*, 3665–3685.
50. Rabilloud, T. *Electrophoresis* **1998**, *19*, 758–760.
51. Klose, J.; Nock, C.; Herrmann, M.; Stuhler, K.; Marcus, K.; Bluggel, M.; Krause, E.; Schalkwyk, L. C.; Rastan, S.; Brown, S. D. et al. *Nat. Genet.* **2002**, *30*, 385–393.
52. Hoving, S.; Voshol, H.; Van Oostrum, J. *Electrophoresis* **2000**, *21*, 2617–2621.
53. Hoving, S.; Gerrits, B.; Voshol, H.; Müller, D.; Roberts, R. C.; Van Oostrum, J. *Proteomics* **2002**, *2*, 127–134.
54. Fey, S. J.; Larsen, P. M. *Curr. Opin. Chem. Biol.* **2001**, *5*, 26–33.
55. Van Oostrum, J.; Hoving, S.; Müller, D.; Schindler, P.; Steinmetz, M.; Towbin, H.; Voshol, H.; Wirth, U. *Chimia* **2001**, *55*, 354–358.
56. Lopez, M. F.; Berggren, K.; Chernokalskaya, E.; Lazarev, A.; Robinson, M.; Patton, W. F. *Electrophoresis* **2000**, *21*, 3673–3683.
57. Lauber, W. M.; Carroll, J. A.; Dufield, D. R.; Kiesel, J. R.; Radabaugh, M. R.; Malone, J. P. *Electrophoresis* **2001**, *22*, 906–918.
58. Lanne, B.; Panfilov, O. *J. Proteome Res.* **2005**, *4*, 175–179.
59. Unlu, M.; Morgan, M. E.; Minden, J. S. *Electrophoresis* **1997**, *18*, 2071–2077.
60. Tonge, R.; Shaw, J.; Middleton, B.; Rowlinson, R. *Proteomics* **2001**, *1*, 377–396.
61. Yan, J. X.; Devenish, A. T.; Wait, R.; Stone, T. *Proteomics* **2002**, *2*, 1682–1698.
62. Karp, N. A.; Lilley, K. S. *Proteomics* **2005**, *5*, 3105–3115.
63. Shaw, J.; Rowlinson, R.; Nickson, J.; Stone, T.; Sweet, A.; Williams, K.; Tonge, R. *Proteomics* **2003**, *3*, 1181–1195.
64. Choe, L. H.; Lee, K. H. *Electrophoresis* **2000**, *21*, 993–1000.
65. Venter, J. C.; Adams, M. D.; Sutton, G. G.; Kerlavage, A. R.; Smith, H. O.; Hunkapiller, M. *Science* **1998**, *280*, 1540–1542.
66. Gygi, S. P.; Rist, B.; Gerber, S. A.; Turecek, F.; Gelb, M. H.; Aebersold, R. *Nat. Biotechnol.* **1999**, *17*, 994–999.
67. Schrimpf, S. P.; Meskenaite, V.; Brunner, E.; Rutishauser, D.; Walther, P.; Eng, J.; Aebersold, R.; Sonderegger, P. *Proteomics* **2005**, *5*, 2531–2541.
68. Molloy, M. P.; Donohoe, S.; Brzezinski, E. E.; Kilby, G. W.; Stevenson, T. I.; Baker, J. D.; Goodlett, D. R.; Gage, D. A. *Proteomics* **2005**, *5*, 1204–1208.
69. Gevaert, K.; Goethals, M.; Martens, L.; Van Damme, J.; Staes, A.; Thomas, G. R.; Vandekerckhove, J. *Nat. Biotechnol.* **2003**, *21*, 566–569.
70. Gevaert, K.; Van Damme, J.; Goethals, M.; Thomas, G. R.; Hoorelbeke, B.; Demol, H.; Martens, L.; Puype, M.; Staes, A.; Vandekerckhove, J. *Mol. Cell. Proteomics* **2002**, *1*, 896–903.
71. Gevaert, K.; Ghesquiere, B.; Staes, A.; Martens, L.; Van Damme, J.; Thomas, G. R.; Vandekerckhove, J. *Proteomics* **2004**, *4*, 897–908.
72. Conrads, T. P.; Alving, K.; Veenstra, T. D.; Belov, M. E.; Anderson, G. A.; Anderson, D. J.; Lipton, M. S.; Pasa-Tolic, L.; Udseth, H. R.; Chrisler, W. B. et al. *Anal. Chem.* **2001**, *73*, 2132–2139.
73. Ong, S. E.; Foster, L. J.; Mann, M. *Methods* **2003**, *29*, 124–130.
74. DeSouza, L.; Diehl, G.; Rodrigues, M. J.; Guo, J.; Romaschin, A. D.; Colgan, T. J.; Siu, K. W. *J. Proteome Res.* **2005**, *4*, 377–386.
75. Stewart, I. I.; Thomson, T.; Figeys, D. *Rapid Commun. Mass Spectrom.* **2001**, *15*, 2456–2465.
76. Gygi, S. P.; Rist, B.; Griffin, T. J.; Eng, J.; Aebersold, R. *J. Proteome Res.* **2002**, *1*, 47–54.
77. Wehmhoner, D.; Dieterich, G.; Fischer, E.; Baumgartner, M.; Wehland, J.; Jansch, L. *Electrophoresis* **2005**, *26*, 2450–2460.
78. Wang, H.; Clouthier, S. G.; Galchev, V.; Misek, D. E.; Duffner, U.; Min, C. K.; Zhao, R.; Tra, J.; Omenn, G. S.; Ferrara, J. L. et al. *Mol. Cell. Proteomics* **2005**, *4*, 618–625.
79. Chaurand, P.; Sanders, M. E.; Jensen, R. A.; Caprioli, R. M. *Am. J. Pathol.* **2004**, *165*, 1057–1068.
80. Caldwell, R. L.; Caprioli, R. M. *Mol. Cell. Proteomics* **2005**, *4*, 394–401.
81. Stoeckli, M.; Chaurand, P.; Hallahan, D. E.; Caprioli, R. M. *Nat. Med.* **2001**, *7*, 493–496.
82. Stoeckli, M.; Staab, D.; Staufenbiel, M.; Wiederhold, K. H.; Signor, L. *Anal. Biochem.* **2002**, *311*, 33–39.
83. Petricoin, E. F.; Ardekani, A. M.; Hitt, B. A.; Levine, P. J.; Fusaro, V. A.; Steinberg, S. M.; Mills, G. B.; Simone, C.; Fishman, D. A.; Kohn, E. C. et al. *Lancet* **2002**, *359*, 572–577.
84. Liotta, L. A.; Ferrari, M.; Petricoin, E. *Nature* **2003**, *425*, 905.
85. Diamandis, E. P. *Mol. Cell. Proteomics* **2004**, *3*, 367–378.
86. Hu, J; Coombes, K. R.; Morris, J. S.; Baggerly, K. A. *Brief Funct. Genomic. Proteomic.* **2005**, *3*, 322–331.
87. Wellmann, A.; Wollscheid, V.; Lu, H.; Ma, Z. L.; Albers, P.; Schutze, K.; Rohde, V.; Behrens, P.; Dreschers, S.; Ko, Y. et al. *Int. J. Mol. Med.* **2002**, *9*, 341–347.
88. Bodovitz, S.; Joos, T.; Bachmann, J. *Drug Disc. Today* **2005**, *10*, 283–287.
89. Sheehan, K. M.; Calvert, V. S.; Kay, E. W.; Lu, Y.; Fishman, D.; Espina, V.; Aquino, J.; Speer, R.; Araujo, R.; Mills, G. B. et al. *Mol. Cell. Proteomics* **2005**, *4*, 346–355.
90. Pawlak, M.; Schick, E.; Bopp, M. A.; Schneider, M. J.; Oroszlan, P.; Ehrat, M. *Proteomics* **2002**, *2*, 383–393.
91. LaBaer, J.; Ramachandran, N. *Curr. Opin. Chem. Biol.* **2005**, *9*, 14–19.
92. Mann, M.; Ong, S. E.; Gronborg, M.; Steen, H.; Jensen, O. N.; Pandey, A. *Trends Biotechnol.* **2002**, *20*, 261–268.
93. Muller, D. R.; Schindler, P; Coulot, M.; Voshol, H.; van Oostrum, J. *J. Mass Spectrom.* **1999**, *34*, 336–345.
94. Towbin, H.; Bair, K. W.; DeCaprio, J. A.; Eck, M. J.; Kim, S.; Kinder, F. R.; Morollo, A.; Mueller, D. R.; Schindler, P.; Song, H. K. et al. *J. Biol. Chem.* **2003**, *278*, 52964–52971.
95. Zaia, J. *Mass Spectrom. Rev.* **2004**, *23*, 161–227.
96. Tackett, A. J.; Dilworth, D. J.; Davey, M. J.; O'Donnell, M.; Aitchison, J. D.; Rout, M. P.; Chait, B. T. *J. Cell Biol.* **2005**, *169*, 35–47.
97. Von Mering, C.; Krause, R.; Snel, B.; Cornell, M.; Oliver, S. G.; Fields, S.; Bork, P. *Nature* **2002**, *417*, 399–403.
98. Bouwmeester, T.; Bauch, A.; Ruffner, H.; Angrand, P. O.; Bergamini, G.; Croughton, K.; Cruciat, C.; Eberhard, D.; Gagneur, J.; Ghidelli, S. et al. *Nat. Cell Biol.* **2004**, *6*, 97–105.
99. Muller, D. R.; Schindler, P.; Towbin, H.; Wirth, U.; Voshol, H.; Hoving, S.; Steinmetz, M. O. *Anal. Chem.* **2001**, *73*, 1927–1934.
100. Gendreizig, S.; Kindermann, M.; Johnsson, K. *J. Am. Chem. Soc.* **2003**, *125*, 14970–14971.
101. Need, A. C.; Motulsky, A. G.; Goldstein, D. B. *Nat. Genet.* **2005**, *37*, 671–681.
102. Merrick, B. A.; Bruno, M. E. *Curr. Opin. Mol. Ther.* **2004**, *6*, 600–607.
103. Anderson, N. L.; Esquer-Blasco, R.; Hofmann, J. P.; Anderson, N. G. *Electrophoresis* **1991**, *12*, 907–930.
104. Veenstra, T. D.; Conrads, T. P.; Hood, B. L.; Avellino, A. M.; Ellenbogen, R. G.; Morrison, R. S. *Mol. Cell. Proteomics* **2005**, *4*, 409–418.
105. Voshol, H.; Glucksman, M. J.; van Oostrum, J. *Curr. Mol. Med.* **2003**, *3*, 447–458.
106. Anderson, L. *J. Physiol.* **2005**, *563*, 23–60.
107. Posadas, E. M.; Simpkins, F.; Liotta, L. A.; MacDonald, C.; Kohn, E. C. *Ann. Oncol.* **2005**, *16*, 16–22.
108. Godl, K.; Wissing, J.; Kurtenbach, A.; Habenberger, P.; Blencke, S.; Gutbrod, H.; Salassidis, K.; Stein-Gerlach, M.; Missio, A.; Cotton, M. et al. *Proc. Natl. Acad. Sci. USA* **2003**, *100*, 15434–15439.

109. Daub, H.; Godl, K.; Brehmer, D.; Klebl, B.; Muller, G. *Assay Drug Dev. Technol.* **2004**, *2*, 215–224.
110. Leung, D.; Hardouin, C.; Boger, D. L.; Cravatt, B. F. *Nat. Biotechnol.* **2003**, *21*, 687–691.
111. Anderson, N. L.; Polanski, M.; Pieper, R.; Gatlin, T.; Tirumalai, R. S.; Conrads, T. P.; Veenstra, T. D.; Adkins, J. N.; Pounds, J. G.; Fagan, R. et al. *Mol. Cell. Proteomics* **2004**, *3*, 311–326.
112. Pieper, R.; Gatlin, C. L.; McGrath, A. M.; Makusky, A. J.; Mondal, M.; Seonarain, M.; Field, E.; Schatz, C. R.; Estock, M. A.; Ahmed, N. et al. *Proteomics* **2004**, *4*, 1159–1174.
113. Zhang, J.; Goodlett, D. R.; Peskind, E. R.; Quinn, J. F.; Zhou, Y.; Wang, Q.; Pan, C.; Yi, E.; Eng, J.; Aebersold, R. H. et al. *Neurobiol. Aging* **2005**, *2*, 207–227.
114. Voshol, H.; Brendlen, N.; Muller, D.; Inverardi, B.; Augustin, A.; Pally, C.; Wieczorek, G.; Morris, R. E.; Raulf, F.; van Oostrum, J. *J. Proteome Res.* **2005**, *4*, 1192–1199.
115. Lamerz, J.; Selle, H.; Scapozza, L.; Crameri, R.; Schulz-Knappe, P.; Mohring, T.; Kellmann, M.; Khamenia, V.; Zucht, H. D. *Proteomics* **2005**, *5*, 2789–2798.
116. Fliser, D.; Wittke, S.; Mischak, H. *Electrophoresis* **2005**, *26*, 2708–2716.
117. Jellum, E.; Thorsrud, A. K. *Clin. Chem.* **1982**, *28*, 876–883.
118. Kuster, B.; Schirle, M.; Mallick, P.; Aebersold, R. *Nat. Rev. Mol. Cell Biol.* **2005**, *6*, 577–583.
119. Neuhoff, V.; Stamm, R.; Eibl, H. *Electrophoresis* **1985**, *6*, 427–448.
120. Neuhoff, V.; Arold, N; Taube, D.; Ehrhardt, W. *Electrophoresis* **1988**, *9*, 255–262.
121. Blum, H.; Beier, H.; Gross, H. J. *Electrophoresis* **1987**, *8*, 93–99.
122. Hochstrasser, D. F.; Harrington, M. G.; Hochstrasser, A. C.; Miller, M. J. *Anal. Biochem.* **1988**, *173*, 424–435.
123. Sitek, B.; Luttges, J.; Marcus, K.; Kloppel, G.; Schmiegel, W.; Meyer, H. E.; Hahn, S. A.; Stuhler, K. *Proteomics* **2005**, *5*, 2665–2679.
124. www.expasy.uniprot.org (accessed Aug 2006).

Biographies

Hans Voshol holds an MSc in Biochemistry and a PhD in Bio-Organic Chemistry from Utrecht University, the Netherlands. After a postdoctoral fellowship at the Institute of Neurobiology of the Federal Institute of Technology (Zurich, Switzerland), he joined Novartis Basel in 1996. His current position is that of project leader in the Proteomics Group of the Genome and Proteome Sciences Department.

Sjouke Hoving was born in the Netherlands in 1966 and received his MSc in the Life Sciences from the Weizmann Institute of Science, Rehovot, Israel in 1993. In 1998, he obtained his PhD in the Natural Sciences from the Swiss Federal Institute of Technology (ETH), Zürich, Switzerland, where he worked on novel methods – the field of proteomics was just emerging – for protein analysis by mass spectrometry. Afterwards, he joined Novartis Research as postdoc in the Proteomics group where he was involved in a project to develop methods to improve the resolution on 2D gels by the use of narrow-range immobilized pH gradient (IPG) strips. Since 2000 he is Research Investigator in the Novartis Institutes for BioMedical Research (Genome and Proteome Sciences Department) and in this function responsible for the 2D gel core facilities in the Proteomics group. The Proteomics group implements, integrates, and applies highly efficient proteomics technologies for target identification, selection, and validation. A variety of proteomics approaches are used to identify proteins involved in disease through the differential analysis of protein expression.

Jan van Oostrum obtained his PhD in Biochemistry and Molecular Biophysics from Columbia University, New York, and joined Ciba-Geigy, Basel, in 1986 in the Department of Biotechnology. In 1991, he changed as a project leader to the Ciba Pharmaceutical Division in Summit, NJ, Department of Enzymology and Molecular Biology. In 1993, he returned to Basel to join the Core Technology Area as a project leader in the Protein Structure Function Unit. Since the merger between Ciba and Sandoz he is heading the Proteomics activities in Novartis Research and is associated with the Genome and Proteome Sciences Department of Novartis.

Comprehensive Medicinal Chemistry II
ISBN (set): 0-08-044513-6

ISBN (Volume 3) 0-08-044516-0; pp. 27–50

3.03 Pharmacogenomics

K Lindpaintner, F. Hoffman-La Roche, Basel, Switzerland

3.03.1	**Introduction**	**51**
3.03.2	**Definition of Terms**	**53**
3.03.2.1	Pharmacogenetics	53
3.03.2.2	Pharmacogenomics	53
3.03.3	**Long-Term Timeframe: Causative Targets – Addressing Deranged Function Directly**	**54**
3.03.3.1	Palliative and Causative-Acting Drugs	54
3.03.3.2	The Twofold Challenge of Common Complex Disease	54
3.03.3.3	Applicability of 'Enetic' Targets/Medicines	55
3.03.3.4	Acceleration of Drug Discovery/Development through 'Better' Targets?	55
3.03.4	**Midterm Timeframe: Pharmacogenomics/Toxicogenomics – Finding New Medicines Quicker and More Efficiently**	**56**
3.03.5	**Short-Term Timeframe: Pharmacogenetics – More Targeted, More Effective Medicines**	**56**
3.03.5.1	Genes and Environment	56
3.03.5.2	An Attempt at a Systematic Classification of Pharmacogenetics	57
3.03.5.3	'Classical Pharmacogenetics'	57
3.03.5.4	Pharmacogenetics as a Consequence of Molecular Differential Diagnosis	61
3.03.5.5	Different Classes of Markers	61
3.03.5.6	Complexity is to be Expected	61
3.03.6	**Incorporating Pharmacogenetics into Drug Development Strategy**	**62**
3.03.6.1	Diagnostics First, Therapeutics Second	62
3.03.6.2	Probability, Not Certainty	62
3.03.7	**Regulatory Aspects**	**63**
3.03.8	**Pharmacogenetic Testing for Drug Efficacy versus Safety**	**64**
3.03.8.1	Greater Efficacy: Likely	64
3.03.8.2	Avoidance of Serious Adverse Effects: Less Likely – With Exceptions	64
3.03.9	**Challenge: Genetics and Society – Ethical–Legal–Societal Issues**	**65**
3.03.9.1	Data Protection Needs to be Matched by 'Person Protection'	65
3.03.9.2	Public Education and Information	66
3.03.9.3	Ethical–Societal Aspects of Pharmacogenetics	66
3.03.10	**Summary**	**67**
References		**67**

3.03.1 Introduction

Advances made over the last 30 years in molecular biology, molecular genetics, and genomics, and in the development and refinement of associated methods and technologies, have had a major impact on our understanding of biology, including the action of drugs and other biologically active xenobiotics. The tools that have been developed to allow these advances, and the knowledge of fundamental principles underlying cellular function thus derived, have become quintessential and indeed indispensable for almost any kind and field of biological research, including future progress in biomedicine and healthcare.

One aspect in particular of the broad scope across which progress in biology has been achieved, namely our understanding of genetics, and, especially, our sequencing of the human genome, has uniquely captured the imagination of both scientists and the public. Given the austere beauty of Mendel's laws of inheritance, the compelling aesthetics of the double-helix structure, and the awe-inspiring accomplishment of cataloguing billions of basepairs, and, last but not least, a public relations campaign unprecedented in its scope in the history of scientific achievement, this reaction is quite understandable. However, the high expectations raised regarding the degree and timeframe of impact that these technologies will have on the practice of healthcare are almost certainly unrealistic. Situated at the interface between pharmacology and genetics/genomics, 'pharmacogenetics and pharmacogenomics' (usually without any further definition of what these terms mean) are commonly touted as heralding a 'revolution' in medicine.

It is important to realize that, with regard to pharmacology and drug discovery, accomplishments in basic biology – starting some time in the last third or quarter of the last century – have indeed already led to what may well be considered a rather fundamental, perhaps paradigmatic, shift from the 'chemical paradigm' to a 'biological paradigm.' Historically, drug discovery was driven by medicinal chemistry, with biology serving an almost secondary, ancillary role that examined new molecules for biological function. The ability to comprehend cell biology and function, based on a newly developed set of tools to investigate the physiological effects of biomolecules and pathways on their basic, molecular level, has since reversed this directionality: the biologist is now driving the process, requesting from the chemist compounds that modulate the function of these biomolecules or pathways, with the expectation of a more predictable impact on physiological function and the correction of its pathological derailments.

Indeed, as pointed out above, the major change in how we discover drugs – from the chemical to the biological paradigm – has already occurred some time ago; what the current advances, in due time, promise to allow us to do is to move from a physiology-based to a (molecular) pathology-based approach toward drug discovery, promising the advancement from a largely palliative to a more cause/contribution-targeting pharmacopoeia.

This communication is intended to provide a – necessarily somewhat subjective – view of what the disciplines of genetics and genomics stand to contribute, and how they have actually contributed for many years, to drug discovery and development, and more broadly to the practice of healthcare. Particular emphasis will be placed on examining the role of genetics, whether acquired or inherited variations at the level of DNA-encoded information, in 'real life.' With regard to common complex disease, a realistic understanding of this role is absolutely essential for a balanced assessment of the impact of genetics' on healthcare in the future. Definitions for some of the terms that are in wide and often unreflected use today – almost always sorely missing from both academic and public policy-related documents on the topic – will be provided, with an understanding that much of the field is still in flux, and that these may well change.

With regard to the actual opportunities and challenges genetics and genomics provide in the field of healthcare, a perspective staged by time windows is helpful. Thus:

1. The aspect that holds the greatest promise, successfully targeting newly recognized, causally relevant targets with innovative drugs based on a more fundamental and functional understanding of disease causation or contribution on the molecular level, is also the one that lies farthest in the future, and commensurately little can be said about it or we will to lose ourselves in pure speculation.
2. The more midterm impact of these technologies, applicable to the earlier stages of drug discovery, will be covered in some detail, as here, too, the true relevance of various applications of genomic technologies has yet to be fully established.
3. The most imminent application to medicines that are either on the market or in clinical trials, i.e., pharmacogenetics, will receive the major emphasis, for the obvious reasons that these applications are in part already being implemented today in clinical trials. Here, a more systematic classification than is generally found will be attempted.

It is important to remain mindful that what will be discussed is, to a large extent, still uncharted territory, so by necessity many of the positions taken, reasoned on today's understanding and knowledge, must be viewed as somewhat tentative in nature. Where appropriate and possible, select examples will be provided, although it should be pointed out that much of the literature in the area of genetic epidemiology and pharmacogenetics lacks the stringent standards normally applied to peer-reviewed research, and replicate data are generally absent.

3.03.2 Definition of Terms

There is widespread indiscriminate use of, and thus confusion about, the terms 'pharmacogenetics' and 'pharmacogenomics.' While no universally accepted definition exists, there is an emerging consensus on the differential meaning and use of the two terms (**Table 1**).

3.03.2.1 Pharmacogenetics

The term 'genetics' relates etymologically to the presence of individual properties, and interindividual differences in these properties, as a consequence of having inherited (or acquired) them. Thus, the term 'pharmacogenetics' describes the interactions between a drug and an individual's (or perhaps more accurately, groups' of individuals) characteristics as they relate to differences in the DNA-based information. Pharmacogenetics, therefore, refers to the assessment of clinical efficacy and/or the safety and tolerability profile – the pharmacological, or reponse phenotype – of a drug in groups of individuals who differ with regard to certain DNA-encoded characteristics. It tests the hypothesis that these differences, if indeed, they are associated with a differential response phenotype, may allow prediction of individual drug response. The DNA-encoded characteristics are most commonly assessed based on the presence or absence of polymorphisms at the level of the nuclear DNA, but may be assessed at different levels where such DNA variation translates into different characteristics, such as differential mRNA expression or splicing, protein levels, or functional characteristics, or even physiological phenotypes – all of which would be seen as surrogate, or more integrated markers of the underlying genetic variant. (It should be noted that some authors continue to subsume all applications of expression profiling under the term 'pharmacogenomics,' in a definition of the terms that is more driven by the technology used than by functional context).

3.03.2.2 Pharmacogenomics

In contrast, the terms 'pharmacogenomics,' and its close relative, 'toxicogenomics,' are etymologically linked to 'genomics,' the study of the genome and of the entirety of expressed and nonexpressed genes in any given physiologic state. These two fields of study are concerned with a comprehensive, genome-wide assessment of the effects of pharmacological agents, including toxins/toxicants on gene expression patterns. Pharmacogenomic studies are thus used to evaluate the differential effects of a number of chemical compounds – in the process of drug discovery, commonly applied to lead selection – with regard to inducing or suppressing the expression of transcription of genes in an experimental setting. Except for situations in which pharmacogenetic considerations are 'front-loaded' into the discovery process, interindividual variations in gene sequence are not usually taken into account in this process. In contrast to pharmacogenetics, pharmacogenomics therefore does not focus on differences among individuals with regard to the drug's effects, but rather examines differences among several (prospective) drugs or compounds with regard to their biological effects using a 'generic' set of expressed or nonexpressed genes. The basis of comparison is quantitative measures of expression, using a number of more or less comprehensive gene expression-profiling methods, commonly based on microarray formats. By extrapolation from the experimental results to theoretically desirable patterns of activation or inactivation of expression of genes in the setting of integrative pathophysiology, this approach is hoped to provide a faster, more comprehensive, and perhaps even more reliable way of assessing the likelihood of finding an ultimately successful drug than previously available schemes involving mostly in vivo animal experimentation.

Table 1 Terminology

Pharmacogenetics

- Differential effects of a drug in vivo in different patients, depending on the presence of inherited gene variants
- Assessed primarily genetic (SNP) and genomic (expression) approaches
- A concept to provide more patient-/disease-specific healthcare
- One drug, many genomes (i.e., different patients)
- Focus: patient variability

Pharmacogenomics

- Differential effects of compounds – in vivo or in vitro – on gene expression, among the entirety of expressed genes
- Assessed by expression profiling
- A tool for compound selection/drug discovery
- *Many 'drugs' (i.e., early-stage compounds) – one genome (i.e., 'normative' genome (database, technology platform))*
- *Focus: compound variability*

Thus, although both pharmacogenetics and pharmacogenomics refer to the evaluation of drug effects using (primarily) nucleic acid markers and technology, the directionalities of their approaches are distinctly different: pharmacogenetics represents the study of differences among a number of individuals with regard to clinical response to a particular drug (one drug, many genomes), whereas pharmacogenomics represents the study of differences among a number of compounds with regard to gene expression response in a single (normative) genome/expressome (many drugs, one genome). Accordingly, the fields of intended use are distinct: the former will help in the clinical setting to find the medicine most likely to be optimal for a patient (or the patients most likely to respond to a drug), while the latter will aid in the setting of pharmaceutical research to find the 'best' drug candidate from a given series of compounds under evaluation.

3.03.3 Long-Term Timeframe: Causative Targets – Addressing Deranged Function Directly

3.03.3.1 Palliative and Causative-Acting Drugs

By far the largest fraction of today's pharmacopoeia does not target disease at its cause – as these causes are largely unknown – but by modulating a pathway that affects the disease-relevant phenotype or function. We refer to such drugs as symptomatic or palliative agents. The pathways they target are known from more than a century of physiological, biochemical, and pharmacological research. The pathways they modulate are disease phenotype-relevant (albeit not disease cause-relevant), and while they are not dysfunctional, their modulation can effectively be used to counterbalance the effect of a dysfunctional, disease-causing pathway. Thus signs and symptoms of the disease can be alleviated, often with striking success, notwithstanding the fact that the real cause of the disease remains untouched. A classical example of such an approach is the acute treatment of thyrotoxicity with β-adrenergic-blocking agents: even though in this case the sympathetic nervous system does not contribute causally to tachycardia and hypertension, dampening even its baseline tonus through this class of rapidly acting drugs can quickly and successfully relieve the cardiovascular symptoms and signs of this condition, and may well prevent a heart attack if the patient has underlying coronary disease, before the causal treatment (in this case, available through partial chemical ablation of the hyperactive thyroid gland) can take effect.

3.03.3.2 The Twofold Challenge of Common Complex Disease

It stands to reason, of course, that a drug that addresses the actual cause of the disease should provide superior treatment. However, finding these 'deranged functions' is not trivial, even with the aid of all the molecular biological, genetic, and genomic tools, which we command today:

There is an emerging consensus that all common complex diseases, i.e., the health problems that are by far the main contributors to society's disease burden as well as to public and private health spending, are multifactorial in nature, i.e., they are brought upon by the coincidence of certain intrinsic (inborn or acquired) predispositions and susceptibilities on the one hand, and extrinsic, environment-derived influences on the other, with the relative importance of these two influences varying across a broad spectrum. In some diseases external factors appear to be more important, while in others intrinsic predispositions prevail.

However, it is important to note – as it is commonly neglected in discussions on this topic – that in the majority of common complex diseases, genetic factors are less important than environmental and lifestyle factors, as exemplified by heritability coefficients generally between 0.2 and 0.5 for these conditions. Thus, it must be recognized from the outset that by targeting the genetic aspects of these diseases we can at best hope to make a minor contribution to curing or preventing them.

The complex nature of these diseases renders the discovery of genetic variants that contribute causally to any of them a major challenge. The complexity confronted occurs on two levels: Several inherent predisposing susceptibility traits and generally more than one environmental or lifestyle risk factor coincide in any one individual for the disease to occur; thus, any of the genetic variants present provides only a modest contribution to overall disease causation, and thus investigational hurdles of discovering it. This intraindividual complexity is further accentuated by inter-individual diversity based on the fact that any one clinical diagnosis is bound to be etiologically heterogeneous at the level of molecular pathology. Thus, consensus exist that the same 'conventional' clinical diagnosis given to different individuals is quite likely to reflect the outcome of different constellations of inborn susceptibility factors and/or of environmental and lifestyle-related risks. So we may expect that both on the level of an individual patient, and even more so on the population level – where all relevant genetic-epidemiological studies need to be conducted – the disease-causing (or, better, disease-contributing) effect of any one intrinsic, genetically encoded characteristic will be, by and large, quite modest, and likely drowned out by noise, unless very large (and very expensive) studies are conducted. Carrying out such studies, which will attach function and clinical outcomes to the human genome sequence and its variations, is a huge task that looms many times larger than the sequencing effort itself. However, it is the sine qua non without which the whole genome-sequencing effort will essentially remain inconsequential.

Since large-scale efforts based primarily on the sib-pair linkage study design have failed, with few exceptions, to yield disease gene variants in common complex disease, it appears that populations with large, genealogically well-characterized families and a certain degree of 'genetic isolation' may provide the most likely successful approach toward characterization of these disease susceptibility gene variants. Not many such populations exist; examples are the Utah Mormons, the Icelandic nation, and the French settlers in northern Quebec, Canada, where initial work has proved to be promising. Of course, results obtained in such isolated populations will need to be validated in other, more 'mainstream' populations, if the new target is to be pursued strictly as a causative target.

Common, complex diseases, and thus the vast majority of what is to be 'clinically applied genetics,' behave almost fundamentally different from rare, classic, monogenic, 'Mendelian' diseases. While in the latter the impact of the genetic variant is typically categorical in nature, i.e., deterministic, in the former case the presence of a disease-associated genetic variant is merely of probabilistic value, raising (or lowering) the likelihood of disease occurrence to some extent, but never predicting it in a black-and-white fashion.

Communicating this difference to a public that has long been misled into a perception of everything 'genetic' being of deterministic, Mendelian quality represents a second, no less important and difficult challenge. Unless we succeed, by engaging in a true dialogue with all stakeholders, in providing the basis for informed discourse and sensible decision making on the societal level, the full potential of our deepening understanding of biology and of these technological advances will not be recognized.

3.03.3.3 Applicability of 'Enetic' Targets/Medicines

Will such newly discovered causative targets, and the medicines that may eventually be developed to modulate them, be indeed applicable only to that fraction of the population with the clinical diagnosis in whom the targeted mechanism contributes materially to the disease? It is too early to tell. Clearly, in the latter group such medicines may be expected to be particularly effective, and sometimes they will be exclusively effective in these patients; however, by uncovering what essentially will commonly be a new, previously not recognized mechanistic pathway, such drugs may be of value as palliative medicines also in those individuals in whom the mechanism in question is actually not dysfunctional. The former case is illustrated by Herceptin (see below); an example for the latter is the finding that glucokinase activators, which correct the molecular defect in the rare mature-onset diabetes of the young (MODY) type 2 patients in whom the enzyme is dysfunctional, can raise the activity of normal glucokinase and thus lower glucose in just about everyone.

3.03.3.4 Acceleration of Drug Discovery/Development through 'Better' Targets?

Hopes have been nurtured by some that the implementation of genetics and genomics would 'smarten' up the drug discovery process and thus potentially accelerate it and reduce its cost. Quite to the contrary, it appears for now that the opposite is happening: latencies from inception of a project to the launch of a new drug have lengthened to almost 15 years, and the average cost per successful launch has gone up to almost US $900 million.[28] Interestingly, about two-thirds of the costs incurred are now apparently spent in the preclinical phase, according to this report. While it is unclear to what extent these data truly reflect reality, it stands to reason that the preclinical phase, in particular, would be more lengthy and expensive if one tackles a completely novel target, about which – other than the association with disease that the genetic approach provides us with – virtually nothing is known. This may be complicated by the fact that such targets may not belong to the classical few 'druggable' target families: such a target may not be chemically tractable at all, or may encounter additional hurdles due to the bias of most chemical libraries in favor of 'conventional' target families. These formidable challenges are somewhat counterbalanced by the expectation that targets selected based on causative disease contribution may, overall, have a somewhat higher likelihood of success; thus, if today's attrition rate of 9 per 10 compounds that are introduced into clinical testing could be reduced by as little as to 8 of 10, this would translate, over time, into a doubling of productivity.

3.03.4 Midterm Timeframe: Pharmacogenomics/Toxicogenomics – Finding New Medicines Quicker and More Efficiently

Once a screen (assay) has been set up in a drug discovery project, and lead compounds are identified, the major task becomes the identification of an optimized clinical candidate molecule among the many compounds synthesized by medicinal chemists. Conventionally, such compounds are screened in a number of animal or cell models for efficacy and toxicity, experiments that, while having the advantage of being conducted in the in vivo setting, commonly take significant amounts of time and depend entirely on the similarity between the experimental animal condition/setting and its human counterpart, i.e., the validity of the model.

Although such experiments will never be entirely replaced by expression profiling on either the nucleic acid (genomics) or the protein (proteomics) level, these techniques offer powerful advantages and complementary information. First, efficacy and profile of induced changes can be assessed in a comprehensive fashion (within the limitations – primarily sensitivity and completeness of transcript representation – of the technology platform used). Second, these assessments of differential efficacy can be carried out much more expeditiously than in conventionally used, (patho-)physiology-based animal models. Third, the complex pattern of expression changes revealed by such experiments may provide new insights into possible biological interactions between the actual drug target and other biomolecules, and thus reveal new elements or branch-points of a biological pathway that may be useful as surrogate markers, novel diagnostic analytes, or additional drug targets. Fourth, increasingly important, these tools serve to determine specificity of action among members of gene families that may be highly important for both efficacy and safety of a new drug. It must be borne in mind that any and all such experiments are limited by the coefficient of correlation with which the expression patterns determined are linked to the desired in vivo physiological action of the compound.

A word of caution regarding microarray-based expression profiling would appear to be in order: the power of comprehensive (almost) genome-wide assessment of expression patterns has led to what may justly be described as somewhat of an infatuation with this technology that at times leaves a degree of critical skepticism to be desired. In particular, the pairwise comparison algorithms used in much of this work (competition staining of a case and a control sample on the same physical array) raise a number of questions regarding selection bias which take on particular significance since the overall sample sizes are commonly (very) small. Biostatistical analytical approaches are commonly less than sophisticated, if used at all. Additionally, it is important to remain aware of the fact that all microarray expression data are of only associative character, and must be interpreted mindful of this limitation.

As a subcategory of this approach, toxicogenomics is increasingly evolving as a powerful adjuvant to classic toxicological testing. As pertinent databases are being created from experiments with known toxicants, revealing expression patterns that may potentially be predictive of longer-term toxic liabilities of compounds, future drug discovery efforts should benefit by insights allowing earlier 'killing' of compounds likely to cause such complications.

When using these approaches in drug discovery, even if implemented with proper biostatistics and analytical rigor, it is imperative to understand the probabilistic nature of such experiments. A promising profile on pharmacogenomic and toxicogenomic screens will enhance the likelihood of having selected an ultimately successful compound, and will achieve this goal quicker than conventional animal experimentation, but will do so only with a certain likelihood of success. The less reductionist approach of the animal experiment will still be needed. It is to be anticipated, however, that such approaches will constitute an important time- and resource-saving first evaluation or screening step that will help to focus and reduce the number of animal experiments that will ultimately need to be conducted.

3.03.5 Short-Term Timeframe: Pharmacogenetics – More Targeted, More Effective Medicines

3.03.5.1 Genes and Environment

It is common knowledge that today's pharmacopoeia, in as much as it represents enormous progress compared with what our physicians had only 15 or 20 years ago, is far from perfect. Many patients respond only partially, or fail to respond altogether, to the drugs they are given, and others suffer adverse events that range from unpleasant to serious and life-threatening.

If we regard a pharmacological agent as one of the extrinsic environmental factors in a common complex disease scenario, with the potential to affect the health status of the individual to whom it is administered, then individually differing responses to such an agent would, under the multifactorial and heterogeneous paradigm of common complex disease elaborated upon earlier, be regarded as the expression of differences in the 'intrinsic' characteristics of these patients. This is true as long as we can exclude variation in the exposure to the drug. This is important, as in clinical practice nonadherence to prescribed regimens of administration, or drug–drug interactions interfering with bioavailability of the drug, is by far the most likely culprit when such differences in response phenotype are observed. The influence of such intrinsic variation on drug response may be predicted to be more easily recognizable and more relevant the steeper the dose–response curve of a given drug is.

The argument for the particular likelihood of observing environmental factor–gene interactions with drugs among all other environmental influences goes along the same lines. Among all these environmental factors to which we are exposed, drugs might be particularly likely to interact specifically and selectively with the genetic properties of a given individual, as their potency and, compared, say, to foodstuffs, narrow therapeutic window make interactions more likely with innate individual susceptibilities that affect the interaction with drugs.

Clearly, a better, more fundamental and mechanistic understanding of the molecular pathology of disease in general and of the role of intrinsic, biological properties regarding the predisposition to contract such diseases, as well as of drug action on the molecular level, will be essential for future progress in healthcare. Current progress in molecular biology and genetics has indeed provided us with some of the prerequisite tools that should help us reach the goal of such a more refined understanding.

3.03.5.2 An Attempt at a Systematic Classification of Pharmacogenetics

Two conceptually quite different scenarios of interindividually differential drug response may be distinguished on the basis of the underlying biological variance (**Table 2**):

1. In the first case, the underlying biological variation is in itself not disease-causing or -contributing, and only becomes clinically relevant in response to the exposure to the drug in question (classical pharmacogenetics).
2. In the second case, the biological variation is directly disease-related, is per se of pathological importance, and represents a subgroup of the overall clinical disease/diagnostic entity. The differential response to a drug is thus related to how well this drug addresses, or is matched to, the presence or relative importance of the pathomechanism it targets, in different patients, i.e., the 'molecular differential diagnosis' of the patient ('disease mechanism-related pharmacogenetics').

Although these two scenarios are conceptually rather different, they result in similar practical consequences with regard to the administration of a drug, namely stratification based on a particular DNA-encoded marker. It seems therefore legitimate to subsume both under the umbrella of 'pharmacogenetics.'

3.03.5.3 'Classical Pharmacogenetics'

This category includes differential pharmacokinetics and pharmacodynamics.

Table 2 Pharmacogenetics systematic classification

'Classical' pharmacogenetics
Pharmacokinetics
Absorption
Metabolism
Activation of prodrugs
Deactivation
Generation of biologically active metabolites
Distribution
Elimination
Pharmacodynamics
Palliative drug action (modulation of disease-symptoms or disease-signs by targeting physiologically relevant systems, without addressing those mechanism that cause or causally contribute to the disease)
'Molecular differential-diagnosis-related' pharmacogenetics

Causative drug action (modulation of actual causative of contributory mechanisms by drug).

Pharmacokinetic effects due to interindividual differences in absorption, distribution, metabolism (with regard to both activation of prodrugs, inactivation of the active molecule, and generation of derivative molecules with biological activity), or excretion of the drug. In any of these cases, differential effects observed are due to the presence at the intended site of action of either inappropriate concentrations of the pharmaceutical agent, or of inappropriate metabolites, or of both, resulting in either lack of efficacy or toxic effects. Pharmacogenetics, as it relates to pharmacokinetics, has been recognized as an entity for more than 100 years, going back to the observation, commonly credited to Archibald Garrod, that a subset of psychiatric patients treated with the hypnotic, sulfonal, developed porphyria. We have since then come to understand the underlying genetic causes for many of the previously known differences in enzymatic activity, most prominently with regard to the P450 enzyme family, and these have been the subject of recent reviews (**Table 3**).[1,2] However, such pharmacokinetic effects are also seen with membrane transporters, such as in the case of differential activity of genetic variants of multidrug-resistant transporter (MDR1) that affects the effective intracellular concentration of antiretrovirals,[24] or of the purine analog-metabolizing enzyme, thiomethyl purine transferase.[25]

Notably, despite the widespread recognition of isoenzymes with differential metabolizing potential since the middle of the twentieth century, the practical application and implementation of this knowledge have been minimal so far. This may be the consequence, on the one hand, of the irrelevance of such differences in the presence of relatively flat dose–effect curves (i.e., a sufficiently wide therapeutic window), as well as, on the other hand, the fact that many drugs are subject to complex, parallel metabolizing pathways, where in the case of underperformance of one enzyme, another one may compensate. Such compensatory pathways may well have somewhat different substrate affinities, but allow plasma levels to remain within therapeutic concentrations. Thus, the number of such polymorphisms that have found practical applicability is rather limited and, by and large, restricted to determinations of the presence of functionally deficient variants of the enzyme thiopurine methyltransferase, in patients prior to treatment with purine analog chemotherapeutics.

In contrast, pharmacodynamic effects may lead to interindividual differences in a drug's effects despite the presence of appropriate concentrations of the intended active (or activated) drug compound at the intended site of action. Here, DNA-based variation in how the target molecule or another (downstream) member of the target molecule's mechanistic pathway can respond to the medicine modulates the effects of the drug. This will apply primarily to palliatively working medicines, as discussed above.

A schematic (**Figure 1**) is provided to help clarify these somewhat complex concepts, in which a hypothetical case of a complex trait/disease is depicted where excessive, dysregulated function of one of the trait-controlling/contributing pathways (**Figure 1A and B**) causes symptomatic disease. The example used refers to blood pressure as the trait, and hypertension as the disease in question, respectively (for the case of a defective or diminished function of a pathway, an analogous schematic could be constructed, and again for a deviant function). A palliative treatment would be one that addresses one of the pathways that, while not dysregulated, contributes to the overall deviant physiology (**Figure 1F**), while the respective pharmacogenetic pharmacodynamic scenario would occur if this particular pathway was, due to a genetic variant, not responsive to the drug chosen (**Figure 1G**). A palliative treatment may also be ineffective if the particular mechanism targeted by the palliative drug due to the presence of a molecular variant provides less than the physiologically expected baseline contribution to the relevant phenotype (**Figure 1H**). In such a case, modulating an a priori unimportant pathway in the disease scenario will not yield successful palliative treatment results (**Figure 1I and J**).

Several of the most persuasive examples we have accumulated to date for such palliative-drug-related pharmacogenetic effects have been observed in the field of asthma. The treatment of asthma relies on an array of drugs aimed at modulating different 'generic' pathways, thus mediating bronchodilation or antiinflammatory effects, often without regard to the possible causative contribution of the targeted mechanism to the disease. One of the mainstays of the treatment of asthma is activation of the β_2-adrenoceptor by specific agonists, which leads to relaxation of bronchial smooth muscles and, consequently, bronchodilation. Recently, several molecular variants of the β_2-adrenoceptor have been shown associated with differential treatment response to such β_2-agonists.[13,14] Individuals carrying one or two copies of a variant allele that contains a glycine in place of arginine in position 16 were found to have a three- and fivefold reduced response to the agonist, respectively.

This was shown in both in vitro[15,16] and in vivo[16] studies to correlate with an enhanced rate of agonist-induced receptor downregulation, but not with any difference in transcriptional or translational activity of the gene, or with agonist binding. In contrast, a second polymorphism affecting position 19 of the β upstream peptide was shown to affect translation (but not transcription) of the receptor itself, with a 50% decrease in receptor numbers associated with the variant allele – which happens to be in strong linkage disequilibrium with a variant allele position 16 in the receptor. The simultaneous presence of both mutations would thus be predicted to result in low expression and enhanced

Table 3 Pharmacogenetics: chronology and systematics

Pharmacogenetic phenotype	Described	Underlying gene/ mutation	Identified
Sulfonal porphyria	*c.* 1890	Porphobilinogen-deaminase?	1985
Suxamethonium hypersensitivity	1957–1960	Pseudocholinesterase	1990–1992
Primaquin hypersensitivity; favism	1958	G-6-PD	1988
Long QT syndrome	1957–1960	*hERG*, etc.	1991–1997
Isoniazid slow/fast acetylation	1959–1960	*N*-acetyltranferase	1989–1993
Malignant hyperthermia	1960–1962	Ryanodine receptor	1991–1997
Fructose intolerance	1963	Aldolase B	1988–1995
Vasopressin insensitivity	1969	Vasopressin receptor 2	1992
Alcohol susceptibility	1969	Aldehyde dehydrogenase	1988
Debrisoquine hypersensitivity	1977	CYP2D6	1988–1993
Retinoic acid resistance	1970	PML-RARA fusion gene	1991–1993
6-Mercaptopurin toxicity	1980		
Thiopurinemethyltransferase	1995		
Mephenytoin resistance	1984	CYP2C19	1993–1994
Insulin-insensitivity	1988	Insulin receptor	1988–1993
Phase I enzyme	**Testing substance**		
Aldehyde dehydrogenase	Acetaldehyde		
Alcohol dehydrogenase	Ethanol		
CYP1A2	Caffeine		
CYP2A6	Nicotine, coumarin		
CYP2C9	Warfarin		
CYP2C19	Mephenytoin, omeprazole		
CYP2D6	Dextromethorphan, debrisoquine, sparteine		
CYP2E1	Chloroxazone, caffeine		
CYP3A4	Erythromycin		
CYP3A5	Midazolam		
Serum cholinesterase	Benzoylcholine, butyrylcholine		
Paraoxonase/arylesterase	Paraoxon		
Phase II enzyme			
Acetyltransferase (NAT1)	*para*-aminosalicylic acid		
Acetyltransferase (NAT2)	Isoniazid, sulfamethazine, caffeine		
Dihydropyrimidin dehydrogenase	5-fluorouracil		
Glutathione transferase (GST-M 1)	*trans*-stilbene oxide		
Thiomethyltransferase	2-mercaptoethanol, D-penicillamine, captopril		
Thiopurine methyltransferase	6-mercaptopurine, 6-thioguanine, 8-azathioprine		
UDP-glucuronosyltransferase (UGT1A)	Bilirubin		
UDP-glucuronosyltransferase (UGT2B7)	Oxazepam, ketoprofen, estradiol, morphine		

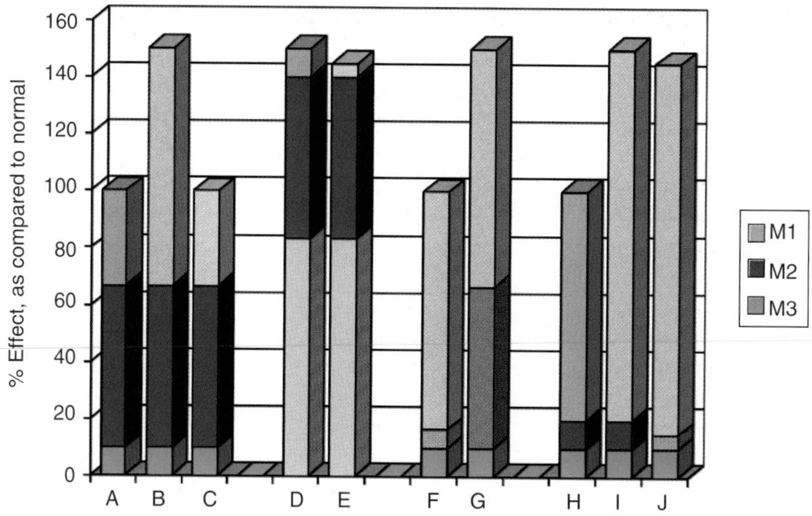

Figure 1 A: normal physiology: three molecular mechanisms (M1, M2, M3) contribute to a trait (such as blood sugar or blood pressure, or cell proliferation); B: diseased physiology D1: derailment (cause/contribution) of molecular mechanism 1 (M1); C: diseased physiology D1: causal treatment T1 (aimed at M1); D: diseased physiology D3: derailment (cause/contribution) of molecular mechanism 3 (M3); E: diseased physiology D3, treatment T1: treatment does not address cause; F: diseased physiology D1, palliative treatment T2 (aimed at M2); G: diseased physiology D1, palliative treatment T2; T2-refractory gene variant in M2; H: normal physiology variant: differential contribution of M1 and M2 to normal trait; I: diseased physiology D1-variant: derailment of mechanism M1; J: diseased physiology D1-variant: treatment with T2. Solid colors indicate normal function; stippling indicates pathologic dysfunction; hatching indicates therapeutic modulation.

downregulation of an otherwise functionally normal receptor, depriving patients carrying such alleles of the benefits of effective bronchodilation as a palliative (i.e., noncausal) countermeasure to their pathological airway hyperreactivity. Importantly, there is no evidence that any of the allelic variants encountered are associated with the prevalence or incidence, and thus potentially the etiology of the underlying disease.[17,18] This would reflect the scenario depicted in **Figure 1H**.

Inhibition of leukotriene synthesis, another palliative approach toward the treatment of asthma, proved clinically ineffective in a small fraction of patients who carried only non-wildtype alleles of the 5-lipoxygenase promoter region.[12] These allelic variants had previously been shown to be associated with decreased transcriptional activity of the gene.[5] It stands to reason – consistent with the clinical observations – that in the presence of already reduced 5-lipoxygenase activity, pharmacological inhibition may be less effective (**Figure 1H–J**). Of note, again, there is no evidence for a primary, disease-causing or -contributing role of any 5-lipoxygenase variants; all of them were observed at equal frequencies in disease-affected and nonaffected individuals.[5]

Pharmacogenetic effects may not only account for differential efficacy, but also contribute to differential occurrence of adverse effects. An example for this scenario is provided by the well-documented 'pharmacogenetic' association between molecular sequence variants of the 12S rRNA, a mitochondrion-encoded gene, and aminoglycoside-induced ototoxicity.[19] Intriguingly, the mutation that is associated with susceptibility to ototoxicity renders the sequence of the human 12S rRNA similar to that of the bacterial 12S rRNA gene, and thus effectively turns the human 12S rRNA into the (bacterial) target for aminoglycoside drug action, presumably mimicking the structure of the bacterial binding site of the drug.[20] As in the other examples, the presence of the 12S rRNA mutation per se has no primary, drug treatment-independent pathologic effect per se.

One may speculate that, analogously, such 'molecular mimicry' may occur within one species: adverse events may arise if the selectivity of a drug is lost because a gene that belongs to the same gene family as the primary target loses its identity vis-à-vis the drug and attains, based on its structural similarity with the principal target, similar or at least increased affinity to the drug. Depending on the biological role of the 'imposter' molecule, adverse events may occur – even though the variant molecule, again, may be quite silent with regard to any contribution to disease causation. Although we currently have no obvious examples for this scenario, it is certainly imaginable for various classes of receptors and enzymes.

3.03.5.4 Pharmacogenetics as a Consequence of Molecular Differential Diagnosis

As alluded to earlier, there is general agreement today that any of the major clinical diagnoses in the field of common complex disease, such as diabetes, hypertension, or cancer, are comprised of a number of etiologically (i.e., at the molecular level) more or less distinct subentities. In the case of a causally acting drug this may imply that the agent will only be appropriate, or will work best, in that fraction of all the patients who carry the (all-inclusive and imprecise) clinical diagnosis in whom the dominant molecular etiology, or at least one of the contributing etiological factors, matches the biological mechanism of action that the drug in question modulates (**Figure 1C**). If the mechanism of action of the drug addresses a pathway that is not disease-relevant – perhaps already downregulated as an appropriate physiologic response to the disease, then the drug may logically be expected not to show efficacy (**Figure 1D and E**).

Thus, unrecognized and undiagnosed disease heterogeneity, disclosed indirectly by the presence or absence of response to a drug targeting a mechanism that contributes to only one of several molecular subgroups of the disease, provides an important explanation for differential drug response and likely represents a substantial fraction of what we today somewhat indiscriminately subsume under the term 'pharmacogenetics.'

Currently, the most frequently cited example for this category of 'pharmacogenetics' is trastuzamab (Herceptin), a humanized monoclonal antibody directed against the *her-2* oncogene. This breast cancer treatment is prescribed based on the level of *her-2* oncogene expression in the patient's tumor tissue. Differential diagnosis at the molecular level not only provides an added level of diagnostic sophistication, but also actually represents the prerequisite for choosing the appropriate therapy. Because trastuzamab specifically inhibits a 'gain-of-function' variant of the oncogene, it is ineffective in the two-thirds of patients who do not overexpress the drug's target, whereas it significantly improves survival in the one-third of patients who constitute the subentity of the broader diagnosis 'breast cancer' in whom the gene is expressed.[3] Parenthetically, some have argued against this being an example of 'pharmacogenetics,' because the parameter for patient stratification (i.e., for differential diagnosis) is the somatic gene expression level rather than particular 'genotype' data.[4] This is a difficult argument to follow, since in the case of a treatment effect-modifying germline mutation it would obviously not be the nuclear gene variant per se, but rather its specific impact on either structure/function or on expression of the respective gene/gene product that would represent the actual physiological corollary underlying the differential drug action. Conversely, an a priori observed expression difference is highly likely to reflect a potentially as yet undiscovered sequence variant. Indeed, as pointed out earlier, there are a number of examples in the field of pharmacogenomics where the connection between genotypic variant and altered expression has already been demonstrated.[5,6]

Another example, although still hypothetical, of how proper molecular diagnosis of relevant pathomechanisms will significantly influence drug efficacy, is in the evolving class of anti-acquired immunodeficiency syndrome (AIDS)/ human immunodeficiency virus (HIV) drugs that target the CCR5 cell surface receptor.[7–9] These drugs would be predicted to be ineffective in those rare patients who carry the δ32 variant, but who nevertheless have contracted AIDS or test HIV-positive (most likely due to infection with an SI (simian immunodeficiency) virus phenotype that utilizes CXCR4).[10,11]

It should be noted that the pharmacogenetically relevant molecular variant need not affect the primary drug target, but may equally well be located in another molecule belonging to the system or pathway in question, either up- or downstream in the biological cascade with respect to the primary drug target.

3.03.5.5 Different Classes of Markers

Pharmacogenetic phenomena, as pointed out previously, need not be restricted to the observation of a direct association between allelic sequence variation and phenotype, but may extend to a broad variety of indirect manifestations of underlying, but often (as yet) unrecognized sequence variation. Thus, differential methylation of the promoter region of O6-methylguanine DNA methylase has recently been reported to be associated with differential efficacy of chemotherapy with alkylating agents. If methylation is present, expression of the enzyme that rapidly reverses alkylation and induces drug resistance is inhibited, and therapeutic efficacy is greatly enhanced.[21]

3.03.5.6 Complexity is to be Expected

In the real world, it is likely that not only one of the scenarios depicted, but a combination of several ones, may affect how well a patient responds to a given treatment, or how likely it is that he or she will suffer an adverse event. Thus, a fast-metabolizing patient with poor-responder pharmacodynamics may be particularly unlikely to gain any benefit from taking the drug in question, while a slow-metabolizing status may counterbalance in another patient the same inopportune pharmacodynamics, while a third patient, who is a slow metabolizer and displaying normal pharmacodynamics, may be more likely to suffer adverse events. In all of them, both the pharmacokinetic and pharmacodynamics properties may result from the interaction of several of the mechanisms described above. In addition, we know of course that coadministration of other drugs or even the consumption of certain foods may affect and further complicate the picture for any given treatment.

3.03.6 Incorporating Pharmacogenetics into Drug Development Strategy

3.03.6.1 Diagnostics First, Therapeutics Second

It is important to note that, despite the public hyperbole and the high-strung expectations surrounding the use of pharmacogenetics to provide 'personalized care,' these approaches are likely to be applicable to only a fraction of medicines that are being developed. Further, if and when such approaches will be used, they will represent no radical new direction or concept in drug development but simply a stratification strategy as we have been using it all along.

An increasingly sophisticated and precise diagnosis of disease, arising from a deeper, more differentiated understanding of pathology at the molecular level, that will increasingly subdivide today's clinical diagnoses into molecular subtypes, will foster medical advances which, if considered from the viewpoint of today's clinical diagnosis, will appear as 'pharmacogenetic' phenomena, as described above. However, the sequence of events that is today often presented as characteristic for a 'pharmacogenetic scenario,' namely, exposing patients to the drug, recognizing a differential (i.e., (quasi-)bimodal) response pattern, discovering a marker that predicts this response, and creating a diagnostic product to be co-marketed with the drug henceforth, is likely to be reversed. Rather, in the case of 'pharmacogenetics' due to a match between drug action and dysregulation of a disease-contributing mechanism, we will likely search for a new drug specifically, and a priori, based on a new mechanistic understanding of disease causation or contribution (i.e., a newly found ability to diagnose a molecular subentity of a previously more encompassing, broader, and less precise clinical disease definition). Thus, pharmacogenetics will not be so much about finding the 'right medicine for the right patient,' but about finding the 'right medicine for the disease (subtype),' as we have aspired to do all along throughout the history of medical progress. This is, in fact, good news: the conventional 'pharmacogenetic scenario' would invariably present major challenges from both a regulatory and a business development and marketing standpoint, as it would confront development teams with a critical change in the drug's profile at a very late point during the development process. In addition, the timely development of an approvable diagnostic in this situation is difficult at best, and its marketing as an 'add-on' to the drug is a less than attractive proposition to diagnostics business. Thus, the 'practice' of pharmacogenetics will, in many instances, be marked by progress along the very same path that has been one of the main avenues of medical progress for the last several hundred years: differential diagnosis first, followed by the development of appropriate, more specific treatment modalities.

Thus, the sequence of events in this case may well involve, first, the development of an in vitro diagnostic test as a stand-alone product that may be marketed on its own merits, allowing the physician to establish an accurate, state-of-the-art diagnosis of the molecular subtype of the patient's disease. Sometimes such a diagnostic may prove helpful even in the absence of specific therapy by guiding the choice of existing medicines and/or of nondrug treatment modalities such as specific changes in diet or lifestyle. Availability of such a diagnostic, as part of the more sophisticated understanding of disease, will undoubtedly foster and stimulate the search for new, more specific drugs; and once such drugs are found, availability of the specific diagnostic will be important for carrying out the appropriate clinical trials. This will allow a prospectively planned, much more systematic approach toward clinical and business development, with a commensurate greater chance of actual realization and success.

3.03.6.2 Probability, Not Certainty

In practice, some extent of guesswork will remain, due to the nature of common complex disease. First, all diagnostic approaches, including those based on DNA analysis in common complex disease, as stressed above, will ultimately only provide a measure of probability, not of certainty. Thus, although the variances of drug response among patients who do (or do not) carry the drug-specific subdiagnosis will be smaller, there will still be a distribution of differential responses. Although by and large the drug will work better in the 'responder' group, there will be some patients among this subgroup who will respond less or not at all, and conversely, not everyone belonging to the 'nonresponder' group will completely fail to respond, depending perhaps on the relative magnitude with which the particular mechanism contributes to the disease. It is important to bear in mind, therefore, that even in the case of fairly obvious bimodality,

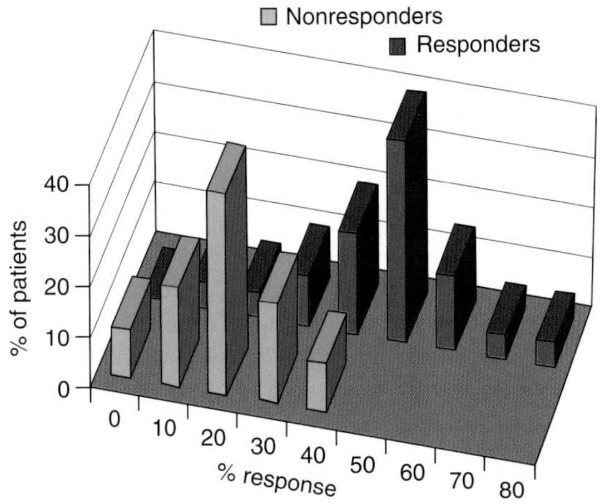

Figure 2 'Responder' status is a quantitative, not qualitative variable.

patient responses will still show distribution patterns, and that all predictions as to responder or nonresponder status will only have a certain likelihood of being accurate (**Figure 2**). The terms 'responder' and 'nonresponder' as applied to groups of patients stratified based on a DNA marker represent, therefore, Mendelian thinking-inspired misnomers that should be replaced by more appropriate terms that reflect the probabilistic nature of any such classification, e.g., 'likely (non)responder.'

In addition, based on our current understanding of the polygenic and heterogeneous nature of these disorders, and even in an ideal world where we would know about all possible susceptibility gene variants for a given disease and have treatments for them, we will only be able to exclude, in any one patient, those that do not appear to contribute to the disease, and therefore deselect certain treatments. We will, however, most likely find ourselves left with a small number – two to four, perhaps – of potentially disease-contributing gene variants whose relative contribution to the disease will be very difficult, if not impossible, to rank in an individual patient. It is likely then, that trial and error, and this great intangible quantity, 'physician experience,' will still play an important role, albeit on a more limited and subselective basis.

The alternative scenario, where differential drug response and/or safety occur as a consequence of a pathologically not relevant, purely drug response-related pharmacogenetics scenario, is more likely to present greater difficulty in planning and executing a clinical development program because, presumably, it will be more difficult to anticipate or predict differential responses a priori. When such a differential response occurs, it will also potentially be more difficult to find the relevant marker(s), unless it happens to be among the 'obvious' candidate genes implicated in the disease physiopathology or the treatment's mode of action. Although screening for molecular variants of these genes, and testing for their possible associations with differential drug response, is a logical first step, if unsuccessful, it may be necessary to embark on an unbiased genome-wide screen for such a marker or markers. Despite recent progress in high-throughput genotyping, the obstacles that will have to be overcome on the technical, data analysis, and cost levels are formidable. They will limit the deployment of such programs, at least for the foreseeable future, to select cases in which there are very solid indications for doing so, based on clinical data showing a near-categorical (e.g., bimodal) distribution of treatment outcomes. Even then, we may expect to encounter for every success – that will be owed to a favorably strong linkage disequilibrium across considerable genomic distance in the relevant chromosomal region – as many or more failures, in cases where the culpable gene variant cannot be found due to the higher recombination rate or other characteristics of the stretch of genome on which it is located.

3.03.7 Regulatory Aspects

At the time of writing, regulatory agencies in both Europe and the US are beginning to show a keen interest in the potential role that pharmacogenetic approaches may play in the development and clinical use of new drugs, and the potential challenges that such approaches may present to the regulatory approval process. While no formal guidelines have been issued, the pharmaceutical industry has already been reproached – albeit in a rather nonspecific manner – for not being more proactive in the use of pharmacogenetic markers. It will be of key importance for all concerned to engage in an intensive dialogue at the end of which, it is hoped, will emerge a joint understanding that stratification according to DNA-based markers is fundamentally nothing new, and not different from stratification according to any other clinical or demographic parameter, as has been used all along.

Still, based on the (in the case of common complex diseases scientifically unjustified) perception that DNA-based markers represent a different class of stratification parameters, a number of important questions will need to be addressed and answered – hopefully always in analogy to 'conventional' stratification parameters, including those referring to ethical aspects. Among the most important ones are questions concerning:

- the need and/or ethical justification (or lack thereof) to include likely nonresponders in a trial for the sake of meeting safety criteria, which, given the restricted indication of the drug, may indeed be excessively broad
- the need to carry out conventional-size safety trials in the disease stratum eligible for the drug (if the stratum represents a relatively small fraction of all patients with the clinical diagnosis, it may be difficult to amass sufficient numbers and/or discourage companies from pursuing such drugs to the disadvantage of patients)
- the need to use active controls if the patient/disease stratum is different from that in which the active control was originally tested
- the strategies to develop and gain approval for the applicable first-generation diagnostic, as well as for the regulatory approval of subsequent generations of tests to be used to determine eligibility for prescription of the drug
- a number of ethical–legal questions relating to the unique requirements regarding privacy and confidentiality for 'genetic testing' that may raise novel problems with regard to regulatory audits of patient data (see below)

A concerted effort to avoid what has been termed 'genetic exceptionalism' – the differential treatment of DNA-based markers as compared with other personal medical data – should be made so as not to complicate further unnecessarily the already very difficult process of obtaining regulatory approval. This seems justified based on the recognized fact that, in the field of common complex disease, DNA-based markers are not at all different from 'conventional' medical data in all relevant aspects – namely specificity, sensitivity, and predictive value.

3.03.8 Pharmacogenetic Testing for Drug Efficacy versus Safety

3.03.8.1 Greater Efficacy: Likely

In principle, pharmacogenetic approaches may be useful both to raise efficacy and to avoid adverse events, by stratifying patient eligibility for a drug according to appropriate markers. In both cases, clinical decisions and recommendations must be supported by data that have undergone rigorous biostatistical scrutiny. Based on the substantially different prerequisites for and opportunities to acquire such data, and to apply them to clinical decision making, we expect the use of pharmacogenetics for enhanced efficacy to be considerably more common than for the avoidance of adverse events.

The likelihood that adequate data on efficacy in a subgroup may be generated is reasonably high, given the fact that, unless the drug is viable in a reasonably sizeable number of patients, it will probably not be developed for lack of a viable business case, or at least only under the protected environment of orphan drug guidelines. Implementation of pharmacogenetic testing to stratify for efficacy, provided that safety in the nonresponder group is not an issue, will primarily be a matter of physician preference and sophistication, and potentially of third-party payer directives, but would appear less likely to become a matter of regulatory mandate, unless a drug has been developed selectively in a particular stratum of the overall indication (in which case a contraindication label for other strata is likely to be issued). Indeed, an argument can be made against depriving those who carry the 'likely nonresponder' genotype regarding eligibility for the drug, but who individually, of course, may respond to the drug with a certain, albeit lower, probability. From a regulatory aspect, use of pharmacogenetics for efficacy, if adequate safety data exist, appears largely unproblematic – the worst-case scenario (a genotypically inappropriate patient receiving the drug) would result in treatment without expected beneficial effect, but with no increased odds to suffer adverse consequences, i.e., much of what one would expect under conventional paradigms.

3.03.8.2 Avoidance of Serious Adverse Effects: Less Likely – With Exceptions

The utility and clinical application of pharmacogenetic approaches toward improving safety, in particular with regard to serious adverse events, will meet with considerably greater hurdles and are therefore less likely expected to become reality. A number of reasons are cited for this: first, in the event of serious adverse events associated with the use of a widely prescribed medicine, withdrawal of the drug from the market is usually based almost entirely on anecdotal evidence from a rather small number of cases – in accordance with the Hippocratic mandate 'primum non nocere.' If the sample size is insufficient to demonstrate a statistically significant association between drug exposure and event, as is typically the case, it will most certainly be insufficient to allow meaningful testing for genotype–phenotype correlations; the biostatistical hurdles become progressively more difficult as many markers are tested and the number of degrees of freedom applicable to the analysis for association continues to rise. Therefore, the fraction of attributable risk shown to be associated with a given at-risk (combination of) genotype(s) would have to be very substantial for regulators to accept such data. Indeed, the low prior probability of the adverse event, by definition, can be expected to yield an equally low positive (or negative) predictive value. Second, the very nature of safety issues raises the hurdles substantially because in this situation the worst-case scenario – administration of the drug to the 'wrong' patient – will result in higher odds of harming the patient.

Therefore, it is likely that the practical application of successfully investigating and applying pharmacogenetics toward limiting adverse events will likely be restricted to the more exceptional case of diseases with dire prognosis, where a high medical need exists, where the drug in question offers unique potential advantages (usually bearing the characteristics of a 'life-saving' drug), and where, therefore, the tolerance even for relatively severe side effects is much greater than for other drugs. This applies primarily to areas like oncology or HIV/AIDS, for which the recently reported highly specific and acceptably sensitive association between the major histocompatibility complex gene variant, *HLA B5701*, and occurrence of a severe hypersensitivity reaction is a prime example.[26,27]

In most other indications, the sobering biostatistical and regulatory considerations discussed represent barriers that are unlikely to be overcome easily. The proposed, conceptually highly attractive, routine deployment of pharmacogenetics as generalized drug surveillance or pharmacovigilance practice following the introduction of a new pharmaceutical agent[22] faces these scientific as well as formidable economic hurdles.

3.03.9 Challenge: Genetics and Society – Ethical–Legal–Societal Issues

3.03.9.1 Data Protection Needs to be Matched by 'Person Protection'

Whereas public attitudes toward genetics and genomics span a broad spectrum from enthusiastic approval to total rejection, it is fair to say that, among the most outspoken voices from the 'lay'-community, there are many who are skeptical, critical, or outright negative about the pursuit of genetic research and the possible implementations of some of its outcomes in society. There is a widespread public sentiment of fear of the consequences which centers primarily on two issues: genetic engineering and the lack of control over one's private medical-genetic data. Both concerns are thoroughly understandable, if not always justified, since any new and powerful technology carries the innate risk of being abused by unscrupulous individuals, to the detriment of others. In the following, as is appropriate for the scope of this discussion, we shall focus on the latter of these concerns, regarding the fate of 'genetic information.'

Much of the discussion about ethical and legal issues relating to pharmacogenetics is centered on the issue of genetic testing, a topic that has recently also been the focus of a number of guidelines, advisories, and white papers issued by a number of committees in both Europe and the US. It is interesting to note that the one characteristic that virtually all these documents share is an almost studious avoidance of defining what exactly is a 'genetic test.' Where definitions are given, they tend to be very broad, including not only the analysis of DNA but also of transcription and translation products affected by inherited variation. In as much as the most sensible solution to this dilemma will ultimately hopefully be a consensus to treat all personal medical data in a similar fashion regardless of the degree to which DNA-encoded information affects it (noting that there really are not any medical data that are not to some extent affected by intrinsic patient properties), it may, for the time being, be helpful to let the definition of what constitutes 'genetic data' be guided by the public perception of 'genetic data' – in as much as the whole discussion of this topic is prompted by these public perceptions.

In the public eye, 'genetic test' is usually understood either as any kind of test that establishes the diagnosis of or predisposition for one of the classic monogenic, heritable diseases, or as any kind of test based on structural nucleic acid analysis (sequence). This includes the (non-DNA-based) Guthrie test for phenylketonuria and forensic and paternity testing, as well as a DNA-based test for lipoprotein (a) (Lp(a)), but not the plasma protein-based test for the same marker (even though the information derived is identical). Since monogenic disease is, in effect, excluded from this discussion, it stands to reason to restrict the definition of 'genetic testing' to the analysis of (human) DNA sequence.

It is clear that DNA-based structural data ('genetic' data) must never be procured without the explicit permission and informed consent of the individual. It is equally clear that, once such information is created, it should be guarded by the most reliable data protection systems that are available to prevent any of this information getting into the hands of third parties not intended to have access. At the same time, to provide the expected benefits in terms of guiding medical management, this information will need to be shared with a more or less extensive number of participants in the patient's healthcare – some of which may potentially use it in ways contrary to the patient's interests. This is true for any drug approved in conjunction with a DNA-based test: mere prescription of such a medicine discloses implicitly the outcome of the test to a wide number of stakeholders, among them certainly the patient's health insurer, who may gain from this information added knowledge about future health risks of this patient and take certain actions; whether these are scientifically or actuarially justified is irrelevant in this discussion.

Since data protection, in real life, is therefore always limited or compromised, additional measures are needed to protect the individual from disadvantages based on his or her data. Thus, personal protection, in addition to data protection, must be provided. This requires a framework of regulations or laws that govern the use of 'personal medical information' – the latter term is preferred to restricting such protection to DNA-derived data because many 'conventional' medical data carry similar or larger information content, particularly in the area of common complex disease. Such a framework can, in democratically governed systems, only arise as a consensus among all stakeholders involved – among them patients, physicians, insurers, and employers – that represents the optimal compromise which maximizes the benefits for both the individual and society. Such a framework will define which uses of the information are endorsed by society as lawful, and which are shunned as illegal. Given such a framework, much of the anxiety that abounds today with regard to the possible leakage of private medical data will be laid to rest, because it is primarily the potential (ab)use of information to their detriment that patients fear (which is not to trivialize the loss of autonomy that is of course also of concern).

3.03.9.2 Public Education and Information

The essential requirement to reach such a consensus that will allow the use of genetics and genomics in the best interest of all concerned is an informed dialogue among the various stakeholders. Such a consensus can only begin to take place once (mis-)perceptions are replaced by objective and neutral information as the basis to form informed opinions. Progress in the fields of genetic and genomics has been rapid and substantial over the last few decades, and it has been accompanied by a great deal of hyperbole in the popular press. Meanwhile, geneticists have continued to cultivate an arcane and forbidding vernacular that adds only to the appearance of purposeful secrecy to which the public is reacting, instead of having made extra efforts to reach out to the public in a concerted educational campaign. As part of the Human Genome Project substantial amounts of funding have been provided to work in the area of bioethics and much progress has been achieved there. Similar or even greater efforts need to be undertaken in the area of public information and education. This will surely go a long way toward resolving some of the fears as well as unrealistic hopes the public currently associates with genetics and genomics. The author of this communication and his colleagues have assembled an interactive CD-ROM-based educational program that is distributed freely upon request.[29]

3.03.9.3 Ethical–Societal Aspects of Pharmacogenetics

Based on the perceived particular sensitivity of 'genetic' data, institutional review boards commonly apply a specific set of rules to granting permission to test for DNA-based markers in the course of drug trials or other clinical research, including (variably) separate informed consent forms, the anonymization of samples and data, specific stipulations about availability of genetic counseling, and provision to be able to withdraw samples at any time in the future.

Arguments have been advanced[22] that genotype determinations for pharmacogenetic characterization, in contrast to 'genetic' testing for primary disease risk assessment, are less likely to raise potentially sensitive issues with regard to patient confidentiality, the misuse of genotyping data or other nucleic acid-derived information, and the possibility of stigmatization. While this is certainly true when pharmacogenetic testing is compared to predictive genotyping for highly penetrant Mendelian disorders, it is not apparent why in common complex disorders issues surrounding predictors of primary disease risk would be any more or less sensitive than those pertaining to predictors of likely treatment success/failure. Both can be expected to provide, in most cases, a modicum of better probabilistic assessment, based on the modest degree of sensitivity, specificity, and positive/negative predictive value we are likely to see with tests for pharmacogenetic interactions. If, however misguided this would be given the anticipated quite limited information content of such tests, such information was to be used 'against' the patient, then two lines of reasoning may indicate an increased potential for ethical issues and complex confrontations among the various stakeholders to arise from pharmacogenetic data.

First, while access to genotyping and other nucleic acid-derived data related to disease susceptibility can be strictly limited, the very nature of pharmacogenetic data calls for a rather more liberal position regarding use. If this information is to serve its intended purpose, i.e., improving the patient's chance for successful treatment, then it is essential that it is shared among at least a somewhat wider circle of participants in the healthcare process. Thus, the prescription for a drug that is limited to a group of patients with a particular genotype will inevitably disclose the receiving patient's genotype to any one of a large number of individuals involved in the patient's care at the medical and administrative level. The only way to limit this quasipublic disclosure of this patient's genotype data would be if he or she were to sacrifice the benefits of the indicated treatment for the sake of data confidentiality.

Second, patients profiled to carry a high disease probability along with a high likelihood for treatment response may be viewed, from the standpoint of, e.g., insurance risk, as quite comparable to patients displaying the opposite profile, i.e., a low risk to develop the disease, but a high likelihood not to respond to medical treatment, if the disease indeed occurs. For any given disease risk, then, patients less likely to respond to treatment would be seen as a more unfavorable insurance risk, particularly if nonresponder status is associated with chronic, costly illness rather than with early mortality: the first case has much more far-reaching economic consequences. The pharmacogenetic profile may thus, under certain circumstances, even become a more important (financial) risk assessment parameter than primary disease susceptibility, and would be expected – in as much as it represents but one stone in the complex disease mosaic – to be treated with similar weight, or lack thereof, as other genetic and environmental risk factors.

Practically speaking, the critical issue is not only, and perhaps not even predominantly, the real or perceived sensitive nature of the information, and how it is, if at all, disseminated and disclosed, but how and to what end it is used. Obviously, the generation and acquisition of personal medical information must always be contingent on the individual's free choice and consent, as must be all applications of such data for specific purposes. Beyond this, however, there is today an urgent need for the requisite dialogue and discourse among all stakeholders within society to develop and endorse a set of criteria by which the use of genetic, indeed of all personal, medical information should occur. It will be critically important that society as a whole endorses, in an act of solidarity with those less fortunate, i.e., at higher risk of developing disease, or less likely to respond to treatment, rules that guarantee the beneficial and legitimate use of the data in the patient's interest while at the same time prohibiting their use in ways that may harm the individual personally, financially, or otherwise. As long as we trust our political decision processes to reflect societal consensus, and as long as such consensus reflects the principles of justice and equality, the resulting set of principles should ensure such proper use of medical information. Indeed, both aspects – data protection and patient/subject protection – are seminal components of the mandates included in the World Health Organization's *Proposed International Guidelines on Ethical Issues in Medical Genetics and Genetic Services*[23] which mandate autonomy, beneficence, nonmaleficence, and justice.

3.03.10 Summary

Genetics and genomics, in their various implementations, will represent an important new avenue toward understanding disease pathology and drug action, and will offer new opportunities of stratifying patients to achieve better treatment success. As such, these approaches represent a logical, consequent step in the history of medicine – evolutionary, rather than revolutionary. The implementation of genetic approaches will take time, and will not apply to all diseases and all treatments equally. Pharmacogenetic information will be probabilistic and relative, not deterministic or absolute. Application of genetics and genomics to the drug discovery and development process, as well as to medical practice, will provide help, but no simple solutions, and will not be a panacea. Importantly, society at large will need to find ways to sanction the proper use of private medical information, thus allowing and protecting its unencumbered use for the benefit of patients while safeguarding them from unintended use. Increased efforts to inform and educate the general public, leading to a more realistic assessment of the actual potential of genetic approaches to provide benefit or cause harm, will be key to quell both the exalted hopes and exaggerated fears that are so often associated with the topic.

References

1. Dickins, M.; Tucker, G. *Int. J. Pharm. Med.* **2001**, *15*, 70–73.
2. Evans, W. E.; Relling, M. V. *Science* **1999**, *206*, 487–491.
3. Baselga, J.; Tripathy, D.; Mendelsohn, J.; Baughman, S.; Benz, C. C.; Dantis, L.; Sklarin, N. T.; Seidman, A. D.; Hudis, C.; Moore, J. et al. *J. Clin. Oncol.* **1996**, *14*, 737–744.
4. Haseltine, W. A. *Nat. Biotechnol.* **1998**, *16*, 1295.
5. In, K. H.; Asano, K.; Beier, D.; Grobholz, J.; Finn, P. W.; Silverman, E. K.; Silverman, E. S.; Collins, T.; Fischer, A. R.; Keith, T. P. et al. *J. Clin. Invest.* **1997**, *99*, 1130–1137.
6. McGraw, D. W.; Forbes, S. L.; Kramer, L. A.; Liggett, S. B. *J. Clin. Invest.* **1998**, *102*, 1927–1932.

7. Huang, Y.; Paxton, W. A.; Wolinsky, S. M.; Neumann, A. U.; Zhang, L.; He, T.; Kang, S.; Ceradini, D.; Jin, Z.; Yazdanbakhsh, K. et al. *Nat. Med.* **1996**, *2*, 1240–1243.
8. Dean, M.; Carrington, M.; Winkler, C.; Huttley, G. A.; Smith, M. W.; Allikmets, R.; Goedert, J. J.; Buchbinder, S. P.; Vittinghoff, E.; Gomperts, E. et al. *Science* **1996**, *273*, 1856–1862.
9. Samson, M.; Libert, F.; Doranz, B. J.; Rucker, J.; Liesnard, C.; Farber, C. M.; Saragosti, S.; Lapaumeroulie, C.; Cognaux, J.; Forceille, C. et al. *Nature* **1996**, *382*, 722–725.
10. O'Brien, T. R.; Winkler, C.; Dean, M.; Nelson, J. A. E.; Carrington, M.; Michael, N. L.; White, G. C., II *Lancet* **1997**, *349*, 1219.
11. Theodorou, I.; Meyer, L.; Magierowska, M.; Katlama, C.; Rouzious, C; Seroco Study Group. *Lancet* **1997**, *349*, 1219–1220.
12. Drazen, J. M.; Yandava, C. N.; Dube, L.; Szczerback, N.; Hippensteel, R.; Pillari, A.; Israel, E.; Schork, N.; Silverman, E. S.; Katz, D. A. et al. *Nat. Genet.* **1999**, *22*, 168–170.
13. Martinez, F. D.; Graves, P. E.; Baldini, M.; Solomon, S.; Erickson, R. *J. Clin. Invest.* **1997**, *100*, 3184–3188.
14. Tan, S.; Hall, I. P.; Dewar, J.; Dow, E.; Lipworth, B. *Lancet* **1997**, *350*, 995–999.
15. Green, S. A.; Turki, J.; Innis, M.; Liggett, S. B. *Biochemistry* **1994**, *33*, 9414–9419.
16. Green, S. A.; Turki, J.; Bejarano, P.; Hall, I. P.; Liggett, S. B. *Am. J. Respir. Cell Mol. Biol.* **1995**, *13*, 25–33.
17. Reihsaus, E.; Innis, M.; MacIntyre, N.; Liggett, S. B. *Am. J. Respir. Cell Mol. Biol.* **1993**, *8*, 334–349.
18. Dewar, J. C.; Wheatley, A. P.; Venn, A.; Morrison, J. F. J.; Britton, J.; Hall, I. P. *Clin. Exp. All.* **1998**, *28*, 442–448.
19. Fischel-Ghodsian, N. *Ann. N. Y. Acad. Sci.* **1999**, *884*, 99–109.
20. Hutchin, T.; Cortopassi, G. *Antimicrob. Agents Chemother.* **1994**, *38*, 2517–2520.
21. Esteller, M.; Garcia-Foncillas, J.; Andion, E.; Goodman, S. N.; Hidalgo, O. F.; Vanaclocha, V.; Baylin, S. B.; Herman, J. G. *N. Engl. J. Med.* **2000**, *343*, 1350–1354.
22. Roses, A. *Lancet* **2000**, *355*, 1358–1361.
23. Proposed International Guidelines on Ethical Issues in Medical Genetics and Genetic Services. http://www.who.int/ (accessed Sept 2006).
24. Fellay, J.; Marzolini, C.; Meaden, E. R.; Back, D. J.; Buclin, T.; Chave, J. P.; Decosterd, L. A.; Furrer, H. J.; Opravil, M.; Pantaleo, G. et al. *Lancet* **2002**, *359*, 30–36.
25. Dubinsky, M.; Lamothe, S.; Yang, H. Y.; Targan, S. R.; Sinnett, D.; Theoret, Y.; Seidman, E. G. *Gastroenterology* **2000**, *1*, 705–713.
26. Mallal, S.; Nolan, D.; Witt, C.; Masel, G.; Martin, A. M.; Moore, C.; Sayer, D.; Castley, A.; Mamotte, C.; Maxwell, D. et al. *Lancet* **2002**, *359*, 727–732.
27. Hetherington, S.; Hughes, A. R.; Mosteller, M. et al. *Lancet* **2002**, *359*, 1121–1122.
28. http://www.bcg.com/publications/files/eng_genomicsgenetics_rep_11_01.pdf (accessed Aug 2006).
29. Roche *Genetics* Educational Program. Available upon request from http://www.rochegenetics.com (accessed Aug 2006).

Comprehensive Medicinal Chemistry II
ISBN (set): 0-08-044513-6

ISBN (Volume 3) 0-08-044516-0; pp. 51–68

3.04 Biomarkers

S E DePrimo, Pfizer Global Research and Development, San Diego, CA, USA

3.04.1	**Introduction**	**69**
3.04.2	**Definition and a Brief History of Biomarker Research**	**70**
3.04.3	**Biomarkers in Target Characterization**	**70**
3.04.3.1	The Need for Application of Biomarkers	70
3.04.3.2	Criteria for Assessing the Utility of Biomarkers in Target Discovery and Validation	71
3.04.4	**Technologies Applied in Biomarker Applications**	**72**
3.04.4.1	Biochemical and Molecular Approaches	72
3.04.4.1.1	Protein detection methods	72
3.04.4.1.2	Protein biomarkers of target modulation, signal transduction, and downstream biological effects	73
3.04.4.2	Applications of 'Omics' Technologies in Biomarker Discovery	74
3.04.4.2.1	Gene expression profiling, genomic alterations, and tissue microarrays	74
3.04.4.2.2	Proteomics in biomarker discovery	75
3.04.4.2.3	Metabonomics and beyond	76
3.04.4.3	Functional Imaging in Preclinical Models	76
3.04.4.3.1	Positron emission tomography and magnetic resonance imaging	77
3.04.4.3.2	Optical imaging with fluorescent and luminescent probes	77
3.04.5	**Biomarker Approaches in the Discovery of Early Markers of Drug Toxicity**	**78**
3.04.6	**Translational Applications: Biomarkers in the Clinic**	**79**
3.04.6.1	Selection of Biomarker Endpoints and Assays	79
3.04.6.2	Biomarker as Surrogates for Clinical Activity	80
3.04.6.3	Biomarker Endpoints Applied in Pharmacokinetic–Pharmacodynamic Modeling	80
3.04.6.4	Regulatory Implications	81
3.04.7	**Conclusions**	**82**
References		**82**

3.04.1 Introduction

The characterization and application of biological markers, or biomarkers, and the associated enabling technologies have been steadily increasing in prominence within the field of drug discovery and development. Application of biomarkers has the potential to accelerate the time frame at multiple stages in the drug discovery and development pipeline, as well as to allow early identification of targets with relevance to disease processes or early deselection of drug candidates with less than optimal pharmacological properties and target. Ultimately, in the pharmaceutical and biotechnology industries the most important outcome will likely be cost savings that would be derived from faster, more efficient development from lead identification through clinical trials; alongside this will come the benefit of increased mechanistic understanding of therapeutics and targets. Biomarkers can be applied across most disease areas and can be measured via a wide range of techniques, from direct biological and morphological observations to quantitative biochemical assays and functional imaging modalities, dependent on the nature of the signal and study in question. This chapter will present a broad overview of biomarker concepts and related technology applications, with an emphasis on biomarker approaches that are of value in the early validation of drug targets and candidate therapeutics. The emergence of biomarkers as a specific field is relatively recent and the field encompasses a variety of directions and technologies; this chapter will describe many facets of biomarker applications, particularly molecular and biochemical approaches, and give relevant examples that can help guide experimental designs. Issues relevant to the discovery and selection of biomarkers and to the technical and biological validation of biomarker endpoints will also be discussed, as will clinical implications and the role of biomarkers in evaluating drug safety.

3.04.2 Definition and a Brief History of Biomarker Research

A reasonable starting point for any discussion of biomarkers applications is with a general definition of the term 'biomarkers.' In recent years a formal definition has been crafted by a committee termed the Biomarkers and Surrogate Endpoint Working Group; the definition of 'biological marker (biomarker)' they have put forth reads as follows: a characteristic that is objectively measured and evaluated as an indicator of normal biological processes, pathogenic processes, or pharmacologic responses to a therapeutic intervention.[1] This definition is broad by necessity, as the field has impacts across biological systems and disease areas and can incorporate any of the technologies that are applicable in basic biological or clinical research. Although there has been an increase in attention to and formalization of the biomarkers field, in practice biomarkers have been an integral component of biomedical research for decades. The primary focus has historically been on biomarkers of disease; indeed, it has been proposed that the initial application of biomarkers may date to the practice of culturing infections to improve cure rates when antibiotics were first introduced during World War II.[2] In the broadest sense, all of the symptoms that are typically associated with the presence of a disease state may be considered as disease biomarkers. With advancements in biochemical and molecular techniques, more specific, quantifiable, and mechanistically relevant types of markers have been characterized (herein referred to as molecular biomarkers). A few examples of clinically useful molecular biomarkers are: glucose and hemoglobin A1c levels in diabetes, circulating viral load in viral infections, and cholesterol, low-density lipoproteins (LDL), and high-density lipoproteins (HDL) levels in cardiovascular disease. Each of these endpoints is diagnostic or prognostic of a disease state, and also can be measured to indicate an effect of a therapeutic intervention, thus fitting the criteria for the biomarkers definition as listed above. Clinically relevant and disease-associated biomarkers such as these are very useful in the clinical testing of drug candidates, as their monitoring can simultaneously confer information about both the mechanistic or biological effect(s) of an intervention (pharmacodynamic effect) as well changes in the status of disease.

The potential for biomarkers to enhance biological understanding of disease processes and drug targets, coupled with their potential to improve the efficiency of decision making during the drug discovery and development process, has led to heightened interest in biomarker applications in the current environment. The impact of this interest may be noted by the observation that, in early 2005, a search of the World Wide Web using the Google search engine yields approximately 1 450 000 matches for the term 'biomarkers.' Similarly, a search for the text word 'biomarker' or 'biomarkers' in the Pubmed biomedical literature database at this time yielded a total of more than 12 000 citations, dating back to the late 1970s. Of the 12 000, more than 8300 were dated in the year 2000 or later. The increased interest can be linked to breakthroughs in molecular technologies in areas such as genomics and proteomics as well as bioinformatics, and also to the increasing sophistication of drug discovery processes and continued emphasis on the rational targeting of specific molecular pathways (e.g., in the area of cancer treatment in particular, a rapidly emerging theme is the development of so-called 'molecularly targeted' therapeutics). Applications of biomarkers are now becoming standard in many phases of drug discovery and target characterization, and are leading to a more comprehensive understanding of drug effects; this ultimately can enable more effective and efficient drug development.

3.04.3 Biomarkers in Target Characterization

3.04.3.1 The Need for Application of Biomarkers

The ability to assess rapidly and accurately whether pharmacological intervention at a specific target point will have a desired biochemical or biological effect can be facilitated by application of appropriate biomarker measurements. Biomarkers can reveal whether a given target enzyme is being impacted by a drug candidate, but can also indicate whether impacting a novel potential target leads to a similar effect as that seen for a known target (whether this be a positive, such as efficacy, or a negative, such as toxicity, effect). An example of a situation of this type is the following: if the goal is identification of a target protein that is involved in preventing programmed cell death (apoptosis), then the biomarker assay of choice would be one indicative of apoptosis induction, such as chromosomal DNA laddering or

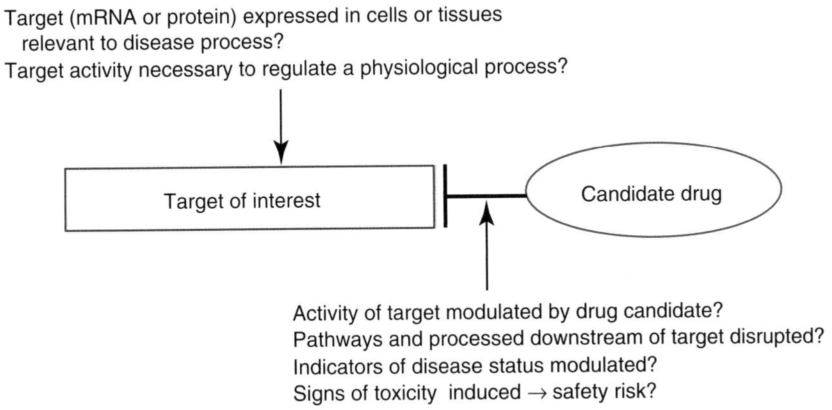

Figure 1 Schematic illustration of key questions that biomarkers can help address at the interface between targets and drug candidates.

caspase-3 cleavage. A clear lack of apoptosis induction in cells in which the potential target is clearly being impacted pharmacologically or by other means (e.g., by small interfering ribonucleic acid (siRNA) targeting) would suggest that this target might not be an appropriate choice for further investigation. Of course, it would be imperative that the key experiments are conducted in the relevant context (e.g., evaluation of a number of distinct cell lines or other models to determine the prevalence of the apoptotic effect in different backgrounds); also very useful would be availability of biomarker assays that would confirm that the target has indeed been impacted (for siRNA, this could be messenger RNA or protein levels of the target; for pharmacological intervention, this could be biochemical change(s) in a protein modification state or in the pathway effected by the candidate target). This last point highlights the importance of being able to determine whether a drug is active both at the level of pharmacological modulation of target and at the level of relevant biological impact subsequently resulting from that pharmacological modulation (such as impact upon downstream signaling events or upon physiological processes).

The example above illustrates some of the key roles biomarkers can play in determining target validation; **Figure 1** expands this concept to outline the full scope of questions that biomarker parameters can address in terms of target and drug candidate selection. The interplay between target and drug in complex cellular systems is at the center of the biomarker interface; beyond enabling the basic characterization of target biology, biomarkers can ultimately also provide insight into both therapeutic value (i.e., efficacy) and toxicity risk of potential targets, as well as of drugs designed to modulate the functionality of those targets.

3.04.3.2 Criteria for Assessing the Utility of Biomarkers in Target Discovery and Validation

In order for a biomarker analysis to be of most utility in characterizing the biology of a target or biological results of pharmacological intervention, a number of criteria should be considered that relate both to the target itself and to the methodology used in biomarker analysis. **Table 1** lists some of the key criteria, and these refer to the specificity with which the effect being observed is linked to activity of the target protein; the extent to which pharmacological intervention disrupts the effect being observed (or induces a different effect); the sensitivity and robustness of the assay; and the degree of difficulty with which the biomarker can be reliably measured in different biological systems and contexts. This last point speaks to a key issue which impacts whether an approach can ultimately be implemented in a clinical setting (e.g., whether the biomarker can be 'translated'). Other considerations, such as correlations of a biomarker with toxicity or disease state, should be weighed in a context-dependent manner. This chapter will primarily refer to applications in target characterization and preclinical drug candidate evaluation, with relevant examples of commonly applied approaches and technologies; however, aspects of biomarker applications in clinical and toxicology settings will also be referenced (often the same or similar applications will be used in more than one setting, and indeed methods primarily applied in, for example, a clinical format may be 'reverse-engineered' to be utilized in a target characterization assay).

Table 1 Concepts for establishing criteria for assessing the utility of putative biomarkers

Target-based considerations	*Technical considerations*
The biomarker is coexpressed with the target (if the abundance or activation of the target is changed by drug effect, then the target itself can be a biomarker)[a]	Robust detection method(s) exist or can be developed (e.g., antibodies are available, specific, and applicable in various assay formats)
The biomarker is regulated by the target (regulation can be in terms of abundance, activation state, etc.)	Detection methods are sufficiently sensitive to detect the biomarker both in the presence and absence of drug
Regulation of the biomarker is relatively target-specific (i.e., not also regulated by several other pathways)	Dynamic range is broad enough to allow measurement of both subtle and dramatic changes in biomarker signal (of course, dramatic changes reduce ambiguity in interpretation)
Regulation of the biomarker follows a similar pattern in all or most organs and cell types (results not confounded by contradictory effects in different compartments in vivo)	Biomarker assay is reasonably high-throughput or can be scaled up as appropriate (this is especially true for biomarker-based screening assays or for analysis of large numbers of samples, as from clinical trials or in vivo studies)
Drug effect induces a significant and reproducible effect on the biomarker (predictable dose–response correlation)	Intra- and interassay variability is low enough to allow clear measurement of drug-induced change in biomarker signal
Changes in the biomarker are correlated with target-dependent changes in disease status or other physiological processes[b]	Suitable positive and negative controls are available (e.g., cells that do not express the biomarker for negative control; other reagents that mimic the drug effect for positive)

[a] For instance, phosphorylation of the target, or amount of mRNA encoding the target.
[b] Note that correlation with disease status is not a prerequisite for a biomarker to be useful as a pharmacodynamic endpoint.

3.04.4 Technologies Applied in Biomarker Applications

3.04.4.1 Biochemical and Molecular Approaches

3.04.4.1.1 Protein detection methods

Perhaps the most common category of biomarkers is that of macromolecules (most typically, proteins or messenger RNA molecules) that can be measured in cells, cell extracts, or biological fluids. Advances made in biochemistry and molecular biology over the last few decades have greatly expanded the number and sensitivity of techniques that can measure specific macromolecules or panels of macromolecules. As knowledge of signal transduction pathways has expanded, so too have the tools with which to assay signaling events and disruptions thereof. Analysis of posttranslational modifications of cellular proteins is perhaps the most commonly applied approach to assessing the role of a potential drug target in mediating a given signaling process. The development of antibody reagents that can detect specific modifications in proteins has been critical in these efforts. This is exemplified by antibodies that can detect protein epitopes that have been phosphorylated on specific amino acid residues; a breakthrough was the development of an antibody specific for phosphotyrosine residues.[3,4] This enabled immunoaffinity purification of many protein kinase substrates, and also indirect monitoring of protein kinase activity in cells via immunoblotting (Western blotting) techniques. When coupled with immunoprecipitation of specific kinase substrate proteins, assessment of phosphotyrosine abundance remains an effective method for determining kinase signaling activity, and is thus a useful biomarker technique for profiling targets that have kinase activity.[5] Further refinements in antibody reagents have led to antibodies that can detect not only specific types of protein modifications but modifications in a specific epitope in a particular protein, thus eliminating the need for prior immunoprecipitation to interrogate a particular protein. Early examples of this are antibodies against phosphoserine sites in the retinoblastoma and p53 tumor suppressor proteins,[6,7] and the list of site-specific phosphoserine and phosphotyrosine antibodies has been growing continuously. Some key practical considerations that should be taken into account with phospho-specific antibodies, such as potential differences between detection of endogenous and recombinant versions of proteins and specificity differences between polyclonal and monoclonal antibodies targeting the same phosphoepitope, are discussed by Craig *et al.*[8] If specificity of detection can be clearly established for such site-specific antibodies via the use of appropriate positive and negative controls, they present a considerable advantage in efficiency over more laborious immunoprecipitation methods.

It should be noted that although antibodies are the primary tool in the measurement of specific proteins and protein epitopes, other types of specific protein capture agents have emerged in recent years. Examples include affibodies, which are small protein domain proteins, derived from combinatorial engineering of the Z domain of Staphylococcal protein A, that can be selected via phage display libraries and can bind with high affinity to specific targets[9,10]; and aptamers, which are synthetic single-stranded DNA or RNA molecules that fold into structures that bind to specific proteins with high affinity.[11] When conjugated with a photoreactive 5-bromo-deoxyuridine moiety, aptamers become photoaptamers, which can be photoactivated to form covalent cross-links with target proteins. While affibodies and aptamers may present some advantage in terms of scalability, flexibility in multiplexing applications (due to lower potential for cross-reactivity), and perhaps enhanced binding affinity, it remains to be determined whether or not either or these types of capture agents will supplant antibodies as broadly applicable reagents for detection of specific proteins in complex biological matrices. Meanwhile, if measurement of protein–protein interactions (or disruption thereof) is the objective, application of surface plasmon resonance technology enables affinity-based monitoring of protein interactions without the need for any labeling or detection agents (as the method detects differences in refractive index at the surface of an immobilized capture molecule).[12,13] Though perhaps most often employed in hit-to-lead screening or screening of antibody interactions, biosensor chips coupled with mass spectrometry can be generated that may become useful in characterizing interactions of novel targets or in biomarker discovery.[14,15]

3.04.4.1.2 Protein biomarkers of target modulation, signal transduction, and downstream biological effects

Changes in the abundance and specific modifications of proteins can provide useful information in two key areas: assessment of target modulation and characterization of the downstream effects of target activity or pharmacological modulation thereof. Target modulation is most readily characterized when pharmacological antagonism (or agonism) of a protein's activity results in changes in modifications of the protein itself; for example, in cases of receptor tyrosine kinase that exhibit autophosphorylation activity, where inhibition results in reduction in the amount of phosphorylated amino acid residues.[5,16–18] Alternatively or in addition, assessment of target modulation can be made via the analysis of proteins that are part of the signaling pathway affected by the target protein (hence, 'downstream' of the point of intervention). In general, in order for a downstream molecule to be a reliable biomarker, its linkage to the target protein of interest should be solidly established, at least in the particular cellular model system wherein the target characterization and drug screening efforts will be focused. Thus, a protein that is, for example, a direct substrate of a kinase or phosphatase target of interest would be a useful biomarker, provided the availability of an assay to detect changes in phosphorylation status of the substrate. If a pathway is well characterized, for example a phosphorylation cascade such as that induced by epidermal growth factor receptor activation, then profiling of modifications to multiple downstream components (such as mitogen-activated protein kinase) is a reasonable approach[19,20]; this has the advantage of reducing the likelihood of a spurious finding due to technical or biological variability in a single endpoint, and thus provides greater confidence that a desired pharmacological effect is being triggered and gives a broader indication of the extent of the effect. Similarly to changes in protein modifications, downstream changes in the abundance of specific proteins, due to changes in the rate of turnover or of gene expression, can also be considered as potential biomarkers, as would disruption of protein complex formation; examples of these would be changes in secreted cytokine levels during inflammatory reactions or induction of receptor homo- or heterodimerization triggered by growth factor binding.[21,22] Proteins that are well correlated with functional changes in cell physiology, such as the Ki67 nuclear antigen associated with cell proliferation or the caspase-3 fragments generated in cells undergoing apoptosis,[19] represent biomarkers of biological consequences further downstream of signaling cascades.

There are several format options for antibody-based assays that can be employed in the assessment of target modulation via protein modifications or downstream consequences. For in vitro characterizations in experiments of moderate throughput, immunoblotting is perhaps the most straightforward approach and, since it features electrophoretic resolution based on molecular weight, provides visual evidence of molecular weight variants or of cross-reactivity with other protein species. Measurements of protein abundance typically rely on enzyme-linked chemiluminscent readout although fluorescence-based detection can be done with suitable gel readers and appropriately labeled detection antibodies. However, quantitation of signal strength via immunoblotting is often less robust than that enabled by the enzyme-linked immunosorbent (ELISA) 'sandwich' assay format,[23] and ELISA can confer greater specificity due to the option of utilizing two distinct antibodies, the so-called 'capture' and 'detection' antibodies (although similar specificity can be gained in the immunoblotting approach by incorporating an immunoprecipitation step, sometimes referred to as 'pull-down,' of a given protein of interest prior to gel separation and membrane probing). The ELISA method, like the related enzyme immunometric assay (EIA) methods, is particularly useful in the analysis of levels of secreted or otherwise soluble factors that are present in fluid matrices (conditioned cell culture media, serum and plasma) and at low concentrations. Further, recent advances in multiplexing technology have led to the ability to measure dozens or even hundreds of different proteins simultaneously in the same sample; this is most commonly performed via antibody-tagged cytometric beads in solution or on planar antibody microarrays.[24] Cytometric beads are versatile in that beads can be coded based either on color or size, and currently allow higher multiplexing of capture/detection antibodies within a single sample, due to lesser effect of antibody cross-reactivity issues as compared to planar sandwich arrays; a comparison of commercially available cytokine multiplex detection kits provides some practical guidance for assay selection and addresses comparison between individual ELISA assays and measurements of the same analyte when performed in multiplex.[25] Another approach to antibody arrays, which can be applied in a highly multiplexed fashion, involves dual-labeling of two different biological samples in which differential protein abundance is being queried; proteins in each samples are directly tagged with, e.g., distinct fluorophores and incubated on the same antibody microarray.[26] This approach eliminates the need for detection antibodies and the associated need to screen for cross-reactivity, but lacks the added specificity provided by the detection antibody and cannot easily discriminate whether a difference in signal is due to different amounts of a single protein or to the fact that the protein forms larger multiprotein complexes in one case (as all proteins in the complex would be tagged with fluorophore).

For detection of protein biomarkers in the full context of organs and tissues (in situ), histological approaches such as immunofluorescence or immunohistochemistry are typically utilized; these can provide indication as to which specific cells or structures in a given tissue specimen are expressing a particular protein species of interest, albeit generally with more limited quantitative rigor than immunoblot or ELISA methods (histological methods are typically most useful in later stages of target validation or in drug development[19,27]). Another option is to use flow cytometry methods to measure protein biomarkers in individual cells that are grown in suspension culture or that have been disaggregated from solid substrates or tissues.[28] Also, it should be noted that nonantibody-based methods are emerging for kinetic analysis of proteins and protein modifications in cells; for instance, quantitative mass spectrometry approaches have been developed which rely upon stable isotope-labeled peptide standards for the monitoring of phosphorylation changes in specific proteins from cell lysates.[29,30] The choice of which of these detection methods are applied will depend on many factors such as the type of specimen (e.g., cultured cells, tissue, biological fluids) and collection method (e.g., fresh or flash-frozen versus formalin fixed); degree of sensitivity and quantitation required; and specificity of antibodies (this can be more of a concern in in situ assays or flow cytometry, which lack the electrophoretic mobility information that is captured in gel-based assays or the dual capture/detection antibodies that are utilized in ELISAs). For any of the commonly used protein detection methods, a paramount concern is that the method of choice in a specific evaluation must be capable of reliably detecting relative differences between treatment groups or between longitudinal samples collected in the same background; this is particularly the case for discovery-oriented experiments, whereas absolute measurements can assume greater importance in clinical studies, depending on the questions addressed and the nature of assays being employed.

3.04.4.2 Applications of 'Omics' Technologies in Biomarker Discovery

The discovery of new biomarkers for a given pathway or drug class is an integral component of successful methods for measuring pharmacological and biological consequences. As new proteins continue to be surveyed as potential drug targets, the need for markers that are tightly linked to their functions or downstream consequences continues to increase. The need is particularly high for novel targets where the associated biochemical pathways are not well characterized, but many validated targets could also benefit from expanded portfolios of biomarker endpoints to track during drug discovery and development efforts. The most powerful approaches toward discovery of molecular biomarkers are derived from recent advances in the fields of proteomics and genomics. Although these fields are covered elsewhere in this volume (*see* 3.01 Genomics; 3.02 Proteomics; 3.05 Microarrays), a description of some examples of biomarker discovery approaches that utilize such technologies is of relevance to the topic of this chapter.

3.04.4.2.1 Gene expression profiling, genomic alterations, and tissue microarrays

Among the battery of emerging technologies often referred to as 'omics,' microarray-based gene expression profiling and other genomic/transcriptomic technologies have thus far been applied most extensively and, arguably, most successfully. Such large-scale profiling has shifted the emphasis of some research efforts from hypothesis testing to hypothesis generation; or, at the very least, biomarker generation. Indeed, individual gene transcripts as well as whole clusters of coregulated transcripts that are modulated by disruption of a given signaling pathway are candidate biomarkers (as well as potential drug targets in their own right if critically important to the same pathway). Microarray transcript profiling can be applied in preclinical models with the aims of better understanding of the pharmacodynamic effects of drug candidates (or other bioactive agents) and ultimately of identification of biomarkers that can be applied during clinical testing. Several examples of this approach have been described, including the use of tumor xenograft models to probe for gene expression biomarkers modulated by in vivo exposure to an antiangiogenic small-molecule kinase inhibitor[31]; exposure of bone marrow-derived macrophages to Gram-positive bacteria to identify gene expression change[32]; animal models for evaluating bladder tissue responses to inflammation-inducing agents[33]; and treatment of cancer cell lines with chemotherapy drugs and signal transduction inhibitors to probe for biomarkers of combination therapy.[34] In one of these studies,[31] a transcript whose level was modulated by the kinase inhibitor in the preclinical model was shown to encode a protein product which was modulated in posttreatment biopsies in some human patients that were treated with the same kinase inhibitor; this demonstrates the potential utility of biomarker discovery studies in translation to clinical applicability (a topic to be discussed later in this chapter).

These are only a few examples of how application of microarray-based or other large-scale expression profiling methods provide a powerful means to identify putative biomarkers, particularly in the context of well-controlled laboratory experiments (as care must be taken to minimize false-positive risk in experiments of this type, given the multidimensional complexity of the data output). However, it must be noted that many potential biomarkers have also been identified in the direct profiling of well-annotated human clinical specimens as opposed to laboratory experiments in model systems,[35] and findings from either realm can be readily implemented in both to probe for biological or diagnostic relevance. An insightful commentary by Liu[36] summarizes the interface between experimental and clinical gene expression biomarker discovery efforts, and contrasts purely correlative, empirical discovery approaches against bottom–up approaches that build on mechanistic insights into cellular or physiological processes. The latter approach is likely to be of the most relevance for target-based research efforts, at least in the near term. In any case, candidate biomarkers identified via transcriptome-wide survey methods are most efficiently probed in subsequent experiments via more narrowly focused assays, particularly if a relatively small number of markers are sufficient to serve as biological readouts (there are at present no hard and fast rules on what comprises an ideal number of analytes to measure in a biomarker panel, but lower is usually better in terms of cost and reproducibility so long as sufficient decision-enabling information is captured[37]). The most common assay format to routinely measure levels of specific transcripts would be variants of quantitative reverse transcription polymerase chain reaction (qRT-PCR) technology,[38,39] which is a sensitive, specific, and versatile platform; however, if transcript level changes are well correlated with changes in the level of the corresponding protein (this may often not be the case[40]), then protein detection methods described earlier can also be employed.

In addition to gene expression changes, another type of biomarker of relevance to target selection, particularly in heterogeneous, mutationally driven diseases such as cancer, is change in protein expression due to changes in gene copy number or other mechanisms. Genes that are selectively lost (genomic deletion) or selectively increased in copy number (genomic amplification) in disease tissue are candidates for causative agents in cancer, and genome-level survey approaches are a means to identify such candidates via comparative genomic hybridization or related genomic methods.[41,42] Confirmation that expression of a given protein is frequently altered in a large number of cases of a particular type of cell or tissue can be determined in a high-throughput fashion through profiling of micro- or macroarrays containing many representative specimens. Tissue arrays usually consist of microscope slides mounted with dozens of small tissue specimens representing a survey of similar disease type (e.g., breast cancer biopsies) or of various normal organs and tissues (for instance, to probe the distribution of expression of a novel target, with regard to potential toxicity).[43,44] Tissue arrays are typically probed for proteins via immunohistochemistry methods, and for gene transcripts or copy number via in situ hybridization. Applications for which tissue arrays are invaluable include confirmation of whether a potential target is overexpressed (or, conversely, never expressed) in a certain disease indication, or likewise that a gene is frequently deleted or amplified, as well as the aforementioned query of the normal tissue distribution of targets in multiple species. An application conceptually related to tissue arrays is that of reverse phase protein microarrays; these consist of arrays of protein lysates rather than intact tissue specimens, and allow for immunoblot probing of protein expression (or activation state, in the case of phosphorylated proteins, for instance) across hundreds of lysates, including serial dilutions to aid in quantitation.[45,46] Reverse phase microarrays are particularly useful for profiling cell lines or tissues that are too limited in quantity for histological handling.

3.04.4.2.2 Proteomics in biomarker discovery

Along with genomic approaches, proteomics has also enabled much progress to be made toward biomarker discovery. The application of multiplex protein profiling has been previously mentioned, and most other proteomic discovery efforts rely on the quantitation of relative differences in abundance of specific proteins. The detection methods are numerous and will not be described in detail in this chapter; indeed, many of the analytic approaches are still in the initial testing phase and will require additional testing. However, the potential utility of proteomic profiling in identification of potential drug targets as well as putative biomarkers is illustrated in innovative applications of proteomic mapping. For example, in an effort to identify proteins induced at the tissue–blood interface in vivo (and thus potentially accessible to targeted antibodies or other biological therapeutics), multidimensional protein identification technology (MudPIT; a 'shotgun' proteomics approach that utilizes chromatographic separation followed by tandem mass spectrometry for identification of peptide sequences to confirm protein identity) was applied to luminal endothelial cell plasma membranes isolated from rat lungs and from cultured microvascular endothelial cells.[47] A large set of proteins was identified, and subsequent validation experiments established specific proteins as putative targets (as well as endothelium-specific biomarkers) in lung and tumor tissues.[48] Such open-ended profiling using proteomic technologies holds much promise for both target and biomarker discovery, but careful application, cautious interpretation, and confirmatory follow-up experiments (using both the initial proteomic screening assay as well as measurements of specific proteins via immunoassays) are essential to optimal success.

3.04.4.2.3 Metabonomics and beyond

As proteomics technologies continue to emerge as powerful biomarker discovery tools, so also does the field of metabolomics (also known as metabonomics). Metabonomic research is focused on the measurement of low-weight components, primarily in biological fluids such as urine and serum, and thus is not of direct relevance in searches for protein or nucleic acid biomarkers that may also be drug targets. However, metabonomics profiles may indeed be of utility in monitoring the biological consequences of modulating the activity of a target. The basic approach relies on ^{1}H nuclear magnetic resonance (NMR) spectroscopy to generate spectrometric readouts of the relative abundance of low molecular weight metabolites.[49] The method can be cost-effective and relatively high throughput with the availability of suitable NMR capability, and is most reliable in comparing metabolite patterns before and after a specific stimulus in samples collected from the same subject. To date, the most common application of metabonomics profiling has been in the toxicology field with the objective of identifying metabolite signatures of toxic stresses[50] (more on this subject below); however, examples of applications in the idenfication of efficacy and mechanistic markers can be found, such as a study of a diabetes drug (the peroxisome proliferator-activated receptor gamma agonist rosiglitazone) in a mouse model.[51] In this study, a subset of specific metabolites – the lipid metabolome – was profiled via thin-layer chromatography coupled with capillary gas chromatography, and a number of alterations were detected. For broader metabolome-wide profiling, it is more typical for a pattern of peaks, or 'fingerprint' (similar to those obtained in mass spectrometry-based proteomic profiling), to be the primary output of metabolome data analysis. However, identification of specific peaks of interest can be done via a variety of means, such as structural similarity matching against databases of known molecules or by further isolation of the peak fraction and subsequent direct analysis.[49] As with genomics and proteomics, metabonomic methods hold much promise but also warrant much rigor in their application (a cautionary perspective on the importance of appropriate methodological and statistical rigor in marker discovery studies is provided by Ransohoff,[52] with a key message being the necessity of performing independent confirmatory experiments to firmly establish linkage between multianalyte changes and the phenomenon in question). The convergences of all of these methodologies along with bioinformatics comprise the nascent field known as systems biology. As this field matures, its impact on biomarker discovery applications will likely continue to increase.

3.04.4.3 Functional Imaging in Preclinical Models

The ability to apply imaging capabilities to cell and small animal models has enabled great strides in the biological sciences and expands the scope of biomarker assays of utility in drug discovery and development. Advances in imaging technologies, many of which were first developed for use in human healthcare rather than for laboratory science, can allow for visualization and quantification of molecular and cellular events noninvasively and in real time. Assessments of physiological endpoints such as vascular blood flow in angiogenesis models or tumor size and anatomy in cancer models can now be performed over time in the same animal (to monitor drug effect, for example), and thus the need for sacrifice prior to histological analysis is reduced or eliminated. Applications in small animal preclinical models have been enabled by miniaturization of imaging equipment and by expansion of the number of reporter probes and contrast agents.[53] It is not within the scope of this chapter to provide an in-depth discussion of imaging technologies (*see* 3.31 Imaging); rather, examples of how imaging applications can be used in biomarker studies will be discussed in brief.

3.04.4.3.1 Positron emission tomography and magnetic resonance imaging

Positron emission tomography (PET) and magnetic resonance imaging (MRI) are powerful technologies that have revolutionized clinical medicine and are now making inroads into the preclinical arena, expanding biomarker methodologies in the process. Cancer treatment and research have perhaps benefited the most from PET and MRI, as some aspects of tumor physiology and anatomy can now be monitored noninvasively and serially, in the same animal.[54,55] PET imaging scanners are capable of measuring the presence and concentration of positron emitting isotopes in living tissues, and thus tracer probes can be synthetically generated for monitoring of molecular transport or enzymatic processes dynamically. The most mature tracer that utilizes PET imaging is a glucose derivative, 2-[^{18}F]fluoro-2-deoxy-D-glucose (FDG), which is transported into metabolically active cells and tissues where it becomes trapped after phosphorylation by the hexokinase enzyme. Since tumors are usually highly metabolically active, FDG–PET imaging can be used to detect and subsequently monitor changes in tumor tissue metabolic activity. Likewise, another PET probe relevant to monitoring changes in tumor physiology is 3′-deoxy-3′[^{18}F]fluorothymidine (FLT), which is a thymidine analog that is preferentially processed by cells that are actively replicating genomic DNA; thus antiproliferative effects induced by cancer targeted agents can be monitored. The application of PET in monitoring specific inducible gene expression events in vivo is rapidly growing via the generation of a number of radiolabeled molecular probes that are captured by specific enzymes or receptors, e.g., herpes tymidine kinase, dopamine D2 receptor, or somatostatin receptor,[56,57] that are placed under the transcriptional control of regulatory sequences in transgenic models. Degradation of a specific protein in cells can also be monitored, as illustrated by the application of a ^{68}Ga-labeled F(ab′)2 fragment of the anti-HER2 therapeutic antibody trastuzumab (Herceptin). This was used to pharmacodynamically monitor heat shock protein inhibitor-induced HER2 protein degradation and subsequent recovery.[58] With the advent of additional PET tracers and small animal scanners (micro-PET systems) with higher resolution (coupled with lowered costs), the range of biomarker studies that can be pursued preclinically will continue to expand.

Magnetic resonance imaging does not rely on reporter probes and is a versatile technology for measuring biophysical magnetic properties in tissues. The primary impact thus far in animal studies has been in imaging tumor anatomy and changes in tumor mass. However, MRI has also enabled great strides in the noninvasive imaging of tissue vasculature (especially tumor vascular changes). A variant of MRI called dynamic contrast enhanced (DCE) MRI relies on tracking time-dependent intensity changes of an injected contrast agent and thus provides information on several physiological parameters in tumor vasculature such as blood flow and vessel permeability. The resolution of MRI allows for imaging even in tissues deep within the body and thus DCE-MRI can be applied to various animal models as well as humans. The DCE-MRI method has been used to assess the activity of antiangiogenesis agents in animal models[59] as well as in the clinic[60] and thus can be added to the repertoire of biomarker approaches that are relevant in characterizing targets with known or predicted roles in angiogenesis.

3.04.4.3.2 Optical imaging with fluorescent and luminescent probes

Measurement of specific molecular targets and processes in living cells and tissues, both microscopically and macroscopically, has been greatly enabled in the last decade by advances in molecular probes that can be optically tracked in real time and engineered to serve as reporters. The engineering of the *Aequorea victoria* green fluorescent protein (GFP) as a protein-based fluorescent probe in vivo was a key event in optical molecular imaging; subsequent genetic engineering has led to the GFP variants that emit at different wavelengths, and additional fluorescent proteins (such as more readily detectable, higher-wavelength red fluorescent proteins) have been isolated from other organisms.[61,62] A key application of relevance to target biomarkers is the ability to generate specific fusion proteins containing a fluorescent protein tag; this can be used to track the subcellular localization of the fusion target of interest, and to follow spatial patterns dynamically. As an approach to target discovery, fluorescent proteins can be used in a variant of shotgun cloning, in which random fusions generated with expression libraries are expressed in a cell population of interest, allowing for visualization and subsequent cloning of the fusion partner motif that is driving a specific localization pattern of interest.[63] Another application for fluorescent probes is in fluorescence resonance energy transfer (FRET), which provides a readout of close molecular interaction between two distinct fluorophores, often linked together in the same engineered polypeptide chain. FRET probes can provide real-time monitoring of enzymatic processes such as proteolysis[64,65] and protein kinase activity.[66,67]

A limitation of utilizing fluorescent proteins in whole-body in vivo imaging, at least in mammalian species, is the relatively shallow depth of penetration (2 mm or less) afforded by the excitation/emission wavelengths of these proteins. Imaging based on bioluminescence has emerged as a complementary approach, particularly for whole-body imaging with aims of tracking the movement of cells throughout the body or monitoring tumor response after treatment with experimental agents. Bioluminescent imaging typically relies on monitoring enzymatic activity of the bioluminescent enzyme luciferase, from the firefly. Transfected cells that express the luciferase transgene can be visualized based on light transmitted; the method has been used, for example, to monitor clearance of engrafted tumor cells[68] or to visualize the effects of antibacterial agents against labeled bacteria in vivo.[69] As is the case with PET and fluorescent protein imaging, gene expression regulation can be visualized through the use of luciferase reporter genes.[70] The bioluminescent method is fairly straightforward although somewhat limited in that absolute signal quantification is not feasible and enzyme substrate must be administered exogenously. Both fluorescent and bioluminescent in vivo imaging approaches utilize reporter gene transfection and thus as biomarker assays they are somewhat artificial. However the unique ability to monitor cell signaling and physiology noninvasively in real time ensures that these and other functional imaging modalities will continue to be powerful tools in cell-based assays in target or lead identification as well as in pharmacology studies in animals.

3.04.5 Biomarker Approaches in the Discovery of Early Markers of Drug Toxicity

The evaluation of drug safety and toxicology has benefited in recent years from biomarker applications, particularly in the area of biomarker discovery. As the number of drugs and drug targets grows, there is a commensurate need for readily measurable markers that can provide an early indicator of undesirable side effects. Indeed, the battery of established markers that can currently be measured in preclinical toxicology studies and in the clinic can all be considered broadly as biomarkers of toxicity[71]; however, there remains a need for additional biomarkers that can serve as indicators of potential toxicity at the earliest possible time, and in a manner that is specific to a given target, pathway, or drug chemotype. The emphasis on toxicity biomarker identification has led to a convergence of a number of disciplines focused on this area; key among these are genomics,[72,73] proteomics,[74] metabonomics,[49,50] and bioinformatics.[75] The ability of these approaches to efficiently evaluate vast numbers of endpoints simultaneously and to sort and correlate large amounts of information can impact toxicology studies in the same manner as in applications in target characterization or pharmacology studies. An emerging paradigm has been the investigation of well-characterized agents with known toxicant profiles in specific, well-established systems; molecular profiles generated by such agents in the appropriate organ or other biological compartment can then be used as comparators for the evaluation of previously uncharacterized agents, to query whether profiles match or to what extent. This is a challenging field given the daunting number of existing agents and of mechanisms of toxicity; however, as databases expand and profiling technologies are refined there is reason to expect considerable progress to be made in coming years.

Some examples of toxicity biomarker discovery efforts are worth noting as cases in point. An early gene expression profiling effort in toxicology applied oligonucleotide microarrays in a rat study of a panel of well-studied compounds and led to identification of response profiles of drug metabolism, metabolic effect, and stress response genes; some of these were previously characterized or otherwise verified by qRT-PCR.[76] In an interesting twist of profiling for toxicant or drug metabolic response transcripts, a variant approach has been described as a means to select for potential targets that are specific to drug effect in a target tissue and not coincidentally up- or downregulated as part of a drug absorption, distribution, metabolism, and excretion (ADME) pharmacokinetic response pathway; by restricting screening of transcripts to those that are not expressed in cells derived from ADME-related tissues (liver, colon, and kidney), this may provide a means to more selective potential targets or target-specific biomarkers.[77] At the same time, the catalog of transcripts that are expressed in ADME-related tissues may be of utility in potential toxicity marker discovery. Meanwhile, increasing emphasis is being placed on the use of surrogate tissues to allow monitoring of toxic effects in accessible compartments such as the peripheral blood or hair follicle rather than in internal organs.[78] Metabonomic analysis of an accessible biological fluid has been applied effectively in an investigation of urine markers of vasculitis; here the effect of a phosphodiesterase known to induce vasculitis in rats was assessed in urine samples, followed by principal components analysis which led to a pattern that could separate cases with vasculitis from those without.[79] Imaging technologies can also be used to assess biomarkers of toxicity; rather than detecting endogenous biomarkers, noninvasive assays utilizing transgenic reporter molecules can be applied in vivo. An example is an approach wherein a transfected luciferase reporter gene under control of the heme oxygenase-1 promoter serves as a bioluminescent reporter probe for toxicant effects in the liver and other tissues.[80] Reiterative testing in preclinical model systems as well as validation studies in both preclinical species and ultimately in the clinic will be critical in establishing the true utility and applicability of biomarkers of drug toxicity as well as in the refinement of large numbers of putative biomarkers into the simplest panels that are sufficiently informative and robust.

3.04.6 Translational Applications: Biomarkers in the Clinic

Roles of biomarker applications in early phases of drug discovery have been outlined, along with reference to their utility in toxicology; however, it is in drug development and the clinic where biomarkers can make perhaps the greatest and longest lasting impact. The expansion of biomarker-related research in the pharmaceutical and biotechnology industry has been driven partly by the promise that application of biomarkers will allow for decision making on a drug candidate during the earliest phases of clinical testing (which would in turn drive cost savings by reducing the expense of large phase II and III clinical trials which might otherwise be required in the absence of clear biomarker activity readout).[81] At the same time, the number of disease biomarkers and potential diagnostic and prognostics markers has been growing rapidly through advances in biochemical and molecular technologies. Thus there has been increased emphasis on the field often referred to as translational biology, and it is in this milieu where clinical biomarker applications are most often focused. The central tenet of translational research is the establishment of relevance and applicability for biological endpoints and assays in both experimental laboratory models and in the clinic; in the biomedical sciences this concept is often referred to as 'bench-to-bedside' (and sometimes, 'bedside-to-bench').[82] In an ideal scenario, the same biomarker(s) that provides evidence of pharmacological intervention in a cell-based assay would also be measured in a clinical study of patient or healthy volunteers, albeit perhaps with a very different analytical assay in terms of sensitivity and reagents.

3.04.6.1 Selection of Biomarker Endpoints and Assays

It is apparent that there is much more latitude in the types of specimens and scope of experimentation that is feasible in laboratory models as opposed to clinical studies in humans. A number of additional variables are encountered in the clinic: e.g., clinical study populations are typically heterogeneous in terms of age, health status, etc., particularly in patient studies as opposed to healthy volunteers; and specimens are often analyzed in laboratories remote from the collection sites, presenting logistical challenges in specimen handling, preservation, storage, and transport.[83] For such reasons, the criteria for biomarker assay implementation in clinical studies is much more stringent. However, the overriding goal of monitoring pharmacodynamic effect is the same in both contexts. Of course, there are molecular biomarkers that are primarily useful as predictive markers rather than pharmacodynamic markers. These are only measured in patient specimens prior to initiation of a therapeutic regimen. An example is the HER2/neu protein, which is highly expressed in approximately 20% of advanced breast tumors and has been exploited as a therapeutic target and led to the development of a humanized monoclonal antibody therapeutic known as trastuzumab.[84] As profiling of genomic DNA sequences to identify polymorphisms associated with particular drug response becomes more extensive, along with increased application of nucleic acid and protein detection, it is likely that the number of predictive biomarkers will increase; this should lead in turn to robust molecular diagnostic assays that can routinely be applied in patient preselection.[85,86] A case in point is that of the epidermal growth factor receptor (EGFR) inhibitor gefitinib (Iressa); recently it has been discovered that somatic mutations in the EGFR gene in lung tumors are strongly correlated with clinical benefit in cancer patients being treated with this drug.[87,88] These conclusions are based on relatively small data sets and additional testing is required to confirm the true relationship between EGFR mutations and clinical benefit. In the establishment of suitability of predictive biomarkers for their intended purpose, similar principles as for consideration of pharmacodynamic markers should always be applied.

An important step in the practical application of biomarker endpoints is validation that a biomarker assay is fit for its intended purpose. The absolute requirements around when an assay is sufficiently validated so as to be useful may vary based on the application as well as the technology platform. For example, an assay that is used as a decision making criterion for further clinical development of a drug candidate would require more rigorous validation than a panel of analytes that are used for gathering information in preclinical experiments. Similarly, results obtained from biomarker discovery efforts with emerging platforms such as the 'omics' technologies may not require the same rigor as a physician-prescribed clinical diagnostic test. However, it will become increasingly important to assess the robustness of biomarker assays as biomarkers become more widely used, particularly in the clinical setting. Some detailed guidelines for this process of validation of biomarker endpoints have been described recently.[89,90] The two key features of assay characterization are: analytical validation (evaluation of assay performance, technology, and specifications) and clinical/biological validation (evaluation of the statistical rigor with which a biomarker signal can be linked to a pharmacological effect or disease status). As with all bioanalytical assays, parameters such as specificity, sensitivity, stability (both specimen and analyte), and precision must be considered, as well as accuracy (unless the objective is primarily to detect a relative difference in longitudinal measurements, such as baseline versus posttreatment levels of a marker; for such assessments of pharmacodynamic activity, this is often sufficient to establish whether a drug effect has occurred, however quantitation is always a useful objective to strive for in clinical biomarker studies). The establishment of guidelines for validation can be challenging, given (1) the multifaceted nature of biomarker applications (ranging from relatively quantitative assays such as ELISAs to more qualitative approaches such as molecular imaging or immunostaining), (2) the frequently heterogeneous nature of the macromolecules being measured, and (3) the biological diversity that is inherent in clinical populations. However, it is imperative that good scientific principles and careful interpretation be applied in all biomarker studies; further, as discussed earlier, multiple studies are useful and often necessary to allow confirmation and cross-validation of initial results.[52]

3.04.6.2 Biomarker as Surrogates for Clinical Activity

As briefly referred to earlier in this chapter, the greatest historical value of biomarkers in drug development has been in the clinic as markers of disease status and response to therapy. Biomarkers that have been strongly linked to improvement in a particular disease are often referred to as surrogate endpoints or surrogate markers. Surrogate endpoints have been defined as: a biomarker that is intended to substitute for a clinical endpoint. A surrogate endpoint is expected to predict clinical benefit (or harm, or lack of benefit or harm) based on epidemiologic, therapeutic, pathophysiologic, or other scientific evidence.[1] After extensive testing, often requiring years of large and well-controlled clinical studies, well-correlated surrogate endpoints can sometimes be used as the basis of regulatory approval in the absence of a large body of evidence based on actual clinical endpoints, which are defined as: characteristics or variables that reflect how a patient feels, functions, or survives.[1] Examples of surrogate endpoints include anatomic measurements of tumor mass shrinkage in certain cancers, or human immunodeficiency virus (HIV) mRNA viral load and CD4 cell count responses to antiretroviral therapy. The burden of proof for a biomarker to attain surrogate endpoint status is sufficiently large such that most biomarkers will not be classified as surrogates (however, all surrogate endpoints can be considered to be biomarkers, whether molecular or otherwise). A case in point is the prostate-specific antigen (PSA) which has long been considered a useful surrogate measure of prostate cancer occurrence and progression, but whose true utility has come under question as the clinical knowledge base has expanded (this also raises the issue of the need for more than one biochemical/molecular biomarker for higher confidence in assessing disease status).[91] A more detailed consideration of the criteria for and implications of surrogate endpoints can be found elsewhere[92]; the road from target identification and early biomarker characterization to establishment of clinical utility as a surrogate endpoint point is likely to be very long and winding, but carefully designed (and statistically well-powered) preclinical experiments will provide a good starting point.

3.04.6.3 Biomarker Endpoints Applied in Pharmacokinetic–Pharmacodynamic Modeling

Perhaps the broadest impact made by biomarker assays has been in the assessment of pharmacodynamic effects of drug candidates both preclinically and in the clinic; the data generated by such pharmacodynamic profiling can allow for extensive modeling of drug concentration–effect relationships to be performed. In colloquial terms, pharmacokinetics (PK) describes what the body does to a drug, while pharmacodynamics (PD) describes what a drug does to the body. The mathematical modeling of quantitative relationships between observed plasma or tissue drug concentrations (pharmacokinetics) and the measured pharmacological effect(s) (pharmacodynamics) is referred to as PK/PD modeling.[93] Application of PK/PD modeling and simulations is useful in preclinical as well as clinical phases of drug testing, and can provide a detailed kinetic view of drug action over time, which ultimately can lead to dose optimization provided that the pharmacodynamic endpoint is well measured and tightly linked mechanistically to drug activity (indeed, the quality of information generated by modeling can only be as good as the input data).[94] A well-characterized biomarker that is linked to target activity or is directly downstream of target signaling is a valid choice for pharmacodynamic input, so long as the bioanalytical assay for quantification of the biomarker is robust and, particularly in the clinical setting, can be performed under conditions that meet Good Laboratory Practice (GLP) conditions or are at least GLP-like[83,90]; of course, the analytical method that measures pharmacokinetic endpoints must also be reliable and quantitative, though this is routinely the case for synthetic drugs.

When surrogate endpoints or biomarkers with at least a preliminary linkage to clinical efficacy are used in the PD component, it may be possible to predict and subsequently test for a drug dose that provides sufficient biological activity but with a lowered risk for adverse side effects; this concept has been termed the optimal biological dose[95] (in contrast with the more traditional and empirically determined maximum tolerated dose concept) and holds great promise for the refinement of therapeutic dosing regimens. The key prerequisite for broad practical application of this concept in the clinic is that the correlation between the pharmacodynamic biomarker endpoint, drug exposure, and

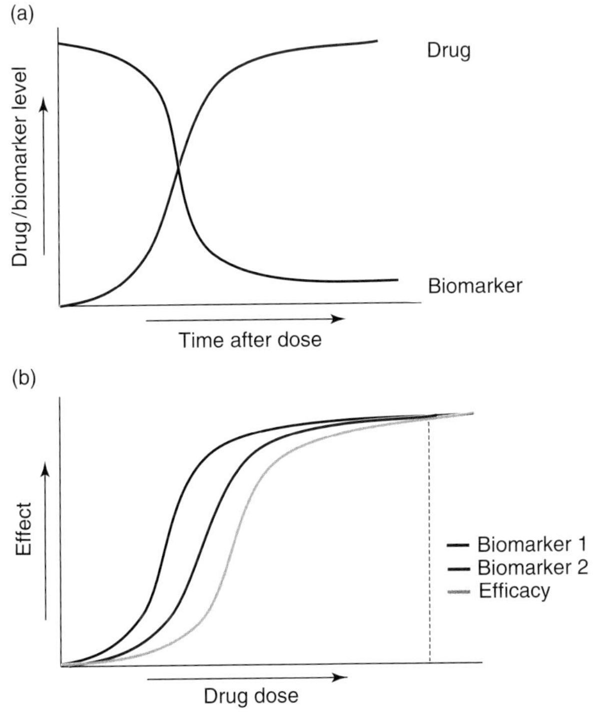

Figure 2 Graphical illustrations of two general concepts in PK/PD modeling. (a) An idealized scenario in which a biomarker signal is shown to decrease as levels of drug are increased over time. (b) The interplay between three endpoints, one of which is a measure of clinical efficacy and the other two are biomarkers; one of the biomarkers (biomarker 2) shows a pharmacodynamic pattern that closely follows that of the efficacy signal, and thus biomarker 2 may be of use in determining an optimal biological dose (indicated by the blue line).

clinical efficacy must be firmly established, ideally in more than one clinical study of reasonable statistical power. Another facet where biomarkers are useful is in defining the PK/PD relationship for any metabolic products of a given drug, i.e., in order to test the pharmacological and biological activity of major metabolites in comparison to the parent compound (this is best done in preclinical models where each metabolite can be first tested in isolation). **Figure 2** illustrates, in a generalized, hypothetical format, two concepts that are integral to PK/PD relationships; one is the temporal relationship between drug levels and changes in the level of a biomarker endpoint (in this illustration, the biomarker decreases over time as the drug level rises), while the other is the dose-dependent relationship between an efficacy endpoint and two pharmacodynamic biomarkers, one of which (biomarker 2) shows a dose–response pattern that is closer to that of the efficacy endpoint (and hence is more relevant than the other biomarker for determination of the optimal biological dose).

3.04.6.4 Regulatory Implications

As the molecular underpinnings behind disease processes and drug mechanisms begin to become better characterized, regulatory agencies are increasingly focusing attention toward biomarkers and their potential role in drug approvals or usage guidelines. As previously mentioned, a subset of biomarkers known as surrogate endpoints have already been established, in select indications, to be predictors of clinical benefit suitable for regulatory approval decision making prior to definitive assessment of clinical outcome per se (e.g., survival). Further, there have been in recent years a few examples of therapeutics which are prescribed on the basis of presence or absence of specific molecular determinants in the disease tissue; a prime case in point is the aforementioned trastuzumab, which is prescribed only for breast cancer patients whose tumor cells express the HER2 protein. Additional cases where molecular or pharmacogenomic information (e.g., in polymorphisms in metabolic enzymes) is described in drug labels are likely to emerge in coming years. However, regulatory agencies are rightly taking a cautious approach toward reliance on either pharmacodynamic or predictive biomarkers in their deliberations. The limiting factors are the difficulties in establishing the true clinical validity of potential surrogate endpoints, and, on a broader scale, the substantial gaps that exist in the scientific understanding of both human disease processes as well as of the mechanistic impact of all potential drug targets on disease (and normal) processes.[96] Thus, empirical assessments of obvious clinical benefit remain the primary driver of drug approvals. Nevertheless, the great potential of biomarker applications and other emerging biomedical technologies has been recognized by regulatory agencies, as exemplified by the recent initiative from the US Food and Drug Administration (FDA) referred to as the Critical Path Initiative.[97] The initiative emphasizes the need for modernization and acceleration of the medical product development process and suggests that new scientific and technical capabilities (including biomarker research) will be pivotal in driving this forward.

3.04.7 Conclusions

The careful and consistent application of biomarker assays throughout target and drug discovery and on into the clinic is becoming an integral part of the drug development process. The benefits of biomarker applications include more complete characterization and understanding of target biological activity and drug mechanism of action; enhanced assessments of drug pharmacological properties and PK/PD relationships; potential early detection of drug toxicities; and the possibility of more efficient and effective clinical testing via enabling early decisions as to whether an agent is exhibiting appropriate activity against its target(s). While the general concept of biomarkers is not new, it has taken on additional import as the number of potential drug targets is expanding and the need for innovative and more rationally targeted agents has grown. Molecular biomarkers will be of most benefit if they can be linked to target activity or direct downstream effects at an early phase of investigation; a number of methodological approaches can be applied to this end, many of which will rely upon well-established protein detection methods or other macromolecular detection assays (e.g., RNA-based gene expression profiling). Biomarker endpoints that can be applied from in vitro laboratory models all the way into the clinic are often referred to as translational biomarkers (or perhaps translatable biomarkers), and are often the most useful due to their impact at many stages and in many contexts. The importance of biomarker discovery efforts should not be overlooked, as this is an expanding field that has been greatly enabled by breakthroughs in the 'omics' technologies; a widening number of potential 'omic' approaches can now be applied toward biomarker discovery, and this is perhaps of greatest utility in the characterization of new targets and at relatively early stages in the process. At the same time, biomarker discovery approaches are being applied in the clinic as well, and advances in the discovery of potential toxicity biomarkers have been furthered by studies of this type. Careful implementation and interpretation of such studies is imperative, and all biomarker assays should undergo extensive testing to establish their validity, particularly when applied in the clinical setting. The status of surrogate endpoints is reserved for those (relatively few at this stage) biomarker endpoints that are ultimately validated as markers for change in disease status and that can serve as predictors of clinical outcome. In this regard in particular, biomarkers will likely continue to be of interest to regulatory agencies in their evaluation of therapeutic agents.

References

1. Biomarkers and Surrogate Endpoint Working Group. *Clin. Pharmacol. Ther.* **2001**, *69*, 89–95.
2. Colburn, W. A. *J. Clin. Pharmacol.* **2003**, *43*, 329–341.
3. Frackelton, A. R., Jr.; Ross, A. H.; Eisen, H. N. *Mol. Cell Biol.* **1983**, *3*, 1343–1352.
4. Huhn, R. D.; Posner, M. R.; Rayter, S. I.; Foulkes, J. G.; Frackelton, A. R., Jr. *Proc. Natl. Acad. Sci. USA* **1987**, *84*, 4408–4412.
5. O'Farrell, A. M.; Foran, J. M.; Fiedler, W.; Serve, H.; Paquette, R. L.; Cooper, M. A.; Yuen, H. A.; Louie, S. G.; Kim, H.; Nicholas, S. et al. *Clin. Cancer Res.* **2003**, *9*, 5465–5476.
6. Kitagawa, M.; Higashi, H.; Jung, H. K.; Suzuki-Takahashi, I.; Ikeda, M.; Tamai, K.; Kato, J.; Segawa, K.; Yoshida, E.; Nishimura, S. et al. *EMBO J.* **1996**, *15*, 7060–7069.
7. Lu, H.; Taya, Y.; Ikeda, M.; Levine, A. J. *Proc. Natl. Acad. Sci. USA* **1998**, *95*, 6399–6402.
8. Craig, A. L.; Bray, S. E.; Finlan, L. E.; Kernohan, N. M.; Hupp, T. R. *Methods Mol. Biol.* **2003**, *234*, 171–202.
9. Renberg, B.; Nygren, P. K.; Eklund, M.; Karlstrom, A. E. *Anal. Biochem.* **2004**, *334*, 72–80.
10. Nord, K.; Gunneriusson, E.; Ringdahl, J.; Stahl, S.; Uhlen, M.; Nygren, P. A. *Nat. Biotechnol.* **1997**, *15*, 772–777.
11. Bock, C.; Coleman, M.; Collins, B.; Davis, J.; Foulds, G.; Gold, L.; Greef, C.; Heil, J.; Heilig, J. S.; Hicke, B. et al. *Proteomics* **2004**, *4*, 609–618.
12. Fivash, M.; Towler, E. M.; Fisher, R. J. *Curr. Opin. Biotechnol.* **1998**, *9*, 97–101.
13. Wilson, S.; Howell, S. *Biochem. Soc. Trans.* **2002**, *30*, 794–797.
14. Nelson, R. W.; Nedelkov, D.; Tubbs, K. A. *Electrophoresis* **2000**, *21*, 1155–1163.
15. Nedelkov, D.; Nelson, R. W. *Trends Biotechnol.* **2003**, *21*, 301–305.
16. Beran, M.; Cao, X.; Estrov, Z.; Jeha, S.; Jin, G.; O'Brien, S.; Talpaz, M.; Arlinghaus, R. B.; Lydon, N. B.; Kantarjian, H. *Clin. Cancer Res.* **1998**, *4*, 1661–1672.
17. Pryer, N. K.; Lee, L. B.; Zadovaskaya, R.; Yu, X.; Sukbuntherng, J.; Cherrington, J. M.; London, C. A. *Clin. Cancer Res.* **2003**, *9*, 5729–5734.
18. Wang, W. L.; Healy, M. E.; Sattler, M.; Verma, S.; Lin, J.; Maulik, G.; Stiles, C. D.; Griffin, J. D.; Johnson, B. E.; Salgia, R. *Oncogene* **2000**, *19*, 3521–3528.

19. Albanell, J.; Rojo, F.; Averbuch, S.; Feyereislova, A.; Mascaro, J. M.; Herbst, R.; LoRusso, P.; Rischin, D.; Sauleda, S.; Gee, J. et al. *J. Clin. Oncol.* **2002**, *20*, 110–124.
20. Xia, W.; Mullin, R. J.; Keith, B. R.; Liu, L. H.; Ma, H.; Rusnak, D. W.; Owens, G.; Alligood, K. J.; Spector, N. L. *Oncogene* **2002**, *21*, 6255–6263.
21. Anido, J.; Matar, P.; Albanell, J.; Guzman, M.; Rojo, F.; Arribas, J.; Averbuch, S.; Baselga, J. *Clin. Cancer Res.* **2003**, *9*, 1274–1283.
22. Ray, C. A.; Bowsher, R. R.; Smith, W. C.; Devanarayan, V.; Willey, M. B.; Brandt, J. T.; Dean, R. A. *J. Pharm. Biomed. Anal.* **2005**, *36*, 1037–1044.
23. Pollack, V. A.; Savage, D. M.; Baker, D. A.; Tsaparikos, K. E.; Sloan, D. E.; Moyer, J. D.; Barbacci, E. G.; Pustilnik, L. R.; Smolarek, T. A.; Davis, J. A. et al. *J. Pharmacol. Exp. Ther.* **1999**, *291*, 739–748.
24. Templin, M. F.; Stoll, D.; Bachmann, J.; Joos, T. O. *Comb. Chem. High Throughput Screen.* **2004**, *7*, 223–229.
25. Khan, S. S.; Smith, M. S.; Reda, D.; Suffredini, A. F.; McCoy, J. P., Jr. *Cytometry B Clin. Cytom.* **2004**, *61*, 35–39.
26. Haab, B. B.; Dunham, M. J.; Brown, P. O. *Genome. Biol.* **2001**, *2*, RESEARCH0004.
27. Piffanelli, A.; Dittadi, R.; Catozzi, L.; Gion, M.; Capitanio, G.; Gelli, M. C.; Brazzale, A.; Malagutti, R.; Pelizzola, D.; Menegon, A. et al. *Breast Cancer Res. Treat.* **1996**, *37*, 267–276.
28. Chow, S.; Patel, H.; Hedley, D. W. *Cytometry* **2001**, *46*, 72–78.
29. Gerber, S. A.; Rush, J.; Stemman, O.; Kirschner, M. W.; Gygi, S. P. *Proc. Natl. Acad. Sci. USA* **2003**, *100*, 6940–6945.
30. Ballif, B. A.; Roux, P. P.; Gerber, S. A.; MacKeigan, J. P.; Blenis, J.; Gygi, S. *Proc. Natl. Acad. Sci. USA* **2005**, *102*, 667–672.
31. Morimoto, A. M.; Tan, N.; West, K.; McArthur, G.; Toner, G. C.; Manning, W. C.; Smolich, B. D.; Cherrington, J. M. *Oncogene* **2004**, *23*, 1618–1626.
32. McCaffrey, R. L.; Fawcett, P.; O'Riordan, M.; Lee, K. D.; Havell, E. A.; Brown, P. O.; Portnoy, D. A. *Proc. Natl. Acad. Sci. USA* **2004**, *101*, 11386–11391.
33. Saban, M. R.; Nguyen, N. B.; Hammond, T. G.; Saban, R. *Am. J. Pathol.* **2002**, *160*, 2095–2110.
34. Taxman, D. J.; MacKeigan, J. P.; Clements, C.; Bergstralh, D. T.; Ting, J. P. *Cancer Res.* **2003**, *63*, 5095–5104.
35. Ramaswamy, S.; Golub, T. R. *J. Clin. Oncol.* **2002**, *20*, 1932–1941.
36. Liu, E. T. *Proc. Natl. Acad. Sci. USA* **2005**, *102*, 3531–3532.
37. Lossos, I. S.; Czerwinski, D. K.; Alizadeh, A. A.; Wechser, M. A.; Tibshirani, R.; Botstein, D.; Levy, R. *N. Engl. J. Med.* **2004**, *350*, 1828–1837.
38. Iyer, V. R.; Eisen, M. B.; Ross, D. T.; Schuler, G.; Moore, T.; Lee, J. C.; Trent, J. M.; Staudt, L. M.; Hudson, J. Jr.; Boguski, M. S. et al. *Science* **1999**, *283*, 83–87.
39. DePrimo, S. E.; Wong, L. M.; Khatry, D. B.; Nicholas, S. L.; Manning, W. C.; Smolich, B. D.; O'Farrell, A. M.; Cherrington, J. M. *BMC Cancer* **2003**, *3*, 3.
40. Griffin, T. J.; Gygi, S. P.; Ideker, T.; Rist, B.; Eng, J.; Hood, L.; Aebersold, R. *Mol. Cell. Proteomics* **2002**, *1*, 323–333.
41. Li, J.; Yang, Y.; Peng, Y.; Austin, R. J.; van Eyndhoven, W. G.; Nguyen, K. C.; Gabriele, T.; McCurrach, M. E.; Marks, J. R.; Hoey, T. et al. *Nat. Genet.* **2002**, *31*, 133–134.
42. Mu, D.; Chen, L.; Zhang, X.; See, L. H.; Koch, C. M.; Yen, C.; Tong, J. J.; Spiegel, L.; Nguyen, K. C.; Servoss, A. et al. *Cancer Cell* **2003**, *3*, 297–302.
43. Kononen, J.; Bubendorf, L.; Kallioniemi, A.; Barlund, M.; Schraml, P.; Leighton, S.; Torhorst, J.; Mihatsch, M. J.; Sauter, G.; Kallioniemi, O. P. *Nat. Med.* **1998**, *4*, 844–847.
44. Schraml, P.; Kononen, J.; Bubendorf, L.; Moch, H.; Bissig, H.; Nocito, A.; Mihatsch, M. J.; Kallioniemi, O. P.; Sauter, G. *Clin. Cancer Res.* **1999**, *5*, 1966–1975.
45. Paweletz, C. P.; Charboneau, L.; Bichsel, V. E.; Simone, N. L.; Chen, T.; Gillespie, J. W.; Emmert-Buck, M. R.; Roth, M. J.; Petricoin, I. E.; Liotta, L. A. *Oncogene* **2001**, *20*, 1981–1989.
46. Nishizuka, S.; Charboneau, L.; Young, L.; Major, S.; Reinhold, W. C.; Waltham, M.; Kouros-Mehr, H.; Bussey, K. J.; Lee, J. K.; Espina, V. et al. *Proc. Natl. Acad. Sci. USA* **2003**, *100*, 14229–14234.
47. Durr, E.; Yu, J.; Krasinska, K. M.; Carver, L. A.; Yates, J. R.; Testa, J. E.; Oh, P.; Schnitzer, J. E. *Nat. Biotechnol.* **2004**, *22*, 985–992.
48. Oh, P.; Li, Y.; Yu, J.; Durr, E.; Krasinska, K. M.; Carver, L. A.; Testa, J. E.; Schnitzer, J. E. *Nature* **2004**, *429*, 629–635.
49. Shockcor, J. P.; Holmes, E. *Curr. Topics Med. Chem.* **2002**, *2*, 35–51.
50. Griffin, J. L.; Bollard, M. E. *Curr. Drug Metab.* **2004**, *5*, 389–398.
51. Watkins, S. M.; Reifsnyder, P. R.; Pan, H. J.; German, J. B.; Leiter, E. H. *J. Lipid Res.* **2002**, *43*, 1809–1817.
52. Ransohoff, D. F. *Nat. Rev. Cancer* **2004**, *4*, 309–314.
53. Weissleder, R. *Nat. Rev. Cancer* **2002**, *2*, 11–18.
54. Lewis, J. S.; Achilefu, S.; Garbow, J. R.; Laforest, R.; Welch, M. J. *Eur. J. Cancer* **2002**, *38*, 2173–2188.
55. Herschman, H. R. *Curr. Opin. Immunol.* **2003**, *15*, 378–384.
56. Gambhir, S. S.; Barrio, J. R.; Phelps, M. E.; Iyer, M.; Namavari, M.; Satyamurthy, N.; Wu, L.; Green, L. A.; Bauer, E.; MacLaren, D. C. et al. *Proc. Natl. Acad. Sci. USA* **1999**, *96*, 2333–2338.
57. Blasberg, R. *Eur. J. Cancer* **2002**, *38*, 2137–2146.
58. Smith-Jones, P. M.; Solit, D. B.; Akhurst, T.; Afroze, F.; Rosen, N.; Larson, S. M. *Nat. Biotechnol.* **2004**, *22*, 701–706.
59. Drevs, J.; Muller-Driver, R.; Wittig, C.; Fuxius, S.; Esser, N.; Hugenschmidt, H.; Konerding, M. A.; Allegrini, P. R.; Wood, J.; Hennig, J.; Unger, C. et al. *Cancer Res.* **2002**, *62*, 4015–4022.
60. Morgan, B.; Thomas, A. L.; Drevs, J.; Hennig, J.; Buchert, M.; Jivan, A.; Horsfield, M. A.; Mross, K.; Ball, H. A.; Lee, L. et al. *J. Clin. Oncol.* **2003**, *21*, 3955–3964.
61. Zhang, J.; Campbell, R. E.; Ting, A. Y.; Tsien, R. Y. *Nat. Rev. Mol. Cell Biol.* **2002**, *3*, 906–918.
62. Weissleder, R.; Ntziachristos, V. *Nat. Med.* **2003**, *9*, 123–128.
63. Bejarano, L. A.; Gonzalez, C. *J. Cell Sci.* **1999**, *112* (Pt 23), 4207–4211.
64. Heim, R.; Tsien, R. Y. *Curr. Biol.* **1996**, *6*, 178–182.
65. Rehm, M.; Dussmann, H.; Janicke, R. U.; Tavare, J. M.; Kogel, D.; Prehn, J. H. *J. Biol. Chem.* **2002**, *277*, 24506–24514.
66. Ting, A. Y.; Kain, K. H.; Klemke, R. L.; Tsien, R. Y. *Proc. Natl. Acad. Sci. USA* **2001**, *98*, 15003–15008.
67. Sato, M.; Ozawa, T.; Inukai, K.; Asano, T.; Umezawa, Y. *Nat. Biotechnol.* **2002**, *20*, 287–294.
68. Sweeney, T. J.; Mailander, V.; Tucker, A. A.; Olomu, A. B.; Zhang, W.; Cao, Y.; Negrin, R. S.; Contag, C. H. *Proc. Natl. Acad. Sci. USA* **1999**, *96*, 12044–12049.

69. Rocchetta, H. L.; Boylan, C. J.; Foley, J. W.; Iversen, P. W.; LeTourneau, D. L.; McMillian, C. L.; Contag, P. R.; Jenkins, D. E.; Parr, T. R., Jr. *Antimicrob. Agents Chemother.* **2001**, *45*, 129–137.
70. Contag, C. H.; Spilman, S. D.; Contag, P. R.; Oshiro, M.; Eames, B.; Dennery, P.; Stevenson, D. K.; Benaron, D. A. *Photochem. Photobiol.* **1997**, *66*, 523–531.
71. Amacher, D. E. *Hum. Exp. Toxicol.* **2002**, *21*, 253–262.
72. Kramer, J. A.; Kolaja, K. L. *Expert Opin. Drug. Saf.* **2002**, *1*, 275–286.
73. Suter, L.; Babiss, L. E.; Wheeldon, E. B. *Chem. Biol.* **2004**, *11*, 161–171.
74. Merrick, B. A.; Bruno, M. E. *Curr. Opin. Mol. Ther.* **2004**, *6*, 600–607.
75. Fielden, M. R.; Matthews, J. B.; Fertuck, K. C.; Halgren, R. G.; Zacharewski, T. R. *Crit. Rev. Toxicol.* **2002**, *32*, 67–112.
76. Gerhold, D.; Lu, M.; Xu, J.; Austin, C.; Caskey, C. T.; Rushmore, T. *Physiol. Genomics* **2001**, *5*, 161–170.
77. Dooley, T. P.; Curto, E. V.; Reddy, S. P.; Davis, R. L.; Lambert, G.; Wilborn, T. W. *Biochem. Biophys. Res. Commun.* **2003**, *303*, 828–841.
78. Rockett, J. C.; Burczynski, M. E.; Fornace, A. J.; Herrmann, P. C.; Krawetz, S. A.; Dix, D. J. *Toxicol. Appl. Pharmacol.* **2004**, *194*, 189–199.
79. Robertson, D. G.; Reily, M. D.; Albassam, M.; Dethloff, L. A. *Cardiovasc. Toxicol.* **2001**, *1*, 7–19.
80. Malstrom, S. E.; Jekic-McMullen, D.; Sambucetti, L.; Ang, A.; Reeves, R.; Purchio, A. F.; Contag, P. R.; West, D. B. *Toxicol. Appl. Pharmacol.* **2004**, *200*, 219–228.
81. Frank, R.; Hargreaves, R. *Nat. Rev. Drug Disc.* **2003**, *2*, 566–580.
82. Marincola, F. M. *J. Transl. Med.* **2003**, *1*, 1.
83. Swanson, B. N. *Dis. Markers* **2002**, *18*, 47–56.
84. Leyland-Jones, B. *Lancet Oncol.* **2002**, *3*, 137–144.
85. Ross, J. S.; Ginsburg, G. S. *Drug Disc. Today* **2002**, *7*, 859–864.
86. Chan-Hui, P. Y. *Drug Disc. Today* **2003**, *8*, 829–831.
87. Paez, J. G.; Janne, P. A.; Lee, J. C.; Tracy, S.; Greulich, H.; Gabriel, S.; Herman, P.; Kaye, F. J.; Lindeman, N.; Boggon, T. J. et al. *Science* **2004**, *304*, 1497–1500.
88. Lynch, T. J.; Bell, D. W.; Sordella, R.; Gurubhagavatula, S.; Okimoto, R. A.; Brannigan, B. W.; Harris, P. L.; Haserlat, S. M.; Supko, J. G.; Haluska, F. G. et al. *N. Engl. J. Med.* **2004**, *350*, 2129–2139.
89. Barker, P. E. *Ann. N. Y. Acad. Sci.* **2003**, *983*, 142–150.
90. Colburn, W. A.; Lee, J. W. *Clin. Pharmacokinet.* **2003**, *42*, 997–1022.
91. Stamey, T. A.; Caldwell, M.; McNeal, J. E.; Nolley, R.; Hemenez, M.; Downs, J. *J. Urol.* **2004**, *172*, 1297–1301.
92. Colburn, W. A. *J. Clin. Pharmacol.* **1997**, *37*, 355–362.
93. Derendorf, H.; Lesko, L. J.; Chaikin, P.; Colburn, W. A.; Lee, P.; Miller, R.; Powell, R.; Rhodes, G.; Stanski, D.; Venitz, J. *J. Clin. Pharmacol.* **2000**, *40*, 1399–1418.
94. Colburn, W. A. *J. Clin. Pharmacol.* **2000**, *40*, 1419–1427.
95. Kleinerman, E. S.; Murray, J. L.; Snyder, J. S.; Cunningham, J. E.; Fidler, I. J. *Cancer Res.* **1989**, *49*, 4665–4670.
96. Katz, R. *Neurorx* **2004**, *1*, 189–195.
97. The critical path to new medical products http://www.fda.gov/oc/initiatives/criticalpath/ (accessed April 2006).

Biography

Samuel E DePrimo received a BSc in biology from King's College in Wilkes-Barre, PA, and in 1996 received his PhD from the University of Cincinnati College of Medicine, where he developed and studied transgenic mouse models for the detection of somatic mutations in the laboratory of James Stringer. This was followed by postdoctoral work in George Stark's laboratory at the Cleveland Clinic Foundation's Lerner Institute, studying p53 signaling pathways and forward genetics applications, and with James Brooks at the Stanford University School of Medicine, in collaboration with Patrick Brown, applying gene expression profiling in studying the biology of prostate cancer cells. Dr DePrimo joined SUGEN, Inc. in 2001, as a scientist in the Biomarkers group, and focused on translational research and analysis of molecular biomarkers in oncology clinical trials of receptor tyrosine kinase inhibitors. He joined Pfizer Global Research and Development at the La Jolla campus in 2003, where as a scientist in the Translational Medicine group he continues to focus on biomarker applications and translational biology, primarily in oncology drug development.

Comprehensive Medicinal Chemistry II
ISBN (set): 0-08-044513-6

ISBN (Volume 3) 0-08-044516-0; pp. 69–85

3.05 Microarrays

D Amaratunga, Johnson & Johnson Pharmaceutical Research & Development LLC, Raritan, NJ, USA
H Göhlmann and P J Peeters, Johnson & Johnson Pharmaceutical Research & Development, Beerse, Belgium

3.05.1	**Introduction**	**87**
3.05.2	**Deoxyribonucleic Acid Microarray Experiments**	**88**
3.05.2.1	Microarray Technologies	88
3.05.2.2	Experimental Design Considerations	90
3.05.3	**Data Analysis Considerations**	**91**
3.05.3.1	Data Preprocessing	91
3.05.3.2	Visual Inspection of the Data	91
3.05.3.3	Identifying and Studying Differentially Expressing Genes	92
3.05.3.4	Sample Classification	93
3.05.4	**Case Studies**	**93**
3.05.4.1	Case Study 1	93
3.05.4.2	Case Study 2	97
3.05.4.3	Case Study 3	100
3.05.4.4	Conclusions from the Case Studies	103
3.05.5	**Discussion and a Look to the Future**	**103**
References		**104**

3.05.1 Introduction

For the pharmaceutical industry, undoubtedly the most enticing prospect stemming from the modern genomics revolution is the promise it offers of improved drug target identification. A drug target is a molecule in the body, usually a protein, that is intrinsically associated with a particular disease process and that could be addressed by a drug to produce a desired therapeutic effect. Typical examples for such targets are receptors on the surface of a cell that can be triggered to give a signal upon stimulation with a drug or enzymes whose activity of converting one substance to another is modulated via a drug. The identification, characterization, and validation of a drug target is a long and difficult exercise, demanding, as it does, a profound understanding of a disease's etiology and the biological processes associated with it, coupled with a fair amount of trial and error experimentation. In fact, up to now, drugs have been developed to address only a few hundred known drug targets. On the other hand, the anticipation is that today's research in genomics and proteomics will yield several thousand new drug targets.

Furthermore, it is likely that these drug targets will lead to drugs that are therapeutically more precise and more effective than those available today. Many of today's drugs were identified by happenstance and show a multitude of side effects. While some lack of specificity may be desirable in certain circumstances, it is still preferable to have substances that are as specific toward their respective targets as possible to avoid undesirable adverse events. Using our ever-increasing understanding of how proteins react and interact with each other, it is expected that, for a given disease, the best possible molecular intervention point should be precisely identifiable. In order to get there, there is a need for experimental procedures that will capture the behavior of all involved interaction partners simultaneously. Such procedures are currently emerging in the genomics area with technologies such as deoxyribonucleic acid (DNA) microarrays for gene activity (= mRNA) profiling, and in related areas such as proteomics, metabolomics, and lipidomics.

The advent of DNA microarrays was heralded by the seminal paper of Schena *et al.*,[1] who introduced the concept of immobilizing microscopic amounts of DNA fragments on glass substrates, and the subsequent success story of Golub *et al.*,[2] who demonstrated that such microarrays could be used both to distinguish among clinically similar types of cancer, acute myeloid leukemia, and acute lymphoblastic leukemia, as well as to generate a list of genes associated with the classification.

The power of DNA microarray technology is that it provides a high throughput instrument for simultaneously screening thousands of genes for differential expression across a series of different conditions. Despite the fact that it may not have sufficient sensitivity to detect subtle expression changes such as those of regulatory genes, a well-designed and properly executed microarray experiment should be able to detect the downstream effects of such changes, which generally tend to be more dramatic. Since these effects are the ones most likely to be associated with druggable targets, DNA microarrays are an important tool for identifying drug targets.

The remainder of this chapter is organized as follows. Section 3.05.2 outlines DNA microarray technology and experiments. Section 3.05.3 describes the various data analysis considerations that are necessary to make sense out of the enormous datasets generated by these experiments. Section 3.05.4 presents a series of case studies from the neuroscience area, although the concepts illustrated therein apply much more globally. Finally, Section 3.05.5 offers some concluding remarks as well as a quick look to the future.

3.05.2 Deoxyribonucleic Acid Microarray Experiments

DNA microarray technology enables scientists to simultaneously assess the relative transcription levels of several (if not all) of the genes within a cell across a series of different conditions: for example, across different pathological conditions, such as healthy versus diseased, different disease states, across time during the progression of a disease or infection, or during therapy and so on. Under the premise that some genes would be differentially expressed as a consequence of a disease process, monitoring expression patterns in this way provides scientists with potentially valuable clues and insights for scrutinizing and understanding the biological and cellular processes underlying that disease.

DNA microarrays are small solid supports (such as glass microscope slides, silicon chips, beads, or nylon membranes) onto which nucleic acid sequences (probes) have been robotically attached in a rectangular array format. Typically, the number of probes arrayed is very large and would correspond to a significant representation of the genome under study, which is in keeping with the concept that this technology is essentially a high throughput screen of the transcriptome. The probes are usually made of small single-stranded DNA sequences (either oligonucleotides or denatured polymerase chain reaction (PCR) products), which are expected to be exactly matching (complementary) to the single-stranded fluorescently labeled target that is finally applied to the microarray. The selection of such short probes is driven by the annotation of the genome of a given organism. The information at the start and the end of a gene plus the untranslated regions of the corresponding mRNA is of high importance for the actual selection of the probe sequence. Owing to the working mechanism of one key enzyme (the reverse transcriptase), the probe sequences tend to be preferentially selected from the 3′-end or the untranslated region of the mRNA.

The labeled target is prepared from mRNA extracted from a biological sample. The mRNA is reverse transcribed using oligo dT as primer, linearly amplified, fluorescently labeled by incorporating Cy5- or Cy3-deoxynucleotides or biotinylated deoxynucleotides to which fluorochromes can attach, solubilized in hybridization buffer, and then dispersed over the surface of the microarray. The microarray slide is then inserted into a sealed hybridization chamber to allow hybridizations between corresponding sequences on the microarray (i.e., the probes) and in the sample (i.e., the labeled target) to occur. Following incubation, the slide is washed using stringent and nonstringent buffers to remove excess sample from the array and to reduce cross-hybridization. The occurrences of any true hybridization for any of the probes are indicated by the remaining fluorescent intensities as measured by a confocal laser microscope or similar instrument. **Figure 1** shows schematically the key steps involved in a typical microarray experiment.

A standard microarray experiment would comprise several samples run through the above process. Comparing the resulting fluorescent intensities across the different samples would indicate which genes are differentially expressed across them.

3.05.2.1 Microarray Technologies

Several different microarray technologies are commercially available from a number of suppliers, such as Affymetrix and GE Healthcare, or they may be produced by an in-house laboratory. A few of the more common microarray technologies are described below.

In a cDNA microarray, the probes are long DNA fragments (a few hundred bases to several kilobases in length) with sequences complementary to the sequences of the genes of interest or representative subsequences thereof.[1] Regions of low complexity are usually excluded as they could diminish hybridization specificity. cDNA probes can be obtained

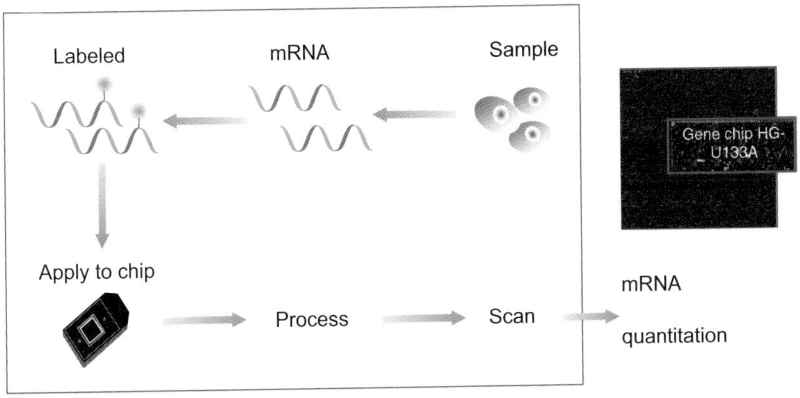

Figure 1 A schematic display of the key steps of a microarray experiment.

from commercial cDNA libraries or, alternatively, PCR can be used to amplify specific genes from genomic DNA to generate cDNA probes. Once obtained, the double-stranded DNA probes are denatured to generate single-stranded probes, which are then robotically spotted onto a glass slide.

Experiments involving cDNA microarrays may be single- or multichannel. Two-channel experiments are the most common. In a two-channel experiment, two samples (e.g., one from control tissue and one from diseased tissue) are prepared for hybridization to the array with different samples labeled with different fluorescent dyes. They are then combined and hybridized to the microarray together. The two samples will hybridize competitively to the probes on the array with the sample containing more transcript for a particular probe prevailing. The two sets of fluorescent intensities are recorded in two scans.

Another type of DNA microarray technology is the oligonucleotide microarray. Affymetrix manufactures a particular type of oligonucleotide microarray[3] in which a gene is represented by a set of 11–20 25-mer oligonucleotide probes called perfect match (PM) probes. The PM probes are carefully selected to have little cross-reactivity with other genes so that nonspecific hybridization is minimized. Nevertheless, some nonspecific hybridization will tend to occur. A second probe called a mismatch (MM) probe, which is identical to the PM probe except for a mismatched base at its center, is placed adjoining to the PM to combat this effect. The idea behind the mismatch probe is to subtract any background hybridization as measured by the MM probe signal from the PM probe signal. Affymetrix synthesizes the probes in situ onto silicon chips using a proprietary photolithographic process.[4] A technological variant of this in situ synthesis process is carried out by Affymetrix' partner NimbleGen. They use a maskless approach based on almost a million tiny mirrors for the photolithographic synthesis of the oligonucleotides. Advantages of this approach are flexibility in the array design (custom chips) and much shorter time frames between the design of a chip and its actual manufacturing as there is no need to generate masks.[5]

Other platforms also use oligonucleotides instead of cDNAs. There is a tendency in these platforms toward using medium-length oligonucleotides as they are thought to provide more of a balance between sensitivity and specificity. Longer oligonucleotides have higher sensitivity as they emit stronger signals and therefore have the ability to detect less abundant transcripts. On the other hand, shorter oligonucleotides have higher specificity as the risk of cross hybridization also increases with length. This makes it challenging to select a set of oligonucleotides to be arrayed such that cross hybridization among known similar sequences is minimized as much as possible and yet sufficiently strong and clear signals are obtained. The most suitable compromise is still an issue for debate. In these platforms, the oligonucleotides are presynthesized as usual rather than synthesizing them in situ. In addition, a substrate other than glass or silicon may be used for immobilizing them.

For instance, in GE Healthcare's Codelink microarrays, presynthesized 30-mer oligonucleotides are inkjet spotted onto a three-dimensional polyacrylamide gel matrix. The polyacrylamide gel can bind considerably more oligonucleotide molecules than a conventional glass surface and the three-dimensional nature of the slide surface enhances the sensitivity of the assay. Applied Biosystems' expression array system uses chemiluminescence hybridization signaling chemistry of 60-mer oligonucleotide probes that are immobilized on a three-dimensional porous nylon substrate. Agilent manufactures two-channel 60-mer oligonucleotide arrays using a proprietary dispensing process similar to the inkjet technology known from current computer printers.

Owing to the diversity of microarray technologies, processes, experimental designs, and even data formats, the Microarray Gene Expression Data (MGED) Society developed the MIAME (minimum information about a microarray experiment) guidelines as a standardized platform for storing and annotating microarray data with a view toward encouraging the generation, storage, and exhange of microarray data in a controlled manner.[3]

Each microarray platform has its own advantages and disadvantages.[6] The greater length of the probes arrayed on cDNA microarrays ensures that they hybridize to sample sequences with higher stringency and are less likely to be affected by minor mutations, but they exhibit lower hybridization sensitivity as longer sequences have higher likelihood for nonspecific binding. They are less expensive to produce than commercial off-the-shelf arrays when produced in quantity and thereby offer some flexibility along with the ability to be customized if manufactured in-house. Then again, they tend to demand much more time and labor-intensive effort in manufacturing, clone maintenance, PCR work, and quality control. On the other hand, commercial oligonucleotide microarrays tend to yield more reproducible results that often also show higher specificity.[7] The disadvantage of using such platforms is certainly the price, and often there are organizational problems caused by having to rely on central facilities. Owing to the necessity of utilizing comparably expensive specialized equipment, many research institutions have had to set up core teams dedicated to performing the microarray work. Regrettably, this removes part of the control a researcher has over his whole experiment.

Other applications of DNA microarrays in drug target research have recently gained more attention but will not be addressed in this chapter further as they require their own experimental design and analysis techniques. Some of these new applications are comparative genomic hybridization (CGH)[8] for the detection of deletions and/or amplifications of longer DNA stretches in the genome, resequencing of specific DNA sequences in a high throughput fashion, and single nucleotide polymorphism (SNP) analysis[9] for the study of inter-individual differences in single nucleotides within the genome, to name a few.

3.05.2.2 Experimental Design Considerations

It is imperative to design a microarray experiment so that its objectives can be met. The objective of many microarray experiments is to identify which genes are differentially expressed among several sets of samples. In such comparative experiments, it is important to be able to generate statistics about what could be considered a differentially expressed gene. This means that replicate samples must be run in order to generate enough data for a statistical analysis to have adequate power to detect gene expression changes. As microarray experiments tend to be expensive, researchers are sometimes reluctant to do too many replicates. However, both natural biological variability and the series of highly technical steps that constitutes a microarray experiment cause the data to be subject to various sources of variation. Minor differences in extraction, amplification, labeling, hybridization, and even the skill and experience of the experimenter could result in substantial differences among what would otherwise be true replicates.

Only with sufficient replication can this variability be overcome and statistical significance assessed adequately. There are two types of replication. One type is 'biological replication,' which involves extracting mRNA from several biological samples and running each sample on a different microarray in order to compensate for intrinsic biological variability in gene expression. The other type is 'technical replication,' which involves running the same sample on multiple arrays in order to compensate for the technical variability of the experimental process.

Sometimes an experiment might turn out to be too large or too complex to be completed in a single session. In that case, it is important to allocate, organize, and run the arrays in such a way that the effects of interest are not confounded with extraneous effects, such as operator effects, manufacturing batch effects, and hybridization day effects.

In general, it is crucial to pay careful attention to quality issues in order to ensure validity of the conclusions reached from the study. The mRNA source and the preparation of the samples should be controlled and their quality assessed. The most common techniques used nowadays are classical agarose gel electrophoresis to assess potential mRNA degradation, spectrophotometrical analysis using a very small quantity of total RNA in a Nanodrop, and miniaturized electrophoresis technologies such as the Agilent Bioanalyzer, which can produce data on mRNA degradation as well as mRNA quantity. The whole issue of mRNA quality is largely dependent on the source from which the mRNA is prepared. Cells cultivated in in vitro cultures tend to be much more homogeneous and generally yield much more consistent mRNA quality and quantity as compared to mRNA from tissues. Furthermore, it is much easier to extract mRNA from some tissues than others: While it is fairly easy to extract mRNA from liver, tissues such as heart or spleen require much more experience to obtain the high-quality mRNA necessary for microarray analysis.

In addition, several control probes are often arrayed on the microarrays for various quality control purposes. The most common are negative control probes, which are used to assess background and nonspecific hybridization, and positive control probes, which are used to measure the abundance of nucleic acid sequences spiked in to the labeled target as a way of assessing labeling and hybridization efficiencies.

3.05.3 Data Analysis Considerations

Owing to the volume and complexity of the data generated by DNA microarray experiments, the data analysis aspect of these studies acquires greater importance than other areas of biomedical research. In order to derive substantive biological information from these studies, careful experimental design and proper statistical and bioinformatics analysis are vital. Typically, a data analysis would consist of several stages. The first stage is always a data preprocessing step, which will be described next. The subsequent stages depend on the objective of the experiment.

3.05.3.1 Data Preprocessing

Preprocessing the raw intensity data enhances the sensitivity of the downstream comparisons. Some preprocessing steps, in the sequence in which they would normally be performed, are:

1. Array quality check: While a certain amount of variation in the intensity values across even identical arrays is inevitable and not a cause for concern, it is not uncommon, given the technical complexity of the experimental process, that a few arrays or a few samples in a study turn out to be defective. It is essential, however, to detect and eliminate them so that they do not unduly affect the data analysis. Arrays or groups of arrays that are substantially different from the rest can be identified by calculating Spearman correlation coefficients between all pairs of arrays. The resulting correlation matrix can be displayed on a heatmap by coloring each correlation on the basis of its value. This will highlight any unusual arrays, samples of poor quality, and bad hybridizations, which will generally tend to have low correlations with most, if not all, of the other arrays and will therefore stand out in the plot.
2. Data re-expression: Microarray data are generally very heavily skewed. Furthermore, the distribution of the intensity values could be quite different for different genes, most noticeably between low-expressing genes and high-expressing genes. The within-gene distribution can be symmetrized and the heterogeneity of variances across the genes greatly reduced by transforming the data. More often than not this can be achieved via a simple logarithmic transformation (logarithms of base 2 are often used). However, sometimes, some other transformation, such as a variance stabilizing transformation, a hybrid linear log transformation, or a started log transformation will do better.
3. Normalization: Spot intensity data will often need to be normalized to reduce monotonic nonlinear array effects. Care must be taken when doing so to ensure that gene-specific effects are at most only slightly dampened, otherwise detecting differential expressed genes will become impossible. Quantile normalization is a popular normalization procedure. A reference array, to which all the arrays are normalized, is defined, usually as an average microarray, such as by calculating gene-wise either the mean or the median across all the arrays. The distributions of the transformed spot intensities are coerced to be as similar as possible to that of the transformed spot intensities of the reference array. Quantile normalization is useful for normalizing a series of arrays where it is believed that a small but indeterminate number of genes may be differentially expressed across the arrays, yet it can be assumed that the distribution of spot intensities does not vary substantially from array to array. This assumption is often valid, and when it is not, the normalization would have to be customized to the particular data being analyzed.
4. Final quality checks: Inevitably, the data will contain a few unusual spot intensity values (i.e., outliers). By identifying and either removing or downweighting such aberrant values, the adverse impact that they would otherwise have on subsequent analyses can be reduced. This can be done by comparing replicates, which can be expected to be very similar to one another. In addition, a final array check can be done via a principal component analysis (PCA) or spectral map analysis as described in (*see* Section 3.05.4.2).

Details of these preprocessing steps and others are described in [10].

3.05.3.2 Visual Inspection of the Data

Gene expression data is typically organized into a gene expression matrix in which the columns represent the samples or experiments and the rows represent the expression vectors for the genes being interrogated by the microarray. The matrix elements are either the spot intensities preprocessed as above or fold changes based on the spot intensities. In trying to expose patterns in this data, it is helpful to somehow render this matrix visually.

A commonly used simple approach is to display the gene expression matrix as a heatmap by coloring each matrix element on the basis of its value. It is likely that the heatmap will initially appear to be devoid of any apparent pattern or order. However, by reordering the rows and/or columns on the basis of a clustering of the genes and/or samples, potentially interpretable patterns of gene expression will often emerge, as groups of co-regulated genes should occupy adjacent or nearby rows and similar samples should occupy adjacent or nearby columns in the display.

PCA is another method for visually presenting the data with a view toward revealing structures in the gene expression vectors. PCA is a technique for summarizing a multidimensional dataset in a few dimensions. Noting that the gene expression matrix has as many dimensions as it has genes, some serious dimension reduction is necessary without too much reduction in information in order for this to produce a meaningful representation. Methodologically, PCA finds a new coordinate system such that the first coordinate, a linear combination of the columns of the data matrix, has maximal variance, the second coordinate has maximal variance subject to being orthogonal to the first, and so on. Plotting the first few coordinates often reveals interesting patterns. For instance, samples sharing similar profiles will tend to lie close to each other. A rough visual estimation of the number of clusters represented in the data should also be possible. In addition, PCA may reveal oddities in the data, such as unusual arrays, and it therefore provides a useful supplementary array quality check as stated in (*see* Section 3.05.3.1).

A number of methods similar to PCA have been developed and project different views of the data. Factor analysis, correspondence factor analysis, multidimensional scaling, and projection pursuit are some of the better known ones.

The biplot and spectral map analysis are extensions of these ideas. The PCA described above is PCA applied to the samples (i.e., to the columns of the gene expression matrix). PCA can also be applied to the genes (i.e., to the rows of the gene expression matrix). When a form of PCA, called the singular value decomposition, is applied to both samples and genes, the simultaneous representation of both samples and genes in a single plot is called a biplot. Thus, a biplot will consist of both sample-related points and gene-related points. As in a PCA plot, the distances between the sample-related points are related to the dissimilarities between the samples; the closer a set of sample-related points, the more similar the samples to which they correspond. Most gene-related points will clump together in the center of the biplot; these would correspond to genes that are not differentially expressing across the samples. Any gene-related points that lie substantially away from this central clump in the direction of a cluster of sample-related points would correspond to genes whose expression profiles are associated with that cluster of samples. Thus, a biplot could be used to reveal not only clusters of samples but also the genes associated with them. As a variation, prior to constructing a biplot, the rows and columns of the gene expression matrix can be centered and weighted in specific ways to allow certain features of the biplot to be highlighted. This is spectral map analysis.[11]

3.05.3.3 Identifying and Studying Differentially Expressing Genes

Identifying genes whose changes in expression are most associated with differences between the samples is one of the fundamental questions addressed in the analysis of gene expression data. Thus, a comparative analysis of the gene expression levels from a diseased cell versus those of a normal cell will help in the identification of the genes and perhaps the biological processes and pathways affected by the disease process. Researchers can then use this information to synthesize drugs that influence and intercede in this process. In order to address this question, both the magnitude and the consistency of the expression changes must be assessed and a valid and reliable metric of differential gene expression derived from the preprocessed spot intensity data.

When comparing two sets of samples, the significance of any gene expression change can be assessed by calculating a t statistic for each gene. The t statistic is essentially a signal-to-noise ratio, in which the difference in mean log spot intensities (essentially the mean fold change on a log scale) is the signal and the standard error is the noise. Note that the fold change by itself, although used occasionally, is unreliable because it does not take into account the fact that not all fold changes can be treated equally which itself is because different genes have different levels of variability. Of course, the t test must be used carefully too. For instance, when the sample sizes are small, as is common in many microarray experiments, the value of the t test statistic tends to be overly dependent on the estimate of noise, to combat which modified versions of the t test (e.g., significance analysis of microarray data-t by [12] and Conditional t by [10]) have been proposed.

When comparing gene expression levels across multiple sets of samples, the t test statistic would be replaced by the F test statistic. In the event that the experimental situation is more complex, such as a temporal or dose–response study, appropriate F test statistics associated with linear ANOVA (analysis of variance) models could be used.

Genes that yield statistically significant t (or F) statistics should be examined for corroborating evidence of biological relevance. Thus, a potentially useful supplementary exercise is to superimpose inter-gene functional relationships onto the t test (or F test) results.[13] Accordingly, a list of genes with statistically significant expression changes can be examined for the repeated occurrence of various functional annotations, such as gene ontology classifications, biochemical pathways, or protein–protein interactions. Via a simple statistical analysis, it can be ascertained whether the set of significant genes contains an over-representation of genes in certain functional categories. There is at present one obstacle to doing this well and that is the scarcity and the questionable quality of the annotation data available for genes and their functional interactions. Nevertheless, as annotation databases improve in coverage, quality, and complexity, their integration into microarray results will be more viable and become an increasingly valuable exercise.

Another potentially useful supplementary exercise is to cluster genes from multiple sets of samples into clusters with similar expression patterns. Hierarchical and nonhierarchical algorithms such as agglomerative nesting, *k*-means, and self-organizing maps have been implemented to cluster similar gene expression patterns. Presuming that genes performing similar functions or belonging to the same biochemical pathway will exhibit a tendency to co-regulate, inferences could be made regarding the unknown genes that share a cluster with genes whose functions or pathways are known. Further inferences can be made by interrogating these clusters as above for over-representation in certain functional categories.[13]

3.05.3.4 Sample Classification

When the samples are heterogeneous, they could be clustered using one of the methods listed above. A cluster analysis will reveal groups of similar samples. If the sample types are known, this exercise can be used as validation that the experiment was successful at separating the samples. In addition, clustering may indicate the existence of new subclasses of existing classes of a phenotype or a disease, thereby leading to a better understanding of the situation at the gene level.

When the samples are known to come from a number of different classes, another primary objective of many DNA microarray experiments is the classification of the samples into the separate classes. This has two objectives. The first is to identify the features (in this case, the genes or certain combinations of genes) that are most associated with the classification. The second is to predict the classification of new samples. Accordingly, classification addresses the question: given a set of samples, how can one assign each sample into the set of known classes with minimum errors? To do this, a set of feature variables is defined and a decision rule (i.e., a classifier) based on these feature variables is derived such that, given the measured values for any sample, the classifier would map it into one of the classes. A number of methods can be used for classification, including linear discriminant analysis and variants, *k* nearest neighbors, random forests, neural networks, and support vector machines. Readers should refer to the References for examples of various microarray classification methods.[10,14]

3.05.4 Case Studies

In an ideal situation the identification of novel drug targets in a given disease is performed using diseased and control tissue. However, biological samples derived from a precisely defined patient population and matched controls are scarce. An exception to this rule is the field of oncology research, where profiling experiments can be performed on tumor tissue and healthy control tissue derived from the same patient. In contrast, at the other end of the spectrum, are the diseases of the central nervous system for which it is much more complicated to obtain affected tissue. As an alternative, post-mortem samples are mostly used in diseases such as bipolar disorder, major depression, and schizophrenia. This complication might be resolved in the future if reports on the possible correlation between expression patterns observed in the brains and lymphoblastoid cells in patients with bipolar disorder, are confirmed.[15]

As an alternative to patient samples, scientists have turned to animal and cellular models of disease to study changes in gene expression related to the etiology and progression of disease states. Both animal and cellular systems can also be used for mode of action studies. In the following case studies, we will address experimental design, microarray experiments, and data analysis in real-life examples.

3.05.4.1 Case Study 1

The aim of the first case study was to elucidate pathways downstream of the corticotropin-releasing factor (CRF) receptor CRF_1 in order to identify novel modulators of this receptor-signaling pathway.[16] CRF is a 41-amino-acid polypeptide that plays a central role in the regulation of the hypothalamic-pituitary-adrenal (HPA) axis, mediating neuroendocrine, autonomic, and behavioral responses to various stressors.[17] Hypothalamic neurons release CRF in response to stress, stimulating the secretion of adrenocorticotropic hormone (ACTH) from the pituitary, which in turn leads to increased release of glucocorticoids from the adrenal glands. Alterations in the CRF system activity have been linked to a number of psychiatric disorders, including anxiety and depression. Since the pituitary is the main target organ for CRF, the study was performed on a mouse pituitary-derived cell line called AtT-20 that expresses the CRF_1 receptor. Experimenting on freshly isolated pituitary cells probably would be closer to the in vivo situation compared to the artificial AtT-20 cell line. However, the pituitary is a heterogeneous organ, comprising many different cell types, of which only a subset (the corticotrophs) respond to CRF. Therefore, changes in gene expression profile in this subset of cells might be diluted when studying the whole organ. In that sense, not withstanding the fact that AtT-20 cells are derived from a mouse pituitary cancer, they are considered a good cellular model of corticotrophs. The experimental setup of this case was as follows: AtT-20 cells were exposed either to CRF, CRF in combination with the CRF_1 antagonist R121919,[37] antagonist alone, or vehicle (DMSO). The transcriptional response to these four treatments was followed over time starting from 0 to 24 h after start of the treatment. Thus, in total 25 samples were processed: time 0 and 6 other time points for each of the four treatments. A downside of this design is the lack of replicates of a specific treatment at each time point. After processing the RNA samples on Affymetrix microarrays, data were preprocessed as described above. In this case data were transformed using a logarithmic transformation with base 2. After transformation, data were normalized using quantile normalization,[18] taking into account the fact that Affymetrix arrays have a limited array-to-array variation. After this preprocessing, data were visually inspected using spectral map analysis. As described above (*see* Section 3.05.3.2), this summarizing technique allows similarities between samples to be identified in an unbiased way based on overall gene expression patterns. In addition, relations between samples and genes can be identified on the biplot. In this specific case the spectral map allowed the identification of both an effect of progressing time and CRF exposure on gene expression. As shown in **Figure 2**, in the biplot of the first two principal components, samples are clustered clock-wise according to time exposed to a certain treatment, irrespective of the nature of this treatment. Interestingly, when plotting the first versus the third principal component differences between the treatments are visualized. This biplot was further applied to identify genes that contribute to the differences between the samples. Genes that do not contribute to this difference end up in the middle of the plot around the so-called centroid. Genes that co-cluster with samples at the periphery of the biplot are genes that show a striking upregulation in that sample. In this manner several genes were found to be strongly upregulated by CRF exposure. The alternative approach that was followed consisted of analyzing expression measurements in the different treatments relative to those observed in the corresponding time point in DMSO-treated control samples. Regulated genes were defined as those showing a greater than two fold change in transcript levels at any one time point. Eighty-eight genes showed a difference in expression after treatment with CRF compared to treatment with the antagonist. Based upon the time point where the response was maximal, genes were classified into 'early responders,' 'intermediate responders,' and 'late responders.' Among the responders were known players in the pathways downstream of the CRF_1 such as the transcription factors Nurr1, Nurr77, and Jun-B, indicating that the setup could confirm previous data obtained by other means. Of interest is the observation that 50 of the 88 genes that were identified using this arbitrary fold change criterion were also identified in the unsupervised spectral map analysis demonstrating the power of this graphical visualization tool.

Confirmation of array data is often done using quantitative PCR techniques such as Taqman technology. In this case study 21 of the most regulated genes were confirmed using quantitative PCR on an ABIPrism 7700 cycler with commercially available assays from Applied Biosystems. Data obtained in this way on all of the tested genes confirmed expression profiles previously observed on the array. Overall, the magnitudes of changes observed with microarray technology were lower than those obtained using quantitative PCR. Ultimately, confirmation needs to be done using a non-RNA-dependent method. This could be based on the protein or on protein function. An example of this is the use of RNA interference (RNAi) technology. By knocking down expression of specific genes identified in the array experiments using RNAi, one could immediately test hypotheses generated by array data analysis.

Figure 2 Spectral map analysis of the microarray data of AtT-20 cells. (a) The first two principal components (PC) of the weighted spectral map analysis applied to the normalized microarray data for all time points and all treatments. On the spectral map, squares depict different samples whereas circles depict different genes (the size of the circle corresponds to intensity on an Affymetrix array). The distances between squares are related to the dissimilarities between the samples; the closer a set of squares, the more similar the samples to which they correspond. Most circles will clump together in the center of the biplot around the centroid (depicted by a cross); these correspond to genes that are not differentially expressed across the samples. Any circle that lies substantially away from this central clump in the direction of a cluster of squares corresponds to a gene whose expression profile is associated with that cluster of samples. Those genes contributing significantly (measured by their distance from the centroid) to difference between samples are annotated with their Affymetrix identifier.[36] The first two PCs identified time as the major discriminator between the samples, distributing the samples clockwise over the plot starting at the T_0 time point and ending at the T_{24} time point. Inserted heat-maps (insets) show representative genes (indicated in bold in biplot) that are maximally induced either after 4 and 8 h or at 0.5, 1, and 24 h, corresponding to their position on the biplot. (b) Biplot of first and third PCs. The third PC identifies the specific CRF effect on AtT-20 cells over time. (c) Summary views on the first three dimensions of the spectral map, showing how this technique identified time and CRF effects in the microarray dataset. uvc,unit column-variance scaling; RW, row weight; CW, column weight. Closure = none; center = double; norm. = global; scale = uvc; RW, mean; CW, constant. (Reproduced from Peeters, P. J.; Gohlmann, H. W.; Van, D. W. I.; Swagemakers, S. M.; Bijnens, L.; Kass, S. U.; Steckler, T. *Mol. Pharmacol.* **2004**, *66*, 1083–1092.)

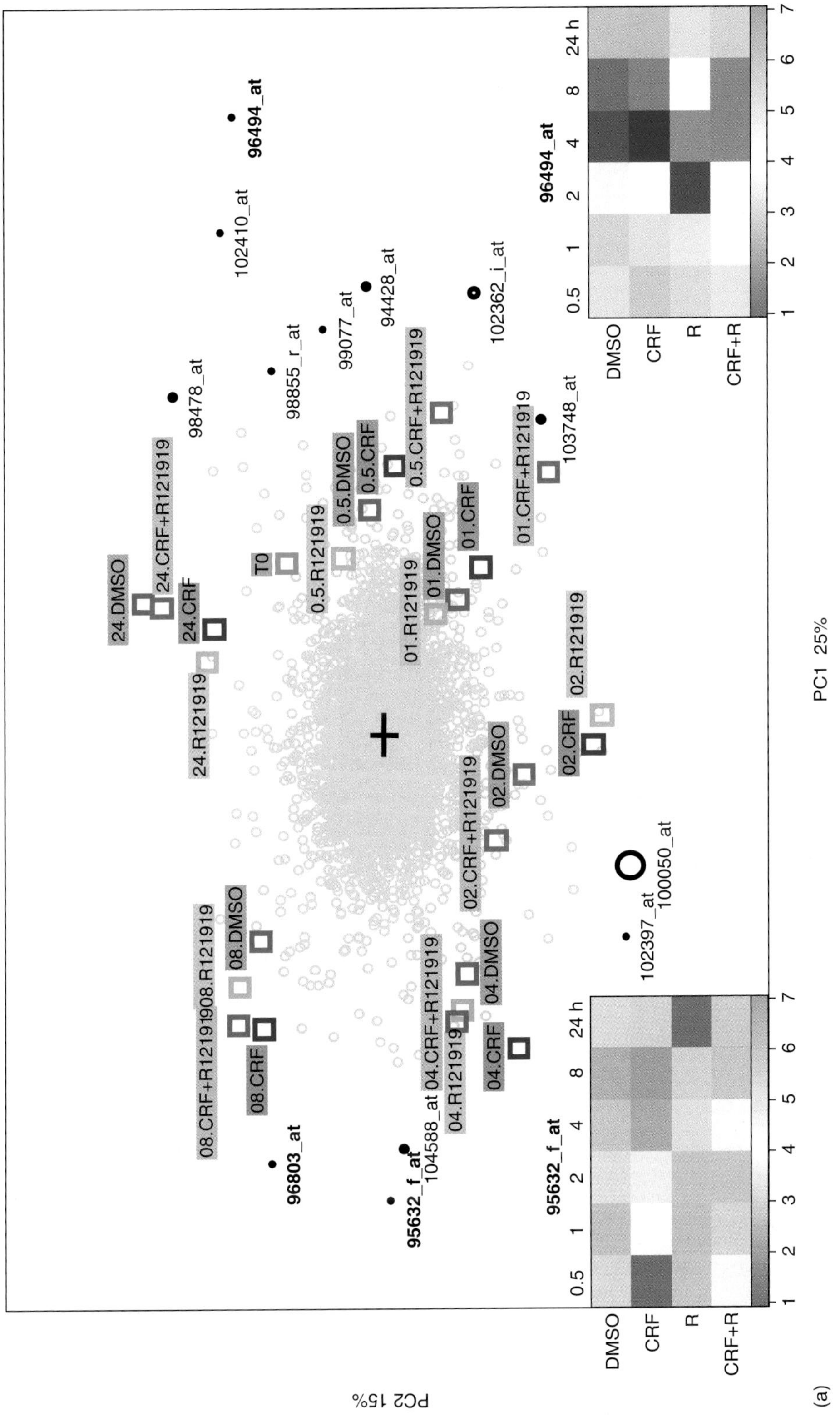
PC1 25%
PC2 15%
96494_at
102410_at
98478_at
98855_r_at
99077_at
94428_at
102362_i_at
103748_at
100050_at
102397_at
96803_at
95632_f_at
104588_at
24.DMSO
24.CRF+R121919
24.CRF
24.R121919
T0
0.5.R121919
0.5.DMSO
0.5.CRF
0.5.CRF+R121919
01.R121919
01.DMSO
01.CRF
01.CRF+R121919
02.R121919
02.CRF
02.DMSO
02.CRF+R121919
04.DMSO
04.CRF+R121919
04.R121919
04.CRF
08.DMSO
08.CRF+R121919
08.R121919
08.CRF
96494_at
0.5 1 2 4 8 24 h
DMSO
CRF
R
CRF+R
1 2 3 4 5 6 7
95632_f_at
0.5 1 2 4 8 24 h
DMSO
CRF
R
CRF+R
1 2 3 4 5 6 7

(a)

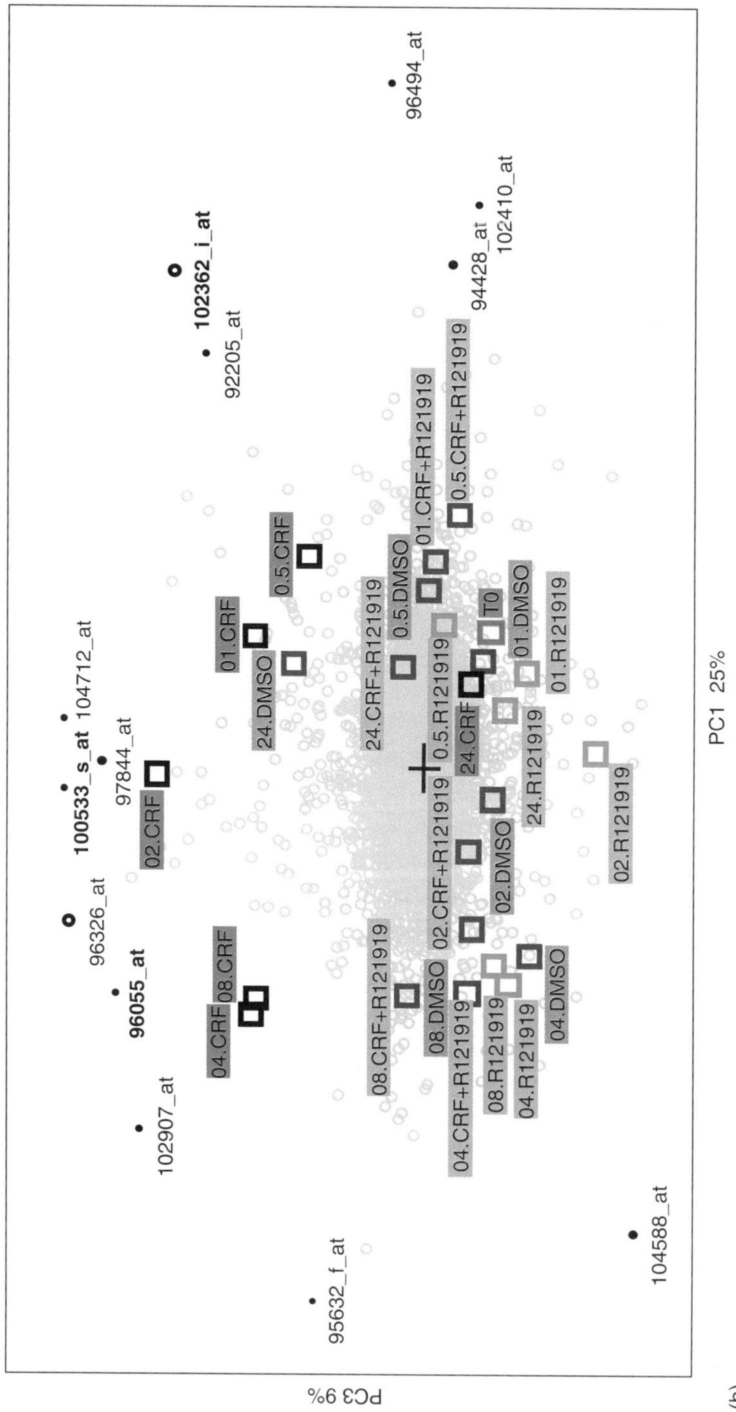

(b)

Figure 2 Continued

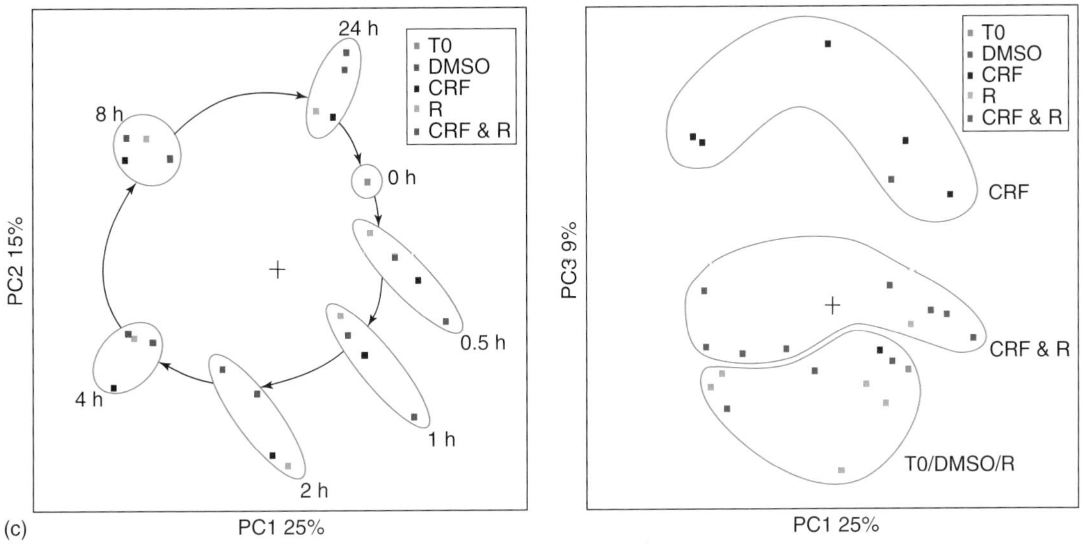

(c)

Figure 2 Continued

3.05.4.2 Case Study 2

In a second case study profiling experiments were done on transgenic animals rather than cell cultures. The aim of this study was to elucidate molecular signatures present in different brain regions of transgenic mice overexpressing CRF (CRF-OE).[19] These signatures are hypothesized to underlie homeostatic mechanisms induced by lifelong elevated CRF levels in the brain. In line with a role of brain CRF in the mediation of endocrine, autonomic, and behavioral responses to stress, transgenic mice overexpressing CRF (CRF-OE) have been reported to show increased anxiety-related behavior, cognitive impairments, and an increased HPA axis activity in response to stress.[20,21] These animals therefore are considered to constitute a good animal model for chronic stress and as an extension thereof, depression. At least part of the behavioral and cognitive changes observed in CRF-OE mice can be attenuated by central administration of a CRF antagonist.

The experimental design of this study aimed at elucidating differences between CRF-OE animals and their wild-type littermates in addition to the effects on gene expression profiles of administrating $10\,mg\,kg^{-1}$ of R121919, a CRF_1 receptor antagonist. To this end six experimental groups were defined. Half of the groups consisted of CRF-OE animals, the other half of wild-type littermates. Animals were either untreated, vehicle treated, or compound treated twice a day for five consecutive days. In experiments using transgenic animals the number of animals is often the limiting factor, which is certainly the case with CRF-OE mice, since these mice are notoriously difficult to breed. As a result each experimental group only consisted of three to four age-matched male animals.

Rather than analyzing whole brain samples, dissection of brain was performed in order to get regional expression data. In total, expression patterns in six regions were studied, including cerebellum, hippocampus, frontal cortex, temporal area, nucleus accumbens, and pituitary. The importance of performing these dissections in a standardized way is exemplified in **Figure 3**. As shown in the spectral map biplot each mouse brain region is characterized by a specific gene expression pattern. Deviations in removing certain brain areas resulting in more or less contribution of surrounding tissue will result in an overall increase of the noise. Therefore consistency is required in removing the tissue (removal of tissue was performed by one operator).

RNA isolated from different animals was processed separately per individual animal and per brain region on Affymetrix murine U74 arrays. As in the previous case study, genes that were called absent (nonreliable detection) in all samples according to Affymetrix's MAS 5.0 software were removed from further analysis. Raw fluorescence intensities from each array were $\log_2$ transformed and data were quantile normalized. Following the group-wise quantile normalization per treatment and genotype, a second quantile normalization was carried out across the data of all samples of a given brain area.[18]

Visual inspection of overall changes in gene expression profile among different treatment groups was done using spectral map analysis per brain region. This analysis revealed overt differences in gene expression pattern predominantly at the level of the pituitary and less so at the level of the other brain areas. Prolonged elevated levels of CRF as present in CRF-OE are expected to induce major changes in the expression profiles in this organ since it is the

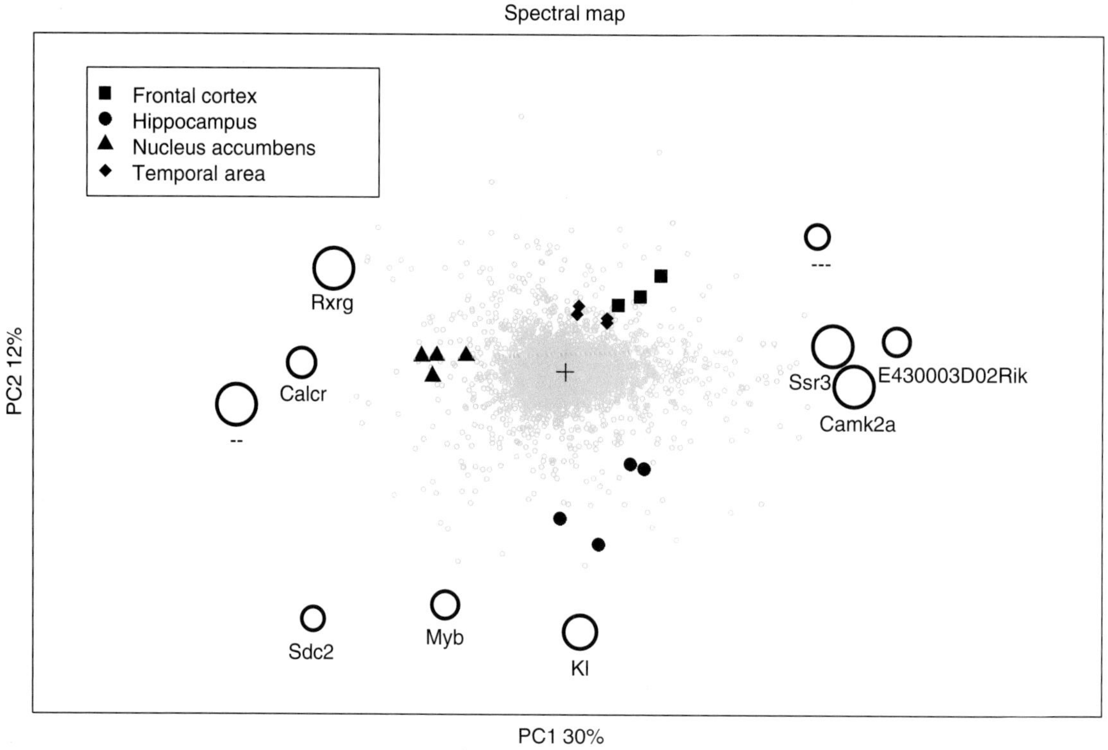

Figure 3 Spectral map analysis of brain regions from mice. The first two components of this spectral map show clustering of samples according to the brain region they are derived from. Genes contributing most are depicted by their gene symbol if applicable. The gene for the calcitonin receptor (Calcr) is, for example, more expressed in the nucleus accumbens than in other brain regions. In contrast, the gene encoding the calcium/calmodulin-dependent protein kinase II-alpha (Camk2a) is more abundant in the other brain regions than in the nucleus accumbens. uvc, unit column-variance scaling; RW, row weight; CW, column weight. Closure = none; Center = double; Norm. = global; Scale = uvc; RW = mean; CW = mean.

primary target of CRF. On the pituitary data spectral map the CRF-OE and wild-type samples (depicted by squares in **Figure 4**) cluster at opposite sites in the biplot. Genes at the periphery of the biplot (contributing significantly to the differences in expression profiles between samples) included kallikreins (Klk9, Klk13, Klk16, and Klk26), a family of serine proteases involved in the processing of biologically active peptides. Upregulation of these enzymes might be an indication for an overactive pituitary. Treatment with the CRF_1 antagonist R121919 or treatment with vehicle did not induce gross changes in expression profile and did not normalize expression patterns in CRF-OE relative to those observed in wild-type animals, as illustrated by the spectral map. An underpowered study as a result of the limited number of animals in each group might account for this negative result. Another more likely explanation is that whereas changes in behavior of CRF-OE mice are the immediate consequence of CRF_1 antagonism, the changes in gene expression are the consequence of prolonged HPA axis activation that are not easily overcome by treatment for 5 days with the CRF_1 antagonist R121919 as in the above experimental setup. A prolonged treatment and/or a larger group of animals might overcome these issues, once again stressing the importance of good experimental design.

Since spectral map analysis only revealed differences between transgenic animals and their wild-type littermates, post hoc analysis was performed on untreated CRF-OE and their corresponding wild-type littermates. Given the low number of animals per group the significance analysis of microarray data algorithm[12] was chosen as an adapted t test for multiple observations. Significance analysis of microarray data assigns a score to each gene on the basis of change in gene expression relative to the standard deviation of repeated measurements. For genes with scores greater than an adjustable threshold, significance analysis of microarray data uses permutations of the repeated measurements to estimate the percentage of genes identified by chance, i.e., the false discovery rate (FDR).[22] Thus, the FDR is the expected proportion of false positives among the tests found to be significant. An extension of this FDR is the so-called q-value.[16] This q-value is similar to the well-known p-value. It gives each hypothesis test a measure of significance in terms of a certain error rate. The p-value of a test measures the minimum false-positive rate that is incurred when calling that test significant. Likewise, the q-value of a test measures the minimum false discovery rate that is incurred

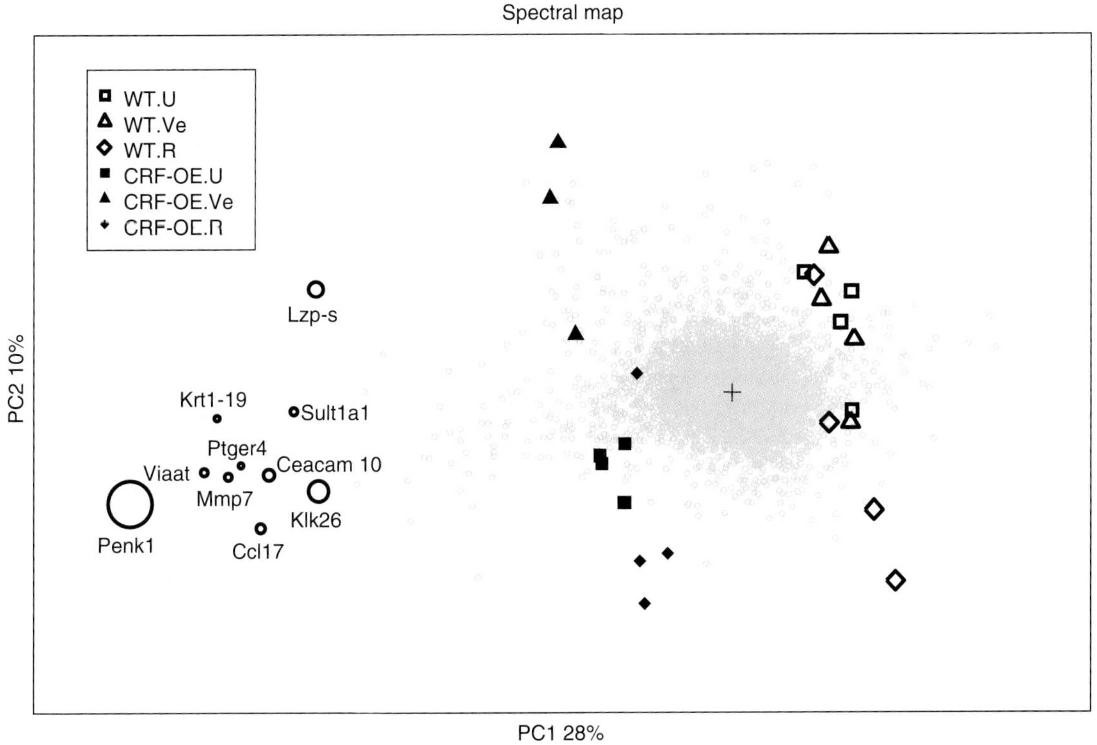

Figure 4 Spectral map analysis of pituitary-derived gene expression profiles of CRF overexpressing mice (CRF-OE) compared to wild-type (WT) mice. Spectral map analysis of microarray data obtained in the pituitary showing the projection of both genes and samples in two dimensions. Positioning of the samples derived from CRF-OE on the opposite side of the centroid compared to the WT samples along the x-axis (PC1) indicates that 28% of the variation in gene expression levels is explained by the genotype of the animals. Genes that contribute largely to the difference between WT and CRF-OE (indicated by their positioning at the extremities of the graph) are highlighted and depicted by their gene symbol. uvc, unit column-variance scaling; RW, row weight; CW, column weight. Closure = none; Center = double; Norm. = global; Scale = uvc; RW = mean; CW = mean.

when calling that test significant. Whereas the *p*-value is commonly used for performing a single significance test, the *q*-value is useful for assigning a measure of significance to each of many tests performed simultaneously, as in microarray experiments. A 10% threshold is accepted practice for array data analysis (as applied for pituitary data). However, deviation of this rule of thumb can be made when the number of significantly changed genes is limited. Therefore, a *q*-value below 20% was used in all other datasets.

In agreement with the spectral map, only a limited number of genes were significantly altered in most brain areas with only 10 genes being downregulated in the hippocampus. In the nucleus accumbens 50 genes were downregulated whereas 11 genes were upregulated. Recurring changes in expression patterns in several brain areas could be clustered into a few pathways such as the glucocorticoid signaling pathway (exemplified by downregulation of 11β-hydroxysteroid dehydrogenase type 1 and upregulation of the immunophilin Fkbp5). Adaptations in the glucocorticoid pathway fit nicely with the CRF changes in CRF signaling, since continued production of CRF in the CRF-OE animals will lead to an excess of glucocorticoids. This excess of glucocorticoids is counteracted by changing the levels of important players in that pathway such as 11β-hydroxysteroid dehydrogenase type 1 and Fkbp5. These changes in individual gene expression profiles in different brain areas could be confirmed by quantitative PCR.

In agreement with results of the spectral map analysis, at the level of the pituitary many more genes (114 genes) differed significantly in their expression between untreated wild-type and CRF-OE animals and were more than two times up- or downregulated. When comparing wild-type to CRF-OE animals, 102 genes had a *q*-value below 10% and were more than 1.5-fold downregulated in CRF-OE. Similarly, 180 genes had a *q*-value below 10% and were more than 1.5-fold upregulated in CRF-OE. In agreement with the spectral map, significance analysis of microarray data identified kallikrein genes *Klk9*, *Klk13*, *Klk16*, and *Klk26*, but in addition also identified *Klk5* and *Klk8* to be significantly upregulated in CRF-OE.

A recurring theme revealed by microarray analysis was the downregulation of neurotensin receptor 2 (*Ntsr2*) mRNA levels in several brain regions including the hippocampus (−54%), the nucleus accumbens (−38%), and the frontal cortex (−59%), but not in the pituitary. Although levels of *Ntsr1* mRNA were below the detection limit in microarray analysis, the observation of *Ntsr2* mRNA downregulation prompted investigators to study the expression levels of all members of the neurotensin (NT) receptor family by quantitative PCR. *Ntsr3* mRNA did not show any change in expression in any area tested. *Ntsr1* mRNA was significantly downregulated in CRF-OE, with the most pronounced effect in the hippocampus. Quantitative PCR also demonstrated a 67% downregulation of *Ntsr2* mRNA in the hippocampus, confirming the microarray data. In order to evaluate the changes in Ntsr1 and Ntsr2 mRNA at the protein level, the presence of receptor (Ntsr1 to Ntsr3) was assessed by autoradiography of $[^{125}I]$NT binding on brain sections. In agreement with array and quantitative PCR results, autoradiography data demonstrated an overall (Ntsr1 to Ntsr3), genetically determined downregulation of the $[^{125}I]$NT binding capacity in CRF-OE versus wild-type animals. Blocking of the Ntsr2 receptors by a saturating concentration of levocabastine, unmasked prominent differences in expression of Ntsr1. A pronounced downregulation of the $[^{125}I]$NT binding in the CRF-OE animals was observed in specific brain regions such as the stratum radiatum ($-63.8 \pm 6.5\%$, $p < 0.001$), the retroslenial granular cortex ($-49.4 \pm 7.5\%$), and the temporal cortex ($-34.8 \pm 3.5\%$). Also, the autoradiography data demonstrated no effect of subchronic administration of R121919 as seen in the gene expression profiling data. These data clearly demonstrate the utility of non-RNA-based technologies such as radioligand binding studies to further validate and investigate microarray data. In addition to the confirmation of quantitative information provided by microarray and PCR this radioligand binding technique supplies further spatial information on where in the brain changes have occurred.

3.05.4.3 Case Study 3

This final case study deals with the complexity of an animal model for irritable bowel syndrome (IBS) in combination with the fact that changes in gene expression are believed only to occur in a small subset of cells. The aim of this study was to investigate long-term changes in gene expression in viscera-specific neurons of both nodose ganglia (NG) and dorsal root ganglia (DRG) in a postinfectious mouse model of IBS.[23] The pathogenesis of IBS is heterogeneous, but at least in a subpopulation of patients emotional stress and enteric infection have been implicated.[24] Therefore, the mouse model of IBS that was chosen for this study consisted of a transient inflammation induced by the nematode *Nippostrongylus brasiliensis* (Nb) combined with exposure to stress. Gene expression profiles were measured both in NG and DRG visceral sensory neurons because the gastrointestinal tract receives dual extrinsic sensory innervation. Vagal afferents have their cell bodies in the NG and project centrally to make synaptic connections in the brainstem, mainly at the level of the nucleus tractus solitarius while spinal afferents arise from the DRG and project into the dorsal horn of the spinal cord (**Figure 5**).[25] Currently, there is a common view that vagal and spinal afferents have different functional roles: spinal afferents play a major role in nociception, while vagal afferents mediate physiological responses and behavioral regulation, particularly in relation to food intake, satiety, anorexia, and emesis. However, there is some overlap, and vagal and spinal afferents share a number of features in common. Both NG and DRG neurons have been shown to become sensitized following inflammatory insult, demonstrating plasticity in the mechanisms that regulate neuronal excitability, which has implications for pain processing.

In this way, extrinsic afferent neurons supplying the gut are prime targets for new treatments of chronic visceral pain disorders such as IBS. However, fiber-tracing experiments measuring the extent to which abdominal viscera-projecting neurons[23] contribute to the total pool of sensory neurons in the DRG and NG, showed that only 3% are visceral sensory neurons. This gives an estimate of the extent to which changes in gene expression occurring in this subpopulation is likely to be diluted when whole ganglia expression is assessed. To circumvent these issues, viscera-specific sensory neurons were labeled using retrograde labeling with fluorescently tagged cholera toxin beta subunit. Labeled neurons in DRG (T10 to T13) and NG were isolated using laser-capture microdissection. This laser-captured material was linearly amplified using a three-round amplification and labeling protocol. To check for linearity of amplification, laser-captured material was spiked with four poly-adenylated prokaryotic genes at a fixed ratio. For these genes probes are present on Affymetrix's microarrays, allowing the efficiency and linearity of the amplification procedure to be monitored.

As mentioned above, the mouse model of IBS consisted of exposing animals to a combination of stress and infection with the nematode *N. brasiliensis*, leading to a transient jejunitis.[26] To assess inflammation at different time points after Nb infection IgE serum levels and jejunal mast cell counts as well as histological analysis were determined. At 3 weeks postinfection, the acute phase of inflammation was finished. Cortisol levels were significantly increased in stressed animals, regardless of infection, at 21 days postinfection. Based on these results, the assessment of mRNA expression levels in viscera-specific neurons was done at 21 days postinfection. At this time-point both NG and DRG visceral

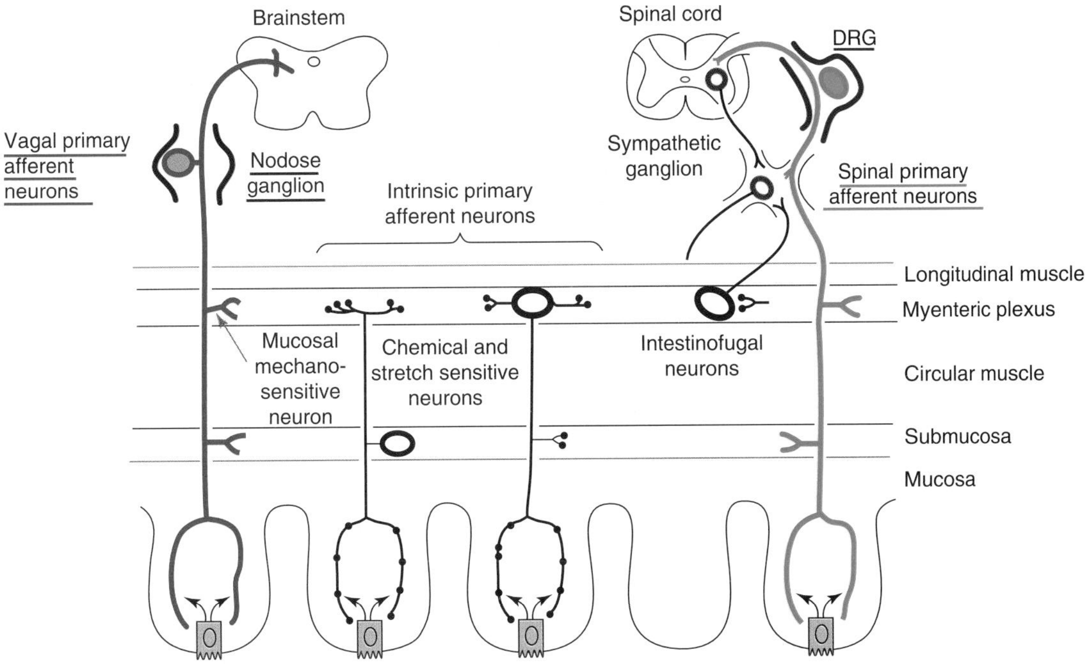

Figure 5 Vagal and spinal afferents in the gastrointestinal system. This simplified diagram of the enteric nervous system indicates that visceral sensory information is relayed from the gut wall to the brain via two main routes. Vagal afferents have their cell bodies in the nodose ganglia and spinal afferents have their cell bodies in the dorsal root ganglia. (Reproduced from Furness, J. B.; Kunze, W. A.; Clerc, N. *Am. J. Physiol.* **1999**, *277*, G922–G928, with permission from the American Physiological Society.)

neurons derived from infected animals showed hyperexcitability in patch clamp electrophysiology experiments. Rather than firing one action potential (as observed in neurons derived from noninfected animals), infected neurons fired a train of action potentials when a current was injected.

In order to evaluate both the effects of infection and stress, four experimental groups were defined each containing 10–12 animals. The first group was exposed to a sham infection and was housed in no stress conditions (SH-NS). The second group was exposed to Nb infection and housed in no stress conditions (INF-NS). A third group was housed in stress conditions but was not exposed to infection (SH-ST). The last group was housed in the same stress conditions but was also infected (INF-ST).

Spectral map analysis of the gene expression in viscera-specific spinal neurons (DRG) revealed no differences among the four experimental groups. Significance analysis of microarray data analysis found no significant differentially expressed genes between the extreme groups (SH-NS versus INF-ST). Upon spectral map analysis of viscera-specific vagal neurons (NG), the animals of groups SH-NS and INF-ST were separated by the first two principal components (**Figure 6**). Animals of groups SH-ST and INF-NS were plotted at the center in the middle, suggesting that their contribution to the variation in the dataset is relatively small compared to the two extreme groups.

A gene-specific two-way ANOVA model was used to analyze the microarray data in more detail:

$$\log_2(Y_{ijk}) = \mu + \alpha_i + \beta_j + (\alpha\beta)_{ij} + \varepsilon_{ijk}$$

where α is the infection effect ($i = 1, 2$), β the stress effect ($j = 1, 2$), and $\alpha\beta$ the interaction.

An overall F test was performed to identify significant genes with any effect resulting in a list of 284 genes. A subsequent test was done to look for significant genes with interaction, infection, and stress effect at the same time. False discovery rate was controlled at 10% at each stage using the algorithm of Benjamini and Hochberg.[27] In this way genes could be classified as affected by stress (59 genes) or infection (13 genes), affected by both (139 genes) and in addition genes were identified that were affected by interaction between stress and infection (73 genes). Of interest in this study is the identification of genes such as the serotonin receptor subtype $5HT_{3A}$, for which antagonists are already on the market for treatment of IBS (**Figure 7**).[28] It is this type of corroborating clinical evidence that further strengthens the data obtained in an animal model. Furthermore, it allows the search for genes showing a similar

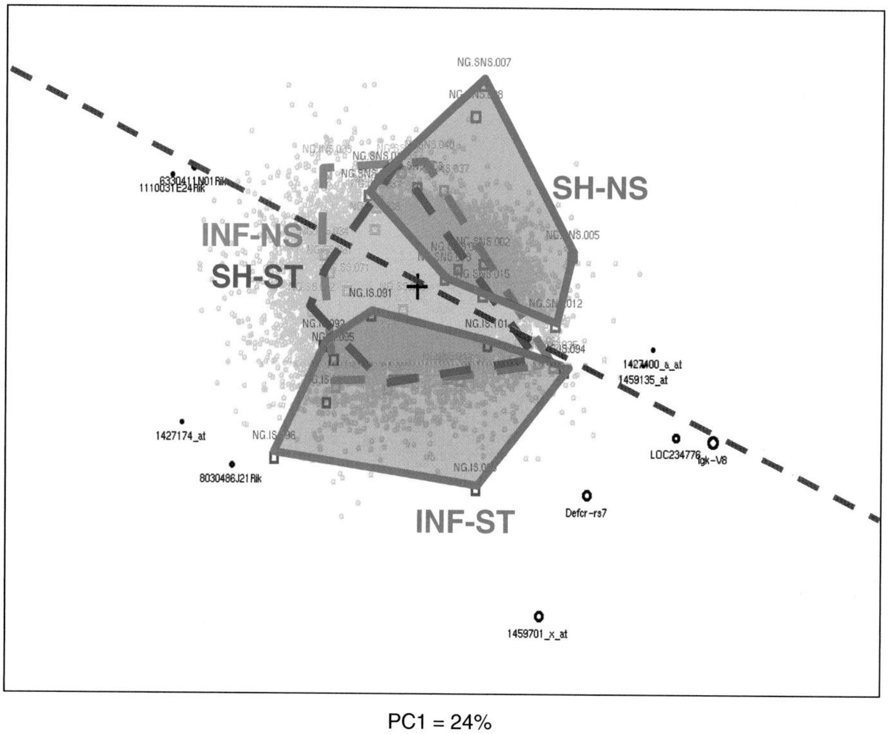

Figure 6 Spectral map analysis of specific gene expression profiles of visceral sensory neurons in nodose ganglia obtained in a mouse model for irritable bowel syndrome (IBS). The spectral map indicates that the first two principal components allow discrimination of the sham/nonstressed animals (SH-NS) from the infected/stressed animals (INF-ST). Samples derived from infected/nonstressed (INF-NS) and sham/stressed (SH-ST) animals end up more in the middle of the biplot in between the two other groups indicating that expression profiles differ the most in the INF-ST versus the SH-NS animals.

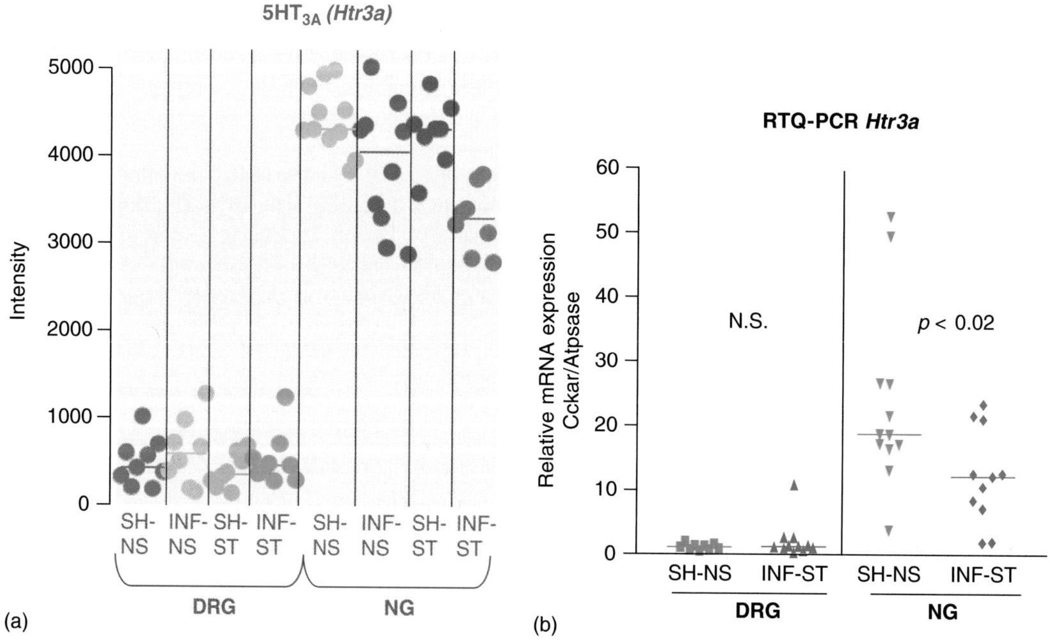

Figure 7 Expression of the $5HT_{3A}$ receptor in a mouse model for IBS. Expression levels of the $5HT_{3A}$ receptor mRNA (*Htr3a*) as assessed both by microarray (a) and quantitative PCR (b) in vagal sensory afferents in mice exposed to stress and *Nippostrongylus brasiliensis* infection compared to control animals.

expression pattern. It is clear that the identification of genes for which pharmacological tools are readily available allows for a straightforward hypothesis testing in animal models. In this way the loop can be closed in the sense that the animal model is first used for target identification followed by validation of that novel target in the same model. Further on, when compounds modulating the target have been identified, the model can again be used for lead generation and optimization.

3.05.4.4 Conclusions from the Case Studies

Although the above-mentioned examples relate mainly to neuroscience research it is clear that the setup of the different studies is generally applicable. Over the past few years a plethora of studies have been performed in several disease areas with the aim of finding novel drug targets.[29–32] In addition to hypothesis generation, microarrays are increasingly used for hypothesis testing. Moreover, with more data becoming available, comparisons between different animal models for the same disease will provide a more powerful approach to the understanding of disease etiology and progression.

3.05.5 Discussion and a Look to the Future

The prevalence of DNA microarray technology testifies to how invaluable a tool it has become in the armamentarium of the research molecular biologist, whether in an academic laboratory or in the pharmaceutical industry. With a well-designed experiment and rigorous standard operating procedures, microarrays will provide valid assessments of the relative transcription levels for thousands of genes simultaneously.

As microarray technology has matured over the past few years and microarray-specific data analysis techniques have been developed, it has become increasingly apparent that microarrays are very comprehensive in characterizing a complex biological sample. Even though the technology itself can surely be further optimized (e.g., with regards to sensitivity and reproducibility), the data generated today is already of high quality. This is reflected in the articles on microarray studies published most recently.[6,33] Use of technical replicates has decreased while use of biological replicates has increased. The main reason can be found in comparably higher variability due to biological variance between individuals rather than technological sources of variance.

A logical consequence emerging from this finding is the need to emphasize a proper design of experiment. Owing to the complex and thereby comprehensive characterization of the samples on the mRNA level, flaws in the experimental design and/or the execution of the biological experiment can often be detected. Combined with the still substantial costs surrounding this genomic technology it becomes apparent that successful microarray studies require not only the input of a statistician determining the right tools for the data analysis but also some level of guidance from a scientist experienced in conducting microarray and/or mRNA experiments. While the former will avoid wasting time after the data is generated (the design of the experiment will define the analysis technique used in the end), the latter will avoid wasting money on microarrays and wasting time on having to repeat experiments. Some of the common avoidable problems include:

- The wealth of data generated by microarrays tends to persuade people to look at answers to biological questions that were not incorporated in the experimental design. Especially when the primary question does not lead to conclusive answers, a feeling of 'but can we not get something out of the data' creeps in. This rarely results in conclusive answers to such secondary questions and tends to be quite time consuming as the data analyst has to apply analysis techniques for which the experiment was not set up. In turn, this results in nonconclusive answers and can spark even more attempts by the scientist interested in the biological interpretation of the results to look even further for information in the data for which the experiment was not designed.
- Trying to incorporate too many different groups into one single experiment usually makes the analysis more difficult and also tends to sacrifice sufficient number of biological replicates per treatment group. In the extreme scenario people will have done a big experiment but might not be able to draw any conclusions because of lack of statistical power. Then the experiment turns out to be a very expensive pilot study. This is even more of an issue if it turns out that due to a large experimental setup that was supposed to cover many different aspects of a certain biological question or phenomenon, the person conducting the biological experiment introduces more technical variance due to time constraints in conducting the experiment.
- There is a lack of experience in working with RNA. Some people are tempted to base their experimental setup on findings, e.g., from behavioral assays or protein work. They then ignore the difference in timing that takes place. While a certain effect may take 4 h to become apparent in a Western blot or a behavioral assay, the effect on the mRNA might have taken place after just 20 min.

What is the future outlook for microarrays in molecular biology research? As the data analysis and the biological interpretation of gene expression data is regarded as the most time-consuming part of a typical microarray, the prospect of future DNA microarrays that would increase even further the amount of data retrieved per sample stresses the importance to continue investing heavily in the development of suitable statistical algorithms and bioinformatics tools to avoid drowning in data. Suppliers such as Affymetrix have already started shipping tiling arrays that cover whole genomes with probes in equal distances rather than with probes for individual genes to discover transcriptional activity outside the current definition of where genes are located. Another approach that is being evaluated and that will also substantially increase the amount of data is the splice variants array, where probes are designed to cover all known gene splice variants.

Recent advances in RNA research have unveiled the presence of so-called microRNAs (miRNAs), small noncoding RNA molecules present in the genome of all metazoa, which function as regulators of gene expression.[34] The few hundred miRNAs that have been discovered so far have been suggested to play an important role in animal development and physiology. Moreover, profiling of expression of some 200 miRNAs in cancer patients proved to be more predictive for classification than mRNA profiling.[35] As insight into the function of these miRNAs continues to grow and, as is likely, their number increases, profiling of noncoding RNAs might become more important in the near future.

Finally, we note that, besides drug target identification, DNA microarrays and associated technologies can be used to identify biomarkers of various kinds to aid in drug development. Efficacy biomarkers would provide gene-level evidence of the efficacy of a drug. Subgroup biomarkers would identify subgroups of patients who would benefit most from a given therapy, giving rise to the optimism of 'personalized medicine.' Toxicity biomarkers would predict adverse events and would be used for compound screening. And so on. Continuing innovations in genomics and analytical technologies are, by the day, offering exciting new strategies for researchers as they search for safe and efficacious drugs.

References

1. Schena, M.; Shalon, D.; Davis, R. W.; Brown, P. O. *Science* **1995**, *270*, 467–470.
2. Golub, T. R.; Slonim, D. K.; Tamayo, P.; Huard, C.; Gaasenbeek, M.; Mesirov, J. P.; Coller, H.; Loh, M. L.; Downing, J. R.; Caligiuri, M. A. et al. *Science* **1999**, *286*, 531–537.
3. Fodor, S. P.; Read, J. L.; Pirrung, M. C.; Stryer, L.; Lu, A. T.; Solas, D. *Science* **1991**, *251*, 767–773.
4. Lockhart, D. J.; Dong, H.; Byrne, M. C.; Follettie, M. T.; Gallo, M. V.; Chee, M. S.; Mittmann, M.; Wang, C.; Kobayashi, M.; Horton, H. et al. *Nat. Biotechnol.* **1996**, *14*, 1675–1680.
5. Singh-Gasson, S.; Green, R. D.; Yue, Y.; Nelson, C.; Blattner, F.; Sussman, M. R.; Cerrina, F. *Nat. Biotechnol.* **1999**, *17*, 974–978.
6. Yauk, C. L.; Berndt, M. L.; Williams, A.; Douglas, G. R. *Nucleic Acids Res.* **2004**, *32*, e124.
7. Wick, I.; Hardiman, G. *Curr. Opin. Drug Disc. Dev.* **2005**, *8*, 347–354.
8. Harding, M. A.; Arden, K. C.; Gildea, J. W.; Gildea, J. J.; Perlman, E. J.; Viars, C.; Theodorescu, D. *Cancer Res.* **2002**, *62*, 6981–6989.
9. Chee, M.; Yang, R.; Hubbell, E.; Berno, A.; Huang, X. C.; Stern, D.; Winkler, J.; Lockhart, D. J.; Morris, M. S.; Fodor, S. P. *Science* **1996**, *274*, 610–614.
10. Amaratunga, D.; Cabrera, J. *Exploration and Analysis of DNA Microarray and Protein Array Data*; Wiley-Interscience: New York, 2003.
11. Wouters, L.; Göhlmann, H. W.; Bijnens, L.; Kass, S. U.; Molenberghs, G.; Lewi, P. J. *Biometrics* **2003**, *59*, 1133–1141.
12. Tusher, V. G.; Tibshirani, R.; Chu, G. *Proc. Natl. Acad. Sci. USA* **2001**, *98*, 5116–5121.
13. Raghavan, N.; Amaratunga, D.; Cabrera, J.; Nie, A.; Qin, J.; McMillian, M. *J. Comput. Biol.* **2006**, submitted for publication.
14. Speed, T. *Statistical Analysis of Gene Expression Microarray Data*; Chapman & Hall/CRC: Boca Raton, 2003.
15. Iwamoto, K.; Kakiuchi, C.; Bundo, M.; Ikeda, K.; Kato, T. *Mol. Psychiatry* **2004**, *9*, 406–416.
16. Peeters, P. J.; Gohlmann, H. W.; Van, D. W. I.; Swagemakers, S. M.; Bijnens, L.; Kass, S. U.; Steckler, T. *Mol. Pharmacol.* **2004**, *66*, 1083–1092.
17. Steckler, T. *Behav. Pharmacol.* **2001**, *12*, 381–427.
18. Amaratunga, D.; Cabrera, J. *J. Am. Stat. Assoc.* **2001**, *96*, 1161–1170.
19. Peeters, P. J.; Fierens, F. L.; Van, D. W. I.; Goehlmann, H. W.; Swagemakers, S. M.; Kass, S. U.; Langlois, X.; Pullan, S.; Stenzel-Poore, M. P.; Steckler, T. *Brain Res. Mol. Brain Res.* **2004**, *129*, 135–150.
20. Stenzel-Poore, M. P.; Cameron, V. A.; Vaughan, J.; Sawchenko, P. E.; Vale, W. *Endocrinology* **1992**, *130*, 3378–3386.
21. Stenzel-Poore, M. P.; Heinrichs, S. C.; Rivest, S.; Koob, G. F.; Vale, W. W. *J. Neurosci.* **1994**, *14*, 2579–2584.
22. Benjamini, Y.; Hochberg, Y. *J. Royal Stat. Soc., Series B* **1995**, *57*, 289–300.
23. Peeters, P. J.; Aerssens, J.; De Hoogt, R.; Gohlmann, H. W.; Meulemans, A.; Hillsley, K.; Grundy, D.; Stead, R. H.; Coulie, B. *Physiol. Genomics* **2006**. 10.1152/Physio/genomics.00169.2005.
24. Spiller, R. C. *Br. Med. Bull.* **2005**, *72*, 15–29.
25. Furness, J. B.; Kunze, W. A.; Clerc, N. *Am. J. Physiol.* **1999**, *277*, G922–G928.
26. Stead, R. H. *Ann. NY Acad. Sci.* **1992**, *664*, 443–455.
27. Benjamini, Y.; Hochberg, Y. *J. Royal Stat. Soc.* **2005**, *57*, 289–300.
28. Camilleri, M. *Br. J. Pharmacol.* **2004**, *141*, 1237–1248.
29. Corton, J. C.; Apte, U.; Anderson, S. P.; Limaye, P.; Yoon, L.; Latendresse, J.; Dunn, C.; Everitt, J. I.; Voss, K. A.; Swanson, C. et al. *J. Biol. Chem.* **2004**, *279*, 46204–46212.
30. Weisberg, S. P.; McCann, D.; Desai, M.; Rosenbaum, M.; Leibel, R. L.; Ferrante, A. W., Jr. *J. Clin. Invest.* **2003**, *112*, 1796–1808.
31. Evans, S. J.; Choudary, P. V.; Neal, C. R.; Li, J. Z.; Vawter, M. P.; Tomita, H.; Lopez, J. F.; Thompson, R. C.; Meng, F.; Stead, J. D. et al. *Proc. Natl. Acad. Sci. USA* **2004**, *101*, 15506–15511.

32. Afar, D. E.; Bhaskar, V.; Ibsen, E.; Breinberg, D.; Henshall, S. M.; Kench, J. G.; Drobnjak, M.; Powers, R.; Wong, M.; Evangelista, F. et al. *Mol. Cancer Ther.* **2004**, *3*, 921–932.
33. Owens, J. *Nat. Rev. Drug Disc.* **2005**, *4*, 459.
34. Ambros, V. *Nature* **2004**, *431*, 350–355.
35. Lu, J.; Getz, G.; Miska, E. A.; Alvarez-Saavedra, E.; Lamb, J.; Peck, D.; Sweet-Cordero, A.; Ebert, B. L.; Mak, R. H.; Ferrando, A. A. et al. *Nature* **2005**, *435*, 834–838.
36. Affymetrix. http://www.affymetrix.com (accessed April 2006).
37. Heinrichs, S. C.; De Souza, E. B.; Schulteis, G.; Lapsansky, J. L.; Grigoriadis, D. E. *Neuropsychopharmacology* **2002**, *27*, 194–202.

Biographies

Dhammika Amaratunga is Senior Research Fellow in Nonclinical Biostatistics at Johnson & Johnson Pharmaceutical Research & Development LLC, where he has been since 1989, working with researchers in drug discovery. In recent years, his primary focus has been on gene expression data analysis. He and his collaborators have written a book (in fact, the first fully authored book on statistical analysis of DNA microarray data) and numerous publications, taught courses and given over 50 (national and international) invited presentations. Over the years, he has been actively involved in a number of committees, including currently PhRMA's Statistics Expert Team on Pharmacogenomics. He has a BSc in Mathematics and Statistics from the University of Colombo, Sri Lanka, and a PhD in Statistics from Princeton University, USA, which he received working under the supervision of Prof John W Tukey.

Hinrnich Göhlmann was born in 1970 in Osnabrück, Germany. He studied Biology at the Technische Hochschule Darmstadt, Germany, and received his Diploma in 1995. His thesis was based on glycolysis research with the yeast *Saccharomyces cerevisiae* in the group of Prof F K Zimmermann. In 1995 he joined the group of Prof R Herrmann at the Zentrum für Molekulare Biologie of the University of Heidelberg, Germany. He received his doctoral degree on his work regarding a whole genome expression analysis of *Mycoplasma pneumoniae* in 1999. Following this in 1999, he joined the department of Functional Genomics of Johnson & Johnson Pharmaceutical Research and Development in Beerse,

Belgium. He is currently a senior scientist responsible for all aspects of gene expression microarray work including data analysis. In 2005, he was awarded Johnson & Johnson's Philip B. Hofmann Research Scientist Award for his contribution in the elucidation of the mechanism of action of a novel drug for the treatment of tuberculosis.

Pieter J Peeters was born in 1972 in Turnhout, Belgium. In 1995, he obtained the degree of bioengineer with a specialization in cellular and genetic biotechnology at the Catholic University of Leuven, Belgium. His first work in the genomics field was in the yeast *Saccharomyces cerevisiae* on the functional analysis of open reading frames, in collaboration with Prof P Philippsen from the *Biozentrum* in Basel, Switzerland. He obtained his PhD in the lab of Prof P Marynen at the Center for Human Genetics at the Catholic University of Leuven in 2000. During his PhD he studied the role of the ETS-variant gene 6 (ETV6) in different mechanisms for leukemogenesis. In 1999, he joined the Johnson & Johnson Pharmaceutical Research and Development organization in Beerse, Belgium as a Research Scientist. There he has been involved in target identification and validation in different disease areas, including affective spectrum disorders such as depression, Alzheimer's disease, and visceral pain, using small model organisms such as *C. elegans*, DNA microarray, and RNA interference technology. He is the author of several peer-reviewed articles.

Comprehensive Medicinal Chemistry II
ISBN (set): 0-08-044513-6
ISBN (Volume 3) 0-08-044516-0; pp. 87–106

3.06 Recombinant Deoxyribonucleic Acid and Protein Expression

F Bernhard, C Klammt, and H Rüterjans, University of Frankfurt/Main, Germany

3.06.1	**Design and Generation of Vectors for High-Level Protein Expression in Bacteria**	**107**
3.06.1.1	Promoters and Control of Gene Expression	108
3.06.1.2	Regulatory Deoxyribonucleic Acid Sequences Important for Protein Expression	109
3.06.1.3	Codon Usage	109
3.06.1.4	Translational Fusion Constructs for Optimized Protein Expression	109
3.06.2	**Preparative-Scale Protein Expression Techniques**	**111**
3.06.2.1	Protein Expression in *Escherichia coli*	111
3.06.2.1.1	Inclusion body formation and coexpression of chaperones	112
3.06.2.1.2	Solubility tags and stable expression of recombinant proteins	113
3.06.2.1.3	Targeting of recombinant proteins	114
3.06.2.1.4	Expression of membrane proteins	114
3.06.2.1.5	Proteolytic degradation of recombinant proteins	114
3.06.2.1.6	Expression of disulfide-bonded proteins	116
3.06.2.2	Protein Expression in Bacterial Hosts other than *Escherichia coli*	116
3.06.2.3	Yeast Expression Systems	116
3.06.2.4	Mammalian Expression Systems	117
3.06.2.4.1	Vector design for gene transfer and expression in mammalian cells	117
3.06.2.4.2	Modified mammalian cell expression systems	118
3.06.2.5	Viral Expression Systems	118
3.06.2.5.1	The baculovirus expression vector system	119
3.06.2.5.2	Semliki forest virus expression systems	119
3.06.2.6	Cell-Free Protein Expression	120
3.06.2.6.1	Design of cell-free expression systems	121
3.06.2.6.2	Components of the cell-free reaction	121
3.06.2.6.3	Preparation of cell-free extracts	122
3.06.2.6.4	Cell-free synthesis of membrane proteins	122
3.06.2.6.5	Cell-free synthesis of disulfide-bridged proteins	123
References		**123**

3.06.1 Design and Generation of Vectors for High-Level Protein Expression in Bacteria

Vectors are generally defined as the basic vehicles that transport and deliver the target genes to be expressed into a suitable host cell. They can be relatively complicated autonomously replicating elements as plasmids, bacteriophages, or viruses or they might consist of less complex DNA molecules that integrate into the host chromosome or that are even not stable in the cellular background and provide only a transient expression of the target gene. However, common characteristics are always a few basic requirements like selection marker, copy number, host range, and origin of replication. The copy number is a key feature as it is linearly related to the gene dosage and thus clearly affects the expression yields. In general, replicating vectors are present in multiple copies up to several hundred molecules per single cell in case of common bacterial expression plasmids. The copy number together with the host range of a DNA molecule is determined by the origin of replication, a sequence motif that is recognized by a specific DNA polymerase. In addition, plasmids are classified into separate incompatibility groups based on their origin of replication. Members of the same group are not compatible and cannot be permanently maintained in the same cell. Integration vectors are devoid of any origins of replication and only present in one or few copies, although for specific cases techniques have been developed to increase the copy number by selective amplification.[1,2] Vectors usually contain selection markers that are indispensable to identify successfully transformed cells. Antibiotic resistance genes encoding, e.g. for the enzyme β-lactamase, are frequently used especially for selection in bacteria. Genes encoding for key enzymes in essential anabolic pathways are also suitable as selection markers if used in combination with specifically engineered auxotrophic host cells. A variety of comprehensive reviews of expression vectors suitable for the most commonly used bacterial host *Escherichia coli* is available.[3–6]

3.06.1.1 Promoters and Control of Gene Expression

Expression cassettes containing strong inducible promoters, a convenient multiple cloning site, and an efficient terminator of transcription, and located on a high copy vector are needed in order to obtain high yields of a recombinant protein. Modifications are expression cassettes that provide dual promoters or multisystem expression vectors containing promoters that are active in bacterial as well as in mammalian cells.[7] The possibility of efficiently inducing the expression at a certain time point during the fermentation process is important as it minimizes metabolic burdens and potential toxic effects of the protein product. Fine-tuning the induction conditions can significantly optimize the expression of foreign proteins.[8] Common inducers employ low-cost chemicals or sudden modifications of the growth conditions. In fact, many promoters are suitable for producing proteins at high levels but they are difficult to completely switch them off.[9,10] Prototypes are the *E. coli* P_{lac} promoter and its cAMP-independent P_{lacUV5} derivative that are frequently in use for protein expression. Induction is performed by the nonhydrolyzable lactose analog isopropyl-β-D-1-thiogalactopyranoside (IPTG) which releases the specific promoter-bound LacI repressor. Considerably increased expression levels, e.g., of up to 30% of the total cellular protein, can be obtained with the related artificial P_{tac} and P_{trc} promoters consisting of synthetic fusions of the P_{lac} with the P_{trp} promoter.[6,9] The intrinsic leaky expression from *lac*-derived promoters in noninduced cells can become cumbersome, sometimes causing a complete loss of protein expression or a reduced viability of the culture.[7] The tight repression of weaker promoters can be achieved in host strains containing the *lacI^Q^* allele which ensures enhanced synthesis rates of the LacI repressor. Furthermore, the *lacI* or *lacI^Q^* genes can be placed on the expression vector in order to additionally increase the LacI copy numbers. Alternatively, an expression system under control of the promoters P_{T7} from phage T7, P_{BAD} from the *E. coli* arabinose operon, or P_L and P_R from the λ phage might be preferred. The P_L and P_R promoters are among the strongest known in *E. coli*. They are tightly regulated and can be induced by temperature switches from 30 °C to 42 °C by virtue of a temperature-sensitive version of the λcI repressor (cI857) and they allow a free choice of host strains.[11,12]

The T7-promoter included in the pET system expression plasmids (Merck, Darmstadt, Germany) has become very popular.[13,14] This promoter is specifically recognized only by the T7-RNA polymerase encoded by the T7 gene 1 which can be provided either on a phage or plasmid vector or as integrated chromosomal copy in engineered host strains carrying the prophage λDE3 under control of the IPTG inducible P_{lacUV5} promoter.[7,13,15] A variety of DE3 derivatives in different cellular backgrounds (e.g., BL21, HMS174) is available (Merck, Darmstadt, Germany; Invitrogen, Carlsbad, CA). Virtually no background expression from the P_{T7} promoter can be observed in *E. coli* cells devoid of any T7-polymerase. Residual T7-RNA polymerase produced by λDE3 host strains due to the leakiness of P_{lacUV5} can be neutralized by coexpression of T7 lysozyme that efficiently binds to T7-RNA polymerase from compatible plasmids like pLysS and pLysE (Merck, Darmstadt, Germany). Other mechanisms for the control of leaky expression have also been proposed.[7] T7-RNA polymerase is extremely effective in the initiation of transcription and has a very processive transcription elongation rate. This could result in the accumulation of large amounts of messenger RNA (mRNA) causing a shortage of the cellular ribosomes. The unprotected mRNA is highly susceptible to degradation and can be stabilized by using the RNaseE mutant strain BL21 (DE3) Star (Merck, Darmstadt, Germany) for expression.

The activity of many strong promoters cannot be modulated after induction. Recombinant proteins can therefore accumulate very fast in the cell and the functional folding of the proteins might then not be able to keep pace. This causes a high risk of unfolded protein precipitating in the host cell as nonsoluble inclusion bodies. In order to solve this problem, a possible option could be the use of the P_{BAD} promoter in combination with its regulator protein AraC. It shows a very fast and tuneable response to changing concentrations of its inducer L-arabinose while heterogeneous cell populations might also account for that effect.[16,17] This allows the modulation of the stability and folding properties of a recombinant protein by controlling the rate of expression.[18] Corresponding vectors are commercially available (Invitrogen, Carlsbad, CA). The P_{BAD} promoter is subject to catabolite repression and a very tight repression can be achieved in the presence of glucose, fructose, and similar metabolites. The rapid kinetics of regulation allows a fast and efficient switch of protein expression from on to off.

3.06.1.2 Regulatory Deoxyribonucleic Acid Sequences Important for Protein Expression

Regulatory sequences especially in the 5′ untranslated regions (UTR) are very different in *E. coli* if compared with eukaryotic control regions. A ribosome binding site (RBS, Shine–Dalgarno sequence) with the highly conserved consensus sequence 5′-UAAGGAGG-3′ needs to be present at 9 ± 3 bp upstream of the first codon that defines the initiation of translation. The RBS is complementary to the 3′ end of the 16S ribosomal RNA (rRNA) and an interaction is crucial for an efficient initiation of translation. Stable mRNA secondary structures covering the RBS or the initiation codon are usually detrimental to gene expression as they can interfere with ribosome binding. Expression constructs therefore usually represent transcriptional fusions where the regulatory sequences of promoters or terminators are provided by the expression cassette of the vector and only the coding sequence from the target gene has to be inserted.

Mammalian genes especially almost routinely require extensive modification of additional regulatory regions in their primary sequences prior to a high-level expression in *E. coli*. Strong translation initiation signals are an indispensable prerequisite for high-level expression as the initiation rate can dramatically determine the final expression level.[19] The sequence context immediately surrounding the translational start codon (up to about the first ten codons) is often critical.[20–22] This area should preferably be free from any rigid secondary structure formation since those could completely abolish any expression of the heterologous protein. Raising the numbers of adenosine residues in that area generally reduces the probability of secondary structure formation. A detailed analysis and systematic modification of the translation initiation region can significantly improve the expression. A practical approach for circumventing any problems with a suboptimal initiation of translation would be to produce a heterologous protein as a fusion C-terminal to a small leader peptide that is already codon-optimized. One possibility would be the T7 tag present in the cloning regions of many commonly used expression vectors of the pET series (Merck, Darmstadt, Germany).

3.06.1.3 Codon Usage

A problem that frequently affects the yield of an expressed protein is the different codon usage of individual species.[23] A strong codon bias is most evident when prokaryotic and eukaryotic systems are compared. Codons that translate into proline and arginine are particularly affected in expression of human genes in *E. coli*.[24] The transfer RNAs (tRNAs) corresponding to rare codons are only found in minor populations in *E. coli* cells and coding sequences containing several rare codons (especially in iterative arrangements) are likely to become poorly expressed. Besides premature termination the misincorporation of a lysine residue in place of arginine at rare codon sites is a well-recognized problem in the structural analysis of proteins resulting in undesired heterogeneous populations in recombinant protein samples.[24–28] One option for addressing this problem is the genetic manipulation of coding sequences according to the codon preferences of the desired host cell. This approach certainly should be considered in the case of the de novo synthesis of a gene in vitro.[29] However, the subsequent introduction of silent mutations in an already cloned gene can be time-consuming as it probably requires multiple steps of mutagenesis. Alternatively, a eukaryotic gene can be expressed in *E. coli* in the presence of a second compatible plasmid encoding for an additional set of tRNAs complementary to rare codons.[30,31] Specialized strains that are used for the high-level expression of rare codon containing genes are BL21 CodonPlus (Stratagene, La Jolla, CA) and the Rosetta-2 derivatives of BL21 (Merck, Darmstadt, Germany). The two strains harbor a plasmid containing extra copies of several rare codons.

3.06.1.4 Translational Fusion Constructs for Optimized Protein Expression

In many cases it is preferable to express a target protein as a single polypeptide chain covalently linked with a second already well-defined protein. Fusion strategies provide the advantage that the engineered fusion products combine the properties of the individual proteins and thus permit high-level production of otherwise poorly synthesized recombinant proteins (**Table 1**). Target proteins with almost or completely unknown functions can be linked with the beneficial binding characteristics or optimal expression and stability properties of a fusion partner. The potential benefits of a fusion system include significantly enhanced expression levels, fast purification by means of affinity chromatography based on the binding properties of the fusion partner, increased probability of proper folding of the attached target protein, prevention of inclusion body formation, and protection from proteolytic degradation.[34] In addition, fusion partners can enable the immobilization of a protein to, e.g., biosensors for functional characterizations. The majority of the common fusion systems place the target protein at the C-terminus of a fusion partner that most likely ensures the transfer of better expression and stability characteristics. However, it should be borne in mind that interference of the fusion partner with the activity of the target protein can occur in some cases and its localization

Table 1 Characteristics of popular fusion tags for the optimized expression of recombinant proteins

Fusion tag	*Size*[a]	*Purification*	*Solubilization*	*Stabilization*	*Source*
$(His)_x$-tag	6–12	+	–	–	Various suppliers
Strep-tag	6–12	+	–	–	IBA GmbH[b]
Protein D	110	–	±	+	Ref. 32
Protein G(B1)	55	–	±	+	Ref. 33
Glutathione *S*-transferase (GST)	230	+	±	+	Various suppliers
Thioredoxin	109	+	±	+	Various suppliers
DsbA	208	–	+	+	Merck[c]
DsbC	235	–	+	+	Merck[c]
MBP	395	+	+	+	New England Biolabs[d]
NusA	495	–	+	+	Merck[c]

[a] Size in amino acid residues; the exact size depends on the individual cloning strategy.
[b] IBA GmbH, Göttingen, Germany.
[c] Merck, Darmstadt, Germany.
[d] New England Biolabs, Frankfurt, Germany.

could make a difference. If possible, the effects of an attachment to either end of the target protein should therefore be compared.[35]

Fusion partners can be classified into small peptide tags and into larger proteins that can have additional benefits for the solubility or the stability of the target protein (**Table 1**). Small peptide tags can principally become attached to either end of a target protein while their accessibility can certainly differ depending on the specific tertiary structure of the target protein. Those tags usually cover about ten amino acid residues and they are efficiently used for the fast and convenient purification and/or detection of target proteins.[36] An advantage of using small peptide tags is that they usually do not affect the activities of the recombinant protein and they need not be removed upon a structural and functional characterization of the target protein. Extended processing and purification steps can therefore be avoided. Frequently employed purification tags include poly$(His)_x$ tag consisting mostly of six consecutive histidine residues and various forms of the biotin binding streptavidin peptide. Proteins containing poly$(His)_x$ tag derivatives usually exhibit a high affinity for Cu^{2+}, Ni^{2+}, Co^{2+}, or Zn^{2+} ions and can be purified from crude extract preparations by one-step affinity chromatography using metal-chelate resins even under denaturating conditions.[37] Also the exceptional strong interaction between biotin and streptavidin (K_D 10^{-15}M) ensures the highly selective binding of a fusion protein. Specific antibodies directed against poly$(His)_x$ tags and Strep-tags are available (Qiagen, Hilden, Germany; IBA GmbH, Göttingen, Germany) and the tags can thus be used for the purification as well as for the detection of a protein. An additional popular peptide tag used for the rapid identification of expressed recombinant proteins by an antibody reaction is the T7 tag inserted into many pET vectors (Merck, Darmstadt, Germany). Small peptide tags are furthermore often combined with larger fusion partners in order to optimize the rapid purification of the recombinant protein.[38]

Fusion proteins like glutathione *S*-transferase (GST), thioredoxin, ubiquitin, and others can significantly increase the expression yield of the attached target proteins (**Table 1**). Thioredoxin is a small monomer encoded by the *trxA* gene which facilitates the soluble expression of a number of mammalian proteins including growth factors and cytokines.[39,40] An affinity resin for the fast purification of thioredoxin fusions is commercially available (Invitrogen, Carlsbad, CA). The ubiquitin of baker's yeast *Saccharomyces cerevisiae* has been successfully used as a fusion tag for small proteins in *E. coli*.[41] A further advantage of that system might be the use of the specific yeast ubiquitin hydrolase in order to remove the tag after isolation of the fusion protein.[42] GST is popular as a purification tag because of its high affinity to glutathione sepharose, and it is additionally effective as an enhancer for the translation efficiency and for the solubility of the target protein.[35,43,44] The *E. coli* maltose binding protein (MBP) has an exceptional and relatively general ability to promote the solubility and stability of the fused recombinant proteins.[38,45,46] The fusion proteins can be easily purified by virtue of the MBP affinity to immobilized amylose resin.[47] The MBP fusion system is available as a kit (New England Biolabs, Frankfurt, Germany) which provides vectors for cytoplasmic or even periplasmic expression by taking advantage of the native MBP leader peptide.

Table 2 Frequently used restriction proteases suitable for the cleavage of fusion proteins

Restriction protease	*Recognition site*[a]	*Source*
Enterokinase	D-D-D-D-K-↓	New England Biolabs[b]
Factor Xa	I-E/D-G-R-↓	Various providers
Thrombin	L-V-P-R-↓-G-S	Various providers
Tobacco etch virus (TEV) protease	E-N-L-Y-F-Q-↓-G	Invitrogen[c]
PreScission protease	L-E-V-L-F-Q-↓-G-P	Amersham Biosciences[d]

[a] The recognition sequence in the one letter code is given; the arrow indicates the site of proteolytic cleavage.
[b] New England Biolabs, Frankfurt, Germany.
[c] Invitrogen, Carlsbad, CA.
[d] Amersham Biosciences, Freiburg, Germany.

Large fusion partners are usually not desired if a structural analysis by nuclear magnetic resonance (NMR) spectroscopy of the expressed protein is under consideration, as the increased number of signals would result in spectra that are too crowded. The relatively small monomeric 11.6 kDa bacteriophage λ head protein D might be considered for this purpose. It shows excellent expression properties, has high thermal stability, and mediates optimal initiation of translation while reducing the risk of inclusion-body formation and protein degradation.[32] Also the small protein G (B1 domain) has been presented as solubility-enhancement tag especially for NMR purposes.[33]

It is often desired to remove the fusion partner after purification in order to obtain a nonmodified target protein. For that purpose, the recognition site for a highly specific 'restriction protease' can be introduced in the linker between the two proteins (**Table 2**). An elegant way that eliminates any additional incubation steps after purification is the intracellular cleavage of the fusion protein directly after translation by virtue of the low-level coexpression of a suitable protease.[48,49] It should be considered that some proteases cleave in between their recognition sites and thus still leave residues attached to the N-terminal end of the target protein (**Table 2**). In addition, it is important to realize that the removal of the fusion partners can reintroduce the initial stability problems, and it may happen that the proteolytic processing is accompanied by unfolding and precipitation of the target protein. The production of fusion proteins also must not necessarily result in functional proteins. However, they still can provide a convenient means of generating antibodies that are specific for the protein under study.

3.06.2 Preparative-Scale Protein Expression Techniques

The design and set-up of an ideal expression system depends on multiple factors and each step can influence the final success. Most system components can be considered as more or less independent modules that can be selected according to the specific requirements (**Table 3**). High level of production of a protein often requires combinatorial approaches where several individual parameters of an expression system have to be optimized. The expression in cellular backgrounds especially by using *E. coli* strains is probably the first method of choice in most cases: no other system is comparable in speediness, simplicity, and cost-effectiveness. However, a valuable toolbox of microbial and eukaryotic expression systems exists and the individual characteristics should be carefully considered in order to make the optimal choice (**Table 4**). Multiple parallel approaches in order to identify the optimal expression technique for a given protein might be the quickest route to success.[50–54] Specific applications like the production of modified eukaryotic proteins or the expression of difficult protein groups like membrane proteins or disulfide-bonded proteins often require special adaptations and the most powerful and highly promising techniques have to be discussed.

3.06.2.1 Protein Expression in *Escherichia coli*

Protein expression in *E. coli* is easy to handle and often reveals a cost-effective and high-level production of heterologous proteins.[51,54] The cells can be grown to very high densities in inexpensive complex media within a short time and yields of recombinant protein up to 1 g L^{-1} have been obtained.[17,55] Numerous elaborated protocols for the fast and efficient manipulation of *E. coli* have been established, a variety of options are possible in order to adapt the expression system to the specific requirements (**Table 3**), and a wide collection of expression vectors with well-defined and strong promoters is available. Many different *E. coli* host strains optimized for a high level of expression of

Table 3 Rational strategies for the design of a procaryotic expression system

Parameters	*General options*
System components	
Host species	*Escherichia coli* (general), *Bacillus subtilis* (e.g., export proteins), *Lactobacillus lactis* (e.g., membrane proteins)
Host strain	Protease deficiency, oxidizing cytoplasm, engineered genotype (e.g., DE3, lacI^q)
Expression vector	Copy number, selection marker, host range, compatibility
Expression cassette	Promoter, induction conditions, available cloning site, peptide tags, or fusion partners
Fermentation conditions	Medium components, addition of cofactors, incubation temperature, oxygen supply
Target gene sequence	
Initation of translation	Optimized environment of the translational start codon
Codon usage	Silent mutagenesis, coexpression of rare codon tRNAs
Targeting	Addition of export signal sequences
Translational fusions	Addition of peptide tags or fusion proteins for better purification and/or stability
Protein characteristics	
Low solubility/stability	Prevention of inclusion bodies (e.g. coexpression of chaperones, addition of solubility tags)
Membrane associated	Construction of fusion proteins, specific host strains
Proteolytic degradation	Construction of fusion proteins, specific host strains
Disulfide bridges	Coexpression of chaperones, engineered host strains, periplasmic expression
Modification	Coexpression of modifying enzymes, addition of cofactors

recombinant proteins are available from various suppliers (Invitrogen, Carlsbad, CA; Merck, Darmstadt, Germany; Stratagene, La Jolla, CA). While it is true that some groups of proteins are more likely to become produced in *E. coli* than others, a wealth of published experience exists that this organism is of importance for almost any application. Depending on the desired protein product, a variety of proven optimization strategies are documented that can considerably accelerate the high-level production of a recombinant protein in the desired form.[56]

3.06.2.1.1 Inclusion body formation and coexpression of chaperones

Unfortunately a frequent observation in *E. coli*, especially in production of eukaryotic proteins, is the formation of inclusion bodies, which are precipitates composed of unfolded and inactive recombinant protein.[4] Protein folding in prokaryotes and eukaryotes follows different pathways. The biosynthesis of proteins in prokaryotes usually proceeds very fast and can thus cause the precipitation of proteins that require a long time to achieve their functional folding. In addition, specific cofactors, chaperones, or accessory proteins essential in order to complete the folding process of a eukaryotic protein might be absent in the prokaryotic environment. A shortage of ligands or substrates could additionally increase the instability of an overproduced protein. Heterologous proteins expressed in *E. coli* therefore do not fold spontaneously into a native and functional state even though they are produced at significant levels.

Several protocols exist that result in the in vitro refolding of proteins out of isolated inclusion bodies.[57,58] However it is generally intended to produce recombinant proteins in a folded and active state. Several pathways for the proper folding of proteins in prokaryotes as well as in eukaryotes are well characterized and key enzymes have been identified and cloned. Best studied are the prokaryotic GroEL/GroES and DnaK/DnaJ/GrpE systems.[59,61] GroEL/ES forms large multisubunit cylinders that actively promote the folding of many proteins through an ATP-dependent cycle.[61,62] The DnaJ activated DnaK interacts with hydrophobic regions in nascent peptide chains and prevents intra- and

Table 4 Comparative evaluation[a] of expression systems

Parameter	*Expression system*				
	Bacterial	*Yeast*	*Mammalian*	*Viral*	*Cell-free*
System characteristics					
Cost-effectiveness	+ + +	+ +	+	+	+[b]
Robustness	+ + +	+ + +	+	+	+
Variety of vectors, host strains	+ + +	+ +	+ +	+ +	+
Speediness of protein production	+ + +	+ +	+	+	+ + +
Set-up of reaction	+ + +	+ +	+	+	+ +
High expression levels	+ + +	+ + +	+	+ +	+ +
Inclusion body formation	+ +	+	+	+	+
Stabilization of recombinant proteins[c]	+	+	+	+	+ +
Specific applications					
Posttranslational modifications	−	+ +	+ +	+ +	+
Secretion of proteins	+ +	+ +	+	+ +	−
Labeling of proteins	+ +	+	±	±	+ + +
Membrane proteins	+	+	+	±	+ + +
Disulfide-bridged proteins	+	+ +	+ +	+ +	+ +
Toxins	±	−	−	+ +	+ + +
Proteins requiring cofactors	+	+	+	+	+ +

[a]The rating ranges from '+ + +', best choice to '−', not recommended.
[b]Commercial systems are relatively expensive.
[c]Stabilization by low intrinsic protease levels, by coexpression of additional proteins, or by the addition of stabilizing compounds.

intermolecular aggregations before the freshly synthesized proteins have adopted their native structure.[60] The nucleotide exchange factor GrpE mediates the final complex dissociation and proteins are released for a folding cycle. The coexpression of cloned chaperones or other beneficial 'helper' proteins can be advantageous in enhancing the stability or the general expression of recombinant proteins.[63–66] A variety of plasmids have been constructed that could be used in order to express the most prevalent chaperone systems in combination with a target protein.[67,68] Boosting the intracellular level of chaperones in *E. coli* can significantly enhance the expression of soluble recombinant proteins.[69]

Coexpression can be addressed principally either by bicistronic or by compatible dual-vector approaches.[70] Bicistronic vectors could contain several reading frames arranged as an operon that is controlled by a single promoter. Unfortunately, genes that are located more distantly to the promoter are commonly much less expressed.[71] A single dual-promoter expression vector that provides two cloning sites each having a separate promoter could alternatively be considered.[72] The use of different induction mechanisms would additionally allow the fine-tuning of the individual expression levels. Coexpression by a dual-vector system usually employs two different compatible vectors for each cloned target protein. However, one vector often becomes dominant because of differences in the copy numbers.[73] Such a bias in copy number may be balanced if different selection pressures are maintained.[74] Dual-vector systems can further be used to express several individual subunits of a heteromeric protein complex in a single host cell.[73,74]

3.06.2.1.2 Solubility tags and stable expression of recombinant proteins

The expression of fusion proteins with a highly stable fusion partner at the N-terminal end of the target protein can often provide a successful approach in order to produce soluble and folded heterologous proteins in *E. coli*.[44,75] A fusion partner with a good probability for transmitting a higher solubility to covalently attached target proteins is the *E. coli* MBP.[44] Precursors of mammalian proteinases show enhanced solubility even with C-terminal attached MBP.[76] Hydrophobic areas at the surface of MBP are supposed to confer a rather unspecific chaperone-like assistance in the folding or refolding of aggregation-prone proteins in *cis*. Other fusion partners like thioredoxin, GST, phage λ protein D, or derivatives of DsbA can act in a similar way to prevent formation of inclusion bodies or to enhance the solubility of specific target proteins, but they seem to be not nearly as generally efficient as MBP.[37,38,44,46,75]

The fermentation temperature may have a pronounced effect on yield, and on the folding and stability of recombinant proteins.[77] Temperatures below 30 °C can significantly improve the solubility of a protein. A reduction of the temperature down to 25 °C about 1 h before induction of expression can generally be recommended for the production of critical proteins. The lack of essential cofactors such as hemes or flavins or simply ions like Mg^{2+} can further prevent the folding process of a recombinant protein upon overexpression. Hence, the accumulation of unstable folding intermediates may finally result in formation of inclusion bodies. The yields of soluble protein should be increased by either enhancing the endogenous cofactor production or by an exogenous supply to the culture.

3.06.2.1.3 Targeting of recombinant proteins

In general, three compartments can be considered in *E. coli* for the targeting of the expressed protein: the cytoplasm, the periplasm, and the export of the protein into the extracellular medium. Production of proteins in the periplasm has the advantage that correct disulfide bonds can be formed due to its oxidizing environment and due to the presence of disulfide isomerases.[78,79] The microenvironment of the periplasm can be manipulated much more easily with regard to pH, redox state, and composition in order to optimize the folding of a recombinant protein or to prevent degradation.[80,81] The majority of proteins exported to the periplasm or to the outer membrane of *E. coli* use the general secretory pathway consisting of an array of secretory (Sec) proteins located in the cytoplasm and in the inner membrane.[82–84] One major targeting mechanism in *E. coli* is exemplified by the attachment of specific[20–25] amino-acid-long hydrophobic extensions known as signal or leader sequences to the N-terminus of a protein. The leader will be cleaved off upon translocation thus leaving recombinant proteins with their natural N-terminal ends.[34] Direction of a recombinant protein into the periplasm can simply be anticipated by fusing a naturally occurring leader sequence of an efficiently exported protein like OmpA, β-lactamase, or alkaline phosphatase to its N-terminus. However, the translocation process of the recombinant protein often interferes with the translocation machinery due to premature folding, aggregation, or sterical blocking by bulky secondary structures. Successful and efficient export can therefore hardly be predicted. It is generally difficult to facilitate the secretion of large amounts of protein and a decrease in the yield down to at least 10% to that of the cytoplasmatic expression should be calculated. Secretion of proteins is also usually not very efficient in *E. coli*. Approaches to direct heterologous proteins to the extracellular space rely on gene fusion technologies, on the use of dedicated translocators like hemolysin,[85] or on coexpression of proteins that damage the outer membrane like Kil,[86] bacteriocin release proteins,[87,88] and the membrane protein TolAIII.[89]

3.06.2.1.4 Expression of membrane proteins

Difficulties in obtaining sufficient amounts of recombinant protein are most pronounced with membrane proteins (**Figure 1**). Mechanisms of membrane targeting and processes leading to the membrane insertion of proteins significantly differ between prokaryotes and eukaryotes. Attachment and targeting by the signal recognition protein, a prevalent mechanism in eukaryotes, does not occur in prokaryotes. Moreover, toxic effects of the overproduced proteins frequently cause a reduced growth or even the death of the host cells. Selection strategies have resulted in the isolation of the specially adapted mutant strains C41 (DE3) and C43 (DE3) from the common host strain BL21 (DE3), which better tolerate the overproduction of membrane proteins.[90] These strains contain proliferated membranes, and recombinant proteins are further directed at elevated levels into inclusion bodies. Culture conditions and genetic manipulations can therefore be crucial for obtaining high yields of a recombinant membrane protein.[91,92] Furthermore, the assembly of membrane proteins in *E. coli* requires complex translocation machineries that may provide promising targets for the optimization of expression.[93]

3.06.2.1.5 Proteolytic degradation of recombinant proteins

Abundant amounts of heterologous proteins are often prone to proteolytic degradation as they are recognized as abnormal by the proteolytic machinery of the host cell. Early degradation is mainly connected with the five major ATP-dependent systems: Lon/La, FtsH/HflB, ClpAP, ClpXP, and ClpYQ/HslUV.[94] Expression in protease-deficient strains could thus be advantageous in order to improve the yields of a recombinant protein, but those mutations often

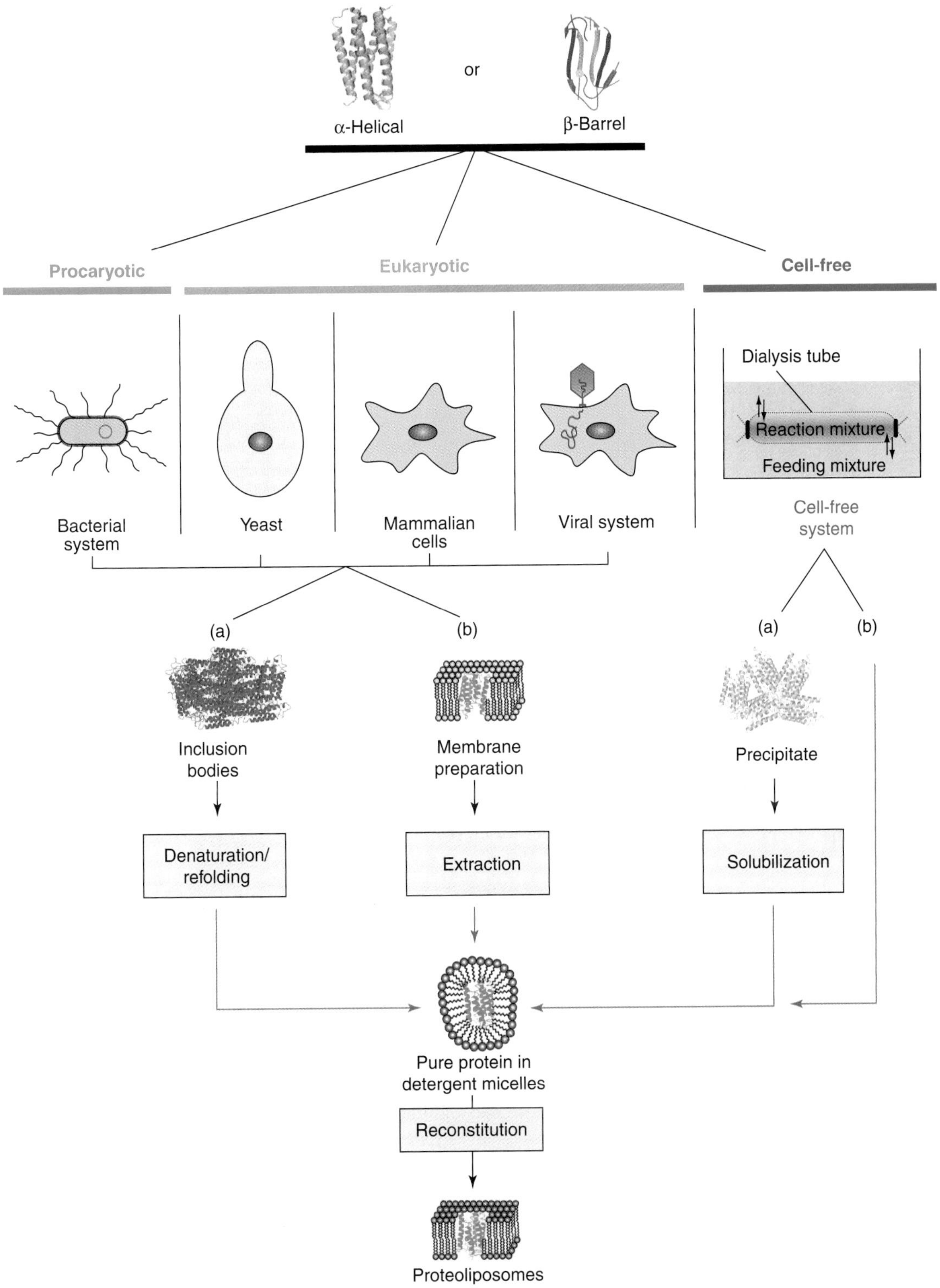

Figure 1 Pathways for the production of membrane proteins in different expression systems. (a) Production as insoluble protein, followed by solubilization and reconstitution. The solubilization from inclusion bodies often requires the initial denaturation of the protein. (b) Isolation or production as soluble protein. Blue arrows indicate the individual purification steps that result in pure membrane protein solubilized in micelles.

cause growth deficiencies.[95] Protease-sensitive proteins could be targeted to the periplasmic space or into inclusion bodies that generally are protected from proteolysis. A preferred strategy in order to prevent degradation is the production of hybrid proteins by the generation of translational fusions.[96] A heterologous protein can thus be shielded by a covalently attached protein that is normally abundant and very stable in the cell.

3.06.2.1.6 Expression of disulfide-bonded proteins

The reducing environment of the *E. coli* cytoplasm and the presence of thioredoxins (TrxA) and glutaredoxins (GrxA-C) that continuously reduce transiently formed disulfide bridges usually prevent the high-level production of oxidized proteins.[79] Consequently, efficient oxidation of recombinant proteins can be achieved by using strains that are defective for some of the major reductases.[97] The BL21 Origami derivatives (Merck, Darmstadt, Germany) have been engineered for a superior expression of disulfide-bonded proteins. These strains carry chromosomal mutations in the thioredoxin reductase (*trxB*) and glutaredoxin reductase (*gor*) genes which are both necessary for the maintenance of a reducing environment in the cytoplasm. Fermentation at reduced temperatures can also favor the oxidation of proteins. An alternative approach is the export of recombinant proteins into the oxidizing environment of the periplasm. The presence of disulfide oxidoreductase and oxidoisomerase (DsbA-C) systems that catalyze the formation and reshuffling of disulfide bonds can considerably facilitate the oxidative folding steps of a recombinant protein in the periplasmic compartment.[64,78,98,99] A further strategy would be the coexpression of oxidizing and disulfide-bridge shuffling chaperones like TrxA, DsbA, DsbC, or the human protein disulfide isomerase (PDI).[100,101]

3.06.2.2 Protein Expression in Bacterial Hosts other than *Escherichia coli*

A variety of bacterial expression hosts other than *E. coli* has been described, and high-level production of recombinant proteins can be achieved with some of them. Limitations in the availability of genetic tools and of expression vectors usually restrict more extensive applications. Nevertheless, obvious advantages of choosing non-*E. coli* hosts include: enhanced expression rates due to a different codon usage, more industrial applications of recombinant proteins after their production in a considered-as-safe organism, and increased efficiencies in the secretion of the target protein. Several proteins could be produced in food-grade microorganisms such as lactic acid bacteria or *Corynebacterium glutamicum*.[102–104] *Lactococcus lactis* has recently been proposed as a promising host for the overproduction of functional membrane proteins of prokaryotic and eukaryotic origin and may represent an interesting alternative to other systems (**Figure 1**).[105] Growth rates are similar to that of *E. coli* and essential techniques for genetic manipulations are available. Expression vectors for lactic acid bacteria are often based on broad-range replicons that also replicate in *E. coli* and a selection of inducible strong promoters is available.[104–107] The efficient secretion systems of *Bacillus subtilis* and probably also of some *Streptomyces* species make these organisms suitable as expression hosts especially for the production of exported proteins.[108–110]

3.06.2.3 Yeast Expression Systems

Eukaryotic expression systems are generally more difficult to handle. Slow growth rates, higher costs, and lower efficiencies are often attributes of eukaryotic expression systems and they are therefore usually consulted when the expression of a recombinant protein in *E. coli* is unlikely for some reason or already has failed. Important and obvious advantages of eukaryotic expression systems are the possibility of obtaining native-like modification patterns of the recombinant protein like glycosylation, phosphorylation, and lipidation. The folding pathway of eukaryotic proteins may in addition depend on specific chaperone systems that do not occur in prokaryotes. The activation of proteins by an attachment of cofactors is also often not possible in *E. coli*. An interesting alternative might be the use of expression systems based on yeast cells. They resemble bacterial systems in many features but still contain important characteristics of eukaryotic expression systems. Many basic mechanisms for the processing and targeting of proteins are very similar to those in mammalian cells. Yeast cells contain an endoplasmatic reticulum and a membrane topology similar to mammalian cells. Efficient pathways for the posttranslational modification of proteins are present and they are frequently essential in order to maintain the stability or the activity of a protein. However, it should be considered that the type of glycosylation in yeast cells sometimes differs from the pattern obtained in mammalian cells. The yeast system can also be very powerful especially with regard to the production of large quantities of secreted proteins.[111–113]

Several yeast species can be used for the production of recombinant proteins. Most popular are baker's yeast *S. cerevisiae*[114,115] the species *Pichia pastoris*,[116,117] and *Hansenula polymorpha*.[118] Other species like *Klyveromyces* sp., *Yarrowia lipolytica*,[119] and *Zygosaccharomyces bailii*[120] might gain increasing interest in the next future. An evident advantage is the unicellular organization of yeasts, which enables the direct transfer of many techniques and manipulations commonly applied for bacteria. Comprehensive protocols exist for the transformation of yeast with foreign DNA. Furthermore, yeasts can be grown in relatively simple and inexpensive media and they reach high cell densities in reasonable times.

An elaborate collection of expression vectors for yeasts is available. Common features are regulatory regions for replication in *E. coli* and a prokaryotic selection marker. This enables the construction and propagation in bacterial cells. The vectors are classified based on their mode of replication into the groups of episomally replicating vectors and chromosomal integration vectors. Stable maintenance of high copy number requires regulatory sequences from the endogenous yeast 2 micron plasmid, while origins of replication from yeast autonomous replicating sequence (ARS) elements are much less stable with loss rates of up to 10% per generation. Integrative vectors are devoid of any origins of replication and they can only be distributed to daughter cells if they integrate into the yeast chromosome by homologous recombination. The efficiency of integration can thus considerably be increased if sequences homologous to chromosomal areas are provided and multiple integrations can be frequent.[117,121] Transformed yeast cells can be screened by a number of selection markers mainly based on auxotrophic deficiencies like HIS3, LEU2, LYS2, TRP1, and URA3.[114] A set of inducible strong promoters is available. Commonly used are the methanol inducible alcohol oxidase 1 (AOX1) and alcohol dehydrogenase promoters (ADH2), the metallothioneine promoter, and galactose inducible promoters (GAL1, GAL7, Gal10).[117,122] Shuttle vector systems suitable for the dual protein expression in *P. pastoris* as well as in *E. coli* are available for the convenient evaluation and rapid comparison of different expression hosts without time-consuming subcloning steps.[123]

Optimized fermenter cultures of *P. pastoris* are able to produce more than $1\,g\,L^{-1}$ of recombinant protein and the ability to grow on defined minimal media makes this species suitable for the efficient labeling of expressed proteins for a structural analysis.[117,122,124] Interesting applications of yeast expression systems are the high secretion levels of recombinant protein,[125,126] the production of disulfide-bonded proteins,[127] and their potential for the high-level production of membrane proteins (**Figure 1**).[128]

3.06.2.4 Mammalian Expression Systems

Expression in mammalian cells provides an environment mostly closely related to the native origin of many eukaryotic proteins. A variety of genetically and phenotypically diverse mammalian cell lines have been used for protein production.[129] Frequently used are baby hamster kidney (BHK) cells,[130] Chinese hamster ovary (CHO) cells,[131] human embryonic kidney (HEK) 293 cells,[132] mouse L-cells,[133] and various myeloma cell lines, which are used to establish relatively stable protein expression systems. Further cell lines for more specific applications are NIH 3T3 cells,[134] murine erythroleukemia (MEL) cells,[135] or lymphoblastoid cell lines like Namalwa[136] and RPMI 1788.[137] Main parameters for the high-level expression of recombinant proteins are the selected cell lines and the specific vector characteristics.[138] Stable productive cell lines are established principally by selecting homogenous cell populations from heterogeneous cell pools. The yield of recombinant protein can often be increased by selective amplification with agents like methotrexate (MTX).[139]

DNA can be transferred into mammalian cells by transfection of coated or encapsulated DNA particles which become incorporated into the cells through endocytosis. Coprecipitation with calcium phosphate[140] or diethylamino-ethyl dextran[141] is also relatively simple and highly effective. Lipofection of compact liposome/nucleic acid complexes and electroporation of cells are further powerful techniques.[142,143]

3.06.2.4.1 Vector design for gene transfer and expression in mammalian cells

Transcriptional and translational control elements, RNA processing, gene copy numbers, stability of mRNA, the site of chromosomal integration, and the impact of recombinant proteins in the host cell are important parameters for efficient gene expression in mammalian cells. A large number of vectors are available and key elements are strong promoters and enhancers of the promoter activity.[144,145] Conserved and essential sequences of eukaryotic promoters are the TATA box and the CAAT box that are located approximately 30 and 80 bp upstream of the mRNA initiation site, respectively. Frequently used constitutive promoters are the adenovirus major late promoter, the human cytomegalovirus immediate early promoter or the simian virus 40 (SV40) and Rous sarcoma virus (RSV) promoters. However, similar to bacterial expression systems, an inducible system is desired in most cases as the produced proteins might become toxic to the host cell.[146] The well-known bacterial *lac* promoter–repressor system was therefore adapted for its use in mammalian cells yielding up to a 10 000–20 000-fold induction of gene expression.[147–149] Selection markers for successfully transfected mammalian cells include the genes encoding for dihydrofolate reductase (DHFR), for aminoglycoside phosphotransferase providing neomycine or G418 resistance,[150] or for enzymes providing resistance against puromycin, hygromycin, or placitidine.[144,151]

Optimized splicing of introns present in most genes of higher eukaryotes is important. The presence of introns in mRNAs may lead to a 10- to 20-fold increased expression rate.[152] Eukaryotic mRNAs have poly(A) tails attached to their 3′ end which modulate their stability and efficiency of translation.[153,154] Effective poly(A) attachment signals in mammalian expression vectors are derived from the genes of mouse β-globulin,[155] herpes simplex virus thymidine kinase,[156] bovine growth hormone,[157] or the SV40 early transcription unit.[158] The initiation of translation depends on a conserved sequence enclosing the AUG start codon (CC(A/G)CCaugG), named the Kozak sequence.[159] Stable hairpin structures of GC-rich regions in the 5′ UTR can further negatively influence the efficiency of transcription.[160]

3.06.2.4.2 Modified mammalian cell expression systems

The copy number of a heterologous gene may significantly increase if it is physically linked to a chromosomal DNA region that is subject of extensive amplification processes. The associated target gene will then become coamplified, resulting in higher yields of recombinant protein due to the increased gene copy numbers. Spontaneous gene amplification in mammalian cells is unfortunately only a rare event, but a variety of artificial gene amplification techniques have been developed that may ultimately lead to considerable gene amplification frequencies.[161] A broad selection of agents and treatments have been found to be active in inducing the amplification of genomic regions. Reagents like hydroxyurea, aphidicolin, and carcinogens, or hypoxia, or the treatment by Ultraviolet- or γ-irradiation have been recommended.[162,163] Frequently used gene amplification systems take advantage of DHFR as a selectable marker present on the expression vector. DHFR catalyzes the reduction of 5,6-dihydrofolate to 5,6,7, 8-tetrahydrofolate, a biocatalyst for the essential synthesis of glycine and thymidine monophosphate, and for purine biosynthesis. CHO cells carrying a DHFR deficiency due to chemical mutagenesis are not able to grow in a medium depleted of nucleosides unless they carry a functional DHFR gene through transfection.[139] The folic acid antagonist MTX binds and inhibits DHFR stoichiometrically. Therefore, stepwise-increased MTX concentrations in the medium can be used to effectively select transfectants that amplify the DHFR gene, resulting in cell populations containing several hundreds of gene copies.[164,165] High levels of recombinant heterologous protein can be achieved by cotransfecting and cointegrating plasmids carrying the DHFR marker and the gene of interest at the same site in the genome.[164] Also an expression cassette including a promotorless DHFR gene can be used. The inserted gene of interest is in that case transcribed from a strong promoter upstream of the DHFR open reading frame, producing a bicistronic message and thus protein production dependent on MTX resistance.[166,167]

A sophisticated technique for antibody production is the fusion of unrestricted and fast-growing myeloma cells with other cells. Fused myeloma and plasma cells create antigen-specific hybridoma cell lines. These fast-growing secretory cell lines are well suited for the expression of recombinant transfected genes and predestinated for the synthesis of monoclonal antibodies.[168,169] A further positive aspect of hybridomas is their natural growth in suspension cultures that saves time-consuming adaptation procedures in the case of large-scale production. The mouse myeloma cell line Sp2/0 Ag14 is a preferred choice for recombinant protein production. It contains deficiencies for synthesizing and secreting endogenous immunoglobulin, is easy to transfect and can proliferate in large-scale serum-free cultures.[170]

The use of powerful viral transcription/translation machineries is an elegant technique for the protein synthesis in a mammalian expression system. COS cell (African-green-monkey kidney cell) lines are most exclusively used for the transient expression of heterologous genes dependent on expression of SV40 large T-antigen, although several other cell lines are suitable.[171–175] Three cell lines (COS-1, COS-3, and COS-7) have been established by transforming the African green monkey cell line CV-1, permissive for the lytic growth of SV40, with an origin-defective SV40 genome and thus creating a cell line harboring the SV40 large T-antigen expression. Transfection with an expression plasmid containing a functional SV40 origin of replication and the gene of interest enables the interaction between the SV40 large T-antigen and the SV40 origin of replication and can finally yield high recombinant protein titers.[176–178] The plasmid replication in COS cells is highest at about 48 h after transfection and the system is not able to produce high amounts of protein over a prolonged period of time.[129] This problem can be solved by using the COS system in extended and modified batch modes that allow multiple harvests and that can result in the cumulative production of preparative amounts of recombinant protein.[179,180]

3.06.2.5 Viral Expression Systems

Viral expression systems are elaborate ways to synthesize heterologous proteins. The systems take advantage of the viral nature to introduce foreign DNA with high efficiency into host cells by infection combined with high replication rates due to strong viral promoters. The most popular viral systems include the already mentioned monkey tumor virus SV40,[175] the baculovirus/insect cells system,[181] and the Semliki Forest virus (SFV) that is used with a wide range of mammalian host cells.[182] Further less prevalent systems use the Epstein–Barr virus,[183] the cytomegalovirus (CMV), RSV, or the SFV-related Sindbis virus.[184] Besides their application in expression systems viral vectors are widely used for vaccines and gene therapies.[185]

3.06.2.5.1 The baculovirus expression vector system

Protein production in insect cells in combination with the baculovious expression vector system (BEVS) presents several advantages like the ease of culture, a high tolerance to by-product concentrations, and high expression levels.[181] Baculoviruses belong to a family of double-stranded DNA viruses with large circular genomes of 120–180 kbp.[186] They are invertebrate-specific and the host range is restricted mainly to arthropods with about 600 insect species reported to be targets of infection.[129] The characteristic production of large amounts of protein during the late phase of infection has implicated their adaptation as vectors for heterologous gene expression. Probably best characterized is the *Autographa californica* nuclear polyhydrosis virus (AcNPV). The two very late gene products, polyhedron and p10, are expressed from strong promoters at levels up to 50% of the total protein content of the infected cell.[187] These proteins are not essential for the formation of viral particles and are thus predestinated to become replaced by heterologous gene products. Diverse expression vectors have been developed that depend on specific host cell lines for infection and propagation.[188–191] The most frequently used insect cell lines are derived from *Spodoptera frugiperda* (SF_9 and SF_{21}) and from *Trichoplusia ni* (Hi_5 and MG_I), having generation times up to 24 h and requiring complex culture media.[187] Various approaches of large-scale protein synthesis in insect cells have been reviewed.[181] Of great interest is the capability of insect cells to perform posttranslational modifications. Cell lines optimized in order to obtain the most favorable glycosylation patterns have been developed.[192,193] The demonstrated high level of expression of several functional G protein-coupled receptors (GPCRs) makes baculovirus-based expression systems very attractive for the preparative production of targets from this important class of membrane proteins (**Figure 1**).[194,195] Activation of the 'late' polyhedron promoter lags significantly behind the infection process and makes the insect/baculovirus system also potentially useful for the expression of cytotoxic recombinant proteins. Recombinant proteins can even effectively become secreted into the medium by adding a signal sequence like the honey bee (*Apis mellifera*) prepromelittin secretory sequence.[196]

3.06.2.5.2 Semliki forest virus expression systems

An effective and well-studied expression system for a rapid and high-level gene delivery is based on replication-deficient alphavirus vectors. These systems benefit from an enormous RNA replication capacity in the cytoplasm that consequently results in extreme expression levels. Most common are vectors of SFV, but similar expression systems have been engineered for the Sindbis virus[197] or the Venezuelan equine encephalitis virus.[198] The pathogenic properties of wild-type SFV, an enveloped single-stranded RNA virus, include induction of neuronal apoptosis, demyelination in the central nervous system, and teratogenesis.[199] These viruses are capable of infecting a broad range of mammalian, amphibian, reptilian, insect, avian, and fish cell lines. Mutant strains with drastically reduced virulence have been made in order to use these pathogenic vectors for experimental gene delivery approaches.

A typical expression system consists of two plasmids based on complementary (cDNA) copies of the SFV genome.[200] The expression vector, containing the SFV nonstructural genes (nsP1-4) and the strong SFV 26S promoter, has a multilinker cloning region into which foreign genes can be inserted. The second helper vector contains the genes for capsid and envelope proteins. The cotransfection of both vectors into BHK cells generates a high titer (10^9–10^{10} particles mL^{-1}) of recombinant and infectious SFV particles.[201] The use of the two-vector system guarantees the production of replication-deficient particles since the RNA packaging signal is only present in the recombinant RNA of the expression vector and no helper RNA will be packed. The infection of host cells will therefore lead to a rapid and high-level expression of recombinant protein without the generation of virus succession. A further advantage is the broad host range of SFV as parallel expression studies can be performed in different cell lines.

The SFV system suffers from safety risks and cytotoxicity, and novel less cytotoxic and temperature-sensitive mutant vectors that are inactive at the blood temperature of 36–37 °C have been developed.[182,202] The addition of a translation enhancement signal to the capsid gene resulted in five- to tenfold increased expression levels.[203] High amounts of topologically different proteins have successfully been expressed with SFV vectors. Besides β-galactosidase with a total cellular protein yield of approximately 25%, more than 50 GPCRs and several ion channels have been expressed at high levels (**Figure 1**).[201]

3.06.2.6 Cell-Free Protein Expression

The described conventional in vivo technologies for protein production depend on the cellular integrity and are only suitable for the production of proteins that do not affect the physiology of the host cell.[204,205] As already discussed, these methods are limited for the expression of many proteins, e.g., by the formation of inclusion bodies,[206,207] by protein instability due to proteolysis,[205,208] or by too low yields in case of most membrane proteins. High-level cell-free (CF) expression systems are a promising new tool for the preparative production of difficult proteins (**Figure 2**). CF systems are principally independent of the cell physiology and they allow direct and immediate control of the reaction at any time. A wide range of critical reaction parameters such as pH, redox potential, and ionic strength can be chosen and adjusted according to the requirements of the specific target protein. Furthermore, any additives that might help to stabilize the recombinant protein after translation can be supplied directly into the reaction and no transport problems through cellular membranes have to be considered. Metabolic conversions or even the breakdown of added substances is usually very low or even not detectable. Options of possible beneficial compounds can include cofactors, ligands, inhibitors, ions, chaperones, and even detergents. The complete and time-independent control in combination with the high flexibility of the reaction conditions provides a challenging opportunity for the preparative production of formerly highly problematic proteins such as membrane proteins, toxins, or unstable proteins.

A unique characteristic of CF expression techniques is the possibility of quickly and easily introducing specific labels into a protein (**Figure 2**). Labeling of proteins with stable isotopes is indispensable for the structural and functional analysis by NMR techniques and for drug screening and ligand interaction studies. Labeling of proteins with spectrally enhanced amino acids can further be very helpful for the analysis of protein interaction studies.[209] The composition and concentration of all low molecular weight substances in the CF reaction is fully defined and the operator has therefore complete control over the amino acid pool of the reaction. Any type of amino acid can thus easily be replaced by a labeled derivative and a 100% label incorporation into the recombinant protein is ensured. The kinetics and

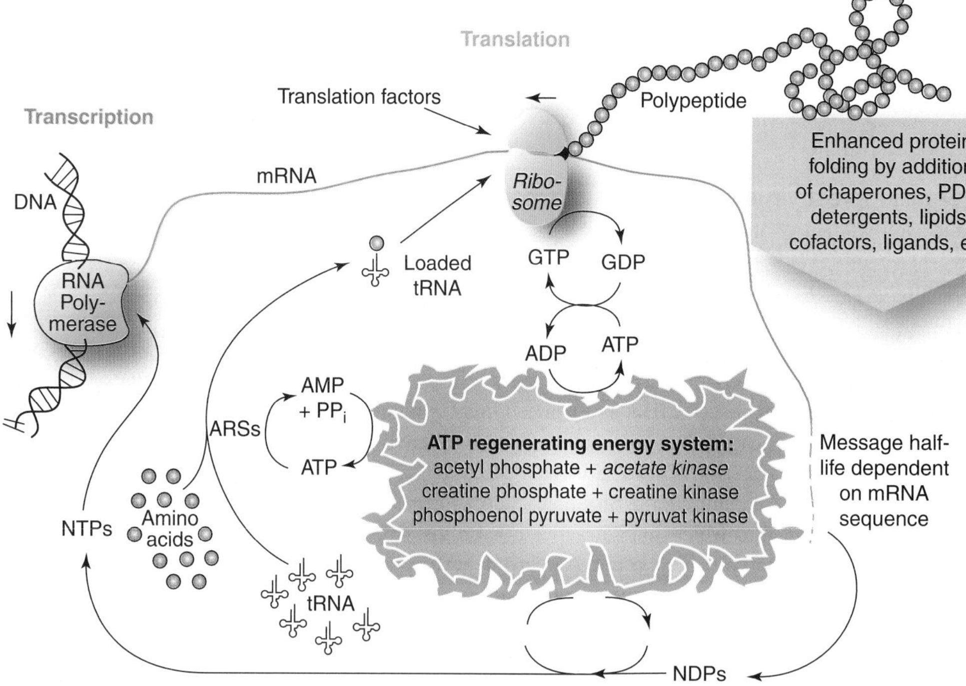

Figure 2 Schematic illustration of cell-free protein synthesis in a bacterial coupled transcription and translation system. The cell extract contains ribosomes, translation factors acetate kinase, and aminoacyl-tRNA synthetases (ARSs). T7 RNA polymerase and substrates like amino acids, the energy regenerating system components, nucleotide triphosphates (NTPs), tRNAs, and salts are supplied and protein synthesis is initiated by adding template DNA. The incorporation of selected isotopic labeled amino acids (red) can easily be achieved in the cell-free system, leading to a selectively isotopic labeled protein. Regeneration of NTPs is accomplished by an ATP regenerating energy system (yellow) based on the hydrolysis of high-energy substrates in the presence of their cognate kinases. To assist stability and folding of the target proteins, detergents, chaperones, and other supplements may be added to the reaction mixture.

efficiency of the production of labeled proteins equals those of the nonlabeled reactions and no laborious optimizations have to be carried out. In contrast to conventional in vivo labeling systems, no switch to auxotrophic host strains and to fermentations in minimal media is necessary. It should be emphasized that there is virtually no background labeling as the target protein is the only protein that is synthesized in substantial amounts during CF synthesis. High-throughput applications attract increasing attention due to the demands of proteomics associated research. CF synthesis also offers a powerful approach as linear DNA templates simply generated by conventional polymerase chain reaction (PCR) can directly be used for the expression of proteins.[210]

3.06.2.6.1 Design of cell-free expression systems

CF expression systems can be set up as pure translation systems with purified mRNA as a template, or alternatively as coupled transcription–translation systems by adding plasmid or linear DNA as a template.[211,212] The simplest design of CF expression is a batch mode reaction with one compartment holding a fixed volume of reaction mixture (RM) in a test-tube. While batch systems are easy to set up and highly suitable for high-throughput applications, they are limited by the rapid accumulation of deleterious by-products like free phosphates from nucleotide consumption that apparently form complexes with magnesium ions. In addition, substrates like nucleotide triphosphates (NTPs) and high-energy phosphate donors are rapidly consumed even in the absence of protein synthesis.[213] The results are obtained in relatively short reaction times that normally do not exceed 1 h. Recent modifications of the batch system by using a novel NTP regeneration system can help to prevent the accumulation of inorganic phosphate,[213] and the supplementation of additional compounds can significantly extend the reaction time and increase the yield of protein synthesis.[214–219] Optimized CF batch systems have the potential to reach a high productivity and may become a powerful technique in the future.

Extended reaction times in CF expression can be achieved by using a continuous-flow CF (CFCF) translation device.[220] A key feature is the continuous supply of energy and substrates concomitant with the removal of reaction by-products. The immobilized RM is continuously perfused by a feeding mixture (FM) containing all low molecular weight precursors and substrates and removing deleterious by-products of the protein production. The RM can alternatively be separated from the FM by a membrane with a variable molecular weight cut-off between 10 and 300 kDa.[221] However, reduced exchange rates by blocked membranes can cause significant problems. The protein synthesis in a CFCF system can continue for more than 20 h and yields of up to 1 mg recombinant protein per mL RM are possible. The CFCF system can either be operated in the translating mode by adding mRNA,[222,223] or in the coupled transcription–translation approach by the addition of phage RNA polymerases.[221,222,224] Many successfully synthesized proteins like the bacteriophage MS2 coat protein,[220] the brome mosaic virus coat protein,[220] globin,[225] calcitonin,[222] DHFR,[222,226–228] chloramphenicol acetyltransferase,[222,221,229–231] interleukin-2,[232] and interleukin-6[223] demonstrate the enormous potential of CFCF protein synthesis.

The relatively complex reaction set-up of the continuous flow mode is simplified in continuous-exchange CF (CECF) expression systems.[230,233] The RM and FM compartments have fixed volumes and are separated by a dialysis membrane. The simplest device for a CECF set-up is a dialysis bag holding the RM that is placed in a suitable container with the FM,[233] but commercially available dialyer device, such as the MicroDialyzer and DispoDialyzer (Spectrum Laboratories, Breda, The Netherlands), can also be used successfully. An additional advantage of the CECF set-up is the accumulation of the synthesized recombinant protein in the RM. High-level synthesis of up to 6 mg recombinant protein per mL RM has been reported.[230,234] Yields of 1–4 mg per mL RM for several functionally active proteins like the green fluorescent protein, DHFR, luciferase, and RNA replicase of tobacco mosaic virus have been obtained with eukaryotic wheat germ extracts.[235] The CECF system is commercially available as a kit (Roche Diagnostics, Mannheim, Germany) but can also be set up individually.[233,236–238]

3.06.2.6.2 Components of the cell-free reaction

All elements involved in gene expression and protein synthesis have to be added to the RM where transcription and translation takes place (**Figure 2**). Components like DNA, a highly processive RNA polymerase like the enzyme encoded by the T7 bacteriophage, NTPs, mRNA, tRNA, aminoacyl tRNA synthetases (ARSs), ribosomes, transcription and translation factors as well as amino acids have to be combined in an optimal pH and salt environment. The required high amounts of free energy for the transcription and translation processes are provided by hydrolysis of the triphosphates ATP and GTP. Crucial for each CF system is therefore an efficient ATP regenerating energy system in order to maintain the NTP concentrations over a long period of time. Conventional energy systems are based on high-energy phosphate donors such as phosphoenol pyruvate in combination with pyruvate kinase,[211,239] creatine phosphate and creatine kinase,[234] or acetyl phosphate with acetate kinase.[240]

3.06.2.6.3 Preparation of cell-free extracts

The quality of the cell extract is crucial for the success of a CF system. The extract represents usually a crude cell lysate which contains most of the essential high molecular weight components for translation. Only T7 RNA polymerase and certain enzymes for the energy regeneration have to be supplied. While many organisms could potentially serve as an extract source, lysates based on the cell types of *E. coli*, wheat germ, and rabbit reticulocytes have been well established.

The bacterial source of choice for the preparation of CF extracts are RNAse-deficient *E. coli* strains, e.g., A19[236] or BL21.[237,241] Extracts of *E. coli* S30 represent the soluble fraction after centrifugation of crude lysates at 30 000 *g*. Endogenous mRNA is removed during a preincubation step of the cell extract either with high salt[236] or with added nucleotides and amino acids. This 'runoff' step releases endogenous mRNA from the ribosomes that will then subsequently become destroyed by endogenous ribonucleases.[242] Alternatively, isolated ribosomes can be added to an S100 extract, centrifuged at 100 000 *g*.[243,244] Extracts of *E. coli* are used in coupled transcription–translation CF systems due to the favorable use of T7 RNA polymerase. Extracts of *E. coli* work well in a wide temperature range between 24 and 38 °C with an optimum at 37 °C.[242,244,245] Due to the relatively simple extract preparation procedure combined with high productivity, the *E. coli* CF system is the most commonly used in vitro protein expression technique and up to 6 mg of protein per ml RM can be synthesized.[234]

A well-defined system using 31 individually purified enzymes isolated with conventional expression systems together with purified ribosomes is termed the 'protein synthesis using recombinant elements' (PURE) system.[246] The advantage of the PURE system is the absence of any inhibitory substances such as nucleases, proteases, and enzymes that hydrolyze nucleoside triphosphates.

The most convenient and promising eukaryotic CF translation system is based on wheat germs isolated from dry wheat seeds.[247] Recent modifications resulted in extracts with a high degree of stability and activity.[235] Important for an enhanced translation efficiency in the wheat germ CF system are the 5′ and 3′ UTRs of eukaryotic mRNAs that play a crucial role in the regulation of gene expression by controlling mRNA translational efficiency, stability, and localization. An optimal 5′ UTR that should therefore be added to mRNAs is the so-called omega sequence (Ω71) of tobacco mosaic virus. For the 3′ UTR the length is more important for an efficient translation than the sequence.[248] Wheat germ extracts possess only low levels of endogenously expressed mRNAs and therefore can be directly used for the expression of templates.[249] The optimal reaction temperature is in the range of 20–27 °C,[250,251] but can be increased to up to 32 °C for higher expression of some templates.[252] The reaction continues up to 60 h and amounts of 1–4 mg of recombinant protein per mL RM can be obtained.[235]

Lysates of rabbit reticulocytes are obtained from blood cells of anemic rabbits that provide a high number of reticulocytes or proerythrocytes. Endogenous mRNA is removed by treatment with micrococcal Ca^{2+}-dependent RNase.[253] The yields of recombinant protein can also be in the range of mg per mL RM, whereas the expression yields of the wheat germ system are usually higher.[254] This system works in an optimal temperature range of 30–38 °C.[255]

In principle, the choice of extract source, either prokaryotic or eukaryotic, for CF synthesis should be chosen according to the origin and biochemical nature of the protein of interest. In general *E. coli*-based systems gain in terms of their higher translation rates, better compatibility with combined transcription–translation formats, easier preparation of extracts as well as reaction set-up, and the availability of mutants with reduced degradative activities. On the other hand they suffer from high degradation of genetic messages, shorter lifetimes, and a great tendency of protein aggregation. Eukaryotic extracts are mainly limited by lower translation rates and the complexity of the genetic constructs that are required for an effective expression. Positive aspects of eukaryotic CF systems are their higher stability and longer lifetime in addition to a better compatibility with eukaryotic mRNAs and the synthesis of eukaryotic proteins.

CECF expression kits based on *E. coli* and wheat germ extracts are commercially available (Roche Diagnostics, Mannheim, Germany). The successful production of more than 40 proteins of different origins, including various enzymes, receptors, hormones, antibodies, and regulatory proteins has been shown by various laboratories.[256,257] Other commercial systems like Expressway (Invitrogen, Carlsbad, CA) focus on the batch mode with expression amounts of up to 1 $mg\,mL^{-1}$.

3.06.2.6.4 Cell-free synthesis of membrane proteins

Membrane proteins today represent less than 1% of the available three-dimensional protein structures, although this is in contrast to their immense medical importance as an estimate of 60–70% of current drug targets are based on membrane proteins. The key bottleneck has been the lack of reliable technologies that ensure the production of a broad variety of recombinant membrane proteins in the required amounts.[258] Toxicity to the host cell, protein aggregation, and miss-folding of overproduced membrane proteins very often result in low yields. Furthermore, the overexpression of integral membrane proteins often causes cell death by overloading the cytoplasmic membranes or by disrupting the membrane integrity. Two protein groups of outstanding pharmaceutical relevance are multidrug resistance transporters of bacterial pathogens and GPCRs as the basic elements of the eukaryotic signal transduction machinery. Preparative-scale expression of GPCRs has not been obtained in most cases, and it is always subject of tedious and long-lasting optimizations.[194,195,259] A promising perspective is that the newly developed CF translation systems offer a powerful alternative for overcoming the tremendous expression barriers for membrane proteins (**Figure 1**). It has recently been shown that functionally active integral membrane proteins, especially small multidrug transporters, GPCR proteins, a light-harvesting membrane protein, and ion channels were expressed in high yields of mg amounts per mL RM in an *E. coli*-based CECF system.[236,238,260–262] The proteins could also be functionally reconstituted into proteoliposomes and isotopically labeled for NMR investigations.[236] Even the addition of mild detergents does not interfere with the translation activity of the CECF systems and results directly in the soluble and functional expression of several integral membrane proteins.[238,260,261] The combination of isotopic labeling and membrane protein expression demonstrates the high potential of the CF method for functional and structural membrane protein research. Finally it should be mentioned that the preparation of membrane protein samples ready for a structural analysis by, e.g., NMR techniques is possible in less than 2 days by using CF expression systems and the structural characterization of even very difficult protein families like the GPCR proteins becomes now feasible.

3.06.2.6.5 Cell-free synthesis of disulfide-bridged proteins

Disulfide-bonded proteins are rarely expressed in traditional expression systems due to their requirements for oxidizing conditions which in eukaryotes are only found in the lumen of the endoplasmatic reticulum (or in the periplasm of prokaryotes). The open character of CF systems allows the direct addition of purified chaperones like the already discussed eukaryotic protein disulfide isomerase (PDI) or bacterial Dsb derivatives with success of the functional expression of single chain antibodies.[263] CF expression reactions are usually operated in the presence of reducing agents like dithiothreitol that stabilize the protein transcription–translation machinery but provide less favorable conditions for the synthesis of disulfide-bonded proteins. However, the use of a dithiothreitol-deficient wheat germ extract in the presence of the PDI chaperone efficiently synthesizes a single-chain antibody variable fragment with dual disulfide bonds.[264] More recently, the problem of the CF expression of disulfide-bonded proteins was overcome by using a combination of iodoacetamide-treated extract, a suitable glutathione redox buffer, and the addition of disulfide bond shuffling chaperones like Skp and DsbC.[265] A recombinant plasminogen activator protein with nine disulfide bonds could productively be expressed with this approach.

References

1. Peredelchuk, M. Y.; Bennett, G. N. *Gene* **1997**, *187*, 231–238.
2. Olson, P.; Zhang, Y.; Olsen, D.; Owens, A.; Cohen, P.; Nguyen, K.; Ye, J. J.; Bass, S.; Mascarenhas, D. *Protein Expr. Purif.* **1998**, *14*, 160–166.
3. Das, A. *Methods Enzymol.* **1990**, *182*, 93–112.
4. Makrides, S. C. *Microbiol. Rev.* **1996**, *60*, 512–538.
5. Hannig, G.; Makrides, S. C. *Trends Biotechnol.* **1998**, *16*, 54–60.
6. Baneyx, F. *Curr. Opin. Biotechnol.* **1999**, *10*, 411–421.
7. Mertens, N.; Remaut, E.; Fiers, W. *Gene* **1995**, *164*, 9–15.
8. Donovan, R. S.; Robinson, C. W.; Glick, B. R. *J. Ind. Microbiol.* **1996**, *16*, 145–154.
9. DeBoer, H. A.; Comstock, L. J.; Vasser, M. *Proc. Natl. Acad. Sci. USA* **1983**, *80*, 21–25.
10. Sawers, G.; Jarsch, M. *Appl. Microbiol. Biotechnol.* **1996**, *46*, 1–9.
11. O'Connor, C. D.; Timmis, K. N. *J. Bacteriol.* **1987**, *169*, 4457–4462.
12. Elvin, C. M.; Thompson, P. R.; Argall, M. E.; Hendry, P.; Stamford, N. P.; Lilley, P. E.; Dixon, N. E. *Gene* **1990**, *87*, 123–126.
13. Tabor, S.; Richardson, C. C. *Proc. Natl. Acad. Sci. USA* **1985**, *82*, 1074–1078.
14. Studier, F. W.; Rosenberg, A. H.; Dunn, J. J.; Dubendorff, J. W. *Methods Enzymol.* **1990**, *185*, 60–89.
15. Studier, F. W.; Moffatt, B. *J. Mol. Biol.* **1986**, *189*, 113–130.
16. Guzman, L. M.; Belin, D.; Carson, M. J.; Beckwith, J. *J. Bacteriol.* **1995**, *177*, 4121–4130.
17. Siegele, D. A.; Hu, J. C. *Proc. Natl. Acad. Sci. USA* **1997**, *94*, 8168–8172.
18. Wycuff, D. R.; Matthews, K. S. *Anal. Biochem.* **2000**, *277*, 67–73.
19. Gold, L. *Methods Enzymol.* **1990**, *185*, 11–14.
20. Zabeau, M.; Stanley, K. K. *EMBO J.* **1982**, *1*, 1217–1224.
21. Stanssens, P.; Remaut, E.; Fiers, W. *Gene* **1985**, *36*, 211–223.
22. Grisshammer, R.; Tate, C. G. *Q. Rev. Biophys.* **1995**, *28*, 315–422.
23. Wada, K.; Wada, Y.; Ishibashi, F.; Gojobori, T.; Ikemura, T. *Nucleic Acids Res.* **1992**, *20*, 2111–2118.
24. Kane, J. F. *Curr. Opin. Biotechnol.* **1995**, *6*, 494–500.
25. Calderone, T. L.; Stevens, R. D.; Oas, T. G. *J. Mol. Biol.* **1996**, *262*, 407–412.
26. Day, A. J.; Aplin, R. T.; Willis, A. C. *Protein Expr. Purif.* **1996**, *8*, 1–16.

27. Zhan, K. *J. Bacteriol.* **1996**, *178*, 2926–2933.
28. Forman, M. D.; Stack, R. F.; Masters, P. S.; Hauer, C. R.; Baxter, S. M. *Protein Sci.* **1998**, 7, 500–503.
29. Gustafsson, C.; Govindarajan, S.; Minshull, J. *Trends Biotechnol.* **2004**, *22*, 346–353.
30. Brinkmann, U.; Mattes, R. E.; Buckel, P. *Gene* **1989**, *85*, 109–114.
31. Baca, A. M.; Hol, W. G. *Int. J. Parasitol.* **2000**, *30*, 113–118.
32. Forrer, P.; Jaussi, R. *Gene* **1998**, *224*, 45–52.
33. Zhou, P.; Lugovskoy, A. A.; Wagner, G. *J. Biomol. NMR* **2001**, *20*, 11–14.
34. LaVallie, E. R.; McCoy, J. M. *Curr. Opin. Biotechnol.* **1995**, *6*, 501–506.
35. Sharrocks, A. D. *Gene* **1994**, *138*, 105–108.
36. Nilsson, J.; Stahl, S.; Lundeberg, J.; Uhlen, M.; Nygren, P. A. *Protein Expr. Purif.* **1997**, *11*, 1–16.
37. Hochuli, E.; Döbeli, H.; Schacher, A. *J. Chromatogr.* **1987**, *411*, 177–184.
38. Pryor, K. D.; Leiting, B. *Protein Expr. Purif.* **1997**, *10*, 309–319.
39. Frorath, B.; Abney, C. C.; Berthold, M.; Scanarini, M.; Northemann, W. *Biotechniques* **1992**, *12*, 558–563.
40. LaVallie, E. R.; DiBlasio, E. A.; Kovacic, S.; Grant, K. L.; Schendel, P. F.; McCoy, J. M. *Biotechnology* **1993**, *11*, 187–193.
41. Kohne, T.; Kusunoki, H.; Sato, K.; Wakamutsu, K. *J. Biomol. NMR* **1998**, *12*, 109–121.
42. Koken, M. H.; Odijk, H. H.; VanDuin, M.; Fornerod, M.; Hoeijmakers, J. H. *Biochem. Biophys. Res. Commun.* **1993**, *195*, 643–653.
43. Smith, D. B.; Johnson, K. S. *Gene* **1988**, *67*, 31–40.
44. Nygren, P. A.; Stahl, S.; Uhlen, M. *Trends Biotechnol.* **1994**, *12*, 184–188.
45. Di Guan, C.; Li, P.; Riggs, P. D.; Inouye, H. *Gene* **1988**, *67*, 21–30.
46. Kapust, R. B.; Waugh, D. S. *Protein Sci.* **1999**, *8*, 1668–1674.
47. Riggs, P. *Mol. Biotechnol.* **2000**, *15*, 51–63.
48. Kapust, R. B.; Waugh, D. S. *Protein Expr. Purif.* **2000**, *19*, 312–318.
49. Shih, Y. P.; Wu, H. C.; Hu, S. M.; Wang, T. F.; Wang, A. H. *Protein Sci.* **2005**, *14*, 936–941.
50. Friedberg, T.; Wolf, C. R. *Adv. Drug Rev.* **1996**, *22*, 187–213.
51. Swartz, J. R. *Curr. Opin. Biotechnol.* **2001**, *12*, 195–201.
52. Andersen, D. C.; Krummen, L. *Curr. Opin. Biotechnol.* **2002**, *13*, 117–123.
53. Hunt, I. *Protein Expr. Purif.* **2005**, *40*, 1–22.
54. Sorensen, H. P.; Mortensen, K. K. *J. Biotechnol.* **2005**, *115*, 113–128.
55. Lesley, S. A. *Protein Expr. Purif.* **2001**, *22*, 159–164.
56. Weickert, M. J.; Doherty, D. H.; Best, E. A.; Olins, P. O. *Curr. Opin. Biotechnol.* **1996**, 7, 494–499.
57. Misawa, S.; Kumagai, I. *Biopolymers* **1999**, *51*, 297–307.
58. Carrió, M. M.; Villaverde, A. *J. Biotechnol.* **2002**, *96*, 3–12.
59. Bukau, B.; Horwich, A. L. *Cell* **1998**, *92*, 351–366.
60. Feldman, D. E.; Frydman, J. *Curr. Opin. Struct. Biol.* **2000**, *10*, 26–33.
61. Sigler, P. B.; Xu, Z.; Rye, H. S.; Burston, S. G.; Fenton, W. A.; Horwich, A. L. *Annu. Rev. Biochem.* **1998**, *67*, 581–608.
62. Wang, J. D.; Weissman, J. S. *Nat. Struct. Biol.* **1999**, *6*, 597–600.
63. Georgiou, G.; Valax, P. *Curr. Opin. Biotechnol.* **1996**, 7, 190–197.
64. Thomas, J. G.; Ayling, A.; Baneyx, F. *Appl. Biochem. Biotechnol.* **1997**, *66*, 197–238.
65. Schlieker, C.; Bukau, B.; Mogk, A. *J. Biotechnol.* **2002**, *96*, 13–21.
66. DeMarco, A.; DeMarco, V. *J. Biotechnol.* **2004**, *109*, 45–52.
67. Nishihara, K.; Kanemori, M.; Kitagawa, M.; Yanaga, H.; Yura, T. *Appl. Environ. Microbiol.* **1998**, *64*, 1694–1699.
68. Nishihara, K.; Kanemori, M.; Yanagi, H.; Yura, T. *Appl. Environ. Microbiol.* **2000**, *66*, 884–889.
69. Cole, P. A. *Structure* **1996**, *4*, 239–242.
70. Castanié, M. P.; Bergès, H.; Oreglia, J.; Prère, M. F.; Fayet, O. *Anal. Biochem.* **1997**, *254*, 150–152.
71. Rucker, P.; Torti, F. M.; Torti, S. V. *Protein Eng.* **1997**, *10*, 967–973.
72. Kim, K. J.; Kim, H. E.; Lee, K. H.; Han, W.; Yi, M. J.; Jeong, J.; Oh, B. H. *Protein Sci.* **2004**, *13*, 1698–1703.
73. Johnston, K.; Clements, A.; Venkatamanrani, R. N.; Trievel, R. C.; Marmorstein, R. *Protein Expr. Purif.* **2000**, *20*, 435–443.
74. Yang, W.; Zhang, L.; Lu, Z.; Tao, W.; Zhai, Z. *Protein Expr. Purif.* **2001**, *22*, 472–478.
75. Zhang, Y.; Olsen, D. R.; Nguyen, K. B.; Olson, P. S.; Rhodes, E. T.; Mascarenhas, D. *Protein Expr. Purif.* **1998**, *12*, 159–165.
76. Sachdev, D.; Chirgwin, J. M. *Protein Expr. Purif.* **1998**, *12*, 122–132.
77. Baneyx, F. In Vivo Folding of Recombinant Proteins in *Escherichia coli.* In *Manual of Industrial Microbiology and Biotechnology*, 2nd ed.; Davies, J. E., Demain, A. L., Cohen, G., Hershberger, C. L., Forney, L. J., Holland, I. B., Hu, W. S., Wu, J. -H. D., Sherman, D. H., Wilson, R. C., Eds.; ASM Press: Washington, DC, 1999, pp 551–565.
78. Missiakas, D.; Raina, S. *J. Bacteriol.* **1997**, *179*, 2465–2471.
79. Aslund, F.; Beckwith, J. *J. Bacteriol.* **1999**, *181*, 1375–1379.
80. Bowden, G. A.; Georgiou, G. *Biotechnol. Prog.* **1988**, *4*, 97–101.
81. Baneyx, F.; Ayling, A.; Palumbo, T.; Thomas, D.; Georgiou, G. *Appl. Microbiol. Biotechnol.* **1991**, *36*, 14–20.
82. Pugsley, A. P. *Microbiol. Rev.* **1993**, *57*, 50–108.
83. Sandkvist, M.; Bagdasarian, M. *Curr. Opin. Biotechnol.* **1996**, 7, 505–511.
84. Driessen, A. J.; Fekkes, P.; vanderWolk, J. P. *Curr. Opin. Microbiol.* **1998**, *1*, 216–222.
85. Blight, M. A.; Holland, I. B. *Trends Biotechnol.* **1994**, *12*, 450–455.
86. Robbens, J.; Raeymaekers, A.; Steidler, L.; Fiers, W.; Remaut, E. *Protein Expr. Purif.* **1995**, *6*, 481–486.
87. VanderWal, F. J.; Luirink, J.; Oudega, B. *FEMS Microbiol. Rev.* **1995**, *17*, 381–399.
88. VanderWal, F. J.; Koningstein, G.; TenHagen, C. M.; Oudega, B.; Luirink, J. *Appl. Environ. Microbiol.* **1998**, *64*, 392–398.
89. Wan, E. W.; Baneyx, F. *Protein Expr. Purif.* **1998**, *14*, 13–22.
90. Miroux, B.; Walker, J. E. *J. Mol. Biol.* **1996**, *260*, 289–298.
91. Wang, D. N.; Safferling, M.; Lemieux, M. J.; Griffith, H.; Chen, Y.; Li, X. D. *Biochim. Biophys. Acta* **2003**, *1610*, 23–36.
92. Montigny, C.; Penin, F.; Lethias, C.; Falson, P. *Biochim. Biophys. Acta* **2004**, *1660*, 53–65.
93. Drew, D.; Fröderberg, L.; Baares, L.; deGier, J. W. L. *Biochim. Biophys. Acta* **2003**, *1610*, 3–10.
94. Gottesman, S. *Annu. Rev. Genet.* **1996**, *30*, 465–506.

95. Roman, L. J.; Sheta, E. A.; Martasek, P.; Gross, S. S.; Liu, Q.; Siler-Masters, B. S. *Proc. Natl. Acad. Sci. USA* **1995**, *92*, 8428–8432.
96. Murby, M.; Uhlén, M.; Stahl, S. *Protein Expr. Purif.* **1996**, 7, 129–136.
97. Bessette, P. H.; Aslund, F.; Beckwith, J.; Georgiou, G. *Proc. Natl. Acad. Sci. USA* **1999**, *96*, 13703–13708.
98. Bardwell, J. C. A. *Mol. Microbiol.* **1994**, *14*, 199–205.
99. Wülfing, C.; Plückthun, A. *Mol. Microbiol.* **1994**, *12*, 685–692.
100. Humphreys, D. L.; Weir, N.; Mountain, A.; Luind, P. A. *J. Biol. Chem.* **1995**, *270*, 28210–28215.
101. Yasukawa, T.; Kanei-Ishii, C.; Maekawa, T.; Fujimoto, J.; Yamamoto, T.; Ishii, S. *J. Biol. Chem.* **1995**, *270*, 25328–25331.
102. Billman-Jacobe, H. *Curr. Opin. Biotechnol.* **1996**, 7, 500–504.
103. Kuipers, O. P.; deRuyter, P. G. G. A.; Kleerebezem, M.; deVos, W. M. *Trends Biotechnol.* **1997**, *15*, 135–140.
104. DeVos, W. M. *Curr. Opin. Microbiol.* **1999**, *2*, 289–295.
105. Kunji, E. R. S.; Slotboom, D. J.; Poolman, B. *Biochim. Biophys. Acta* **2003**, *1610*, 97–108.
106. Sorvig, E.; Grönqvist, S.; Naterstad, K.; Mathiesen, G.; Eijsink, V. G. H.; Axelsson, L. *FEMS Microbiol. Lett.* **2003**, *229*, 119–126.
107. Miyoshi, A.; Jamet, E.; Commissaire, J.; Renault, P.; Langella, P.; Azevedo, V. *FEMS Microbiol. Lett.* **2004**, *239*, 205–212.
108. Wong, S. L. *Curr. Opin. Biotechnol.* **1995**, *6*, 517–522.
109. Binnie, C.; Cossar, J. D.; Stewart, D. I. H. *Trends Biotechnol.* **1997**, *15*, 315–320.
110. Lam, K. H.; Chow, K. C.; Wong, W. K. *J. Biotechnol.* **1998**, *63*, 167–177.
111. Van Heeke, G.; Ott, T. L.; Strauss, A.; Ammaturo, D.; Bazer, F. W. *J. Interferon Cytokine Res.* **1996**, *16*, 119–126.
112. Austin, A. J.; Jones, C. E.; van Heeke, G. *Protein Expr. Purif.* **1998**, *13*, 136–142.
113. Boetner, M.; Prinz, B.; Holz, C.; Stahl, U.; Lang, C. *J. Biotechnol.* **2002**, *99*, 51–62.
114. Romanos, M. A.; Scorer, C. A.; Clare, J. J. *Yeast* **1992**, *8*, 423–488.
115. Schuster, M.; Einhauer, A.; Wasserbauer, E.; Süßenbacher, F.; Ortner, C.; Paumann, M.; Werner, G.; Jungbauer, A. *J. Biotechnol.* **2000**, *84*, 237–248.
116. Hollenberg, C. P.; Gellissen, G. *Curr. Opin. Biotechnol.* **1997**, *8*, 554–560.
117. Cereghino, J. L.; Cregg, J. M. *FEMS Microbiol. Rev.* **2000**, *24*, 45–66.
118. VanDijk, R.; Faber, K. N.; Kiel, J. A. K. W.; Veenhuis, M.; vanderKlei, I. *Enzyme Microb. Technol.* **2000**, *26*, 793–800.
119. Madzak, C.; Gaillardin, C.; Beckerich, J. M. *J. Biotechnol.* **2004**, *109*, 63–81.
120. Branduardi, P.; Valli, M.; Brambilla, L.; Sauer, M.; Alberghino, L.; Porro, D. *FEMS Yeast Res.* **2004**, *4*, 493–504.
121. Sreekrishna, K.; Brankamp, R. G.; Kropp, K. E.; Blankenship, D. T.; Tsay, J. T.; Smith, P. L.; Wierschke, J. D.; Subramaniam, A.; Birkenberger, L. A. *Gene* **1997**, *190*, 55–62.
122. Hensing, M. C. M.; Rouwenhorst, R. J.; Heijnen, J. J.; vanDijken, J. P.; Pronk, J. T. *Antonie van Leeuwenhoek* **1995**, *67*, 261–279.
123. Lueking, A.; Holz, C.; Gotthold, C.; Lehrach, H.; Cahill, D. *Protein Expr. Purif.* **2000**, *20*, 372–378.
124. Vad, R.; Nafstad, E.; Dahl, L. A.; Gabrielsen, O. S. *J. Biotechnol.* **2005**, *116*, 251–260.
125. Parekh, R. N.; Wittrup, K. D. *Biotechnol. Prog.* **1997**, *13*, 117–122.
126. Cereghino, G. P. L.; Cereghino, J. L.; Ilgen, C.; Cregg, J. M. *Curr. Opin. Biotechnol.* **2002**, *13*, 329–332.
127. White, C. E.; Kempi, N. M.; Komives, E. A. *Structure* **1994**, *2*, 1003–1005.
128. Reiländer, H.; Wei, H. M. *Curr. Opin. Biotechnol.* **1998**, *9*, 510–517.
129. Geisse, S.; Gram, H.; Kleuser, B.; Kocher, H. P. *Protein Expr. Purif.* **1996**, *8*, 271–282.
130. Wirth, M.; Bode, J.; Zettlmeissl, G.; Hauser, H. *Gene* **1988**, *73*, 419–426.
131. Page, M. J.; Sydenham, M. A. *Biotechnology* **1991**, *9*, 64–68.
132. Berg, D. T.; McClure, D. B.; Grinnell, B. W. *Biotechniques* **1993**, *14*, 972–978.
133. Gopal, T. V.; Polte, T.; Arthur, P.; Seidman, M. *In Vitro Cell Dev. Biol.* **1989**, *25*, 1147–1154.
134. Kane, S. E.; Reinhard, D. H.; Fordis, C. M.; Pastan, I.; Gottesman, M. M. *Gene* **1989**, *84*, 439–446.
135. Needham, M.; Gooding, C.; Hudson, K.; Antoniou, M.; Grosveld, F.; Hollis, M. *Nucleic Acids Res.* **1992**, *20*, 997–1003.
136. Okamoto, M.; Nakayama, C.; Nakai, M.; Yanagi, H. *Biotechnology* **1990**, *8*, 550–553.
137. Lopez, C.; Chesnay, A. D.; Tournamille, C.; Ghanem, A. B.; Prigent, S.; Drouet, X.; Lambin, P.; Cartron, J. P. *Gene* **1994**, *148*, 285–291.
138. Wenger, R. H.; Moreau, H.; Nielsen, P. J. *Anal. Biochem.* **1994**, *221*, 416–418.
139. Urlaub, G.; Chasin, L. A. *Proc. Natl. Acad. Sci. USA* **1980**, 77, 4216–4220.
140. Graham, F. L.; van der Eb, A. *J. Virology* **1973**, *52*, 456–467.
141. Vaheri, A.; Pagano, J. S. *Virology* **1965**, *27*, 434–436.
142. Wong, T. K.; Neumann, E. *Biochem. Biophys. Res. Commun.* **1982**, *107*, 584–587.
143. Felgner, P. L.; Gadek, T. R.; Holm, M.; Roman, R.; Chan, H. W.; Wenz, M.; Northrop, J. P.; Ringold, G. M.; Danielsen, M. *Proc. Natl. Acad. Sci. USA* **1987**, *84*, 7413–7417.
144. Makrides, S. C. *Protein Expr. Purif.* **1999**, *17*, 183–202.
145. Novina, C. D.; Roy, A. L. *Trends Genet.* **1996**, *12*, 351–355.
146. Angrand, P. O.; Woodfoofe, C. P.; Buchholz, F.; Stewart, A. F. *Nucleic Acids Res.* **1998**, *26*, 3263–3269.
147. Hu, M. C.; Davidson, N. *Cell* **1987**, *48*, 555–566.
148. Figge, J.; Wright, C.; Collins, C. J.; Roberts, T. M.; Livingston, D. M. *Cell* **1988**, *52*, 713–722.
149. Ward, G. A.; Stover, C. K.; Moss, B.; Fuerst, T. R. *Proc. Natl. Acad. Sci. USA* **1995**, *92*, 6773–6777.
150. Colbere-Grapin, F.; Horodniceanu, F.; Kourilsky, P.; Garapin, A. C. *J. Mol. Biol.* **1981**, *150*, 1–14.
151. Karreman, C. *Gene* **1998**, *218*, 57–61.
152. Buchman, A. R.; Berg, P. *Mol. Cell Biol.* **1988**, *8*, 4395–4405.
153. Jackson, R. J.; Standart, N. *Cell* **1990**, *62*, 15–24.
154. Gray, N. K.; Wickens, M. *Annu. Rev. Cell Dev. Biol.* **1998**, *14*, 399–458.
155. Pandey, N. B.; Chodchoy, N.; Liu, T. J.; Marzluff, W. F. *Nucleic Acids Res.* **1990**, *18*, 3161–3170.
156. Schmidt, E. V.; Christoph, G.; Zeller, R.; Leder, P. *Mol. Cell. Biol.* **1990**, *10*, 4406–4411.
157. Goodwin, E. C.; Rottman, F. M. *J. Biol. Chem.* **1992**, *267*, 16330–16334.
158. Van den Hoff, M. J.; Labruyère, W. T.; Moorman, A. F.; Lamers, W. H. *Nucleic Acids Res.* **1993**, *21*, 4987–4988.
159. Kozak, M. J. *Mol. Biol.* **1987**, *196*, 947–950.
160. Grens, A.; Scheffler, I. E. *J. Biol. Chem.* **1990**, *265*, 11810–11816.
161. Wright, J. A.; Smith, H. S.; Watt, F. M.; Hancock, M. C.; Hudson, D. L.; Stark, G. R. *Proc. Natl. Acad. Sci. USA* **1990**, *87*, 1791–1795.

162. Schimke, R. T. *J. Biol. Chem.* **1988**, *263*, 5989–5992.
163. Stark, G. R.; Debatisse, M.; Giulotto, E.; Wahl, G. M. *Cell* **1989**, *57*, 901–908.
164. Kaufman, R. J.; Wasley, L. C.; Spiliotes, A. J.; Gossels, S. D.; Latt, S. A.; Larsen, G. R.; Kay, R. M. *Mol. Cell. Biol.* **1985**, *5*, 1750–1759.
165. Kaufman, R . J. *Methods Enzymol.* **1990**, *185*, 537–566.
166. Kaufman, R. J.; Sharp, P. A. *J. Mol. Biol.* **1982**, *159*, 601–621.
167. Kaufman, R. J.; Murtha, P.; Davies, M. V. *EMBO J.* **1987**, *6*, 187–193.
168. Traunecker, A.; Oliveri, F.; Karjalainen, K. *Trends Biotechnol.* **1991**, *9*, 109–113.
169. Greene, G. L.; Fitch, F. W.; Jensen, E. V. *Proc. Natl. Acad. Sci. USA* **1980**, 77, 157–161.
170. Shulman, M.; Wilde, C. D.; Kohler, G. *Nature* **1978**, *276*, 269–270.
171. Boast, S.; LaMantia, G.; Lania, L.; Blasi, F. *EMBO J.* **1983**, *2*, 2327–2331.
172. Gerard, R. D.; Gluzman, Y. *Mol. Cell. Biol.* **1985**, *5*, 3231–3340.
173. Asselbergs, F. A. *J. Biotechnol.* **1992**, *26*, 301–313.
174. de Chasseval, R.; de Villartay, J. P. *Nucleic Acids Res.* **1992**, *20*, 245–250.
175. Gluzman, Y. *Cell* **1981**, *23*, 175–182.
176. Mellon, P.; Parker, V.; Gluzman, Y.; Maniatis, T. *Cell* **1981**, *27*, 279–288.
177. Edwards, C. P.; Aruffo, A. *Curr. Opin. Biotechnol.* **1993**, *4*, 558–563.
178. Trill, J. J.; Shatzman, A. R.; Ganguly, S. *Curr. Opin. Biotechnol.* **1995**, *6*, 553–560.
179. Ridder, R.; Geisse, S.; Kleuser, B.; Kawalleck, P.; Gram, H. *Gene* **1995**, *166*, 273–276.
180. Wurm, F.; Bernard, A. *Curr. Opin. Biotechnol.* **1999**, *10*, 156–159.
181. Ikonomou, L.; Schneider, Y. J.; Agathos, S. N. *Appl. Microbiol. Biotechnol.* **2003**, *62*, 1–20.
182. Lundstrom, K.; Rotmann, D.; Hermann, D.; Schneider, E. M.; Ehrengruber, M. U. *Histochem. Cell Biol.* **2001**, *115*, 83–91.
183. Brickell, P. M.; Patel, M. S. *Mol. Biotechnol.* **1995**, *3*, 199–205.
184. Yamanaka, R.; Xanthopoulos, K. G. *DNA Cell Biol.* **2004**, *23*, 75–80.
185. Polo, J. M.; Gardner, J. P.; Ji, Y.; Belli, B. A.; Driver, D. A.; Sherrill, S.; Perri, S.; Liu, M. A.; Dubensky, T. W. *Dev. Biol.* **2000**, *104*, 181–185.
186. Ayres, M. D.; Howard, S. C.; Kuzio, J.; Lopez-Ferber, M.; Possee, R. D. *Virology* **1994**, *202*, 586–605.
187. Sarramegna, V.; Talmont, F.; Demange, P.; Milon, A. *Cell Mol. Life Sci.* **2003**, *60*, 1529–1546.
188. Kitts, P. A.; Possee, R. D. *Biotechniques* **1993**, *14*, 810–817.
189. Luckow, V. A. *Curr. Opin. Biotechnol.* **1993**, *4*, 564–572.
190. Davies, A. H. *Biotechnology* **1994**, *12*, 47–50.
191. Jones, I.; Morikawa, Y. *Curr. Opin. Biotechnol.* **1996**, 7, 512–516.
192. Ogonah, O. W.; Freedman, R. B.; Jenkins, N.; Patel, K.; Rooney, B. C. *Biotechnology* **1996**, *14*, 197–202.
193. Palomares, L. A.; Joosten, C. E.; Hughes, P. R.; Granados, R. R.; Shuler, M. L. *Biotechnol. Prog.* **2003**, *19*, 185–192.
194. Massotte, D. *Biochim. Biophys. Acta* **2003**, *1610*, 77–89.
195. Tate, C. G.; Haase, J.; Baker, C.; Boorsma, M.; Magnani, F.; Vallis, Y.; Williams, D. C. *Biochim. Biophys. Acta* **2003**, *1610*, 141–153.
196. Chai, H.; Vasudevan, S. G.; Porter, A. G.; Chua, K. L.; Oh, S.; Yap, M. *Biotechnol. Appl. Biochem.* **1993**, *18*, 259–273.
197. Xiong, C.; Levis, R.; Shen, P.; Schlesinger, S.; Rice, C. M.; Huang, H. V. *Science* **1989**, *243*, 1188–1191.
198. Davis, N. L.; Willis, L. V.; Smith, J. F.; Johnston, R. E. *Virology* **1989**, *171*, 189–204.
199. Atkins, G. J.; Sheahan, B. J.; Liljestrom, P. *J. Gen. Virol.* **1999**, *80*, 2287–2297.
200. Liljestrom, P.; Garoff, H. *Biotechnology* **1991**, *9*, 1356–1361.
201. Lundstrom, K. *Biochim. Biophys. Acta* **2003**, *1610*, 90–96.
202. Lundstrom, K.; Schweitzer, C.; Rotmann, D.; Hermann, D.; Schneider, E. M.; Ehrengruber, M. U. *FEBS Lett.* **2001**, *504*, 99–103.
203. Sjoberg, E. M.; Suomalainen, M.; Garoff, H. *Biotechnology* **1994**, *12*, 1127–1131.
204. Henrich, B.; Lubitz, W.; Plapp, R. *Mol. Gen. Genet.* **1982**, *185*, 493–497.
205. Goff, S.; Goldberg, A. L. *J. Biol. Chem.* **1987**, *262*, 4508–4515.
206. Charles, I. G.; Chubb, A.; Grill, R.; Clare, J.; Lowe, P. N.; Holmes, L. S.; Page, M.; Keeling, J. G.; Moncada, S.; Riveros-Moreno, V. *Biochem. Biophys. Res. Commun.* **1993**, *191*, 1481–1489.
207. Chrunky, B. A.; Evans, J.; Lillquist, J.; Young, E.; Wetzel, R. *J. Biol. Chem.* **1993**, *268*, 18053–18061.
208. Maurizi, M. R. *J. Biol. Chem.* **1987**, *262*, 2696–2703.
209. Sengupta, K.; Klammt, C.; Bernhard, F.; Ruterjans, H. Incorporation of Fluorescence Labels into Cell-Free Produced Proteins. In *Cell-Free Protein Expression*; Swartz, J. R., Ed.; Springer Verlag: Berlin, Germany, 2003, pp 81–88.
210. Ohuchi, S.; Nakano, H.; Yamane, T. *Nucleic Acids Res.* **1998**, *26*, 4339–4346.
211. De Vries, J. K.; Zubay, G. *Proc. Natl. Acad. Sci. USA* **1967**, *57*, 1010–1012.
212. Lederman, M.; Zubay, G. *Biochim. Biophys. Acta.* **1967**, *149*, 253–258.
213. Kim, D. M.; Swartz, J. R. *Biotechnol. Bioeng.* **1999**, *66*, 180–188.
214. Kim, D. M.; Swartz, J. R. *Biotechnol. Prog.* **2000**, *16*, 385–390.
215. Kim, D. M.; Swartz, J. R. *Biotechnol. Lett.* **2000**, *22*, 1537–1542.
216. Kim, D. M.; Swartz, J. R. *Biotechnol. Bioeng.* **2001**, *74*, 309–316.
217. Jewett, M. C.; Swartz, J. R. *Biotechnol. Prog.* **2004**, *20*, 102–109.
218. Jewett, M. C.; Swartz, J. R. *Biotechnol. Bioeng.* **2004**, *86*, 19–26.
219. Jewett, M. C.; Swartz, J. R. *Biotechnol. Bioeng.* **2004**, *87*, 465–472.
220. Spirin, A. S.; Baranov, V. I.; Ryabova, L. A.; Ovodov, S. Y.; Alakhov, Y. B. *Science* **1988**, *242*, 1162–1164.
221. Kigawa, T.; Yokoyama, S. *J. Biochem.* **1991**, *110*, 166–168.
222. Spirin, A. S. Cell-Free Protein Synthesis Bioreactor. In *Frontiers in Bioprocessing II*; Todd, P., Sikdar, S. K., Beer, M., Eds.; American Chemical Society: Washington, DC, 1991, pp 31–43.
223. Volyanik, E. V.; Dalley, A.; McKay, I. A.; Leigh, I.; Williams, N. S.; Bustin, S. A. *Anal. Biochem.* **1993**, *214*, 289–294.
224. Baranov, V. I.; Spirin, A. S. *Methods Enzymol.* **1993**, *217*, 123–142.
225. Ryabova, L. A.; Ortleb, S. A.; Baranov, V. I. *Nucleic Acids Res.* **1989**, *17*, 4412.
226. Baranov, V. I.; Morozov, I. Y.; Ortlepp, S. A.; Spirin, A., S. *Gene* **1989**, *84*, 463–466.
227. Endo, Y.; Otsuzuki, S.; Ito, K.; Miura, K. *J. Biotechnol.* **1992**, *25*, 221–230.
228. Kudlicki, W.; Kramer, G.; Hardesty, B. *Anal. Biochem.* **1992**, *206*, 389–393.

229. Ryabova, L. A.; Volyanik, E.; Kurnasov, P.; Spirin, A. S.; Wu, Y.; Kramer, F. R. *J. Biol. Chem.* **1994**, *269*, 1501–1505.
230. Kim, D. M.; Choi, C. Y. *Biotechnol. Prog.* **1996**, *12*, 645–649.
231. Kitaoka, Y.; Nishimura, N.; Niwano, M. *J. Biotechnol.* **1996**, *48*, 1–8.
232. Kolosov, M. I.; Kolosova, I. M.; Alokhov, V. Y.; Ovodov, S. Y.; Alokhov, Y. B. *Biotechnol. Appl. Biochem.* **1992**, *16*, 125–133.
233. Alakov, Y. B.; Baranov, V. I.; Ovodov, S. J.; Ryabova, L. A.; Spirin, A. S.; Morozov, I. J. Method of Preparing Polypeptides in Cell-Free Translation System. U.S. Patent 5,478,730, 1995.
234. Kigawa, T.; Yabuki, T.; Yoshida, Y.; Tsutsui, M.; Ito, Y.; Shibata, T.; Yokoyama, S. *FEBS Lett.* **1999**, *442*, 15–19.
235. Madin, K.; Sawasaki, T.; Ogasawara, T.; Endo, Y. *Proc. Natl. Acad. Sci. USA* **2000**, *97*, 559–564.
236. Klammt, C.; Lohr, F.; Schafer, B.; Haase, W.; Dotsch, V.; Ruterjans, H.; Glaubitz, C.; Bernhard, F. *Eur. J. Biochem.* **2004**, *271*, 568–580.
237. Torizawa, T.; Shimizu, M.; Taoka, M.; Miyano, H.; Kainosho, M. *J. Biomol. NMR* **2004**, *30*, 311–325.
238. Ishihara, G.; Goto, M.; Saeki, M.; Ito, K.; Hori, T.; Kigawa, T.; Shirouzu, M.; Yokoyama, S. *Protein Expr. Purif.* **2005**, *41*, 27–37.
239. Nirenberg, N. W.; Matthaei, J. H. *Proc. Natl. Acad. Sci. USA* **1961**, *47*, 1588–1602.
240. Ryabova, L. A.; Vinokurov, L. M.; Shekhovtsova, E. A.; Alakhov, Y. B.; Spirin, A. S. *Anal. Biochem.* **1995**, *226*, 184–186.
241. Kigawa, T.; Yabuki, T.; Matsuda, N.; Matsuda, T.; Nakajima, R.; Tanaka, A.; Yokoyama, S. *J. Struct. Funct. Genomics* **2004**, *5*, 63–68.
242. Zubay, G. *Annu. Rev. Genet.* **1973**, 7, 267–287.
243. Gold, L. M.; Schweiger, M. *Proc. Natl. Acad. Sci. USA* **1969**, *62*, 892–898.
244. Gold, L. M.; Schweiger, M. *Methods Enzymol.* **1971**, *20*, 537–542.
245. Kim, D. M.; Kigawa, T.; Choi, C. Y.; Yokoyama, S. *Eur. J. Biochem.* **1996**, *239*, 881–886.
246. Shimizu, Y.; Inoue, A.; Tomari, Y.; Suzuki, T.; Yokogawa, T.; Nishikawa, K.; Ueda, T. *Nat. Biotechnol.* **2001**, *19*, 751–755.
247. Johnston, F. B.; Stern, H. *Nature* **1957**, *179*, 160–161.
248. Sawasaki, T.; Ogasawara, T.; Morishita, R.; Endo, Y. *Proc. Natl. Acad. Sci. USA* **2002**, *99*, 14652–14657.
249. Roberts, B. E.; Paterson, B. M. *Proc. Natl. Acad. Sci. USA* **1973**, *70*, 2330–2334.
250. Anderson, C. W.; Straus, J. W.; Dudock, B. S. *Methods Enzymol.* **1983**, *101*, 635–644.
251. Kawarasaki, Y.; Nakano, H.; Yamane, T. *Biosci. Biotechnol. Biochem.* **1994**, *58*, 1911–1913.
252. Tulin, E. E.; Ken-Ichi, T.; Shin-Ichiro, E. *Biotechnol. Bioeng.* **1995**, *45*, 511–516.
253. Pelham, H. R. B.; Jackson, R. J. *Eur. J. Biochem.* **1976**, *67*, 247–256.
254. Katzen, F.; Chang, G.; Kudlicki, W. *Trends Biotechnol.* **2005**, *23*, 150–156.
255. Woodward, W. R.; Ivey, J. L.; Herbert, E. *Methods Enzymol.* **1974**, *30*, 724–731.
256. Betton, J. M. *Curr. Protein Pept. Sci.* **2003**, *4*, 73–80.
257. Spirin, A. S. *Trends Biotechnol.* **2004**, *22*, 538–545.
258. Bannwarth, M.; Schulz, G. E. *Biochim. Biophys. Acta* **2003**, *1610*, 37–45.
259. Tate, C. G. *FEBS Lett.* **2001**, *504*, 94–98.
260. Berrier, C.; Park, K. H.; Abes, S.; Bibonne, A.; Betton, J. M.; Ghazi, A. *Biochemistry* **2004**, *43*, 12585–12591.
261. Elbaz, Y.; Steiner-Mordoch, S.; Danieli, T.; Schuldiner, S. *Proc. Natl. Acad. Sci. USA* **2004**, *101*, 1519–1524.
262. Shimada, Y.; Wang, Z. Y.; Mochizuki, Y.; Kobayashi, M.; Nozawa, T. *Biosci. Biotechnol. Biochem.* **2004**, *68*, 1942–1948.
263. Ryabova, L. A.; Desplancq, D.; Spirin, A. S.; Pluckthun, A. *Nat. Biotechnol.* **1997**, *15*, 79–84.
264. Kawasaki, T.; Gouda, M. D.; Sawasaki, T.; Takai, K.; Endo, Y. *Eur. J. Biochem.* **2003**, *270*, 4780–4786.
265. Yin, G.; Swartz, J. R. *Biotechnol. Bioeng.* **2004**, *86*, 188–195.

Biographies

From left to right: Heinz Rüterjans, Christian Klammt, Frank Bernhard

Heinz Rüterjans studied chemistry at the University of Muenster, Germany. Since 1979 he has been full professor of biophysical chemistry at the J. W. Goethe University of Frankfurt, Germany. From 1981 to 1982 and from 1992 to 1993 he was Dean of the Faculty of Biochemistry at the University of Frankfurt. In 1995, he became head of the European Large Scale Facility for Biomolecular NMR at the University of Frankfurt. In the same year, he was awarded the Humboldt Prize of the French Government. In 2002, he established the Centre for Biomolecular Magnetic Resonance at the University of Frankfurt. He further acted as president of the European Biophysics Societies Association from 2001 to 2003. Heinz Rüterjans was leading scientist of a variety of research and technology development (RTD) projects of the European Union. His research interests cover the structural and functional analysis of biological macromolecules by high-resolution NMR spectroscopy, the dynamics of protein structures, protein/nucleic acid and protein/lipid interactions, and various issues of the cellular metabolism.

Christian Klammt studied biochemistry at the University of Frankfurt, Germany. Since 2002 he has worked on the structural analysis of integral membrane proteins at the Institute of Biophysical Chemistry at the University of Frankfurt. Christian Klammt is associated with the Centre for Membrane Proteomics in Frankfurt and an expert for the preparative scale cell-free expression of proteins. This new techniques, based on cell-free extracts, for the high-level production of difficult membrane proteins like transporters or G-protein coupled receptors have led to his recently award from the Federation of European Biochemical Societies for the best scientific contribution by a young scientist in the year 2004.

Frank Bernhard studied biology and chemistry at the University of Heidelberg, Germany. During his PhD studies at the Max-Planck Institute for Medical Research in Heidelberg and at Columbus State University in Ohio, USA he analyzed signal transduction mechanisms of bacterial pathogens. After finishing his PhD in 1991 he worked as a group leader at the Technical University in Berlin, Germany on the characterization and genetic engineering of peptide antibiotic synthetases. In 1995 he became laboratory leader at the Institute of Crystallography of the Free University in Berlin where he focused on the structural analysis of proteins. Since 2000 he has been leader of the protein expression unit at the Institute of Biophysical Chemistry at the University of Frankfurt, Germany. Frank Bernhard has more than ten years of experience in the large-scale expression of diverse varieties of proteins for structural research. In recent years, he has concentrated on strategies for efficient labeling of proteins for structural analysis by NMR techniques and for ligand-binding studies.

Comprehensive Medicinal Chemistry II
ISBN (set): 0-08-044513-6

ISBN (Volume 3) 0-08-044516-0; pp. 107–128

3.07 Chemical Biology

R Flaumenhaft, Harvard Medical School, Boston, MA, USA

3.07.1	**Chemical Biology**	**129**
3.07.1.1	High-Throughput Screening in Chemical Biology	130
3.07.1.2	Forward and Reverse Chemical Genetics	130
3.07.2	**Small Molecule Libraries in Chemical Biology**	**132**
3.07.2.1	Compound Collections	132
3.07.2.1.1	Commercially available collections	132
3.07.2.1.2	Collections from federal sources	132
3.07.2.2	Natural Products	133
3.07.2.3	Combinatorial Libraries	133
3.07.2.3.1	Diversity-oriented libraries	133
3.07.2.3.2	Privileged structures	133
3.07.3	**Phenotype-Oriented Screens**	**135**
3.07.3.1	Goals and Scope	135
3.07.3.2	Model Organisms	135
3.07.3.2.1	Cell-based assays	136
3.07.3.2.2	Assays using multicellular organisms	137
3.07.3.3	Detection Methods for Phenotypic Screens	138
3.07.3.3.1	Assays using a uniform readout	138
3.07.3.3.2	High-content screening	138
3.07.3.4	Target Identification	138
3.07.3.4.1	Affinity-based approaches	138
3.07.3.4.2	Genetic approaches	139
3.07.3.5	Specificity	140
3.07.4	**Target-Oriented Screens**	**141**
3.07.4.1	Goals and Scope	141
3.07.4.2	Assays of Direct Ligand Binding	141
3.07.4.2.1	Affinity selection	141
3.07.4.2.2	Small molecule microarrays	142
3.07.4.3	Assays of Protein–Protein Interactions	142
3.07.4.4	Assays of Enzymatic Activity	142
3.07.5	**Profiling and Computation Analyses in Chemical Biology**	**144**
3.07.5.1	Genetic Profiling of Small Molecule Activity	144
3.07.5.2	Cytologic Profiling of Small Molecule Activity	144
3.07.6	**Chemical Biology and Drug Discovery**	**146**
References		**147**

3.07.1 Chemical Biology

This chapter will focus on chemical biology in the framework of target discovery. In this context, chemical biology relies on small molecules to probe complex biological systems. Novel drug targets can be identified by this approach. This discipline has also been referred to as chemical genetics or chemical genomics. While many of the high-throughput approaches in chemical genetics were derived from techniques developed in the biopharmaceutical industry, there is increased interest in and resources dedicated to the use of these platforms in academia. A federal mandate to enhance the translation of knowledge gained by basic research into novel therapeutics has led to initiatives that support the use of small molecules to study disease-oriented biology in the public sector. The Molecular Libraries Roadmap, a component of the US National Institutes of Health (NIH) New Pathways to Discovery, includes initiatives to develop novel high-throughput screening assays and libraries. This chapter will address emerging concepts, technologies, and experimental strategies using small molecules to study proteins, cells, and multicellular organisms.

3.07.1.1 High-Throughput Screening in Chemical Biology

The screening of compounds in biological assays using high-throughput platforms has been performed for more than 40 years. Historically, the goal of these undertakings has been to identify drug candidates with potential for commercial development. A newer development has been the use of high-throughput screening of small molecules with the goal of identifying novel chemical probes to study complex cellular processes. Interrogation of basic biological mechanisms using high-throughput screening can lead to the discovery of drugs with clinical significance. The emphasis of screening in this context, however, is the identification of new probes that will enable selective inhibition or augmentation of biologically important macromolecules.

The term chemical genetics is based on the conceptual similarity of this approach to classic genetics.[1–3] The strategy of deleting, overexpressing, or modifying specific gene products in cellular and animal models has been among the most widely used and successful strategies in molecular biology. Highly selective small molecules that alter protein function can be used in an analogous manner. Such chemical probes have several advantages. The use of these small molecules in both cells and living animals is exceedingly convenient. If the compound is soluble and cell-permeable, it is simply added to the biological system of interest without any manipulations such as transfection of cells or genetic engineering of mice. Small molecules cause conditional alterations of protein function, analogous to conditional mutations, since the investigator determines when they are added to the system. Thus, proteins involved in development can be studied in normal adult organisms. In addition, effects of these probes are titratable, enabling correlations between dose and function. Small molecules can target lipids, carbohydrates, ions, and other biologically important molecules not amenable to direct genetic manipulation. An additional advantage of these small molecules is that they can be used as lead compounds for the development of therapeutics.

Several relatively recent advances from various fields have converged to create the opportunity to analyze complex biological systems using small molecules in high-throughput assays. Relatively large libraries of small molecules containing drug-like compounds have become available for use by individual investigators. Such libraries include commercial collections as well as collections accrued and maintained by the NIH. In addition, advances in combinatorial chemistry have led to the creation of small molecule libraries with enhanced diversity and complexity for probing biological systems.[4,5] The adaptation of commonly used laboratory instruments to high-throughput formats has enabled screening platforms capable of assaying large numbers of compounds for informative phenotypes.[6] In addition, software has been developed that simplifies the analysis of the volumes of data derived from high-throughput assays. These advances have contributed to unprecedented interest and activity in the field of chemical biology.

The development of high-throughput screening facilities in academic centers has hastened the emergence of chemical genetics in academia. Examples of such facilities include the Skaggs Institute at The Scripps Research Institute established in 1996 and the Institute for Chemistry and Cell Biology at Harvard Medical School established in 1998. Since the establishment of these institutes, dozens of high-throughput screening facilities have been created at academic centers throughout the world. A list of academic institutions with screening facilities is available on the website of The Society for Biomolecular Screening.[101] These facilities are open to the academic community. Within these centers, investigators from diverse fields have been able to modify novel assays into high-throughput formats. An advantage of this arrangement is the large number and variety of biological assays that can be tested at such facilities. A second advantage of these facilities is the promotion of direct interaction of synthetic chemists who are pioneering new methods of small compound library development with investigators who have developed the biological assays. Such interactions have the potential to accelerate the development of synthetic strategies that lead to the generation of small molecules with enhanced biologically activity. The initial success of these centers has prompted planning for the development of similar facilities at other academic institutions.

3.07.1.2 Forward and Reverse Chemical Genetics

Forward genetics (from phenotype to gene) is a strategy in which complex biological systems are studied by randomly mutating genes, screening for informative phenotypes, and then identifying the gene mutation responsible for the phenotype of interest (**Figure 1**).[2] By analogy, the term forward chemical genetics has been used to describe a strategy

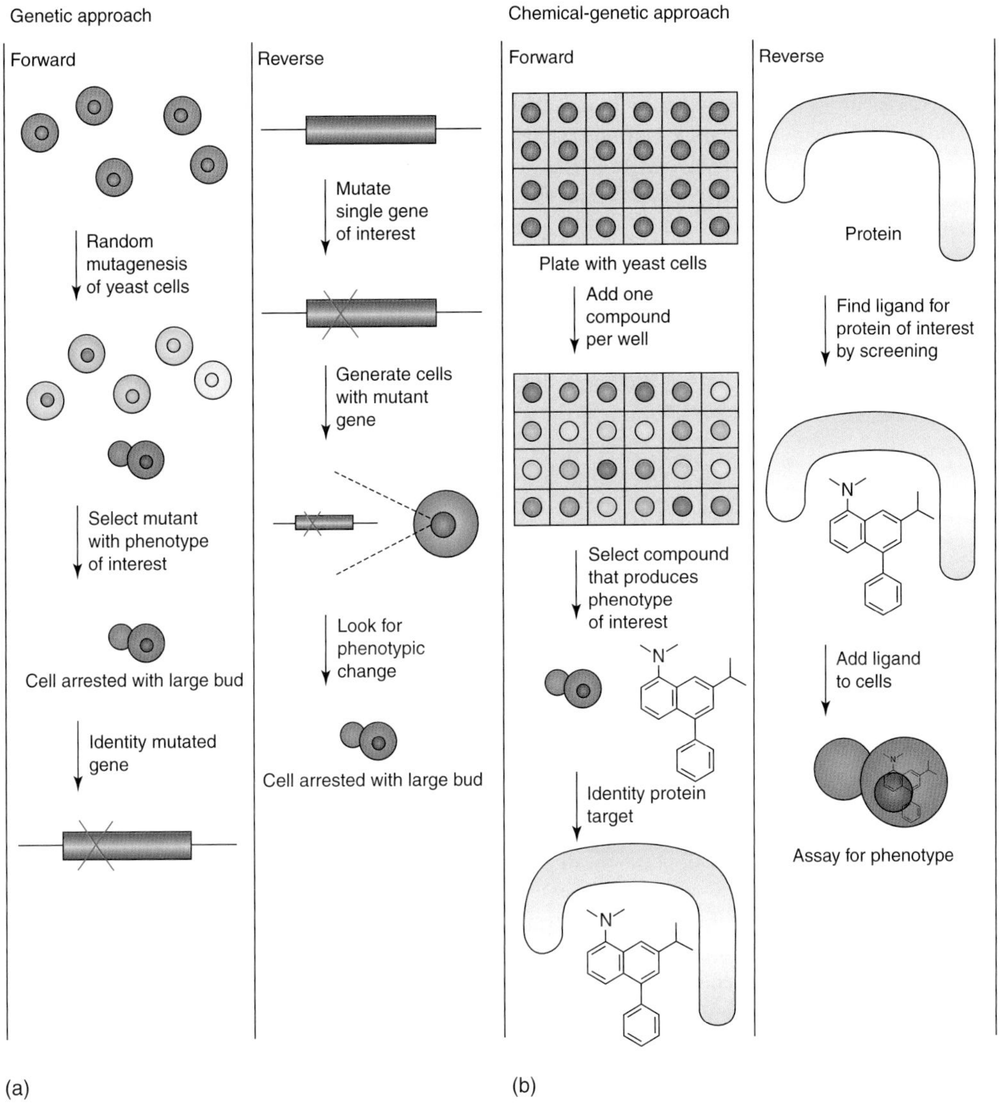

Figure 1 Genetic and chemical-genetic approaches identify genes and proteins, respectively, that regulate biological processes. (a) Forward genetics entails introducing random mutations into cells, screening mutant cells for a phenotype of interest, and identifying mutated genes in affected cells. In the example shown, yeast cells are randomly mutated, cells showing a large-bud phenotype are selected, and genes mutated in these cells are identified. Reverse genetics entails introducing a mutation into a specific gene of interest and studying the phenotypic consequences of the mutation in a cellular or organismal context. In the example shown, a single mutated gene is introduced into yeast cells and a large-bud phenotype is observed. (b) Forward chemical-genetics entails screening exogenous ligands in cells, selecting a ligand that induces a phenotype of interest, and identifying the protein target of this ligand. In the example shown, one compound that induces a large-bud phenotype is selected and the protein target of this ligand is subsequently identified. Reverse chemical-genetics entails overexpressing a protein of interest, screening for a ligand for the protein, and using the ligand to determine the phenotypic consequences of altering the function of this protein in a cellular context. In the example shown, a ligand for a specific protein is found to induce a large-bud phenotype. (Reproduced with permission from Stockwell, B. R. *Nat. Rev. Gen.* **2000**, *1*, 116–125. Nature © Macmillan Publishers Ltd.)

in which small molecules are used much like mutations are used in classical genetics to perturb gene function.[2] Small molecules that impart a desired phenotype in the biological system of interest are selected by high-throughput screening. Following the identification of active compounds, a variety of strategies are used to determine the target of the small molecule. The specificity of the small molecule is also evaluated. Once the target of a small molecule is identified and its specificity proven, the molecule serves as a versatile probe that can cause conditional perturbations of diverse biological systems. The incentive to achieve widely applicable chemical genetic strategies for analyzing varied complex biological systems is great. Yet, there are disadvantages to this approach. Systematic strategies for identifying the targets of active compounds have not become routine. Once the target is identified, it is difficult to prove that the compound acts specifically on that target. Methods for evaluating the molecular targets and specificity of compounds identified by high-throughput screening continue to be developed.

Reverse genetics (from gene to phenotype) refers to the manipulation of a known gene, creation of a cell or organism harboring the genetic mutation or deletion, and thorough phenotyping of the altered cell or organism (**Figure 1**).[2] This strategy is used to determine the role of a specific gene in the physiology of a cell or organism. By analogy, reverse chemical genetics is a process whereby the selectivity of a small molecule for a specific molecular target is established, the small molecule is introduced into a complex biological system such as an organism or cell, and the alteration in phenotype is determined.[2] An important component of reverse chemical genetics is thorough phenotyping. New approaches to profiling biological systems (such as transcriptional profiling and cytologic profiling) provide a means of global phenotyping. These phenotyping strategies aid in the assessment of compound specificity, an important consideration in reverse chemical genetics.

3.07.2 Small Molecule Libraries in Chemical Biology

Availability of a large quantity of high-quality small molecules is essential for a successful chemical biology program. Access to quality compounds outside of the vast proprietary small molecule libraries owned by large pharmaceutical companies has increased substantially over the past decade. Use of compound collections from both commercial and federal sources is becoming more widespread.

3.07.2.1 Compound Collections

3.07.2.1.1 Commercially available collections

A large number of commercial suppliers offer compound libraries. These companies include specialty suppliers who focus on small molecule accrual and synthesis as well as divisions of large companies that offer a wide range of drug discovery products and services. Several pertinent differences exist between suppliers. The source of compounds may be from a combinatorial synthetic program maintained by the company. Alternatively, some companies offer 'handcrafted' compounds collected from synthetic chemists around the world. Large numbers of natural extracts are available commercially and, recently, specialty companies offering purified, characterized compounds derived from natural extracts have emerged. When using a commercial library for high-throughput screening, it is important to know the availability of follow-up compounds that will be needed for detailed evaluation of small molecules selected by the primary screen. Many companies have significant stocks of compounds and offer routine supplies of essentially any compound in the library at reasonably low costs. Other suppliers have programs for custom synthesis of hits identified in initial screening. Obtaining compounds produced by custom synthesis is substantially more expensive and time consuming. The ability to obtain analogs of active compounds to determine structure–activity relationships of active compounds also differs between companies. Some commercial libraries consist of representative compounds selected from larger libraries. Stocks of analogs are often available. Other companies offer lead optimization programs and computational services. The availability of such follow-up services brings to individual investigators some of the chemistry support previously available only at large pharmaceutical companies. The quality and degree of support by individual suppliers is highly variable, however, and should be considered by investigators before screening a commercial library.

3.07.2.1.2 Collections from federal sources

The NIH has several small molecule libraries, a number of which are available to individual investigators. The Molecular Libraries Screening Center Network was established by the NIH to encourage and facilitate high-throughput screening in academia. The NIH Chemical Genomics Center (NCGC) is part of this effort. In addition to providing screening facilities, the NCGC plans to acquire and maintain a collection of approximately 500 000 to 1 million compounds. Small grants to enable investigators to screen specific assays at core sites will be available on a competitive basis. The National Cancer Institute (NCI) maintains several repositories including synthetic and natural compounds. The National Institute of Neurological Disorders and Stroke (NINDS) maintains a 1040 compound library including FDA-approved drugs and natural products that is available for screening. The NCI and NINDS libraries have been used by academic screening centers across the US.

3.07.2.2 Natural Products

Several characteristics of natural products render them well suited for chemical genetic screens. Natural products demonstrate substantial steric complexity. Compared to synthetic compounds, natural products are more likely to contain a larger number of rings and chiral centers.[7] This steric complexity is consistent with the fact that these compounds are synthesized by enzymes that are inherently three-dimensional and chiral.[7] In addition, evolutionary pressures have selected for compounds that interact with targets that are three-dimensional, complex, and biologically relevant. Diversity of chemical structure enhances the likelihood of finding active compounds when assaying complex biological systems that contain multiple potential targets. Despite these considerations that favor natural products and their strong history of yielding drug candidates, natural products have several disadvantages. Once a desired activity is found in a natural extract, the active component of the extract must be purified and the target of the extract identified. Purification and target identification programs can be time-consuming, difficult, and expensive. There has been a trend toward generating natural product libraries that are partially or fully purified rather than or in addition to offering complex extracts.[8–10] Use of purified, partially characterized natural product collections has the appeal of reducing the time-consuming and expensive steps of isolating the active agent from crude extracts.

The Natural Products Branch of the NCI maintains a collection of over 600 000 extracts from microbes, plants, and marine organisms collected from around the world. Many of these extracts are available to academic centers for testing in novel assays. Commercial suppliers also offer natural products. Several vendors have inherited natural product libraries from defunct natural product programs at large pharmaceutical companies. Other suppliers have in house purification programs and offer partially or fully purified products. An innovative method for generation of new natural products is genetic reengineering of the organisms that produce the products.[11–13] Engineered natural products may represent a new source of compounds for chemical genetic screens.

3.07.2.3 Combinatorial Libraries

The increasing demand for diverse small molecules to test in biological assays has motivated synthetic chemists to pioneer new strategies for generating complex compounds via combinatorial chemistry. Combinatorial chemistry enables synthesis of large numbers of chemically defined entities. Compounds can be synthesized by either solution phase or solid phase synthesis using reproducible synthetic pathways. An increasing number of academic institutions have begun combinatorial programs. The logic that more diverse libraries will have a greater chance of yielding hits when screened in a wide range of unrelated biological assays has prompted the advent of diversity-oriented synthesis.

3.07.2.3.1 Diversity-oriented libraries

Diversity-oriented synthesis (DOS) is directed at high-throughput synthesis of small molecules that are diverse and complex in order to perturb biological pathways.[4] Solid-phase rather than solution-phase synthesis is typically used. Both parallel and split-and-mix strategies have been employed. One goal of DOS is to produce compounds that are chemically complex. Compounds produced by DOS are rich in stereochemistry and chiral functional groups (**Figure 2**).[4,14] A second goal of DOS is to produce compounds that are chemically diverse. These compounds are used as chemical probes for perturbing and ultimately understanding cellular processes. As such, they are generally not subject to target-biased methods used in target-oriented synthesis. Instead, the goal of DOS is to generate a collection of compounds that is diverse so as to enhance the likelihood of generating compounds that interact with relevant macromolecules within the biological system being tested.[4] The most commonly used strategy to generate diversity is to attach a combination of building blocks to a single core chemical skeleton. A novel approach has been the use of appendages, termed sigma elements, added to skeletal substrates. These elements encode predictable transformations to create products that cover a substantially greater amount of chemical space than their unmodified substrates.[15,16] The use of libraries that include diversity in stereochemistry, building blocks, and skeletons may ultimately enable systematic study of the role of three-dimensional chemical space in generating effective inhibitors of biological systems.

3.07.2.3.2 Privileged structures

Another strategy for generating combinatorial libraries with interesting biological activities is to use privileged structures for scaffolds. Privileged structures are chemical motifs capable of interacting with a variety of unrelated molecular targets. Identifying and basing library design around a core privileged structure can enhance the likelihood of generating compounds with useful biological activity. For example, Nicolaou and co-workers have created a 10 000-membered

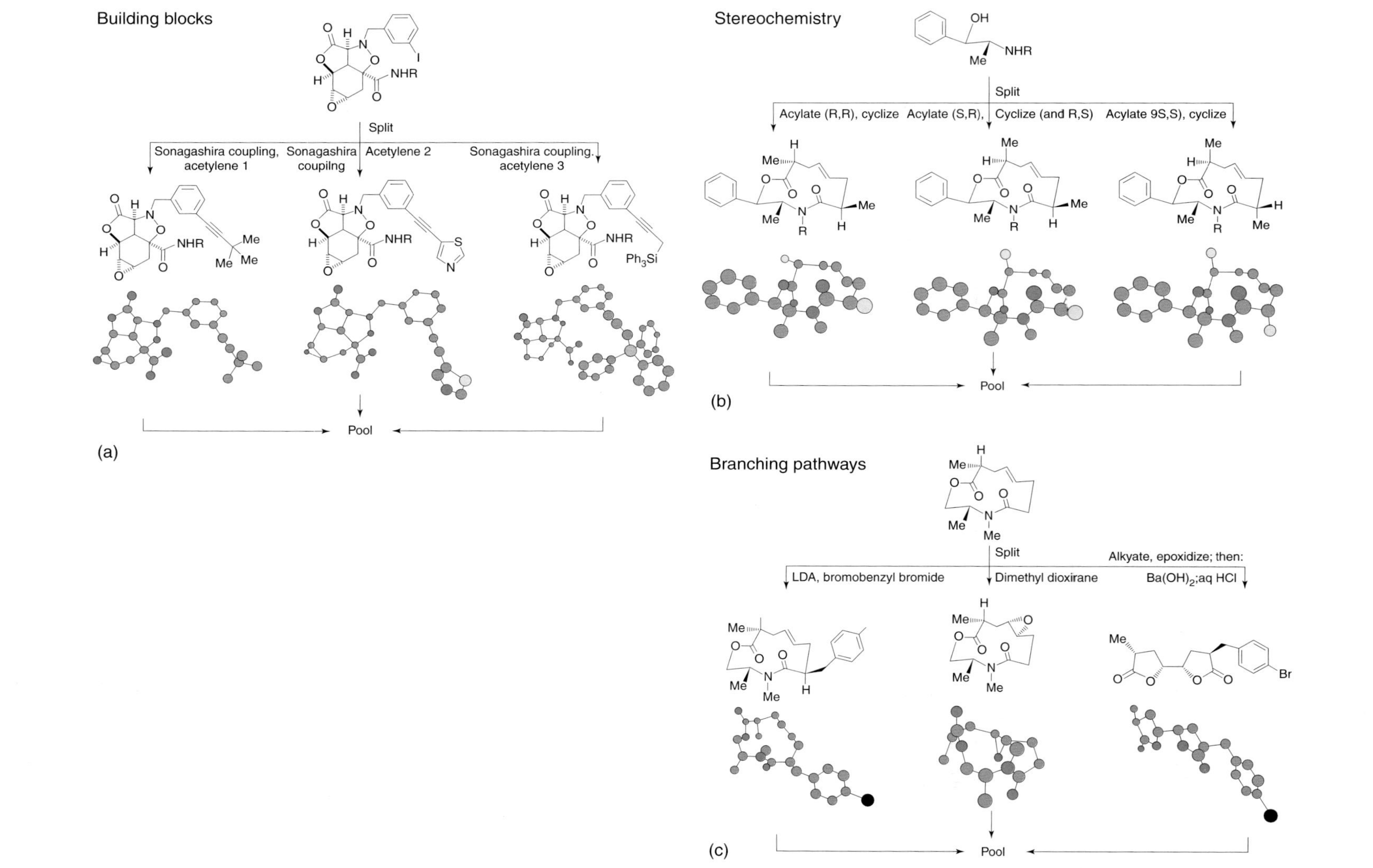

Figure 2 Strategies to increase the structural diversity of products in split-pool syntheses. (a) Alter building blocks[97]; (b) alter stereochemistry[98] (note methyl groups indicated in yellow in ball-and-stick models); and (c) use branching reaction pathways that produce diverse arrays of skeletal atoms (scaffolds), upon which building blocks can be attached.[98] In (a) and (b), indicated or related compounds have been attached to solid supports through their R groups; compounds in (c) have not been attached to solid supports. (Reprinted with permission from Schreiber, S. L. *Science* **2000**, *287*, 1964. Copyright (2000) AAAS.)

natural product-like library based on the 2,2-dimethylbenzopyran template.[17] This library included one compound, termed fexaramine, that bound the farnesoid X receptor 100-fold more tightly than its natural ligand, chenodeoxycholic acid.[18] This example demonstrates the importance of considering biological activity as well as structural attributes when selecting a skeleton for diversity-oriented synthesis.

3.07.3 Phenotype-Oriented Screens

3.07.3.1 Goals and Scope

The objectives of a phenotype-oriented screen are to identify small molecules that impart a desired phenotype in a cell or multicellular organism, identify the molecular target of active compounds, and establish the selectivity of the active compounds. Identification of active small molecules is achieved using high-throughput screening. Assays are considered to be high-throughput when hundreds to thousands of compounds are screened per day. In chemical biology, high-throughput screening often involves adapting a previously established biological assay to a high-throughput format. Required modifications typically include miniaturization of the assay and substitution of rapid, automated procedures for manual manipulations.

A success of modern chemical biology has been the variety of biological assays that have been modified for high-throughput screening. As discussed below, biological systems varying from yeast to fresh human blood cells to zebrafish have been assayed in high-throughput formats. The complexity of the desired phenotype can vary, but often involves the activation (gain-of-signal) or inhibition (loss-of-signal) of a particular function. The nature of the readout (e.g., fluorescence, luminescence, morphology) is limited primarily by the availability of instruments that can perform in a high-throughput setting. The key to successful high-throughput screening is to develop an assay that has a favorable ratio of signal-to-noise and low variance. Variance in particular can be introduced into an assay when converting it into a high-throughput format because of errors resulting from liquid handling equipment. A phenotypic assay should have a signal-to-noise ratio that is several-fold greater than its variance. Once compounds are identified by high-throughput screening, they are subsequently tested in secondary assays to confirm their activity. Secondary assays are typically performed in a low-throughput format. Compounds that are confirmed to have activity in secondary assays are then tested for potency. Potency is typically established by determining the EC_{50} or IC_{50} of the compound in the assay of interest. Small molecules that are active in the low micromolar range or less are considered for further evaluation.

Once potent compounds are identified that impart a desired phenotype in a biological assay, the target of the compound must be determined. If the active compounds are from a complex extract, the active agents of the extract must be purified and identified. More typically, active compounds are from a library of known small molecules. In this case, one can perform database searches to determine whether an active compound has previously been tested in biological assays. Several chemical structure databases and structure-based search engines are available. Structure-based search engines allow investigators to assess analogs of the compound as well as the compound itself. One can determine whether the compound or its analogs have previously been determined to have activity in biological assays. If a candidate target is identified, the ability of the compound to affect the target in an in vitro assay is assessed. Some purified proteins are available from commercial suppliers. In addition, a number of companies offer in vitro assay services. These services will test the compound of interest in assays using purified proteins. Of course, frequently one will not be able to identify the target of an active compound using these relatively straightforward methods. Since the goal of a chemical genetic project is to identify novel probes for biological assays, active compounds with no previously known biological activities are of special interest. For these compounds target identification can become a significant project. Proving that the compound is specific for a proposed target is similarly challenging. New strategies that have been developed to address these challenges in target identification and evaluating specificity are discussed below (*see* Sections 3.07.3.4 and 3.07.3.5).

3.07.3.2 Model Organisms

Any cell or multicellular organism amenable to high-throughput analysis can be studied using forward chemical genetics. Most screens have used cell-based assays. Cultured mammalian cells, microbes, and human cells derived directly from donors have all been used. Multicellular organisms such as zebrafish, fruit flies, and worms have been studied. The breadth of assays amenable to screening and the inventiveness of investigators in adapting assays to a high-throughput format have been strengths of the chemical genetic approach.

3.07.3.2.1 Cell-based assays

Chemical genetics has been used to study cell division in cultured cells. One of the early successes of chemical genetics was the discovery of monastrol as an inhibitor of mitotic spindle bipolarity. Monastrol was selected from a 16 320 compound library that was screened for inhibitors of phosphorylation of nucleolin, a protein that is phosphorylated in cells entering mitosis.[19] Of the 139 initial hits, compounds that directly affected tubulin polymerization were selected against. The effect of the 86 remaining compounds on microtubule, actin, and chromatin distribution in BS-C-1 African green monkey cells was assessed by fluorescence microscopy. One compound elicited a monoastral microtubular array surrounded by a ring of chromosomes and was termed monastrol (**Figure 3**).[19] Monastrol was subsequently found to inhibit Eg5, a mitotic kinesin responsible for assembly and maintenance of the mitotic spindle.[19] Since its initial identification and characterization in BS-C-1 cells, monastrol has been used in laboratories around the world to study spindle assembly in neuronal cells,[20,21] cancer cell lines,[22,23] *Xenopus laevis* egg extract spindles,[24] mouse oocytes,[25] and others. Compounds with greater efficacy have been developed.[22] Eg5 may be a clinically important target for cancer therapy. Eg5 inhibitors are now in clinical trials as antineoplastic agents.[26]

Chemical genetic analyses have also been used to study cell differentiation. A pluripotent mouse cell line termed P19 that resembles embryonic stem (ES) cells was used to identify molecules that affect differentiation. The primary assay involved screening of P19 cells transfected with a luciferase reporter that was constructed using the regulatory region of neuronal Tα1 tubulin, an early neuronal specific marker.[27] Active compounds identified in the primary screen were tested in secondary assays for the ability to induce multiple other neuronal markers. A compound termed TWS119 induced normal neuronal morphology and differentiation markers. The target of TWS119 was subsequently found to be glycogen synthase kinase-3β (GSK-3β).[27] GSK-3β is a serine/threonine kinase that has previously been shown to function in embryogenesis and cell fate determination.[28] This same group has also used progenitor cells to identify compounds that promote differentiation into cardiac myocytes and osteoblasts.[29,30] A compound that may cause dedifferentiation of myoblasts has also been identified (**Figure 4**).[30,31] These examples demonstrate the potency of chemical genetics in identifying small molecules that affect complex cellular phenotypes.

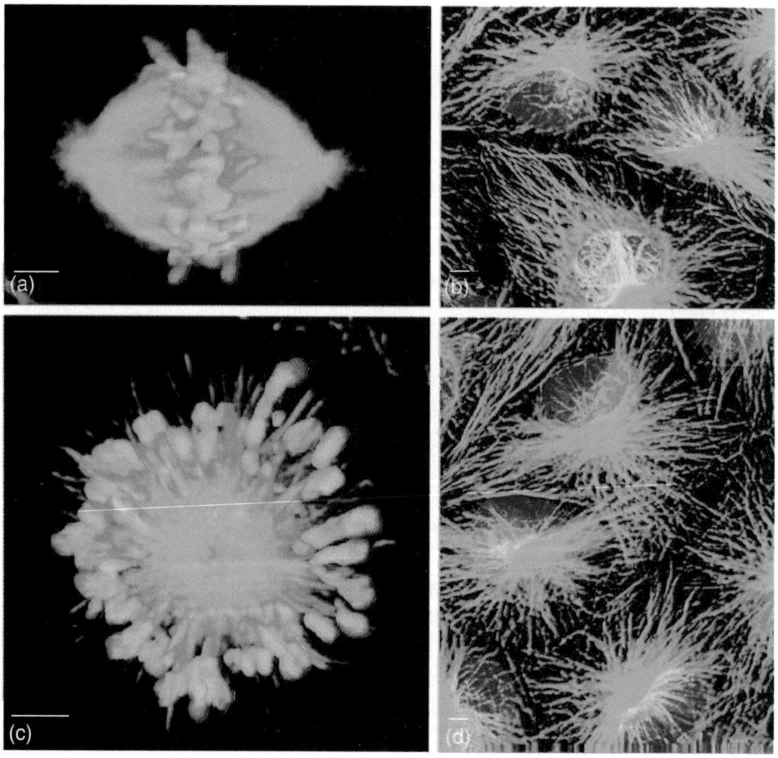

Figure 3 Monastrol causes monoastral spindles in mitotic cells. Immunofluorescence staining (α alpha-tubulin (green), chromatin (blue)) of BS-C-1 cells treated for 4 h with 0.4% DMSO (control) (a and b) or 68 μM monastrol (c and d). No difference in distribution of microtubules and chromatin in interphase cells was observed (b and d). Monastrol treatment of mitotic cells replaces the normal bipolar spindle (a) with a rosette-like microtubule array surrounded by chromosomes (c). Scale = 5 μm. (Reprinted with permission from Mayer, T. U.; Kapoor, T. M.; Haggarty, S. J.; King, R. W.; Schreiber, S. L.; Mitchison, T. J. *Science* **1999**, *286*, 971–974. Copyright (1999) AAAS.)

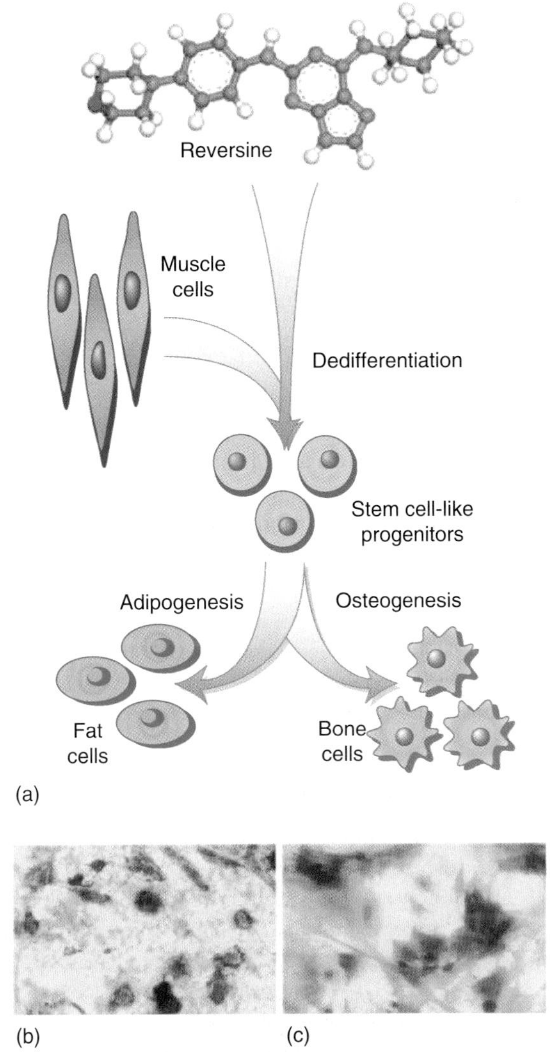

Figure 4 Reversine, a 2,6-disubstituted purine, dedifferentiates lineage-committed myoblasts to multipotent mesenchymal progenitor cells. (a) Dedifferentiation scheme. (b,c) Dedifferentiated C2C12 myoblasts can differentiate into adipocytes under adipogenesis-inducing conditions (b) or into osteoblasts under osteogenesis-inducing conditions (c). (Reproduced with permission from Ding, S.; Schultz, P. G. *Nat. Biotechnol.* **2004**, *22*, 833. © Nature Publishing Group. http://www.nature.com/)

Chemical genetic screens can also be performed on freshly obtained human material. For example, high-throughput screening has been performed on blood platelets. Platelets are particularly well suited for a chemical genetic approach since they are anucleate and therefore not amenable to traditional genetic manipulation.[32] Platelets used for this assay were isolated from human donors. Compounds were screened to identify inhibitors of platelet activation. This screen led to identification of eight novel antiplatelet agents. One of these was found to specifically inhibit phosphodiesterase 3A with an IC_{50} of 15 nM.[33] This compound, termed JF959602, was subsequently tested in a murine model of arterial thrombus formation. The compound was found to potently inhibit thrombus formation initiated by laser-induced endothelial cell injury.[33] This example demonstrates the potential for ex vivo high-throughput screening using human material for identification of biologically active small molecules with therapeutic potential.

3.07.3.2.2 Assays using multicellular organisms

High-throughput screening with small molecules has also been used to study multicellular organisms. Zebrafish development is an example of the use of small molecules to study vertebrate biology.[34] An advantage of this approach is that the small molecule can be applied at any time during development and for a duration determined by the investigator. For example, a compound that inhibits zebrafish otolith development only when incubated with embryos between 14 and 26 h postfertilization was identified.[34] This result showed that the compound acted on a developmentally regulated gene product. In a second zebrafish model, mutant zebrafish were used to identify small molecules capable of reversing phenotypic abnormalities.[35] The zebrafish mutant *gridlock* exhibits an aortic abnormality that mimics aortic coarctation. Fluorescence microangiography was used to screen a 5000-compound library for small molecules that reversed the defect. The screen identified two small molecules that suppressed the *gridlock* phenotype. Both compounds were subsequently found to enhance vascular endothelial growth factor (VEGF) expression. In addition, overexpression of VEGF in *gridlock* zebrafish suppressed the aortic abnormality.[35] These results raise the possibility that chemical genetics can be used in model organisms to study complex phenotypes and identify novel therapeutic strategies for inherited diseases.

3.07.3.3 Detection Methods for Phenotypic Screens

3.07.3.3.1 Assays using a uniform readout

The majority of phenotypic screens in chemical biology use a plate reader to quantify signals from individual wells of an assay plate. The term uniform readout is used to distinguish these assays from high-content screens (described below). Detection methods for such assays include fluorescence, luminescence, and light absorbance (e.g., by a chromophore), autoradiography, fluorescence resonance energy transfer (FRET), surface plasmon resonance (SPR) spectroscopy, and scintillation proximity assay (SPA). Absorbance can be used to assess growth of microbes such as bacteria or yeast. Colorimetric assays are popular for enzyme studies. Cytoblot assays performed in a high-throughput format enable immunochemiluminescence-based detection of DNA synthesis, posttranslational modifications, or neo-antigen expression.[36] Reporter systems using transcriptional activation of a GFP-linked protein are easily amenable to high-throughput formats. These methods are similar in concept to their low-throughput counterparts but have been modified for automation and optimized for detection in high-density arrays. The use of techniques that use homogeneous readouts has been reviewed elsewhere.[37–40]

3.07.3.3.2 High-content screening

High-content screening involves the use of simultaneous recording of multiple parameters in cell-based assays. Automated microscopy is the most frequently used approach to high-content screening. Information such as fluorescence intensity and subcellular localization is readily attained. Several detection methods have been applied in automated microscopy. Cells can be stained with fluorescent probes such as those that detect DNA or actin. In addition, immunologic detection of diverse antigens can be achieved in permeabilized, fixed cells. These methods have been used in high-throughput screening assays to study cell division, cytokinesis, mitotic spindle formation, and centrosome duplication as well as cell migration.[41] Ninety-six-well and 384-well formats have been used. For most of these screens, the microscopic images were scored by visual inspection. An experienced investigator can analyze images from a 384-well plate in approximately 10 min. More recently, automated analyses of microscopic images from high-throughput screens have been performed.[42,43] For example, tracking of GFP-coupled endogenous protein can be accomplished using automated microscopy.[44,45] Algorithms for analyzing images continue to improve and software for detecting boundaries, tracking cells over time, and scoring morphology are becoming increasingly sophisticated.

3.07.3.4 Target Identification

The ability to alter the phenotype of living cells and organisms with a synthetic chemical is a versatile and powerful tool for the biologist. The utility of this strategy for dissecting complex biological systems, however, requires that the chemical has a known molecular target and that it acts selectively on that target. Identifying the target of active small molecules remains a substantial challenge in chemical biology. Approaches based on affinity and genetics have been used.

3.07.3.4.1 Affinity-based approaches

3.07.3.4.1.1 Affinity chromatography

The oldest approach to target identification by affinity is chromatography. The small molecule is immobilized to a suitable solid-phase support via chemical linkage and a lysate of cells is exposed to the immobilized compound. The target that binds the immobilized compound can then be eluted and identified. Although conceptually simple, this strategy has several practical limitations. The majority of compounds identified by high-throughput phenotypic screens are hydrophobic.[46] These compounds will interact nonspecifically with high-abundance proteins in cell lysates. Furthermore, the binding affinity of most small molecules for their targets is in the low micromolar range. These relatively low binding constants are frequently inadequate to enable purification of compounds by chromatography.[46] Chromatography is a better option if a compound binds in the low- to mid-nanomolar range. For example, GSK-3β was identified as the target of TWS119, the compound that induces P19 stem cells to adopt a neuronal phenotype, using affinity chromatography.[27] TWS119 was subsequently found to bind GSK-3β with a K_D of 126 nM. Compounds that bind with such high affinity are not the norm, however. An additional factor in affinity approaches is the concentration of the target in cell lysates.[46] A chromatography approach has a higher likelihood of success if the target protein is relatively abundant. Information about binding constants and target abundance is typically not available a priori. Thus, the utility of affinity chromatography for a particular compound – target pair is difficult to anticipate.

Identifying reactive groups on small molecules for cross-linking represents a second category of concerns in using affinity purification procedures. Appropriate reactive groups may not be available to perform the chemistry required to covalently attach the compound to a support. Even if appropriate reactive groups are available, these groups may be critical to the activity of the compound. Thus, their disruption will inhibit the activity of the compound. Structure–activity relationships can be used to determine which groups are required for activity and can help to direct the linkage strategy. Chemical linkage strategies, however, will not be available for every compound. Limitations of standard affinity chromatography have compelled investigators to develop alternative strategies for target identification.

3.07.3.4.1.2 Tagged libraries

The tagged library approach uses a library of small molecules containing a chemical tag to enable subsequent linkage to the solid-phase support. This strategy circumvents the difficulty of identifying reactive groups for cross-linking to supports. For example, Khersonsky and co-workers synthesized a 1536-compound library based on a triazine scaffold in which each compound contained a triethyleneglycol linker.[47] A compound that inhibited brain/eye morphogenesis in zebrafish was identified from this library. The active compound was immobilized on agarose beads via its linker and used to isolate a protein-binding partner in zebrafish extracts. This strategy is broadly applicable to combinatorial libraries.

3.07.3.4.1.3 Protein microarrays

A newer affinity approach to target identification is the use of protein microarrays. Protein microarrays (or protein chips) are slides onto which purified proteins are printed at high density.[48] A variety of attachment strategies, both covalent and noncovalent, have been used.[49] An advantage of a protein array is that the compound of interest is exposed to relatively purified proteins that are at a high local concentration. This strategy circumvents the problem inherent in lysates in which abundant proteins can inhibit the interaction of the small molecule with a target protein that is in low abundance. A significant challenge in the development of protein microarrays is the generation of proteins for printing. Protein expression collections have been generated by cloning cDNA libraries from human tissue into expression vectors that are transformed into *E. coli*.[50] The yeast proteome (5800 unique proteins representing approximately 80% of proteins in yeast) has been expressed and printed onto glass slides.[51] The slides are chemically derivatized to enable covalent attachment of purified proteins derived from cloned genes. Transfected cell microarrays represent another method to synthesize and localize proteins in an array.[52] To detect binding of a small molecule to a target protein on the protein microarray, the small molecule is typically covalently attached to a carrier that will facilitate visualization of the small molecule. For example, the small molecule can be attached to fluorescently labeled albumin. Alternatively, the small molecule can be attached to biotin. Fluorescently labeled streptavidin can then be used to identify proteins to which the small molecule is bound. The slide is then scanned and the fluorescence associated with each protein is quantified. The identity of each protein is determined by its position on the slide. Protein microarrays have been constructed by several individual laboratories in academia and discovery programs in the biotechnology sector. Such microarrays are increasingly available through commercial sources.

3.07.3.4.2 **Genetic approaches**

3.07.3.4.2.1 Drug-induced haploinsufficiency

One genetic approach that has been used to determine targets of bioactive small molecules is the generation and characterization of drug-resistant or drug-sensitive mutants. Such genetic approaches to target identification have been well developed in yeast. The Saccharomyces Genome Deletion Project has generated single gene deletions for nearly 6000 genes in *Saccharomyces cerevisiae*.[53] Mutant strains are tagged with DNA sequences, termed barcodes, that uniquely identify each mutation. For studies of drug-induced haploinsufficiency, heterozygous deletion mutants are grown under competitive conditions in the presence of drug. Growth differences between heterozygotes and wild-type strains are determined. The deletion in mutants that are either resistant to or sensitive to drug are then identified (**Figure 5**).[54] Multiple studies demonstrate the utility of drug-induced haploinsufficiency in identifying the targets of small molecules.[54–57]

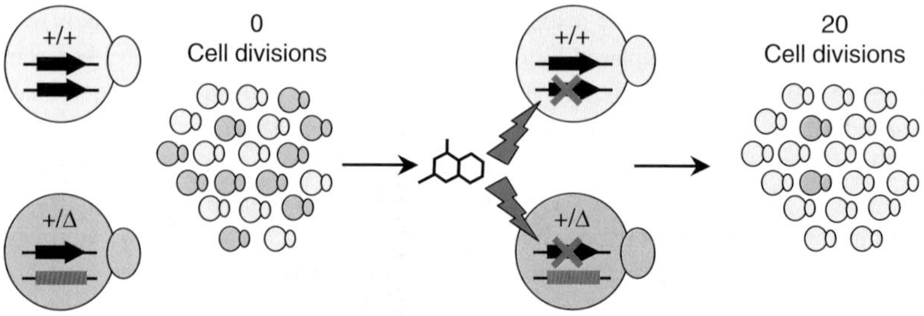

Figure 5 Drug-induced haploinsufficiency. Reduction of target gene copy number can sensitize diploid cells to a drug. Competitive growth experiments can exploit the drug-dependent growth differences between heterozygotes and wild-type cells. (Reproduced with permission from Armour, C. D.; Lum, P. Y. *Curr. Opin. Chem. Biol.* **2005**, *9*, 20. © Elsevier.)

3.07.3.4.2.2 Three-hybrid reporter gene system

Mammalian targets can be screened using a three-hybrid reporter gene approach. This system is based on the observation that cross-linking two small molecules can permit dimerization or oligomerization of two separate target proteins.[58] The three-hybrid system uses a small molecule (ligand A) capable of avidly binding a known receptor that is engineered to include a DNA binding domain.[58] Ligand A is covalently attached to the test compound (ligand B). cDNA from mammalian cells are expressed so that the gene products include a transactivating domain. The binding of ligand B to a receptor – transactivation domain hybrid enables the transcription of a reporter gene.[59] This technique has been used to find targets of kinase inhibitors involved in cell cycle control.[60] For example, several novel targets of Purvalanol B were identified using the three-hybrid reporter system. These targets were subsequently confirmed using affinity approaches and enzymatic assays. This technique could be applied to identify protein targets of compounds of unknown function.

3.07.3.5 Specificity

A major disadvantage of using chemical probes in cells and multicellular organisms is that their specificity is difficult to establish. The ability of a compound to inhibit the function of a specific protein does not indicate that the compound fails to inhibit other proteins. The interpretation of most biological studies using chemical probes relies on its specificity. A small molecule that inhibits the activity of multiple targets will not yield reliable information in complex biological systems. Indeed, the application of small molecules in biological systems has been tarnished historically by inappropriate use of probes that are not specific. Despite several advantages of small molecules over genetic manipulation in terms of convenience and versatility, small molecules have not yet generally approached the specificity of genetic manipulations.

Approaches to assessing the specificity of small molecules have been used. One method to assess the specificity of a compound is to test its activity against purified proteins other than the proposed targets. Many scientific papers have been published demonstrating that small molecules commonly used to inhibit a specific protein also inhibit the function of unrelated proteins.[61–63] For example, two papers evaluating the specificity of commonly used protein kinase inhibitors tested 42 compounds and found that only a handful were specific for their established targets.[61,62] In some instances, the compounds were more potent inhibitors of proteins unrelated to the established target.[61,62] A shortcoming of this approach is that one cannot test all potential targets. In addition, results from in vitro testing rely heavily on assay conditions and may not accurately reflect the in vivo situation. These considerations have compelled researchers to seek alternative strategies.

One strategy that has been used to assess compound specificity is the targetless mutant. In this strategy, a mutant organism such as yeast that lacks the proposed target of a compound is exposed to the compound and the activity of the compound on the mutant yeast tested. If the compound is specific, it should not affect the phenotype of the targetless mutant. If the compound does have an effect on a mutant that lacks the proposed target, then it affects additional targets. This analysis can definitively demonstrate lack of specificity. One must be careful about using this approach to prove specificity. For example, the finding that a compound has no effects in a targetless mutant yeast does not ensure that the compound does not inhibit other targets in mammals. A second consideration in using a targetless mutant is that the mutant must be thoroughly phenotyped to demonstrate that the effect of the drug is identical to that of the mutation. The availability of global analysis such as expression profiling using DNA microarrays enables global phenotyping. Marton and co-workers used yeast mutants to evaluate the specificity of FK506 for calcineurin.[64]

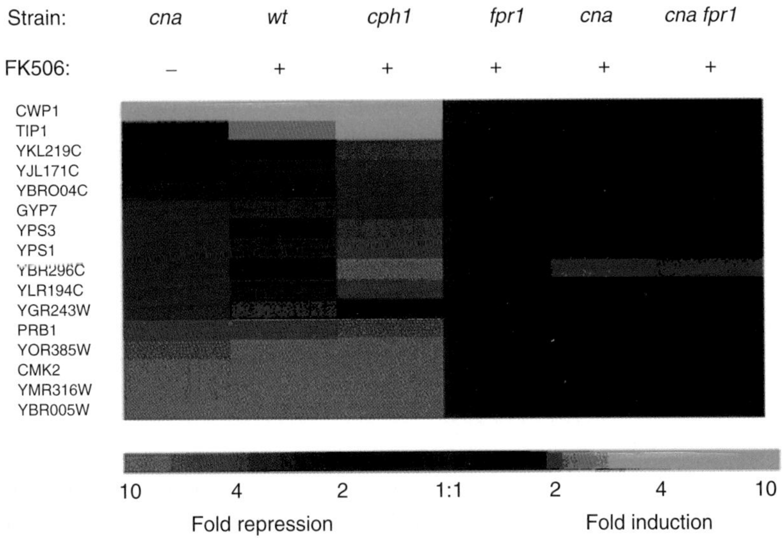

Figure 6 Response of FK506 signature genes in strains with deletions in different genes. Genes with expression ratios greater than a factor of 1.8 in response to treatment with 1 $\mu g\,mL^{-1}$ FK506 are listed (left-hand side) and their expression ratios in the indicated strain are shown on the green (induction)–red (repression) color scale. Calcineurin (*cna*) mutant and FK506 treatment signature genes are in the first two columns. Almost all FK506 signature genes have expression ratios near unity in deletion strains involved in pathways affected by FK506 (*calcineurin*, *fpr1* and *cna fpr1* mutants) but not in deletion strains in unrelated pathways (*cph1*). (Reproduced with permission from Marton, M. J.; DeRisi, J. L.; Bennett, H. A.; Iyer, V. R.; Meyer, M. R.; Roberts, C. J.; Stoughton, R.; Burchard, J.; Slade, D.; Dai, H. *et al. Nat. Med.* **1998**, *4*, 1293. © Nature Publishing Group. http://www.nature.com/)

They found that mutant yeast that lacked calcineurin (*cna* mutants) demonstrated a similar expression pattern to yeast treated with 1 $\mu g\,mL^{-1}$ FK506 (**Figure 6**).[64] This concentration of FK506 elicited little change in expression pattern in calcineurin-deficient yeast. Although developed in yeast, analysis of drug specificity using targetless mutants could also be performed in mammals such as knockout mice.

3.07.4 Target-Oriented Screens

3.07.4.1 Goals and Scope

The goal of target-oriented screens is to identify tools that modify biologically important molecules. In this manner, target-oriented screens are a component of reverse chemical genetics in which the target of the compound is known and the effect of the compound on the phenotype is analyzed. Target-oriented screens involve high-throughput analysis of purified targets. Proteins are the most common target molecules, but lipids, carbohydrates, nucleic acids, or specific ions may also be targets. Small molecules are selected on the basis of their ability to bind the target or perturb its functional activity. The biological function of the target is then assessed by introducing the compound into a complex biological system and carefully evaluating its effects. Unlike phenotype-oriented screens, target-oriented screens cannot identify novel targets. Compounds with novel and unanticipated activities, such as the ability to inhibit only a subset of the downstream functions of a particular protein, have been identified.[65]

3.07.4.2 Assays of Direct Ligand Binding

3.07.4.2.1 Affinity selection

One method to discover small molecules that perturb protein function is to identify compounds that directly bind to the protein of interest. Essentially any protein–ligand interaction is amenable to this type of target-oriented screen provided a purified protein is available. A commonly used strategy is to incubate the protein with mixtures of compounds and then deconvolute the mixtures that demonstrate binding in order to identify the active compound.[66] Deconvolution can be achieved by iterative resynthesis and screening of progressively smaller subpools.[67] This approach, however, is laborious and time consuming. Fractionation of active mixtures by reverse phase, size exclusion, or ion exchange chromatography followed by characterization of active fractions can also be used. Affinity selection is a more efficient method for identification of protein ligands with useful biological characteristics. Selection can be performed using immobilized protein.[68] Alternatively, compounds from a library can be bound to purified protein in a solution-phase and subsequently separated from unbound compounds by size exclusion chromatography. Mass spectroscopy can then be used to identify the bound compound.[69] This generic method of affinity selection requires no label and can be widely applied to evaluate interactions of proteins with the majority of relevant molecular entities.

3.07.4.2.2 Small molecule microarrays

An alternative affinity-based method to identify small molecule ligands for a particular protein is to use an immobilized compound library. Small molecules can be immobilized in high-density spotting arrays. An increasing number of chemistries and photo-labeling techniques have been applied to covalently attach small molecules to surfaces. Attachment can be achieved either by including a tag on the small molecule that binds an immobilized capturing agent or by covalent attachment to a functional group on the surface.[70] Alternative methods of immobilization have been developed (**Figure 7**).[71] A labeled target protein can then be incubated with the array, washed, and detected using an automated array scanner. Potential detection methods include colorimetric, fluorescent, autoradiography, luminescence, chemiluminescence, FRET, SPR spectroscopy, and scintillation proximity assays.[70]

3.07.4.3 Assays of Protein–Protein Interactions

Assays that select compounds on the basis of their ability to directly bind a protein may modulate biological systems by inhibiting protein–protein interactions. Many chemical genetic screens have been designed to select compounds on the basis of their ability to interfere with specific protein–protein interactions. For such assays, the binding of the ligand to protein is not monitored directly. Rather, the ability of the ligand to disrupt protein–protein interactions is monitored. Although protein–protein interactions are the most common molecular interactions evaluated using this strategy, other molecular interactions (such as protein–DNA, protein–lipid, and protein–carbohydrate interactions) can be evaluated.

Fluorescence polarization is a popular technique to evaluate protein–protein interactions. The technique is based on the fact that the molecular spin of a molecule in solution is proportional to its size. Smaller molecules spin more rapidly. In using fluorescence polarization to evaluate protein–protein interactions, a fluorescent probe (a labeled protein) is excited with plane-polarized light. Emission from the probe is measured in the presence and absence of a binding partner. Differences in polarization will occur depending on whether the probe is bound to the binding partner (tumbling slowly) or free in solution (tumbling rapidly). A small molecule that acts as a competitive inhibitor prevents the decrease in rate of spin that occurs upon interaction of the binding partner. Fluorescence polarization was used to identify inhibitors that block the interaction between the Bak BH3 domain and the anti-apoptotic protein, Bcl-x_L.[72] Fluorescence polarization has also been used to identify inhibitors of protein–DNA interactions. For example, the interaction of Tn5 transposase with its recognition sequence has been targeted using this approach.[73]

3.07.4.4 Assays of Enzymatic Activity

Many successful chemical genetic assays have relied on enzymatic activity rather than direct binding to screen for biologically active small molecules. Evaluation of enzymatic activity by high-throughput screening generally requires a method to measure loss of substrate or generation of product. Several strategies have been used. Assays using chromogenic or fluorogenic substrates or sensors are generally amenable to a high-throughput format. Such probes have been used for decades to identify inhibitors of enzymes and have been important in chemical genetic screens. Radiolabeled compounds have been used in high-throughput assays for enzyme systems that are not easily evaluated using chromogenic or fluorogenic probes. For example, radiolabeled fatty acids have been used in a high-throughput format to identify novel inhibitors of fatty acid chain elongation.[74] ELISA-based screens using antibodies that recognize specific protein modifications have been used to detect phosphorylation,[36] acetylation,[36] methylation,[75] and others. These techniques are well established and conceptually analogous to their low-throughput counterparts.

Alternatives to these standard approaches are increasingly being used within the chemical biology community. Fluorescence polarization is popular in this regard. Such studies rely on the generation of a fluorescent probe that retains the activity of the substrate. Atomic resolution structural information of the substrate–enzyme interaction facilitates the placement of the fluorophore so as not to interfere with enzyme binding.[76,77] Another strategy that has been adapted to high-throughput screens of enzyme function is scintillation proximity. Scintillation proximity assays use a scintillant that is incorporated into a bead. The scintillant will only emit light when a radiolabel is in close proximity (i.e., bound to the bead). The emitted light can be detected using a scintillation counter. The typical strategy is to bind an enzyme to a bead and use a radiolabeled substrate.[78] Scintillation proximity is a widely applicable

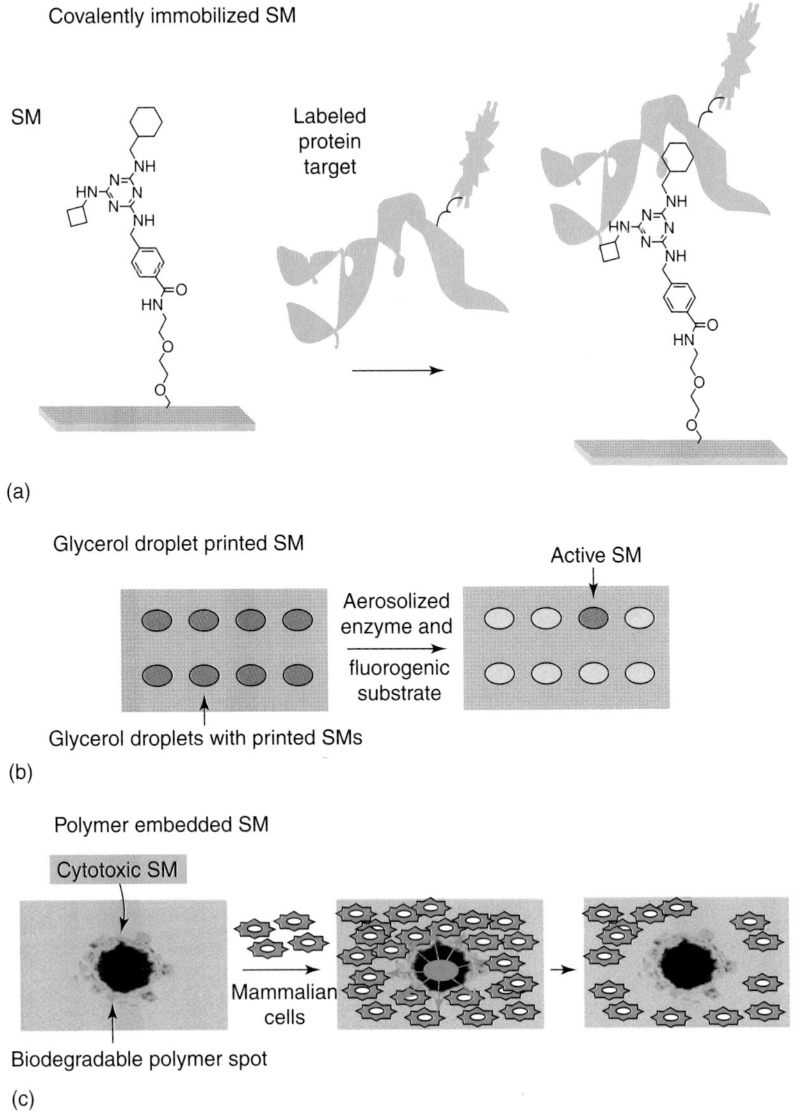

Figure 7 Methods for small molecule (SM) immobilization and target presentation. (a) Covalent attachment of small molecules to the microarray surface. The target, either in the form of a labeled purified protein or as mixture of proteins in a cell lysate, can be probed for small molecule microarray binding. The unbound material is washed away and active small molecules on the arrays are detected by the presence of the label. (b) Nanoliter droplets of glycerol can be used to dissolve and spatially separate the small molecules.[99] The method has been used to probe enzymatic reactions and identify enzyme inhibitors. The enzyme and a specific fluorogenic substrate are aerosolized over the array slide containing the glycerol-immobilized small molecules. Droplets that contain an inhibitor of the enzymatic reaction will prevent the cleavage of the substrate and consequent release of the fluorophor. (c) A biodegradable polymer can be used to embed and print the small molecules on arrays.[100] The polymer solution is printed at 1 nM/drop and dries to form a 200-μm diameter spot in a doughnut-like shape. Picomole amounts of each small molecule are encased into the polymer and then covered with a layer of fibronectin to allow for the attachment of cells onto the slide. The small molecule diffuses radially from the polymer and affects cells growing within several hundred micrometers of the spot. (Reproduced with permission from Chiosis, G.; Brodsky, J. L. *Trends Biotechnol.* **2005**, *23*, 271. © Elsevier.)

technique for target-oriented screens. Technologic advances in mass spectroscopy have enabled the development of high-throughput mass spectroscopy amenable to target-oriented screens. Desorption/ionization on silicon mass spectroscopy (DIOS-MS) has been used in assays to monitor enzyme-catalyzed reactions. Unlike matrix-assisted laser desorption/ionization (MALDI), DIOS-MS uses UV laser light to desorb intact analytes from a silicon surface in the absence of matrix assistance.[79,80] This technology offers a generic approach to evaluate enzyme activity in a high-throughput screen.

3.07.5 Profiling and Computation Analyses in Chemical Biology

An evolving trend in chemical biology is the use of computational analyses to study global effects of small molecules on biological systems. These approaches frequently involve profiling multiple parameters and presenting the data in such a manner as to represent the comprehensive response of the system to the small molecule. These data-rich analyses provide substantially more information regarding drug mechanism and specificity than single parameter approaches. Small molecule profiling can thus be useful for anticipating small molecule effects on complex systems, identifying small molecule targets, assessing small molecule specificity, and anticipating toxicity.

3.07.5.1 Genetic Profiling of Small Molecule Activity

The most widely used profiling technique is genetic profiling. Whole-genome expression profiling using DNA microarrays has enabled global analysis of the transcription response of cells to environmental perturbations. Yeast has served as a robust model for profiling studies. One approach has been to generate a compendium of transcriptional profiles from well-characterized deletion mutants or yeast exposed to drugs with known cellular targets. Each transcriptional profile serves as a fingerprint to help identify pathways that are affected by the mutations or known drugs. The transcriptional profile induced by the small molecule with unknown activity is obtained and compared to the compendium. If exposure to the uncharacterized drug demonstrates a profile similar to that of a known deletion mutant, then its target may be the deleted protein (**Figure 8**).[81] Similarly, uncharacterized drugs that elicit the same expression profile as a drug with an established target may have the same target.[82] The compendium approach has also been used to study the effects of small molecules on pathogens such as *Mycobacterium tuberculosis* and *Bacillus subtilis*.[83,84] Transcriptional profiling can predict responsiveness of cultured cells to small molecules. For example, statistical analyses of transcriptional data have been used to evaluate responses of neuronal cell cultures to psychoactive drugs.[85] Computational analysis of transcription profiles has also been used to correlate genetic profile with chemosensitivity of malignant cells. These studies have been used to identify genes that render cancer cells susceptible to particular small molecules.[86,87] The results imply that transcription profiling could be used to determine the mechanism of action of compounds with no previously described mechanism.

3.07.5.2 Cytologic Profiling of Small Molecule Activity

Simultaneous profiling of a large number of transcriptional events instead of monitoring only a few at a time has led to a revolution in the way that transcriptional activation is studied. This technology has offered new methods to evaluate

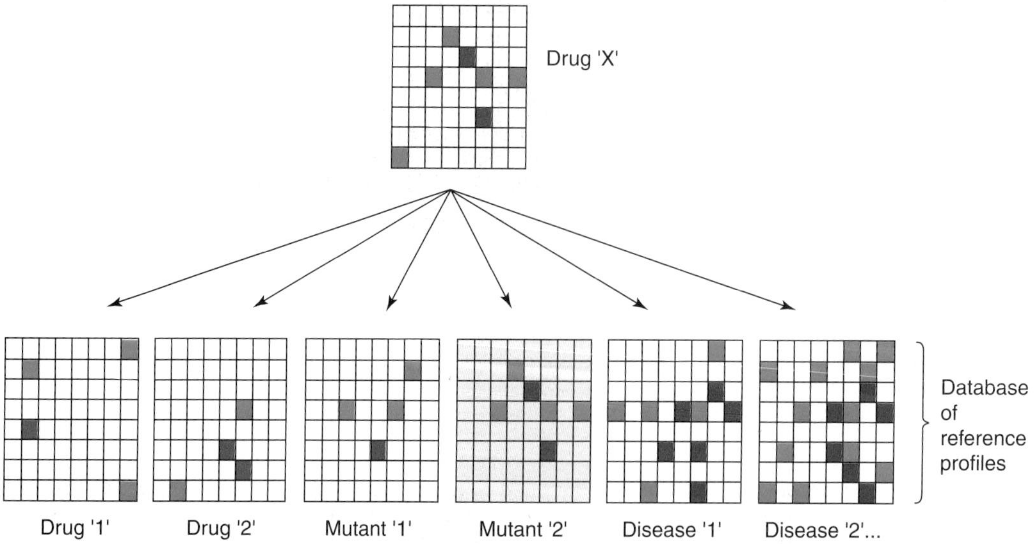

Figure 8 Schematic illustration of how pattern matching an expression profile to a database of reference profiles is used to characterize drug effects, mutations, and diseases. Boxes represent genes; green and red boxes represent induction or repression, respectively. A matching pattern (bold arrow; yellow highlight) infers similarity of samples in other respects. (Reproduced with permission from Hughes, T. R.; Shoemaker, D. D. *Curr. Opin. Chem. Biol.* **2005**, *5*, 21. © Elsevier.)

drug effects on organisms. Success in this field has prompted investigators to consider other parameters that can be evaluated by global profiling. Such consideration has led to the use of automated microscopy and imaging software to perform cytologic profiling of cells (**Figure 9**).[42,88] This strategy enables global analysis of phenotypic responses of cells to small molecules. The approach involves simultaneous evaluation of multiple phenotypic parameters representing a

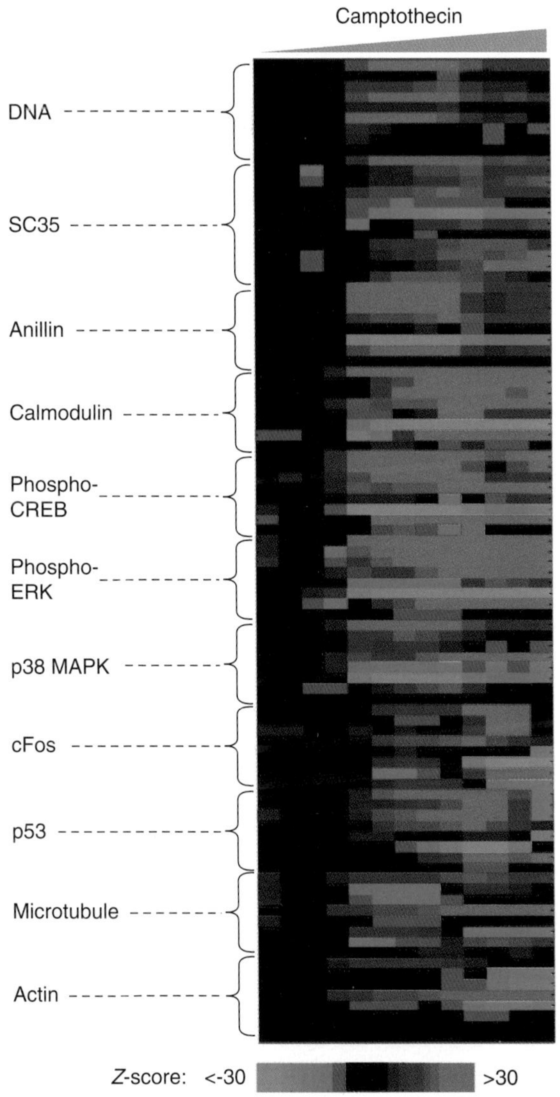

Figure 9 Dose – response profile for camptothecin from a cytological-profiling experiment. This heat plot profiles the response of HeLa cells to increasing concentrations of the topoisomerase inhibitor camptothecin. On the y-axis are a series of descriptors measured on a cell-by-cell basis. The text describes the macromolecule that the probes bind to. Fluorescent small molecules were used to detect DNA (DAPI) and actin (rhodamine-phalloidin). Antibodies labeled with fluorochromes were used in an indirect immunofluorescence protocol to detect the other proteins. Phospho-x refers to an antibody that binds specifically to a phosphorylated epitope on the protein. On the x-axis are increasing drug concentrations (13 concentrations in a 3x dilution series from 66 pM to 35 μM). The color intensity represents the Z-score from a statistical test comparing descriptor values from cells in a treated well to cells in control wells on the same plate. Red indicates a positive deviation from control values, and green a negative deviation. Black indicates no statistically significant difference from controls. Each pixel in the heat plot represents the average Z-score for single wells from two duplicate experiments. For each well, nine nonoverlapping images were collected, and descriptors measured for ~8000 cells. Each well was stained with DAPI (blue channel) plus two other probes (green and red channels), so a single experiment required five sets of plates to cover all 11 probes. Note that at low drug concentrations most descriptors are not different from controls (Z-score ~0), while at high drug concentration many descriptors are different from controls. The approximate EC_{50} value (potency) of the drug is that at which many descriptors change from like control to different. (Reproduced with permission from Mitchison, T. J. *ChemBioChem* **2005**, *6*, 33. © Wiley-VCH Verlag GmbH & Co.)

broad array of cell physiology. The parameters that can be tested are limited only by the probes that are available for cytologic studies. Cell division, signaling, shape change, membrane and protein trafficking, apoptosis, protein synthesis and degradation, and others can be monitored.

3.07.6 Chemical Biology and Drug Discovery

The multiple examples of studies using high-throughput screening to evaluate biological process and new approaches to analyzing global responses of cells and organisms to small molecules demonstrate the breadth of chemical biology. Many of the techniques and tools of chemical biology are the same as those used in drug discovery. While the two endeavors differ somewhat in emphasis, there is unmistakable overlap between the goal of studying complex biological systems using chemical probes (chemical biology) and the objective of identifying new therapeutic targets and commercial drugs (drug discovery). Compared with the well-established drug discovery programs in industry, chemical biology in academia is a relatively new endeavor. There is evidence, however, that this field will grow rapidly. With the success of the NIH-sponsored human genome project, the NIH has sought new programs in genomic research that will build upon the foundation laid by the human genome project and demonstrate tangible health benefits. One of the grand challenges cited in the vision of the US National Human Genome Research Institute is to develop chemical genomics in order to find therapeutics that modulate the activities of specific gene products.[89] The Molecular Libraries Initiative (MLI) component of the NIH Roadmap for Medical Research has been developed with this vision in mind and will expand the availability of small-molecule chemical probes for basic research in the academic community.[90] An increasing number of individual academic institutions are also dedicating resources and faculty to the development of high-throughput facilities. A survey of the literature shows that an increasing number of new approaches to target identification, novel drug targets, and biologically active drugs are being identified in academia.

These observations raise the question of how academia and industry will interact at the interface of chemical biology and drug discovery. Certainly, there is ample opportunity for collaboration. Leadership at the NIH envisions a scenario in which library development and early stage drug discovery performed in academia will lead to the identification of biological targets and useful chemical probes.[90] Chemical probes so identified will then be disseminated within the academic community and made available to biopharmaceutical companies for lead optimization and further development.[90] This scheme has the potential to offer a new source of low-cost opportunities in early drug development to the biopharmaceutical industry. The degree to which biopharmaceutical companies will be interested in developing drugs from leads identified in academia is difficult to predict. Policies regarding the handling of intellectual property derived from federally funded drug discovery remain an area of debate.[91] Whether NIH-funded initiatives represent competition to organizations involved in drug discovery is a second area of uncertainty. For example, the National Center for Biotechnology Information launched PubChem, a public database for storage of data of small molecular compounds. The database consists of fewer than a million compounds. The American Chemical Society (ACS) has argued that this library will compete with the Chemical Abstracts Service (CAS), a 25-million chemical library that is widely used by the biopharmaceutical industry.[92,93] Such policy issues will need to be resolved for chemical genomics to fulfill its potential as a discipline to study complex biological systems using small molecules and provide leads that can be developed into novel therapeutics. At the same time, chemical genomics will require that within academia there is close collaboration between chemists and biologists. Such collaboration is not the cultural norm at most institutions.

Yet, there are forces in both academia and industry that could foster increased collaboration in the areas of chemical biology and drug discovery. The comprehensive sequencing of the approximately 30 000 genes in the human genome has heightened expectations for improved drug development. There are an estimated 5000–10 000 potential drug targets among disease-relevant gene products.[94] Only about 500 of these are targeted by presently available therapeutics. Such estimations suggest vast opportunity in the area of drug discovery. Indeed, there is ever mounting pressure on the pharmaceutical industry to produce drugs with novel mechanisms. The number of new drug targets, however, has decreased in recent years concomitant with a decrease in the number of new chemical entities entering development.[95] Commercial considerations place constraints on the types of targets considered by biopharmaceutical companies. Gene products involved in diseases that are rare or afflict primarily developing nations are not targeted.[96] In the public sector, the aforementioned pressure on federal funded institutions to bring scientific advances to bear on the health of the populous has fostered interest in pursuing small molecule strategies that have historically been in the purview of industry. Partitioning some of the early, high-risk chemical interrogation of less well-characterized gene products to academia may prove an efficient means to enhance fundamental knowledge of basic biology and to produce leads for the private sector. Experimental strategies and technologies developed in an academic setting may also enhance drug discovery in the biopharmaceutical industry.

References

1. Mitchison, T. J. *Chem. Biol.* **1994**, *1*, 3–6.
2. Stockwell, B. R. *Nat. Rev. Genet.* **2000**, *1*, 116–125.
3. Lokey, R. S. *Curr. Opin. Chem. Biol.* **2003**, 7, 91–96.
4. Schreiber, S. L. *Science* **2000**, *287*, 1964–1969.
5. Khersonsky, S. M.; Chang, Y. T. *Comb. Chem. High Throughput Screen.* **2004**, 7, 645–652.
6. Walling, L. A.; Peters, N. R.; Horn, E. J.; King, R. W. *J. Cell. Biochem.* **2001** (Suppl.), 7–12.
7. Ortholand, J. Y.; Ganesan, A. *Curr. Opin. Chem. Biol.* **2004**, *8*, 271–280.
8. Abel, U.; Koch, C.; Speitling, M.; Hansske, F. G. *Curr. Opin. Chem. Biol.* **2002**, *6*, 453–458.
9. Eldridge, G. R.; Vervoort, H. C.; Lee, C. M.; Cremin, P. A.; Williams, C. T.; Hart, S. M.; Goering, M. G.; O'Neil-Johnson, M.; Zeng, L. *Anal. Chem.* **2002**, *74*, 3963–3971.
10. Bindseil, K. U.; Jakupovic, J.; Wolf, D.; Lavayre, J.; Leboul, J.; van der Pyl, D. *Drug Disc. Today* **2001**, *6*, 840–847.
11. Walsh, C. T. *Science* **2004**, *303*, 1805–1810.
12. Watanabe, K.; Rude, M. A.; Walsh, C. T.; Khosla, C. *Proc. Natl. Acad. Sci. USA* **2003**, *100*, 9774–9778.
13. Patel, K.; Piagentini, M.; Rascher, A.; Tian, Z. Q.; Buchanan, G. O.; Regentin, R.; Hu, Z.; Hutchinson, C. R.; McDaniel, R. *Chem. Biol.* **2004**, *11*, 1625–1633.
14. Arya, P.; Joseph, R.; Gan, Z.; Rakic, B. *Chem. Biol.* **2005**, *12*, 163–180.
15. Burke, M. D.; Berger, E. M.; Schreiber, S. L. *Science* **2003**, *302*, 613–618.
16. Burke, M. D.; Berger, E. M.; Schreiber, S. L. *J. Am. Chem. Soc.* **2004**, *126*, 14095–14104.
17. Nicolaou, K. C.; Evans, R. M.; Roecker, A. J.; Hughes, R.; Downes, M.; Pfefferkorn, J. A. *Org. Biomol. Chem.* **2003**, *1*, 908–920.
18. Downes, M.; Verdecia, M. A.; Roecker, A. J.; Hughes, R.; Hogenesch, J. B.; Kast-Woelbern, H. R.; Bowman, M. E.; Ferrer, J. L.; Anisfeld, A. M.; Edwards, P. A. *Mol. Cell.* **2003**, *11*, 1079–1092.
19. Mayer, T. U.; Kapoor, T. M.; Haggarty, S. J.; King, R. W.; Schreiber, S. L.; Mitchison, T. J. *Science* **1999**, *286*, 971–974.
20. Yoon, S. Y.; Choi, J. E.; Huh, J. W.; Hwang, O.; Lee, H. S.; Hong, H. N.; Kim, D. *Cell Motil Cytoskeleton* **2005**, *60*, 181–190.
21. Haque, S. A.; Hasaka, T. P.; Brooks, A. D.; Lobanov, P. V.; Baas, P. W. *Cell Motil. Cytoskeleton* **2004**, *58*, 10–16.
22. Marcus, A. I.; Peters, U.; Thomas, S. L.; Garrett, S.; Zelnak, A.; Kapoor, T. M.; Giannakakou, P. *J. Biol. Chem.* **2005**, *280*, 11569–11577.
23. Leizerman, I.; Avunie-Masala, R.; Elkabets, M.; Fich, A.; Gheber, L. *Cell Mol. Life Sci.* **2004**, *61*, 2060–2070.
24. Miyamoto, D. T.; Perlman, Z. E.; Burbank, K. S.; Groen, A. C.; Mitchison, T. J. *J. Cell Biol.* **2004**, *167*, 813–818.
25. Mailhes, J. B.; Mastromatteo, C.; Fuseler, J. W. *Mutat. Res.* **2004**, *559*, 153–167.
26. Garber, K. *J. Natl. Cancer Inst.* **2003**, *95*, 1740–1741.
27. Ding, S.; Wu, T. Y.; Brinker, A.; Peters, E. C.; Hur, W.; Gray, N. S.; Schultz, P. G. *Proc. Natl. Acad. Sci. USA* **2003**, *100*, 7632–7637.
28. Li, R. *Curr. Biol.* **2005**, *15*, R198–R200.
29. Wu, X.; Ding, S.; Ding, Q.; Gray, N. S.; Schultz, P. G. *J. Am. Chem. Soc.* **2004**, *126*, 1590–1591.
30. Ding, S.; Schultz, P. G. *Nat. Biotechnol.* **2004**, *22*, 833–840.
31. Chen, S.; Zhang, Q.; Wu, X.; Schultz, P. G.; Ding, S. *J. Am. Chem. Soc.* **2004**, *126*, 410–411.
32. Flaumenhaft, R.; Sim, D. S. *Chem. Biol.* **2003**, *10*, 481–486.
33. Sim, D. S.; Merrill-Skoloff, G.; Furie, B. C.; Furie, B.; Flaumenhaft, R. *Blood* **2004**, *103*, 2127–2134.
34. Peterson, R. T.; Link, B. A.; Dowling, J. E.; Schreiber, S. L. *Proc. Natl. Acad. Sci. USA* **2000**, *97*, 12965–12969.
35. Peterson, R. T.; Shaw, S. Y.; Peterson, T. A.; Milan, D. J.; Zhong, T. P.; Schreiber, S. L.; MacRae, C. A.; Fishman, M. C. *Nat. Biotechnol.* **2004**, *22*, 595–599.
36. Stockwell, B. R.; Haggarty, S. J.; Schreiber, S. L. *Chem. Biol.* **1999**, *6*, 71–83.
37. Liu, B.; Li, S.; Hu, J. *Am. J. Pharmacogenomics* **2004**, *4*, 263–276.
38. Jager, S.; Brand, L.; Eggeling, C. *Curr. Pharm. Biotechnol.* **2003**, *4*, 463–476.
39. Silverman, L.; Campbell, R.; Broach, J. R. *Curr. Opin. Chem. Biol.* **1998**, *2*, 397–403.
40. Sittampalam, G. S.; Kahl, S. D.; Janzen, W. P. *Curr. Opin. Chem. Biol.* **1997**, *1*, 384–391.
41. Mitchison, T. J. *ChemBioChem* **2005**, *6*, 33–39.
42. Perlman, Z. E.; Mitchison, T. J.; Mayer, T. U. *ChemBioChem* **2005**, *6*, 218.
43. Li, Z.; Yan, Y.; Powers, E. A.; Ying, X.; Janjua, K.; Garyantes, T.; Baron, B. *J. Biomol. Screen.* **2003**, *8*, 489–499.
44. Feng, Y.; Yu, S.; Lasell, T. K.; Jadhav, A. P.; Macia, E.; Chardin, P.; Melancon, P.; Roth, M.; Mitchison, T.; Kirchhausen, T. *Proc. Natl. Acad. Sci. USA* **2003**, *100*, 6469–6474.
45. Venkatesh, N.; Feng, Y.; DeDecker, B.; Yacono, P.; Golan, D.; Mitchison, T.; McKeon, F. *Proc. Natl. Acad. Sci. USA* **2004**, *101*, 8969–8974.
46. Burdine, L.; Kodadek, T. *Chem. Biol.* **2004**, *11*, 593–597.
47. Khersonsky, S. M.; Jung, D. W.; Kang, T. W.; Walsh, D. P.; Moon, H. S.; Jo, H.; Jacobson, E. M.; Shetty, V.; Neubert, T. A.; Chang, Y. T. *J. Am. Chem. Soc.* **2003**, *125*, 11804–11805.
48. MacBeath, G.; Schreiber, S. L. *Science* **2000**, *289*, 1760–1763.
49. Predki, P. F. *Curr. Opin. Chem. Biol.* **2004**, *8*, 8–13.
50. Lueking, A.; Possling, A.; Huber, O.; Beveridge, A.; Horn, M.; Eickhoff, H.; Schuchardt, J.; Lehrach, H.; Cahill, D. *J. Mol. Cell. Proteomics* **2003**, *2*, 1342–1349.
51. Zhu, H.; Bilgin, M.; Bangham, R.; Hall, D.; Casamayor, A.; Bertone, P.; Lan, N.; Jansen, R.; Bidlingmaier, S.; Houfek, T. et al. *Science* **2001**, *293*, 2101–2105.
52. Ziauddin, J.; Sabatini, D. M. *Nature* **2001**, *411*, 107–110.
53. Giaever, G.; Chu, A. M.; Ni, L.; Connelly, C.; Riles, L.; Veronneau, S.; Dow, S.; Lucau-Danila, A.; Anderson, K.; Andre, B. et al. *Nature* **2002**, *418*, 387–391.
54. Armour, C. D.; Lum, P. Y. *Curr. Opin. Chem. Biol.* **2005**, *9*, 20–24.
55. Lum, P. Y.; Armour, C. D.; Stepaniants, S. B.; Cavet, G.; Wolf, M. K.; Butler, J. S.; Hinshaw, J. C.; Garnier, P.; Prestwich, G. D.; Leonardson, A. et al. *Cell* **2004**, *116*, 121–137.
56. Giaever, G.; Flaherty, P.; Kumm, J.; Proctor, M.; Nislow, C.; Jaramillo, D. F.; Chu, A. M.; Jordan, M. I.; Arkin, A. P.; Davis, R. W. *Proc. Natl. Acad. Sci. USA* **2004**, *101*, 793–798.

57. Baetz, K.; McHardy, L.; Gable, K.; Tarling, T.; Reberioux, D.; Bryan, J.; Andersen, R. J.; Dunn, T.; Hieter, P.; Roberge, M. *Proc. Natl. Acad. Sci. USA* **2004**, *101*, 4525–4530.
58. Spencer, D. M.; Wandless, T. J.; Schreiber, S. L.; Crabtree, G. R. *Science* **1993**, *262*, 1019–1024.
59. Licitra, E. J.; Liu, J. O. *Proc. Natl. Acad. Sci. USA* **1996**, *93*, 12817–12821.
60. Becker, F.; Murthi, K.; Smith, C.; Come, J.; Costa-Roldan, N.; Kaufmann, C.; Hanke, U.; Degenhart, C.; Baumann, S.; Wallner, W. et al. *Chem. Biol.* **2004**, *11*, 211–223.
61. Bain, J.; McLauchlan, H.; Elliott, M.; Cohen, P. *Biochem. J.* **2003**, *371*, 199–204.
62. Davies, S. P.; Reddy, H.; Caivano, M.; Cohen, P. *Biochem. J.* **2000**, *351*, 95–105.
63. Fabian, M. A.; Biggs, W. H., 3rd; Treiber, D. K.; Atteridge, C. E.; Azimioara, M. D.; Benedetti, M. G.; Carter, T. A.; Ciceri, P.; Edeen, P. T.; Floyd, M. et al. *Nat. Biotechnol.* **2005**, *23*, 329–336.
64. Marton, M. J.; DeRisi, J. L.; Bennett, H. A.; Iyer, V. R.; Meyer, M. R.; Roberts, C. J.; Stoughton, R.; Burchard, J.; Slade, D.; Dai, H. et al. *Nat. Med.* **1998**, *4*, 1293–1301.
65. Kuruvilla, F. G.; Shamji, A. F.; Sternson, S. M.; Hergenrother, P. J.; Schreiber, S. L. *Nature* **2002**, *416*, 653–657.
66. Blom, K. F.; Larsen, B. S.; McEwen, C. N. *J. Comb. Chem.* **1999**, *1*, 82–90.
67. Zuckermann, R. N.; Martin, E. J.; Spellmeyer, D. C.; Stauber, G. B.; Shoemaker, K. R.; Kerr, J. M.; Figliozzi, G. M.; Goff, D. A.; Siani, M. A.; Simon, R. J. et al. *J. Med. Chem.* **1994**, *37*, 2678–2685.
68. Kelly, M. A.; Liang, H.; Sytwu, II; Vlattas, I.; Lyons, N. L.; Bowen, B. R.; Wennogle, L. P. *Biochemistry* **1996**, *35*, 11747–11755.
69. Moy, F. J.; Haraki, K.; Mobilio, D.; Walker, G.; Powers, R.; Tabei, K.; Tong, H.; Siegel, M. M. *Anal. Chem.* **2001**, *73*, 571–581.
70. Xu, Q.; Lam, K. S. *J. Biomed. Biotechnol.* **2003**, *5*, 257–266.
71. Chiosis, G.; Brodsky, J. L. *Trends Biotechnol.* **2005**, *23*, 271–274.
72. Degterev, A.; Lugovskoy, A.; Cardone, M.; Mulley, B.; Wagner, G.; Mitchison, T.; Yuan, J. *Nat. Cell. Biol.* **2001**, *3*, 173–182.
73. Ason, B.; Knauss, D. J.; Balke, A. M.; Merkel, G.; Skalka, A. M.; Reznikoff, W. S. *Antimicrob. Agents Chemother.* **2005**, *49*, 2035–2043.
74. Kodali, S.; Galgoci, A.; Young, K.; Painter, R.; Silver, L. L.; Herath, K. B.; Singh, S. B.; Cully, D.; Barrett, J. F.; Schmatz, D. et al. *J. Biol. Chem.* **2005**, *280*, 1669–1677.
75. Cheng, D.; Yadav, N.; King, R. W.; Swanson, M. S.; Weinstein, E. J.; Bedford, M. T. *J. Biol. Chem.* **2004**, *279*, 23892–23899.
76. Hu, Y.; Helm, J. S.; Chen, L.; Ginsberg, C.; Gross, B.; Kraybill, B.; Tiyanont, K.; Fang, X.; Wu, T.; Walker, S. *Chem. Biol.* **2004**, *11*, 703–711.
77. Soltero-Higgin, M.; Carlson, E. E.; Phillips, J. H.; Kiessling, L. L. *J. Am. Chem. Soc.* **2004**, *126*, 10532–10533.
78. Bembenek, M. E.; Roy, R.; Li, P.; Chee, L.; Jain, S.; Parsons, T. *Assay Drug Dev. Technol.* **2004**, *2*, 300–307.
79. Thomas, J. J.; Shen, Z.; Crowell, J. E.; Finn, M. G.; Siuzdak, G. *Proc. Natl. Acad. Sci. USA* **2001**, *98*, 4932–4937.
80. Shen, Z.; Go, E. P.; Gamez, A.; Apon, J. V.; Fokin, V.; Greig, M.; Ventura, M.; Crowell, J. E.; Blixt, O.; Paulson, J. C. et al. *ChemBioChem* **2004**, *5*, 921–927.
81. Hughes, T. R.; Shoemaker, D. D. *Curr. Opin. Chem. Biol.* **2001**, *5*, 21–25.
82. Hughes, T. R.; Marton, M. J.; Jones, A. R.; Roberts, C. J.; Stoughton, R.; Armour, C. D.; Bennett, H. A.; Coffey, E.; Dai, H.; He, Y. D. et al. *Cell* **2000**, *102*, 109–126.
83. Boshoff, H. I.; Myers, T. G.; Copp, B. R.; McNeil, M. R.; Wilson, M. A.; Barry, C. E., 3rd *J. Biol. Chem.* **2004**, *279*, 40174–40184.
84. Hutter, B.; Schaab, C.; Albrecht, S.; Borgmann, M.; Brunner, N. A.; Freiberg, C.; Ziegelbauer, K.; Rock, C. O.; Ivanov, I.; Loferer, H. *Antimicrob. Agents Chemother.* **2004**, *48*, 2838–2844.
85. Gunther, E. C.; Stone, D. J.; Gerwien, R. W.; Bento, P.; Heyes, M. P. *Proc. Natl. Acad. Sci. USA* **2003**, *100*, 9608–9613.
86. Scherf, U.; Ross, D. T.; Waltham, M.; Smith, L. H.; Lee, J. K.; Tanabe, L.; Kohn, K. W.; Reinhold, W. C.; Myers, T. G.; Andrews, D. T. et al. *Nat. Genet.* **2000**, *24*, 236–244.
87. Staunton, J. E.; Slonim, D. K.; Coller, H. A.; Tamayo, P.; Angelo, M. J.; Park, J.; Scherf, U.; Lee, J. K.; Reinhold, W. O.; Weinstein, J. N. et al. *Proc. Natl. Acad. Sci. USA* **2001**, *98*, 10787–10792.
88. Perlman, Z. E.; Slack, M. D.; Feng, Y.; Mitchison, T. J.; Wu, L. F.; Altschuler, S. J. *Science* **2004**, *306*, 1194–1198.
89. Collins, F. S.; Green, E. D.; Guttmacher, A. E.; Guyer, M. S. *Nature* **2003**, *422*, 835–847.
90. Austin, C. P.; Brady, L. S.; Insel, T. R.; Collins, F. S. *Science* **2004**, *306*, 1138–1139.
91. Kaiser, J. *Science* **2004**, *304*, 1728.
92. Kaiser, J. *Science* **2005**, *308*, 774.
93. Marris, E. *Nature* **2005**, *435*, 718–719.
94. Drews, J. *Science* **2000**, *287*, 1960–1964.
95. Drews, J. *Drug Disc. Today* **2003**, *8*, 411–420.
96. Couzin, J. *Science* **2003**, *302*, 218–221.
97. Tan, D. S.; Foley, M. A.; Shair, M. D.; Schreiber, S. L. *J. Am. Chem. Soc.* **1998**, *120*, 8565–8566.
98. Lee, D.; Sello, J. K.; Schreiber, S. L. *J. Am. Chem. Soc.* **1999**, *121*, 10648–10649.
99. Gosalia, D. N.; Diamond, S. L. *Proc. Natl. Acad. Sci. USA* **2003**, *100*, 8721–8726.
100. Bailey, S. N.; Sabatini, D. M.; Stockwell, B. R. *Proc. Natl. Acad. Sci. USA* **2004**, *101*, 16144–16149.
101. The Society for Biomolecular Screening. www.sbsonline.org (accessed Aug 2006).

Biography

Robert Flaumenhaft, MD, PhD, MMSc, is an assistant professor at Beth Israel Deaconess Medical Center and Harvard Medical School. He received his MD and PhD in Cell Biology at New York University and his MMSc from Harvard Medical School. His laboratory work involves the study of platelet biology. These studies focus on the role of platelet activation in thrombus formation and the molecular mechanisms of platelet secretion. He has performed chemical genetic screens in collaboration with investigators at the Institute of Chemistry and Cell Biology at Harvard Medical School to identify novel signaling pathways of platelet activation. He is a past recipient of a Postdoctoral Research Fellowship for Physicians from the Howard Hughes Medical Institute, the Burroughs Wellcome Fund Career Award, the American Society of Hematology Junior Faculty Scholar Award, a Special Project Award from Bayer HealthCare, and a Grant-In-Aid from the American Heart Association.

Comprehensive Medicinal Chemistry II
ISBN (set): 0-08-044513-6
ISBN (Volume 3) 0-08-044516-0; pp. 129–149

3.08 Genetically Engineered Animals

B Bolon, GEMpath Inc., Cedar City, UT, USA

3.08.1	**Introduction**	**151**
3.08.2	**Fundamental Principles for Using Engineered Rodents in Drug Discovery and Development**	**152**
3.08.2.1	Available Methods	152
3.08.2.2	Analysis of Engineered Rodent Models in Drug Discovery and Development	153
3.08.2.2.1	Genotypic analysis	153
3.08.2.2.2	Phenotypic analysis	154
3.08.2.2.3	Interpretation of positive phenotypes from novel constructs	156
3.08.2.2.4	Interpretation of negative phenotypes from novel constructs	157
3.08.3	**Uses for Engineered Animals in Drug Discovery and Development**	**158**
3.08.3.1	General Considerations	158
3.08.3.2	Basic Research	159
3.08.3.3	Applied Research	160
3.08.3.3.1	Genotoxicity	160
3.08.3.3.2	Carcinogenicity	161
3.08.4	**Caveats to Using Engineered Animals in Drug Discovery and Development**	**163**
3.08.5	**Conclusion**	**165**
References		**165**

3.08.1 Introduction

The molecular biology revolution of the past three decades has supplied the foundation for spectacular advancement in our understanding of disease and its treatment. In particular, the advent of dependable genetic engineering techniques that allow deliberate engineering of novel animal models for human diseases has greatly accelerated the pace of biomedical research, including the rate at which pharmaceutical and biopharmaceutical firms can discover and validate new drug targets as well as develop and ultimately manufacture novel therapeutic agents. Engineered animals of many species have been employed in drug discovery and development (DDD) programs, with the choice of species varying with the stage of the project. In discovery-stage tasks, basic research to define and verify aberrant molecular mechanisms that initiate and promote disease is typically conducted using mice (GEM) and/or rats (GER) that have been specifically engineered. In preclinical development efforts to elucidate efficacy or toxicity of new therapeutic candidates, the use of engineered animals depends on the entity being tested; small molecules (*see* 1.09 Traditional Medicines) typically are evaluated using GEM models available from mutant mouse resource centers or commercial vendors, while biomolecules (*see* 1.10 Biological Macromolecules) usually are examined in company-specific GEM or GER models. Finally, a few therapeutic biomolecules actually are fabricated using engineered livestock as bioreactors.

The use of engineered animals in DDD programs usually is limited to inventing and testing novel agents rather than to producing drugs, so this chapter will confine its scope to those DDD functions that use GEM and GER. Inclusion of engineered rodents in DDD is based on the premise that the activity of a modified gene in an animal model can be used to predict the function of a faulty human gene that bears a similar genetic anomaly.[1] The truth of this assertion has been amply proven in the last decade by the demonstration that the biological effects in humans of the 100 top-selling drugs correlate well with the predicted physiological outcome observed in GEM lacking the molecular targets for these agents.[2] Subsequent validation of other new targets and novel agents in various DDD pipelines using GEM suggests that this trend will continue and even accelerate in the future.[3] Thus, the cumulative experience using GEM and GER is that such models add considerable value to both early and late stages of DDD.

3.08.2 Fundamental Principles for Using Engineered Rodents in Drug Discovery and Development

3.08.2.1 Available Methods

Regardless of the species, the growing inventory of procedures that comprise the field of genetic engineering affords a means by which foreign genetic material is introduced reliably and efficiently into the genome of an organism. For DDD, GEM and GER models are constructed using two standard techniques: random introduction of a foreign gene (transgenic) and precise targeting to eradicate (knock out) an endogenous gene.[4–6] The biological impact of a null mutation is to eliminate a gene's function, thus leading to some functional and/or structural deficit. A twist on the knockout strategy is to fabricate the targeting construct in such a manner that it contains a second functional gene (a 'knock in' strategy for testing whether or not the second gene can restore all or part of the deleted gene's functions). In contrast, the biological impact of a transgene depends on the character of the encoded molecule. For example, transgene-associated functional deficits may result from functional antagonism of an endogenous gene's 'positive' action by a transgene's 'dominant negative' action or from physical binding and inactivation of an endogenous gene product by the transgenic product.[7–9] Considerable effort may be required to elucidate the mechanism by which a given phenotype develops in an engineered animal, but such detailed knowledge is a prerequisite for any rational use of new GEM or GER models in DDD.

Transgenic animals usually are constructed either by direct microinjection[10] of the modified genetic material directly into a zygote (one-celled embryo) or by systemic gene delivery (gene therapy) into multicelled embryos or adult animals using chemical[11] or viral[12,13] vectors. Gene-targeted animals are created by inserting the engineered gene in vitro into embryonic stem (ES) cells, and then either injecting into[14,15] or coaggregating[16,17] the ES cells with a blastocyst (multicelled, cavitated embryo). Endogenous genes can also be inactivated by anti-sense oligonucleotides,[18–20] small interfering RNA (*see* 3.09 Small Interfering Ribonucleic Acids),[19,21,22] or ribozymes,[23] all of which interfere with the production of protein from endogenous genes. In contrast to the permanent heritable mutations in knockouts created by gene targeting, these latter three methods generate transient nonheritable reductions in gene activity via an epigenetic mechanism. Furthermore, the transitory nature of the epigenetic modulation typically elicits a partial 'knockdown' effect rather than a complete knockout.[24] At present, all these methods can be applied to create both GEM and GER models except for the conventional gene targeting technique, which is limited only to mice due to the lack of well-defined lines of rat ES cells.

The selection of a genetic engineering strategy for DDD programs of the future will be predicated on the growing number of innovative techniques to specify more exactly the nature and activity of the modified gene. While earlier DDD projects focused on the addition or deletion of a single altered gene, some recent GEM models incorporate several different transgenes and/or null mutations so that complex metabolic interactions can be appraised during xenobiotic exposure.[25–30] Another major new approach is the generation of 'conditional transgenic' GEM. Such animals bear engineered genes that include a tissue-specific promoter[31] and are controlled by a ligand-inducible enhancing or repressing element which limits the impact of the modified gene to a given developmental stage or organ,[29,32–34] or even to a given region or cell type within an organ.[35] A similar novel tactic is the production of GEM with time- and tissue-specific gene deletions.[36–38] This method requires the construction of two GEM lines, a knockout line bearing a functional version of the gene of interest customized by the addition of two excision sites in flanking positions,[39,40] and a second transgenic line incorporating a recombinase gene specific for the excision sites under the control of a suitable (time- and/or tissue-specific) promoter. The two parent lines are normal, but crossing them results in offspring in which the flanked gene is deleted in cells that express recombinase. For added specificity, the recombinase-bearing GEM line can be made using conditional transgenic technology.[41] Even more power might be obtained by evaluating GEM bearing multiple tissue-specific, conditional gene deletions,[42] or greater speed gained by melding RNA interference (RNAi) and conditional transgenic technologies.[43]

Another recent improvement of great importance to DDD is the ability to replace a mouse gene with its human homolog[33,44–47] or to implant functional human cells to repopulate a failing organ in a mutant rodent.[48] Such GEM afford the opportunity early in DDD projects to evaluate the effects of xenobiotics on human proteins of interest in vivo without the need for conducting human clinical trials. This strategy addresses in part the difficulty in defining the extent to which mouse physiological responses reproduce those of humans, thus ameliorating several of the major limitations in animal modeling experiments. A rodent engineered to express human proteins is still an animal, so caution must be used in accepting the GEM phenotype – or lack thereof – produced by insertion of an engineered human gene as a strict analog of naturally occurring genetic events that occur in humans.[49–51] Nevertheless, in many cases such 'humanized' animals will provide a better proxy during preclinical bioassays than will the unmodified animals employed in routine batteries of toxicity tests. This assertion will be particularly true in those instances where the novel therapeutic candidate is a biomolecule of human origin that only interacts with a human protein (*see* 1.10 Biological Macromolecules). In such cases, the use of a humanized GEM as one of the two animal species employed during preclinical efficacy and toxicity testing likely will provide the most effective means of acquiring in vivo data regarding potential mechanism-based effects of the putative new drug.

Additional procedures that should profoundly impact the future of DDD programs are schemes designed to increase the rate at which engineered models can be produced and evaluated. One such strategy is to create GEM lines that contain multiple genetic alterations (usually transgenic insertions) at once to allow several genes to undergo phenotypic screening at one time. Discovery of a phenotype in such a 'multimutant' animal is followed immediately by breeding and analysis of lines that contain the individual transgenes to determine which modified gene elicited the change; loss of the condition in single-transgenic lines suggests that the 'multimutant' phenotype resulted from interactive effects between two (or more) of the original transgenes. Another possibility to greatly decrease the time required for initial characterization of GEM and GER in DDD is to analyze founder mice for phenotypes rather than assessing their progeny.[52] This tactic presupposes that phenotypes will be robust and that the occurrence of a strong phenotype in multiple founders means that the same phenotype can be recaptured merely by creating new founders. The advantage to this approach is that it limits animal husbandry efforts to those lines with potential DDD utility, while the primary disadvantage is that a phenotype might be lost on occasion. To avoid losing phenotypes, some DDD groups have implemented a modified version of this 'speed phenotyping' paradigm, usually by either breeding each founder one time before it is analyzed or by assessing only female founders (as males can be employed in breeding to produce more progeny in a shorter time).

I anticipate that all these options will have great utility in DDD programs in the future. However, techniques that will enjoy particular favor in the near term will be conditional and tissue-specific transgenic and knockout lines, 'humanized' GEM models, and the use of high-throughput analysis of founder animals. Together these methods will allow researchers much greater control and speed when building GEM and GER models for DDD research.

3.08.2.2 Analysis of Engineered Rodent Models in Drug Discovery and Development

The utility of GEM and GER models in DDD programs depends greatly on the extent to which their genotypic (genetic) and phenotypic (functional and structural) characteristics have been mapped. Ultimately, phenotypes must not only be examined in minute detail, but they must be established to arise from the modified gene. In discovery-stage DDD projects, almost all engineered lines are newly created and thus require substantial analysis before they can be used for anything more than basic mechanistic experiments. In contrast, the GEM models employed in later preclinical DDD have been well investigated, and the reproducible nature of their genotypic and phenotypic features allows them to be used to assess both fundamental mechanisms and clinically relevant endpoints such as efficacy and toxicity. This section will briefly discuss the strategies and tactics used to analyze GEM and GER models in sufficient depth to allow for their use in DDD.

3.08.2.2.1 Genotypic analysis

The first critical piece of data required to evaluate the utility of an engineered model for DDD is information regarding the presence and activity of the modified genetic material. An engineered gene in a GEM or GER can carry out its function only if the gene is expressed, a process which involves intranuclear transcription of the gene into messenger RNA (mRNA) followed by intracytoplasmic translation of mRNA into functional protein. Expression of the engineered gene may be confirmed by measuring either the mRNA or the protein, although in certain cases interpretation is complicated by discrepancies between the mRNA and protein readings. For example, distribution of mRNA and protein for a given gene are not identical in many tissues (such as the nervous system, where both mRNA and protein for molecules in peripheral synapses are still produced in the centrally located cell bodies). Thus, measurement of the protein is typically preferred if reagents can be fashioned to detect it.

Genotypic expression analysis to confirm that an engineered gene is present and working correctly usually is performed in tissue homogenates by binding either a labeled nucleotide (for mRNA) or antibody (for protein). In DDD the analytical techniques of choice for measuring RNA are the Northern blot assay, reverse-transcriptase polymerase chain reaction (RT-PCR), and the ribonuclease protection assay (RPA).[53,54] The RT-PCR and RPA methods are more sensitive and specific because their reactions occur in solution (which allows the probe to approach the target molecule in three dimensions) rather than on a membrane (access only in two dimensions). Typically, several oligonucleotide probes are made that react with sequences inside the normal and engineered genes, within the foreign promoter that drives the engineered gene, and/or in the flanking sequences from the microbial vector; the use of multiple probes increases the specificity. Common methods for protein quantification in DDD are Western analysis and enzyme biochemical procedures. The Western technique confirms only the presence of the molecule in question, while the latter measures both the presence and activity of functional protein derived from the engineered gene. A special case adopted for many gene targeting experiments is to use a vector in which the engineered genetic material also contains a nonmammalian chromogenic marker, such as beta-galactosidase[55,56] or green fluorescent protein,[57] because such markers can be used simultaneously to both localize the distribution of the original endogenous gene and confirm that it has been removed. Secreted molecules also may be evaluated in many body fluids using enzyme-linked immunosorbent assays (ELISA)[54] or enzyme kinetic assays[58,59] to detect the physical or functional presence, respectively, of the protein. Regardless of the method, a good analysis of the gene expression pattern in multiple organs is an essential first step in evaluating a new GEM or GER model.

Once the distribution of an engineered gene product in various organs has been defined, the next step is usually to perform a qualitative localization of mRNA and/or protein to various organ compartments or cell types. Such assessments may be conducted in either intact organs or embryos ('whole mount' assays) or in tissue sections (slide-based assays). If feasible, organs are first preserved by chemical fixation (usually by immersion in a dilute solution of a buffered aldehyde) to preserve the overall tissue structure for later microscopic analysis. However, in certain cases (e.g., proteins with delicate antigenic sites, enzymes) fresh tissues are embedded in viscous mounting medium and then flash-frozen in supercooled isopentane (2-methyl butane) to prevent degradation of the molecule while retaining acceptable tissue morphology. The analytical techniques used in DDD to localize sites of transgene expression are comparable to those described above for homogenized tissues. The main method for demonstrating mRNA is in situ hybridization,[53,56] while protein distribution and function are assessed by immunohistochemistry[56,60,61] and enzyme histochemistry,[55,62] respectively. Commercial antibodies[63,64] can sometimes be found if the engineered gene encodes a new variant of a well-characterized protein, although many such reagents are manufactured in murine systems and will thus cross-react with GEM tissues. In most instances, however, antibodies do not exist for proteins derived from new genes discovered in the course of DDD. The arduous labor required to make and validate antibodies coupled with the likelihood that the reagent will be rendered obsolete if the novel gene is dropped from a high-throughput DDD program usually dictates that the in vivo tissue distribution of the engineered gene be inferred only from the mRNA data.

3.08.2.2.2 Phenotypic analysis

Even if expression of the engineered gene is confirmed, the GEM or GER model will be useless for DDD if a distinct phenotype is not induced by the action of the modified gene. Expression of an engineered gene may provoke a functional change, a structural alteration, both of these effects, or neither. These endpoints may be examined either in vivo in living animals or in vitro in isolated tissues, cells, or even embryos. Newly engineered GEM and GER lines are assessed using a tiered strategy (**Table 1**), in which a limited series of tests is first used to reveal the presence of a possible phenotype before a comprehensive battery of assays is deployed to fully characterize the nature and mechanisms of the phenotype.[65,66] Most DDD groups emphasize conventional procedures (clinical evaluation, pathology) in their initial evaluation of GEM and GER, although ancillary tests may be added as well based on the investigator's primary research interests (e.g., behavioral testing for neuroscience programs, radiography for bone biology groups). Some companies also mandate that their discovery group be prepared to employ whatever assay seems warranted to confirm the existence of a phenotype. However, in practice many DDD programs contract with expert academic laboratories to perform such specialized assessments (*see* 1.16 Industry–Academic Relationships). If feasible, the goal is always to define one or more biomarkers that can be readily measured using noninvasive techniques to monitor the progression of disease over time – especially during in vivo treatment of GEM, GER, and ideally human patients with new drug candidates (*see* 3.04 Biomarkers).

In high-throughput DDD programs, novel GEM or GER are usually phenotyped as young adults (2 months of age or so). Subjects used for transgenic projects include the hemizygous (+) engineered animals and age-matched controls (often nontransgenic littermates). For DDD, gene targeting projects typically include homozygous knockout (−/−) animals and age-matched, wild-type (+/+) littermates in the initial screening project (Tier I); heterozygous (+/−) littermates are also often included in the mechanistic evaluation (Tier II) because the reduced gene dosage may result in partial penetrance with diluted intensity of the transgene-induced phenotype.[67,68] Most companies examine both males and females in their initial screens because gene expression and phenotype severity may be sexually dimorphic.[69,70] In the ideal case, the genetic background of GEM and GER will be simplified using congenic (an engineered line in which genetic homogeneity has been attained by repeated backcrossing[71]) or consomic (a line engineered to contain an intact chromosome from a genetically distinct rodent strain[71,72]) technology prior to evaluation as the presence of a truly homogeneous genome eliminates one major source of phenotypic variability. The need for speed, though, usually frustrates this ideal in the high-throughput DDD setting.

Table 1 Sample tiered strategy for phenotyping engineered animals in drug discovery and development

Tier I: Initial phenotypic screen – are there effects of the modified gene?	
Genotypes	Transgenic: hemizygous (+) and wild-type Gene targeted: homozygous knockout (−/−) and wild-type (+/+)
Age	Young adult (8 to 12 weeks old)
Generation	Founder (F_0) or their offspring (F_1 or F_2)
Group size	3 to 6 per genotype (ideally 2 to 3 per sex)
Tests (basic)	Function: activity, appearance, body weight, hematology, serum chemistry, ±hormone levels Structure: appearance of organs, organ weights, histopathology, radiography
Tier II: Mechanistic characterization – what are the effects of the modified gene, and how are they induced?	
Genotypes	Transgenic: hemizygous (+) and wild-type Gene targeted: homozygous knockout (−/−), heterozygous (+/−), and wild-type (+/+)
Age	Young adult (8 to 12 weeks old), and if needed mature (6 to 12 months old) or prenatal
Generation	Offspring (F_1 or later)
Group size	3 to 6 per genotype (ideally 2 to 3 per sex)
Tests (advanced)	Function: behavior, electrophysiology, noninvasive imaging, pharmacologic challenge, telemetry Structure: immunohistochemistry, in situ hybridization, laser capture microdissection, morphometry, ultrastructure

Most DDD groups emphasize conventional procedures in their characterization of GEM and GER. In my experience, initial screening for the presence of a new phenotype can be performed effectively with a series of straightforward functional and structural tests. A typical menu for screening young adult GEM or GER in many DDD programs includes daily observations for clinical abnormalities by the laboratory animal caretakers, in-life experiments by the research staff (planned to address the firm's areas of therapeutic interest), and finally a full necropsy that includes blood collection (to assess blood cell lineages and serum chemistry analytes) and evaluation of at least 30 organs by gross and microscopic methods. These first assessments are performed using standard clinical[73–76] and pathology[77–80] techniques for adult rodents. Occasionally, though, a disparity in the predicted Mendelian ratio (usually reflecting a lack of hemizygous transgenic or homozygous knockout conceptuses) suggests the existence of an embryolethal phenotype; in such instances, the time of intrauterine death must be estimated from the appearance of the dead conceptuses so that clinical, molecular, and microscopic analyses can be used to characterize the phenotype in younger, viable embryos and placentae.[81]

Detection of structural abnormalities during the pathology examination is often used as the 'gold standard' for confirming the presence of a phenotype in Tier I. This preference is dictated by the consistent architecture of normal tissues in mice and rats, which allows biologically relevant changes to be unearthed using only a few animals (usually three to five per group, including animals of both sexes). This ability substantially eases GEM and GER production constraints in the high-throughput discovery phase of DDD. A superior understanding of normal rodent anatomy and physiology is required for proper interpretation of potential phenotypes. Such knowledge may be obtained by consulting appropriate reference materials for adult[82–84] or developing[85,86] mice or adult rats.[87,88] Personnel doing the analysis also should be well experienced with interpreting GEM and GER data (particularly if the specimens are embryos or fetuses), including familiarity with common background changes in strains commonly used in production of transgenic (FVB mouse[89]) and knockout (C57BL/6 × 129 mouse[90,91]) animals. Insufficient expertise with evaluating GEM and GER data can result in such serious errors as misreading normal structures[92] or cyclic variations in organ morphology[93] as gene-induced lesions, or attributing phenotypic variation to physiological processes that do not even occur in rodents.[94]

Phenotypic screening in Tier I is expanded to assess other functional endpoints in some DDD programs, and all firms incorporate specialized tests in Tier II to further characterize newly discovered phenotypes. Examples of such options include behavior,[95–98] electrophysiology,[99–102] metabolic phenotyping ('metabolomics' or 'metabonomics'; *see* 5.43 Metabonomics),[103] noninvasive imaging (*see* 3.31 Imaging),[104–111] telemetry,[99] and pharmacological challenge.[112,113] This latter strategy, which can uncover very subtle phenotypes in apparently normal animals, is of particular importance in research programs where the effects of ablating a receptor gene (total 'knockout') often do not predict with accuracy the impact of receptor inhibition using an antagonist (partial 'knockdown').[114–116] However, the wide baseline variation in functional responses between animals often necessitates large group numbers (typically 10–20) for each genotype and sometimes for each sex, which may preclude the design of studies with adequate statistical power during early DDD projects. (One means by which such large studies can sometimes be avoided is to use each animal as its own comparator by acquiring suitable baseline values.)

In all cases, data acquired for engineered animals should be compared to the findings for 'normal' animals of the same species and strain, ideally by using age- and sex-matched wild-type littermates as controls. Laboratory-specific databases of historical control data (particularly for functional tests) can, over time, serve as a valuable ancillary resource for such comparisons.

3.08.2.2.3 Interpretation of positive phenotypes from novel constructs

The defining aspect that determines whether or not a newly engineered GEM or GER model has utility is the production of a usable and stable phenotype. This evaluation must be made for all proprietary DDD applications, most of which are clustered in the early discovery arena. Many novel GEM and GER have obvious functional and/or structural phenotypes that can be attributed to the influence of the modified gene, even if the phenotypes have no or only partial similarity to a human disease.[117] Ideally, phenotypes will be robust, meaning that the same genetic rearrangement performed by several independent research groups will generate a similar spectrum of lesions. Illustrations of such robustness are found in the anxiety-like syndrome that develops in mice in which the serotonin 1A receptor (5HT1A) has been extirpated[118–120] and the presence of autonomic neural defects and renal agenesis in mice with engineered deletions of glial cell-derived neurotrophic factor (GDNF).[121–123] Interestingly, phenotypes of differing severity can arise from the same transgene merely as a result of variations in the level of transgene expression.[124] Nevertheless, many phenotypes are quite subtle,[125–128] and unanticipated patterns of gene expression[129,130] or phenotypes (e.g., illogical alterations based on our present comprehension of mammalian physiology) are not unknown.[114,127,131–133] The considerable variation in phenotypic strength and stability is often the source of considerable concern for decision makers using engineered animals in a DDD program.

The upshot of this concern is that positive phenotypes in engineered models cannot be presumed to be true consequences of the modified gene without rigorous molecular and physiological testing. Real utility for GEM or GER in DDD (or any field, for that matter) is evident only if good concordance exists between phenotypes resulting from changes to molecules that function, respectively, as the target (e.g., receptor or enzyme involved in a signal transduction cascade) or the effector (e.g., ligand) in a given biological system. For example, mice with null mutations of *RANK* (receptor activator of nuclear factor kappa B (NFκB)) and *RANKL* (RANK ligand), the two main extracellular elements that mediate the primary signaling pathway for osteoclast activation in vivo, develop extensive osteopetrosis[134–136] (**Figure 1**). Similarly, transgenic mice that overexpress *OPG* (osteo-protegerin, a soluble decoy receptor for *RANKL*) also exhibit profound osteopetrosis,[137] while mice that lack *OPG* have severe osteoporosis[138,139] (**Figure 1**). The utility of long-term OPG therapy as a nontoxic means of treating osteoporosis in mice has been shown by systemic transgene delivery[13,140] (**Figure 1**). Taken together, these GEM-based mechanistic studies provide in vivo evidence that OPG supplementation preserves bone without inducing adverse effects. Subsequent studies have confirmed the efficacy and safety of injected OPG as a therapeutic agent in bone-eroding disease, both in other animal models of bone loss[9,141,142] and in human clinical trials in normal postmenopausal women[143] and patients with widespread osteolytic neoplasia.[144] Thus, the typical DDD program will make a series of GEM or GER models as a matter of course to evaluate both branches (target and effector) of a molecular pathway of interest for a consistent and reproducible pattern of positive phenotypes.

Considerable care is required to confirm that a positive phenotype really results from the action of the engineered gene and not just from one or more of the many confounding factors that can influence or even mimic the induction of a true phenotype (**Table 2**). For example, transgene incorporation into the genome occurs at random, so gene expression will depend on the nature of the nearby regulatory elements[145,146] and even the exact position on a chromosome.[147] In like manner, interruption of a codon in an essential gene by introduction of a transgene may produce an unintentional null mutation (insertional mutagenesis) in which the effects of the chance knockout are mingled with the effects (if any) induced by the transgene.[132,148–151] In such cases, the insertion-related phenotype is distinguished from the transgene-related one because insertional events occur in only a single GEM or GER line, while true transgenic phenotypes are similar across many lines. The severity of a phenotype is strongly affected by inherent genetic variations implicit in different strains[90,152–156] or substrains,[157] between sexes,[154] and across generations.[158–160] Epigenetic factors that can impact the development of a phenotype include such husbandry factors as diet[161] and environment,[162] the location of the institution,[163,164] maternal care,[165,166] stress,[167] infections,[168] and an abnormal phenotype in a parent.[169] The most desirable way to reduce most variability is to standardize the genetic constitution of engineered animals – ideally by using the same strains at all institutions and by backcrossing to place each phenotype on a constant genetic background. Unfortunately, this scheme is usually ignored in high-throughput DDD efforts due to time constraints and individual preferences regarding desirable strains for engineering.

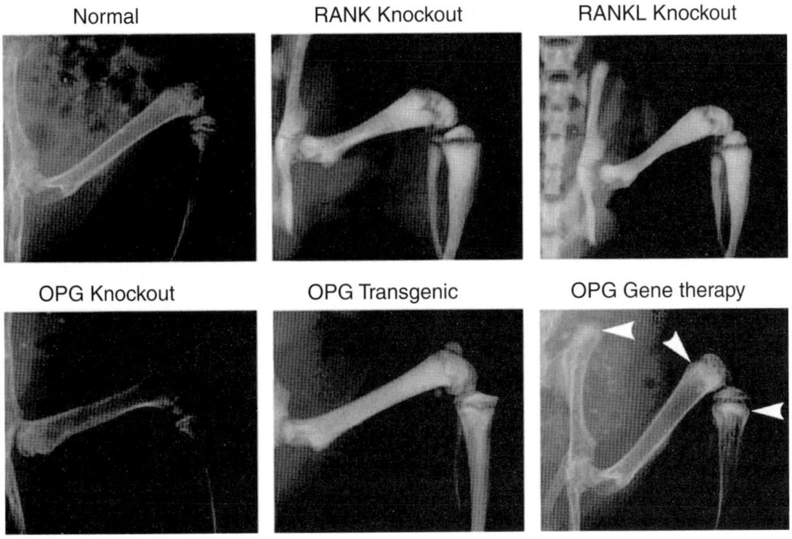

Figure 1 Concordant phenotypes are produced by engineering multiple elements of the same molecular pathway. This example demonstrates the distinct structural phenotypes that develop in mice with engineered mutations in the main signaling pathway that regulates osteoclast function. In this pathway, interaction of the cell-bound receptor RANK (receptor activator of nuclear factor kappa B) and its ligand RANKL enhances osteoclast function unless the soluble decoy receptor OPG (osteoprotegerin) intervenes first to sequester RANKL from RANK. Lifelong deletion of *RANK* or *RANKL* or overexpression of OPG induces widespread osteoclast inactivation and leads to generalized osteopetrosis (increased skeletal radiodensity), while a lifelong null mutation in *OPG* results in osteoclast activation and systemic osteoporosis (decreased bone density). Gene therapy with OPG in wild-type mice induces localized increases in bone radiodensity adjacent to growth plates (arrows). Radiographs were acquired using a bench-top system (Faxitron x-ray, Buffalo Grove, IL, USA) set at 0.3 mA and 55 kV for 49 s. (Adapted from Bolon, B.; Shalhoub, V.; Kostenuik, P. J.; Campagnuolo, G.; Morony, S.; Boyle, W. J.; Zack, D.; Feige, U. *Arthritis Rheum.* **2002**, *46*, 3121–3135, with the permission of John Wiley and Sons.)

A final consideration for experiments in which adult animals are made transgenic using a gene therapy protocol is to show that a potential phenotype is truly associated with the activity of the transgene and not with incidental toxicity driven by the delivery system. For example, chronic overexpression of OPG in mice using an adenoviral vector is associated with chronic hepatic inflammation and hepatocyte apoptosis.[13] However, the hepatic damage resulted from the immune response to the adenovirus and not OPG, a fact demonstrated by the production of identical liver lesions after adenovirus was used to introduce a nontoxic marker protein[13] and by the absence of hepatic injury when long-term OPG overexpression is launched using a different viral vector.[140]

3.08.2.2.4 Interpretation of negative phenotypes from novel constructs

Successful introduction of foreign DNA often elicits no perceptible effect in novel GEM or GER,[114,170,171] even if a transgene is highly expressed or a targeted gene is completely knocked out as confirmed by genotypic (DNA) and expression (mRNA and/or protein) analyses. In some instances, the absence of a phenotype represents the desired conclusion of a DDD program. For example, the production of viable and fertile GEM lines with null mutations for a potential target protein (e.g., beta-secretase knockout mice[172,173]) is an attractive outcome since it implies that chronic drug-mediated inhibition in vivo will induce no substantial side effects. Similarly, the lack of an overt in vivo phenotype could credibly be employed in combination with abundant evidence of in vitro activity to support the selection of a promising 'no observable effect level' (NOEL) for use in preclinical pharmacology and toxicology studies. In most instances, however, the lack of a definite phenotype in a new engineered line is a serious hindrance to a high-throughput DDD project tasked with identifying the activities of novel genes as rapidly as possible.

Researchers in DDD face two main predicaments in assessing the absence of a phenotype in newly generated GEM and GER. The first question is to define an explanation for the missing phenotype. Variations in genetic background are known to diminish a phenotype[174] unless the first engineered mutation is bolstered by the concomitant introduction of a second event.[175] However, the most frequent explanation for the absence of an anticipated phenotype is that other unknown endogenous molecules compensate for the effects elicited by the engineered mutation.[42,114,127,171,175–179] A known case of this phenomenon is the drastic alterations in hippocampal long-term potentiation in mice lacking both neuronal and endothelial variants of nitric oxide synthase (NOS), while animals missing only one NOS form exhibit no

Table 2 Selected factors that influence phenotypic expression in engineered rodents

Factor	*References*
Age	126, 240, 241
Breeding history	159, 242
Generation	158–160, 174
Genetic background	10, 90, 152–157, 235, 243–248
Gene dosage	67, 68, 120, 133, 170, 241, 249, 250
Gene expression level	124
Gene function (pleiotropy)	114, 120, 251
Gene penetrance	239, 252, 253
Genetic instability	50, 158, 160, 210, 236
Localization	129, 130
Hormone levels	166, 167
Laboratory location	163, 164
Maternal abnormalities	169
Maternal care	165, 166
Maternal stress	167
Microbes	162, 168
Nutrition	161, 167
Sex	69, 70, 167

impairment.[180] In many cases, though, homeostatic compensation is proffered as an explanation without any actual evidence to support the contention. A divergent viewpoint is that most 'negative' GEMs actually do have phenotypes, but that the understated functional and structural manifestations were overlooked because the phenotypic analysis was conducted too rapidly on the incorrect tissues using the wrong methods.[181] By this view, inconsistencies between interlaboratory transgenic procedures (e.g., use of different mouse strains) and phenotypic screening practices (e.g., inconsistent use of pharmacological challenges) could account for the apparently unsuccessful outcomes of many GEM experiments. To my knowledge, the truth of this latter contention has yet to be evaluated in a large number of novel GEM and GER in a systematic manner.

Nevertheless, in DDD the explanation for why a phenotype is lacking is usually not the question of chief interest. Instead, the more relevant query in the context of DDD is to make a reasoned and rapid decision regarding the ensuing course of the given project. Generation and characterization of a single GEM or GER line, let alone the several lines needed to validate a new model or the multiple models required to power a high-throughput DDD effort, necessitates significant expense and labor. In this setting, novel GEM or GER models lacking explicit phenotypes are typically presumed to have none because the mutated gene is not critical to the evolution of a given disease or normal homeostatic function. As noted above,[181] it is possible that increasing the number of tests used in phenotypic screening could uncover some additional phenotypes. However, I submit that the absence of a robust phenotype in an organ of interest warrants the conclusion that the novel gene encodes a molecule that cannot be used to create a successful drug.[182]

3.08.3 Uses for Engineered Animals in Drug Discovery and Development

3.08.3.1 General Considerations

In DDD, GEM and GER are employed to answer one of three basic research questions regarding a molecule. First, what is the role of a protein in health and disease? Second, if we make a drug to modify the activity of the molecule, will it have efficacy? And finally, if desirable efficacy is achieved, will the drug also elicit toxicity? Cost considerations are generally secondary to these three fundamental queries.

In functional genomics groups within the DDD setting, genes often are probed first by creating a simple transgenic model[183] as this construct can be completed more rapidly and at much less cost than can gene-targeted or conditional transgenic models for the same gene. Such a high-throughput transgenic program can be used to screen many genes in a short period of time, which allows for the more timely investigation of both potential targets and (for biomolecule DDD programs) prospective therapeutic agents. That said the potential payoff from such transgenic screening efforts is often low. In my experience, 10% or less of transgenic GEM contrived from expressed sequence tag (EST, a transcribed ribonucleic acid molecule representing the full sequence of all DNA codons encoding an unknown protein) exhibit a distinct and pronounced clinical phenotype.[184] This low sensitivity of transgenic analysis has led many firms to focus their current animal engineering efforts on gene targeting experiments as total ablation of a molecular pathway is more likely to yield a prominent phenotype suitable for DDD work than will transgenic overexpression. In my experience, 30–40% of knockout models exhibit some recognizable phenotype – although many simple knockouts yield embryolethal outcomes of no immediate use as models for DDD research. I anticipate that many future GEM models in DDD will be constructed as conditional knockouts to avoid this dilemma, once the speed at which this procedure can be implemented is increased.

The recent boost in utilizing GEM and GER in DDD is instituted in one of two primary experimental strategies. The first and most common paradigm, especially in early discovery projects, is to generate models de novo by engineering a novel gene of interest in hopes of achieving a useful phenotype. The focus for GEM and GER in such programs is to either define the role of a gene in the pathogenesis of a disease process (a hypothesis-driven approach) or to evaluate the functions of previously uncharacterized genes (a biology-driven approach). Alternatively, an existing model for which a sufficient phenotypic analysis already has been defined may be further evaluated. For example, GEM have been used to assess the mechanisms,[26,47,112,177] mutagenic[185–188] and carcinogenic[189–192] properties, and specificity[193] of xenobiotics. At present, GEM models are more readily available than GER constructs for both strategies. However, the number of GER models is increasing based on such desirable aspects of rat biology as more human-like physiological responses for some disease processes (e.g., cancer,[194] hypertension[195]), an extensive behavioral database, and larger size[196] (better suited to surgical manipulation and repeated blood sampling).

In addition to such experimental questions, the choice between engineering a new model and using an existing one in DDD is also founded on financial and time constraints. Most DDD firms use novel GEM and GER in discovery projects to gain and protect proprietary intellectual property concerning molecular targets of interest, and they will continue to do so regardless of the time and expense. Many fundamental mechanistic questions can be answered rapidly using a relatively uncharacterized GEM or GER line, and such information may be enough to allow an early 'go' or 'no go' decision for the whole research project. However, in most instances additional data will be needed before rendering a final verdict, in which case further characterization of novel GEM and GER using the methods described above will be essential; the time required for such a complete assessment usually is measured in months to years. In contrast, later preclinical experiments to define efficacy and toxicity more often than not use commercially available GEM models. This DDD choice is predicated on the ready accessibility of well-characterized animals, which substantially reduces the time required to fully develop a new in-house model, as well as the growing literature regarding their responses to xenobiotic challenge and their increasing acceptance at some regulatory agencies. I predict that biopharmaceutical and pharmaceutical companies will continue to apply this dual strategy when selecting which GEM and GER models to employ for various stages of DDD in the future.

3.08.3.2 Basic Research

Modern DDD programs utilize GEM and GER to address a wide spectrum of basic research questions. Typical applications are to characterize and validate targets, and to confirm that a proposed therapeutic strategy will be effective.[4–6,193,197] Such topics are examined as a component of the basic research engine that drives early-stage discovery efforts, and thus are typically performed using novel engineered models of a proprietary nature. Thus, many GEM and GER models are created in-house, but a growing number (especially GEM) are also available through academic collaborations (*see* 1.16 Industry–Academic Relationships) or from commercial vendors.[198]

A primary use for GEM and GER in basic research is to explore the essential molecular mechanisms whereby a gene regulates the balance between health and disease. In most firms such investigations are performed using newly generated proprietary models. The first prerequisite to this approach is to acquire a detailed knowledge of the physiological machinery that engenders a positive phenotype, as such understanding is necessary before a model can be used to make predictions suitable for guiding further DDD work. As exemplified above using the RANK pathway of osteoclast activation (**Figure 1**), a basic DDD program for a given entity usually begins by confirming that positive GEM and GER phenotypes for paired effector and target molecules are complementary,[134,135,137,138] and that the anticipated effects of pharmacologic intervention to regulate the pathway predict the likelihood of a useful therapeutic strategy for a disease of interest (compare results of systemic transgene delivery[13,140] to parenteral protein injection[9,141–144]). The inability of GEM and GER to provide such corroboration often leads to early termination of the DDD project.

A second purpose for using GEM and GER models in basic research is to probe the impact of pharmacologic intervention. Engineered models are used in this setting to address two basic needs: will chronic therapy elicit undesirable side effects, and what will be the nature of such effects? An instance in which GEM have been used in DDD to evaluate the impact of long-term treatment is provided by mice engineered to lack the beta-secretase, the brain enzyme proposed as the primary molecule responsible for the generation of neurotoxic amyloid beta (A beta) fibrils in patients with Alzheimer's disease.[199] Ablation of beta-secretase has been shown to effectively halt A beta production in knockout mice,[172] but the extensive distribution of this enzyme in extraneural tissues[200] raises the possibility that chronic beta-secretase inhibition will lead to undesirable systemic side effects. This concern was eased by determining that older knockout animals did not develop clinical or anatomic abnormalities either in the brain or in peripheral systems.[173] A DDD example in which GEM studies first revealed and later cataloged a range of side effects is found in the study of mice engineered to overexpress artemin, a GDNF-like neurotrophic factor in the peripheral nervous system put forward as a potential therapy for neuropathies.[201,202] After injection of recombinant artemin was associated with multifocal hyperplasia in the adrenal medulla in one of three preclinical efficacy studies, *artemin*-transgenic mice were generated by embryonic microinjection and systemic gene therapy to determine whether or not this neurotrophic factor actually resulted in adrenal proliferative lesions. In both GEM models artemin overexpression was found to elicit marked neural crest dysplasia (in embryos) and hyperplasia (adults) of the autonomic nervous system, including the adrenal medulla,[203] indicating that the autonomic nervous system represents a likely target for objectionable side effects. Such GEM work thus serves as a powerful means of driving early DDD decisions regarding the utility of potential targets and molecules in a development pipeline.

Many additional reasons have been drawn on to deploy GEM and GER in basic DDD research. Some examples of standard uses include defining the properties of endogenous ligands,[177,204] explaining the importance of cell signaling abnormalities mediated by receptor[177,205–207] or transduction cascade[207] anomalies, and evaluating the impact of metabolism upon toxicity.[26,208–211] Another notable function is to discriminate between xenobiotic-dependent and xenobiotic-independent effects. A hypothetical example of this latter point would be to examine whether or not recurring contact with a therapeutic aspirant (compound A, which is designed to restrain enzyme B) is the mechanism by which a given lesion is induced. If mice engineered to lack enzyme B develop the change after exposure to compound A, the finding does not result from the interaction of A with B and therefore does not embody a mechanism-based toxic effect of A. Finally, in the future GEM and GER will undoubtedly be frequent participants in toxicogenomic experiments to help illuminate disease mechanisms and to create compound- or class-specific signatures (biomarkers of exposure and/or effect; *see* 3.04 Biomarkers) resulting from exposure to new chemical entities.[212] Over time, inclusion of GEM and GER in such basic DDD research applications will allow for more effective bridging between data of early discovery and later preclinical efficacy and toxicity experiments, which in turn will provide a better way of performing risk-to-benefit assessments on novel drug candidates.

3.08.3.3 Applied Research

Modern DDD programs for small-molecule drugs and chemicals exploit several commercially available GEM (and one GER) models during late-stage preclinical research especially to assess the capacity for xenobiotic-induced genetic damage. The two principal applications are abbreviated bioassays for mutagenicity and carcinogenicity. In my experience, carcinogenicity studies are more common.

3.08.3.3.1 Genotoxicity

Several GEM and one GER models can be purchased from commercial vendors to evaluate the mutagenic activity of xenobiotics in vivo. The high cost and time required for such in-life tests in terms of labor, money, and time suggests that these rodent bioassays will be used to supplement rather than replace the rapid bacteria-based screening assays. Nevertheless, genotoxicity assessments made using GEM or GER afford distinctive insights that cannot be evaluated in vitro, including acquisition of data concerning the susceptibility of particular targets (e.g., spinal cord motor neurons) as well as the importance of metabolic, pharmacodynamic, and pharmacokinetic processes to the production of xenobiotic-induced genetic damage.[213] The supplementary mechanistic data obtained from such

GEM- and GER-based mutagenicity assays can be employed, together with the results from conventional and/or GEM-based carcinogenicity bioassays, to significantly enhance the assurance regarding genotoxic risk assessments for new xenobiotics.

The standard GEM and GER models used for in vivo mutagenicity testing in DDD target either endogenous genes for enzymes that participate in DNA repair[214,215] or bacterial transgenes that encode marker proteins.[186–188] The first scheme uses knockout GEM in which the purine salvage pathway required for DNA repair has been disrupted by eliminating one copy of the gene encoding either adenine phosphoribosyltransferase (Aprt[216,217]) or hypoxanthine phosphoribosyltransferase (Hprt[214]). In the absence of xenobiotic exposure, the spontaneous mutation rate for such endogenous genes is low, being affected chiefly by the animal's age.[218] Current commercial models that support the second paradigm are transgenic GEM or GER that bear several dozen copies of either bacterial *lacI* (Big Blue Mouse and Big Blue Rat; Stratagene, La Jolla, CA, USA) or bacterial *lacZ* (Muta Mouse; Covance Research Products, Denver, PA, USA) as stable additions to the native genome. Such bacterial inserts also have a low spontaneous mutation frequency (10^{-6} to 10^{-5}) similar to[213,216,219] or only slightly greater[220] than those of endogenous genes, and in most but not all cases[221] they are at greater risk for developing point mutations or small deletions after exposure to mutagens than is the animal's own genetic material. The tests based on the bacterial genes are particularly accepted as the ex vivo colorimetric readout is amenable to simple and rapid evaluation by both qualitative and quantitative means (**Figure 2**).

In vivo mutagenicity experiments may be readily customized by varying the study design with respect to the length and route of exposure and by judicious selection of a positive control mutagen. Such studies may be further tailored because the GEM models are available on several different genetic backgrounds (B6C3F1 or C57BL/6 strains for the Big Blue Mouse, Fischer 344 for the Big Blue Rat, and (BALB/c × DBA/2) CD2 F_1 for Muta Mouse). This latter feature is desirable because in many instances the same strain can be used for both the mutagenicity and carcinogenicity bioassays; the Big Blue Mouse is a popular choice in my experience as many laboratories perform their 2-year mouse carcinogenicity tests in B6C3F1 or C57BL/6 mice. Employing the same genetic background for both mutagenicity and carcinogenicity screens removes a variable from the risk assessment equation and provides an ancillary avenue for obtaining fundamental data regarding the potential genotoxic activity of new drug candidates.

3.08.3.3.2 **Carcinogenicity**

The GEM models now used for carcinogenicity testing in DDD bear induced mutations in which a tumor suppressor gene is repressed, an oncogene is overexpressed, or both a suppressor gene and an oncogene are altered. These engineered defects provide a built-in predisposition to xenobiotic-induced genetic damage in the form of a preexisting 'first hit'[222] in all tissues. Thus the rationale for employing GEM models in carcinogenicity risk assessment is that the nature of their induced mutations resembles the main spontaneous alterations that form the genetic basis for many human neoplasms.[192,223] Indeed, such GEM are more vulnerable to mutagens (**Figure 3**) than are wild-type mice with the same genetic background as indicated by a substantially reduced latency before the appearance of neoplasms.[224] Therefore, in the DDD setting, GEM provide a readout of a drug candidate's carcinogenic potential of greater sensitivity but in much less time (typically after 6 months of exposure) relative to the 'gold standard' 2-year bioassay in wild-type mice.[191] An added advantage of GEM carcinogenicity models is the ability to harvest mechanistic information regarding the interplay between specific tumor-related genes and new chemical entities.[223]

In the current literature three main GEM are currently used for DDD. The first two models, the Tg.Ac[225] and rasH2[226] GEM, are based on overexpression of the oncogene Ha-*ras*, an early event in the evolution of many human neoplasms that leads to a constitutive increase in cell proliferation.[227] The third model is founded on deletion of one allele of the tumor suppressor gene *p53*, a transcription factor engaged in cell cycle control which is commonly disrupted in spontaneous human and rodent neoplasms.[228] Heterozygous GEM missing one copy of *p53* ($p53^{+/-}$) have a longer lifespan than do knockout GEM ($p53^{-/-}$) and yet develop neoplasms months earlier than do wild-type mice ($p53^{+/+}$)[224] (**Figure 3**). Care must be taken in selecting such models as the genetic background on which the engineered mutation is carried can substantially influence the incidence of spontaneous tumors.[90] Other models that will be of interest to future DDD projects include GEM lacking both copies of the *Xpa* (xeroderma pigmentosum) gene, which encodes a factor needed for nucleotide excision repair of damaged DNA,[28] and the $Xpa^{-/-}/p53^{+/-}$ mouse,[28] in which one *p53* allele and both *Xpa* alleles are missing. The presence of dual mutations that independently predispose cells to neoplastic transformation is anticipated to yield an even more florid response to the genotoxic effects of xenobiotics in carcinogenicity bioassays using $Xpa^{-/-}/p53^{+/-}$ mice than would be apparent in GEM bearing either mutation alone, which would thus increase the sensitivity and possibility reduce the length of the assay.

The ability of these GEM to detect a carcinogenic predisposition depends on the nature of the xenobiotic being tested. Models in which the ability to repair damaged DNA is impacted ($p53^{+/-}$, $Xpa^{-/-}/p53^{+/-}$) are only vulnerable to the actions of genotoxic carcinogens, while those in which cell proliferation has been constitutively activated (rasH2,

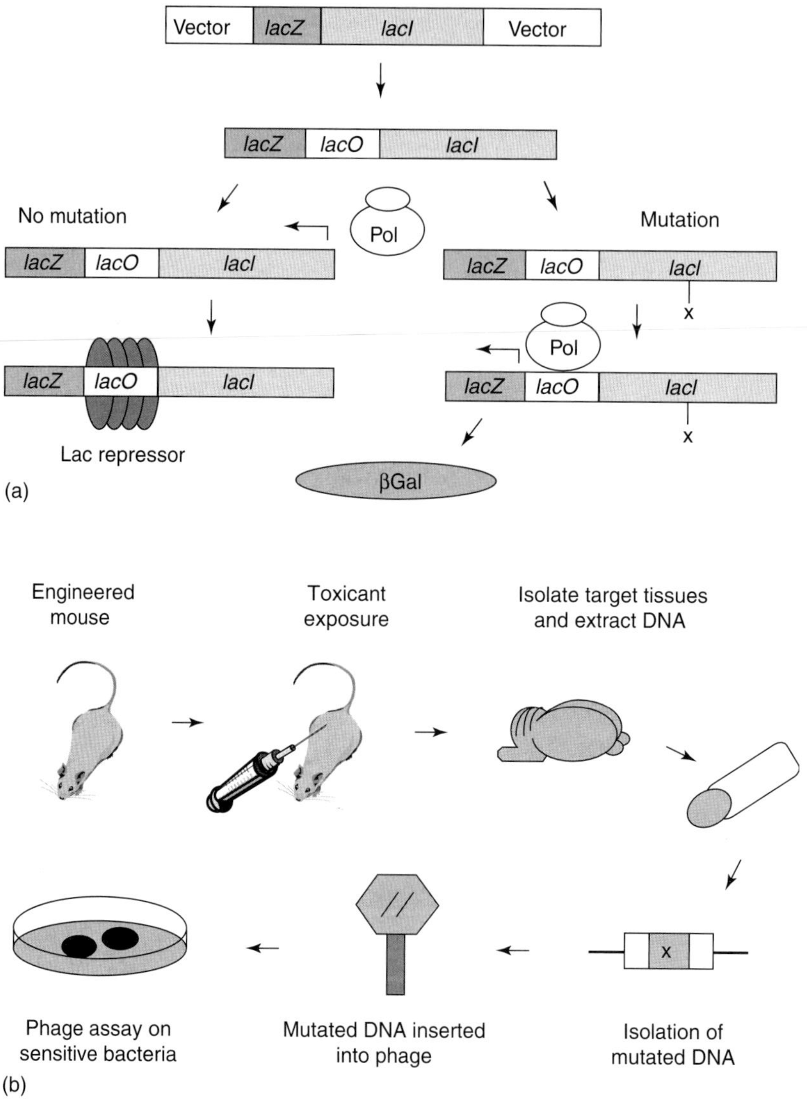

Figure 2 Schematic diagram demonstrating the Big Blue mutagenicity bioassay in GEM. Panel A shows the molecular basis of the test. The top bar shows the conformation of one copy of the transgene as it resides in the mouse genome; the two bacteria-derived elements in the transgene are *lacI*, which encodes a repressor protein, and *lacZ*, which encodes the enzyme beta-galactosidase (βGal). The second bar shows the gene arrangement in the bacterial genetic material; *lacO* is a gene encoding the operator protein, which is the site at which the *lacI*-derived repressor protein binds. Exposure to a genotoxic agent may (right column) or may not (left column) incite an inactivating mutation (denoted by an 'X') in the *lacI* gene. Isolated DNA is transferred into bacteria (see Panel B), where the genes are translated in order by DNA polymerase (Pol). In the absence of a drug-induced *lacI* mutation (left column), Lac repressor binds to *lacO* and prevents the transcription of *lacZ*. In contrast, drug-induced mutations in *lacI* (right column) thwart the production of Lac repressor so that the polymerase can bind to *lacO* and subsequently transcribe *lacZ*. If an appropriate substrate is available, lacZ converts it to a blue entity. Panel B demonstrates the technical aspects of the Big Blue assay. A transgenic mouse bearing the Big Blue construct is given a genotoxic drug (typically for 7 to 28 days). Next, potential target organs are removed and homogenized, and the DNA is isolated and packaged into bacteriophage so that it can be introduced into bacteria. Under appropriate culture conditions, bacteria that receive mutated DNA will make LacZ, and the colonies derived from them will appear on the plates as colored colonies.

Tg.Ac) are susceptible to both genotoxic and nongenotoxic agents. Thus, full characterization of a new drug's carcinogenic potential may require a tiered testing scheme in multiple GEM models to fully assess the risk (**Figure 4**). The low false-positive rate[229–232] and relatively standardized design of GEM-based carcinogenicity protocols – including the ability to reduce the starting-group sizes due to the low frequency of spontaneous tumors[233] – indicates that these models will see increasing use in future DDD projects.

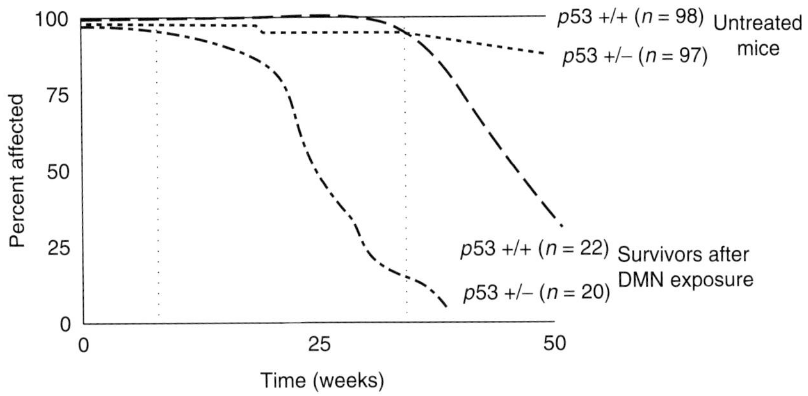

Figure 3 Susceptibility to carcinogens for models such as *p53*-deficient (*p53*$^{+/-}$) mice greatly exceeds that of wild-type (*p53*$^{+/+}$) animals as shown by a higher cancer incidence following a shorter latency. Untreated mice did not develop neoplasms (upper two lines), while animals given 0.0005% dimethylnitrosamine (DMN) in drinking water starting at 8 weeks of age did (lower two curves). Only *p53*$^{+/-}$ mice formed tumors during the 6-month exposure period (bordered by dotted vertical lines), which is the usual length of a GEM carcinogenicity study. All mice had an identical genetic background: 75% C57BL/6 and 25% 129/Sv. (Adapted from Goldsworthy, T. L.; Recio, L.; Brown, K.; Donehower, L. A.; Mirsalis, J. C.; Tennant, R. W.; Purchase, I. F. H. *Fundam. Appl. Toxicol.* **1994**, *22*, 8–19, with the permission of Oxford University Press and Bolon, B. *Basic Clin. Pharmacol. Toxicol.* **2004**, *95*, 154–161, with permission from Blackwell Publishing.)

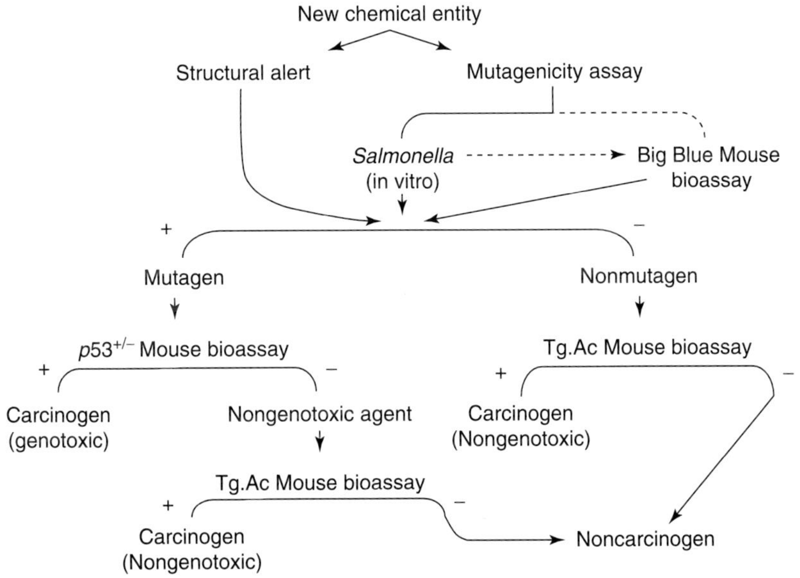

Figure 4 One possible decision tree approach for including various GEM models in the preclinical risk assessment strategy for evaluating the mutagenic and carcinogenic properties of potential therapeutics. (Adapted from Tennant, R. W.; French, J. E.; Spalding, J. W. *Environ. Health Perspect.* **1995**, *103*, 942–950, with the permission of the United States government and Bolon, B. *Basic Clin. Pharmacol. Toxicol.* **2004**, *95*, 154–161, with permission from Blackwell Publishing.)

3.08.4 Caveats to Using Engineered Animals in Drug Discovery and Development

As with any tool used to make DDD decisions, data from GEM and GER must be interpreted in light of side issues that can affect the experimental outcome. First, it must be remembered that each GEM or GER is a unique 'species' unto itself. This truth chiefly reflects insurmountable constraints imposed by current engineering technology, namely the random nature of transgene insertion[50] and the subtle distinctions in flanking minisatellite sequences that follow knockout genes through the process of homologous recombination.[167,234] This situation is regularly complicated when competing research groups place their mutation on strains with dissimilar genetic backgrounds.[157,235] However, additional factors also impact the expression of an engineered gene in its host, especially intrauterine variations in hormone exposure resulting from differences in sex and gestational position as well the degree of maternal stress.[167] While the typical DDD assumption that all GEM or GER bearing a given gene modification will exhibit an identical phenotype is commonly borne out in practice, this correlation is by no means obligatory. For example, knockout GEM in which the engineered gene is identical but the exact nature of the genetic modification is different can have divergent phenotypes, a fact exemplified by mice missing cytochrome P450 1A2 (CYP1A2; the liver enzyme in humans and mice responsible for *N*-hydroxylation of numerous drugs and environmental contaminants but few endogenous molecules). In one laboratory, disruption but not deletion of a single CYP1A2 exon generated distinct knockout phenotypes in early (fatal pulmonary deficits in many but not all pups) and late (no phenotype without pharmacologic challenge) generations,[174] while in another facility CYP1A2 knockout GEM lacking all or part of four exons exhibited only the latter phenotype.[113] Another illustration of this phenomenon is the inability of some but not all hemizygous Tg.Ac mice to develop neoplasms after exposure to 12-*O*-tetradecanoylphorbol 13-acetate, a well-known tumor promoter.[236] The two main messages to be drawn from such instances are that not all GEM or GER in which the same gene has been manipulated will be identical in either genotypic or phenotypic terms, and that unexpected phenotypic drift can occur over time.[210] These facts point to the need for careful characterization of GEM and GER before their use to make critical DDD decisions. In particular, assessments must be made with the clear understanding that 'humanized' GEM and GER are still rodents[49–51] even though they bear human genetic material[44,47] or human cells.[48] Data derived in such models at best only points the way forward in broad strokes, and cannot be used in a formal sense as a substitute for well-designed human clinical trials.

Other aspects also must be considered when interpreting GEM and GER data for DDD, but such matters are comparable to those that must be addressed for experiments conducted using wild-type animals. For example, certain mouse strains favored in GEM production have incidental lesions that might, under some circumstances, mask the changes induced by a true phenotype. Spontaneous conditions illustrating this point include line-specific epilepsy in FVB mice[237] and a high incidence of teratoma in certain 129 strains.[90,157] Various 129 substrains also exhibit attenuation or agenesis of the corpus callosum,[238] although the penetrance of this defect is incomplete[239]; to my knowledge, the relationship of this anatomic aberration with altered cognitive ability over time[155] has not been investigated in detail. Divergence in genetic background between strains also can have a pronounced impact on phenotypic expression (e.g., different variety and latency of neoplasms among parental mouse strains with distinct genetic backgrounds), indicating that background lesions for a given strain used for engineering projects must be well characterized so that spurious findings are not accorded too much weight in making DDD decisions. The ultimate message is that considerable care must be taken when choosing the rodent strain before a genetic engineering project is initiated.

Finally, care must be taken that the use of GEM and GER in DDD is undertaken as part of a well-designed program of mechanism-based research. Many companies have extrapolated GEM data derived from one or two screening experiments performed rapidly in an insufficiently characterized new model into proof that a given target or potential therapeutic has been validated – and wasted millions of dollars in further pursuit of a fruitless project. In my experience, DDD programs make two main blunders in this regard. The first is the overaggressive reading of a clinical abnormality as a gene-associated phenotype when, in fact, the condition results from a confounding factor. One example of such a syndrome is neonatal GEM lacking a neuron-specific protein which developed hydrocephalus and ataxia due to beta-hemolytic *Streptococcus* septicemia and encephalitis secondary to premature tailing for DNA analysis (unpublished data). The second mistake is to miss a real phenotype because the phenotypic screen did not use appropriate tests on animals of the correct age to find the phenotype in an uninteresting (to the firm's DDD areas) organ system. Both these dilemmas may be easily alleviated by preparing a robust database for the molecule of interest before a final verdict is made. At present, the typical GEM- or GER-derived data set from most industrial DDD programs contains two distinct types of information. The first sort is gained from a series of discovery-stage studies using transgenic and/or knockout models (usually generated in-house) to define in detail the relationship among a given target receptor, its ligand, and one or more molecules from the associated signal transduction cascade.[182,203] The second class is obtained later in preclinical development by exposing well-defined commercial GEM models to candidate therapeutic agents to test their mutagenic or carcinogenic potential; while the separation is not absolute, information of this nature is usually only gathered for small molecules, and thus represents the province of chemical and pharmaceutical companies rather than biopharmaceutical firms (*see* 1.09 Traditional Medicines; 2.08 Pharma versus Biotech: Contracts, Collaborations, and Licensing). In the ideal DDD scenario, the phenotypes obtained from the discovery stage mechanistic work would foretell the outcome elicited by administering a drug to modulate the molecular pathway in question.[2] However, candidate therapeutics often do not recapitulate the effects of gene knockout, presumably because the lifelong gene ablation in the knockout animal represents a fundamentally different biological condition from the intermittent and transient inhibition that can be attained using a drug.[115] I anticipate that future DDD work will increasingly use conditional genetic modifications[33,35] and intermittent epigenetic interference[19] to more effectively address such discrepancies and speed pipeline development.

3.08.5 Conclusion

Genetic engineering techniques have matured greatly in recent years, and now function as a major component of modern DDD programs. Numerous well-defined GEM and a burgeoning cohort of GER are readily available as models for both basic and applied research. Use of such models will enable us to accelerate the rate at which we dissect elemental biological mechanisms of health and disease, and develop new, rationally designed drugs to target a host of previously incurable conditions. The increasing utilization of GEM and GER throughout the DDD process likely will lead to the inclusion of bioassays featuring engineered rodents as accepted practice in regulatory submissions within the next few years.

References

1. Sharp, J. J.; Mobraaten, L. E.; Bedigian, H. G. In *Pathology of Genetically Engineered Mice*; Ward, J. M., Mahler, J. F., Maronpot, R. R., Sundberg, J. P., Frederickson, R. M., Eds.; Iowa State University Press: Ames, IA, 2000, pp 3–9.
2. Zambrowicz, B. P.; Sands, A. T. *Nat. Rev. Drug Disc.* **2003**, *2*, 38–51.
3. Zambrowicz, B. P.; Turner, C. A.; Sands, A. T. *Curr. Opin. Pharmacol.* **2003**, *3*, 563–570.
4. Wei, L.-N. *Annu. Rev. Pharmacol. Toxicol.* **1997**, *37*, 119–141.
5. Rudolph, U.; Moehler, H. *Eur. J. Pharmacol.* **1999**, *375*, 327–337.
6. Bolon, B.; Galbreath, E.; Sargent, L.; Weiss, J. In *The Laboratory Rat*; Krinke, G., Ed.; Academic Press: London, 2000, pp 603–634.
7. Beerli, R. R.; Wels, W.; Hynes, N. E. *Biochem. Biophys. Res. Comm.* **1994**, *204*, 666–672.
8. Celli, G.; LaRochelle, W. J.; Mackem, S.; Sharp, R.; Merlino, G. *EMBO J.* **1998**, *6*, 1642–1655.
9. Min, H.; Morony, S.; Sarosi, I.; Dunstan, C. R.; Capparelli, C.; Scully, S.; Van, G.; Kaufman, S.; Kostenuik, P. J.; Lacey, D. L. et al. *J. Exp. Med.* **2000**, *192*, 463–474.
10. Brinster, R. L.; Chen, H. Y.; Trumbauer, M. E.; Yagle, M. K.; Palmiter, R. D. *Proc. Natl. Acad. Sci. USA* **1985**, *82*, 4438–4442.
11. Simoes, S.; Slepuskin, V.; Gaspar, R.; de Lima, M. C. P.; Duzgunes, N. *Gene Ther.* **1998**, *5*, 955–964.
12. Robbins, P. D.; Tahara, H.; Ghivizzani, S. C. *Trends Biotechnol.* **1998**, *16*, 35–40.
13. Bolon, B.; Carter, C.; Daris, M.; Morony, S.; Capparelli, C.; Hsieh, A.; Mao, M.; Kostenuik, P.; Dunstan, C. R.; Lacey, D. L. et al. *Mol. Ther.* **2001**, *3*, 197–205.
14. Capecchi, M. R. *Science* **1989**, *244*, 1288–1292.
15. Koller, B. H.; Smithies, O. *Annu. Rev. Immunol.* **1992**, *10*, 705–730.
16. Wood, S. A.; Allen, N. D.; Rossant, J.; Auerbach, A.; Nagy, A. *Nature* **1993**, *365*, 87–89.
17. Khillan, J. S.; Bao, Y. *Biotechniques* **1997**, *22*, 544–549.
18. Driver, S. E.; Robinson, G. S.; Flanagan, J.; Shen, W.; Smith, L. E.; Thomas, D. W.; Roberts, P. C. *Nat. Biotechnol.* **1999**, *17*, 1184–1187.
19. Lindsay, M. A. *Nat. Rev. Drug Disc.* **2003**, *2*, 831–838.
20. Schiavone, N.; Donnini, M.; Nicolin, A.; Capaccioli, S. *Curr. Pharm. Des.* **2004**, *10*, 769–784.
21. Voorhoeve, P. M.; Agami, R. *Trends Biotechnol.* **2003**, *21*, 2–4.
22. Kissler, S.; Van Parijs, L. *Expert Rev. Mol. Diagn.* **2004**, *4*, 645–651.
23. Kashani-Sabet, M.; Liu, Y.; Fong, S.; Desprez, P. Y.; Liu, S.; Tu, G.; Nosrati, M.; Handumrongkul, C.; Liggitt, D.; Thor, A. D. et al. *Proc. Natl. Acad. Sci. USA* **2002**, *99*, 3878–3883.
24. Wiley, S. R. *Curr. Pharm. Des.* **1998**, *4*, 417–422.
25. te Riele, H.; Brouwers, C.; Dekker, M. *Methods Mol. Biol.* **2001**, *158*, 251–262.
26. Johnson, D. R.; Finch, R. A.; Lin, Z. P.; Zeiss, C. J.; Sartorelli, A. C. *Cancer Res.* **2001**, *61*, 1469–1476.
27. Nakae, J.; Kido, Y.; Accili, D. *Endocrinol. Rev.* **2001**, *22*, 818–835.
28. van Steeg, H.; de Vries, A.; van Oostrom, C. T.; van Benthem, J.; Beems, R. B.; van Kreijl, C. F. *Toxicol. Pathol.* **2001**, *29*, 109–116.
29. Belteki, G.; Haigh, J.; Kabacs, N.; Haigh, K.; Sison, K.; Costantini, F.; Whitsett, J.; Quaggin, S. E.; Nagy, A. *Nucleic Acids Res.* **2005**, *33*, e51. [Erratum in *Nucleic Acids Res.* **2005**, *33*, 2765.]
30. Cox, V.; Clarke, S.; Czyzyk, T.; Ansonoff, M.; Nitsche, J.; Hsu, M. S.; Borsodi, A.; Tomboly, C.; Toth, G.; Hill, R. et al. *Neuropharmacology* **2005**, *48*, 228–235.
31. Törnell, J.; Snaith, M. *Drug Disc. Today* **2002**, 7, 461–470.
32. Schwenk, F.; Kühn, R.; Angrand, P. O.; Rajewsky, K.; Stewart, A. F. *Nucleic Acids Res.* **1998**, *26*, 1427–1432.
33. Shastry, B. S. *Mol. Cell. Biochem.* **1998**, *181*, 163–179.
34. Wang, Y.; DeMayo, F. J.; Tsai, S. Y.; O'Malley, B. W. *Nature Biotechnol.* **1997**, *15*, 239–243.
35. Tsien, J. Z.; Chen, D. F.; Gerber, D.; Tom, C.; Mercer, E. H.; Anderson, D. J.; Mayford, M.; Kandel, E. R.; Tonegawa, S. *Cell* **1996**, *87*, 1317–1326.
36. Rajewsky, K.; Gu, H.; Kuehn, R.; Betz, U. A.; Mueller, W.; Roes, J.; Schwenk, F. *J. Clin. Invest.* **1996**, *98*, 600–603.
37. Dymecki, S. M. *Proc. Natl. Acad. Sci. USA* **1996**, *93*, 6191–6196.
38. Akagi, K.; Sandig, V.; Vooijs, M.; Van der Valk, M.; Giovannini, M.; Strauss, M.; Berns, A. *Nucleic Acids Res.* **1997**, *25*, 1766–1773.
39. Gu, H.; Marth, J. D.; Orban, P. C.; Mossmann, H.; Rajewsky, K. *Science* **1994**, *265*, 103–106.
40. Kühn, R.; Schwenk, F.; Aguet, M.; Rajewsky, K. *Science* **1995**, *269*, 1427–1429.
41. Metzger, D.; Clifford, J.; Chiba, H.; Chambon, P. *Proc. Natl. Acad. Sci. USA* **1995**, *92*, 6991–6995.

42. Furuta, Y.; Behringer, R. R. *Birth Defects Res. (Pt C)* **2005**, *75*, 43–57.
43. Chang, H. S.; Lin, C. H.; Chen, Y. C.; Yu, W. C. Y. *Am. J. Pathol.* **2004**, *165*, 1535–1541.
44. Nebert, D. W.; Dalton, T. P.; Stuart, G. W.; Carvan, M., Jr. *Ann. NY Acad. Sci.* **2000**, *919*, 148–170.
45. Ueno, T.; Tamura, S.; Frels, W. I.; Shou, M.; Gonzalez, F. J.; Kimura, S. *Biochem. Pharmacol.* **2000**, *60*, 857–863.
46. Xie, W.; Barwick, J. L.; Downes, M.; Blumberg, B.; Simon, C. M.; Nelson, M. C.; Neuschwander-Tetri, B. A.; Brunt, E. M.; Guzelian, P. S.; Evans, R. M. *Nature* **2000**, *406*, 435–439.
47. Xie, W.; Evans, M. R. *Drug Disc. Today* **2002**, *7*, 509–515.
48. Tateno, C.; Yoshizane, Y.; Saito, N.; Kataoka, M.; Utoh, R.; Yamasaki, C.; Tachibana, A.; Soeno, Y.; Asahina, K.; Hino, H. et al. *Am. J. Pathol.* **2004**, *165*, 901–912.
49. Cordaro, C. J. *Risk Anal.* **1989**, *9*, 157–168.
50. Liggitt, H. D.; Reddington, G. M. *Xenobiotica* **1992**, *22*, 1043–1054.
51. Setola, V.; Roth, B. L. *Mol. Pharmacol.* **2003**, *64*, 1277–1278.
52. Nagy, A.; Rossant, J. *Int. J. Dev. Biol.* **2001**, *45*, 577–582.
53. Faerman, A.; Shani, M. *Methods. Cell Biol.* **1998**, *52*, 373–403.
54. Shibata, N.; Oda, H.; Hirano, A.; Kato, Y.; Kawaguchi, M.; Dal Canto, M. C.; Uchida, K.; Sawada, T.; Kobayashi, M. *Neuropathology* **2002**, *22*, 337–349.
55. Mercer, E. http://www.rodentia.com/wmc/ (accessed Oct 2006).
56. Franco, D.; de Boer, P. A.; de Gier-de Vries, C.; Lamers, W. H.; Moorman, A. F. *Eur. J. Morphol.* **2001**, *39*, 169–191.
57. Jung, S.; Aliberti, J.; Graemmel, P.; Sunshine, M. J.; Kreutzberg, G. W.; Sher, A.; Littman, D. R. *Mol. Cell. Biol.* **2000**, *20*, 4106–4114.
58. Isola, L. M.; Gordon, J. W. *Dev. Genet.* **1988**, *9*, 181–191.
59. Ganten, D.; Wagner, J.; Zeh, K.; Bader, M.; Michel, J. B.; Paul, M.; Zimmermann, F.; Ruf, P.; Hilgenfeldt, U.; Ganten, U. et al. *Proc. Natl. Acad. Sci. USA* **1992**, *89*, 7806–7810.
60. Larsson, L.-I. *Appl. Immunohistochem.* **1993**, *1*, 2–16.
61. Ramos-Vara, J. A. *Vet. Pathol.* **2005**, *42*, 405–426.
62. Chayen, J.; Bitensky, L. In *Cell Biology: A Laboratory Handbook*, 2nd ed.; Celis, J. E., Ed.; Academic Press: San Diego, CA, 1998; Vol. 3, pp 238–248.
63. Linscott, W. D. *Linscott's Directory of Immunological and Biological Reagents*, 10th ed.; Linscott: Santa Rosa, CA, 1998.
64. Weimer, R. V. *MSRS Catalog of Primary Antibodies*, 3rd ed.; Aerie: Birmingham, MI, 1996.
65. Rossant, J.; McKerlie, C. *Trends Mol. Med.* **2001**, *7*, 502–507.
66. Murray, K. A. *Lab. Anim.* **2002**, *31*, 25–29.
67. Lee, M. A.; Boehm, M.; Kim, S.; Bachmann, S.; Bachmann, J.; Bader, M.; Ganten, D. *Hypertension* **1995**, *25*, 570–580.
68. Sereda, M.; Griffiths, I.; Pülhofer, A.; Stewart, H.; Rossner, M. J.; Zimmermann, F.; Magyar, J. P.; Schneider, A.; Hund, E.; Meinck, H.-M. et al. *Neuron* **1996**, *16*, 1049–1060.
69. Veniant, M.; Menard, J.; Bruneval, P.; Morley, S.; Gonzales, M. F.; Mullins, J. *J. Clin. Invest.* **1996**, *98*, 1966–1970.
70. Cranston, A.; Bocker, T.; Reitmar, A.; Palazzo, J.; Wilson, T.; Mak, T.; Fishel, R. *Nat. Genet.* **1997**, *17*, 114–118.
71. Kwitek-Black, A. E.; Jacob, H. J. *Curr. Hypertens. Rep.* **2001**, *3*, 12–18.
72. Singer, J. B.; Hill, A. E.; Burrage, L. C.; Olszens, K. R.; Song, J.; Justice, M.; O'Brien, W. E.; Conti, D. V.; Witte, J. S.; Lander, E. S. et al. *Science* **2004**, *304*, 445–448.
73. Nachtman, R. G.; Dunn, C. D. R.; Driscoll, T. B.; Leach, C. S. *Lab. Anim. Sci.* **1985**, *35*, 505–508.
74. Ullman-Culleré, M. H.; Foltz, C. J. *Lab. Anim. Sci.* **1999**, *49*, 319–323.
75. Hoff, J. *Lab. Anim.* **2000**, *29*, 47–53.
76. Chew, J. L.; Chua, K. Y. *Lab. Anim.* **2003**, *32*, 48–50.
77. Relyea, M. J.; Miller, J.; Boggess, D.; Sundberg, J. P. In *Systematic Approach to Evaluation of Mouse Mutations*; Sundberg, J. P., Boggess, D., Eds.; CRC Press: Boca Raton, FL, 2000, pp 57–89.
78. Brayton, C.; Justice, M.; Montgomery, C. A. *Vet. Pathol.* **2001**, *38*, 1–19. [Erratum in *Vet. Pathol.* **2001**, *38*, 568.]
79. Bronson, R. T. In *Methods in Molecular Biology: Gene Knockout Protocols*; Tymms, M. J., Kola, I., Eds.; Humana: Totowa, NJ, 2001, pp 155–180.
80. Car, B. D.; Eng, V. M. *Vet. Pathol.* **2001**, *38*, 20–30.
81. Papaioannou, V. E.; Behringer, R. R. *Mouse Phenotypes: A Handbook of Mutation Analysis*; Cold Spring Harbor Laboratory Press: Cold Spring Harbor, NY, 2005.
82. Mohr, U.; Dungworth, D. L.; Capen, C. C.; Carlton, W. W.; Sundberg, J. P.; Ward, J. M. *Pathobiology of the Aging Mouse*; ILSI Press: Washington, DC, 1996.
83. Maronpot, R. R. *Pathology of the Mouse: Reference and Atlas*; Cache River Press: Vienna, IL, 1999.
84. Ward, J. M.; Mahler, J. F.; Maronpot, R. R.; Sundberg, J. P. *Pathology of Genetically Engineered Mice*; Iowa State University Press: Ames, IA, 2000.
85. Kaufman, M. H. *The Atlas of Mouse Development*, 2nd ed.; Academic Press: San Diego, CA, 1992.
86. Kaufman, M. H.; Bard, J. B. L. *The Anatomical Basis of Mouse Development*; Academic Press: San Diego, CA, 1999.
87. Boorman, G. A.; Eustis, S. L.; Elwell, M. R.; Montgomery, C. A. J.; MacKenzie, W. F. *Pathology of the Fischer Rat: Reference and Atlas*; Academic Press: San Diego, CA, 1990.
88. Krinke, G. *The Laboratory Rat*; Academic Press: San Diego, CA, 2000.
89. Mahler, J. F.; Stokes, W.; Mann, P. C.; Takaoka, M.; Maronpot, R. R. *Toxicol. Pathol.* **1996**, *24*, 710–716.
90. Harvey, M.; McArthur, M. J.; Montgomery, C. A. J.; Bradley, A.; Donehower, L. A. *FASEB J.* **1993**, *7*, 938–943.
91. Haines, D. C.; Chattopadhyay, S.; Ward, J. M. *Toxicol. Pathol.* **2001**, *29*, 653–661.
92. Fu, L.; Pelicano, H.; Liu, J.; Huang, P.; Lee, C. *Cell* **2002**, *111*, 41–50. [Erratum in *Cell* **2002**, *111*, 1055.]
93. Peschon, J. J.; Slack, J. L.; Reddy, P.; Stocking, K. L.; Sunnarborg, S. W.; Lee, D.; Russell, W. E.; Castner, B. J.; Johnson, R. S.; Fitzner, J. N. et al. *Science* **1998**, *282*, 1281–1284.
94. Zheng, F.; Plati, A. R.; Potier, M.; Schulman, Y.; Berho, M.; Banerjee, A.; Leclercq, B.; Zisman, A.; Striker, L. J.; Striker, G. E. *Am. J. Pathol.* **2003**, *162*, 1339–1348.
95. Rogers, D. C.; Fisher, E. M.; Brown, S. D.; Peters, J.; Hunter, A. J.; Martin, J. E. *Mammal. Genome* **1997**, *8*, 711–713.
96. Crawley, J. N. *Brain Res.* **1999**, *835*, 18–26.
97. Crawley, J. N. *What's Wrong With My Mouse?: Behavioral Phenotyping of Transgenic and Knockout Mice*; John Wiley: New York, 2000.
98. Moser, V. C. *Neurotoxicology* **2000**, *21*, 989–996.
99. Brockway, B. P.; Mills, P.; Kramer, K. *Lab. Anim.* **1998**, *27*, 40–46.

100. Hartley, C. J.; Taffet, G. E.; Reddy, A. K.; Entman, M. L.; Michael, L. H. *ILAR J.* **2002**, *43*, 147–158.
101. Olgin, J. E.; Verheule, S. *Cardiovasc. Res.* **2002**, *54*, 280–286.
102. Yang, Y.; Frankel, W. N. *Adv. Exp. Med. Biol.* **2004**, *548*, 1–11.
103. Lindon, J. C.; Holmes, E.; Nicholson, J. K. *Curr. Opin. Mol. Ther.* **2004**, *6*, 265–272.
104. Herschman, H. R.; MacLaren, D. C.; Iyer, M.; Namavari, M.; Bobinski, K.; Green, L. A.; Wu, L.; Berk, A. J.; Toyokuni, T.; Barrio, J. R. et al. *J. Neurosci. Res.* **2000**, *59*, 699–705.
105. Louie, A. Y.; Hueber, M. M.; Ahrens, E. T.; Rothbaecher, U.; Moats, R.; Jacobs, R. E.; Fraser, S. E.; Meade, T. J. *Nat. Biotechnol.* **2000**, *18*, 321–325.
106. Mor-Avi, V.; Korcarz, C.; Fentzke, R. C.; Lin, H.; Leiden, J. M.; Lang, R. M. *J. Am. Soc. Echocardiog.* **1999**, *12*, 209–214.
107. Reiman, E. M.; Uecker, A.; Gonzalez-Lima, F.; Minear, D.; Chen, K.; Callaway, N. L.; Berndt, J. D.; Games, D. *Neuroreport* **2000**, *11*, 987–991.
108. Rudin, M.; Beckmann, N.; Porszasz, R.; Reese, T.; Bochelen, D.; Sauter, A. *NMR Biomed.* **1999**, *12*, 69–97.
109. Yu, Y.; Annala, A. J.; Barrio, J. R.; Toyokuni, T.; Satyamurthy, N.; Namavari, M.; Cherry, S. R.; Phelps, M. E.; Herschman, H. R.; Gambhir, S. S. *Nat. Med.* **2000**, *6*, 933–937.
110. Eckelman, W. C. *Drug Disc. Today* **2003**, *8*, 404–410.
111. Heckl, S.; Pipkorn, R.; Nagele, T.; Vogel, U.; Kuker, W.; Voight, K. *Histol. Histopathol.* **2004**, *19*, 651–668.
112. Gonzalez, F. J.; Kimura, S. *Cancer Lett.* **1999**, *143*, 199–204.
113. Liang, H. C.; Li, H.; McKinnon, R. A.; Duffy, J. J.; Potter, S. S.; Puga, A.; Nebert, D. W. *Proc. Natl. Acad. Sci. USA* **1996**, *93*, 1671–1676.
114. Gingrich, J. A.; Hen, R. *Curr. Opin. Neurobiol.* **2000**, *10*, 146–152.
115. Pich, E. M.; Epping-Jordan, M. P. *Ann. Med.* **1998**, *30*, 390–396.
116. Scearce-Levie, K.; Chen, J. P.; Gardner, E.; Hen, R. *Ann. NY Acad. Sci.* **1999**, *868*, 701–715.
117. Grubb, B. R.; Boucher, R. C. *Physiol. Rev.* **1999**, *79*, S193–S214.
118. Heisler, L. K.; Chu, H. M.; Brennan, T. J.; Danao, J. A.; Bajwa, P.; Parsons, L. H.; Tecott, L. H. *Proc. Natl. Acad. Sci. USA* **1998**, *95*, 15049–15054.
119. Parks, C. L.; Robinson, P. S.; Sibille, E.; Shenk, T.; Toth, M. *Proc. Natl. Acad. Sci. USA* **1998**, *95*, 10734–10739.
120. Ramboz, S.; Oosting, R.; Amara, D. A.; Kung, H. F.; Blier, P.; Mendelsohn, M.; Mann, J. J.; Brunner, D.; Hen, R. *Proc. Natl. Acad. Sci. USA* **1998**, *95*, 14476–14481.
121. Moore, M. W.; Klein, R. D.; Farinas, I.; Sauer, H.; Armanini, M.; Phillips, H.; Reichardt, L. F.; Ryan, A. M.; Carver-Moore, K.; Rosenthal, A. *Nature* **1996**, *382*, 76–79.
122. Pichel, J. G.; Shen, L.; Sheng, H. Z.; Granholm, A. C.; Drago, J.; Grinberg, A.; Lee, E. J.; Huang, S. P.; Saarma, M.; Hoffer, B. J.; Sariola, H.; Westphal, H. *Cold Spring Harb. Symp. Quant. Biol.* **1996**, *61*, 445–457.
123. Sanchez, M. P.; Silos-Santiago, I.; Frisen, J.; He, B.; Lira, S. A.; Barbacid, M. *Nature* **1996**, *382*, 70–73.
124. Campbell, I. L. *Int. J. Dev. Neurosci.* **1995**, *13*, 275–284.
125. Domen, J.; van der Lugt, N. M.; Laird, P. W.; Saris, C. J.; Clarke, A. R.; Hooper, M. L.; Berns, A. *Blood* **1993**, *82*, 1445–1452.
126. Krege, J. H.; Hodgin, J. B.; Couse, J. F.; Enmark, E.; Warner, M.; Mahler, J. F.; Sar, M.; Korach, K. S.; Gustafsson, J. A.; Smithies, O. *Proc. Natl. Acad. Sci. USA* **1998**, *95*, 15677–15682.
127. Laird, P. W.; van der Lugt, N. M.; Clarke, A.; Domen, J.; Linders, K.; McWhir, J.; Berns, A.; Hooper, M. *Nucleic Acids Res.* **1993**, *21*, 4750–4755.
128. Sasaki, T.; Irie-Sasaki, J.; Jones, R. G.; Oliveira-dos-Santos, A. J.; Stanford, W. L.; Bolon, B.; Wakeham, A.; Itie, A.; Bouchard, D.; Kozieradzki, I.; Joza, N.; Mak, T. W.; Ohashi, P. S.; Suzuki, A.; Penninger, J. M. *Science* **2000**, *287*, 1040–1046.
129. Feng, G. P.; Mellor, R. H.; Bernstein, M.; Keller-Peck, C.; Nguyen, Q. T.; Wallace, M.; Nerbonne, J. M.; Lichtman, J. W.; Sanes, J. R. *Neuron* **2000**, *28*, 41–51.
130. Su, M.; Hu, H.; Lee, Y.; d'Azzo, A.; Messing, A.; Brenner, M. *Neurochem. Res.* **2004**, *29*, 2075–2093.
131. Ishida, J.; Sugiyama, F.; Tanimoto, K.; Taniguchi, K.; Syouji, M.; Takimoto, E.; Horiguchi, H.; Murakami, K.; Yagami, K.; Fukamizu, A. *Biochem. Biophys. Res. Commun.* **1998**, *252*, 610–616.
132. McNeish, J. D.; Scott, W. J.; Potter, S. S. *Science* **1988**, *241*, 837–839.
133. Tohyama, J.; Vanier, M. T.; Suzuki, K.; Ezoe, T.; Matsuda, J.; Suzuki, K. *Hum. Mol. Genet.* **2000**, *9*, 1699–1707.
134. Dougall, W. C.; Glaccum, M.; Charrier, K.; Rohrbach, K.; Brasel, K.; De Smedt, T.; Daro, E.; Smith, J.; Tometsko, M. E.; Maliszewski, C. R. et al. *Genes Dev.* **1999**, *13*, 2412–2424.
135. Kong, Y. Y.; Yoshida, H.; Sarosi, I.; Tan, H. L.; Timms, E.; Capparelli, C.; Morony, S.; Oliveira dos Santos, A. J.; Van, G.; Itie, A. et al. *Nature* **1999**, *397*, 315–323.
136. Li, J.; Sarosi, I.; Yan, X.-Q.; Morony, S.; Capparelli, C.; Tan, H. L.; McCabe, S.; Elliott, R.; Scully, S.; Van, G. et al. *Proc. Natl. Acad. Sci. USA* **2000**, *97*, 1566–1571.
137. Simonet, W. S.; Lacey, D. L.; Dunstan, C. R.; Kelley, M.; Chang, M.-S.; Luethy, R.; Nguyen, H. Q.; Wooden, S.; Bennett, L.; Boone, T. et al. *Cell* **1997**, *89*, 309–319.
138. Bucay, N.; Sarosi, I.; Dunstan, C. R.; Morony, S.; Tarpley, J.; Capparelli, C.; Scully, S.; Tan, H. L.; Xu, W.; Lacey, D. L. et al. *Genes Dev.* **1998**, *12*, 1260–1268.
139. Mizuno, A.; Amizuka, N.; Irie, K.; Murakami, A.; Fujise, N.; Kanno, T.; Sato, Y.; Nakagawa, N.; Yasuda, H.; Mochizuki, S. et al. *Biochem. Biophys. Res. Commun.* **1998**, *247*, 610–615.
140. Kostenuik, P.; Bolon, B.; Morony, S.; Daris, M.; Geng, Z.; Carter, C.; Sheng, J. *Bone* **2004**, *34*, 656–664.
141. Capparelli, C.; Kostenuik, P. J.; Morony, S.; Starnes, C.; Weimann, B.; Van, G.; Scully, S.; Qi, M. Y.; Lacey, D. L.; Dunstan, C. R. *Cancer Res.* **2000**, *60*, 783–787.
142. Kong, Y. Y.; Feige, U.; Sarosi, I.; Bolon, B.; Tafuri, A.; Morony, S.; Capparelli, C.; Li, J.; Elliott, R.; McCabe, S. et al. *Nature* **1999**, *402*, 304–309.
143. Bekker, P. J.; Holloway, D.; Nakanishi, A.; Arrighi, H. M.; Leese, P. T.; Dunstan, C. R. *J. Bone Miner. Res.* **2001**, *16*, 348–360.
144. Body, J. J.; Greipp, P.; Coleman, R. E.; Facon, T.; Geurs, F.; Fermand, J. P.; Harousseau, J. L.; Lipton, A.; Mariette, X.; Williams, C. D. et al. *Cancer* **2003**, *97*, 887–892.
145. Blackwood, E. M.; Kadonga, J. T. *Science* **1998**, *281*, 60–63.
146. Ogbourne, S.; Antalis, T. M. *Biochem. J.* **1998**, *331*, 1–14.
147. Clark, A. J.; Bissinger, P.; Bullock, D. W.; Damak, S.; Wallace, R.; Whitelaw, C. B. A.; Yull, F. *Reprod. Fertil. Dev.* **1994**, *6*, 589–598.
148. Pham, C. T.; MacIvor, D. M.; Hug, B. A.; Heusel, J. W.; Ley, T. J. *Proc. Natl. Acad. Sci. USA* **1996**, *93*, 13090–13095.
149. Rijli, F. M.; Dolle, P.; Fraulob, V.; LeMeur, M.; Chambon, P. *Dev. Dynam.* **1994**, *201*, 366–377.
150. Robertson, E.; Bradley, A.; Kuehn, M.; Evans, M. *Nature* **1986**, *323*, 445–448.

151. Woychik, R. P.; Stewart, T. A.; Davis, L. G.; D'Eustachio, P.; Leder, P. *Nature* **1985**, *318*, 36–40.
152. Threadgill, D. W.; Dlugosz, A. A.; Hansen, L. A.; Tennenbaum, T.; Lichti, U.; Yee, D.; LaMantia, C.; Mourton, T.; Herrup, K.; Harris, R. C. et al. *Science* **1995**, *269*, 230–234.
153. Crawley, J. N.; Belknap, J. K.; Collins, A.; Crabbe, J. C.; Frankel, W.; Henderson, N.; Hitzemann, R. J.; Maxson, S. C.; Miner, L. L.; Silva, A. J. et al. *Psychopharmacology* **1997**, *132*, 107–124.
154. Cranston, A.; Fishel, R. *Mammal. Genome* **1999**, *10*, 1020–1022.
155. Hengemihle, J. M.; Long, J. M.; Betkey, J.; Jucker, M.; Ingram, D. K. *Neurobiol. Aging* **1999**, *20*, 9–18.
156. Sibilia, M.; Steinbach, J. P.; Stingl, L.; Aguzzi, A.; Wagner, E. F. *EMBO J.* **1998**, *17*, 719–731.
157. Simpson, E. M.; Linder, C. C.; Sargent, E. E.; Davisson, M. T.; Mobraaten, L. E.; Sharp, J. J. *Nat. Genet.* **1997**, *16*, 19–27.
158. Phillips, T. J.; Hen, R.; Crabbe, J. C. *Psychopharmacology* **1999**, *147*, 5–7.
159. Thomas, H.; Hanby, A. M.; Smith, R. A.; Hagger, P.; Patel, K.; Raikundalia, B.; Camplejohn, R. S.; Balkwill, F. R. *Br. J. Cancer* **1996**, *73*, 65–72.
160. Thompson, K. L.; Rosenzweig, B. A.; Sistare, F. D. *Toxicol. Pathol.* **1998**, *26*, 548–555.
161. Zhu, H.; Guo, Q.; Mattson, M. P. *Brain Res.* **1999**, *842*, 224–229.
162. Taurog, J. D.; Richardson, J. A.; Croft, J. T.; Simmons, W. A.; Zhou, M.; Fernandez-Sueiro, J. L.; Balsih, E.; Hammer, R. E. *J. Exp. Med.* **1994**, *180*, 2359–2364.
163. Crabbe, J. C.; Wahlsten, D.; Dudek, B. C. *Science* **1999**, *284*, 1670–1672.
164. Wahlsten, D.; Metten, P.; Phillips, T. J.; Boehm, S. L. N.; Burkhart-Kasch, S.; Dorow, J.; Doerksen, S.; Downing, C.; Fogarty, J.; Rodd-Henricks, K. et al. *J. Neurobiol.* **2003**, *54*, 283–311.
165. Liu, D.; Diorio, J.; Day, J. C.; Francis, D. D.; Meaney, M. J. *Nat. Neurosci.* **2000**, *3*, 799–806.
166. Liu, D.; Diorio, J.; Tannenbaum, B.; Caldji, C.; Francis, D.; Freedman, A.; Sharma, S.; Pearson, D.; Plotsky, P. M.; Meaney, M. J. *Science* **1997**, *277*, 1659–1662.
167. Lathe, R. *Genes Brain Behav.* **2004**, *3*, 317–327.
168. Zeiss, C. J. *Lab. Anim.* **2002**, *31*, 34–39.
169. Fata, J. E.; Kong, Y. Y.; Li, J.; Sasaki, T.; Irie-Sasaki, J.; Moorehead, R. A.; Elliott, R.; Scully, S.; Voura, E. B.; Lacey, D. L. et al. *Cell* **2000**, *103*, 41–50.
170. Langheinrich, M.; Lee, M. A.; Boehm, M.; Pinto, Y. M.; Ganten, D.; Paul, M. *Am. J. Hypertension* **1996**, *9*, 506–512.
171. MacIvor, D. M.; Shapiro, S. D.; Pham, C. T.; Belaaouaj, A.; Abraham, S. N.; Ley, T. J. *Blood* **1999**, *94*, 4282–4293.
172. Luo, Y.; Bolon, B.; Kahn, S.; Bennett, B. D.; Babu-Khan, S.; Denis, P.; Fan, W.; Kha, H.; Zhang, J.; Gong, Y. et al. *Nat. Neurosci.* **2001**, *4*, 213–232.
173. Luo, Y.; Bolon, B.; Damore, M. A.; Fitzpatrick, D.; Liu, H. T.; Zhang, J. H.; Yan, Q.; Vassar, R.; Citron, M. *Neurobiol. Dis.* **2003**, *14*, 81–88.
174. Pineau, T.; Fernandez Salgeuro, P.; Lee, S. S. T.; McPhail, T.; Ward, J. M.; Gonzalez, F. J. *Proc. Natl. Acad. Sci. USA* **1995**, *92*, 5134–5138.
175. Favier, B.; Rijli, F. M.; Fromental-Ramain, C.; Fraulob, V.; Chambon, P.; Dolle, P. *Development* **1996**, *122*, 449–460.
176. Arnold, H. H.; Braun, T. *Int. J. Dev. Biol.* **1996**, *40*, 345–353.
177. Ryffel, B. *Crit. Rev. Toxicol.* **1997**, *27*, 135–154.
178. Shastry, B. S. *Experientia* **1995**, *51*, 1028–1039.
179. Holschneider, D. P.; Shih, J. C. *Int. J. Dev. Neurosci.* **2000**, *18*, 615–618.
180. Son, H.; Hawkins, R.; Martin, K.; Kiebler, M.; Huang, P. L.; Fishman, M. C.; Kandel, E. R. *Cell* **1996**, *87*, 1015–1023.
181. Doetschman, T. *Lab. Anim. Sci.* **1999**, *49*, 137–143.
182. Bolon, B.; Galbreath, E. J. *Int. J. Toxicol.* **2002**, *21*, 55–64.
183. Sung, Y. H.; Song, J.; Lee, H. W. *J. Biochem. Mol. Biol.* **2004**, *37*, 122–132.
184. Bolon, B.; De Rose, M.; Xu, W.; Pisegna, M.; Asuncion, F.; McCabe, J.; Christensen, K.; Hill, D.; Ross, L.; Duryea, D. et al. In *Application of Genomics to Animal Models for Pharmaceutical Studies*, Proceedings of the Cambridge Healthtech Institute's 2nd Annual Meeting, Boston, MA, Nov 15–16, 2000.
185. Ashby, J.; Tinwell, H. *Mutagenesis* **1994**, *9*, 179–181.
186. Friedberg, E. C.; Meira, L. B.; Cheo, D. L. *Mutat. Res.* **1998**, *407*, 217–226.
187. Mirsalis, J. C.; Monforte, J. A.; Winegar, R. A. *Crit. Rev. Toxicol.* **1994**, *24*, 255–280.
188. Suzuki, T.; Hayashi, M.; Sofuni, T. *Mutat. Res.* **1994**, *307*, 489–494.
189. Eastin, W. C.; Haseman, J. K.; Mahler, J. F.; Bucher, J. R. *Toxicol. Pathol.* **1998**, *26*, 461–473.
190. Gulezian, D.; Jacobson-Kram, D.; McCullough, C. B.; Olson, H.; Recio, L.; Robinson, D.; Storer, R.; Tennant, R.; Ward, J. M.; Neumann, D. A. *Toxicol. Pathol.* **2000**, *28*, 482–499.
191. Maronpot, R. R. *Toxicol. Pathol.* **2000**, *28*, 450–453.
192. Tennant, R. W.; French, J. E.; Spalding, J. W. *Environ. Health Perspect.* **1995**, *103*, 942–950.
193. Piwnica-Worms, D.; Marmion, M. *J. Clin. Pharmacol.* **1999**, 30S–33S.
194. Dycaico, M. J.; Provost, G. S.; Kretz, P. L.; Ransom, S. L.; Moores, J. C.; Short, J. M. *Mutat. Res.* **1994**, *307*, 461–478.
195. Paul, M.; Wagner, J.; Hoffman, S.; Urata, H.; Ganten, D. *Annu. Rev. Physiol.* **1994**, *56*, 811–829.
196. Gill, T., Jr.; Smith, G. J.; Wissler, R. W.; Kuntz, H. W. *Science* **1989**, *245*, 269–276.
197. Wolf, C. R.; Henderson, C. J. *J. Pharm. Pharmacol.* **1998**, *50*, 567–574.
198. Donnelly, T. M. *Lab. Anim.* **2004**, *33*, 43–45.
199. Vassar, R.; Citron, M. *Neuron* **2000**, *27*, 419–422.
200. Vassar, R.; Bennett, B. D.; Babu-Khan, S.; Kahn, S.; Mendiaz, E. A.; Denis, P.; Teplow, D. B.; Ross, S.; Amarante, P.; Loeloff, R. et al. *Science* **1999**, *286*, 735–741.
201. Baloh, R. H.; Tansey, M. G.; Lampe, P. A.; Fahrner, T. J.; Enomoto, H.; Simburger, K. S.; Leitner, M. L.; Araki, T.; Johnson, E. M., Jr.; Milbrandt, J. *Neuron* **1998**, *21*, 1291–1302.
202. Masure, S.; Geerts, H.; Cik, M.; Hoefnagel, E.; Van den Kieboom, G.; Tuytelaars, A.; Harris, S.; Lesage, A. S. J.; Leysen, J. E.; van der Helm, L.; Verhasselt, P.; Yon, J.; Gordon, R. D. *Eur. J. Biochem.* **1999**, *266*, 892–902.
203. Bolon, B.; Jing, S.; Asuncion, F.; Scully, S.; Pisegna, M.; Van, G.; Hu, Z.; Yu, Y. B.; Min, H.; Wild, K.; Rosenfeld, R. D.; Tarpley, J.; Carnahan, J.; Duryea, D.; Hill, D.; Kaufman, S.; Yan, X.-Q.; Juan, T.; Christensen, K.; McCabe, J.; Simonet, W. S. *Toxicol. Pathol.* **2004**, *32*, 275–294.
204. Campbell, I. L.; Stalder, A. K.; Akwa, Y.; Pagenstecher, A.; Asensio, V. C. *Neuroimmunomodulation* **1998**, *5*, 126–135.
205. Gonzalez, F. J.; Fernandez-Salguero, P.; Ward, J. M. *J. Toxicol. Sci.* **1996**, *21*, 273–277.

206. Korach, K. S.; Couse, J. F.; Curtis, S. W.; Washburn, T. F.; Lindzey, J.; Kimbro, K. S.; Eddy, E. M.; Migliaccio, S.; Snedeker, S. M.; Lubahn, D. B.; Schomberg, D. W.; Smith, E. P. *Rec. Prog. Hormone Res.* **1996**, *51*, 159–186 [discussion, 186–188].
207. Bolon, B.; Shalhoub, V.; Kostenuik, P. J.; Campagnuolo, G.; Morony, S.; Boyle, W. J.; Zack, D.; Feige, U. *Arthritis Rheum.* **2002**, *46*, 3121–3135.
208. Lee, S. S.; Pineau, T.; Drago, J.; Lee, E. J.; Owens, J. W.; Kroetz, D. L.; Fernandez Salgeuro, P. M.; Westphal, H.; Gonzalez, F. J. *Mol. Cell. Biol.* **1995**, *15*, 3012–3022.
209. Valentine, J. L.; Lee, S. S.; Seaton, M. J.; Asgharian, B.; Farris, G.; Corton, J. C.; Gonzalez, F. J.; Medinsky, M. A. *Toxicol. Appl. Pharmacol.* **1996**, *141*, 205–213.
210. Nebert, D. W.; Duffy, J. J. *Biochem. Pharmacol.* **1997**, *53*, 249–254.
211. Gonzalez, F. J. *Toxicol. Lett.* **2001**, *120*, 199–208.
212. Waters, M.; Boorman, G.; Bushel, P.; Cunningham, M.; Irwin, R.; Merrick, A.; Olden, K.; Paules, R.; Selkirk, J.; Stasiewicz, S.; Weis, B.; Van Houten, B.; Walker, N.; Tennant, R. *EHP Toxicogenomics* **2003**, *111*, 15–28.
213. Provost, G. S.; Kretz, P. L.; Hamner, R. T.; Matthews, C. D.; Rogers, B. J.; Lundberg, K. S.; Dycaico, M. J.; Short, J. M. *Mutat. Res.* **1993**, *288*, 133–149.
214. Deubel, W.; Bassukas, I.; Schlereth, W.; Lorenz, R.; Hempel, K. *Mutat. Res.* **1996**, *351*, 67–77.
215. Wijnhoven, S. W.; van Steeg, H. *Toxicology* **2003**, *193*, 171–187.
216. Van Sloun, P. P.; Wijnhoven, S. W.; Kool, H. J.; Slater, R.; Weeda, G.; van Zeeland, A. A.; Lohman, P. H.; Vrieling, H. *Nucleic Acids Res.* **1998**, *26*, 4888–4894.
217. Wijnhoven, S. W.; Van Sloun, P. P.; Kool, H. J.; Weeda, G.; Slater, R.; Lohman, P. H.; van Zeeland, A. A.; Vrieling, H. *Proc. Natl. Acad. Sci. USA* **1998**, *95*, 13759–13764.
218. Zhang, X. B.; Urlando, C.; Tao, K. S.; Heddle, J. A. *Mutat. Res.* **1995**, *338*, 189–201.
219. Gorelick, N. J.; O'Kelly, J. A.; Walker, V. E.; DeBoer, J. G.; Glickman, B. W. *Toxicol. Pathol.* **1995**, *23*, 823–824.
220. Cosentino, L.; Heddle, J. A. *Mutat. Res.* **2000**, *454*, 1–10.
221. Walker, V. E.; Andrews, J. L.; Upton, P. B.; Skopek, T. R.; deBoer, J. G.; Walker, D. M.; Shi, X.; Sussman, H. E.; Gorelick, N. J. *Environ. Mol. Mutagen.* **1999**, *34*, 167–181.
222. Knudson, A. G. *J. Cancer Res. Clin. Oncol.* **1996**, *122*, 135–140.
223. Contrera, J. F.; DeGeorge, J. J. *Environ. Health Perspect.* **1998**, *106*, 71–80.
224. Goldsworthy, T. L.; Recio, L.; Brown, K.; Donehower, L. A.; Mirsalis, J. C.; Tennant, R. W.; Purchase, I. F. H. *Fundam. Appl. Toxicol.* **1994**, *22*, 8–19.
225. Tennant, R. W.; Stasiewicz, S.; Eastin, W. C.; Mennear, J. H.; Spalding, J. W. *Toxicol. Pathol.* **2001**, *29*, 51–59.
226. Tamaoki, N. *Toxicol. Pathol.* **2001**, *29*, 81–89.
227. Suarez, H. G. *Anticancer Res.* **1989**, *9*, 1331–1343.
228. French, J.; Storer, R. D.; Donehower, L. A. *Toxicol. Pathol.* **2001**, *29*, 24–29.
229. Eastin, W. C.; Mennear, J. H.; Tennant, R. W.; Stoll, R. E.; Branstetter, D. G.; Bucher, J. R.; McCullough, B.; Binder, R. L.; Spalding, J. W.; Mahler, J. F. *Toxicol. Pathol.* **2001**, *29*, 60–80.
230. Storer, R. D.; French, J. E.; Haseman, J.; Hajian, G.; LeGrand, E. K.; Long, G. G.; Mixson, L. A.; Ochoa, R.; Sagartz, J. E.; Soper, K. A. *Toxicol. Pathol.* **2001**, *29*, 30–50.
231. Usui, T.; Mutai, M.; Hisada, S.; Takoaka, M.; Soper, K. A.; McCullough, B.; Alden, C. *Toxicol. Pathol.* **2001**, *29*, 90–108.
232. van Kreijl, C. F.; McAnulty, P. A.; Beems, R. B.; Vynckier, A.; van Steeg, H.; Fransson-Steen, R.; Alden, C. L.; Forster, R.; van der Laan, J.-W.; Vandenberghe, J. *Toxicol. Pathol.* **2001**, *29*, 117–127.
233. Jacobson-Kram, D.; Sistare, F. D.; Jacobs, A. C. *Toxicol. Pathol.* **2004**, *32*, 49–52.
234. Wolfer, D. P.; Crusio, W. E.; Lipp, H. P. *Trends Neurosci.* **2002**, 7, 336–340.
235. Linder, C. C. *Lab. Anim.* **2001**, *30*, 34–39.
236. Weaver, J. L.; Contrera, J. F.; Rosenzweig, B. A.; Thompson, K. L.; Faustino, P. J.; Strong, J. M.; Ellison, C. D.; Anderson, L. W.; Prasanna, H. R.; Long-Bradley, P. E.; Lin, K. K.; Zhang, J.; Sistare, F. D. *Toxicol. Pathol.* **1998**, *26*, 532–540.
237. Goelz, M. F.; Mahler, J.; Harry, J.; Myers, P.; Clark, J.; Thigpen, J. E.; Forsythe, D. B. *Lab. Anim. Sci.* **1998**, *48*, 34–37.
238. Livy, D. J.; Wahlsten, D. *J. Hered.* **1991**, *82*, 459–464.
239. Wahlsten, D.; Metten, P.; Crabbe, J. C. *Brain Res.* **2003**, *971*, 47–54.
240. Jucker, M.; Walker, L. C.; Schwarb, P.; Hengemihle, J.; Kuo, H.; Snow, A. D.; Bamert, F.; Ingram, D. K. *Neuroscience* **1994**, *60*, 875–889.
241. Campbell, I. L.; Krucker, T.; Steffensen, S.; Akwa, Y.; Powell, H. C.; Lane, T.; Carr, D. J.; Gold, L. H.; Henriksen, S. J.; Siggins, G. R. *Brain Res.* **1999**, *835*, 46–61.
242. Wang, M.-H.; vom Saal, F. S. *Nature* **2000**, *407*, 469–470.
243. Taketo, M.; Schroeder, A. C.; Mobraaten, L. E.; Gunning, K. B.; Hanten, G.; Fox, R. R.; Roderick, T. H.; Stewart, C. L.; Lilly, F.; Hansen, C. T.; Overbeek, P. A. *Proc. Natl. Acad. Sci. USA* **1991**, *88*, 2065–2069.
244. Ingram, D. K.; Jucker, M. *Neurobiol. Aging* **1999**, *20*, 137–145.
245. Kobayashi, S.; Yoshida, K.; Ward, J. M.; Letterio, J. J.; Longenecker, G.; Yaswen, L.; Mittleman, B.; Mozes, E.; Roberts, A. B.; Karlsson, S.; Kulkarni, A. B. *J. Immunol.* **1999**, *163*, 4013–4019.
246. Kuzmin, A.; Johansson, B. *Pharmacol. Biochem. Behav.* **2000**, *65*, 399–406.
247. Geisert, E. E. J.; Williams, R. W.; Geisert, G. R.; Fan, L.; Asbury, A. M.; Maecker, H. T.; Deng, J.; Levy, S. *J. Comp. Neurol.* **2002**, *453*, 22–32.
248. Holmes, A.; Lit, Q.; Murphy, D. L.; Gold, E.; Crawley, J. N. *Genes Brain Behav.* **2003**, *25*, 365–380.
249. Pandey, K. N.; Oliver, P. M.; Maeda, N.; Smithies, O. *Endocrinology* **1999**, *140*, 5112–5119.
250. Oh, S. P.; Li, E. *Dev. Dyn.* **2002**, *224*, 279–290.
251. Rinchik, E. M.; Carpenter, D. A.; Handel, M. A. *Proc. Natl. Acad. Sci. USA* **1995**, *92*, 6394–6398.
252. Shen, L.; Pichel, J. G.; Mayeli, T.; Sariola, H.; Lu, B.; Westphal, H. *Am. J. Hum. Genet.* **2002**, *70*, 435–447.
253. Meester-Smoor, M. A.; Vermeij, M.; van Helmond, M. J.; Molijn, A. C.; van Wely, K. H.; Hekman, A. C.; Vermey-Keers, C.; Riegman, P. H.; Zwarthoff, E. C. *Mol. Cell. Biol.* **2005**, *25*, 4229–4236.
254. Bolon, B. *Basic Clin. Pharmacol. Toxicol.* **2004**, *95*, 154–161.

Biography

Brad Bolon earned his BS (1983, Agriculture), DVM (1986), and MS (1986, Veterinary Biosciences) degrees in 6 years at the University of Missouri, 'enjoyed' an anatomic pathology residency (1986–89) at the University of Florida, and acquired a PhD (1993, Pathology) from Duke University while completing postdoctoral training at the Chemical Industry Institute of Toxicology (1989–93). He worked for Pathology Associates International as associate director of the Molecular and Immunopathology Division (1993–94) and later as staff pathologist at the National Center for Toxicological Research (1994–96) before moving to Wyeth-Ayerst Research as a senior scientist (1996–97). From 1997 to 2004, Brad was an experimental pathologist at biotechnology giant Amgen responsible for evaluating genetically engineered mice and the efficacy of novel biopharmaceutical candidates in animal models of disease. He founded GEMpath Inc. (for 'Genetically Engineered Mouse pathology') in 2004 to spend more time in discovery pathology research and writing.

Brad is a Diplomate of the American College of Veterinary Pathology (DACVP, in anatomic pathology, 1991) and the American Board of Toxicology (DABT, 1996; recertified in 2001 and 2006). He has authored or coauthored over seven dozen papers and book chapters and is regularly invited to discuss pathology of engineered animals at national and international meetings.

Comprehensive Medicinal Chemistry II
ISBN (set): 0-08-044513-6

ISBN (Volume 3) 0-08-044516-0; pp. 151–170

3.09 Small Interfering Ribonucleic Acids

L Gianellini and J Moll, Nerviano Medical Sciences S.r.l, Nerviano, Italy

3.09.1	**Introduction**	**171**
3.09.2	**The History of Ribonucleic Acid Interference**	**172**
3.09.3	**Molecular Mechanism of Ribonucleic Acid Interference**	**173**
3.09.3.1	RNase III Family Member (DICER)	173
3.09.3.2	Ribonucleic Acid-Induced Silencing Complex and Activated Ribonucleic Acid-Induced Silencing Complex	173
3.09.3.3	Microribonucleic Acid	175
3.09.4	**Technical Considerations and Limitations**	**176**
3.09.4.1	Critical Factors for Small Interfering Ribonucleic Acid Experiments	176
3.09.4.1.1	Choice of target site	176
3.09.4.1.2	Small interfering ribonucleic acid design criteria	176
3.09.4.2	Delivery into Cells	177
3.09.4.2.1	Getting small interfering ribonucleic acid into the cell	177
3.09.4.2.2	Intracellular expression of short hairpin ribonucleic acid	177
3.09.4.3	Experimental Controls	177
3.09.4.4	Detection of Gene Silencing	178
3.09.4.5	Off-Target Effects	178
3.09.5	**Applications in Target Identification and Validation**	**178**
3.09.5.1	Genome-Wide Ribonucleic Acid Interference Screens	179
3.09.5.1.1	Worms: *Caenorhabditis elegans*	179
3.09.5.1.2	Flies: *Drosophila melanogaster*	179
3.09.5.1.3	Humans: *Homo sapiens*	180
3.09.5.2	Link to Disease	180
3.09.5.3	Validation of Drug Action	181
3.09.5.4	Target Validation in Animal Models	181
3.09.5.4.1	Topical and systemic administration of synthetic small interfering ribonucleic acid	181
3.09.5.4.2	In vivo silencing using vectors	182
3.09.6	**Major Hurdles and Solutions toward Therapeutics**	**182**
3.09.7	**Perspectives**	**183**
References		**184**

3.09.1 Introduction

If the number of primary research papers published each year for a given topic is a measure of the relevance of a technology, then the process of RNA interference (RNAi) and its mediators, small interfering RNAs (siRNA), are the shooting star of the last few years in the field of life sciences. The technology is now established as a standard procedure in most molecular and cellular biology laboratories, and is a determinant factor in the exponential rise in number of genes for which gene function and mechanism of action have been studied at the cellular level. Even new, emerging fields such as systems biology owe their success at least in part to this technology. Rapid progress has also been made in understanding the molecular mechanisms associated with the RNAi process per se, and natural processes involving similar mechanisms either in disease or normal physiology have been discovered and have opened an exciting new field of gene regulation by intracellular noncoding microRNAs (miRNAs).[1] In mammals the estimated number of miRNAs accounts for approximately 1% of genes. The RNAi pathway functions as an ancient cellular defense mechanism to counteract foreign nucleic acids, such as those resulting from viral infection, which often is associated with the generation of double-stranded RNA (dsRNA), a property shared with miRNAs. Since a typical mRNA produces approximately 5000 copies of a protein, targeting mRNA rather than the protein itself should be a very efficient approach to block gene function. Although the impact of RNAi-based approaches for drug development has already been seen at different stages of drug development, the most complex and potentially far-reaching application of this technology will be the development of RNAi-based therapeutics. Since this is an early field and most approaches are in their infancy, this review will touch only briefly upon the usability of siRNA as therapeutics.

As of today, RNAi has its major merits in the area of target identification and validation. By using an appropriate screening design with multiple readouts, target identification and initial validation can be achieved practically simultaneously. The possibility of performing genome-wide or more focused screens, for example kinase restricted kinome-targeted libraries, in different model organisms allows direct comparison of results in different species, thus adding confidence in potential new targets. When applicable, RNAi is nowadays considered a must for any high-quality target validation package for a molecular targeted therapy. Functional validation of the mechanism of action of small-molecule inhibitors by comparing their effects with those of RNAi can also help in understanding an unknown or unexpected drug action and helps in making stop–go decisions on further drug development. The identification and validation of biomarkers to support clinical development of small molecules is another area that benefits from this technology. Although the current hype surrounding potential applications of RNAi is high, and consequently the expectations of companies and investors for siRNA drugs are even higher, it is worthwhile noting that the true potential for such therapeutics is uncertain bearing in mind the pitfalls that have been experienced with DNA-based antisense drugs. The hope is that learning from these failures could result in faster development of siRNA-based drugs, and indeed it appears that initially the same therapeutic niches are being exploited in an attempt to circumvent the current weaknesses of this type of approach.

3.09.2 The History of Ribonucleic Acid Interference

In 1990, Jorgensen and colleagues recognized the first hints of a gene-silencing phenomenon, when attempting to create transgenic petunia plants with an increased flower pigmentation by introduction of the chalcone synthase gene. Surprisingly, instead of increasing pigmentation in petunias, the transgenic plants were found to have no, or low pigmentation, due to lowered levels of endogenous chalcone synthase mRNA.[2] This phenomenon was termed cosuppression without knowledge of the underlying mechanism. It was in 1998 when Mello first used the term RNA interference[3] and a groundbreaking article by Fire and colleagues in 1998[4] demonstrated the dependency of this effect on dsRNA and the usability of RNAi in the worm *Caenorhabditis elegans* as a potent and specific means of genetic interference by dsRNA. However, another three years passed before this method was applicable to mammalian cells. Initial attempts similar to those employed in the nematode failed, since in mammalian cells, the dsRNAs used were found to induce an interferon response, leading to widespread changes in gene expression by global posttranscriptional silencing as a result of inhibition of translation. A breakthrough came in 2001 when Elbashir[5] and colleagues used smaller dsRNA duplexes of ± 21 nucleotides in length, which are not long enough to induce an interferon response, but which were found to efficiently downregulate the mRNA levels of specific genes in several mammalian cell lines. This opened the field to a wide range of applications and it is not surprising that RNAi was selected as the breakthrough of the year in 2002 by *Science* magazine.[6]

At the same time, progress was also being made in the understanding of the molecular mechanism of the RNAi process, and major components such as ribonucleic acid-induced silencing complex (RISC)[7] and RNase III family member (DICER)[8] were identified. It was only a question of time until the first siRNA libraries were created and used to screen for specific cellular phenotypes. As will be discussed later in more detail, genome-wide screens were performed initially in the worm *C. elegans* in 2003 and later in mammals in 2004 using either synthetic siRNAs or siRNAs generated from expression vectors producing longer precursors, so-called short hairpin RNAs (shRNAs) (*see* Section 3.09.4.2.2), which are further processed into siRNAs within cells. In 2002 a simpler method was published[9] making use of complementary DNA (cDNA) which is transcribed into long dsRNA and is digested with exoribonuclease to generate siRNA. Such a library was produced and screened in 2004.[10]

Further major steps were the first successful in vivo applications in mammalian animal models[11] and in 2005, a first clinical phase I trial started (Sirna-27, SIRNA). It is expected that the success or failure of this first trial will have a strong impact on the development of additional siRNA therapeutics in the near future.

3.09.3 Molecular Mechanism of Ribonucleic Acid Interference

The scope of this chapter is to describe briefly the principal mechanisms involved in RNA silencing and its biological role. For a more detailed description some excellent recent reviews have been published.[12–15] RNA interference is an ancient defense pathway of eukaryotic cells and a common denominator for the posttranscriptional gene silencing (PTGS) phenomenon observed in a variety of species such as plants, fungi, and animals. One of the major questions is for which physiological function is this pathway necessary? In part at least, eukaryotic cells appear to use this mechanism to regulate gene expression, by endogenously expressing noncoding regulatory miRNAs, which interfere specifically with endogenous mRNAs. However, RNAi also plays a role when viruses infect eukaryotic cells, or when transposons or transgenes integrate into host genomes, phenomena that are often accompanied by the formation of dsRNA. In these cases, cells are able to sense this challenge and try to counteract the expressed, putatively dangerous genes by destroying their mRNA. This defense mechanism is highly specific and conserved, and is composed of a complex pathway involving multiple protein complexes with different functions. In mammalian cells nonsequence-specific responses are triggered by dsRNAs, which are often generated after viral infection. Key effector proteins of these responses are a protein kinase, which is dsRNA dependent (PKR), and 2′-5′ oligoadenylate synthetase. The degree to which these molecules are fully activated by dsRNAs is at least in part influenced by the length and concentration of the dsRNA.[16]

Interference with chromatin state or promoter activity of a gene represents another potential mechanism for inhibiting gene expression by transcriptional silencing[17] and indeed it has been found that in addition to direct effects on mRNAs, siRNAs are also able to induce DNA or chromatin modifications at their homologous genomic loci. In fission yeast chromatin modifications are only directed by a RNAi mechanism if the homologous DNA sequences are transcribed, coupling RNAi-directed chromatin modification to transcription.[18]

3.09.3.1 RNase III Family Member (DICER)

Biochemical and genetic studies have revealed the molecular mechanism by which dsRNA causes the degradation of target messenger RNA and this process can be divided into four major steps (**Figure 1**). The first step in the RNAi pathway involves the processing of large dsRNAs into small siRNA molecules 21–23 nucleotides long. siRNAs usually bear a 3′ hydroxyl and a 5′ phosphate groups and a 3′ overhang of two unpaired nucleotides on each strand.[5,19] A specific RNase III enzyme, named DICER, was found to be responsible for this processing.[8] DICER has four distinct domains: an N-terminal DExH/DEAH RNA helicase domain, a PAZ (Piwi–Argo–Zwille/Pinhead) domain,[20] a tandem repeat of RNase III catalytic domain sequences, and a C-terminal dsRNA-binding motif. Initially, the PAZ domain was suggested to function as a protein interaction domain. Recent biochemical and structural studies, however, have converged on the view that PAZ is an RNA-binding domain that specifically recognizes the ends of the base-paired helix of siRNA and miRNA duplexes, including the characteristic two-nucleotide 3′ overhangs.[21] This siRNA/miRNA-duplex-specific interaction with PAZ ensures the safe transition of small RNAs into RISC by minimizing the possibility of unrelated RNA processing or RNA turnover products entering the RNA silencing pathway. Sequence homology and functional studies led to the identification of DICER homologs in plants,[22] fission yeast,[23] worms,[24] mice,[25] and humans.[26,27]

The length of a siRNA is species-specific and may reflect differences in the spacing or structure of the RNase III domains of DICER homologs. Among RNase III enzymes, DICER is unique in that cleavage of dsRNA requires ATP, a finding that has been attributed to the presence of an ATP-dependent RNA 'helicase' domain at its N-terminus.[8,19,28] The helicase domain may be a RNA translocase, using ATP-derived energy to drag DICER down the dsRNA, and it has also been suggested that DICER binding to RNA or catalytic activity might be ATP-regulated.[29]

3.09.3.2 Ribonucleic Acid-Induced Silencing Complex and Activated Ribonucleic Acid-Induced Silencing Complex

In the second step, the siRNA duplexes are incorporated into the RISC present in the cytosol. The phosphorylation of the siRNA 5′ terminal is required for entry into RISC[30] and RISC is guided to its target RNA by siRNA. Nykanen[28] found that a second ATP-dependent step was involved in the pathway and showed that following unwinding of the siRNA duplex, RISC was converted to the active form of RISC as indicated with a star (RISC*). In a separate seminal study, RISC* was found to be associated only with the antisense strand of the siRNA.[31] Although the siRNA needs to be double-stranded in order to be efficiently recognized and bound to RISC, the two siRNA strands must unwind before RISC becomes active. Accordingly, it was concluded that either the RISC complex has ATP-dependent helicase activity or a helicase enzyme is associated with RISC.

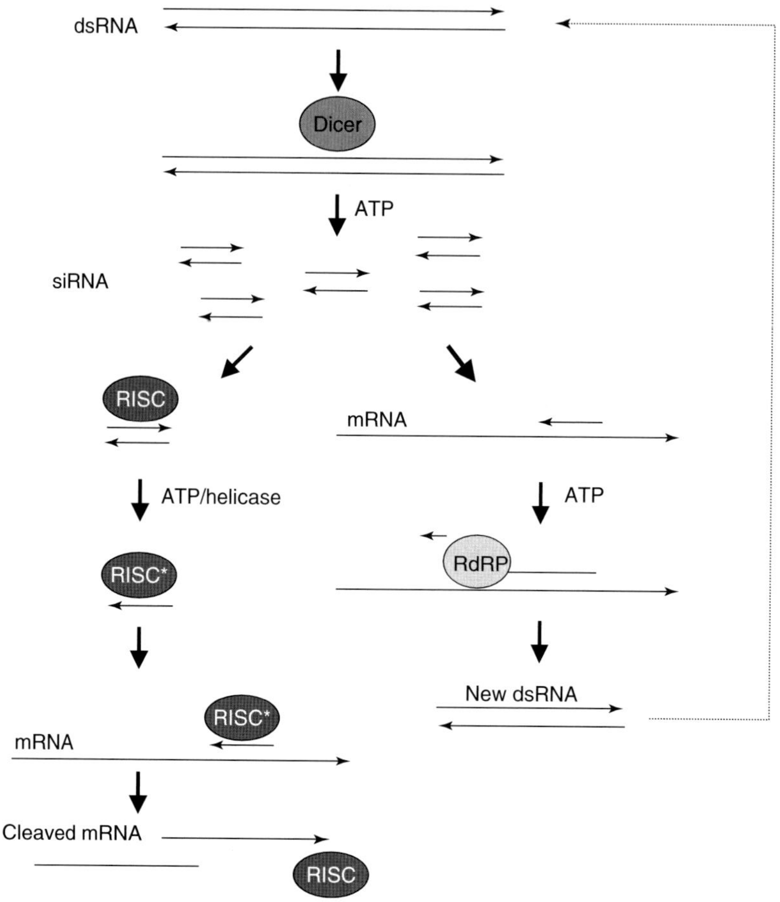

Figure 1 RNAi is initiated by cleavage of long dsRNA into siRNA duplexes of 21–25 nucleotides in length by DICER, in an ATP-dependent manner. These siRNA duplexes are then incorporated into a protein complex called RISC. ATP-dependent unwinding of the siRNA duplex remodels the complex to generate an active RNA-induced silencing complex (RISC*). Finally the RISC* can recognize and cleave a target RNA complementary to the guide strand of the siRNA. Alternatively, in the cells of some organisms (such as *Caenorhabditis elegans*) the siRNA might also unwind and bind to the target mRNA and serve there as a primer for an RdRP (RNA-dependent RNA polymerase). RdRP extends the primer, so that the mRNA becomes double-stranded and hence initiates a new cycle of siRNA production.

The exact composition of RISC is not yet fully elucidated and its components are still being functionally verified. For instance, the components of RISC identified to date in *Drosophila* include fragile X protein (dFXR), Vasa intronic gene protein (VIG),[32] Tudor SN (staphylococcal nuclease),[33] Argonaute2 (AGO2),[34] and siRNA.[34] Recently, the crystal structure of an archaebacterial Ago protein revealed striking similarity with members of the RNase H family.[35] As RNase H cleaves the RNA strand of RNA/DNA duplexes, it was proposed that Ago proteins act by cleaving target RNA in target RNA/siRNA hybrids.

In the last step of the process, the active RISC targets the homologous transcript by base pairing interactions and cleaves the mRNA between the 10th and 11th nucleotide from the 5′ terminus of the siRNAs.[30,36]

In nonmammalian cells, there is evidence for an alternative branch of the RNAi pathway that results in the amplification of the original message, which increases the efficiency of gene silencing.[37] In this case, the unwound siRNA no longer acts as a guide to bring RISC to the target mRNA but merely as a primer for an RNA-dependent RNA polymerase (RdRP), which uses the target mRNA as a template to produce new dsRNA. Subsequently, this dsRNA can be recognized and cleaved by DICER, thus re-entering the RNAi pathway and initiating a new round of silencing. Several RdRPs participating in RNAi have been identified in fungi, plants, and invertebrates. Whether a similar amplification mechanism exists in mammalian cells has to date not been determined.

3.09.3.3 Microribonucleic Acid

As mentioned above, the siRNA antisense product can also be derived from endogenous miRNAs, which are a family of small RNAs of 21–25 nucleotides which negatively regulate gene expression at posttranscriptional level (**Figure 2**). The function of miRNAs was originally defined by *lin-4* and *let-7* RNA, which were identified by genetic analysis of the timing of *C. elegans* development.[38,39] These were initially called small temporal RNAs (stRNAs) because of their temporal expression pattern and their role in temporal gene regulation.

miRNAs are transcribed as longer precursors, termed pri-miRNAs.[40] Upon transcription, pri-miRNAs undergo nuclear cleavage by the RNase III endonuclease Drosha, producing the 60–70-nucleotide spanning stem–loop precursor miRNA (pre-miRNA) with a 5′ phosphate and a two-nucleotide 3′ overhang.[41] The pre-miRNA is actively exported to the cytoplasm where DICER processing trims the hairpin stem and removes the loop and sense strand to create the final 21–23-nucleotide antisense RNAi effector. The antisense strand is subsequently incorporated into the RISC.[31]

The antisense strand of miRNAs often contains mismatches to one or more sites in the 3′ untranslated region (UTR) of the target mRNA. This results in translational repression of the target mRNA, which is a different mechanism compared to the siRNA-mediated mRNA degradation as discussed above. miRNAs are encoded in the genome of large DNA viruses such as EBV,[42] SV40,[43] or HCMV[44] and probably play a role during infection and immune escape. In humans more than 300 miRNAs have been reported so far and it has been extrapolated to a total number of at least 800.[45] The function of most miRNAs still needs to be elucidated, as up to now only a few have been associated with regulatory functions in development or disease.

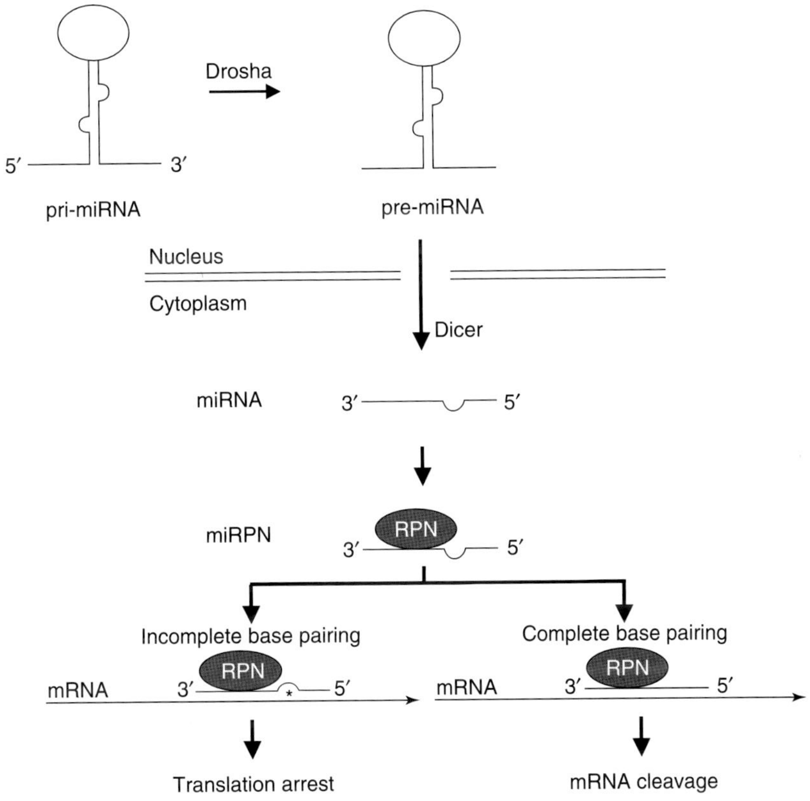

Figure 2 Transcription of endogenous miRNA generates pri-miRNAs that are processed in the nucleus by Drosha ribonuclease III into pre-miRNAs. pre-miRNAs are exported to the cytoplasm and are processed by DICER. Unlike siRNAs, the miRNA are single-stranded and are incorporated into the miRNA–ribonuclear protein complex (miRPN). The miRNAs bind to sites that have partial sequence complementarity typically in the 3′ untranslated region (UTR) of their target mRNA, causing either mRNA degradation (perfect matches with target mRNA) or repression of translation (mismatches with target mRNA). The asteriks indicates a mismatch.

3.09.4 Technical Considerations and Limitations

Despite the high potential of the technology its limits and caveats have to be understood in order to design proper experiments and in order to critically evaluate the results. Artifacts related to off-target effects have generated the need to set up stricter rules for acceptance of siRNA experiments as defined in many peer-reviewed journals. RNAi induced by long dsRNA was originally used to study gene function in plants, worms, and flies. However, when dsRNAs of more than 30 bp in length were used in mammalian cells it was noted that there was an inhibition of protein translation within the cell due to the activation of an interferon response.[26] These nonspecific responses to dsRNA are not triggered by dsRNAs shorter than 30 base pairs, including the siRNA duplexes. Moreover, studies in worms and flies have clearly demonstrated that synthetic siRNAs can produce effects similar to those of the long dsRNAs.[46,47] Based on these experiments, siRNAs are now being optimized for systematic exploration of gene function in a variety of organisms.

In the following sections some of the critical factors are summarized, which may influence the efficiency of RNAi or which should be considered when performing or planning an RNAi experiment.

3.09.4.1 Critical Factors for Small Interfering Ribonucleic Acid Experiments

3.09.4.1.1 Choice of target site

Only a subset of siRNAs designed for different regions of a target mRNA sequence is sufficiently effective. Even though there exist several useful algorithms to increase the success rate, a number of siRNAs must usually be tested against different sites of a particular mRNA to get satisfactory downregulation of the mRNA. It is recommended to avoid the first 75–100 nucleotides of any mRNA as potential target sites, since they may contain protein-binding regulatory sequences (5′ UTR) that could interfere with the action of siRNA.[48] It is also important to ensure that the sequence is specific to the target gene by performing a BLAST database[49] search against all possible genes in the species the siRNA is designed for, in order to avoid cross-reaction with unwanted genes. However, the best-scoring sequences are not necessarily the best choice, since they do not necessarily predict secondary targets and might have other sequence liabilities. Also genetic polymorphisms and alternative splicing of genes have to be taken in consideration to reduce the probability of choosing an inappropriate target site.

3.09.4.1.2 Small interfering ribonucleic acid design criteria

Based on the experience from many siRNA experiments, a few guidelines have been defined for the synthesis of siRNAs. A general rule is that the sequence of one strand should be AA(N19)TT, where N is any nucleotide and these siRNAs should have a two-nucleotide 3′ overhang. Furthermore, siRNAs should have a 5′-phosphate and a 3′-hydroxyl group for best efficiency. A number of additional factors have been proposed to play a role in the activity of siRNAs, including base length, internal thermodynamic stability, and base composition of the siRNA duplex: the use of RNA duplexes longer than 30 nucleotides has been shown to induce a nonspecific degradation pathway of all mRNAs in mammals.[50] A duplex of 19–21 nucleotides appears to be the optimal length for RNAi activity. Since only one strand of the siRNA duplex is incorporated into RISC*[51] it would be desirable to design the siRNA duplex so that the antisense strand is the one which is used preferentially. It has been shown that when siRNA are thermodynamically unstable (AU-rich) or even contain mismatches toward the 5′ end of the antisense strand, this strand is preferentially used by RISC resulting in more efficient silencing.[30,52] Moreover, lower thermal stability around the central region of the duplex facilitates cleavage of mRNA by RISC and higher thermal stability at the 3′ antisense end of the duplex prevents unwinding and incorporation of the sense strand.

To prevent intramolecular folding, palindromic sequences, tandem repeats, and GC stretches of >7–8 nucleotides should be avoided, and a number of algorithms have been developed to assist researchers in the identification of efficient and utilizable siRNA sequences. A collection of free siRNA design tools is available online.[53]

In the majority of cases siRNAs are synthesized chemically. An alternative is the independent transcription of the sense and antisense RNA strands in vitro from DNA oligonucleotide templates. The sense or antisense strand can be transcribed for example by using the T7 phage RNA polymerase.[54] This polymerase produces individual siRNA sense and antisense strands which, once annealed, form siRNAs. Extra nucleotides required by the T7 promoter are removed by RNase digestion.

In the human genome more than half of the genes undergo alternative splicing, which adds additional complexity for siRNA design. Polymorphisms must also be taken into consideration when designing a siRNA and all these factors should be considered during the interpretation of experimental results. Single nucleotide polymorphisms (SNPs) can be found on average once every 300–500 bases in the human genome and it is estimated that the average gene contains around four coding sequence SNPs, for genes with allele frequencies of at least a few percent in the human population.[55] All these factors can be the source of conflicting results when compared to other validation methods.

Another, often neglected property of eukaryotic cells is genetic redundancy. Many pathways can be triggered by different signaling proteins and when one mediator is lacking the intracellular protein network can compensate. This is most evident when considering genetic deletion of genes in mice and it is surprising that many genes originally judged essential for a specific function do not yield the expected phenotype upon deletion. In fact, there are many knockout mice strains for which up to now no obvious phenotype could be assigned, and in most cases this is probably due to genetic redundancy. For example, mice lacking single members of the cyclin-dependent kinase (CDK) family, which are considered essential components for cells to progress through the cell cycle, are viable with no obvious proliferation defects of cells and only combined knockout of several CDKs results in more apparent phenotypes, hinting at redundancy in the system.[56]

siRNA experiments have only just started to be used to examine gene redundancy and an example in *Drosophila* has been recently published, in which the functional redundancies of seven highly related genes belonging to the enhancer of split gene family were studied.[57]

3.09.4.2 Delivery into Cells

3.09.4.2.1 Getting small interfering ribonucleic acid into the cell

In order to understand a gene's function by using chemically synthesized siRNA it is necessary to optimize protocols for efficient delivery into cells. Several transfection reagents exist, the most commonly used being liposomal or amine-based. In some cases electroporation may be used, but cell toxicity can be high with this technique. Cell lines show varying responses to different transfection reagents, and it may be necessary to try more than one reagent or approach. Transfection efficiency is optimized by titrating cell density, transfection time, and the ratio of siRNA to transfection reagent. The cell passage number and the type of antibiotics used can also affect the efficiency of transfection.

An important parameter to be taken into account is the turnover rate of the protein to be targeted. Even when the mRNA level of a specific target is reduced, the protein might still be present and functional. siRNA-mediated RNAi lasts only for three to five cell-doubling times, probably due to gradual dilution of siRNA through cell division. Therefore, multiple transfections might be necessary in cases where the protein is particularly stable or the cells need to be grown for a long time to follow a delayed phenotype.

3.09.4.2.2 Intracellular expression of short hairpin ribonucleic acid

The high costs of synthesizing siRNAs when performing genome-wide screens and the need for more efficient delivery into different cell types has resulted in the development of alternative strategies to generate siRNAs.

One of these strategies is to produce the siRNAs in the cells by RNA polymerase III promoter-based DNA plasmids or expression cassettes.[58] The most commonly used are the U6 and the H1 promoters. These constructs produce small inverted repeats, separated by a spacer of three to nine nucleotides, termed short hairpin RNAs (shRNAs), which are processed by DICER into siRNAs.[59] Transcription begins at a specific initiation sequence, determined by the promoter used and the transcripts terminate with a series of 3′ uridine residues, a feature that seems to favor efficiency of RNAi.[60] In a slightly different approach the sense and antisense strands are transcribed separately by two individual promoters. In this case the two strands are annealed within the cell to form a siRNA,[61] but it has been reported that this method of expressing siRNA results in less potent RNAi effects than expression of a corresponding shRNA.[62]

Using an RNA polymerase II promoter instead of an RNA polymerase III could be advantageous when a more regulated expression of siRNA is required, and a variety of inducible promoters are available to control the timing of the expression. A commonly used promoter is for example the human cytomegalovirus early promoter.[63]

These vectors provide advantages over chemically synthesized siRNAs, but they also have numerous disadvantages, including the often low or variable transfection efficiency, which is highly dependent on the cell type. To overcome these problems and to have a more prolonged expression, viral vectors are employed to deliver shRNA expression cassettes. Retroviral vectors are most widely used and murine retrovirus-based vectors have been shown to be efficient in delivery of shRNA.[64,65] Lentivirus-based vectors have been also tested and appear to be promising vehicles for RNAi because they are effective in infecting noncycling cells, stem cells, and zygotes.[66] Adenoviral vectors are highly effective, but allow only transient expression of siRNA.[67]

3.09.4.3 Experimental Controls

Early studies suggested that due to the high specificity of RNAi, a siRNA with one or two nucleotide sequence mismatches could serve as a negative control. However, with the identification of off-target effects this rule might be too stringent for some sequences. For example, the position and the type of base pairing for the mismatch influence silencing efficiency.[68] The use of scrambled oligonucleotides is not a good alternative since these might not be recognized by the RISC complex and are not sufficiently homologous to the target sequence to function as an adequate control. A more convincing functional control can be to demonstrate the rescue of the target gene function following artificial overexpression of the target gene, although nonphysiological overexpression of a protein might result in additional artifacts. Observation of the same effect(s) upon transfection of different siRNAs, which target different regions within the mRNA of interest, certainly can increase confidence in experimental results.

3.09.4.4 Detection of Gene Silencing

The effect of RNAi should be quantified at both the mRNA and the protein level. In fact, a reduction in protein levels not accompanied by a decrease in mRNA might indicate that other mechanisms are at work, such as the induction of an interferon response. Many different assays to measure mRNA levels, including Northern blot, reverse-transcriptase polymerase chain reaction (RT-PCR) based methods, and microarray analysis, can be utilized to examine the effects of RNAi on the transcript levels. In most cases, assays performed between 24 and 72 h after initiation of RNAi should give appropriate results. Unrelated proteins with known, stable expression are basic controls to be included.

Protein knockdowns are typically confirmed by Western blot analysis, immunofluorescence, or flow cytometry. Although, RNAi generally occurs within 24 h of transfection, both onset and duration of RNAi depend on the turnover rate of the protein of interest, as well as the rate of dilution and longevity of the siRNAs.

3.09.4.5 Off-Target Effects

Off-target effects resulting in the inhibition of genes not targeted by the specific siRNA are related to several cellular mechanisms, and represent one of the key limitations of the technology, potentially giving rise to misleading results. One potential pitfall, for example, is that a transcript with a high sequence homology to the target gene is degraded. Additionally, when there are some mismatches with a close homolog, siRNAs can sometimes act via a less stringent translational repression mechanism. As discussed above, an interferon or stress response can result in more general suppression of gene expression.

Off-target effects have been described for siRNAs with low homology to the nontargeted sequence, sharing as few as 11 contiguous nucleotides.[69] In this study, the effects on target expression of 16 different siRNAs designed against the same gene were compared by microarray analysis. Some of these siRNAs showed strong differences in the expression profiles of genes with uniquely regulated genes for individual siRNAs. However, on the other hand, it has been also reported that just one mismatch between a siRNA and its target mRNA can abrogate silencing.[5] The position of the mismatch is certainly a critical factor, and secondary structures might also contribute to the different responses. A systematic evaluation of 10 siRNAs targeting the menin gene suggests that induction of a stress response may result in off-target effects. Some but not all siRNAs caused a change in p53 and p21 RNA and protein levels, which were unrelated to silencing of the target gene.[70]

Off-target effects are more likely to be seen at high doses of siRNA oligonucleotides (>100 nM), so a careful titration is recommended for any new siRNA experiment. The use of minimal amounts of siRNA may reduce the probability of inducing a nonspecific dsRNA response, of which most mammalian cells are capable.

3.09.5 Applications in Target Identification and Validation

There is no standard definition of target validation and different pharmaceutical companies apply various criteria according to their indication-specific needs.[71] The intervention point for a validated drug target should play an essential role in a disease-relevant process and this link to disease is the most important parameter to be taken into account. A major challenge is to establish cellular or animal models, which predict disease and which also allow a sufficiently high-throughput for testing of compounds.

Target validation is needed not only before starting a project in drug development but often also in later stages, since the characterization of new targets tends to be an open-ended process. siRNAs are useful tools allowing target validation at high-throughput in tissue culture cells and validation of specific targets in animal models of disease. The impact on target identification and validation by using siRNA is still increasing and in this chapter some selected examples will be discussed. It is no coincidence that most of the examples published as of today are related to target identification and validation in cancer. The readouts such as proliferation or apoptosis are directly linked to the disease and are easy to follow; and application of RNAi often simultaneously allows target identification and validation. In any case, alternative methods of target validation should be followed in parallel to increase confidence in a new target. This could mean application of small-molecule agonists or antagonists, dominant negative or constitutive active expression constructs, or the use of genetically modified animal models to name a few. In contrast to knockout animal models where the gene is deleted completely, siRNA is usually a transient knockdown technology with varying degrees of loss of function, dependent on the type of protein. Since the level of silencing hardly ever reaches 100%, careful interpretation of results is needed in case of controversial data using the different techniques.

3.09.5.1 Genome-Wide Ribonucleic Acid Interference Screens

Several genome-wide screens, although often covering partial genomes, have been performed in different species. Lower species such as *C. elegans* have the advantage of being less complex, while higher species are more complex and closer to humans, which is an important factor for a new target for drug development. The majority of the initial screens in different species are related to gross cellular phenotypes affecting cell growth or viability and one of the best-studied examples is cytokinesis (the separation of cells at the end of mitosis), which will be discussed in more detail below. Since cytokinesis is a highly conserved process it allows the best comparison of the various methods and results in different species. It is expected that the available siRNA or shRNA libraries will be much more widely used in the future keeping in mind that we are just looking at the first wave of applications.

3.09.5.1.1 Worms: *Caenorhabditis elegans*

The advantages of *C. elegans* as a model organism for the screen of mutant phenotypes is facilitated by the fact that this worm has a short generation time, consists of only 959 well-defined cells, and is the first metazoan for which most of its genes have been subjected to loss-of function analysis. Kamath[72] and colleagues described a siRNA library, which targets 86% of the predicted 19 427 genes of *C. elegans* and consists of 16 757 bacterial clones. Delivery is simple since the worms can simply be fed with *E. coli* expressing dsRNA complementary to the specific gene of interest.

Mutant phenotypes were identified for 1722 genes, for which the majority had not previously been associated with a phenotype, and 929 of these showed embryonic lethality. The different criteria employed as a readout comprised sterility, embryonic or larval lethality, slow postembryonic growth, or postembryonic defects including morphological and behavioral defects. For this and other genome-wide screens described in the following paragraphs, the specific method of detection has a strong impact on the experimental results and obviously the more robust and simple readouts are applied first.

One of the most systematic screens for growth related phenotypes was performed by Sonnichsen and colleagues[73] with the aim of identifying all genes required for cell division in *C. elegans* embryos. The screen covered the first two embryonic cell divisions using differential interference contrast time-lapse microscopy as a readout, and 20 326 dsRNAs targeting 98% of the *C. elegans* genome were synthesized and screened individually. A total of 8.5% of the predicted genes showed a phenotype when analyzed for 45 distinct defect categories, showing the power of time-lapse microscopy when associated with the appropriate image analysis software. An attempt was made to link these categories to biochemical pathways and protein complexes. For each siRNA, a time-lapse video was analyzed and affiliated according to the specific defect created in the different stages of the cell cycle and as a result previously uncharacterized genes were functionally annotated. The outcome of this screen is publicly available and a searchable database has been established.[74] This is an excellent example of how target identification and validation can go hand in hand in one experimental setting and how intelligently designed readouts can lead to the identification of subtle phenotypes that otherwise might have been overlooked.

3.09.5.1.2 Flies: *Drosophila melanogaster*

Screens in *Drosophila* cells benefit from the early availability of a library consisting of 19 470 dsRNAs with an average length of 408 base pairs, which cover more than 90% of annotated *Drosophila* genes.[75] *Drosophila* cells in tissue culture can take up long RNA fragments from the culture medium and are able to process these long dsRNAs into siRNAs without triggering an interferon response, simplifying the procedure and interpretation of the results. The first application for the 19 740 dsRNA library aimed to identify essential genes for cellular growth and viability[75]; 438 essential genes were identified among which 80% were so far unknown. This was surprising, given the high number of mutants and their associated phenotypes that had already been characterized in *Drosophila*.

The initial proof of concept that siRNA can be useful to identify and to dissect different phenotypes of aberrant cytokinesis was shown by targeting the few known or suspected genes involved in this process.[76] Echard and colleagues[77] extended this concept by making use of the 7216 most conserved genes of the above *Drosophila* library in order to identify genes involved in the different steps in cytokinesis. Insect cells were incubated with single clones of the siRNA library and video microscopy was used as a readout to distinguish defects in terminal stages of cytokinesis.

This screen led to the identification of genes essential for stabilization of the intercellular bridge at the end of mitosis. Later the screen was further extended using the full library and performing a parallel chemical and genetic genome-wide screen to identify both new targets, as well as small-molecule inhibitors of cytokinesis.[78] A total of 214 genes relevant for cytokinesis and 50 inhibitors were identified from a library of 51 000 small molecules, the latter consisting of a mixture of commercial druglike molecules and natural compounds. Most important, the comparison of the two data sets let to the definition of two subphenotypes, which were found in the chemical and the genetic screen, validating the approach. Based on a detailed phenotypic analysis, the putative target of one of the small molecules was identified and the molecular pathway assigned. This work is an impressive demonstration of how genetic and small-molecule screens performed in parallel can lead in a short time to validated small-molecules with good confidence in the relevance of the pathway that the molecular target is associated with.

3.09.5.1.3 Humans: *Homo sapiens*

Two major strategies have been followed for screening using siRNA in mammalian cells: screens have attempted to cover partial or full genomes, or alternatively, selected screens have focused on gene families, such as the systematic screening of all human phosphatases and kinases.[79] A shRNA expression library of a set of 23 742 retroviral vectors that target 7914 human genes was screened for modulators of p53-dependent proliferation arrest.[80] A similar approach targeting 9610 human genes was used to characterize genes important for proteasome function.[81] Both of these libraries utilized a DNA barcode for easy identification of individual siRNA expression vectors associated with a specific phenotype.

A large-scale RNAi library prepared by endonuclease digestion was generated and used to screen for genes essential for cell division in human cells.[10] This approach is claimed to be highly specific, as well as simpler and more cost-effective compared to transfection of chemically synthesized siRNAs or viral expression vectors. The technology makes use of a sequence-verified complementary DNA library from which around 5000 genes were processed in a 96-well format. These cDNA clones were the starting point for the in vitro generation of endoribonuclease-generated short interfering RNAs (esiRNAs). The esiRNAs were transfected into human cervical carcinoma cells (HeLa) followed by a primary screen for viability and a secondary screen using videomicroscopy to identify defects in cell division.[10] This allowed a classification of the identified genes into different phenotypes including defects in mitotic arrest, aberrant cytokinesis, or cell death upon entry into mitosis. This led to the identification of 37 genes necessary for cell division, most of which had not previously been associated with this process. Interestingly, a comparison with the genome-wide RNAi screen as described above in *Drosophila*[75] revealed that a high number (72%) of the human orthologs to essential *Drosophila* genes were also found to be essential for viability of human HeLa cells. A more focused screen in order to identify regulators of apoptosis and chemoresistance has been performed utilizing the human kinase and phosphatase genome.[82] HeLa cells were transfected with siRNAs targeting each of the 650 known and putative kinases in the human genome and their impact on apoptosis or chemoresistance was determined. From this screen 73 kinases were identified as ‘survival kinases’ as defined by their ability to increase apoptosis after downregulation. The same library was rescreened in the presence of low-dose chemotherapeutics including Taxol, Cisplatin, or Etoposide. This led to the identification of additional kinases which increased cellular sensitivity to apoptosis-inducing stimuli, and which may potentially represent new targets in cancer therapy. The percentage of phosphatases involved in cell survival is even higher than that of kinases, considering that 72 of the estimated 222 known and putative phosphatases in the human genome were found to play a role in survival signaling. It will be interesting to see how many of these kinases or phosphatases will turn out to be relevant for survival in a wider range of tumor cell lines or primary tumors.

This strategy can be extended on any gene family with more success stories to be expected in the future. The choice of which screening approach to follow has an analogy in classical small-molecule drug screening, with the option of either conducting a medium- to low-throughput approach using targeted libraries, or entering a classical high-throughput screening (HTS) with millions of compounds.

3.09.5.2 Link to Disease

The identification of target genes of miRNAs has in some cases allowed a link to be made to disease. This was successful in case of miR-375, which targets the pancreatic endocrine protein myotrophin and inhibits glucose-stimulated insulin secretion and exocytosis, linking it to diabetes.[83] Other examples are miR-143, which increases during differentiation of adipocytes, and consequently inhibition of miR-143 reduces adipocyte differentiation making it an interesting target for obesity.[84] Even a potential oncogene function has been associated with the miR-17-92 polycistronic cluster of miRNAs, a region that is commonly amplified in B cell lymphomas.[85] Overexpression of this cluster in cooperation with the proto-oncogene c-myc was able to accelerate tumor development in a murine B cell lymphoma model. The transcription factor c-myc on its own is able to activate expression of six miRNAs and two of these, miR-17-5p and miR-20a suppress E2F1, another known target of c-myc associated with regulation of proliferation. The ability of miRNAs to control the balance between differentiation and proliferation, such as during cardiogenesis by miR-1,[86] or to transfer to germline stem cells the ability to bypass a cell cycle checkpoint[87] point toward other putative therapeutic areas in which miRNAs play a role and certainly, as of today these examples represent only the tip of the iceberg, keeping in mind the high number of miRNA that to date are not functionally annotated.

3.09.5.3 Validation of Drug Action

When comparing inhibition of a target by siRNA to that of a small-molecule inhibitor, differing phenotypes can sometimes be observed. The reasons for these discrepancies are of different origin. First, inhibition using a small molecule usually acts on a small pocket within a protein and, with a few exceptions, the overall structure of the protein is not affected. Since proteins usually interact with multiple other proteins in complexes (in yeast an average of five interactions per protein is estimated[88]), many have scaffold functions, and are important for complex stability. Thus the lack of a protein might destabilize a multiprotein complex to which it normally contributes, creating a different phenotype compared to a small molecule, which typically inhibits an enzymatic activity or a specific protein–protein interaction.

Second, different selectivity profiles for siRNAs due to off-target effects and those of small molecules due to cross-reactivities with other proteins are the rule rather than the exception. Third, regulation by either small molecules or siRNA can lead to a stress response with modulation of stress-related genes including the stress-activated kinases, p53, or heat shock proteins, which themselves are known modulators of gene expression. As outlined above (*see* Section 3.09.4.5), p53 was found to be modulated by some, but not all of the siRNAs designed against different regions within the menin gene.[70] Why some sequences seem to be better inducers of a stress response compared to others is still unknown, but secondary structures might play a role. Stress responses can also be induced by small-molecule drugs and it is not surprising that microarray analysis of some highly selective drugs shows a wider than expected range of changes in gene expression. The use of dominant-negative, enzymatically inactive proteins, or controls of inactive, structurally similar small molecules of the same chemical class should be included as controls and when available, active compounds from different chemical classes should be tested. For example conflicting results are obtained when using different methods for validation of Aurora-B kinase, a mitotic kinase emerging in cancer therapy. When Aurora-B kinase is inhibited by either a small-molecule inhibitor or by overexpression of a dominant-negative kinase mutant, phosphorylation of its substrate histone H3 is reduced. In contrast, RNAi-mediated downregulation has been reported not to show this effect.[89] Probably in this case, residual Aurora-B protein was still able to phosphorylate its substrates making the degree of downregulation a critical factor, since it is in principle possible to see a reduction of histone H3 phosphorylation after siRNA-mediated downregulation of Aurora-B.[90]

3.09.5.4 Target Validation in Animal Models

Despite all these putative pitfalls, siRNA experiments in the appropriate biological system should be a standard component of the target validation data package required for the decision-making on further development of any molecular target. Many reports from pharmaceutical companies using siRNA for target validation deal with targets which are already in essence validated, and siRNA experiments are used to add another piece in the puzzle toward full validation. The list of target validation data in tissue culture is long,[91,92] but the information gained that is relevant for target validation beyond what is already known is sometimes minor. We focus on selected examples of in vivo applications, due to their much higher relevance, and in order to demonstrate some basic concepts.

3.09.5.4.1 Topical and systemic administration of synthetic small interfering ribonucleic acid

Local or topical administration has shown some success in terms of downregulation of targets in different organs. In the eye, ocular neovascularization could be inhibited by targeting vascular endothelial growth factor (VEGF) in a mouse model[93] and in the lung, as will be discussed in more details later, targeting of heme-oxygenase-1 by siRNA enhanced ischemia reperfusion induced apoptosis.[94] However, systemic administration with unmodified siRNA oligonucleotides has major hurdles due to the high clearance and the limited tissue distribution. Plasma stability of unmodified siRNAs is typically in the range of a few minutes. In preclinical animal models these issues have been overcome by intravenous injection of siRNA oligonucleotides at very high volumes into the tail vein of mice using a hydrodynamic transfection method originally set up for expression of plasmids.[95] When applied for siRNA targeting Fas, a receptor associated with induction of apoptosis, mice were successfully protected from induced liver degeneration.[96] One method to induce hepatitis is by injecting agonistic Fas specific antibodies, which leads to the death of animals within 3 days. When animals were treated with the Fas-specific siRNA, 80% of the treated group survived for more than 10 days. Interestingly, mice suffering from established disease also improved after the Fas-targeted siRNA treatment. Although this is an encouraging result and the method for delivery can be used for target validation in preclinical animal models, it is not applicable in the clinic due to the extremely high injection volumes used, which approach the total blood volume of the mice. Hydrodynamic delivery has served as a proof of concept for targets in vivo, resulting in a reduction of target protein levels of 60–90% for up to 10 days.

The respiratory system holds promise for local delivery by aerosols, and first proof of concepts have been demonstrated. Intranasal delivery of a siRNA targeting the mRNA of heme-oxygenase-1, a cytoprotective enzyme, has been shown to enhance ischemia reperfusion induced lung apoptosis.[94] Further, prophylactic treatment of mice inhibited expression or upregulation of the protein after induction by ischemia reperfusion. Downregulation of the gene resulted in increased apoptosis after ischemia reperfusion injury. As expected the gene was specifically downregulated in lung airway and parenchyma, but not in other organs expressing the gene such as liver or kidney. The cellular uptake of the siRNA might be promoted by natural host defense mechanisms in this tissue environment.

3.09.5.4.2 In vivo silencing using vectors

In vivo gene silencing involving RNAi is very challenging, and preclinically different chemically synthesized siRNAs or expression of shRNA from viral or endogenous promoters have been used. The first RNAi-mediated repression of a transcript in vivo was shown for a luciferase reporter gene after hydrodynamic transfection of synthetic siRNA or shRNA expression plasmids injected into mouse liver.[11] In this case long dsRNA lacking essential 3′ and 5′ structures necessary for processing and nuclear export were expressed under the control of a polymerase II promoter. Nuclear localization prevented an undesirable interferon response triggered by long dsRNA in the cytoplasm of mammalian cells. When dsRNA suppressing the transcriptional co-repressor Ski was expressed the phenotypes were very similar to those observed in embryos in which the gene was deleted by homologous recombination techniques.[97]

The creation of standard transgenic animals expressing shRNA directed against a variety of targets with well-known phenotypes yielded disappointing results, with none of the strains showing a distinct or reproducible phenotype.[98] The same group was more successful by using embryonic stem (ES) cells for which the expression and activity of the shRNA were confirmed before generation of genetically modified mice. In this case, germline transmission of the shRNA expression constructs was found and the reduction in mRNA and protein levels were similar as in the ES cells under tissue culture conditions. This opens up the possibility of achieving tissue-specific, inducible, and reversible suppression of gene expression in mice with shorter time-lines as compared to knockout of genes by homologous recombination.

An alternative delivery method is the expression of shRNA from viral vectors. Viral vectors have the advantage of generating a more prolonged downregulation of the target beyond the 5–7 days typically seen with synthetic siRNAs. Furthermore, cell lines, which are difficult to transfect, can be targeted. shRNA expression from viral vectors as a therapeutic has serious limitations, not only due to the disappointing results in clinical trials related to gene therapy in general, but also due to the difficulty of controlling the effective dose and the high variability of shRNA-mediated degradation of mRNA, which can lead to off-target effects.

As vectors adenoviruses or lentiviruses are widely used, and first proof of concepts have been published. In mice, for example, Hommel and colleagues[99] reduced expression levels in the midbrain of tyrosine hydroxylase, an enzyme needed for the synthesis of dopamine, by using a virus-based expression vector.

3.09.6 Major Hurdles and Solutions toward Therapeutics

The major hurdles for siRNA therapeutics are poor cellular uptake, limited serum stability, restricted tissue distribution, and induction of a nonspecific immune stimulation. Current attempts to address these weaknesses consist mainly of chemical modifications, or improvement in formulation, and in this chapter some of the approaches are outlined. Owing to its early stage, gene therapy will not be discussed and the interested reader is referred to some excellent reviews.[100,101]

Modifications of siRNAs to improve druglike properties benefit from years of experience in DNA-based antisense technologies. However, this experience also teaches us that these modifications often lead to the solution of one issue at the cost of creating another, for example the introduction of 2′-*O*-allyl modifications at the 5′ end results in increased stability, but at the cost of lower activity.[102]

Another possible way of modifying siRNAs is represented by poly-2′-*O*-2,4-dinitrophenyl derivatives, which have in the past been used to modify longer RNAs in order to improve cellular permeability in the absence of transfection reagents, and to increase resistance against ribonuclease degradation.[103] A comparison of modified with unmodified siRNA of the same sequence showed that this chemical modification and its associated improvements can be successfully transferred to siRNA.

An alternative modification is to spike siRNAs with locked nucleic acids (LNAs). LNAs are conformational locked nucleotide analogs, which typically contain a methylene bridge connecting the 2′-oxygen with the 4′-carbon of the ribose ring, locking the ribose ring in the 3′-endo conformation, a characteristic of RNA. siRNA oligonucleotides containing LNAs show increased thermal stability and a systematic screen for modifications in different positions has recently been performed.[104] Introduction of LNA modifications in the 3′ ends led to an increase in serum stability ex vivo and additional LNA modifications within the siRNA sequence were tolerated for some but not all positions. Also, a reduction in off-target effects has been suggested, which is probably due to the fact that the strand that displays the weakest binding energy at its closing 5′ base pair is incorporated preferentially and this strand preference can be influenced by LNA. A LNA modification in the 5′ position of the sense strand, therefore, has altered strand bias of the RISC complex in favor of incorporation of the antisense. In 2005 a first clinical trial started using an LNA oligonucleotide, which targets the antiapoptotic protein Bcl-2 as an anticancer intervention for chronic lymphocytic leukemia (Santaris Pharma). Although, this particular LNA oligonucleotide is not working via a RNAi mechanism but rather by an antisense mechanism, the results of this trial might elucidate the safety profile of this type of modification.

A systematic assessment of 2′-*O*-methyl modifications further demonstrated that siRNAs with internal 2′-*O*-methyl modifications, but not those with such terminal modifications, are protected against serum-derived nucleases and those modifications in specific positions are tolerated without loss of activity.[105] As for most other modifications it is important to balance the positive effects on stability with the negative effects on activity or increased toxicity. The key might be to use mixed-backbone oligonucleotides for which the position of modifications is carefully chosen and to attempt to combine different modifications without compromising potency.

Attachment of siRNA to ligands for tissue-specific receptors is one possible means of directing the oligonucleotide to the target cell, and in fact tumor selective delivery has been achieved with ligand-targeted stabilized nanoparticles.[106] Nanoparticles with siRNA were produced by designing molecular conjugates with the following functional requirements: self-assembly, formation of a steric polymer protective surface layer, and exposition of a ligand. This was achieved by means of poly(ethyleneimine) (PEI) that was PEGylated with an RGD-peptide ligand attached at the distal end of the polyethylene glycol as a method to target integrins localized on the surface of endothelial cells of tumors. The siRNA sequence was specific for vascular endothelial growth factor receptor-2 (VEGFR-2) a validated target in cancer therapy. Intravenous administration of these nanoparticles resulted in selective tumor endothelial uptake, target modulation in vivo, inhibition of angiogenesis, and as a consequence inhibition of tumor growth. The high cost of production should be taken into account when considering the use of modified nanoparticles or complex siRNA modifications as therapeutic agents.

Another example of how a combination of modifications can lead to major improvements has recently been published. Soutschek and colleagues[107] showed siRNA activity in vivo following a clinically acceptable way of systemic administration of a modified siRNA, which targets apolipoprotein B (ApoB). In this case, cholesterol was attached to the 3′ end of the sense strand by means of a pyrrolidine linker, a modification which has been shown to enhance cellular uptake of conjugated siRNA in liver cells.[108] Additionally, the sugar backbone was stabilized by introducing phosphorothioate and 2′-*O*-methyl modifications in both the sense and antisense strands, leading to increased resistance toward degradation by exo and endonucleases. After in vivo administration the elimination half-life in rat plasma was 95 min compared to 6 min for unmodified oligonucleotides. Significant levels of this modified ApoB–siRNA were also detected in additional tissues such as kidney, heart, adipose tissue, or lung. In mice, treatment with this modified ApoB–siRNA resulted in a reduction of endogenous mRNA of up to 57% in the liver, and 73% in jejunum, compared to a control group. This was accompanied by a 68% inhibition of ApoB protein levels in plasma. In this study the mechanism of action was proven and the cleavage of the modified ApoB–mRNA occurred at the predicted site. This work certainly shows how, for a specific indication in the appropriate tissue, tailored solutions can be identified to overcome the restrictions of the technology.

Another way for cell specific delivery is linking siRNAs to antibodies recognizing cell-surface receptors.[109] Following this strategy, siRNAs targeting the human immunodeficiency virus (HIV)-1 capsid gene Gag were used to inhibit virus replication in HIV-infected primary T cells. In this case, cell-specific delivery could also be observed when the antibody conjugate was injected into mice.

3.09.7 Perspectives

The potential of the technology and all its facets such as the emerging field of miRNAs awaits further investigation. It is expected that more miRNAs will be functionally annotated and relations to disease will be found beyond the first examples described so far. In parallel, our understanding of the molecular events associated with the RNAi pathway and cross-talk to other pathways will improve further. For example, a recently emerging new function in which RNA interference is involved is regulation of gene expression by DNA methylation.[110]

After the first wave of functional readouts for genes using siRNA-based screens, more subtle screens with more sophisticated readouts will lead to the linkage of new functions to genes and probably also to diseases. Once single gene functions are defined, combination of siRNA knockdowns is expected to explore their interplays or redundancies. For example, an attractive application would be the identification of synthetic lethal genes with the aim of specifically targeting cancer cells (a synthetic lethal gene is defined by a situation in which two nonessential genes become essential when deleted in combination in the same cell). Combining siRNA and microarray technologies will speed up the current increase of knowledge and will allow us to delineate complex networks of gene functions. This creates the need for supporting computational analysis to make best use of these data. These networks will be also extremely valuable to accelerate development of classical small molecules with an unknown mechanism of action since the molecular target for such agents will be more readily identifiable by looking for the changes in gene expression pattern after exposure of cells. The feasibility of this concept has already been demonstrated.[111]

Certainly, additional efforts to improve druglikeness of siRNAs will be made to overcome the hurdles discussed in the previous section and more niches for therapeutic intervention in diseases will be identified. The potential of using siRNAs in therapy is high, keeping in mind the high number of nondruggable targets and any solution for stability, delivery, or formulation found for one siRNA should in principle be transferable to other siRNAs and their targets, which is a big advantage compared to small molecules, for which these kinds of issues have to be solved individually.

There is a strong precedence for a DNA-based oligonucleotide drug as a therapeutic for cytomegalovirus induced retinitis (Vitravene) and the registration of the first siRNA drug is only a question of time. The first clinical trial for a siRNA-based therapeutics has been already initiated. Sirna-27 is a chemically modified siRNA that targets the mRNA of VEGFR-1. This siRNA is used to inhibit ocular neovascularization associated with age-related macular degeneration and the compound is administered by intraocular injection. A more difficult question to answer at present is whether applications will go beyond these niches.

Treatment of genetic diseases and allele-specific destruction of mRNA are other areas where siRNAs hold promise. Selective degradation of a disease-associated mutant transcript but not the wild-type transcript differing only in one nucleotide is possible and proof of concept in cells was shown for the oncogenic Val-12 mutant of K-ras,[65] mutant forms of the Alzheimer's disease associated Tau and amyloid precursor proteins,[112] as well as for the mutant form of the superoxide dismutase SOD1 which gives rise to amyotrophic lateral sclerosis.[113] The field is extremely dynamic and the surrounding excitement and expectations are high, but there is still a long way to go.

References

1. He, L.; Hannon, G. J. *Nat. Rev. Genet.* **2004**, *5*, 522–531.
2. Napoli, C.; Lemieux, C.; Jorgensen, R. *Plant Cell* **1990**, *2*, 279–289.
3. Tabara, H.; Grishok, A.; Mello, C. C. *Science* **1998**, *282*, 430–431.
4. Fire, A.; Xu, S.; Montgomery, M. K.; Kostas, S. A.; Driver, S. E.; Mello, C. C. *Nature* **1998**, *391*, 806–811.
5. Elbashir, S. M.; Harborth, J.; Lendeckel, W.; Yalcin, A.; Weber, K.; Tuschl, T. *Nature* **2001**, *411*, 494–498.
6. Couzin, J. *Science* **2002**, *298*, 2296–2297.
7. Hammond, S. M.; Bernstein, E.; Beach, D.; Hannon, G. J. *Nature* **2000**, *404*, 293–296.
8. Bernstein, E.; Caudy, A. A.; Hammond, S. M.; Hannon, G. J. *Nature* **2001**, *409*, 363–366.
9. Yang, D.; Buchholz, F.; Huang, Z.; Goga, A.; Chen, C. Y.; Brodsky, F. M.; Bishop, J. M. *Proc. Natl. Acad. Sci. USA* **2002**, *99*, 9942–9947.
10. Kittler, R.; Putz, G.; Pelletier, L.; Poser, I.; Heninger, A. K.; Drechsel, D.; Fischer, S.; Konstantinova, I.; Habermann, B.; Grabner, H. et al. *Nature* **2004**, *432*, 1036–1040.
11. McCaffrey, A. P.; Meuse, L.; Pham, T. T.; Conklin, D. S.; Hannon, G. J.; Kay, M. A. *Nature* **2002**, *418*, 38–39.
12. Huppi, K.; Martin, S. E.; Caplen, N. J. *Mol. Cell* **2005**, *17*, 1–10.
13. Scherer, L. J.; Rossi, J. J. *Nat. Biotechnol.* **2003**, *21*, 1457–1465.
14. Denli, A. M.; Hannon, G. J. *Trends Biochem. Sci.* **2003**, *28*, 196–201.
15. Mello, C. C.; Conte, D., Jr. *Nature* **2004**, *431*, 338–342.
16. Nanduri, S.; Carpick, B. W.; Yang, Y.; Williams, B. R.; Qin, J. *EMBO J.* **1998**, *17*, 5458–5465.
17. Schramke, V.; Allshire, R. *Curr. Opin. Genet. Dev.* **2004**, *14*, 174–180.
18. Schramke, V.; Sheedy, D. M.; Denli, A. M.; Bonila, C.; Ekwall, K.; Hannon, G. J.; Allshire, R. C. *Nature* **2005**, *435*, 1275–1279.
19. Zamore, P. D.; Tuschl, T.; Sharp, P. A.; Bartel, D. P. *Cell* **2000**, *101*, 25–33.
20. Cerutti, L.; Mian, N.; Bateman, A. *Trends Biochem. Sci.* **2000**, *25*, 481–482.
21. Lingel, A.; Simon, B.; Izaurralde, E.; Sattler, M. *Nat. Struct. Mol. Biol.* **2004**, *11*, 576–577.
22. Golden, T. A.; Schauer, S. E.; Lang, J. D.; Pien, S.; Mushegian, A. R.; Grossniklaus, U.; Meinke, D. W.; Ray, A. *Plant Physiol.* **2002**, *130*, 808–822.
23. Provost, P.; Silverstein, R. A.; Dishart, D.; Walfridsson, J.; Djupedal, I.; Kniola, B.; Wright, A.; Samuelsson, B.; Radmark, O.; Ekwall, K. *Proc. Natl. Acad. Sci. USA* **2002**, *99*, 16648–16653.
24. Ketting, R. F.; Fischer, S. E.; Bernstein, E.; Sijen, T.; Hannon, G. J.; Plasterk, R. H. *Genes Dev.* **2001**, *15*, 2654–2659.

25. Nicholson, R. H.; Nicholson, A. W. *Mamm. Genome* **2002**, *13*, 67–73.
26. Provost, P.; Dishart, D.; Doucet, J.; Frendewey, D.; Samuelsson, B.; Radmark, O. *EMBO J.* **2002**, *21*, 5864–5874.
27. Zhang, H.; Kolb, F. A.; Brondani, V.; Billy, E.; Filipowicz, W. *EMBO J.* **2002**, *21*, 5875–5885.
28. Nykanen, A.; Haley, B.; Zamore, P. D. *Cell* **2001**, *107*, 309–321.
29. Hutvagner, G.; McLachlan, J.; Pasquinelli, A. E.; Balint, E.; Tuschl, T.; Zamore, P. D. *Science* **2001**, *293*, 834–838.
30. Khvorova, A.; Reynolds, A.; Jayasena, S. D. *Cell* **2003**, *115*, 209–216.
31. Martinez, J.; Patkaniowska, A.; Urlaub, H.; Luhrmann, R.; Tuschl, T. *Cell* **2002**, *110*, 563–574.
32. Caudy, A. A.; Myers, M.; Hannon, G. J.; Hammond, S. M. *Genes Dev.* **2002**, *16*, 2491–2496.
33. Caudy, A. A.; Ketting, R. F.; Hammond, S. M.; Denli, A. M.; Bathoorn, A. M.; Tops, B. B.; Silva, J. M.; Myers, M. M.; Hannon, G. J.; Plasterk, R. H. *Nature* **2003**, *425*, 411–414.
34. Hammond, S. M.; Boettcher, S.; Caudy, A. A.; Kobayashi, R.; Hannon, G. J. *Science* **2001**, *293*, 1146–1150.
35. Song, J. J.; Smith, S. K.; Hannon, G. J.; Joshua-Tor, L. *Science* **2004**, *305*, 1434–1437.
36. Ahlquist, P. *Science* **2002**, *296*, 1270–1273.
37. Sijen, T.; Fleenor, J.; Simmer, F.; Thijssen, K. L.; Parrish, S.; Timmons, L.; Plasterk, R. H.; Fire, A. *Cell* **2001**, *107*, 465–476.
38. Lee, R. C.; Feinbaum, R. L.; Ambros, V. *Cell* **1993**, *75*, 843–854.
39. Reinhart, B. J.; Slack, F. J.; Basson, M.; Pasquinelli, A. E.; Bettinger, J. C.; Rougvie, A. E.; Horvitz, H. R.; Ruvkun, G. *Nature* **2000**, *403*, 901–906.
40. Lee, Y.; Jeon, K.; Lee, J. T.; Kim, S.; Kim, V. N. *EMBO J.* **2002**, *21*, 4663–4670.
41. Lee, Y.; Ahn, C.; Han, J.; Choi, H.; Kim, J.; Yim, J.; Lee, J.; Provost, P.; Radmark, O.; Kim, S.; Kim, V. N. *Nature* **2003**, *425*, 415–419.
42. Pfeffer, S.; Zavolan, M.; Grasser, F. A.; Chien, M.; Russo, J. J.; Ju, J.; John, B.; Enright, A. J.; Marks, D.; Sander, C.; Tuschl, T. *Science* **2004**, *304*, 734–736.
43. Sullivan, C. S.; Grundhoff, A. T.; Tevethia, S.; Pipas, J. M.; Ganem, D. *Nature* **2005**, *435*, 682–686.
44. Pfeffer, S.; Sewer, A.; Lagos-Quintana, M.; Sheridan, R.; Sander, C.; Grasser, F. A.; van Dyk, L. F.; Ho, C. K.; Shuman, S.; Chien, M. et al. *Nat. Methods* **2005**, *2*, 269–276.
45. Bentwich, I.; Avniel, A.; Karov, Y.; Aharonov, R.; Gilad, S.; Barad, O.; Barzilai, A.; Einat, P.; Einav, U.; Meiri, E. et al. *Nat. Genet.* **2005**, *37*, 766–770.
46. Elbashir, S. M.; Lendeckel, W.; Tuschl, T. *Genes Dev.* **2001**, *15*, 188–200.
47. Yang, D.; Lu, H.; Erickson, J. W. *Curr. Biol.* **2000**, *10*, 1191–1200.
48. Elbashir, S. M.; Harborth, J.; Weber, K.; Tuschl, T. *Methods* **2002**, *26*, 199–213.
49. National Center for Biotechnology Information. http://www.ncbi.nlm.nih.gov/BLAST (accessed April 2006).
50. Bass, B. L. *Nature* **2001**, *411*, 428–429.
51. Schwarz, D. S.; Hutvagner, G.; Du, T.; Xu, Z.; Aronin, N.; Zamore, P. D. *Cell* **2003**, *115*, 199–208.
52. Reynolds, A.; Leake, D.; Boese, Q.; Scaringe, S.; Marshall, W. S.; Khvorova, A. *Nat. Biotechnol.* **2004**, *22*, 326–330.
53. THE RNAi WEB. http://www.rnaiweb.com/RNAi/RNAi_Web_Resources/RNAi_Tools___Software/Online_siRNA_Design_Tools/ (accessed April 2006).
54. Donze, O.; Picard, D. *Nucleic Acids Res.* **2002**, *30*, e46.
55. Cargill, M.; Altshuler, D.; Ireland, J.; Sklar, P.; Ardlie, K.; Patil, N.; Shaw, N.; Lane, C. R.; Lim, E. P.; Kalyanaraman, N. et al. *Nat. Genet.* **1999**, *22*, 231–238.
56. Malumbres, M.; Sotillo, R.; Santamaria, D.; Galan, J.; Cerezo, A.; Ortega, S.; Dubus, P.; Barbacid, M. *Cell* **2004**, *118*, 493–504.
57. Kan, L.; Kessler, J. A. *BioEssays* **2005**, *27*, 14–16.
58. Sui, G.; Soohoo, C.; Affar, E. B.; Gay, F.; Shi, Y.; Forrester, W. C.; Shi, Y. *Proc. Natl. Acad. Sci. USA* **2002**, *99*, 5515–5520.
59. Tabara, H.; Sarkissian, M.; Kelly, W. G.; Fleenor, J.; Grishok, A.; Timmons, L.; Fire, A.; Mello, C. C. *Cell* **1999**, *99*, 123–132.
60. Brummelkamp, T. R.; Bernards, R.; Agami, R. *Science* **2002**, *296*, 550–553.
61. Lee, N. S.; Dohjima, T.; Bauer, G.; Li, H.; Li, M. J.; Ehsani, A.; Salvaterra, P.; Rossi, J. *Nat. Biotechnol.* **2002**, *20*, 500–505.
62. Miyagishi, M.; Taira, K. *Oligonucleotides* **2003**, *13*, 325–333.
63. Zeng, Y.; Wagner, E. J.; Cullen, B. R. *Mol. Cell* **2002**, *9*, 1327–1333.
64. Barton, G. M.; Medzhitov, R. *Proc. Natl. Acad. Sci. USA* **2002**, *99*, 14943–14945.
65. Brummelkamp, T. R.; Bernards, R.; Agami, R. *Cancer Cell* **2002**, *2*, 243–247.
66. Rubinson, D. A.; Dillon, C. P.; Kwiatkowski, A. V.; Sievers, C.; Yang, L.; Kopinja, J.; Rooney, D. L.; Ihrig, M. M.; McManus, M. T.; Gertler, F. B. et al. *Nat. Genet.* **2003**, *33*, 401–406.
67. Shen, C.; Buck, A. K.; Liu, X.; Winkler, M.; Reske, S. N. *FEBS Lett.* **2003**, *539*, 111–114.
68. Du, Q.; Thonberg, H.; Wang, J.; Wahlestedt, C.; Liang, Z. *Nucleic Acids Res.* **2005**, *33*, 1671–1677.
69. Jackson, A. L.; Bartz, S. R.; Schelter, J.; Kobayashi, S. V.; Burchard, J.; Mao, M.; Li, B.; Cavet, G.; Linsley, P. S. *Nat. Biotechnol.* **2003**, *21*, 635–637.
70. Scacheri, P. C.; Rozenblatt-Rosen, O.; Caplen, N. J.; Wolfsberg, T. G.; Umayam, L.; Lee, J. C.; Hughes, C. M.; Shanmugam, K. S.; Bhattacharjee, A.; Meyerson, M. et al. *Proc. Natl. Acad. Sci. USA* **2004**, *101*, 1892–1897.
71. Lindsay, M. A. *Nat. Rev. Drug Disc.* **2003**, *2*, 831–838.
72. Kamath, R. S.; Fraser, A. G.; Dong, Y.; Poulin, G.; Durbin, R.; Gotta, M.; Kanapin, A.; Le Bot, N.; Moreno, S.; Sohrmann, M. et al. *Nature* **2003**, *421*, 231–237.
73. Sonnichsen, B.; Koski, L. B.; Walsh, A.; Marschall, P.; Neumann, B.; Brehm, M.; Alleaume, A. M.; Artelt, J.; Bettencourt, P.; Cassin, E. et al. *Nature* **2005**, *434*, 462–469.
74. PhenoBank Database. http://www.worm.mpi-cbg.de/phenobank2/cgi-bin/MenuPage.py (accessed April 2006).
75. Boutros, M.; Kiger, A. A.; Armknecht, S.; Kerr, K.; Hild, M.; Koch, B.; Haas, S. A.; Consortium, H. F.; Paro, R.; Perrimon, N. *Science* **2004**, *303*, 832–835.
76. Somma, M. P.; Fasulo, B.; Cenci, G.; Cundari, E.; Gatti, M. *Mol. Biol. Cell* **2002**, *13*, 2448–2460.
77. Echard, A.; Hickson, G. R.; Foley, E.; O'Farrell, P. H. *Curr. Biol.* **2004**, *14*, 1685–1693.
78. Eggert, U. S.; Kiger, A. A.; Richter, C.; Perlman, Z. E.; Perrimon, N.; Mitchison, T. J.; Field, C. M. *PLoS. Biol.* **2004**, *2*, e379.
79. Jackson, A. L.; Bartz, S. R.; Schelter, J.; Kobayashi, S. V.; Burchard, J.; Mao, M.; Li, B.; Cavet, G.; Linsley, P. S. *Nat. Biotechnol.* **2003**, *21*, 635–637.
80. Berns, K.; Hijmans, E. M.; Mullenders, J.; Brummelkamp, T. R.; Velds, A.; Heimerikx, M.; Kerkhoven, R. M.; Madiredjo, M.; Nijkamp, W.; Weigelt, B. et al. *Nature* **2004**, *428*, 431–437.

81. Paddison, P. J.; Silva, J. M.; Conklin, D. S.; Schlabach, M.; Li, M.; Aruleba, S.; Balija, V.; O'Shaughnessy, A.; Gnoj, L.; Scobie, K. et al. *Nature* **2004**, *428*, 427–431.
82. MacKeigan, J. P.; Murphy, L. O.; Blenis, J. *Nat. Cell Biol.* **2005**, 7, 591–600.
83. Poy, M. N.; Eliasson, L.; Krutzfeldt, J.; Kuwajima, S.; Ma, X.; Macdonald, P. E.; Pfeffer, S.; Tuschl, T.; Rajewsky, N.; Rorsman, P. et al. *Nature* **2004**, *432*, 226–230.
84. Esau, C.; Kang, X.; Peralta, E.; Hanson, E.; Marcusson, E. G.; Ravichandran, L. V.; Sun, Y.; Koo, S.; Perera, R. J.; Jain, R. et al. *J. Biol. Chem.* **2004**, *279*, 52361–52365.
85. He, L.; Thomson, J. M.; Hemann, M. T.; Hernando-Monge, E.; Mu, D.; Goodson, S.; Powers, S.; Cordon-Cardo, C.; Lowe, S. W.; Hannon, G. J.; Hammond, S. M. *Nature* **2005**, *435*, 828–833.
86. Zhao, Y.; Samal, E.; Srivastava, D. *Nature* **2005**, *436*, 214–220.
87. Hatfield, S. D.; Shcherbata, H. R.; Fischer, K. A.; Nakahara, K.; Carthew, R. W.; Ruohola-Baker, H. *Nature* **2005**, *435*, 974–978.
88. Grigoriev, A. *Nucleic Acids Res.* **2003**, *31*, 4157–4161.
89. Keen, N.; Taylor, S. *Nat. Rev. Cancer* **2004**, *4*, 927–936.
90. Carpinelli, P.; Gianellini, L.; Soncini, C.; Giorgini, M.L.; Cappella, P.; Sola, F.; Ceruti, R.; Marsiglio, A.; Storici, P.; Rusconi, L. et al. 1st Annual AACR Meeting, Anaheim, CA, 2005; Abs 663.
91. Milhavet, O.; Gary, D. S.; Mattson, M. P. *Pharmacol. Rev.* **2003**, *55*, 629–648.
92. Dykxhoorn, D. M.; Novina, C. D.; Sharp, P. A. *Nat. Rev. Mol. Cell Biol.* **2003**, *4*, 457–467.
93. Reich, S. J.; Fosnot, J.; Kuroki, A.; Tang, W.; Yang, X.; Maguire, A. M.; Bennett, J.; Tolentino, M. J. *Mol. Vis.* **2003**, *9*, 210–216.
94. Zhang, X.; Shan, P.; Jiang, D.; Noble, P. W.; Abraham, N. G.; Kappas, A.; Lee, P. J. *J. Biol. Chem.* **2004**, *279*, 10677–10684.
95. Zhang, G.; Budker, V.; Wolff, J. A. *Hum. Gene Ther.* **1999**, *10*, 1735–1737.
96. Song, E.; Lee, S. K.; Wang, J.; Ince, N.; Ouyang, N.; Min, J.; Chen, J.; Shankar, P.; Lieberman, J. *Nat. Med.* **2003**, *9*, 347–351.
97. Shinagawa, T.; Ishii, S. *Genes Dev.* **2003**, *17*, 1340–1345.
98. Carmell, M. A.; Zhang, L.; Conklin, D. S.; Hannon, G. J.; Rosenquist, T. A. *Nat. Struct. Biol.* **2003**, *10*, 91–92.
99. Hommel, J. D.; Sears, R. M.; Georgescu, D.; Simmons, D. L.; DiLeone, R. J. *Nat. Med.* **2003**, *9*, 1539–1544.
100. Ryther, R. C.; Flynt, A. S.; Phillips, J. A., III; Patton, J. G. *Gene. Ther.* **2005**, *12*, 5–11.
101. Hannon, G. J.; Rossi, J. J. *Nature* **2004**, *431*, 371–378.
102. Amarzguioui, M.; Holen, T.; Babaie, E.; Prydz, H. *Nucleic Acids Res.* **2003**, *31*, 589–595.
103. Chen, X.; Dudgeon, N.; Shen, L.; Wang, J. H. *Drug Disc. Today* **2005**, *10*, 587–593.
104. Elmen, J.; Thonberg, H.; Ljungberg, K.; Frieden, M.; Westergaard, M.; Xu, Y.; Wahren, B.; Liang, Z.; Orum, H.; Koch, T.; Wahlestedt, C. *Nucleic Acids Res.* **2005**, *33*, 439–447.
105. Czauderna, F.; Fechtner, M.; Dames, S.; Aygun, H.; Klippel, A.; Pronk, G. J.; Giese, K.; Kaufmann, J. *Nucleic Acids Res.* **2003**, *31*, 2705–2716.
106. Schiffelers, R. M.; Ansari, A.; Xu, J.; Zhou, Q.; Tang, Q.; Storm, G.; Molema, G.; Lu, P. Y.; Scaria, P. V.; Woodle, M. C. *Nucleic Acids Res.* **2004**, *32*, e149.
107. Soutschek, J.; Akinc, A.; Bramlage, B.; Charisse, K.; Constien, R.; Donoghue, M.; Elbashir, S.; Geick, A.; Hadwiger, P.; Harborth, J. et al. *Nature* **2004**, *432*, 173–178.
108. Lorenz, C.; Hadwiger, P.; John, M.; Vornlocher, H. P.; Unverzagt, C. *Bioorg. Med. Chem. Lett.* **2004**, *14*, 4975–4977.
109. Song, E.; Zhu, P.; Lee, S. K.; Chowdhury, D.; Kussman, S.; Dykxhoorn, D. M.; Feng, Y.; Palliser, D.; Weiner, D. B.; Shankar, P. et al. *Nat. Biotechnol.* **2005**, *23*, 709–717.
110. Chan, S. W.; Zilberman, D.; Xie, Z.; Johansen, L. K.; Carrington, J. C.; Jacobsen, S. E. *Science* **2004**, *303*, 1336.
111. di Bernardo, D.; Thompson, M. J.; Gardner, T. S.; Chobot, S. E.; Eastwood, E. L.; Wojtovich, A. P.; Elliott, S. J.; Schaus, S. E.; Collins, J. J. *Nat. Biotechnol.* **2005**, *23*, 377–383.
112. Miller, V. M.; Gouvion, C. M.; Davidson, B. L.; Paulson, H. L. *Nucleic Acids Res.* **2004**, *32*, 661–668.
113. Ding, H.; Schwarz, D. S.; Keene, A.; Affar, E. B.; Fenton, L.; Xia, X.; Shi, Y.; Zamore, P. D.; Xu, Z. *Aging Cell* **2003**, *2*, 209–217.

Biographies

Laura Gianellini obtained an education in industrial chemistry from the l'Istituto Tecnico Industriale 'E. Molinari' in Milan, Italy, and has several years of experience in the pharmaceutical industry where she was involved in target validation in immunology and oncology at Pharmitalia, Pharmacia/Pfizer, and Nerviano Medical Sciences S.r.l (Italy). She is responsible for target validation of new targets in cancer therapy using siRNA technologies.

Jürgen Moll studied microbiology at the University of Bayreuth, Germany, and got his PhD in biochemistry at the University of Basel, Switzerland, working on transgenic animal models for cancer. After a postdoctoral training and holding a post as group leader at the Institute for Genetics (Research Center Karlsruhe, Germany), where he investigated extracellular matrix receptors in metastasis, he joined Pharmacia/Pfizer (Italy) as a project leader. At present he is at Nerviano Medical Sciences S.r.l (Italy) where, together with his team, he brought the first Aurora kinase inhibitor for anticancer therapy into the clinic.

Comprehensive Medicinal Chemistry II
ISBN (set): 0-08-044513-6

ISBN (Volume 3) 0-08-044516-0; pp. 171–187

3.10 Signaling Chains

M-H Teiten, R Blasius, F Morceau, and M Diederich, Hôpital Kirchberg, Luxembourg City, Luxembourg
M Dicato, Centre Hospitalier de Luxembourg, Luxembourg City, Luxembourg

3.10.1	**Introduction: Cell-Signaling Pathways**	**190**
3.10.2	**Ubiquitine-Proteasome and Signal Transduction Pathways**	**190**
3.10.2.1	Introduction	190
3.10.2.2	The Ubiquitine Proteasome Pathway	190
3.10.2.3	Targeting Proteins for Ubiquitination by Stress-Responsive Kinases	192
3.10.2.3.1	NFκB pathways in inflammation and immune responses	192
3.10.2.3.2	Protein dephosphorylation targeting ubiquitination	192
3.10.3	**Activity of Bortezomib in Proteasome Inhibition**	**194**
3.10.4	**Proinflammatory Cell-Signaling Pathways**	**195**
3.10.4.1	NFκB Activation Pathways	195
3.10.4.2	NFκB-Signaling Pathways and Apoptosis	195
3.10.4.2.1	Antiapoptotic pathways activated by NFκB	196
3.10.4.2.2	Proapoptotic role	197
3.10.4.3	Inflammatory Signaling Pathways Triggered by NFκB	197
3.10.5	**Vascular Endothelial Growth Factor Cell-Signaling Pathways**	**197**
3.10.5.1	Introduction	197
3.10.5.2	Vascular Endothelial Growth Factor Receptors (VEGFRs)	198
3.10.5.3	Signaling Pathways Activated by Vascular Endothelial Factor Receptor-2	200
3.10.5.4	Regulation of Vascular Endothelial Growth Factor Gene Expression	200
3.10.5.5	Physiological versus Pathological Angiogenesis Linked to Vascular Endothelial Growth Factor Expression	200
3.10.5.5.1	Vascular endothelial growth factor in inflammatory disorders	202
3.10.5.5.2	Pathology of the female reproductive tract	202
3.10.5.5.3	Intraocular neovascular syndromes	202
3.10.5.5.4	Vascular endothelial growth factor and cancer	202
3.10.5.6	Vascular Endothelial Growth Factor as a Therapeutic Tool	203
3.10.6	**Signal Transducer and Activator of Transcription (STAT)**	**203**
3.10.6.1	Chromosomal Localization and Structure of the Signal Transducers and Activators of Transcription	203
3.10.6.2	Signal Transducer and Activator of Transcription Proteins in Signal Transduction/Activation Mechanisms of Signal Transducer and Activator of Transcription Proteins	204
3.10.6.2.1	STAT1 signaling	205
3.10.6.2.2	STAT2 signaling	206
3.10.6.2.3	STAT3 signaling	206
3.10.6.2.4	STAT4 signaling	207
3.10.6.2.5	STAT5 signaling	207
3.10.6.2.6	STAT6 signaling	207
3.10.6.3	Pathological Variations of Signal Transducer and Activator of Transcription Expression	207
3.10.6.3.1	Signal transducers and activators of transcription in tumorigenesis	207

3.10.6.3.2	Signal transducers and activators of transcription in allergic inflammation and autoimmune diseases	208
3.10.6.4	Future Directions	208
3.10.7	**Conclusion**	**208**
References		**208**

3.10.1 Introduction: Cell-Signaling Pathways

Cell-signaling mechanisms describe the molecular events activated within cells to mediate growth, proliferation, differentiation, and survival. Those mechanisms are complex and only partially elucidated. Nevertheless protein domains and interactions were identified and deregulation of those mechanisms was described leading to increased proliferation capacity, sustained angiogenesis, metastasis, as well as resistance to apoptosis. Most diseases present aberrations of cell-signaling processes and compounds that target disease-specific alterations of cell-signaling mechanisms are considered interesting compounds for future therapies. New clinical trials have already generated a multitude of agents targeting cell-signaling pathways that are becoming increasingly complex to understand because of the highly tissue-specific nature of the signaling pathways.

Regulatory pathways are activated by extracellular factors, including hormones, growth factors, or cytokines. These extracellular factors activate intracellular cascades of protein networks within the cell. Whereas cell-signaling cascades demonstrate a high level of specificity, it becomes clear that many pathways are redundant and that inhibition of one specific pathway most often leads to the activation of parallel signaling pathways inducing activation of common targets involved in the development of diseases. Targeting highly specific pathways by single compounds inhibiting only one protein generally leads to resistance to this therapeutic agent.

3.10.2 Ubiquitine-Proteasome and Signal Transduction Pathways

3.10.2.1 Introduction

Lysosomal proteases are mainly responsible for the degradation of proteins taken up from the cell surface and are involved in 20% of normal protein turnover. Therefore, a nonlysosomal proteolytic enzyme complex, the proteasome, degrades most cellular proteins. Proteasomes, which are localized in both the cytosol and nucleus, represent up to 1% of all cellular proteins in eukaryotes.[1] It specifically degrades cytoplasmic and nuclear proteins previously tagged with Ub in an adenosine triphosphate (ATP)-dependent manner. The Ub-proteasome pathway (UPP) is involved in processing and functional modifications of proteins implicated in essential cellular processes,[2] such as survival, function, and cell cycling of normal cells. Indeed, intracellular proteolysis is necessary for physiological regulation of transcription, cell cycle, antigen processing, and signal transduction.[3] Thus, a deregulation may contribute to tumor progression, neurodegenerative or autoimmune diseases, drug resistance, and metabolic disorders that have been clearly demonstrated in humans,[4–7] making this catalytic process inhibition a selective therapeutic target in cancer.

Here we will focus on the role of kinases and protein phosphorylation/dephosphorylation turnovers in ubiquitination targeting proteins as well as the action of the proteasome inhibitors on signal transduction pathways.

3.10.2.2 The Ubiquitine Proteasome Pathway

Proteasome consists of the multicatalytic 26S multisubunit protease complex which comprises a 20S cylinder, capped at both ends by a 19S complex.[8] The 20S complex functions as the catalytic core, whereas 19S subunits constitute the regulatory complexes of the 26S proteasome (**Figure 1**). To be recognized and subsequently degraded by 26S proteasome, cellular proteins targeted to undergo processing and modifications have to be tagged by covalent binding of multiple Ub monomers. Ub is a 76-amino-acid polypeptide that binds to target proteins in an ATP- and a multistep-dependent manner.[9] Indeed, this process, called ubiquitination, requires three classes of enzymes: E1, E2, and E3. The Ub-activating enzyme (E1) catalyzes the transfer of Ub to the Ub-conjugating enzyme (E2). In this first reaction, Ub activation requires energy from ATP hydrolysis and phosphorylation to allow its binding to the enzyme E1 by a high-energy thiol ester bond. Next, Ub can be transferred to E2 by forming an additional thiol ester bond. In the following step, E2 catalyzes the conjugation of an activated Ub moiety to the Ub-ligase enzyme E3 that catalyzes ligation of Ub on specific protein substrates. The specificity of protein ubiquitination for each substrate is determined by E3.[10] Alternatively, transfer of Ub from E2 to the substrate can be directly mediated by E3. Finally, multiple cycles of ubiquitination lead to the formation of a poly-Ub chain as a recognition signal for the 26S proteasome and subsequent

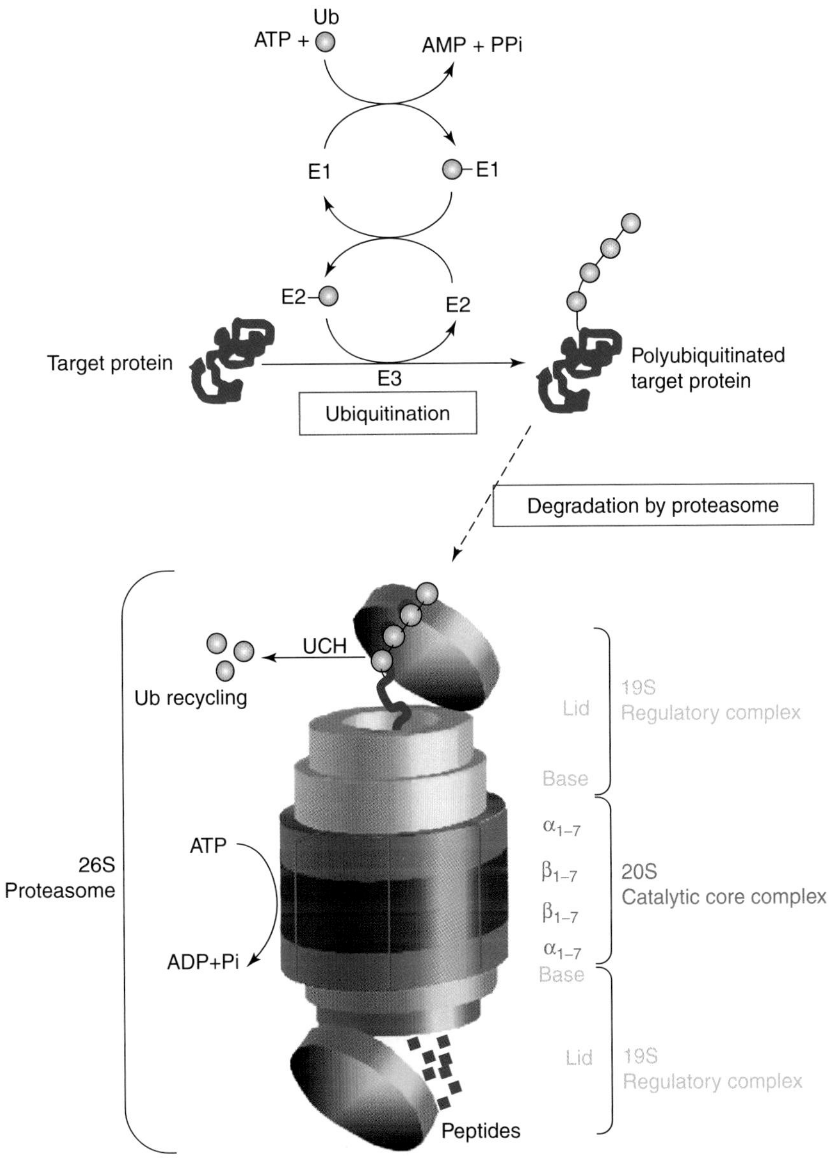

Figure 1 Schematic representation of the ubiquitin-proteasome pathway. Ubiquitin (Ub) binding to target protein is catalyzed in three major enzymatic steps requiring the ubiquitin-activating enzyme (E1) and ATP, the ubiquitin-conjugating enzyme (E2), and the ubiquitin-ligase enzyme (E3). Multiple cycles of this successive enzymatic actions result in the synthesis and attachment of polyubiquitin chains to the target protein. The ubiquitinated protein is then degraded by the 26S proteasome consisting of 20S cylindrical catalytic core (red structure) and two 19S regulatory complexes (yellow structures). The polyubiquitinated protein enters the 19S regulatory complex and, following ubiquitin carboxy-terminal hydrolase (UCH)-mediated deubiquitination and unfolding, it is translocated into the central cavity of the 20S catalytic core complex for degradation by different hydrolytic activities.

degradation.[11] In fact, recognition of ubiquitinated proteins is realized by the 19S regulatory complex which in addition is in charge of deubiquitination, unfolding, and translocation of substrate proteins.[12–14] Then, the ATP-dependent cleavage of peptide bonds occurs in the central catalytic cylinder of the 20S proteasome by the β1, β2, and β5 subunits,[15,16] exhibiting chymotrypsin-like, trypsin-like, and caspase-like activities. These different substrate specificities allow the cleavage of peptides after hydrophobic, basic, and acidic residues respectively.[17]

Regulation of protein turnover by UPP requires the selectivity of the Ub-targeting specific protein substrate. Targeting proteins for ubiquitination is a dynamic process that is necessary for regulating protein stability[18] and a variety of extra- and intracellular signals can initiate the proteolysis process.[19] Indeed, numerous data are in favor of a direct relationship between the targeting of the regulatory proteins for ubiquitination and the phosphorylation/dephosphorylation cascade of signal transduction pathways.[20] Thus, modification of duration and magnitude of transcriptional activity of transcription factors can be initiated by signal transduction cascades in response to stress, damage, and mitogenic stimuli.

Understanding of the Ub-targeting mechanisms has mainly been provided by studies on signal transduction pathways induced in cells subjected to stress. Stress-responsive kinases such as Jun NH_2-kinase (JNK), inhibitor of NFκB (IκB), mitogen-activated protein kinase (MAPK), Janus kinase (JAK), protein kinase C (PKC), phosphatidylinositol-3-kinase (PI3K), and DNA protein kinase (DNA-PK) catalyze phosphorylation of various kinds of cellular proteins affecting their conformation, stability, subcellular localization, and transcriptional activation.

3.10.2.3 Targeting Proteins for Ubiquitination by Stress-Responsive Kinases

Most of the protein kinases are stress-responsive and contribute to the regulation of protein targeting for ubiquitination, such as transcription factors, leading to change in the expression of genes involved in proliferation and apoptosis. The phosphorylation status of targeted proteins for ubiquitination is particularly dependent on the environmental context of the cells. Interestingly, this status is known to be involved in the modification of spatial conformation of the proteins. The adequate conformation of the substrates for the specific E3 enzymes affinity can correspond to the phosphorylated form or not. Thus, protein phosphorylation may lead to ubiquitination or not, determining the stability of the proteins in a particular context. The next section describes significant cases where protein phosphorylation allows or alternatively prevents ubiquitination and degradation by the proteasome in different physiological contexts.

3.10.2.3.1 NFκB pathways in inflammation and immune responses

The UPP plays a crucial role in the transcriptional activity of the Rel family of dimeric transcription factors NFκB. NFκB is involved in the regulation of multiple gene expressions that are critical for the regulation of apoptosis, viral replication, tumorigenesis, inflammation, and various autoimmune diseases. The activation of NFκB results from a variety of stimuli, including growth factors, cytokines (interleukin-1 (IL1), tumor necrosis factor-α (TNF-α)), bacterial lipopolysaccharide (LPS), lymphokines, ultraviolet light, pharmacological agents, and oxidative stress. These stimuli can activate two distinct pathways: (1) the canonical pathway and (2) the alternative pathway, depending on the cell type and the stimulated receptor, that lead to the activation of different NFκB dimers. The UPP is involved in the process of NFκB transcription factor activation whatever the stimulated pathway. In the canonical pathway (**Figure 2**), activated signalosome targets IκB for ubiquitination and proteasome-dependent degradation[21,22] by mediating the N-terminal phosphorylation on serine residues 32 and 36. Thus, IκB phosphorylation provides a specific targeting signal that is recognized by the specific E3 IκB-Ub ligase.[23] Ubiquitination and proteasome-dependent degradation of IκB in the cytoplasm are required for the TNF-induced release and translocation of NFκBp50/p65 toward the nucleus, where it interacts with its target genes. On the other hand, the NFκBp100/RelB cytoplasmic sequestration is IκB-independent. The p100 protein (and also p105) contains a C-terminal IκB-homologous inhibitory region and the dimers containing this NFκB protein are retained in the cytoplasm by virtue of the function of this IκB-homologous region. In fact, the activation process occurs through the cleavage of NFκB inducing kinase (NIK)-mediated phosphorylation of the p100 subunit in a p52 protein. It has been shown that, in the presence of the proteasome inhibitor MG132, the CD27 receptor induced accumulation of polyubiquitinated p100 proteins in cells of the human lymphoblastoid line Ramos, testifying to ubiquitination and proteasomal degradation of the phosphorylated p100, resulting in the generation of a p52 subunit.[24] Thus, the transcriptional activity of NFκB, resulting in inflammatory or protective immune responses, is then dependent on the degradation of specific proteins (IκB, p100, p105) by the UPP, when they exhibit adequate conformation provided by their phosphorylated status.

3.10.2.3.2 Protein dephosphorylation targeting ubiquitination

3.10.2.3.2.1 JNK-activated signaling pathway prevents c-Jun-associated protein degradation

The stress-activated protein kinases (SAPK), also called JNK, are among the stress-responsive kinases capable of targeting ubiquitination. As a cell response to stress stimuli, the kinase phosphorylation cascade MKK1/MKK4/JNK/c-Jun is activated, while JNK-associated inhibitors, such as c-Jun amino-terminal kinase interacting protein-1 (JIP), are inactivated. The phosphorylation of c-Jun by JNK requires its association to the JNK-delta domain, which provides the docking site for trans-signal targeting of this transcription factor.[25] The hypothesis that c-Jun ubiquitination is targeted by its associated protein JNK was supported by the findings that the delta domain was necessary for ubiquitination and proteolytic degradation through the UPP in vivo and in vitro. In contrast, JNK does not target ubiquitination of the nonassociated proteins ELK1 and Jun-D.[26–28] However, it was clearly demonstrated that JNK targeting of c-Jun ubiquitination is limited

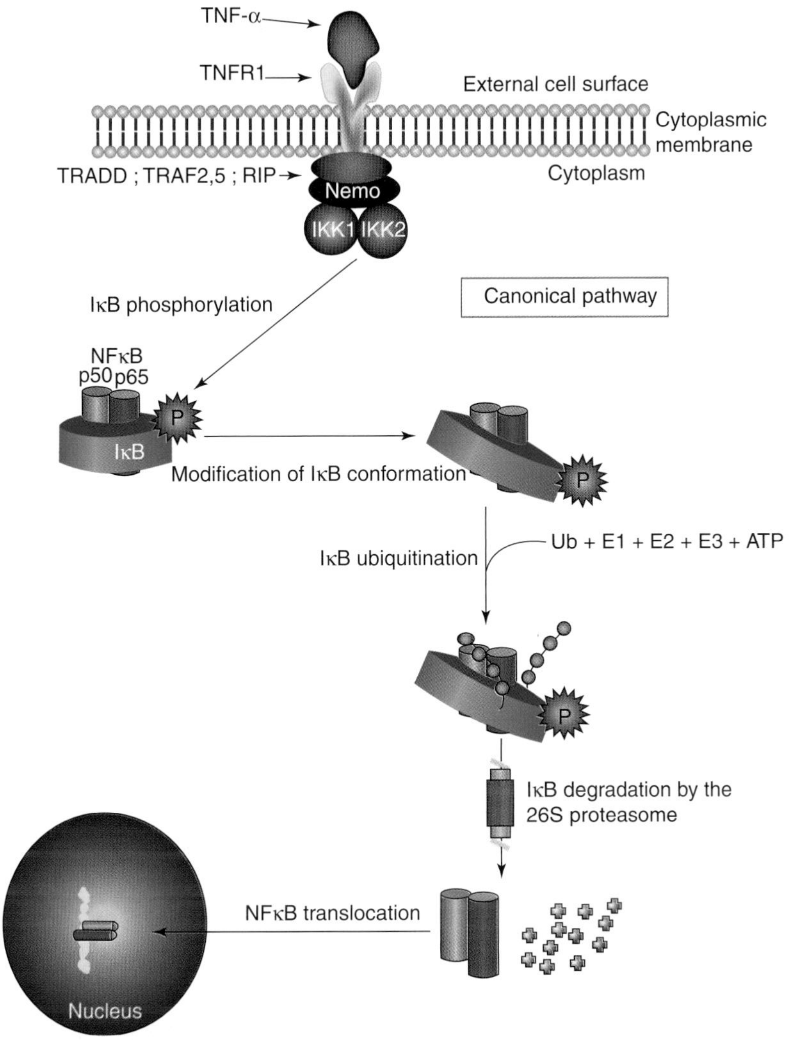

Figure 2 Requirement of IκB phosphorylation for targeting ubiquitination in TNF-α-stimulated cells. Tumor necrosis factor receptor (TNFR) 1 is stimulated by specific association with TNF-α, resulting in the NFκB canonical pathway. The TNFR-associated death domain (TRADD) is first recruited by TNFR1. As an adaptor, TRADD assembles signaling complexes consisting in the receptor-interacting protein (RIP) and TNFR-associated factors (TRAF) that activate the signalosome (NEMO/IKK1/IKK2). The inhibitor of NFκB (IκB) is then phosphorylated/activated, leading to conformational modifications and recognition by the Ub-ligase (E3). IκB is then ubiquitinated and degraded by 26S proteasome, thus releasing NFκB p50/p65, which translocates toward the nucleus, where it interacts with target genes.

to the nonphosphorylated form of the substrate. Indeed, after JNK-mediated phosphorylation of c-Jun, it can no longer be targeted for ubiquitination. Then, while the association of JNK to c-Jun as well as to other proteins (ATF2 and Jun-B) is a prerequisite for their targeting ubiquitination, the phosphorylation in stress conditions of the associated protein prevents this event, possibly by modifying protein conformation, abrogating the action of the Ub-ligase E3 (**Figure 3**).

3.10.2.3.2.2 JNK-activated signaling pathway prevents p53-associated protein degradation

Similarly, JNK plays an important role in the regulation of the targeting ubiquitination of the tumor suppressor protein p53. Indeed, JNK is capable of association with p53 in the nonstressed cells. Conversely, the JNK-signaling pathway via MEKK1 is activated in response to stress, leading to a significant decrease in the amount of JNK-p53 association while the total p53 level increases in cells. This inverse correlation suggests that JNK can target p53 degradation in nonstressed cells. Phosphorylation of p53 by stress-activated kinases such as JNK or DNA-PK results in the dissociation of p53-targeting molecules and stabilizes p53 (**Figure 3**). The stability of p53 is also well known to be controlled by the E3 Ub ligase mdm2. Similarly to JNK-p53 association, the complex formation between mdm2 and p53 occurs in cells

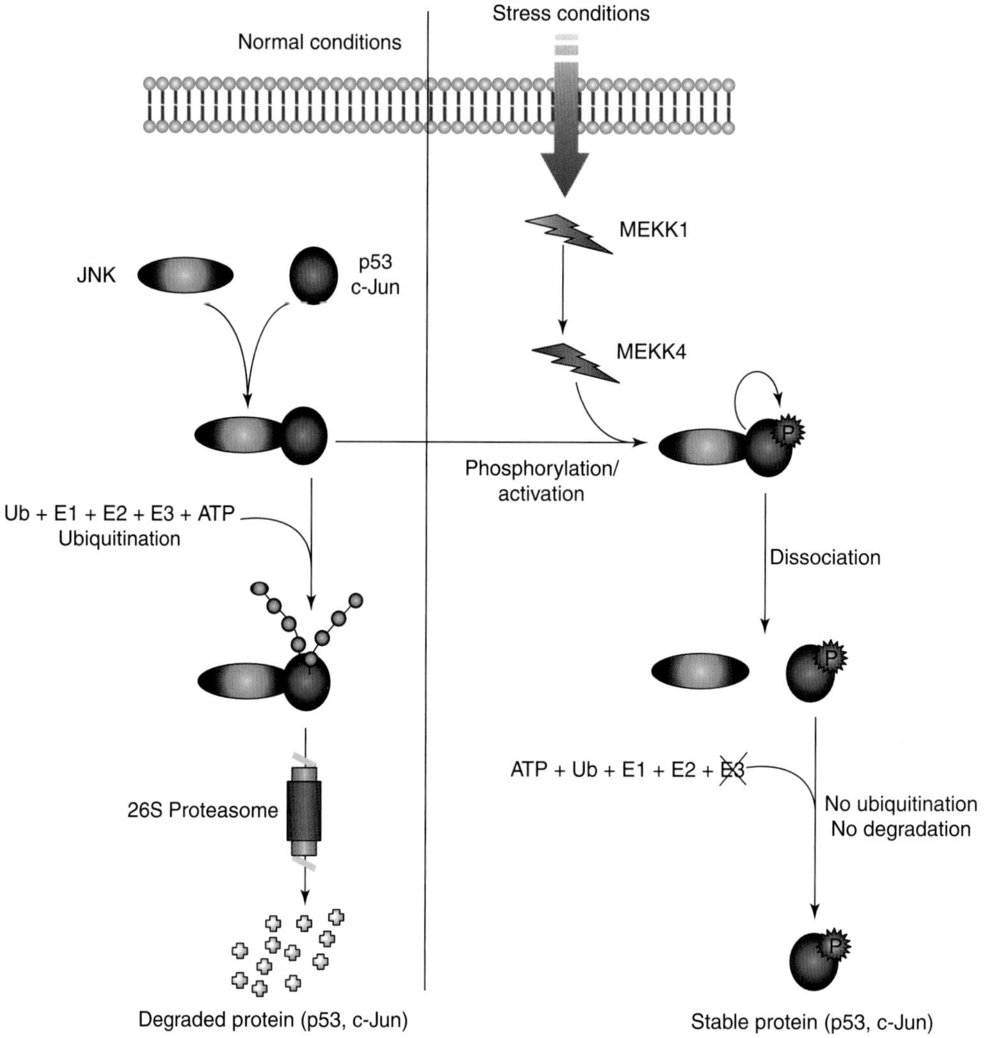

Figure 3 c-Jun and p53 ubiquitination in nonstressed normal growing cells. JNK associates to target proteins such as p53 and c-Jun. In stress conditions, phosphorylation of JNK involves activation of upstream kinases MEKK1 and MKK4. Activated JNK phosphorylates its associated protein, allowing dissociation, and phosphorylated p53 or c-Jun protein half-life is increased due to protection from E3 ligase-mediated ubiquitination and consequently from degradation by the 26S proteasome. In the absence of stress (normal conditions), JNK is not activated and its association to p53 or c-Jun is maintained through the delta domain, which provides the docking site for a trans-signal targeting protein for ubiquitination. Unphosphorylated p53 or c-Jun then undergoes ubiquitination and proteolytic degradation by a 26S proteasome.

growing in normal conditions, allowing ubiquitination and degradation of p53. The JNK-mediated degradation of p53 is still independent of mdm2 since the two complexes occur at different phases of the cell cycle.[20]

JNK-signaling pathway stimulation leads to activation and stabilization of associated proteins in cells (e.g., c-Jun, p53) exposed to stress, whereas JNK promotes ubiquitination in nonstressed normally growing cells where associated proteins are not phosphorylated.

3.10.3 Activity of Bortezomib in Proteasome Inhibition

UPP regulates the stability of multiple proteins, affecting the expression and function of regulatory proteins (e.g., JNK), tumor suppressors (e.g., p53), transcription factors (e.g., NFκB), and proto-oncogenes (e.g., c-Jun). Inhibition of proteasome action can lead to apoptosis by inhibiting NFκB activity, increasing p53 and Bax protein expressions, and accumulating cyclin-dependent kinase inhibitors p27 and p21.[29]

Inhibition of NFκB activity by proteasome inhibitors takes place by blocking the degradation of IκB.[30] This inhibition leads to decreased levels in the proapoptotic proteins Bcl-2 and A1/Bfl-1, triggering cytochrome *c* release, caspase-9 activation, and apoptosis.[31]

Among UPP inhibitors, bortezomib (Velcade) was the first to enter clinical studies. On the basis of data resulting from clinical trials, the US Food and Drug Administration (FDA) granted accelerated approval for this drug for the treatment of patients with multiple myeloma on May 13, 2003. The promising preclinical and clinical activity exhibited by bortezomib on multiple myeloma and other malignancies has confirmed the proteasome as a relevant and important target in the treatment of cancers (for a review, see Rajkumar *et al.*[32]). NFκB inhibition is probably one of the main mechanisms by which bortezomib induces apoptosis and overcomes drug resistance in multiple myeloma.[30] In addition, JNK activation seems to be an important pathway for bortezomib-induced multiple myeloma cell apoptosis. Bortezomib activates JNK, leading to Fas upregulation and caspase-8 and caspase-3 activation. This caspase-8-mediated apoptotic pathway is independent of the caspase-9-mediated pathway. On the other hand, induction of caspase-3 leads to mdm2 degradation and p53 phosphorylation, thereby increasing p53 activity and apoptosis. Bortezomib also induces FasL expression, probably because of increased c-*myc* expression that occurs as a result of proteasome inhibition.[31]

Furthermore, the inhibition of growth in multiple myeloma cell lines and primary multiple myeloma cells is a consequence of the inhibiting effect of bortezomib on IL6-induced Ras/Raf/MAP kinase pathway activation.[30] Bortezomib has no effect on IL6-induced signaling through the JAK/STAT3 pathway. However, the specific effects of proteasome inhibition in malignancy and the precise mechanism of action of bortezomib require further investigation to be fully understood in the future.

3.10.4 Proinflammatory Cell-Signaling Pathways

NFκB transcription factor is a dimeric complex of subunits belonging to the Rel family (p105/50, p100/52, p65 (Rel A), Rel B, and c-Rel). NFκB proteins share the Rel homology domain (RHD), allowing DNA binding, dimerization, and nuclear localization. NFκB proteins are sequestered in the cytoplasm by a family of IκB proteins through interactions between inhibitor ankyrin repeats and RHD. Upon stimulation by various activators such as cytokines, LPS, growth factors, stress inducers, or chemotherapeutic agents, IκB is phosphorylated on two serine residues, which triggers its ubiquitinylation and degradation by the 26S proteasome (*see* Section 3.10.2.3.1). NFκB is then free to enter the nucleus and to activate the transcription of target genes by binding to its cognate decameric DNA sequence 5′-GGGRNNYYCC-3′, where R indicates A or G, Y indicates C or T, and N can be any base. NFκB is involved in the transcription of many proinflammatory as well as antiapoptotic genes, and is thus a key player in the progression of carcinogenesis and inflammatory diseases such as rheumatoid arthritis, inflammatory bowel disease, and asthma.[33] Therapies aiming to suppress NFκB-induced survival genes are thus an interesting approach to fight and cure these diseases.

3.10.4.1 NFκB Activation Pathways

As previously stated, activation of NFκB through IκB phosphorylation and degradation depends on the activation of IκB kinases (IKKs) (**Figure 4**). The IKK complex is composed of three subunits, the catalytic subunits IKKα and IKKβ and the regulatory subunit IKKγ (or NEMO, NFκB essential modulator).[21] The different components of this complex were identified and characterized[34–38] and knockout experiments have determined their role in NFκB activation. From these studies, it appears that IKKγ is required for IKK activity and classical NFκB activation through IκB phosphorylation and degradation pathway,[39,40] but not for the alternative pathway leading to p52/RelB dimer translocation.[41] IKKβ gene disruption in mice leads to death between days 12 and 15 post gestation.

IKKβ and IKKγ are required for the canonical pathway of IκB phosphorylation and degradation in response to proinflammatory stimuli, while IKKα is dispensable for IκB phosphorylation, but is involved and necessary for inducible p100 processing. RelB-p100 forms an inactive dimer and the IKKα-dependent degradation of the C-terminal part of p100 allows nuclear translocation of the so-formed RelB-p52 complex. Activation of this dimer is important for lymphoid organ development and adaptive immune response but not for other NFκB-dependent functions such as apoptosis inhibition and innate immunity.[42,43]

3.10.4.2 NFκB-Signaling Pathways and Apoptosis

Apoptotic cell death is characterized by morphological changes and involves the activation of a family of cysteine aspartate proteases, called caspases. These proteases exist in unstimulated cells as inactive zymogens and are activated by proteolytic cleavage to form a heterotetramer consisting of two large and two small subunits. Two major pathways of

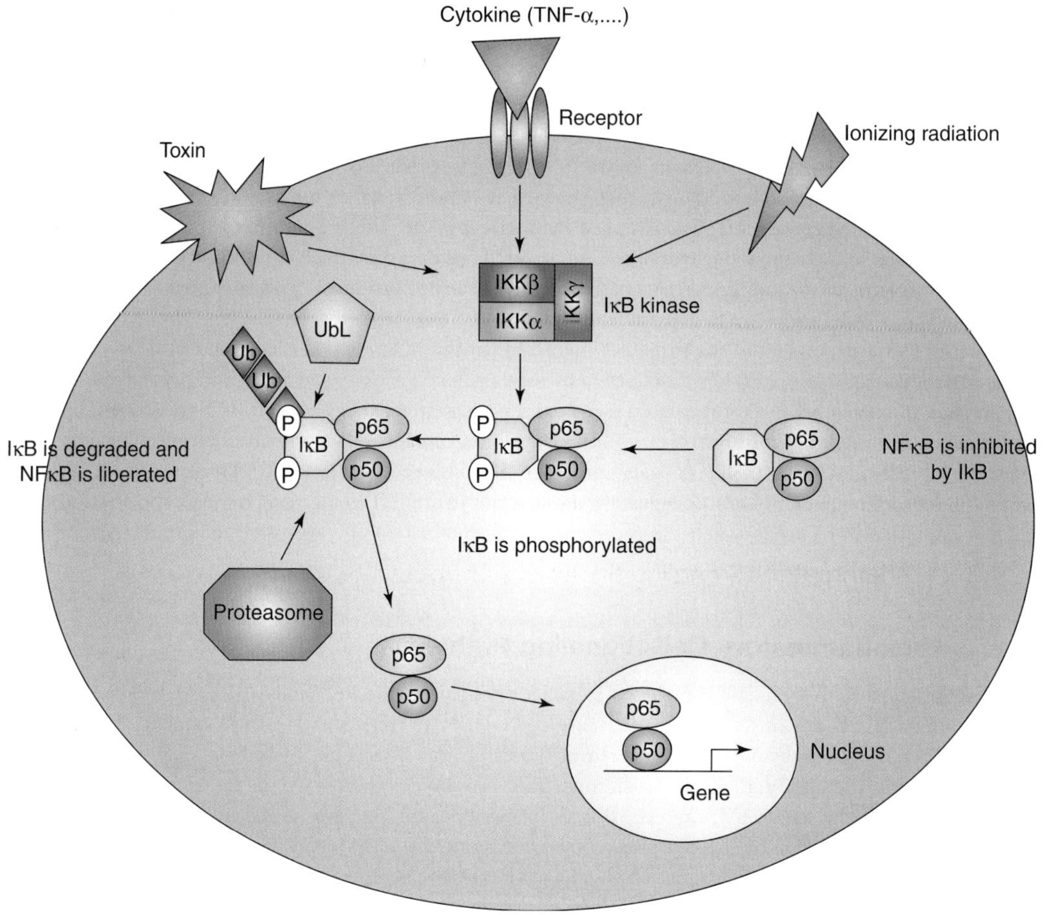

Figure 4 Regulation of proinflammatory cell-signaling pathways.

caspase activation are well characterized to date: (1) the death receptors (DRs) or extrinsic pathway; and (2) the mitochondrial or intrinsic pathway.[44]

Apoptotic signaling through the extrinsic pathway is triggered by the engagement of members of the TNF-receptor family (TNFR1, Fas/CD95, DR3, DR4, DR5, and DR6) by their ligands (TNF-α, lymphotoxin α, FasL, Apo3L, and TNF-related apoptosis-inducing ligand (TRAIL)), resulting in receptor trimerization and activation of the procaspase-8 and -10. DNA damage or cell stress leading to activation of p53 or Bcl-2 family proapoptotic proteins such as Bax and Bak initiate the intrinsic pathway and induce the mitochondrial release of apoptogenic molecules such as cytochrome *c*, apoptosis-inducing factor (AIF) or Smac/DIABLO (second mitochondria-derivated activator of caspases/direct inhibitors of apoptosis (IAP)-binding protein with low pI). Cytochrome *c* binds to the apoptotic protease-activating factor (Apaf)-1 and this dimer forms the apoptosome complex with procaspase-9.[45] Extrinsic and intrinsic pathways are interconnected: caspase-8 proteolysis through the extrinsic pathway leads to cleavage and activation of Bid, a proapoptotic member of the Bcl-2 family. Truncated Bid incorporates into the mitochondrial membrane and induces cytochrome *c* release, which in turn leads to apoptosome formation and thus intrinsic apoptotic pathway.[46]

3.10.4.2.1 Antiapoptotic pathways activated by NFκB

Transcription factors of the NFκB family play a role in the regulation of the apoptotic program in different cell lines.[47] The first evidence of the NFκB cytoprotective role came from the analysis of $p65^{-/-}$ mice that died on embryonic day 15 from massive liver apoptosis, which could be reestablished by p65 reexpression.[48,49] NFκB transcriptional activity leads to the expression of many antiapoptotic genes. Among those gene products, some are directly implicated in inhibiting the apoptotic cascade, such as IAPs, c-FLIP (caspase-8/FADD-like-IL1β-converting-enzyme (caspase-8/FLICE) inhibitory protein), or Bcl-2 family proteins. IAPs are a family of antiapoptotic proteins highly conserved throughout evolution that have the ability to inhibit the activation of various caspases by direct binding and interaction through their baculoviral-inhibitory repeat (BIR).[50] NFκB inhibits TNF-α-induced apoptosis through transcriptional activation of IAP1, IAP2, TNF-receptor-associated factor (TRAF) 1 and TRAF2, thus inhibiting caspase-8 activity.[51] c-FLIP is a proteolytically inactive analog of caspase-8 and competes with procaspases-8 and -10 for binding to FADD, thus preventing their activation and subsequent apoptosis.[52,53] Bcl-2 family proteins such as Bfl-1/A1, Bcl-X_L, or Bcl-2 prevent apoptosis by inhibiting cytochrome *c* leakage from the mitochondria, thus avoiding apoptosome formation and apoptotic cascade. NFκB activation also leads to repression of several proapoptotic genes. For example, two genes coding for the transcription factors Forkhead and growth arrest and DNA-damage-inducible gene (GADD) 153/C/EBP-homologous protein (CHOP),[54,55] as well as the proapoptotic gene *bax*,[56,57] are repressed by the activity of NFκB. NFκB activation also interferes with p53 proapoptotic function, since both transcription factors can inhibit each other's transcriptional activity by competing for a limiting pool of CBP/p300 complexes.[57]

3.10.4.2.2 Proapoptotic role

NFκB may have a proapoptotic activity in some cell types under certain conditions. It regulates the expression of genes whose products can induce apoptosis, such as TNF-receptor superfamily members DR4, DR5, DR6, Fas, and the FasL ligand.[58–61] Moreover, NFκB activity can sometimes be correlated to apoptosis. NFκB is activated during serum starvation-induced human embryonic kidney cell apoptosis, and its inhibition partially protects cells from this death.[62]

3.10.4.3 Inflammatory Signaling Pathways Triggered by NFκB

Numerous proinflammatory mediators, such as IL1 and TNF-α, induce nuclear translocation of NFκB, which in turn activates transcription of gene products involved in inflammatory response, angiogenesis, and cell adhesion. NFκB promotes the expression of cell adhesion molecules (intercellular adhesion molecule (ICAM)-1, vascular cell adhesion molecule (VCAM)-1, E selectin, tenascin C), VEGF, and of matrix metalloproteases (MMP)-2 and -9, responsible for degradation of the extracellular matrix.[63] This transcription factor is also involved in transcription of genes coding for enzymes such as inducible nitric oxide synthase (iNOS), cyclooxygenase (COX)-2, 5- and 12-lipooxygenase (LOX), chemokines, and cytokines (IL1 and TNF-α). COX-2 is upregulated in aggressive colorectal cancers and has been found to promote angiogenesis. Mediators such as nitric oxide, prostaglandins, IL1, or TNF-α are involved in the regulation of blood pressure, platelet aggregation, and body temperature. Moreover, NFκB is also important in the transcription of genes coding for several acute-phase proteins.

NFκB constitutive activity seems to play a role in chronic inflammatory diseases such as inflammatory bowel disease, Crohn's disease, ulcerative colitis, rheumatoid arthritis, and asthma.[33] NFκB activation in chronic inflammation may have a role in tumor initiation, since the antiapoptotic genes activated by this transcription factor may contribute to the survival of altered cells that would otherwise be committed to apoptosis and thereby allow formation of precancerous lesions. For instance, *Helicobacter pylori* infections or constitutive activation of the IL1 gene, both triggering NFκB activation, are risk factors for the development of gastric cancer.[64,65] Constitutive NFκB activation may also be responsible for the development of colitis-associated cancer by patients suffering from inflammatory bowel disease, ulcerative colitis, or Crohn's disease.[66–68]

NFκB and its target genes are potent antiapoptotic and proinflammatory mediators and are thus involved in various pathologies such as chronic inflammatory diseases and cancers. Inhibition of NFκB activation is a very promising therapeutic approach to fight those pathologies. Some inhibitors are already potent NFκB modulators in clinical use, while others remain benchtop compounds still in process. Nevertheless, a better understanding of the different steps leading to NFκB transcriptional activation in different cancer types will permit increased accuracy in the design of various inhibitors, allowing precise targeting of the chosen step.

3.10.5 Vascular Endothelial Growth Factor Cell-Signaling Pathways

3.10.5.1 Introduction

Angiogenesis is a process of new blood vessel development from preexisting vasculature. It plays an essential role in several physiological processes (e.g., embryonic development, normal growth of tissues, wound healing, and in the female reproductive cycle) as well as a major role in many diseases (tumor growth and metastasis spread, cutaneous affections, ophthalmic disorders, and ischemic diseases) depending on its level of expression (**Figure 5**).[69] This process is mediated by several angiogenic growth factors such as acidic and basic fibroblast growth factors, TNF-α, angiogenin, and VEGF.[70,71]

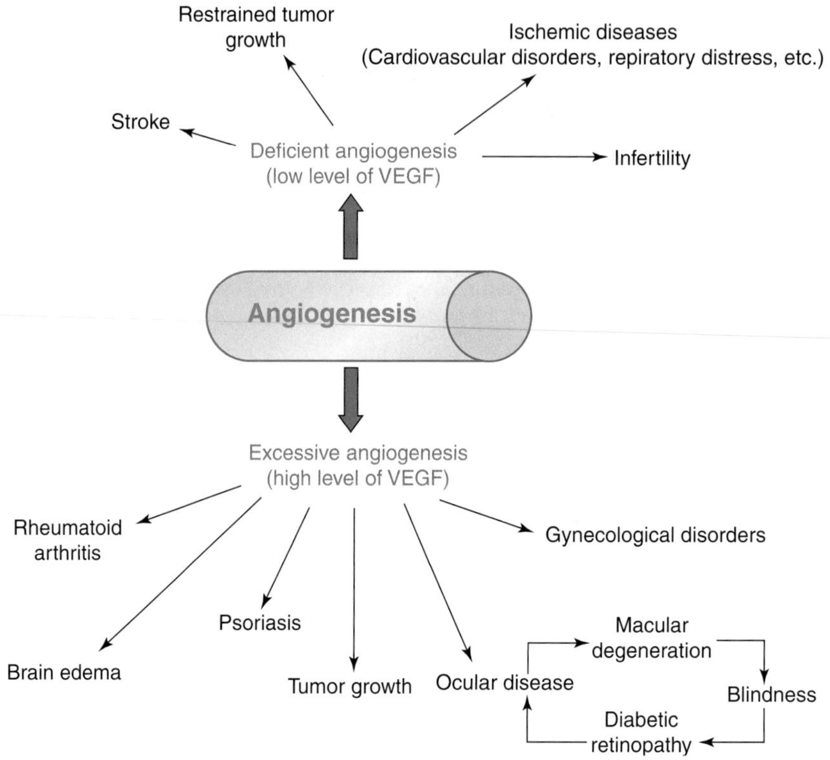

Figure 5 Implications of vascular endothelial growth factor receptors (VEGFRs) level of expression in pathologic angiogenesis.

3.10.5.2 Vascular Endothelial Growth Factor Receptors (VEGFRs)

The VEGF family, composed of VEGF-A, VEGF-B,[72] VEGF-C,[73] VEGF-D,[74] and placental growth factor (PlGF),[75] is highly implicated in the growth and survival of the vascular endothelium. VEGF-A is a key regulator of blood vessel growth, PlGF is known to mediate arteriogenesis, whereas VEGF-C and VEGF-D are shown to regulate lymphatic angiogenesis. Here we will mainly focus on the biology, signaling pathways, and clinical implications of the VEGF-A member, also called VEGF.

VEGF is a heparin-binding glycoprotein secreted as a homodimer of 45 kDa.[76,77] The human VEGF gene is organized as eight exons separated by seven introns. Alternative exon splicing results in the generation of different VEGF isoforms composed of 121–206 amino acids that differ in their expression patterns and in their biochemical and biological properties depending on the vascular endothelial growth factor receptor (VEGFR) they are binding to.[78]

The VEGFR family is composed of three signaling tyrosine kinase (TK) receptors. VEGFR-1/Flt-1 and VEGFR-2/Flk-1 are expressed at the cell surface of most blood endothelial cells, while VEGFR-3/Flt-4 is largely restricted to lymphatic endothelial cells.[79] These VEGFRs consist of seven immunoglobulin G (IgG)-like extracellular domains, a single transmembrane region, and a consensus TK sequence that is interrupted by a kinase insert domain (**Figure 6**). They are distinguished from each others by an inserted sequence.[80] Neuropilins, heparan-sulfated proteoglycans, cadherins, and $\alpha V\beta 3$ integrin serve as coreceptors for some, but not all, VEGF proteins.[81]

VEGFR-1 and VEGFR-2, located on vascular endothelium, are the two high-affinity binding sites of VEGF. VEGFR-2 mediates most of the downstream effects of VEGF in angiogenesis, including endothelial cell proliferation, invasion, survival, and microvascular permeability; whereas VEGFR-1 does not mediate an effective mitogenic signal in endothelial cells. This receptor is critical for physiologic and developmental angiogenesis and it also acts predominantly as a ligand-binding molecule, sequestering VEGF from VEGFR-2 signaling.

As for a variety of receptor TKs, the activation of these growth factor receptors is preceded by the formation of receptor dimers and subsequent receptor phosphorylation.[82] The recognition site for VEGF is located on the first three N-terminal Ig-like loops, whereas dimerization of the receptor is stabilized due to an additional domain located on

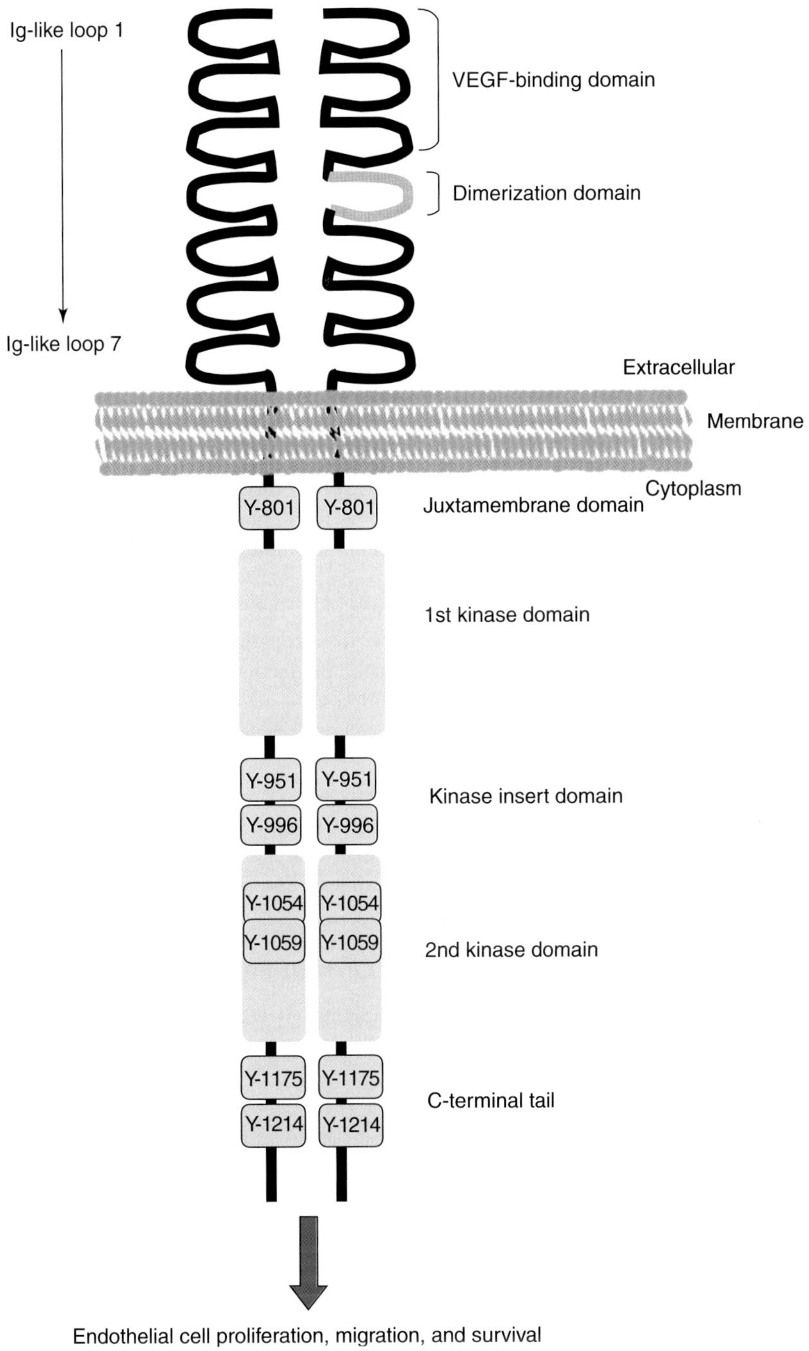

Figure 6 Schematic representation of the VEGFR-2 tyrosine kinase receptor. The VEGFR-2 tyrosine kinase receptor consists of seven immunoglobulin-like structures in the extracellular domain, a single transmembrane region, and a consensus tyrosine kinase domain interrupted by a kinase insert domain.

Ig-like loop 4.[83] Therefore, the binding of VEGF to VEGFR-2 causes receptor dimerization, kinase activation, and autophosphorylation of specific tyrosine residues within the dimeric complex. VEGFR-2 consists of Tyr801, Tyr951, and Tyr996 present in the kinase-insert domain, Tyr1054 and Tyr1059 located in the kinase domain, and Tyr1175 and Tyr1214 in the C-terminal tail; all have been identified as autophosphorylation sites.[84]

3.10.5.3 Signaling Pathways Activated by Vascular Endothelial Factor Receptor-2

Several proteins have been found to associate, via their Src homology-2 (SH2) domain, with these specific autophosphorylated tyrosine residues, so that they are activated to take part in signal transduction cascades[84,85] (**Figure 7**).

In this way, Tyr951 creates a binding site for the VEGFR-associated protein, VRAP,[86] which in turn activates PI3K and phospholipase Cγ1(PLC-γ1), whereas Tyr1175 creates one for ScK (Shc-related adaptor protein)[87] and PLC-γ1.[88] VRAP and Sck are adaptor proteins that facilitate and regulate the interaction of KDR (kinase insert domain-containing receptor) with cytoplasmic effector proteins important to endothelial cell survival and proliferation. Binding of PLC-γ1 activates PKC that stimulates the Ras/Raf/MEK pathway,[89] leading to the subsequent induction of the extracellular kinase (Erk) pathway (p42/44 mitogen-activated protein kinase). Erk can then translocate to the nucleus, where it phosphorylates and activates transcription factors, including c-Jun and the ternary complex factor, which in turn induce immediate transcription of the *c-fos* gene[90–92] leading to cell proliferation. Tyr1059 is responsible for ligand-mediated intracellular Ca^{2+} mobilization, MAPK activation, and endothelial cell proliferation.[93]

Phosphorylated Tyr1214 presents a binding site for the focal adhesion kinase (FAK) that will then interact with PI3K and paxillin in order to ensure focal adhesion and cell migration.[94] Such VEGF-induced cytoskeletal reorganization and cell migration are also established by the interaction of VEGFR-2 with p38MAPK.[95]

VEGFR-2 also activates PI3K, resulting in an increase of the lipid phosphatidylinositol, (3,4,5)P_3, leading to the activation of protein kinase B (Akt/PKB)[96] and the small guanosine triphosphate (GTP)-binding protein Rac. These proteins are known to be implicated in the regulation of vascular permeability and cellular migration.[97] On the one hand Akt/PKB induces Bcl-2 antagonist of cell death (BAD) and inhibits Bcl-2-associated death promoter homolog and caspase-9, thereby promoting cell survival.[98] On the other hand, the Akt pathway activates prostacyclins (PGI_2) and the endothelial nitric oxide synthase (eNOS), with the subsequent production of nitric oxide. These two intracellular mediators are predicted to have a vascular protective effect but also to mediate angiogenic and vascular permeability, increasing the effects of VEGF.[99] Several other important intracellular signaling elements, such as Src, are also activated by VEGFR-2.[100] VEGF is also implicated in the permeabilization of the extracellular membrane of endothelial cells, corresponding to the initial step of angiogenesis. In fact, VEGF induces a variety of enzymes and proteins, such as matrix-degrading metalloproteinases, metalloproteinase interstitial collagenase, and urokinase-type plasminogen (uPA), leading to this process of membrane degradation.[101] Moreover, uPA itself has been shown to increase the production of a variety of different angiogenic factors, including VEGF, suggesting that an autocrine regulatory loop may exist.

3.10.5.4 Regulation of Vascular Endothelial Growth Factor Gene Expression

Many cytokines and growth factors such as PDGF, TNF-α, epidermal growth factor (EGF), transforming growth factor (TGF), interleukins, and also MAP kinases and certain oncogenes such as *ras*, *V-src*, and *HER2* are known to upregulate VEGF mRNA expression or to induce VEGF release.[92,102–105]

Hypoxia has been shown to be one of the major inducers of VEGF expression through both increased transcription and stabilization of VEGF.[106,107] In fact, in hypoxic conditions, the two subunits, HIF-1α and HIF-1α/ARNT, of the hypoxic inducible factor-1 (HIF-1) bind to a hypoxia-responsive element (HRE) located in the 5′ flanking region of the VEGF promoter gene in order to enhance its transcription and stabilization.[108,109] This induction could be inhibited by the inhibitory PAS (Per/ARNT/Sim)[110] and has been shown to implicate PI3K, which is a downstream activator of Ras.[90]

Moreover, a posttranscriptional regulation of the VEGF expression is enhanced by the VHL (von Hippel–Lindau) suppressor gene product.[111]

3.10.5.5 Physiological versus Pathological Angiogenesis Linked to Vascular Endothelial Growth Factor Expression

VEGF is implicated not only in physiological but also in pathological angiogenesis.[112,113] As previously described, it takes part in many intracellular functions, such as endothelial cell permeability, migration, proliferation, and survival. Therefore the development of pharmacological treatments for disorders characterized by inadequate tissue perfusion has been the central point of interest and investigation of many researchers and pharmaceutical firms over the last decades.

VEGF plays an important role in embryonic and early postnatal development,[114,115] skeletal growth, endochondral bone formation,[98] and in ovarian angiogenesis,[116] as well as a partial inhibition of VEGF-achieved impaired organ

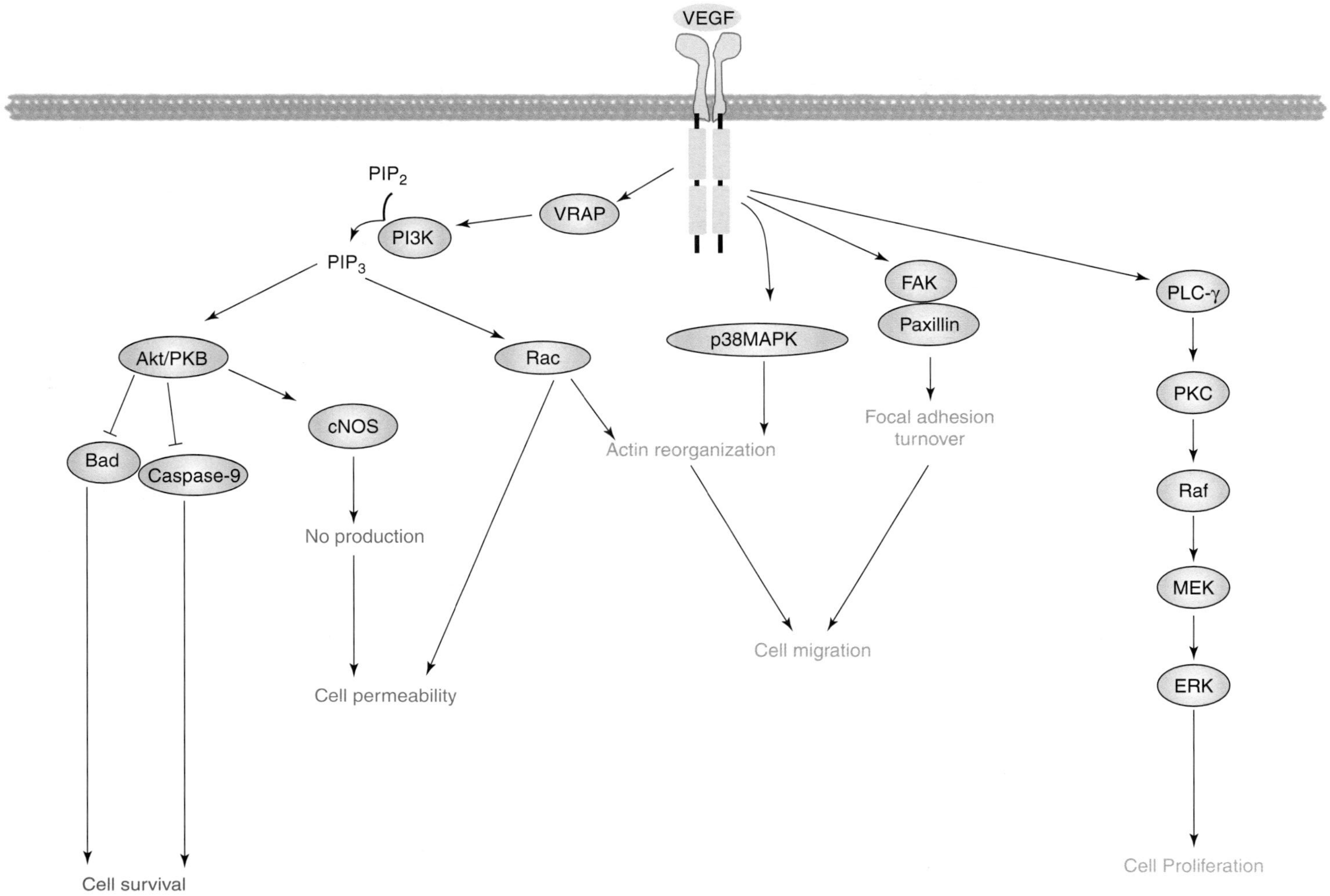

Figure 7 VEGFR-2 intracellular signaling. VEGF binding to the extracellular domain of VEGFR-2 induces the dimerization and autophosphorylation of specific tyrosine residues present in the catalytic domain of the receptor. Several proteins associate by their SH2 domain to these tyrosine residues and are thus activated. Downstream signal transduction molecules then lead to several endothelial cell functions such as vasculopermeability, proliferation, migration, and survival. Arrows indicate positive stimulation and bold lines show inhibition.

development and an increased mortality. Disorders of the vascular system are known to be linked to the overexpression but also to the downregulation of VEGF (**Figure 4**). We will thereafter give details of the implications of VEGF in some diseases: however, the following list is not exhaustive.

3.10.5.5.1 Vascular endothelial growth factor in inflammatory disorders

Psoriasis is a chronic inflammatory skin disorder characterized by dermal angiogenesis and overexpression of VEGF and VEGFR,[117,118] supposedly stimulated by TGF-α and EGF, since receptors of these factors are overexpressed in psoriatic skin.[119] In the case of severe skin lesions, neovastat (AE-941, Aeterna) has shown a promising therapeutic outcome for the treatment of psoriasis.[120]

3.10.5.5.2 Pathology of the female reproductive tract

High levels of VEGF are also involved in pathologic angiogenesis and are responsible for several gynecological disorders, such as endometriosis, dysfunctional uterine bleeding, endometrial hyperplasia, and polycystic syndrome, which is known to be a leading cause of infertility.[121–123] Antiangiogenic compounds are under investigation in order to provide novel therapeutic approaches for such diseases.[124,125]

3.10.5.5.3 Intraocular neovascular syndromes

Retinal hypoxia and excessive secretion of VEGF can lead to an inappropriate retinal vascularization[126] and hemorrhages contributing to visual loss observed in retinopathy of prematurity, diabetic retinopathy, as well as age-related macular degeneration.[127] Antiangiogenic therapies are now under investigation in order to prevent such retinal neovascularization. In the case of wet age-related macular degeneration, such treatment consists of lucentis, an antibody fragment designed to bind all forms of VEGF (ranibizumab, rhuFabV2, Genentech) and in the angiostatic steroid, Retaane (anecortave acetate, Alcon Laboratories),[128] which are both in phase III of clinical trials,[129] whereas Macugen (pegaptanib sodium injection, Eyetech Pharmaceuticals), an aptamer specific for the $VEGF_{165}$ isoform, has recently received US FDA approval.[130] Various others drugs for the treatment of age-related macular degeneration, such as AdPEDF (Genvec),[131] squalamine (Genaera),[132] and combrestatin A4 prodrug (Oxigene),[133] are in early stages of investigation.

3.10.5.5.4 Vascular endothelial growth factor and cancer

Angiogenesis is essential for the growth of most primary tumors and their subsequent metastatic spread. In fact, tumors can absorb sufficient nutriments and oxygen by simple diffusion up to a size of 1–2 mm, at which point their further growth requires connections with the existing blood vessels in order to permit vascular supply. Consequently, VEGF expression is increased and takes part in the angiogenesis process,[112] leading to cell survival by preventing endothelial cell apoptosis.[134] This action is mediated by the PI3K/Akt pathway,[96,98,135] but also by the induction of the expression of antiapoptotic proteins such as Bcl-2[98] and Bcl-A1, by an increased phosphorylation of FAK, and by the stimulation of prostaglandin I_2 and nitric oxide production by endothelial cells.

Moreover, consistent with its role in tumor angiogenesis, the expression of VEGF is upregulated by the common genetic events leading to malignant transformation. These include the activation of oncogenes such as *ras*, *V-src*, *fos*, and *HER2*,[92] but also the loss of tumor suppressor genes, such as *p53*, known to cause cell cycle arrest, degradation of HIF-1α and to inhibit the production of VEGF.[59,136,137]

VEGF expression is increased in the majority of cancers, including hematological malignancies,[138] colon and rectal cancers,[139] lung,[140] breast,[141] kidney and bladder cancers,[142] ovary and uterine cervix carcinomas,[143] intracranial tumors,[144] and others.

As tumors require new vasculature to grow and survive, the inhibition of angiogenesis[145] could be an effective strategy to eradicate cancer. First, monoclonal antibodies can be used to neutralize the ligands VEGF or receptors VEGFR-2 that are key molecular targets. These neutralizing antibodies block the binding of VEGF to its receptor and cause apoptosis of the endothelial cells, taking part of newly formed immature vessels that are dependent on VEGF to maintain cell adhesion. Bevacizumab (Avastin, rhu-Mab-VEGF, Genentech), a specific VEGF monoclonal antibody, is the first antiangiogenic agent to be approved by the FDA for the treatment of colorectal cancer in conjunction with chemotherapy and is still under trial for other cancer types.[113,146] In addition, another VEGF antibody strategy using VEGF trap (Regeneron) has shown promising results in the case of solid tumors in advanced stage, as this compound exhibits a higher affinity for VEGF than other monoclonal antibodies tested.[147,148]

Indirect strategies, such as the use of IMC-1C11[149] and neovastat (AE941, Aeterna) are also tested to target VEGF and its receptors. Neovastat is in phase III clinical trials for the treatment of renal cell carcinoma and non-small-cell lung cancer. It prevents the binding of VEGF to its receptors but also inhibits metalloproteinases and induces endothelial cell apoptosis.[150]

Small molecules that block or prevent the activation of VEGFR-2 TK are also used to inhibit VEGF function and the subsequent signaling pathways.[151,152] The 'target' of these TK inhibitors consists of the ATP-binding site located within the kinase domain of these receptors. Such inhibitor compounds include PTK787 (Vatalanib, Novartis),[153] ZD6474 (AstraZeneca),[154,155] SU11248 (sunitinib malate, Pfizer),[156] SU5416 (semaxinib, Sugen),[157] VGA1102,[158] SU6668 (Sugen),[159] AZD2171(AstraZeneca),[154] and many others.

Finally, VEGF and VEGFR can also be targeted by antisense therapy. This strategy consists of short sequences of oligonucleotides that are designed to be complementary to a region of mRNA that encodes VEGF[160] or VEGFR,[161] as well as blocking the subsequent translation of mRNA into protein and thus the proliferation of endothelial cells in tumor angiogenesis. It can also consist of catalytic antisense RNAs, called ribozymes such as RPI4610 (Ribozyme Pharmaceuticals) that cleave RNA substrates in a sequence-specific manner.[162,163]

The number and variety of novel targeted agents as well as the ongoing clinical evaluation offer a realistic hope for significant advances in cancer treatment. To date, the most successful results of antiangiogenic therapy have been obtained when they are used in combination with certain conventional chemotherapies.[164] Moreover, the combined inhibition of VEGF/EGFR has shown encouraging antitumoral activity.[140]

3.10.5.6 Vascular Endothelial Growth Factor as a Therapeutic Tool

In the previous cited diseases, VEGF was responsible for vascular system disorders leading to acute angiogenesis. But in some others, VEGF may be useful for attempts to increase the collateral vessel formation in order to overcome inadequate tissue perfusion or ischemia, observed in cardiovascular diseases but also in neonatal respiratory distress syndrome[165] and in amyotrophic lateral syndrome.[166,167] Several trials, such as the administration of VEGF recombinant protein[168] or gene transfer using nonviral delivery vector,[139,169] are under investigation and aim to avoid the potential risk of subsequent pathological angiogenesis.

VEGF and its receptors play an important role in the development and regulation of angiogenesis by initiating several signal transduction cascades. In this way, they are highly implicated in several pathological disorders, and various therapeutic approaches aiming to inhibit the function of VEGF/VEGFR are currently under investigation. These antiangiogenic strategies consist of neutralizing antibodies, aptamer and antisense therapy directed against VEGF and VEGFR, and also small-molecular-weight compounds preventing VEGFR TK activity. Currently, clinical trials have a promising outcome and generating hope for the treatment of cancer and ocular disease.

3.10.6 Signal Transducer and Activator of Transcription (STAT)

STAT proteins form a family of transcription factors that transduce signals from the extracellular milieu of cells to the nucleus. These proteins were first described in 1993[170–172] and are involved in a large number of diverse biologic processes, such as fetal development,[173,174] cell growth,[175] transformation,[176] differentiation,[176–178] immune response,[179] inflammation,[179,180] and apoptosis.[181,182] The STAT transcription factor family is composed of seven different members: STAT1, STAT2, STAT3, STAT4, STAT5a, STAT5b, and STAT6.[183]

3.10.6.1 Chromosomal Localization and Structure of the Signal Transducers and Activators of Transcription

The chromosomal localizations of the human STATs were identified on three different chromosomal clusters, as STAT1 and STAT4 are situated on chromosome 2, STAT2 and STAT6 on chromosome 12 and STAT3, STAT5a and STAT5b on chromosome 17 (**Table 1**). Independently of the varying chromosomal localizations, the seven STAT proteins share the same overall structure that is organized into distinct functional domains (**Figure 8**). The highly conserved N-terminal domain, or oligomerization domain, is important for protein–protein as well as dimer–dimer interactions in order to form tetramer and oligomer STAT molecules.[184,185] The DNA-binding domain determines the DNA-binding specificity of the different STAT proteins,[186] while the SH2 domain is essential for the recruitment of the STAT proteins to the phosphorylated receptors. Furthermore, an interaction between the SH2 domain of one STAT monomer and the phosphorylated tyrosine of a second monomer is required for the formation of the STAT dimers.[187] The position of the essential tyrosine residue for the STAT activation and dimerization is located in the transactivation domain and is specific for each family member (**Table 1**). Finally, the C-terminal transcriptional activation domain (also called transactivation domain) carries a conserved serine residue, except for STAT2 and STAT6. This phophorylation site has been described as regulating the transcriptional activity of the concerned STATs.[188,189]

Table 1 Signal transducers and activators of transcription

	Chromosomal localization	*Molecular weight (kDa)*	*Amino acids*	*Phosphorylation sites*	*Dimerization partners*
STAT1	2q32.2	91	750	Y701 S727	1, 2, 3
STAT2	12q13.3	113	851	Y690 /	1
STAT3	17q21.2	89	770	Y705 S727	1, 3
STAT4	2q32.2	89	748	Y693 S721	4
STAT5a	17q21.2	96	794	Y694 S726	5a, 5b
STAT5b	17q21.2	94	787	Y699 S731	5a, 5b
STAT6	12q13.3	94	847	Y641 /	6

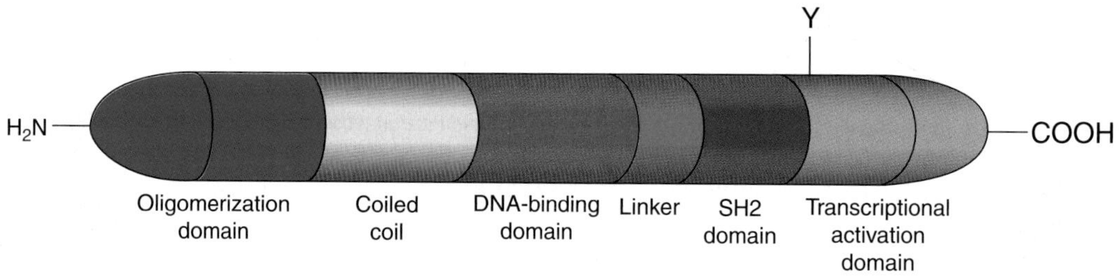

Figure 8 Structure and functional domains of STAT proteins.

Following activation, all STAT proteins can associate as homodimers except for STAT2, which can only form a STAT1-2 heterodimer. Furthermore, STAT1-3 and STAT5a-5b heterodimers may be observed after STAT activation. The central role for this activation and dimerization is played by the tyrosine phosphorylation, as replacement of this residue results in an inactive STAT incapable of nuclear translocation or transactivation.

In addition to these full-length STAT isoforms, shortened STAT proteins lacking regions of the C-terminal transcriptional activation domain were identified in the cases of STAT1, STAT3 and STAT5. These truncated isoforms, termed STATβ, present different transcriptional activities compared to the full-length STATα isoforms, and can act as dominant negative regulators.[190,191] Two different mechanisms were described for the production of C-terminally truncated STAT isoforms. Alternative mRNA splicing could be observed in the case of STAT1,[192] STAT3,[193] and STAT5,[191] while proteolytic cleavage was demonstrated for STAT3[194] and STAT5.[195]

3.10.6.2 Signal Transducer and Activator of Transcription Proteins in Signal Transduction/Activation Mechanisms of Signal Transducer and Activator of Transcription Proteins

The activation of STAT proteins can be induced via the binding of cytokines, growth factors, or hormones to their cell surface receptors (**Figure 9**). The connection of these specific extracellular signaling proteins induces a receptor dimerization, which activates the Janus family of TKs (Jaks) by autophosphorylation (the Jaks family is composed of four members: Jak1, Jak2, Jak3, and Tyk2). In turn, the activated Jaks phosphorylate a tyrosine residue on the cytoplasmic part of the receptor, tyrosine residue, which then becomes the docking site for the STAT proteins.[196] Tyrosine phosphorylation, by the Jaks, in the carboxy-terminal region of the STATs, induces the formation of STAT homo- or heterodimers via the interaction between the phosphotyrosine residue of one STAT and the SH2 domain of the other. Subsequently, the so-formed dimers translocate to the nucleus of the cell, where they bind to specific response elements, induce and regulate target gene transcription.

Alternatively, growth factor receptors possessing intrinsic TK activity, as EGF receptor or platelet-derived growth factor (PDGF) receptor, can also autophosphorylate their receptor cytoplasmic tail,[197,198] leading to STAT activation without the implication of the Jaks. Besides, STAT proteins can be activated by nonreceptor TKs of the Src family.[199]

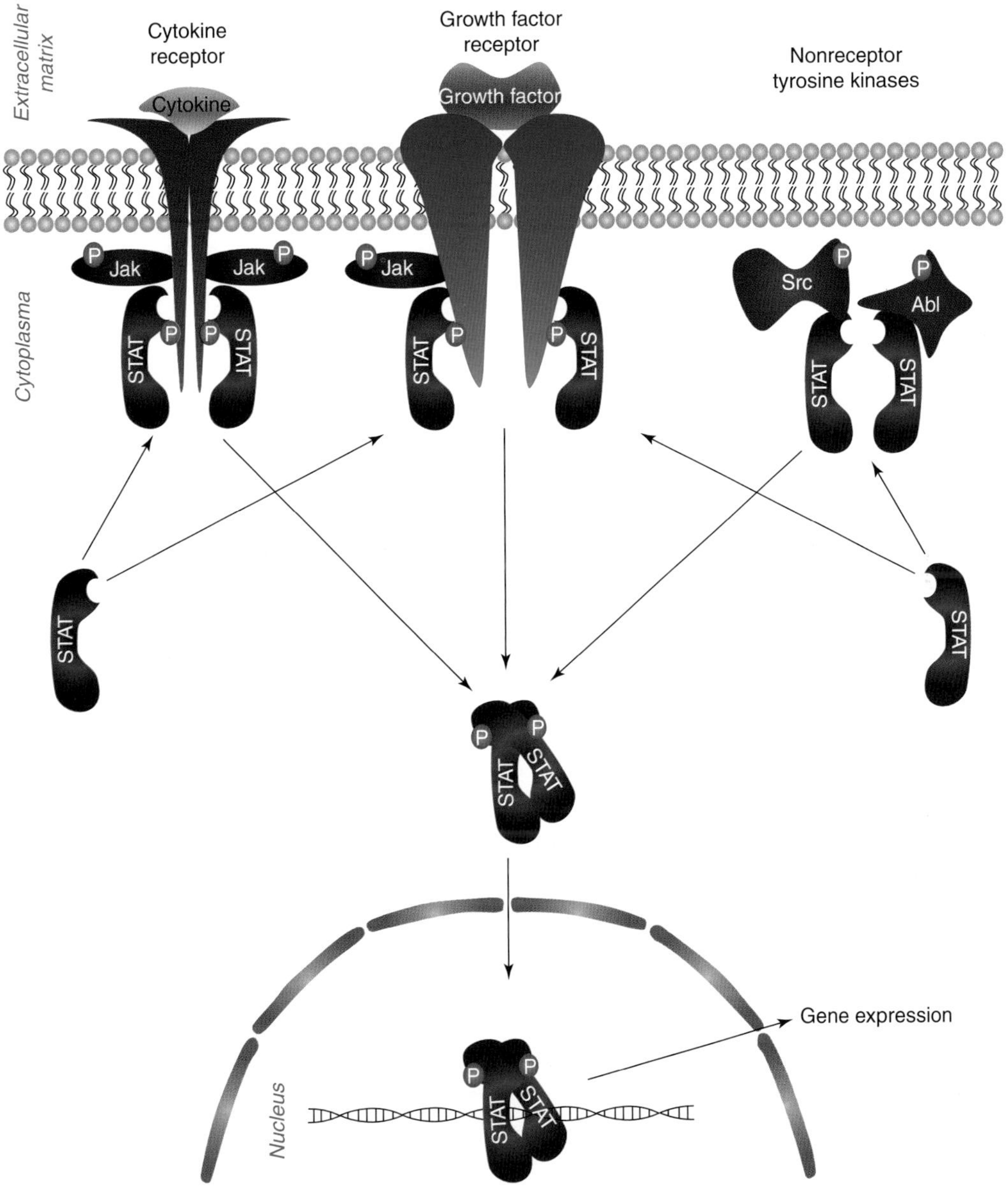

Figure 9 The STAT activation pathway. The binding of a cytokine or growth factor leads to intrinsic receptor tyrosine kinase activity or to the activation of the Janus family kinases associated with the receptor. Subsequent phosphorylation of the cytoplasmic tail of the receptor leads to the formation of docking sites for two STAT monomers. Tyrosine phosphorylation on the STAT proteins induces the dimerization of the latter, which induces the nuclear translocation of the activated STAT dimer. In addition, STAT proteins can also be activated by nonreceptor tyrosine kinases as Bcr or Abl. Once in the nucleus, the STAT dimers can bind to specific DNA response elements and activate transcription of target genes.

These different activation mechanisms control the activity of the STAT proteins and subsequently affect the numerous biologic processes influenced by these pathways. The numerous cytokines activating the Jak/STAT pathways highlight the central role of these proteins (**Table 2**).

3.10.6.2.1 STAT1 signaling

STAT1 is not only implicated in interferon (IFN-α, IFN-β, and IFN-γ) signaling, but it is also activated by IL6, IL10, IL11, IL21, or EGF (**Table 2**). STAT1 knockout mice proved to be viable and displayed no developmental defects.

Table 2 STAT-activating cytokines[247–249] and STAT knockout mice

STAT	*Activating cytokines*	*Phenotype of knockout mice*
STAT1	IFN-α, IFN-β, IFN-γ, IL6, IL10, IL11, IL21, EGF	Defective IFN-dependent immune responses
		High sensitivity to viral and bacterial infections
STAT2	IFN-α, IFN-β	Defective type I-dependent immune responses
STAT3	IL2, IL6, IL7, IL9, IL10, IL11, IL12, IL15, IL21, EGF, CSF1, G-CSF, PDGF, GH, OSM	Embryonic lethality
STAT4	IL12	Impaired Th1-cell development
STAT5a and/or	IL2, IL3, IL4, IL5, IL7, IL9, IL15, IL21, EPO	Deficient mammary gland development and lactogenesis
STAT5b	GH, PRL, EGF, PDGF, GM-CSF	Loss of sexually dimorphic growth and liver gene expression in males
		Female infertility
		Slowed cell-growth after GM-CSF stimulation
		Impaired IL2-induced splenocyte proliferation
		Reduced number of NK cells and impaired IL2-induced T-cell proliferation
		Fetal anemia
		Loss of tolerance and autoimmunity affecting multiple organs
STAT6	IL4, IL13	Impaired Th2 differentiation and defective IgE class switch

However, these mice presented damaged responses to IFNs and their physiological functions related to the IFNs were absent.[200,201] These defective IFN-dependent immune responses led to a high sensitivity toward viral infections and microbial pathogens. Interestingly, STAT1-deficient mice showed normal responses to EGF and other cytokines such as IL10, showing that even if STAT1 is activated by these cytokines, it is only essential for IFN-mediated signaling. More recently, STAT1-deficient mice were proven to be highly susceptible to pulmonary mycobacterial infection.[202]

3.10.6.2.2 STAT2 signaling

Until now, STAT2 activation has almost exclusively been described with type I IFNs (IFN-α and IFN-β). As observed with STAT1, STAT2 knockout mice are viable and develop normally. However, these mice present no response to type I IFNs and are thus defective in type I IFN-dependent immune responses, leading to a high susceptibility to viral infections.[203,204]

3.10.6.2.3 STAT3 signaling

Targeted disruption of STAT3 exposed that this protein is crucial for early embryonic development, as STAT3 knockout mice revealed to be embryonic lethal.[205] Therefore, STAT3 knockout mice cannot assess the functions of STAT3 in adult tissues. In order to gain information about the different roles of STAT3, this protein was knocked out in a tissue- or cell-specific manner. STAT3-deficient T cells showed a reduced response to IL2 and IL6 treatment.[206] Studies in STAT3-deficient macrophages and neutrophils indicated that IL10-induced STAT3 activation is important for antiinflammatory responses in these cells.[207] Mammary glands lacking STAT3 expression presented a delay in involution, which was associated with decreased epithelial apoptosis,[208] while keratinocyte-specific ablation of STAT3 led to compromised hair cycle and wound-healing processes, without affecting skin morphogenesis.[209] Studies on mice with a cardiomyocyte-restricted knockout of STAT3 showed that this protein has a critical role in the protection of inflammation-induced heart damage. Furthermore, a dramatic increase in cardiac fibrosis could be observed in aged mice, although no signs of heart failure could be detected in young STAT3-deficient mice.[210] STAT3 gene disruption in insulin-producing pancreatic beta cells in mice led to glucose intolerance and later to obesity.[211] All these different studies tend to show that STAT3 deletion leads to very diverse effects, which seem to be strongly cell lineage-specific.

Furthermore, STAT3 was also shown to be strongly implicated in interleukin signaling, as an activation by IL2, IL6, IL7, IL9, IL10, IL11, IL12, IL15, and IL21 could be observed. Finally, epidermal growth factor (EGF), colony stimulating factor-1 (CSF-1), granulocyte-CSF (G-CSF), platelet derived growth factor (PDGF), growth hormone (GH), and oncostatin M (OSM) were described as STAT3 activators.

3.10.6.2.4 STAT4 signaling

STAT4 is primarily activated in response to IL12. The phenotype of STAT4 knockout mice is similar to that of IL12-deficient mice and the disruption of STAT4 overcomes the IL12 controlled T-helper cell differentiation along the Th1 pathway.[177,212] The importance of STAT4 in the differentiation of CD4+ T cells into Th1 effector cells was recently confirmed in STAT4-deficient CD4+ T cells.[213] Furthermore, experiments on STAT4-deficient mice revealed that the disruption of STAT4 prevents the development of spontaneous diabetes in nonobese diabetic mice, and reveals the important role of STAT4 in autoimmune diabetes pathogenesis.[214]

3.10.6.2.5 STAT5 signaling

Both STAT5a and STAT5b are known to become activated by a large number of cytokines. Among these, numerous interleukins (IL2, IL3, IL4, IL5, IL7, IL9, IL15, and IL21) implicate STAT5 signaling. Additionally, erythropoietin, GH, prolactin, and growth factor signaling (EGF, PDGF, granulocyte–macrophage colony-stimulating factor (GM-CSF)) display an action on STAT5 activation. First studies with STAT5 knockout mice showed that the deletion of STAT5a leads to deficient prolactin-dependent mammary gland development and lactogenesis in female mice,[215] while STAT5b-deficient males lose the sexual dimorphism of body growth and liver gene expression as a result of impaired GH response.[216] Furthermore, deletion of both STAT5a and STAT5b in female mice resulted in infertility.[217]

Experiments on STAT5a null mice also showed the implication of this protein in the signaling due to GM-CSF, as these mice displayed slowed cell growth after GM-CSF stimulation.[218] Knockout mice models were also used to demonstrate the immunological role of STAT5. During these studies, an impaired IL2-induced splenocyte proliferation could be observed in STAT5a,[219] STAT5b,[220] and STAT5a/5b knockout mice.[221] Furthermore, STAT5b null mice present a defective natural killer (NK) cell development,[220] and thus a reduced number of NK cells, as well as an impaired IL2-induced T-cell proliferation.[221] The loss of tolerance and autoimmunity affecting multiple organs in STAT5a/5b-deficient mice was also related to the signaling through IL2R.[222] In embryos, STAT5a/5b deficiency was shown to lead to fetal anemia and apoptosis of red cell progenitors.[223]

3.10.6.2.6 STAT6 signaling

As STAT6 is essentially activated by IL4 and the related cytokine IL13,[224,225] STAT6 knockout mice lack the physiological functions associated with these interleukins.[224,226,227] Deficient IL4 signaling leads to impaired Th2 differentiation[228] and defective IgE class switching.[226]

These results highlight the importance of STATs in the cellular signaling pathways and show the central role of these proteins in the normal functioning of the cells. However, STAT deregulation as well as aberrant STAT activation can have dramatic consequences, as increasing evidence is underlining the important role of STATs in oncogenesis,[229] tumorigenesis,[230] allergic inflammation, and autoimmune diseases.[180]

3.10.6.3 Pathological Variations of Signal Transducer and Activator of Transcription Expression

3.10.6.3.1 Signal transducers and activators of transcription in tumorigenesis

One decade ago, first studies indicated that STAT signaling is often activated by oncogenes and in tumor cells. First, STAT3 was identified to be activated by the oncogenic Src TK in rodent fibroblast cell lines.[101] Further studies showed that a great number of oncogenes, including *v-Abl, BCR-Abl, v-Eyk*, and *v-Fps* can activate STAT molecules.[231] It is also notable that constitutive activation of essentially STAT1, STAT3, and STAT5 has been demonstrated to be associated with malignant transformation induced by oncoproteins.[232] Constitutive activation of one or more of these STATs was observed in many tumor cell lines and in primary tumors leading to the idea that the STATs play an important role in the malignant progression of human tumors.[197,230]

Different mechanisms were described to assign the transforming activity of constitutively activated STAT proteins. This activity could stem from the activation of antiapoptotic pathways and subsequent upregulation of apoptosis inhibitors such as Bcl-2, Bcl-xL, and Mcl-1, as described for example in breast cancer,[233] multiple myeloma,[197,234] head and neck cancer cell lines,[235] as well as non-Hodgkin's lymphoma.[236] STAT-induced upregulation of genes encoding for cell cycle regulators, such as cyclin D1,[237] cyclin D2,[238] or c-myc,[232] was also described as playing a role in cell transformation. Furthermore, it was shown that activated STATs can increase the transcription of the VEGF gene, an inducer of angiogenesis.[101,239] Thus it seems that activated STAT proteins can contribute to oncogenesis via its control of cell cycle progression and/or apoptosis by inducing genes coding for inducers of angiogenesis, cell cycle regulators, and apoptosis inhibitors. Far more detailed information about the role of STAT proteins in tumorigenesis can be found in several excellent reviews.[197,229,230,233]

3.10.6.3.2 Signal transducers and activators of transcription in allergic inflammation and autoimmune diseases

Asthma is a chronic allergic inflammation whose inflammatory process stems from an unsuitable immune response coordinated by Th2 cells. With regard to this mechanism, it is not surprising that STAT6 signaling, which plays an important role in Th2 differentiation, was investigated in asthma pathogenesis. The implication of STAT6 was shown by the deletion of the genes coding for its expression, as this left the mice resistant to the induction of experimental allergic asthma.[240,241] Increased STAT6 expression was also monitored in subjects with allergic rhinitis after allergen challenge.[242] Furthermore, STAT1, implicated in immune response, was related to asthma, as epithelial STAT1 was observed to be activated in asthmatic compared to normal control subjects.[243]

Supporting evidence for a STAT3 causal role in rheumatoid arthritis was affirmed, as activated STAT3 overexpression was reported in synovial tissues from patients with rheumatoid arthritis.[244] Confirming information came from the observation that a mutation causing hyperactivation of STAT3 led to spontaneous development of autoimmune arthritis.[245] In contrast, the function of STAT1 in rheumatoid arthritis is still controversial as inflammatory and protective roles of this protein have been described in the literature.[246] It may thus be that the STAT1 function depends on the cell type and/or the stage of the inflammatory disease.

3.10.6.4 Future Directions

The various studies on STATs have demonstrated that these proteins are implicated in a large and very diverse number of cellular processes, which also means that a STAT deregulation can potentially affect or corrupt numerous cellular functions. A clear understanding of the molecular mechanisms implicated in the STAT-signaling pathway is thus of capital importance. The harmful effects of STAT deregulation can, for example, be observed in many tumors, where the STATs contribute to cellular dysfunctions in cell cycle regulation, apoptosis inhibitors, and angiogenesis, leading to uncontrolled cell cycle progression and apoptosis. The fact that STAT proteins present a point of convergence for TK signaling positions them as promising targets for future clinical treatment. Indeed, they represent a very limited number of targets, as compared to the multitude of kinases, and are additionally implicated in such diverse functions as cell growth, apoptosis, inflammation, and many others.

3.10.7 Conclusion

Aberrant cell-signaling cascades generally lead to the generation of dominant expression of single regulatory protein, increasing the proliferation of the cell as well as the potential of the capacity for tissue invasion. The equilibrium between apoptosis and survival leads to resistance to classical chemotherapy as well as radiotherapy. A better knowledge of cell-signaling pathways leading to highly specific compounds that can be used with traditional therapies is therefore of utmost importance.

References

1. Gerards, W. L.; de Jong, W. W.; Boelens, W.; Bloemendal, H. *Cell Mol. Life Sci.* **1998**, *54*, 253–262.
2. Roos-Mattjus, P.; Sistonen, L. *Ann. Med.* **2004**, *36*, 285–295.
3. Goldberg, A. L.; Stein, R.; Adams, J. *Chem. Biol.* **1995**, *2*, 503–508.
4. Spataro, V.; Norbury, C.; Harris, A. L. *Br. J. Cancer* **1998**, *77*, 448–455.
5. Naujokat, C.; Hoffmann, S. *Lab. Invest.* **2002**, *82*, 965–980.
6. Schwartz, A. L.; Ciechanover, A. *Annu. Rev. Med.* **1999**, *50*, 57–74.
7. Ciechanover, A. *Cell* **1994**, *79*, 13–21.
8. Hendil, K. B.; Hartmann-Petersen, R. *Curr. Protein Pept. Sci.* **2004**, *5*, 135–151.
9. Hershko, A.; Heller, H.; Elias, S.; Ciechanover, A. *J. Biol. Chem.* **1983**, *258*, 8206–8214.
10. Joazeiro, C. A.; Weissman, A. M. *Cell* **2000**, *102*, 549–552.
11. Weissman, A. M. *Immunol. Today* **1997**, *18*, 189–198.
12. Braun, B. C.; Glickman, M.; Kraft, R.; Dahlmann, B.; Kloetzel, P. M.; Finley, D.; Schmidt, M. *Nat. Cell Biol.* **1999**, *1*, 221–226.

13. Xie, Y.; Varshavsky, A. *Proc. Natl. Acad. Sci. USA* **2000**, *97*, 2497–2502.
14. Navon, A.; Goldberg, A. L. *Mol. Cell* **2001**, *8*, 1339–1349.
15. DeMartino, G. N.; Slaughter, C. A. *J. Biol. Chem.* **1999**, *274*, 22123–22126.
16. Coux, O.; Tanaka, K.; Goldberg, A. L. *Annu. Rev. Biochem.* **1996**, *65*, 801–847.
17. Orlowski, M.; Wilk, S. *Arch. Biochem. Biophys.* **2000**, *383*, 1–16.
18. Isaksson, A.; Musti, A. M.; Bohmann, D. *Biochem. Biophys. Acta* **1996**, *1288*, F21–F29.
19. Hochstrasser, M. *Curr. Opin. Cell Biol.* **1995**, 7, 215–223.
20. Fuchs, S. Y.; Adler, V.; Buschmann, T.; Yin, Z.; Wu, X.; Jones, S. N.; Ronai, Z. *Genes Dev.* **1998**, *12*, 2658–2663.
21. Chen, Z. J.; Parent, L.; Maniatis, T. *Cell* **1996**, *84*, 853–862.
22. Brown, K.; Gerstberger, S.; Carlson, L.; Franzoso, G.; Siebenlist, U. *Science* **1995**, *267*, 1485–1488.
23. Yaron, A.; Gonen, H.; Alkalay, I.; Hatzubai, A.; Jung, S.; Beyth, S.; Mercurio, F.; Manning, A. M.; Ciechanover, A.; Ben-Neriah, Y. *EMBO J.* **1997**, *16*, 6486–6494.
24. Ramakrishnan, P.; Wang, W.; Wallach, D. *Immunity* **2004**, *21*, 477–489.
25. Adler, V.; Franklin, C. C.; Kraft, A. S. *Proc. Natl. Acad. Sci. USA* **1992**, *89*, 5341–5345.
26. Treier, M.; Staszewski, L. M.; Bohmann, D. *Cell* **1994**, *78*, 787–798.
27. Fuchs, S. Y.; Dolan, L.; Davis, R. J.; Ronai, Z. *Oncogene* **1996**, *13*, 1531–1535.
28. Musti, A. M.; Treier, M.; Bohmann, D. *Science* **1997**, *275*, 400–402.
29. Almond, J. B.; Cohen, G. M. *Leukemia* **2002**, *16*, 433–443.
30. Hideshima, T.; Chauhan, D.; Richardson, P.; Mitsiades, C.; Mitsiades, N.; Hayashi, T.; Munshi, N.; Dang, L.; Castro, A.; Palombella, V. et al. *J. Biol. Chem.* **2002**, *277*, 16639–16647.
31. Mitsiades, N.; Mitsiades, C. S.; Poulaki, V.; Chauhan, D.; Fanourakis, G.; Gu, X.; Bailey, C.; Joseph, M.; Libermann, T. A.; Treon, S. P. et al. *Proc. Natl. Acad. Sci. USA* **2002**, *99*, 14374–14379.
32. Rajkumar, S. V.; Richardson, P. G.; Hideshima, T.; Anderson, K. C. *J. Clin. Oncol.* **2005**, *23*, 630–639.
33. Barnes, P. J.; Karin, M. *N. Engl. J. Med.* **1997**, *336*, 1066–1071.
34. DiDonato, J. A.; Hayakawa, M.; Rothwarf, D. M.; Zandi, E.; Karin, M. *Nature* **1997**, *388*, 548–554.
35. Mercurio, F.; Zhu, H.; Murray, B. W.; Shevchenko, A.; Bennett, B. L.; Li, J.; Young, D. B.; Barbosa, M.; Mann, M.; Manning, A. et al. *Science* **1997**, *278*, 860–866.
36. Zandi, E.; Rothwarf, D. M.; Delhase, M.; Hayakawa, M.; Karin, M. *Cell* **1997**, *91*, 243–252.
37. Woronicz, J. D.; Gao, X.; Cao, Z.; Rothe, M.; Goeddel, D. V. *Science* **1997**, *278*, 866–869.
38. Regnier, C. H.; Song, H. Y.; Gao, X.; Goeddel, D. V.; Cao, Z.; Rothe, M. *Cell* **1997**, *90*, 373–383.
39. Rothwarf, D. M.; Zandi, E.; Natoli, G.; Karin, M. *Nature* **1998**, *395*, 297–300.
40. Makris, C.; Godfrey, V. L.; Krahn-Senftleben, G.; Takahashi, T.; Roberts, J. L.; Schwarz, T.; Feng, L.; Johnson, R. S.; Karin, M. *Mol. Cell* **2000**, *5*, 969–979.
41. Dejardin, E.; Droin, N. M.; Delhase, M.; Haas, E.; Cao, Y.; Makris, C.; Li, Z. W.; Karin, M.; Ware, C. F.; Green, D. R. *Immunity* **2002**, *17*, 525–535.
42. Senftleben, U.; Cao, Y.; Xiao, G.; Greten, F. R.; Krahn, G.; Bonizzi, G.; Chen, Y.; Hu, Y.; Fong, A.; Sun, S. C.; Karin, M. *Science* **2001**, *293*, 1495–1499.
43. Ghosh, S.; Karin, M. *Cell* **2002**, *109*, S81–S96.
44. Earnshaw, W. C.; Martins, L. M.; Kaufmann, S. H. *Annu. Rev. Biochem.* **1999**, *68*, 383–424.
45. Strasser, A.; O'Connor, L.; Dixit, V. M. *Annu. Rev. Biochem.* **2000**, *69*, 217–245.
46. Gross, A.; McDonnell, J. M.; Korsmeyer, S. J. *Genes Dev.* **1999**, *13*, 1899–1911.
47. Barkett, M.; Gilmore, T. D. *Oncogene* **1999**, *18*, 6910–6924.
48. Beg, A. A.; Sha, W. C.; Bronson, R. T.; Ghosh, S.; Baltimore, D. *Nature* **1995**, *376*, 167–170.
49. Beg, A. A.; Baltimore, D. *Science* **1996**, *274*, 782–784.
50. Deveraux, Q. L.; Reed, J. C. *Genes Dev.* **1999**, *13*, 239–252.
51. Wang, C. Y.; Mayo, M. W.; Korneluk, R. G.; Goeddel, D. V.; Baldwin, A. S., Jr. *Science* **1998**, *281*, 1680–1683.
52. Irmler, M.; Thome, M.; Hahne, M.; Schneider, P.; Hofmann, K.; Steiner, V.; Bodmer, J. L.; Schroter, M.; Burns, K.; Mattmann, C.; Rimoldi, D.; French, L. E.; Tschopp, J. *Nature* **1997**, *388*, 190–195.
53. Goltsev, Y. V.; Kovalenko, A. V.; Arnold, E.; Varfolomeev, E. E.; Brodianskii, V. M.; Wallach, D. *J. Biol. Chem.* **1997**, *272*, 19641–19644.
54. Tran, H.; Brunet, A.; Griffith, E. C.; Greenberg, M. E. *Sci. STKE* **2003**, *2003*, RE5.
55. Nozaki, S.; Sledge, G. W., Jr.; Nakshatri, H. *Oncogene* **2001**, *20*, 2178–2185.
56. Bentires-Alj, M.; Dejardin, E.; Viatour, P.; Van Lint, C.; Froesch, B.; Reed, J. C.; Merville, M. P.; Bours, V. *Oncogene* **2001**, *20*, 2805–2813.
57. Webster, G. A.; Perkins, N. D. *Mol. Cell Biol.* **1999**, *19*, 3485–3495.
58. Matsui, K.; Fine, A.; Zhu, B.; Marshak-Rothstein, A.; Ju, S. T. *J. Immunol.* **1998**, *161*, 3469–3473.
59. Ravi, R.; Bedi, G. C.; Engstrom, L. W.; Zeng, Q.; Mookerjee, B.; Gelinas, C.; Fuchs, E. J.; Bedi, A. *Nat. Cell Biol.* **2001**, *3*, 409–416.
60. Zheng, Y.; Ouaaz, F.; Bruzzo, P.; Singh, V.; Gerondakis, S.; Beg, A. A. *J. Immunol.* **2001**, *166*, 4949–4957.
61. Kasof, G. M.; Lu, J. J.; Liu, D.; Speer, B.; Mongan, K. N.; Gomes, B. C.; Lorenzi, M. V. *Oncogene* **2001**, *20*, 7965–7975.
62. Grimm, S.; Bauer, M. K.; Baeuerle, P. A.; Schulze-Osthoff, K. *J. Cell Biol.* **1996**, *134*, 13–23.
63. Pahl, H. L. *Oncogene* **1999**, *18*, 6853–6866.
64. El-Omar, E. M.; Carrington, M.; Chow, W. H.; McColl, K. E.; Bream, J. H.; Young, H. A.; Herrera, J.; Lissowska, J.; Yuan, C. C.; Rothman, N. et al. *Nature* **2000**, *404*, 398–402.
65. Peek, R. M., Jr.; Blaser, M. J. *Nat. Rev. Cancer* **2002**, *2*, 28–37.
66. Neurath, M. F.; Pettersson, S.; Meyer zum Buschenfelde, K. H.; Strober, W. *Nat. Med.* **1996**, *2*, 998–1004.
67. Rogler, G.; Brand, K.; Vogl, D.; Page, S.; Hofmeister, R.; Andus, T.; Knuechel, R.; Baeuerle, P. A.; Scholmerich, J.; Gross, V. *Gastroenterology* **1998**, *115*, 357–369.
68. Ekbom, A.; Helmick, C.; Zack, M.; Adami, H. O. *N. Engl. J. Med.* **1990**, *323*, 1228–1233.
69. Carmeliet, P. *Nat. Med.* **2003**, *9*, 653–660.
70. Liekens, S.; De Clercq, E.; Neyts, J. *Biochem. Pharmacol.* **2001**, *61*, 253–270.
71. Gospodarowicz, D.; Ferrara, N.; Schweigerer, L.; Neufeld, G. *Endocrinol. Rev.* **1987**, *8*, 95–114.
72. Olofsson, B.; Pajusola, K.; Kaipainen, A.; von Euler, G.; Joukov, V.; Saksela, O.; Orpana, A.; Pettersson, R. F.; Alitalo, K.; Eriksson, U. *Proc. Natl. Acad. Sci. USA* **1996**, *93*, 2576–2581.

73. Joukov, V.; Pajusola, K.; Kaipainen, A.; Chilov, D.; Lahtinen, I.; Kukk, E.; Saksela, O.; Kalkkinen, N.; Alitalo, K. *EMBO J.* **1996**, *15*, 1751.
74. Achen, M. G.; Jeltsch, M.; Kukk, E.; Makinen, T.; Vitali, A.; Wilks, A. F.; Alitalo, K.; Stacker, S. A. *Proc. Natl. Acad. Sci. USA* **1998**, *95*, 548–553.
75. Maglione, D.; Guerriero, V.; Viglietto, G.; Delli-Bovi, P.; Persico, M. G. *Proc. Natl. Acad. Sci. USA* **1991**, *88*, 9267–9271.
76. Ferrara, N.; Henzel, W. J. *Biochem. Biophys. Res. Commun.* **1989**, *161*, 851–858.
77. Muller, Y. A.; Christinger, H. W.; Keyt, B. A.; de Vos, A. M. *Structure* **1997**, *5*, 1325–1338.
78. Tischer, E.; Mitchell, R.; Hartman, T.; Silva, M.; Gospodarowicz, D.; Fiddes, J. C.; Abraham, J. A. *J. Biol. Chem.* **1991**, *266*, 11947–11954.
79. Tammela, T.; Enholm, B.; Alitalo, K.; Paavonen, K. *Cardiovasc. Res.* **2005**, *65*, 550–563.
80. Ortega, N.; Hutchings, H.; Plouet, J. *Front Biosci.* **1999**, *4*, D141–D152.
81. Robinson, C. J.; Stringer, S. E. *J. Cell Sci.* **2001**, *114*, 853–865.
82. Heldin, C. H. *Cell* **1995**, *80*, 213–223.
83. Barleon, B.; Totzke, F.; Herzog, C.; Blanke, S.; Kremmer, E.; Siemeister, G.; Marme, D.; Martiny-Baron, G. *J. Biol. Chem.* **1997**, *272*, 10382–10388.
84. Claesson-Welsh, L. *Biochem. Soc. Trans.* **2003**, *31*, 20–24.
85. Zachary, I.; Gliki, G. *Cardiovasc. Res.* **2001**, *49*, 568–581.
86. Wu, G. S.; Petersson, S.; Spiik, A. K.; Korsgren, O.; Tibell, A. *Transplant. Proc.* **2001**, *33*, 360.
87. Warner, A. J.; Lopez-Dee, J.; Knight, E. L.; Feramisco, J. R.; Prigent, S. A. *Biochem. J.* **2000**, *347*, 501–509.
88. Takahashi, T.; Yamaguchi, S.; Chida, K.; Shibuya, M. *EMBO J.* **2001**, *20*, 2768–2778.
89. Meadows, K. N.; Bryant, P.; Pumiglia, K. *J. Biol. Chem.* **2001**, *276*, 49289–49298.
90. Mazure, N. M.; Chen, E. Y.; Laderoute, K. R.; Giaccia, A. J. *Blood* **1997**, *90*, 3322–3331.
91. Arsham, A. M.; Plas, D. R.; Thompson, C. B.; Simon, M. C. *J. Biol. Chem.* **2002**, *277*, 15162–15170.
92. Rak, J.; Mitsuhashi, Y.; Sheehan, C.; Tamir, A.; Viloria-Petit, A.; Filmus, J.; Mansour, S. J.; Ahn, N. G.; Kerbel, R. S. *Cancer Res.* **2000**, *60*, 490–498.
93. Zeng, H.; Sanyal, S.; Mukhopadhyay, D. *J. Biol. Chem.* **2001**, *276*, 32714–32719.
94. Qi, J. H.; Claesson-Welsh, L. *Exp. Cell Res.* **2001**, *263*, 173–182.
95. Rousseau, S.; Houle, F.; Landry, J.; Huot, J. *Oncogene* **1997**, *15*, 2169–2177.
96. Fujio, Y.; Walsh, K. *J. Biol. Chem.* **1999**, *274*, 16349–16354.
97. Cross, M. J.; Dixelius, J.; Matsumoto, T.; Claesson-Welsh, L. *Trends Biochem. Sci.* **2003**, *28*, 488–494.
98. Gerber, H. P.; Dixit, V.; Ferrara, N. *J. Biol. Chem.* **1998**, *273*, 13313–13316.
99. Fulton, D.; Gratton, J. P.; McCabe, T. J.; Fontana, J.; Fujio, Y.; Walsh, K.; Franke, T. F.; Papapetropoulos, A.; Sessa, W. C. *Nature* **1999**, *399*, 597–601.
100. Eliceiri, B. P.; Paul, R.; Schwartzberg, P. L.; Hood, J. D.; Leng, J.; Cheresh, D. A. *Mol. Cell* **1999**, *4*, 915–924.
101. Yu, C. L.; Meyer, D. J.; Campbell, G. S.; Larner, A. C.; Carter-Su, C.; Schwartz, J.; Jove, R. *Science* **1995**, *269*, 81–83.
102. Berra, E.; Milanini, J.; Richard, D. E.; Le Gall, M.; Vinals, F.; Gothie, E.; Roux, D.; Pages, G.; Pouyssegur, J. *Biochem. Pharmacol.* **2000**, *60*, 1171–1178.
103. Berra, E.; Pages, G.; Pouyssegur, J. *Cancer Metastasis Rev.* **2000**, *19*, 139–145.
104. Pertovaara, L.; Kaipainen, A.; Mustonen, T.; Orpana, A.; Ferrara, N.; Saksela, O.; Alitalo, K. *J. Biol. Chem.* **1994**, *269*, 6271–6274.
105. Cohen, T.; Nahari, D.; Cerem, L. W.; Neufeld, G.; Levi, B. Z. *J. Biol. Chem.* **1996**, *271*, 736–741.
106. Forsythe, J. A.; Jiang, B. H.; Iyer, N. V.; Agani, F.; Leung, S. W.; Koos, R. D.; Semenza, G. L. *Mol. Cell Biol.* **1996**, *16*, 4604–4613.
107. Minchenko, A.; Bauer, T.; Salceda, S.; Caro, J. *Lab. Invest.* **1994**, *71*, 374–379.
108. Jones, A.; Fujiyama, C.; Blanche, C.; Moore, J. W.; Fuggle, S.; Cranston, D.; Bicknell, R.; Harris, A. L. *Clin. Cancer Res.* **2001**, *7*, 1263–1272.
109. Pugh, C. W.; Ratcliffe, P. J. *Nat. Med.* **2003**, *9*, 677–684.
110. Makino, Y.; Cao, R.; Svensson, K.; Bertilsson, G.; Asman, M.; Tanaka, H.; Cao, Y.; Berkenstam, A.; Poellinger, L. *Nature* **2001**, *414*, 550–554.
111. Gnarra, J. R.; Zhou, S.; Merrill, M. J.; Wagner, J. R.; Krumm, A.; Papavassiliou, E.; Oldfield, E. H.; Klausner, R. D.; Linehan, W. M. *Proc. Natl. Acad. Sci. USA* **1996**, *93*, 10589–10594.
112. Hoeben, A.; Landuyt, B.; Highley, M. S.; Wildiers, H.; Van Oosterom, A. T.; De Bruijn, E. A. *Pharmacol. Rev.* **2004**, *56*, 549–580.
113. Ferrara, N. *Endocrinol. Rev.* **2004**, *25*, 581–611.
114. Carmeliet, P.; Ferreira, V.; Breier, G.; Pollefeyt, S.; Kieckens, L.; Gertsenstein, M.; Fahrig, M.; Vandenhoeck, A.; Harpal, K.; Eberhardt, C. et al. *Nature* **1996**, *380*, 435–439.
115. Ferrara, N.; Carver-Moore, K.; Chen, H.; Dowd, M.; Lu, L.; O'Shea, K. S.; Powell-Braxton, L.; Hillan, K. J.; Moore, M. W. *Nature* **1996**, *380*, 439–442.
116. Ferrara, N.; Chen, H.; Davis-Smyth, T.; Gerber, H. P.; Nguyen, T. N.; Peers, D.; Chisholm, V.; Hillan, K. J.; Schwall, R. H. *Nat. Med.* **1998**, *4*, 336–340.
117. Detmar, M. *J. Invest. Dermatol.* **2004**, *122*, xiv–xv.
118. Bhushan, M.; McLaughlin, B.; Weiss, J. B.; Griffiths, C. E. *Br. J. Dermatol.* **1999**, *141*, 1054–1060.
119. Detmar, M.; Brown, L. F.; Claffey, K. P.; Yeo, K. T.; Kocher, O.; Jackman, R. W.; Berse, B.; Dvorak, H. F. *J. Exp. Med.* **1994**, *180*, 1141–1146.
120. Sauder, D. N.; Dekoven, J.; Champagne, P.; Croteau, D.; Dupont, E. *J. Am. Acad. Dermatol.* **2002**, *47*, 535–541.
121. Reynolds, L. P.; Grazul-Bilska, A. T.; Redmer, D. A. *Int. J. Exp. Pathol.* **2002**, *83*, 151–163.
122. Ferrara, N.; Gerber, H. P.; LeCouter, J. *Nat. Med.* **2003**, *9*, 669–676.
123. McLaren, J.; Prentice, A.; Charnock-Jones, D. S.; Millican, S. A.; Muller, K. H.; Sharkey, A. M.; Smith, S. K. *J. Clin. Invest.* **1996**, *98*, 482–489.
124. Hull, M. L.; Charnock-Jones, D. S.; Chan, C. L.; Bruner-Tran, K. L.; Osteen, K. G.; Tom, B. D.; Fan, T. P.; Smith, S. K. *J. Clin. Endocrinol. Metab.* **2003**, *88*, 2889–2899.
125. Wulff, C.; Wilson, H.; Rudge, J. S.; Wiegand, S. J.; Lunn, S. F.; Fraser, H. M. *J. Clin. Endocrinol. Metab.* **2001**, *86*, 3377–3386.
126. Aiello, L. P.; Avery, R. L.; Arrigg, P. G.; Keyt, B. A.; Jampel, H. D.; Shah, S. T.; Pasquale, L. R.; Thieme, H.; Iwamoto, M. A.; Park, J. E. et al. *N. Engl. J. Med.* **1994**, *331*, 1480–1487.
127. Witmer, A. N.; Vrensen, G. F.; Van Noorden, C. J.; Schlingemann, R. O. *Prog. Retin Eye Res.* **2003**, *22*, 1–29.
128. D'Amico, D. J.; Goldberg, M. F.; Hudson, H.; Jerdan, J. A.; Krueger, D. S.; Luna, S. P.; Robertson, S. M.; Russell, S.; Singerman, L.; Slakter, J. S.; et al., P. *Ophthalmology* **2003**, *110*, 2372–2383; discussion 2384–2385.
129. Gaudreault, J.; Fei, D.; Rusit, J.; Suboc, P.; Shiu, V. *Invest. Ophthalmol. Vis. Sci.* **2005**, *46*, 726–733.
130. Moshfeghi, A. A.; Puliafito, C. A. *Expert. Opin. Investig. Drugs* **2005**, *14*, 671–682.
131. Rasmussen, H.; Chu, K. W.; Campochiaro, P.; Gehlbach, P. L.; Haller, J. A.; Handa, J. T.; Nguyen, Q. D.; Sung, J. U. *Hum. Gene Ther.* **2001**, *12*, 2029–2032.

132. Higgins, R. D.; Sanders, R. J.; Yan, Y.; Zasloff, M.; Williams, J. I. *Invest. Ophthalmol. Vis. Sci.* **2000**, *41*, 1507–1512.
133. Nambu, H.; Nambu, R.; Melia, M.; Campochiaro, P. A. *Invest. Ophthalmol. Vis. Sci.* **2003**, *44*, 3650–3655.
134. Kerbel, R. S. *Carcinogenesis* **2000**, *21*, 505–515.
135. Cai, J.; Ahmad, S.; Jiang, W. G.; Huang, J.; Kontos, C. D.; Boulton, M.; Ahmed, A. *Diabetes* **2003**, *52*, 2959–2968.
136. Mukhopadhyay, D.; Tsiokas, L.; Sukhatme, V. P. *Cancer Res.* **1995**, *55*, 6161–6165.
137. Mukhopadhyay, D.; Tsiokas, L.; Zhou, X. M.; Foster, D.; Brugge, J. S.; Sukhatme, V. P. *Nature* **1995**, *375*, 577–581.
138. Giles, F. J. *Oncologist* **2001**, *6*, 32–39.
139. Takahashi, T.; Ueno, H.; Shibuya, M. *Oncogene* **1999**, *18*, 2221–2230.
140. Herbst, R. S.; Johnson, D. H.; Mininberg, E.; Carbone, D. P.; Henderson, T.; Kim, E. S.; Blumenschein, G., Jr.,; Lee, J. J.; Liu, D. D.; Truong, M. T. et al. *J. Clin. Oncol.* **2005**, *23*, 2544–2555.
141. Jin, Q.; Hemminki, K.; Enquist, K.; Lenner, P.; Grzybowska, E.; Klaes, R.; Henriksson, R.; Chen, B.; Pamula, J.; Pekala, W. et al. *Clin. Cancer Res.* **2005**, *11*, 3647–3653.
142. Shao, Z. M.; Nguyen, M. *Front Biosci.* **2002**, *7*, e33–e35.
143. Abulafia, O.; Triest, W. E.; Sherer, D. M. *Gynecol. Oncol.* **1999**, *72*, 220–231.
144. Ke, L. D.; Shi, Y. X.; Im, S. A.; Chen, X.; Yung, W. K. *Clin. Cancer Res.* **2000**, *6*, 2562–2572.
145. Hudis, C. A. *Oncology (Huntingt.)* **2005**, *19*, 26–31.
146. Midgley, R.; Kerr, D. *Ann. Oncol.* **2005**, *16*, 999–1004.
147. Konner, J.; Dupont, J. *Clin. Colorectal Cancer* **2004**, *4*, S81–S85.
148. Holash, J.; Davis, S.; Papadopoulos, N.; Croll, S. D.; Ho, L.; Russell, M.; Boland, P.; Leidich, R.; Hylton, D.; Burova, E. et al. *Proc. Natl. Acad. Sci. USA* **2002**, *99*, 11393–11398.
149. Posey, J. A.; Ng, T. C.; Yang, B.; Khazaeli, M. B.; Carpenter, M. D.; Fox, F.; Needle, M.; Waksal, H.; LoBuglio, A. F. *Clin. Cancer Res.* **2003**, *9*, 1323–1332.
150. Beliveau, R.; Gingras, D.; Kruger, E. A.; Lamy, S.; Sirois, P.; Simard, B.; Sirois, M. G.; Tranqui, L.; Baffert, F.; Beaulieu, E. et al. *Clin. Cancer Res.* **2002**, *8*, 1242–1250.
151. Drevs, J.; Medinger, M.; Schmidt-Gersbach, C.; Weber, R.; Unger, C. *Curr. Drug Targets* **2003**, *4*, 113–121.
152. Levitzki, A. *Eur. J. Cancer* **2002**, *38*, S11–S18.
153. Goldbrunner, R. H.; Bendszus, M.; Wood, J.; Kiderlen, M.; Sasaki, M.; Tonn, J. C. *Neurosurgery* **2004**, *55*, 426–432; discussion 432.
154. Wedge, S. R.; Kendrew, J.; Hennequin, L. F.; Valentine, P. J.; Barry, S. T.; Brave, S. R.; Smith, N. R.; James, N. H.; Dukes, M.; Curwen, J. O. et al. *Cancer Res.* **2005**, *65*, 4389–4400.
155. Ryan, A. J.; Wedge, S. R. *Br. J. Cancer* **2005**, *92*, S6–S13.
156. Mendel, D. B.; Laird, A. D.; Xin, X.; Louie, S. G.; Christensen, J. G.; Li, G.; Schreck, R. E.; Abrams, T. J.; Ngai, T. J.; Lee, L. B. et al. *Clin. Cancer Res.* **2003**, *9*, 327–337.
157. Litz, J.; Sakuntala Warshamana-Greene, G.; Sulanke, G.; Lipson, K. E.; Krystal, G. W. *Lung Cancer* **2004**, *46*, 283–291.
158. Ueda, Y.; Yamagishi, T.; Samata, K.; Ikeya, H.; Hirayama, N.; Okazaki, T.; Nishihara, S.; Arai, K.; Yamaguchi, S.; Shibuya, M. et al. *Cancer Chemother. Pharmacol.* **2004**, *54*, 16–24.
159. Laird, A. D.; Vajkoczy, P.; Shawver, L. K.; Thurnher, A.; Liang, C.; Mohammadi, M.; Schlessinger, J.; Ullrich, A.; Hubbard, S. R.; Blake, R. A. et al. *Cancer Res.* **2000**, *60*, 4152–4160.
160. Hotz, H. G.; Hines, O. J.; Masood, R.; Hotz, B.; Foitzik, T.; Buhr, H. J.; Gill, P. S.; Reber, H. A. *Surgery* **2005**, *137*, 192–199.
161. Kamiyama, M.; Ichikawa, Y.; Ishikawa, T.; Chishima, T.; Hasegawa, S.; Hamaguchi, Y.; Nagashima, Y.; Miyagi, Y.; Mitsuhashi, M.; Hyndman, D. et al. *Cancer Gene Ther.* **2002**, *9*, 197–201.
162. Oshika, Y.; Nakamura, M.; Tokunaga, T.; Ohnishi, Y.; Abe, Y.; Tsuchida, T.; Tomii, Y.; Kijima, H.; Yamazaki, H.; Ozeki, Y. et al. *Eur. J. Cancer* **2000**, *36*, 2390–2396.
163. Ciafre, S. A.; Niola, F.; Wannenes, F.; Farace, M. G. *J. Vasc. Res.* **2004**, *41*, 220–228.
164. Hurwitz, H.; Fehrenbacher, L.; Novotny, W.; Cartwright, T.; Hainsworth, J.; Heim, W.; Berlin, J.; Baron, A.; Griffing, S.; Holmgren, E. et al. *N. Engl. J. Med.* **2004**, *350*, 2335–2342.
165. Compernolle, V.; Brusselmans, K.; Acker, T.; Hoet, P.; Tjwa, M.; Beck, H.; Plaisance, S.; Dor, Y.; Keshet, E.; Lupu, F. et al. *Nat. Med.* **2002**, *8*, 702–710.
166. Van Den Bosch, L.; Storkebaum, E.; Vleminckx, V.; Moons, L.; Vanopdenbosch, L.; Scheveneels, W.; Carmeliet, P.; Robberecht, W. *Neurobiol. Dis.* **2004**, *17*, 21–28.
167. Storkebaum, E.; Lambrechts, D.; Dewerchin, M.; Moreno-Murciano, M. P.; Appelmans, S.; Oh, H.; Van Damme, P.; Rutten, B.; Man, W. Y.; De Mol, M. et al. *Nat. Neurosci.* **2005**, *8*, 85–92.
168. Lopez, J. J.; Laham, R. J.; Stamler, A.; Pearlman, J. D.; Bunting, S.; Kaplan, A.; Carrozza, J. P.; Sellke, F. W.; Simons, M. *Cardiovasc. Res.* **1998**, *40*, 272–281.
169. Reilly, J. P.; Grise, M. A.; Fortuin, F. D.; Vale, P. R.; Schaer, G. L.; Lopez, J.; Van Camp, J. R.; Henry, T.; Richenbacher, W. E.; Losordo, D. W. et al. *Interv. Cardiol.* **2005**, *18*, 27–31.
170. Shuai, K.; Stark, G. R.; Kerr, I. M.; Darnell, J. E., Jr. *Science* **1993**, *261*, 1744–1746.
171. Sadowski, H. B.; Shuai, K.; Darnell, J. E., Jr.; Gilman, M. Z. *Science* **1993**, *261*, 1739–1744.
172. Darnell, J. E., Jr.; Kerr, I. M.; Stark, G. R. *Science* **1994**, *264*, 1415–1421.
173. Ghatpande, S.; Goswami, S.; Mascareno, E.; Siddiqui, M. A. *Mol. Cell Biochem.* **1999**, *196*, 93–97.
174. Cattaneo, E.; Conti, L.; De-Fraja, C. *Trends Neurosci.* **1999**, *22*, 365–369.
175. Mitchell, T. J.; John, S. *Immunology* **2005**, *114*, 301–312.
176. Weber-Nordt, R. M.; Mertelsmann, R.; Finke, J. *Leuk. Lymphoma* **1998**, *28*, 459–467.
177. Kaplan, M. H.; Grusby, M. J. *J. Leukoc. Biol.* **1998**, *64*, 2–5.
178. Barahmand-pour, F.; Meinke, A.; Kieslinger, M.; Eilers, A.; Decker, T. *Curr. Top. Microbiol. Immunol.* **1996**, *211*, 121–128.
179. Alexander, W. S.; Hilton, D. J. *Annu. Rev. Immunol.* **2004**, *22*, 503–529.
180. Pfitzner, E.; Kliem, S.; Baus, D.; Litterst, C. M. *Curr. Pharm. Des.* **2004**, *10*, 2839–2850.
181. Stephanou, A.; Latchman, D. S. *Int. J. Exp. Pathol.* **2003**, *84*, 239–244.
182. Groner, B.; Hennighausen, L. *Breast Cancer Res.* **2000**, *2*, 149–153.
183. Darnell, J. E., Jr. *Science* **1997**, *277*, 1630–1635.
184. Vinkemeier, U.; Cohen, S. L.; Moarefi, I.; Chait, B. T.; Kuriyan, J.; Darnell, J. E., Jr. *Embo J.* **1996**, *15*, 5616–5626.

185. Xu, X.; Sun, Y. L.; Hoey, T. *Science* **1996**, *273*, 794–797.
186. Horvath, C. M. *Trends Biochem. Sci.* **2000**, *25*, 496–502.
187. Shuai, K. *Prog. Biophys. Mol. Biol.* **1999**, *71*, 405–422.
188. Wen, Z.; Zhong, Z.; Darnell, J. E., Jr. *Cell* **1995**, *82*, 241–250.
189. Zhang, X.; Blenis, J.; Li, H. C.; Schindler, C.; Chen-Kiang, S. *Science* **1995**, *267*, 1990–1994.
190. Caldenhoven, E.; van Dijk, T. B.; Solari, R.; Armstrong, J.; Raaijmakers, J. A.; Lammers, J. W.; Koenderman, L.; de Groot, R. P. *J. Biol. Chem.* **1996**, *271*, 13221–13227.
191. Wang, D.; Stravopodis, D.; Teglund, S.; Kitazawa, J.; Ihle, J. N. *Mol. Cell Biol.* **1996**, *16*, 6141–6148.
192. Yan, R.; Qureshi, S.; Zhong, Z.; Wen, Z.; Darnell, J. E., Jr. *Nucleic Acids Res.* **1995**, *23*, 459–463.
193. Schaefer, T. S.; Sanders, L. K.; Nathans, D. *Proc. Natl. Acad. Sci. USA* **1995**, *92*, 9097–9101.
194. Xia, Z.; Salzler, R. R.; Kunz, D. P.; Baer, M. R.; Kazim, L.; Baumann, H.; Wetzler, M. *Cancer Res.* **2001**, *61*, 1747–1753.
195. Azam, M.; Lee, C.; Strehlow, I.; Schindler, C. *Immunity* **1997**, *6*, 691–701.
196. Imada, K.; Leonard, W. J. *Mol. Immunol.* **2000**, *37*, 1–11.
197. Catlett-Falcone, R.; Dalton, W. S.; Jove, R. *Curr. Opin. Oncol.* **1999**, *11*, 490–496.
198. Catlett-Falcone, R.; Landowski, T. H.; Oshiro, M. M.; Turkson, J.; Levitzki, A.; Savino, R.; Ciliberto, G.; Moscinski, L.; Fernandez-Luna, J. L.; Nunez, G. et al. *Immunity* **1999**, *10*, 105–115.
199. Reddy, E. P.; Korapati, A.; Chaturvedi, P.; Rane, S. *Oncogene* **2000**, *19*, 2532–2547.
200. Durbin, J. E.; Hackenmiller, R.; Simon, M. C.; Levy, D. E. *Cell* **1996**, *84*, 443–450.
201. Meraz, M. A.; White, J. M.; Sheehan, K. C.; Bach, E. A.; Rodig, S. J.; Dighe, A. S.; Kaplan, D. H.; Riley, J. K.; Greenlund, A. C.; Campbell, D. et al. *Cell* **1996**, *84*, 431–442.
202. Sugawara, I.; Yamada, H.; Mizuno, S. *Tohoku J. Exp. Med.* **2004**, *202*, 41–50.
203. Park, C.; Li, S.; Cha, E.; Schindler, C. *Immunity* **2000**, *13*, 795–804.
204. Levy, D. E. *Cell Mol. Life Sci.* **1999**, *55*, 1559–1567.
205. Takeda, K.; Noguchi, K.; Shi, W.; Tanaka, T.; Matsumoto, M.; Yoshida, N.; Kishimoto, T.; Akira, S. *Proc. Natl. Acad. Sci. USA* **1997**, *94*, 3801–3804.
206. Takeda, K.; Kaisho, T.; Yoshida, N.; Takeda, J.; Kishimoto, T.; Akira, S. *J. Immunol.* **1998**, *161*, 4652–4660.
207. Takeda, K.; Clausen, B. E.; Kaisho, T.; Tsujimura, T.; Terada, N.; Forster, I.; Akira, S. *Immunity* **1999**, *10*, 39–49.
208. Chapman, R. S.; Lourenco, P. C.; Tonner, E.; Flint, D. J.; Selbert, S.; Takeda, K.; Akira, S.; Clarke, A. R.; Watson, C. J. *Genes Dev.* **1999**, *13*, 2604–2616.
209. Sano, S.; Itami, S.; Takeda, K.; Tarutani, M.; Yamaguchi, Y.; Miura, H.; Yoshikawa, K.; Akira, S.; Takeda, J. *EMBO J.* **1999**, *18*, 4657–4668.
210. Jacoby, J. J.; Kalinowski, A.; Liu, M. G.; Zhang, S. S.; Gao, Q.; Chai, G. X.; Ji, L.; Iwamoto, Y.; Li, E.; Schneider, M. et al. *Proc. Natl. Acad. Sci. USA* **2003**, *100*, 12929–12934.
211. Gorogawa, S.; Fujitani, Y.; Kaneto, H.; Hazama, Y.; Watada, H.; Miyamoto, Y.; Takeda, K.; Akira, S.; Magnuson, M. A.; Yamasaki, Y. et al. *Biochem. Biophys. Res. Commun.* **2004**, *319*, 1159–1170.
212. Thierfelder, W. E.; van Deursen, J. M.; Yamamoto, K.; Tripp, R. A.; Sarawar, S. R.; Carson, R. T.; Sangster, M. Y.; Vignali, D. A.; Doherty, P. C.; Grosveld, G. C. et al. *Nature* **1996**, *382*, 171–174.
213. Sanchez-Guajardo, V.; Borghans, J. A.; Marquez, M. E.; Garcia, S.; Freitas, A. A. *J. Immunol.* **2005**, *174*, 1178–1187.
214. Yang, Z.; Chen, M.; Ellett, J. D.; Fialkow, L. B.; Carter, J. D.; McDuffie, M.; Nadler, J. L. *J. Autoimmun.* **2004**, *22*, 191–200.
215. Liu, X.; Robinson, G. W.; Wagner, K. U.; Garrett, L.; Wynshaw-Boris, A.; Hennighausen, L. *Genes Dev.* **1997**, *11*, 179–186.
216. Udy, G. B.; Towers, R. P.; Snell, R. G.; Wilkins, R. J.; Park, S. H.; Ram, P. A.; Waxman, D. J.; Davey, H. W. *Proc. Natl. Acad. Sci. USA* **1997**, *94*, 7239–7244.
217. Teglund, S.; McKay, C.; Schuetz, E.; van Deursen, J. M.; Stravopodis, D.; Wang, D.; Brown, M.; Bodner, S.; Grosveld, G.; Ihle, J. N. *Cell* **1998**, *93*, 841–850.
218. Feldman, G. M.; Rosenthal, L. A.; Liu, X.; Hayes, M. P.; Wynshaw-Boris, A.; Leonard, W. J.; Hennighausen, L.; Finbloom, D. S. *Blood* **1997**, *90*, 1768–1776.
219. Nakajima, H.; Liu, X. W.; Wynshaw-Boris, A.; Rosenthal, L. A.; Imada, K.; Finbloom, D. S.; Hennighausen, L.; Leonard, W. J. *Immunity* **1997**, 7, 691–701.
220. Imada, K.; Bloom, E. T.; Nakajima, H.; Horvath-Arcidiacono, J. A.; Udy, G. B.; Davey, H. W.; Leonard, W. J. *J. Exp. Med.* **1998**, *188*, 2067–2074.
221. Moriggl, R.; Topham, D. J.; Teglund, S.; Sexl, V.; McKay, C.; Wang, D.; Hoffmeyer, A.; van Deursen, J.; Sangster, M. Y.; Bunting, K. D. et al. *Immunity* **1999**, *10*, 249–259.
222. Snow, J. W.; Abraham, N.; Ma, M. C.; Herndier, B. G.; Pastuszak, A. W.; Goldsmith, M. A. *J. Immunol.* **2003**, *171*, 5042–5050.
223. Socolovsky, M.; Fallon, A. E.; Wang, S.; Brugnara, C.; Lodish, H. F. *Cell* **1999**, *98*, 181–191.
224. Takeda, K.; Tanaka, T.; Shi, W.; Matsumoto, M.; Minami, M.; Kashiwamura, S.; Nakanishi, K.; Yoshida, N.; Kishimoto, T.; Akira, S. *Nature* **1996**, *380*, 627–630.
225. Palmer-Crocker, R. L.; Hughes, C. C.; Pober, J. S. *J. Clin. Invest.* **1996**, *98*, 604–609.
226. Shimoda, K.; van Deursen, J.; Sangster, M. Y.; Sarawar, S. R.; Carson, R. T.; Tripp, R. A.; Chu, C.; Quelle, F. W.; Nosaka, T.; Vignali, D. A. et al. *Nature* **1996**, *380*, 630–633.
227. Takeda, K.; Kamanaka, M.; Tanaka, T.; Kishimoto, T.; Akira, S. *J. Immunol.* **1996**, *157*, 3220–3222.
228. Kaplan, M. H.; Schindler, U.; Smiley, S. T.; Grusby, M. J. *Immunity* **1996**, *4*, 313–319.
229. Bromberg, J.; Bromberg, J. *J. Clin. Invest.* **2002**, *109*, 1139–1142.
230. Calo, V.; Migliavacca, M.; Bazan, V.; Macaluso, M.; Buscemi, M.; Gebbia, N.; Russo, A. *J. Cell Physiol.* **2003**, *197*, 157–168.
231. Garcia, R.; Jove, R. *J. Biomed. Sci.* **1998**, *5*, 79–85.
232. Bowman, T.; Broome, M. A.; Sinibaldi, D.; Wharton, W.; Pledger, W. J.; Sedivy, J. M.; Irby, R.; Yeatman, T.; Courtneidge, S. A.; Jove, R. *Proc. Natl. Acad. Sci. USA* **2001**, *98*, 7319–7324.
233. Garcia, R.; Bowman, T. L.; Niu, G.; Yu, H.; Minton, S.; Muro-Cacho, C. A.; Cox, C. E.; Falcone, R.; Fairclough, R.; Parsons, S. et al. *Oncogene* **2001**, *20*, 2499–2513.
234. Gomez-Bougie, P.; Bataille, R.; Amiot, M. *Eur. J. Immunol.* **2004**, *34*, 3156–3164.
235. Song, J. I.; Grandis, J. R. *Oncogene* **2000**, *19*, 2489–2495.
236. Jazirehi, A. R.; Bonavida, B. *Oncogene* **2005**, *24*, 2121–2143.
237. Sinibaldi, D.; Wharton, W.; Turkson, J.; Bowman, T.; Pledger, W. J.; Jove, R. *Oncogene* **2000**, *19*, 5419–5427.

238. Martino, A.; Holmes, J. H. T.; Lord, J. D.; Moon, J. J.; Nelson, B. H. *J. Immunol.* **2001**, *166*, 1723–1729.
239. Schaefer, L. K.; Ren, Z.; Fuller, G. N.; Schaefer, T. S. *Oncogene* **2002**, *21*, 2058–2065.
240. Pernis, A. B.; Rothman, P. B. *J. Clin. Invest.* **2002**, *109*, 1279–1283.
241. Chatila, T. A. *Trends Mol. Med.* **2004**, *10*, 493–499.
242. Ghaffar, O.; Christodoulopoulos, P.; Lamkhioued, B.; Wright, E.; Ihaku, D.; Nakamura, Y.; Frenkiel, S.; Hamid, Q. *Clin. Exp. Allergy* **2000**, *30*, 86–93.
243. Sampath, D.; Castro, M.; Look, D. C.; Holtzman, M. J. *J. Clin. Invest.* **1999**, *103*, 1353–1361.
244. Shouda, T.; Yoshida, T.; Hanada, T.; Wakioka, T.; Oishi, M.; Miyoshi, K.; Komiya, S.; Kosai, K.; Hanakawa, Y.; Hashimoto, K. et al. *J. Clin. Invest.* **2001**, *108*, 1781–1788.
245. Atsumi, T.; Ishihara, K.; Kamimura, D.; Ikushima, H.; Ohtani, T.; Hirota, S.; Kobayashi, H.; Park, S. J.; Saeki, Y.; Kitamura, Y.; Hirano, T. *J. Exp. Med.* **2002**, *196*, 979–990.
246. Ivashkiv, L. B.; Hu, X. *Arthritis Rheum.* **2003**, *48*, 2092–2096.
247. Luo, C.; Laaja, P. *Drug Disc. Today* **2004**, *9*, 268–275.
248. Leszczyniecka, M.; Roberts, T.; Dent, P.; Grant, S.; Fisher, P. B. *Pharmacol. Ther.* **2001**, *90*, 105–156.
249. Qin, J. Z.; Kamarashev, J.; Zhang, C. L.; Dummer, R.; Burg, G.; Dobbeling, U. *J. Invest. Dermatol.* **2001**, *117*, 583–589.

Biographies

Marie-Hélène Teiten received her PhD in Biology in 2003 at the Université Henri Poincaré in Nancy (France). Her research project, performed in the Photodynamic Therapy Unit of the Alexis Vautrin Cancer Center (Nancy, France) under the direction of Prof François Guillemin, focused on the subcellular localization and the phodommages induced by photosensitizers used in photodynamic therapy.

Romain Blasius was born in 1976 in Luxembourg. In 1995 he moved to the Université Libre de Bruxelles (Brussels, Belgium), where he received his BSc degree in chemistry in 1999. He then went on to complete his PhD studies under the supervision of A Kirsch and C Moucheron in the area of photochemistry and photophysics of transition metal complexes with DNA and received his PhD in 2003. He subsequently moved to the Laboratoire de Biologie Moléculaire et Cellulaire du Cancer in Luxembourg, where he is currently carrying out his research on the resistance of cancer cells toward chemotherapeutic drugs and the inhibition of this resistance via the use of natural compounds.

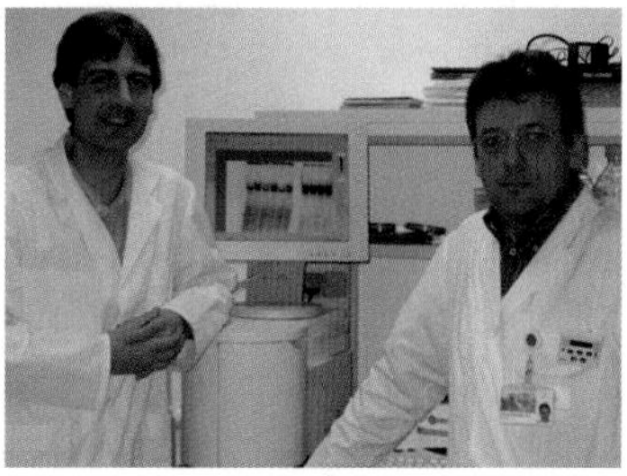

Franck Morceau (right) received his PhD in molecular pharmacology in 1996 at the University of Reims (France). His research focalizes on molecular mechanisms of erythroid differentiation. He is a staff scientist at the Laboratoire de Biologie Moléculaire et Cellulaire du Cancer in Luxembourg.

Marc Diederich (left) received his PhD in molecular pharmacology in 1994 at the University of Nancy (France). He is leading the Laboratoire de Biologie Moléculaire et Cellulaire du Cancer in Luxembourg. Research in this laboratory is mainly focusing on the inhibition of glutathione-based drug resistance mechanism by natural compounds as well as erythroid differentiation mechanisms.

Mario Dicato is Head of Internal Medicine and of the Service of Haematology-Oncology at Luxembourg Medical Centre. Much of his postgraduate training was at the University of Pittsburgh, Pittsburgh, PA, USA and at Harvard University, Boston, MA, USA.

Comprehensive Medicinal Chemistry II
ISBN (set): 0-08-044513-6
ISBN (Volume 3) 0-08-044516-0; pp. 189–214

3.11 Orthogonal Ligand–Receptor Pairs

F C Acher, Université René Descartes – Paris V, Paris, France

3.11.1	**Introduction**	**215**
3.11.2	**The Orthogonal Ligand–Receptor Pair Concept**	**215**
3.11.3	**Orthogonal Ligand–Receptor Pair Applications**	**216**
3.11.3.1	The GTPase Family	216
3.11.3.2	The Cyclophilin–Cyclosporin Pair	217
3.11.3.3	Protein Kinases: The Analog-Sensitive Kinase Alleles (ASKA)	218
3.11.3.4	Kinesin and Myosin	220
3.11.3.5	Nuclear Hormone Receptors	221
3.11.3.6	G Protein-Coupled Receptors (GPCRs)	229
3.11.4	**Conclusion**	**232**
References		**232**

3.11.1 Introduction

The trigger of biological activities relies in many cases on the stablization of a small molecule–protein complex. The affinity of the ligand or substrate for its receptor or enzyme is dependent on complementary noncovalent interactions. Although the interaction types are limited in number (hydrophobic, electrostatic, van der Waals' interactions, and hydrogen bonds), a subtle combination of these delineate the selective affinity of a small molecule for its target protein. Moreover, such a molecular recognition is often the start of a biological process as in signaling pathways. The fine adjustment between the small molecule and the protein may be compared to the fitting of a piece of a jigsaw puzzle, as shown in **Figure 1a**. A minor modification in the ligand or protein will disrupt some of these interactions resulting in a blockade of the activity since a remodeled piece cannot adjust any longer to the jigsaw puzzle unless the complementary piece is also remodeled (**Figure 1**). Similarly, a complementary engineering of the ligand or protein can restore a stabilizing interaction and consequently rescue the impaired functionalities. The concept has been named 'orthogonal ligand–receptor pairs.' In addition to the elegant work, this approach has been successfully used for investigation of mechanisms and target validation. Target validation has indeed become a crucial issue as genomics has brought a large number of new potential therapeutic targets. The orthogonal ligand–receptor pairs approach is a powerful tool when studying complex biological systems, particularly those that are controlled by molecules with multiple cellular targets. The present chapter will define the concept and review several notorious examples of its application.

3.11.2 The Orthogonal Ligand–Receptor Pair Concept

Ligand–protein pairs are characterized by a highly specific mutual recognition. The stability of the complex results from additive noncovalent interactions that are present at the interface between the ligand and the protein. These interactions derive from hydrophobic or van der Waals' contacts, or from polar interactions through hydrogen bonds or electrostatic attraction (**Figure 2a**). They may be accurately visualized and analyzed in 3D structures of the small molecule bound to the protein in x-ray structures or homology models. A close examination of the ligand binding site in those 3D structures suggests how the ligand or the protein might be modified in order to prevent their initial match (**Figures 2** and **3**). The small molecule may thus be chemically modified in order to create a steric hindrance (**Figure 2b**) or an electrostatic repulsion (**Figure 3b**), or a disruption of a hydrogen bond network with specific residues of the binding site. While other positive interactions may still be maintained, the stability of the complex is decreased to such a point that biological activation is abolished. However, further examination of the binding site suggests subsequent compensatory mutations of specific residues in order to restore the initial activity of the ligand–protein complex (**Figures 2d** and **3d**). Similarly, the protein may be first engineered to alter the fine tuning between ligand and protein (**Figures 2c** and **3c**) and then the ligand may be manipulated to restore complementary interactions on a rational basis (**Figure 2d,c**).

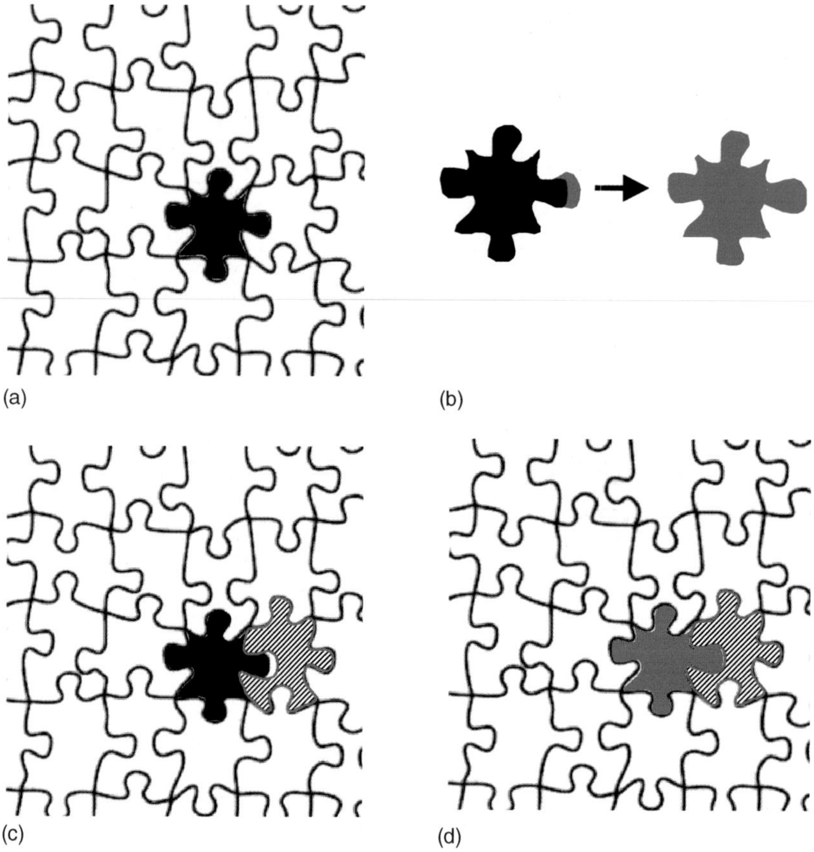

Figure 1 (a) The fine adjustment between a small molecule (black piece) and its cognate protein (white pieces) may be compared to the fitting of a piece of a jigsaw puzzle. (b) The black piece is remodeled to a red piece. (c) Its complementary piece is remodeled to a gray piece and the black piece no longer fits in the puzzle. (d) Matching is recovered between red and gray pieces.

3.11.3 Orthogonal Ligand–Receptor Pair Applications

The strategy of using small molecule–protein reengineered pairs provides powerful tools for probing the biological function of the original pair. One of the major advantages of such an approach is that only the modified system will be triggered by the altered ligand and not the natural pathway. This is of particular interest in cases where the same endogenous ligand is involved in multiple processes. This strategy may be a valuable technique for dissecting complex biological systems and validating potential therapeutic targets.

The orthogonal ligand–receptor pair concept has been successfully investigated and reviewed in several articles.[1–6] This chapter will focus on the application of the concept for target validation. A wide range of biological systems has been studied using this approach, and a selection of these is reported here.

3.11.3.1 The GTPase Family

The orthogonal small molecule–protein strategy was first applied to enzymes of the GTPase family.[7] By switching between GDP-bound inactive and GTP-bound active states, GTPases control many cellular processes, including regulation of the actin cytoskeleton, cell polarity, microtubule dynamics, membrane transport pathways, and transcription factor activity. Because of the pivotal role of GTPases in signaling pathways, the study of factors involved in their activation has been the focus of much attention. The chemical genetic approach was used to demonstrate the substrate specificity. GTPase was turned into XTPase by mutating one single residue that disrupted a hydrogen bond between GTP and elongation factor Tu (EF-Tu), a GTPase involved in protein biosynthesis.[2,3,7] The guanine moiety of GTP is anchored to the EF-Tu enzyme by several hydrogen bonds (N135, K136, D138, and S173) and its bicycle

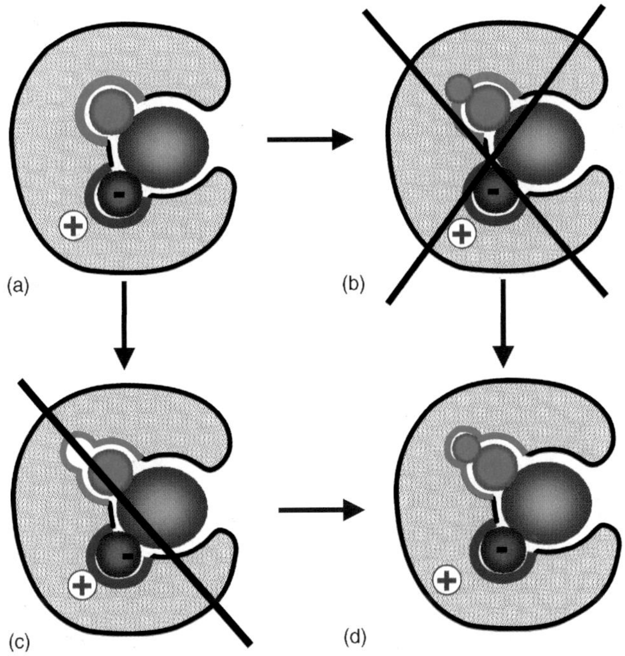

Figure 2 (a) Schematic representation of a small molecule–protein pair with specific mutual recognition based on complementary interactions. Hydrophobic interactions are represented by the contact between a hydrophobic substituent on the small molecule (green sphere) and hydrophobic side chains of the protein residues (green strip). Electrostatic interactions are represented by a negatively charged substituent on the small molecule (red sphere) and positively charged side chains of the protein residues (blue strip). (b) A modified small molecule holding a larger hydrophobic group is no longer able to bind to the protein because of steric hindrance. (c) The initial small molecule shows weaker binding to an engineered binding site where the hydrophobic pocket has been enlarged. (d) Binding between the modified small molecule and the engineered protein is recovered if both modifications are complementary.

sandwiched between two hydrophobic residues (K136 and L175) (**Figure 4a**). If D138 is mutated to an asparagine, the two NH_2 groups of N138 and guanine would be in steric clash in order to maintain the coplanarity of the amide and the purine ring (**Figure 4b**). The complex is thus destabilized and a large decrease in affinity is noted. Nevertheless, when the guanine is replaced by a xanthine entity, perfect matching and affinity are recovered (**Figure 4c**). It is remarkable that such a minor complementary change on substrate and protein leads to a very effective orthogonal ligand–receptor pair. The residue that was mutated (D138) is conserved in all GTPases. Therefore, a similar mutation of that particular aspartate into an asparagine converted a series of other GTPases into XTPases allowing the function of the initial ATPase in complex systems to be specifically investigated.[2]

3.11.3.2 The Cyclophilin–Cyclosporin Pair

As described above (see **Figure 2**), one way of remodeling the ligand–receptor interface is to create a 'bump' on the small molecule and a 'hole' on the protein side. This approach was first successfully applied to the cyclophilin–cyclosporin A (Cyp–CsA) pair by S. Schreiber and co-workers.[8,9] On the basis of the crystal structure of the Cyp–CsA complex, residue 11 (MeVal) of the cyclic peptide CsA was swapped to a MeIle or an α-cyclopentylsarcosine (CpSar11) to generate a bump on the ligand (CsA*, **Figure 5**). On the protein side, three residues of the binding pocket were mutated (S99T, F113G, C115M) to create a hole matching the shape of the ligand bump in CsA* (**Figure 5b**). Cyclosporin A is a well-known immunosuppressive drug used to avoid graft rejection after transplantation. It binds to cyclophilin, and the resulting complex in turn inhibits protein phosphatase calcineurin, which is involved in several cellular functions. The modified CsA ligand (CsA*) had little or no effect on cells expressing the wild-type cyclophilin Cyp; however, it potently inhibited calcineurin signaling in cells expressing the mutated cyclophilin Cyp*. Thus, since calcineurin is ubiquitously expressed, the engineered Cyp*–CsA* pair proved to be an excellent tool to investigate its many cellular roles.

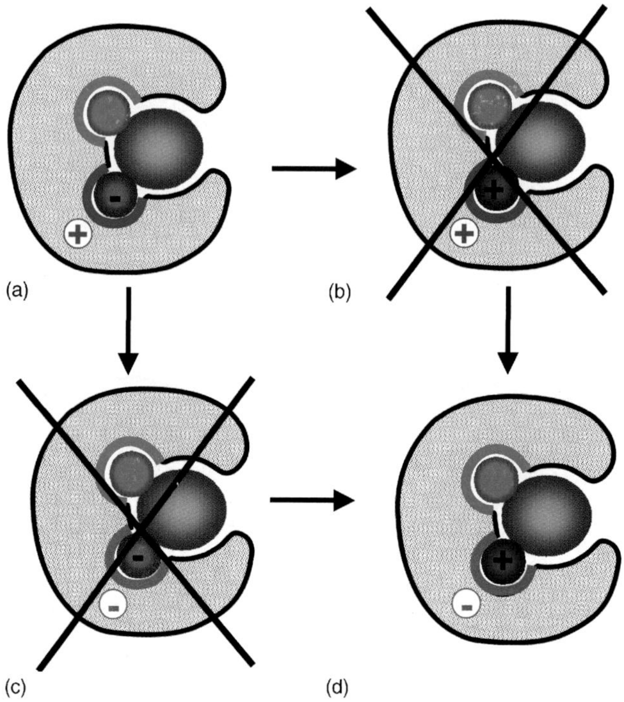

Figure 3 (a) Same as **Figure 2a**. (b) A modified small molecule holding a positively charged group is no longer able to bind to the protein because of electrostatic repulsion. (c) The initial small molecule shows weaker or no binding to an engineered binding site where the positively charged pocket has been changed to negative and results in electrostatic repulsion. (d) Binding between the modified small molecule and the engineered protein is recovered if both modifications are complementary.

3.11.3.3 Protein Kinases: The Analog-Sensitive Kinase Alleles (ASKA)

One of the most impressive proofs of concept in terms of target validation using the orthogonal small molecule–protein pairs was achieved with the ASKA studies by Shokat and co-workers.[4,10] Protein kinases represent the largest superfamily of signaling proteins in all cells. They regulate numerous pathways; consequently, they emerge as therapeutic targets for a large number of diseases that are related to their dysregulation. Protein kinases catalyze the phosphorylation of serine, threonine, and tyrosine residues by adenosine triphosphate (ATP). Because of the high similarity of the ATP binding sites among these enzymes, the identification of specific substrates or inhibitors was a difficult challenge. The ASKA technology proved to be very successful for that purpose. To selectively target one kinase among all others, complementary remodeling is simultaneously introduced in the initial substrate–enzyme pair as described in **Figure 2**.

As a prototype, specific residues of the ATP binding site of a tyrosine kinase v-Src were mutated on the basis of a homology model (**Figure 6a**).[11] The engineered enzyme (V323A, I338A) was shown to only display catalytic activity with an N^6-cyclopentyl-ATP derivative with a similar efficacy to the wild-type kinase v-Src (**Figure 6b**). No affinity of the ATP analog for the wild-type enzyme was detected.[11] Later, other N^6-substituted ATP derivatives were assayed with the I338A or G mutant of v-Src, and proved to be better substrates with improved kinetic constants, for example, N^6-benzyl-ATP and N^6-(2-phenetyl)-ATP.[2,12,13] A closer look at the 3D models of the binding site revealed that the hydrophobic 'hole' in the engineered protein was already present in the wild-type but its access was prevented by a bulky side chain at position 338 (I338 in v-Src) playing the role of a gatekeeper (**Figure 6**). Mutation of that residue allows N^6-ATP-substitutents (bump) to reach this hydrophobic pocket providing a larger interaction surface and consequently a more stable complex (**Figure 6b**).[2,12,13] The strategy was applied to other kinases such as protein kinase Raf-1[14] and used to identify several direct kinase substrates.[4] The identification of substrates of CDK1 and insights into the role of CDK1 illustrates the power of the chemical genetic approach for target validation.[15]

To explore further the role of individual kinases, the ASKA approach was extended to protein kinase inhibitors. PP1 (4-amino-3-phenyl-1-*t*-butyl)pyrazolo[3,4-*d*]pyrimidine) derivatives are known Src family tyrosine kinase inhibitors[16] that were postulated to bind to the active site in a similar orientation to ATP[17] (**Figure 7a**). Chemical modifications of PP1 inhibitors led to potent inhibitors that only bind the Src kinases mutated at position 338 with a smaller residue G or A (**Figure 7**).[17,18] Those

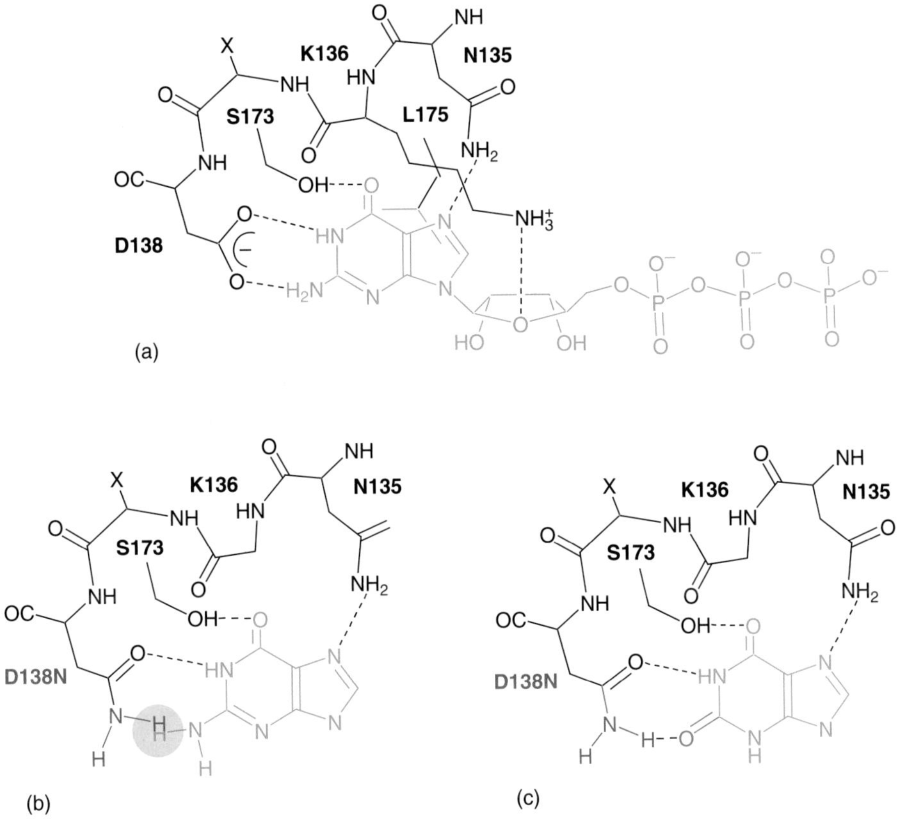

Figure 4 (a) Schematic representation of GTP (blue) bound at EF-Tu GTPase (black); only residues of the binding pocket that surround the guanine moiety are displayed (based on 1EFT of the Protein Data Bank, PDB).[58] The sequence NKXD is conserved in all GTPases with X being a variable residue. (b) Steric impairment shown as a yellow circle when D138 is mutated to N (modified atoms colored in pink); only the guanine part of GTP is displayed. (c) Perfect matching recovered when the guanine moiety is switched to xanthine (modified atom colored in green); only the xanthine part of XTP is displayed (GTP = guanosine-5′-triphosphate, XTP = xanthosine-5′-triphosphate).

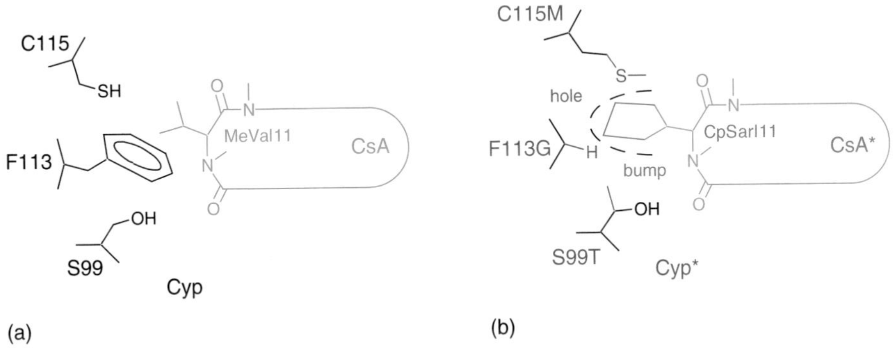

Figure 5 (a) Schematic representation of cyclosporin A (CsA, blue) bound to cyclophilin (Cyp, black); only residues of the binding pocket that surround residue 11 (MeVal) of CsA are displayed (based on 2RMA of the PDB). (b) CsA*, the modified cyclopeptide CsA at position 11 by CpSar, bound to Cyp*, the mutated Cyp (S99T, F113G, C115M); modified atoms of CsA* are colored in green and those of Cyp* in pink.

engineered kinases allow access of a naphthyl substituent to a hydrophobic pocket that was not accessible in the wild-type because of the gatekeeper residue at position 338 (**Figure 7b,c**).[18,19] Such an orthogonal inhibitor–enzyme pair approach has now been widely used to assign specific signaling roles to individual tyrosine and serine/threonine kinases.[10,20–22] While a large number of kinases of the mammalian proteome can be transformed into these analog-sensitive variants, some, nevertheless, do

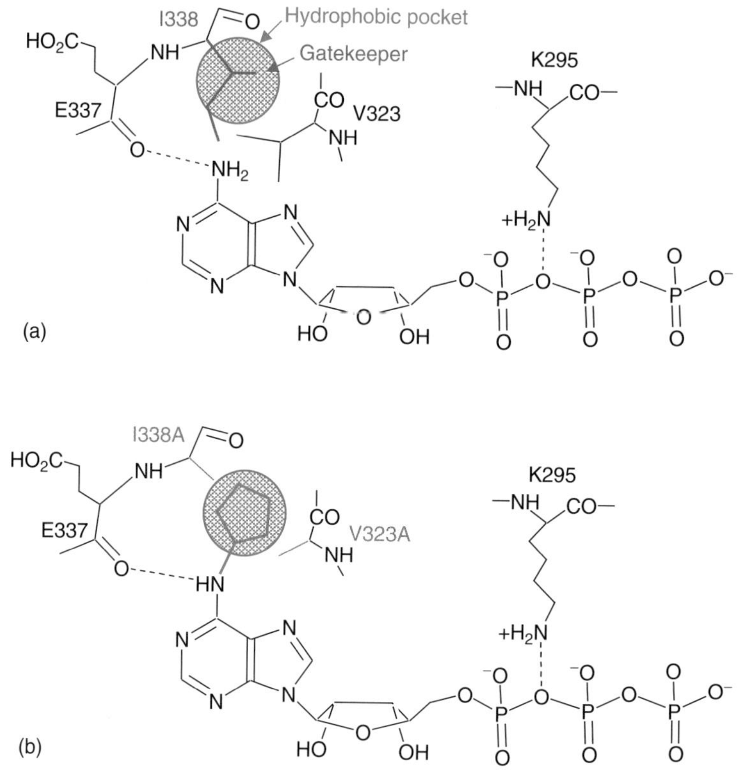

Figure 6 (a) Schematic representation of ATP bound to tyrosine kinase v-Src (based on 1ATP of the PDB and [11]). Only residues in the vicinity of the NH_2 group of the adenine part of the substrate are displayed; residue I338 preventing access to the hydrophobic pocket of kinase (filled light green circle) and named the gatekeeper is shown in red. (b) N^6-cyclopentyl-ATP derivative bound to I338A/V323A v-Src kinase mutant; the cyclopentyl substituent of the adenine part is displayed in green, modified atoms of mutated residues at the binding site are colored in pink; these mutations allow access of the cyclopentyl ring to the hydrophobic pocket (green circle).

not tolerate the gatekeeper mutation because of impaired enzymatic activity. Recently, Shokat and colleagues identified second site suppressor mutations that rescue the enzymatic activity for several of the divergent kinases, enlarging the number of possible applications of the ASKA approach.[23,24]

Taken together these studies demonstrated that a particular kinase could be distinguished from many other cellular kinases while no other techniques allowed such a fine dissection. The approach opened the way to multiple investigations on the role of specific kinases in cellular signaling pathways and provided invaluable tools for drug target validation.

3.11.3.4 Kinesin and Myosin

A similar approach to the ASKA one was applied to myosins and kinesins, which are also ATP-dependent signaling proteins.[2,6] Both types of enzymes belong to the motor protein family. As for kinases, the high binding site similarity among the various isozymes is a major obstacle to the discovery of selective inhibitors and the elucidation of the role of individual isozymes in cellular processes. It was thus anticipated that the chemical genetic methodology might provide a possible clue. Myosin-Iβ was mutated (Y61G) at the ATP-binding pocket in such a manner that a space was created around the N^6 position of the nucleotide part of ATP.[25] A large collection of N^6-modified adenine nucleotides was then screened. N^6-benzyl-ATP, N^6-(2-phenethyl)-ATP, and N^6-(2-methyl)butyl were found to be the most selective inhibitors of ATP hydrolysis by the Y61G mutant over wild-type, while hydrolysis of ATP remained similar in both cases. Similar data were observed with the ADP analogs bearing the same N^6 substituents. These inhibitors locked Y61G myosin-Iβ tightly to actin resulting in actin filament motility being blocked.[25] Gillespie and co-workers designed a remarkable application of this approach to investigate the role of myosin-Iβ in adaptation in sensory cells of the inner ear.[26] A similar application for myosinVb revealed its specific function in membrane transport.[27] In the kinesin family, the core of the ATPase domain is structurally quite similar to that of myosins. Kapoor and Mitchinson[28] mutated residues surrounding N^6 of the adenine moiety of ATP/ADP in order to increase the size of the binding pocket, as

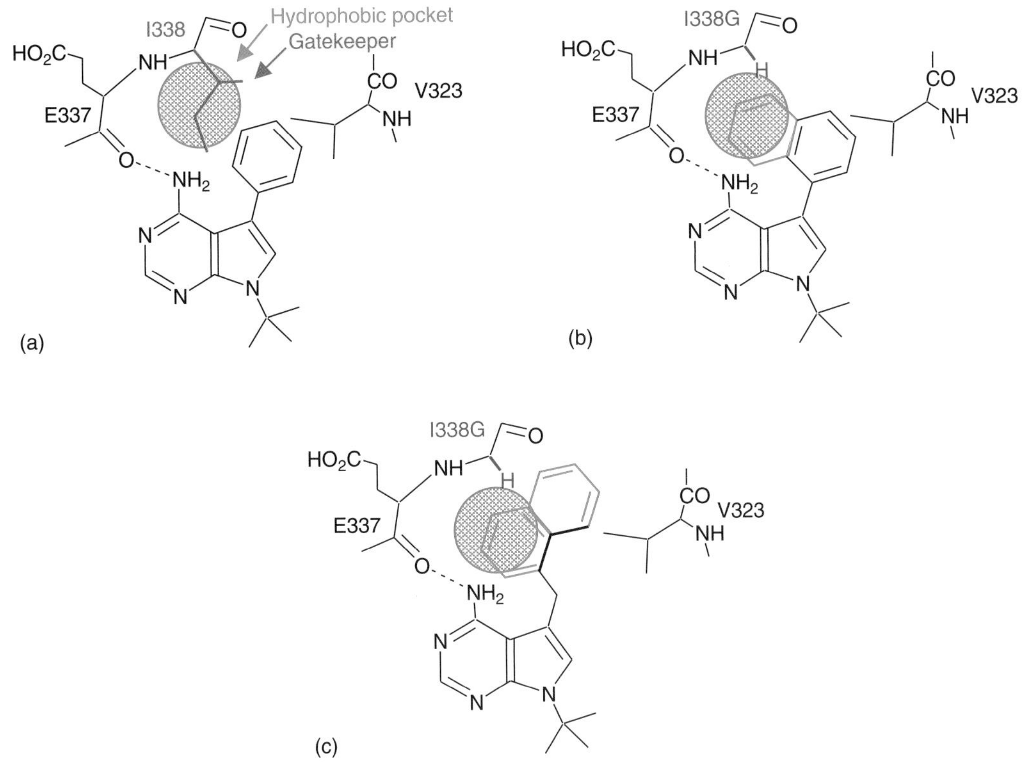

Figure 7 (a) Kinase inhibitor PP1 (4-amino-5-phenyl-7-(*t*-butyl)pyrrolo[2,3–*d*]pyrimidine) bound to v-Src kinase in a similar manner to the adenine part of ATP in **Figure 6a** (same color code). (b) 4-Amino-5-naphthyl-7-(*t*-butyl)pyrrolo[2,3-*d*]pyrimidine, a derivative of PP1, only binds to the mutant I338G of v-Src kinase (and not to wild-type) as the gatekeeper has been removed. (c) 4-Amino-5-methylnaphthyl-7-(*t*-butyl)pyrrolo[2,3-*d*]pyrimidine with similar binding to that in (b).

described for kinases and myosins.[2] The resulting mutated kinesin displays a greatly reduced activity with unmodified ATP but is specifically activated by a modified ATP analog holding a cyclopentyl substituent at the N^6 position of ATP (same substrate as in **Figure 6b**). When the triphosphate moiety of that analog is replaced by a similar group with a nonhydrolyzable bond, the compound is turned into a specific inhibitor of the mutated kinesin.[28] This orthogonal pair could be used to decipher the specific roles of kinesins in cellular processes.

3.11.3.5 Nuclear Hormone Receptors

The nuclear hormone receptor (NHR) superfamily[29] is the largest group of transcription factors that regulate development and metabolism through control of gene expression. Members of that family have been classified according to the chemical similarity of their ligands: the steroid receptors including estrogen receptors (ER); the nonsteroidal receptors including thyroid hormone receptor (TR), vitamin D receptor (VDR), and retinoid X receptors (binding 9-*cis* retinoic acid) (RXR); the retinoic acid receptors (binding all-*trans* retinoic acid) (RAR); and ecdysone receptors (EcR) found in insects. Activation of these receptors depends on both DNA and ligand binding in two different structural domains: the DNA-binding domain (DBD) and ligand-binding domain (LBD). The availability of crystal structures of ligands bound to their LBD have allowed the design of several orthogonal ligand–receptor pairs, which may be powerful tools to regulate gene expression and provide insights into related diseases. Some of the most representative studies in the NHR family targeting RXR,[30–32] RAR,[33] ER,[34–38] TR,[39–42] VDR,[42–44] and EcR[45] are reviewed below.

Corey and co-workers[30] were the first to apply the chemical genetic approach to nuclear receptors. The retinoid X receptor (RXR) activates transcription of target genes in response to its natural ligand 9-*cis* retinoic acid (9cRA) binding. Guided by a 3D structure of 9cRA bound to RXRα binding domain, two residues F313 and L436 were selected for mutations[30] (**Figure 8a**). The single and double mutants resulted in altered ligand specificity. A shift in affinity was observed from natural hormone to synthetic ligands but also the reverse way according to the mutants. In a following study, the authors found that LG335, a compound previously synthesized for a structure–activity relationship study and

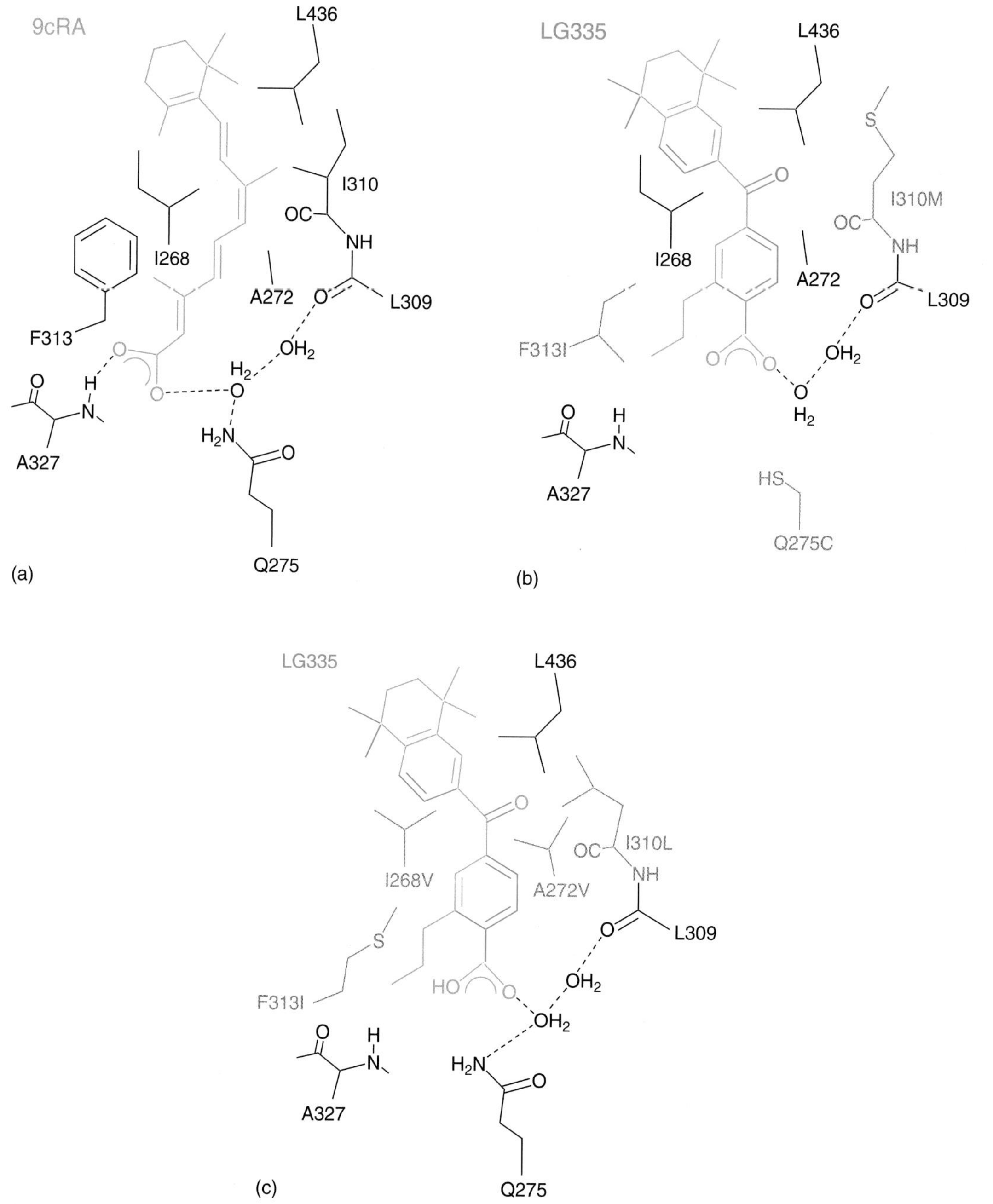

Figure 8 (a) Schematic representation of natural ligand 9-*cis* retinoic acid (9cRA, blue) bound to human retinoid X receptor alpha (RXR, black); hydrophobic and polar contacts are shown (based on 1FBY of the PDB). (b) Synthetic ligand LG335 bound to triple mutant Q275C/I310M/F313I (pink) of RXR. Common atoms between 9cRA and LG335 are colored in blue; additional LG335 atoms are colored in green. (c) Same as in (b) with quadruple mutant I268V/A272V/I310L/F313M (pink).

which was inactive at wild-type RXR, could effectively activate the triple mutant Q275C/I310M/F313I[31] (**Figure 8b**). The LG335–triple mutant pair was thus a powerful tool to investigate the cellular roles of RXR. The orthogonal ligand–receptor pair was further improved with the quadruple mutant I268V/A272V/I310L/F313M[32] (**Figure 8b**). This remodeled receptor displayed a strong selective affinity for the modified ligand LG335. Its EC_{50} was over three orders of magnitude lower than that of 9cRA and in the nanomolar range. This optimized orthogonal ligand–receptor pair allows its use in many potential applications to better characterize transcriptional control by RXR.

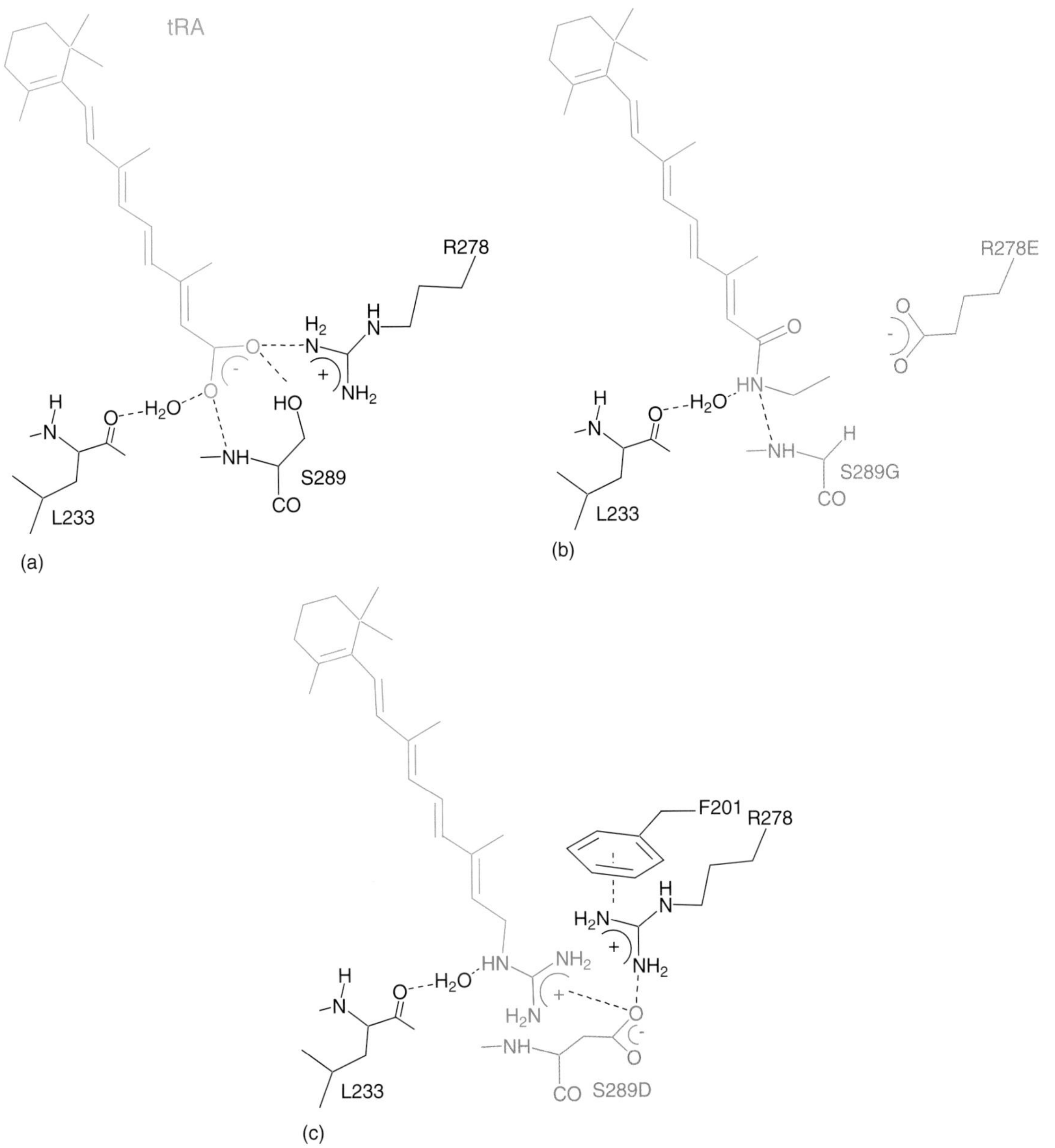

Figure 9 (a) Schematic representation of natural ligand all-*trans* retinoic acid (tRA, blue) bound to human retinoic acid receptors γ (RAR); only polar contacts with the acidic group of tRA are shown (based on 2LBD of the PDB). (b) *N*-ethyl amide of tRA, a neutral derivative of tRA (tRA, blue; *N*-ethyl amide function, green), bound to R278E/S289G RAR mutant (pink). (c) Cationic analog of tRA (blue) where the acidic function has been replaced by a guanidinium function (green), bound to S289D mutant (pink). The positive charge of R278 is neutralized by a putative interaction with D289 and a cation π-interaction with F201.

Doyle and Corey[30–32] demonstrated that the hydrophobic environment of a ligand of a specific type of nuclear receptors can be reshaped in order to tightly bind a chemically modified ligand, as described in **Figure 2**. On the other hand, Koh and his group showed that the polar and electrostatic interactions of the ligand–receptor interface of another close nuclear receptor could be remodeled, as described in **Figure 3**.[33] *Cis* and *trans* retinoic acid (9cRA and tRA, **Figures 8a** and **9a**) are made of a large hydrophobic moiety and a polar carboxylic acid group. In RXR, this acidic function is bound to the protein by means of a hydrogen bond network including a backbone NH, a glutamine residue, and water molecules, while in RAR, the binding of that functional group also involves an electrostatic interaction besides a hydrogen bond network (**Figures 8a** and **9a**). Each of the two polar residues binding tRA in RARγ were

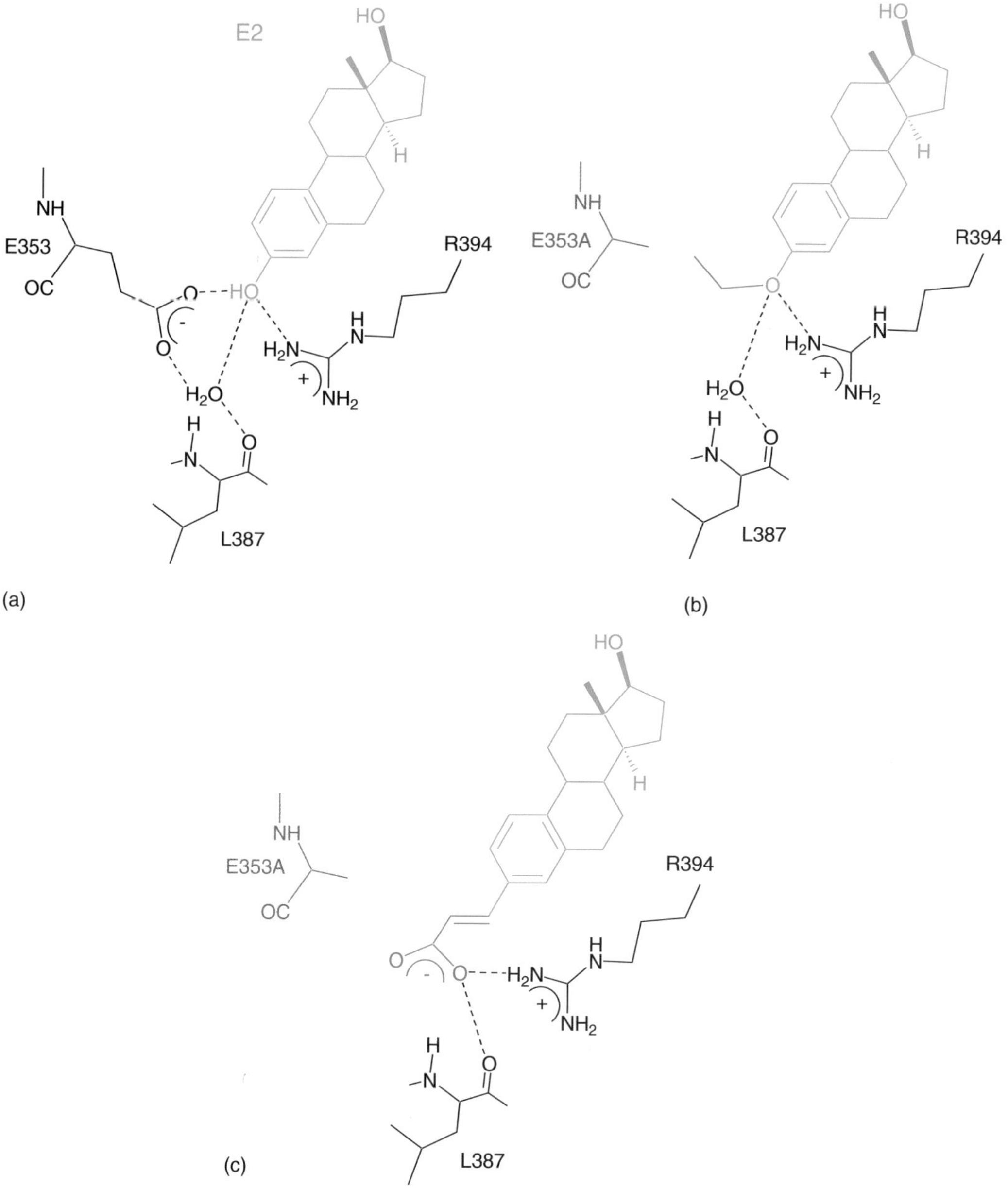

Figure 10 (a) Polar interactions between the phenol function of the estradiol (E2, blue) A ring and estrogen receptor ERα (based on 1ERE of the PDB). (b) Engineered hydrophobic interaction between ethylether (green) of E2 analog and E353A mutant (pink). (c) Intramolecular salt bridge between E353 and R394 in wild-type ERα replaced by an intermolecular electrostatic attraction between the acrylate substituent (green) of E2 and R294 of E353A (pink) mutant of ERα.

changed to an acidic residue in S289D and R278E/S289G mutants.[33] These modified receptors could be activated with significant selectivity over the wild-type receptor with two synthetic ligands bearing either a neutral or a basic functional group in place of the original acidic group of tRA (**Figure 9b,c**).[33] In the case of the S289D mutant, a basic residue R278 remains present at the binding site close to the guanidinium function of the modified tRA; its positive charge is probably neutralized by the protein environment, preventing an electrostatic repulsion with the ligand. However, this charge-reversed ligand–receptor pair displays reduced activity compared to the wild-type ligand–receptor pair. In general, it is observed that engineering ligand–receptor pairs by reversing charges (**Figure 3**) is a more difficult task than remodeling the hydrophobic interactions (**Figure 2**).[3,33]

Coordinated alteration of ligand–receptor pairs was also applied to estrogen receptors (ER).

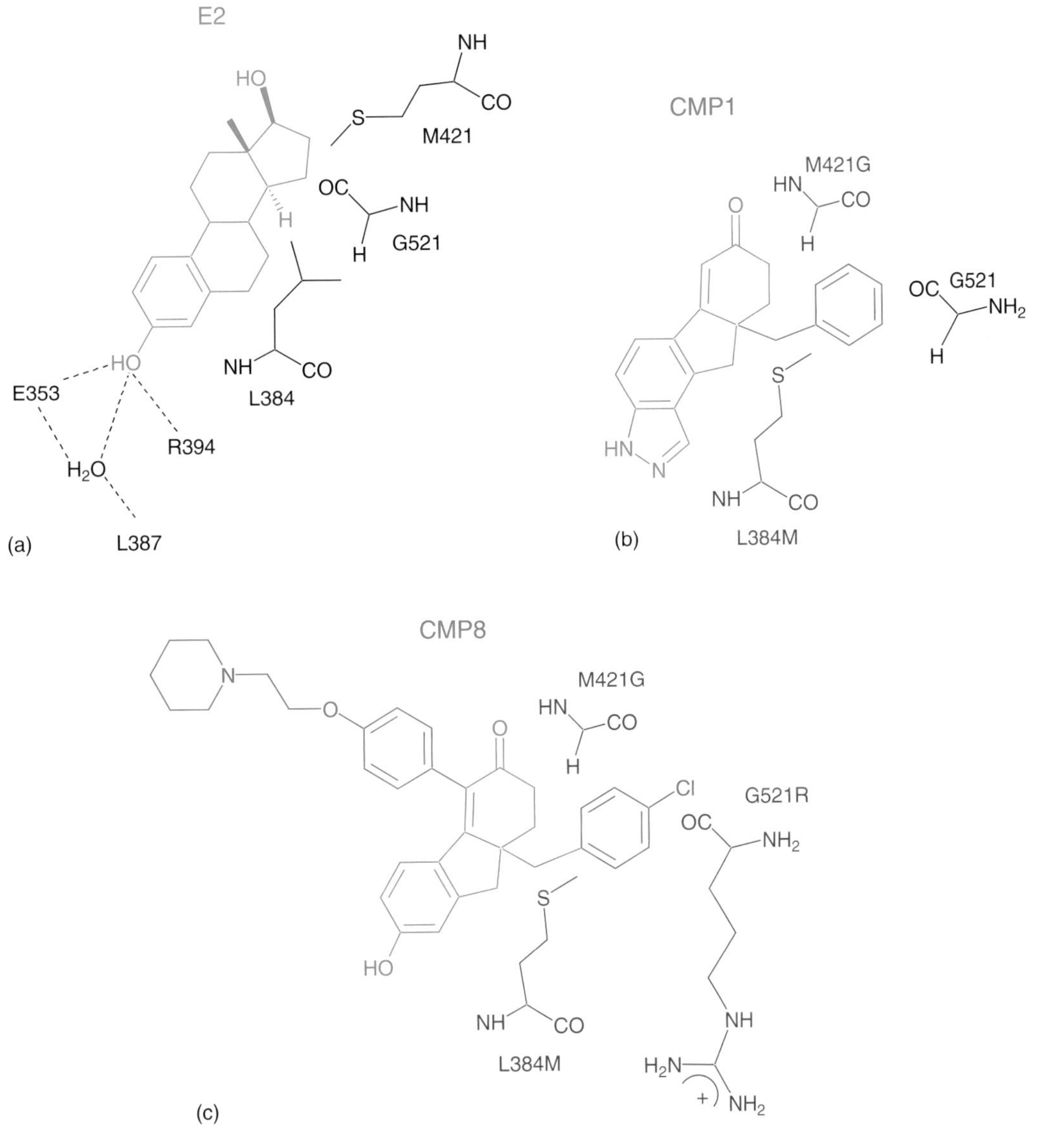

Figure 11 (a) Hydrophobic residues of ERα at the interface with estradiol (E2,blue); polar bonds from Figure 10a are recalled (based on 1ERE of the PDB). (b) 9a-Benzyl tetrahydrofluorenone (CMP1) bound to mutated ERα (L384M/M421G, pink). Tetrahydrofluorenone (blue) binds to wild-type ERα but not its 9a-benzyl (green) derivative; binding is made possible by creating a 'hole' in the protein with the L384M/M421G mutant. (c) Optimized orthogonal pair for ERα: CMP8 (central core, blue; substituents, green) bound to triple mutant L384M/M421G/G521R (pink).

Katzenellenbogen and co-workers reported very significant mismatches when binding residues or binding functional groups of the ligand E2 were each singly modified.[34] Yet the affinity of the modified ligands for their rematched receptors remained weaker than that of estradiol for the wild-type receptor ERα. The best coordinated changes were obtained when replacing the polar interaction between the phenol of estradiol and E353 (**Figure 10a**) by a hydrophobic interaction between A353 and an ethylether group of the new ligand (**Figure 10b**). However, Koh showed that the polar interaction network of the same complex (**Figure 10a**) could be remodeled by exchanging the polar/charged groups at the interface between ligand and binding site residues providing a very efficient rematch for both subtypes ERα and ERβ[35,36] (**Figure 10c**). While the polar group exchange appears to be associated with greater selectivity than neutral remodeling, it may also be unsuccessful as seen with C3-amine functionalized estrogen analogs.[36] Because of the important role of nuclear hormone receptors as transcriptional regulators, the design of orthogonal ligand–receptor pairs may allow some of these receptors to be specifically targeted and avoid interference

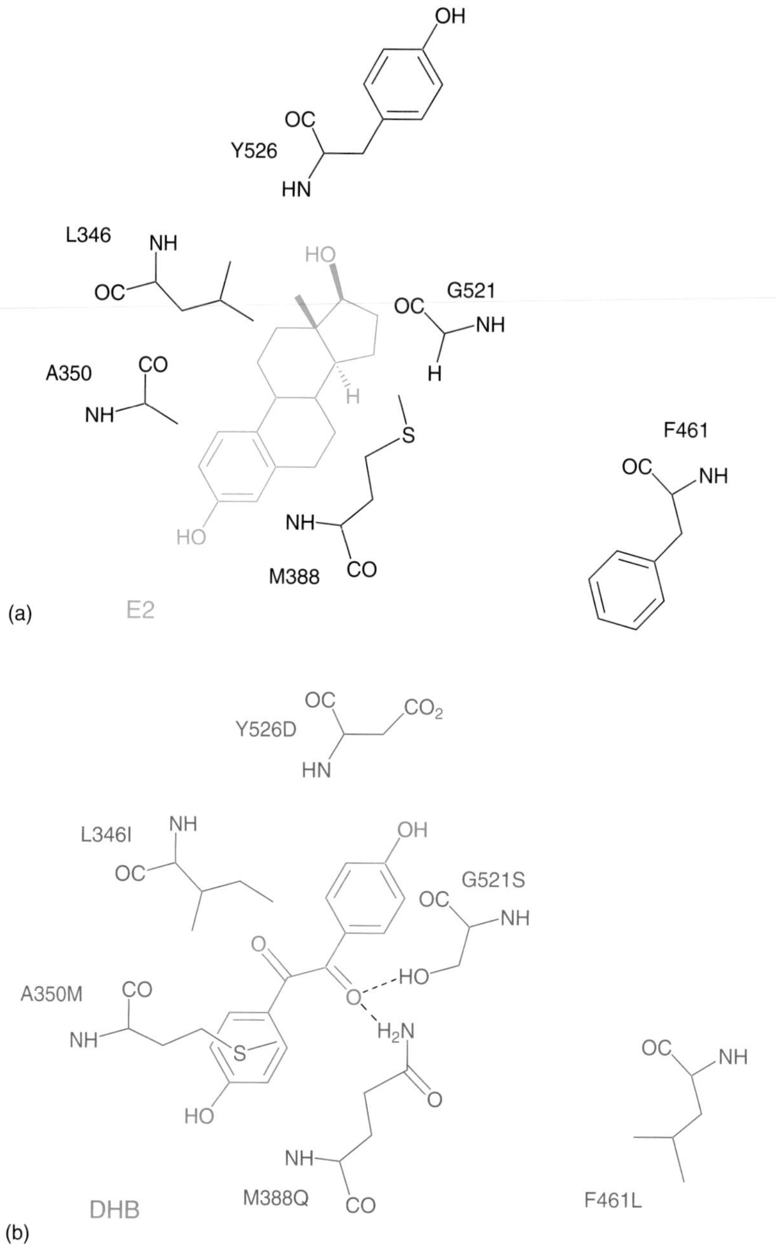

Figure 12 (a) Hydrophobic residues of ERα in the vicinity of estradiol (E2, blue) (based on 1GWR of the PDB). (b) Mutations (pink) of ERα providing the highest specificity for 4,4′-dihydroxybenzil (DHB, green).

with the numerous pathways activated by the same ligands. Thus, optimization of the concept for ER to make ligand-dependent transcription systems is still under investigation as illustrated by recent studies.[37,38] In order to modify the ERα/β-ligand complex to generate such a pair, Gallinari and collaborators[38] started their investigations with a selective ERβ agonist derivative (CMP1) that is inactive at both ERα and ERβ. While terahydrofluorenone binds tightly to ERβ, the benzyl substituent at position C9a of CMP1 prevents it from binding. A molecular modeling examination of the 3D-structure of CMP1 positioned at the ERα binding site allowed definition of a set of residues that may be responsible for the loss of binding. Subsequently, the authors generated a library of human ERα mutants of these residues and identified a double mutant (L384M/M421G) that restores CMP1 binding by creating a 'hole' that accommodates the benzyl 'bump' (**Figure 11a,b**). Yet, the binding of E_2 to that mutant was not totally impaired. A third mutation (G521R) abolished E_2 binding, but an additional substitution of the ligand (CMP8) was needed to

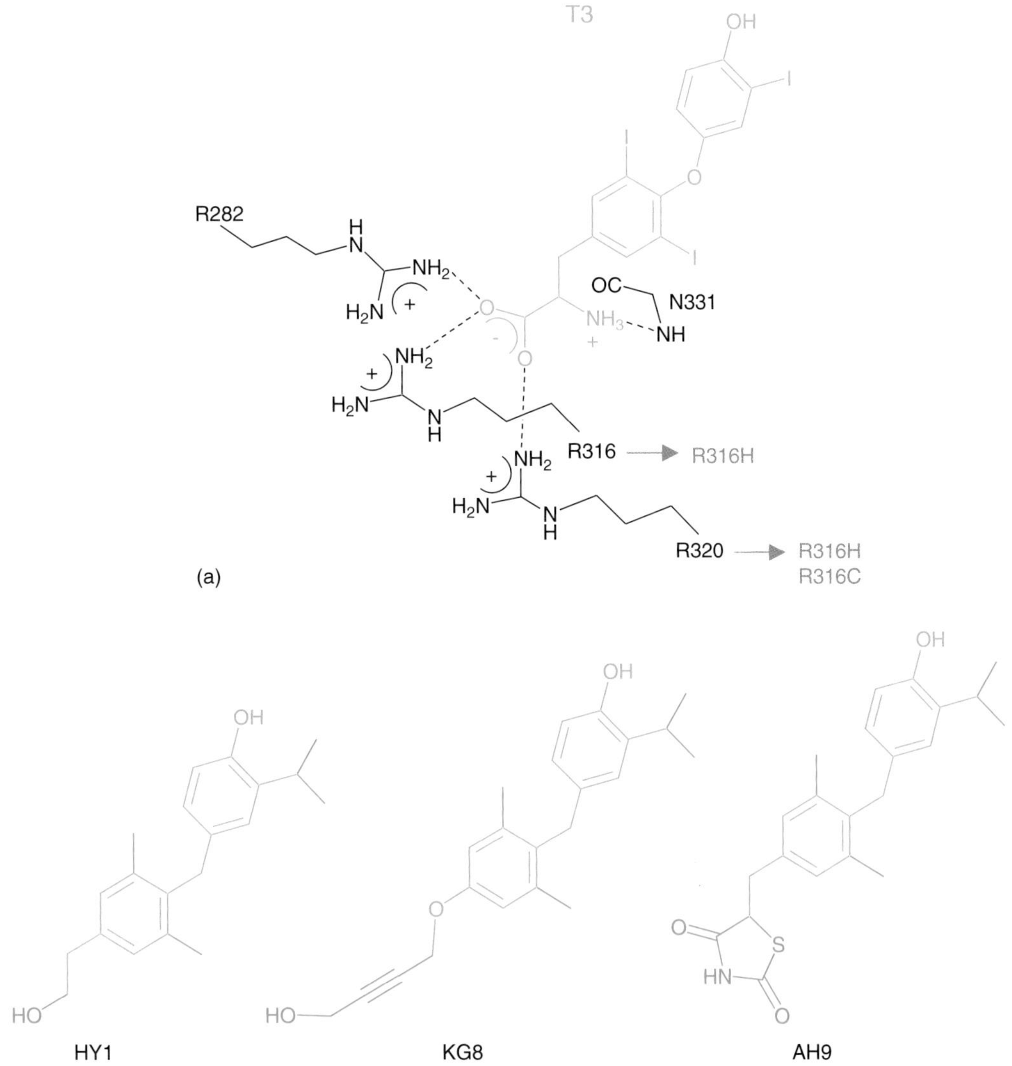

Figure 13 (a) Polar network around the amino acid part of triiodothyronine (T3, blue) bound to thyroid hormone receptor (TR) (based on 1BSX of the PDB). Three of the most common genetic mutations are indicated in pink. (b) Structure of T3 analogs (substituents colored in green) that preferentially activate these mutated receptors over wild-type TR.

confer the desired properties to the engineered pair (**Figure 11c**). The triple mutant was able to bind CMP8 with nanomolar affinity and not estradiol E_2 at physiological concentrations.[38] This orthogonal ligand–receptor pair was finally used to induce a selective transcription of a reporter gene in mammalian cells.[38]

Chockalingam and co-workers recently designed a strategy to engineer ERα to respond specifically to a selected synthetic ligand 4,4′-dihydroxybenzil (DHB).[37] As in previous investigations, the choice of residues to be mutated was initially guided by a 3D model of DHB positioned at the ERα binding site. A series of 14 residues contacting the ligand was selected to generate mutant libraries, which were tested with DHB in an automated screening system. Most selective mutants toward the target ligand and relative to the natural ligand E_2 were chosen to be further engineered by random mutagenesis to improve the selectivity. The two best mutants found (4S and 5E) included five (A350M, L346I, M388Q, G521S, Y526D) and seven (A350M, L346I, M388Q, G521S, Y526D, F461L, V560M) mutations (**Figure 12**). It should be noted that only four of these residues were part of the initial set of contacting residues (A350M, L346I, M388Q, G521S). These engineered receptors display an amazing specificity for DHB with nanomolar affinities and essentially no response with E_2.[37] The last two cases[37,38] that were presented show that access to large libraries and automated assays make possible the discovery of highly optimized orthogonal ligand–receptor pairs.

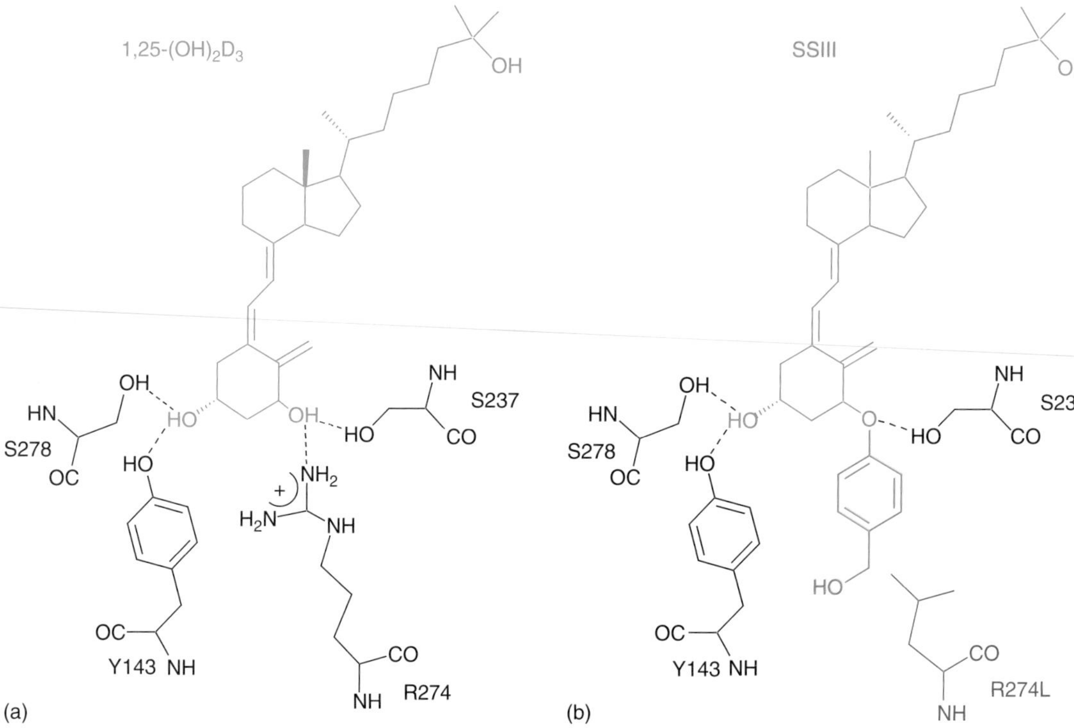

Figure 14 (a) Vitamin D_3 (1,25-$(OH)_2D_3$, blue) bound to vitamin D_3 receptor VDR (based on 1DB1 of the PDB). (b) R274L (pink) is a common mutation found in genetic diseases; the 4-hydroxymethylphenyl ether derivative of 1,25-$(OH)_2D_3$ allows a new hydrophobic interaction with the mutated residue and selective affinity of the altered ligand for the mutated receptor.

The above approach focused on finding a remodeled protein that would specifically bind to a particular synthetic ligand. In another approach, a complementary ligand searches for a given mutated protein. Koh and co-workers undertook this type of study using vitamin D receptors (VDRs)[42–44] and thyroid hormone receptors (TRs)[39–42] of the NHR family. Several genetic diseases are associated with genetic mutations that directly affect the binding of hormones to their NHRs. In those cases, the discovery of synthetic ligands that specifically bind and activate these receptors would allow the function of the impaired systems to be restored. Here, the orthogonal ligand–receptor pair concept finds a highly valuable therapeutic application.[42]

At the TR binding site, the endogenous ligand triiodothyronine (T3) is surrounded by hydrophobic residues in the central part and a polar network in the amino acid part (**Figure 13a**). The acidic function of T3 is anchored to the protein by interactions with three arginine residues (R282, R316, R320). The genetic mutations of these residues are associated with resistance to thyroid hormone (RTH), an autosomal dominant human genetic disease. The affinity of T3 for those mutants is reduced by 100- to 250-fold. While the mutants may still be activated by a higher concentration of T3, supraphysiological concentrations should be avoided because of undesirable cardiac side effects. Accordingly, thyromimics that activate those mutants and not other T3-activated systems should be promising therapeutic drugs. Based on 3D structure models, several T3 analogs have been designed and assayed in order to restore the thyroid function of the RTH mutants. In these thyromimics, the amino acid function is first replaced by a neutral alcohol function. Some of these alcohols (HY1 and KG8, **Figure 13b**) fully restore wild-type efficacy for three of the most common mutations associated with RTH (R316H, R320C, R320H). These T3 analogs are less potent than T3 for the wild-type receptor, but preferentially activate the mutated receptors.[39,41] When the amino acid moiety of T3 is replaced by a biosteric substitute of a carboxylate, more potent but less selective agonists (AH9) are obtained (**Figure 13b**).[40] This result shows that a negatively charged function on the ligand increases its affinity for both wild-type and mutants compared to a neutral function, and further remodeling of the ligand may lead to potent and mutant-selective agonists.

The engineering of the VDR–1,25$(OH)_2D_3$ pair provides another potential therapeutic application of the orthogonal ligand–receptor pair concept. Some genetic mutations of VDR residues at the binding site have been shown to be responsible for genetic diseases such as vitamin D_3-resistant rickets affecting bone development. Mutation of R274 to leucine is one of them (**Figure 14**). It results in a 1000-fold reduction of 1,25$(OH)_2D_3$ activity measured by an increase in EC_{50} from 2 nM to over 2000 nM in cellular assays.[43] The R274L mutation disrupts a hydrogen bond network that involves the guanidinium and phenol functions of R274 and $1,25(OH)_2D_3$ (**Figure 14**). The switch from a highly polar (R274) to a hydrophobic residue (L274) prompted Koh and co-workers to create a complementary hydrophobic 'bump' on the ligand side in order to restore a stabilizing interaction with L274 and consequently rescue the impaired functionality of the mutant. Among a series of substituted $1,25(OH)_2D_3$, SSIII best fulfilled the required properties. The EC_{50} of SSIII in cells expressing the VDR(R274) mutant was reduced to 6.8 nM compared to over 2000 nM for the endogenous ligand $1,25(OH)_2D_3$. Moreover, this synthetic analog did not affect other $1,25(OH)_2D_3$-activated systems in cells.[43] The discovery of mutation-compensating hormone analogs opens the way to possible therapeutic treatments and thus, reinforces interest in the orthogonal ligand–receptor pair concept.

The concept was also successfully applied to the ecdysteroid–ecdysone receptor (EcR) pair.[45] EcRs belonging to the superfamily of nuclear hormone receptors regulate the development and reproduction of insects. A single point mutation (A110P) abolished steroid binding; however, two nonsteroidal ligands were able to activate that mutant selectively. This remodeled pair may find applications in the investigation of insect cellular pathways and disease models.

3.11.3.6 G Protein-Coupled Receptors (GPCRs)

GPCRs make up one of the largest membrane-bound receptor families.[46] GPCRs trigger signaling pathways in response to the binding of numerous ligands. The diversity of these ligands ranges from photons, ions, odorants, small molecules as biogenic amines and amino acids, nucleotides, and peptides up to proteins. The transduction mechanism of GPCRs involves the catalysis of the GDP–GTP exchange on heterotrimeric G proteins and results in many different physiological responses. GPCRs represent the largest family targeted by drugs on the market and new tools to investigate their activation mechanism and in vivo function are still greatly desired. The orthogonal ligand–receptor

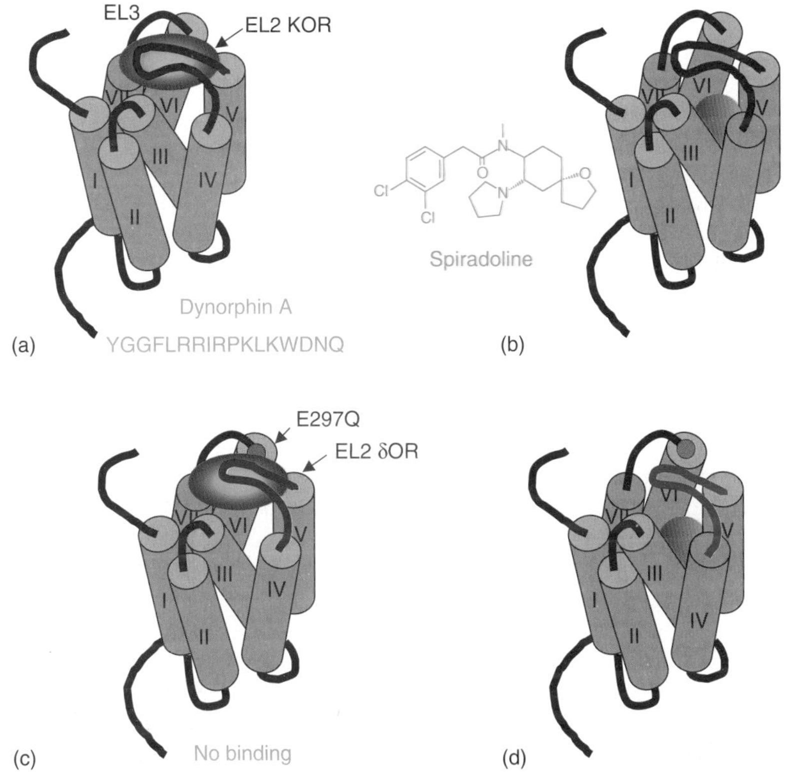

Figure 15 (a) Endogenous dynorphin A (sequence displayed, blue) bound to kappa opioid receptor (KOR). The seven transmembrane helices are represented as cylinders and numbered; extracellular loops EL2 and EL3 are indicated. (b) Synthetic agonist spiradoline (green) bound to KOR more deeply into the membrane than dynorphin A. (c) Exchange of EL2 from KOR with the one of δOR and E297Q mutant affords a protein that does not bind dynorphin. (d) However, spiradolin binds tightly to that mutant (named RASSL for 'receptor activated solely by synthetic ligands').

pair concept may prove to be quite useful in dissecting the multiple GPCR pathways activated by the same ligand, as seen with the other systems reviewed previously. The first attempt at applying the concept to GPCRs was performed by Conklin and co-workers using κ opioid receptors (KOR).[47] These authors later used the approach to control a specific physiological response in heart[48,49] and to differentiate receptors that have the same ligands and effectors, but differ only in desensitization and internalization.[50] The κ opioid receptors may be activated by endogenous peptide ligands such as dynorphin and also by small-molecule ligands such as spiradoline, which are structurally distinct from these (**Figure 15**). Moreover, it was shown that they bind to a different pocket of the seven transmembrane domain, which is common to all GPCRs (**Figure 15a,b**). Thus, engineering the KOR at residues in contact with dynorphin but

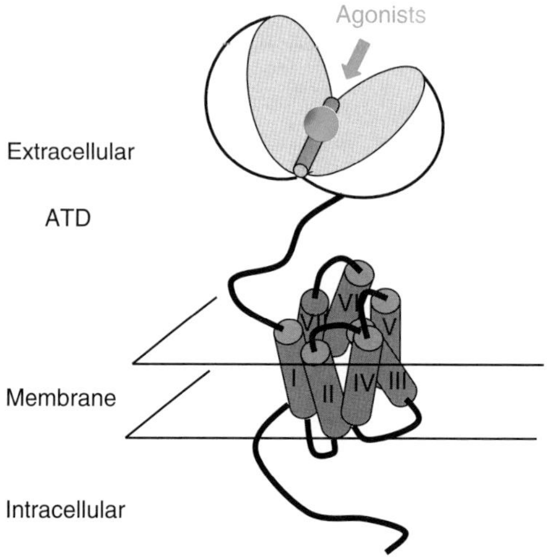

Figure 16 Schematic representation of family C GPCRs. The large extracellular ATD folds into two lobes linked by a hinge. Agonists bind to that domain. The seven transmembrane domain is represented by numbered cylinders.

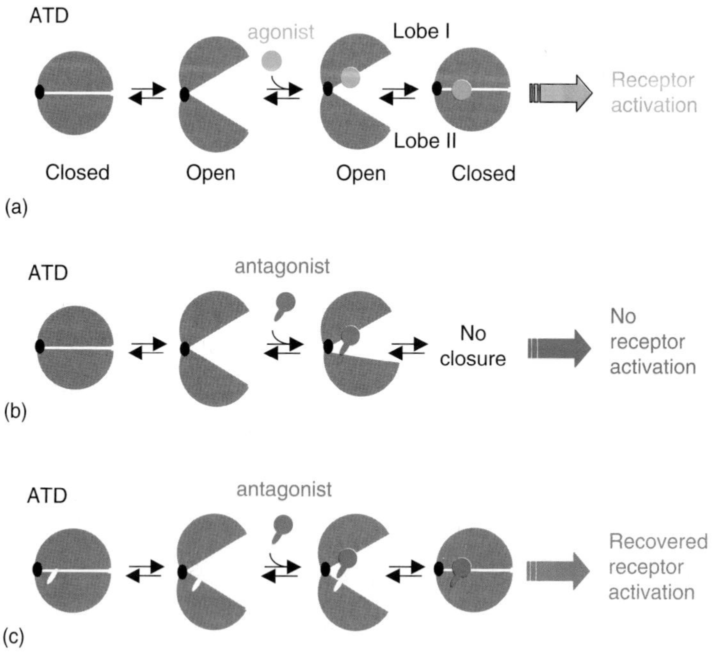

Figure 17 (a) Equilibrium between open and closed conformations of the ATD of family C GPCRs. Binding of the agonist stabilizes the closed conformation, which is supposed to trigger receptor activation. (b) Competitive antagonists prevent ATD closing and subsequent activation. (c) When the ATD is engineered in order to recover a complete closure of the domain, activation of the receptor is observed with the initially competitive antagonists.

not spiradoline (the second extracellular loop of KOR swapped with that of delta opioid receptor and E297Q mutation, **Figure 15c,d**) resulted in a receptor with 1/2000 of the response to dynorphin and unaltered response to spiradoline.[47] Conklin named that modified receptor as 'receptor activated solely by synthetic ligands' (RASSLs).[47] Because synthetic ligands activate both wild-type and mutated receptors, there is not exactly an orthogonal ligand–receptor pair situation where the synthetic ligand should theoretically have no effect on wild-type protein. Yet, an analogous situation is met when using KOR knockout mice where this receptor has been eliminated and thus cannot be activated by synthetic ligands.[48,49] When the RASSLs were expressed in specific tissues, the controlled response allowed the physiological role of the initial receptor in those tissues to be clarified.[48,49] The concept of RASSL has been adapted to serotonin receptors(5HT$_4$),[51] α_{2A}-adrenoceptors,[52] and histamine H$_1$ receptors.[53]

The difficulty in engineering proteins from family1 and 2 of a GPCRs[54] results from the lack of a crystal structure of the transmembrane domain and consequently of accurate 3D models. Bovine rhodopsin is the only structure available, being in the inactive state of the domain. However, for family 3/C of GPCRs,[55] the binding domain of endogenous

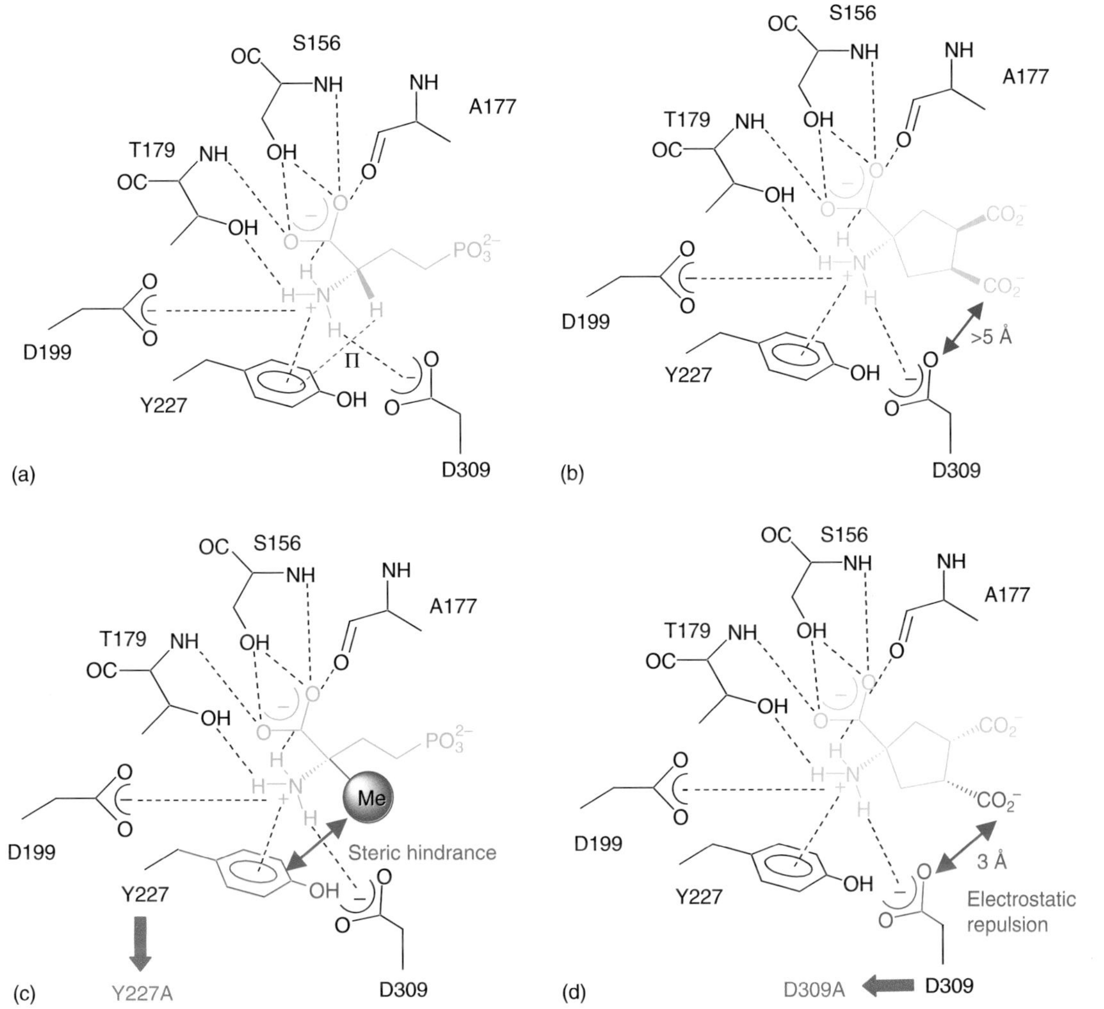

Figure 18 (a) The fine tuning of (*S*)-AP4 (2-amino-4-phosphonobutyric acid, blue) binding at mGlu8 binding site in the closed conformation. The closing angle of the ATD is such that a CH-π interaction takes place between the α-proton of the amino acid and Tyr227. (b) Similar binding of ACPT-I (1-aminocyclopentane-1,3,4-tricarboxylic acid-I) at mGlu8 binding site. The C4-carboxylate of ACPT-I is at such a distance from D309 that no repulsion is observed. (c) When the alpha-proton of (*S*)-AP4 is replaced by a methyl group (MAP4), a steric hindrance with Tyr227 prevents the ATD from closing and no activation of the receptor is detected. If Y227 is mutated to A227, steric hindrance is abolished and activation of mGlu8 is triggered by MAP4. (d) The C4-carboxylate of ACPT-II is in the opposite configuration compared to ACPT-I; consequently, an electrostatic repulsion with D309 takes place upon closing of the domain. If D309 is mutated to A309, repulsion is abolished and activation by ACPT-II is observed.

ligands is not found in the transmembrane domain but in the large extracellular amino terminal domain (ATD), which folds into two lobes connected by a hinge, as demonstrated by an x-ray structure of that domain (illustrated in **Figure 16**).[55] Activation of GPCR family C receptors was proposed to be triggered by the closed conformation of the ATD trapping the endogenous agonists. Competitive antagonists would block that activation by preventing closure of the two lobes of the ATD (**Figure 17a,b**). A close examination of the 3D models of that domain may indicate which residues prevent the ATD from closing and subsequent protein engineering may allow to recover the domain closure (**Figure 17c**). This was actually achieved with metabotropic glutamate receptors (mGluRs) of familyC GPCRs, for which homology models of all subtype binding domains are available.[56] Two types of antagonists were chosen for this study: one for which the inactivation was due to steric hindrance, as shown in **Figure 2b**; and one for which the inactivation was due to a negative charge repulsion, as shown in **Figure 3c**.[57] All mGluRs are activated by endogenous glutamate; however, several subtype-selective synthetic agonists/antagonists are known. (*S*)-AP4 and ACPT-I are competitive agonists of the mGlu4/6/7/8 receptors while (*S*)-MAP4 and ACPT-II are competitive antagonists of these receptors. Careful examination of homology models of the mGlu8R binding site docked with (*S*)-AP4, ACPT-I, and other agonists reveals that the glycine moiety of all agonists is finely fitted to that site to allow the bilobate domain to close at a specific angle (**Figure 18a,b**). It should be noted in particular that the α-proton of that moiety makes a CH-π interaction with a conserved tyrosine Y227. When this proton is substituted by a more bulky group such as a methyl group, closure to the required angle cannot be reached because of steric hindrance between the substituent and the tyrosine residue (**Figure 18c**). Yet if the tyrosine is mutated to an alanine (Y227A), the steric hindrance is abolished and complete closure may be achieved.[57] It should also be noted that the carboxylate at position 4 in the cyclopentyl ring of ACPT-I does not interact with the negative charge of D309, a conserved aspartate that binds to the amino function of agonists (**Figure 18b**). However, with ACPT-II, an ACPT-I diastereoisomer with opposite configuration at C3 and C4, this acidic group at C4 in ACPT-II comes in electrostatic repulsion with D309 upon closing of the ATD lobes and activation of mGlu8 is not observed (**Figure 18d**). If this repulsion is abolished by mutation of D309 to alanine, activation of the mutated mGlu8 receptor is recovered.[57] These results demonstrated unambiguously that a closed conformation of the ATD of family C GPCRs is required for receptor activation. Thus, in this case, the orthogonal ligand–receptor pair concept was a key tool for investigating mechanism of action.

3.11.4 Conclusion

The orthogonal ligand–receptor pair approach has been successfully used in a large number of biological systems. In most cases, the design of complementary remodeling of both the small molecule and the protein relied on crystal structure or homology models of the bound system. New tools such as library screening, now open larger prospects as seen with recent studies. The approach finds unique applications when investigating the role of a family of enzymes/receptors that are activated by the same substrate/ligand. It has helped greatly to gain new insights into various signaling pathways and is, indeed, one of the useful tools in the field of drug discovery. Moreover, an important application may be found with diseases that are linked to genetic mutations that alter the binding of a substrate or ligand to its protein. In such cases, a modification of the small molecule may allow selective activation of the defective protein.

References

1. Clackson, T. *Curr. Opin. Struct. Biol.* **1998**, *8*, 451–458.
2. Bishop, A.; Buzko, O.; Heyeck-Dumas, S.; Jung, I.; Kraybill, B.; Liu, Y.; Shah, K.; Ulrich, S.; Witucki, L.; Yang, F. et al. *Annu. Rev. Biophys. Biomol. Struct.* **2000**, *29*, 577–606.
3. Koh, J. T. *Chem. Biol.* **2002**, *9*, 17–23.
4. Shokat, K.; Velleca, M. *Drug Disc. Today* **2002**, 7, 872–879.
5. Chockalingam, K.; Zhao, H. *Trends Biotechnol.* **2005**, *23*, 333–335.
6. Shah, K. *IUBMB Life* **2005**, *57*, 397–405.
7. Hwang, Y. W.; Miller, D. L. *J. Biol. Chem.* **1987**, *262*, 13081–13085.
8. Belshaw, P. J.; Schoepfer, J. G.; Liu, K.-Q.; Morrison, K. L.; Schreiber, S. L. *Angew. Chem. Int. Ed.* **1995**, *34*, 2129–2132.
9. Belshaw, P. J.; Schreiber, S. L. *J. Am. Chem. Soc.* **1997**, *119*, 1805–1806.
10. Bishop, A. C.; Buzko, O.; Shokat, K. M. *Trends Cell Biol.* **2001**, *11*, 167–172.
11. Shah, K.; Liu, Y.; Deirmengian, C.; Shokat, K. M. *Proc. Natl. Acad. Sci. USA* **1997**, *94*, 3565–3570.
12. Liu, Y.; Shah, K.; Yang, F.; Witucki, L.; Shokat, K. M. *Chem. Biol.* **1998**, *5*, 91–101.
13. Liu, Y.; Shah, K.; Yang, F.; Witucki, L.; Shokat, K. M. *Bioorg. Med. Chem.* **1998**, *6*, 1219–1226.
14. Hindley, A. D. *FEBS Lett.* **2004**, *556*, 26–34.
15. Ubersax, J. A.; Woodbury, E. L.; Quang, P. N.; Paraz, M.; Blethrow, J. D.; Shah, K.; Shokat, K. M.; Morgan, D. O. *Nature* **2003**, *425*, 859–864.

16. Hanke, J. H.; Gardner, J. P.; Dow, R. L.; Changelian, P. S.; Brissette, W. H.; Weringer, E. J.; Pollok, B. A.; Connelly, P. A. *J. Biol. Chem.* **1996**, *271*, 695–701.
17. Bishop, A. C.; Shah, K.; Liu, Y.; Witucki, L.; Kung, C.; Shokat, K. M. *Curr. Biol.* **1998**, *8*, 257–266.
18. Bishop, A. C.; Kung, C.-y.; Shah, K.; Witucki, L.; Shokat, K. M.; Liu, Y. *J. Am. Chem. Soc.* **1999**, *121*, 627–631.
19. Liu, Y.; Bishop, A.; Witucki, L.; Kraybill, B.; Shimizu, E.; Tsien, J.; Ubersax, J.; Blethrow, J.; Morgan, D. O.; Shokat, K. M. *Chem. Biol.* **1999**, *6*, 671–678.
20. Bishop, A. C.; Ubersax, J. A.; Petsch, D. T.; Matheos, D. P.; Gray, N. S.; Blethrow, J.; Shimizu, E.; Tsien, J. Z.; Schultz, P. G.; Rose, M. D. et al. *Nature* **2000**, *407*, 395–401.
21. Chen, X.; Ye, H.; Kuruvilla, R.; Ramanan, N.; Scangos, K. W.; Zhang, C.; Johnson, N. M.; England, P. M.; Shokat, K. M.; Ginty, D. D. *Neuron* **2005**, *46*, 13–21.
22. Gallion, S. L.; Qian, D. *Curr. Opin. Drug Disc. Devel.* **2005**, *8*, 638–645.
23. Alaimo, P. J.; Knight, Z. A.; Shokat, K. M. *Bioorg. Med. Chem.* **2005**, *13*, 2825–2836.
24. Zhang, C.; Kenski, D. M.; Paulson, J. L.; Bonshtien, A.; Sessa, G.; Cross, J. V.; Templeton, D. J.; Shokat, K. M. *Nat. Methods* **2005**, *2*, 435–441.
25. Gillespie, P. G.; Gillespie, S. K.; Mercer, J. A.; Shah, K.; Shokat, K. M. *J. Biol. Chem.* **1999**, *274*, 31373–31381.
26. Holt, J. R.; Gillespie, S. K.; Provance, D. W.; Shah, K.; Shokat, K. M.; Corey, D. P.; Mercer, J. A.; Gillespie, P. G. *Cell* **2002**, *108*, 371–381.
27. Provance, D. W. J.; Gourley, C. R.; Silan, C. M.; Cameron, L. C.; Shokat, K. M.; Goldenring, J. R.; Shah, K.; Gillespie, P. G.; Mercer, J. A. *Proc. Natl. Acad. Sci. USA* **2004**, *101*, 1868–1873.
28. Kapoor, T. M.; Mitchison, T. J. *Proc. Natl. Acad. Sci. USA* **1999**, *96*, 9106–9111.
29. Weatherman, R. V.; Fletterick, R. J.; Scanlan, T. S. *Annu. Rev. Biochem.* **1999**, *68*, 559–581.
30. Peet, D. J.; Doyle, D. F.; Corey, D. R.; Mangelsdorf, D. J. *Chem. Biol.* **1998**, *5*, 13–21.
31. Doyle, D. F.; Braasch, D. A.; Jackson, L. K.; Weiss, H. E.; Boehm, M. F.; Mangelsdorf, D. J.; Corey, D. R. *J. Am. Chem. Soc.* **2001**, *123*, 11367–11371.
32. Schwimmer, L. J.; Rohatgi, P.; Azizi, B.; Seley, K. L.; Doyle, D. F. *Proc. Natl. Acad. Sci. USA* **2004**, *101*, 14707–14712.
33. Koh, J. T.; Putnam, M.; Tomic-Canic, M.; McDaniel, C. M. *J. Am. Chem. Soc.* **1999**, *121*, 1984–1985.
34. Tedesco, R.; Thomas, J. A.; Katzenellenbogen, B. S.; Katzenellenbogen, J. A. *Chem. Biol.* **2001**, *8*, 277–287.
35. Shi, Y.; Koh, J. T. *Chem. Biol.* **2001**, *8*, 501–510.
36. Shi, Y.; Koh, J. T. *J. Am. Chem. Soc.* **2002**, *124*, 6921–6928.
37. Chockalingam, K.; Chen, Z.; Katzenellenbogen, J. A.; Zhao, H. *Proc. Natl. Acad. Sci. USA* **2005**, *102*, 5691–5696.
38. Gallinari, P.; Lahm, A.; Koch, U.; Paolini, C.; Nardi, M. C.; Roscilli, G.; Kinzel, O.; Fattori, D.; Muraglia, E.; Toniatti, C. et al. *Chem. Biol.* **2005**, *12*, 883–893.
39. Ye, H. F.; O'Reilly, K. E.; Koh, J. T. *J. Am. Chem. Soc.* **2001**, *123*, 1521–1522.
40. Hashimoto, A.; Shi, Y.; Drake, K.; Koh, J. T. *Bioorg. Med. Chem.* **2005**, *13*, 3627–3639.
41. Shi, Y.; Ye, H.; Link, K. H.; Putnam, M. C.; Hubner, I.; Dowdell, S.; Koh, J. T. *Biochemistry* **2005**, *44*, 4612–4626.
42. Koh, J. T.; Biggins, J. B. *Curr. Top. Med. Chem.* **2005**, *5*, 413–420.
43. Swann, S. L.; Bergh, J. J.; Farach-Carson, M. C.; Koh, J. T. *Org. Lett.* **2002**, *4*, 3863–3866.
44. Swann, S. L.; Bergh, J.; Farach-Carson, M. C.; Ocasio, C. A.; Koh, J. T. *J. Am. Chem. Soc.* **2002**, *124*, 13795–13805.
45. Kumar, M. B.; Fujimoto, T.; Potter, D. W.; Deng, Q.; Palli, S. R. *Proc. Natl. Acad. Sci. USA* **2002**, *99*, 14710–14715.
46. Kristiansen, K. *Pharmacol. Ther.* **2004**, *103*, 21–80.
47. Coward, P.; Wada, H. G.; Falk, M. S.; Chan, S. D.; Meng, F.; Akil, H.; Conklin, B. R. *Proc. Natl. Acad. Sci. USA* **1998**, *95*, 352–357.
48. Redfern, C. H.; Coward, P.; Degtyarev, M. Y.; Lee, E. K.; Kwa, A. T.; Hennighausen, L.; Bujard, H.; Fishman, G. I.; Conklin, B. R. *Nat. Biotechnol.* **1999**, *17*, 165–169.
49. Scearce-Levie, K.; Coward, P.; Redfern, C. H.; Conklin, B. R. *Trends Pharmacol. Sci.* **2001**, *22*, 414–420.
50. Scearce-Levie, K.; Lieberman, M. D.; Elliott, H. H.; Conklin, B. R. *BMC Biol.* **2005**, *3*, 3.
51. Claeysen, S.; Joubert, L.; Sebben, M.; Bockaert, J.; Dumuis, A. *J. Biol. Chem.* **2003**, *278*, 699–702.
52. Pauwels, P. J. *Trends Pharmacol. Sci.* **2003**, *24*, 504–507.
53. Bruysters, M.; Jongejan, A.; Akdemir, A.; Bakker, R. A.; Leurs, R. *J. Biol. Chem.* **2005**, *280*, 34741–34746.
54. Bockaert, J.; Pin, J.-P. *EMBO J.* **1999**, *18*, 1723–1729.
55. Pin, J.-P.; Galvez, T.; Prézeau, L. *Pharmacol. Ther.* **2003**, *98*, 325–354.
56. Bertrand, H.-O.; Bessis, A.-S.; Pin, J.-P.; Acher, F. *J. Med. Chem.* **2002**, *45*, 3171–3183.
57. Bessis, A.-S.; Rondard, P.; Gaven, F.; Brabet, I.; Triballeau, N.; Prézeau, L.; Acher, F.; Pin, J.-P. *Proc. Natl. Acad. Sci. USA* **2002**, *99*, 11097–11102.
58. Protein Data Bank (PDB). http://www.rcsb.org/pdb/ (accessed Mar 2006).

Biography

Francine C Acher is currently a CNRS Research Director in the Biomedical Institute of the University Paris-V (France). She received her PhD in Chemistry from the University of Paris-VI in 1984 and carried out her postdoctoral research in the Department of Chemistry at University of California–Berkeley with Prof P A Bartlett. Her research focused on structure/function studies and drug design at the interface of chemistry and biology. Specific issues were addressed using chemical tools (synthetic chemistry, molecular modeling), molecular biology, and pharmacology. Her interdisciplinary research was made possible by close collaborations with Accelrys Inc., Dr J-P Pin's research group (Montpellier, France), and more recently Prof J Ngai's group (University of California–Berkeley).

Comprehensive Medicinal Chemistry II
ISBN (set): 0-08-044513-6

ISBN (Volume 3) 0-08-044516-0; pp. 215–234

3.12 Chemoinformatics

V J Gillet, University of Sheffield, Sheffield, UK
A R Leach, GlaxoSmithKline, Stevenage, UK

3.12.1	**Introduction**	**235**
3.12.2	**Representations of Molecules**	**236**
3.12.2.1	Structure and Substructure Searching	236
3.12.3	**Three-Dimensional Structures**	**240**
3.12.3.1	Three-Dimensional Searching	241
3.12.3.2	Pharmacophore Identification	242
3.12.4	**Virtual Screening**	**244**
3.12.4.1	Similarity Searching	245
3.12.4.1.1	Two-dimensional similarity measures	246
3.12.4.1.2	Three-dimensional similarity methods	247
3.12.4.1.3	Evaluation of similarity methods	248
3.12.4.2	Machine Learning Methods	249
3.12.4.3	Structure-Based Virtual Screening	250
3.12.5	**Diversity Analysis**	**252**
3.12.5.1	Dissimilarity-Based Compound Selection	253
3.12.5.2	Cluster-Based Compound Selection	253
3.12.5.3	Partitioning Methods	255
3.12.5.3.1	Pharmacophore fingerprints	255
3.12.5.4	Optimization Methods	256
3.12.6	**Combinatorial Library Design**	**256**
3.12.7	**Prediction of Absorption, Distribution, Metabolism, and Excretion Properties**	**257**
3.12.7.1	Computational Filters	258
3.12.7.2	Prediction of Physicochemical Properties	259
3.12.8	**Conclusions**	**260**
References		**260**

3.12.1 Introduction

The term chemoinformatics first appeared in the literature in 1998 when Brown described it as "The mixing of information resources to transform data into information, and information into knowledge, for the intended purpose of making better decisions faster in the arena of drug lead identification and optimisation."[1] The term was introduced following the explosion of data about compounds and their properties that began to occur in the early to mid 1990s due to the introduction of automation techniques both for the parallel synthesis of compounds and for high-throughput screening (HTS). However, despite the recent emergence of the term, many of the techniques that are encompassed by it have been available for many years, as summarized by Hann and Green in their paper entitled "Chemoinformatics – a new name for an old problem."[2]

Chemoinformatics is now considered to be well established as a discipline, even though the exact term used is still under dispute (the terms 'cheminformatics' and 'chemical informatics' are also in common use). The first university masters course in the subject appeared in 2000 at the University of Sheffield, UK and was closely followed by a similar course at the University of Manchester Institute of Science and Technology (UMIST) and at Indiana University in the

US[3]; several text books have now appeared with chemoinformatics in the title[4–7]; and most of the major pharmaceutical companies now have chemoinformatics departments.

The precise interpretation of the discipline varies but core topics include techniques for the management of databases of chemical structures; storage and retrieval of chemical structure information; structure property prediction methods; virtual screening methods, i.e., the prioritization of compounds for high-throughput screening; similarity and diversity analysis; and the design of combinatorial libraries. Additional topics include reaction databases, patent literature, database design, web applications, and even laboratory information management systems (LIMS).[8] Therefore, a chemoinformatics specialist working in the pharmaceutical sector requires knowledge of techniques from a variety of disciplines, for example, data mining and statistical methods are widely applied, some knowledge of the biological sciences and bioinformatics is desirable, and database and web technologies are also important. However, the definition of chemoinformatics in its broadest sense now extends beyond the context of drug discovery and can include the application of computational methods to all kinds of chemical data such as agrochemicals, foodstuffs, catalysts, and specialist materials.

The focus in this chapter is on techniques that are commonly applied in the drug and agrochemical discovery processes and, in particular, those methods that are suitable for the processing of large databases of compounds. The initial focus is on the representation of structures in databases and searching techniques since many methods in chemoinformatics build on these basic techniques. Subsequent sections of the chapter focus on the analysis of chemical information in order to derive knowledge that can be used to guide drug discovery.

3.12.2 Representations of Molecules

Chemists communicate about chemical structures using structure diagrams. While many software packages exist for entering structures so that they can be viewed on the screen and subsequently imported into documents, the images created by these drawing packages are not suitable for database applications. Various systems for chemical nomenclature have been developed that allow structures to be stored as text; however, the naming conventions tend to be rather complex and are therefore difficult to apply. **Figure 1a** shows the structure of aspirin and **Figure 1b** illustrates a systematic name for aspirin. Line notations have also been developed to provide a concise representation of structures based on alphanumeric characters and can be used to communicate structures, to enter structures into databases or structure drawing packages, or to transfer compounds between systems. The most well known line representation is SMILES (simplified molecular input line entry specification).[9,10] A SMILES representation of aspirin is shown in **Figure 1c**. Although line notations are widely used they must be converted into chemically aware computer-readable representations to allow structural searches to be carried out.

The specialized nature of chemical structural data means that specialized software is required in order to be able to store and search for it within databases. Such computer-based methods for processing databases of structures were first developed in the 1960s. Nowadays, a typical corporate database could consist of 10^6 compounds and the largest publicly available database contains over 26 million small organic and inorganic compounds.[11] Current systems are able to search for structures and data in these databases and have the results returned within seconds. These databases contain compounds that are real in the sense that they have already been synthesized. A more recent development in chemoinformatics involves the creation of databases containing 'virtual' molecules. These are compounds that do not exist yet but which could be synthesized readily, often using combinatorial chemistry techniques. The sizes of these virtual collections can run to billions of structures.

3.12.2.1 Structure and Substructure Searching

The simplest type of search that a chemist may wish to carry out is to find information about a particular compound. For example, has the compound been synthesized already, either within the company or elsewhere, what quantities (if any) are available within the company stores, and what properties have been measured for it. This type of query is known as an exact search. Substructure searching, on the other hand, provides a more general type of search in which a chemist may search for related compounds of interest. It is used to find all compounds within a database that contain a given substructure, for example, all compounds that contain a penicillin ring system.

Most database search systems encode chemical structures as molecular graphs and make use of graph theory to provide tools for searching the databases. A graph is a mathematical object consisting of nodes that are connected by edges. In a molecular graph, the nodes correspond to the atoms and the edges correspond to the bonds. The graphs are typically labeled: the nodes are labeled by atom type, for example, carbon, oxygen, nitrogen etc.; and the edges are

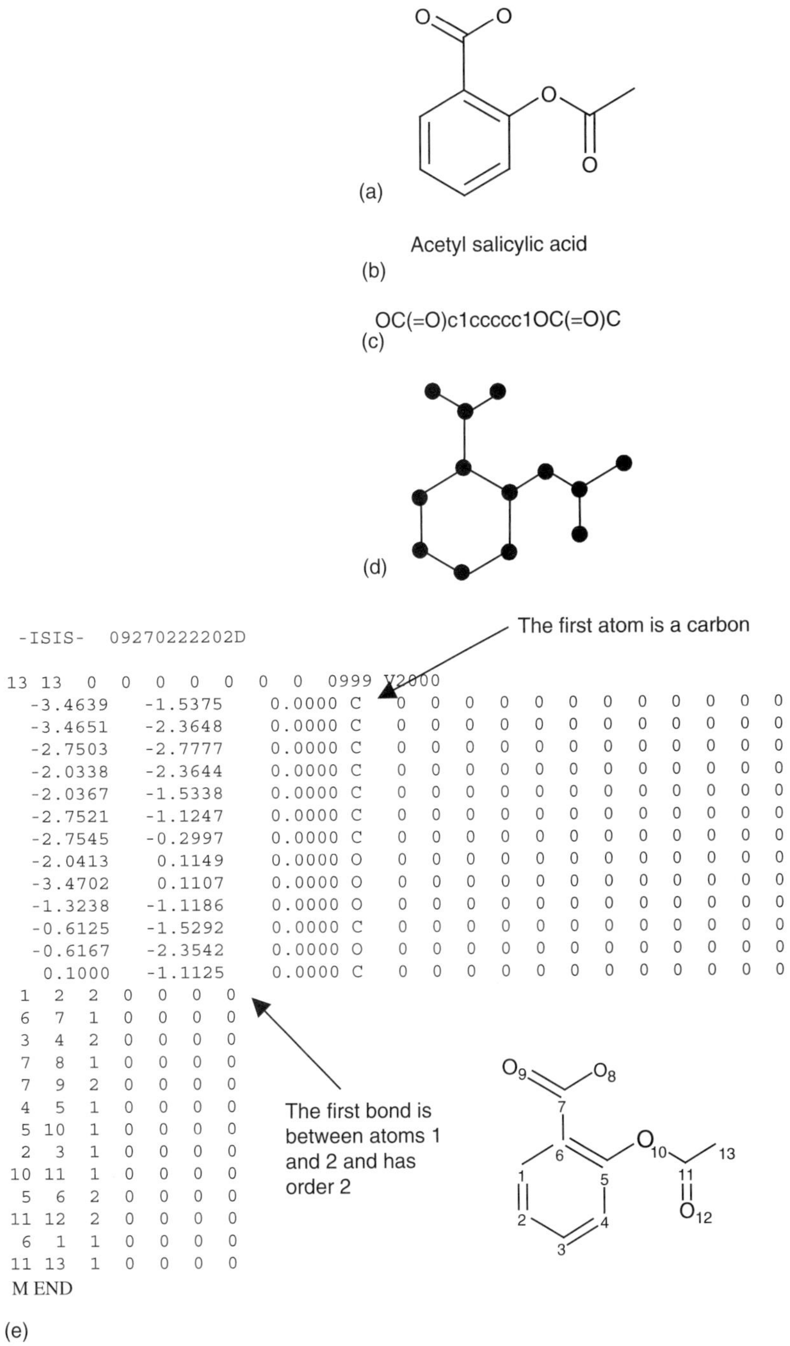

Figure 1 Different representations of aspirin: (a) structure diagram; (b) systematic name; (c) SMILES notation; (d) graph representation; and (e) connection table in MDL format.

labeled by bond type, for example, single, double, triple, or aromatic. The graph representation of aspirin is shown in **Figure 1d**.

A molecular graph is represented internally within a computer as a connection table that records the atoms present in the molecule, together with any associated properties, and the bonds between the atoms. Connection tables are usually hydrogen suppressed, which means that the hydrogen atoms are omitted; they can, however, be easily inferred from chemical valency rules. They can also be redundant, whereby each bond is listed twice, for example, the bond between atoms a and b is associated with both atom a and again with atom b, or nonredundant in which each bond is given once only. Connection tables can include x, y coordinates of the atoms, which enable the graph to be drawn on a page or screen as the familiar chemical structure diagram. Many different connection table formats exist. **Figure 1e** illustrates a connection table representation for aspirin in MDL format.[12]

Exact structure searching is made complex by the fact that there are many different ways in which the atoms in a connection table can be ordered. For a structure consisting of N atoms, there are $N!$ different numberings that are possible; therefore, comparing two connection tables to see if they represent the same structure is a nontrivial task. However, an algorithm to generate a unique numbering of a structure, also known as a canonical representation, was described by Morgan as long ago as 1965.[13] The algorithm is based on calculating the connectivities of atoms in an iterative fashion in an attempt to discriminate the atoms as far as possible. In the first iteration, the connectivity of an atom is the sum of its immediate neighbors. In subsequent iterations, a new connectivity value is calculated for each atom by summing the connectivities of its neighbors. This process continues until the number of distinct connectivity values has reached a maximum. The atom with maximum connectivity is then listed first in the connection table. Its neighbors are listed next in order of decreasing connectivity, then their neighbors are listed, and so on. Various rules have been devised to resolve ties when two atoms have the same connectivity values and subsequent algorithms have been developed to deal with stereochemistry.[14]

If all of the structures in a database are stored in their canonical representations, then an exact structure search can be carried out by converting the query structure to its canonical form and comparing its connection table with those in the database. Hashing algorithms have also been developed that allow structures to be retrieved directly. A hash key is a number that is calculated from a canonical connection table and is used to determine the physical location on disk where a structure is stored. For example, the Augmented Connectivity Molecular Formula used in the Chemical Abstracts Service Registry System[15] is based on a procedure similar in concept to the Morgan algorithm and is used to calculate a hash key. Ideally, each structure will produce a unique hash key, however, there is always a chance that two different structures will produce the same key. Thus, it is necessary to compare the connection table representations for the structures mapped to the same hash key in order to resolve such clashes.

Substructure searching, which aims to identify all structures within a database that contain a structural fragment embedded within a larger structure, is a more complex problem. In graph theoretic terms it is equivalent to searching for a subgraph isomorphism between two graphs. While efficient subgraph isomorphism algorithms do exist, they are not sufficiently fast to allow searching through the large databases that typically exist today. Subgraph isomorphism belongs to a class of problems that are known as NP-complete, which means the time required to find a solution varies exponentially with the size of the problem (in this case, with the number of nodes in the graphs). Thus, substructure search is implemented as a two-stage process with a fast screening stage used to eliminate structures that cannot possibly match the query so that the more expensive graph matching algorithm need only be applied to a subset of the database.

Screening systems are based on fragment bitstrings, which are represented as binary vectors. In a dictionary-based screening system, each position in the bitstring corresponds to a particular fragment, such as a carbonyl group or an amino group. For a given molecule, a bit is set to on or '1' if the molecule contains the fragment it represents, otherwise it is set to off or '0' (see **Figure 2**). The screening stage of substructure search involves generating a bitstring for the query substructure and comparing it with bitstring representations of all the molecules in the database. If a database structure contains a '0' at a position set to '1' in the query this indicates that a fragment present in the query is absent from the database structure, which, therefore, cannot possibly match the query. The database structure can then be eliminated from further consideration. Only those structures that pass the screening stage proceed to the graph matching stage. Ideally, a well-designed screening system should screen out as much as 95% of the database, although this obviously depends on the level of specificity of the query; for example, for a benzene ring as query the screen out is likely to be much lower since a large proportion of the structures in the database will contain it.

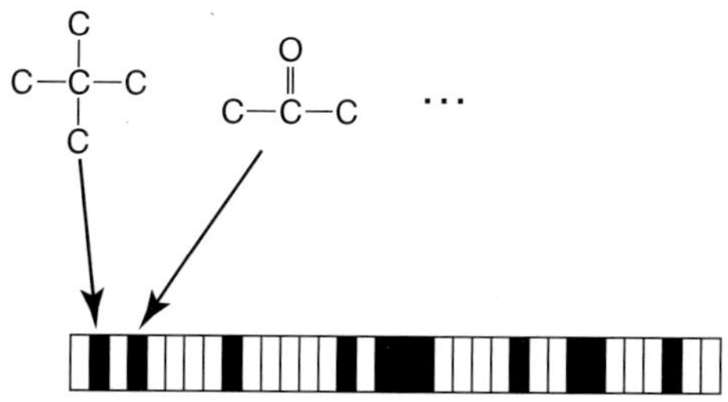

Figure 2 Fragment bitstring.

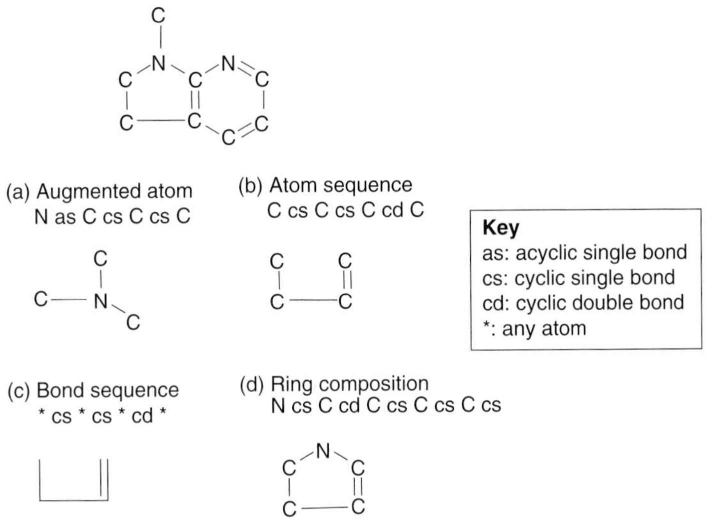

Figure 3 Examples of substructural fragment types included in dictionary-based screening systems.

Examples of the different types of substructural fragments that are included in screening systems are illustrated in **Figure 3**.[16,17] Typically, they include: augmented atoms, which consist of a central atom and its neighboring atoms; atom sequences, which consist of a linear path of atoms with or without the bonds differentiated; bond sequences in which the atom types are ignored; and various types of ring fragments such as ring sizes and bond patterns around rings. The most widely used dictionary-based systems are the MACCS and ISIS systems from MDL Information Systems[12] and the BCI fingerprints from Barnard Chemical Information.[18]

To be effective at screen out, a fragment dictionary needs to be carefully designed since fragments that occur in most molecules are unlikely to be discriminating and, conversely, fragments that occur infrequently are unlikely to be useful since they will rarely occur in queries. The optimum dictionary is, therefore, data set dependent and requires a statistical analysis of the database to be performed.[19] This presents difficulties when new compounds are added to the database, especially if they represent new or previously underrepresented chemistries.

Hashed fingerprints provide an alternative approach to generating fragment bitstrings. These methods involve the exhaustive enumeration of the fragments in a molecule and do not rely on a dictionary of fragments (they therefore avoid the problem of data set dependency). For example, the Daylight fingerprint is constructed by enumerating all linear paths up to a specified length (typically from 2 to 7 atoms).[20] Each path is then hashed to generate a small number of bits (typically four or five), which are set to '1' in the fingerprint bitstring. Since the paths are mapped into a fixed length bitstring collisions can occur whereby a given bit is set by more than one path and, in contrast to the dictionary-based bitstrings, there is no longer a one-to-one mapping between bit position and fragment.

The second stage of substructure searching involves the application of a subgraph isomorphism algorithm to the database structures that pass the screening stage. The brute force approach to substructure search would involve investigating every possible way of mapping the query atoms onto the database structure atoms. For a query substructure consisting of N_q atoms and a database structure consisting of N_d atoms there are $N_q!/N_d! - N_q!$ possible mappings, which is clearly computationally infeasible for all but the smallest of structures. More efficient methods have therefore been developed that make use of heuristics (i.e., rules) aimed at either improving the chances of finding a match early on or efficiently rejecting candidates that cannot give rise to a match.[21] For example, back-tracking methods attempt to reject potential mappings as early as possible.[22] The search begins by mapping a node in the query substructure to a node of the same atom type in the molecule graph. An attempt is then made to extend the mapping by matching neighboring nodes from the query onto corresponding neighboring nodes in the molecule graph. This process continues until either all the nodes have been successfully matched or until it proves impossible to find a match, at which point the algorithm returns (or back-tracks) to the last successful match and attempts an alternative mapping, by, for example, trying a different assignment of the neighboring nodes. This back-tracking procedure continues until a match of all query nodes has been found or until all possible mappings of the initial node have been tried, at which point the algorithm terminates with a mismatch recorded.

The Ullmann algorithm has been found to be particularly efficient for chemical graph matching. The algorithm is a combination of a back-tracking procedure and a refinement or relaxation step in which neighboring atoms are taken into account when considering possible mappings (i.e., an atom in the query is not permitted to match an atom in the molecule unless each of their neighboring atoms also match). A matching matrix is constructed between the query and database structure, which has one row for each query atom and a column for each database structure atom. The elements of the matching matrix take the value '1' if a match is possible between the corresponding atoms (based on atom type and connectivity where the connectivity of the database atom must be equal to or greater than the connectivity of the query atom), otherwise the value is '0.' The aim is to find a matrix in which there is one element per row that is equal to '1' with each mapping occurring in a different column; this then specifies a unique mapping for every query atom onto a database structure atom. The search begins by assigning the first query atom to the first potentially matching molecule atom; all other entries in the row are set to '0.' The refinement step is then applied, which involves checking that the neighbors of the query atom and database atom match. If it fails then the algorithm back-tracks and the next potential match is attempted. If it succeeds then the mapping is extended to the neighboring atoms.

While graph theoretical techniques are widely applied in the processing of chemical structures, the analogy between a graph and a chemical structure is not perfect and special procedures need to be implemented to handle some features of molecules such as aromaticity, tautomerism, and multicentre bonds, etc.[23]

3.12.3 Three-Dimensional Structures

Techniques for the manipulation of three-dimensional (3D) representations of molecules are also widely used. Conformational information about molecules in the form of atomic coordinates can originate either from experimental data or from the application of computational methods. The major sources of experimental data are the Cambridge Structural Database (CSD) and the Protein Data Bank (PDB). The CSD contains data for over 250 K small organic and organometallic molecules, which have been derived using x-ray crystallographic techniques.[24] The PDB, on the other hand, contains a much smaller number of macromolecular structures ($\sim$34 K) including proteins, protein–ligand complexes, and nucleic acids. The structures have usually been derived from either x-ray crystallography or nuclear magnetic resonance.[25]

These experimental databases are very valuable sources of information, however, even the CSD represents only a small proportion of the total number of compounds known; for example, the Chemical Abstracts Registry currently contains around 26 million small organic and inorganic compounds.[11] Furthermore, experimental databases typically contain a single minimum energy conformation of a structure, whereas, it is known that there can be many conformations accessible to a molecule and that the bioactive conformer is not necessarily a minimum energy conformation.[26] Additionally, it is often of interest to consider the conformations of molecules that could potentially be made, i.e., virtual compounds. Thus, computational methods have been developed to generate a single low-energy conformation from a 2D structure representation (a technique often referred to as structure generation) and also to explore the conformational space available to a molecule (a technique known as conformational analysis).

The most widely used methods for structure generation are the CONCORD[27] and CORINA[28] programs. Both programs involve fragmenting the input structure into rings and acyclic components. The acyclic fragments are converted to 3D using standard bond lengths and angles, which are usually extracted from tables. Ring systems are more difficult to process: the conformations of small rings are usually handled via stored ring templates with ring systems being constructed by combining the individual ring conformations together. More complex procedures have been developed to handle larger rings. A comparison of methods has shown conversion rates of between 91 and 99% for the generation of conformations contained in the CSD.[29]

Structure generation programs typically aim to generate a single low-energy conformation, whereas, the goal of conformational analysis is the identification of all accessible minimum-energy conformers of a molecule. However, even for simple molecules there may be a large number of such structures, making conformational analysis a difficult problem. The lowest energy minimum is usually referred to as the global minimum-energy conformation, however, it is important to note that this may not necessarily be the biologically active structure.[26,30] The usual strategy in conformational analysis is to use a search algorithm to generate a series of initial conformations. Each of these in turn is then subjected to energy minimization in order to derive the associated minimum-energy structure. The search algorithms include: systematic conformational search in which the torsion angles of the rotatable bonds are varied systematically; random search, which involves iteratively selecting a previously generated conformer and modifying it at random; and the use of distance geometry and molecular dynamics techniques. Detailed discussion of these techniques is provided by Leach[31] and is beyond the scope of this chapter.

3.12.3.1 Three-Dimensional Searching

The graph analogy of 2D chemical structures is easily extended to the representation of structures in 3D thus allowing graph-based approaches to be used for searching 3D structures. The nodes of a 3D graph represent the atoms of the structure and the edges represent the interatomic distances between the atoms, rather than the bonds as in a 2D graph. The 3D graph representation of a chemical structure is a fully connected graph since all atoms are connected to all other atoms by edges that are labeled with the interatomic distance between the atoms.

The connection table can encode 3D information simply by including the x, y, z coordinates of the atoms. For processing purposes, a 3D structure is usually represented by a distance matrix with the interatomic distances being easily calculated from the coordinates. A distance matrix for a molecule consisting of N atoms is represented by an N by N matrix, with the elements of the matrix filled by the interatomic distances between the atoms.

A major use of 3D searching is to identify compounds that possess 3D properties believed to be important for binding to a particular biological target. These requirements are usually expressed as a 3D pharmacophore, which specifies a set of features together with their relative spatial orientation. Typical features include hydrogen bond donors and acceptors, positively and negatively charged groups, hydrophobic regions, and aromatic rings. The spatial relationships between the features are usually specified as distances or distance ranges or by defining the (xyz) locations of the features together with some distance tolerance (typically given as spherical tolerance regions). The compounds that match a 3D query are those which can adopt the pharmacophore and which could therefore be considered for biological testing. An example of a simple 3D pharmacophore based on interfeature distances generated for three $5HT_{1D}$ receptor agonists is shown in **Figure 4**. It may also be possible to include other geometric features such as centroids, planes, and angles in a 3D pharmacophore together with regions of space that should not be occupied by the ligand (excluded volumes). Methods for elucidating the pharmacophore from a series of known actives are described later.

As for 2D substructure searching, substructure searching in 3D involves a two-stage search. An initial screening stage is used to eliminate rapidly structures that cannot possibly match the query. Those compounds that pass the first stage are then processed by a more time-consuming subgraph isomorphism algorithm. The screening is typically based on bitstrings with each bit position corresponding to a distance range between a pair of atoms or a group of atoms, for example, a bit may correspond to a carbonyl oxygen separated from an amine nitrogen by between 2 and 3 Å. Distances are binned with different bits assigned to different distance ranges, for example, another bit may represent the same pair of atoms separated by 3–4 Å, etc.[32] As for dictionary-based 2D fragment bitstrings, care should be taken when devising binning schemes. For example, it may be more effective to use small distance intervals to represent frequently

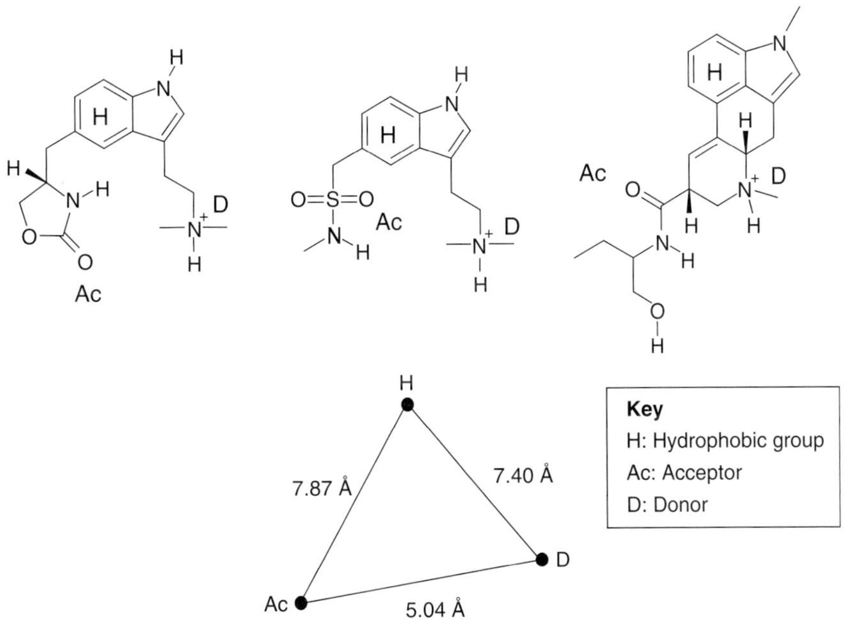

Figure 4 A pharmacophore hypothesis generated for the three $5HT_{1D}$ receptor agonists.

occurring distances with larger intervals used to represent less frequent distances. A bitstring for a single conformation of a molecule is constructed by calculating the distances between all pairs of atoms in the molecule and setting the appropriate bits to '1.' It is also possible to include angular and torsional information within a bitstring based on sets of three and four atoms, respectively, to further improve screening efficiency.[33,34]

The subgraph isomorphism stage can be carried out using an adaptation of the Ullmann algorithm. The query graph is specified in terms of a set of atoms and the distances between them, which are usually specified as distance ranges. Thus, a database structure will match the query if the distances between corresponding atoms in the structure fall within the bounds of the query distance ranges.[35,36]

The above descriptions apply to rigid searching, for example, based on a minimum energy conformation of each molecule. However, in general, molecules are flexible and, to be effective, 3D searching methods should take account of conformational flexibility. There are basically two ways in which this can be done. One is to store several conformers for each molecule within the database and to search each conformer in turn; however, this approach is highly dependent on the resolution of conformational sampling used.[30] The alternative approach is to explore conformational space within the search process itself. The distance between two atoms will depend on the particular conformation adopted and the conformational space available to a flexible molecule can be approximated by assigning interatomic distance ranges, which represent the minimum and maximum distances that are possible geometrically. Distance ranges can be computed using techniques from distance geometry. The search algorithms can then be easily modified to handle distance ranges in the database structure rather than the fixed distances appropriate to a single conformation. This approach can lead to false hits and so a final search process is required to ensure that the conformations retrieved are geometrically feasible. Another technique is known as the directed tweak method.[37] Given a set of distance constraints between pairs of atoms, it is possible to calculate the values that the torsion angles of the intervening rotatable bonds should adopt in order to enable the constraints to be satisfied. This then provides a mechanism for generating a conformation that matches the query.

3.12.3.2 Pharmacophore Identification

The identification of an accurate pharmacophore is a key objective in many drug discovery efforts. As discussed above, a pharmacophore can be used in a database search to identify other molecules that may bind to the receptor of interest. A pharmacophore can also be used as an aid to visualize the potential interactions between the ligand and receptor and as a template for generating alignments for 3D-quantitative structure–activity relationship (QSAR) analysis.[38]

Pharmacophore identification methods are usually applied to a set of active compounds when the structure of the receptor is unknown. Common features amongst the actives are used to hypothesize about the corresponding features present in the receptor. Some methods also make use of inactive compounds, for example, the HypoGen module of Catalyst looks for features that are common to the actives but not to the inactives. Some methods, known as receptor-based methods, also make use of the target structure, for example, the LigandScout program;[39] however, the focus here is on methods used when the structure of the receptor is unknown.

The two major issues in pharmacophore identification are the correct representation of the chemical features, so that bioequivalent features are mapped together, and the appropriate sampling of conformational space so that the bioactive conformation of each compound is found. The definitions of bioequivalence vary in different implementations; however, most methods are based on definitions of hydrogen bond donors, hydrogen bond acceptors, charged centers, and hydrophobic groups. The features are often represented by points outside of the ligand molecular volume in the direction of the hypothetical hydrogen bonds, usually referred to as site points. This is to account for the fact that different ligands can make interactions with the same protein atom via atoms that are in different locations within the binding site. Usually, the requirement is that both the ligand atoms and the corresponding site points are overlaid, with larger tolerances being applied to the ligand atoms. **Figure 5** shows the alignment of the three $5HT_{1D}$ receptor agonists that gives rise to the pharmacophore hypothesis in **Figure 4**.

As already discussed, most molecules can exist in a large number of conformations and ensuring that the bioactive conformations of the compounds are sampled is a major challenge. The handling of conformational flexibility is one of the main ways in which pharmacophore elucidation programs differ. As for database searching, there are two basic methods. In the ensemble approach, a set of conformers of each molecule is precomputed before the pharmacophore search takes place. The search then consists of a series of rigid body alignments in which every combination of conformers is processed in turn.[40,41] Since the conformer generation process is external to the pharmacophore search itself, the method used to generate the conformers may be chosen independently. A limitation of the ensemble approach is that the effectiveness of the pharmacophore search depends on the number of conformers generated per molecule, the range of conformational space sampled, and the conformational search method used.[42]

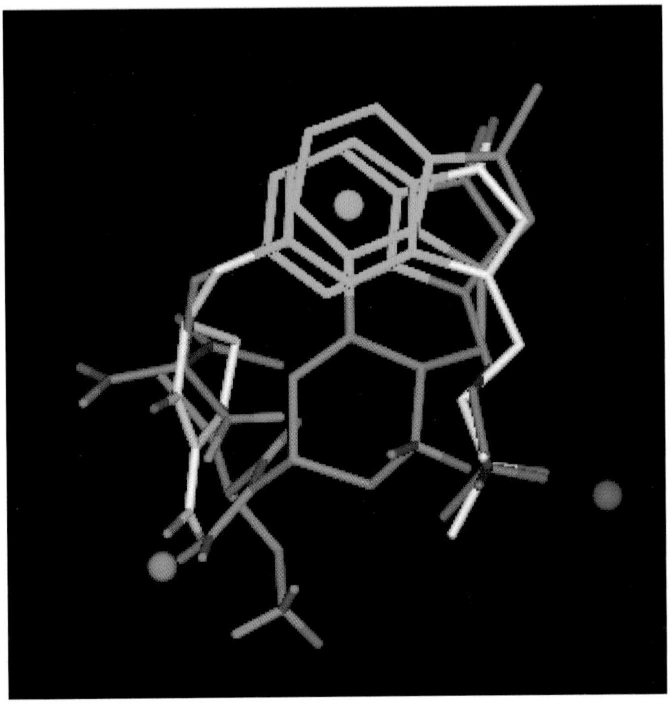

Figure 5 The alignment of the three $5HT_{1D}$ receptor agonists that correspond with the pharmacophore shown in Figure 4. Note that the pharmacophore features correspond to the centroids of the aromatic rings, and site points that correspond to the donor and acceptor atoms.

Rigid body alignments are typically carried out by searching for the maximum common subgraph (MCS) between a pair of graphs using the graph matching technique of clique detection.[41] A clique in a graph is a subset of the nodes such that there is an edge between every pair of nodes in the subgraph, that is, it is a completely connected subgraph and the maximum clique is the largest clique present in the graph. The MCS problem is reduced to finding the maximum clique in what is known as a compatibility or correspondence graph generated from the two graphs being compared. The correspondence graph consists of pairs of matching nodes (that is, nodes of the same type) with one node from each graph. An edge is created between two nodes in the correspondence graph if the corresponding nodes in the original graphs are at the same separation (within some tolerance).

For example, consider the two graphs G_1 and G_2 shown in **Figure 6** each consisting of two acceptors and two donors. The correspondence graph consists of the eight node pairs: A_1a_1, A_1a_2, A_2a_1, A_2a_2, D_1d_1, D_1d_2, D_2d_1, and D_2d_2. The edges in the correspondence graph are as shown, with an edge existing between nodes A_1a_1 and D_2d_1 because the distance between A_1 and D_2 is the same as that between a_1 and d_1, and so on. The maximum clique is shown in bold with the MCS in the original graphs also shown in bold. For pharmacophore identification, the molecular graph representations only contain the features that may potentially be mapped in the pharmacophore. A limitation of clique-detection is that it does not take into account chirality, since a given pattern of features has exactly the same set of interfeature distances as its mirror image. Thus, a postprocessing stage is required to detect and discard such results.

When processing several molecules, as is typically the case in pharmacophore elucidation, the molecule with the smallest number of conformations is chosen as reference and each of its conformations is considered in turn as the reference conformation. All the conformations of the other molecules are then compared to the reference conformation in a pairwise manner and the cliques identified. Any clique that is common to all of the molecules, such that it is matched by at least one conformation of each molecule in the set, corresponds to a common pharmacophore.

For most data sets, there are usually several pharmacophore hypotheses consistent with the data set and in some cases, particularly if the molecules are highly flexible or feature-rich, there may be too many to allow manual inspection of them all. In these situations, an automated means of scoring or ranking the hypotheses generated is necessary. Ranking the hypotheses is usually performed at the end of the search and can be based on various criteria such as the number of pharmacophore points mapped, the root mean square deviation of the overlaid features, the volume overlap of the molecules with the reference molecule, and the conformational energy of the ligands. In the maximum likelihood method, ranking is based on a combination of how well the molecules map onto a hypothesis and the 'rarity' of the

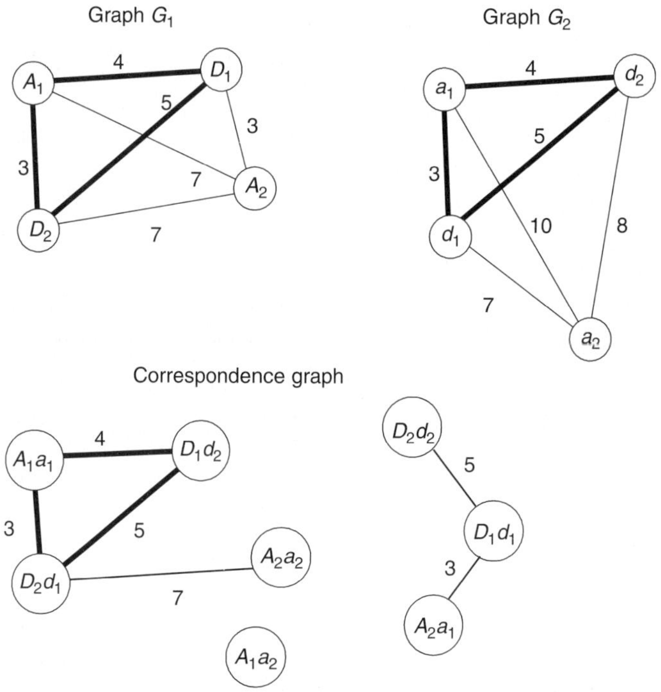

Figure 6 Illustration of the clique detection method. Consider two molecules, each of which contains two donors and two acceptors. The two graphs at the top illustrate the distances between the features in the two conformations being considered. The correspondence graph (bottom) consists of eight nodes but only four edges. The maximum clique in the correspondence graph is shown in bold with the MCS in the original graphs also shown in bold.

hypothesis.[40] The general strategy is to score more highly those hypotheses that are well matched by the active molecules in the set but that are less likely to be matched by a large set of arbitrary molecules.

The alternative approach to handling conformational flexibility via the ensemble approach is to explore conformation on-the-fly. For example, a genetic algorithm (GA) may be used to generate trial conformations and trial mappings of features with a least-squares fitting procedure used to minimize the root mean squared (RMS) distance between pairs of mapped points in the conformers.[43] An alignment is then scored on the goodness of fit of the features to the pharmacophore points, the common volume of the overlay, and the conformational energies of the conformers. The GA continues until convergence, when the best combined score has been found. Theoretically, at least, this approach allows a more thorough exploration of conformational space than the ensemble methods; however, since the optimization method is designed to find the solution with the best score it is unlikely that the full range of plausible hypotheses is identified.

The choice of compounds to input to a pharmacophore generation program can greatly affect the chances of finding the correct solution. The compounds chosen should be structurally diverse since structural analogs are likely to have features in common that are not relevant to the pharmacophore. Also, if they are all flexible, it is usually difficult to deduce any information about their active conformations since there are likely to be many sets of conformations for which the features can be overlaid. For compounds that are structurally diverse, the geometric uncertainty in the pharmacophore is reduced since there will generally be a smaller number of conformations for which the mapped features can be overlaid. Also it is much less likely that a feature exists in all molecules that can be overlaid but does not form an interaction with the protein. A recent comparison of some of the commonly used programs for pharmacophore identification is provided by Patel and coworkers[44] and a review of the current status of pharmacophore methods is provided by van Drie.[45] Recent work in this area has focused on using multiobjective optimization techniques that aim to find multiple solutions that represent alternative but equally plausible hypotheses.[46]

3.12.4 Virtual Screening

Virtual screening is a recently introduced term that appeared following the widespread adoption of automation techniques for synthesizing and testing compounds. Virtual screening refers to the application of computational techniques to the selection compounds for biological screening, either from in-house databases, externally available compound collections, or from virtual libraries, that is, sets of compounds that could potentially be synthesized. Thus, virtual screening can be thought of as the computational equivalent of biological screening.[47,48] In the early days of HTS, the emphasis was on the quantity of compounds being tested. However, the vast size of chemistry space (it has been estimated that more than 10^{60} drug-like compounds could potentially exist[49,50]), the low success rates of early HTS experiments, and the real costs associated with testing large numbers of samples have meant that the emphasis is now on quality rather than quantity with screening sets being carefully designed in order to increase the chances of finding good hits.

Virtual screening encompasses a wide variety of computational techniques with the particular technique that is applied depending on the intended use of the screening set and the information that is available concerning the biological endpoint. The techniques can be divided into ligand-based virtual screening, which is used when the 3D structure of the biological target is unknown, and structure-based virtual screening when the 3D structure of the biological target is used to identify compounds that could potentially bind to the target.

Wilton and colleagues[51] describe three different ligand-based virtual screening scenarios. Early in a drug discovery program, when little is known about the target, but when an active compound is available, such as a compound from the literature, then similarity searching is the most commonly used technique. When several actives are known but no inactives are available then pharmacophore elucidation followed by database searching can be used. When several actives and inactives are known, as might be the case following a round of high-throughput screening, then pattern recognition or machine learning techniques can be used in an attempt to generate a model that is able to distinguish between the two classes. Assuming such a model can be found it can then be applied to compounds that have not yet been tested biologically in order to predict their likelihood of being active. Similarity searching and applications of machine learning methods are described below, with structure-based virtual screening methods discussed in the next section. Pharmacophore methods were described earlier in the context of database searching.

3.12.4.1 Similarity Searching

Similarity methods were first described two decades ago by two groups working independently (*see* 4.08 Compound Selection Using Measures of Similarity and Dissimilarity).[52,53] Since then a large number of different methods have been developed with the recent growth in the size of databases (both real and virtual) resulting in a resurgence of interest in these techniques. Similarity methods are most useful early in a drug discovery project when a limited amount of information is available. Traditionally, similarity searching has been based on a single active compound, which could be derived, for example, from the literature or could be a competitor compound that has been protected by a patent. The rationale for similarity searching is based on the similar property principle,[54] which states that structurally similar compounds tend to have similar properties. Thus, compounds that are similar to a known active compound are likely to exhibit similar activity and are therefore good candidates for testing. Given a way of quantifying the similarity between a pair of compounds then similarity searching involves comparing the query structure of interest with each compound in the database in turn and ranking the database in order of decreasing similarity to the query. The top ranking compounds can then be carried forward for biological testing. An effective similarity search is one that leads to more hits than if compounds were selected at random.

There are two main components required to calculate the similarity of a pair of molecules.[55] First, the molecules should be represented by numerical descriptors. Second, a similarity coefficient is required to quantify the degree of similarity based on the chosen descriptors. Some studies have also investigated the use of weighting schemes to bias the numerical descriptors but most similarity measures assume that the descriptors are equally weighted.

The original similarity methods were based on the fragment screens that had been developed for speeding up substructure searching. For example, Willett and coworkers showed that the similarity between two molecules A and B can be quantified by taking into account the number of fragments they have in common using the Tanimoto coefficient as shown:

$$S_{AB} = \frac{c}{a+b-c}$$

where c is the number of fragments in common between molecule A and molecule B, a is the number of fragments in molecule A and b is the number of fragments in molecule B. When applied to binary bitstrings the Tanimoto coefficient returns a value in the range 0–1, where 0 indicates that there are no fragments in common and 1 indicates that the molecules are identical in terms of their fragment bitstrings (note that this does not necessarily mean they have identical structures).

The Tanimoto coefficient is also applicable to vectors of continuous variables such as physicochemical properties or topological indices (these are various indices that can be derived from the 2D graph representations of molecules[56]), with the general form of the equation as shown:

$$S_{AB} = \frac{\sum_{i=1}^{N} x_{iA}x_{iB}}{\sum_{i=1}^{N}(x_{iA})^2 + \sum_{i=1}^{N}(x_{iB})^2 - \sum_{i=1}^{N} x_{iA}x_{iB}}$$

where molecules A and B are represented by vectors of length N and X_{JA} is the value of the Jth variable for molecule A and X_{JB} is the value of the Jth variable for molecule B. When measuring similarity using continuous variables, the variables should be scaled to be in the same range prior to calculating similarity in order to avoid bias toward particular variables. While the Tanimoto coefficient is the most widely used similarity coefficient, many others exist and are reviewed in Willett and coworkers.[55]

Fragment bitstrings have proved to be remarkably successful for similarity searching considering they were originally developed for a different purpose, that is substructure search. However, they are most effective at identifying close analogs of the target structure and are less effective at identifying nonobvious similarities, that is, compounds that share the same activity but that are based on different chemical scaffolds. Consequently, much of the recent effort in developing new similarity methods has been directed toward new descriptors including both 2D and 3D descriptors, as summarized below.

3.12.4.1.1 Two-dimensional similarity measures

3.12.4.1.1.1 Fragment methods

About the same time as the work of Willett *et al.*, Carhart *et al.* described an approach to similarity searching based on atom-pair descriptors.[52] An atom-pair is composed of two nonhydrogen atoms and their through-bond distance, that is, the shortest bond-by-bond path between them (also known as the topological distance). Each atom is described by element type, the number of nonhydrogen atoms it is bonded to and its number of π-bonding electrons. An example of a simple atom-pair is CX2–(2)–CX2, which represents the sequence -CH_2-CH_2-, with X2 indicating the presence of two nonhydrogen neighboring atoms. Atom-pairs encode more distant information than is the case for the 2D fingerprints described earlier. The similarity between two molecules represented by lists of atom-pairs is calculated using a modified form of the city block distance to obtain a number in the range 0..1. Nilakantan and coworkers introduced related descriptors called topological torsions, which encode sequences of four connected atoms together with their atom types, number of nonhydrogen connections, and number of π-electrons.[57] These were found to complement the atom-pairs descriptors in similarity searches, i.e., they tended to retrieve a different set of actives to those found using the atom-pairs.

Later, Kearsley and colleagues modified the atom-pair definition to represent the binding properties of atoms rather than their specific element types.[58] Thus, atoms are identified as belonging to one of seven binding property classes: cations, anions, neutral hydrogen bond donors and acceptors, atoms that are both donor and acceptor, hydrophobic atoms, and all others. The representation of atoms by their properties rather than element types has been adopted in several more recent descriptors. For example, the CATS (Chemically Advanced Template Search) descriptors[59] are based on counts of atom-pairs up to 10 bonds apart where the atoms are classified as lipophilic, positive, negative, donor, and acceptor. Similog keys are based on a 'DABE' atom-typing scheme in which atoms are described by the presence or absence of the following four properties: donor, acceptor, bulkiness, and electropositivity.[60] A Similog key for a molecule consists of a list of triplets of atoms comprising their DABE codes and the topological distances between the atoms mapped to four intervals. The similarity between two molecules is then calculated using the Tanimoto coefficient applied to the Similog keys.

Descriptors based on the extended connectivities of atoms have recently shown promise for similarity searching. Examples include the Extended Connectivity Fingerprints (ECFPs) and Functional Connectivity Fingerprints (FCFPs).[61] The extended connectivity of an atom is calculated using a modified version of the Morgan algorithm.[13] An initial code is assigned to each atom based on its properties (eg., atom type for ECFPs and generalized atom type for FCFPs) and connectivity. Each atom code is then combined with the codes of its immediate neighbors to produce the next order code for the atom. This process is repeated to produce a code at the required level of description. Recent studies have shown these descriptors to outperform the more conventional fingerprints in searches carried out in the MDDR database.[62]

3.12.4.1.1.2 Graph-based similarity

An alternative approach to 2D fingerprints is to use graph matching techniques to identify the largest subgraph in common between two graphs.[63] The use of clique detection for identifying the MCS between two graphs has already been described in the context of pharmacophore identification. When considering 2D representations of chemical structures the MCS corresponds to the largest substructure in common to the structures. The similarity between two structures can then be quantified by applying a Tanimoto-like coefficient that considers the number of atoms and bonds in the MCS relative to the numbers of atoms and bonds in the structures being matched.

An advantage of graph matching approaches relative to fingerprint methods is that a mapping is generated between the two structures. However, the problem of determining the MCS between two graphs belongs to the class of NP-complete problems, in common with other graph searching techniques. This is a potential problem for virtual screening applications where typically the requirement is to handle very large data sets. However, Raymond and Willett have recently introduced an algorithm called RASCAL that is able to perform thousands of comparisons a second,[64,65] and can therefore be used for virtual screening. The efficiency is achieved through the use of chemically relevant heuristics, a fast implementation of the clique detection process that underlies MCS detection in which the graphs are first converted to line graphs, and the use of very efficient screens that prevent many of the more costly MCS comparisons from being performed.

3.12.4.1.1.3 Reduced graph similarity

Various forms of reduced graph representations of 2D structures have been used for similarity calculations. Reduced graphs were first introduced to provide an intermediate level of search between fragment screening and full atom-by-atom search in Markush structures, which represent families of related compounds as found in chemical patents.[66] A reduced graph is constructed by grouping together connected atoms within the original structure to form a single node in the reduced graph. For example, all atoms within a ring system may be grouped to form a ring node in the reduced graph. The nodes in the reduced graph are connected by edges that correspond to bonds in the original structure.

In similarity searching, the aim is to produce a generalized representation that emphasizes molecular features that could form interactions with a receptor[67] while maintaining the topology between the features. Thus, nodes are typically based on functional groups such as hydrogen bond donors, hydrogen bond acceptors, ring systems, etc. Example reduced graphs are shown in **Figure 7**. The similarity between two molecules can be calculated by converting the reduced graphs into fingerprints as for the more traditional graph representations of structures and then applying a similarity coefficient[68,69] or by applying graph matching techniques to identify an MCS between two reduced graphs.[70,71]

The Feature Trees introduced by Rarey and Dixon are also based on reduced representations of molecules and provide a particularly interesting approach, which, when combined with fast mapping algorithms, allow similarity searching to be performed against large databases.[72,73]

3.12.4.1.2 Three-dimensional similarity methods

The binding of a small molecule to a receptor is governed by its 3D properties such as shape and electrostatics and so it is natural that several methods have been developed for calculating similarity in 3D. As with other methods for analyzing structures in 3D, a major difficulty is the consideration of conformational flexibility. Ensuring adequate

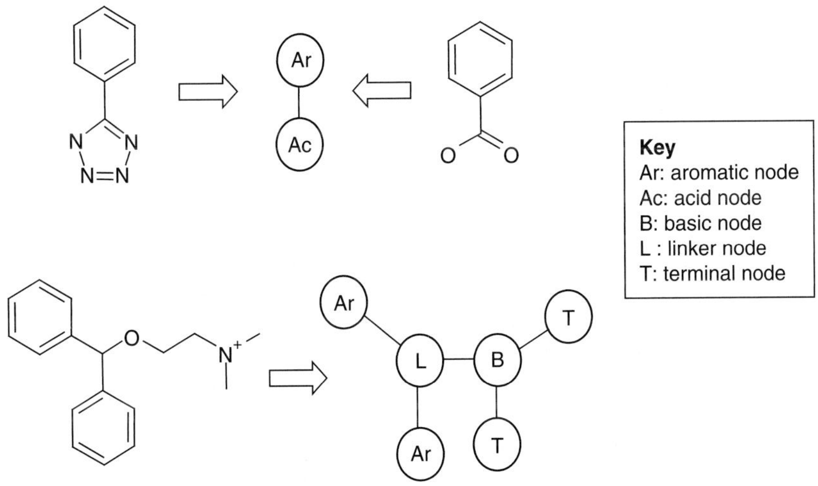

Figure 7 Example reduced graphs.

coverage of conformational space is both difficult to achieve and computationally intensive. 3D similarity methods can be divided into those that require the molecules to be aligned prior to calculation of similarity and methods that are independent of the relative orientation of the molecules. The latter approaches are described first.

3.12.4.1.2.1 Three-dimensional vector-based approaches

The 3D screens initially developed for 3D substructure searching have also been used for 3D similarity searching, paralleling the use of 2D fragment screens in 2D similarity searching. More recent approaches have been based on pharmacophoric fingerprints.[74] Rather than considering all of the atoms in molecules, pharmacophoric fingerprints record the spatial arrangement of pharmacophoric features such as hydrogen bond donors, hydrogen bond acceptors, cations, anions, and aromatic and hydrophobic centers. Each bit in a 3-point pharmacophoric fingerprint represents a particular triplet of features at specified distance ranges. The fingerprint for the 3D conformation of a molecule is constructed by identifying the triplets of features and distances that exist in the conformer and setting the appropriate bits to 'on' in the fingerprint. All other bits are set to 'off.' Conformational flexibility is usually handled by generating an ensemble of conformers for a structure and ORing the fingerprints generated for each conformer. The resulting ensemble fingerprint then represents the conformational space available to the molecule.

Pharmacophore fingerprints have been widely used in diversity studies, which are described later, however, they can also be used in similarity searches as for the 2D fingerprints, although the different characteristics of pharmacophore fingerprints compared to 2D fragment-based fingerprints should be taken into account when devising the similarity calculation.[75] For example, pharmacophoric fingerprints are typically much longer than 2D fingerprints (which usually consist of around 1000 bits) with 5916 3-point pharmacophores possible when considering six different pharmacophoric features and just six different distance bins.[74,76] They also tend to be relatively sparse compared to 2D fingerprints and small changes in the pharmacophoric features present in one molecule relative to another can lead to large changes in the fingerprint.

3.12.4.1.2.2 Alignment-based methods

Alignment-based similarity methods are more computationally demanding than 3D vector-based methods since, as well as taking into account conformational flexibility, they also involve finding an optimum alignment of the conformers before calculating similarity based on the alignment. Some examples of 3D alignment-based methods are given below. A comprehensive review of molecular alignment techniques has been published by Lemmen and Lengauer.[77]

In FlexS the query compound is kept rigid and the database molecules are superimposed on the query.[78] A database structure is first fragmented into relatively rigid fragments, which are then reassembled onto the query in a stepwise manner with the conformational space of the database structure being explored by joining the fragments in all possible ways. Intermediate solutions are evaluated using a scoring function that includes energy-like matching terms for paired intermolecular interactions.

Other approaches are based on aligning field-based properties such as steric, electrostatics, and hydrophobic fields. The properties are calculated at the vertices of a 3D grid that encloses a molecule. For example, the electrostatic potential at a grid vertex, r, can be calculated from the charges on each atom using the following equation:

$$P_{\mathrm{r}} = \sum_{i=1}^{N} \frac{q_i}{|\mathbf{r} - \mathbf{R}_i|}$$

where q_i is the point charge on atom i, at position $\mathbf{R}_i$ from the grid point $\mathbf{r}$. The similarity of two molecules in a given alignment can be calculated by comparing the values of corresponding vertices in each grid using a similarity coefficient. However, the resolution of the grids that is normally used (for example, $1000 \times 1000 \times 1000$) makes a direct grid comparison computationally infeasible. Good and coworkers recognized that such field representations could be approximated by atom-centerd gaussians with similarity calculated using the Carbo coefficient applied to the gaussians.[79] The computationally expensive grid-based calculation is therefore replaced by a fast analytic approach with negligible loss in accuracy. Since the work of Good and coworkers, several groups have further developed this approach. For example, the Field-Based Similarity Searching (FBSS) program uses a genetic algorithm to optimize the alignment of two molecules based on similarities of steric, electrostatic, and hydrophobic fields calculated using gaussian approximations.[80] Gaussian functions have also been used to provide representations of molecular shape that are more realistic than more traditional 'hard sphere' models.[81]

3.12.4.1.3 Evaluation of similarity methods

Given the wide variety of similarity methods that has been developed, it is important to be able to compare the relative effectiveness of one method against another. Similarity methods are usually evaluated via retrospective property prediction or simulated virtual screening experiments.[82,83] Property prediction experiments are typically based on a physicochemical property such as the partition coefficient between *n*-octanol and water (log P) and a data set for which measured values are available. A compound is selected from the data set and its similarity to all other compounds is calculated. The property of the query compound is then predicted from the values of its nearest neighbors. For example, it may be taken as the mean value of the five nearest neighbors. The predicted value is then compared with the known value. This process is repeated for multiple query compounds and the overall effectiveness of the similarity measure is determined from the errors in prediction.

Simulated screening experiments are carried out in a data set where the activities of compounds are known. Usually these studies have involved the use of publicly available data sets such as the MDDR[12] or the NCI Aids and Cancer data sets.[84] The MDDR consists of $\sim$120 K compounds, which are labeled by activity class. A query compound is selected and the database is ranked on similarity to it. A cutoff point is then selected, such as the top 1 or 5% of the database, and the number of compounds in the subset that exhibit the same activity as the query is calculated. A variety of different measures are available to quantify the effectiveness of simulated screening experiments.[85] For example, an enrichment factor can be calculated as the number of active compounds retrieved compared to the number that would have been retrieved if the actives were distributed at random throughout the ranked list.

Several such evaluation studies have been described in the literature (see for example[82,83]). In general, the results have shown that 2D measures tend to outperform 3D methods, which is, at first sight, a surprising result since ligand–receptor binding is based on 3D properties. However, this is widely believed to be due to the difficulties of handling conformational flexibility and the bias toward close analogs in the data sets tested.

Recently, it has been suggested that performance should also be evaluated on ability to suggest novel scaffolds or chemotypes. Compounds that exhibit the same activity as the target and belong to a different chemical series are more valuable than close analogs for a number of reasons. For example, they allow a drug discovery program to move out of the patent space of the initial query compound, they provide alternative lead series should one fail due to poor ADME (adsorption, distribution, metabolism, and excretion) properties (see later) and they provide an alternative route when a lead compound has difficult chemistry. Descriptors vary in their ability to be effective in this respect. As mentioned above, 2D fragment-based fingerprints usually perform well in terms of enrichments; however, they tend to perform less well in the identification of hits with different scaffolds. Some of the newer 2D descriptors have been shown to be better suited to these tasks, for example, reduced graphs, Similog and CATS keys,[59,67,69,86] and comparisons of 3D descriptors with 2D methods have shown that although the 3D descriptors tend to give poorer enrichments, they are more effective at discovering diverse hits.[86] Most evaluations have been based on retrospective studies where the activities are already known; however, a recent study describes the successful application of the 3D shape-based method, rapid overlay of chemical structures (ROCS), to the discovery of novel weakly binding inhibitors of the bacterial protein ZipA. The compounds identified have scaffolds that are significantly different from the compound used as query.[87]

The main difficulty in choosing a similarity method to apply in a given project is that the relative performance of different measures can vary from one problem to another, often in ways that are unpredictable.[88] Thus, a recent focus in similarity searching has been the use of data fusion techniques that attempt to combine the results from different similarity searches to improve on the results obtained using a single measure. The basic approach is to generate a series of rankings of a database using the same query structure and different similarity coefficients or descriptors. The rankings are then combined using, for example, the SUM or MAX rules. In SUM the database is re-ranked based on the sum of the individual rankings for each compound and in MAX it is re-ranked by taking the minimum of the individual ranks for each compound. Data fusion techniques can also be used to combine the ranked lists generated using multiple query compounds with significant improvements being achieved compared to using a single active compound.[89] Alternative approaches have investigated combining different types of descriptors. For example, Xue and coworkers described a fingerprint that combines fragment type descriptors with physicochemical properties.[90]

3.12.4.2 Machine Learning Methods

Machine learning methods are applied when both actives and inactives are known and attempt to learn a model of activity that can then be used to predict about unknown compounds. For example, the model could be used to select additional compounds for testing, to design combinatorial libraries, or to select compounds for purchase from external vendors. When the amount of available data is small, homogeneous, and of high quality then linear model building methods such as multiple linear regression and partial least squares can be used to derive a QSAR. High-throughput screening data, however, is characterized by its high volume, the presence of false positives and false negatives, the diverse nature of the chemical classes involved, and by the possible presence of multiple binding modes. These characteristics mean that linear methods are not applicable.

Typically, HTS data is analyzed using classification techniques with the molecules assigned as 'active' or 'inactive' or into a small number of activity classes (e.g., 'high,' 'medium,' 'low') rather than using quantitative activity values. An early example of this type of approach is substructural analysis described by Cramer and coworkers,[91] which is applied to the fragment bitstrings designed for substructure search. In substructural analysis the assumption is made that each fragment makes a constant contribution to the activity, independent of the other fragments in the molecule. A weight is derived for each fragment that reflects its differential occurrence in active molecules relative to inactive molecules. A fragment that occurs more frequently in actives is given a higher weight than one that is equifrequent in the two classes. A molecule is scored by summing the weights for all of the fragments contained within it. Once a set of weights has been derived, it can be applied to score and rank a new set of molecules in decreasing probability of activity. Many different weighting schemes are possible, for example, the substructure activity frequency (SAF) weighting scheme is:

$$w_i = \frac{act_i}{act_i + inact_i}$$

where w_i is the weight of a fragment *I*, act_i is the number of active molecules that contain the *i*th fragment, and $inact_i$ is the number of inactive molecules that contain the *i*th fragment.

Recursive partitioning is a more recent technique that involves constructing a decision tree by dividing a data set into subsets according to descriptors, or rules, that discriminate between different classes of molecules.[92] For example, a rule may correspond to the presence or absence of a particular feature or to the value of a descriptor. The basic approach is to start with the entire data set as the root node and then identify the descriptor or variable that gives the 'best' split of the data into two or more subsets, the best split being the one that leads to the greatest separation of the compounds into the classes of interest measured using a modified t-test. The same procedure is then applied recursively to each of the subsets, and so on until maximum separation of the classes has been achieved. At this point the terminal nodes are classified as active or inactive according to their content. A decision tree can be used to predict a previously unseen molecule by starting at the root node and following the edge appropriate to the first rule. This continues until a terminal node is reached, at which point the molecule is assigned to the appropriate activity class. An advantage of decision trees is that they are interpretable since they consist of a set of 'rules' that enable specific molecular features and/or descriptor values to be associated with the activity or property of interest. Also, they can identify multiple binding modes. However, as with all machine learning methods, care must be taken to prevent overfitting, which occurs when a model provides an excellent fit to the training data but is poor in prediction. Various mechanisms exist to prevent overfitting from occurring.

Standard recursive partitioning results in a single decision tree. However, it has been shown that more accurate models can be generated through the use of ensembles of trees. For example, Svetnik and coworkers generated ensembles of trees that they called Random Forest.[93] The multiple trees are created using bootstrap samples of the training data with a randomly selected subset of features evaluated at each split. Predictions are made on unseen data by combining the predictions of the ensemble by, for example, majority voting or averaging.

The support vector machine (SVM) is currently a popular binary classification technique that attempts to position a boundary or hyperplane that separates two classes of compounds such as actives and inactives. The descriptors used to represent the compounds are presented as vectors. The hyperplane is positioned using support vectors (i.e., examples in the training set) and, as such, examples in the training that are outside of the boundary are ignored. The use of a subset of the training data reduces the likelihood of over training of the data while maintaining a good degree of generalizability. When the training examples cannot be separated linearly, then kernel functions can be used to transform the data to higher dimensions where it becomes linearly separable. A disadvantage of the SVM is that, as for neural networks, it is a black box making it more difficult to interpret the results than for other methods such as decision trees. Despite this limitation, SVMs have become a popular machine learning method to apply to drug discovery problems. For example, they have been used within an active learning setting to select compounds for successive rounds of screening.[94] More recently they have been applied to the prediction of activity against a range of different G protein-coupled receptors (GPCRs) where they were shown to be effective in lead hopping.[95]

3.12.4.3 Structure-Based Virtual Screening

When the 3D structure of the target compound is known then docking techniques can be used to prioritize compounds for testing, in what is known as structure-based virtual screening. Docking refers to the process of fitting a small molecule into the receptor site of a protein in order to predict the intermolecular interactions that could exist in forming a protein–ligand complex. The number of therapeutic targets for which structural information is available is increasing rapidly due to advances in structural genomics and in structure determination methods.[96] Thus, there is widespread interest in the use of structure-based virtual screening techniques. There are two main components to docking; searching and scoring.

The aim of the search component is to find the relative orientation of the ligand and protein that corresponds to the minimum of the free energy of binding. This is known as finding the correct pose and involves exploring the six degrees of freedom associated with the translation and rotation of one molecule relative to another together with the conformational degrees of freedom of the ligand and the protein. Programs vary in the extent to which they take conformational flexibility into account as described below. There are generally two goals for scoring functions. One is to score the correct pose of a given protein–ligand complex more favorably than other poses explored during the search; the other, more difficult goal is to rank different ligands against the same protein so that the ranking correlates with the actual binding affinities.

The first automated docking program is the DOCK program, which was introduced in 1982.[97,98] The original algorithm was restricted to rigid-body docking and was based on identifying a high degree of shape complementarity between the ligand and the protein binding site. Since then, many different docking programs have been developed that take into account chemical complementarity and conformational flexibility of the ligand.[99–101] As for pharmacophore elucidation methods, there are two different ways in which conformational flexibility is handled. Several docking programs use the ensemble approach in which conformers for each ligand are precomputed and each conformer is docked in turn as a rigid body.[102] The alternative approach is to vary conformation on-the-fly either using a stochastic algorithm such as Monte Carlo, simulated annealing or a GA, or by using a fragment approach.[103–105] The stochastic approaches require the use of a scoring function to guide the search, which is usually a combination of the conformational energy of the ligand and the intermolecular interaction energy, both of which are typically calculated using molecular mechanics forcefields. The fragmentation approach involves breaking the ligand into smaller fragments, usually via its rotatable bonds. A base fragment is chosen and docked into the binding site and the ligand is rebuilt incrementally within the binding site. Several different starting positions are usually considered and a systematic conformational search is carried out in each building step so that many different conformers and poses are considered.[106,107]

The current status of the search component of docking programs is to allow for partial flexibility in the protein. This is typically done by allowing the protein side chains to rotate[108] or by docking to an ensemble of protein conformations.[109–111] However, full treatment of receptor flexibility remains beyond the scope of current methods.[112,113]

Scoring functions can be divided into three different categories: force-field methods, regression-based methods, and knowledge-based methods. The force-field methods quantify binding by considering various contributions calculated in molecular mechanics.[107] Regression-based methods involve summing weighted interaction terms with the weights being derived by fitting to experimental data. The first such scoring function for docking was developed by Böhm who constructed a linear function containing terms that account for hydrogen bonding interactions, ionic interactions, lipophilic interactions, and the loss of internal conformational freedom of the ligand.[114] Weights for the terms were determined using multiple linear regression on experimental binding data for 45 protein–ligand complexes. More recently, a number of alternative functions have been derived using the same approach that differ in the experimental data used to derive the model and in the terms included in the expression.[115–117] The main limitation of these approaches is that they are unlikely to be able to predict the affinity of complexes that have characteristics that differ from the complexes in the training set.

Knowledge-based approaches involve summing protein–ligand atom pair interactions. They are based on a statistical analysis of observed patterns of atom contacts in known protein–ligand complexes.[118–120] The distributions of atom contacts are used to calculate a potential of mean force, which gives the free energy of interaction of a pair of atoms as a function of their separation. An advantage of these approaches is that physical effects that are difficult for other methods to model are automatically taken into account. However, as with the empirical methods, knowledge-based methods are limited by the amount of protein–ligand data that are available, since the number and variety of protein–ligand complexes that exist in the PDB is still rather limited. Also, data is typically only available for high-affinity ligands, whereas, virtual screening invariably identifies relatively weakly binding ligands.

The typical way in which a docking program is evaluated is to extract a protein–ligand complex from the PDB, remove the ligand, minimize its energy, and then attempt to dock it back into the protein. The success, or otherwise, is determined by measuring the root mean squared deviation between the predicted pose and the experimentally derived pose. In general, current docking programs are reasonably successful at identifying the correct pose of a ligand,[108,121,122] however, this is not always the top scoring pose.[123,124] Virtual screening experiments usually only consider the top scoring pose and enrichment rates derived in this way can give misleading results if this pose differs from the known binding mode. When the top scoring pose is used to measure success, performance can be considerably reduced; for example, Wu and Vieth[125] report success rates of around 20–40% for three series of protein–ligand complexes covering thrombin, CDK2, and HIV proteases. Several comparisons of different docking programs have been reported in the literature and a review of these is provided by Cole and coworkers who highlight some of the difficulties associated with carrying out such studies.[126]

Scoring functions have been studied intensely over the last few years, but although the functions are reasonably good at predicting the correct pose for a ligand they are much more limited in their ability to achieve accurate rankings across different ligands. This involves estimating the free energy of binding of the protein–ligand complex, which requires consideration of the electrostatic, inductive and nonpolar intermolecular interactions, conformational energy, and entropy as they relate to the protein, ligand, and solvent effects. While progress has been made in addressing the full thermodynamic cycle involved in binding, these methods require time-consuming simulations to be performed and are therefore only suitable for a small number of compounds. In virtual screening applications, some of the most reliable rankings have been achieved by using consensus scoring techniques that combine the results from different scoring functions.[127–129] These methods are similar to the use of data fusion techniques described earlier for combining the results of different similarity searches.

The limited success that has been achieved in predicting binding affinities has led to the development of pre- and postprocessing filters that aim to bias the docking toward specific targets or a family of targets. For example, the FlexPharm program[130] uses pharmacophore constraints to post-process the results of docking and also to direct the search toward particular poses. In the SIFt method, the interactions between a docked ligand and each of the residues in the active site are recorded in a bitstring or interaction fingerprint for the pose.[131] The fingerprints can then be used to analyze the output from a docking run; for example, dockings that are consistent with known interaction patterns can be identified by comparing the fingerprint of a docked ligand with the fingerprint generated from a known protein–ligand complex. A similar approach has been developed by Kelly and Mancera.[132] Another popular approach has been to develop target-biased scoring functions where docking scores are biased toward interaction patterns seen in known inhibitors.[123]

The results of docking depend on molecules being presented in the correct form and hence careful preparation of the data is necessary. For example, the correct protonation state, tautomeric form, and charge states should be assigned both to the ligands and the protein and where ambiguity exists then alternative forms should be presented.

Until recently, having access to a large database of small molecules suitable for docking has been a barrier to carrying out virtual screening, especially within academic groups. However, the ZINC database now provides a publicly available database of 700 K compounds that have been prepared for docking and which can be downloaded as 2D structures or as precalculated 3D conformers.[133] Various search tools are also available to allow users to extract subsets of compounds based on substructures or Lipinski-type properties (see later). The database has been compiled from supplier catalogs with the vendor information available to facilitate purchase of compounds.

3.12.5 Diversity Analysis

The dramatic increase in the numbers of compounds that can be tested for activity either from existing databases or via combinatorial synthesis has led to considerations of the diversity and coverage of chemical space. Corporate databases typically contain in the order of 10^6 compounds and, therefore, represent a tiny fraction of the chemical space occupied by potentially drug-like molecules (estimates vary from 10^{40} to 10^{60}).[49,50] Furthermore, they are likely to be biased collections containing families of closely related compounds that reflect the focus on analog chemistry in traditional drug discovery programs. Consequently, most pharmaceutical companies are actively engaged in compound acquisition programs with the aim of filling the gaps in chemistry space that exist in their in-house collections. Many suppliers now exist that provide compounds for purchase with the source of the compounds including both traditional and combinatorial synthesis. Compound acquisition programs require methods for comparing data sets to enable the purchase of compounds that are diverse with respect to the compounds that are already available internally.

Diversity analysis is also an important concept when designing screening sets, both when the aim is to screen compounds against a range of biological targets and when screening against a single target for which little is known. More biased compounds sets are appropriate when the compounds are to be screened against a family of targets with related properties, for example, kinases or GPCRs. This can be achieved by placing restrictions on the chemistry space that the compounds should occupy; however, it is still important to select a diverse subset of compounds from within the allowed space. The rationale for diversity stems from the similar property principle described earlier in the context of similarity searching.[54] If structurally similar compounds are likely to share the same activity then increasing the structural diversity of a set of compounds should result in an increase in coverage of biological activity space. Furthermore, it should be possible to design a diverse subset of compounds that covers the same biological space as the larger set from which it is derived while minimizing redundancy.

Compound selection methods usually involve selecting a relatively small set of a few tens or hundreds of compounds from a large database, which could consist of hundreds of thousands or even millions of compounds. Identifying the n most dissimilar compounds in a database containing N compounds, when typically $n << N$, is computationally infeasible since it requires consideration of all possible n-member subsets of the database; therefore, a number of different approximate methods have been developed as described below. Prior to applying a compound selection method, computational filters (described later) are usually used to eliminate undesirable compounds from further consideration.

While diversity is related to similarity, it is the property of a collection of compounds rather than of a pair of molecules. A variety of different ways have been developed to assess data set diversity.[134,135] Some are based on combining pairwise (dis)similarities, whereas others involve quantifying the amount of chemistry space that is covered by the compounds. The dissimilarity of a pair of molecules can be considered as the complement of their similarity. For example, when the similarity of two molecules, S_{AB}, is calculated using the Tanimoto coefficient and binary fingerprint representations of molecules, a similarity value in the range zero to one is returned. Their dissimilarity can be quantified as $1 - S_{AB}$ so that a similarity of one corresponds to dissimilarity of zero, and vice versa. When molecules are represented by their physicochemical properties, such as molecular weight, log$P$, molar refractivity, etc., then pairwise dissimilarity, or distance, can be measured using Euclidean distance (following standardization of the properties). A key consideration in assessing diversity is the choice of descriptors used since this effectively defines the chemistry space that is considered. The main compound selection methods include dissimilarity-based compound selection, clustering, partitioning, and the use of optimization techniques and are described below (*see* 4.08 Compound Selection Using Measures of Similarity and Dissimilarity).

3.12.5.1 Dissimilarity-Based Compound Selection

Dissimilarity-based compound selection (DBCS) methods involve selecting a subset of compounds directly based on their pairwise dissimilarities. The basic algorithm for DBCS is outlined below[54]:

1. Select a compound and place it in the subset.
2. Calculate the dissimilarity between each compound remaining in the database and the compounds in the subset.
3. Choose the next compound as that which is most dissimilar to the compounds in the subset.
4. If there are fewer than n compounds in the subset (n being the desired size of the final subset), return to step 2.

The first compound can be selected either at random or as the one that is most dissimilar to all others in the database. The algorithm then proceeds in iterations with each iteration involving the selection of the compound remaining in the database that is most dissimilar to those already selected. Measuring the dissimilarity between one compound and a set of compounds is typically carried out using either the MaxMin or the MaxSum methods.[136] In MaxMin, each database compound is compared with each compound in the subset and its nearest neighbor identified. The database compound that has the maximum dissimilarity to its nearest neighbor in the subset is selected. In MaxSum, the database compound with the maximum sum of dissimilarities to the compounds in the subset is selected. The order in which compounds might be selected in a DBCS algorithm is shown in **Figure 8a**.

Sphere exclusion algorithms are closely related to DBCS methods. They use a dissimilarity threshold as the radius of an exclusion hypersphere in the descriptor space (see **Figure 8b**). The basic algorithm begins by selecting a compound and then excluding from consideration all the compounds within a sphere centered on the chosen compound. Again, variations on the basic method exist. In one method, the first compound selected is the one that is most dissimilar to all others in the database with subsequent compounds chosen based on dissimilarity to the compounds already chosen. An alternative implementation involves choosing the compounds at random. The use of random numbers results in a different subset being selected each time the algorithm is run.[137] In both cases, the basic algorithm continues until all compounds are either selected or excluded and, hence, in contrast to DBCS, it is not possible to specify the final size of the subset.[138]

3.12.5.2 Cluster-Based Compound Selection

Clustering is the process of dividing a collection of objects into groups (or clusters) such that the objects within a cluster are highly similar whereas objects in different clusters are dissimilar (see **Figure 9a**).[139] Clustering therefore

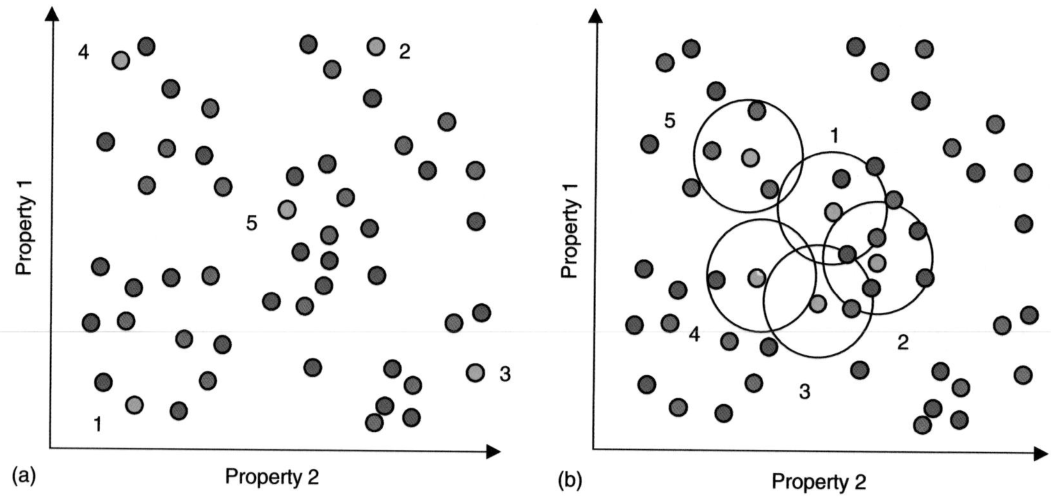

Figure 8 (a) Dissimilarity-based compound selection using the MaxMin function with the order in which the compounds would be selected (green). (b) A sphere exclusion algorithm; again the order in which compounds would be selected is shown.

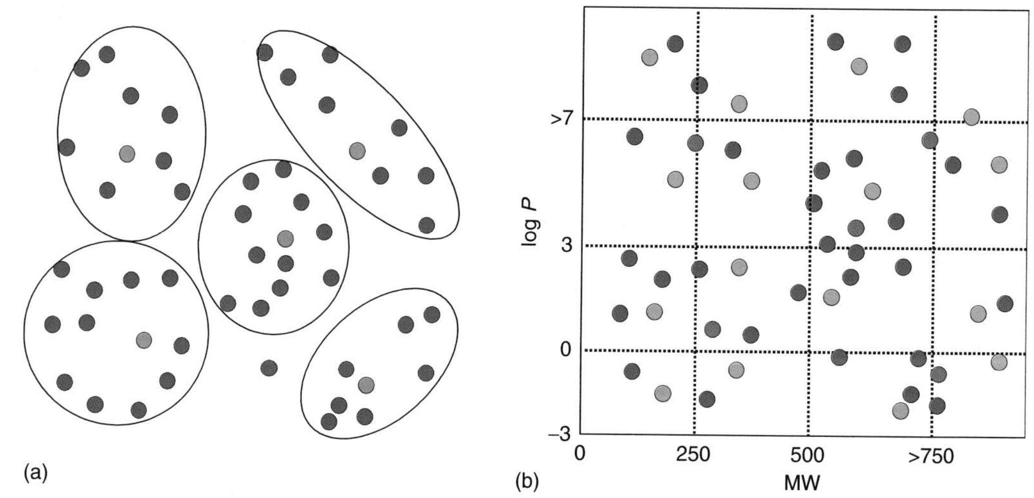

Figure 9 (a) The grouping of the compounds that might be achieved using a clustering algorithm. Here there are five clusters and one singleton. One compound has been selected from each cluster. (b) The compounds are mapped to a portioning scheme based on the properties of log *P* and molecular weight (MW). Each axis has been divided into four bins as shown giving a total of 16 cells. One compound has been selected from each cell.

requires calculation of the pairwise similarities of the compounds in the data set, which is often based on 2D fingerprint representations. Once a database has been clustered then a diverse subset can be selected by choosing one or more compounds from each cluster. For a focused screening set, the compounds could be selected from clusters containing known actives. Many different clustering algorithms have been developed with the methods that are most commonly applied to compound selection including the Jarvis-Patrick method, and Ward's clustering.

Jarvis-Patrick clustering[140] involves first generating a nearest neighbor list for each compound in the database. Compounds are then placed into the same cluster if they share a number of near neighbors; for example, two compounds may be placed in the same cluster if eight out of their fourteen nearest neighbors are in common. Jarvis-Patrick is a relatively fast clustering method, however, the basic algorithm can result in rather skewed clusters with a small number of very large clusters and a large number of singletons.

Ward's clustering is a hierarchical method in which smaller clusters of very similar molecules are embedded within larger clusters.[141] The method begins by placing each compound in its own cluster and then proceeds by merging the most similar clusters together in an iterative manner. Thus, in the first step the closest two compounds are merged into a single cluster; in the next step, the closest two clusters are merged, and so on. The process continues until all compounds are in a single cluster. The next step is to choose a clustering level and various methods are available for automatically determining an appropriate clustering level.[142]

Clustering methods are based solely on intermolecular similarities; hence, they provide a relative measure of the space covered by a data set, rather than an absolute measure. This makes it difficult to compare data sets such as an in-house database and a database offered by a commercial vendor. Such a comparison would require that the databases are combined and then clustered together as a combined unit. The degree of overlap in the two databases could then be assessed by examining the contents of each cluster. If a cluster is mainly occupied by compounds from one of the databases then this indicates a region of space where the databases differ. Thus, a company may decide to augment its internal database by purchasing vendor compounds from clusters that are sparsely occupied by their own in-house compounds.

3.12.5.3 Partitioning Methods

Clustering and dissimilarity-based compound selection methods are based on calculating pairwise (dis)similarities and therefore provide relative measures of diversity. In contrast, partitioning or cell-based methods provide an absolute measure of diversity and involve assessing the extent to which a data set covers some predefined chemistry space. Partitioning methods require the definition of a small number of properties that can be used to define a low-dimensional chemistry space. For example, an early example of the use of partitioning was based on physicochemical properties such as molecular weight, calculated log *P*, number of hydrogen bond donors, etc.[143] Each property or descriptor defines an axis of the chemistry space. The range of values for each property is divided into a set of bins, and the combinatorial product of all bins then defines the set of cells or partitions that comprise the space. A database is mapped onto the space by assigning each molecule to a cell according to its properties (see **Figure 9b**). A diverse subset of compounds can be selected by taking one or more molecules from each cell, and, alternatively, a focused subset can be selected by choosing compounds from a limited number of cells, e.g., from the cells adjacent to a cell occupied by a known active. Since a partitioning scheme is defined independently of the compounds that are mapped onto it, the space occupied by different data sets can be compared easily. Partitioning methods also allow voids or underrepresented regions of the space to be identified and can therefore be used for compound acquisition from external vendors. However, partitioning methods are limited to low dimensional descriptors.

Partitioning schemes have been developed from sets of descriptors known as BCUTs. These are calculated from matrix representations of a molecule's connection table.[144] The diagonals of a matrix represent a property of each of the atoms such as atomic charge, atomic polarizability and atomic hydrogen bonding ability. The off-diagonals are assigned the value 0.1 times the bond type if the atoms are bonded, and 0.001 if the atoms are not bonded. The highest and lowest eigenvalues of each matrix are then extracted for use as descriptors. The three types of matrix described above would thus give rise to six descriptors that describe a six-dimensional space. An initial procedure can be used to identify the most appropriate BCUTs to use for a given data set in order to tailor the method for data sets with different characteristics. This feature is particularly useful when designing focused screening sets.

3.12.5.3.1 Pharmacophore fingerprints

The 3-point pharmacophoric fingerprints described earlier in the context of similarity searching have also been widely used for selecting diverse sets of compounds.[74] The pharmacophore fingerprints for a set of molecules can be combined into an ensemble pharmacophore, which is the union of the individual fingerprints. The resulting fingerprint can then be used to measure total pharmacophore coverage of the data set; to identify pharmacophores that are not represented in the set of molecules; and to compare different sets of molecules. Thus, if the aim were to select a diverse set of compounds this would correspond to maximizing the coverage of pharmacophore triplets over all compounds in the subset. The 3-point pharmacophore fingerprint described by Mason and Pickett[74] is based on six features and six distance ranges and consists of 5196 bits. Discrimination can be increased by using a finer distance resolution and by extending the fingerprint to encode 4-point pharmacophores or quartets of atoms. For example, systems have been described that consist of over 350 million bits for 4-point pharmacophores.[76] Furthermore, 4-point pharmacophores are able to distinguish chiral centers, which is not possible when using 3-point pharmacophores. Pharmacophore fingerprints are also widely used in the design of focused screening sets, for example, the analysis of known active compounds can lead to the identification of privileged substructures. These are features that occur frequently within the known actives and a focused subset of compounds would be one that is enriched in privileged features.

3.12.5.4 Optimization Methods

Optimization techniques can provide effective ways of sampling large search spaces and hence several such methods have been applied to compound selection, for example, the use of Monte Carlo methods combined with simulated annealing.[145,146] In the Monte Carlo method, the selection of a diverse subset proceeds as follows. An initial subset is chosen at random and its diversity is calculated. A new subset is then generated from the first by replacing some of the compounds with others chosen at random. If the new subset is more diverse than the previous subset it is accepted for use in the next iteration; if it is less diverse, then the probability that it is accepted depends on the Boltzmann factor. The process continues for a fixed number of iterations or until no further improvement is observed in the diversity function. A variety of different diversity measures can be used with optimization methods. For example, diversity could be measured as the sum of pairwise dissimilarities, the number of occupied cells in a partitioning scheme, or the number of 3-point pharmacophores covered by the subset. An interesting diversity function that has been used for compound selection is based on computing the minimum spanning tree for the set of molecules.[145] A spanning tree is a set of edges that connect a set of nodes without forming any cycles. The nodes are the molecules in the subset and each edge is labeled by the dissimilarity between the two molecules it connects. The minimum spanning tree is the spanning tree that connects all molecules in the subset with the minimum sum of pairwise dissimilarities. The diversity of the subset then equals the sum of the intermolecular similarities along the edges on the minimum spanning tree.

3.12.6 Combinatorial Library Design

Combinatorial chemistry is the process whereby large numbers of compounds are synthesized in parallel. The technology has developed as a means of providing large numbers of compounds for HTS and the techniques are now sufficiently well developed that it is easy to plan synthetic schemes that could potentially generate massive numbers of compounds. Indeed, for most combinatorial reaction schemes there are many more monomers available than can be handled in practice. For example, consider a benzodiazepine library constructed from three monomer pools, as shown in **Figure 10**. If a search for suitable monomers is carried out in a database of available reagents such as the Available Chemicals Directory[147] then so many possibilities exist that the number of potential products could easily contain millions, if not billions, of compounds. Thus, the key issue in combinatorial library design is the selection of appropriate monomers so that the resulting library is of a manageable size and has the desired properties. The initial emphasis in combinatorial chemistry was on the synthesis of large diverse libraries. However, more recently two trends have emerged. The first is toward the design of smaller, more focused libraries that incorporate as much information about the therapeutic target as possible, as for the selection of screening sets described earlier. The second trend is based on the realization that compounds within combinatorial libraries, whether they are diverse or focused, should have 'drug-like' physicochemical properties so that they constitute good starting points for further optimization.

The two main strategies for combinatorial library design are known as monomer- (or reactant-) based selection and product-based selection. In monomer-based selection optimized subsets of monomers are selected without consideration of the products that will result. Any of the subset selection methods identified earlier, such as clustering or partitioning, can be applied. An early example of monomer-based design using experimental design techniques for the selection of diverse monomers in the synthesis of peptoid libraries is described by Martin and coworkers.[148]

Product-based selection takes the properties of the resulting product molecules into account when selecting the monomers and is appropriate when designing focused libraries, which require consideration of the properties of the resulting products, and also when the aim is to optimize the properties of a library as a whole, such as its diversity or the physicochemical property space it covers. Product-based selection is computationally more demanding than

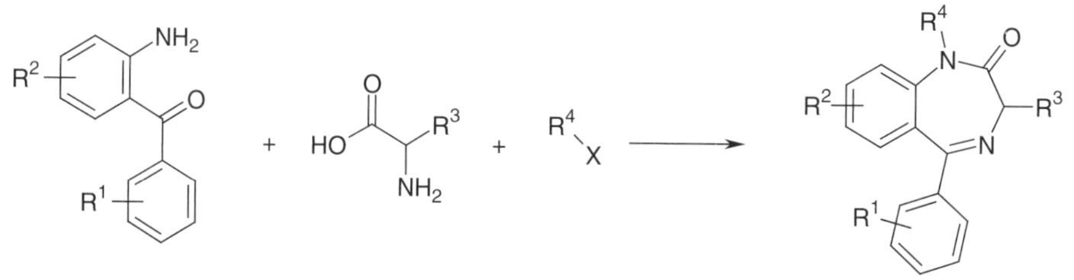

Figure 10 A benzodiazepine library where there are four positions of variability around the ring system and which is built from three monomer pools.

monomer-based selection since it usually requires enumeration of the entire virtual library that could potentially be made, followed by calculation of descriptors for the products. One approach to enumeration is the reaction transform approach. Here, the reaction mechanism itself is encoded as the transformation of the atoms and bonds of the monomers required in order to create the product molecules with the transformation being described by assigning labels to corresponding atoms in the monomers and the products, in a process known as atom mapping.[149]

Recently, Downs and Barnard[150] have described an efficient method of generating fingerprint descriptors of the molecules in a combinatorial library without the need for enumeration of the products. This method is based on earlier technology for handling Markush structures. An alternative approach is to use random sampling techniques to derive a statistical model of the property under consideration.[151]

Product-based selection is typically implemented using an optimization technique such as a GA or simulated annealing,[152–154] which aims to identify a combinatorial subset directly. For example, a GA has been described in which each chromosome encodes one possible combinatorial subset of a predefined size.[153] Thus, consider the design of a two-component combinatorial subset of size $n_A \times n_B$ selected from a possible $N_A \times N_B$ virtual library. Each chromosome consists of $n_A + n_B$ elements, with each element specifying one possible monomer selected from the appropriate monomer pool. The fitness function quantifies the 'goodness' of the combinatorial subset encoded in the chromosome and the GA evolves new potential subsets in an attempt to maximize this quantity. The fitness function could be designed to maximize the diversity of the subset, using MaxSum, MaxMin, or via a partitioning scheme, or it could be designed to focus the subset around a known target compound, for example, by maximizing the sum of similarities to the target compound.

Alternative approaches to product-based library design have been developed that do not directly take the combinatorial constraint into account. These methods have been termed molecule-based methods to distinguish them from library-based methods.[155] They are based on identifying a subset of product molecules with desired properties initially without consideration of the combinatorial chemistry. Once a subset of product molecules has been selected, the molecules are examined to identify monomers that occur frequently within them and a combinatorial subset can be constructed based on the frequently occurring monomers.

A recent trend in library design is to optimize libraries over a number of properties simultaneously, for example, as well as being diverse or focused, it is usually desirable that a library is cheap to synthesize and that the compounds contained within it have 'drug-like' physicochemical properties. Most approaches to designing libraries based on multiple objectives have involved combining different properties via a weighted-sum fitness function. However, recently, the limitations of this approach have been recognized.[156] For example, it can be difficult to choose appropriate weights especially when the objectives are in competition and a single somewhat arbitrary compromise solution is produced. These limitations have been addressed through the implementation of a multiobjective genetic algorithm (MOGA).[156,157] In this approach, multiple objectives are handled independently and a family of equivalent solutions is found, where each solution represents a different compromise in the objectives. This approach allows the relationships between the various objectives to be investigated so that the library designer can make an informed choice on an appropriate compromise solution. This approach has also been extended to allow library size and configuration to be optimized along with other properties.

3.12.7 Prediction of Absorption, Distribution, Metabolism, and Excretion Properties

The development of methods for the accurate prediction of absorption, distribution, metabolism, and excretion (ADME) properties has become a very active area of research.[158–160] The interest in these techniques is due to the high rates of attrition of compounds that have been seen in the later stages of drug discovery: it has been reported that only 1 in 10 compounds entering the clinical development phase of drug discovery actually makes it to the marketplace with many of the failures being due to poor pharmacokinetics.[161] Given these high failure rates, it is now considered important to take ADME properties into account as early as possible in the drug discovery process, to identify and address any issues sooner rather than later and therefore prevent time and resources being spent on compounds with undesirable properties.[162,163]

Experimental methods for measuring ADME properties do not provide sufficient throughput for the analysis of large numbers of compounds and they are costly in terms of resources. Furthermore, they cannot be applied to compounds that have not yet been synthesized, i.e., to virtual compounds. Therefore, there is a huge amount of interest in developing computational models. Computational techniques include the use of computational filters to eliminate compounds with undesirable properties and methods for the prediction of general 'drug-likeness.' These general filters are appropriate in the early stages of drug discovery. The lead optimization stage usually calls for more specific methods that attempt to model properties such as oral absorption, solubility, blood–brain barrier penetration, etc. A variety of different model building approaches have been developed for the prediction of specific ADME properties. Many are based on traditional QSAR techniques such as multiple linear regression and partial least squares; however, more sophisticated machine learning techniques such as neural nets and support vector machines are also being employed.[160] Computational filters are described first then methods for the prediction of more specific ADME properties are discussed.

3.12.7.1 Computational Filters

Computational filters can be applied to virtual compounds to prevent the synthesis of undesirable compounds or to libraries of available compounds to prevent the screening of compounds that are likely to be rejected later in the project. The simplest computational filters consist of substructural searches that are used to eliminate compounds that contain undesirable functional groups such as compounds with toxic functional groups or substructures that are reactive and likely to interfere with the synthesis. Other commonly used filters are based on counts of structural features such as numbers of rotatable bonds, and on physicochemical properties such as molecular weight and log *P*. These criteria are in widespread use following the publication of the 'rule-of-five' by Lipinski and colleagues.[164] The rule-of-five suggests that oral absorption is unlikely for a molecule that satisfies two of the following criteria:

- molecular weight >500;
- number of hydrogen bond donors >5;
- number of hydrogen bond acceptors >10; and
- calculated log $P>5.0$.

The rules were derived from an analysis of just over 2000 compounds that had entered clinical trials. More recent studies using larger data sets have broadly reinforced these results.[165] However, when applying filters such as these it is important to realize that there are always exceptions to the rules. Computational filters are often incorporated within in-house programs, for example, the REOS[166] and ADEPT[149] programs allow several different types of filters to be applied via simple web interfaces.

More sophisticated approaches have been developed that aim to classify compounds as drug-like or non-drug-like.[167] These methods generally involve the use of a training set of known drugs and nondrugs. Several databases have been used to represent drugs, for example, the World Drug Index, the MDL Drug Data Report (MDDR),[12] and the Comprehensive Medicinal Chemistry database. The nondrugs have been selected from databases such as the Available Chemicals Database (ACD)[147] and the SPRESI database[168] with the assumption being made that the compounds contained within them are inactive. The classification methods have included genetic algorithms, neural networks, and decision trees. The algorithms 'learn' classification rules from the input data in the training set with the rules being based on the molecular descriptors used to represent the compounds. Once the algorithms have been trained, then predictions can be made about previously unseen compounds.

An example is the method developed by Gillet and coworkers in which a genetic algorithm is used to devise a scoring scheme for drug-likeness based on a number of physicochemical properties.[169] The properties include counts of hydrogen bond donors, hydrogen bond acceptors, rotatable bonds, and aromatic rings together with molecular weight and a shape index known as the kappa α^2 index. Each property is binned according to ranges of values and a weight is associated with each bin. The GA attempts to find the set of weights that maximizes the discrimination between compounds classed as drugs and those classed as nondrugs. A set of weights is evaluated by using it to assign scores to all of the compounds in the training set with the score for a molecule calculated by summing the appropriate weights across each property. A ranked list is then generated and the distribution of drugs and nondrugs in the ranked list is evaluated. The GA was trained using a sample of the SPRESI database[168] to represent non-drug-like compounds and a sample of the World Drugs Index to represent drug-like compounds. The resulting model was surprisingly effective at distinguishing between the two classes of compounds and the method has been used to filter compound sets prior to HTS.[170] Other related approaches have been described by Ajay and coworkers[171] and Wagener and colleagues[172] amongst others.

A recent focus has been on the prediction of lead-likeness rather than drug-likeness especially when selecting screening sets,[173,174] the reasoning being that HTS aims to identify lead compounds that can subsequently be optimized into drug candidates. The optimization stage usually involves adding functionality to the molecules, which typically results in an increase in properties such as molecular weight and hydrogen bonding groups.

3.12.7.2 Prediction of Physicochemical Properties

The lead optimization stage of drug discovery usually calls for more specific methods that attempt to model properties such as oral absorption, blood–brain barrier penetration, etc. Many ADME models include physicochemical properties as descriptors, such as hydrophobicity and solubility and hence the calculation of these properties has also been widely studied. The aim here is usually to make quantitative predictions and hence many methods are based on techniques originally applied in QSAR studies, for example, multiple linear regression and partial least squares (PLS).

Hydrophobicity is an important property in determining the activity and transport of drugs.[175,176] For example, it is a key factor in membrane permeability and hence drug absorption and also in solubility. It is usually expressed as the logarithm of the partition coefficient between n-octanol and water (log P). Experimental determination of log P can be difficult, particularly for zwitterionic and very lipophilic or polar compounds, so that data is currently available for approximately 30 000 compounds only[177] and, of course, there is no data on compounds yet to be synthesized, such as those contained in virtual libraries.

Many programs have been developed for calculating log P with several of them based on summing contributions over the fragments or atoms in a molecule. In the fragment-based methods, fragment values are obtained from accurately measured experimental log P data for a small set of simple molecules.[178,179] A molecule is broken into fragments and the appropriate values are summed (with correction factors being applied to account for interactions between fragments). A limitation of these methods has been the occurrence of fragments for which data is not available (missing fragments); however, a recent development of the ClogP program has been the implementation of a method to estimate the values of missing fragments.[178,179] Atom-based methods involve summing the contributions from individual atoms.[180–182] The atom contributions are typically determined from a regression analysis using a training set of compounds for which experimental partition coefficients are available.

Recent reviews of methods for predicting solubility are provided by Delaney[183] and Jorgensen and Duffy.[184] Some methods are based on modeling the underlying processes involving crystal effects and affinity for water. Such approaches can be time consuming and are limited in the numbers of compounds for which estimates can be made in reasonable time. Group and atom contribution methods have been developed that are similar to those described for predicting log P. Alternative approaches have been to derive models using statistical and machine learning techniques based on a training set of compounds. A wide variety of model building techniques have been applied including multiple linear regression, partial least squares, neural networks, support vector machines, and genetic algorithms and they have been based on a variety of different molecular descriptors including whole molecule properties such as log P and hydrogen bond counts, topological descriptors, and 3D descriptors. For example, Lind and Maltseva describe a method for predicting solubility that is based on a support vector machine with molecular fingerprints as descriptors.[185]

Polar surface area (PSA) is another molecular property that is widely used in ADME models, especially for the prediction of oral absorption and brain penetration. Polar surface area is defined as the amount of molecular surface arising from polar atoms (nitrogen and oxygen atoms together with their attached hydrogen atoms; some definitions also include sulfur atoms and their attached hydrogen atoms). It has been shown that a single conformation is sufficient to calculate PSA.[186] A fast fragment-based method has also been developed that allows PSA descriptors to be used in virtual screening applications.[187]

The prediction of blood–brain barrier (BBB) penetration has also been studied. For a compound to be able to reach its site of action in the body it will usually have to pass through one or more physiological barriers such as the intestinal cell membrane or the BBB barrier. Thus, drugs acting on the central nervous system need to pass through the BBB barrier.[188] Most early models for predicting BBB penetration are based on multiple linear regression and use physiochemical properties.[189] More recent approaches are based on other multivariate techniques. For example, partial least squares has been used with newly developed descriptors called Volsurf that were specifically developed for the prediction of ADME properties.[190] Volsurf descriptors are based on 3D molecular fields similar to those used for the calculation of molecular similarity. The GRID program[191] is used to calculate molecular interaction fields by placing various probes at each of the vertices of a grid surrounding a molecule and the interaction energy between the probe and the molecule is calculated. The molecular fields are then transformed into a set of numerical descriptors that quantify the molecule's overall size and shape and the balance between hydrophilicity, hydrophobicity, and hydrogen bonding.

Some of the challenges associated with prediction of ADME properties have been described recently by Davis and Riley.[161] For example, QSAR methods were originally developed for the analysis of small homologous sets of molecules and prediction outside of the domain of the model was not required. However, the data sets that are used to derive ADME models are more heterogeneous and may be extracted from corporate databases or may be compiled from published data sets. This presents new challenges for QSAR, which has led to renewed interest in validation studies[192–194] and in assessing the domain of applicability of models. For example, a model cannot be expected to make predictions for compounds that are outside of the descriptor space occupied by the training set.[195] A further issue is that data collected from the literature will often have been generated under different experimental conditions in different laboratories so that the reliability of the data is questionable. Thus, the potential for the success of these methods is currently limited by the lack of good data with which to build models and ultimately by the fact that the processes that determine the fate of a compound in vivo are still poorly understood. Finally, the prediction of properties such as distribution, metabolism, and toxicity is making slow progress, which is largely due to a lack of published data.[163]

3.12.8 Conclusions

This chapter has described some of the major techniques that have been developed in the field of chemoinformatics. As has been discussed, computer methods for the handling of chemical data first appeared about four decades ago. The recent emergence of chemoinformatics as a discipline is largely due to the explosion of data that has become available over the last decade due to the introduction of automation techniques within the chemical industries. Fortunately, progress in computer hardware has permitted the development of techniques that are able to handle the large databases of chemical structures that are now common. While an attempt has been made to summarize the main techniques that are widely applied in the drug discovery process, it is acknowledged that there are many topics that have been excluded. Indeed, it would not be possible to cover all aspects of the field in a single chapter. Omissions include reaction databases, chemical patents, representational issues such as stereochemistry and tautomerism, and data visualization amongst others. For more comprehensive overviews of the field the reader is referred to the textbooks identified in the introduction.

References

1. Brown, F. K. *Annu. Rep. Med. Chem.* **1998**, *33*, 375–384.
2. Hann, M.; Green, R. *Curr. Opin. Chem. Biol.* **1999**, *3*, 379–383.
3. Schofield, H.; Wiggins, G.; Willett, P. *Drug Disc. Today* **2001**, *6*, 931–934.
4. Leach, A. R.; Gillet, V. J. *An Introduction to Chemoinformatics*; Kluwer: Dordrecht, The Netherlands, 2003.
5. Gasteiger, J.; Engel, T. *Chemoinformatics: A Textbook*; Wiley-VCH: Weinheim, Germany, 2003.
6. Bajorath, J. *Chemoinformatics: Concepts, Methods and Tools for Drug Discovery*; Humana Press: New Jersey, NJ, 2004.
7. Oprea, T. I. *Chemoinformatics in Drug Discovery*; Wiley-VCH: Weinheim, Germany, 2005.
8. Bajorath, J. *Drug Disc. Today* **2004**, *9*, 13–14.
9. Weininger, D. *J. Chem. Inf. Comput. Sci.* **1988**, *28*, 31–36.
10. Weininger, D.; Weininger, A.; Weininger, J. L. *J. Chem. Inf. Comput. Sci.* **1989**, *29*, 97–101.
11. CAS Registry Chemical Abstracts Service. http://www.cas.org/EO/resys.html (accessed May 2006).
12. MDL Information Systems, Inc. 14600 Catalina Street, San Leandro, CA 94577, USA. http://www.mdli.com (accessed May 2006).
13. Morgan, H. *J. Chem. Doc.* **1965**, *5*, 107–113.
14. Wipke, W.; Dyott, T. *J. Am. Chem. Soc.* **1974**, *96*, 4825–4834.
15. Freeland, R.; Funk, S.; O'Korn, L.; Wilson, G. *J. Chem. Inf. Comput. Sci.* **1979**, *19*, 94–98.
16. Hodes, L. J. *Chem. Inf. Comput. Sci.* **1976**, *16*, 88–93.
17. Adamson, G. W.; Cowell, J.; McLure, A. H. W.; Town, W. G.; Yapp, A. M.; Lynch, M. F. *J. Chem. Doc.* **1973**, *13*, 153–157.
18. Barnard Chemical Information Ltd. (BCI), 46 Uppergate Road, Stannington, Sheffield S6 6BX, UK. http://www.bci.gb.com (accessed May 2006).
19. Willett, P. *J. Chem. Inf. Comput. Sci.* **1979**, *19*, 159–162.
20. Daylight Chemical Information Systems, Inc., 120 Vantis – Suite 550, Aliso Viejo, CA 92656, USA. http://www.daylight.com (accessed May 2006).
21. Barnard, J. M. *J. Chem. Inf. Comput. Sci.* **1993**, *33*, 532–538.
22. Ray, L. C.; Kirsch, R. A. *Science* **1957**, *126*, 814–819.
23. Barnard, J. M. Representations of Molecular Structures – Overview. In *Handbook of Chemoinformatics*; Gasteiger, J., Ed.; Wiley-VCH: Weinheim, Germany, 2003; Vol. 1, pp 27–50.
24. Allen, F. H. *Acta Crystallogr. Sect. B* **2002**, *58*, 380–388.
25. Berman, H. M.; Westbrook, J.; Feng, Z.; Gilliland, G.; Bhat, T. N.; Weissig, H.; Shindyalov, I. N.; Bourne, P. E. *Nucleic Acids Res.* **2000**, *28*, 235–242.
26. Perola, E.; Charifson, P. S. *J. Med. Chem.* **2004**, *47*, 2499–2510.
27. Rusinko, A.; Sheridan, R. P.; Nilakantan, R.; Haraki, K. S.; Bauman, N.; Venkataraghavan, R. *J. Chem. Inf. Comput. Sci.* **1989**, *29*, 251–255.
28. Gasteiger, J.; Rudolph, C.; Sadowski, J. *Tetrahedron Comput. Methodol.* **1990**, *3*, 537–547.
29. Sadowski, J. 3D Structure Generation. In *Handbook of Chemoinformatics*; Gasteiger, J., Ed.; Wiley-VCH: Weinheim, Germany, 2003; Vol. 1, pp 231–261.
30. Bostrom, J. *J. Comput.-Aided Mol. Des.* **2001**, *15*, 1137–1152.
31. Leach, A. R. *Molecular Modelling Principles and Applications,* 2nd ed.; Pearson Education: Harlow, UK, 2001.
32. Jakes, S. E.; Willett, P. *J. Mol. Graphics* **1986**, *4*, 12–20.

33. Poirrette, A. R.; Willett, P.; Allen, F. H. *J. Mol. Graphics* **1993**, *11*, 2–14.
34. Poirrette, A. R.; Willett, P.; Allen, F. H. *J. Mol. Graphics* **1991**, *9*, 203–217.
35. Jakes, S. E.; Watts, N.; Willett, P.; Bawden, D.; Fisher, J. D. *J Mol Graphics* **1987**, *5*, 41–48.
36. Sheridan, R. P.; Nilakantan, R.; Rusinko, A.; Bauman, N.; Haraki, K. S.; Venkataraghavan, R. *J. Chem. Inf. Comput. Sci.* **1989**, *29*, 255–260.
37. Hurst, T. *J. Chem. Inf. Comput. Sci.* **1994**, *34*, 190–196.
38. Cramer, R. D.; Patterson, D. E.; Bunce, J. D. *J. Am. Chem. Soc.* **1988**, *110*, 5959–5967.
39. Wolber, G.; Langer, T. *J. Chem. Inf. Model.* **2005**, *45*, 160–169.
40. Hahn, M. *J. Chem. Inf. Comput. Sci.* **1997**, *37*, 80–86.
41. Martin, Y. C.; Bures, M. G.; Danaher, E. A.; Delazzer, J.; Lico, I.; Pavlik, P. A. *J. Comput.-Aided Mol. Des.* **1993**, 7, 83–102.
42. Kristam, R.; Gillet, V. J.; Lewis, R. A.; Thorner, D. *J. Chem. Inf. Model.* **2005**, *45*, 461–476.
43. Jones, G.; Willett, P.; Glen, R. C. *J. Comput.-Aided Mol. Des.* **1995**, *9*, 532–549.
44. Patel, Y.; Gillet, V. J.; Bravi, G.; Leach, A. R. *J. Comput.-Aided Mol. Des.* **2002**, *16*, 653–681.
45. van Drie, J. H. *Current Pharm. Des.* **2003**, *9*, 1649–1664.
46. Cottrell, S. J.; Gillet, V. J.; Taylor, R.; Wilton, D. J. *J. Comput.-Aided Mol. Des.* **2004**, *18*, 665–682.
47. Oprea, T. I.; Matter, H. *Curr. Opin. Chem. Biol.* **2004**, *8*, 349–358.
48. Stahura, F. L.; Bajorath, J. *Comb. Chem. High Throughput Screening* **2004**, 7, 259–269.
49. Valler, M. J.; Green, D. *Drug Disc. Today* **2000**, *5*, 286–293.
50. Hann, M. M.; Leach, A. R.; Green, D. V. S. Computational chemistry, molecular complexity and screening set design. In *Chemoinformatics in Drug Discovery*; Oprea, T. I., Ed.; Wiley-VCH: Weinheim, Germany, 2004, pp 43–57.
51. Wilton, D.; Willett, P.; Lawson, K.; Mullier, G. *J. Chem. Inf. Comput. Sci.* **2003**, *43*, 469–474.
52. Carhart, R. E.; Smith, D. H.; Venkataraghavan, R. *J. Chem. Inf. Comput. Sci.* **1985**, *25*, 64–73.
53. Willett, P.; Winterman, V.; Bawden, D. *J. Chem. Inf. Comput. Sci.* **1986**, *26*, 36–41.
54. Lajiness, M. Molecular Similarity-Based Methods for Selecting Compounds for Screening. In *Computational Chemical Graph Theory*; Rouvray, D., Ed.; Nova Science Publishers: New York, 1990, pp 299–316.
55. Willett, P.; Barnard, J. M.; Downs, G. M. *J. Chem. Inf. Comput. Sci.* **1998**, *38*, 983–996.
56. Randić, M. *J. Mol. Graphics Model.* **2001**, *20*, 19–35.
57. Nilakantan, R.; Bauman, N.; Dixon, J. S.; Venkataraghavan, R. *J. Chem. Inf. Comput. Sci.* **1987**, *27*, 82–85.
58. Kearsley, S. K.; Sallamack, S.; Fluder, E. M.; Andose, J. D.; Mosley, R. T.; Sheridan, R. P. *J. Chem. Inf. Comput. Sci.* **1996**, *36*, 118–127.
59. Schneider, G.; Neidhart, W.; Giller, T.; Schmid, G. *Angew. Chem. Int. Ed. Engl.* **1999**, *38*, 2894–2896.
60. Schuffenhauer, A.; Floersheim, P.; Acklin, P.; Jacoby, E. *J. Chem. Inf. Comput. Sci.* **2003**, *43*, 391–405.
61. Scitegic, 9665 Chesapeake Drive, Suite 401, San Diego, CA 92123–1365, USA. http://www.scitegic.com (accessed May 2006).
62. Hert, J.; Willett, P.; Wilton, D. J.; Acklin, P.; Azzaoui, K.; Jacoby, E.; Schuffenhauer, A. *Org. Biomol. Chem.* **2004**, *2*, 3256–3266.
63. Hagadone, T. R. *J. Chem. Inf. Comput. Sci.* **1992**, *32*, 515–521.
64. Raymond, J. W.; Gardiner, E. J.; Willett, P. *J. Chem. Inf. Comput. Sci.* **2002**, *42*, 305–316.
65. Raymond, J. W.; Willett, P. *J. Chem. Inf. Comput. Sci.* **2003**, *43*, 908–916.
66. Gillet, V. J.; Downs, G. M.; Ling, A.; Lynch, M. F.; Venkataram, P.; Wood, J. V.; Dethlefsen, W. *J. Chem. Inf. Comput. Sci.* **1987**, *27*, 126–137.
67. Gillet, V. J.; Willett, P.; Bradshaw, J. *J. Chem. Inf. Comput. Sci.* **2003**, *43*, 338–345.
68. Barker, E. J.; Gardiner, E. J.; Gillet, V. J.; Kitts, P.; Morris, J. *J. Chem. Inf. Comput. Sci.* **2003**, *43*, 346–356.
69. Harper, G.; Bravi, G. S.; Pickett, S. D.; Hussain, J.; Green, D. V. S. *J. Chem. Inf. Comput. Sci.* **2004**, *44*, 2145–2156.
70. Takahashi, Y.; Sukekawa, M.; Sasaki, S. *J. Chem. Inf. Comput. Sci.* **1992**, *32*, 639–643.
71. Barker, E. J.; Cosgrove, D. A.; Gardiner, E. J.; Gillet, V. J.; Kitts, P.; Willett, P. *J. Chem. Inf. Model.*, submitted for publication.
72. Rarey, M.; Dixon, J. S. *J. Comput.-Aided Mol. Des.* **1998**, *12*, 471–490.
73. Rarey, M.; Stahl, M. *J. Comput.-Aided Mol. Des.* **2001**, *15*, 497–520.
74. Mason, J. S.; Pickett, S. D. *Perspect. Drug Disc. Des.* **1997**, *7–8*, 85–114.
75. Good, A. C.; Cho, S. J.; Mason, J. S. *J. Comput.-Aided Mol. Des.* **2004**, *18*, 523–527.
76. Mason, J. S.; Morize, I.; Menard, P. R.; Cheney, D. L.; Hulme, C.; Labaudiniere, R. F. *J. Med. Chem.* **1999**, *42*, 3251–3264.
77. Lemmen, C.; Lengauer, T. *J. Comput.-Aided Mol. Des.* **2000**, *14*, 215–232.
78. Lemmen, C.; Lengauer, T.; Klebe, G. *J. Med. Chem.* **1998**. *41*, 4502–4520.
79. Good, A. C.; Hodgkin, E. E.; Richards, W. G. *J. Chem. Inf. Comput. Sci.* **1992**, *32*, 188–191.
80. Wild, D. J.; Willett, P. *J. Chem. Inf. Comput. Sci.* **1996**, *36*, 159–167.
81. Grant, J. A.; Gallardo, M. A.; Pickup, B. T. *J. Comput. Chem.* **1996**, *17*, 1653–1666.
82. Brown, R. D.; Martin, Y. C. *J. Chem. Inf. Comput. Sci.* **1997**, *37*, 1–9.
83. Brown, R. D.; Martin, Y. C. *J. Chem. Inf. Comput. Sci.* **1996**, *36*, 572–584.
84. National Cancer Institute. http://dtp.nci.nih.gov/webdata.html (accessed May 2006).
85. Edgar, S. J.; Holliday, J. D.; Willett, P. *J. Mol. Graphics Model.* **2000**, *18*, 343–357.
86. Schuffenhauer, A.; Gillet, V. J.; Willett, P. *J. Chem. Inf. Comput. Sci.* **2000**, *40*, 295–307.
87. Rush, T. S.; Grant, J. A.; Mosyak, L.; Nicholls, A. *J. Med. Chem.* **2005**, *48*, 1489–1495.
88. Sheridan, R. P.; Kearsley, S. K. *Drug Disc. Today* **2002**, 7, 903–911.
89. Whittle, M.; Gillet, V. J.; Willett, P.; Alex, A.; Loesel, J. *J. Chem. Inf. Comput. Sci.* **2004**, *44*, 1840–1848.
90. Xue, L.; Godden, J. W.; Stahura, F. L.; Bajorath, J. *J. Chem. Inf. Comput. Sci.* **2003**, *43*, 1151–1157.
91. Cramer, R. D.; Redl, G.; Berkoff, C. E. *J. Med. Chem.* **1974**, *17*, 533–535.
92. Hawkins, D. M.; Young, S. S.; Rusinko, A. *Quant Struct.-Act Relat* **1997**, *16*, 296–302.
93. Svetnik, V.; Liaw, A.; Tong, C.; Culberson, J. C.; Sheridan, R. P.; Feuston, B. P. *J. Chem. Inf. Comput. Sci.* **2003**, *43*, 1947–1958.
94. Warmuth, M. K.; Liao, J.; Ratsch, G.; Mathieson, M.; Putta, S.; Lemmen, C. *J. Chem. Inf. Comput. Sci.* **2003**, *43*, 667–673.
95. Saeh, J. C.; Lyne, P. D.; Takasaki, B. K.; Cosgrove, D. A. *J. Chem. Inf. Model.* **2005**, *45*, 1122–1133.
96. Congreve, M.; Murray, C. W.; Blundell, T. L. *Drug Disc. Today* **2005**, *10*, 895–907.
97. Kuntz, I. D. *Science* **1992**, *257*, 1078–1082.
98. Kuntz, I. D.; Blaney, J. M.; Oatley, S. J.; Langridge, R.; Ferrin, T. E. *J. Mol. Biol.* **1982**, *161*, 269–288.
99. Abagyan, R.; Totrov, M. *Curr. Opin. Chem. Biol.* **2001**, *5*, 375–382.
100. Taylor, R. D.; Jewsbury, P. J.; Essex, J. W. *J. Comput.-Aided Mol. Des.* **2002**, *16*, 151–166.

101. Halperin, I.; Ma, B. Y.; Wolfson, H.; Nussinov, R. *Proteins Struct. Funct. Genet.* **2002**, *47*, 409–443.
102. Friesner, R. A.; Banks, J. L.; Murphy, R. B.; Halgren, T. A.; Klicic, J. J.; Mainz, D. T.; Repasky, M. P.; Knoll, E. H.; Shelley, M.; Perry, J. K. *J. Med. Chem.* **2004**, *47*, 1739–1749.
103. Goodsell, D. S.; Olson, A. J. *Proteins Struct. Funct. Genet.* **1990**, *8*, 195–202.
104. Jones, G.; Willett, P.; Glen, R. C. *J. Mol. Biol.* **1995**, *245*, 43–53.
105. Rarey, M.; Kramer, B.; Lengauer, T.; Klebe, G. *J. Mol. Biol.* **1996**, *261*, 470–489.
106. Rarey, M.; Wefing, S.; Lengauer, T. *J. Comput.-Aided Mol. Des.* **1996**, *10*, 41–54.
107. Ewing, T. J. A.; Makino, S.; Skillman, A. G.; Kuntz, I. D. *J. Comput.-Aided Mol. Des.* **2001**, *15*, 411–428.
108. Jones, G.; Willett, P.; Glen, R. C.; Leach, A. R.; Taylor, R. *J. Mol. Biol.* **1997**, *267*, 727–748.
109. Claussen, H.; Buning, C.; Rarey, M.; Lengauer, T. *J. Mol. Biol.* **2001**, *308*, 377–395.
110. Knegtel, R. M. A.; Kuntz, I. D.; Oshiro, C. M. *J. Mol. Biol* **1997**, *266*, 424–440.
111. Osterberg, F.; Morris, G. M.; Sanner, M. F.; Olson, A. J.; Goodsell, D. S. *Proteins Struct. Funct. Genet.* **2002**, *46*, 34–40.
112. Carlson, H. A. *Curr. Opin. Chem. Biol.* **2002**, *6*, 447–452.
113. Shoichet, B. K.; McGovern, S. L.; Wei, B. Q.; Irwin, J. J. *Curr. Opin. Chem. Biol.* **2002**, *6*, 439–446.
114. Böhm, H.-J. *J. Comput.-Aided Mol. Des.* **1994**, *8*, 243–256.
115. Head, R. D.; Smythe, M. L.; Oprea, T. I.; Waller, C. L.; Green, S. M.; Marshall, G. R. *J. Am. Chem. Soc.* **1996**, *118*, 3959–3969.
116. Murray, C. W.; Auton, T. R.; Eldridge, M. D. *J. Comput.-Aided Mol. Des.* **1998**, *12*, 503–519.
117. Böhm, H.-J. *J. Comput.-Aided Mol. Des.* **1998**, *12*, 309–323.
118. Muegge, I.; Martin, Y. C. *J. Med. Chem.* **1999**, *42*, 791–804.
119. Mitchell, J. B. O.; Laskowski, R. A.; Alex, A.; Forster, M. J.; Thornton, J. M. *J. Comput. Chem.* **1999**, *20*, 1177–1185.
120. Gohlke, H.; Klebe, G. *Curr. Opin. Struct. Biol.* **2001**, *11*, 231–235.
121. Kramer, B.; Rarey, M.; Lengauer, T. *Proteins Struct. Funct. Genet.* **1999**, *37*, 228–241.
122. Nissink, J. W. M.; Murray, C.; Hartshorn, M.; Verdonk, M. L.; Cole, J. C.; Taylor, R. *Proteins Struct. Funct. Genet.* **2002**, *49*, 457–471.
123. Jansen, J. M.; Martin, E. J. *Curr. Opin Chem. Biol.* **2004**, *8*, 359–364.
124. Kontoyianni, M.; McClellan, L. M.; Sokol, G. S. *J. Med. Chem.* **2004**, *47*, 558–565.
125. Wu, G. S.; Vieth, M. *J. Med. Chem.* **2004**, *47*, 3142–3148.
126. Cole, J. C.; Murray, C. W.; Nissink, J. W. M.; Taylor, R. D.; Taylor, R. *Proteins Struct. Funct. Bioinform.* **2005**, *60*, 325–332.
127. Charifson, P. S.; Corkery, J. J.; Murcko, M. A.; Walters, W. P. *J. Med. Chem.* **1999**, *42*, 5100–5109.
128. Clark, R. D.; Strizhev, A.; Leonard, J. M.; Blake, J. F.; Matthew, J. B. *J. Mol. Graphics Model.* **2002**, *20*, 281–295.
129. Bissantz, C.; Folkers, G.; Rognan, D. *J. Med. Chem.* **2000**, *43*, 4759–4767.
130. Hindle, S. A.; Rarey, M.; Buning, C.; Lengauer, T. *J. Comput.-Aided Mol. Des.* **2002**, *16*, 129–149.
131. Deng, Z.; Chuaqui, C.; Singh, J. *J. Med. Chem.* **2004**, *47*, 337–344.
132. Kelly, M. D.; Mancera, R. L. *J. Chem. Inf. Comput. Sci.* **2004**, *44*, 1942–1951.
133. Irwin, J. J.; Shoichet, B. K. *J. Chem. Inf. Model.* **2005**, *45*, 177–182.
134. Willett, P. *Computational Methods for the Analysis of Molecular Diversity*; Kluwer: Dordrecht, The Netherlands, 1997.
135. Lewis, R. A.; Pickett, S. D.; Clark, D. E. Computer-Aided Molecular Diversity Analysis and Combinatorial Library Design. In *Reviews in Computational Chemistry*; Lipkowitz, K. B., Boyd, D. B., Eds.; VCH Publishers: New York, 2000; Vol. 16, pp 1–51.
136. Snarey, M.; Terrett, N. K.; Willett, P.; Wilton, D. J. *J. Mol. Graphics Model.* **1997**, *15*, 372–385.
137. Pearlman, R. S.; Smith, K. M. *Perspect. Drug Disc. Des.* **1998**, *9–11*, 339–353.
138. Hudson, B. D.; Hyde, R. M.; Rahr, E.; Wood, J. *Quant. Struct.-Act. Relat.* **1996**, *15*, 285–289.
139. Downs, G. M.; Barnard, J. M. Clustering Methods and their Uses in Computational Chemistry. In *Reviews in Computational Chemistry*; Lipkowitz, K. B., Boyd, D. B., Eds.; VCH Publishers: New York, 2002; Vol. 18, pp 1–40.
140. Jarvis, R. A.; Patrick, E. A. *IEEE Trans Comput.* **1973**, *C-22*, 1025–1034.
141. Ward, J. H. *J. Am. Statist. Assoc.* **1963**, *58*, 236–244.
142. Wild, D. J.; Blankley, C. J. *J. Chem. Inf. Comput. Sci.* **2000**, *40*, 155–162.
143. Lewis, R. A.; Mason, J. S.; McLay, I. M. *J. Chem. Inf. Comput. Sci.* **1997**, *37*, 599–614.
144. Pearlman, R. S.; Smith, K. M. *J. Chem. Inf. Comput. Sci.* **1999**, *39*, 28–35.
145. Waldman, M.; Li, H.; Hassan, M. *J. Mol. Graphics Model.* **2000**, *18*, 412–426.
146. Agrafiotis, D. K. *J. Comput.-Aided Mol. Des.* **2002**, *16*, 335–356.
147. Available Chemicals Directory. MDL Information Systems, Inc. http://www.mdli.com (accessed May 2006).
148. Martin, E. J.; Blaney, J. M.; Siani, M. A.; Spellmeyer, D. C.; Wong, A. K.; Moos, W. H. *J. Med. Chem.* **1995**, *38*, 1431–1436.
149. Leach, A. R.; Bradshaw, J.; Green, D. V. S.; Hann, M. M.; Delany, J. J. *J. Chem. Inf. Comput. Sci.* **1999**, *39*, 1161–1172.
150. Downs, G. M.; Barnard, J. M. *J. Chem. Inf. Comput. Sci.* **1997**, *37*, 59–61.
151. Beroza, P.; Bradley, E. K.; Eksterowicz, J. E.; Feinstein, R.; Greene, J.; Grootenhuis, P. D. J.; Henne, R. M.; Mount, J.; Shirley, W. A.; Smellie, A. *J. Mol. Graphics Model.* **2000**, *18*, 335–342.
152. Brown, R. D.; Martin, Y. C. *J. Med. Chem.* **1997**, *40*, 2304–2313.
153. Gillet, V. J.; Willett, P.; Bradshaw, J.; Green, D. V. S. *J. Chem. Inf. Comput. Sci.* **1999**, *39*, 169–177.
154. Zheng, W. F.; Cho, S. J.; Waller, C. L.; Tropsha, A. *J. Chem. Inf. Comput. Sci.* **1999**, *39*, 738–746.
155. Sheridan, R. P.; Kearsley, S. K. *J. Chem. Inf. Comput. Sci.* **1995**, *35*, 310–320.
156. Gillet, V. J.; Khatib, W.; Willett, P.; Fleming, P. J.; Green, D. V. S. *J. Chem. Inf. Comput. Sci.* **2002**, *42*, 375–385.
157. Gillet, V. J.; Willett, P.; Fleming, P. J.; Green, D. V. S. *J. Mol. Graphics Model.* **2002**, *20*, 491–498.
158. Stahura, F. L.; Bajorath, J. *Drug Disc. Today* **2002**, 7, S41 – S47.
159. Clark, D. E.; Grootenhuis, P. D. J. *Curr. Opin. Drug Disc. Dev.* **2002**, *5*, 382–390.
160. Penzotti, J. E.; Landrum, G. A.; Putta, S. *Curr. Opin. Drug Disc. Dev.* **2004**, 7, 49–61.
161. Davis, A. M.; Riley, R. J. *Curr. Opin. Chem. Biol.* **2004**, *8*, 378–386.
162. Hodgson, J. *Nat. Biotechnol.* **2001**, *19*, 722–726.
163. van de Waterbeemd, H.; Gifford, E. *Nat. Rev. Drug Disc.* **2003**, *2*, 192–204.
164. Lipinski, C. A.; Lombardo, F.; Dominy, B. W.; Feeney, P. J. *Adv. Drug Deliv. Rev.* **1997**, *23*, 3–25.
165. Oprea, T. I. *J. Comput.-Aided Mol. Des.* **2000**, *14*, 251–264.
166. Walters, W. P.; Murcko, M. A. *Adv. Drug Deliv. Rev.* **2002**, *54*, 255–271.

167. Egan, W. J.; Walters, W. P.; Murcko, M. A. *Curr. Opin. Drug Disc. Dev.* **2002**, *5*, 540–549.
168. The SPRESI database. Daylight Chemical Information Systems. http://www.daylight.com (accessed May 2006).
169. Gillet, V. J.; Willett, P.; Bradshaw, J. *J. Chem. Inf. Comput. Sci.* **1998**, *38*, 165–179.
170. Hann, M.; Hudson, B.; Lewell, X.; Lifely, R.; Miller, L.; Ramsden, N. *J. Chem. Inf. Comput. Sci.* **1999**, *39*, 897–902.
171. Ajay, Walters, W. P., Murcko, M. A., *J. Med. Chem.* **1998**, *41*, 3314–3324.
172. Wagener, M.; van Geerestein, V. J. *J. Chem. Inf. Comput. Sci.* **2000**, *40*, 280–292.
173. Oprea, T. I.; Davis, A. M.; Teague, S. J.; Leeson, P. D. *J. Chem. Inf. Comput. Sci.* **2001**, *41*, 1308–1315.
174. Hann, M. M.; Leach, A. R.; Harper, G. *J. Chem. Inf. Comput. Sci.* **2001**, *41*, 856–864.
175. Martin, Y. C.; DeWitte, R. S. *Perspect. Drug Disc. Des.* **2000**, *18*, A5 – A6.
176. Martin, Y. C.; DeWitte, R. S. *Perspect. Drug Disc. Des.* **1999**, *17*, U1 – U2.
177. Mannhold, R.; van de Waterbeemd, H. *J. Comput.-Aided Mol. Des.* **2001**, *15*, 337–354.
178. Leo, A. J. *Chem. Rev.* **1993**, *93*, 1281–1306.
179. Leo, A. J.; Hoekman, D. *Perspect. Drug Disc. Des.* **2000**, *18*, 19–38.
180. Ghose, A. K.; Viswanadhan, V. N.; Wendoloski, J. J. *J. Phys. Chem. A* **1998**, *102*, 3762–3772.
181. Ghose, A. K.; Crippen, G. M. *J. Comput. Chem.* **1986**, 7, 565–577.
182. Wildman, S. A.; Crippen, G. M. *J. Chem. Inf. Comput. Sci.* **1999**, *39*, 868–873.
183. Delaney, J. S. *Drug Disc. Today* **2005**, *10*, 289–295.
184. Jorgensen, W. L.; Duffy, E. M. *Adv. Drug Deliv. Rev.* **2002**, *54*, 355–366.
185. Lind, P.; Maltseva, T. *J. Chem. Inf. Comput. Sci.* **2003**, *43*, 1855–1859.
186. Clark, D. E. *J. Pharm. Sci.* **1999**, *88*, 807–814.
187. Ertl, P.; Rhode, B.; Selzer, P. *J. Med. Chem.* **2000**, *43*, 3714–3717.
188. Norinder, U.; Haeberlein, M. *Adv. Drug Deliv. Rev.* **2002**, *54*, 291–313.
189. Abraham, M. H.; Ibrahim, A.; Zissimos, A. M.; Zhao, Y. H.; Comer, J.; Reynolds, D. P. *Drug Disc. Today* **2002**, 7, 1056–1063.
190. Crivori, P.; Cruciani, G.; Carrupt, P. A.; Testa, B. *J. Med. Chem.* **2000**, *43*, 2204–2216.
191. Goodford, P. J. *J. Med. Chem.* **1985**, *28*, 849–857.
192. Tropsha, A.; Gramatica, P.; Gombar, V. K. *QSAR Comb. Sci.* **2003**, *22*, 69–77.
193. Golbraikh, A.; Shen, M.; Xiao, Z. Y.; Xiao, Y. D.; Lee, K. H.; Tropsha, A. *J. Comput.-Aided Mol. Des.* **2003**, *17*, 241–253.
194. Sheridan, R. P.; Feuston, B. P.; Maiorov, V. N.; Kearsley, S. K. *J. Chem. Inf. Comput. Sci.* **2004**, *44*, 1912–1928.
195. Stouch, T. R.; Kenyon, J. R.; Johnson, S. R.; Chen, X. Q.; Doweyko, A.; Li, Y. *J. Comput.-Aided Mol. Des.* **2003**, *17*, 83–92.

Biographies

Val J Gillet is Senior Lecturer and Head of the Chemoinformatics Research Group in the Department of Information Studies at the University of Sheffield. She holds MSc and PhD degrees from the University of Sheffield and a first degree in Natural Sciences from Cambridge University. Her research interests are in chemoinformatics and computational approaches to drug design including: combinatorial library design and diversity analysis; the development of novel molecular descriptors; the application of evolutionary algorithms to computational chemistry; the identification of structure–activity relationships; and de novo design. She is an organizer of triennial conferences and annual short courses to industry on chemoinformatics. Dr Gillet and Andrew Leach are joint authors of the first textbook on chemoinformatics (Leach, A. R.; Gillet, V. J. *An Introduction to Chemoinformatics*; Kluwer, 2003).

Andrew R Leach joined Glaxo (now GlaxoSmithKline) in 1994 following postdoctoral research at the University of California – San Francisco and holds an academic position at the University of Southampton (UK). During his career he has been heavily involved in the development of new computational and chemoinformatics methodologies to support drug discovery in areas such as library design, compound acquisition, and structure-based design. He has BA and DPhil degrees in Chemistry from the University of Oxford and is the author of a widely used textbook on molecular modeling. He is currently Director of the GlaxoSmithKline Computational Chemistry department in the UK.

Comprehensive Medicinal Chemistry II
ISBN (set): 0-08-044513-6

ISBN (Volume 3) 0-08-044516-0; pp. 235–264

3.13 Chemical Information Systems and Databases

T Engel, Chemical Computing Group AG, Köln, Germany
E Zass, Informationszentrum Chemie Biologie Pharmazie, Zürich, Switzerland

3.13.1	**Introduction into Information Systems**	**265**
3.13.2	**Database Systems**	**267**
3.13.2.1	Hierarchical Database System	267
3.13.2.2	Network Model	268
3.13.2.3	Relational Model	268
3.13.2.4	Object-Based Model	268
3.13.3	**Classification of Databases**	**269**
3.13.3.1	The Chemical Literature	270
3.13.3.1.1	Primary literature	270
3.13.3.1.2	Secondary literature	270
3.13.3.1.3	Tertiary literature	273
3.13.3.2	Literature Databases	273
3.13.3.2.1	Bibliographic databases	273
3.13.3.2.2	Specific literature databases	275
3.13.3.2.3	Patent databases	275
3.13.3.2.4	Full-text databases	275
3.13.3.3	Factual Databases	276
3.13.3.3.1	Text databases	277
3.13.3.3.2	Metadata databases	277
3.13.3.3.3	Numerical databases	277
3.13.3.3.4	Catalog databases	279
3.13.3.4	Structure and Reaction Databases	279
3.13.3.4.1	General structure databases	280
3.13.3.4.2	Patent databases with structures	281
3.13.3.4.3	Reactions	281
3.13.3.4.4	Sequence databases	283
3.13.3.4.5	Three-dimensional structure databases	283
3.13.3.5	Access to Databases	283
3.13.3.5.1	Databases and interfaces	283
3.13.3.5.2	Hosts and in-house databases	284
3.13.3.5.3	Integration and reference linking	284
3.13.3.5.4	Chemical information on the internet	285
3.13.3.6	Information Sources	285
3.13.4	**Outlook**	**286**
References		**286**

3.13.1 Introduction into Information Systems

The multifaceted information available on compounds or drugs (e.g., literature, physicochemical and physiological properties, spectra) and on reactions or metabolism can be handled in a comprehensive manner only by electronic methods. Such systems for storing and retrieving data are generally called 'information systems' (**Figure 1**). These systems comprise application programs (e.g., search engines, database management systems) and an ordered data

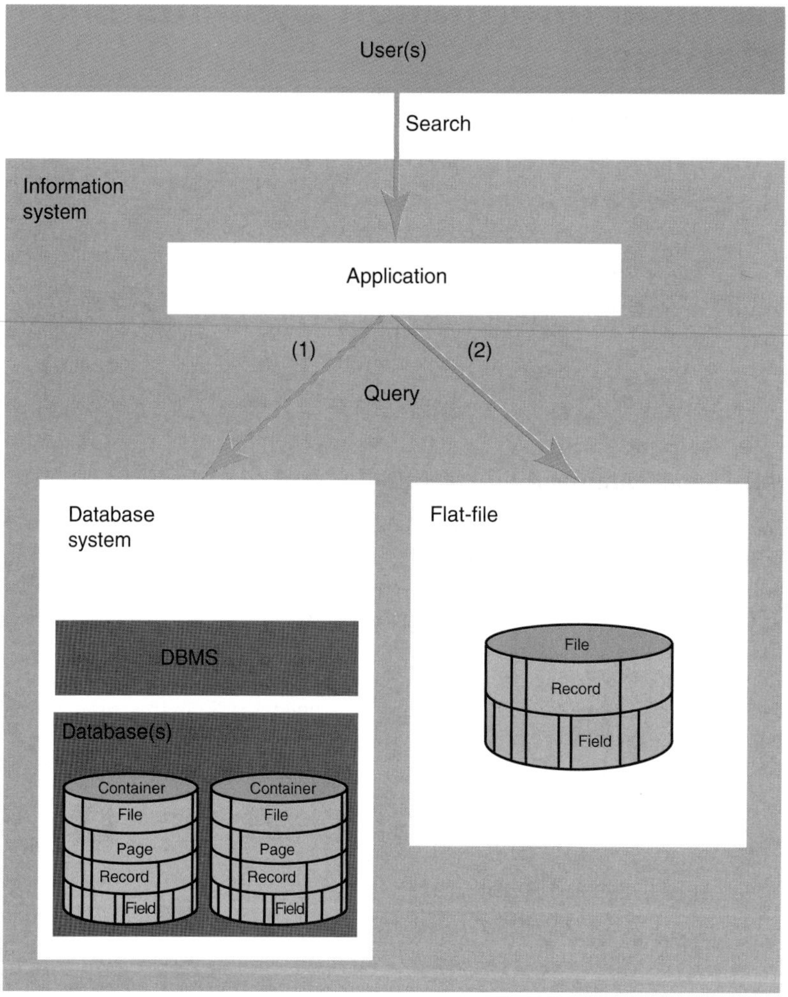

Figure 1 Different organization forms of an Information System. (1) The database(s) (DB), also called container, with organized data and meta-data are part of the database system (DBS), which is managed by the database management system (DBMS). (2) If the user has direct access on the database it is called a flat-file.

collection (database). In chemistry, the term 'database' is often (mis)used for the entire information system, a database system, or a data file itself.[1,2]

In principle there are two different approaches used within information systems to provide information from a database (see **Figure 1**). In one approach the data are made available by a database system (DBS). The DBS is a tool for the efficient computer-supported organization, generation, manipulation, and management of huge data collections. The database itself is only one part of the system, where data can be stored easily, quickly, and reliably. If the data are organized in a DBS the data store is called a 'container.' A second important piece of software is the DataBase Management System (DBMS) (e.g., Oracle, DBase, MS Access) (see **Figure 1**). This software allows the storage of data according to a database structure, which facilitates the retrieval or manipulation of data, user management, security, backup, load balancing, etc.

In the second approach the database is not integrated in a database system, and in this case it is called a 'flat file.' As the name indicates, the data are stored in a file that can be accessed directly by the user. The database itself is defined as a self-describing collection of integrated records. The organization of the database (tables, objects, indices, etc.) is described by meta-data (data about data), which is stored within the database as a data dictionary (system catalog).

The principle of the organization of a database or container is illustrated in **Figure 1**. Fields with unique types of information or characters (numerical, graphical, etc.) consist of bits (0 or 1), which are components of bytes (8 bits = 1 byte). Fields build records (data sets including different attributes) of variable length, which describe corresponding object properties (e.g., name, CAS Registry Number). Some or all of these records are put into relationship with each other and are stored in a file. In order to optimally transfer records between the hard disk and the memory, one or more records are put into a page. A file that contains various pages with congeneric data is called a container.[1] In a flat-file system the database is called a file.

3.13.2 Database Systems

Each database handles data of different origin, nature, or designation in a structured, organized manner. Therefore, the most important task in conceiving an effective database is to structure the imported data. Different conceptual models exist for organizing data in a structured manner.[3] The precursor of the database, the file system, was developed in the 1950s, and this was sufficient for most applications (e.g., library catalogues). The first real database systems became available in the 1970s as hierarchical and networked database systems. About a decade later, relational database systems came into use, and in the 1990s object-oriented database systems were developed.

3.13.2.1 Hierarchical Database System

A hierarchical database system is the simplest type of database system. In systems of this type the various data types (entities) are systematically assigned to various levels (**Figure 2**). The hierarchical system is represented as an upside-down tree with one root segment and ordered nodes. Each parent object can have one or more children (objects), but each child has only one parent. If an object has more than one parent the entity has to be duplicated in another place in the database hierarchy.

In order to trace (find, change, add, or delete) a segment in the database, the sequence in which the data are read is important. Thus, the sequence of the hierarchical path is: parent > child > siblings. The assignment of the data entities uses pointers. In our example, the hierarchical path to K is traced in **Figure 2**.

Typical examples of hierarchical database systems are the file systems used in personal computers. In the domain of chemistry, the in-house compound and reaction database systems MACCS (Molecule Access System) and REACCS (Reaction Access System) were hierarchical database systems, as was the first version of its successor, ISIS.

The primary advantage of hierarchical databases is that the relationship between the data at different levels is simple. This simplicity and efficiency of the data model is a great advantage of the hierarchical DBS (e.g., IMS: Information Management System by IBM). Large data sets, such as a series of measurements where the data values are dependent on different parameters such as boiling point, temperature, or pressure, can be implemented with an acceptable response time.

The disadvantage of hierarchical databases is that the implementation and management of the database requires a good knowledge of the organization (physical level) of the data storage. In addition, it is difficult to administer the structure of the database, as new relations or nodes result in a complex system of management tasks. Therefore, a modification of the logical data-independent data structure in a DBS of this type, which has limited flexibility, may demand significant modifications to the application programs. Furthermore, the hierarchical model suffers from the problem that a child cannot be related to multiple parents, and from the redundancies necessary to remedy this.

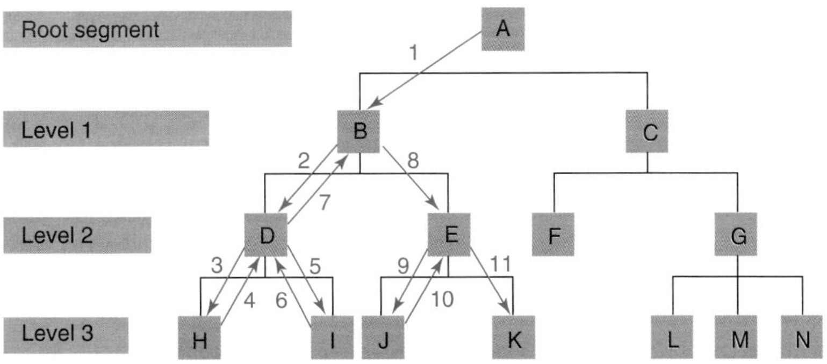

Figure 2 Hierarchical structure of a database. For example, object E in level 2 is the parent of the child objects J and K. The red arrows indicate the hierarchical path to trace the sequence to object K.

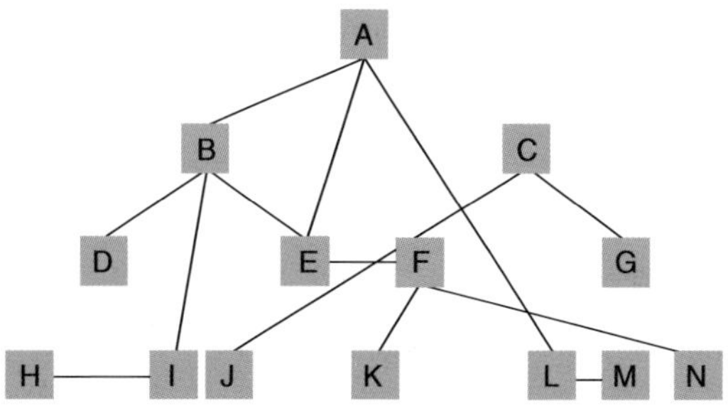

Figure 3 A network model of a database.

3.13.2.2 Network Model

The network model of a database is an improvement over the hierarchical model.[4] In the network database model, a single object can point to many other objects, with the reverse also being true (e.g., an object (child) E can have several parents – A and B; **Figure 3**). While these complex relationships improve the access to desired records, the inherent clarity of the hierarchical system is lost through the complexity of the database organization. As a consequence, design and navigational data access are more complicated. Whereas objects in the hierarchical model have a clear top-down relationship, they are interconnected to others in the network model. Thus, it is possible that a linked object may be deleted during database management, resulting in other objects that are still available but no longer have any relationships.

This database system is implemented in only a few instances due to its complexity and its liability to errors. It is also a model for the World Wide Web.

3.13.2.3 Relational Model

The characteristic of the relational database model is the organization of data in different tables that have relationships between each other. A table is a two-dimensional construct of rows and columns. All entries in one column have a particular correlation (e.g., name, molecular weight), and represent a specific attribute of the objects (records) of the table (file) (**Figure 4**). The sequence or order of individual rows and columns in the tables is irrelevant. Different tables (e.g., different objects with different attributes) in the same database can be related through at least one common attribute. Thus, it is possible to relate objects within tables indirectly by using a key. The range of values of an attribute is called the 'domain,' which is defined by constraints. Schemes define and store such meta-data for the database and the tables.

Relational database models utilize memory very efficiently, avoiding redundancies (e.g., repeated data in a hierarchical system). It is possible to extract both individual elements and combinations of data elements from a table. The main advantage of this structure is that it offers the possibility of changing the structure of the database (adding or deleting tables) without changing the application programs, which are based on prior database structures (data independence).

Among the many approaches to manipulating and querying a relational database the most prevalent is a language called SQL (Structured Query Language).[2] The relational database model was developed by E. F. Codd at IBM in 1970.[5] Oracle provided the first implementation in 1979. IBM replaced the hierarchical database IMS by the relational system DB2, which is also a relational database management system (RDBMS). There are many other DBMSs, such as SQL/DS, XDB, and Ingres.

3.13.2.4 Object-Based Model

For some applications, such as computer-aided engineering systems, software development, or hypermedia, the relational database model is insufficient. In a relational database management system (RDBMS) it is difficult to model very complex objects and environments: the various extensive tables become very complicated, integrity is problematic to observe, and the performance of the system is reduced. This led to two sophisticated object-based models, the object-oriented model and the object-relational model, which are mentioned only very briefly here (for further details, see Lausen and Vossen[6] and Embley[7]). The main difference from the relational DBS is that the data are now stored in object-types with a unique identity number (ID), attributes, and operations. Therefore, the relationship between objects is completely different from that in an RDBMS.

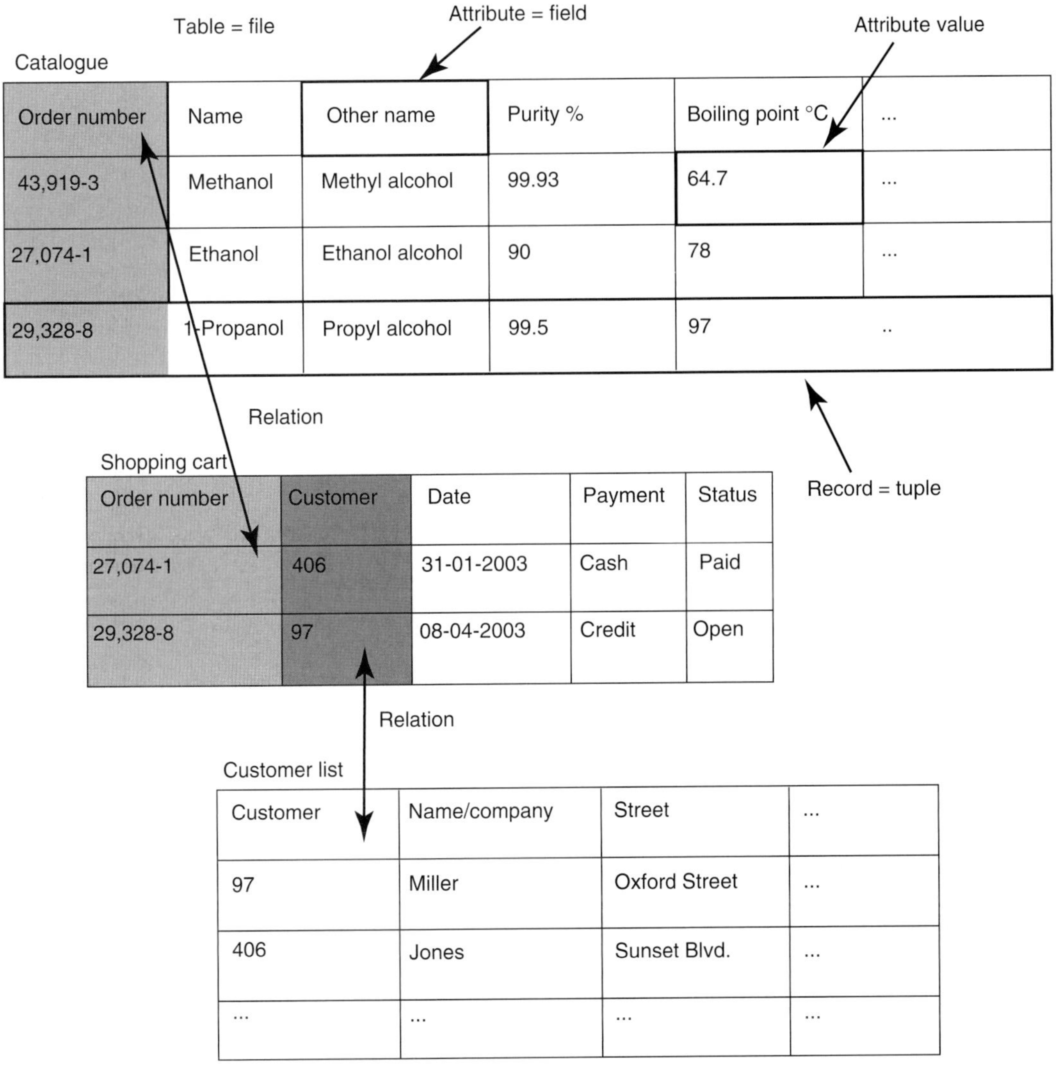

Figure 4 A relational model of a database. The records of each individual table with different attributes are related through at least one common attribute.

3.13.3 Classification of Databases

Most practicing chemists are more interested in the content of databases than in their technical organization, which will be discussed in Section 3.13.3.2. There is now available an almost bewildering variety of databases in chemistry and other sciences, which renders a further classification of databases by content at least useful, if not indispensable, for the evaluation and selection of information sources. In this chapter we present a formal classification of databases by their content type into three broad categories (**Figure 5**)[8,8a]:

- literature (textual)
- factual (alphanumeric)
- structure (topological).

Strict separation of databases into these categories and their subtypes is impossible, as many important chemistry databases cover several types of content: the Beilstein[9–9b] database, for example, contains structures, reactions, numerous physical properties of compounds, and related literature references. In the following classification, the

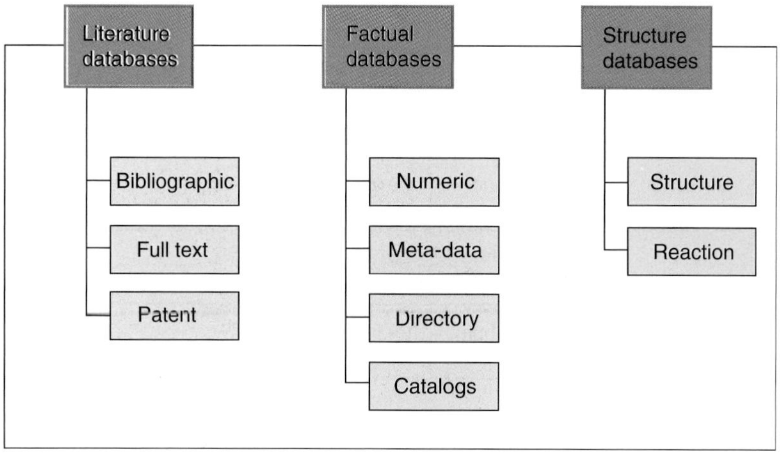

Figure 5 Classification of databases

'assignment' of databases is based on the type of records they contain, or the content that is most characteristic of the database at hand. In the Beilstein example, the content is physical and chemical properties of organic compounds, and hence we classify it as a factual database of the numeric type.

3.13.3.1 The Chemical Literature

The use of computers to handle the structures of chemical compounds, bibliographic, and factual data is indispensable in chemical information[10] because of the volume of this information. Although the number of databases of chemical information not available in another medium or another source is increasing, most chemical databases contain information taken from the published chemical literature. Important chemical databases were, at least in their beginnings, or are even still now, 'electronic versions' of previously established printed sources, reflecting not only their content, but also their organization (data structure). Important examples are all journals that are available as electronic versions (in addition to the printed version), as well as Chemical Abstracts,[11–11b] Biological Abstracts, Medline (Index Medicus), Embase (Excerpta Medica), Science Citation Index,[12] Beilstein,[9–9b] Gmelin,[13] ChemInform,[14–14b] and Science of Synthesis[15] (**Table 1**). Before we categorize these and other databases, we need to put them into the context of chemical information in general.

3.13.3.1.1 Primary literature

Chemical information is usually published for the first time (sic!) and in full detail in the primary literature. This encompasses preprints, articles in journals, patents, research reports, conference proceedings, and theses produced for academic degrees. Information published for the first time on the web, such as manuscripts or lectures, must also be considered as 'primary literature.' Traditionally produced on paper, a growing number of primary sources are available in electronic form (*see* Section 3.13.3.2.4), but the electronic versions seem still a long way from replacing the paper; 'electronic-only' publications are still the exception.

By definition, the primary literature is the key to most information needs. However, both its enormous volume – there are about 10 000 different journals dealing with chemistry – and its diversity have made general searches of primary sources virtually impossible in printed form. So far, electronic availability has not significantly changed this situation – full-text searching is usually too imprecise, structure searching not available,[16] and there are far too many places (e.g., web pages of journal publishers with full-text database interfaces, *see* Section 3.13.3.2.4) to search at.

3.13.3.1.2 Secondary literature

The difficulty of searching the primary scientific literature has been addressed in chemistry since the appearance of the *Pharmaceutisches*, later *Chemisches Zentralblatt*, in 1830. Producers of secondary literature acquire and digest the primary literature, condense the important content into abstracts, and make the content searchable by indexing (e.g., descriptions of topics, compounds, reactions, data) from primary publications in a standardized way.

Table 1 Literature databases

Database	*Content Size (records)*	*Producer Sources*	*Access (host)*
CA (Chemical Abstracts)	Chemistry, biochemistry, chemical engineering >28 million publication records	CAS[23] 9500 journals, 50 patents, reports, conference proceedings, theses, books (since 1907, some earlier publications)	STN: http://www.stn-international.de/stndatabases/databases/ca.html Other hosts: CD-ROM[116](*see* Section 3.13.3.5.1)
CAplus	Same as CA, plus additional publication types, records in process (faster updates)	Same as CA	SciFinder[29–29b] (Scholar[30]): http://www.cas.org/CASFILES/capluscontent.html STN: http://www.stn-international.de/stndatabases/databases/caplus.html (*see* Section 3.13.3.5.1)
BIOSIS (Biological Abstracts)[137]	Biosciences/biomedicine 15 million publication records	Thomson:[34] http://www.biosis.org/products 5000 journals, selected US patents (since 1969)	STN: http://www.stn-international.de/stndatabases/databases/biosis.html Dialog: http://library.dialog.com/bluesheets/html/bl0005.html Other hosts: Web of Knowledge[104]
Medline	Medicine, life sciences *c.* 14 million publication records	US National Library of Medicine: http://www.nlm.nih.gov/pubs/factsheets/medline.html 4300 journals (since 1950)	PubMed: http://www.ncbi.nlm.nih.gov/entrez/query.fcgi?DB = pubmed SciFinder[29–29b] (Scholar[30,30a]): http://www.cas.org/SCIFINDER/medlinecontent.html Many hosts
Embase (Excerpta Medica Database)	Medicine, life sciences *c.* 14 million publication records	Elsevier:[53] http://www.elsevier.com/wps/find/bibliographicdatabasedescription.cws_home/523328/description 4600 journals (since 1974)	Web version: http://www.embase.com Several hosts (e.g., STN: http://www.stn-international.de/stndatabases/databases/embase.html) Dialog: http://library.dialog.com/bluesheets/html/bl0072.html
Life Sciences Collections	Biology, medicine, biochemistry, ecology *c.* 2.4 million publication records	Cambridge Scientific Abstracts: http://www.csa.com Journals, patents, reports, conference proceedings, books (since 1978)	STN: http://www.stn-international.de/stndatabases/databases/lifesci.html
SCI (Science Citation Index)	Publications and citations from science, technology, medicine >30 million publication records	ISI,[32] 3700 journals, since 1974: http://www.isinet.com/products/citation/sci Expanded version, 5800 journals, since 1945/1900: http://www.isinet.com/products/citation/scie)	ISI Web of Knowledge[104] ISI Web of Science [123,123a]STN: http://www.stn-international.de/stndatabases/databases/scisearc.html Dialog: http://library.dialog.com/bluesheets/html/bl0034.html Other hosts: CD-ROM
Scopus	Full-text journal articles with citations (science, technology, medicine, social science) 27 million publication records	Elsevier,[53] 14 000 journals (since 1966, citations since 1996)	http://www.info.scopus.com/detail/what/
Analytical Abstracts	Analytical publications with specific indexing >355 000 publication records	Royal Society of Chemistry: http://www.rsc.org/Publishing/CurrentAwareness/AA/About.asp *c.* 100 journals, reports, conference proceedings, books, standards (since 1980)	STN: http://www.stn-international.de/stndatabases/databases/anabstr.html Dialog: http://library.dialog.com/bluesheets/html/bl0305.html

continued

Table 1 Continued

Database	*Content Size (records)*	*Producer Sources*	*Access (host)*
CEABA-VTB (Chemical Engineering and Biotechnology Abstracts – Verfahrenstechnische Berichte)	Chemical engineering, biotechnology >650 000 publication records	DECHEMA: http://www.dechema.de/ceabavtb.html Journals, technical reports, books (since 1966)	CD-ROM STN: http://www.stn-international.de/stndatabases/databases/ceabavtb.html Dialog: http://library.dialog.com/bluesheets/html/bl0315.html
Derwent Biotechnology Resource	Biotechnology *c.* 350 000 publication records	Thomson Derwent:[46] Journals, patents, conference reports (since 1982)	STN: http://www.stn-international.de/stndatabases/databases/biotecha.html Dialog: http://library.dialog.com/bluesheets/html/bl0357.html
Biotechnology and Bioengineering Abstracts	Biotechnology *c.* 1 million publication records	Cambridge Scientific Abstracts: http://www.csa.com/factsheets/biotclust-set-c.php Journals, conference proceedings (since 1982)	STN: http://www.stn-international.de/stndatabases/databases/bioeng.html
CBNB (Chemical Business News Base)	All aspects of chemical business 1 million publication records	Elsevier Engineering Information: http://www.ei.org/cbnb.html Journals, newspapers, reports, etc. (since 1984)	STN: http://www.stn-international.de/stndatabases/databases/cbnb.html Dialog: http://library.dialog.com/bluesheets/html/bl0319.html
TOXCENTER	Toxicology *c.* 6.5 million publication records	CAS:[23] compilation of several databases (since 1907)	STN: http://www.stn-international.de/stndatabases/databases/toxcenter.html
Cancerlit	All aspects of cancer *c.* 1.9 million publication records (1963–2002)	US National Cancer Institute: http://www.cancer.gov/. 3500 journals, reports, conference proceedings, theses, books	STN: http://www.stn-international.de/stndatabases/databases/cancerli.html Dialog: http://library.dialog.com/bluesheets/html/bl0159.html
Derwent Drug File	Drug development and manufacture *c.* 1.5 million publication records	Thomson Derwent:[46] http://thomsonderwent.com/products/lr/drugfile/ 1150 journals (since 1964)	STN: http://www.stn-international.de/stndatabases/databases/ddfu.html Dialog: http://library.dialog.com/bluesheets/html/bl0377.html
Pharmline	Use of drugs *c.* 168 000 publication records	UK National Health Service: 100 journals (since 1978)	Dialog: http://library.dialog.com/bluesheets/html/bl0174.html
IPA (International Pharmaceutical Abstracts)	Pharmaceutical literature > 401 000 publication records	Thomson:[33] http://scientific.thomson.com/products/ipa Journals (since 1970)	STN: http://www.stn-international.de/stndatabases/databases/ipa.html Dialog: http://library.dialog.com/bluesheets/html/bl0074.html

Secondary literature is thus defined as nonoriginal publications, which exist in two major categories: abstracting and indexing services (e.g., Chemical Abstracts,[11–11b] Science Citation Index[12]) and handbooks (Beilstein,[9–9b] Gmelin,[13] Landolt–Börnstein,[17] Theilheimer,[18–18d] Houben-Weyl,[15] etc.). Secondary sources were the first to appear as databases, starting in 1972 with Chemical Abstracts[11–11b] and Medline.

3.13.3.1.3 Tertiary literature

The classification of the literature into primary, secondary, and tertiary sources models a concentration and abstraction process increasing in this order. While secondary literature directly refers information from original publications in a 1:1 fashion, tertiary literature summarizes a topic with information processed from many different primary sources. Accordingly, tertiary literature is the most refined and digested form, with added value relative to original publications: monographs (books), reference works, and encyclopedias, such as Ullmann's *Encyclopedia of Industrial Chemistry*,[19] the Kirk-Othmer *Encyclopedia of Chemical Technology*,[20] and many more specialized similar publications (e.g., EROS,[21] the *Encyclopedia of Computational Chemistry*[22]).

Handbooks such as those by Gmelin[13] and Beilstein[9–9b] are sometimes classified as tertiary literature, because they do not refer directly to individual primary publications as abstracting publications, but collect information on individual compounds from several publications. However, we prefer to classify these as secondary sources, because their referral function is still dominating. For other handbooks, such as those by Houben-Weyl[15] or Landolt-Börnstein,[17] however, where the original information is more processed and augmented, this is certainly less correct.

Articles in journals are categorized as primary sources, but review articles that summarize a topic based on information already published elsewhere belong to the realm of tertiary literature. The many series named 'Advances in …', 'Annual Report …', 'Progress in …' also fall in this category. Printed in book format, many of these are now also available as e-books.

3.13.3.2 Literature Databases

In literature databases the records and their fields describe individual publications from the primary literature as objects, using character strings, e.g., letters, numbers, and special characters. Bibliographic and content description (indexing) of publications, such as author names, titles of articles, titles of journals or books, publication year, keywords, descriptors, or abstracts, can be retrieved from these databases in a specific way. Literature databases are divided into bibliographic and full-text databases. They may be further characterized by their width of coverage: whether they are of a general type, limited to a scientific subfield, or a specific type of the primary literature (see **Table 1**).

3.13.3.2.1 Bibliographic databases

In contrast to full-text databases (*see* Section 3.13.3.2.4), bibliographic databases do not contain the published information itself, but only references to publications. Thus, a search in a bibliographic database gives us only literature references, whereas in a full-text database complete articles are retrieved. Typical bibliographic databases are Chemical Abstracts[11–11b] and Medline.

The Chemical Abstracts literature database CA (or its augmented version CAplus, see **Table 1**) is the largest single source for chemistry and related fields. It covers all types of primary publications (*see* Section 3.13.3.1.1), and is offered under different interfaces by the producer CAS[23] and all major hosts. Originally starting in 1967, it was recently extended back to 1907 (with some earlier literature) by retroconversion from printed Chemical Abstracts.[11–11b] Besides bibliographic data and informative abstracts, it contains standardized indexing of the content of primary publications. The real strength of CA is this indexing, which was taken over from printed indexes into the database. Topics (phenomena, methods, processes, classes of compounds, types or reactions, biological species, organs, etc.) are described by a limited set of standardized 'general subject headings', which have changed over the course of time, reflecting new developments, but were strictly adhered to within a given time period. This highly standardized general description is further specified by a 'modifying phrase', which is free text making use of a lot of (again standardized[24]) abbreviations to save space.

Compounds have been indexed by their systematic names (before 1972 many trivial names were used). When electronic data processing was introduced by CAS[23] in the 1960s to cope with the increasing amount of primary literature, individual naming of compounds and the use of card indexes were replaced by a central computerized compound registry, the CAS Registry System.[25–25c] This has been available since 1980 as a structure searchable database (*see* Section 3.13.3.4.1.1). Instead of systematic compound names, the indexing in the CA literature database contains, for every compound, its CAS Registry Number, which is a substance identifier (not a code!) used far beyond CAS[23] databases in other sources. The context of a substance in a publication is again described by a 'modifying phrase', and by a set of 81 standardized 'roles'[26] (e.g., preparation, analytical study, occurrence, drug mechanism of action) available from 1967 onwards. Indexing of common compounds, which occur very often in the literature, is subdivided into 22 standardized qualifiers and categories: preparation, biological activities, derivatives, polymers, etc. Printed abstracts are arranged into 80 sections[27,27a] corresponding to major areas of chemistry (e.g., General Biochemistry, Heterocycles, Inorganic Analytical Chemistry). These sections are also assigned to the records in the database.

All this information in the CA records is searchable alone or in combination. The utilization of the powerful, but complex, indexing in searching is supported by an array of tools produced by CAS,[23] from printed CA Index Guides, published lists of abbreviations,[24] roles,[26] a CA Lexicon ('general subject headings' and their changes over time[28]), etc., to the natural language interface provided in SciFinder[29–29b] and SciFinder Scholar.[30,30a]

While CA is the 'central' literature database for chemistry and related fields, biology has been covered since 1969 by BIOSIS, the database version of the printed Biological Abstracts (see **Table 1**). This contains a thorough indexing tailored for biological purposes, particularly for biological species. In the field of medicine, two major databases exist, both of which originally appeared in print, Medline and Embase (see **Table 1**). Medline is not only available for a fee with most major hosts and under the user interfaces SciFinder and SciFinder Scholar, but also for free on the web as PubMed. This database is indexed with MeSH (medical subject headings),[31] a hierarchical thesaurus, which is probably one of the best indexing systems around.

Although usually not seen in this context, all electronic library catalogs (OPACs – online public access catalogs) are bibliographic databases that contain as records books, journal titles (sometimes even individual journal issues, but not articles), etc. In the context of other bibliographic databases, they are very important, because they bridge the gap between references from CA, Medline, etc., and the referenced publications themselves in print or electronic form (*see* Section 3.13.3.5.3).

3.13.3.2.1.1 Special bibliographic databases

The Science Citation Index (SCI)[12] differs from the other bibliographic databases mentioned so far in two important aspects – it is an interdisciplinary science database, at the price of covering a lesser number of journals for a specific discipline than Chemical or Biological Abstracts; and it contains citation data, i.e. for every publication abstracted it gives not only the bibliographic data, but all (earlier) the references cited in the publication, and with that implicitly also all (later) publications citing that publication. CAS[23] recently augmented its literature database with citation data from 1997 onwards, which until then was virtually a domain of the Science Citation Index. However, the SCI lacks the thorough indexing of the discipline databases. The Science Citation Index (SCI) is produced by the Institute for Scientific Information (ISI,[32] now belonging to Thompson[33]).

When searching for a compound, we are well served by the structure databases discussed in Section 3.13.3.4 using the common and precise 'language of chemistry'. Searches for topics in bibliographic databases, however, are more problematic: they depend on matching with the chemist's query the exact representation of a topic in the database (a character string!). This is either the author's own terminology, or the more strictly controlled terms in the indexing of the database producer. This matching is the key problem in keyword searches for a topic. Instead of relying on descriptions by author or indexer to retrieve a relevant publication, one can do a citation search[34,35] based on the fact that scientific publications are not isolated documents, but do cite related earlier publications. Starting from one or several relevant publications on a topic, one uses the SCI[12] to find later publications assumed to be also relevant due to the fact that they cite the earlier relevant publication(s). Both approaches have their inherent problems: the term-matching (character string) problem in a keyword search is replaced by the dependence on the quality of correct citations of the relevant literature in a citation search. Fortunately, practice has shown that both approaches are to a significant extent complementary.

An alternative to the SCI[12] for citation searches is the recently launched Scopus[36] database. This is not a database version of a printed abstracting (secondary) publication, but a new source built directly from journals (primary publications).[36] Like the SCI[12] it is interdisciplinary, and permits both bibliographic, simple topical (no indexing!) and, particularly, citation searches.

Some bibliographic databases, such as Current Contents,[37] were built specifically for current awareness. Current Contents offers the tables of contents of major journals in the areas of Life Science, Physical Sciences, Chemical & Earth Sciences, Clinical Medicine, etc. Also produced by ISI,[32] Current Contents first appeared in print, then on floppy disks, CD-ROM, and now as a database with a web-based interface.

Such specialized current-awareness sources have lost some of their original importance, as almost all bibliographic databases and their interfaces now offer current awareness services, which can be tailored for very specific topics by the user. Most journal publishers also offer alert services directly from their full-text (primary) journal databases.

3.13.3.2.2 Specific literature databases

We discuss here examples of databases that are different from general literature bibliographic databases, being limited either in content by specializing in a subdiscipline, or restricted formally to a subset of the primary literature. The strength of such a limited approach is that the structure of database records, particularly the indexing and search facilities, can be tailored to the user's needs in a particular area. Limiting to a subdiscipline also eliminates false hits when search terms are homonyms (e.g., the term 'mould' means a fungus in a biological context, but something completely different in a metallurgy context).

Examples of such databases are Toxnet,[38] Napralert (Natural Products Alert),[39] business databases such as CIN (Chemical Industry Notes)[40] or CBNB, biotechnology sources (see **Table 1**), and Analytical Abstracts,[41] which has precise search features for analytical procedures that surpass those found in Chemical Abstracts.[11–11b] While Medline and Embase cover all fields of medicine, there are number of specialized sources for AIDS, cancer, pharmaceutical sciences, drug development, etc. (see **Table 1**).

Other special databases cover only one particular type of the primary literature, such as theses (PhD and Master's theses in Dissertation Abstracts[42]), research reports (NTIS[43]), or patents (*see* Section 3.13.3.2.3).

3.13.3.2.3 Patent databases

Patents[44] are both legal documents and primary scientific publications. Therefore, they contain information elements not present elsewhere: patent assignee, priority and other publication data, claims, and legal status data. While other primary publications are indexed only by the database producers, patents are classified (indexed) by the patent-issuing authorities according to the International Patent Classification (IPC),[45] the most extensive existing classification system.

Although general databases such as Chemical Abstracts[11–11b] and Beilstein[9–9b] do cover patents, there is a strong demand for specific bibliographic patent databases. One of the most important is the Derwent[46] World Patent Index (WPI), which was founded in 1963 as a printed information service (**Table 2**). Another major source is INPADOC[47] from the European Patent Office.[48] This lacks special indexing and coding for compounds, and the carefully prepared abstracts that are the hallmark of the WPI, but includes more patent-issuing authorities and, particularly, legal status information that is critical for industry. With (March 2006) over 34 million records, this bibliographic database is even larger than CA (over 26 million) or SCI (*c.* 30 million).

Full-text patent databases are of more recent origin. They emerged from the electronic production of patent documents, and in particular from joint efforts of the EPO[48] and the US and Japanese Patent Offices to convert millions of printed patent documents from their examiner's collections into electronic form. Based on these large patent sets, many commercial full-text patent databases became available at hosts, or for a fee via special patent service and document providers such as Delphion[49] or MicroPatent.[50] In order to promote the use of patents in small and medium enterprises, patent offices made these collections available for free[51] from the EPO,[48] and US patents back to 1790 in the USPTO database (see **Table 2**). These databases serve two purposes: they permit retrieval of full-text patent documents based on patent numbers retrieved in other (bibliographic) sources; and they allow for (simple) searches for inventors, patent assignees, keywords, and the IPC.[45]

3.13.3.2.4 Full-text databases

Until the advent of electronic journals, patents and other primary publications after about 1995, full-text databases were very rare. This type of database is now dominated by the offerings of the patent offices (*see* Section 3.13.3.2.3), the large publishers, and scientific societies. The major journal publishers not only provide their current journals electronically, but they have also converted earlier printed issues into electronic form (backfiles).

Searching in these full-text databases for textual information (keywords, names of authors) is possible with web interfaces, e.g., for e-journals in ACS Journals (American Chemical Society),[52] Elsevier[53] ScienceDirect,[54] etc. But these large full-text databases are virtually unstructured, at least compared with the highly structured bibliographic databases. Therefore, keyword searches can give a high number of irrelevant hits. Another disadvantage is the multitude of publishers, each with their own proprietary search interface and facilities, that needs to be searched (despite the concentration processes in scientific publishing), in contrast to the 'one-stop' approach when searching in bibliographic databases such as Chemical Abstracts.[11–11b] In addition, searching for compounds (structures) and data, as discussed in the following chapters, is virtually impossible.

Daily or weekly full-text news services about all aspects of the healthcare industry, such as the Adis Newsletter,[55] PHARMA MarketLetter,[56,56a] Prous Science Daily Essentials,[57] and Pharmaceutical and Healthcare Industries News Current,[58] are also important for medicinal chemistry.

Table 2 Patent databases

Database	*Content Size (records)*	*Producer Sources*	*Access (host)*
WPI (World Patent Index)	Patents (science, technology) 14.5 million patent families	Thomson Derwent[46] http://thomsonderwent.com/products/patentresearch/dwpi	Derwent Innovations Index: http://scientific.thomson.com/products/dii STN: http://www.stn-international.de/stndatabases/databases/wpidswpx.html
		Published patent documents from 41 patent offices/authorities (since 1963)	Dialog: http://library.dialog.com/bluesheets/html/bl0351.html Questel-Orbit: http://www.questel.orbit.com/EN/customersupport/Userdoc/Fctsht/dwpi.PDF
INPADOC (International Patent Documentation Center Database)	International patents (patent families, legal status data) *c.* 34 million patent families with over 50 million patents issued from 77 organizations	EPO:[48] http://www.european-patent-office.org/inpadoc/index.htm Published patent documents from 65 patent offices/authorities (since 1968)	STN: http://www.stn-international.de/stndatabases/databases/inpadoc.html Dialog: http://library.dialog.com/bluesheets/html/bl0345.html
PCI (Patents Citation Index)	International patents (science, technology) with citations 6.5 million patent families with >66 million citations	Thomson Derwent:[46] (since 1973)	STN: http://www.stn-international.de/stndatabases/databases/dpci.html Dialog: http://library.dialog.com/bluesheets/html/bl0342.html
USPTO	US patents full text 4 million patents	USPTO: http://www.uspto.gov (since 1790)	Free on the web: http://www.uspto.gov/patft/index.html
esp@acenet[51]	Patents full text 50 million patents	EPO:[48] complete patent collections of major countries, time coverage of others variable	Free on the web: http://www.espacenet.com/
Examples of specialized bibliographic and full-text patent databases that cover only certain countries			
JAPIO	Japanese patents 8.5 million patents	Japanese Patent Office Japanese patent applications (since 1973)	STN: http://www.stn-international.de/stndatabases/databases/japio.html Dialog: http://library.dialog.com/bluesheets/html/bl0347.html
IFIPAT	Chemical, mechanical and electrical US patents with special indexing >5.2 million patents	IFI Claims US patents (since 1950)	Dialog: http://library.dialog.com/bluesheets/html/bl0340.html STN: http://www.stn-international.de/stndatabases/databases/ifipat.html
PATDPAFULL	German patents, full-text 1.8 million patents	DPMA German patent documents (since 1987)	STN: http://www.stn-international.de/stndatabases/databases/patdpafull.html

3.13.3.3 Factual Databases

In contrast to bibliographic databases, which refer only to full publications in the primary literature, factual databases immediately provide the required textual or alphanumeric information: physical properties of chemical compounds, spectra, descriptions of research projects, legal information, etc. Although these databases often provide literature references to the origin of the data presented, one does not need to make recourse to the primary literature in every instance as with bibliographic databases (nevertheless, it may be a wise policy to check the validity of the data at the source!). One of the most prolific areas regarding factual databases is molecular biology; a compilation in the annual special database issue of *Nucleic Acids Research* lists over 700 freely available databases.[59,59a]

3.13.3.3.1 Text databases

Printed encyclopedias are typical examples of tertiary literature and have always been important for obtaining fast, concise information about a particular topic. Electronic versions of highly reputed sources, such as Römpp Chemie-Lexikon Online,[60] Merck Index,[61] Ullmann *Encyclopedia of Industrial Chemistry*,[19] Kirk-Othmer *Encyclopedia of Chemical Technology*,[20] *Pharmaceutical Substances* (Kleemann-Engel),[62] and the Chapman & Hall compound dictionaries (CCD),[63] are even more useful because of enhanced search facilities and faster updates compared with their printed equivalents.

Pharmacopoeias are available both in print and electronic forms (CD-ROM or web-based access), and important tertiary sources such as drug handbooks[64] are now also offered electronically.

Another, quite different type in this category are research project databases, which contain information not (yet) available elsewhere, e.g., the general FEDRIP (US Federal Research in Progress),[65] or specialized sources for current pharmaceutical research and clinical tests (**Table 3**).

The CHEMLIST[66] database produced by CAS[23] contains legal and regulatory information about chemical substances, registered by the US Environmental Protection Agency (EPA) and many other regulatory agencies around the world.

3.13.3.3.2 Metadata databases

Meta-databases provide information about the content of databases. Examples are the Gale Directory of Databases,[67] the STNGuide,[68] and the Dialog Dialindex.[69] While the first of these covers all databases regardless of producer or vendor, the latter two are limited to databases held at the respective hosts. This is also true for specialized meta-sources such as the STN Numeriguide,[70] which can be used to identify databases that contain a specific physical property, and the DIALOG Journal Name Finder,[71] which lists databases that abstract a certain journal.

In particular for the vast number of literature, full-text and factual databases associated with drug development, these meta-databases are indispensable. A search in the Dialindex[72] category 'Drug Development Pipeline', for example, gave nine important sources by the producers Adis,[73] IMS Health,[74] and Prous Science,[75] which specialize in this type of information.

Besides such meta-databases, hosts offer further features for selecting appropriate sources: 'clusters' of databases, both as simple lists by topic (e.g., STN,[76] Dialog[77]), and with search facilities in such database clusters, which provide for a query the number of hits in the individual databases in the cluster.

3.13.3.3.3 Numerical databases

Numerical databases primarily contain numeric data about chemical compounds or reactions. As many of these physicochemical values are associated with units, and are dependent on parameters (e.g., boiling point, pressure), they need special representations and search features: range searching (open and closed ranges) for numerical values, and specific searches for values and their associated parameters are essential, while facilities for the conversion of different units are desirable. In addition to numeric data, such as solubility or refraction indexes, textual information (e.g., compound names) and property descriptions (color, or semi-quantitative facts such as 'sparingly soluble') must also be handled in such databases. Bibliographic information leading to the original sources of the data is also needed, unless the database is the original source for the facts presented.

3.13.3.3.3.1 Beilstein and Gmelin

The Beilstein and Gmelin databases both originated from the well-known handbooks that started in the early and mid-nineteenth century, covering organic (Beilstein[9–9b]) and inorganic and organometallic compounds (Gmelin[13]). They are the largest factual databases, the second largest (after the CAS Registry System[25–25c]; *see* Section 3.13.3.4.1.1) structure databases (with presently about 9 million and 2 million compounds, respectively), and are among the largest reaction databases (*see* Section 3.13.3.4.3). Both databases are available via the host STN International,[76] but preferentially via the MDL CrossFire[78–78b,79,79a] database system. Both contain compound descriptions by systematic and trivial names, molecular formula, and structure, and partially also by CAS Registry Numbers, etc. Physical properties are stored and searchable in a large hierarchical data structure with about 400 (Beilstein) and more than 200 (Gmelin) different data fields, grouped in categories such as Spectra, Thermochemical Data, etc., which are subdivided further into Enthalpy, Heat Capacity, etc. Even the large number of data fields (with additional subfields for parameters) does not cover all published measured properties of chemical compounds. In order not to make the data structure too unwieldy, individual numerical property fields were assigned only to those properties occurring relatively commonly in the literature (e.g., the most common enthalpies), while other enthalpies and entropies were grouped together in a field 'Other Thermochemical Data' as keywords without numerical data. The Beilstein database[79,79a] also includes

Table 3 Factual databases

Database	*Content Size (records)*	*Producer Sources*	*Access (host)*
Gmelin	Inorganic and organometallic compounds and their physical and chemical properties 2.3 million substances, 1.8 million reactions, 1.2 million references	MDL: http://www.mdl.com/products/knowledge/crossfire_gmelin/index.jsp Gmelin Handbook, between 110 and 60 journals (since 1791)	MDL CrossFire[78] (updated quarterly) STN: http://www.stn-international.de/stndatabases/databases/gmelin.html (version 1997, no longer updated)
Beilstein	Organic compounds and their physical and chemical properties >9 million substances, >9 million reactions, *c.* 2 million references	MDL: http://www.mdl.com/products/knowledge/crossfire_beilstein/index.jsp/ Beilstein Institut zur Förderung der chemischen Wissenschaften: http://www.beilstein-institut.de/englisch/1024/chemie/index.php3 180 journals, Beilstein Handbook (since 1771)	MDL CrossFire[78,79] STN: http://www.stn-international.de/stndatabases/databases/beilstei.html Dialog: http://library.dialog.com/bluesheets/html/bl0390.html; http://library.dialog.com/bluesheets/html/bl0391.html
DETHERM	500 thermochemical properties (5.6 million datasets) of 25 000 pure compounds and 115 000 mixtures *c.* 20 000 substances with *c.* 440 000 data tables and *c.* 57 000 references	DECHEMA: http://www.dechema.de/Detherm.html/ Journals, patents, reports, etc. (since 1819)	STN: http://www.stn-international.de/stndatabases/databases/detherm.html see also FIZ Chemie Infotherm (properties of 27 000 mixtures, 7100 pure compounds): http://www.fiz-chemie.de/infotherm/servlet/infothermSearch
DIPPR	Physical properties of compounds *c.* 1800 records	Brigham Young University, Chemical Engineering Dep./AIChE Design Institute for Physical Properties: http://www.aiche.org/dippr/ Measured and calculated data (since 1982)	TDS: http://www.tds-tds.com/dipfact.htm
NIST WebBook	Physical and spectroscopic data for compounds	NIST: http://www.nist.gov Selected content from NIST databases	Free on the web: http://webbook.nist.gov/chemistry
NIST Databases	collection of >70 general and specialized databases	NIST: http://www.nist.gov	See: http://www.nist.gov/srd/onlinelist.htm
TOXNET	Toxicology and safety data and literature Cluster of 9 databases	US National Library of Medicine	http://toxnet.nlm.nih.gov Individual databases also available with hosts, on CD-ROM, etc.
Pharmaprojects	Drugs under development >34 000 drugs	PJB Publications: http://www.pjbpubs.com published and unpublished sources (since 1980)	STN: http://www.stn-international.de/stndatabases/databases/phar.html Dialog: http://library.dialog.com/bluesheets/html/bl0128.html
IMS World Drug Monographs	Marketed pharmaceuticals >466 000 drugs	IMS HEALTH Global Services: http://www.imshealth.com	STN: http://www.stn-international.de/stndatabases/databases/drugmono.html
IMS New Product Focus	New drug launches >202 000 drugs	IMS HEALTH Global Services: http://www.imshealth.com (since 1982)	STN: http://www.stn-international.de/stndatabases/databases/imsproduct.html Dialog: http://library.dialog.com/bluesheets/html/bl0446.html

Table 3 Continued

Database	*Content Size (records)*	*Producer Sources*	*Access (host)*
Prous Drug Data Report	Preliminary information on bioactive compounds >177 000 compounds	Prous Science: http://www.prous.com/home Published and unpublished sources (since 1971)	STN: http://www.stn-international.de/stndatabases/databases/prousddr.html Dialog: http://library.dialog.com/bluesheets/html/bl0452.html
Adis R&D Insight	Drugs under development >19 000 drugs	Adis International: http://www.adis.com Journals and unpublished sources (since 1986)	STN: http://www.stn-international.de/stndatabases/databases/adisinsi.html Dialog: http://library.dialog.com/bluesheets/html/bl0107.html

ecological, pharmacological, and toxicological data, starting in their main in 1980, while other properties in Beilstein and Gmelin[80] extend back in time to the beginnings of chemistry in the late eighteenth century. This extensive time coverage is a major advantage of these sources.

3.13.3.3.3.2 Other numerical databases

Besides the general databases Beilstein and Gmelin, there are many specialized numerical databases available, e.g., in the fields of thermodynamics (see **Table 3**), safety and toxicity (RTECS,[81,81a] EINECS[82–82d] HSDB,[38,83,83a] etc.), or biochemical sources such as the enzyme database BRENDA.[84,84a]

The number of other factual databases is so large and still growing that the reader is referred to the meta-databases mentioned in Section 3.13.3.3.2, or the catalogs and web database listings[85,86] of hosts.

3.13.3.3.3.3 Spectroscopic databases

Chemical Abstracts does not index routine spectra from the primary literature, and, while Beilstein and Gmelin do so, they offer only minimal information, and refer otherwise to the primary literature. Given the importance of spectroscopic methods, dedicated spectroscopic databases[87] are needed with the following features:

- display of complete spectra with instrumental and other parameters of measurement
- search for complete spectra or individual signals
- (sub)structure search for compounds, also in combination with spectral search
- comparison of spectra (measured unknown vs database)
- simulation of spectra based on structures, and other tools for structure elucidation and spectrum interpretation.

The spectroscopic database systems SpecInfo[88] and KnowItAll provide such features (**Table 4**).

While spectroscopic databases are obviously preferable for spectrum interpretation and structure elucidation, the number of spectra available in these sources is by about at least a factor of 10 smaller than the mere spectra references in Beilstein or Gmelin.

3.13.3.3.4 Catalog databases

Catalog databases of chemical compounds, including the catalogs of many different chemical suppliers (including package sizes, purity information, prices, and ordering information), are not spectacular, but very much needed examples of factual databases. The most important catalog databases are the Available Chemicals Directory[89–89b] and the Screening Compounds Directory (formerly ACD-SC)[89–89b] (both from MDL), CHEMCATS,[90,90a] and ChemSources.[91–91b] All these are available for a fee, and are usable in combination with searches in large structure or reaction databases. Supplier and other compound-related information (calculated 3D coordinates, biological activities, etc.) is also becoming available for free on the Web, e.g., eMolecules[138] (former Chmoogle), ZINC.[139]

3.13.3.4 Structure and Reaction Databases

Structure and reaction databases contain information on chemical structures as individual compounds or as participants in reactions. The structural diagrams are not stored as graphics (i.e. pictures, which are not structure searchable) in

Table 4 Spectroscopic database systems

Database	*Content Size (records)*	*Producer*	*Access (host)*
SpecInfo	150 000 organic compounds: 26 collections of NMR (^{13}C, ^{1}H, ^{19}F, ^{15}N, ^{17}O, ^{31}P), IR, and mass spectra (between *c.* 1000 and *c.* 100 000 spectra in each collection)	Wiley-VCH (Chemical Concepts): http://www3.interscience.wiley.com/cgi-bin/mrwhome/109609148/HOME	STN: http://www.cas.org/ONLINE/DBSS/specinfoss.html Web version: http://specinfo.wiley.com/specsurf/welcome.html In-house version
KnowItAll (HaveItAll)	Organic and inorganic compounds, polymers: 220 000 IR, 360 000 ^{13}C-NMR, 30 000 ^{1}H-NMR, 198 000 mass spectra, 4200 Raman spectra	Bio-Rad (Sadtler): http://www.bio-rad.com	In-house web version
SDBS	32 000 organic compounds: 23 000 mass spectra, 14 000 ^{1}H-NMR, 12 000 ^{13}C-NMR, 50 000 IR, 3500 Raman spectra, 2500 ESR	National Institute of Advanced Industrial Science and Technology (AIST), Japan	Free on the web: http://www.aist.go.jp/RIODB/SDBS/cgi-bin/cre_index.cgi

such databases, but are represented as connection tables. This representation for two- and three-dimensional structures includes the topological arrangement of atoms and their connections, as well as their stereochemistry (arrangement in space).[10]

In reaction databases, structures of reaction participants are stored in the same manner. In addition, the role of each compound in the reaction (starting material, product, solvent, reagent, catalyst) and reaction-center information is stored (formal, not mechanistic): atoms added/eliminated, bonds formed/broken in the reaction. Reaction centers are determined from an 'atom mapping': which atom/bond in the product corresponds to which atom/bond in the starting material? In an exact reaction search with fully defined structures for starting materials and products, a reaction is already precisely defined by structures and their roles in a reaction, e.g., 2-butene to 2,3-dimethyloxirane. Searching with '2-butene' and '2,3-dimethyloxirane' as partial structures in the same reaction roles, a reaction where the 2-butene unit is simply hydrogenated (not epoxidized) would, formally correctly, also be retrieved, provided that the starting material already contained an epoxide not changed in the reaction, and therefore also as a 'hit' partial structure in the product. Only the definition of the reaction centers (butene double bond to single bond, two new C–O bonds at the former sp^2 carbon atoms) delivers the desired precise result.

3.13.3.4.1 General structure databases

3.13.3.4.1.1 CAS Registry

This database was created in the course of changing the production of Chemical Abstracts[11–11b] (*see* Section 3.13.3.2.1) to electronic data processing. It is the largest and single most comprehensive structure database, having at present[92] more than 85 million records, among them 28 million organic and inorganic compounds, the remainder being sequences of peptides, proteins and nucleic acids.

The CAS Registry[25–25c] database contains, besides the structures (as connection tables) or sequences, and their modifications (with codes for amino acids or nucleotides):

- systematic names and all other trivial or trade names that occur in the primary literature (it is also the largest nomenclature database)
- molecular formula, molecular weight
- special description of rings (useful for generic structure searches)
- calculated (and some measured) physical data, among them pK_a, $\log P$, $\log D$, number of hydrogen donors/acceptors important for medicinal chemistry

- locator field with the names of all databases at STN International,[76] where the particular compound is indexed by its CAS Registry Number (a useful kind of meta-information when searching for literature and properties of compounds)

Structures, partial structures (substructures), and sequences or subsequences are fully searchable, also in combination with the nonstructural information present. The CAS Registry[25–25c] is the core of the CAS[23] database family, and is closely linked by the CAS Registry Number to the bibliographic, reaction, and factual databases produced by CAS, and in a wider sense to many other databases containing that number.

3.13.3.4.1.2 National Cancer Institute (NCI) database

The second largest structure databases, Beilstein and Gmelin, have been discussed in Section 3.13.3.3.3.1. These databases are only accessible for a fee.

The National Cancer Institute (NCI) database[93] is a collection of more than half a million structures, assembled by the NCI's Developmental Therapeutics Program (DTP) or its predecessors in the course of NCI's anticancer screening. Approximately half of this database (*c.* 250 000 structures) is freely publicly available, and is therefore called the 'Open NCI Database'. It contains connection tables for structures, data such as cancer or AIDS screening results, approximately 127 000 CAS Registry Numbers, and 3600 measured log P values. The more recent NIH PubChem[140] project provides freely available information on structure and biological activities of about 5 million small molecules.

3.13.3.4.2 **Patent databases with structures**

As in the case of bibliographic databases, there are special structure databases for patents. While other primary publications contain definite structures with associated information, the legal nature of patents allows Markush structures[94,94a] with generic elements such as 'any halogen,' 'a six-membered heterocycle,' or 'an alkyl group with 5–25 C atoms,' which can represent a large or sometimes even indeterminable number of individual compounds. In order to search and receive meaningful results for both completely defined and Markush structures,[94,94a] special search methods are necessary, and also an appropriate data structure. There are two such special Markush databases: MARPAT (**Table 5**) and Merged Markush Service[95] by Questel-Orbit.[96]

3.13.3.4.3 **Reactions**

In order to search for reactions, one must be able to search for structures and substructures with their roles in the reaction (starting material, product, etc.) and for reaction centers.[97,97a] In addition, numeric searches for yields, reaction conditions (e.g., temperature, pressure, pH), and also text searches for other conditions, types of reactions (name reactions), or compound classes involved, are needed. Reaction searches and reaction databases are thus one of the most demanding tasks in chemical information.

In reaction databases, in the strict sense of the term, the records are individual reactions, and all the information types listed are searchable alone or in any combination.

3.13.3.4.3.1 Databases with reactions

The demanding requirements for 'true' reaction databases are not completely fulfilled by the largest databases with reaction information:

- CrossFire Beilstein (9.7 million reactions) and CrossFire Gmelin (1.7 million reactions, the only inorganic reaction database), with reactions covered since the eighteenth century, are structure-searchable only for starting materials and products, and contain compounds as records
- CASREACT[98,98a] records are publications (at present,[92] *c.* 500 000) with searchable and displayable single- and multi-step reactions (*c.* 11 million), where all reaction participants except solvents are structure-searchable. The time coverage goes back to about 1974, with a relatively small number of older reactions dating back to 1899.

3.13.3.4.3.2 Reaction databases

A large reaction database in the true sense of the definition is SPRESI (ChemReact),[99,99a] which is available under several interfaces (see **Table 5**); this is also a large structure database. Other reaction databases do not try to cover most reactions in the primary literature, either indiscriminately or limited by formal criteria such as in CASREACT[98,98a]

Table 5 Structure databases

Database	*Content size (records)*	*Producer sources*	*Access (host)*
REGISTRY	28 million compounds, 58 million sequences of biopolymers	CAS[23]9000 journals, 37 patent authorities, other primary sources (since 1907)	SciFinder[29–29b] (Scholar[30,30a]): http://www.cas.org/CASFILES/registrycontent.html STN: http://www.stn-international.de/stndatabases/databases/registry.html
NCI	Structures of organic and bioorganic molecules 250 000 structures	US National Cancer Institute/US National Institute of Health	Free on the web: http://cactus.nci.nih.gov
MARPAT	Markush structures from patents >248 000 patents with >640 000 Markush structures	CAS:[23] http://www.cas.org/CASFILES/marpat.html Patents from 44 patent offices/authorities (since 1988)	STN: http://www.stn-international.de/stndatabases/databases/marpat.html
CASREACT	Organic reactions >532 000 documents with >10 million reactions	CAS[23] Journals, patents (since about 1974, some older information, patents since 1991)	SciFinder[29–29b] (Scholar[30,30a]): http://www.cas.org/CASFILES/casreact.html STN: http://www.stn-international.de/stndatabases/databases/casreact.html
ChemInform RX	Organic reactions *c.* 900 000 reactions (125 000 documents)	FIZ Chemie: http://www.fiz-chemie.de/en/katalog/organic/cheminformrx.html 100 core journals (since 1991)	MDL ISIS:[100] http://www.litlink.com/products/knowledge/chem_reaction_lib/index.jsp STN: http://www.stn-international.de/stndatabases/databases/cheminfo.html
SPRESI (ChemReact)	Organic reactions and compounds 4.5 million reactions, 4.5 million compounds, 565 000 journal references, 156 000 patents	InfoChem Journals, patents (1974-2002)	Web version: http://www.spresi.de/faqcontent.htm Daylight: http://www.daylight.com/products/spresi.html see also database subsets at CambridgeSoft: http://chemfinder.cambridgesoft.com/reactions/chemreact.asp, http://chemfinder.cambridgesoft.com/reactions/chemsynth.asp, http://scistore.cambridgesoft.com/software/product.cfm?pid=5018

or Beilstein, but make an intellectual selection of reactions covered according to their synthetic utility. Such selective databases are therefore smaller by at least an order of magnitude. The ChemInform reaction database, based on the printed ChemInform[14–14b] weekly current awareness service, is a typical example, containing about one million reactions since 1991. It is available at STN International,[76] and in-house under the MDL ISIS[100] database system, or the recent web interface DiscoveryGate.[101] Particularly under the interface MDL ISIS,[100] it offers the full spectrum of reaction search facilities, including post-processing of search results such as 'clustering' of retrieved reactions by reagent, solvent, etc. (see **Table 5**).

ISI,[32] better known as a producer of literature databases (*see* Section 3.13.3.2.1.1), also offers structure (Index Chemicus[102–102b]) and reaction databases (Current Chemical Reactions (CCR)[103]), integrated under the interface Web of Knowledge,[104] with literature, citation (SCI[12]), and even patent databases. The Reaction Citation Index is a specific combination of the SCI with CCR,[105,105a] combining structure searches for synthetic methods with literature searches for their applications (citations).

All the reaction databases mentioned so far are general synthesis reaction databases. In addition, dedicated reaction databases exist, for example, for protecting-group chemistry,[106] reactions using enzymes or whole organisms,[107] and metabolic reactions.[108,108a]

3.13.3.4.4 Sequence databases

Besides sequences of peptides and nucleotides in the general structure database CAS Registry,[25–25c] there are specific sequence databases available[59,59a]: GenBank,[109–109b] EMBL,[110,110a] and Swiss-Prot.[111–111b]

3.13.3.4.5 Three-dimensional structure databases

Sources for calculated and measured three-dimensional structures are available. The CAS Registry System[25–25c] contains the largest collection of predicted three-dimensional structures. Another, much smaller, collection is the three-dimensional version of the Available Chemicals Directory (*see* Section 3.13.3.3.4).[89] Such three-dimensional data are often used in drug design.

In contrast to the data calculated by molecular mechanics methods, 'real' three-dimensional structures from x-ray crystallographic structure analyses are provided by crystal structure databases:

- Inorganic Crystal Structure Database (ICSD)[112–112c]
- Cambridge Structural Database (CSD)[113–113b]
- Protein Data Bank (PDB).[114,115,115a]

3.13.3.5 Access to Databases

The history of database searching started around 1970 with a 'textual' approach using retrieval languages. The searching was limited by the terminal–host systems then available, which were relatively cheap, 'dumb,' text terminals, consisting of a keyboard to enter commands, a black-and-white screen to display search output, and a network connection to a large central host computer, which in the beginning was exclusively via slow, relatively expensive phone lines. Chemical structures were displayed, if at all, in 'typewriter style.'

In the 1980s terminals with graphic capabilities became available. Instead of terminals as dedicated hardware, more and more terminal emulations (software) on the personal computers (PCs) becoming available at that time were used. This paved the way for an important change in paradigm from terminal–host to the present client–server systems. Databases are still located and searched centrally on the host's server, but extensive pre- and post-processing by the client software on a local PC is available. This includes not only the graphics so characteristic in chemical information, but all the components of what is considered 'user-friendly,' such as desktop metaphors, mouse and windows, pull-down menus, context-sensitive help, etc. (From our long experience in supporting and training users, we object to the term 'user-friendly'; the term 'usage-friendly' (in German, 'bedienungsfreundlich,' not 'benutzerfreundlich') is more realistic and much more appropriate.)

3.13.3.5.1 Databases and interfaces

Terminology concerning databases and interfaces is often used in a rather sloppy way that is not helpful when discussing information sources or problem solving. For example, SciFinder Scholar[30,30a] is equalized to 'Chemical Abstracts,' or the interface ISI Web of Knowledge[104] is mentioned when in fact the Science Citation Index[12] is meant. We must see this problem in the context of the almost babylonic variety of databases, and their different versions and interfaces. The Chemical Abstracts literature database, for example, is available in six major versions, which differ by content and/or time coverage:

1. CA on CD[116]
2. CA (see **Table 1**)
3. CAplus (see **Table 1**)
4. CAOLD[117,117a]
5. CA versions at hosts other than STN International[76] (less content and time coverage)
6. CA Student edition.[118]

All versions can be considered subsets of the most comprehensive version, CAplus, with the exception of CAOLD.[117,117a] This database, available exclusively from STN International,[76] contains information from the conversion of the printed Chemical Abstracts[11–11b] into a database (*see* Section 3.13.3.2.1) not present in CAplus.

Versions 1 and 6 have specific interfaces, but CAplus is available under SciFinder,[29–29b] SciFinder Scholar,[30,30a] and via STN International.[76] This host offers no less than four different interfaces for CAplus, CA, and other literature databases: the native retrieval language (STN Messenger),[119] support by the front-end software STN Express,[120] STN on the Web,[121] and STN Easy.[122] To top this, the CA database is available in three 'sub-subversions,' which have identical content and searchability, but different price structures for searching.

Web of Knowledge[104]/Web of Science,[123,123a] the Science Citation Index[12] at hosts STN International[76] or Dialog,[77] plus the Science Citation Index on CD-ROM, can also serve as an example of this kind of variety.

Of course, the searchability of a database is above all determined by its content. But within this limitation, different interfaces do have a significant influence on searchability (accessibility) of information. The structure (compound) database CAS Registry,[25–25c] for example, can be searched via molecular formula under both interfaces, SciFinder[29–29b] (Scholar[30,30a]) and STN Messenger,[119] but a search for elemental composition without stoichiometric factors (e.g., 'all compounds which contain Ti and N in any ratio but no other elements') is possible only via the latter interface. Any meaningful discussion must distinguish first between interfaces and databases, and then between their different versions.

3.13.3.5.2 Hosts and in-house databases

In the beginning, databases were only available from hosts such as Dialog[77] or Orbit (now Questel-Orbit[96]), which used a strictly text-based retrieval (command) language for their databases. Many different databases available at one point of access with a powerful common retrieval language, a pay-as-you-go price structure, and support staff, are still the mainstay of hosts. On the negative side, the common retrieval language is a kind of 'least-common denominator,' and less 'user-friendly' (see our comment above regarding this term) than other interfaces, despite more recent web interfaces from hosts, or support from front-end software.[120]

In about 1985, compound and reaction databases with dedicated, graphical interfaces started to become available by a different route. Instead of accessing them at a host's central computer, databases had to be licensed at a fixed annual price, providing unlimited usage for a group of users, and installed in-house on a local database server. Databases under MDL REACCS, its successor MDL ISIS,[100] and CrossFire[78–78b] are available under this access and license model using modern client–server technology. More recently, CAS SciFinder[29–29b]/SciFinder Scholar[30,30a] and MDL DiscoveryGate[101] use a modified access model where the server is again operated centrally at the producer's site, and no longer locally.

Database licenses at fixed cost for, in principle, unlimited searching are economically advantageous for heavily used databases. For special, not routinely used databases, however, the traditional cost model of hosts is preferable. Regarding the general greater 'usage friendliness' (see our comment above regarding this term) of dedicated interfaces as provided with SciFinder[29–29b] (Scholar[30,30a]), CrossFire,[78–78b] or Web of Science,[123,123a] one must not forget that this comes at a cost. The retrieval language STN Messenger[119] permits searches for compounds by their composition, type of ring systems, or biomolecule sequences, which is not possible in exactly the same database CAS Registry[25–25c] under SciFinder[29–29b] (Scholar[30,30a]).

3.13.3.5.3 Integration and reference linking

Chemical information retrieval, or, to phrase it more traditionally, searching the chemical literature, is a stepwise procedure:

1. define the information need (question)
2. select appropriate, available information source(s)
3. rephrase the question according to the selected source(s): query
4. search for compounds, topics, reactions, author names, data, etc., in the secondary or tertiary literature sources
5. identify relevant references to the primary literature in the search results (unless the search has been done in a primary source already)
6. locate/procure the publications for the references
7. retrieve the desired primary information.

This process was particularly tedious in steps 6 and 7 in the early years of database searching, when primary information was only accessible in print. Even the availability of electronic journals and patents did not at first change that, as these existed in separate, differently structured databases produced and offered by different organizations relative to the secondary databases.

In 1997 CAS developed the ChemPort[124] gateway as a link to the primary literature, both in electronic and in print form, from all major interfaces to CAS[23] databases. MDL LitLink[125] and the so-called 'Hop-out' feature in CrossFire[78–78b] offer similar functionality. Even wider reaching is add-on software to OPACs, such as SFX Linking,[126] which links to e-journals as well as the still significant print-only holdings of libraries.

Such integration between databases with literature references and the full text of the corresponding primary publications is certainly the most important type of database integration. A promising approach still in its infancy is Dymond[16] linking, where structures from publications in *Tetrahedron* and *Tetrahedron Letters* are linked directly to the compound records in the factual databases MDL CrossFire Beilstein.[79,79b]

All this involves secondary and primary sources (*see* Section 3.13.3.1), and we thus call it 'vertical integration.' 'Horizontal integration' exists between primary publications (e.g., the CrossRef[127] linking of publishers) as well as between secondary sources: e.g., structure, reaction, literature, and factual databases linked under the interface SciFinder[29–29b] (Scholar[30,30a]), or an even more heterogeneous database collection under Web of Knowledge.[104] Another example is the integrated Major Reference Works[128] under MDL ISIS,[100] where electronic versions of reaction handbooks (Houben-Weyl/Science of Synthesis,[15] EROS[21]) are linked to reaction databases, providing additional information about reaction types for the individual reactions retrieved. Such 'seamless' integrations are very much appreciated by users. Even before the dedicated user interfaces mentioned became available, hosts such as Dialog[77] and STN International[76] provided similar integration under the name of 'cross-file searching' in the form of the transfer of queries and search results between different databases.

3.13.3.5.4 Chemical information on the internet

As the discussion about chemistry on the internet/worldwide web[129] is sometimes at least as chaotic as the web itself, we take recourse again to a classification.

First, almost the entire chemical information available in electronic form, be it in e-journals or databases, is now accessible via the internet, or local intranets using the same protocols for communication. Thus, a discussion about chemical information sources different from the ones mentioned above must concentrate on the worldwide web, which uses the internet, but, contrary to some sloppy use of terminology, is not identical to it.

By definition, the web is certainly not a database, but nevertheless is an important, easily accessible source for chemical and related information. This is offered in three categories:

1. Most of the databases discussed here have web-based interfaces, i.e., they can be accessed on the worldwide web using a standard web browser, sometimes enhanced with special plug-ins or Java to handle chemical structures and other 'nonstandard' information: e.g., STN Easy,[122] Web of Knowledge,[104] DiscoveryGate,[101] PubMed (see **Table 1**), Entrez.[130,130a]
2. Databases available on the Web only: e.g., PubChem,[140] NIST Webbook,[131] Biodegradation Database,[132] ChemFinder,[133] ChemIndustry,[134] MetaXchem.[135]
3. Individual web pages with information not available elsewhere.

Almost everything in 3, and a lot in 2, is free (e.g., all databases listed above), but only few (PubMed, Entrez) in category 1.

Information in categories 1 and 2 is structured, and entirely (1), or mostly (2) reliable, because the sources are well known. On the other hand, this information in the 'deep web'[136] is only accessible via the individual database interfaces, not via the usual search engines. Access is often controlled, demanding at least registration, often payment (almost entirely so for 1). Almost everything in 3, and a lot in 2, is free (e.g., all databases listed above). However, much of the information on the internet (almost everything in category 3) is not reviewed or verified, as is the case in the 'established' literature and databases. Thus, the quality of this information is extremely variable, and continuing access, one of the hallmarks of the traditional publication system, is by no means guaranteed – we all keep getting these infamous 404 messages for information displaced or removed from the web altogether.

3.13.3.6 Information Sources

Every chemist trying to retrieve and use chemical information must be able to select and use appropriate source(s) for the problem at hand. This needs to be done using the plethora of chemical information sources available, which currently undergo quantitative and qualitative changes faster than at any other time. This is the reason why we have concentrated here on presenting a structured overview with references, instead of a more detailed description of the most important sources.

Unfortunately, this diversity of sources is diminished in its utility by a severe lack of easily available meta-information about these sources, describing content, coverage, and organization of the source. Concomitantly, we notice a lack of use of precise terminology and clearly defined criteria to evaluate such sources.

The content and accessibility of every chemical information source, from traditional printed journals to handbooks, databases, and web pages, is a combination of several factors. For a meaningful presentation and discussion of chemical information, a systematic approach, a clear terminology, and meta-information are essential. Such an approach distinguishes clearly between:

- the processing and abstraction state: primary, secondary, tertiary literature/source (*see* Section 3.13.3.1)
- the organization of the information: e.g., databases with records and data fields, web pages with links, journals with articles and their subdivisions (introduction, experimental part, references), patents with well-defined sections
- the interfaces to access and present this information: e.g., printed journal issues with tables of contents, retrieval languages of hosts, dedicated database client software, web browsers (*see* Section 3.13.3.5.1)
- the media (paper, micro-media, electronic) and their implementations to store the information (e.g., loose-leaf collections, microfiche, CD-ROMs)
- the channels to carry the information, e.g., libraries, document delivery services, the internet/worldwide web (*see* Section 3.13.3.5.4), hosts, in-house database servers (*see* Section 3.13.3.5.2), etc.

3.13.4 **Outlook**

Given the enormous number of resources for chemical information that are available, many researchers do not have the time to learn the detail about the various systems and end up searching in only a few resources with which they are familiar. This is a dangerous approach! Knowing that both fee and nonfee resources are available on the internet and that both hold the desired information, it is prudent to search nonfee systems first and then use proprietary databases to fill data gaps.

If users are inexperienced in searching for information they should first consult search engines, meta-databases, or portals. The large search engines generally provide a larger number of hits, but often from commercial, and possibly dubious, sources. Yet if information on a new or rare compound is needed these can be recommended as a first choice. The smaller subject engines provide more reliable data, but vary considerably in their results.

Searchers that are familiar with databases may directly consult known databases (bibliographic databases, numerical databases, etc.), being aware of the fact that they might miss new data sources. A highly important step to getting out of the information labyrinth of information on chemical substances is the evaluation of the contents of the data sources. The reliability and quality of data are only given in peer-reviewed data sources.

Another approach to obtain an overview of chemical information or information related to specified topics in chemistry is to use websites that contain link lists. These link lists are usually provided by universities and private individuals and are classified into subject areas.

All the methods presented above of obtaining information via the internet bear one risk – dead links. Although a search term could be found by a search engine in its own website–meta-data database, the original link to the website could be broken and the information is lost. A more stable source of information is represented by databases.

References

1. Silberschatz, A.; Korth, H. F.; Sudarshan, S. *Database Systems Concepts,* 3rd ed.; McGraw-Hill: New York, 1997.
2. Ulmann, J. D.; Widom, J. *A First Course in Database Systems*; Prentice-Hall: Upper Saddle River, NJ, 1997.
3. Date, C. J. *An Introduction to Database Systems,* 6th ed.; Addison-Wesely: New York, 1995.
4. CODASYL Data Description Language Committee. Report. *Information Systems* **1987**, *3,* 247–320.
5. Codd, E. F. *Commun. ACM* **1970**, *13,* 377–387.
6. Lausen, G.; Vossen, G. *Models and Languages for Object-Oriented Databases*; Addison-Wesley: Harlow, 1998.
7. Embley, D. W. *Object Database Development – Concepts and Principles*; Addison-Welsey: Reading, MA, 1998.
8. Barth, A. Bibliographic Databases. In *Handbook of Chemoinformatics. From Data to Knowledge*; Gasteiger, J., Ed.; Wiley-VCH: Weinheim, 2003; Vol. 2, pp 507–522.

8a. Barth, A. Online Databases in Chemistry. In *Encyclopedia of Computational Chemistry*; Schleyer, P. v. R., Ed.; Wiley: Chichester, 1998; Vol. 3, pp 1968–1981.

9. Beilstein; *Handbuch der organischen Chemie*; 1881–1999; now only produced as database.

9a. Lawson, A. J. The Beilstein Database. In *Handbook of Chemoinformatics. From Data to Knowledge*; Gasteiger, J., Ed.; Wiley-VCH: Weinheim, 2003; Vol. 2, pp 608–628.

9b. Heller, S. R., Ed. *The Beilstein System: Strategies for Effective Searching*; American Chemical Society: Washington, DC, 1998.

10. An introduction into principles of chemical information is provided by Gasteiger, J.; Engel, T., Eds.; *Chemoinformatics, A Textbook*; Wiley-VCH: Weinheim, 2003.
11. Chemical Abstracts printed version. http://www.cas.org/PRINTED/printca.html (accessed June 2006).

11a. Fisanick, W.; Shively, E. R. The CAS Information System: Applying Scientific Knowledge and Technology for Better Information. In *Handbook of Chemoinformatics. From Data to Knowledge*; Gasteiger, J., Ed.; Wiley-VCH: Weinheim, 2003; Vol. 2, pp 556–607.
11b. Fisanick, W.; Amaral, N. J.; Metanomski, V.; Shively, E. R.; Soukup, K. M.; Stobaugh, R. A. Chemical Abstracts Service Information System. In *Encyclopedia of Computational Chemistry*; Schleyer, P. v. R., Ed.; Wiley: Chichester, 1998; Vol. 1, pp 277–315.
12. Synge, R. L. M. *J. Chem. Inf. Comput. Sci.* **1990**, *30*, 33–35.
13. Gmelin *Handbuch der anorganischen Chemie*, print production terminated 1997. Nebel, A.; Toelle, U.; Maass, R.; Olbrich, G.; Deplanque, R.; Lister, P. *Anal. Chim. Acta* **1992**, *265*, 305–312; see also 80.
14. ChemInform (Chemischer Informationsdienst, printed current awareness publication). http://www.fiz-chemie.de/en/katalog/organic/cheminform.html (accessed June 2006).
14a. Wiley-VCH. http://www.wiley-vch.de/publish/en/journals/alphabeticIndex/2073 (accessed June 2006).
14b. Parlow, A.; Weiske, C.; Gasteiger, J. *J. Chem. Inf. Comput. Sci.* **1990**, *30*, 400–402.
15. Houben-Weyl. *Methoden der organischen Chemie*, 1st ed.; 1909. http://www.houben-weyl.com/index.shtml (accessed June 2006). Database version, *Science of Synthesis*. http://www.houben-weyl.com/thieme-chemistry/sos/info/index.shtml (accessed June 2006).
16. For an exception, see Dymond linking. http://www1.elsevier.com/homepage/saa/dymond (accessed June 2006).
17. Landolt-Börnstein, *Numerical Data and Functional Relationships in Science and Technology*. http://www.springer.com/west/home/laboe?SGWID=4-10113-12-95859-0 (accessed June 2006). Landolt-Börnstein is the only of the 'classic' handbooks (Beilstein,[9] Gmelin,[13] Houben-Weyl,[15] Theilheimer[18–18d]) that is not yet available as a true database.
18. Theilheimer, *Synthetic Methods of Organic Chemistry*; printed handbook published since 1946. http://content.karger.com/ProdukteDB/produkte.asp?Aktion=showproducts&ProduktNr=223991&searchWhat=bookseries (accessed June 2006).
18a. Printed, *Journal of Synthetic Methods* (published since 1975). http://www.thomsonderwent.com/products/pca/jsyntheticmethods (accessed June 2006).
18b. MDL ISIS database version. http://www.mdli.com/products/knowledge/journal_synthetic/index.jsp (accessed June 2006).
18c. STN database version. http://www.stn-international.de/stndatabases/databases/djsmonli.html (accessed June 2006).
18d. Finch, A. F. *J. Chem. Inf. Comput. Sci.* **1986**, *26*, 17–22. See also Petersen, A.; Herlan, G. *PharmaChem* **2004**, *3*, 26–29.
19. Ullmann *Encyclopedia of Industrial Chemistry*, Wiley-VCH: Weinheim. http://www.wiley-vch.de/vch/software/ullmann/index.html; http://www3.interscience.wiley.com/cgi-bin/mrwhome/104554801/HOME (accessed June 2006).
20. Kirk-Othmer *Encyclopedia of Chemical Technology*, Wiley-Interscience: New York. http://www3.interscience.wiley.com/cgi-bin/mrwhome/104554789/HOME (accessed June 2006).
21. EROS (*Encyclopedia of Reagents for Organic Synthesis*), Wiley: New York, electronic version. http://www.4ulr.com/products/currentprotocols/interscience/dataencyorganic.html (accessed June 2006).
22. *Encyclopedia of Computational Chemistry*, Wiley-Interscience: New York. http://www3.interscience.wiley.com/cgi-bin/mrwhome/104554772/HOME (accessed June 2006).
23. CAS (Chemical Abstracts Service). http://www.cas.org (accessed June 2006).
24. CAS Standard Abbreviations. http://www.cas.org/ONLINE/standards.html (accessed June 2006).
25. Buntrock, R. E. *J. Chem. Inf. Comput. Sci.* **2001**, *41*, 259–263.
25a. Weisgerber, D. W. *J. Am. Soc. Inf. Sci.* **1997**, *48*, 349–360.
25b. Stobaugh, R. E. *J. Chem. Inf. Comput. Sci.* **1988**, *28*, 180–187.
25c. Dittmar, P. G.; Stobaugh, R. E.; Watson, C. E. *J. Chem. Inf. Comput. Sci.* **1976**, *16*, 111–121.
26. CAS Roles. http://www.cas.org/ONLINE/STN/STNOTES/stnotes5.pdf (accessed June 2006).
27. CA Sections. http://www.cas.org/PRINTED/sects.html (accessed June 2006).
27a. CA Section Thesaurus. http://www.cas.org/ONLINE/QR/casecthes.pdf (accessed June 2006).
28. http://www.cas.org/ONLINE/STN/STNOTES/stnote25.html (accessed June 2006).
29. http://www.cas.org/SCIFINDER/scicover2.html (accessed June 2006).
29a. Ridley, D. D. *Information Retrieval: Scifinder and Scifinder Scholar*; Wiley: Chichester, 2002.
29b. Ridley, D. D. *J. Chem. Inf. Comput. Sci.* **2000**, *40*, 1077–1084.
30. http://www.cas.org/SCIFINDER/SCHOLAR/index.html (accessed June 2006).
30a. Ridley, D. D. *J. Chem. Educ.* **2001**, *78*, 557–558.
31. http://www.nlm.nih.gov/mesh/meshhome.html (accessed June 2006).
32. ISI (Institute for Scientific Information, now Thomson[33]). http://www.isinet.com (accessed June 2006).
33. Thomson (parent company of ISI,[32] Derwent,[46] Dialog,[77] Delphion,[49] BIOSIS[137]). http://www.thomson.com (accessed June 2006).
34. Ridley, D. D. *TrAC Trends Anal. Chem.* **2001**, *20*, 1–10.
35. Whitley, K. M. *J. Am. Soc. Inf. Sci. Technol.* **2002**, *53*, 1210–1215.
36. http://www.info.scopus.com/detail/what (accessed June 2006). This is, to our knowledge, the first database in the domain of the 'secondary literature' not based on a traditional printed version (or an extension of this), but directly from the primary literature. This might herald a lot of more new sources to come.
37. http://www.isinet.com/products/cap/ccc (accessed June 2006).
38. Wexler, P. *Toxicol.* **2004**, *198*, 161–168.
39. http://www.napralert.org/ (accessed June 2006); STN Napralert. http://www.stn-international.de/stndatabases/databases/napraler.html (accessed June 2006). Loub, W. D.; Farnsworth, N. R.; Soejarto, D. D.; Quinn, M. L. *J. Chem. Inf. Comput. Sci.* **1985**, *25*, 99–103.
40. STN. http://www.stn-international.de/stndatabases/databases/cin.html (accessed June 2006); Dialog. http://library.dialog.com/bluesheets/html/bl0019.html (accessed June 2006).
41. Diospatonyi, I.; Horvai, G.; Braun, T. *J. Chem. Inf. Comput. Sci.* **2000**, *40*, 1085–1092.
42. STN. http://www.stn-international.de/stndatabases/databases/dissabs.html (accessed June 2006); Dialog: http://library.dialog.com/bluesheets/html/bl0035.html (accessed June 2006).
43. NTIS (National Technical Information Service database with research report about projects federally funded in the USA). http://www.ntis.gov/products/types/databases/ntisdb.asp?loc1/44-4-3 (accessed June 2006); STN. http://www.stn-international.de/stndatabases/databases/ntis.html; Dialog. http://library.dialog.com/bluesheets/html/bl0006.html (accessed June 2006).
44. Vogt, J. Patent Databases. In *Handbook of Chemoinformatics. From Data to Knowledge*; Gasteiger, J., Ed.; Wiley-VCH: Weinheim, 2003; Vol. 2, pp 743–755.
45. http://www.wipo.int/classifications/en (accessed June 2006).

46. Thomson Derwent (producer of WPI). http://www.derwent.com (accessed June 2006).
47. Lingua, D. G. *World Pat. Inf.* **2005**, *27*, 105–111.
48. EPO (European Patent Office). http://www.european-patent-office.org/index.en.php (accessed June 2006).
49. http://www.delphion.com (accessed June 2006).
50. http://www.micropat.com/static/index.htm (accessed June 2006).
51. http://www.espacenet.com/ (accessed June 2006).
51a. Schwander, P. *World Pat. Inf.* **2000**, *22*, 147–165.
52. http://pubs.acs.org/about.html (accessed June 2006).
53. Elsevier (largest publisher of science journals). http://www.elsevier.com/wps/find/journal_browse.cws_home (accessed June 2006).
54. http://www.sciencedirect.com (accessed June 2006).
55. http://www.stn-international.de/stndatabases/databases/adisnews.html (accessed June 2006).
56. http://www.marketletter.com (accessed June 2006).
56a. STN PHARMAML. http://www.stn-international.de/stndatabases/databases/pharmaml.html (accessed June 2006).
57. http://library.dialog.com/bluesheets/html/bl0458.html (accessed June 2006).
58. http://www.stn-international.de/stndatabases/databases/phic.html (accessed June 2006).
59. von Homeyer, A.; Reitz, M. Databases in Biochemistry and Molecular Biology. In *Handbook of Chemoinformatics From Data to Knowledge*; Gasteiger, J., Ed.; Wiley-VCH: Weinheim, 2003; Vol. 2, pp 756–793.
59a. Galperin, M. Y. *Nucl. Acids Res.* **2005**, *33*, D5–D24; see also 2006 updates. http://nar.oxfordjournals.org/cgi/content/full/34/suppl_1/D3/DC1 (accessed June 2006).
60. Römpp Chemie-Lexikon online. http://www.roempp.com/index.shtml (accessed June 2006).
61. http://www.merckbooks.com/mindex; Internet and CD-ROM. http://scistore.cambridgesoft.com/software/product.cfm?pid=35 (accessed June 2006); STN MRCK. http://www.stn-international.de/stndatabases/databases/mrck.html (accessed June 2006); Dialog. http://library.dialog.com/bluesheets/html/bl0304.html (accessed June 2006).
62. STN Pharmaceutical Substances (Kleemann–Engel). http://www.stn-international.de/stndatabases/databases/ps.html (accessed June 2006).
63. CCD (Combined Chemical Dictionary): 276 000 organic compounds, 188 000 natural products, 46 000 drugs, 103 000 inorganic/organometallic compounds, 14 000 analytical reagents; CHEMnetBASE. http://www.chemnetbase.com/default.html (accessed June 2006); Dialog. http://library.dialog.com/bluesheets/html/bl0303.html (accessed June 2006).
64. Dialog Drug Information Fulltext. http://library.dialog.com/bluesheets/html/bl0229.html (accessed June 2006).
65. http://grc.ntis.gov/fedrip.htm (accessed June 2006); Dialog. http://library.dialog.com/bluesheets/html/bl0266.html (accessed June 2006).
66. http://www.cas.org/Support/substance.html (accessed June 2006); STN CHEMLIST. http://www.stn-international.de/stndatabases/databases/chemlist.html (accessed June 2006).
67. http://www.galegroup.com/pdf/facts/gdod.pdf (accessed July 2006); Dialog. http://library.dialog.com/bluesheets/html/bl0230.html (accessed June 2006).
68. http://www.stn-international.de/stndatabases/databases/stnguide.html (accessed June 2006).
69. http://library.dialog.com/bluesheets/html/bl0411.html (accessed June 2006).
70. http://www.stn-international.de/stndatabases/databases/numerigu.html (accessed June 2006).
71. http://library.dialog.com/bluesheets/html/bl0414.html (accessed June 2006).
72. http://library.dialog.com/bluesheets/html/bloD.html#DRUGDEV (accessed June 2006).
73. Adis International. http://www.adis.com (accessed June 2006).
74. IMS Health. http://www.imshealth.com (accessed June 2006).
75. Prous Science. http://www.prous.com/home (accessed June 2006).
76. STN (Scientific & Technical Information Network) International (important host for chemical databases). http://www.stn-international.de (accessed June 2006); Databases. http://www.stn-international.de/stndatabases/databases/onlin_db.html (accessed June 2006); Search aids. http://www.stn-international.de/training_center/mat_sea_stn.html (accessed June 2006).
77. Host Thomson Dialog. http://www.dialog.com (accessed June 2006); Databases. http://support.dialog.com/publications/dbcat (accessed June 2006); Search aids. http://support.dialog.com/searchaids/dialog (accessed June 2006).
78. Server. http://www.mdl.com/products/knowledge/crossfire_server/key_features.jsp (accessed June 2006); Client. http://www.mdl.com/products/knowledge/crossfire_commander (accessed June 2006).
78a. Cooke, H.; Ridley, D. D. *Austr. J. Chem.* **2004**, *57*, 387–392.
78b. Meehan, P.; Schofield, H. *Online Inf. Rev.* **2001**, *25*, 241–249.
79. MDL CrossFire Beilstein: Nquyen, H; Ilchmann, G. In *Proceedings of the International Chemical Information Conference*, Annecy, France, Oct. 25–28, 1999; Collier, H., Ed.; Infornortics: Tetbury, 1999, pp 105–111.
79a. Cooke, F.; Kopelev, N.; Schofield, H.; Boyce, G.; Dunne, S. *J. Chem. Inf. Comput. Sci.* **2002**, *42*, 1016–1027.
80. Vogt, J.; Vogt, N.; Schunk, A. Databases in Inorganic Chemistry. In *Handbook of Chemoinformatics. From Data to Knowledge*; Gasteiger, J., Ed.; Wiley-VCH: Weinheim, 2003; Vol. 2, pp 629–643.
81. Registry of Toxic Effect of Chemical Substances. http://www.cdc.gov/niosh/rtecs/default.html (accessed June 2006).
81a. Sweet, D. V.; Anderson, V. P.; Fang, J. C. F. *Chem. Health Safety* **1999**, *6*, 12–16.
82. http://ecb.jrc.it; http://ecb.jrc.it/existing-chemicals/ (accessed June 2006).
82a. Heidorn, C. J. A.; Rasmussen, K.; Hansen, B. G.; Norager, O.; Allanou, R.; Seynaeve, R.; Scheer, S.; Kappes, D.; Bernasconi, R. *J. Chem. Inf. Comput. Sci.* **2003**, *43*, 779–786.
82b. Heidorn, C. J. A.; Hansen, B. G.; Norager, O. *J. Chem. Inf. Comput. Sci.* **1996**, *36*, 949–954.
82c. Stephan, U.; Strobel, U. *Nachr. Chem.* **2003**, *51*, 1052–1053.
82d. Geiss, F.; Del Bino, G.; Blech, G.; Norager, O.; Orthmann, E.; Mosselmans, G.; Powell, J.; Roy, R.; Smyrniotis, T.; Town, W. G. *Toxicol. Environm. Chem.* **1992**, *37*, 21–33.
83. Hazardous Substances Databank. http://www.nlm.nih.gov/pubs/factsheets/hsdbfs.html (accessed June 2006).
83a. Fonger, G. C. *Toxicol* **1995**, *103*, 137–145.
84. http://www.brenda.uni-koeln.de (accessed June 2006).
84a. Schomburg, I.; Chang, A.; Ebeling, C.; Gremse, M.; Heldt, C.; Huhn, G. R.; Schomburg, D. *Nucl. Acids Res.* **2004**, *32*, D431–D433.
85. STN database. Cluster medicine. http://www.stn-international.de/stndatabases/clusters/medicine.html (accessed June 2006). Pharmaceutical sciences. http://www.stn-international.de/stndatabases/clusters/pharmaco.html (accessed June 2006).

86. Dialog databases. Medicine & biosciences. http://library.dialog.com/bluesheets/html/bls0019.html#SB0019 (accessed June 2006); Pharmaceuticals. http://library.dialog.com/bluesheets/html/bls0020.html#SB0020 (accessed June 2006).
87. Neudert, R.; Davies, A. N. Spectroscopic Databases. In *Handbook of Chemoinformatics. From Data to Knowledge*; Gasteiger, J., Ed.; Wiley-VCH: Weinheim, 2003; Vol. 2, pp 700–721.
88. Barth, A. J. *Chem. Inf. Comput. Sci.* **1993**, *33*, 52–58.
89. MDL. http://www.mdli.com/products/experiment/index.jsp (accessed June 2006).
89a. CambridgeSoft. http://chemfinder.cambridgesoft.com/chemicals/chemacxpro.asp (accessed June 2006).
89b. Daylight. http://www.daylight.com/products/acd.html (accessed June 2006).
90. http://www.cas.org/CASFILES/chemcats.html (accessed June 2006).
90a. STN CHEMCATS. http://www.stn-international.de/stndatabases/databases/chemcats.html (accessed June 2006).
91. http://www.chemsources.com/chemonline.html (accessed June 2006).
91a. STN CSCHEM. http://www.stn-international.de/stndatabases/databases/cschem.html (accessed June 2006).
91b. STN CSCORP. http://www.stn-international.de/stndatabases/databases/cscorp.html (accessed June 2006).
92. Statistics for CAS databases. http://www.cas.org/cgi-bin/regreport.pl (accessed June 2006).
93. Ihlenfeldt, W.-D.; Voigt, J. H.; Bienfait, B.; Oellien, F.; Nicklaus, M. C. *J. Chem. Inf. Comput. Sci.* **2002**, *42*, 46–57.
94. Berks, A. H.; Barnard, J. M.; O'Hara, M. P. Markush Structure Searching in Patents. In *Encyclopedia of Computational Chemistry*; Schleyer, P. v. R., Ed.; Wiley: Chichester, 1998; Vol. 3, pp 1552–1559.
94a. Berks, A. H. Current State of the Art of Markush Topological Search Systems. In *Handbook of Chemoinformatics. From Data to Knowledge*; Gasteiger, J., Ed.; Wiley-VCH: Weinheim, 2003; Vol. 2, pp 885–903.
95. http://thomsonderwent.com/products/patentresearch/mergedmarkush (accessed June 2006).
96. http://www.questel.orbit.com; Databases. http://www.questel.orbit.com/EN/customersupport/Userdoc/DocPDF/Databasecatalog.pdf (accessed June 2006).
97. Zass, E. Databases of Chemical Reactions. In *Handbook of Chemoinformatics. From Data to Knowledge*; Gasteiger, J., Ed.; Wiley-VCH: Weinheim, 2003; Vol. 2, pp 667–699.
97a. Zass, E. Reaction Databases. In *Encyclopedia of Computational Chemistry*; Schleyer, P. v. R., Ed.; Wiley: Chichester, 1998; Vol. 4, pp 2402–2420.
98. Blake, J. E.; Dana, R. C. J. *Chem. Inf. Comput. Sci.* **1990**, *30*, 394–399.
98a. Blower, P. E., Jr.; Myatt, G. J.; Petras, M. W. *J. Chem. Inf. Comput. Sci.* **1997**, *37*, 54–58.
99. Schinzer, D. *Nachr. Chem. Tech. Lab.* **1993**, *41*, 826–828.
99a. Griepke, G. *J. Chem. Inf. Comput. Sci.* **1997**, *37*, 154–155.
100. http://www.mdl.com/products/framework/isis/index.jsp (accessed June 2006).
101. http://www.litlink.com/products/knowledge/discoverygate/index.jsp (accessed June 2006).
102. http://www.isinet.com/products/litres/indexchemicus (accessed June 2006).
102a. http://www.garfield.library.upenn.edu/essays/V1p063y1962-73.pdf (accessed June 2006).
102b. Antony, A.; Stevens, J. *J. Chem. Inf. Comput. Sci.* **1980**, *20*, 101–105.
103. http://scientific.thomson.com/products/ccr (accessed June 2006).
104. http://www.thomsonisi.com/webofknowledge/index.html (accessed June 2006).
105. http://www.isinet.com/products/citation/rci (accessed June 2006).
105a. Kimberley, R.; Lowe, R.; Banik, G.; Lawlor, B. Implementation of an Alternative Approach to Searching Chemical Reaction Databases. In *Proceedings of the International Chemical Information Conference, Annecy*, Oct. 17–19, 1994; Collier, H., Ed.; Infonortics: Calne, 1994, pp 42–52.
106. Accelrys Protecting Groups. http://www.accelrys.com/products/chem_databases/databases/protecting_groups.html (accessed June 2006).
107. Accelrys Biocatalysis. http://www.accelrys.com/products/chem_databases/databases/biocatalysis.html (accessed June 2006).
108. MDL Metabolite. http://www.litlink.com/products/predictive/metabolite/index.jsp (accessed June 2006).
108a. Accelrys Metabolism. http://www.accelrys.com/products/chem_databases/databases/metabolism.html (accessed June 2006).
109. http://www.ncbi.nlm.nih.gov/Genbank (accessed June 2006).
109a. http://www.ebi.ac.uk/embl (accessed June 2006).
109b. Benson, D. A.; Karsch-Mizrachi, I.; Lipman, D. J.; Ostell, J.; Wheeler, D. L. *Nucl. Acids Res.* **2005**, *33*, D34–D38.
110. http://www.ebi.ac.uk/embl (accessed June 2006).
110a. Kanz, C.; Aldebert, P.; Althorpe, N.; Baker, W.; Baldwin, A.; Bates, K.; Browne, P.; van den Broek, A.; Castro, M.; Cochrane, G. et al. *Nucl. Acids Res.* **2005**, *33*, D29–D33.
111. http://www.ebi.ac.uk/swissprot (accessed June 2006).
111a. http://www.expasy.org/sprot (accessed June 2006).
111b. Bairoch, A.; Boeckmann, B. *Nucl. Acids Res.* **1993**, *21*, 3093–3096.
112. ICSD. http://icsd.ill.fr/icsd (accessed June 2006).
112a. STN ICSD. http://www.stn-international.de/stndatabases/databases/icsd.html (accessed June 2006).
112b. Hellenbrandt, M. *Crystallogr. Rev.* **2004**, *10*, 17–22.
112c. Bergerhoff, G. Inorganic Three-Dimensional Strcuture Databases. In *Encyclopedia of Computational Chemistry*; Schleyer, P. v. R., Ed.; Wiley: Chichester, 1998; Vol. 2, pp 1325–1337.
113. http://www.ccdc.cam.ac.uk/products/csd (accessed June 2006).
113a. Allen, F. H.; Hoy, V. J. Cambridge Structural Database. In *Encyclopedia of Computational Chemistry*; Schleyer, P. v. R., Ed.; Wiley: Chichester, 1998; Vol. 1, pp 154–167.
113b. Allen, F. H.; Lipscomb, K. J.; Battle, G. The Cambridge Structural Database (CSD) of Small Molecule Crystal Structures. In *Handbook of Chemoinformatics. From Data to Knowledge*; Gasteiger, J., Ed.; Wiley-VCH: Weinheim, 2003; Vol. 2, pp 645–666.
114. http://www.rcsb.org/pdb (accessed June 2006).
115. Berman, H. M.; Westbrook, J.; Zardecki, C.; Bourne, P. E. *Protein Struct.* **2003**, 389–405.
115a. Berman, H. M.; Battistuz, T.; Bhat, T. N.; Bluhm, W. F.; Bourne, P. E.; Burkhardt, K.; Feng, Z.; Gilliland, G. L.; Iype, L.; Jain, S.; Fagan, P.; Marvin, J.; Padilla, D.; Ravichandran, V.; Schneider, B.; Thanki, N.; Weissig, H.; Westbrook, J. D.; Zardecki, C. *Acta Crystallogr. D* **2002**, *D58*, 899–907.
116. Chemical Abstracts CD-ROM version. http://www.cas.org/ONLINE/CD/CACD/cover.html (accessed June 2006); http://www.cas.org/ONLINE/CD/collectives.html (accessed June 2006).
117. http://www.stn-international.de/stndatabases/databases/caold.html (accessed June 2006).

117a. Hamill, K. A.; Nelson, R. D.; Vander Stouw, G. G.; Stobaugh, R. E. *J. Chem. Inf. Comput. Sci.* **1988**, *28*, 175–179.
118. CA Student Edition (subset of Chemical Abstracts database). http://www.cas.org/New1/student.html (accessed June 2006).
119. http://www.cas.org/training/basics.pdf (accessed June 2006).
120. http://www.stn-international.de/stninterfaces/stnexpress/stn_exp.html (accessed June 2006).
121. http://www.stn-international.de/stninterfaces/stnow/stn_ow.html (accessed June 2006).
122. http://www.stn-international.de/stninterfaces/stneasy/stn_easy.html (accessed June 2006).
123. http://www.isinet.com/products/citation/wos (accessed June 2006).
123a. Oxley, H. *Database* **1998**, *21*, 37–40.
124. http://www.cas.org/chemport/index.html (accessed June 2006).
125. http://www.mdl.com/products/knowledge/litlink_direct/index.jsp (accessed June 2006).
126. http://www.exlibrisgroup.com/sfx.htm (accessed June 2006).
127. http://www.crossref.org (accessed June 2006).
128. http://mdl.com/products/knowledge/reference_works/index.jsp (accessed June 2006).
129. Tarkhov, A. Chemistry on the Internet. In *Handbook of Chemoinformatics. From Data to Knowledge*; Gasteiger, J., Ed.; Wiley-VCH: Weinheim, 2003; Vol. 2, pp 794–843.
130. http://www.ncbi.nlm.nih.gov/gquery/gquery.fcgi (accessed June 2006).
130a. Maglott, D.; Ostell, J.; Pruitt, K. D.; Tatusova, T. *Nucl. Acids Res.* **2005**, *33*, D54–D58.
131. Linstrom, P. J.; Mallard, W. G. *J. Chem. Eng. Data* **2001**, *46*, 1059–1063.
132. University of Minnesota Biocatalysis/Biodegradation Database. http://umbbd.ahc.umn.edu (accessed June 2006).
133. http://chemfinder.cambridgesoft.com (accessed June 2006).
134. http://www.chemindustry.com (accessed June 2006).
135. http://www.chemie.de/metaxchem/index.jsp (accessed June 2006).
136. http://en.wikipedia.org/wiki/Deep_web (accessed June 2006).
137. Biological Abstracts (database BIOSIS). http://scientific.thomson.com/products/bp/ (accessed June 2006).
138. http://www.emolecules.com/ (accessed July 2006).
139. http://blaster.docking.org/zinc/ (accessed July 2006).
140. http://pubchem.ncbi.nlm.nih.gov/ (accessed July 2006).

Biographies

Thomas Engel studied chemistry and obtained his PhD in 1996 in physical chemistry in Würzburg, Germany. He was assistant head of a laboratory before he joined the research group headed by Johann Gasteiger at the Computer Chemistry Centre at the University of Erlangen-Nuremberg. Thomas has specialized in chemoinformatics and is co-editor of a textbook of chemoinformatics. In addition, he gives advanced training courses in chemo- and bioinformatics and teaches these topics at the university. Since 2005, Thomas has been an application scientist at the Chemical Computing Group Inc. in Cologne, Germany.

Engelbert Zass After studying chemistry at Cologne University (Diploma 1972) and receiving a PhD in organic chemistry at ETH Zürich in 1977, Zass specialized in chemical information. He has been searching public online databases since 1979. In 1995, he helped to start cooperation among Swiss university chemistry departments for running in-house chemistry databases. Based on this practical experience, he has been teaching chemical information courses at ETH and other institutions since 1980. He is a 'Lehrbeauftragter' (lecturer) at ETH as well as at the Universities of Zürich and Berne. As head of the ETH Chemistry, Biology, Pharmacy Information Center he manages a team of scientists and librarians, who combine traditional library services with database server operations, and offer information retrieval and delivery, user support, and chemical information education and training. Presently, he is engaged in projects concerning web-based teaching and development of multimedia teaching modules.

Comprehensive Medicinal Chemistry II
ISBN (set): 0-08-044513-6

ISBN (Volume 3) 0-08-044516-0; pp. 265–291

3.14 Bioactivity Databases

M Olah and T I Oprea, University of New Mexico School of Medicine, Albuquerque, NM, USA

3.14.1	**Introduction**	**293**
3.14.2	**Databases and Management Systems**	**294**
3.14.2.1	Structured Query Language (SQL)	294
3.14.2.2	Database Options	294
3.14.2.3	Open-Source Database Management System Solutions	294
3.14.2.4	Database Management Systems for Chemistry and Biology	295
3.14.2.5	Hierarchical versus Relational Databases	295
3.14.3	**Bioactivity Databases: Some Design Guidelines**	**296**
3.14.3.1	Chemical Information	297
3.14.3.1.1	One-dimensional representation	298
3.14.3.1.2	Two-dimensional representations	298
3.14.3.1.3	Three-dimensional representations	299
3.14.3.2	Target and Protocol Information	299
3.14.3.3	The Bibliographic References	300
3.14.3.4	The Biological Activity Data	301
3.14.3.5	Integration with Other Databases	302
3.14.4	**Biological and Bioactivity Information Databases Examples**	**303**
3.14.4.1	Biological Information Databases	304
3.14.4.2	Bioactivity Information Databases	305
3.14.4.2.1	Open-access databases	305
3.14.4.2.2	Commercially available databases	307
3.14.5	**Conclusions**	**309**
References		**310**

3.14.1 Introduction

The current drug discovery paradigm allocates a short period of time (3–12 months) for the process of lead identification. Thus, medicinal chemists have a rather short amount of time to familiarize themselves with prior art, i.e., background information related to the biological target and to chemotypes relevant on the intended, or related targets. Gathering such background information is enabled by chemical databases such as chemical abstracts via SciFinder,[1] Beilstein,[2] and Spresi,[3,4] by medicinal chemistry related patent databases such as the MDL Drug Data Report (MDDR),[5,6] the World Drug Index (WDI),[4,7] Current Patents Fast Alert,[8] and by collections of bioactive compounds such as Comprehensive Medicinal Chemistry,[9] and the Physician Desk Reference (PDR).[10]

As pharmaceutical drug discovery relies on chemogenomics, the average end-user has learned to expect database systems with built-in search engines that seamlessly mine chemical and biological data within an information-rich, integrated resource. The integration process itself requires hierarchical classification schemes, such that knowledge related to target-focused chemical libraries and biological target families can be mined simultaneously.[11] Such tools being under development, we address bioactivity databases in general with focus on bioinformatics and cheminformatics resources. Technical details are omitted for clarity. Readers are encouraged to consult the general references, as well as software documentation. The first section, 'Databases and Management Systems', introduces databases in general. The second part, 'Bioactivity Databases: Some Design Guidelines', provides guiding principles for designing and maintaining bioactivity databases. The third section, 'Biological and Bioactivity Information Databases Examples', provides a brief overview of currently available bioactivity databases.

3.14.2 Databases and Management Systems

A database is a collection of persistent (factual or virtual) data stored in a computer in a systematic way, such that a computer program can consult them to answer queries. Databases are designed to offer an organized mechanism for information storage, management, and retrieval. The particular software used for this purpose is known as a database management system (DBMS).[12] DBMSs range in complexity: differences include the capability of ensuring data integrity, multiple user access, and what computations they provide given a set of data.[13,14] The first DBMSs were developed in the 1960s. Two key data models arose at this time: the network model, followed by the hierarchical model.[15] These were later replaced by the relational model[16] and the object-oriented database model.[17,18]

3.14.2.1 Structured Query Language (SQL)

Most large-scale databases use the Structured Query Language (SQL),[19] a database language that defines all user and administrator interactions. Designed especially for access to relational database systems, SQL offers a flexible interface for a variety of databases. All database transactions are made in SQL, even if there are a large number of graphical user interfaces that simplify database administration tasks. Initiated by IBM in the late 1970s, the first SQL standard was published by the American National Standards Institute (ANSI) and the International Organization for Standardization (ISO) in 1986. The last revision and addition to this standard was published in 2003 (SQL-2003). Although SQL comes in many flavors, all of them are based upon ANSI SQL.[20] Standard conformity is an important consideration when choosing a relational database management system (RDBMS). Currently, database manufacturers move beyond the ANSI/SQL standard in order to give their products a competitive advantage and to meet customer demand. Such products can claim ANSI/SQL compliance if certain conditions are met.[21]

SQL commands can be divided into two main sublanguages. The Data Definition Language (DDL) contains the commands used to create and destroy databases and database objects. After the database structure is defined with DDL, the Data Manipulation Language (DML) is used to insert, retrieve, and modify the data contained within.[12] One does not need to learn SQL to be a database user. Graphical front-end tools are more than sufficient in most cases. However, database administrators and developers rely upon custom SQL code to ensure that their transactions meet user requirements efficiently.

3.14.2.2 Database Options

Perhaps everyone has worked with at least one desktop database, for example using MS (Microsoft)-Access,[22] or FileMaker Pro.[23] These products are relatively inexpensive and useful for single-user or noninteractive web applications.

For heavy-duty, multiuser environment database application, MS SQL Server,[24] Oracle,[25] and IBM DB2[26] are key among the for-fee server databases. Some open-source systems like MySQL[27] and PostgreSQL[28] are discussed in Section 3.14.2.3.

Currently, almost all database applications call for web interactions. Web-enabled databases are viewable over the internet or intranet with a web browser. Since no special software is needed to access the data, this is a very effective way to distribute data.[29,30]

3.14.2.3 Open-Source Database Management System Solutions

Open-source DBMS alternatives[31] are briefly outlined below: MySQL[27] is focused on maximizing speed, and less on having every possible feature. Its architecture makes it fast, customizable, compact, and stable. It runs on more than 20 platforms. Many companies currently use MySQL to power their web pages.[32,33] Available on a wide number of UNIX platforms, PostgreSQL has a solid engine, and is fairly popular in organizations seeking stability and high volume for large databases.[28,34,35] While not as popular as MySQL or PostgreSQL, Firebird has a dedicated user support, and has many advanced features. It is quite mature for a relatively new open-source framework.[36] Berkeley DB provides fast and reliable database management for special-purpose computing and high-end core internet servers. Extremely portable, it runs under almost all Unix and Windows platforms, as well as a number of embedded, real-time operating systems.[37] Ingres is a mature, high-performance relational database framework. Reliable and cost effective, it supports mission-critical applications in small- to medium-size businesses and high-volume deployments in large-scale enterprises.[38] The above packages come with an open-source license, i.e., they are free to download and test, in order to decide which one fits user requirements; low cost makes them attractive for certain environments.

3.14.2.4 Database Management Systems for Chemistry and Biology

Since it can be stored in textual and graphic format, biological information can be captured in general-purpose databases. Even more complex data, e.g., images or sounds, can be stored without difficulty in binary format. However, this is not the case when dealing with chemical information. Standard DBMSs lack support (i.e., data structures) for handling chemical structures. Special systems or extension modules need to be designed and implemented to enable storage, search, and retrieval for chemical (sub)structures. These can be combined with other data types, as needed.

One such system is MDL ISIS/Base from Elsevier-MDL.[39] This desktop package can be used to manage local (hierarchical) databases for storing, searching, and retrieving chemical structures and associated data. Its form-based searching is customizable and straightforward to use, and combines chemical structure searches with text and numeric queries. MDL Isis/Base is also available as a client to the new MDL Isentris framework (MDL ISIS/Host in previous versions), another ISIS (Integrated Scientific Information System) family member. Isentris and ISIS/Base are enterprise solution chemical DBMSs that can be connected to Oracle[25] using the MDL cartridge. These systems require MDL/Draw (or ISIS/Draw an older product) for structure drawing. The incorporated development toolkit allows the creation of dialog boxes, tool bars, buttons, and special forms. Custom modules can be written using the ISIS programming language for automating procedures such as data registration. ISIS/Base/Draw is a good solution for Windows-based front-end systems.[40]

Another stable and powerful enterprise solution is available from Daylight CIS,[4] which offers THOR, a client-server distributed chemical information database system for (non-Microsoft) network environments.[41] THOR features extremely fast data retrieval time, independent of database size or complexity. The primary key used in the database is the molecular structure stored in SMILES[42,43] format: this chemical linear notation was designed by Daylight. This feature distinguishes Daylight components from competition, as they require minimal storage and provide space efficiency and very fast retrieval times; various toolkits are available for high-end customization, tailored to data visualization, query, and storage. DayCart[44] is a fully integrated cartridge which extends Oracle clients and servers with chemical intelligence.

ChemFinder, from CambridgeSoft Corporation,[45] is a small-enterprise DBMS that can be used standalone or connected to Oracle and MS Access. Extension modules for MS Word and MS Excel are available. Accessible through a web browser, the ChemOffice WebServer is a solution platform for chemical and biological data storage and sharing. Its Software Developer's Kit allows customization for user-defined functionality. ChemOffice WebServer Software Developer's Kit extends the Microsoft and Oracle platforms, allowing information scientists to use the most powerful development tools. ChemDraw is the equivalent MDL's ISIS/Draw tool for drawing chemical structures.

The Java-based[46] JChem Base, from ChemAxon Ltd., is another small-enterprise solution that allows the query of mixed structural and nonstructural data. It can integrate a variety of database systems (Oracle, MS SQL Server, IBM DB2, MS Access) with web interfaces and offers a fast similarity, exact-structure and (sub)structure search engine. Using the JChem Cartridge for Oracle the user can acquire additional functionalities from within Oracle's SQL. The system includes Marvin, a Java-based chemical editor and viewer.[47]

Focused on small-enterprise solutions for the desktop, ChemoSoft from Chemical Diversity Labs Inc., offers a low-cost, reliable, and efficient solution for most of the above tasks: ChemoSoft has an interface with standard SQL servers (Oracle, MS SQL Server, Borland Interbase[48]) and ChemWebServer. The SQL Link Library is intended to connect ChemoSoft to the SQL server, where the user can export, import, browse, and edit data. The ChemWebServer is intended to exposure chemical databases via the internet.[49]

Two products handle two-dimensional (2D) and three-dimensional (3D) data equally well: Catalyst, an integrated DBMS and pharmacophore modeling environment from Accelrys, enables seamless access to pharmacophore-based alignment of molecules, shape based 3D searching, and automated generation of pharmacophore hypotheses based on structure–activity relationship data.[50] The second 3D database is UNITY, a DBMS package accessible through SYBYL, the molecular modeling system from Tripos. UNITY combines database searching with molecular design and analysis tools.[51] UNITY allows building of structural queries based on molecules, molecular fragments, pharmacophore models, receptor sites, and other geometric features and relationships. The UNITY relational database interface within SYBYL provides access to Oracle data associated with structures.[52]

3.14.2.5 Hierarchical versus Relational Databases

Somewhat outdated, the hierarchical database model was replaced by the relational model. Since it is still available in several database applications, e.g., MDL ISIS/Base,[40] we are discussing some inherent limitations of this earlier approach. A hierarchical database links records in a data tree, in a parent–child relationship, such that each record type has only one owner. Hierarchical relationships between different data can enable the system to answer certain queries with ease, yet others are difficult to answer. The condition for a one-to-many relationship (a relational database concept) between parent and child often results in redundant data, a situation that frequently makes it difficult to expand or modify hierarchical databases. When fields are added or removed, applications that access the database need updates. Given such restrictions, hierarchical models are somewhat obsolete, as they cannot be easily used to relate timely, real-world information.

In the relational database model, data tables relate to each other through common values. As in spreadsheets, data are stored in tables, independent of the way data are physically laid up. A row, corresponding to a record, represents a collection of information about a separate item, whereas a column represents the characteristics of an item. The relationship is a logical link between tables. The relational DBMS uses matching values in multiple tables to relate information in one table with information in other tables. The relational database model provides flexibility, allowing changes to the database structure without having to update any applications that rely on that structure. This data model permits the designer to create a consistent logical model of information, to be refined through database normalization. Basically, the rules of table schema normalization are enforced by eliminating redundancy and inconsistent dependency in the table designs. Codd provided a set of rules that define the basic characteristics of a relational database, rules that a DBMS must meet in order to be considered relational.[53,54] These rules brought clarity and rigor to the database field. Actually, the relational model is based on the information principle: All information is represented by data values in relations.

3.14.3 Bioactivity Databases: Some Design Guidelines

Computer-based systems for information capture, storage, and retrieval are of critical importance in understanding and mining the chemistry–biology interface. Such information is pertinent to new target discovery, and to understanding disease models, as well as to the study of active chemotypes and privileged structures. A variety of chemical, e.g., SciFinder,[1] Beilstein,[2] and Spresi,[3,4] or medicinal chemistry databases, e.g., MDDR,[5,6] or drug databases, e.g., the Physician Desk Reference (PDR),[10] are available. These do not capture the biological endpoint in numerical form, i.e., there is no searchable field to identify in a quantitative manner what is the target-related activity of a particular compound. Such information is important if one considers that (a) not all chemotypes indexed in patent databases are active – some are claims with no factual basis; and that (b) not all chemotypes disclosed as active are equally potent for the target of choice. Should one decide to pursue a certain interaction 'hotspot' in a given ligand–receptor structure (assuming that good in silico models are available), it would be convenient to mine bioactivity databases for similar chemotypes, to be used as bioisosteric replacements.

Furthermore, biological research produces large amounts of data that need to be organized, queried and reduced to scientific information and knowledge. Management of biological data involves acquisition, modeling, storage, integration, analysis, and interpretation of diverse data types.[55] In this chapter, biological activity refers to experimentally measured data for a set of chemical compounds on a given biological target (as well as cell, organ, and organism), using predefined experimental protocols. After curation and standardization, these measured values together with related information can be indexed in a bioactivity database. In the largest context, databases need to handle data in a structured and organized way. Consequently, the key task when designing an effective bioactivity database is to structure the information properly. To accomplish this, several models might be imagined. The model discussed below has a two-level structural design: the internal level corresponds to the database itself, while the external level provides cross-referencing support (stored identifiers) for accessing external records from other databases. Such a database model needs to provide a set of unique and stable identifiers for linking to external levels of other databases. Those databases will perceive this one as external, hence the cross-linking through external levels is bidirectional. A good design should allow the database to be integrated with others in a databases network, e.g., CABINET (*see* Section 3.14.3.5 below).[56]

The internal Bioactivity component lies on top of three major autonomous components:

- Chemical information: combines chemical structures with experimental or calculated chemical and physical properties (*see* Section 3.14.3.1 for details);
- Target/Protocol: biological target and experimental protocol data (*see* Section 3.14.3.2 for details);
- Reference: bibliographic information for all units in the database (*see* Section 3.14.3.3 for details).

3.14.3.1 Chemical Information

Chemical databases contain chemical structures encoded in machine-readable formats and associated molecular data. Chemical structures define the atomic connectivity, expressed in connection tables that store two- and three-dimensional atomic coordinates. An abstract model and a simple table schema in a relational database model concept for storing chemical information are shown in **Figures 1** and **2**.

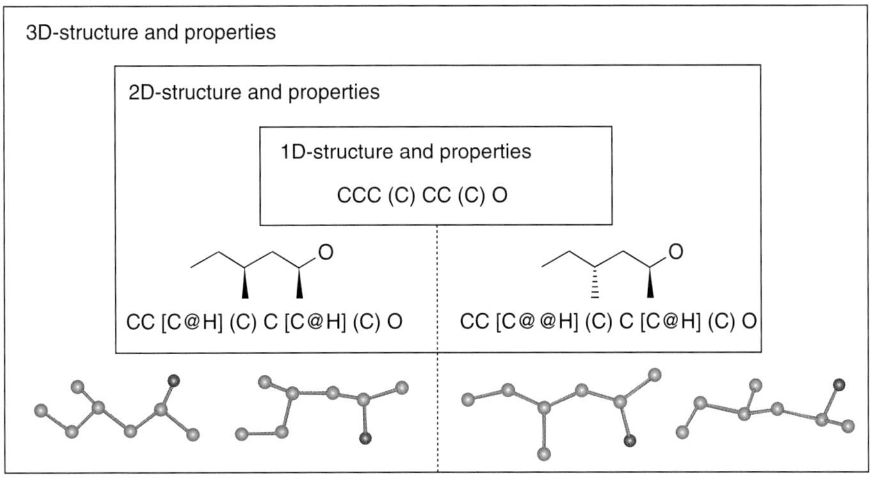

Figure 1 Hierarchical chemical information representation example, for 4-methyl-hexan-2-ol 1D, 2D, and 3D refer to one-, two-, and three-dimensional structure depictions.

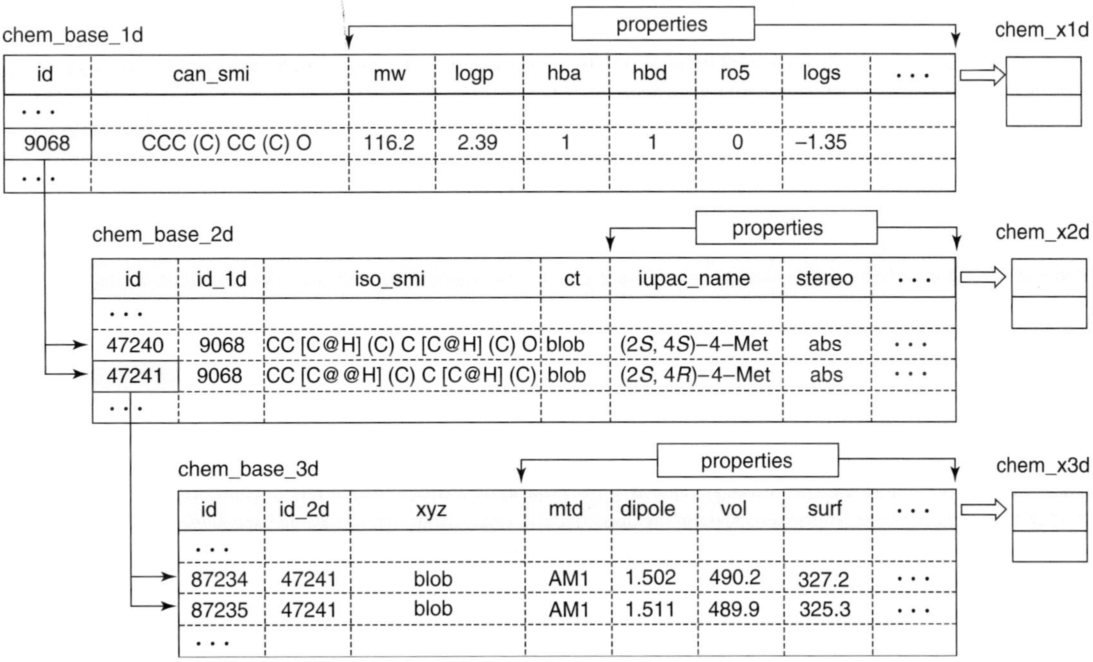

Figure 2 Table relationships for the hierarchical chemical information model. Auxiliary relationships to subset tables (chem_x1d, _x2d, _x3d) can be derived for each level. id – unique identifier for relating with the bioactivity unit; can/iso_smi – canonical/isomeric SMILES; mw, logp, hba, hbd, ro5 – Lipinski's[57] rule-of-five components; logs – logarithm of water solubility; ct/xyz – connection table/atomic coordinates vector, values stored as BLOBs (binary long objects); stereo – stereochemical information; mtd – 3D optimization method; vol, surf – molecular volume and molecular surface area, respectively.

3.14.3.1.1 One-dimensional representation

One-dimensional (1D) representation stores chemical structures as canonical nonisomeric SMILES,[43] without stereochemical information. Unique SMILES are registered in the chem_base_1d table along experimental and calculated properties that depend only on connectivity (without atomic coordinates). Canonical SMILES may serve as index in the table to speed up the search; this philosophy is used by Daylight.[41] The SMILES string is small, taking approximately 1.5 ASCII characters per atom, so it can be manipulated with relative ease. Instead of querying graph isomorphisms to find an identical compound, the canonical SMILES index can be queried using a very fast string match. Often, the entire chemical database can fit in memory, which totally eliminates slow disk I/O.[58]

Other chemical formats may assist with data processing: InChI, the International Chemical Identifier, is a relatively new invariant character string representation of chemical structures. The conversion of structural information uses IUPAC structure conventions and rules for normalization and canonicalization of an input structure representation. This enables an automatic conversion of graphical representations for chemical substances into the unique InChI labels, which can later be restored by any chemical structure drawing software.[59] Unlike SMILES technologies from Daylight, InChI is a nonproprietary format.

An important class of 1D descriptors relate to molecular similarity. Molecular fingerprints are binary representations for chemical structures characterization, typically stored in the database. Similarity searching involves matching the target structure with all of the structures in a database. MDL fingerprints, also termed keys or keybits, use a predefined set of definitions and derive fingerprints based on pattern matching of the structure to the defined 'keyset.' [60] The 320 set (which is public), optimized for drug discovery, can be generated with the Fingerprint module from Mesa Analytics and Computing LLC, using SMILES as input.[61] Daylight software generates hashed fingerprints[62] with a variable number of bits (typically between 512 and 2048). Most similarity searching systems are based on the comparison of pairs of molecular fingerprints and use Tanimoto's symmetric distance-between-patterns[63] or other (sometimes asymmetric) coefficients.[64] Computed storable descriptors may also include: size-, hydrophobicity-, and electronic effects-related, as well as hydrogen bonding descriptors and topological indices.[65]

Additional information from upper levels (2.5D: *R*/*S* and *E*/*Z*;[66] 3D: atomic coordinates) is not likely to influence these properties and query results. If this happens, those properties should be stored to their corresponding level(s). In accordance to the normalization rules,[54] such properties should be stored here as distinct records. Depending on the number of duplicate structures, a considerable amount of redundancy can be eliminated when using the relational database model.

3.14.3.1.2 Two-dimensional representations

Two-dimensional (2D) representations extend the 1D with stereochemical information. The chemical structure can be stored in the isomeric SMILES format, along other required formats. The current version of SMILES does not support atomic coordinates, but future Daylight releases (e.g., version 5.0) are likely to make this feature available. If plain coordinates (usually for hand-drawn structures) are required, an alternative SMILES-derived solution is represented by the TDT (Thor Data Tree) format.[41] It combines the classic SMILES with related information, which is in this case a vector of atomic coordinates.

Many chemical data formats are available today, in both text (ASCII) and binary formats. We note that text formats are redundant: for example MDL's MOL format needs about 100 characters per atom.[67] If such a format is required, it is preferable to store data in a compressed format. Our studies using an in-house dictionary-based compression algorithm[69] indicate that for each molecule compressed individually, a compression ratio of 6 or higher can be achieved without significant real-time computations. This becomes more important when one considers that data transfer between client and server takes network bandwidth, which may considerably slow the search/retrieval process if the data is not compressed. Data compression (slower process) is typically performed once, when registering records into the database, whereas decompression (faster process) occurs much more often, i.e., at each data retrieval step.

Binary formats are more compact and comparable in size with the compressed binary already mentioned. Using binary formats may lead to loss in flexibility: storing MDL,[39] CambridgeSoft,[45] or OpenEye[70] binaries requires native proprietary software modules in order to process information, whereas ASCII (text) is more flexible and easier to process, e.g., using in-house software.

One relatively recent format, initially designed for representing chemical information on the internet is the Chemical Markup Language (CML).[71,72] Developed using internet tools like XML[73,74] and Java,[46] CML is capable of holding extremely complex information structures and can act as an information-interchange mechanism. It interfaces with relational and object-oriented database architectures. Even though a large amount of generic XML software is available, only a few applications are actually using CML. For newly developed database applications, it is advisable to store chemical structures also in compressed CML format. Due to the existing huge volume of data and available computational tools, many of the above formats will continue to coexist. Conversion software and import–export options are available, but the user should be aware that sometimes results are unexpected, due to implementation errors that typically arise from a lack of a clear documentation for particular formats.

If no 2D atomic coordinates are stored, toolkits for generating flat representation of molecules can be used, such as those from Daylight[4] or OpenEye.[70] If coordinates are deposited, the task of showing a structure in a standalone or web-oriented application becomes easier; many packages, some open-source,[75] are available.[47,76] Along the structure, at this level one can store properties that depend on stereochemistry, as well as flags that indicate absolute/relative stereochemistry, racemic/diastereoisomeric mixture and *cis*/*trans* double-bond geometry. Generic (trivial, commercial), IUPAC names,[77] CAS registry numbers,[78] and other external identifiers and optional keywords may be also registered at this level. From the chem_base_2d table, a set of supplementary relationships can be derived to other tables which store special information for only a subset of compounds, such as data related to natural products, drugs, or compound mixtures. If 3D information is required for certain subsets, e.g., bioactive conformers, a relationship from the chem_base_3d table (see below) can be derived as well.

3.14.3.1.3 Three-dimensional representations

Three-dimensional (3D) representations extend the 2D sets by introducing atomic coordinates (e.g., Cartesian). This level is dedicated for storing automatically generated conformers from 2D structures, 3D molecular models, bioactive conformers extracted from experimentally determined structures, etc. To preserve the natural 2D→3D hierarchy representation, stereochemistry needs to be conserved: all chiral centers and double bond geometries must remain the same as in the associated 2D structure. The 3D level is designated for structures with real coordinates, and for properties and features that actually depend on 3D structures. Properties like partial charges, heat of formation, or other energy values from computation, volumes and surfaces, VolSurf descriptors,[68,79] etc., can be registered here.

Cheminformatics involves the application of computational techniques to the analysis of large volumes of molecular structure data. Many docking and searching systems for databases of 2D and 3D chemical structures are available.[80] In order to take advantage of these algorithms and others, each level of the Chemical Information entity should provide the maximum amount of structured information, while maintaining schematic flexibility.

3.14.3.2 Target and Protocol Information

Bioactivity is typically measured for a series of compounds on a given target. A bioactivity database cannot exist without both biological and chemical information. Having discussed cheminformatics aspects, let us briefly discuss bioinformatics databases: many target and gene oriented databases are available for free on the internet. These databases have some degree of disparity because their purpose and classification criteria differ. However, a rich and organized set of hyperlinks enables the user to navigate across many of them. A common characteristic for these databases is the existence of unique, constant identifiers for every entry. Another aspect is the ability to access these entries using uniform resource locators[81] (URLs) that can be easily built from identifiers. This leads to an excellent cross-linking support for accessing these resources in a programmatic manner (i.e., from one web page to another).

Taking into account these considerations, the designer should store in the Target unit some basic data and many external identifiers. The basic information should refer to: an internal identifier valid through the whole bioactivity database; some internal target description, e.g., name, target function, classification, and species; some searchable (dictionary-defined) keywords; perhaps some dataset-dependent (contextual) comments. The internal identifier is used for relating records within the bioactivity database. Many classifications may be applied. For example, using functional criteria, a protein may be one of enzyme, receptor, ion channel, transporter, or 'other' (unspecified). Furthermore, an enzyme falls in one of six major categories: oxidoreductase, transferase, ..., ligase,[82] while receptors are divided in nuclear hormone receptor (NHR),[83] integrins or G protein-coupled receptor (GPCR),[84] etc.

Keywords may help refine the result of the query/retrieval process. It is advisable to use a relatively small number of keywords. These keywords should be defined such that they are meaningful for a significant number of entries. Keywords that are too general will apply to almost every entry, while those that are too specific will apply to too few entries (hard to find). Statistically speaking, there should be some variance for each keyword (i.e., zero or almost zero variance produces sets that have almost the same keyword representation).

For targets, one has to consider synonyms: a significant number of targets have two or more names, which can be confusing to nonbiologists. A dictionary can keep track of name equivalents, in order to provide a high success rate for queries. Standardization is a desire; this problem is partially solved for protein-oriented databases such as Swiss-Prot.[85] Current data management technology is confronted with database instability (e.g., once-free and web-based databases vanish and become commercial products), their evolving nature (e.g., some unique protein identifiers in Swiss-Prot were redefined in release 46 compared to release 45), diversity and scientific context that characterize biological data.[55] Thus, standardization is far from trivial.

Additional text fields may be defined to register full-text search capability (if the DBMS supports this, acting like a web search engine) and error data. If some inconsistency between miscellaneous sources is detected, the user needs to be notified. Because molecular biology has an extremely dynamic character, it is easy to find discrepancies between sources. For example, one Target table record stores five external identifiers, but one of them may point to an outdated link; that identifier should be flagged, and an alert should be generated. Updating on a weekly/monthly basis the Target unit ensures that information stays as current as possible.

Experimental protocol data are recorded as well. Protocols are usually standardized and published, but they are also subject to change. Newly discovered targets and an increase in technical accuracy are two reasons for such change. Examples of protocol information that can be stored include specific/nonspecific ligand, radioligand, substrate, temperature, pH, buffer, incubation time, etc. The protocol, target, and biology are strongly related to each other, explaining why these components are grouped in the Target/Protocol unit.

Due to a rather small amount of physically stored information, a database schema with a small number of tables should be sufficient. The idea is to keep advantage of the ‘heavy’ information already existent in other databases by registering external identifiers and building (hyper)links, not by copying and locally storing that data, except for the minimum amount of (critical) data. In time, such data becomes deprecated, hard to modify, and even useless. Thus, as much as possible, one should store pointers (i.e., identifiers) to data, not data themselves.

Section 3.14.4.1 illustrates some of the most representative biological-oriented databases at this time. An exhaustive discussion of bioinformatics database systems is beyond the goal of this work. Besides, by simply performing a web search, one is likely to identify many, some novel, resources. This is another reason why the Target/Protocol unit must be updated periodically: some databases appear, some disappear, and most get updated. To paraphrase Herakleitus of Ephesos, one feels that one cannot hyperlink to the same database twice.[86]

3.14.3.3 The Bibliographic References

Literature databases are divided into bibliographic databases, which contain only references to information, and full-text databases, which hold the original published work. Widely used bibliographic databases with relevance to chemical and biological research include Chemical Abstracts Service,[87] MEDLINE (via PubMed),[88] and BIOSIS.[89] The bibliographic references component supplies data describing briefly the content of a document. Its function is to locate an article, patent, book, or other literature where information related to the query was published.

In its simplest form, bibliographic reference data may be stored in one single text field, as in the references section of every publication: authors, optional title, journal or book name, volume or chapter, and inclusive page numbers. This is satisfactory for small systems, when retrieval for use as quotation in scientific papers is required. However, using reference information in the free format is unproductive and sometimes unfeasible. This issue is addressed by storing reference data in several searchable fields. Because of the disparity of reference formats (e.g., article and patent data types differ), a two-level hierarchy can be used: the first level identifies the document type while the second stores details for the given document. Some fields such as author names, journal names, base URLs, and keywords may be stored in global tables as common resources. Adjacent to these standard fields, it is proper to register extensions like web hyperlinks to the full-text publication, cross-linking identifiers, and abstracts. The increase of existing information will increase the hit rate when looking for a specific topic. Whereas hyperlinks may be useless when online subscriptions are not available, the digital object identifiers (DOIs)[90] or PubMed[88] identifiers (both external) are likely to work – at least access to the abstract is ensured, if it is not already indexed. A suitable structured Reference unit will allow the execution of complex literature queries. Searches by author, document type, publication name and year, or by topic keywords, as well as bibliometric and citation analyses become easy. A more detailed schema is illustrated and described for the Reference unit in **Figure 3**. We consider this a simple and intuitive unit in comparison to all other units, where complex, purpose-dependent schemata are needed. This example is not a database schema, which implies tables, field names and types, relationships, and constraints. It is only a possible model for how to divide and restructure bibliographic information.

The entry point in this schema is represented by the Directory. The document type element (doc_type) indicates the reference description which should be further consulted (e.g., Journal or Patent). It is also possible that other reference types may be available, e.g., in-house reports, book chapters, etc. – but for the purpose of illustration we discuss patents and journal articles. The Directory also contains a unique reference identifier for linking to other units (ref_id) and an associated Digital Object Identifier (DOI). Currently, the International DOI Foundation plans to

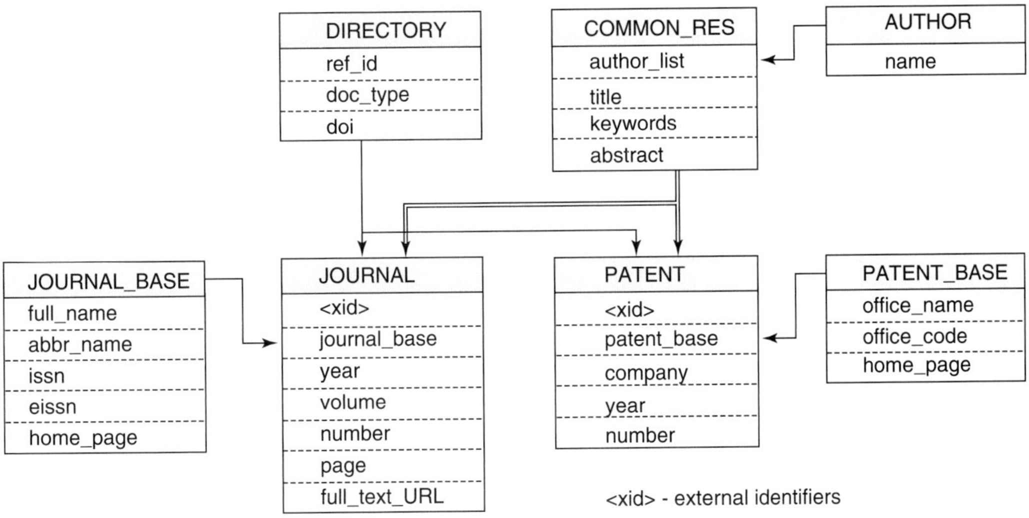

Figure 3 A general structure for the Reference unit.

extend the DOI assignation to other document types, including patents.[91] This is the reason why DOI is included at the entry level, in fact serving as a second unique identifier. Not every entry has a DOI entry, as not all publications have allocated such values.

Usually, authors publish more than one paper; thus author names should be stored as a separate set of unique names. An enumeration of names from this set corresponds to all authors in the publication (author_list). Author list, beside title, keywords, and abstract constitute the common resources which apply to all document types (Common_Res).

To keep redundancy at a low level, journal and patent descriptions should be kept separately (Journal_Base and Patent_Base). For example, from a relational model perspective, it is detrimental to store journal names with every reference when a simple pointer to this information is sufficient. Base description besides the rest of the reference data (year, page, etc.) form the full reference. External identifiers (xid, e.g., PubMed or CAS accession number) may be recorded as well for each reference. Additional document types may be added to this model. Because each type has its own specific elements, corresponding subschemata must be designed (such as Journal_Base/Journal entities).

Given the ubiquity of the internet, some comments about web hyperlinks and digital object identifiers as bibliographic references are required. Even if most publishers do no agree to this, internet references are a necessity due to the value and amount of information that can be found using a simple web browser. The internet is a dynamic system where hyperlinks, and implicitly pointed web pages, appear, get relocated, or disappear. Some web servers maintain consistent address and service for a long time, and even if the URL[81] changes, a redirect mechanism is available to track the moved resource. It is good practice to store additional reference information next to the hyperlink, just in case the URL becomes invalid. This will enable the end-user to track the referred data, given the additional information.

DOI is a system for identifying content objects in a digital environment. DOIs are uniquely assigned persistent identifiers which provide current information to existing data on the internet. Information about a digital object may change over time, including its URL, but its DOI does not change over time. Using DOIs as identifiers makes managing intellectual property in a networked environment much easier and more convenient, and allows the construction of automated services and transactions.[90] The referred object can be easily accessed using the hyperlink resulted from concatenation of the base URL 'http://dx.doi.org' with the DOI value.[91]

3.14.3.4 The Biological Activity Data

The Bioactivity unit lays on the previously described units (**Figure 4**). It relates biological activity data (primarily activity type and value) with unique identifiers for the chemical compound, the biological target (or cell, etc.), the experimental protocol, and the bibliographic references.

The most used activity types are:

- IC_{50} – Inhibitory Concentration 50%, the molar concentration of an antagonist/inhibitor that reduces the response to an agonist by 50% (IC_x – other percentage values can be specified)

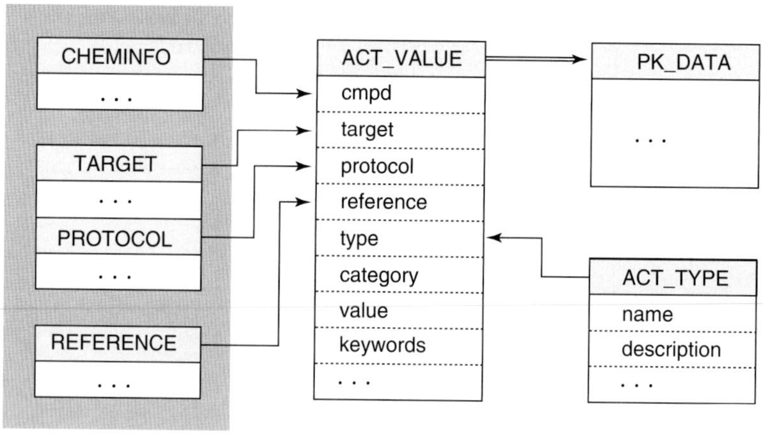

Figure 4 A general structure for the Bioactivity unit.

- A_2 – the molar concentration of an antagonist that requires double concentration of the agonist to elicit the same submaximal response, obtained in the absence of antagonist
- EC_{50} – Effective Concentration 50%, the molar concentration of an agonist/substrate that produces 50% of the maximal possible effect (or reaction velocity) of that agonist (substrate)
- ED_{50} – Effective Dose 50%, the dose of a drug that produces, on average, a specified all-or-none response in 50% of a test population or, if the response is graded, the dose that produces 50% of the maximal response to that drug
- K_i, K_d – inhibition respectively direct binding experiment equilibrium dissociation constants.[92]

The activity value corresponds to that molar concentration or dose for which a certain condition is satisfied. Typically, numerical activity values range from picomolar to molar, depending on how active the tested compound was on a particular target. To enable further analyses and facilitate data mining, all activity values should be converted to the same scale: the negative base 10 logarithm ($\log_{10}$) is the most suitable. Sometimes, multiple experiments are performed under the same conditions for the same target/compound pair, which enables the estimation of the confidence interval. The standard error of mean (SEM, sometimes incorrectly identified with the standard deviation, SD) can be stored together with the activity. SEM values need to be checked with respect to the mean value: for example, an activity of 10 ± 15 nM is incorrectly described, since the given range, between -5 nM and 25 nM, returns errors when converted to $\log_{10}$ (logarithms are undefined for negative input).

Not every compound shows activity on a given target: some experiments cannot precisely determine the exact activity value, only activities less than or greater than an estimated threshold; other experiments can only determine the inhibition percent for a given ligand concentration. These are special cases which should be still stored in the database because there is some useful information which may help in further activity data mining. Next to primary activity data, one needs to store additional keywords, experimental observations, errors, and sometimes images (e.g., Schild plots). Keywords such as 'agonist,' 'antagonist,' 'partial/inverse/competitive,' and 'inhibitor' allow queries to retrieve bioactivity values selected by effect, while keywords like 'active,' 'inactive' may help rapidly discriminate between these categories.

A compound may have activities recorded for more than one target, i.e., a series of chemical compounds tested on a set of biological targets (sometimes with different protocols), leading thus to many-to-many relationships in a relational DBMS. However, the Bioactivity schema can range from very simple to very complex, depending on how many kinds of data should be stored and how related or unrelated these are. Pharmacokinetic (PK) data such as oral availability, plasma protein binding, volume of distribution, half-life, total blood clearance, etc., or experimental values from clinical trials can be registered in a dedicated subunit linked to the standard activity subunit.[93]

3.14.3.5 Integration with Other Databases

Given the multidisciplinary nature of research, a single, isolated database can no longer meet user needs. Many open-access and commercial databases are accessible via the internet, but taking advantage of the information stored in heterogeneous databases can sometimes be taxing. The sheer size of the data volume, the growing number of databases, the excess of data types and formats, database heterogeneity, and even the interdisciplinary nature of the chemistry–biology–informatics arena are some of the reasons. One may query a database, visually inspect the results, find the data of interest, and subsequently use the data to query other databases (i.e., queries within queries). Such manual multidatabase queries can be time-consuming, especially when large amounts of data are involved. Automated data integration from multiple sources has become a critical need.

Heterogeneous databases have different structures, data types, stored formats, and interpretations for the data. Given the variability of these databases, it is not possible to manage them with a single DBMS. Bringing together different information from different sources into a single data model is difficult, even for the same kinds of data, because of the disparity of purposes between various information sources. A multidatabase is a distributed system that acts as a front end to multiple local DBMSs or is structured as a global layer on top of local DBMSs.[94] The global system provides full database functionality and interacts with local DBMSs at their external interface. The global system provides some resources (federation schema or multidatabase language) for resolving the differences in data representation and function between local DBMSs. A multidatabase provides the ability to solve the communication diversity, helps data interchange between local heterogeneous databases, and reinforces data accessibility without affecting the local databases' autonomy. Distributed database systems should be capable to manage collections of autonomous databases without a global schema and control.[94]

Federated DBMSs provide an evolutionary approach to integrating multiple databases over time: they allow continued operation of existing applications, support controlled integration of existing databases, and facilitate incorporation of new applications and new databases.[18] Technologies such as CORBA,[95] ICE,[96] Java,[46] XML,[73] and HTML[97,98] provide a powerful and flexible method of integrating data from different databases. Each information source can be organized to represent its kind of data in the most natural and expressive manner for that specific kind of information. Given query languages that are adequately expressive, a federation of databases can appear to be a single integrated information system.[99] CABINET, a federation of chemical and biological information databases, was recently released by Metaphorics LLC.[100] CABINET servers provide the ability to navigate easily, using a web browser, through diverse chemical and biological information sources and to explore the relationships between them. Each CABINET server has the ability to query other CABINET servers for related information. Similarity relationships which can be explored include chemical structure similarity, protein sequence similarity, similarities in enzyme function, and various textual links.[56]

An effective and widely used integration technique is to embed external identifiers corresponding to related information from other databases in the database. One can observe in almost every current web-enabled database links to other web resources, constructed from two major components: the URL[81] which acts as entry point into the database, and a text, numeric, or alphanumeric identifier for selecting a specified resource within. One natural constraint is to keep unchanged at least the unique identifiers (i.e., accession codes or accession numbers) designed to be used for cross-linking. Such identifiers can be removed (an alert is returned in response to the query), but under no circumstances should they be reassigned. If this is not strictly enforced, the link may point to completely unexpected information. For technical reasons, sometimes modified identifiers become available, e.g., in the Swiss-Prot[85] database some of the Entry Name values changed. For example, the gonadotropin releasing hormone receptor, GRHR_HUMAN, associated to the unmodified Primary Accession Number P30968, was changed to GNRHR_HUMAN. However, an IDTracker[101] tool is provided in order to keep track of such changes. This example does not question the quality of Swiss-Prot, an invaluable resource for scientists worldwide; it merely illustrates the dynamic nature of this field. High-quality database systems offer linking support and full documentation. This should be consulted before constructing applications that access the resource. If documentation is unavailable, simple tests executed directly in the web browser can help guess the underlying rules.

As a final identifier-based linking specification, one can create supplementary linking tables between internal (usually local primary keys) and external identifiers. This is preferred when the core data must remain unmodified or when one/many-to-many relationships are imposed. More tables lead to more complex database schemata and more complex SQL statements. The resulting schema needs to remain flexible and extensible. The embedded identifier linking method is illustrated in **Figure 5** for the main units: Chemical Information and Target/Protocol. Several external resources are shown, but one can find and use many others. Note that the identifiers may be directly used in SQL statements (command line or program-embedded).

3.14.4 Biological and Bioactivity Information Databases Examples

The volume of data related to genomics and proteomics that becomes available each day makes it difficult for any one person, or group, to maintain pace. This generally creates specialized scientific fields (e.g., serine proteases or NHRs),

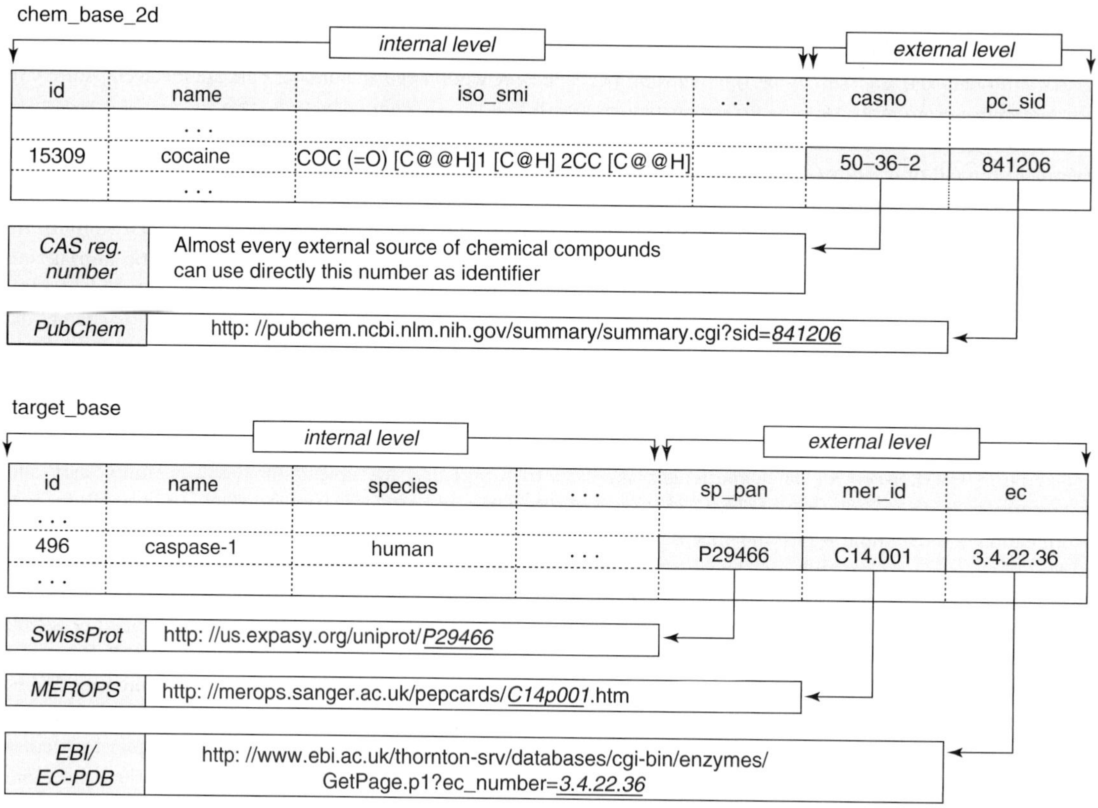

Figure 5 Linking a Chemical Structure table (chem_base_2d) and a Target/Protocol table (target_base) to some external resources: PubChem,[102] Chemical Abstracts Service,[103] Swiss-Prot,[85] MEROPS,[104] and EC-PDB[105] databases. Based on the general rule, URLs are resolved based on the stored identifiers.

as discovery scientists focus on a class or family of macromolecules, their biological functions, and related therapeutic impact. Consequently, a need for class-specific, specialized databases, has emerged. In the field of bioinformatics, these databases collect and organize data around a single class of macromolecules (e.g., GPCRs), or around a particular topic of interest (e.g., cancer), whereas in the field of cheminformatics, they often collect target-focused small molecules (e.g., kinase inhibitors). When curated well, these resources allow scientists to query a single source in order to obtain a significant fraction of the most relevant data. What follows is a succint presentation of certain bioinformatics and cheminformatics databases relevant for bioactivity.

3.14.4.1 Biological Information Databases

An invaluable network of databases is available from the European Bioinformatics Institute (EBI). Some of them are described below. The EBI mission is to ensure that the growing body of information from molecular biology and genome research is placed in the public domain and is freely accessible to all facets of the scientific community in ways that promote scientific progress.[106] One of the most popular databases, maintained collaboratively by the Swiss Institute for Bioinformatics (SIB) and the European Bioinformatics Institute (EBI), is the UniProtKB/Swiss-Prot Protein Knowledgebase (Swiss-Prot). This is a curated protein sequence database that provides a high level of annotation and a minimal level of redundancy. Together with UniProtKB/TrEMBL, it constitutes the UniProt (Universal Protein Resource) Knowledgebase, an access point to all publicly available information about protein sequences.[107] Two classes of data can be distinguished in Swiss-Prot: the core data and the annotation. For each sequence entry the core data consists of the sequence data, the citation information, and the taxonomic data. The annotation consists of descriptions like: function of the protein, domains and sites, secondary and quaternary structure, similarities to other proteins, etc. A high degree of integration with other biomolecular databases is provided: Swiss-Prot is currently cross-referenced with about 60 different databases. Cross-references are provided in the form of pointers to information related to Swiss-Prot entries.[85] Another EBI database, Enzyme Structures Database EC-PDB[105] contains the known enzyme structures that have been deposited in the Protein Data Bank (PDB).[108] Currently, about 15 000 entries are available. A related database, applied to enzymes and nuclear receptors, Representativity of Target Families in the Protein Data Bank (PDBRTF) is available from IMIM.[109]

Two databases with pharmacological relevance are GPCRDB[84] and NucleaRDB[83], which collect and disseminate data related to GPCRs and NHRs, respectively. They store experimental data (sequence, structure, mutation and ligand-binding data) as well as computationally derived data (multiple sequence alignments, cDNA–protein alignments, phylogenetic trees, correlated mutation analysis, and 3D models). The data organization is based on the pharmacological classification of receptors and access to the data is obtained via a hierarchical list of known families in agreement with this classification. For specific families, one can access individual sequences, the multiple alignments, the profiles used to perform the latter, 2D visualization and the phylogenetic tree. A maximum degree of integration with other biomolecular databases is provided for each protein. Entries for GPCRs and NHRs from other databases (e.g., Swiss-Prot) point to these cross-reference tables.[83]

MEROPS[104] is a database for peptidases and the proteins that inhibit them. It uses hierarchical, structure-based schemes for the classification of peptidases and their inhibitors. Each one is assigned to a family on the basis of statistically significant similarities in amino acid sequence, and families that are thought to be homologous are grouped together in a 'clan.' Currently, about 3000 individual peptidases and inhibitors under 180 families and 49 clans are included in the database (release 7.00). The system provides navigation using indexes of name, MEROPS identifier, and source organism as well as links to supplementary pages showing sequence identifiers, structure, and literature references.[104]

The TCDB[110] database details a comprehensive IUBMB approved classification system for membrane transport proteins, known as the Transporter Classification (TC) system.[111] The TC system is analogous to the Enzyme Commission (EC) system[82] for classification of enzymes, but incorporates phylogenetic information as well. Transport systems are classified hierarchically on the basis of five criteria: transporter class, subclass, family or superfamily, subfamily, and substrate or range of substrates transported. TCDB is a curated database of factual information from over 10 000 published references and contains about 3000 protein sequences classified into over 550 transporter families based on the TC system.[110]

TransportDB is a relational database designed for describing the predicted cellular membrane transport proteins in organisms whose complete genome sequences are available. For each organism, the complete set of membrane transport systems is classified into different types and families according to the TC classification system.[111] Web pages were created to provide user-friendly interfaces for access, query and data download. Additional features, e.g., a BLAST[112] search tool against known transporter protein sequences, comparison of transport systems from different organisms, and phylogenetic trees of individual transporter families are also provided.[113]

3.14.4.2 Bioactivity Information Databases

3.14.4.2.1 Open-access databases

3.14.4.2.1.1 BIDD Databases

The National University of Singapore Bioinformatics and Drug Design group[114] develops methods, software, and databases for drug discovery. Their bioinformatics databases provide information about drugs, natural products, protein targets, ADME/Tox, drug-protein binding, and other relevant information. The following databases are searchable online: Therapeutic Target Database (1174 target entries, 1251 drug/ligand entries),[115] Drug Adverse Reaction Target database,[116] Drug ADME Associated Protein Database (321 protein entries),[117] Therapeutically Relevant Multiple Pathways Database,[118] Computed Ligand Binding Energy database (14 731 entries),[119] and Kinetic Data of Biomolecular Interaction (20 803 entries).[120] These databases contain cross-references to other relevant databases, associated references, and ligand structures.

3.14.4.2.1.2 Blueprint

The Blueprint Initiative is a free resource for biomolecular data focused on public databases and other software.[121] SMID (Small Molecule Interaction Database) is a relational online database for small molecule/domain interactions, determined from the Molecular Modeling Database (MMDB,[122] hosted and maintained by NCBI). SMID links small-molecules to their protein partners and their families, giving a comprehensive picture of small-molecule binding.[123] SMID further offers physicochemical and biochemical details about the participating small and macromolecules. Currently, 182 301 interactions for 4283 small molecules and 2807 domains are indexed. BIND (Biomolecular Interaction Network Database) is a collection of records documenting molecular interactions.[124] BIND includes high-throughput data submissions and hand-curated information from literature. BIND stores full descriptions of molecular interactions (181 295 records), complexes (3520 records), and pathways (8 records). The SeqHound database integrates biological sequence, taxonomy, annotation, and 3D structures.[125] Its annotated links include Genbank,[126] MMDB,[122] MedLine,[88] and BIND.[124] SeqHound is a resource for programmers, hence it features a simple web interface with limited functionality.

3.14.4.2.1.3 ChemBank

ChemBank,[127] implemented at Harvard University and supported by the National Cancer Institute,[128] is a free database about small molecules and resources for studying their effects on biology. ChemBank aims to assist biologists who seek molecular probes that can be used to perturb a biological system, and to enable chemists to design novel compounds or libraries. ChemBank features a chemical structures database for over 1 million compounds, including 150 000 commercially available compounds with vendor catalog numbers and a reference set of over 700 000 other druglike compounds; a small-molecule bioactives database for over 6300 known bioactive compounds; and biological assay data from high-throughput screens.[127] Powered by Daylight software, ChemBank can be queried by compound name, substructure, or similarity. Each query result is further expandable to information that includes compound name, chemical data (SMILES, 2D depiction, molecular weight, and formula), identifiers (ICCB[129] and CAS numbers[103]), vendors, classification, and ICCB plate locations. Along with basic data, a list of characterized activities and observed effects are presented as well as cross-references to the PubMed[88] database. Downloadable files of compound structures from the bioactives database and other useful information are available in both SDF and XML formats. Supplementary resources relevant to small molecules, cheminformatics, and high-throughput screening are incorporated.

3.14.4.2.1.4 *Ki*Bank

Launched by the University of Tokyo, *Ki*Bank[130,131] is a free online database that targets in silico drug-design scientists. *Ki*Bank contains chemical structures with associated binding affinity for given targets. The July 2005 update has over 11 700 K_i values, 50 target proteins, and 4400 chemical structures.[132] Two search modes are available: by protein name/function, and by chemical name respectively. Target-oriented queries retrieve a list with all compounds that bind to the queried target. Structure visualizations (2D and 3D) are accessible for most compounds. In addition to chemical information, *Ki*Bank stores K_i values, species, limited experimental protocol information, and PubMed[88] bibliographic citations as hyperlinks. Chemistry-oriented queries retrieve a list of all targets for which bioactivity data is available on that chemical. These two lists are cross-referenced, so the end-user can easily switch from chemical queries to target queries and vice versa. The bioactivity data is downloadable in comma separated file format.

3.14.4.2.1.5 Ligand Info

Ligand Info,[133] the Small-Molecule Meta-Database, compiles some of the publicly available small molecule databases (about 1.1 million entries). Compound records contain 3D structures and, occasionally, biological activity information. Some molecules have Food and Drug Administrative (FDA)[134] approval status or anti-human immunodeficiency virus (HIV) activity data. Available in the SDF format, this resource is intended for virtual screening locally, or online (using the provided Java-based tool). Ligand Info is based on the similarity tenet, i.e., small molecules with similar structure have similar binding properties. Thus, the system enables a fast and sensitive structural index (based on averaging related molecules) for compound similarity searching. The index profile helps to focus on frequent, common features of a compound family. This tool can interactively cluster sets of molecules and create index profiles on the user side, and automatically download similar molecules from other databases.

3.14.4.2.1.6 PDSP K_i

The PDSP K_i database[135] is a public domain resource that provides information related to drugs and their binding properties to an expanding number of molecular targets. Designed as a data warehouse for both published and in-house K_i values or affinity, this database covers a large number of drugs and drug candidates binding to GPCRs, ion channels, transporters, and enzymes. The user interface provides tools for customized data mining (K_i graphs, receptor and ligand selectivity mining). Searchable fields include: receptor name, species name, tissue source, radiolabeled and tested ligand, bibliographic references as well as K_i value range. The system retrieves a list of all matching entries, cross-linked with corresponding entries in PubChem[102] and PubMed[88] (if available). Currently, the database has 34 940 K_i values. This database can be downloaded as a compressed ASCII file. Additional K_i data can be easily entered to the now-local database using the provided input web page.

3.14.4.2.1.7 PubChem

Matching one of the key goals of the National Institutes of Health (NIH) Roadmap,[136] the Molecular Libraries Initiative (MLI) has a dedicated cheminformatics resource, PubChem,[102] hosted by the National Library of Medicine (NLM). Organized as three databases within NLM's Entrez,[137] PubChem provides a high volume of information on the biological activities of small molecules, as it links chemical structures to other Entrez databases, seeking live information update on biological properties. PubChem links to PubMed scientific literature and NCBI's protein 3D structure resource. Powered by Openeye software[70] and CACTUS,[138] PubChem may be searched online on the basis of descriptive terms, chemical properties, and structural similarity. The three databases are:

1. PubChem Substances Database, which contains descriptions of chemical samples from a variety of sources, and links to PubMed citations, protein 3D structures, and biological screening results that are available in PubChem BioAssay.
2. PubChem Compounds Database, which contains validated chemical depiction information provided to substances from PubChem Substance; structures stored within PubChem Compounds are pre-clustered and cross-referenced by identity and similarity. Additionally, calculated properties and descriptors are available for searching and filtering of chemical structures.
3. PubChem BioAssay Database, which contains bioactivity screens of chemical substances described in PubChem Substance; it provides searchable descriptions of each bioassay, including descriptions of the conditions and readouts specific to that screening procedure. All the data are freely available through a PubChem FTP site.

3.14.4.2.1.8 ZINC

ZINC is a free online resource of commercially available compounds dedicated to virtual screening practitioners. It contains over 3.3 million compounds in ready-to-dock 3D formats.[139] The database or subsets are downloadable in several common file formats: SMILES, mol2,[140] SDF, and DOCK flexibase.[141] Compounds from the ZINC database may also be purchased directly from more than 20 vendors. The query can search molecular (sub)structures and several properties such as Lipinski's rule-of-five[57] criteria. Structures can be entered using the Java Molecular Editor,[142] which generates SMILES, or directly as SMILES or SMARTS. The result lists all the entries matching the query in 2D format, next to vendor identifiers and the (pre)calculated properties; 3D structures can be displayed by request. An upload service is available for adding new entries in the database or to filter user datasets according to ZINC-defined filters.

3.14.4.2.1.9 eMolecules

Started by eMolecules, Inc., this[160] is a new free open-access search engine for chemistry-related information. Its mission is to discover, curate, and index all of the public chemical information in the world, and make it available to the public. eMolecules distinguishes itself by extremely fast searches, an appealing presentation of results, and high-quality chemical drawings. Millions of molecules from hundreds of sources are merged into a single, searchable chemical database. It lets the user run searches by entering text or by drawing molecular structures via Java Molecular Editor. eMolecules also provides code that users can embed into their own web sites for direct access to it, as well as hosted cheminformatics systems and full web sites for chemical suppliers, pharmaceutical, and other chemical industries.

3.14.4.2.2 Commercially available databases

3.14.4.2.2.1 Accelrys

The Accelrys[143] bioactivity databases include BIOSTER, Biotransformations, and Metabolism.[144] BIOSTER is a compilation of over 9 500 active molecules that include drugs, agrochemicals, and enzyme inhibitors from literature. This database provides keywords indicating the mode of action and cross-references to reports for each active compound. The Biotransformations database covers the metabolism of drugs, agrochemicals, food additives, and industrial and environmental chemicals in vertebrates. It is indexed with original literature, test systems and a set of generic keywords. The Metabolism database provides a reference source for entry into the primary literature on the metabolic fate of organic molecules. It covers biotransformations of organic molecules in a wide variety of species. The BIOSTER and Metabolism databases have been designed for use with the MDL–ISIS[40] system, while Biotransformations is a standalone database. Accelrys further distributes Kinase ChemBioBase, a database produced by Jubilant Biosys.[145] This database provides a collection of small-molecule inhibitors of hundreds of kinase targets in a readily available 3D multiconformation Catalyst[50] format.

3.14.4.2.2.2 Aureus Pharma

AurSCOPE is a collection of annotated structure databases that capture biological and chemical information related to a given therapeutic or biopharmaceutical topic from literature, mostly patents and journals.[146] These databases capture in vitro and in vivo biological data, together with chemical information and structure–activity relationships. Complete descriptions of the biological test methods are provided.[147] AurSCOPE GPCR (500 000 bioactivities for 106 000 ligands and 2300 targets) contains biological and chemical data relating to GPCR chemistry, pharmacology, and physiology; AurSCOPE ADME/Drug–Drug Interactions (97 000 bioactivities for 4520 molecules, 1770 metabolites, and 420, targets) contains biological and chemical information related to metabolic properties of drugs, which enables the identification of potential drug–drug interactions. AurSCOPE Ion Channel is focused on drugs described as ion channel blockers, openers, or activators, that captures all ion channels (calcium, chloride, potassium, sodium) and transmitter-gated ion channels; AurSCOPE hERG Channel database (7750 bioactivities for 1155 ligands) contains chemical and biological information relating to the human ether-a-go-go related gene (hERG) potassium channel.[147]

3.14.4.2.2.3 CambridgeSoft

CambridgeSoft[45] and GVK Biosciences[148] released the MediChem database, a medicinal chemistry database indexing over 500 000 compounds in more than 650 000 records selected from the top 25 medicinal chemistry journals.[149] The database consists of chemical information, literature reference (standard and PubMed ID), and biological activity data (bioassay, target, activity). Assays include absorption, distribution, metabolism, excretion (ADME), binding information for a target and its mutants, functional assays (e.g., cell based or in vivo), toxicity, etc. Records can be queried by target platform, e.g., kinase, GPCR, NHR, etc. Ashgate Drugs Synonyms and Properties is a database of over 8000 drug substances currently in common use worldwide, while Traditional Chinese Medicines (TCM, also available from Daylight CIS) contains information on 10 458 compounds isolated from 4 636 TCM natural sources, comprised mostly of plants, a small number of animals, and a few minerals. The Merck Index is a structure-searchable encyclopedia of chemicals, drugs, and biologic active compounds. More than 10 000 monographs on single substances and related groups of compounds cover chemical, generic, and brand names. Searchable fields include structures and stereochemistry, registry numbers, physical properties, toxicity information, therapeutic uses and literature.[149]

3.14.4.2.2.4 Eidogen-Sertanty

Eidogen-Sertanty[150] offers the Kinase Knowledgebase, which captures published information for therapeutically relevant kinases; this provides a comprehensive view of the explored chemistry space around a target of interest and its relevant anti-targets. Inhibitor structural data enable the rapid grouping of known inhibitors into scaffolds and provide a layout around patentable chemotypes.[151] Kinase Knowledgebase covers more than 300 000 unique molecules ($>$140 000 assayed molecules) for over 300 annotated kinase targets, captured from over 1700 journal articles and patents. Quantitative structure–activity relationships (QSAR) models based on in vitro data, and advanced activity models based on cellular activity and toxicity can further be selected. The curation process captures chemical synthesis steps with detailed experimental procedures. Pertinent chemistry information is organized in protocols of generic reaction sequences, and incorporates synthetically feasible reagents contextualized around patent claims. This generates all the specific examples from a given patent, and a comprehensive ensemble of structures that can possibly be made by the reported synthetic methodology that could be potentially relevant for the relevant target. Synthetic pathways leading to these molecules are linked to biological information.[151]

3.14.4.2.2.5 Elsevier MDL

MDL Comprehensive Medicinal Chemistry stores 3D models and various properties, e.g., drug class, measured $\log P$, and pK_a values for over 8400 pharmaceutical compounds.[152] Produced in cooperation with Prous Science,[6] MDL Drug Data Report (MDDR) covers patent literature, journals, meetings, and congresses. MDDR contains over 132 000 biologically relevant compounds, including launched and candidate drugs, and well-defined derivatives, with about 10 000 entries annual update. The National Cancer Institute Database (NCI,[128] also available from Daylight CIS) v2001.1 contains over 213 000 structures with CORINA generated[153] 3D models. Publicly available NCI databases include: NCI-127 K, Plated Compounds, AIDS, and Cancer. These databases can be searched by compound and property, using MDL ISIS/Base[40] or MDL Database Browser via DiscoveryGate.[154] DiscoveryGate, a web-enabled discovery environment, integrates, indexes, and links scientific information to give immediate access to compounds and related data, reactions, original journal articles and patents, and authoritative reference works on synthetic methodologies. The MDL Patent Chemistry Database indexes chemical reactions, substances, and substance-related information from organic chemistry and life sciences patent publications. The first release contains approximately 1.5 million reactions, along with at least 1.5 million organic, inorganic, organometallic, and polymeric compounds, with associated data.[152]

3.14.4.2.2.6 GVK Biosciences

GVK Biosciences[148] databases capture chemical structures, biological activities, toxicity, and pharmacological data for a large number of compounds curated from patents and over 125 journals.[155] Over 1 million compounds and 3 million SAR points are indexed in Oracle, XML, and ISIS/Base formats. MediChem Database, codistributed by CambridgeSoft, was described in Section 3.14.4.2.2.3. Target Inhibitor Databases are focused on specific protein families: kinases, phosphatases, proteases, GPCRs, NHRs, transporters, and ion channel blockers. Natural Product Database contains compounds derived from natural plant, animal, marine, and microbial sources. Target Information Database details commercially explored targets, including function, transcription factors, pathway, protein–protein interactions, products in the pipeline, etc. Toxicity Databases contain in vitro and in vivo toxicity data, mechanistic terms, and details of compounds, while Reaction Database registers reactions reported in medicinal chemistry journals. PK Databases (Drug, Clinical Candidate, and Preclinical) contain pharmacokinetic (PK) parameters of compounds at various stages of drug discovery and development.[155]

3.14.4.2.2.7 Jubilant Biosys

PathArt is a pathway database that dynamically builds molecular interaction networks from curated databases. This product has comprehensive information on over 900 regulatory as well as signaling pathways that allows users to upload and map microarray expression data onto the pathways. GPCR Annotator captures a wide range of therapeutically relevant areas related to GPCRs; the annotator module allows users to classify the GPCR family hierarchy from sequence input.[156] Kinase ChemBioBase, codistributed by Accelrys (*see* Section 3.14.4.2.2.1) is a comprehensive database, currently containing over 217 000 compounds active on more than 350 kinases. Quality-checked SAR points with additional information are curated from about 1100 patents and over 500 journal articles. BioMarker is a collection of curated data from public domain databases and literature, on over 1000 disease biomarkers. Drug Database captures over 1500 approved drugs. GPCR Ligand Database captures over 400 000 small-molecule GPCR agonists/antagonists from 12 000 journal articles and 5000 patents, covering 60 GPCR receptor classes. Protease Inhibitors Database includes over 200 000 small-molecule protease inhibitors active on approximately 200 proteases, captured from 1000 journal articles and 1500 patents, while the Nitrilase and Nitrile Hydratase Knowledgebase comprises information on these enzymes.[156]

3.14.4.2.2.8 Sunset Molecular Discovery

Sunset Molecular Discovery LLC[157] integrates knowledge from target-driven medicinal chemistry with clinical PK data in the WOMBAT-PK database,[93] and provides up-to-date coverage of the medicinal chemistry literature in the WOMBAT database, as it appears in peer-reviewed journals.[158] WOMBAT (WOrld of Molecular BioAcTivity) release 2005.2 contains over 135 000 entries with 268 000 biological activities on more than 1100 unique targets (GPCRs, ion channels, enzymes, and proteins). The information is curated from more than 5800 papers published in medicinal chemistry journals. Additional experimental properties and calculated descriptors are available, as well as a comprehensive set of keywords related to biology and experimental protocols. Approximately 87% of the targets have Swiss-Prot Primary Accession Numbers[85]; all indexed papers contain the DOI[90] identifiers and/or have a direct cross-reference to the URL providing the original paper in PDF format. WOMBAT-PK 2005.2 (the WOMBAT Database for Clinical PK) captures 657 drugs with 4721 clinical PK measurements. Clinical data and physicochemical properties for launched drugs are captured from multiple literature sources. Integrated to WOMBAT, WOMBAT-PK captures in vitro and in vivo data for drugs, as captured in WOMBAT. Both databases are available in the MDL-ISIS/Base format, and are available in the RDF format, as well as the Oracle/Daycart[44] (Daylight) format.

3.14.5 Conclusions

Integrated resources, where bioinformatics and cheminformatics data are seamlessly converging into a comprehensive picture are quickly becoming reality. Annotated bioactivity databases (discussed in Section 3.14.4.2) are de facto becoming the second-generation chemical databases.[159] As hierarchical classification schemes for biological and chemical entities mature, extracting knowledge from such databases becomes easier.[11] The major task for discovery scientists remains data analysis and interpretation, resulting in knowledge creation. This requires familiarity with fundamental principles in both chemistry and biology, and certainly skills with complex queries. Learning how to proper query such disjoint sources might be an issue, but the major hurdles of the past, i.e., data collection and, recently, integration, are fading. The age of informatics-driven pharmaceutical discovery has arrived.

Acknowledgment

This work was supported by New Mexico Tobacco Settlement funds for Biocomputing.

References

1. American Chemical Society, CAS Online/SciFinder. http://www.cas.org/SCIFINDER/ (accessed June 2006).
2. Elsevier MDL, *CrossFire Beilstein Database*. http://www.beilstein.com/ (accessed June 2006).
3. InfoChem GmbH, München, *Spresi Database*. http://www.spresiweb.de/ (accessed June 2006).
4. Daylight Chemical Information System, Inc., Santa Fe, USA. http://www.daylight.com/ (accessed June 2006).
5. Elsevier MDL, *MDDR – MDL Drug Data Report*. http://www.mdli.com/products/knowledge/drug_data_report/ (accessed June 2006).
6. Prous Science, Barcelona, Spain. http://www.prous.com/ (accessed June 2006).
7. Derwent Publications Ltd., *WDI: World Drug Index*. http://thomsonderwent.com/products/lr/wdi/ (accessed June 2006).
8. Thomson Scientific, *The Current Patents Fast Alert Database*. http://scientific.thomson.com/currentpatents/ (accessed June 2006).
9. Elsevier MDL, *The Comprehensive Medicinal Chemistry Database*. http://www.mdli.com/products/knowledge/medicinal_chem/ (accessed June 2006).
10. Thomson Healthcare, *The Physician Desk Reference*, 57th ed.; Thomson: Philadelphia, PA, 2003. Available online at http://www.pdr.net/ (accessed June 2006).
11. Cases, M.; Garcia-Serna, R.; Hettne, K.; Weeber, M.; van der Lei, J.; Boyer, S.; Mestres, J. *Curr. Top. Med. Chem.* **2005**, *5*, 763–772.
12. Date, C. J. *An Introduction to Database Systems*, 7th ed.; Addison-Wesley: New York, 2000; pp 2–57.
13. Wikipedia, *The Free Encyclopedia*. http://www.wikipedia.org/ (accessed June 2006).
14. Elmasri, R.; Navathe, S. *Fundamentals of Database Systems*, 3rd ed.; Addison-Wesley: New York, 1999.
15. Bagui, S.; Earp, R. *Database Design Using Entity-Relationship Diagrams*; CRC Press: Boca Raton, FL, 2003, pp 1–22.
16. Date, C. J. *An Introduction to Database Systems*, 7th ed.; Addison-Wesley: New York, 2000; Part II, pp 109–326.
17. Bertino, E.; Martino, L. *Object-Oriented Database Systems*; Addison-Wesley: New York, 1993.
18. Larson, J. A. *Database Directions: From Relational to Distributed, Multimedia and Object-Oriented Database Systems*; Prentice-Hall PTR: Upper Saddle River, NJ, 1995.
19. Date, C. J.; Darwen, H. *A Guide to the SQL Standard: A User's Guide to the Standard Database Language SQL*, 4th ed.; Addison-Wesley: New York, 1997.
20. SQL standards are available from http://www.ansi.org/ (accessed June 2006) and http://www.iso.org/ (accessed June 2006) respectively: ANSI X3.135-1986, ISO 9075-1987 (SQL-86); ANSI X3.135-1989, ISO/IEC 9075:1989 (SQL-89); ANSI X3.135-1992, ISO/IEC 9075:1992 (SQL-92); ISO/IEC 9075-*:1999 (SQL-99); ISO/IEC 9075-*:2003 (SQL-2003).
21. Doll, S. Is SQL a standard anymore? http://builder.com.com/5103-6388-1046268.html (accessed June 2006).
22. Microsoft Corporation, *MS Office*. http://office.microsoft.com/ (accessed June 2006).
23. FileMaker, Inc., *FileMaker Pro*. http://www.filemaker.com/products/fm/ (accessed June 2006).
24. Microsoft Corporation, *MS SQL server*. http://www.microsoft.com/sql/ (accessed June 2006).
25. Oracle Corporation, *Oracle*. http://www.oracle.com/ (accessed June 2006).
26. IBM Corporation, *DB2 Universal Database Management System*. http://www-306.ibm.com/software/data/db2/ (accessed June 2006).
27. MySQL AB, *MySQL*. http://www.mysql.com/ (accessed June 2006).
28. PostgreSQL Global Development Group, *PostgreSQL*. http://www.postgresql.org/ (accessed June 2006).
29. Khurana, G. S.; Khurana, B. S. *Web Database Construction Kit*; Waite Group Press: Corte Madera, CA, 1996.
30. Hall, M. *Core Servlets and JavaServer Pages*; Sun Microsystems Press/Prentice-Hall PTR Book: Santa Clara, CA, 2004.
31. Linux SQL Databases and Tools. http://linas.org/linux/db.html (accessed June 2006).
32. MySQL AB, *MySQL Technical Reference Manual*. http://dev.mysql.com/doc/ (accessed June 2006).
33. DuBois, P. *MySQL*, 1st ed.; New Riders Publishing: Indianapolis, IN, 1999.
34. Momjian, B. *PostgreSQL: Introduction and Concepts*, 1st ed.; Addison-Wesley: Boston, MA, 2001.
35. Douglas, K.; Douglas, S. *PostgreSQL*, 1st ed.; Sams: Indianapolis, IN, 2003.
36. FirebirdSQL Foundation, *Firebird – Relational Database for the New Millenium*. http://firebird.sourceforge.net/ (accessed June 2006).
37. Sleepycat Software Inc., *Berkeley DB*, 1st ed.; Pearson Education: Berkeley, CA, 2001.
38. Computer Associates International, Inc. http://opensource.ca.com/projects/ingres/ (accessed June 2006).
39. Elsevier MDL. http://www.mdli.com/ (accessed June 2006).
40. Elsevier MDL, *ISIS – Integrated Scientific Information System*. http://www.mdl.com/products/framework/isis/ (accessed June 2006).
41. Daylight CIS, *THOR v4.8*. http://www.daylight.com/products/thor.html (accessed June 2006).
42. Weininger, D. *J. Chem. Inf. Comput. Sci.* **1988**, *28*, 31–36.
43. Weininger, D.; Weininger, A.; Weininger, J. L. *J. Chem. Inf. Comput. Sci.* **1989**, *29*, 97–101.
44. Daylight CIS, *DayCart - Daylight Chemistry Cartridge for Oracle*. http://www.daylight.com/products/f_daycart.html (accessed June 2006).
45. CambridgeSoft Corporation. http://www.cambridgesoft.com/ (accessed June 2006).
46. Sun Microsystems, Inc., *Sun Developer Network - Java Technology*. http://java.sun.com/ (accessed June 2006).
47. ChemAxon Ltd., *Marvin, JChem Base, JChem Cartridge*. http://www.chemaxon.com/ (accessed June 2006).
48. Borland Software Corporation, *InterBase 7.5*. http://www.borland.com/us/products/interbase/ (accessed June 2006).
49. Chemical Diversity Labs, Inc., *ChemoSoft*. http://chemosoft.com/modules/db/ (accessed June 2006).
50. Accelrys Software Inc., *Catalyst*. http://www.accelrys.com/products/catalyst/ (accessed June 2006).
51. Tripos, Inc *Support for Tripos Software*. http://www.tripos.com/ (accessed June 2006).

52. *UNITY.* http://www.tripos.com/ (accessed June 2006).
53. Codd, E. F. *Commun. ACM* **1970**, *13*, 377–387.
54. Date, C. J. *An Introduction to Database Systems*, 7th ed.; Addison-Wesley: New York, 2000, pp 348–388.
55. Topaloglou, T. In *Biological Data Management: Research, Practice and Opportunities*, Proceedings of the 30th VLDB Conference, Toronto, Canada, 2004.
56. Metaphorics LLC, *CABINET - Chemical and Biological Informatics NETwork.* http://www.metaphorics.com/products/cabinet.html (accessed June 2006).
57. Lipinski, C. A.; Lombardo, F.; Dominy, B. W.; Feeney, P. J. *Adv. Drug Delivery Rev.* **1997**, *23*, 3–25.
58. Dalke, A. *Computers – History of Chemical Nomenclature*, 2003. http://www.dalkescientific.com/writings/diary/archive/2003/10/15/Computers.html (accessed June 2006).
59. IUPAC, *The IUPAC International Chemical Identifier (InChI).* http://www.iupac.org/inchi/ (accessed June 2006).
60. Durant, J. L.; Leland, B. A.; Henry, D. R.; Nourse, J. G. *J. Chem. Inf. Comput. Sci.* **2002**, *42*, 1273.
61. Mesa Analytics and Computing LLC, *Measures Software.* http://www.mesaac.com/ (accessed June 2006).
62. Daylight CIS, *Smi2fp_ascii fingerprint module.* http://www.daylight.com/ (accessed June 2006).
63. Tanimoto, T. T. *Trans. NY Acad. Sci. Ser.* **1961**, *2*, 576–580.
64. Willett, P.; Barnard, J. M.; Downs, G. M. *J. Chem. Inf. Comput. Sci.* **1998**, *38*, 983–996.
65. Olah, M.; Bologa, C.; Oprea, T. I. *J. Comput.-Aided Mol. Des.* **2004**, *18*, 437–449.
66. Pearlman, R. S. *3D QSAR in Drug Design: Theory, Methods, and Applications*; Kubinyi, H., Ed.; ESCOM Science Publishers: Leiden, 1993; Vol 1, pp 41–79.
67. Elsevier MDL, *MOL File Format.* http://www.mdl.com/downloads/public/ctfile/ctfile.pdf (accessed June 2006).
68. VolSurf software is available from Molecular Discovery Ltd. http://moldiscovery.com/ (accessed June 2006).
69. Salomon, D. *Data Compression*; Springer-Verlag: New York, 1998; pp 103–162.
70. OpenEye Scientific Software. http://www.eyesopen.com/ (accessed June 2006).
71. Murray-Rust, P.; Rzepa, H. S. *J. Chem. Inf. Comput. Sci.* **2003**, *43*, 757–772.
72. The Chemical Markup Language. http://www.xml-cml.org/, http://cml.sourceforge.net/ (accessed June 2006).
73. World Wide Web Consortium (W3C), *Extensible Markup Language (XML).* http://www.w3.org/XML/ (accessed June 2006).
74. DuCharme, B. *XML: The Annotated Specification*; Prentice Hall PTR: Upper Saddle River, NJ, 1999.
75. SourceForge, *The JMol Project.* http://jmol.sourceforge.net/ (accessed June 2006).
76. Elsevier MDL, *MDL Chime.* http://www.mdli.com/products/framework/chime/index.jsp (accessed June 2006).
77. Wisniewski, J. L. *Handbook of Cheminformatics*; Gasteiger, J., Ed.; Wiley-VCH: Weinheim, Germany, 2003; Vol 2, pp 51–79.
78. Fisanick, W.; Shively, E. R. *Handbook of Cheminformatics*; Gasteiger, J., Ed.; Wiley-VCH: Weinheim, Germany, 2003; Vol 2, pp 556–607.
79. Cruciani, G.; Crivori, P.; Carrupt, P. A.; Testa, B. *J. Mol. Struct. (Theochem)* **2000**, *503*, 17–30.
80. Willett, P. *J. Med. Chem.* **2005**, *48*, 4183–4199.
81. Uniform Resource Identifiers (URI), *Generic Syntax – Draft Standard RFC 2396*, 1998. http://www.ietf.org/rfc/rfc2396.txt (accessed June 2006).
82. International Union of Biochemistry and Molecular Biology (IUBMB), *Enzyme Classification.* http://www.chem.qmul.ac.uk/iubmb/ (accessed June 2006).
83. Horn, F.; Vriend, G.; Cohen, F. E. *Nucleic Acids Res.* **2001**, *29*, 346–349.
84. Horn, F.; Weare, J.; Beukers, M. W.; Hörsch, S.; Bairoch, A.; Chen, W.; Edvardsen, O.; Campagne, F.; Vriend, G. *Nucleic Acids Res.* **1998**, *26*, 277–281.
85. Swiss Institute of Bioinformatics, *ExPASy Proteomics Server/Swiss-Prot Protein Knowledgebase.* http://www.expasy.org/sprot/ (accessed June 2006).
86. Herakleitus is quoted saying that "You cannot step into the same river twice." This is his Doctrine of Flux – see http://www.utm.edu/research/iep/h/heraclit.htm (accessed June 2006) for details.
87. American Chemical Society, *Chemical Abstracts Service.* http://www.cas.org/ (accessed June 2006).
88. National Center for Biotechnology Information/National Library of Medicine, *Entrez PubMed.* http://www.ncbi.nlm.nih.gov/entrez/ (accessed June 2006).
89. Thomson Scientific, *BIOSIS Bibliographic Database.* http://www.biosis.org/, http://www.cas.org/ONLINE/DBSS/biosisss.html (accessed June 2006).
90. International DOI Foundation, *The Digital Object Identifier System.* http://www.doi.org/ (accessed June 2006).
91. International DOI Foundation, *DOI Handbook*, 2005. http://dx.doi.org/10.1000/186 (accessed June 2006).
92. Neubig, R. R.; Spedding, M.; Kenakin, T; Christopoulos, A. *Pharmacol Rev.* **2003**, *55*, 597–606.
93. Oprea, T. I.; Benedetti, P.; Berellini, G; Olah, M; Fejgin, K.; Boyer, S. *Molecular Interaction Fields*; Cruciani, G., Ed.; Wiley-VCH: Weinheim, Germany, 2005; Vol. 24, pp 249–272.
94. Bright, M. W.; Hurson, A. R.; Pakzad, S. H. *Computer* **1992**, *25*, 50–60.
95. Object Management Group, Inc., *CORBA – Common Object Request Broker Architecture.* http://www.corba.org/, http://www.omg.org/ (accessed June 2006).
96. ZeroC, Inc., *ICE – The Internet Communication Engine.* http://www.zeroc.com/ (accessed June 2006).
97. World Wide Web Consortium (W3C), *HyperText Markup Language (HTML).* http://www.w3.org/MarkUp/ (accessed June 2006).
98. Powell, T. A. *HTML: The Complete Reference*, 2nd ed.; Osborne/McGraw-Hill: Berkeley, CA, 1999.
99. Povolna, V.; Dixon, S.; Weininger, D. *Cheminformatics in Drug Discovery*; Oprea, T. I., Ed.; Wiley-VCH: Weinheim, Germany, 2004; Vol. 23 pp 241–269.
100. Metaphorics LLC. http://www.metaphorics.com/ (accessed June 2006).
101. Swiss Institute of Bioinformatics, *ExPASy/Swiss-Prot IDTracker.* http://www.expasy.org/cgi-bin/idtracker (accessed June 2006).
102. National Center for Biotechnology Information, *PubChem.* http://pubchem.ncbi.nlm.nih.gov/ (accessed June 2006).
103. American Chemical Society, *Chemical Abstracts Service, CAS Registry.* http://www.cas.org/EO/regsys.html (accessed June 2006).
104. Rawlings, N. D.; Tolle, D. P.; Barrett, A. J. *Nucleic Acids Res.* **2004**, *32*, D160–D164.
105. European Bioinformatics Institute, *Enzyme Structures Database.* http://www.ebi.ac.uk/thornton-srv/databases/enzymes/ (accessed June 2006).
106. EBI–European Bioinformatics Institute. http://www.ebi.ac.uk/; an exhaustive list of available databases can be found at: http://www.ebi.ac.uk/Databases/ (accessed June 2006).
107. Swiss Institute for Bioinformatics and the European Bioinformatics Institute, *UniProtKB/Swiss-Prot Protein Knowledgebase Database.* http://www.ebi.ac.uk/swissprot/ (accessed June 2006).

108. Berman, H. M.; Westbrook, J.; Feng, Z.; Gilliland, G.; Bhat, T. N.; Weissig, H.; Shindyalov, I. N.; Bourne, P. E. *Nucleic Acids Res.* **2000**, *28*, 235–242.
109. Chemogenomics Lab Research Unit, Institute Municipal D'Investigacio Medica, Barcelona, *PDBRTF Database*. http://cgl.imim.es/pdbrtf/ (accessed June 2006).
110. Saier Lab Bioinformatics Group, *TCDB – Transport Classification Database*. http://www.tcdb.org/ (accessed June 2006).
111. Nomenclature Committee of the International Union of Biochemistry and Molecular Biology (NC-IUBMB), *Membrane Transport Proteins*; http://www.chem.qmul.ac.uk/iubmb/mtp/ (accessed June 2006).
112. National Center for Biotechnology Information, *Basic Local Alignment Search Tool (BLAST)*; http://www.ncbi.nlm.nih.gov/BLAST/ (accessed June 2006).
113. Ren, Q.; Kang, K. H.; Paulsen, I. T. *Nucleic Acids Res.* **2004**, *32*, D284–D288.
114. Bioinformatics and Drug Design Group, Computational Science Department, National University of Singapore. http://bidd.nus.edu.sg/ (accessed June 2006).
115. Chen, X.; Ji, Z. L.; Chen, Y. Z. *Nucleic Acids Res.* **2002**, *30*, 412–415.
116. Ji, Z. L.; Han, L. Y.; Yap, C. W.; Sun, L. Z.; Chen, X.; Chen, Y. Z. *Drug Safety* **2003**, *26*, 685–690.
117. Sun, L. Z.; Ji, Z. L.; Chen, X.; Wang, J. F.; Chen, Y. Z. *Clin. Pharmacol. Ther.* **2002**, *71*, 405–416.
118. Zheng, C. J.; Zhou, H.; Xie, B.; Han, L. Y.; Yap, C. W.; Chen, Y. Z. *Bioinformatics* **2004**, *20*, 2236–2241.
119. Chen, X.; Ji, Z. L.; Zhi, D. G.; Chen, Y. Z. *Comp. Chem.* **2002**, *26*, 661–666.
120. Ji, Z. L.; Chen, X.; Zheng, C. J.; Yao, L. X.; Han, L. Y.; Yeo, W. K.; Chung, P. C.; Puy, H. S.; Tay, Y. T.; Muhammad, A.; Chen, Y. Z. *Nucleic Acids Res.* **2003**, *31*, 255–257.
121. Samuel Lunenfeld Research Institute, Toronto, *The Blueprint Initiative*. http://www.blueprint.org/ (accessed June 2006).
122. National Center for Biotechnology Information, The NCBI Structure Group, *MMDB – The Molecular Modeling Database*. http://www.ncbi.nlm.nih.gov/Structure/ (accessed June 2006).
123. The Blueprint Initiative, *SMID – Small Molecule Interaction Database*. http://smid.blueprint.org/ (accessed June 2006).
124. Alfarano, C.; Andrade, C. E.; Anthony, K.; Bahroos, N.; Bajec, M.; Bantoft, K.; Betel, D.; Bobechko, B.; Boutilier, K.; Burgess, E. et al. *Nucleic Acids Res.* **2005**, *33*, D418–D424.
125. The Blueprint Initiative, *SeqHound Database*. http://www.blueprint.org/seqhound/index.html (accessed June 2006).
126. Benson, D. A.; Karsch-Mizrachi, I.; Lipman, D. J.; Ostell, J.; Wheeler, D. L. *Nucleic Acids Res.* **2004**, *32*, D23–D26.
127. Broad Institute, Cambridge, *ChemBank Project*. http://chembank.broad.harvard.edu/ (accessed June 2006).
128. National Institutes of Health, National Cancer Institute. http://www.cancer.gov/ (accessed June 2006).
129. Institute of Chemistry and Cell Biology, Harvard Medical School. http://iccb.med.harvard.edu/ (accessed June 2006).
130. Aizawa, M.; Onodera, K.; Zhang, J.-W.; Amari, S.; Iwasawa, Y.; Nakano, T.; Nakata, K. *Yakugaku Zasshi* **2004**, *124*, 613–619.
131. Zhang, J.-W.; Aizawa, M.; Amari, S.; Iwasawa, Y.; Nakano, T.; Nakata, K. *Comput. Biol. Chem.* **2004**, *28*, 401–407.
132. Quantum Molecular Interaction Analysis Group, Institute of Industrial Science, University of Tokyo, *KiBank*. http://kibank.iis.u-tokyo.ac.jp/ (accessed June 2006).
133. von Grotthuss, M.; Koczyk, G.; Pas, J.; Wyrwicz, L. S.; Rychlewski, L. *Comb. Chem. High Throughput Screen* **2004**, 7, 757–761.
134. US Food and Drug Administration, Department of Health and Human Services. http://www.fda.gov/ (accessed June 2006).
135. Roth, B. L.; Kroeze, W. K.; Patel, S.; Lopez, E. *Neuroscientist* **2000**, *6*, 252–262.
136. Austin, C. P.; Brady, L. S.; Insel, T. R.; Collins, F. S. *Science* **2004**, *306*, 1138–1139.
137. National Center for Biotechnology Information, *Entrez, The Life Sciences Search Engine*. http://www.ncbi.nlm.nih.gov/gquery/gquery.fcgi (accessed June 2006).
138. University of Erlangen-Nürnberg, Erlangen, Germany, *The CACTVS System Home Page*. http://www2.ccc.uni-erlangen.de/software/cactvs/ (accessed June 2006).
139. Irwin, J. J.; Shoichet, B. K. *J. Chem. Inf. Model.* **2005**, *45*, 177–182.
140. Tripos, Inc., *MOL2 File Format*. http://www.tripos.com/ (accessed June 2006).
141. Kearsley, S. K.; Underwood, D. J.; Sheridan, R. P.; Miller, M. D. *J. Comput.-Aided Mol. Des.* **1994**, *8*, 565–582.
142. Molinspiration Cheminformatics, *Java Molecular Editor*. http://www.molinspiration.com/jme/ (accessed June 2006).
143. Accelrys Software Inc. http://www.accelrys.com/ (accessed June 2006).
144. Accelrys Software Inc., *Chemical Database Product Listing*. http://www.accelrys.com/products/chem_databases/ (accessed June 2006).
145. Jubilant Biosys Ltd., Bangalore, India. http://www.jubilantbiosys.com/ (accessed June 2006).
146. Aureus Pharma, Paris, France. http://www.aureus-pharma.com/ (accessed June 2006).
147. Aureus Pharma, *AurSCOPE*. http://www.aureus-pharma.com/Pages/Products/Aurscope.php (accessed June 2006).
148. GVK Biosciences Private Ltd., Hyderabad, India. http://www.gvkbio.com/ (accessed June 2006).
149. CambridgeSoft Corporation, *Chemical Database*. http://www.cambridgesoft.com/databases/ (accessed June 2006).
150. Eidogen-Sertanty. http://www.eidogen-sertanty.com/ (accessed June 2006).
151. Eidogen-Sertanty, *Kinase Knowledgebase*. http://www.eidogen-sertanty.com/products_kinasekb.html (accessed June 2006).
152. Elsevier MDL, *MDL Discovery Knowledge Product Listing*. http://www.mdli.com/products/knowledge/ (accessed June 2006).
153. Molecular Networks GmbH, *CORINA*. http://www.mol-net.de/software/corina/ (accessed June 2006).
154. Elsevier MDL, *DiscoveryGate*. https://www.discoverygate.com/ (accessed June 2006).
155. GVK Biosciences, *Database Products*. http://www.gvkbio.com/informatics/dbprod.htm (accessed June 2006).
156. Jubilant Biosys Ltd., *Products*. http://www.jubilantbiosys.com/products.htm (accessed June 2006).
157. Sunset Molecular Discovery LLC. http://www.sunsetmolecular.com/ (accessed June 2006).
158. Sunset Molecular Discovery LLC, *Products*. http://www.sunsetmolecular.com/products/ (accessed June 2006).
159. Savchuck, N. P.; Balakin, K. V.; Tkachenko, S. E. *Curr. Opin. Chem. Biol.* **2004**, *8*, 412–417.
160. eMolecules Chemical Search. http://www.emolecules.com/ (accessed June 2006).

Biographies

Marius Olah, MS, PhD, is Research Scientist in the Division of Biocomputing at the University of New Mexico School of Medicine (USA). Born in Lugoj (Romania), he earned his doctoral degree (in physical chemistry and quantitative structure–activity relationships) under the supervision of Zeno Simon. He worked at the West University of Timisoara and at the Institute of Chemistry of the Romanian Academy for 3 years before moving to New Mexico in 2003.

Tudor I Oprea, MD, PhD, is Professor of Biochemistry and Molecular Biology and Chief of the Division of Biocomputing at the University of New Mexico School of Medicine (USA). Born in Timisoara (Romania), he earned his doctoral degree (in molecular physiology) under the supervision of Francisc Schneider. He later worked as a postdoctoral fellow at Washington University with Garland Marshall, and Los Alamos National Laboratory with Angel Garcia. After 6 years at AstraZeneca Mölndal (Sweden), he moved to New Mexico in 2002. He received the Hansch Award from the QSAR and Modeling Society in 2002, and is the Chair of the same Society (2006–10).

Comprehensive Medicinal Chemistry II
ISBN (set): 0-08-044513-6
ISBN (Volume 3) 0-08-044516-0; pp. 293–313

3.15 Bioinformatics

T Lengauer and C Hartmann, Max Planck Institute for Informatics, Saarbrücken, Germany

3.15.1	**Introduction**	**316**
3.15.2	**Sequencing Genomes**	**317**
3.15.3	**Molecular Sequence Analysis**	**318**
3.15.3.1	Sequence Alignment	318
3.15.3.2	Phylogeny Construction	320
3.15.3.3	Finding Genes	321
3.15.3.4	Analyzing Regulatory Regions	322
3.15.3.5	Finding Repetitive Elements	322
3.15.3.6	Analyzing Genome Rearrangements	324
3.15.4	**Molecular Structure Prediction**	**324**
3.15.4.1	Protein Structure Prediction	324
3.15.4.1.1	Secondary structure prediction of proteins	325
3.15.4.1.2	Homology-based protein structure prediction	325
3.15.4.1.3	De novo protein structure prediction	326
3.15.4.2	Ribonucleic Acid Secondary Structure	327
3.15.5	**Molecular Networks**	**327**
3.15.5.1	Metabolic Networks	328
3.15.5.2	Regulatory Networks	330
3.15.5.3	The Virtual Cell	330
3.15.6	**Analysis of Expression Data**	**330**
3.15.6.1	The Complementary Deoxyribonucleic Acid Microarray	331
3.15.6.2	Configuration of Experiments and Low-Level Analysis	331
3.15.6.3	Classification of Samples	332
3.15.6.4	Classification of Probes	332
3.15.6.5	Beyond cDNA	333
3.15.7	**Protein Function Prediction**	**333**
3.15.7.1	What is Protein Function?	333
3.15.7.2	Function from Sequence	334
3.15.7.3	Protein Interaction Networks	335
3.15.7.4	Genomic Context Methods	335
3.15.7.4.1	Gene neighborhood and gene order	336
3.15.7.4.2	Domain fusion	336
3.15.7.4.3	Phylogenetic profiles	336
3.15.7.5	Function from Structure	336
3.15.7.6	Text Mining	337
3.15.7.7	Information Integration	338
3.15.8	**Analysis of Genetic Variations**	**338**
3.15.8.1	Genetic Variations in Humans: Analyzing Predispositions	338
3.15.8.2	Genetic Variations in Pathogens: Analyzing Resistance	339
3.15.9	**Outlook**	**340**
References		**341**

3.15.1 Introduction

The process of drug development can roughly be segmented into two phases. In the first phase, a suitable protein is searched for that is hypothesized to be the target for the drug to be developed. In the second phase, the drug molecule is selected or developed. The second phase has a much longer tradition than the first. For hundreds of years, people have searched for effective drugs. Originally, the drug that was responsible for the therapeutic effect could not be identified from the cocktail in which it was administered. For example, it was known that extracts from certain plants had therapeutic or soothing effects when administered in a certain fashion, but no one knew why. However, in the last 150 years, research has focused increasingly on the drug molecule itself. With the advent of systematic experimental procedures in analytic and synthetic chemistry, the effect of the therapy could begin to be attributed to specific molecules, and those molecules could be modified and refined to enhance their actions and limit adverse side effects. However, in most cases, the target protein was still not known. As the field increased in maturity over the last century, target proteins were identified for increasing numbers of drugs. While certain target proteins were known, it was still not possible to search for new and more effective drug targets for a given disease. Only in the mid-1990s did the technology for systematically searching for drug targets become available, with the advent of whole-genome sequencing (WGS), the related screening procedures for biological data, and the accompanying developments in bioinformatics that are targeted toward configuring experiments and analyzing and interpreting the resulting voluminous and complex data. However, even today, we do not know the target proteins for many drugs on the marketplace.

The computational side of the field mirrors the development on the experimental side. Basically, we can distinguish three subfields: drug screening, docking, and bioinformatics. The first two fields belong to what is usually called computational chemistry. Drug screening aims at the computational comparison between drugs or druglike molecules. This procedure does not require knowledge of the structure of the target protein (or even its identity, for that matter). Drug screening is based on the idea that one can transfer information on the biological activity of a molecule from another, if the two molecules are sufficiently similar in make-up. Thus, one can attempt to find lead candidates for a drug for a certain protein on the basis of a molecular comparison with another drug targeting the same protein, or with a substrate for this protein. (The protein itself does not necessarily have to be known, though this helps greatly for validation purposes.) In docking, the protein structure has to be available, and a suitable drug can be developed or selected from a large compound database based on a structural analysis of how tightly the drug molecule binds to the active site of the protein. Molecular analyses based on docking can be much more specific than those based on drug screening without incorporating the structure of the binding site of the protein.

These two approaches to selecting or developing drugs are the topics of other chapter in this book (*see* 4.12 Docking and Scoring; 4.19 Virtual Screening). Here, we will concentrate on the much more recent, and still immature, area of the systematic search for target proteins. A target protein has to meet a multitude of requirements. First and foremost, it has to have a biological function whose modulation (often blocking) is effective for therapy. The activation of cellular receptors by suitable agonists or their deactivation by antagonists and the blocking of an enzyme by an inhibitor are examples. Usually, finding a protein that plays an appropriate role within a disease process affords a more or less comprehensive biological analysis that aims at understanding at least the relevant parts of the molecular basis of the disease. The required data have only recently become available through large-scale, often whole-genome, screening procedures. Uncovering essential parts of the biological map of the disease requires a multitude of bioinformatical procedures, which are the subject of this chapter.

To date, over 260 genomes have been completely sequenced (**Table 1** lists genomes whose completion marked the progress in genome sequencing technology). Most genomes are those of small organisms, among them a large number of pathogens. However, there is also an increasing amount of sequence data from mammals: a dozen mammals will be completely sequenced before long. Genome sequence data do not only afford us with nature's blueprints of the relevant organisms: the comparison between different organisms also enables us to look into the past, and provides insight into the evolutionary development of species. This kind of information is essential for uncovering the biological function of molecules and unraveling the structure of biochemical networks.

While sequence data are the historical basis of genome-wide biomolecular information, they are not sufficient to analyze the inner workings of cells. Since the genome is the same in all cells of an organism, we need additional information to distinguish between different tissues and cell states. This information is afforded by gene and protein expression data. Furthermore, data on protein–protein interactions and the concentrations or fluxes of metabolites enhance the datascape on cellular processes. All these data can and must be brought together and analyzed in concert, in order to enlighten us about cellular processes in general, and disease processes in particular.

This chapter is organized as follows, with each section building on the previous one. In Section 3.15.2 we discuss how one arrives at a complete genome sequence; in particular, what bioinformatics support is needed along the way.

Table 1 Landmark genome sequencing projects

Organism	*Colloquial name*	*Genome length (kb)*	*Date sequenced*
Haemophilus influenzae	Bacterium (ear infections)	1 830	1995
Escherichia coli (K12)	Bacterium in the intestine	4 639	1997
Saccharomyces cerevisiae	Baker's yeast	12 069	1997
Caenorhabditis elegans	Nematode, worm	97 000	1998
Drosphila melanogaster	Fruitfly	13 7000	2000
Arabidopsis thaliana	Mouse-ear cress	1 15 428	2000
Mus musculus	Mouse	2 497 000	2002
Homo sapiens	Human	2 866 000	2003

In Section 3.15.3 we discuss the bioinformatics procedure for analyzing molecular sequences. Section 3.15.4 concentrates on the analysis and prediction of the structure of biomolecules. All of these procedures aim at better understanding the building blocks of the cell. The analysis of molecular interactions (docking) is not discussed in this chapter, since this topic is discussed in detail (*see* 4.12 Docking and Scoring). In Section 3.15.5, we embark on the road toward understanding how the building blocks act in concert to motorize cellular processes. In Section 3.15.6, we distinguish different cell types and cell states. Section 3.15.7 discusses the central notion of protein function, and surveys computational procedures that uncover elements of protein function by analyzing certain types of molecular data and, eventually, integrating these sources of information. In Section 3.15.8 we move away from the generic view of a species and discuss ventures into distinguishing individuals within a species, a step that needs to be taken if one wants to arrive at therapies that take the individual genome of the patient (and the relevant pathogen) into account. Finally, Section 3.15.9 gives an outlook on future developments.

3.15.2 Sequencing Genomes

Sequencing a genome means experimentally reading off the bases along the genomic DNA sequence. The main problem with today's sequencing techniques is that the result of sequencing becomes increasingly inaccurate, as the sequencer proceeds along a continuous piece of DNA sequence. As a result, the longest DNA sequences that can be sequenced continuously are less than 1000 base pairs long. These experimental restraints require the genome sequence to be split into fragments whose ends can be read up to this length from either side. The sequence 'reads' have to cover the complete genome if the total genome sequence is to be recovered. In order to assemble the complete genome from the reads, algorithms specially tailored to this problem have to be used. All of these algorithms have in common that they derive the information on the positions of the fragments in the genome sequence from overlaps between different fragments. Fragmentation is done either in a deterministic fashion, using restriction enzymes that cut the DNA sequences in specific places, or – more commonly today – randomly (shotgun sequencing). One can show that, in the random approach, statistically one needs to cover each piece of the genome several times with fragments in order to recover the complete genome sequence using assembly algorithms (e.g., for the human genome a coverage of sevenfold was required[1]).

The success of assembly algorithms depends on the quality of the reads as well as on some characteristics of the sequenced genome. Reading errors may cause the alignment of overlapping fragment ends to fail, and in this way corrupt the local assembly. This drawback can be compensated for by increasing the fold coverage of the genome with fragments. Repeats in the genome sequence pose another danger, because they confound the alignment of fragments. Repeats can be handled by splitting the genome into fragments of two or three defined lengths. Then, the end reads of such fragments can be linked with quite accurate distance information. This linking information can be used to generate scaffolds reaching across large sections of the genome even if the respective part of the sequence is not completed covered with reads. More information on sequencing genomes can be obtained from Green.[2] The correct assembly of the sequence reads usually provides a first draft of the original genome sequence. Generally, this draft contains many gaps, mostly in highly repetitious regions or in regions that have not been covered with enough fragments. To complete the genome sequence from this first draft, gaps have to be closed, the assembly has to be validated, and, in general, the sequence quality has to be improved.[3]

The most popular application of fragment assembly algorithms arose during the sequencing of the human genome. Within the Human Genome Project, a public consortium of scientists was formed to integrate all efforts toward this monumental quest. The project started in 1990, and was completed in 2003. The strategy of sequencing the whole genome that has been applied by the Human Genome Project is often referred to as hierarchical shotgun sequencing (HS), or sometimes the map-based, BAC-based (bacterial artificial chromosome), or clone-by-clone sequencing approach.[4] First, the whole genome is cloned, and the clones are cut into fragments. The composition of all these fragments ensures multiple coverage of the whole genome with sufficient overlap between the fragments for later reassembly. This can be done by special restriction enzymes that are able to cut the DNA at specific positions defined by characteristic local sequence patterns. The resulting sequence fragments are called BAC clones (sometimes referred to as contigs), with fragment lengths between 150 and 200 kbp, and are again cloned to build up so-called BAC libraries. The BAC clones are organized in a so-called physical map, which locates each single BAC library along the whole genome sequence. Each BAC is then cloned and fragmented randomly with the shotgun approach and then reassembled using computers.

In 1997, Weber and Myers proposed application of the shotgun sequencing method to the whole human genome directly (i.e., WGS) without the intervening step of the BAC libraries. Previously, the WGS approach had only been applied to bacterial genomes, which are comparatively short and have few repetitions. Weber and Myers also introduced the idea of paired reads at both ends of larger fragments of determined size. A year later, Venter and his group at Celera Genomics announced the sequencing of the human genome using the WGS method, and claimed that they would finish significantly before the planned completion date of the Human Genome Project. This announcement sent shock waves through the research community; a particular concern was that if a private company finished sequencing the genome first, this would lead to serious patenting issues. Thus, a race started between the two groups for the first draft of the human genome sequence. This competition accelerated the whole process of sequencing the human genome as well as the development of better fragment assembly algorithms. The additional information on the distance of paired reads is utilized by the assembly algorithms of Myers and his bioinformatics group.[5,6] In February 2001 both groups presented their drafts of the human genome sequence, 4 years before the original target date for completing the sequence of the human genome. The draft assembly of the public consortium was computed by Kent at the University of California, Santa Cruz.[7] The consortium required 2 more years for the completion of their human genome sequence, finishing their work in April 2003.

The usability of both strategies depends upon the characteristics of the genome to be sequenced. WGS works fine on bacterial genomes. For long genome sequences with many repeats the WGS approach is able to provide a draft of approximately 80% completeness. However, finishing a repetitive genome is difficult with the WGS approach.

In the slipstream of the Human Genome Project, the WGS approach was applied to the dog genome,[8] and to probes from the Sargasso Sea containing many organisms.[9] A hybrid approach that utilizes aspects of both competing methods was used to finish sequencing the mouse genome[10] and the human genome. Completed drafts of a total of 12 mammalian genomes are expected in the near future.[11,12]

Since these mammalian genomes are evolutionary quite close to the human, they harbor precious information for further studies of the human genome.

3.15.3 Molecular Sequence Analysis

Biological, pharmaceutical, and medical applications do not only require sequences, they need a functional description of organismic, especially human, life. The completion of genome sequences or even creation of rough drafts is a crucial step toward this goal, but there is still much more work to be done: the genome has to be mined for genes and for regions that regulate gene expression. Heritage and genetic variations between single individuals have to be analyzed, to assist the study of the relationship between diseases and genes. There are many computational tools that provide varied support for these efforts. Basic software tools that concentrate on the computational analysis of genome sequences are compiled in utilities such as the Staden Package[13] or the Wisconsin Package (GCG).[14]

3.15.3.1 Sequence Alignment

Sequence alignment is at the basis of analyzing the evolutionary and functional relationships between molecular sequences. An alignment of two sequences is an arrangement of both sequences, one written below the other. Matching sequence positions are written one on top of the other. (If two characters in such a column are different, we speak of a 'mismatch'; if one sequence is missing a character, a gap character ('–') is used. **Figure 1a** illustrates a

```
Hemoglobin alpha-1    1  MVLSPADKTNVKAAWGRVGAHAGEYGAEALERMPLSFPTTKTYFPHF-D   48
Hemoglobin beta       1  MVHLTPEEKSABTALNGRV--NVDEVGGEALGRLLVVYPNTQRFFESFGD   48

Hemoglobin alpha-1   49  LSH-----GSAQVKGHGKKVADALTNAVAHVDDMPNALSALSDLHAHKLR   93
Hemoglobin beta      49  LSTPDAVMGNPKVKAHGKKVLGAFSDGLAHLDNLKGTFATLSELHCDKLH   98

Hemoglobin alpha-1   94  VDPVNFKLLSHCLLVTLAAHLPAEFTPAVHASLDKFLASVSTVLTSKYR  142
Hemoglobin beta      99  VDPENFRLLGNVLVCVLAHHFGKEFTPPVQAAYQKVVAGVANALAHKYH  147
```

(a)

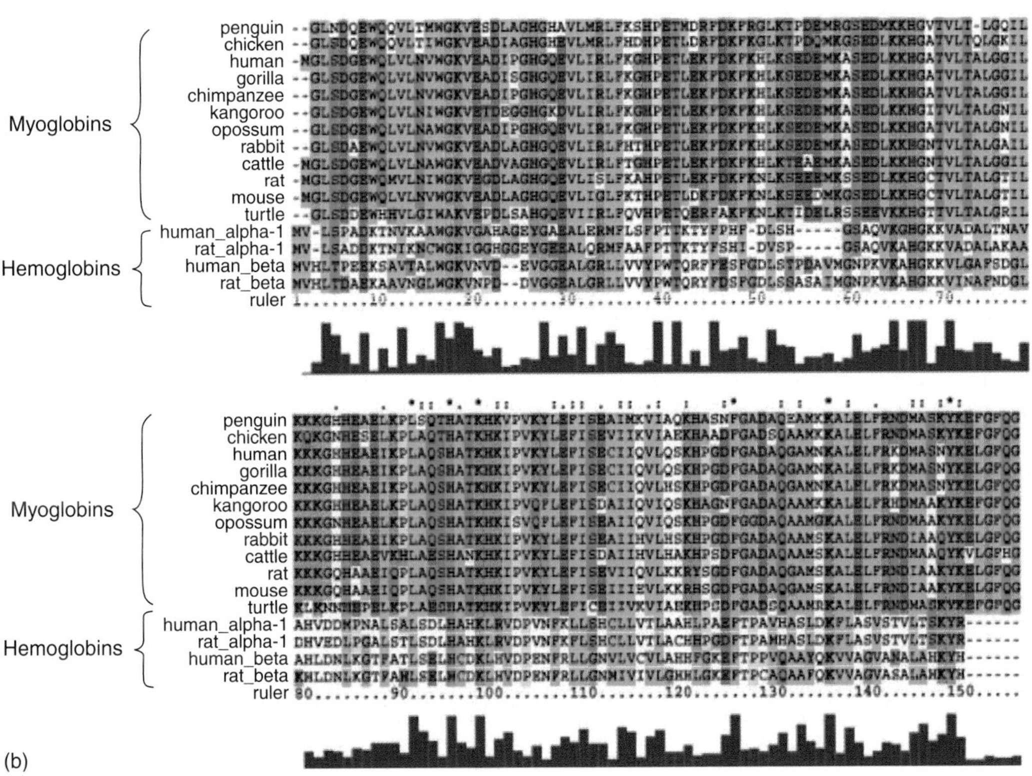

(b)

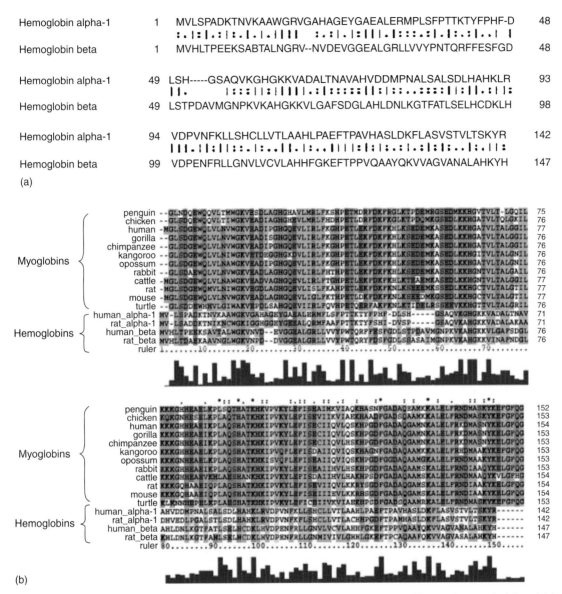

Figure 1 (a) Global alignment of the human hemoglobin α_1 and hemoglobin β sequences. The numbers on the left and right sides of the alignment represent sequence positions at the end of each row. The signs in the middle row signify similarities of matched amino acid residues: |, identical; :, similar; ., not similar. (b) Multiple alignment of several myoglobin and hemoglobin sequences. The alignment was generated with ClustalW. The colors represent chemical groups of amino acid residues. The bar graphs represent the degree of conservation of the alignment columns.

sample pairwise alignment of the human hemoglobin α_1 and β chains. **Figure 1b** shows a multiple alignment of myoglobin and hemoglobin sequences from different organisms. Depending on the context, matches/mismatches can have varying meanings. They can indicate evolutionary, structural, or other correspondence. In evolutionary correspondence, the two sequence positions can be referred back to the same sequence position in a (lowest) common ancestor. In structural correspondence, the two residues lie on top of each other in the structural superposition of the two proteins. A gap signifies an insertion into the sequence that does not have the gap character or a deletion from the sequence that has the gap character. We cannot tell which of the two is the case since the alignment model is symmetric with respect to the relationship between the sequences. Thus, alignment columns with gap characters are also called 'indels.'

Sequence alignment algorithms are based on probabilistic models for the occurrence of positional (mis-)matches.[15] The probability that the two sequences stem from a common origin in the way that the alignment represents is the product over all alignment columns of the probabilities of seeing the (mis-)matches or indels in these columns. The match probabilities are derived from some database of match events. In the evolutionary context, the database contains manually curated multiple alignments that are supposedly biologically correct[16,17]. Usually it is assumed that the (mis-) match probabilities only depend on the nature of the affected nucleotides or residues and are independent of their neighbors. A string of adjacent indels (called gaps) is treated as a unit. The alignment that has the highest probability is assumed to be the biologically correct one according to this model. Early approaches arrange the costs for (mis-)matches into so-called substitution matrices. For protein sequences, the PAM[18] and/or BLOSUM[19] matrices are the most widely used. The number that is part of the matrix name is an indication of the targeted evolutionary distance. For example, PAM-250 is a mutation matrix for sequences that have an evolutionary distance that allows for 250 mutations (PAM, point accepted mutations) in every 100 sequence positions, on average. Since the number of mutations increases with rising evolutionary distance, the correct choice of the substitution matrix is essential for the quality of an alignment. Recently, a method for scoring sequence alignments has been presented that adaptively determines the evolutionary distance of the aligned sequences.[20] Usually, the cost of deletions and insertions are assumed to depend linearly on the gap length, and their values are determined by heuristics or fitted to a data set of trusted alignments.[21]

Using such a cost theme, algorithms are able to solve the optimization task very efficiently via the dynamic programming paradigm.[22] They can be used to either align complete sequences (global alignment) or only some parts of both sequences (local alignment). In the case of pairwise alignments, the runtime of these algorithms grows linearly in the product of the lengths of the two sequences. This is acceptable for single alignment queries, but unacceptable for similarity searches in large databases. For database searches there are other algorithms, of which BLAST is the most popular.[23] This heuristic algorithm is based on gapless alignment, and tries to optimize a local similarity measure. Its popularity does not only arise from its speed but also from the fact that it provides a significance value (p value). BLAST is a rare example of a major bioinformatics tool that provides a confidence estimate for its output that is based solid statistical theory.[24]

Things become more complicated when many sequences must be aligned. Extending the cost scheme for pairs of sequences to an all-pairs cost for aligning multiple sequences leads to a straightforward extension of the known dynamic programming algorithms. Unfortunately, the runtime of such an algorithm grows exponentially with the number of sequences to be aligned, which disqualifies it for most applications. Again, one has to rely on heuristics. One such approach is to compute all pairwise alignments and then assemble the alignment incrementally, starting with the most similar sequences. In each step we take a cluster of already multiply aligned sequences and align them with a simple modification of the pairwise sequence algorithm that allows for two multiple alignments to be aligned. Here, in effect, each sequence position does not present a unique nucleotide (in DNA/RNA), amino acid (in proteins), or gap character, but a profile of choices, derived from the respective multiple-alignment column. Aside from that scoring and algorithm are essentially the same as in pairwise sequence alignment. This version of multiple-sequence alignment is called 'progressive' alignment. ClustalW[25,26] is a common multiple-alignment program that works this way.

Another approach to multiple alignment maps the common pattern of a sequence family to the parameters of a so-called hidden Markov model (HMM).[27] An HMM can be trained to a given set of sequences. A new sequence can be aligned to the trained model. Doing this for a number of sequences, in turn, directly retrieves the multiple alignment – in linear time in the number of sequences. In essence, the process is tantamount to computing a set of pairwise alignments (sequenced against the HMM), where the scores depend on the sequence position. Such a scoring system is based on a 'position-specific scoring matrix' (PSSM). The training process learns the characteristics of each sequence position, and codes them into the parameters of the HMM.

T-COFFEE,[28] DIALIGN,[29,30] and MUSCLE[31] are other popular multiple-alignment programs.

Sequence alignment methods provide the algorithmic basis for a great variety of sequence analysis methods that are used for, for example, the discovery of genes, the analysis of regulatory regions,[32] or the construction of phylogenies. They can be applied to proteins to discern their evolutionary[18,33,34] or structural relationships,[35,36] or to classify them into families.[37–39]

3.15.3.2 Phylogeny Construction

A phylogeny describes the evolutionary relationships between molecular sequences using a treelike structure. In general, the 'leaves' are labeled with known sequences of some of the observed organisms. Edges denote ancestral relationships between the nodes of the tree, in the ideal case their length measuring the ancestral distance in terms of time. Interior nodes represent assumed common ancestors whose sequences are unknown. **Figure 2a** shows a phylogenetic tree of the myoglobin sequences used in **Figure 1b**. Phylogenetic trees come in undirected and directed versions. Most phylogenies are undirected, meaning that we cannot tell which of two interior tree nodes joined by an edge is the ancestor and which the descendant. Among many others, PHYLIP and PAUP* are two widely used software libraries for constructing phylogenies.

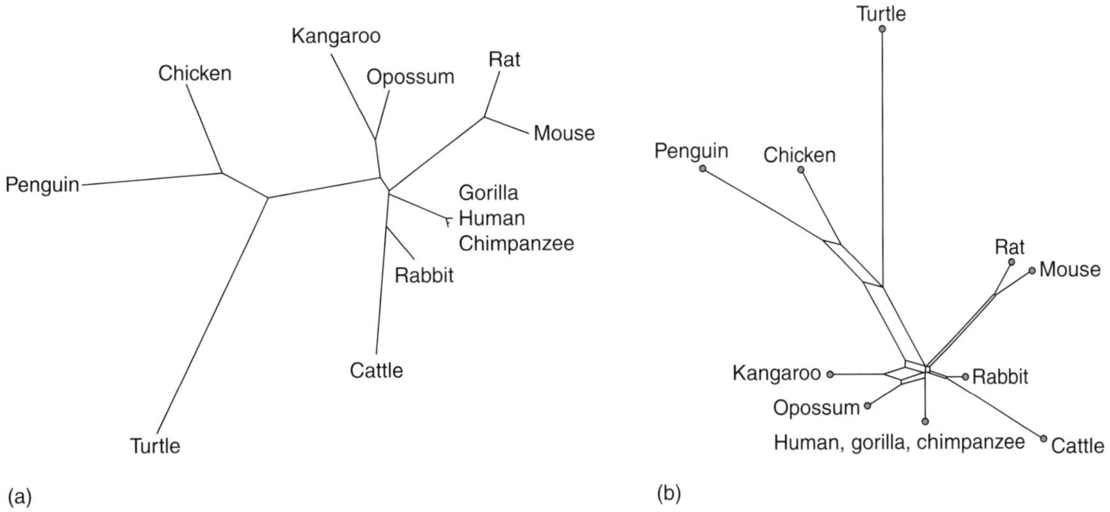

Figure 2 (a) A phylogenetic tree of the myoglobin sequences aligned in **Figure 1b**. (b) A split diagram of the myoglobin sequences aligned in **Figure 1b**.

Methods for inferring phylogenies from sequence data can de divided roughly into distance-based methods, maximum-parsimony methods, and maximum-likelihood methods.

The distance-based approach deduces the evolutionary distance from the pairwise similarity of the sequences, either measured by the score of a pairwise alignment of all involved sequences or by other techniques (e.g., based on the maximum-likelihood approach).[22] The scores are arrayed in a distance matrix. Although, in many cases, these distances cannot be mapped accurately onto a tree, the tree generation is guided by this matrix. The tree that fits the distances best is hard to find – the problem is 'NP-hard.' Thus, we have to search for proper approximations of the optimal solution. Like others, the program ClustalW (*see* Section 3.15.3.1) constructs a multiple alignment and builds a phylogenetic tree bottom-up by successively clustering the most closely related sequences.

The maximum-parsimony approach is based on the reasonable assumption that evolution introduces as few changes as possible to generate the observed sequence data. In this model, every mismatch between two directly related sequences (connected by a single edge) incurs some costs. The generation of a tree then aims at minimizing these costs.

The maximum-likelihood approach is based on a probabilistic model of evolution. Here, the phylogeny with the highest probability is searched for, given the observed sequence data. This approach is much more compute-intensive than the parsimony-based approach, but has the advantage of being founded on a statistically more sound and generalizable theoretical basis.[40]

All three approaches have in common that they have rather limited abilities to model the highly complex mechanisms of evolution. Specifically, the hypothesis that the phylogeny can be modeled in a tree structure is not valid in many biological settings because, for example, gene duplications, horizontal gene transfer, or recombination events can take place. The split-decomposition method takes account of the fact that the distance measure derived from the sequence alignments need not be treelike.[41] The method generates so-called split diagrams that are more treelike if the underlying distance measure allows for this, but which exhibit deviations from the tree shape where the data suggest. **Figure 2b** shows a split diagram of the myoglobin sequences from **Figure 1b**. There are plenty of articles elucidating and discussing the current state of the field of phylogeny construction.[42–44]

3.15.3.3 Finding Genes

The overtly most interesting parts of a genome sequence are the regions that code for proteins, our genes. Genes make up only 1.5% of the human genome. **Figure 3** illustrates the complex structure of a eukaryotic gene. All parts of this structure have to be identified accurately.

There are two basic approaches to identifying genes: 'contents sensors' and 'signal sensors.'[45] Contents sensors classify DNA segments into types such as 'coding' and 'noncoding.' The homology-based (or extrinsic) content sensor does this with a comparison to known sequences using sequence alignment. With the availability of a rising number of genomes, one can search for regions that are more conserved than others, and thus may be more likely to a have a

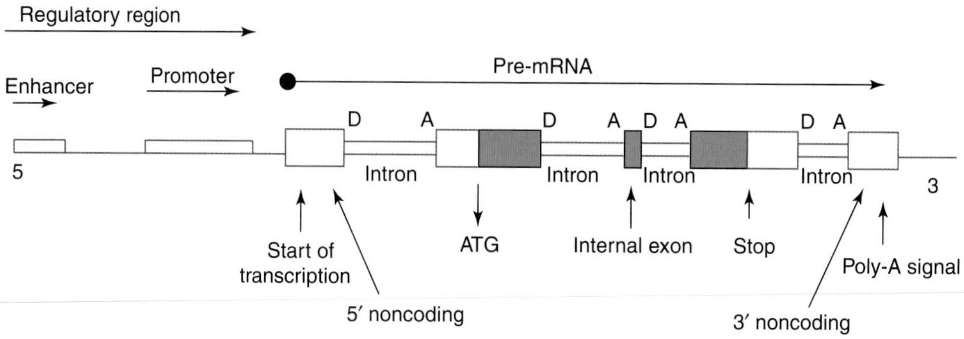

Figure 3 The structure of a eukaryotic gene. Only the shaded parts represent coding regions.

function or be a coding region.[46] Attempts to even capture the evolutionary development of genes have been made.[47] Without the knowledge of related sequences, the genes have to be identified with ab initio (intrinsic) approaches. Here, the target sequence is scanned for specific patterns such as start codons, CpG islands, or unusual codon frequencies. With these patterns, either statistical classifiers,[48,49] HMMs,[50,51] or neural networks[52] can be used to uncover the structure of a gene.

Using the best of these methods, the accuracy for predicting coding nucleotides in mammalian sequences reaches 93%.[53] The detection of exons is even better, but the prediction of their correct boundaries is much more difficult, and as the exons become shorter the situation gets worse. The accuracy at the gene prediction level is even lower – below 80%. With increasing length of introns, the number of exons that form a gene, and the length of the noncoding regions, the accuracy of gene prediction drops further – below 50%.[54] Obviously, there is still room for improvement.

3.15.3.4 Analyzing Regulatory Regions

Regulatory regions are regions in the genome that do not code for proteins but instead control the expression of other, coding, regions. Regular regions occur in the vicinity of coding regions. The promoter regions, located upstream of the genes (**Figure 4**), are the most prominent type. They are usually 10–15 nucleotides long. They harbor binding sites for the transcription factors (TFs), proteins that initiate DNA transcription. Other regulatory regions include binding sites for coactivators, coexpressors, or factors that regulate epigenetic modifications of the DNA.[55] They span only a small stretch of DNA (10–15 nucleotides) but hold certain motifs.

Methods for finding and delineating regulatory regions suffer from the same drawbacks as gene-finding strategies. The situation is even worse here: because regulatory regions are not expressed, there are no negative images such as the mRNA for the genes that could imply their boundaries. Similar to the gene-finding problem, intrinsic methods are applicable. Additionally, the overrepresentation of certain motifs in the regulatory region of genes that are known to be regulated by the same transcription factors provides an opportunity to use comparative methods.[56,57] Roughly speaking, most of the methods currently used in detecting such motifs are either alignment based or consensus based. The consensus-based approaches usually enumerate all possible motifs of a given length, and rank their frequencies in the given set of regulatory sequences. Current tools have been estimated to find about half of the existing promoters, but their specificity is still low, implying a significant number of false positives.[58]

Enhancers, silencers, and matrix-attached regions are other types of regulatory regions for which predictions have been attempted with similar technology. Compared with protein-based bioinformatics, the analysis of regulatory regions is still at an early stage of development. It is especially difficult because the processes of molecular recognition based on proteins binding to DNA are very intricate and cannot yet be analyzed at a structural level. Thus, we have to derive all information from the rather indirect methods of sequence analysis and comparison.

3.15.3.5 Finding Repetitive Elements

As mentioned in Section 3.15.2, repeats in the DNA sequence are a stumbling block for genomic sequence assembly. Despite the fact that a significant part of the human genome consists of repeats, much of their origin and function is still uncertain. (There are theories that many of these regions are the result of endogenous viruses and selfish genes.[59]) It is assumed that up to 50% of the human genome has a repetitive nature.[60]

There are numerous tools that detect and mark repeats in strings; some target biomolecular sequences explicitly. When they are used as a first step in genome analysis, the detected repeats are usually masked (i.e., excluded from

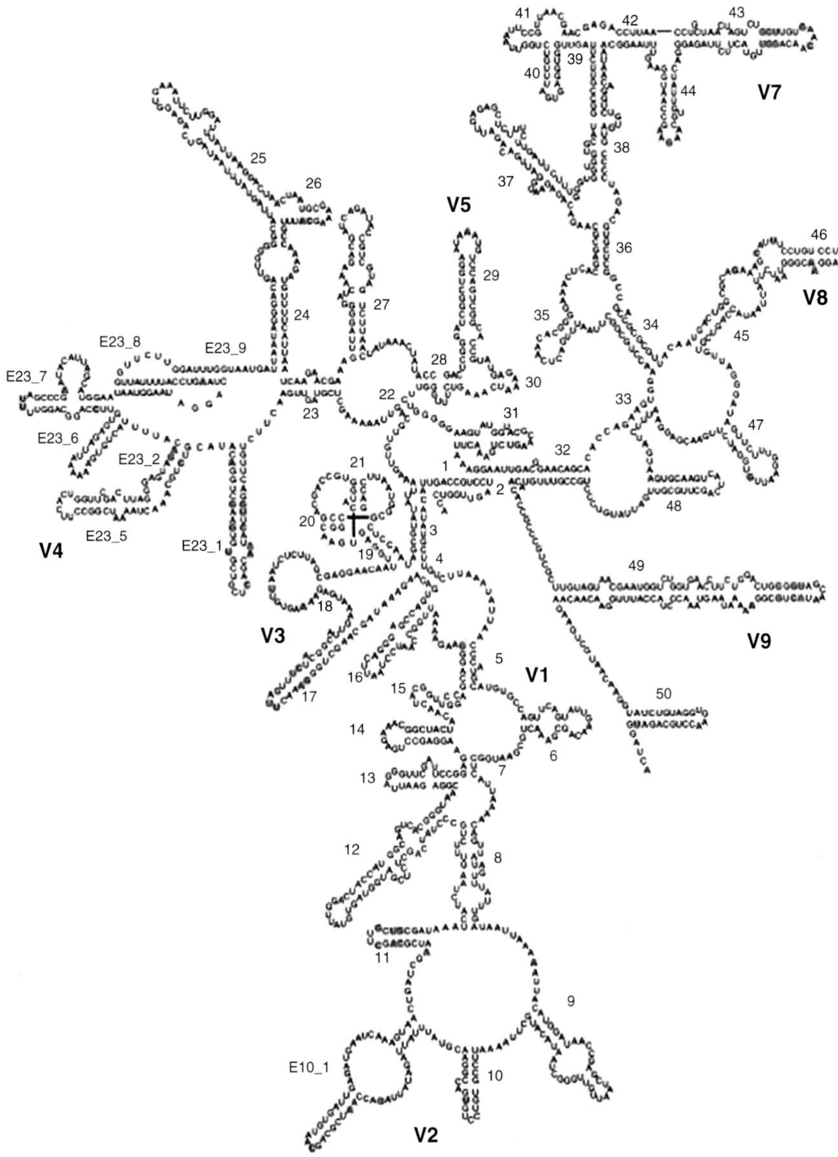

Figure 4 Secondary structure of the RNA from the small ribosomal subunit of *Tetrahymena bergeri*. The fat lines denote pseudoknots.

further analysis). A cautious use of automation is still vital, because repeats can lie in functional regions. RepeatMasker and similar tools scan the target sequence for exact or approximate matches to patterns in a dictionary of known repeats. RepeatMasker scores the similarity between match and pattern, and uses a heuristically determined threshold to tell repeats apart from unique fragments. This technique can only rediscover known repeat patterns, and the use of threshold bears the risk of false predictions.

The task of finding repeats in a sequence is much harder when there is no supporting information or prior knowledge (e.g., the length of the repeat, or motifs). Tools with this ability usually search for exact repeats first, and then for so-called degenerated repeats (slightly changed), using the exact repeats as query sequences. A popular algorithmic technique (used by REPuter[61] and MUMmer, for example)[62] organizes the data in a structure called a suffix tree, allowing fast detection of exact repeats.

Some programs specialize in finding an important type of repeat – the tandem repeat.[63] This repeat does not allow for insertions between the original sequence and its copy, and is thus easy to find. The visualization of the detected repeats and their relation is an important feature for the functional study of repeats.

Repeat-finding programs can be used for a multitude of tasks:

- Checking genome assemblies. Assembly programs (*see* Section 3.15.2) are imperfect, and thus the sequence assemblies they produce may contain errors. Repeat-finding programs have detected palindromic repeats in the human genome sequence that were due to wrong assembly.
- Identifying low copy repeats in human diseases. Several human diseases are associated with deletions or duplications of specific genomic regions. Such repeat patterns can be identified by a repeat-finding program.
- Checking the uniqueness of hybridization probes. This application is technological. On a microarray we want to deposit sequence probes that are unique in the genome under consideration. This uniqueness can be ensured with a repeat finder.
- Comparative genomics. Concatenating two different genomes and looking for repeats with varying rates of divergence allows comparison of the two genomes at different grades of similarity.

3.15.3.6 Analyzing Genome Rearrangements

In addition to mutations and recombinations that shape a species, there are also large-scale rearrangement operations in genomes that largely act at a chromosomal level and lead to differences between species. As long as only one chromosome is affected, these are translocation (integrating some sequence fragment at another location) and reversal (the same as translocation, but with integration in the oppositely directed DNA strand). Fission and fusion operations enable the interchange of segment fragments between different chromosomes.

Genome rearrangements can be analyzed by comparing two or more genomes computationally. The knowledge of correspondences between parts in both genomes can help reveal the time sequence of rearrangement operations. This scenario is similar to that faced by sequence alignment algorithms (*see* Section 3.15.3.1), with the difference that the mutation, insertion, and deletion sequence operations have been changed to genome rearrangement operations. Again analogously, some costs are attributed to the genome rearrangement operations. Then, the sequence of operations with the lowest cost that transforms one genome sequence into the other is searched for. The respective algorithms make the assumption that rearrangements occur at the level of large conserved blocks.[64,65] This assumption may not be justified, however: recent comparative investigations between the genomes of the human and mouse have revealed that genome rearrangements occur on different scales. Intra-chromosomal micro-rearrangements affect sequence segments of less than 1 Mb in length; macro-rearrangements affect much longer sequence segments, and can occur inside a chromosome as well as between chromosomes.[64] The rearrangement problem is algorithmically more complex than that of sequence alignment.[65,66]

As the number of available mammalian genomes has increased rapidly in recent years, the idea to infer phylogenies from their pairwise rearrangement history has gained popularity.[67,68] Additionally, useful insights into the relationship between genome rearrangement and disease can be gained.[69]

3.15.4 Molecular Structure Prediction

The function of a molecule is tightly connected to its three-dimensional structure: proteins, RNA, and DNA interact with other molecules by binding to them in a predictable and often unique structural arrangement. Therefore, the knowledge of the three-dimensional structure of a molecule can lead to detailed understanding of its biological function. While DNA has a uniform double-helix structure, RNA and proteins can fold into much more diverse arrangements.

The technology of experimentally resolving molecular structures (*see* 3.20 Protein Crystallization; 3.21 Protein Crystallography) lags substantially behind that of uncovering molecular sequences. Therefore, the demand for reliable molecular structure prediction methods is high, especially for proteins. Unfortunately, current-day structure prediction methods fall short of our needs, especially with respect to their reliability. Methods for protein structure prediction are at a more advanced stage than those for RNA structure prediction. We will discuss both fields in turn.

3.15.4.1 Protein Structure Prediction

The prediction of the native structure of proteins from their amino acid sequence is one of the most popular topics in bioinformatics. As well as the methods that try to predict the complete three-dimensional structure of a protein from its sequence, there are many methods that aim at computing special structural features and properties of a protein. Every 2 years the Critical Assessment of Structure Prediction (CASP) international experiment is held, predicting protein structures that are unknown at the time of prediction but are being resolved.[70] Furthermore, there are servers on the Internet that continually assess and compare the performance of protein structure prediction methods.[71,72]

3.15.4.1.1 Secondary structure prediction of proteins

The three-dimensional structure of a protein contains three main motifs of secondary structure: α helices (H), β strands (E), and so-called coil regions (C), which belong to neither H nor E. Different types of helices may be distinguished (α helix, 3_{10} helix, etc.), just as several types of turns can be differentiated from the coil regions. Because these variants do not change the nature of the problem, they will not be discussed here.

Initial approaches to secondary structure prediction of proteins (from the 1960s) tried to predict the secondary structure of a single residue from its amino acid type alone. More advanced methods (from the 1970s) included the amino acid types of neighboring residues, but the prediction accuracy was still low (in the low 60% range). The most promising approach today is as follows. We first search for evolutionarily related proteins (homologs) in a sequence database. These sequences, including the query sequence, are then organized in a multiple alignment, from which a profile of the amino acid preferences at each alignment position is derived. In the final stage, the sequence positions of this profile are classified – neural networks[73,74] are a popular method of classification. In the early 1990s, the introduction of evolutionary information was the key component that boosted structure prediction accuracy above 70%.[74] Accuracy can now be increased further, to just below 80%, by running several methods and forming a consensus between them.[75–77] Secondary structure is used frequently as the input to tertiary structure prediction methods, or as a tool for validating tertiary structure prediction methods.

A somewhat simpler task is to assign the secondary structure motifs to residues of a resolved three-dimensional protein structure. Techniques that are able to do so can also be used to evaluate the accuracy of the secondary structure prediction methods. Due to differing assignments by tools such as DSSP[78] and STRIDE,[79] one cannot expect prediction accuracy to increase much above 90%.

3.15.4.1.2 Homology-based protein structure prediction

This approach makes use of the growing number of resolved protein structures. It is based on the observation that the three-dimensional structures of two proteins with similar amino acid sequences are also similar. It was observed in the early 1990s[80] that two proteins whose sequence similarity exceeds 20% have a very similar secondary structure (and, in fact, also a common fold) in the overwhelming number of cases. (There are notable exceptions, where the deletion of a single amino acid critically changes the fold.[81]) This observation suggests modeling the structure of a target protein by searching for a structure 'template' among similar protein sequences within the set of structurally resolved proteins. Then, the target sequence can be structurally aligned to the sequence of the template, that is, aligned in such a way that (mis-)matches pair residues with matching positions in the protein structures. The resulting alignment serves as a recipe for modeling the structure of the target protein.

The database of templates can be the complete PDB (database of all structurally resolved proteins (*see* 3.17 The Research Collaboratory for Structural Bioinformatics Protein Data Bank)) or, if runtime is to be saved, a subset of that set (e.g., a set of protein structures whose sequence have mutual similarities that do not exceed 40%). This approach has resulted in a database of 5700 template structures to date.

The template structure is found by aligning the target sequence to all sequences in the template database. The similarity of two sequences is measured by the score of a pairwise alignment (*see* Section 3.15.3.1 for more on sequence alignment). The scores for mutations are defined by a PSSM, which reflects the preferences of amino acids to occur in certain structural environments. This matrix is derived from a statistical analysis of the protein structure database. Since the selection of an appropriate template structure is most critical for the following modeling steps, any additional information that may improve the quality of the scoring function (and thus the selection of a protein template) is helpful.[82] Such information may stem from evolutionarily related proteins on both the query side and the template side from which a sequence profile can be derived or from secondary structure predictions. This alignment technique is also known as protein threading, since the sequence of the query protein is mapped to the backbone structure of the template, like pearls threaded onto a string.[83]

Aside from algorithmic aspects, the quality of the alignment also depends on the availability of sufficiently similar templates. The alignment process becomes increasingly unreliable if the similarity between the target and template sequences falls below about 40%. However, it is very accurate and robust when the similarity rises above 70%. The scoring function is of great importance: a faulty alignment leads invariably to a wrong protein model. Since the quality of the alignment and the applied scoring function are the most crucial items for the quality of the whole model, much effort has been spent on their improvement. This concerns both the functional form of the scoring function and its parameterization. As noted in Section 3.15.3.1, the scoring function is a sum of terms. More terms can afford more accurate scoring, in general. However, the terms are usually determined by a statistical fitting procedure that is based on the available structure data. If the number of terms to be fitted is too large for the volume of data on which the fitting procedure is based, unreliable scoring functions result. As the number of protein structure increases, refitting will yield more accurate scoring parameters. Thus, the quality of the alignment depends not only on the appropriateness of the algorithm but also on the availability of experimental data.

The higher the score the more attractive the template for use as a model in the structure prediction of the target. The structural alignment affords the mapping of the target residues onto three-dimensional coordinates of the template residues. In a subsequent step, the side chains of the template structure have to be exchanged by those of the target sequence. In the case of a match between amino acids, modeling is started with the conformer of the side chain in the template structure; in the case of a mismatch, the side chain of the amino acid of the template structure is replaced by a structural prototype of the side chain on the target side. This prototype can stem either from a rotamer library or be modeled by energy minimization, or by a combination of the two. There are homology-based protein modeling tools that combine these steps with a broad spectrum of algorithmic strategies such as discrete optimization, nonlinear optimization, and constraint programming.[84–86]

The interpretation of gaps in the alignment depends on the sequence in which they appear: if the template sequence contains gaps, the aligned residues of the query sequence cannot be matched to the template structure. This occurs frequently when loops of the query protein have no counterpart in the template protein. Conversely, if gaps appear in the query sequence, the aligned residues of the template structure will not be present in the final model, and so they rip apart the template structure.

The resulting structure model will contain inconsistencies such as wrong lengths and angles of bonds as well as clashes. A final refinement procedure performing energy minimization yields the resulting structure model. The final model can be submitted to plausibility checks for validation (*see* 3.23 Protein Three-Dimensional Structure Validation).

The whole modeling procedure critically depends on the quality of the threading step. Thus, it is this step that requires most rigorous evaluation. Here, the knowledge of the secondary structure can help: if the similarity between the secondary structure of the template and the query sequence is not high enough, the threading is questionable. Another important target for evaluation is the score of the alignment. This score itself is not very informative. We need a value that rates the confidence that we can have in the alignment. Often, confidence values are based on statistical significance, which rates how unlikely it is to obtain the alignment by chance (p values). But other, more heuristic choices of confidence values have proven effective as well.[87] Another method of measuring the quality of protein threading methods is to assess them in fold recognition benchmarks.[71,72] In this test the template structure that is most similar to the query sequence must be identified. The performance is measured by the number of correctly assigned folds or by rating the quality of the alignment on which the fold assignment is based.

Homology modeling methods are quite successful in protein structure predictions. Today, we can reach an accuracy of above 70% of correctly assigned folds over the whole protein structure database. The main limitation is that such methods can only rediscover already known folds and structural motifs. As we can assume to have uncovered only about a third of all protein structures used by nature,[88] this approach has definite limitations. This has stimulated efforts to enlarge the number of known protein structures.

Structural genomics initiatives[89] have been launched that aim at structurally resolving, in particular, those proteins that are theorized to fold into as yet unknown structures (*see* 3.25 Structural Genomics). The most ambitious structural genomics initiatives aim at complete coverage of the structural protein space. So far, the throughput of these methods is not yet high enough to consider high-throughput structure determination a reality.[90]

3.15.4.1.3 De novo protein structure prediction

In de novo structure prediction we attempt to model the protein without any template structure. Originally, approaches to this problem followed the physical route. Thermodynamically, the native structure of a protein is that in which an ensemble of protein molecules solved in water achieves the minimum free energy. The energy landscape spanned by the conformational space of the protein is exceedingly complex, and its global minimum is concealed among a large set of local minima with close to the same energy. Since, in addition, we do not have an accurate method of estimating the energy of a protein conformation, direct energy minimization is not practical. To date, only small stretches of chains (a couple of dozen residues) in small proteins have been modeled satisfactorily in isolated cases using methods based solely on energy minimization.

A second approach was developed in the late 1990s, when Baker and his group assembled native protein folds from structural templates of small protein fragments (nonapeptides) that stem from already resolved protein structures.[91,92] The assembly process is supported by an effective scoring function.[93] This ROSETTA procedure has advanced the field considerably, and has led to significant progress in recent CASP events.[94] It has also been applied to improving secondary structure predictions.[95]

3.15.4.2 Ribonucleic Acid Secondary Structure

As in proteins, the knowledge of the three-dimensional structure of an RNA molecule is of great benefit for the study of its interactions and function, and, again, the structure is hard to resolve experimentally. The prediction problem for the three-dimensional structure of RNA is quite similar to that of proteins: the sequences of RNA molecules are magnitudes easier to determine than their structures. However, because of the much smaller number of resolved RNA structures than resolved protein structures (less than 800 for RNA, more than 30 000 for proteins), homology-based modeling techniques are hampered by a lack of data. Thus, RNA structure prediction is still dominated by the prediction of its secondary structure.

The differences in the chemical nature of RNA and proteins result in different secondary structure motifs: the secondary structure elements of RNA are helical base parings (see **Figure 4**). These base pairings do not always follow the known Watson–Crick base pairs (A-U, C-G), but other pairings such as (U · · G, A · · G) or even (U · · U) are also observed. Base-pairing patterns usually assume a treelike secondary structure, as shown in **Figure 4**. In some cases the treelike structure is disturbed by a few pairings, the so-called pseudo-knots.[96,97]

The approaches to secondary prediction in RNA use two main strategies. In comparative sequence analysis, homologous RNA sequences are searched for and a multiple alignment is created. From the alignment, correlated mutations that preserve the presence of putative base pairings can be identified. Based on this analysis, we can confirm base pairings in the secondary structure of the target RNA. This technique was traditionally executed manually, but, recently, computer algorithms have been developed to extract the secondary structure from the multiple alignment.[98–100]

The other approach is based on energy rules. Local features such as pairings of Watson–Crick base pairs or different kinds of unpaired stretches of sequences called loops add terms to a sum that estimates the free energy. The respective energetic contributions of base pairs and loops are deduced from experiments and simple model assumptions. Then, a combinatorial optimization algorithm searches for the RNA secondary structure with lowest energy. The optimization can be done efficiently using dynamic programming, if we assume that the RNA structure is treelike, as described above. This method has been pioneered by Zuker and Stiegler,[101] and it is quite popular, since it lends itself to many generalizations. One can select different trade-offs between the complexity of the energy model and the runtime of the minimization algorithm.[102] One can compute not only the optimal but also a set of near-optimal secondary structures, and display them together in a so-called energy dot plot in order to identify the putatively most stable regions of the structure.[103] Finally, since the energy contributions from local structure elements can be chosen to be temperature-dependent, one can assemble a 'movie' of the energetically most favorable RNA structures as the temperature rises (or falls). Also, thermodynamic quantities such as the heat capacity can be computed with the same efficient dynamic programming approach.[104] All of these methods disregard pseudo-knots, however. These have to be factored in by special algorithms that make the computation more expensive.[105,106]

3.15.5 Molecular Networks

The analysis of molecular interactions (protein–small molecule, protein–protein, protein–DNA, etc.) is a central issue in bioinformatics. Molecular interactions form the basic elements in a complex biochemical circuitry that motorizes the living cell. It is important to understand the single interactions. However, a coherent picture of the workings of a cell can only come about through the understanding of the networks formed by many of these interactions. The development of appropriate models for biochemical networks is one of the tasks of the emerging field of systems biology.[107] Understanding the molecular networks in a cell requires the integration of a variety and large volume of experimental data into a unified framework.

This section will focus on two network classes, metabolic and regulatory networks. This classification is reasonable since both kinds of networks have their own databases and analysis procedures. A more general view of protein interaction networks is afforded by cell-wide protein interaction screening experiments. This topic is taken up in Section 3.15.7.3.

3.15.5.1 Metabolic Networks

The metabolic network is made up by all chemical reactions that involve the metabolism of small molecules (called metabolites) with the help of catalytic proteins (called enzymes). A metabolic pathway is a series of successive reactions that changes one or more educts into one or more products. Whenever an educt and a product of a pathway are the same kind of molecule, this pathway is called a metabolic cycle. Along a pathway, the metabolites are either combined to form larger molecules (anabolism) or they are degraded to smaller molecules (catabolism). Metabolic pathways can also serve to store energy or to release usable energy.

Metabolic networks are composed of quite homogeneous building blocks such as the one shown in **Figure 5**. The educt and product are represented by vertices, and the connecting reaction by an edge that is labeled with the catalyzing enzyme, and possibly augmented with cofactors. This representation only describes the connectivity (topology) of a reaction. For simulation we need to know the reaction kinetics, that is, the concentration gradients of the involved molecular partners in terms of their starting concentrations. While the connectivity of many reactions is known today, their kinetics are sorely missing data for most reactions.

Reactions are the building blocks for complex metabolic networks, such as the one shown in **Figure 6**. Since the structure of each reaction is like any other, these networks are homogeneous. This greatly simplifies their analysis. The application of general network analysis techniques to these networks affords insights into the organizational principles of the cell.[108]

Metabolic network databases include organism-specific databases such as EcoCyc[109] (for *E. coli*), HinCyc[110] (for *H. influenzae*), and PseudoCyc[111] (for *Pseudomonas aeruginosa*), and databases embracing a number of various organisms, even ranging over different kingdoms of life, such as KEGG,[112] MPW,[113] and MetaCyc.[114] The metabolic pathway is the basic descriptive unit of these databases. The databases provide data records and images that describe such pathways and their interconnections. Increasingly, the images can be processed by computer. Another group of databases concentrates on a single reaction, such as BRENDA,[115] ENZYME,[116] and the integrated effort IntEnz.[117] All metabolic databases provide their own sets of tools for querying their data, enabling one to navigate through the search results and to visualize them. These databases are comprehensive enough to afford comparisons of the topologies of

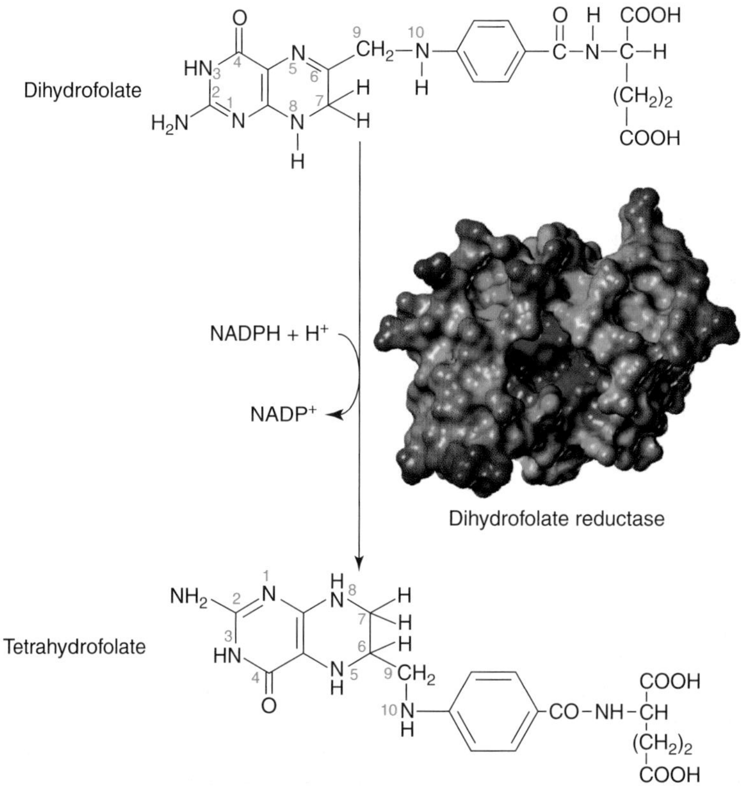

Figure 5 A metabolic reaction. Educt: dihydrofolate. Product: tetrahydrofolate. Enzyme: dihydrofolate reductase (DHFR). This is the reaction blocked by the inhibitor bound to DHFR in **Figure 6**.

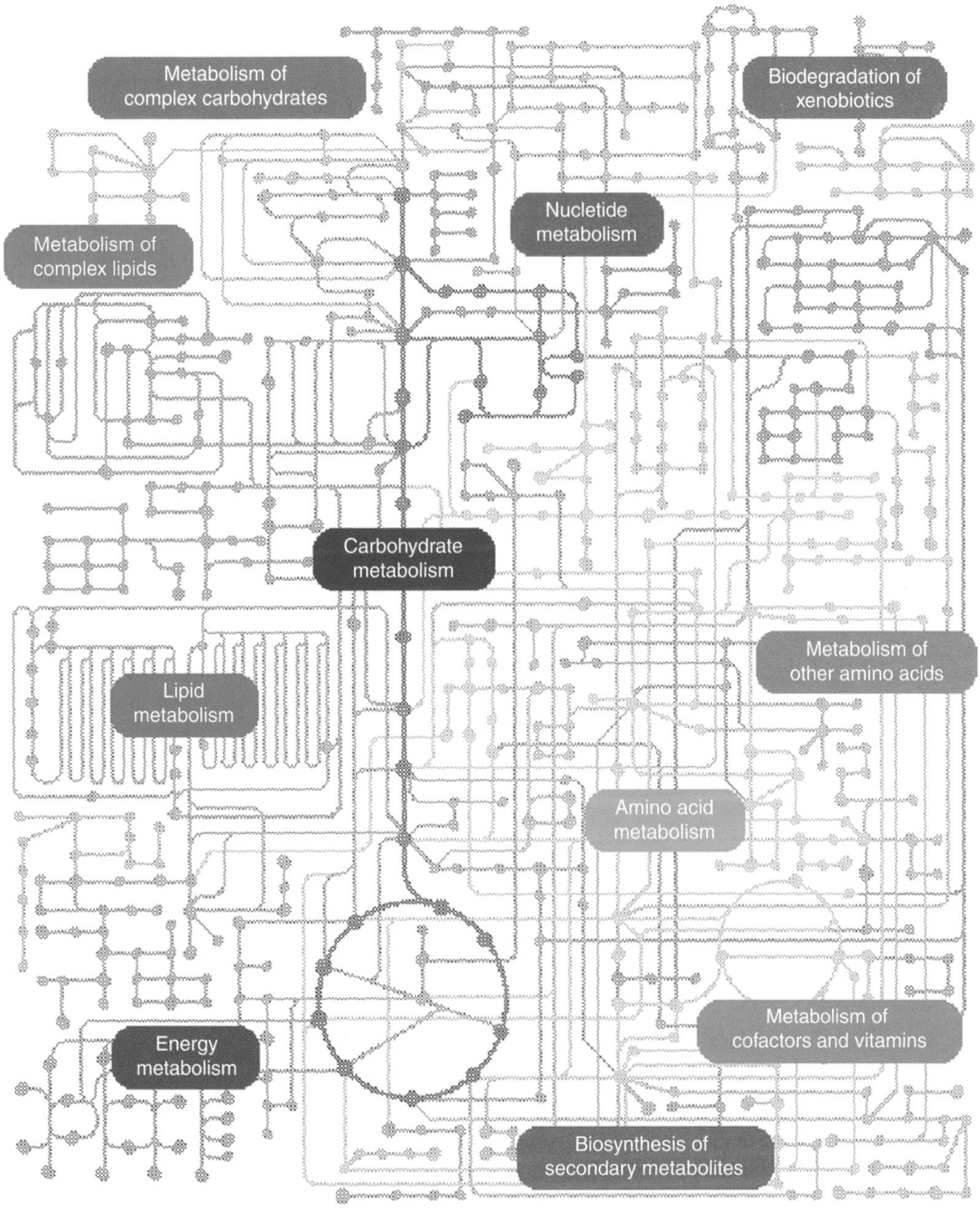

Figure 6 Schematic diagram of part of the metabolic network of *E. coli*.

metabolic pathways in different species. If the genome sequence is complete, we can even reason about the absence of enzymes and thus gaps in pathways.[118]

One computational method of analyzing the steady state properties of metabolic networks is the flux balance analysis method. This approach uses linear optimization to assign flux values to the edges of a metabolic network that maximize a given objective function.[119] A function of interest is growth per substrate uptake, for example.[120,121] In the metabolic flux analysis method, the pathways are decomposed into the so-called elementary flux modes.[122,123] These are the minimal components of a pathway that can act coherently and independently under steady state conditions. The programs METATOOL[124] and GEPASI[125] can perform such analyses on medium-sized pathways. Recently, cell-wide steady state models of metabolic networks for *E. coli* and *S. cerevisiae* have been published.[126]

The steady state of a cell applies to only a restricted set of simulation scenarios. Whenever it is dropped, one has to rely on more general analysis methods that are based on ordinary differential equations that model the reaction kinetics in the pathway under consideration. Much effort has been expended on the development of appropriate simulation tools.[127–129] These tools usually require the network kinetics to be completely described, but the relevant data are missing in many cases. Large metabolite screening projects aim at measuring the concentrations of metabolites in a cell-wide fashion.[130–132]

3.15.5.2 Regulatory Networks

The regulatory network is responsible for intra- and intercellular communication. Like the metabolic network, it is based on molecular interactions, but here its role is not to turn around matter but to control cellular processes or transmit information (*see* 3.10 Signaling Chains). The signal transduction pathways and protein interaction networks that control the transcription of genes make up two crucial parts of the regulatory network.[133] Generally, its structure is more inhomogeneous, since there are many important ways in which a molecule can affect another (agonistic, antagonistic, activating, deactivating, inhibitory, etc.). This results in very complex notation when representing regulatory relationships, for which a general standard is not yet in sight. The greater part of our knowledge on regulatory pathways is still buried in the literature, that is, it is not in a computer-digestible form. Several attempts are being made to systematically mine the literature with automated methods in order to extract this knowledge. The results of these efforts suffer from the unsystematic notation and ambiguities that arise when text generated manually is parsed; thus, they should be used with the greatest caution (*see* Section 3.15.7.6).

There are a number of databases of regulatory networks, but they are less comprehensive than the well-developed metabolic databases. Whereas regulatory databases contain only a few thousand genes, metabolic databases contain well over a 100 000 genes. TRANSPATH[134] is a prominent database of signal transduction pathways. The database provides tools for querying and as well as an integrated tool for the analysis of microarray data in the context of the regulatory networks (*see* Section 3.15.6). The databases DIP[135] and BIND[136] center on general protein interactions and also describe molecules in terms of pathways and provide reaction maps.

One way to analyze regulatory networks is to use the same framework of differential equations that is used for the nonstationary analysis of metabolic networks. These methods may produce accurate results, presuming that sufficient data are available. As in metabolic networks, this approach lacks the model hierarchy that biologists usually attribute to biochemical networks. The database DOQCS[137] provides models of signal pathways that also hold the kinetics of the associated chemical reactions at various levels of detail. Qualitative simulation methods reduce the general differential equations model to simpler functional forms[138] or even to discrete models.[139] A more challenging approach is to build regulatory networks with (partially) unknown relationships and pathways directly from experimental data. One promising route for this is to use mRNA expression data (*see* Section 3.15.6).

3.15.5.3 The Virtual Cell

The vision of systems biology is to synthesize all cellular molecular networks together in unified model of the whole cell, and thereby predict the behavior of cells.[140] The E-Cell Project provides a software suite allowing for the simulation of in silico models of the cell.[141,142] The project is embedded into a large metabolite screening project that is responsible for generating the necessary data for the simulation.[143] Another virtual cell project, Genomic Object Net, is based on a combination of discrete and continuous simulation methods.[144]

3.15.6 Analysis of Expression Data

The measurement of the profile of transcribed genes in a cell provides information that exceeds the information from the genome sequence. Whereas all cells have the same genome, cells differ in their expression profile, since the expression of genes changes in response to the needs of the cell and its functional role in the whole organism. Thus, the expression profile affords a distinction between different cell states within the same tissue (e.g., healthy/diseased) and between different tissues. When the first technologies for measuring transcript expression levels were developed in the early 1990s, they bore the hope of being the source for completely new approaches toward diagnostics and therapy of diseases. Since then, it has become clear that things are not as simple as they first seemed.

In principle, expression profiles can be measured at different levels, either with the mRNA transcript (gene expression – transcriptomics) or at the level of the synthesized product, the posttranslationally modified protein

(protein expression – proteomics). The former is much easier to do, since mRNA transcripts can be backtranslated into DNA (cDNA), which is very uniform and easy to handle by laboratory procedures. In contrast, proteins are highly diverse and harder to handle uniformly in a protein expression measurement. On the other hand, the measurement of the mRNA expression level only affords indirect information on the abundance of the finished protein, which, in the end, is performing the molecular function. The correlation between mRNA expression levels and protein expression levels is reduced by the regulation of protein translation and posttranslational modification as well as the degradation of proteins in the cell. Still, mRNA expression data afford an interesting molecular census of the cell that holds important information for classifying cell types and learning about gene function. Expression data were first measured on the basis of expressed sequence tags (ESTs),[145] but now mainly at the level of cDNA microarrays.[146]

The advent of cDNA microarrays in the late 1990s has created a large surge in statistics and bioinformatics activity targeted to both configuring the arrays and interpreting the resulting data. Despite and partly because of this progress, the limits of microarray measurements are becoming more and more apparent. The high hopes of making protein function accessible directly through microarray measurements could not be fulfilled. However, expression measurements do support diagnostic procedures, and can, in concert with other biological data, shed light on the biological role of genes and proteins.

3.15.6.1 The Complementary Deoxyribonucleic Acid Microarray

In this short summary, we focus on the technical aspects of the cDNA microarray (or DNA chip), insofar as they are relevant to bioinformatics (*see* 3.01 Genomics; 3.05 Microarrays). A cDNA microarray consists of an arrangement of complementary sequence probes for relevant genes, arrayed and bound to a glass or nylon plate. The number of nucleotides in the probes ranges from 25 across and about 70 up to full-length genes. The cell under examination is lysed, and its mRNA is extracted. Then, the mRNA is translated back to cDNA, and labeled radioactively or fluorescently for detection. The cDNA is then run over a microarray, and hybridizes to the respective probes that are affixed to the microarray. Depending on the labeling method, a suitable device measures the quantity of hybridized cDNA in each array cell. The detected signal levels provide the raw data for analysis and interpretation.

Current cDNA microarrays can hold as many as 10 000 probes. Thus, one can screen for genes that are most likely responsible for a common disease (heart chip, cancer chip, etc.). In addition to the one-channel arrays described here,[147] there are also two-channel arrays that allow for hybridizing cDNA from two different cell lysates, each colored differently, in the same experiment.[148] The measurement produces a ratio profile of the two cells. Several companies have proprietary microarray technologies. A widely used protocol for making two-channel arrays is provided for free by the Pat Brown Lab at Stanford University.

3.15.6.2 Configuration of Experiments and Low-Level Analysis

The interpretability and general worth of microarray data sensitively depends on the appropriate configuration of the microarray experiment. This demand had been underestimated in the early years of this technology, but, nowadays, the importance and difficulty of the consistent preparation of a microarray experiment has gained widespread attention.[149] Strict quality control of the purified RNA populations, the proper selection and configuration of the probes on the chip,[150] an adequate number of replications of the experiment,[151] and careful logging of the experiment are vital requirements. For this purpose, the MIAME standard (minimum information about a microarray experiment) has been defined.[152] The public database ArrayExpress of gene expression data, which follows this standard, contains more than 650 experiments.[153] The National Center for Computational Biology provides another database of expression data, the Gene Expression Omnibus (GEO).[154]

The configuration of a microarray experiment has an immediate effect on the subsequent low-level analysis of the generated data.[155,156] The purpose of this procedure is to turn the raw data (luminance values of radioactivity or fluorescence) into scalar quantities that represent the expression levels of the measured transcripts. For example, the expression level may be defined as the number of RNA copies per cell or in terms of fold changes, based on some reference profile. This normalization step sounds easier than it actually is, since the errors from technological and biological sources introduce variance in expression levels that has to be incorporated and appropriately modeled. The choice of the appropriate statistical method contributes critically to the quality of the low-level analysis, and quite a few normalization methods tailored to low-level processing of expression data have been published. Overviews of basic normalization methods[157,158] and more advanced methods[159,160] are available.

The low-level analysis produces an expression data matrix of normalized expression levels with the dimensions No. of probes $\times$ No. of experimental samples. Here, the number of genes (probes) usually exceeds the number of experimental samples by far, roughly several tens of thousands of probes as opposed to only a few dozen samples. This off-square shape of the expression data matrix hampers the analysis of expression data.

On the basis of the different expression levels measured in a set of experiments, the low-level analysis finally has to determine a set of genes whose expression changes significantly between the experiments – more so than expected by chance. The question of what is significant in this context has to be answered in terms of a model of background variation specific to the gene and the experiment.

3.15.6.3 Classification of Samples

The interpretation of the data in the expression data matrix can be decomposed roughly into classifying columns and rows. Each column represents a sample. Thus, classifying columns means clustering the samples into groups that represent cell conditions, disease types, and the like. Clearly, this procedure supports diagnosis and prognosis if we are talking about disease chips. As soon as microarray technology matured, it was applied to many clinically relevant diseases, especially cancer. An early study attempted to distinguish two types of leukemia, acute myeloid leukaemia and acute lymphoblastic leukemia, because the associated tumors are hard to classify with histological methods.[161] Since then, there have been many other studies on different types of cancer.

Many sample classification methods for expression profiles are available, and the choice depends on the availability of labeled data. These data are expression profiles from samples that have been confirmed to originate from cells that are in a known disease state. If these data are available, a statistical classifier is trained on them, such that future samples can be reliably assigned to disease states. Such a classifier may be based on decision trees, neural networks, support vector machines, or other statistical learning technologies.[162,163] The success of this approach depends on the disease under investigation: For certain diseases, the classification accuracy can reach 90% or more. This has been substantiated quite impressively on a wide variety of cancers.[164] The large number of genes in expression profiles cannot be handled effectively by supervised learning methods. One has to reduce the expression profile by special feature selection techniques, that is, select a subset of all genes that is especially indicative of the type of disease. Owing to the various correlations in gene expression, there are usually a few dozen genes that allow for much more efficient classification with nearly unchanged accuracy. Often the accuracy even increases, since the selected subset of genes masks some variance in the data set. The role in classification also makes these genes especially suspicious for being important actors in the disease process. Since feature selection concentrates on a set of genes, it is a more general approach for determining genes that are relevant for a disease than detecting differentially expressed genes, which takes only single genes into account. If no labeled data are accessible, for example when discovering new types or states of diseases, unsupervised methods must be used. These methods cluster the expression profiles of the samples into groups that may correspond to that of known disease types or to new ones.[165]

3.15.6.4 Classification of Probes

The classification of probes aims at finding those genes that are relevant for the course of a disease. In contrast to the former classification problem, this task is much more difficult and error prone when only starting from differential expression profiles. The number of candidate genes is usually much larger than the number of available samples; thus, it is very hard to acquire detailed information on the role of a gene in the disease. However, there are approaches that try to address this challenge.

One early approach tried to cluster similar expression differentials of genes. It was hoped that such co-expressed genes might have functional association.[166] However, it soon became apparent that such a procedure is quite inaccurate. Hence, additional information was searched for to improve this approach. The combined analysis of gene expression data together with the upstream regulatory regions of the genes is promising, since these regions control transcriptional regulation. If co-expressed genes share functional sites in their upstream regulatory regions, then their transcription should be controlled by the same regulatory regime, which points to a functional relationship between these genes.[167–170] Recently, this idea has been extended by including an explicit model of the involved regulators.[171]

In addition to regulatory information, other observations on genes can provide hypotheses on functional dependencies that may be confirmed by the expression data. The mutual membership of genes in a metabolic or regulatory pathway is one such source. One can create and assess the differential expression of all genes in a pathway simultaneously. If the change in expression level is significant in the statistical sense, this can be an indication that the disease process interacts with this pathway in some way. This model can then be used to test other genes with unknown function. If the test result is significant, it indicates a functional relationship of the genes with the pathway.[172,173]

There are other data sources that can provide useful information on functional relationships for this approach: choices are the vast scientific literature (*see* Section 3.15.7.6) or protein interaction data (*see* Section 3.15.7.3). Protein interaction data have been analyzed together with expression data to reveal molecular pathways.[174,175] Here, it is assumed that the proteins associated with genes that share a similar expression profile are likely to interact. All methods presented here make use of statistical learning techniques. Bayesian networks are a general tool for integrating heterogeneous sources of information, but Boolean networks and graph models are also applicable. The relevant statistical learning procedures can produce regulatory networks that may be quite abstract and not closely linked to the molecular regulation process.[176] Therefore, their interpretation presents certain problems.

3.15.6.5 Beyond cDNA

We mentioned above that protein expression measurements may afford more direct information on the functional role of the involved proteins inside the cell. Protein expression can be measured with a combination of a separation technique (gels, columns) and mass spectrometry to identify the expressed proteins and their abundance[177] (*see* 3.02 Proteomics). Peptide and protein arrays aim at detecting proteins, their expression levels, or some aspects of their functionality, directly. These chips are quite similar to cDNA arrays, but functional proteins are bound to the chip instead of nucleotide probes. The chip is exposed to a certain ligand, and the proteins are tested with respect to their ability to bind to this ligand. The same principle is applied conversely, by affixing a set of ligands to the chip and then exposing them to a protein solution. These techniques are more difficult to apply than cDNA analysis, since proteins may denature while affixing them to the array, and the binding ability of ligands may suffer from the loss in flexibility resulting from being fixed to the array.[178–180] The new emerging tissue, cell, and antibody microarrays all come with their own special experimental procedural requirements.[181,182] Eventually, all these microarray experiments will produce matrices of signal strengths, and all bioinformatics techniques used in cDNA microarray analysis should be extendible to the low-level output of different microarrays.

3.15.7 Protein Function Prediction

The knowledge of the function of proteins is the most fundamental prerequisite for understanding the biological processes in a cell or organism. Whereas the genomes of a growing number of species have been or are being completely sequenced (around 1000 today), information on the function of their proteins is still limited, often partial, inaccurate, or even plainly wrong. This section concentrates on the various bioinformatics approaches that aim at narrowing the gap between the knowledge on protein sequences and protein function. There are several recent reviews focusing on this topic.[183,184]

3.15.7.1 What is Protein Function?

What we perceive to be the function of a protein depends on the particular context. From the biochemist's point of view, the function of a protein refers to its binding partners, the respective modes of binding, and the reactions it may catalyze. From the biologist's point of view, the role of the protein in a complex biological process (cell cycle, apoptosis, etc.) or even its localization inside the cell (cytosolic, membrane standing, extracellular, etc.) is an attribute of its function. The different meanings that the notion 'protein function' can acquire in different contexts calls for a suitable ontology, that is, a structured vocabulary for talking about protein function.[185]

Here we present some of the work done to establish a unifying ontology for protein function.[186] The first approach is the classification of enzymes by the Enzyme Commission.[187] This notational scheme classifies enzymes hierarchically by the reactions that they catalyze. The location in the hierarchy is represented by the so-called EC number, a four-level numerical code (similar to IP numbers on the internet) in which each successive number homes in on the catalyzed reaction to a growing level of detail. Invented more recently, the MIPS Functional Catalog contains a larger variety of protein function categories.[188] Originally defined for handling functional annotations of the yeast genome, the MIPS Functional Catalog has been broadened to apply a wide range of organisms. Like the EC numbers, the MIPS categories are organized in a hierarchical treelike structure. However, the MIPS catalog has six levels, whereas the EC uses only four levels. One of the most extensive efforts in establishing an ontology for protein function was undertaken by the international Gene Ontology Consortium, which maintains the Gene Ontology (GO).[189] Similar to the approaches described above, the GO is organized hierarchically, but, here, there are three separate hierarchies representing three fundamentally different views on protein function: cellular component, molecular function, and biological process. Furthermore, the GO categories can belong to several supercategories, for instance, the category 'cell growth and/or cell maintenance' refines the two categories 'cellular process' and 'physiological processes.' This generalizes the treelike structure of the hierarchy to that of a directed acyclic graph.

Such ontologies are valuable tools for categorizing proteins by their function. They are under continuing development, and, to a growing extent, functional annotations of genes and proteins are being linked to the ontologies in databases or in the literature. These annotations provide an essential basis for today's protein function prediction programs. But when dealing with phenomena such as functional relationships between proteins, multiple protein functions, and function variability of a protein under different physiological conditions, a protein function annotation guide is necessary, in addition to the largely context-free hierarchical ontology. Although the GO provides a basic annotation guide with its ontology, the need for a highly expressive, consistent, and standardized framework for annotating protein function still remains. As the first larger regulatory networks are being understood,[190] initial efforts are being undertaken to develop the expressive tools for this purpose. An example of an application in a supracellular scenario has also been presented.[191]

3.15.7.2 Function from Sequence

The structure of a protein is mainly determined by the sequence of its coding gene. Since the structure of a protein determines its function, the function of a protein must be encoded largely by the sequence of its gene. This observation directly leads to the conclusion that two genes that are similar in sequence should code for proteins with similar function. In many cases, this conclusion is applicable to the prediction of protein function from sequence, especially to those parts of the protein that are most relevant to its function. On the other hand, there are pairs of proteins whose sequences are very similar but their function is significantly different. As a rule, sequence similarity originates from a common evolutionary origin. Evolutionarily related genes or proteins are also called homologous. Homologous genes can occur in different or the same species. As a gene is handed down from generation to generation, its sequence may change slightly, whereas the function of the gene remains the same. Such ancestor–descendant relationships between genes are denoted by the term 'orthology.' Thus, orthologous genes are found in different species that have a common ancestor in which the genes originate from the same ancestral gene. In general, gene function is maintained in orthologous genes. However, sequence similarity can also arise due to the duplication of a gene within the same organism. A gene duplication results in two identical copies of the gene within the same genome. One of the copies is submitted to the selective pressure to maintain the original function of the gene. The second copy can acquire a new function. Such genes that have a common evolutionary origin but are separated by a gene duplication are called paralogous. It is hard to determine from sequence similarity alone whether two genes are homologous or paralogous, and, in fact, the exact definitions of orthology and paralogy are still subject to debate.[192] Since orthologous proteins are likely to have the same function, the detection of orthology is one of the mechanisms used to infer protein function from sequence. The reporting of function due to paralogous gene pairs leads to false predictions in this context. It is obvious that a correct computational categorization of homology in orthology and paralogy requires the application of phylogenetics. There have been approaches to identify gene duplications in the ancestral history of the involved species, but they are mathematically complex and computationally demanding.[193,194] More practical is the use of heuristics, such as the program Orthostrapper, which only analyzes single pairs of genes and does not aspire to map complete phylogenies.[195]

As an approximation to phylogenetic analysis of orthology, one can use whole-genome sequence data, and reduce the concept of orthology to a pure sequence similarity criterion. Specifically, if two proteins in two different genomes retrieve each other as the top hit in a sequence database search (e.g., using BLAST) of the respective genome, then there is a good chance that they are orthologous. One can extend this analysis to larger groups of genes. A set of genes, each of them in a different species, which point to each other as the top hit in a BLAST search, forms a so-called cluster of orthologous genes (COGs). A database of COGs has been developed.[196,197] Since such an approach foregoes all phylogenetic analysis it is likely to produce quite a number of false-positive hits. Hence, the resulting COG database has to be carefully curated manually. Another protein database that provides orthology information based on this model is SMART.[198]

The application of the concept of orthology is not limited to complete gene or protein sequences. Only a fraction of the sequence of a gene actually codes for those elements of a protein that are characteristic for a specific function. This affords an approach to function prediction by simply searching the protein for such short patterns of sequence that are indicative of a certain function, the so-called sequence motifs. For example, a well-known motif for an ATP-binding site is the P loop. Its characterizing sequence fragment has length 8, respecting the pattern [AG]-x(4)-G-K-[ST]. This motif pattern defines positional constraints for an amino acid sequence: the single upper-case letters denote unique residues according to the single-letter code for amino acids. The letters in square brackets designate the respective choices of one amino acid from two (or more, in general) available alternatives, and x(4) represents an insertion of four arbitrary adjacent amino acid residues at the respective position. Such motifs are typically retrieved from multiple sequence alignments (*see* Section 3.15.3.1) of protein sequences that are known to share a common function. From such an alignment, the highly conserved regions of the sequences are selected to define the characteristic positional constraints of the motif by consensus. The first database utilizing sequence motifs for the representation of functional sites of proteins was PROSITE.[199,200] Whereas motif generation started exclusively manually, methods have been developed to derive them automatically.[201,202] The weighting of sensitivity against specificity is a central issue in motif generation: a strict motif is likely to reject true positives – proteins that have the function represented by the motif. On the other hand, an imprecise motif may accept many false positives: protein sequences that contain the motif but do not have the respective function. In general, different motifs can be combined for a search, in order to ensure that most of the true positives comprise at least one of these motifs. The EMOTIF Database[203] contains motifs generated with this approach.

PSSMs or HMMs (*see* Sections 3.15.3.1, 3.15.3.3, 3.15.3.4, and 3.15.4.1.2) are other approaches for characterizing conserved regions in protein sequences. They are more suitable for longer sequence fragments that span segments up to complete protein domains, and they can be generated easily on the basis of multiple alignments.[204] Several databases that make use of PSSMs and HMMs in this context exist. Several motif and domain databases have been integrated in the domain database InterPro.[205]

Motif and domain databases are complemented by supervised learning methods, primarily based on neural networks. They are able to predict many aspects of protein function; for example, cellular localization (through the analysis of signal peptides and the prediction of transmembrane helices) and posttranslational modification features (glycosylation and phosphorylation sites). The Center for Biological Sequence Analysis at the Technical University of Denmark, Lyngby, maintains a server that offers many such methods, and integrates them in the ProFun method. This method classifies proteins according to their predicted function (e.g., enzyme class or participation in a biological process such as amino acid biosynthesis or energy metabolism). The overall functional prediction is made on the basis of a large number of methods analyzing or predicting features of the protein that can be derived directly from the sequence, simple ones that can be computed directly (e.g., sequence length, charge and amino acid composition), and more complex ones that need to be predicted (e.g., secondary structure, signal peptides, and sites for posttranslational modifications). The predictors for some function classes reach high rates of accuracy, in some cases up to 90%.[206]

3.15.7.3 Protein Interaction Networks

Protein–protein interactions make up an important part of all intracellular interactions. In the late 1990s, assays for measuring protein interaction data were developed by scaling of the yeast-two-hybrid (Y2H) method[207] to cover substantial parts of a genome.[208] In principle, these procedures generate a binary matrix with the dimensions No. of proteins × No. of proteins defining, for each pair of proteins, whether they interact or not. Another procedure, named tandem-affinity purification, can to latch onto a protein and pull out with it a whole attached complex of proteins. The result is a set of protein complexes instead of a binary matrix.[209] Bioinformatics can utilize these data for the analysis of cell-wide protein networks.[210,211]

Both procedures generate interaction data that are contaminated by a significant number of false negatives and false positives. One explanation for this is that laboratory procedures inadequately imitate the natural cellular conditions, which biases protein binding. For instance, laboratory procedures are notorious for being unable to differentiate between binding events in different physiological states to distinguish between transient binding partners and those that bind to each other over longer time periods. Still, the resulting data are used as a basis for the construction and bioinformatics analysis of protein interaction networks,[212–215] These methods mainly follow the unsupervised learning paradigm (i.e., they cluster the data). The resulting clusters represent putative groups of functionally related proteins.[216] This 'guilt by association' approach can also point to interesting drug targets.[217] The global topology of the constructed interaction networks is a target of fundamental studies.[218,219] In the face of the noise present in protein interaction data, the combined analysis of protein interaction data and other protein function data appears most promising (*see* Section 3.15.7.7). The amount of publicly available protein interaction data in repositories is growing significantly.[220]

3.15.7.4 Genomic Context Methods

The presence of a growing number of completely sequenced genomes has afforded entirely new ways of approaching the quest for elucidating protein function. The new power afforded by complete genomes is that one can reason not only about the presence of a protein but also about its absence. Analysis of patterns of groups of genes occurring in different organisms by so-called genomic context methods is a powerful tool for obtaining information on protein function. In order to discern functional associations in proteins on the basis of genome sequences, approximately 30 completely sequenced genomes are required. Today, this number has been reached for prokaryotes, but not for higher organisms such as mammals. The database STRING provides predictions of the functional association between proteins that has been derived with genomic context methods among others.[221]

Genomic context methods usually reveal functional associations between proteins that are more general than physical binding. Additional functional associations include taking part in the same biological process (cell cycle, apoptosis, etc.) or cooperating in producing a certain phenotype (genetic association). Several genomic context methods are briefly discussed below.

3.15.7.4.1 Gene neighborhood and gene order

When two genes in two species occur in close proximity in both genomes, they are likely to be functionally associated. This evidence becomes stronger if the number of involved species increases.[222] The respective order of the genes along the genome may also be used for the analysis.[223] There are approaches that generalize this method by revealing general conservation patterns that present cycles of association that alternate between homology and neighborhood.[224,225]

3.15.7.4.2 Domain fusion

This method utilizes the observation that two protein are likely to be binding partners if their genes are fused in another species.[226,227]

3.15.7.4.3 Phylogenetic profiles

The concurrent presence or absence of two proteins in a species may provide evidence for their functional association.[228] The corresponding comparison of proteins from different species has been based originally on a binary classification of protein pairs in orthologous and not orthologous. This approach has been improved through the incorporation of gradual levels of evolutionary distance. STRING and PLEX are servers that provide analysis of protein function based on phylogenetic profiles.

In general, the effectiveness of genomic context methods is much higher for prokaryotes than for eukaryotes. This follows not only from fewer fully sequenced eukaryotic genomes but also from the organization of genes in prokaryotes (e.g., their collection in operons), which is more suitable for the application of genomic context methods.

3.15.7.5 Function from Structure

The structure of a protein determines the interactions that it can perform with other molecules. These interactions are the basis of the functional role of the protein. Methods that deduce protein function directly from structure are still evolving. Automating this process is quite difficult: on the one hand, the same fold – originating from gene duplication or convergent evolution – may realize different functions. On the other hand, the same function (e.g., catalysis of a specific reaction) can be attained with various protein folds.[229] This ambiguity can be eliminated with the help of detailed orthology analysis (*see* Section 3.15.7.2). Furthermore, remote homologies can suggest a functional relationship. For example, two remotely homologous enzymes can share the same reaction mechanism, while differing in substrate specificity.[230]

The comparison of protein structures provides the methodical basis for analyzing the similarities and evolutionary relationships of proteins in terms of structure, similar to alignment methods that analyze the evolutionary relationship of proteins in terms of sequence. The conventional approach to structurally comparing proteins is to superpose their (rigid) structures. At first, the two protein chains are aligned structurally (i.e., pairs of amino acids, each from one of the proteins, are matched if they occupy corresponding spatial locations in the protein structures. In principle, the algorithms used for sequence alignment (*see* Section 3.15.3.1) are also applicable to this task. However, the scoring function has to be adapted, since now it no longer measures sequence evolution but structural similarity. The simplifying assumption of sequence alignment, dealing with the residues as independent from their neighbors in the sequence, cannot be made either, since residues that are adjacent in space influence each other.

There are several automatic structural superposition methods that were used to create structural protein classification databases. The database CATH[231] provides one of the most popular structural classifications of proteins; it has been derived automatically and uses the structural superposition method SSAP.[232,233] The only structural protein classification of comparable popularity is SCOP. This database actually achieves a higher classification consistency, because it is curated manually.[234] Other protein structure superposition methods include DALI/FSSP,[235,236] CE,[237,238] and FATCAT.[239] New algorithms also take the structural flexibility of proteins into account.[240] As in sequence alignment, it is also possible to multiply align protein structures.[241] Recent reviews provide comprehensive overviews of the available methods.[242,243]

The structural similarity of functionally related proteins allows for the derivation of structural motifs from their (structural) alignments; the most prominent motifs are specific ligand-binding sites. Various algorithmic methods for detecting such motifs have been developed.[244–248] Attempts to calculate the statistical significance for the occurrence of a motif have also been made, since this is a critical issue in structural comparison, as it is in sequence search.[246,249]

Since the functional sites must be accessible for interaction partners, they are usually located at the protein surface. Therefore, the analysis of the protein surface may help identify functional sites – a significant contribution toward elucidating protein function. Early methods for searching for functional sites identify geometric features such as clefts and cavities along the protein surface.[250] This approach has been extended to considering the physicochemical properties of the protein surface in order to improve prediction accuracy.[251–253] Other approaches utilize the evolutionary conservation of functional residues,[254–256] and Jones and Thornton provide an overview of recent developments.[257] The rating of residue conservation among homologous proteins requires an accurate molecular clock measuring the speed of evolution.[258] The ConSurf server provides an interface for an online annotation of the protein structure with the residue conversation that is calculated among the protein and its homologous.[259]

3.15.7.6 Text Mining

The literature is the most comprehensive source of information on protein function. However, this information is not easily accessible by computer. Text mining is the computer-based approach used to harvest this information. The two major classes of text-mining methods address the problems of information retrieval and information extraction, respectively. Information retrieval calls for the selection of documents according to some user-defined criteria. In contrast, information extraction accesses pieces of information comprising prespecified types of events, entities, or relationships from documents. Information extraction is much harder than information retrieval. Early experience in text mining has been gathered mainly in the newswire domain, but the insights gained there are not directly applicable to text mining in the protein function context. The main reason for this is that articles published in the newswire domain usually target a general audience, whereas biomedical articles usually aim at a small community of domain experts, and, thus, are more difficult to interpret. Current efforts mainly aim at adapting established text mining methods to the characteristics of the biomedical literature. The available approaches employ natural language-processing techniques, ranging from direct pattern-matching approaches[260] to customization of established natural language-processing systems.[261]

The reliable extraction of relationships between proteins, genes, diseases, and other biological entities is a central goal of text-mining methods in the biological context. There are two main classes of problems that make progress in this field difficult. The first comprises general natural language problems that include tracking references to the same object consistently throughout the text and detecting precisely hypothesis and their negations. The second covers domain-specific matters that concern the highly variable nomenclature for genes and proteins, for instance. Nevertheless, recent approaches that specialize in certain tasks have proved to be quite effective. Such methods focus on the extraction of facts concerning the subcellular localization of proteins[262] and protein–protein interactions,[261] for example. One approach that relies on machine-learning techniques has recently been reported, combining information retrieval and extraction for the determination of protein–protein interactions.[263] The system described there helps maintain the BIND database of protein interactions, among other applications. Further information can be found in recent reviews.[264–267]

Whereas information retrieval methods have reached performance levels that justify their immediate use,[263] the results provided by information extraction methods must be used with caution. Method improvement is necessary. Furthermore, the field would benefit greatly from an evaluation standard. Such a standard would also help in determining the algorithms most suitable for retrieval and extraction tasks. The pace of idea exchange and the development of an evaluation standard can be accelerated by the launch of critical assessment contests.[268] The established TREC conference on text-mining methods supports these efforts, and has started a genomics track. Launched in 2003, the BioCreAtIvE (Critical Assessment for Information extraction Systems in Biology) competition assesses text-mining methods in the biomedical domain.

3.15.7.7 Information Integration

The protein function prediction methods described in this section are based on a variety of data, including sequence data, structure data, protein interaction data, literature data, mRNA expression data, and genomic context data. Each kind of data gives a different type of hint about protein function, and, typically, the reliability and specificity of the prediction based on a single kind of data is quite limited. This suggests integrating the weak or partial hints coming from each data source in order to arrive at a more comprehensive, specific, and reliable prediction of protein function. The situation is rather like that of having to identify a criminal: hints from all kinds of backgrounds, such as circumstantial evidence, witnesses, family background, and personal history, have to be integrated in order to rank list suspects.

The integration of information on protein function has become an important field in computational biology. Methods have been reported that evaluate protein interaction data in combination with mRNA expression data.[269,270] Marcotte *et al.*, presented the first study integrating several genomic context methods with primary experimental data and mRNA expression data.[271] The Bork group followed with two studies, one concentrating on genomic context methods[272] and the other on protein interaction data.[273] More recently, these researchers provided the database STRING, containing precomputed functional associations based on genomic context data.[274] Phylogenetic profiles seem to be the most powerful genomic context method. Domain fusion analysis contributes strong signals, but it is applicable only to a small subset of proteins. Surprisingly, mRNA expression data and protein interaction data carry comparatively little signal. The STRING server has been used to recover functional modules in *E. coli.*[275] Protein interaction data have been put into the context of other sources of signals for protein function.[273,276,277] All of these studies indicate that, when integrated, the methods are much stronger than any single method by itself.[278–280]

3.15.8 Analysis of Genetic Variations

The generic genome of a species is responsible for all the common traits shared by the members of the species. In this section, we focus on the slight but important differences between these members that promote them to genomic individuals. In humans, these differences are responsible for distinctions in appearance, and, more importantly, for individual predispositions to diseases and responses to certain drugs. In the same context, the individual genomic differences of the infectious agents (bacteria and viruses) play an important role, since they hold the key to understanding the phenomena of resistance to drug treatment. The combined analysis of the genomic variations of pathogen and host leads to an understanding of the problems or benefits of immune response. Therefore, we will briefly remark on genetic variations in both contexts.

3.15.8.1 Genetic Variations in Humans: Analyzing Predispositions

The commonly observed rate of variation in human genomes is about one difference in 1200 base pairs. (This is remarkable small, five times as small as in the chimpanzee and 12 times as small as in the fruitfly.) The length of these differences is usually 1 base pair, and thus they are referred to as single nucleotide polymorphisms (SNPs). However, SNPs come in blocks that are handed down through the generations together, and separated by break-points introduced via recombination. These so-called haplotype blocks are the basis for the analysis of genetic predispositions for diseases and individuals' responses to drug therapy. The International HapMap Project provides a database that contains haplotype blocks.[281] About 1000 diseases are known to be caused by alterations occurring in a single gene, so-called monogenetic diseases, among them sickle cell anemia (caused by an SNP in the hemoglobin gene), cystic fibrosis (caused by a mutation in an ion transporter), and hemophilia (in which a mutation disables a protein facilitating blood clotting). Several of these diseases have been studied for decades; the OMIM (Online Mendelian Inheritance in Man) database can be queried for information on disease-relevant mutations and the related phenotypes.[282]

In the case of monogenetic diseases, the genes responsible are identified by statistical genetics methods. Such analyses are based on molecular markers – short, polymorphic stretches of repetitive DNA or SNPs, usually without any apparent function. Such polymorphisms can be located along the genome via genetic comparison, both for healthy people and for those who carry the disease. Statistical procedures are used for investigations into the observed instances of the polymorphisms and if they occur with preference in one of the two groups (healthy and diseased, respectively). If this procedure finds a significant correlation between the disease and one of the markers, this suggests that the region of genome sequence around the marker is involved in the disease, that is, the genes in this region become suspect for playing a role in the disease. Breaking this suggestion down to single genes, the probability of being related to the disease for a gene decreases inversely proportional to the distance of the gene from the marker. This probability is mapped to a statistical score, the so-called LOD (log odds) score, which quantifies the likelihood of a gene being involved in the disease. If a high-scoring genomic region can be found, it is screened with experimental or bioinformatics procedures to identify candidate genes and provide insight into their function.

Methods from statistical genetics can be categorized by the sort of data they analyze: linkage analysis works on familial data (i.e., pedigrees of families in which the disease has occurred with high frequency). This analysis can greatly help to elucidate the genetic basis of the disease, but the required familial data are usually sparse and hard to obtain. Association analysis deals with larger genetic data sets that are retrieved on a population scale, without knowledge of familial relationships between the individuals in the population.[283–285] Both kinds of analysis have helped to reveal important parts of the molecular basis of many monogenetic diseases. Even though none of the investigated diseases is curable at present, this knowledge has implications for therapy.

Diseases caused by a defect in a single gene are notoriously rare because, in general, they confer such a severe disadvantage on the affected individual that they are barely propagated in the population. Our common diseases are caused by alterations of multiple genes, influences from the environment, and the behavior of the affected individual, or any combination of these. Therefore, they are referred to as complex diseases. Current genomic research based on the analysis of the genetic variations between individuals mainly focuses on these diseases. The identification of the relevant and responsible genes is much harder in complex diseases than in monogenetic diseases.[286,287]

The Iceland genotyping program is one of the most ambitious current genotyping efforts, since it aims at genotyping the complete Icelandic population. In concert with the extensive pedigree information that has been collected in Iceland for almost 1000 years, this provides a unique database that allows for detailed studies of the genetic basis of complex diseases.[288] The project is carried out by the company deCODE Genetics, Inc.,[289] which has been founded for this purpose, and the pharmaceutical company Hoffman-LaRoche. The project provoked controversy,[290–294] and required special legislation in Iceland.[295] To date, the project has identified 15 genes that are in some way responsible for 12 common diseases.

Genes that are shown to be linked to a disease can be suitable targets for the design of drugs against the disease, or they can provide stepping stones to finding such drug targets. The field of pharmacogenomics uses genomics, be it on the basis of individual genetic variants or not, to search for drug targets and the corresponding drugs (*see* 3.03 Pharmacogenomics).[296,297] Often, the term 'pharmacogenetics'[298] is used synonymously with 'pharmacogenomics,' but it can also point specifically to research that deals with individual genomics differences. (There is some controversy about the meaning of 'pharmacogenomics' and 'pharmacogenetics' in the literature.[299,300]) Pharmacogenomics can benefit from all the technological achievements presented in this chapter, be it experimental or bioinformatics technologies; mRNA microarrays are of special interest (*see* 3.03 Pharmacogenomics).[301] Experimental target finding and drug design is often easier when it is performed in some model organism. Comparative genomics analysis can thus help to transfer the knowledge gained on models to human research.[302,303]

The special challenges that pharmacogenomics poses for bioinformatics are reviewed by Altman and Klein.[304] This group maintains and collects data relevant for pharmacogenomics (genomic, phenotypical, and clinical information) for the PharmGKB database.[305] The associated project addresses interlinking and querying the data, visualizing the query results, and maintaining both the confidentiality and privacy of critical data. Pharmacogenetics also covers the bioinformatics analysis of structural and functional consequences of the so-called 'nonsynonymous' SNPs, which occur in the coding region of a gene and change a residue, and therefore have an effect on the structure and function of the gene product.[306] The transcriptional effects of SNPs in the regulatory regions of genes are more indirect than those of nonsynonymous SNPs in coding regions of genes, and these effects are therefore more difficult to analyze with bioinformatics methods.

3.15.8.2 Genetic Variations in Pathogens: Analyzing Resistance

The analysis of the genome of a pathogen and its interaction with the host can help reveal the molecular basis of pathogenicity. Analyzing the genomic variation in the pathogen that is the result of selective pressure imposed on the pathogen by the immune system of the host or by a given drug therapy can help reveal mechanisms by which the pathogen acquires resistance. We will discuss this issue using the example of the drug treatment resistance of HIV/AIDS (for further reading see recent reviews[307–311]).

Upon infection with a pathogen, the immune system of the host mounts an attack on the intruder, with the goal of eradicating it. A drug can influence the subsequent battle in two ways: it either targets the pathogen directly or it assists the immune system. As the therapy continues, the pathogen experiences a selective pressure from the drug treatment, and the pathogen reacts to it by evolving into resistant variants. This defensive action of the pathogen has to be analyzed at the genomic level, at best also involving the genotype of the host. Bioinformatics can support this analysis.

HIV is a rapidly mutating virus that attacks the immune system of the host. Currently, more than 20 drugs are available that target viral proteins, but the high mutational rate of the virus enables it to rapidly become resistant to each drug. Therefore, according to the highly active antiretroviral therapy (HAART) scheme, several drugs with different target proteins and different modes of activity are administered simultaneously. The development of a resistant strain of the virus usually requires months to a year or two, and then the drug combination must be adapted to the new dominant strain. Since the virus develops multiple resistances over time during such a treatment, every drug combination should be effective against the current viral strain and should also lower the chance for the virus to become resistant again. The combination of drugs can be selected with the help of bioinformatics methods. This requires two kinds of data: (1) genotypic resistance data relating the genotype of the dominant viral strain inside the patient to clinical disease measures (here the so-called viral load, the number of free virus particles in 1 ml of blood serum) and (2) phenotypic resistance data originating from laboratory experiments, in which viral strains are exposed to HIV drugs, and the so-called resistance factor is measured. The resistance factor is the quotient of the drug concentration that halves the growth of the tested virus strain divided by the respective concentration for the wild type. Several servers provide resistance and mutation data information.[312,313] These data can be analyzed with bioinformatics methods that use (parts of) the sequences of the viral protein targets as input, and predict the effectiveness of single drugs[314,315] and drug combinations as well as a the possible shortest mutational path to resistance.[316]

3.15.9 Outlook

This chapter summarizes the state of the art in a wide variety of areas in bioinformatics, all of which can contribute to providing useful information for the analysis of the molecular basis of diseases, the search for target proteins, and the analysis of the effect of drug therapies. We mentioned repeatedly that the field is just emerging. Therefore, a coherent picture of how all of these components will act in concert, for example to systematically search for protein targets, cannot be given yet.

For a protein to be a suitable target, its appropriate role in the disease process is only one factor. The protein has to be suitable for drug design. This implies an appropriate shape of the binding pocket to be able to bind to small-molecule drugs, for instance. Also, the protein must allow for the development of specific drugs, that is, there should be no other proteins with similar binding pockets to which the drug may bind and cause side effects. To our knowledge, there is no integrated computational approach that covers all of these aspects, to date.

When will the field be mature enough to provide the first successes in pharmaceutical and clinical practice? This question is hard to answer; also, we have not undertaken a comprehensive survey of successes that can already be reported. However, in our opinion, the general picture is as follows.

Given a protein target, computational support for developing effective drugs has already shown numerous successes. These successes cannot be attributed to a specific computational method. Rather, the pharmaceutical industry tends to use different methods in turn or in concert. Computational methods for drug development had their first industrial applications in the early 1990s. Since the drug development process takes about a decade, drugs developed with the aid of computers only arrived on the market a few years ago.

We know of no case in which a protein target has been identified with substantial bioinformatics support and, for this protein, a drug developed and put on the market, successfully. We believe that there has simply not been enough time for this to happen. We expect such successes to be reported within the next decade.

However, bioinformatics has been instrumental in the advance of disease diagnosis and prognosis as well as therapy selection. In all of these cases, our informal inquiries have yielded substantial support for the fact that bioinformatics procedures have entered clinical practice. We expect bioinformatics procedures such as the computational analysis of gene expression data for the determination of the type and state of a disease or the computational analysis of the resistance of a pathogenic strain to drug therapy to become increasingly widespread in clinical practice over the next few years. We anticipate that this development will occur before the advent of 'computer drugs' having a substantial share of the pharmaceutical marketplace.

In general, we are well aware that the quest for unraveling the molecular processes of a disease to an extent that affords the effective development and selection of drug therapies is a great challenge. However, we are encouraged by promising recent developments that incremental progress will be made along this route and that important intermediate results will find their way into pharmaceutical and medical practice not only in the long term but also in the more immediate future.

References

1. Venter, J. C.; Adams, M. D.; Myers, E. W.; Li, P. W.; Mural, R. J.; Sutton, G. G.; Smith, H. O.; Yandell, M.; Evans, C. A.; Holt, R. A. et al. *Science* **2001**, *291*, 1304–1351.
2. Green, E. D. *Nat. Rev. Genet.* **2001**, *2*, 573–583.
3. Celniker, S. E.; Wheeler, D. A.; Kronmiller, B.; Carlson, J. W.; Halpern, A.; Patel, S.; Adams, M.; Champe, M.; Dugan, S. P.; Frise, E. et al. *Genome Biol.* **2002**, *3*, 1–14.
4. Lander, E. S.; Linton, L. M.; Birren, B.; Nusbaum, C.; Zody, M. C.; Baldwin, J.; Devon, K.; Dewar, K.; Doyle, M.; FitzHugh, W. et al. *Nature* **2001**, *409*, 860–921.
5. Huson, D. H. *J. ACM* **2002**, *49*, 603–615.
6. Huson, D. H.; Reinert, K.; Kravitz, S. A.; Remington, K. A.; Delcher, A. L.; Dew, I. M.; Flanigan, M.; Halpern, A. L.; Lai, Z.; Mobarry, C. M. et al. *Bioinformatics* **2001**, *17*, S132–S139.
7. Kent, W. J.; Haussler, D. *Genome Res.* **2001**, *11*, 1541–1548.
8. Kirkness, E. F.; Bafna, V.; Halpern, A. L.; Levy, S.; Remington, K.; Rusch, D. B.; Delcher, A. L.; Pop, M.; Wang, W.; Fraser, C. M. et al. *Science* **2003**, *301*, 1898–1903.
9. Venter, J. C.; Remington, K.; Heidelberg, J. F.; Halpern, A. L.; Rusch, D.; Eisen, J. A.; Wu, D.; Paulsen, I.; Nelson, K. E.; Nelson, W. et al. *Science* **2004**, *304*, 66–74.
10. Waterston, R. H.; Lindblad-Toh, K.; Birney, E.; Rogers, J.; Abril, J. F.; Agarwal, P.; Agarwala, R.; Ainscough, R.; Alexandersson, M.; An, P. et al. *Nature* **2002**, *420*, 520–562.
11. Gibbs, R. A.; Weinstock, G. M.; Metzker, M. L.; Muzny, D. M.; Sodergren, E. J.; Scherer, S.; Scott, G.; Steffen, D.; Worley, K. C.; Burch, P. E. et al. *Nature* **2004**, *428*, 493–521.
12. Gagneux, P. *Trends Ecol. Evol.* **2004**, *19*, 571–576.
13. Staden, R. *Mol. Biotech.* **1996**, *5*, 233–241.
14. Womble, D. D. *Methods Mol. Biol.* **2000**, *132*, 3–22.
15. Durbin, R.; Eddy, S. R.; Krogh, A. *Biol. Sequence Anal.* **1998**.
16. Raghava, G.; Searle, S. M.; Audley, P. C.; Barber, J. D.; Barton, G. J. *BMC Bioinform.* **2003**, *4*, 47.
17. Thompson, J. D.; Plewniak, F.; Poch, O. *Bioinformatics* **1999**, *15*, 87–88.
18. Dayhoff, M. O. *Protein Segment Dictionary 78: From the Atlas of Protein Sequence and Structure*; National Biomedical Research Foundation, Georgetown University Medical Center: Silver Spring, MD, 1978; Vol. 5, Suppl. 1–3.
19. Henikoff, S.; Henikoff, J. G. *Adv. Protein Chem.* **2000**, *54*, 73–97.
20. Muller, T.; Vingron, M. *J. Comput. Biol.* **2000**, 7, 761–776.
21. Zien, A.; Zimmer, R.; Lengauer, T. *J. Comput. Biol.* **2000**, 7, 483–501.
22. Gusfield, D. *Algorithms on Strings, Trees, and Sequences: Computer Science and Computational Biology*; Cambridge University Press: Cambridge, UK, 1997.
23. Altschul, S. F.; Gish, W.; Miller, W.; Myers, E. W.; Lipman, D. J. *J. Mol. Biol.* **1990**, *215*, 403–410.
24. Karlin, S.; Altschul, S. F. *Proc. Natl. Acad. Sci. USA* **1990**, *87*, 2264–2268.
25. Chenna, R.; Sugawara, H.; Koike, T.; Lopez, R.; Gibson, T. J.; Higgins, D. G.; Thompson, J. D. *Nucleic Acids Res.* **2003**, *31*, 3497–3500.
26. Higgins, D. G.; Sharp, P. M. *Gene* **1988**, *73*, 237–244.
27. Eddy, S., R. *Bioinformatics* **1998**, *14*, 755–763.
28. Notredame, C.; Higgins, D. G.; Heringa, J. *J. Mol. Biol.* **2000**, *302*, 205–217.
29. Morgenstern, B. *Nucleic Acids Res.* **2004**, *32*, W33–W36.
30. Subramanian, A. R.; Weyer-Menkhoff, J.; Kaufmann, M.; Morgenstern, B. *BMC Bioinformatics* **2005**, *6*, 66.
31. Edgar, R. C. *Nucleic Acids Res.* **2004**, *32*, 1792–1797.
32. Pavlidis, P.; Furey, T. S.; Liberto, M.; Haussler, D.; Grundy, W. N. *Pac. Symp. Biocomput.* **2001**, *6*, 151–163.
33. Gonnet, G. H.; Cohen, M. A.; Benner, S. A. *Science* **1992**, *256*, 1443–1445.
34. Henikoff, S.; Henikoff, J. G. *Proc. Natl. Acad. Sci. USA* **1992**, *89*, 10915–10919.
35. Bowie, J. U.; Luthy, R.; Eisenberg, D. *Science* **1991**, *253*, 164–170.
36. Kelley, L. A.; MacCallum, R. M.; Sternberg, M. J. *J. Mol. Biol.* **2000**, *299*, 499–520.
37. Bateman, A.; Birney, E.; Cerruti, L.; Durbin, R.; Etwiller, L.; Eddy, S. R.; Griffiths-Jones, S.; Howe, K. L.; Marshall, M.; Sonnhammer, E. L. *Nucleic Acids Res.* **2002**, *30*, 276–280.
38. Haft, D. H.; Selengut, J. D.; White, O. *Nucleic Acids Res.* **2003**, *31*, 371–373.
39. Letunic, I.; Goodstadt, L.; Dickens, N. J.; Doerks, T.; Schultz, J.; Mott, R.; Ciccarelli, F.; Copley, R. R.; Ponting, C. P.; Bork, P. *Nucleic Acids Res.* **2002**, *30*, 242–244.
40. Yang, Z. *J. Mol. Evol.* **1996**, *42*, 294–307.
41. Bandelt, H. J.; Dress, A. W. *Mol. Phylogenet. Evol.* **1992**, *1*, 242–252.
42. Brocchieri, L. *Theor. Popul. Biol.* **2001**, *59*, 27–40.
43. Felsenstein, J. *Inferring Phylogenies*, 1st ed.; Sinauer Associates: Sunderland, MA, USA, 2003.
44. Sawa, G.; Dicks, J.; Roberts, I. N. *Brief Bioinform.* **2003**, *4*, 63–74.
45. Mathe, C.; Sagot, M. F.; Schiex, T.; Rouze, P. *Nucleic Acids Res.* **2002**, *30*, 4103–4117.
46. Alexandersson, M.; Cawley, S.; Pachter, L. *Genome Res.* **2003**, *13*, 496–502.
47. Pedersen, J. S.; Hein, J. *Bioinformatics* **2003**, *19*, 219–227.
48. Solovyev, V.; Salamov, A. *Proc. Int. Conf. Intell. Syst. Mol. Biol.* **1997**, *5*, 294–302.
49. Snyder, E. E.; Stormo, G. D. *J. Mol. Biol.* **1995**, *248*, 1–18.
50. Burge, C.; Karlin, S. *J. Mol. Biol.* **1997**, *268*, 78–94.
51. Reese, M. G.; Eeckman, F. H.; Kulp, D.; Haussler, D. *J. Comput. Biol.* **1997**, *4*, 311–323.
52. Brunak, S.; Engelbrecht, J.; Knudsen, S. *J. Mol. Biol.* **1991**, *220*, 49–65.
53. Rogic, S.; Mackworth, A. K.; Ouellette, F. B. *Genome Res.* **2001**, *11*, 817–832.
54. Rogic, S.; Ouellette, B. F.; Mackworth, A. K. *Bioinformatics* **2002**, *18*, 1034–1045.
55. Nardone, J.; Lee, D. U.; Ansel, K. M.; Rao, A. *Nat. Immunol.* **2004**, *5*, 768–774.
56. Buhler, J.; Tompa, M. *J. Comput. Biol.* **2002**, *9*, 225–242.

57. Pevzner, P. A.; Sze, S. H. *Proc. Int. Conf. Intell. Syst. Mol. Biol.* **2000**, *8*, 269–278.
58. Qiu, P. *Biochem. Biophys. Res. Commun.* **2003**, *309*, 495–501.
59. Bromham, L. *Trends Ecol. Evol.* **2002**, *17*, 91–97.
60. Smit, A. F. A.; Hubley, R.; Green, P. RepeatMasker Open-3.0. 1996–2004. See http://www.repeatmasker.org (accessed April 2006).
61. Kurtz, S.; Schleiermacher, C. *Bioinformatics* **1999**, *15*, 426–427.
62. Delcher, A. L.; Kasif, S.; Fleischmann, R. D.; Peterson, J.; White, O.; Salzberg, S. L. *Nucleic Acids Res.* **1999**, *27*, 2369–2376.
63. Benson, G. *Nucleic Acids Res.* **1999**, *27*, 573–580.
64. Pevzner, P.; Tesler, G. *Genome Res.* **2003**, *13*, 37–45.
65. Pevzner, P. A. *Computational Molecular Biology An Algorithmic Approach*; MIT Press: Cambridge, MA, 2000.
66. Tesler, G. *J. Comput. Syst. Sci.* **2002**, *65*, 587–609.
67. Murphy, W. J.; Pevzner, P. A.; O'Brien, S. J. *Trends Genet.* **2004**, *20*, 631–639.
68. Eichler, E. E.; Sankoff, D. *Science* **2003**, *301*, 793–797.
69. Raphael, B. J.; Volik, S., Collins, C.; Pevzner, P. A. *Bioinformatics* **2003**, *19*, I162–II171.
70. Venclovas, C.; Zemla, A.; Fidelis, K.; Moult, J. *Proteins* **2003**, *53*, 585–595.
71. Bujnicki, J. M.; Elofsson, A.; Fischer, D.; Rychlewski, L. *Proteins Struct. Funct. Genet.* **2001**, *45*, 184–191.
72. Koh, I. Y.; Eyrich, V. A.; Marti-Renom, M. A.; Przybylski, D.; Madhusudhan, M. S.; Eswar, N.; Grana, O.; Pazos, F.; Valencia, A.; Sali, A. et al. *Nucleic Acids Res.* **2003**, *31*, 3311–3315.
73. Jones, D. T. *J. Mol. Biol.* **1999**, *292*, 195–202.
74. Rost, B.; Sander, C. *J. Mol. Biol.* **1993**, *232*, 584–599.
75. Albrecht, M.; Tosatto, S. C.; Lengauer, T.; Valle, G. *Protein Eng.* **2003**, *16*, 459–462.
76. Cuff, J. A.; Clamp, M. E.; Siddiqui, A. S.; Finlay, M.; Barton, G. J. *Bioinformatics* **1998**, *14*, 892–893.
77. Selbig, J.; Mevissen, T.; Lengauer, T. *Bioinformatics* **1999**, *15*, 1039–1046.
78. Kabsch, W.; Sander, C. *Biopolymers* **1983**, *22*, 2577–2637.
79. Frishman, D.; Argos, P. *Proteins* **1995**, *23*, 566–579.
80. Sander, C.; Schneider, R. *Proteins Struct. Funct. Genet.* **1991**, *9*, 56–68.
81. Qu, B. H.; Strickland, E.; Thomas, P. J. *J. Bioenerg. Biomembr.* **1997**, *29*, 483–490.
82. McGuffin, L. J.; Jones, D. T. *Bioinformatics* **2003**, *19*, 874–881.
83. Jones, D. T.; Miller, R. T.; Thornton, J. M. *Proteins* **1995**, *23*, 387.
84. Dunbrack, R. L., Jr. *Proteins Struct. Funct. Genet.* **1999**, *43*, 81–87.
85. Sanchez, R.; Sali, A. *Proteins Struct. Funct. Genet.* **1997**, *41*, 50–58.
86. Schwede, T.; Kopp, J.; Guex, N.; Peitsch, M. C. *Nucleic Acids Res.* **2003**, *31*, 3381–3385.
87. Sommer, I.; Zien, A.; Von Öhsen, N.; Zimmer, R.; Lengauer, T. *Bioinformatics* **2002**, *18*, 802–812.
88. Wolf, Y. I.; Brenner, S. E.; Bash, P. A.; Koonin, E. V. *Genome Res.* **1999**, *9*, 17–26.
89. Rodrigues, A.; Hubbard, R. E. *Brief Bioinform.* **2003**, *4*, 150–167.
90. O'Toole, N.; Grabowski, M.; Otwinowski, Z.; Minor, W.; Cygler, M. *Proteins Struct. Funct. Bioinformatics* **2004**, *56*, 201–210.
91. Simons, K. T.; Kooperberg, C.; Huang, E.; Baker, D. *J. Mol. Biol.* **1997**, *268*, 209–225.
92. Bonneau, R.; Strauss, C. E.; Rohl, C. A.; Chivian, D.; Bradley, P.; Malmstrom, L.; Robertson, T.; Baker, D. *J. Mol. Biol.* **2002**, *322*, 65.
93. Simons, K. T.; Ruczinski, I.; Kooperberg, C.; Fox, B. A.; Bystroff, C.; Baker, D. *Proteins* **1999**, *34*, 82–95.
94. Bradley, P.; Chivian, D.; Meiler, J.; Misura, K. M.; Rohl, C. A.; Schief, W. R.; Wedemeyer, W. J.; Schueler-Furman, O.; Murphy, P.; Schonbrun, J. et al. *Proteins* **2003**, *53*, 457–668.
95. Meiler, J.; Baker, D. *Proc. Natl. Acad. Sci. USA* **2003**, *100*, 12105–12110.
96. Han, K.; Byun, Y. *Nucleic Acids Res.* **2003**, *31*, 3432–3440.
97. van Batenburg, F. H.; Gultyaev, A. P.; Pleij, C. W. *Nucleic Acids Res.* **2001**, *29*, 194–195.
98. Grate, L. *Proc. Int. Conf. Intell. Syst. Mol. Biol.* **1995**, *3*, 136–144.
99. Hofacker, I. L.; Stadler, P. F. *Comput. Chem.* **1999**, *23*, 401–414.
100. Parsch, J.; Braverman, J. M.; Stephan, W. *Genetics* **2000**, *154*, 909–921.
101. Zucker, M.; Stiegler, P. *Nucleic Acids Res.* **1981**, *9*, 133–148.
102. Lyngso, R. B.; Zuker, M.; Pedersen, C. N. *Bioinformatics* **1999**, *15*, 440–445.
103. Zuker, M. *Science* 1989, *244*, 48–52.
104. McCaskill, J. S. *Biopolymers* **1990**, *29*, 1105–1119.
105. Cai, L.; Malmberg, R. L.; Wu, Y. *Bioinformatics* **2003**, *19*, I66–I73.
106. Lyngso, R. B.; Pedersen, C. N. *J. Comput. Biol.* **2000**, *7*, 409–427.
107. Ideker, T.; Galitski, T.; Hood, L. *Annu. Rev. Genomics Hum. Genet.* **2001**, *2*, 272–343.
108. Jeong, H.; Tombor, B.; Albert, R.; Oltval, Z. N.; Barabasi, A. L. *Nature* **2000**, *407*, 651–654.
109. Karp, P. D.; Riley, M.; Saier, M.; Paulsen, I. T.; Collado-Vides, J.; Paley, S. M.; Pellegrini-Toole, A.; Bonavides, C.; Gama-Castro, S. *Nucleic Acids Res.* **2002**, *30*, 56–58.
110. Karp, P. D.; Ouzounis, C.; Paley, S. *Proc. Int. Conf. Intell. Syst. Mol. Biol.* **1996**, *4*, 116–124.
111. Romero, P.; Karp, P. *J. Mol. Microbiol. Biotechnol.* **2003**, *5*, 230–239.
112. Kanehisa, M.; Goto, S.; Kawashima, S.; Nakaya, A. *Nucleic Acids Res.* **2002**, *30*, 42–46.
113. Selkov, E., Jr.; Grechkin, Y.; Mikhailova, N.; Selkov, E. *Nucleic Acids Res.* **1998**, *26*, 43–45.
114. Karp, P. D.; Riley, M.; Paley, S. M.; Pellegrini-Toole, A. *Nucleic Acids Res.* **2002**, *30*, 59–61.
115. Schomburg, I.; Chang, A.; Ebeling, C.; Gremse, M.; Heldt, C.; Huhn, G.; Schomburg, D. *Nucleic Acids Res.* **2004**, *32*, D431–D433.
116. Bairoch, A. *Nucleic Acids Res.* **1999**, *27*, 310–311.
117. Fleischmann, A.; Darsow, M.; Degtyarenko, K.; Fleischmann, W.; Boyce, S.; Axelsen, K. B.; Bairoch, A.; Schomburg, D.; Tipton, K. F.; Apweiler, R. *Nucleic Acids Res.* **2004**, *32*, D434–D437.
118. Osterman, A.; Overbeek, R. *Curr. Opin. Chem. Biol.* **2003**, *7*, 238–251.
119. Holzhutter, H. G. *Eur. J. Biochem.* **2004**, *271*, 2905–2922.
120. Schilling, C. H.; Edwards, J. S.; Letscher, D.; Palsson, B. *Biotechnol. Bioeng.* **2000**, *71*, 286–306.
121. Kauffman, K. J.; Prakash, P.; Edwards, J. S. *Curr. Opin. Biotechnol.* **2003**, *14*, 491–496.

122. Schuster, S.; Dandekar, T.; Fell, D. A. *Trends Biotechnol.* **1999**, *17*, 53–60.
123. Stelling, J.; Klamt, S.; Bettenbrock, K.; Schuster, S.; Gilles, E. D. *Nature* **2002**, *420*, 190–193.
124. Pfeiffer, T.; Sanchez-Valdenebro, I.; Nuno, J. C.; Montero, F.; Schuster, S. *Bioinformatics* **1999**, *15*, 251–257.
125. Mendes, P. *Trends Biochem Sci.* **1997**, *22*, 361–363.
126. Reed, J. L.; Vo, T. D.; Schilling, C. H.; Palsson, B. O. *Genome Biol.* **2003**, *4*, 54.
127. Goryanin, I.; Hodgman, T. C.; Selkov, E. *Bioinformatics* **1999**, *15*, 749–758.
128. Voit, E. O. *Computational Analysis of Biochemical Systems: A Practical Guide for Biochemists and Molecular Biologists*; Cambridge University Press: Cambridge, UK, 2000; Vol. 21, pp 692–696.
129. Wiechert, W. *J. Biotechnol.* **2002**, *94*, 37–63.
130. Allen, J.; Davey, H. M.; Broadhurst, D.; Heald, J. K.; Rowland, J. J.; Oliver, S. G.; Kell, D. B. *Nat. Biotechnol.* **2003**, *21*, 692–696.
131. Nielsen, K. F.; Smedsgaard, J. *J. Chromatogr. A* **2003**, *1002*, 111–136.
132. Soga, T.; Ohashi, Y.; Ueno, Y.; Naraoka, H.; Tomita, M.; Nishioka, T. *J. Proteome Res.* **2003**, *2*, 488–494.
133. Wei, G. H.; Liu, D. P.; Liang, C. C. *Biochem. J.* **2004**, *381*, 1–12.
134. Krull, M.; Voss, N.; Choi, C.; Pistor, S.; Potapov, A.; Wingender, E. *Nucleic Acids Res.* **2003**, *31*, 97–100.
135. Xenarios, I.; Salwinski, L.; Duan, X. J.; Higney, P.; Kim, S. M.; Eisenberg, D. *Nucleic Acids Res.* **2002**, *30*, 303–305.
136. Bader, G. D.; Betel, D.; Hogue, C. W. *Nucleic Acids Res.* **2003**, *31*, 248–250.
137. Sivakumaran, S.; Hariharaputran, S.; Mishra, J.; Bhalla, U. S. *Bioinformatics* **2003**, *19*, 408–415.
138. de Jong, H.; Geiselmann, J.; Hernandez, C.; Page, M. *Bioinformatics* **2003**, *19*, 336–344.
139. Peleg, M.; Yeh, I.; Altman, R. B. *Bioinformatics* **2002**, *18*, 825–837.
140. Snoep, J. L.; Westerhoff, H. V. *Cur. Genomics* **2004**, *5*, 687–697.
141. Takahashi, K.; Yugi, K.; Hashimoto, K.; Yamada, Y.; Pickett, C. J. F.; Tomita, M. *IEEE Intell. Syst.* **2002**, *17*, 64–71.
142. Takahashi, K.; Ishikawa, N.; Sadamoto, Y.; Sasamoto, H.; Ohta, S.; Shiozawa, A.; Miyoshi, F.; Naito, Y.; Nakayama, Y.; Tomita, M. *Bioinformatics* **2003**, *19*, 1727–1729.
143. Triendl, R. *Nature* **2002**, *417*, 7.
144. Matsuno, H.; Tanaka, Y.; Aoshima, H.; Doi, A.; Matsui, M.; Miyano, S. *In Silico Biol.* **2003**, *3*, 389–404.
145. Adams, M. D.; Kelley, J. M.; Gocayne, J. D.; Dubnick, M.; Polymeropoulos, M. H.; Xiao, H.; Merril, C. R.; Wu, A.; Olde, B.; Moreno, R. F. *Science* **1991**, *252*, 1651–1656.
146. Schena, M.; Shalon, D.; Davis, R. W.; Brown, P. O. *Science* **1995**, *270*, 467–470.
147. Lipshutz, R. J.; Fodor, S. P.; Gingeras, T. R.; Lockhart, D. J. *Nat. Genet.* **1999**, *21*, 20–24.
148. Brown, P. O.; Botstein, D. *Nat. Genet.* **1999**, *21*, 33–37.
149. Forster, T.; Roy, D.; Ghazal, P. *J. Endocrinol.* **2003**, *178*, 195–204.
150. Tobler, J. B.; Molla, M. N.; Nuwaysir, E. F.; Green, R. D.; Shavlik, J. W. *Bioinformatics* **2002**, *18*, S164–S171.
151. Zien, A.; Fluck, J.; Zimmer, R.; Lengauer, T. *J. Comput. Biol.* **2003**, *10*, 653–667.
152. Brazma, A.; Hingamp, P.; Quackenbush, J.; Sherlock, G.; Spellman, P.; Stoeckert, C.; Aach, J.; Ansorge, W.; Ball, C. A.; Causton, H. C. et al. *Nat. Genet.* **2001**, *29*, 365–371.
153. Brazma, A.; Parkinson, H.; Sarkans, U.; Shojatalab, M.; Vilo, J.; Abeygunawardena, N.; Holloway, E.; Kapushesky, M.; Kemmeren, P.; Lara, G. G. et al. *Nucleic Acids Res.* **2003**, *31*, 68–71.
154. Edgar, R.; Domrachev, M.; Lash, A. E. *Nucleic Acids Res.* **2002**, *30*, 207–210.
155. Krajewski, P.; Bocianowski, J. *J. Appl. Genet.* **2002**, *43*, 269–278.
156. Yang, Y. H.; Speed, T. *Nat. Rev. Genet.* **2002**, *3*, 579–588.
157. Quackenbush, J. *Nat. Genet.* **2002**, *32*, 496–501.
158. Park, T.; Yi, S. G.; Kang, S. H.; Lee, S.; Lee, Y. S.; Simon, R. *BMC Bioinform.* **2003**, *4*, 33.
159. Zien, A.; Aigner, T.; Zimmer, R.; Lengauer, T. *Bioinformatics* **2001**, *17*, S323–S331.
160. Huber, W.; von Heydebreck, A.; Sueltmann, H.; Poustka, A.; Vingron, M. *Stat. Appl. Genet. Mol. Biol.* **2003**, *2*, Article 3.
161. Golub, T. R.; Slonim, D. K.; Tamayo, P.; Huard, C.; Gaasenbeek, M.; Mesirov, J. P.; Coller, H.; Loh, M. L.; Downing, J. R.; Caligiuri, M. A. et al. *Science* **1999**, *286*, 531–537.
162. Hastie, T.; Tibshirani, R.; Friedman, J. *The Elements of Statistical Learning*; Springer-Verlag: New York, 2001.
163. Friedman, N.; Kaminski, N. *Ernst Schering Res. Found. Workshop* **2002**, *38*, 109–131.
164. Russo, G.; Zegar, C.; Giordano, A. *Oncogene* **2003**, *22*, 6497–6507.
165. Sharan, R.; Elkon, R.; Shamir, R. *Ernst Schering Res. Found. Workshop* **2002**, *38*, 83–108.
166. Iyer, V. R.; Eisen, M. B.; Ross, D. T.; Schuler, G.; Moore, T.; Lee, J. C.; Trent, J. M.; Staudt, L. M.; Hudson, J.; Boguski, M. S. et al. *Science* **1999**, *283*, 83–87.
167. Lyons, T. J.; Gasch, A. P.; Gaither, L. A.; Botstein, D.; Brown, P. O.; Eide, D. J. *Proc. Natl. Acad. Sci. USA* **2000**, *97*, 7957–7962.
168. Werner, T. *Biomol. Eng.* **2001**, *17*, 87–94.
169. Palin, K.; Ukkonen, E.; Brazma, A.; Vilo, J. *Bioinformatics* **2002**, *18*, S172–S180.
170. Yu, H.; Luscombe, N. M.; Qian, J.; Gerstein, M. *Trends Genet.* **2003**, *19*, 422–427.
171. Segal, E.; Yelensky, R.; Koller, D. *Bioinformatics* **2003**, *19*, I273–I282.
172. Hanisch, D.; Zien, A.; Zimmer, R.; Lengauer, T. *Bioinformatics* **2002**, *18*, 145S–154S.
173. Rahnenführer, J.; Domingues, F. S.; Maydt, J.; Lengauer, T. *Stat. Appl. Genet. Mol. Biol.* **2004**, *3*.
174. Ideker, T.; Thorsson, V.; Ranish, J. A.; Christmas, R.; Buhler, J.; Eng, J. K.; Bumgarner, R.; Goodlett, D. R.; Aebersold, R.; Hood, L. *Science* **2001**, *292*, 929–934.
175. Segal, E.; Wang, H.; Koller, D. *Bioinformatics* **2003**, *19*, I264–I272.
176. de Jong, H. *J. Comput. Biol.* **2002**, *9*, 67–103.
177. Binz, P. A.; Muller, M.; Walther, D.; Bienvenut, W. V.; Gras, R.; Hoogland, C.; Bouchet, G.; Gasteiger, E.; Fabbretti, R.; Gay, S. et al. *Anal. Chem.* **1999**, *71*, 4981–4988.
178. Howbrook, D. N.; van der Valk, A. M.; O'Shaughnessy, M. C.; Sarker, D. K.; Baker, S. C.; Lloyd, A. W. *Drug Disc. Today* **2003**, *8*, 642–651.
179. Wilson, D. S.; Nock, S. *Angew. Chem. Int. Ed. Engl.* **2003**, *42*, 494–500.
180. Templin, M. F.; Stoll, D.; Schrenk, M.; Traub, P. C.; Vohringer, C. F.; Joos, T. O. *Trends Biotechnol.* **2002**, *20*, 160–166.

181. Ko, I. K.; Kato, K.; Iwata, H. *Biomaterials* **2005**, *26*, 687–696.
182. Wulfing, P.; Diallo, R.; Muller, C.; Wulfing, C.; Poremba, C.; Heinecke, A.; Rody, A.; Greb, R. R.; Bocker, W.; Kiesel, L. *J. Cancer Res. Clin. Oncol.* **2003**, *129*, 375–382.
183. Domingues, F. S.; Lengauer, T. *Appl. Bioinform.* **2003**, *2*, 3–12.
184. Rost, B.; Liu, J.; Nair, R.; Wrzeszczynski, K. O.; Ofran, Y. *Cell. Mol. Life Sci.* **2003**, *60*, 2637–2650.
185. Lan, N.; Montelione, G. T.; Gerstein, M. *Curr. Opin. Chem. Biol.* **2003**, *7*, 44–54.
186. Ouzounis, C. A.; Coulson, R. M. R.; Enright, A. J.; Kunin, V.; Pereira-Leal, J. B. *Nat. Rev. Genet.* **2003**, *4*, 508–519.
187. Webb, E. C. *Enzyme Nomenclature 1992. Recommendations of the Nomenclature Committee of the International Union of Biochemistry and Molecular Biology*; Academic Press: New York, 1992.
188. Mewes, H. W.; Frishman, D.; Gruber, C.; Geier, B.; Haase, D.; Kaps, A.; Lemcke, K.; Mannhaupt, G.; Pfeiffer, F.; Schuller, C. et al. *Nucleic Acids Res.* **2000**, *28*, 37–40.
189. Ashburner, M.; Ball, C. A.; Blake, J. A.; Botstein, D.; Butler, H.; Cherry, J. M.; Davis, A. P.; Dolinski, K.; Dwight, S. S.; Eppig, J. T. et al. *Nat. Genet.* **2000**, *25*, 25–29.
190. Davidson, E. H.; Rast, J. P.; Oliveri, P.; Ransick, A.; Calestani, C.; Yuh, C. H.; Minokawa, T.; Amore, G.; Hinman, V.; Arenas-Mena, C. et al. *Science* **2002**, *295*, 1669–1678.
191. Kam, N.; Harel, D.; Kugler, H.; Marelly, R.; Pnueli, A.; Hubbard, E. J. A.; Stern, M. J. *Modeling of C. elegans Development: A Scenario-Based Approach.* International Workshop on Computational Methods in Systems Biology (CMSB 2003), 2003.
192. Fitch, W., M. *Trends Genet.* **2000**, *16*, 227–231.
193. Eulenstein, O.; Mirkin, B.; Vingron, M. *J. Comput. Biol.* **1998**, *5*, 135–148.
194. Yuan, Y. P.; Eulenstein, O.; Vingron, M.; Bork, P. *Bioinformatics* **1998**, *14*, 285–289.
195. Storm, C. E.; Sonnhammer, E. L. *Bioinformatics* **2002**, *18*, 92–99.
196. Tatusov, R. L.; Koonin, E. V.; Lipman, D. J. *Science* **1997**, *278*, 631–637.
197. Tatusov, R. L.; Natale, D. A.; Garkavtsev, I. V.; Tatusova, T. A.; Shankavaram, U. T.; Rao, B. S.; Kiryutin, B.; Galperin, M. Y.; Fedorova, N. D.; Koonin, E. V. *Nucleic Acids Res.* **2001**, *29*, 22–28.
198. Letunic, I.; Copley, R. R.; Schmidt, S.; Ciccarelli, F. D.; Doerks, T.; Schultz, J.; Ponting, C. P.; Bork, P. *Nucleic Acids Res.* **2004**, *32*, D142–D144.
199. Bucher, P.; Bairoch, A. *Proc. Int. Conf. Intell. Syst. Mol. Biol.* **1994**, *2*, 53–61.
200. Falquet, L.; Pagni, M.; Bucher, P.; Hulo, N.; Sigrist, C. J.; Hofmann, K.; Bairoch, A. *Nucleic Acids Res.* **2002**, *30*, 235–238.
201. Nevill-Manning, C. G.; Wu, T. D.; Brutlag, D. L. *Proc. Natl. Acad. Sci. USA* **1998**, *95*, 5865–5871.
202. Jonassen, I. *Comput. Appl. Biosci.* **1997**, *13*, 509–522.
203. Huang, J. Y.; Brutlag, D. L. *Nucleic Acids Res.* **2001**, *29*, 202–204.
204. Attwood, T. K. *Brief Bioinform.* **2000**, *1*, 45–59.
205. Mulder, N. J.; Apweiler, R.; Attwood, T. K.; Bairoch, A.; Barrell, D.; Bateman, A.; Binns, D.; Biswas, M.; Bradley, P.; Bork, P. et al. *Nucleic Acids Res.* **2003**, *31*, 315–318.
206. Jensen, L. J.; Gupta, R.; Blom, N.; Devos, D.; Tamames, J.; Kesmir, C.; Nielsen, H.; Staerfeldt, H. H.; Rapacki, K.; Workman, C. et al. *J. Mol. Biol.* **2002**, *319*, 1257–1265.
207. Fields, S.; Song, O. *Nature* **1989**, *340*, 245–246.
208. Uetz, P.; Giot, L.; Cagney, G.; Mansfield, T. A.; Judson, R. S.; Knight, J. R.; Lockshon, D.; Narayan, V.; Srinivasan, M.; Pochart, P. et al. *Nature* **2000**, *403*, 623–627.
209. Gavin, A. C.; Bosche, M.; Krause, R.; Grandi, P.; Marzioch, M.; Bauer, A.; Schultz, J.; Rick, J. M.; Michon, A. M.; Cruciat, C. M.; Remor, M. et al. *Nature* **2002**, *415*, 141–147.
210. Schachter, V. *Drug Disc. Today* **2002**, *7*, S48–S54.
211. Cho, S. Y.; Park, S. G.; Lee, D. H.; Park, B. C. *J. Biochem. Mol. Biol.* **2004**, *37*, 45–52.
212. Schwikowski, B.; Uetz, P.; Fields, S. *Nat. Biotechnol.* **2000**, *18*, 1257–1261.
213. Saito, R.; Suzuki, H.; Hayashizaki, Y. *Bioinformatics* **2003**, *19*, 756–763.
214. Bader, G. D.; Hogue, C. W. *BMC Bioinform.* **2003**, *4*, 2.
215. Krause, R.; von Mering, C.; Bork, P. *Bioinformatics* **2003**, *19*, 1901–1908.
216. Brun, C.; Chevenet, F.; Martin, D.; Wojcik, J.; Guenoche, A.; Jacq, B. *Genome Biol.* **2003**, *5*, R6.
217. Legrain, P.; Wojcik, J.; Gauthier, J. M. *Trends Genet.* **2001**, *17*, 346–352.
218. Wuchty, S. *Proteomics* **2002**, *2*, 1715–1723.
219. Barabasi, A. L.; Oltvai, Z. N. *Nat. Rev. Genet.* **2004**, *5*, 101–113.
220. Hermjakob, H.; Montecchi-Palazzi, L.; Lewington, C.; Mudali, S.; Kerrien, S.; Orchard, S.; Vingron, M.; Roechert, B.; Roepstorff, P.; Valencia, A. et al. *Nucleic. Acids. Res.* **2004**, *32*, D452–D455.
221. von Mering, C.; Jensen, L. J.; Snel, B.; Hooper, S. D.; Krupp, M.; Foglierini, M.; Jouffre, N.; Huynen, M. A.; Bork, P. *Nucleic Acids Res.* **2005**, *33*, D433–D437.
222. Overbeek, R.; Fonstein, M.; D'Souza, M.; Pusch, G. D.; Maltsev, N. *Proc. Natl. Acad. Sci. USA* **1999**, *96*, 2896–2901.
223. Dandekar, T.; Snel, B.; Huynen, M.; Bork, P. *Trends Biochem. Sci.* **1998**, *23*, 324–328.
224. Kolesov, G.; Mewes, H. W.; Frishman, D. *J. Mol. Biol.* **2001**, *311*, 639–656.
225. Kolesov, G.; Mewes, H. W.; Frishman, D. *Bioinformatics* **2002**, *18*, 1017–1019.
226. Enright, A. J.; Iliopoulos, I.; Kyrpides, N. C.; Ouzounis, C. A. *Nature* **1999**, *402*, 86–90.
227. Marcotte, E. M.; Pellegrini, M.; Ng, H. L.; Rice, D. W.; Yeates, T. O.; Eisenberg, D. *Science* **1999**, *285*, 751–753.
228. Pellegrini, M.; Marcotte, E. M.; Thompson, M. J.; Eisenberg, D.; Yeates, T. O. *Proc. Natl. Acad. Sci. USA* **1999**, *96*, 4285–4288.
229. Thornton, J. M.; Todd, A. E.; Milburn, D.; Borkakoti, N.; Orengo, C. A. *Nat. Struct. Biol.* **2000**, *7*, 991–994.
230. Todd, A. E.; Orengo, C. A.; Thornton, J. M. *J. Mol. Biol.* **2001**, *307*, 1113–1143.
231. Pearl, F. M.; Bennett, C. F.; Bray, J. E.; Harrison, A. P.; Martin, N.; Shepherd, A.; Sillitoe, I.; Thornton, J.; Orengo, C. A. *Nucleic Acids Res.* **2003**, *31*, 452–455.
232. Taylor, W. R.; Orengo, C. A. *J. Mol. Biol.* **1989**, *208*, 1–22.
233. Orengo, C. A.; Taylor, W. R. SSAP: Sequential Structure Alignment Program for Protein Structure Comparison. In *Computer Methods For Macromolecular Sequence Analysis*; Doolittle, R., Abelson, J., Simon, M., Eds.; Academic Press: San Diego, CA, 1996; Vol. 266, pp 617–635.
234. Lo Conte, L.; Brenner, S. E.; Hubbard, T. J.; Chothia, C.; Murzin, A. G. *Nucleic Acids Res.* **2002**, *30*, 264–267.
235. Holm, L.; Sander, C. *J. Mol. Biol.* **1993**, *233*, 123–138.

236. Holm, L.; Sander, C. *Nucleic Acids Res.* **1997**, *25*, 231–234.
237. Shindyalov, I. N.; Bourne, P. E. *Protein Eng.* **1998**, *11*, 739–747.
238. Shindyalov, I. N.; Bourne, P. E. *Nucleic Acids Res.* **2001**, *29*, 228–229.
239. Ye, Y.; Godzik, A. *Nucleic Acids Res.* **2004**, *32*, W582–W585.
240. Ye, Y. Z.; Godzik, A. *Protein Sci.* **2004**, *13*, 1841–1850.
241. Guda, C.; Lu, S. F.; Scheeff, E. D.; Bourne, P. E.; Shindyalov, I. N. *Nucleic Acids Res.* **2004**, *32*, W100–W103.
242. Kolodny, R.; Koehl, P.; Levitt, M. *J. Mol. Biol.* **2005**, *346*, 1173–1188.
243. Novotny, M.; Madsen, D.; Kleywegt, G. J. *Proteins* **2004**, *54*, 260–270.
244. Russell, R. *J. Mol. Biol.* **1998**, *279*, 1211–1227.
245. Kleywegt, G. *J. Mol. Biol.* **1999**, *285*, 1887–1897.
246. Bradley, P.; Kim, P. S.; Berger, B. *Proc. Natl. Acad. Sci. USA* **2002**, *99*, 8500–8505.
247. Barker, J. A.; Thornton, J. M. *Bioinformatics* **2003**, *19*, 1644–1649.
248. Spriggs, R. V.; Artymiuk, P. J.; Willett, P. *J. Chem. Inf. Comput. Sci.* **2003**, *43*, 412–421.
249. Stark, A.; Sunyaev, S.; Russell, R. B. *J. Mol. Biol.* **2003**, *326*, 1307–1316.
250. Liang, J.; Edelsbrunner, H.; Woodward, C. *Protein Sci.* **1998**, 7, 1884–1897.
251. Wei, L.; Altman, R. B. *Pac. Symp. Biocomput.* 1998, 497–508.
252. Stahl, M.; Taroni, C.; Schneider, G. *Protein Eng.* **2000**, *13*, 83–88.
253. Liang, M. P.; Banatao, D. R.; Klein, T. E.; Brutlag, D. L.; Altman, R. B. *Nucleic Acids Res.* **2003**, *31*, 3324–3327.
254. Lichtarge, O.; Bourne, H. R.; Cohen, F. E. *J. Mol. Biol.* **1996**, *257*, 342–358.
255. Armon, A.; Graur, D.; Ben-Tal, N. *J. Mol. Biol.* **2001**, *307*, 447–463.
256. Chelliah, V.; Chen, L.; Blundell, T. L.; Lovell, S. C. *J. Mol. Biol.* **2004**, *342*, 1487–1504.
257. Jones, S.; Thornton, J. M. *Curr. Opin. Chem. Biol.* **2004**, *8*, 3–7.
258. Pupko, T.; Bell, R. E.; Mayrose, I.; Glaser, F.; Ben-Tal, N. *Bioinformatics* **2002**, *18*, 71S–77S.
259. Glaser, F.; Pupko, T.; Paz, I.; Bell, R. E.; Bechor-Shental, D.; Martz, E.; Ben-Tal, N. *Bioinformatics* **2003**, *19*, 163–164.
260. Blaschke, C.; Andrade, M. A.; Ouzounis, C.; Valencia, A. *Proc. Int. Conf. Intell. Syst. Mol. Biol.* **1999**, 60–67.
261. Friedman, C.; Kra, P.; Yu, H.; Krauthammer, M.; Rzhetsky, A. *Bioinformatics* **2001**, *17*, S74–S82.
262. Craven, M.; Kumlien, J. *Proc. Int. Conf. Intell. Syst. Mol. Biol.* **1999**, 77–86.
263. Donaldson, I.; Martin, J.; de Bruijn, B.; Wolting, C.; Lay, V.; Tuekam, B.; Zhang, S.; Baskin, B.; Bader, G. D.; Michalickova, K. et al. *BMC Bioinform.* **2003**, *4*, 11.
264. de Bruijn, B.; Martin, J. *Int. J. Med. Inf.* **2002**, *67*, 7–18.
265. Hirschman, L.; Park, J. C.; Tsujii, J.; Wong, L.; Wu, C. H. *Bioinformatics* **2002**, *18*, 1553–1561.
266. Shatkay, H.; Feldman, R. *J. Comput. Biol.* **2003**, *10*, 821–855.
267. Krallinger, M.; Erhardt, R. A. A.; Valencia, A. *Drug Disc. Today* **2005**, *10*, 439–445.
268. Yeh, A. S.; Hirschman, L.; Morgan, A. A. *Bioinformatics* **2003**, *19*, I331–I339.
269. Kemmeren, P.; van Berkum, N. L.; Vilo, J.; Bijma, T.; Donders, R.; Brazma, A.; Holstege, F. C. *Mol. Cell* **2002**, *9*, 1133–1143.
270. Kemmeren, P.; Holstege, F. C. *Biochem. Soc. Trans.* **2003**, *31*, 1484–1487.
271. Marcotte, E. M.; Pellegrini, M.; Thompson, M. J.; Yeates, T. O.; Eisenberg, D. *Nature* **1999**, *402*, 83–86.
272. Huynen, M.; Snel, B.; Lathe, W., III; Bork, P. *Genome Res.* **2000**, *10*, 1204–1210.
273. von Mering, C.; Krause, R.; Snel, B.; Cornell, M.; Oliver, S. G.; Fields, S.; Bork, P. *Nature* **2002**, *417*, 399–403.
274. von Mering, C.; Huynen, M.; Jaeggi, D.; Schmidt, S.; Bork, P.; Snel, B. *Nucleic Acids Res.* **2003**, *31*, 258–261.
275. Von Mering, C.; Zdobnov, E. M.; Tsoka, S.; Ciccarelli, F. D.; Pereira-Leal, J. B.; Ouzounis, C. A.; Bork, P. *Proc. Natl. Acad. Sci. USA* **2003**, *100*, 15428–15433.
276. Tucker, C. L.; Gera, J. F.; Uetz, P. *Trends Cell Biol.* **2001**, *11*, 102–106.
277. Ideker, T.; Ozier, O.; Schwikowski, B.; Siegel, A. F. *Bioinformatics* **2002**, *18*, 233S–240S.
278. Pellegrini, M. *Curr. Opin. Chem. Biol.* **2001**, *5*, 46–50.
279. Valencia, A.; Pazos, F. *Curr. Opin. Struct. Biol.* **2002**, *12*, 368–373.
280. Huynen, M. A.; Snel, B.; von Mering, C.; Bork, P. *Curr. Opin. Cell Biol.* **2003**, *15*, 191–198.
281. Gibbs, R. A.; Belmont, J. W.; Hardenbol, P.; Willis, T. D.; Yu, F. L.; Yang, H. M.; Ch'ang, L. Y.; Huang, W.; Liu, B.; Shen, Y. et al. *Nature* **2003**, *426*, 789–796.
282. Hamosh, A.; Scott, A. F.; Amberger, J.; Bocchini, C.; Valle, D.; McKusick, V. A. *Nucleic Acids Res.* **2002**, *30*, 52–55.
283. Elston, R. C. *Genet. Epidemiol.* **1998**, *15*, 565–576.
284. Liu, B.-H. *Statistical Genomics: Linkage, Mapping, and QTL Analysis*; CRC Press: Boca Raton FL, 1998.
285. March, R. E. *Mol. Biotechnol.* **1999**, *13*, 113–122.
286. Page, G. P.; George, V.; Go, R. C.; Page, P. Z.; Allison, D. B. *Am. J. Hum. Genet.* **2003**, *73*, 711–719.
287. Brookes, A. J. *Trends Mol. Med.* **2001**, 7, 512–516.
288. Gulcher, J.; Stefansson, K. *Clin. Chem. Lab. Med.* **1998**, *36*, 523–527.
289. Hakonarson, H.; Gulcher, J. R.; Stefansson, K. *Pharmacogenomics* **2003**, *4*, 209–215.
290. Abbott, A. *Nature* **1999**, *400*, 3.
291. Gulcher, J. R.; Stefansson, K. *Nature* **1999**, *400*, 307–308.
292. Palsson, B.; Thorgeirsson, S. *Nat. Biotechnol.* **1999**, *17*, 407.
293. Gulcher, J.; Stefansson, K. *Nat. Biotechnol.* **1999**, *17*, 620.
294. Abbott, A. *Nature* **1999**, *400*, 602.
295. Hodgson, J. *Nat. Biotechnol.* **1999**, *17*, 127.
296. Evans, W. E.; Relling, M. V. *Science* **1999**, *286*, 487–491.
297. Lindpaintner, K. *J. Mol. Med.* **2003**, *81*, 141–153.
298. Schmith, V. D.; Campbell, D. A.; Sehgal, S.; Anderson, W. H.; Burns, D. K.; Middleton, L. T.; Roses, A. D. *Cell Mol. Life Sci.* **2003**, *60*, 1636–1646.
299. Tsai, Y. J.; Hoyme, H. E. *Clin. Genet.* **2002**, *62*, 257–264.
300. Goldstein, D. B.; Tate, S. K.; Sisodiya, S. M. *Nat. Rev. Genet.* **2003**, *4*, 937–947.
301. Gerhold, D. L.; Jensen, R. V.; Gullans, S. R. *Nat. Genet.* **2002**, *32*, 547–551.

302. Stoll, M.; Kwitek-Black, A. E.; Cowley, A. W., Jr.; Harris, E. L.; Harrap, S. B.; Krieger, J. E.; Printz, M. P.; Provoost, A. P.; Sassard, J.; Jacob, H. J. *Genome Res.* **2000**, *10*, 473–482.
303. Hoopengardner, B.; Bhalla, T.; Staber, C.; Reenan, R. *Science* **2003**, *301*, 832–836.
304. Altman, R. B.; Klein, T. E. *Annu. Rev. Pharmacol. Toxicol.* **2002**, *42*, 113–133.
305. Hewett, M.; Oliver, D. E.; Rubin, D. L.; Easton, K. L.; Stuart, J. M.; Altman, R. B.; Klein, T. E. *Nucleic Acids Res.* **2002**, *30*, 163–165.
306. Sreekumar, K. R.; Aravind, L.; Koonin, E. V. *Curr. Opin. Genet. Dev.* **2001**, *11*, 247–257.
307. Francois, I. *Drug News Perspect.* **2001**, *14*, 46–49.
308. McGrath, K. M.; Hoffman, N. G.; Resch, W.; Nelson, J. A.; Swanstrom, R. *Virus Res.* **2001**, *76*, 137–160.
309. Paine, K.; Flower, D. R. *J. Mol. Microbiol. Biotechnol.* **2002**, *4*, 357–365.
310. Sassetti, C.; Rubin, E. J. *Curr. Opin. Microbiol.* **2002**, *5*, 27–32.
311. Wilson, J. W.; Schurr, M. J.; LeBlanc, C. L.; Ramamurthy, R.; Buchanan, K. L.; Nickerson, C. A. *Postgrad. Med. J.* **2002**, *78*, 216–224.
312. Shafer, R. W.; Jung, D. R.; Betts, B. J.; Xi, Y.; Gonzales, M. J. *Nucleic Acids Res.* **2000**, *28*, 346–348.
313. Beerenwinkel, N.; Schmidt, B.; Walter, H.; Kaiser, R.; Lengauer, T.; Hoffmann, D.; Korn, K.; Selbig, J. *Proc. Natl. Acad. Sci. USA* **2002**, *99*, 8271–8276.
314. Draghici, S.; Potter, R. B. *Bioinformatics* **2003**, *19*, 98–107.
315. Beerenwinkel, N.; Daumer, M.; Oette, M.; Korn, K.; Hoffmann, D.; Kaiser, R.; Lengauer, T.; Selbig, J.; Walter, H. *Nucleic Acids Res.* **2003**, *31*, 3850–3855.
316. Beerenwinkel, N.; Lengauer, T.; Daumer, M.; Kaiser, R.; Walter, H.; Korn, K.; Hoffmann, D.; Selbig, J. *Bioinformatics* **2003**, *19*, I16–I25.

Biographies

Thomas Lengauer, PhD is Director at the Max Planck Institute for Informatics in Saarbrucken, Germany. His background is in Math (PhD, Berlin, Germany 1976) and Computer Science (PhD, Stanford 1979). In the 1970s he performed research in theoretical computer science, in the 1980s on design methods for integrated circuits. He has been engaged in research in computational biology since the beginning of the 1990s. His major focuses of research are protein bioinformatics, computational drug screening, and design and bioinformatics for understanding and curing diseases. Previously, he held the positions of a full professor at the University of Paderborn, Germany (1984–1992) and of a Director of the Institute for Algorithms and Scientific Computing at the German National Research Center for Computer Science in Sankt Augustin, near Bonn Germany (1992–2001). Lengauer is a founding member of the International Society for Computational Biology (ISCB), a member of the steering board of the international conference series RECOMB, and he headed the steering board of the European bioinformatics conference series ECCB since its foundation in 2002 until 2005. In 2001 he co-founded the BioSolveIT GmbH, Sankt Augustin, Germany, which develops and distributes Cheminformatics software. In 2003 he received the Konrad Zuse Medal of the German Informatics Society and the Karl Heinz Beckurts Award. He is a member of the German Academy of Sciences Leopoldina.

Christoph Hartmann is research scientist at the Max Planck Institute for Informatics in Saarbrucken, Germany. He received his diploma in computer science at the University of Bonn (2004). Currently he is pursuing his PhD studies in the Department of Computational Biology and Applied Algorithmics headed by Prof Thomas Lengauer. His research interests include computational docking strategies for protein–ligand complexes, molecular simulations, optimization heuristics as well as scoring functions for protein structure assessment and protein structure prediction.

Comprehensive Medicinal Chemistry II
ISBN (set): 0-08-044513-6

ISBN (Volume 3) 0-08-044516-0; pp. 315–347

3.16 Gene and Protein Sequence Databases

M-J Martin, T Kulikova, M Pruess, and R Apweiler, EMBL Outstation European Bioinformatics Institute, Hinxton, Cambridge, UK

3.16.1	**Introduction**	**349**
3.16.1.1	Primary and Secondary Databases	350
3.16.2	**Nucleotide Sequence Databases**	**351**
3.16.2.1	Data Bank of Japan/European Molecular Biology Laboratory/GenBank	351
3.16.2.1.1	Database formats	352
3.16.2.1.2	Editorial control and data redundancy	352
3.16.2.1.3	Third party annotation in International Nucleotide Sequence Databases	354
3.16.2.2	National Center for Biotechnology Information Reference Sequences	355
3.16.2.3	Genome Reviews	356
3.16.2.3.1	Data enhancements	356
3.16.2.3.2	Evidence tags	356
3.16.2.3.3	Statistics and further developments	357
3.16.3	**Protein Sequence Databases**	**357**
3.16.3.1	GenPept	357
3.16.3.2	Entrez Protein	358
3.16.3.3	Reference Sequences	358
3.16.3.4	The Universal Protein Resource	358
3.16.3.4.1	UniProt Archive	358
3.16.3.4.2	UniProt Knowledgebase	359
3.16.3.4.3	UniProt Reference Clusters	362
3.16.4	**Integrated and Comparative Databases**	**363**
3.16.4.1	InterPro	363
3.16.4.1.1	Integrating the signatures	363
3.16.4.1.2	InterProScan	364
3.16.4.2	The International Protein Index	364
3.16.4.2.1	Nonredundant complete UniProt proteome sets	365
3.16.4.2.2	Building the International Protein Index	366
3.16.4.3	Integr8	366
3.16.4.3.1	Proteome analysis	367
3.16.4.3.2	The Integr8 browser	368
3.16.5	**Summary and Conclusion**	**368**
References		**369**

3.16.1 Introduction

The understanding of genetics has advanced remarkably in the last 30 years. The completion of sequencing of the human genome and numerous model organism genomes has provided a fertile ground for biologists to learn more about genes, proteins, structures, and metabolic processes that drive the engine of life. For many years, large amounts of sequence data have been collected along with functional information and other biological features in biosequence databases, allowing for the integration of biological knowledge. However, the increasingly larger data sets present a challenge in data management, and, to this end, much effort has gone into the design and maintenance of biological sequence databases. Although their extensive use is relatively recent, biosequence databases have a long history.

In the early 1960s, Dayhoff and her colleagues at the Protein Information Resource (PIR) collected all the protein sequences known at that time. This collection was published in the *Atlas of Protein Sequence and Structure*.[1] At this particular point in the history of biology the focus was on sequencing proteins through traditional techniques such as the Edman degradation. As the atlas evolved, it included text-based descriptions as well as information on the evolution of many protein families. This work was the first annotated database of macromolecular sequences, distributed in its early years in printed form. By the early 1970s, the data had increased sufficiently so that researchers needed it in electronic form for computer searching and analysis. The contents of the atlas were distributed electronically by PIR on magnetic tape, and the distribution included some basic programs that could be used to search and evaluate the data.

The advent of DNA sequence databases in 1982, initiated by the European Molecular Biology Laboratory (EMBL) and joined shortly thereafter by GenBank, led to the next phase in the history of sequence databases: the veritable explosion in the amount of nucleotide sequence databases available to researchers. Both EMBL (then based in Heidelberg) and the National Center for Biotechnology Information (NCBI, part of the National Library of Medicine at the National Institutes of Health in the USA) were contributing to the input activity, which consisted of transcribing and interpreting what was published in print journals to an electronic format more appropriate for use with computers. The DNA Databank of Japan (DDBJ) joined the data-collecting collaboration a few years later. In 1988, after a meeting of these three groups (now referred to as the International Nucleotide Sequence Database Collaboration), there was an agreement to use a common format for data elements within a unit record and to have each database update only the records that were directly submitted to it. Now, all three centers (the National Institute of Genetics in Mishima, Japan; the European Bioinformatics Institute (EBI) in Hinxton, UK; and the NCBI in Bethesda, MD, USA) are collecting direct submissions and distributing them so that each center has copies of all the sequences, meaning that they each act as a primary distribution center for these sequences. However, each record is managed by the database that created it, and can be updated only by that database. DDBJ/EMBL/GenBank records are updated automatically every 24 h at all three sites; therefore, all sequences present in DDBJ also will be present in EMBL and GenBank, and so forth.

On a parallel track, the foundations for the Swiss-Prot protein sequence database were also laid in the early 1980s, when Bairoch at the University of Geneva converted the PIR atlas to a format similar to that used by EMBL for its nucleotide database. In this initial release, called PIR+, additional information about each of the proteins was added, increasing its value as a curated, well-annotated source of information on proteins. In the summer of 1986, Bairoch began distributing PIR+ on the US BIONET (a precursor to the Internet), renaming it Swiss-Prot. At that time, it contained the grand total of 3900 protein sequences; this was seen as an overwhelming amount of data, in stark contrast to today's standards. Because Swiss-Prot and EMBL followed similar formats, a natural collaboration developed between these two European groups; these collaborative efforts strengthened when both the EMBL and Swiss-Prot operations were moved to the EMBL outstation (the EBI) in Hinxton, UK. One of the first collaborative projects undertaken was to create a new supplement to Swiss-Prot. Maintaining the high quality of Swiss-Prot entries is a time-consuming process involving extensive sequence analysis and detailed curation by expert annotators.[2] So as to allow the quick release of protein sequence data not yet annotated to the stringent standards of Swiss-Prot, a new database called TrEMBL (for 'translation of EMBL nucleotide sequences') was created. This supplement to Swiss-Prot initially consisted of computationally annotated entries derived from the translation of all coding sequences (CDS) found in DDBJ/EMBl/GenBank, including only data that were not already present in Swiss-Prot.

3.16.1.1 Primary and Secondary Databases

The principal contribution of sequence databases to the biological community is making the sequences themselves accessible. It is important to distinguish between primary (archival) and secondary (curated) databases. The primary databases mainly contain experimental results with some interpretation but with no curated review. The secondary databases carry additional information through curated reviews. The nucleotide sequences in DDBJ/EMBL/GenBank are derived from the sequencing of a biological molecule that exists in a test tube somewhere in a laboratory. Each DNA and RNA sequence will be annotated to describe the analysis from experimental results that indicate why that sequence was determined in the first place. The great majority of the protein sequences available in public databases have not been determined experimentally, which may have downstream consequences when analyses are performed. For example, the assignment of a product name or function qualifier based on a subjective interpretation of similarity analysis (such as BLAST) can be very useful. Therefore, the DNA, RNA, or protein sequences are the computable items to be analyzed, and they represent the most valuable component of primary databases.

3.16.2 Nucleotide Sequence Databases

Nucleotide sequence databases store and distribute to the public the data on nucleic acid sequences. The nucleotide sequence databases can be divided into primary databases, storing and distributing sequences coming from the genome sequencing projects and from smaller sequencing efforts, and secondary databases, largely utilizing the nucleotide content of the primary databases and adding/superimposing annotation to it. DDBJ, EMBL, and GenBank, also known as INSD (International Nucleotide Sequence Databases), are primary databases and are described below.

3.16.2.1 Data Bank of Japan/European Molecular Biology Laboratory/GenBank

The vast majority of the publicly available nucleotide sequence data are collected, organized, and distributed by the INSD Collaboration, a joint effort of the DDBJ,[3] the EMBL Nucleotide Sequence Database,[4] and GenBank.[5] The DDBJ/EMBL/GenBank databases are data repositories, accepting nucleic acid sequence data from the scientific community and making it freely available. In most cases the sequence data represent supporting material for publications in scientific journals, as printing the nucleotide sequence data together with the articles became impractical a long time ago. At the point of writing the nucleotide content of the INSD databases is 116 billion nucleotides – an amount of data that clearly cannot be printed, and should not, because it will be of no practical use. The arrangement between the journals (which recommend the submission of sequence data to the INSD prior to acceptance of the article discussing the results) and the INSD (which accepts the submissions and assign accession numbers for the data) proved to be beneficial for everybody involved: for the authors, who want to make their sequence data publicly accessible and to provide all necessary supporting material for their scientific findings; for the journals, which cannot print sequences for practical reasons but have to provide a reference to supporting information; for the collaborative databases, which strive for completeness, with the aim of recording and making available every publicly known nucleic acid sequence; and for the users, who know that if they need to search for a certain nucleotide sequence, the most likely place to find it will be INSD. **Figure 1** shows the dynamics of the database nucleotide content (as a percentage of the whole database) for four organisms, traditionally popular for nucleotide sequence studies: *Homo sapiens*, *Mus musculus*, *Rattus norvegicus*, and *Caenorhabditis elegans*.

The collaborative databases exchange new/updated entries daily, so all sequences present in DDBJ are also present in EMBL and GenBank, and vice versa, with only a day's worth of contents possibly missing or not up to date. To ensure that this is the case, the databases exchange various checklists on a regular basis. Full releases of EMBL, GenBank, and DDBJ are produced every 2–3 months (nonsynchronously), with incremental daily updates available by anonymous FTP. Since their conception in the 1980s, the nucleic acid sequence databases have experienced exponential growth,

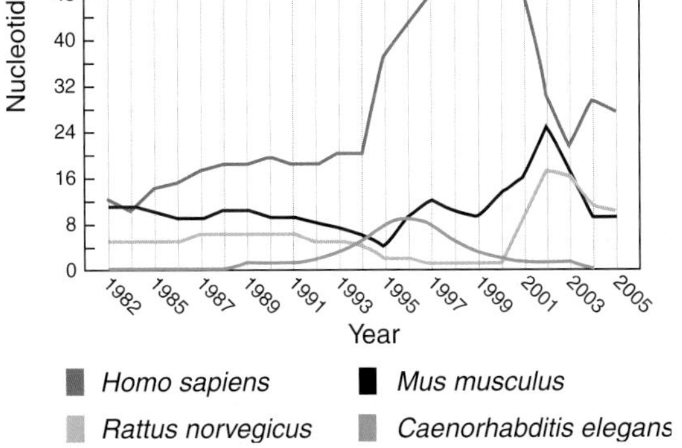

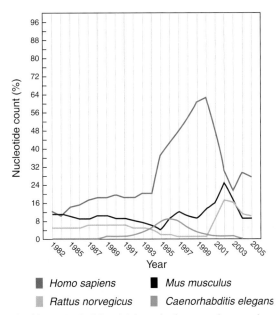

Figure 1 Percentage of the nucleotide content of the database for four popular organisms.

which is not surprising, considering the huge increase in sequence data available publicly due to technological advances. At the time of writing, the DDBJ/EMBL/GenBank nucleotide sequence databases contain around 116 billion nucleotides in more than 64 million individual entries; the daily inter-database exchange volume is 30 000 entries on average (varying between a few thousand and a million entries).

3.16.2.1.1 Database formats

The flatfile is the elementary unit of information in INSD. At the time of writing, it is the format of exchange from GenBank to the DDBJ and EMBL databases, and vice versa. The DDBJ/EMBL/GenBank flatfiles are the most commonly used formats in the representation of nucleotide sequences. The formats of the DDBJ flatfile and the GenBank flatfile are similar, but not identical; the EMBL format (**Figure 2**) is very different from the other two, but the essential functionality of the formats is the same for all three. The EMBL format, unlike GenBank and DDBJ, uses line-type prefixes, which indicate the type of information present in each line of the record. The feature section (see below), prefixed with FT in the EMBL database, is identical in content in all nucleotide sequence databases; the feature table format is described in the 'The DDBJ/EMBL/GenBank Feature Table: Definition' document, which is shared between the databases and regularly updated by them.[50]

The flatfile in all three formats can be separated into three parts: the header, which contains the information (descriptors) applicable to the whole record; the feature table, which contains the annotation for the sequence linked to the locations; and the nucleotide sequence itself. All nucleotide sequence database flatfiles in all formats end on the last line.

3.16.2.1.1.1 The header

The header is the most database-specific part of the record. The collaborating databases are not obliged to carry the same information in this segment. Minor variations of the content therefore exist, but considerable effort is made to ensure that the same information is carried from one to the other. This part contains the identifiers associated with the entry, the short description of the contents, the references (both to the literature and to the submission, at least one reference of any type per entry), the taxonomic information, cross-references, keywords, and comments.

3.16.2.1.1.2 The feature table

The middle segment of the INSD record, the feature table (FT), is the representation of the biological information in the record as provided by the submitter(s). The FT format design is based on a tabular approach, and consists of the feature key (a single word or abbreviation indicating the described biological property), the location information (instructions for finding the feature), and qualifiers (auxiliary information about a feature). The DDBJ/EMBL/GenBank feature table documentation describes in great detail the allowed features and what qualifiers are permitted with them. The FT format and wording use common biological research terminology whenever possible. For example, an item in the feature table such as shown in **Figure 3** might be read as: 'The feature CDS is a coding sequence beginning at base 23 and ending at base 400, has a product called 'alcohol dehydrogenase' and is coded for by a gene with the gene symbol 'adhI'.'

3.16.2.1.1.3 The sequence

The feature table is followed by the nucleotide sequence itself, in the format which can be seen in the database entry example (see **Figure 2**).

3.16.2.1.2 Editorial control and data redundancy

The final editorial control over each database entry in INSD rests with the submitters of individual records. Because of the commitment by the databases to store nucleotide sequence data supporting publications, a certain level of redundancy in INSD is unavoidable, especially where popular study subjects are concerned. However, while true in some respects, this statement is at the same time misleading. A study that is supposedly the same as another (e.g., when two independent groups sequence and submit to the INSD the same gene of the same organism), in practice either supports the results of the other or contradicts it to a certain degree, therefore the data are not redundant as such. As to the redundancy as in '100% sequence identity,' it is difficult to estimate across the whole of the database, due to the fact that most of the genomic sequences are open ended (i.e., unfinished), as far as biological reality is concerned. Very few complete sequences of large genomic components (such as complete chromosomes of eukaryotic organisms) are available at the time of writing, and those that are available do not have 'duplicates'; so, no comparison is possible. For medium-sized genomic components, such as complete microbial genomes, there are a few complete

```
ID    TRBG361    standard; mRNA; PLN; 1859 BP.
XX
AC    X56734; S46826;
XX
SV    X56734.1
XX
DT    12-SEP-1991 (Rel. 29, Created)
DT    15-MAR-1999 (Rel. 59, Last updated, Version. 9)
XX
DE    Trifolium repens mRNA for non-cyanogenic beta-glucosidase
XX
KW    beta-glucosidase.
OS    Trifolium repens   (white clover)
OC    Eukaryota; Viridiplantae; Streptophyta; Embryophyta;
      Tracheophyta;
OC    Spermatophyta; Magnoliophyta;  eudicotyledons;  core
      eudicots;  rosids;
OC    eurosids I; Fabales;  Fabaceae;  Papilionoideae; Trifolieae;
      Trifolium.
XX
RN    [5]
RP    1-1859
RX    MEDLINE;   91322517.
RX    PUBMED;  1907511.
RA    Oxtoby E., Dunn M.A. , Pancoro A., Hughes M.A.  ;
RT    "Nucleotide  and derived amino acid sequence of the
      cyanogenic
RT    beta-glucosidase (linamarase) from white clover
      (Trifolium repens L.).";
RL    Plant Mol.  Biol.   17(2):209-219 (1991).
XX
RN    [6]
RP    1-1859
RA    Hughes M.A.;
RT     ;
RL    Submitted (19-NOV-1990) to the EMBL/GenBank/DDBJ
      databases.
RL    M.A. Hughes, UNIVERSITY OF NEWCASTLE UPON TYNE, MEDICAL
      SCHOOL, NEW CASTLE
RL    UPON TYNE, NE2 4HH, UK
XX
FH     Key                  Location/Qualifiers
FH
FT     Source                     1..1859
FT                          /db_xref="taxon:3899"
FT                          /mol_type="mRNA"
FT                          /organism="Trifolium repens"
FT                          /tissue_type="leaves"
FT                          /clone_lib="lambda gt10"
FT                          /clone="TRE361"
FT     CDS                        14.. 1495
FT                          /db_xref= "GOA: P26204"
FT                          /db_xref= "HSSP: P26205"
FT                          /db_xref= "InterPro: IPR001360"
FT                          /db_xref= "UniProt/Swiss-Prot: P26204"
FT                          /note="non-cyanogenic"
FT                          /EC_number= "3.2.1.21"
FT                          /product= "beta-glucosidase"
FT                          /protein_id= "CAA40058.1"
FT
                              /translation="MDFIVAIFALFVISSFTITSTNAVEASTLLDIGNLSRSSPRGFI
                              FGAGSSAYQFEGAVNEGGRGPSIWDTFTHKYPEKIRDGSNADITVDQYHRYKEDVGIM
                              KDQNMDSYRFSISWPRILPKGKLSGGINHEGIKYYNNLINELLANGIQPFVTLFHWDL
                              PQVLEDEYGGFLNSGVINDFRDYTDLCFKEFGDRVRYWSTLNEPWVFSNSGYALGTNA
                              PGRCSASNVAKPGDSGTGPYIVTHNQILAHAEAVHVYKTKYQAYQKGKIGITLVSNWL
                              MPLDDNSIPDIKAAERSLDFQFGLFMEQLTTGDYSKSMRRIVKNRLPKFSKFESSLVN
                              GSFDFIGINYYSSSYISNAPSHGNAKPSYSTNPMTNISFEKHGIPLGPRAASIWIYVY
                              PYMFIQEDFEIFCYILKINITILQFSITENGMNEFNDATLPVEEALLNTYRIDYYYRH
                              LYYIRSAIRAGSNVKGFYAWSFLDCNEWFAGFTVRFGLNFVD"
"

FT     mRNA                 1..1859
FT                                /evidence=EXPERMENTAL
XX
SQ     Sequence 1859 BP;  609 A;  314 C; 355 G;  581 T;  0 other;
       aaacaaacca aatatggatt ttattgtagc catatttgct ctgtttgtta ttagctcatt         60
< >
```

Figure 2 Example of an EMBL nucleotide sequence database flatfile.

sequences of the supposedly same genomic elements (genomes of the same strain, sequenced by different groups, for instance) available for comparison, and for these elements one thing can be said – they are never 100% identical.

However, a level of 100% sequence identity can be estimated for some subsets of the data. For example, such estimation can be done for EMBLCDS: a subset of the nucleotide sequence data, showing each CDS feature annotated

```
Key          Location/ Qualifiers
<...>

CDS           23..400
             /product="alcohol dehydrogenase"
             /gene="adhI"
```

Figure 3 An example item in the INSD feature table.

in EMBL as a separate entry. Recently, CRC32 (Cyclic Redundancy Code, a checksum with a length of 32 bits), estimated for the nucleotide sequence of the CDS features, was added to all EMBLCDS entries. Although CRC32 can miss some of the sequence differences, that is, two different sequences can have exactly the same CRC32 checksum, the introduction of checksums allows the degree of 100% nucleotide sequence identity for the CDS features to be estimated reasonably well: at the time of writing, for nearly 3.5 million CDS features annotated in the EMBL database, 3.0 million unique checksums have been calculated, which means that 500 000 CDSs are exactly the same, sequence-wise, as one or more of the others.

In the data acquisition area, electronic bulk submission of sequences from the major sequencing centers overshadows all other input into the databases in volume; however, the relatively low number of submissions from smaller sequencing groups (direct submissions) is an important part of the databases, due to being usually more heavily annotated when the submission is received and heavily manually curated after submission. DDBJ/EMBL/GenBank records are often the primary source from which records in other databases such as the Reference Sequence collection (RefSeq)[6] and the Universal Protein Resource (UniProt)[7] are derived.

As mentioned before, the archival nature of the databases dictates the policy that the data are supplied by and is kept up-to-date (or not, as the case might be) by the submitters and their co-authors. One of the exceptions from this 'submitter supplies all the data' principle is the addition of cross-references. It must be noted that cross-references are not in the collaboratively controlled area of data in the nucleotide sequence databases, and the policies that apply to cross-referencing by the databases can be and are different. The EMBL Nucleotide Sequence Database, for example, inserts the cross-references into the entries on the basis of the information provided or extracted directly from the databases to which referencing is done. This ensures the consistent use of a particular type of cross-reference throughout the EMBL database. Sometimes the cross-referencing is done using the resources of more than one database. For example, cross-referencing to InterPro[8] is done via UniProt and UniParc[9]: UniParc is used to check for 100% protein sequence identity between the contents of/translation qualifier in the EMBL CDS feature and the sequence in the UniProt entry, and cross-references to InterPro are only inherited from those UniProt entries where these sequences are identical. The relative ease of doing cross-referencing this way is due to the databases involved being completely or partially in-house to the EBI.[10]

However, regardless of how many cross-references are added and how good and stable they are, they only imply the annotation. The main body of the annotation in INSD cannot be updated, corrected, or amended without the permission of the original submitter. This fact and the redundancy (in some senses) of the sequence data in INSD led to the creation of the Third Party Annotation section of EMBL, GenBank, and DDBJ, and to the secondary nucleotide sequence databases RefSeq, Ensembl,[11] and EBI Genome Reviews.[12] Third party annotation, RefSeq, and Genome Reviews are discussed later in this chapter.

3.16.2.1.3 Third party annotation in International Nucleotide Sequence Databases

Until 2002, the DDBJ/EMBL/GenBank databases collected and distributed only primary nucleotide sequence and annotation data resulting from direct sequencing of cDNAs, ESTs (expressed sequence tags), genomic DNA, etc. Primary data, as already mentioned above, are defined as a sequence that has been determined by a submitter, accompanied by the annotation attached to the sequence by the same group. Primary database entries remain in the ownership of the original submitter and the co-authors of the submission publication(s). The owners of database entries have privileges to update the data; only relatively minor changes are made at the level of the database. For example, when the format of the database changes, the old data have to be fitted into the new format; this is done by the databases, however, no new annotation as such is attached to the data by the database staff (with the exception of the cross-references, see above).

In response to demand from the research community, INSD created the third party annotation data set. The first database entries were released to the public in January 2002. The types of data that make up the third party

```
AH    TPA SPAN   PRIMARY_IEDENTIFIER  PRIMARY_SPAN     COMP
AS    1-251      BE529226.1           1-251
AS    68-450     BE524624.1           1-383
AS    394-1086   AJ420881.1           1-693
AS    826-1211   AV561543.1           1-386
```

Figure 4 Flatfile extract from BN000024, showing the new line types.

annotation data set include reannotations of existing entries, combinations of novel sequence and existing primary entries, and annotation of trace archive data and whole genome shotgun (WGS) data. Third party annotation submitters are required to provide DDBJ/EMBL/GenBank accession and version numbers and nucleotide locations for all primary entries to which their third party annotation entry relates. For third party annotation sequences created from the trace archive data, the trace archive identifier (e.g., TI123445566), and corresponding nucleotide locations must be provided.

Third party annotation entries can easily be distinguished from their primary counterparts. 'Third party annotation' appears at the beginning of each description (DE) line, and the keywords 'Third Party Annotation' appear in the keyword (KW) line.

The flatfile extract shown in **Figure 4** illustrates the two new line types that have been created for third party annotation entries. The assembly header (AH) line holds column headings for the assembly information. The assembly (AS) lines provide information on the composition of the third party annotation sequence by listing base span(s) of the third party annotation sequence, together with identifiers and base span(s) of contributing sequences; the last column (COMP) shows whether the participating base span is on the complement strand in the source sequence.

In order to ensure sequence annotation of the highest quality, entries that are yet to be discussed in peer reviewed publications are held confidential and are not visible to database users. This is an important difference from the policy of data release for primary entries.

Third party annotation sequences, like other types of nucleotide data, are exchanged among the DDBJ/EMBL/GenBank database collaborators on a daily basis. At the time of writing, 4525 third party annotation entries are publicly available, of which 2156 are derived from *Drosophila melanogaster*, 893 entries from *H. sapiens*, and 598 entries from *Oryza sativa*. These figures do not correspond proportionally to the representation of the same organisms in the main body of the database (which continues to be firmly dominated by human sequences), but all three organisms are very popular study subjects; also, it must be taken into account that the number of entries in the third party annotation section of the database is relatively low, and that the statistics are not yet very demonstrative.

3.16.2.2 National Center for Biotechnology Information Reference Sequences

Some sequences are represented more than once in DDBJ/EMBL/GenBank, which leads to a certain degree of redundancy (here used in the sense of 'same study subject sequenced and annotated many times over by different groups') in the DDBJ/EMBL/GenBank data. The NCBI RefSeq collection is a secondary database that aims to provide a comprehensive, integrated, nonredundant set of sequences, including genomic DNA, transcripts, and protein products, for certain organisms of major research interest. RefSeq is a public database of nucleotide and protein sequences with corresponding feature and bibliographic annotation, aiming to provide a stable reference for gene identification and characterization, mutation analysis, expression studies, polymorphism discovery, and comparative analyses. Sequences are annotated to include coding regions, conserved domains, variation, references, names, database cross-references, and other features, using a combined approach of collaboration and other input from the scientific community, automated annotation, propagation from GenBank and curation by NCBI staff. The main features of the RefSeq collection include:

- nonredundancy;
- explicitly linked nucleotide and protein sequences;
- updates to reflect current knowledge of sequence data and biology;
- data validation and format consistency;
- distinct accession series; and
- ongoing curation by NCBI staff and collaborators, with reviewed records indicated.

One of the main features of the RefSeq collection is ongoing curation, using a combined approach of collaboration and other input from the scientific community, automated annotation, propagation from GenBank and curation by

NCBI staff, with review status indicated in each record. In addition, the annotated RefSeq records and/or supplementary information are provided by collaborations established with nomenclature groups, model organism databases, and other groups in the scientific community. RefSeq records indicate the source INSD data, include references and annotations relevant to the gene, transcript, and protein, and indicate curation, with attribution to the group that performed it. RefSeq represents a nearly nonredundant collection, which is a summary of available information, and is an attempt to present a 'current' view of the sequence information, names, and other annotations. The latest release of RefSeq at the time of writing (release 10 March 2005) included about 190 000 genomic sequences and 355 000 RNA sequences from a total of 2827 organisms.

3.16.2.3 Genome Reviews

Primary databases represent results of experiments in the laboratory, supplemented with the interpretation of the results supplied by the research group. Secondary databases add value to the archival data by offering curated reviews of the primary data. Secondary databases are therefore a better data source for more up-to-date and summary data, while primary databases can be used to monitor ongoing research and the 'history' of the research.

One of the secondary nucleotide sequence databases is Genome Reviews,[51] a database in a EMBL-like format in which annotation has been corrected and enhanced (compared with the current status of the original submission) through the integration of data from many sources. The Genome Reviews database aims to overcome some of the archival database problems (such as lack of updates) by providing edited versions of entries representing complete genome sequences in the collaborative nucleotide sequence databases.

3.16.2.3.1 Data enhancements

Each Genome Review represents an edited version of the original flatfile (in EMBL database flatfile compatible format, see below), maintaining the link with the primary submission via the use of identifiers, with additional annotation imported from other data sources such as the UniProt Knowledgebase (UniProtKB), InterPro (see below), and others. For each particular data type, a preferred source of information (database) was chosen, and the annotation from the chosen source imported into the Genome Review, to either supplement or correct the annotation in the original submission. UniProtKB, which is a well-annotated resource where redundant information is merged and literature-based curation is used to extract experimental data, proved to be especially useful for the information-extracting purposes.

Also, annotations applied inconsistently among the original submissions have been standardized, and deleted in some cases, usually where the annotation coverage is low or varies significantly between genomes. For example, in some cases CDS features that were identified by the UniProt curators as 'false' (i.e., unlikely to encode a real protein) were removed. Further plans for annotation enhancement in Genome Reviews include introducing features (e.g., CDS or tRNA features) that are missing in the original submissions to the flatfiles.

New types of features and feature qualifiers have been introduced to describe additional data not previously present in EMBL (e.g., 'biological_process' and '/cellular_component' qualifiers were introduced to contain information that was either missing in the original entries or was contained in nonspecific qualifiers, such as '/note'). Despite this, the overall number of feature types and qualifier types used in Genome Reviews has decreased in comparison to the original submissions due to standardization of annotation and to removal of redundant data.

New features have been added to Genome Reviews by mapping features annotated on protein sequences onto corresponding regions of DNA. For example, regions of DNA encoding the mature peptides produced after cleavage of the primary translations, 'mat_peptide' features, were introduced. The number of cross-references (described by 'db_xref' qualifiers) was increased substantially in comparison with the original submissions.

At the time of writing, the latest release of Genome Reviews (1 March 2005) contained the following new features: for the first time annotations for tRNA genes have been added to all Genome Reviews entries where the original EMBL submission did not contain any such annotations, and annotations for protein-coding sequences corresponding to UniProt entries that were not originally annotated have been added by mapping the protein sequence against the corresponding genome.

3.16.2.3.2 Evidence tags

The most notable change to feature qualifiers in comparison to the original EMBL format has been the introduction of evidence tags. Evidence tags are attached to most feature qualifiers, indicating the primary source of the information. The evidence tags show the database that was used as a source of the information, and the identifiers within the database, where appropriate. The presence of an exclamation mark before the database identifier indicates that a deduction has been made from the absence of this identifier from the database in question.

```
FT   CDS             complement(19..870)
FT                   /codon_start=1
FT                   /locus_tag="Xfb0001  {UniProt/Swiss-Prot:Q9PHK6}"
FT                   /product="Putative replication protein Xfb0001
FT                   {Swiss-Prot:Q9PHK6}"
FT                    /biological_process="DNA replication {GO:0006260}"
FT                    /protein_id="AAF85568.1 {EMBL:AE003850}"
FT                    /db_xref="GO:0006260 {UniProt/Swiss-Prot:Q9PHK6}"
FT                    /db_xref="UniPort/Swiss-Prot:Q9PHK6 {EMBL:AE003850}"
FT                    /db_xref="UniParc:UPI00000223E {EMBL:AAF85568}"
FT                    /transl_table=11
FT
                  /translation="MPVITVYRHGGKGGVAPMNSSHIRTPRGEVQGWSPGAVRRNTEFL
                  MSVREDQLTGAGLALTLTVRDCPPTAQEWQKIRRAWEARMRRAGMIRVHWVTEWQRRGV
                  PHLHCAIWFSGTVYDVLLCVDAWLAVASSCGAGLRGQHGRIIDGVVGWFQYVSKHAARG
                  VRHYQRCSENLPEGWKGLTGRVWGKGGYWPVSDALRIDLQDHRERGDGGYFAYRRLVRS
                  WRVSDARSSGDRYRLRSARRMLTCSDTSRSRAIGFMEWVPLEVMLAFCANLAGRGYSVT
                  SE"
```

Figure 5 Excerpt of a Genome Reviews sample record, with evidence tags in curly braces.

Evidence tags are always located at the end of the qualifier value. They are contained within curly braces and preceded by a space. **Figure 5** shows an excerpt of a Genome Reviews sample record, specifically an example of CDS annotation in one of the Genome Reviews, where the application of evidence tags can be seen clearly.

3.16.2.3.3 Statistics and further developments

At the time of writing, the last release of the EBI Genome Reviews (March 2005) contained a total of 366 files for each chromosome and plasmid from 218 prokaryotes with completely sequenced genomes. In total, this Genome Reviews release files contained 691 million nucleotides and 643 000 annotated protein-coding sequences. In later releases of Genome Reviews, data will also be available for eukaryotic organisms.

The Genome Reviews data are synchronized with UniProtKB, and are distributed in flatfiles in EMBL-like format to ensure compatibility with existing tools. The main difference between the Genome Reviews and EMBL formats is the presence of evidence tags; if the evidence tags cause problems for a software package, they can be removed from the files by the supporting software.

3.16.3 Protein Sequence Databases

With the availability of hundreds of complete genome sequences from both eukaryotic and prokaryotic organisms, efforts are now focused on the identification and functional analysis of the proteins encoded by these genomes. The large-scale analysis of these proteins has started to generate huge amounts of data, particularly from the range of newly developed technologies in protein science. This increasing information has strengthened the central role that the protein databases play as a resource for protein data, freely available to the scientific community. There are a number of protein sequence databases that gather information about protein sequences. Understanding the differences between the various resources regarding data content, management, and distribution is essential to exploit them fully.

3.16.3.1 GenPept

An example of a protein sequence repository is the GenBank Gene Products Data Bank (GenPept), produced by the NCBI. The entries in this database are derived from translations of the sequences contained in the DDBJ/EMBL/GenBank nucleotide sequence database, and contain minimal annotation, which has been extracted primarily from the corresponding nucleotide entry. The entries lack additional annotation, and the database does not contain proteins derived from amino acid sequencing. Also, multiple records may represent a certain protein, with no attempt made to group these records into a single database entry.

3.16.3.2 Entrez Protein

Entrez Protein, produced by the NCBI, is another example of a sequence repository. The database contains sequence data translated from the nucleotide sequences of the DDBJ/EMBL/GenBank database as well as protein sequences from Swiss-Prot,[13] PIR-PSD,[14] RefSeq,[6] and the Protein Data Bank (PDB).[15] The database differs from GenPept in that many of the entries contain additional information derived from their curated database sources such as Swiss-Prot and PIR-PSD. As with GenPept, the sequence collection is redundant.

3.16.3.3 Reference Sequences

The NCBI RefSeq collection is a secondary database that provides a nonredundant data set of sequences representing genomic DNA, transcripts (RNA), and protein products, for major research organisms. The database, as of release 9 of January 2005, incorporates data from over 2700 organisms, and includes over a million proteins, comprising relevant taxonomic diversity spanning prokaryotes, eukaryotes, and viruses. One of the main features of the RefSeq collection is that NCBI staff and collaborators provide supplementary information for a subset of organisms. The curated protein records are created via a combined process of automated computational methods, collaboration, and manual data review. The level of curation is indicated on RefSeq records as a COMMENT feature, ranging from PREDICTED to REVIEWED, when sequence level curation has taken place. The RefSeq database, with less than 4% of records manually reviewed, can still be considered a sequence repository rather than a curated database.

3.16.3.4 The Universal Protein Resource

While repositories are essential in providing the user with sequences as quickly as possible, the addition of information to the sequence greatly increases the value of the database for users. In curated databases, expert biologists enrich the sequence data by combining experimental verified information extracted from the scientific literature with that obtained from rigorous sequence analysis. These databases are valuable sources of biochemical annotation, cellular and physiological properties, and literature references for a certain gene product.

An important step forward with regard to protein sequence databases was the decision by the National Institutes of Health in 2002 to award a grant to combine the curated Swiss-Prot, TrEMBL, and PIR-PSD databases into a single resource, UniProt.[7] Created to centralize protein database information, UniProt consists of three major databases:

- the UniProt Archive (UniParc), providing a nonredundant sequence collection of all publicly available protein sequence data;
- the UniProt Knowledgebase (UniProtKB), combining the work originally done with the expertly curated Swiss-Prot, TrEMBL, and PIR-PSD databases; and
- the UniProt Reference Clusters (UniRef), offering nonredundant views of the data contained in UniProtKB and UniParc.

3.16.3.4.1 UniProt Archive

The aim of UniParc is to provide a complete collection of protein sequences from predominantly publicly available resources. Most protein sequence data are derived from the translation of DDBJ/EMBL/GenBank CDS. However, a great deal of additional protein data are derived from other sources. Primary protein sequence data resulting from the direct sequencing of proteins is submitted directly to the Swiss-Prot and TrEMBL[13] and PIR-PSD databases. Furthermore, a large number of protein sequences are published in patent applications, are derived from structures published in PDB,[15] can be translated from gene predictions in resources such as Ensembl,[11] or are available in specialized databases. UniParc[9] was designed to capture all the available protein sequence data and contains all protein sequences from Swiss-Prot and TrEMBL, PIR-PSD, DDBJ/EMBL/GenBank CDS translations, Ensembl, the International Protein Index (IPI),[16] PDB, RefSeq, FlyBase,[17] WormBase,[18] and patent offices in Europe, the USA, and Japan. The combination of these sources makes UniParc the most comprehensive publicly accessible nonredundant protein sequence database available. UniParc represents each unique sequence only once, assigning it a unique UniParc identifier. Identical sequences are combined in a single UniParc entry regardless whether they are from the same or different species. UniParc cross-references the accession numbers of the source databases, and provides sequence versions that are incremented each time the underlying sequence changes, making it possible to observe sequence changes in all source databases. UniParc records display flags to indicate the status of the entry in the original source database, with 'active' meaning that the sequence is still present in the source database and 'obsolete' indicating that the sequence no longer exists in the source database. The main use of UniParc is for sequence similarity searches, since all known protein sequences are represented. UniParc records contain no annotations since an annotation will only be relevant for the actual context of the sequence. Provision of this context-dependent information is the purpose of the UniProtKB. UniParc of April 2005 contained 5 141 316 unique sequences from 12 166 358 different source database records.

3.16.3.4.2 UniProt Knowledgebase

UniProtKB is the centerpiece of the UniProt resource, and aims to provide all known information on a certain protein. UniProtKB consists of two sections: UniProtKB/Swiss-Prot, a manually curated database providing extensive annotation extracted from literature information and curator-evaluated computational analysis, and UniProtKB/TrEMBL, a section with computationally analyzed records awaiting for manual annotation. One of the first steps in the creation of UniProtKB was to incorporate all suitable PIR-PSD sequences missing from Swiss-Prot and TrEMBL. Bidirectional cross-references between Swiss-Prot and TrEMBL records and PIR-PSD entries were created for easy tracking of the now historic and no longer updated PIR-PSD records to the corresponding UniProt entries.

UniProtKB is, by design, nonredundant. The concept of nonredundancy in UniProtKB is very different from that in UniParc. In UniParc, all sequences that are 100% identical over their entire length are merged regardless of species. In contrast, UniProtKB aims to describe, in a single record, all protein products derived from a certain gene (or genes if the translation from different genes in a genome leads to indistinguishable proteins) from a certain species. An accession number is assigned to the whole record of gene products, and each protein form derived by alternative splicing, proteolytic cleavage, and post-translational modification (PTM) is described with position-specific annotation in the feature table. Furthermore, unique and stable feature identifiers (FTId) are created to distinguish between these different protein forms, since they have different functions and roles or only exist during specific developmental stages or under certain environmental conditions, even when they are all derived from a single gene. Currently, these identifiers are systematically attributed to the FT VARIANT lines of human UniProtKB entries, to alternative splicing events (VARSPLIC), and to certain glycosylation sites (CARBOHYD), but will ultimately be assigned to all types of FT lines. The tool VARSPLIC,[19] which is freely available, enables the recreation of all annotated splice variants from the feature table of a UniProtKB entry, or for the complete database. A FASTA-formatted file containing all splice variants annotated in the UniProtKB can be downloaded for use with similarity search programs.

UniProtKB only represents a subset of the sequences given in UniParc. The intensive curation process of merging sequences that are not 100% identical over the whole length due to alternative splicing, polymorphisms, and sequencing fragments or errors explains why distinct sequences in UniParc refer to a single UniProtKB record. Furthermore, a lot of the UniParc data are not included in UniProtKB, such as translations of eukaryotic gene predictions that are highly unstable and quickly replaced with new predictions, or notoriously bad annotated proteins like patent data, which will dilute the quality of the database. Most of the data in UniProtKB is obtained from DDBJ/EMBL/GenBank CDS translations, the sequences of the PDB structures, and the sequence data derived from amino acid sequencing submitted directly to UniProt. However, some data from these sources, for example CDS translations leading to small fragments or not coding for real proteins, synthetic sequences, nongermline immunoglobulins and T cell receptors, most patent application sequences, and highly overrepresented data such as GP120 sequences, are actively excluded from UniProtKB. UniParc sequences from other UniParc source databases, identified by the UniProt curators as important sequences missing in UniProtKB, are included. In this way, UniProtKB contains the relevant sequences available in the protein sequence repositories and, at the same time, minimizes the amount of unstable and low-quality data.

An important feature that makes UniProtKB a central hub for protein-related information is the extensive cross-references to external data collections, such as underlying DNA sequence entries in the DDBJ/EMBL/GenBank nucleotide sequence databases, two-dimensional polyacrylamide gel electrophoresis and three-dimensional protein structure databases, various protein domain and family characterization databases, PTM databases, species-specific data collections, variant databases, and disease databases. This interconnectivity is achieved almost exclusively via DR (database cross-reference) lines in the database flatfiles. The main annotation principles of UniProtKB are following the established procedures used to annotate Swiss-Prot, TrEMBL, and PIR-PSD, and this will be explained in some detail in Section 3.16.3.4.2.3 on the UniProtKB flatfile.

3.16.3.4.2.1 UniProt Knowledgebase Swiss-Prot

The Swiss-Prot section of UniProtKB is the major source for curated protein sequences. Each entry in Swiss-Prot is thoroughly analyzed and annotated by expert biologists. Literature-based curation is used to extract experimental data. This experimental knowledge is supplemented by manually confirmed results from various sequence analysis programs. Annotation includes the description of properties such as the function of a protein, PTMs, domains and sites, secondary and quaternary structure, similarities to other proteins, diseases associated with deficiencies in a protein, developmental stages in which the protein is expressed, in which tissues the protein is found, pathways in which the protein is involved, and sequence conflicts and variants. This information is stored mainly in the description (DE) and gene (GN) lines, the comment (CC) lines, the feature table (FT) lines, and the keyword (KW) lines. In March 2005 there were 178 940 UniProtKB/Swiss-Prot records, containing over 600 000 comments and nearly a million sequence features. The addition of a number of qualifiers in the comment and feature table lines during the annotation process allows users to distinguish experimentally verified data.[20] Gene names are standardized through the use of authoritative sources such as the HUGO Gene Nomenclature Committee's Genew project,[21] FlyBase,[17] and the Mouse Genome Database (MGD).[22]

3.16.3.4.2.2 UniProt Knowledgebase/Translation of European Molecular Biology Laboratory

The annotation process of a fully curated Swiss-Prot entry is a very labor-intensive and time-consuming process. To deal with the large numbers of new protein sequences, the UniProtKB/TrEMBL section aims to make these sequences directly available as computer-annotated entries, while waiting for manual annotation.

The life of a UniProtKB/TrEMBL record begins with the translation of a new CDS as soon as this appears in the DDBJ/EMBL/GenBank nucleotide sequence databases. At this point, all annotation in the UniProtKB/TrEMBL entry derives from the corresponding nucleotide entry. UniProtKB/TrEMBL also uses PDB as a primary source of protein data, as well as sequences derived from amino acid sequencing directly submitted to UniProt. UniParc sequences identified by the UniProt curators as important proteins missing in UniProtKB are also included. As soon as a new UniProtKB/TrEMBL record is created, redundancy is removed by merging multiple redundant records.[23] The automated enhancement of the information content in UniProtKB/TrEMBL follows,[24] which is based on various systems of standardized transfer of annotation from well-characterized proteins in UniProtKB/Swiss-Prot to unannotated UniProtKB/TrEMBL entries belonging to defined functional groups.[25,26] The definition of such protein groups uses resources such as InterPro,[8] PROSITE,[27] and Pfam,[28] among others. This process of adding accurate, high-quality information to UniProtKB/TrEMBL entries brings the standard of annotation in TrEMBL closer to that found in Swiss-Prot.

3.16.3.4.2.3 The UniProt Knowledgebase flatfile

A sample UniProtKB/Swiss-Prot flatfile entry corresponding to the DDBJ/EMBL/GenBank entry used as an example before is shown in **Figures 6–11**. The UniProtKB flatfile looks quite similar to the EMBL Nucleotide Sequence Database flatfile, as both are made up of different line types, each of them beginning with a two-character line code indicative of the type of data stored in the line. Some line types may occur more than once, and some do not appear at all. **Figure 6** shows the first lines of the sample entry.

The identification line (ID) is the first line in every UniProtKB entry, and contains the entry name. In our UniProtKB/Swiss-Prot example, ROA1_HUMAN is the entry name for the human heterogeneous nuclear ribonucleoprotein A1. The accession number (e.g., P09651) is found on the AC (accession) line(s). UniProt/TrEMBL records use the accession number and also the species identification code as the entry name (e.g., Q13174_HUMAN). The accession number should be always used to cite a specific entry, since it provides a stable way of identifying a particular entry. The three date (DT) lines show when the entry was created, when the sequence was updated last, and when the most recent annotation was added. The DE (description) line(s) lists all the names known for a particular protein. The GN (gene name) line follows, listing all known gene symbols for the gene. This line can be absent if no gene name has been given. The organism species (OS) and the organism classification (OC) provide the scientific name of the source organism, its common name where available, and its taxonomic classification. The OX (organism taxonomy cross-reference) line stores the identifier of a specific organism in the NCBI's taxonomy database. An OG (organelle) line is present if the sequence comes from a particular organelle (e.g., chloroplast) or extrachromosomal element. Thereafter, the reference block comprises the RN (reference number), RP (reference position), RC (reference comment), RX (reference cross-reference), RG (reference group), RA (reference author), RT (reference title), and RL (reference location) lines. These mainly include information from the published literature on sequencing work, but also on three-dimensional structure, PTMs, and isoforms. Some sequence data from mega-sequencing projects are no longer published in the traditional sense. Information submitted to UniProt is also included in the reference block. The next blocks of information in a UniProt record are the CC lines (**Figure 7**).

The CC (comments) lines contain biological information relating to the sequence. The DR (database cross-references) lines interconnect UniProt to other biomolecular databases. These cross-links allow users to easily retrieve additional information on a certain protein (**Figure 8**).

```
ID     ROA1_HUMAN    STANDARD;    PRT;    371 AA.
AC     P09651;
DT     01-MAR-1989 (Rel. 10, Created)
DT     01-AUG-1990 (Rel. 15, Last sequence update)
DT     15-MAR-2004 (Rel. 43, Last annotation update)
DE     Heterogeneous nuclear ribonucleoprotein A1 (Helix-destabilizing
DE     protein) (Single-strand binding protein) (hnRNP core protein A1).
GN     HNRPA1.
OS     Homo sapiens (Human).
OC     Eukaryota; Metazoa; Chordata; Craniata; Vertebrata; Euteleostomi;
OC     Mammalia; Eutheria; Primates; Catarrhini; Hominidae; Homo.
OX     NCBI_TaxID=9606;
```

Figure 6 The first lines of a UniProtKB/Swiss-Prot flatfile entry (description in the text).

```
CC       -!-     FUNCTION: Involved in the packaging of pre-mRNA into hnRNP
CC               particles, transport of poly(A) mRNA from the nucleus to the
CC               cytoplasm and may modulate splice site selection.
CC       -!-     SUBCELLULAR LOCATION: Nuclear. Shuttles continuously between the
CC               nucleus and the cytoplasm along with mRNA. Component of
CC               ribonucleosomes.
CC       -!-     ALTERNATIVE PRODUCTS:
CC               Event=Alternative splicing; Named isoforms=2;
CC               Name=A1-B;
CC               IsoID=P09651-1; Sequence=Displayed;
CC               Name=A1-A;
CC               IsoID=P09651-2; Sequence=VSP_005824;
CC               Note=Is twenty times more abundant than isoform A1-B;
CC       -!-     SIMILARITY: BELONGS TO THE A/B GROUP OF HNRNP, WHICH ARE BASIC AND
CC               GLY-RICH PROTEINS.
CC       -!-     SIMILARITY: Contains 2 RNA recognition motif (RRM) domains.
```

Figure 7 The CC lines of a UniProtKB/Swiss-Prot flatfile entry.

```
DR     EMBL; X12671; CAA31191.1; -.
DR     EMBL; X06747; CAA29922.1; ALT_SEQ.
DR     EMBL; X04347; CAA27874.1; -.
DR     EMBL; X79536; CAA56072.1; -.
DR     PIR; S02061; S02061.
DR     PDB; 1HA1; 15-MAY-97.
DR     PDB; 1L3K; 17-APR-02.
DR     PDB; 1UP1; 17-SEP-97.
DR     PDB; 2UP1; 22-JUN-99.
DR     SWISS-2DPAGE; P09651; HUMAN.
DR     Aarhus/Ghent-2DPAGE; 207; NEPHGE.
DR     Aarhus/Ghent-2DPAGE; 2114; NEPHGE.
DR     Aarhus/Ghent-2DPAGE; 3612; NEPHGE.
DR     Genew; HGNC:5031; HNRPA1.
DR     GK; P09651; -.
DR     MIM; 164017;-.
DR     GO; GO:0005737; C: cytoplasm; TAS.
DR     GO; GO:0030530; C: heterogeneous nuclear ribonucleoprotein com...; TAS.
DR     GO; GO:0005654; C: nucleoplasm; TAS.
DR     GO; GO:0003723; F: RNA binding; TAS.
DR     GO; GO:0006397; P: mRNA processing; TAS.
DR     GO; GO:0006405; F: RNA-nucleus export; TAS.
DR     InterPro; IPR000504; RNA_rec_mot.
DR     Pfam; PF00076; rrm; 2.
DR     SMART; SM00360; RRM; 2.
DR     PROSITE; PS50102; RRM; 2.
DR     PROSITE; PS00030; RRM_RNP_1; 2.
```

Figure 8 The DR lines of a UniProtKB/Swiss-Prot flatfile entry.

```
KW     Nuclear protein; RNA-binding; Repeat; Ribonucleoprotein; Methylation;
KW     Alternative splicing; 3D-structure; Polymorphism.
```

Figure 9 The KW lines of a UniProtKB/Swiss-Prot flatfile entry.

```
FT   INIT_MET     0     0
FT   DOMAIN       3    93    GLOBULAR A DOMAIN.
FT   DOMAIN      94   184    GLOBULAR B DOMAIN.
FT   DOMAIN      13    96    RNA-binding (RRM) 1.
FT   DOMAIN     104   183    RNA-binding (RRM) 2.
FT   DOMAIN     217   239    RNA-BINDING RGG-BOX.
FT   DOMAIN     194   371    Gly-rich.
FT   DOMAIN     319   356    NUCLEAR TARGETING SEQUENCE (M9).
FT   MOD_RES    193   193    METHYLATION (BY SIMILARITY).
FT   VARSPLIC   251   302    Missing (in isoform A1-A).
                             /FTId=VSP_005824.
FT   VARIANT     72    72    N→S (in dbSNP:6533).
                             /FTId=VAR_014711.
FT   MUTAGEN    325   325    G→A: NO NUCLEAR IMPORT NOR EXPORT.
FT   MUTAGEN    326   326    P→A: NO NUCLEAR IMPORT NOR EXPORT.
FT   MUTAGEN    333   334    GG→LL: NORMAL NUCLEAR IMPORT AND EXPORT.
FT   CONFLICT   139   139    R→P (in Ref. 4).
FT   HELIX       10    13
FT   STRAND      14    18
FT   TURN        22    23
. .  25 FT lines omitted
. .
```

Figure 10 The FT lines of a UniProtKB/Swiss-Prot flatfile entry.

```
SQ   SEQUENCE    371 AA;    38715    MW;    B3EEFA5AE1DB7C26    CRC64;
     SKSESPKEPE  QLRKLFIGGL  SFETTDESLR  SHFEQWGTLT  DCVVMRDPNT  KRSRGFGFVT
     YATVEEVDAA  MNARPHKVDG  RVVEPKRAVS  REDSQRPGAH  LTVKKIFVGG  IKEDTEEHHL
     RDYFEQYGKI  EVIEIMTDRG  SGKKRGFAFV  TFDDHDSVDK  IVIQKYHTVN  GHNCEVRKAL
     SKQEMASASS  SQRGRSGSGN  FGGGRGGGFG  GNDNFGRGGN  FSGRGGFGGS  RGGGGYGGSG
     DGYNGFGNDG  GYGGGGPGYS  GGSRGYGSGG  QGYGNQGSGY  GGSGSYDSYN  NGGGRGFGGG
     SGSNFGGGGS  YNDFGNYNNQ  SSNFGPMKGG  NFGGRSSGPY  GGGGQYFAKP  RNQGGYGGSS
     SSSSYGSGRR  F
/ /
```

Figure 11 The SQ lines of a UniProtKB/Swiss-Prot flatfile entry.

The DR lines are followed by the KW (keyword) lines, which list relevant keywords that can be used to retrieve a specific subset of protein entries from the database (**Figure 9**), and the FT (feature) lines describe regions or sites of interest in the sequence (**Figure 10**).

The feature table provides detailed information on PTMs, binding sites, active sites of an enzyme, secondary structure, sequence conflicts and variations, signal sequences, transit peptides, propeptides, transmembrane regions, and other sequence-related characteristics. The SQ (sequence header) line holds the sequence itself, and every record ends with the usual '//' characters (**Figure 11**).

3.16.3.4.3 UniProt Reference Clusters

The concept of the UniProt nonredundant reference clusters UniRef100, UniRef90, and UniRef50 is based on the same principles as UniProtKB, but with some relevant differences. UniRef merges sequences automatically across different species, and also includes some data from UniParc that is not part of UniProtKB, such as translations from highly unstable gene predictions. UniRef100 is based on all UniProtKB records, as well as UniParc records overrepresented in UniProtKB, Ensembl protein translations, PDB, and RefSeq sequences. The production of UniRef100 begins with the clustering of all these records by sequence identity. Identical sequences and subfragments are presented as a single UniRef100 entry, containing the accession numbers of all the merged entries, the protein sequence, bibliography, links to the corresponding UniProtKB and UniParc records, as well as to close sequence neighbors having at least 95% sequence identity.

UniRef90 and UniRef50 are built from UniRef100, to provide nonredundant sequence collections for the scientific user community to use in performing faster homology searches. All records having more than 90% or 50% sequence identity are merged into a single UniRef90 or UniRef50 entry, respectively. UniRef90 and UniRef50 yield a size reduction of approximately 40% and 65%, respectively.

3.16.3.4.3.1 Specialized protein databases

In addition to the protein databases described above, there are numerous specialized databases available to the life sciences community. Some are devoted to one particular aspect of proteins while others seek to consolidate and exploit already existing resources to their full potential. One example of the former type of database is PDB, archiving three-dimensional structural data. GOA produces a dynamic controlled vocabulary that can be applied to all organisms, even while knowledge of gene and protein roles in cells is still accumulating and changing. IntAct stores protein interaction data. Other examples are MEROPS,[29] a catalog of peptidases, and model organism databases such as SGD,[30] for *Saccharomyces* proteins. In contrast, InterPro is an example of an integrated protein resource combining a number of databases that use different methodologies and a varying degree of biological information on well-characterized proteins to derive protein signatures for protein families, domains, and sites. This database and its member databases are described below.

3.16.4 Integrated and Comparative Databases

So far in this chapter, mainly primary sequence databases have been introduced, which provide detailed information on nucleotide or protein sequences. Despite a high level of links to other specialized databases, the need for better integration of data from different resources is growing. Such integrated resources can facilitate database searches considerably, and thus help the user to quickly gather comprehensive information, as well as compare data from different sources in context. Some of these resources are centered on proteins, and can be used to study a protein in the context of other protein family members, within and between species, whereas others aim at combining genomic and proteomic information. The user can undertake species-specific searches, or searches for other biological entities, or learn specifically about protein functions or whole proteomes.

3.16.4.1 InterPro

The identification of possible DNA-coding regions can be deduced by similarity to previously characterized genes. Inferring biological function for a coding region can be a complicated process, which cannot always be achieved by sequence similarity searches. Protein sequence comparisons often provide the first clues to the structure and function of novel proteins, as functional constraints are known to persist in evolution. Protein domain signature databases are available for identifying distant relationships in novel sequences with a known protein family. InterPro[8] is an integrated resource of protein families, domains, and functional sites that amalgamates information from its member databases, which are currently PROSITE,[27] PRINTS,[31] Pfam,[28] ProDom,[32] SMART,[33] TIGRFAMS,[34] PIR SuperFamily (PIRSF),[35] SUPERFAMILY,[36] PANTHER,[37] and Gene3D.[38] InterProScan[39] is a tool that combines the different protein recognition methods and scanning tools of each method into one powerful search resource unifying the strength of the individual signature database methods to ensure the best prediction of protein domains for a query translation. In the absence of biochemical characterization of a protein, domain predictions can be a good guide to protein function.

Release 9.0 of InterPro (17 February 2005) contains 11 605 entries, representing 2982 domains, 8373 families, 222 repeats, 27 active sites, 21 binding sites, and 20 PTM sites.

3.16.4.1.1 Integrating the signatures

The crucial feature of InterPro is its ability to group proteins in families, and protein sequences into domains, integrating the methods of the member databases. To diagnose a domain or characteristic region of a protein family in a protein sequence, the member databases employ variations of the handful of signature methods available, which include patterns, profiles and hidden Markov models (HMMs): PROSITE uses regular expressions and profiles; PRINTS uses position-specific scoring matrix (PSSM)-based fingerprints; ProDom uses automatic sequence clustering; and Pfam, SMART, TIGRFAMs, PIRSF, SUPERFAMILY, PANTHER, and Gene3D use HMMs. All these different methods have their individual strengths and weaknesses – regular expressions, for example, are ideal for finding short motifs, but not so reliable in the identification of members of highly divergent families, fingerprints are excellent for determining specific subfamily memberships, but less reliable in finding short motifs, and methods based on HMMs and profiles are the best for identifying members of divergent superfamilies, but not for determining specific subfamily memberships. InterPro integrates all of these methods to enable the best possible classification of proteins.

In InterPro, signatures from the member databases are integrated manually as they are developed. For this, the signatures must overlap, at least in part; in position to the protein sequence, they should have at least 75% overlap; and they must all describe the same biological entity (family, domain, etc.). New signatures from member databases are either integrated into existing InterPro entries, or assigned unique InterPro accession numbers. Each InterPro entry is described by one or more of these signatures, and corresponds to a biologically meaningful family, domain, repeat, or site (e.g., PTM). Not every entry, though, contains a signature from each member database; only those signatures that correspond to each other are united. Entries are assigned a type to describe what they represent, which may be family,

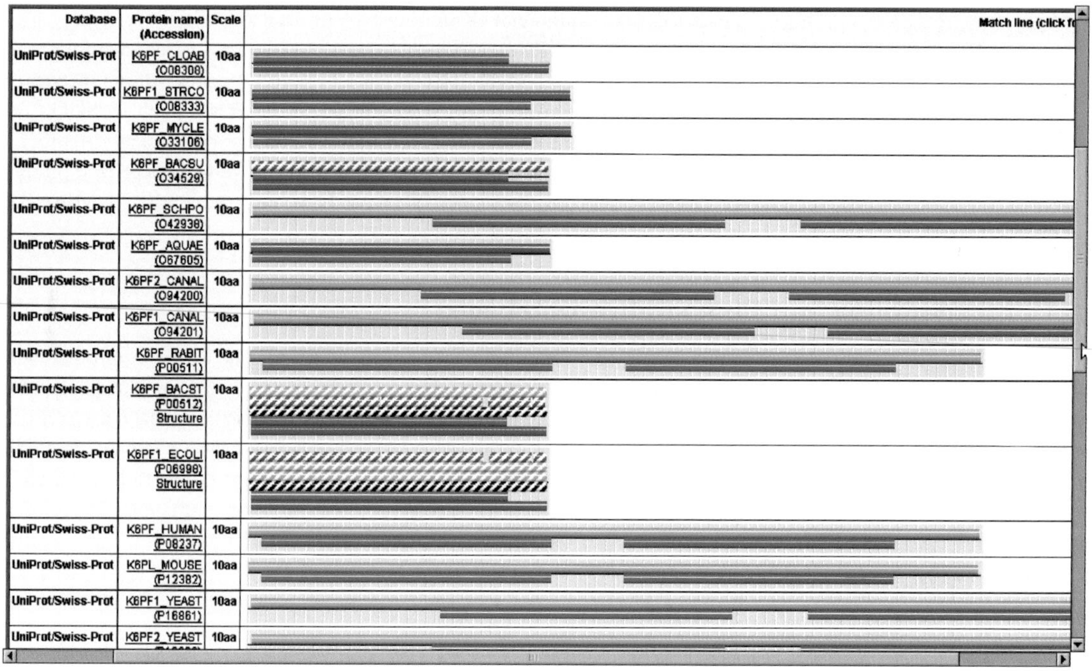

Figure 12 Overview matches for InterPro entry IPR000023 (the phosphofructokinase family). Each protein has one match line, which contains matches colored by their InterPro entry. The vertical lines are drawn at 10 amino acid (aa) intervals.

domain, repeat, PTM, active site, or binding site. They may be related to each other through two different relationships: the parent/child and contains/found in relationship. Parent/child relationships are used to describe a common ancestry between entries, whereas the contains/found in relationship generally refers to the presence of genetically mobile domains. InterPro entries are annotated with a name, an abstract, mapping to gene ontology (GO)[40] terms, and links to specialized databases. InterPro groups all protein sequences matching related signatures into entries. All hits of the protein signatures in InterPro against a composite of the UniProt components UniProt/Swiss-Prot and UniProt/TrEMBL are precomputed. The matches are available for viewing in each InterPro entry in different formats (**Figure 12**).

3.16.4.1.2 InterProScan

Protein matches are calculated using the InterProScan software package. This sequence search tool combines the different protein signature recognition methods native to the InterPro member databases into one resource. After submission of a protein sequence to InterProScan, an output with all results in a single format is provided, which can be HTML, text, or XML. The search is sequence based, and uses the tools provided by the member databases (i.e., PfScan for PROSITE patterns and profiles; hmmpfam for Pfam, SMART, and TIGRFAMs HMMs; fingerPRINTScan for PRINTS fingerprints;[41] BlastProDom for ProDom patterns; and PSI-BLAST for SUPERFAMILY PSSMs). Each of the applications returns lists of hits to the individual databases. The hits are combined, and the InterPro families to which they belong are returned to the user. InterProScan is more than a simple wrapping of sequence analysis applications, though, since it has to perform a considerable number of data lookups for some database and program outputs. Some postprocessing of data is linked to the software package, too; some of the outputs are filtered through family-specific thresholds for increased accuracy of the results. As a result of a sequence search, matches to the member databases and the corresponding InterPro entries are shown. The position of the signatures within the sequence and a graphical and a table view of the matches, as well as links to the corresponding results in the member databases, are provided (**Figures 13** and **14**).

3.16.4.2 The International Protein Index

The genome sequences of many organisms are now known, including those of several higher eukaryotes. There is, however, not much known about the corresponding proteomes of these organisms. The information available on the encoded proteins, gained either experimentally or computationally, is stored in a variety of databases. This lack of a single resource that could offer full coverage of known and predicted proteins makes it difficult to obtain sufficient

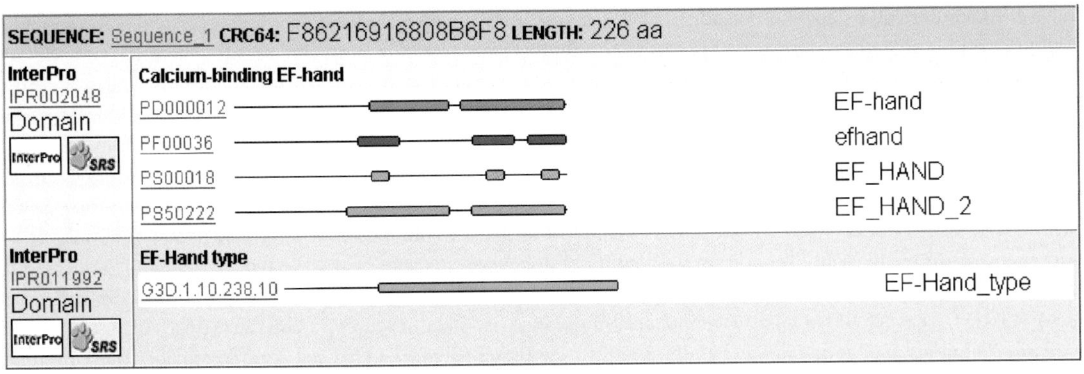

Figure 13 InterProScan output for the human calcium-binding protein 1 (CABP1), graphical view. The signatures are color coded, and the width of the colored bands represents the boundaries of the signatures. The codes on the left-hand side of the bands are the accession numbers of and links to the source databases (PD, ProDom; PF, Pfam; PS, PROSITE; G3D, Gene3D) and to InterPro (IPR). On the right-hand side, the identification codes from the source databases are shown.

SEQUENCE: Sequence_1 CRC64: F86216916808B6F8 LENGTH: 226 aa	
InterPro IPR002048 Domain	**Calcium-binding EF-hand** PRODOM PD000012 *EF-hand* 1.9999999999999998E-26 [93-146]T 3.0E-34 [154-224]T PFAM PF00036 *efhand* 2.8E-6 [85-113]T 6.3E-8 [162-190]T 0.0020 [199-225]T PROFILE PS00018 *EF_HAND* 0.0 [94-106]T 0.0 [171-183]T 0.0 [208-220]T PROFILE PS50222 *EF_HAND_2* 19.362 [77-146]T 23.54 [161-224]T
Parent	IPR011992
Children	IPR000261 IPR001751
Found in	IPR001125 IPR003299 IPR008080 IPR012008
Contains	no entries
GO terms	Molecular Function: calcium ion binding (GO:0005509)
InterPro IPR011992 Domain	**EF-Hand type** CATH G3D.1.10.238.10 *EF-Hand_type* 2.101408E-31 [65-226]T
Parent	no parent
Children	IPR002048
Found in	IPR001217 IPR003153 IPR007736
Contains	no entries
GO terms	Molecular Function: calcium ion binding (GO:0005509)

Figure 14 InterProScan output for human CABP1, table view. Information on the positions of each match on the sequence is provided. Links to the source databases (and to GO) entries are shown (see **Figure 13** for definitions).

information on proteomes. The IPI[16] is an attempt to overcome this obstacle, by providing nonredundant data sets for certain eukaryotic species. The sets are built from the UniProt/Swiss-Prot, UniProt/TrEMBL, Ensembl, and RefSeq databases, which all provide proteome information.

3.16.4.2.1 Nonredundant complete UniProt proteome sets

The UniProt protein sequence resource constructs biweekly complete nonredundant proteome sets that can be downloaded via the EBI's Integr8 browser[12] (see below), which is a further development of the Proteome Analysis database.[42] Each set and its analysis is made available shortly after the appearance of a new complete genome sequence in the nucleotide sequence databases. A standard procedure is used to create proteome sets for bacterial, archaeal, and nonmetazoan eukaryotic genomes from UniProtKB. Proteome sets for certain metazoan genomes, currently *H. sapiens*, *M. musculus*, *R. norvegicus*, *Brachydanio rerio*, and *Arabidopsis thaliana*, are produced by a separate procedure, since the UniProtKB proteome sets for metazoan organisms may be incomplete, as a full and accurate set of coding sequence predictions are not yet available in the nucleotide sequence databases. Therefore, another data source is used, the Ensembl genome browser.[11] Ensembl is a joint project between the EMBL-EBI and the Wellcome Trust Sanger Institute that aims at developing a system that maintains automatic annotation of large eukaryotic genomes. It is a comprehensive source of stable annotation with confirmed gene predictions that have been integrated from external data sources. Ensembl annotates known genes and predicts new ones, with functional annotation from InterPro (see above), OMIM, the NCBI's Online Mendelian Inheritance in Man resource,[52] SAGE (Serial Analysis of Gene Expression) libraries, and gene families. UniProt and Ensembl jointly offer additional sets, comprising the UniProtKB nonredundant proteome sets for these organisms plus additional nonredundant proteins from Ensembl.

3.16.4.2.2 Building the International Protein Index

In addition to the UniProt and Ensembl proteome sets, the EBI also produces monthly a third data set for *H. sapiens*, *M. musculus*, *R. norvegicus*, *B. rerio*, and *A. thaliana*, aimed at users who want data sets offering complete coverage of all predicted coding sequences. IPI[53] is, as mentioned before, derived from UniProtKB, Ensembl, and the NCBI's RefSeq project. RefSeq is another database that stores protein sequences linked to higher eukaryotic genomes, and it contains, in different sections, protein sequences linked to experimentally determined mRNAs, and predicted sequences derived automatically from the genome. Thus, IPI represents a top-level guide to the main resources that describe the human, mouse, rat, zebrafish, and mouse-ear cress proteomes. There is no consensus on the gene number or on the identity and structure of each gene for the genomes of these organisms, although they are well studied. The data in the different databases are the result of different gene prediction algorithms and experimentally determined sequences. IPI attempts a semantic resolution of the different data in the databases, and its aim is to provide maximum coverage with minimal redundancy, stable identifiers to allow the tracking of sequences between releases, and interdatabase cross-references. Each IPI entry represents a cluster of entries from the source databases believed to represent the same protein. One difficulty in creating IPI is that there is no absolute way of telling whether two entries in molecular biology databases represent different biological entities or the same entity rendered differently owing to in silico or experimental artifacts. To assemble IPI data sets, an automatic and pragmatic approach is chosen to build clusters through combining knowledge already present in the primary data sources (and in the cross-references between them) with the results of protein sequence similarity comparisons. After a cluster is assembled, a master entry from among the cluster members is chosen, which supplies the IPI entry with its sequence and annotation. Finally, an identifier is chosen for each cluster.

3.16.4.3 Integr8

High-standard protein sequence databases (e.g., UniProt/Swiss-Prot) and the model organism databases (e.g., FlyBase,[17] WormBase,[18] ZFIN,[43] or MGD,[22] to name but a few) try to link complete genomes data and experimental results through manual curation, but due to the amount of available data this process still has a growing backlog of unprocessed data. To overcome this obstacle, the Integr8 project has been developed, as a joint effort between different European data providers. It aims at integrating genomic and proteomic data into one single resource, allowing the choice of individual views on different types of information.

Traditionally, different types of data describing genomes and proteomes are stored in different databases. These data of interest comprise, for example, gene and protein sequences, experimentally obtained feature information, provided by microarray or protein identification and interaction assays, or information drawn by in silico comparison with known sequences. Integr8 is an approach that aims to make these different types of data searchable, by storing the relationships of biological entities (e.g., 'gene,' 'protein,' and 'species') to each other; apart from this information, only the core data necessary to identify the biological objects are integrated (a completely automated process). For more detailed analyses, links to the source databases are provided. Such a framework of cross-correlated data allows for new types of data to be attached to existing database entries (e.g., protein–protein interactions). Also, complex queries between these different resources and entity-centric views of complete genomes and proteomes are possible. The integration is done against a reference genomic framework, taken from EMBL or Ensembl (depending on the organism). Other source databases are scanned regularly for updates and new entries, and mappings are calculated independently when necessary. Stable identifiers for the biological entities are provided.

To achieve this, some existing tools for data integration based around the EMBL nucleotide and the UniProt protein sequence database have been used. These are the Proteome Analysis database, the CluSTr database, IPI, UniParc, and Genome Reviews. The BioMart tool allows customized queries on the complete proteome data, based on individual selections of organisms and protein attributes. IPI, UniParc, and Genome Reviews have already been described in this chapter, so in the following we will focus on the analysis of proteomes, and how Integr8 can be browsed.

3.16.4.3.1 Proteome analysis

The statistical analysis of proteomes is an important part of the Integr8 web portal. The tool for this, formerly known as the Proteome Analysis database, is a highly integrative resource itself. It complements Genome Reviews, in which whole genomes are annotated, by providing comparative information about the proteins encoded by the genomes, the so-called 'proteomes'. The proteome sets are built from UniProtKB, as described above. Analysis of the proteomes is carried out by combining information from different resources. InterPro provides information on protein families, domains, and sites. A specifically designed application enables InterPro proteome comparisons for any one proteome against one or more of the proteomes in the database, using different types of analyses (an example is shown in **Figure 15**). Furthermore, users can obtain statistics describing each complete proteome and download associated data underpinning these statistics. This approach offers a view of this data from the perspective of organisms with completely deciphered genomes. It is by its nature very protein-centric; however, enclosing it in the larger context of Integr8 helps to provide more powerful interfaces for interrogating these data, and to link them more closely to the underlying genomic data.

Another feature of proteome analysis, the automatic clustering of proteins, is provided by the CluSTr database.[44] The CluSTr database offers an automatic classification of UniProtKB proteins into groups of related proteins. The clustering is based on the analysis of all pairwise comparisons between protein sequences using the Smith–Waterman algorithm.[45] Statistical significance is estimated using Monte Carlo simulation, resulting in a Z score.[46] Analysis carried out at different levels of protein similarity yields a hierarchical organization of clusters. By working with clusters at different levels of similarity, biologically meaningful clusters can be selected for different groups of proteins, making use of the flexibility of the database. Clusters for mammalian proteins, plant proteins, and for 11 complete eukaryotic

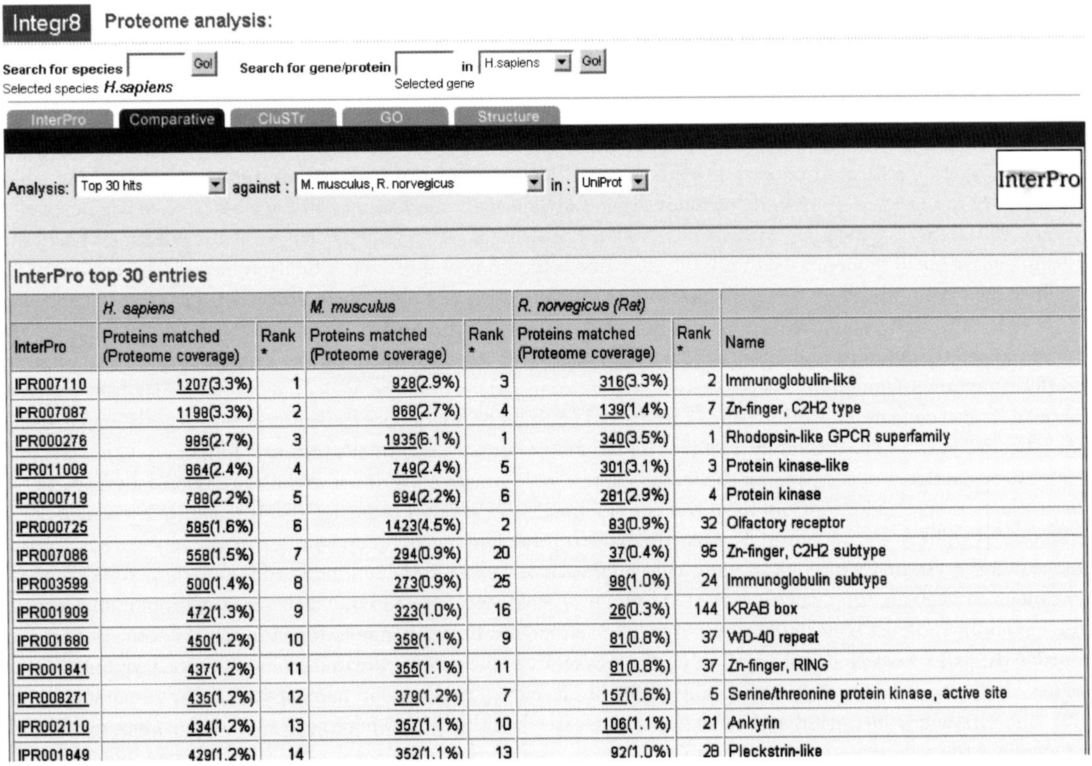

InterPro	*H. sapiens* Proteins matched (Proteome coverage)	Rank *	*M. musculus* Proteins matched (Proteome coverage)	Rank *	*R. norvegicus (Rat)* Proteins matched (Proteome coverage)	Rank *	Name
IPR007110	1207(3.3%)	1	928(2.9%)	3	316(3.3%)	2	Immunoglobulin-like
IPR007087	1198(3.3%)	2	868(2.7%)	4	139(1.4%)	7	Zn-finger, C2H2 type
IPR000276	985(2.7%)	3	1935(6.1%)	1	340(3.5%)	1	Rhodopsin-like GPCR superfamily
IPR011009	864(2.4%)	4	749(2.4%)	5	301(3.1%)	3	Protein kinase-like
IPR000719	788(2.2%)	5	694(2.2%)	6	281(2.9%)	4	Protein kinase
IPR000725	585(1.6%)	6	1423(4.5%)	2	83(0.9%)	32	Olfactory receptor
IPR007086	558(1.5%)	7	294(0.9%)	20	37(0.4%)	95	Zn-finger, C2H2 subtype
IPR003599	500(1.4%)	8	273(0.9%)	25	98(1.0%)	24	Immunoglobulin subtype
IPR001909	472(1.3%)	9	323(1.0%)	16	26(0.3%)	144	KRAB box
IPR001680	450(1.2%)	10	358(1.1%)	9	81(0.8%)	37	WD-40 repeat
IPR001841	437(1.2%)	11	355(1.1%)	11	81(0.8%)	37	Zn-finger, RING
IPR008271	435(1.2%)	12	379(1.2%)	7	157(1.6%)	5	Serine/threonine protein kinase, active site
IPR002110	434(1.2%)	13	357(1.1%)	10	106(1.1%)	21	Ankyrin
IPR001849	429(1.2%)	14	352(1.1%)	13	92(1.0%)	28	Pleckstrin-like

Figure 15 The Integr8 view of the top 30 entries for *H. sapiens*, in comparison with those for *M. Musculus* and *R. norvegicus*, obtained by InterPro analysis.

genomes (including *H. sapiens*) have been built. In the proteome analysis part of the Integr8 browser, CluSTr currently covers these 11 complete eukaryotic genomes, with more to come. It also provides links to InterPro, with its wealth of information on protein families, domains, and functional sites.

A functional classification of proteomes is performed according to the assignment of proteins with a special selection of GO terms. This dynamic controlled vocabulary can be applied to all organisms, even while knowledge of gene and protein roles in cells is still accumulating and changing.[40] For proteome analysis, the whole range of GO terms is not used, which is very comprehensive with currently more than 17 700 terms in total, a number that is constantly growing. Instead, for assigning proteins to GO terms, only a carefully chosen selection of high-level terms from each of the three GO sections (molecular function, biological process, and cellular component) called GO Slim is used. (A number of such 'GO Slims' exist, each custom built for specific projects.) A functional classification of the proteins within each proteome set has been generated to show the percentage of proteins involved in certain functions. All organisms are linked to the NEWT taxonomy browser,[54] and each entry contains links to related resources.

To complete the information about the proteomes, links to structural information databases such as the Homology derived Secondary Structure of Proteins (HSSP) database,[47] PDP,[15] and the Structural Classification of Proteins (SCOP) database[48] are provided, for individual proteins from each of the proteomes.

3.16.4.3.2 The Integr8 browser

The tool used to view the contents of Integr8 and the proteome analyses is BioMart, a development of EnsMart.[49] A user can customize an analysis. First, after having chosen the 'Focus: UniProt Proteomes' option (BioMart also offers the choice of data from the European Macromolecular Structure Database, VEGA, and dbSNP), the user makes an individual selection. It is possible to choose one or more species, and then more filters can be applied, such as certain genomic regions, the availability of external identifiers (e.g., InterPro or GO IDs, or gene names or keywords), and certain GO terms. Next, different protein attributes are chosen. The user will then be provided with a list of proteins that match the requested criteria. Currently (as of April 2005), proteome analysis in Integr8 is available for 252 complete genomes, 19 of which are eukaryota, 203 bacteria, and 20 archaea.

3.16.5 Summary and Conclusion

Genomics and proteomics are rapidly developing fields, and the databases handling the resulting data are similarly advancing. Owing to substantial progress in informatics (i.e., the development of database models and search algorithms), it is not only possible to store huge amounts of biological data but also to query the data in quick and meaningful user-friendly ways.

In this chapter, we have introduced different kinds of molecular biological databases. Both nucleic acid and amino acid sequence resources, from plain repositories to carefully manually annotated knowledge bases, have been explained in detail. The INSD collaboration consists of the DDBJ, EMBL, and GenBank databases. These are data repositories, accepting nucleic acid sequence data from the scientific community and making it freely available. The flatfiles of the individual databases, which are the elementary units of information in INSD, show distinctive features, but this does not prevent data from being exchanged on a regular basis, ensuring the identity of the nucleotide sequences. The situation is slightly different for protein sequences. A number of high-quality resources exist in parallel, each focusing on different features. GenPept is a database that provides translations from the nucleotide sequence databases, without a lot of additional annotation. UniParc simply stores all known protein sequences, even those that have been withdrawn from other databases – thus providing a 'history' of all protein sequences, without further information other than links to the source databases and reports on the status ('active' or 'not active') of the entry in the source databases. Entrez Protein contains some additional information, derived from other protein sequence databases. Other resources (e.g., RefSeq and UniProt) aim at providing a comprehensive, integrated, nonredundant set of sequences. The protein sequences are presented with a lot of additional information, derived from automatic and, most importantly, manual annotation. In RefSeq, the majority of the protein records are automatically generated with minimal manual intervention. In UniProtKB, every entry has undergone a process of manual annotation. Additional information is given about PTMs, pathways a protein is involved it, diseases that can be associated with certain mutations, and many other features of a protein. All these databases link to further resources that provide more specialized information.

We also introduced integrated and comparative databases, with their growing importance in providing related data from widespread resources quickly and reliably: the more data there are, the greater the need for data integration. No single resource can at present handle all the available data on a certain biological subject (be it a gene or even a whole genome, a protein or an organism), so thorough cross-links (based on stable identifiers) are of crucial importance.

The trend is not necessarily to gather all the data into one place, but to store only core data, and link outwards to specialized resources. It is also very important to provide intuitive, user-friendly interfaces. Database browsers also need to be easy to use to encourage users to undertake more complicated queries. InterPro is an integrated resource of protein families, domains, and functional sites that amalgamates the information in its 10 member databases. InterPro integrates the different methods used by the component databases to diagnose a protein domain or characteristic region of a protein family in a protein sequence, combining their individual strengths and overcoming their weaknesses. This integrated database provides a valuable tool for predicting protein function. Another integrated resource is Integr8, which provides various genomic and proteomic data. The relationships of biological entities to each other are stored, but apart from this information, only the core data necessary to identify the biological objects are integrated. Statistical analysis and functional characterization for more than 250 complete proteomes is available.

Many lessons have been learned, often the hard way, since the 1980s, when the sequence databases as the core resources in genomics and proteomics were established. The further development of existing biomolecular databases and the creation of new ones will certainly remain an area of growing importance for the life sciences. Both advanced database techniques and careful assessments of the kind of data that need to be stored are crucial for progress in this field. Concerning the data, keeping the databases up to date is a task that needs the active involvement of the whole scientific community. The quality of DDBJ/EMBL/GenBank, UniProt, InterPro, Integr8, and other community resources depends on the submission of new data and updates to these databases. It needs to be kept in mind that most errors, inaccuracies, and omissions in the databases are not introduced by the curators of the database but by the data submitters. Also, most corrections in databases are initiated by curators, and not by update requests from users. If database users can be encouraged to become more actively involved, data quality can be improved further.

References

1. Dayhoff, M. O. *Atlas of Protein Sequence and Structure*. National Biomedical Research Foundation: Washington, DC, 1978; Vol. 5.
2. Apweiler, R. *Brief. Bioinform.* **2001**, *2*, 9–18.
3. Tateno, Y.; Saitou, N.; Okubo, K.; Sugawara, H.; Gojobori, T. *Nucleic Acids Res.* **2005**, *33*, D25–D28.
4. Kanz, C.; Aldebert, P.; Althorpe, N.; Baker, W.; Baldwin, A.; Bates, K.; Browne, P.; van den Broek, A.; Castro, M.; Cochrane, G. et al. *Nucleic Acids Res.* **2005**, *33*, D29–D33.
5. Benson, D. A.; Karsch-Mizrachi, I.; Lipman, D. J.; Ostell, J.; Wheeler, D. L. *Nucleic Acids Res.* **2005**, *33*, D34–D38.
6. Pruitt, K. D.; Tatusova, T.; Maglott, D. R. *Nucleic Acids Res.* **2005**, *33*, D501–D504.
7. Bairoch, A.; Apweiler, R.; Wu, C. H.; Barker, W. C.; Boeckmann, B.; Ferro, S.; Gasteiger, E.; Huang, H.; Lopez, R.; Magrane, M. et al. *Nucleic Acids Res.* **2005**, *33*, D154–D159.
8. Mulder, N. J.; Apweiler, R.; Attwood, T. K.; Bairoch, A.; Bateman, A.; Binns, D.; Bradley, P.; Bork, P.; Bucher, P.; Cerutti, L. et al. *Nucleic Acids Res.* **2005**, *33*, D201–D205.
9. Leinonen, R.; Diez, F. G.; Binns, D.; Fleischmann, W.; Lopez, R.; Apweiler, R. *Bioinformatics* **2004**, *20*, 3236–3237.
10. Brooksbank, C.; Cameron, G.; Thornton, J. *Nucleic Acids Res.* **2005**, *33*, D46–D53.
11. Hubbard, T.; Andrews, D.; Caccamo, M.; Cameron, G.; Chen, Y.; Clamp, M.; Clarke, L.; Coates, G.; Cox, T.; Cunningham, F. et al. *Nucleic Acids Res.* **2005**, *33*, D447–D453.
12. Kersey, P.; Bower, L.; Morris, L.; Horne, A.; Petryszak, R.; Kanz, C.; Kanapin, A.; Das, U.; Michoud, K.; Phan, I. et al. *Nucleic Acids Res.* **2005**, *33*, D297–D302.
13. Boeckmann, B.; Bairoch, A.; Apweiler, R.; Blatter, M.; Estreicher, A.; Gasteiger, E.; Martin, M. J.; Michoud, K.; O'Donovan, C.; Phan, I. et al. *Nucleic Acids Res.* **2003**, *31*, 365–370.
14. Wu, C. H.; Yeh, L. S.; Huang, H.; Arminski, L.; Castro-Alvear, J.; Chen, Y.; Hu, Z.; Kourtesis, P.; Ledley, R. S.; Suzek, B. E. et al. *Nucleic Acids Res.* **2003**, *31*, 345–347.
15. Berman, H.; Henrick, K.; Nakamura, H. *Nat. Struct. Biol.* **2003**, *10*, 980.
16. Kersey, P. J.; Duarte, J.; Williams, A.; Karavidopoulou, Y.; Birney, E.; Apweiler, R. *Proteomics* **2004**, *4*, 1985–1988.
17. Drysdale, R. A.; Crosby, M. A.; Gelbart, W.; Campbell, K.; Emmert, D.; Matthews, B.; Russo, S.; Schroeder, A.; Smutniak, F.; Zhang, P. et al. *Nucleic Acids Res.* **2005**, *33*, D390–D395.
18. Chen, N.; Harris, T. W.; Antoshechkin, I.; Bastiani, C.; Bieri, T.; Blasiar, D.; Bradnam, K.; Canaran, P.; Chan, J.; Chen, C. K. et al. *Nucleic Acids Res.* **2005**, *33*, D383–D389.
19. Kersey, P.; Hermjakob, H.; Apweiler, R. *Bioinformatics* **2000**, *16*, 1048–1049.
20. Junker, V.; Apweiler, R.; Bairoch, A. *Bioinformatics* **1999**, *15*, 1066–1067.
21. Wain, H. M.; Lush, M.; Ducluzeau, F.; Povey, S. *Nucleic Acids Res.* **2003**, *30*, 169–171.
22. Blake, J. A.; Richardson, J. E.; Bult, C. J.; Kadin, J. A.; Eppig, J. T. Members of the Mouse Genome Database Group. *Nucleic Acids Res.* **2003**, *31*, 193–195.
23. O'Donovan, C.; Martin, M. J.; Glemet, E.; Codani, J.; Apweiler, R. *Bioinformatics* **1999**, *15*, 258–259.
24. Apweiler, R. *Brief. Bioinf.* **2001**, *2*, 9–18.
25. Fleischmann, W.; Möeller, S.; Gateau, A.; Apweiler, R. In Proceedings of the German Conference on Bioinformatics (GCB'98); Zimmermann, O., Schomburg, D., eds.; Cologne: Germany, 1998.
26. Kretschmann, E.; Fleischmann, W.; Apweiler, R. *Bioinformatics* **2001**, *17*, 920–926.
27. Hulo, N.; Sigrist, C. J. A.; Le Saux, V.; Langendijk-Genevaux, P. S.; Bordoli, L.; Gattiker, A.; De Castro, E.; Bucher, P.; Bairoch, A. *Nucleic Acids Res.* **2004**, *32*, D134–D137.

28. Bateman, A.; Coin, L.; Durbin, R.; Finn, R. D.; Hollich, V.; Griffiths-Jones, S.; Khanna, A.; Marshall, M.; Moxon, S.; Sonnhammer, E. L. L. et al. *Nucleic Acids Res.* **2004**, *32*, D138–D141.
29. Rawlings, N. D.; Tolle, D. P.; Barrett, A. J. *Nucleic Acids Res.* **2004**, *32*, D160–D164.
30. Dwight, S. S.; Balakrishnan, R.; Christie, K. R.; Costanzo, M. C.; Dolinski, K.; Engel, S. R.; Feierbach, B.; Fisk, D. G.; Hirschman, J.; Hong, E. L. et al. *Brief. Bioinform.* **2004**, *5*, 9–22.
31. Attwood, T. K.; Bradley, P.; Flower, D. R.; Gaulton, A.; Maudling, N.; Mitchell, A. L.; Moulton, G.; Nordle, A.; Paine, K. et al. *Nucleic Acids Res.* **2003**, *31*, 400–402.
32. Bru, C.; Courcelle, E.; Carrere, S.; Beausse, Y.; Dalmar, S.; Kahn, D. *Nucleic Acids Res.* **2005**, *33*, D212–D215.
33. Letunic, I.; Copley, R. R.; Schmidt, S.; Ciccarelli, F. D.; Doerks, T.; Schultz, J.; Ponting, C. P.; Bork, P. *Nucleic Acids Res.* **2004**, *32*, D142–D144.
34. Haft, D. H.; Selengut, J. D.; White, O. *Nucleic Acids Res.* **2003**, *31*, 371–373.
35. Wu, C. H.; Nikolskaya, A.; Huang, H.; Yeh, L. S.; Natale, D. A.; Vinayaka, C. R.; Hu, Z. Z.; Mazumder, R.; Kumar, S.; Kourtesis, P. et al. *Nucleic Acids Res.* **2004**, *32*, D112–D114.
36. Gough, J.; Karplus, K.; Hughey, R.; Chothia, C. *J. Mol. Biol.* **2001**, *313*, 903–919.
37. Mi, H.; Lazareva-Ulitsky, B.; Loo, R.; Kejariwal, A.; Vandergriff, J.; Rabkin, S.; Guo, N.; Muruganujan, A.; Doremieux, O.; Campbell, M. J. et al. *Nucleic Acids Res.* **2005**, *33*, D284–D288.
38. Pearl, F.; Todd, A.; Sillitoe, I.; Dibley, M.; Redfern, O.; Lewis, T.; Bennett, C.; Marsden, R.; Grant, A.; Lee, D. et al. *Nucleic Acids Res.* **2005**, *33*, D247–D251.
39. Zdobnov, E. M.; Apweiler, R. *Bioinformatics* **2001**, *17*, 847–848.
40. Harris, M. A.; Clark, J.; Ireland, A.; Lomax, J.; Ashburner, M.; Foulger, R.; Eilbeck, K.; Lewis, S.; Marshall, B.; Mungall, C. et al. *Nucleic Acids Res.* **2004**, *32*, D258–D261.
41. Scordis, P.; Flower, D. R.; Attwood, T. K. *Bioinformatics* **1999**, *15*, 799–806.
42. Pruess, M.; Fleischmann, W.; Kanapin, A.; Karavidopoulou, Y.; Kersey, P.; Kriventseva, E.; Mittard, V.; Mulder, N.; Phan, I.; Servant, F. et al. *Nucleic Acids Res.* **2003**, *31*, 414–417.
43. Sprague, J.; Clements, D.; Conlin, T.; Edwards, P.; Frazer, K.; Schaper, K.; Segerdell, E.; Song, P.; Sprunger, B.; Westerfield, M. *Nucleic Acids Res.* **2003**, *31*, 241–243.
44. Kriventseva, E. V.; Servant, F.; Apweiler, R. *Nucleic Acids Res.* **2003**, *31*, 388–389.
45. Smith, T. F.; Waterman, M. S.; Fitch, W. M. *J. Mol. Evol.* **1981**, *18*, 38–46.
46. Comet, J. P.; Aude, J. C.; Glemet, E.; Risler, J. L.; Henaut, A.; Slonimski, P. P.; Codani, J. J. *Comput. Chem.* **1999**, *23*, 317–331.
47. Dodge, C.; Schneider, R.; Sander, C. *Nucleic Acids Res.* **1998**, *26*, 313–315.
48. Lo Conte, L.; Brenner, S. E.; Hubbard, T. J.; Chothia, C.; Murzin, A. G. *Nucleic Acids Res.* **2002**, *30*, 264–267.
49. Kasprzyk, A.; Keefe, D.; Smedley, D.; London, D.; Spooner, W.; Melsopp, C.; Hammond, M.; Rocca-Serra, P.; Cox, T.; Birney, E. *Genome Res.* **2004**, *14*, 160–169.
50. The DOBJ/EMBL/GenBank Feature Table: Definition. http://www.ebi.ac.uk/embl/Documentation/FT_definitions/feature_table.html (accessed March 2006).
51. Genome Reviews. http://www.ebi.ac.uk/GenomeReviews (accessed March 2006).
52. OMIM – Online Mendelian Inheritance in Man. http://www.ncbi.nlm.nih.gov/entrez/query.fcgi?db=OMIM (accessed March 2006).
53. IPI – International Protein Index. http://www.ebi.ac.uk/IPI/IPIhelp.html (accessed March 2006).
54. NEWT UniProt Taxonomy Browser. http://www.ebi.ac.uk/newt/display (accessed March 2006).

Biographies

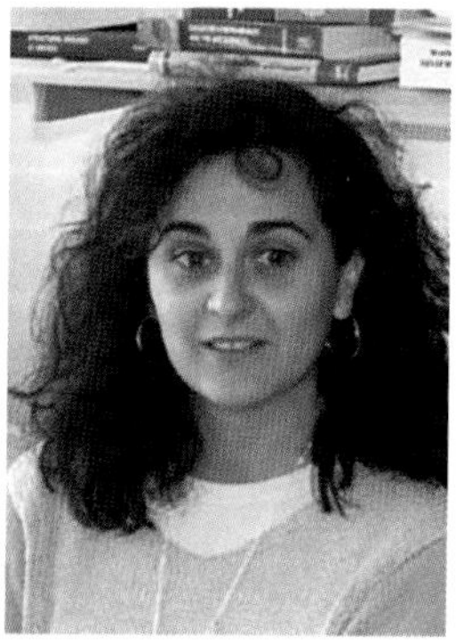

Maria-Jesus Martin, PhD, is a Sequence Database Group Coordinator at the EMBL Outstation European Bioinformatics Institute (EBI) in Cambridge, UK. She has worked for many years on protein sequence databases, and she is responsible for the technical aspects of the Universal Protein Resource (UniProt) at EBI. She supervises production software development, database design and workflows, data management, and data analysis systems for the UniProt databases.

Tamara Kulikova, MSc, has been a Sequence Database Group Coordinator at the EMBL Outstation European Bioinformatics Institute (EBI) in Cambridge, UK, since 1998. She is responsible for the maintenance and development of the EMBL Nucleotide Sequence Database at the EBI.

Manuela Pruess, PhD, has held postdoctoral positions at the Institute of Technical and Business Information Systems of the University of Magdeburg, Germany, and at the German Research Centre for Biotechnology (GBF) in Braunschweig. From there, she moved to the company BIOBASE in Wolfenbuettel, where she worked on biological databases, and in 2001 she joined the Sequence Database Group at the EMBL Outstation European Bioinformatics Institute (EBI) in Cambridge, UK, where she is responsible for quality assurance and project management.

Rolf Apweiler, PhD, has been working on the Swiss-Prot database since 1987, and in 1994 he became leader of the Swiss-Prot group at the EMBL Outstation European Bioinformatics Institute (EBI) in Cambridge, UK. Since 2001 he has been leading the Sequence Database Group at EBI. He is a member of several committees, review panels, and advisory boards, including the Nomenclature Committee of the IUBMB (the 'Enzyme Commission'), council of the Human Proteome Organization (HUPO), the expert committee on 'Bibliometric Mapping of Excellence in the Area of Life Sciences' of the European Commission, the FlyBase advisory board, and the Bioinformatics review panel of the German Ministry of Research.

Comprehensive Medicinal Chemistry II
ISBN (set): 0-08-044513-6

ISBN (Volume 3) 0-08-044516-0; pp. 349–372

3.17 The Research Collaboratory for Structural Bioinformatics Protein Data Bank

P E Bourne, W F Bluhm, N Deshpande, and Q Zhang, University of California, La Jolla, CA, USA
H M Berman and J L Flippen-Anderson, Rutgers – The State University of New Jersey, Piscataway, NJ, USA

3.17.1	**History of the Research Collaboratory for Structural Bioinformatics (RCSB) Protein Data Bank (PDB)**	**373**
3.17.2	**Data Deposition, Standards, and Annotation**	**376**
3.17.3	**Problem Solving With the Research Collaboratory for Structural Bioinformatics Protein Data Bank**	**376**
3.17.3.1	Overall Searching and Browsing Capabilities	376
3.17.3.2	Finding Ligands and Exploring Protein–Ligand Interactions	377
3.17.3.3	Finding Disease-Related Structures	377
3.17.3.4	Exploring Genetic and Induced Mutations	380
3.17.4	**Using Structural Information to Understand Disease**	**381**
3.17.4.1	Acquired Immune Deficiency Syndrome (AIDS)	381
3.17.4.2	Diabetes	381
3.17.4.3	The Common Cold	382
3.17.5	**The Future**	**382**
References		**384**

3.17.1 History of the Research Collaboratory for Structural Bioinformatics (RCSB) Protein Data Bank (PDB)

Once it was demonstrated that crystals could diffract x-rays and possibly provide information to derive the structure of molecules, early scientific pioneers took the challenge of using this powerful technology to try to uncover the structures of biological molecules. In 1934, Crowfoot and Bernal obtained diffraction photographs from pepsin.[1] More than 20 years later, in 1957, Kendrew and Perutz determined the first three-dimensional structure of a protein–myoglobin.[2,3] Over the next several years the structures of hemoglobin,[4,5] lysozyme,[6,7] and ribonuclease[8,9] were revealed, each providing rich information about the intricacies and beauty of protein architecture and the relationships between shape and function. By the early 1970s, there were at least a dozen examples of protein structures that had been determined by x-ray crystallography, and it was clear to some that there would be much to learn by comparative analyses of the data these structures provided. In 1971, the PDB was born of such interest, and it was managed by Brookhaven National Laboratory[10] until 1998, when its management was assumed by the Research Collaboratory for Structural Bioinformatics (RCSB).[11]

Over the years the number of structures grew, and by early 2006 there were more than 35 000 individual structures (**Figure 1**), of which approximately 6000 are considered nonredundant (at 30% sequence identity). In 2005, 5439 structures were deposited, which is more than double the number of depositions in 1999. The numbers promise to become even larger as structural genomics takes off and the challenge of solving protein structures on a genomic scale becomes a reality. In addition to the rising number of structures, the complexity of the types of structures that are being solved and deposited is much greater. More than 100 structures have molecular weights greater than 500 000.[12] Among these structures are many ribosomal subunits, viruses, and large multienzyme complexes. The methods for determining such large structures include both x-ray crystallography and cryo-electron microscopy.

As the methods for determining structures have improved, it is now possible to answer biological questions with structural biology. The structures involved in various pathways can now be analyzed and used to understand

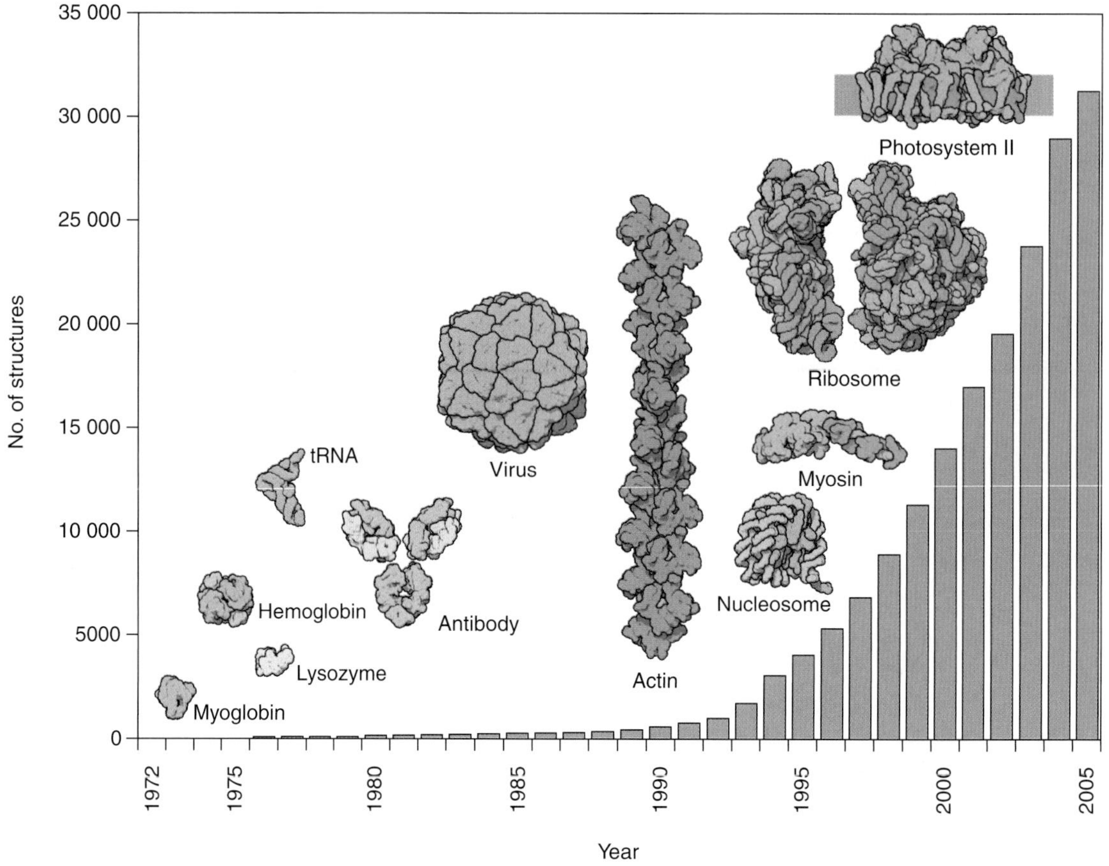

Figure 1 Growth in the number of structures available in the PDB archives each year with examples showing the increasing complexity of structures available. (Reproduced by courtesy of David S. Goodsell of The Scripps Research Institute/RCSB PDB.)

mechanisms. For example, the structures involved in glycolysis that help define a single molecular pathway have been analyzed by many different groups (**Figure 2**). In the future, it is likely that structural biologists will study whole systems, and not just individual structures, at the molecular level at a time.

The realization that structure is central to our understanding of biology can be seen in the growing usage of the PDB. In the earliest days, it served as a kind of safe deposit box for crystallographers wishing to ensure that their data were not lost. As the value of these data became more apparent, many researchers, such as chemists, bio- and chemo-informaticians, modelers, biochemists, crystallographers, and computational biologists, began to download coordinate sets to perform classification and comparative analyses. This work gave rise to a variety of secondary databases that are now widely used, such as two of the most prominent structural classifications CATH (Class, Architecture, Topology and Homologous superfamily)[13] and SCOP (Structural Classification of Proteins).[14] As the Internet became more popular, the PDB also began to see a rise in usage by educators who use the data and resources found at the PDB for teaching. Now, with the availability of simple as well as complex queries, biologists use the data and resources found in the PDB in their own research. On any given day, about 10 000 individual users access the PDB online, and download one file every second of the day, seven days a week. These numbers can only become larger as time goes by.

From the beginning, the PDB was a project in which groups from all over the world contributed structures and multiple international sites distributed data. Most recently, these arrangements have been formalized in the worldwide PDB (wwPDB; wwpdb.org),[15] whose mission is to keep the PDB a single archive of structures available to the global community. The members of wwPDB are the RCSB PDB, a consortium consisting of groups from Rutgers, the State University of New Jersey and the San Diego Supercomputer Center at the University of California, San Diego; the Macromolecular Structure Database at the European Bioinformatics Institute (MSD-EBI), and the Protein Data Bank Japan (PDBj) at the Institute for Protein Research at Osaka University. Data are deposited and processed at each

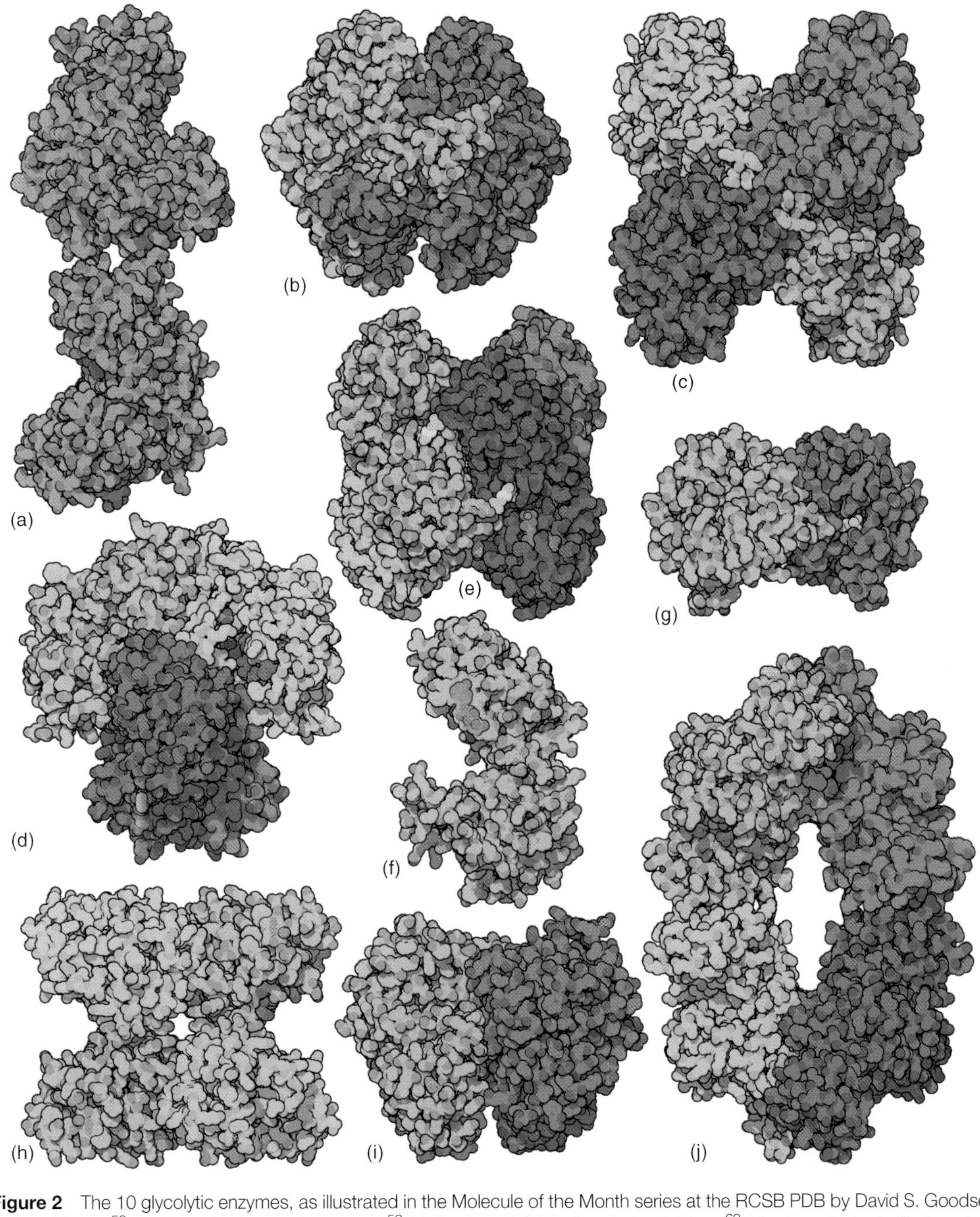

Figure 2 The 10 glycolytic enzymes, as illustrated in the Molecule of the Month series at the RCSB PDB by David S. Goodsell: (a) hexokinase,[58] (b) phosphoglucose isomerase,[59] (c) fructose 1,6-bisphosphate aldolase,[60] (d) glyceraldehyde-3-phosphate dehydrogenase,[61] (e) phosphofructokinase,[62] (f) phosphoglycerate kinase,[63] (g) triose phosphate isomerase,[64] (h) phosphoglycerate mutase,[65] (i) enolase,[66] and (j) pyruvate kinase.[67]

wwPDB site. The final released entries follow the same format and content guidelines, and are stored in a single copy of the PDB archives. Each wwPDB site develops and presents its own unique views to access the data in the archives.

In the following sections, the procedures and resources described involving the PDB archives are particular to the RCSB PDB.

3.17.2 Data Deposition, Standards, and Annotation

Structural data for proteins, nucleic acids, protein–nucleic acid complexes, viruses, and large macromolecular machines such as the ribosome are archived in the PDB. The structures have been determined using x-ray crystallography, nuclear magnetic resonance (NMR), and cryo-electron microscopy. The data represented for each structure must be consistent for all entries, so that searches will retrieve all relevant and unique information. The PDB Exchange Dictionary[16] currently contains more than 3000 terms, and is used to define every data item in a PDB file for consistency across the archive.

Several different types of information about each structure are archived, including atomic coordinates, sequences, chemical descriptions of small-molecule ligands, and details about experimental procedures that were used to derive the structure. During the annotation process, all the information that has been deposited is checked for accuracy and consistency. Macromolecular sequences in the entry are compared with existing sequence databases, to ensure that the reported sequence is correct, even if some residues could not be completely characterized by the experimental data. These checks also provide information about the name of the molecule and its oligomerization state. Any ligands present in the entry are compared with pre-existing ligands in the PDB Chemical Component Dictionary using Ligand Depot, a small-molecule resource developed by the RCSB.[17] The dictionary currently contains more than 6000 ligands, and a significant project is underway within the RCSB to verify their stereochemistry and to provide a consistent system for ligand nomenclature across the archive. The same checks are performed on new ligands as they are added. Information about secondary structure and the biologically active unit for the macromolecule are also added at this point. For accurate descriptions of the experiment, the RCSB PDB has made software tools available to help researchers working on the structures (depositors) collect and save as much data as possible during the various stages of a structure determination study.[18]

The RCSB PDB Validation Server is used to assess the quality of a structure.[19] The chemistry and geometry of all the components of the structure are checked. Incorrect chirality, close intramolecular contacts, poor valence geometry, misplaced solvent molecules, and missing residues are evaluated. The Validation Server uses programs developed by the RCSB and others[20–22] to produce reports detailing the results of the checking procedures. These programs and reports are used by both the depositors and the RCSB PDB staff who work together to create data files that are as complete and accurate as possible. Depositors are encouraged to use the Validation Server prior to deposition. Users of the PDB are encouraged to examine entries so that they can determine for themselves whether or not a structure meets their research needs for structure function studies, molecular modeling or drug design. A detailed description of related tools and resources available from the RCSB PDB can be found in *Current Protocols in Bioinformatics*.[23]

Each entry is assigned a unique identifier, known as the PDB ID, at the completion of the deposition process. This PDB ID is often listed in publications describing the structure, and can be used as a 'tag' to search for information about the structure in the database. The structures in the PDB can be visualized and searched online, or data can be downloaded for offline visualization and analysis.

3.17.3 Problem Solving With the Research Collaboratory for Structural Bioinformatics Protein Data Bank

3.17.3.1 Overall Searching and Browsing Capabilities

An interesting way of exploring structures relevant to medicinal chemistry is to enter the RCSB PDB via the Molecule of the Month column, linked to from the home page of the PDB website. This feature illustrates important biological molecules and how they function through descriptive text and pictures, with links to specific PDB entries and other resources. To dig deeper, various search interfaces are available on the RCSB PDB website. From the home page, users can search by PDB ID (if known, such as from a scientific publication), author name, or a keyword search using general terms such as 'diabetes' or 'insulin.' A number of different reports can be generated to compare various features of the selected structures. Selecting any of the PDB IDs listed on the results page will bring up a structure summary page for that individual structure. This latter page provides summary information about the entry, an illustration, links to coordinates, and links to detailed reports (such as biology and chemistry, or materials and methods) available within the RCSB PDB, as well as links to external web resources.

A number of additional search options, referred to as Query-by-Example, are available from the structure summary page. They offer simple ways of retrieving search results that share particular attributes. For example, all structures associated with a particular author can be quickly retrieved by simply choosing a single author name from the page.

Derived features including SCOP[14] and/or CATH[13] structure classifications, and Gene Ontology (GO) terms[24] describing molecular function, biochemical process, and cellular location can also be searched in this way. For example, choosing 'hormone activity' under GO terms and 'molecular function' on the structure summary page for the insulin structure 1APH[25] retrieves all structures that have been classified under the same GO branch.

From this page, it is also possible to retrieve the PubMed abstract, and, using terms from it, to search MEDLINE for all structures with abstracts that contain the same terms.

There is also a more advanced and specialized search interface that allows a user to search for structures that have common characteristics in their experimental details, geometry, biology, chemistry, SCOP and CATH classifications, and citations.

Search Sequence provides several ways of finding structures that contain similarities to a given amino acid or nucleic acid sequence. Search Unreleased looks for structures that have been submitted to the PDB but have not yet been released. Search Ligands provides an interface for looking at all the small molecules covalently or noncovalently associated with macromolecules.

Using the 'PDB ID or keyword' option on the home page also queries the static web pages on the RCSB PDB site. For example, a search for insulin finds a number of pages, including the corresponding Molecule of the Month edition.

An alternative to searching is browsing, which is particularly useful in situations where queries cannot be quantitatively defined. A number of browsers, such as Biological Process, Molecular Function, or Disease, are available. Each browser (e.g., Browse Database→Disease) offers a hierarchical classification that can be expanded (e.g., to Cancers→Colon Cancer) to show all structures associated with a particular disease.

Finally, a number of tools are available for evaluating and refining search results. Tabular reports (e.g., Structure Summary, Ligands, or Primary Citation) can be produced from the results obtained from various searches. These reports can also be customized from a large number of available attributes describing various aspects of the structures, to aid the user in evaluating the relevance of their search results.

No single search strategy can be recommended as the best practice for all possible scenarios. In general, however, users might be advised to start with simple keyword searches for one or more search terms of interest. Based on the number and relevance of the results returned by these simple searches, the queries can then be refined based on additional search terms, or more specific searches can be composed (e.g., using the Advanced Search option). Alternatively, browsing can provide some good first impressions of the content of the PDB.

3.17.3.2 Finding Ligands and Exploring Protein–Ligand Interactions

Particularly relevant to medicinal chemistry is the study of chemical entities bound to proteins, DNA, and RNA. These entities include commercial drugs, known inhibitors, toxic agents, and small molecules found in the cell such as ATP and GTP. The PDB website represents these entities by their common names, a three-letter code, a two-dimensional (2D) chemical diagram, and a SMILES string (actually used to construct the 2D diagram).[26] Entities may be downloaded as a MOL file (defined by Molecular Design Ltd as a file that describes chemical structure) or as part of the PDB structure file.

There are several ways to search for and display ligands. If a particular structure has ligands, they are displayed in the Chemical Component section on the Structure Summary web page (**Figure 3**). Users can retrieve a list of PDB structures with this ligand, view the ligand structure itself, or view ligand interactions with the macromolecule. The ligand structure view provides a chemical structure for the ligand, a 2D MarvinView sketch, a SMILES string, and a link to the MOL file. The SMILES string opens up the Search Ligands form, to let the user modify the ligand and run a similarity/substructure search.

The ligand interaction view launches the LigPro Ligand Explorer,[27] a three-dimensional (3D) interactive tool that is specifically designed for inspecting protein–ligand interactions, such as hydrophilic, hydrophobic, and other van der Waals interactions. Ligand Explorer dynamically generates the associated ligand–macromolecule contact list, centers the view at the user-selected ligand, calculates interactions within a user-specified range, and provides one-click inspections of different types of interactions. Ligand Explorer provides two-way communication between the macromolecular sequence and the structure viewers – a click on any residue in the sequence viewer will highlight that residue in the structure viewer; a click on any fragment/residue in the structure viewer will highlight the corresponding sequence in the sequence viewer. The ATP-binding site of a protein kinase (PDB ID 1ATP[28]) is examined with Ligand Explorer in **Figure 4a**.

3.17.3.3 Finding Disease-Related Structures

The disease browser offers an ideal starting point for a user interested in structures that have been solved for proteins implicated in human disease. This browser is accessible from the search menu tab under Browse Database → Disease.

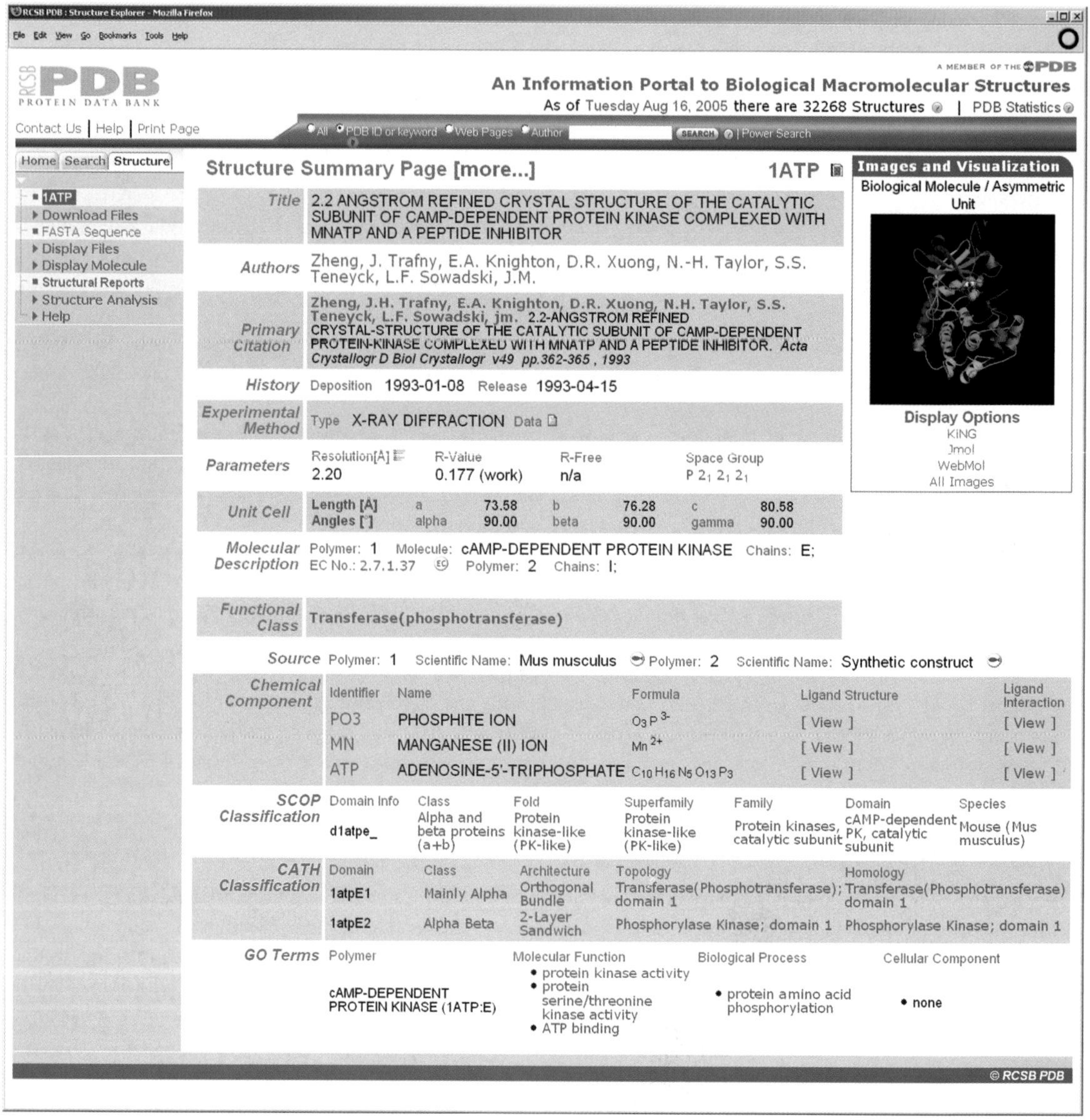

Figure 3 The Structure Summary page for 1ATP.[28] The Chemical Component section in the middle gives information about ligands and provides links to view the ligand structure and interaction.

Figure 4 Ligand Explorer. (a) The view for 1ATP. When the user selects 'ATP_1' from the left side bar and 'Hydrophilic Interactions,' and clicks on the apply button, LigPro computes the view centered at ligand ATP_1, displays the protein residues that have hydrophilic (H bond) interactions with ATP in the structure viewer, and highlights these residues in red in the sequence viewer at the top. The green dashed lines connect putative H bond donors and acceptors, with distances (in angstroms) displayed at the midpoint. Selecting an interacting protein residue in the sequence viewer highlights this residue in yellow in the structure viewer (SER53 here). The number of calculated interactions is displayed in the status bar at the bottom. Clicking on a noninteracting protein residue in the sequence viewer turns on the all-atom display for this residue. Through the analysis menu on the top, the user can measure the distance between any two atoms, the angle between any three atoms, and the dihedral angles between the planes made by any four atoms. The image or selected interactions can be saved (under the File menu) for further analysis or publication. (b) View of the hydrogen bond interaction of the drug indinavir with the critical ASP25 residue in each of the two chains of the HIV-1 protease 1HSG.[38]

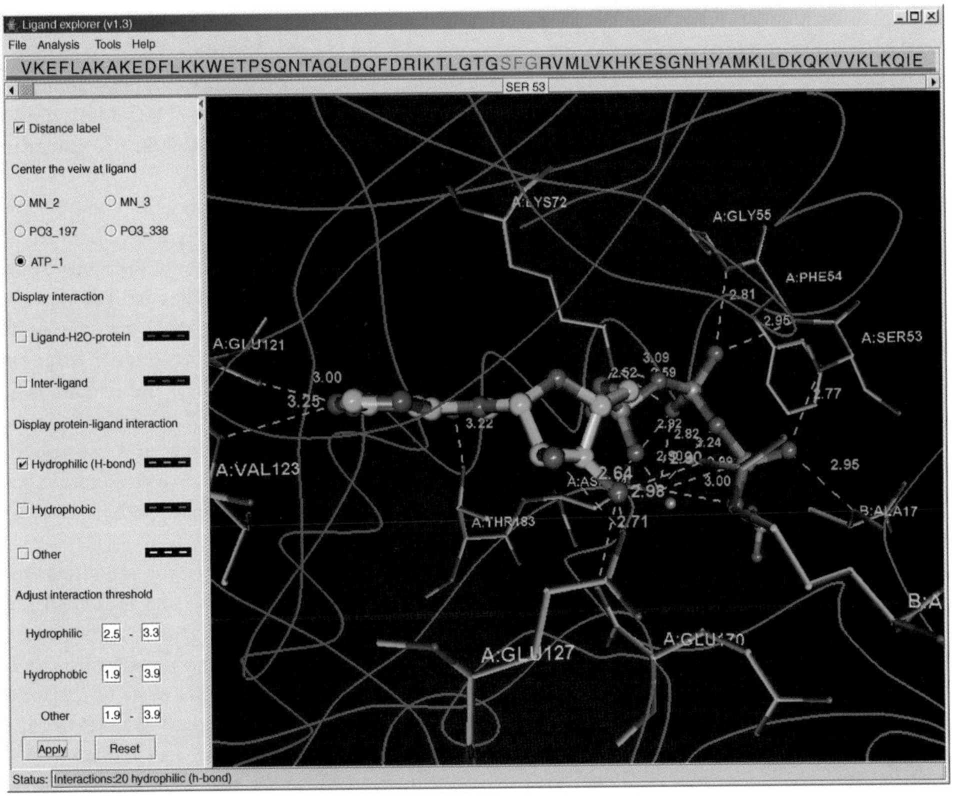
Ligand explorer (v1.3)
File Analysis Tools Help
VKEFLAKAKEDFLKKWETPSQNTAQLDQFDRIKTLGTGSFGRVMLVKHKESGNHYAMKILDKQKVVKLKQIE
SER 53
Distance label
Center the veiw at ligand
MN_2
MN_3
PO3_197
PO3_338
ATP_1
Display interaction
Ligand-H2O-protein
Inter-ligand
Display protein-ligand interaction
Hydrophilic (H-bond)
Hydrophobic
Other
Adjust interaction threshold
Hydrophilic 2.5 - 3.3
Hydrophobic 1.9 - 3.9
Other 1.9 - 3.9
Apply
Reset
Status: Interactions:20 hydrophilic (h-bond)
A:LYS72
A:GLY55
A:PHE54
A:SER53
A:GLU121
A:VAL123
A:THR183
A:GLU127
B:ALA17
2.81
2.95
3.09
2.52
3.00
3.25
3.22
2.77
2.82
3.24
2.64
2.98
2.71
3.00
2.95

(a)

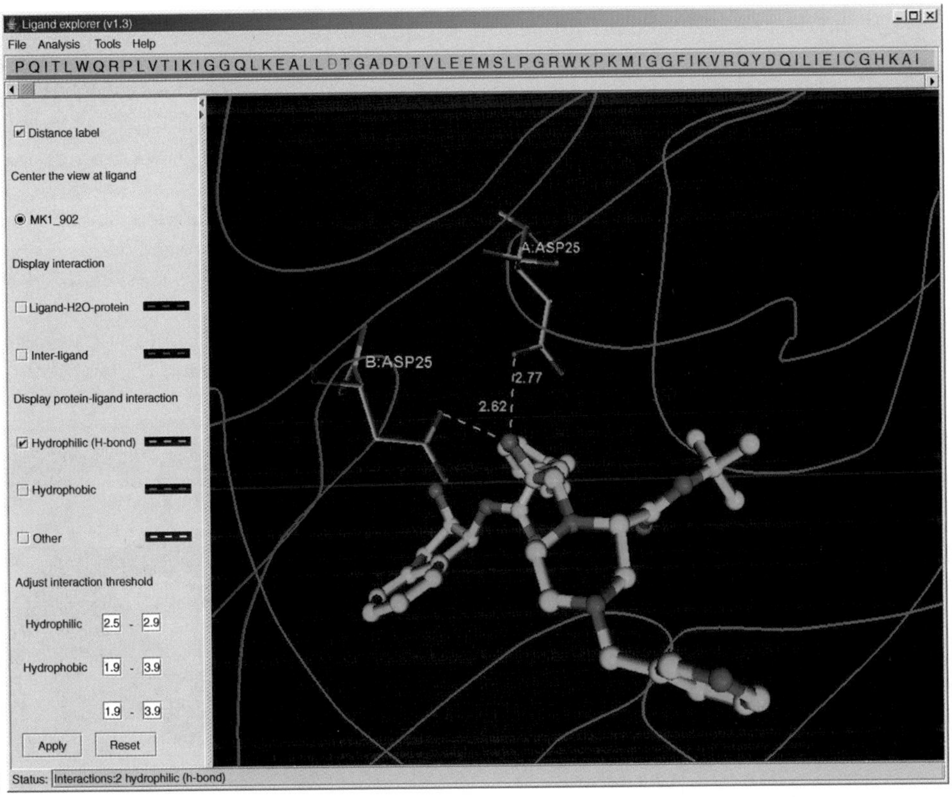
Ligand explorer (v1.3)
File Analysis Tools Help
PQITLWQRPLVTIKIGGQLKEALLDTGADDTVLEEMSLPGRWKPKMIGGFIKVRQYDQILIEICGHKAI
Distance label
Center the view at ligand
MK1_902
Display interaction
Ligand-H2O-protein
Inter-ligand
Display protein-ligand interaction
Hydrophilic (H-bond)
Hydrophobic
Other
Adjust interaction threshold
Hydrophilic 2.5 - 2.9
Hydrophobic 1.9 - 3.9
1.9 - 3.9
Apply
Reset
Status: Interactions:2 hydrophilic (h-bond)
A:ASP25
B:ASP25
2.77
2.62

(b)

The hierarchy in the browser is based on the chapters and sections in the e-book *Genes and Disease*.[29] The focus of the e-book is the set of inherited diseases caused by a mutation on a single gene. Identification of mutations in multiple genes whose products interact, disease-causing mutations whose phenotype is influenced by environmental conditions, and mutations in regulatory elements causing inherited diseases are challenges that will be addressed in the future.

Structures of proteins associated with any given disease are identified following a mapping to one or more OMIM (Online Mendelian Inheritance in Man) numbers.

A user interested in retrieving structures linked to a disease (e.g., colon cancer) can enter the term and perform a search across the hierarchy in the disease browser or run a keyword search: 'colon' and 'cancer.' There are 61 structures (at the time of writing) of proteins that are associated with colon cancer, as retrieved from the disease browser. Structures of proteins with similar sequences can be eliminated using the homology reduction function available from the query results page. This will allow the user to focus on just the proteins that differ from each other at the sequence level. Upon removing homologous sequences (menu item Narrow Query → Remove Similar Structures → 90% identity), five structures are returned. A number of tabular reports can be generated for the structures listed on the results page. The Summary Reports → Biological Details report lists EC (Enzyme Classification) numbers and GO terms and ID associated with each of the structures. Depending on the user's focus, any of these structures may be further explored. For example, the GO term definitions for structure 1CTQ[29a] suggest that this protein participates in molecular signaling processes within the cell and plays a role in modulating the cell cycle, both of which are generally implicated in the process of transformation of healthy tissues to carcinomas.

The structure summary page for 1CTQ provides the user with further insight into the structure. An interesting report to view from this page would be the biology and chemistry report (found under the Structural Reports menu item). This report lists the OMIM numbers and OMIM clinical synopses associated with 1CTQ. Clicking on the OMIM 190020 link leads to the OMIM summary for the gene *HRAS*, coding for p21. The summary lists research on this gene and mutations leading to its transition to a transforming gene. A point mutation at codon 12 replaces the glycine residue at that position, drastically impeding GTP hydrolysis to GDP by p21. The decrease in GTP to GDP hydrolysis results in p21 remaining in its active state, leading to uncontrolled cell proliferation and transformation. An interesting question arising from this information could be: what structural differences exist between the mutant and the wild type and could those differences explain the functional differences? The user may now be able to look for structures of mutants of p21 and compare them with the structure of 1CTQ (**Figure 5**), as described in our next section.

3.17.3.4 Exploring Genetic and Induced Mutations

A user interested in looking for the effects of mutations on 3D structure can start with a known structure of a nonmutant protein. We can look for structures with a similar sequence by using the Search Database menu (Search Database→ Sequence), then entering '1CTQ' in the PDB ID box and running the sequence similarity search using either BLAST[30] or FASTA.[31] At the time of writing, 203 structures are retrieved by this search. We can refine the results by looking for the text word 'mutant' (menu item Refine this Search; then entering 'mutant' as the keyword in the text box for Keyword – advanced). At the time of writing, this refinement returns 46 structures.

Many of these structures have point mutations at the 12th codon. Looking at each of these structures provides insights into the phenotype of the mutation. For example, a G12D transforming mutant of *HRAS* (1AGP)[32] was found

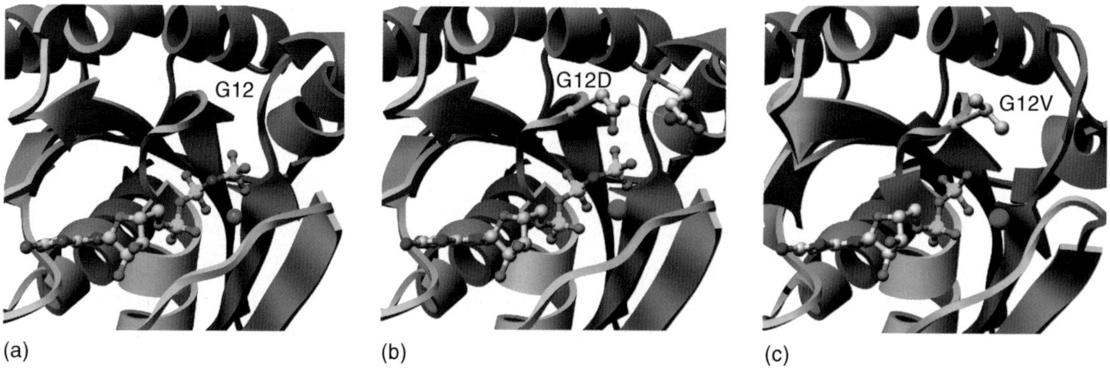

Figure 5 Structures of (a) p21 (1CTQ)[29a] and two transforming mutants of p21: (b) 1AGP[32] and (c) 2Q21.[68] The images show the location of the transforming mutation at position 12 in the polypeptide chain. The image was created with Chimera.[69]

to crystallize in a space group different from the wild type. Also, the structure of 1AGP around the active site was found to be different from that of the wild type. On the other hand, a G12P non-transforming mutant had a structure very similar to that of the wild type in the active site. Structure 2Q21[33] is a G12V transforming mutant where the valine side chain interferes with GTP hydrolysis to GDP. The user can also explore a number of structures of mutants with point mutations at other positions in the sequence. The structure summary pages for all structures have links to the PubMed abstracts, where available. By following this link, the user can access an abstract of the analysis of the structure, along with the information necessary to retrieve the complete article.

3.17.4 Using Structural Information to Understand Disease

Disease, defined here as the abnormal functioning of an organism, is brought about in a number of ways, such as a genetic defect leading to the production of an incorrectly functioning protein, environmental stress, or the intrusion of a foreign body such as a virus. Understanding the cause of a disease requires a detailed knowledge of events occurring at the molecular level. Thus, macromolecular structure is very important to the understanding and treatment of diseased states. This section explores three specific and somewhat different outcomes of the impact of understanding structure on our view and treatment of disease.

3.17.4.1 Acquired Immune Deficiency Syndrome (AIDS)

AIDS is the bubonic plague of the late-twentieth and twenty-first centuries. At the end of 2003, an estimated 40 million people worldwide were infected with the human immunodeficiency virus (HIV); AIDS is estimated to have taken 20 million lives, and an estimated 5 million people contract the disease annually.[34] At the early stage of the HIV virus life cycle, the protein of the virus exists as a single polypeptide chain. As the virus develops, the enzyme HIV-1 protease is responsible for cutting the chain into segments that will mature and infect new cells. Disrupting this event prevents the virus from reproducing. Structure studies have revealed the detailed mechanism by which the enzyme excises the protein and have provided the key for designing inhibitors that bind more tightly to the enzyme than the protein chain they are charged with cutting.

The earliest HIV-related crystal structures found in the PDB were published in the early 1990s: a complex between a synthetic protease of HIV-1 and a substrate-based hydroxyethylamine inhibitor (7HVP)[35] and an HIV protease complex with L-700,417 (4PHV).[36] The first NMR structure of an HIV zinc finger-like domain was published in 1990 (2ZNF).[37] Only 16 years later, an HIV keyword search of the PDB returns almost 400 structures, including several genetic strains of the enzyme, complexes of the enzyme with many different drugs and inhibitors, and dozens of mutant enzymes. Hundreds more are most likely stored in the proprietary databases of pharmaceutical companies, where they are used to test and refine new drug candidates.

At least six protease inhibitors that attack HIV-1 have already been approved to treat people infected with the virus, and several others are in late stages of clinical development. Structures of all six drugs currently in clinical use are available in complexes in the PDB: indinavir, which is also discussed below (1HSG),[38] saquinavir (1HXB),[39] ritonavir (1HXW),[40] nelfinavir (1OHR),[41] amprenavir (1T7J),[42] and Kaletra, which is a combination of lopinavir (1MUI)[43] and ritonavir.

The details of how inhibition of HIV protease may take place is highlighted by the structure of the complex of HIV-1 protease and indinavir (1HSG),[38] marketed today under the tradename Crixivan as part of the so-called protease cocktail treatment. The interaction of the inhibitor and the protein is shown in **Figure 4b** using the Ligand Explorer software. Highlighted is the critical hydrogen-bonding interaction of the hydroxyaminepentane amide moiety interacting with the hydroxyl groups of the critical ASP 25 carboxyl groups from both of the polypeptide chains of this symmetric protein. Ligand Viewer can be used to explore all the hydrophilic and hydrophobic interactions between this viral inhibitor and the virus protease. Additional information on the structures of HIV protease can be found in the HIV-1 Protease Protein Structures Database.

3.17.4.2 Diabetes

Insulin is a hormone that carries messages describing the amount of sugar that is available in the blood at any one time. It is synthesized in the pancreas in response to food intake. It then informs liver, muscle, and fat cells to take glucose from the blood and store it for subsequent use. Insufficient production of insulin causes glucose levels to rise in the blood, leading to the disease diabetes mellitus. Diabetes is most often found in adults, but it can occur in children as well, and it is one of the major chronic diseases of the modern world. Early treatment of diabetes consisted of injections of insulin from either pigs or cows. Now the insulin given is produced using recombinant methods.

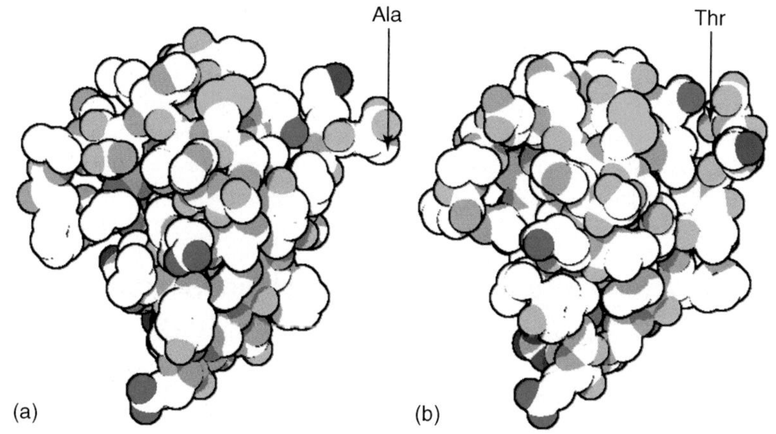

Figure 6 The structure of insulin (from the Molecule of the Month). (a) Insulin from pig (from the PDB entry 4INS[46]). (b) Insulin from human (from the PDB entry 2HIU[47]). The two structures are very similar, with a nondisruptive single amino acid substitution.

Insulin has historically been much studied, since it is a small protein that proved to be relatively easy to isolate from natural sources, such as from pig and beef pancreases. Dorothy Hodgkin took the first x-ray diffraction photographs of insulin in 1935. The structure was not solved until 34 years later, when the structure of 2-Zn insulin was reported by Hodgkin and her co-workers in August 1969.[44] She was awarded the 1964 Nobel Prize in Chemistry for her work on insulin. The earliest insulin structure that can be found in the PDB is that of 4-Zn insulin, published in *Nature* by Hodgkin *et al.* in 1976 – 1ZNI.[45] Since that time, almost 200 structures of insulin or closely related molecules have been deposited in the PDB archives. Approximately one-quarter of these structures were studied using NMR techniques.

The structure of insulin shown in **Figure 6**, taken from the RCSB PDB Molecule of the Month feature, illustrates that there is only a single amino acid difference between pig[46] and human[47] insulin – a threonine in human at the end of the chain is replaced by alanine in pig. Its surface location as revealed by the structure and the conserved chemical nature of the substitution reveals details of how insulin binds to the appropriate cell receptor.

3.17.4.3 The Common Cold

A cure for the common cold is still illusive, but there are moderately effective treatments that attack the problem indirectly. The rhinovirus responsible for the common cold is the heart of the problem in the majority of cases. Currently, there are more than 75 structures of rhinovirus in the PDB. The structure of the virus itself, HRV1A (1R1A),[48] complexed with various antiviral ligands,[49,50] and ligand binding in drug-resistant mutants[51] have been published. For a review of this work, see the article by Bella and Rossmann.[52]

While we now understand the basic structure of the protein viral capsid, we have been unable to find a lasting cure, since the virus continues to evolve. The basic action of the rhinovirus is similar to that of HIV, but less deadly for most. The virus attaches itself to the cell surface, and its RNA enters the human cell, which then directs the cell to replicate the virus. Unlike HIV, which focuses on the immune system, rhinoviruses mainly attack the respiratory tract. **Figure 7** illustrates one such capsid of the rhinovirus. The capsid itself is made up of 60 individual building blocks arranged in an icosohedral arrangement. **Figure 7a**[53] illustrates a rhinovirus bound to a receptor protein on the cell surface, shown in blue. In **Figure 7b**,[54] antibodies against the infection can attach themselves to the virus in the same position to prevent the virus from binding to the cell surface and causing infection. Work toward finding pharmaceutical agents effective against the common cold continues.

3.17.5 The Future

As technologies improve, macromolecular structures are being determined at an ever-increasing rate. This increase is also due in part to the structural genomics projects worldwide.[55] While only one project has as its direct goal the elucidation, treatment, and eradication of a specific disease (tuberculosis), the large-scale determination of structures based on complete genomes, specific pathways, and unique protein folds will bring forth a lot of useful information specific to understanding structure–disease relationships. **Figure 8** attempts to provide some insight as to this potential impact. The method of classification of PDB structures by their disease involvement, focusing on genetic

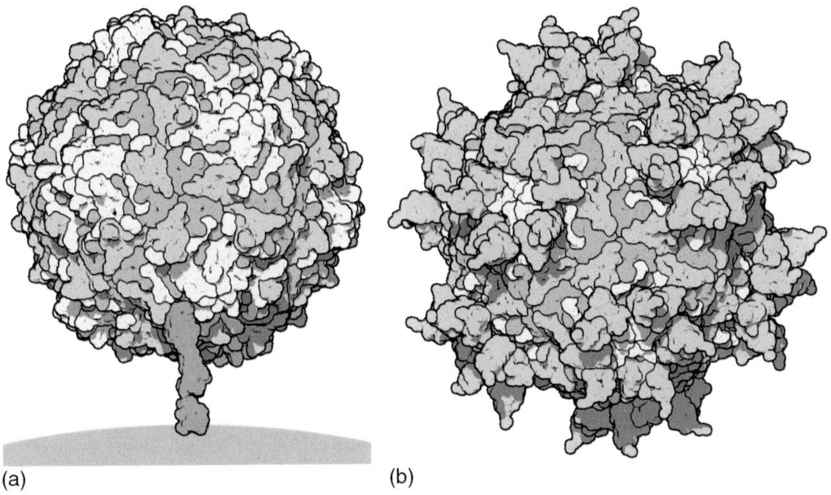

Figure 7 The structure of rhinovirus (from the Molecule of the Month). (a) Rhinovirus (from the PDB entry 1DGI)[53] is bound to a receptor protein on the cell surface, shown in blue. (b) Fragments of antibodies (in light blue) bound to rhinovirus (from PDB entry 1RVF[54]).

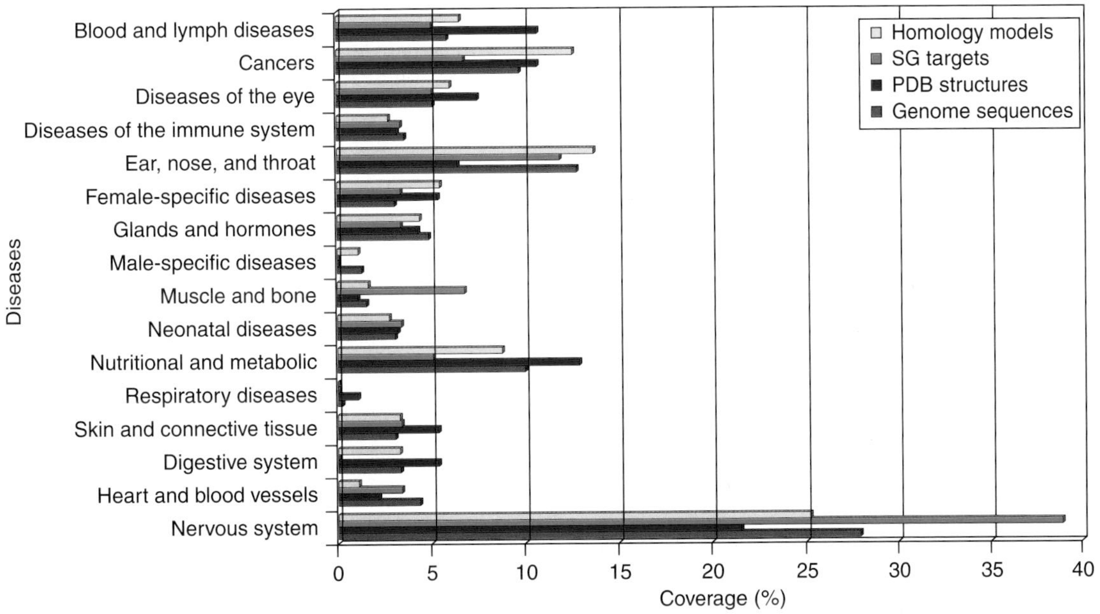

Figure 8 Structural coverage of well-classified diseases, as reported from the RCSB PDB structural genomics portal. The abscissa refers to the normalized distribution of structures/proteins/targets/homology models across the disease classes (number of occurrences per disease class/total occurrences for all disease classes) × 100. The occurrence is based on nonredundant sequence clusters with 40% sequence identity. Blue depicts the normalized distribution of PDB structures associated with that disease class. Red depicts the normalized distribution of proteins identified in the human genome associated with that disease. Green identifies those sequence targets from structural genomics that are associated with the disease. Yellow indicates the normalized distribution for which structures can be assigned by homology modeling (based on data from the Superfamily database).[70]

diseases, has been given above, and is shown graphically in blue in **Figure 8**. The red bar is based on human genome annotation, and after normalization indicates the relative number of proteins associated with a given disease class. As such, it is an approximate measure of how structure determination has been focused relative to the importance of specific disease classes. So, for example, diseases of the central nervous system are underrepresented by structure, whereas diseases associated with nutrition and metabolism are overrepresented – at this time. The various structure genomics projects worldwide deposit their targets – the sequences of the structures they are trying to determine – in a public database managed by the RCSB PDB (TargetDB).[56] Based upon the sequences that can be annotated, it can be estimated that structural genomics will allow for the detailed study for most classes of disease (green bar). The yellow bar indicates how well we can model structures associated with particular diseases.

In addition to understanding the structure of single proteins, it will be necessary to understand in detail how proteins interact with one another and with potential therapeutics. At the present time, we have several examples of multimolecular systems complexed with drugs. An outstanding case in point is the recently published structure of the 50S ribosome particle complexed with a series of antibiotics.[57] It is likely that more examples will follow.

The RCSB PDB will continue to provide resources for archiving and understanding these molecules in order to facilitate drug design.

Acknowledgments

The RCSB PDB is supported by funds from the National Science Foundation, the National Institute of General Medical Sciences, the Office of Science, Department of Energy, the National Library of Medicine, the National Cancer Institute, the National Center for Research Resources, the National Institute of Biomedical Imaging and Bioengineering, and the National Institute of Neurological Disorders and Stroke.

References

1. Bernal, J. D.; Crowfoot, D. M. *Nature* **1934**, *133*, 794–795.
2. Watson, H. C. *Prog. Stereochem.* **1969**, *4*, 299–333.
3. Kendrew, J. C.; Bodo, G.; Dintzis, H. M.; Parrish, R. G.; Wyckoff, H. *Nature* **1958**, *181*, 662–666.
4. Bolton, W.; Perutz, M. F. *Nature* **1970**, *228*, 551–552.
5. Perutz, M. F.; Rossmann, M. G.; Cullis, A. F.; Muirhead, G.; Will, G. *Nature* **1960**, *185*, 416–422.
6. Blake, C. C. F.; Koenig, D. F.; Mair, G. A.; North, A. C. T.; Phillips, D. C.; Sarma, V. R. *Nature* **1965**, *206*, 757–761.
7. Blake, C. C. F.; Johnson, L. N.; Mair, G. A.; North, A. C. T.; Phillips, D. C.; Sarma, V. R. *Proc. R. Soc. Lond. Ser. B* **1967**, *167*, 378–388.
8. Kartha, G.; Bello, J.; Harker, D. *Nature* **1967**, *213*, 862–865.
9. Wyckoff, H. W.; Hardman, K. D.; Allewell, N. M.; Inagami, T.; Tsernoglou, D.; Johnson, L. N.; Richards, F. M. *J. Biol. Chem.* **1967**, *242*, 3749–3753.
10. Bernstein, F. C.; Koetzle, T. F.; Williams, G. J. B.; Meyer, E. F., Jr.; Brice, M. D.; Rodgers, J. R.; Kennard, O.; Shimanouchi, T.; Tasumi, M. *J. Mol. Biol.* **1977**, *112*, 535–542.
11. Berman, H. M.; Westbrook, J.; Feng, Z.; Gilliland, G.; Bhat, T. N.; Weissig, H.; Shindyalov, I. N.; Bourne, P. E. *Nucleic Acids Res.* **2000**, *28*, 235–242.
12. Dutta, S.; Berman, H. M. *Structure* **2005**, *13*, 381–388.
13. Orengo, C. A.; Michie, A. D.; Jones, S.; Jones, D. T.; Swindells, M. B.; Thornton, J. M. *Structure* **1997**, *5*, 1093–1108.
14. Conte, L.; Bart, A.; Hubbard, T.; Brenner, S.; Murzin, A.; Chothia, C. *Nucleic Acids Res.* **2000**, *28*, 257–259.
15. Berman, H. M.; Henrick, K.; Nakamura, H. *Nature Struct. Biol.* **2003**, *10*, 980.
16. Westbrook, J., Henrick, K., Ulrich, E. L., Berman, H. M. The Protein Data Bank Exchange Dictionary. In *International Tables for Crystallography*; Hall, S., McMohan, B., Eds.; Kluwer: Dordrecht, Germany, 2005; Vol. G.
17. Feng, Z.; Chen, L.; Maddula, H.; Akcan, O.; Oughtred, R.; Berman, H. M.; Westbrook, J. *Bioinformatics* **2004**, *20*, 2153–2155.
18. Yang, H.; Guranovic, V.; Dutta, S.; Feng, Z.; Berman, H. M.; Westbrook, J. *Acta Crystallogr. D* **2004**, *60*, 1833–1839.
19. Westbrook, J.; Feng, Z.; Burkhardt, K.; Berman, H. M. *Methods Enzymol.* **2003**, *374*, 370–385.
20. Laskowski, R. A.; McArthur, M. W.; Moss, D. S.; Thornton, J. M. *J. Appl. Crystallogr.* **1993**, *26*, 283–291.
21. Lovell, S. C.; Davis, I. W.; Arendall, W. B., 3rd; de Bakker, P. I.; Word, J. M.; Prisant, M. G.; Richardson, J. S.; Richardson, D. C. *Proteins* **2003**, *50*, 437–450.
22. Vaguine, A. A.; Richelle, J.; Wodak, S. J. *Acta Crystallogr. D* **1999**, *55*, 191–205.
23. Dutta, S.; Burkhardt, K.; Bluhm, W. F.; Berman, H. M. *Curr. Protocols Bioinform.* **2005**, 1.9.1–1.9.40.
24. The Gene Ontology Consortium. *Nature Genetics* **2000**, *25*, 25–29.
25. Gursky, O.; Badger, J.; Li, Y.; Caspar, D. L. *Biophys. J.* **1992**, *63*, 1210–1220.
26. Weininger, D. *J. Chem. Inf. Comput. Sci.* **1988**, *28*, 31–36.
27. Moreland, J. L.; Gramada, A.; Buzko, O. V.; Zhang, Q.; Bourne, P. E. *BMC Bioinform.* **2005**, *6*, 21.
28. Zheng, J.; Trafny, E. A.; Knighton, D. R.; Xuong, N. H.; Taylor, S. S.; Ten Eyck, L. F.; Sowadski, J. M. *Acta Crystallogr. D* **1993**, *49*, 362–365.
29. Scheidig, A. J.; Sanchez-Llorente, A.; Lautwein, A.; Pai, E. F.; Corrie, J. E.; Reid, G. P.; Wittinghofer, A.; Goody, R. S. *Acta Crystallogr. D* **1994**, *50*, 512–520; NCBI. *Genes and Disease*, e-book (http://www.ncbi.nlm.nih.gov). National Center for Biotechnology Information: Bethesda, MD (accessed April 2006).
29a. Scheidig, A. J.; Burmester, C.; Goody, R. S. *Struct. Fold. Des.* **1999**, 7, 1311–1324.
30. Altschul, S. F.; Gish, W.; Miller, W.; Myers, E. W.; Lipman, D. J. *J. Mol. Biol.* **1990**, *215*, 403–410.
31. Pearson, W. R.; Lipman, D. J. *Proc. Natl. Acad. Sci. USA* **1988**, *24*, 2444–2448.
32. Franken, S. M.; Scheidig, A. J.; Krengel, U.; Rensland, H.; Lautwein, A.; Geyer, M.; Scheffzek, K.; Goody, R. S.; Kalbitzer, H. R.; Pai, E. F. et al. *Biochemistry* **1993**, *32*, 8411–8420.
33. Tong, L. A.; de Vos, A. M.; Milburn, M. V.; Kim, S. H. *J. Mol. Biol.* **1991**, *217*, 503–516.
34. Joint United Nation Programme on HIV/AIDS (UNAIDS). *2004 Report on the Global AIDS epidemic*; UN, Geneva, Switzerland, 2004.
35. Swain, A. L.; Miller, M. M.; Green, J.; Rich, D. H.; Schneider, J.; Kent, S. B.; Wlodawer, A. *Proc. Natl. Acad. Sci. USA* **1990**, *87*, 8805–8809.
36. Bone, R.; Vacca, J. P.; Anderson, P. S.; Holloway, M. K. *J. Am. Chem. Soc.* **1991**, *113*, 9382–9384.

37. Summers, M. F.; South, T. L.; Kim, B.; Hare, D. R. *Biochemistry* **1990**, *29*, 329–340.
38. Chen, Z.; Li, Y.; Chen, E.; Hall, D. L.; Darke, P. L.; Culberson, C.; Shafer, J. A.; Kuo, L. C. *J. Biol. Chem.* **1994**, *269*, 26344–26348.
39. Krohn, A.; Redshaw, S.; Ritchie, J. C.; Graves, B. J.; Hatada, M. H. *J. Med. Chem.* **1991**, *34*, 3340–3342.
40. Kempf, D. J.; Marsh, K. C.; Denissen, J. F.; McDonald, E.; Vasavanonda, S.; Flentge, C. A.; Green, B. E.; Fino, L.; Park, C. H.; Kong, X. P. et al. *Proc. Natl. Acad. Sci. USA* **1995**, *92*, 2484–2488.
41. Kaldor, S. W.; Kalish, V. J.; Davies, J. F., II; Shetty, B. V.; Fritz, J. E.; Appelt, K.; Burgess, J. A.; Campanale, K. M.; Chirgadze, N. Y.; Clawson, D. K. et al. *J. Med. Chem.* **1997**, *40*, 3979–3985.
42. Surleraux, D. L. N. G.; Tahri, A.; Verschueren, W. G.; Pille, G. M. E.; De Kock, H. A.; Jonckers, T. H. M.; Peeters, A.; De Meyer, S.; Azijn, H.; Pauwels, R. et al. *J. Med. Chem.* **2005**, *48*, 1813–1822.
43. Stoll, V.; Qin, W.; Stewart, K. D.; Jakob, C.; Park, C.; Walter, K.; Simmer, R. L.; Helfrich, R.; Bussiere, D.; Kao, J. et al. *Bioorg. Med. Chem.* **2002**, *10*, 2803–2806.
44. Adams, M. J.; Blundell, T. L.; Dodson, E. J.; Dodson, G. G. *Nature* **1969**, *224*, 491–494.
45. Bentley, G.; Dodson, E.; Dodson, G.; Hodgkin, D.; Mercola, D. *Nature* **1976**, *261*, 166–168.
46. Baker, E. N.; Blundell, T. L.; Cutfield, J. F.; Cutfield, S. M.; Dodson, E. J.; Dodson, G. G.; Hodgkin, D. M.; Hubbard, R. E.; Isaacs, N. W.; Reynolds, C. D. et al. *Philos. Trans. R. Soc. Lond. B* **1988**, *319*, 369–456.
47. Hua, Q. X.; Gozani, S. N.; Chance, R. E.; Hoffmann, J. A.; Frank, B. H.; Weiss, M. A. *Nature Struct. Biol.* **1995**, *2*, 129–138.
48. Kim, S. S.; Smith, T. J.; Chapman, M. S.; Rossmann, M. C.; Pevear, D. C.; Dutko, F. J.; Felock, P. J.; Diana, G. D.; McKinlay, M. A. *J. Mol. Biol.* **1989**, *210*, 91–111.
49. Kolatkar, P. R.; Bella, J.; Olson, N. H.; Bator, C. M.; Baker, T. S.; Rossmann, M. G. *EMBO J.* **1999**, *18*, 6249–6259.
50. Abel, K.; Yoder, M. D.; Hilgenfeld, R.; Jurnak, F. *Structure* **1996**, *4*, 1153–1159.
51. Matthews, D. A.; Dragovich, P. S.; Webber, S. E.; Fuhrman, S. A.; Patick, A. K.; Zalman, L. S.; Hendrickson, T. F.; Love, R. A.; Prins, T. J.; Marakovits, J. T. et al. *Proc. Natl. Acad. Sci. USA* **1999**, *96*, 11000–11007.
52. Bella, J.; Rossmann, M. G. *J. Struct. Biol.* **1999**, *128*, 69–74.
53. He, Y.; Bowman, V. D.; Mueller, S.; Bator, C. M.; Bella, J.; Peng, X.; Baker, T. S.; Wimmer, E.; Kuhn, R. J.; Rossmann, M. G. *Proc. Natl. Acad. Sci. USA* **2000**, *97*, 79–84.
54. Smith, T. J.; Chase, E. S.; Schmidt, T. J.; Olson, N. H.; Baker, T. S. *Nature* **1996**, *383*, 350–354.
55. Gerstein, M.; Edwards, A.; Arrowsmith, C. H.; Montelione, G. T. *Science* **2003**, *299*, 1663–1664.
56. Chen, L.; Oughtred, R.; Berman, H. M.; Westbrook, J. *Bioinformatics* **2004**, *20*, 2860–2862.
57. Tu, D.; Blaha, G.; Moore, P. B.; Steitz, T. A. *Cell* **2005**, *121*, 257–270.
58. Aleshin, A. E.; Kirby, C.; Liu, X.; Bourenkov, G. P.; Bartunik, H. D.; Fromm, H. J.; Honzatko, R. B. *J. Mol. Biol.* **2000**, *296*, 1001–1015.
59. Lee, J. H.; Chang, K. Z.; Patel, V.; Jeffrey, C. J. *Biochemistry* **2001**, *40*, 7799–7805.
60. Dalby, A.; Dauter, Z.; Littlechild, J. A. *Protein Sci.* **1999**, *8*, 291–297.
61. Mercer, W. D.; Winn, S. I.; Watson, H. C. *J. Mol. Biol.* **1976**, *104*, 277–283.
62. Evans, P. R.; Farrants, G. W.; Hudson, P. J. *Phil. Trans. R. Soc. Lond. B* **1981**, *293*, 53–62.
63. Watson, H. C.; Walker, N. P.; Shaw, P. J.; Bryant, T. N.; Wendell, P. L.; Fothergill, L. A.; Perkins, R. E.; Conroy, S. C.; Dobson, M. J.; Tuite, M. F. et al. *EMBO J.* **1982**, *1*, 1635–1640.
64. Lolis, E.; Petsko, G. A. *Biochemistry* **1990**, *29*, 6619–6625.
65. Winn, S. I.; Watson, H. C.; Harkins, R. N.; Fothergill, L. A. *Phil. Trans. R. Soc. Lond. B* **1981**, *293*, 121–130.
66. Zhang, E.; Brewer, J. M.; Minor, W.; Carreira, L. A.; Lebioda, L. *Biochemistry* **1997**, *36*, 12526–12534.
67. Valentini, G.; Chiarelli, L.; Fortin, R.; Speranza, M. L.; Galizzi, A.; Mattevi, A. *J. Biol. Chem.* **2000**, *275*, 18145–18152.
68. Koole, L. H.; Neidle, S.; Crawford, M. D.; Krayevski, A. A.; Gurskaya, G. V.; Sandstrom, A.; Wu, J.-C.; Tong, W.; Chattopadhyaya, J. *J. Org. Chem.* **1991**, *56*, 6884–6892.
69. Pettersen, E. F.; Goddard, T. D.; Huang, C. C.; Couch, G. S.; Greenblatt, D. M.; Meng, E. C.; Ferrin, T. E. *J. Comput. Chem.* **2004**, *25*, 1605–1612.
70. Madera, M.; Vogel, C.; Kummerfeld, S. K.; Chothia, C.; Gough, J. *Nucleic Acids Res.* **2004**, *32*, D235–D239.

Biographies

Philip E Bourne received a BSc and a PhD from The Flinders University in South Australia; this was followed by a postdoctoral term at Sheffield University in Sheffield, UK. His professional interests focus on bioinformatics, and structural bioinformatics in particular. This implies algorithms, metalanguages, biological databases, biological query languages, and visualization with special interest in evolution, cell signaling and apoptosis. He is currently a professor in the Department of Pharmacology at the University of California at San Diego, an adjunct professor at the Burnham Institute and the Keck Graduate Institute, and, since 1998, he has been a co-director of the Research Collaboratory for Structural Bioinformatics Protein Data Bank.

Wolfgang F Bluhm received a Diploma in physics from the University of Siegen, Germany, and an MSc in engineering sciences and a PhD in bioengineering from the University of California at San Diego. He did his postdoctoral training in the Department of Medicine at the University of California at San Diego, working on a variety of topics in cardiac physiology. He joined the staff of the Research Collaboratory for Structural Bioinformatics Protein Data Bank in 2000. Since 2001, he has been the Production Manager for the query and distribution systems of the Protein Data Bank. His responsibilities include new feature development, data distribution, and scientific outreach.

Nita Deshpande received an MSc in biotechnology from Madurai Kamaraj University in Madurai, India, followed by a PhD in biology from the University of Texas at Austin. She then did a postdoctoral fellowship in cardiology in Dr Kathy Griendling's laboratory at Emory University in Atlanta, while working on an MSc in computer science from Southern Polytechnic State University in Marietta, GA. Since September 2000, she has worked as a Database Applications Programmer at the Research Collaboratory for Structural Bioinformatics Protein Data Bank, focusing on creating and maintaining the query interface and the integration of biological data from external databases.

Qing Zhang received a BSc in chemistry from Beijing University in China. She then obtained her PhD in chemistry and structural biology from the University of California at Berkeley. She worked as a bioinformatics scientist on the human genome project at Lawrence Berkeley National Laboratory before joining the PDB project in 2004. She is currently involved in the efforts to include information on protein–protein interactions on the RCSB PDB website.

Helen M Berman received an AB in chemistry from Barnard College in New York City, followed by a PhD and post-doctoral work at the University of Pittsburgh in Pittsburgh. She is currently a Board of Governors Professor of Chemistry and Chemical Biology at Rutgers, The State University of New Jersey. Berman, with her collaborators, examines the structural properties of nucleic acid-containing molecules using x-ray crystallography and computational approaches. A major focus of her work has been to establish methods to collect and archive structural data for analysis. She is both the founder and director of the Nucleic Acid Database and is also one of the founders of the PDB. Since 1998, Berman has been the overall director of the Research Collaboratory for Structural Bioinformatics Protein Data Bank.

Judith L Flippen-Anderson received a BA in chemistry from Northeastern University in Boston, followed by an MS in chemistry from Arizona State University in Tempe. She then joined the staff at the Naval Research Laboratory in

Washington, DC, where her research interests centered on x-ray crystallographic studies of small molecules related to drug design and dense energetic materials. Since January 2003, she has directed the education and outreach activities of the Research Collaboratory for Structural Bioinformatics Protein Data Bank.

Comprehensive Medicinal Chemistry II
ISBN (set): 0-08-044513-6

ISBN (Volume 3) 0-08-044516-0; pp. 373–388

3.18 The Cambridge Crystallographic Database

F H Allen, G M Battle, and S Robertson, Cambridge Crystallographic Data Centre, Cambridge, UK

3.18.1	**Introduction**	**389**
3.18.2	**Overview of the Cambridge Structural Database**	**390**
3.18.2.1	Information Content of the Cambridge Structural Database	390
3.18.2.2	Data Acquisition for the Cambridge Structural Database	391
3.18.2.3	Statistical Overview of the Cambridge Structural Database	392
3.18.3	**The Cambridge Structural Database System**	**393**
3.18.3.1	Searching, Visualizing, and Analyzing Cambridge Structural Database Information	393
3.18.3.1.1	The ConQuest search program	393
3.18.3.1.2	The mercury visualizer	395
3.18.3.1.3	Data analysis using Vista	395
3.18.3.2	Knowledge-Based Libraries of Structural Information	395
3.18.3.2.1	Mogul: a knowledge base of intramolecular geometry	396
3.18.3.2.2	IsoStar: a knowledge base of intermolecular interactions	398
3.18.4	**Applications of the Cambridge Structural Database System in Medicinal Chemistry**	**398**
3.18.4.1	Cambridge Structural Database-Based Research: An Overview	398
3.18.4.2	The Relevance of Crystal Structure Data in Medicinal Chemistry Research	399
3.18.4.3	Intramolecular Geometries and Conformational Analysis	400
3.18.4.4	Metal Coordination	401
3.18.4.5	Intermolecular Interactions	401
3.18.4.5.1	Locating nonbonded interactions using the Cambridge Structural Database	402
3.18.4.5.2	Strong hydrogen bonds	402
3.18.4.5.3	Weak hydrogen bonds	404
3.18.4.5.4	Interactions not mediated by hydrogen	404
3.18.4.5.5	Applications of the IsoStar library of intermolecular interactions	405
3.18.4.5.6	Predicting hydrogen-bonded contacts	406
3.18.5	**Applications Software that uses Knowledge Derived from the Cambridge Structural Database**	**406**
3.18.5.1	Programs for Predicting Intermolecular Interactions	406
3.18.5.2	Protein–Ligand Docking Programs	406
3.18.6	**Conclusions**	**407**
References		**407**

3.18.1 Introduction

Crystal structure analyses are remarkable for the richness of structural information that they provide. Because this information yields both the geometric structures of molecules and also characterizes the nature and geometry of their interactions with other molecules and ions, crystal structure data are crucially important to a very wide range of scientific activities. In medicinal chemistry, this experimental structural and supramolecular information is an essential adjunct to (1) rational molecular design, via conformational analysis and validation, and the prediction of protein-ligand interactions, and (2) pharmaceutical materials design and drug delivery, through studies in crystal engineering, crystal growth and polymorphic systems, all of which contribute to our ability to infer and predict the crystal structures of novel molecules.

Since the late 1960s, the results of published crystal structure analyses, together with some directly deposited data, have been collected in five databases, which together cover the complete spectrum of chemical compounds.[1] This chapter concentrates on the Cambridge Structural Database (CSD) of small organic and metal–organic molecules, curated by the Cambridge Crystallographic Data Centre (CCDC) in the UK.[2] We begin by discussing its information content and the facilities provided by the associated software[3] in the distributed CSD system. This software now includes knowledge-based systems[4,5] that provide click-of-a-button access to extensive libraries of molecular and supramolecular information.

A principal purpose of the chapter is to illustrate the scientific value of locating and analyzing the crystallographic results for many chemical structures or substructures taken together. Techniques for data visualization and data analysis then permit, for example the determination of mean values for geometrical parameters, the observation of preferred conformations or coordination sphere geometries (and the mapping of their interconversion pathways in both cases), and the analysis and visualization of intermolecular interactions. Thus, we describe (1) how the CSD can be used for basic research in some of the areas listed above, (2) how CSD data have been converted into rapidly accessible electronic libraries of structural knowledge, and (3) how these libraries can, in their turn, be used as knowledge engines that underpin further software applications designed to solve problems in structural chemistry, rational drug design, and crystallography. Use of small-molecule CSD data in conjunction with macromolecular information from the Protein Data Bank (PDB)[6] is also highlighted.

3.18.2 Overview of the Cambridge Structural Database

3.18.2.1 Information Content of the Cambridge Structural Database

Each individual crystal structure forms an entry in the CSD, and is identified by a CSD reference code (refcode) (e.g., BAGFIT02). Six letters identify each chemical compound, and two supplementary digits identify additional structure determinations (studies of other polymorphs, studies under different experimental conditions, studies by different scientists, etc.). The information content of each entry is summarized in **Table 1** and **Figure 1**.

The most important information item added by CCDC staff is the 2D chemical structure representation, which forms the basis for 2D and 3D substructure searching[2,3] at the molecular and supramolecular levels. To increase the

Table 1 Information content of the CSD

Bibliographic and chemical text and comment	Compound name(s) – systematic and trivial Amino acid sequence for peptides Chemical formula Author's name Journal name and literature citation Text indicating special experimental conditions or results (neutron study, powder study, polymorph, nonambient temperature or pressure, absolute configuration determined, etc.) Chemical class (alkaloid, steroid, etc.) Text comment concerning disorder and errors located during validation
Chemical diagram and connection table	Formal two-dimensional chemical structure diagram in terms of atom and bond properties. Bond types used in the CSD are single, double, triple, quadruple (metal–metal), aromatic, delocalized double, and π bonds Bit-encoded screen records (see text)
Crystal structure data	Cell dimensions and standard uncertainties Space group and symmetry operators Reduced-cell parameters Z' (number of chemical entities per asymmetric unit) Calculated density Structural precision indicators Atomic coordinates and standard uncertainties

AZHPXA

S.J. Cline, D.J. Hodgson
J.Am.Chem.Soc., 102, 6285, 1980
2-Phenyl-7-methyl-8-azahypoxanthine

Formula:	C11 H9 N5 01
Colour:	Colorless
Extra information:	Antiallergenic drug
Spacegroup:	C2/c
R-factor:	3.8

(a)

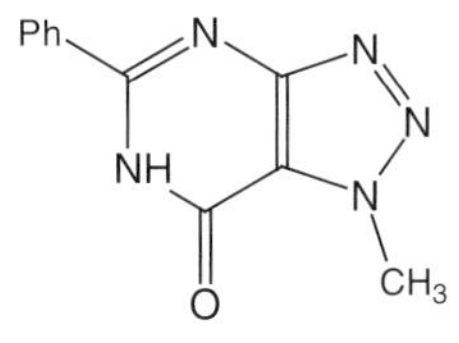

(b)

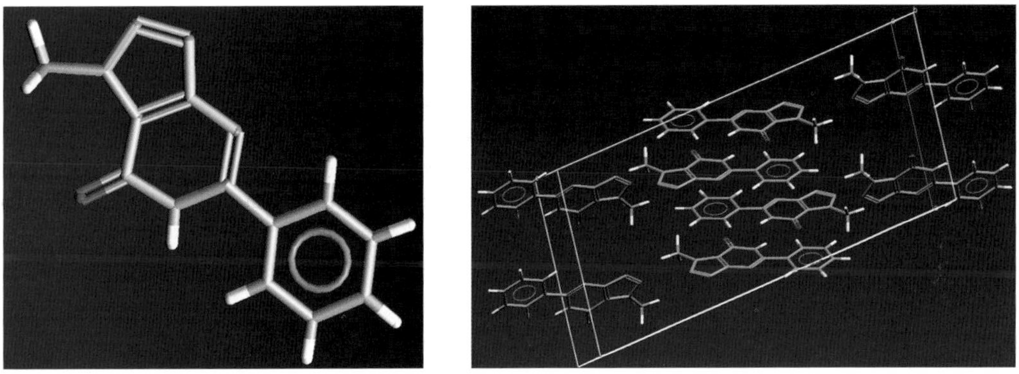

(c) (d)

Figure 1 Schematic view of the information content of a CSD entry: (a) bibliographic, chemical, and crystallographic text; (b) two-dimensional (2D) chemical structural formula (chemical connectivity), together with (c) three-dimensional (3D) molecular structure and (d) 3D crystal structure derived from stored atomic coordinates and crystal data.

speed of these searches, each connection table is analyzed in order to assign cyclic/acyclic flags to bonds and to generate a bitmap or 'screen' record. This contains codified yes/no information indicating the presence or absence of specific substructural features. The screens then act as search heuristics to minimize calls on computer-intensive code for atom-by-atom and bond-by-bond substructure matching. Bit-screens are also generated from other CSD information fields, including text fields, for similar reasons.

Complete information, crystallographic and chemical, is then validated using the CCDC program PreQuest. The principal checks carried out are: (1) self-consistency of crystal data (i.e., cell parameters, Z value, and density) and chemical constitution; (2) a cross-check that the atomic connectivity derived from the coordinate data and standard covalent radii matches that in the encoded 2D chemical diagram; (3) self-consistency of the geometry calculated from the published atomic coordinates with the geometry presented in the paper; and (4) that there are no unreasonably short intermolecular contacts, nor any large unexplained voids in the extended crystal structure, perhaps indicative, respectively, of an incorrect space group or missing solvent molecule(s). Any unresolved errors are referred back to the original author(s) for clarification. CCDC scientific editors also resolve issues connected with crystallographic disorder, and add text remarks concerning any special aspects of the crystallographic experiment or its results.

3.18.2.2 Data Acquisition for the Cambridge Structural Database

Until the early 1990s, most of the information entering the CSD was retyped from coordinate listings in published papers or associated deposition documents. This situation changed radically with the universal acceptance of the Crystallographic Information File (CIF),[7] adopted as an international standard for the electronic interchange of crystal structure data by the International Union of Crystallography in 1991. Most major journals now require electronic data deposition in CIF format, and electronic data capture has increased rapidly, from a mere 30% in 1997 to 98.6% in 2004, and most major journals now have direct deposition arrangements with the CCDC. The vast majority (99%) of CSD entries arise from published work abstracted from more than 1000 literature sources. However, the CCDC also encourages personal communications of data that would otherwise be lost to the scientific community, thus increasing the value of the CSD. In order to assist scientists to prepare format and syntax compliant CIFs for deposition, the CCDC makes the enCIFer program[8] freely available to researchers via its website.

3.18.2.3 Statistical Overview of the Cambridge Structural Database

On January 1, 2005 the CSD contained 335276 structures, with 28916 new entries having been added during 2004. Summary statistics are given in **Table 2**, and more detailed statistical information is regularly updated on the CCDC website. The CSD is fully retrospective, with its earliest reference dating back to 1923, and its earliest coordinate sets dating from 1936. **Figure 2** shows the cumulative rate of growth of the CSD for the period 1970–2004.

These data show that (1) the time taken for the CSD to double its number of entries has increased from 3.6 years in the 1980s to settle at around 9.0 years by 2004, and (2) continuation of current trends indicates that the CSD will archive its 500 000th structure during 2009. However, this projection does not take account of increased publication rates arising from improved crystallographic and computing technology, or of any changes that may accelerate current practices for placing crystal structure data into the public domain. One problem arising in an era of high-throughput crystallography is that the time required to publish a crystal structure is now comparable to, or even longer than, the time required to carry out the diffraction analysis itself. This means that many thousands of structures remain in the archives of individual laboratories, and much valuable data is being lost to science. This problem has been recognized,[9] and a number of initiatives are in progress to mitigate this situation.

Table 2 Summary statistics for the CSD on January 1, 2005

	Structures	*Percent of the CSD*
Total No. of structures	335 276	100.0
No. of different compounds	303 733	–
No. of literature sources	1 083	–
Organic structures	145 559	43.4
Transition metal present	172 240	51.4
Li to Fr or Be to Ra present	17 832	5.5
Main group metal present	21 856	6.5
3D coordinates present	301 942	90.1
Error-free coordinates	295 600	97.9[a]
Neutron studies	1 219	0.4
Powder diffraction studies	835	0.3
Low/high temperature studies	96 843	28.9
Absolute configuration determined	5 826	1.7
Disorder present in structure	65 775	19.6
Polymorphic structures	10 853	3.2
R factor < 0.100	306 284	91.3
R factor < 0.070	262 194	78.2
R factor < 0.050	168 310	48.7
R factor < 0.030	29 090	8.7
N atoms with 3D coordinates	21 882 879	–

[a] Taken as a percentage of structures for which coordinates are present in the CSD.

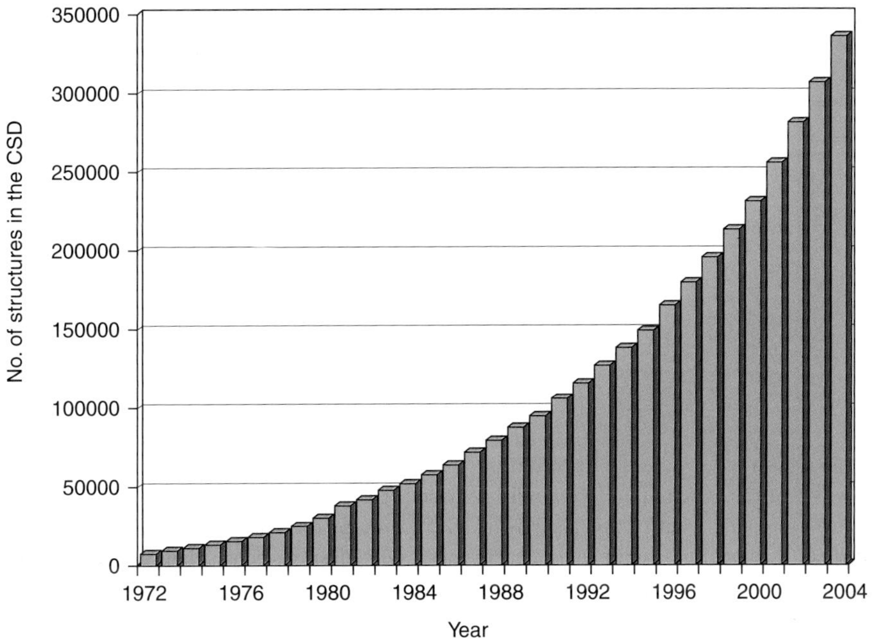

Figure 2 Histogram showing the cumulative growth rate of the CSD for the period 1970–2004.

3.18.3 The Cambridge Structural Database System

The distributed CSD system comprises the CSD itself together with six major software components. One of these is PreQuest, the software used by the CCDC's scientific editors to create value-added and fully checked CSD entries. This software is released so that users of the CSD system can create their own in-house databases of proprietary structures in CSD format. These private databases can be searched separately or together with the main CSD. The other five software components fall into two categories: (1) software for CSD access, structure visualization, and data analysis, and (2) knowledge-based libraries that provide click-of-a-button access to many millions of geometrical data items derived principally from the CSD, but incorporating some information from the PDB[6] as well.

3.18.3.1 Searching, Visualizing, and Analyzing Cambridge Structural Database Information

3.18.3.1.1 The ConQuest search program

ConQuest[3] provides search, retrieval, and display facilities for the CSD. Individual queries can be entered to interrogate the bibliographic, chemical text and numerical fields listed in **Table 1**. Most importantly (**Figure 3**), ConQuest provides extensive graphical facilities for defining 2D and 3D substructure searches. The 2D searches interrogate the chemical connection tables alone, while the internal mapping of atomic coordinates and connection tables forms the basis for:

- Systematic 3D substructure searching. This can be (1) intramolecular, for example to locate 3D pharmacophoric patterns or to retrieve substructures having specific conformations by use of torsion angle constraints, or (2) intermolecular, where searching is applied to extended crystal structures, for example to locate hydrogen bonds or other non-bonded interactions, again using appropriate chemical and geometrical constraints.
- The retrieval of calculated 3D geometrical parameters for each occurrence of the defined substructure, as shown in **Figure 3**. These data can then be used in further analyses (e.g., using the Vista program described below).

ConQuest has facilities for combining individual queries, including 2D and 3D substructure queries, using Boolean logic. The program also displays the information content of CSD entries, selected from either the main database, or from the subset of entries resulting from a search. Display panes show bibliographic and chemical text, crystal data, 2D

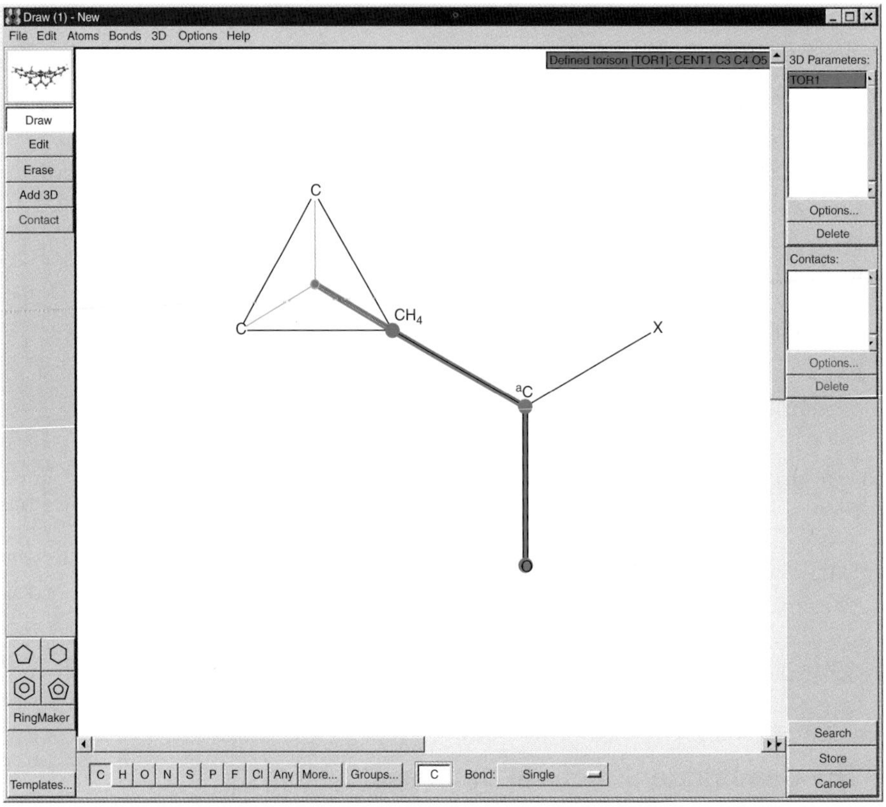

(a)

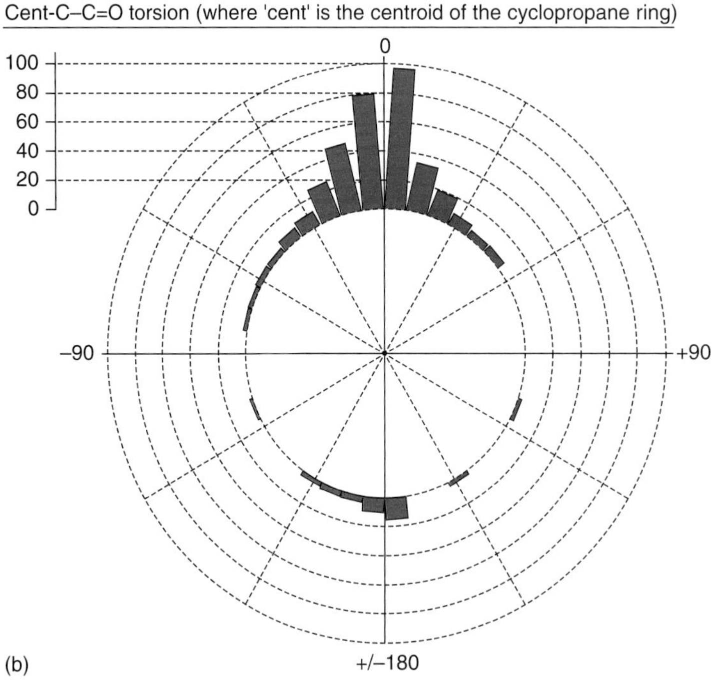

(b)

Figure 3 (a) ConQuest query of the cent-C–C=O torsion angle of cyclopropyl–carbonyl and (b) the corresponding Vista-generated histogram of the torsion angle (cent is the centroid of the cyclopropyl ring).

chemical diagrams, and 3D molecular or crystal structures. ConQuest can access Mercury or Vista directly (see below), to provide more extensive 3D structure visualization or data analysis facilities. ConQuest can also output information for search hits in a variety of formats (e.g. cif and mol2), and transfer data to other programs. Currently (2005), ConQuest is being upgraded to provide links from CSD entries to the electronic literature.

3.18.3.1.2 The mercury visualizer

Mercury[3] provides general and advanced functionality for viewing 3D molecular and crystal structures, as summarized in **Table 3**. A unique feature of Mercury applied to CSD entries is its ability to import chemical bond types from the 2D connection tables and display them on 3D images, as shown in **Figure 4a**.

However, the most important functionality in Mercury, and one which is vital in supramolecular studies, is the ability to locate, build, and display networks of intermolecular and intramolecular hydrogen bonds, short nonbonded contacts, and user-specified contact types. Mercury will use distance criteria relative to van der Waals radii sums, or direct (ångstrom) values. An example hydrogen-bonded network, constructed and viewed in Mercury, is shown in **Figure 4b**. The facilities for displaying a slice through a crystal in any direction are illustrated in **Figure 4c**, and such displays can be valuable in rationalizing crystal morphology and predicting how to control it. Finally, Mercury has the ability to read several structures into the same visualization window and manipulate and overlay them separately. This is especially valuable when comparing conformations, examining differences between polymorphic forms, etc.

3.18.3.1.3 Data analysis using Vista

Vista displays molecular geometry and other parameters relating to a molecular or supramolecular substructure in a spreadsheet format. These data are normally retrieved from the CSD using ConQuest according to user-supplied specifications (e.g. see **Figure 3**). Vista performs a variety of analysis and display functions, including generation of:

- histograms and scattergrams of parameter distributions referred to Cartesian or polar axes;
- simple descriptive statistics for parameter distributions;
- statistical analyses, including linear regression and principal-component analysis;
- hyperlinking from spreadsheet data back to the original CSD entry; and
- preparation of plots for reports and publications.

3.18.3.2 Knowledge-Based Libraries of Structural Information

The software facilities described in Section 3.18.3.1 permit CSD information to be searched and analyzed in a very comprehensive manner. However, while efficient in themselves, they can be time-consuming for scientists who wish to access standard geometrical data, such as the mean value of a particular type of bond, the conformational preferences exhibited by a specific substructure, or the spatial characteristics and metrics of a common hydrogen bond. For this reason, the CCDC has compiled two structural knowledge bases designed to provide instant click-of-a-button access to a very wide range of information on geometrical structure, both intramolecular, via Mogul,[4] and intermolecular, via IsoStar.[5]

Table 3 Principal facilities of the Mercury visualizer

- Browse the entire CSD, load hit lists from ConQuest searches, or read in crystal structure data in other common formats (mol2, pdb, cif, mol)
- Rotate, translate and scale the 3D crystal structure display and view down cell axes, reciprocal cell axes, and normals to planes
- Range of visualization options (different display styles, coloring and labeling options, ability to hide and then redisplay atoms and molecules, etc.)
- Measure distances, angles, and torsion angles
- Create and display centroids, least-squares mean planes, and Miller planes
- Display anisotropic displacement parameters as ellipsoids
- Display unit cell axes and the contents of any number of unit cells in any direction (including fractions of unit cells)
- Locate, display, and build networks of hydrogen bonds or other nonbonded contacts
- Display a slice through the crystal in any direction
- View, and superpose, two or more structures in the same window
- Display CSD entry information, including 2D chemical diagrams
- Save images in a variety of formats

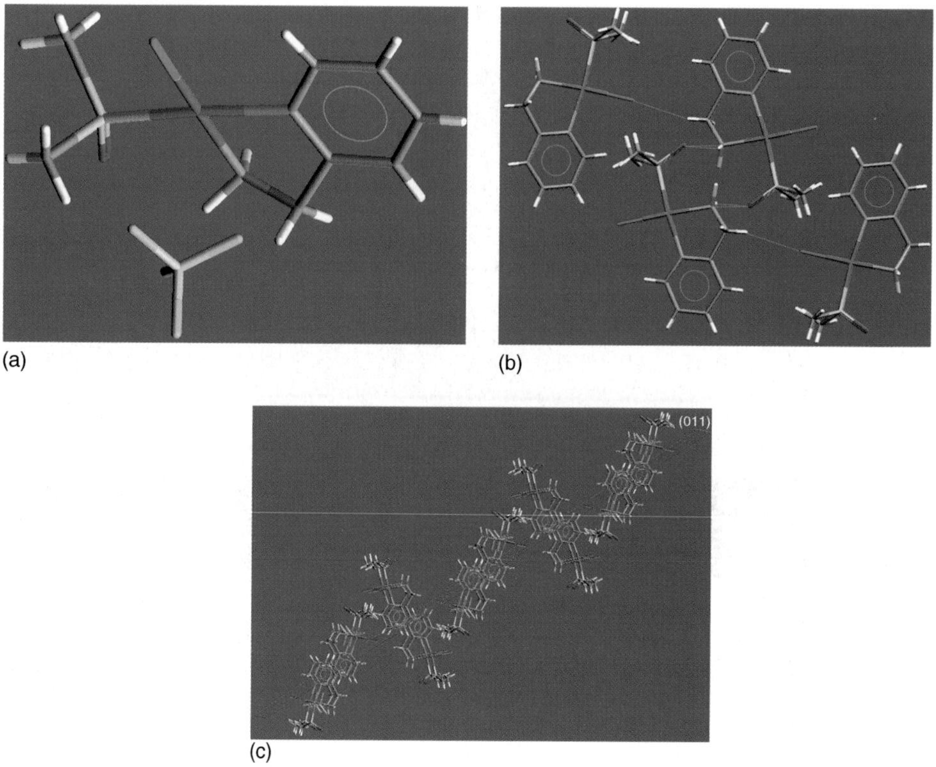

Figure 4 Mercury plots of the CSD entry LECNEM (an aminomethylpyridine-chloro-dimethylsulfoxide-palladium complex thought to have antitumor properties) showing (a) the 3D structure, (b) short-range interactions (within the range of van der Waals radii), and (c) a slice through the unit cell.

3.18.3.2.1 Mogul: a knowledge base of intramolecular geometry

Mogul[4] operates by searching precomputed libraries of bond lengths, valence angles, and torsion angles derived from the CSD. The libraries are built in the following manner: (1) bond, angle, and torsional fragments (acyclic torsions only, at present) are constructed from every entry in the CSD; (2) these fragments are then classified by the evaluation of their 'key components' (i.e., their atom-based and bond-based properties); and (3) the fragments are then grouped on the basis of these components, together with the value of the geometric feature. This process generates a search tree, which enables very fast searching based only on the 'key components' of a query fragment,[4] thus obviating the need for the computer-intensive atom-by-atom, bond-by-bond searching inherent in normal substructure searching methodologies.

To view data concerning a bond, angle, or torsion angle, a query molecule (2D or 3D) is either imported into the Mogul interface (from the CSD or using a variety of input formats (mol2, cif, res, pdb, etc.)) or sketched. The geometric feature of interest is selected by the user (**Figure 5a**), the software computes the 'key components' of the chemical fragment, and the search is carried out. A histogram of the required distribution is obtained in seconds, together with descriptive statistics (**Figure 5b**). Occasionally, a histogram may contain rather few hits due to the low frequency of occurrence of the search fragment environment in the CSD. Although these hits are exact with respect to the query fragment, there are insufficient entries in the histogram for it to be useful. In these situations, it is possible to generalize the search (i.e., relax the level of fragment environment specification), so that additional chemically related hits are obtained. The results of a generalized search can be added to the histogram obtained from the exact search, and all hits are listed by their similarity/relevance to the query fragment. The user may choose the level of fragment similarity to be included in the final distribution. Complete geometry searches (i.e., to locate CSD distributions for all bonds, angles, and torsions in the target molecule) may also be carried out.

It is possible to hyperlink from bar(s) in the Mogul histograms back to the CSD entries that generated those values; individual 2D and 3D structures can be viewed, together with data such as the CSD reference code, publication details,

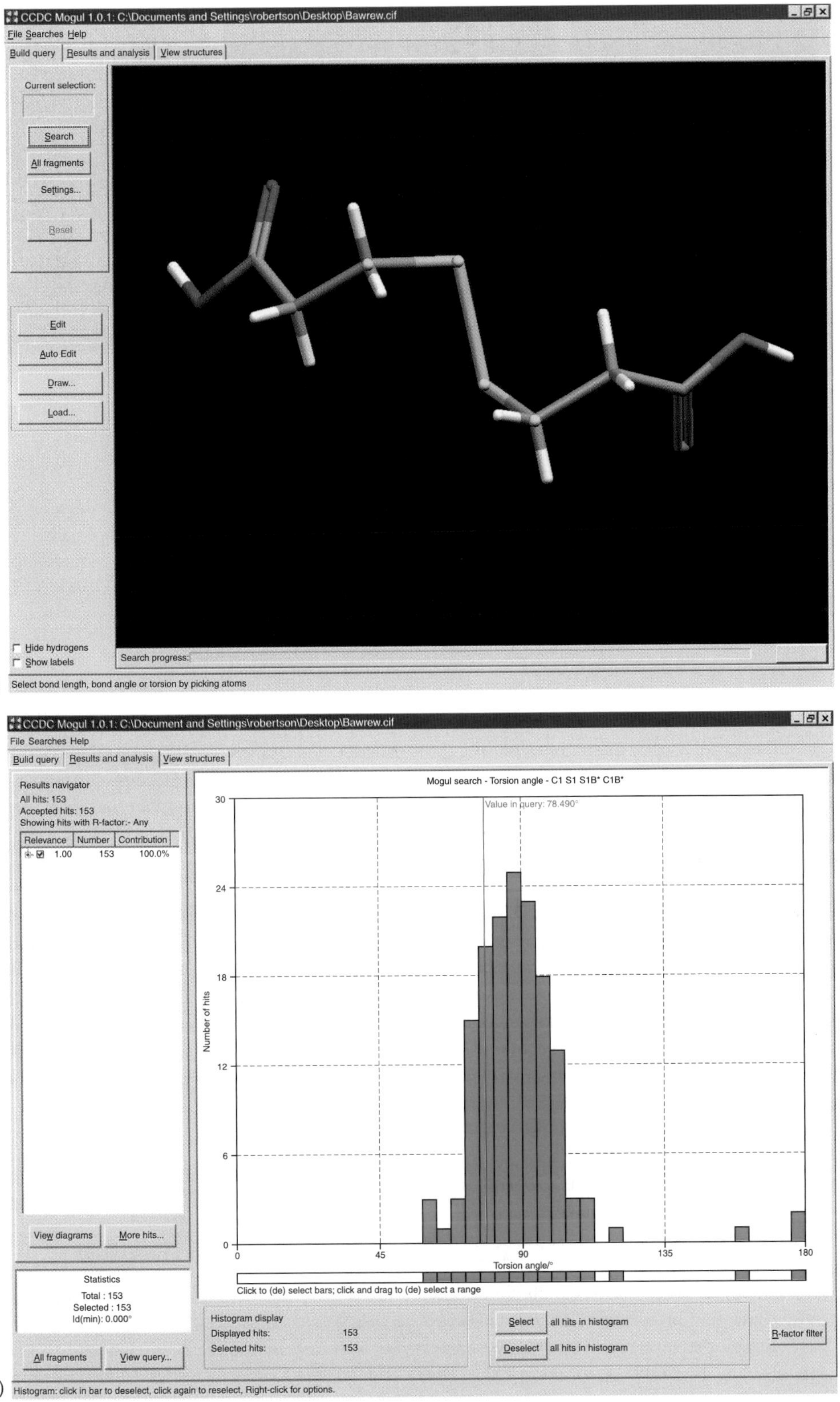

(a) (b)

Figure 5 Results of a Mogul search for a C–S–S–C torsion. (a) The search was carried out by selecting the relevant atoms that make up the torsion in the query molecule. (b) The results are presented in a distribution histogram, illustrating that the preferred C–S–S–C torsion is around 90°.

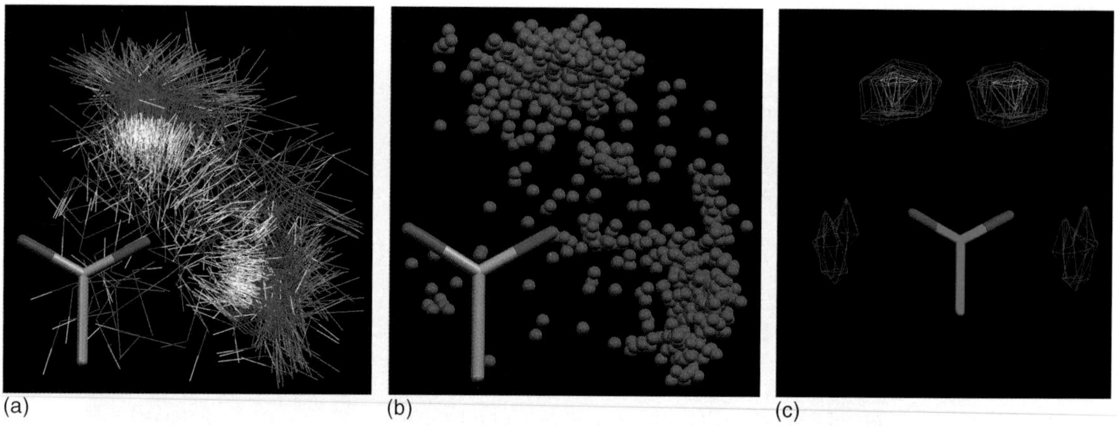

Figure 6 IsoStar scatterplots of a charged carboxylate central group and an OH contact group taken from (a) the CSD and (b) the PDB. (c) A CSD plot contoured on the donor oxygen atoms.

compound name, formula, and R-factor. The hyperlinking feature is particularly useful in examining outliers, and in determining any chemical effects that may be responsible for any multimodality of, especially, torsion angle distributions. Data can also be retrieved from the precomputed libraries via an instruction file interface, and this facility makes it possible to integrate Mogul quite readily with third-party software, as discussed in Section 3.18.4.3.

3.18.3.2.2 IsoStar: a knowledge base of intermolecular interactions

IsoStar[5] gathers together a vast amount of information on intermolecular interactions in a readily accessible form. For a given contact between a central group (A) and a contact group (B), the CSD search results for an interaction A···B are transformed into an easily visualized form by overlaying the A moieties. This results in a 3D scatterplot (**Figure 6a** and **b**), showing the experimental distribution of the B moieties around the static central group (A); these scatterplots can also be presented in contoured form (**Figure 6c**). IsoStar contains data retrieved from the CSD and from protein–ligand complexes stored in the PDB.[6] IsoStar also contains over 1500 potential energy minima calculated using distributed multipole analysis and intermolecular perturbation theory.[10]

Version 1.7 of IsoStar, released in December 2004, covers 300 central groups and 48 contact groups, and contains around 20 000 CSD-based scatterplots and 5500 PDB-based scatterplots. The user may interact with the basic scatterplots to generate contoured surfaces, change the A···B distance limit for data presentation, control the display style, etc. As with Mogul, the scatterplot data are hyperlinked to the master CSD and PDB files, so that the structural origin and environment of a specific interaction can be examined in detail. As with the hyperlinking feature in Mogul, hyperlinking in IsoStar can be used to investigate outliers in any scatterplot, and to investigate the full chemical nature of the contacting atoms. IsoStar therefore contains a vast amount of information of use in supramolecular chemistry, crystal engineering, and organic crystal chemistry, and also provides ready access to information that is invaluable in studying protein–ligand interactions as part of the rational drug design process.

3.18.4 Applications of the Cambridge Structural Database System in Medicinal Chemistry

3.18.4.1 Cambridge Structural Database-Based Research: An Overview

Research applications of the CSD began to appear in the literature from the mid-1970s, once basic software for searching, visualizing and analyzing CSD data began to be distributed. Indeed, CSD software developments and research methodologies have always been closely linked, with new research applications demanding extensions to software functionality. By 2005, more than 1200 papers had appeared in which the CSD system formed the computational infrastructure for research projects. The CCDC maintains a freely available, classified and searchable database, WebCite,[11] containing references and short synopses for all CSD applications studies that we can locate in the open literature. It is impossible to cover all aspects of CSD usage in a chapter of this length, and the WebCite database gives a comprehensive view of the wide range of application areas for CSD information.

Principal areas of activity, with key references, have been:

- mean molecular dimensions[12,13];
- structure correlation and reaction pathways[14];
- conformational analysis[15,16];
- hydrogen bond geometry and directionality for both strong[17,18] and weak[19] hydrogen bonds;
- nonbonded interactions not mediated by hydrogen[20,21];
- crystal engineering[22];
- protein–ligand interactions and protein–ligand docking[23,24]; and
- metal coordination sphere geometry and related applications.[25]

All of these topics and more are covered in three recent reviews of CSD applications.[26–28]

The CSD has a number of attractions as a source of data for structure-based research: (1) crystal structure analyses provide precise experimentally determined information about 3D molecular structure, and this precision has improved dramatically with the general availability, and subsequent technical improvement, of automated diffractometers that have taken place since the mid-1970s; (2) the technique provides equally precise information about intermolecular interactions; and (3) the technique is now the method of choice for geometry characterization, hence the CSD exhibits exceptional chemical diversity which is continually being extended as the world's output of crystal structures continues to increase.

In rational drug design and development, the CSD has some obvious, and very straightforward applications: (1) in providing structural information on individual molecules, to check stereochemistries or to provide starting geometries for molecular modeling; (2) in validating computational methodologies (e.g., the generation of 3D structures from 2D representations, or in the maintenance of conformational validity of protein–ligand docking results); and (3) in direct exploitation of the chemical diversity of the CSD as a source of potential new lead molecules.[29–31] However, beyond the individual structure level, many applications of the CSD in medicinal chemistry rely upon structural information derived from many crystal structures taken together, and rely on knowledge-mining techniques that range from the simple to the sophisticated.

3.18.4.2 The Relevance of Crystal Structure Data in Medicinal Chemistry Research

Before we address these more advanced applications of the CSD, we consider briefly the appropriateness of transferring knowledge gained from crystal structure data in the condensed state to the in vivo situation that is the concern of medicinal chemistry and rational drug design. An important underlying principle here is the principle of structure correlation, enunciated by Bürgi and Dunitz[14] in the late 1970s and through the 1980s, beginning with their classic studies of reaction pathways in a series of aminoketones located using the CSD. In the Bürgi–Dunitz hypothesis, the static distortions exhibited by a specific molecular fragment in a wide variety of crystalline environments are assumed to map the distortions that the fragment would undergo along a reaction or interconversion pathway, that is, the various static fragments are considered to form a series of structural 'snapshots' along the pathway, and the observed structures tend to concentrate in low-lying regions of the potential energy hypersurface. Thus, there is a direct correlation between structural behavior and energetics in crystals and in solution.

The issue of the relevance of CSD data has been considered in some depth by Taylor[27] in relation to the use of data on (1) molecular conformations, (2) metal coordination, and (3) intermolecular interactions. With respect to conformational information, the Bürgi–Dunitz hypothesis has been tested[16] by comparing CSD torsion angle distributions for 12 common molecular fragments with potential energy curves obtained from high-level ab initio calculations on appropriate model compounds. Each fragment was able to adopt two conformers (*anti* and *gauche*), and the complementarity of the CSD and ab initio results was striking, with the relative frequencies of the two conformers being strongly correlated with their potential energy differences across the full range of examples. These results suggest that torsional distributions from small-molecule crystal structures generally serve as good guides to torsional distributions in other phases. In his review, Taylor[27] provides evidence that crystal structure conformations may be more representative of in vivo situations than the in vacuo results provided by ab initio methods. Finally, we note that a comparison[32] of CSD torsion angle distributions with those for corresponding fragments in protein-bound ligands also showed clear similarities within the experimental limits imposed by the resolution of the protein structures.

Similar computational and protein-bound ligand comparisons are cited by Taylor[27] in his analysis of the relevance of CSD data on intermolecular interactions. In particular, he notes that although there are few statistically significant differences between the geometries of nonbonded contacts in CSD and PDB structures,[33] the same is not true for nonbonded contact frequencies.[34] These discrepancies seem to be due, however, to the much higher frequency of hydrophobic contacts in PDB structures than in small-molecule structures in the CSD. Taylor[27] also notes that there are occasional situations in which systematic crystal packing effects (e.g., the formation of common motifs) can occasionally bias CSD surveys.

3.18.4.3 Intramolecular Geometries and Conformational Analysis

A simple and obvious use of CSD data has been to generate mean values for standard geometrical parameters, such as bond lengths and valence angles, to act as benchmarks against which new data may be compared, or to act as restraints during the refinement of novel structures. Two major printed compilations of mean bond lengths were generated during the late 1980s for organic molecules,[12] and for metal–organic complexes of the d and f block metals.[13] Another notable compilation is that of Engh and Huber,[35] who derived mean bond lengths and valence angles for peptidic structures in the CSD, basing their classification on 31 carbon, nitrogen, and oxygen atom types that are most appropriate to the protein environment. These data continue to be used extensively in the determination, refinement, and validation of novel protein crystal structures, and are built into many key computer programs in structural biology.

A common use of CSD system software in rational drug design is to observe conformational preferences about freely rotatable bonds or in ring systems in crystal structures, so as to validate the output of computational procedures, or as input to conformation generators. Here, the chemical diversity of the experimental CSD data is particularly valuable, especially in situations where computational procedures are less reliable or unavailable.

Conformational analysis using ConQuest and Vista for a single rotatable bond is illustrated in **Figure 3**, where a simple histogram conveys the significant and valuable information – that the *cis*- and *trans*-bisected conformations of a carbonyl substituent with respect to a cyclopropane ring are the preferred arrangements. Extension of this procedure to study two contiguous torsions is exemplified in **Figure 7**.

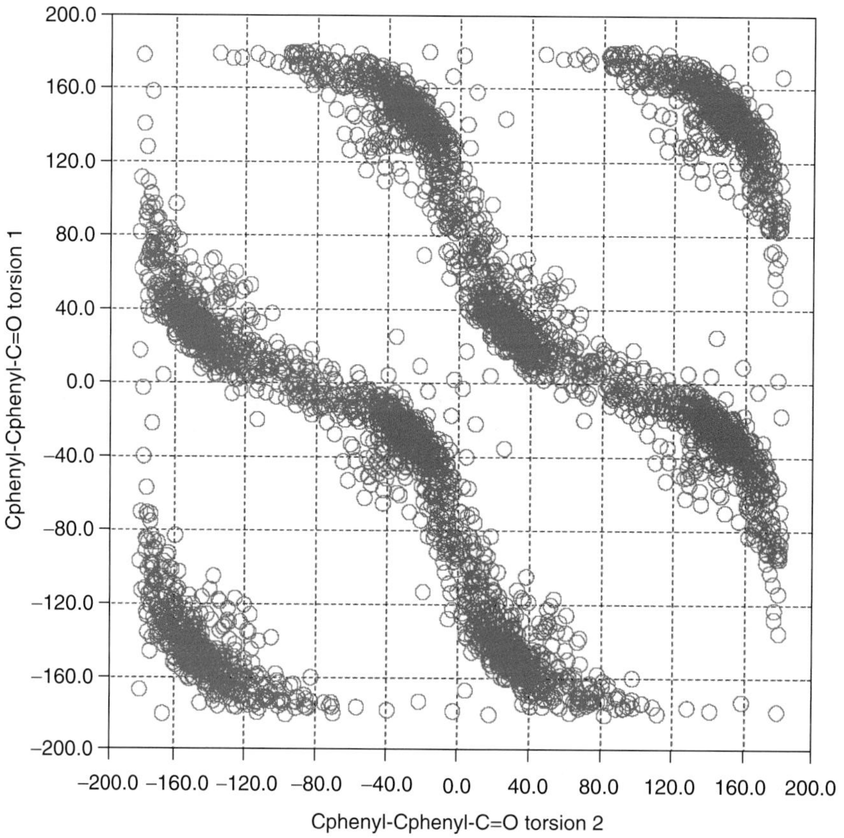

Figure 7 Symmetry expanded Ramachandran-like plot of the $O{=}C{-}C_{ar}{-}C_{ar}$ torsion angles in benzophenone structures retrieved from the CSD using ConQuest.

Here, the CSD has been used[36] to study conformational variations in benzophenones by generating a scatterplot of the two torsion angles (TOR1 and TOR2) that quantify the conformations of the two independent phenyl rings with respect to the >C=O group. In the original paper,[36] **Figure 7** was superimposed on the contoured calculated potential energy hypersurface, which has energy minima at TOR1, TOR2 = $+30, -30°$, and its symmetry equivalents, and where these global minima are connected by low-energy valleys that correspond to the conformational interconversion pathways depicted so clearly in **Figure 7**, again in agreement with the structure correlation principle. Further extension of CSD analysis to higher dimensions in torsional space (e.g., in the conformational analysis of n-membered rings) normally requires use of multivariate statistical and numerical techniques, such as principal component analysis (included in Vista) or cluster analysis. CSD applications of this type have been applied to rings of size 5–8, and these results are reviewed elsewhere.[26]

It is unsurprising that a number of electronic conformational libraries have been derived from the CSD for use in software for rational drug design. Thus, the conformation generator MIMUMBA[37] incorporates a CSD-based torsional library of 216 molecular fragments that exhibit a rotatable bond. More recently, the *et* program[38] extends the MIMUMBA philosophy by taking account of correlations between the torsion angles of adjacent rotatable bonds as observed in 18 000 diverse organic molecules retrieved from the CSD. These correlations can be very strong and can therefore impose important restrictions on conformational space, as illustrated in **Figure 7**. The *et* program has been validated by generating conformers for 113 molecules whose protein-bound conformations are available in the PDB. By comparison with a distance geometry-based approach,[39] *et* was found to generate a conformer within 1.5 Å RMSD for about 90 of the ligands, as opposed to <80 for the distance–geometry-based program.

As with printed compilations, one-off electronic libraries provide a snapshot reflecting knowledge at the time the compilation was made. They can also lack generality, in being tied to one specific software application. The CCDC's Mogul library, described above, avoids these two difficulties, since it is regularly updated (at least annually) from the latest version of the CSD, and can be readily integrated into a variety of applications programs. The first significant integration has been with the CRYSTALS package[40] for structure solution and refinement from x-ray diffraction data. Here, direct interaction between CRYSTALS and Mogul permits the former program to perform reality checks on the developing molecular structure during refinement, by comparing observed geometry in the novel structure with the 'historical' geometrical knowledge stored in Mogul. Poor agreement can indicate problems, such as the misassignment of element types or untreated disorder.

3.18.4.4 Metal Coordination

Knowledge about ligand binding to metal ions is important to our structural understanding of metalloenzymes of pharmaceutical importance, and our ability to predict binding modes. Crystallography is the method of choice for the characterization of novel metal–organic species, and the CSD contains a wealth of data on metal coordination. These data have been used, for example, in the compilation of tables of standard bond lengths for all metals[13] and for in-depth analysis of more specific subsets,[41] which concentrate on bond lengths for those metals most commonly found in enzymes: calcium, magnesium, manganese, iron, copper, and zinc.

More importantly, the CSD has been used to provide data on metal coordination numbers and the polyhedral geometries of their coordination spheres. Thus, Glusker[42] has used the CSD to determine the most likely coordination number(s) of magnesium, sodium, calcium, potassium, zinc, cadmium, iron, cobalt, copper, and molybdenum, and lists the types of atoms that are most likely to bind to specific metals. In this work and elsewhere[41] the geometry of metal coordination with respect to the donor group (e.g. carboxylate) has also been studied, and information of this sort can be valuable in identifying metal binding sites in proteins.[43]

Other studies have used statistical methodologies to map coordination spheres so as to identify geometrical preferences and to locate their interconversion pathways,[44] while applications of the CSD to study conformational and other problems in structural inorganic chemistry are reviewed by Orpen.[25] The use of the CSD in studying metal coordination assumes considerable importance in the modeling environment, since computational approaches in this area can pose considerable difficulties, as exemplified by a comparative study of CSD and ab initio results for phosphate groups coordinated by sodium ions.[45]

3.18.4.5 Intermolecular Interactions

Crystal structures have always been the primary source of experimental information on hydrogen bonding and on other nonbonded interactions not mediated by hydrogen. Since its inception, the CSD system has played a key role in accelerating and improving the focus of these studies, by making it possible to retrieve, analyze, and visualize intermolecular interactions in a chemically systematic manner. In studying nonbonded interactions, four basic questions are of interest: (1) What types of interactions occur? (2) What is the geometry of the interaction? (3) Is the interaction directional? (4) What is the strength of the interaction? The CSD is ideally suited to answering the first three questions, but crystal structure data can only provide some relativistic information about, for example, the strengths of different hydrogen bonds. For this reason, CSD analyses are often coupled with ab initio calculations, for example using the IMPT methodology.[10] Here, the CSD is used to locate heavily populated areas of interaction space, and the computer-intensive calculations can then concentrate on these very restricted areas using simple molecules that model the interaction type.

Two scientific developments have given particular impetus to this area of research: (1) the fundamental importance of computational paradigms for protein–ligand docking in high-throughput screening methodologies, and (2) the increased status gained by the subfield now commonly referred to as crystal engineering – the design of solids having predictable structures and properties. Individual papers, reviews, and monographs now proliferate,[11,14–16] and in this short survey, we concentrate only on selected aspects that are relevant to studies in medicinal chemistry and to drug development processes. Thus, to the four basic objectives above, we may add two more questions, both of which depend on CSD information: (5) What are the relative abilities of functional groups to form non-bonded interactions? (6) What types of extended motifs are formed by different types of interactions? This latter question is of particular importance in crystal engineering and drug development, and often explains differences between the structures of polymorphic forms.

3.18.4.5.1 Locating nonbonded interactions using the Cambridge Structural Database

The Mercury visualizer[3] is an important tool for surveying crystal structures to locate short interactions of all types. A typical example is illustrated in **Figure 8**.

Here, the expected hydrogen bonds are formed between the various acceptors and donors, but a short dipolar interaction is also exhibited between pairs of carbonyl groups. Once a focus interaction has been decided, the chemical substructures of the two interacting groups can be drawn into the ConQuest window, and the limiting geometry of the interaction entered as search constraints (*see* Section 3.18.3). The primary search constraint will be the distance limit for the 'nonbond' connecting the functional groups, but other limits (e.g., those based on the directionality of an hydrogen bond at the acceptor or at the donor hydrogen atom), may also be employed. ConQuest will systematically output user-defined geometrical descriptors for each occurrence of the interaction, and their distributions can be displayed and analyzed using Vista.

3.18.4.5.2 Strong hydrogen bonds

The basic procedures of Section 3.18.4.5.1 are exemplified for N–H···O=C hydrogen bonds in **Figure 9**.

CSD analysis combined with ab initio calculations is exemplified by a study[46] of the resonance-induced hydrogen bonding of N–H or O–H donors to sulfur acceptors in $(R^1R^2)C{=}S$ systems. The >C=S bond is not a natural dipole

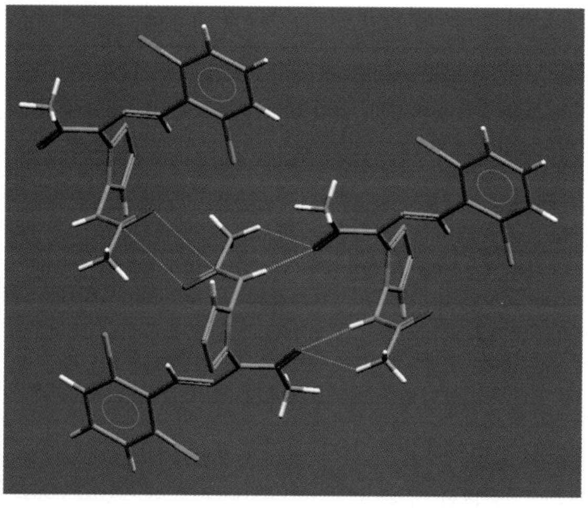

Figure 8 Mercury plot of the CSD entry AABHTZ showing hydrogen-bonded interactions between donor and acceptor groups (C=O and N–H) and short-range (less than the sum of van der Waals radii) contacts between C=O groups.

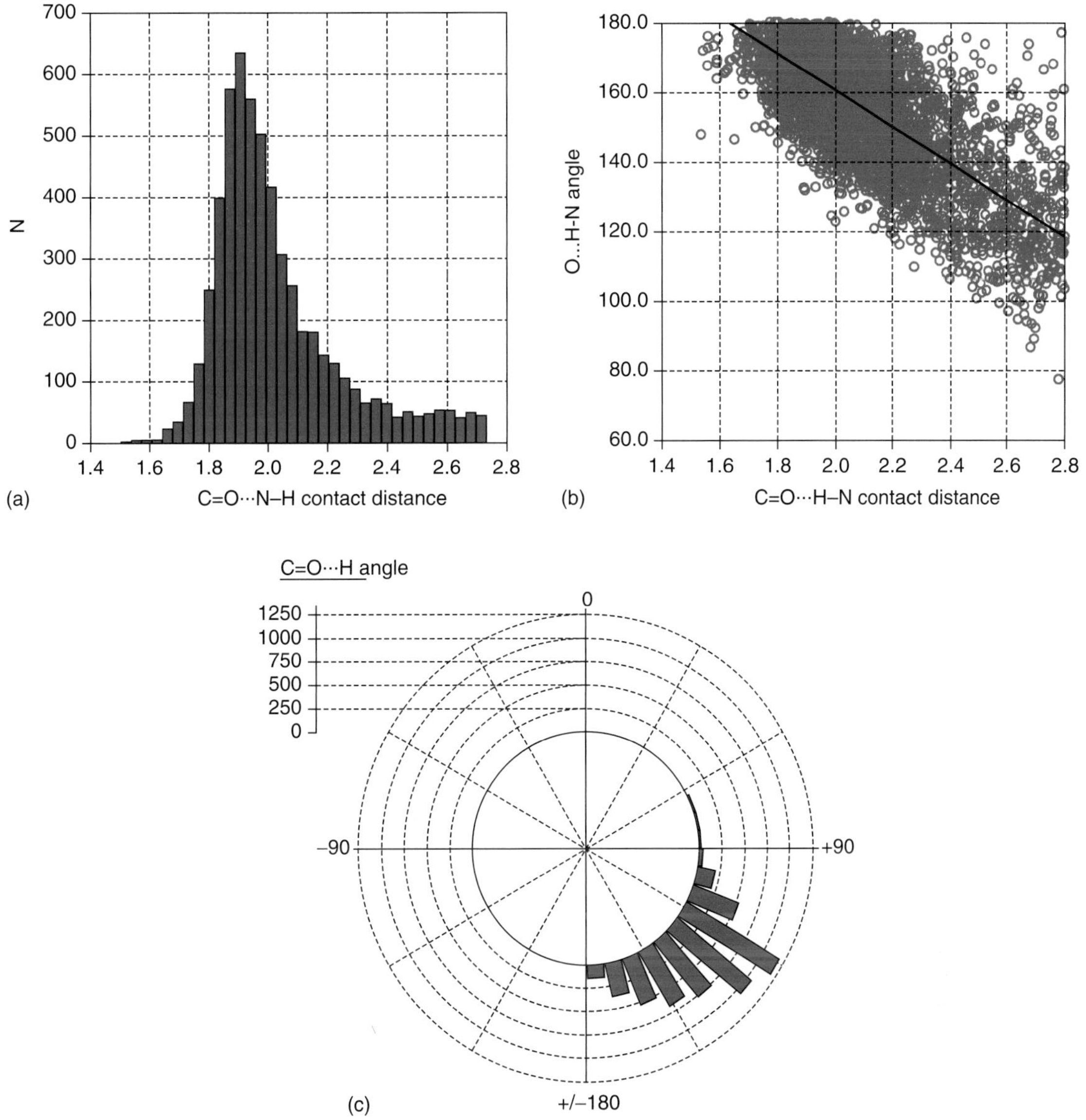

Figure 9 Vista plots of a C=O· · ·N–H contact search carried out using ConQuest. (a) The histogram shows the unimodal distribution of the H· · ·O contact distance, while (b) the scattergram shows the dependence of contact distance on the angle of approach of the contact atom. (c) The polar plot illustrates the lone pair directionality in the hydrogen approach to the oxygen atom.

due to the almost equal electronegativities of carbon and sulfur, by contrast to the situation in >C=O bonds, where the electronegativity of oxygen makes it a strong acceptor. Nevertheless, the structure of thiourea is dominated by N–H···S=C(R^1R^2) bonding. The CSD analysis[46] of all (N, O)–H···S=C(R^1R^2) substructures showed that only those systems in which one or both of R^1, R^2 were electron-donating substituents (e.g., the amine groups of thiourea) formed hydrogen bonds. Here, the effective electronegativity of sulfur is significantly increased by resonance effects (**Figure 10**), so that it now becomes an effective acceptor, and the geometrical distributions show typical hydrogen-bonding behavior of the sort illustrated in **Figure 9**. Importantly also, further analysis revealed: (1) a significant preference for the donor hydrogen atom to approach the acceptor sulfur atom in the >C=S plane, and at a C=S···H angle of *c.* 105° (**Figure 10a**), which clearly shows the sulfur lone pair directionality; and (2) an interaction energy of $-20\,\mathrm{kJ\,mol^{-1}}$ computed using IMPT[10] with an O–H donor in the >C=S plane at d(S···H) = 2.40 Å and with a C=S···H angle of 95° (**Figure 10b**). This value is somewhat less attractive than interaction energies computed for >C=O···H–O bonds (*c.* $-28\,\mathrm{kJ\,mol^{-1}}$).

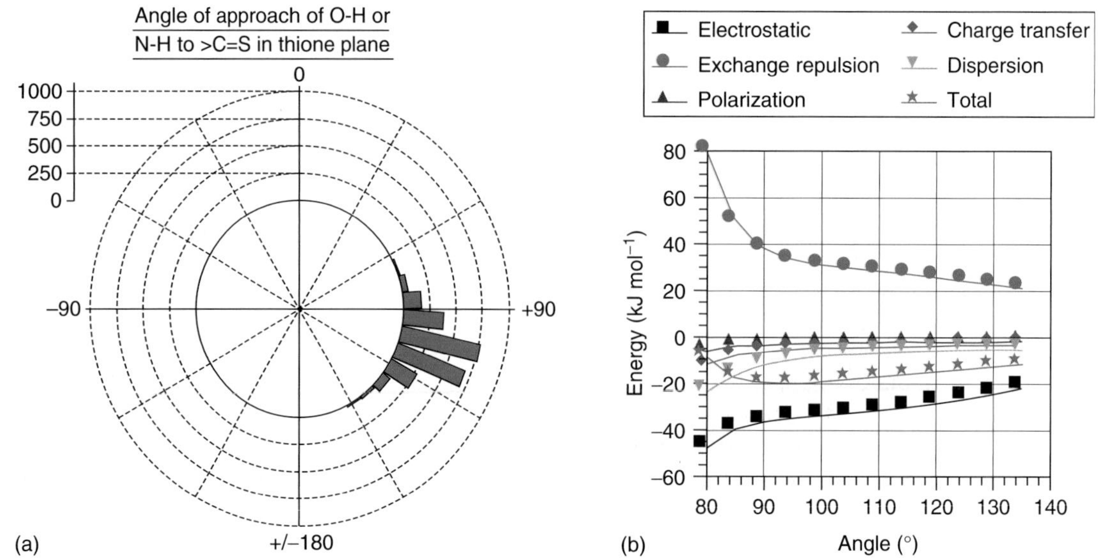

Figure 10 Directionality of [N or O]–H···S=C(R^1R^2) hydrogen bonds at the acceptor sulfur atom in thiones: (a) polar histogram of the H···S=C angle, and (b) interaction energies calculated using the IMPT procedure,[10] using thiourea and methanol as model molecules, restricting the donor hydrogen atom to lie in the thione plane, and varying the H···S=C angle.

3.18.4.5.3 **Weak hydrogen bonds**

Perhaps the most important contribution of CSD analysis to hydrogen bond research has been to establish the existence of a wide range of weaker hydrogen bonds[19] involving (1) weak donors and strong acceptors (e.g., C–H···O, C–H···N); (2) strong donors and weak acceptors (e.g., (O, N)–H···Cl, (O, N)–H···π); and (3) weak donors and weak acceptors (e.g., C–H···Cl, C–H···π). Of particular importance was the clear identification of short C–H···O and C–H···N contacts as true hydrogen bonds in the early 1980s.[47] This much cited work, based on neutron diffraction studies retrieved from the CSD, finally ended all speculation as to the nature of these short interactions involving acidic C–H hydrogen atoms, and put an end to the 'dark ages'[19] that had existed since the late 1960s in which such interactions had to be described using the most circumspect phraseology that did not include the phrase 'hydrogen bond'! Since that time, the importance of weaker interactions is now being recognized in stabilizing protein secondary structures, in protein–ligand binding, and in influencing the solid state conformations of cyclic and acyclic peptides.[19,48]

3.18.4.5.4 **Interactions not mediated by hydrogen**

A review of supramolecular synthons[49] illustrates the structural importance of a wide range of attractive nonbonded interactions that are not mediated by hydrogen, and notes the value of CSD analyses in identifying and characterizing these interactions. In practice, the combination of CSD analysis and ab initio calculations has again proved valuable, so that the relative robustness of these interactions can be compared with one another and with the more well-understood hydrogen-bonded interactions.

The marked tendency of the halogens (X=Cl, Br, I) to form short contacts with each other and electronegative nitrogen and oxygen atoms is well known. A combined CSD/IMPT analysis of C–X···O=C< systems[20] showed a marked preference for the shortest X···O interactions to form along the extension of the C–X bond and with interaction energies ranging from -7 to $-10\,\mathrm{kJ\,mol^{-1}}$, depending on the nature of X and the bonding environment of O=C<. These energies are comparable to the strengths of C–H···O hydrogen bonds. The importance of group–group interactions has also been highlighted[50] during a CSD analysis designed to locate isosteric replacements in modeling protein–ligand interactions. A later paper[51] presented an in-depth study of carbonyl–carbonyl interactions, and showed that dipolar >C=O···O=C< interactions most commonly form in a slightly sheared antiparallel arrangement having interaction energies of about $-20\,\mathrm{kJ\,mol^{-1}}$, comparable to the energies exhibited by medium-strength hydrogen bonds, for example the (N, O)–H···S bonds discussed above. Carbonyl–carbonyl interactions have also been shown to be significant in stabilizing certain protein secondary structure motifs[52] and in stabilizing the partially allowed Ramachandran conformations of asparagine and aspartic acid.[53]

3.18.4.5.5 Applications of the IsoStar library of intermolecular interactions

During the many studies on hydrogen bonding summarized in the WebCite database,[11] a key piece of methodology emerged: the idea of converting hydrogen bond data retrieved from the CSD into scatterplots, and thence into contoured surfaces.[54] It is this idea that underpins the IsoStar library (*see* Section 3.18.3.2.2), which gathers together the intrinsic spatial features observed in most of the existing studies of intermolecular interactions, and extends them systematically within a coherent methodology. Structure-based drug design relies, of course, on information about noncovalent interactions, and such information can assist significantly in the design of novel ligands. IsoStar presents a wealth of information about noncovalent interactions in a visual form, and can answer questions of relevance in drug design in a few button clicks. Not only will IsoStar indicate when specific interactions have a high frequency of occurrence, it will also indicate when interactions are unlikely to occur, thus directing chemical attention away from certain atoms and functional groups in the design process.

This 'unlikelihood' is exemplified often in IsoStar. For example, organically bound fluorine is very unlikely to accept hydrogen bonds,[55] and aromatic oxygen is a very poor hydrogen bond acceptor.[56] Other examples concern the relative hydrogen-bonding abilities of two different acceptors in a chemical functional group. Thus, the distribution of donor hydrogen atoms around esters is shown in **Figure 11a** and **b** in standard and contoured forms.

It is clear that hydrogen atom acceptance by the carbonyl oxygen atom is strongly preferred, and the lone pair directionality at the oxygen atom of **Figure 9** is clearly exhibited in the contoured plot of **Figure 11b**. By contrast, hydrogen bond density around the ester oxygen atom is very small, especially when viewed in the contoured plot. An explanation for this can be found in the potential energy minima, also available in IsoStar: the $C=O\cdots H$ bonds in this system are much stronger, by about $9\,kJ\,mol^{-1}$, than they are to the ester oxygen atom, and the frequencies of hydrogen bond formation in **Figure 11a** and **b** result simply from this competition effect. Other cases of competition arise in heterocyclic systems such as oxazole, and isoxazole.[57] The IsoStar plots of **Figure 11c** and **d** show quite clearly the strong preference of donor hydrogen atoms for the ring nitrogen in both cases, and IMPT calculations show that these $N–H\cdots O$ bonds are stronger by up to $6\,kJ\,mol^{-1}$ than the hydrogen bonds that form to the ring oxygen atoms.

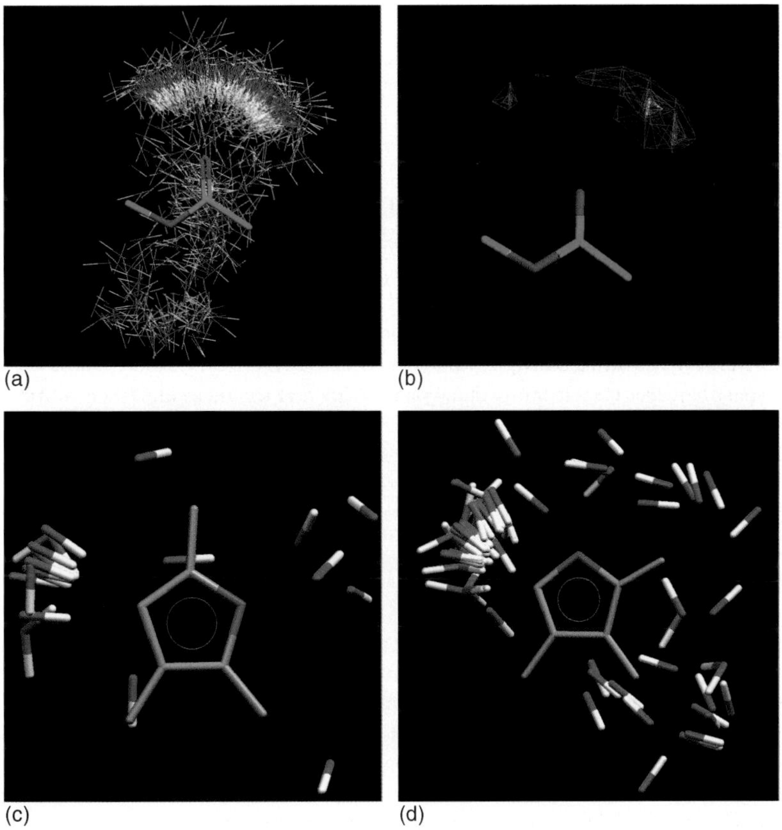

Figure 11 IsoStar plots showing (a) the distribution of OH groups around esters, taken from the CSD; (b) the contour plot of (a), produced by contouring on the O contact atoms; and (c) the distribution of OH groups around oxazole and (d) around isoxazole.

3.18.4.5.6 Predicting hydrogen-bonded contacts

More recently, Infantes and Motherwell[58] have taken the study of hydrogen bond competition effects a step further, through the creation of a database, CSDContact, of chemical groups and their contacts, derived from >40 000 organic structures in the CSD. The database records the atom types of each donor and acceptor pair, the functional groups from which they are derived, and the accessible surface of each acceptor. They find that the accessible surface and the ratio of the number of donors to the number of acceptors in a structure are useful parameters in predicting the probable number of hydrogen-bonded contacts to a given acceptor. Such information is valuable for assisting in the prediction of packing arrangements in molecular crystals, and, therefore, in selecting the most likely form from a set of polymorphic structures predicted by computational methodologies, thus contributing valuable insights to the drug development process.

3.18.5 Applications Software that uses Knowledge Derived from the Cambridge Structural Database

Previous sections have shown that crystal structure data have contributed profoundly to many aspects of medicinal chemistry and the life sciences. The improved accessibility of knowledge from the CSD in the form of derived libraries of structural information, such as Mogul and IsoStar, now opens many interesting possibilities for knowledge-based applications software, of which the CRYSTALS–Mogul link[40] highlighted in Section 3.18.4.3 is an obvious example in crystallography itself. The role of CSD data in two application areas related to the life sciences is outlined below.

3.18.5.1 Programs for Predicting Intermolecular Interactions

A number of different approaches have been taken to this problem. GRID[59] identifies interaction 'hot spots' in protein-binding sites using empirical energy functions that are parameterized, in part, to reproduce the geometric distributions of nonbonded contacts taken from the CSD. SuperStar[23,60] and X-SITE[61] utilize a purely knowledge-based approach. These programs generate interaction maps by estimating the probability of an interaction between the protein and a probe (a small functional grouping such as methyl or carbonyl) based on how often the interaction has been observed in crystal structures.

SuperStar partitions the protein-binding site into its constituent chemical functional groups, the partitioning being done in such a way that each group corresponds to one of the central groups in the IsoStar library. The crystallographically observed probability distributions of the chosen probe atom around each of the chemical groups present in the protein binding site are then retrieved from IsoStar. Each IsoStar scatterplot is then overlaid on all parts of the protein-binding site that it matches. Overlapping distributions are combined and normalized to the same scale. This generates a contoured surface that highlights 'hot spots' within the binding site, that is, regions where the protein–probe interaction is particularly favorable. **Figure 12** shows the SuperStar map of a carboxypeptidase-binding site, indicating where aromatic CH groups (blue) and carbonyl (yellow) groups are favored.

The crystallographically observed ligand (glycyl-L-tyrosine) is shown for comparison. These maps, in turn, can be used to hypothesize possible 3D arrangements of functional groups that should interact well with the binding site, and therefore confer binding affinity for guiding drug design. The CSD system itself can then be used to locate any examples of these pharmacophores that exist in crystal structures.

3.18.5.2 Protein–Ligand Docking Programs

The use of protein–ligand docking programs for high-throughput virtual screening is becoming increasingly important, and many of the leading programs exploit crystallographic data. Both GOLD[24] and FlexX[62] utilize torsional angle distributions extracted from the CSD. In GOLD, these are used to restrict the conformational space sampled during docking. This increases the chances of finding the correct answer by biasing the ligand conformational search toward torsion angle values that are commonly observed in crystal structures. FlexX works by breaking down the ligand into its constituent molecular fragments. These rigid fragments are then reassembled within the binding site to generate the docked conformation. During this process, CSD information is used to determine the torsion angles around the bonds linking individual fragments.

Additionally, both programs use small-molecule crystal structure data to ensure that protein–ligand hydrogen bonds have energetically favorable geometries. GOLD uses CSD information on intermolecular interactions to characterize acceptors according to their preferred hydrogen-bonding geometries. Acceptors can either form hydrogen bonds along

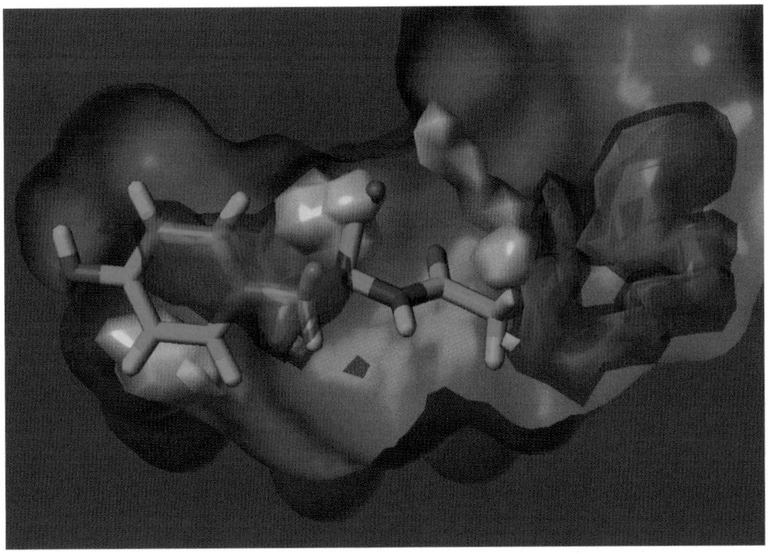

Figure 12 SuperStar (CSD) map of the 3CPA binding site using aromatic CH (blue) and C=O (yellow) probes. The ligand is glycyl-L-tyrosine.

the lone pair direction (e.g., the oxygen atoms of nitro groups), within the plane of the lone pairs (e.g., ether oxygen atoms), or they may show no strong directional preferences (e.g., phosphate oxygen atoms). Using this information during docking ensures that geometrically reasonably hydrogen bonds are formed between the ligand and protein.

In the program LUDI[63,64] the ligand is both designed and built algorithmically within the protein-binding site. Molecular fragments from a preassembled library are placed at positions where they are expected to interact favorably with the protein. Individual fragments are then connected (by single bonds, bridges, fusions, etc.) to form the complete ligand. Once again, CSD data are used to guide the initial placement of fragments via searches of the CSD performed to determine the angular and dihedral ranges in which particular atoms could form hydrogen bonds.

3.18.6 Conclusions

While small-molecule crystal structure analysis may be regarded as a mature technique, the results of every structure determination are valuable not just for the reasons that prompted the original analysis but also to provide structural information for other scientists in a very wide range of disciplines. This chapter has tracked the unique journey of crystal structures from the creation of a database and its access software, through the acquisition of knowledge and the creation of knowledge bases of structural information, to the use of this stored knowledge to solve problems in the life sciences. When taken together in this way, the rapidly growing reservoir of information that is embodied in the CSD, and in the other crystallographic databases, will continue to provide the precise experimental data that are crucial to the processes of lead discovery and rational molecular design. Indeed, the projected growth of the CSD, now rapidly approaching its half-millionth crystal structure, can only increase its already considerable chemical diversity, and make it an even more potent research tool in the future.

References

1. Allen, F. H.; Glusker, J. P. *Acta Crystallogr. B* **2002**, *58*, preface. http://journals.iucr.org/b/issues/2002/03/01/me0172/index.html (accessed May 2006).
2. Allen, F. H. *Acta Crystallogr. B* **2002**, *58*, 380–388.
3. Bruno, I. J.; Cole, J. C.; Edgington, P. R.; Kessler, M.; Macrae, C. F.; McCabe, P.; Pearson, J.; Taylor, R. *Acta Crystallogr. B* **2002**, *58*, 389–397.
4. Bruno, I. J.; Cole, J. C.; Lommerse, J. P. M.; Rowland, R. S.; Taylor, R.; Verdonk, M. L. *J. Comput.-Aided Mol. Design* **1997**, *11*, 525–537.
5. Bruno, I. J.; Cole, J. C.; Kessler, M.; Jie Luo, Motherwell, W. D. S.; Purkis, L. H.; Smith, B. R.; Taylor, R.; Cooper, R. J.; Harris, S. E. et al. *J. Chem. Inf. Comput. Sci.* **2004**, *44*, 2133–2144.
6. Berman, H. M.; Battistuz, T.; Bhat, T. N.; Bluhm, W. F.; Bourne, P. E.; Burkhardt, K.; Feng, Z.; Gilliland, G. L.; Iype, L.; Jain, S. et al. *Acta Crystallogr. D* **2002**, *58*, 899–907.
7. Hall, S. R.; Allen, F. H.; Brown, I. D. *Acta Crystallogr. A* **1991**, *47*, 655–685.
8. Allen, F. H.; Johnson, O.; Shields, G. P.; Smith, B. R.; Towler, M. *J. Appl. Crystallogr.* **2004**, *37*, 335–338.

9. Allen, F. H. *Crystallogr. Rev.* **2004**, *10*, 3–15.
10. Hayes, I. C.; Stone, A. *J. Mol. Phys.* **1984**, *53*, 83–105.
11. CCDC WebCite database, available at: http://www.ccdc.cam.ac.uk (accessed May 2006).
12. Allen, F. H.; Kennard, O.; Watson, D. G.; Brammer, L.; Orpen, A. G.; Taylor, R. *J. Chem. Soc. Perkin Trans.* **1987**, *2*, S1–S19.
13. Orpen, A. G.; Brammer, L.; Allen, F. H.; Kennard, O.; Watson, D. G.; Taylor, R. *J. Chem. Soc. Dalton Trans.* **1989**, S1–S83.
14. Bürgi, H.-B.; Dunitz, J. D. *Structure Correlation*; VCH: Weinheim, Germany, 1997.
15. Allen, F. H.; Doyle, M. J.; Auf der Heyde, T. P. E. *Acta Crystallogr. B* **1991**, *47*, 412–428.
16. Allen, F. H.; Harris, S. E.; Taylor, R. *J. Comput.-Aided Mol. Design* **1996**, *10*, 247–254.
17. Taylor, R.; Kennard, O. *Acc. Chem. Res.* **1984**, *17*, 320–326.
18. Jeffrey, G. A.; Saenger, W. *Hydrogen Bonding in Biological Structures*; Springer-Verlag: Berlin, Germany, 1991.
19. Desiraju, G. R.; Steiner, T. *The Weak Hydrogen Bond in Structural Chemistry and Biology*; Oxford University Press: Oxford, UK, 1999.
20. Lommerse, J. P. M.; Stone, A. J.; Taylor, R.; Allen, F. H. *J. Am. Chem. Soc.* **1996**, *118*, 3108–3116.
21. Allen, F. H.; Baalham, C. A.; Lommerse, J. P. M.; Raithby, P. R. *Acta Crystallogr. B* **1998**, *54*, 320–329.
22. Nangia, A. *Cryst. Eng. Comm.* **2002**, 93–101.
23. Verdonk, M. L.; Cole, J. C.; Taylor, R. *J. Mol. Biol.* **1999**, *289*, 1093–1108.
24. Jones, G.; Willett, P.; Glen, R. C.; Leach, A. R.; Taylor, R. *J. Mol. Biol.* **1997**, *267*, 727–748.
25. Orpen, A. G. *Acta Crystallogr. B* **2002**, *58*, 398–406.
26. Allen, F. H.; Motherwell, W. D. S. *Acta Crystallogr. B* **2002**, *58*, 407–422.
27. Taylor, R. *Acta Crystallogr. D* **2002**, *58*, 879–888.
28. Allen, F. H.; Taylor, R. *Chem. Soc. Rev.* **2004**, *33*, 463–475.
29. DesJarlais, R. L.; Seibel, G. L.; Kuntz, I. D.; Furth, P. S.; Alvarez, J. C.; Ortiz de Montellano, P. R.; DeCamp, D. L.; Babe, L. M.; Craik, C. S. *Proc. Natl. Acad. Sci. USA* **1990**, *87*, 6644–6648.
30. Lam, P. Y. S.; Prabhakar, K. J.; Eyermann, C. J.; Hodge, C. N.; Ru, Y.; Bacheler, L. T.; Meek, J. L.; Otto, M. J.; Rayner, M. M.; Wong, Y. N. et al. *Science* **1994**, *263*, 380–384.
31. Chowdhury, S. F.; Di Lucrezia, R.; Guerrero, R. H.; Brun, R.; Goodman, J.; RuizPerez, L. M.; Pacanowska, D. G.; Gilbert, I. H. *Bioorg. Med. Chem. Lett.* **2001**, *11*, 977–980.
32. Böhm, H.-J.; Klebe, G. *Angew. Chem. Int. Ed. Engl.* **1996**, *35*, 2588–2614.
33. Boer, D. R.; Kroon, J.; Cole, J. C.; Smith, B.; Verdonk, M. L. *J. Mol. Biol.* **2001**, *312*, 275–287.
34. Verdonk, M. L.; Cole, J. C.; Taylor, R. *J. Mol. Biol.* **1999**, *289*, 1093–1108.
35. Engh, R. A.; Huber, R. *Acta Crystallogr. A* **1991**, *47*, 392–398.
36. Rappoport, Z.; Biali, S. E.; Kaftory, M. *J. Am. Chem. Soc.* **1990**, *112*, 7742–7750.
37. Klebe, G.; Mietzner, T. *J. Comput.-Aided Mol. Design* **1994**, *8*, 583–606.
38. Feuston, B. P.; Miller, M. D.; Culberson, J. C.; Nachbar, R. B.; Kearsley, S. K. *J. Chem. Inf. Comput. Sci.* **2001**, *41*, 754–763.
39. Blaney, J. M.; Crippen, G. M.; Dearing, A.; Dixon, J. S. *DGEOM, &hash;590, Quantum Chemistry Program Exchange*; Indiana University: Bloomington, 1990.
40. Betteridge, P. W.; Carruthers, J. R.; Cooper, R. I.; Prout, K.; Watkin, D. J. *J. Appl. Crystallogr.* **2003**, *36*, 1487.
41. Harding, M. M. *Acta Crystallogr. D* **1999**, *55*, 1432–1443.
42. Glusker, J. P. *Adv. Protein Chem.* **1991**, *42*, 1–76.
43. Carrell, H. L.; Glusker, J. P.; Burger, V.; Manfre, F.; Tritsch, D.; Biellmann, J.-F. *Proc. Natl. Acad. Sci. USA* **1989**, *86*, 4440–4444.
44. Allen, F. H.; Mondal, R.; Pitchford, N. A.; Howard, J. A. K. *Helv. Chim. Acta* **2003**, *86*, 1129–1139.
45. Schneider, B.; Kabelac, M.; Hobza, P. *J. Am. Chem. Soc.* **1996**, *118*, 12207–12217.
46. Allen, F. H.; Bird, C. M.; Rowland, R. S.; Raithby, P. R. *Acta Crystallogr. B* **1997**, *53*, 696–701.
47. Taylor, R.; Kennard, O. *J. Am. Chem. Soc.* **1982**, *104*, 5063–5070.
48. Umezawa, Y.; Tsuboyama, S.; Takahashi, H.; Uzawa, J.; Nishio, M. *Bioorg. Med. Chem.* **1999**, 7, 2021–2026.
49. Desiraju, G. R. *Angew. Chem. Int. Ed.* **1995**, *34*, 2311–2327.
50. Taylor, R.; Mullaley, A.; Mullier, G. W. *Pesticide Sci.* **1990**, *29*, 197–213.
51. Allen, F. H.; Baalham, C. A.; Lommerse, J. P. M.; Raithby, P. R. *Acta Crystallogr. B* **1998**, *54*, 320–329.
52. Maccallum, P.; Poet, R.; Milner-White, E. J. *J. Mol. Biol.* **1995**, *248*, 374–384.
53. Deane, C. M.; Allen, F. H.; Taylor, R.; Blundell, T. L. *Protein Eng.* **1999**, *12*, 1025–1028.
54. Rosenfield, R. E., Jr.; Swanson, S. M.; Meyer, E. F., Jr.; Carrell, H. L.; Murray-Rust, P. *J. Mol. Graphics* **1984**, *2*, 43–46.
55. Dunitz, J. D.; Taylor, R. *Chem. Eur. J.* **1997**, *3*, 89–98.
56. Böhm, H.-J.; Brode, S.; Hesse, U.; Klebe, G. *Chem. Eur. J.* **1996**, *2*, 1509–1513.
57. Nobeli, I.; Price, S. L.; Lommerse, J. P. M.; Taylor, R. *J. Comp. Chem.* **1997**, *18*, 2060–2074.
58. Infantes, L.; Motherwell, W. D. S. *Chem. Commun.* **2004**, 1166–1167.
59. Goodford, P. J. *J. Med. Chem.* **1985**, *28*, 849–857.
60. Verdonk, M. L.; Cole, J. C.; Watson, P.; Gillet, V.; Willett, P. *J. Mol. Biol.* **2001**, *307*, 841–859.
61. Laskowski, R. A.; Thornton, J. M.; Humblet, C.; Singh, J. *J. Mol. Biol.* **1996**, *259*, 175–201.
62. Rarey, M.; Kramer, B.; Lengauer, T.; Klebe, G. *J. Mol. Biol.* **1996**, *261*, 470–489.
63. Böhm, H.-J. *J. Comput.-Aided Mol. Design* **1992**, *6*, 61–78.
64. Böhm, H.-J. *J. Comput.-Aided Mol. Design* **1992**, *6*, 593–606.

Biographies

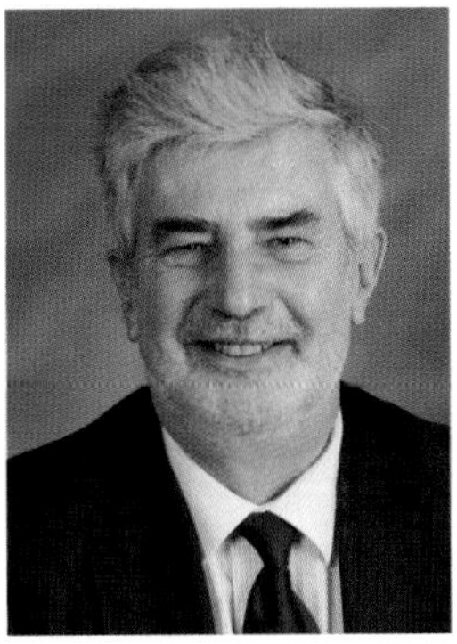

Frank H Allen is Executive Director of the Cambridge Crystallographic Data Centre, where he has worked since 1970, following undergraduate and graduate studies in chemistry and crystallography at Imperial College, London, and postdoctoral work at the University of British Columbia, Vancouver. At the Cambridge Crystallographic Data Centre he has been involved in creating the Cambridge Structural Database (CSD), software development, and in research applications of CSD data. He has authored or co-authored more than 200 publications, and has edited 16 reference books and conference proceedings. He became a Fellow of the Royal Society of Chemistry in 1992, was awarded the RSC Silver Medal and Prize for Structural Chemistry in 1994, and the Herman Skolnik Award of the American Chemical Society Division of Chemical Information in 2003. He was appointed Visiting Professor of Chemistry at the University of Bristol in 2002. He has held a range of senior positions in the British and European Crystallographic Associations and in the International Union of Crystallography (IUCr), including the Editorship of *Acta Crystallographica*, Section B from 1993 to 2002.

Gary M Battle graduated with a BSc in chemistry from the University of Essex in 1998. He moved to the University of Warwick, where he completed a PhD in synthetic organic chemistry with Dr Andrew Clark, in 2002. He then joined the Cambridge Crystallographic Data Centre as an Applications Scientist in the Applications & Marketing Group.

Susan Robertson graduated from Heriot-Watt University, Edinburgh, in 1999 with a BSc in chemistry with French. She then completed a PhD in carborane chemistry with Prof Alan Welch in 2002. She subsequently joined the CCDC as an Applications Scientist in the Applications & Marketing Group.

Comprehensive Medicinal Chemistry II
ISBN (set): 0-08-044513-6

ISBN (Volume 3) 0-08-044516-0; pp. 389–410

3.19 Protein Production for Three-Dimensional Structural Analysis

M M T Bauer and G Schnapp, Boehringer Ingelheim Pharma, Biberach, Germany

3.19.1	**Introduction**	**412**
3.19.2	**Purification from Natural Sources**	**412**
3.19.3	**Choice of Clone**	**413**
3.19.3.1	Search for Homologous or Closely Related Proteins in the Protein Data Bank (PDB) Database	413
3.19.3.2	Sequence Alignments and Analysis	413
3.19.3.3	Limited Proteolysis Studies	413
3.19.4	**Cloning**	**413**
3.19.4.1	Polymerase Chain Reaction (PCR)	414
3.19.4.2	Gateway System	414
3.19.5	**Expression Systems**	**415**
3.19.5.1	Cell-Free Protein Expression for Structural Research	415
3.19.5.2	Bacteria	416
3.19.5.3	Yeast	416
3.19.5.4	Insect Cells (Baculovirus System)	416
3.19.5.5	Mammalian Cells	418
3.19.6	**Fermentation**	**418**
3.19.6.1	Shaking Flasks	418
3.19.6.2	Spinner Flasks or Roller Bottles	418
3.19.6.3	Fermenters	419
3.19.6.4	Wave Bioreactor	419
3.19.7	**Purification**	**419**
3.19.7.1	Cell Lysis	420
3.19.7.2	Refolding	420
3.19.7.2.1	Inclusion bodies purification	420
3.19.7.2.2	Solubilization of inclusion bodies protein	421
3.19.7.2.3	Refolding conditions and screens	421
3.19.7.3	Capture Step with Affinity Tags	421
3.19.7.4	Fractionated Precipitation	421
3.19.7.5	Low-Pressure Column Chromatography	422
3.19.7.5.1	Affinity chromatography	422
3.19.7.5.2	Ion exchange chromatography	422
3.19.7.5.3	Hydrophobic interaction chromatography	423
3.19.7.5.4	Size exclusion chromatography (SEC)	423
3.19.7.6	Buffer Exchange	423
3.19.7.6.1	Dialysis	423
3.19.7.6.2	Size exclusion chromatography	424
3.19.7.6.3	Ultrafiltration	424
3.19.7.7	Concentration	424
3.19.7.7.1	Concentration with polyethyleneglycol	424
3.19.7.7.2	Ultrafiltration	424
3.19.8	**Analysis**	**424**

3.19.8.1 Electrophoresis 424
3.19.8.1.1 Sodium dodecyl sulfate–polyacrylamide gel electrophoresis 424
3.19.8.1.2 Native and isoelectric focusing polyacrylamide gel electrophoresis 425
3.19.8.1.3 Western blotting: identification 425
3.19.8.2 Protein Concentration Determination Methods 425
3.19.8.2.1 Colorimetric measurements 426
3.19.8.2.2 Ultraviolet absorption 426
3.19.8.3 Particle Size/Aggregation Status 426
3.19.8.3.1 Size exclusion chromatography 426
3.19.8.3.2 Dynamic light scattering 426
3.19.8.3.3 Analytical ultracentrifugation 426
3.19.8.4 Mass Spectrometry 427
3.19.8.5 Activity Assays 427
3.19.8.6 Circular Dichroism Spectroscopy and One-Dimensional (1D)-Nuclear Magnetic Resonance Spectroscopy 427
3.19.8.7 Sequence Analysis 428
3.19.9 Deglycosylation 428
3.19.10 Labeling for Nuclear Magnetic Resonance Studies 428
3.19.10.1 Cell-Free System 428
3.19.10.2 *Escherichia coli* 428
3.19.10.3 Insect Cells (Selective Labeling of Amino Acids) 429
3.19.11 Membrane Proteins 429
References 429

3.19.1 Introduction

The two most commonly used protein structure solution methods, x-ray crystallography and NMR spectroscopy, require support from large quantities, measured in milligrams, of suitably purified proteins of interest. The protein is required in the performance of crystallization trials for x-ray crystallography or for obtaining NMR measurements in solution. High standards are set and have to be met, regarding the individual protein's purity, folding status, and activity, during its production by specialized laboratories. The specifications which the protein has to meet vary according to whether it is used in activity assays or for biological testing purposes. For this reason, many structural research groups have access to a protein laboratory dedicated to protein production for structural research. The main focus of such a laboratory is described in this chapter.

Depending on the research focus (in pharmaceutical research: on the therapeutic areas), different classes and families of proteins may be of interest, e.g., membrane proteins such as membrane channels, or G protein-coupled receptors (GPCRs) or proteolytic enzymes. Due to the difficulties in expressing, purifying, and crystallizing membrane proteins, structural research work and, in consequence, the production of protein for structural research mostly focus on water-soluble proteins.

This chapter presents an overview of the necessary steps and process requirements for a laboratory intending to produce protein for structural research studies. There are general textbooks available that deal with protein expression and purification.[1–6]

3.19.2 Purification from Natural Sources

The supply of proteins in milligram to gram amounts for structural research can be achieved by accessing either natural sources such as microbial organisms, organ tissue, or blood plasma; or from genetically engineered expression systems.

In pharmaceutical research, the main focus is usually on human proteins. Since resources for human blood, plasma, or tissue are very limited and cost-prohibitive, access to, and extraction of, proteins from those sources is correspondingly restricted, though possible. Blood or plasma proteins such as thrombin and coagulation factors X and IX are well-known examples.[7–9] These proteins occur in sufficient amounts and show acceptable stability during the purification process. The purification of human proteins from natural sources is difficult for the following reasons:

- The collection of human tissue or blood is complicated and/or expensive.
- Human fluids and tissues contain an extensive range of proteins which first have to be separated from the selected protein of interest.
- Human fluids and tissues contain proteases that may act on the protein of interest during the purification process.
- Purification tools such as affinity tags are not available.

Most of the cytoplasmatic, mitochondrial, or membrane-associated proteins obtained from human tissue have to be cloned in suitable expression systems, to allow the production of protein amounts sufficient for structural biology purposes.

The principles of protein purification from natural sources are similar to those used with genetically engineered organisms (*see* Section 3.19.7).

3.19.3 Choice of Clone

Prior to starting the cloning process, sequence analysis and bioinformatics tools are required to enable the most effective search for the best appropriate construct, with respect to its solubility, folding, nucleotide and codon usage.

3.19.3.1 Search for Homologous or Closely Related Proteins in the Protein Data Bank (PDB) Database

A strong indication of where suitable sequences for cloning are located is the finding of closely related proteins in the structures deposited in the PDB.[10] With a sequence homology of $>80\%$ between the protein of interest and the PDB protein and its domain borders, start and stop points for cloning and domain contacts can be predicted.

3.19.3.2 Sequence Alignments and Analysis

Sequence alignments of any protein of interest with any related proteins with a known structure can help to predict secondary structure elements: hydrophobic and hydrophilic parts of the protein surface or stabilizing disulfide bonds. These items of information are necessary for plotting length and mutation planning. Strongly hydrophilic areas on the protein surface should be avoided, as well as the destruction of intramolecular contacts in α-helices or β-sheets caused by choosing cloning borders incorrectly. Inserting point mutations can help to increase solubility.

Determination of where in the protein sequence solubility patches and orthologs of increased solubility are to be found may improve expression success.

For structural studies on membrane proteins and multidomain complexes, concentration on one or two domains and extramembranal areas is useful and facilitates crystallization.

3.19.3.3 Limited Proteolysis Studies

If the protein of interest has been expressed and purified but failed to crystallize, limited proteolysis studies followed by activity and a mass spectrometry (MS) analysis of stable fragments can help to find other soluble and active domains to support crystallization. In that case, the full length of the protein of interest has to undergo proteolytic degradation with different proteases (serine, metallo-, cysteine, and aspartic proteases). The fragments are then to be tested for activity and separated by high-performance liquid chromatography (HPLC) or sodium dodecyl sulfate–polyacrylamide gel electrophoresis (SDS-PAGE) and Western blot. With MS and/or Edman sequencing, the active fragments are then analyzed and can be cloned for expression.

3.19.4 Cloning

After the identification of one or multiple suitable protein sequence constructs, as described in Section 3.19.3, cloning of the protein has to be performed. To shorten timelines from cloning to crystallization or NMR studies, parallelization in the cloning, expression testing, and first purification steps are recommended.

Which cloning strategy to follow depends on the expression system used. Vectors containing tags for protein purification and/or detection shorten the protein production process. If it is unclear as to what is the best expression system, a parallel cloning strategy into an *Escherichia coli*, baculovirus, or mammalian expression system should be followed. This can be achieved by using the Gateway system (Invitrogen), thus allowing the shuttling of the gene of interest from an entry clone into a different expression vector.

3.19.4.1 Polymerase Chain Reaction (PCR)

The genes of interest can be cloned into expression vectors by conventional restriction digestion and ligation, or by amplification with PCR, using gene-specific primers. The gene-specific primers can be inserted to indicate restriction sites, protease cleavage sites or tags (e.g., C-terminal His-Tags). When using PCR, it is necessary to work with a high-fidelity, proof-reading Taq polymerase to avoid the occurrence of point or frameshift mutations.

Linear templates generated by PCR can also be used directly in the cell-free translation system. This avoids further time-consuming cloning steps.

3.19.4.2 Gateway System

Gateway cloning technology (Invitrogen[11]) is a universal system for the cloning and subcloning of DNA sequences. It facilitates both gene function analysis and protein expression. DNA segments are transferred between vectors using phage lambda-based site-specific recombination instead of restriction endonucleases and ligase.[12] This powerful system can easily transfer one or more DNA sequences into multiple vectors in parallel reactions, while maintaining orientation and reading frame.

The gene of interest can be moved into an entry vector via PCR, restriction endonuclease digestion, and ligation, or site-specific recombination from a cDNA library constructed in a Gateway-compatible vector (**Figure 1**). A gene in the entry clone can then be transferred simultaneously into destination vectors. This is done by combining the entry clone

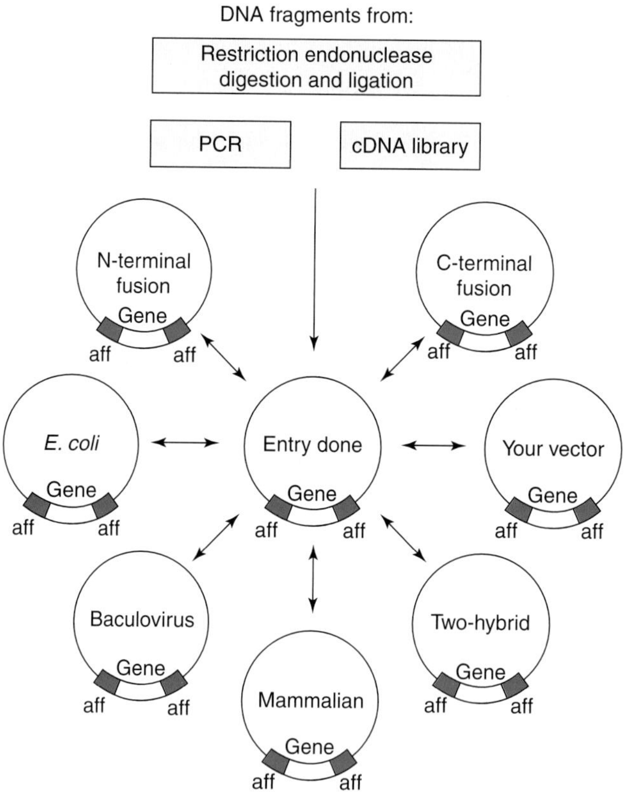

Figure 1 Overview on the Gateway technology (Invitrogen): the gene of interest can be moved into an entry vector via PCR, restriction endonuclease digestion, and ligation, or site-specific recombination from a cDNA library. A gene in the entry clone can then be transferred simultaneously into destination vectors.

with a Gateway destination vector and clonase enzyme mix in a single tube, incubating it for 1 h, transforming with *E. coli*, and plating.

In the entry clones for protein expression, the placement of the translation signals is determined whether the protein is expressed as a native or as a fusion protein. With native protein expression or C-terminal fusion, the translation signal must be included downstream of the attB1 recombination site. In this case the attB1 site will reside in the 5′ untranslated region of the mRNA. In N-terminal fusion the translation signals and the fusion protein sequences are provided by the destination vector and will be upstream of the attB1 site. Consequently the 25-bp attB1 site becomes part of the coding sequence and inserts 8 amino acids between the fusion domain and the protein encoded by a gene. To get rid of the fusion domain and the 8-amino-acid sequence from the attB1 site, a protease cleavage site should be introduced between the attB1 site and the gene.

The attB1 sequence has not been observed to affect protein yield in *E. coli*, insect, or mammalian cells.

Once a gene is configured as an entry clone it can be easily moved into any destination vector using the LR reaction (cleavage and re-ligation of genes between attachment sites of gene and vector). The currently available destination vectors[11]concentrate on protein expression applications. Vectors for *E. coli*, insect, and mammalian expression system are available with N-terminal His- or GST tag, or for native protein expression. However, it is possible to convert any vector into a Gateway destination vector using the Gateway conversion system.

3.19.5 Expression Systems

Different expression systems have been described as being suitable for protein production destined for structural research studies. Sufficient protein yields are obtained from bacteria, yeast, insect, and mammalian cells, and NMR labeling has been shown to be possible in bacterial, yeast, and insect cells.

Bacterial expression strains are quickly cloned and tested for expression. Transformation and growth of bacteria can be performed with few significant safety or sterility problems and are very cheap. Proteolytic degradation of the protein of interest during its expression is very rare; most high-level expression hosts on the market are deficient in proteases. Bacteria do not perform posttranslational modifications such as glycosylation and phosphorylation. For that reason, protein batches from a bacterial expression host are more homogeneous, do not contain microheterogeneities, and therefore facilitate crystallization. On the other hand, however, the lack of posttranslational modifications may sometimes lead to a decrease in protein solubility. Many proteins, however, are either toxic for bacteria, or the prokaryotes express unfolded protein in inclusion bodies.

Cell-free expression indicates the degree of solubility and native folding of the protein construct within a short time. Cellular influences and problems of overexpression in different hosts do not play a role in this system. Labeling for NMR studies is easy to perform in this system. On the other hand, upscaling for large protein quantities is difficult.

Yeast expression hosts produce glycoproteins but the glycosylation pattern is different from mammalian ones. Secreted expression facilitates purification while purification of proteins expressed into the cytosol needs special cell disruption techniques, such as French Press or cell mills.

During the last decade, insect cell expression has become the workhorse in protein production for structural studies. Expression yields are usually good and protein folding and all posttranslational modifications can be performed.

Insect cells and mammalian cells require not only more effort to achieve sterility (laminar flow benches) but also need further steps in the cloning process.

Expression hosts such as *Leishmania tarentolae*[13] provide several advantages such as easy handling and eukaryotic expression pattern, but are not widely used.

On the whole, the most productive, fastest way to obtain protein crystals is to test two or more expression systems in parallel. This will provide at least one suitable system for the least work.

Parallelization[14–17] and automation make test expressions possible with different expression temperatures and induction conditions; heat shock before expression, different strains, different media, and the addition of chaperones.[18–22]

3.19.5.1 Cell-Free Protein Expression for Structural Research

One of the major bottlenecks in structural proteomics is the production of soluble and folded proteins in the quantities required for structural analyses. Platforms for NMR spectroscopy must support the production of proteins labeled with stable isotopes (^{15}N, ^{13}C, ^{2}H), and those for x-ray crystallography must enable incorporation of selenomethionine or other appropriately labeled amino acids.

Cell-free methods for protein synthesis[23] offer an alternative to *E. coli* cell-based platforms. They have been used in the production of various kinds of protein, including membrane proteins[24] and proteins that are toxic to cells.[25] The

RIKEN Structural Genomics Center[26] has pioneered the use of cell-free protein production for structural proteomics through a coupled transcription–translation system employing *E. coli* extracts.[27–30] The Center for Eukaryotic Structural Genomics[31] has developed an automated platform based on wheatgerm, cell-free technology to produce labeled proteins for NMR spectroscopy.[32]

Cell-free methods appear to introduce a new degree of complexity as the cells must first be grown and the cell extracts prepared. Today, a number of cell-free systems based on *E. coli*, reticulocytes, wheatgerm and insect cells are commercially available from Ambion, CellFree Sciences, Invitrogen, Novagen, Promega, Qiagen, Roche, and Shimadzu, and therefore this obstacle has been removed. The cell-free approach offers many potential advantages. A clear advantage of in vitro translation over in vivo expression is the relative ease with which single components of the system can be altered or exchanged. Certainly it can allow the synthesis of proteins toxic to cell division. It also allows most of the metabolic resources to be focused only on product synthesis. More importantly, it provides enormous flexibility when manipulating protein synthesis and folding. The major disadvantage of the cell-free approach, compared to the *E. coli* cell-based one, is low protein yield. The cell-free yields are sufficient to make this approach attractive for NMR spectroscopy.[33] The required amount is 0.3 mL of 0.25 mM protein, which, for a 20 kDa target, represents 1.5 mg. The requirement for entering crystallization trials is between 5 and 10 mg of each protein and, therefore, the cell-free approach is still too expensive for protein crystallization purposes. An interim approach for x-ray crystallography might be to use cell-free screening in place of *E. coli* screening, to determine which proteins are to be produced on a large scale in *E. coli*. A significant advantage of the cell-free approach for screening is the fact that linear templates generated by PCR can be used directly for in vitro expression of the encoded protein. This avoids certain time-consuming cloning steps.[34–36]

3.19.5.2 Bacteria

Bacterial expression is mostly done in *E. coli*,[37] although other prokaryotic expression systems may also give good expression rates. A variety of *E. coli* strains with different features is commercially available (Novagen/Merck Bioscience,[38] Invitrogen,[11] Stratagene,[39] etc.). The most common strain for high expression levels is BL21(DE3). Due to lack of proteases, the highly controlled induction of the Lac-promoter with isopropyl-β-D-thiogalactopyranosid (IPTG)[40] and strong transcription under T7 RNA polymerase control (DE3 site), the strain shows fast growth and very high expression (up to 100 mg L^{-1}) of the protein of interest (**Figure 2**). These high levels of foreign protein sometimes result in inclusion-body formation or heat shock protein overexpression.[41] These problems may be solved by adjusting expression levels by varying the amounts of IPTG in expression induction, or, by choosing a growth temperature of less than 30 °C.

For expression of proteins toxic for *E. coli*, host strains containing a pLysS plasmid increase the stringency of the control of T7 RNA polymerase, with the help of T7 lysozyme. Some eukaryotic proteins raise problems in BL21(DE3) expression due to the usage of codons for rare t-RNAs in *E. coli*. Codon plus (Stratagene[39]) or Rosetta (Novagen[38]) strains contain a plasmid coding for these rare codon t-RNAs for AUA, AGG, AGA, CUA, CCC, GGA. These strains may help to overcome problems when expressing eukaryotic proteins in *E. coli*. Proteins which need disulfide bond formation for folding can be expressed in special strains that bear mutations in either thioredoxin or glutathione reductase.

Autoinduction has become available with the Novagen Overnight express autoinduction system[42,43] (for further information, see Novagen,[38] Stratagene[39]).

3.19.5.3 Yeast

Expression in *Saccharomyces cerevisiae* or *Pichia pastoris* is less common than bacterial or insect cell expression but also delivers good expression results.[11,44,45] Yeast is a eukaryotic expression system that is both less costly and easier to use than mammalian or insect cells. Yeast grows fast in shaking flasks and fermenters; it produces acceptable amounts of protein and needs very little tending. Expression of the protein of interest is induced by adding methanol to the medium. The protein-processing and folding mechanisms are identical to mammalian ones; posttranslational modifications are very similar (glycosylation is mostly N-linked of the high mannose type). Protein expression in the extracellular medium, with the help of a signal peptide, facilitates purification (ammonium sulfate precipitation) without the need for cell disruption.

3.19.5.4 Insect Cells (Baculovirus System)

Since 1983, when the baculovirus expression vector system (BEVS) technology was introduced, the baculovirus system has become one of the most versatile and powerful eukaryotic vector systems for recombinant protein expression.[46]

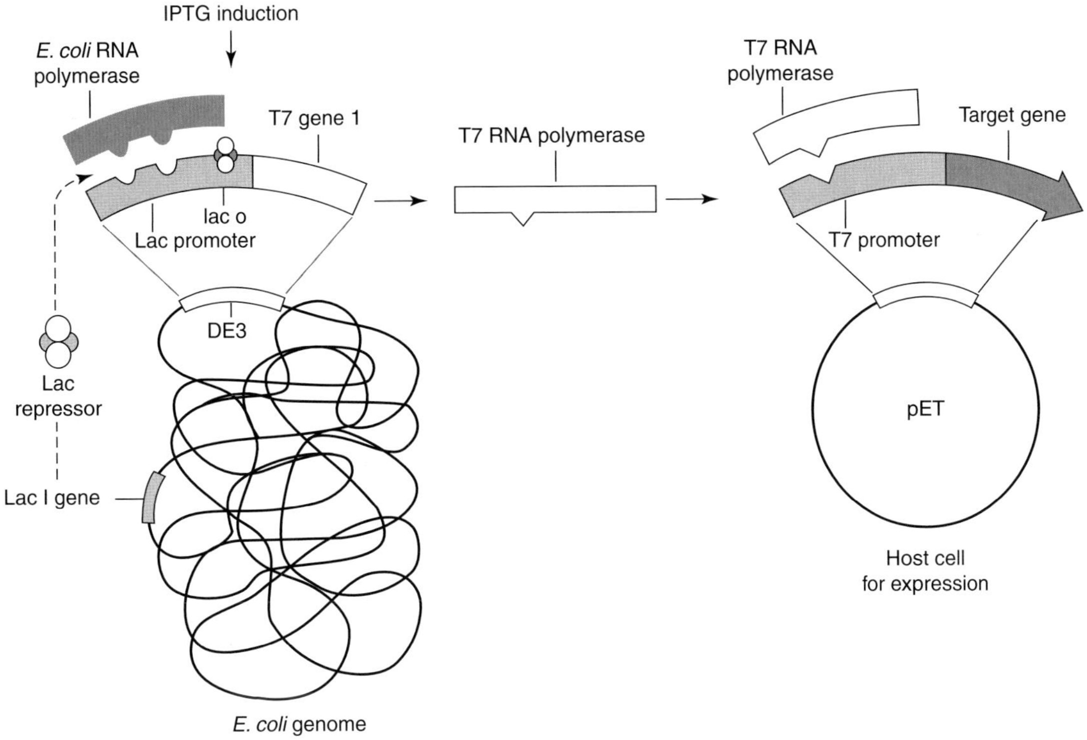

Figure 2 Schematic illustration of the gene translation control in the pET plasmid system (Novagen) for *E. coli* expression. (Reproduced with permission from Merck Biosciences.[38])

Table 1 Insect cell culture media commonly used in BEVS applications

Serum/hemolymph-dependent media	*Serum-free media*
Grace's supplemented (TNM-FH)	Sf-900 II SFM
IPL-41	Express-Five SFM
TC-100	Insect X-Press

BEVS is based on the infection of insect cells with recombinant baculovirus carrying the gene of interest and subsequent expression, by insect cells, of the corresponding recombinant protein. The advantages of the baculovirus system are:

- Safety: baculoviruses are essentially nonpathogenic to mammals and have a restricted host range limited to invertebrate species.
- Scale-up: large-scale production of recombinant proteins is possible.
- High levels of recombinant gene expression: in many cases the recombinant proteins are soluble and easily recovered from infected cells late in infection when host protein synthesis is diminished.
- Posttranslational modifications: insect cells posttranslationally modify proteins in a manner similar to that of mammalian cells.
- Use of cell lines ideal for suspension culture: cells from the fall army worm, *Spodoptera frugiperda*, or from the cabbage looper, *Trichoplusia ni*, grow well in suspension cultures, allowing production in large-scale bioreactors.

The most common cell lines used for BEVS applications are Sf9, Sf21, Tn-368, and High-Five BTI-TN-5B1-4. Of these, it is Sf9, a clonal isolate of the *S. frugiperda* cell line IPLB-SF21-AE, which is probably the most widely used. Commonly used insect cell culture media are listed in **Table 1**.

Traditionally, Grace's supplemented (TNM-FH) medium has been the medium of choice for insect cell culture. However, other serum/hemolymph-dependent and serum-free formulations have evolved since Grace's medium was introduced.

The baculovirus expression system typically utilizes site-specific transposition in *E. coli* (Bac-to-Bac system, Invitrogen[11]), or homologous recombination in insect cells (e.g., BaculoGold, Pharmingen, BD Biosciences, Missisauga, Ontario, Canada) to generate recombinant viruses. These approaches require a great deal of hands-on time over a period of 10 days to several weeks. Faster methods such as the BaculoDirect system have been developed.[11]

3.19.5.5 Mammalian Cells

Human proteins expressed in mammalian cells have the same glycosylation pattern as if they had been produced in human cells, the same specific activity and often better solubility and stability. These factors play an important role in the production of biopharmaceuticals. Either Chinese hamster ovary (CHO) cells or HEK293 are traditionally used for protein production. These cells grow in suspension and are adapted to chemically defined serum-free medium. This saves production costs and facilitates the purification of the recombinant protein from the cell culture medium.

Transient transfection has traditionally provided a quick way of producing proteins in mammalian cells within 2 weeks of the cloning of their corresponding cDNA. Many published works have recently focused on production processes that are simple and cost-effective. With the use of the human embryonic kidney 293-EBNA1 (HEK293) cell line in combination with polyethylenimine and the oriP-based pTT vector, Durocher *et al.*[47] have brought reports of a simple, rapid, and scalable transfection method, which yields high amounts of recombinant protein.

Stable transfection is performed with expression vectors (e.g., IRES Bicistronic expression vectors, BD Biosciences) that allow rapid selection of positive clones that express the target protein. These vectors include a single cassette that expresses both the selected gene of interest and the selection marker of the reporter gene from the same promoter, so that virtually all transfected cells expressing the selection marker also express the protein in question. With these vectors, fewer colonies are needed to locate clones that are expressing high levels of the target protein.

Recently, recombinant baculoviruses modified to contain mammalian cell-active expression cassettes have been demonstrated to direct transient gene expression in a wide range of cell lines and in primary cells (BacMam system). Gene transfer via baculoviruses provides a unique system in which the viruses are readily generated and amplified in insect cells but do not replicate in the transduced mammalian cells, thus providing a high degree of biosafety.

3.19.6 Fermentation

To produce protein amounts necessary for structural research, fermentation in a 5–50 L range is necessary. Depending on expression levels and number of purification steps, protein yields of 1–30 $mg\,L^{-1}$ of expression culture can be achieved. The upscaling from shaking or roller flask small-scale cultures to fermenters has to be monitored thoroughly and, in some cases, screening for media and culture conditions is necessary. High cell density culture of *E. coli* protein often leads to expression of heat shock proteins, inclusion bodies, or low yields of the desired protein.

3.19.6.1 Shaking Flasks

For *E. coli*, yeast, and insect cells, small-scale fermentation for up to 1 L cell culture in shaking flasks is very popular. For good air saturation, baffled culture flasks have to be used in *E. coli* and yeast cultures. Fernbach flasks are used for expression in insect and mammalian cells. As no high cell densities are achievable in shaking flask, media with low nutrient content such as Luria broth (LB) medium[48] are sufficient. To avoid inclusion body or heat shock protein production in *E. coli* by expression at low temperatures, shakers with cooling and heating function are necessary. Insect cells need low-speed shakers with cooling (27 °C).

3.19.6.2 Spinner Flasks or Roller Bottles

Cells that grow in suspension, such as insect or mammalian cells (CHO), can be cultured in spinner flasks. These contain a Teflon-coated magnetic stir bar suspended from a glass or stainless-steel rod that is driven from below by a magnetic stirrer. Microcarrier spinner flasks have a large plastic paddle that is perpendicular to the stir bar. In these vessels, more media is stirred at a given speed than in vessels without the paddle. This increased aeration favors cell growth, particularly when larger volumes of media are used. Mammalian cells usually grow well in spinner flasks, whereas insect cells tend to aggregate. For the maintenance of insect cell suspensions, we prefer Fernbach flasks.

Roller bottles can be used for both cells in suspension and adherent cells. The rotation at the longitudinal axis keeps the cells in motion, whereas adherent cells slowly attach themselves to the surface. Advantages are: (1) increased culture surface; (2) slow, constant movement of the medium; and (3) increased aeration of the cells. However, the whole system is time-consuming and a special roller system is required.

3.19.6.3 Fermenters

For large-scale expression, fermenters with controlled culture conditions (temperature, pH value, air or O_2 saturation, glucose or other nutrient concentration) are necessary. Different fermenters are required for different cell cultures: *E. coli* needs air control; yeast cultures require high concentrations of oxygen. Both need fermenters with an antifoam pump, a feeder pump, and high stirring speeds while insect and mammalian cells need O_2/N_2 control, higher sterility, low temperatures, and gentle stirring.

Media used for fermentation runs with *E. coli* differ from those for LB medium for low cell densities (OD_{600} at end of run: 3–4), 2YT, or for a glucose-enriched medium up to fed batch fermentation[49,50] for very high cell densities (OD_{600} up to 100). Duration of fermentation is from 5 h (low cell densities) up to 2 days (fed batch). Of course, stable isotope-labeled full media can be used in fermenter expression for NMR spectroscopy purposes. If a minimal medium has to be used for single amino acid labeling, then that expression strain requires several days for adaptation to the medium.

Yeast needs high oxygen levels and a rich medium to obtain high cell densities in the fermenter.

3.19.6.4 Wave Bioreactor

The Wave Bioreactor (Wave Biotech, Somerset, NJ, USA) is a patented bioreactor for cell culture (**Figure 3**).[51] In this device, cell culture is performed in presterile plastic bags. The bags, called cellbags, are single-use bioreactors. This eliminates the need for cleaning and sterilization and reduces the risk of contamination. The cellbag is filled with media and cells and inflated to form a rigid gas-impermeable chamber. It is then placed on a Wave Bioreactor rocking platform. The gentle wave motion provides excellent oxygenation and mixing, with minimal shearing forces. Air is continuously passed through via the sterilizing filters provided on each cellbag. This oxygenation technique provides an optimal environment for cell growth and expression, and cell densities of over 6×10^6 cells mL^{-1} have been achieved. On completion of the cultivation, the culture is harvested and the used cellbag is discarded. Wave bioreactors are available for 0.1–500 L culture volume. They can be used with a variety of cells, including CHO, hybridomas, HEK293, insect cells, T cells, plant and primary human cell lines.

3.19.7 Purification

After producing the proteins of interest in suitable expression systems or after gaining sufficient amounts from natural sources, the protein of interest must then be separated from any contaminating substances such as sugars, lipids, and/or

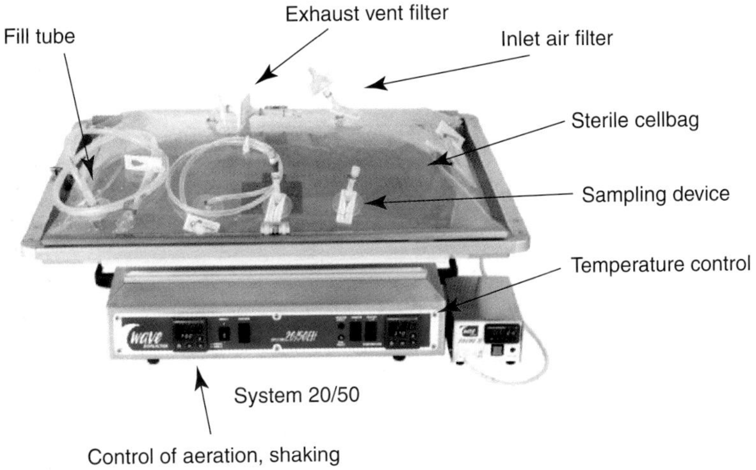

Figure 3 Wave Bioreactor System (Wave Biotech AG): the figure shows a cell bag filled with medium and cells on a wave bioreactor rocking platform.

other proteins. There is no general procedure for protein preparation and a wide range of literature is available on that subject (*see* Section 3.19.1). In most cases, it consists of the following steps:

- cell lysis
- centrifugation
- a least one (column) chromatography step for purification
- concentration
- buffer exchange

Mostly, the purification procedure consists of three steps:

1. capture phase for quick isolation and stabilization of the protein
2. intermediate purification procedure for removal of bulk impurities
3. polishing to achieve over 90% purity levels when separating the protein of interest from trace contaminants and/or closely related substances

Over the last decade, developments in recombination techniques have revolutionized protein production in many ways. Engineered proteins have been designed with improved solubility and protein properties and simplified purification with the help of affinity tags. It has to be mentioned that these protein mutations may interfere with folding, function, or crystallizability of the protein of interest.

In bacterial expression systems, refolding of inclusion bodies protein sometimes requires additional time-consuming steps, together with a reliable and robust test of molecular activity of the protein of interest.

3.19.7.1 Cell Lysis

All proteins that are produced intracellularly require cell disruption for release. Depending on the expression system used, different methods of breaking up the expression hosts are necessary. When proteins are secreted into the medium by the usage of secretion signals, no cell lysis is necessary. Fractionated precipitation or ultrafiltration can be performed to concentrate the protein of interest out of the large volume of cell medium.

Proteins expressed in the periplasm can be isolated by mild osmotic shock.

Cells from mammalian tissue origin are destroyed by mechanical procedures such as homogenizers or potters. Mammalian and insect cells from cell cultures have to undergo freeze/thaw cycles or mild detergent treatment.

Bacterial cell membrane protection is more stable: treatment with lysozyme destabilizes the cell by hydrolysis of the 1,4 glycosidic bonds in the peptidoglycan cell wall. Subsequent detergent treatment releases most of the soluble content of the cells. Sonication, French Press, Constant Cell Disruption System (Constant System Ltd., Daventry, Northants, NN114SD, UK), or bead mill treatment are also very effective cell disruption methods.

Yeast cells have to go through harsh treatment (French Press or cell mill with glass beads) to undergo lysis.

Independently of the chosen lysis procedure, the addition of protease inhibitors, reducing agents, and glycerol to the lysis buffer helps to stabilize the protein and protects it against degradation.

After lysis, cell membrane fragments and cells with other insoluble or large particles have to be separated from the soluble protein by centrifugation (40 000 **g** for 30–60 min) and filtering it through 0.2 μm pore size filter.

3.19.7.2 Refolding

Proteins produced in a microbial host at high expression levels are often produced in an unfolded or improperly folded form and then accumulate in intracellular vesicles called inclusion bodies.[52] Inclusion bodies consist of more than 90% of the protein of interest. After gaining this nearly pure protein fraction, refolding has to be performed to obtain active protein suitable for structural studies.

The process of refolding consists of three steps: (1) total denaturation of the protein; (2) slow renaturation (duration: hours to weeks.); and (3) the gain of the correctly folded protein (yields: 5–40% of the total inclusion bodies protein) by column chromatography (for a review on refolding, see Rudolph *et al.*[53]).

3.19.7.2.1 Inclusion bodies purification

To separate inclusion bodies protein from soluble protein and cell components, extensive washing steps (resuspension of lysis pellet in different solutions followed by centrifugation to pellet the inclusion bodies protein) are necessary. Suitable buffers for washing contain detergent (solubilizes membrane fragments), high salt concentrations (salt solubilizes proteins), and up to 1 M urea.

3.19.7.2.2 Solubilization of inclusion bodies protein

After washing, the pelleted inclusion bodies protein has to be solubilized and totally denatured. The pellet is homogenized in a buffer containing a chaotropic agent such as 6 M guanidinium hydrochloride or 8 M urea and dithiothreitol (DTT) or β-mercaptoethanol (β-ME), which will reduce disulfide bonds (for proteins containing more than one cysteine residue) and is then incubated for 1 h at 37 °C. After that procedure, insoluble parts are to be removed by centrifugation.

3.19.7.2.3 Refolding conditions and screens

Fast dilution of the denatured protein in, or slow dialysis against, an appropriate refolding buffer reduces the concentration of the chaotropic denaturant. This procedure takes several hours to several weeks and produces a mixture of different folding conditions of the protein of interest: A larger amount will fold incorrectly, precipitate during the refolding procedure, and can then be easily removed by means of centrifugation or expanded bed chromatography. Another amount will exhibit incorrect folding but will stay in solution and will then require separation from the correctly folded protein by means of chromatography.

Different proposals for refolding screens are to be found in the literature; the main components of refolding buffers are arginine or sucrose (which act as stabilizing agents); sodium chloride or potassium chloride as ionic molecules; detergents for solubilizing hydrophobic patterns of the protein; low concentrations of urea or guaninidium hydrochloride as chaotropes; reducing/oxidizing agent mixture for correct formation of disulfide bonds (oxidized and reduced glutathion); pH and temperature variations.[54–57]

After refolding, the insoluble, incorrectly folded proteins are to be removed by centrifugation. Any remaining impurities and soluble but incorrectly folded protein molecules are to be removed by column chromatography.

3.19.7.3 Capture Step with Affinity Tags

In the last decade, a broad variety of fusion proteins or peptides for simplified protein purification and detection have been described. The fusion partner is encoded on the plasmid or viral DNA and either C- or N-terminally linked to the protein of interest. For larger fusion tags the introduction of an appropriate protease cleavage site between fusion protein and the protein of interest is considered to remove the tag prior to crystallization. Careful removal of the protease before starting crystallization experiments is very important so as to avoid cleavage of the protein of interest during the crystallization procedure.

Plasmids containing DNA coding for proteins like glutathione-*S*-transferase,[58,59] maltose-binding protein, NusA, calmodulin-binding peptide, intein, thioredoxin, cellulose-binding protein, and Fc fragments of antibodies[17] are commercially available, together with the appropriate protein purification tools. Many of these large fusion partners increase the solubility of the expressed fusion protein (see, for example, Kapust and Waugh[60]). Sometimes fusion partners simulate solubility of the expressed protein. After removal of the tag, the protein of interest precipitates irreversibly.

Small affinity tags such as Poly-His-tag and Strep-tag[61,62] are widely used as they do not change protein properties and do not need to be cleaved off. When using Poly-His-tags for protein purification, it has to be taken into account that *E. coli* strains produce protein SlyD (20 kDa) containing a potential metal-binding site domain. This protein binds tightly to metal chelate chromatography and is often detected as a contamination band.[63]

Affinity tag chromatography can be performed in batch mode, by centrifugal or syringe filter devices, or on chromatographic equipment with prepacked columns according to the manufacturer's instructions.[58,62,64]

In some cases, removal of the elution agent is necessary to maintain protein stability: Fc-fragment fusion proteins are eluted in extreme acid conditions: pH has to be adjusted to neutral values immediately after elution. Using imidazole in high concentrations in the elution of Poly-His-tagged proteins destabilizes proteins and it has to be removed by dialysis.

3.19.7.4 Fractionated Precipitation

Fractionated precipitation with ammonium sulfate or alcohol has been a widely used method for tissue or plasma protein purification.[65] Solubility of proteins depends on different parameters, such as temperature, pH, salt, and protein concentration, and is lowest at pH values of around the isoelectric point of the protein. Therefore, different proteins precipitate at different conditions. This can be used as a fast and quantitative protein purification step, especially for proteins secreted to the medium.

Precipitation of proteins with 1–4 M ammonium sulfate or 10–40% (w/v) polyethyleneglycol (PEG) under pH and temperature control is a convenient method for fast increase of protein concentration. The precipitate is dissolved in a small amount of solubilizing buffer and dialyzed against the same buffer to regain solubility and perform the next purification step.

3.19.7.5 Low-Pressure Column Chromatography

Column chromatography covers protein separation steps performed on chromatography columns filled with a variety of column matrices (mostly derivatives of agarose) using different protein characteristics for separation. The columns require pumps to be run and an ultraviolet detector for protein determination. Complete equipment for column chromatography from either GE Healthcare or BioRad provides additional pH and conductivity monitors and fraction collectors.

Protein characteristics used for separation are size, charge, affinity toward substrates or inhibitors, hydrophobicity, lectin, or antibody binding. The most prominent column chromatography methods are described below.

A variety of simplified affinity and ion exchanger-binding tools are available on the market.[58,62,64,66] These are tools that do not require chromatography equipment but are used in a batch mode or as centrifugal devices. The separation results and protein-binding capacities are not comparable to chromatography methods, but give quick, low-cost results for pre-studies.

Before starting column chromatography, clearance of the protein sample and the use of centrifugation and filtration to remove all dust and cellular particles are indispensable.

3.19.7.5.1 Affinity chromatography

The most efficient purification method is column chromatography with an affinity medium, although it is not applicable in all cases. Similar to the affinity-tag purification capture step mentioned above (Section 3.19.7.3), affinity chromatography binds proteins with respect to their specific and natural properties.

The user must have comprehensive knowledge of the protein of interest and how it functions or reacts enzymatically, any possible cofactors or substrates, and/or about its sequence and modification pattern.

A very specific, but expensive, method of affinity chromatography is purification by specific antibodies coupled to column matrices. The binding to these matrices is performed at neutral pH values while elution of the bound protein of interest requires pH values of around 2–3, which is not suitable for every protein, but very effective when possible.

The same holds for capturing enzymes to specific inhibitors, cofactors, or substrate analogs coupled via activated groups to agarose: elution can be performed by the addition of the coupled substance in a free form. Commercially available examples such as adenosine triphosphate or NADPH agarose for capturing kinases, lectin matrices for the capture of glycoproteins, or heparin agarose for binding of blood coagulation factors are widely used.

Capturing the protein of interest by its binding to specific inhibitors, peptides, or substrate analogs chemically bound to activated agaroses is very efficient. These affinity chromatographies are very helpful, particularly for removing randomly folded protein from correctly folded ones, after the refolding procedure. Proteins and peptides can be coupled via amino groups to cyanogen bromide or 6-aminocaproic acid *N*-hydroxysuccinimide ester-activated-Sepharose, or via carboxyl groups to aminohexyl Sepharose. Many other matrices with a variety of functional groups for protein and substrate coupling are commercially available.[4]

3.19.7.5.2 Ion exchange chromatography

Ion exchange chromatography matrices capture proteins from a solution by ionic interaction. The overall isoelectric point (IP) can be calculated from the amount and K_i values of all charged amino acid residues in a protein. At buffer pH values above this IP, the protein is negatively charged (anionic); at pH values below that, the protein is positively charged (cationic).

For the binding of anionic proteins, anion exchanger matrices such as quaternary amines or diethylaminoethyl groups coupled via a linker to a cellulose matrix are available.

For the binding of cationic proteins, sulfopropyl or carboxymethyl groups can be used. The proteins are eluted by a sodium chloride gradient providing increasing ionic strength to break the interaction of protein with charged matrix.

The proteins bind at suitable pH values in the absence of ionic substances such as salts. Usually, a linear salt gradient from 0 to 1 M is used for stepwise elution of contaminating proteins and proteins of interest. Depending on the diversity and amount of contaminants, one-step ion exchange chromatography may lead to purity that is sufficient for structural studies. Microheterogeneities of the protein of interest (due to posttranslational modifications such as glycosylation or phosphorylation or to aggregation status) result in multiple-peak elutions. These different protein homologs require separation from each other prior to crystallization.

3.19.7.5.3 Hydrophobic interaction chromatography

Protein mixtures can be separated by their different hydophobic/hydrophilic natures through hydrophobic interaction chromatography. The proteins are applied on the column matrix carrying hydrophobic ligands (phenyl-, ethyl, or alkyl residues) in a solution containing a high percentage of hydrophilic substances (e.g., >1.5 M ammonium sulfate), to decrease the water solubility of the proteins to a minimum. Elution is performed by the stepwise decrease of the hydrophilic substance.

3.19.7.5.4 Size exclusion chromatography (SEC)

For the separation of proteins with a high molecular mass from those with a low one, SEC is the method of choice. As SEC columns do not have high separation capacities, this method is not suitable as a capture, but only for the last, polishing step in the protein purification procedure. A high degree of caution is necessary in the choice of running buffer: its pH value has to differ by at least 1 pH unit from the IP of the protein. Additionally, the NaCl concentration needs to be above 150 mM to avoid unspecific interference to the column matrix. Unless it is necessary for protein solubilization, detergent should be avoided so as to prevent partial denaturation of the protein.

SEC columns are restricted in the volume of the sample to be applied, depending on the column diameter and length. For this reason, concentration of the protein sample prior to application on the SEC column is recommended. The smaller the sample volume and flow rate are, the better the separation results.

Before starting crystallization experiments, SEC is recommended to determine the aggregation status of the protein (**Figure 4**). Monomeric proteins or protein aggregations of defined size crystallize more easily than undefined aggregate mixtures.

3.19.7.6 Buffer Exchange

Before starting the affinity process, hydrophobic or ion exchange chromatography, or crystallization experiments, the protein sample has to be converted to defined buffer conditions, depending on the nature of the protein in hand and the method to be performed. Different methods are available for changing the buffer but it is imperative not to dilute the protein sample too far. The most effective, mildest, but also most time-consuming method is dialysis.

3.19.7.6.1 Dialysis

A variety of dialysis tools are on the market: regenerated cellulose tubings, Slide-A-Lyzer,[67] tube-O-Dialyzer,[68] and dialysis buttons,[69] all of which work on the same principle: the protein sample is transferred to a container with a semipermeable membrane (usually cellulose) closure. By means of this membrane, small molecules such as buffer

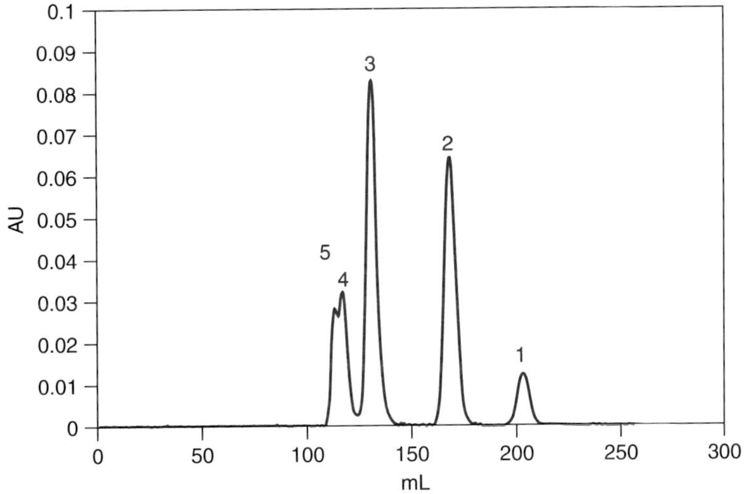

Figure 4 Size exclusion chromatography on HiLoad 26/60 Superdex 75 pg on fast performance liquid chromatography (FPLC) (GE Healthcare Amersham Biosciences[58]). Peak 1, cytochrome *c*, 1 mg mL^{-1}, Sigma C-3006, molecular weight 12.4 kDa. Peak 2, carbonic anhydrase, 1 mg mL^{-1}, Sigma C-7025, molecular weight 29 kDa. Peak 3, bovine serum albumin, 1 mg mL^{-1} Sigma P0914, molecular weight 67 kDa. Peak 4, bovine serum albumin dimer, 1 mg mL^{-1} Sigma P0914, molecular weight 134 kDa. Peak 5, bovine serum albumin tetramer, 1 mg mL^{-1} Sigma P0914, molecular weight 268 kDa.

ingredients, salts and detergents permeate, protein, or other molecules larger than the molecular weight cut-off of the dialysis membrane will remain in the container. Defined buffer conditions are obtained after approximately 6 h time and two exchanges of the dialysis buffer (volume 100 × sample volume).

3.19.7.6.2 Size exclusion chromatography

SEC for buffer exchange can either be performed as described above (in Section 3.19.7.5.4) at conventional column chromatography workstations or with low-cost and low-tech variations of the same method: PD10/NAP-5,[58] zip-tips for MS,[70] bio-spin chromatography columns.[71] Caution has to be taken as described above (in Section 3.19.7.5.4) to choose the right buffer to avoid unspecific interactions with the column materials. The protein sample will then, afterwards, have been diluted.

3.19.7.6.3 Ultrafiltration

The fastest method of exchanging buffers and simultaneously concentrating protein samples is ultrafiltration. Again, many tools with different molecular weight cut-offs (1–100 kDa) are offered on the market: Centrifugal devices of different sizes; N_2 pressure-driven stirring cells or pump driven crossflow filtration systems.[66,70,72] The buffer of the protein sample is pressed through a membrane permeable for low-molecular-weight components: the proteins remain in the sample chamber. After several refills, buffer exchange is more than 99% complete.

3.19.7.7 Concentration

Protein crystallization and NMR spectroscopy experiments need high protein concentrations (usually around 10 mg mL^{-1}). For that reason, concentration of the purified protein prior to structure analysis methods is necessary. Concentration steps have to be performed with caution, as many proteins tend to precipitate at very high concentrations (> 20 mg mL^{-1}), in low-ionic-strength solutions.

3.19.7.7.1 Concentration with polyethyleneglycol

PEG has a very high water-binding capacity. Embedding dialysis tubing filled with protein solution, in solid PEG (molecular weight of the PEG larger than the cut-off of the dialysis membrane) draws the buffer out of the sample and concentrates the protein in a very slow and nonreactive manner.

3.19.7.7.2 Ultrafiltration

Ultrafiltration procedures, as described above (in Section 3.19.7.6.3), with centrifugal, nitrogen, or pump-driven devices, are the most common methods used to increase protein concentration.

3.19.8 Analysis

Analysis of the purified protein prior to crystallization or NMR spectroscopy is essential. During the purification procedure, different stress factors act upon the protein of interest. Proteases may degrade the protein; intermolecular disulfide bridges and protein aggregation may disturb structure studies. For those reasons, the degree of purity, correct folding, sequence, and aggregation status together with the activity of the enzyme must all be tested. Programs calculating physicochemical properties of proteins can be found at the Expasy program[73a] or the Swissprot websites.[73b]

3.19.8.1 Electrophoresis

Protein separation techniques in an electric field offer a variety of analytical data on the protein of interest.[74,75]

Equipment for standard gel electrophoresis processing consists of a power supply and a gel chamber. Automated electrophoresis with medium to high throughput gives faster and more reproducible results by using microfluidic separation.[76,77]

3.19.8.1.1 Sodium dodecyl sulfate–polyacrylamide gel electrophoresis

After total denaturation of a protein probe with anionic detergent (sodium dodecyl sulfate), reducing agent (β-mercaptoethanol or dithiotreitol), and heat, the absolute molecular weight of the components can be determined. Gels with different polyacrylamide content (6–30%) are used to separate proteins of molecular weights from 200 to 5 kDa (**Figure 5**). The proteins are visualized by different staining methods. Silver staining is more sensitive than Coomassie blue staining.

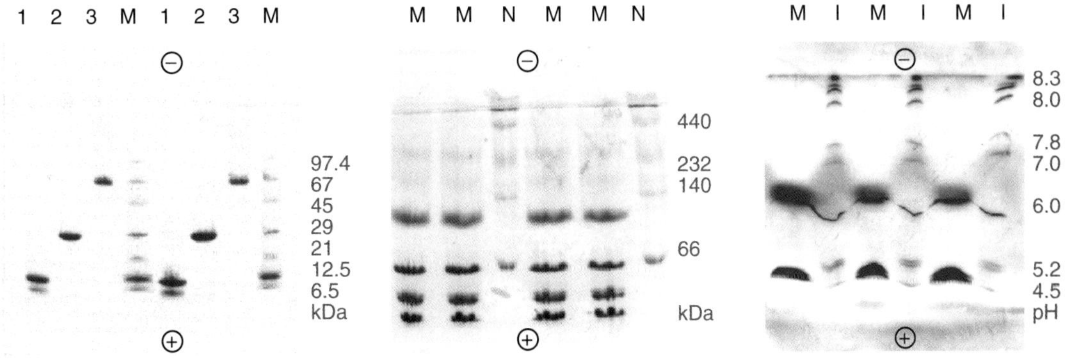

(a) SDS-page Gradient: 8–25 % Polyacrylamide

(b) Native page Gradient: 8–25 % Polyacrylamide

(c) IEF page Gradient: pH 3–9

Figure 5 Comparision of polyacrylamide gel electrophoresis (PAGE) results. PAGEs for PHAST System from GE Healthcare Amersham Biosciences.[58] (a) Lane 1, cytochrome *c*, 1 mg mL^{-1}, Sigma C-3006, molecular weight 12.4 kDa. Lane 2, carbonic anhydrase, 1 mg mL^{-1}, Sigma C-7025, molecular weight 29 kDa. Lane 3, bovine serum albumin, 1 mg mL^{-1} Sigma P0914, molecular weight 67 kDa. Proteins are denatured in Laemmli buffer and separated on SDS-PAGE 8–25% polyacrylamide according to their absolute molecular weight. Lane M, protein test mixture 6, Serva 39207 (molecular weight 97.4, 67, 45, 29, 21, 12.5, 6.5 kDa). (b) Lane M, proteins of protein test mixture 6, Serva 39207 (compare to lane M on PAGE in part (a)) dissolved in H_2O are separated on native PAGE 8–25% polyacrylamide according to their molecular weight and aggregation status. Lane N, high-molecular-weight native calibration kit, Amersham Biosciences 17044501 (molecular weight 669, 440, 232, 140, 66 kDa). (c) Lane M, proteins of protein test mixture 6, Serva 39207 (compare to lane M on PAGE in part (A)) dissolved in H_2O are separated on isoelectric focusing (IEF)-PAGE pH 3–9 according to their isoelectric point. Lane I, test mixture for PI determination pH 3–10, Serva 39211.

3.19.8.1.2 Native and isoelectric focusing polyacrylamide gel electrophoresis

Polyacrylamide gels with nondenaturated protein samples deliver additional knowledge: in the absence of SDS and β-ME or DTT, the proteins become separated according to a combination of molecular weight and overall charge: the larger the protein, the shorter the sample propagation; and the more negatively charged the protein is, the faster it moves toward the cathode. Proteins with IP higher than the pH of the buffer system of the electrophoresis will not move toward the cathode. Proteins that occur with mono-, di-, or an even higher aggregated state will show up as different bands on a native gel. Separation of proteins in an electric field under native conditions gives an approximation of the uniformity of a protein's aggregation status. The protein moves in an electric field which, depending on whether its net charge is toward the anode or cathode, will affect the separation distance, depending on the size of the molecule. After staining with Coomassie, the proteins forming dimers or larger aggregates may show a ladder of bands related to the different molecular weights of the aggregates (**Figure 5**).

Isoelectric focusing gels also consist of polyacrylamide, but also contain ampholyte (bivalently charged molecules) mixtures (varieties from pH 2 to 9.5 are available), to separate the proteins according to their isoelectric points (**Figure 5**). The determination of the isoelectric point of the purified protein is necessary to decide which ion exchange chromatography method to use for separation (with a cation or anion exchanger) and to analyze the posttranslational modifications of the protein of interest: phosphorylation and glycosylation can shift the IP of a protein far away from the one originally calculated.

3.19.8.1.3 Western blotting: identification

The most specific method for detection of single proteins includes the binding of specific antibodies (immunoprecipitation and Western blotting). Similar to Northern/Southern blotting, the bands separated on an electrophoresis are transposed by a second electrophoresis on to a nitrocellulose membrane. Protein-specific antibodies and antibodies carrying a detection group (e.g., horseradish peroxidase, alkaline phosphatase) are applied. A large range of specific antibodies for the specific detection of many different proteins are commercially available.

3.19.8.2 Protein Concentration Determination Methods

Knowing the exact concentration of the purified protein sample is of special interest for crystallization and NMR experiment setups. For both methods, protein concentrations from 5 to 50 mg mL^{-1} are necessary. A number of different methods are used to determine the concentration of a highly purified protein in solution.

3.19.8.2.1 Colorimetric measurements

The Bradford method[78] or Lowry method[79] use the specific binding of dyes to proteins followed by the determination of the color intensity at a defined wavelength, by a spectrophotometer. These methods refer to a calibration curve usually done with bovine serum albumin. The standard deviations for these measurements are very high (up to 20% from the correct value) as a number of buffer ingredients interfere with the measurement and different proteins have different dye-binding capacities.

3.19.8.2.2 Ultraviolet absorption

The more exact protein concentration determination is the ultraviolet absorption value at 280 nm measured in a spectrophotometer in quartz cuvettes. This method requires the knowledge of the exact protein sequence and a very high degree of purity (>95%). The molar extinction coefficient ε' of the protein at 280 nm is defined by the π/π^* absorption of aromatic amino acids by the formula:

$$\varepsilon' = a \times 5.7 \times 10^3 + b \times 1.28 \times 10^3 \ \ \mathrm{M^{-1}\ cm^{-1}}$$

where a = number of tryptophanes in the AA sequence of the protein of interest and b = number of tyrosines in the AA sequence of the protein of interest.

The relationship between the absorption at 280 nm of the protein solution (E), and the concentration of the protein of interest (c) in the solution, is as follows:

$$E = \varepsilon \times c \times d$$

where $\varepsilon = \varepsilon'$/molecular weight of the protein of interest and d = diameter of cuvette.

When nucleic acids are present in the protein solution, the value has to be corrected according to the formula[2]: Protein concentration in mg mL^{-1} = $1.55A_{280} - 0.76A_{260}$.

3.19.8.3 Particle Size/Aggregation Status

Protein used for crystallization and NMR studies has to be in a defined aggregation status and correspond to the native status in vivo. As proteins sometimes form dimers or larger aggregates, especially after the refolding process, the size of the particles in a purified protein sample has to be determined. In measuring the aggregation status of a protein, it must be remembered that proteins behave differently in different environments: highly ionic solutions or the presence of detergents usually prevent aggregation, while high protein concentrations or pH values around the isoelectric point of a protein enhance it.

3.19.8.3.1 Size exclusion chromatography

SEC has already been described in Section 3.19.7.5.4. The method needs a chromatographic system with ultraviolet detection and a suitable gel filtration column (for example, Superdex 75 or 200 for proteins of molecular weight 10–60 kDa or 30–300 kDa respectively, from GE Healthcare).[58] Calibration of the column with a number of proteins of known molecular weight (e.g., bovine serum albumin (BSA), cytochrome c, carbonic anhydrase) is necessary (**Figure 4**).

3.19.8.3.2 Dynamic light scattering

Dynamic light scattering measures the intensity of laser light when scattered by molecules in solution. The scattering is related to the hydrodynamic radius (Rh), as in the Stokes–Einstein equation. Particles of different size and aggregation status are recognized by their increasing Rh value.[80] Proteins in solution have the tendency to aggregate, depending on the buffer conditions and the protein concentration. A protein solution containing a mixture of different aggregation states of a protein is inapplicable for crystallization and NMR studies. For that reason, appropriate protein buffer conditions have to be screened for, with the help of dynamic light scattering, to start structural determination experiments with a homogenous particle size distribution.

3.19.8.3.3 Analytical ultracentrifugation

Analytical ultracentrifugation determines the duration of sedimentation of particles in a solution undergoing high centrifugal forces. The equipment is a considerable financial investment (e.g., Optima XL-I analytical ultracentrifuge, Beckman Coulter[81]) and requires operator expertise. However it does deliver highly reliable measurements.

3.19.8.4 Mass Spectrometry

MS methods give the absolute molecular weight of a protein.[82] The method is absolutely necessary to determine the correct size and sequence of a purified protein and its correct cleavage product after removal of affinity tags by protease cleavage. For proteins with posttranslational modifications such as glycosylation or phosphorylation, MS data detect the modification performed by the expression host. In some cases, limited proteolysis followed by MS detection of resulting peptides can help to detect unexpected modifications. For proteins labeled with stable isotopes for NMR measurements, MS data show the degree of labeling.

Modern ionization techniques in MS, such as electrospray ionization and matrix-assisted laser desorption ionization in combination with, for example, time-of-flight or Fourier transform ion cyclotron resonance analyzers, allow the determination of the molecular weight of biopolymers to within 1 or 2 mass units. This precision is generally sufficient to confirm the correct sequence of an expressed protein following its purification and removal of affinity tags. Posttranslational modifications such as phosphorylation and, to a lesser extent, glycosylations which are dependent on their heterogeneity are also easily detected by mass measurement of the whole protein. The mass shift of a protein before and after stable isotope labeling allows the degree of labeling to be determined.

Identification of an unknown protein is generally performed by creating protein fingerprints by digestion with specific endoproteases (e.g., using trypsin). The subsequent mass determination of the respective peptides and then their comparison with the calculated peptide masses, in accordance with the sequences of protein databases, leads to their identification.

As an extension, collision-induced dissociation experiments on several peptides are applied to create sequence tags which often enable unambiguous identification of the protein. The site of a posttranslational modification may also be determined.

3.19.8.5 Activity Assays

Protein structure research in pharmaceutical companies mainly results in the provision of tools for ligand-binding studies and molecular modeling. Many of the proteins of interest are enzymes or receptors. This means that knowledge of the activity or binding capacity of the protein used for crystallization or NMR studies is helpful. In most cases, an activity or binding assay is available and should be used to test the activity of the purified protein sample for structural research purposes.

3.19.8.6 Circular Dichroism Spectroscopy and One-Dimensional (1D)-Nuclear Magnetic Resonance Spectroscopy

Determination of the secondary structure by circular dichroism spectroscopy or 1D-NMR is a good indicator of the correct folding of a protein. Some proteins only show proper folding status in protein complex form, or in the presence of cofactors or substrates. Proteins being refolded following an inclusion body preparation have to be tested for their native state. For this reason, a routine measurement on a circular dichroism spectrometer or 1D-NMR, prior to crystallization experiments or multidimensional NMR studies, prevents undesirable loss of time and material on unnecessary crystallization or NMR efforts on unfolded protein.

Circular dichroism spectra in the wavelength range of 190–250 nm reflect the overall protein secondary structure, while measurements between 250 and 300 nm show the contribution of aromatic side chains.[83] Conformational and structural changes to protein resulting from mutagenesis or ligand binding and refolding success, as well as folding/unfolding transitions in globular proteins, can be followed by circular dichroism spectroscopy.[84] If the percentage of β-sheets, α-helix, β-turns, and random coil regions of the purified protein, determined by circular dichroism spectroscopy, matches the predicted secondary structure, the protein can be used for structural studies.

1D ^{1}H NMR spectra exhibit a characteristic spread of signals called dispersion, which is dependent on the folding state of a protein. As the dispersion is caused by the formation of secondary and tertiary structure, a completely unfolded protein has the proton signals from the amino acids confined to narrow ppm regions. Partial and/or complete folding results in clearly higher or lower spreading of the signals. For example, in the region between 1 and -1 ppm, where signals of the aliphatic side chains of the amino acids are found, an unfolded protein will generally have signal values of above 0 ppm, whereas a folded protein will also show signals with values of under 0 ppm. A similar picture can be seen in the region of the amide protons between 6 and 10 ppm; the unfolded state then results in the collapse of signals at around 8 ppm. With isotopically labeled protein, the values of ^{13}C chemical shifts can also be used to monitor its folding state. As the folded state is stabilized by hydrogen bonds from amide to carbonyl groups, thus trapping the amide proton, hydrogen deuterium exchange can also be used to determine regions of stable secondary structure and allow assessment of the regions of stable secondary structure versus less stable or unfolded regions.

3.19.8.7 Sequence Analysis

After the PCR reaction, DNA sequence analysis and cloning procedure have to be performed to prove the correct sequence of the expression vector. Access to a DNA sequencer or to a DNA-sequencing provider (e.g., MWG, Ebersberg, Germany[85]) is absolutely fundamental for the cloning of proteins for structural studies.

Protein sequence analysis after the purification process has become facultative in most cases as MS measurements are accurate enough to detect minor variances in protein sequence. In cases where the MS molecular mass differs from the calculated mass, N-terminal sequencing or sequencing of cleavage products after limited proteolysis has to be performed to identify and characterize the isolated protein.

3.19.9 Deglycosylation

Extracellular mammalian proteins contain glycosylation sites. Eukaryotic expression systems will recognize those glycosylation sites and synthesize glycoproteins. Bacteria do not produce glycoproteins and are therefore the crystallizer's preferred expression hosts for those proteins. Most carbohydrate chains are not stabilized by intramolecular contacts in proteins and are therefore very motile. Due to this, crystallization contacts are often disturbed and crystallization is inhibited. To overcome this problem, removal of the carbohydrate chains can be necessary. Depending on the nature of the glycosylation (N- or O-) and on the expression host, a number of glycosidases are available (e.g., Roche,[86] Prozyme,[87] or Calbiochem[38]).

3.19.10 Labeling for Nuclear Magnetic Resonance Studies

3.19.10.1 Cell-Free System

One method of obtaining uniformly ^{13}C, ^{15}N-labeled protein samples through in vitro synthesis is to add each of the 20 ^{13}C, ^{15}N-labeled amino acids to the starting lysate.[88] This is often a prohibitive cost item compared with cell-based expression, and is usually only considered if cell-based expression is poor. Another, more cost-effective alternative is to add a commercial mix of ^{13}C, ^{15}N-labeled amino acids derived from algal sources. Nowadays, mixtures containing the labeled forms of all the amino acids are commercially available.

Information on specific residues and specific sites in a protein can also be obtained by selective incorporation of only one or few amino acids labeled with ^{13}C or ^{15}N. This selective labeling strategy dramatically simplifies the NMR spectrum. Yokoyama and coworkers recognized early on that there are significant advantages in producing specifically labeled proteins for NMR studies by in vitro methods.[89–91] One of the main problems with producing specifically labeled proteins using cell-based systems is that the label is scrambled, due to the metabolism of amino acid. With cell-free synthesis, metabolic scrambling is significantly reduced, thus giving the researcher much more control over labeling schemes. In addition, isotope incorporation is much more efficient. For the same amount of protein, in vitro synthesis requires less than one-fourth the amount of labeled amino acid than is typically used in cell-based expression.[90] This result is of particular importance when it is necessary to incorporate costly labeled amino acids. Specifically labeled proteins can be produced by supplementing the cell-free reaction mixture with one labeled and 19 unlabeled amino acids.

A method of obtaining sequence-pecific resonance assignments is to incorporate a single labeled residue into a protein through the use of amber or nonsense codons and matching aminoacylated tRNAs within a cell-free expression system.[91,92]

3.19.10.2 *Escherichia coli*

For protein studies with NMR spectroscopy, labeling with stable isotopes such as ^{15}N, ^{13}C, or ^{2}H, or combinations thereof, is necessary. Specific labeling of single amino acids is performed by premixed medium containing 19 unlabeled and one labeled amino acid. Complete labeling of the protein of interest has to be done in medium, where all nitrogen atoms are exchanged to ^{15}N. *E. coli* expression in labeled full medium gives good results and does not need adaption of the bacteria to the labeled medium. Commercially available labeled full media (^{15}N, ^{13}C, ^{2}H, or double/triple labeled from Silantes[93] or Eurisotop[94] and Spectra Stable Isotopes[95]) give excellent protein expression yields comparable to unlabeled full media. Minimal media prepared from M9 salts[48] enriched with ^{13}C carbon or ^{15}N nitrogen source require some adaptation time for the *E. coli*. Expression rates, especially in triple-labeled minimal media, are very low.

3.19.10.3 Insect Cells (Selective Labeling of Amino Acids)

A suitable labeling medium is the crucial item for successful isotope labeling of recombinant proteins in insect cells. Culture media for uniform ^{15}N-labeling of recombinant proteins with insect cells suitable for getting a ^{1}H–^{15}N heteronuclear single quantum coherence (HSQC) spectrum are commercially available (e.g., Bioexpress-2000, Cambridge Isotope Laboratories[94]). For insect cells, as for other animal cells, all 20 (or at least the 10 essential) amino acids have to be present in labeled form in the expression medium for uniform isotope labeling. They cannot be synthesized from ^{13}C-glucose and NH_4^+ (and D_2O), as in *E. coli*. Many of the 20 amino acids are cost-prohibitive and quantities of 100–2000 mg L^{-1} are required for maximum protein expression. Moreover, for frequently used culture media (e.g., Sf 900 II), the medium formulations are not subject to public disclosure. However, the composition of some insect media, such as IPL-41, is known.[68] Strauss *et al.*[96] have used uniformly ^{15}N-labeled Abl with recently BV-infected Sf9 cells using labeling medium BE2000-N (CIL[94]) and have obtained 91% label incorporation. The reported ^{1}H-^{15}N HSQC spectrum for the ^{15}N-labeled Abl was of good quality. The labeling protocol used by the authors is usually applicable for other recombinant proteins and extends the range of recombinant proteins suitable for NMR studies. However, the cost of the labeled amino acids can often restrict this labeling protocol to a limited number of experiments.

For mammalian cell culture, a method for the uniform isotope labeling of recombinant proteins has already been described.[97] It has not, however, been generally applied.

3.19.11 Membrane Proteins

While approximately one-third of the projects in the pharmaceutical industry are GPCRs and membrane channels, structural research studies on those proteins are rare. The root cause for this lies in low expression rates and difficulties in the purification and crystallization of membrane proteins. Different scientific consortia and biotech companies continue trying to overcome these problems, but with hitherto low success rates.

References

1. Deutscher, M. P. Guide to Protein Purification. In *Methods of Enzymology* 182; Academic Press: San Diego, 1990.
2. Harris, E. L. V.; Angal, S.; *Protein Purification Application*; IRL Press at Oxford University Press: Oxford, 1990.
3. Harris, E. L. V.; Angal, S.; *Protein Purification Methods*; IRL Press at Oxford University Press: Oxford, 1989.
4. Janson, J.-C., Ryden, L., Eds. *Protein Purification*; Wiley-VCH: Weinheim, 1998.
5. Pingoud, A.; Urbanke, C. *Arbeitsmethoden der Biochemie*; W. de Gruyter: Berlin, New York, 1997.
6. Cooper, T. G. *The Tools of Biochemistry*; Wiley: New York, 1977.
7. Bode, W.; Turk, D.; Karshikov, A. *Protein Sci.* **1992**, *1*, 426–471.
8. Nar, H.; Bauer, M.; Schmid, A.; Stassen, J.-M.; Wienen, W.; Priepke, H. W. M.; Kauffmann, I. K.; Ries, U. J.; Hauel, N. H. *Structure* **2001**, *9*, 29–37.
9. Brandstetter, H.; Bauer, M.; Huber, R.; Lollar, P.; Bode, W. *Proc. Natl. Acad. Sci. USA* **1995**, *92*, 9796–9800.
10. RCSB Protein Data Bank, Rutgers, the State University of New Jersey Department of Chemistry and Chemical Biology, 610 Taylor Road, Piscataway, NJ 08854-8087, USA. www.rcsb.org/pbd (accessed Aug 2006).
11. Invitrogen Corporation, 1600 Faraday Avenue, Carlsbad, CA 92008, USA. www.invitrogen.com (accessed Aug 2006).
12. Landy, A. *Annu. Rev. Biochem.* **1989**, *58*, 918–941.
13. Breitling, R.; Klingner, S.; Callewaert, N.; Pietrucha, R.; Geyer, A.; Ehrlich, G.; Hartung, R.; Müller, A.; Contreras, R.; Beverley, S. M. et al. *Protein Expr. Purif.* **2002**, *25*, 209–218.
14. Vincentelli, R.; Bignon, C.; Gruez, A.; Canaan, S.; Sulzenbacher, G.; Tegoni, M.; Campanacci, V.; Cambillau, C. *Acc. Chem. Res.* **2003**, *36*, 165–172.
15. Stevens, R. C. *Structure* **2000**, *8*, R177–R185.
16. Chambers, S. *DDT* **2002**, 7, 759–765.
17. Braun, P.; LaBaer, J. *Trends Biotechnol.* **2003**, *21*, 383–388.
18. Schäffner, J.; Winter, J.; Rudolph, R.; Schwarz, E. *Appl. Environ. Microbiol.* **2001**, *67*, 3994–4000.
19. Lund, P. A. *Adv. Microb. Physiol.* **2001**, *44*, 93–140.
20. Hoffmann, F.; Rinas, U. *Adv. Biochem. Eng. Biotechnol.* **2004**, *89*, 143–161.
21. Ellis, R. J.; Hartl, F. U. *Curr. Opin. Struct. Biol.* **2000**, *10*, 13–15.
22. Lee, D. H.; Kim, M. D.; Lee, W. H.; Kweon, D. H.; Seo, J. H. *Appl. Microbiol. Biotechnol.* **2004**, *63*, 549–552.
23. Jermutus, L.; Ryabova, L. A.; Plückthun, A. *Curr. Opin. Biotechnol.* **1998**, *9*, 534–548.
24. Klammt, C.; Löhr, F.; Schäfer, B.; Haase, W.; Dötsch, V.; Rüterjans, H.; Glaubitz, C.; Bernhard, F. *Eur. J. Biochem.* **2004**, *271*, 568–580.
25. Chrunyk, B. A.; Evans, J.; Lillquist, J.; Young, P.; Wetzel, R. *J. Biol. Chem.* **1993**, *268*, 18053–18061.
26. RIKEN Structural Genomics/Proteomics Initiative 230-0045 1-7-22 Suehiro, Tsurumi, Yokohama 230-0045, Japan. www.rsgi.riken.go.jp/rsgi_e/index.html (accessed Aug 2006).
27. Kim, D. M.; Kigawa, T.; Choi, C. Y.; Yokoyama, S. *Eur. J. Biochem.* **1996**, *239*, 881–886.
28. Kigawa, T.; Yabuki, T.; Yoshida, Y.; Tsutsui, M.; Ito, Y.; Shibata, T.; Yokoyama, S. *FEBS Lett.* **1999**, *442*, 15–19.

29. Yokoyama, S.; Hirota, H.; Kigawa, T.; Yabuki, T.; Shirouzu, M.; Terada, T.; Ito, Y.; Matsuo, Y.; Kuroda, Y.; Nishimura, Y. et al. *Nat. Struct. Biol.* **2000**, 7, 943–945.
30. Center for Eukaryotic Structural Genomics, University of Wisconsin-Madison, USA 53706-1549. www.uwstructuralgenomics.org (accessed Aug 2006).
31. Yokoyama, S. *Curr. Opin. Chem. Biol.* **2003**, 7, 39–43.
32. Vinarov, D. A.; Markley, J. L. *Exp. Rev. Proteomics* **2005**, *2*, 49–55.
33. Tyler, R. C.; Aceti, D. J.; Bingman, C. A.; Cornilescu, C. C.; Fox, B. G.; Frederick, R. O.; Jeon, W. B.; Lee, M. S.; Newman, C. S.; Peterson, F. C. et al. *Proteins* **2005**, *59*, 633–643.
34. Lesley, S. A.; Brow, M. A. D.; Burgess, R. R. *J. Biol. Chem.* **1991**, *266*, 2632–2638.
35. Martemyanov, K. A.; Spirin, A. S.; Gudkov, A. T. *FEBS Lett.* **1997**, *414*, 268–270.
36. Ohuchi, S.; Nakano, H.; Yamane, T. *Nucleic Acids Res.* **1998**, *26*, 4339–4346.
37. Goulding, C. W.; Perry, L. J. *J. Struct. Biol.* **2003**, *142*, 133–143.
38. Merck Biosciences Ltd., Beeston, Nottingham NG9 2JR, UK. www.emdbiosciences.com (accessed Aug 2006).
39. Stratagene, 11011 N. Torrey Pines Road, La Jolla, CA 92037, USA. www.stratagene.com/homepage (accessed Aug 2006).
40. Jacob, F.; Monod, J. *J. Mol. Biol.* **1961**, *3*, 318–356.
41. Hoffmann, F.; Rinas, U. *Adv. Biochem. Eng. Biotechnol.* **2004**, *89*, 73–92.
42. Grabinsky, A.; Mehler, M.; Drott, D. *Innovations* **2003**, *17*, 4–6.
43. Studier, F. W.; Rosenberg, A. H.; Dunn, J. J.; Dubendorff, J. W. *Methods Enzymol.* **1990**, *185*, 60–89.
44. Cregg, J. M.; Cereghino, J. L.; Shi, J.; Higgins, D. R. *Mol. Biotechnol.* **2000**, *16*, 23–52.
45. Romanos, M. A.; Scorer, C. A.; Clare, J. J. *Yeast* **1992**, *8*, 423–488.
46. O'Reilly, D. R.; Miller, L. R.; Luckow, V. A. *Baculovirus Expression Vectors: A Laboratory Manual*; Oxford University Press: Oxford, New York, 1994.
47. Durocher, Y.; Perret, S.; Kamen, A. *Nucleic Acid Res.* **2002**, *30*, 9.
48. Sambrook, J.; Fritsch, E. F.; Maniatis, T. *Molecular Cloning*; Cold Spring Harbor Laboratory Press: Cold Spring Harbor, NY, 1989.
49. Buck, K. K. S.; Subramanian, V.; Block, D. E. *Biotechnol. Prog.* **2002**, *18*, 1366–1376.
50. Sandén, A. M.; Prytz, I.; Tubulekas, I.; Förberg, C.; Le, H.; Hektor, A.; Neubauer, P.; Pragai, Z.; Harwood, C.; Ward, A. et al. *Biotechnol. Bioeng.* **2002**, *81*, 158–166.
51. Wave Biotech, LLC, 300 Franklin Square Drive, Somerset, NJ 08873, USA. www.wavebiotech.com (accessed Aug 2006).
52. Lilie, H.; Schwarz, E.; Rudolph, R. *Curr. Opin. Biotechnol.* **1998**, *6*, 497–501.
53. Rudolph, R.; Böhm, G.; Lilie, H.; Jaenicke, R. Folding Proteins. In *Protein Function – A Practical Approach*; Creighton, T. E., Ed.; Oxford University Press: Oxford, 1997.
54. Chen, G.-Q.; Gouaux, E. *Proc. Natl. Acad. Sci. USA* **1997**, *94*, 13431–13436.
55. Rudolph, R.; Lilie, H. *FASEB J.* **1996**, *10*, 49–56.
56. Armstrong, N.; de Lencastre, A.; Gouaux, E. *Protein Sci.* **1999**, *8*, 1475–1483.
57. Vincentelli, R.; Canaan, S.; Campanacci, V.; Valencia, C.; Maurin, D.; Frassinetti, F.; Scappucini-Calvo, L.; Bourne, Y.; Cambillau, C.; Bignon, C. *Protein Sci.* **2004**, *13*, 2782–2792.
58. GE Healthcare Biosciences AB, SE-751 84 Uppsala, Sweden. www.amershambiosciences.com (accessed Aug 2006).
59. Merck Biosciences Ltd., Beeston, Nottingham NG9 2JR, UK. www.merckbiosciences.co.uk (accessed Aug 2006).
60. Kapust, R. B.; Waugh, D. S. *Protein Sci.* **1999**, *8*, 1668–1674.
61. Clonetech, 1290 Terra Bella Avenue, Mountain View, CA 94043, USA. www.clontech.com/clontech (accessed Aug 2006).
62. IBA GmbH, Rudolf-Wissell-Str. 28, 37079 Göttingen, Germany. www.iba-go.com (accessed Aug 2006).
63. Wülfing, C.; Lombardero, J.; Plückthun, A. *J. Biol. Chem.* **1994**, *269*, 2895–2901.
64. QIAGEN GmbH, QIAGEN Strasse 1, 40724 Hilden, Germany. www.qiagen.com (accessed Aug 2006).
65. Brandstetter, H.; Turk, D.; Hoeffken, H. W.; Grosse, D.; Stürzebecher, J.; Martin, P. D.; Edwards, B. F. P.; Bode, W. *J. Mol. Biol.* **1992**, *226*, 1085–1099.
66. Vivascience, now: Sartorius AG, Weender Landstrasse 94 – 108, D-37075 Goettingen, Germany. www.vivascience.com (accessed Aug 2006).
67. Pierce Biotechnology, Inc. Rockford, IL 61105, USA. www.piercenet.com (accessed Aug 2006).
68. CHEMICON International, Inc., 28820 Single Oak Drive, Temecula, CA 92590, USA. www.chemicon.com (accessed Aug 2006).
69. Hampton Research, 34 Journey, Aliso Viejo, CA 92656-3317, USA. www.hamptonresearch.com (accessed Aug 2006).
70. Millipore Corporate Headquarters, 290 Concord Rd., Billerica, MA 01821, USA. www.millipore.com (accessed Aug 2006).
71. Bio-Rad Laboratories, 1000, Alfred Nobel Drive, Hercules, CA 94547, USA. www.bio-rad.com (accessed Aug 2006).
72. Pall Corp., 2200 Northern Boulevard, East Hills, NY 11548, USA. www.pall.com (accessed Aug 2006).
73a. Expasy Proteomics Web Server, Département Biologie Structurale et Bioinformatique, Centre Médical Universitaire 1, rue Michel-Servet CH-1211 Genève, Switzerland. www.expasy.org (accessed Aug 2006).
73b. Swiss Institute of Bioinformatics, Biozentrum – University of Basel, Klingelbergstrasse 50-70, CH-4056 Basel, Switzerland. www.ebi.ac.uk/swissprot (accessed Aug 2006).
74. Westermeier, R. *Electrophoresis in Practice*; VCH: Weinheim, 1993.
75. Gersten, D. M. *Gel Electrophoresis: Proteins*; John Wiley: Chichester, 1996.
76. Agilent Technologies, Inc., 395 Page Mill Rd., Palo Alto, CA 94306, USA. www.chem.agilent.com (accessed Aug 2006).
77. Caliper Life Sciences, 68 Elm Street, Hopkinton, MA 01748, USA. www.caliperLS.com (accessed Aug 2006).
78. Bradford, M. M. *Anal. Biochem.* **1976**, *72*, 248–254.
79. Lowry, O. H.; Rosebrough, N. J.; Farr, A. L.; Randall, R. J. *J. Biol. Chem.* **1951**, *193*, 265–275.
80. Bergfors, T. Dynamic Light Scattering. In *Protein Crystallization, IUL Biotechnology Series*; Bergfors, T. M., Ed.; International University Line: La Jolla, CA, 1999.
81. Beckman Coulter, Inc., 4300 N. Harbor Boulevard, Fullerton, CA 92834-3100, USA. www.beckmancoulter.com (accessed Aug 2006).
82. De Hoffman, E.; Stroobant, V. Mass Spectrometry–Principles and Applications, 2nd ed.; Wiley: New York, 2001.
83. Venyaminov, S. Y.; Yang, J. T. Determination of Protein Secondary Structure. In *Circular Dichroism and the Conformational Analysis of Biomolecules*; Fasman, G. D., Ed.; Plenum Press: New York, 1990.
84. Sreerama, N.; Woody, R. W. Circular Dichroism of Peptides and Proteins. In *Circular Dichroism*; Berova, N., Nakanishi, K., Woody, R. W., Eds.; Wiley-VCH: New York, 2000.
85. MWG-BIOTECH AG, Anzingerstr. 7a, D-85560 Ebersberg, Germany. www.mwgbiotech.com/html/all/index.php (accessed Aug 2006).

86. ROCHE Diagnostics - Applied Science, Sandhofer Strasse 116, D-68305 Mannheim, Germany. www.roche-applied-science.com (accessed Aug 2006).
87. Prozyme, 1933 Davis Street, San Leandro, CA 94577-1258, USA. www.prozyme.com/glyko/index.html (accessed Aug 2006).
88. Kigawa, T.; Yabuki, T.; Yoshida, Y.; Tsutsui, M.; Ito, Y.; Shibata, T.; Yokoyama, S. *FEBS Lett.* **1999**, *442*, 15–19.
89. Kigawa, T.; Muto, Y.; Yokoyama, S. *J. Biomol. NMR* **1995**, *6*, 129–134.
90. Yabuki, T.; Kigawa, T.; Dohmae, N.; Takio, K.; Terada, T.; Ito, Y.; Laue, E. D.; Cooper, J. A.; Kainosho, M.; Yokoyama, S. *J. Biomol. NMR* **1998**, *11*, 295–306.
91. Noren, C. J.; Anthony-Cahill, S. J.; Griffith, M. C.; Schultz, P. G. *Science* **1989**, *244*, 182–188.
92. Weiss, S. A.; Smith, G. C.; Kalter, S. S.; Vaughn, J. L. *In Vitro* **1981**, *17*, 495–502.
93. Silantes GmbH, Gollierstr. 70 c, D-80339 München, Germany. www.silantes.com (accessed Aug 2006).
94. Cambridge Isotope Laboratories, 50 Frontage Road, Andover, MA 01810-5413, USA. www.isotope.com (accessed Aug 2006).
95. Spectra Stable Isotopes, 9108A Guilford Rd., Columbia, MD 21046, USA. www.SpectraStableIsotopes.com (accessed Aug 2006).
96. Strauss, A.; Bitsch, F.; Fendrich, G.; Graff, P.; Knecht, R.; Meyhack, B.; Jahncke, W. *J. Biomol. NMR* **2005**, *31*, 343–349.
97. Hansen, A. P.; Petros, A. M.; Mazar, A. P.; Pederson, T. M.; Rueter, A.; Fesik, S. W. *Biochemistry* **1992**, *51*, 12713–12718.

Biographies

Margit M T Bauer studied chemistry and biology at the University of Regensburg. From 1988 to 1991 she was working on her PhD at the Max-Planck-Institute for Biochemistry, Martinsried, in Prof Robert Huber's group, on protein purification and crystallization for x-ray studies. From 1992 to 1994 her postdoctoral research was carried out at the Max-Planck-Institute for Biochemistry, Martinsried, in the group of Prof Wolfram Bode. Her work was focused on isolation and x-ray studies of blood coagulation factors. Since 1995 she has been employed at Boehringer Ingelheim Pharma, in the Department for Integrated Lead Discovery, working in the Structural Research Group of Dr Herbert Nar.

Gisela Schnapp studied chemistry at the University of Würzburg from 1985 to 1991. Her PhD on rDNA transcription in the mouse was completed from 1991 to 1994 at the German Cancer Research Center, Heidelberg, in the group of Prof Ingrid Grummt. In 1995 she carried out postdoctoral research on expression and functional analysis of bone morphogenetic factors at the University of Ulm, in the group of Prof Knöchel. From 1995 to 1997 she worked for Dr

Rettig at Boehringer Ingelheim Pharma, in the Department of Oncology Research, on purification and functional analysis of telomerase. From 1997 to 2002 Dr Schnapp was at the Department for Oncology Research, Boehringer Ingelheim Pharma, and transferred to the Department for Integrated Lead Discovery, Structural Research Group, under Dr Herbert Nar, in 2002.

Comprehensive Medicinal Chemistry II
ISBN (set): 0-08-044513-6

ISBN (Volume 3) 0-08-044516-0; pp. 411–432

3.20 Protein Crystallization

G E Schulz, Albert-Ludwigs-Universität, Freiburg im Breisgau, Germany

3.20.1	**History of Protein Crystals**	**433**
3.20.2	**Crystallization of Native Proteins**	**434**
3.20.2.1	Soluble Proteins	434
3.20.2.2	Integral Membrane Proteins	434
3.20.2.3	Monotopic Membrane Proteins	435
3.20.2.4	Mobile Parts and Associations	436
3.20.3	**Crystallization of Modified Proteins**	**436**
3.20.3.1	Heterogeneous Poisoning	436
3.20.3.2	Crystal Engineering	438
3.20.4	**Crystallization Procedures**	**438**
3.20.4.1	Theory	438
3.20.4.2	Conditions	438
3.20.4.3	Apparatus	439
3.20.4.4	Automation	440
3.20.5	**Properties and Modifications of Crystals**	**440**
3.20.5.1	Size and Shape	440
3.20.5.2	Atomic Order	441
3.20.5.3	Crystal Soaking	442
3.20.5.4	Cocrystallization and Protein Labeling	442
3.20.6	**Crystal Statistics**	**443**
3.20.6.1	Space Groups and Packing Densities	443
3.20.6.2	Packing Arrangements	443
3.20.7	**Conclusion**	**445**
References		**445**

3.20.1 History of Protein Crystals

Biochemical studies at the molecular level were first carried out in the mid 19th century. Blood was the first material to be studied in this way because it was readily available. In 1864, Hoppe-Seyler from Tübingen reported that he had purified and crystallized hemoglobin from dogs and some rodents using ethanol as precipitant.[1] The crystals Hoppe-Seyler produced shrinked appreciably on drying in a vacuum and contained a globin (he called this hemoglobin) so he concluded that these were protein crystals. This observation was made a long time before its importance was appreciated. Sixty-two years later protein crystals became important when Sumner demonstrated that the enzyme urease was a protein and could be purified to form crystals.[2] This observation contradicted Willstätter's concept of enzymes being less well-defined colloids that occurred at very low concentrations in cells, which prevailed at that time. A couple of years later, the enzymes pepsin, trypsin, chymotrypsin, lysozyme, and Warburg's yellow enzyme were also crystallized, thus overriding the colloid concept completely.

In 1934 Bernal and Crowfoot observed an x-ray reflection pattern from a pepsin crystal that had been kept in a closed capillary to prevent it from drying out.[3] This pattern demonstrated that pepsin possesses a unique structure, which, in principle, could be determined. Moreover, it showed that dried protein crystals became disordered and lost their reflection pattern. The complexity of this pattern, however, prevented any advances in solving the crystal structure because the available methods sufficed only for much smaller unit cells. This impediment was removed just

20 years later in 1954 when the Perutz group[4] used the isomorphous heavy atom replacement method to phase a hemoglobin crystal diffraction pattern. This method had already been used by Cork for alums,[5] by Robertson and Woodward for phthalocyanines,[6] and by the Bijvoet group[7] for strychnine. In 1960, this method was used to yield the first protein structure after Kendrew drove the myoglobin analysis project to completion.[8] The structures of numerous natural proteins have now been elucidated,[9] and it is thought that only a few types of polypeptide chainfold remain unknown.[10]

3.20.2 Crystallization of Native Proteins

3.20.2.1 Soluble Proteins

The first protein to be structurally elucidated was myoglobin from whale.[8] Myoglobin is a small (18 kDa) protein for storing dioxygen that occurs at very high concentrations in the muscles of diving mammals such as whales. Myoglobin is a rather rigid protein, which is suitable for a protein that merely stores a gas. Since it was available in large quantities, the purification was easy as it was possible to cut very narrow bands in every chromatography step. After myoglobin and hemoglobin, the next structures to be elucidated were those of extracellular enzymes such as lysozyme,[11] ribonuclease, and chymotrypsin, which are hydrolases for saccharides, nucleic acids, and peptides, respectively. Hydrolysis does not involve any great movement within the enzymes because material is degraded and nothing needs to be shielded. Moreover, these enzymes work outside the cell and need to be resistant to attack by all kinds of agents. Consequently, they are as rigid as myoglobin. Therefore, it can be concluded from these first structural analyses that crystallization and handling of a rigid protein is easier than that of proteins with mobile parts.

The reason for this observation is a minimization of the entropy reduction required for crystallization. A rigid protein crystallizes from a solution because this process increases the total entropy of the crystallization drop: while the protein reduces its entropy by crystal formation, the solvent entropy increases after protein removal and thus compensates for the loss. If a mobile protein has to adopt a defined conformation in a crystal, however, the required entropy reduction becomes excessive and can no longer be compensated by an entropy increase in the solvent: As a consequence, the mobile protein fails to crystallize. Still, numerous crystals of proteins are known that possess mobile parts that are not involved in packing contacts and therefore may remain mobile in the crystal so that they do not require entropy reduction. The rare cases where mobile parts are rigidified in contacts can be recognized by the fact that a given protein adopts different conformations in different packing contacts. An analysis of such differences often leads to very interesting data on protein deformations, for example, induced-fit motions of an enzyme during catalysis. However, it should be kept in mind that numerous proteins are completely mobile for functional reasons[12,13] and are unlikely to crystallize on their own. It is likely that they adopt defined structures only in complexes.[14]

3.20.2.2 Integral Membrane Proteins

In simple terms, cells consist of a cytosol containing water-soluble proteins surrounded by a membrane filled with integral membrane proteins. Attempts to grow 2-dimensional crystals of membrane proteins within the membrane[15] and to analyze them by electron diffraction using an electron microscope have been largely abandoned because of the limited rate of success. As an alternative, membrane proteins were crystallized in cubic and other lipid phases but this yielded only small crystals that were difficult to find and isolate from the highly viscous medium.[16,17] Presumably, the viscosity reduces the diffusion rate too much for obtaining large crystals. In other experiments, the membrane proteins were removed from their natural environment by covering their nonpolar surfaces with detergents and then dissolving them in water.[18] Such a soft micellar environment requires that the membrane protein is intrinsically stable so that it does not disintegrate when the natural pressure of the surrounding membrane is removed.

Looking back at the first membrane protein crystals solved by x-ray analysis helps to explain the situation. The first structure came from a bacterial photoreaction center,[19] which only channels electrons and no massive particles and therefore requires a solid protein scaffold. The second structure to be solved was that of a porin from the outer membrane of bacteria, which has to be intrinsically stable in order to endure all kinds of attack.[20] The subsequently established membrane protein structures were from porin-like proteins.[21] Structures of regulated ion channels and pumps that have to perform motions were established much later.[22,23] Usually, these mobile membrane proteins have to be rigidified by a bound inhibitor before they crystallize.[24] In summary, membrane protein crystallization follows the same rigidity rules as soluble proteins but it is generally more difficult to arrest a membrane protein in a rigid state than a soluble protein.

3.20.2.3 Monotopic Membrane Proteins

Monotopic membrane proteins constitute a third category of proteins besides cytosolic and integral membrane proteins. Among them are enzymes that work on nonpolar substrates dissolved in the membrane such as prostaglandins, terpenoids, and carotenoids. In order to accept their substrate, these enzymes generally use a nonpolar patch on their surface to dip halfway into the membrane.[25–27] In the cytosol, these enzymes may associate with each other by burying their nonpolar patches in an oligomer, which is usually not well defined. This type of contact is stabilized by water exclusion rather than by the weak direct nonpolar interactions and is thus poorly defined. Consequently, the oligomer lacks a uniform structure and therefore does not crystallize. These monotopic membrane proteins are usually only crystallizable in the presence of detergents that cover the nonpolar patches so that the proteins can associate in a defined manner with contacts between the polar surfaces.

Interestingly, the nonpolar patches are usually close together in the crystals and most likely covered by a common detergent micelle.[28] Limiting the number of micelles in a crystal of course diminishes the required entropy reduction on crystallization and thus furthers crystal growth. The described effects were particularly evident in the crystallization of a monomeric carotenoid-processing enzyme.[27] During preparation this enzyme behaved like a soluble protein that even appeared as a reasonable monomer band in the size-exclusion chromatography (**Figure 1**) but did not crystallize. Crystals grew only after applying a detergent, which changed the size-exclusion run drastically because all protein molecules went into micelles (**Figure 1**). Consequently, the monomer band in the size-exclusion run did not mean that the protein was monomeric in solution. Rather, it is in an association/dissociation equilibrium with conversion rates fast enough to result in a single peak during chromatography, but too slow to yield crystals.

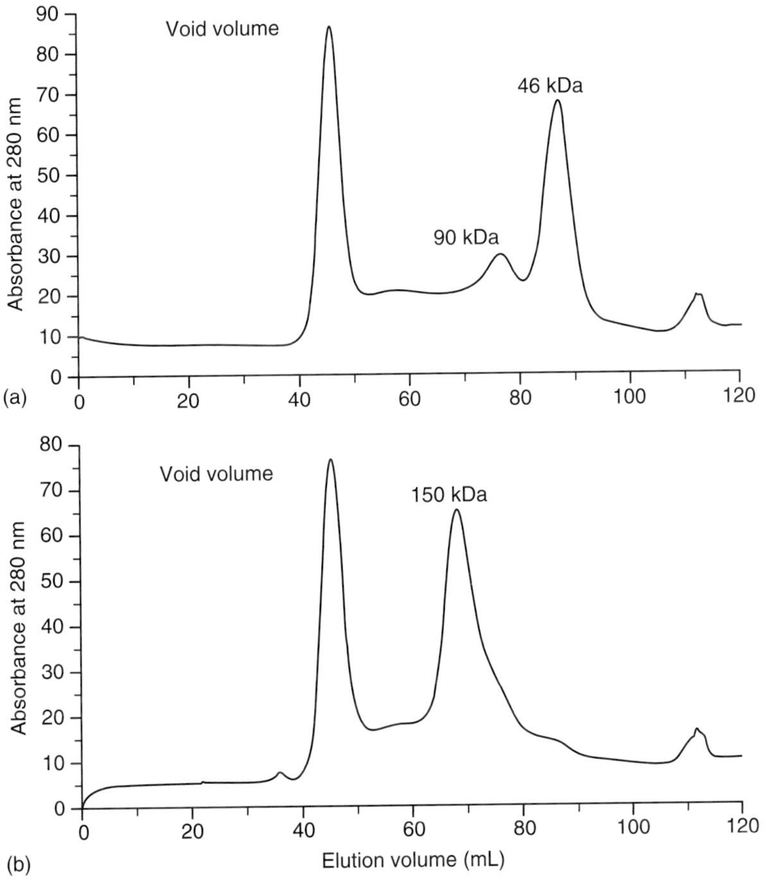

Figure 1 Size-exclusion chromatography with an apocarotenoid-15,15′-oxygenase using a Superdex 200 column.[27] (a) Run without detergent under normal protein preparation conditions (50 mM Tris-HCl at pH 8.0, 200 mM NaCl, 1 mM EDTA, 2 mM DTT). (b) Run under the same conditions but with 0.25% C_8E_{4-8} (octylpolyoxyethylene). In SDS-PAGE, all fractions below 100 mL contained exclusively the prepared enzyme with a polypeptide mass of 54 kDa. The run without detergent shows monomer (46 kDa), dimer (90 kDa), and aggregates (void). The run with detergent shows micelles (150 kDa) and aggregates (void). (Reproduced with permission from Science – Supplementary Material.)

3.20.2.4 Mobile Parts and Associations

Experience has indicated that mobile parts hinder crystallization, which is backed up by the physics. It is therefore advisable to check a given protein for mobile parts. This could happen by accident as occurred with the elongation factor Tu, which only yielded suitable crystals after it was degraded by a mold in the crystallization setup. This degradation was then repeated with a defined trypsin digestion.[29] Elongation factor Tu undergoes large motions when fulfilling its function and the proteolytic cut essentially killed the protein, allowing the resulting 'dead rock' to form crystals. Such solid remainders can often be identified by limited proteolytic digestion of a protein.[30] It is now common practice to isolate single, rigid domains from mobile multidomain eukaryotic proteins and crystallize them separately.[31]

Mobile parts of a protein can also be detected by ^{1}H-nuclear magnetic resonance (NMR) where they show a strong amide signal and no signal beyond the amide signal at low field.[32] Such amide shifts are caused by chains in a rigid environment. The same applies for signals beyond the methyl signal at high field. ^{1}H-NMR may therefore help to guide attempts to remove mobile parts. Furthermore, mobile parts can be recognized by hydrogen-deuterium exchange experiments. The amide protons have a pK_a value of around 12 and are rather quickly exchanged at pH 8–9. In this pH range the exchange rate follows the amide availability from the solvent, which is high in mobile parts but low in solid parts. In these experiments, the exchange is stopped by lowering the pH to 2, and the amount of exchange is measured either by pepsin digestion and mass spectroscopy or directly by NMR.[33]

Crystallization is not only hindered by partial mobility but also by undefined protein association. During evolution, proteins were developed to avoid undefined associations in the cytosol as this could compromise their function. The most famous example of a pathogenic association of rigid proteins is sickle cell hemoglobin. This association was enforced during evolution by the fight of the organism against a parasite.[34] A further example of medical importance is the association of denatured proteins in amyloid fibers.[35] To achieve solubility, proteins tend to carry mobile charged side chains such as lysines and glutamates that protrude into the solvent and carry hydration shells. Both mobility and hydration shell hinder association.

Association does not always mean that large visible precipitates are formed. Equally bad are undefined oligomeric aggregates. These can be detected in dynamic light scattering experiments because they show a distribution of diminished diffusion coefficients.[36] A specialized apparatus (ProteinSolutions, Charlotteville/VA, USA) provides data for a dissolved protein within a reasonable period of time and has become a popular means of checking protein preparation for monodispersity. The beauty of this method, however, does not warrant its actual application because the crystallization setups contain commercial screens with numerous given conditions, each of which needs to be tested by dynamic light scattering. This is very time-consuming and wastes precious protein in support analysis. Consequently, this method is only used in extreme cases where crystals are not found in any screen.

3.20.3 Crystallization of Modified Proteins

3.20.3.1 Heterogeneous Poisoning

It is common knowledge that proteins have to be highly purified before they can be crystallized. Impurities hinder growth because they may attach to growing crystal faces and prevent further protein molecules from binding. A more serious situation arises, however, when a part of the protein surface is modified. As shown in **Figure 2**, such a protein will bind perfectly with its intact surface to a growing crystal face, but will offer the impaired part of its surface for the association of further molecules. While impurities do not bind well and therefore need to be present at high concentrations to stop a growing crystal face, the partially modified proteins bind well and therefore poison a growing face at low concentrations. Consequently, the conventional purity of a protein preparation is generally less important than the structural homogeneity of the protein itself.

It is very difficult to prove that a protein preparation is homogeneous. In some cases, such as incomplete posttranslational modifications (e.g., phosphorylations or glycosylations), the heterogeneity can be recognized by additional lines in the mass spectrum of the protein. The more serious cases, however, such as incomplete disulfide bond formations, asparagine and glutamine deaminations, and conformational changes, are more difficult to detect, the latter in particular. In order to avoid such mishaps, the protein should be produced in large amounts so that only narrow bands around the major peaks of the chromatographic runs have to be collected. All preparation steps should be under mild conditions and the total time required for the preparation should not exceed a couple of days. Longer purification campaigns tend to increase the conformational heterogeneity resulting in the problems described above.

The problems associated with long preparation times became obvious around 1980. Only a few, very abundant proteins would crystallize because they could be purified in sufficient amounts within a short time period. However, most of the proteins of interest occur only in small amounts in native material; purification procedures for these took

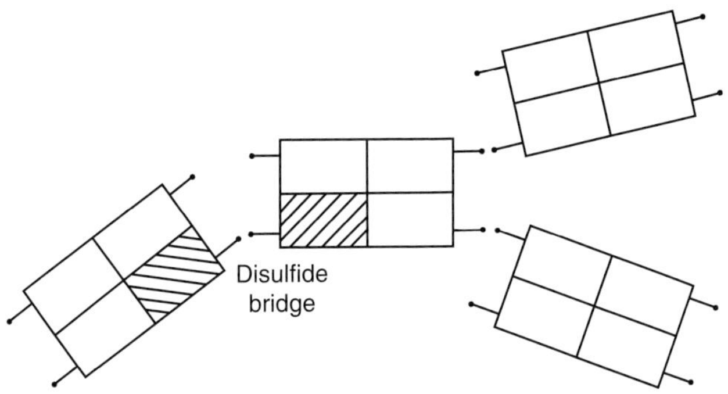

Figure 2 Sketch of a tetrameric protein with one exposed cysteine per protomer. In the case of histidine ammonium-lyase[39] the protein failed to crystallize in hanging drops despite the presence of reducing agents in both drop and reservoir. Only the exchange of one particular cysteine for serine resulted in suitable crystals. As indicated, a solvent-accessible cysteine in an oligomer gives rise to undefined aggregates. In contrast, a single solvent-accessible cysteine in a monomer (marked by striation) results in dimerization. Even very small amounts of such dimers function as poisons because as soon as a growing crystal face incorporates half of these dimers, the next layer cannot attach because the other half is fixed in an undefined manner.

more than a month to complete, which was too long to retain conformational homogeneity. Fortunately, the situation changed drastically with the emergence of recombinant proteins, which opened up possibilities for even the least abundant proteins. Without this development, x-ray structure analysis would probably have come to an end only a quarter of a century after determination of the first structure.

Conformational heterogeneity preventing crystallization is usually present in fusion proteins, for example, in cases where a protein of interest is expressed as the C-terminal extension of a glutathione *S*-transferase (GST). Such a fusion increases the chances of expressing a eukaryotic protein in a bacterium and also facilitates the purification procedure. Since only very few cases are known in which a fusion protein yielded suitable crystals, it is advisable to remove the helping protein by a specific protease after production and purification. For this purpose, the target sequence of the applied protease is expressed in the linker between GST and the protein of interest. The remaining part of the linker usually contains only a few unstructured residues that rarely disturb crystallization. A similar situation arises with proteins carrying an N- or C-terminal His_6-tag, which is most helpful for purification. Such tags are unpopular because they tend to hinder crystallization. However, many cases are known where His-tagged proteins yielded suitable crystals. Some fusion proteins involving the maltose-binding protein (MBP) as the N-terminal part were engineered for an extremely short linker. This resulted in a couple of rigid fusion proteins yielding suitable crystals.[37]

A further well-recognized heterogeneity arises in proteins that contain noncovalently bound prosthetic groups or cofactors. Care has to be taken that the binding places are fully occupied to avoid a mixture of *holo* and *apo* protein molecules with slightly different conformations, because such a mixture may cause crystal growth poisoning. Usually, precursors of these cofactors are added to the culture medium and the incorporation is measured photometrically, for instance, by following the A_{450}:A_{280} absorption ratio in an FAD-binding protein that relates the FAD contents (A_{450}) to the fixed number of tryptophans and tyrosines (A_{280}) in the polypeptide.

The cytosol is a reducing medium. In most cells the redox potential is regulated by the equilibrium between reduced and oxidized glutathione. Therefore, most cytosolic proteins contain cysteines in their interior and on their surfaces without risking disulfide formation. However, surface cysteines of cytosolic proteins turned out to cause persistent crystallization problems even when the crystallization reservoir was full of reducing agents such as beta-mercaptoethanol. First, such agents may themselves add to the thiols and cause surface heterogeneity. Second, the crystallization setups are usually not gas-tight and have to endure weeks of oxygen intrusion before the crystals are grown. This leads to irreproducible redox conditions causing erratic crystal growth. From our experience, it is advisable to exchange exposed cysteines for serines at the beginning,[38,39] because even minute amounts of disulfide dimers can poison crystal growth (**Figure 2**).

A single exposed thiol in a monomeric protein may cause disulfide dimer formation. Larger aggregates occur with a single exposed cysteine in an oligomeric protein or with more than one exposed cysteine in monomeric proteins. Such disulfide bridges can be detected by comparing SDS-PAGE runs performed in oxidized and reduced environments. The cysteines concerned can usually be identified in a given amino acid sequence by predicting the solvent exposure of the residues.[40] The respective programs exhibit a rather high rate of success. Each Cys→Ser mutant based on such a prediction must then be checked by appropriate SDS-PAGE runs. Thus, disulfide heterogeneity can often be circumvented without investing too much effort.

3.20.3.2 Crystal Engineering

Many proteins do not crystallize at all, while others yield unsuitable crystals. In both cases, it is advisable to mutate surface residues thus providing novel packing contact opportunities.[41,42] Since only a limited number of residues are exposed on the surface, these should be identified beforehand. In many recent cases, the general folding pattern of a given protein was derived from structurally established homologs that were recognized as such from the amino acid sequences. If this is not the case, one may look at the sequence variability of closely related homologs with more than about 70% sequence identity, because the exposed regions are usually revealed as the regions of the highest variations.[43] Completely new proteins without detectable homologs should be run through surface prediction programs[40] and mutated accordingly. In some cases, such as the beta-barrel outer membrane proteins OmpA and OmpX from *Escherichia coli*, the solvent-exposed regions were identified by biochemical studies.[41,44]

After locating a surface position, the question arises as to which mutations should be applied. It seems clear that long flexible side chains should be avoided, in particular if they are highly hydrated like lysine or glutamate. Arginine is much more rigid than lysines and is preferable provided a positive charge is suitable.[45] Tyrosines are a good choice[44] because they are comparatively rigid, amphiphilic, and most abundant in antibody-binding sites where presumably they function like viscous glue. In general, it seems advisable to apply changes that move the isoelectric point toward pH 7.

In some cases the obtained crystals diffract merely to medium resolution, whereas higher resolution is required for analytical results of sufficient significance. If the packing pattern is complicated and contains more than one molecule per asymmetric unit, it would be advisable to start a general mutational attempt with new screenings of the crystallization conditions in order to find another crystal form.[38] With only one asymmetric molecule, however, the actual crystal contacts should be analyzed for possible mutational improvements and these should be introduced. In such a case, crystals can usually be reproduced under conditions close to the original ones.

It is also conceivable that the crystal order will improve after the addition of further packing contacts. An attempt with a 100 kDa enzyme, however, did not improve the order but increased the crystallization speed by a factor of forty.[46] Most likely it facilitated nucleation and growth by increasing the binding strength of a contact but failed to improve the precision of the contacts. Unfortunately, every mutation requires a new preparation and in most cases also a new screening so that crystal engineering remains a field for patient researchers with endurance.

3.20.4 Crystallization Procedures

3.20.4.1 Theory

Proteins crystallize when water is continuously removed from a protein solution so that the remaining water becomes less mobile. If the proteins manage to associate in an orderly manner in a crystal adopting a state of lower entropy, the protein-embedding water is released and increases the entropy of the solvent. Crystal growth will continue as long as the net entropy change is positive. Crystal nucleation requires a certain degree of supersaturation, which increases the driving force for splitting the solution into a low and a high entropy region. Nucleation is a highly improbable event because a number of molecules have to associate in a productive manner.[47,48] Since only a small number of large crystals are required for an x-ray analysis, a few nuclei are sufficient. Therefore, the supersaturation should be limited to allow long periods for the nuclei to develop. As soon as the first nucleus is formed and starts to grow, it depletes the solution, which, fortunately, tends to inhibit further nucleation.

It seems clear that water removal should be slow and should stop at a supersaturation level at which nuclei can form within days. In general, water removal from the protein is slowed by adding a precipitant so that the water is not actually removed but displaced from the protein to the precipitant.[49] Convenient precipitants are salts, in particular ammonium sulfate, or small molecular mass alcohols such as methylpentanediol that have the same vapor pressure as water, or inert polymers such as polyethylene glycol that always remain in a state of high entropy.

3.20.4.2 Conditions

Protein crystallization involves the adjustment of numerous parameters giving rise to a multitude of possible conditions. For membrane proteins, detergent has to be added, which increases the number of important parameters even more.[50] Each experiment under any given condition requires a certain mixture, the production of which is tedious. No wonder that the provision of these mixtures has now become a profitable commercial enterprise. They are offered in sets of around 100 conditions. For the crystallization of a normal protein about half a dozen of these screening sets are usually tried. The screens reflect general experience gathered over many years of protein crystallization.[51] However, they cannot cover every conceivable condition but instead sample the parameter space rather coarsely. If a crystal is found under a given condition, this condition has to be refined by performing fine-tuned sampling around it.

The applied protein concentrations are rarely below 1 mg mL^{-1} or above 10 mg mL^{-1}. The pH ranges between 4 and 10 and is usually, but not always, fixed by a buffer. The most successful precipitants are polyethylene glycols with molecular masses of between 400 and 8000 that are chains of 9–180 oxyethylenes, at concentrations of 2–30%. Common detergents are compounds similar to octylglucoside and dodecylmaltoside as well as compounds similar to octyltetraoxyethylene and dodecyltetraoxyethylene. They are usually used at concentrations slightly above the critical micelle concentration (CMC). Detergents with higher polarity such as lauryldimethylaminoxide (LDAO) or even charged detergents such as dodecylsulfate turned out to be unsuccessful precipitants.[50]

Ammonium sulfate is a popular salt that allows high ionic strength and fits the water structure as it is located at the corresponding end of the Hofmeister series. It is used at concentrations of 0.5–4.0 M. Other suitable salts are phosphates, chlorides, citrates, malonates, acetates, and formates used in the 0.1–3.0 M range. Conditions with divalent cations such as Ca^{2+} or with chaotropic cations such as Li^{+} at the other end of the Hofmeister series are frequent. Many conditions include small organic compounds as precipitates. The most popular ones are methylpentanediol (MPD), iso-propanol, ethanol, butanediol, pentanediol, hexanediol, ethylene glycol, and glycerol. The applied concentration range is 10–60%. The rather nonpolar organic solvents, however, tend to denature proteins.

3.20.4.3 Apparatus

Initially, scientists set up supersaturated protein solutions in a batch and waited for crystal formation. These batch crystallizations were usually difficult to follow under the microscope; they required a lot of protein and they failed if the adjusted supersaturation was not suitable. Therefore, these batches were superseded by the hanging drop method, where a drop of a protein solution near the saturation level hangs from a glass plate and is dried by vapor diffusion over a reservoir adjusted to a given supersaturation level (**Figure 3a**). The equilibrium between drop and reservoir is reached after a couple of days,[46] which means that the protein solution is driven into the supersaturation region rather slowly but continuously, so that enough time is available for nucleation. Moreover, the drops can be easily followed under the microscope. In general, the glass plates are chemically modified to exhibit a nonpolar surface diminishing the number of nucleation points. Unfortunately, setting up of hanging drops is very labor intensive, and becomes tiresome for screens conducted under several hundred conditions. This major disadvantage led to screens being developed where the drops sit on a bridge over a reservoir, which is produced from a polymer with an inert surface (**Figure 3b**). A simple cover is used to seal, for example, 96 reservoirs in a single manual operation. These 96-well plates can be loaded by applying eight drops at a time using special pipettes. Development within the drops can be monitored under a microscope. In summary, screens are now performed in sitting drops with limited handling requirements, whereas the more time-consuming hanging drops are restricted to the crystallization refinements around a suitable condition detected in a screen.

The size of drop used varies appreciably. A 1 μL drop of a 10 mg mL^{-1} protein solution contains 10 μg protein and potentially forms an 8 nL crystal, or a (200 μm)^3 cube. This is far above the limit needed for x-ray analysis. Today, the lower limit for x-ray grade crystals is about 20 pL or 100 × 20 × 10 μm^3. This limit is reached with a 25 nL drop of a 1 mg mL^{-1} protein solution. Since the screens need not yield x-ray grade crystals in the first place, the drop size can be

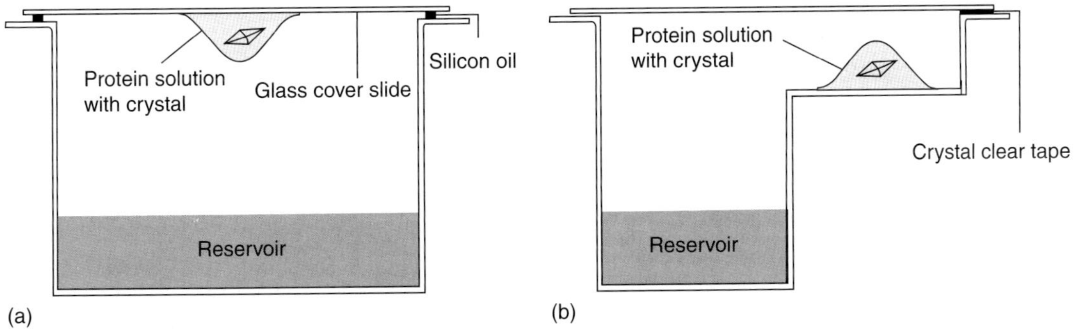

Figure 3 Crystallization setups using small drops. (a) A drop hangs from a glass plate and is dried for a couple of days to meet the precipitant concentration of the reservoir. (b) The drop is sitting on a pedestal instead of hanging. Sitting drop arrangements can be produced much more easily than hanging drops and are generally used in crystallization screenings.

reduced to the lower nanoliter range to save protein. The handling of such small drops, however, is very difficult. At the other end of the scale, the large crystals needed for neutron diffraction analyses require drops of 20–100 μL just to supply the protein. To grow large crystals it is advisable to replenish the protein in the drop or to use macroseeding, which is the placing of a freshly grown crystal into a fresh protein solution.

Apart from hanging and sitting drops numerous other setups have been recommended and used. Among them is a novel mini-batch procedure that has proved to be very successful in one laboratory.[52] Over the years numerous arrangements involving diffusion through a gel have been proposed[53]; however, the setup for this is rather tedious. Recently, the use of a thin agarose gel to increase the viscosity of a drop was proposed.[54] Increasing the viscosity using glycerol has been practiced for some time, and has removed a serious twinning problem in at least one instance.[55]

All the apparatus described so far works under normal gravity conditions. Crystallization at high gravity in a centrifuge has been tried[56] but was abandoned because it involved too much work. In contrast, crystallization at low gravity in a spacecraft became extremely popular in the years following publication of a study in 1984.[57] However, crystallization in space was certainly no breakthrough. The lack of gravitational force removes the convection around the growing crystals so that the added protein has to be supplied through the slow diffusion process, which limits the growth rate. Crystallization at low gravity has never been proven to be superior or inferior to crystallization under normal gravitational conditions. It is no surprise that this expensive method was abandoned after the national subsidies declined.

3.20.4.4 Automation

The mechanics of mini-batch crystallization and sitting drops are simple and therefore readily automated. Commercial equipment is available for this.[58] Crystallization robots are most useful where enough protein is available to run them continuously. Intermittent use tends to create problems because solutions run dry, etc., so that automation can be less efficient than if done manually. Robots are particularly useful if they are run with nanoliter drops that are difficult to handle. These small drops are necessary when only small amounts of protein can be produced, which is very often the case for human proteins of pharmaceutical interest.

Setting up the crystallization wells constitutes less than half the required workload. The remainder has to be invested in repeated visual checking of the wells. These checks have as yet not been fully automated. It is now common practice to photograph the wells automatically and to store the pictures digitally. In order to spot crystals, however, the operator has to look through the pictures on a monitor. In principle, the visual detection of crystals could be done with pattern recognition programs. However, at present the failure rate of such programs is high, largely due to the fact that they need to detect crystals covered with precipitate. Further progress in this respect is to be expected in the future.

Besides crystallization, current structural genomics projects invest heavily in the automated production of a large number of proteins. In particular, protein expression and solubility problems have to be solved. This is performed by engineering[59] and directed-evolution approaches.[60,61]

3.20.5 Properties and Modifications of Crystals

3.20.5.1 Size and Shape

The intensity of an x-ray reflection is proportional to the number of unit cells, which is the crystal volume divided by the volume of the unit cell. The accuracy of an x-ray intensity determination depends on the number of scattered and detected quanta according to the Poisson law: for instance 10 000 detected quanta are required to reduce this basic error to 1%. The real error is larger, mostly because the background radiation has to be subtracted but also because of the commonly required digital-to-analog conversion in modern detectors. Since the ratio between signal and background improves with the volume, a lower limit for crystal sizes exists. This limit depends on the experimental setup. Minimum values at modern synchrotrons approach 20 pL or $(25\,\mu m)^3$ for unit cells around $(100\,Å)^3$. At the other end of the scale, large crystals cause appreciable errors because the x-ray absorption varies for each reflection. The absorption effect can be alleviated by using a well-penetrating radiation with about 1 Å wavelength, which is available at a synchrotron. Absorption becomes an important issue, however, if a long wavelength around 2.0 Å is required for collecting data near a given atomic absorption edge, for instance, that of iron and sulfur.

The situation is completely different for neutron diffraction studies, which are most useful for locating hydrogen atoms that are of central importance for explaining enzyme mechanisms. The neutron sources are still so weak that

Figure 4 Crystals of phenylalanine ammonia-lyase from parsley with a molecular mass of 4 × 78 kDa.[62] These feather-like entities are single crystals that were used to solve the structure at 1.7 Å resolution. After 2–3 days the crystals reached their maximum size (about 500 μm in length). The crystal solvent content was as low as 44%.

protein crystals of about 100 nL or $(500\,\mu m)^3$ are required for an analysis. Growing such large single crystals is often a prohibitive task calling for new experimental approaches.

Apart from its size, the shape of a crystal matters if one or two dimensions are very short giving rise to plates or needles. Thin plates tend to break into two or more pieces, the diffraction patterns of which may no longer be separable from each other. The length of needles can easily exceed the cross-section of the x-ray beam so that the crystal is no longer bathed in the beam, which may cause diffraction intensity errors. Therefore, globular crystals are preferable if they can be obtained. A particular case of an eccentric shape was observed with the phenylalanine ammonia-lyase from parsley, which formed feather-like single crystals diffracting x-rays to 1.7 Å resolution.[62] An example of these attractive crystals is shown in **Figure 4**.

3.20.5.2 Atomic Order

A crystal contains an array of unit cells that cause interference when scattering x-rays. The interference focuses the generally diffuse scatter to localized reflections of high intensity. This focusing effect depends very much on the atomic order. As an example, a 0.5 Å missetting of a given carbon atom within the framework of a 100 Å unit cell decreases its contribution to the intensity of a reflection at 2 Å resolution by a factor of 10. To avoid such missettings, the protein has to be a solid block because a 0.5 Å displacement is already reached in a 1% compression (or elongation) of the protein in one direction. Furthermore, the crystal packing contacts have to form with the same accuracy, which requires a solid protein surface. Accordingly, the quality of the protein material determines the ordering in the crystals and consequently the resolution obtained. Resolutions of 2 Å are now common for protein crystals. In contrast, crystals of globular ribonucleic acid (RNA) molecules like transfer-RNA, ribozymes, or ribosomes usually reach resolutions no better than 3 Å. This demonstrates that the protein material is generally more rigid and better defined than RNA material: it is not surprising that the original 'RNA world' was superseded by a better defined and thus more efficient 'protein world.'

On a larger scale, several types of crystalline disorders are known. The simplest situation is a poly- or oligocrystal, which can in general be visually detected. Even-shaped oligocrystals can be detected under crossed polarization filters in a microscope, because the birefringence may not be uniform over the whole crystal. This check is not possible for the rare protein crystals belonging to the cubic crystal system that show no birefringence at all.[63] Owing to the difference between the visual wavelengths around 5000 Å and x-rays at 1 Å, a birefringent crystal need not show any x-ray diffraction at all, because it may be well ordered at the dimensions of the wavelength of light but disordered at the atomic level. Curiously enough, however, cases exist where crystals show a symmetry violation in the birefringence pattern that is not apparent in the x-ray pattern. This disorder was presumably caused by small but regular domain displacements.[64]

Many crystals show anisotropic x-ray diffraction patterns, for instance, they reach 2 Å resolution along the *a*- and *b*-axes but only 4 Å resolution along the *c*-axis. This means that the ordering depends on the direction indicating that either the order within the protein is anisotropic or, more likely, the crystal packing contacts in one or two directions are less well defined than in others. Such anisotropy gives rise to a protein structure model that is inaccurate in the direction of the low-resolution diffraction pattern.

Crystal agglomerates formed by a small number of single crystals may appear visually as single crystals because the constituents are attached via slightly different but almost equivalent faces. This effect is called twinning although it is not restricted to twins but may also occur with triplets or even with quadruplets.[65] If this twinning is coarse-grained, it may be recognizable in the crystal shape or by birefringence under the polarizing microscope and avoided by splitting the crystal and using only one of the resulting parts. Finer-grained twinnings can often be detected in the geometry of the diffraction pattern. However, so-called merohedral fine-grained twinnings do exist and they do not change the geometry of the diffraction pattern but only the intensity distribution within this pattern. Such twinnings may only be detected rather late in a structure analysis. If detected and properly dealt with, merohedrally twinned crystals may still yield high-quality structures.[38]

3.20.5.3 Crystal Soaking

The phase problem arising in a structure analysis of protein crystals was initially solved by soaking the crystals in solutions of heavy atom compounds.[4,8,11] This is possible because the crystals contain about 50% solvent in interconnected channels so that molecules with masses below about 1000 Da tend to reach every available protein surface patch. Heavy atoms sitting at defined places on the proteins can be located in so-called difference-Patterson maps and subsequently used for the phase determination. Useful heavy atoms should have more than about 50 electrons. Various mercury and platinum compounds are most frequently applied. Other popular compounds contain uranium, lead, thallium, iridium, osmium, tungsten, tantalum, numerous lanthanides, barium, xenon, and iodine atoms. In general, phase determination through heavy atom soaking is a most tedious task because binding of these compounds tends to modify slightly the packing scheme of the native protein and also its conformation. Consequently, the derivatized crystals tend to become nonisomorphous with respect to the native crystal, which is an outcome that destroys the phase information.

As a further general experience, most heavy atom compounds fail to bind at a defined, recurring position in the crystal and thus cannot contribute to phasing. Alternatively, many of the compounds break the crystals, presumably because they bind at packing contacts. A somewhat special heavy atom agent is xenon, which may fill holes in the nonpolar cores of proteins without too much destruction. However, its application requires a special apparatus that bathes the crystal in 10–20 bar of this noble gas.[66] Holes and thiols in the core of a protein may also be reached by methylmercury compounds. In another approach, crystals were bathed in highly concentrated bromide salt solutions and shock-frozen to 100 K, which, surprisingly, left some of the bromides at defined positions on the protein surface.[67] These bromides were then applied for phase determination. In summary, the defined binding of heavy atoms in crystals remains an undertaking with unpredictable outcome and usually involves many unsuccessful experiments.

Once a protein structure is established, the analysis may change to address functional questions. For this purpose, protein crystals are often soaked with known ligands of the protein under scrutiny. Such an analysis may have important commercial consequences, for instance, for the pharmaceutical industry,[68] or it may help clarifying chemical reactions involving the protein.[69] Once the basic structure of the crystallized protein is known, such ligand-soaking experiments are no longer disturbed by nonisomorphism because each analysis is independent of all others. However, the experiments may become tiresome because in most cases soaking breaks the crystals, in particular if ligand binding induces conformational changes that deteriorate the packing contacts.

3.20.5.4 Cocrystallization and Protein Labeling

Since crystal nonisomorphism and crystal destruction experienced after soaking the crystals in heavy atom compounds often prohibits phasing for a long time, one may try to circumvent these problems by reacting the protein with the heavy atom compound prior to crystallization and cocrystallize the complex. This detour worked out well in some cases but, in general, it gives rise to new crystal forms that cannot be combined with the results of the native crystals and therefore fail to yield initial phases.

A very promising method of ligand analysis is the binding of the ligand to the protein in solution and subsequent cocrystallization. If the protein structure has already been established, the obtained crystal form is irrelevant because the structure of any new crystal can be determined with the molecular replacement method. It should be noted that many proteins and in particular membrane channels can only be crystallized after they are intrinsically stabilized by a bound ligand.[24,70]

A well-established modern method of phase determination involves a change of the protein to be crystallized rather than a modification of the crystal. As soon as a protein can be expressed in *E. coli*, it is usually possible to obtain the same protein with all methionine sulfurs being replaced by selenium atoms.[71] With its 34 electrons, selenium is not a traditional heavy atom. However, the anomalous scattering of its K-shell electrons can be used for phasing with the so-called multiwavelengths anomalous diffraction method if the appropriate x-ray wavelengths are available. Since the percentage of analyses of proteins expressed in *E. coli* are overwhelming and since almost every synchrotron supplies the radiation required for the multiwavelengths anomalous diffraction method, selenium labeling has become very popular.

In general, the crystals of selenium-labeled proteins grow under the same conditions as the native proteins yielding isomorphous crystals. Moreover, isomorphism is not even necessary for this method because it need not be related to a native crystal. Many eukaryotic proteins cannot be expressed in sufficient amounts in *E. coli*. In such cases, the structure analysis may proceed via the molecular replacement method based on a homolog that can be expresssed in *E. coli* and structurally elucidated. Alternatively, *E. coli* expression can be dramatically improved by back-translating a known amino acid sequence to an *E. coli*-adjusted gene sequence, which can then be synthesized without excessive costs and used for the expression.[72]

3.20.6 Crystal Statistics

3.20.6.1 Space Groups and Packing Densities

The first two images of an x-ray diffraction pattern of a given single crystal may reveal the geometry of the unit cell and thus the crystal system, except for a residual ambiguity between trigonal and hexagonal. After collecting a complete data set, the symmetry of the diffraction pattern indicates the basic point symmetry of the unit cell, and the systematically missing reflections narrow down to the space group. Since all biological material is chiral, none of the possible space groups contains an inversion or a mirror element, leaving a total of 65 biocompatible ones among the total of 230 space groups. At this point, only enantiomeric space groups like $P3_1$ and $P3_2$ cannot be distinguished from each other, or the special case of $I222$ and $I2_12_12_1$. These remaining ambiguities are resolved during phase determination. In summary, the determination of the space group is generally a fast procedure.

As the crystallized protein is usually well characterized, the packing density V_M of a crystal can be derived from the volume of the unit cell V_{uc} and the molecular mass M_r of the protomer set αβ, which is $M_r = M_{r\alpha} + M_{r\beta}$. The set $\alpha\beta$, of course, stands for any composition of the crystallized protein. The packing density relates V_{uc} to the total protein mass in the unit cell, which is a multiple z of the protomer set in the asymmetric unit multiplied by the number n of asymmetric units in the unit cell:

$$V_M = V_{\text{uc}}/(n \cdot z \cdot M_r)$$

While n is determined by the space group, for instance $n = 2 \cdot 4 \cdot 2 = 16$ for space group $I4_122$, the z value is usually known only after the crystal structure has been established. In principle, z can be determined from the densities of the crystal ρ_c and the bathing solution ρ_s as:

$$z = N_A \cdot V_{\text{uc}} \cdot \rho_p \cdot (\rho_c - \rho_s)/(n \cdot Mr \cdot (\rho_c - \rho_s))$$

using the general value $\rho_p = 1.35\,\text{g}\,\text{cm}^3$ for the protein density. Since the measurement of ρ_c is difficult and requires large crystals, z is generally determined during the structure analysis. However, in many cases only $z = 1$ is possible because the volume $z \cdot M_r/\rho_p$ has to fit into the asymmetric unit. A statistical analysis of the V_M values of all established crystal structures resulted in a rather narrow distribution.[73] **Figure 5** shows a particular selection of about 2300 crystals analyzed at resolutions in the range of 2.0–2.4 Å. This distribution has been converted to a distribution of the solvent volume fraction x according to:

$$x = (V_{\text{uc}} - V_{\text{protein}})/V_{\text{uc}} = 1 - 1/(\rho_p \cdot V_M) = 1 - 1.23/V_M$$

with V_M given in the common units (Å^3/Dalton). While the V_M distribution is skewed, the x distribution is closer to a Gaussian and therefore obviously a more natural representation (**Figure 5**).

3.20.6.2 Packing Arrangements

In crystallized oligomers the packing contacts have to be distinguished from the oligomer interfaces. This is generally possible by comparing the solvent-accessible surface areas buried in a contact. Note that the buried area is twice the

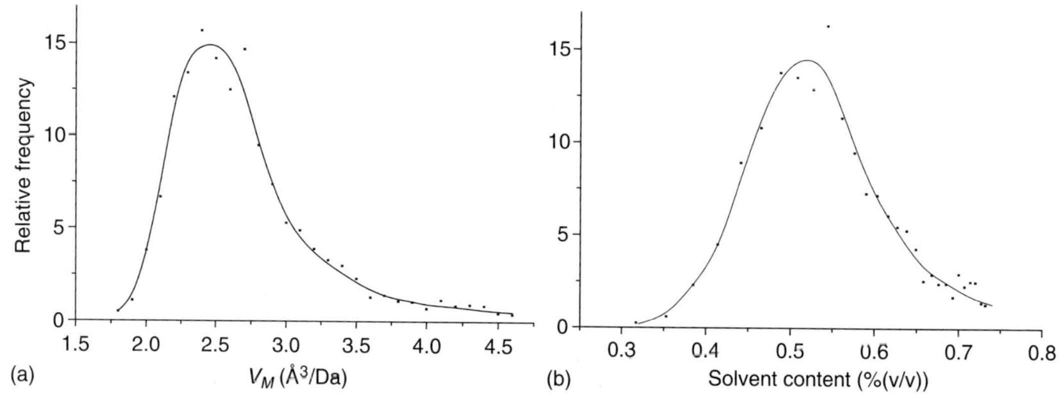

Figure 5 Statistics of about 2300 protein crystals diffracting to resolutions of between 2.4 and 2.0 Å (a) Statistics of the packing parameter V_M (see text),[73] which is 1.23 Å^3 per Dalton for hypothetical crystals without solvent. (b) The same distribution plotted over the solvent content *x* is close to a Gaussian distribution with a tail of highly solvated crystals. The average solvent content is 52%.

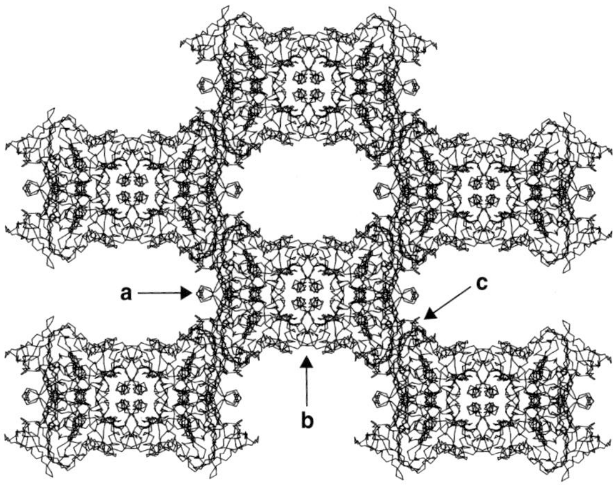

Figure 6 Crystal packing of NADH peroxidase with a molecular mass of 4 × 50 kDa from *Streptococcus faecalis*. The solvent content is 67%, which is in the highly solvated tail of **Figure 5**. The crystals contain a tight large contact (**a**, 2900 Å^2) and a small contact (**b**, 460 Å^2), which together define the tetramer found in solution.[77] Surprisingly, the third packing contact (**c**, 620 Å^2) is larger than the smaller tetramer contact. This feature and the fact that the crystal contains just three large packing contacts and large channels, indicates that this crystal may have evolved to form a dioxygen-destroying organelle in this anaerobic bacterium, in correspondence with a recent report on a similar organelle.[78]

area of the resulting interface. The interfaces of packing contacts range between 50 and 800 Å^2 whereas oligomeric interfaces are between 500 and 3000 Å^2.[74–76] Oligomeric interfaces usually contain rotation axes that may coincide with crystal axes. A closer inspection of the contacts in the overlapping region around 800 Å^2 has shown that oligomeric interfaces are usually much more tightly packed than crystal contacts. Therefore, the assignment of the oligomers in a crystal structure is usually straightforward.

One particular case of a large crystal interface seems worth mentioning. The enzyme NADH peroxidase protects an anaerobic bacterium by converting the hazardous compound dioxygen to H_2O_2, which is then further reduced to form water. The enzyme was established as a tetramer by equilibrium centrifugation.[77] The crystals were constructed using only three packing contacts with interfaces of 2900, 620, and 460 Å^2 and contained as much as 67% solvent (see **Figure 6**). Owing to the space group *I222*, the 2900 and 460 Å^2 interfaces clearly formed the tetramer whereas the larger 620 Å^2 contact formed the crystal. The small number of contacts, the inverted contact sizes, and the high percentage of solvent, forming channels with diameters of up to 40 Å, indicated that this packing is probably natural. Most likely, the bacterium uses these crystals as organelles for dioxygen destruction. NADH, dioxygen, and H_2O_2 can diffuse quickly through the spacious channels and the active centers are easily available. A similar case for another enzyme is now known.[78] However, natural protein crystals are rare. Clear cases are insulin crystals as a special storage form in the pancreas. Such insulin crystals are currently used as a slowly released depot form in treating diabetis mellitus.[79]

All symmetry elements of a space group run through the whole crystal. In addition, many crystals contain local symmetry elements called noncrystallographic symmetry. A simple case of noncrystallographic symmetry occurs whenever a C_2-symmetric dimeric molecule crystallizes without the molecular twofold axis becoming crystallographic. A more complex case was observed with cubic crystals of an isosahedral virus, in which the twofold and the threefold axes of the icosahedral symmetry are crystallographic whereas the fivefold axes are local because they cannot be crystallographic as such.[80] The percentage of established crystal structures harboring noncrystallographic symmetry elements is increasing because the analytical methods are improving on the experimental as well as on the computational side. The most complicated noncrystallographic symmetry occurs in virus crystals.[81] However, enzyme crystals with as many as 20 protomers per asymmetric unit have been reported.[82]

A survey of the crystal structures with $z > 1$, that means with more than one protomer set per asymmetric unit, showed that local symmetry elements are abundant.[83] In many crystals the noncrystallographic symmetry relates intrinsically symmetric oligomeric proteins to each other. The enzyme crystal with 20 protomers, for instance, contained five C_4-symmetric tetramers in the asymmetric unit.[82] Surprisingly, these complicated associations are commonly not asymmetric themselves but contain plenty of local symmetry. This local symmetry reduces the number of crystal packing contact types and therefore increases the probability that a dissolved protein could find its correct place in a growing crystal face. One may therefore predict that crystals with, say, more than four molecules in the asymmetric unit should show a regular noncrystallographic symmetry.

3.20.7 Conclusion

The x-ray analysis of proteins has certainly come of age. Many more structures are presently known than can be registered and sorted in a reasonable way. The ongoing structural genomics campaigns will produce numerous unpublished structure factor and coordinate files that will serve as raw material for future analyses. Fortunately, deposition in the Protein Data Bank requires structure analysis to be of a high standard, thus securing raw data of high quality. The production of homogeneous proteins and their crystallization has become the limiting factor for structure analyses since high-quality x-ray beams are easily available at the synchrotrons and phasing procedures have become much easier than 20 years ago, either through selenium labeling or molecular replacement using one of the numerous available protein structures deposited in the Protein Data Bank. Although we may reach a situation in the not too distant future where most essential proteins are known with respect to their polypeptide chainfolds, protein crystallization will remain of central interest because all important biological processes require fine structural details and involve complexes between proteins and other cell constituents. The structural elucidation of such complexes will keep researchers in this area busy for a long time to come.

References

1. Hoppe-Seyler, F. Über die chemischen und optischen Eigenschaften des Blutfarbstoffs. *Archiv für pathologische Anatomie und Physiologie und für klinische Medizin*; Virchow, R., Eds.; Verlag Reimer: Berlin, Germany, 1864; Vol. 29, pp 233–235.
2. Sumner, J. B. *J. Biol. Chem.* **1926**, *69*, 435–441.
3. Bernal, J. D.; Crowfoot, D. C. *Nature* **1934**, *133*, 794–795.
4. Green, D. W.; Ingram, V. M.; Perutz, M. F. *Proc. Roy. Soc. A* **1954**, *225*, 287–307.
5. Cork, J. M. *Philos. Mag., Ser.* 7 **1927**, *4*, 688–698.
6. Robertson, J. M.; Woodward, I. *J. Chem. Soc. (London)* **1937**, *Abstract 37*, 219–230.
7. Bokhoven, C.; Schoone, J. C.; Bijvoet, J. M. *Acta Crystallogr.* **1951**, *4*, 275–280.
8. Kendrew, J. C.; Dickerson, R. E.; Strandberg, B. E.; Hart, R. G.; Davies, D. R.; Phillips, D. C.; Shore, V. C. *Nature* **1960**, *185*, 422–427.
9. Schulz, G. E. *Angew. Chem. Int. Ed. Engl.* **1981**, *20*, 143–151.
10. Schmidt, A.; Lamzin, V. S. *Curr. Opin. Struct. Biol.* **2002**, *12*, 698–703.
11. Blake, C. C. F.; Koenig, D. F.; Mair, G. A.; North, A. C. T.; Phillips, D. C.; Sarma, V. R. *Nature* **1965**, *206*, 757–763.
12. Dunker, A. K.; Obradovic, Z. *Nature Biotech.* **2001**, *19*, 805–806.
13. Uversky, V. N.; Gillespie, J. R.; Fink, A. L. *Proteins: Struct. Funct. Genet.* **2000**, *41*, 415–427.
14. Schulz, G. E. Nucleotide Binding Proteins. In *Molecular Mechanisms of Biological Recognition*; Balaban, M., Ed.; Elsevier/North Holland Biomedical Press: Amsterdam, the Netherlands, 1979, pp 79–94.
15. Unger, V. M.; Hargrave, P. A.; Baldwin, J. M.; Schertler, G. F. X. *Nature* **1997**, *389*, 203–206.
16. Nollert, P.; Royant, A.; Pebay-Peyroula, E.; Landau, E. M. *FEBS Lett.* **1999**, *457*, 205–208.
17. Misquitta, L. V.; Misquitta, Y.; Cherezov, V.; Slattery, O.; Mohan, J. M.; Hart, D.; Zhalnina, M.; Cramer, W. A.; Caffrey, M. *Structure* **2004**, *12*, 2113–2124.
18. Michel, H. *J. Mol. Biol.* **1982**, *158*, 567–572.
19. Deisenhofer, J.; Epp, O.; Miki, K.; Huber, R.; Michel, H. *Nature* **1985**, *318*, 618–624.

20. Weiss, M. S.; Wacker, T.; Weckesser, J.; Welte, W.; Schulz, G. E. *FEBS Lett.* **1990**, *267*, 268–272.
21. Schulz, G. E. *Biochim. Biophys. Acta* **2002**, *1565*, 308–317.
22. Dutzler, R.; Campbell, E. B.; Cadene, M.; Chait, B. T.; MacKinnon, R. *Nature* **2002**, *415*, 287–294.
23. Dong, J.; Yang, G.; Mchaourab, H. S. *Science* **2005**, *308*, 1023–1028.
24. Toyoshima, C.; Nomura, H. *Nature* **2002**, *418*, 605–611.
25. Loll, P. J.; Picot, D.; Garavito, R. M. *Nature Struct. Biol.* **1995**, *2*, 637–643.
26. Wendt, K. U.; Poralla, K.; Schulz, G. E. *Science* **1997**, *277*, 1811–1815.
27. Kloer, D. P.; Ruch, S.; Al-Babili, S.; Beyer, P.; Schulz, G. E. *Science* **2005**, *308*, 267–269.
28. Wendt, K. U.; Lenhart, A.; Schulz, G. E. *J. Mol. Biol.* **1999**, *286*, 175–187.
29. Kabsch, W.; Gast, W. H.; Schulz, G. E.; Leberman, R. *J. Mol. Biol.* **1977**, *117*, 999–1012.
30. Fontana, A.; Polverino der Laureto, P.; De Fillipis, V.; Scaramella, E.; Zambonin, M. *Folding Des.* **1997**, *2*, R17 – R26.
31. Kamada, K.; De Angelis, J.; Roeder, R. G.; Burley, S. K. *Proc. Natl. Acad. Sci. USA* **2001**, *98*, 3115–3120.
32. Rehm, T.; Huber, R.; Holak, T. A. *Structure* **2002**, *10*, 1613–1618.
33. Spraggon, G.; Pantazatos, D.; Klock, H. E.; Wilson, I. A.; Woods, V. L., Jr.; Lesley, S. A. *Protein Sci.* **2004**, *13*, 3187–3199.
34. Benesch, R. E.; Kwong, S.; Benesch, R. *Nature* **1982**, *299*, 321–324.
35. Nelson, R.; Sawaya, M. R.; Balbirnie, M.; Madsen, A. O.; Riekel, C.; Grothe, R.; Eisenberg, D. *Nature* **2005**, *435*, 773–778.
36. D'Arcy, A. *Acta Crystallogr., Sect. D* **1994**, *50*, 469–471.
37. Smyth, D. R.; Mrozkiewicz, M. K.; McGrath, W. J.; Listwan, P.; Kobe, B. *Protein Sci.* **2003**, *12*, 1313–1322.
38. Claus, M. T.; Zocher, G. E.; Maier, T. H. P.; Schulz, G. E. *Biochemistry* **2005**, *44*, 8620–8626.
39. Schwede, T. F.; Rétey, J.; Schulz, G. E. *Biochemistry* **1999**, *38*, 5355–5361.
40. Rost, B.; Sander, C. *Protein: Struct. Funct. Genet.* **1994**, *20*, 216–226.
41. Pautsch, A.; Vogt, J.; Model, K.; Siebold, C.; Schulz, G. E. *Proteins: Struct. Funct. Genet.* **1999**, *34*, 167–172.
42. Derewenda, Z. S. *Structure* **2004**, *12*, 529–535.
43. Faller, M.; Niederweis, M.; Schulz, G. E. *Science* **2004**, *303*, 1189–1192.
44. Pautsch, A.; Schulz, G. E. *Nat. Struct. Biol.* **1998**, *5*, 1013–1017.
45. Czepas, J.; Devedjiev, Y.; Krowarsch, D.; Derewenda, U.; Otlewski, J.; Derewenda, Z. S. *Acta Crystallogr., Ser. D* **2004**, *60*, 275–280.
46. Mittl, P. R. E.; Berry, A.; Scrutton, N. S.; Perham, R. N.; Schulz, G. E. *Acta Crystallogr., Ser. D* **1994**, *50*, 228–231.
47. Aizenberg, J.; Black, A. J.; Whitesides, G. M. *Nature* **1999**, *398*, 495–498.
48. Yau, S.-T.; Vekilov, P. G. *Nature* **2000**, *406*, 494–497.
49. Majeed, S.; Ofek, G.; Belachew, A.; Huang, C.-c.; Zhou, T.; Kwong, P. D. *Structure* **2003**, *11*, 1061–1070.
50. Bannwarth, M.; Schulz, G. E. *Biochim. Biophys. Acta* **2003**, *1610*, 37–45.
51. Gilliland, G. L.; Tung, M.; Blakeslee, D. M.; Ladner, J. E. *Acta Crystallogr., Ser. D* **1994**, *50*, 408–413.
52. Rayment, I. *Structure* **2002**, *10*, 147–151.
53. Drenth, J. *Principles of Protein X-Ray Crystallography*; Springer Verlag: New York, 1999, pp 4–16.
54. Zhu, D.-W.; Lorber, B.; Sauter, C.; Ng, J. D.; Bénas, P.; Le Grimellec, C.; Giegé, R. *Acta Crystallogr., Ser. D* **2001**, *57*, 552–558.
55. Muller, Y. A.; Schulz, G. E. *Science* **1993**, *259*, 965–967.
56. Pitts, J. E. *Nature* **1992**, *355*, 117.
57. Littke, W.; John, C. *Science* **1984**, *225*, 203–204.
58. Mayo, C. J.; Diprose, J. M.; Walter, T. S.; Berry, I. M.; Wilson, J.; Owens, R. J.; Jones, E. Y.; Harlos, K.; Stuart, D. I.; Esnouf, R. M. *Structure* **2005**, *13*, 175–182.
59. Pédelacq, J.-D.; Piltch, E.; Liong, E. C.; Berendzen, J.; Kim, C.-Y.; Rho, B.-S.; Park, M. S.; Terwilliger, T. C.; Waldo, G. S. *Nat. Biotech.* **2002**, *20*, 927–932.
60. Waldo, G. S. *Curr. Opin. Chem. Biol.* **2003**, 7, 33–38.
61. Yang, J. K.; Park, M. S.; Waldo, G. S.; Suh, S. W. *Proc. Natl. Acad. Sci. USA* **2003**, *100*, 455–460.
62. Ritter, H.; Schulz, G. E. *Plant Cell* **2004**, *16*, 3426–3436.
63. Dreyer, M. K.; Schulz, G. E. *J. Mol. Biol.* **1996**, *259*, 458–466.
64. Dreyer, M. K.; Schulz, G. E. *Acta Crystallogr., Ser. D* **1996**, *52*, 1082–1091.
65. Rosendal, K. R.; Sinning, I.; Wild, K. *Acta Crystallogr., Ser. D* **2004**, *60*, 104–143.
66. Schiltz, M.; Fourme, R.; Broutin, I.; Prange, T. *Structure* **1995**, *3*, 309–316.
67. Devedjiev, Y.; Dauter, Z.; Kuznetsov, S. R.; Jones, T. L. Z.; Derewenda, Z. S. *Structure* **2000**, *8*, 1137–1146.
68. Kubinyi, H. *Curr. Opin. Drug Disc. Dev.* **1998**, *1*, 4–15.
69. Reinert, D. J.; Balliano, G.; Schulz, G. E. *Chem. Biol.* **2004**, *11*, 121–126.
70. Mosbacher, T. G.; Bechthold, A.; Schulz, G. E. *J. Mol. Biol.* **2005**, *345*, 535–545.
71. Hendrickson, W. A.; Horton, J. R.; LeMaster, D. M. *EMBO J.* **1990**, *9*, 1665–1672.
72. Baedeker, M.; Schulz, G. E. *FEBS Lett.* **1999**, *457*, 57–60.
73. Kantardjieff, K. A.; Rupp, B. *Protein Sci.* **2003**, *12*, 1865–1871.
74. Carugo, O.; Argos, P. *Protein Sci.* **1997**, *6*, 2261–2263.
75. Dasgupta, S.; Iyer, G. H.; Bryant, S. H.; Lawrence, C. E.; Bell, J. A. *Proteins: Struct. Funct. Genet.* **1997**, *28*, 494–514.
76. Janin, J. *Nat. Struct. Biol.* **1997**, *4*, 973–974.
77. Stehle, T.; Ahmed, S. A.; Claiborne, A.; Schulz, G. E. *J. Mol. Biol.* **1991**, *221*, 1325–1344.
78. Kerfeld, C. A.; Sawaya, M. R.; Tanaka, S.; Nguyen, C. V.; Phillips, M.; Beeby, M.; Yeates, T. O. *Science* **2005**, *309*, 936–938.
79. Mühlig, P.; Klupsch, T.; Kaulmann, U.; Hilgenfeld, R. *J. Struct. Biol.* **2003**, *142*, 47–55.
80. Harrison, S. C.; Olson, A. J.; Schutt, C. E.; Winkler, F. K. *Nature* **1978**, *276*, 368–373.
81. Stehle, T.; Gamblin, S. J.; Yan, Y.; Harrison, S. C. *Structure* **1996**, *4*, 165–182.
82. Kroemer, M.; Schulz, G. E. *Acta Crystallogr., Ser. D* **2002**, *58*, 824–832.
83. Vonrhein, C.; Schulz, G. E. *Acta Crystallogr., Ser. D* **1999**, *55*, 225–229.

Biography

Georg E Schulz was born in Berlin. He grew up in West Berlin and began his studies at the Technische Universität. In 1962 he moved to Heidelberg and received his PhD degree from the Ruperto-Carola Universität in 1966. Following a postdoc year with H W Wyckoff at Yale University (USA), he joined the Max-Planck-Institut für Medizinische Forschung in Heidelberg. Since 1984 he has been teaching Biochemistry at the Albert-Ludwigs-Universität in Freiburg im Breisgau. He has been a member of the Deutsche Akademie für Naturforscher Leopoldina since 1998 and joined the Board of Reviewers of Science Magazine in 2001. Together with R H Schirmer he wrote the monograph 'Principles of Protein Structure' (Springer Verlag, New York) which was published in 1979.

Comprehensive Medicinal Chemistry II
ISBN (set): 0-08-044513-6
ISBN (Volume 3) 0-08-044516-0; pp. 433–447

3.21 Protein Crystallography

M T Stubbs II, Martin Luther University, Halle, Germany

3.21.1	**Introduction**	**449**
3.21.1.1	A Nobel History	449
3.21.2	**Diffraction Theory**	**451**
3.21.2.1	The Nature of Crystals	451
3.21.2.2	The Interaction of X-rays with Crystals	452
3.21.2.3	The Solution of the Phase Problem	457
3.21.2.3.1	Isomorphous replacement	457
3.21.2.3.2	Anomalous dispersion	459
3.21.2.3.3	Molecular replacement	460
3.21.2.4	From Map to Model	462
3.21.3	**Practical Crystallography**	**465**
3.21.3.1	The Generation of X-rays	465
3.21.3.2	Handling of Crystals	465
3.21.3.3	The Detection of X-rays	468
3.21.3.4	Data Collection	468
3.21.4	**Quality Control in Protein Crystallography**	**470**
3.21.5	**A Look into the Future?**	**470**
References		**471**

3.21.1 Introduction

X-ray crystallography is one of the most widespread sources of information for the functional analysis of biological molecules at an atomic level. Advances in scientific instrumentation, x-ray sources and detectors, computing, and in particular molecular biology have lead to a situation where experimental three-dimensional structure data of a drug target are no longer an academic dream but a realistic goal. The crystal structure of a protein thereby represents a 'gold standard' of modern drug design. It is the aim of this chapter to provide a theoretical and practical background to this important method, which is based on my lecture course and recurring questions raised by my graduate students. While it has now become possible to solve a protein crystal structure without any knowledge of the underlying principles, crystallographic theory is ignored at the risk of failure to elucidate important structural details if problems occur. In this short chapter, it is not possible to cover all details pertaining to x-ray crystallography; for these, the interested reader is directed to Rossmann and Arnold's excellent recent reference work.[1] Before covering current methods for protein structure determination, it is instructive to follow the development of x-ray crystallography over the last century.

3.21.1.1 A Nobel History

The discovery in Würzburg of radiation that was able to penetrate solid matter by Wilhelm Conrad Röntgen (1845–1923) in 1895 attracted worldwide attention.[2] While this was reflected in his award of the first Nobel Prize for Physics in 1901 "in recognition of the extraordinary services he has rendered by the discovery of the remarkable rays subsequently named after him," there existed considerable speculation as to the nature of this radiation. Although generally accepted as being electromagnetic in nature, no direct proof could be found, as the rays failed to be refracted in any measurable sense. Working in Munich in the department of the theoretician Sommerfeld, Max von Laue (1879–1960) postulated that if x-rays were electromagnetic, then they should be of extremely short wavelength. If the wavelength was of the order of interatomic distances, then a crystal should act as a three-dimensional diffraction grating for x-rays. His two assistants Friedrich and Knipping carried out the essential experiment in 1912[3]; the diffraction patterns obtained from a number of minerals proved the electromagnetic wave hypothesis, and von Laue was awarded the Nobel Prize for Physics in 1914 "for his discovery of the diffraction of x-rays by crystals." It was not long before von Laue's observations found practical application; father-and-son team Sir William Henry Bragg (1862–1942) and William Lawrence Bragg (1890–1971) showed that the diffraction data from crystals could be used to derive the atomic structure therein,[4] and thus were also awarded the Nobel Prize in Physics in 1915 "for their services in the analysis of crystal structure by means of x-rays."

At a time when the chemical identity of proteins and other biological macromolecues was not yet established, W. L. Bragg championed x-ray methods for the study of living systems. Although Sumner, Northrop, and Stanley had shown that enzymes and viruses could be crystallized (an achievement for which they were awarded the Nobel Prize in Chemistry in 1946), all attempts to obtain diffraction images from protein crystals failed. A decisive breakthrough was made in Cambridge in 1934, when J. D. Bernal (1901–1971) and Dorothy Crowfoot Hodgkin (1910–1994) noted that crystals of pepsin lose their birefringence on exposure to air.[5] Bernal and Crowfoot correctly surmised that the protein crystals must contain a high degree of solvent; by mounting the crystals in capillaries and keeping them moist, they were able to obtain diffraction patterns from a protein crystal for the first time. Although this was at a time when even the structure determination of molecules with only few atoms was a major undertaking, the way was paved for the use of x-ray analysis to explore biological macromolecules at the atomic level.

Working initially in Bragg's laboratory in the Royal Institution in London and then later in Leeds, W. T. Astbury (1889–1961) obtained fiber diffraction patterns from keratin as well as a number of other biological specimens. Keratin diffraction patterns measured at varying humidity indicated two distinct, characteristic spacings corresponding to 5.4 Å and 3.4 Å, which Astbury termed α and β, respectively.[6] Further experiments revealed these spacings to be present in the diffraction patterns of other proteins as well. Based on precise measurements concerning the stereochemistry of amino acids, in particular the planarity of the peptide bond, Linus Pauling (1901–1994) of CalTech was able to explain Astbury's observations by proposing models for the α-helix and the β-sheet.[7] Pauling received the Nobel Prize for Chemistry in 1954 "for his research into the nature of the chemical bond and its application to the elucidation of the structure of complex substances."

The 1950s represented a golden age for structural biology. Upon hearing of Pauling's model, Max F. Perutz (1914–2002), who had been working on hemoglobin crystals in Bragg's laboratory in Cambridge since 1937, made x-ray measurements from his crystal data that confirmed the existence of the α-helix.[8] Furthermore, he set his PhD student Francis H. C. Crick (1916–2004) the task of providing a mathematical treatment for the diffraction from a helix.[9] The subsequent collaboration between Francis Crick and James D. Watson (1928–), culminating in the proposal of a double helical structure for DNA,[10] has been well documented, as well as the role played by the x-ray fiber diffraction patterns of Rosalind Franklin (1920–1958)[11] and Maurice Wilkins (1916–).[12]

Meanwhile, Perutz was working on ways of extracting structural information from protein crystal data. While the intensities of the individual reflections in the diffraction pattern could be measured with some accuracy, the vital information for calculating an electron density, the phase information, was missing (*see* Section 3.21.2.2). Perutz suggested that the introduction of heavy atoms into the large solvent channels of hemoglobin crystals could give rise to measurable changes in the diffraction pattern, which could in turn be used to solve the phase problem.[13] In 1958, the first ever protein structure, that of myoglobin, saw the light of day at a resolution of 6 Å.[14] Within 2 years, the structure had been further improved to include data to 2 Å,[15] and was joined by that of hemoglobin.[16] In 1962, Max Perutz and John C. Kendrew (1917–1997) were awarded the Nobel Prize for Chemistry "for their studies of the structures of globular proteins" and Francis Crick, James D. Watson, and Maurice Wilkins the Noble Prize for Medicine "for their discoveries concerning the molecular structure of nucleic acids and its significance for information transfer in living material." Two years later in 1964, Dorothy Crowfoot Hodgkin, who launched the field of protein crystallography by recognizing that protein crystals must be kept in a humid environment, became a Nobel Laureate for Chemistry "for her determinations by x-ray techniques of the structures of important biochemical substances."

Over the next quarter century, the methods established by Perutz were applied to a wide variety of biological macromolecules, including enzymes, antibodies, and entire viruses. As the number of structures solved increased, so it became necessary to catalog and characterize them. Thus the Protein Data Bank (PDB) was born,[17,18] a depository for the coordinates of all published structures. The realization that many structures possess similar folds lead to a new method for structure determination: molecular replacement or the Patterson search method.[19,20] The advent of synchrotrons as a source of high-intensity x-rays facilitated data collection from weaker and/or smaller crystals; the ability to tune the wavelength allowed the use of multiple wavelength anomalous dispersion (MAD) to solve the phase problem.[21] Advances in molecular biology meant that it was no longer necessary to restrict crystallographic analysis to protein of high natural abundance; it was now possible to choose a target protein and have a realistic chance of solving its structure.

Yet there was one class of proteins whose structure determination remained elusive: the membrane proteins. The first structure of an integral membrane protein was revealed in 1985: that of the bacterial photosynthetic reaction center.[22] For this, the crystallographers Johann Deisenhofer (1943–), Robert Huber (1937–), and Hartmut Michel (1948–) of the Max-Planck-Institut für Biochemie in Martinsried shared the Nobel Prize in Chemistry in 1988 "for the determination of the three-dimensional structure of a photosynthetic reaction center". In part for his work on the structure determination of ATP synthase,[23] John E. Walker (1941–) was made a Nobel Laureate in Chemistry in 1997 "for the elucidation of the enzymatic mechanism underlying the synthesis of adenosine triphosphate (ATP)." Finally, Roderick MacKinnon (1956–) provided groundbreaking structures of ion channels,[24] for which he was awarded the Nobel Prize in Chemistry in 2003 "for fundamental studies on ion channels."

Thus x-ray crystallography boasts an illustrious career. Protein crystallography has progressed from a discipline driven by technology to a field driven by biological, physiological, or pharmaceutical questions. The recent structure determinations of ribosomes and their subunits[25–27] show that even vast multimolecular complexes are amenable to crystallographic analysis, and have provided detailed information as to the workings of, e.g., antibiotics.[28]

Today, there are some 34 000 structures deposited in the PDB,[18] of which 29 000 have been solved using x-ray methods, and the numbers are increasing at an exponential rate. This resource provides a basis for theoretical analyses of biological macromolecular structure. Experience with experimental structures shows, however, that at least for the time being, our theories are not sufficiently advanced to describe the intricacies of biology in detail at an atomic level. For this, it will remain necessary to probe the nature of proteins using experimental biophysical methods; the aim of this chapter is to provide an understanding for the processes involved in protein crystallography.

3.21.2 Diffraction Theory

3.21.2.1 The Nature of Crystals

The word 'crystal' is derived from the Greek κρυσταλλοζ, meaning 'ice.' In contrast to a glass, a crystal is formed through the orderly repetition of molecules in three dimensions (one-dimensional (e.g., fibrous molecules) and two-dimensional (e.g., molecules embedded in a membrane) also exist; only three-dimensional crystal arrangements will be dealt with in this chapter). The macroscopic crystal can be divided into a number of small crystallites; the smallest possible fragment that can be repeated to form the crystal is termed the unit cell (**Figure 1**). In the most general case, the unit cell is defined by a parallelepiped with three cell axes a, b, and c and angles between the axes of α, β, and γ. End-to-end stacking of unit cells in the directions of the cell axes, termed the crystal lattice, then describes a perfect macroscopic crystal. As we shall see later, the large number of molecules present in a regular array within the crystal allows the amplification of the weak interaction between x-rays and matter, making possible the visualization of proteins at an atomic level.

In the simplest of cases, the unit cell consists of a single molecular entity. More often than not, however, crystals are formed from unit cells containing multiple molecular copies. Should these copies be so arranged in the unit cell that rotation, translation, or inversion of the cell allows superposition on the original coordinates, then the resulting crystal exhibits higher-order symmetry. The asymmetric unit defines the molecular arrangement from which the unit cell (and thereby the entire crystal) can be constructed based on the given symmetry elements (**Figure 2**). Apart from the identity operation, only twofold, threefold, fourfold, and sixfold rotations are allowed for crystallographic axes, as it is not possible to tessellate building blocks based on any other rotational symmetry. While for example fivefold or sevenfold axes cannot form crystallographic axes, it is important to remember that the asymmetric unit can contain any arrangement of molecules. If the asymmetric unit contains more than one identical copy, we speak of noncrystallographic symmetry relating the individual subunits.

The presence of symmetry elements leads to restrictions in the unit cell axes; for example, the introduction of a crystallographic twofold axis along b requires the cell angles α and γ to be 90°, while a threefold axis requires that $a=b$, $\alpha=\beta=90°$, and $\gamma=120°$. These restrictions give rise to the known crystal classes: triclinic (no symmetry), monoclinic (one twofold axis), orthorhombic (three mutually perpendicular twofold axes), trigonal (one threefold axis), tetragonal (one fourfold axis), hexagonal (one sixfold axis), rhombohedral (one threefold axis, all axes equal), and cubic (twofold and threefold axes). In addition to rotational symmetries, the three-dimensional translational repetition of unit cells within a lattice allows for further crystallographic symmetries, namely screw axes, glide planes, and cell centering. As an example, a threefold screw axis involves a one-third revolution of the asymmetric unit together with a one-third translation along the crystallographic c axis (**Figure 2**). Two more repetitions of this operation result in an asymmetric unit that has been shifted by one complete unit cell, by definition equivalent to the original unit cell. In this example,

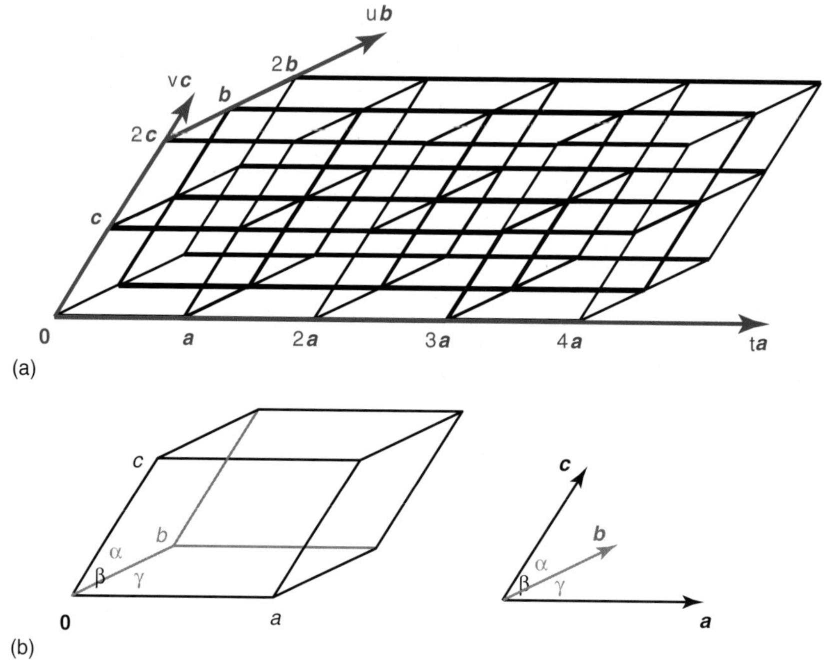

Figure 1 (a) A macroscopic crystal is defined by an orderly repetition of molecules in three dimensions. Conceptually, the crystal can be divided up into a lattice of smaller crystallites; the smallest possible repeating unit is termed the unit cell (b). Both the crystal and the unit cell are described by their unit cell axes *a*, *b*, and *c* and included angles of α, β, and γ. The crystal can then be reconstructed by translations of the unit cell along the axis vectors ***a***, ***b***, and ***c***.

if the 120° rotation is clockwise and the translation is in the positive direction, the screw axis is denoted 3_1; on the other hand, if the translation is in the negative direction, the screw axis is denoted 3_2. In the absence of centering, the cells are termed primitive (symbol P); centering operations can be in a single face (C-centering), in all six faces of the cell (F-centering), or in the body of the cell (I-centering, from the German *innenzentriert*).

Combining all possible symmetry operations results in 230 possible three-dimensional space groups; these are detailed in the *International Tables for Crystallography Volume A*.[29] As proteins and other biological macromolecules are chiral, crystal symmetries involving inversion, mirror, and glide planes cannot occur, so that only 65 space groups are applicable for these molecules.

3.21.2.2 The Interaction of X-rays with Crystals

As shown by von Laue's ingenious experiment, x-rays are electromagnetic waves of high energy (or short wavelength). Waves have both amplitude and phase, important concepts for understanding x-ray crystallography (**Figure 3**). X-rays are scattered by electrons; other diffraction phenomena important for the biosciences are neutron crystallography (neutrons are scattered by atomic nuclei, i.e., by neutrons and protons) and electron crystallography (electrons are scattered by the charge distribution, i.e., by electrons and protons). X-ray diffraction is by far the most established of these techniques, and many procedures in x-ray crystallography have become more or less routine.

When an x-ray photon impinges on an electron, the electron begins to oscillate and thereby acts as a secondary source of x-radiation. The ensuing waves disperse radially from the scattering center, with the same wavelength as the incoming wave (assuming elastic scattering, i.e., no energy loss) but an amplitude corresponding to the scattering power of an electron and a phase shift of 180° (**Figure 4**). A second electron nearby would also produce the same scatter, but due to its different position, the resulting waves will have a phase different to those from the first electron. If a detector is placed at some distance away, it will record the resultant diffraction pattern arising from addition of the individually scattered waves. Due to their different phases, the amplitude will be modulated as a function of position along the detector.

Of course, electrons are not found in isolation in macromolecules; rather, they are associated with atoms and bonds. An approximation used in biological crystallography is that the charge distribution is centered on the atomic nuclei; in this case, scattering by individual electrons is replaced by that of electron density distributions. From solution of the

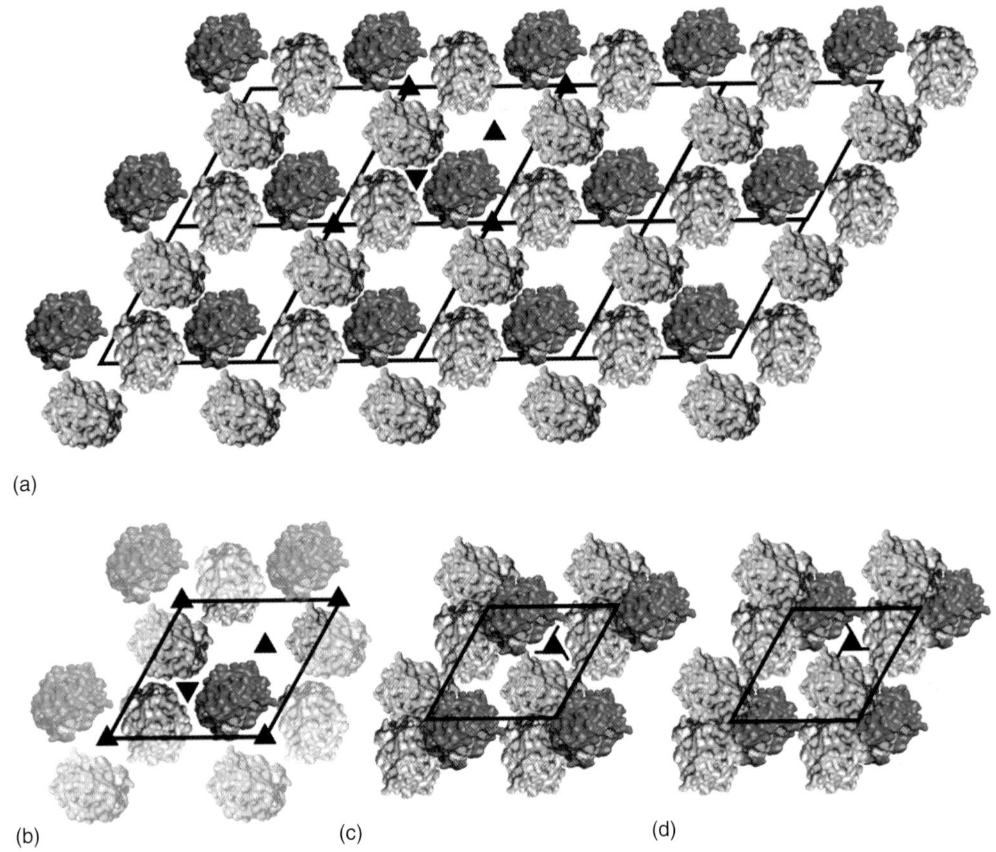

Figure 2 Schematic diagram depicting a single layer of a crystal with trigonal symmetry; triangles represent threefold axes. (a) If the three colored molecules are identical, the threefold axis represents a crystallographic symmetry, as a 120° rotation superimposes individual unit cells onto one another. This restricts the cell axes to $a=b$, $\alpha=\beta=90°$, and $\gamma=120°$. Each threefold axis represents an alternative origin of equal validity. Note the large solvent channels between the molecules. (b) The unit cell corresponding to (a) contains three molecules in a trigonal P3 space group (note that two of the molecules of the complete lattice are made up of two 'halves' in the unit cell). Each equivalent molecule represents an alternative choice of asymmetric unit. The next layer of the crystal would stack directly on top at a distance *c* corresponding to the crystallographic ***c*** axis. (c, d) In contrast to (b), each orientation of the molecule represents a single layer of the crystal, each layer separated by one-third of the crystallographic *c* axis. The transformation between the colored molecules represents a screw axis, clockwise in (c) and anticlockwise in (d) which correspond to the space groups $P3_1$ and $P3_2$, respectively. These arrangements exhibit a tighter packing density than (b); note that the crystal contacts in each are fundamentally different.

Schrödinger equation, it is possible to derive the electron density distribution and thereby the degree of scattering of any particular atom at a given wavelength (the so-called scattering or form factor). The atomic form factor f falls off with increasing angle of diffraction; maximum scatter is in the forward (undeflected) direction and is directly related to the atomic number Z of the scattering atom (**Figure 4**). This belies a major problem of biological crystallography: biomolecules are composed predominantly of atoms with low electron density – hydrogen, carbon, nitrogen and oxygen, with $Z=1$, 6, 7, and 8, respectively – and therefore scatter x-rays only weakly. This is compounded by the fact that atoms are in motion; this motion leads to an exponential reduction in the intensities at higher diffraction angle and thereby a reduction in resolution (see below). The degree of damping of the scattering factor is governed by the temperature- or B-factor – the higher the B-factor, the more rapid the falloff in intensity with scattering angle (**Figure 4**).

Addition of the diffracted waves from each atom in our object results in a defined diffraction pattern. Mathematically, the diffraction pattern is a Fourier transform of the diffracting object, each direction within the diffraction pattern associated with an amplitude F and a phase ϕ. If we have an electron density distribution $\rho(\mathrm{r})$, then the diffraction $\mathbf{F}$ in a given direction θ is given by

$$\mathbf{F}(\mathbf{S}) = \int \rho(\mathbf{r}) \exp[2\pi i\mathbf{r}.\mathbf{S}]d^3\mathbf{r} \qquad [1]$$

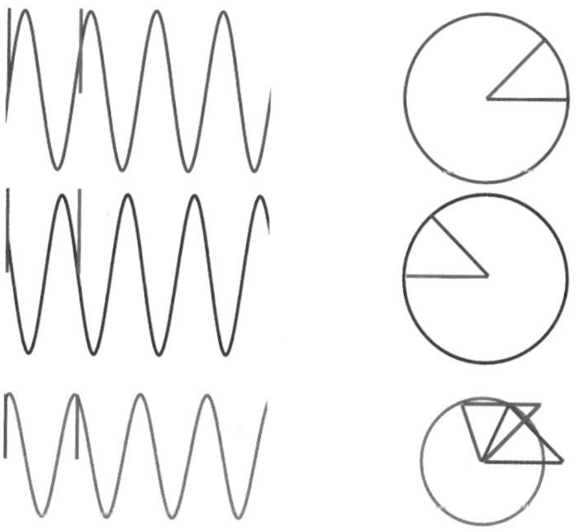

Figure 3 The anatomy of a wave. (a) A wave is defined by its wavelength, its amplitude, and its phase. The red and the blue sine waves have the same wavelength and amplitude but differ in their phases. Addition of the two waves results in the pink wave, also of the same wavelength; due to the phase difference of the two waves, however, the resultant wave exhibits a new amplitude and phase. Addition of the waves is simplified by using the Argand representation in (b) amplitudes are given by the radii of the circles, while the phases are denoted by the angle subtended by the wave vector from the horizontal axis (straight lines). Every point of the sine wave from (a) is found as the projection of the height on the vertical axis of (b) (compare the blue and green elements of the wave cycle). The amplitude and phase of the resultant wave is then given by the simple vector addition of the two components (bottom).

where the magnitude of the diffraction vector $|\mathbf{S}| = 2\sin\theta/\lambda$. A special property of the Fourier transform is that it can be inverted: the electron density is simply given by the inverse Fourier transform:

$$\rho(\mathbf{r}) = \int \mathbf{F}(\mathbf{S}) \exp[-2\pi i\mathbf{r}.\mathbf{S}]d^3\mathbf{S} \qquad [2]$$

Note here however that **F** is a vector quantity, with magnitude and phase.

The phenomenon of diffraction is not restricted to x-radiation; the same occurs for example with visible light. With the latter, however, it is possible to recombine the diffracted waves to obtain an image of the diffracted object. This is achieved through the use of a lens; the lens ensures that both the phase and the amplitude information present in the diffraction pattern are maintained and recombined up to the point of image formation. For x-radiation, however, there exist at present no suitable lenses; an alternative means to image formation is therefore necessary. In principle, construction of the image is possible providing the phases of every part of the diffraction pattern are known – this is a property of Fourier transforms. Unfortunately, however, current diffraction experiments can measure accurately the intensities of the scattered rays, but all phase information is lost. While knowledge of the atomic positions within the molecule (i.e., the structure) can be used to completely reconstruct amplitudes, phases, and intensities in the diffraction pattern, the absence of experimental phase information means that the reverse process, from intensities to structure, is not possible directly. It is the burden of the crystallographer to determine these phases, i.e., to solve the phase problem.

A special situation arises when the object is a crystal. Under special conditions, all the unit cells in the crystal diffract in phase, leading to an enormous amplification in the scattering, so that individual reflections can be measured. While not strictly correct physically, Bragg provided a graphic illustration of the conditions for crystal diffraction (**Figure 5**).[4] He considered that diffraction was caused by reflection of the incoming x-rays by imaginary planes within the crystal. For constructive interference to occur, all reflected waves from the crystal must be in phase. Bragg's law states that, for a wavelength λ, a glancing angle θ, and an interplane spacing d, constructive interference occurs when

$$n\lambda = 2d\sin\theta \qquad [3]$$

where n is an integer. If the diffraction pattern fades away at an angle $2\theta_{max}$, then this defines a minimum distance $d_{min} = \lambda/2\sin\theta_{max}$ that can be resolved from the crystal – this is termed the resolution of the diffraction data.

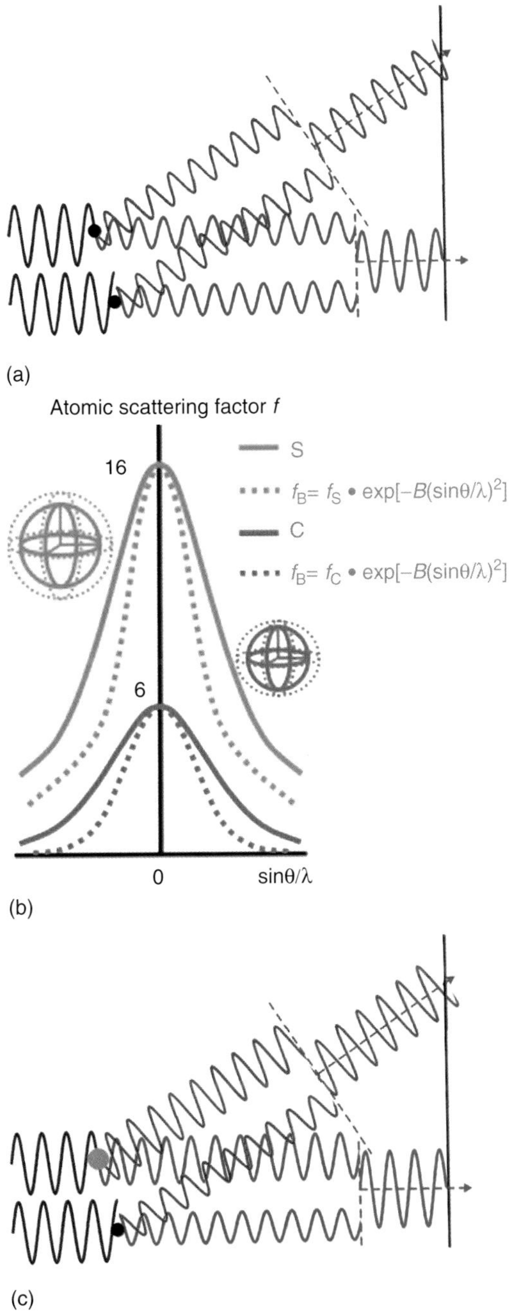

Figure 4 Diffraction by two atoms. (a) When an incoming x-ray photon (blue) strikes an atom, the latter acts as a center for secondary waves (red and green). Here two atoms are separated by a distance similar to the x-ray wavelength – the resulting secondary waves interfere with one another, leading to a diffraction pattern that travels to the detector (right). Depending on the positions of the atoms (the structure), the resultant amplitude will vary as a function of scattering angle, being maximum in the undeflected direction. Not only do the diffracted waves arrive at the detector with different amplitudes; they also arrive at different phases of their cycle. As intensities are measured during x-ray diffraction experiments, however, the phase information is lost. (b) The scattering amplitude *f* afforded by an atom is a function of its atomic number, and shows a fall off with increasing scattering angle. Thermal displacement of the atom about its equilibrium position results in a further reduction in scattered intensity, characterised by the B-factor. (c) Substitution of a 'light' atom with a 'heavy' atom results in increased scatter from that atom; if the position is the same, then the phases for the secondary waves are not altered. As a result of the increased amplitude, however, the resultant diffracted waves exhibit both different amplitudes and phases compared to the nonsubstituted structure (cf. (a)). While the phase is immeasurable, the change in intensity is measurable, and can be used to solve the phase problem.

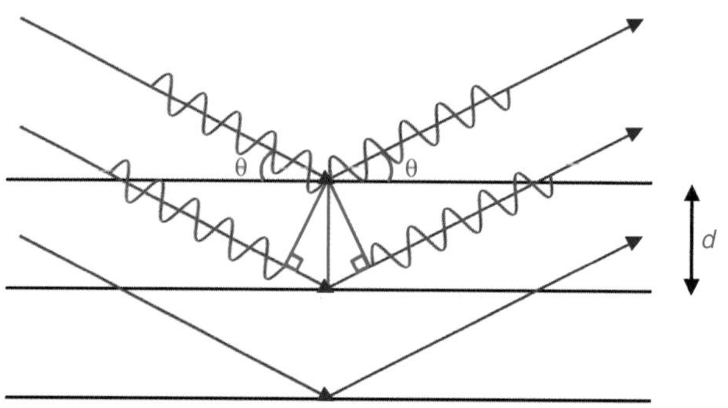

Figure 5 Bragg's law.[4] The diffraction phenomenon is treated as a reflection of incident x-rays of wavelength λ at a grazing angle of θ from imaginary planes within the crystal separated by a distance *d*. For constructive interference to occur, the path difference experienced by reflected waves from consecutive planes must equal an integer number *n* of waves. As the path difference is $2 \cdot (d\sin\theta)$, the well-known relationship $n\lambda = 2d\sin\theta$ can be easily derived.

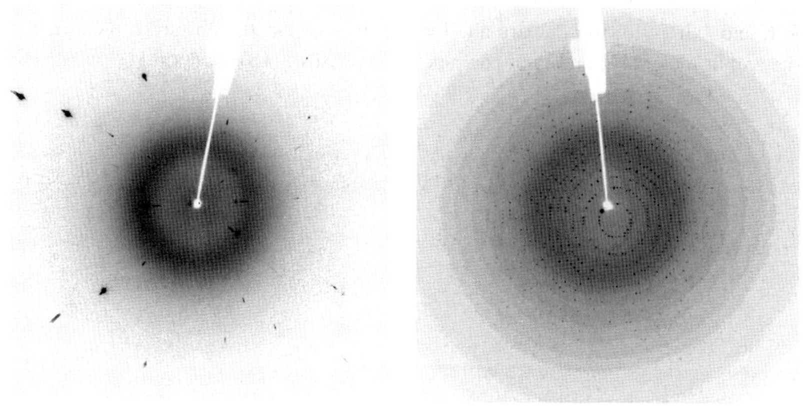

(a) (b)

Figure 6 Typical diffraction patterns obtained from an in-house x-ray source. (a) The strong, widely spaced diffraction spots indicate that the crystal is defined by short unit cell axes – the hallmark of a salt crystal. Note the slight smearing of the reflections, which is due to scattering of the (weak) Cu Kβ radiation. The dark diffuse background ring corresponds to disordered solvent (the average distance between water molecules in solution is 3.5 Å); the white vertical feature is the shadow of the beamstop (see **Figure 12**). (b) The regular arrangement of the reflections in the horizontal and vertical directions of this orthorhombic trypsin crystal[61] reveals two directions of the reciprocal lattice; the third dimension is projected as a series of concentric rings. Knowing the crystal-to-detector distance, the spacing of the reflections yields the unit cell dimensions; their proximity to one another confirms that this is a protein crystal.

The resolution of the diffraction data is a function of the crystal – a poorly ordered crystal exhibits low-resolution scattering. One of the most frustrating moments in a crystallographer's work is the discovery of a beautifully formed crystal that fails to diffract (or only to low resolution). There can be many reasons for such behavior. If the crystalline periodicity is not maintained for all of the individual unit cells, then this will result in a reduction of the resolution and an increase in the background – a corollary of this is that in some cases, microcrystals have been shown to diffract to higher resolution than (composite) larger crystals. Similarly, differences in the contents of the individual unit cells (through, e.g., mobility or multiple conformations of part of the molecule) will also tend to lower resolution. In particular, the high solvent content of protein crystals (up to 95% has been recorded)[30] results in a large proportion of the cell being occupied by disordered solvent as well as reduced intermolecular contacts to stabilize the crystal, so that protein crystals generally diffract less than their small-molecule counterparts.

A further consequence of Bragg's law is the concept of the reciprocal lattice. Rearranging eqn [3], we can see that scattering occurs at angles of 2θ when $\sin\theta = n\lambda/2d$. Thus the scattering angle shows an inverse relationship to the interlayer spacing – for large distances within the crystal, reflections are close together, while short distances reveal themselves in widely spaced reflections (**Figure 6**). The fundamental repeating units of the crystal are the unit cell axes ***a***, ***b***, and ***c***, so that reflections will be observed at angles corresponding to these three directions. In addition, reflection can be obtained along all possible diagonals of the crystal, so that a three-dimensional crystal gives rise to a three-dimensional diffraction pattern. If we define a scattering vector **S** with magnitude $|2\sin\theta/\lambda|$ and direction perpendicular to the imaginary scattering plane, then it can be shown that all scattering vectors **S** lie on a three dimensional lattice defined by $\mathbf{S} = h\mathbf{a}^* + k\mathbf{b}^* + l\mathbf{c}^*$, where h, k, and l are integers and $\mathbf{a}^*$, $\mathbf{b}^*$, and $\mathbf{c}^*$, the reciprocal lattice vectors, are inversely related to the crystal lattice parameters $\mathbf{a}$, $\mathbf{b}$, and $\mathbf{c}$. Thus the positions of reflections within the diffraction pattern are solely dependent on the nature of the crystal; the phases $\phi(hkl)$ and amplitudes $F(hkl)$ (and therefore the measured intensities $I(hkl)$) are determined however by the contents of the unit cell. For a crystal, eqn [2] can be rewritten

$$\rho(xyz) = \Sigma_{hkl}\, F(hkl)\, \exp[i\phi(hkl)]\, \exp[-2\pi i(hx + ky + lz)] \qquad [4]$$

where x, y, and z are fractional coordinates within the unit cell and Σ_{hkl} represents the sum over all reflections hkl. It is convenient to describe the atomic positions, electron density, and crystal as belonging to real or direct space, while the amplitudes, phases, intensities and diffraction pattern are in reciprocal space.

3.21.2.3 The Solution of the Phase Problem

As mentioned above, the primary data from the x-ray diffraction experiment consist of the measured intensities $I(hkl)$ for all reflections h, k, and l. In order to calculate an electron density and thereby solve the structure, however, it is necessary to have the amplitudes $F(hkl)$ and their corresponding phases $\phi(hkl)$ (see eqn [1]). While the amplitudes can be obtained directly from the intensities (as $I(hkl) = \mathrm{F}^*(hkl) \cdot \mathrm{F}(hkl)$), the phase information (which contains the information necessary to solve the structure) is lost in most x-ray crystallographic experiments. It is therefore necessary to elucidate these phases using other means. Four fundamental methods exist for solving the phase problem: (1) heavy atom derivatives; (2) anomalous dispersion; (3) molecular replacement; and (4) direct methods. While direct methods are widely used for the solution of small-molecule (fewer than 1000 atoms) crystal structures, they are dependent on the availability of near-atomic resolution data, so that they are rarely used for biological macromolecule structure solution and will not be discussed here further.

3.21.2.3.1 Isomorphous replacement

The heavy atom method is the oldest of these techniques, first having been used for proteins in the study of hemoglobin.[13] The technique makes use of the fact that protein crystals contain large solvent channels (**Figure 2**) through which small-molecule compounds can diffuse and bind to the protein surface. If a heavy atom (defined as possessing significantly high electron density) binds to a distinct location within the unit cell, it will give rise to distinct changes in the phases of all reflections (**Figure 4**). While it is not possible to observe the phase change directly, the measurable change in the intensities of the diffraction pattern – some reflections become stronger, some weaker – can be related to the position of the heavy atom. This is achieved by way of the Patterson map, the key to the solution of the phase problem.

Patterson[31] showed that Fourier transformation of the experimental intensity data – with no prior phase information – yields a map in direct space that corresponds to an autocorrelation of the electron density map. This means that every peak in the Patterson map corresponds to a cross-vector between two atoms in the electron density map (**Figure 7**). For simple (few atom) molecules, it is possible to use the Patterson map directly to elucidate the structure; indeed, this procedure was how the first nontrivial crystallographic structures were determined. For a molecule of N atoms, however, the Patterson map will have $N(N-1)/2$ cross-vectors, so that the Patterson map becomes extremely crowded for large structures. Further complications of the Patterson map are a strong origin peak (as every atom has a 'cross-vector' to itself) and an ambiguity in direction – for every cross-vector from one atom to another, there is an equivalent cross vector in the opposite direction, and therefore the Patterson map is centrosymmetric.

The Patterson map can however be simplified considerably given a 'derivative' (i.e., heavy atom modified) data set that is isomorphous to the 'native' data set (i.e., the crystal form and cell constants of the respective crystals are identical). In such a case, the native and derivative Patterson maps can be subtracted from one another, revealing cross-vectors for the heavy atoms. In general, the number of heavy atom positions is sufficiently small that their absolute positions can be deduced from cross-vectors. As the Patterson map contains multiple peaks for each heavy atom position, however, it is important that all heavy atoms correspond to the same origin. This is particularly important when multiple heavy atom derivative data sets are being used to determine the phases.

As noted earlier, knowledge of the structure allows the calculation of both amplitudes and phases for the resulting diffraction pattern. Thus determination of the heavy atom positions results in a partial structure that can be used to calculate the heavy atom contributions to phases and amplitudes of the derivative data sets. This in turn is utilised to provide estimates of the native and derivative phases using the Harker construction (**Figure 8**).[32] This vector

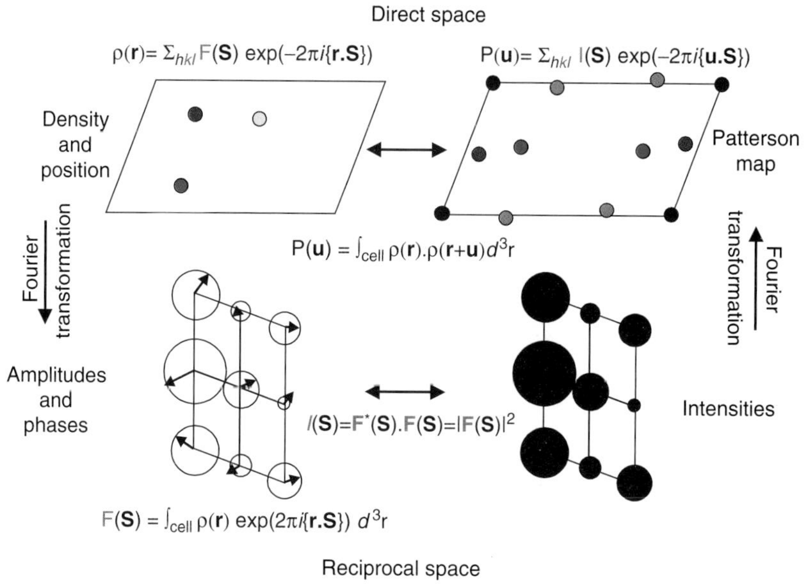

Figure 7 Relationships in diffraction space. The diffraction pattern of an object is the Fourier transform of the electron density, consisting of both amplitudes and phases. During the x-ray diffraction experiment, diffracted intensities are measured, however, the phase information is lost. The Fourier transform of the intensities results in the Patterson map,[31] which is related to the electron density as follows. For any two atoms in the structure, the vector between them, centered at the origin, has a value corresponding to the product of their densities. Thus, the red atom and the yellow atom result in the orange cross-vectors, the red and blue atoms result in the magenta cross-vectors, and the yellow and blue atoms result in the green cross-vectors. As each atom has a 'cross-vector' to itself, a large peak is found at the origin; the Patterson map is centrosymmetric. (Reproduced with permission from Stubbs, M. T. In *International Tables for Crystallography, Volume F: Crystallography of Biological Macromolecules*; Rossmann, M. G., Aruold, E., Eds.; Kluwer Academic Publishers: Dordrecht, The Netherlands, 2001, pp 256–260, with kind permission of Springer Science and Business Media.[62])

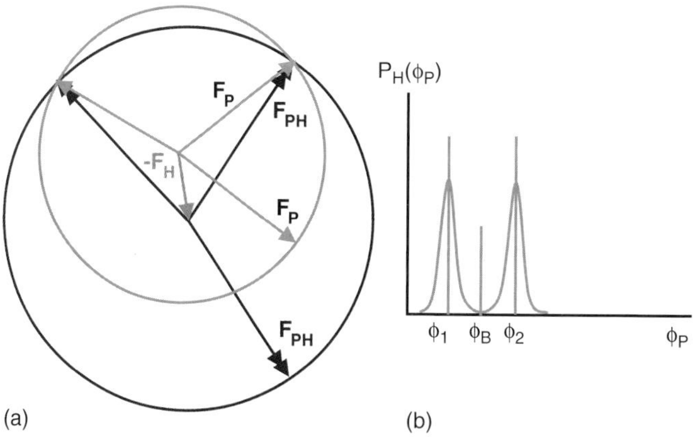

Figure 8 The Harker construction and phase ambiguity for single isomorphous replacement (SIR). (a) Solution of the heavy atom structure by Patterson methods allows the contribution of the heavy atoms $\mathbf{F}_H(hkl)$ to the scattered amplitude and phase to be estimated for any reflection (red Argand vector). Assuming that the derivative and native data sets differ only as a result of the substitution of the heavy atoms, the relationship $\mathbf{F}_{PH}(hkl) = \mathbf{F}_P(hkl) + \mathbf{F}_H(hkl)$ implies that two circles of radius F_{PH} and F_{PH} (the experimentally determined magnitudes) whose centers are separated by the vector $\mathbf{F}_H(hkl)$ should yield phases for $\mathbf{F}_P(hkl)$ and $\mathbf{F}_{PH}(hkl)$ at their intersection. SIR therefore results in two possible phases; due to errors in the data, these give rise to a phase probability (b). While ϕ_1 and ϕ_2 represent the most probable phases, the 'best phase' ϕ_B corresponds to the minimum phase error[33]; the distribution of the phase probability is recorded in the figure of merit (FoM). The phase ambiguity may be resolved using a second derivative (MIR) or anomalous scattering (SIRAS).

construction uses the known relationship between the measured amplitudes of the native and derivative reflections and the phased heavy atom contribution.

While this gives an estimate for the native phase, for each reflection there remains also an alternative phase that is equally probable (the so-called phase ambiguity). This ambiguity can be resolved through the use of two or more separate derivatives, with different heavy atom contributions and derivative amplitudes but with a common native phase (multiple isomorphous replacement or MIR). A phase probability can be calculated from these multiple measurements; the centroid of this probability distribution provides the so-called best phase.[33] The spread of the distribution is reflected in the figure of merit (FoM): a value of 1 indicates zero phase error, whilst a value of 0 marks a random phase distribution. As phases can be calculated for all reflections, the overall FoM indicates the quality of the phasing; an overall FoM of 0.5 corresponds to an average phase deviation of 60°. The variation of the FoM with resolution can be used to determine to which resolution the first experimental maps should be calculated – although the diffraction data may extend to high resolution, should there be large errors in the phases, inclusion of high-resolution data will result in noisy maps and obscuration of features necessary for structure interpretation. These high-resolution data can be used at a later stage however, for example during density modification or structure refinement (*see* Section 3.21.2.4).

Although MIR yields a set of phases that can be used to generate an electron density map, there still remains an ambiguity with regards to the handedness. This is a consequence of Friedel's law, which states that reflections $\mathbf{F}(hkl)$ and $\mathbf{F}(-h-k-l)$ have equal amplitudes but opposite phases; as only the intensities can be measured, $I(hkl) = I(-h-k-l)$, and the diffracted intensities are centrosymmetric (in short, that the diffraction pattern is the same regardless of whether it is measured in the 'forward' or 'reverse' direction). The handedness of the map can be determined by inspection of secondary structure elements such as α-helices: if the helices appear to be left-handed, then the (equally probable) wrong phase set has been chosen, and the opposite phases must be taken.

The absolute configuration of the data can be determined if an anomalous scatterer (*see* Section 3.21.2.3.2) is present. In this case, Friedel's law no longer applies, and small but significant differences between $I(hkl)$ and $I(-h-k-l)$ can be utilized to establish the correct hand. Anomalous data can also be used to resolve the aforementioned phase ambiguity in the case of a single derivative – this so-called single isomorphous replacement with anomalous scattering or SIRAS is gaining increasing popularity,[34] particularly since improved x-ray sources and detectors yield superior data of sufficient quality to provide accurate measurements of the small anomalous scattering contribution.

3.21.2.3.2 Anomalous dispersion

While it has been a tacit assumption that all atoms are composed of free electrons loosely associated around the nuclei, a special case arises when the frequency of the incoming x-rays (related to the wavelength by $\nu = c/\lambda$) approaches the resonance frequency of atoms within the crystal. Under these conditions, the secondary wave emanating from the resonant atom undergoes a phase shift compared to the other (nonabsorbing) atoms, modifying the diffraction pattern in both amplitude and phase, and the atom is said to be an anomalous scatterer. The form factor f (*see* Section 3.21.2.2, **Figure 4**) of the atom in question becomes modified by

$$f_{\text{anom}} = f + f' + if'' \qquad [5]$$

where f' represents the real component of the absorption (and is strongly negative near the absorbance edge), i is the imaginary number $\sqrt{(-1)}$, and f'' is the imaginary component of the form factor (and is 90° out of phase with f). It is customary to express both f' and f'' in units of electrons; they are strongly dependent on the wavelength, yet even at their maxima (resonance frequency), they are only fractions of the nonanomalous component f (see **Figure 13**).

As mentioned above, the phase shift results in a breakdown in Friedel's law, so that the intensity $I(hkl)$ measured for a reflection going through the crystal in one direction is (marginally) different to the corresponding $I(-h-k-l)$ reflection measured in the opposite direction. Similar to the methods used for the determination of heavy atom positions, the anomalous scatterers may be located using an anomalous Patterson map derived from the differences between $I(hkl)$ and $I(-h-k-l)$. Initial phasing then proceeds in a manner analogous to that used for isomorphous replacement. The primary advantage over the latter technique is that it is possible to derive a data set from a single crystal, practically eliminating any problems of nonisomorphism.

Although anomalous dispersion can be measured for a variety of heavy atoms using a conventional x-ray source, the advent of synchrotron radiation revolutionized the field of structural biology. The ability to tune the wavelength at synchrotrons gave birth to the method of Multiple wavelength Anomalous Dispersion or MAD.[21] The position of the adsorption edge of the atom in question is determined by measuring the x-ray fluorescence of the crystal as a function of x-ray energy (it is customary at synchrotrons to express the energy E of the x-rays in terms of eV – this can be converted into wavelength λ using the relationship $E = h\nu = hc/\lambda$, where Planck's constant $h = 4.1356 \times 10^{-15}$ eV s and the velocity of light $c = 2.9979 \times 10^{8}$ m s^{-1}). The peak of the fluorescence scan corresponds to the maximum absorption (f' is maximum), whilst the inflection of the scan corresponds to the maximum dispersion (f'' is maximum). Data sets are then collected from the crystal at wavelengths equivalent to the peak and the inflection, as well as a remote wavelength (far removed from the absorption edge; this can be at an energy above or below the edge or both). Provided that radiation damage is negligible, the three or four resulting data sets then differ only in the anomalous scattering contribution (if they can be collected from a single crystal). Maps resulting from MAD experiments are often of such high quality that they can be traced in a single session.

The choice of anomalous scatterer has also been heavily influenced by the possibilities opened through synchrotron radiation. While many 'heavy' atoms exhibit a strong anomalous signal at the corresponding wavelength, selenium, with an absorption K edge at 0.9800 Å (12 660 eV), has become one of the most popular atoms for ab initio structure determinations.[35] The reason for this lies in the fact that today most proteins are produced recombinantly. As many proteins also contain methionine residues, it is relatively straightforward to introduce Se atoms into the protein itself by feeding the recombinant organisms with selenomethionine (SeMet) while suppressing methionine biosynthesis, replacing a finite number of sulfur atoms with selenium. In the majority of cases, the SeMet protein is ostensibly identical to the native protein, and usually crystallizes isomorphously. Thus the use of SeMet MAD is one of the least 'intrusive' methods for determining the structure of biological macromolecules.

With improvements in synchrotrons, detectors and crystallographic hardware and software, the current trend appears to be moving away from MAD and toward single wavelength anomalous dispersion (SAD). As the name suggests, anomalous scattering data are collected at a single wavelength, usually at the peak. This has the advantage of reducing the exposure time as well as the risk of radiation damage. Again, SeMet crystal data collections currently predominate; in recent times, however, the use of halide soaks such as sodium bromide have gained ground[36] – these anomalous scatterers can be soaked at high concentration for a short time and then measured at the appropriate wavelength to provide high-quality phased data sets. Such a procedure obviates the somewhat time-consuming task of producing, purifying, and crystallizing SeMet labeled protein, allowing native and derivative data to be collected from a single crystal.

Finally, SAD can be carried out on in-house x-ray sources such as copper anodes. Increasingly, phasing can be done at home on crystals of sufficient size for data collection containing suitable anomalous scatterers. Particular promise has been shown for halides such as potassium iodide or cesium chloride, as well as phasing using the inherent anomalous signal from sulfur, present in cysteines and methionines.[34,37] The first steps have also been taken in using alternative anode materials such as chromium[38,39] – it remains to be seen whether this will take the place of other methods described here, however.

3.21.2.3.3 Molecular replacement

As more and more structures are being solved and deposited in the PDB, it is increasingly the case that a similar structure exists to that in your crystal. In such a situation, it can be easier to use the existing structure to obtain initial phase information and thereby solve the unknown structure. This technique, known as molecular replacement,[20] is also based on the Patterson map, and is therefore also sometimes known as the Patterson Search method.[19] Although the method is essentially an attempt to solve a six-dimensional search problem, with three rotational and three translational parameters defining the transformation of the known coordinates into the unknown crystal system, the Patterson map is used to reduce this into two consecutive three-dimensional operations, saving considerably on computation time.

Prior to commencing molecular replacement, it is necessary to know the number of molecules in the asymmetric unit. This is most often calculated using the so-called Matthews coefficient V_M[40]: given a unit cell volume V (Å^3) containing N asymmetric units (a function of the space group), a monomer molecular weight of M (Da) and n molecules in the asymmetric unit, the specific volume V_M is evaluated as

$$V_M = V/(N \cdot n \cdot M) \quad [6]$$

For proteins, V_M typically varies between 1.9 Å^3 Da^{-1} and more than 2.9 Å^3 Da^{-1}, with a minimum of 1.7 Å^3 Da^{-1} (corresponding to a solvent content of around 22%) and an average value of 2.4 Å^3 Da^{-1} Substitution of n into eqn [6] (remembering in the case of a macromolecular complex that M refers to total molecular weight) to yield a V_M in the desired range usually provides a good estimate of the number of molecules in the asymmetric unit. In general, a small value of V_M indicates a low solvent content and therefore high resolution and vice versa; thus if your crystal only diffracts to low resolution, you should expect a higher than average V_M.

The first stage in molecular replacement is to generate the molecular transform of the search model. This is carried out computationally by placing the model in a large triclinic unit cell (so large that no complications are caused by

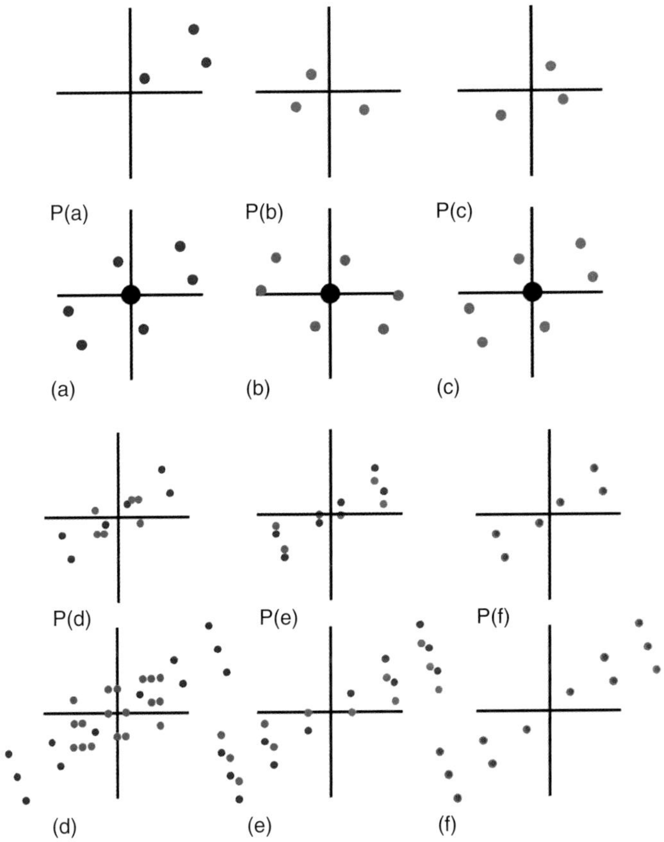

Figure 9 Molecular replacement: use of the Patterson function to position a known molecule (red) in an unknown unit cell (blue). (a–c) The rotation function allows the orientation of the molecule to be found by use of an intramolecular vector set. The Patterson map **P(a)** of the unknown molecule is synthesized from Fourier transformation of the measured intensities. Using the coordinates of the known structure b, a Patterson function is calculated **P(b)** and rotated around the origin until a high correlation is found with the experimental Patterson (**P(c)**). Applying the rotation matrix thus found to the original coordinates yields the correct orientation of the molecule (c). Note that the Patterson functions **P(a)** and **P(c)** coincide despite the different positions (a) and (c). (d–f) The translation function positions the correctly oriented molecule relative to the crystallographic origin (in this example P2) using cross-vectors between symmetry-related molecules. The oriented molecule is placed at various positions within the unit cell, the Patterson function calculated and compared with the experimental map. At maximum correlation, the molecule has been correctly positioned, and phasing may commence.

interference between 'neighboring' molecules in the synthetic crystal) followed by Fourier transformation and model Patterson map synthesis. The model Patterson is then rotated against the experimental Patterson map (**Figure 9**) and the correlation coefficient between the two maps is evaluated for each angle. The resulting rotation function should yield a high peak when the model Patterson is in the correct orientation. For multiple identical molecules in the asymmetric unit, the rotation function correlating the experimental Patterson map to itself after rotation can also be used to determine the orientation of noncrystallographic symmetry axes.

Once the orientation has been determined, the translation function must be evaluated to place the model structure in the correct position of the new unit cell. To this end, the correctly oriented model can be translated through the unit cell, a model Patterson calculated, and the correlation to the experimental Patterson determined (**Figure 9**). For computational reasons, however, the translation function is normally calculated in reciprocal space. It is useful to carry out the translation function in all possible space groups that match the Laue symmetry – e.g., if the crystal form is $P3_1$, then $P3_2$ and P3 should also be tested. If the solution is correct, then there should be a significant difference for the translation function results using the correct space group. Should there be several molecules in the asymmetric unit, then each corresponding rotation peak must initially be searched individually. The molecular replacement search is only complete however when all molecules have been assigned to the same origin (as was the case for the Patterson solution for heavy atoms; *see* Section 3.21.2.3.1). This involves keeping one (reference) molecule fixed and carrying out the translation function for the other molecules relative to this. The correct solution should then give a significant increase in the correlation coefficient.

While the separation of the search into a rotation and translation component reduces computation considerably, this can also lead to slight deviations from the correct orientation(s). For this reason, it is advisable to carry out a rigid body refinement of the resulting model before proceeding further. In this case, each model is allowed to be rotated and translated in small increments from its start position. As a measure of the quality of the correctly oriented and positioned model(s), the crystallographic R-factor R_{fac} is minimized:

$$R_{fac} = 100 \cdot \Sigma_{hkl}\{||F_{obs}(hkl)|-\kappa|F_{calc}(hkl)||\}/\Sigma_{hkl}|F_{obs}(hkl)| \quad [7]$$

where $F_{obs}(hkl)$ is the measured amplitude for an individual reflection *hkl*, $F_{calc}(hkl)$ the amplitude calculated from the model for the same reflection, κ a factor that scales the calculated amplitudes to the observed data (independent of *h*, *k* and *l*), and Σ_{hkl} represents the sum over all reflections in the resolution range chosen for the search. The absolute value of R_{fac} will depend on the degree of similarity of the search model to the final structure; values between 40% and 50% are usually indicative of a correct solution (values under 20% are generally taken to represent a fully refined structure (see below); a random arrangement of atoms gives rise to a value of around 55%).

Before embarking on molecular replacement, it is important to realize that the method depends more on structural similarity than on sequence similarity. It is therefore recommended to carry out all searches at low (around 3 Å) resolution, where differences between the search target molecules are less significant. If the desired structure exhibits low sequence identity to the search model (less than say 50%), then it is prudent to try a number of different trial models that may show less sequence identity. Similarly, if the target protein consists of multiple domains and it is suspected that these can act independently of one another, or that parts of the molecule could exhibit high flexibility, then it might be better to take smaller fragments as search molecules – with the proviso that as the fraction of the asymmetric unit covered by the search molecule is reduced, so the chance of finding a significant rotation and translation function solution decreases.

3.21.2.4 From Map to Model

Having obtained initial phases as described in the previous section, an electron density map can be calculated by Fourier transformation. It is not always (!) possible to recognize features in a first electron density map, however (**Figure 10a**). In order to improve the map (phases), it is necessary to use information about protein crystals and their electron densities. One of the most universally applied forms of initial phase improvement is solvent flattening. Protein crystals contain large amounts of solvent; this will in general be disordered, and so will not contribute to the crystal diffraction.

By knowing the protein content of the crystal (from calculating the Matthews coefficient V_M; see above), it is therefore possible to estimate the threshold density below which is noise; points with density below the threshold are set to a suitable average value. This method is particularly useful for locating molecular boundaries (**Figure 10**). If the asymmetric unit contains more than one molecule, then real space averaging can lead to a dramatic improvement in the map. Provided that the transformation between individual molecules can be identified (e.g., by the coordinates of heavy atoms or anomalous scatterers, or through location of a noncrystallographic symmetry axis), then the densities of the various copies can be equivalenced to yield an averaged density.

These density modification procedures (which also include so-called histogram matching and map skeletonization) become particularly powerful when carried out in a cyclic process (**Figure 10**). Repeated density modification, phase calculation and combination, and map calculation until convergence results in most cases in an electron density map of superior quality, and can even be used for phase extension, i.e., determination of phases to a higher resolution than those at which the experimental phases are reliable. Once again, the FoM can be used to ascertain the quality of the phases.

With reasonable phases and sufficiently high resolution, the initial interpretation of the electron density map should pose few problems; at resolutions better than about 2 Å (**Figure 11**), modern programs can even build a model completely automatically. In many cases, however, it will be necessary to build at least part of the model manually. The general procedure is (1) identification of secondary structure elements, (2) interpretation of possible side chains, assigning them as large or small, interior or exterior, and (3) matching of the protein sequence to the density. The ease with which this process can be carried out is dependent on the quality of the phases and the resolution of the data (**Figure 11**). It is important to remember that the connectivity of the electron density (i.e., the continuity of the density for connected atoms in the protein) can be disrupted by errors in the phases, mobility of the residues involved

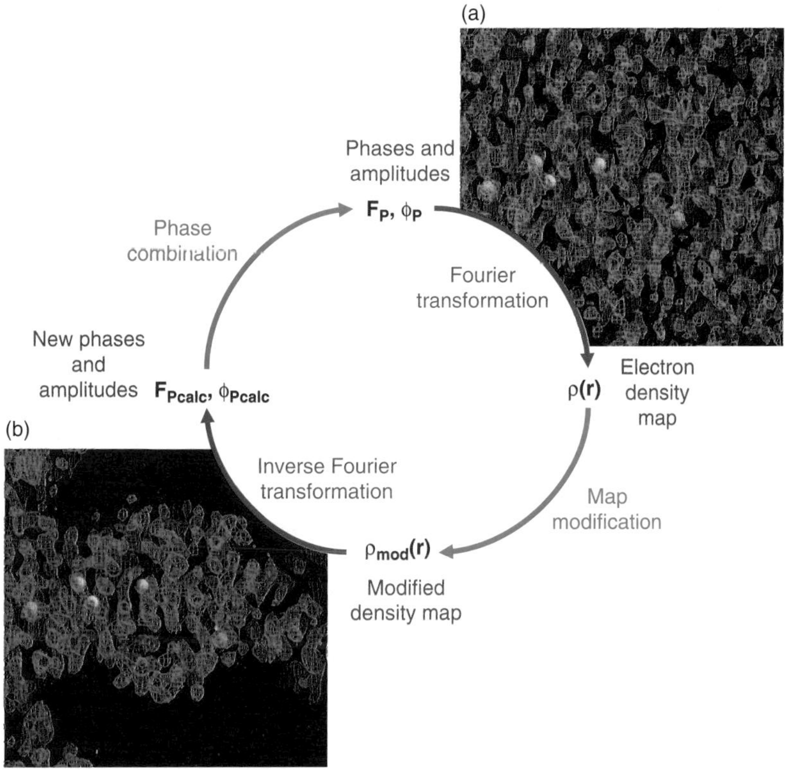

Figure 10 Density modification. (a) The initial electron density map from a SeMet MAD experiment was of insufficient quality for model building (yellow spheres reveal the positions of the Se atoms used for phasing). (b) Repeated cycles of solvent flattening and phase recombination resulted in clear differentiation between protein and solvent and allowed interpretation of secondary structure (β-strands are clearly visible). Interestingly, although there are two molecules in the asymmetric unit, electron density averaging failed to converge; subsequent completion of the structure revealed significant conformational differences between the two monomers. Data taken from the structure determination of surfactin thioesterase domain SrfTE.[63]

(leading to reduced or missing density), alternative conformations or ambiguous density – one example is the presence of disulfide bridges, which can cause problems at lower resolutions. Clues as to the possible interpretation of individual side chains can come from neighboring heavy atom positions used in the phase determination (in particular, SeMet phasing reveals the positions of methionines at an early stage of the interpretation), or to already interpreted and well-defined surrounding atoms (such as charged groups, where charge compensation is usually necessary).

In nearly all cases, the first building session will result in a partial model with errors. To this end, the model must be refined – the *xyz* positions of the atoms of the model are allowed to move (and their B-factors allowed to vary) according to a modified force field that resembles a molecular dynamics simulation, with a 'pseudoenergy' related to the R_{fac} described in eqn [7]. Unless working at atomic resolution (more the exception than the rule for biological macromolecules), refinement of the structure requires that geometrical parameters such as bond lengths, bond angles, torsion angles etc. are restrained to minimize deviation from ideal values. Simulated annealing,[41] in which the model is heated in silico to high temperatures and allowed to cool down slowly, provides a method to overcome local energy barriers and is often valuable in the early stages of refinement. At low resolution, the parameter to observable ratio can be improved by applying constraints, where geometrical parameters are fixed to ideal values. In place of free refinement of the x, y, and z coordinates for each atom, torsion angle dynamics can be particularly useful in reducing the number of variables.

During the molecular dynamics run, the pseudoenergy term is used to decrease the R_{fac} while maintaining the standard geometries of amino acids and proteins. If the model is for the most part correct, then the reduction in the R_{fac} will reflect an improvement in the phases, and a new map calculated from the ensuing phases should indicate clearer electron density for previously uninterpretable regions. At early stages of the structure solution, it is wise to combine the model phases with any experimental phase information at hand (from, e.g., MIR measurements) to avoid model bias. The latter is a very real problem for x-ray crystallography, in particular for molecular replacement – the phases tend to be dominated by the model, so that subsequent electron densities have a tendency of reproducing

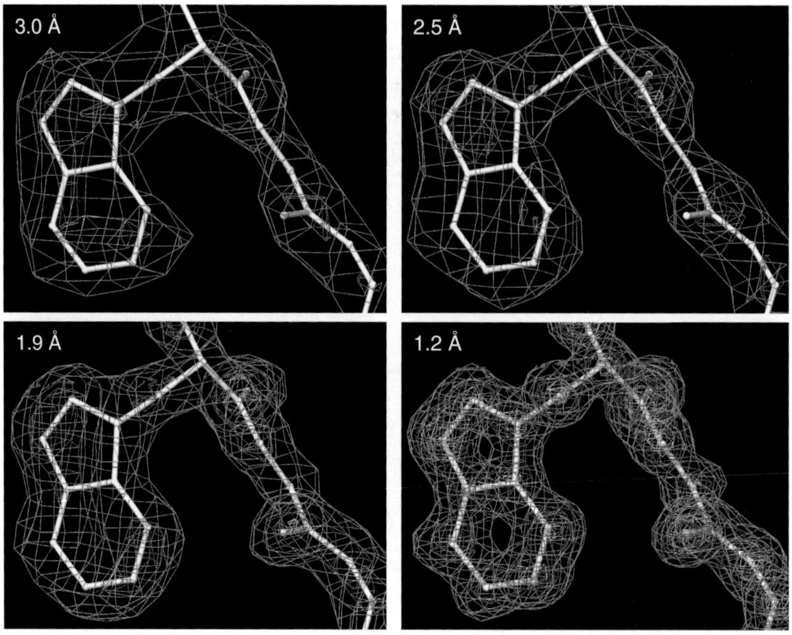

Figure 11 Information content of $2F_{obs} - F_{calc}$ maps at increasing resolution; each map contoured at 1.0 σ (light blue), 2.5 σ (magenta), and 3.7 σ (yellow). Density shown for a Trp-Gly dipeptide from trypsin phased at the highest resolution.[64] Given adequate phases, the overall shape and Cα trace can be evaluated at resolutions between 5 and 3.5 Å. Side chains can be discerned at 3 Å, carbonyl oyxgens at 2.8 Å, while side chains and peptide bond planes are well resolved at resolutions better than 2.5 Å. Holes in the aromatic rings of Phe and Tyr at resolutions of 1.5 Å are indicators of correct phases, and individual atoms can be observed at 1.2 Å; at high resolution, hydrogen atoms can even be visualized.

model density, even when this is incorrect. For this reason it is essential during density interpretation to only build model fragments where there is a degree of certainty (this of course is relative, but a conservative interpretation is generally advisable). In particular, solvent and substrate molecules should only be built into the density at late stages of the refinement.

To avoid model bias (or to be able to recognise when it occurs), the free R-factor R_{free} is generally monitored during refinement.[42] R_{free} is calculated in exactly the same way as R_{fac} (eqn [5]), but for only a fraction (between 5% and 10%) of the reflections (the test set). The essential difference is that reflections in the test set are not used for refinement (for this, the remaining 90–95% reflections of the working set are used) and are therefore independent of the model. The R_{free} calculated in this way should decrease with refinement as the R_{fac} (R_{free} is generally 2–5% higher than R_{fac}) if the model is correct; an increase in R_{free}, however, indicates that something is wrong with the model and it is necessary to backtrack.

Recalling eqn [4], if we had perfect phases, then

$$\rho_{obs}(xyz) = \Sigma_{hkl} F_{obs}(hkl)\ \exp\{i\phi_{calc}(hkl)\}\ \exp\{-2\pi i(hx + ky + lz)\} \qquad [8]$$

The model phases are of course imperfect; if they are close to the correct solution, however, then a difference Fourier map calculated according to

$$\Delta\rho_{obs}(xyz) = \Sigma_{hkl}\{F_{obs}(hkl) - F_{calc}(hkl)\}\ \exp\{i\phi_{calc}(hkl)\}\ \exp\{-2\pi i(hx + ky + lz)\} \qquad [9]$$

should reveal errors or omissions in the model. Positive difference density $\Delta\rho_{obs}$ indicates that atoms are missing from the model, while negative difference density shows that an atom should not be at this position, or that its temperature factor is too low. Thus a so-called ($F_{obs} - F_{calc}$) map (corresponding to $\Delta\rho_{obs}$) is particularly useful in completing the structure. As ρ_{obs} suffers from model bias, it is customary to work with maps relating to ($\rho_{obs} + \Delta\rho_{obs}$) or ($2F_{obs} - F_{calc}$) maps: electron density for missing atoms is upweighted, while that for wrongly placed atoms is downweighted. Model bias can be reduced in part by calculating so-called omit maps, where a portion of the structure is removed from the refinement and phasing calculations prior to difference map synthesis. Electron density maps are commonly scaled according to their standard deviations σ; it is usual to contour ($2F_{obs} - F_{calc}$) maps at 0.8–1.0 σ and ($F_{obs} - F_{calc}$) maps at 2.5–3.0 σ (**Figure 11**).

Most structure determinations involve multiple cycles of building and refinement before the R_{fac} (and R_{free}) converge to reasonable values (around or below 20%), and so patience is required. The problem of model bias should always be borne in mind, as should the fact that the phases of all reflections contribute to the synthesis of an electron density map, and that all atoms of the model contribute to the phases of all reflections. This latter point is often neglected; a consequence thereof, however, is that a poorly or wrongly interpreted portion of the molecule remote from the point of interest (e.g., the binding of a ligand at the active site of an enzyme) may yet influence the electron density at this region. Particular care should be taken in the placement of solvent molecules, which are not constrained and laborious to insert and check by hand.

3.21.3 Practical Crystallography

3.21.3.1 The Generation of X-rays

X-rays are generated when charged particles interact with an electromagnetic field. For the purposes of x-ray crystallography, x-ray generation is provided by the acceleration of electrons. The first x-ray tubes involved a heated cathode and an anode sealed in a vacuum. Electrons generated at the cathode are accelerated toward the anode target; the ensuing deceleration in the target gives rise to a continuum of x-rays ('white radiation', also known as *Bremsstrahlung*) (**Figure 12**). The minimum wavelength as well as the intensity distribution of the rays is dependent upon the accelerating voltage. In addition, a series of intense spectral lines is superimposed upon the 'white' background that is characteristic of the nature of the anode target used. These lines correspond to the excitation of inner electrons in the target by bombarding electrons of sufficient energy; relaxation of the excited electrons back to their ground state results in quantized x-ray production of defined wavelength. The most commonly used target for protein crystallography is copper, whose strongest spectral line (Cu K_α radiation, resulting from the L to K energy level transition) has a wavelength of 1.5405 Å. Molybdenum, whose short Mo K_α wavelength of 0.7093 Å interacts only weakly with biological macromolecules, is more often used by small-molecule crystallographers, while chromium has attracted interest in recent years as the Cr K_α wavelength of 2.29 Å is suitable for a wider range of anomalous scattering techniques (*see* Section 3.21.2.3.2).[38,39]

Clearly, bombardment of an anode with high-energy electrons leads to serious problems of heat dissipation. While so-called sealed tube generators with stationary targets are relatively easy to maintain and to align, the heating problem presents obvious limitations to the maximum intensity obtainable. As described above, the interaction of x-rays with biological matter is particularly weak. In order to overcome this, rotating anodes have been developed, which now represent the workhorse of most in-house sources used for structural biology. For particularly small or weakly scattering crystals, however, the use of synchrotron radiation has become indispensable. In a synchrotron, relativistic electrons (or positrons) are held in a closed orbit by way of a series of bending magnets. The centripetal acceleration experienced by the electrons toward the center of the ring results in an intense beam of white electromagnetic radiation, ranging from infrared to hard x-rays (**Figure 13**). Higher intensities can be obtained using so-called insertion devices such as wigglers and undulators: multipole magnets inserted into straight sections of the electron beam path between bending magnets that allow a controlled deflection of the electrons.

For nearly all single crystal projects, monochromatic radiation is required. In former times, this was achieved by placing a filter in the beam path: for Cu radiation, for example, a nickel filter will absorb most of the Cu K_β radiation, leaving a relatively clean Cu K_α beam. According to Bragg's law, a well-defined and precisely oriented crystal will allow only a single wavelength to pass through. Applying a slight curvature to a crystal mirror allows focusing and collimation of the x-ray beam to a certain extent, so that most modern x-ray sources are fitted with perpendicular mirror optics to focus the beam in the horizontal and vertical directions. More recently, mirrors coated with graded multilayers have become available that can provide higher flux or lower beam divergence. Continuing advances in x-ray optics have opened the way to microfocus x-ray sources with reasonable x-ray flux yet lower running costs. Monochromators are particularly important for MAD and SAD experiments carried out at synchrotrons – to be able to collect a reliable anomalous data set at the absorption edge, it is necessary to have a precision in wavelength of the order of 0.0001 Å (corresponding to 2 eV; **Figure 13**).

3.21.3.2 Handling of Crystals

To be able to measure crystal diffraction, it is necessary to remove the crystal from the drop in which it has crystallized and introduce it to the x-ray camera. As first observed by Bernal and Crowfoot, protein crystals tend to be fragile and are particularly sensitive to humidity changes. It is therefore necessary to maintain sufficient moisture within the crystal

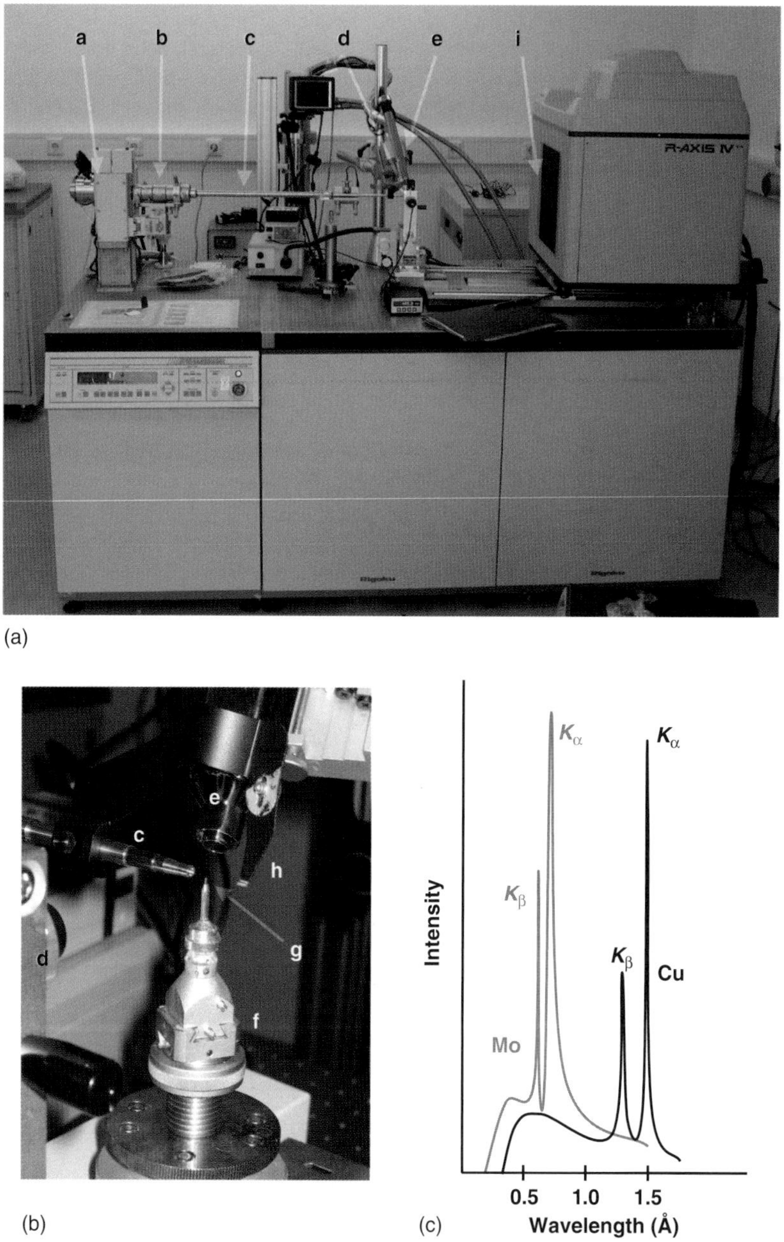

Figure 12 (a) A typical in-house x-ray crystallographic facility: (**a**) rotating anode; (**b**) confocal mirror optics; (**c**) x-ray collimator; (**d**) microscope for aligning crystal in x-ray beam; (**e**) cryo-cooling equipment; (**i**) image plate detector. (b) close up of crystal: (**c**) x-ray collimator exit; (**d**) microscope; (**e**) cryohead; (**f**) goniometer head for positioning crystal; (**g**) loop containing frozen crystal; (**h**) beamstop to prevent primary x-ray beam reaching detector. (c) Characteristic x-ray spectra for Cu and Mo anodes; note the sharp inner shell transition lines superimposed on the white *Bremsstrahlung* radiation.

during data collection. While it is sometimes possible to remove the crystal from the crystal plate and bring it directly to the camera, this can be risky, particularly if several crystals are present in the drop. In most cases, it is wise to prepare a harvesting buffer. This differs from the mother liquor in which the crystal grew in one crucial respect – the mother liquor contains saturating concentrations of dissolved protein in equilibrium with crystalline material. To compensate for this lack of protein, the harvesting buffer contains a higher concentration of the precipitating agent used for crystallization, so that transfer should not lead to dissolution of the crystal (which can be the case if the reservoir solution of the crystallization setup is used to mount the crystal). The harvesting buffer is also used for soaking

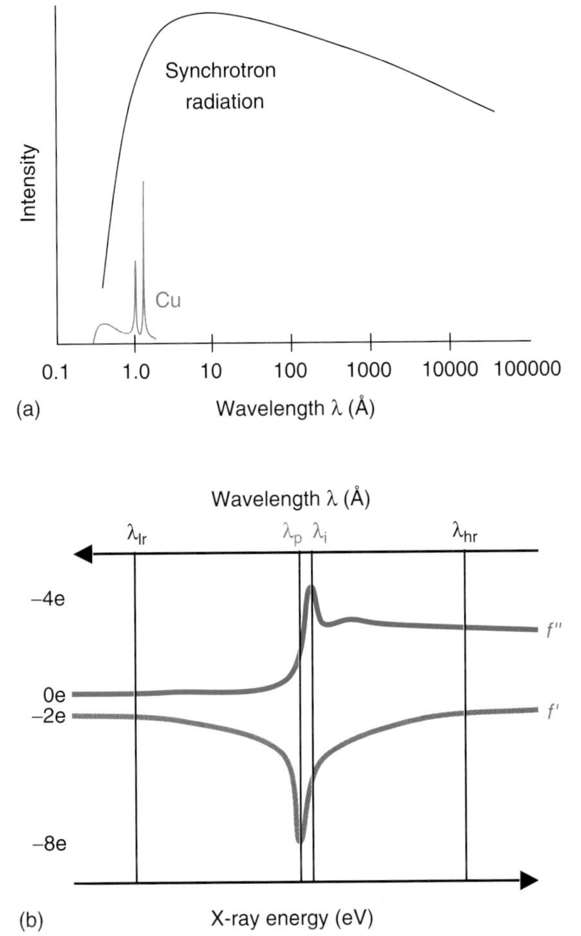

Figure 13 (a) Comparison of the radiation spectrum from a synchrotron source and a conventional rotating anode; not only is there a tremendous increase in intensity, but there is also a wide spectral range. (b) A typical anomalous scattering curve for Se (atomic number 34). Theoretically, the absorption maximum occurs at 12.6578 keV (peak $\lambda_p = 0.9795$ Å) and the dispersive maximum at inflection $\lambda_i = 0.97935$ Å; these values can vary due to the atomic environment or the synchrotron and should therefore always be measured experimentally through an x-ray fluorescence scan. For MAD experiments, additional data can be collected below (low remote, λ_{lr}) or above (high remote, λ_{hr}) the absorption edge.

experiments, where a crystal is placed in a buffer in which heavy atoms, halides, substrates, ligands, or inhibitors are dissolved, allowing diffusion of the compound into the crystal prior to subsequent measurement.

To handle crystals, it is helpful to have a polarizing stereo microscope, a set of micro tools (e.g., fine mounted needles such as used by dentists, siliconized capillaries, wax, and/or fine nylon loops) and a steady hand. Once the crystallization vessel has been opened, it is necessary to work quickly, as the drops can dry out and extinguish all hopes of obtaining a reasonable diffraction pattern, so all materials should be at hand before mounting. Crystals often tend to stick to the crystallization plate base or cover slips, but can be prised off gently using a needle. It is also sometimes necessary to cut crystals that have grown together – this can usually be achieved by applying gentle pressure to the interface between the two crystals with a fine needle.

For measurements at room temperature, the crystal can be drawn into x-ray capillaries. Although the crystal must be maintained in a humid environment, there is a high risk of crystal slippage during data collection if too much liquid is present, which is detrimental to obtaining a useful data set. Moisture must therefore be removed carefully from the crystal (using drawn-out glass capillaries or thin strips of filter paper) until only a thin film of liquid surrounds it that adheres it to the capillary wall. A drop of harvesting buffer is then introduced to one end of the capillary, which is then sealed at both ends using wax, making sure to avoid heating of the crystal. The crystal can then be placed in the x-ray beam by positioning the capillary on the goniometer head, the interface between crystal and x-ray camera (**Figure 12**).

Today, most crystals are measured using Cryocrystallography[43]: due to reduced radiation damage at low temperatures, more data can be collected from a single crystal. As a result of the high solvent content of protein crystals, it is usually necessary to prepare harvesting buffers containing a cryoprotectant to prevent the formation of ice crystals, which would lead to crystal deterioration and cracking. Common cryoprotectants are glycerol (5–30%), ethylene glycol (20–40%), polyethylene glycol (PEG)400 (10–30%), and paraffin oil (100%).[44] The choice of protectant and concentration is dependent on the harvesting buffer composition and on crystal compatibility; it is therefore advisable to prepare a range of these. Prior to crystal mounting, the minimum concentration of cryoprotectant should be tested at which the solution freezes as a glass. Cryo conditions are achieved by introducing the solution into a small nylon loop of dimension slightly larger than the crystal to be examined, and mounting it in a carefully controlled stream of nitrogen gas at about 100 K (**Figure 12**). Once the appropriate cryobuffer has been chosen, the crystal can be fished out of its mother liquor using the loop, introduced to the cryobuffer for a few seconds, then mounted on the goniometer head into the cryostream and x-ray beam, ready for measurement.

Freezing of crystals can lead to a deterioration of the diffraction pattern, although this can only be ascertained if a separate measurement is made at room temperature. The resolution of the diffraction pattern can however be improved by 'crystal annealing' – the frozen crystal is removed from the cryostream, allowed to thaw, and then returned to the stream. While this does not work for every crystal, it is always worth attempting when desperation sets in. Unfortunately, this process is difficult to control; a more promising method seems to be the free mounting system, where the crystal is placed in a stream of air of controlled humidity.[45] The humidity can then be manipulated while monitoring the x-ray diffraction pattern, allowing possible transitions in crystal quality to be followed.

3.21.3.3 The Detection of X-rays

Traditionally, x-rays were detected using photographic film; film has a large dynamic range and small grain pixel size, affording high spatial resolution of diffraction spots. On the other hand, film is relatively insensitive, requiring long exposure times, and data evaluation (involving offline developing, fixing, and densitometry of the films) is a lengthy and tedious process. Early attempts to use electronic detection devices (such as scintillation counters used for the structure determination of lysozyme)[46] were useful for crystals with small unit cells, but became less suitable as the number of reflections increased. Today, all measurements are made using area detectors, which collect a large number of reflections for a single crystal orientation. Detectors are divided into photon counters and integrating detectors. As the name suggests, photon counters measure the individual x-ray photons that contribute to a single reflection; it is therefore possible to follow the image building up with time. The most widely distributed photon counters are multiwire proportional counters, which measure the x-ray induced ionization of xenon gas between a fine anode wire and a planar cathode.[47,48] Such detectors are however expensive to maintain and show limitations at high count rates.

More widespread are integrating detectors in the form of image plates (**Figure 12**) and charge coupled devices (CCDs). In contrast to photon counters, integrating detectors count the total number of photons during the exposure period. Reminiscent of photographic film, the detection of x-rays in an image plates proceeds via the excitation of a BaFBr: Eu phosphor to a metastable energetic state.[49] Instead of chemical development, however, the phosphor can be read out using a red HeNe laser, resulting in luminescence of the stored image which can be detected by a photomultiplier. The image plate is then erased by application of a strong light pulse, preparing it for exposure of a subsequent image. Image plate systems are suitable for long exposures; the lengthy readout time is however a disadvantage when working with high fluxes, such as at a synchrotron. In this case, CCDs are preferred. X-rays are converted to visible light by way of a phosphor screen, which is then recorded directly using a CCD camera.[50] As they provide a very rapid readout, CCDs are ideal for strongly scattering crystals or high intensity sources. However, they tend to be very expensive, and the buildup of background noise with long exposure times make them less suitable for weakly scattering crystals on home sources. A particular problem of CCDs is their small active detection area – in order to cover the whole of reciprocal space, it is usual to have the CCD mounted on a camera that is able to rotate the crystal around several axes (so-called kappa, three-circle and four-circle geometries) with the ability to swing out the detector in 2θ.

3.21.3.4 Data Collection

Regardless of the detector used, the steps necessary to collect a data set are as follows: (1) crystal characterization and indexing; (2) data collection strategy; (3) integration of diffracted intensities; and (4) scaling of the data set. Visual inspection of the first diffraction pattern can already provide much information to the experienced observer: the resolution is easily calculated from the experimental geometry, the quality of the spots can be ascertained (splitting or smearing of reflections indicates either a cracked crystal, disorder, or a particularly long unit cell axis) and the mosaicity (which represents the angular deviation from an ideal crystal and determines over how many images individual reflections are spread) estimated. It is useful to take at least two images at 90° to each other to access any possible anisotropy of the diffraction data. If there are problems separating neighboring reflections, this might be due to a long cell axis; in this case, the detector should be moved back to a longer distance from the crystal to spatially resolve the axis – but it should be remembered that this will be at the expense of the resolution of the diffraction pattern.

Most detector software is able to index the diffraction pattern on the basis of one or two images. Indexing starts with selecting a number of strong reflections on the image and attempts to fit these positions to a single set of unit cell parameters a, b, c, α, β, γ and their orientation to the x-ray camera system. Success in indexing requires prior knowledge of the camera geometry, in particular the crystal-to-detector distance, the position of the beam center on the detector, and the x-ray wavelength. If a suitable solution for the unit cell axes can be found, various suggestions for the crystal form are presented – for example, if $a=b$ and $\alpha=\beta=\gamma=90°$, then a tetragonal space group is possible. This information is then used to determine a data collection strategy – the higher the crystal symmetry, the less data need be collected for a complete data set. The orientation of the crystal axes are used to define the best start and end positions (angles) of the data collection.

The data collection strategy followed depends on the type of experiment being carried out. For simple ligand soaks, for example, a complete data set of relatively low redundancy is sufficient. Completeness is defined as the ratio of measured to expected reflections, and two values are generally quoted: the overall completeness and that for the highest resolution shell. Redundancy corresponds to the total number of single measurements for an individual reflection $I(hkl)$, and is also quoted as overall and highest shell values. On the other hand, in-house SIRAS measurements ask for high redundancy, achieved by making multiple measurements of all reflections, by which the weak anomalous signal can be detected through averaging of the reflections. A third strategy is recommended for MAD measurements – after collecting the first complete data set for a particular wavelength, the crystal is rotated by 180° and the corresponding Friedel pairs are collected. Other criteria may also come into play, such as availability of beam time, radiation damage, or detector geometry.

It should be noted that the strategy is dependent on the space group assignment. Should the actual space group exhibit a lower symmetry (and this happens surprisingly often), then more data must be collected to achieve a complete data set. The correct space group can only be assigned following integration and scaling – it is therefore good practice to integrate and scale a subset of the data while the measurement is still running.

Integration involves determining the (partial) intensity for each reflection as it appears on individual images. As the data have been indexed, it is possible to predict in advance when and where individual reflections will appear. The intensity of each reflection hkl is integrated, subtracted from the local background and assigned both an intensity value $I(hkl)$ and an error estimate $\sigma(I(hkl))$. Reflections are divided into fully and partially recorded – partials are distributed over several consecutive images, while fully recorded reflections appear on only one image and can be measured more accurately.

To obtain the final data set, it is necessary to merge each partial measurement to a single value $I(hkl)$ and deviation $\sigma(I(hkl))$. Furthermore, multiple- and symmetry-related measurements of the same reflection are scaled to a single average value $\langle I(hkl)\rangle$, resulting in a unique data set – in the case of anomalous scattering, it is important to maintain separation of the Friedel pairs $\langle I(hkl)\rangle$ and $\langle I(-h-k-l)\rangle$. In addition to completeness and redundancy, a variety of criteria are used to assess the quality of the data set. The average value of $\langle I/\sigma(I)\rangle$ is a good indicator for judging the resolution of the data set; by dividing the data into resolution shells (as discussed for the completeness and redundancy), the shell in which $\langle I/\sigma(I)\rangle$ is still greater than 2 for more than half the reflections is chosen as the highest resolution limit. The scaling R-factor R_{sym} indicates the deviation of individual measurements of multiple (symmetry related) measurements from the final averaged values:

$$R_{sym} = \Sigma_{hkl}\{|I(hkl) - \langle I(hkl)\rangle|\}/\Sigma_{hkl}\, I(hkl) \qquad [10]$$

Obviously, R_{sym} is also dependent on the redundancy of the data – the higher the redundancy, the more the number of individual measurements and therefore the higher will be R_{sym}; on the other hand, a larger number of measurements means a more accurate determination of the correct intensity $\langle I(hkl)\rangle$. R_{sym} is also quoted for all data and for the highest resolution shell; overall values between 0.05 and 0.10 and highest shell values between 0.30 and 0.50 are generally deemed acceptable.

It is common to place the final data set on an absolute scale using Wilson statistics,[51] where plotting the logarithm of the average intensity as a function of $\sin^2\theta/\lambda^2$ yields a scale factor corresponding to the total number of electrons in the cell and the overall B-factor for the crystal. The final intensities and sigma values are then converted to amplitudes $F(hkl)$ and $\sigma(F(hkl))$ for further analysis.

An issue not dealt with above is that of twinning, where two or more lattices are present in the one macroscopic crystal.[52] If the two lattices do not superimpose (epitaxial twinning), then the diffraction pattern reveals split reflections, and it may be possible to index on a single lattice. Merohedral twinning on the other hand occurs when two or more lattices match each other perfectly – this can be the case for two trigonal lattices with coincident *c* axes, for example, so that the crystal may masquerade as hexagonal. Twinning poses special problems in x-ray analysis, and if undetected may even prevent a successful structure determination.

3.21.4 Quality Control in Protein Crystallography

Finally, some brief comments are provided to the criteria by which a successful crystal structure determination can be judged. The final structure is of course dependent on the quality of the measured diffraction data. While it is important to collect data at the highest possible resolution, care should be taken that the strength of the reflection intensities are significant ($I/\sigma(I)$ criterion), that the data are complete (this can be a particular problem if the space group exhibits lower symmetry than that suggested from the unit cell axes), and that the data are well scaled (R_{sym} test). It is better to work with a data set at lower resolution with strong intensities and low R_{sym} values than high resolution and poorer data. A valuable maxim to follow is that the crystal currently being measured might be the last one that diffracts, so it is imperative that sufficient data are collected for the subsequent structure determination – the story of 'the one that got away' is probably common to every crystallographic laboratory. Nevertheless, careful data collection cannot rule out pathological cases such as merohedral twinning, which is often only recognized when a structure solution fails to materialize.

As should be evident from this chapter, by far the most important factor for a successful structure determination is the quality of the phases – features of a well-phased low-resolution map are more reliable than those of a poorly phased high-resolution map. Model bias can be a particular problem when applying molecular replacement techniques; if this is suspected, then experimental phase information from isomorphous replacement or anomalous dispersion should be strived for. The quality of experimental phases can be estimated from the figure of merit FoM; the crystallographic R-factors R_{fac} and R_{free} provide a measure of the agreement between the final model and the diffraction data. Nevertheless, a low R_{fac} is only indicative of a correct structure if the model satisfies known geometrical constraints, such as bond lengths and angles and the Ramachandran plot for proteins.

Despite enormous technological and computational progress in the determination of x-ray structures, there remain a number of pitfalls for the practicing crystallographer. In all but the most perfectly phased high-resolution structure determinations, there remains a degree of subjectivity in the interpretation of the available electron density. While many crystallographic investigations continue to reveal unexpected results (for example, the existence of a DAsp residue in porcine pancreatic trypsin),[53,54] such interpretations should be reserved for the final stages of structure solution. Ambiguous density may afford an incorrect tracing of the model, as can multiple conformations or weak density. Particular problems involved in the elucidation of protein–ligand binding modes have been dealt with elsewhere.[55]

3.21.5 A Look into the Future?

The past half century has seen the field of biocrystallography develop from an academic endeavor into an indispensable tool for the analysis of biochemical processes. Modern technological advances have made it possible to understand the molecular action of a (potential) drug in a degree of detail unthinkable even a decade ago. Developments in in-house data collection facilities assure the confidentiality necessary for today's drug design process. Nevertheless, progress continues to be made in all aspects of structural biology. Crystal improvement strategies[56] allow the measurement of crystals previously deemed unsuitable for analysis. Structural genomics projects[57] have fueled the implementation of high-throughput robotics systems for crystallization, data collection, and phasing, and has seen the advent of Fed-Ex or 'mail-in' crystallography.[58] Dedicated synchrotron microfocus beamlines are now in place capable of collecting high-quality data from crystals only a few micrometers in size.[59] Membrane protein structural biology has progressed from the realm of fantasy into that of feasibility; all prognoses indicate that this field will evolve in due course to the status enjoyed by conventional structural biology.[60] It is anticipated that the emphasis of biomolecular structure determination will move toward the analysis of multimolecular complexes. The future of x-ray crystallography, with its many chances and challenges, is bright.

References

1. Rossmann, M. G.; Arnold, E.; *International Tables for Crystallography, Volume F: Crystallography of Biological Macromolecules*; Kluwer Academic Publishers: Dordrecht, The Netherlands, 2001.
2. Röntgen, W. C. *Eine neue Art von Strahlen*; Verlag und Druck der Stahel'schen K. Hof- und Kunsthandlung: Würzburg, Germany, 1895.
3. Friedrich, W., Knipping, P., Laue, M., *Proc. Bavarian Acad. Sci.* 1912, 303-322.
4. Bragg, W. L.; Bragg, W. H. *Proc. R. Soc. London Ser. A 88*, 428–438.
5. Bernal, J. D.; Crowfoot, D. *Nature* **1934**, *133*, 794.
6. Astbury, W. T.; Dalgleish, C. E.; Darmon, S. E.; Sutherland, G. B. B. M. *Nature* **1948**, *162*, 596.
7. Pauling, L.; Corey, R. B. *J. Am. Chem. Soc.* **1950**, *72*, 5349.
8. Perutz, M. F. *Nature* **1951**, *168*, 653–654.
9. Cochran, W.; Crick, F. H.; Vand, V. *Acta Crystallogr.* **1952**, *5*, 581–586.
10. Watson, J. D.; Crick, F. H. *Nature* **1953**, *171*, 737–738.
11. Franklin, R.; Gosling, R. G. *Nature* **1953**, *171*, 740–741.
12. Wilkins, M. H. F.; Stokes, A. R.; Wilson, H. R. *Nature* **1953**, *171*, 738–740.
13. Green, D. W.; Ingram, V. M.; Perutz, M. F. *Proc. R. Soc. London Ser. A* **1954**, *225*, 287.
14. Kendrew, J. C.; Bodo, G.; Dintzis, H. M.; Parrish, R. G.; Wyckoff, H.; Phillips, D. C. *Nature* **1958**, *181*, 662–666.
15. Kendrew, J. C.; Dickerson, R. E.; Strandberg, B. E.; Hart, R. G.; Davies, D. R.; Phillips, D. C.; Shore, V. C. *Nature* **1960**, *185*, 422.
16. Perutz, M. F.; Rossmann, M. G.; Cullis, A. F.; Muirhead, H.; Will, G.; North, A. C. T. *Nature* **1960**, *185*, 416.
17. Bernstein, F. C.; Koetzle, T. F.; Williams, G. J.; Meyer, E. E.; Brice, M. D.; Rodgers, J. R.; Kennard, O.; Shimanouchi, T.; Tasumi, M. *J. Mol. Biol.* **1977**, *112*, 535–542.
18. Berman, H. M.; Westbrook, J.; Feng, Z.; Gilliland, G.; Bhat, T. N.; Weissig, H.; Shindyalov, I. N.; Bourne, P. E. *Nucleic Acids Res.* **2000**, *28*, 235–242.
19. Hoppe, W. *Acta Crystallogr.* **1957**, *10*, 750–751.
20. Rossmann, M. G., Ed. *The Molecular Replacement Method*; Gordon & Breach: New York, 1972.
21. Guss, J. M.; Merritt, E. A.; Phizackerley, R. P.; Hedman, B.; Murata, M.; Hodgson, K. O.; Freeman, H. C. *Science* **1988**, *241*, 806–811.
22. Deisenhofer, J.; Epp, O.; Miki, K.; Huber, R.; Michel, H. *Nature* **1985**, *318*, 618–624.
23. Abrahams, J. P.; Leslie, A. G.; Lutter, R.; Walker, J. E. *Nature* **1994**, *370*, 621–628.
24. Doyle, D. A.; Morais Cabral, J.; Pfuetzner, R. A.; Kuo, A.; Gulbis, J. M.; Cohen, S. L.; Chait, B. T.; MacKinnon, R. *Science* **1998**, *280*, 69–77.
25. Ban, N.; Nissen, P.; Hansen, J.; Moore, P. B.; Steitz, T. A. *Science* **2000**, *289*, 905–920.
26. Schluenzen, F.; Tocilj, A.; Zarivach, R.; Harms, J.; Gluehmann, M.; Janell, D.; Bashan, A.; Bartels, H.; Agmon, I.; Franceschi, F. et al. *Cell* **2000**, *102*, 615–623.
27. Wimberly, B. T.; Brodersen, D. E.; Clemons, W. M., Jr.; Morgan-Warren, R. J.; Carter, A. P.; Vonrhein, C.; Hartsch, T.; Ramakrishnan, V. *Nature* **2000**, *407*, 327–339.
28. Schluenzen, F.; Zarivach, R.; Harms, J.; Bashan, A.; Tocilj, A.; Albrecht, R.; Yonath, A.; Franceschi, F. *Nature* **2001**, *413*, 814–821.
29. Hahn, T., Eds., *International Tables for Crystallography*; Volume A, Space-Group Symmetry 4th ed; Kluwer Academic Publishers: Dordrecht, The Netherlands, 1995.
30. Phillips, G. N., Jr.; Lattman, E. E.; Cummins, P.; Lee, K. Y.; Cohen, C. *Nature* **1979**, *278*, 413–417.
31. Patterson, A. L. *Phys. Rev.* **1934**, *46*, 372–376.
32. Harker, D. *Acta Crystallogr.* **1956**, *9*, 1–9.
33. Blow, D. M.; Crick, F. H. C. *Acta Crystallogr.* **1959**, *12*, 794–802.
34. Uson, I.; Schmidt, B.; von Bulow, R.; Grimme, S.; von Figura, K.; Dauter, M.; Rajashankar, K. R.; Dauter, Z.; Sheldrick, G. M. *Acta Crystallogr. Sect. D* **2003**, *59*, 57–66.
35. Hendrickson, W. A.; Horton, J. R.; LeMaster, D. M. *EMBO J.* **1990**, *9*, 1665–1672.
36. Devedjiev, Y.; Dauter, Z.; Kuznetsov, S. R.; Jones, T. L.; Derewenda, Z. S. *Structure* **2000**, *8*, 1137–1146.
37. Debreczeni, J. E.; Bunkoczi, G.; Ma, Q.; Blaser, H.; Sheldrick, G. M. *Acta Crystallogr., Sect. D* **2003**, *59*, 688–696.
38. Kwiatkowski, W.; Noel, J. P.; Choe, S. *J. Appl. Crystallogr.* **2000**, *33*, 876–881.
39. Yang, C.; Pflugrath, J. W.; Courville, D. A.; Stence, C. N.; Ferrara, J. D. *Acta Crystallogr., Sect. D* **2003**, *59*, 1943–1957.
40. Matthews, B. W. *J. Mol. Biol.* **1968**, *33*, 491–497.
41. Brunger, A. T.; Adams, P. D.; Clore, G. M.; DeLano, W. L.; Gros, P.; Grosse-Kunstleve, R. W.; Jiang, J. S.; Kuszewski, J.; Nilges, M.; Pannu, N. S. et al. *Acta Crystallogr., Sect. D* **1998**, *54*, 905–921.
42. Brünger, A. T. *Acta Crystallogr., Sect. D* **1993**, *49*, 24–36.
43. Hope, H. *Acta Crystallogr., Sect. B* **1988**, *44*, 22–26.
44. Garman, E. F.; Schneider, T. R. *J. Appl. Crystallogr.* **1997**, *30*, 211–237.
45. Kiefersauer, R.; Than, M. E.; Dobbek, H.; Gremer, L.; Melero, M.; Strobl, S.; Dias, J. M.; Soulimane, T.; Huber, R. *J. Appl. Crystallogr.* **2000**, *33*, 1223–1230.
46. Blake, C. C. F.; Koenig, D. F.; Mair, G. A.; North, A. C. T.; Phillips, D. C.; Sarma, V. R. *Nature* **1965**, *206*, 757–761.
47. Hamlin, R.; Cork, C.; Howard, A.; Nielsen, C.; Vernon, W.; Matthews, D.; Xuong, N. H. *J. Appl. Crystallogr.* **1981**, *14*, 85–93.
48. Blum, M.; Metcalf, P.; Harrison, S. C.; Wiley, D. C. *J. Appl. Crystallogr.* **1987**, *20*, 235–242.
49. Amemiya, Y.; Matsushita, T.; Nakagawa, A.; Satow, Y.; Miyahara, J.; Chikawa, J.-I. *Nucl. Instrum. Methods Phys. Res. A* **1988**, *266*, 645–653.
50. Milch, J. R.; Gruner, S. M.; Reynolds, G. T. *Nucl. Instrum. Methods* **1982**, *201*, 43–52.
51. Wilson, A. J. C. *Nature* **1942**, *150*, 151–152.
52. Yeates, T. O. *Methods Enzymol.* **1997**, *276*, 344–358.
53. Stubbs, M. T.; Morenweiser, R.; Sturzebecher, J.; Bauer, M.; Bode, W.; Huber, R.; Piechottka, G. P.; Matschiner, G.; Sommerhoff, C. P.; Fritz, H. et al. *J. Biol. Chem.* **1997**, *272*, 19931–19937.
54. Di Marco, S.; Priestley, J. P. *Structure* **1997**, *5*, 1465–1474.
55. Stubbs, M. T. Protein–Ligand Interactions Studied by X-Ray. In *Encyclopedic Reference of Genomics and Proteomics in Molecular Medicine*; Ganten, D., Ruckpaul, K., Eds.; Springer-Verlag: Heidelberg, Germany 2006; in press.
56. Heras, B.; Martin, J. L. *Acta Crystallogr., Sect. D* **2005**, *61*, 1173–1180.
57. Dry, S.; McCarthy, S.; Harris, T. *Nat. Struct. Biol.* **2000**, 7, 946–949.

58. Smith Schmidt, T. *Nature* **2003**, *423*, 799–800.
59. Nelson, R.; Sawaya, M. R.; Balbirnie, M.; Madsen, A. O.; Riekel, C.; Grothe, R.; Eisenberg, D. *Nature* **2005**, *435*, 773–778.
60. White, S. H. *Prot. Sci.* **2004**, *13*, 1948–1949.
61. Rauh, D.; Klebe, G.; Stubbs, M. T. *J. Mol. Biol.* **2004**, *335*, 1325–1341.
62. Stubbs, M. T. In *International Tables for Crystallography, Volume F: Crystallography of Biological Macromolecules*; Rossmann, M. G., Arnold, E., Eds.; Kluwer Academic Publishers: Dordrecht, The Netherlands, 2001, pp 256–260.
63. Bruner, S. D.; Weber, T.; Kohli, R. M.; Schwarzer, D.; Marahiel, M. A.; Walsh, C. T.; Stubbs, M. T. *Structure* **2002**, *10*, 301–310.
64. Tziridis, A.; Rauh, D.; Neumann, P.; Steinmetzer, P.; Stürzebecher, J.; Stubbs, M.T. *J. Mol. Biol.* **2006**, in press.

Biography

Milton T Stubbs II was born in 1962 in New York, USA, and raised and educated in the UK. After receiving his BSc in Physics at the University of Durham, he commenced his DPhil under the supervision of Professors Andrew Miller and Sir David Phillips at the Laboratory of Molecular Biophysics of the University of Oxford. During his doctoral work, he used x-ray small-angle scattering to study viruses in solution, including influenza and foot and mouth disease virus, at the Synchrotron Radiation Source in Daresbury, and completed his thesis in 1986 in the Department of Biochemistry of the University of Edinburgh. In 1987, Dr Stubbs joined the laboratory of Prof Robert Huber at the Max-Planck-Institut für Biochemie in Martinsried, Germany, where he worked on a variety of crystallographic projects, including structural analyses of blood coagulation proteins. From 1992 to 1993 he was Högskolelektor at the Centrum för strukturbiokemi of the Karolinska Institutet in Stockholm, before returning to Martinsried. He headed the x-ray crystallography facility of the Institut für Pharmazeutische Chemie at the Philipps-Universität Marburg from 1996 to 2002, where he worked with Prof Gerhard Klebe. In 2002 he was appointed Professor of Physical Biotechnology at the Martin-Luther-Universität Halle-Wittenberg.

Dr Stubbs's research interests lie in the structural biology of proteins of therapeutic relevance, with a particular focus on macromolecular conformational transitions, including the influence of protein flexibility on ligand affinity.

Comprehensive Medicinal Chemistry II
ISBN (set): 0-08-044513-6
ISBN (Volume 3) 0-08-044516-0; pp. 449–472

3.22 Bio-Nuclear Magnetic Resonance

T Carlomagno, M Baldus, and C Griesinger, Max Planck Institute for Biophysical Chemistry, Göttingen, Germany

3.22.1	**Introduction**	**473**
3.22.1.1	Samples	475
3.22.1.2	Assignment	476
3.22.1.3	Molecular Weight Considerations	479
3.22.2	**Structural NMR Parameters**	**481**
3.22.2.1	Distances from Dipolar Couplings via Relaxation or Dipolar Recoupling	481
3.22.2.2	Paramagnetic Relaxation Enhancement	482
3.22.2.3	Dipolar Couplings	483
3.22.2.4	Scalar Couplings	485
3.22.2.5	Cross-Correlation between Interactions	486
3.22.2.6	Chemical Shifts	486
3.22.3	**Structure Calculation**	**488**
3.22.3.1	Examples	490
3.22.3.2	Fast Acquisition Schemes	491
3.22.4	**Multidomain Proteins, Protein–Protein, and Protein–Ligand Complexes**	**492**
3.22.4.1	Preparation	492
3.22.4.2	Binding Affinities	492
3.22.4.2.1	Tight binding	492
3.22.4.2.2	Weak binding	493
3.22.5	**Dynamics**	**495**
3.22.5.1	Sub-Correlation Time Dynamics	496
3.22.5.2	μs to ms Dynamics on the Scale	497
3.22.5.3	ns to μs Dynamics on the Scale	499
3.22.5.4	Domain Motion in Multidomain Proteins	500
3.22.5.5	Dynamics in Solid-State Nuclear Magnetic Resonance	501
3.22.6	**Conclusion**	**502**
References		**502**

3.22.1 Introduction

Structure determination of proteins and protein/ligand complexes by nuclear magnetic resonance (NMR) spectroscopy has reached a level of high precision and breadth of applicability that we will cover in part in this chapter. The possibility of determining structures and the structure of complexes relies on NMR parameters of individual nuclei or pairwise nuclear interactions. These parameters depend on the structure and therefore inversely structural information can be extracted from them.

While the structures of proteins have been determined by liquid-state NMR spectroscopy for more than 20 years, the possibility of also determining structures from solid-state NMR using magic angle sample spinning is rather novel and only a few examples have been presented so far. NMR is a complementary method to x-ray crystallography in the sense that it provides information on biomolecules in solution or in the native membrane environment. It is also able to yield information on biomolecular systems that are too heterogeneous to crystallize, such as glycosylated proteins, protein aggregates, or unfolded proteins in solution. We will give examples of liquid- and solid-state NMR approaches toward structure determination from the experimental side, describe the similiarities and differences of the information that can be obtained, and illustrate the state of the art with a few representative examples.

Despite great differences in the approaches that different researchers take, the workflow for structure determination by NMR is fairly similar. Since the structural information is encoded in NMR parameters of nuclei or pairs of nuclei, assignment of the nuclei is a prerequisite. The assignment process and the measurement of the structurally relevant NMR parameters may be done using the same experiments, such as NOESY and COSY[1] in the pre-triple resonance era of protein structure determination. However, the strategy of the past 15 years since the introduction of three-dimensional (3D) NMR spectroscopy[2] and ^{13}C and ^{15}N labeling[3,4] has been to use for assignment only those experiments that rely exclusively on NMR interactions between directly bonded nuclei. In a second step experiments that report on the structural NMR parameters are recorded and then translated into structures (**Figure 1**). We will therefore first describe experiments that are used for assignment and then switch to experiments that are used to record the NMR parameters necessary for the determination of the structure.

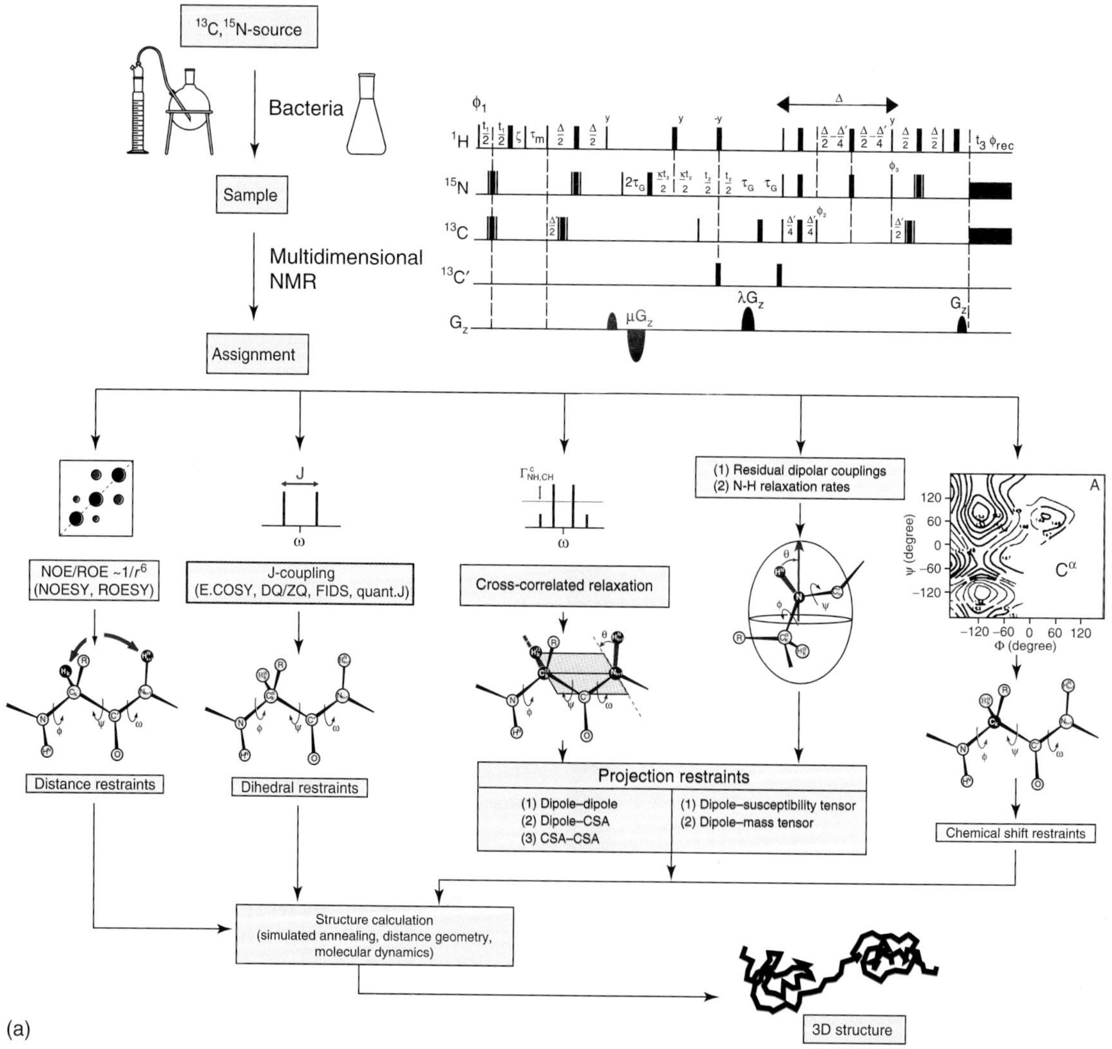

Figure 1 (a) Flowchart for structure determination of proteins with liquid-state NMR. A labeled sample is subjected to a number of pulse sequences that yield assignment. Then structural parameters (distances, J-couplings, cross-correlated relaxation rates, dipolar couplings, chemical shifts) are measured and together with molecular dynamics used to derive a structure. (b) the same procedure for solid-state NMR. Here orientation restraints are not available for powder samples under magic angle sample spinning.

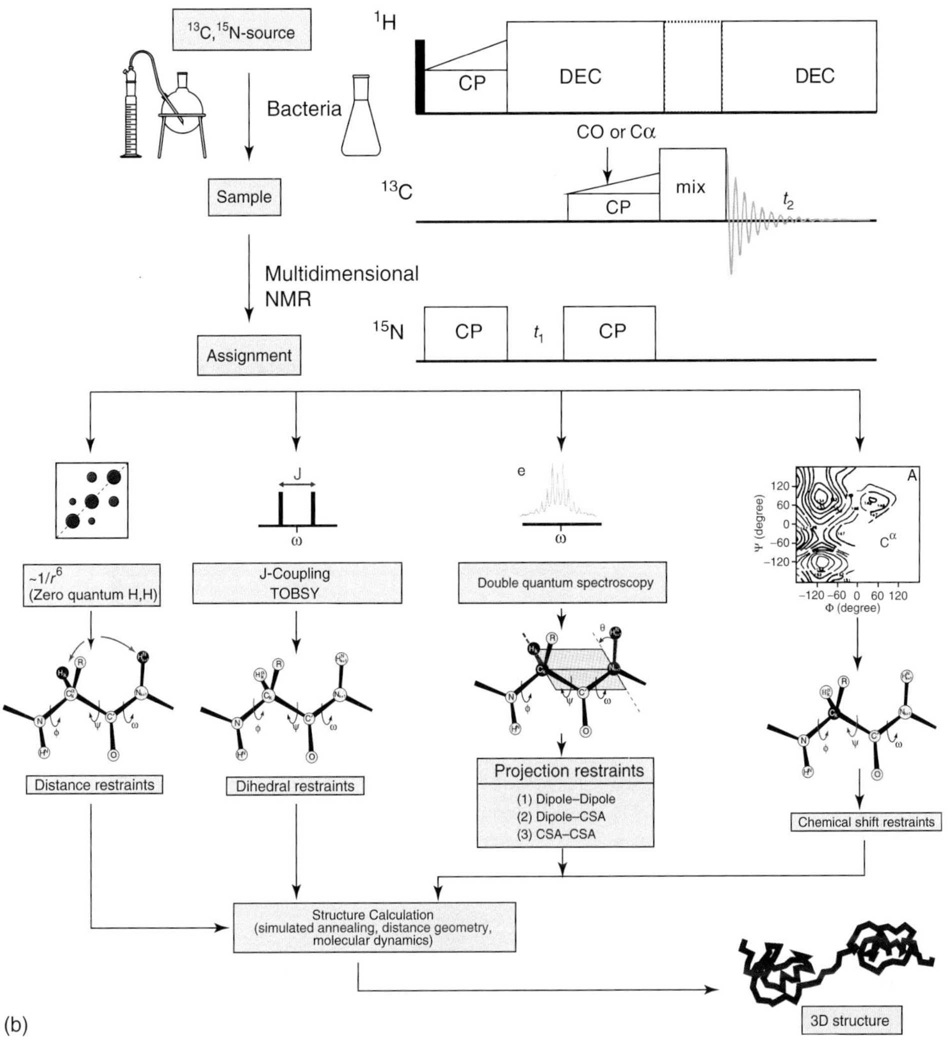

(b)

Figure 1 continued

3.22.1.1 Samples

NMR spectroscopy can be done in solution on soluble proteins or oligonucleotides provided that concentrations of the solutions in the range of 100 μM to 2 mM can be achieved. Measurement in 300 μL is conventionally performed for liquid-state NMR while rotors of 30–80 μL are used for solid-state NMR measurements. While 'proton-only' spectroscopy is possible for full structure determination of soluble proteins up to approximately 100–150 amino acids, because it is possible to label expressable proteins easily, fully labeled compounds are now used even for proteins as small as 60 amino acids. This is because in addition to the conventional proton–proton distances measured by the nuclear Overhauser effect (NOE) more NMR parameters such as chemical shifts of heteronuclear spins, dipolar couplings, cross-correlated relaxation (CCR) rates, and couplings are easily accessible. Also, dynamical properties that are uniquely accessible with NMR spectroscopy can be measured much more easily from proteins or oligonucleotides that include heteronuclear spins as will be explained in more detail below.

An overview over the samples accessible for NMR in solution or in solid state is given in **Figure 2**. From an NMR spectroscopic view, the large anisotropic interactions are averaged out either by intrinsic fast rotation of the molecule (solution) or by magic angle sample (MAS) spinning and heteronuclear decoupling, which remove the anisotropic interactions such that only the isotropic parts remain (**Figure 3**). The anisotropic interactions have an angular dependence with respect to the magnetic field of $3\cos^2\theta - 1$. At the magic angle of 54.47°, $3\cos^2\theta - 1$ assumes the value 0. Thus, both in solution and in the solid state, spectra arise that have sharp lines and reflect the isotropic values of the magnetic interactions.

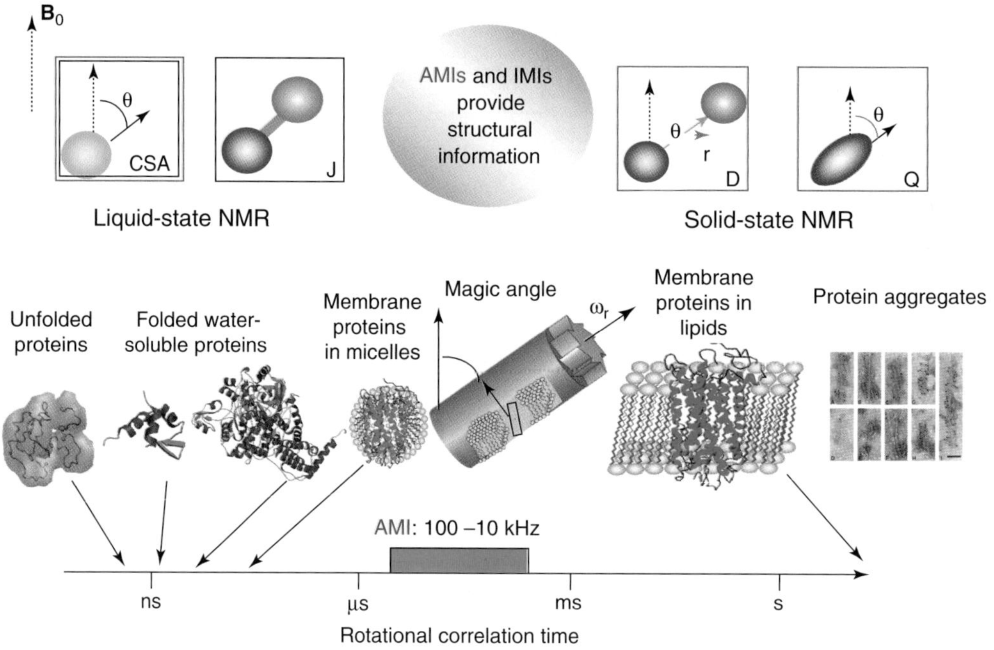

Figure 2 Liquid- and solid-state NMR. Above, the different interactions that pertain in NMR are shown. Anisotropic magnetic interactions (AMIs) are indicated by red boxes, isotropic magnetic interactions (IMIs) by black boxes. The chemical shift, dipolar coupling, and quadrupolar coupling are the anisotropic interactions whose leading term depends on the orientation with respect to the magnetic field with $3\cos^2\theta - 1$. Only the scalar coupling J does not depend on the orientation of the molecule with respect to the external magnetic field. Samples suitable for liquid-state NMR are folded or unfolded proteins, their complexes, and membrane proteins reconstituted in micelles. With solid-state NMR, membrane proteins reconstituted in liposomes or heterogeneous fibrils can be investigated.

3.22.1.2 Assignment

Proteins are built from amino acids and they have a repetitive backbone. The difference lies in the side chains on the level of constitution and in addition on the different conformations the various amino acids adopt in a protein structure. A prerequisite for the structure determination of a protein is the difference of resonance frequencies due to different chemical environments of the nuclei, i.e., the chemical shift. A comparison of a proton NMR spectrum of lysozyme and a ^{13}C NMR spectrum recorded under MAS conditions and proton decoupling of fully labeled ubiquitin is shown in **Figure 4a** and **b**, respectively. Due to the repetitiveness of the amino acid backbone the scalar couplings between the directly bonded nuclei are rather uniform and can therefore be used for assignment purposes. Similarly, the dipolar couplings depend on the distance between nuclei and therefore are also uniform along the backbone of the amino acids. Consequently, they are ideal for assignment purposes. Both approaches of using one-bond couplings, ^{1}J, require isotopic labeling of the protein. Labeling strategies will be discussed below. The most successful strategies rely on the usage of uniformly labeled proteins, meaning that all amino acids are labeled similarly, e.g., fully with ^{13}C and ^{15}N in all positions. The scalar couplings as well as the dipolar couplings are indicated in **Figure 5** for the backbone of the proteinaceous amino acids.

The assignment strategy is similar for liquid- and solid-state NMR spectroscopy. A sufficient number of chemical shifts need to be correlated with each other in order to obtain the required spectral resolution. For liquid-state NMR this is conventionally three-dimensional correlation spectroscopy that relies on transfer between the involved nuclei via the scalar couplings.[5] For solid-state NMR, 2D as well as 3D NMR experiments are recorded that rely on dipolar couplings between directly bonded nuclei.[6] Due to the r^{-3} dependence of the dipolar couplings, dipolar couplings across a C–C or a C–N bond are at least fivefold smaller than across a single bond. While in liquid-state NMR detection is normally done on the amide protons, in solid-state NMR detection occurs on ^{13}C. This is due to the fact that it is difficult to spin out the proton–proton dipolar couplings completely. In liquid-state NMR, however, the proton linewidths are narrow enough.

For molecules below approximately 25 kDa the following strategy is applied for liquid-state NMR spectroscopy. Backbone assignment is obtained by the combination of a CBCA(CO)NH[7] experiment in which the C_β and the C_α

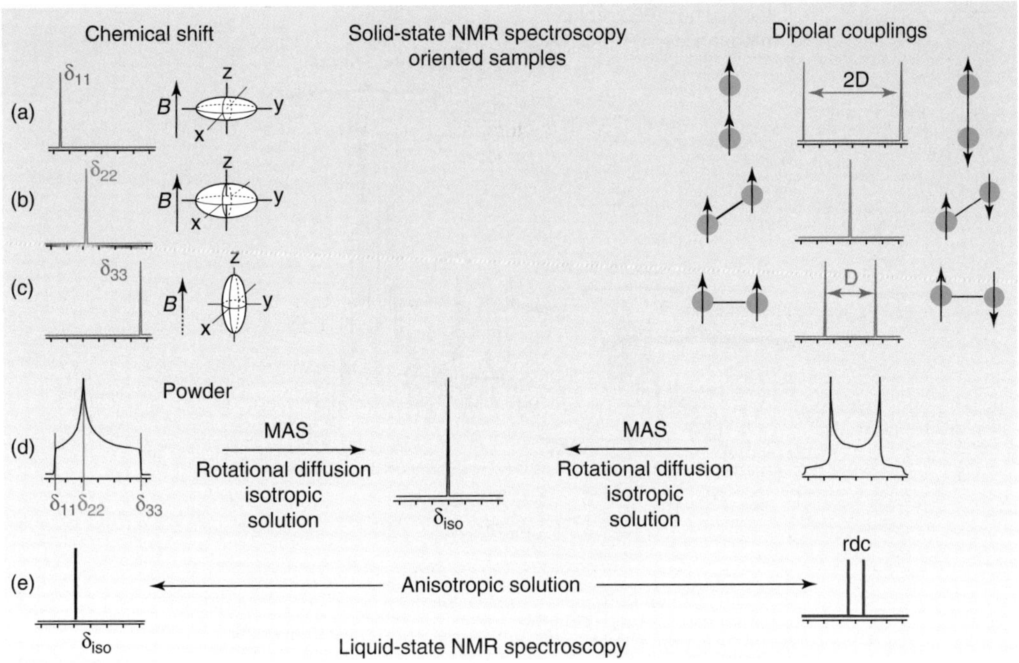

Figure 3 Averaging of the anisotropic interactions exemplified with chemical shift (left) and the dipolar couplings (right). Due to their tensorial nature, both interactions depend on the relative orientation of the chemical shift anisotropy tensor or the tensor of the dipolar coupling with respect to the external field. Extreme orientations of the magnetic field along the principal components of the tensors are indicated (a, b, c). Averaging over all orientations yields the powder spectrum which is a so-called Pake pattern (d). Magic angle sample (MAS) spinning or rotation in isotropic solution yield the isotropic spectrum (middle of d). In anisotropic solution residual anisotropic interactions are recovered (e).

nuclei resonate in the ω_1 dimension, the ^{15}N in the ω_2 dimension, and the amide proton H^N in the ω_3 dimension. Due to the coupling topology it is easy to transfer the magnetization exclusively on this pathway in which the C_β and the C_α of the given amino acid are correlated with the amide moiety of the following amino acid. The coupling constants on this pathway are rather large: 130 Hz for the $H_\beta C_\beta$ and 140 Hz for the $H_\alpha C_\alpha$ coupling, 35 Hz for the $C_\beta C_\alpha$ and 55 Hz for the $C_\alpha C'$ coupling, 15 Hz for the C′N coupling, and 90 Hz for the NH coupling.

For the second experiment, CBCANH,[8] there are two transfer pathways, either via the $^1J(N_iC_{\alpha i}) = 11$ Hz coupling within the same amino acid or via the $^2J(N_{i+1}C_{\alpha i}) = 7$ Hz to the following amino acid. The cross-peak amplitudes for the first transfer depend on $\sin^2\pi^1J(N_iC_{\alpha i})\tau\cos^2\pi^2J(N_{i+1}C_{\alpha i})\tau$ while it depends on $\cos^2\pi^1J(N_iC_{\alpha i})\tau\sin^2\pi^2J(N_{i+1}C_{\alpha i})\tau$ for the second transfer. Thus, adjustment of the delay to values that are optimized for the transfer via the single bond or the transfer via two bonds are feasible. Both experiments use detection on the amide proton. However, the magnetization flow can either start on the aliphatic H_β and H_α protons, or else an out-and-back experiment is done in which the magnetization flows from the amide proton via the nitrogen to the aliphatic carbons and then back again. In fact, for larger proteins, the out-and-back version is superior for the CBCANH experiment.

The approach taken for solid-state NMR is very similar to the one taken for liquid-state NMR assignment with respect to the pathway of coherence transfers. However, it differs in that carbons and not protons are detected. One-bond correlations can be obtained by transfer via the dipolar couplings that are large between directly bonded spins. Here, the magnetization of the protons is transferred to the nitrogen via cross-polarization then from the N to the C′ via specific cross-polarization, and then further on via nonspecific cross-polarization for detection on the C_β and C_α resonances.[9] For the NCACX[9] experiment, cross-polarization between the N and aliphatic carbons leads to transfer to the C_α and all other carbons due to the limited spectral dispersion. In both the liquid-state and the solid-state experiments, selective transfers are performed. While selective pulses are applied in the liquid-state INEPT-type (INEPT = insensitive nucleus enhancement by polarization transfer) transfers, selective Hartmann Hahn conditions for the transfer via dipolar couplings are applied for the solid-state experiments.[6] The transfer schemes are indicated in **Figure 6** and a representative example is given for the CaM/C20W complex in solution and an SH3 in the solid state in **Figure 7**.

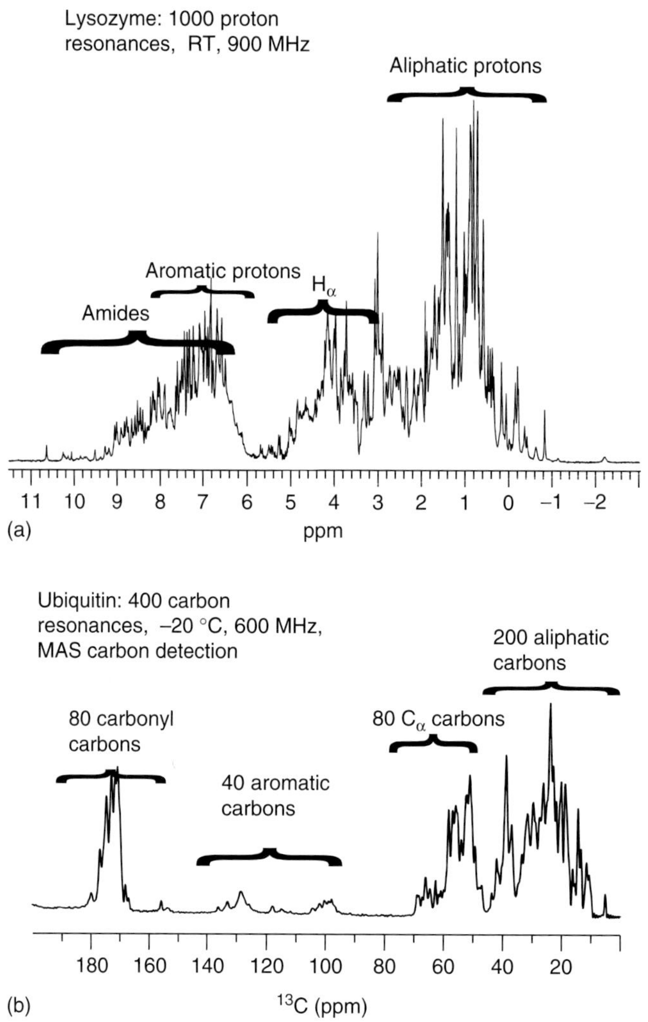

Figure 4 (a) 900 MHz ^{1}H spectrum of lysozyme; the spectral ranges of the individual proton resonances are indicated. (b) 600 MHz ^{13}C spectrum of ubiquitin; the ranges of chemical shifts are indicated.

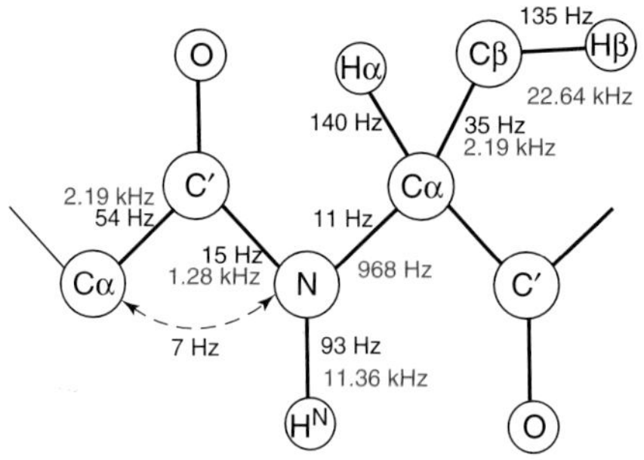

Figure 5 Scalar and maximal dipolar couplings between different nuclear pairs in the protein backbone. The dipolar couplings are in red, the scalar couplings in black.

Relying on the assignment of the C_β and C_α resonances, further assignment is now obtained by correlation of the side chain carbons and protons with these C_β and C_α resonances. In the liquid state, this is achieved with HCCH-TOCSY (total correlation spectroscopy) experiments or HCC(CO)NH experiments. In these experiments, proton magnetization is transferred to carbon resonances by INEPT transfer and then further transferred via a broadband TOCSY experiment to other aliphatic carbons and back again for proton detection via INEPT.[5]

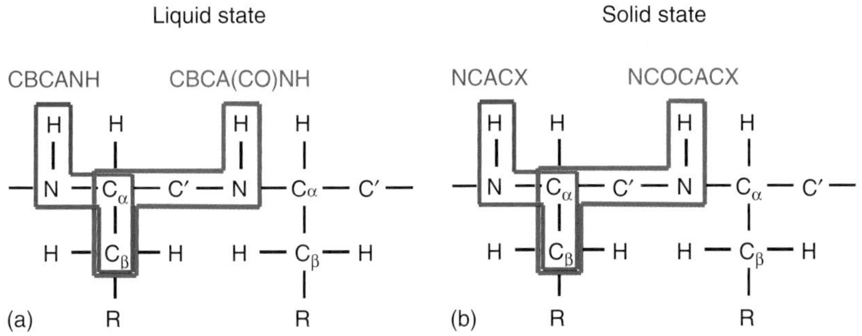

Figure 6 Correlation scheme for liquid- and solid-state NMR assignment of proteins when ^{13}C and ^{15}N labeling has been performed. (a) The CBCA(CO)NH correlates the C_β and C_α with the N and H of the following amino acid while the CBCANH experiment correlates mainly with the NH of the same amino acid. (b) In the NCACX and the NCOCACX experiments, magnetization is transferred from the amide protons to the indicated carbons which are then detected. The magnetization flow is essentially the same as in (a).

For solid-state NMR, cross-polarization from proton to carbon followed by a ^{13}C,^{13}C- spin diffusion step leads to carbon–carbon correlations.[10,11] Since the magnetization transfer is proportional to the inverse third power of their distances, r^{-3}, only single bond correlations are observed for small mixing times of approximately 4 ms. For larger mixing times, correlations between carbons via relayed spins are also observed (spin diffusion). Here spin diffusion refers to coherent transfer between carbon nuclei due to the large dipolar couplings between the ^{13}C,^{13}C pairs. This mechanism is different from the same notion used in liquid-state NMR that refers to incoherent transfer due to relaxation via an intermediate spin for large molecules (spin diffusion limit).

3.22.1.3 Molecular Weight Considerations

For solid-state NMR, the molecular weight of a sample does not affect the linewidth, in contrast to liquid-state NMR in which the rotational correlation time governs the achievable linewidth. Roughly speaking this is due to the fact that the tensorial interactions are 'spun out' in solution with size-dependent speed, namely the correlation time of rotational diffusion.

The linewidth obtained is directly proportional to the square of the relevant anisotropic interaction multiplied with the correlation time. Some relevant interactions are given in **Table 1** and have been indicated in **Figure 7**. The quadratic dependence of the line width Γ from the interaction size b_{kl} originates from the Bloembergen and Redfield relaxation theory. The double commutator of the respective interactions $A^q_{kl}(I_k, I_l)$ multiplied with the spectral density function $J(\omega_q)$ determines the linewidth[12,13]:

$$\Gamma = b_{kl}b_{kl}\sum_{q=-2}^{2} J(\omega_q)\left[A^q_{kl}(I_k, I_l), [A^{-q}_{kl}(I_k, I_l),]\right] = b^2_{kl}\tau_c/5 \quad [1]$$

with $J(\omega_q) = \frac{4}{5}\frac{\tau_c}{1+\omega_q^2\tau_c^2}$.

For solid-state NMR, there is no overall tumbling of the molecules in solution. This motion is frozen out. Therefore, local motion and the details of MAS and the decoupling determine the linewidths and integrals in solid-state NMR spectra. In addition, cross-talk between different interactions such as chemical shift anisotropy (CSA) and heteronuclear couplings may influence the integral of the lines and the linewidth.[14] For regular spinning speeds of the order of 15 kHz, however, the proton–proton dipolar couplings cannot be spun so that ^{13}C detection is mandatory in solid-state NMR experiments.

For liquid-state NMR proton detection is compromised by the fast relaxation of transverse magnetization of the ^1H,^{13}C and ^1H,^{15}N dipolar couplings. Due to interference of different relaxation mechanisms or the inefficiency of relaxation mechanisms for certain coherences it is possible to obtain narrow linewidths even for large molecules with long correlation times.[15] These so-called relaxation optimized sequences rely on the choice of coherence that relaxes the slowest during a certain period of time that is needed for transfer of magnetization from one nucleus to the next. For example in the case of the amide ^{15}N,^1H- pair the relaxation via the CSA of both ^1H and ^{15}N is of equal magnitude at 900 MHz as the dipolar coupling between the two mentioned nuclei. Since the principal axes of these two interactions are almost perfectly parallel to each other, one of the resonances in the ^{15}N or also the

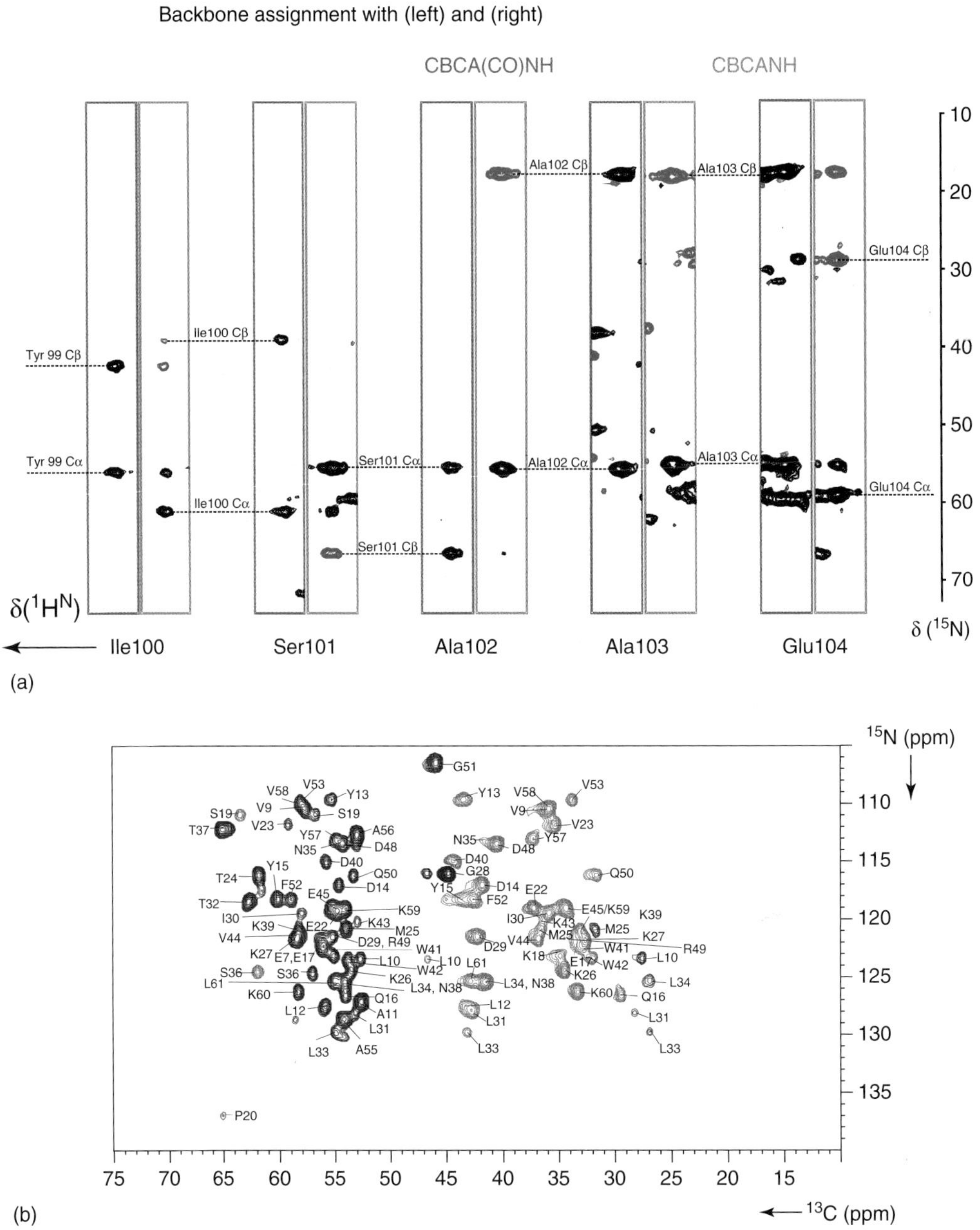

Figure 7 (a) Strips from a 3D CBCA(CO)NH and CBCANH experiment. Sequential amino acids are connected based on the same C_β and C_α chemical shifts. (b) Assigned NCACX experiment for fully labeled α-spectrin SH3 domain.

Table 1 Interaction sizes for dipolar interactions

	Dipolar Interaction $b_{kl}/2\pi = -\mu_0 \frac{\gamma_k \gamma_l \hbar}{4\pi r_{kl}^3}/2\pi$
H,H: 1.78, 2.2 Å	21.38, 11.32 kHz
C,H: 1.09 Å	22.64 kHz
N,H: 1.09 Å	11.36 kHz

^{1}H- doublet is only weakly affected by relaxation, while the other line is. This so-called TROSY (transverse relaxation optimized spectroscopy) effect can be used to narrow the lines during detection of amide protons as well as during times in which the ^{15}N is tranverse in order to transfer magnetization to neighboring carbons. The same effect is used for experiments on ^{13}C-labeled RNA bases in which the CSA of the carbons and the dipolar coupling are similar at approximately 600 MHz.[16]

If CSA is much smaller than the dipolar coupling, zero and double quantum coherences turn out to be optimal with respect to relaxation. This strategy has been used for aliphatic ^{1}H,^{13}C- pairs in side chains in proteins and for ^{1}H,^{13}C-pairs in sugars of RNAs.[17] Finally, enhancement schemes have been designed for CH_2 groups also under the assumption of negligible CSA.[18,19]

Finally, ^{1}H,^{1}H- dipolar relaxation broadens the proton lines very efficiently. This relaxation is difficult to switch off except for dilution of the protons. The most successful approaches are the use of 100% deuterated proteins in which the amide hydrogens are back-exchanged for protons during purification.[20] In complete analogy to liquid-state NMR, such samples can also be studied with proton detection in solid-state NMR. Here, the average distance to the nearest protons is increased to a level where the dipolar couplings can now be spun out using fast MAS.[21]

Introduction of protons only into amide hydrogens suffers from the lack of protons to define side chain contacts especially in the hydrophobic core. Therefore, the labeling scheme that reintroduces protons into the methyl groups (Ile, Leu, Val, Ala) has been shown to be successful even for large proteins.[22] It turns out that the simple HMQC experiment that also employs double and zero quantum coherence in ^{1}H,^{13}C- pairs is already relaxation optimized. In addition, methyl groups profit from the thrice-enhanced signal due to the presence of three protons with identical chemical shifts. Structurally, they are important since they are abundant in the hydrophobic core of any protein. Therefore, the combination of full deuteration except for amides and methyl groups has made structural information of proteins up to 80 kDa accessible and has allowed following functional transitions in proteins exceeding 300 kDa.[23]

3.22.2 Structural NMR Parameters

The following structural parameters are in use in both liquid- and solid-state NMR (see **Figure 1**).

- Dipolar couplings to measure distances from relaxation in liquid-state NMR and from recoupling or spin diffusion experiments in solid-state NMR.
- Dipolar couplings in liquid-state NMR upon alignment of biomolecules either by dissolving them in anisotropic media or using or introducing intrinsic magnetic anisotropies.
- Chemical shifts to derive local structure.
- J-couplings to define dihedral angles.
- Cross-correlated relaxation rates to define intervector angles.

3.22.2.1 Distances from Dipolar Couplings via Relaxation or Dipolar Recoupling

As indicated in **Table 1**, dipolar coupling depends on the distance between the two involved spins and is a powerful structural parameter. Imagine the situation of 3, 4, 5, or 6 spins and the possibility to measure all distances between them, namely 3, 6, 10, and 15 distances. Then it is obvious that the 3D structure can be derived just from these internuclear distances since there are 3, 6, 9, and 12 internal coordinates to be determined which are fewer unknowns compared to the number of knowns. Distances are measured in the liquid state NOEs for assignment purposes mostly between protons because they provide the largest dipolar couplings. Due to the overall motion of the protein, the dipolar couplings are stochastically modulated, which leads to the NOE. The stochastic modulation of the dipolar coupling between spins k and l leads to magnetization flow from one nucleus to the next one by the so-called cross-relaxation rate σ^l_{kl}:

$$\begin{aligned}\sigma^l_{kl} &= \frac{1}{10} b^2_{kl}[6J(\omega_k + \omega_l) - J(\omega_k - \omega_l)] \\ &= -\frac{4}{50} b^2_{kl}\tau_c = -\frac{4}{50}\left(\frac{\gamma_k \gamma_l \mu_0 \hbar}{4\pi r^3_{kl}}\right)^2 \tau_c\end{aligned} \qquad [2]$$

The last very simple expression for the cross-relaxation rate[24] is obtained for large proteins in the spin diffusion limit. The $J(\omega)$ were introduced with eqn [1] and are the so-called spectral density functions that reflect the overall tumbling time (rotational correlation time τ_c). From the last expression of eqn [2] it is apparent that the cross-relaxation rate depends only on the internuclear distance r_{kd} and the correlation time τ_c.

Thus, during the time of magnetization flow, the mixing time of the NOESY experiment τ_m, the amount $\sigma^l\tau_m$ of magnetization of one spin is transferred to magnetization of a second spin. Provided there is a calibration distance from which the respective cross-relaxation rates can be determined at small mixing times ($\sigma^l\tau_m \ll 1$) one can obtain unknown distances from the comparison of the cross-relaxation rate for the two protons with known distance according to:

$$\frac{\sigma^l_{ij}}{\sigma^l_{\text{calib}}} = \frac{r_{ij}^{-6}}{r_{\text{calib}}^{-6}} \qquad [3]$$

More sophisticated analyses of NOEs taking into account the spin diffusion at longer mixing times in programs such as MARDIGRAS[25] and IRMA[26] as well as the scaling of the cross-relaxation rates due to dynamics[27] are rarely taken into account in structure calculations. By contrast, rather crude extraction routines for the NOEs are conventionally used for structure elucidation. Given the abundance of parameters described in this article for soluble proteins, a more sophisticated treatment of NOEs is required and first steps toward a joint calculation of dynamical structures are under way.[28]

In magic angle solid-state NMR, proton distance restraints cannot be extracted from relaxation effects since they are much too slow compared to the other interactions. Since the strong dipolar interactions are not averaged out by the molecular tumbling, relaxation effects due to motion are negligible compared to the direct effect of the dipolar couplings. They can therefore be used directly for distance measurements. However, for so-called zero quantum mixing between proton spins, spin diffusion rates are observed that behave very similar to the NOE build-up in the NOESY or ROESY experiment:

$$\sigma^{0Q}_{kl} = b^2_{kl} J^{0Q}(\omega_k - \omega_l) = \left(\frac{\gamma_k\gamma_l\mu_0\hbar}{4\pi r^3_{kl}}\right)^2 J^{0Q}(\omega_k - \omega_l) \qquad [4]$$

The transfer rate[29] depends only on the distance with r_{kl}^{-6} and the zero-quantum line shape function $J^{0Q}(\omega_k - \omega_l)$. Thus calibration of distances can be done similarly to liquid-state NMR by referencing to a known distance. However, in solid-state NMR this is even simpler: due to the fact that there is no correlation time in eqn [4] it is possible to calibrate for a certain pulse sequence the distances on a simple compound and then transfer this information to more complicated molecules in which the calibration might not be possible. The spin diffusion rate between the protons is measured by correlating the two protons involved with their respective ^{15}N or ^{13}C nucleus that is directly bonded.[30] From this, the result of CHHC and NHHC experiments in which the strong dipolar coupling between CH and/or NH is used for polarization transfer and then the strong dipolar couplings between the protons is used to transfer magnetization within the proton network. Then, via another H,C-cross-polarization transfer step this magnetization is transferred back to the carbons for usual solid-state NMR detection. It turns out that the H,H spin diffusion transfer is governed by an r^{-6} rather than an r^{-3} dependence such that the CHHC or NHHC experiment performed at small mixing times yields distance information very similar to a small mixing time NOESY experiment.

Alternatively, NMR experiments are successfully employed in which the size of dipolar couplings are selectively reintroduced in constant times or evolution periods between one pair or a number of equivalent pairs of nuclei in a protein by recoupling. This is easily possible for example for well-resolved carbon resonances such as methyl groups and carbonyls or between heteronuclei such as amide nitrogens and methyl groups. The pairs of nuclei have to be chosen such that the desired dipolar couplings are not superseded by strong dipolar couplings in similar chemical shift ranges (dipolar truncation). This problem can be avoided by performing the described CHHC or NHHC experiments. Alternatively, removal of the strong one-bond dipolar couplings can be achieved by labeling the amino acids such that only every second carbon is ^{13}C. Then the long-range distances are no longer obscured by the strong one-bond dipolar couplings between neighboring ^{13}C^{13}C- pairs.[31]

A representative NOESY from a protein in solution shown in 2D and 3D (**Figure 8a**) as well as a 0Q–H,H correlation experiment of ubiquitin in which only correlations between CH_2 groups are shown (**Figure 8b**).

3.22.2.2 Paramagnetic Relaxation Enhancement

Distance measurement is an essential tool for structure determination from NMR spectroscopy. However, the range of distances that can be observed between protons is limited to 5 Å under good conditions 7 Å. This is due to the coupling power of two interacting spins that is given by the product of the squares of their gyromagnetic ratios in eqn [2]. By replacing one of the spins by an electron spin, the relaxation can be enhanced over the proton–proton relaxation by a factor of 660^2 resulting in an extension of the distances to be accessible by a factor of $660^{1/3} = 8.7$. Therefore, distances of 30 to 40 Å become accessible. Paramagnetic tagging has been successfully used for the characterization of large proteins[32] and membrane proteins investigated in micelles[33] as well as partially folded proteins in solution.[34]

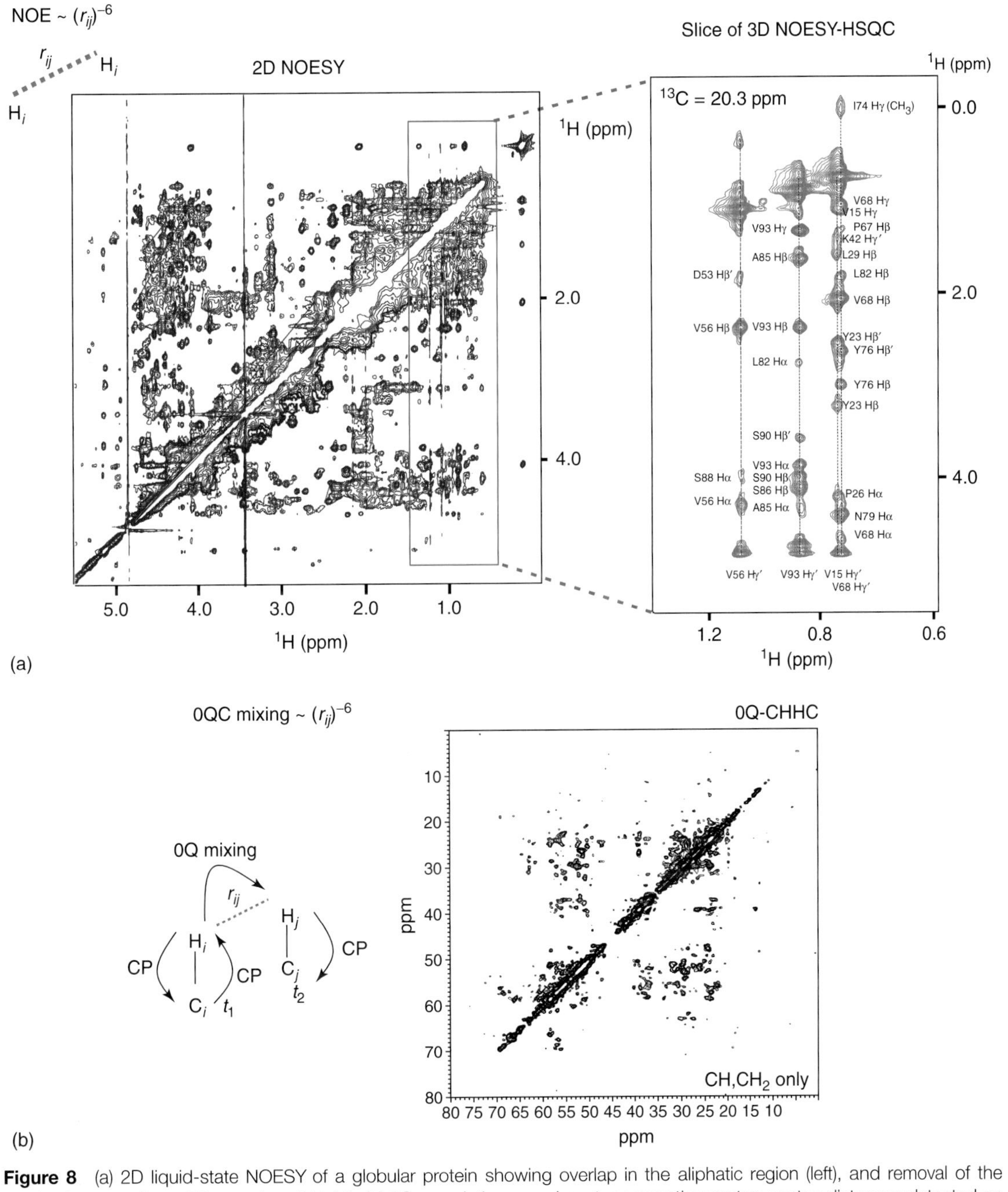

Figure 8 (a) 2D liquid-state NOESY of a globular protein showing overlap in the aliphatic region (left), and removal of the overlap by recording a 3D experiment (right). (b) 0Q–correlation experiment representing proton–proton distances detected on the carbons bound to the involved protons, respectively.

3.22.2.3 Dipolar Couplings

Dipolar couplings are averaged out in isotropic solution, but can be reintroduced if the isotropic orientation distribution is disturbed. Technically speaking, the anisotropy of the solution must have a rank two component (American football shape) because the dipolar coupling is also a rank two interaction between two nuclei. The anisotropy of the orientation distribution can be visualized by an ellipsoid that is fixed in the molecular frame. The lengths of the principal axes reflect the probability of finding that axis along the magnetic field. Alignment of proteins in water can be achieved by using intrinsic or engineered anisotropic magnetic susceptibilities[35] or by dissolving the protein in an anisotropic medium.[36] To scale the dipolar couplings to a value that is smaller than the respective J coupling, alignment of the order of 10^{-3} is desired. Except for the scaling, the spectrum of the aligned protein is then identical to the spectrum that this protein would yield in a single crystal that is switched fast between the three main axes of the alignment tensor with the populations according to the lengths of the axes.[37] Dipolar couplings are mainly applied to improve the precision of structures, to investigate protein–protein complexes when few NOEs are measured and to investigate dynamics from dipolar couplings. These applications are summarized in recent reviews.[38]

It is remarkable that the orientation information that is reintroduced by alignment in solution is not easy to recover for powder samples in solid-state NMR. At least in micro- or nanocrystalline samples, texture effects known from powder pattern x-ray analysis could in principle be used to achieve a nonuniform distribution of orientations of the molecules with respect to the magnetic field. However, by MAS spinning the dipolar couplings are averaged out for each molecular orientation individually, which prevents small texture effects from being used in solid-state NMR spectroscopy under MAS spinning.

Alternatively, membrane peptides or proteins can be macroscopically oriented on glass plates or in the magnetic field.[39,40] There are two methods for mechanically aligning lipid bilayers[39]: deposition from organic solvents followed by evaporation and lipid hydration, and fusion of unilamellar reconstituted lipid vesicles with the glass or polymer surface. Maintaining a constant hydration level of the sample is critical. For this reason, stacked glass plates are in general placed in thin polymer films that achieve heat-sealing and hence stable sample hydration. In addition, bicelles have been used to study membrane proteins by NMR.[41] They represent molecular aggregates composed of long-chain phospholipids (such as 1,2-dimyristoyl-*sn*3-glycerophosphocholine) and either short-chain lipids or surfactants. The long-chain lipids are organized into planar bilayers with the short-chain lipids arranged in a rim surrounding the bilayer edges. Bicellar solutions are lyotropic liquid crystalline solutions and can form a nematic phase that aligns in the magnetic field. The orientation of these bicelles has been shown to be affected by the addition of lanthanide ions.

While residual dipolar couplings D_{kl}^{res} reintroduced in solution carry distance information r_{kl}^{-3}, they also reflect information on the orientation of the respective internuclear vector with respect to the alignment tensor measured with the polar angles θ_{kl} and ϕ_{kl}. Da is the axial component of the alignment tensor while R is its rhombicity. The axial component can be understood as the difference between the length and the diameter of the 'American football' mentioned above. The rhombicity indicates a deviation from the axial symmetry of the football.

$$D_{kl}^{\mathrm{res}} = -\frac{B_0^2}{15kT}\frac{\gamma_k\gamma_l h}{16\pi^3 r_{kl}^3} Da\left(\left(3\cos^2\theta_{kl} - 1\right) + \frac{3}{2}R\sin^2\theta_{kl}\cos 2\varphi_{kl} \right) \qquad [5]$$

Most dipolar couplings used focus on the orientation information inherent in the dipolar coupling and therefore choose internuclear vectors with fixed distances.

The most prominent of them are NH, $C_\alpha C'$, $H^N C'$, $C_\alpha C_\beta$, and CH_3 The dipolar couplings are overlapped with the J couplings that exist between all these nuclei. By measurement of two spectra, one in isotropic solution for the J couplings and one in anisotropic medium for J + D, the size and sign of the dipolar couplings can be extracted (**Figure 9**). In most experiments for the determination of dipolar couplings, they are measured from a splitting that does not carry information on the sign of the interaction that leads to the splitting. In this situation, the absolute sign of the dipolar coupling can be determined, provided the absolute signs of the J coupling are known and the absolute value of the dipolar coupling does not exceed the absolute value of the J coupling. It is precisely for this reason that the scaling of the alignment is desired to be in the 10^{-3} range.

Dipolar couplings between nuclei that are separated by more than two bonds carry angular as well as distance information. Such couplings have been measured successfully mainly between protons. Distances of up to 10 Å have been measured in this way by selective experiments, focusing only on dipolar couplings, e.g., of amide protons in proteins or H1′ protons in oligonucleotides.[42]

Dipolar couplings are used as restraints for structure calculation together with the restraints derived from other NMR parameters such as NOEs. Due to their angular dependence, in particular the dipolar couplings between directly bonded nuclei depend strongly on the details of the structure. While this allows them to restrict a structure very tightly, it also makes the pseudoenergy involved very rugged. For example, distance restraints vary only a little when an NH vector rotates by 90° while the dipolar coupling can change from one extreme to the other extreme value. Because of this ruggedness of the energy landscape introduced by the dipolar couplings, it has turned out that they should be used only in later steps of structure calculation when a first fold has already been established.[43] Another implementation translates the restraints from dipolar couplings into allowed and disallowed regions or relative angles of two vectors[44] and thus ensures better convergence.

Alignment can be achieved via so-called external alignment media excellently reviewed recently.[38] Suffice to say, that there are approximately 25 different alignment conditions that rely on lipids forming anisotropic disks, compressed

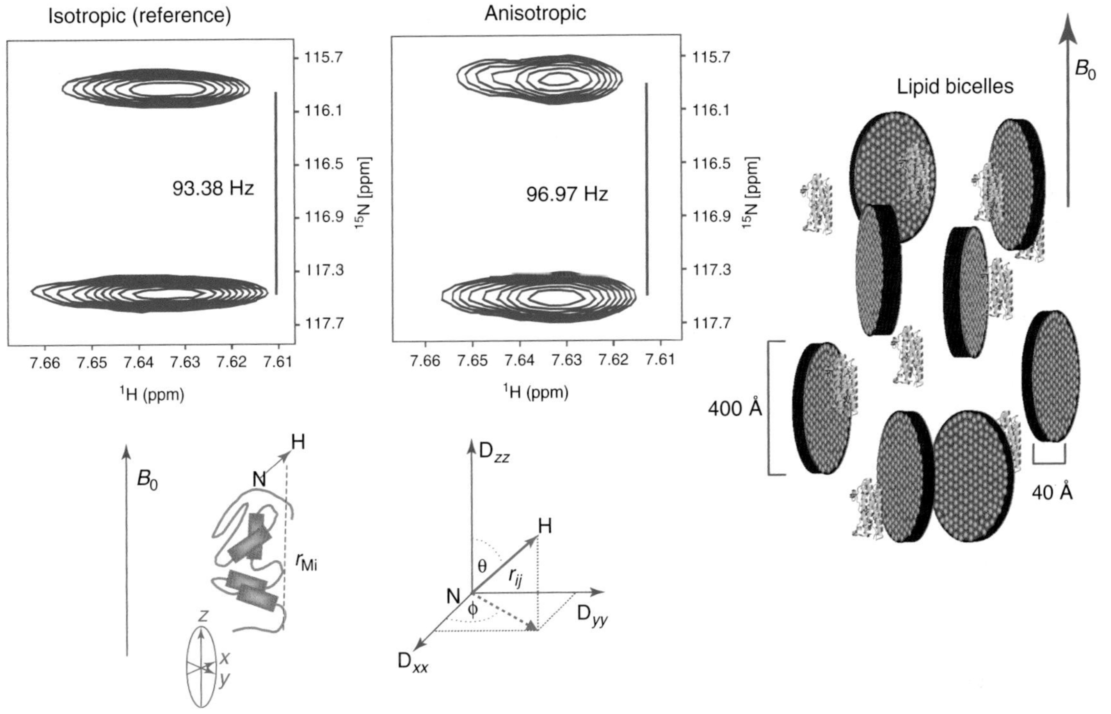

Figure 9 Alignment of proteins under various conditions (bicelles shown here as an example). Due to the complete alignment of the bicelles, a protein with an anisotropic shape can no longer assume all orientations with equal probability. This creates anisotropy of the solution with an order parameter in the 10^{-3} range. The NH dipolar coupling depends on the orientation, e.g., of the NH vector in the alignment frame, depicted as an orthogonal coordinate system. The dipolar coupling is extracted by measuring the splitting due to the NH coupling in isotropic phase (J) and in anisotropic phase (J + D). From the difference of the splittings in the two spectra, the dipolar coupling D can be extracted.

gels that have anisotropic holes, and solutions of anisotropic particles such as filamentous viruses. Most of these alignment media induce complete alignment for the lipids or phages in the magnetic field due to cooperative effects. However, the interaction with the alignment medium should be small (no binding of the solute to the alignment medium). Alternatively, alignment can also be achieved from paramagnetic parts of the molecule, e.g., in paramagnetic metalloproteins, or by attaching paramagnetic tags to proteins. Provided the paramagnetism is anisotropic,[45] dipolar couplings as well as pseudocontact shifts can be observed and used for the determination of structure as well as dynamics (as will be discussed later).

Dipolar couplings between directly bonded nuclei can be measured with highly sensitive sequences such as HSQC,[35] HNCO,[46] or CBCA(CO)NH[47] derived sequences that rely on polarization transfer steps via a single bond. Spectra in isotropic and anisotropic media are taken, and the extracted dipolar couplings are then obtained from the splitting of the anisotropic spectrum subtracted from the splitting of the isotropic spectrum. The precision of dipolar couplings is high and depends mainly on the signal-to-noise ratio. For NH dipolar couplings, errors as low as 0.01 Hz have been reported, though normally errors in the range of 0.2 to 0.5 Hz are observed. Dipolar couplings are average over the structure. Due to the small size of the dipolar couplings, they average dynamics between very fast timescales (ps) to slow timescales of the order of ms (see below).

3.22.2.4 Scalar Couplings

Scalar couplings are mainly used for assignment experiments. Here, the small variance of scalar couplings via a single bond under changes of the conformation is used. Scalar couplings over two and three bonds, however, report on torsional angles in biomolecules. In paricular, the almost universal Karplus type curve that relates the scalar coupling between two nuclei separated by three bonds with the torsional angle about the middle bond can be used to measure this torsional angle very accurately.[48] The only drawback of scalar couplings is the necessity of calibration of the coupling constant. However, recently it has become possible to calculate scalar couplings quite accurately from density functional theory (DFT) calculations.[49] The most used scalar coupling constants are the $^3J(H_i^N, H_i^\alpha)$, the $^3J(H_i^\beta, H_i^\alpha)$ and the heteronuclear couplings $^3J(H_i^\beta, C_i')$, and $^3J(H_i^\beta, N_{i-1})$. $^1J(H_i^\alpha, C_i^\alpha)$ are also used for the definition of the angle ψ in proteins.[50]

Scalar couplings exist only between nuclei that are bonded to each other. Therefore, the existence of bonds can be established by the observation of scalar couplings. This is especially true for hydrogen bonds that define structure, be it in proteins in secondary structure elements such as alpha-helices or beta-sheets, side chains of amino acids such as histidines, or oligonucleotides between the bases.[51] Hydrogen bridges have been established from the observation mainly of N,N and N,C couplings that reach a size of up to 7 Hz and 2 Hz, respectively (**Figure 10**).[52]

In liquid-state NMR spectra, scalar couplings are mostly the dominating interactions, being larger than the dipolar couplings induced by weak alignment: in solid-state NMR experiments, the situation is the reverse. Here the dipolar couplings dominate the situation and only after careful decoupling of the dipolar couplings can scalar couplings be used. In this case, TOBSY[53] spectra (analogous to TOCSY in liquid state)[35] are implemented. Scalar couplings become the dominating interactions in cases of flexible parts of proteins in solid-state spectra. In this situation, dipolar couplings are motionally averaged and only the scalar couplings that are not affected by orientational motion are observed.

TOBSY multiple-pulse schemes[53] then insure that both isotropic chemical shifts and residual anisotropic chemical shift and dipolar interactions that may be present under MAS conditions are suppressed. The resulting 2D and 3D TOBSY-based correlation experiments can therefore be interpreted solely based on scalar coupling transfers and minimize artifacts due to residual dipolar interactions. Scalar couplings are averaged by conformational variations. The timescales of averaging cover the range from ps to ms similar to dipolar couplings.

3.22.2.5 Cross-Correlation between Interactions

Scalar couplings can be used to measure torsional angles. A more general approach to the measurement of intervector angles in biomolecules can be realized by the analysis of double quantum coherences between distant nuclei. For the two extreme situations of two CH vectors that are either parallel of orthogonal to each other, a double quantum spectrum under MAS spinning will reflect a triplet-Pake pattern, since for any given orientation the two CH dipolar couplings have the same value. However, for the perpendicular situation, the dipolar couplings are never identical and therefore a different Pake pattern is obtained, which only in the case where both vectors are perpendicular to the magnetic field is a triplet. MAS spinning during the double quantum evolution time will introduce side bands that are depicted in (**Figure 11a**) whose intensities depend on the intervector angle. Cross-correlation effects are observed not only for double quantum coherences but also for single quantum carbon resonances, e.g., by cross-correlation between a CSA and a dipolar coupling.[54]

It is one of the beauties of NMR spectroscopy that the liquid-state spectrum can also be derived by averaging of the individual Pake patterns and transforming the width into the linewidth of the liquid-state lines. Differential broadening is then observed for the individual multiplet lines[55] according to:

$$\Gamma^c_{\mathrm{NH,CH}} = \frac{\gamma_H\gamma_N \; \gamma_H\gamma_C}{(r_{\mathrm{NH}})^3 (r_{\mathrm{CH}})^3} \left(\frac{\mu_0}{4\pi}\hbar\right)^2 \frac{1}{5}(3\cos^2\theta_{\mathrm{NH,CH}} - 1)\tau_c \qquad [6]$$

Thus the cross-correlation of interactions yields the intervector angle. This technique has been mainly applied to measure the backbone angle ψ in proteins, the sugar pucker in oligonucleotides,[56] and to distinguish the sign of the angle ϕ in proteins.[57] An example of the measurement of the angle between the $C_{\alpha i-1}H_{\alpha i-1}$ and the N_iH_i vector is given in **Figure 11b**. The traces through the double quantum spectrum, in which the $C_{\alpha i-1}$ and the N_i are active, are given and the extracted intervector angles $\theta\cdot$

Much rarer has been the use of CCR for the measurement of side chain angles in proteins.[58] Cross-correlated relaxation is a unique method for measuring correlated motion which we will not be able to cover in this review.[59]

3.22.2.6 Chemical Shifts

Chemical shifts are the basis of spectral assignment. However, they also encode information about the conformation of a molecule and are therefore important structural parameters. They can be used in order to define the backbone geometry of proteins by the observation of the C_α, C_β and C' chemical shifts that can be directly translated into dihedral restraints[60] (**Figure 12**), but also from ^{31}P spectra in oligonucleotides that define the backbone angles α and ζ.[61] Chemical shifts can in addition be predicted from DFT calculations quite accurately.

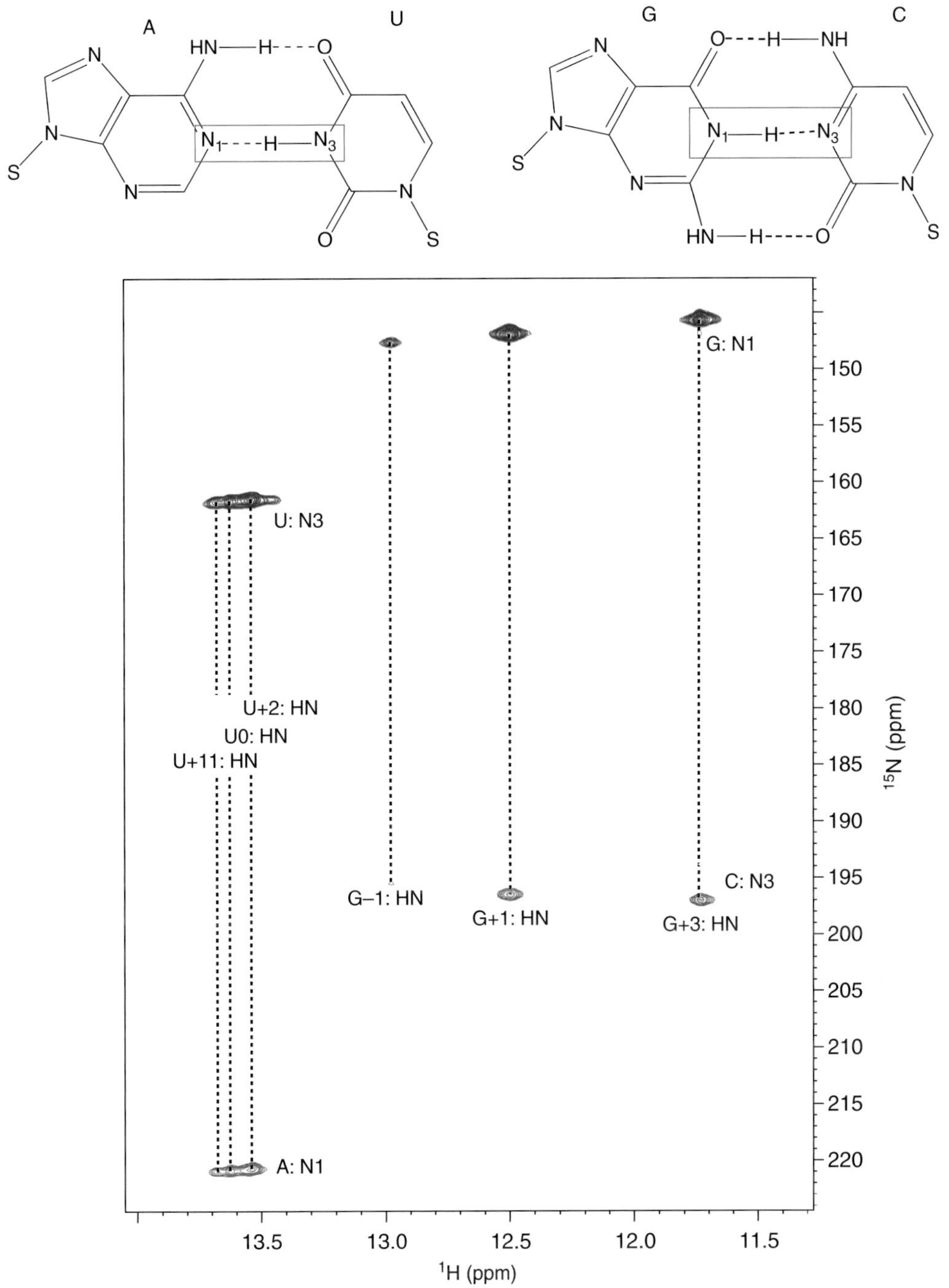

Figure 10 Scalar couplings between nitrogens in base-paired oligonucleotides. The observation of a cross peak between the bases can be interpreted as a proof for the existence of the hydrogen bond.

For solids, the nature of chemical shifts as tensors becomes obvious. Indeed, they have three principal values along orthogonal axes and are therefore dependent on the orientation of the molecule with respect to the external magnetic field. MAS spinning removes the anisotropic part of the chemical shift and the isotropic part is observed similar to the isotropic chemical shift obtained from the fast averaging by rotational diffusion of molecules in solution (**Figure 3**). Since these isotropic chemical shifts obtained in liquid- and solid-state spectra are in nature the same, they can be used for the same purposes for samples in both types of NMR spectroscopy. This is extremely helpful in determining the conformation of the backbone with very little spectroscopic information.

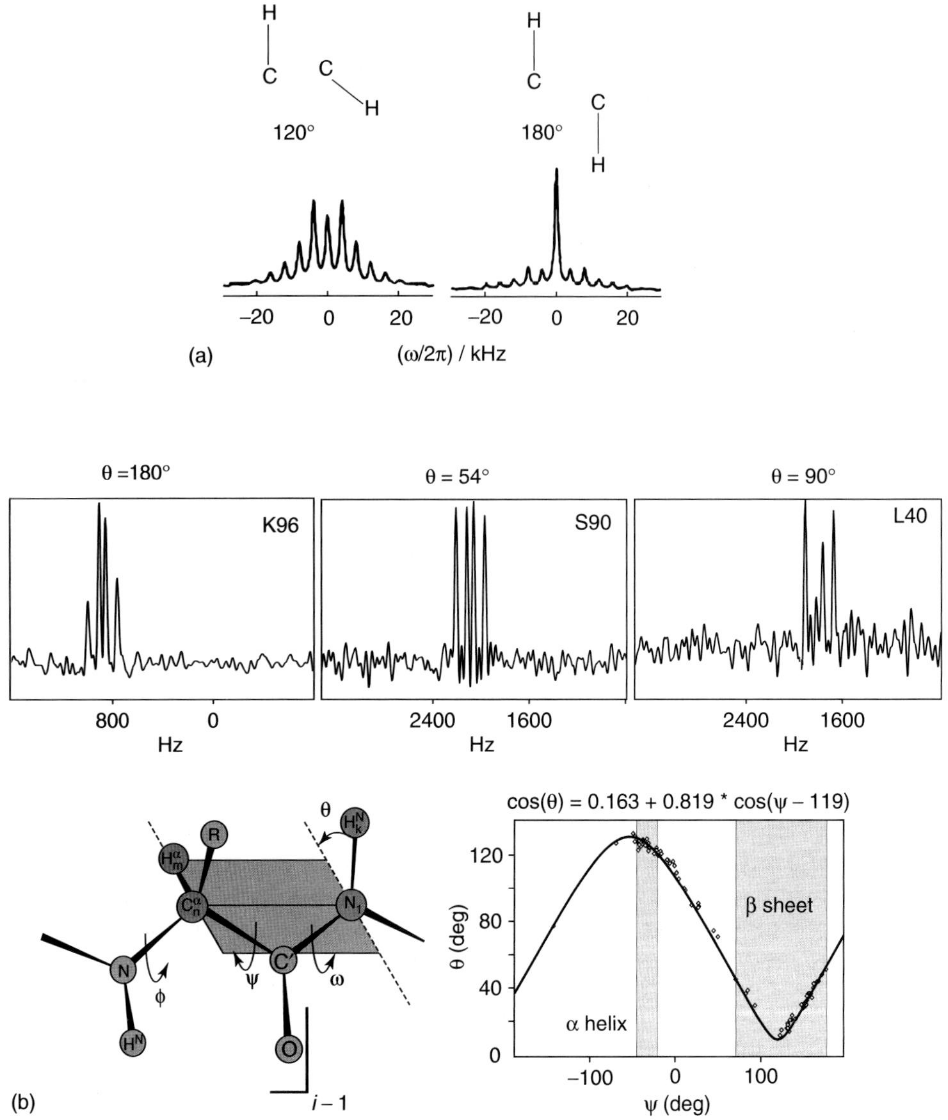

Figure 11 (a) Magic angle sample (MAS) spinning ^{13}C,^{13}C double quantum spectrum. Depending on the relative orientation of the two CH vectors, the spectrum looks different. This can be used to determine projection angles in the solid state. (b) Liquid-state analog of the solid-state experiment in (a) in which double quantum coherence is now excited between the N and C in a peptide chain. The projection angle θ can be essentially read of from the appearance of the multiplets.

3.22.3 Structure Calculation

Structure determination via NMR relies on the assignment of the resonances of the protein and then the measurement of a set of the above-discussed NMR parameters. While it is obvious that the precision of the structure can be increased the more parameters can be measured, the sensitivity of the experiments that make them accessible is quite different and therefore not all of them are always used. For example, chemical shifts have to be measured to obtain assignment and from the assigned resonances, one can very easily derive secondary structural information. In addition, NOEs as well as dipolar couplings of the backbone are very easily accessible in solution and are therefore the first ones to be used. Distances are derived from NOEs by assigning each NOE to a range of distances and enforcing this distance range to each of the individual protons. Flat bottom potentials with zero energy toward distances that are smaller than the restraint are mostly used. Such a potential avoids so-called non-NOEs, namely enforcing a minimal distance, e.g., of at least 5 Å of protons that do not show an NOE.

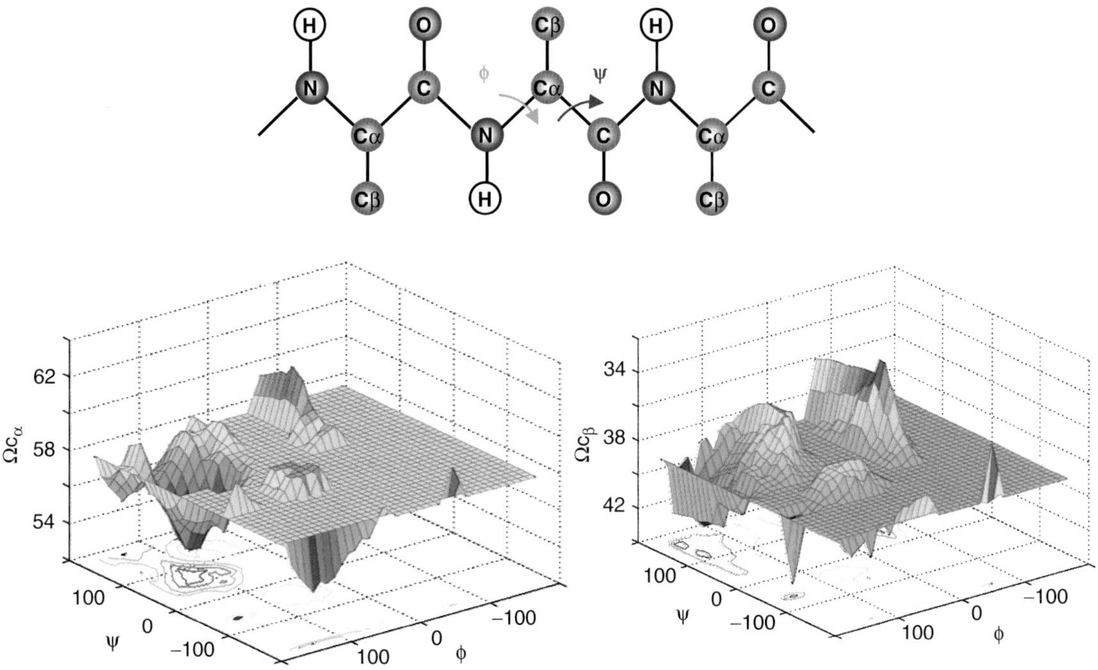

Figure 12 Dependence of the C_β and C_α chemical shifts in proteins from the dihedral angles in the backbone.

Scalar couplings are either directly converted into a range of dihedral angles that is allowed or a potential is included that calculates the deviation of the experimental from the expected coupling into an energy.

Chemical shifts are directly used in the TALOS routine as restraints.

Several protocols exist for the incorporation of dipolar couplings into structure calculation. The first one extracts from the distribution of the dipolar couplings the size of the tensor Da and R (of eqn [5])[62] and calculates an energy penalty from the difference of experimental and expected dipolar coupling.[43] The problem is the ruggedness of this restraint that changes sign even for very minute changes of the structure. A less rugged restraint can be formulated by transforming the dipolar restraints into intervector angle restraints.[44] In addition, the tensor size that is derived may have to be adjusted during the calculation by variation of the alignment tensor during the calculation.[63] Structure calculation has even been done based exclusively on dipolar couplings.[64]

For solid-state NMR, backbone assignment and the measurement of chemical shifts is also the first information to be obtained.[40,65,66] Due to the lack of orientational restraints in isotropic solid-state samples distance information has to be acquired in order to determine higher resolution structures. Prime information are the H,H distances measured in a NOESY type fashion by CHHC and NHHC experiments as described above.

The calculation protocol used most for structure calculation relies on restrained molecular dynamics[67] and the so-called simulated annealing protocol.[68] The molecule is heated to an elevated temperature (normally 1000 K for proteins and 2000 K for oligonucleotides) that is accomplished by defining velocities for the atoms that obey the Maxwell distribution for the given temperature. The molecule is then able to cross high barriers between different conformations. Cooling of the molecule in successive steps allows the molecule to overcome less and less over high energy barriers and forces it to settle into energy minima. The procedure is repeated several times from different nonstructured conformations and from the family of structures. Normally, a subset of structures with the lowest energies and the least violations of experimental restraints is selected. From this subset the root mean standard deviation (RMSD) or the atomic positions is calculated as one quality factor of the family of structures. Additionally, other protocols can be used to check, e.g., the distribution of ϕ and ψ angles that should fall into the allowed regions of the Ramachandran space for most amino acids, chirality, etc.[69]

New directions when using sparse data are explored in which NOEs between the side chains that are difficult to measure and are not required any more. Along these lines, the determination of a structure without any NOEs has been demonstrated on several examples,[70] either relying only on several sets of dipolar couplings for the backbone or by also taking paramagnetic restraints such as relaxation and pseudocontact shifts into account.[71]

Another direction is the use of NOESY restraints that need not be assigned, as the most tedious step in structure determination of proteins is to assign the proton–proton NOEs. However, provided proton assignment is available, the programs CANDID and ARIA use all proton–proton distances that are compatible with a cross peak as an ambiguous distance restraint.[72] Integration of the use of ambiguous NOEs with fully automated assignment protocols is under way and it can be foreseen that in the near future structure elucidation of small proteins will be done to high resolution with little if any intervention by skilled trained personel.[73]

For solid-state derived structures that rely on similar restraints (distances, chemical shifts, dihedral restraints) similar calculation protocols are applied. For fully labeled samples, however, there may be also intermolecular distances in addition to the intramolecular distances because of crystal packing. However, in the context of the calculations of symmetrical dimers in solution, protocols are being developed that allow the separation of intra- and intermolecular distance restraints as well.[74]

3.22.3.1 Examples

Liquid-state NMR derived structures have been solved for more than 20 years. We present here as an example the structure of the periplasmic domain of DcuS (**Figure 13a**), which has been solved recently in our laboratory.[75] DcuS is a bacterial two-component sensor, which resides in the membrane and is sensitive to fumarate. It should be noted that neither the structure of DcuS without fumarate nor with fumarate could be obtained by x-ray crystallography. However, the structure of a homologous protein, CitA, in the presence of its cognate small molecule, citrate, could be obtained by x-ray crystallography and is shown for comparison (**Figure 13b**).[76] Interestingly, the CitA structure could not be solved in the absence of citrate by x-ray crystallography.

The DcuS structure has been solved using all the parameters indicated above. From the comparison of the two structures a possible mechanism of transmission of the signal can be inferred, namely through the C-terminal helix, which shows markedly different conformations in the two structures. The modeled extensions of the two helices into the membrane are indicated on the CitA structure.

The structure of a beta-barrel protein, PagP, is shown in **Figure 14a** that has been obtained from partially deuterated samples in which only the methyl groups and the amides were protonated. The sample has been reconstituted in micelles. Here, the molecular weight is approximately 80 kDa. This structure constitutes one of the largest that has been solved ab initio from NMR spectroscopy.[77]

The residual structure of alpha-synuclein, a partially unfolded protein, is demonstrated that has been derived mainly from dipolar couplings as well as paramagnetic relaxation restraints (**Figure 14b**).[34] The monomeric form adopts a residual tertiary structure in the absence of any secondary structure. NMR is unique in the description of such unfolded proteins in solution with atomic resolution.

Solid-state NMR has a much shorter history of structure elucidation. Probably, the first molecular structure determined purely by MAS-based NMR was reported by Terao and coworkers and relates to the dipeptide glycyl isoleucine.[78] In this case, chemical shift selective transfer methods were applied to determine ($^{13}C^{13}C$) distance constraints. Three years later, the molecular structure of the tripeptide *N*-formyl-L-Met-L-Leu-L-Phe-OH was published.[79] Here, NMR experiments were conducted on a uniformly ($^{13}C^{15}N$) labeled sample. Three differently ($^{13}C^{15}N$) labeled protein samples were employed to determine the 3D fold of a microcrystalline sample of the alpha-spectrin SH3 domain (1M8M)[31,80] and of the 11-amino acid stretch of transthyretin (TTR(105-115).[81]

Using indirectly detected proton–proton interactions, the 3D structure of uniformly ($^{13}C^{15}N$) labeled polyethylene glycol (PEG)-precipitated ubiquitin was determined in the solid state.[82] No attempts were made to obtain microcrystalline material. An ensemble of 10 structures of ubiquitin selected according to the lowest overall energy and aligned along the backbone atoms of residues M1 to V70 with the aid of MOLMOL is shown in **Figure 14c**. This

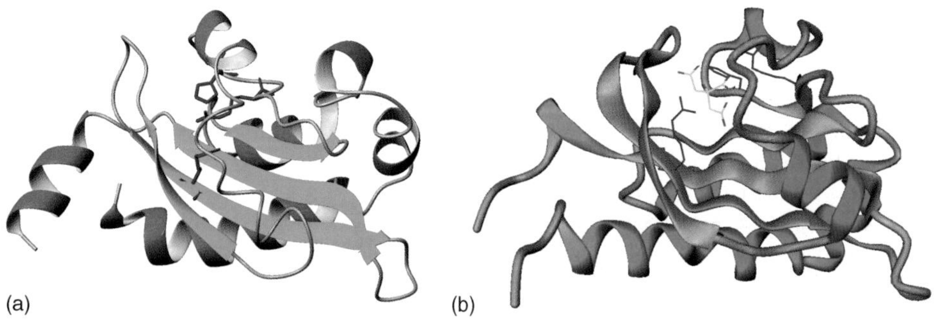

Figure 13 Comparison of the structure of the periplasmic domains of DcuS (a) solved by NMR and CitA in the presence of citrate, (b) solved by x-ray crystallography. Refer to the text for more details.

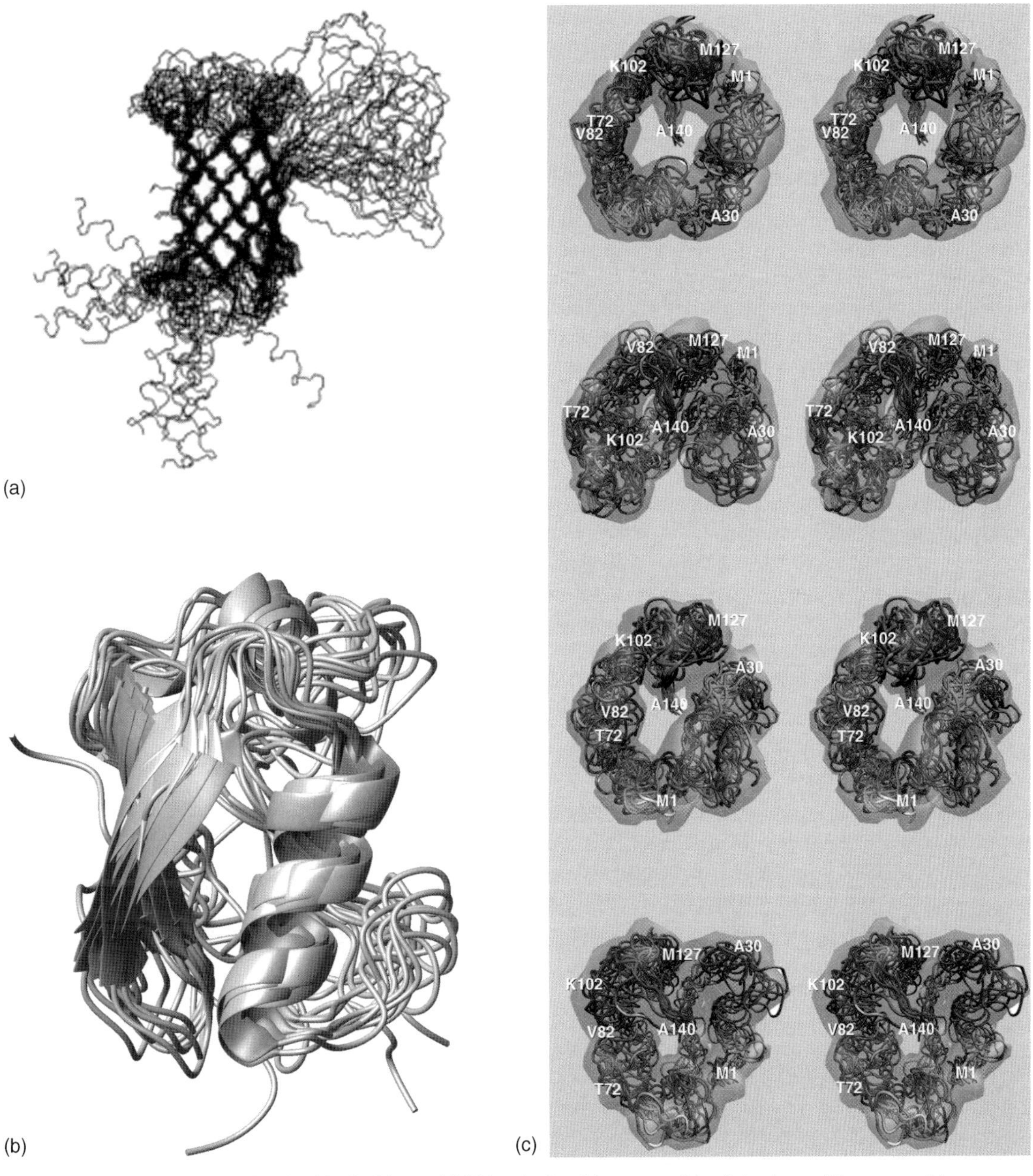

Figure 14 (a) structure of PagP solved by liquid-state NMR in micelles; (b) an ensemble of structures of the α-synuclein monomer that does not adopt secondary but residual tertiary structure; and (c) structure of ubiquitin solved by solid-state NMR in the powder.

structure is based on the analysis of C/NHHC data and secondary chemical shift information. The overall fold and the characteristic secondary structure elements are reproduced well, suggesting that single crystal and microcrystalline structure must be closely related. The refinement of this structure is currently ongoing. Note that this result was obtained without the use of the recently published strategy of block-labeling proteins.[83]

3.22.3.2 Fast Acquisition Schemes

Structure determination by liquid-state NMR spectroscopy has become more and more automated and defined with respect to the acquisition of the data as well as their interpretation. For some samples, where neither availability nor solubility pose a problem conventional 3D and 4D experiments can be replaced by faster versions such as a set of projection reconstruction spectra or using the so-called GFT approach (G-matrix Fourier transform). In these experiments, evolution times are incremented simultaneously in a concerted fashion such that weighted sums and differences of chemical shifts are recorded in certain frequency dimensions. Due to this reduction in dimensionality, less measurement time is required without compromising the number of chemical shifts that are reflected in the spectra.[84] These pulse sequences are useful for structural genomics projects that are undertaken mainly in the USA and Japan.

3.22.4 Multidomain Proteins, Protein–Protein, and Protein–Ligand Complexes

Many proteins are constituted of different globular domains that are interconnected by rather loose linker domains. These domains then can have different amounts of mobility between them. They are expected to move almost unhindered if, e.g., in a two-domain protein these domains do not interact except for the covalent linker (weak binding). However, there may also be a tight binding between them that would hold even if there was no covalent linkage between the two domains. Then a protein–protein complex can be understood as a two- or multidomain protein where the linker is missing and similarly tight and weak binding can be observed. Essentially ligand–protein complexes can also be understood as a noncovalent interaction between two molecules.

3.22.4.1 Preparation

With respect to labeling, there are differences between multidomain proteins and protein–protein complexes. Unless chemical ligation is used[85] based on inteins,[86] the individual domains in a multidomain protein are labeled in the same way. However, using intein technology, block labeling can be done in which different parts of the proteins can be labeled differently. For protein–protein complexes, a similar situation exists. Two proteins may not be expressable separately, e.g., because one is toxic and needs to be inhibited by its binding partner or one is unfolded and is degraded during expression without its binding partner. Then the protein complex has to be expressed and the components will be labeled in the same way. If the complex can be reconstituted, however, then the individual domains can be labeled in a different way that makes spectroscopy considerably simpler.

3.22.4.2 Binding Affinities

Interactions between molecules can be studied by NMR spectroscopy in a large variety of binding constants. Although NMR was used for a long time just as a tool to determine the structure, the unique ability of NMR to detect molecular interactions has recently established its importance in the field of drug screening and optimization.[87] For example, the complex between calmodulin (CaM) and its cognate peptide from the Ca-ATPase (C20W) is a tight complex with a binding constant in the nM range (see below). However, much weaker binding in the mM range can also easily be detected by NMR.

3.22.4.2.1 Tight binding

Tight binding of molecules to other molecules requires structural analysis very much along the lines as has been described for protein structure determination in the liquid or also the solid state (DcuS-periplasmic domain for liquid-state and ubiquitin for solid-state NMR). In this case, the dissociation constant of the complex is so high that the rate of dissociation is normally found to be on the order of several ms or even longer. The induced chemical shifts in simple NMR experiments such as ^{15}N,^{1}H-HSQC experiments haven been used for structure–activity relationships (SARs) studies by NMR.[88]

As an example we show the complex complex between calmodulin and a cognate peptide C20W[89] that is part of the plasma membrane Ca^{2+}-ATPase which is responsible for pumping calcium ions out of the cell. The peptide binds exclusively to the C-terminal domain with a subnanomolar dissociation constant while contacts to the N-terminal domain are not observed. The structure of the complex is shown in **Figure 15**. It has been determined from 1634 NOE restraints (791 intraresidual, 383 sequential, 307 medium range and 153 long range) that could be identified within CaM, corresponding to an average of 15.6 NOEs per residue. For the peptide C20W, 163 intramolecular NOE restraints could be assigned (102 intraresidual, 32 sequential, 29 medium range); and 48 intermolecular NOEs, detected between the peptide and the protein, defined the interface between the two molecules (**Figure 15b**). Based on the temperature dependence of amide proton chemical shifts, 60 intramolecular hydrogen bond restraints were incorporated for CaM and nine for the peptide C20W. The acceptor of the hydrogen bond was identified by backcalculation of structures derived only from NOEs. In addition, 24 distance restraints between the four Ca^{2+} ions and protein groups derived from the crystal structure of Ca^{2+} loaded CaM were used in the structure calculation.

Since solid-state NMR is not adversely affected by large molecular weight, protein–protein complexes can be investigated without big changes in the approach. This is an especially attractive feature. Until recently, no structural information of a high-affinity ligand bound to a G protein-coupled receptor (GPCR) was available. Unlike rhodopsin,

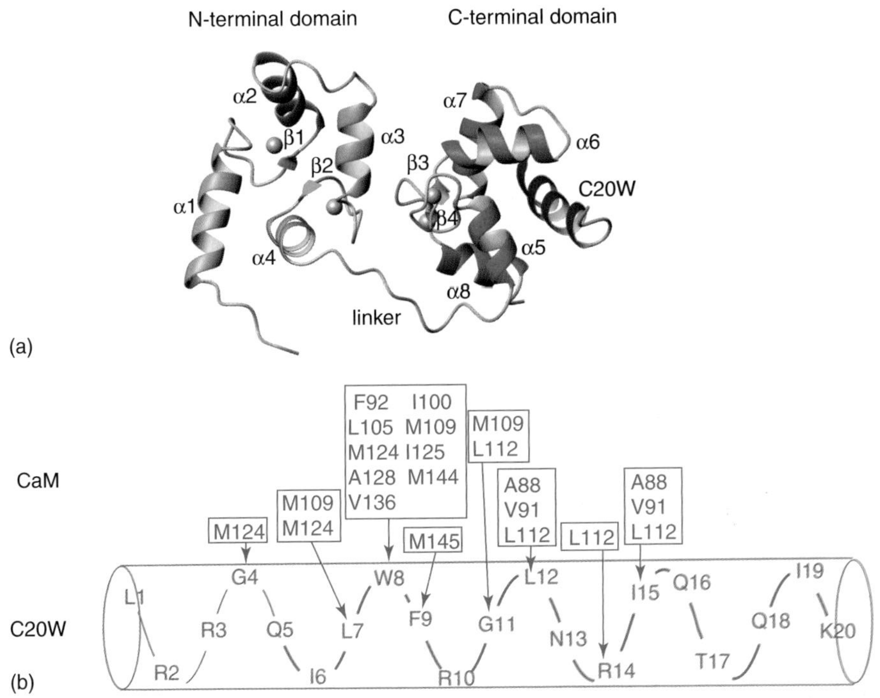

Figure 15 (a) Structure of the complex between calmodulin (CaM) and the cognate peptide C20W; The peptide binds only to the C-terminal domain of CaM. (b) NOEs between the peptide and the protein.

the recombinant expression of GPCR in large quantities is usually difficult and must involve carefully optimized biochemical procedures. Restrictions regarding the availability of functional receptors also affect ligand quantities that can be studied. Moreover, the chemical environment including lipids and receptor protein can hamper the unambiguous spectral identification of a bound ligand in a solid-state NMR experiment. Recently, Luca *et al.*[90] were able to show how 2D double-quantum solid-state NMR experiments can be utilized to detect microgram quantities of bound neurotensin, a 13-residue neuropeptide that binds in high affinity (subnanomolar) to the NTS-1 (101 kDa) receptor. For the 6-residue, biologically active, C-terminal sequence of neurotensin, a homonuclear double quantum–single quantum correlation spectrum is sufficient to assign all C_α and C_β resonances of the uniformly ^{13}C,^{15}N labeled ligand. These chemical shift assignments can be used to construct the backbone model of the ligand in complex with the receptor (**Figure 16**). The solid-state NMR data indicate that the resulting beta-strand conformation is only adopted in the presence of the receptor and may serve as a structural template for future pharmacological studies. The structure of this peptide could be used for correct predictions of the binding affinity of peptide mimetics.[91]

A similar solid-state NMR study was recently performed on a kaliotoxin (KTX), a toxin that binds in high affinity to the chimeric KcsA–Kv1.3 potassium channel.[92] Using indirectly detected proton–proton interactions in addition to chemical shifts, the 3D structure of a uniformly (^{13}C,^{15}N) labeled version of kaliotoxin was determined in the solid state.[93] Proton–proton distance and dihedral angle restraints derived from conformation-dependent chemical shifts resulted in 3D structure with an average backbone RMSD of 0.81 Å (PDB entry: 1XSW) (**Figure 17a**). In **Figure 17b**, an ensemble of structures representing the toxin is shown when bound to the potassium channel KcsA–Kv1.3 potassium channel that has been reconstituted functionally in proteoliposomes. Residues that are perturbed by complex formation with a chimeric KcsA–Kv1.3 channel are indicated in red in a surface representation (**Figure 17c**).

3.22.4.2.2 Weak binding

If the binding constant of the drug to the target is of the order of μM to mM, techniques can be used for the detection of the binding as well as for the determination of the structure that rely on the fact that the drug in its target-bound form as compared to the free form has largely different translational as well as rotational diffusion times (correlation time), which will affect the relaxation times of the drug. The different relaxation of the bound versus the free form is

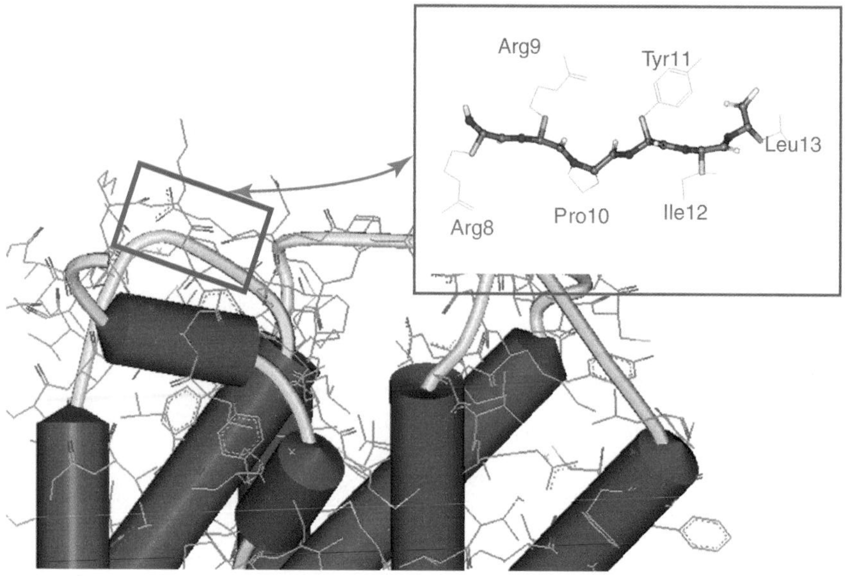

Figure 16 Structure of neurotensin 8-13 when bound to the neurotensin receptor. The structure of the GPCR is a model. The backbone structure of the peptide has been determined experimentally mainly from chemical shifts.

reflected in broader linewidths and cross-correlated relaxation rates as well as an increase in the cross-relaxation rates (NOE). The fast exchange between bound and free form leads to an averaging of the properties of the bound and the free form weighted with the populations and the correlation times, which are proportional to the molecular weight. For weak binding, the experiments are performed in such a way that the drug is present in excess (normally mM concentration) compared to the target protein (normally μM concentration). Observing the drug resonances then yields a strong signal due to the rather large concentration of the drug. Frequently, the bound form is present to less than 10% in solution. However, still the relaxation rates of the bound form dominate the cross-relaxation[94] and transfer cross-correlated relaxation[95,96] properties:

$$\sigma_{ij}^{\mathrm{av}} = \sigma_{ij}^{\mathrm{bound}} p^{\mathrm{bound}} + \sigma_{ij}^{\mathrm{free}} p^{\mathrm{free}}, \quad \Gamma_{ijkl}^{\mathrm{av}} = \Gamma_{ijkl}^{\mathrm{bound}} p^{\mathrm{bound}} + \Gamma_{ijkl}^{\mathrm{free}} p^{\mathrm{free}} \qquad [7]$$

provided the molecular mass of the drug M^{free} is much less than that of the target protein M^{bound} fulfilling the following equation:

$$M^{\mathrm{bound}} p^{\mathrm{bound}} > M^{\mathrm{free}} p^{\mathrm{free}} \qquad [8]$$

This approach has been used for the potential anticancer drug class, epothilones. Epothilones exhibit extraordinary antiproliferative activity in vitro and they efficiently induce cell death in paclitaxel-resistant tumor cell lines at up to 5000-fold lower concentrations than paclitaxel.[97–99] Epothilone binds to tubulin and stabilizes microtubules inducing apoptosis in these cells. The complex between tubulin and epothilone is difficult to investigate since so far it has not been possible to label tubulin with stable isotopes for NMR spectroscopic investigation. However, since the binding constant is weak, transfer NOESY (trNOE)[100] and transfer cross-correlated relaxation (trCCR)[101,102] could be successfully applied to this compound. The tubulin-bound conformation of epothilone A was calculated from 46 interproton distance restraints and seven torsion angle restraints measured for a 0.5 mM solution of epothilone A in water in the presence of 5 μM tubulin.[103] The distance restraints were derived from transferred NOE experiments. To filter out spin diffusion-mediated peaks, only those signals were taken into account than were present with opposite sign to the diagonal peaks in a transferred ROESY experiment. The dihedral angle restraints were obtained by measuring CH–CH dipolar–dipolar transferred CCR rates in CH–CH and CH–C–CH moieties[104] as well as CH–CO dipolar–CSA transferred CCR rates for 60–70% ^{13}C labeled epothilone A in complex with tubulin. Two main changes are observed between the epothilon free compared to bound to tubulin.[103] The thiazole ring rotates by 180° and a local conformational change in the region of O1–C6 moves the 3OH group by 3.8 Å. Epothilones derivates, which have the

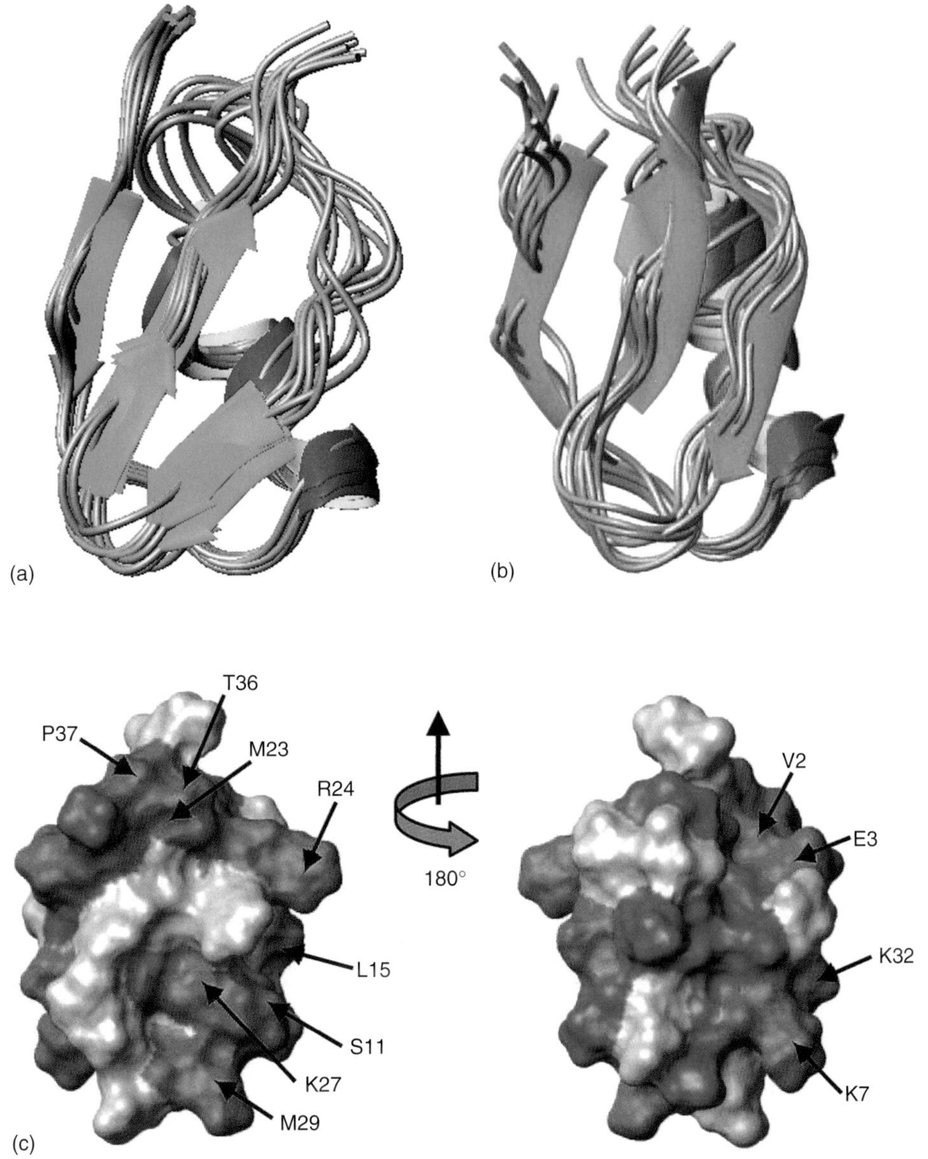

Figure 17 (a) Structure of kaliotoxin (KTX) without and (b) with the potassium channel KcsA-Kv1.3 bound. One beta-strand of the KTX becomes considerably longer upon binding. The amino acids that show contacts between the toxin and the ion channel are indicated in red (c).

thiazole locked in the correct conformation by cyclization, corroborate the relevance of the found conformation. The comparison of the structures is given in **Figure 18**. Approaches toward the intercorrelation of binding modes of several weak binders are under way.[105]

3.22.5 Dynamics

Structural NMR parameters are modified due to dynamics on all different timescales, which is sometimes a nuisance in the process of structure determination but it is a rich source of dynamical information to be extracted from NMR spectra. In this chapter, we cannot review all the aspects of motion to be picked up by NMR measurements but will focus on some selected aspects.

The timescales of motion that are accessible are depicted in **Figure 19**. Each of the indicated timescales can be addressed by different parameters. It should be mentioned that for conformational conversions up to the ms range, only

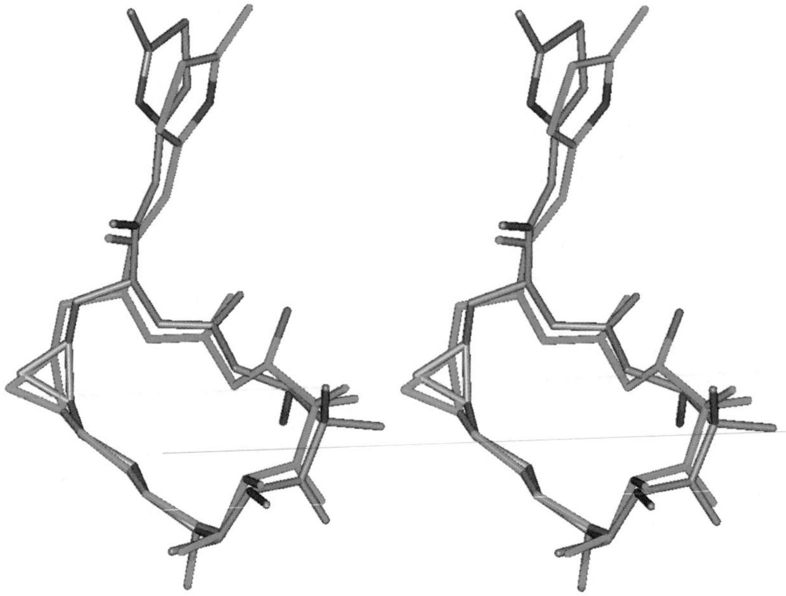

Figure 18 Stereoview of epothilone when bound to tubulin (green) and without tubulin (grey). The conformation of eopthilone changes quite dramatically when it is bound by the tubulin.

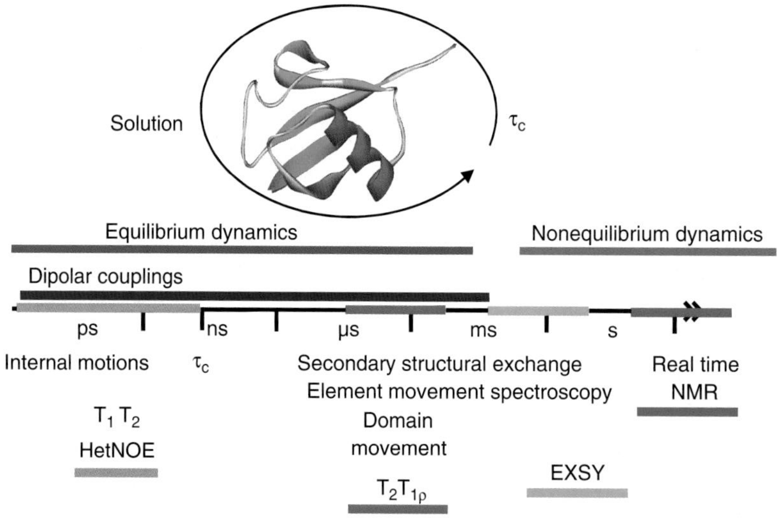

Figure 19 Timescales for NMR spectroscopy with the various NMR parameters that address them.

equilibrium dynamics is accessible because the free induction decay (FID) requires several ms in order to resolve the spectrum. Nonequilibrium processes are accessible by so-called real time NMR.[106] Fascinating real applications are feasible using ultrafast 2D acquisition schemes in which a 2D experiment can be acquired in a single scan.[107] Nevertheless, the sensitivity has still been a limitation in such techniques so far.

3.22.5.1 Sub-Correlation Time Dynamics

The most common approach is the measurement of relaxation parameters of the backbone nitrogens in proteins that report on motion on timescales faster than the correlation time. The nitrogen has a dipolar coupling to the amide proton and it has a rather well-defined CSA that is moderately asymmetric and almost parallel to the direction of the NH bond. Three parameters can be measured rather easily: T_1, the longitudinal relaxation time of the nitrogens, and T_2, the transverse relaxation time of the nitrogens as well as the steady state NOE of the ^{15}N upon irradiation into the amide proton.[108]

$$\begin{aligned}
1/T_1 =& \frac{1}{3}(\sigma_\parallel - \sigma_\perp)^2 \gamma_N^2 B_0^2 J(\omega_N) \\
&+ \left(\frac{\hbar\mu_0\gamma_H\gamma_N}{8\pi r_{HN}^3}\right)^2 (J(\omega_N - \omega_H) + 3J(\omega_N) + 6J(\omega_N + \omega_H)) \\
1/T_2 =& \frac{1}{18}(\sigma_\parallel - \sigma_\perp)^2 \gamma_N^2 B_0^2 (4J(0) + 3J(\omega_N)) \\
&+ \left(\frac{\hbar\mu_0\gamma_H\gamma_N}{8\pi r_{HN}^3}\right)^2 \left(2J(0) + \frac{1}{2}J(\omega_N - \omega_H)\right. \\
&\left. + \frac{3}{2}J(\omega_N) + 3J(\omega_H)3J(\omega_N + \omega_H)\right) \\
\mathrm{NOE} =& \frac{\gamma_H}{\gamma_N} \frac{\left(\frac{\hbar\mu_0\gamma_H\gamma_N}{8\pi r_{HN}^3}\right)^2 (-J(\omega_N - \omega_H) + 6J(\omega_N + \omega_H))}{\frac{1}{3}(\sigma_\parallel - \sigma_\perp)^2 \gamma_N^2 B_0^2 J(\omega_N) + \left(\frac{\hbar\mu_0\gamma_H\gamma_N}{8\pi r_{HN}^3}\right)^2 (J(\omega_N - \omega_H) + 3J(\omega_N) + 6J(\omega_N + \omega_H))}
\end{aligned} \quad [9]$$

Since the T_2 is also modulated by exchange processes that affect the chemical shifts, some researchers measure in addition the cross-correlated relaxation between the CSA of the ^{15}N and the dipolar coupling that is not affected by exchange processes. The interpretation of the rates is done with different models, of which the simplest one and the one requiring many assumptions is the most commonly used, namely the Lipari Szabo model-free approach.[109] It requires the uncoupling of the overall rotational diffusion from internal motion. Then all motions are cast into one internal correlation time and the data analysed accordingly. In this model, an overall correlation time τ_c, an order parameter S^2 and an internal correlation time τ_e can be extracted in an almost automatic way.[110]

$$J(\omega_q) = \frac{4}{5}\left(\frac{S^2\tau_c}{1 + \omega_q^2\tau_c^2} + \frac{(1 - S^2)\tau_e}{1 + \omega_q^2\tau_e^2}\right) \quad [10]$$

Anisotropies of the rotational diffusion are taken into account in this software.

Field-dependent measurements often reveal that the simple model may not be valid in all situations such that various attempts to obtain different models of motion are being developed. For example, the uncoupling of overall and internal motion is not a requirement in the model of Meirovitch and Freed, which however requires much more data to be measured.[111]

3.22.5.2 μs to ms Dynamics on the Scale

Motion slower than the correlation time is not picked up by T_1 and heteronuclear NOE and requires different ways of measurement. One approach is the so-called relaxation dispersion, which measures the modulation of chemical shifts due to interconversion of conformations. Provided conformations have different chemical shifts, then an exchange process will average the two chemical shifts involved. For an exchange process that is fast on the chemical shift frequency scale: $k \gg \Delta\upsilon$, the linewidth is given by[112]:

$$1/T_2^{\mathrm{exchange}} = p_A p_B \Delta\Omega^2 \frac{\tau_{ex}}{1 + (\omega_{eff}\tau_{ex})^2} \quad [11]$$

The linewidth is measured in the time domain by variation of a delay during which the transverse coherence decays. This also allows one to apply a transverse magnetic field of adjustable field strength, which then reports also on the timescale of the chemical shift exchange. The result of these measurements is then a rate of interconversion and in addition the product of the populations of the involved species and the difference in chemical shifts. Relaxation dispersion is broadly applied in protein structural biology since it has been shown that functional dynamics, e.g., of enzymes or of binding events relates to the internal dynamics detected by relaxation dispersion.

As an example the three relaxation rates $1/T_1$, $1/T_2$, and the heteronuclear NOE are plotted for calmodulin and the calmodulin/C20W complex. The rates adopt more uniform values upon binding of the peptide. The linker region

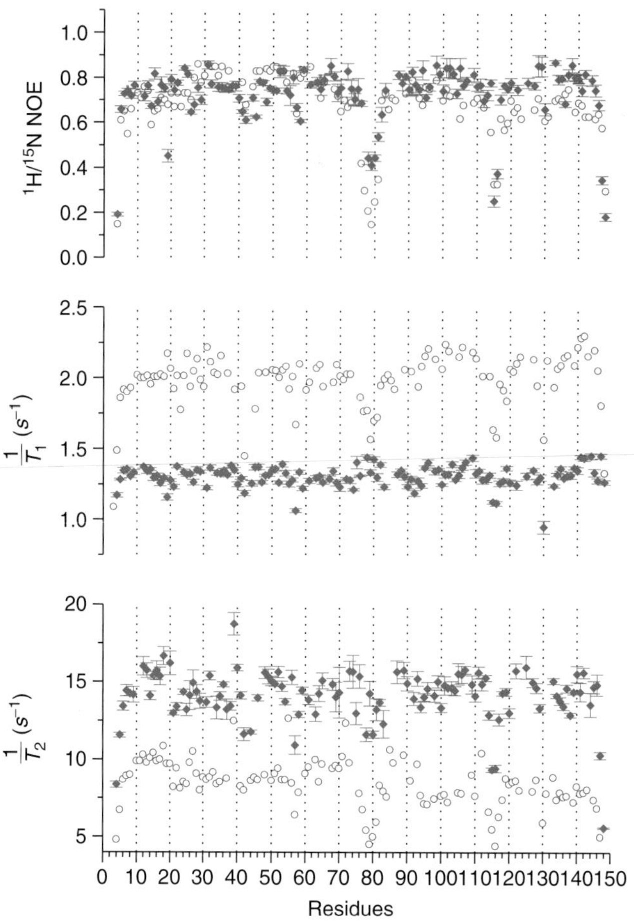

Figure 20 Various relaxation parameters that have been measured for the complex between calmodulin and C20W (red) and calmodulin alone (blue), represented along the protein sequence. Information on the interpretation is given in the text.

around amino acid 80 becomes less flexible after the peptide is bound. This is true for both fast and slow timescales (**Figure 20**).

A fascinating application of relaxation dispersion is the detection of functional dynamics in enzymatic reactions. For enzymes to reach a certain turnover they need to bind the product substrate, catalyze the reaction, and release the product in a given time. If protein dynamics is detected that occurs on the same timescale, it is highly probable that it is the dynamics the protein has to show in order to catalyze the reaction. Such functional dynamics has been observed for the first time on the reaction of binase.[113] In this study, the authors show that the active site of binase is flanked by loops that are flexible at the 300 μs timescale (**Figure 21**). One of the catalytic residues, His101, is located on such a flexible loop. In contrast, the other catalytic residue, Glu72, is located on a beta-sheet, and is static. The residues Phe55, part of the guanine base recognition site, and Tyr-102, stabilizing the base, are the most dynamic. The findings suggest that binase possesses an active site that has a well-defined bottom, but which has sides that are flexible to facilitate substrate access and egress, and to deliver one of the catalytic residues. The motion in these loops does not change on complexation with the inhibitor d(CGAG) and compares well with the maximum *k*cat (1500 s^{-1}) of these ribonucleases. This observation indicates that the NMR-measured loop motions reflect the opening necessary for product release, which is apparently rate-limiting for the overall turnover (**Figure 21**). The backbone dynamics is reflected with the thickness of the tube and color coding. The dynamics is found to have a timescale of 300 μs, which is nicely compatible with the maximum reaction rate of 1500 Hz.The dynamics does not depend on the absence or presence of the substrate, indicating that the rate-limiting step is motion of the loop of which the catalytic residues His101 and Tyr102 are a part. Several further studies along these lines have been published recently.[114]

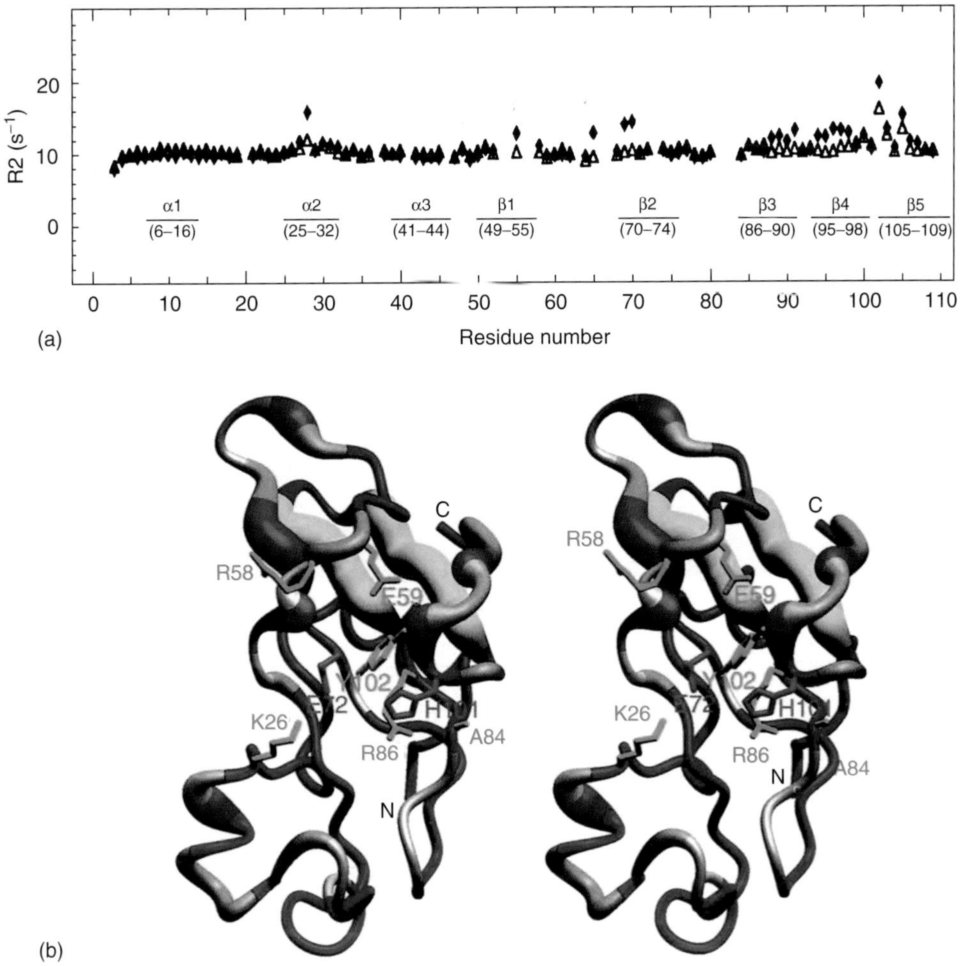

Figure 21 (a) T_2 relaxation measured for binase with a sequence in which the term from the chemical exchange is removed (open triangles) not removed (closed triangles). Those amino acids that have different values for the two spectra undergo chemical exchange. (b) Mapping of the catalytic residues of binase onto the structure; while H101 is part of a mobile loop, the other important residue E72 is rather rigid. More information on the interpretation is given in the text.

3.22.5.3 ns to μs Dynamics on the Scale

The gap between the timescale accessible by relaxation measurements and by relaxation dispersion is between the correlation time and approximately 100 μs. The rates here are too fast to access by relaxation dispersion unless nuclei whose chemical shifts are strongly dependent on conformation are used, such as ^{19}F, and they are slower than the correlation time. Here, the measurement of dipolar couplings can be undertaken. Dipolar couplings are averaged over all timescales until exchange processes between different conformations become so slow that individual chemical shifts can be observed. If a vector between two dipolar coupled spins moves the average dipolar coupling is smaller. While this can be measured directly in solid-state spectra, e.g., from the Pake pattern in solid-state spectra, in liquid-state spectra it requires the measurement of dipolar couplings in five linearly independent alignment media. From such measurements, however, the size of the dipolar coupling can be reconstructed and motion can be extracted. Such measurements have been performed on protein G and ubiquitin and showed marked dynamics in the time range between the correlation time and 50 μs where relaxation dispersion measurements start to be active.[115] A result from these studies is given in **Figure 22** in which the order parameter of this type of motion is color-coded. Residues whose side chains point toward the outside of the protein have side chains that are more mobile than residues whose side chains point toward the interior of the protein. This finding was recently corroborated from a careful analysis of protein x-ray structures with a resolution of less than 1 Å.[116]

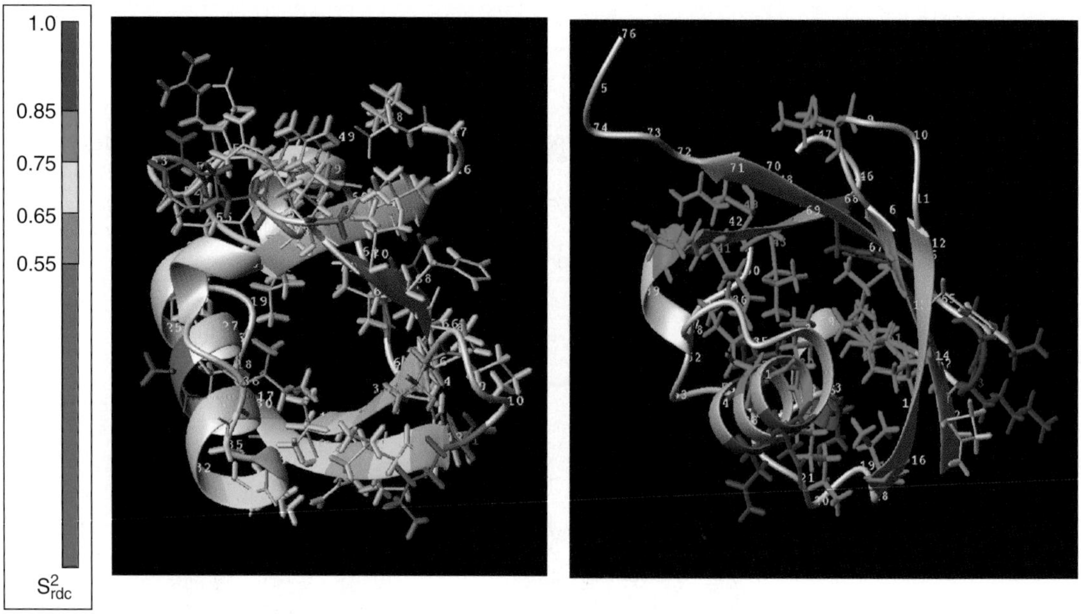

Figure 22 Backbone dynamics on a timescale between the correlation time and 50 μs. A color scale is given on the left. Those residues that have mobile backbones have side chains that point predominantly toward the water and vice versa.

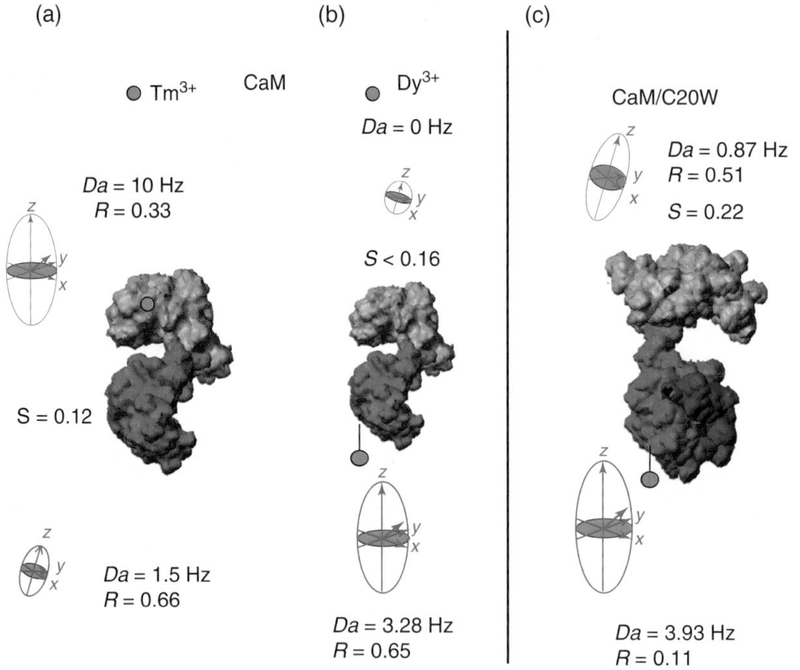

Figure 23 Domain motion of calmodulin and the calmodulin C20W complex revealed by paramagnetic tagging. While the two domains move in the absence of the peptide with an order parameter $S = 0.12$, they become less mobile ($S = 0.22$) upon binding of the peptide.

3.22.5.4 Domain Motion in Multidomain Proteins

The definition of domain orientation in a multidomain protein is possible. If for example the two domains are rigidly connected dipolar couplings can be measured in a suitable alignment medium and the relative orientation is obtained by enforcing the alignment tensors to be parallel.

However, when the domains become dynamic with respect to each other, external alignment media are not able to detect the amount of mobility of these domains with respect to each other. We consider again the case of calmodulin and the complex of calmodulin with the cognate peptide C20W. A cysteine can be mutated into calmodulin such that a paramagnetic tag can be added. Then the alignment of the tagged domain can be compared to the alignment of the domain that is not tagged. The reduction of the axial component *Da* is a measure for the order parameter, while the change in rhombicity *R* is a measure for the anisotropy of the motion of the nontagged domain versus the tagged one. As can be appreciated from **Figure 23**, the order parameter of the motion of the N-terminal domain with respect to the C-terminal one is found to be around 0.12.[117,118] When the peptide C20W in blue is bound to the calmodulin, the amount of motion of the two domains with respect to each other is reduced and the order parameter increases to 0.22.

3.22.5.5 Dynamics in Solid-State Nuclear Magnetic Resonance

The timescale separations in solid and liquid state are different because there is no rotational correlation time in the solid state. Therefore, all interactions in solid-state NMR behave as the chemical shifts in liquid-state NMR. Thus, a motion that is faster than the timescale of the interaction reduces the size of the interaction even down to zero, to extreme line broadening and for dynamics slower than the reciprocal magnitude of the interactions to a doubling of lines. Such fast motion was for example detected in the protein phospholamban when it was reconstituted in liposomes (**Figure 24**). While the dipolar couplings were not averaged out for the N-terminal hydrophobic part, they were

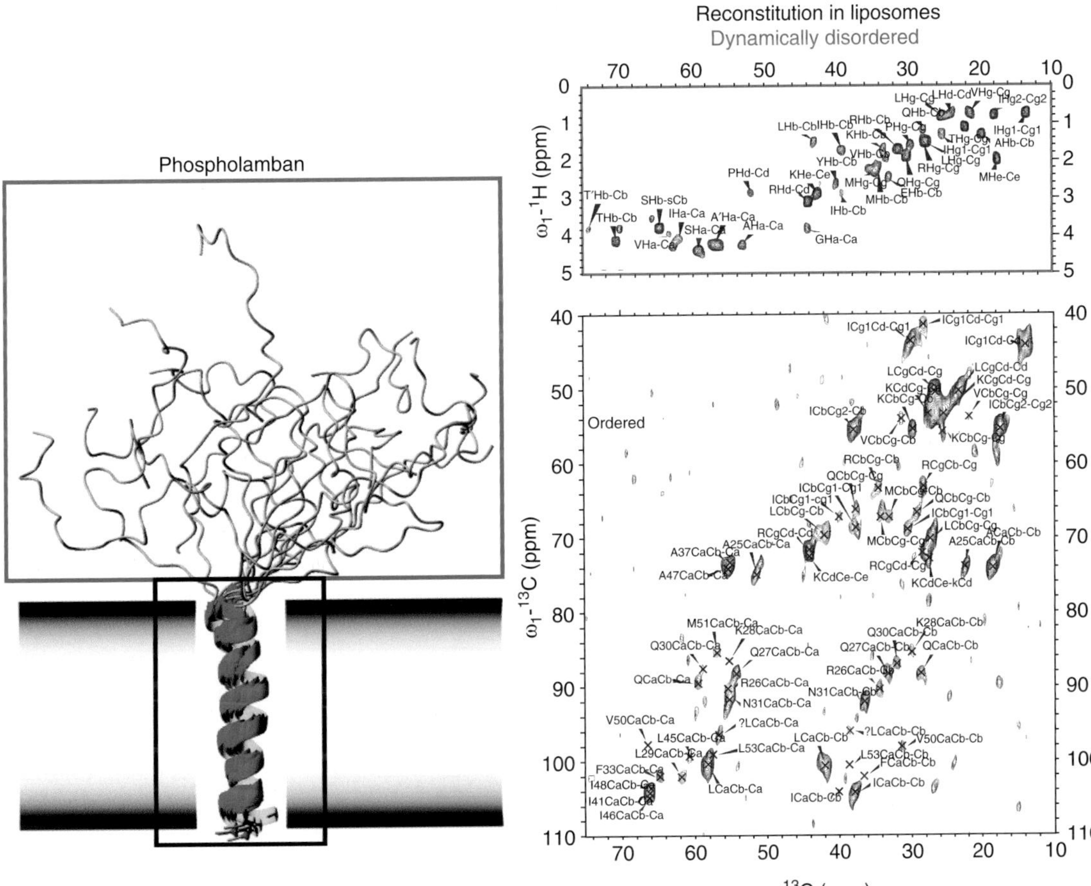

Figure 24 NMR spectra of phospholamban reconstituted in liposomes. The C-terminal part of the peptide is fully ordered and immobile in the membrane and therefore exhibits large dipolar couplings that can be used for correlations (black box). The N-terminal part, however, is not structured and fully flexible. The dipolar couplings are averaged out but due to the fast motion (faster than 50 μs), a J(^{1}H,^{13}C) correlation experiment can be recorded.

completely averaged out for the C-terminal part and only scalar couplings were observable, indicating motion on a timescale that is considerably faster than the ^{15}NH dipolar coupling and must therefore occur on a timescale of shorter than approximately 50 μs (**Figure 23**).[119]

3.22.6 Conclusion

We have reviewed the most promising applications of NMR spectroscopy both in solution as well as in the solid state. The goal of the article was to show the similarity of the principles that underly liquid- and solid-state NMR and to show that NMR cannot only provide structure but is also very successful in describing the dynamical features of proteins. The breadth of applications including folded and weakly folded proteins as well as membrane proteins and complexes has been shown with examples. Exciting novel developments are ongoing both in elucidation of structures of larger and larger proteins as well as the understanding of biomolecular function by the detailed study of the dynamics revealed by NMR.

References

1. Wüthrich, K. *NMR of Proteins and Nucleic Acids*; Wiley: New York, 1986.
2. Vuister, G. W.; Boelens, R.; Kaptein, R. *J. Magn. Reson.* **1988**, *80*, 176–185.
3. Oh, B. H.; Westler, W. M.; Derba, P.; Markley, J. L. *Science* **1988**, *240*, 908–911.
4. Ikura, M.; Kay, L. E.; Bax, A. *Biochemistry* **1990**, *29*, 4659–4667.
5. Sattler, M.; Schleucher, J.; Griesinger, C. *Prog. NMR Spectrosc.* **1999**, *34*, 93–158.
6. Baldus, M. *Prog. NMR Spectrosc.* **2002**, *41*, 1–47.
7. Grzesiek, S.; Bax, A. *J. Am. Chem. Soc.* **1992**, *114*, 6291–6293.
8. Grzesiek, S.; Bax, A. *J. Magn. Reson.* **1992**, *99*, 201–207.
9. Pauli, J.; Baldus, M.; van Rossum, B.; de Groot, H.; Oschkinat, H. *ChemBioChem* **2001**, *2*, 272–281.
10. Bloembergen, N. *Physica* **1949**, *15*, 386–426.
11. Seidel, K.; Lange, A.; Becker, S.; Hughes, C. E.; Heise, H.; Baldus, M. *Phys. Chem. Chem. Phys.* **2004**, *6*, 5090–5093.
12. Bloembergen, N.; Purcell, E. M.; Pound, R. V. *Phys. Rev.* **1948**, *73*, 679–712.
13. Redfield, A. G. *IBM J. Res. Dev.* **1957**, *1*, 19–31.
14. Sachleben, J.; Gaba, J.; Emsley, L. *Solid State Nucl. Magn. Reson.* **2006**, *29*, 30–51.
15. Pervushin, K.; Riek, R.; Wider, G.; Wüthrich, K. *Proc. Natl. Acad. Sci. USA* **1997**, *94*, 12366–12371.
16. Pervushin, K.; Riek, R.; Wider, G.; Wüthrich, K. *J. Am. Chem. Soc.* **1998**, *120*, 6394–6400.
17. Marino, J. P.; Diener, J.; Moore, P. B.; Griesinger, C. *J. Am. Chem. Soc.* **1997**, *119*, 7361–7366.
18. Peti, W.; Carlomagno, T.; Griesinger, C. *J. Biomol. NMR* **2000**, *17*, 99–109.
19. Miclet, E.; Williams, D. C.; Clore, G. M.; Bryce, D. L.; Boisbouvier, J.; Bax, A. *J. Am. Chem. Soc.* **2004**, *126*, 10560–10570.
20. Gardner, K. H.; Kay, L. E. *Annu. Rev. Biophys. Biomol. Struct.* **1998**, *27*, 357–406.
21. Reif, B.; Griffin, R. G. *J. Magn. Reson.* **2003**, *160*, 78–83.
22. Gardner, K. H.; Rosen, M. K.; Kay, L. E. *Biochemistry* **1997**, *36*, 1389–1401.
23. Kanelis, V.; Forman-Kay, J. D.; Kay, L. E. *IUBMB Life* **2001**, *52*, 291–302.
24. Jeener, J.; Meier, B. H.; Bachmann, P.; Ernst, R. R. *J. Chem. Phys.* **1979**, *71*, 4546–4553.
25. Liu, H.; Spielmann, H. P.; Ulyanov, N. B.; Wemmer, D. E.; James, T. L. *J. Biomol. NMR* **1995**, *6*, 390–402.
26. Boelens, R.; Koning, T. M. G.; Kaptein, R. *J. Mol. Struct.* **1988**, *173*, 299–311.
27. Withka, J. M.; Swaminathan, S.; Bolton, P. H. *J. Magn. Reson.* **1990**, *89*, 386–390.
28. Lindorff Larssen, K.; Best, R. B.; De Pristo, M. A.; Dobson, C. M.; Vendruscolo, M. *Nature* **2005**, *433*, 128–132.
29. Lange, A.; Seidel, K.; Verdier, L.; Luca, S.; Baldus, M. *J. Am. Chem. Soc.* **2003**, *125*, 12640–12648.
30. Lange, A.; Luca, S.; Baldus, M. *J. Am. Chem. Soc.* **2002**, *124*, 9704–9705.
31. Castellani, F.; van Rossum, B.; Diehl, A.; Schubert, M.; Rehbein, K.; Oschkinat, H. *Nature* **2002**, *420*, 98–102.
32. Battiste, J. L.; Wagner, G. *Biochemistry* **2000**, *39*, 5355–5365.
33. Roosild, T. P.; Greenwald, J.; Vega, M.; Castronovo, S.; Riek, R.; Choe, S. *Science* **2005**, *307*, 1317–1321.
34. Dedmon, M. M.; Lindorff-Larsen, K.; Christodoulou, J.; Vendruscolo, M.; Dobson, C. M. *J. Am. Chem. Soc.* **2005**, *127*, 476–477.
35. Tolman, J. R.; Flanagan, J. M.; Kennedy, M. Am.; Prestegard, J. H. *Proc. Natl. Acad. Sci. USA* **1995**, *92*, 9279–9283.
36. Tjandra, N.; Bax, A. *Science* **1997**, *278*, 1111–1114.
37. Saupe, A. *Angew. Chem. Int. Ed. Engl.* **1968**, *7*, 97–102.
38. Blackledge, M. *Prog. Nucl. Magn. Res. Spectrosc.* **2005**, *46*, 23–61.
39. De Angelis, A. A.; Jones, D. H.; Grant, C. V.; Park, S. H.; Mesleh, M. F.; Opella, S. J. In *Nuclear Magnetic Resonance of Biological Macromolecules, Part C: Methods in Enzymology*; Thomas, L. J., Ed.; Academic Press: San Diego, CA, 2005; Vol. 394, pp 350–382.
40. Opella, S. J.; Marassi, F. M. *Chem. Rev.* **2004**, *104*, 3587–3606.
41. Sanders, C. R.; Hare, B. J.; Howard, K. P.; Prestegard, J. H. *Prog. NMR Spectrosc.* **1994**, *26*, 421–444.
42. Boisbouvier, J.; Delaglio, F.; Bax, A. *Proc. Natl. Acad. Sci. USA* **2003**, *100*, 11333–11338.
43. Schwieters, C. D.; Kuszewski, J. J.; Tjandra, N.; Clore, G. M. *J. Magn. Reson.* **2003**, *160*, 65–73.
44. Meiler, J.; Blomberg, N.; Nilges, M.; Griesinger, C. *J. Biomol. NMR* **2000**, *16*, 245–252.
45. Bertini, I.; Luchinat, C.; Parigi, G.; Pierattelli, R. *ChemBioChem* **2005**, *6*, 1536–1549.
46. Chou, J. J.; Delaglio, F.; Bax, A. *J. Biomol. NMR* **2000**, *18*, 101–105.
47. Chou, J.; Bax, A. *J. Am. Chem. Soc.* **2001**, *123*, 3844–3845.

48. Karplus, M. *J. Chem. Phys.* **1959**, *30*, 11–15.
49. Dingley, A. J.; Masse, E.; Peterson, R. D.; Barfield, M.; Feigon, J.; Grzesiek, S. *J. Am. Chem. Soc.* **1999**, *121*, 6019–6027.
50. Vuister, G. W.; Delaglio, F.; Bax, A. *J. Am. Chem. Soc.* **1992**, *114*, 9674–9675.
51. Dingley, A. J.; Grzesiek, S. *J. Am. Chem. Soc.* **1998**, *120*, 8293–8297.
52. Hennig, M.; Geierstanger, B. H. *J. Am. Chem. Soc.* **1999**, *121*, 5123–5126.
53. Baldus, M.; Meier, B. H. *J. Magn. Reson. Ser. A* **1996**, *121*, 65–69.
54. Duma, L.; Hediger, S.; Lesage, A.; Sakellariou, D.; Emsley, L. *J. Magn. Reson.* **2003**, *162*, 90–101.
55. Reif, B.; Hennig, M.; Griesinger, C. *Science* **1997**, *276*, 1230–1233.
56. Schwalbe, H.; Carlomagno, T.; Junker, J.; Hennig, M.; Reif, B.; Richter, C.; Griesinger, C. *Methods Enzymol.* **2001**, *338*, 35–81.
57. Carlomagno, T.; Bermel, W.; Griesinger, C. *J. Biomol. NMR* **2003**, *27*, 151–157.
58. Crowley, P.; Ubbink, M.; Otting, G. *J. Am. Chem. Soc.* **2000**, *122*, 2968–2969.
59. Pelupessy, P.; Ravindranathan, S.; Bodenhausen, G. *J. Biomol. NMR* **2003**, *25*, 265–280.
60. Case, D. A. *J. Biomol. NMR* **2001**, *21*, 321–333.
61. Gorenstein, D. G. *Chem. Rev.* **1994**, *94*, 1315–1338.
62. Clore, G. M.; Gronenborn, A. M.; Bax, A. *J. Magn. Reson.* **1998**, *133*, 216–221.
63. Sass, H. J.; Musco, G.; Stahl, S. J.; Wingfield, P. T.; Grzesiek, S. *J. Biomol. NMR* **2001**, *21*, 275–280.
64. Hus, J. C.; Marion, D.; Blackledge, M. *J. Am. Chem. Soc.* **2001**, *123*, 1541–1542.
65. Saito, H. *Magn. Reson. Chem.* **1986**, *24*, 835–852.
66. Luca, S.; Heise, H.; Baldus, M. *Accounts Chem. Res.* **2003**, *36*, 858–865.
67. Zuiderweg, E. R. P.; Scheek, R. M.; Boelens, R.; van Gunsteren, W. F.; Kaptein, R. *Biochimie* **1985**, *67*, 707–715.
68. Nilges, M.; O'Donoghue, S. I. *Prog. NMR Spectrosc.* **1998**, *32*, 107–139.
69. Laskowski, R. A.; Rullmann, J. A. C.; Mac Arthur, M. W.; Kaptein, R.; Thornton, J. M. *J. Biomol. NMR* **1996**, *8*, 477–486.
70. Beraud, S.; Bersch, B.; Brutscher, B.; Gans, P.; Barras, F.; Blackledge, M. *J. Am. Chem. Soc.* **2002**, *124*, 13709–13715.
71. Hus, J. C.; Marion, D.; Blackledge, M. *J. Mol. Biol.* **2000**, *298*, 927–936.
72. Linge, J. P.; Habeck, M.; Rieping, W.; Nilges, M. *Bioinformatics* **2003**, *19*, 315–316.
73. Güntert, P. *Prog NMR Spectrosc.* **2003**, *43*, 105–125.
74. Fossi, M.; Castellani, T.; Nilges, M.; Oschkinat, H.; van Rossum, B. J. *Angew. Chem. Int. Ed.* **2005**, *44*, 6151–6154.
75. Pappalardo, L.; Janausch, I. G.; Vijayan, V.; Zientz, E.; Junker, J.; Peti, W.; Zweckstetter, M.; Unden, G.; Griesinger, C. *J. Biol. Chem.* **2003**, *278*, 39185–39188.
76. Reinelt, S.; Hofmann, E.; Gerharz, T.; Bott, M.; Madden, D. R. *J. Biol. Chem.* **2003**, *278*, 39189–39196.
77. Hwang, P. M.; Choy, W.-Yi.; Lo, E. I.; Chen, L.; Forman-Kay, J. D.; Raetz, C. R. H.; Prive, G. G.; Bishop, R. E.; Kay, L. E. *Proc. Natl. Acad. Sci. USA* **2002**, *99*, 13560–13565.
78. Nomura, K.; Takegoshi, K.; Terao, T.; Uchida, K.; Kainosho, M. *J. Am. Chem. Soc.* **1999**, *121*, 4064–4065.
79. Rienstra, C. M.; Tucker-Kellogg, L.; Jaroniec, C. P.; Hohwy, M.; Reif, B.; McMahon, M. T.; Tidor, B.; Lozano-Perez, T.; Griffin, R. G. *Proc. Natl. Acad. Sci. USA* **2002**, *99*, 10260–10265.
80. Castellani, F.; van Rossum, B. J.; Diehl, A.; Rehbein, K.; Oschkinat, H. *Biochemistry* **2003**, *42*, 11476–11483.
81. Jaroniec, C. P.; MacPhee, C. E.; Bajaj, V. S.; McMahon, M. T.; Dobson, C. M.; Griffin, R. G. *Proc. Natl. Acad. Sci. USA* **2004**, *101*, 711–716.
82. Seidel, K.; Etzkorn, M.; Heise, H.; Becker, S.; Baldus, M. *ChemBioChem* **2005**, *6*, 1638–1647.
83. Zech, S. G.; Wand, A. J.; McDermott, A. E. *J. Am. Chem. Soc.* **2005**, *127*, 8618–8626.
84. Liu, G. H.; Shen, Y.; Atreya, H. S.; Parish, D.; Shao, Y.; Sukumaran, D. K.; Xiao, R.; Yee, Y.; Lemak, A.; Bhattacharya, A. et al. *Proc. Natl. Acad. Sci. USA* **2005**, *102*, 10487–10492.
85. Dawson, P. E.; Muir, T. W.; Clark-Lewis, I.; Kent, S. B. H. *Science* **1994**, *266*, 776–779.
86. Xu, R.; Ayers, B.; Cowburn, D.; Muir, T. W. *Proc. Natl. Acad. Sci. USA* **1999**, *96*, 388–393.
87. Roberts, G. C. K. *Drug Disc. Today* **1999**, *5*, 230–240.
88. Hajduk, P. J.; Sheppard, G.; Nettesheim, D. G. *J. Am. Chem. Soc.* **1997**, *119*, 5818–5827.
89. Elshorst, B.; Hennig, M.; Försterling, H.; Diener, A.; Maurer, M.; Schulte, P.; Schwalbe, H.; Griesinger, C.; Krebs, J.; Schmid, H. et al. *Biochemistry* **1999**, *38*, 12330–12332.
90. Luca, S.; White, J. F.; Sohal, A. K.; Filippov, D. V.; van Boom, J. H.; Grisshammer, R.; Baldus, M. *Proc. Natl. Acad. Sci. USA* **2003**, *100*, 10706–10711.
91. Bittermann, H.; Einsiedel, J.; Hübner, H.; Gmeiner, P. *J. Med. Chem.* **2004**, *47*, 5587–5590.
92. Lange, A.; Giller, K.; Hornig, S.; Martin-Eauclaire, M. F.; Pongs, O.; Becker, S.; Baldus, M. *Nature* **2006**, *440*, 959–962.
93. Lange, A.; Becker, S.; Seidel, K.; Pongs, O.; Baldus, M. *Angew. Chem. Int. Ed. Engl.* **2005**, *44*, 2089–2092.
94. Ni, F. *Prog. NMR Spectrosc.* **1994**, *26*, 517–606.
95. Carlomagno, T.; Felli, I. C.; Czech, M.; Fischer, R.; Sprinzl, M.; Griesinger, C. *J. Am. Chem. Soc.* **1999**, *121*, 1945–1948.
96. Blommers, M. J. J.; Stark, W.; Jones, C. E.; Head, D.; Owen, C. E.; Jahnke, W. *J. Am. Chem. Soc.* **1999**, *121*, 1949–1953.
97. Bollag, D. M.; McQueney, P. A.; Zhu, J.; Hensens, O.; Koupal, L.; Liesch, J.; Goetz, M.; Lazarides, E.; Woods, C. M. *Cancer Res.* **1995**, *55*, 2325–2333.
98. Altmann, K.-H.; Wartmann, M.; O'Reilly, T. *BBA-Rev. Cancer* **2000**, *1470*, M79–M91.
99. Kowalski, J.; Giannakakou, P.; Hamel, E. *J. Biol. Chem.* **1997**, *272*, 2534–2541.
100. Ni, F. *Prog. Nucl. Mag. Res.* **1994**, *26*, 517–606.
101. Carlomagno, T.; Felli, I. C.; Czech, M.; Fischer, R.; Sprinzl, M.; Griesinger, C. *J. Am. Chem. Soc.* **1999**, *121*, 1945–1948.
102. Blommers, M. J. J.; Stark, W.; Jones, C. E.; Head, D.; Owen, C. E.; Jahnke, W. *J. Am. Chem. Soc.* **1999**, *121*, 1949–1953.
103. Carlomagno, T.; Blommers, M. J. J.; Meiler, J.; Jahnke, W.; Schupp, T.; Petersen, F.; Schinzer, D.; Altmann, K.-H.; Griesinger, C. *Angew. Chem. Int. Ed. Engl.* **2003**, *42*, 2511–2515.
104. Carlomagno, T.; Sánchez, V.; Blommers, M. J. J.; Griesinger, C. *Angew. Chem. Int. Ed. Engl.* **2003**, *42*, 2515–2517.
105. Víctor, M.; Sánchez-Pedregal, M.; Reese, J.; Meiler, M.; Blommers, J. J.; Griesinger, C.; Carlomagno, T. *Angew. Chem. Int. Ed. Engl.* **2005**, *44*, 2–4.
106. Balbach, J.; Forge, V.; van Nuland, N. A. M.; Winder, S. L.; Hore, P. J.; Dobson, C. M. *Nat. Struct. Biol.* **1995**, *2*, 865–870.
107. Gal, M.; Mishkovsky, M.; Frydman, L. *J. Am. Chem. Soc.* **2006**, *128*, 951–956.
108. Cavanagh, J.; Fairbrother, W. J.; Palmer, A. G.; Skelton, N. J. *Protein NMR Spectroscopy*; Academic Press: Orlando, FL, 1996.

109. Lipari, G.; Szabo, A. *J. Am. Chem. Soc.* **1982**, *104*, 4546–4559.
110. http://cpmcnet.columbia.edu/ (accessed Aug 2006).
111. Tugarinov, V.; Liang, Z. C.; Shapiro, Y. E.; Freed, J. H.; Meirovitch, E. *J. Am. Chem. Soc.* **2001**, *123*, 3055–3063.
112. Fischer, M. W. F.; Majumdar, A.; Zuiderweg, E. R. P. *Prog. Nucl. Magn. Reson.* **1998**, *33*, 207–272.
113. Wang, L.; Pang, Y.; Holder, T.; Brender, J. R.; Kurochkin, A. V.; Zuiderweg, E. R. P. *Proc. Natl. Acad. Sci. USA* **2002**, *98*, 7684–7689.
114. Eisenmesser, E. Z.; Millet, O.; Labeikovsky, W.; Korzhnev, D. M.; Wolf-Watz, M.; Bosco, D. A.; Skalicky, J. J.; Kay, L. E.; Kern, D. *Nature* **2005**, *438*, 117–121.
115. Lakomek, N. A.; Farès, C.; Becker, S.; Carlomagno, T.; Meiler, J.; Griesinger, C. *Angew. Chem. Int. Ed. Engl.* **2005**, *44*, 7776–7778.
116. Davis, I. W.; Arendall, W. B., III; Richardson, D. C.; Richardson, J. S. *Structure* **2006**, *14*, 265–274.
117. Bertini, I.; Del Bianco, C.; Gelis, I.; Katsaros, N.; Luchinat, C.; Parigi, G.; Peana, M.; Provenzani, A.; Zoroddu, M. A. *Proc. Natl. Acad. Sci. USA* **2004**, *101*, 6841–6846.
118. Rodriguez-Castaneda, F.; Haberz, P.; Leonov, A.; Griesinger, C. *Magn. Reson. Chem.* **2006**, *8*, 1275–1278.
119. Andronesi, O. C.; Becker, S.; Seidel, K.; Heise, H.; Young, H. S.; Baldus, M. *J. Am. Chem. Soc.* **2005**, *127*, 12965–12974.

Biographies

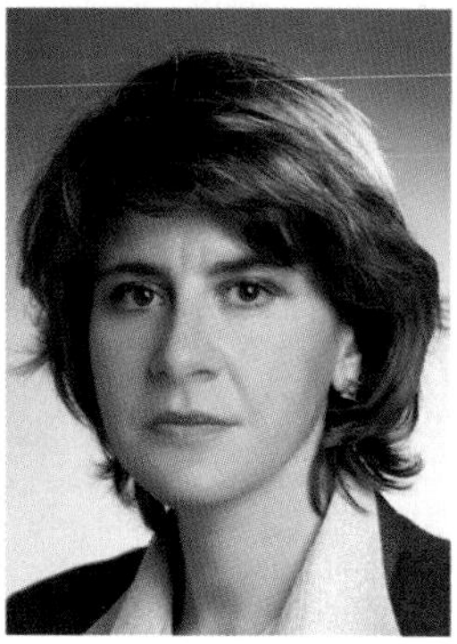

Teresa Carlomagno worked at the University of Naples and at the IRBM, Rome with Renzo Bazzo; she received her PhD from the University of Naples, Italy in 1996. From 1997 to 1999, she worked as a postdoctoral fellow with C Griesinger at the University of Frankfurt, Germany. In 2000 she joined the group of J Williamson at the Scripps Research Institute in La Jolla, USA, where she remained until 2001. Since 2002 she leads a research group of liquid-state NMR applied to protein and nucleic acids complexes at the MPI for Biophysical Chemistry in Göttingen, Germany.

Marc Baldus worked in Zurich/Switzerland and in Nijmegen/The Netherlands; he received his PhD from the ETH in Zurich with Richard Ernst and Beat Meier in 1996. From 1997 to 1999, he worked as a postdoctoral fellow with Robert G Griffin at MIT, Boston, USA. Following a period as lecturer at the University of Leiden, he joined the MPI for Biophysical Chemistry in 2000, where he currently directs the group in solid-state NMR.

Christian Griesinger studied Chemistry and Physics in Frankfurt, Germany; he received his PhD from the university of Frankfurt with Horst Kessler in 1986. From 1986 to 1989, he worked as a postdoctoral fellow with Richard Ernst at the ETH Zürich, Switzerland. He then became a full professor at the University of Frankfurt and since 1999 became a director at the Max Planck Institute for Biophysical Chemistry and directs the department of NMR-based structural Biology.

Comprehensive Medicinal Chemistry II
ISBN (set): 0-08-044513-6

ISBN (Volume 3) 0-08-044516-0; pp. 473–505

3.23 Protein Three-Dimensional Structure Validation

R P Joosten, Centre for Molecular and Biomolecular Informatics, Nijmegen, The Netherlands
G Chinea, Center for Genetic Engineering and Biotechnology, Havana, Cuba
G J Kleywegt, University of Uppsala, Uppsala, Sweden
G Vriend, Centre for Molecular and Biomolecular Informatics, Nijmegen, The Netherlands

3.23.1	**Introduction**	**507**
3.23.1.1	Why	508
3.23.1.2	Sources of Errors	509
3.23.1.3	The Good, the Bad, and the Ugly	509
3.23.1.4	Working with Errors	509
3.23.1.5	Correlation of the Structure with Experimental Data	510
3.23.1.6	The Data Problem	511
3.23.1.7	Overview	511
3.23.2	**Different Types of Errors**	**511**
3.23.2.1	Administrative Errors	511
3.23.2.2	What's in a Name?	513
3.23.2.3	Hetero Compounds	514
3.23.2.4	Geometry	515
3.23.2.5	Contact Analysis	517
3.23.2.6	Hydrogen Bonds and Water	518
3.23.2.7	Using Experimental X-ray Data	519
3.23.3	**Time Flies When You Are Validating Structures**	**520**
3.23.3.1	The Quick and the Dead	520
3.23.3.2	Database Generation	520
3.23.3.3	Errors over the Years	521
3.23.3.4	The Future of Validation	523
3.23.4	**Conclusion**	**526**
References		**527**

3.23.1 Introduction

The use of (experimental) three-dimensional (3D) models of protein and nucleic acid structures is now commonplace in biochemistry and medicine alike. These models give us more insight into the inner workings of the cell, which is of paramount importance in the ongoing fight against disease. They form the basis of structure-based (or, rather audaciously, rational) drug design,[1] which is one of many techniques used for the discovery and optimization of new medicines.

The total number of macromolecular structures publicly available in the Protein Data Bank (PDB)[2] is increasing rapidly (**Figure 1**). Of the more than 30 000 protein and nucleic acid structures currently in the PDB, only a relatively small number is unique (**Table 1**).[3] This redundancy introduces a problem of convenience for users of this structural data: which of the available 3D structure models should one use?

To help make this very important choice, a vast number of tools have been developed with a widely varying level of sophistication. These tools supply information about the quality of the experimental data, the correlation between the measured data and the resulting 3D model, the quality of the structure model with respect to our general knowledge of

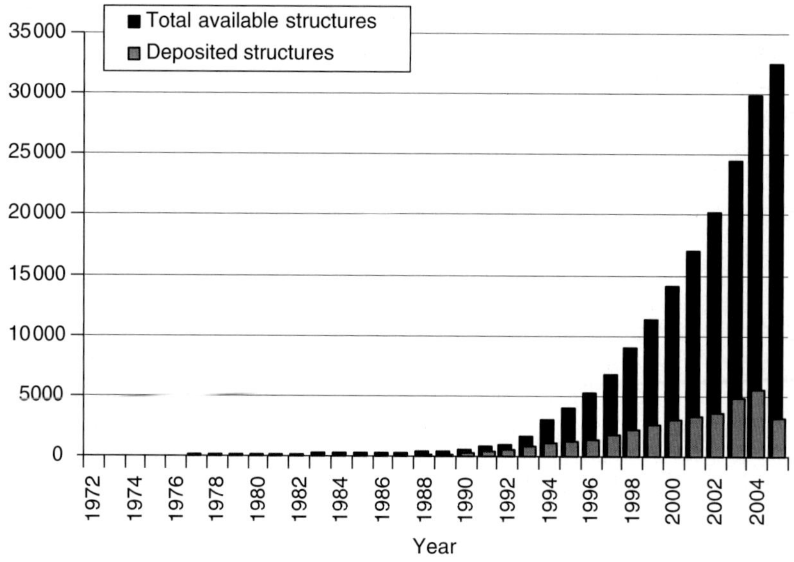

Figure 1 Growth history of the Protein Data Bank (PDB),[2] which has grown exponentially since its modest start in 1972. It currently (July 2005) holds 32 545 structures determined using x-ray diffraction, NMR, and several other methods. At the current rate of growth more than 100 structures are added every week. The black columns represent the total number of entries in the PDB for each single year, and the grey columns show the number of structures added per year. Data from http://www.rcsb.org.

Table 1 Unique entries in the PDB

Sequence identity	*Number of entries*[a]
⩽25%	3977
⩽50%	6957
⩽75%	8303
⩽99%	13 970
⩽100%	29 081

[a]Only entries with 40 to 10 000 amino acid residues are considered.

physics and chemistry, and the correlation between the model and our ever-increasing knowledge of the specific features of protein and nucleic acid structures. The goal of all these tests is the same: to provide insight into the quality of the final structure model so that the 'best' structure can be used for further research.

3.23.1.1 Why

The conclusions drawn from a 3D structure model are, at best, as good as the model itself. Structure validation helps us to identify anomalies and errors in structure models. This enables 3D coordinate users to select the best structure models and to take into account any abnormalities in those structures. Some errors can even be fixed before the 3D coordinates are used.

A striking example of the importance of using the correct structure model occurred in 1989. Immediately after the coordinates of the human immunodeficiency virus protease (HIV-1 protease) became available, they were used to direct the search for inhibitors.[4] Because the initial structure model included a mistake in the dimer interface region, understanding the molecular mechanisms of the release of HIV-1 protease became difficult.[5] Based on a comparison with the homologous retroviral Rous sarcoma virus (RSV) protease, Blundell and Pearl[5] suggested that there might be something wrong with the proposed HIV-1 protease structure. This idea was reinforced by a homology model of the HIV-1 protease, built using RSV as a template, that showed a small but important difference in structure.[6] Finally, the discrepancy was solved by Wlodawer *et al.*,[7] who determined the structure of a synthetic HIV-1 protease, but now with the correct chain tracing of the N-terminal strands at the dimer interface. The development of HIV protease inhibitors in the fight against the human immunodeficiency virus was a major breakthrough for structure-based drug design. It has shown the importance of validating protein models to extract the right structural information.

3.23.1.2 Sources of Errors

The processes of obtaining structure coordinates of macromolecules by means of x-ray crystallography, nuclear magnetic resonance (NMR), or any other method consist of many complicated steps, each introducing uncertainties and errors. Crystallization and data recording introduce measurement errors and noise. Human interpretation of data is often required when no straightforward answer is available. This may result in a significant degree of subjectivity in the model building process.[8] The computer software used during refinement is not without weaknesses either. The cumulative effect of errors and human interpretation is almost impossible to assess but Chapter 3.24 gives a very thorough description of the problems one can encounter.

To complicate matters even further, users of 3D structure coordinates are not necessarily trained in the methods that were used to obtain those coordinates. Fortunately for them, many tools have been developed to provide insight into the ambiguities, anomalies, and errors of those coordinates.

3.23.1.3 The Good, the Bad, and the Ugly

Telling right from wrong is no black or white matter. First of all, we have to establish what is right, or better, what is normal. This is derived from sources like (a subset of) the PDB, that we assume to be representative for the proteins we study. Any 'normal' value has to be accompanied by a standard deviation to give some perspective on the distribution of the values one could encounter. Using these two values to separate right from wrong requires a third value: the cut-off. Given an expected value, a standard deviation, and a cut-off we can separate the good from the bad. However, not every anomaly is bad, some are just ugly. For any large set with a Gaussian distribution a certain fraction of outliers is expected.

To minimize the risk of false positives (reported outliers that are in fact correct), a suitable deviation from the expected value has to be allowed. **Table 2** shows the relation between the Z-score (the number of standard deviations an observed value differs from the expected value) and the probability that a flagged outlier may be a false positive.[9]

So with a 4σ cut-off, about 1 in 10 000 flagged outliers is a false positive. Therefore, the PDBREPORT database[10] with its 11.6 million errors (July 2005) contains about 1100 flagged errors that are actually correct. The errors in **Figure 2**, however, are all definitely real errors.

3.23.1.4 Working with Errors

When an error is identified in a structure model, the user must assess its impact on the overall quality of the work for which the structure model will be used. For instance, an unlikely short inter atomic distance between two atoms (a bump) at some uninteresting spot on the protein surface may be less problematic than a similar bump in the active site. Critical errors must be fixed, preferably using experimental data. This, however, requires expert knowledge of the method used to solve the structure and is therefore not preferred by most structure users.

Alternatively, the structure can be improved by using standard conformation libraries.[11] It is important to realize that by doing so, one is normalizing the structure, fixing it to use the most reasonable conformations, and as a consequence making it pass conformation tests, while not necessarily improving the correlation with the experimental data.

Table 2 Relation between Z-score and the probability an outlier is a false positive

Z-score	*Probability (%)*
1	31.74
2	4.54
3	0.26
4	0.01

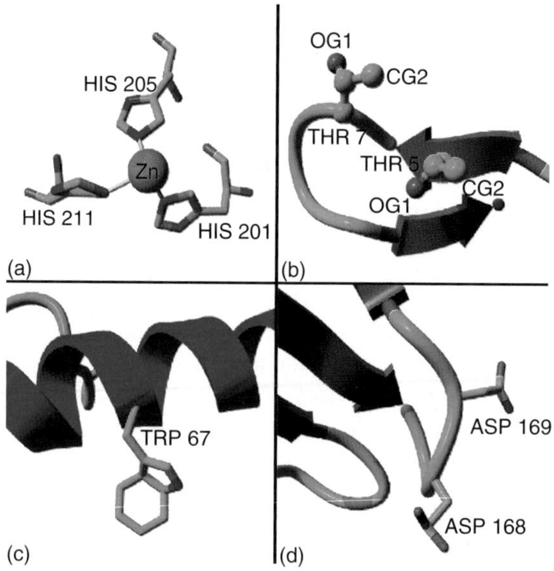

Figure 2 Some examples of errors in the PDB. (a) Histidine 211 that coordinates the zinc ion has a 23σ deviation from normal planarity whereas the other two histidines are planar (PDB entry 1BIW). (b) The C_β of threonine 5 has inversed chirality. The $O_{\gamma 1}$ and $C_{\gamma 2}$ atoms are swapped; threonine 7 has correct chirality (PDB entry 5RXN). (c) The two rings in tryptophan 67 make an angle of 106 degrees. This is a 37σ deviation from the correct planar configuration (PDB entry 7GPB). (d) Aspartate residue 168 in PDB entry 1DLP has a 64σ deviation from planarity. Its direct neighbor, aspartate 169 is 'normal.'

There is a risk involved in using standard libraries: circularity. For instance, using side chain conformation libraries[11–15] to reposition side chains to the preferred rotamer has a reinforcing effect on the database distribution of the preferred rotamer. In other words, the most popular rotamer becomes even more popular. Residue conformations purely based on conformational libraries should be flagged so that they can be excluded when generating new libraries. Often, this is done implicitly or explicitly by crystallographers. If a side chain cannot be properly seen in the experimental electron density, it will have a very high B-factor (if no B-factor restraints are used) or an atomic occupancy of (nearly) zero. Some crystallographers prefer leaving out the ill-defined atoms completely, submitting structures with partial amino acid residues. In any of these cases, the user of the 3D coordinates has to be aware of the problem. Validation programs like WHAT_CHECK[10] warn the user if such residues are encountered.

Some errors cannot be resolved either because the problem is too global in nature, such as bad Ramachandran statistics,[16] or no straightforward solution is at hand due to a lack of (experimental) data. In those cases one can either choose another structure, if such an alternative is available, or just 'learn to live with it.' The latter is not as bad as it sounds. A known error is much less of a problem than an error that goes undetected. If the drug designer or molecular biologist knows that certain residues are likely to be imprecise or wrong, he or she can take appropriate action in the design of experiments.

3.23.1.5 Correlation of the Structure with Experimental Data

The best-known quality score for x-ray structures is the so-called R-factor. It describes the correlation between the measured data and the final macromolecular structure. The lower the R-factor, the better. Or so it seemed until one of us showed that a purposely misthreaded structure can still have the perfectly acceptable R-factor of 21.4%.[17] This proved that the R-factor by itself is not reliable enough. The free R-factor, R_{free}, is calculated from crystallographic reflection data that is not used during the building and refinement of the protein structure.[18] It is therefore an independent measure of the quality of the structure. R_{free}, or better the difference between the R-factor and R_{free}, can also be used to recognize over-refinement of 3D structure.[17,19] Similar independent quality factors have been developed for NMR, also relying on experimental data that is not used during refinement.[20]

Apart from these global correlation tests, there are numerous local correlation validation tests. For x-ray structures, the Electron Density Server[21] gives a summary of important validation values including the Real Space R-factor (RSR)[11] and the real space correlation factor, which give information of the fit to the data at a residue level. We shall discuss working with experimental x-ray data in more detail in Section 3.23.2.7.

The number of distance or angular restraints per residue is a commonly used quality indication for NMR-derived 3D coordinates. The QUEEN program[22] supplies statistics on the actual information content of distance restraints, correcting for the redundancy that is usually encountered in sets of restraints. The correlation between data and structure on a per residue basis is best described by the number of restraint violations, that is, the number of interatomic distances longer than prescribed by the restraint distance. The size of each violation has to be taken into account as well. For drug designers, a 1.0 Å violation is worse than five 0.2 Å violations.

3.23.1.6 The Data Problem

In order to validate the data used to derive the protein structure model, it has to be made available by the depositor of the 3D coordinates. At the time of writing (July 2005) only 55% of all NMR-derived and 64% of all x-ray-derived entries in the PDB have experimental data available and in both cases the spread in standards used upon data deposition makes it hard to use these experimental data in a fully automated way.[21,23] Because ever more scientific journals require the submission of experimental data, and because deposition of experimental data is now an integral part of the deposition of PDB entries, these percentages are increasing rapidly (87% for this year's x-ray structures so far). This, however, does not solve the problem for the older PDB entries.

Fortunately, here are numerous validation tools that do not require experimental data. Instead, they use geometrical parameters like bond lengths, angles and planarity,[24–27] protein backbone torsion angles,[16] and the relative orientation of amino acids. These tools supply the user with a wealth of information about the protein structure but no details are given about the correlation with the experimental data. Therefore, any anomalies or errors flagged during the validation of the model in question cannot be traced back to the original experiment or the interpretation of the experimental data. This also limits the ability of the user to correct the errors.

3.23.1.7 Overview

In this chapter, we shall discuss different types of errors that are of special interest to those who use the structures in the PDB to collect information about the workings of proteins and those who use 3D coordinates to design drugs. Where possible, we shall elaborate on the errors with examples of difficult structures or interesting errors. We shall pay special attention to the hydrogen-bonding network that is thoroughly explored and validated in the WHAT_CHECK package. This package is available, free of charge, to both academic and commercial users[10] and can be used for all types of protein structures, independent of the method by which they are derived. Apart from validating the entire PDB, we encourage the validation of protein (homology) models for drug design.

We will also give a historic overview of the quality of published structures. Have they improved or not? Which errors are less common and which are not? Finally, we will take a look at work to be done in the field of structure validation. Many more types of errors are still waiting to be discovered.

3.23.2 Different Types of Errors

A larger number of different classes of errors have been identified, some more important than others. Discussing all those classes is beyond the scope of this chapter. Instead, we shall focus on the types of errors that are of special significance to those who want to use 3D structural data of proteins and nucleic acids for biological and drug design purposes.

3.23.2.1 Administrative Errors

Errors with the format of PDB files, for example, missing atoms or wrong atomic occupancies, are easy to make and seem innocuous, but can have serious implications when left undetected. A missing side chain can give unexpected results in electrostatics and hydrophobicity calculations. What does a drug docking program do with atoms with the wrong names?

The standard for PDB files is well defined but not all programs that produce PDB files strictly adhere to this standard. Even simply reading in a file and writing it out again with the same software can cause many deviations from the standard that make a PDB file invalid.

CRYST1, MTRIX, and SCALE records describe the crystal parameters and may be important when symmetry-derived molecules are considered. For instance, in drug docking care must be taken that the target site is not influenced by symmetry-related molecules. Some flexible residues may seem buried and relatively rigid due to crystal packing. This can only be checked if the molecular symmetry is properly described.[28]

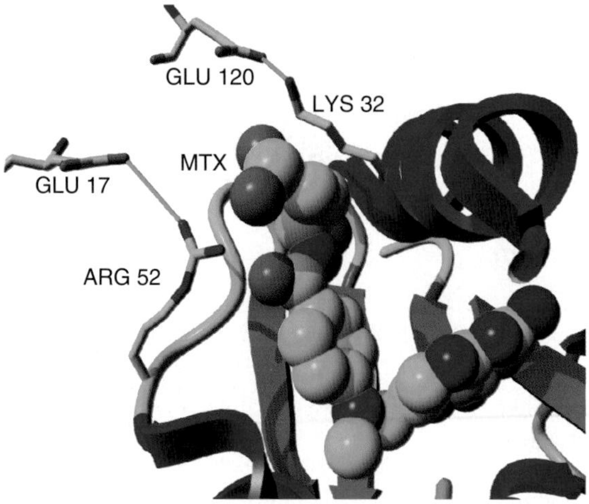

Figure 3 The binding site of methorexate is strongly influenced by intersymmetry interactions (PDB entry 4DFR). Lysine 32 and arginine 52, both positively charged, coordinate the charged group in methorexate (MTX, in space-filling representation). Their conformation is restricted by two residues from an adjacent copy of the same protein. It is not included in the PDB entry because this copy can be generated by applying crystal symmetry operators. Glutamate 17 forms a salt bridge with arginine 52. Lysine 32 is pulled towards glutamate 120 forming a hydrogen bond. As a result, the binding of methorexate cannot be exactly the same as in the native protein. The two glutamate residues that influence the binding site are produced by a symmetry operation and cannot be seen without applying this operation. Therefore, their influence can only be assessed when this symmetry is properly administrated.

We find an example in PDB entry 4DFR. The binding site for the hetero compound is influenced by side chains from a symmetry-related molecule that is not found directly in the PDB file (**Figure 3**). This influences the way the compound is bound. This should be taken into account when structures like this are used to benchmark docking algorithms. Without proper administration of the crystal symmetry, such slight distortions of binding sites would go undetected.

In PDB entries the MTRIX records describe noncrystallographic symmetry transformations between related molecules. The determinant of this matrix should be exactly 1 or one molecule may be 'inflated' with respect to the other. This means that a molecule produced by application of this matrix will have systematically longer (or shorter) bonds than the original molecule. If this deviation from 1 is too large it may be wise to validate noncrystallographic symmetry-related molecules separately and only use the best one.

Atoms with an occupancy <1 should be inspected carefully. They can have an alternative position in which case all positions must be taken into account in calculations, or they can be flagged as 'not-there.' This means that there is no experimental data to support their position. Users must treat such atoms with caution. In PDB entry 4DFR, for example, we find many atoms with occupancy <1. Water molecules especially have a wide variety of occupancies without any specified alternate positions.

Atomic occupancies >1 are also encountered. Sometimes the reason for this is obvious: a few PDB entries (e.g., 4HHB) use the number of electrons instead of the occupancy. Some occupancies >1 are caused by occupancy refinement, in which the occupancy of atoms is adapted to ensure the optimal fit with the electron density. This is, again, encountered in the water molecules in PDB entry 4DFR. At any rate, occupancies >1 should be treated with a healthy dose of suspicion.

An exceptionally complicated situation is found in PDB entry 1CBQ (**Figure 4**). It has a phosphate group on a threefold symmetry axis with the P-O1 bond exactly on the axis. Only one of the other oxygen atoms, O2, is present in the file because O3 and O4 are symmetry related to O2. The phosphorous atom has its occupancy set to 0.33, which makes it a full atom when the threefold symmetry is applied. Unfortunately, oxygen atom O1 has occupancy 1.00, which makes it three full atoms on the same position after the symmetry operation. This in turn causes WHAT_CHECK to detect a very serious atomic bump. Upon closer inspection of the PDB file, the cause of the error is easily identified as a typo: O1 and O2 have their occupancies switched.

This example shows that seemingly serious errors can be caused by simple administrative mistakes. We observe a form of error inheritance: one error can cause a number of anomalies that are flagged by validation programs. On the other hand, the atoms O2, O3, and O4 now have occupancy 0.33, and as WHAT_CHECK does not issue warnings for atoms with occupancy <0.5, any errors related to these three oxygen atoms will remain undetected.

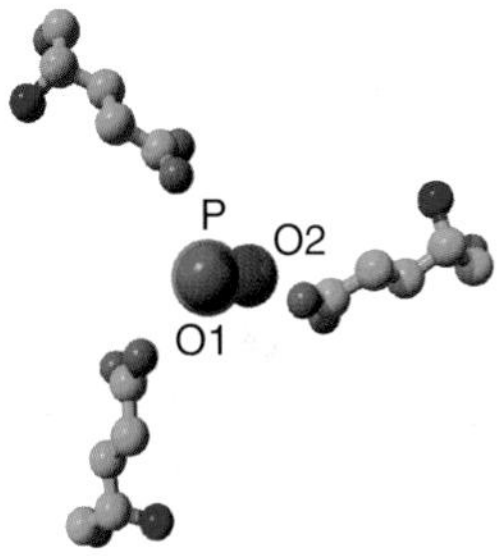

Figure 4 The phosphate group on a threefold symmetry axis in PDB entry 1CBQ has an occupancy problem. The atoms O3 and O4 are symmetry related to O2 and are formed after applying the proper symmetry. The phosphorus atom (in green) has occupancy = 0.33 so that one whole atom is formed after applying symmetry. O1 on the same symmetry axis has occupancy = 1.00. This means that, taking symmetry into account, there are three atoms at one site. This in turn causes error messages for close atomic contacts (bumps).

Missing atoms, either purposely left out or just lost when switching from one program to the other, can also cause a lot of problems when using a protein structure. A missing surface side chain can make the difference between a docking calculation that works and a docking without a usable result. In such cases, a badly positioned side chain is often more realistic than no side chain at all. It does not work the other way around. Extra side chain atoms are not welcome. It is therefore an error that PDB entry 1VNS has a C_β on glycine 126.

3.23.2.2 What's in a Name?

There are numerous things that can go wrong with naming schemes in the PDB. We shall only discuss a few relevant types of errors here. The PDB provides a very wide variety of nomenclature-related errors, some insignificant to drug designers, many annoying, and some catastrophic.

It is important to realize that there might be discrepancies between the protein sequence in the SEQRES record in a PDB entry, the corresponding ATOM records, and the sequence in sequence databases like UNIPROT.[29,30] Residues for which the side chain cannot be identified in the electron density are sometimes renamed to alanine during the refinement process. We find an example of this in PDB entry 1WCM in which a poly-alanine chain replaces the amino acid residues in the original sequence found in UNIPROT entry RPB4_YEAST. Discrepancies like this one should be intercepted upon deposition of the structure to the PDB. Before using the 3D coordinates, one has to ensure that you are looking at the protein with the right sequence.

Chain identifiers in PDB records have to be unique. However, hetero compounds associated with a certain protein chain should have the same chain ID. This is useful for drug docking as the smallest unique unit (that is, the drug target) can be extracted with all ligands and ions included, merely by selecting a chain identifier. This is particularly true for hetero compounds that are covalently bound to the protein chain. PDB entry 1MYP is an example of what can go wrong (**Figure 5**). The *N*-acetyl-D-glucosamine groups have a different chain ID than the asparagine residues to which they are bound.

ATOM records in PDB files have four positions reserved for atom names. The first two are reserved for the chemical (or Mendeleev) symbol, the third for a remoteness indicator (α, β, γ, δ, etc.), and the fourth for a branch number. This means that CA- - (dashes denote spaces here) is a calcium atom whereas -CA- is a C_α atom. In that sense, the chlorine atom found in PDB record 1NDE named 3CL3, should have been called CL33 to avoid it being recognized as a carbon atom (**Figure 6**).

Another fine example of badly chosen atom names can be found in entry 1E6Y, which has an atom named OXT in the middle of a hetero compound. This can cause errors in molecular graphics and dynamics programs because the atom name OXT is reserved for the 'second' C-terminal oxygen atom of amino acid chains.

The naming of hydrogen atoms is more complicated than usual. First of all, an exception to the atom-naming scheme was necessary for protons as often two branch indicators are needed. In such cases, the second one is put in the first 'empty' space so 1H12 is a correct PDB name for the atom H121. Secondly, because NMR software is poorly standardized, many naming schemes are used. Fortunately, a much needed 'Rosetta stone' is available at the BioMagResBank.[31] Stereochemistry is an important factor in naming hydrogen atoms.[32] One example can be found in glycine. The two C_α atoms have unique names depending on their prochirality (Rectus or Sinister). The proper names for hydrogen atoms are important for calculation of bond and torsion angles in calculations like simulated annealing and drug docking.

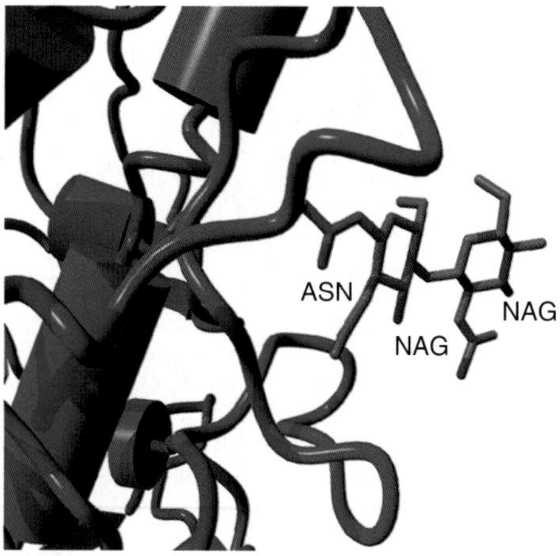

Figure 5 Detail of PDB entry 1MYP. Assigning the right chain ID can help when selecting a protein molecule with all the (covalently) bound hetero compounds. In this case, the protein is designated chain D (in blue) and the *N*-acetyl-D-glucosamine groups (NAG, in red) have chain identifier B. This causes a single molecule with only covalent bonds to have two different chain identifiers.

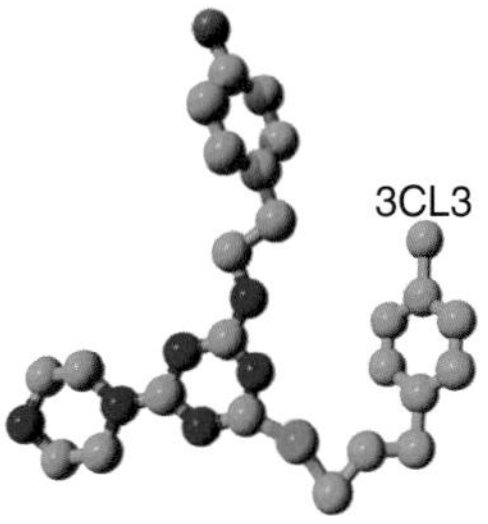

Figure 6 Choosing nonstandard atom names can have serious consequences when working with hetero compounds. The atom 3CL3 in the ligand of PDB entry 1NDE is recognized as a carbon because the first two positions in its name, which denote the atom type, are a digit and the letter C. The right name would have been CL33. That way, the first two letters would be CL, which stands for chloride, the right atom type.

3.23.2.3 Hetero Compounds

The refinement and validation of hetero compounds is limited due to a lack of statistical data and probably also due to a lack of quantum chemical expertise of many crystallographers and NMR spectroscopists.[33] New compounds do not have a predefined naming scheme and many programs either crash because of unexpected input or ignore the compound completely. The geometry of new compounds can only be validated by adding its topology and target geometry as an extra library file. Generating such libraries used to be hard work but two web servers can now help. The PRODRG[34,35] server can produce topology files for nine popular programs from various types of input. This input includes PDB coordinates and simple text drawings. Optional energy minimization can produce an ideal geometry.

The HIC-Up (Hetero-Compound Information Center – Uppsala)[13] database contains a collection of hetero compounds extracted from the PDB. It has both experimental and idealized coordinates for each molecule and a complete list of all the PDB entries that contain this molecule.

A nice example of problems that can occur with hetero compounds can be found in PDB entry 1JX6. The LuxP protein in this coordinate file contains an unknown compound named AI-2 (**Figure 7**). Because of the good electron density, the composition of this molecule could be derived by fitting atoms to the electron density and using potential hydrogen bonds to tell the difference between carbon and oxygen.[36] An orthocarbonate moiety was found in AI-2. The proposed molecular structure, however, could not be stable enough to exist in a protein. This is not picked up by any protein validation tool because such complicated chemistry is not (yet) implemented. Software packages like Logic and Heuristics Applied to Synthetic Analysis (LHASA) [37] that are used for optimizing (the synthesis of) small compounds are able to detect such instable compounds. Careful study showed that the compound was in fact a borate diester: the number 2 atom in one of the rings (**Figure 7**) was a boron atom instead of carbon. Unfortunately, this solution

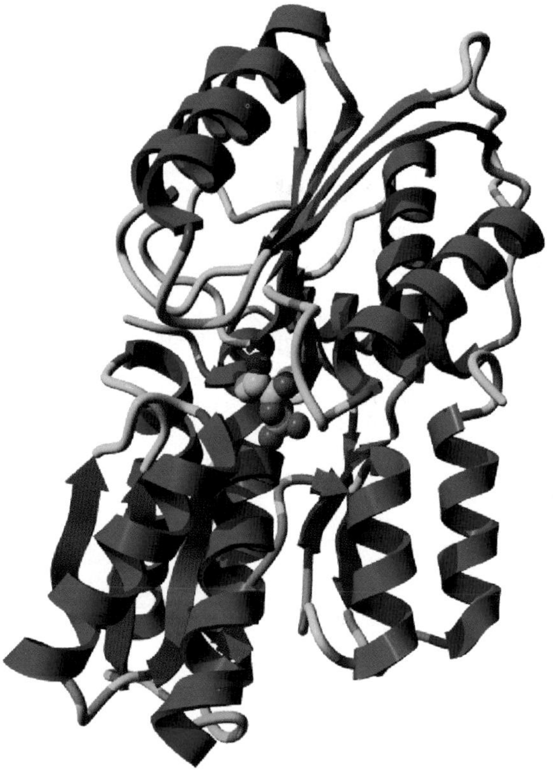

Figure 7 Cartoon representation of PDB entry 1JX6. The previously unknown AI-2 compound in space-filling representation contains one boron atom (in green). Because boron is very uncommon in protein structures, geometrical restraints cannot be produced automatically with tools like PRODRG.[34,35] From a crystallographic point of view it is hard to tell whether this atom should be a carbon or a boron atom. The (physico-) chemical knowledge that can be used to make that distinction is not yet implemented in protein validation software.

introduces new problems as boron is not yet supported by many programs, including PRODRG, because it is seldom encountered in protein structures.

Extra complications occur when hetero compounds are covalently linked to each other. When two sugars are bound through a condensation reaction, one oxygen atom is lost. However, there are no strict rules prescribing which one. In a lot of cases we do not even know which one is lost in the actual chemical reaction. This can cause numerous missing atom errors, even when there is no real problem.

Another interesting problem occurs in mannose. α-D-mannose and β-D-mannose have the same topology and covalent geometry apart from the oxygen atom connected to C_1, which can be either above or below the plane. Their names (MAN and BMA, respectively) should reflect this so that the proper stereochemistry can be used when validating the structure. Unfortunately, this does not happen all the time. In the PDB, a lot of mannose groups called MAN have the wrong name (e.g., PDB entry 1HGD). This is reflected by HIC-Up where the MAN record is used for both α-D-mannose and β-D-mannose, whereas the BMA record only describes β-D-mannose.

3.23.2.4 Geometry

The geometry of amino acids is without doubt the most extensively validated set of parameters in protein structures. Root mean square deviations from standard bond lengths and angles[24,25] are typically reported when a structure is published. Even though these angles and bond lengths are usually restrained during refinement, and are therefore not independent values, they still provide interesting information about the structure model. Not just obvious outliers, like 1C0P (**Figure 8**), but also structures with too tightly restrained geometry should be flagged. The latter because tight restraints may inhibit maximal correlation with the experimental data, delivering models that suboptimally represent the biological structures. When enough experimental data is available, the restraints should be relaxed accordingly.

A systematic deviation of bond lengths (i.e., all bonds are a bit too long or too short) may imply an error in the cell dimension or x-ray wavelength.[38] This should have been intercepted in the early stages of structure determination[39]

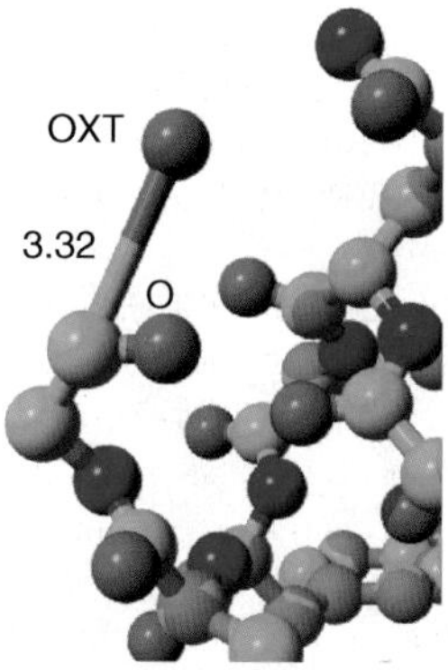

Figure 8 Detail of the C-terminus of PDB entry 1COP. The terminal glycine residue has an extremely long bond between the carboxylate carbon and terminal oxygen OXT. The bond length is 3.32 Å.

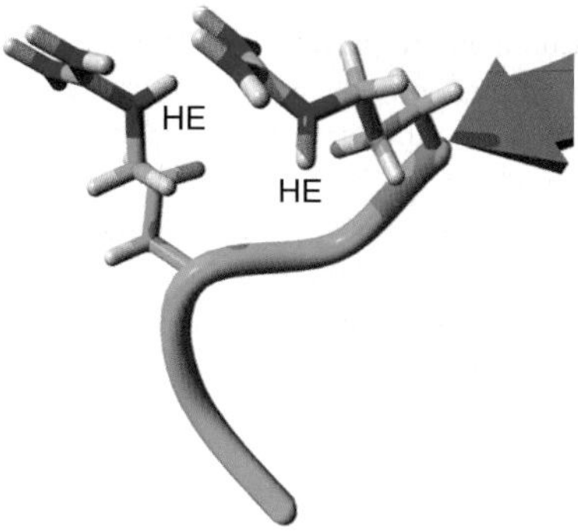

Figure 9 Detail of PDB entry 1D3Z. Two arginine residues near the C-terminus have their H_ε-atoms sticking far out of the guanidine plane as a result of under-restrained geometry. Wrong hydrogen positions can have serious consequences for the calculation of hydrogen bonds, pK_a values, and electrostatic potentials.

and indeed modern x-ray reflection and indexing software does so. These systematic deviations can be resolved to a reasonable extent by refinement with improved crystal parameters but early validation of the crystal parameters is preferable.[38]

Proline puckering[26] and the planarity of amino acid side chains[27] are validated by programs like WHAT_CHECK and PROCHECK.[40] It is not just the position of the heavy atoms that should be planar; the hydrogen atoms should be planar in certain residues as well.[32] The NMR structure ensemble 1D3Z has a number of arginine residues with strongly deviating H_ε atoms (**Figure 9**). This error can be traced back to a refinement problem: the H_ε atoms were not constrained to the guanidine plane. Because these atoms can make a hydrogen bond it is important to position them correctly. The calculation of pK_a values can be greatly influenced by the local hydrogen bonding network.[41] Proper electrostatics and pK_a values are of paramount importance for the design of protein inhibitors and activators.

Another aspect of protein geometry is the large number of torsion angles in the protein backbone and the residue side chains. These angles are not validated against a single value. Instead, they are compared against a distribution of values that is considered normal. This is done because there is no ideal value with a well-defined standard deviation. Many different conformations are possible, but some are more likely than others. If many unlikely conformations are found in a PDB entry, or a few unlikely rotamers are found close together, further inspection of the structure is necessary.

We describe two groups of torsion angles: on the one hand we have the φ- and ψ-angles that were initially described by theory,[16] on the other we have the side chain χ-angles that are usually compared to a database distribution derived from high-quality PDB entries.[14,15,42] The well-known Ramachandran plot is a scatter plot of φ and ψ-angles against a background of expected angle combinations. Certain zones in the plot are typical for the secondary structure types, while other zones are deemed 'forbidden.' Because backbone torsion angles are usually not restrained during refinement, the Ramachandran plot is seen as an independent validation tool. Unfortunately, there remains a problem with the interpretation of such plots: PROCHECK divides the plot into four sections (most favored, allowed, generously allowed, and disallowed) but a binary approach was proposed later.[43] WHAT_CHECK uses a database approach in which the scoring system is not discrete but continuous.[44] This continuum of scores is separately defined for each residue type. Therefore, WHAT_CHECK can express the quality of the Ramachandran plot as a real Z-score, rather than a percentage in a certain region. MOLPROBITY uses a method described by Lovell *et al.*[45] that takes into account the C_β positions of the residues.

Because of the global nature of the Ramachandran plot, large deviations from acceptable values imply that a structure is not very good as a whole and should preferably not be used for drug design or modeling. Individual outliers must be inspected because they may contain valuable information.[46]

Side chain χ-angles have a specific distribution, which becomes more apparent at high resolution.[42] There is also some context specificity in this distribution[47]: for example, the rotamer distribution for histidine at the beginning of an α-helix is different from that at the end of a helix.[48] Deviating rotamers should be inspected because they may be at an interesting site for drug design. If experimental data is available, alternative rotamers can be built if sufficient evidence is found to allow it.

Omega angles in the peptide bond form a different kind of validation set because they can typically have just two values: 0° or 180°. This corresponds to the *cis*- and *trans*-conformation, respectively. An in-depth evaluation of ω-angles was performed by MacArthur and Thornton, who describe the distribution of these angles based on small molecules as well as protein structures.[49]

cis-Peptide bonds are relatively uncommon in protein structures because they are energetically unfavorable. However, about 5% of all peptide bonds between any residue X and a proline residue (X-Pro) are found in the *cis*-conformation. This fraction of observed *cis*-peptide bonds increases with improved resolution because a lot of *cis*-peptides are not recognized at low resolution.[50] An equal relation is observed for *cis*-peptides with no prolines involved. However, the *cis*-peptide fraction is only 0.03% for bonds without proline. WHAT-CHECK has a routine that flags potential *cis*-peptides. Because the refinement of the flagged peptides may have been performed with the wrong peptide conformation, the local backbone may be distorted. Therefore, this section has to be carefully inspected using the experimental data, so that the correct peptide conformation can be found. Additional refinement may be necessary.

3.23.2.5 Contact Analysis

The simplest form of contact analysis is bump detection. No two atoms can fill the same space, so a minimum separation between all atoms must exist. This separation is usually chosen as the sum of the two Van der Waals radii minus some tolerated overlap. Because bumps are energetically unfavorable, they can have a profound impact on molecular dynamics and docking studies. It is therefore important that they are flagged and inspected during structure validation. Inter-symmetry bumps, i.e., bumps between symmetry related molecules, are a special case because they are easily missed when symmetry is not taken into account. Proper administration of crystal symmetry is therefore indispensable.

When we look at the type of inter-atomic contact, packing becomes less straightforward but also more valuable as a validation tool. It has been shown that the packing quality of proteins increases with decreasing R-factor.[51] This implies that better versions of a protein have better packing. It has also been shown that correctly folded proteins make twice as many hydrophobic contacts as misfolded proteins.[52]

There are numerous approaches to packing analysis but the principle remains the same: is a certain atom (or group of atoms) likely to be found near another atom (or group)? A very comprehensive study was performed by Singh and Thornton, resulting in the Atlas of protein side-chain interactions.[53] This wealth of information can be used to create packing validation software.

There are many software tools that validate protein packing. PROSA-II expresses residues packing as an energy potential to give a packing score.[54] Profile3D applies the knowledge derived from the PDB.[55] This program uses the model's sequence, which is compared to a 3D structure profile calculated from high-quality protein structures. The program ERRAT compares the distribution of atoms of a certain type (C, N, or O) near the atom that is to be evaluated, to the expected number of atomic contacts from a database.[56] It is based on the knowledge that the distribution of atoms at a given distance from a central atom is nonrandom. A similar but more sophisticated approach is the directional atomic contact analysis (DACA).[51] This is based on the interaction between a rigid group of atoms (e.g., three consecutive atoms in a lysine side chain or the phenyl group plus C_β of phenylalanine) and atoms of a certain type (e.g., alanine C_β or methionine S_δ). The secondary structure of the fragment is taken into account, as is the spatial orientation of the contacting atoms. This leads to a vast number of possible contacts, each with a specific score. Comparison of the environment of a fragment and the probability of such an environment, summed over all interactions, gives a correlation value that can be used for validation purposes. However, one must bear in mind that an improbable packing environment can be the result of more than just bad model building. A residue with an abnormal packing environment can in fact be very important for the function of a protein. Also, a badly packed residue can be an interesting target for mutation studies and protein design in both medical and industrial applications.[57]

3.23.2.6 Hydrogen Bonds and Water

In x-ray crystallography, except in ultra high-resolution structures, hydrogen atoms cannot be seen. In NMR it is completely the other way round: the protons dominate the structure calculation process. In any case, hydrogen atoms, especially those that form hydrogen bonds, are extremely important for drug design. Therefore, hydrogen-related validation tests are indispensable.

The simplest tests entail the validation of hydrogen nomenclature. The PDB has a strict definition for hydrogen names. Unfortunately, structure calculation programs use different naming schemes, as we have mentioned in Section 3.23.2.2. Many NMR ensembles do not even contain all the hydrogen atoms that are to be expected, but only the ones that can be seen in the experiment.[32] There are many geometrical considerations when validating hydrogen atoms. We have already discussed the planarity in Section 3.23.2.4 but also dihedral angles are important as staggered conformations are energetically more favorable than eclipsed conformations.

It becomes much more complicated, but also a lot more interesting when hydrogen bonds are taken into consideration. Individual hydrogen bonds are of course important for protein structure in general, but a lot more information can be derived from the hydrogen bond network.[58] First, the position of missing hydrogens can be ascertained; second, the optimal orientation of histidine, glutamine, and asparagine can be found. This is of particular interest for x-ray structures because the difference between the NH_2 group and the oxygen atom for asparagine and glutamine, and the nitrogen and carbon atoms in histidine is not visible at normal resolutions. The planar end of the side chains of these amino acids is therefore easily flipped into the wrong conformation.[59] Such errors can be very costly in electrostatics and pK_a calculations. Deviations of up to 2 pK_a units as a result of a single flip have been reported.[60] This brings us to the third advantage of an optimized hydrogen bonding network: proper calculation of the ionization states of charged residues and histidine. The final advantage is the validation of water molecules. In crystallography, adding water molecules to the model has a significant lowering effect on the R-factor. It is therefore very tempting to fill peaks in the electron density with water. However, not all peaks are water; some may be ions like Na^+.[61,62] In this sense, water molecules without any hydrogen bonds are rather suspect. Clusters of water molecules that do not make any hydrogen bonds (directly or indirectly) to the protein or another compound may be something more interesting than water. On the other hand, they may also be islands of spurious density kept alive by 'feeding' it water during refinement.

An optimized hydrogen bond network has many merits for the design of potential drugs. Without it, a proper description of the binding site is not possible in terms of electro-statics, local pH, and pK_a or hydrogen bonding potential. H-bonds are also important for validation purposes: we have already mentioned the histidine, asparagine, and glutamine flips as well as validation of crystal water, but hydrogen bonds are important for describing secondary structure too.[63] The number of buried unsatisfied H-bond donors in a structure is also a validation criterion because good structures tend to have very few of them.[64]

WHAT_CHECK uses an empirical hydrogen bond force field to perform a very thorough validation of the H-bond network. It considers the full hydrogen bond network at once, taking into account potential side chain flips, crystal symmetry, and H-bonds geometry (also for bifurcated bonds). The main problem lies in the number of proton conformations that can be generated. For instance, using 10 degree steps, there are 36 options for hydroxyl groups in serine, threonine, and tyrosine, and there are 366 orientations for each water molecule. This creates a mind-boggling number of possible networks, which needs to be reduced significantly before calculation becomes feasible. First, a list of all possible hydrogen bonds is made and any H-bond that can be unambiguously assigned is flagged. This way, the network is subdivided into independent clusters of ambiguous conformations. The problem is now greatly reduced and subnetworks that have fewer than a fixed number of possible structures can be fully assigned. We are left only with the larger subsections of the H-bond network for which a separate approach is used. A randomly selected ambiguous conformation is selected and set to a fixed value and the subnetwork is evaluated. Threshold acceptance[65] is used to create an energy cut-off. This way, any change can be either accepted or rejected. After a series of permutations, the energy threshold is lowered. These steps are repeated until the threshold energy is reduced to zero. The whole procedure is repeated a few times and the best solution is used. Finally, a local optimization for each hydrogen bond is performed.

The calculation of the complete hydrogen bond network is very computer intensive but can be performed in a reasonable amount of time on a normal workstation. The optimization of the H-bond network can have such a large effect on drug docking experiments that it is worth every CPU cycle.

3.23.2.7 Using Experimental X-ray Data

Validation of the actual experimental data is typically the task of the crystallographer or NMR spectroscopist as this requires specialized software and expertise. Nevertheless, a few 'sanity checks' can sometimes be carried out by nonexperts using either specialized software (e.g., SFCHECK[66] for x-ray data) or web-accessible databases (e.g., EDS[21]). For instance, one can obtain information about the resolution and completeness of the crystallographic data (higher resolution and completeness usually leads to better models), or about any problems with the crystals or the data that may have gone undetected by the crystallographers (e.g., twinning or anisotropic diffraction of the crystals). Finally, the Wilson B-factor that is calculated from the data can be compared to the average B-factor of the model and it should be of similar magnitude. If, for example, the average B-factor of the model is 20 Å^2 and the Wilson B-factor is 60 Å^2, then this suggests that the model's B-factors are unreliable.

However, a much more important advantage of the availability of experimental data is that it allows electron-density maps to be calculated. This, in turn, enables even nonexperts to inspect the quality of the density on which the model is based, both overall and for parts of the model that are of specific interest, such as ligands, cofactors, catalytically active residues, etc. In addition, statistics that measure the fit of the model and the density can be derived.

A global statistic that measures the fit between model and data is the R-value (*see* Section 3.23.1.5), but for most applications knowledge of the local quality of the fit between model and density is of greater interest. For this reason, Jones[11] introduced the real-space fit, which compares experimental electron density with the density distribution calculated directly from the structure model. If the model is a faithful representation of the data, these two densities should be very similar. The real-space fit is usually calculated for one amino acid residue, nucleotide, or small molecule at a time and thus provides a local quality score. The fit can be expressed as an R-value (lower values imply a better fit) or as a correlation coefficient (values closer to one imply a better fit). These real-space fit statistics are readily available from EDS for the large majority of crystal structures in the PDB for which the experimental data was deposited (about 15 000 entries in EDS in July 2005). In addition, since EDS contains real-space fit statistics for millions of amino acid residues and nucleotides, typical values (average and sample standard deviation) for each type of residue or nucleotide at various resolutions are easy to compute. These values, in turn, can be used to flag residues that have unusually high real-space R-values (given their type and the resolution). A cut-off of two standard deviations above the mean is used in EDS, which tends to flag those residues that have very poor density indeed.

An example of the importance of the use of electron-density maps to select between alternative, seemingly equivalent entries in the PDB is provided in **Figure 10**. PDB entries 268D and 1D63 are both complexes of a small fragment of DNA with the drug berenil. Both structures were determined at 2.0 Å resolution, with similar R-factors (16% and 17%, respectively) and similar average B-factors (29 and 31 Å^2, respectively). The overall completeness of the data for 1D63 was only 73%, whereas that for 268D was 99%.

The latter statistic could be used to make a case for using 268D rather than 1D63 for subsequent modeling or design studies. However, this would be an unfortunate choice, as the density for the two berenil molecules reveals. The

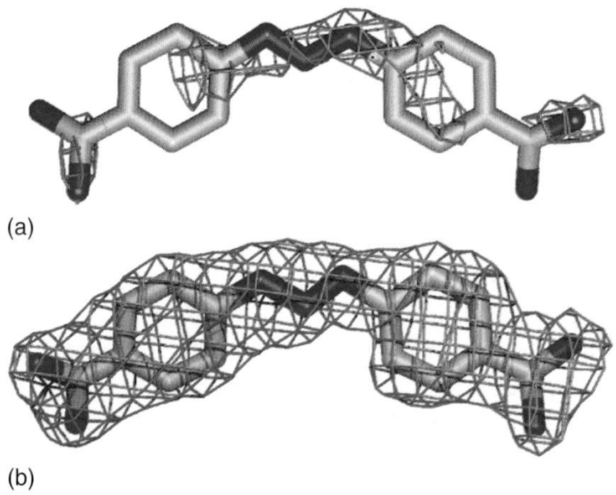

Figure 10 Electron-density maps (taken from EDS)[21] for the drug berenil in complex with DNA from two different PDB entries, both determined at 2.0 Å resolution. (a) Density for berenil in PDB entry 268D, and (b) in PDB entry 1D63.

berenil molecule in 1D63 has excellent density as one would expect for a fully occupied site in a 2 Å electron-density map, and this is reflected in a real-space R-value of 0.11, close to the average value of 0.10 for the entire model (minus the waters). The berenil molecule in 268D, on the other hand, has a real-space R-value of no less than 0.44 (compared to an average of 0.12 for the entire model).

A similar trend is visible in the behavior of the real-space correlation coefficient (1D63: 0.93 for berenil, 0.96 overall; 268D: only 0.82 for berenil, compared to 0.95 for the entire model) and the B-factors (1D63: 35 $Å^2$ compared to 31 $Å^2$ overall; 268D: 56 $Å^2$ for berenil, which is about twice the average of 29 $Å^2$ for the entire model). In this particular case, the appearance of the density maps and the statistics suggest that the berenil site in 268D probably had a very low occupancy, making 1D63 the obvious model of choice for any further computational or experimental studies.

3.23.3 Time Flies When You Are Validating Structures

Structure validation is not new. Many years of research have led to the validation tools that are available today. But the development of structure validation and structure determination is not standing still; many new tools and methods are being developed. We therefore want to put validation in a temporal perspective. In this section we shall look at the past, present, and future of validation.

3.23.3.1 The Quick and the Dead

We can categorize validation checks into two types: living and dead. Some validation methods are completely based on definitions or on solid experimental data. No one will dispute that 'P 1-' is a nonstandard space group in protein crystallography and will therefore cause problems with many software packages. Nor will anyone say that a nonbonded interatomic distance can be as short as 1.281 Å in a protein. The typical bond lengths and angles as determined by Engh and Huber[24] have been proven to be correct when extrapolated to macromolecules.[38] These undisputed values in validation need no further development and are therefore considered dead. Small updates have occurred,[67] but the changes were small enough not to significantly modify the results of the validation of structures. The dead validation parameters are accepted and have found their way into a lot of refinement protocols.

However, not all validation parameters are as well determined as the bond lengths and angles. There are no perfect side chain torsion angles or ideal residue packing environments. Our definition of right and wrong, or better normal and abnormal, is based on information extracted from the best entries in the PDB. But the protein data bank is continuously updated and our perception of what is 'the best' needs to be updated as well. The databases from which we extract validation parameters grow and/or change in composition. This makes tools that rely on these databases' living tests.

This is not without consequences for the validation results of older structures. For certain validation tests, the average resolution of the database is an important factor. The average values of certain validation parameters are significantly resolution dependent.[42] This may result in bias towards structures that have the same resolution as the structures in the database even to the extent that higher resolution structures sometimes score worse than lower resolution structures if the latter have resolutions close to those of the files in the validation database. Furthermore, because the average quality of structures increases with date of deposition,[68,69] database derived tools become sterner (see **Table 3**). As a result of this, older structures have the disadvantage that they can look quite bad even though they had the best possible refinement at that time. Unfortunately, this is unavoidable. After all, drug designers and other users of protein 3D coordinates should use the best data available by our current standards.

3.23.3.2 Database Generation

Many validation tools are based on the comparison of some parameter with the average value for that parameter in a set of 'good-quality' PDB entries. There are many ways to generate such a set of PDB entries but they all follow a basic principle. Here, we describe parameters that are common to most data set generation protocols.[3,27,70–72]

First of all, the method with which the structure models were derived is considered. Usually, only crystal structures are used, but pure NMR databases may be useful for specific purposes. Provided that x-ray structures are used, the resolution becomes the second selection criterion. Low-resolution structures are usually excluded. Even high-resolution structures can be bad so the R-factor has to be taken into account. High R-factor structures have less correlation with the experimental data. In this sense, the difference between R-factor and R_{free} should also be reasonable (less than 5%) because a large difference may imply overinterpretation of the experimental data.

Table 3 Quality indicators for PDB entry 1CRN based on databases for 2000, 2002, and 2005

Structure Z-score[a]	*2000*	*2002*	*2005*
First-generation packing quality	0.163	0.163	0.163
Second-generation packing quality	−0.547	−1.509	−0.531
Ramachandran plot appearance	−0.230	−0.244	−0.754
Chi-1/chi-2 rotamer normality	−0.195	−0.738	−1.079
Backbone conformation	0.470	1.089	0.841

[a] Positive is better than average; for more information see PDBREPORT.

Table 4 Date at which a nonredudant, 500 entry database was available for different resolutions

Resolution cut-off (Å)	*Date*
1.8	October 2002
1.7	December 2003
1.6	March 2004
1.5	June 2005
1.0	2011[a]

[a] At current rate of growth of the number of 1.0 Å (or better) entries in the PDB.

Normally, the structure database contains only native proteins, no mutants. This is especially important for packing analysis, as mutants may have distorted packing environments. The same holds for the amino acid sequence. No nonnatural amino acids should be included. Incomplete residues (e.g., with missing side chain atoms) can distort many calculations so they must not be included in validation databases.

There are many other possible criteria like chain length, average B-factors, occupancy, and hetero compounds. Any database created with the parameters above may still be significantly biased. It is very important that the database is filtered resulting in minimal pairwise sequence identity. To ensure that data redundancy does not bias the statistics, individual proteins in the database should only have low levels of mutual sequence identity. WHAT_CHECK uses a cut-off of 30%. **Table 4** shows the resolution cut-off that can be used to create a database of more than 500 nonredundant PDB entries over the years. At the current rate (assuming exponential growth of the PDB) we can make a 1 Å database in 2011.

When updating a database to a new version, there is a final criterion that is as important as it is straightforward: any new structure in the database should have passed the validation based on the previous version of the database. Because of the steady growth of the PDB, databases can be updated at regular intervals resulting in a good representation of our current knowledge. This keeps the so-called living tests alive.

3.23.3.3 Errors over the Years

In 1996 we reported that there were more than 1 million 'errors' in the PDB.[10] This was received with skepticism at the time.[73] Surely, those errors must have been caused by the old structures in the PDB? Apparently not, as we now see 11.6 million outliers in the protein data bank. The number of newly introduced tests does not account for that increase alone. Newly submitted structures still contain errors, even ones that are easy to detect.[74]

To see which errors have become less frequent and which errors are more common now than they used to be, we have selected a few dead tests that are independent of the development of the PDB in terms of average quality and resolution (**Figure 11**). From the beginning of the PDB until the early 1990s the number of deposited structures per year were so small that the graphs are quite noisy. However, it is clear that most errors are as old as the PDB itself. An interesting exception is errors with nonunique chain identifiers, which only started occurring in 1990 (**Figure 11b**). This is caused by the limitations of the PDB format that only allows a chain identifier of one character. The large structures that are now resolved can be so large that the number of chains exceeds the number of possible chain identifiers.

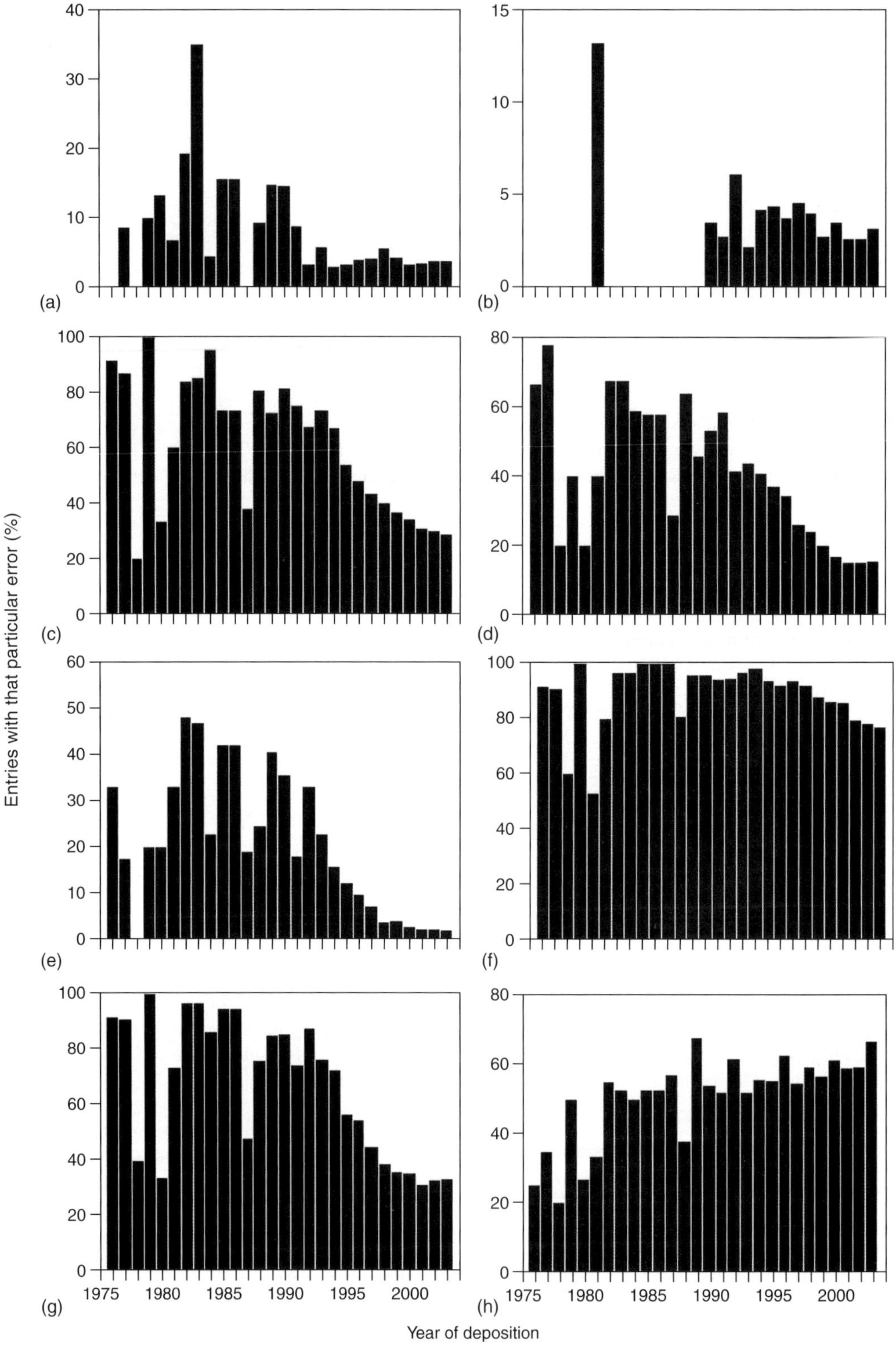

Figure 11 The development of errors over the years. All PDB entries were grouped by year of deposition and evaluated by WHAT_CHECK[10] for eight different dead checks. (a) Atoms too close to the symmetry axis; (b) chain names not unique; (c) chirality deviations; (d) side-chain planarity problems; (e) high bond angle deviation; (f) unusual bond angles; (g) unusual bond lengths; and (h) water molecules without hydrogen bonds.

The advent of modern refinement software in the early 1990s had a very clear effect when we look at geometrical parameters (**Figure 11c–g**). The percentage of PDB entries with these geometry errors has declined significantly in the last 10 years. Errors regarding parameters that are not refined (**Figure 11a** and **11b**) have been reasonably steady in the same period. It is surprising that the number of PDB entries with nonhydrogen-bonded water molecules is slowly increasing even though proper validation of hydrogen bonds has been available for many years (**Figure 11h**).[58] We assume that the growth of the average number of water molecules per PDB entry surpasses the effect of hydrogen bond validation. After all, any structure with a single water molecule that is flagged as having no hydrogen bonds is used to calculate the percentages in **Figure 11h**.

In **Figure 12a** we show the development of a few structural normality Z-scores over the last 15 years. To avoid bias caused by a higher fraction of high-resolution structures in the last few years, we have only taken into account x-ray structures with a resolution between 1.8 and 2.2 Å.

Side-chain rotamer libraries, which are now commonly used in structure building, have had a large effect on the average χ_1–χ_2 distribution Z-score, whereas backbone libraries, which have also been available for a long time, have had a much smaller effect. The average residue packing quality, for which no such libraries exist, has stayed stable over the last 15 years. The average quality of the Ramachandran plot has steadily increased, showing that the quality of structures in general has indeed risen.

Figure 12b shows how these normality score are a function of resolution. This figure shows that there is a good correlation between validation scores and the real structure quality, strengthening the conclusions drawn from **Figure 12a**.

3.23.3.4 The Future of Validation

Over the years we have written software to find errors, funnies, and anomalies in PDB files, and we made a website that lists all these observations.[10] The previous section shows that these validation efforts, and those of several other groups,[75–87] combined with all efforts by the x-ray and NMR communities to improve the structure determination process, are paying off. Structures have been getting better over the last 10 years. So where should structure validation go from here? We see several major validation topics that will help continue the observed trend of getting better structures with the progress of time.

The first topic is simply making the existing checks more accurate. We warn, for example, against missing C-terminal oxygen atoms. This is an important topic for drug design because a C-terminus normally is charged and thus influences electrostatic calculations. C-termini are seldom located near the active site, but if they are, missing atoms will mean missing interactions with the studied ligand. But what if the last residue in the chain is not the C-terminus because the real C-terminal residue(s) are too mobile to be observed in the electron density map? In that case, we issue a warning that is not relevant. It will require harmonization of the PDB with UniProt[30] and a lot of artificial intelligence to read the PDB file-header to resolve this problem. If one day we succeed in doing so, we can issue error messages for missing C-terminal oxygen atoms rather than the warnings we issue today.

The second topic is designing new checks for errors that hitherto remained unnoticed. Alternate atom positions are often observable in high-resolution x-ray structures. Information about alternate atom positions is important for rational drug design because their presence is a strong indicator for local mobility that could modify the shape of the ligand-binding pocket. However, a very large number of alternate atom positions are administratively fouled up beyond all recognition. **Figure 13** shows two examples.

The third topic is a better understanding of the physicochemical properties of proteins. If we want to call something wrong, we first must know what is right. For example, loops often can adjust their conformation to accommodate crystal packing. What can we say about these loops? In an attempt to better understand this flexibility, we studied intermolecular contacts in covalently identical proteins related by noncrystallographic symmetry. Noncrystallographic symmetry related loops, by definition, have a different structural environment but are guaranteed to be crystallized under the same conditions. Nobody will be surprised by the observation that residues in loops are more often involved in intermolecular contacts than one would expect from their frequency in the protein. It is also not surprising that loops that are involved in intermolecular contacts show on average the largest backbone displacements (for penta-peptides in loops, strands, and helices the average C_α-displacements are 0.42, 0.27, and 0.25 Å, respectively). In a data set of 270 crystal structures with noncrystallographic symmetry (all solved at better than 2.0 Å resolution) we could find only a few examples of loops that made contacts in one monomer but not in the noncrystallographic symmetry-related one. We looked at all loops in the first monomer of the PDB file and compared the contact patterns with the noncrystallographic symmetry-related loops in the second monomer. **Table 5** shows that there is a strong tendency for these pairs of loops to both be involved in contacts or to both not be involved. Part of this observed relation is caused by

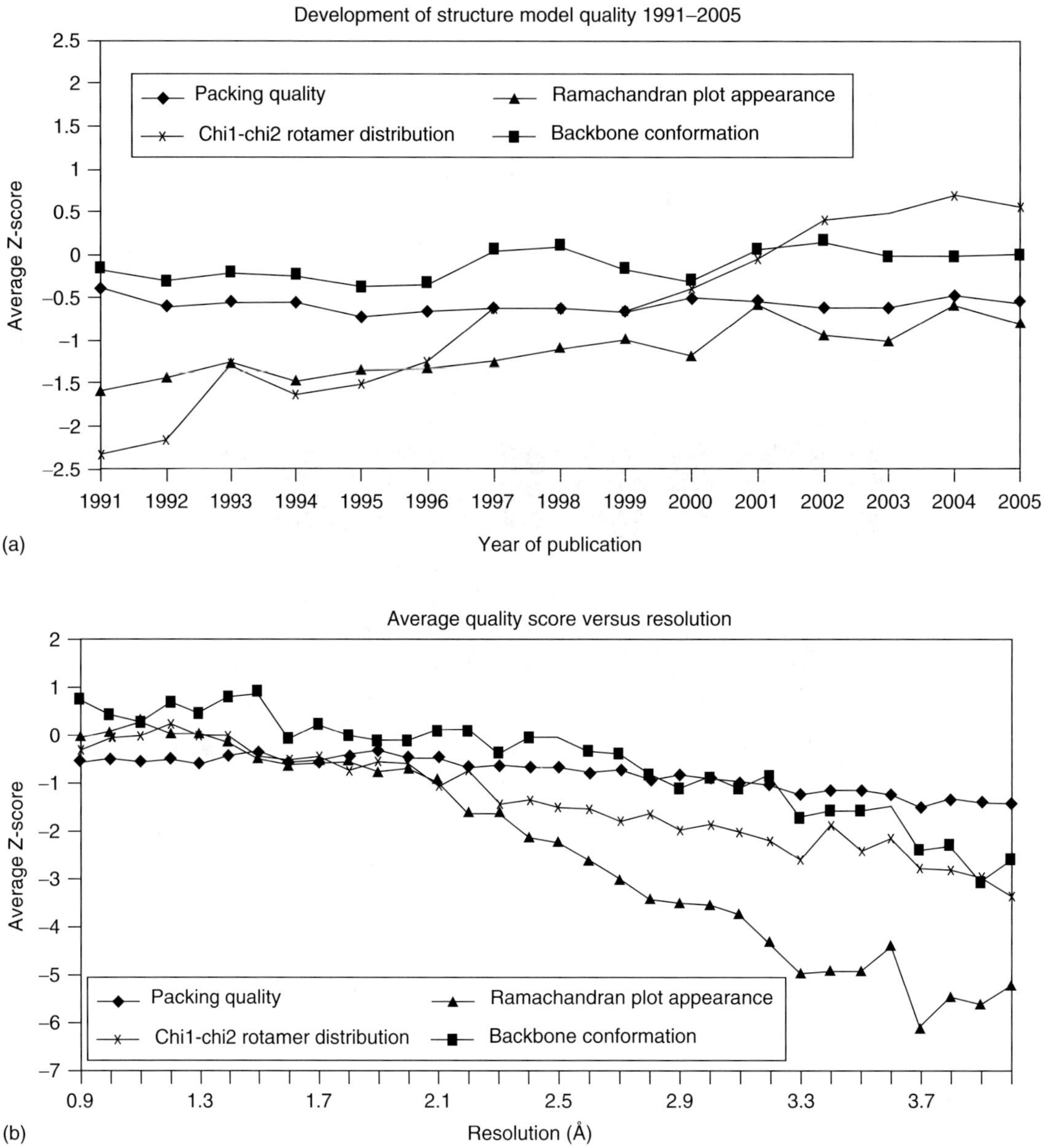

Figure 12 (a) The development of four structure quality indicators over the last 15 years for structures with resolution between 1.8 and 2.2 Å. There is a clear improvement of quality over time. All structures had at least 100 residues and a maximum of 1000 residues. Data sets for were truncated at 100 structures per year. (b) The resolution dependence of the same four quality indicators. This dependence shows that high-resolution structures perform better than low-resolution structures. All structures had at least 100 residues and a maximum of 1000 residues. Data sets for every 0.1 Å resolution bin were truncated at 50 structures.

the fact that dimers in the asymmetric unit prefer to pack in such a way that they are related by an internal pseudo twofold axis. However, **Table 5b**, which is based on the same data as **Table 5a** but with the intra asymmetric unit contacts and the contacts with ligands neglected, shows a similar tendency.

There probably exist many possible explanations for this phenomenon. We like to think that "certain loops seem predestined to make contacts," but that is not an explanation. The most likely explanation for this predestination is the capability to alter their structure, a capability we think is essential for the formation of crystals. After all, how big is the chance that any given protein would crystallize if it could not adjust its structure to fit well in the crystal? We could, however, find no correlation between the RMS deviation between pairs of noncrystallographic symmetry-related loops and parameters such as the number of glycines in the loop, the end-to-end distance, the internally buried hydrophobic surface, the number of salt bridges or hydrogen bonds, or a whole series of other parameters.

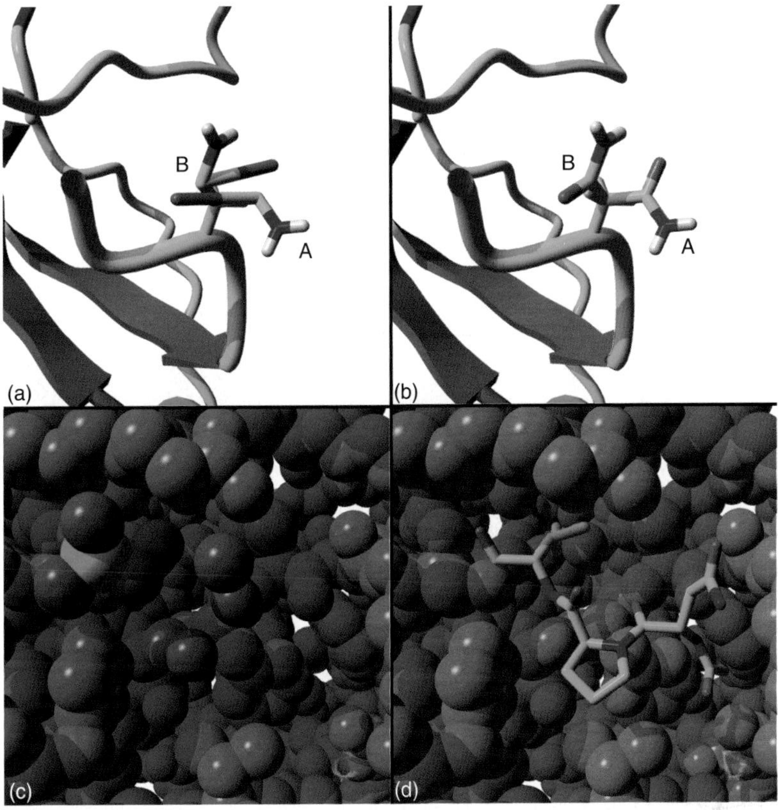

Figure 13 Examples of problems with alternate atom administration. (a) Asparagine 143 in PDB entry 1HJ8 has two alternate conformations, labeled A and B. Owing to an administrative error, the $O_{\delta 1}$ atoms in the two conformation have been swapped. This results in two malformed side chains. (b) The same PDB entry but now with the $O_{\delta 1}$ atoms swapped back to form two good side chain conformations. (c) The sulfate and water molecules in PDB entry 1GVK occupy the same space as the peptide in (d); they are therefore alternates. However, they are not administrated as such in the coordinates. Instead, a very thorough explanation is given in the entry header, which makes human interpretation straightforward. Unfortunately, automated processing of this PDB entry is problematic.

Table 5 Comparison of noncrystallographic symmetry-related penta-peptides in loops[a]

Pep1 makes contacts	*Pep2 makes contacts*	*Preference parameter*
(a) Including intra-asymmetric unit contacts		
No	No	1.75
No	Yes	−0.83
Yes	No	−0.83
Yes	Yes	0.07
(b) Not using intra-asymmetric unit contacts		
No	No	0.73
No	Yes	−0.88
Yes	No	−0.90
Yes	Yes	0.28

[a]The data set contained 8289 penta-peptide pairs. The preference parameters are relative to a model in which all peptides have an equal change of making a contact. Pep1 and Pep2 are noncrystallographic symmetry related penta-peptide pairs. In (a) all contacts are counted, in (b) contacts with ligands with contacts within the asymmetric unit are excluded. Similar trends are observed if only helical, strand, or loop penta-peptides are studied.

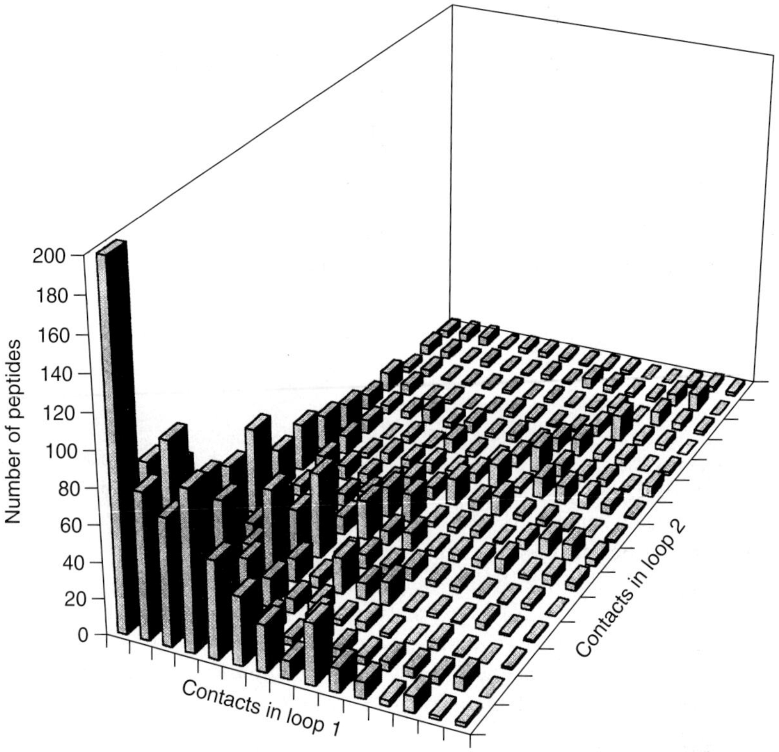

Figure 14 Frequency of occurrence versus number of contacts made in pairs of noncrystallographic symmetry related penta-peptides. Every interatomic contact is counted, so a pair of large, tightly packed residues can easily make 100 contacts. Each bin represents 20 atomic contacts. The bar representing 0–20 contacts for both peptides is 677 high and has been truncated for clarity.

We went one step further and analyzed the number of contacts made by penta-peptides in loops that are involved in crystal contacts. **Figure 13** shows on the x-axis the number of contacts made by a penta-peptide, on the y-axis the number of contacts made by the noncrystallographic symmetry-related peptide, and the height of the bars indicates the frequency of occurrence of that event in the set of 270 proteins. In this figure we only show those pairs of penta-peptides for which the sets of contacted residues overlap for less than 33%. That is, at least two-thirds of the contact partners of the one peptide are different from the contact partner of the noncrystallographic symmetry-related peptide (contact partners that are related by noncrystallographic symmetry or crystallographic symmetry are considered the same). **Figure 14** shows that on top of the already concluded 'predestination to make contacts' there is an even more mysterious predestination to make roughly equally many contacts even when the residues contacted are radically different.

If we want to make progress in rational drug design, we must not only increase the quality of the protein structures we work with, but we must also increase our understanding of those protein structures. What can move, and what cannot. Which interactions are favorable and which are not. The story about contacts between noncrystallographic symmetry-related molecules is highly confusing, but we must figure out what is going on before we can start validating loops that make noncrystallographic symmetry contacts, or any crystallographic contact at all for that matter.

3.23.4 Conclusion

In building macromolecular structures, as in science in general, anything that can go wrong will go wrong. Fortunately, most problems are intercepted in time. The PDB contains a wealth of data for (structural) biologists, biochemists, and drug designers but there are many pitfalls in using 3D coordinates.

Structure validation is indispensable for the identification of errors in protein structures. We have shown a variety of errors and discussed their implication for those who use the PDB entries that contain them. Many problems can be resolved and even those that cannot are a smaller threat when they are identified in time.

The methods used to obtain 3D protein models are still improving. Together with validation they have resulted in better structures over the years. Unfortunately, several problems are now more common than they used to be.

A constant effort from all parties in this process is required to maintain this overall trend of improving structural data. With the ever-increasing thirst for protein and nucleic acid structures and the pressure to publish before the competitor, the risk of a trade-off between speed and quality exists. We believe that the use of validation software throughout the process of structure determination can help avoiding this trade-off.[88] By intercepting errors before they can cause problems in the next refinement round structures can be solved faster and at the same time more reliably.

Validators continue to develop newer and better tools to recognize problems in structures, both known errors and newly discovered types of error. Because of the increasing size of the PDB more statistical data is available to help this development. We are still a long way from perfect structures but we try to move closer one step at a time. The future will point out how big these steps are.

Acknowledgments

The authors acknowledge financial support from: the EMBRACE project that is funded by the European Commission within its FP6 Programme, under the thematic area 'Life sciences, genomics and biotechnology for health,' contract number LHSG-CT-2004–512092, and the BioRange programme of the Netherlands Bioinformatics Centre (NBIC), which is supported by a BSIK grant through the Netherlands Genomics Initiative (NGI). GJK is a Research Fellow of the Royal Swedish Academy of Sciences (KVA), supported through a grant from the Knut and Alice Wallenberg Foundation. The figures were created with YASARA.[89]

In this chapter, we have mentioned a lot of PDB entries without a reference to the authors. This was done because we did not want to single them out as bad scientists. The examples we used are just that: examples. We could have picked a wide selection of other PDB entries. And after all, any person who shares his or her data with the (scientific) community is a much better scientist than all those who do not.

References

1. Hol, W. G. J. *Angew. Chem., Int. Ed. Engl.* **1986**, *25*, 768–778.
2. Bernstein, F. C.; Koetzle, T. F.; Williams, G. J.; Meyer, E. F., Jr.; Brice, M. D.; Rodgers, J. R.; Kennard, O.; Shimanouchi, T.; Tasumi, M. *J. Mol. Biol.* **1977**, *112*, 535–542.
3. Wang, G.; Dunbrack, R. L., Jr. *Bioinformatics* **2003**, *19*, 1589–1591.
4. Navia, M. A.; Fitzgerald, P. M.; McKeever, B. M.; Leu, C. T.; Heimbach, J. C.; Herber, W. K.; Sigal, I. S.; Darke, P. L.; Springer, J. P. *Nature* **1989**, *337*, 615–620.
5. Blundell, T.; Pearl, L. *Nature* **1989**, *337*, 596–597.
6. Weber, I. T.; Miller, M.; Jaskolski, M.; Leis, J.; Skalka, A. M.; Wlodawer, A. *Science* **1989**, *243*, 928–931.
7. Wlodawer, A.; Miller, M.; Jaskolski, M.; Sathyanarayana, B. K.; Baldwin, E.; Weber, I. T.; Selk, L. M.; Clawson, L.; Schneider, J.; Kent, S. B. *Science* **1989**, *245*, 616–621.
8. Brändén, C.-I.; Jones, T. A. *Nature* **1990**, *343*, 687–689.
9. Nabuurs, S. B.; Krieger, E.; Spronk, C. A. E. M.; Hooft, R. W. W.; Vriend, G. In *Computational Medicinal Chemistry and Drug Discovery*; Bultnick, P., de Winter, H., Langenaeker, W., Tollenaere, J. P., Eds.; Marcel Dekker Inc.: New York, 2003, pp 387–403.
10. Hooft, R. W. W.; Vriend, G.; Sander, C.; Abola, E. E. *Nature* **1996**, *381*, 272.
11. Jones, T. A.; Zou, J. Y.; Cowan, S. W.; Kjeldgaard, M. *Acta Cryst. A* **1991**, *47*, 110–119.
12. Ponder, J. W.; Richards, F. M. *J. Mol. Biol.* **1987**, *193*, 775–791.
13. Kleywegt, G. J.; Jones, T. A. *Acta Cryst. D* **1998**, *54*, 1119–1131.
14. Chinea, G.; Padron, G.; Hooft, R. W.; Sander, C.; Vriend, G. *Proteins* **1995**, *23*, 415–421.
15. Lovell, S. C.; Word, J. M.; Richardson, J. S.; Richardson, D. C. *Proteins: Struct. Funct. Genet.* **2000**, *40*, 389–408.
16. Ramachandran, G. N.; Ramakrishnan, C.; Sasisekharan, V. *J. Mol. Biol.* **1963**, *7*, 95–99.
17. Kleywegt, G. J.; Jones, T. A. *Structure* **1995**, *3*, 535–540.
18. Brunger, A. T. *Nature* **1992**, *355*, 472–475.
19. Kleywegt, G. J.; Brunger, A. T. *Structure* **1996**, *4*, 897–904.
20. Clore, G. M. ; Garrett *D. S. J. Am. Chem. Soc.* **1999**, *121*, 9008–9012.
21. Kleywegt, G. J.; Harris, M. R.; Zou, J.; Taylor, T. C.; Wählby, A.; Jones, T. A. *Acta Crystallogr. Sect. D* **2004**, *60*, 2240–2249.
22. Nabuurs, S. B.; Spronk, C. A. E. M.; Krieger, E.; Maassen, H.; Vriend, G.; Vuister, G. W. *J. Am. Chem. Soc.* **2003**, *125*, 12026–12034.
23. Doreleijers, J. F.; Rullmann, J. A.; Kaptein, R. *J. Mol. Biol.* **1998**, *281*, 149–164.
24. Engh, R.; Huber, R. *Acta Crystallogr. Sect. A* **1991**, *47*, 392–400.
25. Parkinson, G.; Voitechovsky, J.; Clowney, L.; Brünger, A. T.; Berman, H. *Acta Crystallogr. Sect. D* **1996**, *52*, 57–64.
26. Cremer, D.; Pople, J. A. *J. Am. Chem. Soc.* **1975**, *97*, 1354–1358.
27. Hooft, R. W. W.; Sander, C.; Vriend, G. *J. Appl. Crystallogr.* **1996**, *29*, 714–716.
28. Hooft, R. W. W.; Sander, C.; Vriend, G. *J. Appl. Crystallogr.* **1994**, *27*, 1006–1009.
29. Lesk, A. M.; Boswell, D. R.; Lesk, V. I.; Lesk, V. E.; Bairoch, A. *Protein Seq. Data Anal.* **1989**, *2*, 295–308.
30. Apweiler, R.; Bairoch, A.; Wu, C. H.; Barker, W. C.; Boeckmann, B.; Ferro, S.; Gasteiger, E.; Huang, H.; Lopez, R.; Magrane, M. et al. *Nucleic Acids Res.* **2004**, *32*, 115–119.
31. Seavey, B. R.; Farr, E. A.; Westler, W. M.; Markley, J. L. *J. Biomol. NMR* **1991**, *1*, 217–236.

32. Doreleijers, J. F.; Vriend, G.; Raves, M. L.; Kaptein, R. *Proteins* **1999**, *37*, 404–416.
33. Kleywegt, G. J.; Henrick, K.; Dodson, E. J.; van Aalten, D. M. F. *Structure* **2003**, *11*, 1051–1059.
34. van Aalten, D. M. F.; Bywater, R.; Findlay, J. B.; Hendlich, M.; Hooft, R. W.; Vriend, G. *J. Comput.-Aided Mol. Des.* **1996**, *10*, 255–262.
35. Schuettelkopf, A. W.; van Aalten, D. M. F. *Acta Crystallogr. Sect. D* **2004**, *60*, 1355–1363.
36. Chen, X.; Schauder, S.; Potier, N.; Van Dorsselaer, A.; Pelczer, I.; Bassler, B. L.; Hughson, F. M. *Nature* **2002**, *415*, 545–549.
37. Corey, E. J.; Howe, W. J.; Orf, H. W.; Pensak, D. A.; Petersson, G. *J. Am. Chem. Soc.* **1975**, *97*, 6116–6124.
38. EU 3-D Validation Network. *J. Mol. Biol.* **1998**, *276*, 417–436.
39. Vriend, G.; Rossmann, M. G.; Arnold, E.; Luo, M.; Griffith, J. P.; Moffat, K. *J. Appl. Crystallogr.* **1986**, *19*, 134–139.
40. Laskowski, R. A.; MacArthur, M. W.; Moss, D. S.; Thornton, J. M. *J. Appl. Crystallogr.* **1993**, *26*, 283–291.
41. Nielsen, J. E.; Vriend, G. *Proteins* **2001**, *43*, 403–412.
42. MacArthur, M. W.; Thornton, J. M. *Acta Crystallogr. Section D* **1999**, *55*, 994–1004.
43. Kleywegt, G. J.; Jones, T. A. *Structure* **1996**, *4*, 1395–1400.
44. Hooft, R. W.; Sander, C.; Vriend, G. *Comput. Appl. Biosci.* **1997**, *13*, 425–430.
45. Lovell, S. C.; Davis, I. W.; Arendall, W. B., III; de Bakker, P. I.; Word, J. M.; Prisant, M. G.; Richardson, J. S.; Richardson, D. C. *Proteins* **2003**, *50*, 437–450.
46. Herzberg, O.; Moult, J. *Proteins: Struct. Funct. Genet.* **1991**, *11*, 223–229.
47. Dunbrack, R. L.; Karplus, M. *J. Mol. Biol.* **1993**, *230*, 543–574.
48. Vriend, G.; Sander, C.; Stouten, P. F. *Protein Eng.* **1994**, *7*, 23–29.
49. MacArthur, M. W.; Thornton, J. M. *J. Mol. Biol.* **1996**, *264*, 1180–1195.
50. Jabs, A.; Weiss, M. S.; Hilgenfeld, R. *J. Mol. Biol.* **1999**, *286*, 291–304.
51. Vriend, G.; Sander, C. *J. Appl. Cryst.* **1993**, *26*, 47–60.
52. Bryant, S. H.; Amzel, L. M. *Int. J. Pept. Protein Res.* **1987**, *29*, 46–52.
53. Singh, J.; Thornton, J. M. *Atlas of Protein Side Chain Interactions*; IRL Press: Oxford, UK, 1992; Vols. I & II.
54. Sippl, M. J. *Proteins* **1993**, *17*, 355–362.
55. Luthy, R.; Bowie, J. U.; Eisenberg, D. *Nature* **1992**, *356*, 83–85.
56. Colovos, C.; Yeates, T. O. *Protein Sci.* **1993**, *2*, 1511–1519.
57. Eijsink, V. G.; Vriend, G.; Van der Zee, J. R.; Van den Burg, B.; Venema, G. *Biochem. J.* **1992**, *285*, 625–628.
58. Hooft, R. W. W.; Sander, C.; Vriend, G. *Proteins* **1996**, *26*, 363–376.
59. McDonald, I. K.; Thornton, J. M. *Protein Eng.* **1995**, *8*, 217–224.
60. Nielsen, J. E.; Andersen, K. V.; Honig, B.; Hooft, R. W.; Klebe, G.; Vriend, G.; Wade, R. C. *Protein Eng.* **1999**, *12*, 657–662.
61. Nayal, M.; Di Cera, E. *J. Mol. Biol.* **1996**, *256*, 228–234.
62. Shui, X.; McFail-Isom, L.; Hu, G. G.; Williams, L. D. *Biochemistry* **1998**, *37*, 8341–8355.
63. Kabsch, W.; Sander, C. *Biopolymers* **1983**, *22*, 2577–2637.
64. McDonald, I. K.; Thornton, J. M. *J. Mol. Biol.* **1994**, *238*, 777–793.
65. Dueck, G.; Scheuer, T. *J. Comput. Phys.* **1990**, *90*, 161–175.
66. Vaguine, A. A.; Richelle, J.; Wodak, S. J. *Acta Crystallogr. Sect. D* **1999**, *55*, 191–205.
67. Engh, R. A.; Huber, R.; , International Tables for Crystallography Volume F Crystallography of Biological Macromolecules, International Tables for Crystallography Volume F Crystallography of Biological Macromolecules; Kluwer Academic Publishers: Dordrecht, The Netherlands, 2001, pp 382–392.
68. Weissig, H.; Bourne, P. E. *Bioinformatics* **1999**, *15*, 807–831.
69. Kleywegt, G. J.; Jones, T. A. *Structure* **2002**, *10*, 465–472.
70. Hobohm, U.; Scharf, M.; Schneider, R.; Sander, C. *Protein Sci.* **1992**, *1*, 409–417.
71. Noguchi, T.; Onizuka, K.; Akiyama, Y.; Saito, M. *Proc. Int. Conf. Intell. Syst. Mol. Biol.* **1997**, *5*, 214–217.
72. Heringa, J.; Sommerfeldt, H.; Higgins, D.; Argos, P. *Comput. Appl. Biosci.* **1992**, *8*, 599–600.
73. Jones, T. A.; Kleywegt, G. J.; Brunger, A. T. *Nature* **1996**, *383*, 18–19.
74. Badger, J.; Hendle, J. *Acta Crystallogr. Sect. D* **2002**, *58*, 284–291.
75. Vila, J.; Williams, R. L.; Vasquez, M.; Scheraga, H. A. *Proteins* **1991**, *10*, 199–218.
76. Ohlendorf, D. H. *Acta Crystallogr. Sect. D* **1994**, *50*, 808–812.
77. Schultze, P.; Feigon, J. *Nature* **1997**, *387*, 668.
78. Hendlich, M.; Lackner, P.; Weitckus, S.; Floeckner, H.; Froschauer, R.; Gottsbacher, K.; Casari, G.; Sippl, M. J. *J. Mol. Biol.* **1990**, *216*, 167–180.
79. Gregoret, L. M.; Cohen, F. E. *J. Mol. Biol.* **1990**, *211*, 959–974.
80. Holm, L.; Sander, C. *J. Mol. Biol.* **1992**, *225*, 93–105.
81. Novotny, J.; Bruccoleri, R.; Karplus, M. *J. Mol. Biol.* **1984**, *177*, 787–818.
82. Morris, A. L.; MacArthur, M. W.; Hutchinson, E. G.; Thornton, J. M. *Proteins* **1992**, *12*, 345–364.
83. Miyazawa, S.; Jernigan, R. L. *J. Mol. Biol.* **1996**, *256*, 623–644.
84. Bohm, G.; Jaenicke, R. *Protein Sci.* **1992**, *1*, 1269–1278.
85. Delarue, M.; Koehl, P. *J. Mol. Biol.* **1995**, *249*, 675–690.
86. Allain, F. H.; Varani, G. *J. Mol. Biol.* **1997**, *267*, 338–351.
87. Chung, S. Y.; Subbiah, S. *Proteins* **1999**, *35*, 184–194.
88. Kleywegt, G. J.; Jones, T. A. *Methods Enzymol.* **1997**, *277*, 208–230.
89. YASARA. http://www.yasara.org (accessed May 2006).

Biographies

Robbie P Joosten studied natural science at the Radboud University, Nijmegen (the Netherlands), from 1999 to 2005. He is currently working on his PhD project at the Centre for Molecular and Biomolecular Informatics in Nijmegen, focusing on the validation and refinement of x-ray structures.

Glay Chinea studied chemistry in Hungary in the early 1980s and received his PhD in mass spectrometry after which he started working at the CIGB in Havana, Cuba. During a sabbatical at the EMBL in 1991, he started working on protein structure modeling and validation, the field he has ever since been in.

Gerard J Kleywegt obtained a degree in chemistry from the University of Leiden (the Netherlands) in 1986. In 1991, he obtained his doctorate from the University of Utrecht (the Netherlands), where he had worked under the guidance of Profs Rob Kaptein and Rolf Boelens on automating the interpretation of homonuclear 2D and 3D NMR spectra of proteins. After a short period of employment with a commercial software company, he moved to Uppsala

(Sweden) to join the laboratory of Prof Alwyn Jones. In Uppsala, he became a protein crystallographer with a penchant for methods and software development, structure validation, and structural bioinformatics. He is currently working as an independent investigator, supported by a research fellowship from the Royal Swedish Academy of Sciences. He was the coordinator of the Swedish Structural Biology Network from its inception in 1994 until 2004, and has been its programme director since 2004. He is a member of the scientific advisory boards of the MSD database at EBI and of the international wwPDB consortium, and a co-editor of the journal *Acta Crystallographica, Section F.*

Gert Vriend studied biochemistry at the University of Utrecht (the Netherlands) and got his PhD in 1983 at the Agricultural University of Wageningen (the Netherlands). His PhD project was the study of the assembly process of plant viruses using NMR, EPR, etc. He did post-doctorate from Purdue, Indiana, USA, on the x-ray structure of the common cold virus, and in Groningen, the Netherlands, where he worked on several structures while starting the WHAT IF project. Since 1989 he works at the EMBL in Heidelberg, Germany, where he continued working on (and with) the WHAT IF program. In summer 1999, he took up a position at the University of Nijmegen, the Netherlands, where he continued the projects started at the EMBL, and added several new projects, mainly in the area of data-mining. His main topics of present interest are (1) model building by homology, (2) structure verification, (3) specialized databases, and (4) application of computers in wet biology.

Comprehensive Medicinal Chemistry II
ISBN (set): 0-08-044513-6

ISBN (Volume 3) 0-08-044516-0; pp. 507–530

3.24 Problems of Protein Three-Dimensional Structures

R A Laskowski and G J Swaminathan, European Bioinformatics Institute, Wellcome Trust Genome Campus, Hinxton, Cambridge, UK

3.24.1	**Introduction**	**532**
3.24.2	**Problems with X-ray Structures**	**532**
3.24.2.1	Severe Errors	532
3.24.2.1.1	Totally incorrect structures	532
3.24.2.1.2	Topologically incorrect structures	532
3.24.2.1.3	Errors resulting from incorrect determination of the space group	533
3.24.2.1.4	Sequence-register errors	534
3.24.2.1.5	Overinterpretation of experimental data	535
3.24.2.1.6	Deliberately incorrect structures	535
3.24.2.2	Obsolete Structures	535
3.24.2.3	Errors in Current Structures	536
3.24.2.3.1	Reliability of structures improving	537
3.24.2.3.2	Significant errors that remain	537
3.24.2.3.3	Sequence-register shift errors	537
3.24.2.3.4	Reliability of the atomic coordinates	538
3.24.2.3.5	Ligand errors	538
3.24.2.3.6	Water molecules	539
3.24.2.4	Incomplete Quaternary Structure	540
3.24.3	**Problems with Nuclear Magnetic Resonance Structures**	**540**
3.24.3.1	Dealing with Ensembles of Models	541
3.24.3.2	Quality of Nuclear Magnetic Resonance Structures	542
3.24.3.3	Re-Refined Nuclear Magnetic Resonance Structures	542
3.24.3.4	Errors in Nuclear Overhauser Effect (NOE) Assignments	542
3.24.4	**Problems with Electron Microscopy (EM) Structures**	**542**
3.24.5	**Problems with Very Large Three-Dimensional Structures**	**543**
3.24.5.1	Limitations of the Old Protein Data Bank Format	543
3.24.5.2	One Structure, Several Entries	543
3.24.5.3	Virus Structures	544
3.24.6	**Problems with Homology Models**	**544**
3.24.6.1	Homology Modeling	544
3.24.6.2	Reliability of Homology Models	544
3.24.6.3	Model Databases	544
3.24.7	**Problems with Predicted Structures**	**545**
3.24.8	**Some Advice on Coping with the Problems**	**545**
3.24.8.1	Global Measures of Quality	545
3.24.8.1.1	The resolution and *R*-factor	545
3.24.8.1.2	Stereochemical parameters	546
3.24.8.2	Local Measures of Quality	546
3.24.8.2.1	Estimates of coordinate errors	546
3.24.8.2.2	Atomic *B*-factors	546
3.24.8.2.3	The uppsala electron density server (EDS)	547
3.24.8.2.4	Stereochemical checks	547
3.24.9	**Conclusion**	**548**
References		**548**

3.24.1 Introduction

As the previous chapter has demonstrated, experimentally determined protein structures are prey to a number of sources of error. The validation techniques described there are aimed at assisting the experimenter in reducing these errors as far as possible, particularly where they may be the result of misinterpretation of the data. However, even where these techniques have been followed rigorously, no three-dimensional (3D) model of a protein structure can be regarded as completely error-free; it is the result of experiment, and any experimental measurement, however precisely made, contains errors – both systematic and random. The question is: how large are these errors and what is their likely impact on the use to which the model is to be put – be it to understand the details of a particular catalytic reaction, or the recognition of a specific substrate or the binding of a candidate drug molecule?

Errors in protein models vary in their severity, from gross errors – where the model is essentially incorrect – to minor errors such as the uncertainties associated with the reported *x*-, *y*-, *z*-coordinates of each atom.

There have been a number of well-publicized cases of grossly incorrect models, and some of these will be described below. In 1996 a correspondence to *Nature*[1] suggested that there were over a million errors in the structure models deposited at the Protein Data Bank (PDB), which contained a mere 3442 structures at the time (*see* 3.17 The Research Collaboratory for Structural Bioinformatics Protein Data Bank). This was perhaps overstating the case, as was pointed out in a subsequent response.[2] The majority of the quoted errors were the result of structures refined using older dictionaries being compared against values in newer dictionaries. Nevertheless, this very public discussion made the point that the information, including protein coordinates, in the PDB cannot be relied on unthinkingly and uncritically.

In recent years, experimental methods have undergone a number of significant improvements (*see* 3.21 Protein Crystallography; 3.22 Bio-Nuclear Magnetic Resonance) and validation of structure models is now a routine part of the deposition process.[3] However, even as recently as 2004 a structure solved in 2000 was shown to contain a serious error (see below). And a paper published in 2002 suggested that about 3% of the amino acids in structures deposited at that time were incorrectly modeled (again, see below).

In other words, even recently released models may occasionally contain severe errors although such cases are exceedingly rare. However, the problems of protein 3D structures are not just a matter of errors in the deposited coordinates. There are a wide variety of problems that one needs to be aware of when looking at or using these structures. Some are specific to the method of the structure's determination, whereas others are more general. In this chapter we describe the problems and give some tips on how to deal with them.

3.24.2 Problems with X-ray Structures

Structures determined by x-ray crystallography are generally regarded as the most accurate and reliable of all experimentally determined structures, particularly when the data approaches atomic resolution. They represent around 85% of the entries in the PDB and, as the techniques have improved, so x-ray crystallography has enabled the determination of larger and larger macromolecular complexes. However, this is not to say that x-ray structures come free of problems.

3.24.2.1 Severe Errors

In the early 1990s it became clear that several published models had contained serious errors. Brändén and Jones[4] listed five examples of protein models that had been replaced by more accurate representations. Other examples came to light soon after.

3.24.2.1.1 Totally incorrect structures

The worst type of error is where the model is completely wrong. The most commonly cited example is that of photoactive yellow protein from *Ectothiorhodospira halophila*, solved in 1989.[5] The incorrect model, PDB code **1phy**, is shown in **Figure 1a** and bears practically no resemblance to the corrected model,[6] **2phy**, shown in **Figure 1b**. The folds in the two models are completely different, and when they are superposed the root-mean-squared distance (RMSD) between equivalent C_α atoms is 15 Å. This kind of error is extremely rare. In this case it resulted from poor phasing of low-resolution data (2.4 Å) leading to a gross misinterpretation of the electron density.

3.24.2.1.2 Topologically incorrect structures

A different type of severe error is where the connectivity of the chain between secondary structure elements is wrong. This is exemplified by the original model of D-alanyl-D-alanine peptidase from *Streptomyces R61*,[7] **1pte**, shown in **Figure 2a**. It was

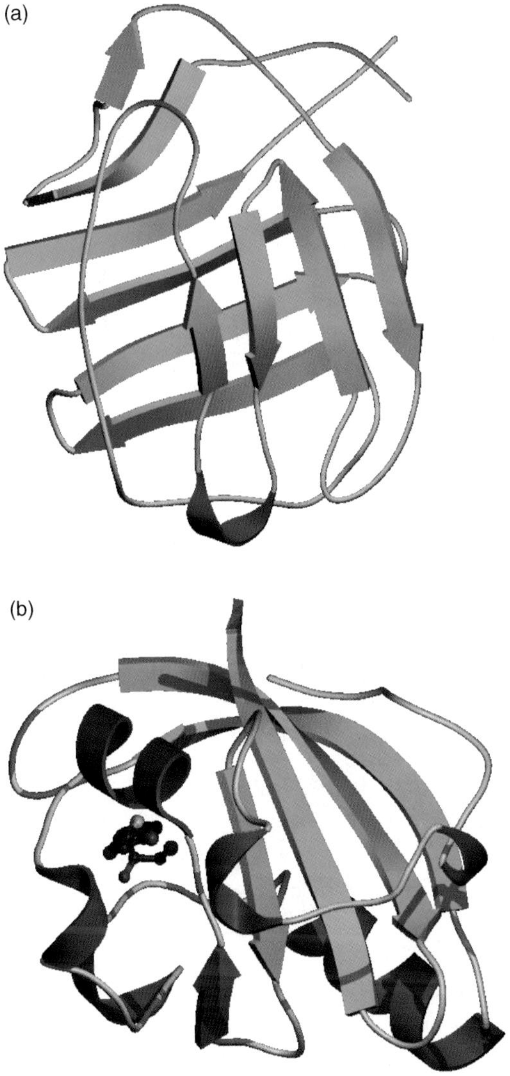

Figure 1 An example of a completely incorrect structure. (a) The incorrect model of photoactive yellow protein, PDB entry **1phy**, has a β-clam fold; (b) The corrected model, PDB entry **2phy**, has a very different, α/β fold. The two models bear practically no resemblance to one another. The first model resulted from a gross misinterpretation of the electron density due to poor phasing of low-resolution data (2.4 Å) while the second model benefited from data to 1.4 Å resolution.

solved at low resolution (2.8 Å) at a time when the protein's sequence was unknown. This had made tracing the chain through the electron density especially difficult as the sequence had to be inferred from the blobs of electron density rather than serving as a guide to the fitting known side chains to those blobs. When the structure was solved again, with the known sequence **Figure 2b** (PDB code **3pte**, resolution 1.6 Å),[8] it turned out that the chain had been completely mistraced in the earlier model and, as a consequence, the inferred sequence was also severely in error. Interestingly, most of the secondary structure elements are in the right places as these would have been the easiest to see in the electron density; conversely, the loop regions, being more flexible and disordered, would have had more poorly defined electron density and would have been more difficult to fit correctly. As a result of the mistracing (in one case the chain goes the wrong way down a β-strand), most of the protein's residues are in the wrong place in the 3D model.

3.24.2.1.3 Errors resulting from incorrect determination of the space group

Where the space group of the crystal is incorrectly determined the model of the structure can be completely wrong. Such problems usually arise when the phasing information is insufficient to discriminate between enantiomorphic space groups

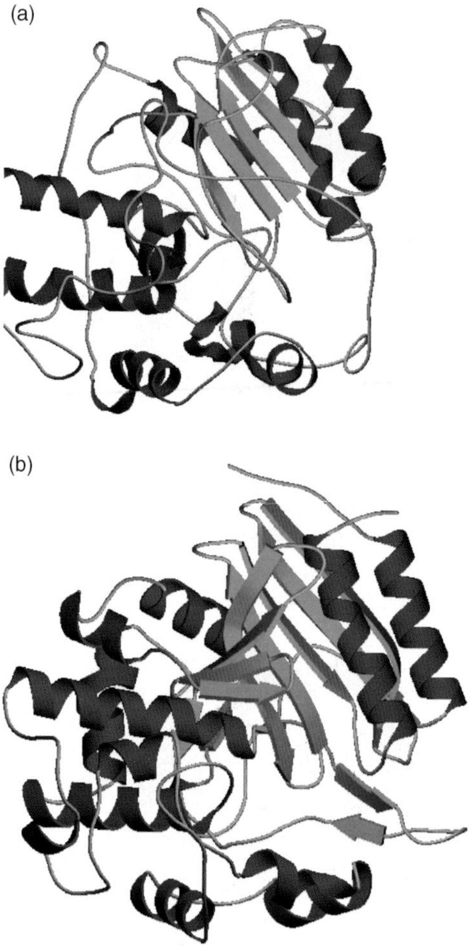

Figure 2 A topologically incorrect model. (a) The initial model of D-alanyl-D-alanine peptidase determined at low resolution, PDB entry **1pte**; (b) The improved model at higher resolution, PDB entry **3pte**. By comparing the α-helices (colored red) and β-sheets (colored green) in the two models it can be seen that the various secondary structure elements are fairly similar, but the connectivity between the various elements differs. As a consequence, one of the strands in the incorrect model has been built in the wrong direction and most of the residues in the incorrect model are in totally wrong positions.

such as $P4_1$ and $P4_3$ or $P4_12_12$ and $P4_32_12$. One example is that of chloromuconate cycloisomerase, PDB entry **1chr**. The crystal structure of this enzyme was first reported at 3 Å resolution in space group *I*4.[9] The crystallographic *R*-factor for the refined model and the Ramachandran plot statistics were also reasonable for a model built on such low-resolution data. However, a re-evaluation of the same structure using the original data deposited with the PDB suggested that the crystal actually belonged to space group *I*422. Rebuilding the model based on the new space group identified a 25-residue stretch at the N-terminal end of the protein which had been built out of register (see the following section). With these errors corrected, the crystallographic *R*-factor dropped from 19.5% to 18.9% in the new model (PDB entry **2chr**)[10] while the number of residues in the disallowed regions of the Ramachandran plot dropped from 10 to 1.

3.24.2.1.4 Sequence-register errors

Occasionally, particularly at low resolution, a residue might be placed into the electron density belonging to the next residue. This most frequently occurs at loops in the structure and results in a sequence-register shift error, which persists until a compensating error is made further down the chain when two residues are fitted into the density belonging to a single residue.[11] Such errors can often result in a significant part of the model being incorrect and, if in a region also containing the active site, have potentially serious implications in understanding the molecular function of a protein. Various techniques for identifying frameshift errors during model building and refinement have been documented in the literature[12,13] but once they make their way into the deposited coordinates are difficult to pick up without access to the original experimental data.

One such error was recently picked up purely by chance.[14] It was a 2-residue shift in chain M of PDB entry **1kc9** – a 3.1 Å resolution model of the large ribosomal subunit from *Deinococcus radiodurans*. Several of this structure's 30 chains were being used as part of the LiveBench[15] protein structure prediction benchmarking experiment which tests the ability of fold recognition servers to identify the fold of any given protein sequence. The structure's sequences had been made public for use in this test, but the 3D model itself had not yet been released. All the servers correctly identified the sequence of chain M as having the same fold as chain M of PDB entry **1jj2** and all generated a model of the structure of chain M based on their match to this structure. However, when **1kc9** was released it was found that the predicted models of chain M consistently differed from the experimentally determined model. This unusual result prompted a check against the experimental data by the structure's depositors and revealed the register error in chain M. Indeed, other chains in the model also exhibited serious sequence register errors. A new version of the structure was deposited as PDB entry **1kpj**, which itself was subsequently superseded first by entry **1lnr** and then by **1nkw**.

3.24.2.1.5 Overinterpretation of experimental data

When the experimental data are of poor quality, or are ambiguous, there is a danger of overinterpretation; this often leads to a part of the model being wrong. One fairly recent example concerns the binding of a peptide to botulinum neurotoxin B light chain (BoNT/B). A model of this protein, complexed with a 38-residue peptide of its natural substrate synaptobrevin II, was reported at 2.0 Å resolution in the year 2000 (PDB entry **1f83**).[16] It was proposed that the binding of the peptide revealed how the full-length synaptobrevin II binds to the protein and is catalyzed. However, questions were raised about the validity of the peptide's binding as the average *B*-factors of its atoms were over four times the average for the rest of the structure – see **Figure 3a**.[17] Furthermore, very few of its residues were in favorable regions of the Ramachandran plot and a re-examination of the electron density maps showed there was actually little evidence for the peptide being in the proposed binding site.

In response, the authors of the original model explained these observations by suggesting they were due to the peptide's low occupancy and high flexibility.[18] However, when a model of the related botulinum neurotoxin A light chain (BoNT/A) in a product-bound state was published (PDB entry **1e1h**),[19] it suggested the mode of substrate binding to be markedly different from that predicted by the initial BoNT/B model (**Figure 3b**). Recently, an independent experimental determination of this second protein has confirmed this mode of binding (PDB entry **1xtg**).[20]

3.24.2.1.6 Deliberately incorrect structures

In this final section on severe errors we describe structure models that are deliberately incorrect. In fact, we know of only two such models – for why would anyone deliberately and with premeditation misinterpret their experimental data to produce a totally erroneous protein model?

The reason, in both cases, was to prove a point about the unreliability of the structure validation criteria in use at the time (the early 1990s), specifically the *R*-factor which was the key measure of the agreement between the final model and the experimental data. Both structures were deliberately traced backwards through the electron density. The first was cellular retinoic-acid-binding protein type II (CRABP) complexed with a bound all-*trans*-retinoic acid and the person making the point was Gerard Kleywegt.[21] Despite the model being obviously totally wrong, the final *R*-factor attained was a perfectly respectable 0.214 using data to 2.9 Å resolution. Had the model been submitted to the PDB it would probably have been accepted. Tellingly, other measures give the game away. The R_{free} is a pretty woeful 0.617, and the Ramachandran plot has 11 residues in the disallowed regions and only 53 residues (43%) in the most favored regions.

Although never submitted to the PDB, the erroneous model can be seen in the PDBsum database[22] under PDB id **fake**. Even the secondary structure looks unusual in that only 12 of the protein's 137 residues (8.76%) are in regular secondary structures whereas one typically finds 50–60% of the residues to be in a helix or strand – except in very small proteins. Surprisingly, and perhaps worryingly, the PDB still contains protein structures that look in even worse shape than **fake** does – although we will not name and shame them here.

The second backward-traced structure was of α_{2U}-globulin,[23] the purpose here being to test the use of R_{free} when the crystal structure exhibits noncrystallographic symmetry.

3.24.2.2 Obsolete Structures

Cases of severe errors such as those described above are rare. Many of the erroneous models have been withdrawn from the PDB but can still be found in the Archive of Obsolete PDB Entries.[24] This holds superseded and withdrawn models and, as of July 2005, contained 1516 superseded structures, of which 799 had been withdrawn from the PDB.

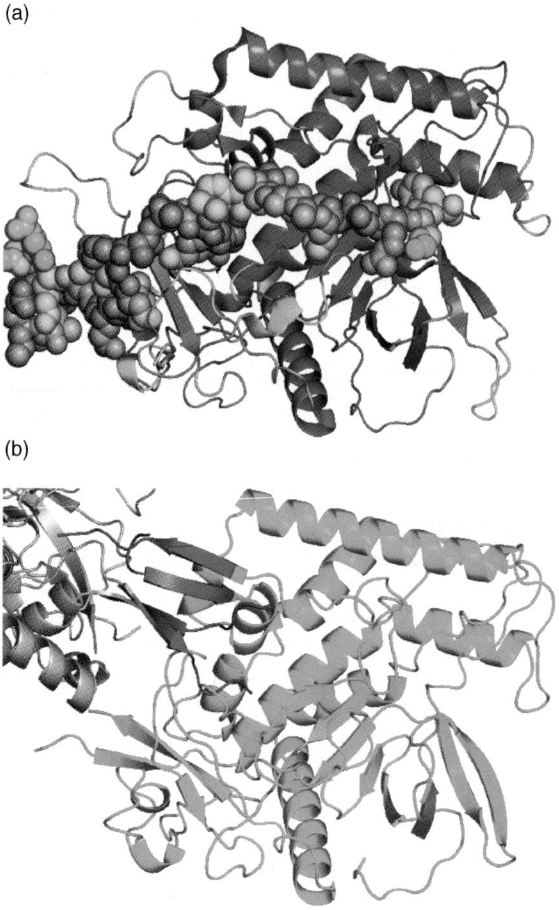

Figure 3 An example of overinterpretation of data. (a) A model of neurotoxin type B from *Clostridium botulinum* (BoNT/B) in complex with a 38-residue peptide of its natural substrate synaptobrevin II, PDB entry **1f83**. The model is colored by *B*-factor (blue lowest, via orange, to green highest), showing that the peptide, with its greens and oranges, has considerably higher *B*-factors than the rest of the protein – which is largely blue. This strongly hints at the peptide being poorly defined at this location. (b) A model of the related botulinum neurotoxin type A, PDB entry **1e1 h**. The model consists of two molecules, one colored green and the other, off the screen, colored gray. The green molecule is shown in the same orientation as the model in a. The protein is self-cleaving. The red loops – one from the green molecule and the other from the gray one – are where the cleavage has occurred. The breaks in the loops can be clearly seen. This model strongly suggests that the location of the active site is here, rather than the location proposed by the depositors of the model in a.

The archive's website provides a graphical history of each model, some of which have gone through several incarnations (e.g., PDB entry **1fd1** was successively replaced by **2fd1**, **3fd1**, **4fd1**, and **5fd1** over a period of 12 years between 1981 and 1993). Note that not all superseded models contain serious structural errors; in many cases PDB entries have become obsolete simply due to the availability of better-quality data for the same protein.

3.24.2.3 Errors in Current Structures

Most of the really serious errors, described above, are largely a thing of the past as experimental and computational techniques have improved sufficiently to make such errors less likely. The models identified as being seriously in error have now either been superseded or withdrawn from the PDB. Of course, the PDB contains many vintage models and it is possible that there still lurk among them ones that have serious errors that are yet to be detected. But the vast majority of experimentally determined structures in the PDB are essentially correct. The remaining errors will largely consist of the random errors associated with any experimental measurement, or errors of the kind detailed in the previous chapter: bent or flipped side chains, mislabeled atoms, unfeasible bonds, clashing atoms, unusual torsion angles, misplaced waters, mispuckered prolines, and so on. Such errors, where they occur in biologically important regions of the structure, can have a significant impact on, say, the biochemical interpretation of the protein's function or on attempts to design a viable drug molecule.

So how reliable are the models in the current PDB and are newly released models more sound than those of, say, 10 or 20 years ago?

3.24.2.3.1 Reliability of structures improving

An analysis in 1999 of all models in the PDB at the time[25] suggested that the overall quality of the models did appear to be improving with date of deposition; more recent models were deemed more reliable than older ones. This was put down to: firstly, improvements in crystallization and data collection techniques, higher data quality and consequently better-resolved structures, and secondly, the use of more accurate target values during refinement and good validation programs to check on final geometry.

A different view was put forward by a more recent analysis which claimed that the average quality of crystal structures, as measured by analyzing their Ramachandran plots, did not show an improvement with time for the period 1991 to 2000. On the other hand, it did appear that the fit of the models to the experimental data had improved over that period, as evidenced by a decrease in the discrepancy between the R_{free} and R-factor.[26]

3.24.2.3.2 Significant errors that remain

Another recent analysis[27] suggested that significant errors that are relatively simple to detect and fix still remain in models newly deposited at the PDB. The analysis used a set of global and local quality control criteria incorporating checks between the model and the computed electron density as well as stereochemical quality indicators. The checks were applied to a small set of 28 x-ray crystal structures, being the models released by the PDB on a single day (6 March 2001) for which experimental data were available. The types of errors found included residues in the disallowed regions of the Ramachandran plot, severe deviations of bond lengths and bond angles from their ideal values, His/Asn/Gln side chain flips, and main chain or side chain atoms with low correlation to regions of high electron density. The authors found the average error rate (i.e., number of amino acids with probable errors) to be 3.6%, with the 'worst' of the models having an error rate of 10.6%.

The majority of the errors were His/Asn/Gln side chain flips – which are easy to detect and rectify. These side chains are symmetrical in terms of shape and so fit their electron density equally well when rotated by 180°. However, they are not symmetrical when the chemical nature of the side chain atoms is taken into account. The correct orientation is usually taken to be the one that optimizes the local hydrogen-bonding network, as conveniently computed by WHAT_CHECK[28] (*see* 3.23 Protein Three-Dimensional Structure Validation). And this orientation can have a crucial bearing on the interactions the side chain can make, particularly if it happens to be in a site of interest (e.g., catalytic or binding site).

The second most common type of error were poor fits between the main chain or side chain atoms and the electron density at those locations. Mostly these were disordered, rather than misplaced, side chains for which the density was poor in any case. Such limitations of the experimental data are quite common, but are treated in different ways by different crystallographers. In some cases, the coordinates of the side chains or segments of the main chain for which there is no, or very little, electron density are omitted entirely from the PDB entry. In other cases, the side chains, or indeed whole residues, for which essentially no electron density is visible are assigned plausible coordinates in the PDB file but flagged as having 'zero occupancy'. When using such a model one needs to be particularly vigilant to be aware that certain parts have no experimental basis whatsoever. In the situation where such 'invisible' residues are omitted from the coordinate files the gaps are easy to spot when viewing or using the model.

3.24.2.3.3 Sequence-register shift errors

A second recent study focused on sequence-register shift errors and discovered that these types of error, already described above, are far from being a thing of the past. The study was limited to a particular class of protein structure: the oligosaccharide-binding (OB)-fold,[29] but could equally have been any other fold. The OB-fold is characterized by five antiparallel β-strands forming a closed or partly open barrel.

The principal method used for finding the errors was ProsaII[30] which computes the 'energy' of local regions within the model using empirically derived potentials of mean force. It can also provide an energy for the model as a whole. The authors of the study found that 12 (1.4%) of the 842 protein chains they looked at contained sequence-register shift errors. The 12 originated from five separate PDB entries as some of the entries were structures of multimers. The incorrect models ranged in resolution from 2.2 to 3.2 Å, with deposition dates of 1983 to 2000. The authors suggest that, on the basis of their analysis, it is reasonable to estimate that around 1% of protein sequences in the PDB could contain such errors.

3.24.2.3.4 Reliability of the atomic coordinates

Another recent study concluded that the errors in atomic coordinates in x-ray crystal structures are actually somewhat larger than is generally appreciated.[31] This can have a significant impact in any analysis of the atomic positions.

The study involved automatically re-refining three models in the PDB (resolutions 1.3, 1.8, and 2.3 Å). Crucially, each was refined between 10 and 20 times, giving 10 to 20 different final conformers. The net result, therefore, was an ensemble of models, each consistent with the experimental data – much as one gets in nuclear magnetic resonance (NMR) spectroscopy. Then, from each ensemble were selected the five models giving the lowest R_{free}. The variation in the atomic coordinates across the five-model ensembles provided an estimate of the positional uncertainties in the structures. The authors found that the RMSDs of the main chain atoms varied between 0.1 and 0.5 Å while for all protein atoms this variation was 0.3–0.75 Å. This is higher than usual error estimates give. For example, all-atom errors are traditionally estimated by theoretical methods such as the Luzzati plot, or Read's σ_A plot (*see* 3.21 Protein Crystallography), and are typically in the range 0.1–0.3 Å, much lower than the errors found here.

One particularly interesting aspect of the study was the authors' conclusions as to why the errors should be so large. Protein molecules are not the static, rigid entities that one finds in the beautiful illustrations in the literature. Rather, any single molecule will exhibit a number of large-scale motions over a range of timescales, while its individual atoms will jiggle and wiggle with varying anisotropic thermal motions.[32] This dynamic nature is only approximated in experimentally determined protein structures. In crystal structures, the dynamic motion, together with the structural heterogeneity that comes from differences in conformation from one location in the crystal to another, are modeled by the individual atoms' isotropic *B*-factors and occupancies. Only at atomic resolution is there sufficient data to allow anisotropic atomic motions to be modeled by anisotropic *B*-factors. DePristo *et al.* suggest that this deficiency in modeling the full heterogeneity of protein molecules leads to the multiplicity of quite different solutions in their study.[31] In other words, no single conformation with isotropic *B*-factors is able to explain the data sufficiently, and other conformations may be equally consistent with the experimental data. Thus, particularly at low resolutions, it might be more appropriate to deposit ensembles of models, much as is done for NMR structures.

3.24.2.3.5 Ligand errors

As of July 2005, over 18 000 of the 32 000 structures in the PDB were of proteins with small molecule ligands bound. The ligands spanned a variety of types, including biological substrates, enzyme cofactors, transition state analogs, protein inhibitors, or merely molecules from the buffer solution used in the experimental determination of the structure. The molecules covered nearly 7000 different chemical entities. A protein structure with bound ligand can provide information on how the protein interacts with the bound, or similar, molecule and, in many cases, can reveal how the protein performs its biological function. In drug design, where a lead compound needs to be optimized to improve its binding, the determination of the structure with bound ligand can help explain how the binding is achieved, the factors determining the strength and specificity of binding, and how the molecule might be modified to alter these.

Yet, despite their smaller size, the models of the small molecules in the PDB are often less reliable than those of the proteins or nucleic acids to which they are bound.[33] The primary reason for this is that the standard geometrical libraries used for restraining bond lengths and bond angles during structure refinement have been fine-tuned over the years for protein and nucleic acid structures. Indeed, as there are only 20 amino acid types and four nucleotides, there is a standard set of target values for their geometrical and stereochemical parameters. In contrast, the small molecules that can bind to these macromolecules come in a wide range of sizes and chemical types and so no set of 'ideal' bond lengths and angles is likely to be known for any given molecule. Often, these need to be individually compiled for each new molecule, taking into account its atom types and bond orders, using tables derived from crystal structures of small molecules.[34,35] Because a thorough understanding of each molecule's chemistry is required, errors do occur.

3.24.2.3.5.1 Errors in target values

An example of the types of errors one gets is provided by a recent analysis of bond lengths, bond angles, and torsion angles, in 139 adenosine triphosphate (ATP) molecules in the PDB.[33] The analysis showed that the variation in bond lengths of the ATP molecules was around 0.2–0.3 Å, while some torsion angles varied by as much as 60° across the models. These values were much larger than necessary and suggest use of inappropriate target values during refinement – and indeed some of the values of the improper torsion angles revealed that, for some of the models, some chemically important restraints must have been omitted, been far too weak or had incorrect target values.[33]

3.24.2.3.5.2 Doubtful ligand structures

Another picture of the quality of ligands in the PDB is given by a study that was not specifically looking for errors.[36] Its aim was to derive a large, diverse test set of reliable protein–ligand complexes as a data set for use in validating ligand docking programs. To ensure reliability of the complexes, the ligands were subjected to several checks to identify those that were unsuitable for the test data set. The starting set consisted of 305 complexes. Of these, 61 (i.e., 20%) were rejected because the ligands failed one or more of the checks. Most of these (38 out of 61) failed the check for clashes with crystallographically related chains. Around one-third (21 of the 61) exhibited clashes between the protein and ligand atoms. In two cases the ligand had been incorrectly defined in the PDB file and in a further two the ligand had dubious geometry. The final check was one to ensure that the ligand structure was consistent with the electron density computed from the deposited diffraction data. This check could only be made for 70 of the 305 entries as only for these were structure factors available. In these 70 there were six cases where the placement of the ligand could not be reconciled with the electron density map.

A separate, but particularly striking, example of a ligand's location not matching the experimental data is where the authors of a crystal structure re-examined their model after it had been published and decided that the original placement of the ligand in the model could not be supported by the electron density. Consequently, they withdrew their structural conclusions in a short correction to the original paper.[37] However, the model, complete with bound ligand, remains unmodified in the PDB (entry **1fqh**).

Even where a ligand sits beautifully in the electron density there may still be crucial uncertainties concerning its chemistry. If the ligand has any branches that are geometrically symmetrical but chemically disparate – like the protein side chains His, Asn, and Gln – the correct conformation of these branches cannot be inferred from the data unless, perhaps, the data have been collected to atomic resolution. If the ligand contains any tautomeric groups, determination of its tautomeric state requires that the hydrogen atoms be resolved in electron density maps – which is not generally possible. Similarly, the data provide no information on the state of ionization of a ligand or protein as the pK_a of side chains may be strongly influenced by the local microenvironment in which they find themselves.

3.24.2.3.5.3 Inconsistencies in ligand nomenclature

Quite apart from being of doubtful quality, many ligands in the PDB cause problems as a result of inconsistent nomenclature. Even in such commonplace ligands as ATP the atom names of equivalent atoms may differ from one PDB entry to another. A rarer cause of trouble, but equally inconveniencing, is in the naming of the ligands themselves. They are assigned three-character 'residue names'. Unfortunately, the names are not always unique and there are over 60 instances where the same name has been used for totally different chemical entities. One example is MAL which corresponds to maltose ($C_{12}H_{22}O_{11}$) in 44 PDB entries, malonate in four PDB entries (where its formula is given as $C_3H_4O_4$ – which is actually malonic acid), and L-malate ($C_4H_4O_5$) in another four PDB entries. The Research Collaboratory for Structural Bioinformatics (RCSB) is making efforts to address such inconsistencies in the PDB data.[38]

Another problem is where the ligand is assigned the wrong name. A recent study of carbohydrate structures in the PDB found that about 30% of these contained errors, the most common being the misassignment of the α-/β-isoforms. Other errors included missing or surplus atoms or connections and interchanged coordinates.[39]

3.24.2.3.6 **Water molecules**

Water plays a crucial role in all living systems, and globular proteins are, for the most part, immersed in an aqueous solution. Whenever a ligand binds to a protein it usually needs to displace one or more waters at the site of binding, and this will govern the strength and favorability of the binding. It is thus important to know where on the protein's surface the waters are likely to be found. In crystal structures many waters are well ordered – i.e., consistently bind at the same site on the protein's surface throughout the crystal. Where they are not well ordered, the electron density of the water positions becomes difficult to distinguish from noise and there is a degree of subjectivity in identifying which density corresponds to water molecules and which does not.

In other words, if two different crystallographers are given the same data from a diffraction experiment, they may agree on the majority of the water positions, but are likely to disagree on a significant minority. This assertion has been put to the test[40] when the same synchrotron data for poplar leaf plastocyanin to 1.6Å resolution were refined independently in two separate laboratories. The final models contained 171 and 189 water molecules, of which 159 were common to both.

The danger of waters is when they are overzealously added to a model – as has happened on a number of occasions. This can hide errors in either the experimental data or in other parts of the model.[21,41] Each extra water adds four adjustable parameters to the refinement (the x-, y-, and z-coordinates of the water molecule, plus its temperature factor). These extra parameters can help artificially reduce the apparent discrepancy between the observed and calculated structure factors, as measured by the R-factor, and make a poor model appear better than it is.

3.24.2.4 Incomplete Quaternary Structure

One aspect of x-ray structures not often appreciated is that the deposited coordinates are not necessarily of the biologically relevant unit. So, for example, what may look like a monomer from the PDB entry, is, in real life, a dimer. Indeed it may also be a dimer in the crystal itself, but only the coordinates of one molecule will have been deposited; the other member of the dimer needs to be generated using the appropriate crystallographic symmetry transformations. The reason for this is that the data from an x-ray crystallographic experiment provide information on what is in the crystal's asymmetric unit. If the molecule forms a symmetrical dimer, then the asymmetric unit is defined by a single copy of the molecule rather than the more biologically meaningful pair.

Similarly this is the case for trimers, tetramers, and so on, right up to large multiprotein complexes such as the virus coat proteins (which will be mentioned later). In each of these cases, the coordinates of the monomer may be all that is deposited. Conversely, proteins that act as monomers in vivo, may pack together in the crystal in pairs or triplets, etc., so that the deposited coordinates are of a nonbiological dimer, trimer, and so on.

Thus the PDB entry can often be most misleading because it does not show the true quaternary structure. Sometimes this is fairly obvious because a protein looks rather odd in some way; for example, the model of aspartate aminotransferase (PDB entry **1aaw**) has a fairly odd-looking protrusion coming out of it. When the biological unit – in this case a dimer – is regenerated, it becomes clear that the protrusion wraps around the other protein molecule of the dimer, which, in turn, has an identical protrusion interacting with it. What is more, because the symmetry-related partner is absent in the PDB entry, the ligand that binds between the two molecules is apparently missing some of its interactions – a crucial issue when the binding of a ligand molecule to the protein is being studied. Similarly, for the active site. In **1aaw** the active site residues are reported as being Lys258, Arg292, and Arg386. However, if you view the model you will find that the Arg292 residue appears to be on completely the wrong side of the protein, far from the site where the structure's ligand is bound. This peculiarity is explained as soon as the full dimer is generated: the Arg292 complements the Lys258 and Arg386 active site residues on the other protein in the dimer while that protein's Arg292 reciprocates by complementing the Lys258 and Arg386 residues for this one.

A second example is that of PDB entry **1buu** which is shown in **Figure 4**. Again, the model as deposited in the PDB looks somewhat odd with a single, and quite long, helix protruding from the rest of the structure (**Figure 4a**). In this case the biological unit is a trimer (**Figure 4b**), with the protruding helices packing together to form a coiled-coil which is inserted into a cellular membrane, the remainder of the structure forming a lectin domain that recognizes carbohydrates.

Even being aware of the problem of incomplete quaternary structure may not always help as, for many proteins, the true biological unit is not known. Happily, there are computational methods for determining what the most likely unit is. These work by applying the crystal symmetry operations to the molecules in the asymmetric unit and generating all symmetry-related copies surrounding it. Then, the resultant protein–protein interfaces and interactions across them are examined. From the quantity and nature of these interactions it is often possible to tell which interfaces are likely to be the true, biological interfaces and which are merely artifacts of crystal packing.

The Protein Quaternary Structure (PQS) server[42,43] performs such analyses for all x-ray structures deposited in the PDB. It suggests the most likely quaternary structure and provides the complete coordinates of this model. For any serious analysis use of the true quaternary structure of a protein, or protein complex, is preferable as otherwise crucial information relating to the protein's function may be lost. It should be borne in mind, however, that PQS is not always correct. In most cases it gives the right answer, but in some it does not.

For new models, not in the PDB, the PITA (Protein InTerfaces and Assemblies) program performs a similar analysis.[44] Its web server[45] takes a standard PDB file, including the crystallographic symmetry operators, and suggests the most likely biological unit for the macromolecules in it. Another site, performing a similar service, is PISA (Protein Interfaces, Surfaces and Assemblies).[46]

3.24.3 Problems with Nuclear Magnetic Resonance Structures

The models of biomolecules obtained by NMR spectroscopy are quite different in nature from those determined by x-ray crystallography and present their own set of problems. They represent a significant and important fraction of the models in the PDB; numerically, as of July 2005, they accounted for around 15% of the models in the data bank, covering about the same percentage of the unique sequences in the database. Thus many unique and important proteins have only been solved by NMR. Furthermore, proteins in solution, as determined by NMR, can show quite large conformational differences from their crystal form, as determined by crystallography,[47] and so can provide additional information about mobility.

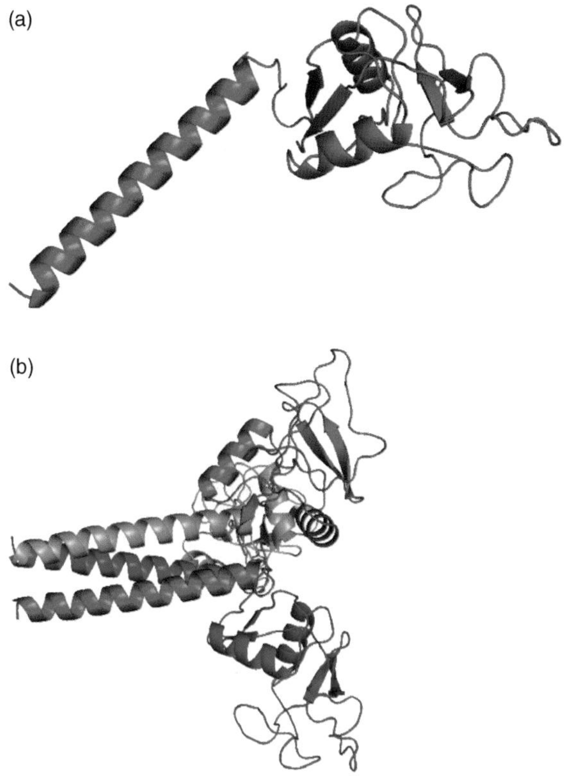

Figure 4 Incomplete quaternary structure. (a) The model of rat mannose-binding protein, as deposited in PDB entry **1buu**, shows a long α-helix apparently suspended in space; (b) in real life this protein forms a trimer, as supported by other experimental evidence. Here the biological trimer has been generated by the Protein Quaternary Structure (PQS) server from the coordinates of the monomer and the appropriate crystallographic symmetry operators. This protein is involved in the opsonization of bacterial cell surfaces with the prong formed by the coiled-coil helices penetrating the bacterial cell wall.

Inevitably, there is some overlap between crystal and NMR structures, which has provided a means of comparing the models obtained by the two methods. In fact, there are currently about 300 proteins solved by both x-ray crystallography and NMR spectroscopy (in some case multiple times by one or other technique). The PDB provides a handy list of these.[48]

3.24.3.1 Dealing with Ensembles of Models

The first problem with using structures solved by NMR is that they are usually deposited as an ensemble of models, all of which are largely consistent with the experimental data. Typically the ensemble comprises around 20 models, but sometimes there can be as many as 80. Often the models can exhibit quite a wide range of variability among themselves, especially in poorly defined or particularly flexible regions of the structure. The problem, therefore, is how to deal with an ensemble of models rather than a single one.

Usually the experimentalists will identify one of the models as being the 'representative' of the whole ensemble, in which case it is convenient to use this representative in any analytical work – although of course the information about the variability of the structure, inherent in the other members of the ensemble, is thus lost. There is no universally agreed method of obtaining an ensemble representative. It may be the model that differs the least from all others; or it may be an 'average, energy minimized' model obtained by calculating an average set of coordinates from all members of the ensemble and then energy minimizing to counteract the unphysical bond lengths and angles that the averaging process introduces. In many cases, the average model is deposited as one PDB entry and the ensemble as another – so the same structure, or rather the outcome of the same experiment, appears as two separate entries in the PDB!

An alternative way of selecting a representative is provided by Olderado (On Line Database of Ensemble Representatives and Domains).[49,50] This identifies the most representative model as follows: it first locates the rigid body domains in the structure, then superposes all models on the largest domain, and finally clusters the models according to their structural similarity. The model closest to the centroid of the largest of the clusters is taken as the most representative of the whole ensemble.

3.24.3.2 Quality of Nuclear Magnetic Resonance Structures

The second typical problem with using NMR structures is concern over their likely quality and reliability. NMR structures have always been regarded as the poor cousins of crystal structures. A number of comparisons between NMR and crystal structures have fuelled such concerns by showing that the former tend to be of a lower quality than the latter.[51–55] The reason for NMR structures generally faring poorly in such comparisons is that the structural information contained in the data is poorer than that given by x-ray diffraction data. Recently, the information content of NMR data has been supplemented by new experimental techniques and improved methods of spectrum interpretation,[56] so more recent models are likely to be more accurate than older ones.

However, this is not easy to prove as there are still no generally accepted criteria for defining the precision and accuracy of a model derived from NMR data.[57] A recent study concluded that the precision of an NMR structure normally exceeds its accuracy.[58] This is as a result of current procedures in structure calculation and selection of representative conformations in the NMR ensemble. When the best 20 or 30 models that have the lowest overall energy and overall intermodel RMSD are chosen, other models that fulfill the accuracy criteria might be overlooked for the sake of precision. There is now a concerted move in the NMR community to establish community-wide standards for defining accuracy and precision of NMR-derived structures.[57] Thus, for the user of an NMR structure, the easiest indicators to rely on are the stereochemical checks described in the previous chapter and available for every PDB entry.

3.24.3.3 Re-Refined Nuclear Magnetic Resonance Structures

Recently, it has been shown that many of the early models can be significantly improved by re-refining them using more up-to-date force fields and refinement protocols. Indeed, there are two web servers giving access to the re-refined coordinates of many early NMR structures. The first is DRESS, a Database of Refined Solution NMR Structures,[59] which, as of July 2005, contained 100 structures.[60] As well as giving the new coordinates, the database shows indicators of how the newer models are an improvement over the old in terms of stereochemical quality measures.

The second database, RECOORD,[61,62] contained 545 re-refined NMR structures as of July 2005. It, too, shows the improvements achieved by the re-refinement procedure. In general, the new models tend to be more accurate but less precise than their earlier counterparts. Thus if the structure you are interested in is an early NMR structure, it is worth checking whether it is included in either of these two databases.

3.24.3.4 Errors in Nuclear Overhauser Effect (NOE) Assignments

One potential problem, even with the re-refined models described above, is the nature of the experimental data deposited with the PDB. NMR experimentalists deposit the distance and dihedral restraints they used in solving their structure (and they do so in a wide variety of formats). However, these do not constitute the original experimental measurements in the same way that the structure factors in x-ray crystallography do. The restraints result from assignment of the experimentally observed nuclear overhauser effects (NOEs), and are thus subject to errors of assignment. Even a single error in the NOE assignments can lead to a serious error in the resultant model. One example is given by PDB entry **1stw**, now obsolete. This model of a protein–DNA complex incorrectly showed a tryptophan residue intercalating between two bases of the DNA. When the error was detected and corrected (PDB entry **2stw**) the orientation of the protein in the complex was found to be opposite to that originally reported and the tryptophan no longer lay between the DNA bases.[63] A recent study has suggested that at least 25% of NMR structures in the PDB have serious NOE assignment errors, whereas over 40% have at least one assignment error.[64]

3.24.4 Problems with Electron Microscopy (EM) Structures

As of July 2005 the PDB contained 97 structure models from electron microscopy (EM) experiments. Perhaps the most obvious problem with these models is their low resolution – typically in the range of 6–15 Å – well below the limits where individual side chains, let alone atoms, can be discerned. Despite this, the PDB entries do contain atomic coordinates. These either consist of the C^{α} backbone atoms only, or come from the coordinates of one or more crystal structures solved at higher resolution which have been fitted into the EM map volumes. The coordinates are, at best, an approximation of the atomic positions of various components making up the sample and are not subject to any rigorous positional or geometric refinement procedures. They should, therefore, be treated with extreme scepticism.

3.24.5 Problems with Very Large Three-Dimensional Structures

As the technology of solving protein structures has improved it has become possible to determine the structures of larger and larger proteins and protein complexes. There are a number of very large structures in the PDB such as viruses, the ribosome, large multienzyme complexes, chaperonins, and structural protein assemblies (*see* 3.17 The Research Collaboratory for Structural Bioinformatics Protein Data Bank).[65] However, these very large structures have brought with them their own unusual problems.

3.24.5.1 Limitations of the Old Protein Data Bank Format

The main problem of very large structures stems from the limitations of the PDB file format, devised when the PDB was established in 1971.[66] The format was based on the 80-column punched cards that were then in common use for storing data and programs, with each data item having a fixed size and location on the card and all text being in upper case.

As larger and larger structures have been solved, so the shortcomings of this format have become manifest. The most serious limitations are that each structure can have a maximum of 99 999 atoms, each chain can have no more than 9999 residues, and there can be no more than 26 chains in all (these are identified by the letters A to Z). In fact, the chain maximum has now been extended to 62 by additionally allowing the numbers 0–9 and the lower case letters *a*–*z*, although some structures have plundered the top row of the keyboard and resorted to the punctuation marks: entry **1gav** has chains labeled !, #, _, -, =, :, <, >, and |.

Internally, the RCSB,[67] who act as keepers of the PDB archive, use a format called the macromolecular Crystallographic Information File (mmCIF) which does not suffer from these restrictions.[68] Indeed, PDB structures can be downloaded in this format, or in XML, as can software for reading and interpreting it.[69] However, the old format is still very popular on account of its simplicity; it is easy to read, edit, and modify using a text editor or via simple operating system commands. What is more, a huge number of computer programs in everyday use rely on the old format. So switching to the new one would be a major rewriting exercise involving a significant part of the bioinformatics community.

3.24.5.2 One Structure, Several Entries

How, then, are the really large protein and protein/RNA complexes handled within the bounds of this restrictive file format? The answer is that the structures are split across two or more PDB entries. For example, the model of the BAFF/BAFF-R complex contains 120 protein chains in total. It has been split across two PDB entries: **1otz** containing the 60 BAFF (B-cell activating factor) chains and **1p0t** containing the 60 BAFF-R (B-cell activating factor) chains. So to view or analyze the whole complex requires first recombining these two entries, and second finding software that can handle the resultant file. As of July 2005 there were over 20 such split structures.

The situation is even worse for NMR and EM structures. NMR structures, as already mentioned, are often deposited as two separate PDB entries: in one is the ensemble of models and in the other a single representative model. For large structures, or large ensembles, the entry for the ensemble has to be split into two or more separate entries to overcome the atom numbering restriction. So there are cases where three or more PDB entries correspond to a single structure determination. One such is the N-terminal domain of enzyme I from *Escherichia coli*. Its energy minimized average model is stored as PDB entry **1eza**, while entries **1ezb**, **1ezc**, and **1ezd** hold the 50 models that constitute its ensemble.

As if this were not enough, one structure determined by EM has been deposited as 11 separate, and somewhat eccentrically named, entries: **1m8q**, **1mvw**, **1o18**, **1o19**, **1o1a**, **1o1b**, **1o1c**, **1o1d**, **1o1e**, **1o1f**, and **1o1g**. Each holds a model of the myosin–actin complex from insect flight muscles. The models were derived by fitting multiple copies of the model of the atomic structure of myosin (PDB entry **2mys**, resolution 2.8 Å) and of actin (PDB entry **1atn**, resolution 2.8 Å) into the maps of the muscle protein complex obtained by electron tomography (resolution 70 Å).

3.24.5.3 Virus Structures

A different set of problems relate to models of virus capsids.[65] These are made up of symmetrical assemblies of repetitions of one or more proteins. The PDB stores the coordinates of the smallest unit, together with the symmetry operators that will regenerate the entire protein coat from it. Typically, the coats have icosahedral symmetry, so 60 copies of the basic unit need to be generated.

However, there are different conventions and complications associated with the representation of the symmetry operators and this can sometimes cause errors when the full virus particle is generated.[65] Many of these problems are resolved in the VIPER (Virus Particle Explorer) database[70,71] which has been specifically set up for simplifying and standardizing the generation of complete icosahedral virus particles.

3.24.6 Problems with Homology Models

3.24.6.1 Homology Modeling

Although seemingly plentiful, the known protein structures constitute a mere drop in the ocean of all known biological proteins. As of July 2005, the PDB held around 32 000 biomolecular structures containing one or more protein chains, corresponding to around 8000 nonredundant (i.e., less than 90% sequence identical) protein sequences. This was but 0.4% of the 2 million protein sequences in the UniProt/SwissProt and UniProt/TrEMBL databases[72] at the same date. Thus, for 99.6% of the currently sequenced proteins there is no 3D model available.

However, it is known that proteins with very similar sequences have very similar 3D structures; specifically, proteins sharing a sequence identity greater than around 35% will have essentially the same fold. This makes it possible to generate a 'homology model' of the 3D structure of a given protein based on the 3D model of a closely related protein. Homology modeling used to be done largely 'by hand', using molecular graphics software, fitting the sequence of the unknown protein onto the model of the known one. Nowadays, the bulk of the modeling can be performed automatically. Indeed there are several servers that generate models automatically from user-submitted sequences. The best known is SWISS-MODEL.[73,74]

Much structure-based drug design in the pharmaceutical industry relies on homology modeling wherein a model of the target protein is generated and used for inhibitor design. Thus if the model is a poor one, the chances of designing a viable inhibitor will be seriously diminished.

Even where the model is a good one, obtained from a very similar protein, one needs to be cautious about taking the model's atomic positions at face value as they do not represent positions based on any experimental data. Indeed, any experimental errors in the template model will be propagated into the homology model, and probably exacerbated in the process.

3.24.6.2 Reliability of Homology Models

The reliability of homology modeling has been evaluated over the past 10 years as part of the ongoing CASP (Critical Assessment of Techniques for Protein Structure Prediction)[75] experiment. To date (July 2005) there have been six such experiments wherein the sequences of proteins whose structures are being solved by crystallography and NMR groups around the world are made public prior to the release of the 3D models. Predictions are then invited in various categories, including homology, or comparative, modeling. Once the structures have been solved the accuracy of the models can be assessed, with the results presented at a specially convened conference and made available on the web.[76]

These experiments have revealed that the crucial factor in obtaining a good model is the initial sequence alignment between the two sequences identifying which are the equivalent residues in the two proteins. At low sequence identity (i.e., at around 30–35%) the chances of there being errors in the alignment increase, and so the danger of serious errors in the model becomes high. Over the course of the CASP experiments there have been significant improvements in the quality of the alignments, and consequently of the resultant models. Nowadays, large parts of the models are approximately correct but in many cases are not sufficiently accurate to reproduce the key functional residues[77] – which is obviously a problem where homology models are used in structure-based drug design.

3.24.6.3 Model Databases

The PDB used to accept and include homology-built models. However, with effect from 2 July 2002, all theoretical models were removed from the standard PDB archive and transferred to a separate ftp site.[78] The rationale behind this policy change was that there should be a clear distinction between experimental and theoretical models and that they should not all be lumped together in a single archive. As of July 2005 there were 882 models on the models site. In addition to homology models they included other types of theoretical models such as molecular dynamics (MD) trajectories. For example, 350 models from an MD simulation of reduced glutaredoxin were formerly stored across 5 PDB entries: **1upy**, **1upz**, **1uq0**, **1uq1**, and **1uq2**.

Another well-known database of homology models is ModBase.[79] This contains automatically generated models for all sequences in the SwissProt and TrEMBL databases that have detectable similarity to an experimentally determined protein structure. Each model has associated with it a measure of its likely reliability. While it is useful to have ready access to such models, it is important to remember not to place too much reliance on their accuracy.

3.24.7 Problems with Predicted Structures

Many proteins do not have relatives of known 3D structure in the PDB, so their structures cannot be approximated even by homology modeling. In these cases, to get an idea of their structure requires resorting to the straw-clutching predictive techniques of secondary structure prediction and fold recognition. Now, whereas homology-built models need to be treated with extreme caution, models derived by these predictive techniques require treatment with extreme prejudice. Occasionally, these methods approximate the right answer – usually for small, single-domain proteins where they may produce topologically near-correct models[75] – but generally, they are wildly wrong.

A number of websites exist for generating predicted models based on fold recognition, or 'threading', and are usually honest enough to preface their predictions with appropriate health warnings. A full list can be found on the LiveBench[15] website[80] which regularly evaluates these servers.

3.24.8 Some Advice on Coping with the Problems

So far in this chapter we have seen that there are many different types of problem associated with the structures of biomacromolecules. And, because we have focused on the problems, this may have cast an unduly pessimistic light on protein and DNA 3D models. This was not the intention, but it is important to keep in the back of one's mind that such problems do exist. The majority of models in the PDB are mostly correct, with only the usual experimental errors to worry about. However, even with such structures, care should be taken in overinterpreting the apparent precision of the atomic coordinates, particularly in important regions such as active sites.

The previous chapter (*see* 3.23 Protein Three-Dimensional Structure Validation) describes the checks one can run and how, for many x-ray structures, it is possible to judge the quality of different parts of the model by actually inspecting the electron density. For most purposes, though, there is sufficient information about the models on the web to guide one as to which are likely to be reasonable for one's purposes and which should be avoided. Here we present a summary of what to look for and where to find it.

3.24.8.1 Global Measures of Quality

3.24.8.1.1 The resolution and *R*-factor

The most obvious starting points are the global measures, such as the resolution and *R*-factor, reported for each model. These are recorded in the original PDB file and consequently reported by any website describing the structure.

The resolution reflects the amount of detail discernible in the electron density map and hence the chances of the final model being a reasonable one. The highest-resolution models are those solved to 'atomic' resolution (0.6–1.2 Å). Here the electron density is so clear that many of the hydrogen atoms become visible, and alternate occupancies of side chains or whole loops become more easily distinguishable. These models are 'truer' to the experimental data than the lower-resolution structures because they require fewer geometrical restraints during refinement. The majority of the PDB's crystal structures have been solved at 'medium' resolution of around 2.0 Å. At this resolution the path of a protein or DNA chain is fairly straightforward to trace correctly through the electron density and the majority of the side chains or nucleotide bases can be confidently fitted. Below around 3.0 Å, the backbone can still be reasonably well traced but the side chains become difficult to place. At 4.0 Å or lower, only the general locations of the regions of regular secondary structure are apparent; so, in these cases, only the C_α coordinates may be deposited. Sometimes, however, a good model can be obtained from poor data. If the crystal has a high degree of noncrystallographic symmetry, it is possible to use density-averaging techniques to produce excellent electron density maps even with data at less than 4.0 Å resolution, e.g., the viral capsids mentioned above.

The other well-known global measure is the reliability, or crystallographic *R*-factor, which indicates how well the final model agrees with the experimental data. Typically its value for protein and nucleic acid structures is around 0.20 (or equivalently, 20%); higher values indicate poorer agreement with the experimental data. Values in the range 0.40 to 0.60 can be obtained from a totally random model, so models with such values are unreliable and probably would never be published. Because the *R*-factor can be willfully, or innocently, manipulated, a better measure is Brünger's free *R*-factor, or R_{free},[81] which is an application of the statistical technique of cross-validation. Its value will tend to be larger than the *R*-factor, although it is not clear what a good value might be. Brünger has suggested that any value above 0.40 should be treated with caution[82] and that the difference between R_{free} and R be small, ideally <0.05.

Global measures for NMR structures are harder to come by. An equivalent of the resolution is the number of experimentally derived restraints per residue (*see* 3.22 Bio-Nuclear Magnetic Resonance). The more restraints the better, although some restraints may be completely redundant and so should not be counted. NMR equivalents of the crystallographic *R*-factor have been introduced in recent years,[83,84] but none has yet become standard. Before these, the most common measures of the agreement of the final models with the data were the numbers of 'restraint violations' exhibited by each model and the RMSD across the ensemble of solutions. None of these measures are consistently computed or universally reported, so it is difficult to get the same instant assessment of the likely quality of an NMR structure that is given by the resolution and *R*-factor for a crystal structure.

3.24.8.1.2 Stereochemical parameters

The next guide to the overall quality of a protein model can be given by global checks on the model's geometry, stereochemistry, and other structural properties. Such checks are readily available on the web. The best of these is the Ramachandran plot of the protein's ψ–ϕ main-chain torsion angles which can readily identify proteins with rather too many outlying residues. A number of websites provide Ramachandran plots for existing protein structures.

Two websites in particular provide this plot plus a number of additional geometrical checks. The PDBsum[85] website,[22] which specializes in structural analyses and pictorial representations of all models in the PDB, gives the full output of the PROCHECK program.[86] The PDBREPORT website,[87] provides detailed and numerous analyses of all PDB models as calculated by the WHAT_CHECK program[1] (*see* 3.23 Protein Three-Dimensional Structure Validation). The analyses include space group and symmetry checks, geometrical checks on bond lengths, bond angles, torsion angles, proline puckers, bad contacts, planarity checks, checks on hydrogen bonds, and more, including an overall summary report.

3.24.8.2 Local Measures of Quality

3.24.8.2.1 Estimates of coordinate errors

Even a good model overall may have poorly defined regions within it. Ideally, one would like to have error bars on all the atomic coordinates. Theoretically, for x-ray structures, it is possible to compute these. However, it is so computationally expensive that only a handful of models have such error estimates. As of July 2005 there were only eight structures (out of roughly 32 000 in the PDB) which had coordinate error estimates. All had been solved at atomic resolution (ranging from 0.89 to 1.3 Å). Of these, the six deposited prior to 2002 were all very small: one was not a protein at all, but a carbohydrate (cycloamylose, PDB code **1c58**), another was a 6-bp fragment of DNA (**362d**), three were marginally differing copies of the same 13-residue enterotoxin (**1etl**, **1etm**, and **1etn**), and the sixth was the crystal structure of the 54-residue rubredoxin (**4rxn**). The two models deposited since then have been considerably larger. One is a model of streptavidin, a bacterial binding protein, comprising a dimer of two 119-residue protein chains solved to 0.96 Å resolution (**1luq**), and the largest is a 259-residue carbonic anhydrase with several ligands bound, solved to 0.95 Å resolution (**1lug**).

For NMR structures one cannot compute error estimates as the theory does not provide a means of obtaining them directly from the experimental data. However, some progress to getting such estimates is being made.[57]

3.24.8.2.2 Atomic *B*-factors

In the absence of error estimates, an atom's *B*-factor is often used as a guide to the precision of the atom's coordinates. *B*-factors are closely related to positional errors but the relationship cannot be easily formulated,[88] as *B*-factors can sometimes soak up features of the structure not included in the model, such as disorder, occupancy, dynamics, errors, invalid restraints, etc. In general, though, the atoms with the largest *B*-values will tend to be those with the largest positional uncertainty. A common rule of thumb is to beware of atoms with *B*-values in excess of 40.0 although it is better to compare individual *B*-factors with the average *B*-factor of the molecule. So, if the majority of the atoms in, say, the active site under study happen to have large *B*-values, it is safe to conclude that the site is not well determined and needs care over what conclusions can be drawn from it. A quick way to view the *B*-factors is to display the model in a molecular graphics program such as RasMol and to color the atoms or residues by temperature factor.

3.24.8.2.3 The uppsala electron density server (EDS)

Another useful source of information on local quality, although only for x-ray structures – and not all at that – is the Uppsala Electron Density Server (EDS).[89,90] The main purpose of this server is to display electron density maps for PDB entries, computed directly from the experimental structure factors. And, indeed, one can use it for this purpose: i.e., examine the electron density maps, although such examination benefits from a little patience and some previous experience. However, the server also provides some useful statistics about the models which can more readily identify the residues that are better, or more poorly, defined by the data. Plots of the real-space *R*-factor (RSR), for example, show how well each residue fits its electron density. Other useful plots include: the occupancy-weighted average temperature factor and a *Z*-score associated with the residue's RSR for the given resolution. In essence, in all these plots, if a residue is indicated by a tall spike then it is either particularly disordered or does not sit well within an appropriate region of electron density. A quick glance at the plots shows which residues, or regions, to beware of.

3.24.8.2.4 Stereochemical checks

Finally, the stereochemical checks mentioned previously can provide local as well as global quality indicators. These show which residues have unusual geometry. Of course, unusual geometry does not always mean 'incorrect' geometry as there have been many cases where unusual conformations in a protein model have been associated in some way with the protein's biological function.[91–93] And the only way of being sure whether the oddity is a feature of the true structure or an error is by reference to the original experimental data (assuming it is of sufficient quality to resolve the issue).

Table 1 Selected websites offering structure quality assessments

Name	*Address*	*Description*
RCSB PDB	http://www.rcsb.org/pdb	Provides detailed geometric information for all PDB entries in the archive
PDBsum	http://www.ebi.ac.uk/thornton-srv/databases/pdbsum	A pictorial atlas of PDB entries which includes quality indicators such as the Ramachandran plot and full PROCHECK analysis
PDBREPORT	http://swift.cmbi.ru.nl/gv/pdbreport/	Detailed structure quality reports for each PDB entry calculated using WHAT_CHECK
MolProbity	http://kinemage.biochem.duke.edu/molprobity/	Graphical contact and geometrical analysis of an uploaded PDB-style file or PDB entry
Verify3D	http://nihserver.mbi.ucla.edu/Verify_3D/	Provides a graphical analysis of quality of an uploaded PDB file
AQUA	http://www.nmr.chem.uu.nl/users/jurgen/Aqua/server	An interactive program to assess the quality of structures determined by NMR spectroscopy using deposited experimental data and PDB files
EDS	http://eds.bmc.uu.se/eds	The Uppsala Electron Density Server offers statistics regarding the accuracy and quality of x-ray crystal structures based on the analysis of electron density maps
Biotech Validation Suite	http://biotech.ebi.ac.uk:8400	A validation suite for protein structures, including full PROCHECK and WHAT_CHECK analyses
JCSG Validation Tools	http://www.jcsg.org/scripts/prod/validation1.cgi	Provides tools for validating structures and x-ray experimental data
VADAR	http://redpoll.pharmacy.ualberta.ca/vadar	Volume, Area, Dihedral Angle Reporter for analyzing protein structure quality
STAN	http://xray.bmc.uu.se/cgi-bin/gerard/rama_server.pl	Assesses protein structure geometry and associated waters

Note: These websites are correct as at July 2005.

In addition to the PROCHECK and PBDREPORT analyses mentioned above, the PDB website[94] provides its own set of geometrical analyses on each entry, consisting of tables of average, minimum and maximum values for the protein's bond lengths, bond angles, and dihedral angles. Any unusual values are highlighted. A backbone representation of the model, colored according to the 'Fold Deviation Score', can be viewed in RasMol; the redder the coloring the more unusual the residue's conformational parameters. **Table 1** lists these, and several other, web servers that provide analyses of protein structural quality for both existing models in the PDB and for user-uploaded ones.

3.24.9 Conclusion

In this chapter we have described some of the many and varied problems that come with protein 3D models. We hope we have not scared the reader from using these models, which are both beautiful and a rich source of biological information. Nor have we intended to instill such a distrust of them that the reader will view every model as totally unreliable and unbelievable. Rather, we have meant to foster a healthy scepticism that will lead to the reader taking great care in ensuring that a given model is sufficiently reliable for his or her scientific needs. Often this will require a considerable investment of effort, but that effort will be amply rewarded if meaningful conclusions are to be drawn.

It is sufficient to remember that all experimentally determined protein structures, like the results of any scientific measurement, are likely to contain errors. These will vary in severity from one structure to another, and from one region of a protein model to another. Thus it is necessary to determine whether the size and location of these errors is relevant. As for models that are not the direct result of experimental data, even greater care is required for these.

References

1. Hooft, R. W.; Vriend, G.; Sander, C.; Abola, E. E. *Nature* **1996**, *381*, 272.
2. Jones, T. A.; Kleywegt, G. J.; Brünger, A. T. *Nature* **1996**, *383*, 18–19.
3. Westbrook, J.; Feng, Z.; Burkhardt, K.; Berman, H. M. *Methods Enzymol.* **2003**, *374*, 370–385.
4. Brändén, C.-I.; Jones, T. A. *Nature* **1990**, *343*, 687–689.
5. McRee, D. E.; Tainer, J. A.; Meyer, T. E.; Beeumen, J. V.; Cusanovich, M. A.; Getzoff, E. D. *Proc. Natl. Acad. Sci. USA* **1989**, *86*, 6533–6537.
6. Borgstahl, G. E. O.; Williams, D. R.; Getzoff, E. D. *Biochemistry* **1995**, *34*, 6278–6287.
7. Kelly, J. A.; Knox, J. R.; Moews, P. C.; Hite, G. J.; Bartolone, J. B.; Zhao, H.; Joris, B.; Frere, J. M.; Ghuysen, J. M. *J. Biol. Chem.* **1985**, *260*, 6449–6458.
8. Kelly, J. A.; Kuzin, A. P. *J. Mol. Biol.* **1995**, *254*, 223–236.
9. Hoier, H.; Schlömann, M.; Hammer, A.; Glusker, J. P.; Carrell, H. L.; Goldman, A.; Stezowski, J. J.; Heinemann, U. *Acta Crystallogr.* **1994**, *D50*, 75–84.
10. Kleywegt, G. J.; Hoier, H.; Jones, T. A. *Acta Crystallogr.* **1996**, *D52*, 858–863.
11. Jones, T. A.; Kjeldgaard, M. *Methods Enzymol.* **1997**, *277*, 173–208.
12. Kleywegt, G. J.; Read, R. J. *Structure* **1997**, *5*, 1557–1569.
13. Kleywegt, G. J. *J. Mol. Biol.* **1997**, *273*, 371–376.
14. Bujnicki, J.; Rychlewski, L.; Fischer, D. *Bioinformatics* **2002**, *18*, 1391–1395.
15. Bujnicki, J. M.; Elofsson, A.; Fischer, D.; Rychlewski, L. *Protein Sci.* **2001**, *10*, 352–361.
16. Hanson, M. A.; Stevens, R. C. *Nat. Struct. Biol.* **2000**, 7, 687–692.
17. Rupp, B.; Segelke, B. *Nat. Struct. Biol.* **2001**, *8*, 663–664.
18. Stevens, R. C.; Hanson, M. A. *Nat. Struct. Biol.* **2001**, *8*, 664.
19. Segelke, B.; Knapp, M.; Kadkhodayan, S.; Balhorn, R.; Rupp, B. *Proc. Natl. Acad. Sci. USA* **2004**, *101*, 6888–6893.
20. Breidenbach, M. A.; Brunger, A. T. *Nature* **2004**, *432*, 925–929.
21. Kleywegt, G. J.; Jones, T. A. *Structure* **1995**, *3*, 535–540.
22. PDB sum. http://www.ebi.ac.uk/thornton-srv/databases/pdbsum (accessed April 2006).
23. Kleywegt, G. J.; Brünger, A. T. *Structure* **1996**, *4*, 897–904.
24. http://pdbobs.sdsc.edu (accessed Oct 2006).
25. Weissig, H.; Bourne, P. E. *Bioinformatics* **1999**, *15*, 807–831.
26. Kleywegt, G. J.; Jones, T. A. *Structure* **2002**, *10*, 465–472.
27. Badger, J.; Hendle, J. *Acta Crystallogr.* **2002**, *D58*, 284–291.
28. Vriend, G. *J. Mol. Graphics* **1990**, *8*, 52–56.
29. Venclovas, Č.; Ginalski, K.; Kang, C. *Protein Sci.* **2004**, *13*, 1594–1602.
30. Sippl, M. J. *Proteins* **1993**, *17*, 355–362.
31. DePristo, M. A.; de Bakker, P. I. W.; Blundell, T. L. *Structure* **2004**, *12*, 831–838.
32. Frauenfelder, H.; Sligar, S. G.; Wolynes, P. G. *Science* **1991**, *254*, 1598–1603.
33. Kleywegt, G. J.; Henrick, K.; Dodson, E. J.; van Aalten, D. M. F. *Structure* **2003**, *11*, 1051–1059.
34. Lide, D. R. *CRC Handbook of Chemistry and Physics*, 75th ed.; CRC Press: Boca Raton, FL, 1995.
35. Allen, F. H.; Kennard, O.; Watson, D. G.; Brammer, L.; Orpen, A. G.; Taylor, R. *J. Chem. Soc. Perkins Trans.* **1987**, *II*, S1–S19.

36. Nissink, J. W.; Murray, C.; Hartshorn, M.; Verdonk, M. L.; Cole, J. C.; Taylor, R. *Proteins* **2002**, *49*, 457–471.
37. Hanson, M. A.; Oost, T. K.; Sukonpan, C.; Rich, D. H.; Stevens, R. C. *J. Am. Chem. Soc.* **2000**, *124*, 10248.
38. Bhat, T. N.; Bourne, P.; Feng, Z. K.; Gilliland, G.; Jain, S.; Ravichandran, V.; Schneider, B.; Schneider, K.; Thanki, N.; Weissig, H. et al. *Nucleic Acids Res.* **2001**, *29*, 214–218.
39. Lutteke, T.; Frank, M.; von der Lieth, C. W. *Carbohydr. Res.* **2004**, *339*, 1015–1020.
40. Fields, B. A.; Bartsch, H. H.; Bartunik, H. D.; Cordes, F.; Guss, J. M.; Freeman, H. C. *Acta Crystallogr.* **1994**, *D50*, 709–730.
41. Kleywegt, G. J. *Acta Crystallogr.* **2000**, *D56*, 249–265.
42. Molecular Structure Database. http://pqs.ebi.ac.uk (accessed April 2006).
43. Henrick, K.; Thornton, J. M. *Trends Biochem Sci.* **1998**, *23*, 358–361.
44. Ponstingl, H.; Kabir, T.; Thornton, J. M. *J. Appl. Crystallogr.* **2003**, *36*, 1116–1122.
45. Protein Interfaces and Assemblies. http://www.ebi.ac.uk/thornton-srv/databases/pita (accessed April 2006).
46. Molecular Structure Database. http://www.ebi.ac.uk/msd-srv/prot_int/pistart.html (accessed April 2006).
47. Swaminathan, G. J.; Holloway, D. E.; Colvin, R. A.; Campanella, G. K.; Papageorgiou, A. C.; Luster, A. D.; Acharya, K. R. *Structure* **2003**, *11*, 521–532.
48. RCSB Protein Data Bank. http://www.rcsb.org/ (accessed April 2006).
49. http://neon.ce.umist.ac.uk/olderado (accessed April 2006).
50. Kelley, L. A.; Gardner, S. A.; Sutcliffe, M. J. *Protein Eng.* **1996**, *9*, 1063–1065.
51. MacArthur, M. W.; Thornton, J. M. *Proteins* **1993**, *17*, 232–251.
52. Abagyan, R. A.; Totrov, M. M. *J. Mol. Biol.* **1997**, *268*, 678–685.
53. Ratnaparkhi, G. S.; Ramachandran, S.; Udgaonkar, J. B.; Varadarajan, R. *Biochemistry* **1998**, *37*, 6958–6966.
54. Doreleijers, J. F.; Rullmann, J. A. C.; Kaptein, R. *J. Mol. Biol.* **1998**, *281*, 149–164.
55. Spronk, C. A. E. M.; Linge, J. P.; Hilbers, C. W.; Vuister, G. W. *J. Biomol. NMR* **2002**, *22*, 281–289.
56. Spronk, C. A. E. M.; Nabuurs, S. B.; Krieger, E.; Vriend, G.; Vuister, G. W. *Prog. Nucl. Magn. Reson. Spectrosc.* **2004**, *45*, 315–337.
57. Snyder, D. A.; Bhattacharya, A.; Huang, Y. P. J.; Montelione, G. T. *Proteins* **2005**, *59*, 655–661.
58. Spronk, C. A. E. M.; Nabuurs, S. B.; Bonvin, A. M. J. J.; Krieger, E.; Vuister, G. W.; Vriend, G. *J. Biomol. NMR* **2003**, *25*, 225–234.
59. ID-PROF. http://www.cmbi.kun.nl/dress (accessed April 2006).
60. Nabuurs, S. B.; Nederveen, A. J.; Vranken, W.; Doreleijers, J. F.; Bonvin, A. M. J. J.; Vuister, G. W.; Vriend, G.; Spronk, C. A. E. M. *Proteins* **2004**, *55*, 483–486.
61. RECOORD. http://www.ebi.ac.uk/msd/recoord (accessed April 2006).
62. Nederveen, A. J.; Doreleijers, J. F.; Vranken, W.; Miller, Z.; Spronk, C. A. E. M.; Nabuurs, S. B.; Guntert, P.; Livny, M.; Markley, J. L. et al. *Proteins* **2005**, *59*, 662–672.
63. Werner, M. H.; Clore, G. M.; Fisher, C. L.; Fisher, R. J.; Trinh, L.; Shiloach, J.; Gronenborn, A. M. *J. Biomol. NMR* **1997**, *10*, 317–328.
64. Zhang, H.; Neal, S.; Wishart, D. S. *J. Biomol. NMR* **2003**, *25*, 173–195.
65. Dutta, S.; Berman, H. M. *Structure* **2005**, *13*, 381–388.
66. Bernstein, F. C.; Koetzle, T. F.; Williams, G. J. B.; Meyer, E. F., Jr.; Brice, M. D.; Rogers, J. R.; Kennard, O.; Shimanouchi, T.; Tasumi, M. *J. Mol. Biol.* **1977**, *112*, 535–542.
67. Berman, H. M.; Westbrook, J.; Feng, Z.; Gilliland, G.; Bhat, T. N.; Weissig, H.; Shindyalov, I. N.; Bourne, P. E. *Nucleic Acids Res.* **2000**, *28*, 235–242.
68. Bourne, P. E.; Berman, H. M.; Watenpaugh, K.; Westbrokk, J. D.; Fitzgerald, P. M. D. *Methods Enzymol.* **1997**, *277*, 571–590.
69. Westbrook, J.; Feng, Z. K.; Jain, S.; Bhat, T. N.; Thanki, N.; Ravichandran, V.; Gilliland, G. L.; Bluhm, W.; Weissig, H.; Greer, D. S. et al. *Nucleic Acids Res.* **2002**, *30*, 245–248.
70. Reddy, V. S.; Natarajan, P.; Okerberg, N.; Li, K.; Damodaran, K. V.; Morton, R. T.; Brooks, C. L., III; Johnson, J. E. *J. Virol.* **2001**, *75*, 11943–11947.
71. Virus Particle Explorer. http://mmtsb.scripps.edu/viper (accessed April 2006).
72. Bairoch, A.; Apweiler, R.; Wu, C. H.; Barker, W. C.; Boeckmann, B.; Ferro, S.; Gasteiger, E.; Huang, H.; Lopez, R.; Magrane, M.; Martin, M. J.; Natale, D. A.; O'Donovan, C.; Redaschi, N.; Yeh, L. S. *Nucleic Acids Res.* **2005**, *33*, D154–D159.
73. Swiss-Model. http://www.expasy.ch/swissmod/SWISS-MODEL.html (accessed April 2006).
74. Schwede, T.; Kopp, J.; Guex, N.; Peitsch, M. C. *Nucleic Acids Res.* **2003**, *31*, 3381–3385.
75. Moult, J. *Curr. Opin. Struct. Biol.* **2005**, *15*, 285–289.
76. Protein Structure Production Centre. http://predictioncenter.org (accessed Oct 2006).
77. DeWeese-Scott, C.; Moult, J. *Proteins* **2004**, *55*, 942–961.
78. ftp://ftp.rcsb.org/pub/pdb/data/structures/models
79. Pieper, U.; Eswar, N.; Braberg, H.; Madhusudhan, M. S.; Davis, F. P.; Stuart, A. C.; Mirkovic, N.; Rossi, A.; Marti-Renom, M. A.; Fiser, A. et al. *Nucleic Acids Res.* **2004**, *32*, D217–D222.
80. BioInfo Bank Institute. http://bioinfo.pl/LiveBench (accessed April 2006).
81. Brünger, A. T. *Nature* **1992**, *355*, 472–475.
82. Brünger, A. T. *Methods Enzymol.* **1997**, *277*, 366–396.
83. Gronwald, W.; Kirchhofer, R.; Gorler, A.; Kremer, W.; Ganslmeier, B.; Neidig, K. P.; Kalbitzer, H. R. *J. Biomol. NMR* **2000**, *17*, 137–151.
84. Huang, Y. J.; Powers, R.; Montelione, G. T. *J. Am. Chem. Soc.* **2005**, *127*, 1665–1674.
85. Laskowski, R. A.; Chistyakov, V. V.; Thornton, J. M. *Nucleic Acids Res.* **2005**, *33*, D266–D268.
86. Laskowski, R. A.; MacArthur, M. W.; Moss, D. S.; Thornton, J. M. *J. Appl. Crystallogr.* **1993**, *26*, 283–291.
87. PDBREPORT Database. http://www.cmbi.kun.nl/gv/pdbreport (accessed April 2006).
88. Tickle, I. J.; Laskowski, R. A.; Moss, D. S. *Acta Crystallogr.* **1998**, *D54*, 243–252.
89. Electron Density Server. http://eds.bmc.uu.se (accessed April 2006).
90. Kleywegt, G. J.; Harris, M. R.; Zou, J.-Y.; Taylor, T. C.; Wählby, A.; Jones, T. A. *Acta Crystallogr.* **2004**, *D60*, 2240–2249.
91. Herzberg, O.; Moult, J. *Proteins* **1991**, *11*, 223–229.
92. Petock, J. M.; Torshin, I. Y.; Weber, I. T.; Harrison, R. W. *Proteins* **2003**, *53*, 872–879.
93. Videau, L. L.; Arendall, W. B.; Richardson, J. S. *Proteins* **2004**, *56*, 298–309.
94. RCSB Protein Data Bank. http://www.rcsb.org/pdb (accessed April 2006).

Biographies

Roman A Laskowski received a BSc in physics from the University of Birmingham, an MSc in astrophysics from Queen Mary College, London, and a PhD in computational molecular biology from Birkbeck College, London. His postdoctoral research at University College London under Prof Janet Thornton involved developing methods for validating protein 3D structures. He is currently a senior researcher at the European Bioinformatics Institute where he has developed the ProFunc web server and further developed the PDBsum web server. He has co-written several programs in widespread use in the field of structural bioinformatics, including PROCHECK, LIGPLOT, and SURFNET.

G Jawahar Swaminathan received a MSc in biotechnology from the University of Roorkee, India followed by a PhD in protein crystallography from the National Institute of Immunology, New Delhi. He then proceeded to a postdoctoral research fellowship at the University of Bath, where his research involved determination of many crystal structures of proteins involved in inflammation. He joined the Macromolecular Structure Database (MSD), a member of the Worldwide Protein Data Bank (wwPDB) in 2002 as a curator working on the annotation of PDB entries. He is currently employed as a senior curator at the MSD, where his responsibilities include development of structure data deposition and structure validation software. He is also the author of biobar, a bioinformatics web browser search toolbar application.

Comprehensive Medicinal Chemistry II
ISBN (set): 0-08-044513-6

ISBN (Volume 3) 0-08-044516-0; pp. 531–550

3.25 Structural Genomics

W Shi and M R Chance, Case Western Reserve University, Cleveland, OH, USA

3.25.1	**Overview**	**551**
3.25.2	**Target Selection**	**553**
3.25.3	**Cloning and Expression of the Protein**	**554**
3.25.4	**Protein Purification**	**555**
3.25.5	**Structure Determination**	**555**
3.25.6	**Metalloproteomics**	**556**
3.25.7	**Comparative Computational Modeling**	**557**
3.25.8	**Structure to Function**	**557**
3.25.9	**Auto-Publishing**	**558**
3.25.10	**Outlook for Protein Structure Initiative-2**	**558**
References		**558**

3.25.1 **Overview**

The field of Structural Genomics has become an international effort to obtain protein structures on a genomic scale. Multi-institutional collaborative structural genomics centers have been initiated in the US and worldwide. The Protein Structure Initiative (PSI), funded by the National Institute of General Medical Science (NIGMS)[1] of the National Institute of Health, aims to acquire structural and functional information on all proteins in the biosphere through a combination of experimental and computational approaches. In phase 1 of the PSI (PSI-1; September 2000–June 2005), nine structural genomics centers (top nine centers in **Table 1**) were selected to develop new approaches and technologies to streamline and automate the steps of protein structure determination, and to incorporate those methods into high-throughput pipelines from DNA sequence information to high-resolution three-dimensional protein structures.[2,3] Each center in the PSI-1 is responsible for developing and testing an integrated structural genomics effort that includes open reading frame (ORF) target selection, cloning, expression, protein purification, structure determination by x-ray crystallography or nuclear magnetic resonance (NMR) spectroscopy, and comparative computational modeling.[4–11] The targets selected for these nine centers included species with minimal genomes (*Mycoplasma genitalium*, *Mycoplasma pneumoniae*), pathogenic microorganisms (*Mycobacterium tuberculosis*) and parasites (*Plasmodium falciparum*), model organisms (*Caenorhabditis elegans*, yeast), and higher organisms (mouse and human), which provided protein diversity and had biomedical relevance. Upon completion of the PSI-1 in June 2005, more than 1100 protein structures had been solved and deposited in the Protein Data Bank (PDB), including over 700 unique structures (structures sharing less than 30% of their sequence with proteins with existing structures).

Other structural genomics initiatives include Structure to Function (S2F), a structural genomics project focusing on proteins from *Haemophilus influenzae* (the first completely sequenced bacterial genome) with unknown functions, and 10 other centers in North America, Europe, and Asia (**Table 1**). Most of these structural genomics centers work with bacterial proteins. Coordination among the centers to maximize the effectiveness of target selection has been facilitated by the development of a target registration data base, TargetDB,[12] maintained by the Research Collaboratory for Structural Bioinformatics (RCSB) Protein Data Bank. The target selection and progress of the production and solutions of structures for structural genomics centers worldwide can be tracked through a query form. This database allows the centers to monitor overlap in target selection and has allowed many scientists outside the centers to access information on proteins of interest.

Following the announcement of 10 new structural genomics centers, PSI advanced to its second phase, Rapid Production Phase, in July 2005. Two types of center were included in PSI-2. Four large-scale centers, proven to be productive during the PSI-1, are expected to generate 3000–4000 structures in total in the next 5 years: Joint Center for Structural Genomics (JCSG), Midwest Center for Structural Genomics (MCSG), New York Structural GenomiX Research (NYSGRC), and Northeast Structural Genomic Consortium (NESG). Six specialized centers are also

Table 1 Structural genomics centers worldwide

Center name	*Target species*	*Country*	*Website*
Berkeley Structural Genomics Center (BSGC)	*Mycoplasma genitalium* and *Mycoplasma pneumoniae*	US	www.strgen.org
Center for Eukaryotic Structural Genomics	*Arabidopsis thaliana*	US	www.uwstructuralgenomics.org
Joint Center for Structural Genomics (JCSG)	*Thermotoga maritima* and mouse	US	www.jcsg.org
Midwest Center for Structural Genomics (MCSG)	Multiple microorganisms	US	www.mcsg.anl.gov
New York Structural Genomics Research Consortium (NYSGRC)	Bacteria and eukaryotes	US	www.nysgrc.org
Northeast Structural Genomics Consortium (NESG)	*Drosophila melanogaster*, *Saccharomyces cerevisiae*, *Caenorhabditis elegans*, mouse, and human	US	www.nesg.org
The Southeast Collaboratory for Structural Genomics (SECSG)	*Pyrococcus furiosus*, *C. elegans*, and human	US	www.secsg.org
Structural Genomics of Pathogenic Protozoa Consortium (SGPP)	*Leishmania*, *Trypanosoma*, and *Plasmodium falciparum*	US	www.sgpp.org
TB Structural Genomics Consortium (TB)	*Mycobacterium tuberculosis*	US	www.doe-mbi.ucla.edu/TB
Structure to Function (S2F)	*Haemophilus influenzae*	US	S2f.carb.nist.gov
Montreal-Kingson Bacterial Structural Genomics Initiative (BSGI)	*E. coli*	Canada	www.bri.nrc.ca/brimsg/bsgi.html
Protein Structure Factory (PSF)	Human	Germany	www.rzpd.de/psf
M. tuberculosis Structural Proteomic Project (XMTB)	*M. tuberculosis*	Germany	xmtb.org
Northwest Structural Genomics Centre (NWSGC)	*M. tuberculosis*	England	www.nwsgc.ac.uk
Oxford Protein Production Facility (OPPF)	Human and human pathogens	England	www.oppf.ox.ac.uk
Bacterial Targets at IGS-CNRS (BIGS)	*E. coli*	France	Igs-server.cnrs-mrs.fr/Str_gen
Marseilles Structural Genomics Programme (MSGP)	*M. tuberculosis*	France	Afmb.cnrc-mrs.fr/rubrique93.html
Yeast Structural Genomics (YSG)	Yeast	France	genomics.eu.org
RIKEN Structural Genomics Initiative (RSGI)	*Thermus thermophilus*	Japan	www.riken.jp/engn/index/html

included to develop new technologies for determining the structures of 'difficult proteins,' such as small protein complexes, membrane proteins, and proteins from higher organisms, including human. The progress of the four large-scale structural genomics centers in PSI-1 is summarized in **Table 2**.

The pipeline of structure determination in structural genomics involves: (1) target selection on the basis of primary amino acid sequence analysis; (2) generation of cDNA clone of an ORF of the selected target by PCR amplification and incorporation of the PCR product into a suitable expression vector; (3) expressing the protein in bacteria or other systems at a high level; (4) purification of the protein in sufficient quantity and quality for structural analysis (x-ray crystallography or NMR spectroscopy); (5) biophysical characterization of the protein prior to structure determination; (6) structure determination by x-ray crystallography or NMR; (7) structure validation and deposition in the Protein Data Bank; (8) comparative computational modeling using the new structure as template; and (9) structural-functional analysis of the target (**Figure 1**). Each step will be discussed in detail in the following sections.

Table 2 Progress of the four large-scale structural genomics centers in PSI-1. The table shows the number of targets processed successfully through each stage for the four large-scale structure genomics centers

Progress	*Centers*			
	JCSG	*MCSG*	*NESG*	*NYSGRC*
Selected	6678	15331	12262	2306
Cloned	3270	5681	5197	1685
Expressed	2951	4255	3291	1375
Purified	1339	2150	1669	1068
Crystallized	1180	841	–	391
Structure (x-ray)	226	281	109	191
Structure (NMR)	–	–	91	–
Deposited in PDB	193	281	200	185
Accurate models	–	–	–	11021

JCSG, Joint Center for Structural Genomics; MCSG, Midwest Center for Structural Genomics; NYSGRC, New York Structural GenomiX Resarch; NESG, Northeast Structural Genomic Consortium.

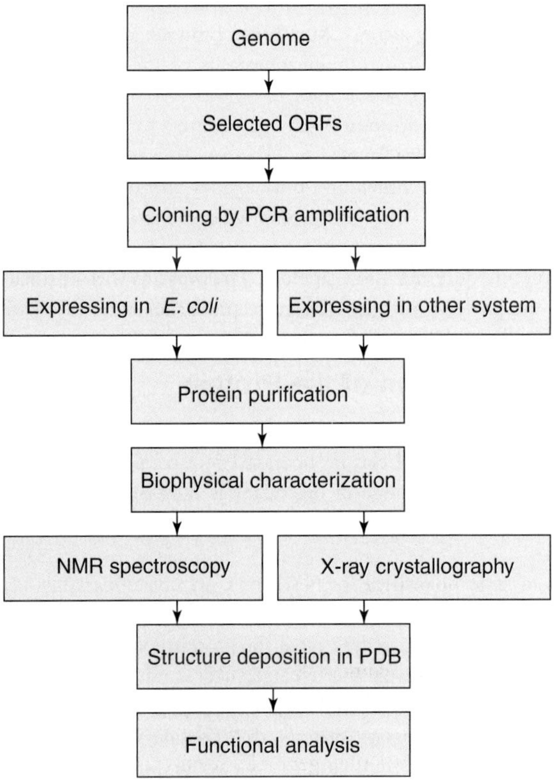

Figure 1 Structure determination pipeline in structural genomics from DNA information to three-dimensional structures.

3.25.2 Target Selection

Target selection is the most important step in structural genomics pilot studies. In PSI-1, target selection is focused on providing structural coverage for all protein families. In accordance with the goal of PSI to maximize diversity of the protein structures, protein targets are generally selected for their expected structural distinction from existing protein structures based on primary amino acid sequence analysis. As a general rule, the ORF targets selected for structural genomics must have <30% sequence identity to proteins that have already been deposited in the PDB. This general rule is based on the observation that reliable models of protein can usually be generated from structural templates that share >30% sequence identity.[13,14]

To achieve comprehensive and uniform coverage, Chandonia and Brenner proposed the Pfam5000 strategy, which involved selecting the 5000 most important families from the Pfam database as sources for targets.[15] The Pfam database is a collection of protein families and domains and contains multiple protein sequence alignments and profiles of these families. Focusing on the largest families and supplementing the strategy with other criteria, such as bacterial sequences or eukaryotic sequences, the strategy can be modified in different ways to reflect various requirements and changing priorities, and can complement the alternative single-genome strategy. Targeting large protein families is generally accepted as an effective strategy because these are often evolutionarily conserved and functionally important. It is estimated that 16 000 experimental structures using an optimal target selection strategy are required to accurately model the majority of all proteins.[14]

Some structural genomics centers in PSI-1 choose proteins of biomedical interest, thus having a direct impact on drug discovery.

The PSI-1 structures are dominated by single-domain, prokaryotic proteins for many reasons. First, the cloning strategies can be easily executed. Bacterial genes lack introns and have well-defined start and stop sites. Second, the gene products from the lower organisms are mostly comprised of small, soluble domains that are easily expressed and purified from bacterial systems. Third, most of the PSI structures are solved by x-ray crystallography utilizing the single or multiple anomalous dispersion methods with selenomethionine (Se-Met) substituted proteins. Labeling proteins with Se-Met is simple with bacterial proteins. On the other hand, protein production and crystallization of eukaryotic proteins and protein complexes have proven to be difficult for many reasons, including their requirement of posttranslational modification or functional partners. Membrane proteins are also eliminated from the list of potential targets, due to the difficulties experienced in obtaining crystals of membrane proteins. These problems will be addressed by the six specialized centers in PSI-2.

A commonly used strategy by most of the structural genomics centers to increase the chances of solving the structure of a target protein is studying protein orthologs from several species. However, once the first protein structure is solved, the effort of solving the structures of the remaining orthologs is abandoned. A data mining system for selecting eukaryotic targets has also been developed.[16] Several protein characteristics including protein length, composition of charged and polar residues, hydrophobicity, presence of a signal sequence, and COG assignment play an important role throughout the stages of the structure determination pipeline. The proteins with optimal combinations of these features are selected for efficiency and the proteins with less favorable properties are treated with special care.[17]

3.25.3 Cloning and Expression of the Protein

The production of a cDNA clone for any particular ORF of interest by PCR amplification and its insertion into a suitable expression vector is an easy task and can be operated in a parallel fashion for high-throughput production. Cloning of almost any protein from a wide range of organisms is now considered to be a routine molecular biology operation. The successful rate for cloning is approximately 95% (lower for eukaryotic targets and higher for prokaryotic targets).[17]

The expression of proteins in large quantities for NMR or x-ray crystallography studies is more challenging than cloning. Expressing recombinant proteins in *Escherichia coli* is relatively successful using established methods. The four large-scale structural genomics centers have success rates for protein expression ranging from 51% to 90%, with an average success rate of 74% (**Table 2**). However, this average success rate is biased by the predominantly prokaryotic target selections; successful expression of proteins from higher eukaryotes is nearer 50% while only 1 in 10 are typically soluble.[18] There are other well-studied expression systems that provide more favorable success rates including *Pichia pastorisis*, baculovirus, and mammalian or insect cell culture systems for expressing proteins from higher organisms. The availability of these different expression systems significantly increases the chances of producing large quantities of protein, particularly for the targets from eukaryotic organisms. A newly developed cell-free wheatgerm expression system is also used to increase the production of proteins from higher organisms.[19–21] It is desirable to design compatible expression vectors to exchange inserts into the different expression systems.

The large-scale structural genomics centers utilize robotic expression platforms as part of their high-throughput technology, making it possible to test many parameters of expression conditions.[7,21,22] A new technology developed to express protein at a high level in *E. coli* for structural genomics is the auto-induction protocol, which allows automatic induction of bacterial protein production.[22] These protocols produce 10 times more protein than traditional methods. Auto-induction protocols also allow many cultures to be inoculated in parallel and induced simply by growth to saturation, making it a powerful tool for screening clones for expression and solubility automatically. Many laboratories have utilized auto-induction protocols and some of the technologies used have been commercialized, such as the Overnight Express Auto Induction System by Novagen.

Other new technologies for protein expression that have been developed in structural genomics studies include a novel SUMO-based protein expression system (US Patent Application 10/188,343) for high-throughput cloning and expression and a mammalian cell expression system based on fluorescent fusions to evaluate whether the target has folded properly and is in soluble form.[22–24] Both methods have been utilized in high-throughput overexpression of protein targets in structural genomics.

3.25.4 Protein Purification

Purification is an important step that provides the large quantity of highly purified proteins for x-ray crystallography and NMR analysis. To separate the protein from the other contents of the cells that were utilized for expression, the C-terminal or N-terminal affinity tags were used for single-step purification in most of the structural genomics centers. The use of hexahistidine engineered into expression vectors (N-terminal or C-terminal) is the most commonly used technique for purifying recombinant proteins in the large-scale structural genomics center.[7,20,22,25] The affinity tags were sometimes cleaved from the purified proteins. Since the protein purity correlates to the success of protein crystallization, additional chromatography steps after histidine-tag affinity purification are incorporated (anion exchange followed by size-exclusion gel filtration) to remove remaining impurity if necessary.[7]

The cloning, expression, and purification of proteins in these large-scale structural genomics centers have become highly automated, utilizing industrial-scale platform.[22] To facilitate parallel purification on the high-throughput scale, a customized purification robotics was developed by the JCSG to process cultures directly from fermentation to affinity purification.[7] This automation integrates many functions necessary to lyze cells and separate the soluble from insoluble fractions. It also purifies the insoluble portions of each sample through detergent extractions for use in refolding studies.

Prior to structure determination, purified proteins are subjected to biophysical characterization studies including sodium dodecyl (lauryl) sulfate–polyacrylamide gel electrophoresis (SDS-PAGE) to confirm purity, mass spectrometry for construct verification, analytical gel filtration, and UV/visible absorbance spectroscopy to identify possible bound cofactors.[22]

3.25.5 Structure Determination

Purified proteins are subjected to structural studies by NMR spectroscopy or x-ray crystallography. Although both methods have made contributions in the past, improvement in crystal growth methods and automation in many steps including data collection and structure solution by advances in instrumentation and computer technology have made x-ray crystallography the primary choice for structure determination in PSI.

Samples for NMR studies are required to be isotopically labeled. This can be done by supplementing the cell growth media with the required labeled compounds. The major limitation for structure determination by NMR is the size of the protein, which must not be greater than 30 kDa. New spectrometers in the range of 800–1000 MHz increase the limit of size of the proteins to 60 kDa. In addition, data collection and processing for NMR is significantly slower than that for x-ray crystallography. Most of the crystallography data collection is done at a synchrotron radiation source, which can be completed in a few minutes to a few hours.

In the past, the major bottleneck in structure determination is obtaining diffraction quality crystals for x-ray crystallography. Improvements in crystallization techniques have long been sought and considerable effort has gone into understanding the mechanism of the crystallization process. New technologies have been developed and applied to generate homogeneous protein for crystallization. A method called limited proteolysis combined with mass spectrometry (LP/MS) has been applied to design new expression constructs for protein targets that failed to produce crystals.[22] The technique was applied to 164 targets and 50% of these targets were totally resistant to proteolysis, 45% were partially degraded to a stable domain, and 5% were degraded to multiple fragments. It has been shown that the targets resistant to proteolysis are good candidates for crystallographic studies with 27% of these targets yielding three-dimensional structures thus far, whereas only 9% of the targets showing partial proteolysis have yielded three-dimensional structures to date. Large-scale subcloning of the stable domains for the targets with partial proteolysis has been carried out in the NYSGRC.

To rapidly identify initial crystallization conditions, a sparse matrix sampling method has been designed that efficiently combines the conditions that have previously been proven to be successful in protein crystallization.[26] Several companies have developed and commercialized crystallization screen kits based on this concept. These screen kits, mainly the result of practical experience, have been proven to be very effective for crystallization of a large assortment of proteins. In future, PSI is likely to generate additional information that can be used to improve efficiency in finding lead conditions for obtaining protein crystals.

Many different crystallization conditions have to be tested to maximize the chances of finding the lead conditions for protein crystallization and this approach needs a large quantity of purified proteins and can be time consuming. To overcome this technical hurdle, many structural genomics centers have adopted automated robotic devices to setup 96-well format crystallization trials that require minimum amounts of sample volume (50 nL).[7,22] It takes about 2 min to set up each plate, with a throughput of 2880 different crystallization conditions per hour. Automated crystallization plate tracking, imaging, and analysis have also been developed to manage the large protein crystallization data output.[27] Once the initial crystallization conditions have been found, they need to be refined to produce diffraction quality crystals for data collection; these fine screens are still carried out by hand in many instances.

There have been significant advances in recent years in methods and instrumentation for macromolecular crystallography data collection at US synchrotron radiation sources, such as the Advanced Photon Source at Argonne National Laboratory, National Synchrotron Light Source at the Brookhaven National Laboratory, Stanford Synchrotron Radiation Laboratory, and Advanced Light Source at the Lawrence Berkeley Laboratory as well as at many synchrotron centers throughout the world. Both the brightness (intensity with fine collimation) and tunability of synchrotron sources contributes significantly to their value to structural biology. In many cases, the high brightness of synchrotron x-radiation sources provide a significant enhancement in the resolution of data collected from both weakly diffracting and small crystals relative to laboratory-based rotating anode generators. It also shortens the exposure time for data collection from such weakly diffracting and small crystals. With the use of synchrotron radiation sources, together with the advanced technologies in charge-coupled device (CCD) detectors and cryo-crystallography, a single crystal with dimensions as small as 10–30 μm can be used to solve protein structures.

Synchrotron radiation is also the most valuable tool for solving the 'phase problem,' which remains one of the rate-limiting steps in de novo macromolecular structure determination by x-ray crystallography. The broad-band nature of synchrotron radiation sources allows for data collection at, and around, the absorption edges of anomalous scatterers maximizing the anomalous contribution of each derivative allowing convenient extraction of phase information.[28] Thus, the incorporation of only a single type of anomalous scatterer allows for highly efficient structure determination by single or multiwavelength anomalous dispersion techniques. Optimization of anomalous signal also significantly enhances the quality of maps calculated by traditional isomorphous replacement methods. Most of the protein structures in structural genomics are determined using single or multiple anomalous dispersion methods with Se-Met substituted proteins.[7,22,25]

Crystal mounting and screening robots have been implemented in many synchrotron beamlines to screen crystals automatically for quality and to facilitate the automated, high-throughput data collection required in structural genomics. The most time-consuming step in x-ray crystallography structure determination is likely to be electron density map interpretation. Automated software for structure determination, including electron density interpretation, refinement, and PDB deposition, has been developed and will enhance high-throughput structure determination.[29]

Once a structure is solved and refined, it is validated and subsequently deposited into the PDB. The coordinates and the experimental data associated with the structure are usually released to the public upon deposition.

3.25.6 Metalloproteomics

Metalloproteomics is a high-throughput method to characterize metalloproteins using x-ray absorption spectroscopy (XAS). The presence of metal ions is detected by a fluorescence scan using a synchrotron radiation source. Only 20% of the purified protein targets in structural genomics yield diffraction quality crystals. Any step to improve crystal quality or aid in crystallographic phasing, even if it affects only a small percentage of the total targets examined, may have a major impact on the speed and cost of the overall structural genomics project.

An initial test of a high-throughput XAS methodology has been carried out at Stanford Synchrotron Radiation Laboratory using purified gene product samples from *Pyrococcus furiosus* from the Southeast Collaboratory for Structural Genomics (SECSG) and distributions of Co, Ni, Cu, and Zn were mapped.[30] A high-throughput XAS analysis of metal binding with a large set of proteins and quantification of six metals (Mn, Fe, Co, Cu, Ni, and Zn) is reported by the NYSGRC.[5,31] Over 10% of 654 analyzed proteins showed the presence of transition metal atoms in stoichiometric amounts; these totals, as well as the abundance distribution, are similar to transition metal content observed in the

PDB. The metalloprotein annotation is posted in several databases such as IceDB,[32] the integrated consortium experimental database in the NYSGRC, MODBAS,[33,34] and BIND.[35] The information can be accessed by other scientists to facilitate the phasing step in structure determination or to aid in computational modeling of these metalloproteins. The ultimate goal of the metalloproteomics project is to develop a metalloprotein annotation database where at least one protein from each of the Pfam families is experimentally analyzed by this technique.

3.25.7 Comparative Computational Modeling

Protein structures can be modeled using the primary amino acid sequence and a structure template (high-resolution three-dimensional structure of a homologous protein). Comparisons of predicted protein models with experimental protein structures of the same protein have shown that reliable protein models can be constructed if the template shares >30% sequence identity. Structural genomics aims to determine the structures by x-ray crystallography or NMR of proteins predicted to be sufficiently dissimilar to those for which the structure has already been determined. The remaining protein sequences will be modeled based on one or more of the experimental structures. The number of modeled structures is expected to be 100 times greater than the number of experimentally determined structures.[36]

The modeling of protein structures involves three steps. The first step is to find the related known structures in the PDB and to perform a fold assignment followed by primary amino acid sequence alignment. The next step is comparative modeling, which relies on alignment and the structures of templates to produce three-dimensional models of the target proteins. Finally, the models are evaluated using structural and energetic criteria in addition to sequence similarity.

What information can be extracted from the model depends on the results of model evaluation. In general, the accuracy of constructed models depends on the sequence similarity between the model sequence and the templates. The models based on >30% sequence identity are suitable for many applications including functional analysis and screening of small molecule databases for potential drug discovery. The accuracy of a model based on a template with >50% sequence identity equals the accuracy of a low-resolution crystal structure (3.0 Å resolution).[36] One hundred and ninety-one structures have been solved in NYSGRC during PSI-1. Comparative computational modeling has generated fold assignments for 127 521 sequences and accurate models for 11 021 sequences based on these experimental structures.[32] A suite of bioinformatics programs and databases can be found at MODBASE,[33,34] a comprehensive database of annotated comparative protein structure models.

3.25.8 Structure to Function

The goal of structural genomics is to obtain the structures for all proteins by experimental methods and computational modeling. This information can be used to determine the molecular and cellular functions of these proteins and to assist in structure-based drug design. The targets selected for structural genomics are often 'hypothetical proteins' with previously unknown functions. The function of these proteins can often be predicted by comparison to other structures. The functional annotation of a hypothetical protein is relatively straightforward if the new structure demonstrates similar fold to that of the proteins whose function are known. Common structure comparison methods including DALI[37] and VAST[38] identify proteins with a similar fold through structural alignment. These methods have led to many discoveries in functional analysis.[39] The biological functions of the proteins were subsequently confirmed by biochemical assays.

There are two drawbacks of these structural comparison methods. First, the targets in structural genomics are selected because of their expected structural dissimilarity to the known structures. The structure comparison methods usually fail to produce any match from the database. Second, if the new structure possesses a popular fold such as TIM barrel, it is difficult to pinpoint the function of the protein. New methods have been developed to find similarity not in the overall fold, but in the functional sites.[40–42] These programs perform searches for small functional patterns by comparing target proteins to a database of three-dimensional templates generated from known enzyme active sites. A functional site comparison method called 'PINT' has been applied to over 200 structures determined through structural genomics.[42] It has shown promising results finding functional centers within an overall similar fold for 11% of proteins and, more importantly, detecting functional similarity across folds for an additional 6% of all structures. However, some similarities were found only by DALI and not by PINT because of the distortion in the active sites. The two approaches complement each other and a combined functional annotation strategy can help to overcome defects inherent to each method.

Sometimes, the unexpected presence of a ligand in the structure of a protein with unknown function gives a clue to its biological function, and the hypothesis based on the information can be tested and confirmed through subsequent biochemical experiment. This has been the case for several structural genomics targets. For example, TM04449 from the JCSG was solved with four flavin adenine dinucleotide (FAD) molecules in the active site and fold comparison search using DALI failed to find any match with significant similarity.[7] Later, the structure was used to confirm that it is a unique member of the thymidylate synthases and suggests an alternative flavin-dependent mechanism for thymidylate synthesis.[43]

3.25.9 Auto-Publishing

Over 1100 structures were determined in PSI-1 and only approximately 10% of these resulted in peer-reviewed publications. To assist scientists in fields other than structural biology in utilizing the large influx of information from structural genomics, an auto-publishing protocol has been developed to construct a structure paper semi-automatically and in a less time-consuming manner. First, the protocol extracts data collection and refinement statistics from the coordinate file in PDB and generates a table based on the information. It plots three figures including a ribbon diagram of the structure, an amino acid sequence alignment with the related proteins, and a molecular surface representation with conserved residues color coded and mapped onto the surface. In addition, it generates a simple description of the structure. However, background information and structural analysis still need to be prepared by the author. The first sample paper on a structural genomics target prepared by the auto-publishing protocol is published.[44] The auto-publishing website is still undergoing improvement and will probably be available for public access in 2006.

3.25.10 Outlook for Protein Structure Initiative-2

The rationales for PSI-2 can be summarized as follows: (1) structural descriptions will help researchers illuminate structure–function relationships and thus formulate better hypotheses and design better experiments; (2) the PSI collection of structures will serve as the starting point for structure-based drug development by permitting faster identification of lead compounds and their optimization; (3) the design of better therapeutics will result from comparisons of the structures of proteins that are from pathogenic and host organisms and from normal and diseased human tissues; (4) the PSI collection of structures will assist biomedical investigators in research studies of key biophysical and biochemical problems, such as protein folding, evolution, structure prediction, and the organization of protein families and folds; and (5) technical developments, the availability of reagents and materials, and experimental outcome data in protein production and crystallization will directly benefit all structural biologists and provide valuable assistance to a broad range of biomedical researchers.

During PSI-1, over 1100 structures were solved; most of these were prokaryotic proteins from a variety of microorganisms, providing scientists with a large bacterial structure database for new antibacterial drug discovery.[45] More importantly, these large-scale structural genomics centers have developed new technologies to automate every step from target selection and protein production to structural and functional analysis in the structure determination pipeline and are ready to produce a large number of structures in PSI-2. The PSI-1 budget was $270 million for the 5-year period. With 1100 structures solved, each structure costs approximately $250 000. In PSI-2, with a $300 million total budget and 4000 structures expected to come through the pipeline, the cost will be reduced to less than $75 000 per structure.

References

1. National Institute of General Medical Sciences. http://www.nigms.nih.gov/psi (accessed April 2006).
2. Editoriall. *Nat. Struct. Mol. Biol.* **2004**, *11*, 201.
3. Burley, S. K. *Nat. Struct. Mol. Biol.* **2000**, 7, 932–934.
4. Chance, M. R.; Bresnick, A. R.; Burley, S. K.; Jiang, J. S.; Lima, C. D.; Sali, A.; Almo, S. C.; Bonanno, J. B.; Buglino, J. A.; Boulton, S. et al. *Protein Sci.* **2002**, *11*, 723–738.
5. Chance, M. R.; Fiser, A.; Sali, A.; Pieper, U.; Eswar, N.; Xu, G.; Fajardo, J. E.; Radhakannan, T.; Marinkovic, N. *Genome Res.* **2004**, *14*, 2145–2153.
6. Gerstein, M.; Edwards, A.; Arrawsmith, C. H.; Montelione, G. T. *Science* **2003**, *299*, 1663.
7. Lesley, S. A.; Kuhn, P.; Godzik, A.; Deacon, A. M.; Mathews, I.; Kreusch, A.; Spraggon, G.; Klock, H. E.; McMullan, D.; Shin, T. et al. *PNAS* **2002**, *99*, 11664–11669.
8. Terwilliger, T. C.; Park, M. S.; Waldo, G. S.; Berendzen, J.; Hung, L. W.; Kim, C. Y.; Smith, C. V.; Sacchettini, J. C.; Bellinzoni, M.; Bossi, R. et al. *Tuberculosis* **2003**, *83*, 223–249.
9. Burley, S. K.; Bonanno, J. B. *Methods Biochem. Anal.* **2003**, *44*, 591–612.
10. Burley, S. K.; Almo, S. C.; Bonanno, J. B.; Capel, M.; Chance, M. R.; Gaasterland, T.; Lin, D.; Sali, A.; Studier, F. W.; Swaminathan, S. *Nat. Genet.* **1999**, *23*, 151–157.
11. Shi, W.; Ostrov, D.; Gerchman, S.; Kycia, H.; Studier, F. W.; Edstrom, W.; Bresnick, A. R.; Ehrlich, J.; Blanchard, J.; Almo, S. C. et al. In *Protein Chips, Biochips, and Proteomics: The Next Phase of Genomics Discovery*; Marcel Dekker: New York, 2003, pp 299–324.

12. PDB Protein Data Bank. http://targetdb.pdb.org (accessed April 2006).
13. Baker, D.; Sali, A. *Science* **2001**, *294*, 93–96.
14. Vitkup, D.; Melamud, E.; Moult, J.; Sander, C. *Nat. Struct. Biol.* **2001**, *8*, 559–566.
15. Chandonia, J. M.; Brenner, S. E. *Proteins* **2005**, *58*, 166–179.
16. Liu, J.; Hegyi, H.; Acton, T. B.; Montelione, G. T.; Rost, B. *Proteins* **2004**, *56*, 188–200.
17. Goh, C. S.; Lan, N.; Douglas, S. M.; Wu, B.; Echols, N.; Smith, A.; Milburn, D.; Montelione, G. T.; Zhao, H.; Gerstein, M. *J. Mol. Biol.* **2004**, *336*, 115–130.
18. Huang, R. Y.; Boulton, S. J.; Vidal, M.; Almo, S. C.; Bresnick, A. R.; Chance, M. R. *Biochem. Biophys. Res. Commun.* **2003**, *307*, 928–934.
19. Ma, L. C.; Sawasaki, T.; Tsuchimonchi, M.; Mazda, S.; Gunsalus, K. C.; Macapagal, D.; Shatry, R.; Ho, C. K.; Acton, T. B.; Endo, Y. et al. *J. Struct. Funct. Genomics*, submitted for publication, 2005.
20. Acton, T. B.; Gunsalus, K. C.; Xiao, R.; Ma, L. C.; Aramini, J.; Baran, M. C.; Chiang, Y. W.; Climent, T.; Cooper, B.; Denissova, N. G. et al. *Methods Enzymol.* **2005**, *394*, 210–243.
21. Morita, E. H.; Sawasaki, T.; Tanaka, R.; Endo, Y.; Kohno, T. A. *Protein Sci.* **2003**, *12*, 1216–1221.
22. Bananno, J. B.; Almo, S. C.; Bresnick, A.; Chance, M. R.; Fiser, A.; Swaminathan, S.; Jiang, J.; Studier, F. W.; Shapiro, L.; Lima, C. D. et al. *J. Struct. Funct. Genomics* **2005**, *6*, 225–232.
23. Mancia, F.; Patel, S. D.; Rajala, M. W.; Scherer, P. E.; Nemes, A.; Ira Schieren, I.; Hendrickson, W. A.; Shapiro, L. *Structure* **2004**, *12*, 1355–1360.
24. Patel, S. D.; Rajala, M. W.; Rosetti, L.; Scherer, P. E.; Shapiro, L. *Science* **2004**, *304*, 1154–1158.
25. Kim, Y.; Dementieva, I.; Zhou, M.; Wu, R.; Lezondra, L.; Quartey, P.; Joachimiak, G.; Korolev, O.; Li, H.; Joachimiak, A. *J. Struct. Funct. Genomics* **2004**, *4*, 1–8.
26. Jancarik, J.; Kim, S. H. *J. Appl. Crystallogr.* **1991**, *24*, 409–411.
27. Mayo, C. J.; Diprose, J. M.; Walter, T. S.; Berry, I. M.; Wilson, J.; Owens, R. J.; Jones, E. Y.; Harlos, K.; Stuart, D. I.; Esnouf, R. M. *Structure* **2005**, *13*, 175–182.
28. Hendrickson, W. A. *Science* **1991**, *254*, 51–58.
29. Jiang, J. S.; Lin, Z. *Acta Crystallogr., Sect. D*, submitted for publication, 2005.
30. Scott, R. A.; Shokes, J. E.; Cosper, N. J.; Jenney, F. E.; Adams, M. W. W. *J. Synchrotron Rad.* **2005**, *12*, 19–22.
31. Shi, W.; Zhan, C.; Manjasetty, B.; Marinkovic, N.; Sullivan, M.; Huang, R.; Chance, M. R. *Structure* **2005**, *13*, 1473–1486.
32. IceDB – Integrated Consortium Experimental Database. http://www.nysgrc.org/nysgrc/icedb.html (accessed April 2006).
33. Pieper, U.; Eswar, N.; Braberg, N. H.; Madhusudhan, M. S.; Davis, F.; Stuart, A. C.; Mirkovic, N.; Rossi, A.; Marti-Renom, M. A.; Fiser, A. et al. *Nucleic Acids Res.* **2004**, *32*, D217–D222.
34. Database of Comparative Protein Structure Models. http://modbase.compbio.ucsf.edu/modbase-cgi-new/index.cgi (accessed April 2006).
35. Unleashed Informatics Limited. http://www.BIND.ca/ (accessed April 2006).
36. Sanchez, R.; Pieper, U.; Melo, F.; Eswar, N.; Marti-Renom, M. A.; Madhusudhan, M. S.; Mirkovic, N.; Sali, A. *Nat. Struct. Mol. Biol.* **2000**, 7, 986–990.
37. Holm, L.; Sander, C. *J. Mol. Biol.* **1993**, *233*, 123–138.
38. Gibrat, J. F.; Madej, T.; Bryant, S. H. *Curr. Opin. Struct. Biol.* **1996**, *6*, 377–385.
39. Zhang, C.; Kim, S. H. *Curr. Opin. Chem. Biol.* **2003**, 7, 28–32.
40. Wallace, A. C.; Borkakoti, N.; Thornton, J. M. *Protein Sci.* **1997**, *6*, 2308–2323.
41. Kleywegt, G. J. *J. Mol. Biol.* **1999**, *285*, 1887–1897.
42. Stark, A.; Shkumatov, A.; Russell, R. B. *Structure* **2004**, *12*, 1405–1412.
43. Myllykallio, H.; Lipowski, G.; Leduc, D.; Filee, J.; Forterre, P.; Liebl, U. *Science* **2002**, *297*, 105–107.
44. Zhan, C.; Fedorov, E. V.; Shi, W.; Ramagopal, U. A.; Thirumuruhan, R.; Manjasetty, B. A.; Almo, S. C.; Fiser, A.; Chance, M. R.; Fedorov, A. A. *Acta Crystallogr. F* **2005**, *61*, 959–963.
45. Schmid, M. B. *Nat. Rev. Microbiol.* **2004**, *2*, 739–746.

Biographies

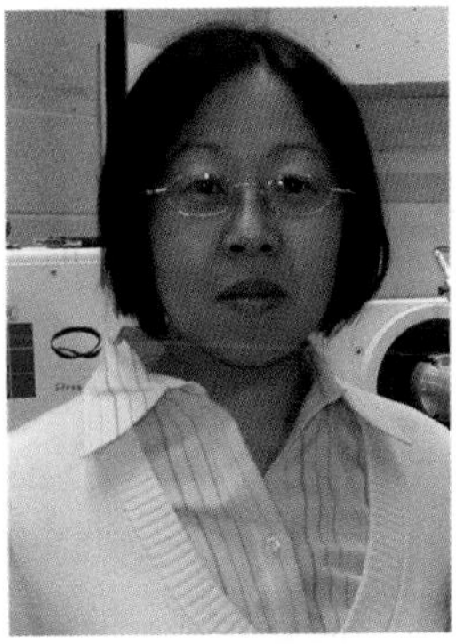

Wuxian Shi obtained her BS from University of Science and Technology of China (1987) and PhD from Pennsylvania State University (1996). She is currently an assistant professor in the Case Center for Proteomics of Case Western Reserve University and Chief Crystallographer of the Center for Synchrotron Biosciences at the National Synchrotron Light Source. Dr Shi is a member of the New York Structural GenomiX Research Consortium.

Mark R Chance received his BA from Wesleyan University (1980) and his PhD from the University of Pennsylvania (1986). After 13 years on the faculty of Albert Einstein College of Medicine, he recently joined the Case Western Reserve University Faculty as professor of physiology & biophysics and Director of the Case Center for Proteomics. Dr Chance is a member of the New York Structural GenomiX Research Consortium.

Comprehensive Medicinal Chemistry II
ISBN (set): 0-08-044513-6
ISBN (Volume 3) 0-08-044516-0; pp. 551–560

3.26 Compound Storage and Management

W W Keighley, T P Wood, and T J Winchester, Pfizer Global Research and Development, Sandwich, UK

3.26.1	**Introduction**	**561**
3.26.1.1	Materials Management: Centralized or Distributed – Where does Centralization Pay and How Can Flexibility be Retained	562
3.26.1.2	Background and History	562
3.26.1.3	Building the Liquid Compound Set	563
3.26.2	**Compound Acquisition and Processing**	**563**
3.26.2.1	Registration	564
3.26.3	**Sample Storage and Retrieval**	**565**
3.26.3.1	Solid Sample Storage and Supply	565
3.26.3.2	Liquid Sample Storage and Supply	566
3.26.3.2.1	First steps in automation	566
3.26.3.2.2	Current practice	566
3.26.3.3	Rationale of Storage Conditions	567
3.26.3.4	Quality Control of Compound Solutions	568
3.26.3.5	Customer Interaction	568
3.26.4	**Processing Compounds for Use**	**569**
3.26.4.1	Center of Emphasis Model and High-Throughput Screening	569
3.26.4.2	Primary Screen Supply: BasePlate	569
3.26.4.2.1	Subsets and compressed sets	570
3.26.4.3	Compounds Synthesized for Activity against a Specific Target	571
3.26.4.4	Hardware Selection and Trade-Offs	572
3.26.5	**Future Development**	**573**
References		**573**

3.26.1 Introduction

Materials management is an essential support function for all drug discovery stages. Encompassing sample acquisition, storage, dispensing, and distribution, it is difficult to think of any process that can happen in chemistry or biology laboratories that is not touched, in some way, by materials management. Every pharmaceutical and biotechnology company has a materials management requirement – in larger companies often supported by a department devoted to that function[1,2] – but also appearing as an additional task for some chemistry or biology teams. Materials management has responsibility not only for efficiently supplying compounds for screening, altering the process and product as required to meet changing customer requirements, but also for maintaining those compounds in good condition and dispensing using compound sparing protocols.

Increasingly, as companies become more globalized, the role involves maintaining an audit trail of sample usage that not only can assist the customer, but will also satisfy Customs and Excise criteria for import and export valuations, and can track the use of the compound collection as a capital asset. These aspects are not dealt with here as we concentrate on the processes involved in securely storing and accurately supplying samples for screening.

3.26.1.1 Materials Management: Centralized or Distributed – Where does Centralization Pay and How Can Flexibility be Retained

Although there is no one 'right' way to organize materials management, it is possible to identify definite efficiency gains in centralizing the function. Alternatively it can be argued that ultimate flexibility is retained in a distributed process. How should we incorporate the best of both approaches?

Where standardized processes exist, or become necessary due to size (or best scientific practice), a centralized process can be very effective in terms of efficient usage of capital equipment and manning. Centralization ensures consistency of products and allows best practice to be put in place quickly and effectively to permit a high degree of compound stewardship, both in terms of maintaining compound integrity and also compound sparing behavior. In the case of events that have to occur at intervals, or can be planned for and anticipated, centralization is very effective. As example, the Pfizer Centers of Emphasis (CoEs) maintain source copies of the entire screening compound file (where 'file' represents the compound collection, library, or set) and these can be aliquoted and distributed to all Discovery sites when replenishment of the screening set is needed. It is just as easy, and nearly as quick, to prepare seven replicates, one for each site, than it would be to prepare a single copy, and because there has been investment in storage and replication facilities, we can be confident that the product is quality-assured and consistent.

Where different sites have very different requirements, in terms of compound plate format, compressed or singleton presentation, solvent, etc., a centralized process struggles to cope and distributed processes become much more desirable because of the flexibility and tailoring to specific needs that they can offer. It is worth asking at this point: why are the requirements different, do they need to be? Almost always, requirements have been driven by the customer and have arisen due to historical processes, having undergone multiple iterations and revisions, rather than having been planned from the outset. While often it would be desirable to start over again and build a whole new process, this rarely happens due to the time and cost involved, and to a basic reluctance in many customer groups to change. Even if requirements can be standardized to a large degree, the presence of a local materials handling function at the point of use, i.e., as the screening plate is prepared for the individual screen, becomes a necessity in order to support the different biological buffers required by individual target systems and to impart the rapid turnaround and interaction that is the hallmark of most screening processes.

At Pfizer, although the bulk supply for primary screening, and the cherry-picked supply for follow-up has been centralized, a highly distributed local sample handling function has been retained, and enhanced, for 'closed loop' work (*see* Section 3.26.4.3). Although the process followed is standardized across sites, the location of the process is variable according to site, in some cases residing in a centralized (for the site) materials management function, in others as a specialism within the screening teams, and in others totally distributed to the level of the individual biologist who uses a standard set of applications within the setting of open access laboratories – shared laboratories offering automation and software on a walk-up basis for all projects. By supporting this level of flexibility, we have been able to minimize costs incurred in developing and supporting the software applications in use while retaining different site based models that work well for historical reasons.

3.26.1.2 Background and History

In the Pfizer organization, prior to merger with Warner Lambert, a completely distributed model was in place. The three Discovery sites (Sandwich, Groton, and Nagoya) each maintained a store of dry samples synthesized at or acquired by that site. At Sandwich and Groton, liquid sets existed, comprising mainly the same samples as available from the dry store, and samples were exchanged on an ad hoc basis between these two larger sites; Nagoya received liquid supplies from Groton and all three sites conducted distribution of samples to screens from within the screening teams themselves. This organization was not planned but grew in response to the advent of high-throughput screening (HTS) and a desire to screen at a rate faster than dry samples could be supplied – hence the development of a liquid store. Although not planned, the system worked well enough; however, the lack of organized sharing of the liquid set and any global inventory meant that it could be years before new compounds synthesized at one site were screened at another.

The merger with Warner Lambert provided a trigger to organize and formalize compound sharing practices in Pfizer. At the time of the merger, three new Discovery sites joined the Pfizer organization: at La Jolla and Ann Arbor in the USA, and at Fresnes in France. To have six sites each maintaining their own liquid set and sharing this with all other sites would have been a recipe for chaos and, instead, it was agreed to centralize materials management by setting up two CoEs, at Groton and Sandwich, who would consolidate and distribute samples worldwide. The initial consolidation phase was fraught with difficulties, generally not associated with sharing the physical samples, but with sharing and consolidating the matching electronic data, since each legacy company had a different set of standard formats and data standards for the representation of the chemical structures and for storage of the inventory information. In parallel with this attempt to consolidate the legacy sets of both companies, Pfizer entered into four major collaborations to enrich its screening set via structures amenable to highly parallel chemistry follow-up. These collaborations, being new, offered the opportunity to set registration and data content and format standards from the outset, and formed the backbone of what was to become the new Pfizer screening set. When a further merger, with Pharmacia, occurred in 2002, the lessons learned from the first global consolidation and exchange were well applied and the entire compound set, now comprising the combined assets of all of the previous mergers of the legacy companies, was distributed to the Discovery sites in short time.

By 2003, each CoE had a complete, replicated copy of the Pfizer liquid compound set, had distributed a complete copy of this to each of the Discovery sites and placed copies in storage ready for replenishment as required, and had begun to populate 'on demand' stores to supply cherry-picked samples for follow-up to the primary screen. This service would be provided by single use tube storage in $-20\,^{\circ}C$ stores supplied by the Swiss manufacturer REMP AG, which came on line at the CoEs during late 2003 and early 2004.

3.26.1.3 Building the Liquid Compound Set

The liquid sample sets that were created at the original Pfizer sites were first compiled in response to a desire, and ability, to screen samples at a rate faster than could be supported by weighing from the dry compound store. In its first iteration, the liquid set consisted of source liquids at approximately 30 mM generated by flicking an approximate milligram of compound into each tube of a 96 format rack and solubilizing with 1 ml dimethyl sulfoxide (DMSO). This was soon superseded by an accurately weighed sample, solvated to an accurate 30 mM via a simple software application used to drive a Tecan robot. The solubilization process has changed very little since, though it is now more controlled, as we have become aware of the potential impact of the process on liquid sample quality (*see* Section 3.26.3.4).

In those early days, liquid samples were used, almost exclusively, for HTS and confirmation of primary screen hits. For IC_{50} and any subsequent screening, an aliquot of dry material was always requested, but this was set to change. As the size of the screening collection grew, so the numbers of hits confirmed from primary screens became too large for the dry sample weighing process to support the increasing number of requests for dry sample. Experimental evidence was collated to support the premise that testing using the liquid samples did indeed produce the same results as the corresponding dry sample, but it remained difficult to persuade customers to abandon ingrained practices. Only with the major increase in the compound collection via parallel chemistry, and the lack of a dry sample for the vast majority of new registrations, did this change. Today, all hit confirmation, plate-based potency, and selectivity studies are conducted entirely using liquid samples. Dry samples are still requested for secondary pharmacology studies, but this too is changing and we can predict a (relatively close) future where dry sample will be reserved for in vivo experimentation, often made to demand since all earlier work will have relied on the original liquid straight from parallel synthesis via autopurification in quantities designed to supply the primary screening stream only. It is interesting to note how similar are the issues encountered, and the solutions employed, for the consolidation of the Astra AB and Zeneca Group PLC compound collections following the formation of AstraZeneca in 1999[3] and for the management of the archive and screening sets of a range of the industry's major players.[4]

3.26.2 Compound Acquisition and Processing

Later sections will discuss the methods used to maintain the purity and integrity of stored compounds, but the ability to track accurately the compounds we own starts earlier, at registration.

The majority of pharmaceutical companies will synthesize in-house, at least some of the compounds that they screen. The information collated at registration of these compounds paves the way to making best use of this material by recording the details that describe it uniquely and unambiguously, and would enable an exact resynthesis. While it can (or should) be relatively easy to control the quality of data submitted for in-house synthesis, it becomes a larger and more difficult task to control the quality of the data submitted by outside suppliers. For large external suppliers, it is reasonable to expect familiarity with standard data items and formats; however, where material is acquired from smaller collaborators, there is often a lack of previous experience in creating, for example, structure definition (SD) files, and this skill has to be taught as part of the collaborative process. Quality-control processes also vary, both in content and rigor, from supplier to supplier and it is important to set clear minimum standards at the outset, e.g., how will structure identity be confirmed? What percentage purity is minimum acceptable? Will solution concentration be checked? How?

What variance from target concentration is acceptable? It's also important that supplier quality control is backed up by an internal random sampling and testing program. The time taken to build up relationships with a supplier, reaching the point where the material and data supplied is of consistent and reliable quality, makes it very attractive to stay with known and trusted collaborators, and places an initial extra hurdle in the way of entering agreements with new suppliers.

Despite this temptation to stay with the old and trusted, venturing beyond traditional 'high cost' suppliers can offer real cost savings provided the relative risks associated with the various 'low cost' operators is well understood. Outsourcing became of interest when internal capacities could not be increased at the rate that programs demanded. Outsourcing also gave a degree of flexibility in capacity and skill sets that chemistry departments had previously not enjoyed. In the first instance, external suppliers were nearly all US- or Europe-based, where the intellectual and manufacturing capabilities were well established, but recent years have seen a burgeoning of capabilities and capacity in the developing nations, with excellent quality chemistry available in, for example, India, China, and other Asian states, and the former Eastern bloc at very keen prices.[5] This expanded capacity and sharp competition has additionally resulted in a much more fluid and competitive market in both the USA and Europe. Together with a renewed emphasis on innovation and advanced techniques within academic centers fuelled by initiatives such as the recent National Institutes of Health (NIH) plan to fund Centres of Excellence in Chemical Methodologies and Library Development with the primary goal of accelerating the production of new high-quality libraries,[6] this will continue to improve upon the variety of chemistries amenable to automation and increase the accessibility to and exploitation of these by the industry partners.

3.26.2.1 Registration

The bare minimum of information for getting a compound collection into Pfizer's Global Compound Management (GCM) database – from the compound registration viewpoint, results from the construction of GCM around the parent–compound–batch model and it is much easier to assimilate any compound collection, from any source, if its data is in this format:

- the structure (MDL molfile form) – just the parent structure, no salt fragments
- a compound parent identifier
- a compound form identifier (compound parent identifier + salt suffix). Salt suffix 'talks,' i.e., each salt code represents a different type of salt, e.g., HCl, HBr, Na, etc.
- a batch identifier.

The following is an illustration using Viagra (sildenafil citrate):

- the structure, just sildenafil represented, not the citrate
- a compound parent identifier – identifies sildenafil itself, does not vary between salt forms, e.g., PF-1,000,000
- a compound form identifier (compound parent identifier + salt suffix), e.g., Viagra = Sildenafil + Citrate, PF-1,000,000-10
- a batch identifier - distinguishes between different batches (synonym – lots) of the compound form and each batch identifier should be unique within the collection, hence as per PF-1,000,000-10-001.

New compounds synthesized in house, or to our specification, are described in this way. Where a collection has been acquired as part of a merger, if a different method of describing the compound has been used, rather than converting the information to the standard format it can be quick to reregister the compounds using our own Global electronic Compound Registration system (GeCR), giving all of the compounds new PF numbers as the primary identifier. This has the added benefit of deduplicating the new additions where identical structures already exist in the Pfizer collection, though stereoisomers can create a problem if the acquired libraries have not been unambiguously described.

Incorporating an existing library into our compound inventory also requires that we load accurate weights available, or volumes and concentrations in the case of solution samples, in order that data driven compound handling processes can properly function further downstream. Where an accurate electronic inventory does not accompany the physical samples, the value of the library is greatly diminished.

3.26.3 Sample Storage and Retrieval

3.26.3.1 Solid Sample Storage and Supply

As the activity profile of compounds emerging from primary and follow-up screens is developed, there is often a requirement for screening scientists to have access to nonsolubilized material to allow them to make fresh solutions or apply the compounds to test programs that use different media to the DMSO solvent used in the liquid stores. There is also a small, but significant, demand for dry material as intermediates for further syntheses.

At Sandwich, our Automated Dry Sample Bank (ADSB) stores over 500 000 dry, (or Neat) samples and stocks each sample in individual 4 dram ($\sim 12\,cm^3$) vials. This store was supplied and installed by REMP AG,[7] and has been in production since mid 2003. The impact was almost immediate with mean order turnround times dropping from over 4 days to just less than 24 h. ADSB is in fact the acronym for two automated stores, an automated device for reweighing and capping vials, a bank of manually operated balances under software control, and the administration software that brings it all together. The physical and data flows associated with operation of the system are shown in **Figure 1**. Empty vials are tare-weighed (cap off) at the decapping, weighing, and capping station (DWCS) and provided to all chemistry laboratories.

When newly synthesized samples are submitted, the weight of sample submitted is checked in the DWCS (again cap off – cap weights can vary considerably and this process allows for the return of the vial without its original cap) before being sent to storage. Where an order for the compound already exists, the vial is diverted to the weighing team prior to storage and, prior to storage, after each weighing episode the check weigh is repeated at the DWCS. This robust checking process ensures that our inventory is very accurate, an important factor in maintaining customer confidence.

While the main bank is designed to hold samples in the 4-dram vial format, the second store, a much smaller version of its big sister, has a capacity of around 26 000 30-mL bottles and is designed to hold 'bulk' dry samples. This store, with a footprint of just $24\,m^2$ is the smallest automated store commissioned by REMP AG. Similar systems for the storage and presentation of dry samples are in place at virtually all of the larger pharmaceutical companies, and systems are available from a number of suppliers in addition to REMP; such products include The Automation Partnership's Haystack,[8] and RTS Life Sciences[9] narrow aisle store.

The operation of ADSB and interaction of its various components is shown in **Figure 1**. Note that automated weighing does not feature in this design. Reports of up to 100 000 weighings per annum have been published using samples stored in Archimedean screw-top vials, and handled robotically[10] and so long as only suitable samples are presented, automated weighing can be valuable: an example of such a system is in place at Pfizer's dry sample bank in Ann Arbor, MI, to which free-flowing powders are routed, and this robot performs productively as an additional weighing team technician. However due to the limited range of samples that may practically be handled in this manner we have assessed that the automation of dry material weighing does not offer sufficient return on investment in terms of throughput and reliability.

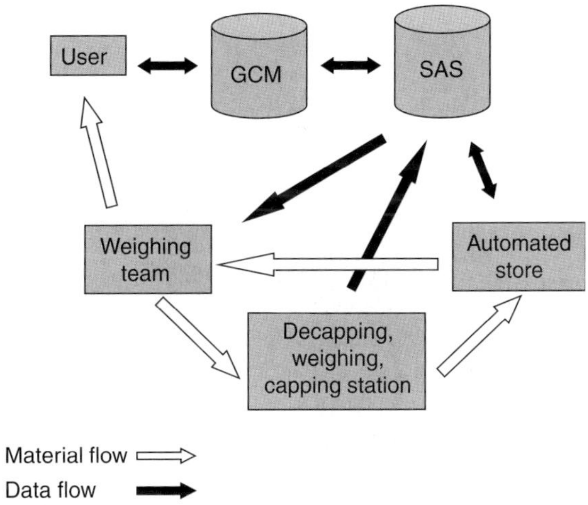

Figure 1 The physical and data flows for dry sample storage and provision. SAS, sample administration system.

3.26.3.2 Liquid Sample Storage and Supply

3.26.3.2.1 First steps in automation

The older of our automated liquid stores, the Automated Liquid Sample Bank (ALSB) commissioned from Manchester-based RTS Thurnall (now RTS Life Sciences), came into production in Spring 2000, and has over the last 5 years cherry-picked some 1.5 million samples from its 2.5 million inventory. The design of this system bears strong similarities to a store constructed at the same time for the Eli Lilly HTS operation Sphinx, both systems being based on paternoster storage at $-20\,^{\circ}C$ with integrated defrost and liquid handling. The ALSB continues to be a valuable tool supporting many drug discovery processes within Pfizer's Global Research and Development lines; however, it is beginning to reflect the limitations of a machine for which the design specifications were set over 8 years ago. While the vast bulk of Pfizer's screening programs utilize 384 or higher-density formats, ALSB dispenses in 96-well configuration, and hence virtually all output must be reconfigured off-line. Samples are contained in multiuse tubes, resulting in exposure to multiple freeze–thaw cycles, which has the potential to be detrimental to compound integrity, possibly leading to compound precipitation (*see* Section 3.26.3.4). Finally, the minimum reliable dispense volume of the system is 10 μL, which, being far more than necessary to run many screening programs, makes the system inefficient in compound use. These limitations, plus the requirement by Pfizer to develop a Global Center of Emphasis model for liquid compound supply, led to the installation at Sandwich of a second automated liquid order fulfilment system.

3.26.3.2.2 Current practice

In April 2004, a second automated liquid store was put into production, supplied by REMP AG, as for the automated dry bank. Again, samples are stored in 100% DMSO at $-20\,^{\circ}C$, but the REMP store works on completely different operating and supply principles to ALSB, being both based on the 384 format, and on a single-use principle.

Each sample in SREMP (Sandwich REMP store), replicated 17 times, is contained in a sealed microtube held in a 16×24 (384) storage rack matrix with the standard SBS microplate footprint (**Figure 2**).

Storage racks are stored in 8-place trays on shelves lining each side of the system's twin aisles. With the shelving units being around 50 levels high, and the aisle length accommodating over 100 units, the store capacity is some 60 000 000 samples. The four robots of the system each select samples from a discrete quadrant of the store; however, should a robot require service attention it can be removed and the remaining robot will control the entire aisle. To further enable convenient servicing of the equipment, chilled areas immediately adjacent to the cold store are provided into which the robot can be removed allowing a more pleasant environment (than $-20\,^{\circ}C$) for the engineers while avoiding raising the temperature of the working parts unnecessarily.

The multiple replicates of each compound are stored at just two volumes, 2 μL at 4 mM, being sufficient to source a single point retest screen, and 10 μL at 4 mM for multipoint IC_{50} screens. As each screen is sourced from a single-use

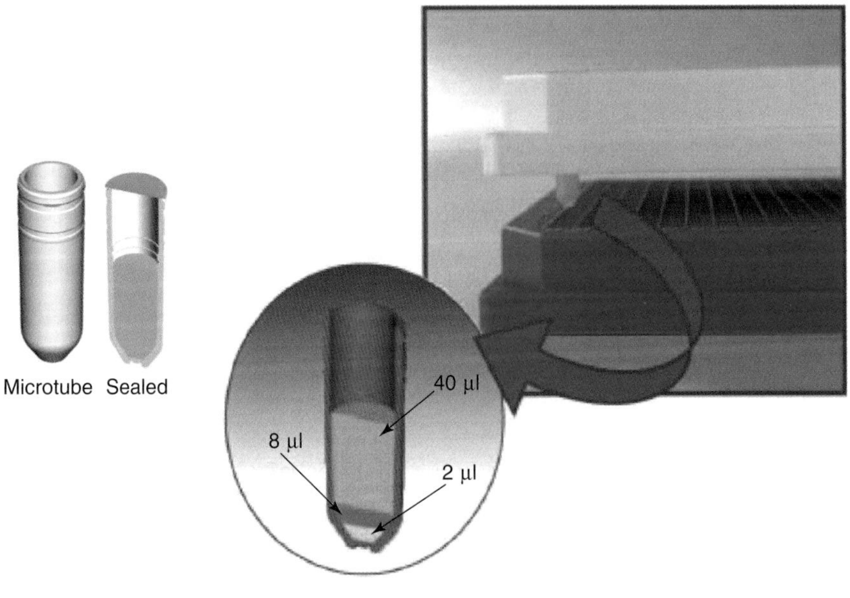

Figure 2 Storage media for single-use liquid samples.

tube, freeze–thaw cycles are no longer an issue. Samples are delivered from SREMP an order of magnitude faster than from ALSB, and in a 384 matrix for rapid reformatting into assay-ready plates.

To maximize productivity, the store is loaded such that replicates of each sample are represented in each of the four quadrants, allowing up to four orders to be processed simultaneously. When stock of a particular compound becomes exhausted in any one quadrant, the system will pass the delivery container between each of the four quadrants within a single order. Should the system be idle, a new order will go to each robot on a 'round robin' principle, such that no single robot performs more than around 25% of the total workload. Selected compounds are arrayed in delivery racks to the customer's chosen format in the following way. Trays are pulled from the appropriate shelf, and the storage tray containing the requested sample picked and placed beneath a puncher, which pushes the sample tube into a delivery rack held immediately below. Once all samples in that order have been punched, the delivery container is delivered to an input/output (I/O) buffer for collection and further processing by the system operator. The combined production of SREMP and the equivalent system in Groton (GREMP) satisfies all cherry-picking requirements for Pfizer's Discovery sites globally, each system requiring the attentions of only one full-time operator.

3.26.3.3 Rationale of Storage Conditions

Many pharmaceutical companies manage liquid sample stores, but not all of these operate at − 20 °C. Historically, compound solutions have been stored at temperatures as low as is practically possible in the laboratory environment and storage has invariably been in laboratory refrigerators and freezers, which tend to maintain temperatures of + 4 °C and − 20 °C respectively. It was this practice that drove the functional design specifications of the ALSB and REMP stores. The liquid sample bank of Novartis, in Switzerland, is somewhat different, operating at + 4 °C and keeping stocks in 90% DMSO.[11] The perceived advantages of a less complex refrigeration plant, and samples that do not require thawing, are balanced against the possibility that solutions might be less stable at relatively higher temperatures. However the Novartis compound management operation does replace liquid stocks over a 5-year usage cycle. At the other extreme liquid samples stored at − 80°C tend to be stock for very long-term storage, with manual access facility, as robotic systems are very difficult to maintain and operate at such extremely low temperatures.

Until recently best storage conditions have been opinion rather than data based since few publications had attempted to describe conclusive comparative experiments which could enable identification of optimum sample storage temperatures. Over the last several years, a flood of publications have addressed the issue of optimal storage conditions. Thus, Kozikowski *et al.*[12,13] reported on compound stability at room temperature suggesting that, at least to the 6-month timescale, the majority of compounds remain viable in DMSO solution. For samples stored at − 20 °C, and subject to multiple freeze–thaw cycles, a gradual deterioration in sample quality was observed particularly in the range of 10–25 freeze–thaw cycles. In these experiments plates were left open exposed to the atmosphere for 2 h during the thaw phase to simulate liquid handling processes. No extra peaks were observed in analysis while solid precipitate was noted in many samples, leading to the proposal that compound loss was due to precipitation rather than to degradation. We concur with this explanation and suggest that the increased precipitation can be ascribed to the altered properties of DMSO following water uptake during solution exposure to the atmosphere, rather than to the freeze–thaw cycles per se. In-house data confirms that samples contained in multiuse containers, and exposed to the atmosphere during aspiration, can display significant deterioration with no such degradation being observed from similar samples stored in single-use containment. Further, in earlier studies we noted no alteration in the concentration of samples subject to in excess of 20 freeze–thaw cycles when sample tubes remained sealed during the thaw phase, sampling taking place through self-sealing silicon septa. Introduction of instrumentation designed to minimize water ingress during compound preparation, processing, and storage (e.g., Tekbench and Tekcel tube store/tube server (Tekcel)[14] comPOUND and compiler (TTP Labtech)[15]) should offer alternative ways of dealing with this problem.

The sample repositories at Pfizer utilize the whole range of storage temperatures, as befits the purpose and content of each system, with long-term samples stored at − 80 °C, cherry-pickable liquids at − 20 °C, diluted 'mother' or intermediate stock samples where the DMSO content is reduced to 5% stored at + 4 °C, and other stocks, destined for short-term use, are stored at room temperature as are our solid stores. Again, most storage media are used, assay solutions in 384-well microplates, liquid samples in 384 and 96 formatted minitubes, and the dry collection in 4-dram vials. More recently, we have taken an active interest in standardizing the containers used at the various Pfizer sites, to simplify exchange of samples and to minimize bespoke configuration of robotics software for handling the various different containers. This has not been without difficulty, particularly where local processes at a single site may rely upon a particular container type. To date, however we have succeeded in reducing the globally supported container list from > 50 items to just 10 items, with the additional benefit of being able to bulk buy these fewer container types for all sites at greatly reduced cost.

3.26.3.4 Quality Control of Compound Solutions

If stored dry and under conditions avoiding extremes of temperature, in an airtight container and excluding light, most solid compounds will be perfectly stable for many years. However, this may not be the case for compounds stored in solution, particularly those solubilized in DMSO. Lipinski[16,17] and others[18] have published widely regarding the effect of even small quantities of water on the ability of dissolved compounds to stay in DMSO solution, particularly when such solutions are exposed to repeated freeze–thaw cycles. Now that the importance of maintaining DMSO in a totally dry state is understood, it is still challenging to maintain since the hygroscopic property of DMSO invariably results in some water uptake by the solvent during compound dissolution and sample distribution. Moreover, the legacy sets of many companies have been constructed with scant regard for water uptake since the impact of this is only now being widely appreciated and this has bearing on the quality of these samples today.

Published observations, particularly those of Popa-Burke[19] and Carmody,[20] indicate that the measured concentration of compounds in DMSO solution is far less than the intended target concentration. However, other studies[12,13,18] while showing a wide variation in measured concentration compared to expected, fail to correlate this observation with marked change in compound integrity, or with the number of freeze–thaw cycles experienced by the solution. Analysis, in-house, of representative batches of compounds from the Pfizer liquid compound set support the observations from Amphora and GSK, and find a wide spread of concentration distributed around a value significantly lower than the original target concentration. A similar spread was detected by analysis of freshly prepared solutions using evaporative light scattering detection (ELSD) (for methodology see Mathews *et al.*[21]) suggesting that a large part of the observed variation from target concentration is introduced as part of the initial solution preparation and not due to inappropriate storage conditions.

Thus variation from intended compound concentration can be ascribed to a combination of sample quality, the initial sample preparation process and the inherent solubility of the compound in DMSO. When pure crystalline material, purchased from a reputable supplier, was weighed and solubilized, close to expected target concentration was reached. However our analysis of compounds routinely submitted for solubilization show the presence of up to 30% of other materials, including trapped solvent, water of crystallization, and column residues from the purification process, findings supported also by work at ChemRx[22] and Novartis.[23] This, combined with recorded variations of up to 3 mg for the same destination tubes going through routine handling, taring and Genevac processes, contribute to the lower than expected actual concentration. Pfizer teams are committed to development of a standard best practice that can be applied at all global sites, which will result in less reliance on tare weights, and more on analytical determination of solution concentration to best ensure accuracy of the resultant concentration and hence confidence of end users in the quality of the compounds stored in a company's liquid repositories. An important part of this best practice will be inclusion of quality control analysis, at various stages of sample preparation, such that solutions of less than acceptable quality, do not enter the collection. For solutions already existing in the set, ELSD analysis at decision-making points (e.g., IC_{50}) in the hit to lead process, will provide additional data to assist Therapeutic Areas to make go/no-go decisions on further development.

While our findings suggest that, when stored under suitable and well-controlled conditions, the quality of a liquid compound file will not change markedly over many years, this is not a universal finding, as others suggest that degradation of compounds over time can mean that screening sets have to be rebuilt on a regular basis.[24] Differences may be ascribed to the structural classes represented in the different collections, but reconstitution would also offer the opportunity to revitalize a screening set with new structures of high purity and predicted attractive physicocochemical properties and is something that would likely take place even in the absence of compound solution quality deterioration.

3.26.3.5 Customer Interaction

Despite the complexity and diversity of Pfizer's compound storage systems, inventory control allows each Pfizer site to monitor quantity and location of any sample in Pfizer's multimillion strong compound collection. This is achievable via the GCM software system giving each Pfizer scientist access to the corporate materials management database. All Pfizer global compound stores are linked to an ordering software application, MatTrack, which directs requests for compounds to the order fulfilment system most convenient to the location of the scientist placing the order, facilitating speed of order turnround. In addition, MatTrack can be accessed to track the status of specific orders, the stock level of any particular compound, and the geographical location of every liquid and solid batch of that compound.

The automated systems described are a complex combination of hardware and software applications that operate by taking in a customer order and delivering what is required without any further interaction. In practical terms intervention with the order process and interrogation of status can be necessary and is often useful and the teams responsible for operating the order fulfilment systems have a professional interest in the level of interaction they can offer customers in the prosecution of their orders. Each of the several thousand scientists from all of Pfizer's sites can place orders against the Sandwich systems, and some may be unaware of the full range of orderable material available to them. In addition, orders can be placed incorrectly by, for example, ordering replicates when only a single sample is required, or by ordering too much or too little for the screening operation. It is in circumstances such as these that the human element of the order fulfilment systems, the operating team, provide an invaluable extra dimension to the automation.

An Active Order Listener, a software tool, allows operators to examine the order profile and using their experience they can often identify erroneous entries in the order list. Communication with the order originator usually allows any mistakes to be rectified, and the order can then be reset for processing. In addition to reactive response to customer orders, the team encourage scientists to contact them to discuss special requirements and conditions associated with their order.

An important aspect of compound storage is the interaction of the storage team with colleagues responsible for preparing the assay ready plates from the store delivered compounds, contributing to the smooth running of the liquid handling devices to dilute and reformat compounds as well as making the final delivery of plates. This interaction of equipment, process, and trained staff has proved to be very efficient, and orders placed in the morning can be completed and delivered to the scientist's laboratory by early afternoon of the same day.

3.26.4 Processing Compounds for Use

Setting aside those compounds obtained as the result of mergers, all compounds reaching materials management fall into two categories: (1) those synthesized to enrich the content of the compound collection, for use in HTS and with no single target in mind, and (2) those synthesized primarily for activity against a particular target and submitted specifically for testing in an appropriate screen(s). We can take these two categories as use cases to describe how a sample logistics process can operate.

In the first case, compounds are processed for distribution to all participating screening sites, there to enter storage until required for HTS. Our CoE model provides for this via the creation, from each input plate of compounds, of an array of products ranging from screening sources, through cherry-pickable tube arrays, to supersources destined for long-term storage, ultimately to replenish the first two products.

3.26.4.1 Center of Emphasis Model and High-Throughput Screening

Figure 3 shows the input and output streams for our CoE model. Development of this model for materials management has proved a major success in the goal to achieve global, corporate, best practice. As a general rule, the Groton CoE supplies liquid sample orders originating from the US mainland sites at La Jolla, St. Louis, Ann Arbor, Cambridge, and Groton itself, while the Sandwich CoE supplies UK and Japanese orders. The ordering software directs orders from originating sites to the appropriate CoE, but where samples have limited availability, the order is automatically redirected to the sister facility for completion. The inventories at each store are mostly replicated, with specific sample nonavailability usually arising as a consequence of shipping delays when transferring new material across the Atlantic to the individual stores. The capacity and performance of each store is such that either could support the whole of Pfizer's global cherry-picking requirement, albeit with a limited performance, during the course of a Groton or Sandwich disaster recovery program.

3.26.4.2 Primary Screen Supply: BasePlate

The SREMP and GREMP stores are principally designed to supply liquid samples to screens as follow-up to hits emerging from primary HTS. Also supplied direct from the CoEs are stock plates containing every sample in the screening set, and from these stock plates each Discovery site conducts it's own supply to primary HTS. The process flow used at Sandwich is illustrated in **Figure 4**. Supply of samples to these primary screens, based on a single point determination for every compound in the Pfizer collection, is sourced at the Sandwich site from BasePlate, a rapid plate replication device supplied by The Automation Partnership, Royston, UK. Stock samples are stored at $-20\,^{\circ}\mathrm{C}$ in 384-well plates at 4 mM concentration and from these copies are diluted to 200 μM. It is from these lower-concentration 'Mother' plates that multiple 'Daughter' plates for HTS are replicated. While the 4 mM stocks are designed to last a number of years, the mother stocks have a shelf-life of a matter of months, supplying primary screening daughter plates to robotic screening systems on a just-in-time basis. The same basic steps are followed at all Pfizer sites though exact detail, and equipment in use, will vary by location.

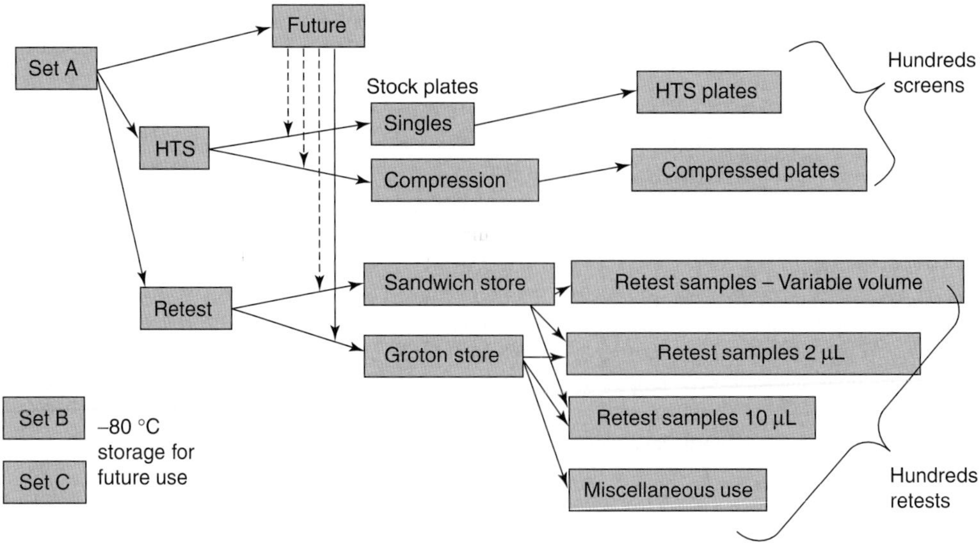

Figure 3 Process flow for the center of emphasis (CoE) model. From initial input of newly synthesized compounds, multiple copies in a variety of formats are created both for immediate use and long-term storage.

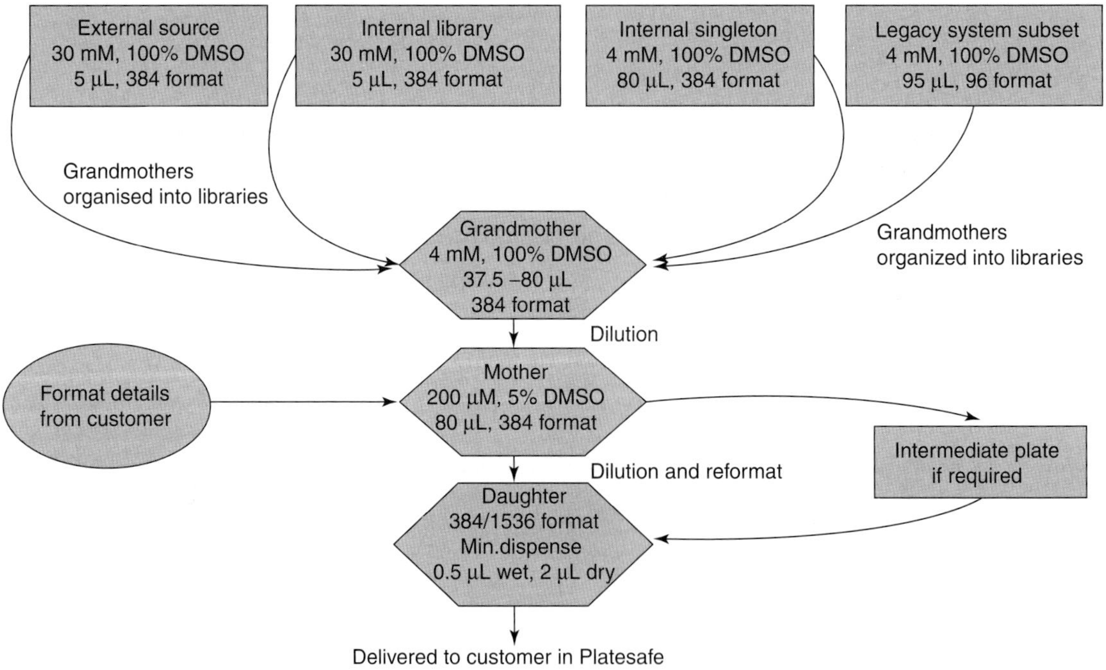

Figure 4 Production workflow for preparation of HTS plates.

3.26.4.2.1 Subsets and compressed sets

With the Pfizer compound library of multiple million samples, screening just a section of this huge collection is an increasingly attractive proposition in an environment of rapid hit seeking and budgetary constraint. While the only guaranteed representation of the collection is the full set, several tools exist to explore a large portion of the file's diversity and representation of its chemical space, by analyzing selections of compounds.

Other sets, compiled for a range of purposes, are available and have been shared across all Pfizer sites. These include a random set, assembled by simply sampling the same well in the entire plated collection, and this is the subset of choice with which to validate a screening protocol. Additionally, specific collections have been assembled, for example directed against G protein-coupled receptors (GPCRs) or kinases, selecting those structures with higher probability of activity against these biological targets, and a 'bug box' selected for known penetration of the cell wall. A recent addition to the subset battery is a collection of diverse, representative compounds covering both parallel chemistry and traditional medicinal chemistry collections. This Global Diverse Representative Subset (GDRS) was defined using computational techniques to maximize coverage of chemical space while proportionately representing the various chemical series and combinatorial libraries existing in the full collection. GDRS has been further subdivided into overlapping tranches representing, for example low-molecular-weight 'fragments,' druglike, Lipinski 'rule of five' compounds, bases, neutrals, etc., permitting a project to select a subset particularly suitable for a specific target where practical information on target specificity is available. Confirmed actives from screens using GDRS are not only followed up per se, but generate an iterative search for similar structures around the hit molecule, and hits from these follow-up compounds generate further iterative searches until suitable lead material has been identified. Review of historical data demonstrates that screening iteratively in this way over 40% of available hits can be identified by screening just 5% of the total compound collection. Alternatively, if the original hit is a member of a chemical library series, that entire library can be screened, compounds being selected from the plate-based store, rather than by individual cherry-picks.

BasePlate works in conjunction with a proprietary software package, Mosaic (Titian software plc),[25] allowing partitioning of the total inventory into libraries. In addition to the subsets described above, which are classed as libraries for this purpose, a library can be, for example, those compounds from a specific supplier, a particular collection from a synthetic chemistry program, or compounds directed toward a particular biological target. Mosaic efficiently combines plates from the same library (as defined) within the same container set (plated compounds are stored, 26 at a time, in Platesafe magazines) allowing subsets of the collection to be rapidly selected by magazine rather than by individual plate. Where individual plates are required, the Mosaic inventory permits BasePlate to locate individual required plates, returning these to the correct library set after sampling. The Platesafe design protects against evaporative loss hence the plates can be stored unsealed, facilitating rapid and automated liquid handling to produce output in 384 or 1536 format.

Subset screening is just one of the ways in which screening costs can be contained, and it will always have the disadvantage that it does not test every structure in the corporate collection. An alternative, in use at several Pfizer sites, is compression: multiple compounds per well, each compound appearing in at least two wells, and formatted using a self-deconvoluting algorithm allowing identification of the single active component in a mixture. Current compression schemes have limited flexibility in terms of screening concentrations (due to compound load) and alternative compression algorithms are in development which may be more applicable to screens requiring higher concentrations of test compounds, or that utilize reagents sensitive to a high compound load. The vast majority of Pfizer screens are developed for use in 384-well microplates, and an increasing percentage ($\sim$50% at Sandwich currently) are further miniaturized to 1536 format, thus economizing on reagent and maintaining a reasonably high daily throughput to maintain the viability of screening against the entire set for appropriate targets. This usage profile mimics, to some extent, that of the pharmaceutical industry as a whole. For larger pharma the favored formats are 384 and 1536, with a bias currently still to 384, while smaller pharma and biotech are still happy to screen in a mix of 96 and 384 formats – testament to the smaller compound libraries that are addressed.[15]

3.26.4.3 Compounds Synthesized for Activity against a Specific Target

Unlike the process described above supporting HTS and follow-up, a very different system is used in support of chemistry-led disease area projects, although this process often is initiated as a result of hit series identified from an HTS or subset screen. At Pfizer, this process in its entirety is known as 'closed loop.' Similar processes, with the same aims in mind, are in place at many pharma locations.[26] The basic assumption behind closed loop is that, if more information on a compound's activities is available at an early stage in series selection, a better decision can be made. This means not only testing the compound activity against the primary target, but also accruing selectivity data, ADME data, safety science data, as appropriate during the first round of screening. To achieve this it has been necessary to increase the throughput of many of these additional screens, which had not previously handled large numbers of primary screen compounds, but in addition it has been important to devise a rapid, reliable, and efficient way of preparing compound plates to supply all of these additional tests so that little extra preparation time by the biologist was required, as illustrated in **Figure 5**.

It has been necessary to modify many of our sample handling methods to cater for this new process flow, and this has been done in an iterative fashion in collaboration with chemistry and biology teams. Designed to be accessible to teams

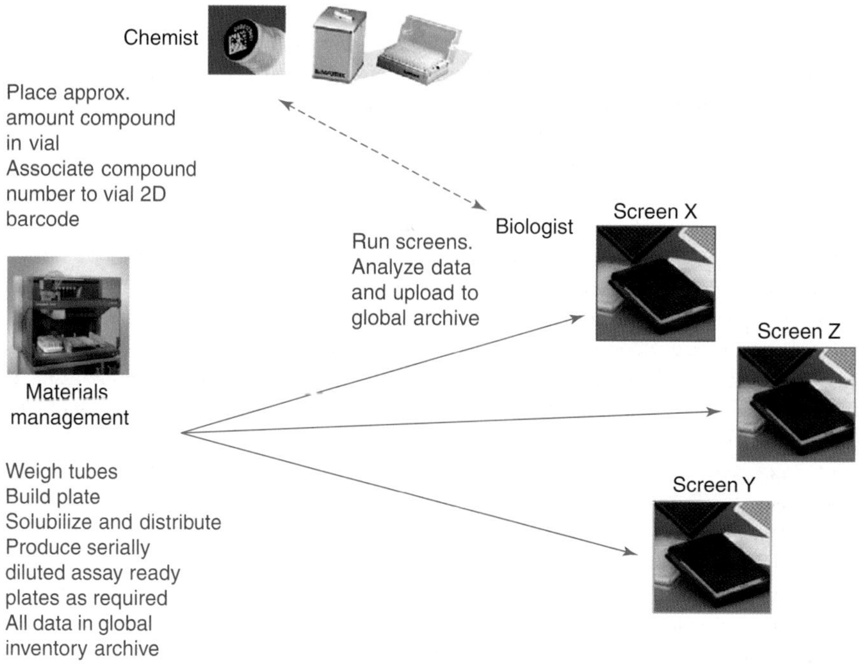

Figure 5 Outline of the sample flow within the 'closed loop.'

with a range of resources and equipment across the Discovery sites, the process accepts a set of standardized inputs and produces a range of standardized outputs both from newly acquired material and also from short-and long-term stores. While the inputs and outputs must change as business needs and technology changes, the intent is that the formats will remain standardized and few in number, with new formats being introduced only where clear business need can be described. The end result has much more in common with a production line than with a traditional chemistry or biology laboratory, and we have learned to use many of the tactics of the production line as a result, e.g., any critical piece of equipment must be duplicated, even if this means some redundancy in the system. Redundancy in high-value capital equipment is not always an obvious choice to make but where reliability of supply is critical and other processes are totally dependent on the supply, the extra spending is justified.

3.26.4.4 Hardware Selection and Trade-Offs

Our initial equipment selection for closed loop sample logistics was based upon a fully integrated robotic platform centred on a Tecan Evolution.[27] Ancillary workstations enabled this system to accept input in either standard microplates or in REMP tube racks from our cherry-picking store. The input compounds could then be sampled and diluted to dilution series in a variety of buffers and at a range of concentrations to feed the various screens selected for each closed loop project, all without user intervention. Although a single system initially had capacity well in excess of what was required, throughput did soon become rate-limiting since customer demand increased rapidly as the benefits of the process became evident. Additionally the system offered a clear single point of failure. In designing the back-up process we opted for a less integrated system which nonetheless offers the potential of greater throughput. By producing the dilution series in 100% DMSO at higher than final concentrations (i.e., treating it as a mother plate), the daughter plates for each screen can be created by dilution in a variety of buffers in plate replication mode. This modification to the process has multiple benefits:

1. Compounds with stability or solubility problems in aqueous media are better handled.
2. Problems with 'sticky' compounds and cross-contamination are minimized.
3. Throughput is much increased since the slow dilution series step is conducted once only for each compound set regardless of the number of screens it will feed.
4. The process no longer relies on one robotic platform, but additionally can run on any combination of our random access liquid handlers and plate replicators.

The importance of appropriate handling of compounds at this dilution stage cannot be overstated. If all problems in achieving accurate concentrations in 100% DMSO source plates (Section 3.26.3.4) are overcome, it is still possible to negate this achievement by precipitation of the compound at the aqueous dilution stage. Properly handled, this issue can be minimized but the absolute aqueous solubility of a compound cannot be changed and so it is equally important to determine the screening concentration achieved. Currently, it is technically challenging to carry out concentration estimates in the 1–100 μM range on a production basis[28] and so estimates of aqueous drug solubility can be a useful pointer to potentially unreliable screening data.[29]

The secondary process, as described, is no longer fully automated, requiring human intervention between random access liquid handler and plate replicator, but due to the greatly increased throughput enabled by the combination of this with the fully automated process we currently have no requirement for additional overnight unattended running. Should capacity again be exceeded by increased demand, the decision to opt for semiautomation would have to be balanced alongside the requirement for an extended working day or shift system. We also have realized, quite early in this venture, that machine time may not be limiting but the ability of a single scientist to organize the multiple, varied outputs for different project teams will be. Relying as we do on a spreadsheet detailing the requirements (plate formats, volumes, concentrations, buffers, etc.) for many screens for each project limits the number of projects we can accommodate. Efficiencies may allow some slight increase, but to offer support to more chemistry-led projects, which is our goal, requires software to assist management of the workflow. Together with software suppliers Titian plc we have developed a workflow application to guide and assist the operator in processing inputs and producing outputs for closed loop. Based on assignment of each compound to a panel of screens associated with the relevant project and target stage, the workflow manager will aggregate compounds by group onto plates and guide the user to the correct processing requirement for that group of compounds on scanning of the plate barcode. The workflow will also accommodate concatenation of compounds from across groups onto contiguous plates destined for generic assays (e.g., human ether a-go-go related gene (hERG)) resulting in availability of a battery of information against appropriate multiple targets for each compound.

3.26.5 Future Development

A principal drawback to automated systems is that they are, in general, relatively inflexible in what they can do and deliver. However, the needs of screening scientists are varied and cannot all be satisfied by the limited deliverable range of the compound stores. We are currently developing processes by which more concentrated solutions, at 30 mM, can be held in the REMP store inventory. While not available to the global screening community, these solutions could be ordered and used by store administrators to replenish depleted 4 mM stocks, and to supply those screens for which 4 mM is either not enough material, or the solvent concentration would be toxic to the assay reaction mixture. Finally, these stocks could act as source material for future subset assembly, reducing reliance on the slower and less efficient ALSB.

We expect to assimilate new storage formats, integrate new output formats, and further develop sampling and dilution protocols where these can benefit our customers.

Among emerging technologies, dispensing well plates[30,31] offer promise as both a storage and rapid, noncontact dispensing device if challenges around evaporation and compound stability can be overcome. Additional quality control measures, e.g., acoustic auditing[32] for volume confirmation and checking of DMSO water content, will be introduced as we further understand the ideal conditions necessary to maintain compound integrity. Real-time concentration and purity assessment of the sample in parallel or just in advance of screening will be a goal but with no single analysis method currently able to detect all compounds at assay-type concentrations this still seems far off. Radio frequency identification tags (RFID),[33] assuming price drops as volume use increases, will confer ultimate traceability for both storage and transfer of samples within and between discovery sites, automatically tracking the contents of sample containers.

A reduction in the amount of each compound synthesized can be anticipated on economy grounds. Bearing in mind the increasing difficulties in obtaining accurate sample concentrations as weight available decreases, we can nevertheless imagine synthesis designed to support the intended raft of primary screening and no more – no long-term storage, no inventory to manage. Since the majority of compounds in early stage screening fall at the first hurdle there are interesting economies to be had if this small-scale synthesis and processing could be achieved reliably.

References

1. Koppal, T. *Drug Disc. Dev.* **2004**, *7*, 48–50.
2. Spencer, P. A. *Eur. Pharm. Rev.* **2004**, *2*, 51–57.
3. Yates, I. *Drug Disc. World* **2003**, *4*, 35–42.

4. Sofia, M.; Stevenson, J. M.; Houston, J. *Pharmaceut. Disc.* **2005**, *5*, 22–31.
5. Wong, J. F. *Gen. Eng. News* **2005**, *25*, 46–47.
6. McGee, P. *Drug Disc. Dev.* **2005**, *8*, 42–44.
7. REMP AG. http://www.remp.com (accessed May 2006).
8. The Automation Partnership. http://www.automationpartnership.com (accessed May 2006).
9. RTS Life Sciences. http://www.rts-group.com (accessed May 2006).
10. Schopfer, U.; Hoehn, F.; Hueber, M. *Eur. Pharm. Rev.* **2005**, *10*, 68–73.
11. Schopfer, U. *Podium Presentation*, 1QPC Meeting on Compound Management and Screening, London, May 2004.
12. Kozikowski, B. A.; Burt, T. M.; Tirey, D. A.; Williams, L. E.; Kuzmak, B. R.; Stanton, D. T.; Morand, K. L.; Nelson, S. L. *J. Biomol. Screen.* **2003**, *8*, 205–209.
13. Kozikowski, B. A.; Burt, T. M.; Tirey, D. A.; Williams, L. E.; Kuzmak, B. R.; Stanton, D. T.; Morand, K. L.; Nelson, S. L. *J. Biomol. Screen.* **2003**, *8*, 210–215.
14. Tekcel. http://www.tekcel.com (accessed May 2006).
15. Comley, J. *Drug Disc. World* **2005**, *6*, 59–78.
16. Lipinski, C. A. *Podium Presentation*, IQPC Meeting on Compound Management and Screening, London, May 2004.
17. Oldenburg, K.; Pooler, D.; Scudder, K.; Lipinski, C. A.; Kelly, M. *Comb. Chem. HTS* **2005**, *8*, 499–512.
18. Cheng, X.; Hochlowski, J.; Tang, H.; Hepp, D.; Beckner, C.; Kantor, S.; Scmitt, R. *J. Biomol. Screen.* **2003**, *8*, 292–304.
19. Poppa-Burke, I. G.; Issakova, O.; Arroway, J. D.; Bernasconi, P.; Chen, M.; Coudurier, L.; Galasinski, S.; Jadhav, A. P.; Janzen, W. P.; Lagasca, D. et al. *Anal. Chem.* **2004**, *76*, 7278–7287.
20. Carmody, C.; Blaxill, Z.; Besley, S.; Farrant, D.; Taylor, N. *Poster Presentation*, ELRIG Meeting: Advances in High Content Screening, High Throughput Screening and Compound Management, Stevenage, UK, June 2004.
21. Mathews, B. T.; Higginson, P. D.; Lyons, R.; Mitchell, J. C.; Sach, N. W.; Snowden, M. J.; Taylor, M. R.; Wright, A. G. *Chromatographia* **2004**, *60*, 625–633.
22. Yan, B.; Fang, L.; Irving, M.; Zhang, S.; Boldi, A. M.; Woolard, F.; Johnson, C. R.; Kshirsagar, T.; Figliozzi, G. M.; Krueger, C. A. et al. *J. Comb. Chem.* **2003**, *5*, 547–559.
23. Letot, E.; Koch, G.; Falchetto, R.; Bovermann, G.; Oberer, L.; Roth, H-J. *J. Comb. Chem.* **2004**, 7, 364–371.
24. Tulsi, B. *Drug Disc. Dev.* **2004**, 7, 39–45.
25. Titian Software plc. http://www.titian.co.uk (accessed May 2006).
26. Brideau, C.; Hunter, J.; Maher, J.; Adam, S.; Fortin, L. J.; Ferentinos, J. *J. Assoc. Lab. Automat.* **2004**, *9*, 123–127.
27. Tecan. http://www.tecan.com (accessed May 2006).
28. Lane, S.; Boughtflower, B.; Mutton, I.; Patterson, C.; Taylor, N.; Blaxill, Z.; Carmody, C.; Farrant, D. *Poster Presentation*, ELRIG meeting: Compound Management, Emerging Themes, Stevenage, UK, April 2005.
29. Pitt, A. *Pharmaceut. Disc.* **2005**, *5*, 46–49.
30. Keating, S. *Drug Disc. Dev.* **2005**, *1*, 46–48.
31. Steger, R.; Bohl, B.; Zengerle, R.; Koltay, P. *J. Assoc. Lab. Automat.* **2004**, *9*, 291–299.
32. Ellson, R. *Podium Presentation*, ELRIG meeting: Compound Management, Emerging Themes, Stevenage, UK, April 2005.
33. Ng, S. *Insights* **2005**, *18*, 16–17.

Biographies

Wilma W Keighley currently leads a department responsible for all aspects of medium-throughput plate based screening, i.e., screening for the hit to CAN process, at Pfizer's UK site (Sandwich). With a remit covering reagent and compound provision, assay development and execution, and data management, the team is around 60 strong and screens on behalf of all six Therapeutic Areas located at the Sandwich site.

Dr Keighley has a background in pharmacology, obtaining her degrees from Glasgow and Southampton Universities. She carried out postdoctoral research in the Department of Clinical Pharmacology, Southampton University, UK, funded by the Parkinson's Disease Society and with Prof Stephen Thesleff at the Institute Farmakologiska, Lund University, Sweden before joining British American Tobacco Co. Ltd. as a research scientist. Several years later, Wilma joined Pfizer, spending time in preclinical safety and drug candidate support, and in the cardiovascular therapeutic zone, before transferring to the New Leads team specializing in support for automated screening where her main focus has been in the integration of the various aspects of a screening effort to ensure that the physical world is reliably reflected in the database. She played a major role in introducing the first fully automated screening platforms to Sandwich and the first automated liquid compound storage and retrieval system to Pfizer.

Over the past 8 years Wilma's team has grown to provide engineering, automation and data support for all compound management and HTS at the Sandwich site. Recent organizational changes, centralizing all Therapeutic Area plate-based screening in one group, have resulted in the opportunity to put into practice for the hit to CAN program those lessons which have been learned in HTS, realizing the efficiencies that standardization can bring to the primary screening program.

Dr Keighley is married with two teenage daughters and lives near to the Pfizer site in South East England.

Terry P Wood is currently team leader of the group responsible for the storage and delivery of the corporate compound collection, at Pfizer's research center at Sandwich, UK, using a number of automated repositories.

With degrees in chemistry and biology, Terry joined a new group at Pfizer in 1987 charged with exploring the emerging science of HTS. The requirement to produce quality data and maintain motivation of the laboratory team in those early days led Terry, together with global colleagues and vendors, to develop and apply a range of semiautomated solutions to the repetitive tasks required to screen a few thousand compounds each week – meagre throughput by today's standards, but a tough challenge when the level of automation was a hand-held repeating pipette. From challenges such as these Terry's keen interest in automation and robotics developed.

Foreseeing that sample supply might become rate limiting to the screening process, Pfizer commissioned one of the pharmaceutical industry's first fully automated liquid stores, a project in which Terry was a key player in the design and testing team, and leading the way for him to commit full time to his current role of managing sample storage and supply at Sandwich.

Toby J Winchester is a materials management specialist within a centralized plate-based screening team supporting all Therapeutic Area targets at Pfizer's Sandwich site. In this role, Toby leads a matrix based team responsible for all compound supply to chemistry led hit to CAN projects.

Toby originally joined Pfizer's Veterinary Medicine Division as a fermentation specialist working in the natural products area, but found that his experience and skills in process control served him well in the new role of sample logistics. Initially focusing on sample logistics specifically for HTS, Toby and his team have gradually brought the same benefits to the less standardized low-and medium-throughput world of Therapeutic Area primary pharmacology.

Comprehensive Medicinal Chemistry II
ISBN (set): 0-08-044513-6

ISBN (Volume 3) 0-08-044516-0; pp. 561–576

3.27 Optical Assays in Drug Discovery

B Schnurr, T Ahrens, and U Regenass, Discovery Partners International AG, Allschwil, Switzerland

3.27.1	**Introduction**	**577**
3.27.2	**Absorption/Emission-Based Assays**	**578**
3.27.2.1	Immunoassays	578
3.27.2.2	Assays Generating Colored End Products	580
3.27.2.2.1	Oxidoreductases and glyoxalase (direct assay format)	581
3.27.2.2.2	Caspase-3 (direct assay format)	581
3.27.2.2.3	Phosphatases (semidirect assay format)	582
3.27.2.2.4	Protein–protein interactions (indirect format)	582
3.27.3	**Radioactive Assays**	**583**
3.27.3.1	Technologies for Radioactive Assays	584
3.27.3.1.1	Filtration assay	584
3.27.3.1.2	Scintillation proximity assay (SPA) technology	584
3.27.3.1.3	Flash plate technology	585
3.27.3.2	Targets for Radioactive Assays	585
3.27.3.2.1	G protein-coupled receptor	585
3.27.3.2.2	Kinases	586
3.27.3.2.3	Scintillation proximity assay bead assay for long-chain fatty acyl coenzyme a synthetase	588
3.27.3.2.4	Scintillation proximity assay bead assay for translation factors	588
3.27.3.2.5	Flash plate assays for deoxyribonucleic acid polymerase, primase, and helicase activities	588
3.27.3.2.6	Flash plate assay for poly(ADP-ribose) polymerase-1 (PARP-1) inhibitors	589
3.27.3.2.7	Image flash plates and imaging beads in peptide–protein binding assays	589
3.27.3.2.8	Scintillation proximity assay for human cytochrome P450s (CYPs)	589
3.27.3.3	Comparison of Radioactive Assays with other Formats	590
3.27.4	**Bioluminescent and Chemiluminescent Assays: Kinases, Proteases, Enzyme-Linked Immunosorbent Assays**	**590**
3.27.5	**Optical Biosensors**	**592**
3.27.5.1	Technology	592
3.27.5.2	Surface Plasmon Resonance	593
3.27.5.3	Absorption, Distribution, Metabolism, and Excretion (ADME) Application with Surface Plasmon Resonance	594
3.27.5.4	High-Throughput Application with Optical Biosensors	594
3.27.6	**Turbidometric Assays: Isomerases**	**595**
3.27.7	**Conclusions**	**595**
References		**596**

3.27.1 Introduction

A critical component of the hit and lead discovery process is the high-throughput screening (HTS) step. A successful HTS campaign is dependent upon the skills of many different scientific groups, their tools, and systems. Biochemists and medicinal chemists who select the chemicals to be tested and assess the drugability of the targets, assay development specialists who design and develop the test systems, and logistic groups who manage the compound collection and produce assay tools need to collaborate closely.

The design of the screening assay is also a crucial part of the hit and lead discovery process and its format and readout can profoundly influence the identification of truly active compounds that modulate the biological target of interest.

Optical readouts are among the formats most often applied in assays used in the HTS process. Optical readouts make use of measurable changes in some characteristic features of light induced by the interaction of assay components. Therefore, changes in the signal from optical readouts in the presence of chemical test compounds are a direct reflection of the interaction with assay components. However, depending on the selection of the optical readout, the optical properties of compounds can influence assay performance in different ways.

Optical assays can be grouped into different classes:

1. UV to visual absorption/emission spectroscopy, which is based on Beer's law. The law states that the absorbance A of a molecule or assay product is directly proportional to the path length of the measured solution and the concentration of the molecule or product of interest. Absorbance assays remain popular, mainly because they are straightforward to set up, homogeneous, and highly cost effective.
2. Luminescence measurements. These are characterized by detectors, which determine the emission of light from a molecule after an excitation step. Luminescence can be divided into:
   - Photoluminescence: the excitation of the system is induced by photons (light). Fluorescence represents a special form of photoluminescence, where the absorbed energy cannot be transmitted to the environment in the form of oscillations. In the case of fluorescence, rigid molecules remain in the excited stage for a short period of time (nanosecond to microsecond scale) and emit the energy spontaneously as photons. In contrast, in the case of phosphorescence, emission occurs over a longer time period.
   - Chemiluminescence: the excitation of the system is induced by a chemical reaction.
   - Bioluminescence: the excitation of the system is induced by a chemical reaction in a viable organism.

The use of fluorescent dyes and associated systems has become a major readout technology in drug discovery. This topic will be reviewed in Chapter 3.28 of this volume.

The use of radioisotopic tracers has a long history in cell physiology and assay technologies. Radioisotopic assays are still the format of choice for many biochemical assays. The measurement of radioisotopic molecules is performed by liquid scintillation counting, which is the most sensitive method. The liquid scintillator plays the role of an energy transducer, converting energy from nuclear decay into light. The light is then measured by photomultiplier tubes, which amplify the signal and display as counts. Owing to their importance in drug discovery and the fact that light plays an essential role in the quantification of isotopic decay, such assays are considered in the present review.

Optical biosensors have not only become valuable tools in drug discovery for analysis of the kinetic properties in drug–target interactions but also for HTS. Optical biosensors represent a set of technologies that make use of the optical characteristics of light to study molecular interactions. They are designed to reveal changes in the characteristics of light coupled with specific features of the sensor surface, where one of the interaction partners is immobilized. Technologies such as plasmon surface resonance, wave guides, and resonant mirrors partly work without the need of a molecular tag or label and therefore significantly complement most other profiling technologies (for a recent review see [1]).

In this chapter, specific assay types will be discussed in detail. They are grouped into the following categories:

1. absorption/emission-based assays;
2. radioactive assays;
3. bioluminescent and chemiluminescent assays;
4. optical biosensors; and
5. turbidometric assays.

This review will focus on biochemical assays. Cellular assays is the topic of Chapter 3.29 in this volume.

3.27.2 Absorption/Emission-Based Assays

3.27.2.1 Immunoassays

Immunoassays are still one of the most important and powerful techniques covering a broad range of different assay types. The assays are based on detecting and quantifying antigens by the use of antibodies. Either isotopic readouts (radioimmunoassay; RIA), where the antibody carries a radioactive label, or nonisotopic readouts such as enzyme-linked immunosorbent assays (ELISA), where antibodies used for quantification are coupled to an enzyme, are applied. The coupled enzyme converts substrates to produce colored end products. Lack of radioactive waste and high sensitivity compared to RIA achieved with novel enzyme substrates, as well as high specificity by antibody characteristics make ELISAs attractive. Amounts of antigens much less than 1 pg mL^{-1} can be detected in most cases and large numbers of high-quality kits and tools are available from many suppliers.

ELISAs are classified into three types: (1) antibody capture assays; (2) sandwich-type assays; and (3) antigen capture assays (**Figure 1**), all of which require multiple wash steps. These three assay types can be performed under different experimental conditions, antibody excess, antigen excess, as an antibody competition assay, or as an antigen competition assay, therefore offering 12 different experimental formats. Additional assay formats are possible, such as combining the primary antibody with a receptor protein. Garrett and co-workers developed an assay to identify compounds competing with binding of estradiol to the human estrogen receptor (hER) by using the ligand-binding domain of the recombinant hER as the primary 'capture molecule' and a specific anti-17β-estradiol antibody to determine free 17β-estradiol in competition to immobilized 17β-estradiol fused with bovine thyroglobulin (17β-estradiol-BTG).[2] The assay represents an 'indirect competition ELISA': binding of the 17β-estradiol-specific antibody to immobilized 17β-estradiol-BTG is competed by soluble 17β-estradiol. When hER, which does not bind to the coated 17β-estradiol-BTG conjugate, is added as the 'capture molecule,' binding of anti-17β-estradiol antibody to the coated conjugate is increased, which was monitored after adding a secondary horseradish peroxidase-labeled anti-IgG antibody metabolizing 3,3′,5,5′-tetramethyl benzidine to a product that absorbs light at 450 nm. Estrogenic compounds interfere with the hER – estradiol interaction, increase free estradiol, and therefore decrease OD_{450nm}. The antibody has

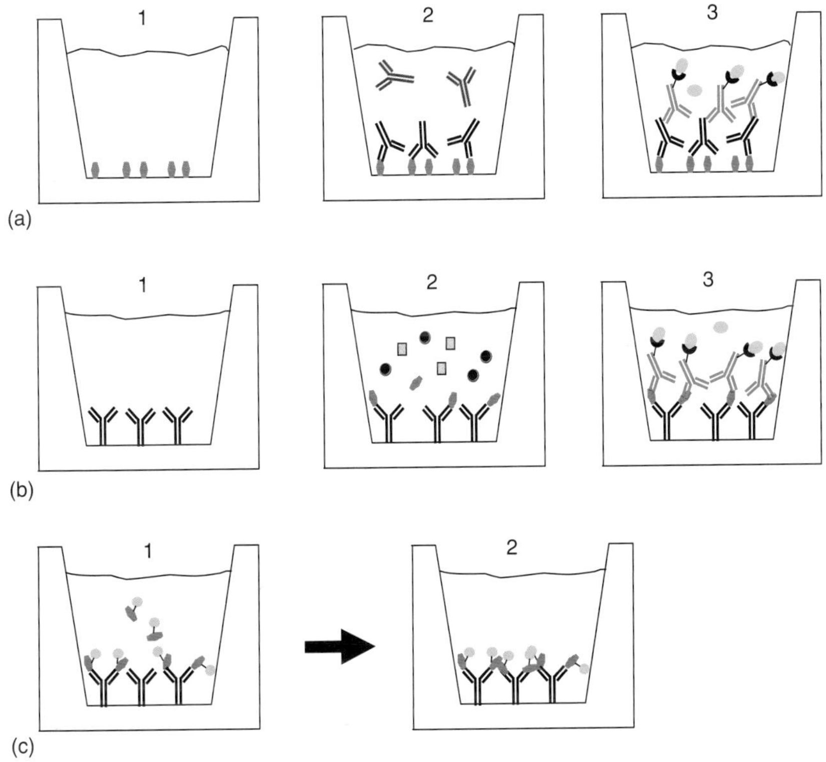

Figure 1 Schematic overview of three principal immunoassay formats.(a) In antibody capture ELISA the antigen is coated to the surface of microtiter plates and only antigen-specific antibodies bind (1,2). This format is typically used for antibody titer determinations. Following incubation with a secondary enzyme-conjugated antibody (3), which targets the primary antibody, quantification is performed by substrate addition resulting in a colored product. (b) Principle of a 'sandwich' ELISA, which quantifies the amount of antigen in a sample captured between two layers of specific antibodies. An antigen-specific antibody is coated to the surface of microtiter plates (1). After antigen binding (2) a second, antigen-specific and enzyme-conjugated antibody is added to the wells (3). After substrate addition the colored product is measured in a spectrophotometer. (c) In antigen-capture ELISA the enzyme-conjugated or directly labeled antigen is incubated on antibody-coated wells (1) of microtiter plates. Bound antigen is quantified by adding an enzyme substrate and following the change in optical properties (2).

to be specific for 17β-estradiol and should not bind estrogenic compounds. The assay is not able to distinguish between agonists and antagonists, but allows initial screening and forms an alternative to isotopic assay variants.

In some cases, an immunoassay represents an alternative to traditional assay systems including the measurement of drug sensitivity of *Plasmodium falciparum* parasites, which uses a labor-intensive in vivo schizont maturation assay defined by the World Health Organization (WHO). Noedl and colleagues[3] examined various approaches to the in vitro assessment of malaria drug sensitivity like an isotopic assay (incorporation of tritium-labeled hypoxanthine) and colorimetric assays based on parasite lactate dehydrogenase (pLDH) and histidine-rich protein 2 (HRP2). The colorimetric assays are performed using hemolyzed culture samples gained from malaria-infected samples. The pLDH levels are measured by double-site enzyme-linked LDH immunodetection (DELI) and the HRP2 levels are measured by a double-site sandwich ELISA. Noedl and colleagues.[3] concluded that for malaria drug sensitivity testing the new ELISA-type assays (DELI and HRP2) will be the methods of choice because of sensitivity, performance, simplicity, dispensability of highly specialized equipment, and applicability to screen for new antimalarial drugs.

ELISA opens up high-throughput alternatives to complementary assay systems such as gel permeation chromatography or fibrinogen affinity gel electrophoresis. For studying sulfated polysaccharide–protein interactions, a new method called SPC-ELISA (sulfated polysaccharide-coating-ELISA) has been described.[4] This assay is based on the efficient coating of sulfated polysaccharides to 96-well plates with high binding capacity (Microlon 600, Greiner, Frickenhausen, Germany), followed by incubation with human fibrinogen and binding detection with a mouse antihuman fibrinogen monoclonal antibody (clone FG-21, Sigma, Deisenhofen, Germany). This paper is an example of selecting the appropriate assay design and type for high throughput. Rosenau and co-workers have transformed complex gel shift assays into HTS suitable chemiluminescent assay systems (*see* Section 3.27.4) to determine deoxyribonucleic acid (DNA) binding of transcription factors.[5] Nuclear factor kappa B (NFκB) and activating protein-1 (AP-1), two transcription factors that play a crucial role in many cellular signaling events, have been used for assay development. After pre-coating of assay plates with streptavidin, biotinylated DNA consensus sequences for either NFκB or AP-1 were immobilized onto the plates. The plates were incubated with cell lysates or nuclear fractions that contained the transcription factors of interest. Bound transcription factors were detected and quantified with specific primary antibodies and a HRP (horseradish peroxidase)-labeled secondary antibody. A chemiluminescent substrate (SuperSignal, Pierce, Rockfort, IL, USA) was used in this case for increased sensitivity and robustness. This type of assay can also be applied to other transcription factors and represents a high-throughput alternative to gel shift assays.

Generally, the greatest challenges in developing an ELISA are choosing the right design and the availability of specific antibodies, which can be polyclonal, affinity-purified polyclonal or monoclonal, enzyme-coupled antibodies, or uncoupled. The optimal concentrations of antibody and antigen have to be determined by titrating the individual components in order to reach the highest possible sensitivity and robustness of the assay. To save assay development time, commercially available ELISA kits can be used, if available. Unfortunately, this is often associated with increased reagent costs. Useful and valuable information for ELISA setup can be downloaded from supplier homepages (e.g., R&D Systems, Minneapolis, MN, USA).

ELISA is mainly used in basic research and preclinical and clinical diagnostic applications with low to medium throughput, but not as often in high-throughput screening because it is not homogeneous and requires several pipetting and washing steps. Typically, the assay procedure involves: (1) coating of microtiter plate with antigen; (2) blocking of unspecific binding sites to reduce background signal; (3) incubation with specific antibodies (primary antibody); (4) incubation with enzyme-linked antibodies directed against the primary antibody; and (5) incubation with a substrate of the enzyme and monitoring the color development. Nevertheless, ELISA is attractive in HTS for several reasons: (1) it is highly sensitive; (2) it is robust and reproducible; (3) it has fairly low-cost special equipment requirements; (4) it has low reagent costs (unless commercial kits are used); (5) different designs are possible; (6) a large set of tools is available; and (7) references for most of the targets are available in the literature. In addition, multiple wash steps reduce the chance for interactions between small molecules and secondary assay elements like primary and secondary antibodies and enzymes coupled to antibodies.

Multiple washing steps are a major disadvantage for achieving high throughput and miniaturization to high-density plate formats (1536-well). ELISAs are performed in 96- or 384-well formats. In summary, immunoassays are not the first choice for HTS, but for some targets they represent a true alternative because of sensitivity, reagent cost, equipment required, robustness, and reliability.

3.27.2.2 Assays Generating Colored End Products

The requirements for an HTS assay are simplicity (few pipetting steps, no washing steps, no isotopic material involved), and the presence of as few assay components as possible to avoid multiple interference points.

Homogeneous assays and direct measurement of end products, due to their optical properties are preferred. Examples include phosphatase assays, ATPase assays, and β-lactamase assays, where results obtained in 384-well and 1536-well formats have been compared.[6] The sensitivity was reduced and signal to background ratios were lower in the 1536-well format, but acceptable Z' values[7] could be reached and the potencies of reference inhibitors were comparable to lower density formats. In addition to the direct homogeneous assays (e.g., metabolized NADPH/NADH or change of absorption properties of a colorless substrate for caspase-3), two additional readout formats are possible: (1) semidirect (determination of free phosphate); and (2) indirect measurements, which require additional pipetting steps but can still be performed in a homogeneous format (coupled enzymes like horseradish peroxidase or alkaline phosphatase). Examples of assays making use of the different readout formats are given in the following sections.

3.27.2.2.1 Oxidoreductases and glyoxalase (direct assay format)

The activity of oxidoreductases and glyoxalases can be monitored by direct measurement of the consumption of substrates or cofactors in a homogeneous assay format. In the case of the oxidoreductases, the reduction of NAD^+ or $NADP^+$ to NADPH or oxidation of NADPH can be quantified by spectrophotometric measurement of optical density (OD) at 340 nm. The enzyme reaction is performed by mixing the assay components including test compounds, incubation, and measurement of the change in NADPH concentration. The main advantages of this assay type are simplicity, inexpensiveness, fast assay development, high throughput using high-density formats, and finally its robustness. Usually, only three pipetting steps are needed (adding compounds, reaction mix containing NADPH and substrate, and finally the enzyme); incubation times range from 30 min to 2 h. With a constant incubation time, assay plates can be measured directly in a spectrophotometer.

The glyoxalase system is part of the methylglyoxal metabolism. Glyoxalase is thought to play a crucial role in insulin-dependent diabetes mellitus (IDDM). Glyoxalase-I and glyoxalase-II are significantly elevated in mononuclear and polymorphonuclear cells from IDDM patients, and assays described by Ratliff and co-workers[8] for both enzymes are easy to adapt for HTS. In the case of glyoxalase-I, methylglyoxal and glutathione are preincubated in order to form hemithioacetal as a substrate for the enzyme. The enzyme activity is determined by detecting the formation of *S*-D-lactoylglutathione at OD 240 nm. The glyoxalase-II activity can be determined by following hydrolysis of *S*-D-lactoylglutathione to D-lactate and glutathione at OD 240 nm. However, at this wavelength interference with small molecules and assay plate material has to be considered. The assays are homogeneous and can be optimized for high-density formats to be used for HTS.

3.27.2.2.2 Caspase-3 (direct assay format)

Caspase-3, also known as apopain, CPP-32, and Yama, is a cysteine protease that is activated early in a sequence of events associated with programed cell death or apoptosis. Usually, assays for identification of inhibitors are those using fluorogenic or colorimetric tetra-peptide substrates in 96-well, 384-well, or even 1536-well microtiter plate format. In addition, radiometric formats can be used and are described in the section for isotopic assays. For spectrophotometric detection, the chromophore *p*-nitroaniline (pNA) is monitored by determining the OD at 405 nm after cleavage from the labeled tetrapeptide DEVD (one letter code for amino acids Asp-Glu-Val-Asp), as shown in **Figure 2**.

In addition to standard microtiter formats, micro fluidic assays with fluorogenic readouts were developed for several caspase isoforms on the Caliper Technologies Labchip platform.[9] The micro fluidic screening technology was compared to plate-based screening in terms of reagent consumption, data quality, and ease of operation. The micro fluidic system has some advantages, but the investment required needs to be taken into account. The methodology of choice is driven mainly by the infrastructure and accessible reader systems available in addition to the extent of usage.

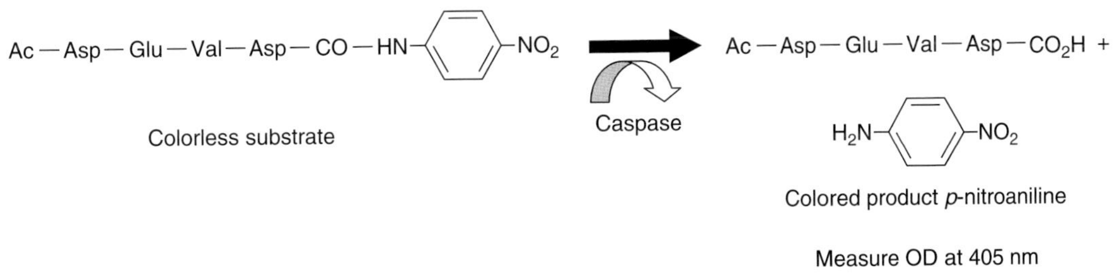

Figure 2 Spectrophotometric detection of caspase activity. Cleavage of the labeled, colorless substrate DEVD by caspase results in the release of the chromophore *p*-nitroaniline, which can be detected by monitoring the OD at 405 nm.

3.27.2.2.3 Phosphatases (semidirect assay format)

Phosphorylation by kinases and dephosphorylation by phosphatases play an important role in cell physiology and regulate metabolic and signaling events. The dephosphorylation of phosphorylated proteins or lipids can be monitored by determining the released inorganic phosphate. Phosphate is captured in a molybdate complex, which changes its absorption properties. The colored complex can be detected and quantified with a microplate reader by measuring optical density between 600 and 680 nm. The colored complex has a broad absorbance peak with a maximum at about 650 nm. The assay principle is outlined in **Figure 3**, and a typical protocol is described in **Table 1**. Usually, only four pipetting steps are needed and the duration of the assay is between 30 min and 1 h.

Classically, inorganic phosphate was determined as a complex of malachite green,[10] and ammonium molybdate[11] for determining the phosphatase activity of calcineurin. This method was further developed and ready-to-use reagents can be purchased (BiomolGreen from Biomol, Plymouth Meeting, PA, USA). This type of assay delivers highly reproducible results (**Table 2**). A disadvantage is the possible interaction of test compounds with the molybdate complex, resulting in false-positive compounds. This requires additional control experiments to exclude unspecific interactions. One of the easiest control experiments is running the assay in thc presence of the potential active compounds in the absence or presence of inorganic phosphate without adding enzyme. In addition, secondary isotopic assays can further help to identify truly active compounds. In cases where inorganic phosphate is released by an enzyme reaction, this type of assay can be applied and is therefore not limited to phosphatases. Other enzyme activities that can be monitored include ATPases, pyrophosphatases, and phosphodiesterases (5′-nucleotidase coupled). Advantages for HTS compared to other methods are: (1) no radioactivity; (2) no excessive mixing; (3) high sensitivity; (4) can be miniaturized; and (5) cost effectiveness. Since the phosphate reagent is highly sensitive, the background might be increased due to free phosphate originating from lab equipment. Therefore, unused plasticware is recommended or items should be rinsed with dH_2O after cleansing with detergents.

3.27.2.2.4 Protein–protein interactions (indirect format)

An assay for screening small molecule libraries was developed by Jin and co-workers[12] to identify active compounds preventing the formation of the helical bundle complex of HIV gp41 molecules, which is important for virus–cell fusion

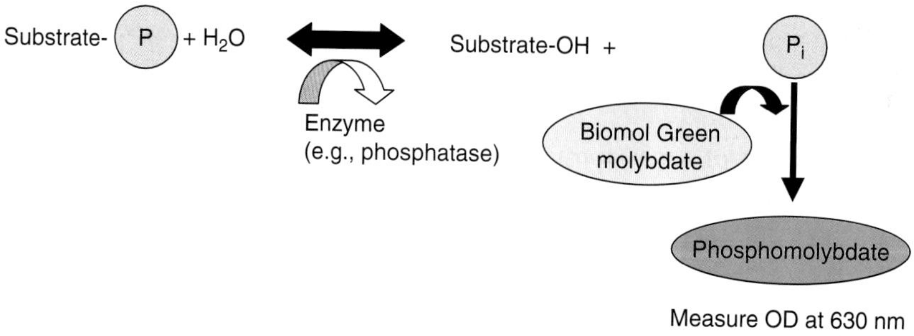

Figure 3 Phosphatase assay with molybdate. Phosphorylated substrate is enzymatically dephosphorylated and inorganic phosphate is released. In a second step, inorganic phosphate is captured in a molybdate complex. The change in absorption properties can be quantified with a microplate reader at 600–680 nm.

Table 1 Phosphatase assay protocol using Biomol Green reagent

Assay plates	*Per well*
Test compound (in reaction buffer)	1 μL
Phosphopeptide(in reaction buffer)	20 μL
Phosphatase(in reaction buffer)	10 μL
Incubate for 10 min at RT	
Biomol Green reagent	60 μL
Incubate for 20–60 min at RT	
Measure OD at 630 nm	

Table 2 Screening statistics of a phosphatase assay using the protocol described in **Table 1**

HTS batches	*Plates (N)*	*Average values of N plates*					
		Z'	S/B	$OD_{100\%}$[a]	$OD_{0\%}$	$CV_{100\%}$	$CV_{0\%}$
HTS batch 1	30	0.7	3.8	0.264	0.071	4.10%	9.29%
	30	0.8	4.4	0.286	0.066	3.18%	4.74%
	60	0.8	4.0	0.279	0.069	4.29%	4.74%
	90	0.8	4.3	0.296	0.070	3.02%	5.22%
	94	0.8	4.3	0.311	0.072	3.18%	5.73%
	70	0.9	5.2	0.366	0.070	2.23%	4.81%
HTS batch 2	100	0.8	4.0	0.288	0.071	2.69%	5.12%
	100	0.9	4.6	0.316	0.069	2.71%	4.38%
	100	0.9	5.2	0.359	0.070	2.35%	5.13%
	100	0.9	4.7	0.325	0.070	2.37%	4.25%
	50	0.9	4.2	0.277	0.067	2.50%	3.81%
	100	0.8	3.7	0.288	0.077	2.78%	5.18%

In total 131 648 compounds were tested in batch 1 and 193 600 compounds in batch 2. The assay was performed in 384-well plates. The results (Average Z′, S/B, CV values) demonstrate the robustness of this assay type. For abbreviations see **Table 3**.
[a] OD = optical density.

after viral gp120-CD4 interaction. For the assay, designated as a modified ELISA, two fusion proteins were constructed. Thioredoxin was fused to the N-terminal helix of gp41 (Trx-N) and horseradish peroxidase fused with glutathione *S*-transferase (HRP-GST) was fused to the C-terminal helix of gp41 (HRP-GST-C). Three N-helices, which form a trimeric coiled-coil, and three C-helices are involved in the formation of the fusion-active helical bundle complex. Trx-N is coated on the surface of the microplate. If the binding of HRP-GST-C is inhibited by peptides or small molecules, no activity of HRP can be recovered after washing and staining with *o*-phenylenediamine dihydrochloride as a substrate. As a signal, OD at 490 nm can be monitored with a standard absorption reader. The authors demonstrated the suitability for HTS reaching an average Z' value of 0.89 and were able to identify active compounds that could be validated in an in vivo model.

Aviezer and colleagues[13] developed an assay that was used for the identification of compounds inhibiting the interaction of Fibroblast Growth Factor and its receptor. Fibroblast growth factor-2 (FGF-2) was immobilized on heparin precoated 96-well plates. A fusion protein (FRAP) of the extracellular FGF receptor-1 (FGFR-1) region with alkaline phosphatase (AP) was used in the assay, and was found at high levels in conditioned medium of transfected NIH 3T3 cells. The binding of FRAP to the immobilized FGF-2 was detected by monitoring the optical density at 405 nm after the added AP substrate *p*-nitrophenyl phosphate was metabolized by the enzyme. The assay was optimized and adapted for running on a robotic system. At least one potent inhibitor was identified and confirmed using an isotopic cellular binding assay (FGFR-1 expressing CHO cells) and a proliferation assay with bovine aortic endothelial cells. This example of a sensitive and robust assay is made possible by: (1) a low number of pipetting steps; (2) nonisotopic principle; and (3) exclusion of influence of colored compounds, which are washed out before the detection solution is dispensed.

3.27.3 **Radioactive Assays**

Even though radioactive assays are associated with several disadvantages, such as (1) safety issues, (2) possible contamination of automation units, (3) handling of radioactive waste, (4) specific infrastructure, (5) difficulties in miniaturization, (6) exposure of personnel, and (7) labor-intensive nature, they still play an important role in drug discovery for several target families (G protein-coupled receptors (GPCRs), kinases, proteases, transporters, and others). Historically, filtration assays formed the standard for isotopic assays, and it was not until the 1990s that true alternatives were developed. These new technologies are homogeneous scintillation proximity type assays provided by Amersham (SPA) (GE Healthcare, Little Chalfont, Bucks, UK) and PerkinElmer, Welleslay, MA, USA (Flash Plate). Reader systems and disposables are offered and innumerable application notes are available from both companies for many different target classes.

One of the major advantages of radioactive assay systems is the direct measurement using radiolabeled ligands or substrates. The label usually has no impact on the shape and structure of a ligand or substrate and has no influence on the affinity and reaction performance. This also makes radioactive assays very attractive as secondary assays for any validation of active compounds identified in a nonisotopic HTS assay format.

3.27.3.1 Technologies for Radioactive Assays

3.27.3.1.1 Filtration assay

Filtration assays, which are usually performed in the 96-well format, mainly because no higher density filtration unit is available, are used for HTS for ligand binding to cell membranes and proteins. The binding reactions are performed in polypropylene or polystyrene plates and the reaction mix is then transferred onto the filter plates via a vacuum system (Millipore, Billerica, MA, USA; TomTec, Hampton, CT, USA; PerkinElmer, Welleslay, MA, USA; Brandel, Gaithersburg, MD, USA). After washing and drying, filter plates have to be sealed on the back side, scintillation liquid has to be added, and the plates have to be sealed on the top before they are measured in a liquid scintillation counter (MicroBeta; TopCount, both PerkinElmer). Recently, new developments have shown that the 384-well format also can be applied either using Millipore's $\text{MultiScreen}_{\text{HTS}}$ 384-well filter plates, as described in a case study in association with Sanofi-Aventis. This case study describes an inhibition filtration assay for identification of inhibitors for FabD enzyme (malonyl-coenzyme A: acyl carrier protein (APC) transacylase). More than 270 000 compounds were screened. Substantial time saving, high reproducibility with Z' values reaching 0.75, and a low background make this type of filtration assay attractive for screening campaigns. As an alternative, 384-well filtration assays can be performed using a specific type of the Brandel Harvester (Brandel, Gaithersburg, MD, USA), which allows the reaction to be run in a 384-well plate and harvests the reaction mix into four 96-well filter plates. The advantage of this system is the usage of any commercially available 96-well filter plate. Systems from Brandel and Millipore can be fully integrated in automation units. Radioactive filtration assays are often the method of choice for certain assays, not only because of simplicity but also because of the quality of data generated, particularly in binding assays, where other labels could interfere with binding. Additionally, the influence of colored compounds or quenching effects is reduced to a minimum, since the test compound is washed out before the addition of scintillation liquid.

3.27.3.1.2 Scintillation proximity assay (SPA) technology

SPA technology uses coated resin beads that contain a scintillant. Upon binding of an isotopically labeled assay component to the beads, there is an increased likelihood that particles from radioactive decay cause light emission from the scintillant compared to unbound components (**Figure 4**). The light output can be quantified by a photo multiplier tube (PMT)-based scintillation counter or by a charge coupled device (CCD)-based image reader. For the two reader principles Amersham developed two different types of beads: the SPA scintillation beads (PMT) and the SPA imaging beads (CCD). Choice of bead type depends on different parameters: (1) instrument, (2) throughput, (3) format, and (4) emission. There are some advantages to using the SPA imaging beads such as higher throughput and adaptation to a 1536-well plate format. The emission lies in the red region (615 nm), which reduces color quenching caused by yellow/orange/red compounds, usually represented in a higher portion than blue compounds in screening compound collections. The characteristics of the radioisotope are important in SPA assays and the shorter the path length of the decay particle the more suitable the radioisotope. Tritium and iodine-125 are most suitable, but SPA has also been successfully applied using carbon-14, sulfur-35, and phosphorus-33. The homogeneous SPA technology is open to many applications (receptor-ligand binding, radioimmunoassay, signal transduction/molecular interactions) and screening of many different target classes (receptors, kinases, transferases, proteases, nucleases, lipid modifying enzymes, DNA/RNA modifying enzymes). A large variety of beads coated with different coupling molecules like WGA (wheat germ agglutinine), glutathione, copper chelate, nickel chelate, and streptavidin and innumerable application notes are available as well as many publications on the subject.[14–17]

Recently, PerkinElmer (Welleslay, MA, USA) developed FlashBlue GPCR beads, which are based on the same principle as that described for the SPA beads from Amersham. Bound radioisotope induces emission of blue light that can be detected by liquid scintillation counters like TopCount and MicroBeta as well as by luminescence counters in 96-well and 384-well formats. FlashBlue GPCR beads are specifically designed for high-throughput homogeneous

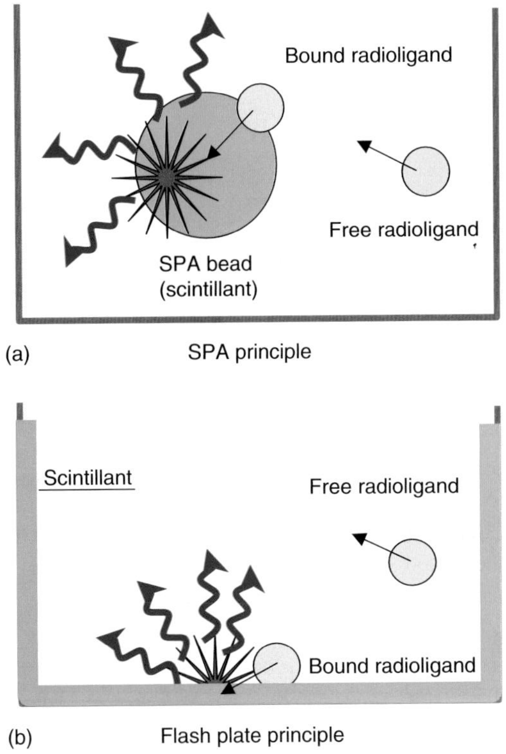

Figure 4 Principles of SPA and FlashPlate technology. (a) Radioligand binds to the coated surface of SPA beads and induces emission of light. (b) Scintillant is embedded in the surface of the assay plate. Radioligand bound to the surface of the assay well leads to the emission of light. Free radioligand, not proximal to the scintillant, causes only minor background emission.

GPCR-radioligand binding assays. They consist of a polystyrene core with a hydrophilic coating, and WGA is covalently attached to capture cellular membranes.

3.27.3.1.3 Flash plate technology

The Flash Plate technology is also based on the proximity of radioligands and scintillant, which in this case is embedded in the surface of an assay plate, the Flash Plate (PerkinElmer, Wellesley, MA, USA). Instead of immobilizing a molecule on the surface of a free-floating bead, the molecule has to be immobilized on the surface of the assay well. This allows additional washing steps, which could be important for increasing the S/B ratio, especially if radioligands are used with longer path length, like phosphorus-33. The possibility of immobilizing molecules on the surface of a Flash Plate has improved since different coatings are available such as glutathione, goat-anti-mouse AB, myelin basic protein, nickel chelate, phospholipid, protein A, sheep-anti-rabbit AB, streptavidin, and WGA. With most coatings, plates are available in the 384-well format. As for SPA, tritium and iodine-125 are most suitable, but with carbon-14, sulfur-35, and phosphorus-33 the Flash Plate technology can also be applied. By analogy to the SPA Image Beads, PerkinElmer (Wellesley, MA, USA) offers the ImageFlash Plate with a special scintillant that emits in the red region of the spectrum reducing the interference of colored compounds. Application notes can be found on PerkinElmer's homepage.

3.27.3.2 Targets for Radioactive Assays

3.27.3.2.1 G protein-coupled receptor

GPCRs still represent one of the most important target classes in drug discovery. High-throughput assays are performed today using stably transfected cell lines overexpressing the receptor of interest. Either whole cells or membranes are used for biochemical and in vitro testing. Assays applied to identify and study GPCR modulators are summarized in **Figure 5**.

Several of these assay techniques make use of radioisotopes. One of the most robust methodologies and usually straightforward to set up is the displacement assay, where a high-affinity radioactively labeled ligand is displaced from

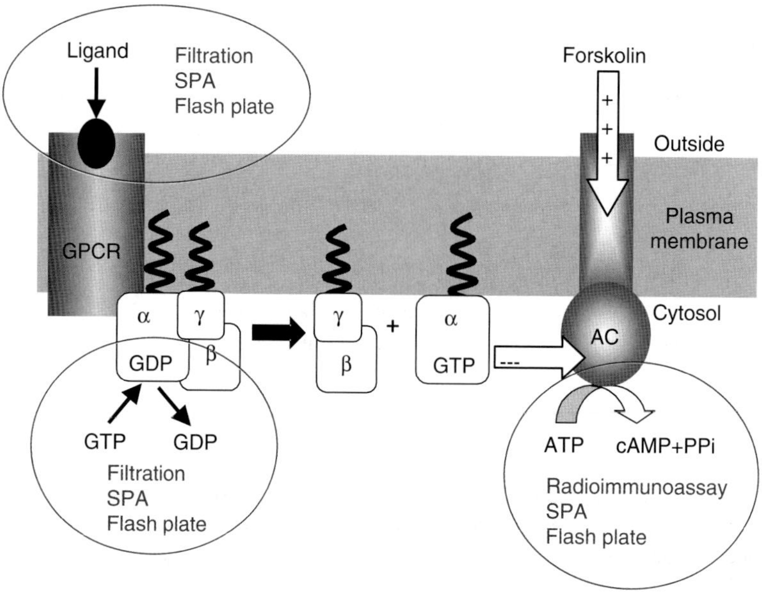

Figure 5 Assays applied to profile compounds interfering with GPCR functionality. Isotopic assay variants for identification of GPCR modulators. For ligand binding either membrane fractions or whole cells can be used for all given technologies. In the case of monitoring GTPγS binding, only membrane fractions can be used. cAMP determination is performed after lysis of cells, which were stimulated with ligand, in the presence of forskolin and small molecules for a distinct period (example of GPCR that inhibits AC).

the receptor by test compounds. Filtration assays are currently the most reliable technique for this purpose, and can be run in 384-well format, as recently published by Millipore. Nevertheless, it is still considered as one of the most labor-intensive assay formats. Alternatives, like SPA or Flash Plate formats are available and reduce the labor cost substantially, since no separation of the bound from the unbound radioactive ligand is needed and the assays can be adapted to higher density formats. However, not all receptors can be easily assayed using SPA or Flash Plate technology. The robust and reliable displacement-screening assay should then be considered as an alternative. In this context, it needs to be considered that even though filtration assays are more labor intensive, material costs are substantially lower.

The successful development of SPA based screening assays was described for histamine receptors[18] and for the adenosine 2a receptor using WGA-coated yttrium silicate (YSi) or red-shifted yttrium oxide (YO) beads.[14] For the A2a-selective radioligand [^{3}H]-SCH 58261, the assay gave identical K_d values compared to a filtration assay. Z' values in the range of 0.7–0.8 were obtained. The homogeneous format makes this assay suitable for HTS, and the use of an imaging system for signal detection increases the productivity. To validate the system, the library of pharmacologically active compounds (LOPAC, Sigma Aldrich, St. Louis, MO, USA) was screened with this newly developed SPA technology and 18 hits could be identified, whereby 12 were either adenosine-based A2a agonists, xanthines, or other reported adenosine receptor antagonists. The other six compounds, which showed activity in the SPA assay, did not show activity in the filtration assay. Reasons for this difference between assays are not clear and have also been described for other targets like kinases.[15]

3.27.3.2.2 Kinases

Despite the availability of several alternatives, isotopic assays played and still play a very important role in the identification and profiling of kinase inhibitors. Filtration assays were among the first used in the early days of kinase inhibitor discovery. The principle of the kinase filtration assay is the separation of phosphorylated [^{32}P]- or [^{33}P]-labeled substrate from the donor of the labeled phosphate, [γ^{32}P]ATP, using filtration units. Histone H1 was commonly used as substrate for serine/threonine kinases[19] and polyGly-Tyr was mainly used for tyrosine kinases as phosphate acceptor. With minor modifications, the filtration assay principle can be applied for all kinases. Parameters that need to be adjusted include incubation time, substrate concentration, choice of substrate, ATP and enzyme concentrations, as well as buffer components. The assay development time is usually very short for radioactive filtration assays in comparison to nonradioactive alternatives. This is particularly true for serine/threonine kinases, where the development of antibodies specifically recognizing the phosphorylated substrates often posed the major obstacle. In contrast, many phosphotyrosine-specific antibodies are commercially available and are used in alternative assay technologies. However, the main advantages of the filtration assay for kinases are (1) the fast setup phase, (2) no influence of secondary effects like interaction with other components, (3) high specificity and sensitivity, (4) allows exact characterization of binding sites, and (5) robust, reliable, and high S/B ratios, therefore reaching very good screening statistics and Z' values >0.8.

Researchers tried to avoid the tedious filtration steps in screening, and the development of alternatives to isotopic filtration assays gained in importance in the 1990s. The first Flash Plate assays for tyrosine and serine/threonine kinases using different substrates and [^{33}P]ATP were described (Flash Plate, Schnurr and Schächtele, application note # 6). For both the protein kinase C and the vascular endothelial growth factor receptor kinase (KDR) assay, stability and high reproducibility could be shown. Results were confirmed with the filtration approach. This assay was used as a basis for other kinases and screening campaigns were run on an automation unit composed of a Biomek2000 (BeckmanCoulter, Fullerton, CA, USA) and a Quadra96 (Tomtec, Hampton, CT, USA).

The Flash Plate technology is still used for HTS and for the identification of small molecule inhibitors of kinases. Sun and colleagues[20] developed an assay for the centrosome-associated serine/threonine kinase STK15/Aurora2 using myelin basic protein (MBP) coated Flash Plates. With a high signal/noise (S/N) ratio and Z' values of 0.69 the robustness of this system could be demonstrated. Flash Plates were coated overnight with MBP, followed by washing prior to addition of a compound reaction mix containing [^{33}P]ATP. The kinase reaction was started by the addition of enzyme and the reaction was stopped by washing. Comparison with the alternative LANCE technology demonstrated higher S/N ratios using the isotopic Flash Plate technology.[20]

The SPA technology can also be applied for kinase assays by immobilizing the substrate on the beads.[15,21–24] Here, a biotinylated peptide substrate was captured by streptavidin-coated SPA beads. This design is not working for all kinases and alternative designs were developed as described for protein kinase A (PKA).[16] In this study avidin-coated polylysine YSi beads were used for capturing a biotinylated PKA substrate peptide. The assay was optimized for the 384-well format with regard to the following parameters: (1) amount of beads; (2) avidin concentration; and (3) ATP concentration. With the optimized conditions, the K_m values for the substrate and ATP were determined and found to be close to the those reported in the literature. Enzyme titration as well as kinetic studies were performed, which showed that the reaction runs in a linear fashion for up to 100 min. Assay development was concluded by determining the effects of EDTA, DMSO, and the stability of the 'stopped reaction mix' on assay performance. In a small screening campaign an S/N ratio between 5 and 9 and Z' values between 0.6 and 0.8 were obtained. In comparison to SPA assays using streptavidin beads, this assay using polylysine YSi beads is more cost effective.

Flash Plate technology currently allows the use of up to a 384-well format. In contrast, the SPA technology was successfully adapted to the high-density 1536-well format, dramatically reducing assay costs while increasing throughput.[17] In brief, a serine/threonine kinase assay was established using streptavidin-coated LEAD seeker imaging beads (GE Healthcare, Little Chalfont, Bucks, UK), biotinylated peptide substrate, and [^{33}P]ATP. The assay was optimized and compared to the 384-well format with respect to the screening process in general, costs, and results. The reaction volume for the 384-well format was 50 μL and for the 1536-well format 7 μL which allowed reagent costs to be reduced by a factor of about 7, and the bead costs by a factor >9. The LEAD seeker imaging system detects luminescence by means of a CCD camera, allowing simultaneous recording of entire plates. This technology dramatically increased the throughput in comparison to PMT-based readers.

Drug discovery efforts have been hampered so far by the lack of lipid kinase assay formats suitable for HTS. Conventionally, PI3K activity has been measured using solution-based assays with phospholipid vesicles. The reaction has to be terminated by acidified organic solvents and subsequent phase separation by extraction or thin-layer chromatography. Assays using solid-phase immobilized phospholipids have been reported for the identification of phospholipid-binding proteins and antibodies. A radiometric, robust in vitro PI3K assay in 384-well format has been described using a solid-phase immobilized phospholipid as substrate for PI3Kγ.[25] Crude lipid extract as a source for phosphatidylinositols (PtdIns) offers an economic advantage over using the physiological substrate PtdIns(4,5)P2. PtdIns, which has been shown to be a substrate for PI3K, was immobilized to Maxisorp plates (Nalge Nunc International, Rochester, NY, USA) and transfer of [^{33}P]phosphate from radiolabeled [γ^{33}P]ATP to PtdIns was analyzed. Known PI3K inhibitors like wortmannin, quercetin, and LY294002 were tested in the MaxiSorp plate assay for PI3Kγ inhibition and IC_{50} values were in good correlation to previously reported values in other assay formats. An HTS with 469 plates resulted in an overall mean S/B of 19.2 and an overall mean $Z'=0.66$. However, high day-to-day and intraday variations were observed (6–15% coefficient of variance (CV) for compound median signal and 15–35% CV for background median signal). This was mainly due to the washing steps required in the protocol.

3.27.3.2.3 Scintillation proximity assay bead assay for long-chain fatty acyl coenzyme a synthetase

Fatty acyl coenzyme A (CoA) synthetases are a class of enzymes involved in the activation of fatty acids through ligated high-energy CoA thioester bonds. This activation step is a prerequisite for intracellular fatty acid uptake and the biosynthesis of glycerolipids, cholesterol esters, and triacylglycerols. Acyl CoA synthetases are involved in cell proliferation and cell signaling processes and represent a class of interesting drug target molecules.

Within the acyl CoA synthetase class there have been five isoenzymes described. The SPA methodology was adapted to screen modulators of acyl CoA synthetase 5 in a homogeneous assay format.[26] The reaction was performed in the presence of coenzyme A and [^{3}H]-palmitate as substrates and binding of the reaction product [^{3}H]-palmitoyl-CoA to YSi-SPA beads was used as readout system for acyl CoA synthetase activity. Owing to the difficulty of assaying this class of lipid-utilizing enzymes this approach might be useful for other family members like fatty acid transporting proteins (FATPs) and other acyl CoA transferases.

3.27.3.2.4 Scintillation proximity assay bead assay for translation factors

In prokaryotes translation initiation is mediated by initiation factors (IFs) IF1, IF2, and IF3. Their function is the formation of a 70S ribosome complex that places the mRNA and the formyl-Met-tRNA (fMet-tRNA) in the correct position for starting the elongation process. New anti-infective drugs are in high demand and strictly bacteria-specific and conserved proteins are ideal targets for new drugs. Owing to the lack of structural information, the identification of specific IF2 inhibitors, for example, is restricted to high-throughput screening of chemical libraries and natural products.

IF2 activity assays have been developed that measure the GTPase or the fMet-tRNA binding activity but require multiple steps. A primary screening assay based on the SPA methodology has been developed.[27] IF2 and the IF2-C domain, which binds fMet-tRNA, were biotinylated to allow specific capture by streptavidin-coated beads. The biotinylated proteins were incubated in the presence of [^{3}H]fMet-tRNA and the resulting complex was bound to SPA beads. The assay was validated in a 96-well format by screening of $\sim$4500 compounds. Inhibiting compounds were defined by a residual activity $<$50% binding and the hit rate was 1.5%. Of these, 24% were verified by retesting and IC_{50} values were determined. The strongest inhibitor showed an IC_{50} value of 11 μM. This homogeneous assay allowed the rapid identification of inhibitors of the fMet-tRNA and IF2 interaction. These inhibitors can form the basis for a lead selection and optimization program.

3.27.3.2.5 Flash plate assays for deoxyribonucleic acid polymerase, primase, and helicase activities

DNA replication proteins represent a class of extremely well established anti-infective drug targets. A number of anti-infective therapies are directed against enzymes involved in DNA replication (e.g., virally encoded DNA polymerases or bacterially encoded DNA gyrases). DNA polymerases, primases, and helicases share the common feature of utilizing a DNA substrate and requiring nucleoside triphosphates for enzymatic activity.

The Flash Plate technology has been adapted to a wide range of proteins in this drug target class using a common set of reagents.[28] The approach has allowed the rapid characterization of DNA polymerase, DNA primase, and DNA helicase activities.[28] In principle, biotinylated DNA oligonucleotide substrates were tethered to streptavidin-coated scintillant-embedded Flash Plates. To test for specific functions adaptations were necessary for each enzyme.

DNA polymerization reactions are conventionally assayed via scintillation counting of acid precipitable [^{3}H]DNA after capture on glass fiber filters, scintillation counting of DEAE-filtermat-captured [^{3}H]DNA products, or bead-based scintillation proximity assay. For DNA polymerase (*Escherichia coli* Klenow fragment) reaction measured by Flash Plate technology, a short complementary oligonucleotide primer was annealed to the tethered oligonucleotide and polymerization was measured by incorporation of [^{3}H]dNTPs at the 3′ end of the primer. When comparing $K_{m,app(dATP)}$ values using the Flash Plate assay and filter assay equivalent data were obtained (43 and 65 nM, respectively). For the DNA polymerase assay an S/B = 6 and $Z' = 0.62$ were obtained when the assay was performed in 96-well Flash Plates.

Analysis of DNA primase activity is normally monitored by denaturing gel autoradiography of [^{33}P]- or [^{32}P]UTP incorporated into short RNA oligonucleotides or by scintillation counting of acid-precipitable product. Sensitivity can be increased indirectly by coupling RNA primer synthesis to DNA polymerization with the addition of exogenous DNA polymerase and [^{3}H]dNTPs. In the Flash Plate DNA primase assay, direct synthesis of short oligoribonucleotides complementary to the tethered oligonucleotide was measured by incorporation of [^{3}H]rNTPs or by subsequent polymerase elongation with [^{3}H]dNTPs. For the direct primase assay an S/B = 2.7 and $Z' = 0.39$ were observed, making this assay not optimal for HTS.

In recent years SPA (Amersham Pharmacia Biosciences, Chalfont St Giles, UK) and homogeneous fluorescence-based helicase assays have been developed that usually require the custom contract synthesis of labeled oligonucleotide duplex substrates. The Flash Plate assay for DNA helicase (*E. coli* DnaBi; *SV40T* antigen) offers the advantage of rapid analysis of a wide range of easily synthesized substrates because unwinding of a [^{33}P]-labeled oligonucleotide complementary to the tethered oligonucleotide was measured. Almost identical IC_{50} values were determined using either gel-shift (23 μM) or Flash Plate (20 μM) methods. An S/B ratio of ~2.8 and Z' values ~0.43 were obtained under screening conditions for the helicase assay. In summary, the streptavidin-coated Flash Plate system provides a flexible platform by which a variety of DNA replication proteins can be rapidly assayed in HTS.

3.27.3.2.6 Flash plate assay for poly(ADP-ribose) polymerase-1 (PARP-1) inhibitors

Poly(ADP-ribose) polymerase-1 (PARP-1) plays a pivotal role in the repair of nuclear DNA damage through the base excision repair pathway. Following DNA damage (by genotoxic agents or endogenous sources such as oxygen radicals), PARP-1 binds DNA at the site of the lesion and, using NAD^+ as substrate, synthesizes polymers of ADP-ribose, which are transferred to acceptor proteins and play a role in DNA repair.

An ELISA has been applied to test for PARP-1 activity[29] but the method involves multiple washing steps and is labor intensive. The principle of the Flash Plate assay is based on activation of PARP-1 by binding to double-stranded DNA oligonucleotide. Using NAD^+ and [^{3}H]NAD^+, activated PARP-1 then synthesizes labeled poly(ADP-ribose) chains and after termination of the reaction by adding acetic acid brings the product into proximity with the 0.015 M sodium carbonate/0.035 M sodium bicarbonate precoated Flash Plate wells.[30] An S/B of 20 and a $Z' = 0.62$ were obtained under screening conditions in 384-well plates. Comparisons of IC_{50} values of known PARP-1 inhibitors were comparable to the values achieved in the Flash Plate assay. The application of the Flash Plate methodology is simple and robust and can be used for other polymer-synthesizing enzymes within the PARP family.

3.27.3.2.7 Image flash plates and imaging beads in peptide–protein binding assays

In addition to SPA imaging beads (Amersham Pharmacia Biosciences, Chalfont St Giles, UK), recently Image Flash Plates (PerkinElmer Life Sciences, Wellesley, MA, USA) have been developed that emit light at 615 nm when exposed to β-radiation. CCD cameras are more sensitive for red-shifted scintillation and therefore such systems are well suited to analyze peptide–protein interactions. Binding of a [^{33}P]-labeled protein and a biotinylated peptide was measured in binding assays using either streptavidin-coated Image Flash Plates or SPA imaging beads. Both detection systems were established and compared for HTS in a 384-well format.[31] The assay in Image Flash Plates needed a wash step and the assay using image beads a centrifugation step for best performance. The wash step resulted in lower scintillation counts in the Image Flash Plate, however, Z' values of 0.64 and 0.87 were obtained in plates and with beads, respectively. A reference inhibitor produced similar IC_{50} values in both systems. Red-shifted scintillation methods offer the advantage of minimizing color-quenching effects caused by colored compounds and, depending on instrumentation, imaging beads or Image Flash Plates might offer possibilities to analyze peptide–protein interactions in HTS.

3.27.3.2.8 Scintillation proximity assay for human cytochrome P450s (CYPs)

Inhibition screens for human CYPs are being used in preclinical drug metabolism to support drug discovery programs. Drug metabolism studies play an important role in lead optimization and require the development of HTS assays for Cytochrome P450 (CYP) inhibition. The most commonly used assay to test for CYP inhibition is fluorometric and uses a probe that is metabolized by CYP to a fluorescent metabolite. However, some probes available for microtiter plate fluorescence assays are not CYP-isoform selective and cannot be used with human liver microsomes. In addition, high-throughput LC/MS/MS assays for CYPs are available that are sensitive and selective but difficult to set up. A homogeneous radiometric inhibition screen for CYP2D6 using the SPA principle has also been developed.[32] [*O*-methyl-^{14}C]dextromethorphan was used as a substrate to test for the dextromethorphan *O*-demethylase activity of CYP2D6. Beads coated with poly(ethyleneimine) (PEI) were used to irreversibly capture the product [^{14}C] labeled HCHO of the enzyme reaction by a possible covalent interaction. Applying human liver microsomes or purified CYP2D6 the reaction was selectively inhibited by quinidine, an inhibitor of CYP2D6. In a second set of experiments it was shown that SPA is able to distinguish potent from weak inhibitors.

SPA-based CYP inhibition assays are easy to use, are amenable to automation, and permit the use of either microsome preparations or recombinant enzymes.

3.27.3.3 Comparison of Radioactive Assays with other Formats

A comparison between different technologies can be made on a more theoretical basis by comparing the assay procedures regarding cost, time, and resources. More importantly, the assay format should be able to identify profiles and deliver an optimum of validated compounds. Screening the same compound collection in parallel in different assay systems can give an idea about the importance of the assay format for drug discovery.[15] Three different assay technologies for a tyrosine kinase assay are described in this study. Thirty thousand compounds in mixtures of five compounds were evaluated using either SPA, homogeneous time-resolved fluorescence resonance energy transfer (HTR-FRET), or fluorescence polarization (FP). All assays were conducted in a 384-well format and the validation experiments carried out in the 96-well format. Results demonstrated that the overlap of the identified active compounds was very low between the three different assays after the primary screening and only one compound could be identified by all three different formats. In the next step, dose–response curves were evaluated for all active compounds in all three assay variants. Interestingly, at this level all 40 compounds identified in the SPA assay were active in the HTR-FRET assay and 35 of these 40 compounds were active in the FP assay, but the rank order of activities was different for all of the three assays, and the hit rate for the SPA assay was relatively low, although the cut-off criteria were set less stringent. Only functional follow-up assays will demonstrate which hits develop into valuable leads. The different hits in the different formats could be due to enzyme and substrate concentrations, as well as different types of substrates. The hit rate could be increased by changing the substrate for the SPA assay. Other aspects to be considered include the fact that: (1) SPA works with immobilized substrate in contrast to the fluorescent assays, which run in solution phase; and (2) assay components interact differently in different formats. The selection of the assay format can therefore determine the outcome of an HTS project. Follow-up assays in different formats can be of outmost importance for lead selection.

3.27.4 Bioluminescent and Chemiluminescent Assays: Kinases, Proteases, Enzyme-Linked Immunosorbent Assays

Bioluminescence and chemiluminescence in drug screening have been recently reviewed.[33,34] Bioluminescence is characterized by exergonic reactions of molecular oxygen with different substrates (luciferins) and enzymes (luciferases) resulting in photons of visible light. Factors affecting the color of the emissions are: (1) the amino acid composition of the luciferase; (2) the structure of the luciferin; (3) the presence of green fluorescent proteins (GFPs); and (4) mechanisms that control intensity and kinetics. Luciferases are the enzymes that catalyze the light-emitting reactions in organisms and in in vitro luciferase/luciferin reactions. The spectrum of a bioluminescence often matches the fluorescence spectrum of the reaction product.

Firefly luciferase catalyzes the condensation of luciferin with ATP followed by the reaction with oxygen and a cyclization process of the peroxide; therefore, firefly luciferase is an ATP-activating enzyme, generating AMP as a leaving group (for review see[35]). Assays to monitor ATP levels have been described for cell viability determinations (CellTiter-Glo luminescent assay, Promega Corporation, Madison, WI, USA) and more recently for the determination of kinase activities. Since kinases are enzymes that transfer a phosphate group from ATP to a whole variety of substrates, such as peptides, proteins, lipids, and sugars, monitoring the change in ATP levels can be used as a general principle to measure kinase activity in a homogeneous format. The assay uses luciferase to monitor the decrease in ATP, which is altered in kinase reactions where phosphate is transferred to the substrate (**Figure 6**).

Promega Corporation (Kinase-Glo, Madison, WI, USA) and Cambrex corporation (PKLight, East Rutherford, NJ, USA) both provide assay kits to monitor ATP levels using luciferase/luciferin reactions. The assays can be performed in high-density microtiter plates in a homogeneous format measuring luminescence. A significant advantage of the assay format is its independence from phosphorylation-specific antibodies and therefore its general usefulness for kinases. Since the luminescence signal is very stable, plates can be collected during a high-throughput screen and measured batch-wise. For assay optimization, the optimal amount of ATP and substrate used in the assay needs to be determined for each kinase.[36–38]

Pommereau and co-workers[39] have compared the bioluminescent kinase assay with a micro fluidic assay format. In this case the luciferin/luciferase kit from Lumitech (now part of Cambrex Corporation, East Rutherford, NJ, USA) was used. Once the optimal linear ATP concentration for the serine/threonine kinase had been determined, a set of about 2000 compounds was tested. Among others, compounds with an apparent 'negative' inhibition have been identified in the bioluminescent format. These could be true activators of the kinase, inhibitors of luciferase or compounds, which quench the luminescence signal. Indeed, all compounds turned out to be luciferase inhibitors. In general, the overlap of positive compounds between the two assay formats was >94%. The only limitation of the bioluminescent assay

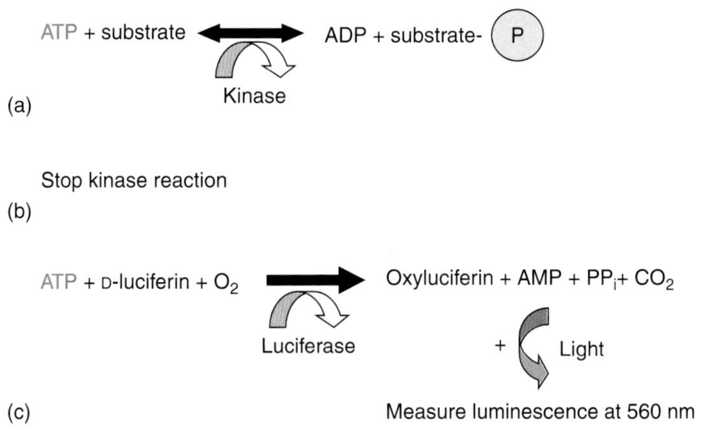

Figure 6 Kinase activity using a luciferase/luciferin system. The assay is based on two enzyme reactions. (a) A kinase-specific substrate is phosphorylated utilizing ATP; (b) The reaction is stopped; and (c) luciferase is added and uses the remaining ATP pool to oxidize luciferin to oxyluciferin under emission of luminescence (560 nm).

Table 3 Kinase assay screening statistics

	Activity controls		
	100%	*0% (staurosporine)*	*0% (no kinase)*
Mean cps (60 min)[a]	603 000	1 915 000	1 981 000
CV (in %)[b]	9.5	3.2	4.0
Number of wells	160	112	112
Z′factor[c]	–	0.74	0.70
S/B[d]	–	3.2	3.3

Kinase activity is determined as a measure of ATP-dependent conversion of luciferin to oxyluciferin and light. 100% control wells (160) were in the presence of kinase whereas 0% control wells (112) were either in the presence of a kinase inhibitor (staurosporine) or without kinase.

[a] Mean cps, luminescence measured as counts per second.

[b] Coefficient of variation $(\mathrm{CV}) = \dfrac{\mathrm{SD}_{(\mathrm{controls})}}{\mathrm{mean}_{(\mathrm{controls})}} \times 100.$

[c] $Z'\mathrm{factor}^{7} = 1 - \dfrac{3 \times \mathrm{SD}_{(0\%\ \mathrm{activity\ control})} + 3 \times \mathrm{SD}_{(100\%\ \mathrm{activity\ control})}}{|\mathrm{mean}_{(0\%\ \mathrm{activity\ control})} - \mathrm{mean}_{(100\%\ \mathrm{activity\ control})}|}.$

[d] Signal/background $(\mathrm{S/B}) = \dfrac{\mathrm{mean}_{(0\%\ \mathrm{activity\ control})}}{\mathrm{mean}_{(100\%\ \mathrm{activity\ control})}}.$

format might be the signal to background ratio dependent upon the ratio between optimal ATP and substrate concentration. The ATP concentration is mainly determined by the K_m of luciferase. The ATP consumption assay is a highly cost-effective and efficient assay for high-throughput use. Typical screening statistics are given in **Table 3**.

The high Z' factor and the S (without kinase or 100% inhibition)/B (full kinase activity) ratio of >3.0 indicate the robustness of the assay. Validation of screening hits in duplicates reinforces the reproducibility of the system (**Figure 7**).

Based on a similar principle making use of the luciferin/luciferase system, a caspase-3 assay has been developed.[40] In this case a caspase-sensitive peptide substrate has been conjugated to luciferin. In the conjugate form, luciferin cannot act as a substrate for luciferase. If the caspase is active and cleaves the peptide, free aminoluciferin is formed and acts as a luciferase substrate leading to bioluminescence. This single step assay has been compared to other caspase assay formats and performed very well. The format should be applicable to other proteases. In conclusion, advantages in using this technology are: (1) assay development is simple and rapid; (2) no label of substrate is necessary; (3) there is no limitation on length of peptide substrate, proteins, lipids, or any other phosphate acceptor; and (4) no expensive reagents are needed.

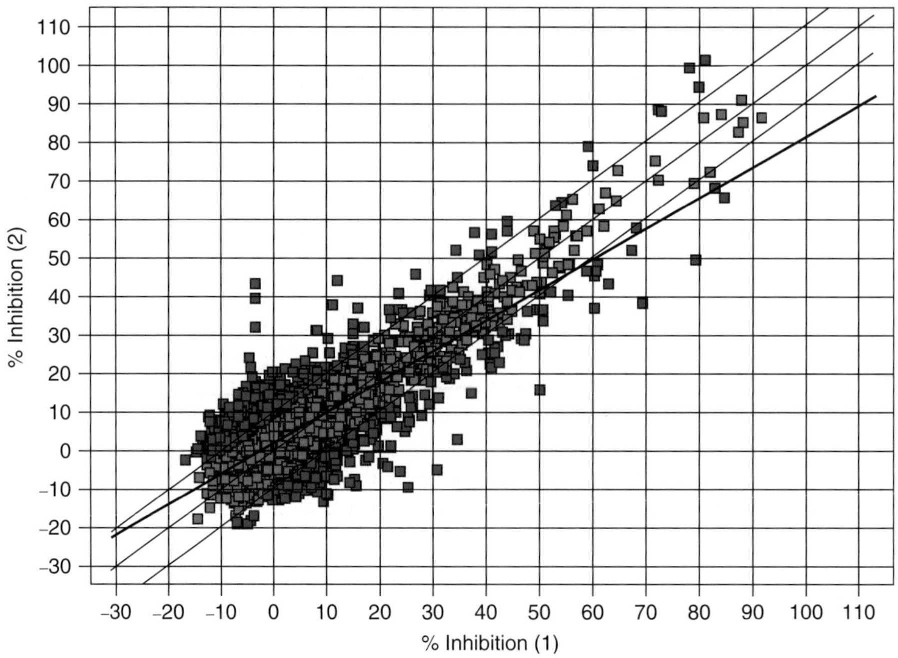

Figure 7 Validation of screening hits in a kinase assay using the luciferase/luciferin system. Primary screening hits were retested in duplicates. Percent inhibition of kinase activity of the duplicate determinations are depicted as a scatter plot. Red squares, inhibition (1) and (2) within 10%; blue squares, inhibition (1) and (2) differ by $>10\%$; $R=0.81$ between % inhibition (1) and (2). (Courtesy of Dr Peter Zbinden.)

ELISAs can be developed using color and fluorescent and chemiluminescent substrates. Chemiluminescence makes use of chemical reactions catalyzed by enzymes coupled to assay components and which, with the proper substrates, generate light. Well-known substrates are luminol oxidized by horseradish peroxidase or dioxethane as a substrate for alkaline phosphatase. ELISA is normally quite cumbersome, requiring several wash steps and can therefore lead to assay errors. In addition they normally need a stop reagent and have a limited dynamic range. Chemiluminescent readouts are not only used with ELISAs, but are the method of choice for branched-chain DNA technology (bDNA, Quantigene, Bayer HealthCare, Tarrytown, NY, USA). This technique is not described here in detail, since it is based on RNA quantification from cellular assays. High-throughput applications have been described.[41]

Electrochemiluminescence[42,43] circumvents many of the negative aspects of ELISA. This technology has been developed mainly by the IGEN corporation (Gaithersburg, MD, USA) under the name Origen Technology. Typically, assay products or constituents of biological samples can be quantified by two antibodies, one labeled with biotin and the other with an ester of ruthenium tris-bipyridine chelate. The complex can be bound to streptavidin-labeled magnetic beads. The bead complex is then carried through a flow cell where it is captured by a magnetic field. Electrochemiluminescence is generated by the conversion of the ruthenium label in the presence of tissue plasminogen activator to active redox states via a series of oxidation-reduction steps by applying an electrical potential. Active states decay by releasing photons at 620 nm. The assay can be performed with an Origen MS-Series Analyzer, which is able to read 384-well formats. High-throughput applications have recently been described for the determination of rat insulin.[44] High Z' values of >0.8 have been obtained and intra- and interassay variation was less than 5%. The assay is homogeneous and therefore involves only one incubation step, and the detection process is fast. The system can be completely automated and the M-Series M8 analyzer can read 384-well plates. This technology offers a robust, reliable, and efficient alternative to classical ELISA.

3.27.5 Optical Biosensors

3.27.5.1 Technology

Optical biosensors offer a significant advantage over many other profiling technologies applied during the drug discovery process, since they do not rely on any type of fluorescent or radiolabeling. The label-free technique leads to a lower number of false positives in drug profiling and also avoids expensive labeling processes. Initially, optical biosensors have been mainly applied in secondary profiling and kinetic analyses of protein–protein or protein–drug interactions, but more recently high-throughput applications have been developed (for review see[1,45]).

Optical biosensors exploit the so-called evanescent wave phenomenon. Evanescent waves are generated when polarized light hits a metal film under total internal reflection. When the total reflection happens in a prism covered with a metal film, the reflected photons create an electrical field on the opposite side of the prism–metal interface. This field is called the evanescent field. When the energy of the photons is such that they can interact with the free electron constellations in the metal film (resonance), the photons are absorbed and converted into surface plasmons (oscillation of free electrons). When the composition of the medium and surface changes, the momentum of the plasmons changes and therefore also the angle of the incident light at which resonance occurs. These changes of the refractive angles are measured in surface plasmon resonance (SPR), e.g., as light of a certain wavelength that is absorbed or reflected by the metal surface. Normally, one of the interaction partners is immobilized on the sensor surface while the other interactant is free in solution and passes over the immobilized interactant. Association and dissociation of the two interaction partners can be measured in real time as a change of the refractive index or resonance angle or change in resonance wavelength. This change is linear to the number of molecules bound. The technology therefore delivers real time and quantitative information on binding events, high degree of sensitivity, and a high dynamic range (millimolar to picomolar sensitivity in affinity constants). Surfaces can be regenerated and used for several experiments.

Optical biosensors and in particular SPR have applications in drug discovery and development from the target to the marketed drug. In this chapter we will focus on applications in early discovery. The reader is referred to recent reviews for in depth discussion on technology and applications.[1,73]

3.27.5.2 Surface Plasmon Resonance

The sequencing of the human genome has revealed a large number of proteins of already known families, such as well-known disease targets like GPCRs, ion channels, and others. For many of these potential drug targets the natural ligands are unknown and no ligands are available, which would enable competition assays for high-throughput ligand-finding approaches to be established. Similarly, many newly identified signal transduction proteins have not been analyzed for potential interaction partners and no small molecules are available for their modulation. Several techniques are available for ligand identification, among them SPR[46,47] where cell lines have been identified expressing a ligand to the orphan transmembrane tyrosine kinase receptor TIE2 using BIAcore (Uppsala, Sweden) sensor chips. The ectodomain of TIE2 was fused to the Fc portion of human IgG and immobilized to the surface of an SPR chip. Control proteins such as fusion proteins with other ectodomains of transmembrane tyrosine kinase receptors were used as controls. Cell lines were identified that had specific binding activities to TIE2. The gene responsible for the binding protein, called *Angiopoietin-1*, was cloned. SPR was used to verify the expression of Angiopoietin-1 in supernatants from transgenic cells using the TIE2-Fc BIAcore chips.

Methods have been developed in recent years to detect bound ligands directly from the chip surface. Mass spectrometry of the bound protein ligand or ligand digests combined with database searches allows the direct identification of the bound protein.[48–50] Recent improvements have made it possible to identify the binding of low-molecular-weight analytes,[45,51,52] broadening the scope for use in drug discovery and development significantly.

The technology has application in the characterization of hits or leads from high-throughput screens or for compounds during the lead optimization cycle. The binding characteristics of compounds can be quantitatively determined and compounds can be clustered according to affinity and structure,[53] therefore going beyond just measuring the effect of compounds on enzymatic activity. The affinity and dissociation rates have been determined for a group of 14 HIV protease inhibitors.[54] While a correlation between binding affinity and K_i could be demonstrated, such a correlation was not found between dissociation half time and affinity. Therefore, the determination of the dissociation half time introduces a new aspect of interaction and compound characteristics. Compounds with relatively low binding affinities or K_i can have favorable association/dissociation characteristics. Hämäläinen and co-workers[55] have expanded the HIV protease inhibitor characterization to 290 structurally diverse compounds. Compounds with low K_i have a high association rate, indicating high recognition of the target, but fast dissociation, indicating poor final fit. This information can be useful to select and optimize lead structures. In addition to the primary target, other proteases were immobilized and a specificity analysis has been performed demonstrating nonspecific compounds with high general protein-binding properties. A kinetic analysis of estrogen receptor–ligand interactions was performed by Rich and colleagues,[56] which indicated that agonists bind to the receptor with association rates much slower than antagonists. This finding supports the evidence that antagonists bind to altered conformations of the receptor. The

SPR analysis allows conclusions to be made regarding the mechanism of action of compounds and can be of help in driving the optimization process.

Stability of the protein during the coupling step can be critical for the development of SPR assays. Casper and co-workers[57,58] have shown that p38 MAP kinase can be coupled to a chip surface in the presence of a reversible, structure-stabilizing kinase-binding molecule. It was again shown that compounds with similar affinity can differ significantly in association and dissociation constants, and that allosteric binders have relatively slow association rate constants. This technology might be useful for other kinases or proteins with known reversible ligands.

3.27.5.3 Absorption, Distribution, Metabolism, and Excretion (ADME) Application with Surface Plasmon Resonance

SPR can be used to provide information concerning ADME (absorption, distribution, metabolism, and excretion) properties of chemical compounds. Human serum proteins, such as albumin or α-acidic glycoprotein, can be immobilized on sensor chips. The analysis consumes a low amount of sample and is rapid and can be automated. As the binding response in an SPR assay is dependent on the molecular mass close to the chip surface, data need to be adjusted to molecular weight of each compound in order to deliver comparable data[59–61]. Danelian and co-workers[62] have used SPR to study the direct interaction between drugs and liposome surfaces. Liposomes of different compositions can be coupled to the dextran matrix. Data indicate that compounds strongly bound to liposomes have a high chance of being highly absorbed in vivo. The technology has the advantage that the liposome surface is very stable and can be used for up to 2 weeks and washed for repeated use.[60,62] The possibility of studying interactions at more physiologically relevant conditions than those under which enzymatic or binding studies are normally performed in vitro represents an important advantage in drug discovery. The methodology is expected to be generally applicable to other proteins provided suitable immobilization procedures can be found (for discussion of generation of surfaces see [63]). The throughput is generally about 100 compounds per 24 h, sufficient for a lead selection or lead optimization program.

3.27.5.4 High-Throughput Application with Optical Biosensors

Several approaches are pursued to increase the throughput. One approach is to organize a large number of flow cells in an array format on a chip. Several companies are engaged in developing protein biochip systems.[64] BIAcore (Uppsala, Sweden) has recently launched its SPR-based array technology and signed the first commercial deal. Applied Biosystems (Foster City, CA, USA) together with HTS Biosystems (Hopkinton, MA, USA) have developed the 8500 Affinity Chip Analyzer, based on grating-coupled surface plasmon resonance. Grating coupling is based on a fine grating on the chip surface, which provides coupling and allows imaging of the entire surface at once.[64] BIAcore has now acquired all assets from HTS Biosystems including the FLEX Chip System. Another approach to increase throughput is based on the use of colloidal gold or silver nanoparticles. Englebienne and co-workers[65] used colloidal gold nanoparticles sensitized with the binding protein. Gold nanoparticles coated with protein are able to directly report ligand association and dissociation by shifts in the SPR band of gold (localized surface plasmon resonance, LSPR). Protein–protein interactions as well as protein–small molecule interactions could be monitored. The same throughput has been calculated to be about 3000 samples per day using a clinical chemistry analyzer. The use of triangular silver nanoparticles in LSPR has been demonstrated with biotin-streptavidin as a model system[66] and more recently applied by Haes and colleagues[67,68] to monitor amyloid β-derived diffusible ligands.

Small molecule microarrays (SMMs) are an expanding field, where small synthetic or semisynthetic molecules are immobilized on a variety of surfaces.[69,70] Arrays are used to probe small molecule–ligand interactions or protein function profiling. The protein-binding partner can either be fluorescently labeled or detected with a labeled antibody to the binding protein. Scientists at Graffinity Pharmaceuticals (Heidelberg, Germany) have developed a SPR-based fragment-screening platform.[71] The microarrays in this case consist of gold-coated micro-structured glass plates. The whole array surface is then coated with a self-assembled monolayer on which chemical compounds are immobilized. A special plasmon reader has been developed to read the >9000-compound-rich sensor slides. A fragment-based affinity fingerprint is generated with what is called 'RAISE' (rapid affinity instructed structure evolution). The technology has been validated by the successful identification of fragments that bind to factor VIIa.[72]

Other label-free optical readout technologies are pursued in high-throughput screening. Gauglitz and colleagues are using reflectometric interference spectroscopy (RIfS).[73] Part of white light is reflected at the interface of a thin layer and another part is reflected at the second interface. The two reflected beams can either be superimposed or form an interference pattern. This pattern can be constructive, where waves add in phase, or destructive, producing smaller peaks than either wave alone. The pattern depends on incidence, wavelengths, and optical density of the layer (given by refractive index and physical thickness of the layer). The properties of the interference pattern are sensitive to changes in or at the layer, e.g., by binding of proteins and ligands. For direct optical detection of binding events, simultaneous imaging with a CCD camera can be applied. Assays can be run in 96- or 384-well formats. A high-throughput assay for the identification of thrombin inhibitors has been developed on the basis of RIfS.[74] In this case plates were coated with a SiO_2 layer onto which a dextran layer was added. A known thrombin inhibitor was covalently attached to the dextran layer, which enabled the detection of thrombin binding and inhibition by small molecules.

A new system called BIND (biomolecular interaction detection system) has been developed by scientists at SRU Biosystems (Woburn, MA, USA).[75] It is based on a new class of optical biosensors, so-called guided-mode resonant filters. They are made from a few homogeneous dielectric layers combined with a grating. This layout exhibits a narrow reflection spectral band due to the grating. The system works such that the incoming light is trapped in the waveguide via evanescent coupling. However coupling becomes leaky due to the grating layer and energy is coupled out of the waveguide into radiation modes. This light interferes destructively with the incoming light, similar to resonance conditions. Outside this resonance the light does not couple and is transmitted or reflected. Since coupling is highly sensitive to the wavelength of light and the angle of incidence, sharp resonant peaks can be observed in the reflected light when these parameters are changed. By applying biomolecules or cells to the surface of the sensor, the resonant coupling of light is modified and so the reflected and transmitted output is tuned. SRU has developed sensors in 96- and 384-well formats. A readout instrument based on a two-dimensional CCD camera determines the peak wavelength values of reflected light at resonant reflection conditions. Binding of biochemical material to the sensor shifts the peak wavelength values to greater wavelengths.[75] As examples for the application of the technology, protein A–IgG interaction and human serum albumin as well as carbonic anhydrase II binding to small molecule drugs were selected. In the case of interaction with small molecules, the protein partner is immobilized on the sensor chip. Equilibrium dissociation constants determined with the BIND system correlate well with published data. The system also allows cellular assays to be performed and can determine cell density, cell death, and detachment. The IC_{50} determined for vinblastine on CHO cells is similar to values determined for other cell lines with other methods. In addition, when the surface of the chip is coated with antibodies directed toward cell surface molecules, the interaction of whole cells with the antibodies could be quantified. This application might be of interest to identify inhibitors of protein–cell or cell–cell interactions. The BIND system offers a sensitive, label-free assay technology with a broad range of applications.

3.27.6 Turbidometric Assays: Isomerases

Turbidometric assays make use of fine suspensions, which are formed during material precipitation such as protein denaturation, either by measuring light scattering (nephelometry) or by determining absorbed light spectrophotometrically. The development of high-throughput assays for certain enzymes has proven difficult. Protein disulfide isomerases represent a protein family with multiple biological functions and with potential implications in disease processes.[76–78] Smith and colleagues[79] have recently described the successful adaptation of a protein disulfide isomerase assay to high-throughput mode using a turbidometric assay. The assay makes use of the catalytic reduction of insulin by the isomerase and the subsequent aggregation of insulin chains in the presence of dithiothreitol. The turbidity can be monitored spectrophotometrically at 650 nm. A signal to background ratio of 3–6 was obtained and the assay was generally linear for 30 min. An important step that enabled a high-throughput mode was the successful conversion of the kinetic assay into an endpoint assay by the addition of H_2O_2, which acts as an oxidant and immediately depletes the reductant dithiothreitol. The assay showed good performance with Z' values above 0.7 in 384-well plates. At least 100 plates could be screened per day making this assay a true high-throughput assay to search for protein disulfide isomerase modulators. The methodology developed should also have broader applications in transforming kinetic turbidimetric assays into endpoint screening assays.

3.27.7 Conclusions

Optical assays play an important role in hit/lead finding and high-throughput screening. ELISA is highly sensitive, robust, and reproducible and in most cases inexpensive. A major disadvantage is inhomogeneity, but a broad range of antibodies and luminescent readout technologies make ELISAs attractive. Assays where colored products are formed represent ideal homogeneous formats. Oxidoreductases, glyoxalases, caspases, and phosphatases form a large group of drug targets that can be assayed directly or indirectly by color changes of components of the assay reaction. Radioactive assays still play an important role in drug discovery. Filtration assays and more recently SPA beads and FlashPlates are important for a broad range of targets: kinases, GPCRs, fatty acid synthases, DNA polymerases, primases, helicases, and CYPs among others. Radioactive assays are easy to set up and robust. A major disadvantage remains, i.e., the risk involved in handling radioactive materials. Optical biosensors represent a more recent technology, which finds increasing applications in drug discovery. Optical biosensors make use of the behavior of light when it hits metal films in a prism. The behavior of light is altered when the metal film is coated with biomolecules. Optical sensors are therefore label-free technologies that develop from low- to high-throughput applications and become highly valuable in all phases of drug discovery and development.

References

1. Cooper, M. A. *Nat. Rev. Drug Disc.* **2002**, *1*, 515–528.
2. Garrett, S. D.; Lee, H. A.; Morgan, M. R. *Nat. Biotechnol.* **1999**, *17*, 1219–1222.
3. Noedl, H.; Wongsrichanalai, C.; Wernsdorfer, W. H. *Trends Parasitol.* **2003**, *19*, 175–181.
4. Alban, S.; Gastpar, R. *J. Biomol. Screen.* **2001**, *6*, 393–400.
5. Rosenau, C.; Emery, D.; Kaboord, B.; Qoronfleh, M. W. *J. Biomol. Screen.* **2004**, *9*, 334–342.
6. Lavery, P.; Brown, M. J. B.; Pope, A. *J. Biomol. Screen.* **2001**, *6*, 3–9.
7. Zhang, J. H.; Chung, T. D. Y.; Oldenburg, K. R. *J. Biomol. Screen.* **1999**, *4*, 67–73.
8. Ratliff, D. M.; Vander Jagt, D. J.; Eaton, R. P.; Vander Jagt, D. L. *J. Clin. Endocrinol. Metab.* **2004**, *81*, 488–492.
9. Wu, G. E.; Irvine, J.; Luft, C.; Pressley, D.; Hodge, C. N.; Janzen, B. *Comb. Chem. High Throughput Screen.* **2003**, *6*, 303–312.
10. Hess, H. H.; Derr, J. E. *Anal. Biochem.* **1975**, *63*, 607–613.
11. Martin, B.; Pallen, C. J.; Wang, J. H.; Graves, D. J. *J. Biol. Chem.* **1985**, *260*, 14932–14937.
12. Jin, B. S.; Lee, W. K.; Ahn, K.; Lee, M. K.; Yu, Y. G. *J. Biomol. Screen.* **2005**, *10*, 13–19.
13. Aviezer, D.; Seddon, A. P.; Wildey, M. J.; Böhlen, P.; Yayon, A. *J. Biomol. Screen.* **2001**, *6*, 171–177.
14. Bryant, R.; McGuinness, D.; Turek-Etienne, T.; Guyer, D.; Yu, L.; Howells, L.; Caravano, J.; Zhai, Y.; Lachowicz, J. *Assay Drug Dev. Technol.* **2004**, *2*, 290–299.
15. Sills, M. A.; Weiss, D.; Pham, Q.; Schweitzer, R.; Wu, X.; Wu, J. J. *J. Biomol. Screen.* **2002**, *7*, 191–199.
16. Mallari, R.; Swearingen, E.; Liu, W.; Ow, A.; Young, S. W.; Huang, S. G. *J. Biomol. Screen.* **2003**, *8*, 198–204.
17. Sorg, G.; Schubert, H. D.; Buttner, F. H.; Heilker, R. *J. Biomol. Screen.* **2002**, *7*, 11–19.
18. Crane, K.; Shih, D. T. *Anal. Biochem.* **2004**, *335*, 42–49.
19. Gopalakrishna, R.; Chen, Z. H.; Gundimeda, U.; Wilson, J. C.; Anderson, W. B. *Anal. Biochem.* **1992**, *206*, 24–35.
20. Sun, C.; Newbatt, Y.; Douglas, L.; Workmann, P.; Aherne, W.; Linardopoulos, S. *J. Biomol. Screen.* **2004**, *9*, 391–397.
21. Park, Y. W.; Cummings, R. T.; Wu, L.; Zheng, S.; Cameron, P. M.; Woods, A.; Zaller, D. M.; Marcy, A. I.; Hermes, J. D. *Anal. Biochem.* **1999**, *269*, 94–104.
22. Beveridge, M.; Park, Y. W.; Hermes, J.; Marenghi, A.; Brophy, G.; Santos, A. *J. Biomol. Screen.* **2000**, *5*, 205–211.
23. McDonald, O. B.; Chen, W. J.; Ellis, B.; Hoffman, C.; Overton, L.; Rink, M.; Smith, A.; Marshall, C. J.; Wood, E. R. *Anal. Biochem.* **1999**, *268*, 318–329.
24. Evans, D. B.; Rank, K. B.; Sharam, S. K. *J. Biochem. Biophys. Methods* **2002**, *50*, 151–161.
25. Fuchikami, K.; Togame, H.; Sagara, A.; Satoh, T.; Gantner, F.; Bacon, K. B.; Reinemer, P. *J. Biomol. Screen.* **2002**, 7, 441–450.
26. Bembenek, M. E.; Roy, R.; Li, P.; Chee, L.; Jain, S.; Parsons, T. *Drug Dev. Technol.* **2004**, *2*, 300–307.
27. Delle Fratte, S.; Piubelli, C.; Domenici, E. *J. Biomol. Screen.* **2002**, 7, 541–546.
28. Earnshaw, D. L.; Pope, A. J. *J. Biomol. Screen.* **2001**, *6*, 39–46.
29. Dillon, K. J.; Smith, G. C.; Martin, N. M. *J. Biomol. Screen.* **2003**, *8*, 347–352.
30. Decker, P.; Miranda, E. A.; de Murica, G.; Muller, S. *Clin. Cancer Res.* **1999**, *5*, 1169–1172.
31. Merk, S. E.; Schubert, H. D.; Moreth, W.; Valler, M. J.; Heilker, R. *Comb. Chem. High Throughput. Screen.* **2004**, 7, 763–770.
32. Delaporte, E.; Slaughter, D. E.; Egan, M. A.; Gatto, G. J.; Santos, A.; Shelley, J.; Price, E.; Howells, L.; Dean, D. C.; Rodrigues, A. D. *J. Biomol. Screen.* **2001**, *6*, 225–231.
33. Roda, A.; Guardigli, M.; Pasini, P.; Mirasoli, M. *Anal. Bioanal. Chem.* **2003**, *377*, 826–833.
34. Roda, A.; Pasini, P.; Mirasoli, M.; Michelini, E.; Guardigli, M. *Trends Biotechnol.* **2004**, *22*, 295–303.
35. Wilson, T.; Woodland Hastings, J. *Annu. Rev. Cell Dev. Biol.* **1998**, *4*, 197–230.
36. Goueli, S.; Larson, B.; Hsiao, K.; Worzella, T.; Gallagher, A.; Matthews, E. *Cell Notes* **2004**, *10*, 21–23.
37. Somberg, R.; Pferdehirt, B.; Kupcho, K. *Promega Notes* **2003**, *83*, 14–17.
38. Promega. *Kinase-Glo^{TM} Luminescent Kinase Assay*, Technical Bulletin 318, 2002
39. Pommereau, A.; Pap, E.; Kannt, A. *J. Biomol. Screen.* **2004**, *9*, 409–416.
40. O'Brien, M. A.; Daily, W. J.; Hesselberth, P. E.; Moravec, R. A.; Scurria, M. A.; Klaubert, D. H.; Bulleit, R. F.; Wood, K. V. *J. Biomol. Screen.* **2005**, *10*, 137–148.
41. Warrior, U.; Fan, Y.; David, C. A.; Wilkins, J. A.; McKeegan, E. M.; Kofron, J. L.; Burns, D. J. *J. Biomol. Screen.* **2000**, *5*, 343–351.
42. Yang, H.; Leland, J. K.; Yost, D.; Massey, R. J. *Bio/Technol.* **1994**, *12*, 193–194.
43. Deaver, D. R. *Nature* **1995**, *377*, 758–760.
44. Golla, R.; Seethala, R. *J. Biomol. Screen.* **2004**, *9*, 62–70.
45. Myszka, D. G.; Rich, R. L. *PSTT* **2000**, *3*, 310–317.
46. Williams, C. *Curr. Opin. Biotechnol.* **2000**, *11*, 42–46.
47. Davis, S.; Aldrich, T. H.; Jones, P. F.; Acheson, A.; Compton, D. L.; Jain, V.; Ryan, T. E.; Bruno, J.; Radziejewski, C.; Maisonpierre, P. C. et al. *Cell* **1996**, *87*, 1161–1169.
48. Williams, C.; Addona, T. A. *Trends Biotechnol.* **2000**, *18*, 45–48.
49. Nedelkov, D.; Nelson, R. W. *J. Mol. Recognit.* **2000**, *13*, 140–145.

50. McDonnell, J. M. *Curr. Opin. Chemi. Biol.* **2001**, *5*, 572–577.
51. Karlsson, R.; Stahlberg, R. *Anal. Biochem.* **1995**, *228*, 274–280.
52. Davis, T. M.; Wilson, W. D. *Anal. Biochem.* **2000**, *284*, 348–353.
53. James, P. *Biacore J.* **2004**, *1*, 4–7.
54. Markgren, P. O.; Hämäläinen, M.; Danielson, U. H. *Anal. Biochem.* **2000**, *279*, 71–78.
55. Hämäläinen, M. D.; Markgren, P. O.; Schaal, W.; Karlén, A.; Classon, B.; Vrang, L.; Samuelsson, B.; Hallberg, A.; Danielson, H. *J. Biomol. Screen.* **2000**, *5*, 353–359.
56. Rich, R. L.; Hoth, L. R.; Geoghegan, K. F.; Brown, T. A.; LeMotte, P. K.; Simons, S. P.; Hensley, P.; Myszka, D. G. *PNAS* **2002**, *99*, 8562–8567.
57. Casper, D.; Bukhtiyarova, M.; Springman, E. B. *Biacore J.* **2003**, *3*, 4–7.
58. Casper, D.; Bukhtiyarova, M.; Springman, E. B. *Anal. Biochem.* **2004**, *325*, 126–136.
59. Frostell-Karlsson, A.; Remaeus, A.; Roos, H.; Andersson, K.; Borg, P.; Hämäläinen, M.; Karlsson, R. *J. Med. Chem.* **2000**, *43*, 1986–1992.
60. McWirther, A.; based on work by: Widegren, H.; Nordin, H.; Hämäläinen, M.; Westerlund, L.; Frostell-Karlsson, A. *Biacore J.* **2003**, *3*, 16–19.
61. Rich, R. L.; Myszka, D. G. *Biacore J.* **2001**, *1*, 8–11.
62. Danelian, E.; Karlén, A.; Karlsson, R.; Winiwarter, S.; Hansson, A.; Löfas, S.; Lennernäs, H.; Hämäläinen, M. D. *J. Med. Chem.* **2000**, *43*, 2083–2086.
63. Lahiri, J.; Isaacs, I.; Tien, J.; Whitesides, G. M. *Anal. Chem.* **1999**, *71*, 777–790.
64. Sage, L. *Am. Chem. Soc.* **2004**, *4*, 137–139.
65. Englebienne, P.; VanHoonacker, A.; Verhas, M. *Analyst* **2001**, *126*, 1645–1651.
66. Haes, A. J.; VanDuyne, R. P. *J. Am. Chem. Soc.* **2002**, *124*, 10596–10604.
67. Haes, A. J.; Chang, L.; Klein, W. L.; VanDuyne, R. P. *J. Am. Chem. Soc.* **2004**, *127*, 2264–2271.
68. Haes, A. J.; Hall, W. P.; Chang, L.; Klein, W. L.; VanDuyne, R. P. *Nano Letters* **2004**, *4*, 1029–1034.
69. Vetter, D. *J. Cell. Biochem.* **2002**, *39*, 79–84.
70. Walsh, D. P.; Chang, Y. T. *Comb. Chem. High Throughput Screen.* **2004**, 7, 557–564.
71. Metz, G.; Ottleben, H.; Vetter, D. *Meth. Princ. Med. Chem.* **2003**, *19*, 213–236.
72. Dickopf, S.; Frank, M.; Junker, H. D.; Maier, S.; Metz, G.; Ottleben, H.; Rau, H.; Schellhaas, N.; Schmidt, K.; Sekul, R. et al. *Anal. Biochem.* **2004**, *335*, 50–57.
73. Gauglitz, G. *Anal. Bioanal. Chem.* **2005**, *381*, 141–155.
74. Birkert, O.; Gauglitz, G. *Anal. Bioanal. Chem.* **2002**, *372*, 141–147.
75. Cunningham, B. T.; Li, P.; Schulz, S.; Lin, B.; Baird, C.; Gerstenmaier, J.; Genick, C.; Wang, F.; Fine, E.; Laing, L. *J. Biomol. Screen.* **2004**, *9*, 481–490.
76. Ferrari, D. M.; Söling, H. *Biochem. Soc.* **1999**, *339*, 1–10.
77. Gallina, A.; Haneley, T. M.; Mandel, R.; Trahey, M.; Broder, C. C.; Viglianti, G. A. *J. Biol. Chem.* **2002**, *277*, 50579–50588.
78. Barbouche, R.; Miquelis, R.; Jones, I. M.; Fenouillet, E. *J. Biol. Chem.* **2003**, *278*, 3131–3136.
79. Smith, A. M.; Chan, J. C.; Oksenberg, D.; Urfer, R.; Wexler, D. S.; Ow, A.; Gao, L.; McAlorum, A.; Huang, S. *J. Biomol. Screen.* **2004**, *9*, 614–620.

Biographies

Bernhard Schnurr is currently Vice President Assay Development & Profiling at Discovery Partners International AG in Switzerland. He joined Discovery Partners International AG in 2001. Previously, Dr Schnurr was Head of HTS Factory at Discovery Technology Ltd, which merged in 2001 with Discovery Partners International AG. From 1996 to 1997 he was Project and Team Leader for Analytics and HTS at Burecco AG. He joined the Tumor Biology Center in Freiburg i. Br., Germany in 1994 as a Team leader for Assay Development and HTS (Institute for Molecular Medicine and Natural Compound Research) until 1996. In 1993 Dr Schnurr obtained his PhD in Biology from the University of Freiburg i. Br., Germany. He performed his thesis at Goedecke AG, Freiburg i. Br. in Germany.

Thomas Ahrens is currently Team Leader Assay Development and Profiling at Discovery Partners International AG in Switzerland. Dr Ahrens joined Discovery Partners International AG in 2004. In 1999 he obtained his PhD in Molecular Biology from the Institute of Genetics, University of Karlsruhe, Germany and completed his postdoctoral work at the Biozentrum, University of Basel, Switzerland in 2004.

Urs Regenass is currently CEO of Discovery Partners International AG. He was global Head of Pharma Knowledge/Information Management and from 1994 to 1998 Global Head of Core Drug Discovery Technologies in Research, first at Ciba and then at Novartis following the merger in 1996. He joined Ciba-Geigy in 1981 in Oncology and served as project and unit head until 1994. He and his department contributed to the discovery and development of signal transduction inhibitors (Gleevec, EGF-R and VEGF-R kinase inhibitors, PKC inhibitors), polyamine biosynthesis inhibitors, and many other projects in the area of oncogenes, suppressor genes, signal transduction, and growth control. Dr Regenass obtained his PhD in Cell Biology and Genetics from the Biocenter, University of Basel and completed his postdoctoral work at the Jackson Laboratory, Bar Harbor, Maine. He was teaching experimental oncology at the University of Basel from 1994 to 2001 and presently serves on the board of directors of two start-up companies in Switzerland.

Comprehensive Medicinal Chemistry II
ISBN (set): 0-08-044513-6

ISBN (Volume 3) 0-08-044516-0; pp. 577–598

3.28 Fluorescence Screening Assays

D Ullmann, Evotec AG, Hamburg, Germany

3.28.1	**Introduction**	**599**
3.28.2	**Homogeneous Soluble Biological Assays**	**600**
3.28.3	**Statistics in High-Throughput Screening**	**600**
3.28.4	**Fluorescence Read-Out Artifacts in High-Throughput Screening**	**601**
3.28.5	**Bulk Fluorescence Techniques**	**602**
3.28.5.1	Overview	602
3.28.5.2	Total Fluorescence Intensity (FLINT)	602
3.28.5.3	Fluorescence Polarization	602
3.28.5.4	Förster Resonance Energy Transfer	604
3.28.5.5	Time-Gated Fluorescence – Homogeneous Time-Resolved Fluorescence (HTRF)	605
3.28.5.6	Fluorescence Lifetime	605
3.28.5.7	Time-Resolved Anisotropy Analysis (TRA)	606
3.28.6	**Fluctuation Techniques – The Way Down to Single Molecules**	**606**
3.28.6.1	Overview	606
3.28.6.2	Instrumentation for Fluorescence Fluctuation Techniques	607
3.28.6.3	Data Acquisition for Fluorescence Fluctuation Techniques	608
3.28.6.4	Fluorescence Correlation Spectroscopy	608
3.28.6.5	Fluorescence Intensity Distribution Analysis	609
3.28.6.6	Two-Dimensional Fluorescence Intensity Distribution Analysis (2D-FIDA)	610
3.28.6.7	Other Fluorescence Intensity Distribution Analysis-Based Methods	610
3.28.7	**Lifetime Techniques Applied to Confocality – Future Developments for High-Throughput Screening Read-Outs**	**610**
3.28.8	**Compound Artifact Correction – Molecular Resolution Applied to High-Throughput Screening**	**611**
3.28.9	**Conclusion**	**614**
References		**614**

3.28.1 Introduction

In high-performance drug discovery, high-throughput screening (HTS) is used to test libraries of 10^4–10^6 potential drug candidates (compounds) for biological activity toward a specific target. The resulting active compounds ('hits') are further evaluated, and a specific drug for the biological target of interest might be designed in subsequent follow-up studies. Possible HTS screening campaigns include the inhibition of binding events, the activation of protein function, and the control of enzymatic reactions. In HTS it is desirable to test as many samples as possible in the shortest possible time with the smallest possible fraction of false classifications with respect to the investigated drug candidates.

When testing a biological target up to 10^6 times, miniaturization is needed, since reagent and test compound savings are an important consideration. On the other hand, HTS requires a precise, efficient, and controllable sample handling to ensure the same assay conditions for every tested compound, to enable screening of all drug candidates in the fastest possible time, and to allow for tracking of each compound. Modern screening systems use sample volumes as small as 1 μL on sample carriers with up to 2080 sample wells, incorporate efficient liquid sample handling, and use databases for compound tracking.[1–9]

Besides the demands on the equipment, HTS requires an efficient read-out mode that enables short data acquisition times down to 1 s and analysis methods with the highest possible statistical accuracy. Failures in the classification of the biological activities of the compounds will result in subsequent validation tests being wasted on nonactive compounds (false positives) or – even worse – potential active drug candidates remaining undiscovered (false negatives).

In recent years, labeling with fluorescent tracers has started to emerge as the method of choice for tagging purposes in life science research and development, where it is replacing radio-labeling. Due to its superior sensitivity to environmental properties and also its multidimensionality (i.e., its ability to provide various simultaneous read-outs such as intensity, lifetime, anisotropy, and spectral characteristics), fluorescence has gained much interest in life science applications. By attaching a dye to the biological molecule of interest, even low biological activities will lead to observable changes in the fluorescence properties of the dye that can be monitored.[10,11]

However, not all fluorescence techniques fulfill the prerequisites for HTS. Using bulk fluorescence techniques such as total fluorescence emission, robust assay systems that remain compatible with HTS can be miniaturized down to volumes of 10 μL, but not further.[12] In contrast, with confocal fluorescence microscopy the measurement volume can be as low as 1 fL, and assay volumes of 1 μL are well established.[1,2] Beside minute assay volumes, confocal techniques offer the key advantage of applying new efficient fluorescence read-out methods, which reach the highest possible sensitivity by tracking biological activities down to the single-molecule level.

The most detailed information on biomolecular processes can be obtained by detecting fluorescence from single molecules. Therefore, fluorescence-based assays using single-molecule detection (SMD) techniques are evolving as a very important tool in science.[8–10,13,15–19] These techniques use samples of highly diluted fluorophores, and include not only the direct detection and analysis of single-molecule events[13,15–21] but also spectroscopic analysis by means of fluctuation methods such as fluorescence correlation spectroscopy (FCS) or fluorescence intensity distribution analysis (FIDA).[22–28] These techniques enable different molecules in a sample (e.g., bound and unbound) to be distinguished and quantified, and, furthermore, allows a number of different read-outs that are inherent to the fluorescence signal to be combined.[29–31] The use of multiple, simultaneously acquired parameters improves the resolution of distinct species significantly, and hence opens the door to the study of complex assemblies of biomolecular interactions. Thus, the major advantages of these techniques in HTS are their high statistical accuracy, even at measurement times of about 1 s, and their very low consumption of precious biological material. As a consequence, they are increasingly applied in HTS.[3–8,13,32,33]

This chapter presents a review of fluorescence detection techniques for HTS on homogeneous soluble biological assays, focusing on the improvements offered by confocal techniques. The advantages and disadvantages of the different methods are clarified by presenting some biological examples, and problems applying the fluorescence read-out in HTS are discussed, such as the intrinsic fluorescence emission of the compounds.

3.28.2 Homogeneous Soluble Biological Assays

Moving the activity of biomolecules such as enzymes, receptors, or other proteins away from its natural cellular environment to a soluble biological assay results in some disadvantages as well as major benefits for HTS applicability.

- (Liquid) sample handling as well as read-out of homogeneous soluble assays is much easier and more straightforward than for cellular assays. This is a major advantage for automation and speed toward efficient HTS, and most HTS approaches make use of this assay design.[1–9,12,32]
- The biological activity might change markedly when moved from the natural cellular environment. The question arises whether the biological characteristics of a target changes and/or whether potential drug candidates behave differently. Thus, there is a possibility of choosing false-positive or false-negative compounds when using homogeneous soluble biological assays. Therefore, HTS has recently started to move toward cellular systems by using sophisticated cellular handling and fluorescence read-out devices such as imaging techniques.[32,33]

This chapter will focus on HTS applications using homogeneous soluble biological assays.

3.28.3 Statistics in High-Throughput Screening

Different approaches exist to judge from the read-out whether a compound shows biological activity.[1–9,12] In any case, one needs to include control samples without any compound, that is, samples that show no biological activity at all (negative controls), as well as control samples that contain a known reagent leading to the desired biological activity (positive controls, such as an inhibitor of enzymatic or binding reaction, which can be the unlabeled ligand). By comparing the read-out of the compound samples to that of the control samples, one can easily determine the biological activity of each compound (eqn [1a]). On the other hand, the statistical accuracy of this determination can be expressed by the statistical screening parameter Z', which takes into account both the standard deviation of the read-out as well as the dynamic range (i.e. the difference between positive and negative control (eqn [1b]).[34] In order to be HTS compatible, an assay must at least express a Z'-value of above 0.5.

$$\text{activity} = \frac{X_{\text{cpd}} - X_{\text{neg}}}{X_{\text{pos}} - X_{\text{neg}}} \qquad [1a]$$

$$Z' = 1 - \frac{3\sigma(\text{high}) + 3\sigma(\text{low})}{\mu(\text{high}) - \mu(\text{low})} \qquad [1b]$$

where 'activity' is the biological activity of the compound, X_{cpd} is the read-out (e.g., total intensity, polarization etc.) of the compound, X_{pos} and X_{neg} are the medians of the same read-out for all positive or negative samples, respectively, σ is the standard deviation, μ is the mean value, and 'high' and 'low' denote high control and low control, repectively.

In a second step, a criterion has to be chosen that rates the activity of the compounds. In a standard approach, a compound is rated as active if its activity is above a certain threshold (e.g., set to three times above the standard deviation of the activity of the negative controls).[1]

The easiest way to judge the activity of any reagent or compound in more detail is to titrate it against the target, that is, to observe the change in the assay read-out on increasing the reagent or compound concentration (compare **Figures 2, 5, 6b, 7,** and **9**) and the subsequent determination of K_D or IC_{50} values. The K_D value represents the binding constant of a reagent to another molecule, while the IC_{50} value characterizes the competition or inhibition of the reagent toward the biological interaction of interest (protein–peptide binding, enzymatic reaction, etc.). The following equations represent the easiest approach for the determination of the K_D value (eqn [2a]) or the IC_{50} value (eqn [2b]) from titration curves, as exemplified in **Figures 5, 6b, 7,** and **9**:

$$Y(X) = \frac{n(Y_{\max} - Y_{\min})X^n}{K_D + X^n} + Y_{\min} \qquad [2a]$$

$$Y(X) = \frac{Y_{\max} - Y_{\min}}{1 + (X/IC_{50})} + Y_{\min} \qquad [2b]$$

where $Y(X)$ is a read-out such as fluorescence intensity or polarization, as a function of the compound or reagent concentration X; $Y_{\max}$ and $Y_{\min}$ represent the maximum and minimum read-out values, respectively; and n is the number of binding sites (the Hill coefficient).[35]

3.28.4 Fluorescence Read-Out Artifacts in High-Throughput Screening

The fluorescence read-out has some disadvantages in HTS due to its sensitivity toward artifacts, which might obscure the fluorescence data and lead to a wrong classification of the compound activity (e.g., a change in the fluorescence read-out is detected although the biological activity is not affected). Some of the potential artifacts are described below, with autofluorescence and quenching being the most prevalent effects:

- Autofluorescence – additional background fluorescence signal might appear upon the addition of the test compounds due to intrinsic compound fluorescence or fluorescence from impurities. This autofluorescence leads to an increase in the total fluorescence signal as well as to a deterioration of other read-outs such as fluorescence polarization, as shown in **Figure 1a**.
- Quenching – a direct interaction of the test compound with the dye label (and not the biology) might lead to a quench of the dye fluorescence emission. This will result in a decrease in the total fluorescence signal and a change of the fluorescence characteristics such as the fluorescence polarization (as shown in **Figure 1b**), although the biological activity is not influenced.
- Artifacts from the instrument – variations in the fluorescence tracer concentration or assay volume might show up due to mis-dispensing by the HTS machinery. This will lead to, for example, a change in the overall fluorescence signal.

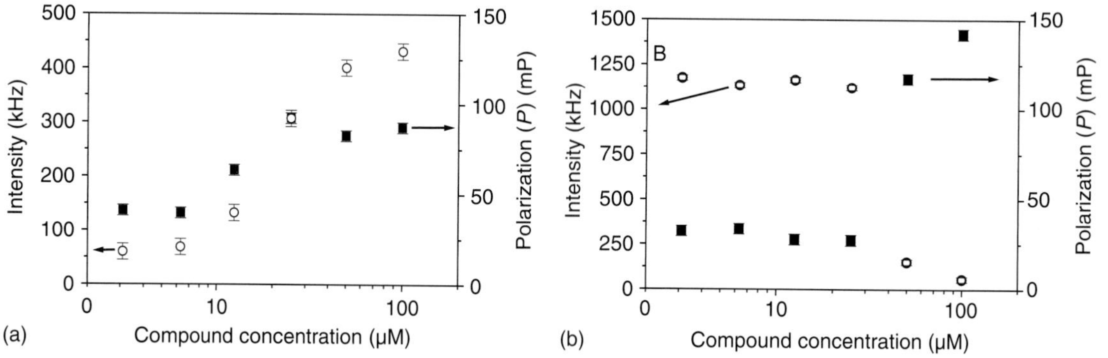

Figure 1 Effects of background fluorescence resulting from (a) a compound or impurities (autofluorescence) and (b) the direct interaction of a compound with the fluorophore (quenching reaction) on the bulk fluorescence intensity and polarization read-out. Two different compounds were titrated in (a) and (b) against rhodamine green (1 nM (a) and 20 nM (b) in aqueous PBS-buffer solution; excitation was at 488 nm). The open and solid symbols refer to intensity and polarization read-outs, respectively.

- Optical artifacts – variations in the adjustment of the excitation or detection volume or in the excitation power will obscure the data.
- Inner-filter effects – exhibited by test compounds, since they are usually used at a relatively high concentration (10–100 μM).

The different fluorescence techniques used in HTS show different sensitivities toward these artifacts, which are described in detail in Section 3.28.8. Fluorescence fluctuation techniques display the greatest flexibility in identifying and correcting for these artifacts.

3.28.5 Bulk Fluorescence Techniques

3.28.5.1 Overview

This section briefly summarizes the most used bulk fluorescence techniques in HTS, and discusses their advantages and limitations – bearing in mind that miniaturization is only possible down to about 10 μL.[12,32]

Bulk fluorescence techniques excite fluorescence in a relatively large volume (above 100 μm^3), far above that used in confocal excitation (diffraction-limited focus of about 1 μm^3). The experimental prerequisites are very simple (e.g., lamp excitation by means of an ordinary lens). Therefore, bulk techniques offer very simple and straightforward approaches for detecting fluorescence in HTS applications.[2,12]

3.28.5.2 Total Fluorescence Intensity (FLINT)

Measuring the integrated fluorescence intensity of a substantial part of a sample well (e.g., from a 1536-well microplate) over a predefined time window (e.g., 100 ms) represents the simplest of all read-out possibilities. The overall fluorescence intensity basically scales linearly with the fluorescence quantum yield (i.e., the efficiency of the fluorescence emission of the dye label – the quantum yield is highly sensitive to any environmental changes such as polarity variations, for example due to a switch from a polar aqueous environment to relatively nonpolar binding sites, or quenching reactions with molecules in close proximity). The major advantages of FLINT are high speed and ease of detection and analysis. It even enables the simultaneous read out of several wells on a plate.

On the other hand, FLINT is rather prone to the artifacts discussed above. However, there are elegant techniques available to circumvent these limits (e.g., resonance energy transfer or fluorescence fluctuation methods – see later). Examples of significant assays using the total fluorescence intensity as the read-out are enzymatic turnover and calcium release assays.[36]

3.28.5.3 Fluorescence Polarization

In HTS, fluorescence polarization is one of the most frequently used techniques. A sample is excited with linearly polarized light, and the fluorescence emission parallel and perpendicular with respect to the polarization of the incident light is detected simultaneously or subsequently.

The polarization, P, is defined as

$$P = \frac{I_{\text{parallel}} - I_{\text{perpendicular}}}{I_{\text{parallel}} + I_{\text{perpendicular}}} \quad [3]$$

where I_{parallel} and $I_{\text{perpendicular}}$ represent the emitted light parallel and perpendicular to the incident polarization direction, respectively.[11]

Instead of P, the fluorescence anisotropy, r, can instead be used:

$$r = \frac{I_{\text{parallel}} - I_{\text{perpendicular}}}{I_{\text{parallel}} + 2\ I_{\text{perpendicular}}} \quad [4]$$

P and r can be converted to each other.[11] The anisotropy, r, is straightforwardly related to other spectroscopic values such as the fluorescence lifetime, τ, via the Perrin equation (eqn [5a]) as well as the Stokes equation (eqn [5b]):

$$\frac{1}{r} = \frac{1}{r_0}\left(1 + \frac{\tau}{\tau_r}\right) \quad [5a]$$

$$\tau_r = \frac{M V \eta}{RT} \quad [5b]$$

where r_0 is the limiting anisotropy (at time zero), τ_r is the rotational correlation time (the time required to rotate through an angle whose cosine is 1/e, or approximately 68.5°), M is the molecular mass (Da), V is the specific volume of the molecule including hydration ($cm^3 g^{-1}$), η is the viscosity of the medium ($g cm^{-1} s^{-1}$), R is the gas constant ($8.314\,J\,mol^{-1}K^{-1}$), and T is the absolute temperature (K).

It follows that r or P is – to a first approximation (under the condition $\tau/\tau_r \gg 1$) – directly proportional to the mass and indirectly proportional to the lifetime:

$$r \propto M \quad [6a]$$

$$r \propto \frac{1}{\tau} \quad [6b]$$

Therefore, biological reactions that result in a change of the mass or lifetime of the labeled biomolecule (binding, cleavage, etc.) will show a change in the overall polarization or anisotropy. Consequently, this read-out is most straightforward for any biological assay designed for HTS application, since a change in mass is easy to predict. However, the polarization read-out assigned to bulk fluorescence has some limitations for HTS:

- Autofluorescence or quench from compounds will lead to a deterioration of the overall polarization read-out, and result in a false classification of the activity of the compound. In contrast to the overall fluorescence intensity, the polarization is insensitive to inner-filter effects or variations in the fluorescence tracer concentration or assay volume, since it is an intrinsic property of the molecule, independent of concentration etc.
- The range of molecular mass is limited. The value of the lifetime itself determines in which range of molecular mass an experiment will be successful. **Figure 2a** demonstrates the relevant dependency in a plot of the anisotropy as a function of the protein mass for different fluorophore lifetimes. The graph clearly shows where the limits are; for example, a protein–protein binding event can barely be detected for protein masses above 50 kDa, if fluorescein or rhodamine is used as the labeling dye ($\tau_f = 4$ ns).

Nevertheless, frequently used dyes for polarization measurements in HTS express lifetimes from 1 ns to 4 ns, which allows the binding reaction of low-molecular-weight molecules (<10 kDa) to proteins (or antibodies, etc.) being detected reliably and with excellent statistics.[9,37] Further applications are protease assays.[38] The determination of the amount of quench upon binding is important for all polarization measurements. If quenching occurs – which is true for most assays – binding as well as competition curves (K_D and IC_{50} determination) will be incorrectly estimated, as outlined in **Figure 2b**.[9] This has to be considered when determining K_D or IC_{50} values.

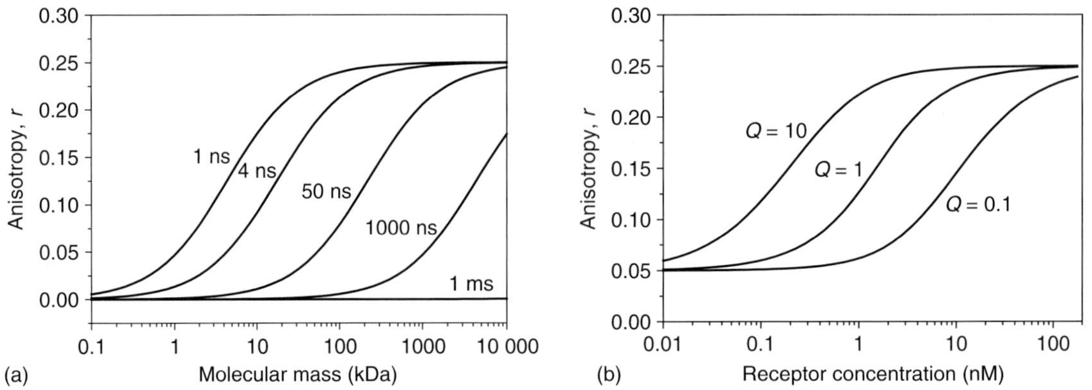

Figure 2 (a) Dependency of the anisotropy, r, on the molecular mass, M, of a dye-labeled molecule calculated for different lifetimes, τ (1 ns, 4 ns, 50 ns, 1000 ns, and 1 ms), using eqn [5] (and assuming typical values for the limiting anisotropy $r_0 = 0.25$ and rotational diffusion time $\tau_r = 0.23\,\text{ns kDa}^{-1}$). (b) Simulated change of the apparent anisotropy, r, following the increased binding of a dye-labeled ligand ($c = 1$ nM) to a receptor upon titration of the receptor. The binding curves shift ($K_D = 10$ nM ($Q = 10$), 1.2 nM ($Q = 1$), 0.2 nM ($Q = 0.1$)) due to quench upon binding. The value Q indicates the amount of quench. The overall anisotropy, r, is calculated from the emission-weighted contributions of each species, that is, unbound and bound ligand ($r = I_1/(I_1 + I_2)r_1 + I_2/(I_1 + I_2)r_2 = (Qr_1 + r_2)/(Q + 1)$ with the intensity and anisotropy, I_i and r_i, respectively, of unbound ($i = 1$, $r_1 = 0.05$) and bound ($i = 2$, $r_2 = 0.25$) ligand and the amount of quench $Q = I_1/I_2$).

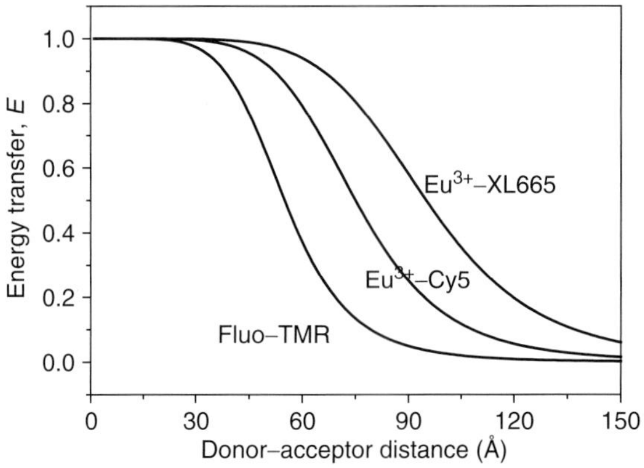

Figure 3 The efficiency of fluorescence resonance energy transfer, E, calculated for different donor–acceptor fluorophore pairs with different R_0 values using eqn [7] (Fluo–TMR, fluorescein–tetramethylrhodamine ($R_0 = 55$ Å); Eu^{3+}–Cy5, europium chelate–Cy5 ($R_0 = 75$ Å); Eu^{3+}–XL665, europium cryptate–XL665 ($R_0 = 95$ Å); values for R_0 are from Pope *et al.*[9]

3.28.5.4 Förster Resonance Energy Transfer

One of the most generic homogeneous assay principles uses Förster resonance energy transfer.[39] Förster resonance energy transfer constitutes two fluorophores – a donor and an acceptor dye – with different excitation and emission wavelengths. Ideally, only the donor dye is excited by light, and transfers a fraction of its energy to an acceptor dye. This leads to a decrease (quench) in the fluorescence emission of the donor, and the appearance of the fluorescence emission of the acceptor. The fraction of transferred energy depends on the overlap of the donor emission and acceptor excitation spectra as well as on the orientation and distance of both dyes. Förster resonance energy transfer is mostly referred to as fluorescence resonance energy transfer (note that the energy from the donor to the acceptor is transferred by nonradiative processes, and thus should not be called fluorescence). In some cases the acceptor is a quencher, which opens up another possibility for a read-out strategy. **Figure 3** shows the fractional energy transfer E of some typical donor–acceptor pairs, as calculated from the Förster theory[39]:

$$E = \frac{R_0^6}{R_0^6 + R^6} \qquad [7]$$

where R_0 is the distance for half-transfer efficiency from the donor to the acceptor, and R is the donor–acceptor distance; note the sixth power dependency.

Owing to the sixth power dependency of eqn [7], dye pairs such as fluorescein–tetramethylrhodamine are limited to applications with acceptor–donor distances below 60 Å, which complicates the design of a Förster resonance energy transfer assay. Nevertheless, Förster resonance energy transfer has been realized in, for example, peptide–protease assays, where the C and N termini of the peptide are labeled with one dye each, and the protease activity leads to a cleavage of the peptide, and, thus, a separation of the donor and the acceptor, followed by a decrease in the Förster resonance energy transfer signal.

The use of Förster resonance energy transfer in HTS is straightforward, and has some advantages over the polarization or the total fluorescence read-out:

- The detection of the acceptor fluorescence is definitely free of any autofluorescence, since excitation occurs for the donor and is spectrally well separated.
- The simultaneous or subsequent detection of the donor and acceptor fluorescence can be used to calibrate the read-out toward artifacts such as variations in fluorescence tracer concentration or assay volume by, for example, calculating the fraction of acceptor fluorescence.

3.28.5.5 Time-Gated Fluorescence – Homogeneous Time-Resolved Fluorescence (HTRF)

Rare earth elements (especially lanthanides) have been used as long-lived tracers for time-resolved fluorescence (TRF) since the 1980s. The first applications were heterogeneous immunoassays using europium chelates,[40] which later were transformed into homogeneous formats. A further development took advantage of europium cryptates, their complexes being more stable than the classical chelates.[41] A number of different Förster resonance energy transfer assays using lanthanide complexes as donors have been developed (e.g., LANCE[42] or HTRF,[43] which are generally referred to as time-resolved energy transfer (TRET) systems. One of the most important advantages of TRET systems (such as europium cryptate–XL665, see **Figure 3**) results from the separation of the donor–acceptor lifetime (microseconds to milliseconds) from the low-lifetime components (<10 ns) originating in background signals such as solvent scattering, autofluorescence from compounds, or fluorescence from impurities. (Consequently, the fluorescence detection is gated, in order to register only light, which is delayed (e.g., >10 ns) with respect to a short excitation pulse.) However, the downsides to TRF are that it requires the use of rare earth elements in assay development, and fluorescence detection devices capable of detecting fluorescence with such a long lifetime.

3.28.5.6 Fluorescence Lifetime

Many scientific applications use the fluorescence lifetime, since it is an intrinsic molecular property of the fluorophore and is able to detect minute changes in the direct environment of the fluorophore.[11] Its sensitivity as well as its intrinsic molecular property, which is independent of the set-up or adjustment of the instrument or the overall fluorophore concentration (in contrast to fluorescence intensity), results in high statistical accuracy, and makes fluorescence lifetime analysis (FLA) a very valuable tool for HTS applications.[6,13,44,45]

Two main methods are generally used for FLA: frequency domain and time domain data acquisition. In the frequency domain, sinusoidal modulated light is used, and the fluorescence lifetime calculated from the phase shift and demodulation of the fluorescence emission.[45,46] In the time domain, the fluorescence lifetime is determined from the time-dependent decay of the fluorescence emission after employing repetitive brief excitation pulses (e.g., applying time-correlated, single-photon counting, TCSPC).[47,48] The fluorescence lifetime analysis using TCSPC is performed using a straightforward deconvolution approach taking into account the 'repetitive excitation' effect of the fluorescence not fully decaying to zero, before the arrival of the following laser pulses.[14,19,49–52] To perform the analysis, the instrumental response function, p, of the equipment has to be measured in a separate experiment with only scattered light reaching the detector. Contributions of any fluorescence to this data set must be strictly prevented. The model function, F_i, in the ith histogram channel is given by the convolution of this instrumental response function, p_i, with the exponential fluorescence lifetime decay. Contributions to F originating in any of the previous laser pulses result in a geometric series, A_τ, which leads to the theoretical expression for F_i given by

$$F_i = a_1\left(\sum_{j=0}^{i} P_j e^{(i-j/\tau_1)} + A_{\tau_1} e^{-i/\tau_1}\right) + a_2\left(\sum_{j=0}^{i} P_j e^{(i-j/\tau_2)} + A_{\tau_2} e^{-i/\tau_2}\right) + \cdots \qquad [8a]$$

where

$$A_\tau = \frac{1}{1-e^{-k/\tau}} \sum_{j=0}^{k} P_j e^{-(k-j/\tau)} \qquad [8b]$$

and a_m and τ_m denote the fractional fluorescence intensities and lifetimes of the different species, m, respectively.

FLA determines the lifetimes of the different fluorescent species present in the sample, and takes into account background and autofluorescence signals. Due to the robustness as well as the possibility to resolve and quantify different components of a sample (e.g., free versus bound ligand), the lifetime read-out displays high statistical accuracy, and is increasingly used in HTS with measurement times below 1 s.[6,13,44] TCSPC has been shown to be superior for HTS applications, compared with frequency domain FLA. However, the applicability of FLA to biological reactions is limited, since changes in the fluorescence lifetime (due to environmental quenching or polarity effects) are not always predictable. Therefore, new approaches in assay development use additional quencher labels to induce a lifetime change upon binding or enzymatic reactions (see the example in **Figure 7**).[6]

3.28.5.7 Time-Resolved Anisotropy Analysis (TRA)

TRA comprises the combination of fluorescence lifetime and fluorescence polarization. Lifetime data using TCSPC are collected simultaneously or subsequently on two detectors monitoring the fluorescence emission parallel and perpendicular with respect to the linear polarization of the incident light. The two time-dependent decays of the fluorescence emission obtained from the repetitive brief excitation pulses (compare the explanation of TCSPC above) are globally analyzed, and the rotational correlation time, τ_r (i.e., the average time of the rotation cycle of a molecule under the Brownian motion regime), as well as the fluorescence lifetime, τ, are determined for every fluorescent species present in the sample. In this approach, eqn [8a] has to be applied to both polarization directions (F_{par} for parallel polarized and F_{perp} for perpendicular polarized fluorescence light), and is modified as follows[11,14,15,44]:

$$F_{i,\mathrm{par}} = \frac{1}{3} I_{\mathrm{tot}} \left[\left(\sum_{j=0}^{i} P_j e^{-(i-j/\tau)} + A_\tau e^{-i/\tau} \right) + \frac{4}{5} \left(\sum_{j=0}^{i} P_j e^{-(i-j/\tau_0)} + A_{\tau_0} e^{-i/\tau_0} \right) \right] \qquad [9a]$$

$$F_{i,\mathrm{perp}} = \frac{1}{3} I_{\mathrm{tot}} \left[\left(\sum_{j=0}^{i} P_j e^{(i-j)/\tau} + A_\tau e^{-i/\tau} \right) - \frac{2}{5} \left(\sum_{j=0}^{i} P_j e^{-(i-j)/\tau_0} + A_{\tau_0} e^{-i/\tau_0} \right) \right] \qquad [9b]$$

with the substitution $1/\tau_0 = 1/\tau + 1/\rho$, and the total fluorescence intensity, I_{tot}, originating in the anisotropy description $I_{tot} = I_{par} + 2I_{perp}$. Eqns [9a] and [9b] describe a simple one-component solution, but can easily be modified for more complex cases.

The rotational correlation time is typically within the several hundred picoseconds to several tens of nanoseconds range, and a mass-dependent change upon biochemical reaction may be deduced in addition to a possible lifetime change. Because of the simultaneous determination of the intrinsic molecular fluorescence lifetime and rotational properties, the statistical accuracy in resolving and distinguishing different fluorescent species is further enhanced compared with FLA, which opens up further perspectives for its application to HTS. Biological changes are detectable with higher sensitivity, since variations in mass (rotation) and lifetime (e.g., quenching) are simultaneously monitored. Deterioration of the fluorescence read-out (e.g., due to autofluorescence) is much easier to distinguish since artifacts influence the lifetime differently than they do the rotational correlation time.[44]

3.28.6 Fluctuation Techniques – The Way Down to Single Molecules

3.28.6.1 Overview

Since HTS applications aim at saving precious biological material, the straightforward approach would be to reduce the assay volume. However, because in macroscopic (bulk) fluorescence techniques the signal is averaged over most of the assay volume, the signal will inevitably degrade as the volume is reduced.[2] In order to overcome this limitation, read-out methods based on the detection of single fluorophores have been developed for HTS. Here, the measurement volume is microscopically small (1 fL); thus, miniaturization does not alter the measurement statistics.

The signal detected from single fluorescent molecules fluctuates due to their diffusion into and out of the excitation or detection volume, each causing a burst in fluorescence emission and, thus, in detected fluorescence photons. In SMD, these bursts from single molecules are directly analyzed. So far, SMD has generated important new results and insights into biological systems, and will play an important role in the future development of detection techniques.[13,15–21,53,54] However, it needs acquisition times of at least several seconds, since a reasonable number of single-molecule events has to be gathered, in order to reach a sufficiently high accuracy. Therefore, applications for HTS purposes use other analysis methods based on the statistical analysis of the fluorescence fluctuations, which offer much lower data acquisition times. Here, more than one fluorophore can be present in the detection volume, as opposed to SMD, where at the most one fluorophore at a time is allowed.

The analysis of fluorescence signal fluctuations opens up the possibility of resolving and quantifying various components of a sample expressing different fluorescence and, hence, molecular characteristics; a feature comparable to FLA and TRA. These characteristics are directly associated with the signal fluctuations; for example, brightly fluorescing particles give rise to high fluorescence emission and detection rates and, therefore, to fluctuations with high amplitudes. Slowly diffusing fluorescing molecules remain in the detection volume and emit fluorescence over a long period of time, thus generating broader fluctuations compared with fast-diffusing fluorophores. Additionally, the fluorescence fluctuations from highly concentrated molecules show much smaller amplitudes than from molecules of low concentration. Since the fluctuating signal is influenced by a large number of molecular properties, the statistical accuracy of the characterization of a biological target will be increased by the simultaneous measurement of a variety of fluorescence parameters. Several different analysis methods have evolved over the last few years that take advantage of this molecular resolution.[22–28] In contrast to SMD, these methods use the whole signal data stream to extract the information, and, therefore, the necessary data acquisition times can be lowered to below 1 s, which led to their widespread application in HTS[1–9,13,15,32,44,55–57] and their integration into the FCS++ read-out portfolio of the EVOscreen HTS platform (Evotec Technologies, Hamburg, Germany). However, as outlined above, the analysis requires fluctuations of a certain amplitude, that is, the concentration of fluorescing molecules has to be close to the single-molecule level, and specialized instruments such as a confocal microscope have to be used. The different fluctuation methods are outlined further below.

3.28.6.2 Instrumentation for Fluorescence Fluctuation Techniques

Fluorescence fluctuation experiments generally require only a small number of fluorescent molecules ($\sim$0.1–10) to be present in the detection volume at any time, a condition that can be fulfilled most easily through the use of a confocal geometry in combination with low fluorophore concentrations.

The use of confocal optics for fluorescence fluctuation techniques has been described in detail elsewhere.[29,58–60] The basic principle of confocal fluorescence microscopy is to monitor only a small volume of the sample (the observation volume) in the focal plane of the microscope at any time. Using an objective lens, the exciting laser light is focused close to the diffraction limit, to a diameter of about 1 μm. Fluorescence emitted by the sample is imaged back to a pinhole by the same objective, and, subsequently, it is detected by an avalanche photodiode (APD). The pinhole may be adjusted to limit the observation volume to about 1 fL. Alternatively, multiphoton excitation (MPE) greatly facilitates the realization of confocal excitation, since the nonlinearity of the absorption process (e.g., quadratic dependence of the absorption on the excitation intensity for two-photon excitation) confines the fluorescence emission to the focal region of the excitation beam,[61] thus making the confocal pinhole redundant.

Using this set-up brings along some features that enable the detection of fluorescence signals from single molecules. Highly focused laser excitation combined with fluorophores expressing high quantum yields and a low tendency for photo-destruction, such as rhodamines, allows for the highest fluorescence emission achievable in practice.[62–64] Furthermore, the tiny detection volume of about 1 fL minimizes the background signal originating from solvent scattering or impurities. Together with high-precision optics and filters, which efficiently separate the fluorescence from the background signal, a high fluorescence- (or signal-) to-background ratio is realized. Finally, a number of other advances have also contributed to the extensive application of fluorescence fluctuation methods today, including the increased availability of stable, user-friendly laser sources with a variety of wavelengths (primarily diode lasers and all-solid-state laser systems), the development of efficient correlator electronics along with improvements in computer processing speed, and the introduction of high quantum efficiency detectors with broad spectral sensitivity (primarily APDs (e.g., SPCM-AQ-131, EG&G Optoelectronics, Vaudreuil, Quebec, Canada) and, more recently, a new generation of photomultiplier tubes (e.g., the H7421 series, Hamamatsu, Japan)). These prerequisites enable not only the observation of a very small number of fluorescence photons, as emitted from a single fluorophore, but also allow the clear distinction of the fluorescence signal from the background level, and, thus, the analysis of the signal fluctuation originating solely from diffusion of the single molecules into and out of the laser focus.

The very tiny detection volume opens up new advantages for HTS, since submicroliter sample volumes can be exploited. This reduces the consumption of precious biological material to the lowest possible amount, and allows the use of high-density sample carriers, such as 1536- or even 2080-well plates, which are now routinely used within the EVOscreen uHTS system (Evotec Technologies).[1,32,33,44] The introduction of the Fluorescence Imaging Plate Reader (FLIPR) (Molecular Devices, Sunnyvale, USA) has introduced new possibilities for the measurements of fast changes in intercellular calcium and membrane potential changes. This instrument has stimulated kinetic cellular approaches in HTS and initiated the development of new generations of instruments specially designed to measure transient cellular fluorescence or luminescence signals.

3.28.6.3 Data Acquisition for Fluorescence Fluctuation Techniques

The data acquisition electronics for the fluorescence fluctuation techniques require the creation of a multichannel scaler (MCS) trace, which is obtained by counting the number of detected photons within consecutive time windows of constant size (e.g., 1 ms in **Figure 4a**) either for each detector separately, or for a set of detectors. This can be done using, for example, a separate electronic PC card.[19,29,65] The MCS trace represents the time-dependent fluorescence signal intensity, $F(t)$, on the macroscopic time scale.

3.28.6.4 Fluorescence Correlation Spectroscopy

FCS analyses the temporal characteristics of fluorescence fluctuations by calculating the correlation function, $G(t_c)$, from an MCS trace, $F(t)$ (small time windows of, for example, 50 ns):

$$G(t_c) = \frac{\langle F(t)F(t+t_c)\rangle}{\langle F(t)\rangle} \qquad [10]$$

where t is the measurement time, t_c is the correlation time, and $\langle \ldots \rangle$ denotes averaging over the time t.

The calculated correlation function (**Figure 4b**) of these fluctuations decays with time constants that are characteristic of the molecular processes causing these fluorescence changes (e.g., diffusion into and out of the detection volume as well as the reaction kinetics).[22,66–70] The amplitude of the decay is related to the molecular concentration or mean number of particles in the detection volume, N, while the inflection point represents the mean diffusion time, τ_{diff}, of the fluorescing molecules through the detection volume, which is solely dependent on the diffusion coefficient:

$$G(t_c) = 1 + \frac{1}{N}\,\frac{1}{1+t_c/\tau_D}\left(\frac{1}{1+(\omega_0/Z_0)^2 t_c/\tau_D}\right)^{1/2} \qquad [11]$$

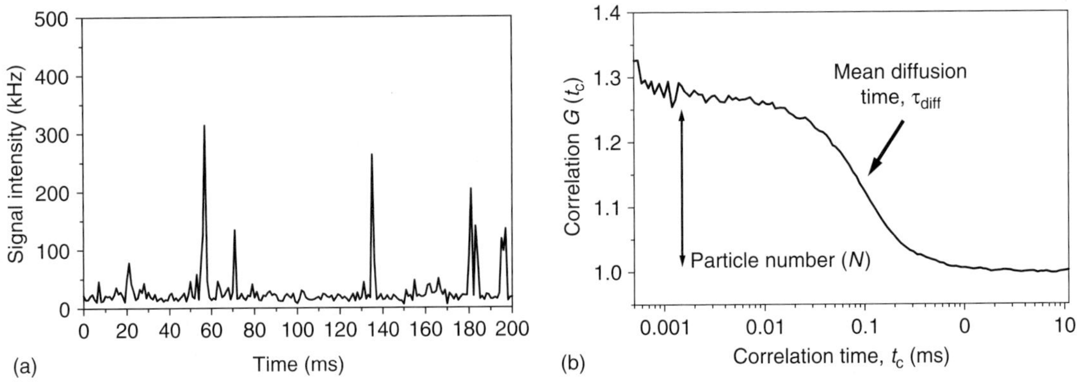

Figure 4 (a) Multichannel scaler trace for rhodamine 6G in water recorded using a confocal fluorescence microscope (the dye concentration was 10^{-12} M, as typically used in single-molecule experiments; the excitation was at 532 nm). The signal intensity, $F(t)$, is plotted as a function of the measurement time, t (number of photons detected in subsequent time windows of 1 ms duration). (b) Typical FCS curve; the correlation function, $G(t_c)$, is calculated for different correlation times, t_c, using eqn [10] The analysis according to eqn [11] yields the particle number (i.e., the mean number of fluorophores in the detection volume), $N = 3.6$, and a mean diffusion time, $\tau_{diff} = 95\,\mu s$. The data were taken from a confocal fluorescence measurement of an aqueous rhodamine 6G solution (the dye concentration was 3 nM).

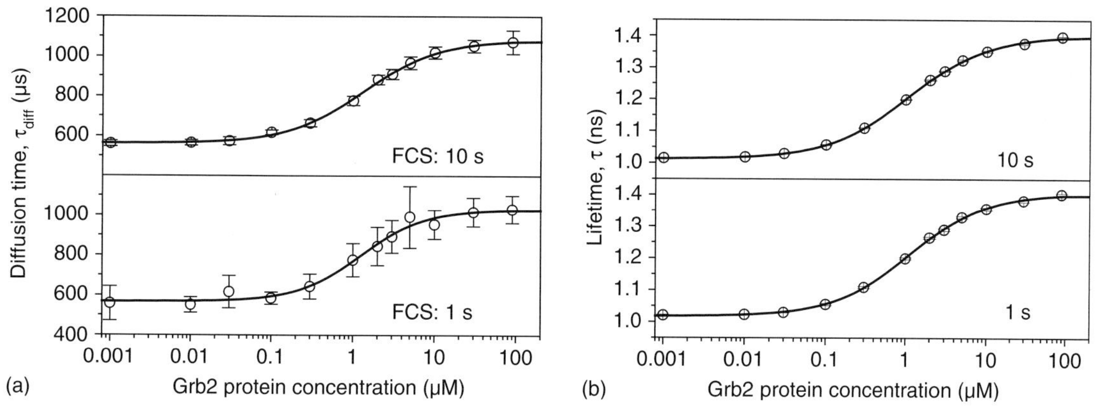

Figure 5 Measurement of the mean diffusion time, τ_{diff}, obtained by (a) FCS and of the mean fluorescence lifetime, τ, obtained by (b) FLA, observing the increased binding of a small Cyanin5-labeled peptide to the SH_2 domain of the Grb2 protein upon titration of the protein (the excitation was at 633 nm; for details on the assay, see Palo *et al.*[30] The error bars result from 10 measurements, and vary for different measurement times (1 s and 10 s). This results in the following Z' values (eqn [1b]); $Z' = 0.70$ (FCS, 10 s), 0.02 (FCS, 1 s), 0.98 (FLA, 10 s), and 0.95 (FLA, 1 s). While FLA can readily be used at measurement times far below 1 s, FCS is not recommended in this case, and is therefore rarely used in HTS.

The shape of the detection volume is in general assumed to be Gaussian in the radial direction and Lorentzian in the axial direction, with characteristic $1/e^2$ radii z_0 and z_0.[22,66,67,71] Hence, FCS is able to resolve components of a sample with different diffusion coefficients due to, for example, different molecular masses. For example, the binding of a dye-labeled peptide to a larger protein is followed by an increase in the mean diffusion time, τ_{diff}. In this way, biological assays can easily be developed for HTS applications.[55,57] However, for measurement times as short as 1 s and for an increasing background signal, the correlation curves become severely noise limited. Thus, for these conditions, FCS loses its accuracy, as shown in **Figure 5**. Due to this problem, further fluctuation methods have been developed for HTS.

3.28.6.5 Fluorescence Intensity Distribution Analysis

FIDA relies on a collection of instantaneous values of the fluctuating fluorescence intensity by building up a frequency histogram of the signal amplitudes throughout a measurement (from the relevant MCS trace $F(t)$).[27,28] The resulting distribution of signal intensities is then analyzed by a theory that relates the specific fluorescence brightness, q (intensity per molecule in kilohertz), to the absolute concentration, c (mean number of molecules in the detection volume), of the molecules under investigation, and accounts for possible background signals due to scattering or impurities. Consequently, FIDA distinguishes the species in a sample according to their different values of specific molecular brightness, q. Thus, the overall signal intensity is split up into its molecular contributions:

$$I_{tot} = I_1 + I_2 + \cdots = c_1 q_1 + c_2 q_2 + \cdots \qquad [12]$$

where I_{tot} is the total signal intensity, I_i ($i = 1, 2, \ldots$) is the fractional signal intensity contributed by the ith species, c_i and q_i are the mean number of molecules in the detection volume and mean brightness (count rate per particle, measured in kilohertz) of the ith species, respectively, and $I = cq$.

The theory behind FIDA has been described extensively.[22,23,28] In principle, two Poissonian distributions have to be taken into account; one expressing the possibility of finding a certain number of molecules in the detection volume, the other recalling the distribution of number photon counts within the MCS time window arising from a certain number of noninteracting molecules in the detection volume, which emit a specific brightness q. Furthermore, the inhomogeneous excitation profile evoked by the strongly focused laser light is taken into account by a spatial brightness profile of the focus. The combination of these parts is efficiently performed by the use of the generating function principle, which allows a very fast calculation of theoretical FIDA distributions and enables fitting times in the submicrosecond range.

Reactions resulting in the quenching of the fluorescence emission of a dye label or in the aggregation of dye-labeled molecules (i.e., a change in the molecule brightness) can easily be monitored applying FIDA. Possible HTS applications have been introduced by enzyme cleavage reactions as well as inhibition assays, where a fluorescently labeled ligand, bound to its vesicular acceptor protein, competes against the unlabeled form.[4,5,8,28,32,56] A very

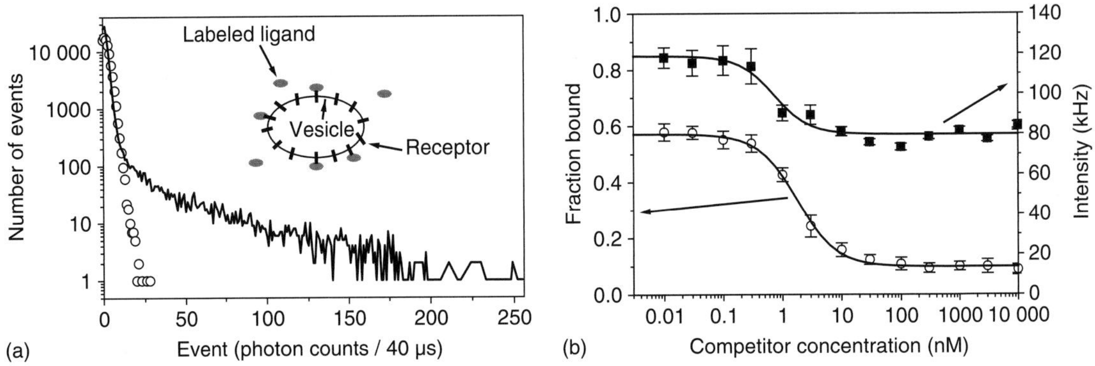

Figure 6 Example assay for FIDA: binding of an MR121-labeled ligand (propranolol) to the β_2-adregenic receptor embedded in vesicular membranes (inhomogeneous size about 100 nm). (a) The assay principle and two exemplary FIDA histograms, resulting from a sample containing only 6 nM free ligand (open circles – the FIDA fit results in brightness and concentration values of $q = 22$ kHz and $c = 6.3$) and a sample containing free ligand (6 nM) and vesicles (0.2 mg mL^{-1}) (black line – the FIDA fit results in brightness and concentration values of $q = 20$ kHz and $c = 6.9$ for the free ligand and $q = 920$ kHz and $c = 0.0012$ for the ligand bound to membrane-embedded receptors). Since several receptors are present on one vesicular particle, the brightness of the vesicles with bound ligands is significantly increased compared with the free ligand, which is obvious by the tail in the FIDA data. (b) The titration of a competitor (unlabeled propranolol) to a mixture of the labeled ligand (3 nM) and vesicles (0.2 mg mL^{-1}). The fraction of bound labeled ligand was determined using FIDA (similar to Schaertl *et al.*,[56] calculated from $c_1q_1/(c_1q_1 + c_2q_2)$ (c_i and q_i are the concentration and brightness of the ith species obtained from FIDA; $i = 1$, bound labeled ligand (vesicles); $i = 2$, unbound labeled ligand). The measurement time was 2 s. A hyperbolic fit to the competition curves results in IC_{50} values of 1.78 ± 0.15 nM for FIDA and 0.72 ± 0.22 nM for the fluorescence intensity. The Z' values (0.71 determined for FIDA and 0.29 calculated for the overall intensity using eqn [1b]) clearly prove that FIDA is much more suited to HTS.

good assay application of FIDA is presented in **Figure 6**, which was later also used for HTS. Since FIDA is based solely on a statistical process (i.e., building up of a histogram) rather than on a numerical operation (i.e., calculating the correlation function, as in the case of FCS), FIDA is able to maintain good statistical accuracy down to acquisition times of 1 s.

3.28.6.6 Two-Dimensional Fluorescence Intensity Distribution Analysis (2D-FIDA)

2D-FIDA is based on the extension of the FIDA technique, and improves its performance. 2D-FIDA is applied to a two-detector set-up monitoring either different polarization or emission bands of the fluorescence signal. In addition to the increase in performance, 2D-FIDA achieves additional molecular resolution by the concurrent determination of two specific brightness values originating from both detection channels, q_1 (channel 1) and q_2 (channel 2).[29] By observing the molecularly resolved fluorescence anisotropy, simple as well as more complex binding events and enzymatic reactions may be followed, as has been shown in Kask *et al.*,[29] and will be presented later for an HTS approach including autofluorescence correction. Alternatively, such events may be resolved by labeling with two dyes expressing different fluorescence emission bands. As a read-out, either two-color excitation by different lasers or one-color excitation employing a Förster resonance energy transfer dye pair can be chosen. Thus, the use of a second detector improves the power of FIDA to distinguish between molecular components, and the technique is therefore increasingly used in HTS.[8,57]

3.28.6.7 Other Fluorescence Intensity Distribution Analysis-Based Methods

Other FIDA-based methods, such as fluorescence intensity multiple distribution analysis (FIMDA)[30] and fluorescence intensity and lifetime distribution analysis (FILDA),[31] combine the FIDA read-out (specific molecular brightness, q) with other fluorescence parameters, such as the mean diffusion time in the case of FIMDA and the fluorescence lifetime in the case of FILDA. Owing to the use of a second molecular parameter, both approaches gain further accuracy to distinguish between molecular components, and have already been used to various biological applications, including HTS.[13,23,30,31]

3.28.7 Lifetime Techniques Applied to Confocality – Future Developments for High-Throughput Screening Read-Outs

Fluorescence lifetime methods such as FLA and TRA (as presented above) can easily be adapted for use with a confocal device, and therefore exploit the advantages of confocality or microscopic fluorescence techniques in HTS.

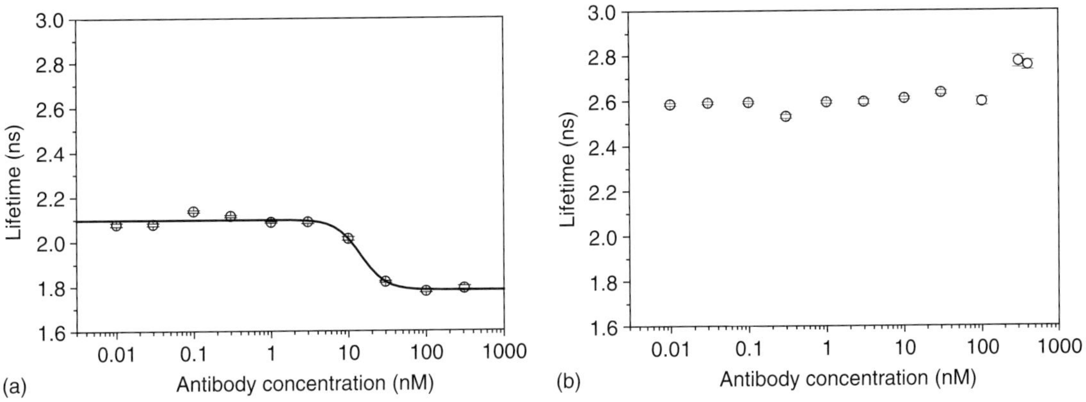

Figure 7 Example assay for cFLA: binding of tetramethylrhodamine (TAMRA)-labeled protein A to MR121 (a) labeled or (b) unlabeled IgG-mouse antibody. The fluorescence lifetime as a function of the antibody concentration is determined using cFLA (fluorescence excitation was at 532 nm and detection at around 570 nm; the measurement time was 1 s). A lifetime change is only observable if the antibody is labeled with MR121, due to a quench of the TAMRA label induced by Förster resonance energy transfer toward the MR121 label. A hyperbolic fit to the data results in a binding constant of $K_D = 14.4 \pm 1.9$ nM. A change in the polarization signal is not observable; the polarization remains constant at around $P = 150$ mP (data not shown), and is therefore useless for HTS.

Confocal FLA (cFLA) has been used several times in SMD experiments[14,16–19,21,49,50,54,72–74] as well as in HTS.[6,13,15,32,44] It offers tremendous accuracy, and fulfills all the demands of HTS; a very low dynamic range (>0.1 ns) between positive and negative control, and measurement times below 1 s are sufficient for HTS.[8,44] As already mentioned for FLA, the applicability of FLA to biological assays is limited, since changes in the fluorescence lifetime (due to environmental quenching or polarity effects) are not always predictable. A possible approach to circumvent this limitation in HTS assay development is the use of an additional quencher label to induce a lifetime change upon binding, as presented in **Figure 7** for a protein–protein interaction. Since an induced Förster resonance energy transfer quenches the donor dye, the binding of a protein to an antibody can conveniently be followed by a decrease in the donor lifetime. In this case, the fluorescence lifetime represents the optimum HTS read-out, since a polarization change is not detectable due to the rather high molecular mass (40 kDa and 150 kDa) of both binding partners (see **Figure 2a**).

Confocal TRA (cTRA) has to take into account that the polarization of the exciting laser beam as well as that of the detected fluorescence is slightly distorted due to diffraction by the objective lens (caused by the small aperture or tight focusing of the objective). This distortion can be well corrected by the introduction of correction factors into eqn [9], as described elsewhere.[14,75] The significant advantage of cTRA over methods exploiting only one fluorescence parameter such as cFLA or polarization is depicted in **Figure 8**. In this example, the hybridization of DNA is best followed by cTRA, reaching a statistical accuracy (Z' value, see eqn [1b]) that cannot be matched by the other methods.

Both cFLA and cTRA offer the major advantage that they do not require fluctuations of the fluorescence signal, and can, therefore, be used in a wide concentration range of the labeled assay components, from the subnanomolar to the micromolar range.

Furthermore, the confocal lifetime read-out can be combined with fluctuation methods such as FIDA, which is realized in FILDA (as presented above).[31] The concurrent determination of the fluorescence lifetime and brightness opens up new opportunities for HTS applications as presented in Eggeling *et al.*[44] However, the advantages of FILDA for HTS are limited, since in most cases the brightness and the lifetime read-out depend significantly on each other (e.g., a quench leads to a decrease in both lifetime and brightness), and do not offer much new information. Therefore, current HTS developments combine independent fluorescence read-out parameters such as fluorescence lifetime and rotation, as realized in cTRA.[44]

3.28.8 Compound Artifact Correction – Molecular Resolution Applied to High-Throughput Screening

Several artifacts that arise when applying the fluorescence read-out to HTS are described in Section 3.28.4. The predominant ones are autofluorescence and quenching effects, caused by the test compounds and/or impurities (see **Figure 1**). In the case of autofluorescence, the additional background signal just adds up to the assay signal. In the

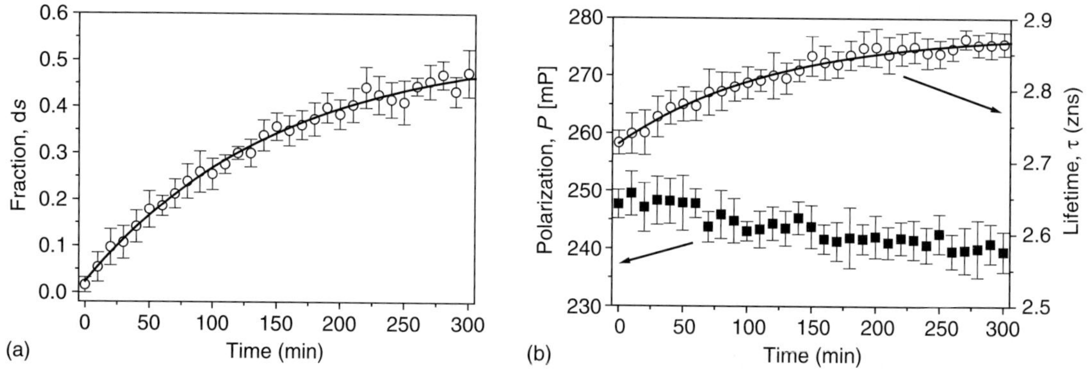

Figure 8 Example assay for cTRA: hybridization of an MR121-labeled DNA (single strand with 22 base pairs) to the relevant unlabeled counter-DNA strand, which results in the formation of double-stranded (ds)DNA (the excitation was at 633 nm, the measurement time was 1 s, error bars result from 10 measurements, and DNA concentrations were 5 nM). The fluorescence lifetime, rotational correlation time, and polarization values were determined to be repectively $\tau = 2.56$ ns, $\tau_c = 2.1$ ns, and $P = 245$ mP for the single strand, and $\tau = 2.87$ ns, $\tau_c = 2.2$ ns, and $P = 238$ mP for the double strand. (a) cTRA analysis. Fixing τ and τ_c in a two-component fit, the fraction of dsDNA is obtained (using eqns [9] corrected for artifacts induced by the confocality, as described in Section 3.28.8), which increases with time due to the hybridization process (a time constant of $t_0 = 152 \pm 11$ min was calculated from a fit of $y_0 + (1 - \exp{-t/t_0})$ to the data). From eqn [1b] a Z' value of 0.59 was determined). (b) Analysis using 2D-FIDA (no change in P) or cFLA. For the latter technique, a time constant of $t_0 = 122 \pm 12$ min was calculated. However, the Z' value for this data set was determined to be 0.04, which demonstrates that cTRA is superior to cFLA in this HTS application.

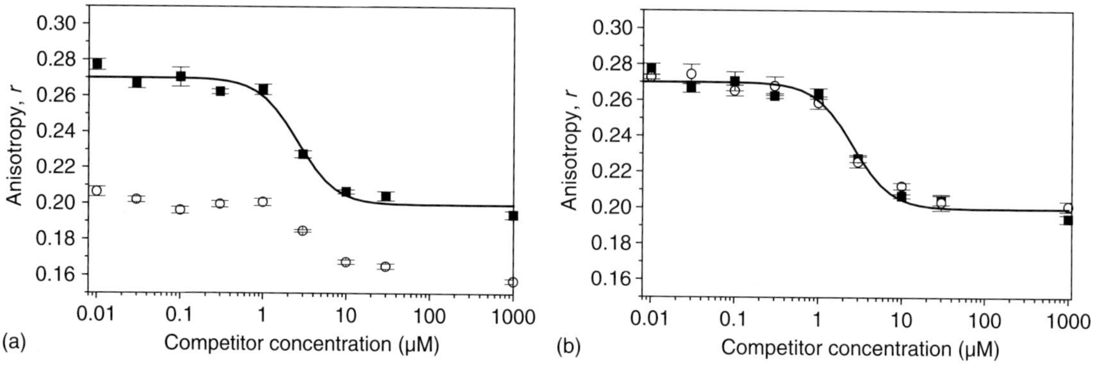

Figure 9 Binding of a small Cyanin5-labeled peptide to the SH_2 domain of the Grb2 protein monitored by a change in the fluorescence anisotropy. Unlabeled peptide (competitor) was titrated with (open circles) and without (solid circles) addition of fluorescent impurities (1 μM rhodamine 800, excitation was at 633 nm). The anisotropy was determined from (a) the overall signal and from (b) 2D-FIDA. The latter analysis is able to separate the fluorescence signal of the labeled peptide from the impurities. A dose–response fit of the data resulted in IC_{50} values of 2.6 ± 0.4 for the pure (solid line (a)) and 2.2 ± 0.3 μM for the impure corrected (solid line (b)) sample. Oligino *et al.*[75] report an IC_{50} of 2 μM for the same peptide motive. For more details on the assay system, see Palo *et al.*[30]

case of quenching, there is a chance that the quenching interaction between the compound and the dye will not affect all dye labels, and thus some assay signal will remain unaffected. As a consequence, a read-out method must be able to distinguish between artifact and assay signal. This prerequisite is fully met by the fluorescence fluctuation methods described above as well as by the lifetime-based methods (c)FLA and (c)TRA, which all distinguish and quantify different fluorescent species of a sample (e.g., autofluorescence and assay components) by a difference in a fluorescence parameter (molecular resolution). By adding an additional fit component in the analysis step, which accounts for the autofluorescent or quenched species, the deterioration of the read-out can be corrected, as presented in **Figure 9** for the polarization read-out using 2D-FIDA.

In many cases, the addition of an extra-fit component leads to a decrease in statistical accuracy, since the fit gains additional freedom. In HTS this will lead to a decreased Z' value (see eqn [1b]) and thus a decreased accuracy in characterizing the compounds. Therefore, the correction step in such cases should only be applied to samples that exhibit the autofluorescence or quenching artifacts. This, however, demands an additional identification step – a procedure used for the HTS example of **Figure 10**. Here, a TAMRA-labeled DNA molecule binds to a protein (p53). The bound and unbound states are distinguished by measuring the fluorescence anisotropy. The following five steps describe the identification and correction procedure:

1. Compounds and assay components are dispensed on 2080-well plates. In addition, some of the wells are spiked with autofluorescent and quenching compounds.
2. Consequently, the read-out will show anisotropy values that are far off with respect to the expected positive and negative controls (here, far above the positive control, **Figure 10a**; positive and negative controls are represented by the upper and lower lines, respectively).
3. Artifact identification step: the total fluorescence intensity is plotted against the fluorescence read-out (e.g., polarization or activity (calculated from eqn [1a])) of all sample wells, including the compound and control wells (**Figure 10b**). From this, a corridor can be plotted, calculated from the median plus three times the standard deviation of the read-out pair (intensity versus anisotropy) of the positive or negative control, respectively, which describes the potential range in which the biological activity is expected to be present. Thus, every value pair outside this corridor must be affected by artifacts.
4. Artifact correction and hit identification step: all wells outside the corridor are again fitted with a model that accounts for the additional autofluorescent or quenching component. The result is presented in **Figure 10c**. As a criterion for the selection of a compound with positive biological activity (hit), the threshold was set to the anisotropy value of the negative control plus three times its standard deviation (gray line).
5. To check for potential failures of the analysis (e.g., of the autofluorescence correction), a second identification step was introduced, which applies the corridor method of step 3 to the value pair anisotropy versus particle number (concentration c obtained by the 2D-FIDA fit) (**Figure 10d**). Value pairs outside the corridor are excluded (bad) or may be measured a second time. As an example, compound 693 was identified as 'bad' and subsequently marked in **Figure 10c**.

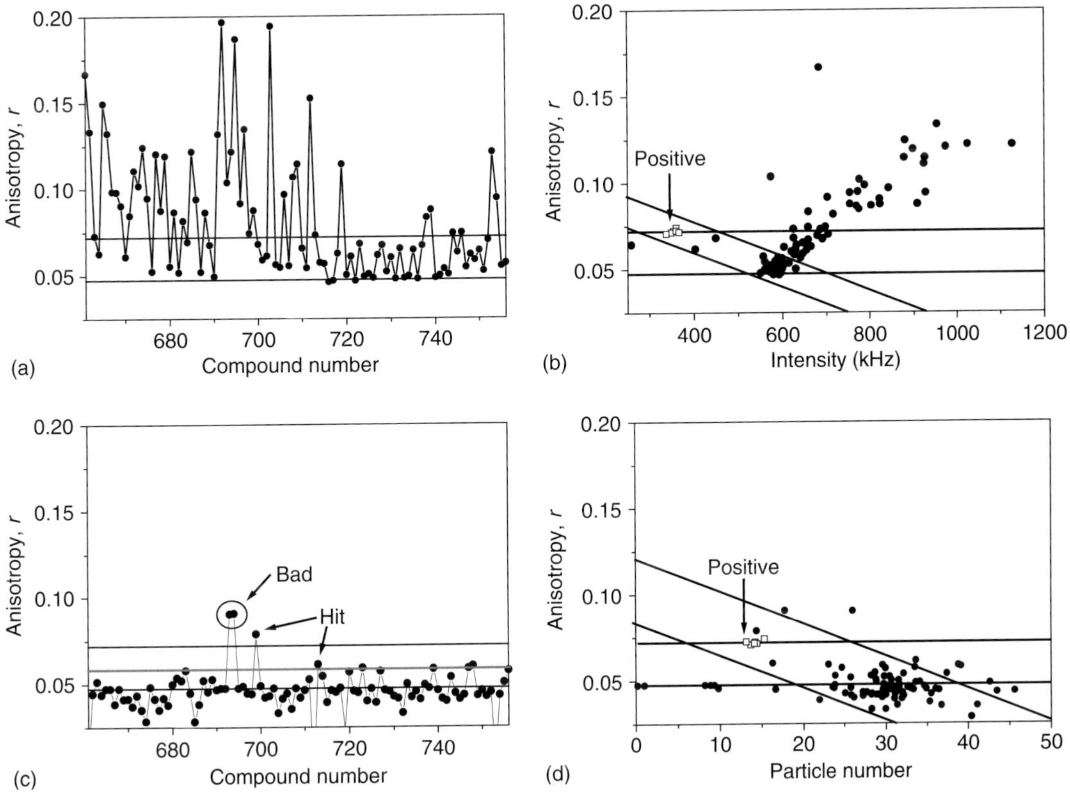

Figure 10 Identification and correction of autofluorescence and quenching artifacts as applied to the polarization read-out of a DNA–protein binding screen. For details, see the text.

The above-described analysis method might fail if the difference in the read-out parameter (e.g. polarization or lifetime) of the assay compared with the autofluorescence signal is very low. Furthermore, the correction might introduce a decrease in statistical accuracy (as mentioned above), which degrades the whole analysis toward small measurement times and HTS. Therefore, new read-out methods are increasingly applied to HTS that apply more than one read-out parameter to achieve an improved molecular resolution (e.g., cTRA and FILDA[32,44]). This reduces the possibility that the autofluorescence read-out coincides with that of the assay, since both have to match more than one fluorescence parameter. Also, the concurrent use of several fluorescence parameters will strengthen the fit and prevent a decrease in statistical accuracy when applying the correction, and thus circumvents the identification step.

It is worth mentioning that microscopic techniques are much more prone to autofluorescence than macroscopic techniques, which apply different fluorophore concentrations (10^{-8} M for the former technique, and up to the micromolar range for the latter). Because compounds are added at a relatively high concentration (micromolar range), the ratio of the assay signal to the autofluorescence signal will be much higher for the macroscopic techniques. However, these techniques do not offer autofluorescence correction steps. An elegant way to use both advantages (high fluorophore concentrations concomitant with low assay volumes) is realized in cFLA and cTRA. Here, the measurement is independent from the fluorophore concentrations, and the detection volume is as small as 1 fL, which saves precious assay material, and the autofluorescence signals are less prominent.

3.28.9 Conclusion

This chapter has outlined the latest developments in fluorescence techniques, with an emphasis on HTS applications. Starting with prominent bulk fluorescence techniques, which cover assay volumes of 10 μL and above, the broad range of single-molecule detection methods currently available for HTS was presented. Here, the detection volume is as small as 1 fL, and assay formats of 1 μL are established in routine use. The advantages and disadvantages of fluctuation techniques versus bulk read-out methods were discussed. Since the latter techniques use higher fluorescence tracer concentrations, artifacts originating in autofluorescence or quenching are less dominant compared with fluctuation techniques, where only nanomolar (0.1–10 nM) tracer concentrations are allowed. On the other hand, bulk techniques do not offer any artifact correction steps, nor do they provide for molecular resolution. The ability to resolve the biochemical reaction at a molecular level offers several advantages. By applying different read-outs simultaneously such as FIDA, polarization, and lifetime techniques, a higher statistical accuracy is achieved, not only in distinguishing the biochemical reaction partners (e.g., bound versus unbound) but also in discriminating autofluorescence or quenching artifacts. Furthermore, the multiplexing possibility opens the door to high-content screening, since different parameters can be looked at in one assay. In addition, the low assay volume saves precious biological samples. Future developments will aim at globally analyzing the data stream, which makes the decision of which read-out to choose redundant.

References

1. Wölcke, J.; Ullmann, D. *Drug Disc. Today* **2001**, *6*, 637–646.
2. Jäger, S.; Brand, L.; Eggeling, C. *Curr. Pharm. Biotechnol.* **2003**, *4*, 463–476.
3. Gribbon, P.; Sewing, A. *Drug Disc. Today* **2003**, *8*, 1035–1043.
4. Gribbon, P.; Schaertl, S.; Wickenden, M.; Williams, G.; Grimley, R.; Stuhmeier, F.; Preckel, H.; Eggeling, C.; Kraemer, J.; Everett, J. et al. *Curr. Drug Disc. Tech.* **2004**, *1*, 27–35.
5. Gribbon, P.; Lyons, R.; Laflin, P.; Bradley, J.; Chambers, C.; Williams, B. S.; Keighley, W.; Sewing, A. *J. Biomol. Screen.* **2005**, *10*, 99–107.
6. Turconi, S.; Bingham, R.; Haupts, U.; Pope, A. *Drug Disc. Today HTS* Suppl. **2001**, 27–39.
7. Rüdiger, M.; Haupts, U.; Moore, K. J.; Pope, A. J. *J. Biomol. Screen.* **2001**, *6*, 29–37.
8. Haupts, U.; Rüdiger, M.; Ashman, S.; Turconi, S.; Bingham, R.; Wharton, C.; Hutchinson, P. J.; Carey, C.; Moore, K. J.; Pope, A. *J. Biomol. Screen.* **2003**, *8*, 19–33.
9. Pope, A. J.; Haupts, U. M.; Moore, K. *J. Drug Disc. Today* **1999**, *4*, 350–362.
10. Eigen, M.; Rigler, R. *Proc. Natl. Acad. Sci. USA* **1994**, *91*, 5740–5747.
11. Lakowicz, J. R. *Principles of Fluorescence Spectroscopy*; Plenum Press: New York, NY, 1999.
12. Lavery, P.; Brown, M. J.; Pope, A. J. *J. Biomol. Screen.* **2001**, *6*, 3–9.
13. Eggeling, C.; Schaffer, J.; Volkmer, A.; Seidel, C. A. M.; Brand, L.; Jäger, S.; Gall, K. *Proceedings of the BioSensor Symposium*; Tuebingen: Germany, 2001.
14. Schaffer, J.; Volkmer, A.; Eggeling, C.; Subramaniam, V.; Striker, G.; Seidel, C. A. M. *J. Phys. Chem. A* **1999**, *103*, 331–336.
15. Hovius, R.; Vallotton, P.; Wohland, T.; Vogel, H. *Trends Pharmacol. Sci.* **1999**, *21*, 266–273.
16. Keller, R. A.; Ambrose, W. P.; Goodwin, P. M.; Jett, J. H.; Martin, J. C.; Wu, M. *Appl. Spectrosc.* **1996**, *50*, 12A–32A.

17. Plakhotnik, T.; Donley, E. A.; Wild, U. P. *Annu. Rev. Phys. Chem.* **1997**, *48*, 181–212.
18. Xie, X. S.; Trautman, J. K. *Annu. Rev. Phys. Chem.* **1998**, *49*, 441–480.
19. Eggeling, C.; Berger, S.; Brand, L.; Fries, J. R.; Schaffer, J.; Volkmer, A.; Seidel, C. A. M. *J. Biotechnol.* **2001**, *86*, 163–180.
20. Tamarat, P.; Maali, A.; Lounis, B.; Orrit, M. *J. Phys. Chem.* **2000**, *104*, 1–16.
21. Ambrose, W. P.; Goodwin, P. M.; Jett, J. H.; van Orden, A.; Werner, H. J.; Keller, R. A. *Chem. Rev.* **1999**, *99*, 2929–2956.
22. Rigler, R.; Elson, E. L. *Fluorescence Correlation Spectroscopy – Theory and Applications*; Springer-Verlag: Berlin, Germany, 2001.
23. Kraayenhof, R.; Visser, A. J. W. G.; Gerritsen, H. C. *Fluorescence Spectroscopy, Imaging and Probes – New Tools in Chemical, Physical and Life Sciences*; Springer-Verlag: Berlin, Germany, 2002.
24. Thompson, N. L.; Lieto, A. M.; Allen, N. W. *Curr. Opin. Struct. Biol.* **2002**, *12*, 634–641.
25. Krichevsky, O.; Bonnet, G. *Rep. Progr. Phys.* **2002**, *65*, 251–297.
26. Hess, S. T.; Huang, S.; Heikal, A. A.; Webb, W. W. *Biochemistry* **2002**, *41*, 697–705.
27. Chen, Y.; Müller, J. D.; So, P. T. C.; Gratton, E. *Biophys. J.* **1999**, *77*, 553–567.
28. Kask, P.; Palo, K.; Ullman, D.; Gall, K. *Proc. Natl. Acad. Sci. USA* **1999**, *96*, 13756–13761.
29. Kask, P.; Palo, K.; Fay, N.; Brand, L.; Mets, Ü.; Ullman, D.; Jungmann, J.; Pschorr, J.; Gall, K. *Biophys. J.* **2000**, *78*, 1703–1713.
30. Palo, K.; Mets, Ü.; Jäger, S.; Kask, P.; Gall, K. *Biophys. J.* **2000**, *79*, 2858–2866.
31. Palo, K.; Brand, L.; Eggeling, C.; Jäger, S.; Kask, P.; Gall, K. *Biophys. J.* **2002**, *83*, 605–618.
32. Eggeling, C.; Brand, L.; Ullmann, D.; Jäger, S. *Drug Disc. Today* **2003**, *8*, 632–641.
33. Jäger, S.; Garbow, N.; Kirsch, A.; Preckel, H.; Gandenberger, F. U.; Herrenknecht, K.; Rüdiger, M.; Hutchinson, P. J.; Bringham, R.; Ramon, F. et al. *J. Biomol. Screen.* **2003**, *8*, 648–659.
34. Zhang, J. H.; Chung, T. D.; Oldenburg, K. R. *J. Biomol. Screen.* **1999**, *4*, 67–73.
35. Hulme, E. C. In *Receptor–Ligand Interactions (A Practical Approach)*; Hulme, E. C., Ed.; Oxford University Press: New York, 1992.
36. Kassack, M. U.; Hofgen, B.; Lehmann, J.; Eckstein, N.; Quillan, J. M.; Sadee, W. *J. Biomol. Screen.* **2002**, *7*, 233–246.
37. Turconi, S.; Shea, K.; Ashman, S.; Fantom, K.; Earnshaw, D. L.; Bingham, R. P.; Haupts, U. M.; Brown, M. J.; Pope, A. J. *J. Biomol. Screen.* **2001**, *6*, 275–290.
38. Flotow, H.; Leong, C. Y.; Buss, A. D. *J. Biomol. Screen.* **2002**, *7*, 367–371.
39. Förster, T. *Ann. Phys.* **1948**, *2*, 55–75.
40. Hemmila, I.; Dakubu, S.; Mukkala, V. M.; Siitari, H.; Lovgren, T. *Anal. Biochem.* **1984**, *137*, 335–343.
41. Mathis, G. *Clin. Chem.* **1993**, *39*, 1953–1959.
42. Karvinen, J.; Hurskainen, P.; Gopalakrishnan, S.; Burns, D.; Warrior, U.; Hemmila, I. *J. Biomol. Screen.* **2002**, *7*, 223–231.
43. Mathis, G. *J. Biomol. Screen.* **1999**, *4*, 309–314.
44. Eggeling, C.; Gall, K.; Palo, K.; Kask, P.; Brand, L. *SPIE Proc.* **2003**, *4962*, 101–109.
45. Clegg, R. M.; Schneider, P. C. In *Fluorescence Microscopy and Fluorescent Probes*; Slavik, J. Ed.; Plenum Press: New York, NY, 1996, pp 15–33.
46. Weber, G. *J. Phys. Chem.* **1981**, *85*, 949–953.
47. Wild, U. P.; Holzwarth, A. R.; Good, H. P. *Rev. Sci. Instrum.* **1977**, *48*, 1621–1627.
48. O'Connor, D. V.; Phillips, D. *Time-Correlated Single Photon Counting*; Academic Press: New York, NY, 1984.
49. Zander, C.; Sauer, M.; Drexhage, K. H.; Ko, D. S.; Schulz, A.; Wolfrum, J.; Brand, L.; Eggeling, C.; Seidel, C. A. M. *Appl. Phys. B* **1996**, *63*, 517–523.
50. Brand, L.; Eggeling, C.; Zander, C.; Drexhage, K. H.; Seidel, C. A. M. *J. Phys. Chem. A* **1997**, *101*, 4313–4321.
51. Fries, J. R.; Brand, L.; Eggeling, C.; Kollner, M.; Seidel, C. A. M. *J. Phys. Chem.* **1998**, *102*, 6601–6613.
52. Eggeling, C.; Fries, J. R.; Brand, L.; Gunther, R.; Seidel, C. A. *Proc. Natl. Acad. Sci. USA* **1998**, *95*, 1556–1561.
53. Diaspro, P. *Confocal and Two-Photon Microscopy: Foundations, Applications and Advances*; Wiley-Liss: New York, NY, 2001.
54. Herten, D. P.; Tinnefeld, P.; Sauer, M. *Appl. Phys. B* **2000**, *71*, 765–771.
55. Auer, M.; Moore, K. J.; Meyer-Almes, F. J.; Günther, R.; Pope, A. J.; Stoeckli, K. A. *Drug Disc. Today* **1998**, *3*, 457–465.
56. Schaertl, S.; Meyer-Almes, F. J.; Lopez-Calle, E.; Siemers, A.; Kramer, J. *J. Biomol. Screen.* **2000**, *5*, 227–237.
57. Ullmann, D.; Busch, M.; Mander, T. *Inn. Pharm. Tech.* **2000**, *99*, 30–40.
58. Koppel, D. E.; Axelrod, D.; Schlessinger, J.; Elson, E. L.; Webb, W. W. *Biophys. J.* **1976**, *16*, 1315–1328.
59. Rigler, R.; Widengren, J. In *BioScience*; Klinge, B.; Owman, C. (Eds.); Lund University Press: Lund, Sweden, 1990, pp 180–183.
60. Rigler, R.; Mets, Ü.; Widengren, J.; Kask, P. *Eur. Biophys. J.* **1993**, *22*, 169–175.
61. Denk, W.; Strickler, J. H.; Webb, W. W. *Science* **1990**, *24*, 73–76.
62. Tsien, R. Y.; Waggoner, A. In *Handbook of Biological Confocal Microscopy*; Pawley, J. B. Ed.; Plenum Press: New York, NY, 1995, pp 267–279.
63. Eggeling, C.; Widengren, J.; Rigler, R.; Seidel, C. A. M. In *Applied Fluorescence in Chemistry, Biology and Medicine*; Rettig, W.; Strehmel, B.; Schrader, M.; Seifert, H. Eds.; Springer-Verlag: Berlin, Germany, 1999, pp 193–240.
64. Eggeling, C.; Widengren, J.; Rigler, R.; Seidel, C. A. M. *Anal. Chem.* **1998**, *70*, 2651–2659.
65. Eggeling, C. Analyse von Photochemischer Kinetik und Moleküldynamik durch mehrdimensionale Einzelmolekül-Fluoreszenzspektroskopie. Thesis/Dissertation, Georg-August Universität Göttingen, 2000.
66. Magde, D.; Elson, E. L.; Webb, W. W. *Phys. Rev. Lett.* **1972**, *29*, 705–708.
67. Ehrenberg, M.; Rigler, R. *Chem. Phys.* **1974**, *4*, 390–401.
68. Thompson, N. L. In *Topics in Fluorescence Spectroscopy* Lakowicz, J. R. Ed.; Plenum Press: New York, NY, 1991; Vol. 1, pp 337–378.
69. Widengren, J.; Rigler, R. *Cell. Mol. Biol.* **1998**, *44*, 857–879.
70. Visser, A. J. W. G.; Hink, M. A. *J. Fluorescence* **1999**, *9*, 81–87.
71. Aragón, S. R.; Pecora, R. *J. Chem. Phys.* **1976**, *64*, 1791–1803.
72. Shera, E. B.; Seitzinger, N. K.; Davis, L. M.; Keller, R. A.; Soper, S. A. *Chem. Phys. Lett.* **1990**, *174*, 553–557.
73. Müller, R.; Zander, C.; Sauer, M.; Deimel, M.; Ko, D. S.; Siebert, S.; Arden-Jacob, J.; Deltau, G.; Marx, N. J.; Drexhage, K. H. et al. *Chem. Phys. Lett.* **1996**, *262*, 716–722.
74. Koshioka, M.; Saski, K.; Masuhara, H. *Appl. Spectrosc.* **1995**, *49*, 224–228.
75. Oligino, L.; Lung, F. D.; Sastry, L.; Bigelow, J.; Cao, T.; Curran, M.; Burke, T. R., Jr.; Wang, S.; Krag, D.; Roller, P. P. et al. *J. Biol. Chem.* **1997**, *272*, 29046–29052.

Biography

Dirk Ullmann obtained his PhD in chemistry and biochemistry from the University of Leipzig, Germany, on protease-catalyzed peptide synthesis, and did his postdoctoral training on protein engineering at Brandeis University, Waltham, MA.

He is currently the director of biology services at Evotec, with global responsibilities for reagent production, assay development and screening. Ullmann joined Evotec in September 1997. Initially, he was responsible for developing and designing new ultra-high-throughput assays based on confocal fluorescence correlation spectroscopy (FCS) and related new proprietary technologies (FCS +). From April to December 1999 he held the position of section head of assay development, responsible for setting up new own proprietary assays for confocal fluorescence read-outs and also for assay development within screening service Evotec offers. Between January 2000 and April 2002 he set up and directed screening operations within the drug discovery services unit, embedding EVOscreen uHTS technology into a screening factory environment.

During his scientific career, Ullmann has published more than 30 papers within the life sciences area, and was presented with the Friedrich-Weygand-Award 1995 in Peptide Chemistry by the Max-Bergmann-Kreis e.V. He is a member of the Association for Laboratory Automation, the German Chemical Society, and the Society of Biomolecular Screening.

Comprehensive Medicinal Chemistry II
ISBN (set): 0-08-044513-6

ISBN (Volume 3) 0-08-044516-0; pp. 599–616

3.29 Cell-Based Screening Assays

A Weissman, J Keefer, A Miagkov, M Sathyamoorthy, S Perschke, and F L Wang,
NovaScreen Biosciences Corporation, Hanover, MD, USA

3.29.1	**Introduction to Cell-Based Assays**	**617**
3.29.2	**Cell Sources and Cell Culture**	**618**
3.29.2.1	Cell Selection	619
3.29.2.2	Cell Culture	622
3.29.3	**Readouts of Cellular Responses**	**623**
3.29.3.1	Electrophysiology and Ion Fluxes	624
3.29.3.2	Fluorescent Dyes for Voltage and Calcium	625
3.29.3.3	Receptor Binding and Active Transport	627
3.29.3.4	Migration and Chemotaxis	627
3.29.3.5	Proliferation	628
3.29.3.6	Absorption, Distribution, Metabolism, and Elimination/Excretion	628
3.29.3.7	Toxicity and Cell Death	629
3.29.3.8	Flow Cytometry	629
3.29.3.9	Multiple Outputs (Multiplexed Assays)	630
3.29.3.10	Calorimetric Enzyme Immunometric Assay and Similar Antibody-Based Techniques	631
3.29.3.11	Reporter Genes	633
3.29.4	**Cell-Based Screening Examples**	**633**
3.29.4.1	Binding versus Functional Readouts for Ion Channels – The $5HT_3$ Ligand-Gated Ion Channel	633
3.29.4.2	Choosing Among Different Readouts for a Targeted G Protein-Coupled Receptor – The μ Opioid Receptor	634
3.29.4.3	A Multitarget and Multioutput Assay – Immunomodulation through the Toll Receptors	638
3.29.4.4	Targeting a Cell-Based Model of Organ Response – Neurotoxicity	639
3.29.5	**Future Directions for Cell-Based Assays**	**641**
References		**644**

3.29.1 Introduction to Cell-Based Assays

Central to the success of any medicinal chemistry-based project are the assays that establish the biological activities of new chemical entities. The results of these assays are then used for a family of compounds to determine their relative potencies and elucidate a chemical structure that is the basis of that activity. These assays illuminate the path forward for the medicinal chemist, serving as the direction for future synthetic work. Because of the pivotal role that bioassays play in routinely guiding these efforts and in measuring the progress of the medicinal chemist, they must routinely provide rapid, relevant, and reliable pharmacological information. The contents of the first edition of this *Comprehensive Medicinal Chemistry* series were a testament to the importance of this fact, containing a significant number of pages devoted to defined pharmacologically relevant assays. At that time, the emphasis was on organ, enzyme, and especially receptor-binding assays, as a relatively new and strongly productive strategy for future drug discovery.

Over the last decade pharmacologists added powerful tools and experimental methods to those offered by standard enzyme or binding inhibition assays. In particular, the determination of effects of chemical compounds on a complex cellular level quickly became an essential component of early drug development. Tissue, organ, or whole-animal experimentation provided much of the basic information on the biological effects of a compound in the past. The established regulatory environment needed to provide information about the efficacy and safety of new chemical entities reinforced this in vivo approach. These regulations required tests in animal models and ultimately in humans. Animal studies do provide predictive information about the possible therapeutic or adverse effects that may occur in humans. However, there are severe limitations in extrapolating data from different kinds of animals to human systems. Central to this concern are the intrinsic species differences as expressed in the anatomy, catabolic/metabolic, and susceptibility/regenerative properties in response to various chemicals. In addition to these problems, conventional animal data are expensive and time-consuming to obtain, raise ethical issues concerning animal usage, and are not often compatible with the current high-throughput requirements of the drug and chemical industries.

There is now a renewed emphasis to develop credible in vitro endpoints in cell models that parallel the known effects of compounds in humans. This information, often obtained in a high-throughput format, can be used at an early stage in drug development, minimizing the amount of animals, time, and expense needed to carry out preclinical studies. Cell use has also been advanced by progress in instrumentation (specifically computers), robotics, intracellular dyes, fiberoptics, and new fluorescence-based techniques such as fluorescence resonance energy transfer (FRET) and time-resolved fluorescence (TRF). Advances in recombinant deoxyribonucleic acid (DNA) technology have also enabled the ready use of human protein targets expressed in almost any cell. The culmination of this work, for drug discovery, has been the increased use of whole cells in screening assays in search of effective therapeutic compounds. A cell-based assay for our purposes can be defined as an analytical procedure that measures a biological activity resulting from the interaction of a test substance on individual or groups of viable cells. The important aspect of these assays is the use of the whole cell that has all the cellular response components intact and thus may more closely approximate what one might observe in vivo.

There are several advantages to use of cell-based assays versus those using an isolated protein. First, membrane-bound targets are in a more natural environment and surrounded by accessory proteins that operate more normally than when prepared as membrane fractions. Second, one can visualize, in real time, cell responses occurring in the seconds to minutes range. This immediate feedback is a boon to rapid assay development and allows troubleshooting to occur during an experiment. Third, most functional assays allow us to distinguish between agonists and antagonists as well as partial agonists and inverse agonists, while binding assays lack this ability. Finally, most response and feedback signal pathways are intact in the cell, allowing the measurement of secondary as well as primary responses to initial drug interactions. With these properties, we believe that cell-based assays constitute a paradigm-changing technology in the process of drug discovery. When properly designed, cell-based assays now permit a medicinal chemistry project to determine and quantify not only the immediate pharmacological interactions of a compound, but also the multiple types of cellular signal transduction-mediated consequences of that initial action. Cell-based assays can thus provide a direct functional readout of a compound's pharmacological activity and enable discovery of new drug targets for the ultimate therapeutic endpoint. Serendipitous results from this type of assay can provide the opportunities for major advances in the field, providing a real advantage over other drug discovery methods.

From our experience as participants in the preclinical drug discovery arena, we conclude that cell-based functional assays will play an increasingly important role in lead discovery and the medicinal chemistry characterization process. In this chapter we address the use and relevance of in vitro-based cellular assays for drug discovery. It is an attempt to provide a primer for the medicinal chemist and covers the basics of cell-based functional assays and the type of information that can be derived from them. We begin by describing what cells are available, how to select a cell type, and how to grow them. This discussion then leads to a description of what techniques and readouts are available for cell-based assays and how they are selected and used. Finally, we consider the in vivo relevance of in vitro cell assays to medicinal chemistry and drug development by considering several practical examples in ion channels, G protein-coupled receptors (GPCRs), tyrosine kinase-linked receptors, and drug toxicity. Additional details can be particularly found in Chapters 3.27, 3.28, and 3.32.

3.29.2 Cell Sources and Cell Culture

Tissue culture was first attempted in a limited way in 1885, but did not find broad relevance until the introduction of antibiotics, which prevented the growth of bacteria in the cultures. It was not until the early part of the twentieth century that culturing of single cells became a common practice. An early pioneer, Carrell, began a cell line, from the heart of a chick embryo. This cell line was propagated successfully for 34 years. In the early twentieth century, those heart cells were applied to the toxicity testing of germicides.[1] Ebeling then used that experience to culture a line of epithelial cells. Later, hybrid cells were constructed by fusing cells of different species as well as different kinds of cells from the same species.[2] Since that time, cells have become a workhorse for biological research. Today, a vast array of cell types, from both nonprofit and commercial sources, is available to researchers.

3.29.2.1 Cell Selection

Selecting an appropriate cell model is one of the first steps in designing and implementing a modern cell-based assay. For this purpose, cell lines are now available from a variety of species among bacteria, insects, amphibians, and mammals. Cells from different organisms have attributes that provide both unique value and critical flaws for specific applications of in vitro research.

Insect cells are relatively easy and inexpensive to grow and can be engineered to express large quantities of a variety of transfected receptors and enzymes. However, posttranslational modification of the protein, particularly glycosylation, may differ from the human cell and could change the expressed protein's pharmacological properties. Accessory proteins, such as G proteins, may be different or absent in the insect cell, therefore limiting the usefulness of this cell class to mimic a particular human system. This problem can be mitigated by artificially introducing the required G protein or other necessary accessory protein into the cell. Care has to be taken to validate insect cell systems due to the evolutionary distance from humans. However, all cells, even human ones, need to be validated to ensure equivalence with the biological target in humans. The expressed phenotype of a cell can also vary as a function of tissue source and culture conditions.

Another source of cells is oocytes (eggs) from the *Xenopus* frog that are routinely used for certain in vitro assays such as ion channel responses.[3] They are large enough to be individually microinjected with gene constructs for a receptor. They can then be easily monitored electrophysiologically for changes resulting from the expression and pharmacological modulation of the gene product. The large cell size also makes handling of these cells easier than most mammalian cells, allowing for intracellular manipulations, including alterations in membrane potential. One disadvantage for oocytes is that they do not replicate. Eggs must be harvested anew from their living source for each of the assay iterations. A second caveat is that, like insect cells, amphibian cell types do not always provide results that correlate well with mammalian cell systems or tissues. For example, there is a reported lower affinity of ether-a-go-go related gene (hERG) blockers when they are tested on ion channels expressed in Chinese hamster ovary cell line (CHO) or human embryonic kidney cell line (HEK).[4] This lower affinity may be attributed to the more highly lipid oocyte cell membrane.

Tumors that have been isolated from specific human or animal cancers are another convenient cell source. These cells often express markers that are characteristic of their tissue of origin and are generally able to proliferate rapidly. Although they rarely express the full complement and comparable levels of cellular markers found in primary cells from the same organ, these cell types often retain many of the specific functions that are relevant to profiling the physiological effects of chemical compounds.[5] Numerous tumor cell lines are commercially available from a large variety of human and animal tissues. This allows researchers a steady access to a variety of cells that have structural and biochemical features similar to normal healthy tissue. Importantly, although immortalized cells and primary cells, derived from the same organ, have similar characteristics, they also have significant differences, mostly related to their less differentiated and proliferative state.[6]

In addition to cell lines derived from tumors, cells from humans and animals can be directly cultured. Primary cells, taken from a specific tissue or organ, can be ideal for testing since many are fully differentiated and express numerous features typical of the entire organ of origin. However, this end-stage differentiation can also limit their usefulness. Most primary cells are limited in their ability to proliferate and thus have a defined lifetime. Hepatocytes and neurons, like oocytes, must be harvested from a donor when more cells are needed. Other primary cells capable of division can be cultured only for a limited number of passages, typically about 10, before their properties change or vitality diminishes. Indeed, all primary cells in culture can lose expression of particular features such as receptors or morphology over time.[7] One challenge has been to find a way to prevent primary cells, with all their special structures and functions, from reverting to a less organ-specific type cell. Possible solutions have been to co-culture them with different types of cells, particularly from the same organ. Studies have shown that, in co-cultures, some cells can survive longer and are better able to maintain their differentiated functions.[8] Even if their phenotype can be successfully maintained, primary cells tend to be difficult and expensive to grow. They often require special media and/or culture plate with surface treatments for optimal growth and attachment. An additional issue is related to the origins of these primary cells. Cells are usually derived from a single individual having traits unique to that donor, but representing only a single example of the tested species. For inbred animal species this variation is less of a problem than when a human donor is the source of the cells. Due to their high cost, short lifespans and heterogeneity, primary cells are of limited value in large-scale assays, even though they may have great similarities to the organ of origin. For these reasons, primary cells are not generally used for high-throughput screening (HTS) primary screens, being better suited for smaller screening projects and secondary assays.

Blood is a readily available source of primary human cells for use in certain cell-based assays. However, the variety of cell types is limited, and, depending on the cell type desired, yield per donor can be small. When considering blood, like any primary tissue source, one must be careful to consider possible effects from factors such as age, sex, ethnicity, and individual genetic variances. The donor's diet, smoking, health status, and particularly medications, can have a dramatic impact on some receptor types, and on the general tone of the cells.[9–11] Psychological state, daily and monthly hormonal cycles, or infections can also impact, depending on the target studied. In some cases, blood cells from more than one donor can be pooled to reduce individual differences, yet this blending may induce immune responses. Ideally the target would be validated to confirm that person-to-person differences are minor. Most providers of blood products routinely test the samples for infectious agents such as hepatitis and human immunodeficiency virus (HIV). This pathological testing is crucial for the safety of those who will handle the samples. Still, exposed workers should receive hepatitis vaccinations as a precaution and be properly trained in safe handling of the cells and disposal of biological wastes.

A technique of immortalization can be employed to overcome the finite replication problem of most primary cells and allow long-term culture. Here, the DNA-encoding viral genes, such as from the Epstein–Barr virus, are transfected into the cells of interest. The transfected gene can be engineered for activation by a triggering chemical (i.e., tetracycline), causing the cell to replicate indefinitely or to differentiate when the gene is turned off.[12] Additionally, hybrid cells can be made by fusing a proliferative cell with another target cell of interest. For instance, monoclonal antibody-producing hybridomas are made by fusing primary, mortal spleen cells from a mouse that has been immunized with the desired antigen, and immortalized myeloma cells. The fusion agent can be a chemical like poly(ethylene glycol) (PEG) or an infectious agent like the Sendai virus. These techniques can provide a continuous supply of identical cells that have the capacity to overexpress a specific phenotype typical of their tissue of origin. This gene expression in either primary or transfected cell lines can be responsive to conditions in the culture such as availability of growth factors, cell confluence, density, pH, cell substrates, temperature, and method of harvest (trypsinization, ethylenediaminetetraacetic acid (EDTA), mechanical). Continuous cell lines can be developed from these transfected, immortalized cells, thus eliminating the need for further animal or human cell donors.

Finally, naturally proliferating cells can be modified using molecular biology techniques to express proteins that can be functionally coupled to a desired biological system. Transfection can yield a transient or stable phenotype. Transient transfections last for days and affect only a portion of the cells, whereas stable transfections permanently express the protein. This technique can be further enhanced by transfecting into the host cell the target and a readout system like a reporter gene. A number of cell types have been found to express few endogenous receptors themselves, yet can be easily transfected to produce exogenous receptors. Commonly used cell lines include CHO, simian monkey fibroblast (COS), and HEK293. While these are useful host cells, they can sometimes impart significant changes in the functional protein they produce. See **Table 1** for a list of the common types of cells that are used as a source of proteins for

Table 1 Typical sources of cells used to produce target human protein

Recombinant protein (transfected into cell type)	*Endogenous protein*
Bacteria (*Escherichia coli*)	Tumor cell lines
Yeast (*Saccharomyces cerevisiae*)	Hybrid cells (hybridomas)
Insect (Sf-9)	Primary cells
Frog (oocytes)	Blood cells
Rodent (CHO)	Stem cells
Monkey (COS)	Immortalized cells
Human (HEK)	
Tumor cell lines	
Primary cells	

screening assays. The high level of similarity between a gene product expressed in a transfected mammalian cell and that expressed in vivo is a great benefit, enabling rational tests of therapeutic strategies. However, the cell line must be validated to determine exceptions to this rule. One study compared the pharmacological binding profile of human angiotensin type 1 (AT1) receptor from three tissue sources. The results revealed a significant difference between all three sources of receptor tested under the same binding conditions. Binding properties (K_i) for most ligands are similar. However, there is a substantial difference in the affinity for certain agonists (**Figure 1**) despite identical DNA coding for the receptor in the three cell types. Specifically, the angiotensin III peptide ligand revealed a 400-fold difference in its affinity in the recombinant receptor ($IC_{50} \cong 4.6$ nM) expressed in CHO cells versus the liver ($IC_{50} > 3000$ nM) or KAN-TS cells ($IC_{50} \cong 2000$ nM). Binding differences can be explained by cell culture conditions, receptor density, accessory proteins, and posttranslational modifications. It is also important to be aware of the level of endogenous receptors in the cell of choice, just as with primary or tumor lines. These endogenous receptors may dimerize with the receptor target or your drug candidate may act at endogenous receptors as well as the transfected target. For example, HEK293 cells are reported to have β-adrenergic, somatostatin, muscarinic, and 10 other GPCR receptors.

An alternative to using these conventional cell lines is to clone the protein of interest into a tumor or other cell line approximating the target tissue. A disadvantage would be that the protein interaction with an uncommon cell would take more effort to validate. With either common or unique cell hosts, expression levels of the protein may also cause variations in a cell-based assay. Generally, cloned receptors are present at levels of 0.5–5 pmol mg^{-1} of protein and in the range of a few thousand to a few hundred thousand receptors per cell. High expression levels can cause changes in desensitization. Low expression levels may be insufficient to initiate a functional response. For instance, expression of the glucagon-like peptide 1 (GLP-1) receptor at 1800–5600 receptors per cell shows no agonist effect

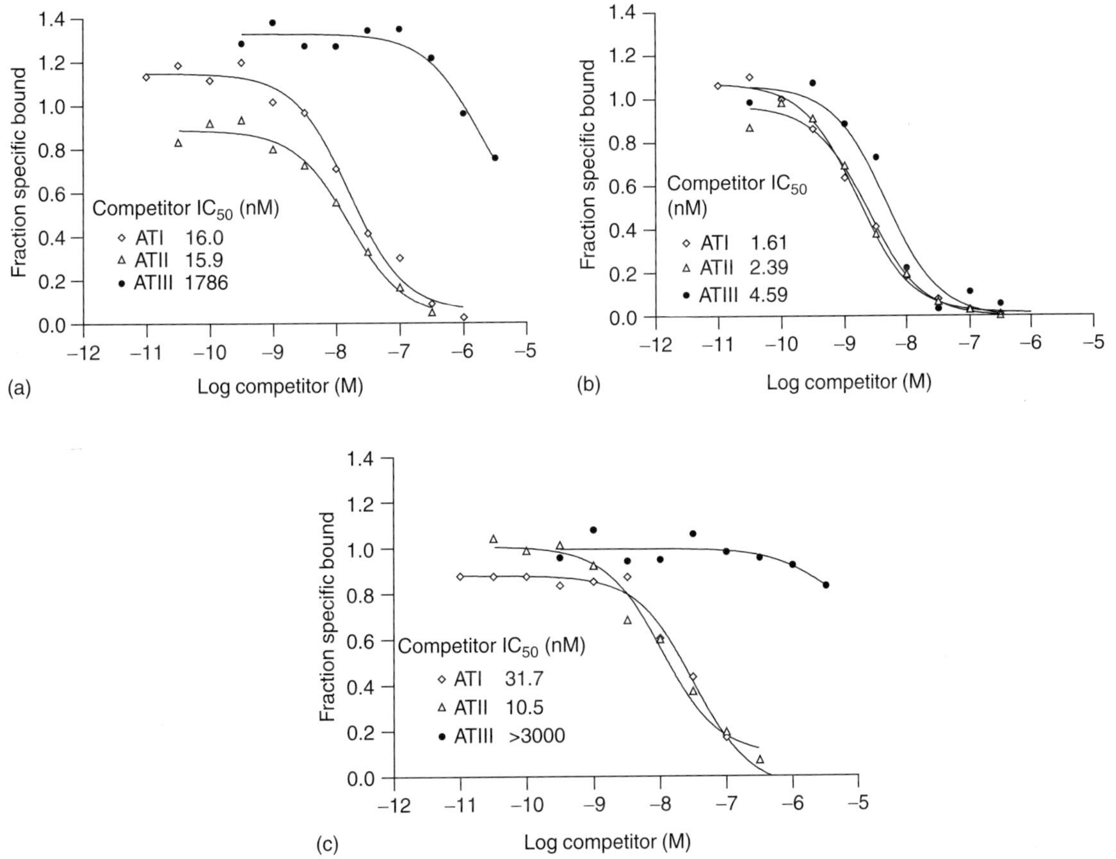

Figure 1 Pharmacological comparison of the binding of three angiotensin peptides to three different sources of AT1 receptor: (a) an endogenously expressing human neuroblastoma cell line, KAN-TS; (b) a transfected CHO cell; and (c) human liver. IC_{50} curves are shown for angiotensin I, II, and III in competition with Sar1, [^{125}I]Tyr4, Ile8-angiotensin II (an AT1-selective radioligand) for binding at the AT1 receptors.

whereas there was a response when the expression levels reached 13 000–270 000.[13] The highest number of receptors showed the largest response, but also a 13-fold increase in agonist EC_{50}. Also, partial agonist effects are particularly sensitive to expression levels. Therefore, cells chosen must be validated to confirm similar or advantageous characteristics with respect to the target tissue.

3.29.2.2 Cell Culture

Once a cell-based assay is developed, cells that are grown or harvested from the organism of interest must be obtained in sufficient number to support this effort. If the cells are to be cultured, an on-site sterile facility is usually required. Alternatively, some cells can be purchased already plated in 96- or 384-well plates at an additional cost. A popular collection of cells is the ATCC (American Type Culture Collection, Manassas, VA, USA). ATCC offers extensive references and descriptions of these lines and supplies them at a low cost. Other sources of both primary and immortalized cell lines are the European Collection of Cell Cultures (Salisbury, UK) and private companies such as Cambrex (East Rutherford, NJ, USA). A useful reference, the Cell Line Database of the Interlab Project (IST, Genova, Italy) includes a listing of cell lines from different catalogs, including ATCC and the European Collection.[14] If the target of interest is on human blood cells, several companies, such as Biological Specialty Corp. (Colmar, PA, USA) and Analytical Biological Services (Wilmington, DE, USA), provide either whole blood or separated cell populations. Some researchers obtain cell samples from individuals or clinical populations on an as-needed basis. However, consent forms and prior institutional review and approval of the project are generally required. Other sources of cells can be the laboratories that have created particular cells and published articles on their work. Upfront licensing agreements and substantial fees may be required, particularly for nonacademic use.

Cultures of cells are obtained either as frozen aliquots or as live actively growing cultures. The cryovial is kept frozen and stored in liquid nitrogen. When needed, it is rapidly thawed and diluted in media to dilute the cryopreservatives. This cell suspension is pelleted by centrifugation and then the resulting pellet is resuspended gently in media. Each cell line has specific media requirements. Prepared media such as HAM's DMEM or RPMI-1640 consist of a balanced salt solution, glucose, supplements such as amino acids, vitamins, fatty acids, and a bicarbonate or *N*-2-hydroxyethylpiperazine-*N*′-propranesulfonic acid (HEPES) buffer. Glutamine is added when a new bottle of medium is opened, since glutamine is relatively unstable long-term in solution and lasts about a month at 4 °C. Antibiotics such as penicillin–streptomycin may be added to lessen the chance of media contamination. Other antibiotics, such as G418, Zeocin, or blasticidin, are often added to cultures of clones to maintain the transfected cells. The purpose of this antibiotic is to limit growth or kill any untransfected cells that do not have the antibiotic resistance gene cloned in along with the target protein gene. Antibiotic use may mask bacterial contamination, so routine use of antibiotics is discouraged. An antibiotic-free culture technique will limit low levels of contamination and the production of antibiotic-resistant bacteria. Other signs of the health of a culture can be assessed visually. Most media contain phenol red as a pH indicator. As cells proliferate and metabolize in culture, the media can become acidic and turn yellow. Media removed from a CO_2 incubator will become basic and turn purple. Bacterial contamination turns the culture cloudy and yellow.

Supplements are often added to the culturing media. Each cell line may have individual requirements such as steroids or assorted growth factors. In general, fetal bovine (or horse serum) is added to media at 5–20% volume to supply often unknown factors important to cell growth and health. Some groups use defined media with specific growth factors in place of serum, to lessen the variable effects due to serum batch-to-batch differences. If pure water systems are available, bulk media can be made from powder to reduce transportation and storage costs. However, extreme care must be taken to use high-quality, sterile water and to filter the media, to prevent contamination of the media. This is generally only done when large quantities of particular media types are required. Cells are typically cultured in T-flasks or roller bottles, which provide plenty of surface area and are capped to lessen the chance of bacterial contamination while allowing gas exchange. Cells are generally diluted with fresh media and replated two to three times a week to maintain growth. The incubation chambers are generally humidified, at 37 °C, with 5% or 10% CO_2 as a pH buffer. Once a cell line is growing, aliquots can be frozen and stored in liquid nitrogen for future use. Cell phenotype may change with time in culture so many labs limit the number of passages of cells used before restarting cells from the frozen stock.

The successful production of cells for assays is dependent on the surface on which the cells are grown. While some cell lines grow in suspension, other cell lines are anchorage-dependent and require special substrates. Extracellular matrices such as collagen, fibronectin, poly-D-lysine, and laminin can optimize cell adherence to plasticware. For example, Caco-2 cells are grown on collagen whereas primary neurons are often grown on poly-D-lysine with laminin. There are also cell factories that supply additional surface area for cell growth in a space-efficient design. Similarly, a microcarrier substrate in a bioreactor can effectively increase that growth surface and increase cell yields. Although cells may grow on many different surfaces, these surfaces can have a profound effect on their proliferation and phenotypic expression. For instance, human bone cells express different chemokines depending on cell culture substrate.[15] The use of a substrate that is similar to the three-dimensional structure of bone tissue favors the differentiation of osteoclasts from peripheral blood mononuclear cells.[16] Attention must also be paid to growth and cell-harvesting conditions. If cells reach confluence, the target expression may be downregulated, yet a high density of cells will produce a stronger response in many fluorescence-based assays. So these factors must be balanced when developing an assay. Removal of cells from plates by the addition of the enzyme trypsin may also digest or otherwise degrade the receptor of interest. Improper harvesting may decrease cell viability. Typically one aims to maintain at least 90% viability. Some cell-based assays may require performance at 37 °C while certain interfering cell functions, such as receptor desensitization and dye transport, may be reduced by performing the assay at room temperature.[17]

Traditional cultures can be modified to form a more complex cell organization that more closely mimics human organ systems. These models can be as simple as the co-culturing of two cell types together, such as liver and heart cells, to measure the effects of hepatic biotransformation of a compound on the heart muscle. A more complicated system can involve some structural modeling and multiple cell types that can be used, for example in transdermal/epidermal permeability and metabolism studies or development of hemopoietic activity in culture simulation of structural bone. In essence, these three-dimensional cell models employ a supporting matrix, which serves as a scaffold to allow multiple cells to interact, such as fibroblasts and epithelial cells, thereby creating a more realistic representation of a human organ system. Presently, a number of these models are in use, including one to measure eye and skin irritation, mucosal absorption, blood–brain barrier, and kidney transport. There are systems that attempt to mimic multiple organ function. One example uses cultures linked by circulating media much in the way the heart recirculates blood through the different organs of the body.[18]

3.29.3 Readouts of Cellular Responses

Medicinal chemists require as much data as possible concerning the physiological effects, metabolic fate, and possible toxicity of putative therapeutic compounds. The earlier in development this is obtained, the more this information can guide their chemistry effort. The ultimate goal is to weed out unsuitable drugs and cultivate the most effective drug with the fewest side-effects. The obvious benefits of cellular assays for this purpose are their high content of information combined with their speed, efficiency, and throughput, resulting in reduced screening costs as compared to traditional animal and human testing. An additional consideration is that these cellular studies can reduce the expenditure of limited compound quantities, as well as the number of animals that would be needed to provide similar information. Where the cell-based assay often fails is in its capacity to model complex processes adequately at the whole organism level and thus an inability to predict reliably effects of the compound in humans. For instance, the interaction of many organs and cellular systems is regulated by the absorption, distribution, metabolism, and elimination/excretion (ADME) of a compound in a highly structured process. The amount of compound or its metabolites that a target cell is exposed to in vivo may ultimately depend on the capacity and efficiency of an individual's organs like the gastrointestinal tract, liver, kidney, and heart to process those compounds. The range of these integrated multiorgan responses can be difficult to predict for a typical human patient. This is especially true when those compounds have novel structures that can produce atypical biological responses not previously observed in human populations. However, innovative approaches to this problem, such as multiple cell-type models of organ function or whole-body function, are responding to this deficiency.[19]

The strengths of in vitro assays are their ability to provide detailed cellular mechanisms of a drug's action by providing a simple system with well-defined elements amenable to experimental manipulation. This system is usually composed of three components: (1) the biological model; (2) endpoint measurements; and (3) the test protocol. The biological model is the system used for evaluation, for example, of liver function. Hepatocytes from animal and human livers are commercially available and routinely used to examine metabolism and toxicity in this organ.[20] The greater the ability of the biological model to mimic in vivo structure and function accurately, the more valuable is the data derived from it in extrapolating to human effects. The endpoint measurement is quantitation of the test drug and its metabolites or some form of cellular readout. Cell readouts include voltage change, a second-messenger response such as calcium flux, a total cellular response such as growth, proliferation, or death, and an engineered output such as a reporter gene linked to luciferase. A partial list of those target readouts is shown in **Table 2**. The test protocol is usually determined experimentally to define the boundaries of the cellular response over factors such as dose, time, temperature, growth phase, or density.

Table 2 Cell-based approaches used for major drug target categories

Major target (approx. no. in family)[69]	*Subtypes*	*Example of target and drug*	*Mechanism (time)*	*Common cell-based screening techniques*
Ion channels (200)[41,66]	Voltage-gated	Calcium channel (amlodipine)	Ion flow change (ms–min)	Planar patch clamp, atomic absorption, dye (voltage/calcium)
	Ligand-gated	Serotonin $5HT_3$ (granisetron)	Ion flow change (ms–min)	Same as voltage-gated
GPCRs nonodorant (350)[67]	$G_{i/o}$	μ opioid (morphine)	Modulates cAMP (5–20 min)	cAMP EIA; calcium dye using cells transfected with $G_q15/16$; aequorin or similar readout
	G_s	Histamine H2 (ranitidine)	Increases camp (5–20 min)	Same as $G_{i/o}$
	G_q	Histamine H1 (loratidine)	Increases Ca^{2+} and inositol phosphates (5 s–2 min)	Calcium dye
Cell surface transmitter transporters (19)[66]	–	Serotonin uptake (fluoxetine)	Movement of molecule (s–min)	Radiolabeled ligand flow into or out of the cell
Protein kinases (500)[68] and phosphatases (200)[68]	Protein kinase	BCR/Abl (Imatinib/Gleevec)	Phosphorylation (min–1 day)	EIA-based detection of phosphorylated substrate, cell proliferation
	Protein phosphatases	PP2B = calcineurin (ciclosporin A)	Dephosphorylation (min–1 day)	EIA-based isolation of dephosphorylated substrate
Nuclear hormone receptors (50)[67]	–	Estrogen (tamoxifen)	Binds to response element (approx. 1 h)	Reporter gene, cell proliferation, EIA

cAMP, cyclic adenosine monophosphate; EIA, enzyme immunometric assay.

Individual components of cells such as receptors, enzymes, and individual proteins are the predominant target of drug development assays. However, an examination of a whole-cell response can often provide more essential information to the medicinal chemist trying to understand the impact of compounds on a biological system. Measurements range from a determination of agonist/antagonist responses to the pattern of changes across a large number of genes or proteins on a microchip. In addition, larger-scale assessments can be made, for example, approximating a whole organ's function in vivo through validated cell models of toxicity and ADME. These whole-cell studies are often used as an intermediate step. They bridge assays establishing an effect on a well-defined protein target, to those observing the interactions among cells, organized into tissues and organ systems, in the intact human body.

3.29.3.1 Electrophysiology and Ion Fluxes

The functional measurement of ion channel activity was one of the earliest techniques that found wide acceptance as a valuable cell-based assay. The use of ion-sensitive dyes, membrane potential dyes (MPDs), atomic absorption, or radioactive methods can assess ion flux across a cell membrane. However, electrophysiology (EP) is the most informative method of studying channels and is widely recognized as the 'gold standard.' With the insertion of a very thin electrode (traditionally a glass pipette) into the cell, the difference in electrical charge, due to ion fluxes across the cell membrane, can be accurately measured while compounds are added to the incubation medium. Voltage clamp is generally used, where the voltage is held constant and the current due to ion movement is measured. The ability to add drugs or ions inside as well as outside the cell is an important additional advantage because many channel blockers work intracellularly. EP's shortcoming is that it has a very low throughput for screening compounds, on the order of 10–15 data points per day, and requires extensive training to gain proficiency. Whole-cell EP has recently been automated by the use of suction of a cell on to a plastic or glass chip with holes that approximate the size of a cell. A single-well format is used by Nanion (Munich, Germany), which advocates its ease of use and a tripling of traditional patch clamp rates. Other companies have adapted this technique to up to 384 wells. Companies include Flyion (Tubingen, Germany), Molecular Devices/Axon (Sunnyvale, CA, USA), and Sophion (Ballerup, Denmark). These tools allow for throughput on the order of up to 1000 or so data points per day. The tradeoff is the limited range of experimental designs that they can perform, relatively high materials cost per well, a significant percentage of failed cell measurements, an inability to choose a cell visually, and a high initial price for the instrument. In addition there are concerns about non-giga ohm physical seals between cells and the instrument chip surface, and adherence of hydrophobic drugs to plastic chips, which could alter IC_{50} values in some systems. Noninstrument-dependent factors such as cell density can also have significant effects on EC_{50} curves.

There are several other cell-based methods to study ion channel activity. Direct measurements of ion flux can be done using substitute labeled ions of similar size and charge to the natural ion. One example is the use of a rubidium isotope ^{86}Rb to assess ion flux through potassium channels. In this method, cells are loaded with the isotope during an incubation step, washed, and potential candidate blockers are added. In the final step, cells are harvested via filtration and the remaining ^{86}Rb radioactivity that has not flowed through the channel is counted. Potassium channel blockers will increase the radioactivity in the cell since the efflux of ^{86}Rb, like potassium, is prevented from flowing through the ion channel along its electrochemical gradient. However, due to the high levels of radioactivity present, this technique is considered a poor vehicle for drug screening. As an alternative, nonradioactive Rb can be used and quantified by atomic absorption. In this method, cells are incubated as above, but nonradiolabeled RbCl replaces KCl. Then the cells are washed, test compound added, the supernatant and cells are separated, and the cells lysed open. Samples are then placed in the atomic absorption instrument where the liquid sample is vaporized in a flame. The amount of light absorbed from the Rb lamp is proportional to the ionic content of a sample. This technique suffers from less sensitivity than patch clamp (about fivefold). More sophisticated instrumentation to measure Rb such as sold by Aurora Biomed (Vancouver, Canada), can speed analysis time, improve IC_{50} accuracy, and reduce sample required.

3.29.3.2 Fluorescent Dyes for Voltage and Calcium

Ion flux can also be measured in whole cells by dyes sensitive to voltage/membrane potential. These dyes are formulated to enter depolarized cells where they bind to intracellular proteins or membranes. Increased depolarization (increased positive charge of the interior of the cell relative to exterior) results in more influx of the anionic dye and thus an increase in fluorescence. In contrast, hyperpolarization (decreased positive charge) is indicated by a decrease in fluorescence. MPD assays are generally less sensitive than any of the above techniques (EC_{50}s for channel blockers are lower), and have the added complication that drugs can interfere with the dye's fluorescence to produce a high rate of false negatives and false positives. This is a particular problem with voltage dyes, as the dye–cation interaction is not selective and is thus prone to artifacts produced by cationic drugs. For voltage dyes, a drug may interact at a site other than the channel and cause a membrane potential change that is not specific to just a single channel type. In the case of one particular potassium channel, hERG, which is linked to a heart arrhythmia side effect of certain drugs, the MPD does not produce the same rank order potency as patch clamping for a series of well-known drugs. Also, a false-positive rate of 12% was found during screening.[21,22] An alternate to the standard voltage dyes such as the MPD (Molecular Devices, Sunnyvale, CA, USA) or $DiSBAC_4(3)$ (Invitrogen, Carlsbad, CA, USA) is Invitrogen's FRET-based system. In this assay coumarin phospholipid is the donor molecule and resides in the outer cell membrane. It transfers energy to the $DiSBAC_2(3)$ acceptor dye, which moves to the outer membrane under depolarizing conditions, increasing FRET-based fluorescence. Conversely, $DiSBAC_2(3)$ moves to the inner membrane when the cell is more negative, making it unavailable for FRET interactions. A ratiometric readout of two emissions, rather than one wavelength, can be derived with FRET. This provides an internal control and reduces experimental errors arising from well-to-well variations in cell number, well addition artifacts, dye loading, plate inconsistencies, temperature fluctuations, and signal intensities. Further applications of FRET are discussed in subsequent sections.

Calcium measurements are particularly well suited to dye techniques since there is a large extracellular level of this ion relative to intracellular, resulting in a strong fluorescent response. In tandem with specific agonists and antagonists for the target receptor, calcium dyes provide an extremely effective technique to measure receptor-mediated effects. Other Invitrogen dyes for K^+ (PBFI), Na^+ (SBFI), Cl^- (MQAE), H^+ (SNARF) are seldom used due to technical difficulties such as dye specificity and relatively small differences between intracellular and extracellular ion concentrations. For example, the difference between extracellular and intracellular potassium is approximately 40-fold, whereas calcium difference is 10 000 fold (1 mM versus 100 nM). The appeal of dye-based cell assays is the relatively low expense and ease of conversion to high-throughput for use as an initial screen. Most of these other ion cell-based assays follow a standard methodology. Cells are washed with physiological salt solution, loaded with dye for 60 min, placed in a fluorometer, and the fluorescence change is measured upon addition of agonist or antagonist. Examples of two cell-based dye assays, calcium increase with the neuronal ganglia nicotinic receptor and membrane potential increase with the hERG potassium channel are shown in **Figures 2** and **3**, respectively. In general, the calcium response (in relative fluorescent units) is much smaller than the voltage and as such may be more vulnerable to experimental noise. A more complete review and comparison of dye techniques can be found in Birch *et al.*[23]

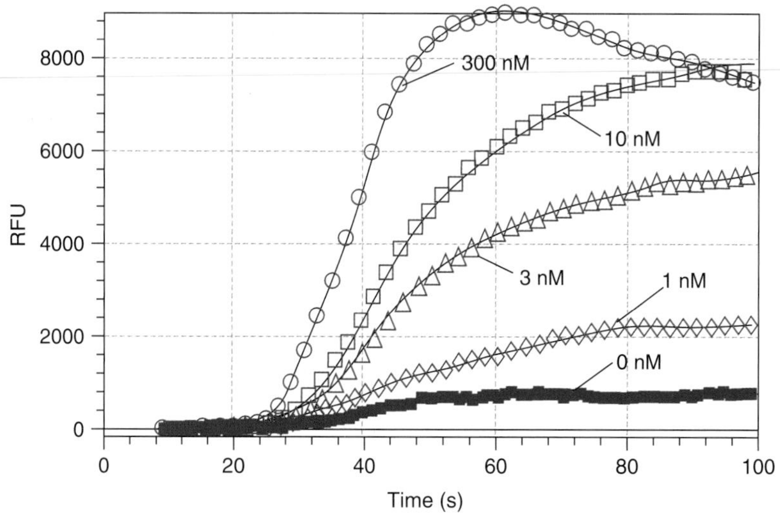

Figure 2 Calcium dye (Molecular Devices) readout of the effect of a nicotinic agonist, epibatidine, at the nicotinic (neuronal autonomic ganglia) receptor in human neuroblastoma cells. Data are a dose–response from five concentrations of epibatidine, 0–300 nM. Scale is in relative fluorescent units (RFU). Drug was added at 30 s and the response peaked between 45 and 90 s. Measurements were performed at 25 °C rather than 37 °C, to slow the reaction, and reduce receptor desensitization. Ex/Em 485/525.

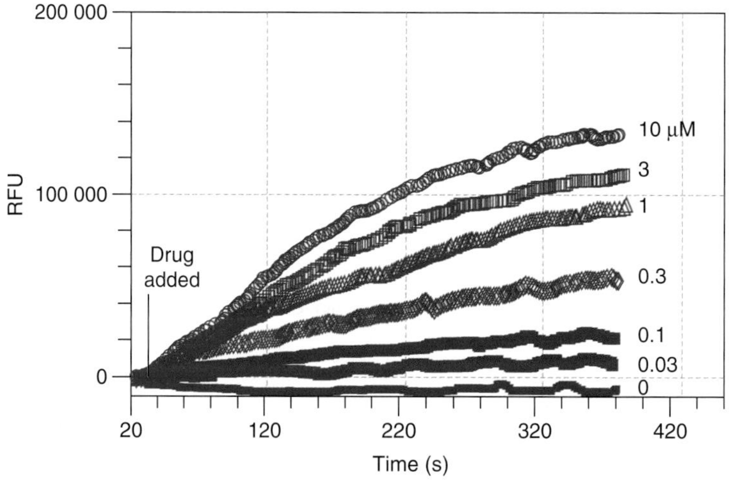

Figure 3 Membrane potential dye (Molecular Devices) readout of the effects of a channel blocker of the hERG potassium channel cloned into CHO cells. Data are a dose–response from seven concentrations of the drug astemizole, 0–10 μM. Scale is in relative fluorescent units. Astemizole was added at 30 s in these figures and the response peaked between in about 340 s. Measurements were performed at 25 °C. The presence of the hERG channels causes the cells to become hyperpolarized relative to an untransfected CHO cell. Upon addition of astemizole, the membrane potential becomes more positive. Ex/Em 530/565 nm.

Effects mediated through the specific G_q, GPCRs, can also be measured by dye-based measurements of intracellular calcium. A secondary calcium response occurs in some cells as activated kinases phosphorylate and open calcium channels. The calcium assay can be used to test both the agonist and antagonist property of a compound in the same well. To accomplish this, cells loaded with dye are tested with the compound alone and 2 min later with the addition of that compound and a known agonist at a concentration that yields a known response level of about 50–90%. Drug inhibition or IC_{50} values can be calculated from the compound's ability to decrease the agonist response. A plate-based reader is mostly used in these assays; however, a variation of these methods can be performed with a fluorescence-activated cell sorter (FACS: BD Biosciences, San Jose, CA, USA) or a biochip (Caliper, Hopkinton, MA, USA) instrument.[24] Both of these techniques detect increased fluorescence in individual, not groups of, responding cells.

3.29.3.3 Receptor Binding and Active Transport

Receptor-binding assays have traditionally been a mainstay of drug development. These assays employ a simple biological preparation of cell membranes and a labeled compound to bind specific protein receptors in those membranes.[25] One major appeal of this assay technique is the ability to combine different batches of cells and to make frozen cell preparations of them for later use. This pooling can eliminate many of the assay-to-assay variations originating from the growing of living cells such as viability, density, and functional state. In one of our typical binding assays for the substance P (SP, neurokinin-1) receptor, previously frozen cell membranes are incubated with 1 nM [^{3}H]-SP at room temperature for 60 min. The tissue is then collected by vacuum filtration on filter paper circles and measured by scintillation counter for radioactivity. Nonspecific binding is assessed by wells containing a high concentration (1 μM) of unlabeled SP that successfully competes with the lower levels of the radioactive SP.

If whole cells are used instead, for these binding assays, interpretation of the results can be complicated by secondary cell uptake of the radioactive or fluorescent compound used to bind to the receptor. This uptake can be mediated through binding to the receptor and subsequent internalization or by an active or passive transport across the cell membrane. An additional problem in using whole cells for binding assays is that they require a relatively high sodium concentration in the buffer to preserve cell integrity. Physiological levels of sodium (150 mM) are known to decrease binding affinity for agonists (but not antagonists) in some GPCRs. For these reasons, binding assays are not routinely performed in whole cells. A receptor-binding assay, using only cell components, is generally useful when one wants to determine the affinity of various drugs at a receptor or to determine the number of receptors per cell. A standard binding assay, however, is not designed to characterize a ligand as an agonist, partial agonist, or antagonist. For these characterizations, a functional assay is needed.

In contrast, for membrane-bound transporter proteins, assays can be formatted to measure both passive binding and active transport. For the assessment of compounds that are transported across cell membranes, such as neurotransmitters, a cell-based method is preferable. A ligand-binding assay, using a cell fraction, will reveal only passive interaction of a compound with a specific transporter protein and will not provide data on whether it facilitates or retards that transport process. With a binding assay, compounds are simply allowed to compete against a specific radioligand, like the uptake inhibitors [^{3}H]-citalopram (serotonin transporter) or [^{3}H]-nisoxetine (norepinephrine transporter), for their respective binding sites on the transporter protein. This assay format does not, for example, reveal the impact of a compound on any secondary or allosteric site on the transporter protein that would influence the actual physical transport. Neither does it reveal if the cell takes up the compound. However, an active cell-based assay can provide such information by examining the uptake of a labeled substrate in either intact cells or, in some cases, vesicles formed from intracellular organelles containing the transporter protein. In this case, a decrease in radioactivity in the presence of another compound indicates interference with the transporter uptake. This interference may be via competition for the site of transport, or at a more distant site. With this method it is also possible to identify compounds that modulate the activity of a transporter but that do not have any effect in the passive binding assay.

3.29.3.4 Migration and Chemotaxis

Movement or migration, especially in response to a chemical signal, is a fundamental feature of certain cell types and can be easily measured in a cell-based assay. These signals can include peptide growth factors, cytokines, and a variety of extracellular matrix proteins. Cell migration is an essential process in embryonic development, angiogenesis, wound healing, immune response, inflammation, and cancer metastasis. Microporous membrane inserts in multiwell plates are generally used for cell migration and the closely related cell invasion assays, the most widely accepted of which is the Boyden Chamber assay. This assay starts with a population of cells in the upper chamber or well insert of a two-compartment microplate. A chemoattraction factor is then added to the bottom chamber or well and the number of cells passing through the membrane from the top to the bottom chamber is assessed.[26] Other systems employ a single compartment system and solid substrate like agar as the medium for cell migration. Cell motility is then quantified either by counting the number of cells that have migrated through the filter or by measuring the distance traveled into the filter by several of the fastest-moving cells. Current methods have replaced the older more onerous process of staining and manually counting translocated cells in these assays. The development of a fluorescence-blocking membrane between chambers now allows bottom measurements of only migrating cells labeled with fluorescent dye, such as calcein-AM, and not those cells still residing in the top chamber.

3.29.3.5 Proliferation

The capacity of certain cells to proliferate in culture is a hallmark of their viability as well as their functionality. Cells respond to a variety of environmental factors by replicating themselves. For instance, lymphocytes are stimulated to proliferate in the presence of specific antigen signals while fibroblasts increase proliferation in response to both mechanical and chemical signals caused by tissue injury. These periods of replication differ from the resting state and may represent a point in the cell's existence when it is most sensitive to intervention such as drug therapy.[27] Establishing conditions to promote proliferation is essential in certain cell-based assays in order accurately to model the in vivo characteristic of certain disease states dependent on cellular division such as cancer.

When proliferation is continuous there is little differentiation, but not all cells have this ability to proliferate continually. Many cells are in end-stage differentiation when removed from a living organism or in confluent cultures, where the close proximity of other cells limits their ability to divide. However, when cells do retain this capacity, two competing processes, namely contact inhibition and cell migration, can influence their rates of proliferation. For instance, the growth of large colonies of epidermal keratinocytes can depend on the presence of epidermal growth factor (EGF) and transforming growth factor-α (TGF-α) to increase the cell proliferation rates significantly. However, this growth can also be dependent on the outward migration of rapidly proliferating cells located in the perimeter of a cell grouping. Thus, the effect of EGF and TGF-α in promoting cell growth relies on both the ability to increase the rate of proliferation and cell migration.[28]

Cell proliferation is always accompanied by an increase in the absolute amount of nucleic acids. Therefore assays based on quantification of nucleic acids provide a high-throughput method to measure proliferation rates. The classic radioactivity-based method uses tritiated thymidine incorporation into newly formed DNA to quantify the amount of cell division. An alternative to radioactivity is the use of fluorescent dyes, such as ethidium bromide. This dye shows strong fluorescence enhancement when bound to cellular nucleic acids. It can thus be used to measure cell proliferation in a cell culture treated with RNAase. The fluorescence measured is directly proportional to the amount of DNA and cell number.

3.29.3.6 Absorption, Distribution, Metabolism, and Elimination/Excretion

Some biokinetic properties of drugs, such as ADME, are not necessarily restricted to in vivo measures. Cell-based models of absorption and metabolism are used routinely in drug discovery. However, distribution and elimination studies are presently used much less, due to the difficulties of modeling in vitro the interactions among the many organs involved. For example, absorption studies of a drug's movement across cell membranes, particularly the intestine, are mostly performed with Caco-2 cells (human colon carcinoma). The top of the Caco-2 monolayer contains microvilli characteristic of the intestinal brush border of the intestine.[29] These cells are grown in a confluent monolayer on porous membrane filters that are mounted in diffusion chambers that mimic the barrier between the lumen of the intestine and the bloodstream. The resulting permeability measurements are based on the rate of appearance of test compound in the bottom fluid reservoir as analyzed by ultraviolet absorbance, high-performance liquid chromatography (HPLC), or mass spectrometry (MS).

Absorption properties of compounds in other organs, such as the kidney and brain, can often be predicted by other models, such as one using Madin-Darby canine kidney (MDCK) cells. MDCK cells or the same cells transfected with human multidrug resistant transporter (MDR1) cDNA, have been used to measure passive transcellular diffusion and P-glycoprotein-mediated transport, respectively.[30] The activity of other proteins responsible for drug transport can also be measured in cell-based assays. Measurements are made by competing unknown compounds with known fluorescent substrates of that transporter and observing the deceased fluorescence of the cell. However, this method lacks specificity and is not often used for ADME. Individual drug transporters can also be expressed in frog oocytes. Some are commercially available (BD Biosciences, Franklin Lakes, NJ, USA).

Cell-based drug metabolism studies are usually performed by incubating a drug with human hepatocytes, commercially available, and then analyzing the drug and metabolite levels with liquid chromatography/MS. Excretion of a drug generally depends on its transport across glomerular and proximal tubule epithelium. Renal transport can be examined in a variety of cellular models. Epithelial culture models recreate human renal tubular function in an in vitro environment. The model consists of primary epithelial cells that were isolated from kidneys that are grown on a supporting extracellular matrix-coated support. This matrix typically sits inside the well of a microplate in much the same way that Caco-2 cells are grown. This culture model provides a two-compartment culture system with a polarized, functional renal tubular epithelium situated as a barrier between them. Compounds can be introduced into either of the compartments, apical or basal, and evaluated for uptake, metabolic transformation, or transcellular transport by the kidney cells.

3.29.3.7 Toxicity and Cell Death

Cells can die in one of two ways. First, they can be killed directly by mechanical damage or exposure to a toxic substance or pathogen. This results in a process called necrosis and involves a series of rapid cellular changes leading to the disruption of the plasma membrane and swelling of intracellular organelles such as mitochondria. Eventually, the cell membrane ruptures and the cell contents leak out into the surrounding tissue. This death can be observed rather easily in a high-throughput format by measuring cell viability directly in cells cultured in multiwell plates. In general these assays involve some measurement of cell membrane integrity. Common colorimetric readouts include: the increased uptake of the dye methylene blue; the decreased uptake and binding of another dye, neutral red; the release of the intracellular enzyme, lactate dehydrogenase; or the release of a reduced tetrazolium salt, MTT, into the culture medium. Additional fluorescent readouts can include the increased uptake and binding of ethidium bromide to nucleic acids in cells with damaged membranes or the uptake and retention of the dye calcein-(AM) that is decreased in cells with nonintact membranes. While all these endpoints exploit the increased permeability of the cell membrane as a measure of viability and/or cell number, they differ in their sensitivity and overall performance.[31]

A second scenario of cell mortality involves a more delayed toxicological process called programmed cell death (PCD) or apoptosis. It is characterized by cell shrinkage, a breakdown of mitochondria to release cytochrome *c* and the development of membrane blebs or bubbles on the cell surface with the eventual degradation of cell chromatin (DNA and protein). This type of cell suicide averts the inflammatory response associated with necrosis. Cells make the decision to commit PCD when there is an imbalance between positive signals needed to maintain survival and negative signals, which induce death. The continual survival of most cells relies on constant signals from other cells and, for many, continued adhesion to their growing surface. PCD is initiated by either signals arising within the cell or death activators binding to their cell surface. Over its development, a series of biochemical features associated with this process include specific receptor activation (i.e., tumor necrosis factor (TNF) or death receptors), protein cleavage at specific locations, translocation of signaling proteins such as Bcl-2 from the mitochondria, increased mitochondrial membrane permeability with the release of cytochrome *c*, activation of a cascade of aspartic acid-specific cysteine proteases (caspases), the appearance of phosphatidylserine on the cell membrane surface, DNA fragmentation, and the final dissolution of the cell.[32] Cell-based assays are currently available to examine all of these intermediate PCD processes and mostly involve fluorescently labeled reagents that can be visualized microscopically, in a plate reader, an automated cytometer, or various other fluorescent-based instruments.

3.29.3.8 Flow Cytometry

Flow cytometry is the general term that refers to a type of cell analysis based on detecting fluorescence levels of single cells. This technique relies on a fairly complex instrument called a flow cytometer that 'pushes' cells in a single file past a laser that excites a fluorochrome. Before loading into the cytometer, the cells are incubated with fluorescently labeled probes (such as antibodies) that recognize extracellular molecules of interest. A set of optics focuses the lasers on passing cells. The data from a photomultiplier tube or photodiode record how much light is emitted by each cell and this in turn is a function of the amount of labeled material on that cell. The data output is a histogram, which can be interpreted to determine what percentage of the analyzed cell sample expresses a particular level of the protein of interest. The cells are also analyzed based on how much light they scatter in two different directions, which can be extrapolated to size and shape information about the cells. The typical cytometer (FACS) has two lasers and can record six cell-specific parameters (four fluorescence and two scatter-related). With the increase in available fluorescent labels and antibodies, the flow cytometers are increasingly used for multiparameter analyses of cellular function. In addition, some instruments incorporate more lasers and detectors so many more fluorochromes can be excited and detected at

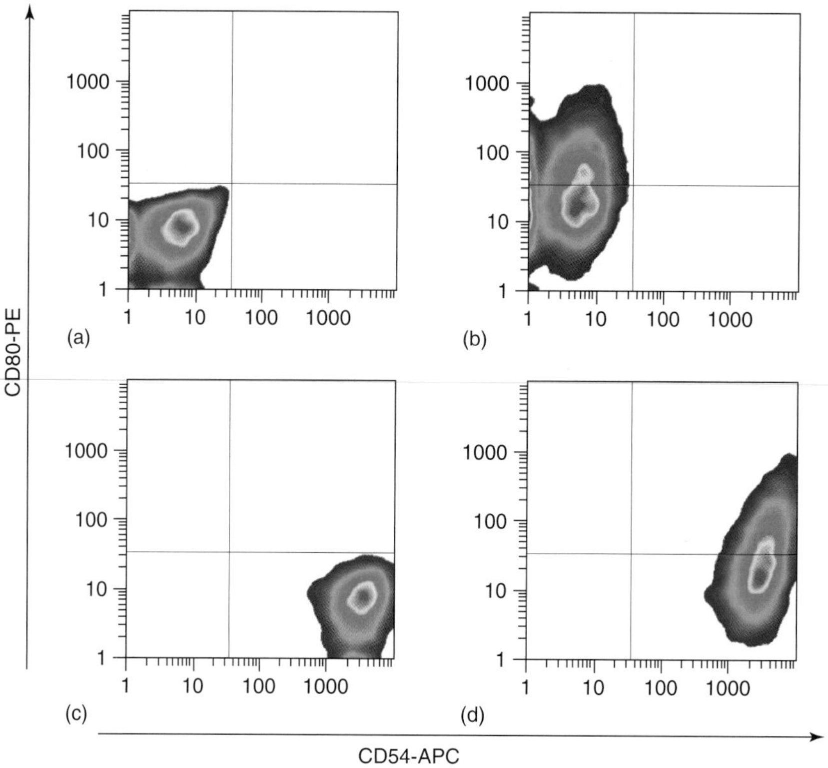

Figure 4 Two-color analysis of expression of CD80 (B7-1) and CD54 (intercellular adhesion molecule (ICAM-1)) by THP-1 cell line (human acute monocytic leukemia cell line). Cells were stimulated with lipopolysaccharide and then stained with the isotype control antibodies (a), antiCD80-PE conjugate as shown by increased fluorescence on the *y*-axis only (b), antiCD54-APC conjugate as shown by increased fluorescence on the *x*-axis only (c), or both CD80-phycoerythrin (PE) and CD54-allophycocyanin (APC) as shown by increased fluorescence along both the *y*-axis and *x*-axis (d). The figure shows how flow cytometry can simultaneously measure expression of two proteins on a cell surface.

once. Typically, one or two lasers intersect a sample to excite two to four fluorescent reporter molecules. More sophisticated instruments now have three to four lasers that commonly allow eight, and with further configuration, 11 different colors to be detected. This multiparameter analysis allows for simultaneous detection of several proteins in the single sample or cell and greatly increases the content and throughput of an assay. A typical flow cytometry readout of two colors is shown in **Figure 4**.

The other extremely powerful application of flow cytometry is intracellular protein staining. This is a method that allows one to take a snapshot in time of the protein content of a single cell. The aim is to make use of the rapid flow cytometric analysis of cells to analyze both surface phenotype and intracellular protein content of single cells within a population of cells responding to stimulation. The technique initially requires fixation of cells in paraformaldehyde, which cross-links proteins and prevents their loss. Secondly, permeabilization of cellular membranes with a detergent generates gaps in the membrane, allowing antibodies to specific protein access to the cell's interior. Similar procedures may be performed to stain for DNA instead of proteins.

3.29.3.9 Multiple Outputs (Multiplexed Assays)

When a ligand binds to a cell receptor and effects a modification of cell function, more than just a single protein is altered as a result of that interaction. There is always a corresponding change in the overall profile of mRNA and protein expression in the cell.[33] Thus to achieve the most accurate representation of the cell's functional state, one often needs to assess the transcription or expression of several proteins under the same assay conditions. This analysis can distinguish responses created along similar cellular pathways that may or may not involve the same initial receptor event. One classic example is changes in the cytokine profile of T lymphocytes in response to different stimuli.[34] These protein profiles are not static, but change in a dynamic way over a timeframe that ranges from 6 h to several days. Changes in protein expression levels are also associated with cell death, cellular metabolism, cell differentiation, migration, and proliferation. Similar types of cellular profile in response to a stimulus can be observed for phosphorylation states of key proteins, such as enzymes or transcription factors.

There are several ways to measure many proteins simultaneously in an assay. Most methods are antibody-based and utilize a pair of specific antibodies (one capture and one detection). The assays primarily differ in the detection and analyte separation techniques and can be divided into two major classes: (1) positional or flat arrays; and (2) microsphere-based or liquid arrays. In the first case, the single analyte-specific capture antibody is spotted at different positions of the flat array. Nitrocellulose, plastic, or glass can be used as the array surface. Then the whole array is incubated with the sample and the proteins are visualized with detection antibodies. A typical output for this assay is a flat surface with dots of varying intensity.[35] Thus, one can determine relative concentration of multiple proteins in a single assay, since the position of the capture antibody on the array is known and the intensity of the dot is proportional to the analyte concentration. The number of analytes per array is only limited by the size of the array surface and so is virtually unlimited. The drawback of this method is the relative high price and rigidity of an array. Once the array is printed, the user cannot change its composition. The data generated using the array can be used directly in a cell-based assay or just to select the optimal readouts and parameters for futher screening campaigns.

In the case of liquid arrays, each analyte-specific capture antibody is immobilized on microbeads with known and unique fluorescence characteristics. Then a sample is incubated with these beads and the protein of interest is detected using fluorophore-labeled detection antibodies. The beads are then analyzed on the flow cytometer, which simultaneously identifies each bead by its color and measures the level of the analyte associated with it by its fluorescent tag. Multiple measurements are taken for each analyte-specific bead, thus further validating the assay results. Using this system one can quantify up to 100 analytes in a single sample. Furthermore, the assay format allows users to configure the multiplex each time the assay is run, making it one of the most flexible assay platforms.

Multiplexed assays can also be used to quantify levels of nucleic acids or the activity of transcription factors. Traditional methods typically used gel-shift or electrophoretic mobility shift assays (EMSA) as well as colorimetric enzyme immunometric assay (EIA)-based assays. The technique was originally developed for DNA-binding proteins, but has been extended to allow detection of RNA-binding proteins due to their interaction with a particular RNA sequence. EMSAs are cumbersome, usually involve radioactivity, and are not amenable to high-throughput applications. In general, purified proteins or crude nuclear cell extracts are incubated with a ^{32}P-radiolabeled DNA or RNA probe, followed by separation of the complexes from the free probe by electrophoresis through a nondenaturing polyacrylamide gel. The DNA–protein or RNA–protein complexes migrate more slowly than the unbound probe of interest. In one example of multiplexing (Marligen Biosciences, Ijamsville, MD, USA), the assay simultaneously measures the binding of multiple transcription factors. Cells are treated with a drug candidate of interest and nuclear extracts are prepared. These extracts are then incubated with a mixture of labeled probes followed by the addition of a digestion reagent. Individual assays rely on the specific binding of transcription factors to cognate recognition sequences to protect labeled probes from digestion. Undigested probes are then captured on to the surface of fluorescently labeled microspheres where each microsphere is color-coded for its unique transcription factor. Signal intensity corresponds to the amount of activated transcription factor present within the test sample. Signals are detected with a compact flow cytometer as each microsphere passes through the Luminex instrument (Luminex Corporation, Austin, TX, USA). The amount of label associated with bead surfaces correlates with the amount of transcription factor present in the nuclear extract. This method claims a faster speed and greater ease of use than EMSA or EIA. In addition there is a lower limit of detection, which allows use of significantly less protein. A single plex use of this newly developed technique, measuring nuclear factor kappa B (NFκB), is described by Shurin *et al.*[36]

3.29.3.10 Calorimetric Enzyme Immunometric Assay and Similar Antibody-Based Techniques

As previously discussed, cell-based assays often rely on the use of specific antibodies, either polyclonal or monoclonal, as a detection method. Antibodies are valued for their high target specificity, high affinity, and high sensitivity, which allows measurement of physiologically relevant quantities of chemicals and proteins with a small number of cells. Assays are generally in a competitive or sandwich format. Competitive assays use an antibody attached to the assay plate and a target chemical attached to an enzyme or other signaling molecule. Sandwich assays use an antibody attached to the plate aimed at one epitope of the target molecule. Then a secondary antibody binds to the attached target molecule, to form the sandwich. Finally, a third antibody attached to a signaling molecule is added to react with the secondary antibody. It is important to note that the sensitivity of each technique is determined by the readout method, detection instrument sensitivity, and the affinity of the antibody.

Initially, the only detection method available was radioactivity and thus radioimmunoassays (RIAs) were widely used. Although radioactivity provides good sensitivity and reproducibility, problems occur with sample handling, safety precautions, high-energy radiation, short half-lives for ^{125}I-labeled compounds, and separations of labeled and nonlabeled assay components.[37] Further refinement of this approach with ^{3}H-labeled assay components solved some, but not all, of these issues. More recently, researchers have generally migrated to EIA technology, which eliminates radiation safety and disposal issues. There are a number of enzyme-based detection methods. Among the more popular are alkaline phosphatase and horseradish peroxidase (HRP) using ultraviolet absorbance and luciferase for luminescence detection. Fluorescence systems are also available. These are described in detail later in this chapter under GPCR methods. The challenge with these methods is the time involved in performing the assay. Once the cell reaction is completed, cells must be lysed, the lysates added to the plate, and then generally 2 h are required for each antibody-binding step, depending on the experimentally determined parameters, plus a wash step for each. A period of time is also required for the enzyme action, which is dependent on time and temperature effects. Still the commercially available reagents are very stable (> 3 months at 4 °C in solution) and produce consistent results.

One of the newer attractive options is offered by FRET-based assays. In this assay, a pair of antibodies recognizing distinct epitopes of the same protein is labeled with either a donor or an acceptor fluorescence dye. Interaction with the protein brings the fluorescent dyes to close proximity with each other, resulting in energy transfer from the donor to the acceptor dye. Fluorescence of the acceptor is then detected and used as the assay readout. Since excitation of the acceptor is directly dependent on formation of the antibody–analyte–antibody complex, the specific fluorescence is proportional to the analyte concentration. For instance, a FRET system for cyclic adenosine monophosphate (cAMP) detection uses a cAMP donor molecule and an anti-cAMP antibody labeled with the acceptor molecule. This technique offers a number of advantages over traditional EIAs or RIAs: there are no long incubation and washing steps, the assay can be performed in a streamlined 'mix and measure' mode, and it can be performed in whole cells, thus eliminating an additional step in the sample preparation. All these qualities make the FRET-based immunoassays highly adaptable for miniaturization and ultra-high-throughput format. Another recently introduced technique is the TRF-based assays. The final step of this assay does not require incubation with a detecting enzyme because the antibody tag, generally europium, is detected directly with very little background fluorescence. Fluorescence polarization has also been used as a detection technique with these antibody-based assays.

The best assay methodology to screen for activity at your target depends on the reagent costs, equipment availability, and the detection sensitivity needed. Early-stage research efforts or studies examining unique targets will make or contract out the production antibody to their target. The next step is to work out empirical parameters such as antibody dilution, time required, antibody selectivity, and antibody stability. Kits containing the antibodies and detection reagents for many targets are available from a wide variety of sources. Vendors include Perkin-Elmer (Boston, MA, USA), Amersham (GE Healthcare, Piscataway, NJ, USA), R&D systems (Minneappolis, MN, USA), CisBio International (Cedex, France), Biomol (Plymouth Meeting, PA, USA), and others. For instance, one study compared several methods for quantifying changes in cAMP and concluded that TRF was the best overall balance of sensitivity and accuracy.[38]

A variation of this theme is the use of a plate coated with DNA sequences to measure production of transcription factors. In this method, cells are treated with a drug candidate, then a homogenate of lysed whole cells or only their nuclei are prepared. The lysate is incubated in the DNA-coated plate, and then washed. Subsequently, as with EIA, an antibody specific to the transcription factor is added, incubated, washed, and then an antibody with a detection molecule is added, incubated, and washed. Finally a substrate is added (if enzyme detection is used) and the response measured. This works in many cases with 50 000 cells, if a luminescence detection method is used. Colorometric methods are generally less sensitive. Kits have the potential to provide results from cell lysates in 3 h. These assays are commercially available as kits by BD Biosciences (Franklin, NJ, USA), Oxford Biomedical (Oxford, MI, USA) and others. Whole-cell or nuclear extraction kits are also commercially available. BD Biosciences currently has kits for over 40 different transcription factor targets.

Another cell-based technique to quantify protein levels is the ELISPOT (enzyme-linked immunospot). This technique was originally introduced to enumerate antibody-secreting B cells.[39] In the current method, cells are deposited on to a membrane coated with one antibody specific for a protein, followed by an appropriate incubation period. Subsequently, the protein of interest is detected in the environment immediately surrounding the cell secreting it, by another antibody specific for a different epitope of the protein. The signal detected by the HRP enzyme/substrate results in a colorimetric footprint of the cells and can be quantitated by visual scoring or specialized plate-readers. For cytokines, ELISPOT may be more sensitive than an EIA/enzyme-linked immunosorbent assay (ELISA), since it takes advantage of the higher concentration of the secreted cytokine close to its source. In addition, ELISPOT allows quantitation of cytokine secretion on a per-cell basis.

3.29.3.11 Reporter Genes

Knowledge about cell communication has increased significantly in the last few years. Transcription factors can be selectively activated or deactivated by other proteins, often as the final step in signal transduction. As stated earlier, several signal transduction pathways contribute to regulation of gene transcription in cells. These pathways act by stimulating the interaction of the transcription factors with sequences known as the response elements, located in the promoter region of the responsive genes. A response element sequence can be attached to a reporter gene. Then researchers can selectively monitor the activation of the particular signal transduction pathway or secondary messenger cascade, leading to regulation of gene expression inside the cell. A number of hormones and growth factors stimulate target cells by activation of second messenger pathways that regulate phosphorylation of nuclear or transcription factors, which in turn modulate or alter gene transcription. For example, extracellular signals or ligands bind to cell surface receptors like GPCRs, ion channel-linked receptors, or receptor tyrosine kinases. This ligand binding causes activation of downstream signal transduction pathways, which in turn can activate kinases to phosphorylate various transcription factors. These activated factors then bind to their specific response elements in the promoter regions of the hormone or growth factor-responsive genes.

To facilitate understanding the activation or inhibition of various pathways on gene expression, these specific response elements or sequences have been fused to genes encoding suitable reporter proteins. A variety of reporter genes are available with broad applications, ranging from characterization of receptors, development of drug screening, and gene delivery systems to temporal and spatial gene expression, specifying the exact times and physical location in the cell where the changes are occurring. Some commonly used reporter genes include chloramphenicol acetyl transferase, luciferase (from firefly), secreted alkaline phosphatase (SEAP), green fluorescent protein (GFP), β-galactosidase, and β-lactamase. These enzymes can have multiple substrate readout mechanisms. For example, in the Live-Blazer (Invitrogen, Carlsbad, CA, USA), CCF4-AM is added to cells. Its coumarin moiety excites at 409 and donates to a fluorescein moiety, which emits at 530 nm. Expression of β-lactamase leads to cleavage of the substrate and loss of FRET, producing a robust blue fluorescent signal detectable at 460 nm. Ultimately, the choice of a reporter system can be based on the ease of use (cell lysis or intact cells), on the specific characteristics of the signal they produce (absorbance, fluorescence, FRET, luminescence), or can be influenced by factors such as toxicity related to product buildup. For example, GFP can build up in a cell over time and cause toxicity.[40]

3.29.4 Cell-Based Screening Examples

The previous sections discussed the choice and treatment of cells and the methods that can be employed to obtain readouts of their functional activity in a cell-based assay. In this section, we provide some background on specific cell targets of interest in drug development and show examples of how the appropriate cells and screening methods are chosen, and how the results can be interpreted. Ion channels, GPCRs, multitarget/multioutput assays (Toll receptors), and drug properties (toxicity) are showcased.

3.29.4.1 Binding versus Functional Readouts for Ion Channels – The $5HT_3$ Ligand-Gated Ion Channel

Cellular responses range from very rapid – <1 ms – to those that can be observed over days. One type of rapid response is the selective flow of ions across cell membranes that can directly or through a secondary signal pathway mediate a compound's effects. Ion channels gate the flow of ions across cell membranes in response to ligand binding, intracellular signals, or changes in voltage. They closely control intracellular levels of chloride, sodium, calcium, and potassium as well as regulating the cell electrical potential that is based on the concentration gradient of those ions. Most of the ion channels perform a unique function, which is defined by the specific ion, their cellular location, and the stimulus that triggers the conformational change that allows for ion flow through them. Ion channels were initially associated with only excitable cells such as neurons and muscle cells, but were later found to be important in all types of cell. There are two main classes, voltage-gated channels and ligand-gated channels, and a number of less common classes (vanilloid, cation, cyclic nucleotide activated, etc.). In voltage-sensitive channels, changes in the potential across a cell membrane cause the channels to open. Channels are usually open for a short time and then return to their normal closed state. Numerous voltage-gated channels exist for the major cations, including 10 for sodium, 73 for potassium, and 11 for calcium, and this list continues to grow.[41]

The other large group of ion channels is the ligand-gated ion channels, also referred to as ionotropic receptors. These channels comprise a group of transmembrane proteins that allow the flow of specific ions across the cell membrane in response to binding of a chemical messenger. One such ligand-gated ion channel that we have worked with in the course of drug discovery is the serotonin $5HT_3$ (5-hydroxytryptamine, subtype 3). It is interesting to note that all the other 12 known 5HT receptors are GPCRs, not ion channels. This channel is prevalent in muscle and brain, existing as either a homopentamer of A subunits or a heteropentamer of A and B subunits. Blockade of $5HT_3$ is used as an antiemetic therapy (granisetron) in humans to lessen postoperative nausea. In our screening effort the cell line N1E-115, a mouse neuroblastoma, was chosen for a cell-based assay, based on the literature reports that these cells had $5HT_3$-mediated function, and a membrane preparation of these cells is used for the $5HT_3$ binding assay already established at our company. The choice of a cell originating from a rodent species, in this case, was guided by its channel expression, our previous experience with this cell, and the concept that initial rodent testing precedes human clinical testing. One school of thought believes that a human source for all targets is preferable since the drug will eventually be given to humans. However, often these human sources are the human recombinant protein expressed in a non-neural cell, either animal or human. Therefore, as we discussed previously, cellular feedback and regulatory systems may not be identical to that found in the human target organ. Other issues, such as patent rights controlling the human $5HT_3$ protein, also guided this particular cell choice.

Since $5HT_3$ is a channel that regulates the flow of cations, the most appropriate functional measurement techniques were either a calcium or MPD. Despite some reports in the literature, we saw no calcium response to $5HT_3$ agonists.[42,43] However, we did observe a large response with MPD, and used this dye for our drug-screening project (**Figure 5** and **Table 3**). Results from binding at Novascreen, on the same cell line (N1E-115), and the average and range values from the National Institute of Mental Health's Psychoactive Drug Screening Program (PDSP) K_i database are compared with the results from the MPD.[44] From these data we can see that there is a good correlation between antagonist binding and functional K_i determinations. There does seem to be a discrepancy between functional EC_{50}s and the higher affinity levels found with binding assays. These disparate results are currently under study.

From this example it appears that a cell-based screen of $5HT_3$ antagonists would work well, but the $5HT_3$ agonist assay may be less reliable. The large range of literature K_i values in the literature-based database indicates large laboratory, tissue, and technique-based variances. These can only be addressed by careful control of variables in the screening laboratory and use of numerous controls during validation. Differences between binding and MPD values at Novascreen bring up a current controversy about binding assays in general. Binding assays use radiolabeled antagonists because they bind generally to a higher-affinity site and do not desensitize the target. However, the affinity site may be different from that occupied by agonists. This would be a target validation issue that may be addressed by also using radiolabeled agonists for binding. Binding may also be altered because channels have multiple states (open, closed, intermediate) that expose different binding regions. Other items that may cause variations between binding and functional data include receptor reserve and differences in buffers.

3.29.4.2 Choosing among Different Readouts for a Targeted G Protein-Coupled Receptor – The μ Opioid Receptor

Protein or small-molecule ligands can bind to a specific receptor, causing a conformational shift leading to activation of intracellular proteins that can induce a cellular response. There are several major types of receptor: nuclear hormone, tyrosine-kinase-linked and GPCRs. These GPCRs are the most studied and are the target for roughly half of current drugs, with emphasis on neuronal communication and immune system function.[45] The receptors consist of seven transmembrane regions. GPCRs respond to ligand binding by interacting with a heterotrimeric G protein consisting of an α, β, and γ subunit (**Figure 6**). Response in an individual cell is particularly dependent on the number of receptors and the types and abundance of G proteins available for coupling. The GPCRs may be grouped by the type of $G\alpha$ subunit, whether $G_{o/i}$, G_s, or G_q, and these couple to cAMP increase, cAMP modulation, and phospholipase C activation respectively. In addition to α, the β and γ subunits may act to increase phospholipase C and modify other second messengers. Following activation, receptors desensitize and are removed from the membrane. Typically the receptor is phosphorylated by a G protein kinase, and then bound to β-arrestin. The receptor may be directly recycled to the surface or destroyed by proteolysis. In this section, we will discuss how the mechanisms of GPCR have been used for drug screening, especially in cell-based assays.

Numerous measurement techniques exist, some universal, some subtype-selective, to measure cellular responses mediated by GPCRs. These were discussed in the cellular readout sections above covering EIA and similar methods. Typically, we use a cAMP EIA assay for screening a series of different G_s-coupled receptors. This assay uses cells that are washed and incubated for a period with compounds to look for agonist activity at the receptor of interest. Following this incubation, the cells are lysed in hydrochloric acid and the cell supernatant is added to a 96-well plate surface coated with goat antirabbit immunoglobulin G (IgG) antibody. In this assay, cAMP is detected by the rabbit antibody to

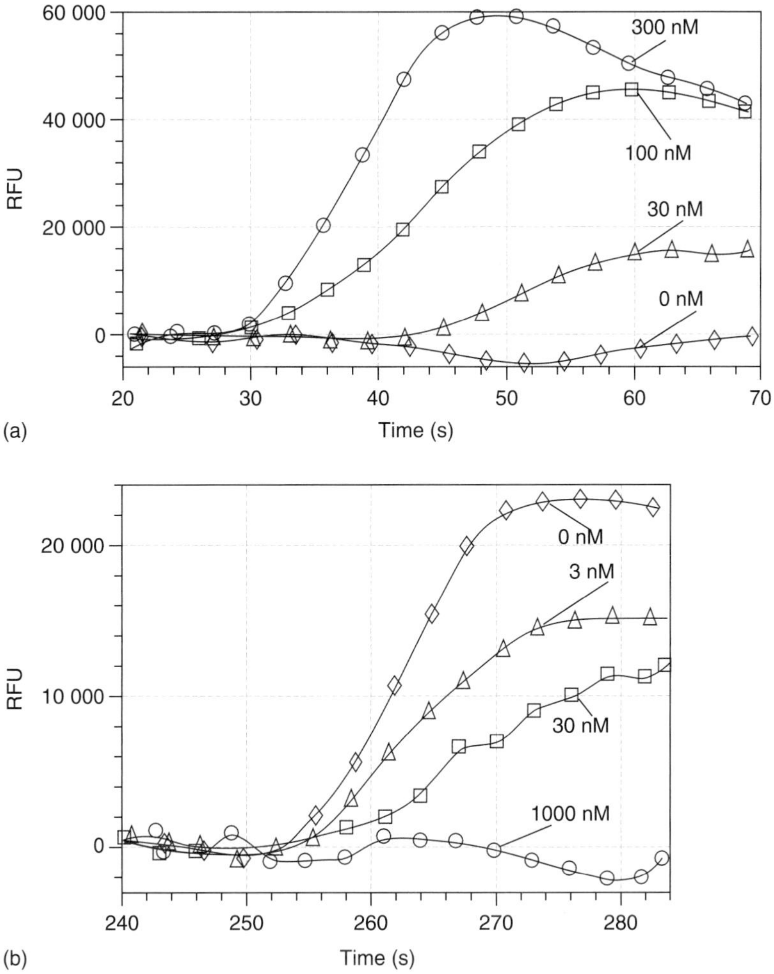

Figure 5 Membrane potential dye dose–response to either a $5HT_3$ agonist or antagonist. Fluorescent response (relative fluorescent units) versus time. (a) Agonist, 1-(*m*-chlorophenyl) biguanide (*m*-CPG), is added at 30 s. (b) For antagonist assay, MDL 72222 is added at 30 s and then 100 nM agonist (*m*-CPG) is added at 250 s. 530 nm excitation (Ex)/565 nm emission (Em), at 25 °C. Note that, for agonists, the highest fluorescence response is at the highest drug dose, whereas for antagonists, the highest fluorescence response is with the 0 nM control.

Table 3 Comparison of $5HT_3$ serotonin receptor affinities

Drug	*N1E-115-binding (Novascreen) K_i (nM)*	*Literature (PDSP K_i database)[44] average K_i (nM) and range*	*N1E-115 MP dye (Novascreen) K_i or EC_{50} (nM)*
Antagonist	K_i	K_i	K_i
MDL-72222	35	12 (0.8–87)	8.3
Metoclopramide	49	390 (0.15–910)	30
Agonist	K_i	K_i	EC_{50}
m-CPBG	3.5	0.94 (N/A)	60
2-*m*-5HT	100	806 (0.16–3500)	1400

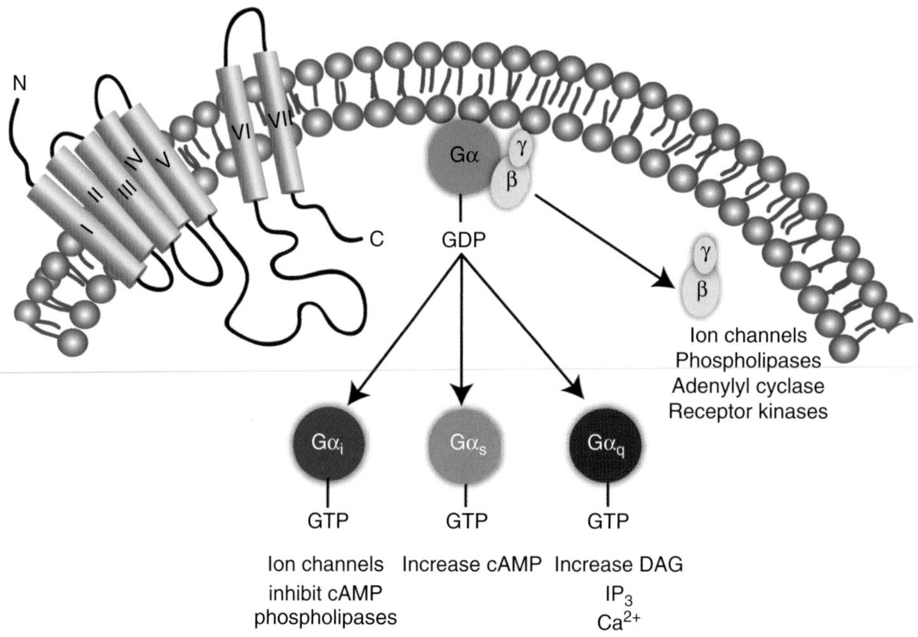

Figure 6 Schematic for the mechanism of GPCR function.

cAMP, linked to a marker enzyme, phosphodiesterase. The assay components react for several hours and the plate is washed and a substrate for the detection enzyme added. After the appropriate time (1 h with phosphodiesterase) the response is read in a detector (absorbance at 405 nM in this case). In this competitive assay, increased absorbance is associated with decreased cAMP. cAMP levels are calculated in comparison with a standard curve. Generally the sensitivity is enough to measure cAMP in 50 000–100 000 cells in a 96-well plate. Sensitivity may be reduced by diluting the supernatant or increased 10-fold by acetylation of the cAMP. Our usual cAMP increases are about threefold for a robust G_s response.

G_i assays are similar to G_s assays except that cells are incubated with forskolin (10 μM) or another cAMP stimulant (such as the β-adrenergic agonist, isoproterenol) prior to agonist addition. Forskolin will typically stimulate cAMP by 5–100-fold and agonists at G_i-coupled receptors will decrease this cAMP increase by 40–80%. G_i receptor functional responses may also be measured using membranes and either europium or ^{35}S-labeled guanosine triphosphate (GTP)γS. Stimulation by agonist causes guanosine diphosphate (GDP) on the α subunit to be replaced with GTP. Then, upon desensitization, the GTP is enzymatically converted back to GDP. Substituting GTPγS, which cannot be hydrolyzed, for GTP, leads to an irreversible increase of bound labeled GTPγS following activation by an agonist. Typically this assay has high (often 50%) background levels, so it works best with clones that express a high number of receptors and hence produce a larger response. In general, it is easier to use this method with G_i-coupled receptors because most cells have a large proportion of G_i's (80% of total G proteins) rather than G_s or G_q. Additionally, radioactive 35S is more sensitive in general than europium, which is measured by TRF. However, 35S is hazardous to work with and has a short half-life.

Effects mediated through G_q-coupled receptors can be measured by dye-based measurements of intracellular calcium. In our method, cells are plated the night before at 50 000 cells per well in 96-well black-sided, clear-bottom plates, incubated with a calcium-specific fluorescent dye such as fluo-4 (Molecular Probes, Carlsbad, CA, USA) or calcium-3 dye (Molecular Devices, Sunnyvale, CA, USA). Cells are incubated with the dye and then placed in a fluorometer such as the Flexstation (Molecular Devices). A response typically peaks at 7–30 s after ligand addition and decreases to basal at 60 s. A secondary calcium response occurs in some cells as the increased calcium opens calcium-sensitive channels or activates kinases that phosphorylate and open calcium channels. A typical readout of a dose–response calcium assay is shown in **Figure 7**. Many researchers use addition of a positive control after addition of the agonist and after the response returns to basal calcium levels. For example, calcium ionophore or carbachol that can activate muscarinic M3 receptor, present on many cells, can be used. Use of a positive control allows the researcher to normalize the response for each well and account for differences in cell density or dye loading. This is particularly useful to control edge effects on multiwell plates.

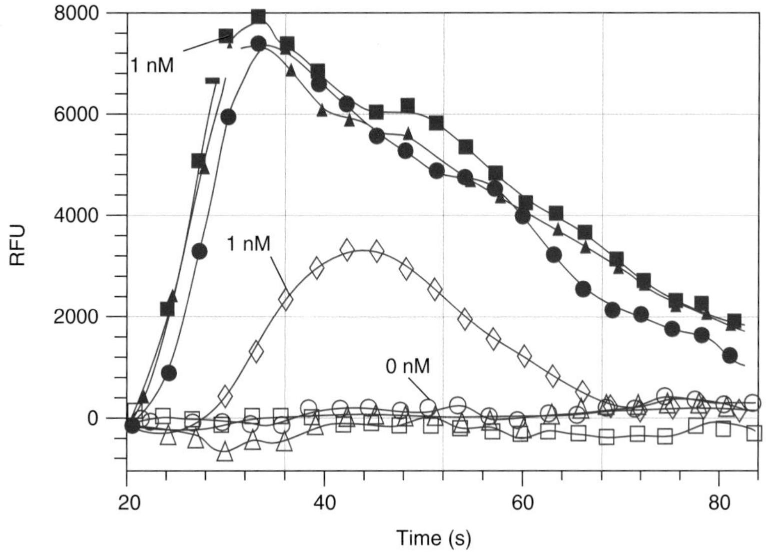

Figure 7 Calcium dye (Molecular Devices) dose–response curve for the μ opioid agonist DAMGO on the μ opioid receptor. Calcium response in fluorescence (relative fluorescent units) versus time. Ex/Em 485/535 nm. Agonist added at 20 s. Note the secondary hump at 50 s, which may be due to opening of calcium channels, whereas the initial response comes from intracellular calcium release.

An advantage of the calcium system is the ability to screen both the agonist and antagonist sequentially in the same well. To do this, cells are loaded with dye and placed in the fluorometer. The compound of interest is added and the response noted. Next the antagonist is added at a concentration needed to obtain a reproducible, robust, but not maximal response. (A maximum response would be difficult to overcome by an antagonist and thus decrease the sensitivity of the response.) An antagonist will inhibit this response and an IC_{50} can be identified. IC_{50}s calculated this way may be converted to the more universal K_i, using a modified Cheng–Prusoff equation of $K_i = IC_{50}/(1 + L/EC_{50})$.[46] L refers to the agonist added, as does the EC_{50}. This equation compensates for the differences in ligand concentrations used in different labs. A less controversial method is to perform a Schild plot, where the slope of one indicates a competitive antagonist and the x intercept is the pA_2. pA_2 is the negative log of the concentration of antagonist, which would produce a twofold shift in the concentration–response curve for an agonist.[47] The pA_2 (derived from functional experiments) will equal the K_d from binding experiments if antagonist and agonist compete for binding to a single class of receptor sites. The ability to identify noncompetitive antagonists is important since noncompetitive sites are less conserved and might be more specific to a targeted receptor and thus a better therapeutic site to modulate receptor activity. Another advantage of a calcium signal readout over time is the ability to pick out false-positive compounds. In many cases, these produce a calcium response that increases too quickly or does not decrease over time, as would an actual hit on the target.

These different ways of measuring GPCR activity can be used to highlight the process involved in choosing the best method to screen activity at a particular target. We compared four separate analysis systems for the μ opioid receptor, which is generally considered to be G_i-linked. We used: (1) intracellular calcium; (2) MPD; (3) cAMP inhibition; and (4) GTPγS shift. Dose–response curves are shown in **Figure 8**. In this case, the G_i-coupled receptor inhibited forskolin-induced cAMP increase and increased GTPγS binding as expected. The calcium and membrane potential appear to be the more sensitive of these assays. The MPD showed a strong signal. However, with this dye, some agonist drugs tested caused nonspecific increases that were probably due to dye interactions and the addition of buffer to the wells often resulted in a significant baseline shift, which interferes with measurement of peak response. The calcium response was interesting as we discovered that the cells needed to be primed for 5 min with 100 μM adenosine triphosphate (ATP) to see the calcium response. The ATP acts at the P_2Y GPCR. The priming may work by priming phospholipase C response to the βγ subunits released upon μ opioid receptor stimulation.[48]

After comparing the techniques, Novascreen generally performs μ opioid receptor functional assays using the calcium response due to ease of use, slightly higher sensitivity, rapid response, immediate feedback, decreased cost, and ability to perform the agonist and antagonist determination in each well. An example of a μ opioid agonist response is shown in **Figure 7**. An example of a small screen for a G_i-coupled receptor using a typical G_i-coupled calcium

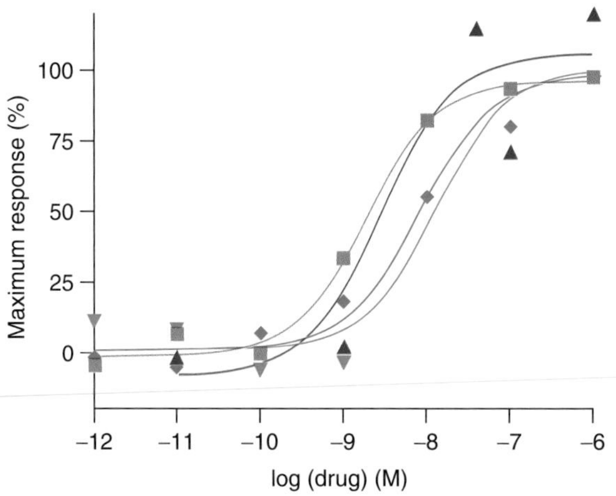

Figure 8 Comparison of EC_{50} values obtained with calcium dye, voltage dye, cAMP EIA, and GTPγS methods for the μ opioid agonist, DAMGO. The calcium, voltage, and GTPγS values are reported as percent maximum response; cAMP values are reported as percent inhibition.

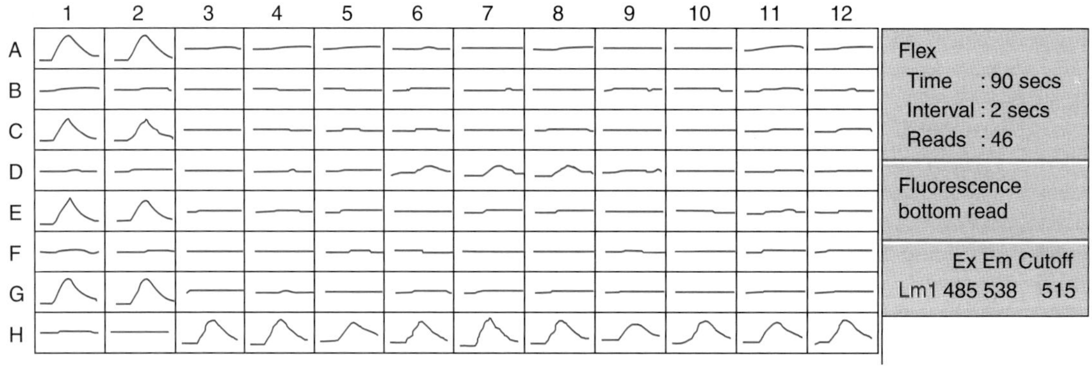

Figure 9 Typical agonist screen for a G_i-coupled receptor using calcium dye (Molecular Devices) in a 96-well plate. Readout is fluorescence with Ex/Em of 485/538 nm. Columns 1 and 2 contain alternating agonist and blank wells as controls. Agonists produce a peak. In the remaining columns, row G contains blank and row H contains agonist, again as controls. Columns 6–8 row D shows an agonist hit of about 50% the maximum response (triplicate points.). Thus, on this plate, 21 drugs were screened for agonist activity and one hit was found (ignore column 12). More controls were run than on a normal assay due to the early development stage of this project and the use of a phospholipid as the agonist.

response is shown in **Figure 9**. Other available systems for GPCR measurement include: (1) reporter gene assay; (2) aequorin technique[49]; (3) use of Gαq15/16 to couple G_i and G_s to calcium increase[50]; (4) bioluminescence resonance energy transfer (BRET) technique[51]; (5) inositol phosphates (CisBio, Bedford, MA); (6) whole-cell voltage clamp – tumor cells or oocytes[52]; (7) frog melanocytes – with cAMP increase the cells appear black[53]; (8) GFP-linked β-arrestin (high content screening (HCS))[54]; (9) cellular dielectric spectroscopy[55]; and (10) cells with a cyclic nucleotide-gated (CNG) channel, where cAMP increase also increases intracellular calcium as measured by a calcium dye (BD Biosciences, Franklin Lakes, NJ, USA).

3.29.4.3 A Multitarget and Multioutput Assay – Immunomodulation through the Toll Receptors

An advantage of cell-based assays is the presence of all the proteins in a target cell as well as all its readouts. Therefore a measured cell response can be the result of the activation or repression of multiple pathways mediated through multiple targets. In the earlier examples of ion channels and GPCR assays, cell responses can be easily linked to a specific target and the single readout measured within a minute or less of test compound addition. However, in complicated pathways, one needs to look at a cellular response in terms of multiple targets and outputs, sometimes over an extended time period.

Biologically, many signal pathways lead to the same outcome, allowing them to be explored simultaneously. In one multitarget/whole-pathway/phenotypic assay, a desired readout can come from several possible sources. Particularly useful results are obtained when the assay uses cells similar to the human target, such as peripheral blood mononuclear cells. For example, Synta Pharmaceutical screened primary human blood cells for inhibition of interleukin-12 (IL12) production, a marker of inflammation. In this phenotypic screen, there was no bias toward selecting the target of the compounds, just the output of the cells.[56] So this was a multitarget, one-output screen. Due to the multiple targets, finding the specific target and pathway for active compounds entailed some extra work. However, the extra work was justified by the relevance of the model to the disease process. These IL12 production screens resulted in a compound that inhibits synthesis of IL12 subunits and is now in phase II trials.

In other cases one can read multiple output parameters arising from interactions with a single known target. This is an important consideration in inflammatory diseases where the desired outcome is a change in the amounts of cytokines produced and the relative levels of cell surface markers. In our example of screening, we have combined both the multitarget and multiplex/multi-readout aspects of a screen. The goal is to produce an adjuvant that would stimulate the immune system better than injection of an antigen alone. To achieve this goal we began an HTS campaign looking for new agonists of the Toll-like receptors. The Toll receptors are newly discovered targets for viral and bacterial components, such as endotoxin, that stimulate cells of the innate immune system. Several of these receptors are expressed on THP-1 monocytes: Toll 1, 2, 4, 5, 6, 7, 8, and perhaps the nucleic acid binding receptors Toll 3 and 9. Their presence was confirmed by Western blotting. Functionality was determined by increase of NFκB DNA-binding activity in response to stimulation with the few known Toll-like receptor specific ligands. However, the stimulation of NFκB is not a productive assay outcome since there are 100 or so chemicals that increase this signaling without producing an adjuvant-like response. Therefore, an alternate endpoint would be preferred to reduce the number of false hits when looking for specific Toll receptor ligands.

Inflammatory mediators produced by a stimulated immune cell are a good source of responses. These include IL8, RANTES (regulated on activation, normal T expressed and secreted), and TNF-α. Markers such as CD40 and CD54 on the cell surface also change. Thus a screen looking at all these responses would be very informative. These assays are being performed by Novascreen. Cells are exposed to test compounds at 10 μM overnight and then the supernatants are used for cytokine analysis and the cells are reacted to antibodies for particular CDs of interest. The study, screening 20 000 compounds, is currently under way and generating hits at the approximate rate of 0.14%. An example of an unexpected hit is the microtubule inhibitor vincristine.

The disadvantages of a multiple-targets assay are trying to find which target is activated and eliciting the response you are interested in. Follow-up receptor-specific assays still need to be performed with those active compound hits to identify the actual molecular target.[57] Additionally, there is the aspect of added time and cost for measuring and analyzing multiple outputs in this type of assay format. One opinion is that several different assay outputs confuse rather than clarify the screen and that only perhaps two simultaneous readouts would be justified. The second assay would thus provide a good control for the first assay and replace a secondary screening effort. Certainly, there is a clear advantage in saving both compound and compound handling time, and the knowledge that all assays were performed on the same well of cells under the same conditions with the same age-related compound impurities present. This type of assay also enhances the chance of a serendipitous discovery of new pathways or routes to the desired result. **Figure 10** shows an example of flow cytometry results for known test compounds, including the Toll 4 agonist, lipopolysaccharide (LPS), and an immunomodulator, interferon-gamma (IFN-γ) as a control. In addition to the readouts for the screen (CD40 and CD54), other cell surface markers, major histocompatibility complex (MHC) I and II, CD80 and CD86, are shown. Although IFN-γ and LPS both increase CD54, the CD40 response is only markedly increased with LPS. IFN-γ response is characterized by MHC II increase.

3.29.4.4 Targeting a Cell-Based Model of Organ Response – Neurotoxicity

The appropriate in vitro models of drug toxicity should be chosen based on target organ specificity of the relevant species and include endpoints relevant to the mechanism in those organs. Building the predictive capabilities of current in vitro toxicity testing requires the use of not one cell readout, but a battery of endpoints derived from mechanism-specific as well as general cytotoxicity assays. One of the current challenges associated with this combination of information from different cell assays is to validate such a data set so that a broad predictive model of toxicity may be developed. Such a model must provide specific target organ information and a possible mechanism of

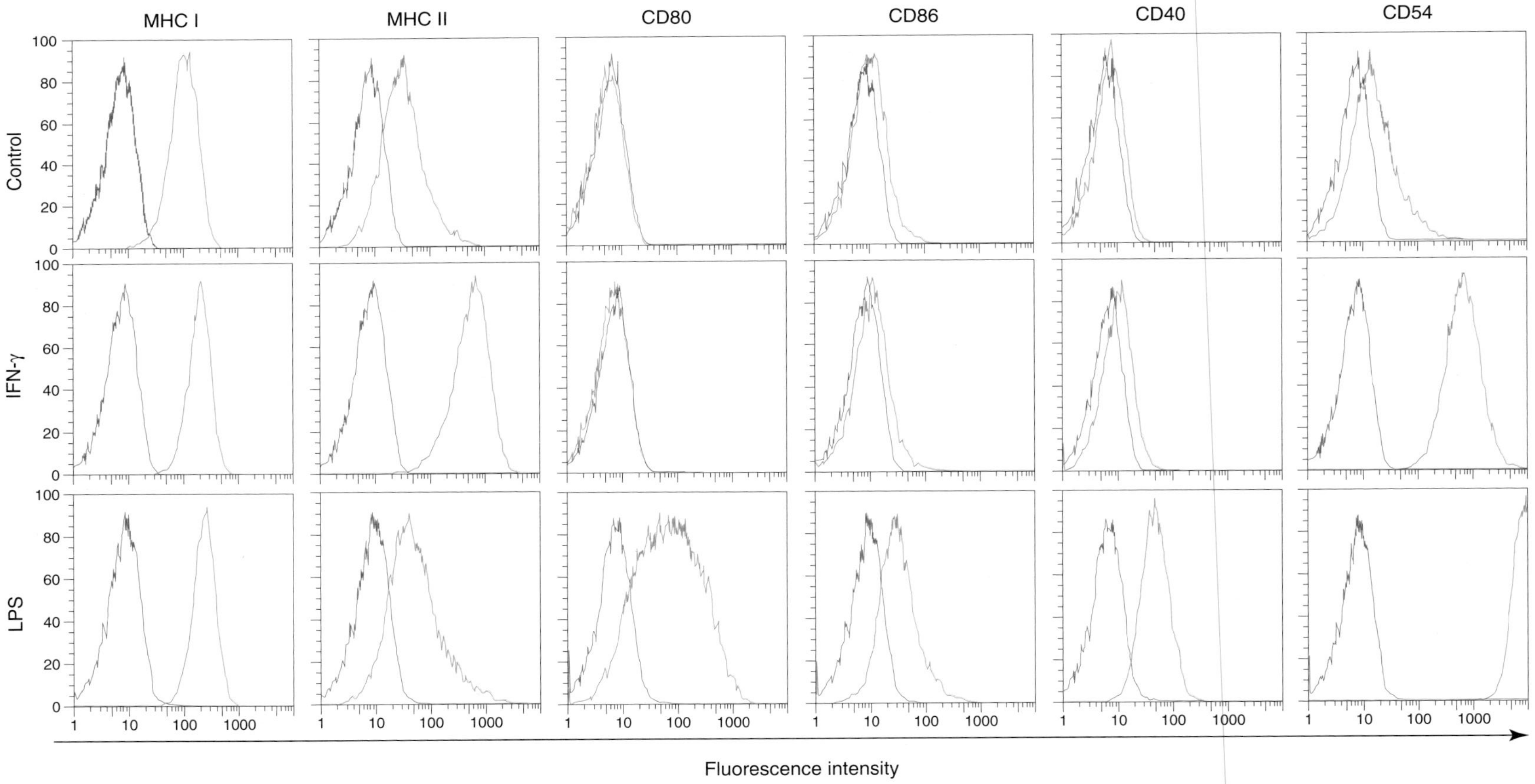

Figure 10 Expression of six immunological activation markers on the surface of THP-1 cells stimulated by LPS. Cells were incubated for 24 h with either media alone (as a negative control), LPS (1 μg mL^{-1}), or interferon-gamma (IFN-γ 500 IU mL^{-1}) (as a positive control for MHC II expression). Cells were stained with a fluorophore-labeled receptor-specific antibody (green) or an antibody isotype control (red).

action across a range of specific cell responses. The ultimate validation of this in vitro model of cellular toxicity can be found in a strong correlative relationship, among chemically diverse compounds, to their in vivo toxicity in humans.

Significant correlations among a select group of cell-based toxicity measures and median lethal dose (LD_{50}) data from rodent as well as human poison data have been obtained in past studies. For example, one concerted effort has been the Multicenter Evaluation of in Vitro Cytotoxicity, where a small group of carefully chosen compounds were tested in a series of cytotoxic assays in many different laboratories and compared to the available data on human toxicity.[58] This effort used 50 chemicals and had 77% accuracy in predicting actual human acute general toxicity. The result was compared to the 65% accuracy in predictions obtained from animal testing. Other groups, using selected groups of organic solvents[59] and other toxic compounds[60] have also been able to show strong correlations with LD_{50} data, which were mostly obtained from rodents. While these types of studies constitute a proof of principle, the correlation of highly toxic chemical effects and in vivo data is not the best predictive model for the medicinal chemist. In fact, many prototype drugs being developed have some adverse effects, but lack a highly toxic profile. The one exception would be cytoxic compounds selected as cancer treatment agents.

In response to these limitations, our company Novascreen, with National Institutes of Health funding and oversight, has produced a cell-based toxicity predictive database designed around a chemical set that is approximately two-thirds pharmaceuticals. These compounds, by definition, have high efficacy with low toxicity. Seven measures of cellular toxicity with these compounds at six concentrations in duplicate across 7 days of treatment were collected in two human-derived cells. These measures included an indicator of apoptosis, mitochondrial function, protein synthesis, glucose utilization, morphology, viability and proliferation in parallel with drug-treated cultures. An example of this type of data is shown in **Figure 11**, where the activity of the enzyme caspase-3, associated with the induction of apoptosis, is measured in SK-N-MC and NT2 human neuronal cells, after exposure to a drug. This multiassay approach resulted in an initial in vitro toxicity database that has been annotated with chemical data, and LD_{50} values from animal studies. Relevant human toxicity data were accessed from online public sources (MSDS, EXTOXNET, Registry of Toxic Effects of Chemical Substances National Library of Medicine, National Cancer Institute National Toxicology Program) and successfully correlated with the Novascreen in vitro toxicity data. Those human correlations were much higher than we obtained with rodent LD_{50} values (**Figure 12**) and provide encouraging indications that the in vitro data have a useful predictive capacity in humans.

3.29.5 Future Directions for Cell-Based Assays

Developments in cell-based assays have increased our ability to explore the interaction of chemistry with complex levels of biological activity. Access to a variety of signal cascades within a single cell allows us to see the impact of compounds beyond their initial interface with that cell. This often reveals drug effects along related cellular pathways, ending in a final set of cellular consequences. New techniques are continually introduced that allow us to track these specific molecular changes against the busy background of other ongoing cellular processes. These methods exploit the scientific progress in aligned fields such as engineering and informatics, allowing measurements that are increasingly sensitive, specific, and perhaps impractical with older methods. We expect to see this progress continue, with improvements in signal-to-noise ratio, reduction in the number of cells needed per assay, adoption of methods with more universal application, and a decline in the cost and expertise needed to perform them. For example, laser-based instruments are being replaced by ones using light-emitting diodes, sensitive charge-coupled device cameras are smaller and more affordable, and image analysis of cells is being completely automated. In addition, miniaturization and microfluidics are reducing the quantity of reagents used and nanotechnology promises to decrease assay requirements further. In fact, single-cell screening, as mentioned in Chapter 3.32 is becoming a reality. Advances in reagents will certainly have a large impact on future cell assays. Currently, new and better dyes for ions are being developed, pairing of multiple dyes using FRET rather than direct fluorescence is increasing assay specificity and sensitivity, and new ways to tag molecules allow us to follow their movement and metabolism in real time.

While advances in instruments and reagents are undoubtedly having an impact on the future directions of cell-based assays, developments in cell biology are revolutionizing the basic substrate for those assays. For example, stem cells are undifferentiated cells that are capable of self-renewal and differentiation into multiple lineages of mature cells. Stem cells can be isolated from embryonic, fetal, or adult tissue. The use of the first two types is limited by federal funding restrictions and ethical issues. Thus the third type, mesenchymal stem cells, appears to have more current potential. They can be isolated from adipose tissue, muscle, and bone marrow and can produce several cell lineages such as osteoblasts, adipocytes, myoblasts, and chondrocytes.[61] As more ground-breaking work is achieved in patient therapy for diseases such as Parkinson's, cystic fibrosis, and Alzheimer's, the technologies derived can be applied to cell-based

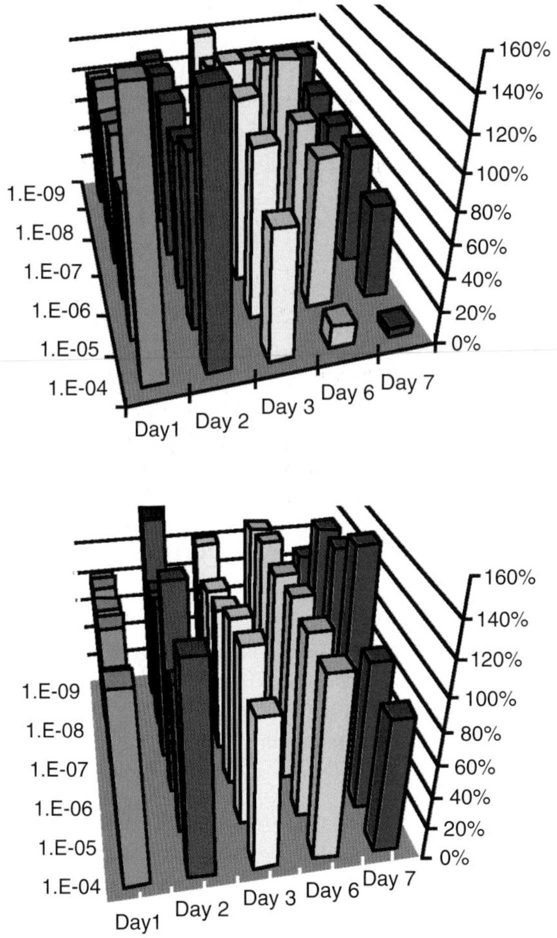

Figure 11 Comparison of the effects of the drug chloroquine on the intracellular activity of caspase-3, an indicator of apoptosis. Two different human-derived neuronal cell lines, (a) SK-N-MC and (b) dNT2, are shown. SNMC is a tumor cell line while the dNT2 is a primary neuronal cell line that is terminally differentiated with retinoic acid so that it resembles the morphology of mature brain cells more closely. This cell-based assay has a fluorescent readout of enzyme activity after 1–7 days of drug treatment at six different molar doses (1 nM to 10 μM). There is a clear difference in the pattern of response between cell lines on days 6 and 7 at high drug dose, indicating a reduction in viable cells following an initial induction of apoptosis at an earlier time. In addition to screening for drug toxicity, such information can be also be used to choose a cell line with the sensitivity appropriate to detecting toxic effects on the target organ, in this case the central nervous system.

screening. Understanding the characteristics of cells that work successfully in transplantation will require knowledge of cell enzymes, receptors, ion channels, and other targets that will fit nicely as screening targets. The cell-based techniques described in this chapter can be envisioned to play a central role in screening both the quality and function of those cells before placement into the patient.

Recent reports indicate that the 'ecological niche' or local microenvironment that a stem cell encounters governs its developmental behavior and differentiated fate. The hope and eventual expectation is that, when exposed to the optimally defined microenvironment, stem cells will differentiate in a manner appropriate to the local region, and integrate seamlessly with the existing circuits of the transplant recipient.[62] For this to occur, the stem or progenitor cell would need to respond to proliferative, developmental, and/or guidance cues provided by the host. For example, adult neural stem cells produced neurons when transplanted into the neurogenic zone of the hippocampus, but produce astrocytes in the environment of the spinal cord. Further investigation has showed that a specific component of the local environment, the regional astrocytes from the hippocampus, was capable of instructing these stem cells to adopt a neuronal fate in vitro.[63] In addition to regional differences within the nervous system, the microenvironment encountered by a stem cell may vary as a function of the age of the host organism.

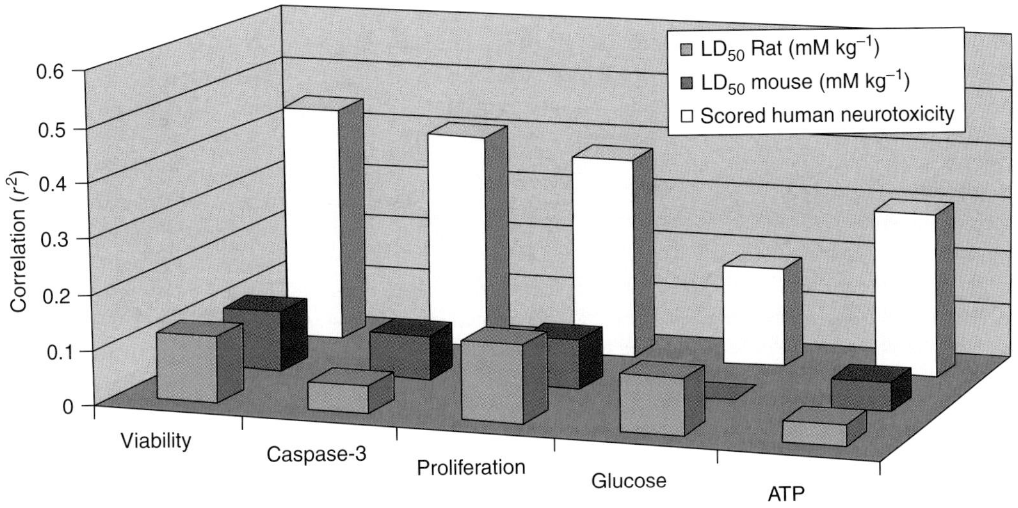

Figure 12 Correlations between each of several Novascreen in vitro toxicity measures, using neuronal cell cultures, and literature values for LD_{50} from rat or mouse or scored human neurotoxicity data. All of the in vitro data shows much better correlation with reported human neurotoxicity than it does with animal LD_{50} values. This may indicate that certain in vitro measures may be more appropriate than rodent LD_{50} values for the ultimate goal of predicting human neurotoxicity. Data are from treatment with approximately 200 different drugs.

Another promising area of research lies in exploring the extent to which precursor cells that appear to be partially differentiated stem cells derived from non-neural tissues can be cultured to generate specific classes of neurons and glial cells. While it is not surprising that embryonic stem cells can differentiate spontaneously into heart, blood, bone, and neuronal cells, unspecialized cells derived from adult sources as diverse as muscle, brain, and bone marrow have recently been shown to 'transdifferentiate' and acquire properties of cells from other lineages. Most intriguingly, cells that appear committed to a particular lineage may dedifferentiate to a more pluripotent state.[64] These aspects of developmental plasticity raise the hope for numerous types, sources, and manipulations of cells that can be employed in cell-based assays. Teasing out, understanding, and then mimicking these guiding in vivo interactions to control in vitro differentiation pose a major challenge. This challenge must be faced for cell biologists to develop realistic and precise cell-based functional assays. The past decade has seen enormous progress in our understanding of the specific requirements of stem cells to proliferate and differentiate along specified lineages. This progress has been made possible by the discovery of myriads of growth factors and substrate conditions, followed by careful testing in culture. However, the excitement of stem cells is tempered by the fact that pluripotent stem cells from bone marrow have been in use for decades with limited curative applications. Similarly, gene therapy's great potential has been slow to achieve clinical success.

As well as their apparent applications in patient therapy, stem cells are certain to play a central role in future clinical testing and research. Developments in stem cell biology present a potent opportunity to refine the drug discovery process further by finetuning the differentiated state of the therapeutically targeted cell type. This will allow the cell-based functional assay to mimic more closely the conditions found in the normal intact organism. Additionally, and perhaps more relevant to cell-based assays, these stem cells may also be used as a source of precisely differentiated cell types with specific relevance to disease states that can be individualized for the variations found within that patient population. It could eliminate many of the false positives that are now confronted by medicinal chemists when in vitro results of lead compounds are not validated by in vivo studies. Stem cell-based innovations may strengthen the link between cell-based functional assays and the real drug discovery target that will produce in vivo efficacy.

Promising developments in instrumentation, reagents, cell types, cell culture, increased information about existing cells, and knowledge about signaling pathway interactions will also greatly aid drug discovery. One effort in this area is from the Alliance for Cellular Signaling. Their overall goal is to understand as completely as possible the relationships between sets of inputs and outputs in signaling cells that vary both temporally and spatially. An aspect of this work is a freely available database of significant proteins on the signaling gateway maintained by the Nature publishing group. There are 3000 proteins in the database, curated by experts on each protein. Also, at this time, they are working to identify all the signaling proteins in several mouse cell types, including a macrophage cell line. For example, one report describes a macrophage, listing 6 G protein subunits, 47 kinases, 40 receptors, 7 ion channels, 40 transcription factors and their cofactors, and seven transporters.[65] Yet the work, which falls into the realm of systems biology, is only in its early stages. Insights gained now via DNA arrays, particularly in cancer, may be extended using protein data to other disease states. As more data are available, the pathway interconnections and relative importance will become more evident. These types of information will provide a robust framework for additional studies and the option of discovery-based rather than traditional hypothesis-driven work.

We hope that this chapter covers the major techniques involved in cell-based assays. This exciting field promises to provide a pipeline of innovations in both techniques and instrumentation. As useful information continues to emerge from cell-based studies, it should be evident to the chemist that these types of assay provide new research opportunities rather than unnecessary complications. As with most research endeavors, the researcher's expertise, comfort level, and the availability and sensitivity of instrumentation are key factors. These factors ultimately determine the appropriate techniques and methodologies that will be applied to the scientific question. However, one should appreciate that cell-based assays better approximate the disease target in its natural cell environment than an isolated protein or membrane. Therefore cells can provide a better understanding of the disease processes in vivo. We believe that this knowledge can lead to the productive development of new and improved compounds to treat human disease.

References

1. Ebeling, A. H. *J. Exp. Med.* **1919**, *30*, 531–537.
2. Schneeberger, E. E.; Harris, H. *J. Cell Sci.* **1966**, *1*, 401–406.
3. Dascal, N. *CRC Crit. Rev. Biochem.* **1987**, *22*, 317–387.
4. Weerapura, M.; Nattel, S.; Cartier, D.; Caballero, R.; Hebert, T. E. *J. Physiol.* **2002**, *540*, 15–27.
5. Good, E.; Perschke, S.; Lopez, R.; Chang, S.; Kinsler, A.; Snowman, A.; LaCombe, J.; Fedock, M.; Zeppetello, R.; Zysk, J. *J. Biomol. Screen.* **1998**, *3*, 231–236.
6. Rheinwald, J. G.; Beckett, M. A. *Cell* **1980**, *22*, 629–632.
7. Block, G. D.; Locker, J.; Bowen, W. C.; Petersen, B. E.; Katyal, S.; Strom, S. C.; Riley, T.; Howard, T. A. *J. Cell Biol.* **1996**, *132*, 1133–1149.
8. Hsieh, F. H.; Sharma, P.; Gibbons, A.; Goggans, T.; Erzurum, S. C.; Haque, S. J. *Proc. Natl. Acad. Sci. USA* **2005**, *02*, 14380–14385.
9. Guo, C. J; Lai, J. P.; Luo, H. M.; Douglas, S. D.; Ho, W. Z. *J. Neuroimmunol.* **2002**, *131*, 160–167.
10. Stefano, G. B.; Burrill, J. D.; Labur, S.; Blake, J.; Cadet, P. *Med. Sci. Monit.* **2005**, *11*, MS35–MS42.
11. Zvara, A.; Szekeres, G.; Janka, Z.; Kelemen, J. Z.; Cimmer, C.; Santha, M.; Puskas, L. G. *Dis. Markers* **2005**, *21*, 61–69.
12. Ryding, A. D.; Sharp, M. G.; Mullins, J. J. *J. Endocrinol.* **2001**, *171*, 1–14.
13. Fehmann, H. C.; Pracht, A.; Goke, B. *Pancreas* **1998**, *17*, 309–314.
14. Cell Line Database. Interlab Project, Biotechnology Dept. IST, Genova, Italy. www.biotech.ist.unige.it (accessed May 2006).
15. Grassi, F.; Piacentini, A.; Cristino, S.; Toneguzzi, S.; Cavallo, C.; Facchini, A.; Lisignoli, G. *Histochem. Cell Biol.* **2003**, *20*, 391–400.
16. Faust, J.; Lacey, D. L.; Hunt, P.; Burgess, T. L.; Scully, S.; Van, G.; Eli, A.; Qian, Y.; Shalhoub, V. *J. Cell Biochem.* **1999**, *72*, 67–80.
17. Boguslavsky, J. *Drug Disc. Mag.* **2004**, 35–39.
18. Lichtenberg, A.; Dumlu, G.; Walles, T.; Maringka, M.; Ringes-Lichtenberg, S.; Ruhparwar, A.; Mertsching, H.; Haverich, A. *Biomaterials* **2005**, *26*, 555–562.
19. Shin, A.; Chin, K. C.; Jamil, M. F.; Kostov, Y.; Rao, G.; Shuler, M. L. *Biotechnol. Prog.* **2004**, *20*, 338–345.
20. LeCluyse, E.; Madan, A.; Hamilton, G.; Carroll, K.; DeHaan, R.; Parkinson, A. *J. Biochem. Mol. Toxicol.* **2000**, *14*, 177–188.
21. Tang, W.; Kang, J.; Wu, X.; Rampe, D.; Wang, L.; Shen, H.; Li, Z.; Dunnington, D.; Garyantes, T. *J. Biomol. Screen.* **2001**, *6*, 325–331.
22. Dorn, A.; Hermann, F.; Ebneth, A.; Bothmann, H.; Trube, G.; Christensen, K.; Apfel, C. *J. Biomol. Screen.* **2005**, *10*, 339–347.
23. Birch, P. J.; Dekker, L. V.; James, I. F.; Southan, A.; Cronk, D. *Drug Disc. Today*, **2004**, *9*, 410–418.
24. Tran, L.; Farinas, J.; Ruslim-Litrus, L.; Conley, P. B.; Muir, C.; Munnelly, K.; Sedlock, D. M.; Cherbavaz, D. B. *Anal. Biochem.* **2005**, *341*, 361–368.
25. Hoolengerg, M. D. Receptor Models and the Action of Neurotransmitters and Hormones: Some New Perspectives. In *Neurotransmitter Receptor Binding*, 2nd ed.; Yamamura, H. I., Enna, S. J., Michael, J. K., Eds.; Raven Press: New York, **1985**, pp 1–39.
26. Chicoine, M. R.; Silbergeld, D. L. *J. Neurooncol.* **1997**, *35*, 249–257.
27. Weissman, A. D.; Caldicott, S. *Clin. Exp. Pharm. Physiol.* **1995**, *22*, 376.
28. Barrandon, Y.; Green, H. *Cell* **1987**, *50*, 1131–1137.
29. Sambuy, Y.; De Angelis, I.; Ranaldi, G.; Scarino, M. L.; Stammati, A.; Zucco, F. *Cell Biol. Toxicol.* **2005**, *21*, 1–26.
30. Braun, A.; Hammerle, S.; Suda, K.; Rothen-Rutishauser, B.; Gunthert, M.; Kramer, S. D.; Wunderli-Allenspach, H. *Eur. J. Pharm. Sci.* **2000**, *11*, S51–S60.
31. Pulliam, L.; Stubblebine, M.; Hyun, W. *Cytometry* **1998**, *32*, 66–69.
32. Brenner, C.; Kroemer, G. *Science* **2000**, *18*, 1150–1151.
33. Nau, G. J.; Richmond, J. F.; Schlesinger, A.; Jennings, E. G.; Lander, E. S.; Young, R. A. *Proc. Natl. Acad. Sci. USA* **2002**, *5*, 1503–1508.
34. Prabhakar, U.; Eirikis, E.; Reddy, M.; Silvestro, E.; Spitz, S.; Pendley, C., 2nd; Davis, H. M.; Miller, B. E. *J. Immunol. Methods* **2004**, *291*, 27–38.
35. Chan, S. M.; Ermann, J.; Su, L.; Fathman, C. G.; Utz, P. J. *Nat. Med.* **2004**, *10*, 1390–1396.
36. Shurin, G. V.; Ferris, R.; Tourkova, I. L.; Perez, L.; Lokshin, A.; Balkir, L.; Collins, B.; Chatta, G. S.; Shurin, M. R. *J. Immunol.* **2005**, *174*, 5490–5498.
37. Armando, I.; Jezova, M.; Juorio, A. V.; Terron, J. A.; Falcon-Neri, A.; Semino-Mora, C.; Imboden, H.; Saavedra, J. M. *Am. J. Physiol. Renal Physiol.* **2002**, *283*, F934–F943.
38. Gabriel, D.; Vernier, M.; Pfeifer, M. J.; Dasen, B.; Tenalillon, L.; Bouhelai, R. *Assay Drug Dev. Technol.* **2003**, *1*, 291–303.
39. Kalyuzhny, A. E. *Methods Mol. Biol.* **2005**, *302*, 15–31.
40. Li, X.; Zhao, X.; Fang, Y.; Jiang, X.; Duong, T.; Fan, C.; Huang, C. C.; Kain, S. R. *J. Biol. Chem.* **1998**, *273*, 34970–34975.
41. Catterall, W. A.; Chandy, K. G.; Gutman, G. A. *The IUPHAR Compendium of Voltage-Gated Ion Channels*; IUPHAR Media: Leeds, UK, 2002.

42. Quirk, P. L.; Rao, S.; Roth, B. L.; Siegel, R. E. *J. Neurosci. Res.* **2004**, 77, 498–506.
43. Hargreaves, A. C.; Lummis, S. C.; Taylor, C. W. *Mol. Pharmacol.* **1994**, *46*, 1120–1128.
44. Roth, B. L.; Kroeze, W. K.; Patel, S.; Lopez, E. *Neuroscientist* **2000**, *6*, 252–262.
45. Filmore, D. *Modern Drug Disc.* **2004**, 7, 55–56.
46. Lazareno, S.; Birdsall, N. J. *Trends Pharmacol. Sci.* **1993**, *14*, 237–239.
47. Arunlakshana, O.; Schild, H. O. *Br. J. Pharmacol.* **1959**, *14*, 48–58.
48. Samways, D. S.; Li, W. H.; Conway, S. J.; Holmes, A. B.; Bootman, M. D.; Henderson, G. *Biochem. J.* **2003**, *375*, 713–720.
49. Dupriez, V. J.; Maes, K.; Le Poul, E.; Burgeon, E.; Detheux, M. *Receptors Channels* **2002**, *8*, 319–330.
50. Liu, A. M.; Ho, M. K.; Wong, C. S.; Chan, J. H.; Pau, A. H.; Wong, Y. H. *J. Biomol. Screen.* **2003**, *8*, 39–49.
51. Heding, A. *Exp. Rev. Mol. Diagn.* **2004**, *4*, 403–411.
52. Yokota, Y.; Sasai, Y.; Tanaka, K.; Fujiwara, T.; Tsuchida, K.; Shigemoto, R.; Kakizuka, A.; Ohkubo, H.; Nakanishi, S. *J. Biol. Chem.* **1989**, *264*, 17649–17652.
53. Menzaghi, F.; Behan, D. P.; Chalmers, D. T. *Curr. Drug Targets CNS Neurol. Disord.* **2002**, *1*, 105–121.
54. Oakley, R. H.; Hudson, C. C.; Cruickshank, R. D.; Meyers, D. M.; Payne, R. E., Jr.; Rhem, S. M.; Loomis, C. R. *Assay Drug Dev. Technol.* **2002**, *1*, 21–30.
55. Ciambrone, G. J.; Liu, V. F.; Lin, D. C.; McGuiness, R. P.; Leung, G. K.; Pitchford, S. *J. Biomol. Screen.* **2004**, *9*, 467–480.
56. Koppal, T. *Drug Disc. Dev.* **2004**, 30–36.
57. Hart, C. P. *Drug Disc. Today* **2005**, *10*, 513–519.
58. Ekwall, B. *Toxicol. In Vitro* **1999**, *13*, 665–673.
59. Naskali, L.; Engelke, M.; Tähtil, H. *ATLA* **1994**, *22*, 175–179.
60. Schmuck, G.; Schluter, G. *Toxicol. Ind. Health* **1996**, *12*, 683–696.
61. Olivier, V.; Faucheux, N.; Hardouin, P. *DDT* **2004**, *9*, 803–811.
62. Shyu, W. C.; Lee, Y. J.; Liu, D. D.; Lin, S. Z.; Li, H. *Front Biosci.* **2006**, *1*, 889–907.
63. Emsley, J. G.; Mitchell, B. D.; Kempermann, G.; Macklis, J. D. *Prog. Neurobiol.* **2005**, *75*, 321–341.
64. Engelhardt, M.; Bogdahn, U.; Aigner, L. *Brain Res.* **2005**, *8*, 98–111.
65. Oda, K.; Kimura, T.; Matsuoka, Y.; Funahashi, A.; Muramatsu, M.; Kitano, H. *AfCS Research Reports [online].* **2004**, *2*, 14. Available from http://www. signaling-gateway. org/reports/v2/DA0014/DA0014.htm (accessed May 2006).
66. Alexander, S. P.; Mathie, A.; Peters, J. *Guide to Receptors and Channels*, 1st ed. (**2005** revision) *Br. J. Pharmacol. 144*, S1–S128, doi:10:1038/sj.bjp.0706158.
67. Gray, P. A; Fu, H.; Luo, P.; Zhao, Q.; Yu, J.; Ferrari, A.; Tenzen, T.; Yuk, D. I.; Tsung, E. F.; Cai, Z. et al. *Science* **2004**, *306*, 2255–2257.
68. Forrest, A. R.; Ravasi, T.; Taylor, D.; Huber, T.; Hume, D. A.; Grimmond, S.; RIKEN GER Group; GSL Members. *Genome Res.* **2003**, *13*, 1443–1454.
69. Venter, J. C.; Adams, M. D.; Myers, E. W.; Li, P. W.; Mural, R. J. *Science* **2001**, *291*, 1304–1351.

Biographies

All authors are currently employed by NovaScreen Biosciences Corporation (NBC), a subsidiary of Caliper Life Sciences. The company routinely performs over 700 receptor-binding, enzyme, cell-based and toxicity assays to support drug discovery. Clients include over 170 pharmaceutical, biotech, academia, and government laboratories worldwide. In addition, NBC performs computational modeling of drug activity based on it own proprietary pharmacological database. The company's emphasis has been in central nervous system pharmacology, toxicology, neuroscience, and, more recently, immunology. Arthur Weissman, PhD (physiology/pharmacology) is the Chief Technical Officer and Vice President of NBC. He received his training at Cornell University, New York Medical College, and UCLA Brain Research Institute. He was formerly a scientist at NIDA/NIH and Director of Research at a Mount Sinai Medical School hospital in New Jersey. James Keefer, PhD (biochemistry), Senior Scientist, worked at Johnson & Johnson, GlaxoSmithKline and trained at University of Chicago, University of Nebraska, and NCI/NIH. Alexei Miagkov,

PhD (immunology/cell biology), Scientist, trained at Moscow State University and Johns Hopkins. Malathi Sathyamoorthy, PhD (biochemistry), Senior Scientist, trained at University of Maryland-Baltimore and NIAID/NIH. Scott Perschke, MS (pharmacology), is Director of Assay Development and has worked for 18 years at NBC with training at Johns Hopkins and Liberty University. F L Wang, MD, PhD (pharmacology), Scientist, trained at Wuhan School of Medicine, University of Maryland School of Medicine, and NINDS/NIH.

Comprehensive Medicinal Chemistry II
ISBN (set): 0-08-044513-6
ISBN (Volume 3) 0-08-044516-0; pp. 617–646

3.30 Small Animal Test Systems for Screening

M Muda, S McKenna, and B G Healey, Serono Research Institute, Rockland, MA, USA

3.30.1	**Introduction**	**647**
3.30.2	***Xenopus laevis* as a Screening Tool**	**649**
3.30.3	***Drosophila melanogaster* as a Screening Tool**	**649**
3.30.3.1	*Drosophila* Reverse Genetics Screens	651
3.30.4	***Caenorhabditis elegans***	**652**
3.30.5	**The Zebrafish as a Screening Tool**	**652**
3.30.6	**The Mouse as Screening Tool**	**653**
References		**655**

3.30.1 Introduction

Since the beginning of the twentieth century biologists have tried to identify and develop experimental systems using simple and tractable model organisms to answer fundamental questions about complex biological processes such as gene regulation, cell proliferation, tissues, and embryonic development. Bacteria and phages were the first organisms that attracted scientists' attention, and many breakthrough discoveries were made by studying biochemistry and the genetics of prokaryotes and their viruses. Using the powerful combination of genetics and biochemical methods, fundamental biological discoveries were made by dissecting at the molecular level metabolic, morphologic, and physiological pathways in bacteria and phages. These studies became the stepping stones of modern molecular biology that paved the way to the understanding of the genetics, biochemistry, and physiology of more complex organisms. Only a few organisms have become widely used as model systems to address biologically relevant questions, both fundamental and applied.

Specifically, simple eukaryote model organisms such as Baker's yeast (*Saccharomyces cerevisiae*), the filamentous fungi *Neurospora crassa*, the amoebae *Dictyostelium discoideum*, the nematode *Caenorhabditis elegans*, and the fruit fly *Drosophila melanogaster* have all been extensively studied because of easy genetic analysis, simple nutritional requirements, rapid growth, and generation time.

More complex organisms such as the vertebrate zebrafish (*Danio rerio*), the frog *Xenopus laevis*, and the mouse *Mus musculus* have also become popular systems for biologists. Recently, several of these model organisms have crossed the borders of basic research laboratories and made their way into pharmaceutical research and development laboratories. In fact, these model systems have started to be exploited by pharmaceutical companies at all stages of the drug discovery process, ranging from the identification of novel drug targets to the discovery of the mechanism of action of a drug to potential toxicity issues. A strong indication of the increased use of model systems for drug discovery is the increasing number of companies directly using model systems in their discovery programs. One of the first companies directly working with model systems was Exelixis (San Francisco, CA), who have applied *Drosophila*, *Caenorhabditis elegans*, and *Danio rerio* for drug discovery. Additionally, Phylonix (Cambridge, MA) and Devgen (Ghent-Zwijnaarde, Belgium) are actively utilizing model systems for both internal and collaborative programs. Each of the model systems has its own strengths, which are linked to the distinct collection of molecular tools available and also to their specific physiology that might better reproduce certain human processes.

Recent methodological and technological advances have supported the completion of genomic sequencing of *Saccharomyces cerevisiae*, *Drosophila melanogaster*, *Caenorhabditis elegans*, zebrafish, and mouse genomes. Full genomic sequence comparisons have confirmed that many important molecular components are conserved throughout evolution (**Table 1**). In fact, these genomic sequence comparisons have reiterated the fact that essential gene and protein sequences are usually highly conserved throughout evolution, including many essential genes involved in cancer and diabetes. Many genetically tractable organisms such as yeast, *Drosophila melanogaster*, and *Caenorhabditis elegans* have been instrumental for the dissection and the molecular organization of biochemical pathways involved in many human diseases. Decades of experimentation have clearly demonstrated the functional conservation of complex biochemical

Table 1 Comparison of small-animal models

Scientific name (common name)	*Drosophila melanogaster (fruit fly)*	*Caenorhabditis elegans (worm)*	*Danio rerio (zebrafish)*	*Mus musculus (mouse)*
	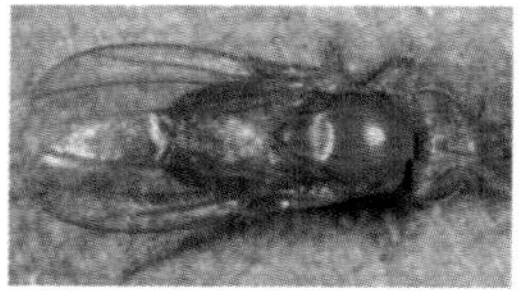	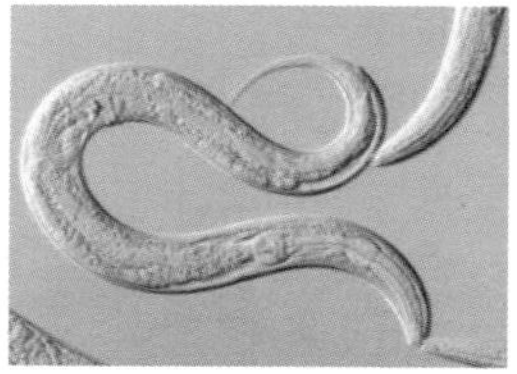	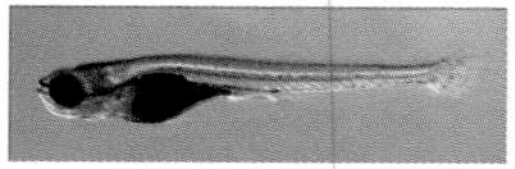	
Genome size (no. of genes)	14 000	20 000	18 000	35 000
Genome sequenced	Yes	Yes	Yes	Yes
Human similarity	62%	43%	80%	96%
Generation time	10 days	3 days	2–3 months	3 months
Cells	10^6	959	10^8	10^9
Transparent	No	Yes	Yes, larvae	No
High-throughput screen capable (96 well)	Yes, larvae	Yes	Yes, larvae	No

pathways among organisms such as yeast, *Drosophila*, and *Caenorhabditi elegans* and humans. Moreover, the functional domains of homologous proteins are usually even more highly conserved than the overall amino acid identity.

This conservation has allowed researchers to integrate many biochemical studies initiated in human cells with the power of genetic studies in tractable model organisms. The vast set of genetic data harvested from full genomic sequencing projects of prokaryotes and eukaryotic organisms has extensively validated the use of model organisms for the study and molecular dissection of human diseases. Using specific molecular genetic methods for each organism, genomic-wide screens have been designed to identify novel genes relevant for human diseases.

The initial impact of these drug discovery efforts was directly related to the identification and validation of disease-related targets. The studies involved either the overexpression or knock down of specific genes to ascertain the function of said gene by observing the phenotypes of the transgenic animals. This functional annotation was called reverse genetics. In the case of overexpression, a plasmid containing the gene of interest is transfected into the animal, and the effect on the animal is determined by analyzing the treated population. For knock-down studies, a gene is rendered obsolete through random chemical mutagenesis, or directly targeted using siRNAs and morpholinos. Morpholinos are chemically stable antisense sequences, made up of 18–25 subunits (six-membered morpholine ring with a covalently attached genetic base: adenine, cytosine, guanine or thymine. attached via phosphorodiamidate chemistry) resulting in substantially better antisense properties than RNA or DNA.

Because of their small size, short generation time, high fecundity and cost-effective maintenance, certain model organisms can also be used for high-throughput screens of small molecules. In fact, in many cases, standard microtiter plates (96 wells) can be used for whole-animal assay to assess toxicity, identify novel therapeutic compounds, and unravel drug mechanisms of action.

3.30.2 *Xenopus laevis* as a Screening Tool

The study of the African frog *Xenopus laevis* has made seminal contributions to various areas of biology such as embryology, cell biology, and biochemistry. *Xenopus laevis* has been widely utilized by embryologists as a model organism to decipher the complex mechanisms underlying early steps of embryo development and organ morphogenesis in vertebrates. However, this frog has also played a role in pharmacology and drug discovery mostly via the use of cells derived from it, including oocytes and melanophores. Together, these cells have been used for the study and the screening of ion channels and G protein-coupled receptors. Owing to their versatility in expressing mammalian membrane proteins, *Xenopus laevis* oocytes have been widely used to express and functionally characterize ion channels and human G protein-coupled receptors. Melanophores, large cells found in fish and amphibian, are filled with granule-like organelles (melanosomes) containing the pigment melanin. Redistribution of melanosomes is regulated by the intracellular concentration of cAMP and diacyl-glycerol produced from the activation of adenylate cyclase and phospholipase C, respectively. In eukaryotes, both adenylate cyclase and phospholipase pathways are common intracellular signaling pathways downstream of GPCR receptors. Hence, the changes in melanosome localization, aggregation, or dispersion, straightforwardly monitored by changes in light transmittance, allows for screening of agonists or antagonists of GPCR receptors. While the tetraploid genome of *Xenopus laevis* has hampered the usage of genetic and genomic analysis, a close relative, *Xenopus tropicalis*, has been recently identified as a valid alternative for this important vertebrate model organism. Because of its relatively small and diploid genome, genomic analysis of *Xenopus tropicalis* will complement previous studies with a much needed genetic/genomic approach.

3.30.3 *Drosophila melanogaster* as a Screening Tool

Modern biology has been permeated to a great extent by genetics, and a number of fundamental discoveries have been made using the fruit fly *Drosophila melanogaster*. This small fly has been one of the preferred model organisms used by biologists since the beginning of last century. The importance of this model organism is underscored by the numerous researchers that have been awarded the Nobel Prize utilizing *Drosophila melanogaster* for their experimental investigations. In 2000, a collaborative effort between the Berkeley *Drosophila* Genome Project (BDGP) and Celera Genomics led to the complete sequencing of the *Drosophila melanogaster* genome. The *Drosophila melanogaster* genome was quickly completed by a whole-genome shotgun sequencing, in which randomly generated pieces of complex genome sequences were reassembled by computer.[1] This became a proof of principle, and this same approach was later successfully used for the sequence completion of the human genome. Remarkably, whole *Drosophila melanogaster* genome comparison with human disease genes to assess the presence of the human gene counterparts in the fly has revealed that 178 out of 287 (62%) genes appear to be conserved.[2] Therefore, *Drosophila melanogaster* seems a good model for the functional study of genes implicated in diverse human diseases such as cancer as well as neurological and metabolic diseases. Historically, the study of *Drosphila melanogaster* has been particularly useful for the identification of gene function due to a vast set of powerful technologies that can be used to manipulate its genome. Many methods for manipulating the genome, such as generation of transgenic animals and the use of transposable elements for generation of genomic wide screens, have been developed in the study of *Drosophila melanogaster*.

Studies using genetic interactions and genetic screens in *Drosophila melanogaster* are furthering our understanding of complex human diseases such as cancer and diabetes. For instance, genetic screens have made great contributions in identifying and arranging components within signaling pathways involved in essential cell functions such as proliferation, regulation of life span, cell size, and cell growth.

Drosophila melanogaster genetics has highlighted the functional role of many molecular components downstream of growth factor receptor tyrosine-kinase and insulin/insulin-like growth factor receptors. For example, the guanine nucleotide exchange factor responsible for the exchange of GDP to GTP on the proto oncogene protein RAS, named Son-of-Sevenless (SOS), was first identified in *Drosophila melanogaster* genetic screens designed to discover genes involved in the differentiation of photoreceptor cells.[3] *SOS* was identified via a genetic screen as a gene working downstream of *Sevenless*, which produces the receptor essential for photoreceptor differentiation, thereby deriving its name. *SOS* was also found to function downstream from *DER*, which produces a second tyrosine kinase receptor that is homologous to the human epidermal growth factor receptor.

The mammalian *SOS* homolog, SOS1, was isolated based on sequence similarity.[4] Like its fly counterpart, the SOS1 protein acts as a positive regulator between a tyrosine kinase and RAS proteins by catalyzing exchange of GDP for GTP. Normal RAS proteins can bind either GDP or GTP, and mutations of RAS that lock the protein conformation in the GTP-bound state are frequently found in human cancer. RAS has been shown to associate and activate RAS activated factor, RAF, and constitutive activation of this pathway drive tumors formation. Importantly, *Drosophila RAS* (*DRAS1*) was identified in a genetic screen as a gene functioning downstream of the tyrosine kinase receptor *Sevenless*.[5] Together with many other examples, this demonstrates that the same components are found in the RAS/extracellular signal-regulated kinase (ERK) pathway, downstream of the receptor tyrosine kinase gene in *Drosophila melanogaster* and humans (**Figure 1**). Importantly, this pathway is conserved not only at the molecular level but has retained similar function during evolution, regulating cell differentiation and proliferation.

More recently, several groups have been using *Drosophila melanogaster* genetic screens to unmask the functions of human tumor suppressor genes. Tumor suppressor genes, when inactivated by genetic mutation, result in uncontrolled

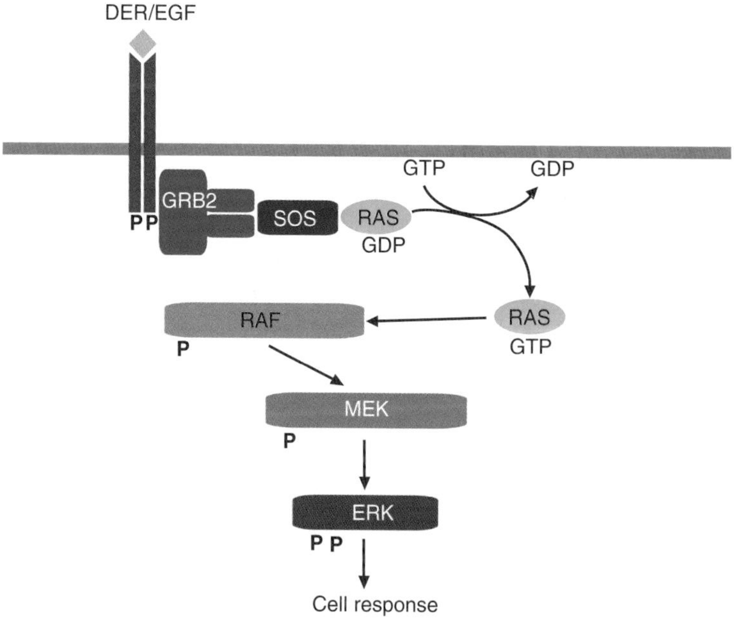

Figure 1 The RAS/ERK signaling pathway. This modular enzymatic cascade is conserved from yeast to humans. The RAS/ERK cascade has been shown to play a pleiotropic role, regulating many diverse functions such as development, memory, stress responses, proliferation, differentiation, and apoptosis. Genetic analysis in *Drosophila melanogaster*, lead to the discovery of SOS, a guanine nucleotide exchange factor for RAS acting downstream of receptor tyrosine kinases.

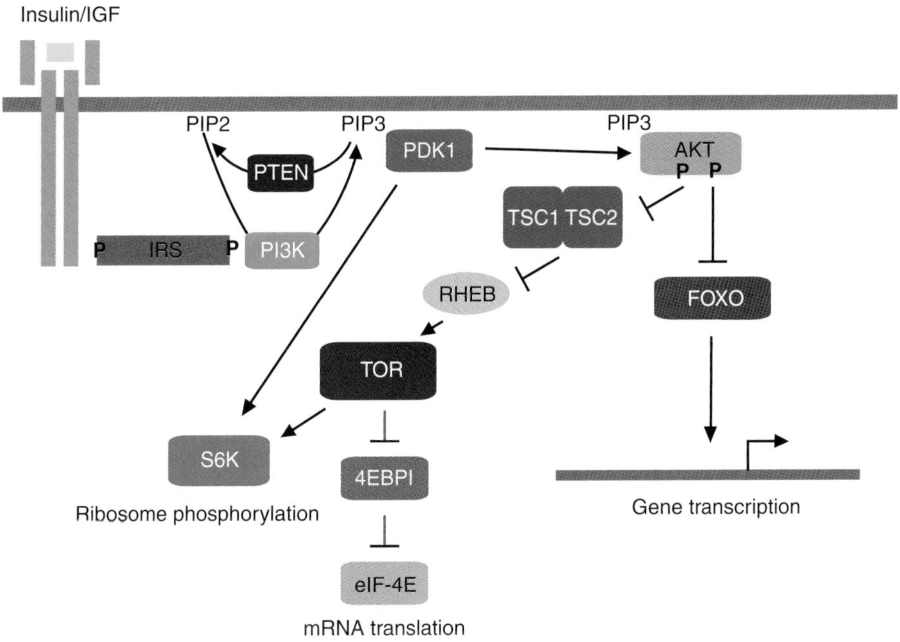

Figure 2 The target of rapamycin (TOR) signaling pathway, which couples nutrient availability (amino acids) with cell growth. Mutation of the *TSC1* and *TSC2* tumor suppressor genes is responsible in humans for the tuberous sclerosis syndrome. The tumor suppressor complex, TSC1/TSC2, antagonizes the TOR signaling pathway. *Drosophila melanogaster* genetic screens were instrumental in illuminating the functions of these human tumor suppressor genes as regulators of the TOR pathway.

cell growth and proliferation. Dissecting the function of tumor suppressor genes in mammals has been complex because it is likely that homozygous loss of function mutations of genes with key roles in the control of cell growth are accompanied by major developmental defects that, in turn, result in lethality. Clever genetic *Drosophila melanogaster* screens have been developed that use advanced genetics tools to generate homozygous mutant tissues in the eyes or the head of otherwise heterozygous fly. In one such type of screen that uses somatic recombination, several groups identified the gene *dTSC1* as a key regulator of cell growth. Inactivation of the *dTSC1* gene results in enhanced tissue growth due to cell-autonomous increase in cell size. The *dTSC1* gene is homologous to the human hamartin gene (also named *TSC1*), and, together with a second *Drosophila melanogaster* protein *dTSC2* (homologous to the tuberin gene, also named *TSC2*), forms a complex homologous to the human tuberous sclerosis complex (**Figure 2**). Tuberous sclerosis is an autosomal dominant disease that affects about 1 in 6000–7000 newborns. Genetic inactivation of the *TSC* genes in humans causes a complex disease that results in benign tumor growth (hamartomas) in the brain and in other vital organs such as kidney, lung, and skin. Tuberous sclerosis complex derives its name from the presence of hamartomas, tuber-like lesions in the brain and other organs. These tumors lesions become calcified and sclerotic with time. The growth of benign tumors in multiple organs in afflicted patients results in numerous pathological symptoms ranging from neurological disorders (epilepsy) to kidney failure and skin tumors.

3.30.3.1 *Drosophila* Reverse Genetics Screens

Classical genetics involves the generation of genomic mutations, analyzing the corresponding phenotypes and then identifying the affected genes. The availability of gene sequences from genome projects has reversed this procedure. Much of today's research is based on reverse genetics, which begins with the knowledge of the gene sequence and then dissects its functions by analyzing the phenotype of an induced targeted mutation. Additionally, the systematic analysis of large gene families has been made possible by the systematic use of gene knock-down methods such as RNA interference (RNAi)[6] and knockout animals.[7] *Drosophila melanogaster* has the additional advantage that cell lines have been established that can be used to perform biochemical experiments that complement genetic studies. Remarkably, RNAi has been extensively implemented using *Drosophila melanogaster* cell lines, and this powerful yet simple technology has been used to dissect signaling pathways and to identify novel genes regulating cytoskeleton dynamics.[8,9]

3.30.4 *Caenorhabditis elegans*

A model system that offers advantages over *Drosophila melanogaster* for the functional study of genes, chosen by Sydney Brenner in 1965, is *Caenorhabditis elegans*.[10] This soil nematode was an ideal model for study because of its transparency, small size (1 mm in length), rapid growth and expansion (hermaphrodite), and its relatively simplicity as an organism (959 total cells). During development, *Caenorhabditis elegans* comprises 1090 cells, of which 131 undergo programmed cell death. The remaining 959 cells of the adult hermaphrodite subdivide into a variety of tissues and primitive organs. Although simple, *Caenorhabditis elegans* has skin, an intestine, neurons, glands, and a reproductive system. The transparency of *Caenorhabditis elegans* coupled with these rudimentary organs and tissues allows the facile study of developmental biology, neuronal signaling, and disease biology, including cancer, depression, Alzheimer's disease, and apoptosis.[11]

The genome sequencing of *Caenorhabditis elegans* in 1998 was the first demonstration that a multicellular organism could be completely sequenced, and was a critical step in the genomics effort.[12] This work led to the development of whole-genome mapping and new sequencing technologies, and was the basis for the human genome project. Comparison between the genomes of *Caenorhabditis elegans* and humans has shown that many disease-related genes are conserved across the species, and indicates that many of the critical signaling pathways can be evaluated in this model.[11] Considerable progress has been made toward the understanding of programmed cell death in mammals using *Caenorhabditis elegans*.[13] Following the work on the *Caenorhabditis elegans* death genes *CED3* and *CED4*,[14] it was shown that transfection of the human *BCL-2* into *Caenorhabditis elegans* improved the survival of *CED1* mutant worms.[15] In addition to being the first demonstration of the expression of a human gene into *Caenorhabditis elegans*, it indicated that human BCL-2 can functionally interact with the nematode's cell death machinery. Later work on the *Caenorhabditis elegans* cell death inhibitory gene *CED-9* supported the hypothesis that the cell death mechanism was conserved between species. In fact, *BCL-2* and *CED-9* are homologous sequences, and the expression of *BCL-2* in the *CED-9*-deleted nematodes could partially restore a normal phenotype.[16] This research was a significant step toward the validation and acceptance of *Caenorhabditis elegans* as a model system for drug discovery.

Following the successful use of *Caenorhabditis elegans* for the study of gene function and target validation, the model was applied to small-molecule screening. In addition to the advantages identified for the study of gene function, the testing of small molecules is facilitated by the fact that the nematodes can be plated in 96-well microtiter plates in the presence of test compounds diluted in aqueous media. The nematodes ingest the compounds through feeding, where they are absorbed in the intestine, distributed throughout the nematode, and interact with the target of interest. Active compounds in the screen are inherently orally bioavailable, enhancing the value of the resulting positives over traditional high-throughput screening positives. Screens using *Caenorhabditis elegans* are typically functional in nature, measuring changes in phenotypes. Screens have been performed for the study of central nervous system disorders using standard fluorescence plate readers such as a pharynx pumping assay for depression.[17] In the area of cancer, an assay was developed in *Caenorhabditis elegans* that measured the inhibitory activity of farnesyl transferase inhibitors on the multi-vulva phenotype in *LET-60* gain-of-function nematodes, the nematode ortholog of human *RAS*.[18] This study demonstrated that this assay could be used to screen for inhibitors of farnesyl transferase and potentially other downstream members of the RAS pathway.

3.30.5 The Zebrafish as a Screening Tool

The newest model system introduced into drug discovery is the zebrafish (*Danio rerio*). Zebrafish are well suited to phenotyping for similar reasons identified for *Caenorhabditis elegans*: transparent (as embryos), small size (embryos can live in 96-well plates), and scaleable (300 eggs per female). A significant benefit of the zebrafish as a model is that it is a vertebrate, allowing the study of specific development and disease processes that do not exist in the previous model systems, and it was the first vertebrate studied in a large forward genetic screen.[19] The majority of these screens have been focused on developmental biology, resulting in the generation of thousands of distinct mutations.[20,21] Additional screens have been carried out that focused on disease biology, including polycystic kidney disease,[22,23] cholesterol processing,[24] tissue regeneration,[25] heart disease,[26–31] anemia,[32–34] cancer,[35–38] and nervous system disorders.[39–44]

The sequencing of the zebrafish genome has been completed by the Sanger Center, and there is greater than 80% similarity between the zebrafish and human genes. Like *Drosophila melanogaster* and *Caenorhabditis elegans*, this conservation between species allows for the study of human diseases in the zebrafish. In a study designed to test the direct role of *BRAF*, a *RAF* kinase family member, gain-of-function mutations on the development of nevi (moles) and melanoma, transgenic zebrafish were created with human wild-type *BRAF* or the constitutive active mutant $BRAF^{V600E}$.[38] Transgenic fish were produced by microinjection of either *BRAF* or $BRAF^{V600E}$ under the melanocyte promoter mitfa. The results showed that the wild-type *BRAF* zebrafish did not develop nevi while the $BRAF^{V600E}$ zebrafish developed nevi in 10% of the animals by 8 weeks. Histology revealed that these nevi were not dysplastic and there was no local tissue invasion. Following up on this result, $BRAF^{V600E}$ transgenic fish were created in a tumor suppressor *P53*-deficient background, resulting in lesions that developed into invasive melanoma. This new transgenic melanoma model demonstrates the role of the $BRAF^{V600E}$ mutation in the development of nevi and melanoma in the zebrafish.

In addition to the use of zebrafish for genetic screens and target validation in an animal model, zebrafish can be used to perform small-molecule screens by measuring phenotypic changes. These phenotypes can be measured with automated microscopy, allowing large numbers of compounds to be tested in a manner not achievable with any other model system. A recent example is a screen for compounds that could rescue a lack of blood flow to the trunk and tail of zebrafish embryo that have a vascular defect (gridlock phenotype) caused by a hypomorphic mutation in the *HEY2* gene.[45] The screen was carried out in 96-well plates (three mutant embryos/well) by testing compounds followed by visual scoring of the circulation on a dissecting microscope. The screen identified a novel class of compounds that allowed the embryos to develop normal vasculature. Similarly, zebrafish screens have also been designed to assess compound toxicity.[46] The current trend in drug discovery is to perform in vitro toxicity assays earlier in the drug discovery process to improve the quality of compounds entering the hit-to-lead phase. Zebrafish toxicology assays have advantages over in vitro assays in that they are more physiologically relevant, and may be able to pick up toxic effects due to metabolites while still being run at the throughput of in vitro assays. Two recent zebrafish toxicology screens have been developed for determining cardiotoxicity. Both screens utilized microscopy to measure the heartbeat of zebrafish embryos treated with QT-prolonging drugs, and were able to identify the induced bradycardia.[31,47]

3.30.6 The Mouse as Screening Tool

Although the relative cost and size of rodents has limited their use as large-scale screening tools for the identification of active molecules early in the drug discovery process, the biological similarities of rodents to humans has made them useful for evaluating limited libraries of small molecules for specific pharmacological properties such as efficacy, toxicity, or pharmacokinetics. Typically, this has followed in vitro prescreening efforts to identify compounds that demonstrate potency and efficacy in cell-based assays, as well as demonstrating potential for bioavailability in predictive models such as the Caco-2 permeability and hepatocyte metabolic assays. Compound screening in mouse models is then used to select the most promising candidate from a family of related compounds as a final step in prioritizing lead molecules for progression beyond the discovery phase.

Pharmacokinetic determination of compounds in mice and rats is often a bottleneck in the discovery process, and is therefore one in which methods for rapid screening have been actively sought. Advances in compound detection in complex biological fluids such as serum has permitted the use of cassette or '*N*-in-one' dosing, in which several compounds are co-administered into mice or rats in a single solution for the parallel assessment of pharmacokinetic behavior. As each compound in the sample can be evaluated individually, this method has permitted up to dozens of compounds being tested in parallel.[48] Although problems may arise when one compound interferes with the metabolic pathways that regulate other compounds in the cassette, leading to false conclusions on the pharmacokinetic properties of the later compounds, in most cases these problems can be minimized though preliminary studies on representative compounds from each class to be included.

Pharmacodynamics, or the determination of the effects of compounds on the organism, are important metrics used to predict efficacy or toxicity based on the use of readily assessable biological changes. The identification of surrogate markers that may be related to but are more easily assessed than the primary endpoints allows pharmacodynamic assessment of compounds in a rapid manner. For example, there is emerging evidence from the clinic that the development of skin rashes in cancer patients shortly after the initiation of treatment with inhibitors of epidermal growth factor receptor (EGFR) is associated with improved antitumor response rates.[49] If this observation is borne out, this marker may be used as an early predictor of drug efficacy. Similarly, in compound screening in mice, surrogate marker identification permits the rapid screening of candidate molecules to select those likely to achieve the ultimate efficacy endpoint. In one mouse xenograft model, inhibition of phosphorylation of EGFR in an implanted fibrosarcoma was correlated to EGFR phosphorylation status in distant epidermal keratinocytes.[50] Hence, utilizing a simple skin biopsy harvested shortly after treatment may be a rapid and accessible method of predicting drug effects in the tumor itself. In the future, the same relatively simple assay may be used to support drug screening in preclinical models as well as predicting clinical outcome in patients.

Toxicogenomics employs microarray technology to evaluate the impact of exogenous compounds on the expression of up to thousands of genes in target organs such as the liver.[51,52] By comparing the gene expression profiles of drug candidates to those of known toxic compounds, an early signal of potential toxicity may be observed. Such technologies may eventually replace the cumbersome and costly pathology-based toxicity studies now required in drug development. Metabonomics is a relatively recent technology that may be used for the rapid in vivo screening of compounds for perturbations of the metabolic status in experimental animals. The utility of this approach is based on the fact that endogenous factors (i.e., disease), as well as exogenous factors (i.e., drugs), can disrupt metabolic homeostasis in various and predictable ways. Metabolic profiling of biological fluids such as serum or urine by mass spectrometry and nuclear magnetic resonance spectroscopy results in a metabolite fingerprint unique to the pertubating factors.[53] While the full potential of this technology is still under evaluation, it is likely that this will represent a powerful new tool to assess drug safety and toxicity of compounds in a rapid manner.

Genetic mouse models are rapidly becoming valuable tools in screening compounds for specific functions. Disease models created by genetically modifying mice to correspond with the genotype and phenotype of human disease allow the assessment of compounds in more relevant model systems. Compounds active in models that are the genetic correlate of the clinical indication would be expected to have a higher rate of success in patients. In an early version of this approach, a transgene formed by the *BCR/ABL* gene fusion, identical to that involved in the etiology of chronic myeloid leukemia in humans, was introduced into mouse bone marrow cells and transferred into mice, resulting in leukemic blast crisis.[54] Importantly, these blasts were sensitive to STI571 (Gleevec), a molecule that would become one of the earliest targeted therapies in cancer. Key disadvantages of genetic models are that they are typically expensive, and, unless a key surrogate biomarker is identified, the throughput is limited and relies heavily on in vitro screening to focus the number of candidate molecules. Other genetic models may be more useful in supporting a higher number of compounds by relying on simpler endpoints. The coupling of highly sensitive luminescence detection methods with transgenic mice expressing the promoter-controlled luciferase gene has opened new opportunities for the evaluation of drug candidates. In one example, the vascular endothelial growth factor receptor 2 (VEGFR2) promoter has been used to drive luciferase expression[55] in transgenic mice. In this model, dexamethasone was found to suppress luciferase expression as well as angiogenesis in a wound-healing model. Using such biophotonics imaging technologies, the parallel determination of compound efficacy, biodistribution, and toxicity may eventually be readily ascertained in living animals.

While the use of mice for drug compound screening has both advantages and disadvantages, the mouse has proven invaluable in screening for novel targets in support of drug discovery. Selective deletion, or knockout, of specific known genes in mice results in phenotypes that are remarkably similar to the effects of drugs which inhibit the protein products of these same genes.[56] Conditional knockout mice, in which the targeted gene is flanked by cleavable *LOXP* sequences, have been used to examine the effect of gene deletion only in specific tissues or at predetermined stages of development, thus avoiding problems such as neonatal lethality caused by traditional knockout technologies. This is accomplished with the cyclization recombinase enzyme (CRE), which recognizes and removes the *LOXP* sites together with the intervening sequences. By controlling *CRE* gene expression with tissue-specific or inducible promoter systems, the deletion of the target gene can be readily controlled in mice.[57] A more recent method for selectively downregulating gene expression in rodents is through the delivery of silencing RNA or RNAi, which interact with and destabilize the RNA transcripts that code for the protein of interest. Transgenic mice created with lentiviral vector-based short-hairpin RNAs have been produced in which the targeted gene is effectively silenced.[58] Unlike traditional knockout approaches, RNA interference typically does not result in the complete inhibition of the target protein, and may therefore more closely mimic the effect of a pharmaceutical agent. The use of conditional promoter systems in such mice may eventually permit the rapid prediction of drug efficacy even before the compounds themselves are synthesized. Technologies bridging genetic and pharmacological approaches are currently being developed to support target validation in mice. ATP analog sensitive kinase alleles code for biologically active variants of wild-type kinases that have been modified to be sensitive to a specific small-molecule inhibitor. In this way, the impact of inhibiting the wild-type version of the kinase can be assessed prior to the initiation of a screen for selective inhibitors of the naturally occurring kinase.[59] Gene traps and chemical mutagenesis have been used to identify previously unknown or unappreciated genes that are involved in disease phenotypes. For example, *N*-ethyl-*N*-nitrosourea has been used to randomly induce point mutations in mouse embryos by inducing single base point mutations.[60] Resulting phenotypes can be correlated to the pertubated genotype for the identification of the affected gene. One advantage of this method over a classical gene knockout strategy is that in many cases the affected gene is only partially reduced in function, avoiding some of the problems of neonatal lethality. A major shortcoming, however, is that preselected genes cannot be targeted for phenotype evaluation.

As a complex biological organism, the mouse represents a tool that spans the evolutionary gap between lower organisms and humans. Mice therefore represent an important tool for the identification of novel drug candidates that may have clinical efficacy. For what mice lack in amenability to true high-throughput screening, they often make up for in predictiveness to human conditions.

References

1. Adams, M. D.; Celniker, S. E.; Holt, R. A.; Evans, C. A.; Gocayne, J. D.; Amanatides, P. G.; Scherer, S. E.; Li, P. W.; Hoskins, R. A.; Galle, R. F. et al. *Science* **2000**, *287*, 2185–2195.
2. Fortini, M. E.; Skupski, M. P.; Boguski, M. S.; Hariharan, I. K. *J. Cell Biol.* **2000**, *150*, 23–30.
3. Olivier, J. P.; Raabe, T.; Henkemeyer, M.; Dickson, B.; Mbamalu, G.; Margolis, B.; Schlessinger, J.; Hafen, E.; Pawson, T. *Cell* **1993**, *73*, 179–191.
4. Bowtell, D.; Fu, P.; Simon, M.; Senior, P. *Proc. Natl. Acad. Sci. USA* **1992**, *89*, 6511–6515.
5. Simon, M. A.; Bowtell, D. D.; Dodson, G. S.; Laverty, T. R.; Rubin, G. M. *Cell* **1991**, *67*, 701–716.
6. Muda, M.; Worby, C. A.; Simonson-Leff, N.; Clemens, J. C.; Dixon, J. E. *Biochem. J.* **2002**, *366*, 73–77.
7. Rong, Y. S.; Golic, K. G. *Science* **2000**, *288*, 2013–2018.
8. Clemens, J. C.; Worby, C. A.; Simonson-Leff, N.; Muda, M.; Maehama, T.; Hemmings, B. A.; Dixon, J. E. *Proc. Natl. Acad. Sci. USA* **2000**, *97*, 6499–6503.
9. Kiger, A. A.; Baum, B.; Jones, S.; Jones, M. R.; Coulson, A.; Echeverri, C.; Perrimon, N. *J. Biol.* **2003**, *2*, 27.
10. Brenner, S. *Genetics* **1974**, *77*, 71–94.
11. Riddle, D.; Blumenthal, T.; Meyer, B.; Priess, J. *C. elegans II*; Cold Spring Harbor Laboratory Press: Plainview, NY, 1997.
12. The *C. elegans* Sequencing Consortium. *Science* **1998**, *282*, 2012–2018.
13. Kaufmann, S. H.; Hengartner, M. O. *Trends Cell. Biol.* **2001**, *11*, 526–534.
14. Ellis, H. M.; Horvitz, H. R. *Cell* **1986**, *44*, 817–829.
15. Vaux, D. L.; Weissman, I. L.; Kim, S. K. *Science* **1992**, *258*, 1955–1957.
16. Hengartner, M. O.; Horvitz, H. R. *Cell* **1994**, *76*, 665–676.
17. Kaletta, T.; Butler, L.; Bogaert, T. *Caenorhabditis elegans* Functional Genomics in Drug Discovery Expanding Paradigms. In *Model Organisms in Drug Discovery*; Carroll, P. M., Fitzgerald, K., Eds.; John Wiley: Chichester UK, 2003, Chapter 3, pp. 41–79.
18. Hara, M.; Han, M. *Proc. Natl. Acad. Sci. USA* **1995**, *92*, 3333–3337.
19. Eisen, J. S. *Cell* **1996**, *87*, 969–977.
20. Driever, W.; Solnica-Krezel, L.; Schier, A. F.; Neuhauss, S. C.; Malicki, J.; Stemple, D. L.; Stainier, D. Y.; Zwartkruis, F.; Abdelilah, S.; Rangini, Z. et al. *Development* **1996**, *123*, 37–46.
21. Haffter, P.; Granato, M.; Brand, M.; Mullins, M. C.; Hammerschmidt, M.; Kane, D. A.; Odenthal, J.; van Eeden, F. J.; Jiang, Y. J.; Heisenberg, C. P. et al. *Development* **1996**, *123*, 1–36.
22. Sun, Z.; Amsterdam, A.; Pazour, G. J.; Cole, D. G.; Miller, M. S.; Hopkins, N. *Development* **2004**, *131*, 4085–4093.
23. Otto, E. A.; Schermer, B.; Obara, T.; O'Toole, J. F.; Hiller, K. S.; Mueller, A. M.; Ruf, R. G.; Hoefele, J.; Beekmann, F.; Landau, D. et al. *Nat. Genet.* **2003**, *34*, 413–420.
24. Farber, S. A.; Pack, M.; Ho, S. Y.; Johnson, I. D.; Wagner, D. S.; Dosch, R.; Mullins, M. C.; Hendrickson, H. S.; Hendrickson, E. K.; Halpern, M. E. *Science* **2001**, *292*, 1385–1388.
25. Poss, K. D.; Keating, M. T.; Nechiporuk, A. *Dev. Dyn.* **2003**, *226*, 202–210.
26. Gerull, B.; Gramlich, M.; Atherton, J.; McNabb, M.; Trombitas, K.; Sasse-Klaassen, S.; Seidman, J. G.; Seidman, C.; Granzier, H.; Labeit, S. et al. *Nat. Genet.* **2002**, *30*, 201–204.
27. Xu, X.; Meiler, S. E.; Zhong, T. P.; Mohideen, M.; Crossley, D. A.; Burggren, W. W.; Fishman, M. C. *Nat. Genet.* **2002**, *30*, 205–209.
28. Li, Q. Y.; Newbury-Ecob, R. A.; Terrett, J. A.; Wilson, D. I.; Curtis, A. R.; Yi, C. H.; Gebuhr, T.; Bullen, P. J.; Robson, S. C.; Strachan, T. et al. *Nat. Genet.* **1997**, *15*, 21–29.
29. Garrity, D. M.; Childs, S.; Fishman, M. C. *Development* **2002**, *129*, 4635–4645.
30. Curran, M. E.; Splawski, I.; Timothy, K. W.; Vincent, G. M.; Green, E. D.; Keating, M. T. *Cell* **1995**, *80*, 795–803.
31. Langheinrich, U.; Vacun, G.; Wagner, T. *Toxicol. Appl. Pharmacol.* **2003**, *193*, 370–382.
32. Donovan, A.; Brownlie, A.; Zhou, Y.; Shepard, J.; Pratt, S. J.; Moynihan, J.; Paw, B. H.; Drejer, A.; Barut, B.; Zapata, A. et al. *Nature* **2000**, *403*, 776–781.
33. Njajou, O. T.; Vaessen, N.; Joosse, M.; Berghuis, B.; van Dongen, J. W.; Breuning, M. H.; Snijders, P. J.; Rutten, W. P.; Sandkuijl, L. A.; Oostra, B. A. et al. *Nat. Genet.* **2001**, *28*, 213–214.
34. Davidson, A. J.; Ernst, P.; Wang, Y.; Dekens, M. P.; Kingsley, P. D.; Palis, J.; Korsmeyer, S. J.; Daley, G. Q.; Zon, L. I. *Nature* **2003**, *425*, 300–306.
35. Amatruda, J. F.; Shepard, J. L.; Stern, H. M.; Zon, L. I. *Cancer Cell* **2002**, *1*, 229–231.
36. Kalev-Zylinska, M. L.; Horsfield, J. A.; Flores, M. V.; Postlethwait, J. H.; Vitas, M. R.; Baas, A. M.; Crosier, P. S.; Crosier, K. E. *Development* **2002**, *129*, 2015–2030.
37. Langenau, D. M.; Traver, D.; Ferrando, A. A.; Kutok, J. L.; Aster, J. C.; Kanki, J. P.; Lin, S.; Prochownik, E.; Trede, N. S.; Zon, L. I. et al. *Science* **2003**, *299*, 887–890.
38. Patton, E. E.; Widlund, H. R.; Kutok, J. L.; Kopani, K. R.; Amatruda, J. F.; Murphey, R. D.; Berghmans, S.; Mayhall, E. A.; Traver, D.; Fletcher, C. D. et al. *Curr. Biol.* **2005**, *15*, 249–254.
39. Li, L.; Dowling, J. E. *Proc. Natl. Acad. Sci. USA* **1997**, *94*, 11645–11650.
40. Peitsaro, N.; Kaslin, J.; Anichtchik, O. V.; Panula, P. *J. Neurochem.* **2003**, *86*, 432–441.
41. Nicolson, T.; Rusch, A.; Friedrich, R. W.; Granato, M.; Ruppersberg, J. P.; Nusslein-Volhard, C. *Neuron* **1998**, *20*, 271–283.
42. Malicki, J.; Schier, A. F.; Solnica-Krezel, L.; Stemple, D. L.; Neuhauss, S. C.; Stainier, D. Y.; Abdelilah, S.; Rangini, Z.; Zwartkruis, F.; Driever, W. *Development* **1996**, *123*, 275–283.
43. Neuhauss, S. C.; Biehlmaier, O.; Seeliger, M. W.; Das, T.; Kohler, K.; Harris, W. A.; Baier, H. *J. Neurosci.* **1999**, *19*, 8603–8615.

44. Darland, T.; Dowling, J. E. *Proc. Natl. Acad. Sci. USA* **2001**, *98*, 11691–11696.
45. Peterson, R. T.; Shaw, S. Y.; Peterson, T. A.; Milan, D. J.; Zhong, T. P.; Schreiber, S. L.; MacRae, C. A.; Fishman, M. C. *Nat. Biotechnol.* **2004**, *22*, 595–599.
46. Spitsbergen, J. M.; Kent, M. L. *Toxicol. Pathol.* **2003**, *31*, 62–87.
47. Milan, D. J.; Peterson, T. A.; Ruskin, J. N.; Peterson, R. T.; MacRae, C. A. *Circulation* **2003**, *107*, 1355–1358.
48. White, R. E.; Manitpisitkul, P. *Drug. Metab. Dispos.* **2001**, *29*, 957–966.
49. Perez-Soler, R. *Oncology (Williston Park)* **2003**, *17*, 23–28.
50. Bucana, C. D.; Fidler, I. J. *Int. J. Oncol.* **2004**, *24*, 19–24.
51. Pennie, W. D.; Kimber, I. *Toxicol. In Vitro* **2002**, *16*, 319–326.
52. Nuwaysir, E. F.; Bittner, M.; Trent, J.; Barrett, J. C.; Afshari, C. A. *Mol. Carcinogen.* **1999**, *24*, 153–159.
53. Keun, H. C. *Pharmacol. Ther.* **2006**, *109*, 92–106.
54. Dash, A. B.; Williams, I. R.; Kutok, J. L.; Tomasson, M. H.; Anastasiadou, E.; Lindahl, K.; Li, S.; Van Etten, R. A.; Borrow, J.; Housman, D. et al. *Proc. Natl. Acad. Sci. USA* **2002**, *99*, 7622–7627.
55. Zhang, N.; Fang, Z.; Contag, P. R.; Purchio, A. F.; West, D. B. *Blood* **2004**, *103*, 617–626.
56. Zambrowicz, B. P.; Sands, A. T. *Nat. Rev. Drug Disc.* **2003**, *2*, 38–51.
57. Kuhn, R.; Schwenk, F.; Aguet, M.; Rajewsky, K. *Science* **1995**, *269*, 1427–1429.
58. Rubinson, D. A.; Dillon, C. P.; Kwiatkowski, A. V.; Sievers, C.; Yang, L.; Kopinja, J.; Rooney, D. L.; Ihrig, M. M.; McManus, M. T.; Gertler, F. B. et al. *Nat. Genet.* **2003**, *33*, 401–406.
59. Denzel, A.; Hare, K. J.; Zhang, C.; Shokat, K.; Jenkinson, E. J.; Anderson, G.; Hayday, A. *J. Immunol.* **2003**, *171*, 519–523.
60. Justice, M. J.; Noveroske, J. K.; Weber, J. S.; Zheng, B.; Bradley, A. *Hum. Mol. Genet.* **1999**, *8*, 1955–1963.

Biographies

Marco Muda obtained 'Dottore in Scienze Biologiche' degree from Universita degli Studi di Milano, Italy, during 1981–86; he did PhD in biology from Centre Medical Universitaire (CMU), University of Geneva, Switzerland from 1993 to 1997. From 1986 to 1989, he was a fellow at Istituto di Ricerche Farmacologiche Mario Negri, Milan, Italy, and from 1989 to 1991, he was a research fellow at the Biology Department, Massachusetts Institute of Technology, Cambridge, MA, USA. Subsequently, during 1992–93, Muda was a research scientist and interim head of the Molecular Biology Laboratory at the Biotechnology Department of 'Dompe, SPA, Consorzio Biolaq,' L'Aquila, Italy.

During 1997–2000, Muda worked as a postdoctoral research fellow in Prof Dixon's laboratory, Department of Biological Chemistry, The University of Michigan, Ann Arbor, MI, USA; from 2000 to 2003, he was a senior principal investigator, Serono Reproductive Biology Institute, MA, USA. Since 2003, he has been working as a group leader at Serono Research Institute MA, USA.

The fellowships and awards secured by him include: the Fellowship Banca Popolare di Milano, Italy (1989); the Productivity Prize, Dompe Spa, Italy (1992); the Long Term Fellowship from Human Frontier Science Program, Strasbourg, France (1998); and the Long Term Fellowship from NATO/CNR, Rome, Italy (1999).

To his credit, he has 24 peer-reviewed journal publication, one journal review, one book chapter, one patent, 12 scientific meeting presentations and posters.

Sean McKenna is Director of Biotherapeutic Discovery for the Serono Research Institute where he is responsible for overseeing molecular biology, protein expression, protein purification, and pharmacology. The department focuses on the discovery of novel protein-based drugs in Serono's key therapeutic areas.

Dr McKenna came to Serono in 1996 from The Jackson Laboratory where he worked on the development of research models for human disease. While at Serono, Sean has served as head of the In Vivo Pharmacology and Cell Pharmacology Groups, and as the Director of Lead Discovery. In addition, he has led several projects including Ovulation Induction and Protein Engineering. Sean has made major contributions to both the small molecule and protein therapeutic drug discovery efforts and is recognized for his solid background in pharmacology and creative contributions to a wide range of projects.

Dr McKenna received his BS from the University of Massachusetts at Amherst, and his PhD in Biomedical Sciences at the University of Connecticut Medical Center.

Brian G Healey is Head of Lead Discovery at the Serono Research Institute supporting the Cancer and Reproductive Health Therapeutic Areas. The department supports the drug discovery process at SRI from target identification through lead declaration for both small molecules and biotherapeutics. The department is comprised of High Throughput Screening (HTS) for 'Hit' identification, In vitro Pharmacology for 'Hit' optimization, Cellular Pharmacology for cell function, and mechanism-of-action (MOA), Cancer Signaling, and Biomarkers for target validation, signaling studies, and the development of bioanalytical assays to measure target modulation & efficacy in vivo. Prior to joining Serono in 2001, Brian worked at Bristol-Myers Squibb leading the Automation Technologies group developing automated screening solutions. Brian received his PhD from Tufts University in 1996 in analytical chemistry.

Comprehensive Medicinal Chemistry II
ISBN (set): 0-08-044513-6
ISBN (Volume 3) 0-08-044516-0; pp. 647–657

3.31 Imaging

P M Smith-Jones, Memorial Sloan Kettering Cancer Center, New York, NY, USA

3.31.1	**Introduction**	**659**
3.31.2	**Imaging Modalities**	**660**
3.31.2.1	Anatomical Imaging	661
3.31.2.2	Functional Imaging	661
3.31.2.3	Molecular Imaging	661
3.31.3	**Molecular Imaging Probes**	**663**
3.31.3.1	Molecular Probes against Cell Receptors	664
3.31.3.2	Molecular Probes against Apoptosis and Angiogenesis	665
3.31.3.2.1	Apoptosis imaging	665
3.31.3.2.2	Angiogenesis imaging	665
3.31.3.3	Internal Cell Processes: Metabolism and General Viability	665
3.31.3.3.1	Metabolic probes	665
3.31.3.3.2	Amino acid probes	665
3.31.3.3.3	Proliferation probes	665
3.31.3.3.4	Reporter gene transfection and imaging	665
3.31.3.3.5	Radiolabeled biomolecular probes	666
3.31.4	**Small-Animal Imaging Applications**	**667**
3.31.4.1	Imaging Receptor Occupancy and Inhibition of a Characterized Radiotracer by Cold Competitors	667
3.31.4.1.1	Central nervous system imaging	667
3.31.4.1.2	Steroid hormone imaging	667
3.31.4.2	Imaging Drug Effects on Cell Surface – Radiolabeled Annexin	667
3.31.4.3	Imaging Cellular Metabolism with [^{18}F]-Fluorodeoxyglucose	668
3.31.4.3.1	General viability	668
3.31.4.3.2	Cytotoxic agents	668
3.31.4.3.3	Photodynamic therapy	669
3.31.4.3.4	Androgen ablation	669
3.31.4.3.5	Infection	670
3.31.4.3.6	Radiotherapy	670
3.31.4.4	Imaging Cellular Proliferation with [^{18}F]Fluorothymidine	671
3.31.4.5	Amino Acid Metabolism	672
3.31.4.6	Measuring Drug Pharmacokinetics	672
3.31.4.6.1	Direct measurements	672
3.31.4.6.2	Indirect measurements	673
3.31.5	**Conclusions**	**674**
References		**675**

3.31.1 Introduction

In vivo imaging techniques have been around in the clinic for at least three decades, and techniques such as magnetic resonance imaging (MRI), computer tomography (CT), ultrasound, single photon emission computed tomography (SPECT), and positron emission tomography (PET) have become indispensable services in general clinical practice for diagnosing and monitoring a variety of disorders. More recently small dedicated animal-imaging systems have been designed and installed in a preclinical work environment. These imaging systems are becoming an important

Table 1 Small-animal imaging methods

Method	*Imaging Anatomical*	*Functional*	*Molecular*	*Probe*	*Depth/resolution*	*Imaging*
Ultrasound	x			Microbubbles	No limit/50 μm	s
MRI/MRS	x	x	x	Gadolinium and dysprosium chelates Iron oxide particles	No limit/10–100 μm	min/h
CT	x	x		Organic iodides	No limit/20 μm	min
PET		x	x	^{11}C, ^{18}F compounds	No limit/1–2 mm	min
SPECT		x	x	^{99m}Tc, ^{111}In, ^{123}I compounds	No limit/0.5–1 mm	min
Optical						
Fluorescence		x	x	GFP, fluorochromes	< 10 mm/1 μm[a]	s/min
Bioluminescence			x	Luciferin	< 10 mm/1 μm[a]	min
Microscopy		x	x	Dyes	< 1 mm/1 μm[a]	s/min

[a] Values for shallow imaging only. Scattered light can reduce resolution to 10 mm at 10 mm depth.
MRS, magnetic resonance spectroscopy; GFP, green fluorescent protein.

cornerstone in both basic research and drug development. The main advantage of these imaging technologies is that they are noninvasive and can be used repeatedly to monitor disease progression or responses to applied therapies. Alternatively they may be used to study directly the pharmacodynamics of a drug or the effects of a competing drug on its target.

Small animals may be, and were, imaged on clinical machines before the advent of dedicated small-animal imaging devices, but because of the difference in size (20 g versus 80 kg), they gave poor resolution images which were hard to quantify. A list of most small-animal imaging modalities is shown in **Table 1**. It can be seen that they fall into two main groups depending on the depth at which organs may be imaged. All of the imaging modalities except ultrasound rely on the use of electromagnetic radiation. Since tissue absorbs most visible radiation, most optical methods can only be used to interrogate tissue up to 5 or 10 mm below the surface, but at these depths there is a loss of resolution due to photon scattering. In contrast, all of the other methodologies can give good resolution throughout the whole animal.

Until recently, a most common initial evaluation of most therapeutic drugs was comprehensive animal testing to determine the pharmacokinetics. These techniques are considered more suitable because they can determine drug effect on a given target and do not require animal sacrifice or tissue collection. In addition some of these methods can be considered quantitative and hence the number of animals required to study a particular drug can be reduced by imaging the same group of animals over a period of time, rather than by conventional methods which require the killing of multiple groups of animals at numerous time points. Another important feature of such high-end technologies being applied with animals is that researchers are becoming more aware of the molecular mechanisms underlying particular diseases and they can now use these methods with transgenic animals (animals genetically manipulated to exhibit disease symptoms so that effective treatment can be studied). The wide spread availability of animal-imaging systems has opened up new approaches both for screening new therapeutic agents and for collecting preclinical data for eventual investigational new drug (IND) submissions.

3.31.2 Imaging Modalities

Small-animal imaging modalities fall into three main categories: (1) anatomical; (2) functional; or (3) molecular imaging (**Table 1**). Anatomical imaging looks at the imaging of gross anatomy and diseases that produce structural abnormalities. Enhanced images can be obtained by using contrast agents that reflect the vascularization of these organs. Functional imaging looks at the particular function of a target tissue and requires higher spatial resolution as well as the introduction of an imaging or contrast agent to derive physiological parameters for that tissue. Molecular imaging goes one step further and looks at what is happening in a particular tissue at the molecular level and is reliant on the introduction of a tracer to monitor physiological processes occurring within the cell.

3.31.2.1 Anatomical Imaging

The three main techniques of anatomical imaging are CT, MRI, and ultrasound.

CT uses an external x-ray source to obtain image data from different angles around a body, and then uses computer processing of this information to show a cross-section of body tissues and organs. CT imaging can show several types of tissue depending on the attenuation of the x-rays by that tissue. Hence lung, bone, and soft tissue can be clearly differentiated. In addition, if a contrast agent is given, then blood vessels and other tissues heavily perfused by blood can also be seen. Conversely, small tumors in heavily perfused organs can also be imaged as cold spots.[1] Some recently developed systems can image at the subcellular level.[2]

Magnetic resonance images reflect the weighted distribution of protons (i.e., hydrogen nuclei) in water, fat, and tissue.[3] The basic components for MRI are a strong homogeneous static magnetic field, a radiofrequency source, a radiofrequency coil, and a gradient system. Protons possess an inherent magnetic momentum (spin) that induces a small magnetic field with a direction represented by the magnetic moment. Normally the magnetic moments are oriented randomly, but when a static field is applied, they align parallel to the direction of the applied field. However this alignment is not perfect and the nuclei experience a torque, which causes them to spin at a precise frequency dependent on the magnetic field strength (i.e., 200 MHz for a 4.7 T field). To obtain information about these spins, the system has to be excited by a radiofrequency pulse of the same frequency. After excitation, the nuclei return to their ground state by emitting electromagnetic radiation or transferring energy to the surrounding nuclei. During relaxation the longitudinal component of the net magnetization returns to its equilibrium value with a characterizational relaxation time of T1 (milliseconds to seconds) and the transverse component has a characteristic relaxation time of T2 (microseconds to milliseconds). A weaker magnetic field may be superimposed on the main magnetic field to create a magnetic field gradient which enables spatial differentiation of the signals emitted from the region under investigation. The duration and orientation of the field may be altered during data acquisition to give two- and three-dimensional images to reflect the characteristics of the tissues being examined. Up until recently, only small Gd compounds have been used as MRI contrast agents; however, recently, more complex molecules carrying superparamagnetic iron oxide particles have been described.[4]

Ultrasound imaging is a method of obtaining images from inside a body through the use of high-frequency sound waves. The reflected sound wave echoes are recorded and displayed as a real-time visual image. Doppler ultrasound is a special technique for examining blood flow and can be used to study blockages to blood flow or the buildup of plaque inside a vessel. Recent developments in ultrasound have included the introduction of microbubbles as high-sensitivity contrast agents.[5] Ultrasound is a useful way of examining many of the body's internal organs in real time, as opposed to CT or MRI, which can take minutes or up to hours for data collection and detailed image processing.

3.31.2.2 Functional Imaging

Functional imaging requires the introduction of a chemical probe into the living animal. The physical amount of the probe is dependent on the detection methods (**Table 1**). CT and MRI can also be used to perform functional imaging. However they require the administration of large (millimolar) amounts of contrast agents, which may perturb physiological functions. Optical imaging methods require the introduction of moderate amounts of probe (fluorescent 50–10 nmolar; bioluminescent 100–10 nmolar). Other methods of functional imaging such as SPECT or PET rely on the use of very small amounts (subnanomolar) of radiotracers.

3.31.2.3 Molecular Imaging

Molecular imaging, as the name implies, studies cellular responses at the molecular level. Molecular imaging can be performed with various optical systems, SPECT or PET. The optical methods are outlined in **Table 1** and they all use a highly sensitive charge-coupled device camera and differ only in the type of probe given. Bioluminescence imaging relies on the incorporation of the luciferase gene into cells and the subsequent administration of substrate to produce light at these cells. Fluorescence imaging relies on the administration of a bioactive molecule tagged with a fluorophore. The fluorophore is excited by an external light source and then decays to the ground state by emitting light of another frequency. Both methods rely on the detection of visible or near-infrared light at the site of interest.[6] These methods are relatively cheap and suited to high-throughput screening, but they are compromised by the fact that the light is attenuated and scattered by tissue, so the methods are not quantitative.

SPECT relies on the external detection of gamma rays which are produced at the site of interest by the physical decay of a radioisotope. The isotope may be incorporated into a biologically active molecule that is used to track a biological event or process. The gamma rays are energetic enough to pass through the animal tissue with minimal

Table 2 Commercially available SPECT radionuclides

Isotope	*Half-life*	*Gamma energy (%)*	*Notes*
^{67}Ga	3.26 days	93.3 (39.2)	Chelator chemistry
		184.6 (21.2)	
		300.2 (16.8)	
		393.5 (4.68)	
^{99m}Tc	6.01 h	140.5 (89)	Generator-produced. Inexpensive
			Optimal gamma energy
			Chelator chemistry
^{123}I	13.27 h	159.0 (83)	Optimal half-life, low-energy gamma
^{125}I	59.4 days	35.5 (6.68)	Long half-life
^{131}I	8.02 days	284.3 (6.14)	High-energy gamma rays
		364.5 (81.7)	
		637 (7.17)	
^{111}In	2.80 days	171.3 (90)	Chelator chemistry
		245.4 (94)	Low-energy gamma rays

attenuation or scatter. However, in order to reduce the manufacturing costs, the gamma detectors do not normally surround the animals and hence they have to be rotated around the subject to produce the three-dimensional images and hence data collection times can be long. However the general method is very versatile[7] and there are numerous commercially available tracers for imaging and monitoring most human diseases. Various common SPECT isotopes are listed in **Table 2** and their half-lives range from hours to several months. ^{99m}Tc is of particular interest because it is a generator-produced inexpensive radionuclide that can be used to label numerous clinically important tracers. Radiohalogens can be used to label most biomolecules directly,[8,9] and radiometals such as ^{99m}Tc, ^{67}Ga, and ^{111}In rely on the use of bifunctional metal chelators (BMCs).[10]

The three-dimensional imaging of the distribution of a PET probe relies on the use of positron-emitting radionuclides. Upon decay, the positron, ejected from the nucleus, travels a short distance before interacting with an electron to create two characteristic annihilation gamma rays. These gamma rays each have an energy of 0.511 MeV and are emitted at approximately 180° to each other. A ring of detectors around the animal detects the almost simultaneous arrival of the gamma rays. The spatial resolution of PET is dependent on both the mean range of the positron before it interacts with the electron and size of the individual detector elements. In most animal systems the resolution can be as low as 0.5 mm for ^{18}F.[11] Various PET isotopes are listed in **Table 3**. Very-short-lived isotopes such as ^{13}N, ^{15}O, and ^{82}Rb can only be used as simple compounds (i.e., $^{13}NH_3$, $H_2^{15}O$) as perfusion agents. Longer-lived isotopes such as ^{11}C, ^{18}F, $^{75/76}Br$, and ^{124}I can be built into organic structures so that the final probe mimics a natural compound or drug. Other longer-lived isotopes, particularly metallic nuclides, can be chelated by specific donor groups built into the probe. Unlike SPECT, there are several PET isotopes of biological importance (e.g., ^{11}C) and they can be incorporated into the molecular probe so that it is chemically indistinguishable from the native compound. This enables the biological behavior of novel drugs to be investigated noninvasively and can speed up the clinical development of new drugs.

All the methods listed in **Table 1** have applications in imaging drug response, but the optimal timing for such responses varies according to the methods used. In general, nuclear magnetic resonance (NMR), CT, and ultrasound can only determine late changes in the target physiology, whereas the molecular imaging methods can detect very early changes before the cell has responded to the drug as well as any cascade of events which lead to responses in the cell's biological pathways. Because of the large range of probes available, PET and SPECT have become the methods of choice to determine drug effects quantitatively for a wide range of oncological and neurological ligands.

Table 3 PET radionuclides

Isotope	*Half-life*	*Positron energy (MeV)*	*Positron yield (%)*	*Notes*
^{11}C	20.3 min	0.961	99.8	Complex organic molecules possible
^{13}N	10 min	1.198	100	Simple compounds only (i.e., NH_3)
^{15}O	122 s	1.732	99.9	Simple compounds only (i.e., H_2O)
^{18}F	109.8 min	0.634	96.7	Complex organic molecules possible
				Commercially available
				Low energy and high yield of positrons
^{55}Co	17.5 h	1.021	25.6	
		1.113	4.26	
		1.498	46	
^{62}Cu	9.74 min	2.92	93.9	Cu chemistry well documented
		1.75	1.8	
		0.87	1.5	
^{64}Cu	12.7 h	0.653	17.4	Chelator chemistry
				Low positron yield
^{68}Ga	67.8 min	0.822	1.1	Generator-produced
		1.899	88	
^{75}Br	96.7 min	1.580	4.8	Halogen labeling
		1.721	52	
		2.008	7	
^{76}Br	16.2 h	0.871	6.3	Halogen labeling
		0.990	5.2	
		3.381	25.8	
		3.941	6	
^{86}Y	14.7 h	1.033	1.9	Halogen labeling
		1.221	11.9	
		1.545	5.6	
		1.988	3.6	
		3.141	2	
^{94m}Tc	52 min	2.438	67.6	Tc chemistry well documented
^{124}I	4.18 days	1.534	11.7	I chemistry well documented
		2.138	10.8	Commercially available

3.31.3 Molecular Imaging Probes

A number of molecular probes have been designed to report on: (1) the presence of unoccupied cell receptors, which in many diseases are upregulated; (2) the expression of unique proteins on the cell, which are indicative of a disease state; (3) internal cell processes, metabolism, and general viability; and (4) cellular signaling pathways. Standard commercially available isotopes for SPECT imaging are shown in **Table 2** and commonly used PET isotopes are listed in **Table 3**. Only ^{18}F, ^{68}Ge (^{68}Ga), and ^{124}I are currently commercially available in the US, but some academic cyclotron centers are supplying limited amounts of ^{64}Cu and ^{76}Br. For some probes, such as [^{18}F]-fluorodeoxyglucose, it is worthwhile noting that the anesthetic used to sedate the animals during imaging may have an effect on the uptake of the probe.[12]

3.31.3.1 Molecular Probes against Cell Receptors

Cell receptors are an important indication of diseased states in oncology and the study of their function is important in understanding neurological disorders. Classically the main drive to develop clinical PET was to map the presence and function of human neurological receptors. Numerous radioligands have been developed for the dopamine,[13–26] benzodiazepine,[27,28] opioid,[29–31] acetylcholine,[32–35] and adrenergic[36] receptor systems (**Table 4**). More recently, numerous tracers have been developed for receptors associated with cancer. These include steroid hormones (i.e., androgen,[37] estrogen,[38] and progestin[39]) and peptide-based ligands (i.e., somatostatin,[40–42] bombesin,[43] alpha-melanocyte-stimulating hormone,[44] and vasoactive intestinal peptide[45]).

Table 4 Central nervous system ligands

Receptor	*Ligand*	*Specificity*	*Ref.*
Dopamine	[^{99m}Tc]TRODAT-1	DAT	13
D1	[^{11}C]SCH-23390	D1	14
	[^{11}C]SCH-39166	D1	15
	[^{11}C]Methylspiperone	D2, 5HT2	16
	[^{18}F]-*N*-Methylspiroperidol		17
D2	[^{76}Br]Bromospiroperidol	D2, 5HT2	18
	[^{18}F]Fallypride	D2	18
	[^{18}F]Spiperone	D2, S2	20
	[^{11}C]Raclopride	D2	21
	[^{76}Br]Bromolisuride	D2	22
	[^{123}I]Iodobenzamide	D2, D3	23
Serotonin	[^{18}F]Setoperone	5HT2	24
	[^{11}C]Lu29-024	5HT2	25
	[^{11}C]WAY 100635	5HT1A	26
Benzodiazepine	[^{11}C]RO 15.1788		27
	[^{11}C]PK11195	Peripheral	28
Opioid	[^{11}C]Carfentanyl	μ	29
	[^{11}C]Diprenorphine	μ, κ	30
	[^{18}F]3-acetylcyclofoxy	μ	31
Acetlycholine	[^{11}C]MQNB	M1, M2	32
(muscarinic)	[^{11}C]Levetimide	M1, M2	33
	[^{11}C]Scopolamine	M1, M2	34
	N-[^{11}C]methyl]-benztropine	M1, M2	35
Adrenergic	[^{11}C]CGP 12177	β1, β2	36

3.31.3.2 Molecular Probes against Apoptosis and Angiogenesis

3.31.3.2.1 Apoptosis imaging

Apoptosis, a common mechanism of cell death, is an energy-dependent, genetically controlled process by which cell death is activated through an internally regulated suicide program. In contrast to necrotic cell death, apoptosis tends to occur during less intense, chronic tissue insult. Once initiated, apoptosis is characterized by a cascade of morphological and biochemical events, including: phosphatidylserine externalization, cytoplasm shrinkage, chromatin and nucleus condensation, DNA degradation, and fragmentation of the cell into smaller apoptotic bodies by a budding process. Finally, these membrane-enclosed bodies are engulfed and phagocytosed by macrophages and neighboring cells, which then remove the cell fragments without inducing any concomitant inflammatory response. Several imaging agents based on annexin V, which binds to phosphatidylserine, have been evaluated to study drug-induced apoptosis.[46–57]

3.31.3.2.2 Angiogenesis imaging

A variety of therapeutic strategies in oncology have focused on the inhibition of tumor-induced angiogenesis. There is a keen interest in methods that allow noninvasive monitoring of molecular targets involved in angiogenesis which would support information for planning and controlling corresponding therapies. Moreover, such techniques would provide an insight into the formation of new sprouting blood vessels, the involved processes and regulatory mechanisms in patients. At the moment, development of radiotracer-based techniques is mainly concentrated on different targets, including peptidic and nonpeptidic $\alpha v\beta 3$ integrin-binding ligands,[58–60] colchicines,[61] vascular endothelial growth factor,[62] and single-chain antifibronectin antibody fragments.[63]

3.31.3.3 Internal Cell Processes: Metabolism and General Viability

3.31.3.3.1 Metabolic probes

[^{18}F]-Fluorodeoxyglucose ([^{18}F]FDG) is the most commonly used PET tracer in clinical practice.[64] [^{18}F]FDG is taken up into the cell by endothelial glucose transport and is converted to [^{18}F]FDG-6-phosphate. Unlike glucose, which is metabolized further, the phosphorylated [^{18}F]FDG cannot undergo further metabolism and is then trapped in the cell. An increased cellular uptake of [^{18}F]FDG and a higher rate of intracellular phosphorylation are the underlying mechanisms for the high uptake and trapping of FDG by cancer cells. [^{18}F]FDG is commonly used to image tumors in small-animal models.[64–87] Other metabolic probes, such as [^{11}C]acetate,[88,89] [^{11}C]palmitate,[88] and [^{123}I]-fatty acids,[90] have been used.

3.31.3.3.2 Amino acid probes

Increased transport and/or protein synthesis are well-established phenomena. Radiolabeled methionine[91,92] is one of the best established radiolabeled amino acid tracers, but it is limited by the short half-life of ^{11}C. To overcome this, several ^{18}F-labeled analogs such as 2-[^{18}F]fluoro-L-tyrosine ([^{18}F]Tyr),[93] l-3-[^{18}F]fluoro-α-methyl tyrosine ([^{18}F]FMT),[94,95] *O*-(2-[^{18}F]fluoroethyl)-L-tyrosine ([^{18}F]FET),[72,96] and 6-[^{18}F]fluoro-L-DOPA ([^{18}F]DOPA)[97] have been produced.

3.31.3.3.3 Proliferation probes

Since [^{18}F]FDG uptake and protein synthesis are only an indirect reflection of cell proliferation, DNA synthesis is a more reliable indicator of this process. The success of [^{3}H]thymidine to measure cell proliferation in vitro led to the development of [^{11}C]thymidine to measure cell proliferation. However, its short biological and physical half-lives together with the problems associated with circulation metabolites hampered its applications. Consequently a number of alternative compounds have been labeled with ^{18}F, $^{125/131}$I, and ^{76}Br. Of these agents 3′-deoxy-3′-[^{18}F]fluoro-thymidine ([^{18}F]FLT) has been shown to be the most useful.[69,72,75,81–84,98–101]

3.31.3.3.4 Reporter gene transfection and imaging

Reporter gene transfection has become an important area of small-animal imaging. In this system genes are transferred into the animal tissue using a vehicle such as an adenovirus, adeno-associated virus, retrovirus, or liposome, or as naked DNA. Both the reporter gene and the therapy gene are simultaneously inserted into the cell. The reporter gene uses a specific probe to track the expression of the therapy gene. The reporter gene causes the expression of the desired enzyme or receptor by the cell, which then takes up and irreversibly metabolizes specific PET probes, which can indicate the presence of the therapy gene. One specific advantage of this system is that one reporter enzyme can trap a number of reporter probes, thus increasing the instrument signal and detection limits. A number of probes have been designed to be metabolized by the enzyme herpes simplex virus thymidine kinase (HSV-1-TK),[102,103] Na/I symporter,[104] norepinephrine transporter,[105] and somatostatin receptor subtype 2.[106]

3.31.3.3.5 Radiolabeled biomolecular probes

Most laboratories do not have access to a cyclotron and radiochemistry facilities to produce and label with ^{11}C and ^{18}F and so they rely on commercially available isotopes such as $^{67/68}Ga$, ^{99m}Tc $^{123/124/125/131}I$, ^{111}In, and $^{186/188}Re$, although in the US, several cyclotron facilities will supply small quantities of ^{76}Br and ^{64}Cu. All these isotopes may be attached to other probes, which may be used to look at various cellular targets, such as antibodies and peptides. The antibodies can either be used intact or as $F(ab')_2$, Fab′, scfv, or various constructs thereof (i.e., diabodies, triabodies, etc.[107]). An intact antibody will have a biological blood half-life of 3–4 days and a hepatobiliary clearance, whereas the smaller fragments will have a much more rapid blood clearance (<1 day) and be excreted via the kidneys. Since intact antibodies have a slow blood clearance and are divalent, their maximal uptake is normally seen 3–5 days post injection, as opposed to scFv antibody fragments, which can have a maximal uptake within a few hours.

These probes can either be directly labeled at a tyrosine residue with $^{123/124/125/131}I$ using the conventional oxidation methods with iodogen or chloramine T or using indirect iodination with the Hunter-Bolton reagent or allied methods at a lysine residue.[10] Radiometals are attached to biomolecules via BMCs. The BMCs are conjugated in advance to the biomolecule, which is then purified and characterized. The radiometal is then added to the conjugate and the chelate allowed to form. The inherent advantages of using radiometals over radioiodines is that, if the probe is internalized by the cell, the radioiodine is invariably lost from the complex and excreted from the cell, whereas the radiometal is normally trapped inside the cell.

^{99m}Tc and $^{186/188}Re$ are normally attached to biomolecules using chelators which have amine, amide, thiol, phosphine, oxime, and isonitrile donors (for a review of labeling methods, see Liu and Edwards[108]). ^{64}Cu, $^{67/68}Ga$, and ^{111}In may be attached to biomolecules using polyaminopolycarboxylate chelators, of which cyclohexyldiethylenetriamine-pentaacetic acid (CHX-DTPA)[109] and 1,4,7,10-tetraazacyclododecane tetraacetic acid (DOTA) derivatives[110,111] seem the most promising. An example of PET imaging with a ^{64}Cu-labeled intact immunoglobulin G (IgG) is shown in **Figure 1**.

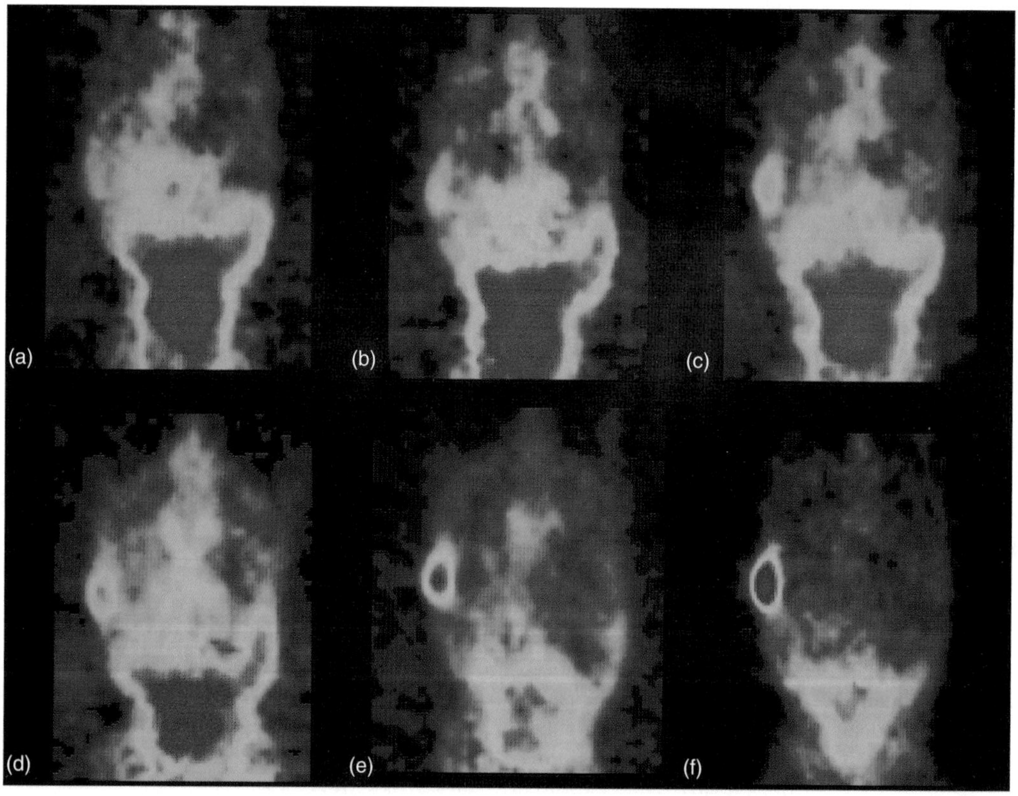

Figure 1 Sequential microPET images of an athymic mouse with a BT474 breast tumor, injected with ^{64}Cu-DOTA-herceptin, at: (a) 1; (b) 2; (c) 4; (d) 6; (e) 17; and (f) 27 h post injection.

3.31.4 Small-Animal Imaging Applications

3.31.4.1 Imaging Receptor Occupancy and Inhibition of a Characterized Radiotracer by Cold Competitors

3.31.4.1.1 Central nervous system imaging

Mukherjee *et al.*[19] have studied [^{18}F]fallypride for evaluating receptor occupancy by the antipsychotic drugs clozapine, risperidone, and haloperidol in both rodents and nonhuman primates. In rodents, clozapine (0.1–100 mg kg^{-1}) competed with [^{18}F]fallypride at all the doses administered. At doses over 40 mg kg^{-1}, clozapine was able to displace all the administered [^{18}F]fallypride. A pseudo biphasic profile of receptor occupancy by clozapine was observed. This behavior was compared with other neuroleptics such as risperidone and haloperidol that exhibited over 90% receptor occupancy at doses over 0.1 mg kg^{-1} and did not exhibit a biphasic nature. Dopamine D2 receptor occupancy in the monkeys was studied using PET after acute subcutaneous doses of the various drugs. At therapeutically relevant doses, clozapine, risperidone, and haloperidol were able to compete significantly with the binding of [^{18}F]fallypride in all brain regions in rhesus monkeys.

[^{11}C]-*N*-Methylspiperone has been used to look at dopamine receptor function in patients with pituitary tumors.[16] The patients were given a baseline study and then treated with the dopamine antagonist haloperidol before a repeat study. The most marked uptake and the largest effect of haloperidol pretreatment were seen in a patient with a hormonally active prolactinoma.

[^{18}F]GBR 13119 has been studied in rats and monkeys.[112] In rats, blockage of specific binding was demonstrated with dopamine reuptake inhibitors, but no effect was observed for pretreatment with serotonin or norepinephrine reuptake inhibitors. The striatum of living primates was also imaged using PET and intravenous administration of [^{18}F]GBR 13119.

[^{18}F]spiperone ([^{18}F]SP) has been evaluated in baboons.[20] Three to six studies were performed in each of five male baboons. Each animal was pretreated with either ketanserin (serotonergic (5HT2)), eticlopride (dopaminergic (D2)), or unlabeled SP to compete with [^{18}F]SP for specific binding sites. Sequential PET scans and arterial blood samples were collected for 3 h after intravenous injection of [^{18}F]SP. There was no detectable in vitro or in vivo specific binding of SP in cerebellum. The specific binding of SP in striatal tissue in vitro was approximately 74% to D2 sites and 26% to 5HT2 sites, whereas ketanserin displaced all specific binding in frontal cortex. In close agreement, specific binding measured in vivo with PET revealed that 68% of apparent striatal binding could be blocked by pretreatment with eticlopride, and 34% by ketanserin.

3.31.4.1.2 Steroid hormone imaging

3.31.4.1.2.1 [^{18}F]-Fluoroestradiol ([^{18}F]FES)

Aliaga *et al.*[68] studied the effect of chemotherapy and hormone therapy on various breast cancer models using [^{18}F]FDG and 16α-[^{18}F]fluoroestradiol ([^{18}F]FES). Tumor metabolic activity was estimated from the relative uptake (% injected dose g^{-1}) of [^{18}F]FDG uptake, whereas estrogen receptor content was determined from [^{18}F]FES retention. ^{18}F activity values were obtained by small-animal PET imaging and confirmed by tissue sampling and radioactivity counting. Reliable uptake measurements could be obtained for tumors of 0.2 ml or over. The Balb/c MC7-L1 and MC4-L2 grew well and showed good uptake of both FDG and FES. Chemotherapy and hormone therapy delayed the growth of MC7-L1 and MC4-L2 tumors and it was concluded that these tumors were suitable models for the monitoring of estrogen receptor+ breast cancer therapy using small-animal PET imaging.

3.31.4.2 Imaging Drug Effects on Cell Surface – Radiolabeled Annexin

Collingridge *et al.*[46] used ^{124}I and ^{125}I-labeled annexin V to study apoptotic cells and tumors. Radiation-induced fibrosarcoma (RIF-1) cells induced to undergo apoptosis in vitro showed a drug concentration-dependent increased binding of [^{125}I]-annexin V and the radiolabeled *N*-succinimidyl-3-benzoic acid annexin derivative, [^{125}I]SIB-annexin V, and an increase in terminal deoxynucleotidyl transferase-mediated nick end labeling (TUNEL)-positive cells and a decrease in clonogenic survival. In RIF-1 tumor-bearing mice, rapid distribution of [^{125}I]SIB-annexin V-derived radioactivity to kidneys was observed and the radiotracer accumulated in urine. The binding of [^{125}I]SIB-annexin V to RIF-1 tumors increased by 2.3-fold at 48 h after a single intraperitoneal injection of 5-fluorouracil (5-FU: 165 mg kg^{-1} body weight), compared to a 4.4-fold increase in TUNEL-positive cells measured by immunostaining. PET images with both radiotracers demonstrated intense localization in the kidneys and bladder. Unlike [^{124}I]SIB-annexin V,

[^{124}I]-annexin V also showed localization in the thyroid region, presumably due to deiodination of the radiolabel. The authors concluded that [^{124}I]SIB-annexin V was an attractive candidate for in vivo imaging of apoptosis by PET.

Mandl *et al.*[53] studied the uptake of [^{99m}Tc]-annexin V in syngeneic orthotopic murine BCL1 lymphoma model using in vivo bioluminescence imaging (BLI) and small-animal SPECT. BCL1 cells labeled for fluorescence and bioluminescence assays (BCL1-gfp/luc) were injected into mice at a dose that leads to progressive disease within 2–3 weeks. Tumor response was followed by BLI and SPECT before and after treatment with a single dose of 10 mg kg^{-1} doxorubicin. Biodistribution analyses revealed a biphasic increase of annexin V uptake within the tumor-bearing tissues of mice. An early peak, occurring before actual tumor cells loss, was observed between 1 and 5 h after treatment, and a second longer sustained rise from 9 to 24 h after therapy. The multimodality imaging revealed the temporal patterns of tumor cell loss and annexin V uptake, giving a better understanding of the timing of radiolabeled annexin V uptake for its development as a marker of therapeutic efficacy.

Subbarayan *et al.*[57] studied the uptake of [^{99m}Tc]-annexin V and evaluated its usefulness in detecting apoptosis that occurs during tumor shrinkage after photodynamic therapy (PDT). RIF-1 tumors were grown in C3H mice and treated with PDT. [^{99m}Tc]-annexin V tumor uptake was investigated at 2, 4, and 7 h after PDT by autoradiography. At all time points, [^{99m}Tc]-annexin V was clearly imaged in the PDT-treated tumors, whereas the untreated tumors showed no uptake of the radiolabeled compound. Histopathology and immunohistochemistry of PDT-treated tumors confirmed the evidence of apoptosis compared with untreated tumors. They concluded that the detection of apoptosis using [^{99m}Tc]-annexin V in tumor tissue in living animals after PDT treatment was a novel technique which could be used as a noninvasive means of detecting and serially imaging tissues undergoing apoptosis after cancer treatment protocols in humans.

Yagle *et al.*[113] used an ^{18}F-labeled annexin V to study cycloheximide-induced liver apoptosis in rats. The pre-treatment of rats with cycloheximide resulted in a three- to ninefold increase in uptake of [^{18}F]-annexin V in the liver of treated animals at 2 h, compared with controls. By morphologic analysis, treated livers showed a three- to sixfold higher level of apoptosis than controls, with higher levels also seen with longer exposure to cycloheximide. TUNEL assays performed on liver slices showed that cycloheximide induced a five- to eightfold increase in the number of TUNEL-positive nuclei. The TUNEL results correlated with the uptake of [^{18}F]-annexin V in dissected liver tissue, with an r^2 value of 0.89. Biodistribution analysis of normal rats showed highest uptake of [^{18}F]-annexin V in the kidneys and urinary bladder, indicating rapid renal clearance of [^{18}F]-annexin V metabolites.

Ke *et al.*[114] studied the uptake of [^{111}In]-DTPA-PEG-annexin V by MDA-MB-468 tumors in mice before and after poly(L-glutamic acid)-paclitaxel and/or C225 antiepidermal growth factor receptor (EGFR) antibody. Tumor apoptotic index increased from 1.67%±0.31% at baseline to 7.60%±0.72% and 11.07%±1.81%, respectively, 4 days after treatment with poly(L-glutamic acid)-paclitaxel or combined poly(L-glutamic acid)-paclitaxel and C225. Tumor uptake (percentage of injected dose per gram of tumor [%ID g^{-1}]) of PEGylated [^{111}In]-DTPA-PEG-annexin 4 days after treatment was significantly higher in tumors treated with poly(L-glutamic acid)-paclitaxel (10.76±1.38%ID g^{-1}; $P=0.001$) and with combined poly(L-glutamic acid)-paclitaxel and C225 (9.84±2.51%ID g^{-1}; $P=0.029$) than in nontreated tumors (6.14±0.67%ID g^{-1}), resulting in enhanced visualization of treated tumors. [^{111}In]-DTPA-PEG-annexin V distributed into the central zone of tumors, whereas [^{111}In]-DTPA-annexin V was largely confined to the tumor periphery. Furthermore, uptake of [^{111}In]-DTPA-PEG-annexin V by tumors correlated with apoptotic index ($r=0.87$, $P=0.02$). Increase in tumor uptake of the nonspecific PEGylated protein [^{111}In]-DTPA-PEG-ovalbumin was also observed after poly(L-glutamic acid)-paclitaxel treatment (55.6%), although this increase was less than that observed for [^{111}In]-DTPA-PEG-annexin V (96.7%).

3.31.4.3 Imaging Cellular Metabolism with [^{18}F]-Fluorodeoxyglucose

3.31.4.3.1 General viability

Palmedo *et al.*[79] compared the uptake of [^{18}F]FDG, [^{99m}Tc]-methoxyisobutylisonitrile (MIBI), and [^{99m}Tc]-dimercaptosuccinic acid (DMSA) in immunosuppressed rats implanted with HH-16 clone 4 mammary tumor cells. Tumor uptake was highest for [^{18}F]FDG > [^{99m}Tc]-DMSA > [^{99m}Tc]-MIBI. The uptake ratios (tumor to muscle) correlated well with the ratios calculated by residue on ignition analysis, determined by imaging. The researchers concluded that [^{18}F]FDG revealed the best uptake and imaging properties and may be the radiopharmaceutical of choice for routine breast cancer imaging.

3.31.4.3.2 Cytotoxic agents

[^{18}F]FDG has been used to monitor the effects of a cytotoxic agent, combretastatin A-4, on liver metastases in B6D2F1 mice.[87] In untreated mice with liver metastases, a strong correlation ($r^2=0.98$) was found between the quantitative estimates of [^{18}F]FDG uptake obtained by analysis of PET images and those obtained from ex vivo assay of liver plus metastases excised immediately after imaging. The effective limit of resolution was in livers containing a number of small metastases (range 8–14) with a single volume equivalent of approximately 200 mm^3. A single intraperitoneal dose of combretastatin A-4 resulted in an average 30% volume destruction of metastatic mass by 24 h after administration.

Spaepen *et al.* studied the effect of the chemotherapy agent Endoxan on mice with 15–20 mm Daudi tumors.[80] At 1–3 days postinjection of Endoxan, reductions in [^{18}F]FDG uptake and viable tumoral cell fraction were observed and these reductions preceded changes in tumor size. By 8–10 days postinjection, [^{18}F]FDG uptake had stabilized despite a further reduction in viable tumoral cell fraction. At these time points a major inflammatory response was observed. At day 15, an increase in viable tumor cells was again observed and this was accurately predicted by an increase in [^{18}F]FDG uptake, while the tumor volume remained unchanged. In contrast to variations in tumor volume, the authors found that [^{18}F]FDG was a good marker for chemotherapy response monitoring, but noted that optimal timing was crucial since a transient increase in stromal reaction may result in overestimation of the fraction of viable cells.

Barthel *et al.*[69] studied the effect on 5-FU on C3H/Hej mice bearing the radiation-induced fibrosarcoma 1 tumors using [^{18}F]FLT and [^{18}F]FDG. Tumor [^{18}F]FLT uptake decreased at 24 and 48 h after 5-FU treatment (47.8 $\pm$ 7.0 and 27.1 $\pm$ 3.7%). The drug-induced reduction in tumor [^{18}F]FLT uptake was found to be significantly more pronounced than that of [^{18}F]FDG. The decrease in tumor [^{18}F]FLT uptake correlated with the proliferating cell nuclear antigen (PCNA)-labeling index ($r = 0.71$, $P = 0.031$) and tumor volume changes after 5-FU treatment ($r = 0.58$, $P = 0.001$), which was explained by changes in catalytic activity but not translation of TK1 protein. In comparison with untreated animals, TK1 levels were lower at 24 h posttreatment (78.2 $\pm$ 5.2%) but higher at 48 h (141.3 $\pm$ 9.1%, $P < 0.001$).

Miyagawa *et al.*[76] studied the effect of the effect of ganciclovir (GCV) treatment on Fischer 344 rats with intracerebral HSV-TK-transduced RG2TK+ xenografts. Measurements were taken of [^{14}C]aminocyclopentane carboxylic acid ([^{14}C]APAC) and [^{67}Ga]-DTPA plasma clearance ($K_{(1)}$), [^{14}C]ACPC transport (partial differential $K_{(1)}$), relative glucose utililization (R), and normalized radioactivity (% dose g^{-1}) in tumor and brain tissues. GCV treatment reduced partial differential $K_{(1)}$ and % dose g^{-1} of [^{14}C]ACPC in RG2TK+ xenografts to approximately 30% of that in nontreated animals (from 34 $\pm$ 9 (mean $\pm$ sd) to 9.5 $\pm$ 2.7 μL min^{-1} g^{-1} and from 0.28 $\pm$ 0.09 to 0.11 $\pm$ 0.04% dose g^{-1}, respectively). GCV had a significant but substantially smaller effect than toxicity on glucose utilization and little or no effect on passive vascular permeability of RG2TK+ xenografts. These differences could not be explained by differences in plasma amino acid or glucose concentration at the time of the study. Histology revealed a large fraction of dead tumor cells and only a sparse distribution of apoptotic cells in GCV-treated tumors. Many CD34-positive endothelial cells in GCV-treated tumors showed only weak or marginal L amino acid transport 1 (LAT1) staining, whereas CD98 staining remained unchanged. Survival was significantly increased by GCV treatment from 18 $\pm$ 4 to 56 $\pm$ 17 days. The researchers concluded that amino acid transport imaging may be a good surrogate paradigm to monitor treatment response of brain tumors.

3.31.4.3.3 Photodynamic therapy

Moore *et al.* investigated studies of PDT with hematoporphyrin ester, and 630 nm light on subcutaneous T50/80 mouse mammary tumors using [^{18}F].[77] [^{18}F]FDG uptake into untreated control tumors was 3.8% of the injected activity and the uptake of [^{18}F]FDG by treated tumors decreased by 0.7% for every 100 mm^3 reduction in remaining viable histological volume. Outcome was further compared with that measured by: (1) T2-weighted proton imaging on a 4.7 T MRI system; and (2) histological analysis of subsequently sectioned tumors. PET using [^{18}F]FDG described the absolute volume of surviving tumor histological mass to the same degree as high-resolution MRI, but quantitatively described, at early times, the extent of tumor destruction by PDT.

Lapointe *et al.*[74] used [^{18}F]FDG with small-animal PET to assess early tumor response after PDT in mice.[74] PDT was performed with Photofrin, which has been approved for clinical use, and disulfonated aluminum phthalocyanine, which is a second-generation drug. The study was performed using a mouse model implanted with two contralateral murine mammary tumors (5 mm diameter × 2.5 mm thickness) on the back. Only one tumor was subjected to PDT, whereas the other tumor served as a control. [^{18}F]FDG tumor uptake after 15 min was a direct measurement of tumor metabolism and demonstrated the relative efficacy of the two PDT drugs to induce tumor necrosis through indirect vascular stasis or direct cell kill.

3.31.4.3.4 Androgen ablation

Haaparanta *et al.*[71] studied the effect of antiestrogen toremifene in rats with 7,12-dimethylbenzanthracene (DMBA)-induced mammary carcinomas. [^{18}F]FDG uptake at 15 min p.i. correlated better with the fractional change in tumor volume ($r = 0.284$ (untreated) and $r = 0.721$ (treated)) and at 240 min ($r = 0.932$ (untreated)), than at 45 min

($r = -0.137$ (untreated) and $r = 0.265$ (treated)). Inverse relations were found for the fraction of unmebolized FDG and change in tumor volume ($r = 0.070$ (45 min) and $r = -0.872$ (240 min) for untreated tumors and ($r = -0.963$ (15 min) and $r = -0.715$ (45 min)) for treated tumors. No significant therapy-induced morphometrical changes were observed.

Carnochan and Brooks[70] studied the effect of androgen ablation on the uptake of various tracers in the hormone-responsive rat mammary carcinoma OES.HR1. From 4 days after tumor growth arrest induced by estrogen ablation, a sustained fall in tumor cell proliferation was demonstrated, which was associated with reduced tumor uptake of [^{131}I] 5-iododeoxyuridine ([^{131}I]-IUdR), [^{3}H]2-deoxy-D-glucose ([^{3}H]DG), and [^{99m}Tc]hexylmethylpropylene amineoxine ([^{99m}Tc]-HMPAO). Whereas reduced levels of tumor [^{3}H]DG could be accounted for by changes in blood flow, this was not the case for [^{131}I]IUdR, which was found to be closely related to percentage S-phase cells within tumor ($r = 0.73$, $P < 0.002$). This group concluded that [^{124}I]IUdR is a promising alternative to [^{18}F]FDG for the early assessment by PET of tumor response to treatments directed at specific targets.

Agus *et al.*[67] studied the effect of androgen withdrawal [^{18}F]FDG uptake in a CWR22 prostate model. They found a significant decrease in [^{3}H]DG accumulation in the tumors at 48 h after androgen withdrawal (62% of baseline: 95% confidence interval: 0.59, 0.65) and a maximum decline at day 10 (38% of baseline). Using PET, parallel changes in tumor metabolism were demonstrated and preceded changes in tumor volume and marked declines in serum prostate-specific antigen. [^{3}H]DG accumulation returned to near baseline on reintroduction of androgen. The decrease in [^{3}H]DG accumulation was associated with a decline in the proportion of tumor cells in active cell cycle from > 60% to < 5% at 7–10 days after androgen withdrawal. No increase in the proportion of tumor cells undergoing apoptosis was observed during this time period, implying an arrest in a G_0/early G_1 state. [^{3}H]DG accumulation in tumor cells, measured directly by PET, correlated with androgen changes in the host.

3.31.4.3.5 Infection

Kaim *et al.*[72] compared the uptake of [^{18}F]FET with that of [^{18}F]FDG in activated inflammatory white blood cells. Unilateral thigh muscle abscesses were induced in 11 rats by intramuscular inoculation of 0.1 ml of a bacterial suspension. In these animals, areas with increased [^{18}F]FDG uptake corresponded to cellular inflammatory infiltrates mainly consisting of granulocytes. The standardized uptake value (SUV) was calculated to be 4.08 ± 0.65 (mean ± sd). The uptake of [^{18}F]FET in activated white blood cells was not increased: the SUV of the abscess wall, at 0.74 ± 0.14, was even below that of contralateral muscle. The authors found that the low uptake of [^{18}F]FET in nonneoplastic inflammatory cells promised a higher specificity for the detection of tumor cells than [^{18}F]FDG, since the immunological host response will not be labeled and inflammation can be excluded.

van Waarde *et al.*[82] studied the uptake of [^{18}F]FLT and [^{18}F]FDG in Wistar rats with both C6 rat glioma tumors and sterile inflammation. At 24 h after turpentine injection, the rats received an intravenous bolus of either radiotracer. Tumor-to-muscle ratios of [^{18}F]FDG at 2 h after injection (13.2 ± 3.0) were higher than those of [^{18}F]FLT (3.8 ± 1.3). [^{18}F]FDG showed high physiologic uptake in brain and heart, whereas [^{18}F]FLT was avidly taken up by bone marrow. [^{18}F]FDG accumulated in the inflamed muscle, with 4.8 times higher uptake in the affected thigh than in the contralateral healthy thigh, in contrast to [^{18}F]FLT, for which this ratio was not significantly different from unity (1.3 ± 0.4). The researchers concluded that [^{18}F]FLT had a higher tumor specificity.

Wyss *et al.*[86] studied the effect of ceftriaxone antibiotic treatment on [^{18}F]FDG uptake in experimental soft-tissue infections in rats. PET scans were performed in treated and nontreated animals at days 3, 5, and 6 after inoculation of the infection. Additional autoradiography was performed at day 7 and in three animals at day 11. The difference of [^{18}F]FDG uptake on day 5 (after 3 days of antibiotic treatment) between both groups proved to be significant (df = 6; $T = 2.52$; $P = 0.045$). [^{18}F]FDG uptake determined on the other days did not reveal significant difference between the two groups. This group concluded that the effect of antibiotic treatment on [^{18}F]FDG uptake is less evident than that reported for therapy monitoring of cancer treatment.

3.31.4.3.6 Radiotherapy

Kubota *et al.*[73] studied the effects of radiotherapy on the uptake of a number of tracers in an AH109A model. Metabolic tracers for glucose, amino acid, nucleic acid metabolism, [^{18}F]FDG, [^{14}C]Met, [^{3}H]Thd, and [^{18}F]FdUrd and conventional ^{67}Ga citrate were compared. [^{18}F]FDG showed a large uptake change and a steady response to radiotherapy, which was similar to ^{67}Ga. [^{18}F]FdUrd showed a rapid decrease, but the range of change in uptake was narrow. [^{3}H]Thd and [^{14}C]Met showed the most rapid response to irradiation and a high sensitivity for monitoring radiotherapy.

Abe *et al.*[65] studied the effect of external 20 Gy radiotherapy [^{18}F]FDG uptake in mouse mammary carcinoma MM48, FM3A, and rat hepatoma AH109A. Following 20 Gy irradiation of the radioresistant tumor (MM48), the

$[^{18}F]$FDG uptake ratio was found to be unchanged, whereas in radiosensitive tumors (FM3A) the $[^{18}F]$FDG uptake ratio was 0.37, the relative tumor volume was 0.31, and the calculated total tumor uptake was 0.11 on the eighth day after irradiation. The total tumor uptake was lower than the relative tumor volume. AH109A began to regrow after 10 Gy irradiation, accompanied by elevated uptake of $[^{18}F]$FDG on the seventh day. These authors concluded that $[^{18}F]$FDG uptake by tumor was a good marker of radiotherapeutic effects as well as relapse of cancers and is more sensitive than morphological methods.

Ohira *et al.*[78] studied the effect of 20 Gy radiotherapy on $[^{99m}Tc]$-MIBI and 2-deoxy-D-[1–^{14}C]-glucose in three animal models of breast cancer. The FM3A, MM48, and Ehrlich tumors have different growth rates and radiosensitivities. Uptake of $[^{18}F]$FDG, but not of $[^{99m}Tc]$-MIBI, correlated significantly with growth rates. Compared with $[^{14}C]$DG, $[^{99m}Tc]$-MIBI accumulated more in cancer cells and less in infiltrating fibroblasts and macrophages in all tumor models. Irradiation significantly decreased $[^{99m}Tc]$-MIBI uptake, but a rapid increase was noted at recurrence on day 7. Changes in $[^{18}F]$FDG uptake were not significant at recurrence. Microvessel density in tumor tissue correlated significantly with $[^{99m}Tc]$-MIBI uptake on double-tracer macroautoradiography. The researchers concluded that the accumulation of $[^{99m}Tc]$-MIBI in cancer cells is preferential and can be used as a sensitive marker to examine the response to radiotherapy and that angiogenesis seems to enhance the accumulation of $[^{99m}Tc]$-MIBI in tumors.

3.31.4.4 Imaging Cellular Proliferation with $[^{18}F]$Fluorothymidine

Waldherr *et al.*[84] compared $[^{18}F]$FDG and $[^{18}F]$FLT uptakes in mice with A431 tumors treated with the ErbB-selective kinase inhibitor PKI-166. They observed a markedly lowered tumor $[^{18}F]$FLT uptake within 48 h of drug exposure; within 1 week $[^{18}F]$FLT uptake decreased by 79%. $[^{18}F]$FLT uptake by the xenografts significantly correlated with the tumor proliferation index as determined by PCNA staining ($r = 0.71$). Changes in $[^{18}F]$FLT uptake did not reflect inhibition of ErbB kinase activity itself but, rather, the effects of kinase inhibition on tumor cell proliferation. Tumor $[^{18}F]$FDG uptake generally paralleled the changes seen for $[^{18}F]$FLT. However, the baseline signal was significantly lower than that for $[^{18}F]$FLT. This group concluded that $[^{18}F]$FLT PET provides noninvasive, quantitative, and repeatable measurements of tumor cell proliferation during treatment with ErbB kinase inhibitors and provides a rationale for using this technology in clinical trials of kinase inhibitors.

Oyama *et al.*[99] studied the effects of androgen ablation uptake of $[^{18}F]$FLT in the androgen-dependent human prostate tumor, CWR22, in athymic mice. The microPET study with dynamic imaging showed that $[^{18}F]$FLT uptake in blood reached its plateau within 1 min and was rapidly cleared, whereas $[^{18}F]$FLT uptake in tumor reached its plateau in 30 min and remained for up to 60 min. MicroPET using $[^{18}F]$FLT successfully imaged the implanted CWR22 tumor in the mice at both 1 and 2 h after injection. At 2 h after injection, the $[^{18}F]$FLT uptake in tumor was 0.69%ID g^{-1} and was the highest activity of all organs measured. There was a marked reduction of $[^{18}F]$FLT uptake in tumor after castration or diethylstilbestrol treatment; however, there were no differences in $[^{18}F]$FLT uptake in the tumor in the control group. These writers concluded that $[^{18}F]$FLT is a useful tracer for detection of prostate cancer in an animal model and that $[^{18}F]$FLT has the potential for monitoring the therapeutic effect of androgen ablation therapy in prostate cancer.

Sugiyama *et al.*[81] studied the effect of 20 Gy x-ray irradiation on C3H/He mice bearing SCCVII tumors as well as PDT on BALB/c nu/nu mice with HeLa tumors, using $[^{18}F]$FLT, $[^{18}F]$FDG, $[^{14}C]$deoxyglucose ($[^{14}C]$DG), and $[^{3}H]$thymidine ($[^{3}H]$Thd). Expression of PCNA was determined in untreated and treated tumors. Tumor volumes decreased to 39.3% ± 22.4% at 7 days after radiotherapy. The PCNA labeling index was reduced in x-ray-irradiated tumors (control, 53.2% ± 8.7%; 6 h, 38.5% ± 5.3%; 24 h after radiotherapy, 36.8% ± 5.3%). $[^{18}F]$FLT uptake in tumor expressed as %ID g^{-1} decreased significantly at 6 h and remained low until 3 days after radiotherapy (control, 9.7 ± 1.2%ID g^{-1}; 6 h, 5.9 ± 0.4%ID g^{-1}; 24 h, 6.1 ± 1.3%ID g^{-1}; 3 days after radiotherapy, 6.4 ± 1.1%ID g^{-1}). $[^{18}F]$FDG uptake tended to gradually decrease but a significant decrease was found only at 3 days (control, 12.1 ± 2.7%ID g^{-1}; 6 h, 13.3 ± 2.3%ID g^{-1}; 24 h, 8.6 ± 1.8%ID g^{-1}; 3 days after radiotherapy, 6.9 ± 1.2%ID g^{-1}). PDT resulted in a reduction of the PCNA labeling index (control, 82.0% ± 8.6%; 24 h after PDT, 13.5% ± 12.7%). Tumor uptake of $[^{18}F]$FLT decreased (control, 11.1 ± 1.3%ID g^{-1}; 24 h after PDT, 4.0 ± 2.2%ID g^{-1}), whereas $[^{18}F]$FDG uptake did not decrease significantly after PDT (control, 3.5 ± 0.6%ID g^{-1}; 24 h after PDT, 2.3 ± 1.1%ID g^{-1}). Changes in the uptake of $[^{18}F]$FLT and $[^{18}F]$FDG were similar to those of $[^{3}H]$Thd and $[^{14}C]$DG, respectively. The authors concluded that changes in $[^{18}F]$FLT uptake after radiotherapy and PDT were correlated with those of $[^{3}H]$Thd and the PCNA labeling index. The decrease in $[^{18}F]$FLT uptake after treatments was more rapid or pronounced than that of $[^{18}F]$FDG.

3.31.4.5 Amino Acid Metabolism

Deehan *et al.*[115] studied the uptake of the radioiodinated amino acid analog l-3-iodo-alpha-methyl tyrosine (IMT) by HSN and OES.HR1 tumors in hooded rats. Maximum tumor uptake was seen at 15 min postinjection, although an improved tumor-to-brain ratio was seen at 24 h. Brain uptake of IMT was also found to be substantially reduced by competition with another large neutral amino acid phenylalanine; however, relatively less effect was seen in tumor, and in skeletal muscle no change in IMT uptake was observed. Similar levels of IMT uptake were found in the OES.HR1 tumor during growth supplemented by exogenous estrogen. Following arrest of tumor growth by removal of the estrogen stimulus, IMT uptake was seen to fall from 1.7% to 1.0% of the injected dose per gram: this was matched by a fall in tumor blood flow as estimated by [^{99m}Tc]-hexamethylpropylene amine oxime distribution. This group concluded that IMT uptake was more strongly influenced by blood flow than cell proliferation and that intratumoral distribution of IMT is principally determined by diffusion.

3.31.4.6 Measuring Drug Pharmacokinetics

3.31.4.6.1 Direct measurements

Brix *et al.*[116] radiolabeled and studied the anticancer drug 5-FU in ACI and Buffalo rats with transplanted MH3924A and TC5123 Morris hepatomas. The potentials and limitations of dynamic ^{18}F PET and metabolic ^{19}F MRI examinations for noninvasive 5-FU monitoring were investigated. Selective 5-[^{19}F]FU and [^{19}F]alpha-fluoro-beta-alanine (FBAL) MR images were acquired 5 and 70 min after 5-FU injection using a chemical shift selective (CHESS) MRI sequence. After administration of 5-[^{18}F]FU, the kinetics of the regional 5-[^{18}F]FU uptake was measured by dynamic PET scanning over 120 min. To allow a comparison between PET and MRI data, SUVs were computed at the same points in time. The TC5123 hepatoma showed a significantly ($P < 0.002$) higher mean SUV at 5 and 70 min post-5-FU injection than the MH3924A cell lines, whereas there were no significant differences between the mean SUV measured in the liver of both animal populations. In contrast to the PET data, no significant differences in the mean 5-[^{19}F]FU and [^{19}F]FBAL MR signal values in the tumor of both models were observed. The MR images, however, yielded the additional information that 5-FU is converted to FBAL only in the liver and not in thc hepatomas.

Baling *et al.*[117] also used ^{18}F to measure the pharmacokinetics of this chemotherapeutic drug. Pretreatment with eniluracil (5-ethynyluracil) was used to prevent catabolism of FU. Anesthetized rats bearing a subcutaneous rat colorectal tumor were given eniluracil or placebo and injected intravenously 1 h later with [^{18}F]FU or [^{3}H]FU. In the ^{18}F studies, dynamic PET image sequences were obtained 0–2 h after injection. Tumors were excised and frozen at 2 h and then analyzed for labeled metabolites by high-performance liquid chromatography. Biodistribution of radiolabel was determined by direct tissue assay. Eniluracil improved tumor visualization in PET images. With eniluracil, tumor-standardized uptake values ((activity/g)/(injected activity/g body weight)) increased from 0.72 ± 0.06 (mean$\pm$sem; $n=6$) to 1.57 ± 0.20 ($n=12$; $P < 0.01$), and tumor uptake increased by factors of 2 or more relative to plasma ($P < 0.05$) and bone, liver, and kidney ($P < 0.01$). Without eniluracil ($n=5$), 57%$\pm$4% of recovered radiolabel in tumor at 2 h was on catabolites, with the rest divided among FU (2%$\pm$1%), anabolites of FU (38%$\pm$7%), and unidentified peaks (4%$\pm$2%). With eniluracil ($n=8$), catabolites, FU, and anabolites comprised 2%$\pm$1%, 41%$\pm$5%, and 57%$\pm$4%, respectively, of the recovered radiolabel in tumors. These authors concluded that eniluracil increased tumor accumulation of ^{18}F relative to host tissues and fundamentally changed the biochemical significance of that accumulation. Since the catabolism was suppressed, the tumor radioactivity reflected the therapeutically relevant aspect of FU pharmacokinetics – namely, uptake and anabolic activation of the drug.

Osman *et al.*[118] studied ^{11}C-labeled analogs of the tricyclic carboxamide *N*-[2-(dimethylamino)ethyl]acridine-4-carboxamide (DACA), a DNA-intercalating agent capable of inhibiting both topoisomerases I and II, in mice bearing U87MG tumors. The acridine DACA, the phenazine [^{11}C](9-methoxyphenazine-1-carboxamide (SN 23490), the pyridoquinoline [^{11}C]2-(4-pyridyl)quinoline-8-carboxamide (SN 23719), and the dibenzodioxin [^{11}C]dibenzo[1,4]diox-in-1-carboxamide (SN 23935) were found to be cytotoxic in in vitro assays with an IC_{50} of 1.4 ± 1.8, 0.4 ± 0.6, 1.3 ± 1.6, and 24 ± 36 μM, respectively, in HT29, U87MG, and A375M cell lines. Ex vivo biodistribution studies with carbon-11 radiolabeled compounds in mice bearing human tumor xenografts showed rapid clearance of ^{11}C-radioactivity (parent drug and metabolites) from blood and the major organs. Rapid hepatobiliary clearance and renal excretion were also observed. There was low (<5%ID g^{-1}) and variable uptake of ^{11}C-radioactivity in three tumor types for all of the compounds. Tumor (U87MG) to blood ^{11}C-radioactivity for [^{11}C]DACA, SN 23490, SN 23719, and SN 23935 at 30 min were 2.9 ± 1.1, 2.3 ± 0.6, 2.6 ± 0.6, and 0.7 ± 0.2, respectively. For SN 23719, the distribution of ^{11}C-radioactivity in normal tissues and tumors determined ex vivo was in broad agreement with that determined in vivo by whole-body PET scanning. [^{11}C]DACA was rapidly and extensively metabolized to several plasma metabolites and a major tumor metabolite. In contrast, [^{11}C]SN 23935, [^{11}C]SN 23490, and [^{11}C]SN 23719 showed less extensive metabolism. In the tumor samples, the parent [^{11}C]DACA and [^{11}C]SN 23935 represented between 0.3 and 1.5%ID g^{-1}, whereas [^{11}C]SN 23490 and [^{11}C]SN 23719 represented between 1.5 and 2.8%ID g^{-1}. The researchers concluded that, using a strategy with ^{11}C-labeling and PET, they could determine the tissue distribution and metabolic stability of novel tricyclic carboxamides with the view of selecting analogs with potentially better in vivo activity against solid tumors.

3.31.4.6.2 Indirect measurements

17-Allylaminogeldanamycin (17-AAG) is the first hsp90 inhibitor to be tested in a clinical trial. It causes the degradation of HER2 and other hsp90 targets, and has antitumor activity in preclinical models. We have exploited the mechanism by which 17-AAG inhibits HER2 expression to image its pharmacodynamic effects.[119] We labeled F(ab′)$_2$ fragments of Herceptin with ^{68}Ga and were able to image tumors every 24 h with no decrease in signal. With this reagent, the effect of 17-AAG on HER2 expression in xenografts was quantitatively imaged (**Figure 2**). We showed that, by 24 h after 17-AAG administration, HER2 expression fell 80% in the tumor (**Figures 3** and **4**). This change correlated well with the results obtained by immunoblotting of tumor tissue. As this technique is noninvasive and can be repeated every 24 h, we have used it to determine the kinetics of loss and recovery of HER2 expression in response

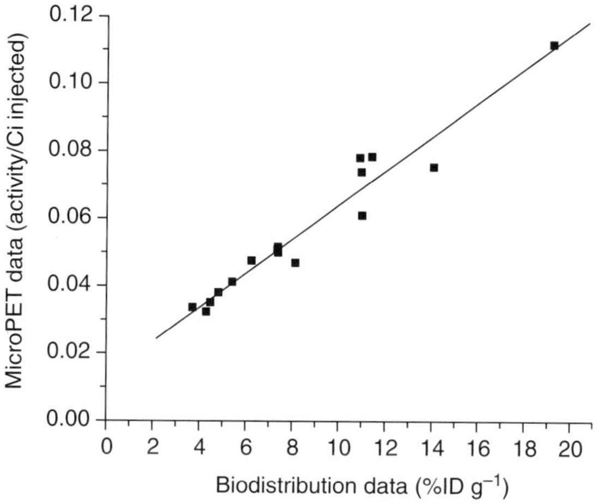

Figure 2 Correlation of microPET ROI analysis and biodistribution data for ^{68}Ga-DOTA-F(ab′)$_2$-herceptin uptake by BT474 tumors at $3\frac{1}{2}$ h post injection.

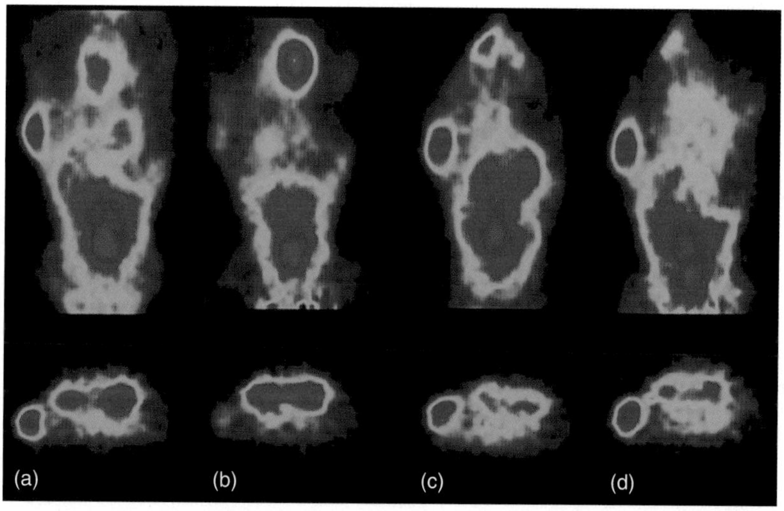

Figure 3 MicroPET images (coronal slice and transverse slice through tumor and kidneys) of the two different nude mice with single BT474 tumors at 3 h post injection with ^{68}Ga-DOTA-F(ab′)$_2$ Herceptin. (a) A mouse prior to 17AAG treatment; (b) the same mouse 24 h later after it had received 150 mg kg^{-1} 17-AAG. (c and d) Comparable images of a control mouse which received two doses of ^{68}Ga-DOTA-F(ab′)$_2$-herceptin 24 h apart.

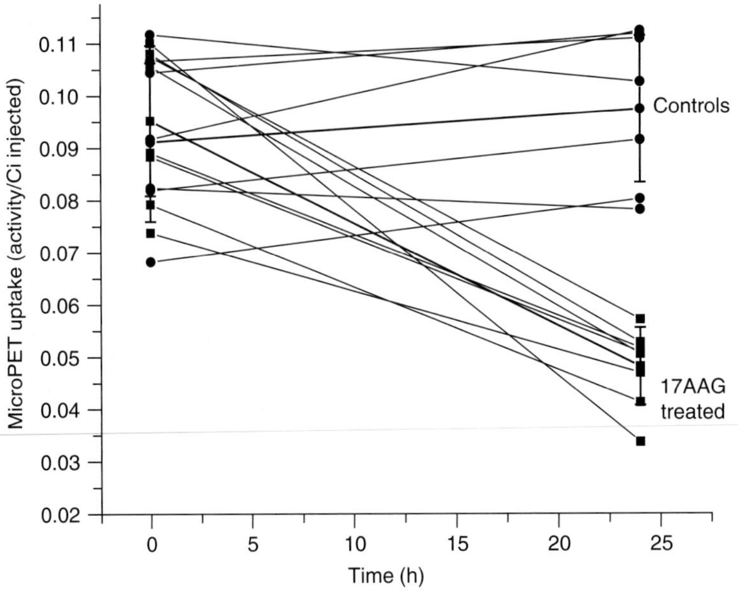

Figure 4 MicroPET ROI analysis of BT474 tumor uptake of ^{68}Ga-DOTA-F(ab′)$_2$-herceptin in control and 17-AAG-treated animals.

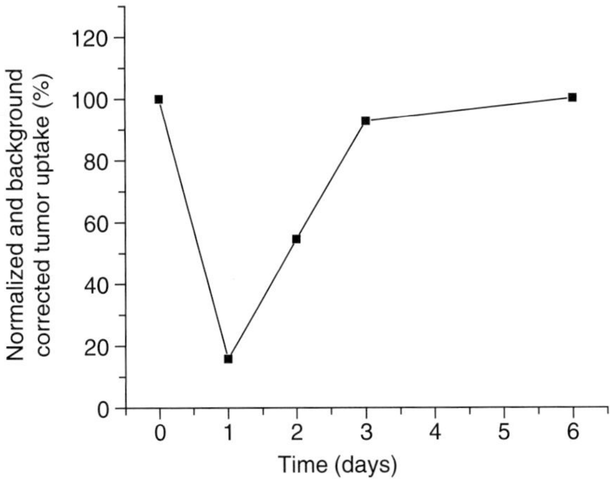

Figure 5 PET ROI analysis of a CWR22-RV1 prostate tumor in an athymic mouse. The mouse was imaged with ^{68}Ga-DOTA-F(ab′)$_2$-herceptin on days 0, 1, 2, 3, and 6 and treated with 17-AAG (100 mg kg^{-1}) after scan on day 0.

to a single drug treatment over 7 days (**Figure 5**). As EGFR antibodies and inhibitors of the HER kinase catalytic domain have also been shown to induce receptor degradation, the technique potentially has a wider use. Further, the ability to image the kinetics of drug action will allow a more rational approach to testing the optimal dose and schedule of combination therapy.

3.31.5 Conclusions

Small-animal imaging is a valuable tool to investigate new drugs and validate their potential in vivo. CT and MRI are good methods for anatomical and functional imaging, but can not be reliably used for molecular imaging since they require potentially pharmacologically active doses of drugs. Optical methods of imaging can be performed at the tracer level using bioluminescence and fluorescent imaging techniques, but they can only yield planar images which can not give quantitative data. Small-animal imaging with PET and SPECT permits the noninvasive study of novel drugs as well as their effects in animals over substantial periods of time. The methods are directly transferable into the clinic and offer a rapid and cost effective way to develop new therapeutic strategies.

References

1. Weber, S. M.; Peterson, K. A.; Durkee, B.; Qi, C.; Longino, M.; Warner, T.; Lee, F. T., Jr.; Weichert, J. P. *J. Surg. Res.* **2004**, *119*, 41–45.
2. Ritman, E. L. *Annu. Rev. Biomed. Eng.* **2004**, *6*, 185–208.
3. Boesch, C. *Mol. Aspects. Med.* **1999**, *20*, 185–318.
4. Weinmann, H. J.; Ebert, W.; Misselwitz, B.; Schmitt-Willich, H. *Eur. J. Radiol.* **2003**, *46*, 33–44.
5. Morawski, A. M.; Lanza, G. A.; Wickline, S. A. *Curr. Opin. Biotechnol.* **2005**, *16*, 89–92.
6. Ntziachristos, V.; Ripoll, J.; Wang, L. V.; Weissleder, R. *Nat. Biotechnol.* **2005**, *23*, 313–320.
7. Acton, P. D.; Kung, H. F. *Nucl. Med. Biol.* **2003**, *30*, 889–895.
8. Wilbur, D. S. *Bioconjug. Chem.* **1992**, *3*, 433–470.
9. Adam, M. J.; Wilbur, D. S. *Chem. Soc. Rev.* **2005**, *34*, 153–163.
10. Buchsbaum, D. J. *Cancer Res.* **1995**, *55*, 5729s–5732s.
11. Weber, S.; Bauer, A. *Eur. J. Nucl. Med. Mol. Imaging* **2004**, *31*, 1545–1555.
12. Toyama, H.; Ichise, M.; Liow, J. S.; Vines, D. C.; Seneca, N. M.; Modell, K. J.; Seidel, J.; Green, M. V.; Innis, R. B. *Nucl. Med. Biol.* **2004**, *31*, 251–256.
13. Kung, H. F.; Kim, H. J.; Kung, M. P.; Meegalla, S. K.; Plossl, K.; Lee, H. K. *Eur. J. Nucl. Med.* **1996**, *23*, 1527–1530.
14. Halldin, C.; Stone-Elander, S.; Farde, L.; Ehrin, E.; Fasth, K. J.; Langstrom, B.; Sedvall, G. *Int. J. Rad. Appl. Instrum. [A]* **1986**, *37*, 1039–1043.
15. Halldin, C.; Farde, L.; Barnett, A.; Sedvall, G. *Int. J. Rad. Appl. Instrum. [A]*, **1991**, *42*, 451–455.
16. Muhr, C.; Bergstrom, M.; Lundberg, P. O.; Bergstrom, K.; Hartvig, P.; Lundqvist, H.; Antoni, G.; Langstrom, B. J. *Comput. Assist. Tomogr.* **1986**, *10*, 175–180.
17. Shiue, C. Y.; Fowler, J. S.; Wolf, A. P.; McPherson, D. W.; Arnett, C. D.; Zecca, L. *J. Nucl. Med.* **1986**, *27*, 226–234.
18. Maziere, B.; Loc'h, C.; Hantraye, P.; Guillon, R.; Duquesnoy, N.; Soussaline, F.; Naquet, R.; Comar, D.; Maziere, M. *Life Sci.* **1984**, *35*, 1349–1356.
19. Mukherjee, J.; Christian, B. T.; Narayanan, T. K.; Shi, B.; Mantil, J. *Neuropsychopharmacology* **2001**, *25*, 476–488.
20. Perlmutter, J. S.; Moerlein, S. M.; Hwang, D. R.; Todd, R. D. *J. Neurosci.* **1991**, *11*, 1381–1389.
21. Farde, L.; Ehrin, E.; Eriksson, L.; Greitz, T.; Hall, H.; Hedstrom, C. G.; Litton, J. E.; Sedvall, G. *Proc. Natl Acad. Sci. USA* **1985**, *82*, 3863–3867.
22. Maziere, B.; Loc'h, C.; Stulzaft, O.; Hantraye, P.; Ottaviani, M.; Comar, D.; Maziere, M. *Eur. J. Pharmacol.* **1986**, *127*, 239–247.
23. Kung, H. F.; Guo, Y. Z.; Billings, J.; Xu, X.; Mach, R. H.; Blau, M.; Ackerhalt, R. E. *Int. J. Rad. Appl. Instrum. [B]* **1988**, *15*, 195–201.
24. Maziere, B.; Crouzel, C.; Venet, M.; Stulzaft, O.; Sanz, G.; Ottaviani, M.; Sejourne, C.; Pascal, O.; Bisserbe, J. C. *Int. J. Rad. Appl. Instrum. [B]* **1988**, *15*, 463–468.
25. Amokhtari, M.; Andersen, K.; Ibazizene, M.; Dhilly, M.; Dauphin, F.; Barre, L. *Nucl. Med. Biol.* **1988**, *25*, 517–522.
26. Hirani, E.; Opacka-Juffry, J.; Gunn, R.; Khan, I.; Sharp, T.; Hume, S. *Synapse* **2000**, *36*, 330–341.
27. Maziere, M.; Hantraye, P.; Prenant, C.; Sastre, J.; Comar, D. *Int. J. Appl. Radiat. Isot.* **1984**, *35*, 973–976.
28. Shah, F.; Hume, S. P.; Pike, V. W.; Ashworth, S.; McDermott, J. *Nucl. Med. Biol.* **1994**, *21*, 573–581.
29. Frost, J. J.; Wagner, H. N., Jr.; Dannals, R. F.; Ravert, H. T.; Links, J. M.; Wilson, A. A.; Burns, H. D.; Wong, D. F.; McPherson, R. W.; Rosenbaum, A. E. et al. *J. Comput. Assist. Tomogr.* **1985**, *9*, 231–236.
30. Jones, A. K.; Luthra, S. K.; Maziere, B.; Pike, V. W.; Loc'h, C.; Crouzel, C.; Syrota, A.; Jones, T. *J. Neurosci. Methods* **1988**, *23*, 121–129.
31. Channing, M. A.; Eckelman, W. C.; Bennett, J. M.; Burke, T. R., Jr.; Rice, K. C. *Int. J. Appl. Radiat. Isot.* **1985**, *36*, 429–433.
32. Syrota, A.; Paillotin, G.; Davy, J. M.; Aumont, M. C. *Life Sci.* **1984**, *35*, 937–945.
33. Dannals, R. F.; Langstrom, B.; Ravert, H. T.; Wilson, A. A.; Wagner, H. N., Jr. *Int. J. Rad. Appl. Instrum. [A]* **1988**, *39*, 291–295.
34. Mulholland, G. K.; Jewett, D. M.; Toorongian, S. A. *Int. J. Rad. Appl. Instrum. [A]* **1988**, *39*, 373–379.
35. Dewey, S. L.; MacGregor, R. R.; Brodie, J. D.; Bendriem, B.; King, P. T.; Volkow, N. D.; Schlyer, D. J.; Fowler, J. S.; Wolf, A. P.; Gatley, S. J. et al. *Synapse* **1990**, *5*, 213–223.
36. Brady, F.; Luthra, S. K.; Tochon-Danguy, H. J.; Steel, C. J.; Waters, S. L.; Kensett, M. J.; Landais, P.; Shah, F.; Jaeggi, K. A.; Drake, A. et al. *Int. J. Rad. Appl. Instrum. [A]* **1991**, *42*, 621–628.
37. Bonasera, T. A.; O'Neil, J. P.; Xu, M.; Dobkin, J. A.; Cutler, P. D.; Lich, L. L.; Choe, Y. S.; Katzenellenbogen, J. A.; Welch, M. J. *J. Nucl. Med.* **1996**, *37*, 1009–1015.
38. VanBrocklin, H. F.; Pomper, M. G.; Carlson, K. E.; Welch, M. J.; Katzenellenbogen, J. A. *Int. J. Radiat. Appl. Instrum. [B]* **1992**, *19*, 363–374.
39. Pomper, M. G.; Katzenellenbogen, J. A.; Welch, M. J.; Brodack, J. W.; Mathias, C. J. *J. Med. Chem.* **1988**, *31*, 1360–1363.
40. Albert, R.; Smith-Jones, P.; Stolz, B.; Simeon, C.; Knecht, H.; Bruns, C.; Pless, J. *Bioorg. Med. Chem. Lett.* **1998**, *8*, 1207–1210.
41. Smith-Jones, P. M.; Bischof, C.; Leimer, M.; Gludovacz, D.; Angelberger, P.; Pangerl, T.; Peck-Radosavljevic, M.; Hamilton, G.; Kaserer, K.; Kofler, A. et al. *Endocrinology* **1999**, *140*, 5136–5148.
42. Smith-Jones, P. M.; Stolz, B.; Bruns, C.; Albert, R.; Reist, H. W.; Fridrich, R.; Macke, H. R. *J. Nucl. Med.* **1994**, *35*, 317–325.
43. Zhang, H.; Chen, J.; Waldherr, C.; Hinni, K.; Waser, B.; Reubi, J. C.; Maecke, H. R. *Cancer Res.* **2004**, *64*, 6707–6715.
44. Eberle, A. N.; Froidevaux, S. J. *Mol. Recognit.* **2003**, *16*, 248–254.
45. Virgolini, I.; Kurtaran, A.; Raderer, M.; Leimer, M.; Angelberger, P.; Havlik, E.; Li, S.; Scheithauer, W.; Niederle, B.; Valent, P. et al. *J. Nucl. Med.* **1995**, *36*, 1732–1739.
46. Collingridge, D. R.; Glaser, M.; Osman, S.; Barthel, H.; Hutchinson, O. C.; Luthra, S. K.; Brady, F.; Bouchier-Hayes, L.; Martin, S. J.; Workman, P. et al. *Br. J. Cancer* **2003**, *89*, 1327–1333.
47. Ohtsuki, K.; Akashi, K.; Aoka, Y.; Blankenberg, F. G.; Kopiwoda, S.; Tait, J. F.; Strauss, H. W. *Eur. J. Nucl. Med.* **1999**, *26*, 1251–1258.
48. Tait, J. F.; Smith, C.; Gibson, D. F. *Bioconjug. Chem.* **2002**, *13*, 1119–1123.
49. Wen, X.; Wu, Q. P.; Ke, S.; Wallace, S.; Charnsangavej, C.; Huang, P.; Liang, D.; Chow, D.; Li, C. *Cancer Biother. Radiopharm.* **2003**, *18*, 819–827.
50. Blankenberg, F. G.; Robbins, R. C.; Stoot, J. H.; Vriens, P. W.; Berry, G. J.; Tait, J. F.; Strauss, H. W. *Chest* **2000**, *117*, 834–840.
51. D'Arceuil, H.; Rhine, W.; de Crespigny, A.; Yenari, M.; Tait, J. F.; Strauss, W. H.; Engelhorn, T.; Kastrup, A.; Moseley, M.; Blankenberg, F. G. *Stroke* **2000**, *31*, 2692–2700.
52. Kolodgie, F. D.; Petrov, A.; Virmani, R.; Narula, N.; Verjans, J. W.; Weber, D. K.; Hartung, D.; Steinmetz, N.; Vanderheyden, J. L.; Vannan, M. A. et al. *Circulation* **2003**, *108*, 3134–3139.
53. Mandl, S. J.; Mari, C.; Edinger, M.; Negrin, R. S.; Tait, J. F.; Contag, C. H.; Blankenberg, F. G. *Mol. Imaging* **2004**, *3*, 1–8.
54. Ntziachristos, V.; Schellenberger, E. A.; Ripoll, J.; Yessayan, D.; Graves, E.; Bogdanov, A., Jr.; Josephson, L.; Weissleder, R. *Proc. Natl Acad. Sci. USA* **2004**, *101*, 12294–12299.

55. Ogura, Y.; Krams, S. M.; Martinez, O. M.; Kopiwoda, S.; Higgins, J. P.; Esquivel, C. O.; Strauss, H. W.; Tait, J. F.; Blankenberg, F. G. *Radiology* **2000**, *14*, 795–800.
56. Post, A. M.; Katsikis, P. D.; Tait, J. F.; Geaghan, S. M.; Strauss, H. W.; Blankenberg, F. G. J. *Nucl. Med.* **2002**, *43*, 1366–1367.
57. Subbarayan, M.; Hafeli, U. O.; Feyes, D. K.; Unnithan, J.; Emancipator, S. N.; Mukhtar, H. *J. Nucl. Med.* **2003**, *44*, 650–656.
58. DeNardo, S. J.; Burke, P. A.; Leigh, B. R.; O'Donnell, R. T.; Miers, L. A.; Kroger, L. A.; Goodman, S. L.; Matzku, S.; Jonczyk, A.; Lamborn, K. R. et al. *Cancer Biother. Radiopharm.* **2000**, *15*, 71–79.
59. Haubner, R.; Wester, H. J.; Weber, W. A.; Mang, C.; Ziegler, S. I.; Goodman, S. L.; Senekowitsch-Schmidtke, R.; Kessler, H.; Schwaiger, M. *Cancer Res.* **2001**, *61*, 1781–1785.
60. Chen, X.; Park, R.; Tohme, M.; Shahinian, A. H.; Bading, J. R.; Conti, P. S. *Bioconjug Chem.* **2004**, *15*, 41–49.
61. Zareneyrizi, F.; Yang, D. J.; Oh, C. S.; Ilgan, S.; Yu, D. F.; Tansey, W.; Liu, C. W.; Kim, E. E.; Podoloff, D. A. *Anticancer Drugs* **1999**, *10*, 685–692.
62. Lu, E.; Wagner, W. R.; Schellenberger, U.; Abraham, J. A.; Klibanov, A. L.; Woulfe, S. R.; Csikari, M. M.; Fischer, D.; Schreiner, G. F.; Brandenburger, G. H. et al. *Circulation* **2003**, *108*, 97–103.
63. Viti, F.; Tarli, L.; Giovannoni, L.; Zardi, L.; Neri, D. *Cancer Res.* **1999**, *59*, 347–352.
64. Avril, N. E.; Weber, W. A. *Radiol. Clin. North Am.* **2005**, *43*, 189–204.
65. Abe, Y.; Matsuzawa, T.; Fujiwara, T.; Fukuda, H.; Itoh, M.; Yamada, K.; Yamaguchi, K.; Sato, T.; Ido, T. *Eur. J. Nucl. Med.* **1986**, *12*, 325–328.
66. Aft, R. L.; Lewis, J. S.; Zhang, F.; Kim, J.; Welch, M. J. *Cancer Res.* **2003**, *63*, 5496–5504.
67. Agus, D. B.; Golde, D. W.; Sgouros, G.; Ballangrud, A.; Cordon-Cardo, C.; Scher, H. I. *Cancer Res.* **1998**, *58*, 3009–3014.
68. Aliaga, A.; Rousseau, J. A.; Ouellette, R.; Cadorette, J.; van Lier, J. E.; Lecomte, R.; Benard, F. *Nucl. Med. Biol.* **2004**, *31*, 761–770.
69. Barthel, H.; Cleij, M. C.; Collingridge, D. R.; Hutchinson, O. C.; Osman, S.; He, Q.; Luthra, S. K.; Brady, F.; Price, P. M.; Aboagye, E. O. *Cancer Res.* **2003**, *63*, 3791–3798.
70. Carnochan, P.; Brooks, R. *Nucl. Med. Biol.* **1999**, *26*, 667–672.
71. Haaparanta, M.; Paul, R.; Huovinen, R.; Kujari, H.; Bergman, J.; Solin, O.; Kangas, L. *Nucl. Med. Biol.* **1995**, *22*, 483–489.
72. Kaim, A. H.; Weber, B.; Kurrer, M. O.; Westera, G.; Schweitzer, A.; Gottschalk, J.; von Schulthess, G. K.; Buck, A. *Eur. J. Nucl. Med. Mol. Imaging* **2002**, *29*, 648–654.
73. Kubota, K.; Ishiwata, K.; Yamada, S.; Kubota, R.; Tada, M.; Sato, T.; Ido, T. *Tohoku J. Exp. Med.* **1992**, *168*, 437–439.
74. Lapointe, D.; Brasseur, N.; Cadorette, J.; La Madeleine, C.; Rodrigue, S.; van Lier, J. E.; Lecomte, R. *J. Nucl. Med.* **1999**, *40*, 876–882.
75. Mach, R. H.; Huang, Y.; Buchheimer, N.; Kuhner, R.; Wu, L.; Morton, T. E.; Wang, L.; Ehrenkaufer, R. L.; Wallen, C. A.; Wheeler, K. T. *Nucl. Med. Biol.* **2001**, *28*, 451–458.
76. Miyagawa, T.; Oku, T.; Sasajima, T.; Dasai, R.; Beattie, B.; Finn, R.; Tjuvajev, J. G.; Blasberg, R. *J. Nucl. Med.* **2003**, *44*, 1845–1854.
77. Moore, J. V.; Waller, M. L.; Zhao, S.; Dodd, N. J.; Acton, P. D.; Jeavons, A. P.; Hastings, D. L. *Eur. J. Nucl. Med.* **1998**, *25*, 1248–1254.
78. Ohira, H.; Kubota, K.; Ohuchi, N.; Harada, Y.; Fukuda, H.; Satomi, S. *J. Nucl. Med.* **2000**, *41*, 1561–1568.
79. Palmedo, H.; Hensel, J.; Reinhardt, M.; Von Mallek, D.; Matthies, A.; Biersack, H. *J. Nucl. Med. Biol.* **2002**, *29*, 809–815.
80. Spaepen, K.; Stroobants, S.; Dupont, P.; Bormans, G.; Balzarini, J.; Verhoef, G.; Mortelmans, L.; Vandenberghe, P.; De Wolf-Peeters, C. *Eur. J. Nucl. Med. Mol. Imaging* **2003**, *30*, 682–688.
81. Sugiyama, M.; Sakahara, H.; Sato, K.; Harada, N.; Fukumoto, D.; Kakiuchi, T.; Hirano, T.; Kohno, E.; Tsukada, H.; Kohno, E. et al. *J. Nucl. Med.* **2004**, *45*, 1754–1758.
82. van Waarde, A.; Cobben, D. C.; Suurmeijer, A. J.; Maas, B.; Vaalburg, W.; de Vries, E. F.; Jager, P. L.; Hoekstra, H. J.; Elsinga, P. H. *J. Nucl. Med.* **2004**, *45*, 695–700.
83. Wagner, M.; Seitz, U.; Buck, A.; Neumaier, B.; Schultheiss, S.; Bangerter, M.; Bommer, M.; Leithauser, F.; Wawra, E.; Munzert, G. et al. *Cancer Res.* **2003**, *63*, 2681–2687.
84. Waldherr, C.; Mellinghoff, I. K.; Tran, C.; Halpern, B. S.; Rozengurt, N.; Safaei, A.; Weber, W. A.; Stout, D.; Satyamurthy, N.; Barrio, J. et al. *J. Nucl. Med.* **2005**, *46*, 114–120.
85. Wyss, M. T.; Weber, B.; Honer, M.; Spath, N.; Ametamey, S. M.; Westera, G.; Bode, B.; Kaim, A. H.; Buck, A. *Eur. J. Nucl. Med. Mol. Imaging* **2004**, *31*, 312–316.
86. Wyss, M. T.; Honer, M.; Spath, N.; Gottschalk, J.; Ametamey, S. M.; Weber, B.; von Schulthess, G. K.; Buck, A.; Kaim, A. H. *Nucl. Med. Biol.* **2004**, *31*, 875–882.
87. Zhao, S.; Moore, J. V.; Waller, M. L.; McGown, A. T.; Hadfield, J. A.; Pettit, G. R.; Hastings, D. L. *Eur. J. Nucl. Med.* **1999**, *26*, 231–238.
88. Schelbert, H. R. *Ann. Biomed. Eng.* **2000**, *28*, 922–929.
89. Oyama, N.; Miller, T. R.; Dehdashti, F.; Siegel, B. A.; Fischer, K. C.; Michalski, J. M.; Kibel, A. S.; Andriole, G. L.; Picus, J.; Welch, M. J. *J. Nucl. Med.* **2003**, *44*, 549–555.
90. Knapp, F. F. J.; Kropp, J. *Eur. J. Nucl. Med.* **1995**, *22*, 361–381.
91. Kubota, K.; Matsuzawa, T.; Takahashi, T.; Fujiwara, T.; Kinomura, S.; Ido, T.; Sato, T.; Kubota, R.; Tada, M.; Ishiwata, K. *J. Nucl. Med.* **1989**, *30*, 2012–2016.
92. Mitterhauser, M.; Wadsak, W.; Krcal, A.; Schmaljohann, J.; Eidherr, H.; Schmid, A.; Viernstein, H.; Dudczak, R.; Kletter, K. *Appl. Radiat. Isot.* **2005**, *62*, 411–415.
93. Ishiwata, K.; Kubota, K.; Murakami, M.; Kubota, R.; Senda, M. *Nucl. Med. Biol.* **1993**, *20*, 895–899.
94. Melega, W. P.; Perlmutter, M. M.; Luxen, A.; Nissenson, C. H.; Grafton, S. T.; Huang, S. C.; Phelps, M. E.; Barrio, J. R. *J. Neurochem.* **1989**, *53*, 311–314.
95. Inoue, T.; Tomiyoshi, K.; Higuichi, T.; Ahmed, K.; Sarwar, M.; Aoyagi, K.; Amano, S.; Alyafei, S.; Zhang, H.; Endo, K. *J. Nucl. Med.* **1998**, *39*, 663–667.
96. Heiss, P.; Mayer, S.; Herz, M.; Wester, H. J.; Schwaiger, M.; Senekowitsch-Schmidtke, R. *J. Nucl. Med.* **1999**, *40*, 1367–1373.
97. Becherer, A.; Karanikas, G.; Szabo, M.; Zettinig, G.; Asenbaum, S.; Marosi, C.; Henk, C.; Wunderbaldinger, P.; Czech, T.; Wadsak, W. et al. *Eur. J. Nucl. Med. Mol. Imaging* **2003**, *30*, 1561–1567.
98. Lu, L.; Samuelsson, L.; Bergstrom, M.; Sato, K.; Fasth, K. J.; Langstrom, B. *J. Nucl. Med.* **2002**, *43*, 1688–1698.
99. Oyama, N.; Ponde, D. E.; Dence, C.; Kim, J.; Tai, Y. C.; Welch, M. J. *J. Nucl. Med.* **2004**, *45*, 519–525.
100. Schwartz, J. L.; Tamura, Y.; Jordan, R.; Grierson, J. R.; Krohn, K. A. *J. Nucl. Med.* **2003**, *44*, 2027–2032.
101. Shields, A. F.; Grierson, J. R.; Dohmen, B. M.; Machulla, H. J.; Stayanoff, J. C.; Lawhorn-Crews, J. M.; Obradovich, J. E.; Muzik, O.; Mangner, T. *J. Nat. Med.* **1998**, *4*, 1334–1336.
102. Tjuvajev, J. G.; Stockhammer, G.; Desai, R.; Uehara, H.; Watanabe, K.; Gansbacher, B.; Blasberg, R. G. *Cancer Res.* **1995**, *55*, 6126–6132.
103. Alauddin, M. M.; Shahinian, A.; Gordon, E. M.; Conti, P. S. *Mol. Imaging* **2004**, *3*, 76–84.

104. Groot-Wassink, T.; Aboagye, E. O.; Wang, Y.; Lemoine, N. R.; Reader, A. J.; Vassaux, G. *Mol. Ther.* **2004**, *9*, 436–442.
105. Anton, M.; Wagner, B.; Haubner, R.; Bodenstein, C.; Essien, B. E.; Bonisch, H.; Schwaiger, M.; Gansbacher, B.; Weber, W. A. *J. Gene Med.* **2004**, *6*, 119–126.
106. Verwijnen, S. M.; Sillevis Smith, P. A.; Hoeben, R. C.; Rabelink, M. J.; Wiebe, L.; Curiel, D. T.; Hemminki, A.; Krenning, E. P.; de Jong, M. *Cancer Biother. Radiopharm.* **2004**, *19*, 111–120.
107. Sundaresan, G.; Yazaki, P. J.; Shively, J. E.; Finn, R. D.; Larson, S. M.; Raubitschek, A. A.; Williams, L. E.; Chatziioannou, A. F.; Gambhir, S. S.; Wu, A. M. *J. Nucl. Med.* **2003**, *44*, 1962–1969.
108. Liu, S.; Edwards, D. S. *Chem. Rev.* **1999**, *99*, 2235–2268.
109. Camera, L.; Kinuya, S.; Garmestani, K.; Wu, C.; Brechbiel, M. W.; Pai, L. H.; McMurry, T. J.; Gansow, O. A.; Pastan, I.; Paik, C. H. et al. *J. Nucl. Med.* **1994**, *35*, 882–889.
110. Smith-Jones, P. M.; Vallabahajosula, S.; Goldsmith, S. J.; Navarro, V.; Hunter, C. J.; Bastidas, D.; Bander, N. H. *Cancer Res.* **2000**, *60*, 5237–5243.
111. Chappell, L. L.; Ma, D.; Milenic, D. E.; Garmestani, K.; Venditto, V.; Beitzel, M. P.; Brechbiel, M. W. *Nucl. Med. Biol.* **2003**, *30*, 581–595.
112. Kilbourn, M. R.; Carey, J. E.; Koeppe, R. A.; Haka, M. S.; Hutchins, G. D.; Sherman, P. S.; Kuhl, D. E. *Int. J. Radiat. Appl. Instrum. [B]* **1989**, *16*, 569–576.
113. Yagle, K. J.; Eary, J. F.; Tait, J. F.; Grierson, J. R.; Link, J. M.; Lewellen, B.; Gibson, D. F.; Krohn, K. A. *J. Nucl. Med.* **2005**, *46*, 658–666.
114. Ke, S.; Wen, X.; Wu, Q. P.; Wallace, S.; Charnsangavej, C.; Stachowiak, A. M.; Stephens, C. L.; Abbruzzese, J. L.; Podoloff, D. A.; Li, C. *J. Nucl. Med.* **2004**, *45*, 108–115.
115. Deehan, B.; Carnochan, P.; Trivedi, M.; Tombs, A. *Eur. J. Nucl. Med.* **1993**, *20*, 101–106.
116. Brix, G.; Bellemann, M. E.; Haberkorn, U.; Gerlach, L.; Lorenz, W. *J. Nucl. Med. Biol.* **1996**, *23*, 897–906.
117. Bading, J. R.; Alauddin, M. M.; Fissekis, J. D.; Shahinian, A. H.; Joung, J.; Spector, T.; Conti, P. S. *J. Nucl. Med.* **2000**, *41*, 1714–1724.
118. Osman, S.; Rowlinson-Busza, G.; Luthra, S. K.; Aboagye, E. O.; Brown, G. D.; Brady, F.; Myers, R.; Gamage, S. A.; Denny, W. A.; Baguley, B. C. et al. *Cancer Res.* **2001**, *61*, 2935–2944.
119. Smith-Jones, P. M.; Solit, D. B.; Akhurst, T.; Afroze, F.; Rosen, N.; Larson, S. M. *Nat. Biotechnol.* **2004**, *22*, 701–706.

Biography

Peter M Smith-Jones was educated at Kings College, University of London, UK and University of Basle, Switzerland. He is currently an Associate Attending Radiochemist in the Department of Radiology at Memorial Sloan Kettering Cancer Center in New York, an Assistant Professor of Radiopharmacy at Cornell University Medical Center in New York, and an Adjunct Assistant Professor at City University of New York. His current areas of research include radiolabeling ligands, peptides, and proteins for in vivo diagnosis and therapy.

Comprehensive Medicinal Chemistry II
ISBN (set): 0-08-044513-6

ISBN (Volume 3) 0-08-044516-0; pp. 659–677

3.32 High-Throughput and High-Content Screening

M Cik, Johnson & Johnson Pharmaceutical Research & Development, Beerse, Belgium
M R Jurzak, Merck KGaA, Darmstadt, Germany

3.32.1	**Introduction**	**679**
3.32.2	**Context of High-Throughput Screening**	**680**
3.32.2.1	Automation of Screening	681
3.32.2.2	General High-Throughput Screening Assay Prerequisites	682
3.32.2.3	Considerations in Single-Parametric Biochemical Assays	683
3.32.2.4	Considerations in Single Parametric Cellular Assays	683
3.32.2.5	Evaluation of Assay Quality	683
3.32.3	**Hit Characterization and the Hit-to-Lead Process**	**684**
3.32.4	**Principles of High-Content Screening and Important Assay Parameters**	**685**
3.32.5	**High-Content Screening Hardware**	**685**
3.32.5.1	Image Detection	687
3.32.5.2	Confocal Systems	687
3.32.6	**Analysis Software Requirements**	**687**
3.32.7	**Labeling Techniques**	**688**
3.32.7.1	Cellular and Subcellular Stains	689
3.32.7.2	Antibody Staining	689
3.32.7.3	Fluorescent Proteins Biosensors	690
3.32.8	**High-Content Screening Assay Examples and Applications**	**690**
3.32.8.1	Translocation Events	690
3.32.8.2	Other Multiparametric Cellular Events	691
3.32.8.2.1	Cellular toxicity	691
3.32.8.2.2	The cell cycle	691
3.32.8.2.3	Neurite outgrowth	692
3.32.8.2.4	Micronucleus and colony formation	692
3.32.9	**Data Analysis and Storage**	**692**
3.32.10	**Multidimensional Cytological Profiling**	**693**
3.32.11	**Applications of High-Content Screening within the Drug Discovery Process**	**693**
References		**694**

3.32.1 Introduction

Horace Walpole coined the term serendipity in 1754 after he had read a poem about three princes of Serendip (Sri Lanka) who were "always making discoveries, by accident and sagacity, of things they were not in quest of." Louis Pasteur clearly recognized the importance of serendipity in the development of science when he stated that chance favors the prepared mind.[1] Today's drug discovery is designed to minimize the serendipitous effect within its processes by seeking for opportunities to get compounds tested and selected meaningfully. High-throughput screening (HTS) can be defined as the screening of large numbers of substances in an efficient and timely manner. The ultimate goal is the discovery of active substances which can serve directly or after optimization as templates for molecules of commercial value. The market will mostly be the pharmaceutical sector but other life science areas, such as the agrochemical industry, also employ HTS as part of their discovery research. Now that the pharmaceutical industry has pioneered the field and fueled the technical developments, other organizations are starting to discover the benefits of HTS. In recent years, academic institutions have built up HTS facilities, with the aim of efficiently discovering chemical molecules as tools for interrogating biological networks.[2,3]

The unifying principle of HTS is the interrogation of substances in a test system, which allows for the efficient differentiation between active and inactive substances. At the end of the nineteenth century, working on the hypothesis that staining activity might be related to specific affinity, Paul Ehrlich screened 100 dyes for their ability to cure mice of *Trypanosoma* infection, resulting in the discovery of nagana red, named after the nagana disease in cattle caused by *Trypanosoma brucei*. Thus Ehrlich was not only the originator of the receptor theory but also a pioneer of the idea that synthetic molecules might be used for therapeutic purposes. According to Sneader,[4] the first random, large-scale screening was performed by Charles Pfizer and Company in 1948–49, when a team of 56 scientists investigated 100 000 soil samples for their antibiotic activity. This endeavor resulted in the discovery of terramycin, which would finally capture half of the broad-spectrum antibiotic market.

In its current definition, HTS refers to screening at a much larger scale and employs usually a comprehensive level of sophisticated automation. The rise of rational drug design in the early 1980s prompted companies to embark on comprehensive or mass screening, which at that time resulted in throughputs of at best 100 compounds per week. To boost productivity, processes and equipment needed to be developed and optimized. As a result of the growing new discipline, the first conference dedicated to HTS technologies was held at SRI International in 1992.[5]

In the following years, HTS became more and more integrated into drug discovery processes and internal dedicated groups have been established within pharmaceutical companies. In 1994 the Society for Biomolecular Screening was founded and has since then provided a forum for scientists interested in HTS.[6] In 1995 came the establishment of the Association for Laboratory Automation,[7] whose mission is to advance the utilization of laboratory robotics and automation. More than 10 years later, HTS is an integral part of the drug discovery process within the industry and many building blocks of this multidisciplinary field have reached a high level of sophistication. In this chapter we will shortly review a number of current developments within the field of HTS and finally focus on high-content screening (HCS), one of the novel approaches in the field.

3.32.2 Context of High-Throughput Screening

Modern HTS laboratories are capable of screening enormous numbers of compounds. Throughputs of 100 000 compounds per day, can be achieved in major pharmaceutical companies. A smaller fraction of HTS laboratories is even capable of reaching throughputs exceeding 1 million screened compounds per week.[8] Therefore, the technology has matured to a level such that throughput will no longer be a bottleneck for most HTS laboratories. Although the number of screened wells per time is an easily measurable and objective performance parameter, it does not directly define the desired output, which is the efficient identification of novel lead structures. The HTS process and the often interlinked combinatorial chemistry approaches have recently been in the focus of critical evaluations of the entire modern drug discovery process. Since early promises of increased productivity have not yet been realized, the question has been raised as to whether the investments in these technologies were justified.[9] However, it is crucial to understand that HTS productivity is embedded into many other drug discovery processes, which determine the overall output. The composition of the chemical library used for screening is one of the main factors in the success of the entire process. During the evolution of the HTS process, screening capabilities, size and selection of the compound library have influenced each other. Initially, natural substances and endogenous ligands largely drove medicinal chemistry. Inspired by the introduction of peptide combinatorial chemistry in the late 1980s, in some pharmaceutical companies high-throughput organic synthesis (or combinatorial chemistry) was used to build large compound collections, in parallel to the development of increased screening capabilities during the 1990s.[10] Today libraries of major pharmaceutical companies have grown to a considerable size. However, since screening large collections of compounds generates substantial costs, there has been a recent development toward increased quality and optimized diversity of libraries rather than increased size. The 'druglikeness' concept has been further developed for the requirements of an optimized screening collection and properties of 'lead-likeness' have been proposed.[11] Essentially, lead-like parameters put even stricter rules on the physicochemical parameters, since the selected molecules have to be amenable for additional modifications.[12,13]

The current size of compound libraries of the leading pharmaceutical companies is generally in the range of 10^6. Recent findings on the poor long-term stability of compounds dissolved in dimethyl sulfoxide (DMSO) have prompted many companies to take measures to improve and to redissolve their compound collections.[14–16]

Therefore, a high degree of sophistication is necessary to compose a screening library that has the desired physicochemical properties, being on the one hand as diverse as possible but allowing for sufficient redundancy to generate some overlap in biological activities necessary for the series analysis of hits. Computational property filters are used to rapidly select drug- and lead-like properties before purchase of external compounds or internal synthesis to increase diversity and usefulness of collections.[17,18] In addition, improved storage conditions, automated compound management and enhanced analytical quality control (QC) measures are employed to increase the stability of the library and provide flexible and rapid compound management (*see* 3.26 Compound Storage and Management).[19,20]

Another critical processes feeding into the HTS operation is the selection of the biological target. Even if the contribution of a target to a certain pathology has been clearly shown, e.g., leptin and obesity,[21] it does not mean that the target is suitable for direct small-molecule intervention. It seems that some protein families are not amenable to modulation by small molecules. This concept has recently been described as the 'druggability' of targets.[22] Druggability of a protein describes its ability to be modulated by a small molecule with physicochemical properties compatible with the intended therapeutic application. Since a high affinity interaction of the ligand and the therapeutic targets is based on complementary structures in terms of volume, topology, and physicochemical properties, such interaction will usually require a small binding pocket on the surface of the target protein.[23] Such structures are often found on enzymes or receptors, which naturally interact with small bioorganic molecules, but might be scarce on other protein families, such as transcription factors, which tend to interact via flat surfaces with other interaction partners.[24] Since such proteins often interact with each other via relatively planar surfaces, protein–protein interactions are considered to be poor targets for libraries of small molecules.[25] Current drugs are estimated to target between 399 and 483 biomolecules.[9,22] Using similarity-based considerations, it has been estimated that approximately 3000 targets are druggable by small molecules. Among those the pathophysiological and therapeutic relevance in a particular indication still needs to be demonstrated. Such considerations reduce the number of targets amenable for small-molecule screening and of pathophysiological relevance from 30 000 genes in the human genome to 600–1500 interesting for HTS.[22] Therefore, the number of well-defined molecular biochemical targets seems to be more limited than initially expected from the sequencing of the human genome. Target validation is, therefore, at present considered to be one of the major hurdles within the drug discovery process.[26]

3.32.2.1 Automation of Screening

A certain level of automation is a prerequisite for an efficient HTS laboratory. Initially, higher productivity and freedom from performing repetitive tasks has been a major motivator to introduce laboratory automation.[27] In the early 1990s screening libraries grew by combinatorial or parallel synthesis approaches, and demanded higher screening capabilities (and vice versa). In the following years it became clear that dependable automation would be able to deliver higher reproducibility and better data quality. With greater sophistication of the equipment, miniaturization of assay volume and increased parallelization became important means to reduce screening costs and to respond to the increasing pressure on program cycle times. The current number of laboratories performing HTS screening is estimated to be around 450 hosted by 100 pharmaceutical and 150 biotech companies, each producing an average of 6–8 million data points per year.[26] The market for HTS detection technologies alone, which covers only one part of HTS operations, has been estimated to be around $241 million in 2005.[28]

In general HTS assays require a certain number of operations such as plate handling, various forms of liquid dispensing or transfer, incubation steps, mixing, and signal detection. Depending on the needs of the assay and organization, the level of automation can range from manual use of a single hand-held (electronical) multichannel pipette to a fully integrated robotic system. If there are many processing units used to complete a given assay, these can be equipped with plate stackers, and laboratory personnel can control the flow of events within the assay by carrying the plate stackers to the next working unit after a task has been completed. Such operation can provide surprisingly high throughput and maximal flexibility for a limited investment in hardware. However, this type of operation requires fully dedicated and alert operators and becomes stressful when large libraries have to be screened. In addition, time scheduling, might not be compatible with certain assays, since the events cannot be exactly controlled at the plate level; in this case assays that require short and timed incubations might be impaired.

An intermediate level of integration is provided by workstations which can perform a number of predefined operations or assays, usually built around a liquid-handling equipment facility combined with some plate storage, dispensing unit(s), and reader functionality. Plate handling is usually performed with grippers and/or short conveyer belts. Typical examples of such workstation platforms are the Tecan FreedomEVO,[29] the Beckman Biomek FX,[30] the MultiPROBE II or PlateTrak,[31] the Sciclone ALH 3000 from Caliper Life Sciences,[32] the TekBench from TekCel,[33] and the Cybi-Well from Cybio.[34] Workstations can be designed to perform dedicated assays efficiently even when unattended. However, due to their limited footprint, normally cannot provide large plate capacities. In addition, the number of components that can be integrated within the limited reach of the plate handling devices of a workstation might not be sufficient to perform complex assays efficiently. To circumvent the footprint limitations certain vendors have designed workstations which can access devices or plates at levels above or below the primary work surface, thus extending their functionality while still keeping the compact footprint (see Tecan FreedomEVO and Tekcel TekBench or Velocity11).

To further extend the throughput and flexibility of a workstation, an open architecture robotic system is the screening platform of choice for most major screening laboratories. Robotic systems usually encompass a robotic arm (rotating or moved by a linear track) or alternatively a conveyor belt system for plate movements. Within reach of the robotic arm (or along the conveyor system) single operating units are mounted which can perform the desired tasks.

Modern systems are often combinations of several established units and provide flexibility for future extensions by integration of new items of equipment, possibly from other suppliers. Due to strong overlaps in technology most workstation vendors will also provide integrated robotic solutions. Examples for more specialized producers of integrated robotic solutions are The Automation Partnership,[35] ThermoLabsystems,[36] and SSI Robotics.[37] An extensive overview of existing automation vendors can be found elsewhere.[7]

Generally, open architecture robotic systems are custom-built from preexisting modules or equipment units connected via a plate transport system. A crucial feature of such complex systems is the sophisticated software, which interfaces unit movement and operation, bar code and plate tracking, and finally manages reader data output. Whereas integration of novel equipment still needs expert support, the PC-based scheduling software often exhibits a user-friendly graphic user interface.

Above the level of hand-held manual operation, each additional step of automation provides some specific advantages. The optimal solution will depend on the size of the library to be screened, the variability of targets and choice of assays employed, the desired number of screening campaigns, the available number and expertise of laboratory personnel, and, not least, the available investment. In practice, fully developed HTS laboratories will often use a combination all these levels of automation to reach an optimal balance between throughput and flexibility. Thus, single stacker-equipped washers can be efficiently used for tasks such as plate-coating; workstations might be best for dedicated assays such as enzyme-linked immunosorbent assays (ELISAs) or filtration assays; and fully integrated robotic units might be best used for homogenous assays. However, even robotic units might sometimes need off-line addition of valuable or unstable controls. If the necessary reader mode is not available on-line or the reader capacity is the throughput-limiting step of the robotic operation, it might need support by off-line reader capacity.

Although technical developments in the field of robotics have reached some level of maturity, today most operators will state that the reliability and performance of robotic systems are directly linked to the level of technical support provided by the vendor and the expertise of the user. However, quality standards for robotic laboratory equipment have recently been proposed.[38] Despite the improvements still needed to reach a robust "walk-away operation" the technical solutions currently available are not considered as the major bottleneck in HTS. In addition a number of interest groups have built local networks, which provide support and sharing of experiences.[39]

3.32.2.2 General High-Throughput Screening Assay Prerequisites

In general, the microplate has established itself as the major assay platform, starting with the 96-well plate initially used in the diagnostics field, which was the first major step toward parallelization. To allow for easier automated handling of various plate types the Society of Biomolecular Screening has proposed standard dimensions of a microplate, including length (127.8 mm) and width (85.5 mm), which is now known as the SBS standard. The pressure to reduce costs of the screening and technical improvement in pipetting equipment and plate manufacturing has led to increased use of higher density plates with 384 and 1536 wells. Despite the obvious advantage in reagent savings, the use of smaller volumes is more demanding on the equipment used for liquid handling and detection. Thus reduction of assay volumes can lead to compromised assay parameters especially due to stronger evaporation effects and changed surface-to-volume ratios. Conventional plate-readers, which scan the plate by measuring single wells, might easily become the rate-limiting step in screening operation when high-density plates are used. Such bottlenecks can be overcome by parallelization of conventional single-channel readers by installating 96-multilens array readers (plate::vision from Zeiss) or charge-coupled device (CCD) camera-based readers, which measure signals simultaneously from entire plates (LEADseeker from GE Healthcare and ViewLux, from PE). HTS laboratories that need to frequently screen large libraries will especially benefit developments toward miniaturization; thus, despite the technical challenges, a wide range of assays has been mastered in 1536-well format.[40–43] Whereas even higher well densities than 1536 have found some applications in specialized laboratories,[44] currently the low-volume version of 384-well plates seem to be the preferred format, providing an attractive balance between low volume (15–25 μL), accessibility by standard equipment, and reliability of hardware and assay performance. Despite the possibility of further savings of valuable reagents offered by moving to the 1536-well (or higher) plate density, only 16% of the liquid handling market in 2003 was occupied by low-volume nanoliter dispensing tools, indicating that assay volumes $<5\,\mu L$, are not the predominant format.[45] In general, coevolution of equipment and assay technologies has occurred, and for most targets a number of technologies and assay formats will be available.

3.32.2.3 Considerations in Single-Parametric Biochemical Assays

The methods of molecular biology can achieve the modification, overexpression, and purification of most drug targets and have opened the possibility of screening approaches on single, isolated biochemical targets. The availability of potentially modified and pure proteins has stimulated the development of a plethora of novel assay technologies. It is beyond the scope of this review to give a comprehensive overview of these technologies (*see* 3.27 Optical Assays in Drug Discovery; 3.28 Fluorescence Screening Assays; 3.29 Cell-Based Screening Assays). Typically development of HTS assays techniques tends toward a simplification of steps, i.e., the favored goal is an add, mix, and read process. In its simplest form, the assay reagents are mixed with the compounds to be tested and after short incubation the activity is measured in some form of a physically detectable signal. Such signals will in most cases be of optical nature, i.e., a change of optical density, fluorescence, or emission of light. For better signal-to-background resolution, such readout signals can be further modulated by filtering of the appropriate wavelengths or can be time-resolved. Heterogenous types of assays require some sort of a separation step. These are typically either radiometric filtration assays or ELISAs, which are tedious and require extensive washing. Unfortunately, filtration is not easy to automate and washing steps require large buffer volumes, making these technologies less attractive for HTS. Therefore, homogenous assay formats have been developed using radiometric and various nonradiometric technologies. Most biochemical assays are based either on a binding event, i.e., of a ligand to its receptor, or on an enzymatic activity, i.e., modification of a substrate to a product. If the analytes participating in the reaction cannot be directly measured, a label can be added either by direct attachment of a radioisotope or a fluorescent label. Alternatively, indirect detection can be performed using either specific antibodies or other tags, such as biotin, which can be used for subsequent detection with specific reaction partners.

3.32.2.4 Considerations in Single Parametric Cellular Assays

Cellular assays provide some obvious technical advantages for example when the isolation of the biochemical target is difficult or cannot be upscaled, as often found with membrane proteins. Initially, reporter gene assays have been one of the few really scalable cellular assays.[46,47] Such assays monitor the transcriptional regulation of a promoter of the gene of interest linked to the coding region of a reporter gene. By coupling the response to the expression of an enzyme, a highly amplified and therefore sensitive signal is obtained.[48,49] However, reporter gene assays might suffer from interference of compounds acting distally to the target. Therefore, their results have to be verified in a number of control assays aiming at filtering the non target-related or cytotoxic effects.[50] Another important cellular screening parameter has been the direct measurement of second messengers such as cAMP and intracellular Ca^{2+}.[51] The introduction of the fluorescence imaging plate reader (FLIPR^{R}) from Molecular Devices Corp. has introduced new possibilities for the measurement of fast changes in intercellular Ca^{2+} and membrane potential changes.[52] This instrument has stimulated cellular approaches in HTS and initiated the development of new generations of instruments specially designed to measure cellular fluorescence or luminescence signals.[53] The reference technology for the measurements of ion current across cellular membranes is patch clamping, and recently automated systems from Molecular Devices Corp. have been introduced to the market. Particularly for the measurement of fast currents automated patch clamp technologies will bring new possibilities for directly addressing the target class of ion channels in rapid screening.[54]

3.32.2.5 Evaluation of Assay Quality

Whereas throughput was the challenge at the beginning of HTS, today assay quality and dependable detection of active compounds are the main problems. One important assay parameter is the definition of controls used to define the assay performance and for calculation of the activities. Whereas uninhibited samples are typically used to serve as the positive control, the negative or blank control is often more difficult to define. In the best case, a pharmacologically relevant standard can be used to define the level of maximal effect or the level of nonspecific signal blank (NSB). If such standards are not available, one of the assay components, such as the enzyme or substrate, can be left out from certain wells to define NSB; but one must ascertain during assay development that such control values are not too artificial to be reached by a pharmacologically active compound. In some cases the positive controls might need the addition of an agonistic compound for stimulation of the system. Assays searching for agonistic activities might have accordingly reversed controls, i.e., the blank value represents the unperturbed control.

Assay performance and its suitability for HTS is usually calculated using the mean (M) and standard deviation (SD) of such control wells. Useful statistical parameters are:

1. Signal to background: $S/B = M_{\text{positive controls}}/M_{\text{NSBcontrols}}$
2. Signal to noise: $S/N = (M_{\text{positive controls}} - M_{\text{NSBcontrols}})/SD_{\text{NSBcontrols}}$
3. Screening window coefficient[55]: $z' = 1 - \dfrac{(3SD_{\text{positive controls}} + 3SD_{\text{NSBcontrols}})}{(M_{\text{positive controls}} - M_{\text{NSBcontrols}})}$

As a rule of thumb, a S/B higher than 3 and S/N greater than 10 is required for HTS. The parameter z' is a statistical tool that reflects the assays dynamic range and the variability associated with the measurements. In recent years the z' value has been accepted as the most relevant parameter describing the assay robustness, and z' values above 0.5 are considered as sufficient for screening campaigns. However, all these parameters only describe to a certain level the robustness of the assay, they do not guarantee that the assay can detect the correct pharmacology and/or is capable of detecting hits with the desired sensitivity. This can be difficult to control beforehand, especially on novel targets lacking necessary reference substances. In addition to the difficult quest for the correct pharmacology of novel targets, target-unrelated effects have been recognized as a major challenge for HTS operation.[56] Such undesired assay interference leads to false positives or false negatives, which can create a significant burden on resources during the hit validation phase. Phenomena such as inner filter effect caused by colored compounds, quenching of fluorescence by various mechanisms, autofluorescence of compounds, light scattering resulting from particles, and photobleaching are common mechanisms of assay interference. Currently, no interference-free technology exists and therefore a balance between assay robustness and HTS feasibility has to be found for each screening laboratory and each single target. Another major disturbing effect is caused by compound precipitation, and, therefore, exceeding the compound solubility limits in HTS is not recommended; but due to the variability of physicochemical properties within the compound collection this cannot always be avoided.[57]

Studies that have screened a subselection of compounds with different technologies have shown only a limited overlap of active compounds, demonstrating a complex interplay of target-, compound-, and assay-related activities.[58,59] The high costs of such parallel screens using different technologies and the fact that the outcomes can only be empirically understood limit their broad applicability. However, such studies reveal the drawbacks of such isolated biochemical approaches. As a result, the technical development of assay technology is working toward more robust techniques, and the so-called 'label-free' methods are currently in a late stage of development.[60] Surface plasmon resonance (SPR) is currently the only established label-free technology. Due to the dynamic readout, it does not only provide the direct measurement of affinity but also important kinetic parameters such as the K_{on} and K_{off} values.[61] The throughput of the new instrument generation Biacore S51, allows its application for hit validation or lead series selection but not in primary screening.

Whereas initially the information from HTS was in the form of digital active/nonactive information, today most HTS laboratories will engage in further hit confirmation, such as rapid repetition of activity and determination of detailed IC_{50} values. The available expertise and technological infrastructure of HTS can also be employed for further compound characterization such as confirmation of the activity by an independent, orthogonal technology to filter for assay technology-related compound interference.

3.32.3 Hit Characterization and the Hit-to-Lead Process

Each company has developed its own strategy to combine available technical and process elements to maximize the results from HTS. However, as a general development, most companies will today acknowledge that potency of compounds within the primary assay is not a sufficient parameter for the selection of a particular compound class for further testing or synthesis activity. Within the 'hit-to-lead' process organizations aim at collecting a broad range of data, allowing the most educated selection of hit series. In addition to the determination of the compound's potency, the chemical structures are validated by analysis and, if necessary, resynthesis. Beyond such data quality steps, other parameters such as the physicochemical properties, including compound solubility, lipophilicity and early absorption, distribution, metabolism, excretion, and toxicity (ADMET) parameters are either calculated in silico or determined experimentally. As a result, a multiparametric data set is created, which should allow hit series with the highest propensity for further development to be selected. Dedicated chemistry efforts are sometimes included to extend series and to monitor the overall properties of the synthesized analogs (hit-to-lead chemistry). Such processes are necessary to protect the post-HTS processes from getting blocked by potentially high numbers of hit series. Navigation through such multiparametric data provides opportunities for a more rational selection, but, because of its complexity, it also presents a challenge for each organization, since it requires novel IT solutions able to deal with multiparametric data sets.

3.32.4 Principles of High-Content Screening and Important Assay Parameters

As already discussed elsewhere (*see* 3.29 Cell-Based Screening Assays), physiologically relevant information obtained from assays performed in the cellular context can improve the entire drug discovery process from target validation to lead optimization. More and more, those assays are being done at cellular and subcellular level using high-content instruments and subsequent image analysis. HCS technologies are being used in the pharmaceutical industry with the aim of improving the quality and efficiency of lead compound generation. Due to steady improvements in the technology, HCS is not only used to improve target validation and lead compound selection, but certain well-characterized assays are already used in primary HTS.[62] Since Lansing Taylor coined the term high-content screening in 1996, many definitions of HCS have appeared in the literature. It is difficult to find a consensus on the definition of HCS, but most scientists would agree that HCS can be defined as a multiparametric analysis of cell populations or subcellular events that measures spatial and temporal changes of phenotypic parameters. Those changes could be movements of fluorescence-labeled proteins, changes in fluorescent intensities, changes in cellular morphology or motility, and other parameters that can be observed under a microscope. An important feature of HCS assays is their potential for multiplexed data acquisition. By capturing multiple fluorescent or bright-field events from the same cellular population, a tremendous amount of biological data can potentially be obtained. In contrast to multiplexed biochemical assays where extracted proteins or messenger RNAs (mRNAs) are used, HCS multiplexing is achieved in a cellular context. The data collected in HCS are obtained from whole cell imaging. However, it is important to note that for quantitative and physiologically relevant imaging, it is very important that the cells, subcellular structures, and biochemical events are not disrupted by fluorescence labeling of the protein of interest.

3.32.5 High-Content Screening Hardware

In HCS, image quality is an important parameter. The first microscope was developed in the seventeenth century, when Anthony van Leeuwenhoek used a microscope to observe and describe bacteria and other life forms in a drop of water. Later, several major improvements were made, and at the end of the nineteenth century, an immersion lens was developed with a numerical aperture of 1.5. A valuable addition to light microscopy was fluorescence microscopy, with first specialized fluorescence microscopes produced in 1960s. Thus cell biologists have been using light microscopy for centuries, and more recently fluorescence microscopy has been used to collect images, usually from a small number of samples. These small sets of data still required extensive human interaction for image acquisition, image analysis, and archiving. More recently, during the last decade and thanks to progresses in automation and robotics, light microscopy is becoming increasingly popular within screening applications. With the need for higher throughput, new instruments and hardware were created that allowed automation of image acquisition at high speed.[63] Thus, high-throughput microscopy was born. It can be defined as the automated acquisition of cellular images at high speed. To move from high-throughput microscopy to HCS, high-speed image acquisition needs to be integrated with a powerful multiparametric analysis of cell populations which measures spatial and temporal changes in the cellular phenotype. As a consequence of the enormous data load HTS applications require a flexible data structure, and also a storage medium capable of managing terabytes of data. Until recently, the focus of the HCS industry was automation of the imaging instrumentation. The first company to introduce a high-resolution, high-throughput cell imaging system was Cellomics,

Table 1 HCS instrumentation specifications

Instrument	*Manufacturer*	*Objectives*	*Plate formats*	*Laser line*	*Camera*[a]	*Other light source*	*Filter positions*
ArrayScan VTI	Cellomics	5–40 ×	96–384		CCD camera	Mercury–xenon lamp	10
KineticScan	Cellomics	5–40 ×	96–384		CCD camera	Mercury–xenon lamp	10
IN Cell 1000	GE Healthcare	4–40 ×	96–384		CCD camera	Xenon lamp	6
						Halogen lamp	
IN Cell 3000	GE Healthcare	40 ×	96–384	364 nm	3 CCD cameras	Red, green, blue LED	
				488 nm			
				647 nm			
Opera	Evotec Technologies	10–60 ×	96–2080	405 nm	4 CCD cameras	Xenon lamp	
				488 nm			
				532 nm			
				635 nm			
Discovery-1	Molecular Devices	2–40 ×	6–1536		CCD camera	Xenon lamp	8
Image express	Molecular Devices	4–40 ×	6–1536		CCD camera	Xenon lamp	10
MIAS-2	MAIA Scientific	2–63 ×	6–384		Inensified camera	Halogen lamp	8
					CCD camera		
Pathway HT	BD Biosciences	4–60 ×	96–384		CCD camera	Mercury lamp white LED	16
Cell lab IC 100	Beckman	2–40 ×	6–1536		CCD camera	Mercury lamp	10
Acumen Explorer	TIP Labtech		96–1536	405 nm	4 PMTs		
				488 nm			
iCyte	Compucyte	4–40 ×	96–384	405 nm	4 PMTs	Mercury lamp	
				488 nm		Halogen lamp	
				633 nm			

[a] CCD, charge-coupled device; PMT, photomultiplier tube.

which in 1997 launched the ArrayScan, a multi-wavelength instrument based on a white light illumination source and a cooled CCD camera.[64] Today, there is a wide range of instrumentation available, as many other manufactures have developed their own instruments. They differ mostly with regard to the illumination technology, image acquisition devices, optics, and autofocusing approaches. **Table 1** summarizes current manufacturers of imaging instrumentation and the main technical specifications of their image readers.

With regard to the light source, instruments can be divided into two groups: those with broad-spectrum white light sources, such as mercury arc lamps or xenon lamps, and those with a laser-based illumination. The advantage of a white light source is the possibility of multicolor applications by the use of appropriate filter sets. Typically, lasers offer a controlled excitation spot and greater power density, but laser-based instruments need to be equipped with multiple lasers to allow multicolor applications. Due to the limited lifetime of the laser sources, their regular replacement can be costly when used extensively. The next sections will describe some technical specifications of today's instruments.

3.32.5.1 Image Detection

HCS instruments can be equipped with a wide range of image detectors. Some instruments are equipped with cooled CCD cameras for good image quality and a wide dynamic range. Image intensifier cameras are used for low light level fluorescence imaging. Photomultiplier tubes (PMTs) provide the highest signal-to-noise ratio and good dynamic range. PMTs are mostly used in combination with a laser scanning cytometer (LSC) and Acumen Explorer. Some instruments use multiple CCD cameras for simultaneous acquisition of cellular phenomena at different wavelengths. The IN Cell 3000[65] has three CCD cameras for simultaneous imaging. The Opera[66] has four CCD cameras for detection of four fluorescent wavelengths. Other instruments such as the ArrayScan VTI[67] and Pathway HT[68] have one CCD camera. Some of the laser-scanning HCS instruments have multiple PMT detectors to allow simultaneous detection of up to four colors. The Acumen Explorer[69] and iCyte[70] are equipped with one to three lasers and up to four PMTs for simultaneous excitation and detection of up to four colors. Currently, there is a growing market trend toward CCD-based instrumentation, while PMT-based instruments are staying at the same level of market penetration.[26]

3.32.5.2 Confocal Systems

Standard fluorescence microscopy involves continuous illumination of a sample for a certain period of time while the image is being recorded. Emitted light is also collected from a layer outside the focal plane, and therefore images include out-of-focus blur. To avoid this problem, confocal microscopy is being increasingly used, as it is capable of producing images free from out-of-focus information. Consequently, confocal microscopy can also provide improved special resolution and better signal-to-noise ratios. In its high-end application mode, confocal microscopy uses multiple sequential scans along the focal plane, allowing three-dimensional reconstitution of samples. There are several commercially available confocal microscopes with exceptional resolution of fixed objects, but their image acquisition is too slow to be used for drug screening applications. However, there are currently three HCS instruments with increased acquisition rates that can be used in drug discovery: the IN Cell 3000, the Opera, and the Pathway HT.[71] To increase throughput, the Pathway HT and the Opera use a multi-pinhole spinning disk, a so-called Nipkow disk (**Figure 1**). This allows almost instant scanning of a large sample area. For the same purpose, the IN Cell 3000 uses line scanning through a confocal slit mask. A different approach to create a confocal impression is by inserting a special device, ApoTome,[72] in the light path of a standard fluorescence microscope. ApoTome uses a combination of a grid projection system and a mathematical correction to display a defined optical section through the specimen. With the ApoTome device, stray light from out-of-focus planes can be removed and images are improved to almost confocal quality.

Focusing is one of the critical steps in automated microscopy. Automated HCS instruments locate the sample plane applying various autofocus technologies. Due to different reaction speeds, the choice of the autofocusing technology also determines the overall image acquisition rate. One approach is to focus on the bottom of a well using a laser and to work with an offset to capture cells adhering to the bottom of the well. Object-based autofocusing is a more sophisticated way to find cells, but it is more difficult as it requires both hardware and software to work in a constant feedback loop.[73]

3.32.6 Analysis Software Requirements

Sophisticated and powerful analysis software combined with high-throughput microscopy hardware is another key elements of HCS. Once good-quality images have been obtained, it is the task of software to extract information out of the images. Therefore, image analysis software tools are critical for data interpretation and one of the key challenges in HCS. Analysis software quantifies changes in spatial distribution and intensities of fluorescent probes or other identified objects. Ideally, the software contains a set of algorithms and tools that can easily be modified by the user and that can be applied to images for various biological applications. Finally it is necessary to translate image information into some kind of a numerical output. Another basic challenge in image analysis is the proprietary file formats of images generated by different instruments and, therefore, the lack of file format standardization which would allow easier algorithm comparison and development.[26,74,75] At the recent Cambridge Healthtech Institute High-Content Analysis 2005 conference in San Francisco, standardization was identified by users as one of the highest priorities that the instrumentation industry must tackle. A standardized and open image analysis work environment would allow one to save or export images to a preferred image analysis package, and this would open up the market for third-party algorithm developers. Definiens[76] and Cytoprint[77] are such companies active in hardware-independent automated image analysis. As an example, **Figure 2** shows an image captured in Tiff file format on the MIAS-2 system and analyzed using three different analysis algorithms.

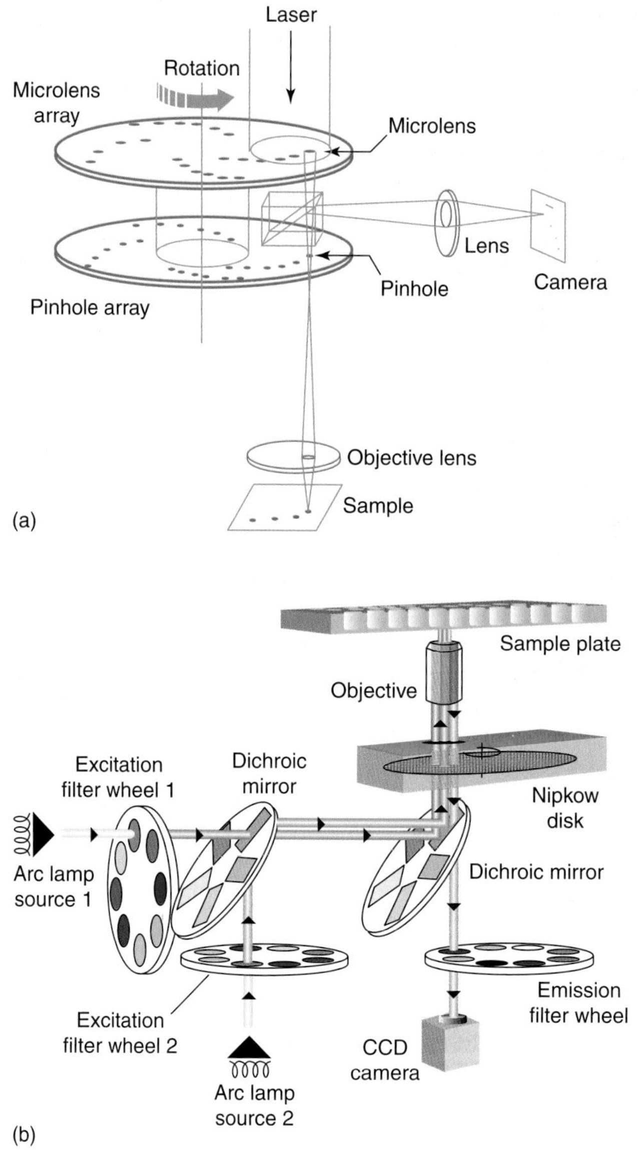

Figure 1 Light paths of (a) the Opera (b) Pathway HT confocal imaging systems. (Reprinted from Zemanova, L.; Schenk, A.; Valler, M. J.; Nienhaus, G. U.; Heilker, R. *Drug Disc. Today* **2004**, *8*, 1085–1093, Copyright (2004) with permission from Elsevier.)

3.32.7 Labeling Techniques

Cells can be visualized directly using bright-field microscopy, or advantage can be taken of the tremendous variety of reagents that have been developed over the years for fluorescent microscopy. Bright-field microscopy allows visualization of cells without any labeling; thus biological responses can be followed in real time without any interference. Such a noninvasive and universal approach can be used to count cells and to follow up cellular growth or colony growth, as recently achieved with the incorporation of the MAIA Scientific microscopy reader technology into the Cello platform from The Automation Partnership.[78] A bright-field application, however, has special technical requirements and is supported by only few suppliers. In contrast, the selection of an optimal combination of fluorescent probes to measure multiple, specific cellular events is generally very important for a successful HCS application. To label cellular targets, fluorescent probes and tools can be divided into the following categories: cellular stains, antibody-based labels, and fluorescent protein biosensors.

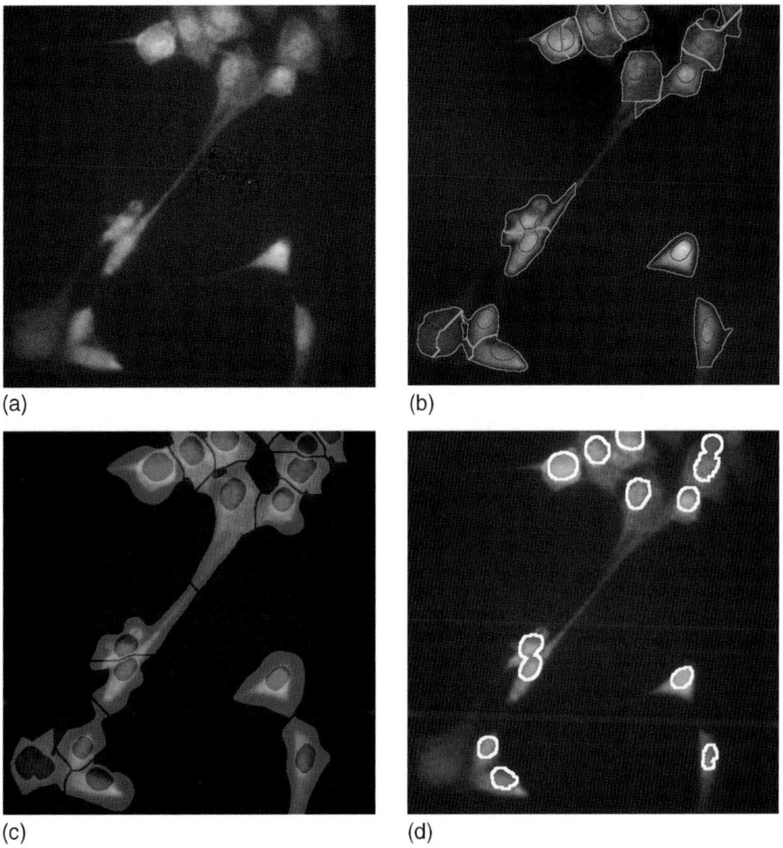

Figure 2 Image analysis with different algorithms. (a) Two mode image was captured using MIAS-2 system. The image was analyzed using different algorithms: (b) DCILabs, (c) MIAS-2 algorithm, and (d) IN Cell 3000. Blue color of the analysis mask shows the nucleus, green or white shows the cell membrane.

3.32.7.1 Cellular and Subcellular Stains

Direct staining of cells and cellular structures by specific dyes is a technique that has been used by microscopist for centuries. More recently, fluorescent probes have been developed to stain whole cells (Calcein-AM), cellular organelles (MitoTracker), DNA, and proteins, as well as biochemical events or second messengers (calcium influx with Fluo-4, membrane potential change with JC-1). Some of the most widely used labels are DNA dyes. These dyes can be used to label the nucleus either as a primary analysis parameter or simply as a topographical marker for individual cells, allowing other biological events to be quantified in reference to the cellular nucleus, providing confidence in image analysis. Dyes most often used for this purpose are the blue-shifted Hoechst 33342 and DAPI dyes[79] or the red-shifted DRAQ5 dye.[80]

3.32.7.2 Antibody Staining

The second large group of labeling techniques is based on the use of primary antibodies which specifically recognize the desired biological target. Visualization is achieved using a secondary antibody tagged with a fluorescent label. Antibodies can be used simply to track protein translocations in endpoint assays or to detect changes in protein expression levels. A particularly interesting group of antibodies are those that recognize modified, i.e., phosphorylated proteins. Staining with such phosphoprotein-specific antibodies is dependent on the activity of kinases and phosphatases and can be used to study signaling pathways within cells. A well-established example of such application is phospho extracellular signal-regulated kinase (ERK) staining. Using an anti phospho-ERK specific antibody, an endpoint assay can be designed to detect a kinase-dependent specific signal in fixed cells.[81] This assay technology is termed cell-based FLISA, and is used in cellular screening of kinase inhibitors. This type of assay and specific antibodies can also be used to measure expression levels of other proteins such as p21.[82] The use of those highly specific immunological reagents has also some disadvantages. For example, each of those antibodies has to be developed and tested for its applicability and specificity in HCS assay. This approach usually also requires cell fixation and permeabilization, so that live cell applications are not possible; and the assay set-up requires as many incubation and washing steps as ELISA.

3.32.7.3 Fluorescent Proteins Biosensors

Fluorescent protein biosensors are biological target molecules tagged with a fluorescent dye or fluorescent protein such as green fluorescent protein (GFP).[83,84] These biosensors can be used for noninvasive, homogenous methods to study fluorescence distribution within the cell with spatial and temporal resolution. Combining tools of molecular and cellular biology provides numerous ways for using these fluorescent protein biosensors. GFP was isolated from the light-emitting organ of the jellyfish *Aequorea victoria* in 1962[85] and it was cloned 30 years later by Prasher *et al.*[86] GFP has a intrinsic ability to generate fluorescence in the absence of any cofactors and it is mostly used as a genetically encoded tag. A number of spectral variants of GFP have been produced by mutagenesis. Those are wavelength-shifted GFP variants such as the blue FP, cyan FP, and yellow FP.[87] There are numerous applications reported in which GFP proteins, enhanced spectral mutants, and novel fluorescent proteins have been used.[88,89] Redistribution assay technology from Bioimage[90] makes use of tracking target movements from one cellular compartment to another in live or fixed cells. Another widely used GFP biosensor is a chimera consisting of β-arrestin fused to GFP.[91,92] This GFP–β-arrestin chimera has already been used to detect G protein-coupled receptor (GPCR) activation at the scale of full library HTS campaigns.[93]

Recently, an alternative antibody and protein labeling technique has emerged: nanoscale quantum dots. This has some advantages over traditional fluorescent chemical conjugates as the dots are more resistant to photobleaching and engineered to emit multiple fluorescent emission spectra after excitation by a single wavelength.[94] A new way to label cell surface proteins using a biophysical probe has been described by Chen *et al.*,[95] who used the *Escherichia coli* enzyme biotin ligase to ligate biotin sequence specifically to a short peptide. Accepting a ketone isostere, the enzyme creates ketone groups that can be tagged with a ketone probe and specifically conjugated to different molecules.[96] As the ketone group is absent in natural proteins, this method can be used to selectively derivatize proteins on the cell surface under physiological conditions.

As described above, there is a tremendous variety of state-of-the-art tools and reagents available for labeling cellular targets in HCS.

3.32.8 High-Content Screening Assay Examples and Applications

3.32.8.1 Translocation Events

Binding of agonistic small molecules or peptide ligands to a GPCR initiates a wide range of cellular events. Initially, the GPCR causes G proteins to activate the second messenger generating enzyme and is subsequently phosphorylated by GPCR-related kinases, which have been activated by the second messenger cascade. The phosphorylated GPCR binds to arrestin and aggregates on the cell surface prior to endocytosis. Once internalized, the GPCR is either targeted to the lysosome for degradation or dephosphorylated and recycled back to the plasma membrane.[97] Based on this cycle of events Xsira Pharmaceuticals have developed a GFP-based approach to monitor GPCR activation.[98] Transfluor technology is used to monitor translocation of arrestin–GFP fusion proteins upon receptor activation.[91,92] This universal GPCR technology is applicable to all GPCRs, and it is compatible with image-based HCS. The translocation can be monitored via HCS imagers by visualizing and quantifying membrane-associated fluorescent pits. High-affinity interactions of GPCRs with GFP–arrestins result in a further internalization event and the appearance of fluorescent vesicles within the cell. The ability to detect and quantify fluorescent pits and vesicles is becoming one of the standard assays to demonstrate the capabilities of HCS instruments and analysis algorithms. The advantage of this technology over conventional GPCR assays is the universality of the method for all different GPCR subtypes and signal transduction pathways and the uniform assay conditions. Another method to study receptor internalization has been developed by GE Healthcare. Their pH-sensitive cyanine dye CypHer-5 increases its fluorescence in the acidic environment of the endosome. This method has broader application as it can be used not only for GPCRs but also for any cell-surface receptor that internalizes upon activation.[97] The receptor of choice has to be modified to contain a VSV tag at the N-terminus. This tag is recognized by an antibody labeled with CypHer-5. Ligand activation will cause internalization of the receptor–antibody–ligand complex and the associated dye will become fluorescent in the acidic environment if excited with the appropriate red-shifted wavelength of 630 nm. Both GFP-tagged arrestins and CypHer techniques can be used as a generic GPCR or receptor assay to monitor agonist-mediated internalization.

Biosensors are very powerful tools in studying processes involved in signal transduction. Bioimage has developed a proprietary technology, called Redistribution, to study effects of compounds on intracellular targets.[62,90,99] This technology is specifically designed for the discovery of compounds that affect intracellular translocation events in response to signaling. Typical targets of Redistribution assays are kinases. Instead of looking at their catalytic activity, labeled kinases can be tracked in their cellular movements thus providing a functional, phenotypic readout of their activation. There are two advantages of this approach. First, targets that cannot be studied using standard drug discovery approaches, i.e., because of an unknown substrate of a particular enzyme, can be studied with this technique, exemplified by protein translocation. Second, kinases are studied in their natural and physiological environment within the cell. Other examples of targets that can be used in such assays are the Ets2 repressor factor, forkhead proteins or STAT3. One of the compounds discovered in a forkhead Redistribution screen is currently progressing toward lead optimization. The nuclear factor kappa B (NFκB) family of transcription factors is also extensively used in cytoplasm to nucleus translocation assays.[100] Activation of the cytoplasmatic complex consisting of NFκB and IκB results in IκB phosphorylation and the dissociation of the protein subunit p65. The dissociation exposes a nuclear localization sequence which allows a subsequent translocation of p65 from cytoplasm to the nucleus and binding to DNA. Such p65 translocation events can be monitored using p65 specific antibodies or, alternatively, GFP fusion proteins and HCS instrumentation.[101]

In addition to nuclear translocation, membrane to cytoplasm translocation is also a widely used readout in HCS screens. An Akt Redistribution assay has been described in which a EGFP–Akt1 fusion protein translocation from the cytoplasm to the plasma membrane was monitored.[102]

3.32.8.2 Other Multiparametric Cellular Events

3.32.8.2.1 Cellular toxicity

An additional broad field of application of HCS technology is in assays that monitor cell survival. Cell death can occur by two distinct mechanisms, necrosis and apoptosis. Necrosis is the pathological process triggered after exposure of cells to physical or chemical insults. Apoptosis or programmed cell death, on the other hand, is the physiological process used to eliminate unwanted cells during development or other natural biological events. Many cell-based assay technologies are available for studying cellular toxicity events such as necrosis or apoptosis. Cellomics offers an apoptosis, viability, and cytotoxicity kit suitable for HCS that measures multiple viability, toxicity, and apoptotic parameters.[67] Molecular Probes also offers number of cell death reagents and ready-to-use kits to measure both apoptosis and necrosis.[103] A number of other reagents are available to measure cytotoxic events. They include probes for mitochondrial membrane potential, cytochrome C release, nuclear condensation, and caspase activity. One of the early markers of apoptosis is an alteration of the plasma membrane where translocation of phosphatidylserine to the outside of a cell occurs. Annexin V is a calcium-dependent phospholipid-binding protein that binds with a high affinity to phosphatidylserine. Annexin V labeled with a fluorescent dye is widely used as a marker of early apoptosis. Another early apoptotic marker is caspase activation. Specific antibodies against cleaved (proteolytically activated) caspases are often used as markers for the detection of apoptosis.

3.32.8.2.2 The cell cycle

Especially for applications in cancer, understanding the mechanism and regulation of cell proliferation status is one of the major challenges in cell biology. Normal cell proliferation is regulated by a complex series of events, classified into distinct cell cycle stages. During the cell cycle, cells replicate their DNA and finally divide. Based on the cellular DNA content, cell cycle analysis separates the cell population into cells in G0/G1, S, G2, and M stages. Using laser scanning cytometry or HCS instruments, the total amount of DNA per cell can be precisely determined to obtain a cell cycle distribution. The overall DNA content information can be combined with the maximal pixel peak value to discriminate mitotic cells. As a typical cell cycle assay requires only one detection channel, it can be easily combined with the measurement of other markers. Very often, the cell cycle assay is combined with the measurements of apoptosis, thus allowing potential toxic effects of investigated compounds to be identified early in the discovery process. A different approach to measuring the cell cycle is used by GE Healthcare, where expression of an EGFP is used as a reporter signal, placed under the control of the cyclin B1 promoter. This promoter limits EGFP production to late S and G2 phases of the cell cycle. Cellular localization of the reporter protein is controlled by the cytoplasmatic retention sequence, while its destruction is controlled by the cyclin B1 D-box. Thus in this assay the cell cycle phase is determined on the basis of the expression pattern and intensity of EGFP fluorescence.[104] Another important cell cycle parameter is the mitotic index, defined as a percentage of cells within the M phase.[96] Histone H3 phosphorylation is a well-characterized biological marker for M phase. An antibody specific for phosphohistone H3 can be used to assess the mitotic index of cell populations after treatment with compounds that have a potential to interfere with the cell cycle. Nucleotide analogs such as 5-bromo-2-deoxyuridine (BrdU) are employed to quantify the duplication of DNA during S phase (BrdU is a thymidine analog, which is incorporated during the synthesis of DNA and can be easily detected by commercially available antibodies). BrdU incorporation is one of the most common methods for detecting DNA synthesis and cell proliferation.[96]

3.32.8.2.3 Neurite outgrowth

Marked morphological changes within the cell population can also be a valuable readout parameter. Neurite outgrowth, for example, is an important event in neuronal development. Stimulation or inhibition of neurite growth are implicated in a wide range of central nervous system disorders. Stimulation of neurite outgrowth is the most relevant parameter for screening of compounds with neurotrophic activity. Thus, such compounds might find their application in disorders such as stroke, Parkinson's disease, Alzheimer's disease, and spinal cord injury. Drug discovery projects have therefore focused on the identification of new compounds that promote neurite growth and branching. Traditionally, counting was done manually on images taken form the microscopic field.[105] Manual methods are time-consuming, subjective, and unsuitable for large-scale compound screening. Therefore, there is a need for an automated neurite measurement that can be used for compound screening at a larger scale. Most current microscopy systems and analysis algorithms are based on fluorescence labeling of neurites. Ramm *et al.* have published a validation of such automatic neurite outgrowth measurement using the IN Cell Analyzer 1000 and fluorescence imaging at 10× magnification.[106] He reported that no difference was observed between manual and automatic scoring for five relevant parameters, i.e., number of neurites, neurite length, total cell area, number of cells, and neurite length per cell. Another widely used neurite outgrowth analysis package is developed by CSIRO.[107] CSIRO's neurite outgrowth detection software provides robust image analysis routines for neurons and neuronlike cells. This analysis algorithm is integrated into Evotec, BD, and Molecular Devices HCS analysis software. An automated neurite outgrowth application module is also described by Cellomics.[67]

3.32.8.2.4 Micronucleus and colony formation

The micronucleus assay is frequently used to assess chromosomal damage as a consequence of mutagen exposure.[108,109] Micronuclei are chromosomes or chromosomal fragments that become separated during mitosis. Scoring of micronuclei is mostly done manually or semiautomatically using image analysis. Although the criteria for scoring have been standardized in order to minimize analysis errors, the manual scoring system remains an important source of variability. In addition, manual counting is also a very time-consuming and tiring process. More than 20 years ago, attempts were made to automate the counting of micronuclei using flow cytometry. The major disadvantages of flow cytometry are that the measurements are performed in suspension and that it is not possible to relate micronuclei to individual cells. To overcome those limitations, either LSCs or image-based analysis could be used.[110] A comparison of results obtained by automated image analysis with those of visual scoring should be used as a very important parameter for assessment of automated systems.[109] Another assay where imaging is proving to be a very useful addition is the colony formation assay. Colony formation assay or clonogenic assay is used to measure cellular response to different compounds, mutagens, and radiation. Mostly, large numbers of colonies and a large area of culture plates need to be analyzed to obtain a statistically significant sample population.[111] As manual counting of colonies is time-consuming and tedious, there is a need for an automated approach. DCILabs have developed an automated routine that delivers results that are comparable to manual counting.[112] Wells are scanned at low resolution and large mosaics are constructed for the image analysis algorithm.

3.32.9 Data Analysis and Storage

Naturally, with the introduction of HCS, the volume of data produced in screening campaigns will increase by orders of magnitude. High-speed microscopy creates large image files and gigabytes (Gb) of data on a daily basis. Typically, a well-based endpoint assay creates approximately 1–10 kB of data for one 384-well plate. The same plate scanned with an HCS instrument could generate an equivalent of 2 GB, which is up to a million-fold increase in data volume. Thus, recent improvements in HCS technology and availability of fluorescent bioprobes have not only provided a new platform for drug discovery, but also generated additional bottlenecks in downstream processes by flooding the IT infrastructures with terabytes of data from a single screening campaign. Therefore, there is a growing need to create a flexible database infrastructure capable of managing terabytes of data when high-throughput, high-resolution, multiparametric HCS is employed. Full integration of this instrumentation and analysis software with data mining software is crucial for the successful implementation of HCS in today's drug discovery processes. Currently, only few manufacturers are developing fully integrated tools from image acquisition to data mining.[113] However, the potential pitfall of such fully integrated approaches is the use of proprietary sets of imaging and analysis tools that cannot be integrated with other instrumentation and software. An alternative approach, without such limitations, is to integrate HCS instrumentation into an open-platform database structure.[74]

3.32.10 Multidimensional Cytological Profiling

The applications described above are all phenotypic experiments in which small molecules are tested in order to find specific effects on cells. Phenotypic experiments can be used in HTS mode to find hits or later to further characterize smaller sets of compounds in secondary screens. A variant of the latter application has recently been extended, and is described as cytological drug-profiling.[114,115] In this study the biological activity of a limited number of small molecules was characterized in depth, testing numerous cellular variables at once after exposure to different compound concentrations. Perlman *et al.* assembled a test set of 100 compounds, 90 of which were drugs with known mechanisms of action, and selected 11 distinct assay readouts covering a broad range of biological phenotypes. In total 93 descriptors were extracted from more than 600 000 images yielding $\sim 10^9$ data points. This immense data set was clustered by software into classes represented by distinct heat plots describing relative changes of descriptors from control values. Using such data reduction approach most of the 61 drugs that showed a strong response in HeLa cells could be clustered into groups based on their heat plot patterns. These groups reflected a common reported target or mechanism of action, as confirmed by blinded compounds included in the test set. Therefore, the described application of multiparametric cytological profiling can be used to suggest the mechanism of action for new drugs. Interestingly, although the assay descriptors were selected in a hypothesis-independent manner, none of them could be omitted from the data set without losing analytical power.[115] This technique for concentration–response profiling is an elegant way to characterize the effects of drugs and can be used in the hit-to-lead phase of drug discovery in which many compounds have to be evaluated. Alternatively, such biological profiles could be used for the selection in stages of lead optimization, as they provide a fast tool for ranking clinical candidates and discriminating them from competitors.

3.32.11 Applications of High-Content Screening within the Drug Discovery Process

As we have already described, HCS technologies are being used more and more in the pharmaceutical industry to improve the efficiency and quality of the lead compound generation process. Many technical developments have been achieved in hardware and software development and the technology is becoming more mature. Rather than just relying on testing large numbers of compounds in simple biochemical assays, HCS uses sophisticated assays and algorithms to increase the chance of finding leads with presumably higher potential of becoming a successful drug. HCS measures multiple parameters of individual cells, providing better quality data and a closer estimate of the biological complexity. As cultured cells are biologically more relevant than isolated proteins and HCS data are much more informative than simple readouts from HTS screens, there is growing support in the field for the hypothesis that HCS might provide greater speed and efficiency in generating new drugs. However, it has to be emphasized that, as discussed in earlier debates in the field, the biochemical and cellular screening approaches still have their intrinsic pros and cons.[116] Although the natural environment of the cell will show the biochemical targets in the biologically relevant context, cellular screening of intracellular targets has the disadvantage that the penetration of the cellular membrane may distort the chemical structure–activity relationship and therefore renders the series analysis of hits and chemical lead optimization difficult. On the other hand, screening directly for compounds already having relevant cellular activity might simplify the hit selection process and shorten lead optimization cycles. Monastrol, for example, an inhibitor of the mitotic kinesin Eg5, was discovered using automated microscopy in a combination of two phenotype-based screens.[117] Potent Eg5 inhibitors are currently in clinical trials for cancer treatment. A recently described Akt1 translocation assay delivered active hits with cellular activity after the screening of a small library.[102] This is remarkable, since despite the fact that Akt1 is a druggable and validated target for cancer, no specific inhibitors from biochemical screens had previously been reported.

Phenotypic phenomena might be a result of multiple biochemical targets triggering cellular networks, and the identification of each single interacting target might not always be successful. Although targeting relevant cellular phenotypes might partially circumvent the target validation bottleneck in drug discovery, a potential 'black box' component of these approaches has to be addressed by intelligent design of control experiments and postscreening target identification. Target-oriented projects have been the mainstream of drug discovery since the beginning of the era of molecular biology. Indeed it is intellectually extremely rewarding to be able to compensate pathophysiological states by influencing the activity of a single target molecule. However, the limitations of such target-based paradigms are more and more widely recognized, and phenotype-oriented approaches are considered as an innovative alternative.[118]

Recent technical developments in the field of HCS technologies have opened new possibilities for a number of steps in the drug discovery process, such as target validation, lead finding, and lead optimization. However, efficient integration of multiparametric, phenotypic approaches into productive programs will still require hardware and software developments, improvements in assay know-how, and a powerful IT environment. Most important, however, will be the integration of these technical possibilities into a novel, biology-centered paradigm in the drug discovery process, which will require an extension of the single target-focused mind set. Recent developments in the field of HCS have been encouraging; however, it will take several more years before HCS will improve drug discovery success rates.

References

1. Sneader, W. In *Comprehensive Medicinal Chemistry, Vol. 1*; Hansch, C., Sammes, P. G., Taylor, J. B., Eds.; Pergamon Press: Oxford, 1990, pp 7–81.
2. Schreiber, S. L. *Bioorg. Med. Chem.* **1998**, *6*, 1127–1152.
3. Austin, C. P.; Brady, L. S.; Insel, T. R.; Collins, F. S. *Science* **2004**, *306*, 1138–1139.
4. Sneader, W. *Drug Discovery: The Evolution of Modern Medicines*; John Wiley: New York, 1985.
5. The First Forum on Data Management Technologies in Biological Screening, SRI International, Menlo Park, CA, April 22–24, 1992.
6. Society for Biomolecular Screening. http://www.sbsonline.org/ (accessed August 2006).
7. Association for Laboratory Automation. http://www.labautomation.org/ (accessed August 2006).
8. Fox, S.; Farr-Jones, S.; Sopchak, L.; Boggs, A.; Comley, J. *J. Biomol. Screen.* **2004**, *9*, 354–358.
9. Drews, J. *Science* **2000**, *287*, 1960–1964.
10. Devlin, J. P. In *Integrated Drug Discovery Technologies*; Mei, H. -Y., Czarnik, A. W., Eds.; Marcel Dekker: New York, 2002, pp 221–246.
11. Lipinski, C. A.; Lombardo, F.; Dominy, B. W.; Feeney, P. J. *Adv. Drug Deliv. Rev.* **1997**, *23*, 3–25.
12. Hann, M. M.; Oprea, T. I. *Curr. Opin. Chem. Biol.* **2004**, *8*, 255–263.
13. Teague, S. J.; Davis, A. M.; Leeson, P. D.; Oprea, T. *Angew. Chem. Int. Ed. Engl.* **1999**, *38*, 3743–3748.
14. Kozikowski, B. A.; Burt, T. M.; Tirey, D. A.; Williams, L. E.; Kuzmak, B. R.; Stanton, D. T.; Morand, K. L.; Nelson, S. L. *J. Biomol. Screen.* **2003**, *8*, 210–215.
15. Kozikowski, B. A.; Burt, T. M.; Tirey, D. A.; Williams, L. E.; Kuzmak, B. R.; Stanton, D. T.; Morand, K. L.; Nelson, S. L. *J. Biomol. Screen.* **2003**, *8*, 205–209.
16. Cheng, X.; Hochlowski, J.; Tang, H.; Hepp, D.; Beckner, C.; Kantor, S.; Schmitt, R. *J. Biomol. Screen.* **2003**, *8*, 292–304.
17. Baurin, N.; Baker, R.; Richardson, C.; Chen, I.; Foloppe, N.; Potter, A.; Jordan, A.; Roughley, S.; Parratt, M.; Greaney, P. et al. *J. Chem. Inf. Comput. Sci.* **2004**, *44*, 643–651.
18. Harper, G.; Pickett, S. D.; Green, D. V. S. *Combin. Chem. High Throughput Screen.* **2004**, 7, 63–71.
19. Chan, J. A.; Hueso-Rodriguez, J. A. *Methods Mol. Biol.* **2002**, *190*, 117–127.
20. Archer, J. R. *Assay Drug Dev. Technol.* **2004**, *2*, 675–681.
21. Friedman, J. M.; Halaas, J. L. *Nature* **1998**, *395*, 763–770.
22. Hopkins, A. L.; Groom, C. R. *Nat. Rev. Drug Disc.* **2002**, *1*, 727–730.
23. Cochran, A. G. *Chem. Biol.* **2000**, 7, R85–R94.
24. Jones, S.; Thornton, J. M. *Proc. Natl. Acad. Sci. USA* **1996**, *93*, 13–20.
25. Arkin, M. R.; Wells, J. A. *Nat. Rev. Drug Disc.* **2004**, *3*, 301–317.
26. Comley, J.; Fox, S. *Drug Disc. World* **2004**, *Spring*, 25–34.
27. Elands, J. P. In *Handbook of Drug Screening*; Seethala, R., Fernandes, P. B., Eds.; Marcel Dekker: New York, 2001, pp 477–492.
28. Comley, J. *SP*^{2} **2004**, June, 24–36.
29. Tecan. http://www.tecan.com/ (accessed August 2006).
30. Beckman. http://www.beckman.com/ (accessed August 2006).
31. Perkin Elmer. http://www.perkinelmer.com/ (accessed August 2006).
32. Caliper Life Sciences. http://caliperls.com/ (accessed August 2006).
33. TekCel. http://www.tekcel.com/ (accessed August 2006).
34. Cybio. http://www.cybio-ypn.com/.
35. The Automation Partnership. http://www.automationpartnership.com/ (accessed August 2006).
36. ThermoLabsystems. http://www.thermo.com/ (accessed August 2006).
37. SSI Robotics. http://www.ssirobotics.com/ (accessed August 2006).
38. Caillet, C.; Pegon, Y.; Le Neel, T.; Morin, D.; Baudiment, C.; Truchaud, A. *J. Assoc. Lab. Autom.* **2005**, *10*, 48–53.
39. Laboratory Robotics Interest Group (LRIG) http://www.lab-robotics.org/ and European Laboratory Robotics Interest Group (ELRIG) http://www.elrig.org/ (accessed August 2006).
40. Beveridge, M.; Park, Y. W.; Hermes, J.; Marenghi, A.; Brophy, G.; Santos, A. *J. Biomol. Screen.* **2000**, *5*, 205–211.
41. Lavery, P.; Brown, M. J. B.; Pope, A. J. *J. Biomol. Screen.* **2001**, *6*, 3–9.
42. Turconi, S.; Shea, K.; Ashman, S.; Fantom, K.; Earnshaw, D. L.; Bingham, R. P.; Haupts, U. M.; Brown, M. J. B.; Pope, A. J. *J. Biomol. Screen.* **2001**, *6*, 275–290.
43. Harris, A.; Cox, S.; Burns, D.; Norey, C. *J. Biomol. Screen.* **2003**, *8*, 410–420.
44. Wolcke, J.; Ullmann, D. *Drug Disc. Today* **2001**, *6*, 637–646.
45. Comley, J. *Drug Disc. World* **2004**, Summer, 43–54.
46. Dhundale, A.; Goddard, C. *J. Biomol. Screen.* **1996**, *1*, 115–118.
47. Suto, C. M.; Ignar, D. M. *J. Biomol. Screen.* **1997**, *2*, 7–9.
48. Maffia, A. M., III; Kariv, I.; Oldenburg, K. R. *J. Biomol. Screen.* **1999**, *4*, 137–142.

49. Fitzgerald, L. R.; Mannan, I. J.; Dytko, G. M.; Wu, H. L.; Nambi, P. *Anal. Biochem.* **1999**, *275*, 54–61.
50. Johnston, P. A. *Drug Disc. Today* **2002**, *7*, 353–363.
51. Williams, C. *Nat. Rev. Drug Disc.* **2004**, *3*, 125–135.
52. Schroeder, K. S. *J. Biomol. Screen.* **1996**, *1*, 75–80.
53. Comley, J. *Drug Disc. World* **2004**, *Winter*, 49–60.
54. Zheng, W.; Spencer, R. H.; Kiss, L. *Assay Drug Dev. Technol.* **2004**, *2*, 543–552.
55. Zhang, J. H.; Chung, T. D. Y.; Oldenburg, K. R. *J. Biomol. Screen.* **1999**, *4*, 67–73.
56. Comley, J. *Drug Disc. World* **2003**, *Summer*, 91–97.
57. McGovern, S. L.; Caselli, E.; Grigorieff, N.; Shoichet, B. K. *J. Med. Chem.* **2002**, *45*, 1712–1722.
58. Sills, M. A.; Weiss, D.; Pham, Q.; Schweitzer, R.; Wu, X.; Wu, J. Z. J. *J. Biomol. Screen.* **2002**, *7*, 191–214.
59. Wu, X.; Glickman, J. F.; Bowen, B. R.; Sills, M. A. *J. Biomol. Screen.* **2003**, *8*, 381–392.
60. Comley, J. *Drug Disc. World* **2005**, *Winter*, 63–74.
61. Lofas, S. *Assay Drug Dev. Technol.* **2004**, *2*, 407–415.
62. Almholt, D. L.; Loechel, F.; Nielsen, S. J.; Krog-Jensen, C.; Terry, R.; Bjorn, S. P.; Pedersen, H. C.; Praestegaard, M.; Moller, S.; Heide, M. et al. *Assay Drug Dev. Technol.* **2004**, *2*, 7–20.
63. Honeysett, J. M. *PharmaGenomics* **2003**, *October*, 33–44.
64. Giuliano, K. A.; DeBiasio, R. L.; Dunlay, R. T.; Gough, A.; Volosky, J. M.; Zock, J.; Pavlakis, G.; Taylor, D. L. *J. Biomol. Screen.* **1997**, *2*, 249–259.
65. IN Cell 3000. http://www.gehealthcare.com/ (accessed August 2006).
66. Evotec Technologies. http://www.evotec-technologies.com/ (accessed August 2006).
67. Cellomics. http://www.cellomics.com/ (accessed August 2006).
68. Pathway HT. http://www.bd.com/ (accessed August 2006).
69. Acumen Explorer. http://www.ttplabtech.com/ (accessed August 2006).
70. Olympus Europa. http://www.olympus-europa.com/ (accessed August 2006).
71. Zemanova, L.; Schenk, A.; Valler, M. J.; Nienhaus, G. U.; Heilker, R. *Drug Disc. Today* **2003**, *8*, 1085–1093.
72. Apo Tome. http://www.zeiss.de/ (accessed August 2006).
73. Geusebroek, J. M.; Cornelissen, F.; Smeulders, A. W.; Geerts, H. *Cytometry* **2000**, *39*, 1–9.
74. Cik, M. *Eur. Pharmaceut. Rev.* **2004**, 47–49.
75. Trask, O. J. Jr.; Large, T. H. *Curr. Drug Disc.* **2001**, *September*, 25–29.
76. Definiens. http://www.definiens.com/ (accessed August 2006).
77. Cytoprint. http://www.cytoprint.com/ (accessed August 2006).
78. The Automation Partnership. http://www.maia-scientific.com/ (accessed August 2006).
79. Hoechst and DAPI dyes. http://www.probes.com/ (accessed August 2006).
80. Biostatus. http://www.biostatus.co.uk/ (accessed August 2006).
81. Vogt, A.; Cooley, K. A.; Brisson, M.; Tarpley, M. G.; Wipf, P.; Lazo, J. S. *Chem. Biol.* **2003**, *10*, 733–742.
82. Grand-Perret, T.; Cik, M.; Arts, J.; Vander, B. A.; Ercken, M.; Valckx, A.; Vermeesen, A.; Roevens, R.; Janicot, M. *Drugs Exp. Clin. Res.* **2004**, *30*, 89–98.
83. Chen, I.; Ting, A. Y. *Curr. Opin. Biotechnol.* **2005**, *16*, 35–40.
84. Giuliano, K. A.; Taylor, D. L. *Trends Biotechnol.* **1998**, *16*, 135–140.
85. Shimomura, O.; Johnson, F. H.; Saiga, Y. *J. Cell. Comp. Physiol.* **1962**, *59*, 223–240.
86. Prasher, D. C.; Eckenrode, V. K.; Ward, W. W.; Prendergast, F. G.; Cormier, M. J. *Gene* **1992**, *111*, 229–233.
87. Pollok, B. A.; Heim, R. *Trends Cell Biol.* **1999**, *9*, 57–60.
88. Misteli, T.; Spector, D. L. *Nat. Biotechnol.* **1997**, *15*, 961–964.
89. Miyawaki, A. *Curr. Opin. Neurobiol.* **2003**, *13*, 591–596.
90. Bioimage. http://www.bioimage.com/ (accessed August 2006).
91. Barak, L. S.; Oakley, R. H.; Shetzline, M. A. *Assay Drug Dev. Technol.* **2003**, *1*, 409–424.
92. Oakley, R. H.; Hudson, C. C.; Cruickshank, R. D.; Meyers, D. M.; Payne, R. E., Jr.; Rhem, S. M.; Loomis, C. R. *Assay Drug Dev. Technol.* **2002**, *1*, 21–30.
93. Garippa, R. J. Cambridge Healthtech Institute 2nd High-Content Analysis Conference, 2005.
94. Quantum dots. http://www.qdots.com/.
95. Chen, I.; Howarth, M.; Lin, W. Y.; Ting, A. Y. *Nat. Methods* **2005**, *2*, 99–104.
96. Gasparri, F.; Mariani, M.; Sola, F.; Galvani, A. *J. Biomol. Screen.* **2004**, *9*, 232–243.
97. Milligan, G. *Drug Disc. Today* **2003**, *8*, 579–585.
98. Xsira Pharmaceuticals. http://www.xsira.com/ (accessed August 2006).
99. Pedersen, H. C.; Pagliaro, L. *Gene. Eng. News* **2004**, *24*, 1–2.
100. Vakkila, J.; DeMarco, R. A.; Lotze, M. T. *J. Immunol. Methods* **2004**, *294*, 123–134.
101. Tenjinbaru, K.; Furuno, T.; Hirashima, N.; Nakanishi, M. *FEBS Lett.* **1999**, *444*, 1–4.
102. Lundholt, B. K.; Linde, V.; Loechel, F.; Pedersen, H. C.; Moller, S.; Praestegaard, M.; Mikkelsen, I.; Scudder, K.; Bjorn, S. P.; Heide, M. et al. *J. Biomol. Screen.* **2005**, *10*, 20–29.
103. Molecular Probes. http://www.probes.com/.
104. GE Healthcare. http://www.gehealthcare.com/ (accessed August 2006).
105. Bilsland, J.; Rigby, M.; Young, L.; Harper, S. *J. Neurosci. Methods* **1999**, *92*, 75–85.
106. Ramm, P.; Alexandrov, Y.; Cholewinski, A.; Cybuch, Y.; Nadon, R.; Soltys, B. J. *J. Biomol. Screen.* **2003**, *8*, 7–18.
107. CSIRO. http://www.cmis.csiro.au/ (accessed August 2006).
108. Fenech, M. *Mutat. Res.* **2000**, *455*, 81–95.
109. Varga, D.; Johannes, T.; Jainta, S.; Schuster, S.; Schwarz-Boeger, U.; Kiechle, M.; Patino, G. B.; Vogel, W. *Mutagenesis* **2004**, *19*, 391–397.
110. Smolewski, P.; Ruan, Q.; Vellon, L.; Darzynkiewicz, Z. *Cytometry* **2001**, *45*, 19–26.
111. Dahle, J.; Kakar, M.; Steen, H. B.; Kaalhus, O. *Cytometry* **2004**, *60A*, 182–188.
112. DCILabs. http://www.users.pandora.be/dcilabs/ (accessed August 2006).
113. Abraham, V. C.; Taylor, D. L.; Haskins, J. R. *Trends Biotechnol.* **2004**, *22*, 15–22.

114. Perlman, Z. E.; Slack, M. D.; Feng, Y.; Mitchison, T. J.; Wu, L. F.; Altschuler, S. J. *Science* **2004**, *306*, 1194–1198.
115. Mitchison, T. J. *Chembiochem* **2005**, *6*, 33–39.
116. Moore, K.; Rees, S. *J. Biomol. Screen.* **2001**, *6*, 69–74.
117. Mayer, T. U.; Kapoor, T. M.; Haggarty, S. J.; King, R. W.; Schreiber, S. L.; Mitchison, T. J. *Science* **1999**, *286*, 971–974.
118. Butcher, E. C. *Nat. Rev. Drug Disc.* **2005**, *4*, 461–467.

Biographies

Miroslav Cik obtained his PhD in molecular neurobiology at the University of London, UK. In 1995, Miroslav joined the Department of Biochemical Pharmacology at Johnson & Johnson Pharmaceutical Research & Development, Belgium, where he became major investigator and a project leader for a central nervous system target. In 2001, Miroslav joined the Assay Development and High-Throughout Screening department. As head of assay development, Miroslav is responsible for the design and development of HTS cell-based assays for receptors and intracellular targets. He is also responsible for evaluation and implementaion of novel technologies and instrumentation, such as automated microscopy platforms for high-content screening.

Mirek R Jurzak received his PhD in biology at the Wolfgang Goethe University after practical thesis work at the Max-Planck Institute for Biophysics (both in Frankfurt, Germany). He had postdoctoral posts at the Max-Planck Institute for Clinical and Physiological Research, the W G Kerckhoff Institute in Bad Nauheim, Germany, and the Department of Psychology and Pharmacology at the University of lowa, lowa City, IA, USA. In 1996 he joined the Janssen Research Foundation in Beerse, Belgium, where he held several positions, first leading an assay development laboratory and later the department of Assay Development and High-Throughput and Screening. Since 2002 he has been head of the Central Assay Development and Screening Department at Merck KGaA, Darmstadt, Germany.

Comprehensive Medicinal Chemistry II
ISBN (set): 0-08-044513-6

ISBN (Volume 3) 0-08-044516-0; pp. 679–696

3.33 Combinatorial Chemistry

P Seneci, Desenzano del Garda, Italy

3.33.1	**Scope and Definitions**	**697**
3.33.2	**History and Trends**	**697**
3.33.2.1	Numbers, Diversity, and Quality	697
3.33.2.2	Combinatorial Chemistry and Natural Products: Friends or Foes?	699
3.33.2.3	Expanding the Scope of Medicinal Chemistry: Target Identification and Validation	700
3.33.3	**Solid-Phase Libraries in Medicinal Chemistry**	**702**
3.33.3.1	Chemical Assessment on Solid Phase: A Gateway toward Medicinal Chemistry Libraries	703
3.33.3.2	Solid-Phase Libraries of Discretes	708
3.33.3.3	Solid-Phase Pool Libraries	726
3.33.4	**Medicinal Chemistry Libraries: Examples**	**738**
3.33.4.1	Chemical Assessment on Solid-Phase	738
3.33.4.2	Solid-Phase Pool Libraries	741
3.33.4.3	Solid-Phase Libraries of Discretes	746
References		**753**

3.33.1 Scope and Definitions

The term 'combinatorial chemistry' has become popular in pharmaceutical research (and in other areas) since the late 1980s, although its meaning has evolved through the years. This chapter will briefly survey past achievements, describe current favorite approaches, and look at future trends of combinatorial chemistry applied to drug discovery projects in general and to medicinal chemistry in particular.

Combinatorial libraries, that is, collections of compounds prepared simultaneously using the same synthetic route, have been prepared either on solid phase (SP),[1,2] or in solution. Both formats have been extensively used in drug discovery, but this chapter will focus only on the former as the latter is covered in detail in other chapters (*see* 3.34 Solution Phase Parallel Chemistry; 3.35 Polymer-Supported Reagents and Scavengers in Synthesis).

SP libraries of discretes (each compound obtained as spatially separated from the other library components, and screened as a discrete on relevant biological targets) and of mixtures (prepared as pools of compounds with predefined composition by split-and-mix synthesis,[3,3a,3b] and screened as such) have both been popular in the past, and are still used widely.

SP libraries of discretes and mixtures have been prepared by medicinal chemists to find new active structures (hit discovery), to optimize their druggability (hit to lead), and to select a preclinical candidate (lead optimization). More recently, medicinal chemists have used SP libraries as tools to help the elucidation of biochemical pathways, and even to identify or validate novel targets in drug discovery. These applications of combinatorial chemistry will be reviewed here, and appropriate references will be provided.

3.33.2 History and Trends

3.33.2.1 Numbers, Diversity, and Quality

Combinatorial chemistry, and combinatorial libraries, were first described in drug discovery in the early 1980s.[4,4a,4b] Although combinatorial libraries of biological[5] and inorganic[5a] nature were reported at the same time, or even earlier, synthetic libraries became popular in pharmaceutical research at that time due to so-called high-throughput screening (HTS).[6,6a] In fact, for some years it had been possible to test many compounds for their activity on a selected target in a short period of time, using automated assays and HTS protocols. Large pharmaceutical companies were screening their historical compound collections, but a method to increase the productivity of medicinal chemists to match the requirements of HTS, and thus to allow the evaluation of many more compounds per biological target, was needed. Combinatorial chemistry was the answer to such a need.

At first, the focus was on creating large libraries, and methods to obtain them were favored. Libraries of peptides and oligonucleotides, whose synthesis is repetitive and could easily be automated, were the first to be reported. SP synthesis (SPS) was ideally suited to these biomolecules, and for the quick purification of intermediate and final supported compounds via multiple solvent washings. The sensation created by the report by Furka[3] on split-and-mix synthesis, and the appearance of two back-to-back relevant papers in *Nature*,[3a,3b] shifted even more early combinatorial efforts toward SP libraries. Using split-and-mix synthesis, large numbers of individual targets are prepared as mixtures by mixing and portioning resin aliquots appropriately. Due to its spatial isolation from other beads, though, each resin bead at the end of the multiple synthetic protocol carries only one final compound, which can be both analytically characterized and biologically tested in a selected assay. Several reviews have extensively covered this methodology and the libraries made by using it.[7,7a,7b] It is fair to say that some academic and private groups at that time targeted huge libraries, to find hits on each possible biological target.

However, the focus of SP libraries began to change over time, the effects of which are visible in the structure and composition of modern libraries, and can be summarized in three words: diversity, rationale, and quality.

Initially, SP libraries comprised peptides and nucleotides. Their usefulness, though, proved to be limited due to several factors. First, as tools to discover new structural hits for targets of interest, they are bound to give bio-oligomeric ligands that suffer from well-known drawbacks (instability to proteases and nucleases, suboptimal physicochemical properties, etc.). Second, their diversity is also limited, as their common scaffold (α-amino acidic, or phosphoester-sugar-nucleobase) is only diversifed by nature and order of the amino acid side chains, or of the nucleobases. Small organic molecule (SOM) libraries, in the contrast, can be produced through any organic chemistry transformation, and sample a much wider portion of the so-called chemical diversity space through diverse scaffolds and functional groups to be decorated.

The first SOM library was reported in 1992,[8] and thereafter the efforts of medicinal chemists have largely focused on this large class of libraries. The nonrepetitive nature of SOM chemistry, and the wide range of experimental conditions needed to prepare different SOM libraries, make the chemical assessment and synthesis of these libraries more complex than for their bio-oligomeric counterparts. A careful assessment, including the rehearsal of building blocks for library synthesis, often leads to a smaller size for the planned SOM library. A smaller size, but a high diversity content, is also driven by computational methods. These allow selection, among large 'virtual' libraries described in silico by their calculated properties, of compound subsets with more probability of spanning the diversity space efficiently (hit discovery libraries); of better interacting with a computational model of the target pharmacophore (hit to lead, lead optimization); or of better complying with the theoretical properties influencing the 'druggability' of a compound (Lipinski rule of five, etc., in lead optimization). The use of virtual libraries in drug discovery to orient the efforts of medicinal chemists from early hit discovery to late lead optimization is now widespread. For a more detailed description, which is not the focus of this chapter, the reader should see recent reviews.[9,9a,9b]

Last but not least, there is the quality aspect. As any medicinal chemist knows, the issue is not how to synthesize large numbers of compounds in short time – for example, we could put an equimolar mixture of 100 amines and 100 carboxylic acids under the right experimental conditions, and obtain a mixture where each of the theoretically obtainable carboxylic amides is surely represented – the problem is how to determine if an active principle is present in a mixture comprising many compounds, and what its structure is.

Split-and-mix libraries address both the high-throughput synthesis and the structure determination requirements. For the latter, two main approaches exist. First, deconvolutive methods,[10,10a,10b] where the complexity of a mixture is simplified to obtain the structure of an active compound via iterative synthesis and biological testing of smaller pools. Second, encoding methods,[10b,11,11a] where the structure of a library component supported on a single bead is determined after identifying which bead(s) supports an active compound. This is obtained by determining the structure of one or more chemical tags also linked to the resin bead (one tag, one library individual: chemical encoding); or via reading a nonchemical code attributed to each library component and to the resin beads supporting it (radiofrequency or laser optical tags: non-chemical encoding).

In reality, encoding methods add complexity to a planned combinatorial synthetic strategy. If chemical tags are involved, these correspond to a more extensive chemical assessment, to more synthetic steps, and to more synthetic restrictions, to ensure the integrity of the tags during the whole synthesis. If any encoding is involved, additional instrumentation and techniques are needed to 'read' the code and/or to perform properly each encoding step.

If deconvolution methods are involved, screening of pools of compounds takes place, often leading to false positives (synergistic effects among pool members, active impurities, etc.) or false negatives (compounds theoretically included into the library structure, but in reality not prepared due to unexpected synthetic hurdles). As important as the above, the synthesis of large split-and-mix libraries would require a huge effort to analytically characterize the whole library. Although high-throughput analysis has made significant improvements recently, only a statistical evaluation of large pool libraries can take place (typically not more than 10–15% of library members), thus leaving a considerable amount of uncertainty about purity and composition of the library.

To summarize, neither medicinal chemists nor biologists have a lot of confidence in large split-and-mix libraries to find hits, or to progress them to leads, or to optimize them. Their preference is for smaller discrete libraries, whose quality can be more thoroughly determined.[12,12a–12c] As for diversity and rationale, a small set of medium-sized discrete libraries of SOMs, derived from larger virtual libraries filtered for activity/druggability, spans much more of a hypothetical drug diversity space than the old, large bio-oligomeric libraries. As for profiling, the smaller size allows preparation of multimilligram amounts of each analytically pure compound, thus allowing its testing not only in primary in vitro efficacy screening but also in secondary assays yielding a fuller druggability profile for each component of the whole library. High-throughput purification techniques are widely used[13,13a,14] (*see* 3.37 High-Throughput Purification).

Most of today's efforts in combinatorial medicinal chemistry are devoted to the synthesis of solution- and SP discrete libraries of small organic molecules. Most, though, does not mean all: SP split-and-mix libraries are still prepared and tested, especially as tools in early steps of the drug discovery process, as we will see later in this chapter.

3.33.2.2 Combinatorial Chemistry and Natural Products: Friends or Foes?

Combinatorial chemistry and natural products were often seen as juxtaposed techniques to access meaningful drug candidates. In late 1980s and early 1990s the hype on combinatorial chemistry and large pool libraries was so great that many people believed that meaningful chemical diversity would predominantly be gained through combinatorial methods. Conversely, the known drawbacks of natural products research, such as its long and tedious structure determination, often ending into the discovery of an already known chemical entity, led many groups to neglect it in favor of other, more trendy projects.

Now, the picture is different. We have already mentioned the reality check that focused the scope of combinatorial chemistry without reducing its relevance in drug discovery in general, and in medicinal chemistry in particular. Meanwhile, the analytical structure characterization of natural products has become a short, routine operation, using recently introduced high-throughput approaches,[15,15a] and the exploitation of largely untapped sources of biodiversity has secured novel structural inputs.[16,16a–16c] Considering that a recent review[17] proves that 61% of new small molecules registered as drugs in the period 1982–2001 are either traceable or inspired by structures of natural origin, we can understand why there is a new wave of natural products-driven projects in pharmaceutical research.

Highs and lows for a discipline are the rule, rather than an exception, and natural products and combinatorial chemistry will most likely ride the rollercoaster further. What I would consider a real, long-lasting trend is a marriage between the two, that is, the use of combinatorial methods in natural products research, or in natural products-inspired chemical diversity. The advantage of accessing pharmaceutically meaningful, unprecedented (and unpredictable!) chemical diversity is the hallmark of natural compounds. I would like to challenge any medicinal chemist in asking if he or she would have been able to rationally design any of the structures reported in **Figure 1** before they were isolated from natural sources. As the vast majority of natural sources is still unexploited,[16,16a–16c] it is reasonable to assume that many novel drugs await discovery.

Once discovered, these structures serve as models to medicinal combinatorial chemists. In fact, focused decoration libraries may be built accessing large quantities of the natural product itself or of suitable intermediates, and decorating them on their available functionalities. Natural products-inspired focused libraries are also obtained by modular synthetic strategies, allowing access to several analogs of the natural structure via a fully synthetic scheme. Both library classes can be more easily appreciated through recent reviews.[18,18a,18b]

Inspiration to the medicinal chemist may also be indirect, but by no means less significant. The complexity embedded in several natural compounds has prompted a number of groups to build natural products-like libraries, defined as medium-large libraries of stereochemically defined, complex structures obtained via diversity-oriented synthesis,[19,19a,19b] often prepared on SP. Although not directly related to natural products, they try to mimic the span of chemical diversity and selectivity exhibited by those structures using chemical transformations not accessible to the organisms producing the natural compounds. Their usefulness in finding novel structural hits, and even in chemically validating therapeutically useful targets (chemical genetics,[20,20a–20c] see next section) has been recently reviewed.[21,21a–21c]

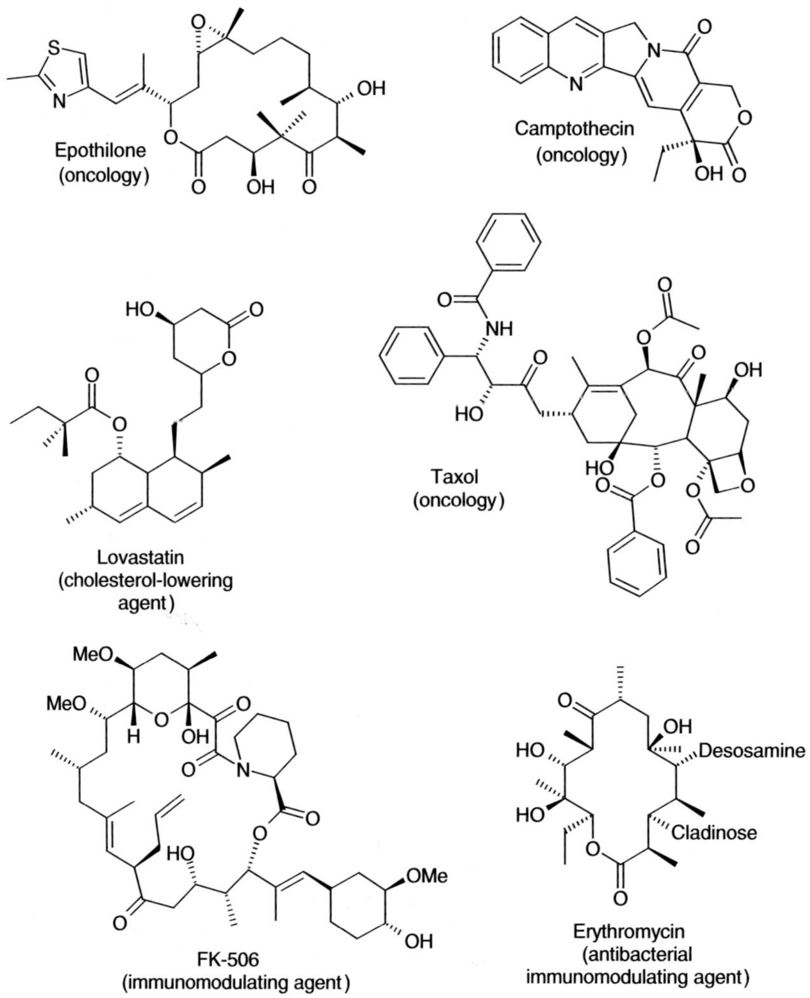

Figure 1 Natural products with pharmacological properties.

3.33.2.3 Expanding the Scope of Medicinal Chemistry: Target Identification and Validation

The impact of genetics, genomics, and postgenomics factors on modern drug discovery has been extensively described in several reviews,[22,22a,22b] and their crucial importance need not to be restated here. The impact on chemistry, and medicinal chemistry in particular, has also been extremely significant. While, historically, biological disciplines were considered the only sources of novel targets and the only tools to validate them, recently some leading medicinal chemists have exploited chemistry and chemicals in the quest for novel drug discovery targets.

Even before the advent of genetics it was common to use chemicals to probe biological systems via specific interactions with natural macromolecules. For example, natural products were often use to elucidate poorly known biological pathways, and sometimes this led to the identification of their molecular target and their validation as a therapeutically relevant objective.[23,23a,23b] Such approaches were somewhat shielded and obscured by the advent of genetic approaches, but no real reason exists to completely neglect them. In fact, the renaissance of chemical identification and validation of drug discovery targets by using modern technologies has led to the introduction of the general term 'chemical genetics.'[20,20a–20c] While this term may sound a bit pompous, there are clearly similarities between the chemical and genetic approaches for target identification and validation. Their comparison is visually represented in **Figure 2**, according to an earlier review.[24]

Let us assume that the specific alteration of the normal phenotype of a living cell is observed in a model cellular system, and that the molecular target responsible for this alteration is sought for. A sound hypothesis must link this alteration to a disease state, and eventually to a positive interference with the development of the disease, leading to

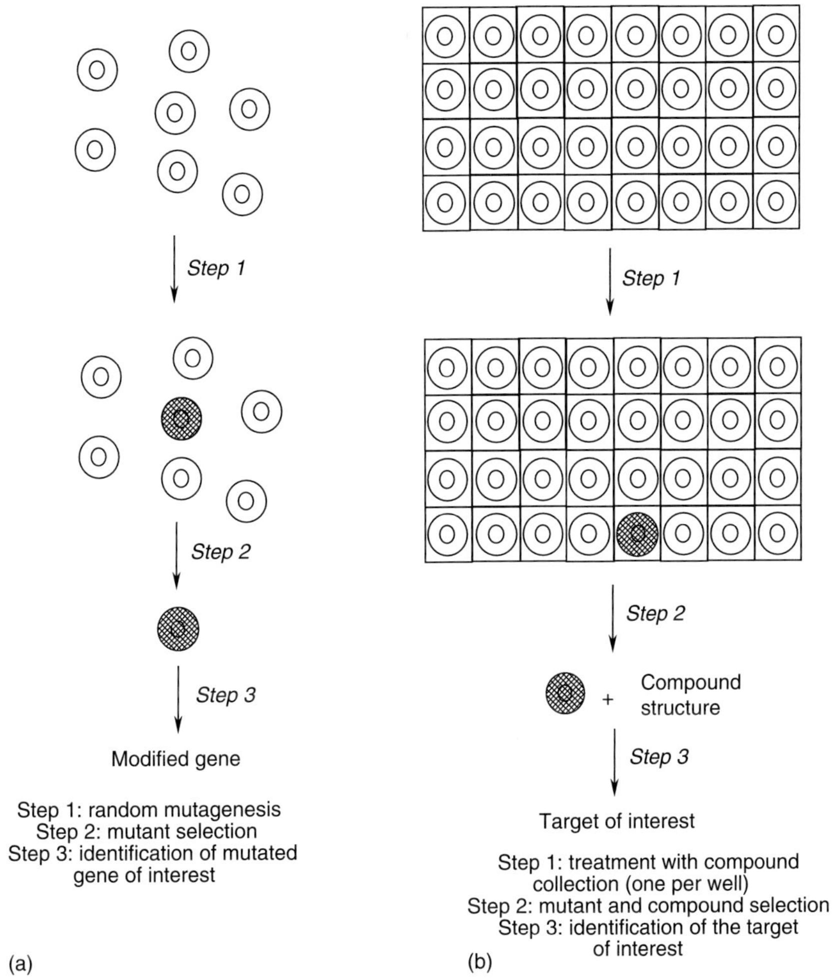

Figure 2 Target identification and validation: (a) genetics and (b) chemical genetics approaches.

significant benefits for patients. No information is available about the gene(s) responsible for the alteration, nor about the gene product(s) that would then represent the biological target for a novel drug discovery project. The classical genetics approach (**Figure 2a**) would consist of:

- random mutations introduced into the cells (step 1);
- selection of mutant cells showing the desired phenotype (step 2, highlighted cell); and
- identification of the mutated gene(s) responsible for the alteration and of its gene product(s) as a target for drug discovery (step 3).

As a consequence, the identified gene product is expressed in suitable amounts, and used to screen chemical collections searching for the desired interaction (inhibition, activation, and so on).

The chemical genetics approach (**Figure 2b**) is as follows:

- random parallel screen of a diverse chemical collection on cells (step 1);
- selection of cells showing the desired phenotype (step 2, highlighted cell), and thus of the compound responsible for the alteration; and
- identification of the molecular target of the compound as a target for drug discovery (step 3).

As a result, both the identified gene product and at least one interacting compound are obtained, paving the way for more focused biological and chemical efforts.

Let us now compare each step of the two processes. The random mutation of cells of any origin, or even of other living organisms, is more and more a routine operation, which combines high reliability, ease, and throughput. Such properties make step 1 of the genetics approach very cost- and effort-effective. The same cannot be said for step 1 of the 'classical' chemical genetics approach. To explore a substantial amount of phenotype variations it is necessary to access some substantial, meaningful chemical diversity. Druglike properties (penetration through cell membranes, reasonable physicochemical properties, good stability, etc.) must be intrinsic to the large chemical collection to be screened. A high degree of diversity (significantly different structures are likely to determine significantly different phenotypic alterations) and of complexity (complex natural products often exert potent biological actions with exquisite levels of specificity) is also necessary. Historically, academic institutions did not have access to large, diverse, druglike collections of chemicals, and large pharmaceutical companies that possessed large collections often had collections with a strong bias toward the history of the company and a low degree of druglike properties. Today, the commercial availability of meaningful collections can assist both industrial and academic institutions to screen the required large chemical diversity,[25,25a] but a more cost-effective approach consists of preparing medium-large libraries of stereochemically defined, complex structures obtained via diversity oriented synthesis[19,19a,19b] as chemical genetics collections.[20c]

As for step 2, this is the same for the two processes. A fast reading method, which can accurately detect the desired phenotypical change, is enough to guarantee the effectiveness of the modified cell selection and, consequently, for chemical genetics, of the active compound identification. Step 3 for the genetics approach is nowadays routine, as the detection of the mutated loci and the subsequent identification of the mutated gene and gene product is easy and efficient. This is not yet the case for chemical genetics. For some examples, an almost immediate correlation was reported between the phenotypic change and the identification of the molecular target, but usually only hints about the pathways involved are obtained. The identification of the target usually requires additional efforts, such as labeling of the chemical entity and its detection as an entity complexed with its target, or affinity chromatography techniques. Some reviews are available for the interested reader.[26,26a,26b]

The overall higher efficiency, and the larger diffusion, often orient researchers and laboratories toward genetics approaches to target identification. Nevertheless, several points should be mentioned:

- Sometimes genetics approaches can not be successful; for example, when a mutation is fatal and the cell or the organism is not viable, or when the studied pathway is simply too complex and poorly understood. A compound interacts with the corresponding gene product reversibly and when the organism is developed and viable, without causing cell death while providing useful albeit crude information on the biological pathway involved.
- The genetics approach irreversibly mutates the organism, and the mutation may originate other compensating effects, thus hiding the original mutation, or may cause other additional effects, thus affecting the screen results. The chemical genetics approach is reversible, as the compound can be washed away from the medium, and thus the biological machinery involved can, in principle, be switched on and off at will.
- Once the target is identified, its chemical genetics validation for drug discovery purposes using the same small molecule inhibitor(s) as a tool will be significantly easier and more valuable than its genetics counterpart.
- More importantly, the small molecule inhibitor(s) identified may even become, through structural optimization, valuable leads, thus speeding up the whole target-to-lead process.

The most likely hurdle on the road to the effective use of chemical genetics is, as recently reported,[27] "the cultural difficulty of getting biologists and chemists to collaborate." The eventual reward is, quoting the same source, the appearance of "multi-skilled individuals holding some of the keys to deciphering the human genome, and help to discover tomorrow's life-saving drugs." Who, I ask you, is better qualified than a modern medicinal chemist for this role?

3.33.3 Solid-Phase Libraries in Medicinal Chemistry

Medicinal chemistry libraries are so numerous that even a review focusing only on SP could not cover them all appropriately. Thus, only some of the libraries reported from the year 2000 were selected on the basis of their biological activity, analytical characterization, and purity. Excellent yearly reviews by Dolle[28,28a–28g] complement this section.

SP discrete and pool libraries will be reported separately, and for each class, libraries spanning target identification to lead optimization will be mentioned. Small organic molecule libraries are singled out, and only a few examples of peptidomimetics or nucleotide analog libraries (in particular those related to pool libraries) are included.

SP publications are reported that have lead to one or more compounds that potentially are of therapeutic relevance, but whose combinatorial exploitation has up to now remained at the chemical assessment stage (less than 25 compounds prepared). Any paper reporting more than 25 simultaneously prepared compounds is classified as a library, and the same is true also for smaller arrays when they have been tested on a biological target. Although this is an arbitrary classification method, it allows a reasonable partition of papers into three categories.

Each section contains a table, where the most relevant descriptors for each library are reported; several figures containing the generic structures of each library, including their point of attachment to the solid support (heteroatoms/functional groups in italic, carbon atoms labeled with an asterisk) and their 'combinatorializable' positions (bold); and a short description for each cited library.

3.33.3.1 Chemical Assessment on Solid Phase: A Gateway toward Medicinal Chemistry Libraries

Natural products were extensively used as medicinal chemistry scaffolds, leading to the SPS of active principles from natural sources, and/or of some analogs (**Table 1**). Tripathy *et al.*[29] presented the synthesis of indenopyrrolocarbazoles (**1**) (**Figure 3**), obtained by decoration of a indene CH_2-unsubstituted structure, as representatives of a class of potent kinase inhibitors with broad specificity and, thus, potential toxicity, which may eventually lead to selective and potent kinase inhibitors. Olsen *et al.*[30] have reported the synthesis of 2,4,6-trisubstituted benzenes (**2**) as constrained and diverse mimetics of hapalosin, a natural multidrug resistance modulator with cytotoxic activity. Paterson and Temal-Laib[31] paved the way toward synthetic polyketide libraries with different sizes and stereochemistry by assessing and fine tuning the reactivity of supported α-chiral aldehydes in asymmetric aldol reactions, leading to diastereomeric tetraketides (**3**). Janda and co-workers[32] reported the SPS of anandamide (arachidonyl-ethanolamide), the most prominent endocannabinoid, with a variety of physiological action, and of its analogs **4** via iterative assembly of tetrayne frameworks. The SPS of clavulone analogs (**5**), with potentially strong cytotoxicity, was reported by Tanaka *et al.*[33] as being expandable toward a natural product-like library for chemical genetics. Braese and co-workers[34] focused on naturally occurring phthalides and isoindolinones (**6**) as ideal scaffolds for library generation, preparing a set of diverse heterocycles with up to four diversity points. Albericio and co-workers[35–37] reported the synthesis of several naturally occurring lamellarins and their analogs with general structures **7** and **8**, using a versatile SP experimental protocol sequence. Klein *et al.*[38] reported polycyclic β-carbolines (**9**) as combinatorializable scaffolds inspired by known natural products with a wide range of pharmacological effects.

Kamal and co-workers[39] reported the SP synthesis of pyrrolo[2,1-*c*][1,4]benzodiazepines (**10**) (**Figure 4**) as analogs of the naturally occurring antitumor antibiotic chicamycin A. Andersen and Stroemgaard[40] and Joensson[41] reported SP synthetic studies aimed at naturally occurring philantotoxins and their analogs (**11**), which may give access to extremely strong effectors on ion channels and central nervous system (CNS) receptors. Albericio and co-workers[42] reported the synthesis of decorated, rigid tentoxin-related compounds (**12**), including the original phytotoxic fungal metabolite, as a diversity-oriented platform for chemical genetics studies. Arnusch and Pieters[43] and Cristau *et al.*[44] have reported biaryl ether-containing macrocycles (**13**), which can be seen as simplified constrained analogs of the natural antibiotic vancomycin, made by Ugi multi-component reactions (MCRs). You *et al.*[45] reported the SPS and stereochemical assignment of thiazole- and oxazole-containing macrolactams tenuecyclamides A–D (**14**) as precursors of interesting natural products-like libraries. Bleomycin analogs (**15**), a class of potential glycopeptide antitumor antibiotics acting through oxidative DNA damage, were successfully targeted by Hecht and co-workers,[46,47] and directly screened on solid support. Taddei and co-workers[48] reported the SPS of a combinatorially significant, natural product-like macrolactam (**16**), via the key intramolecular hydroformylation of a terminal alkene. Rew and Goodman[49] assessed the SPS of amine-bridged cyclic enkephalin analogs (**17**), as potential novel opioid ligands, using a final Fukuyama–Mitsunobu SP cyclization. Finally, several saframycin A analogs (**18**) were prepared as potential antiproliferative alkaloids by Myers and Lanman.[50]

Heterocyclic scaffolds have also been the target of SP assessments, as their pharmaceutical relevance is well known. Shintani *et al.*[51] reported the SPS of tetrasubstituted pyridines (**19**) (**Figure 5**) bearing a 3-CN substitution, in analogy to a known IKK-β inhibitor, obtaining higher yields on SP than in solution for the key MCR. Barluenga and co-workers[52] described the versatile SPS of tetrasubstituted piperidones (**20**) as structural cores toward more complex biologically compounds, such as alkaloids. Xiao *et al.*[53] prepared on SP the well-known sedative and hypnotic drug thalidomide, and some close analogs (**21**), with significant antiangiogenic potential. Another known structural lead, the dipeptidyl peptidase IV inhibitor 2-cyanopyrrolidide **22**, was prepared on SP by Willand *et al.*[54] as a model for fast and efficient analoging to search for novel treatments of type-II diabetes. Hall and co-workers[55] used MCRs to synthesize an array of bicyclic heterocycles in solution, and assessed an SPS MCR protocol, leading to polyfunctionalized piperidines (**23**), with significant advantages compared with their solution counterparts. Komatsu *et al.*[56] reported the SPS of pyrrolidine-based byciclic compounds (**24**) through a traceless linker SP strategy suitable for more diversity-oriented efforts. Arya and co-workers have published several efforts aimied at SPS of tetrahydroquinoline-based polycyclic structures, such as

Table 1 SP chemical assessment

Compound class	*Compounds made*	*Library type*[a]	*Analytical characterization; purification; purity/yield*[b]	*Ref.*
Indenopyrrolocarbazoles, **1**	6	NPD	HPLC; none; 61–98% (yield)	29
Hapalosin mimetics, **2**	12	NPB	HPLC; silica gel; 10–40% (yield)	30
Diasteroisomeric tetraketides, **3**	2	NPB	HPLC, NMR; silica gel; 15–42% (yield)	31
Anandamide analogs, **4**	12	NPB	NMR; none; 5–40% (yield)	32
Clavulones, **5**	6	NPB	HPLC; prep. HPLC; 7–49% (yield)	33
Isoindolines, **6**	20	NPB	NMR; none; 17–90% (yield)	34
Lamellarins, **7**	>20	NPB	HPLC, NMR; prep. HPLC; NG	35
Lamellarins, **8**	6	NPB	HPLC; prep. HPLC; 4–10% (yield)	36,37
β-Carbolines, **9**	15	NPB	HPLC; none; 40–90% (purity)	38
Chicamycin analogs, **10**	8	NPB	HPLC, MS; silica gel; 45–82% (yield)	39
Philantotoxin analogs, **11**	2	NPB	HPLC, MS; none; >75% (purity)	40, 41
Tentoxin analogs, **12**	8	NPB	HPLC; none; NG	42
Vancomycin mimics, **13**	7	NPB	HPLC; prep. HPLC; 30–38% yield	43, 44
Tenuecyclamides, **14**	4	NPB	PLC; none; 33–54% (yield)	45
Bleomycin analogs, **15**	6	NPB	HPLC; none; 17–22% (yield)	46,47
Macrocycles, **16**	1	NPB	HPLC; prep. HPLC; 56% yield	48
Enkephalin analogs, **17**	9	NPB	HPLC, MS; none; 72–98% (purity)	49
Saframycin A analogs, **18**	23	NPB	HPLC, NMR; silica gel; 24–58% (yield)	50
3-Cyano-6-(2-hydroxyphenyl)pyridines, **19**	20	HET	HPLC; none; 40–100% (purity)	51
Polysubstituted piperidones, **20**	15	HET	NMR; none; 42–93% (yield)	52
Thalidomide analogs, **21**	6	HET	NMR; silica gel; 40–98% (yield)	53
2-Cyanopyrrolidides, **22**	1	HET	HPLC, NMR; silica gel; NG	54
Hydroxyalkyl piperidines, **23**	3	HET	HPLC, MS; none; 54–62% (yield)	55
Bicyclic pyrrolidines, **24**	4	HET	NG; none; 45–90% (yield)	56
Tetrahydroquinoline-based polycyclics, **25**	1	HET	NMR, HPLC; silica gel; 45% yield	57
Tetrahydroquinoline-based polycyclics, **26**	2	HET	HPLC, NMR; silica gel; 25–27% (yield)	58
Tetrahydroquinoline-based polycyclics, **27**	1	HET	NMR, MS; silica gel; 40% yield	59
Dibenz-1,5-oxazocines, **28**	17	HET	HPLC; none; 12–75% (yield)	60
[2.2.1]Bicyclic lactams, **29**	18	HET	HPLC, NMR; silica gel; 49–65% (yield)	61
1,3-Thiazines, **30**	20	HET	HPLC, NMR; none; 30–60% (yield)	62
2-Aminothiophenol-derived heterocycles, such as **31**	40	HET	HPLC; silica gel; 40–71 (yield)	63
2N-containing heterocycles, such as **32**	18	HET	HPLC, MS; none; 70–96% (purity)	64
6-Acylamino-4-oxo-1,4-dihydrocinnoline-3-carboxamides, **33**	>9	HET	HPLC; prep. HPLC; 42–88% (yield)	65

Table 1 Continued

Compound class	*Compounds made*	*Library type*[a]	*Analytical characterization; purification; purity/yield*[b]	*Ref.*
Isoxazolinocyclobutenones, **34**	9	HET	NMR; none; 27–38% (yield)	66
Isoxazolinopyrrole-2-carboxylates, **35**	10	HET	NMR; silica gel; 6–24% (yield)	67
4-Hydroxy-4,5-dihydroisoxazole-2-oxides, **36**	3	HET	NG; none; 72 to >95% (yield)	68
Bicyclic aminal lactones, **37**	19	HET	HPLC, NMR; none; 51–98% (purity)	69
Spirocyclic ketal lactones, **38**	12	HET	NMR, HPLC; none; 9–36% (purity)	70
Benzopyranoisoxazoles, **39**	9	HET	HPLC, NMR; none; 33–55% (yield)	71
6*H*-Pyranobenzimidazole-6-ones, **40**	21	HET	HPLC; none; 37–82% (yield)	72
1,3,4-Thiadiazolium-2-amidines, **41**	17	HET	HPLC; silica gel; 38–93% (yield)	73
Oxazole-based macrocycles, **42**	3	HET	HPLC; prep. HPLC; NG	74
Benzofuran-based dimers, **43**	8	HET	NMR, HPLC; none; >80% (purity)	75
Cyclopent-2-enones, **44**	11	OTH	NMR; silica gel; 18–40% (yield)	76
Cyclopropyl phenyl methanones, **45**	2	OTH	HPLC, NMR; none; >90% (purity)	77

[a] HET, heterocyclic; NPD, natural products, decorated; NPB, natural products, built; OTH, miscellaneous libraries.
[b] HPLC, high-performance liquid chromatography; MS, mass spectrometry; NG, not given; NMR, nuclear magnetic resonance; prep., preparative.

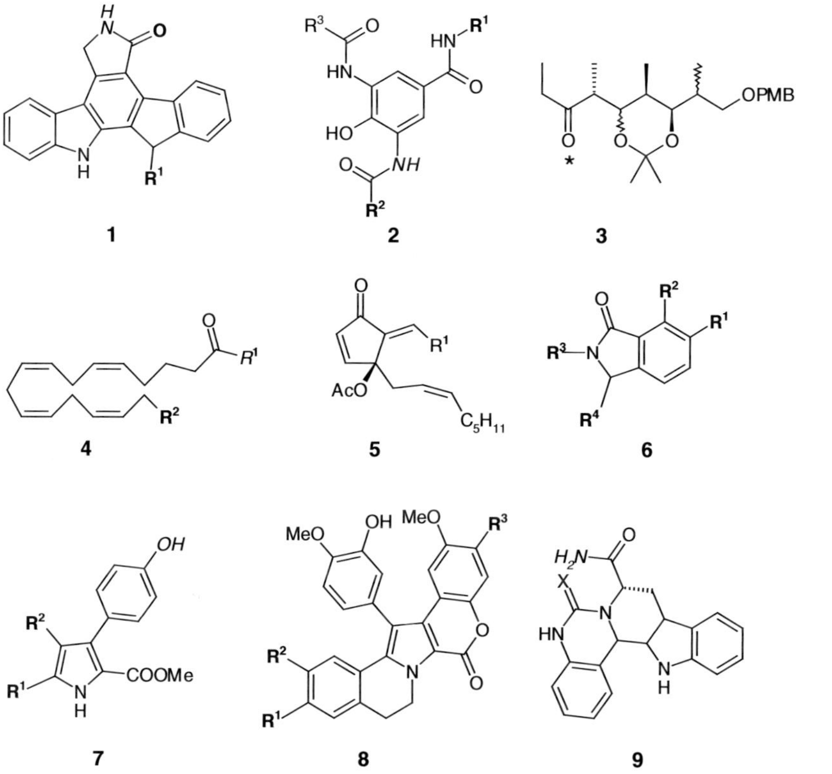

Figure 3 SP chemical assessment: structures **1**–**9**.

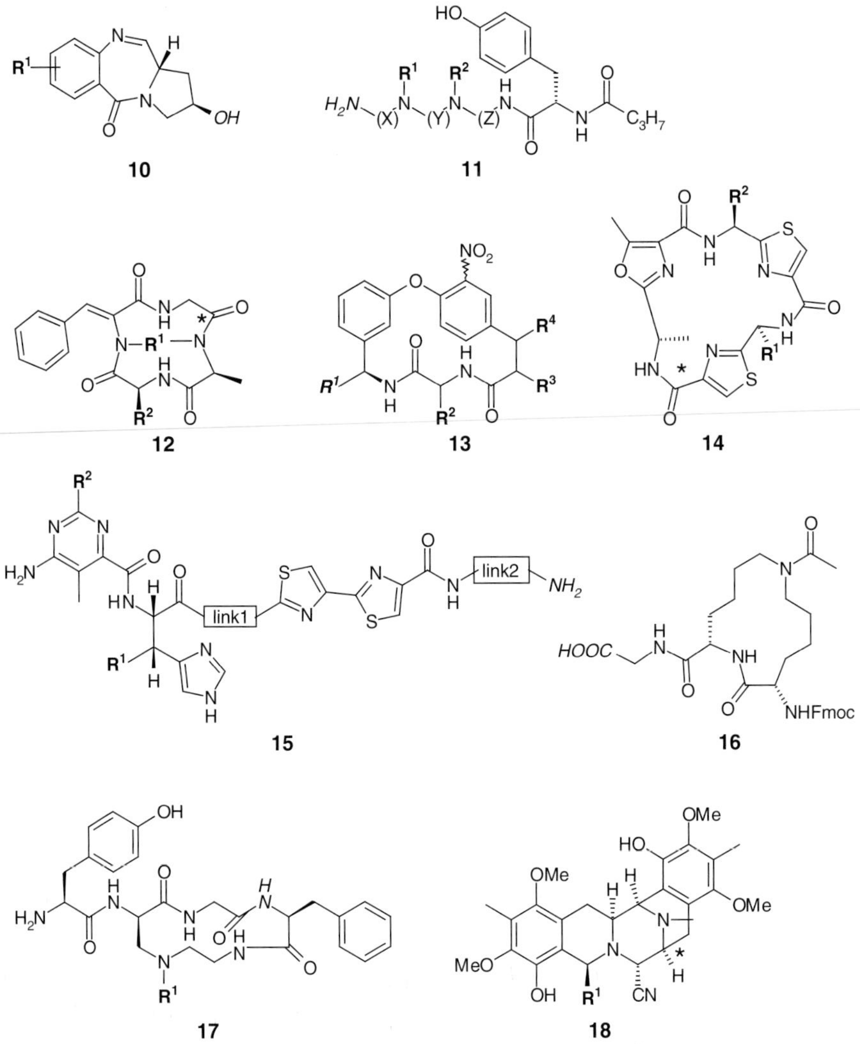

Figure 4 SP chemical assessment: structures **10–18**.

the eight-membered ring-containing lactam **25**,[57] the tricyclic amine derivatives **26**,[58] and the 10-membered ring-containing unsaturated lactam **27**,[59] as valuable structural platforms for chemical genetics studies.

Kiselyov and co-workers[60] focused on dibenz[*b*,*g*]1,5-oxazocines (**28**)(**Figure 6**), due to their proven activity in CNS and oncology. Bicyclic lactam derivatives (**29**) were reported by Savinov and Austin[61] both as diversity-generating scaffolds and as a gateway to polyfunctional ring-opening reaction products. Kappe and co-workers[62] described the access to SPS libraries of biologically relevant, but largely unexploited, 1,3-thiazines via the synthesis of a set of compounds (**30**). Barany and co-workers[63] have reported an SP protocol leading to dihydrobenzothiazines (**31**), to dihydrobenzothiazinones, and to dihydrobenzothiazine-1,1-dioxides as analogs of known anticancer, antinflammatory, antirheumatic, and antibacterial agents. Purandare *et al.*[64] have validated the SP access to libraries of at least five di-nitrogen-containing heterocyclic scaffolds such as **32** via a common SP intermediate. Sereni *et al.*[65] reported the SPS of cinnoline derivatives (**33**) as potential active principles in the CNS, oncology, and parasitology. Isoxazinolocyclobutanones (**34**) and isoxazole-containing congeners were described by Kurth and co-workers[66] as novel and flexible key components of biologically active compounds. The same group reported another isoxazoline-driven SP assessment,[67] leading to pharmaceutically relevant isoxazolinopyrrole esters (**35**). Isoxazole-based scaffolds were also studied by Righi *et al.*,[68] through the SPS of dihydroisoxazole-2-oxides (**36**), which are useful in themselves and as intermediates toward aminopolyhydroxylated derivatives.

Bartlett and co-workers have reported the SP access to two classes of druglike lactones, either characterized by a bicyclic aminal structure (**37**) (**Figure 7**)[69] or by a spirocyclic ketal structure (**38**),[70] both expandable for lead

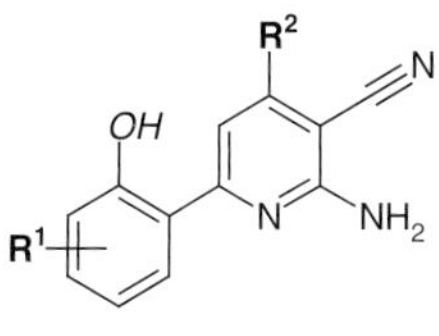
19

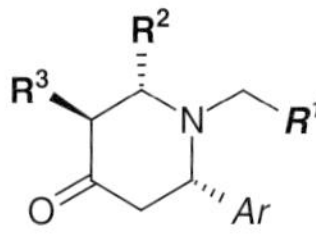
20

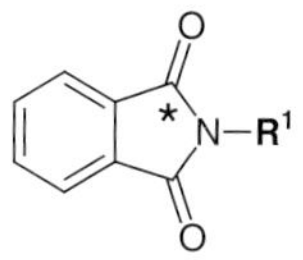
21

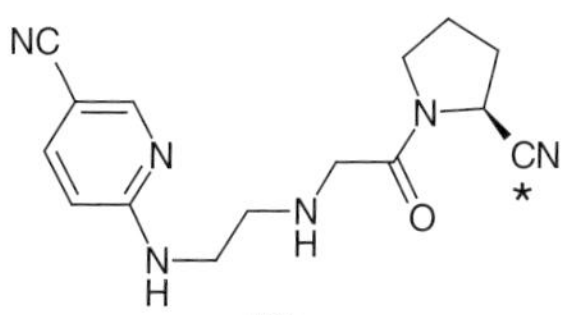
22

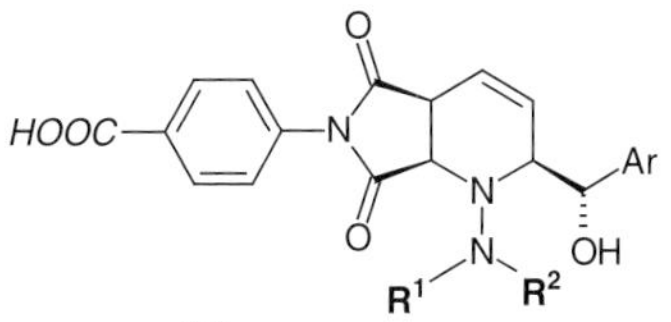
23

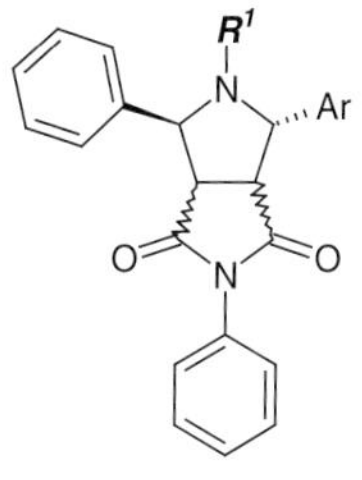
24

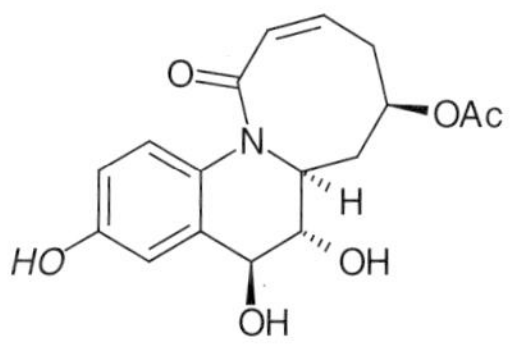
25

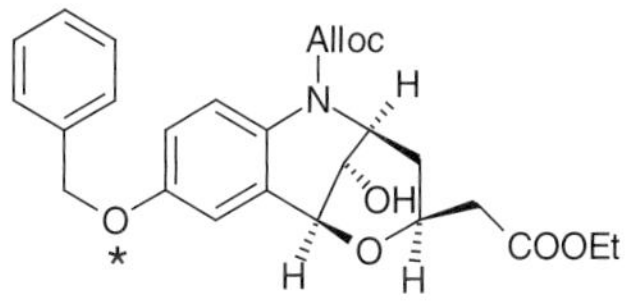
26

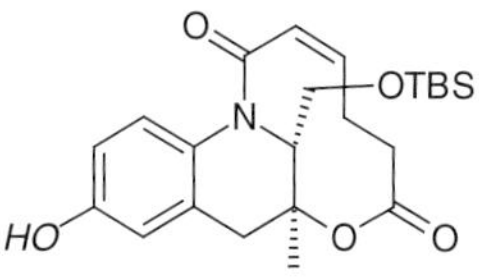
27

Figure 5 SP chemical assessment: structures **19**–**27**.

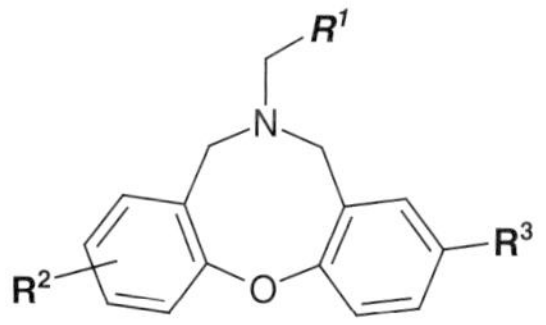
28

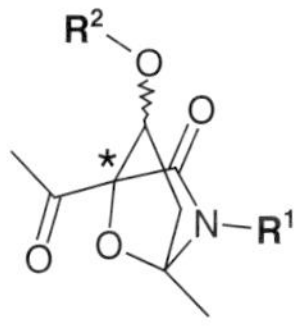
29

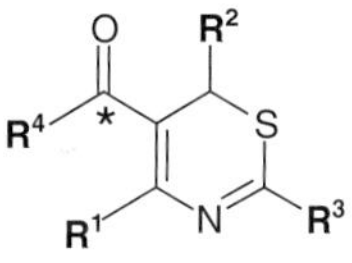
30

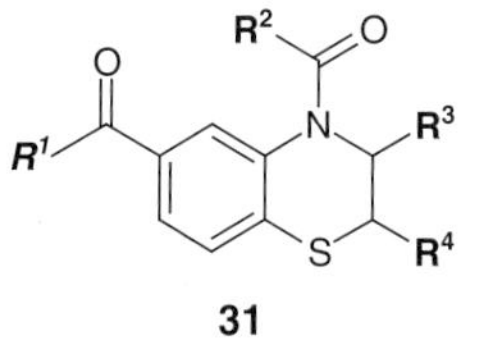
31

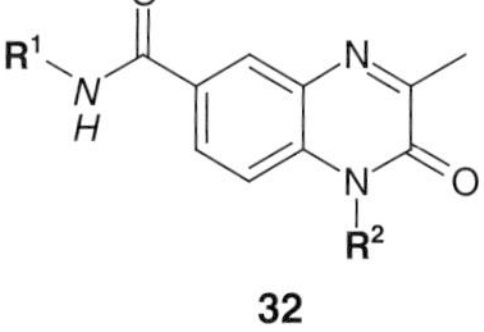
32

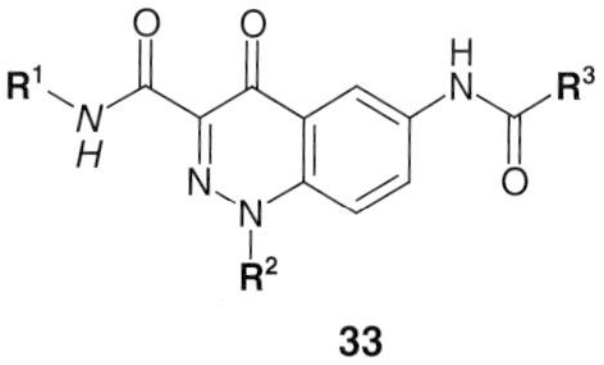
33

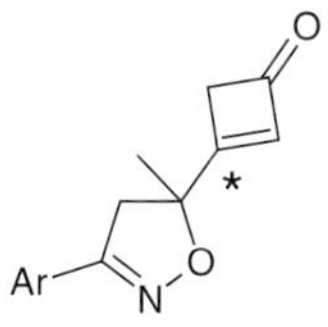
34

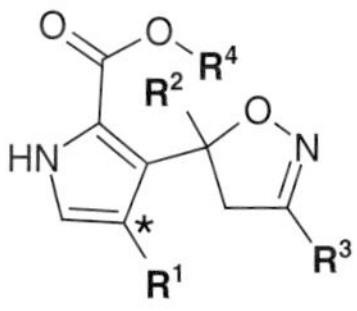
35

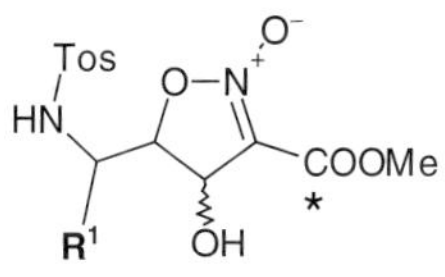
36

Figure 6 SP chemical assessment: structures **28**–**36**.

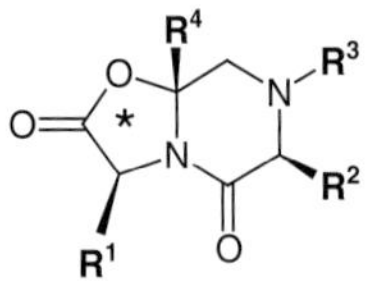

37

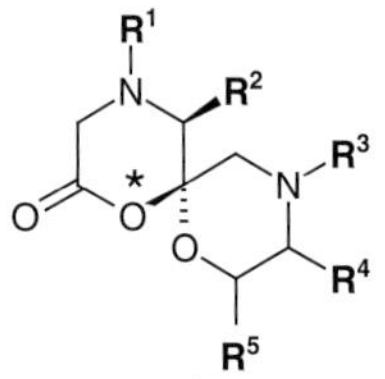

38

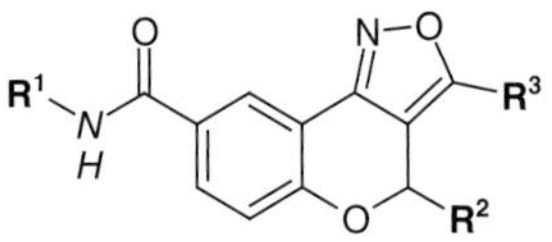

39

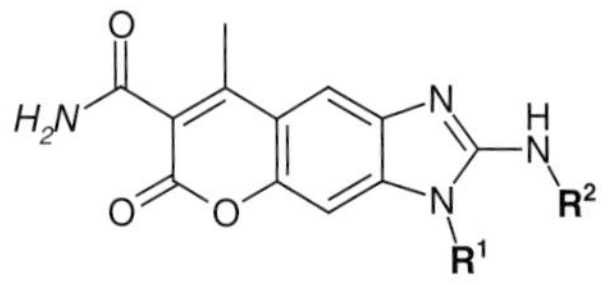

40

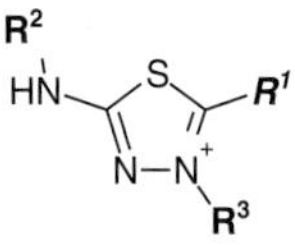

41

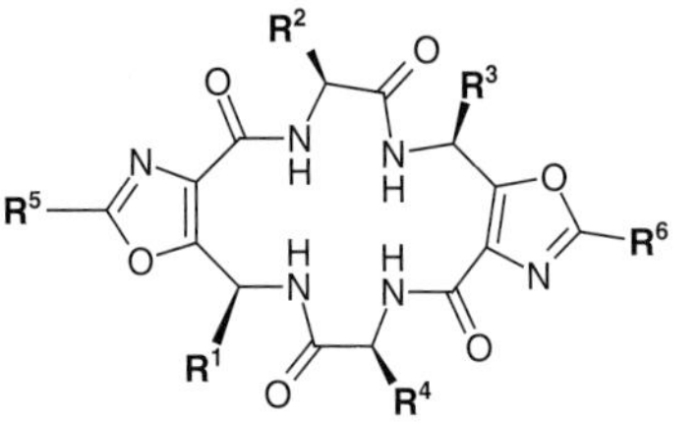

42

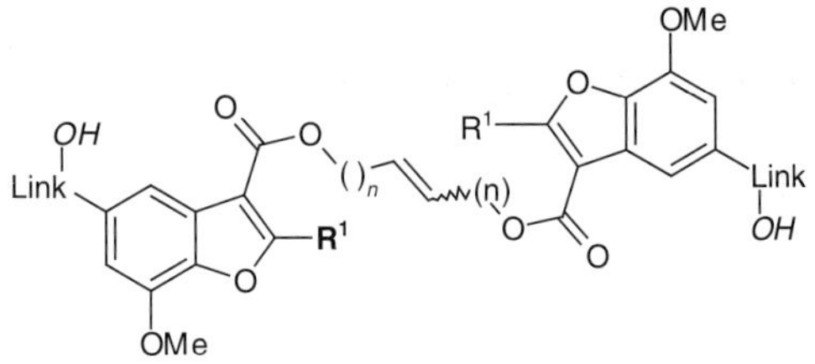

43

Figure 7 SP chemical assessment: structures **37–43**.

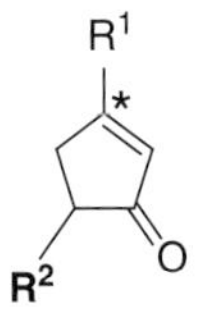

44

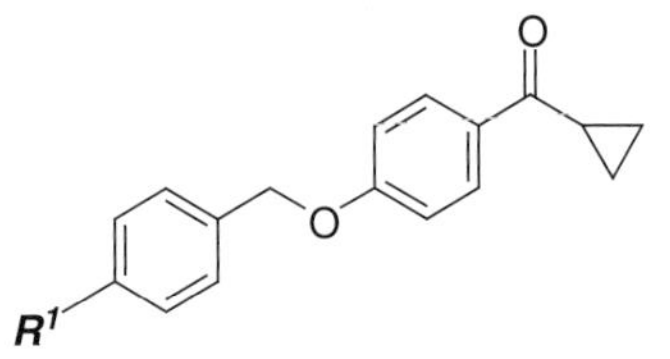

45

Figure 8 SP chemical assessment: structures **44** and **45**.

identification, or even for chemical genetics purposes. Chao *et al.*[71] focused their attention on benzopyranoisoxazoles (**39**) as potential new steroid mimetic templates, with a variety of possible applications. Song and Lam[72] reported the SPS of pyranobenzimidazoles (**40**) as psoralen analogs for photochemotherapy in autoimmune diseases. Thiadiazolium salts (**41**) were prepared in high yields on SP by Barany and co-workers,[73] as novel analogs of antibacterial and antifungal chemotypes. Mann and Kessler[74] reported the SPS of macrocyclic systems (**42**) containing two oxazole rings, as rigid and polyfunctional scaffolds for lead discovery and chemical genetics. Liao *et al.*[75] reported dimeric structures (**43**) based on a core benzofuran nucleus for exploring protein–protein homodimeric interactions among 'difficult' biological targets.

Finally, two papers mention SP assessment studies on scaffolds other than natural products- or heterocycle-related molecules (**Figure 8**). Cheng and Kurth[76] reported the SPS of a model cyclopent-2-enone library (**44**), which, due to the abundance of this motif in pharmacologically active compounds, represents a meaningful target. Grover *et al.*[77] reported the SPS of two multipurpose cyclopropyl phenyl methanones (**45**), and their SP transformation into oximes, alcohols, or alkanes as gateways toward more complex, diversity-oriented scaffolds.

3.33.3.2 Solid-Phase Libraries of Discretes

Natural products are represented in recently prepared SP discrete libraries, either from decoration of an advanced natural product-like intermediate, or from fully synthetic pathways (**Table 2**). Considering the former, Poirier and co-workers[78] reported the SPS of four sublibraries of estradiols such as **46** (**Figure 9**), identifying some strong

Table 2 SP discrete libraries

Compound class	*Library size, type*	*Analytical method; purification; yields/purities*[a]	*Therapeutic area; target; drug discovery phase*[b]	*In vitro activity, best structure*	*Ref.*
Estradiol derivatives, **46**	156, NPD	HPLC; none; >90% (purity), 27–46% (yields)	Oncology; steroid sulfatase; H2L	>95% inhibition at 1 nM R^1 = Bn, R^2 = 3-cyclopentyl-Pr^n	78
Estradiol derivatives, **47**	50, NPD	HPLC; none; >90% (purity), 18–66% (yield)	Oncology; steroid sulfatase; H2L	>95% inhibition at 1 nM R^1 = Bn, R^2 = 4-Bu^tPh	79
Lupeol derivatives, **48**	96, NPD	HPLC; none; 60–90% (purity), 70–85% (yield)	Antimalarial; *Pseudomonas falciparium*; LO	MIC = 13.07 μM *n* = 6, R^1 = OBn	80
Thiazole peptide analogs, **49**	500, NPD	HPLC; prep. HPLC; >70% (purity), 30–50% (yield)	Antibacterial; MICs, various pathogens; LO	MICs <1 μM, various G^+ bacteria R^1 = 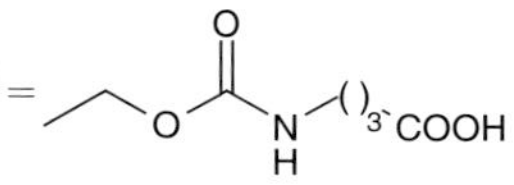	81
Naltrindole derivatives, **50**	40, NPD	HPLC; none; 48->70% (purity)	Immunosuppressant, pain; δ-opioid receptor; H2L	None	82
Saphenamycin analogs, **51**	12, NPD	HPLC; silica gel; 20–100% (yield)	Antibacterial; MIC, *Bacillus subtilis*; LO	MIC = 0.07 μM R^1 = 3-Cl	83
Butenolides, **52**	76, NPB	HPLC; none; 55–99% (purity), 23–99% (yield)	Antibacterial, antifungals, oncology; none; LD	None	84
Quorum sensing effectors, **53**	96, NPB	NMR, MS (4 compounds); none; NG	Antibacterial; virulence factors; LO	Strong activity at 50 μM R^1 = H, R^2 = 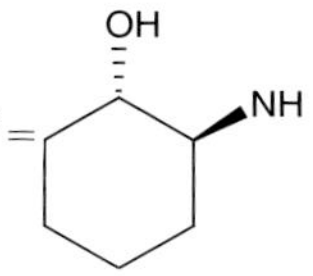	85
Glycine betaine analogs, **54**	7, NPB	HPLC, NMR; none; >90% (yield), >95% (purity)	Antibacterial; growth inhibition; H2L	Lethal, most potent analog R^1 = 4-NO_2, X = Br	86
Philanthotoxin analogs, **55**	18, NPB	HPLC; prep. HPLC; 39–69% (purified)	CNS; nAChR; LO	IC_{50} = 790 nM X = CH_2, R^1 = 4-OH, R^2 = Et	87
Philanthotoxin-12 analogs, **56**	6, NPB	HPLC; prep. HPLC; >98% (purity)	CNS; nAchR; LO	IC_{50} = 580 nM *n* = 11	88

continued

Table 2 Continued

Compound class	*Library size, type*	*Analytical method; purification; yields/purities*[a]	*Therapeutic area; target; drug discovery phase*[b]	*In vitro activity, best structure*	*Ref.*
Negamycin analogs, **57**	180, NPB	NG; none; NG	Antibacterial;- MIC G– pathogens; H2L	MIC from 2 to 128 μM $R^1 = Me$	89
Kojic acid tripeptides, **58**	30, NPB	HPLC; none; 49–95% (yield)	Oncology; tyrosinase; H2L	$IC_{50} = 240$ nM $AA^1 = Phe$, $AA^2 = Trp$, $AA^3 = Tyr$	90
Muramyl dipeptide analogs, **59**	60, NPB	HPLC; none; >75% (purity)	Antibacterial.; none; H2L	None	91
Mureidomycin analogs, **60**	80, NPB	HPLC; none; 15–45% (purity)	Antibacterial; MIC, various G^+ and G– strains - LD	No activity	92
Deglycobleomycin analogs, **61**	108, NPB	HPLC; prep. HPLC; >70% (crude HPLC), 27–85% (purified)	Antibacterial; DNA cleavage; LO	Two analogs better than DB	93
Desferrioxamine B analogs, **62**	122, NPB	HPLC; none; 25–78% (purity)	Blood disorders; Fe overload; LO	47% Fe removal $R^1 = (CH_2)_4$, $NR^2R^3 = NH(CH_2)_3$ $R^4 = (CH_2)_2$, $NR^5R^6 = NH(CH_2)_5$	94
Aurilide analogs, **63**	25, NPB	HPLC; prep. HPLC; 11–25% (yield)	Oncology; none; LO	None	95
α-Substituted prolines, **64**	48, HET	HPLC; prep. HPLC; 50–95% (purity), 3–52% (yield)	Various; none; LD	None	96
N-Sulfonylprolyl amides, **65**	>50, HET	HPLC; SPE; >90% (purity), 54–70% (yield)	CNS; FKBP12 rotamase; LO	$IC_{50} = 110$ nM $R^1 = Bn$, $R^2 = (CH_2)_4Ph$	97
Hydroxyprolines, **66**	10 200, HET	HPLC; none; >80% purity for 80.4% of compounds	Antiviral, CV; proteases; LD	None	98
Alkoxyprolines, **67**	17 000, HET	MS (12.5% sampling); none; NG	Antiviral, CV; proteases; LD	None	99
Hydroxyproline-based peptidomimetics, **68**	33, HET	HPLC; prep. HPLC; 10–75% (crude HPLC)	Antiviral; HCV NS3 protease; LO	$IC_{50} = 800$ nM $Ar\text{-}R^1 =$ 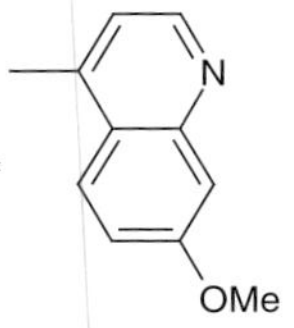	100

Azarene pyrrolidin-2-ones, **69**	56, HET	HPLC; none; >70% purity for >70% compounds	Antithrombin; fXa; H2L	0.1 μM, 77% inhibition $R^1 = CH_2NH(4\text{-}Py)$, Ar = 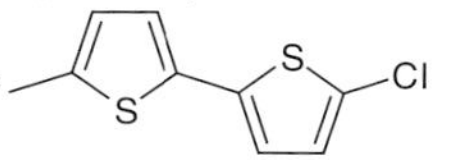	101
Pyrrolidin-2-ones, **70**	12 000, HET	HPLC; none; >80% purity for 85% of compounds	Anti-inflammatory, antiepileptic, oncology; integrins, CCR5; LD	None	102
Piperidin-2,5-diones, **71**	7, HET	HPLC; silica gel; 20–55% (yield)	CNS, oncology; cruzain; H2L	$K_i = 16\,nM$ $R^1 = (CH_2)_3Ph$	103
Fused bicyclic lactams, **72**	7, HET	HPLC; SEC; 40–67% (yield)	Oncology, osteopor.; αvβ3 integrin; H2L	$IC_{50} = 3.7\,nM$ $n = 2$, stereochemistry: 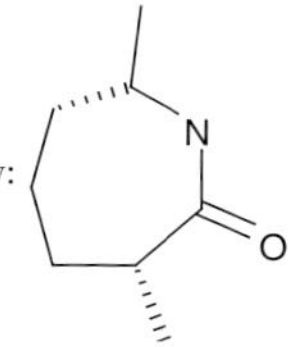	104
Mono- and bicyclic furan-4-ones, **73**	16, HET	HPLC; silica gel; 84–95% (purity)	Various; cathepsin K; LD	$K_i = 1.8\,\mu M$ $R^1 = Bu^i$, $R^2 =$ 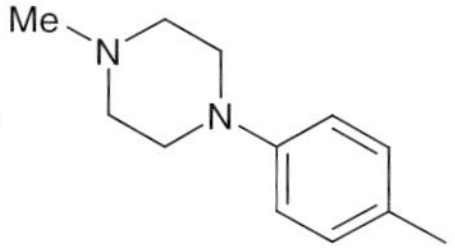	105
Polymorphic scaffolds, **74**	>25, HET	HPLC; none; >80% (purity), >67% (yield)	None; TGT ID/Val, LD	None	106
2-Pyridones, **75**	80, HET	HPLC; none; >85% (purity)	Pain; KOR receptor; LO	$IC_{50} = 5.5\,nM$ $Ar^1 = Ph$, $Ar_2 = 2\text{-MeO,5-FPh}$	107
5-Substituted nicotinic acids, **76**	180, HET	HPLC; none; 75–85% (yield), 71–96% (purity)	None; LD	None	108
3-Arylthio-3-nicotinyl propionic acids, **77**	39, HET	HPLC; none; 80–90% (purity)	Oncology, restenosis; $\alpha_v\beta_3$ receptor; H2L	$IC_{50} = 16\,nM$ $R^1 =$ 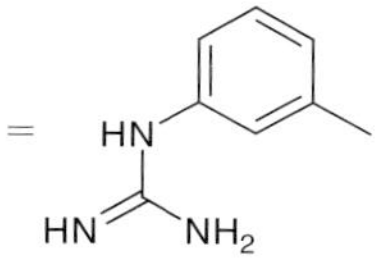	109
2,5-Disubstituted pyrroles, **78**	57, HET	NMR; none; 50% to >80% (purity)	Oncology; B-cell proliferation. inhibitor; LD	$IC_{50} = 10\,nM$ $Ar^1 = 2\text{-F,4-HOOCPh}$, $Ar^2 = 2(4,7\text{-di-Me})benzofuran$	110

continued

Table 2 Continued

Compound class	*Library size, type*	*Analytical method; purification; yields/ purities*[a]	*Therapeutic area; target; drug discovery phase*[b]	*In vitro activity, best structure*	*Ref.*
Thiophene asparty1 ketones, **79**	48, HET	HPLC; prep. HPLC; 2–64% (purified)	Anti-inflammatory, neurodegenerative; caspase-3; LO	$K_i = 20$ nM Cycle:2,5 thiophene, $R^1 =$ 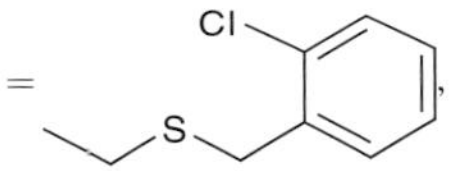 $R^2 =$ 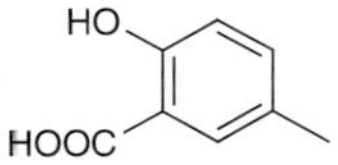	111
Phthalides, **80**	100, HET	HPLC; none; >90% purity, 35% average yield	None; LD	None	112
Trisubstituted benzofurans, **81**	320, HET	HPLC, MS; none; 35% average purity	Osteoporosis; ER-β; H2L	$IC_{50} = 79$ nM $R^1 =$ 1-Naphthyl, $R^2 =$ 4-OH	113
Pentasubstituted benzopyrans, **82**	2000, HET	HPLC; prep. HPLC; 24–85% (yield)	CNS, CV, HIV; K^+ channels; LD	None	114
2,3-Disubstituted indoles, **83**	>>100, HET	HPLC; NG; NG	CNS; $5HT_{2A}$; LD	$IC_{50} = 2.7$ nM Ar = Ph, $NR^1R^2 =$ *N*-piperidine	115
Tetrasubstituted indoles, **84**	33, HET	HPLC; prep. HPLC; 20–82% (purified)	Various; none; LD	None	116
Indole-based peptidomimetics, **85**	>200, HET	HPLC; none; average 80% purity	CV; PAR-1; H2L	$IC_{50} = 240$ nM $AA^1 =$ 3,4-diFPhe, $AA^2 =$ Citrulline, $R^1 =$ 2,6-diCl, $NR^2R^3 = N(CH_2)_4$, $R^4 =$ Bn	117
Phthalimido amides, **86**	45, HET	HPLC; prep. HPLC; 5–95% (yield)	Antiparasitic; hypoxanthine-guanine-xanthyne phosphorybosyl transferase; H2L	$IC_{50} = 1.5$ uM $R^1 =$ *p*-BrPh	118
Quinoline-4-carboxamides, **87**	42, HET	HPLC; none; >80% (purity)	Inflammation; VCAM/VLA4; H2L	$IC_{50} = 1.7$ nM $R^1 =$ 2,6-di-FPh	119
Indolizino[8,7-b]-5-carboxylates, **88**	576, HET	HPLC; none; NG	None; LD	None	120
N-Arylpiperazines and diazepines, **89**	36, HET	HPLC; silica gel; 42–98% (crude HPLC), 1–72% (purified)	CNS, hypertension; none; LD	None	121

4-Phenyl-2-carboxy-piperazines, **90**	160, HET	HPLC; prep. HPLC; >80% (purity), 18–85% (yield)	Various; none; LD	None	122
Pyrrolemethyl piperazines, **91**	5, HET	HPLC; none; 56–94% (yield)	CNS; hD4 receptor; H2L	$IC_{50} = 79$ nM $R^1 =$ 	123
N-Arylpiperazines, **92**	72, HET	HPLC; none; 18–57% (yield)	Psychiatric. disorders; $5HT_{2A}$; LD	$IC_{50} = 1$ nM $R^1 = Cl$, $n = 2$, $R^2 =$ 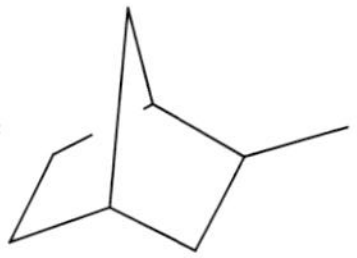	124
2-Oxo-1,4-piperazines, **93**	384, HET	HPLC; none; 89% average purity, 61% average yield	Various; none; LD	None	125
Biaryl diketopiperazines, **94**	>300, HET	HPLC; prep. HPLC; NG	Infertility; FSH receptor; LO	$EC_{50} = 1.2$ nM $R^1 = n\text{-}C_8H_7^1$, $R^2 = Me$, $R^3 = Bu^n$, R^4 = NHMe	126
Aryl pyrimidines, **95**	40, HET	HPLC; none; 80–95% (purity), 40–99% (yield)	Various; none; LD	None	127
Aryl pyrimidines, **96**	96, HET	HPLC; SLE–prep. HPLC; 25–48% (purified)	Various; none; LD	None	128
Tetrasubstituted aryl pyrimidines, **97**	80, HET	TLC–NMR; none; >90% (purity)	Antitubercular drugs; MIC M. tuberculosis; LD	MIC = 25 μM $R^1 = NH(CH_2)_7CH_3$, $R^2 =$ 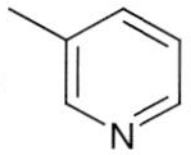	129
Quinazolinones, **98**	27, HET	HPLC; prep. HPLC–silica gel; NG	Neurodegenerative diseases; PARP-1; LO	$IC_{50} = 6$ nM R^1 = p-CN	130
Quinazolinediones, **99**	39, HET	NMR; none; 2–100% (yield)	Oncology; Col2 cell proliferation; LD	Inhibition at 20 mg mL^{-1} = 62% R^1 = 3-OH, R^2 = H, R^3 = 3,4-di-Ome	131
Aminoquinazolinones, **100**	36, HET	HPLC; prep. HPLC; 70–93% (purity)	Various; none; LD	None	132
Diaminoquinazolines, **101**	NG, HET	HPLC, FIA-MS (12.5 sampling); NG	Various; none; LD	None	133
Tetrahydropyrido[2,3-*d*]pyrimidines, **102**	40, HET	HPLC; none; 30–85% (yield), 83–99% (purity)	Various; none; LD	None	134
Tetrahydropyrido[2,3-*d*]pyrimidines, **103**	27, HET	HPLC; none; 55–96% (purity)	Various; none; LD	None	135

continued

Table 2 Continued

Compound class	*Library size, type*	*Analytical method; purification; yields/purities*[a]	*Therapeutic area; target; drug discovery phase*[b]	*In vitro activity, best structure*	*Ref.*
Xanthines, **104**	18, HET	HPLC; none; 44–48% (yield)	Erectile dysfunction; PDE5; LO	93% inhibition at 1 μM R^1 = Me, Ar = 4-BrPh	136
Spiroimidazolidinones, **105**	167, HET	HPLC; none; 75–100% (purità)	CNS; GPCRs; LD	None	137
Spirohydantoins, **106**	≫10, HET	HPLC; prep. HPLC; NG	CNS, respiratory diseases; NK1; LD	$pK_i = 7.34$ $n = 1$, R^1 = CONH(3-MePh)	138
Isoxazolino imidazolidinediones, **107**	990, HET	HPLC; none; >70% purity, 10% samples	Various; none; LD	None	139
4-Substituted imidazoles, **108**	35, HET	HPLC; prep. HPLC; 20–80% purity	Various; 5HTs; LD	None	140
4-Sulfonaminomethyl imidazoles, **109**	≫100, HET	HPLC; prep. HPLC; >90% (purity)	Antifungals; MIC *Candida* spp.; LD	MIC = <0.1->10 μM	141
4-Sulfonaminomethyl imidazoles, **110**	29, HET	HPLC; prep. HPLC; >90% (purity)	Antifungals; MIC *Candida* spp.; H2L	MIC = <0.1->10 μM	142
5-Trifluoromethyl ketoimidazoles, **111**	155, HET	HPLC; prep. HPLC; 16–59% (yield)	Respiratory, antibacterial; COX-2, proteases; LD	None	143
Tetrasubstituted pyrazoles, **112**	96, HET	HPLC; none; 45 to >65% purity	Oncology; ER; H2L	RBA (100% for estradiol) = 23% $R^1 = Bu^t$, R^2 = m-OH, R^3 = *p*-OHPh	144
1,4-Diazepane-2-ones, **113**	90, HET	HPLC; none; 14–72% (purity)	Inflammation, cancer; LFA-1 receptor; LD	IC_{50} = 70 nM $R^1 = CH_2$-2-naphthyl, $R^2 = Bu^i$, $R^3 = CH_2$-3-quinolyl, X = NH_2	145
Tetrahydro-1,4-benzodiazepine-2-ones, **114**	62, HET	HPLC; silica gel; 2–61% (yield)	Various; GPCRs, etc.; LD	None	146
Benzo[*b*][1,4]diazepines, **115**	23, HET	HPLC; silica gel; 7–53% (purified)	CNS; Neuronal Na^+ channels; LD	50% inhibition, 100 mM R^1 = Ph, R^2 = H, R^3 = Me	147
Tetrasubstituted chiral 1,3-oxazolidines, **116**	96, HET	HPLC; none; average >95% purity, 52–99% (yield)	Various; none; LD	None	148
N-Aryloxazolidin-2-ones, **117**	379, HET	HPLC, NMR (15%); none; NG	Antibacterial; MIC G+ pathogens; LO	MIC = 1–64 μM $n = 1$, X = CH, $R^1 = CHF_2$	149
N-Aryloxazolidin-2-ones, **118**	172, HET	HPLC; none; 60->90% (purity)	CNS, antibacterial; MAO, GPIIb/IIa; LO	None	150

2-Substituted oxazolines, **119**	34, HET	HPLC; none; 86–100% (purity), 32–75% (yield)	Antibacterials; none; LD	None	151
Tetrahydrooxazepines, **120**	320, HET	HPLC; prep. HPLC; 60–80% (crude HPLC)	Inflammation; cell permeability; LD	High permeability imparted by library compounds	152
3,5-Disubstituted isoxazoles, **121**	32, HET	HPLC; none; 69–97% (purity), 65–97% (yield)	Various; none; LD	None	153
3,5-Disubstituted isoxazoles, **122**	173, HET	HPLC; none; 74–95% (purity), 60–86% (yield)	Antithrombotics; none; LD, H2L	Thrombosis prot, 70%, at 30 mg kg^{-1} $R^1 = CH_2OH$, $R^2 = 2\text{-}Cl$, $R^3 = 4\text{-}Br$	154
3,4-Disubstituted isoxazoles, **123**	34, HET	HPLC; none; 95% (purity and yield)	Various; none; LD	None	155
3-Hydroxymethyl isoxazoles, **124**	50, HET	HPLC; none; 2–89% (yield)	Various; none; LD	None	156
Aminothiazoles, **125**	20, HET	HPLC; prep. HPLC; 75–98% (purity), 5–60% (yield)	Oncology; CDK-2; LO	$IC_{50} = 3$ nM $R^1 =$ 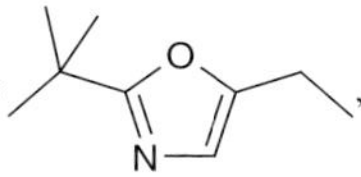 , $R^2 = Bn$	157
1,3,5-Triaminotriazines, **126**	96, HET	HPLC; none; 94% (average purity)	Oncology, antibacterials; none; LD	None	158
1,3,5-Triaminotriazines, **127**	> 100, HET	HPLC; none; > 90% compounds with > 98% purity	Oncology; U937 leukemia cells growth inhibition; LD	$GI_{50} = 1$ μM $R^1 = 3,4\text{-di-OMeBn}$, $R^2 = H$, $R^3 = $ cyclohexyl, X = 4-OmeBnNH	159
1,3,5-Triaminotriazines, **128**	1536, HET	HPLC; none; > 90% compounds with > 98% purity	Oncology; none; TGT ID/Val, LD	Targets identified $R^1 = Pr^i$, $R^2 = H$, $R^3 = 3\text{-OmeBn}$, PS = resin, link = 4-[CONH-$(C_2H_4O)_2C_2H_4$]Ph	160
1,3,5-Triaminotriazines, **128**	1536, HET	HPLC; none; > 90% compounds with > 98% purity	Albinism; pigment induction in zebrafish melan-p1 cells; TGT ID/Val, LD	Most active compound $R^1 = R^3 = CH_2$-cychexyl, $R^2 = H$, link = 4-[CONH-$(C_2H_4O)_2C_2H_4$]Ph	161
Exocyclic amino nucleosides, **129**	1234, HET	HPLC; none; > 60% (purity)	Antiviral, oncology; nucleotide binding targets; LD	None	162
Adenosine analogs, **130**	29, HET	HPLC; none; 91–98% (purity, HPLC)	Hypertension; renal, A_{1A} receptor; H2L	$K_i = 419$ nM $R^1 = H$, $R^2 =$ 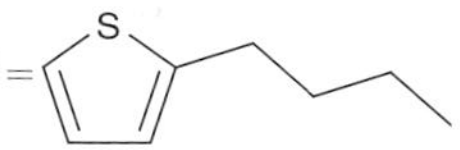	163
Benzo[*a*][1,2,3]triazinones, **131**	42, HET	GC, MS, NMR; none; 10–70% (yield)	Oncology, CNS, antibacterial; 5HTs; LD	None	164

continued

Table 2 Continued

Compound class	*Library size, type*	*Analytical method; purification; yields/purities*[a]	*Therapeutic area; target; drug discovery phase*[b]	*In vitro activity, best structure*	*Ref.*
Triazole-tethered pyrrolidines, **132**	18, HET	HPLC; prep. HPLC; 7–45% (yield)	CV, CNS, pulmonary; ECE-1; LO	$IC_{50} = 150$ nM R^1 = 2-NaphthSO$_2$, R^2 = 4-FBn, R^3 = Me	165
Sulfahydantoins, **133**	80, HET	HPLC; prep. MPLC; 3–38% (yield)	Various; serine proteases; LD	None	166
Furazano[3,4-*b*]pyrazines, **134**	407, HET	HPLC; none; 75% compounds purity >70%	Antibacterials; none; H2L	None	167
Cyclic thioether peptidomimetics, **135**	19, HET	HPLC; prep. HPLC; 4–30% (purified)	Eating disorders; mMC1R; LD	$IC_{50} = 164$ nM $R^1 = (CH_2)_3NH_2$, R^2 = Phe-His, R^3 = H, AA^1 = Trp	168
Tetrazolyl-containing hydantoins, **136**	>14, HET	HPLC; prep. HPLC; 39–85% (yield)	Various; GHS-R; LO	$IC_{50} = 600$ nM Ar = *p*-Ph; $R^1 = (CH_2)_3$piperid, $R^2 = (CH_2)_2Ph$	169
Linear aminoamides, **137**	60, OTH	HPLC; silica gel; 55% to >98% (purity), 41–88% (yield)	Anesthetic, antiarrhythmic; none; H2L	None	170
Ureas, **138**	33, OTH	HPLC; SCX; 66–100% (yield)	Various; none; LD	None	171
Disubstituted guanidines, **139**	70, OTH	HPLC; none; 43–83% (yield), 31–99% (purity)	Various; none; LD	None	172
3,4-Diamino cyclopentanols, **140**	54, OTH	HPLC; none; 21–91% (purity), 45–98% (yield)	Various; prostanoid receptors; LD	None	173
Aryl tertiary amides, **141**	36, OTH	HPLC; none; 0 to >90% (yield)	Various; none; LD	None	174
Oxalic acid arylamides, **142**	60, OTH	HPLC; SCX; 12–92% (yield)	Oncology, allergy; PTP1B; H2L	None	175
Tetrasubstituted benzylguanidines, **143**	5000, OTH	HPLC; none; >70% (purity, selected compounds)	Various; NMDA; LD	None	176
Arylmethyl cinnamic acids, **144**	20, OTH	HPLC; none; 21–95% (yield)	Inflammatory. diseases; EP_3; H2L	$K_i = 20$ nM R^1 = β-naphthyl	177
Trisubstituted benzamides, **145**	784, OTH	NG; NG; NG	Antiviral; HRV 3C protease; H2L	NG, NG	178
Aryloxoisobutyric acids, **146**	480, OTH	HPLC, MS; none; average purity >70%	Dyslipidemia; PPAR-γ/δ; H2L	$IC_{50}\gamma/\delta = 4/19$ nM R^1 = 2,4-di-CF$_3$Bn, R^2 = 4-CF$_3$Bn	179
Arylthioisobutyric acids, **147**	226, OTH	HPLC; none; 70–90% (purity)	Dyslipidemia PPAR-α; LO	$IC_{50} = 1$ nM R^1 = 2,6-di-EtPh, R^2 = 3,4-(OCH$_2$O)Ph(CH$_2$)$_3$	180

Acyl resorcinol carbamates, **148**	112, OTH	HPLC; prep. HPLC; 55–88% (yield)	Oncology; integrin αvβ3; H2L	$IC_{50} = c.$ 10 nM R^1 = neopent.amide R^2 = 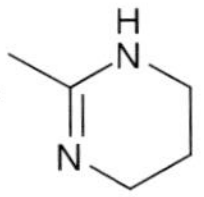	181
Arylsulfonamide hydroxamates, **149**	>260, OTH	HPLC; prep. HPLC; >80% (purity)	Inflammation, respiratory diseases; PCP proteinase; LO	$IC_{50} = 0.024$ nM R^1 = $FmocNHCH_2$, R^2 = 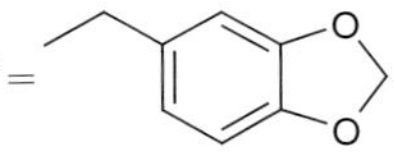	182
Aryl vinyl sulfones, **150**	30, OTH	HPLC; none; 40–95% (purity)	Multiple indications; cysteine proteases; LD	None	183
Diaryl ethers, **151**	32, OTH	HPLC; none; >90% (purity, HPLC) 34–95% (yield)	Antibacterial; none; LD	None	184
Arylmethyl amidines, **152**	37, OTH	HPLC; prep. HPLC; >95% (purity, HPLC)	Oncology, osteoporosis; integrins; H2L	$IC_{50} = 0.7$ nM X^1 = NH, X^2 = Me, R^1 = 4-Ph	185
Aryl diamides, **153**	400, OTH	HPLC; SPE; 1–58% (yield)	Osteoporosis; V-ATPase; H2L	No activity	186
Aromatic polyamine amides, **154**	85, OTH	HPLC; prep. HPLC; NG	Antiparasitic; trypanothione Reductase; LD	$K_i = 76$ nM Ar = 5-Br,3-indolCH_2COOH	187
Biphenylalanine ureas, **155**	500, OTH	HPLC; none; >90% (purity)	Oncology, restenosis; $\alpha_v\beta_3$ receptor; H2L	$K_i = 2.5$ nM R^1 = SO_2-(2,4,6tri-MePh), R^2 = $CONHPr^n$	188
Biphenylalanines, **156**	>180, OTH	HPLC; NG; >95% (purity)	Inflammatory diseases; integrin α4β1; H2L	$IC_{50} = 11.3$ nM R^1 = *o*-NMe_2	189
Bisarylthio *N*-hydroxypropionamides, **157**	44, OTH	HPLC; crystallization; 18–93% (yield)	Oncology, arthritis; MMP-2; H2L	$IC_{50} = 2$ nM R^1 = 2-pyridyl	190
Tertiary arylamines, **158**	1280, OTH	HPLC; none; NG	CV; LXRa; H2L	$EC_{50} = 45$ nM $n = 1$, R^1 = 2-Cl, 3-CF_3	191
Aryl 1,5-enediols, **159**	34, OTH	HPLC; prep. HPLC; 24–38% (crude HPLC)	Pain; μ-opioid receptor; LO	$IC_{50} = 8.8$ nM (*S*,*S*,*S*,*R*) R^1 = $CONH_2$	192

[a] FIA, flow injection analysis; GC, gas chromatography; MPLC, medium-pressure chromatography; SCX, strong cation exchange; TLC, thin-layer chromatography.
[b] CV, cardiovascular; G–, Gram negative; G+, Gram positive; HCV, human cytomegalovirus; H2L, hit-to-lead; LD, lead discovery; LO, lead optimization; nAChR, nicotinic acetylcholine receptor; NG, not given; TGT ID/Val, target identification and validation; TNF, tumor necrosis factor.

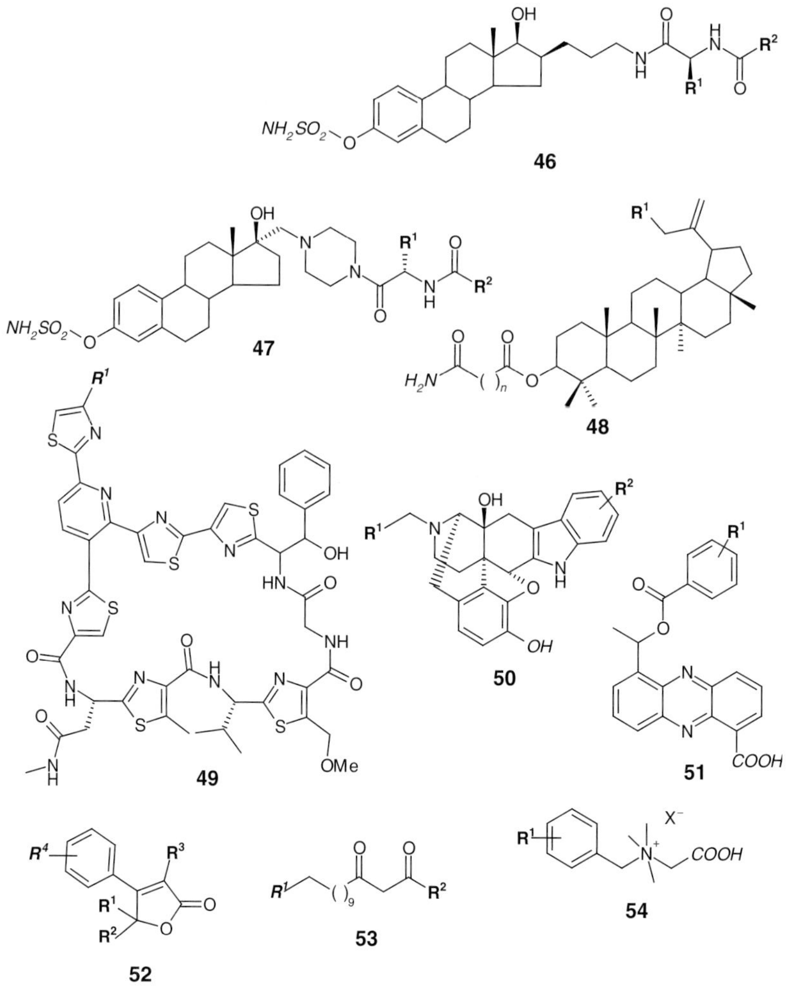

Figure 9 SP discrete libraries: structures **46–54**.

inhibitors of steroid sulfatase when R^2 is a hydrophobic substituent. Ciobanu and Poirier [79] also reported two estradiol sublibraries such as **47**, containing a piperazine group, with similar activity and structure–activity relationship (SAR) on steroid sulfatase. Srinivasan *et al.*[80] introduced lupeol analogs such as **48** as novel antimalarial compounds, and validated their hypothesis by discovering several moderately active compounds. Clough *et al.*[81] focused on the thiazole peptide antibiotic GE2270, producing on SP a set of analogs (**49**) by decoration with amides, esters, acids, alcohols, and amines, and eventually identifying eight potent compounds with better solubility than the parent compound. Naltrindole, a potent δ-opioid ligand, was used as scaffold for nitrogen and indole decoration by Takahashi and co-workers,[82] leading to R^1 and R^2-decorated analogs (**50**). Laursen *et al.*[83] decorated on SP saphenic acid, a fragment of the known antimicrobial agent saphenamycin, to give highly potent analogs (**51**).

Fully synthetic SP libraries of natural products are also abundant. Ma and co-workers[84] reported the parallel SP of five sublibraries of butenolides (**52**), varying according to the nature of R^2 and R^4. Smith *et al.*[85] synthesized on SP a library of acyl-homoserine lactone analogs (**53**) as quorum-sensing inhibitors in *Pseudomonas aeruginosa*, obtaining significant agonist and antagonist structures. Glycine betaine analogs (**54**) were prepared on SP by Cosquer *et al.*,[86] and characterized as strong antibacterial agents using specific transporters to penetrate the cell wall of resistant bacteria.

Stroemgaard *et al.*[87,88] reported two SP libraries (**55** and **56**) modeled on the structures of philantotoxins, a natural family of noncompetitive antagonists of ionotropic receptors (**Figure 10**), yielding submicromolarly active compounds with better potency and selectivity than their parent natural compounds. Raju and co-workers[89] explored the SP N-terminal modifications of negamycin, a natural antibiotic with activity against Gram-negative pathogens, identifying a moderately active *N*-acyl analog out of the library (**57**). Kojic acid, a known inhibitor tyrosinase inhibitor with low

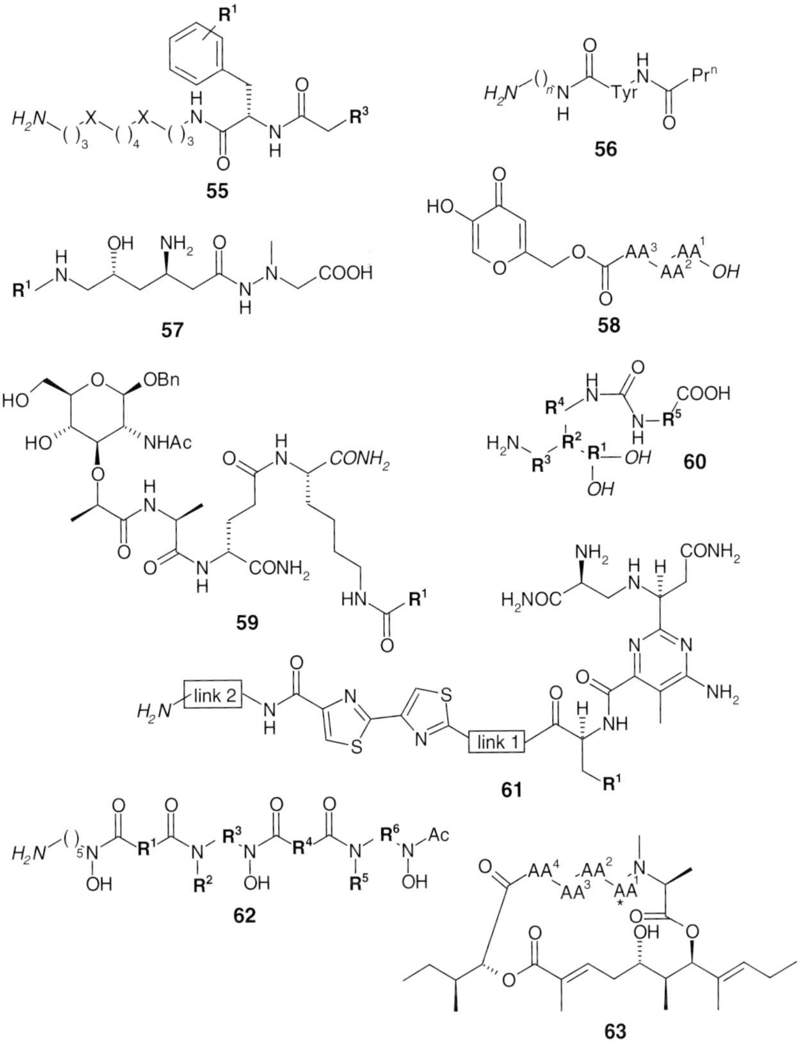

Figure 10 SP discrete libraries: structures **55–63**.

potency and stability, was coupled by Kim *et al.*[90] to amino acids on SP to give the library **58**, obtaining a 100-fold potency increase for several library members, and a significantly higher stability. Muramyl dipeptide amide analogs (**59**) were targeted by Liu and co-workers[91] as therapeutics in several areas, with the aim of preventing known inflammatory side effects. Bozzoli *et al.*[92] targeted mureidomycin, a *P. aeruginosa*-active antibiotic with a modular structure theoretically amenable to SPS and combinatorial exploitation, preparing the library **60**, which lacked antibacterial activity. Two members of an SPS library of deglycobleomycin analogs (**61**) reported by Hecht and co-workers[93] showed a higher potency than the parent compound in a DNA relaxation assay, thus paving the way for sugar-containing bleomycin libraries. Siderophoric iron chelators for thalassemias and anemias were identified by Poreddy *et al.*[94] through SPS of desferrioxamine B analogs (**62**). Finally, a cyclic depsipeptide SP set of libraries (**63**) was prepared by Takahashi and co-workers[95] as analogs of aurilide, a potent cytotoxic agent from Japanese sea hares, representing novel constrained natural products-like derivatives with potential use in various therapeutic areas.

Heterocyclic SP libraries with biological activity have been frequently reported since 2000. Scott *et al.*[96] reported the SPS of α-substituted prolines (**64**) as cores for lactam-based rigid, privileged peptidomimetic scaffolds (**Figure 11**). Wei *et al.*[97] focused on prolyl and pipecolyl amides (**65**) as inhibitors of FKBP12 rotamase activity, identifying two potent analogs active in an in vivo model of Parkinson's disease. Player and co-workers[98] tapped significantly into hydroxyproline-oriented diversity by synthesizing on SP a library (**66**) bearing three decorated functions, as a large lead discovery tool. A similar, and even larger, alkoxyproline library (**67**) was reported by Boldi and co-workers.[99] A more

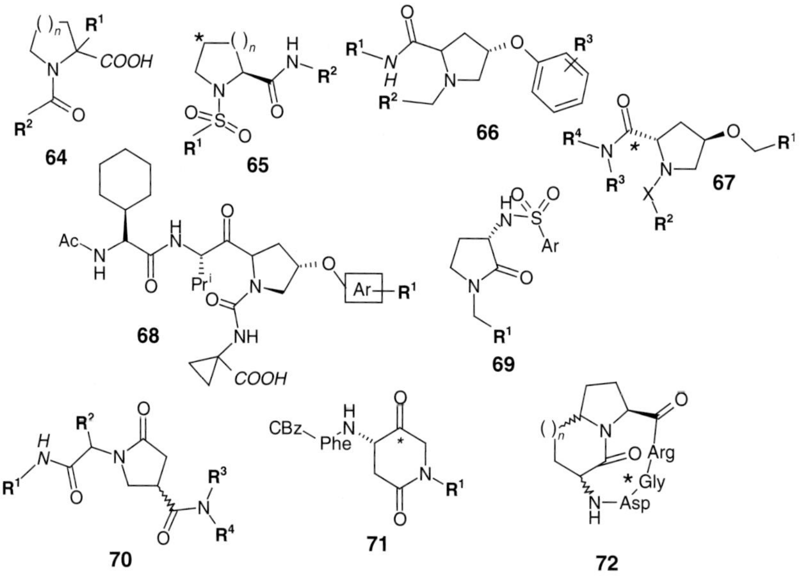

Figure 11 SP discrete libraries: structures **64–72**.

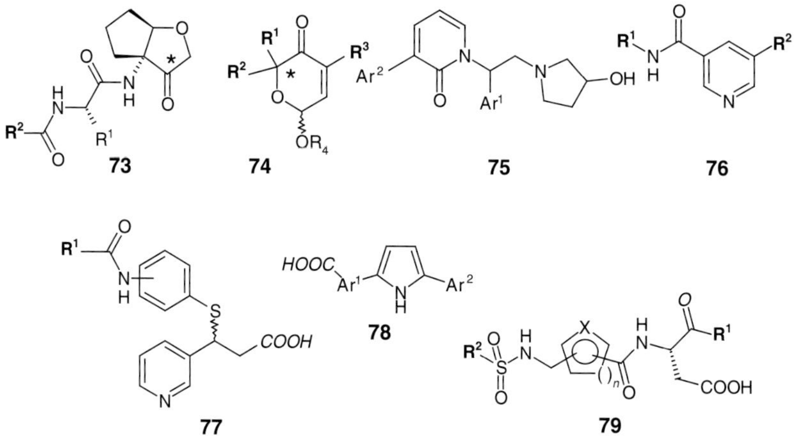

Figure 12 SP discrete libraries: structures **73–79**.

focused effort on hydroxyprolines led to peptidomimetics (**68**) as analogs of a known hepatitis C NS3 protease inhibitor by Poupart *et al.*,[100] who identified submicromolar inhibitor leads. Gong *et al.*[101] reported the SPS of azaarene sulfonamidopyrrolidinones (**69**) as highly potent factor Xa-directed antithrombotics. Player and co-workers[102] exploited the pyrrolidin-2-one scaffold through SPS of γ-lactams (**70**). Huang and Ellman[103] focused on diketopiperazines, and prepared on SP the Cbz-containing cyclic ketones (**71**) as rationally designed, potent, and selective inhibitors of the cysteine protease cruzain. Scolastico and co-workers[104] synthesized the cyclic pseudopeptides (**72**), containing the RGD sequence, as 50-fold more potent integrin antagonists than the original lead structure.

Quibell and co-workers[105] reported a small array of furan 4-ones (**73**) (**Figure 12**), inspired by the structure of known cathepsin K inhibitors, and showed significant selectivity trends for cathepsins K, L, and S. Couladouros and Strongilos[106] reported a modular approach to access SP diversity containing pyranic cores, all accessible from 'polymorphic' scaffolds (**74**) and all qualifying as biologically privileged structures. Aryl pyridones (**75**) were prepared by Semple and co-workers[107] as analogs of known κ opioid receptor agonists, reaching low nanomolar potency. Fernandez and co-workers[108] reported the SPS of substituted nicotinic acid amides (**76**) with good yields and purities. A nicotinic moiety is present also in the propionic acid SP library **77**, reported by Vianello and co-workers[109] as a source of potent and selective integrin αvβ3 antagonists. Kobayashi and co-workers[110] reported a mixed SP/solution phase synthesis leading to disubstituted pyrroles (**78**) as selective retinoid acid receptor agonists having similar agonist potency to retinoic acid and moderate selectivity compared with similar receptors. Choong and co-workers[111] focused on thiophene- and pyridine-based aspartyl ketones (**79**) as caspase-3 inhibitors with low nanomolar potency and selectivity.

Garibay and co-workers[112] reported a previously unreported SP library of phthalides (**80**) (**Figure 13**), obtained with good yields and purities. Smith and co-workers[113] focused on benzopyrans (**81**) as analogs of raloxifene, a known estrogen receptor modulator, increasing its potency in a bone pit assay. Gong and co-workers[114] built a large lead discovery SP library of pentasubstituted benzopyrans (**82**) to target, among others, activators of potassium channels. Smith and co-workers[115] prepared several arrays of 2-aryltryptamines (**83**) as G protein-coupled receptor (GPCR) ligands, finding a potent and selective serotonin ($5HT_{2A}$) antagonist. Indoles were also targeted by Janda and co-workers,[116] who reported the SPS of tetrasubstituted indoles (**84**) as privileged pharmacophores. Zhang and co-workers[117] prepared the indole-based peptide mimetics **85** as thrombin receptor antagonists with more potency and selectivity than the hit which directed their SPS. Aronov *et al.*[118] used rational drug design and parallel SPS to synthesize a library of indole-containing peptidomimetics (**86**) as potent druglike antiparasitics. Chen and co-workers[119] focused on quinazoline carboxamides (**87**) as potential vascular cell adhesion molecule-1/very late antigen 4 (VCAM/VLA-4) antagonists with strong in vitro activity. Grimes and co-workers[120] reported a library of constrained β-turn peptidomimetics based on a tetracyclic structure (**88**).

Ruhland and co-workers[121] reported the SPS of six- and seven-membered rings containing two nitrogen atoms, to give a library of biologically recurrent motifs (**89**) (**Figure 14**). *N*-Arylpiperazines were also targeted by Nilsson *et al.*,[122] Bergauer *et al.*,[123] and Zajdel and co-workers.[124] The first report presented the SPS of 2-carboxycompounds (**90**) as a lead discovery library designed according to the well-known Lipinski's rules. The second introduced some pyrrole-containing compounds (**91**), designed (and found) to be selective dopamine D4 agonists for attention deficit hyperactivity disorders. The last dealt with an SP library (**92**) with potential GPCR-binding properties, out of which several strong and selective $5HT_{1A}$ ligands were found. 2-Oxopiperazines (**93**) were prepared by Gonzalez-Gomez *et al.*[125] as constrained dipeptide mimetics, to ensure a wide range of biological activities. Biaryl diketopiperazines were

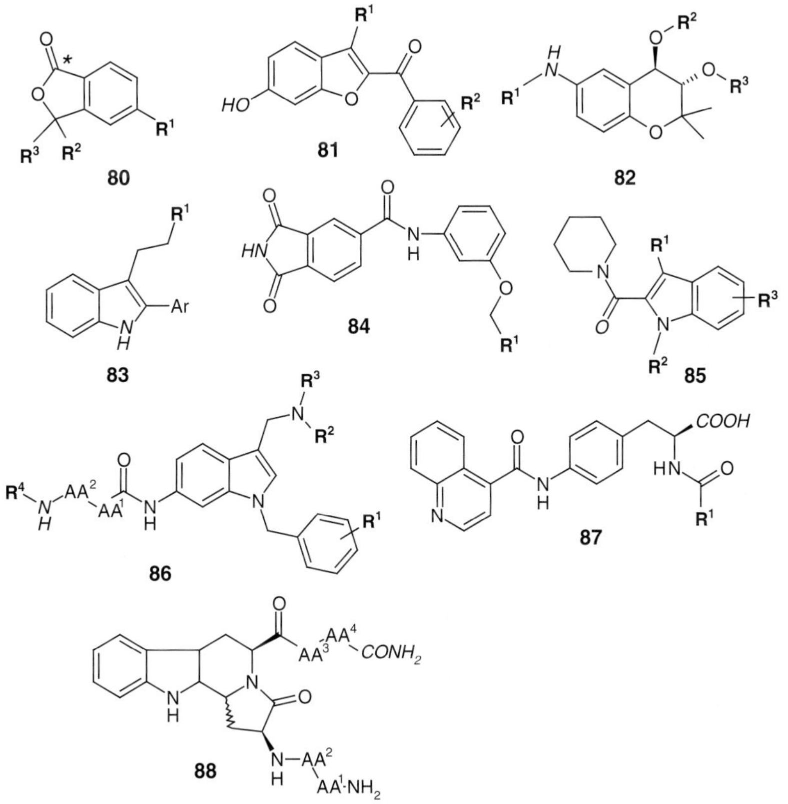

Figure 13 SP discrete libraries: structures **80–88**.

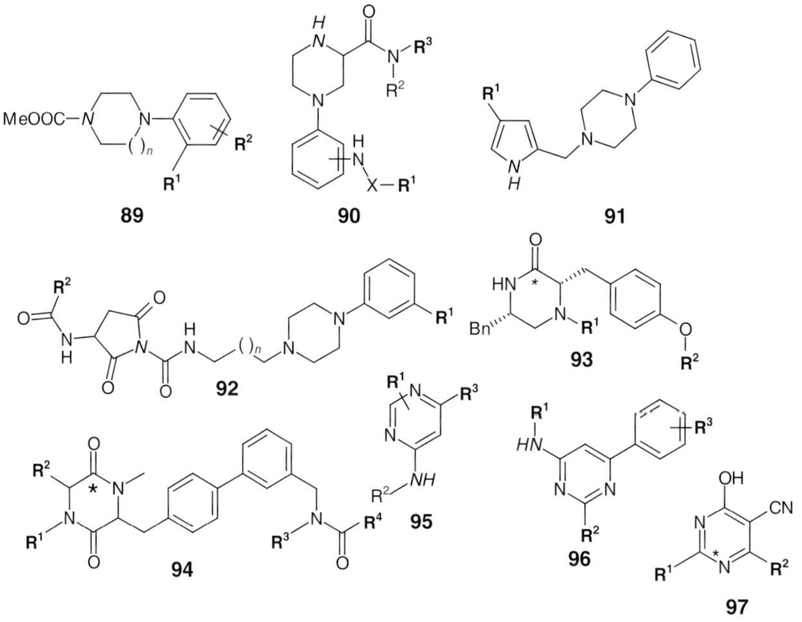

Figure 14 SP discrete libraries: structures **89–97**.

targeted by Guo and co-workers,[126] and three SP libraries (**94**) were designed to expand the SAR around two known FSH agonists, obtaining a 1000-fold activity increase for the best analogs. Ma and co-workers[127] reported the SPS of biaryls and arylpyrimidines (**95**), both being recognized as pharmaceutically privileged structures. Wade and Krueger[128] focused on the same scaffold by synthesizing on SP an easily expandable library (**96**). Tetrasubstituted pyrimidines (**97**) were targeted by Kumar *et al.*,[129] yielding moderately active antitubercular compounds.

Hattori and co-workers[130] synthesized an SP library of quinazolinones (**98**) (**Figure 15**) aiming to achieve more bioavailable PARP-1 inhibitors, and eventually to identify an attractive therapeutic candidate for neurodegenerative disorders. Quinazolinediones (**99**) were the focus of the work by Choo and co-workers,[131] as rationally designed, selective antineoplastics active on colon carcinoma cells. Kesarwani *et al.*[132] focused their attention on quinazolin-4(3H)-ones (**100**) due to their recurrence in anticonvulsant, antibacterial, and antidiabetic structures. Dener and co-workers[133] reported the SPS of a large SP library of 2,4-diaminoquinolines (**101**), evaluating its quality by statistical characterization of several random compounds. Falcò and co-workers[134,135] reported two druglike SP libraries (**102** and **103**) based on a common tetrahydropyrido[2,3-*d*]pyrimidine scaffold. Beer *et al.*[136] reported the SPS of xanthines (**104**) as PDE5-active structural analogs of sildenafil.

Feliu *et al.*[137] prepared spiroimidazolidinones (**105**) (**Figure 16**) on SP, as a GPCR-privileged source of ligands with a constrained structure. Spirohydantoins (**106**) were prepared by Bleicher and co-workers,[138] and their testing as NK-1 ligands showed selected compounds to have high affinity for this receptor. Kurth and co-workers[139] reported an SP library of isoxazolinohydantoins (**107**) as a diverse and druglike lead discovery collection. Gelens *et al.*[140] focused on 4-substituted imidazoles (**108**) as potential bioactives, due to the recurrence of this structural scaffold in active principles. The same scaffold was targeted by Saha and co-workers,[141,142] through the SPS of two sulfonamide libraries (**109** and **110**), whose testing as antifungals eventually produced several yeast-selective, broad-spectrum agents. 5-Substituted imidazoles were also targeted by Hamper and co-workers,[143] and trifluoromethyl substituted compounds (**111**) were prepared. Stauffer and Katzenellenbogen[144] targeted an SP library of tetrasubstituted pyrazoles (**112**) as analogs of a known hit active on the estrogen receptor, eventually identifying potent and selective ERα antagonists. Seven-membered heterocyclic rings containing two nitrogen atoms were the basis of three recent publications. Wattanasin and co-workers[145] prepared on SP a focused library of 1,4-diazepane-2-ones (**113**), from which nanomolar antagonists of the lymphocyte function-associated antigen 1 (LFA-1) were identified. Im *et al.*[146] prepared two β-turn peptidomimetic libraries based on a tetrahydro-1,4-benzodiazepine-2-one scaffold (**114**) as potential GPCR-binding arrays. Lam and co-workers[147] prepared an array of nitrogen-containing heterocycles, including benzo[*b*][1,4]diazepines (**115**), and identified a potent neuronal sodium channel blocker.

Janda and co-workers[148] focused on a chiral 1,3-oxazolidine scaffold (**116**) (**Figure 17**), obtaining an SP library with excellent purities and yields. Singh *et al.*[149] reported the SPS of four libraries of *N*-aryloxazolidinones (**117**) as potential

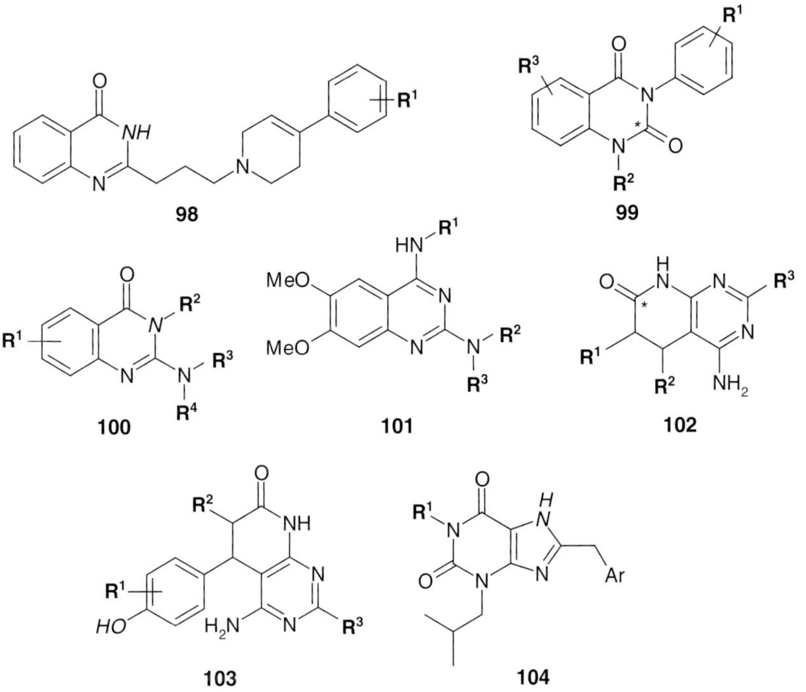

Figure 15 SP discrete libraries: structures **98–104**.

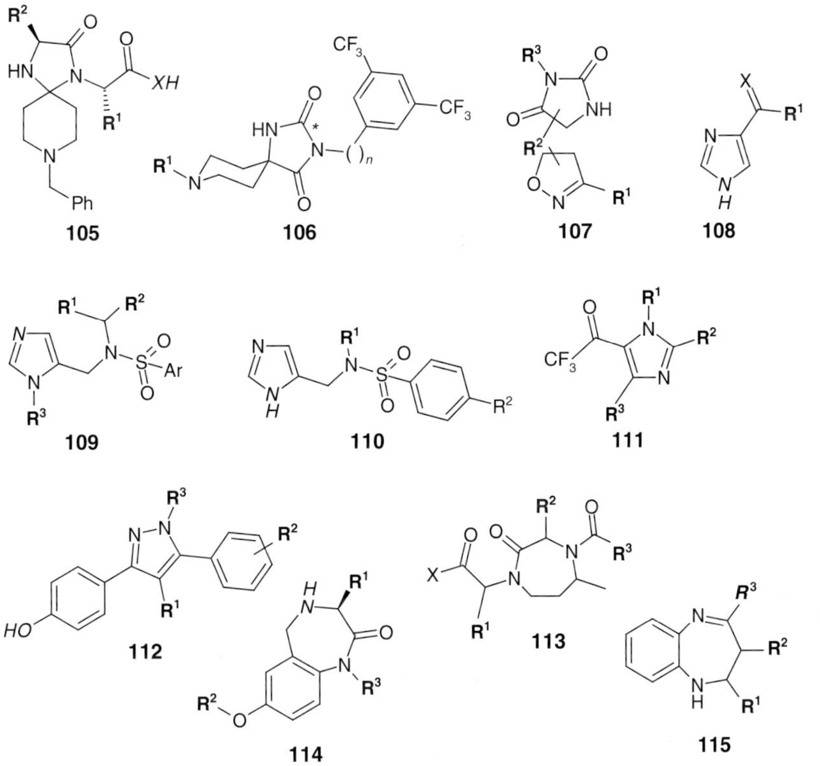

Figure 16 SP discrete libraries: structures **105–115**.

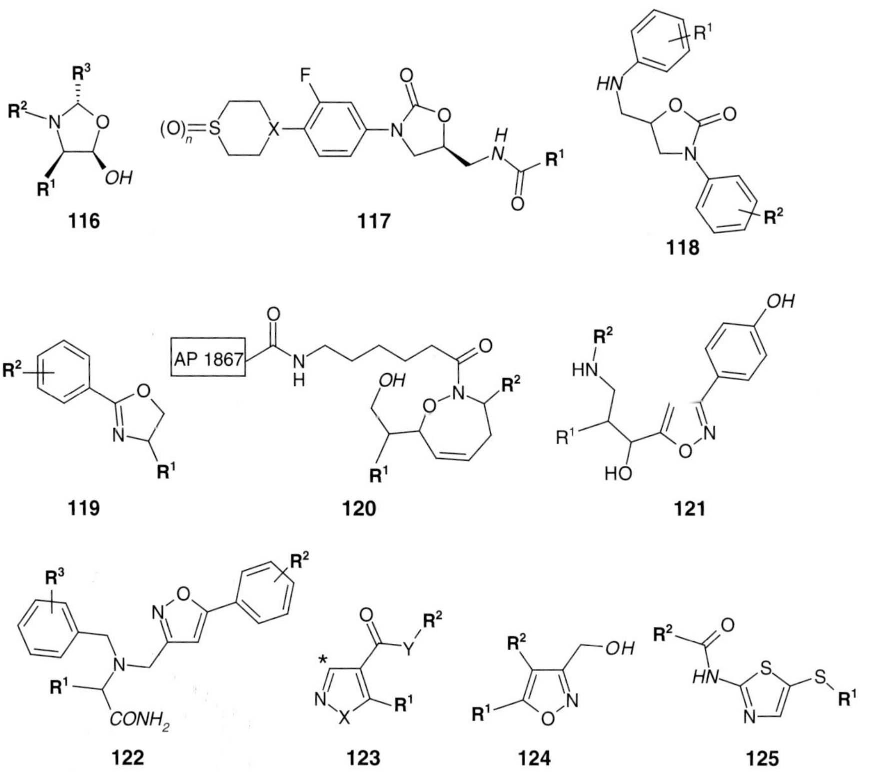

Figure 17 SP discrete libraries: structures **116–125**.

antibacterials, which showed in vivo activity with similar, or better, potency to linezolid. A similar library (**118**) was reported by Buchstaller.[150] Pirrung and Tumey[151] reported an SP library of oxazolines (**119**) as a source of siderophoric antibacterials. Verdine and co-workers[152] reported a library of tetrahydrooxazepines (**120**) as inducers of protein dimerization (AP1867 is a synthetic analog of FK506) and as cell permeabilizers, opening the way to novel therapeutic approaches. Isoxazoles were the common scaffolding of four recent SP libraries. Batra and co-workers[153,154] reported the libraries **121** and **122**, and successfully characterized the latter as a source of antithrombotic agents with in vivo activity. De Luca *et al.*[155] reported the SPS of pyrazoles and isoxazoles (**123**) with exceptionally high yields and purities. Cereda and co-workers[156] focused on 3-hydroxymethylisoxazoles (**124**) as privileged bioactive fragments. A focused SP library of aminothiazoles (**125**) was prepared by Kim and co-workers,[157] identifying in vitro and in vivo potent CDK-2 inhibitors with good selectivity.

Chang and co-workers focused on SP libraries of 1,3,5-trisubstituted triazines (**126–128**) (**Figure 18**) as sources of meaningful collections for chemical genetics studies in target identification and lead discovery. Their first report[158] introduced a sound SP strategy to exploit this heterocyclic scaffold, introducing **126** as a model library. The second[159] introduced **127** as a myoseverin-inspired library used to find novel microtubule-destabilizing agents with better in vitro potency. The last two reports[160,161] discussed the preparation and testing of the library **128** to respectively identify novel biological targets in the zebrafish embryo and to identify targets and binding triazines determining the most common form of albinism. Two nucleoside-inspired libraries were reported by Varaprasad *et al.*,[162] and by Van Calenbergh and co-workers.[163] The former introduced the SPS of 13 sublibraries of general structure **129** as a lead discovery collection for antiviral and oncology targets. The latter focused on adenosine analogs (**130**), leading to a rationalization of SARs in this class and allowing the rational selection of better adenosine A_1 receptor antagonists. Benzo[*a*][1,2,3]triazinones (**131**) were selected by Braese and co-workers[164] as broad-range privileged pharmaceutical scaffolds. Triazole-tethered pyrrolidines (**132**) were targeted as Zn^{2+} metalloprotease inhibitors by Kitas and co-workers,[165] establishing meaningful SARs and identifying compounds with similar potency as the original lead. Sulfahydantoins (**133**) were synthesized on SP by Albericio and co-workers[166] as serine protease inhibitors. Fernandez *et al.*[167] reported the SPS of several arrays of furazano[3,4-*b*]pyrazines (**134**) as diverse antibacterial candidates. A cyclic thioether peptidomimetic scaffold (**135**) was exploited combinatorially by Meldal and co-workers,[168] finding a

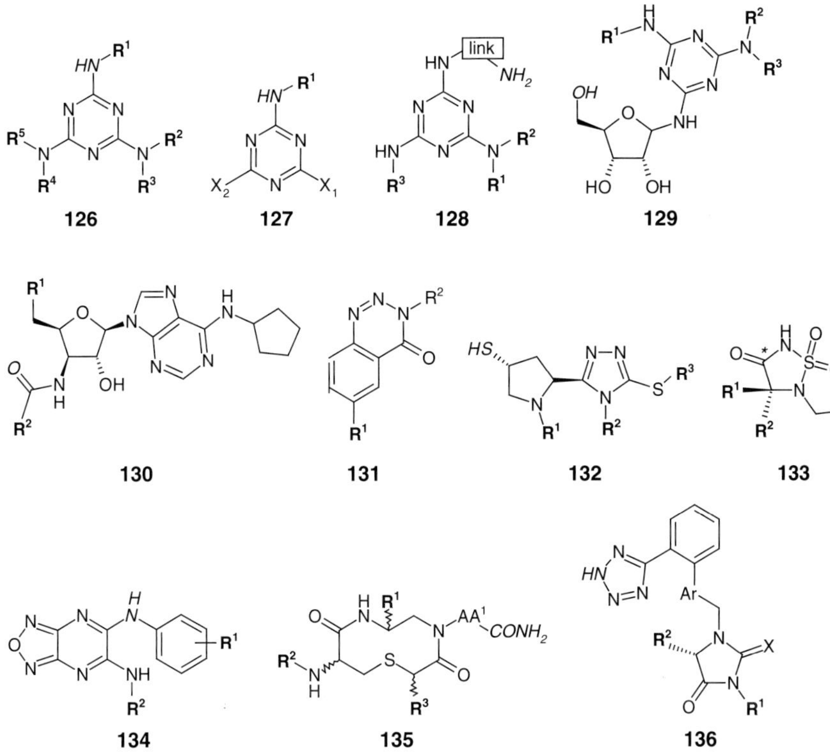

Figure 18 SP discrete libraries: structures **126–136**.

promising melanocortin receptor agonist for eating disorder treatment. Tetrazole-containing hydantoins and thiohydantoins (**136**) were prepared by Severinsen *et al.*[169] as moderately potent growth hormone secretagogue receptor ligands.

A final set of SPS libraries can neither be classified as natural products-inspired nor as heterocyclic. Shannon *et al.*[170] presented an SP library of aminoamidic analogs of lidocaine and procainamide (**137**) (**Figure 19**). Janda and co-workers[171] prepared a library of biologically relevant compounds (**138**) with high yields through a versatile SP method. Guanidines (**139**) were selected by Sandanayake *et al.*[172] as well-known biologically relevant functionalities. Guan *et al.*[173] reported stereochemically defined 3,4-diaminocyclopentanols (**140**), paving the way for the diversification of such scaffold-based SP libraries into druglike, pharmaceutically relevant lead discovery tools. Katritzky and co-workers[174] reported the SPS of tertiary amides (**141**) as simple but pharmaceutically common compounds. Georgiadis *et al.*[175] focused on oxalic acid amides (**142**) as analogs of known PTP1B inhibitors with therapeutic potential. Several sublibraries of trisubstituted benzyl guanidines (**143**) were reported by Hopkins *et al.*[176] A much smaller set of arylmethyl cinnamic acids (**144**) was presented by Juteau and co-workers[177] as a source of potent and selective human EP3 prostanoid receptor inhibitors. An example of rational design of an SP H2L library was reported by Reich and co-workers,[178] through SPS of trisubstituted benzamides (**145**) showing nanomolar potency on human rhinovirus 3C protease. Brown *et al.* focused on peroxisome proliferator-activated receptor (PPAR) agonists for treating dyslipidemia, and reported two SP libraries used to identify hits or leads for these targets. The former[179] (aryloxoisobutyric acids (**146**)) led to the identification of low nanomolar dual PPAR-γ/δ agonists, while the second[180] (two sublibraries of arylthioisobutiric acids (**147**)) led to low nanomolar, selective PPAR-α agonists. Aryl resorcinol carbamates (**148**) were selected by Gopalsamy and co-workers[181] as a source of $\alpha v\beta 3$ vitronectin receptor inhibitors, yielding several moderately active inhibitors.

Dankwardt and co-workers[182] reported an SP library of arylsulfonamide hydroxamates (**149**) (**Figure 20**) targeted against procollagen C proteinase, providing up to 12 500 times more potent inhibitors. Wang and Yao[183] exploited a vinyl sulfone scaffold to prepare the SP library **150** for lead discovery of cysteine protease inhibitors, or for target identification/validation studies. Diaryl ethers (**151**) were prepared on SP by Braese and co-workers[184] as potential antibacterials, or as core structures for natural products-like antiinfectives. An SP library of nonpeptide RGD mimetics,

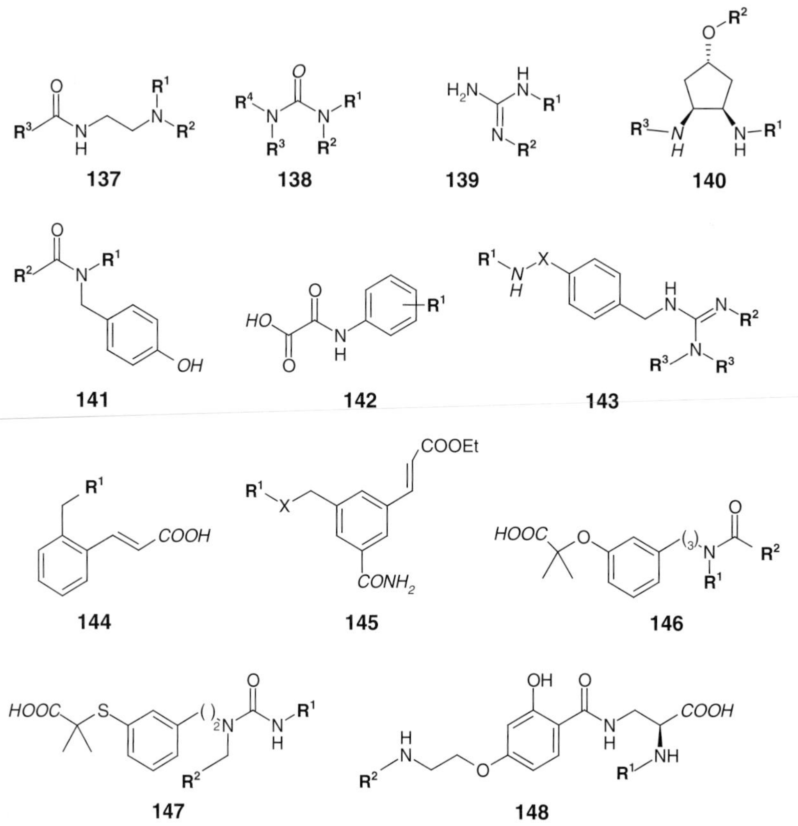

Figure 19 SP discrete libraries: structures **137–148**.

including selective subnanomolar αvβ3 inhibitors (**152**), was published by Kessler and co-workers.[185] Edvinsson and co-workers[186] reported 10 SP libraries of aryl diamides (**153**) as analogs of a potent vacuolar ATPase inhibitor, unfortunately devoid of significant biological activity. Aromatic polyamineamides (**154**) were targeted by Chitkul and Bradley[187] as trypanothione reductase inhibitors, leading to strong enzyme inhibitors. αvβ*x* inhibitors were targeted by Urbahns *et al.*,[188] reporting an SP library of biphenylalanine ureas (**155**), among which several single-digit nanomolar inhibitors of αvβ3 were found; and by Castanedo *et al.*,[189] reporting biphenylalanines (**156**), and leading to similar potencies as dual inhibitors of αvβ1/αvβ7. Chollet *et al.*[190] focused on hydroxamates (**157**) as matrix metalloproteinase (MMP) inhibitors, identifying low nanomolar inhibitors of MMP-2, MMP-9, and MMP-13 with selectivity versus MMP-1. Collins and co-workers[191] identified a potent and selective nonsteroidal liver X receptor (LXR) agonist through screening of polyaromatic tertiary amines (**158**) made by an easy and flexible SP route. Finally, Verdine and co-workers[192] reported the SPS of an exhaustively stereodiversified library of 1,5-enediols (**159**), similar to endomorphin-2 and thus providing low nanomolar μ-opioid receptor ligands.

3.33.3.3 Solid-Phase Pool Libraries

The SP libraries listed in this section were all made by split-and-mix synthesis,[7,7a,7b] and the identity of active individuals was determined (even if activities and structures were often undisclosed) either by deconvolutive[10,10a,10b] or encoding (chemical or nonchemical)[10b,11,11a] methods. A significant number were obtained using directed sorting[10b,11,11a] techniques, which produce spatially separated compounds while employing split-and-mix synthetic methodologies.

Natural products-inspired pool libraries have been well represented since 2000 (**Table 3**). Akritopoulou-Zanze and Sowin[193] reported the split-and-mix SP synthesis of highly diversified, chemically encoded macrolides (**160**) (**Figure 21**), made by decoration of a baccatin intermediate. Jagtap *et al.*[194] disclosed a decorative SP library of taxol analogs (**161**) through nonchemical encoding and directed sorting,[10b,11,11a] obtaining either improved cytotoxicity, or better solubility than taxol itself. Morphinan derivatives were the focus of two reports from Takahashi and co-workers.[195,196] The first

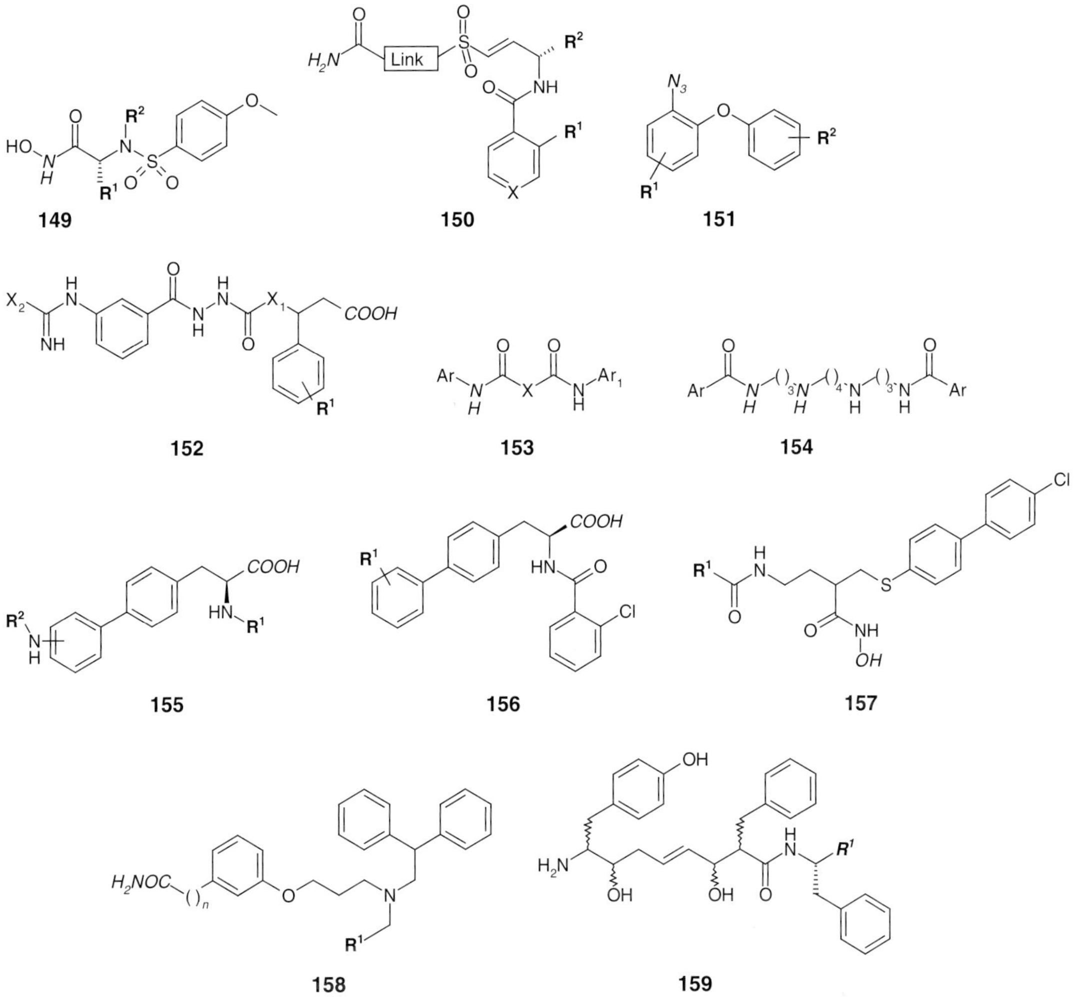

Figure 20 SP discrete libraries: structures **149–159**.

reported a nonchemically encoded SP decoration library of sulfonylaminoderivatives (**162**) as potential opioid receptor ligands, while the second presented some closely related symmetric dimeric norbinaltorphimines (**163**).

Player and co-workers[197] focused their attention on nonchemically encoded[10b,11,11a] tyrphostins (**164**) (**Figure 22**), as synthetic analogs of the naturally occurring kinase inhibitor erbstatin, to be evaluated on a panel of kinase targets. Takahashi and co-workers reported two SP libraries made by nonchemical encoding and directed sorting. The former[198] (**165**) was inspired by phlorizin, a Na^+/glucose co-transporter inhibitor, while the second[199] (**166**) was focused on vitamin D_3 analogs as potential cell regulators for therapeutic purposes. Renault *et al.*[200] and Manku *et al.*[201] reported two polyamine SP pool libraries, both made by nonchemical encoding methods. While the first paper reported a small model library (**167**), the second introduced a large library (**168**) for lead discovery purposes. Maletic and co-workers[202] focused on inhibitors of the Mur pathway enzymes as antibacterials, and reported the SP library of *N*-acetylmuramic acid analogs **169**. Macrocyclic, natural products-like SP libraries were targeted both by Ramaseshan *et al.*[203] and by Jefferson and co-workers.[204] The first paper reported a library of macrolactones (**170**) for lead discovery purposes using nonchemical encoding, while the second introduced a library of macrocyclic peptidomimetics (**171**) from which a potent series of bacterial protein synthesis was identified.

Heterocyclic SP pool libraries are also abundant. Willoughby and co-workers[205] targeted CCR5 and its potential antagonists to fight HIV infection through pyrrolidines (**172**) (**Figure 23**), whose deconvolution[10,10a,10b] yielded the structures of several highly potent antiviral compounds. Aryloxyprolines (**173**) were reported by Jackson and co-workers,[206] and led via deconvolution to moderately active, but structurally novel tumor necrosis factor (TNF-α) signaling inhibitors. Fenwick *et al.*[207] identified a low nanomolar reversible inhibitor of cathepsin K by in vitro screening

Table 3 SP pool libraries

Compound class	*Library size, nature*	*Structure determination method*	*Analytical method; purification; yield/purity*	*Therapeutic area;– target*[b]	*Biological activity, structure of active*	*Ref.*
Macrolides, **160**	70 000, NPD	Chemical encoding	NG; none; NG	Antibacterial, others; MICs against pathogens	None	193
Paclitaxel analogs, **161**	24, NPD	Nonchemical encoding	HPLC; prep. TLC; >80% purity	Oncology; A2780 cell cytotoxic	$IC_{50} = 14\,ng\,mL^{-1}$ $R^1 = H$, $R^2 =$ 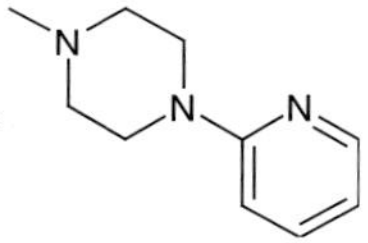	194
Morphinan analogs, **162**	339, NPD	Nonchemical encoding	HPLC; none; >80% (purity)	Pain; κ agonist	None	195
Norbinaltorphimine analogs, **163**	120, NPB	Nonchemical encoding	HPLC; none; 25–75% (yield)	Pain; κ agonist	None	196
Tyrphostin analogs, **164**	4500, NPB	Nonchemical encoding	HPLC; none; 86% average purity	Various; kinases	None	197
Phlorizin analogs, **165**	132, NPB	Nonchemical encoding	HPLC; none; >40% (purity)	Various; Na^+/glucose transporters	None	198
Vitamin D_3 analogs, **166**	72, NPB	Nonchemical encoding	HPLC; GPC; NG	Oncology, immunomodulation; none	None	199
Polyamines, **167**	16, NPB	Lanterns – encoding	NMR; none; 50–85% (yield)	Oncology, antiparasitic; ion channels	None	200
Polyamines, **168**	4913, NPB	Mass encoding	NG; none; NG	Various; DNA and RNA targets	None	201
N-Acetylmuramic acid analogs, **169**	9, NPB	Deconvolution, encoding	HPLC; none; 53% (average yield)	Antibacterial; Mur pathway	None	202
Macrolactones, **170**	59, NPB	Nonchemical encoding	HPLC; none; 10–62% (yield)	Various; none	None	203
Macrocyclic peptidomimetics, **171**	12 000, NPD 204	Nonchemical encoding	HPLC; none; NG	Antibacterial; MICs against pathogens	MICs = 25–100 μM $R^1 = (CH_2)_3$-guanidine, $R^2 = Bn$, $R^3 = NH_2$, $R^4R^5 = $–COCO–	204

Pyrrolidines, **172**	11 700, HET	Deconvolution	HPLC; none; NG	Antiviral; CCR_5	$IC_{50} = 1$ nM $NR^1R^2 =$ 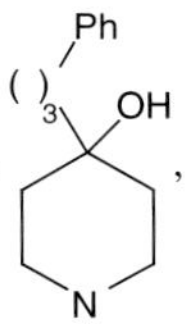 , $R^3 = CH_2$-cychexyl	205
Aryloxyprolines, **173**	1728, HET	Deconvolution	NMR, HPLC (statistical); none; NG	Various; TNF-α inhibition	$IC_{50} = 8.1$ μM $R^1 =$ Bn, $R^2 = n$-pentyl$_2$, $R^3 =$ 3-quinolyl	206
Tetrahydrofuranones, **174**	NG, HET	Nonchemical encoding	HPLC, NMR; none; NG	Osteoporosis; cathepsin K	$K_i = 11$ nM $R^1 =$ (p-Ph)Ph, $R^2 =$ sec-Bu	207
Ethylenediamino piperidines, **175**	180, HET	Nonchemical encoding	HPLC; none; >80% purity	Various; RNA interacting	No activity	208
Tetrahydroquinolinones, **176**	27, HET	Nonchemical encoding	HPLC, MS; none; NG	Various; TGT ID/Val, LD	None	209
Arylindoles, **177**	128 000, HET	Deconvolution	NG; none; NG	Various; GPCRs, $5HT_6$	$IC_{50} = 0.7$ nM $R^1 = R^2 =$ Me, $R^3 = m$-Me, Ar $= m$-BrPh, $n = 2$	210
Dialkoxyindoles, **178**	64, HET	Color encoding	HPLC; none; 51–84% (yield)	Various; none	None	211
2,3-Disubstituted benzofurans, **179**	90, HET	Nonchemical encoding	HPLC–NMR; none; >90% compounds >80% purity	Various; none	None	212
Trisubstituted pyridines, **180**	220, HET	Deconvolution	MS; none; 95% compounds confirmed	Various; none	None	213
Bicyclic guanidines, **181**	75, HET	Tea bags	HPLC; none; 70–80% (purity)	Gastrointestinal; none	None	214
4-Imidazolidinones, **182**	16 000, HET	Positional scanning	HPLC; none; NG	Various; none	None	215
Biheterocyclic imidazolines, **183**	>50 000, HET	Positional scanning	HPLC; none; NG	CV, CNS, oncology; none	None	216
Ureido hydantoins, **184**	>100, HET	Tea bags	HPLC; none; >80% purity for 10% compounds	Various; none	None	217

continued

Table 3 Continued

Compound class	*Library size, nature*	*Structure determination method*	*Analytical method; purification; yield/purity*	*Therapeutic area;– target*[b]	*Biological activity, structure of active*	*Ref.*
4-Imidazolidinones, **185**	34, HET	Tea bags	HPLC; none; 27–84% (yield)	Various; none	None	218
Biheterocyclic imidazolines, **186**	33, HET	Tea bags	HPLC; none; 60–82% (purity)	Antiviral, anti-inflammatory; none	None	219
Dihydroimidazolium salts, **187**	40, HET	Tea bags	HPLC, MS; none; 60–75% (crude yield)	Antiallergic, antihyperglycemic; none	None	220
Biheterocyclic imidazoles, **188**	8649, HET	Chemical encoding	HPLC (statistical); none; NG	Various; suppression of iNOS induction, in rats	$ED_{50} = 1.2\ mg\ kg^{-1}$ $R^1 =$ 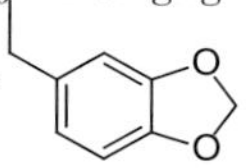 $R^2 = H$, $X = NCOOMe$ (6-piperazine), $n = 2$	221
Benzimidazolones, **189**	48, HET	Nonchemical encoding	HPLC; prep. HPLC; 40–51% (yield)	CNS, CV; phosphodiesterases, factor Xa	None	222
Piperazines, **190**	300, HET	Deconvolution	HPLC; none; NG	Antiviral; HIV protease	$IC_{50} = <0.1\ nM$ $R^1 = CH_2(4\text{-}Ph)\text{-}2\text{-}thiazole$, $R^2 =$ 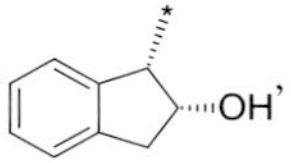, $R^3 = Bn$	223
Piperazines, **191**	902, HET	Deconvolution	HPLC; none; NG	Antiviral; HIV protease	$IC_{50} = 0.1\ nM$ $R^1 = SO_2(2\text{-}thiophene)$, $R^2 =$ 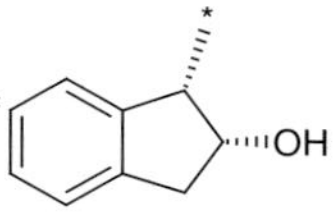	224
Imidazolyl glycopyranosides, **192**	48, HET	Nonchemical encoding	HPLC; none; 55% (avg yield)	Various; none	None	225
Diaminopyrimidines, **193**	162, HET	Nonchemical encoding	HPLC; none; 86% of compounds >80% purity	Various – Protein kinases	None	226

2-Aminopyrimidines, **194**	45, HET	Nonchemical encoding	HPLC, MS; none; 13–90% (yield)	CNS, others – NOS, adenosine receptors	None	227
Quinazolinones, **195**	11, HET	Tea bags	HPLC; none; 58–85% (purity)	CNS, oncology; 5HTs, TNF	None	228
Tricyclic hydroxyindolines, **196**	78, HET	Nonchemical encoding	HPLC; none; NG	Various; TGT ID/Val, LD	None	229
Benzodiazepinones, **197**	10 530, HET	Nonchemical encoding	HPLC (5% compounds); none; >75% compounds >75% purity	Various; none	None	230
Benzodiazepinediones, **198**	200, HET	Tea bags	HPLC; none; 65–92% (purity)	CNS; none	None	231
Bicyclic pyrimidines, **199**	NG, HET	MS	HPLC; none; 0–95% (yield, test cases)	Various; nucleotide binding proteins	None	232
Nucleoside analogs, **200**	25 000, HET	Nonchemical encoding	HPLC (6% of compounds); none; 75% of them >85% pure	Various; nucleotide binding proteins	None	233
Peptidotriazoles, **201**	450 000, HET	On-bead screening	HPLC (statistical); none; NG	Leishmaniasis; cysteine protease B	$K_i = 870$ nM $AA^1 = Gly$, $AA^2 = Leu$, $AA^3 = ClPhe$, $AA^4 = Leu$, $R^1 = $ red Arg	234
Thiadiazole ethers, **202**	96, HET	Nonchemical encoding	HPLC; none; 64 to >98% (purity)	CV prophylaxis; IL8	$IC_{50} = 160$ nM $R^1 = $ 4-FPh, Ar = α-naphthyl, $n = 2$	235
Ureido acids, **203**	9 660, OTH	Deconvolution	NG; none; NG	Inflammatory diseases; VLA-4	$IC_{50} = 3.2$ nM $R^1 = $ 3,5-di-ClPh, $X = SO_2$, $W = NHCH_2$–pPh, $Y = CH(CH_2$-*p*-biaryl)	236
Peptoids, **204**	10 648, OTH	Positional scanning	HPLC; none; NG	Antibacterial; MICs against pathogens	MICs = 31–62 $\mu g\,mL^{-1}$ $R^1 = R^3 = (Ph)_2CH(CH_2)_2$, $R^2 = $ 2-(*N*-Me-pyrrolid)-$(CH_2)_2$	237
Diamines, **205**	>50 000, OTH	Deconvolution	MS; none; 73% of expected compounds found on a statistical set	Antibacterial; MICs against *M. tuberculosis*	MIC = 0.2 $\mu M\,mL^{-1}$ $R^1 = $ 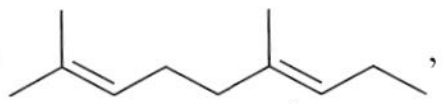 , $R^2 = R^4 = H$, $R^3 = $ 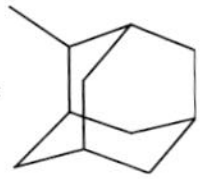	238

continued

Table 3 Continued

Compound class	*Library size, nature*	*Structure determination method*	*Analytical method; purification; yield/purity*	*Therapeutic area;– target*[b]	*Biological activity, structure of active*	*Ref.*
Dipeptide analogs, **206**	300, OTH	Nonchemical encoding	HPLC; none; NG	Antibacterial; PPAT	$IC_{50} = 6$ nM $R^1 = Pr^i$, $R^2 = His$, $R^3 = Glu$, $R^4 = Fmoc$	239
Dansyl peptides, **207**	381, OTH	Positional scanning	MS; none; NG	Pain; neuropeptide FF antagonist	$K_i = 1.4$ μM $AA^1 = Ser$, $AA^2 = Gly$	240
Reduced peptidomimetics, **208**	34 000, OTH	On-bead screening	HPLC; none; NG	Leishmaniasis; cysteine protease B	$K_i = 1$ μM H_2N-D-K-H-F(CH_2NH)-L-L-V-K	241
Tris amine peptidomimetics, **209**	300, OTH	Lanterns – encoded	HPLC; none; average 72% purity	Various; PS translocase activity	Strong activity $AA^1 = Gly$, $AA^2 = Phe$, $R^1 = SO_2(4\text{-}NO_2Ph)$	242
Phenylakylamides, **210**	108, OTH	Deconvolution	GC-NMR; none; 40–70% (yield)	Sleep disorders; hMT_2 receptor	$K_i = 0.7$ nM $R^1 = Pr^n$, $R^2 = m\text{-}OMe$, $n = 3$	243
Aryl-containing peptidomimetics, **211**	158 400, OTH	Chemical encoding	NG; none; NG	None	None	244
Phenoxypropanolamines, **212**	5800, OTH	Deconvolution	MS (5% compounds); >70% confirmed with >70% purity	CV, CNS; none	None	245
Biaryls, **213**	31 372, OTH	Chemical encoding	NG; none; NG	Infertility; follicle-stimulating hormone agonists	$EC_{50} = 260$ nM $R^1 = (CH_2)_2\text{-}p\text{-}ClPh$, $R^2 = meta$, $R^3 = Bu^n$, $R^4 = CONHMe$, X = 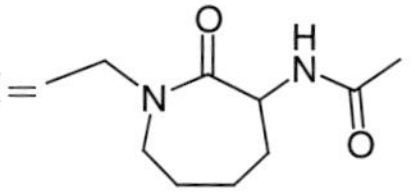	246
Aminoaryls, **214**	NG, OTH	Deconvolution	NG; none; NG	CNS; NPY2	$IC_{50} = 450$ nM $R^1 = Me$, $R^2 = H$, $R^3 = 2\text{-}Bz\text{-}thiophene$, $X = SCH_2$, $n = 1$	247

Dihydrostilbenes, **215**	36, OTH	Nonchemical encoding	HPLC; prep. HPLC; 20–45% (yield)	Various; none	None	248
Alkylaminobenzanilides, **216**	10 800, OTH	Nonchemical encoding	HPLC; none; 84% compounds >80% purity	Oncology, others; none	None	249
Diaryl diamides, **217**	144, OTH	Nonchemical encoding	NMR; none; 17% of compounds tested, 83% of them >90% purity	Oncology; farnesyltransferase	Weak activity	250

[b] iNOS, inducible nitric oxide synthase; NG, not given.

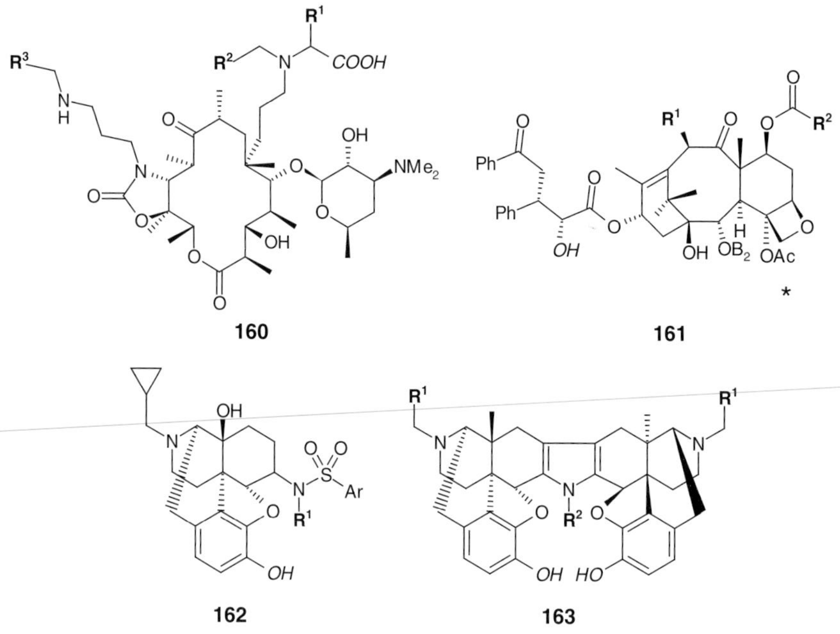

Figure 21 SP pool libraries: structures **160–163**.

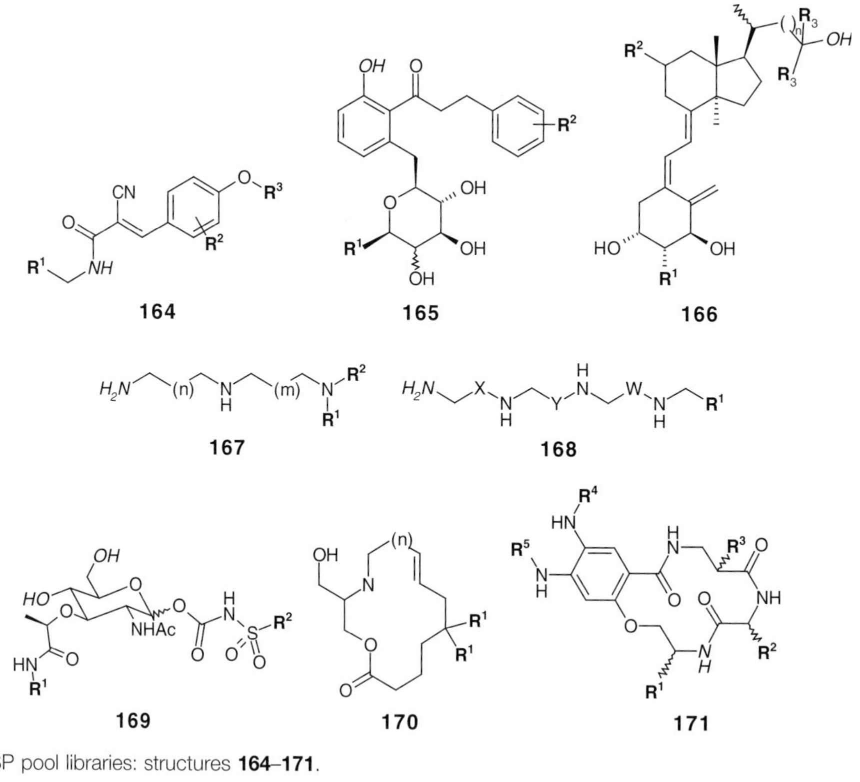

Figure 22 SP pool libraries: structures **164–171**.

of a directed sorting[10b,11,11a] cyclic alkoxyketone library (**174**). Jefferson and co-workers[208] focused their attention on ethylenediamine-functionalized nitrogen-containing heterocycles, such as **175**, without finding any activity on various RNA-targeted functional assays. Tetrahydroquinolines were the structural motif selected by Arya and co-workers[209] for lead discovery and chemical genetics studies, through the directed sorting SP library **176**. Willoughby and co-workers[210]

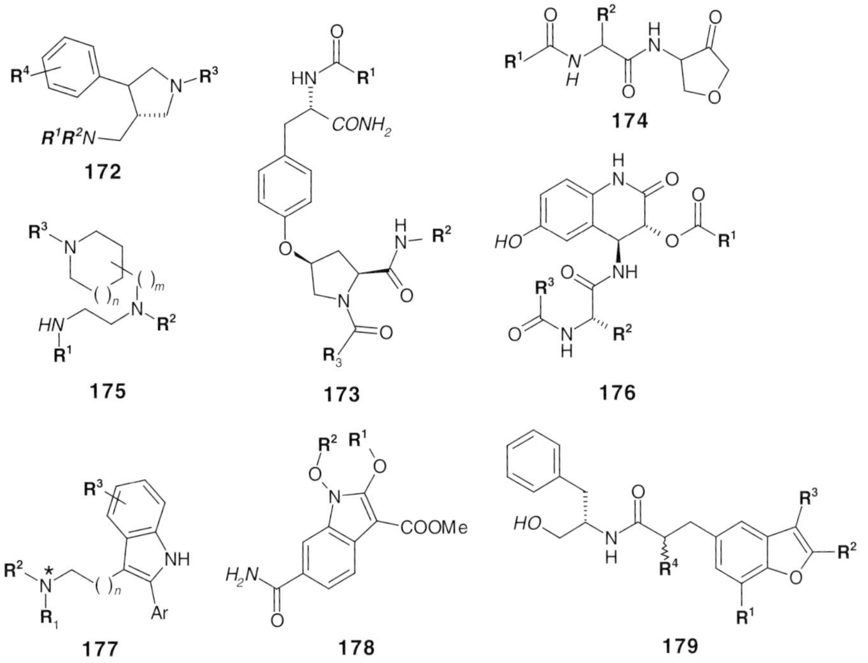

Figure 23 SP pool libraries: structures **172–179**.

reported an SP library of arylindoles (**177**), identifying after deconvolution the structures of subnanomolar inhibitors of various GPCRs. Indoles were also targeted by Wu and Ede,[211] through the 1,2-dialkoxy SP library **178**, using a simple color encoding method.[10b,11,11a] 2,3-Disubstituted benzofurans (**179**) were prepared by directed sorting by Liao and co-workers[212] for chemical genetics studies.

Janda and co-workers[213] prepared a versatile SP pool library of pyridines (**180**) (**Figure 24**), with deconvolution[10,10a,10b] as the planned structure determination method for positives. Houghten and co-workers have reported seven attempts to exploit chemical diversity for pharmaceutical purposes focusing on reduced imidazoles, using positional scanning,[10,10a,10b] or 'tea bags'[10b,11,11a] to produce SP pool libraries. They first[214] reported a library of urea-containing bicyclic guanidines (**181**) as a source of antifungals or antiulcer compounds. Then, a library of 2,3, 5-trisubstituted imidazolones (**182**)[215] was presented as a lead discovery tool, and a library of imidazoline-tethered nitrogen-containing heterocycles (**183**)[216] was prepared as a screening collection for, among others, cardiovascular, CNS, and oncology targets. A library of urea-containing hydantoins and thiohydantoins (**184**)[217] was prepared as a model library leading to larger lead discovery expansions, as was a library of 1,2,5-trisubstituted imidazolones (**185**)[218] with defined stereochemistry. A library of biheterocyclic imidazolines (**186**)[219] was made for viral and inflammatory diseases, and, finally, a library of imidazolidinium salts (**187**)[220] was introduced as a source of antiallergic and antihyperglycemic agents. McMillan and co-workers[221] exploited some chemically encoded[10b,11,11a] bis-heterocyclic, imidazole-containing structures (**188**) to discover potent, in vivo active allosteric inducible nitric oxide synthase inhibitors.

Bianchi *et al.*[222] reported the SPS of a nonchemically encoded, directed sorting[10b,11,11a] library of pharmaceutically relevant benzimidazolones (**189**) (**Figure 25**). Cheng and co-workers,[223] and Raghavan and co-workers[224] focused on indinavir analogs as potential HIV protease inhibitors, through two SP pool libraries of piperazines (**190** and **191**), which were prepared, biologically characterized, and deconvoluted[10,10a,10b] to yield valuable SARs and in vivo active leads. Glycopyranosides (**192**) were targeted by Sofia and co-workers[225] as directed sorting, universal carbohydrate-based, pharmacophore-mapping libraries. Arvanitis *et al.*[226] prepared a pool library of diaminopyrimidines (**193**) as ATP analogs to inhibit protein kinases, via directed sorting. 2-Aminoanilines, pyridines, and pyrimidines (**194**) are similar scaffolds targeted via directed sorting by Zhu and co-workers,[227] as privileged pharmaceutical scaffolds. Houghten *et al.*[228] reported a model SP library of quinazolinones (**195**) using the 'tea bag' synthetic protocol.[10b,11,11a] Triheterocyclic scaffolds (**196**) (directed sorting) were targeted by Arya and co-workers[229] as small molecular probes in chemical genetics studies.

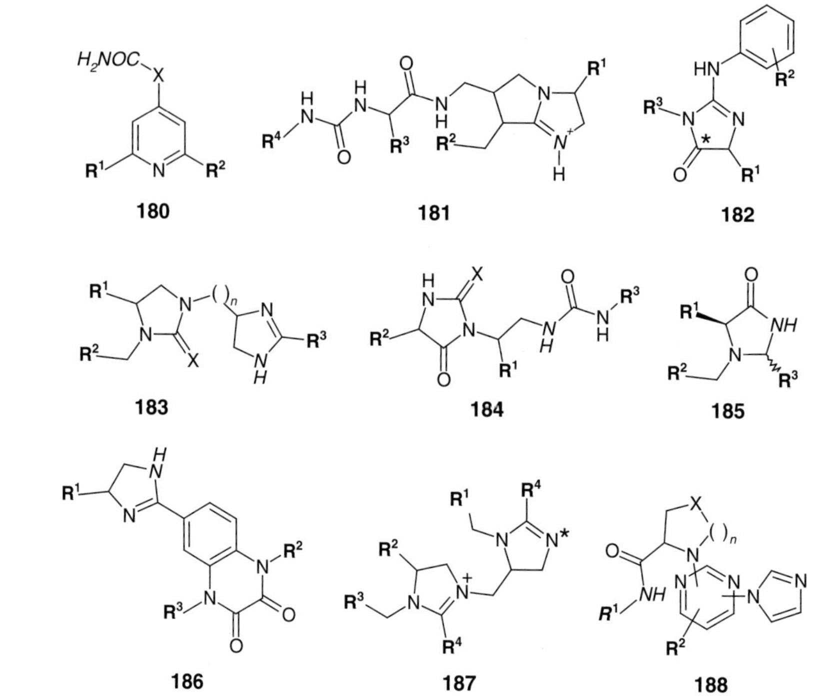

Figure 24 SP pool libraries: structures **180–188**.

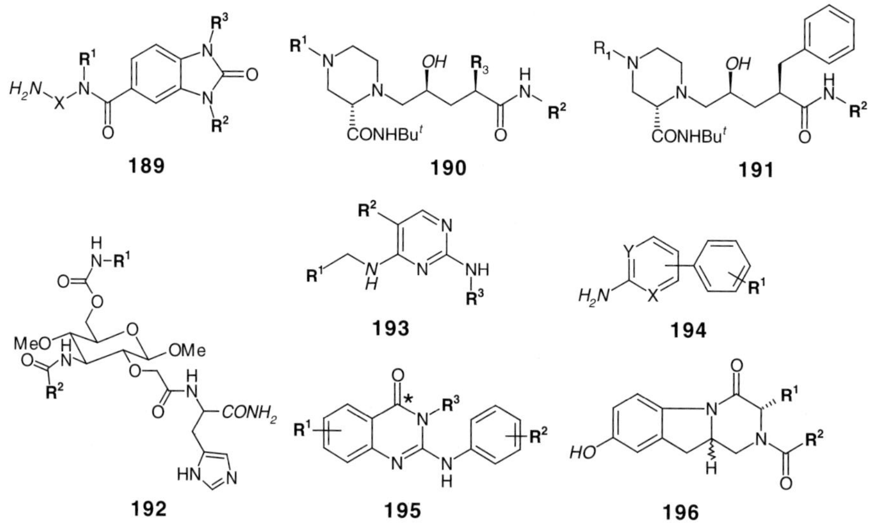

Figure 25 SP pool libraries: structures **189–196**.

Benzodiazepines were targeted by two research groups (**Figure 26**). Herpin and co-workers[230] presented an SP directed sorting[10b,11,11a] library of 1,5-benzodiazepin-2-ones (**197**) for lead discovery, and Houghten and co-workers[231] used 'tea bags'[10b,11,11a] to prepare a CNS-targeted SP library of 4,5-dihydro-1,4-benzodiazepine-2,3-diones (**198**). Nucleoside analogs were targeted both by Makara and co-workers,[232] and by Epple and co-workers.[233] The first paper reported a large pool library of bicyclic pyrimidines (**199**) as ATP analogs for lead finding on nucleotide-binding proteins. The second introduced an SP library of nucleoside analogs (**200**), made by directed sorting and intended for systematic screening on nucleotide-binding targets. Regarding heterocycles containing three heteroatoms, Meldal *et al.*[234] described a large SP library of peptidotriazoles (**201**) made by split-and-mix synthesis and biologically characterized

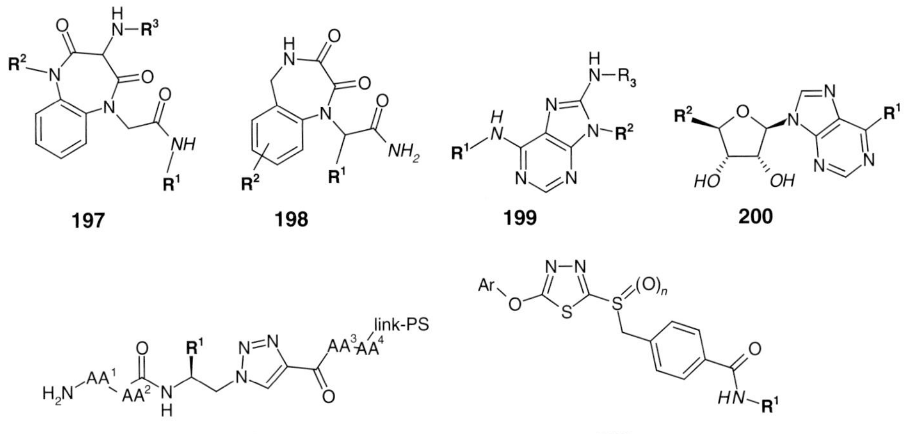

Figure 26 SP pool libraries: structures **197–202**.

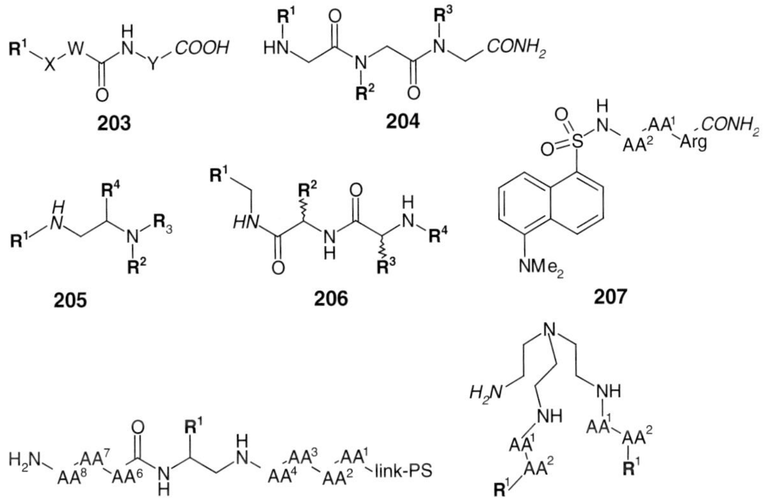

Figure 27 SP pool libraries: structures **203–209**.

by on-bead screening,[7,7a–7c] finding several inhibitors of a *Leishmania* protease. Pernerstorfer *et al.*[235] reported an SP library of thiadiazole ethers (**202**), made by directed sorting, from which a nanomolar IL8 inhibitor was obtained.

SP pool libraries unrelated to natural products or to heterocycles were also reported. De Laszlo *et al.*[236] presented an SP library of ureido acids (**203**) (**Figure 27**), which, after deconvolution,[10,10a,10b] led to several potent VLA-4 antagonists for chronic inflammatory disorders. Humet *et al.*[237] reported a library of peptoids (**204**) as weakly active antibacterials found by positional scanning.[10,10a,10b] 1,2-Diamines (**205**) were prepared by Lee *et al.*,[238] identifying after deconvolution some potent antitubercular hits. Zhao and co-workers[239] focused on phosphopantetheine adenyltransferase as a validated antibacterial target, using a directed sorting[10b,11,11a] SP library of dipeptide analogs (**206**) to identify potent in vitro enzyme inhibitors. Prokai and co-workers[240] used positional scanning to identify a CNS-bioavailable, high-affinity neuropeptide FF antagonist from a library of *N*-dansyl peptides (**207**). Meldal *et al.*[241] identified several strong, in vivo active inhibitors of a *Leishmania* protease, focusing on reduced peptide bond-containing peptidomimetics (**208**), via on-bead screening.[7,7a–7c] Shukla *et al.*[242] identified several strong synthetic phosphatidylserine translocase agents using a nonchemically encoded SP library of tris amine-based peptidomimetics (**209**).

Pegurier *et al.*[243] focused on melatonin receptors as targets for sleep disorders, and reported a SP library of phenylalkylamides (**210**) (**Figure 28**), whose deconvolution[10,10a,10b] provided potent, although unselective, hMT_2

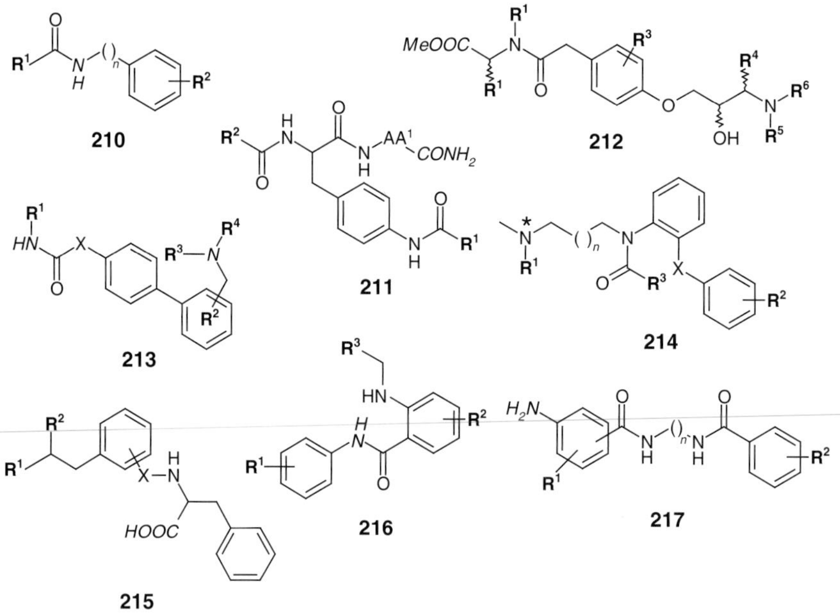

Figure 28 SP pool libraries: structures **210–217**.

receptors. Lam *et al.*[244] reported a large SP library of aryl-containing peptidomimetics (**211**), validating a new peptide-based encoding technique via on-bead screening.[7,7a–7c] Phenoxypropanolamines (**212**), structurally characterized by deconvolution, were reported by Bryan and co-workers[245] as lead discovery collections in CNS, cardiovascular, and other therapeutic areas. Follicle-stimulating hormone receptor agonists were sought for by Guo and co-workers[246] for treatment of infertility via a large, chemically encoded[10b,11,11a] SP library of biaryls (**213**), establishing a SAR which was expanded and refined in further studies. Aminoaryls (**214**) were reported by Andres *et al.*[247] as a source of several moderate NPY_2 ligands. Ferguson *et al.*[248] reported an SP library of dihydrostilbenes (**215**) made by directed sorting,[10b,11,11a] and intended as a preliminary effort around a druglike scaffold. El-Araby *et al.*[249] focused on pure, druglike alkylaminobenzanilides (**216**), made by directed sorting. Finally, Park *et al.*[250] reported an SP library of diaryl diamides (**217**), made by directed sorting, as a source of weak farnesyltransferase inhibitors.

3.33.4 Medicinal Chemistry Libraries: Examples

This section will be structured like the previous one, illustrating the substantial amount of work on SP combinatorial libraries reported by four leading research groups, led by K C Nicolaou (The Scripps Research Institute and the University of California, San Diego, CA), S L Schreiber (Harvard University, Cambridge, MA), P G Schultz (The Scripps Research Institute and The Genomics Institute of the Novartis Research Foundation, San Diego, CA), and H Waldmann (Max-Planck-Institute and Dortmund University, Dortmund, Germany). Each group has used SP libraries in recent years to achieve important scientific and technological goals in the field of drug discovery, from early target identification to lead optimization.

3.33.4.1 Chemical Assessment on Solid-Phase

Nicolaou has reported the assessment of several SP routes, leading to classes of pharmaceutically relevant compounds (**Table 4**). For example, a Dess–Martin periodinane-mediated cascade cyclization of unsaturated anilides was fully assessed in solution and on SP to give the tetracycle **218** (**Figure 29**).[251] Several indoles and indolines, such as **219**, were prepared on SP via supported selenyl bromide reaction with *o*-allyl or *o*-prenyl anilines, followed by decoration and diversification of various functional groups.[252] Several complex scaffolds, including annulated heterocyclic systems such as **220** and enediyne structures, were obtained on SP through the use of a common α-tosyloxyketone intermediate.[253] Nicolaou focused also on a specific class of privileged pharmaceutical compounds, containing the 2,2-dimethylbenzopyran moiety, such as in **221**[254] and **222**.[255] The use of a selenium linker; the assessment of, among

Table 4 SP chemical assessment: examples

Compound class	*Compounds made*	*Analytical characterization; purification; yields*	*Therapeutic area; drug discovery phase*	*Target screened/ identified/ validated*	*Ref.*
Tetracyclic compound, **218**	1	HPLC; silica gel; 28% yield	Various; TGT ID/Val, LD	None	251
Indolines, **219**	30	HPLC; silica gel; 9–64% yield	Various; TGT ID/Val, LD	None	252
Pyrazine-containing macrocycles, **220**	29	HPLC; silica gel; 38–95% yield	Various; TGT ID/Val, LD	None	253
Benzopyran-containing heterocycles, **221**	29	HPLC; silica gel; 18–95% yield	Various; TGT ID/Val, LD	None	254
Benzopyran-containing heterocycles, **222**	24	HPLC; silica gel; 42–95% yield	Various; TGT ID/Val, LD	None	255
Small-molecule reference compounds, **223**	5	HPLC; none; 69–90% (yield)	Tools for DD; SP linker for diversity-oriented libraries	None	256
Functionalized prolines, **224**	1	HPLC; silica gel; 79% (yield)	Various; TGT ID/Val, LD	None	257
3,4-Dihydropyrimidinones, **225**	2	HPLC; none; NG	Various; TGT ID/Val, LD	None	258
Bicyclic diols, **226**	1	NMR; none; >70% (purity)	Various; TGT ID/Val, LD	None	259
Dihydroxyallenes, **227**	1	NMR; none; NG	Various; TGT ID/Val, LD	None	260
Biaryl-containing medium rings, **228**	1	HPLC, NMR; none; 55 (yield)	Various; TGT ID/Val, LD	None	261
Carpanone analogs, **229**	6	HPLC, NMR; silica gel; 77–81% (yield)	Various; TGT ID/Val, LD	None	262
Polycyclic compounds, **230**	1	NMR; none; NG	Various; TGT ID/Val, LD	None	263
Macrocycles, **231**	NG	NG; none; NG	Various; TGT ID/Val, LD	None	264
Dimeric molecules, **232**	11	NMR; none; 5–97% (yield)	Various; TGT ID/Val, LD	None	265
PNA-encoded cathepsin inhibitors, **233**	6	NG; none; NG	Tools for drug discovery; affinity fishing, specificity profiling	Cathepsins	266
2,6,9-Substituted purines, **234**	24	HPLC; none; >85% (purity)	Various; TGT ID/Val, LD	None	267
2,6,9-Substituted purines, **235**	24	HPLC, NMR; none; NG	Various; TGT ID/Val, LD	None	268
Cyclic peptides, **236**	2	HPLC; none; 6–7% (yield)	Various; TGT ID/Val, LD	None	269
Biaryls, **237**	15	HPLC, NMR; none; >90% (purity)	Various; TGT ID/Val, LD	None	270
Reference azides, **238**	2	NG; none; NG	Tools for drug discovery; immobilization of libraries on glass slides	None	271
6,6-Spiroketals, **239**	4	HPLC, NMR, none, 6–16% (yield)	Various; TGT ID/Val, LD	None	272

NG, not given.

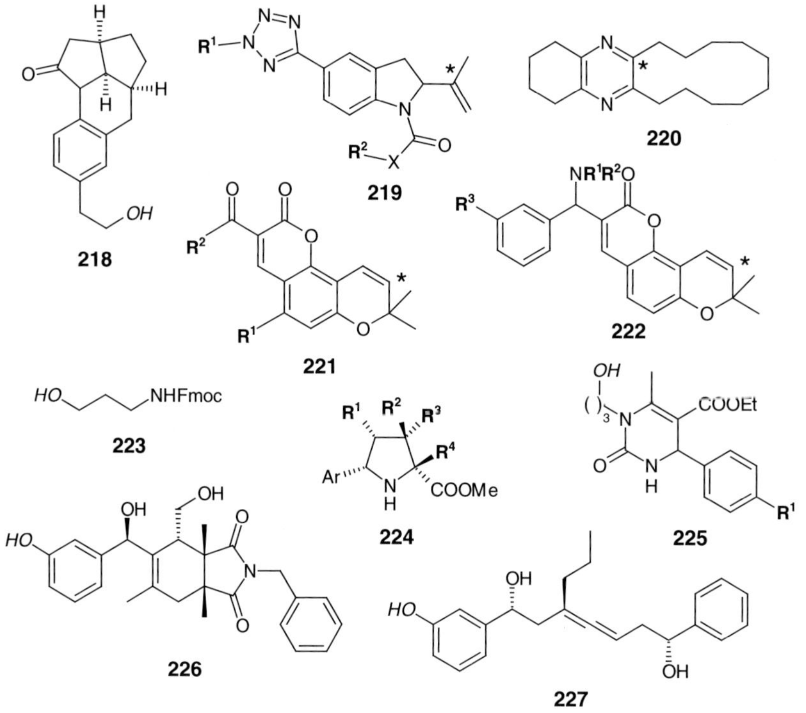

Figure 29 SP chemical assessment, examples: structures **218–227**.

others, annulation, aryl coupling, condensation, and glycosidation reactions; and the use of radiofrequency encoding[10b,11,11a] led to the development of a benzopyran-oriented SP strategy amenable to large and diverse library synthesis, as discussed below.

Schreiber has set up a robust and efficient SP environment, including a high-capacity solid support and an alkyl-tethered diisopropylarylsilane linker, suitable for bead-based library synthesis and screening, as shown by loading on, and releasing from, single beads standard compounds such as **223**.[256] These large resin beads were used in a series of SP chemical assessments leading to diversity-oriented synthesis[19,19a,19b] of medium-large SP libraries. For example, a three-component reaction between a benzaldehyde, an amino acid ester, and an α,β-unsaturated ester based on a catalytic asymmetric [3 + 2] cycloaddition of azomethine ylides, led to stereochemically defined, trisubstituted prolines (**224**).[257] 3,4-Dihydropyrimidinones (**225**) were obtained via a three-component Biginelli reaction involving an aryl aldehyde, a urea, and a β-ketoester.[258] Boronic ester annulation reactions were the common thread of two papers, leading respectively to some stereochemically pure bicyclic diols (**226**) from allylboronic esters[259] and to dihydroxyallenes (**227**) from alkynylboronic esters.[260]

The biaryl-containing medium ring macrocycle **228** (**Figure 30**), an example of a natural products-like scaffold, was prepared via C–C coupling between two aryls using organocuprates on SP.[261] Complex polycycles (**229**) were designed as analogs of carpanone, a well-known, biologically active natural product, and were prepared via oxidative heterocoupling with concomitant stereocontrol on five stereocenters.[262] A four-component Ugi reaction involving an aldehyde, an isonitrile, an unsaturated acid, and an alcohol was successfully assessed on SP to yield the complex tetracycle **230** after intramolecular Diels–Alder and ring-opening/closing metathesis, proving the suitability of complex cascade reactions in a diversity-oriented synthetic pathway.[263] Careful control of substitution patterns on chosen building blocks (i.e., different stereochemistry of one or more stereocentres) in a common synthetic pathway allowed Schreiber to reach different scaffolds, such as **231**,[264] due to alternative reaction pathways, paving the way to single libraries containing different structural elements, as we will see below. A known drawback of SPS (i.e., the possibility of site–site interactions on the same resin bead) was even exploited, to yield homodimeric molecules such as **232** via 'intra-site' cross-olefin metathesis.[265]

Schultz reported a novel chemical encoding method on SP, based on peptide nucleic acids (PNAs) as tags, and validated it in a microarray format used for functional profiling, detecting specific interactions of the encoded tetrapeptide inhibitors **233** with several cathepsins.[266] Turning to classes of pharmaceutically relevant compounds, his group focused on kinase inhibitors derived from natural purine-containing nucleobases, such as **234**[267] and **235**.[268]

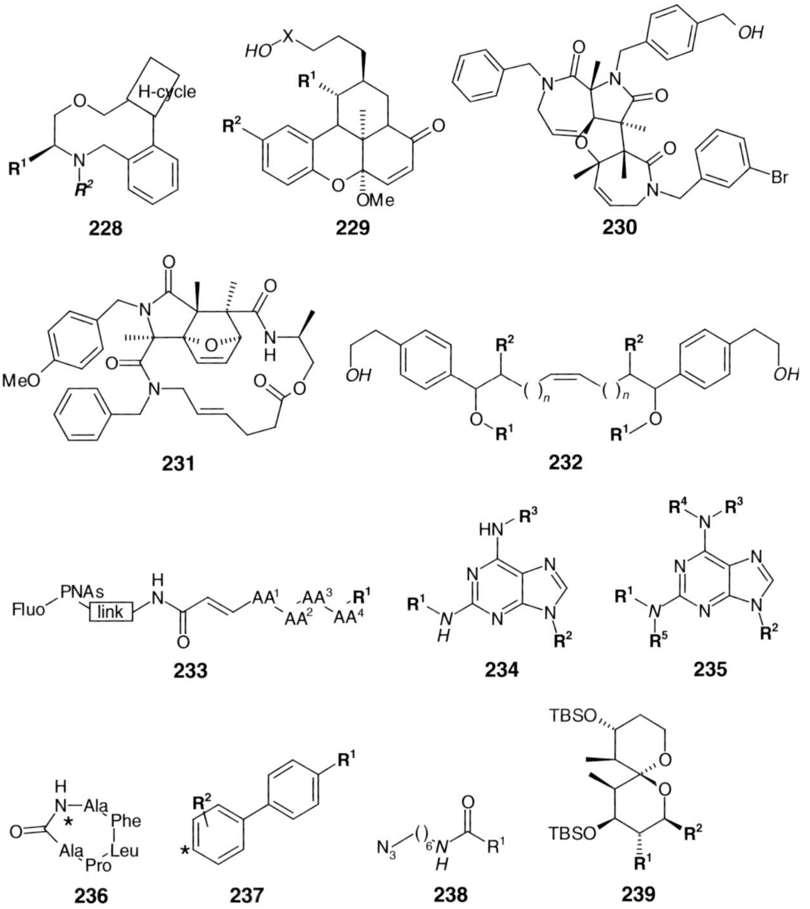

Figure 30 SP chemical assessment, examples: structures **228–239**.

A robust SPS was assessed, including a traceless linker strategy, and allowing the decoration of at least three positions of the scaffolds, and several large SP libraries were later obtained, as discussed below.

Waldmann reported the use of an aryl hydrazide-based traceless linker to prepare cyclic peptides, such as **236**,[269] and various aryl or heteroaryl-based scaffolds such as **237**,[270] all with relevance to medicinal chemistry. Another linker, based on the Staudinger reaction between a supported phosphane group and alkyl azides such as **238**, was developed as a novel anchoring method for small-molecule microarrays with good chemical compatibility and stability.[271] Finally, 6,6-spiroketals (**239**) were targeted as natural products-occurring structural motifs through an asymmetric SPS involving two asymmetric, boron-mediated aldol condensations.[272]

3.33.4.2 Solid-Phase Pool Libraries

Schreiber and his group have reported a number of chemically encoded,[10b,11,11a] bead-based medium-large SP libraries of complex stereochemically defined molecules to appropriately span the druglike chemical diversity space for chemical genetics[20,20a–20c] and lead discovery purposes (**Table 5**). Dihydropyrancarboxamides (**240**) (**Figure 31**) were prepared through stereoselective cycloaddition of supported vinyl ethers with protected heterodiene carboxylates, followed by amidation of the deprotected carboxylic function.[273] The decoding protocol and the product recovery in a 'one bead, one stock solution' format was fully validated,[274] and this library was successfully transposed into a cutting edge technology platform, including microplate storage, analytical instrumentation for decoding, and structure determination procedures.[275] The same library was characterized in terms of compound release from the beads (yield and purity), robotic partitioning into microplates of eluates, and assaying of compounds in a phenotypic assay on human lung carcinoma cells, discovering the structures of several hits.[276] 1,3-Dioxanes (**241**) were also combinatorially exploited on SP, after assessing a route including on-bead nucleophilic opening of a hydroxy epoxide, cyclization of

Table 5 SP pool libraries: examples

Compound class	*Library size, nature*[a]	*Analytical characterization; purification method; yield*[b]	*Theraputic area; drug discovery phase*[b]	*Target*	*Biological activity*	*Ref.*
Dihydropyrancarboxamides, **240**	4320 (BB)	Statistical characterization (2.5% compounds); >98% identified via LC/MS and decoding	Various; TGT ID/Val, LD	A549 lung carcinoma cells	Hits (inhibition, yes/no) $R^1 = $ 3-thiophene, $R^2R^3 = N$-morpholine, link $= NSO_2$-p-OMePh	273
Polyfunctionalized 1,3-dioxanes, **241**	3780 (BB)	HPLC; none; 70% compounds >90% purity (projected)	Various; TGT ID/Val, LD	Microscopy evaluation, HeLa cells	Hit (yes/no) $R^1 = H$, $XR^2 = S$-(p-OHPh), $R^3 = SO_2Ph$	277
Polyfunctionalized 1,3-dioxanes, **241**	3780 (BB)	HPLC; none; 70% compounds >90% purity (projected)	Various; TGT ID/Val, LD	Ure2p binding	$K_d = 7.5\ \mu M$ $R^1 = H$, $XR^2 =$ 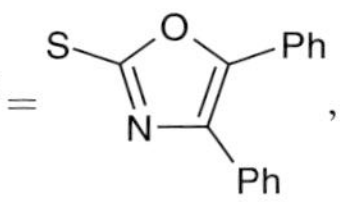 , $n = p - 1$, uretupamine B	278
Polyfunctionalized 1,3-dioxanes, **241**	7200 (BB)	HPLC (1% compounds); all confirmed after cleavage by LC/MS	Oncology; LD	HDAC6 inhibition	$IC_{50} = 1.2\ \mu M$ $R^1 = H$, $XR^2 = S$-(2,3-diphoxazole), $R^3 = p$-NHCO-$(CH_2)_6$CO-NHOH Tubacin	279–283
Polyfunctionalized 1,3-dioxanes, **242**	18 K (BB)	HPLC (1% of compounds); 76% compounds >70% purity	Various; TGT ID/Val, LD	CV malfunction Inducer in zebrafish	Atrioventricular block at $6\ \mu M$ $R^1 = H$, $XR^2 =$ 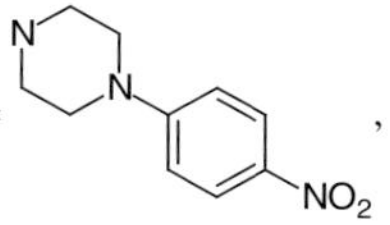 , $m = 0$, $n = 1$	284
Biaryl-containing nine-membered rings, **243**	1412 (BB)	HPLC (10% compounds); none; 62% analyzed >70% purity	Various; TGT ID/Val, LD	Development malfunction inducer in *A. thaliana*	Hit, 100 nM $R^1 = Bn$, $R^2 = R^3 = $ 2,4,6-tri-OMe, $R^4 = Me$	285

Proprietary CD set (**240** + **241** + **243**)	12 396 (BB)	Refs 273, 277, 279, 285	Various; TGT ID/Val, LD	Hap3p-GST protein	$K_d = 5.03\ \mu M$ 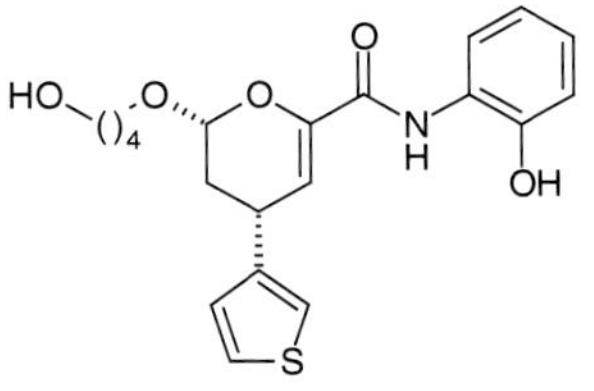 Haptamide B	286
Polycyclic compounds, **244**	29 400 (BB)	Statistical characterization; 100 macrobeads; 88% identified via LC/MS	Various; TGT ID/Val, LD	None	None	287
Polycyclic compounds, **244**	6336 (BB)	Ref. 287	Various; TGT ID/Val, LD	Calmodulin binding	$K_d = 121$ nM $R^1 =$ 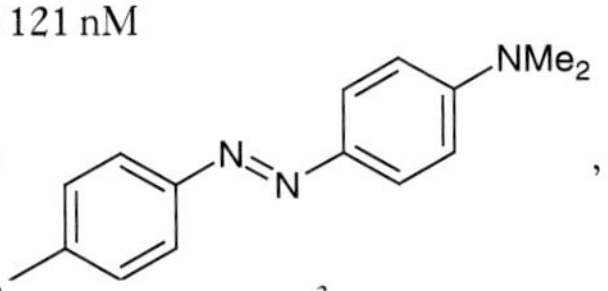, $R^2 = 3$-Cl, 4-FPh, $R^3 = 4$-Cl(1-OH), $R^4 =$ Me, $R^5 =$ H	288
Ketoamides, **245**	80 (BB)	NG; none; NG	Tools for DD-Techniques for small molecule printing on microarrays	Streptavidin	Positive hit $R^1 =$ 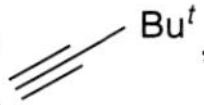, $R^2 =$ allyl, $R^3 = CH_2CH_2O$-array surface	289
Galanthamine analogs, **246**	2527 (BB)	HPLC; none; all compounds confirmed by MS (86% of the original 2946)	Various; TGT ID/Val, LD	Protein trafficking inhibition (Golgi apparatus to plasma membrane)	Strong inhibition at 750 nM $R^1 = CH_2$cycprop, $R^2 = p$-OMeBn, $R^3 =$ H, $R^4 =$ Obn Secramine	290
Tricyclic compounds, **247**	2500 (BB)	HPLC; none; 80% samples (<10% compounds) showed >80% purity	Various; TGT ID/Val, LD	None	None	291

continued

Table 5 Continued

Compound class	*Library size, nature*[a]	*Analytical characterization;*	*Theraputic area; drug discovery phase*[b]	*Target*	*Biological activity*	*Ref.*
Spirooxindoles, **248**	3144 (BB)	HPLC; none; purity >70% (statistical)	Various; TGT ID/Val, LD	Latrunculin B-induced growth arrest enhancers	Hits, yes/no $R^1 = R^2 = H$, $R^3 = CH_2CH_2Ph$, $R^4 = O\text{-}(CH_2)_2OMe$, LINK = *o*-$O(CH_2)_2OH$	292
Trisubstituted tetrahydrofurans, **249**	1260 (BB)	HPLC (10% compounds); 70% of them >70% purity	Various TGT ID/Val, LD	None	None	293, 294
PNA-encoded tetrapeptides, **250**	7 (CE)	NG; none; NG	Tool for DD	Caspase activation upon induction of apoptosis	Best probe AA^1 = Asp, AA^2 = Val, AA^3 = Glu, AA^4 = Asp	295
PNA-encoded tetrapeptides, **250**	4000 (CE)	NG; none; NG	Tools for DD	Activity profiling of dust mite allergens	Best probe AA^1 = Nle, AA^2 = Val, AA^3 = Ala, AA^4 = Lys	296
PNA-encoded tetrapeptides, **251**	192 (CE)	NG; none; NG	Tools for DD	Protease profiling	Best probe, complex lysates AA^1 = Asp, AA^2 = Val, AA^3 = Thr, AA^4 = Asp	297
Benzopyran-based libraries, **252**	>200 (D)	HPLC; silica gel; >95% purity	Various; TGT ID/Val, LD	None	None	298
Benzopyran-based libraries, **253**	10 215 (DS)	TLC, HPLC, MS for 5% compounds, NMR for 1%; 69% compounds >80% purity	Various; TGT ID/Val, LD	None	None	299
Benzopyran-based libraries, **254**	170 (DS)	HPLC; none; >90% purity	Various; TGT ID/Val, LD	None	None	300

NG, not given.
[a] BB, bead-based libraries; CE, chemically encoded; DS, directed sorting.
[b] LC/MS, liquid chromatography/mass spectrometry.

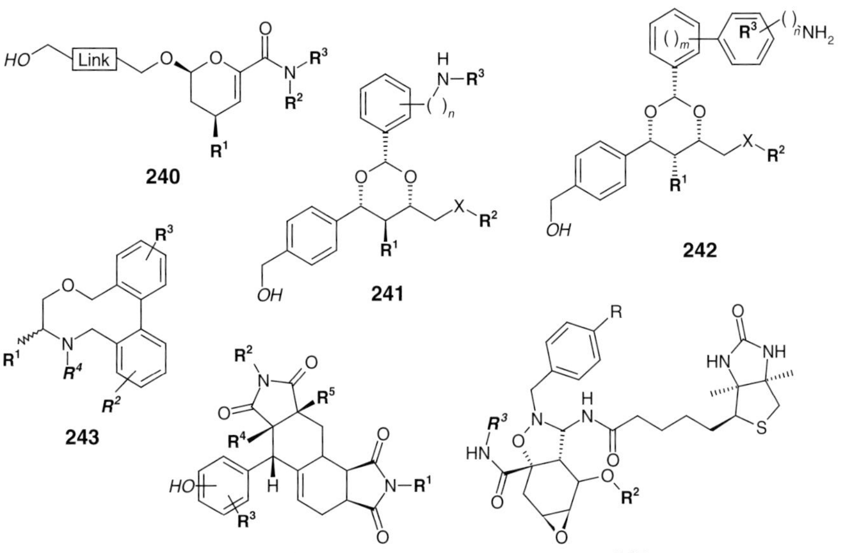

Figure 31 SP pool libraries, examples: structures **240–245**.

the resulting β-diol with a protected amine-containing aryl aldehyde dimethyl acetal, and functionalization of the deprotected amine.[277] The library was tested on a phenotypic assay measuring whole-organism development of zebrafish embryos, finding hits affecting CNS development[277]; and, after postsynthesis printing in a small-molecule microarray format, it was used to find binders of Ure2p, a central repressor of potentially relevant genes for disease treatment.[278] Another 1,3-dioxane-containing larger bead-based library using the same general structure was prepared as a set of structural analogs of histone deacetylase (HDAC) inhibitors, using an SP strategy similar to the one described above.[279] Screening of this library on HDAC-6 identified a compound, named tubacin,[280,281] that acts as a selective inducer of α-tubulin acetylation, and another, named histacin,[281] that acts as a selective inducer of histone acetylation. Both compounds were then used to further characterize the function of HDACs in cellular processes,[282] while a 617-member library subset of active compounds was used to thoroughly map the meaningful chemical space for HDAC selective inhibition.[283] An even larger 1,3-dioxane SP library (**242**) was reported, and a small subset was used in a whole-organism phenotypic assay to discover a single inducer of cardiovascular malfunctions in zebrafish embryos.[284] A library composed of biaryl-containing medium rings (**243**) was created as previously described,[261] and was tested in several phenotypic assays, discovering, among other hits, compounds able to induce development malfunctions in *Arabidopsis thaliana* seeds.[285] A collection of chemical libraries, including structures **240**, **241**, and **243**, was tested after being spotted in a small-molecule microarray, to discover haptamide B, a transcription factor inhibitor.[286] A large library of polycyclic compounds (**244**) was prepared through a six-step stereoselective synthesis and two consecutive Diels–Alder reactions.[287] A subset of this library was printed on a small-molecule microarray using a novel diazobenzylidene linker, and several calmodulin binders were identified.[288] A small SP bead-based library of ketoamides (**245**) was printed on a microarray, and validated for streptavidin-biotin-binding detection, among other methods.[289]

Analogs of galanthamine, a natural product with acetylcholinesterase inhibition, were targeted through library (**246**) (**Figure 32**), leading, after screening on a cell-based phenotypic assay, to the structure of secramine, a potent protein-trafficking inhibitor.[290] Interestingly, galanthamine has no effect in this assay, showing that these natural product-inspired libraries can lead to compounds with different activities to the source structure. Tricyclic compounds (**247**) were prepared via an SPS, including Ferrier and Pauson–Khand reactions on a glycan template, and were targeted toward chemical genetics studies.[291] Spirooxindoles (**248**) were successfully obtained using a three-component coupling reaction between an aromatic aldehyde, a lactone, and a dipolarophile, followed by a Sonogashira coupling between an alkyne and an aryl iodide, and led to the identification of an enhancer of cell growth arrest induced by latrunculin B.[292] Finally, libraries of skeletally diverse small molecules, such as trisubstituted furans (**249**), were obtained from the same SPS by employing differently substituted building blocks, which were able to promote alternative reaction pathways.[293,294] It is apparent how such an approach could enlarge the chemical diversity obtainable from a single small-molecule combinatorial library.

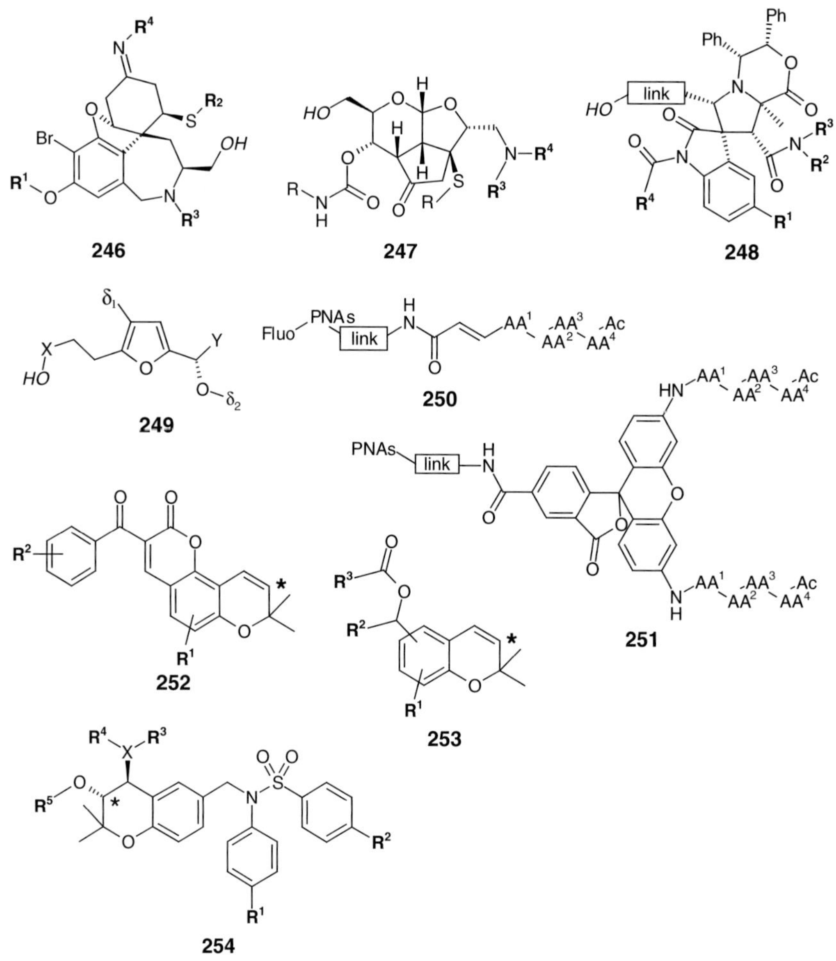

Figure 32 SP pool libraries, examples: structures **246–254**.

Schultz reported the use of PNA-encoded tetrapeptide libraries (**250**[295,296] and **251**[297]), using the synthetic method described above,[266] to profile the function of protein classes in a microarray format. Chemical probes to detect caspase activation upon apoptosis induction were identified.[295] Probes for activity profiling of dust mite allergens,[296] and for protease profiling in crude cell lysates and blood samples,[297] were also identified.

Nicolaou exploited combinatorially the careful SP assessment performed on 2,2-dimethyl benzopyran-containing scaffolds.[254,255] Several model SP libraries (**252**) were prepared using radiofrequency-encoded methods[10b,11,11a] and selenium SP linkers[298]; a large library collection (**253**), containing nine different benzopyran-based scaffolds and using, among others, acylations, sulfurylations, glycosylations, condensations, reductive aminations, and organometallic additions, was prepared[299]; and 'libraries from libraries,' such as **254** (**Figure 32**), were reported simply by functionalizing some subsets of the original library collection and using postcleavage, high-yield transformations leading to a combinatorial explosion of benzopyran-inspired structures.[300]

3.33.4.3 Solid-Phase Libraries of Discretes

Waldmann has reported several SP discrete libraries, either composed of natural product analogs, or inspired by their structures (**Table 6**). A small library (**255**) (**Figure 33**), composed of analogs of teleocidin, a tumor-promoting protein kinase C (PKC) activator, was prepared via on-bead cyclization of an amino acid intermediate, and was tested to give a selective, strong downregulator of the single PKCδ isoform.[301] Pepticinnamicin E analogs (**256**) were prepared via both standard peptide-coupling conditions and organic reactions,[302,303] identifying after in vitro screening several moderately potent farnesyltransferase inhibitors able to induce apoptosis in a Ras-transformed tumor cell line.[304]

Table 6 SP Discrete libraries: examples

Compound class	*Library size*	*Analytical characterization; purification method; yield*	*Therapeutic area; drug discovery phase*	*Target screened*	*Biological activity*	*Ref.*
Teleocidin analogs, **255**	31	HPLC; silica gel; 10–61% (yield)	Oncology; H2L	PKCδ	Strong downregulation at 200 nM $R^1 = Me$, $R^2 = Pr^i$, $R^3 = Ph$	301
Pepticinnamin E analogs, **256**	51	HPLC; prep. HPLC; 89% to >95% (purity), 3–40% (yield)	Oncology; H2L	PFT	$IC_{50} = 1.1\ \mu M$ $R^1 = OCH_2(p\text{-}NO_2)Ph$, R^2 = 5-imidazole, $R^3 = R^4 = H$, + = D	302–304
2-Arylaminothiazoles, **257**	23	HPLC; SP extraction; 81–99% (purity), 19–69% (yield)	Oncology; LD	VEGFR-2	$IC_{50} = 7.4\ \mu M$ $R^1 = H$, $R^2 = CH_2$-2-furyl, $R^3 = R^4 = Ph$	305, 306
Dysidiolide analogs, **258**	8	HPLC; silica gel; 6–27% (purified)	Oncology; H2L	Cdc25C	$IC_{50} = 800\ nM$ $R^1 =$ 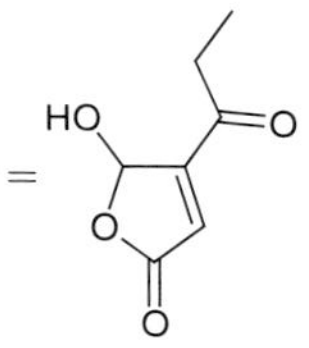	307, 308
Dysidiolide analogs, **259**	147	HPLC; silica gel; 19–82% (yield)	Oncology; LD	Cdc25C	$IC_{50} = 350\ nM$ $R^1 = 2\text{-}(6\text{-}C_{11}H_{23})$ naphthyl	309
Indomethacin analogs, **260**	197	HPLC; silica gel; >98% purity	Oncology; H2L	Tie-2	$IC_{50} = 3\ \mu M$ $R^1 = 3\text{-}SO_3H$, $R^2 = p\text{-}ClPh$, $R^3R^4 = CH_2CH_2S\text{–}CH_2$	310
Sulindac analogs, **261**	239	HPLC; silica gel; 47–98% (yield)	Various; TGT ID/Val, LD	Ras signaling inhibition; MDCKF3 cells	$IC_{50} = 10\ \mu M$ $R^1 = 3\text{-}Br$, $R^2 = Me$, $R^3 = H$ (3-thiophene)	311, 312

continued

Table 6 Continued

Compound class	*Library size*	*Analytical characterization; purification method; yield*	*Therapeutic area; drug discovery phase*	*Target screened*	*Biological activity*	*Ref.*
Kinase-directed heterocyclic libraries, **262**	45 140	Statistical evaluation; 78–95% purity for a few compounds	Various; TGT ID/Val, LD	Sulfotransferase inhibition; β-AST-IV	$IC_{50} = 1.8\ \mu M$ $R^1X = p$-di-Ph-ether, $Y = N$, $R^2 = R^3 = R^5 = H$, $R^4 = CH_2$-1-naphthyl	313, 314
Kinase-directed heterocyclic libraries, **262**	45 140	Statistical evaluation; 78–95% purity for a few compounds	Various; TGT ID/Val, LD	Osteogenesis; inducing activity in mesenchymal progenitor cells	$EC_{50} = 1\ \mu M$ $R^1X = O$-1-naphthyl, $R^2 =$ cyclohexyl, $Y = N$, $R^3 =$ 4-*N*-morpholino-Ph, $R^4 = R^5 = H$ Purmorphamine	315
Kinase-directed heterocyclic libraries, **262**	50 000	Statistical evaluation; 78–95% purity for a few compounds	Various; TGT ID/Val, LD	GSK3β target identified; neuronal differentiation inducers	Strong effect, 1 μM $R^1X = R^2 = H$, $Y = O$, $R^3 = R^5 = 3\text{-}NH_2Ph$, $R^4 =$ absent TWS119	316
Kinase-directed heterocyclic libraries, **262**	50 000	Statistical evaluation; 78–95% purity for a few compounds	Various; TGT ID/Val, LD	Dedifferentiation of cells; C2C12 cells	Full prevention of differentiation at 5 μM $R^1X =$ NH-(4-*N*-morpholino-Ph), $R^2 = R^3 = R^5 = H$, $Y = N$, $R^4 =$ cyclohexyl Reversine	317
Kinase-directed heterocyclic libraries, **262**	139	NG; none; NG	Various; TGT ID/Val, LD	Carbonyl sulfotransferase inhibition	$IC_{50} = 20\ \mu M$ $R^1X =$ NH-*p*-BrBn, $R^2 = C_2H_4OH$, $Y = N$, $R^3 = Bn$, $R^4 = R^5 = H$	318
Kinase-directed heterocyclic libraries, **262**	275	NG; none; NG	Various; TGT ID/Val, LD	Sulfotransferase inhibition; EST	$IC_{50} = 0.5\ \mu M$ $R^1X =$ 1-azepine, $R^2 = Pr^i$, $Y = N$, $R^3 = p$-OMeBn, $R^4 = R^5 = H$	319

Kinase-directed heterocyclic libraries, **262**	275	NG; none; NG	Various; TGT ID/Val, LD	IP3K	$IC_{50} = 10.2\ \mu M$ $R^1X = NHCH_2$-*m*-CF_3Ph, $R^2 = R^4 = R^5 = H$, $Y = N$, $R^3 =$ *p*-NO_2Bn,	320
Kinase-directed heterocyclic libraries, **262**	275	NG; none; NG	Various; TGT ID/Val, LD	Microtubule-binding capacity; disassembly of microtubules	$IC_{50} = 11\ \mu M$ $R^1X = NH$-*p*-OMePh, $R^2 = Pr^i$, $Y = N$, $R^3 =$ *p*-OMeBn, $R^4 = R^5 = H$ Myoseverin	321
Kinase-directed heterocyclic libraries, **262**	1 561	NG; none; NG	Various; TGT ID/Val, LD	NQO1 target identified; microtubule stability regulator; inhibitor of polymerization	Strong at 50 μM $R^1X = NHCH$-$(Pr^i)CH_2OH$, $R^2 = Pr^i$, $Y = N$, $R^3 = S$-(*m*-NH_2Ph), $R^4 = R^5 = H$ Diminutol	322
Kinase-directed heterocyclic libraries, **263**	100 000	Statistical evaluation; 78–95% purity for a few compounds	Various; TGT ID/Val, LD	Differentiation of P19 cells into cardiomyocytotoxin; inducers	$EC_{50} = 100\ nM$ $R^1X = NH$-*p*-OMePh, $R^2 = CH_2CH_2OH$, $R^3 = H$ Cardiogenol D	323
Kinase-directed heterocyclic libraries, **263**	100 000	Statistical evaluation; 78–95% purity for a few compounds	Various; TGT ID/Val, LD	Wnt signalling pathway agonists in *Xenopus*	Strong effect at 10 μM $R^1X = NH_2$, $R^2 =$ 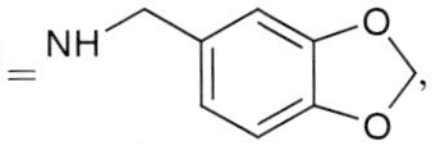 , $R^3 = H$, $R^4 =$ *m*-OmePh	324
2,3,5-Trisubstituted indoles, **264**	38	HPLC; none; >95% (purity)	Various; TGT ID/Val, LD	None	None	325
Arylsulfonamides, **265**	NG	HPLC; prep. HPLC; 70–90% (purity), 30–50% (yield)	Oncology; LD, H2L	XIAP/caspase 3 interaction	80–100% inhibition, 20 μM $R^1 =$ *o*-F, $R^2 =$ *p*-Ph, $R^3 = Et$, $L_1 = (CH_2)_3$, $L_2 =$ 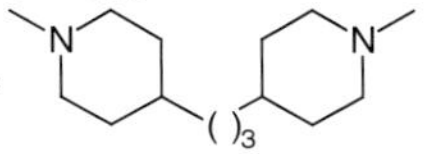	326

continued

Table 6 Continued

Compound class	*Library size*	*Analytical characterization; purification method; yield*	*Therapeutic area; drug discovery phase*	*Target screened*	*Biological activity*	*Ref.*
Ceramide analogs, **266**	528	Statistical (20%), MS-NMR; NG	Various; TGT ID/Val, LD	Apoptosis screen; U937 cells	$IC_{50} = 4\,\mu M$ $R^1 = n\text{-}C_9H_{19}$, $R^2 = H$, $R^3 = CH_2OH$, $R^4 = CH(OH)\text{-}CH=CHC_7H_{15}$	327
Biaryl-containing nine-membered rings, **267**	78	HPLC; prep. HPLC; >90% purity	Various; TGT ID/Val, LD	None	None	328
Macrolactones, **268**	48	HPLC; prep. HPLC; 6–75% (yield)	Various; TGT ID/Val, LD	None	None	329
Bicyclic compounds, **269**	244	HPLC; none; 70–80% (yield)	Various; TGT ID/Val, LD	Multifactor. cellular pathways (40 cell-based assays)	Full matrices analzed	330
Polycycles, **270**	864	HPLC; none; >80% purity (5% compounds)	Various; TGT ID/Val, LD	None	None	331
Dimethylbenzopyrans, **271**	121	HPLC; prep. HPLC; >90% purity	Oncology; H2L	NADH: ubiquinone oxidoreductase	$IC_{50} = 24\,nM$ $Y = R^1 = H$, $R^2 = Ome$, $X = O$	332
Vancomycin analogs, **272**	82	HPLC; prep. HPLC; >90% purity	Antibacterial; LD	MICs against Vanco-resistant strains	MICs = $<0.03\text{–}2\,\mu M$ $R^1 = H$, $R^2 = O(CH_2)_8\text{-}CH=CH_2$	333

NG, not given.

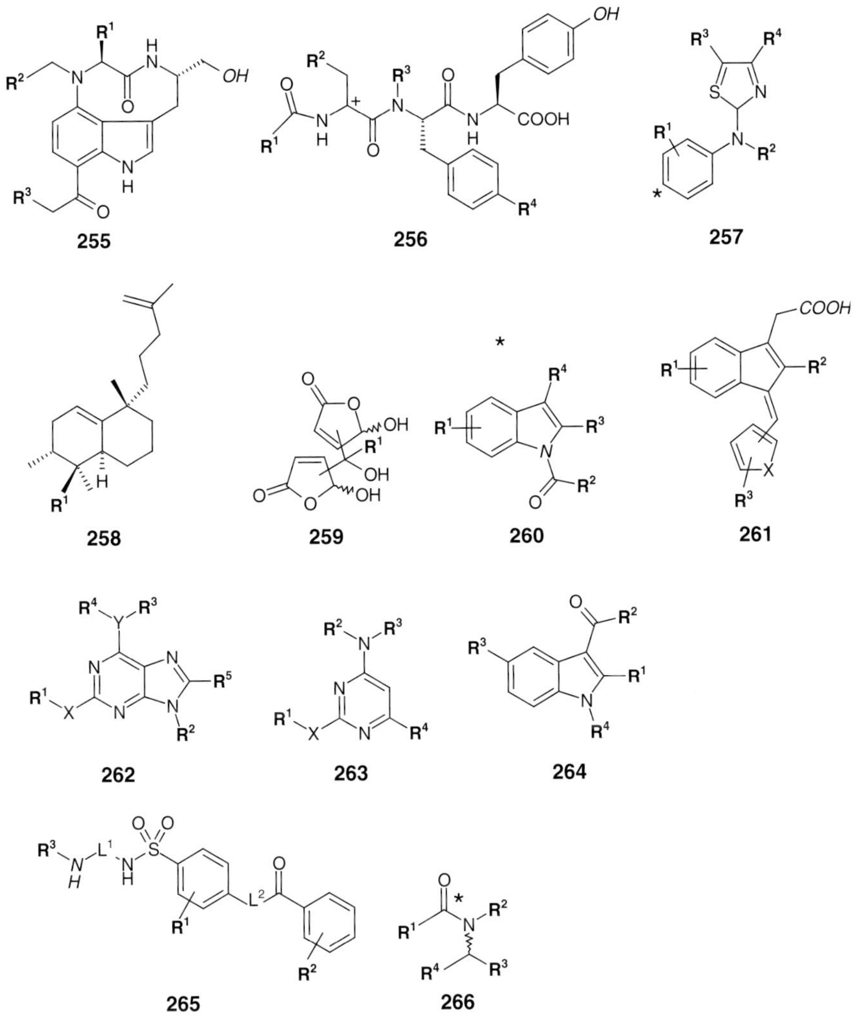

Figure 33 SP discrete libraries, examples: structures **255–266**.

Arylaminothiazoles (**257**) were targeted for their potential kinase inhibition, and were prepared via the previously described traceless aryl hydrazide linker,[269,270] identifying moderately potent inhibitors of two receptor tyrosine kinases, VEGFR-2 and Tie-2.[305,306] Several papers reported the SPS and characterization of dysidiolide analogs as phosphatase inhibitors and as chemical genetics tools. The small model library **258** was prepared via Wittig condensation and an asymmetric Diels–Alder reaction, and proved the potency and selectivity among various phosphatases for each library member,[307,308] reflected in their cytotoxicity on various tumor cell lines. A larger, more diverse library of dysidiolide analogs (**259**) was used to validate the in silico hypothesized structural analogy between cdc25c (the 'natural' target of dysidiolide) and apparently unrelated molecular targets such as acetylcholinesterase, leading to unprecedented novel hits for nonphosphatase targets.[309] Indomethacin analogs (**260**) were made via a mixed SP–solution phase strategy, and were used to identify antiangiogenetic hits acting through the Tie-2 kinase.[310] Finally, the closely related sulindac analogs **261** were made on SP via Knoevenagel condensation, and were used to identify compounds affecting the Ras signaling pathway,[311] and more precisely the Ras–Raf interaction for oncology treatments.[312]

Schultz targeted medium-large nucleotide-like libraries as kinase inhibitors, preferring the extensive use of automation to obtain each library member as a discrete, and using these libraries to clarify the nature of several complex metabolic pathways. The SPS route leading to such libraries, assessed in previously mentioned papers,[267,268] was thoroughly described for the purine library **262**,[313] and validated through an extensive analytical characterization. Its screening on a purified enzyme, β-AST-IV, led to the identification of a 'first in class' potent and highly selective sulfotransferase inhibitor.[314] Screening of this library in multipotent mesenchymal progenitor cells allowed the identification of purmorphamine, an inducer of osteoblast differentiation whose molecular mechanism of action is being identified[315]; in embryonic stem cells it led to the discovery of TWS119, an inducer of neuronal differentiation for which the GSK-3β kinase was identified as the molecular target[316]; and in a myogenic lineage-committed myoblast cell line it led to the discovery of reversine, a compound able to de-differentiate the committed cells and to make them once again multipotent mesenchymal progenitor cells with differentiation potential.[317] Smaller subsets of library **262** were used to identify unprecedented, albeit modestly potent, inhibitors of carbohydrate sulfonyltransferases,[318] and more potent inhibitors of estrogen sulfotransferases[319]; to identify the first ATP-competitive, purine-based inhibitors of the IP3 K kinase as calcium level regulators[320]; to identify myoseverin, a microtubule-binding compound endowed with novel transcriptional effects, including priming the natural regenerative capacity of tissues[321]; and to identify diminutol, an inhibitor of microtubule polymerization targeting NQO1, a quinone oxidoreductase involved in redox regulation.[322] A larger nucleotide-like library also containing pyrimidine-like compounds, such as **263** was used to identify cardiogenol D, an inducer of cardiomyogenesis in embryonic stem cells via unknown mechanisms with potential therapeutical applications[323]; and to identify agonists of the Wnt signaling pathway, which should be of use in elucidating the complexity of this therapeutically relevant process.[324] A small SP library of indoles (**264**) was also reported,[325] possibly as a gateway toward more complex chemical genetics library collections. Arylsulfonamides (**265**) were prepared as XIAP-caspase-3-directed medium-sized libraries, prepared on SP via simple organic reactions, and allowed the identification of the first protein–protein interaction inhibitors of this process, with significant potential for novel oncology treatments.[326] Finally, a ceramide library (**266**) (**Figure 33**) was prepared via on-bead tail–core assembly, and tested in a cellular screen for apoptosis induction, determining a useful first SAR.[327]

Schreiber reported four SP libraries of discretes. Tellimagrandin I analogs (**267**) (**Figure 34**) were prepared via on-bead C–C coupling/cyclization of two aryl bromides.[328] Stereochemically defined, macrolide-like lactones (**268**) were prepared for future chemical genetics studies.[329] Bicycles containing 12-membered rings (**269**) were prepared as a small library for multidimensional screening (40 cellular assays), to develop a so-called 'cellular measurement space' and to determine the consequences of stereochemical and skeletal diversity in cells.[330] Finally, small molecule libraries such as **270** were prepared by joining with convergent SP strategies structural elements reminiscent of different natural products, to determine any additive or synergistic effect in small-molecule screens.[331]

Nicolaou reported a 'hit-to-lead' expansion of his benzopyran-based SP pool libraries, following up on primary screen results from the original large libraries.[299,300] Novel and potent NADH:ubiquinone oxidoreductase inhibitors (**271**)

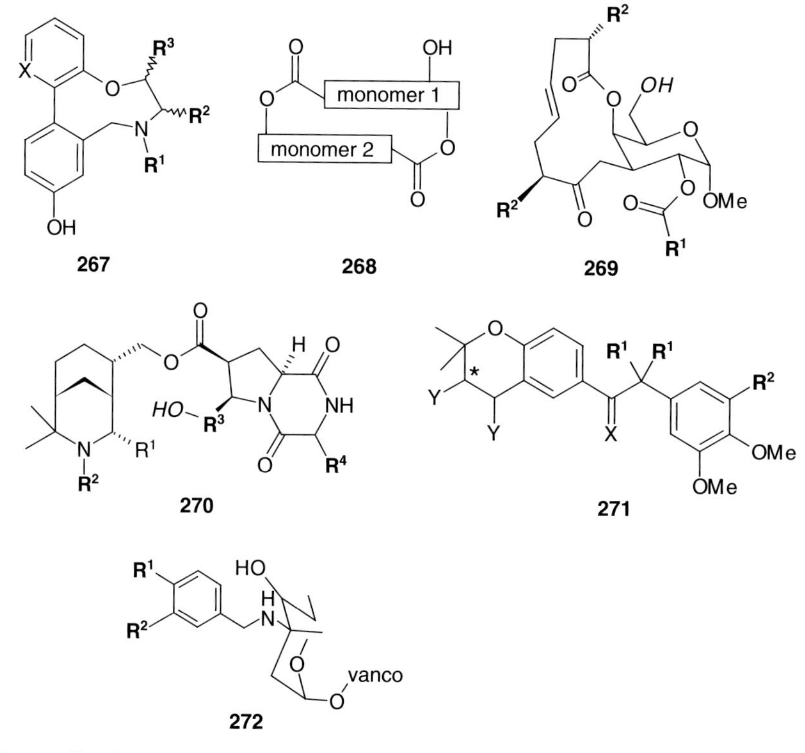

Figure 34 SP discrete libraries, examples: structures **267–272**.

with chemotherapeutic/chemopreventative potential were published.[332] An SP discrete library (**272**) (**Figure 34**), inspired by the structure of the potent antibiotic vancomycin, identified some valuable hits against vancomycin-resistant bacterial strains.[333]

References

1. Seneci, P. *Solid Phase Synthesis and Combinatorial Technologies*; John Wiley: New York, 2000.
2. Zaragoza Dorwald, F. *Organic Synthesis on Solid Phase. Supports, Linkers, Reactions*; Wiley-VCH: Weinheim, Germany, 2000.
3. Furka, A.; Sebestyen, F.; Asgedom, M.; Dibo, G. *Int. J. Pept. Protein Res.* **1991**, *37*, 487–493.

3a. Lam, K. S.; Salmon, S. E.; Hersh, E. M.; Hruby, V. J.; Kazmierski, W. M.; Knapp, R. J. *Nature* **1991**, *354*, 82–84.

3b. Houghten, R.; Pinilla, C.; Blondelle, S. E.; Appel, J. R.; Dooley, C. T.; Cuervo, J. H. *Nature* **1991**, *354*, 84–86.

4. Furka, A. Notarized document file number 36237/1982. Dr. Judit Bokai, state notary public, June 15, 1982, Budapest.

4a. Geysen, H. M.; Meloen, R. H.; Barteling, S. J. *Proc. Natl. Acad. Sci. USA* **1984**, *81*, 3998–4002.

4b. Houghten, R. A. *Proc. Natl. Acad. Sci. USA* **1985**, *82*, 5131–5135.

5. Smith, G. P. *Science* **1985**, *228*, 1315–1317.

5a. Hanak, J. J. *J. Mater. Sci.* **1970**, *5*, 964–971.

6. Burch, R. M. *J. Recept. Res.* **1991**, *11*, 101–113.

6a. Darlison, M. G. *Trends Neurosci.* **1992**, *15*, 469–474.

7. Furka, A. *Drug Devel. Res.* **1995**, *36*, 1–12.

7a. Balkenhohl, F.; von dem Bussche-Hunnefeld, C.; Lansky, A.; Zechel, C. *Angew. Chem. Int. Ed.* **1996**, *35*, 2289–2337.

7b. Lam, K. S.; Lebl, M.; Krchnak, V. *Chem. Rev.* **1997**, *97*, 411–448.

7c. Seneci, P. Synthetic Organic Libraries: Solid-Phase Pool Libraries. In *Solid Phase Synthesis and Combinatorial Technologies*; John Wiley: New York, 2000, pp 264–338.

8. Bunin, B. A.; Ellman, J. *J. Am. Chem. Soc.* **1992**, *114*, 10997–10998.
9. Seneci, P. Synthetic Organic Libraries: Library Design and Properties. In *Solid Phase Synthesis and Combinatorial Technologies*; John Wiley: New York, 2000, pp 165–209.

9a. Toledo- Sherman, L. M.; Chen, D. *Curr. Opin. Drug Disc. Devel.* **2002**, *5*, 414–421.

9b. Langer, T.; Krovat, E. M. *Curr. Opin. Drug Disc. Devel.* **2003**, *6*, 370–376.

10. Konings, D. A. M.; Wyatt, J. R.; Ecker, D. J.; Freier, S. M. *J. Med. Chem.* **1997**, *40*, 4386–4395.

10a. Seneci, P. Direct Deconvolution Techniques for Pool Libraries of Small Organic Molecules. In *Combinatorial Chemistry and Combinatorial Technologies: Methods and Applications*; Miertus, S., Fassina, G., Eds.; Marcel Dekker: New York, 1999, pp 91–125.

10b. Barnes, C.; Balasubramanian, S. *Curr. Opin. Chem. Biol.* **2000**, *4*, 346–350.

11. Seneci P., Encoding Techniques for Pool Libraries of Small Organic Molecules. In *Combinatorial Chemistry and Combinatorial Technologies: Methods and Applications*; Miertus, S., Fassina, G., Eds.; Marcel Dekker: New York, 1999, pp 127–167.

11a. Ede, N. J.; Wu, Z. *Curr. Opin. Chem. Biol.* **2003**, 7, 374–379.

12. Golebiowski, A.; Klopfenstein, S. R.; Portlock, D. E. *Curr. Opin. Chem. Biol.* **2001**, *5*, 273–284.

12a. Hermkens, P. H. H.; Muller, G. *Ernst Schering Res. Found. Workshop* **2003**, *42*, 201–220.

12b. Geysen, H. M.; Schoenen, F.; Wagner, D.; Wagner, R. *Nat. Rev. Drug Disc.* **2003**, *2*, 222–230.

12c. Golebiowski, A.; Klopfenstein, S. R.; Portlock, D. E. *Curr. Opin. Chem. Biol.* **2003**, 7, 308–325.

13. Kyranos, J. N.; Cai, H.; Zhang, B.; Goetzinger, W. K. *Curr. Opin. Drug Disc. Devel.* **2001**, *4*, 719–728.

13a. Edwards, P. J. *Comb. Chem. HTS* **2003**, *6*, 11–27.

14. Pottorf, R. S.; Player, M. R. *Curr. Opin. Drug Disc. Devel.* **2004**, 7, 777–783.
15. Bindseil, K. U.; Jakupovic, J.; Wolf, D.; Lavayre, J.; Leboul, J.; van der Pyl, D. *Drug Disc. Today* **2001**, *6*, 840–847.

15a. Rouhi, A. M. *Chem. Eng. News* **2003**, *81*, 84–86.

16. Haefner, B. *Drug Disc. Today* **2003**, *8*, 536–544.

16a. Berlinck, R. G. S.; Hajdu, E.; da Rocha, R. M.; de Oliveira, J. H. H. L.; Hernández, I. L. C.; Seleghim, M. H. R.; Granato, A. C.; de Almeida, E. V. R.; Nuñez, C. V.; Muricy, G. et al. *J. Nat. Prod.* **2004**, *67*, 510–522.

16b. Tulp, M.; Bohlin, L. *Drug Disc. Today* **2004**, *9*, 450–458.

16c. Clardy, J.; Walsh, C. *Nature* **2004**, *432*, 829–837.

17. Newman, D. J.; Cragg, G. M.; Snader, K. M. *J. Nat. Prod.* **2003**, *66*, 1022–1037.
18. Hall, D. G.; Manku, S.; Wang, F. *J. Comb. Chem.* **2001**, *3*, 125–150.

18a. Breinbauer, R.; Vetter, I. R.; Waldmann, H. *Angew. Chem. Int. Ed.* **2002**, *41*, 2879–2890.

18b. Ortholand, J.-Y.; Ganesan, A. *Curr. Opin. Chem. Biol.* **2004**, *8*, 271–280.

19. Schreiber, S. L. *Science* **2000**, *287*, 1964–1969.

19a. Burke, M. D.; Schreiber, S. L. *Angew. Chem. Int. Ed.* **2004**, *43*, 46–58.

19b. Borman, S. *Chem. Eng. News* **2004**, *82*, 32–40.

20. Schreiber, S. L. *Bioorg. Med. Chem.* **1998**, *6*, 1127–1152.

20a. Crews, C. M.; Splittgerber, U. *Trends Biol. Sci.* **1999**, *24*, 317–320.

20b. Specht, K. M.; Shokat, K. M. *Curr. Opin. Cell Biol.* **2002**, *14*, 155–159.

20c. Khersonsky, S. M.; Chang, Y.-T. *ChemBioChem* **2004**, *5*, 903–908.

21. Arya, P.; Baek, M.-G. *Curr. Opin. Chem. Biol.* **2001**, *5*, 292–301.

21a. Liao, Y.; Hu, Y.; Zhu, Q.; Donovan, M.; Fathi, R.; Yang, Z. *Curr. Med. Chem.* **2003**, *10*, 2285–2316.

21b. Boldi, A. M. *Curr. Opin. Chem. Biol.* **2004**, *8*, 281–286.

21c. Tan, D. S. *Comb. Chem. HTS* **2004**, 7, 631–643.

22. Lindsay, M. A. *Nat. Rev. Drug Disc.* **2003**, *2*, 831–838.

22a. Weinshilboum, R.; Wang, L. *Nat. Rev. Drug Disc.* **2004**, *3*, 739–748.

22b. Kramer, R.; Cohen, D. *Nat. Rev. Drug Disc.* **2004**, *3*, 965–972.

23. Evans, F. J. *J. Ethnopharmacology* **1991**, *32*, 91–101.
23a. Hung, D. T.; Jamison, T. F.; Schreiber, S. L. *Chem. Biol.* **1996**, *3*, 623–629.
23b. Caterina, M. J.; Schumacher, M. A.; Tominaga, M.; Rosen, T. A.; Levine, J. D.; Julius, D. *Nature* **1997**, *389*, 816–824.
24. Stockwell, B. R. *Nat. Rev. Genet.* **2000**, *1*, 116–125.
25. Seneci, P.; Miertus, S. *Mol. Diversity* **2000**, *5*, 75–89.
25a. Schuffenhauer, A.; Popov, M.; Schopfer, U.; Acklin, P.; Stanek, J.; Jacoby, E. *Comb. Chem. HTS* **2004**, 7, 771–781.
26. Mayer, T. U. *Trends Cell Biol.* **2003**, *13*, 270–277.
26a. Burdine, L.; Kodadek, T. *Chem. Biol.* **2004**, *11*, 593–597.
26b. Tochtrop, G. P.; King, R. W. *Comb. Chem. HTS* **2004**, 7, 677–688.
27. Gura, T. *Nature* **2000**, *407*, 282–284.
28. Dolle, R. E. *Mol. Diversity* **1998**, *3*, 199–232.
28a. Dolle, R. E. *Mol. Diversity* **1998**, *3*, 233–256.
28b. Dolle, R. E.; Nelson, K. H., Jr. *J. Comb. Chem.* **1999**, *1*, 235–282.
28c. Dolle, R. E. *J. Comb. Chem.* **2000**, *2*, 383–433.
28d. Dolle, R. E. *J. Comb. Chem.* **2001**, *3*, 477–517.
28e. Dolle, R. E. *J. Comb. Chem.* **2002**, *4*, 369–418.
28f. Dolle, R. E. *J. Comb. Chem.* **2003**, *5*, 693–753.
28g. Dolle, R. E. *J. Comb. Chem.* **2004**, *6*, 623–679.
29. Triparhy, R.; Learn, K. S.; Reddy, D. R.; Iqbal, M.; Singh, J.; Mallamo, J. P. *Tetrahedron Lett.* **2002**, *43*, 217–220.
30. Olsen, J. A.; Jensen, K. J.; Nielsen, J. *J. Comb. Chem.* **2000**, *2*, 143–150.
31. Paterson, I.; Temal-Laib, T. *Org. Lett.* **2002**, *4*, 2473–2476.
32. Qi, L.; Meijler, M. M.; Lee, S.-H.; Sun, C.; Janda, K. D. *Org. Lett.* **2004**, *6*, 1673–1675.
33. Tanaka, H.; Hasegawa, T.; Iwashima, M.; Iguchi, K.; Takahashi, T. *Org. Lett.* **2004**, *6*, 1103–1106.
34. Knepper, K.; Ziegert, R. E.; Braese, S. *Tetrahedron* **2004**, *60*, 8591–8603.
35. Marfil, M.; Albericio, F.; Alvarez, M. *Tetrahedron* **2004**, *60*, 8659–8668.
36. Cironi, P.; Manzanares, I.; Albericio, F.; Alvarez, M. *Org. Lett.* **2003**, *5*, 2959–2962.
37. Cironi, P.; Cuevas, C.; Albericio, F.; Alvarez, M. *Tetrahedron* **2004**, *60*, 8669–8675.
38. Klein, G.; Ostresh, J. M.; Nefzi, A. *Tetrahedron Lett.* **2003**, *44*, 2211–2215.
39. Kamal, A.; Reddy, G. S. K.; Reddy, K. L.; Raghavan, S. *Tetrahedron Lett.* **2002**, *43*, 2103–2106.
40. Andersen, F. T.; Stroemgaard, H. *Tetrahedron Lett.* **2004**, *45*, 7929–7933.
41. Joensson, D. *Tetrahedron Lett.* **2002**, *43*, 4793–4796.
42. Jimenez, J. C.; Chavarria, B.; Lopez-Macia, A.; Royo, M.; Giralt, E.; Albericio, F. *Org. Lett.* **2003**, *5*, 2115–2118.
43. Arnusch, C. J.; Pieters, R. J. *Eur. J. Org. Chem.* **2003**, 3131–3138.
44. Cristau, P.; Vors, J.-P.; Zhu, J. *Tetrahedron* **2003**, *59*, 7859–7870.
45. You, S.-L.; Deechongkit, S.; Kelly, J. W. *Org. Lett.* **2004**, *6*, 2627–2630.
46. Smith, K. L.; Tao, Z.-F.; Hashimoto, S.; Leitheiser, C. J.; Wu, X.; Hecht, S. M. *Org. Lett.* **2002**, *4*, 1079–1082.
47. Thomas, C. J.; Chizhov, A. O.; Leitheiser, C. J.; Rishel, M. J.; Konishi, K.; Tao, Z.-F.; Hecht, S. M. *J. Am. Chem. Soc.* **2002**, *124*, 12926–12927.
48. Dessole, G.; Marchetti, M.; Taddei, M. *J. Comb. Chem.* **2003**, *5*, 198–200.
49. Rew, Y.; Goodman, M. *J. Org. Chem.* **2002**, *67*, 8820–8826.
50. Myers, A. G.; Lanman, B. A. *J. Am. Chem. Soc.* **2002**, *124*, 12969–12971.
51. Shintani, T.; Kadono, H.; Schubert, T.; Shogase, Y.; Shimazaki, M. *Tetrahedron Lett.* **2003**, *44*, 6567–6569.
52. Barluenga, J.; Mateos, C.; Aznar, F.; Valdes, C. *Org. Lett.* **2002**, *4*, 3667–3670.
53. Xiao, Z.; Schaefer, K.; Firestine, S.; Li, P.-K. *J. Comb. Chem.* **2002**, *4*, 149–153.
54. Willand, N.; Joossens, J.; Gesquiere, J.-C.; Tartar, A. L.; Evans, D. M.; Roe, M. B. *Tetrahedron* **2002**, *58*, 5741–5746.
55. Touré, B. B.; Hoveyda, H. R.; Tailor, J.; Ulaczyk-Lesanko, A.; Hall, D. G. *Chem. Eur. J.* **2003**, *9*, 466–474.
56. Komatsu, M.; Okada, H.; Akaki, T.; Minakata, S. *Org. Lett.* **2002**, *4*, 3505–3508.
57. Arya, P.; Couve-Bonnaire, S.; Durieux, P.; Laforce, D.; Kumar, R.; Leek, D. M. *J. Comb. Chem.* **2004**, *6*, 735–745.
58. Arya, P.; Durieux, P.; Chen, Z.-X.; Joseph, R.; Leek, D. M. *J. Comb. Chem.* **2004**, *6*, 54–64.
59. Khadem, S.; Joseph, R.; Rastegar, M.; Leek, D. M.; Oudatchin, K. A.; Arya, P. *J. Comb. Chem.* **2004**, *6*, 724–734.
60. Ouyang, X.; Chen, Z.; Liu, L.; Dominguez, C.; Kiselyov, A. S. *Tetrahedron* **2000**, *56*, 2369–2377.
61. Savinov, S. N.; Austin, D. J. *Org. Lett.* **2002**, *4*, 1419–1422.
62. Strohmeier, G. A.; Haas, W.; Kappe, C. O. *Chem. Eur. J.* **2004**, 2919–2926.
63. Yokum, T. S.; Alsina, J.; Barany, G. *J. Comb. Chem.* **2000**, *2*, 282–292.
64. Purandare, A. V.; Gao, A.; Poss, M. A. *Tetrahedron Lett.* **2002**, *43*, 3903–3906.
65. Sereni, L.; Tatò, M.; Sola, F.; Brill, W. K.-D. *Tetrahedron* **2004**, *60*, 8561–8577.
66. Cheng, W.-C.; Wong, M.; Olmstead, M. M.; Kurth, M. J. *Org. Lett.* **2002**, *4*, 741–744.
67. Hwang, S. H.; Kurth, M. J. *J. Org. Chem.* **2002**, *67*, 6564–6567.
68. Righi, P.; Scardovi, N.; Marotta, E.; ten Holte, P.; Zwanenburg, B. *Org. Lett.* **2002**, *4*, 497–500.
69. Lewis, J. G.; Bartlett, P. A. *J. Comb. Chem.* **2003**, *5*, 278–284.
70. Trump, R. P.; Bartlett, P. A. *J. Comb. Chem.* **2003**, *5*, 285–291.
71. Chao, E. Y.; Minick, D. J.; Sternbach, D. D.; Shearer, B. G.; Collins, J. L. *Org. Lett.* **2002**, *4*, 323–326.
72. Song, A.; Lam, K. S. *Tetrahedron* **2004**, *60*, 8605–8612.
73. Kappel, J. C.; Yokum, T. S.; Barany, G. *J. Comb. Chem.* **2004**, *6*, 746–752.
74. Mann, E.; Kessler, H. *Org. Lett.* **2003**, *5*, 4567–4570.
75. Liao, Y.; Fathi, R.; Yang, Z. *J. Comb. Chem.* **2003**, *5*, 79–82.
76. Cheng, W.-C.; Kurth, M. J. *J. Org. Chem.* **2002**, *67*, 4387–4391.
77. Grover, R. K.; Mishra, R. C.; Kundu, B.; Tripathi, R. P.; Roy, R. *Tetrahedron Lett.* **2004**, *45*, 7331–7334.
78. Poirier, D.; Ciobanu, L. C.; Berube, M. *Bioorg. Med. Chem. Lett.* **2002**, *12*, 2833–2838.
79. Ciobanu, L. C.; Poirier, D. *J. Comb. Chem.* **2003**, *5*, 429–440.
80. Srinivasan, T.; Srivastava, G. K.; Pathak, A.; Batra, S.; Raj, K.; Singh, K.; Puri, S. K.; Kundu, B. *Bioorg. Med. Chem. Lett.* **2002**, *12*, 2803–2806.

81. Clough, J.; Chen, S.; Gordon, E. M.; Hackbarth, C. J.; Lam, S.; Trias, J.; White, R.; Candiani, G.; Donadio, S.; Romanò, G. et al. *Bioorg. Med. Chem. Lett.* **2003**, *13*, 3409–3414.
82. Tanaka, H.; Ohno, H.; Kawamura, K.; Ohtake, A.; Nagase, H.; Takahashi, T. *Org. Lett.* **2003**, *5*, 1159–1162.
83. Laursen, J. B.; de Visser, P. C.; Nielsen, H. K.; Jensen, K. J.; Nielsen, J. *Bioorg. Med. Chem. Lett.* **2002**, *12*, 171–175.
84. Ma, S.; Duan, D.; Wang, Y. *J. Comb. Chem.* **2002**, *4*, 239–247.
85. Smith, K. M.; Bu, Y.; Suga, H. *Chem. Biol.* **2003**, *10*, 563–571.
86. Cosquer, A.; Ficamos, M.; Jebbar, M.; Corbel, J.-C.; Choquet, G.; Fontanelle, C.; Uriac, P.; Bernard, T. *Bioorg. Med. Chem. Lett.* **2004**, *14*, 2061–2065.
87. Stroemgaard, K.; Brier, T. J.; Andersen, K.; Mellor, I. R.; Saghyan, A.; Tikhonov, D.; Usherwood, P. N. R.; Krogsgaard-Larsen, P.; Jaroszewski, J. W. *J. Med. Chem.* **2000**, *43*, 4526–4533.
88. Stroemgaard, K.; Mellor, I. R.; Andersen, K.; Neagoe, I.; Pluteanu, F.; Usherwood, P. N. R.; Krogsgaard-Larsen, P.; Jaroszewski, J. W. *Bioorg. Med. Chem. Lett.* **2002**, *12*, 1159–1162.
89. Raju, B.; Mortell, K. H.; Anandan, S.; O'Dowd, H.; Gao, H.; Gomez, M.; Hackbarth, C. J.; Wu, C.; Wang, W.; Yuan, Z. et al. *Bioorg. Med. Chem. Lett.* **2003**, *13*, 2413–2418.
90. Kim, H.; Choi, J.; Cho, J. K.; Kim, S. Y.; Lee, Y.-S. *Bioorg. Med. Chem. Lett.* **2004**, *14*, 2843–2846.
91. Liu, G.; Zhang, S.-D.; Xia, S.-Q.; Ding, Z.-K. *Bioorg. Med. Chem. Lett.* **2000**, *10*, *Bioorg. Med. Chem. Lett.* **2000**, *10*, 1361–1363.
92. Bozzoli, A.; Kazmierski, W.; Kennedy, G.; Pasquarello, A.; Pecunioso, A. *Bioorg. Med. Chem. Lett.* **2000**, *10*, 2759–2763.
93. Leitheiser, C. J.; Smith, K. L.; Rishel, M. J.; Hashimoto, S.; Konishi, K.; Thomas, C. J.; Li, C.; McCormick, M. M.; Hecht, S. M. *J. Am. Chem. Soc.* **2003**, *125*, 8218–8227.
94. Poreddy, A. R.; Schall, O. F.; Osiek, T. A.; Wheatley, J. R.; Beusen, D. D.; Marshall, G. R.; Slomczynska, U. *J. Comb. Chem.* **2004**, *6*, 239–254.
95. Takahashi, T.; Nagamiya, H.; Doi, T.; Griffiths, P. G.; Bray, A. M. *J. Comb. Chem.* **2003**, *5*, 414–428.
96. Scott, W. L.; Alsina, J.; O'Donnell, M. J. *J. Comb. Chem.* **2003**, *5*, 684–692.
97. Wei, L.; Wu, Y.-Q.; Wilkinson, D. E.; Chen, Y.; Soni, R.; Scott, C.; Ross, D. T.; Guo, H.; Howorth, P.; Valentine, H. et al. *Bioorg. Med. Chem. Lett.* **2002**, *12*, 1429–1433.
98. Vergnon, A. L.; Pottorf, R. S.; Player, M. R. *J. Comb. Chem.* **2004**, *6*, 91–98.
99. Boldi, A. M.; Dener, J. M.; Hopkins, T. P. *J. Comb. Chem.* **2001**, *3*, 367–373.
100. Poupart, M.-A.; Cameron, D. L.; Chabot, C.; Ghiro, E.; Goudreau, N.; Goulet, S.; Poirier, M.; Tsantrizos, Y. S. *J. Org. Chem.* **2001**, *66*, 4743–4751.
101. Gong, Y.; Becker, M.; Choi-Sledeski, Y. M.; Davis, R. S.; Salvino, J. M.; Chu, V.; Brown, K. D.; Pauls, H. W. *Bioorg. Med. Chem. Lett.* **2000**, *10*, 1033–1036.
102. Vergnon, A. L.; Pottorf, R. S.; Winters, M. P.; Player, M. R. *J. Comb. Chem.* **2004**, *6*, 903–910.
103. Huang, L.; Ellman, J. A. *Bioorg. Med. Chem. Lett.* **2002**, *12*, 2993–2996.
104. Belvisi, L.; Bernardi, A.; Checchia, A.; Manzoni, L.; Potenza, D.; Scolastico, C.; Castorina, M.; Capelli, A.; Giannini, G.; Carminati, P. et al. *Org. Lett.* **2001**, *3*, 1001–1004.
105. Watts, J.; Benn, A.; Flinn, N.; Monk, T.; Ramjee, M.; Ray, P.; Wang, Y.; Quibell, M. *Bioorg. Med. Chem.* **2004**, *12*, 2903–2925.
106. Couladouros, E. A.; Strongilos, A. T. *Angew. Chem. Int. Ed.* **2002**, *41*, 3677–3680.
107. Semple, G.; Andersson, B.-M.; Chhajlani, V.; Georgsson, J.; Johansson, M. J.; Rosenquist, A.; Swanson, L. *Bioorg. Med. Chem. Lett.* **2003**, *13*, 1141–1145.
108. Fernandez, J.-C.; Solè-Feu, L.; Fernandez-Forner, D.; de la Figuera, N.; Forns, P.; Albericio, F. *Tetrahedron Lett.* **2005**, *46*, 581–585.
109. Vianello, P.; Cozzi, P.; Galvani, A.; Meroni, M.; Varasi, M.; Volpi, D.; Bandiera, T. *Bioorg. Med. Chem. Lett.* **2004**, *14*, 657–661.
110. Kobayashi, N.; Kaku, Y.; Higurashi, K.; Yamauchi, T.; Ishibashi, A.; Okamoto, Y. *Bioorg. Med. Chem. Lett.* **2002**, *12*, 1747–1750.
111. Choong, I. C.; Lew, W.; Lee, D.; Pham, P.; Burdett, M. T.; Lam, J. W.; Wiesmann, C.; Luong, T. N.; Fahr, B.; DeLano, W. L. et al. *J. Med. Chem.* **2002**, *45*, 5005–5022.
112. Garibay, P.; Vedsoe, P.; Begtrup, M.; Hoeg-Jensen, T. *J. Comb. Chem.* **2001**, *3*, 332–340.
113. Smith, R. A.; Chen, J.; Mader, M. M.; Muegge, I.; Moehler, U.; Katti, S.; Marrero, D.; Stirtan, W. G.; Weaver, D. R.; Xiao, H. et al. *Bioorg. Med. Chem. Lett.* **2002**, *12*, 2875–2878.
114. Gong, Y.-D.; Seo, J.-s.; Chon, Y. S.; Hwang, J. Y.; Park, J. Y.; Yoo, S.-e. *J. Comb. Chem.* **2003**, *5*, 577–589.
115. Smith, A. L.; Stevenson, G. I.; Lewis, S.; Patel, S.; Castro, J. L. *Bioorg. Med. Chem. Lett.* **2000**, *10*, 2693–2696.
116. Lee, S.-H.; Clapham, B.; Koch, G.; Zimmermann, J.; Janda, K. D. *J. Comb. Chem.* **2003**, *5*, 188–196.
117. Zhang, H.-C.; McComsey, D. F.; White, K. B.; Addo, M. F.; Andrade-Gordon, P.; Derian, C. K.; Oksenberg, D.; Maryanoff, B. E. *Bioorg. Med. Chem. Lett.* **2001**, *11*, 2105–2109.
118. Aronov, A. M.; Munagala, N. R.; Ortiz de Montellano, P. R.; Kuntz, I. D.; Wang, C. C. *Biochemistry* **2000**, *39*, 4684–4691.
119. Chen, L.; Trilles, R.; Miklowski, D.; Huang, T.-N.; Fry, D.; Campbell, R.; Rowan, K.; Schwinge, V.; Tilley, J. W. *Bioorg. Med. Chem. Lett.* **2002**, *12*, 1679–1682.
120. Grimes, J. H., Jr.; Angell, Y. M.; Kohn, W. D. *Tetrahedron Lett.* **2003**, *44*, 3835–3838.
121. Ruhland, T.; Bang, K. S.; Andersen, K. *J. Org. Chem.* **2002**, *67*, 5257–5268.
122. Nilsson, J. W.; Thorstensson, F.; Kvarnstroem, I.; Oprea, T.; Samuelsson, B.; Nilsson, I. *J. Comb. Chem.* **2001**, *3*, 546–553.
123. Bergauer, M.; Huebner, H.; Gmeiner, P. *Bioorg. Med. Chem. Lett.* **2002**, *12*, 1937–1940.
124. Zajdel, P.; Subra, G.; Bojarski, A. J.; Duszynska, B.; Pawlowski, M.; Martinez, J. *J. Comb. Chem.* **2004**, *6*, 761–767.
125. Gonzalez-Gomez, J. C.; Uriarte-Villares, E.; Figueroa-Perez, S. *Synlett* **2002**, 1085–1088.
126. Guo, T.; Adang, A. E. P.; Dong, G.; Fitzpatrick, D.; Geng, P.; Ho, K.-K.; Jibilian, C. H.; Kultgen, S. G.; Liu, R.; McDonald, E. et al. *Bioorg. Med. Chem. Lett.* **2004**, *14*, 1717–1720.
127. Ma, Y.; Margarida, L.; Brookes, J.; Makara, G. M.; Berk, S. C. *J. Comb. Chem.* **2004**, *6*, 426–430.
128. Wade, J. V.; Krueger, C. A. *J. Comb. Chem.* **2003**, *5*, 267–272.
129. Kumar, A.; Sinha, S.; Chauhan, P. M. S. *Bioorg. Med. Chem. Lett.* **2002**, *12*, 667–669.
130. Hattori, K.; Kido, Y.; Yamamoto, H.; Ispida, J.; Kamijo, K.; Murano, K.; Ohkubo, M.; Kinoshita, T.; Iwashita, A.; Mihara, K. et al. *J. Med. Chem.* **2004**, *47*, 4151–4154.
131. Choo, H.-Y. P.; Kim, M.; Lee, S. K.; Kim, S. W.; Chung, I. K. *Bioorg. Med. Chem. Lett.* **2002**, *12*, 517–523.
132. Kesarwani, A. P.; Srivastava, G. K.; Rastogi, S. K.; Kundu, B. *Tetrahedron Lett.* **2002**, *43*, 5579–5581.
133. Dener, J. M.; Lease, T. G.; Novack, A. R.; Plunkett, M. J.; Hocker, M. D.; Fantauzzi, P. P. *J. Comb. Chem.* **2001**, *3*, 590–597.

134. Falcò, J. L.; Matallana, J. L.; Barberena, J.; Teixidò, J.; Borrell, J. I. *Mol. Diversity* **2003**, *6*, 3–11.
135. Falcò, J. L.; Borrell, J. I.; Teixidò, J. *Mol. Diversity* **2003**, *6*, 85–92.
136. Beer, D.; Bhalay, G.; Dunstan, A.; Glen, A.; Haberthuer, S.; Moser, H. *Bioorg. Med. Chem. Lett.* **2002**, *12*, 1973–1976.
137. Feliu, L.; Subra, G.; Martinez, J.; Amblard, M. *J. Comb. Chem.* **2003**, *5*, 356–361.
138. Bleicher, K. H.; Wuetrich, Y.; De Boni, M.; Kolczewski, S.; Hoffmann, T.; Sleight, A. J. *Bioorg. Med. Chem. Lett.* **2002**, *12*, 2519–2522.
139. Park, K.-H.; Ehrler, J.; Spoerri, H.; Kurth, M. J. *J. Comb. Chem.* **2001**, *3*, 171–176.
140. Gelens, E.; Koot, W. J.; Menge, W. M. P. B.; Ottenheijm, H. C. J.; Timmerman, H. *Bioorg. Med. Chem. Lett.* **2000**, *10*, 1935–1938.
141. Saha, A. K.; Liu, L.; Simoneaux, R. L.; Kukla, M. J.; Marichal, P.; Odds, F. *Bioorg. Med. Chem. Lett.* **2000**, *10*, 2175–2178.
142. Saha, A. K.; Liu, L.; Marichal, P.; Odds, F. *Bioorg. Med. Chem. Lett.* **2000**, *10*, 2735–2739.
143. Hamper, B. C.; Jerome, K. D.; Yalamanchili, G.; Walzer, D. M.; Chott, R. C.; Mischke, D. A. *Biotechnol. Bioeng.* **2000**, *71*, 28–37.
144. Stauffer, S. R.; Katzenellenbogen, J. A. *J. Comb. Chem.* **2000**, *2*, 318–329.
145. Wattanasin, S.; Albert, R.; Ehrhardt, C.; Roche, D.; Sabio, M.; Hommel, U.; Welzenbach, K.; Weitz-Schmidt, G. *Bioorg. Med. Chem. Lett.* **2003**, *13*, 499–502.
146. Im, I.; Webb, T. R.; Gong, Y.-D.; Kim, J.-I.; Kim, Y.-C. *J. Comb. Chem.* **2004**, *6*, 207–213.
147. Kong, K.-H.; Chen, Y.; Ma, X.; Chui, W. K.; Lam, Y. *J. Comb. Chem.* **2004**, *6*, 928–933.
148. Tremblay, M. R.; Wentworth, P., Jr.; Lee, G. E., Jr.; Janda, K. D. *J. Comb. Chem.* **2000**, *2*, 698–709.
149. Singh, U.; Raju, B.; Zhou, J.; Gadwood, R. C.; Ford, C. W.; Zurenko, G. E.; Schaadt, R. D.; Morin, S. E.; Adams, W. J.; Friis, J. M. et al. *Bioorg. Med. Chem. Lett.* **2003**, *13*, 4209–4212.
150. Buchstaller, H.-P. *J. Comb. Chem.* **2003**, *5*, 789–793.
151. Pirrung, M. C.; Tumey, L. N. *J. Comb. Chem.* **2000**, *2*, 675–680.
152. Koide, K.; Finkelstein, J. M.; Ball, Z.; Verdine, G. L. *J. Am. Chem. Soc.* **2001**, *123*, 398–408.
153. Batra, S.; Rastogi, S. K.; Kundu, B.; Patra, A; Bhaduri, A. P. *Tetrahedron Lett.* **2000**, *41*, 5971–5974.
154. Batra, S.; Srinivasan, T.; Rastogi, S. K.; Kundu, B.; Patra, A.; Bhaduri, A. P.; Dixit, M. *Bioorg. Med. Chem. Lett.* **2002**, *12*, 1905–1908.
155. De Luca, L.; Giacomelli, G.; Porcheddu, A.; Salaris, M.; Taddei, M. *J. Comb. Chem.* **2003**, *5*, 465–471.
156. Cereda, E.; Ezhaya, A.; Quai, M.; Barbaglia, W. *Tetrahedron Lett.* **2001**, *42*, 4951–4953.
157. Kim, K. S.; Kimball, S. D.; Misra, R. N.; Rawlins, D. B.; Hunt, J. T.; Xiao, H.-Y.; Lu, S.; Qian, L.; Han, W.-C.; Shan, W. et al. *J. Med. Chem.* **2002**, *45*, 3905–3927.
158. Bork, J. T.; Lee, J. W.; Khersonsky, S. M.; Moon, H.-S.; Chang, Y.-T. *Org. Lett.* **2003**, *5*, 117–120.
159. Moon, H.-S.; Jacobson, E. M.; Khersonsky, S. M.; Luzung, M. R.; Walsh, D. P.; Xiong, W.; Lee, J. W.; Parikh, P. B.; Lam, J. C.; Kang, T.-W. et al. *J. Am. Chem. Soc.* **2002**, *124*, 11608–11609.
160. Khersonsky, S. M.; Jung, D.-W.; Kang, T.-W.; Walsh, D. P.; Moon, H.-S.; Jo, H.; Jacobson, E. M.; Shetty, V.; Neubert, T. A.; Chang, Y.-T. *J. Am. Chem. Soc.* **2003**, *125*, 11804–11805.
161. Williams, D.; Jung, D.-W.; Khersonsky, S. M.; Heidary, N.; Chang, Y.-T.; Orlow, S. J. *Chem. Biol.* **2004**, *11*, 1251–1259.
162. Varaprasad, C. V.; Habib, Q.; Li, D. Y.; Huang, J.; Abt, J. W.; Rong, F.; Hong, Z.; An, H. *Tetrahedron* **2003**, *59*, 2297–2307.
163. Van Calenbergh, S.; Link, A.; Fujikawa, S.; de Ligt, R. A. F.; Vanheusden, V.; Golisade, A.; Blaton, N. M.; Rozenski, J.; Jzerman, A. P. I.; Herdewjin, P. *J. Med. Chem.* **2002**, *45*, 1845–1852.
164. Gil, C.; Schwoegler, A.; Braese, S. *J. Comb. Chem.* **2004**, *6*, 38–42.
165. Kitas, E. A.; Loeffler, B.-M.; Daetwyler, S.; Dehmlow, H.; Aebi, J. D. *Bioorg. Med. Chem. Lett.* **2002**, *12*, 1727–1730.
166. Albericio, F.; Bryman, L. M.; Garcia, J.; Michelotti, E. L.; Nicolas, E.; Tice, C. M. *J. Comb. Chem.* **2001**, *3*, 290–300.
167. Fernandez, E.; Garcia-Ochoa, S.; Huss, S.; Mallo, A.; Bueno, J. M.; Micheli, F.; Paio, A.; Piga, E.; Zarantonello, P. *Tetrahedron Lett.* **2002**, *43*, 4741–4745.
168. Bondebjerg, J.; Xiang, Z.; Bauzo, R. M.; Haskell-Luevano, C.; Meldal, M. *J. Am. Chem. Soc.* **2002**, *124*, 11046–11055.
169. Severinsen, R.; Lau, J. F.; Bondensgaard, K.; Hansen, B. S.; Begtrup, M.; Ankersen, M. *Bioorg. Med. Chem. Lett.* **2004**, *14*, 317–320.
170. Shannon, S. K.; Peacock, M. J.; Kates, S. A.; Barany, G. *J. Comb. Chem.* **2003**, *5*, 860–868.
171. Lee, S.-H.; Matsushita, H.; Koch, G.; Zimmermann, J.; Clapham, B.; Janda, K. D. *J. Comb. Chem.* **2004**, *6*, 822–827.
172. Sandanayake, S.; Perera, S.; Ede, N. J. *QSAR Comb. Sci.* **2004**, *23*, 655–661.
173. Guan, Y.; Green, M. A.; Bergstrom, D. E. *J. Comb. Chem.* **2000**, *2*, 297–300.
174. Katritzky, A. R.; Rogovoy, B. V; Kirichenko, N.; Vvedensky, V. *Bioorg. Med. Chem. Lett.* **2002**, *12*, 1809–1811.
175. Georgiadis, T. M.; Baindur, N.; Player, M. R. *J. Comb. Chem.* **2004**, *6*, 224–229.
176. Hopkins, T. P.; Dener, J. M.; Boldi, A. M. *J. Comb. Chem.* **2002**, *4*, 167–174.
177. Juteau, H.; Gareau, Y.; Labelle, M.; Lamontagne, S.; Tremblay, N.; Carriere, M.-C.; Sawyer, N.; Denis, D.; Metters, K. M. *Bioorg. Med. Chem. Lett.* **2001**, *11*, 747–749.
178. Reich, S. H.; Johnson, T.; Wallace, M. B.; Kephart, S. E.; Fuhrman, S. A.; Worland, S. T.; Matthews, D. A.; Hendrickson, T. F.; Chan, F.; Meador, J., III et al. *J. Med. Chem.* **2000**, *43*, 1670–1683.
179. Liu, K. G.; Lambert, M. H.; Leesnitzer, L. M.; Oliver, W., Jr.; Ott, R. J.; Plunkett, K. D.; Stuart, L. W.; Brown, P. J.; Willson, T. M.; Sternbach, D. D. *Bioorg. Med. Chem. Lett.* **2001**, *11*, 2059–2062.
180. Brown, P. J.; Stuart, L. W.; Hurley, K. P.; Lewis, M. C.; Winegar, D. A.; Wilson, J. G.; Wilkison, W. O.; Ittoop, O. R.; Willson, T. M. *Bioorg. Med. Chem. Lett.* **2001**, *11*, 1225–1227.
181. Gopalsamy, A.; Yang, H.; Ellingboe, J. W.; Kees, K. L.; Yoon, J.; Murrills, R. *Bioorg. Med. Chem. Lett.* **2000**, *10*, 1715–1718.
182. Dankwardt, S. M.; Abbot, S. C.; Broka, C. A.; Martin, R. L.; Chan, C. S.; Springman, E. B.; Van Wart, H. E.; Walker, K. A. M. *Bioorg. Med. Chem. Lett.* **2002**, *12*, 1233–1235.
183. Wang, G.; Yao, S. Q. *Org. Lett.* **2003**, *5*, 4437–4440.
184. Knepper, K.; Lormann, M. E. P.; Braese, S. *J. Comb. Chem.* **2004**, *6*, 460–463.
185. Sulyok, G. A. G.; Gibson, C.; Goodman, S. L.; Holzemann, G.; Wiesner, M.; Kessler, H. *J. Med. Chem.* **2001**, *44*, 1938–1950.
186. Edvinsson, K. M.; Hersloef, M.; Holm, P.; Kann, N.; Keeling, D. J.; Mattsson, J. P.; Norden, B.; Scherbukhin, V. *Bioorg. Med. Chem. Lett.* **2000**, *10*, 503–507.
187. Chitkul, B.; Bradley, M. *Bioorg. Med. Chem. Lett.* **2000**, *10*, 2367–2369.
188. Urbahns, K.; Haerter, M.; Vaupel, A.; Albers, M.; Schmidt, D.; Brueggemeier, U.; Stelte-Ludwig, B.; Gerdes, C.; Tsujishita, H. *Bioorg. Med. Chem. Lett.* **2003**, *13*, 1071–1074.

189. Castanedo, G. M.; Sailes, F. C.; Dubree, N. J. P.; Nicholas, J. B.; Caris, L.; Clark, K.; Keating, S. M.; Beresini, M. H.; Chiu, H.; Fong, S. et al. *Bioorg. Med. Chem. Lett.* **2002**, *12*, 2913–2917.
190. Chollet, A.-M.; Le Diguarher, T.; Kuharczyk, N.; Loynel, A.; Bertrand, M.; Tucker, G.; Guilbaud, N.; Burbridge, M.; Pastoureau, P.; Fradin, A., et al. *Bioorg. Med. Chem.* **2002**, *10*, 531–544.
191. Collins, J. L.; Fivush, A. M.; Watson, M. A.; Galardi, C. M.; Lewis, M. C.; Moore, L. B.; Parks, D. J.; Wilson, J. G.; Tippin, T. K.; Binz, J. G. et al. *J. Med. Chem.* **2002**, *45*, 1963–1966.
192. Harrison, B. A.; Gierasch, T. M.; Neilan, C.; Pasternak, G. W.; Verdine, G. L. *J. Am. Chem. Soc.* **2002**, *124*, 13352–13353.
193. Akritopoulou-Zanze, I.; Sowin, T. J. *J. Comb. Chem.* **2001**, *3*, 301–311.
194. Jagtap, P. G.; Baloglu, E.; Barron, D. M.; Bane, S.; Kingston, D. G. I. *J. Nat. Prod.* **2002**, *65*, 1136–1142.
195. Ohno, H.; Kawamura, K.; Otake, A.; Nagase, H.; Tanaka, H.; Takahashi, T. *Synlett* **2002**, 93–96.
196. Tanaka, H.; Moriwaki, M.; Takahashi, T. *Org. Lett.* **2003**, *5*, 3807–3809.
197. Guo, G.; Arvanitis, E. A.; Pottorf, R. S.; Player, M. R. *J. Comb. Chem.* **2003**, *5*, 408–413.
198. Tanaka, H.; Zenkoh, T.; Setoi, H.; Takahashi, T. *Synlett* **2002**, 1427–1430.
199. Hijikuro, I.; Doi, T.; Takahashi, T. *J. Am. Chem. Soc.* **2001**, *123*, 3716–3722.
200. Renault, J.; Lebranchu, M.; Lecat, A.; Uriac, P. *Tetrahedron Lett.* **2001**, *42*, 6655–6658.
201. Manku, S.; Wang, F.; Hall, D. G. *J. Comb. Chem.* **2003**, *5*, 379–391.
202. Maletic, M.; Anntonic, J.; Leeman, A.; Santorelli, G.; Waddell, S. *Bioorg. Med. Chem. Lett.* **2003**, *13*, 1125–1128.
203. Ramaseshan, M.; Dory, Y. L.; Deslongchamps, P. *J. Comb. Chem.* **2000**, *2*, 615–623.
204. Jefferson, E. A.; Arakawa, S.; Blyn, L. B.; Miyaji, A.; Osgood, S. A.; Ranken, R.; Risen, L. M.; Swayze, E. E. *J. Med. Chem.* **2002**, *45*, 3430–3439.
205. Willoughby, C. A.; Berk, S. C.; Rosauer, K. G.; Degrado, S.; Chapman, K. T.; Gould, S. L.; Springer, M. S.; Malkowitz, L.; Schleif, W. A.; Hazuda, D. et al. *Bioorg. Med. Chem. Lett.* **2001**, *11*, 3137–3141.
206. Jackson, R. W.; Tabone, J. C.; Howbert, J. J. *Bioorg. Med. Chem. Lett.* **2003**, *13*, 205–208.
207. Fenwick, A. E.; Garnier, B.; Gribble, A. D.; Ife, R. J.; Rawlings, A. D.; Witherington, J. *Bioorg. Med. Chem. Lett.* **2001**, *11*, 195–198.
208. Jefferson, E. A.; Sprankle, K. G.; Swayze, E. E. *J. Comb. Chem.* **2000**, *2*, 441–444.
209. Couve-Bonnaire, S.; Chou, D. T. H.; Gan, Z.; Arya, P. *J. Comb. Chem.* **2004**, *6*, 73–77.
210. Willoughby, C. A.; Hutchins, S. M.; Rosauer, K. G.; Dhar, M. J.; Chapman, K. T.; Chicchi, G. G.; Sadowski, S.; Weinberg, D. H.; Patel, S.; Malkowitz, L. et al. *Bioorg. Med. Chem. Lett.* **2002**, *12*, 93–96.
211. Wu, Z.; Ede, N. J. *Org. Lett.* **2003**, *5*, 2935–2938.
212. Liao, Y.; Reitman, M.; Zhang, Y.; Fathi, R.; Yang, Z. *Org. Lett.* **2002**, *4*, 2607–2609.
213. Fujimori, T.; Wirsching, P.; Janda, K. D. *J. Comb. Chem.* **2003**, *5*, 625–631.
214. Acharya, A. N.; Nefzi, A.; Ostresh, J. M.; Houghten, R. A. *J. Comb. Chem.* **2001**, *3*, 189–195.
215. Yu, Y.; Ostresh, J. M.; Houghten, R. A. *J. Comb. Chem.* **2001**, *3*, 521–523.
216. Acharya, A. N.; Ostresh, J. M.; Houghten, R. A. *J. Comb. Chem.* **2001**, *3*, 612–623.
217. Nefzi, A.; Giulianotti, M.; Truong, L.; Rattan, S.; Ostresh, J. M.; Houghten, R. A. *J. Comb. Chem.* **2002**, *4*, 175–178.
218. Rinnova, M.; Vidal, A.; Nefzi, A.; Houghten, R. A. *J. Comb. Chem.* **2002**, *4*, 209–213.
219. Acharya, A. N.; Ostresh, J. M.; Houghten, R. A. *J. Comb. Chem.* **2002**, *4*, 214–222.
220. Acharya, A. N.; Ostresh, J. M.; Houghten, R. A. *Tetrahedron Lett.* **2002**, *43*, 1157–1160.
221. McMillan, K.; Adler, M.; Auld, D. S.; Baldwin, J. J.; Blasko, E.; Browne, L. J.; Chelsky, D.; Davey, D.; Dolle, R. E.; Eagen, K. A. et al. *Proc. Natl. Acad. Sci. USA* **2000**, *97*, 1506–1511.
222. Bianchi, I.; La Porta, E.; Barlocco, D.; Raveglia, L. F. *J. Comb. Chem.* **2004**, *6*, 835–845.
223. Cheng, Y.; Rano, T. A.; Huening, T. T.; Zhang, F.; Lu, Z.; Schleif, W. A.; Gabryelski, L.; Olsen, D. B.; Stahlhut, M.; Kuo, L. C. et al. *Bioorg. Med. Chem. Lett.* **2002**, *12*, 529–532.
224. Raghavan, S.; Yang, Z.; Mosley, R. T.; Schleif, W. A.; Gabryelski, L.; Olsen, D. B.; Stahlhut, M.; Kuo, L. C.; Emini, E. A.; Chapman, K. T. et al. *Bioorg. Med. Chem. Lett.* **2002**, *12*, 2855–2858.
225. Jain, R.; Kamau, M.; Wang, C.; Ippolito, R.; Wang, H.; Dulina, R.; Anderson, J.; Gange, D.; Sofia, M. J. *Bioorg. Med. Chem. Lett.* **2003**, *13*, 2185–2189.
226. Arvanitis, E. A.; Chadha, N.; Pottorf, R. S.; Player, M. R. *J. Comb. Chem.* **2004**, *6*, 414–419.
227. Zhu, S.; Shi, S.; Gerritz, S. W.; Sofia, M. J. *J. Comb. Chem.* **2003**, *5*, 205–207.
228. Yu, Y.; Ostresh, J. M.; Houghten, R. A. *J. Org. Chem.* **2002**, *67*, 5831–5834.
229. Arya, P.; Wei, C.-Q.; Barnes, M. L.; Daroszewska, M. *J. Comb. Chem.* **2004**, *6*, 65–72.
230. Herpin, T. F.; Van Kirk, K. G.; Salvino, J. M.; Yu, S. T.; Labaudiniere, R. F. *J. Comb. Chem.* **2000**, *2*, 513–521.
231. Nefzi, A.; Ong, N. A.; Houghten, R. A. *Tetrahedron Lett.* **2001**, *42*, 5141–5143.
232. Makara, G. M.; Ewing, W.; Ma, Y.; Wintner, E. *J. Org. Chem.* **2001**, *66*, 5783–5789.
233. Epple, R.; Kudirka, R.; Greenberg, W. A. *J. Comb. Chem.* **2003**, *5*, 292–310.
234. Tornoe, C. W.; Sanderson, S. J.; Mottram, J. C.; Coombs, G. H.; Meldal, M. *J. Comb. Chem.* **2004**, *6*, 312–324.
235. Pernerstorfer, J.; Brands, M.; Schirok, H.; Stelte-Ludwig, B.; Woltering, E. *Tetrahedron* **2004**, *60*, 8627–8632.
236. de Laszlo, S. E.; Li, B.; McCauley, E.; Van Riper, G.; Hagmann, W. K. *Bioorg. Med. Chem. Lett.* **2002**, *12*, 685–688.
237. Humet, M.; Carbonell, T.; Masip, I.; Sanchez-Baeza, F.; Mora, P.; Canton, E.; Gobernado, M.; Abad, C.; Perez-Paya, E.; Messeguer, A. *J. Comb. Chem.* **2003**, *5*, 597–605.
238. Lee, R. E.; Protopopova, M.; Crooks, E.; Slayden, R. A.; Terrot, M.; Barry, C. E., III *J. Comb. Chem.* **2003**, *5*, 172–187.
239. Zhao, L.; Allanson, N. M.; Thomson, S. P.; MacLean, J. K. F.; Barker, J. J.; Primrose, W. U.; Tyler, P. D.; Lewendon, A. *Eur. J. Med. Chem.* **2003**, *38*, 345–349.
240. Prokai, L.; Prokai-Tatrai, K.; Zharikova, A.; Li, X.; Rocca, J. R. *J. Med. Chem.* **2001**, *44*, 1623–1626.
241. St. Hilaire, P. M.; Alves, L. C.; Herrera, F.; Renil, M.; Sanderson, S. J.; Mottram, J. C.; Coombs, G. H.; Juliano, M. A.; Juliano, L.; Arevalo, J. et al. *J. Med. Chem.* **2002**, *45*, 1971–1982.
242. Shukla, R.; Sasaki, Y.; Krchnak, V.; Smith, B. D. *J. Comb. Chem.* **2004**, *6*, 703–709.
243. Pegurier, C.; Curtet, S.; Nicolas, J.-P.; Boutin, J. A.; Delagrange, P.; Renard, P.; Langlois, M. *Bioorg. Med. Chem.* **2000**, *8*, 163–171.
244. Liu, R.; Marik, J.; Lam, K. S. *J. Am. Chem. Soc.* **2002**, *124*, 7678–7680.
245. Bryan, W. M.; Huffman, W. F.; Bhatnagar, P. K. *Tetrahedron Lett.* **2000**, *41*, 6997–7000.

246. Guo, T.; Adang, A. E. P.; Dolle, R. E.; Dong, G.; Fitzpatrick, D.; Geng, P.; Ho, K.-K.; Kultgen, S. G.; Liu, R.; McDonald, E. et al. *Bioorg. Med. Chem. Lett.* **2004**, *14*, 1713–1716.
247. Andres, C. J.; Zimanyi, I. A.; Desphande, M. S.; Iben, L. G.; Grant-Young, K.; Mattson, G. K.; Zhai, W. *Bioorg. Med. Chem. Lett.* **2003**, *13*, 2883–2885.
248. Ferguson, R. D.; Su, N.; Smith, R. A. *Tetrahedron Lett.* **2003**, *44*, 2939–2942.
249. El-Araby, M.; Guo, H.; Pottorf, R. S.; Player, M. R. *J. Comb. Chem.* **2004**, *6*, 789–795.
250. Park, J. G.; Langenwalter, K. J.; Weinbaum, C. A.; Casey, P. J.; Pang, Y.-P. *J. Comb. Chem.* **2004**, *6*, 407–413.
251. Nicolaou, K. C.; Baran, P. S.; Zhong, Y.-L.; Sugita, K. J. *J. Am. Chem. Soc.* **2002**, *124*, 2212–2220.
252. Nicolaou, K. C.; Roecker, A. J.; Hughes, R.; van Summeren, R.; Pfefferkorn, J. A.; Winssinger, N. *Bioorg. Med. Chem.* **2003**, *11*, 465–476.
253. Nicolaou, K. C.; Montagnon, T.; Ulven, T.; Baran, P. S.; Zhong, Y.-L.; Sarabia, F. *J. Am. Chem. Soc.* **2002**, *124*, 5718–5728.
254. Nicolaou, K. C.; Pfefferkorn, J. A.; Cao, G.-Q. *Angew. Chem. Int. Ed.* **2000**, *39*, 734–739.
255. Nicolaou, K. C.; Cao, G.-Q.; Pfefferkorn, J. A. *Angew. Chem. Int. Ed.* **2000**, *39*, 739–743.
256. Tallarico, J. A.; Depew, K. M.; Pelish, H. E.; Westwood, N. J.; Lindsey, C. W.; Shair, M. D.; Schreiber, S. L.; Foley, M. A. *J. Comb. Chem.* **2001**, *3*, 312–318.
257. Chen, C.; Li, X.; Schreiber, S. L. *J. Am. Chem. Soc.* **2003**, *125*, 10174–10175.
258. Lusch, M. J.; Tallarico, J. A. *Org. Lett.* **2004**, *6*, 3237–3240.
259. Micalizio, G. C.; Schreiber, S. L. *Angew. Chem. Int. Ed.* **2002**, *41*, 152–154.
260. Micalizio, G. C.; Schreiber, S. L. *Angew. Chem. Int. Ed.* **2002**, *41*, 3272–3276.
261. Spring, D. R.; Krishnan, S.; Schreiber, S. L. *J. Am. Chem. Soc.* **2000**, *122*, 5656–5657.
262. Lindsley, C. W.; Chan, L. K.; Goess, B. C.; Joseph, R.; Shair, M. D. *J. Am. Chem. Soc.* **2000**, *122*, 422–423.
263. Lee, D.; Sello, J. K.; Schreiber, S. L. *Org. Lett.* **2000**, *2*, 709–712.
264. Sello, J. K.; Andreana, P. R.; Lee, D.; Schreiber, S. L. *Org. Lett.* **2003**, *5*, 4125–4127.
265. Blackwell, H. E.; Clemons, P. A.; Schreiber, S. L. *Org. Lett.* **2001**, *3*, 1185–1188.
266. Winssinger, N.; Harris, J. L.; Backes, B. J.; Schultz, P. G. *Angew. Chem. Int. Ed.* **2001**, *40*, 3152–3155.
267. Ding, S.; Gray, N. S.; Ding, Q.; Schultz, P. G. *J. Org. Chem.* **2001**, *66*, 8273–8276.
268. Ding, S.; Gray, N. S.; Ding, Q.; Wu, X.; Schultz, P. G. *J. Comb. Chem.* **2002**, *4*, 183–186.
269. Rosenbaum, C.; Waldmann, H. *Tetrahedron Lett.* **2001**, *42*, 5677–5680.
270. Stieber, F.; Grether, U.; Waldmann, H. *Chem. Eur. J.* **2003**, *9*, 3270–3281.
271. Koehn, M.; Wacker, R.; Peters, C.; Schroeder, H.; Soulere, L.; Breinbauer, R.; Niemeyer, C. M.; Waldmann, H. *Angew. Chem. Int. Ed.* **2003**, *42*, 5830–5834.
272. Barun, O.; Sommer, S.; Waldmann, H. *Angew. Chem. Int. Ed.* **2004**, *43*, 3195–3199.
273. Stavenger, R. A.; Schreiber, S. L. *Angew. Chem. Int. Ed.* **2001**, *40*, 3417–3421.
274. Blackwell, H. E.; Perez, L.; Schreiber, S. L. *Angew. Chem. Int. Ed.* **2001**, *40*, 3421–3425.
275. Blackwell, H. E.; Perez, L.; Stavenger, R. A.; Tallarico, J. A.; Eatough, E. C.; Foley, M. A.; Schreiber, S. L. *Chem. Biol.* **2001**, *8*, 1167–1182.
276. Clemons, P. A.; Koehler, A. N.; Wagner, B. K.; Sprigings, T. G.; Spring, D. R.; King, R. W.; Schreiber, S. L.; Foley, M. A. *Chem. Biol.* **2001**, *8*, 1183–1195.
277. Sternson, S. M.; Louca, J. B.; Wong, J. C.; Schreiber, S. L. *J. Am. Chem. Soc.* **2001**, *123*, 1740–1747.
278. Kuruvilla, F. G.; Shamji, A. F.; Sternson, S. M.; Hergenrother, P. J.; Schreiber, S. L. *Nature* **2002**, *416*, 653–657.
279. Sternson, S. M.; Wong, J. C.; Grozinger, C. M.; Schreiber, S. L. *Org. Lett.* **2001**, *3*, 4239–4242.
280. Haggarty, S. J.; Koeller, K. M.; Wong, J. C.; Grozinger, C. M.; Schreiber, S. L. *Proc. Natl. Acad. Sci. USA* **2003**, *100*, 4389–4394.
281. Haggarty, S. J.; Koeller, K. M.; Wong, J. C.; Butcher, R. A.; Schreiber, S. L. *Chem. Biol.* **2003**, *10*, 383–396.
282. Wong, J. C.; Hong, R.; Schreiber, S. L. *J. Am. Chem. Soc.* **2003**, *125*, 5586–5587.
283. Haggarty, S. J.; Clemons, P. A.; Wong, J. C.; Schreiber, S. L. *Comb. Chem. HTS* **2004**, 7, 669–676.
284. Wong, J. C.; Sternson, S. M.; Louca, J. B.; Hong, R.; Schreiber, S. L. *Chem. Biol.* **2004**, *11*, 1279–1291.
285. Spring, D. R.; Krishnan, S.; Blackwell, H. E.; Schreiber, S. L. *J. Am. Chem. Soc.* **2002**, *124*, 1354–1363.
286. Koehler, A. N.; Shamji, A. F.; Schreiber, S. L. *J. Am. Chem. Soc.* **2003**, *125*, 8420–8421.
287. Kwon, O.; Park, S. B.; Schreiber, S. L. *J. Am. Chem. Soc.* **2002**, *124*, 13402–13404.
288. Barnes-Seeman, D.; Park, S. B.; Koehler, A. N.; Schreiber, S. L. *Angew. Chem. Int. Ed.* **2003**, *42*, 2376–2379.
289. Hergenrother, P. J.; Depew, K. M.; Schreiber, S. L. *J. Am. Chem. Soc.* **2000**, *122*, 7849–7850.
290. Pelish, H. E.; Westwood, N. J.; Feng, Y.; Kirchhausen, T.; Shair, M. D. *J. Am. Chem. Soc.* **2001**, *123*, 6740–6741.
291. Kubota, H.; Lim, J.; Depew, K. M.; Schreiber, S. L. *Chem. Biol.* **2002**, *9*, 265–276.
292. Lo, M. M.-C.; Neumann, C. S.; Nagayama, S.; Perlstein, E. O.; Schreiber, S. L. *J. Am. Chem. Soc.* **2004**, *126*, 16077–16086.
293. Burke, M. D.; Berger, E. M.; Schreiber, S. L. *Science* **2003**, *302*, 613–618.
294. Burke, M. D.; Berger, E. M.; Schreiber, S. L. *J. Am. Chem. Soc.* **2004**, *126*, 14095–14104.
295. Winssinger, N.; Ficarro, S.; Schultz, P. G.; Harris, J. L. *Proc. Natl. Acad. Sci. USA* **2002**, *99*, 11139–11144.
296. Harris, J.; Mason, D. E.; Li, J.; Burdick, K. W.; Backes, B. J.; Chen, T.; Shipway, A.; Van Heeke, G.; Gough, L.; Ghaemmaghami, A. et al. *Chem. Biol.* **2004**, *11*, 1361–1372.
297. Winssinger, N.; Damoiseaux, R.; Tully, D. C.; Geierstanger, B. H.; Burdick, K.; Harris, J. L. *Chem. Biol.* **2004**, *11*, 1351–1360.
298. Nicolaou, K. C.; Pfefferkorn, J. A.; Roecker, A. J.; Cao, G.-Q.; Barluenga, S.; Mitchell, H. J. *J. Am. Chem. Soc.* **2000**, *122*, 9939–9953.
299. Nicolaou, K. C.; Pfefferkorn, J. A.; Mitchell, H. J.; Roecker, A. J.; Barluenga, S.; Cao, G.-Q.; Affleck, R. L.; Lillig, J. E. *J. Am. Chem. Soc.* **2000**, *122*, 9954–9967.
300. Nicolaou, K. C.; Pfefferkorn, J. A.; Barluenga, S.; Mitchell, H. J.; Roecker, A. J.; Cao, G.-Q. *J. Am. Chem. Soc.* **2000**, *122*, 9968–9976.
301. Meseguer, B.; Alonso-Diaz, D.; Griebenow, N.; Herget, T.; Waldmann, H. *Chem. Eur. J.* **2000**, *6*, 3943–3957.
302. Thutewohl, M.; Kissau, L.; Popkirova, B.; Karaguni, I.-M.; Nowak, T.; Bate, M.; Kuhlmann, J.; Mueller, O.; Waldmann, H. *Angew. Chem. Int. Ed.* **2002**, *41*, 3616–3620.
303. Thutewohl, M.; Waldmann, H. *Bioorg. Med. Chem.* **2003**, *11*, 2591–2615.

304. Thutewohl, M.; Kissau, L.; Popkirova, B.; Karaguni, I.-M.; Nowak, T.; Bate, M.; Kuhlmann, J.; Mueller, O.; Waldmann, H. *Bioorg. Med. Chem.* **2003**, *11*, 2617–2626.
305. Stieber, F.; Mazitschek, R.; Soric, N.; Giannis, A.; Waldmann, H. *Angew. Chem. Int. Ed.* **2002**, *41*, 4757–4761.
306. Stieber, F.; Grether, U.; Mazitschek, R.; Soric, N.; Giannis, A.; Waldmann, H. *Chem. Eur. J.* **2003**, *9*, 3282–3291.
307. Brohm, D.; Metzger, S.; Bhargava, A.; Mueller, O.; Lieb, F.; Waldmann, H. *Angew. Chem. Int. Ed.* **2002**, *41*, 307–311.
308. Brohm, D.; Philippe, N.; Metzger, S.; Bhargava, A.; Mueller, O.; Lieb, F.; Waldmann, H. *J. Am. Chem. Soc.* **2002**, *124*, 13171–13178.
309. Koch, M. A.; Wittenberg, L.-O.; Basu, S.; Jeyarai, D. A.; Gourzoulidou, E.; Reinecke, K.; Odermatt, A.; Waldmann, H. *Proc. Natl. Acad. Sci. USA* **2004**, *101*, 16721–16726.
310. Rosenbaum, C.; Baumhof, P.; Mazitschek, R.; Mueller, O.; Giannis, A.; Waldmann, H. *Angew. Chem. Int. Ed.* **2004**, *43*, 224–228.
311. Mueller, O.; Gourzoulidou, E.; Carpintero, M.; Karaguni, I.-M.; Langerak, A.; Herrmann, C.; Moroy, T.; Klein-Hitpass, L.; Waldmann, H. *Angew. Chem. Int. Ed.* **2004**, *43*, 450–454.
312. Waldmann, H.; Karaguni, I.-M.; Carpintero, M.; Gourzoulidou, E.; Herrmann, C.; Brockmann, C.; Oschkinat, H.; Mueller, O. *Angew. Chem. Int. Ed.* **2004**, *43*, 454–458.
313. Ding, S.; Gray, N. S.; Wu, X.; Ding, Q.; Schultz, P. G. *J. Am. Chem. Soc.* **2002**, *124*, 1594–1596.
314. Chapman, E.; Ding, S.; Schultz, P. G.; Wong, C.-H. *J. Am. Chem. Soc.* **2002**, *124*, 14524–14525.
315. Wu, X.; Ding, S.; Ding, Q.; Gray, N. S.; Schultz, P. G. *J. Am. Chem. Soc.* **2002**, *124*, 14520–14521.
316. Ding, S.; Wu, T. Y. H.; Brinker, A.; Peters, E. C.; Hur, W.; Gray, N. S.; Schultz, P. G. *Proc. Natl. Acad. Sci. USA* **2003**, *100*, 7632–7637.
317. Chen, S.; Zhang, Q.; Wu, X.; Schultz, P. G.; Ding, S. *J. Am. Chem. Soc.* **2004**, *126*, 410–411.
318. Armstrong, J. I.; Portley, A. R.; Chang, Y.-T.; Nierengarten, D. M.; Cook, B. N.; Bowman, K. G.; Bishop, A.; Gray, N. S.; Shokat, K. M.; Schultz, P. G. et al. *Angew. Chem. Int. Ed.* **2000**, *39*, 1303–1306.
319. Verdugo, D. E.; Cancilla, M. T.; Ge, X.; Gray, N. S.; Chang, Y.-T.; Schultz, P. G.; Negishi, M.; Leary, J. A.; Bertozzi, C. R. *J. Med. Chem.* **2001**, *44*, 2683–2686.
320. Chang, Y.-T.; Choi, G.; Bae, Y.-S.; Burdett, M.; Moon, H.-S.; Lee, J. W.; Gray, N. S.; Schultz, P. G.; Meijer, L.; Chung, S.-K. et al. *ChemBioChem* **2002**, 897–901.
321. Rosania, G. R.; Chang, Y.-T.; Perez, O.; Sutherlin, D.; Dong, H.; Lockhart, D. J.; Schultz, P. G. *Nat. Biotech.* **2000**, *18*, 304–308.
322. Wignall, S. M.; Gray, N. S.; Chang, Y.-T.; Juarez, L.; Jacob, R.; Burlingame, A.; Schultz, P. G.; Heald, R. *Chem. Biol.* **2004**, *11*, 135–146.
323. Wu, X.; Ding, S.; Ding, Q.; Gray, N. S.; Schultz, P. G. *J. Am. Chem. Soc.* **2004**, *126*, 1590–1591.
324. Liu, J.; Wu, X.; Mitchell, B.; Kintner, C.; Ding, S.; Schultz, P. *Angew. Chem. Int. Ed.* **2005**, *44*, 2–4.
325. Wu, T. Y. H.; Ding, S.; Gray, N. S.; Schultz, P. G. *Org. Lett.* **2001**, *3*, 3827–3830.
326. Wu, X.; Wagner, K. W.; Bursulaya, B.; Schultz, P. G.; Deveraux, Q. L. *Chem. Biol.* **2003**, *10*, 759–767.
327. Chang, Y.-T.; Choi, G.; Ding, S.; Prieska, E. E.; Baumruker, T.; Lee, J.-M; Chung, S.-K.; Schultz, P. G. *J. Am. Chem. Soc.* **2002**, 1856–1857.
328. Krishnan, S.; Schreiber, S. L. *Org. Lett.* **2004**, *6*, 4021–4024.
329. Schmidt, D. R.; Kwon, O.; Schreiber, S. L. *J. Comb. Chem.* **2004**, *6*, 286–292.
330. Kim, Y.-k.; Arai, M. A.; Arai, T.; Lamenzo, J. O.; Dean, E. F., III; Patterson, N.; Clemons, P. A.; Schreiber, S. L. *J. Am. Chem. Soc.* **2004**, *126*, 14740–14745.
331. Chen, C.; Li, X.; Neumann, C. S.; Lo, M. M.-C.; Schreiber, S. L. *Angew. Chem. Int. Ed.* **2005**, *44*, 2–4.
332. Nicolaou, K. C.; Pfefferkorn, J. A.; Schuler, F.; Roecker, A. J.; Cao, G.-Q.; Casida, J. E. *Chem. Biol.* **2000**, 7, 979–992.
333. Nicolaou, K. C.; Cho, S. Y.; Hughes, R.; Winssinger, N.; Smethurst, C.; Labischinski, H.; Endermann, R. *Chem.-Eur. J.* **2001**, 7, 3798–3823.

Biography

Pierfausto Seneci received his doctorate in 1983 from the University of Pavia in Italy under the direction of Prof Desimoni, following which he undertook postdoctoral research at the University of Milan with Prof C Fuganti (1984–85). In 1986 he joined Pierrel, and then held a series of senior R&D positions at Marion Merrell Dow (1987–95, Gerenzano, Italy, and Strasbourg, France), the Selectide Corporation (1995–96, Tucson, AZ), SmithKline Beecham

(1996–97, Rennes, France), and GlaxoWellcome (1997–2000, Verona, Italy), working in the fields of antibacterials, anti-inflammatories, and the central nervous system. In 2000, he joined a neurodegenerative disease-oriented company, Nucleotide Analog Design AG (NADAG, Munich, Germany), which he led as a Chief Executive Officer to a merger with Sireen AG in 2004 to form Sirenade Pharmaceuticals AG, of which Dr Seneci remained Chief Scientific Officer. In May 2005 he joined NiKem Research srl (Baranzate, Italy) as a Chief Business Officer. He has held the position of associate professor at the University of Milan since 2003. His research interests include medicinal chemistry, natural products, and diversity-oriented synthesis of natural products-like libraries in solution and on solid phase.

Comprehensive Medicinal Chemistry II
ISBN (set): 0-08-044513-6

ISBN (Volume 3) 0-08-044516-0; pp. 697–760

3.34 Solution Phase Parallel Chemistry

M Ashton and B Moloney, Evotec (UK) Ltd, Abingdon, UK

3.34.1	**Introduction**	**761**
3.34.2	**The Evolution of Combinatorial Chemistry**	**762**
3.34.2.1	From a Few Reaction Types to Many	762
3.34.2.2	The Synthesis of Discrete Compounds	763
3.34.2.3	The Continued Need for Large Libraries	763
3.34.2.4	Evolution and Terminology	763
3.34.3	**Comparison of Solid Phase versus Solution Phase Synthesis**	**763**
3.34.3.1	The Applicability of Solid Phase Parallel Synthesis	764
3.34.3.2	The Applicability of Solution Phase Parallel Synthesis	765
3.34.3.3	Selecting the Appropriate Parallel Synthesis Technology	768
3.34.4	**Solution Phase Library Synthesis**	**768**
3.34.4.1	Library Design	768
3.34.4.2	Route Validation	770
3.34.4.3	Synthesis and Purification of Intermediates	771
3.34.4.4	Preproduction: Synthesis of a Rehearsal Library	771
3.34.4.5	Library Production	771
3.34.5	**Reaction Monitoring, Downstream Processing, and Purification**	**775**
3.34.5.1	Reaction Monitoring	775
3.34.5.2	Parallel Work-up of Compounds and Key Intermediates	776
3.34.5.2.1	Liquid–liquid extraction	776
3.34.5.2.2	Purification of intermediates	776
3.34.5.3	Purification of Final Compounds	776
3.34.5.3.1	Resin scavenging	776
3.34.5.3.2	Resin capture and release/solid phase extraction	777
3.34.5.4	Analytical Characterization	778
3.34.6	**Examples of Solution Phase Parallel Synthesis**	**778**
3.34.6.1	Scaffold Derivatization	779
3.34.6.2	Heterocyclic Libraries	779
3.34.6.3	Multicomponent Reactions	784
3.34.6.4	Dynamic Combinatorial Chemistry	785
3.34.6.5	Fluorous Chemistry	785
3.34.7	**New and Future Trends; the Continued Evolution of Combinatorial Chemistry**	**786**
References		**787**

3.34.1 Introduction

A chapter on solution phase parallel chemistry cannot ignore the groundwork laid that enables us to consider this methodology to make small, medium, and large libraries of compounds. Solution phase parallel chemistry, while having its roots in traditional organic synthesis, owes a debt of gratitude to solid phase organic synthesis (SPOS), especially those results reported over the past 10 years. Without the emergence of SPOS as a viable technique to make libraries of compounds for biological screening, we would probably not be in a position today to consider making these numbers of compounds in solution phase. For example, without the significant experience in the synthesis of small molecules on a solid support, we would probably not have the breadth of scavenging resins that we have today to help purify products generated in solution phase. It is chemists' ability to synthesize molecules on solid phase that has led to the synthesis of not only a wide variety of scavenger resins but also a vast number of solid supported reagents that also aid the synthesis of pools of compounds via solution phase. Unfortunately, SPOS was unable to provide the diversity of compounds required to satisfy biologists' thirst for hit molecules against a plethora of new and emerging biological targets. Thus, the chemist was challenged with identifying methods for producing high numbers of compounds with a high degree of structural diversity. Solution phase parallel synthesis was born.

The evolution of combinatorial or parallel synthesis will be discussed first in this chapter; from its roots in the solid phase synthesis of discrete compounds to today's demands for high-quality, well-designed libraries synthesized using solution phase parallel chemistry. The advantages and disadvantages of SPOS will be examined and compared with solution phase methodologies. This is not meant as a comprehensive review or comparison as there are many literature references to such a discussion. It is merely aimed at being used as a lead in to more detailed discussions on solution phase synthetic procedures.

3.34.2 The Evolution of Combinatorial Chemistry

Solid phase chemistry was originally primarily used for the synthesis of peptides. It was only during the last 10 years that it has been adapted for the synthesis of small molecules as a tool to aid drug discovery. The main driver for this change in emphasis was the need to produce large pools of compounds to satisfy the demands of high throughput screening – simply put, traditional or classical chemistry could not make enough compounds in a year to meet the daily throughput of high throughput screening. This was achieved via the synthesis of combinations of compounds around a central scaffold: combinatorial chemistry. However, this could not be achieved using traditional chemical synthesis techniques due to downstream issues such as compound work-up or purification. So it was achieved by carrying out the chemistry on a solid support or resin and cleaving the final molecules from the support using a reagent such as a strong acid. Unfortunately, this technology limits the range of chemistry that can be carried out. The chemistry has to be compatible with the solid support and the final cleavage step. This limits the diversity of the compounds synthesized and thus tested against the biological target.

Combinatorial chemistry was seen by many as the panacea to solve this bottleneck in the lead generation process. With a blinkered focus on numbers to fuel the ever hungrier screening operations, the combinatorial chemist was able to produce a staggering amount of compounds in a short space of time. The synthetic procedures tended to be focused on a key scaffold, sometimes a privileged structure, that was modified by the addition of monomers in a multi-parallel fashion in order to generate a library of compounds. The result was a large number of compounds with limited diversity. Today's combinatorial or parallel synthesis chemist will do fewer syntheses on a higher number of scaffolds in order to enhance diversity.

3.34.2.1 From a Few Reaction Types to Many

Arguably, the early days of combinatorial chemistry were focused more on numbers of compounds than the actual quality of the compounds. The target of the early combinatorial chemist was to build as large a corporate collection as possible. Whether the compounds were 50% pure or existed as mixtures of compounds was irrelevant. To achieve this, skills used in solid phase combinatorial peptide synthesis were applied to the synthesis of small molecules, with some success. The primary reaction in solid phase peptide synthesis was amide bond formation, and this was exploited to the full for the synthesis of small molecules, with the result that the majority of compounds had an amide bond as a key step in the synthetic scheme, which magnified the occurrence of the amide bond in products formed from combinatorial chemistry. That said, amides are not all bad news – 28% of all compounds in the Comprehensive Medicinal Chemistry database are carboxamides.[1] In addition, another well-used reaction in solid phase synthesis is reductive amination, and 30% of compounds in the Comprehensive Medicinal Chemistry database are tertiary amines.[1] However, through the demand for more diverse compounds, more diverse reactions were reported that could be applied to solid phase synthesis. This being said, the solid support was still the key issue. Reaction conditions and cleavage reagents had to be compatible with the support, thus precluding some reactions that could be carried out in solution on particular scaffolds.

3.34.2.2 The Synthesis of Discrete Compounds

In recent years, the pharmaceutical industry's focus has switched from the desire to access sheer numbers of compounds synthesized using combinatorial chemistry to an increase in the quality of the compounds. This shift in focus has been driven by the lack of positive results from combinatorial chemistry when considering the low numbers of leads and development candidates (not even taking into account drugs) resulting from the screening of combinatorial libraries. It is fair to say that many people hoped that combinatorial chemistry would fuel the pipelines of the drug discovery industry. Unfortunately, combinatorial chemistry failed to deliver against this, although it did give us a valuable tool that could be utilized better in the search for new drugs. The reasons for the early failure of combinatorial chemistry can probably be attributed to the low purity of the compounds synthesized together with poor compound design, remembering that in many case it was the solid support that dictated the types of compounds synthesized and not the intuitive design of the medicinal chemist.

In order to keep pace with the shift in focus, combinatorial chemistry has needed to evolve from the early days of chemistry-driven libraries of low to moderate purity to today's high-purity, high-quality 'drug-like' or 'lead-like' libraries. The focus has very much changed from one of chemistry being the driving force to that of the quality and design of the compounds being the primary focus. In the early days the nature of the compounds synthesized was very much a function of the available technology. Today the reverse is becoming true, as it is the nature of the compounds to be synthesized that chooses the particular technology to be used, be it solid phase or solution phase chemistry for example. Just because a compound can be synthesized is no longer the driving force for synthesizing it.

3.34.2.3 The Continued Need for Large Libraries

There is still a need within the pharmaceutical industry for large discovery libraries, for example for 'file enrichment' to populate a screening collection. The screening of large numbers of compounds can quickly lead to early structure–activity relationships (SARs), and may provide a practical starting point for a drug discovery program where little to no information is known about the target. This is especially relevant with the likelihood of a plethora of new targets evolving from genomics activities. Unlike the libraries prepared in the early days of combinatorial chemistry, the libraries that can be prepared now tend not to be so chemistry-driven. In the early libraries, only those monomers/groups of monomers where successful results could be gained quickly were prepared as part of the final library. This resulted in a bias toward compounds that had undesirable properties, few functional groups, a lack of structural diversity, and, in many, cases a low purity. Today, the design of the libraries is much better defined in terms of the drug-like or lead-like properties of the compounds, and has required that the combinatorial chemist be innovative in the chemistry that can be carried out to meet the diverse breakdown of the library to be synthesized. This is now possible due to the significant advances in the chemistry that can be carried out in parallel, in solution and/or solid phase, coupled with improvements in parallel purification. This increase in efficiency has been achieved despite significant increases in expectations with respect to compound purities.

3.34.2.4 Evolution and Terminology

Today, combinatorial chemistry is often considered an unfashionable phrase due to the perceived failure of the early technology. Other terms have evolved to better describe the overall concept of the synthesis of multiple compounds by the most efficient means – multiple parallel synthesis; array synthesis; diversity-oriented synthesis and target-oriented synthesis; high-speed analog chemistry; high-throughput organic synthesis; just-in-time libraries; and the rise in popularity in multi-component reactions, to name but a few.

3.34.3 Comparison of Solid Phase versus Solution Phase Synthesis

Solid phase synthesis is the synthesis of molecules attached to a solid support. The attachment to the solid support is a covalent one, and must be cleaved by a chemical reaction. Solid phase synthesis can today produce libraries in many formats using many different reaction types to prepare the final compounds. But in many cases the reaction required to cleave the compounds from the solid support or the incompatibility of the solid support to certain chemistries still results in many limitations with respect to the types of compounds that may be synthesized using solid phase synthesis. On the other hand, the essentially unlimited flexibility of solution phase synthesis methodologies has resulted in a re-emergence of this classical approach and its application to library synthesis. Solution phase synthesis is the synthesis of molecules primarily in solution (i.e., neither molecule is attached to a solid support during the reaction step). Solid

Table 1 Advantages (✔) and disadvantages (×) of solution phase and solid phase approaches to library synthesis

Solution Phase		*Solid Phase*	
Classical organic synthesis directly transferable and method flexibility	✔	'Classical' chemistry often has to be re-validated for SPOS	×
Amenable to scale up	✔	Less amenable to scale up (route may require redesigning)	×
No additional steps[a]	✔	Need a loading/cleavage strategy	×
No unnecessary functionality	✔	Need a point of attachment to resin	×
Wide choice of solvents available for reaction steps	✔	Limited solvents available due to need for resin swelling/compatibility with support	×
Wide choice of concentrations suitable for reaction steps	✔	Need concentrated solutions for interactions of reagents with resin	×
Wide choice of catalysts suitable for reaction steps (homo- and heterogeneous)	✔	Cannot use heterogeneous catalysts	×
Extensive literature	✔	Limited (but growing) literature	×
Extensive range of noncompeting reaction methodologies	✔	Need to ensure that reagents do not interfere with the solid support (e.g., cannot use an acidic step on an acid-labile resin)	×
Wide range of temperatures can be used	✔	Need to avoid very low and very high temperatures as these may be detrimental to the support	×
Reaction monitoring	✔	Need to cleave to monitor the reaction fully	×
Isolation and purification of intermediates	✔	Only final compounds obtained	×
Isolation and purification of intermediates	×	Only need to isolate and purify final compounds	✔
Parallel purification	×	Ease of purification/reaction work-up	✔
Stoichiometric reagents may not be sufficient	×	Excess reagents to drive reaction to completion	✔
Automation challenging	×	Automation friendly	✔

[a] Unless a protection/deprotection strategy is required.

supports may still be used within the reactions to purify a particular reaction (scavenger resin) or to support a reaction reagent (solid-supported reagents), but, importantly, in solution phase synthesis the reactants are not attached to any supports. There are a few exceptions to this; for example, a reaction product may be attached to a solid support after the reaction is complete so as to be able to remove the by-products and other reagents by simple washes – this technique is termed catch and release, and will not be covered in this chapter. Early restrictions with respect to reaction format, purification of products, reaction work-up, etc., have received much attention and are no longer drawbacks for carrying out reactions in parallel in solution phase.

The advantages and drawbacks of each methodology have been well reported, and are summarized in **Table 1**, although it should be noted that due to extensive research over the years, some of the perceived disadvantages of each methodology have been resolved.

3.34.3.1 The Applicability of Solid Phase Parallel Synthesis

Some of the early justifications for using solid phase synthesis for the synthesis of small molecules (e.g., the use of excess reagents to drive reactions to completion) are still valid, and often assist in the synthesis of compounds that would prove difficult in solution phase. Normally, for solution phase synthesis, stoichiometric amounts of reagents are required. Excess reagents or unreacted reagents will need to be scavenged away from the reaction media as opposed to the simpler approach or washing away, as in solid phase synthesis.

An argument often postulated for solid phase synthesis is the ease of purification of final compounds – the argument goes that by careful selection of cleavage conditions only the compounds desired are cleaved from the resin, leading to highly pure compounds obtained simply by filtering off the solution, washing the resin and evaporation. Unfortunately the often preferred method of cleavage from resin is the use of the strong acid such as trifluoroacetic acid, which can often lead to other side reactions being performed, resulting in impurities present in solution along with the desired compound. In addition to this, intermediate reactions in the sequence may not have proceeded to completion, and these 'incomplete' compounds will probably be cleaved into solution along with the desired compounds. Another issue with using strong acid to cleave a molecule from a solid support is the potential for the strong acid to also cleave off certain parts of the solid support, which may sometimes be undetected by standard analytical techniques. Other cleavage methods are possible, and do lead to purer compounds (e.g., alkylative cleavage from regeneratable Michael (REM) resin typically yields very pure compounds, but often at the expense of yield). Safety-catch resins also typically give a more favorable reaction profile, but again at the expense of yield, as sub-stoichiometric quantities of the activating agent are used to effect cleavage. The use of this activating agent, often a reactive alkyl halide, may limit the reactions that can be performed on the compound itself, and therefore a safety-catch approach may not be feasible.

There are also arguments for solid phase chemistry that can override a more traditional solution phase approach. For instance, the intermediates in the reaction sequence may be unstable and not possible to isolate, but could be stabilized by attaching to a solid support. In a similar vein, the need for a protection/deprotection strategy can be circumvented with a loading/cleavage step, which may also ease handling. This strategy is often used in a 'mixed-phase' approach whereby a number of steps are performed in solution phase followed by loading a late intermediate onto resin, possibly then undergoing a transformation on solid phase, followed by cleavage of the desired compounds. This 'mixed-phase' approach can be very successful, and uses advantages of both solid and solution phase synthesis.

3.34.3.2 The Applicability of Solution Phase Parallel Synthesis

The arguments for solution phase parallel synthesis are also strong. There has been a shift in recent years away from the synthesis of large multi-thousand-member libraries to the efficient preparation of smaller, more focused, or targeted libraries. One has to question as to whether the often lengthy validation time required to ensure that the synthetic scheme is feasible on solid phase is justified in today's climate of smaller more focused libraries. Dolle[2] reports that 79% of the biologically active libraries published in 2003 contained less than 500 members. It should also be noted that this is not an entirely new concept: Lahana and others alluded to this when they commented that if drug discovery is like looking for a needle in a haystack, why increase the size of the haystack[3,4]?

Menon and co-workers[5] recently presented a solution phase parallel synthesis approach to rapidly synthesize approximately 300 analogs of the Ras/Raf protein interaction inhibitor **1** (**Figure 1**).

This study brings together many of the aspects noted above as advantages for solution phase parallel chemistry: small focused library with a short efficient solution phase route envisaged, and with significant diversity; isolation of intermediates; no obvious point of attachment to resin for solid phase chemistry; use of solid-supported reagents to assist in later downstream work-up and purification; use of scavenger resins to assist in the initial purification; and the

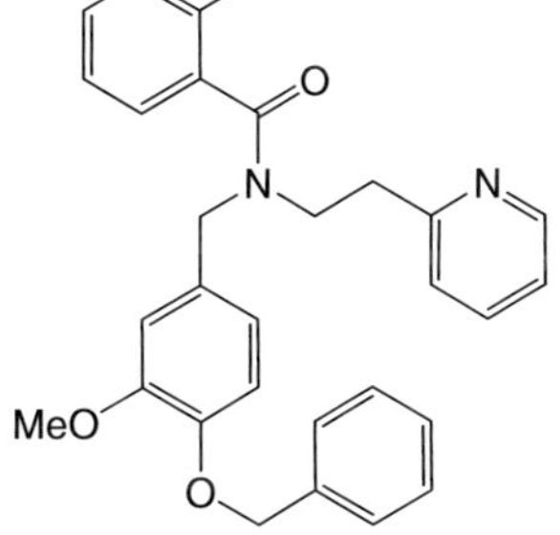

1

Figure 1 Ras/Raf protein interaction inhibitor.

use of ion exchange purification techniques. This is a prime example whereby the types of molecules required to be synthesized very much dictate the synthetic approach; in this case, solution phase synthesis.

Menon envisaged potential structural modifications of **1**, as noted in **Figure 2**. Each compound was selected to answer certain structure–activity questions, and, as such, the route used to synthesize the library had to be able to yield the maximum number of members for the library.

A rapid synthesis of analogs was envisaged by a short two-step solution phase synthetic protocol involving an initial reductive amination, aided by the use of polymer support borohydride as the reducing agent, with subsequent purification using a tandem scavenging protocol. The resulting secondary amines were 'capped' in the second step with a targeted selection of acid chlorides, sulfonyl chlorides, and isocyanates using polymer-supported morpholine as the base and a tandem scavenging approach to assist in the purification of the final compounds, as shown in **Scheme 1**.

Minor quantities of the unwanted tertiary amine formed by the bis-reductive alkylation of the aldehyde on the primary amine in step i of **Scheme 1** were suppressed by using a slight excess of the secondary amine, which was removed by the use of scavenger resins.[6–9] To remove all unwanted reagents and reactants, a dual-scavenging strategy was employed – polymer-supported aldehyde resin was used to remove the excess secondary amine, and polymer-supported tris-(2-aminoethyl)amine was used to sequester any unreacted aldehyde. For step ii, a solid phase reagent/scavenging protocol was also employed: polymer-supported morpholine was used as the base; unreacted secondary amine was scavenged using resin-bound isocyanate; and excess acid chloride, sulfonyl chloride, and isocyanate was sequestered by the use of polymer-supported tris-(2-aminoethyl)amine. The products were purified with ion exchange chromatography[10,11] where necessary to obtain the desired products in >80% purity by combined liquid chromatography/mass spectrometry (LC/MS). The parallel solution phase syntheses were typically performed on a 100–300 mg scale of starting reagents to allow for archiving of the desired compounds. Analogs of **1** synthesized in this study were initially screened at 20 μM final concentration in an Elk1-luciferase reporter assay in HeLa cells, and compounds exhibiting ≥50% inhibition were further profiled in this assay to obtain an IC_{50} measurement. Analogs exhibiting interesting levels of inhibition ($IC_{50} < 20\,\mu M$) in the Elk1-luciferase reporter assay were subsequently evaluated for their antiproliferative activity (WST-1 assay) as well as their ability to inhibit anchorage independent

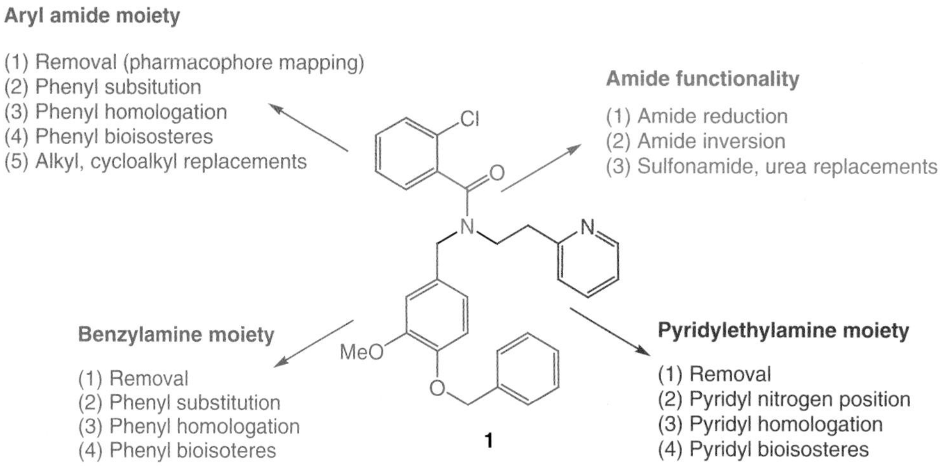

Figure 2 Potential structural modifications of the Ras/Raf inhibitor **1**.

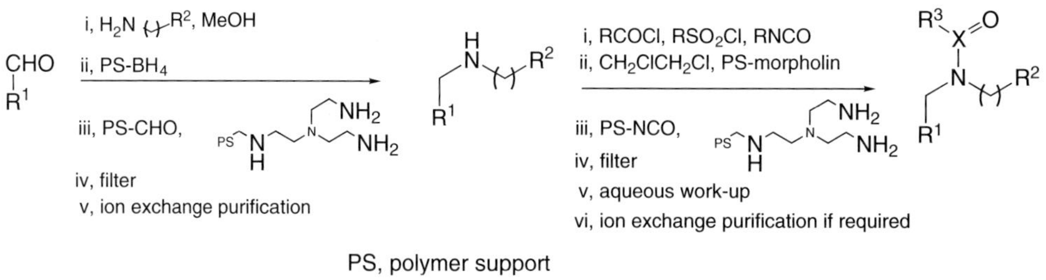

Scheme 1

colony formation (soft agar growth) in HCT-116 cells. Data for selected analogs are presented by Menon that identify the key structural features responsible for the activity (or lack thereof) in the assays. Overall, a modest improvement in potency was observed in the analog series, and a preliminary SAR was elucidated. Evaluation of the hypothesis is underway through the synthesis of additional target compounds; in particular, conformationally restricted molecules. The results of this study are to be published in due course.

Solution phase parallel chemistry does not, in theory, suffer from the long validation times needed to validate a synthetic sequence on solid phase, as the chemistry to be used has probably been well exemplified in the literature and just needs to be applied to the particular set of monomers required. Therefore, when considering a more focused/targeted library of around 500 compounds, it is often advisable to consider a solution phase approach unless there are overriding reasons, as noted above, why a solid phase strategy may be more advantageous. In addition, should the particular compounds/library/template/proposed synthetic sequence be novel to the literature, it will in most cases be more straightforward to design a route for solution phase chemistry than it will be to design one for solid phase, especially if the desired compounds do not lend themselves readily to an obvious attachment point to resin and do not contain suitable functionality to do so.

There is a need in the pharmaceutical industry for the synthesis of large diversity oriented or target focused libraries. This may be for 'file enrichment' either to enhance and supplement the corporate compound collection for future high-throughput screening campaigns, or to replenish the corporate compound collection following a review and subsequent cull of legacy compounds that no longer meet the currently accepted requirements with respect to purity and 'drug-likeness'. The use of solution phase chemistry, once considered impossible by which to prepare a multi-thousand-component library of discrete, pure library compounds, can now effectively be used to achieve this.

When designing a synthetic approach to the synthesis of a library of compounds one should remember that there are some limitations to the use of solution phase parallel chemistry. As noted previously, in cases of slow or low-yielding reactions one cannot just use multiple equivalents of particular reagents to drive the reaction to completion without some thought to the work-up and purification of the target compounds. Unlike solid phase chemistry, one cannot simply wash away the excess equivalents of the reagents. This was always the limiting factor when considering the use of a solution phase approach; however, multiple techniques have now emerged to cope with this issue, thus allowing a more widespread use of solution phase parallel chemistry. For example, one can consider the use of solid-supported reagents. These are reagents that are covalently bound to a solid support, allowing the use of excess equivalents of such reagents in a particular reaction. On completion of the reaction the solid-supported reagents can be separated from the reaction mixture by filtration, leaving just the desired compound in solution, ready for entry into the next step in the synthetic sequence and/or purification. Today, there are many examples of standard reagents bound to a solid support that can be used in this way, such as amide coupling reagents, palladium catalysts, oxidizing agents, and reducing agents, among many others. Alternatively, scavenger resins may be used to remove excess equivalents of reagents used in the reaction; this will be discussed further.

In addition, compound libraries are often designed to rapidly expand the SAR of an active compound obtained from high-throughput screening, or of a compound designed and prepared in a medicinal chemistry laboratory (hit explosion libraries). Although for the purpose of the confirmation of the biological activity, the resynthesis of the active molecule (hit) will be expedited by following the original synthetic route, be that solution phase or solid phase, the synthesis of key small focused libraries required to expand the SAR of the original hit will probably be more facile via a solution phase route. The advantages of using a solution phase approach at this stage is the lack of any need to validate any solid phase approaches for problematic reagents, the applicability to more traditional medicinal chemistry approaches that will be used in lead optimization, and the probable experience of the chemists. Even if the synthesis of a large library is warranted, it may still be prudent to consider a solution phase approach.

As stated earlier, for solution phase parallel synthesis, alternatives to using excess reagents to drive reactions to completion are key to the success of the approach. In this regard, microwave-assisted organic synthesis is a fast-growing area of research with an extensive literature on the topic[12–18] (*see* 3.36 Microwave-Assisted Chemistry). Generally, short reaction times are observed, leading to the rapid validation of problematic chemical steps, with often high yields and purities obtained. By using solution phase parallel chemistry in combination with microwave technology, one can efficiently validate a synthetic step and/or route, followed by the rapid generation of a small focused library.

The use of solution phase parallel synthesis allows for the consideration of purification of intermediates at defined stages, an option not possible with a solid phase approach. For example, the penultimate compounds in the synthetic sequence can be purified by traditional and/or automated means to high purity, say $>95\%$, which gives the chemist the best possible chance of obtaining the final compounds in high purity with a high return from the synthesized library. This has an added benefit in that the intermediates may be stored and used in the synthesis of key follow-up libraries, negating any need to repeat the whole synthetic route.

Once the synthetic sequence is complete and the desired final compounds are in hand, solid-supported material can assist in the clean-up and purification of these compounds. For example, scavenger resins are widely used to remove excess reagents from solution phase reactions, and catch-and-release resins are used as a clean-up step by removing all compounds from the reaction media other than the desired compounds, which can then be obtained in pure form. More details on these approaches will be discussed later.

3.34.3.3 Selecting the Appropriate Parallel Synthesis Technology

With the number of technologies available today to the medicinal or parallel synthesis chemist, there is no longer the need for the technology to dictate what compounds are synthesized. Once compounds are designed and selected for synthesis, the relevant synthetic technology may be selected and used. For example, the relative merits of a solution phase approach compared with a solid phase approach should each be assessed to determine which would be the preferred method for the synthesis of the compounds/library required. A number of factors should be considered initially, which may direct the synthesis to one particular methodology: the size of the library; the number of synthetic transformations in the route; the precedence and ease for each synthetic step; the number of points of diversity required for the final compounds; the number of individual members (or monomers) at each diversity point; the order in which each diversity point is introduced; additional reactive functionalities; value in obtaining intermediates; etc. Often a compromise will be reached. For example, introducing a high degree of diversity (determined by the number of discrete monomers) toward the beginning of a solution phase synthetic route will result in a significant number of intermediates that will need to be isolated and purified prior to their use in subsequent steps. In this case, a preferred approach would be for the chemist to explore an alternative breakdown of the virtual library by either reducing the diversity, and, hence, the number of monomers at the early step and increasing the diversity at a later step or considering an alternative synthetic strategy, including the reordering of the synthetic steps or even the exploration of a solid phase route. In contrast, the often lengthy validation time required for a solid phase route may not be justified for a small array of compounds, and a solution phase strategy would be required.

So there are valid reasons why one approach may be favored over another on a case-by-case basis, and the processes are often similar; however, the following sections will discuss these processes with particular regard to the solution phase parallel synthesis approach and with emphasis on the rapid generation of smaller focused libraries.

3.34.4 Solution Phase Library Synthesis

The synthesis of a compound library using solution phase parallel chemistry can be broken down into several key and discrete stages, namely library design, route validation, monomer screening, synthesis of key intermediates, preproduction, and production of library compounds in a single or multiple batches. Each of these will be discussed in turn in this section. As integral parts of the library design process the concepts of drug-likeness and lead-likeness will be discussed as well as other considerations, for example the application of in silico absorption, distribution, metabolism, excretion, and toxicity (ADMET). For route validation, the concept of design of experiments for assistance in selection of the most appropriate and efficient reaction conditions will be reviewed. Monomer screening is an essential part of the overall library production process in order to maximize the diversity in the library, and to ensure that all possible and preferred monomers can be included in the synthesis of the library by the selection of appropriate conditions for each class of monomers if necessary. A number of lessons can be learnt from preproduction (sometimes referred to as 'rehearsal'), which is the first time that the library compounds are prepared to the exact production protocols (quantities, time frames, etc.) in a parallel format. During preproduction, a subset of scaffolds and monomers are chosen to represent the library as a whole, with up to 20% of the library being prepared in this phase. The lessons learnt can then be applied to the production of the remainder of the library to ensure for the effective and efficient preparation of the target molecules. For all stages the (quantity and quality standards) will be discussed where relevant, as will the use of enabling automation technologies. Other essential parts of the process – reaction monitoring and compound purification – will be discussed in Section 3.34.5.

3.34.4.1 Library Design

From the early days of 'make what can be made' using available technology to synthesize large numbers of compounds with arguably little regard for quality or design, the pharmaceutical industry's focus today is almost entirely on quality: quality with respect to the design of the library compounds by careful selection of the library members with a consideration of computational tools together with quality with respect to the specification (purity and quantity) of the library compounds. Using the 'make what can be made' philosophy, very large libraries were synthesized containing many times the number of compounds required, and only those compounds that were produced in sufficient quantity and with an appropriate purity were progressed. This often meant that the compounds successfully prepared were skewed toward a specific set of monomers, meaning that diversity was often low and that the compounds all possessed similar physical properties.

As previously noted, there is still a need in the pharmaceutical industry for large, diverse discovery libraries, for example for 'file enrichment' to populate a screening collection. The advent of genomics has resulted in many new targets being identified, for which a significant number of well-designed compounds are needed, both from a diversity point of view and from a targeted viewpoint. The screening of large numbers of compounds can quickly lead to some useful SAR information, and may provide a practical starting point for a drug discovery program where little to no information is known about the target. However, unlike the libraries prepared in the early days of combinatorial chemistry, significantly more initial library design is performed, to make sure that all compounds meet appropriate drug-like[19–22] or lead-like[23–27] criteria and that the library's synthetic route is feasible for all monomers. This ensures that the final library is not skewed toward only those monomers that perform better under certain conditions but instead is more in line with monomers selected by the medicinal chemist; if necessary, library production can be broken down into several batches to allow for different conditions to be used for diverse sets of monomers. In this way, the desired compounds dictate the chemistry and not vice versa, so that the library can be tailored to ensure that the compounds have appropriate functional groups, good structural diversity, and high purity, and do not have undesirable physical properties.

There are many types of libraries that can be accessed using combinatorial chemistry. Libraries can be designed to identify new compounds that interact with biological targets, as demonstrated by various screening technologies (hit compounds), or to expand on biologically active molecules using key structure–activity information.

In addition to large multi-thousand-member diverse, or random, libraries, there are more focused or targeted libraries. These libraries are designed to specifically interact with a certain biological target or target class such as kinases or G protein-coupled receptors. Libraries may be designed based on protein structure information gathered using protein crystallography (structure-based design) or by using the structures of known ligands that interact with a specific target or class of targets (ligand-based design) including the use of neural networks to design compounds biased toward certain biological targets. In many cases, where known ligands are used to design libraries, a privileged structure may be extracted and used to design libraries with a higher probability of interacting with a biological target (focused or targeted libraries) Examples have been noted in the literature[28] whereby diverse libraries based around a similar pharmacologically relevant core have been prepared by different investigators and result in diverse biological activity, indicating that the exact nature of the monomers can fine tune the biological activity that may be inherent in the core. By taking this approach, not only is it possible to obtain good biological activity from a chemically tractable starting point, but, potentially, a favorable patent position can also be achieved. One could term these types of libraries 'classical combinatorial libraries'.

In addition to these random and targeted libraries, the combinatorial or medicinal chemist can use combinatorial chemistry (using either a combinatorial or noncombinatorial approach) to prepare smaller libraries of tens to hundreds of compounds based around screening hits in order to enrich any available SAR. The focus of these arrays of compounds is to rapidly explore a defined amount of diversity around the screening hit or lead compound. Parallel synthesis enables the medicinal chemist to accomplish this in a very short space of time. One could term these types of libraries 'nonclassical combinatorial libraries' or 'just-in-time libraries,' where the focus is very much on the speed of synthesis. They should be seen as a crucial resource for the immediate expansion and exploration of SARs around the hit molecules, providing a boost to the initial hit-to-lead activities of a drug discovery program.

The design of the library, in terms of the number of compounds that could potentially be made by the synthetic sequence, is iteratively modified down to the number of compounds it is feasible to prepare based on the need for which the library is being produced and the use of computational tools. A consideration should also be given to the method by which the compounds are to be prepared – solid phase or solution phase, the number of synthetic transformations, the number of points of diversity, where diversity should be random, and where diversity needs to be more focused. Only by considering these factors initially can a sensible synthetic strategy be employed. For example, if maximal diversity is required in the last step in a synthetic sequence, it would be preferable to perform solution phase chemistry by making a key set of intermediate compounds and by purifying these to high purity before commencement of the final step, which could be automated to further increase the efficiency of the overall process. However, if maximal diversity is required in the first step, this would make solution phase synthesis untenable, as there would be too many intermediates to purify in order to ensure a good return from the library unless the order of the synthetic steps in the synthetic sequence could be reversed. Thus, in instances like this, one must either compromise on the diversity of the first step (i.e., limit the number of different monomers to be used) or, preferably, redesign the synthetic sequence so that this step giving maximal diversity is introduced as a later step in the sequence. This philosophy can also be used in reverse, whereby a synthetic step that may be difficult to perform in a multiple parallel sense could be redesigned such that the step could be performed early in the synthetic scheme in fewer numbers, and from which the intermediate products prepared could be readily purified prior to the next step.

Using computational chemistry tools, libraries can now be filtered based on agreed criteria prior to synthesis, not just for diversity or similarity as appropriate but also to remove undesirable functionalities either from a synthetic point of view or, more increasingly, from a consideration of ADMET properties. As stated earlier, diverse discovery libraries are primarily aimed at identifying hit molecules; however, optimization libraries such as the 'just-in-time libraries' are driven by the rapid need for good SAR information. A library designed for 'hit optimization' may target more than one parameter potency and/or selectivity, whereas a library designed for 'lead optimization' may target many more parameters such as potency, selectivity, aqueous solubility, and increased oral bioavailability; this is termed multi-parameter optimization. To this end, technologies such as in silico ADMET tools have become invaluable in supporting the design of compound libraries.

Parallel synthesis has needed to evolve to meet today's needs for high-purity libraries, which are often produced in significant quantities, as these 'quality' well-designed compounds will be around for many years and be part of many screening campaigns.

3.34.4.2 Route Validation

As previously discussed, if the library is small, say 100–500 members, which is becoming more common today, it would not make sense to use a solid phase strategy unless there is an overriding reason why this would be preferred. With this in mind, a careful consideration of the sequence of steps and in particular the diversity inducing steps should be at the forefront when designing the overall library and the synthetic sequence.

Each step in the synthetic scheme should be assessed using a range of scaffolds and monomers that best represent the diversity of the overall library to be prepared. For instance, a set of amine monomers may include primary (e.g., butylamine or benzylamine) and secondary (e.g., dimethylamine or *N*-methylbenzylamine) amines; straight chain (e.g., ethylamine or pentylamine) and branched (e.g., *s*-butylamine or isobutylamine) amines; primary (e.g., cyclohexylamine or cyclopropylamine) and secondary (e.g., *N*-ethylcyclopentylamine or *N*-benzylpyrrolidine) exocyclic amines; secondary amines integral to a ring system (e.g., piperidine or morpholine); and anilines (e.g., aniline or 3-aminopyridine). It is important that suitable examples of all such diverse amines are included in the route validation to help ensure a set of reaction conditions can be identified that is capable of providing products from the diversity of monomers. Sometimes this will require the use of more than a single set of reaction conditions. In parallel to the investigation of the reaction conditions, the purification and analytical strategy will be investigated. Thus, it is important that during this phase a diverse set of final compounds is synthesized in order to investigate the necessary range of analytical and purification methods required for the final library. When one considers the necessary purification of the final compounds, certain considerations need to be made. For example in a reductive amination step using a carbonyl with a diverse set of amine monomers, products may be formed that respond differently to various techniques. The reaction of the carbonyl with a primary amine leads to a secondary amine that may be difficult to separate from any excess primary amine used without careful selection of an appropriate scavenger resin; whereas a carbonyl that is reacted with a secondary amine leads to a tertiary amine. It is now a relatively facile process to scavenge away any excess secondary amine from the reaction media, leaving only pure tertiary amine in solution (*see* Section 3.34.5.3.1).

A result that is sometimes obtained from the route validation step is that different sets of monomers require slightly different reaction conditions to proceed; for example, when considering reactions where one of the reagents is an amine, primary and secondary amines may proceed at room temperature whereas the same reaction with anilines may require more forcing conditions to proceed, such as elevated temperatures or microwave chemistry. Likewise, aldehydes may proceed more readily than ketones in a reductive amination step. Using different sets of reaction conditions in parallel can be tolerated in a solution phase library strategy, but is less likely to be allowed using a solid phase approach. This is partly due to the intolerance of the solid support for different conditions and the fact that the exact monitoring of each different reaction is difficult and time-consuming using solid phase synthesis. Therefore, by preparing the library in solid phase, the bias of the monomers may be toward only those that work well under standard and uniform conditions, and the overall diversity of the library may be compromised; this can be circumvented by using different conditions for different sets of monomers in solution phase, which can often be done in parallel with no loss in efficiency.

Each library prepared may also require subtly different analytical techniques to be used for the effective characterization of the final compounds. The diverse compounds generated in the route validation phase can be used to assess the most appropriate method for the final analysis in advance of the bulk of the library being available for analysis.

An important consideration to be given to the route validation phase is the final specification of the compounds to be produced with respect to purity and quantity. Intermediates and final compounds are fully isolated, purified, and characterized in order to assess the quantities required for the library synthesis.

Solid-supported reagents are being increasingly evaluated at this stage. Some key references from Ley and co-workers[29–31] outline the strategies employed, in one case using resin-bound reagents in all stages of a five-step synthetic sequence.[32]

In addition, suitable methods for the purification of the library compounds may be validated. For example, the use of resin scavenging methodology to remove excess reactants may yield compounds of appropriate purity and no further purification may be required. In a similar way, the use of solid phase extraction (SPE) methods can remove unwanted compounds from the reaction mix, yielding suitably pure compounds. As a final measure, the use of medium-high-throughput preparative liquid chromatography can be used to deliver high-quality compounds. **Table 2** highlights some scavenger resins routinely used for the purification of solution phase parallel chemistry reactions.

Specific methods for the work-up, clean-up, analysis, and purification of library compounds will be discussed in a later section.

3.34.4.3 Synthesis and Purification of Intermediates

By using solution phase methodology, not only can the library be designed to be drug-like or lead-like taking advantage of almost every chemical reaction known, but by purifying at intermediate stages, high purities of the final compounds can be achieved. Purifying at each intermediate stage helps to ensure that the next step can proceed without any unwanted side reactions due to the presence of reaction impurities from unwanted side reactions and/or incomplete reactions. Purifying the penultimate compounds prior to the final step helps to facilitate the best possible return from the production library. In addition, quantities of the intermediates can be stored to ensure that they are available if required for hit follow-up activities.

One can envisage the synthesis of novel intermediates in bulk quantities that can be exploited via solution phase parallel synthesis. D'Alessio *et al.*[33] presented a new class of potent cyclin-dependent kinase 2 (CDK-2) inhibitors via a solution phase synthetic protocol in which a key building block was synthesized and isolated prior to rapid diversity expansion, as shown in **Scheme 2**. The key intermediate (**2**) was prepared in four steps in an overall 77% yield with a single isomer of the protected pyrazole formed. This intermediate was then rapidly diversified, as shown in **Scheme 3**.

D'Alessio prepared a small diverse library of 26 compounds, followed by a small 17-member array with one position fixed, with representative examples from all the structural motifs noted. This diversification would have been extremely difficult to achieve in a solid phase approach.

3.34.4.4 Preproduction: Synthesis of a Rehearsal Library

The next step in the library production process is to prepare a subset, typically up to 20%, of the library on a production scale using the exact conditions and protocols that will be used for the synthesis of the main library. Problems can occur in the parallelization of the synthesis for a variety of reasons, and it is important that these are identified and engineered out of the process before committing the entire library to the final reaction conditions. For example, during the parallel synthesis of the library the reactions are often performed on a reduced scale to that carried out in validation and monomer screening because during the earlier stage it is important to ensure that sufficient quantities of each compound are obtained to enable full compound characterization. Issues with solvent quantities, reagent solubilities and numbers of equivalents, and solution concentrations are magnified on this smaller scale. Timings are often different due to the multitude of reactions that are being performed in parallel. Successful preproduction has only been achieved if the hit rate (i.e., the number of compounds produced to the required purity and quantity), when extrapolated to the whole library, is sufficient to meet the library goals.

3.34.4.5 Library Production

Following successful completion of preproduction, the entire library can now be synthesized. Depending on the size of the final library, this may be carried out in multiple batches. Lessons learned in each production batch can be applied to each subsequent batch to ensure that the hit rate is maintained or even improved upon.

Table 2 A guide for the selection of scavenger and ion exchange resins

NUCLEOPHILES	Recommended Solid-Supported Scavenger
Acidic phenols	Trisamine (see electrophile scavenger table), Carbonate (see ion exchange table)
Alcohols	Benzoyl chloride, Boronic acid (for diols), Tosyl chloride
Alkoxides	Isocyanate, Tosyl chloride
Amines – All	Anhydride, 2-Chlorotrityl chloride, Isocyanate, Isothiocyanate, Tosic acid (see ion exchange table), Tosyl chloride
Amines - Primary	Activated ketone, Benzaldehyde, 4-Phenethyloxybenzaldehyde
Amines - Primary and Secondary	Methylisatoic anhydride
Amines – Secondary	Benzoyl chloride
Amines – Anilines	Isocyanate, Tosic acid (see ion exchange table), Tosyl chloride
Aminothiols	4-Benzyloxybenzaldehyde
Azides	Triphenylphosphine (see electrophile scavenger table)
Enolates	Benzaldehyde
Hydrazines	Anhydride, Benzaldehyde, 4-Benzyloxybenzaldehyde, Isocyanate, Isothiocyanate, 4-Phenethyloxybenzaldehyde, Tosyl chloride
Hydroxamides	Tosyl chloride
Hydroxylamines	4-Benzyloxybenzaldehyde, 4-Phenethyloxybenzaldehyde
Organometallics	Benzaldehyde, Isocyanate, Tosyl chloride
Thiol/Thiolates	2-Chlorotrityl chloride, Isocyanate, Maleimide, 4-Maleamidobutyramidomethyl, Thiophenol (see electrophile scavenger table)
Triphenylphosphine/Triphenylphosphine oxide	Phenylmethyl chloride

ELECTROPHILES	Recommended Solid-Supported Scavenger
Acids - H^+	Dimethylamine, Morpholine, Piperidine, Pyridine
Carboxylic acids, Acid chlorides, Sulfonyl chlorides	AM, Carbonate/Water (see ion exchange table), DETA, DMAP, Dimethylsilyl, EDA, Sulfonyl amide, Trisamine,
Acid anhydrides	AM, Carbonate/Water (see ion exchange table), DETA, EDA, Sulfonyl amide, Trisamine
Alkylating agents	AM, N-(2-mercaptoethyl)amino methyl, Thiol, Thiophenol, Triphenylphosphine
Boronic acids	Cyclohexyldiol, DEAM, Diol
Carbonyls – Aldehydes	AM, Cyclohexyldiol, Diol, DETA, EDA, Sulfonyl amide, Sulfonyl hydrazine, Thiol
Carbonyls – Ketones	Cyclohexyldiol, Diol, Sulfonyl hydrazine, Thiol, Trisamine
Chloroformates	AM, Sulfonyl amide
Epoxide	Thiophenol
Imines	Trisamine
Isocyanates	AM, DETA, EDA, Sulfonyl amide, Trisamine
Isothiocyanates	AM, DETA, EDA, Trisamine
Metals - Sn (IV), Ti (IV)	DEAM
Metals – Palladium	DEAM, N-(2-mercaptoethyl) amino methyl, Thiol, Trimercaptotriazine, Triphenylphosphine
Metals – Mercury	Thiol, Duolite

INTERNAL REFERENCES

	Lab Book	Scavenged
1	OD1045-096	Primary amines
2	OD1069-170	Primary amines
3	OD1246-001	Triflates, tresylates
4	OD1309-001	Primary over secondary amines
5	OD1101-012	Primary amines
6	OD1227-040	Primary over secondary amines
7	OD1218-048	Diols
8	OD1194-048	Diols
9	OD1042-156	Amines
10	OD1135-087	Primary and secondary amines
11	OD1045-096	Secondary amines
12	OD1052-034	Amines
13	OD795-118	Amines
14	OD1227-098	Secondary amines
15	OD1218-048	Secondary over tertiary amines
16	OD1235-106	Secondary amines
17	OD1246-001	Secondary amines
18	OD1309-001	Secondary over tertiary amines
19	OD1643-050	Alcohol
20	OD1433-014	Catch and release of amines
21	OD1643-050	Secondary amine from ketone
22	OD1045-096	Catch and release of amines
23	OD1040-134	Catch and release of amides
24	OD1041-018	Catch and release of amines
25	OD1042-062	Purification of amines
26	OD1121-006	Catch and release of amines
27	OD1193-006	Purification of amides
28	OD1068144	Amines

	Lab Book	Scavenged
29	PSH	Fluoride/TBAF
30	OD1433-001	Catch and release of amines
31	PSH	Acid chlorides
32	OD1169-026	Purification of carboxylic acids
33	PSH	Carboxylic acid from ester
34	PSH	Acid chlorides
35	OD1194-084	MeSO3H
36	OD1507-108	HOBt
37	OD1218-051	MeSO3H
38	PSH	Acid chlorides
39	OD943-056	Purification of carboxylic acids
40	OD1246-001	Triflates, tresylates
41	OD1027-076	Acid chlorides
42	OD1101-012	Isocyanates
43	OD352-182	Isocyanates
44	OD599-126	Sulfonyl chlorides, isocyanates
45	OD1227-040	Aldehyde
46	PSH	Isocyanates
47	PSH	Sulfonyl chlorides
48	PSH	Alkylating agents
49	OD1426-056	Acid chlorides
50	OD1426-110	Acid chlorides
51	OD1643-050	Aldehyde
52	OD1194-020	HCl
53	OD1337-016	Ketones
54	OD1682-040/044	Mercury
55	OD1599-108	Azides
56	OD1052-034	Isothiocyanates

KEY

MP	Macroporous resin
PS	Polystyrene resin
FSG	Functionalised silica
PSH	Parallel Synthesis Handbook

Table 2 Continued

NUCLEOPHILE SCAVENGERS	Resin	Support	Loading (mmol g^{-1})	Supplier	EOAI Reference	Literature Reference	Scavenges
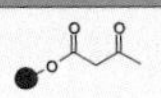	Activated ketone (Ketoester, AAEM)	PS PS PS	3 3 3	Sigma Aldrich Avecia Polymer Labs	1-4	L1	Primary amines
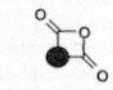	Anhydride	MP	5-7	Novabiochem	5	Cat	Amines, Hydrazines
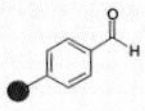	Benzaldehyde	MP MP PS PS PS	1.8 3 1-1.6 3 2.5-3	Polymer Labs Sigma-Aldrich Argonaut Polymer Labs Sigma-Aldrich		L2-3	Primary amines, Enolates, Hydrazines, Organometallics
	4-Benzyloxy-benzaldehyde	PS	2-3	Novabiochem	6	L4	Aminothiols , Hydrazines, Hydroxylamines
	4-Phenethyloxy-benzaldehyde	PS PS MP	2-2.5 1.8-3 0.7-1.2	Sigma-Aldrich Novabiochem Sigma-Aldrich		Cat	Primary amines, Hydrazines, Hydroxylamines
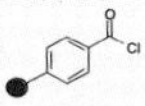	Benzoyl chloride	PS FSG	2.1 1	Sigma-Aldrich Silicycle		L2, L5	Secondary amines
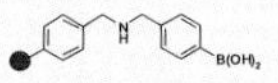	Boronic acid	PS	0.7	EOAI	7-8		Diols
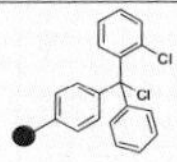	2-Chlorotrityl chloride	PS	1-1.6	Novabiochem		Cat	Amines, Thiols, Thiolates
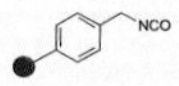	Isocyanate (NCO)	MP MP MP PS PS PS PS FSG	0.7-1.7 1.8 0.5-2 1-1.8 1.5 2 1.4-1.8 1.2	Argonaut Polymer Labs Sigma-Aldrich Argonaut Polymer Labs Sigma-Aldrich Novabiochem Silicycle	9-14	L2, L4, L6	Alkoxides, Amines, Anilines, Hydrazines, Organometallics, Thiols, Thiolates,
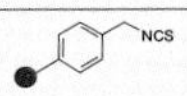	Isothiocyanate (NCS)	PS PS FSG	1 1.5 0.8	Sigma-Aldrich Novabiochem Silicycle	15	L7	Amines, Hydrazines
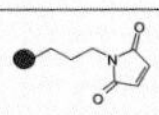	Maleimide	FSG	1	Silicycle		L8	Thiols, Thiolates
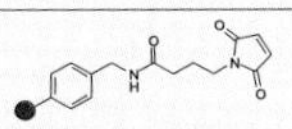	4-Maleamidobutyr-amidomethyl	PS	0.4	Sigma-Aldrich		Cat	Thiols, Thiolates
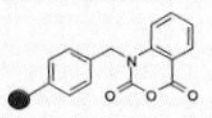	Methylisatoic anhydride (MIA)	PS	2.6	Polymer Labs	14, 16-18	L9	Primary and Secondary amines
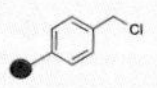	Phenylmethyl chloride (Merrifield, CMS)	PS PS PS FSG	1 5.5 1-1.6 1.2	Polymer Labs Sigma Aldrich Novabiochem Silicycle		L10	Triphenylphosphine/ Triphenylphosphine oxide
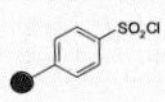	Tosyl chloride (Sulfonyl chloride, TsCl)	MP PS PS PS FSG	2.5-3 2-3 1.6-2.2 2.9 1	Sigma-Aldrich Argonaut Sigma-Aldrich Novabiochem Silicycle	19	L11-12	Alcohols, Alkoxides, Amines, Anilines, Hydrazines, Hydroxamides, Organometallics

ION EXCHANGE RESINS	Resin	Support	Loading (mmol g^{-1})	Supplier	EOAI Reference	Literature Reference	Function and use
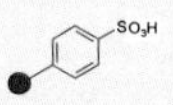	Tosic Acid (Sulfonic Acid, TsOH)	MP MP MP MP PS FSG	3 2-2.5 1.6 3 3.5 0.8	Argonaut Sigma-Aldrich Novabiochem Polymer Labs Polymer Labs Silicycle	20-21	L13-14	Strong acidic cation exchange Catch and release of amines Removal of secondary amine from ketone Removal of amines from amides Removal of fluoride/TBAF
	SCX	FSG	0.3	IST	22-27	Cat	
	Amberlyst A15	MP MP MP	 4.6	Lancaster Acros Fluka	28-29	L15	
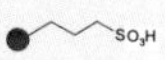	SCX-2	FSG	0.4	IST	23	Cat	Strong acidic cation exchange Purification of amides
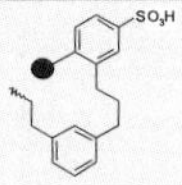	MCX	MP	1	Waters	30	Cat	Mixed mode cation exchange Catch and release of amines
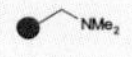	Amberlyst A21	MP MP MP	 4.8	Lancaster Acros Fluka	31	Cat	Weak basic anion exchange Removal of acid chlorides
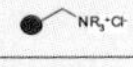	SAX	FSG	0.6	IST		Cat	Strong basic anion exchange
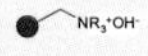	Ambersep 900	MP MP MP	 1.2 	Lancaster Acros Fluka	32-34	Cat	Strong basic anion exchange Removal of carboxylic acids and acid chlorides
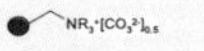	Ambersep 900-CO3	MP		EOAI	35		Strong basic anion exchange Removal of MeSO3H, HOBt, carboxylic acids and acid chlorides
	Carbonate	MP MP FSG	1.9 2.8 0.7	Polymer Labs Argonaut Sigma Aldrich	36-37	Cat	
	Amberlyst A27-CO3	MP	1	EOAI	38		
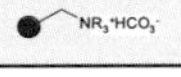	Amersep HCO3	MP MP	5.8 1.8	Novabiochem Polymer Labs	39	Cat	Strong basic anion exchange Removal of carboxylic acids

continued

Table 2 Continued

ELECTROPHILE SCAVENGERS	Resin	Support	Loading (mmol g^{-1})	Supplier	EOAI Reference	Literature Reference	Scavenges
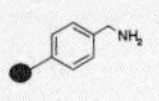	Aminomethyl (AM)	MP PS PS PS FSG PS	2-3.5 2 4 1.3-1.9 1.6 1.2-1.8	Sigma-Aldrich Sigma-Aldrich Sigma-Aldrich Argonaut Silicycle EOAI	40-48	L2, L6, L16-17	Acids (All), Aldehydes, Alkylating agents, Chloroformates, Isocyanates, Isothiocyanates
	Cyclohexyldiol	FSG	0.8	Silicycle		Cat	Aldehydes, Ketones, Boronic acids
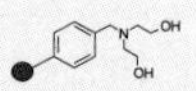	N,N- diethanolamino methyl (DEAM)	PS PS	1.5-2.2 1.7	Argonaut Polymer Labs		L18	Boronic acids, Sn (IV), Ti (IV), Palladium
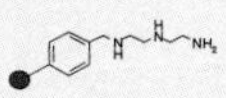	Diethylenetriamine (DETA)	MP MP PS PS PS FSG	4.5 6 6 4-5 2.5-3 3.7	Polymer Labs Sigma-Aldrich Polymer Labs Sigma-Aldrich Sigma-Aldrich Silicycle	49-50	L6, L17	Acids (All), Aldehydes, Alkylating agents, Isocyanates, Isothiocyanates
	Dimethylamine	FSG	1.6	Silicycle		Cat	H+
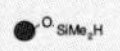	Dimethylsilyl	FSG	1.5	Silicycle		Cat	Acids (All)
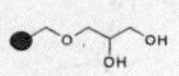	Diol	FSG	1.1	Silicycle		Cat	Aldehydes, Ketones, Boronic acids
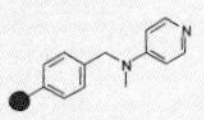	DMAP	PS	0.35	Argonaut		L19	Acid chlorides, Sulfonyl chlorides
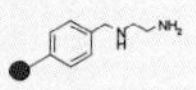	Ethylenediamine (EDA)	MP MP PS PS FSG	3.3 5-6 5 2.5-3 2.7	Polymer Labs Sigma-Aldrich Polymer Labs Sigma-Aldrich Silicycle	51	L16	Acids (All), Aldehydes, Isocyanates, Isothiocyanates
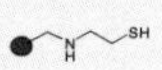	N-(2-mercaptoethyl) amino methyl	PS PS	1.3 1.4	Sigma-Aldrich Novabiochem		L20	Alkylating agents, Palladium
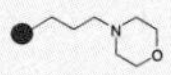	Morpholine (MPH)	MP MP PS PS PS FSG	1.8 0.7-1.5 3 3-4 3.2 1	Polymer Labs Sigma-Aldrich Polymer Labs Sigma-Aldrich Novabiochem Silicycle		L4, L6	H+
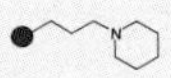	Piperidine (PIP)	MP PS PS PS FSG	1.8 3 3-4 3-4 1	Polymer Labs Polymer Labs Sigma-Aldrich Novabiochem Silicycle	52	L6	H+
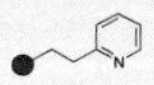	Pyridine	FSG	1.3	Silicycle		Cat	H+
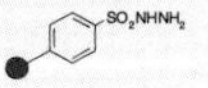	Sulfonyl hydrazine (TsNHNH2, Hydrazine)	PS PS PS FSG	1.8-3.2 3 2.5 2	Argonaut Polymer Labs Sigma-Aldrich Silicycle	53	Cat	Aldehydes, Ketones
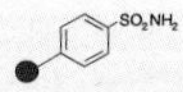	Sulfonyl amide (TsNH2)	PS FSG	1.5-2.5 0.9	Sigma-Aldrich Silicycle		Cat	Acids (All), Aldehydes, Chloroformates, Isocyanates
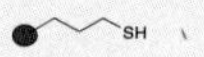	Thiol	FSG	1.2	Silicycle	54	L21	Aldehydes, Alkylating agents, Ketones, Mercury, Palladium
	Duolite (Amberlite GT73)	MP		Sigma-Aldrich	54	Cat	
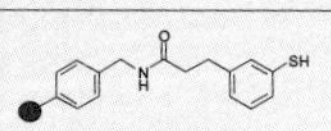	Thiophenyl	PS	1-1.5	Argonaut		L22	Alkylating agents, Epoxide, Thiols, Thiolates
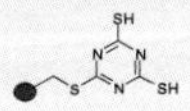	Trimercaptotriazine (TMT)	MP	0.9	Argonaut		Cat	Palladium
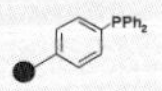	Triphenylphosphine	PS PS PS PS	 1.5 3 1-1.5	Argonaut Polymer Labs Sigma-Aldrich Novabiochem	55	L23-25	Alkylating agents, Azides, Palladium
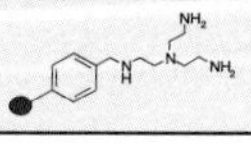	Trisamine	MP PS PS	2-3 3-5 4-5	Argonaut Argonaut Sigma-Aldrich	56	L4, L6, L26	Acids (All), Imines, Isocyanates, Isothiocyanates, Ketones, Acidic phenols

LITERATURE REFERENCES

L1 *Tet Lett*, 2000, 41, 8963	L10 *Org Lett*, 2001, 3, 1869	L19 *J Am Chem Soc*, 1985, 107, 4249
L2 *Tet Lett*, 1996, 37, 7193	L11 *J Comb Chem*, 2000, 2, 675	L20 *Tet Lett*, 1993, 34, 7685
L3 *J Am Chem Soc*, 1971, 93, 492	L12 *Tet Lett 1998*, 39, 975	L21 *Energy and Fuels*, 1999, 13, 1046
L4 *Tetrahedron*, 54, 3983	L13 *Tet Lett*, 1997, 38, 3357	L22 *Tet Lett*, 2000, 41, 2483
L5 *J Org Chem*, 1997, 62, 6797	L14 *Synthesis*, 1997, 553	L23 *J Org Chem*, 1983, 48, 326
L6 *J Am Chem Soc*, 1997, 119, 4882	L15 *Bioorg Med Chem Lett*, 1998, 8, 2391	L24 *Tet Lett*, 1996, 37, 7595
L7 *Bioorg Med Chem Lett*, 2000, 10, 2697	L16 *J Am Chem Soc*, 1997, 119, 4874	L25 *Synthesis*, 1987, 386
L8 *Org Lett*, 2001, 3, 3491	L17 *Tet Lett* 1999, 40, 9195	L26 *Tet Lett*, 1998, 39, 3635
L9 *Tet Lett*, 1998, 39, 8233	L18 *J Comb Chem*, 2000, 2, 228	Cat *Supplier catalogue*is primary source of reference

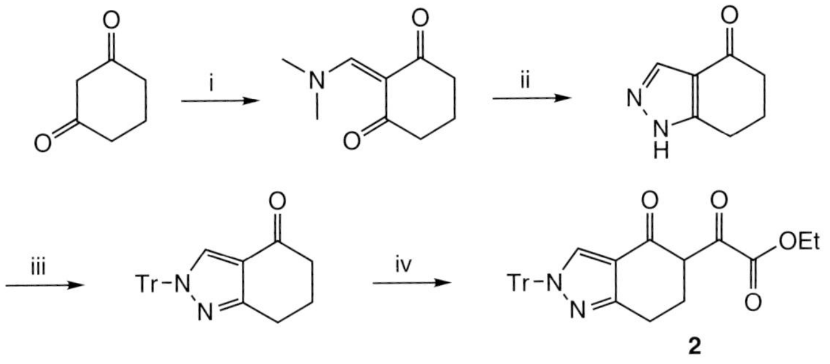

i, $Me_2NCH(OMe)_2$, reflux, ih, 90%; ii, $NH_2NH_2.2HCl$, NaOH/MeOH, reflux, 3 h, 98%; iii, TrCl, Et_3N, CH_2Cl_2, 94%; iv $(CO_2Et)_2$, $(Me_3Si)_2NLi$, 93%

Scheme 2

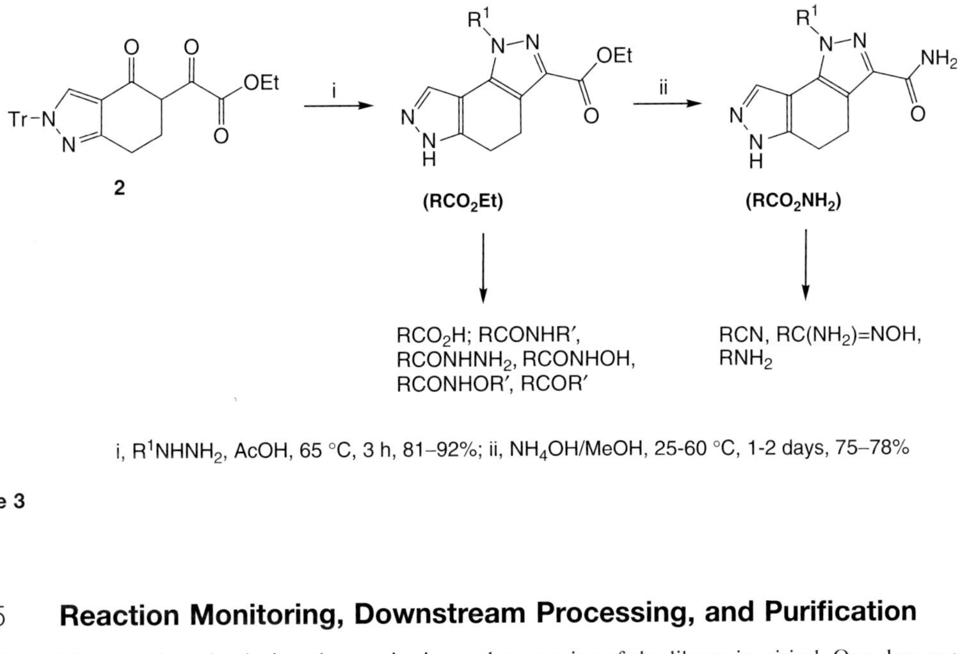

i, R^1NHNH_2, AcOH, 65 °C, 3 h, 81–92%; ii, NH_4OH/MeOH, 25-60 °C, 1-2 days, 75–78%

Scheme 3

3.34.5 Reaction Monitoring, Downstream Processing, and Purification

As in all parallel synthesis technologies, the monitoring and processing of the library is critical. One does not want to expend significant efforts for the synthesis of a library to find at the end of it that the compounds cannot be analyzed or that the analysis shows that the reaction sequence has been unsuccessful. As stated earlier one of the major advantages of solution phase parallel synthesis over solid phase is the monitoring of reactions. Traditional techniques such as thin-layer chromatography (TLC), nuclear magnetic resonance (NMR), and LC/MS can be used, which can only be applied to the analysis of compounds prepared from a solid phase route after cleavage from the resin.

3.34.5.1 Reaction Monitoring

One of the major advantages of solution phase parallel synthesis is that the progress of each reaction can be easily monitored using standard methods (e.g., TLC or LC/MS). In order to perform this in-process check (IPC), a small aliquot is removed from the reaction media and analyzed without detriment to the reaction. If the IPC method shows that the reaction has progressed to completion, the reaction can be stopped, worked up, and the products isolated and purified as appropriate. If the analysis shows that the reaction has not gone to completion, it can be left for a longer period of time, the reaction temperature elevated to help drive the reaction to completion, or additional equivalents of catalysts and/or reagents added. These studies are used extensively in the validation and monomer screening stages, but can also be used in the preproduction and production phases by analyzing representative samples from the library as a whole.

3.34.5.2 Parallel Work-up of Compounds and Key Intermediates

The efficient preparation of the library is by no means the end of the process. In order to maximize the deliverables from the library, techniques have evolved for the parallel work-up and purification of arrays of compounds. These include the development of automated techniques for compound work-up, the use of solid-supported chemicals for SPE for the scavenging of reaction impurities (scavenger resins), and the use of parallel high-performance liquid chromatography (HPLC) purification using either mass or ultraviolet (UV)- or evaporative light-scattering (ELS)-based detection. Excess monomers, for example, can be removed with the aid of scavenger resins. The use of these techniques is now widespread, and has even been adopted by the medicinal chemist as a tool for the effective synthesis of single molecules.

3.34.5.2.1 Liquid–liquid extraction

Depending on the number of reactions that require this technique, liquid–liquid extraction can be used as an effective work-up procedure even for large numbers of compounds. Suitable automation tools (e.g., the LISSY liquid sampling system[34]) can add the second (aqueous) phase to the reaction vessels and can exploit the differences in the conductivity of the bi-layers to effect the separation.

3.34.5.2.2 Purification of intermediates

A typical solution phase parallel synthesis operation will involve a multistep synthetic scheme with an increasing cumulative number of intermediates being produced as monomers that are added to or incorporated in the scaffold in each of the synthetic steps. Each of these intermediates will require isolation, and may require purification to assist in the effective synthesis of the ensuing steps. On planning the synthetic route, one must consider the number of these intermediates, the need to prepare, isolate, and purify each one, the time taken to perform these operations, and the value in the overall number proposed when designing the diversity in the library. The most efficient solution phase parallel library synthesis protocol with respect to number of final compounds versus the time taken in the synthesis of intermediates is one in which the final step introduces the highest diversity. This may be considered a limitation with respect to solution phase parallel synthesis; however, by paying due attention to library design with respect to scaffolds and monomers such that an analysis of the diversity of the final library is appropriate for its purpose, one can efficiently prepare a large library in this way.

It has now become commonplace within parallel synthesis laboratories to use automated equipment to purify reaction intermediates in parallel. Purification stations such as the Horizon[35] or the Flashmaster[36] from Biotage have become invaluable in the parallel purification of reaction intermediates on a scale of grams to tens of grams. Both instruments purify the compounds using chromatography, and can be programmed to separate the reactants and collect the product automatically.

3.34.5.3 Purification of Final Compounds

The quality of the compounds undergoing biological screening has a direct influence on the quality and validity of the results obtained. It is therefore imperative that sound analytical methods are used for the analysis of all members of a compound library, and that the appropriate methods are identified early in the overall library synthesis procedure such that issues with quantification and or qualification can be addressed prior to the analysis of a large array of compounds.

Arguably, the early days of combinatorial chemistry were associated with compounds of questionable or unknown purity being used in high-throughput biological screens. As a direct result of this, the results obtained were of questionable value, generating numerous false positives, for which significant resources will have been spent in the follow-up of these apparent hits; and generating false negatives and other nonreproducible results that have a direct impact on the lead discovery process.

A number of purification methodologies are now utilized for the purification of final compounds from the parallel synthesis of a compound library.

3.34.5.3.1 Resin scavenging

A facile method of purifying the products from a solution phase reaction is by the use of scavenger resins. In a recent example from Contour-Galcéra *et al.*,[37] the synthesis of a targeted library of tertiary amines, which was prepared to optimize the biological activity of a lead compound described as an inhibitor of CDC25 phosphatases, was outlined (**Scheme 4**). The library of tertiary amines was prepared from the reaction of aldehydes with a slight excess of secondary amines. This was followed by reaction scavenging with a resin of suitable functionality to react with the

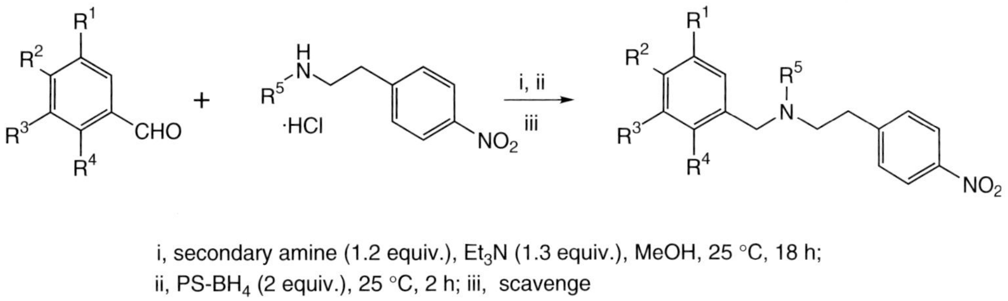

Scheme 4

excess secondary amine. Purification of this reaction required the separation of the excess secondary amine starting material from the tertiary amine product. Thus, the scavenger resin has to be capable of reacting with the secondary amine while being unreactive toward the tertiary amine product. In this case, methylisocyanate polystyrene resin was added to the reaction vessels to covalently bind to the excess secondary amine, effectively removing it from solution. The scavenger resin plus amine starting material was then removed by filtration, leaving the pure tertiary amine in solution.

In the similar manner, secondary amines can be prepared from the reaction of aldehydes and a slight excess of primary amines (**Scheme 4**, $R^5 = H$), followed by scavenging of the excess primary amine. This is efficiently achieved using 4-benzyloxybenzaldehyde polystyrene resin to remove the primary amine in the presence of the desired secondary amine product. This resin will selectively scavenge primary amines in the presence of secondary amines, thus leaving the secondary amines in solution.

3.34.5.3.2 Resin capture and release/solid phase extraction

In tandem with resin scavenging, the use of ion exchange esins to remove unwanted impurities from solution phase reactions is well documented. In a recent example of the generation of a focused library in order to expand the SAR for an mGluR5 antagonist program, Eastman *et al.*[38] noted that the purification of Suzuki reactions with strong cationic exchange resin (SCX) was readily accomplished for compounds containing tertiary amines.[39,40] Utilizing a high-throughput parallel solution phase Suzuki coupling approach in tandem with SCX purification afforded the desired focused library (**Scheme 5**). The library synthesis was performed on the Argonaut Trident and Zinsser workstations. Purification was accomplished by also utilizing the Zinsser workstation.

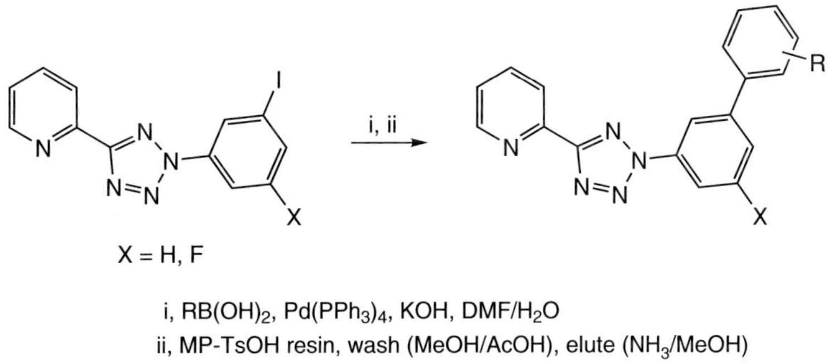

Scheme 5

The Trident liquid handler from Argonaut[41] was used to transfer stock solutions of the iodides, boronic acids and KOH to the reaction vessels, followed by the palladium catalyst. Following overnight reaction at 65 °C and subsequent cooling, the Zinsser liquid handler was used to transfer the reaction mixtures onto suitable columns containing diatomaceous earth, from which the organic components were eluted with dichloromethane. The reaction mixtures were then concentrated, and the resulting residue was dissolved in a methanol/acetic acid mixture, and loaded onto SCX columns (Argonaut MP-TsOH resin). Each column was washed with methanol/acetic acid (4 × 4 mL), and the eluents containing impurities and by-products were discarded; the required products were then eluted from the columns with 1 N NH_3 (4 × 4 mL). The purification relies on the fact that the tertiary amine products are 'captured' by the SCX resin, and all of the nontertiary amine impurities are washed away. The tertiary amine product is then displaced from the SCX resin by the addition of ammonia. The tertiary amine product is then collected and analyzed. Eastman noted that of the 192 reactions carried out, 72% gave purities in excess of 90% by HPLC/MS, with a random sample of compounds analyzed by ^{1}H NMR in order to confirm the final structure and corroborate the HPLC purity.

In another example, scientists at Lundbeck described the use of tablets of functionalized polystyrene beads both alone and in combination with solid-supported reagents for solution phase parallel synthesis, as shown in **Scheme 6.** Accurately preweighed tables were used to carry out solution phase Mitsunobu and acylation reactions.

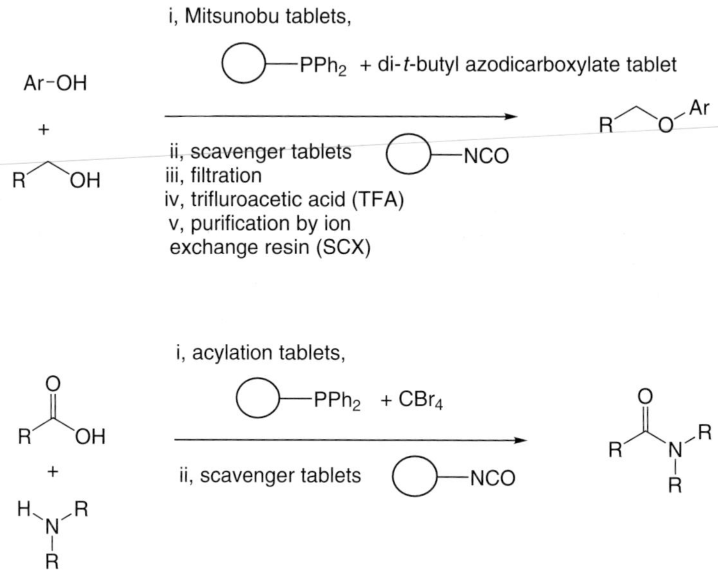

Scheme 6

3.34.5.4 Analytical Characterization

Once the parallel synthesis and the reaction work-ups of a library are complete, the analysis of the library should not become the bottleneck. Traditional methods for analyzing the success of a chemical step involve NMR (for identity and purity), elemental analysis (for purity), HPLC/MS (for identity and purity), and gravimetric analysis (for yield). When considering the number of compounds required to be analyzed from a compound library, one has to compromise on the methods that can be used (e.g., NMR and elemental analysis on a large number of compounds may not be feasible), and exploit those that are amenable to high throughput (e.g., HPLC/MS). Automated weighing is possible for compounds by using pre-tared vials or a microtiter plate format whereby the wells are removable.

When assigning identity and purity by HPLC/MS alone, one should be cognizant of potential limitations; for example some compounds may not have a suitable chromophore to allow detection by UV. In these cases, an alternative detection methodology such as ELS may be used. In some cases, certain compounds may give poor MS responses. In these cases, the incorporation of different co-solvents or alternative ionization methods may be more successful. However, by performing analytical validation at an early stage, the appropriate method can be used for the library, and different methods can be used for different scaffolds/sets of monomers with a library, if appropriate.

3.34.6 Examples of Solution Phase Parallel Synthesis

A wide diversity of chemistries can now be readily accessed using solution phase synthesis carried out in parallel. There are many examples whereby scientists use both solid and solution phase synthesis to investigate different regions of an active molecule to better understand SARs. Examples of this includes libraries designed to improve activity against plasmepsin,[42] factor Xa,[43] caspase-3,[44] dihydrofolate reductase,[45] p56lck,[46] Kv1.5 channel,[47] the farnesoid X receptor,[48] and antibacterials.[49,50]

By using a combination of solid and solution phase synthesis, a wider range of chemistries can be accessed. This demonstrates the recent acceptance of solution phase synthesis not only in the support of lead optimization but also in the synthesis of lead generation libraries for screening.

Solution phase parallel synthesis is a key tool in its own right for the synthesis of libraries of compounds. Many different types of chemistries can now be carried out in solution in parallel. This section illustrates a selection of chemistries that have been used to synthesize libraries of compounds using solution phase parallel synthesis. Solution phase parallel synthesis has been used to derivatize core scaffolds as well as synthesizing libraries of acyclic, heterocyclic, monocyclic, and polycyclic scaffolds.

3.34.6.1 Scaffold Derivatization

There are many examples whereby central core scaffolds have been decorated using solution phase parallel synthesis. Atrash[51] reports on the use of resin plug-bound palladium(0) to catalyze Suzuki reactions in solution to give a variety of biaryl compounds. Workers at Aventis (now Sanofi-Aventis) synthesized a library of functionalized oxindoles as potent Janus kinase 3 (JAK3) inhibitors.[52] From a high-throughput screen, an oxindole compound was identified as having submicromolar activity against the JAK3 enzyme. Following docking studies of the high-throughput screening hit in a homology model of the JAK3 enzyme, a focused library of approximately 700 oxindoles was designed and synthesized as shown in **Scheme 7**. Custom synthesized oxindole cores were condensed with commercially available aldehydes to give a two-dimensional library. The most potent compound identified had an IC_{50} of 27 nM against JAK3, representing a 40-fold improvement in activity compared with the original screening hit.

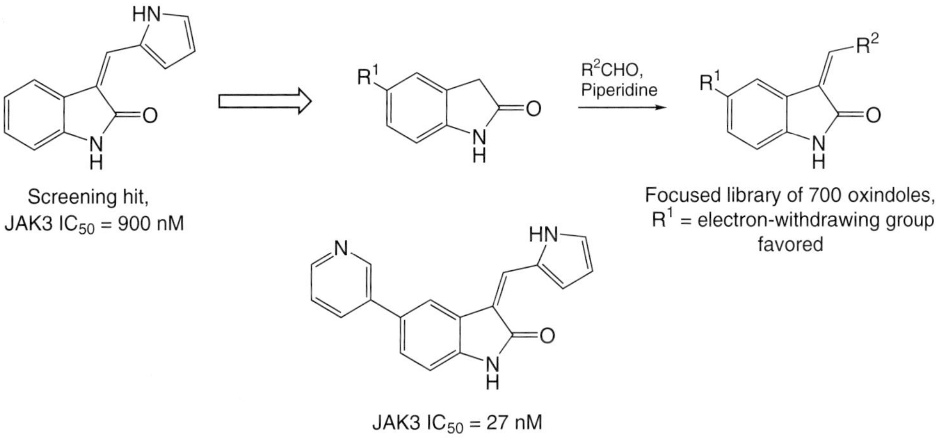

Scheme 7

Aromatic substitutions have been well exemplified using solution phase parallel synthesis. Menichincheri *et al.*[53] describes the sequential displacement of fluorine atoms in a substituted pyrimidine. There are a number of examples whereby a heterocyclic core has been derivatized using solution phase synthesis (**Figure 3**); Fu[54] reports the polymer-assisted acylation of 3-aminopyrazolinones, Touzani[55] reports the condensation of *N*-alkyl heteroarylamines with *N*-hydroxymethyl pyrazoles, Menon[56] reports the derivatization of indazoles, in one of a few examples Sezen[57] describes C–H bond functionalization of imidazoles, and Devocelle[58] describes the polymer-supported synthesis of hydroxamic acids in solution.

3.34.6.2 Heterocyclic Libraries

There are many examples of libraries of heterocycles synthesized using parallel solution phase synthesis. In many examples the heterocyclic core is derivatized using parallel synthesis. There are also examples whereby the heterocyclic core is synthesized using parallel synthesis via some form of cyclization.

Researchers at Bristol-Myers Squibb[59,60] used solution phase parallel synthesis to expand the SARs around p56lck inhibitors (**Scheme 8**). An amino thiazole compound was identified in a random screen as having an IC_{50} of 3.2 μM against p56lck. By investigating substituents on the exocyclic amino group, a compound with an activity of 35 nM was

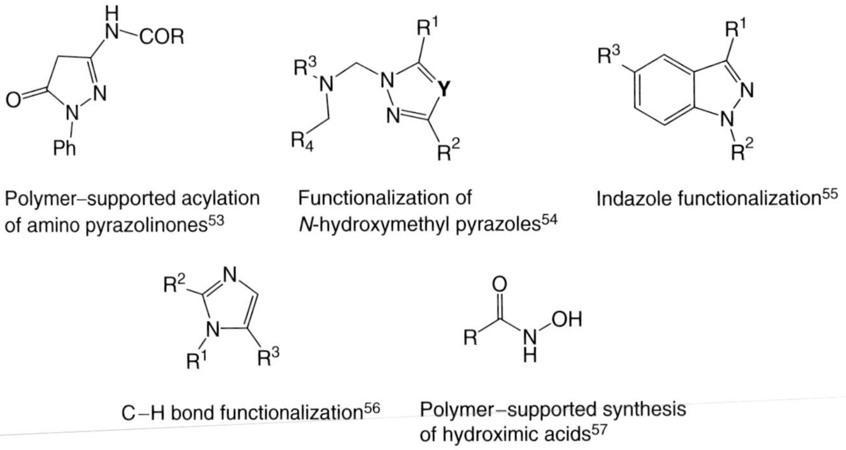

Figure 3 Examples of scaffold derivatization using solution phase parallel synthesis.

quickly identified. In parallel, a library of benzothiazoles was also synthesized with pleasing results. Parallel solution phase synthesis was used to quickly prepare a range of analogs utilizing information gleamed from the screening of the previous analogs. It was apparent that key molecular differences at the exocyclic amino site (cyclopropyl versus isopropyl) were critical for activity against the kinase. Following a number of iterations, a compound with an activity of 4 nM against p56lck was identified, with excellent kinase selectivity.

There are many reports of the synthesis of libraries of benzimidazoles. Beaulieu[61] at Boehringer Ingelheim reported a simple and efficient method for the solution phase synthesis of benzimidazoles (**Scheme 9**). The benzimidazole was synthesized by the addition of oxone to a solution of 1,2-phenylenediamines and an aldehyde in wet dimethylformamide at room temperature. The initially formed benzoimidazoline is quickly oxidized, to give the resulting benzimidazoles. A wide range of diamine substituents can be used in conjunction with aliphatic, aromatic, and heteroaromatic aldehydes. Crude products were isolated by extraction or filtration, and little further purification was required.

Raju and co-workers[62] also reported on the solution phase parallel synthesis of benzimidazoles using phenylenediamines. They describe the solution phase synthesis of a range of *o*-phenylenediamines that were subsequently converted to the corresponding benzimidazoles in excellent yields and purities.

In another reported synthesis of benzimidazoles,[63] multi-substituted 5-aminobenzimidazoles were synthesized by the nucleophilic substitution of the two fluorine atoms of 1,5-diflouro-2,4-dinitrobenzene (**Scheme 10**). Simultaneous reduction of the two nitro groups is followed by condensation with a range of aldehydes to give the benzimidazoles. The pendant aromatic amino group may be further derivatised by a variety of electrophilic reagents.

The parallel solution phase synthesis of more than 2200 7-trifluormethyl-substituted pyrazolo-[1,5-α]-pyrimidines and 4,5,6,7-tetrahydropyrazolo[1,5-α]pyrimidine carboxamides was reported by Dalinger and co-workers.[64] Key reactions included the assembly of the pyrazolopyrimidine ring by condensation of 5-aminopyrazole derivatives with the corresponding trifluoromethyl-β-diketones. The libraries from libraries were obtained in good purities using solution phase acylation and reduction methodologies (**Scheme 11**). The final products were readily crystallized from solution.

Spirohydantoins are considered to be privileged structures, and are thus attractive for libraries designed to have diverse biological activity. Nieto and co-workers[65] report on the synthesis of a library of 168 diverse spirohydantoins through a two-step solution phase parallel synthesis route starting from a range of *N*-substituted piperidinones, as shown in **Scheme 12**. The Strecker reaction was used to generate α-amino nitriles from aniline and trimethylsilyl cyanide (or KCN). Subsequent reaction of the anilido nitrogen with a diverse set of isocyanates gave the library in a high yield and purity.

A library of trisubstituted 1,2,4-triazoles was synthesized by researchers at Bristol-Myers Squibb[66] via Mitsunobu chemistry using parallel solution phase synthesis. as shown in **Scheme 13**. The compounds were found to have low nanomolar antagonist activity against the NPY Y1 receptor.

An efficient diversity-oriented solution phase strategy was used for the parallel synthesis of di- and trisubstituted pyrrole libraries.[67] Methyl esters were converted to 1,2-di- and 1,2,5-trisubstituted pyrroles in three steps, as noted in **Scheme 14**.

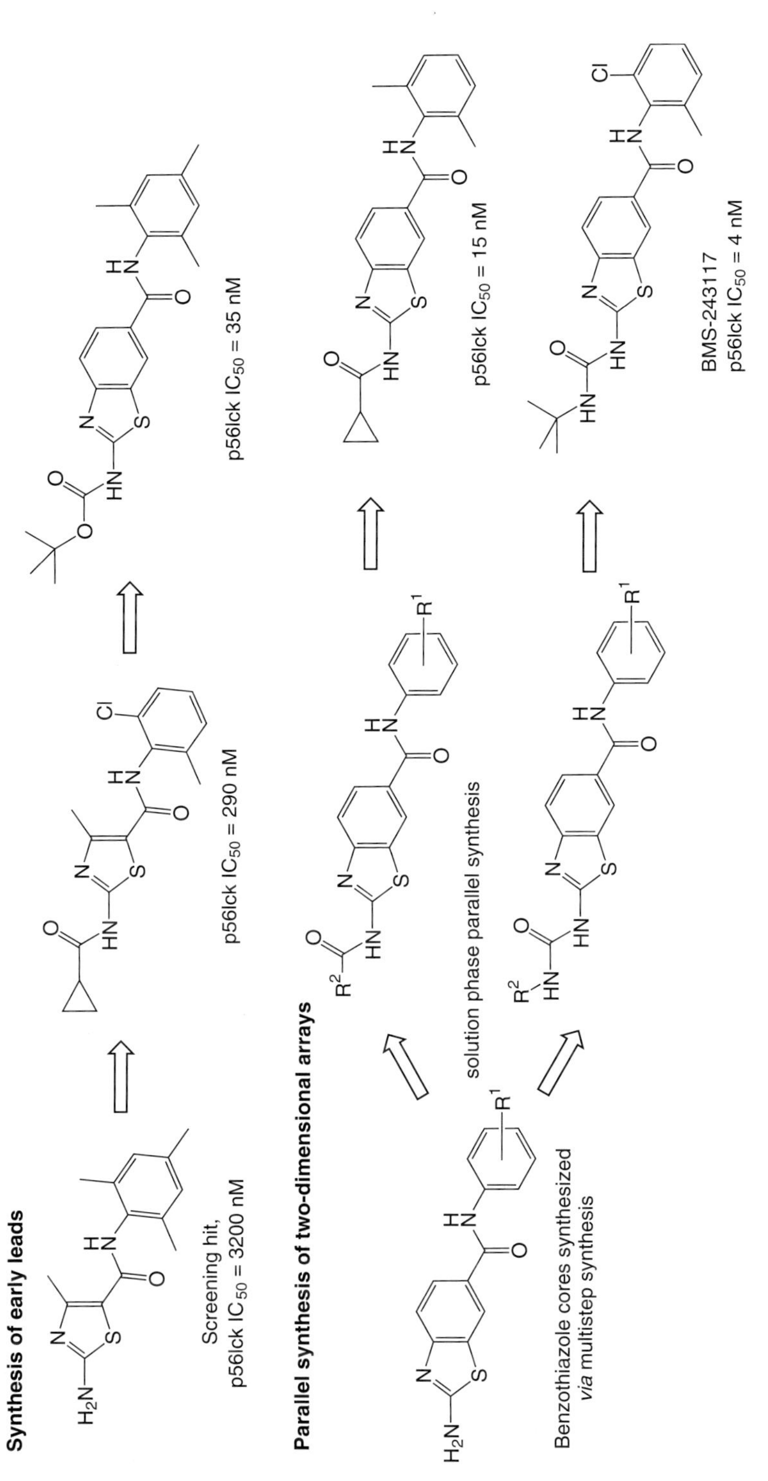

Synthesis of early leads
Screening hit,
p56lck IC$_{50}$ = 3200 nM
p56lck IC$_{50}$ = 290 nM
p56lck IC$_{50}$ = 35 nM
Parallel synthesis of two-dimensional arrays
solution phase parallel synthesis
p56lck IC$_{50}$ = 15 nM
BMS-243117
p56lck IC$_{50}$ = 4 nM
Benzothiazole cores synthesized
via multistep synthesis

Scheme 8

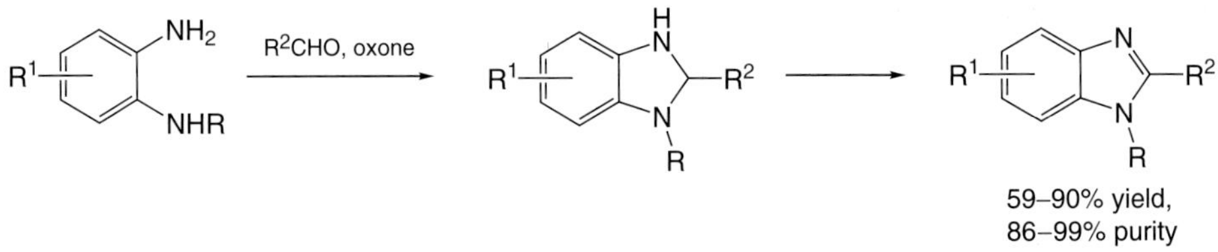

Scheme 9

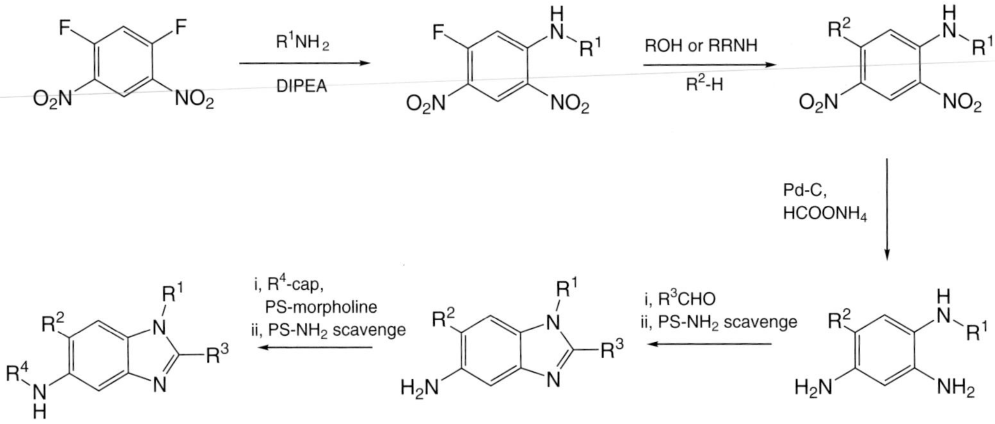

Scheme 10 Solution phase parallel synthesis of 5-aminobenzimadoles.

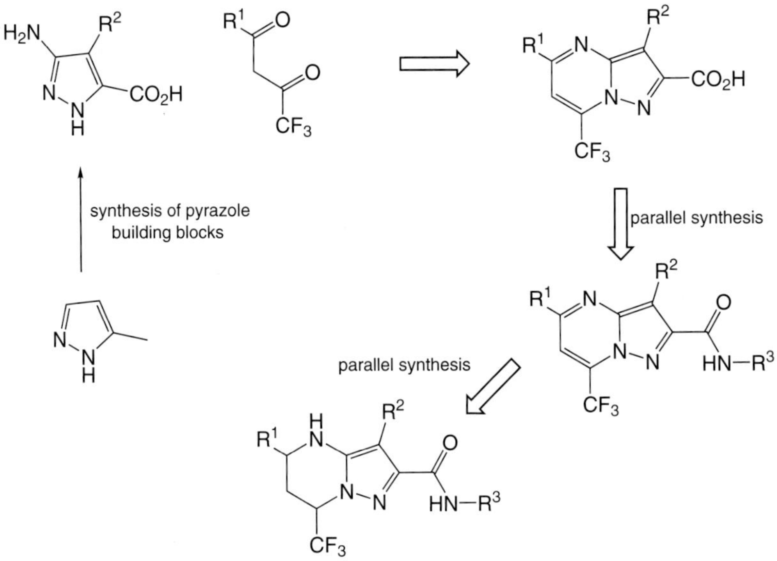

Scheme 11

Starting from phenylalanine derivatives, a solution phase parallel synthesis of benzazepine-3-one is described by Tourwe and co-workers (**Scheme 15**).[68] They show that the two nitrogen atoms of the heterocycle can be selectively derivatized. Common reagents such as aldehydes, carboxylic acids, isocyanates, and sulfonyl chlorides are used to access a wide diversity. The high-throughput nature of this synthesis was demonstrated by the extensive use of

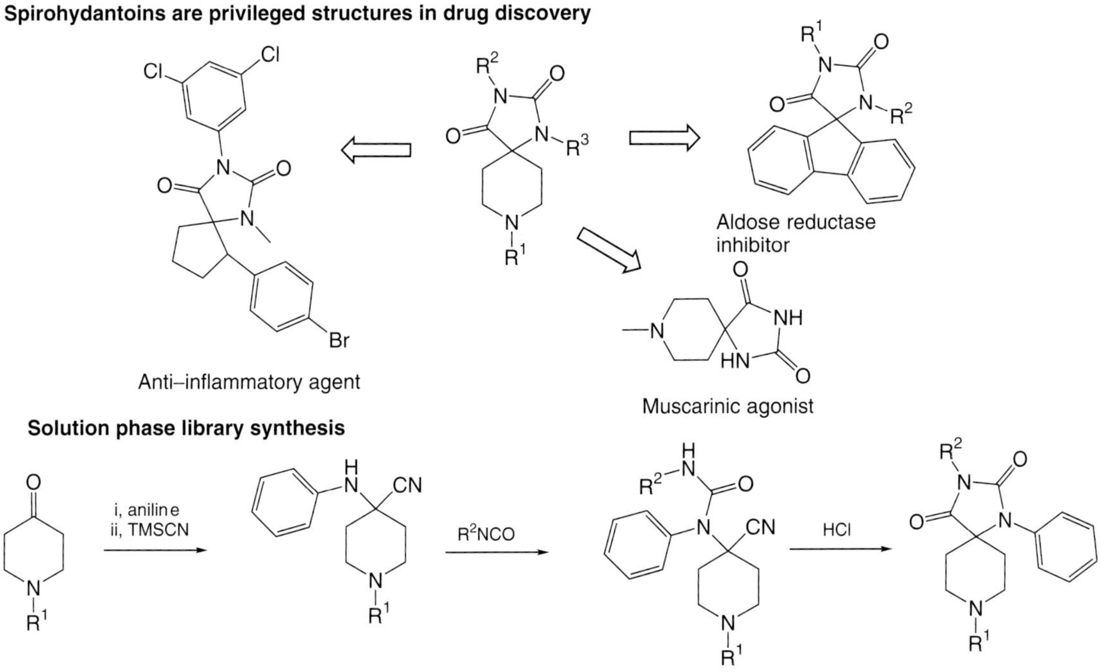

Scheme 12

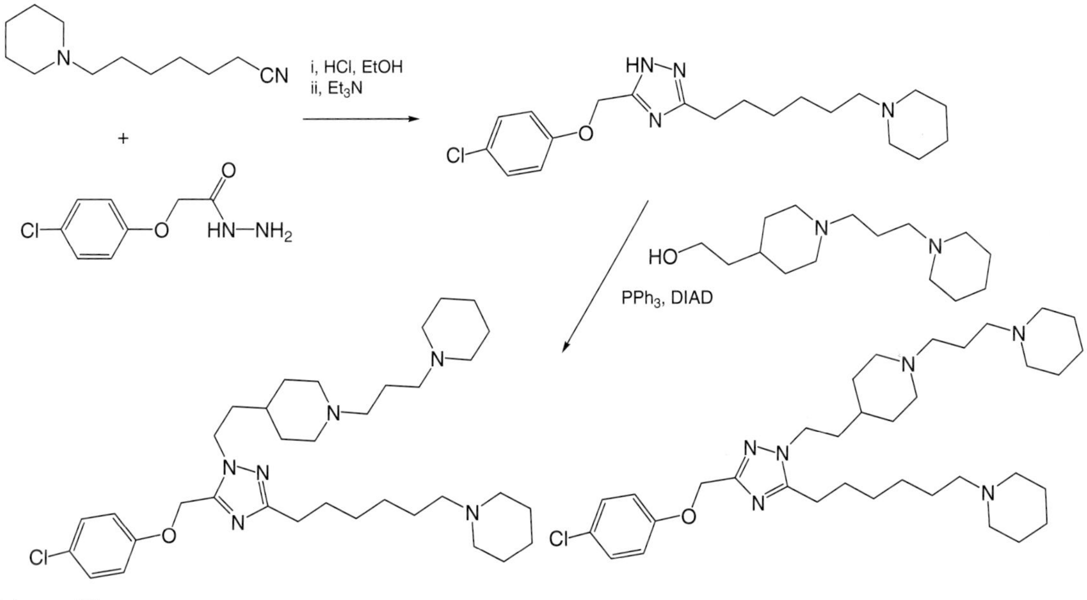

Scheme 13

polymer-supported reagents and scavenger resins. Polymer-bound cyanoborohydride worked very well for the reductive amination step, and scavenging of the excess amine was accomplished using polymer-supported benzaldehyde. The cyclization to give the benzazepinone ring was carried out in the presence of the polymer-supported coupling reagent 1-ethyl-3-(3-dimethylaminopropyl)carbodiimide (EDC). After Boc deprotection, the second nitrogen atom was either acylated using carboxylic acids, sulfonylated, or converted to ureas. The acylation was performed in the presence of the polymer-supported EDC.

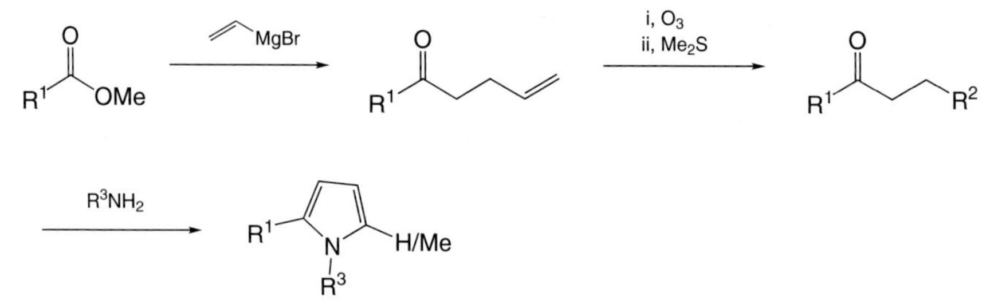

Scheme 14

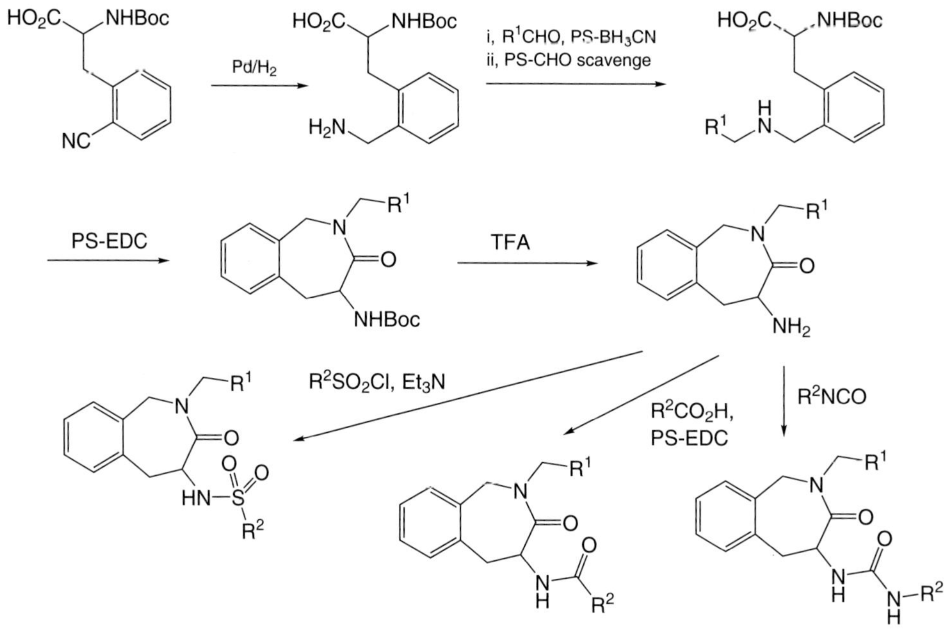

Scheme 15

3.34.6.3 Multicomponent Reactions

One of the original reactions carried out in parallel in solution was the multicomponent reaction. The original attraction of this reaction was the fact that it was a single step, and thus it was hoped that using solution phase parallel synthesis would not create a poor-purity profile. This type of reaction continues to be attractive, based on the structural diversity of the resulting products. A library of 2-substituted quinolines was synthesized[69] using a modification of the Grieco multicomponent reaction, as shown in **Scheme 16**. The authors demonstrated this reaction by the synthesis of a 25-member library in solution.

Multi-component reactions are routinely used to generate libraries of compounds using solution phase technologies.[70–74]

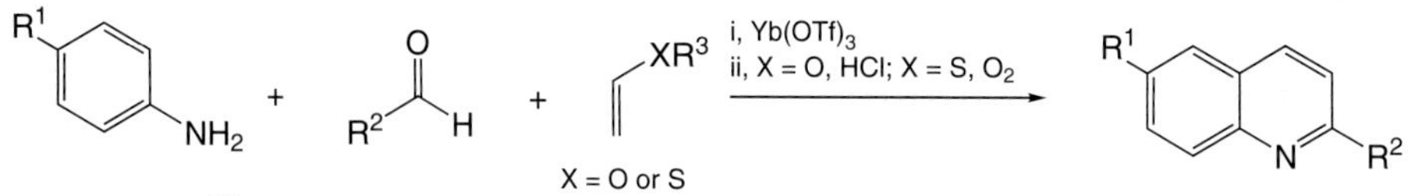

16

Scheme 16

3.34.6.4 Dynamic Combinatorial Chemistry

Using a play-on solution phase parallel synthesis, dynamic combinatorial chemistry (DCC) has been reported.[75,76] DCC is a molecular recognition strategy whereby building blocks react with one another reversibly in solution under thermodynamic control in the presence of a molecular target, enzyme, or receptor. Specific members of the library are amplified on the basis of their preferred target interactions. In its application to the discovery of CDK-2 inhibitors,[77] oxindole and aryl hydrazine building blocks were reacted together in the presence of enzyme crystals (**Scheme 17**). In solution studies in the absence of the enzyme, five oxindoles were reacted with six hydrazines. All of the possible 30 products were detected using LC/MS. The reaction was then repeated in the presence of the enzyme. Resulting electron density maps showed one combination of oxindole and hydrazine to be prevalent. This combination was resynthesized, and the resulting compound was found to be a potent inhibitor of CDK-2, with an IC_{50} of 30 nM.

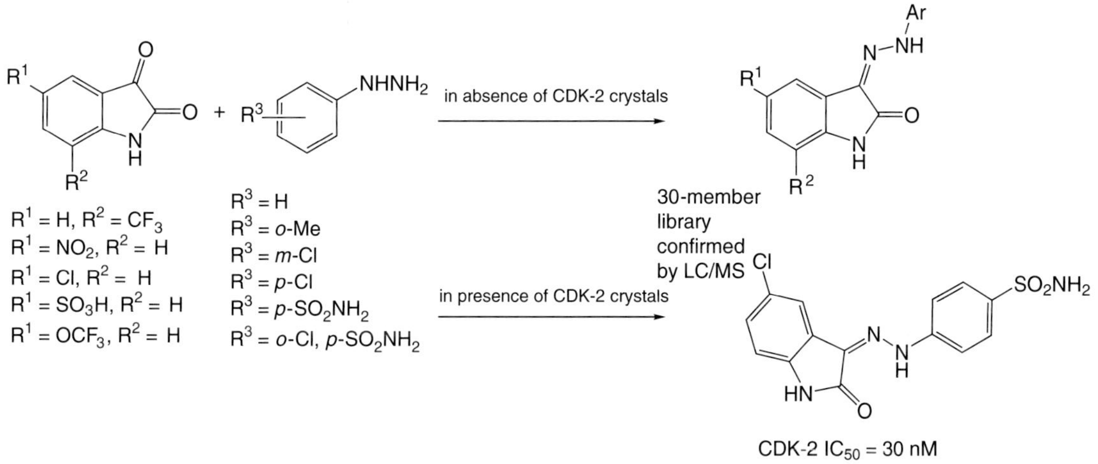

Scheme 17

3.34.6.5 Fluorous Chemistry

Fluorous chemistry was introduced as a means of synthesizing individual compounds in solution by incorporating a fluorous tag with each compound in order to facilitate the purification of the products. Fluorous chromatography is used to separate the reaction products from the impurities. After purification, the fluorous tags are removed from the compounds, to give the final products. In 2002, Curran described the synthesis of a 560-member library of analogs of the natural product mappicine.[78] A seven-component mixture was carried through a four-step synthesis, after which each individual compound was purified using fluorous chromatography (**Scheme 18**).

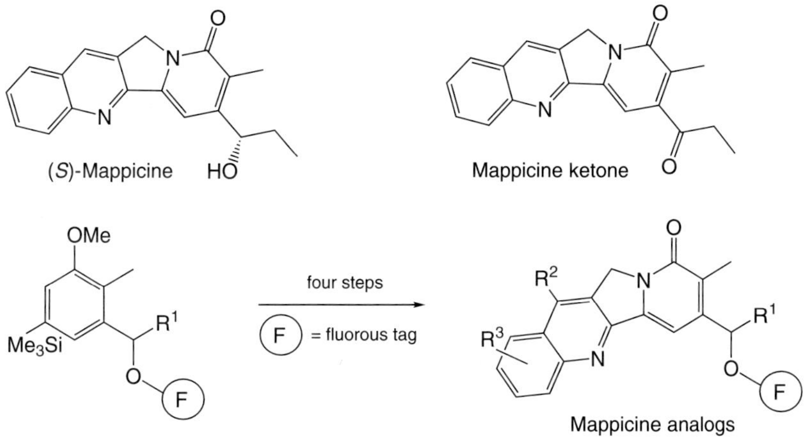

Scheme 18

Fluorous reagents have been used as reagents, scavengers, and protecting groups[79,80] for the parallel synthesis of libraries of compounds in solution. Heterocycles such as hydantoins[81] and pyrimidines[82] have been synthesized using such reagents. For the synthesis of hydantoins, perfluoroalkyl-tagged esters were reacted with isocyanates in solution followed by triethylamine-mediated intramolecular cyclization to urea and concomitant tag release, as shown in **Scheme 19**. Product purification was carried out using SPE. Using an analogous approach, thiohydantoins could be prepared.

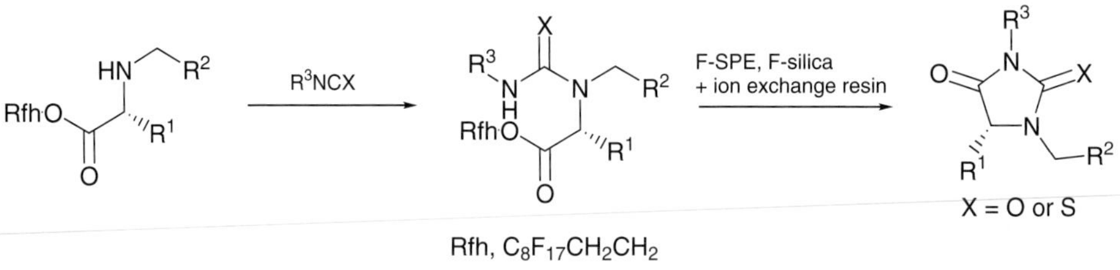

Scheme 19

A fluorous catch-and-release strategy was used for the synthesis of disubstituted pyrimidines, as shown in **Scheme 20**. A fluorous thiol was reacted with a 2,4-dichloro-5-substituted pyrimidine. Following reaction with a nitrogen nucleophile, the fluorous tag is oxidized with oxone to the sulfone. The sulfone is then displaced with a second amine or other nucleophile, releasing the tag. The fluorous tag acts as a phase tag for intermediate and product purification using SPE cartridges.

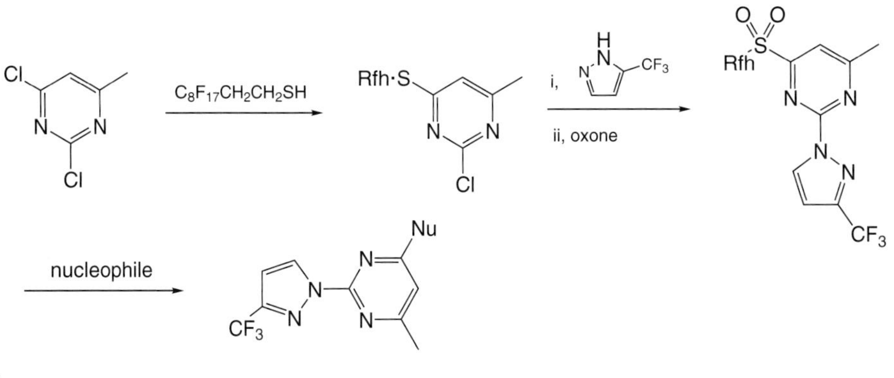

Scheme 20

Other publications describing the synthesis of compounds in solution using fluorous chemistry include reports on fluorous electrophilic scavengers,[83] the synthesis and reactions of fluorous-Cbz-protected amino acids,[84] a fluorous-based quinazoline 2,4-dione synthesis,[85] fluorous dienophiles,[86] a fluorous ruthenium catalyst for ring-closing metathesis,[87] and a fluorous version of Evans chiral auxiliary.[88]

3.34.7 New and Future Trends; the Continued Evolution of Combinatorial Chemistry

Combinatorial chemistry has already yielded several compounds that are currently undergoing clinical development. But it is fair to say that combinatorial chemistry has not lived up to initial expectations in this regard. However, Ashton *et al.*[89] comment that this is more likely to be a result of an incorrect application of the technology rather than a failing in the technology. Combinatorial chemistry should be viewed simply as a technology that can provide arrays of compounds in parallel in a shorter space of time than would be required to synthesize the compounds in a sequential manner. As such, it is a tool that is very much dependent on the quality of the design of the compounds. Should combinatorial chemistry therefore be held responsible for the lack of development candidates?

Today, combinatorial chemistry can be used where large libraries of thousands of compounds need to be routinely prepared. The advances in the design, synthesis, purification, and analysis of large arrays of compounds have been aided by the simultaneous advances in automated technologies that enable the chemist to prepare libraries more efficiently.

The synthesis of arrays of compounds should still be considered as a core competence of the pharmaceutical industry today; however, these compounds need to be prepared in a 'smarter' fashion than once was the case. Even before initial chemistry is performed, one must give thought to the design of the library: does it need a gene target focus or should it be designed purely with diversity in mind? One must use all the available and appropriate computational tools to assist in the selection of suitable scaffolds and monomers. On commencing synthesis, one must prepare the compounds efficiently and effectively by using well-validated procedures where available. Due consideration should be given to the processes for design, synthesis, and downstream activities.

By the sharing of knowledge between the combinatorial chemist and the medicinal chemist, the techniques pioneered and routinely used by the combinatorial chemist have been adopted by medicinal chemists for the preparation of smaller focused and hit-to-lead libraries. Using all of the technologies used in parallel synthesis, such as library design, synthesis, and purification, combined with a knowledge of the biological target, libraries can be prepared efficiently and effectively, to maximize the chance of first finding the hit, secondly developing that hit into a lead, and, ultimately, optimizing the lead to yield a development candidate.

What once was called combinatorial chemistry has given way to new terminology that better describes the focus of today. The generic phrase 'parallel synthesis' is used primarily to describe the synthesis of small- to mediumsized arrays, and has largely superseded 'combinatorial chemistry' and its associated baggage. Other terms give an indication of the need for rapid access to an array of compounds for SAR expansion; for example, high-speed analog chemistry,[90] high-throughput organic synthesis,[91] and just-in-time library synthesis. There are new terms that are focused on the library design element; for example, diversity-oriented synthesis[92–94] and target-oriented synthesis. In addition, the concept of efficient chemical synthesis is captured in the re-emergence of multicomponent reactions.[95–97]

In summary, one should not be too critical of combinatorial chemistry. The industry demanded large numbers of compounds when it was evident that combinatorial chemistry could provide such high numbers. The number of compounds was the primary aim, with compound design and quality coming second. A reality check and the lack of results has now forced the industry to re-evaluate the use of combinatorial chemistry, and to use it as it should be used, as a compound synthesis tool to deliver compounds based on a rationale design and to a high quality. Combinatorial chemistry is not the sole answer to the industry's challenge of increasing its R&D productivity, but it is an invaluable tool within the toolbox of the 'drug hunter.' The real gain in efficiency and productivity will come from knowing when to use the most applicable tool. From the first chemistry-driven libraries prepared either as proof of concept or targeted at high-throughput screens, through the design of 'drug-like' libraries, to today's 'lead-like' libraries, often aimed at a particular biological target, combinatorial chemistry will continue to evolve along with the needs of the pharmaceutical industry, and will continue to be an invaluable tool in the drug discovery process.

References

1. Ghose, A. K.; Viswanadhan, V. N.; Wendoloski, J. J. *J. Comb. Chem.* **1999**, *1*, 55–68.
2. Dolle, R. E. *J. Comb. Chem.* **2004**, *6*, 623–679.
3. Lahana, R. *Drug Disc. Today* **1999**, *4*, 447–448.
4. Gane, P. J.; Dean, P. M. *Curr. Opin. Struct. Biol.* **2000**, *10*, 401–404.
5. Lu, Y.; Sakamuri, S.; Chen, Q.; Keng, Y.; Khazak, V.; Illgen, K.; Schabbert, S.; Weber, L.; Menon, S. R. *Bioorg. Med. Chem. Lett.* **2004**, *14*, 3957–3962.
6. Parlow, J. J.; Devraj, R. V.; South, M. S. *Curr. Opin. Chem. Biol.* **1999**, *3*, 320.
7. Booth, R. J.; Hodges, J. C. *J. Am. Chem. Soc.* **1997**, *119*, 4882.
8. Flynn, D. L.; Crich, J. Z.; Devraj, R. V.; Hockerman, S. L.; Parlow, J. J.; South, M. S.; Woodard, S. *J. Am. Chem. Soc.* **1997**, *119*, 4874.
9. Kaldor, S. W.; Siegel, M. G. *Curr. Opin. Chem. Biol.* **1997**, *1*, 101.
10. Siegel, M. G.; Hahn, P. J.; Dressman, B. A.; Fritz, J. E.; Grunwell, J. R.; Kaldor, S. W. *Tetrahedron Lett.* **1997**, *38*, 3357.
11. Lawrence, M. R.; Biller, S. A.; Fryszman, O. M.; Poss, M. A. *Synthesis* **1997**, 553.
12. Loupy, A., Ed. *Microwaves in Organic Synthesis*; Wiley-VCH: Weinheim, Germany, 2002.
13. Hayes, B. L. *Microwave Synthesis: Chemistry at the Speed of Light*; CEM: Matthews, NC, 2002.
14. Georgsson, J.; Hallberg, A.; Larhed, M. *J. Comb. Chem.* **2003**, *5*, 350–352.
15. Husemoen, G.; Olsson, R.; Andersson, C.-M.; Harvey, S. C.; Hansen, C. H. *J. Comb. Chem.* **2003**, *5*, 606–609.
16. Pottorf, R. S.; Chadha, N. K.; Katkevics, M.; Ozola, V.; Suna, E.; Ghane, H.; Regberg, T.; Player, M. R. *Tetrahedron Lett.* **2003**, *44*, 175–178.
17. Zbruyev, O. I.; Stiani, N.; Kappe, C. O. *J. Comb. Chem.* **2003**, *5*, 145–148.
18. Takvorian, A. G.; Combs, A. P. *J. Comb. Chem.* **2004**, *6*, 171–174.
19. Lipinski, C. A.; Lombardo, F.; Dominey, B. W.; Feeney, P. *J. Adv. Drug Deliv. Rev.* **1997**, *23*, 3–25.
20. Rishton, G. M. *Drug Disc. Today* **1997**, *2*, 382–384.
21. Ajay.; Walters, W. P.; Murcko, M. A.; *J. Med. Chem.* **1997**, *2*, 382–384.

22. Sadowski, J.; Kubinyi, H. A. *J. Med. Chem.* **1998**, *41*, 3325–3329.
23. McGovern, S. L.; Caselli, E.; Grigorieff, N.; Shoichet, B. K. *J. Med. Chem.* **2002**, *45*, 1712–1722.
24. Roche, O.; Schneider, P.; Zuegge, J.; Guba, W.; Kansy, M.; Alanine, A.; Bleicher, K.; Danel, F.; Gutnecht, E. M.; Rogers-Evans, M. et al. *J. Med. Chem.* **2002**, *45*, 137–142.
25. Teague, T. S. J.; Davis, A. M.; Leeson, P. D.; Oprea, T. I. *J. Chem Info. Comput. Sci.* **1999**, *38*, 3743–3748.
26. Oprea, T. I.; Davis, A. M.; Teague, S. J.; Leeson, P. D. *Angew. Chem.* **2001**, *41*, 1308–1315.
27. Rishton, G. M. *Drug Disc. Today* **2003**, *8*, 86–89.
28. Sauer, W. H. B.; Schwarz, M. K. *Chimica* **2003**, *57*, 276–283.
29. Caldarelli, M.; Habermann, J.; Ley, S. V. *Bioorg. Med. Chem. Lett.* **1999**, *9*, 2049–2052.
30. Habermann, J.; Ley, S. V.; Smits, R. *J. Chem. Soc. Perkin Trans.* **1999**, *1*, 2421–2423.
31. Habermann, J.; Ley, S. V.; Scicinski, J.; Scott, J. S.; Smits, R.; Thomas, A. W. *J. Chem. Soc. Perkin Trans.* **1999**, *1*, 2425–2427.
32. Caldarelli, M.; Habermann, J.; Ley, S. V. *J. Chem. Soc. Perkin Trans.* **1999**, *1*, 107–110.
33. D'Alessio, R.; Bargiotti, A.; Metz, S.; Brasca, M. G.; Cameron, A.; Ermoli, A.; Marsiglio, A.; Polucci, P.; Roletto, F.; Tibolla, M. et al. *Bioorg. Med. Chem. Lett.* **2005**, *15*, 1315–1319.
34. Zinsser Analytic: http://www.zinsser-analytic.com/10.asp (accessed Aug 2006).
35. Biotage Horizon purification system. http://www.biotagedcg.com (accessed Aug 2006).
36. Biotage FlashMaster purification system. http://argonaut.biotage.com(accessed Aug 1006).
37. Contour-Galcéra, M.-O.; Lavergne, O.; Brezak, M.-C.; Ducommun, B.; Prévost, G. *Bioorg. Med. Chem. Lett.* **2004**, *14*, 5809–5812.
38. Eastman, B.; Chen, C.; Smith, N. D.; Poon, S.; Chung, J.; Reyes-Manalo, G.; Cosford, N. D. P.; Munoz, B. *Bioorg. Med. Chem. Lett.* **2004**, *14*, 5485–5488.
39. Organ, M.; Arvanitis, E.; Dixon, C.; Lavorato, D. *J. Comb. Chem.* **2001**, *3*, 473.
40. Organ, M.; Mayhew, D.; Cooper, J.; Dixon, C.; Lavorato, D.; Kaldor, S.; Siegel, M. *J. Comb. Chem.* **2001**, *3*, 64.
41. Argonaut. http://www.argonaut.com (accessed Aug 2006).
42. Noteburg, D.; Schaal, W.; Hamelink, E.; Vrang, L.; Larhed, M. *J. Comb. Chem.* **2003**, *5*, 456–464.
43. Lam, P. Y. S.; Adams, J. J.; Clark, C. G.; Calhoun, W. J.; Luettgen, J. M.; Knabb, R. M.; Wexler, R. R. *Bioorg. Med. Chem. Lett.* **2003**, *13*, 1795–1799.
44. Isabel, E.; Black, W. C.; Bayly, C. I.; Grimm, E. L.; Janes, M. K.; McKay, D. J.; Nicholson, D. W.; Rasper, D. M.; Renaud, J.; Roy, S. et al. *Bioorg. Med. Chem. Lett.* **2003**, *13*, 2137–2140.
45. Wu, Z.; Erole, F.; FitzGerald, M.; Perera, S.; Riley, P.; Campbell, R.; Pham, Y.; Rea, P.; Sandanayake, S.; Mathieu, M. N. et al. *J. Comb. Chem.* **2003**, *5*, 166–171.
46. Wityak, J.; Das, J.; Moquin, R. V.; Shen, Z.; Lin, J.; Chen, P.; Doweyko, A. M.; Pitt, S.; Pang, S.; Shen, D. R. et al. *Bioorg. Med. Chem. Lett.* **2003**, *13*, 4007–4010.
47. Peukert, S.; Brendel, J.; Pirard, B.; Bruggermann, A.; Below, P.; Kleemann, H.-W.; Hemmerle, H.; Schmidt, W. *J. Med. Chem.* **2003**, *46*, 486–498.
48. Nicolau, K. C.; Evans, R. M.; Roecker, A. J.; Hughes, R.; Downes, M. *Org. Biomol. Chem.* **2003**, *1*, 908–920.
49. Zhi, C.; Long, Z.-Y.; Gambino, J.; Xu, W.-C.; Brown, N. C.; Barnes, M.; Butler, M.; LaMarr, W.; Wright, G. E. *J. Med. Chem.* **2003**, *46*, 2731–2739.
50. Kaizerman, J. A.; Gross, M. I.; Ge, Y.; White, S.; Hu, W.; Duan, J.-X.; Baird, E. E.; Johnson, K. W.; Tanaka, R. D.; Moser, H. E. et al. *J. Med. Chem.* **2003**, *46*, 3914–3929.
51. Atrash, B.; Reader, J.; Bradley, M. *Tetrahedron Lett.* **2003**, *44*, 4779–4782.
52. Adams, C.; Aldous, D. J.; Amendola, S.; Barnborough, P.; Bright, C.; Crowe, S.; Eastwood, P.; Fenton, G.; Foster, M.; Harrison, T. K. P. et al. *Bioorg. Med. Chem. Lett.* **2003**, *13*, 3105–3110.
53. Menichincheri, M.; Bassini, D. F.; Gude, M.; Angliolini, M. *Tetrahedron Lett.* **2003**, *44*, 519–522.
54. Fu, J.; Shuttleworth, S. J. *Tetrahedron Lett.* **2003**, *44*, 3843–3845.
55. Touzani, R.; Garbacia, S.; Lavastre, O.; Yadav, V. K.; Carboni, B. *J. Comb. Chem.* **2003**, *5*, 375–378.
56. Menon, S.; Vaidya, H.; Pillai, S.; Vidya, R.; Mitscher, L. A. *Comb. Chem. High Throughput Screen.* **2003**, *6*, 471–480.
57. Sezen, B.; Sames, D. *J. Am. Chem. Soc.* **2003**, *125*, 10580–10585.
58. Devocelle, M.; McLoughlin, B. M.; Sharkey, C. T.; Fitzgerald, D. *J. Org. Biomol. Chem.* **2003**, *1*, 850–853.
59. Das, J.; Lin, J.; Moquin, R. V.; Shen, Z.; Spergel, S. H.; Wityak, J.; Doweyko, A. M.; DeFax, H. F.; Fang, Q.; Pang, S. et al. *Bioorg. Med. Chem. Lett.* **2003**, *13*, 2145–2149.
60. Wityak, J.; Das, J.; Moquin, R. V.; Shen, Z.; Lin, J.; Chen, P.; Doweyko, A. M.; Pitt, S.; Pang, S.; Shen, D. R. et al. *Bioorg. Med. Chem. Lett.* **2003**, *13*, 4007–4010.
61. Beaulieu, P. L.; Hache, B.; von Moos, E. *Synthesis* **2003**, *11*, 1683–1692.
62. Raju, B.; Nguyen, N.; Holland, G. W. *J. Comb. Chem.* **2002**, *4*, 320–328.
63. Li, L.; Liu, G.; Wang, Z.; Yuan, Y.; Zhang, C.; Tian, H.; Wu, X. *Zhang. J. Comb. Chem.* **2004**, *6*, 811–821.
64. Dalinger, I. L.; Vatsadse, I. A.; Ivachtchenko, A. V.; Shevelev, S. A.; Ivachtchenko, A. V. *J. Comb. Chem.* **2005**, 7, 236–245.
65. Nieto, M. J.; Philip, A. E.; Poupaert, J. H.; McCurdy, C. R. *J. Comb. Chem.* **2005**, 7, 258–263.
66. Martin, S. W.; Romaine, J. L.; Chen, L.; Mattson, G.; Antal-Zimanyi, I. A.; Poindexter, G. S. *J. Comb. Chem.* **2004**, *6*, 35–57.
67. Hansford, K. A.; Zanzarova, V.; Dörr, A.; Lubell, W. D. *J. Comb. Chem.* **2004**, *6*, 893–898.
68. Van den Eynde, I.; van Rompaey, K.; Lazzaro, F.; Tourwé, D. *J. Comb. Chem.* **2004**, *6*, 468–473.
69. Demaude, T.; Knerr, L.; Pasau, P. *J. Comb. Chem.* **2004**, *6*, 768–775.
70. Mont, N.; Teixido, J.; Borrell, J. I.; Kappe, C. O. *Tetrahedron Lett.* **2003**, *44*, 5385–5387.
71. Zhao, Z.; Leister, W. H.; Strauss, K. A.; Wisnoski, D. D.; Lindsley, C. W. *Tetrahedron Lett.* **2003**, *44*, 1123–1127.
72. Sabitha, G.; Reddy, G. S. K. K.; Reddy, K. B.; Yadav, J. S. *Tetrahedron Lett.* **2003**, *44*, 6497–6499.
73. Adrian, J. C.; Sanpper, M. L. *J. Org. Chem.* **2003**, *68*, 2143–2150.
74. Gedey, S.; Vainiotalo, P.; Zupko, I.; de Witte, P. A. M.; Fulop, F. *J. Heterocycl. Chem.* **2003**, *40*, 951–956.
75. Hochgurtel, M.; Biesinger, R.; Kroth, H.; Piecha, D.; Hofmann, M. W.; Krause, S.; Schaaf, O.; Nicolau, C.; Eliseev, A. V. *J. Med. Chem.* **2003**, *46*, 356–358.
76. Buchstaller, H.-P. *J. Comb. Chem.* **2003**, *5*, 789–793.
77. Congreve, M. S.; Davis, D. J.; Devine, L.; Granata, C.; O'Reilly, M.; Wyatt, P. G.; Jhoti, H. *Angew. Chem. Int. Ed.* **2003**, *42*, 4479–4482.
78. Zhang, W.; Luo, Z.; Hiu-Tung Chen, C.; Curran, D. P. *J. Am. Chem. Soc.* **2002**, *124*, 10443–10450.

79. Zhang, W. *Tetrahedron* **2003**, *59*, 4475–4489.
80. Zhang, W. PharmaChem Directory 2003, 18–20.
81. Zhang, W.; Lu, Y. *Org. Lett.* **2003**, *5*, 2555–2558.
82. Zhang, W. *Org. Lett.* **2003**, *5*, 1011–1014.
83. Zhang, W.; Chen, C. H. T.; Nagashima, T. *Tetrahedron Lett.* **2003**, *44*, 2065–2068.
84. Curran, D. P.; Amatore, M.; Campbell, M.; Go, E.; Guthrie, D.; Luo, Z. *J. Org. Chem.* **2003**, *68*, 4643.
85. Schwinn, D.; Glatz, H.; Bannwarth, W. *Helv. Chim. Acta* **2003**, *86*, 188–195.
86. Werner, S.; Curran, D. P. *Org. Lett.* **2003**, *5*, 3293–3296.
87. Yao, Q.; Zhang, Y. *J. Am. Chem. Soc.* **2003**, *126*, 74–75.
88. Hein, J. E.; Hultin, P. G. *Synlett* **2003**, *5*, 635–638.
89. Ashton, M. R.; Moloney, B. *Curr. Drug Disc.* **2003**, *8*, 9–11.
90. DeSimone, R. W. *Drug Disc. Today* **2003**, *8* (4), 156.
91. Wang, Y.; Miller, R. L.; Sauer, D. R.; Djuric, S. W. *Org. Lett.* **2005**, 7, 925–928.
92. Schreiber, S. *Science* **2000**, *287*, 1964–1969.
93. Burke, M. D.; Schreiber, S. L. *Angew. Chem. Int. Ed.* **2004**, *43*, 46–58.
94. Spring, D. R. *Org. Biomol. Chem.* **2003**, *1*, 3867–3870.
95. Timmer, M. S. M.; Risseeuw, M. D. P.; Verdoes, M.; Filippov, D. V.; Plaisier, J. R.; van der Marel, G. A.; Overkleeft, H. S.; van Boom, J. H. *Tetrahedron Asym.* **2005**, *16*, 177–185.
96. Musonda, C. C.; Taylor, D.; Lehman, J.; Gut, J.; Rosenthal, P. J.; Chibale, K. *Bioorg. Med. Chem. Lett.* **2004**, *14*, 3901–3905.
97. Huang, W.; O'Donnell, M.-M.; Bi, G.; Liu, J.; Yu, L.; Baldino, C. M.; Bell, A. S.; Underwood, T. *J. Tetrahedron Lett.* **2004**, *45*, 8511–8514.

Biographies

Mark Ashton gained his BSc and PhD in chemistry at Bath University in the UK, where he also undertook postdoctoral research. He has worked within the outsourcing industry for the past 10 years, and has both operational and commercial experience of a wide range of pharmaceutical-related projects, including focused library design and synthesis plus medicinal chemistry projects on a range of target classes and therapeutic indications together with technology development and optimization.

Ashton joined Evotec AG in 1995, then named Oxford Diversity. Initially working as a combinatorial chemist and team leader, he helped to initiate and manage the combinatorial chemistry activities of the company. In 1998, Ashton was promoted to a department manager of discovery services, and was responsible for the management of a number of combinatorial and medicinal chemistry projects. In May 2000, he was promoted to the director of discovery services, and in March 2003 to the president of discovery services. As the president of discovery, he was responsible for the discovery division of Evotec: a division of over 250 chemists and biologists involved in projects ranging from high-throughput screening through parallel synthesis and medicinal chemistry to preclinical development. In April 2005, he was promoted to his current role of executive vice president of business development, in which he is responsible for the global, commercial, and marketing activities of Evotec, discussing outsourcing needs with the world's leading pharmaceutical and biotechnology companies. This covers Europe, the USA, and Asia.

Ashton is the author and coauthor of a number of reviewed scientific publications and patents.

Brian Moloney gained his BSc in chemistry at the University of Leicester, and his PhD at UWIST, Cardiff, in the UK. He then undertook a period of postdoctoral research at Stanford University in the USA. Moloney then worked for over 10 years in the agrochemical industry, first with Schering Agrochemicals, then AgrEvo, following a merger with Hoechst, and latterly with Aventis CropScience, following the merger with Rhône-Poulenc Agro. During his time at AgrEvo, Moloney set up and ran the automation and combinatorial synthesis groups for over 5 years, during which time he also worked on secondment to the HMR, now Sanofi-Aventis, Combinatorial Technologies Center in Tucson, AZ.

Moloney joined Evotec AG in 2000, then named Oxford Asymmetry International, as a department manager, and is responsible for the management of the parallel chemistry collaborative projects for large and small pharmaceutical, biotechnology and agrochemical companies in the Discovery Services Division. In addition, he is also responsible for many of the support functions of the division, namely automation, building blocks synthesis, data registration, reformatting, and discovery analysis, and purification.

Moloney is the author and coauthor of a number of reviewed scientific publications and patents, and had a product released to the market in 2006.

Comprehensive Medicinal Chemistry II
ISBN (set): 0-08-044513-6
ISBN (Volume 3) 0-08-044516-0; pp. 761–790

3.35 Polymer-Supported Reagents and Scavengers in Synthesis

S V Ley, I R Baxendale, and R M Myers, University of Cambridge, Cambridge, UK

3.35.1	**Introduction**	**791**
3.35.1.1	Historical Perspective	792
3.35.1.2	Suitable Supports for Immobilized Reagents	792
3.35.1.3	Practical Advantages of Immobilized Reagents	793
3.35.1.4	Site Isolation Effects	794
3.35.2	**The Development of Methods and Protocols**	**795**
3.35.2.1	Immobilized Scavengers	795
3.35.2.2	Catch-and-Release Strategies	795
3.35.2.3	One-Pot Reactions	796
3.35.2.4	Improved Reaction Quenching and Scavenging Techniques	797
3.35.3	**Multistep Synthesis of Biologically Active Compounds**	**800**
3.35.3.1	Synthesis of Commercially Available Drug Substances	801
3.35.3.2	Chemical Library Synthesis Using Immobilized Reagents	804
3.35.3.3	Applications of Encapsulated Catalysts in Library Construction	812
3.35.3.4	Synthesis of Natural Products	814
3.35.3.4.1	The Amaryllidaceae alkaloids oxomaritidine and epimaritidine	815
3.35.3.4.2	The alkaloid epibatidine	816
3.35.3.4.3	The plant lignan carpanone	818
3.35.3.4.4	Plicamine and the related amaryllidaceae alkaloids plicane and obliquine	819
3.35.3.4.5	The epothilones	821
3.35.4	**Flow Chemistry in Organic Synthesis**	**826**
3.35.4.1	Microreactors	826
3.35.4.2	Advantages of Flow Techniques in Practice	827
3.35.4.3	Use of Flow Techniques in Organic Synthesis	828
3.35.5	**Conclusions**	**830**
References		**831**

3.35.1 Introduction

This chapter highlights the latest developments and key advances in preparative techniques that are rapidly being adopted in the modern synthesis laboratories.[1] In particular, we describe the impact on organic synthesis by the judicious use of immobilized species to conduct reactions, quench chemical processes, scavenge by-products, and bring about the isolation of pure products.[2–11] While the focus of the chapter is on the generation of medicinally and biologically relevant structures, broader topics, encompassing new reaction types, cleaner chemical processes, and possibilities for scale-up are also discussed. Where appropriate, we have highlighted the main opportunities and the remaining challenges in these areas, within the context of today's changing commercial, environmental, and ethical climates. Our overall aim is to present a vision for how synthesis chemistry is responding toward a more productive, efficient, and sustainable future.[12,13]

The pharmaceutical industry is without doubt a very demanding sector. Medicinal chemists are expected to make discoveries and prepare compounds at a phenomenal rate with increasing levels of structural diversity. Hence, the new methods adopted need to provide alternatives to the labor-intensive practices of the past, such as manual optimization, aqueous work-ups and extractions, difficult crystallizations or distillations, and chromatographic purifications.

The desire to reduce these repetitive bottleneck operations has led to a noticeable increase in the use of automation,[14] informatics,[15] and other technology-based approaches in the laboratory.[16] Furthermore, by combining the use of supported reagents with new equipment such as microfluidic flow reactors,[17–33] focused microwaves,[34] and the practice of catch-and-release strategies,[35–40] many new opportunities for synthesis are being recognized.

3.35.1.1 Historical Perspective

The pioneering work of Letsinger[41] and Merrifield[42] on solid-phase organic synthesis has become a widely used tool for rapidly constructing large compound libraries. In solid-phase synthesis, the substrate undergoing transformation is covalently attached to the support material and the reagents and/or coupling partners are present in solution. Following a transformation, the material is purified by a sequence of washing steps to elute solution-labile impurities, before a cleavage step is performed to liberate the functionalized product from the support, thereby enabling its isolation in a pure form.

While solid-phase synthesis has been an important advance, particularly for combinatorial applications, there are intrinsic difficulties associated with these methods that often outweigh the benefits, especially for syntheses of complex, biologically active compounds. Therefore, the application of solid-phase reagents, catalysts, and scavenging techniques presents an attractive alternative to linear, substrate-bound synthesis. This approach embraces many of the advantages of conventional solution-phase chemistry such as real-time reaction monitoring, convergency, and rapid optimization, but also enables purification by the simple expedient of filtration to remove the spent reagents. These techniques also readily accommodate multistep processes, parallel methods, batch-splitting, and reaction scale-up.

The first serious application of supported species was in the formation of ocadec-9-enoic acid butyl ester using a sulfonic acid resin in 1946.[43] The 1950s saw further advances in the form of acidic and basic ion exchange techniques, which are still used as scavengers and buffers in water treatment plants today.[44] In 1957 the first review of ion exchange resin catalysts was published.[45]

The 1990s, however, heralded a dramatic change in the use of supported reagents in synthesis. Underlying this was the recognition by the pharmaceutical industry as to how to better make greater numbers of compounds in response to high-throughput screening capability. This led to the concepts of combinatorial chemistry.[46–53] Consequently, renewed interest in immobilized agents was triggered by the development of efficient agents to quench chemical reactions and scavenge out unwanted by-products. The time-saving potential of supported reagents was quickly recognized when it was discovered that clean materials could be obtained that were suitable for direct biological evaluation by a simple filtration of the reaction mixture to remove spent reagents or captured by-products. In combination with sequestering processes, solid-supported reagents also provide powerful synthetic tools for conducting multistep chemical operations. This latter topic forms the basis of a major review that summarizes relevant contributions from some 1500 papers.[9] This review documents early contributions and provides a significant reference source for supported reagents in chemical transformations, and details many scavenging applications. The wide-ranging use of immobilized reagents to individual processes has also been collected in a number of other articles.[2,54–58] Further reviews of interest highlight alternative aspects of this science; for example, soluble polymer species,[59,60] microwave methods with supported reagents,[34,61–67] supported chiral catalysts,[5,68–70] encapsulation and entrapment processes,[71–90] phase-switching methods,[91,92] natural product synthesis using supported reagents and scavengers,[54–58] chemical library generation,[93] and stop-flow and continuous-flow reactor systems.[22,24,94,95]

3.35.1.2 Suitable Supports for Immobilized Reagents

Reagents can be immobilized by tethering them to an insoluble (or semi-soluble) support material; this attachment can be via covalent or electrostatic interaction with the functionalized solid matrix. Commonly, this support is divinyl benzene (DVB) cross-linked form of polystyrene. These polystyrene resins can be either micro- or macroporous, depending on the degree of cross-linking. To date, the majority of the immobilized reagents that have been developed are polystyrene based; as polystyrene supports are not only cheap but are also easy to handle, achieve high loadings, and, importantly are relatively chemically inert.[8]

In general, polystyrene resins swell adequately in most common organic solvents; however, this is dependent upon their cross-linking. Microreticular resins are defined by a low level of cross-linking (1–2% DVB) and swell more easily than their macroreticular counterparts ($>30\%$ DVB). Correspondingly, the bulk characteristics of the supporting polymer in terms of its physical properties and architecture can infer vastly altered reactivity and reaction kinetics for the same immobilized reagent. In this regard, a variety of other polymers such as acrylamides, polyureas, and co-polymer formulations including PEGylated polystyrenes and Jandagel[96] have received considerable attention. With an increasing selection of synthetic polymers becoming available with new physical and functional characteristics, polymers will continue to be useful for the preparation of supported reagents.

However, while polystyrene may be the most commonly used support, it is certainly not the only material that has been used. The encapsulation of catalysts inside polymeric matrices is one class of reagent involving polymers in an innovative way.[79,80,97] Alternative supports such as controlled pore glass, monoliths,[98–104] cellulose,[105,106] zeolites,[107–109] and silicas have also been used.[110–112] Irrespective of which support is employed, it is essential that the bound reactive species remains accessible within the support matrix to the soluble substrates.

3.35.1.3 Practical Advantages of Immobilized Reagents

The simplicity of separation allows supported reagents to be used in excess. Reagent concentrations can thus be used to force reactions to completion, which in turn leads to cleaner conversions. As discussed, another practical advantage is how the reaction mixtures can easily be manipulated; for example, by filtering off the spent supported reagent and evaporating the filtrate, the products can be isolated readily.

By contrast, solution-phase reagents often create unwanted by-products that can be difficult to remove, even with extensive aqueous work-up and conventional chromatographic purification procedures. For example, triphenylphosphine is a common, versatile, and widely used reagent in many synthetic transformations, including the Staudinger,[113] Mitsunobu[114–115] and Wittig reactions.[116] Chemists readily recognize the difficulties associated with removal of the resultant phosphine oxide by-product and excess triphenylphosphine in these reactions. The utility of immobilized reagents to overcome these difficulties is illustrated by the use of a polymer-supported triphenylphosphine equivalent in a Wittig reaction. This enables facile removal of the derived phosphine oxide and excess reagent by filtration (**Scheme 1**).[117–120] This example readily demonstrates the practical advantages achievable using supported reagents.

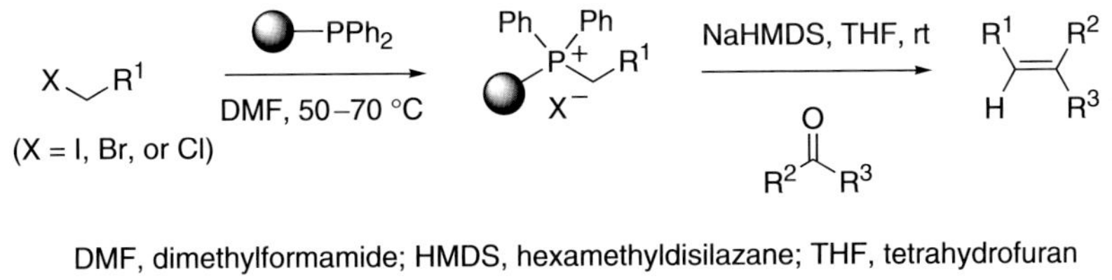

Scheme 1 Immobilized triphenylphosphine used in a Wittig reaction.

Importantly, many of the recovered spent reagents may be recycled a number of times, which is necessary to make them financially viable options. The ease of handling supported reagents is also noteworthy, particularly when dealing with expensive catalysts, since these systems are more likely to be incorporated into automated or flow processes.

Also, the process of immobilizing reagents produces reactive entities that often no longer possess the same safety or environmental concerns that their homogeneous counterparts display. Problems associated with foul stenches, extreme toxicity, and noxious build-up or other hazardous issues such as flammability or explosion are significantly reduced by immobilization. This aspect is nicely exemplified in the thionylation reaction shown in **Scheme 2**, whereby the

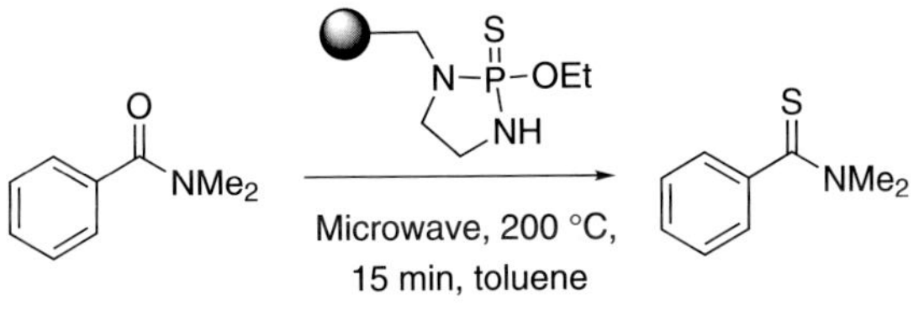

Scheme 2 Immobilized Lawesson's reagent used in the conversion of amides to thioamides.

malodorous sulfur by-products remain immobilized using the heterogeneous equivalent of Lawesson's reagent[121] in the conversion of amides to thioamides.[122] Other examples include the use of microencapsulated osmium tetroxide to perform dihydroxylation reactions to avoid the issues associated with release of volatile metallic osmium.[75]

Supported reagents generally have slower reaction kinetics but the reactivity of supported systems can be greatly enhanced using short bursts of energy provided through focused microwave heating.[123–125] Indeed, this is often preferred over prolonged heating using standard methods such as oil baths, which can result in partial decomposition of many of the commonly used supporting materials. For reaction systems requiring the use of poor microwave-absorbing solvents, doping with ionic liquids can provide a thermocouple for effective heating.[126–138]

3.35.1.4 Site Isolation Effects

Immobilization of a reagent facilitates both reaction work-up and product purification and consequently accelerates the synthesis overall process. This inevitably frees up more time for the creative element of synthesis planning and thereby increases overall productivity and output (**Figures 1** and **2**).

The three-dimensional and steric environment of the support matrix creates a system of porelike structures with unusual topographies; they exist in isolation as a consequence of phase partitioning. Such structural characteristics can

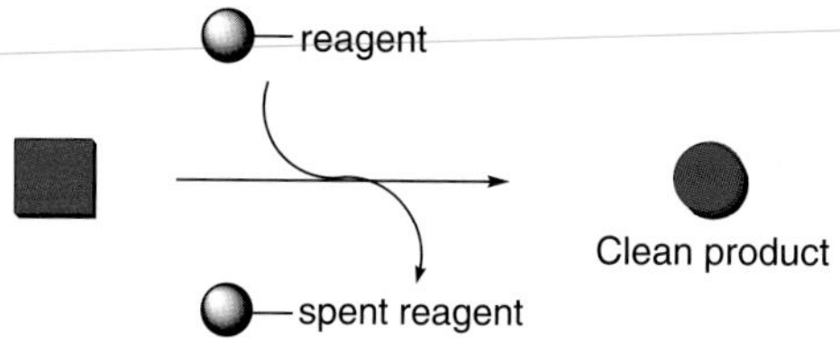

Figure 1 Supported reagents are able to transform substrates into new chemical products, where spent or excess reagents can be readily removed by filtration.

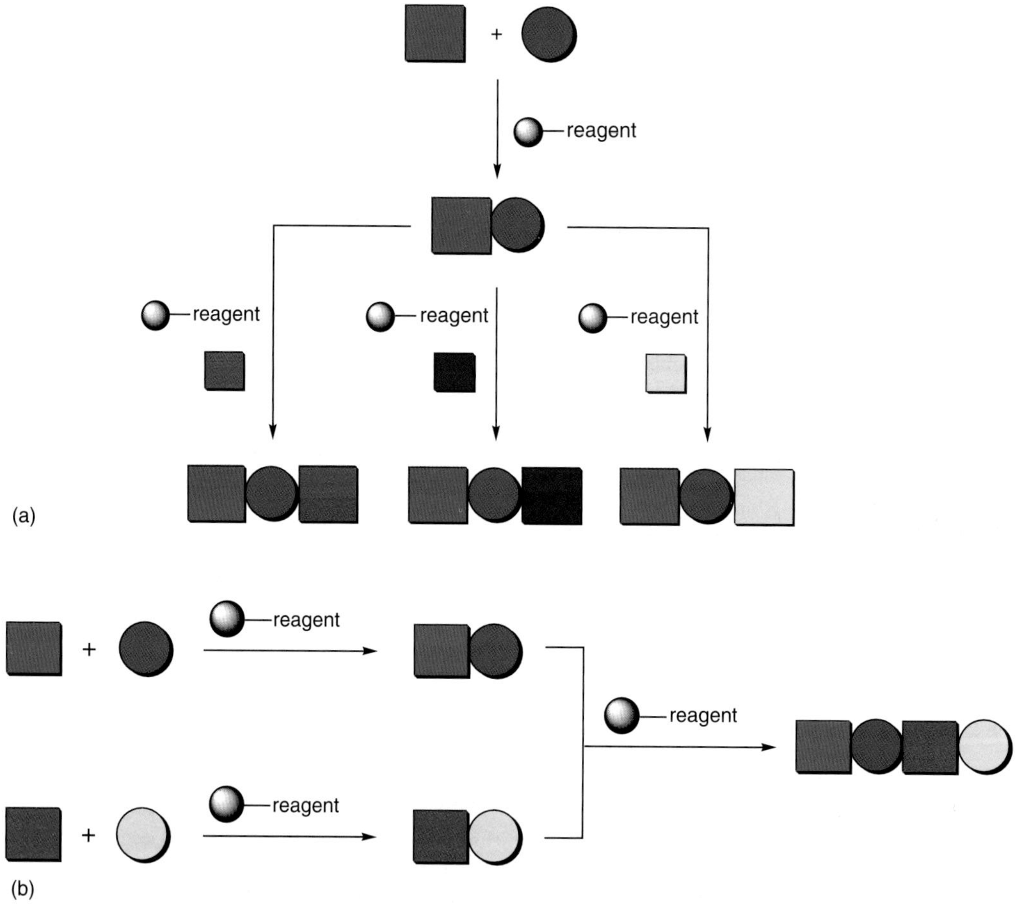

Figure 2 Process arrangements employing supported reagents. (a) Batch-splitting process. (b) Convergent synthesis.

produce reagents with unique reactivities that can help overcome solvent-specific issues common to conventional solution-phase chemistry methods.[138–143]

The support material impacts on the reactivity of the attached functional groups. It can allow certain reactions to proceed at higher than usual concentrations; ones that would normally only be possible under very high dilution conditions. Macrocyclization is one example of a reaction that typically requires very high dilution. The reasoning here is that if the rate of intermolecular coupling is minimized, then it follows that there will be a concomitant increase in the intramolecular coupling reaction. It has been shown that by using an appropriate polymer-supported coupling reagent, for example an immobilized tetrakistriphenylphosphine palladium(0) reagent (**Scheme 3**), that controlled

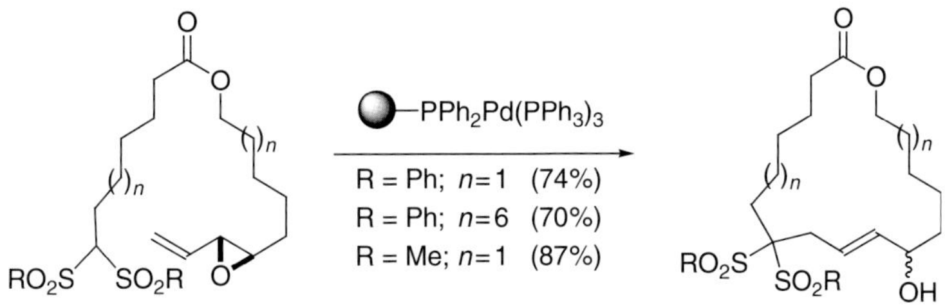

Scheme 3 Macrocyclization reactions immobilized tetrakistriphenylphosphine palladium(0).

concentrations are of less significance.[144] No dimeric or oligomeric side-products were obtained with the immobilized system when compared with the solution-phase equivalent reaction.

These same principles also allow selective monoprotection of equivalently difunctional molecules.[37–39,145–152] As substrate binding is temporary for the duration of the reaction, other pendant functional groups are able to react in preference. As the reagents are anchored to the support materials they are site isolated; this also allows the simultaneous use of multiple reagents in one-pot transformations. This concept creates many new opportunities for organic synthesis.

3.35.2 The Development of Methods and Protocols

3.35.2.1 Immobilized Scavengers

Scavengers are supported compounds that selectively sequester by-products from a reaction and render them insoluble such that they can be readily removed by filtration (**Figure 3**). Their use can be highly effective at improving the purity profile of complex reaction streams without resorting to liquid–liquid extractions or column chromatography. Scavengers can exploit both ionic and covalent interactions, and bind either organic or inorganic impurities. In the simplest of cases, electrophilic or nucleophilic by-products are removed by reciprocally functionalized supports.

3.35.2.2 Catch-and-Release Strategies

A related concept involves a process referred to as a 'catch-and-release' (**Figure 4**). Here, a suitably functionalized support can be designed to react with the desired product in a reaction mixture. After filtration and washing with a solvent to remove by-products or other unwanted materials. The captured compound can then be released in a pure

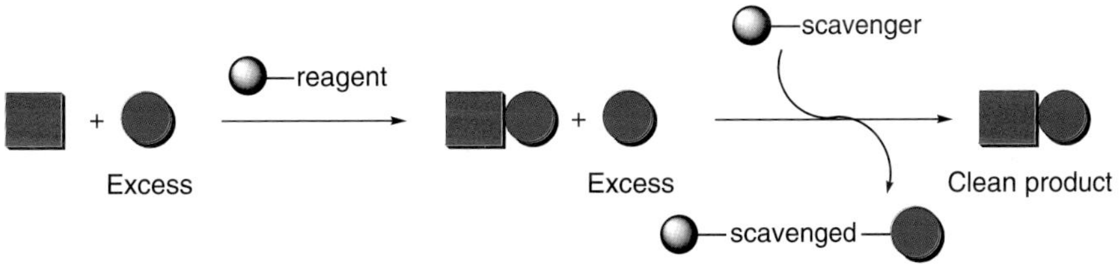

Figure 3 Removal of excess components using supported scavengers.

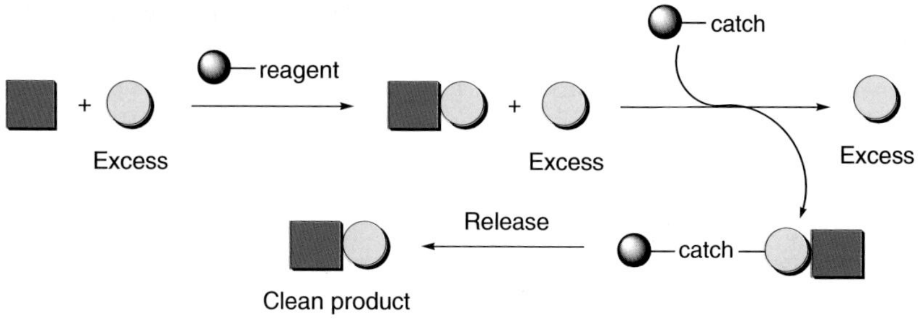

Figure 4 The catch-and-release principle.

form from the support by an appropriate chemical process. The release mechanism can be driven by a variety of reactions such as hydrolysis or acid–base exchange. In more sophisticated examples the concept of 'catch-activate-and-release' can be invoked to provide a different product from the cleavage process. Similarly, the release step may involve a chemical functional group interchange such as a reductive or oxidative conversion.

3.35.2.3 One-Pot Reactions

As many reagents used in traditional solution-phase chemistry are incompatible with one another, the property of site isolation using supported reagents provides yet another opportunity for multistep transformations, since it is possible to use mutually incompatible reagents, such as oxidants and reductants, together in a single reaction vessel. Likewise, the existence of short-lived intermediates can be determined by movement between two independent polymeric trapping agents. This concept was a landmark observation in the 1970s, employing two polymeric agents simultaneously for sequential reactions in one pot.[153–155]

A short while later, combinations of different polymer-supported triphenylphosphine-ligated transition metal complexes were used as catalysts during the functionalization of various alkenes.[156] A so-called 'wolf and lamb' approach employing the site isolation effect showed that enolates could be generated and subsequently *C*-acylated in situ by an immobilized trityllithium base and an acyl transfer polymer.[157] Furthermore, a third and final step was incorporated to lead directly to a 3,5-diphenylpyrazole in a single reaction pot (**Scheme 4**). This work constitutes the first real attempt at a multistep application of supported reagents to synthesize a target compound.

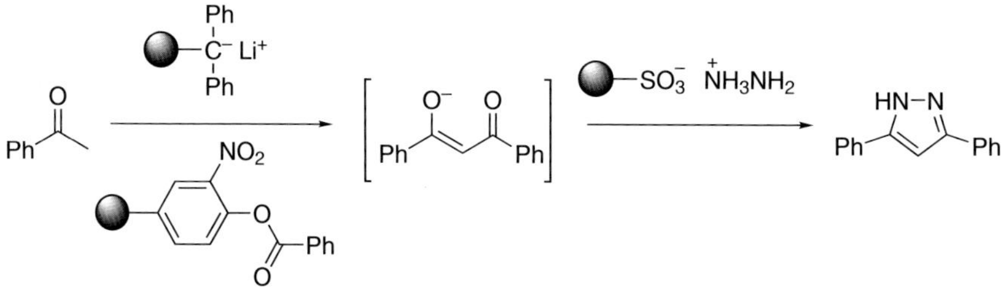

Scheme 4 Preparation of 3,5-diphenylpyrazole in a one-pot process.

Another early example describes the direct conversion of acetals to unsaturated nitriles.[158] In this process the acetal underwent acid catalyzed hydrolysis by an acidic Amberlyst 15 sulfonic acid resin, the unmasked carbonyl was then immediately trapped by a supported phosphonate reagent in a Wadsworth–Horner–Emmons alkenylation reaction (**Scheme 5**). The process was found to be solvent-independent, suggesting that the resin provided a

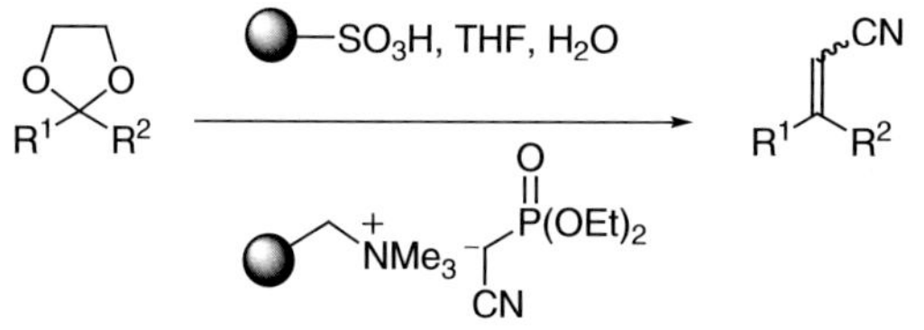

Scheme 5 Direct one-pot transformation of acetals into nitriles.

microenvironment for the reaction away from the external medium. Such chemistries serve to illustrate the potential of using supported reagents for progressing a substrate through a predetermined sequence of chemical transformations, leading to a target compound formed in solution.

Indeed, the benefits of being able to manipulate reactive intermediates in solution are neatly exemplified by the preparation of a series of trihydroxy nucleosides (**Scheme 6**).[159] Here, the oxidative cleavage of the 1,2-diol

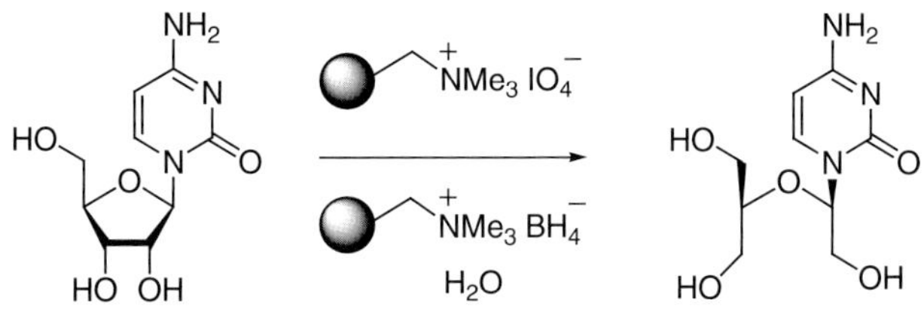

Scheme 6 Simultaneous application of an oxidant and reductant for 1,2-diol cleavage.

functionality generates a highly unstable dialdehyde component, which is intercepted by an in situ supported borohydride reagent to effect reduction. This cascade sequence was achieved by pumping solutions of the substrate through a prepacked column containing an equal stoichiometry of the periodate and the borohydride resins. Evaporation of the solvent from the reaction stream directly furnished the triol product without the need for any further chemical processing and generates the product in better yield than the conventional synthesis approach. This general process was later used for the cleavage of 1,2,5,6-diisopropylidene mannitol during a synthesis of phosphatidylcholines.[160]

Pioneering work by Cardillo *et al.* in 1982 demonstrated how a two-step supported reagent system could be used for the conversion of β-iodoamines into amino alcohols (**Scheme 7**).[161] The interesting aspect of this reaction was the use

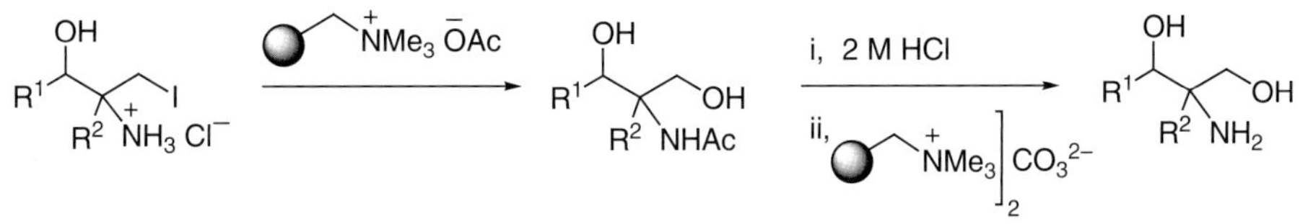

Scheme 7 A three-step preparation of amino alcohols from β-iodoamines.

of a supported acetate resin to promote iodine displacement, presumably via an intermediate aziridine, and subsequently promote acyl transfer to the amine group. The final hydrolysis and conversion to the free base was achieved using hydrochloric acid and a carbonate resin. This reaction is of synthetic interest since amino alcohols are usually very soluble in water, which consequently causes isolation problems during a conventional aqueous work-up procedure.

The overall increase in complexity of transformations was neatly demonstrated in the preparation of the β-adrenergic receptor agonist propanolol. Three immobilized reagents were used to bring about the preparative steps in this route (**Scheme 8**).

The next significant use of immobilized reagents as synthesis tools was beautifully demonstrated by Parlow *et al.* in 1995.[162] This team performed an orchestrated sequence of reactions that led to the preparation of a specifically functionalized pyrazole (**Scheme 9**). The synthetic steps required oxidation of a secondary benzylic alcohol using chlorochromate immobilized on a polyvinyl pyridine, followed by α-bromination and displacement with a supported alkoxide derived from a 3-hydroxy pyrazole. The same product was also obtained (albeit in 48% yield) when all the components were added together at the start in a single pot.

However, the real utility of supported organic synthesis comes from harnessing all the power these systems have to offer. That is, by using supported reagents, scavenging and quenching agents, and catch-and-release methods, together with continuous-flow and other techniques, tremendous opportunities for synthesis then emerge.

3.35.2.4 Improved Reaction Quenching and Scavenging Techniques

The recognition of the importance of reaction quenching and scavenging in organic synthesis occurred in the mid-1990s, when various industrial research groups independently demonstrated their utility for automated chemical library generation. In particular, the research groups of Parlow and Flynn (then at Monsanto), Kaldor and Siegel (at Eli-Lilly),

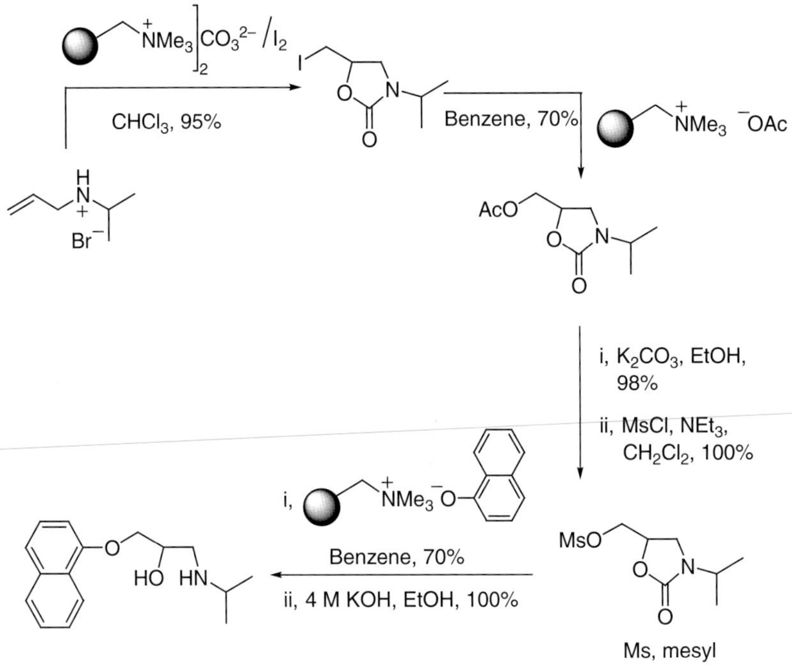

Scheme 8 Synthesis of the β-adrenergic receptor agonist propanolol.

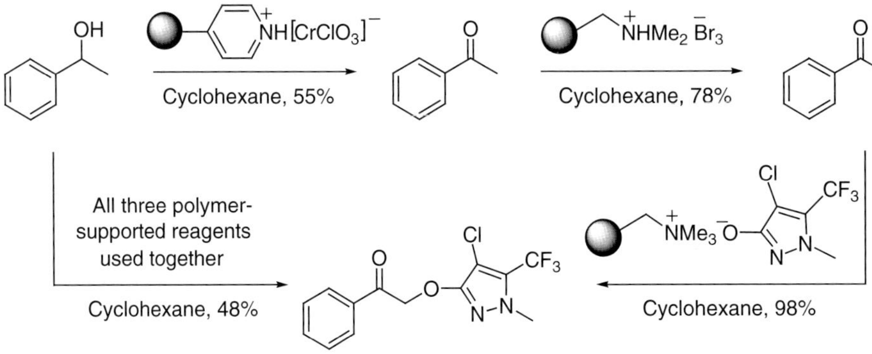

Scheme 9 A one-pot multi-reagent approach versus a sequential approach.

and Hodges (at Parke-Davis) were all key players. Between them, they reported a number of applications focused on reaction quenching and the chemoselective scavenging of by-products, excess substrates, and spent reagents. As a result of these efforts new and improved polymer-supported scavenging systems became commercially available. While it is not possible to do full justice to these early pioneers here, a few examples are discussed that illustrate the potential of these methods.

Kaldor and Siegel[163] were the first to use supported scavengers in the expedient construction of various chemical arrays. **Scheme 10** illustrates how they used supported reagents and scavengers via a two-step process that led to a large collection of amides, sulfonamides, ureas, and thioureas that were then assayed for biological activity. The initial reductive amination (step i) required the use of an excess primary amine to drive the imine formation to completion. Following reduction with an immobilized borohydride (step ii), the excess primary amine was scavenged by an immobilized aldehyde (step iii). The purified secondary amine was then reacted with an excess of a sulfur-containing isocyanate (step iv), which after scavenging the unreacted isocyanate with an aminomethylated polystyrene resin (step v) gave the final urea product. Kaldor went on to use these methods to prepare various antirhinoviral lead compounds with submicromolar potency.[164]

Parlow and Flynn used modified Mitsunobu reagents in the parallel synthesis of a solution-phase library (**Scheme 11**).[165] A disadvantage often encountered in this reaction is the difficult removal of by-products, in this case

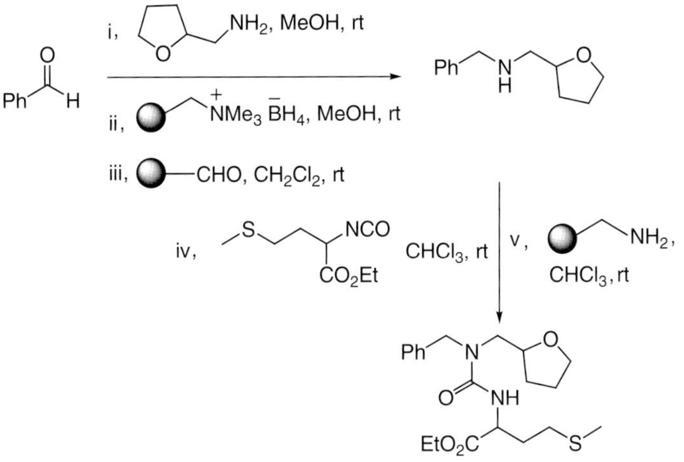

Scheme 10 Construction of parallel arrays of amides, sulfonamides, ureas, and thioureas.

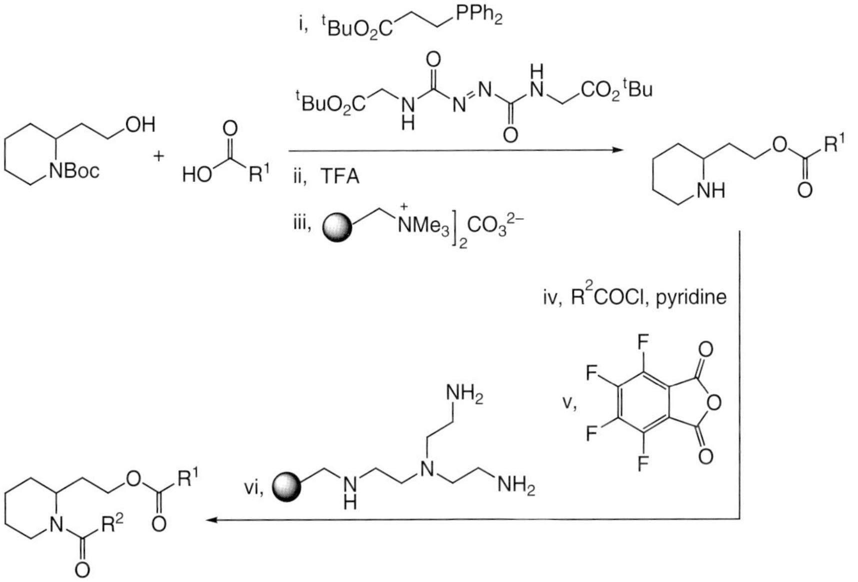

Scheme 11 Chemically tagged Mitsunobu reagents in the solution-phase library synthesis.

dialkyl hydrazinedicarboxylate and triphenylphosphine oxide. In the example, bespoke chemically tagged phosphine and azodicarboxylate reagents (step i) were specially developed to be used in the Mitsunobu[114,115] process. Masked carboxylic acid tags (*t*-butyl esters) were attached to both reagents, and, after the Mitsunobu reaction were deprotected, with trifluoroacetic acid were deprotected scavenging with a carbonate base. The deprotected secondary amine was then progressed by acylation with an acid chloride in pyridine. Any remaining free amine was derivatized using tetrafluorophthalic anhydride[166] in step v and subsequently removed, along with any remaining acid chloride, by a polyamino scavenging resin in step vi.

Recently, a more generic strategy for the parallel synthesis of small-molecule libraries based on the catch-and-release principle[167,168] was reported. In this approach, a method of selectively and reversibly immobilizing intermediates in a conventional solution-phase reaction sequence was devised (**Scheme 12**).[169]

The idea was based on a phase switch concept[91] involving reversible and selective noncovalent interactions between a resin-bound metal and an organic metal-chelating tag linked (through the hydroxyl groups) to the compound of interest.[170] The tagged intermediates were immobilized (or phase switched) when they were sequestered at the end of the reaction by chelation to a resin-bound metal. After washing the resin, the purified intermediates were easily

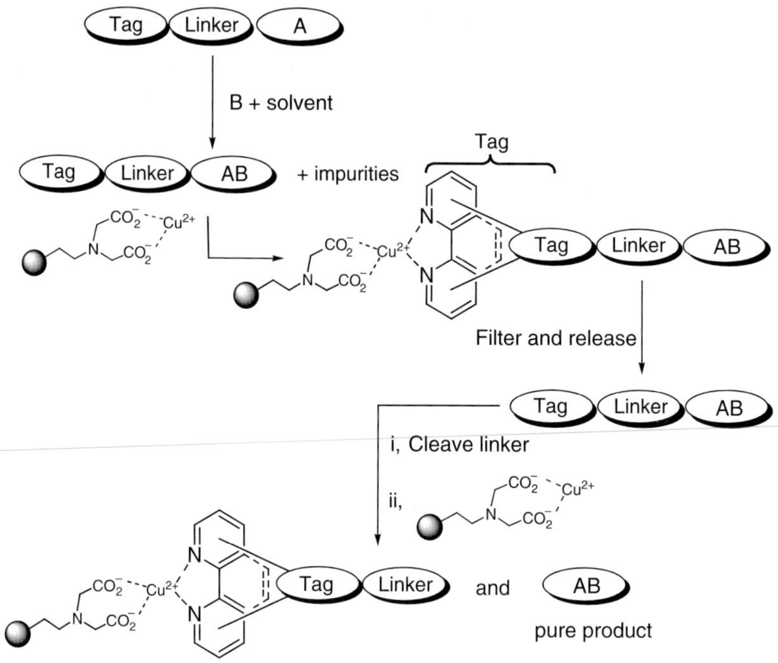

Scheme 12 Phase-switch method using immobilization through metal-chelated tags.

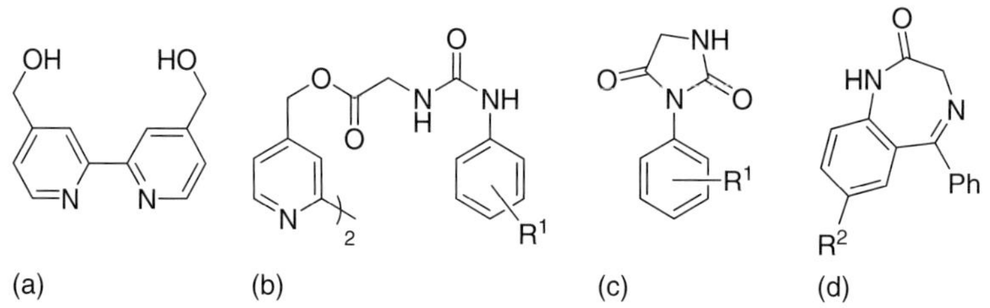

Figure 5 (a) Tag, (b) ureas, (c) hydantoins and (d) benzodiazepines.

obtained by competitive displacement from the resin. The final step required controlled cleavage of the tag component to yield the desired products. To exemplify this principle a series of ureas, hydantoins, and benzodiazepines were prepared (**Figure 5**).[169,171]

In this work, the resin offering the best loading of copper(II) ions was the inexpensive IRC-718 material containing an aminodiacetic acid residue, and the preferred metal-chelating tags were bidentate pyridine-containing ligands such as 4,4′-bis(hydroxymethyl)-2,2′-bipyridine. A range of chemistries were developed from this initial work based on the same tag and linker system (e.g., a purification by sequestration of a Horner–Emmons reaction).[172] In addition, a diverse set of compounds was purified based on the ability to selectively sequester the chemically benign bipyridyl tags. This universal tag offers a versatile handle for developing many more new reagents and scavengers. The concept however is much more general and many more examples can be expected in the future.

3.35.3 Multistep Synthesis of Biologically Active Compounds

In the hope of finding new drug candidates, chemists now have a variety of options that they can select from the new toolbox of opportunities, not just the preparation of any structure by the designed application of specific compounds or limited compound sets.

3.35.3.1 Synthesis of Commercially Available Drug Substances

In order for any new technology to be adopted it must demonstrate advantages over prevailing practices. Consequently, the effectiveness of supported reagents was initially validated by the efficient preparation of commercially available drugs. The compound sildenafil (Viagra), for example, which is used for the treatment of male erectile dysfunction, has become one of the largest-selling globally marketed prescription drugs in recent history. It acts by inhibiting the phosphodiesterase enzyme PDE-5, the main phosphodiesterase present in the smooth muscle of the corpus cavernosum.[173]

A key feature of this synthesis of sildenafil was the convergent strategy that was adopted. The reaction between the resin-bound activated ester (generated from **1** and **2**) and heterocyclic amine **3** formed the important amide bond with any excess amine being scavenged by polymer-supported isocyanate (**Scheme 13**).[174] The supported hydroxybenzotriazole

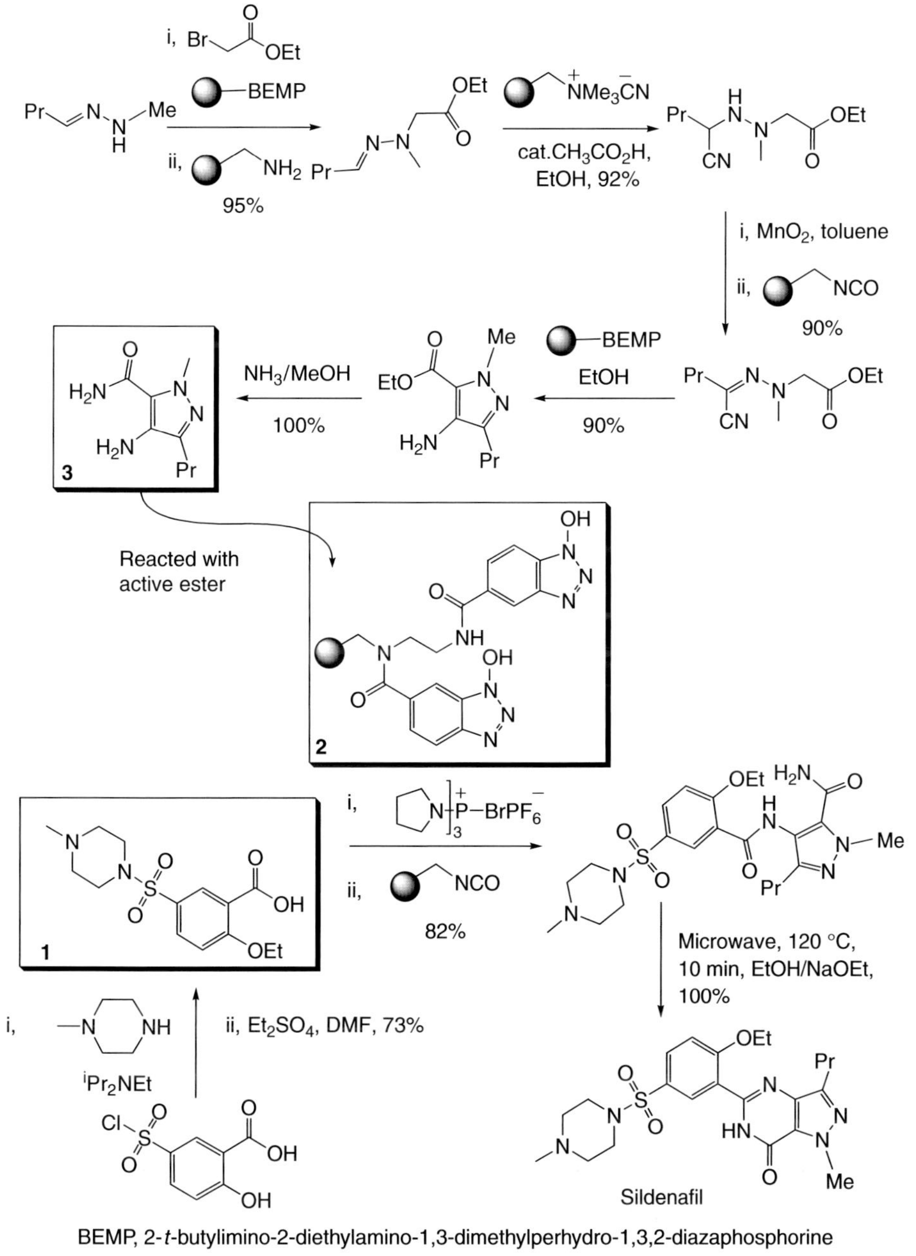

Scheme 13 Synthesis of sildenafil (Viagra).

used in the coupling step performed a dual function; namely, it facilitated the simultaneous activation of the acid **1** and also its purification from by-products accumulated during its formation. Cyclodehydration was then effected using catalytic sodium ethoxide under focused microwave irradiation conditions to produce gram quantities of Sildenafil in excellent overall yield. Clearly this synthesis could be readily adapted for analog preparation.

Adiposity is a major risk factor for insulin resistance and type II diabetes. Peroxisome proliferator-activated receptor gamma (PPAR-γ), an orphan nuclear receptor, is highly expressed in adipose tissue. Rosiglitazone (Avandia), a PPAR-γ agonist, has recently been introduced as an antihyperglycemic thiazolidinedione which is effective in treating noninsulin-dependent diabetes mellitus (type II diabetes).[175] Rosiglitazone was recently synthesized using a multistep supported-reagent approach (**Scheme 14**).[176] Noteworthy is the introduction of the pyridine moiety, which provided a

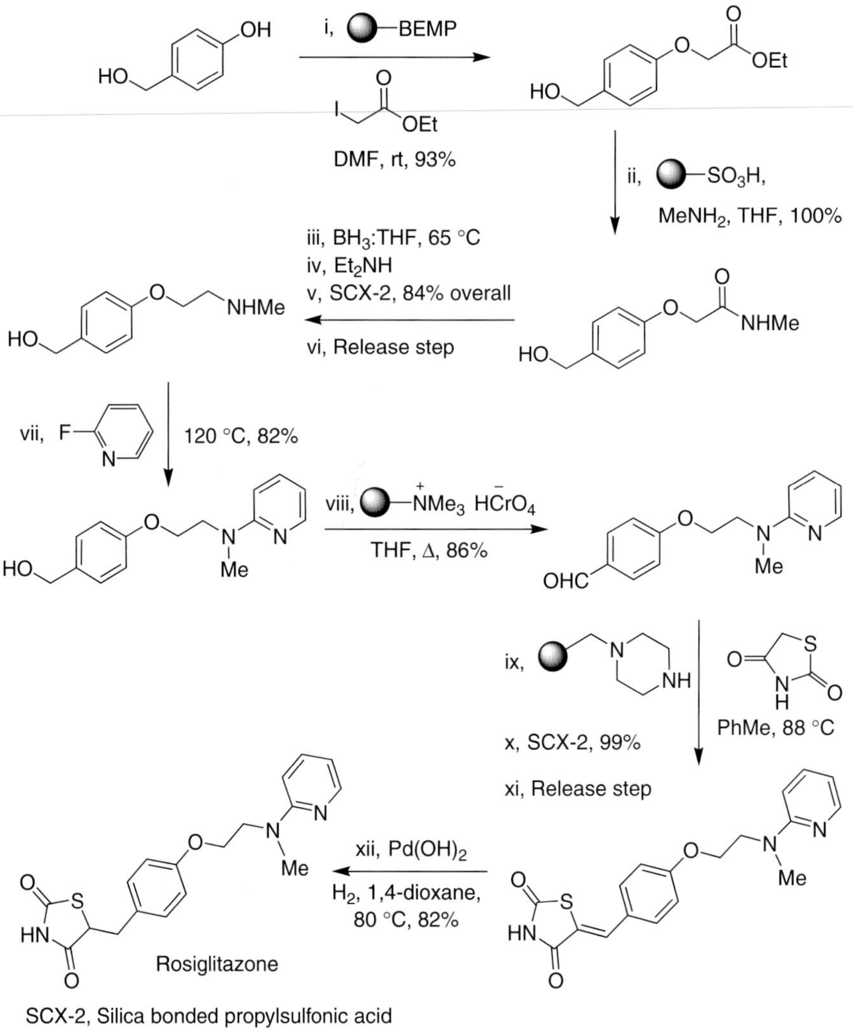

Scheme 14 Synthesis of rosiglitazone (Avandia).

molecular handle to facilitate purification of the molecule in the subsequent steps of the synthesis. It is also interesting to note that the overall yield for this synthesis (46%) was higher than the yield reported in the initial shorter solution-phase synthesis (31%).[177]

Salmeterol (Serevent) is a potent, long-acting β_2 adrenoceptor agonist used as a bronchodilator for preventing bronchospasms in patients with asthma and chronic obstructive pulmonary disease. Although a number of routes for its preparation exist, many of them have had certain drawbacks, including extensive chromatographic separation of diastereomeric mixtures. The route illustrated in **Scheme 15** circumvents these problems and enables salmeterol to be

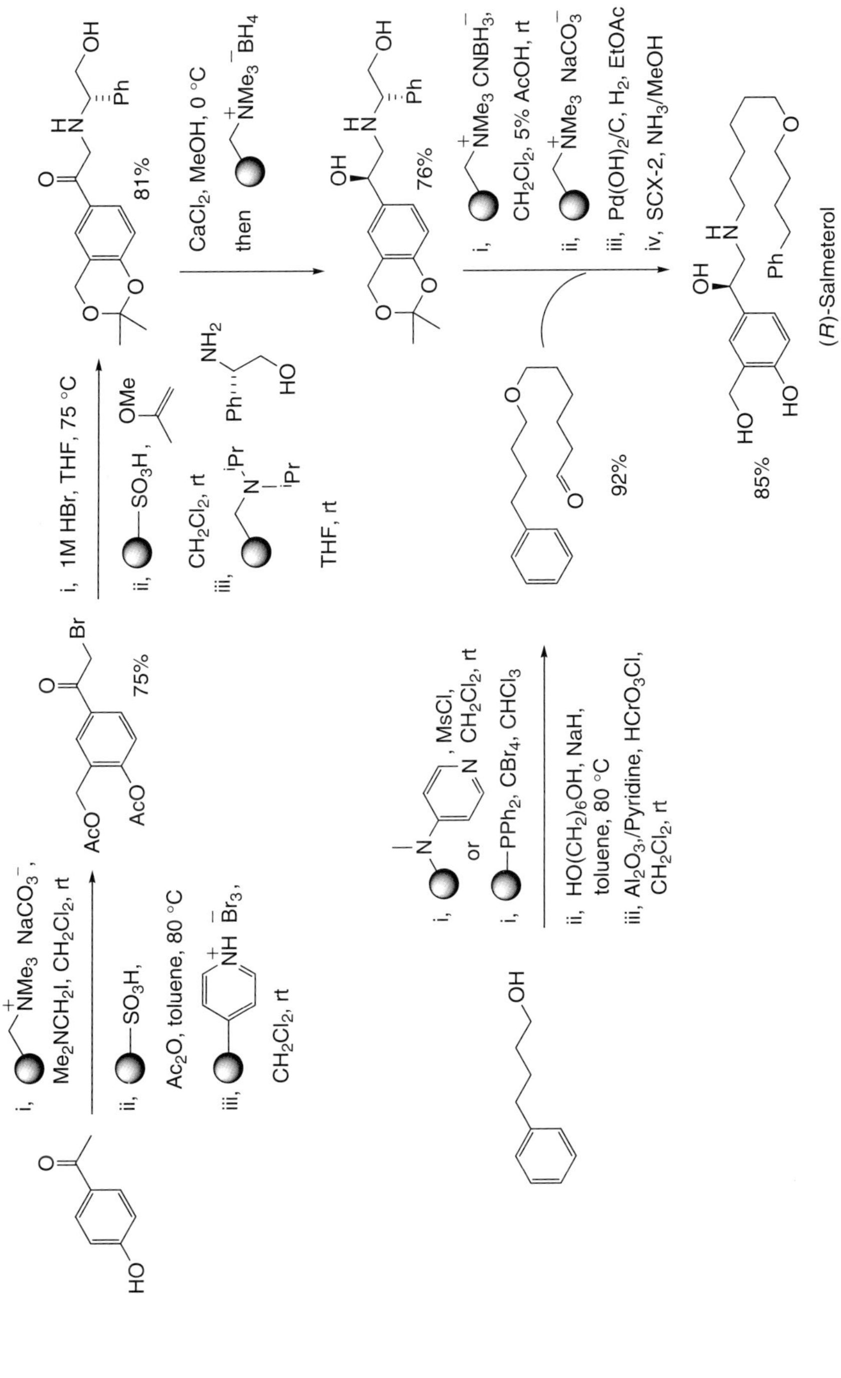

Scheme 15 Synthesis of salmeterol (Serevent).

synthesized without chromatographic purification.[178,179] However, a single recrystallization enhanced the diastereomeric purity of an intermediate, and all remaining transformations and purifications were performed using supported reagents. An interesting transformation in this work is the initial condensation of the starting phenolic material with an immobilized methyleneiminium salt. This reagent, prepared from a polymer-supported carbonate resin and Eschenmoser's amine, was found to be general for the *ortho*-aminomethylation of phenolic compounds.

3.35.3.2 Chemical Library Synthesis Using Immobilized Reagents

Modern medicine, while capable of many efficacious remedies for complex diseases and illnesses, still struggles to find cures for the common cold.[180] In a search for potential antirhinoviral candidates, scavenger resins were used to aid reaction purification during the preparation of a 4000-membered urea library (**Scheme 16**).[164] In these examples,

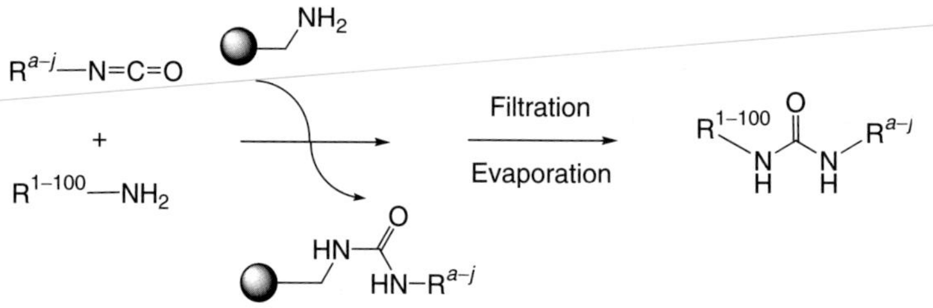

Scheme 16 Library of antirhinoviral candidates for biological screening after excess isocyanates were scavenged using an immobilized amine catalysts.

amines were reacted with an excess of isocyanate to generate the corresponding ureas, with the excess isocyanate impurities being scavenged using aminomethylpolystyrene.

In a program designed to find new triazole analogs to generate molecular diversity in a compound collection, a robust and readily automated synthesis using a 'catch, cyclize, and release' protocol was used to generate a 2500-member combinatorial library; 64 members were 3-thioalkyl-1,2,4-triazoles, which has been reported previously.[181] The synthesis made particular use of the strong polymer-supported base PS-BEMP and involved the condensation of an acyl hydrazide and isothiocyanate to give the diacylhydrazide. This remained associated with the PS-BEMP as the ion pair, which, after washing the resin, was thermally cyclized to the 3-thio-1,2,4-triazole. Subsequent treatment with a substoichiometric amount of alkylating agent released the *S*-alkylated triazole products into solution (**Scheme 17**).

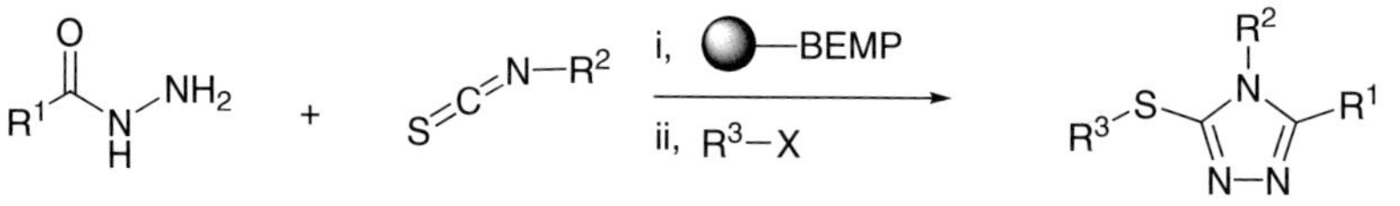

Scheme 17 Library synthesis of 3-thioalkyl-1,2,4-triazole using a catch-cyclize-and-release protocol.

This procedure was used successfully on a number of automated platforms, including the Myriad Core System (Mettler-Toledo Ltd) and the Argonaut Quest 210 (Argonaut Technologies, Inc.).

Benzimidazole derivatives (**4** and **5**) are of interest in medicinal chemistry programs because of their wide-ranging activities. The ring system is present in a range of antiparasitic, fungicidal, anithelemintic, and anti-inflammatory drugs. In addition, compounds containing this structural feature have also been found to possess antiviral and antihistamine activity, as well as the ability to modulate ion channels; they also have a developing role in other chemotherapeutics.[182]

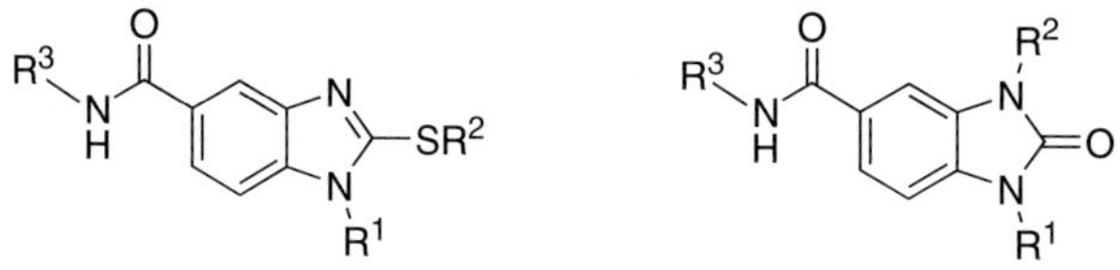

4

2-Alkylthiobenzimidazoles

5

Benzimidazolin-2-ones

It is therefore clear that these are desirable constructs for automated preparative approaches. Toward this aim, a 72-membered library of 2-alkylthiobenzimidazoles and benzimidazolin-2-ones have been prepared both manually and in a fully automated fashion.[183] Automated aqueous work-ups were incorporated with in-line scavenging and catch-and-release protocols to provide the products containing three points of diversity, without requiring any silica gel chromatography.

A Zinsser Sophas robotic synthesizer was used in another automated preparation of benzimidazole derivatives. In just one experiment, 96 thiobenzimidazoles derivatives and a 72-membered library of benzimidazolin-2-ones were rapidly prepared (**Scheme 18**).[184] The robotic system performed all the reactions sequentially, with manual intervention only required during the hydrogenation step.

In another example of automated synthesis, the Quest 205 Organic Synthesiser was used for the large-scale preparation of 2-aminothiazole scaffolds which were used for combinatorial decoration (**Scheme 19**).[185] The reaction

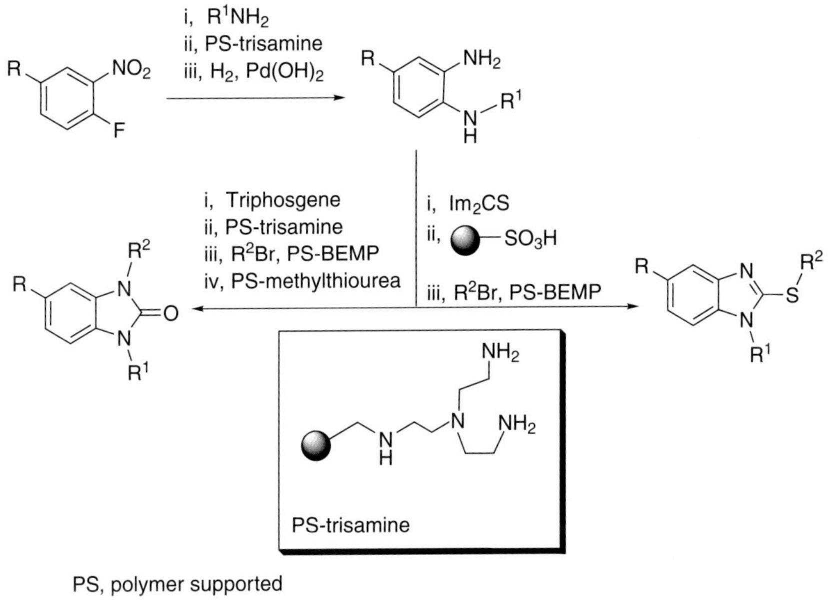

Scheme 18 Automated library synthesis of thiobenzimidazoles and benzimidazolin-2-ones.

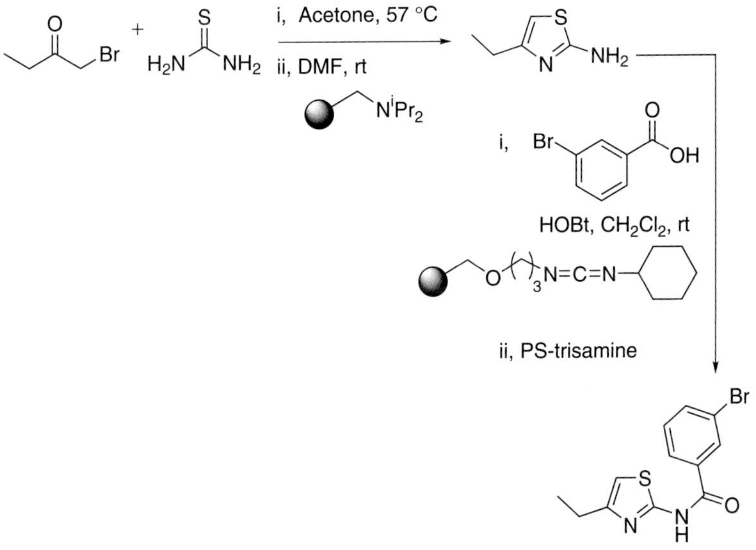

Scheme 19 Automated parallel synthesis of 2-aminothiazole scaffold.

of thioureas and α-bromoketones in acetone generated a thiazole hydrobromide salt that was isolated cleanly by precipitation and neutralization with an immobilized tertiary amine base. The free thiazole was then coupled with a carboxylic acid via the activated ester intermediate prepared using a supported carbodiimide and HOBt. Excess acid remained bound to the resin, and the HOBt was removed using polymer-supported trisamine at the end of the sequence.

In a very early example an ACT 496 automatic synthesizer successfully produced a 96-membered array of compounds in a single automated step. In this array, 12 aromatic aldehydes were reacted with eight aliphatic amines in the presence of polymer-supported cyanoborohydride to produce the corresponding secondary amines with a very high compound success rate (**Scheme 20**).

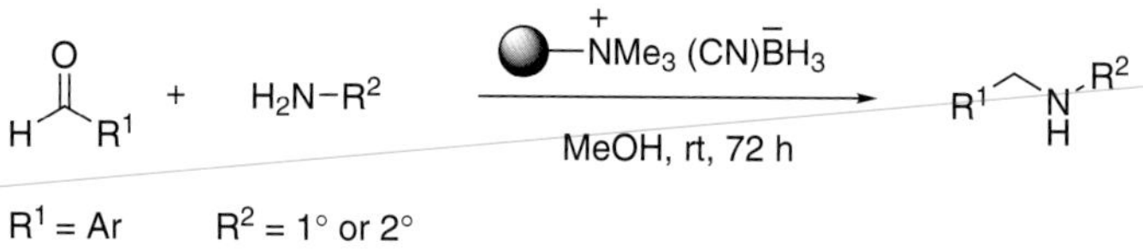

Scheme 20 Polymer-supported cyanoborohydride reductions in the synthesis of secondary amines.

The automated sequential application of polymer-supported perruthenate[186–188] and polymer-supported cyanoborohydride has been used in an oxidation–reductive amination procedure. This allowed simple alcohols to be transformed into more complex amines, which were then further derivatized using immobilized amino sulfonylpyridinium chlorides (**Scheme 21**).[189]

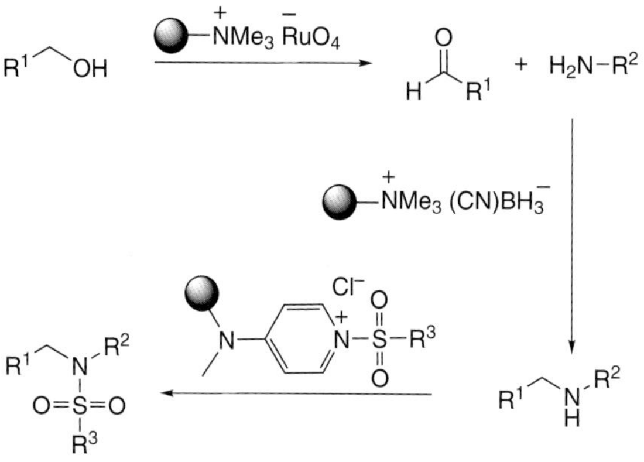

Scheme 21 An oxidation–reductive animation procedure.

Similarly, an extended library of amides was prepared on a Trident automated platform. The synthesis involved combinations of standard solution-phase reagents and supported scavenging agents (**Scheme 22**).[190] First, a collection of 24 secondary amines was generated by a titanium(IV) isopropoxide mediated reductive amination procedure (three ketones, three aldehydes, and four amines) using a Trident platform. The amines were then used in a split array synthesis to create sublibraries of ureas, amides, and sulfonamides. Various immobilized scavenger systems were used to assist in the parallel work-up and purification, including a catch-and-release technique for separating the desired amine from contaminating alcohol by-products following the borohydride reduction step. Although this work involved the formation of relatively simple compounds, it aptly demonstrates the ease and effectiveness of employing supported reagents for the generation of chemical libraries involving automated purification strategies.

An automated, multicomponent, sequential condensation reaction has been used to make a 4080-member library of 5-arylidine 4-thiazolidinones using supported scavenging as the principle method of purification. The one-pot synthesis involved a three-component condensation of mercaptoacetic acid reacting with an amine and a carbonyl compound. Further structural decoration was introduced using the 'libraries from libraries' principle, where the core template was derivatized using aldol chemistry with a second carbonyl unit at the 5-methylene position (**Scheme 23**).[191] After each of the synthetic steps, basic alumina was added to remove trace amounts of impurities by filtration, thereby increasing

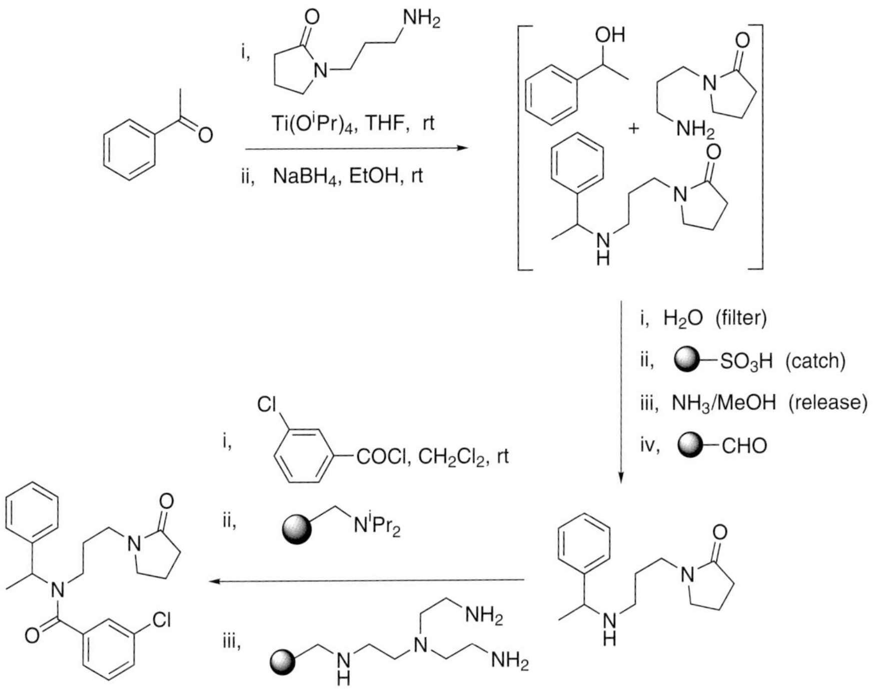

Scheme 22 Automated preparation of amide, urea, and sulphonamide libraries.

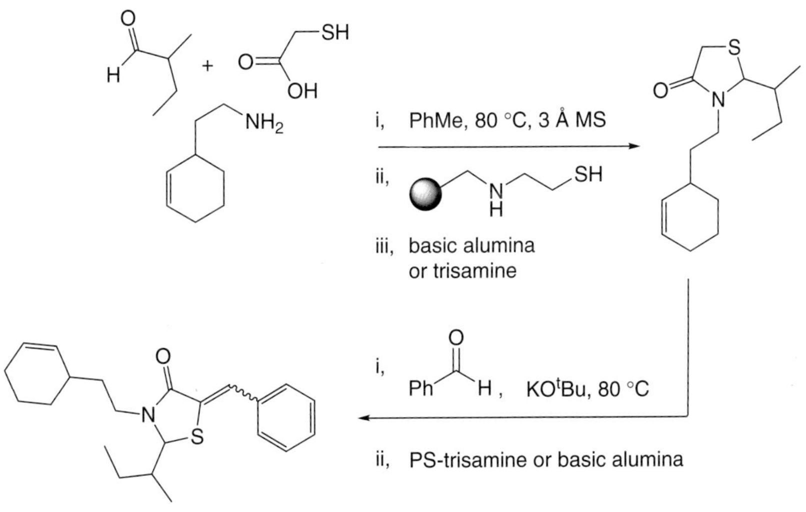

Scheme 23 Three-component condensation reaction using supported scavenger purification.

the overall purity of the products. Alternatively, trisamine resin could be substituted for purification in this step, to give equally high-quality materials for suitable biological evaluation.

Substituted pyrazoles are of interest as potential p38 kinase inhibitors. A recent automated library synthesis of these compounds involved the direct treatment of chalcones with hydrazine monohydrate followed by in situ trapping of the unstable pyrazoline intermediates with a number of electrophiles (**Scheme 24**).[192] The various classes of electrophile included acid chlorides, chloroformates, isocyanates, and sulfonyl chlorides. A supported diisopropylamine base was used in those reactions where reagent combinations generated hydrochloric acid. Trace amounts of pyrazoline and excess electrophiles were easily sequestered using a combination of supported isocyanate and trisamine resins. Overall, more than 1500 derivatives were generated in a 96-well format by this approach.

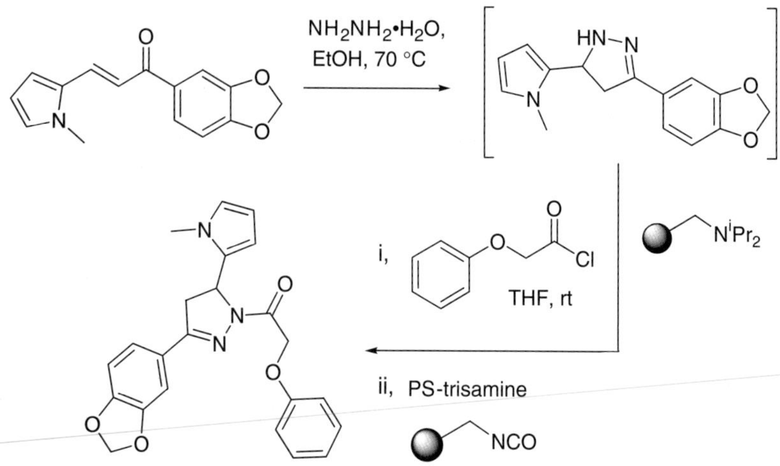

Scheme 24 Automated library preparation of substituted pyrazoles.

The 1,5-biaryl pyrazole structure is found in a number of important pharmaceuticals; for example, the nonsteroidal anti-inflammatory drugs (NSAIDs) such as celecoxib (**6**, Celebrex), a cyclooxygenase-2 (COX-2) inhibitor used to treat arthritis, menstrual cramps, and colonic polyps, and tepoxaline (**7**, Zubrin), which is effective in relieving pain associated with osteoarthritis.[193]

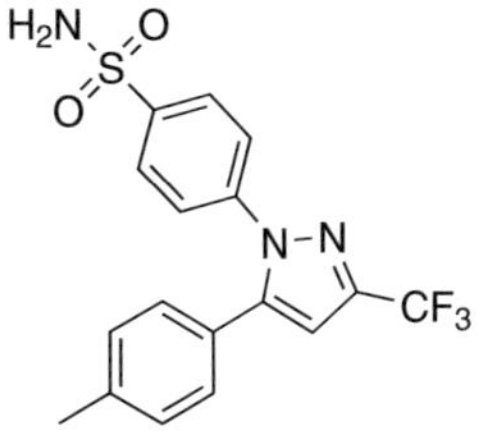

6
Celecoxib

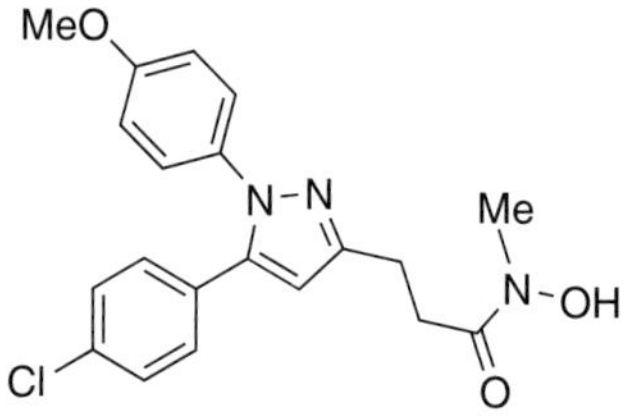

7
Tepoxaline

Toward new and efficient routes to this class of compound, a fully automated synthesis using immobilized reagents has been developed.[194] A 1,5-biaryl pyrazole library was prepared using the route exemplified in **Scheme 25**. The synthesis employed a solution-phase condensation reaction of a diketone and hydrazine to form the pyrazole ring. This was followed by a catch-and-release purification protocol, which also afforded the carboxylic acid. A 192-membered library was obtained in a single run using a top-filtration robotic synthesizer without any manual intervention.

Cathepsin D is a lysosomal aspartic protease which is thought to play a role in the metastatic potential of several types of cancer. A high activated cathepsin D level in breast tumor tissue, for example, is associated with an increased incidence of relapse and metastasis. In addition, high levels of active cathepsin D have also been found in colon, prostate, uterine, and ovarian cancer. Again, this is another interesting target where syntheses of potential drug candidates have been tackled by supported-reagent technologies.

Poly-4-vinylpyridine, an inexpensive base, was used to facilitate the coupling of acyl and sulfonyl chlorides with a range of nitrogen nucleophiles in the preparation of a 300-member library of cathepsin D inhibitors (**Scheme 26**).[195,196] Excess unreacted electrophilic components were scavenged using aminomethylpolystyrene, and any remaining amine fragments were removed using a supported sulfonic acid. Alternatively, the acyl component could be generated from a carboxylic acid by activation using either 1-ethyl-3-(3-dimethyl-aminopropyl)carbodiimide (EDC) or 2,4,6-triisopropylbenzenesulfonyl chloride. Both of these coupling agents and their by-products contain either a basic or an acidic function, which can be used as a handle for later sequestration. These methods gave library members with yields ranging from 60% to 100% and in >90% purity.

Potential *N*-type calcium channel blockers are also of interest. One particular family of structures found to be useful in this therapeutic area are the *N*,*N*-dialkyldipeptidylamines. A collection of 30 of these amines (surrounding a lead compound) has been prepared using multiple parallel synthesis methods.[197] The strategy employed a one-pot

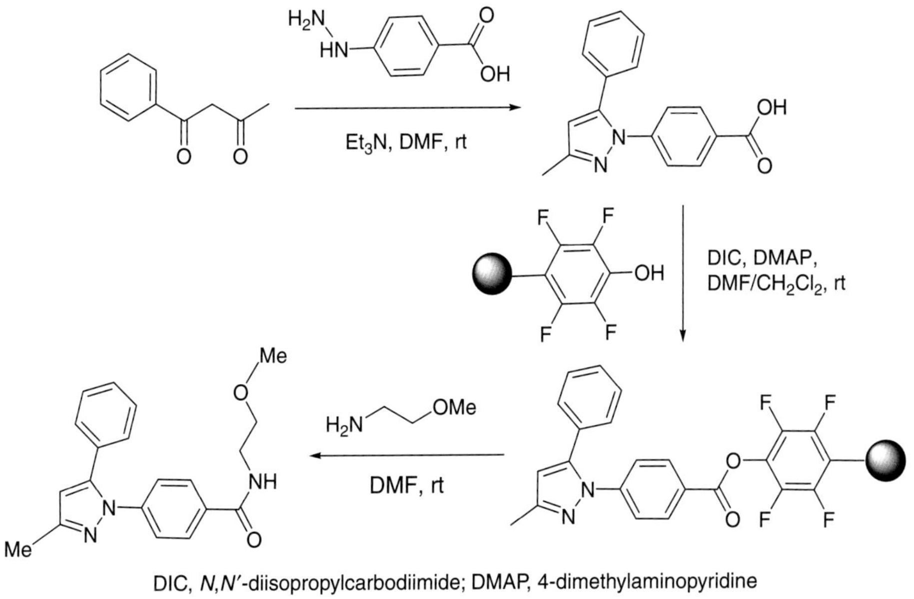

Scheme 25 Synthesis of 1,5-biaryl pyrazole library.

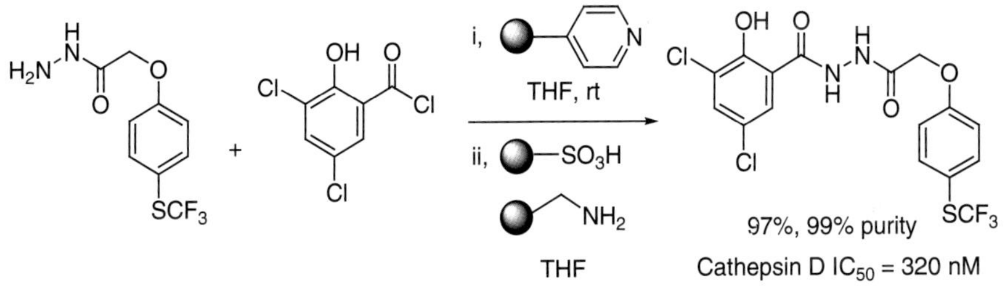

Scheme 26 Synthesis of Cathepsin D inhibitors.

procedure for the coupling of an *N*,*N*-disubstituted leucine acid with a tyrosine amine (**Scheme 27**). The compounds were then screened in an in vitro assay, which led to new leads with excellent potencies. Again, good use of immobilized reagents had been made in this library synthesis of peptidic compounds.

Cardiovascular disease is one of the most common causes of mortality in the western world. However, one of the most frequently prescribed oral antithrombotic therapies, Warfarin, has a narrow therapeutic window and its use is complicated by the need for frequent patient monitoring owing to the drug's significant side effects i.e., abnormal bleeding, drug–drug interactions and dietary impacts. Although other compounds have come on to the market, such as clopidogrel (Plavix), there is still an unmet need to develop new and effective oral anticoagulants in the form of potent and selective tissue factor VIIa inhibitors (tissue factor VIIa is a serine protease inhibitor and a key enzyme in blood coagulation).

The application of polymer-assisted synthesis in a range of chemistry programs has been vigorously explored by Parlow *et al.*, particularly in a search for tissue factor VIIa inhibitors. This work led to investigations into two different compound series. The first approach was based on the utilization of a tripeptide, in which the scissile amide bond had been replaced by an electron-deficient carbonyl group. The design was of the general form D-Phe-L-AA-Arg-α-ketothiazole, which is closely related to the known inhibitor D-Phe-L-Phe-Arg-chloromethylketone, whose crystal structure has recently been determined while bound to active site of tissue factor VIIa.[198,199] This has allowed comparisons to be made with closely related proteases such as thrombin and factor Xa.

In order to rapidly prepare libraries of α-ketothiazole peptidyl protease inhibitors in a parallel format, a multistep polymer-assisted synthesis was developed (**Scheme 28**). Initially, each of the steps in the synthesis was independently validated to identify and optimize the conditions, such that each transformation could be performed in a high-yielding parallel format. In total, a 38-membered array was prepared using this methodology, with overall purity levels ranging from 70% to 99%.

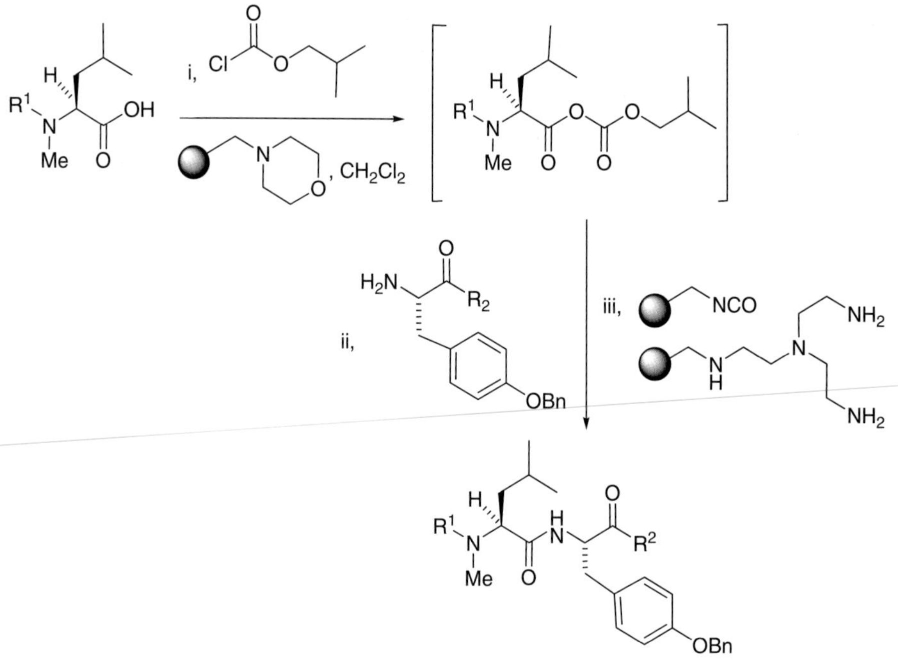

Scheme 27 Synthesis *N*-type calcium channel blockers.

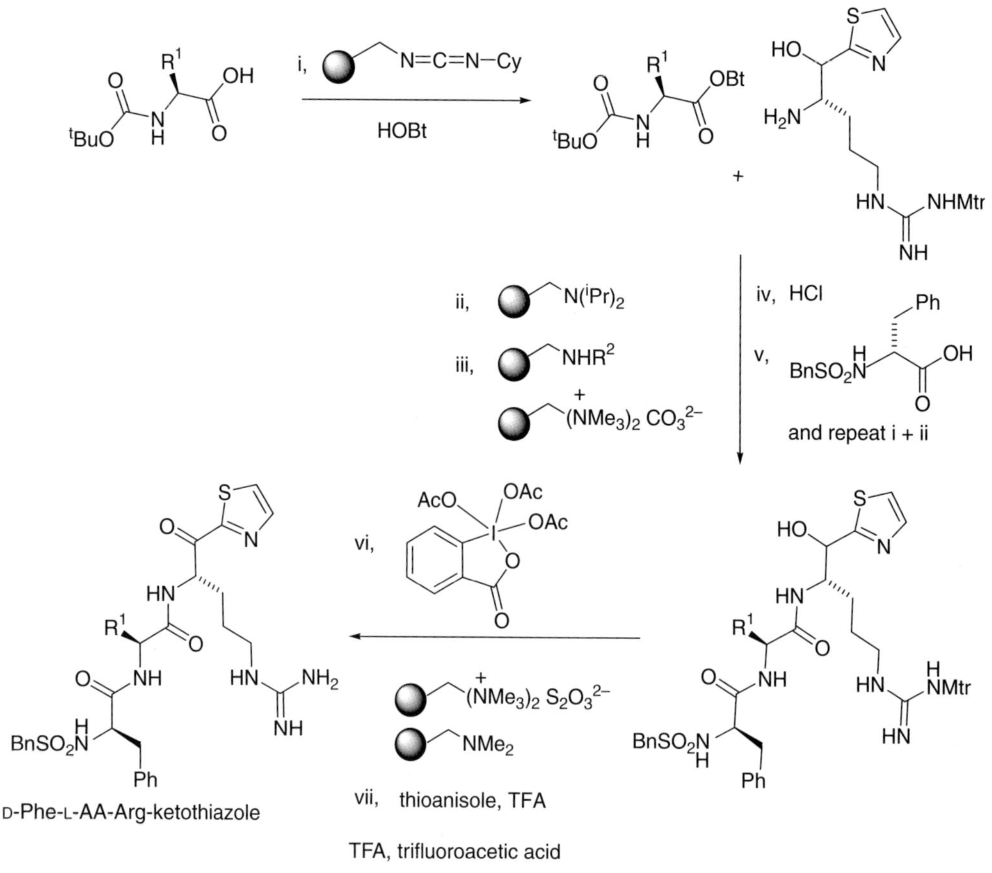

Scheme 28 Polymer-assisted synthesis of α-ketothiazole library.

Metalloproteinases are efficient processing enzymes of many bioactive mediators, such as cytokines, chemokines, growth factors, their receptors, and specific matrix protein anchors for these molecules. These species have proposed pathogenic roles, which include tissue breakdown, metastasis, and tumor angiogenesis. A polymer-supported reagent approach toward the generation of an array of substituted hydroxamic acid derivatives, known to have activity against matrix metalloproteinases, has been described (**Scheme 29**).[200] Five synthetic transformations were required to afford an array of 27 final compounds in greater than 90% purity without the use of any chromatography.

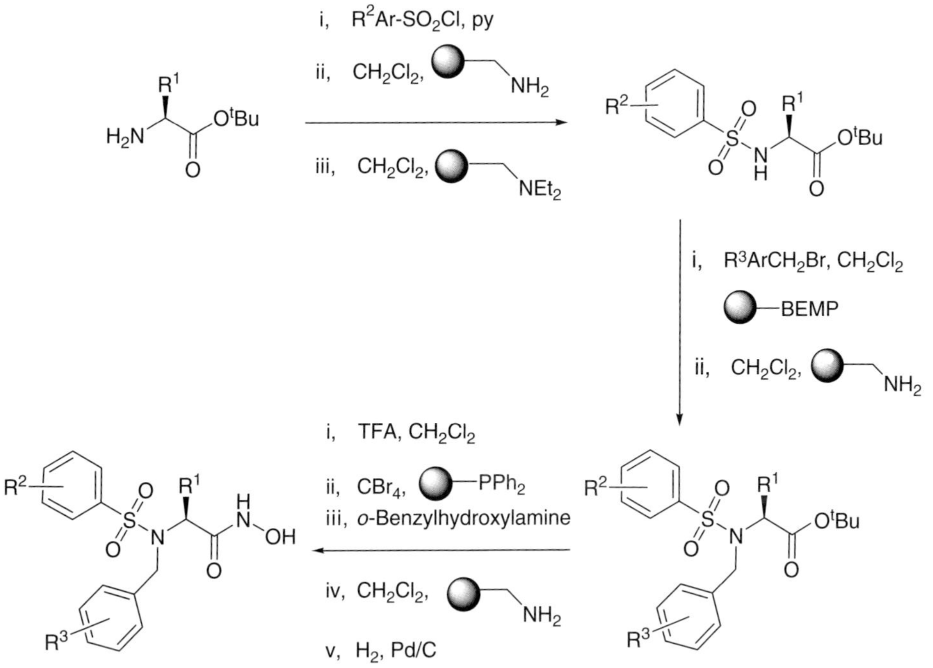

Scheme 29 Generation of substituted hydroxamic acid derivatives as potential matrix metalloproteinase inhibitors.

An extension to the above work was a further synthesis of hydroxamic acid derivatives, this time a fully automated polymer-assisted solution-phase synthesis of histone deacetylase (HDAC) inhibitors.[201,268] A series of immobilized reagents and scavenger agents were used to perform sequential automated synthetic transformations in a common solvent (dimethylformamide), providing a 36-membered array of HDAC inhibitors. The robot platform was preloaded with the necessary resins and reagents, and 180 reactions were allowed to run unattended over a 4-day period. A total of 34 out of 36 target compounds were successfully obtained, with purities ranging between 55% and 80%. The protocols employed in this synthesis included a catch-and-release in-line purification step (**Scheme 30**).

Scaffold decoration is an important aspect of generating chemical diversity, creating novel structures and lead candidates. The use of immobilized reagents is very useful for achieving this kind of chemical manipulation. Heterocyclic derivatives, such as thiomorpholine analogs, have found a wide application base in both medicine and agriculture. An early example of a fast and flexible method for library generation of piperidino-thiomorpholines has been developed.[202] In this synthesis an orchestrated six-step reaction sequence using different supported reagents was accomplished (**Scheme 31**). *N*-Sulfonylation followed by α-bromination of 4-piperidone proved to be facile, and could be performed on a large scale. Nucleophilic substitution with an *N*-Boc-protected aminothiol using Amberlyst 21 as the base, followed by cleavage of the Boc group with Amberlyst 15 (or alternatively trifluoroacetic acid) yielded the corresponding imine, which was reduced with a supported cyanoborohydride. The resulting thiomorpholine derivative was then decorated by reaction with different isocyanates or isothiocyanates, then purified by an acid/base scavenging sequence.

In addition, intermediates from various stages of the reaction sequence were diverted into alternative reaction pathways, allowing access to a larger number of heterocyclic compounds in a relatively short period of time (**Scheme 32**). Being able to change the core scaffold is clearly as important as changing the fingertips of a molecule. This divergent approach also exemplifies the general versatility of a solution-phase synthesis using supported reagents, and exemplifies the opportunities that may be attained in drug discovery programs, particularly in the area of establishing structure–activity relationships. When automated, this route was used to generate over 5000 product variants.

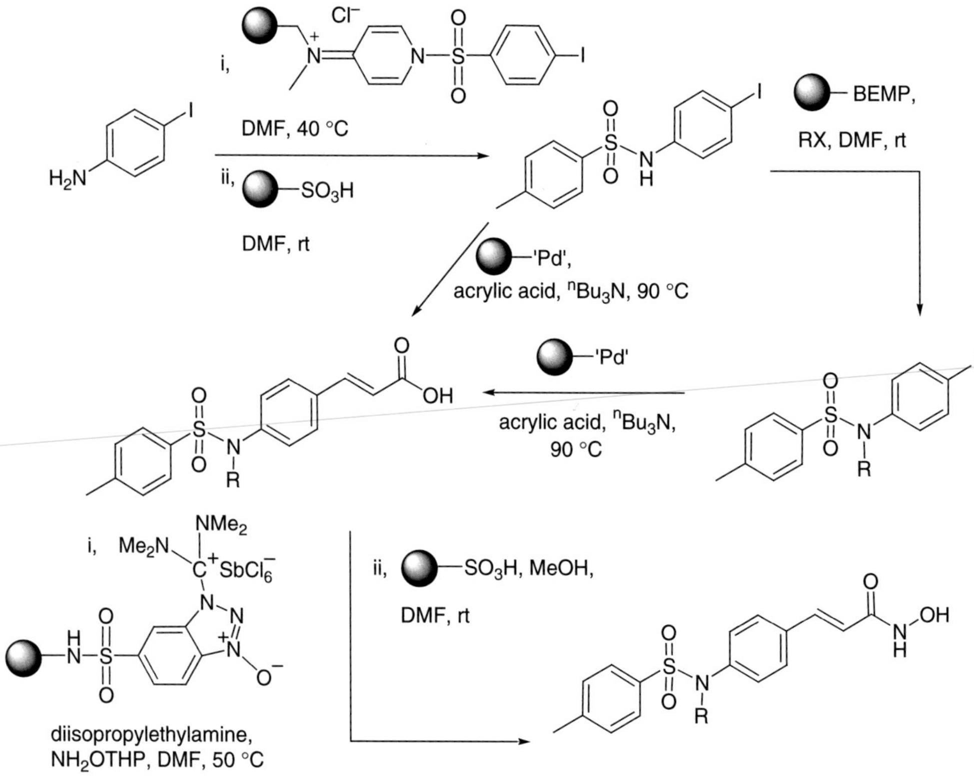

Scheme 30 Synthesis of HDAC inhibitor library.

3.35.3.3 Applications of Encapsulated Catalysts in Library Construction

Catalytic transition metal species are essential elements for the assembly of bioactive molecules. The need for practical and economic translation of these laboratory methods to large-scale operations, coupled with the trend for clean manufacturing processes, has led to the development of new strategies for reagent immobilization that may allow for the recovery and reuse of catalysts.

In typical supported transition metal catalysts the metal is coordinated to a ligand which is covalently bound to a polymer backbone, or it can be adsorbed on an inert surface such as silica or carbon. In this approach synthesis of the polymer-bound ligand can be lengthy and expensive and there can be problems associated with leaching and the reactivity of the catalyst, owing to competitive complexation with the synthesized product. An alternative approach that may offer a solution to these limitations is the use of microencapsulation: a process of entrapping materials in a polymeric coating. This approach has found application in drug delivery systems,[86] radiation therapies,[82] cell entrapment,[85] and agriculture.[84]

Microcapsules can be prepared by an in situ interfacial polymerization.[203] This involves dispersing an organic phase containing polyfunctional monomers and/or oligomers along with the material to be encapsulated into an aqueous phase containing a mixture of emulsifiers and protective colloid stabilizers. This resulting oil-in-water emulsion undergoes in situ interfacial polymerization, with monomers/oligomers reacting spontaneously at the phase interface to form microcapsule walls. The permeability and size of these microcapsules and the coordinating properties of the polymer matrix may be tuned by varying the kinds of monomers/oligomers and other reagents and conditions used in the encapsulation procedure. Efficient entrapment of transition metal-based catalysts requires the design of systems possessing ligating functionality in order to retain the metal species. These systems should be physically robust and chemically inert to reaction conditions while also being cost effective. This area is undergoing considerable growth at present and what follows highlights the applications where microencapsulated catalysts have been employed.

An efficient Suzuki–Miyaura coupling protocol using fibrous polystyrene-based palladium catalysts (FibreCat) and microwave heating has been developed (**Scheme 33**).[204] After the reaction the catalyst was recovered by filtration. The biaryl compounds formed in this reaction are important structural features in many drug substances.

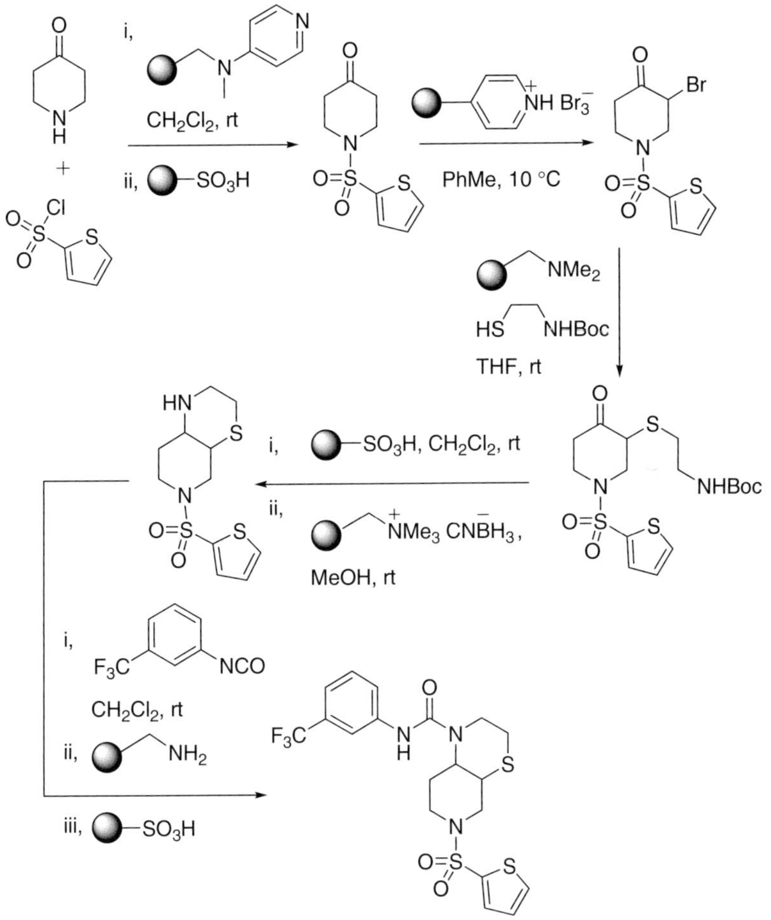

Scheme 31 An orchestrated six-step library preparation of piperidino-thiomorpholine derivatives.

Polyurea microcapsules[205] were also found to be suitable carriers for palladium in Suzuki-type reactions (**Scheme 34**). The urea backbone of these particular microcapsules was ideal for ligating, and thus retaining metal species, dispensing with the need for expensive phosphine ligands. These encapsulated palladium catalysts (Pd-EnCats) were shown to be highly successfully in an initial set of 10 cross-coupling reactions of aryl boronic acids and aryl bromides.[78] The reactions were conducted in a toluene, ethanol, and water solvent system using 5 mol% of catalyst. Furthermore, the catalysts were recovered and reused four times, without any noticeable loss in activity. Analytical studies found that there was an extremely low level of palladium leaching (0.2%), and products contained less than 5 ppm of contaminating palladium by direct filtration. These Suzuki couplings are also readily achieved under continuous flow conditions using in-line focused microwave heating.[270]

The preliminary study was extended to the use of Pd-EnCats in the Heck reaction using both conventional solvent conditions and supercritical carbon dioxide mixtures ($scCO_2$) (**Scheme 35**).[77] A typical reaction involved the treatment of 1-bromo-4-nitrobenzene with *n*-butyl acrylate in isopropyl alcohol in the presence of Pd-EnCats (2.5 mol%) and Bu_4NOAc, where the requisite unsaturated ester was furnished in 91% yield.

Supercritical carbon dioxide is a promising alternative solvent for organic synthesis,[206] as well as being of considerable interest in hydrogenation reaction[207] and palladium-mediated C–C bond-forming reactions.[208,209] Heck reactions have been carried out in $scCO_2$, and under these conditions it was found that yields were generally higher than in isopropyl alcohol, despite a lower catalyst loading. The catalyst was separated by filtration of the ethyl acetate solution into which the reaction mixture had been vented, demonstrating an advantage of using $scCO_2$; as carbon dioxide-soluble products can be extracted from the heterogeneous reaction mixture while the catalyst remains in the reaction cell. After success with the Heck reactions, a range of Suzuki couplings of tolylboronic acid with aryl halides and a series of Stille couplings between aryl halides and trimethylphenyltin were also conducted in $scCO_2$.

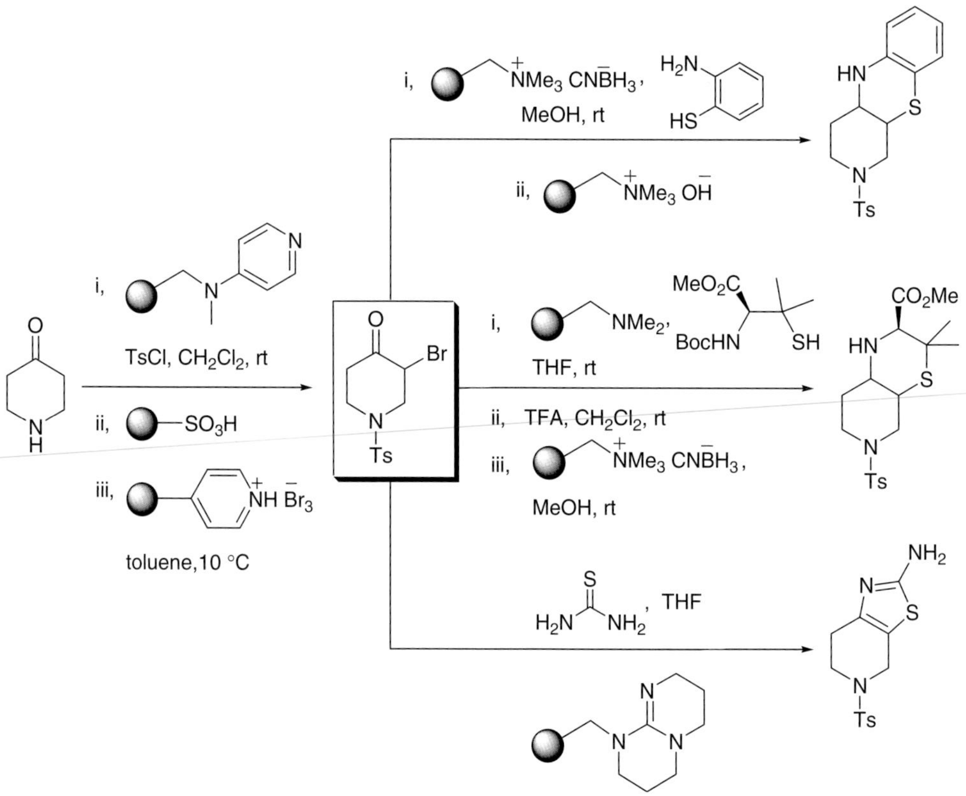

Scheme 32 Additional compound collections from template decoration.

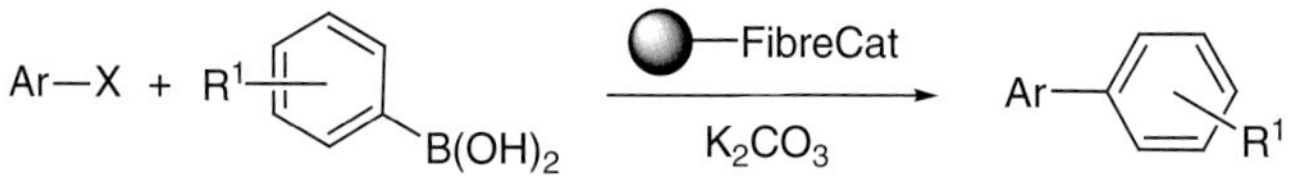

Scheme 33 Suzuki–Miyaura coupling protocol using fibrous polystyrene-based palladium catalysts.

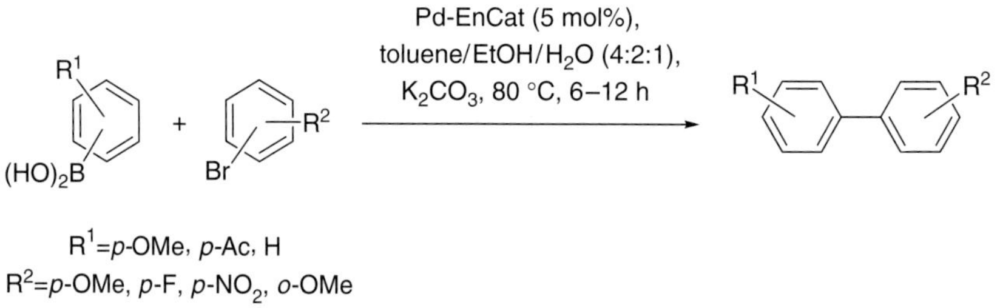

Scheme 34 Suzuki-type cross-coupling reactions employing polyurea encapsulated palladium(II) acetate as catalysts.

Not only have Pd-EnCats demonstrated great utility in cross-coupling reactions in conventional and supercritical media, they have also been used in hydrogenolysis of epoxides[73] (**Scheme 36**) and the chemoselective reduction of aryl ketones (**Scheme 37**).[76]

3.35.3.4 Synthesis of Natural Products

Natural products and their various derivatives have dutifully served the pharmaceutical industry for years. They provide the impetus for many drug discovery programs, which will undoubtedly continue given nature's bountiful supply of

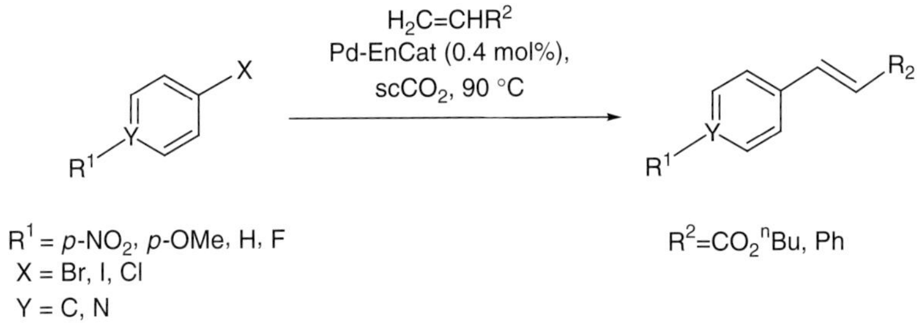

Scheme 35 Heck reactions conducted in supercritical carbon dioxide using Pd-EnCats.

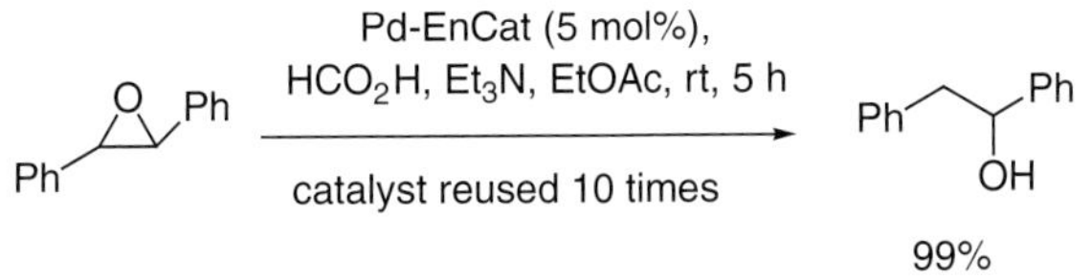

Scheme 36 Recyclable Pd-EnCats as catalysts for epoxides hydrogenolysis.

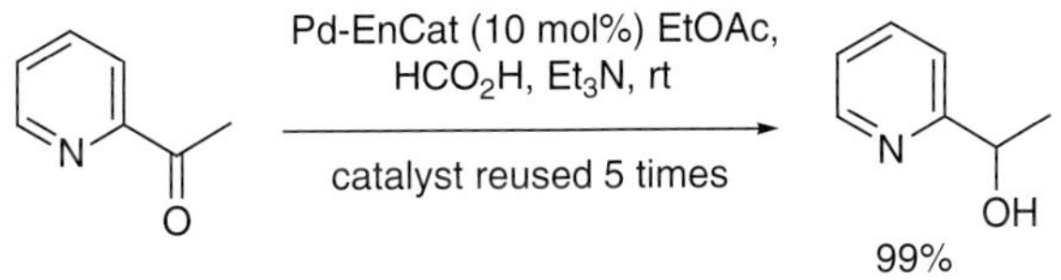

Scheme 37 Transfer Pd-EnCats in the chemoselective reduction of aryl ketones.

exquisite molecular architectures. Moreover, these complex structures present a constant challenge to the frontiers of synthesis methodology and help provide the tools for future discoveries.

The typically complex structures found in nature are also suitable targets for synthesis using immobilized reagents, scavengers, quenching agents, and catch-and-release techniques. This next section highlights a selection of these syntheses, but it should not go unnoticed that the same methods can also be adapted for analog preparation.[55]

3.35.3.4.1 The Amaryllidaceae alkaloids oxomaritidine and epimaritidine

The Amaryllidaceae alkaloids have been studied extensively owing to their structural variation and biological activities.[210] Using an orchestrated array of supported reagents, the first natural products ever to be synthesized by these methods were the alkaloid natural products[57] oxomaritidine (8) and epimaritidine (9) in just five and six steps respectively (**Scheme 38**).[211]

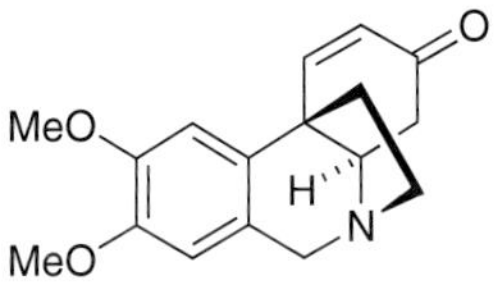

8
Oxomaritidine

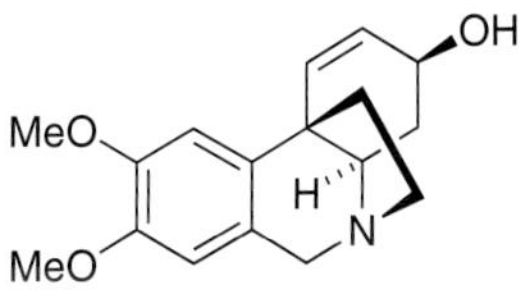

9
Epimaritidine

This landmark synthetic achievement demonstrated how these effective methods are robust as well as applicable to the production of gram quantities of the materials without recourse to chromatography, water washes, crystallization, distillation or other work-up procedures. In fact only filtrations had to be performed to remove the spent reagents, followed by solvent evaporation to yield the intermediates, giving the products in remarkably high yield and purity.

Noteworthy in this synthesis was the specific development of an immobilized hypervalent iodine diacetate reagent[212] (step iv), which was used to bring about phenol oxidative coupling to give the *para–para* coupled

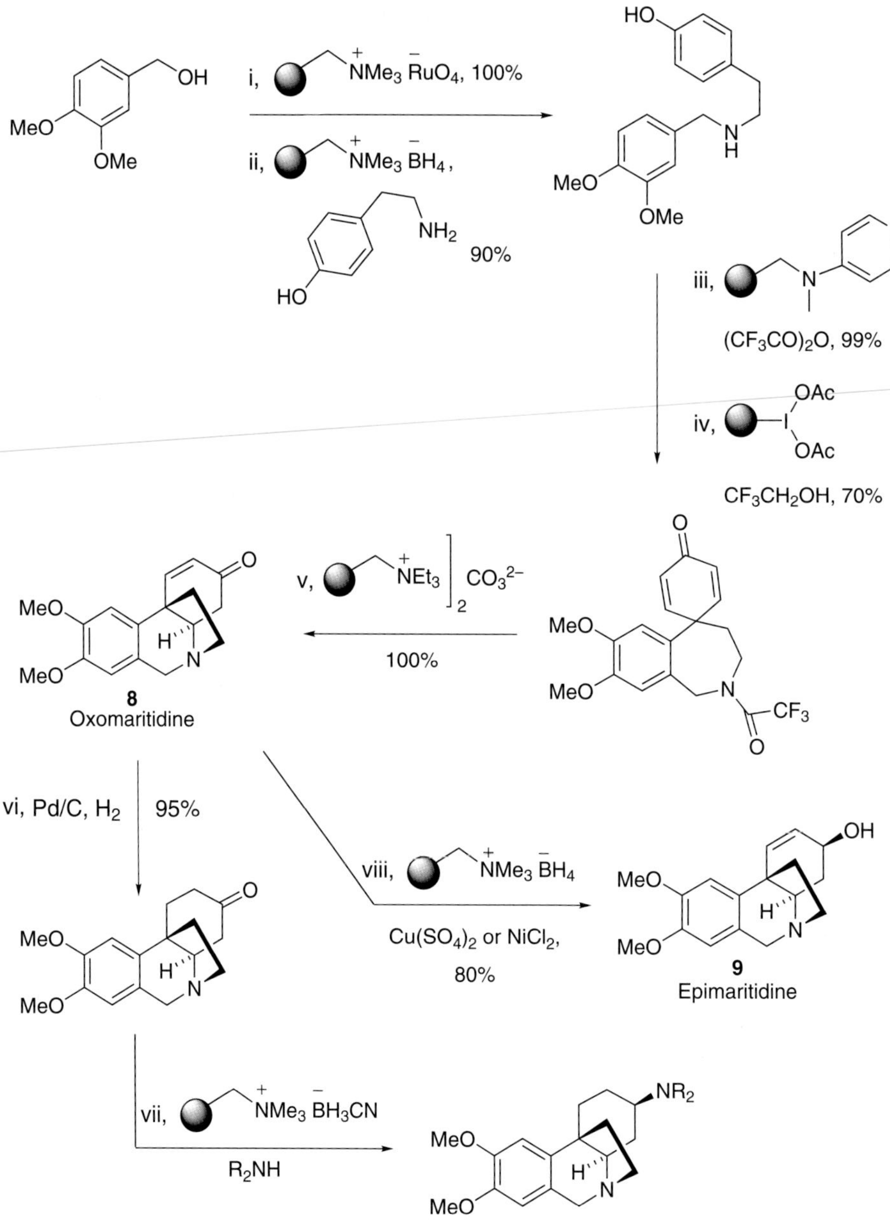

Scheme 38 The synthesis of epimaritidine and oxomaritidine; the first orchestrated application of solid-supported reagents in the synthesis of natural products.

spirodienone in 70% yield. This yield was higher than the corresponding solution-phase variant, and with the added advantage that the spent polymeric reagent could be isolated by filtration and conveniently recycled by treating it with peracetic acid. In addition, several amine analogs of oxomaritidine were prepared via reductive amination and were screened for biological activity (**Scheme 38**).

3.35.3.4.2 The alkaloid epibatidine

An impressive example of the application of supported reagents to natural product synthesis was the preparation of the analgesic epibatidine (**10**),[213–215] a compound isolated from the Ecuadorian poison frog *Epipedobates tricolor*. This otherwise small and innocuous frog is a member of the Dendrobatidae family (commonly known as the poison dart frogs), and it shares with them a number of common characteristics. Although epibatidine is highly toxic at concentrations only slightly higher than its effective dose, clearly limiting its therapeutic potential, it provides a lead structure for the design of new ligands for exploring nicotinic cholinergic receptor subtypes.[216] Over the past few years, considerable efforts have been made to determine the structural features that account for the high binding affinity of epibatidine. It has shown itself to be a potent agonist of ganglionic nicotinic receptors, and elicits cardiorespiratory effects similar to those of the structurally similar nicotine.[217] Due to the therapeutic potential of this compound, analogs of epibatidine are also of interest.[218–220]

This synthesis of epibatidine which uses more than 10 polymer-bound reagents is a comprehensive demonstration of how immobilized reagents, scavengers, and catch-and-release techniques can be used to prepare natural products (**Scheme 39**). Focused microwave heating at 100 W in 1 min bursts over 30 min was also used to promote a

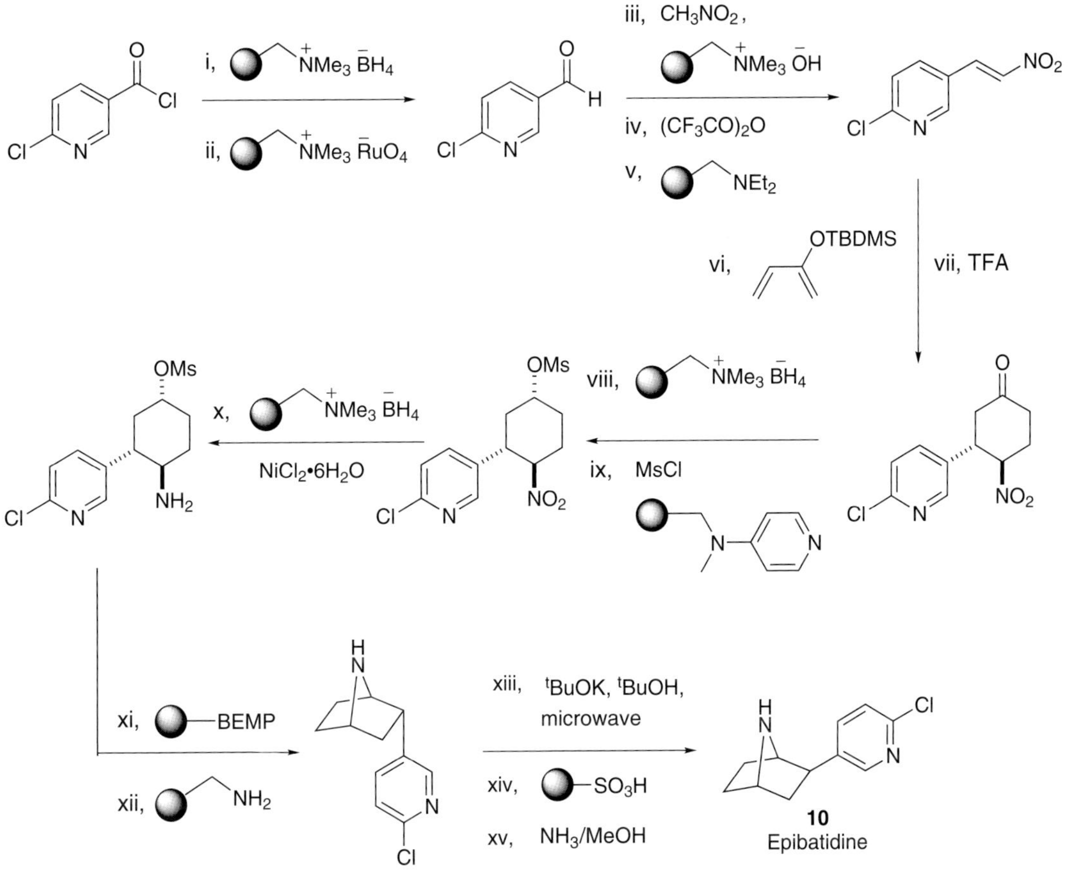

Scheme 39 Synthesis of the alkaloid epibatidine using 10 polymer-supported reagents or scavenger-assisted steps.

base-catalyzed epimerization (step xiii) in the final step leading to the natural product. This was one of the very early reports of the use of focused microwaves in synthesis.

An interesting aspect of this work is the use of a nickel-mediated borohydride reduction of the aromatic nitro compound to the corresponding amine (step x), without complication from competitive reduction of the chloropyridine ring. The use of the polymer-supported base (BEMP) to achieve selective cyclization of the mesylate (shown in step xi) is again an attractive element in this synthesis. Sequentially adding the supported reagents in sealed porous pouches (**Figure 6**) meant that the five steps from the acid chloride starting material to the nitroalkene could be carried out in a single pot (steps i to v). Progress of the reactions was readily monitored by thin-layer chromatography or liquid chromatography–mass spectrometry. When complete, the pouch of spent reagent was lifted out, washed, and the next reagent pouch added.

In recent times a number of other natural product syntheses using these methods have been accomplished. For example, the simpler alkaloid nornicotine (**11**) as well as nicotine (**12**) itself and various structural analogs have been prepared.[221]

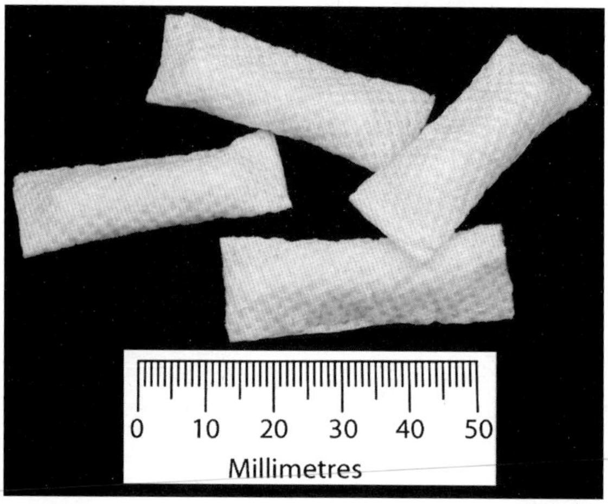

Figure 6 Immobilized reagents in sealed porous pouches.

Another study that was of interest concerns the preparation of various neolignans. The anti-malarial neolignan polysyphorin (**13**), isolated from *Piper polysphorum* C and from the leaves and stems of *Rhaphidopora decursiva*,[222,223] has several family members that also display interesting biological properties; for example, rhaphidecursinol B (**14**), which demonstrates activity against *Plasmodium falciparum*, the parasite responsible for severe forms of malaria,[224,225] and surinamensin, which demonstrates activity against the vector-borne disease leishmaniasis.[226] Polysyphorin,[227] rhaphidecursinol, (−)-surinamensin (**15**), and oxo-surinamensin (**16**) have been made by employing a variety of immobilized reagents, including supported enzyme systems, to effect key transformations.[228]

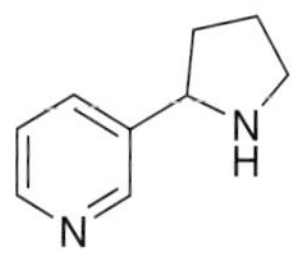

11
Nornicotine

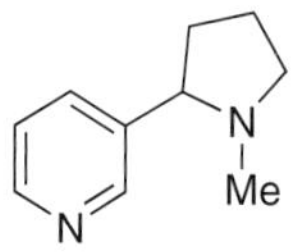

12
Nicotine

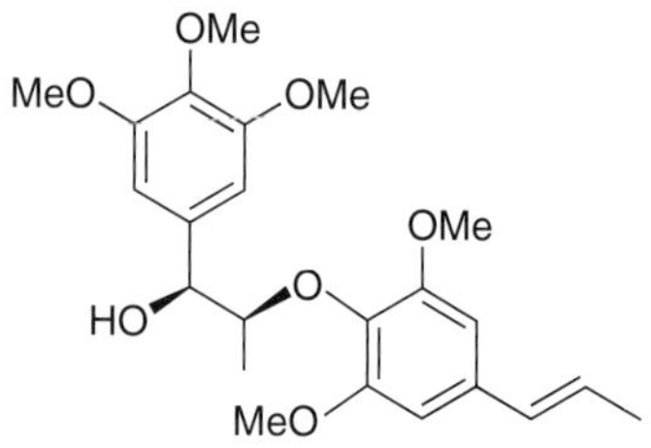

13
Polysyphorin

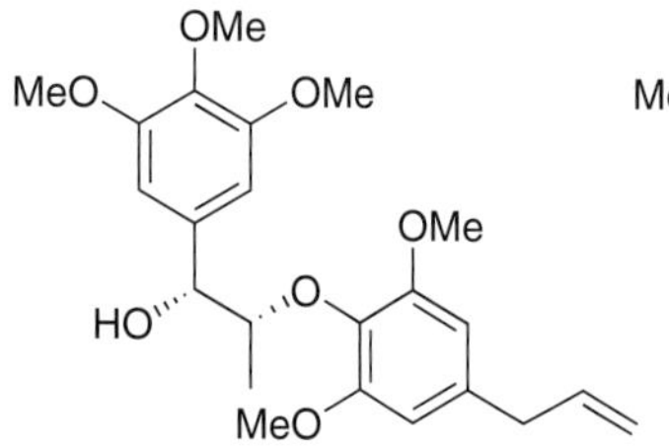

14
Rhaphidecursinol B

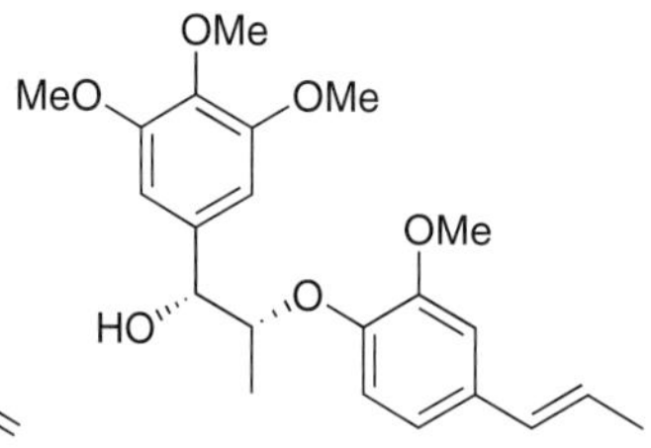

15
(–)-Surinamensin

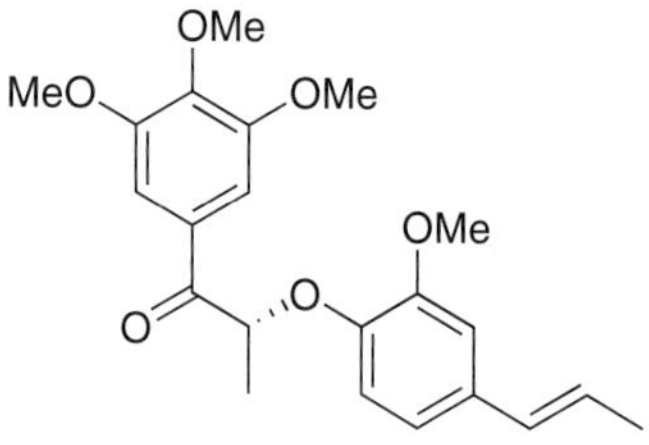

16
Oxo-surinamensin

3.35.3.4.3 The plant lignan carpanone

Lignans are a very common and structurally diverse group of natural products of phenylpropanoid origin, and are found in many plants in various forms and levels of abundance. In plants they perform important functions in defense against fungal and viral infection. There is also mounting evidence that lignans have potential as efficacious therapies in certain human ailments.[229–233] While there are yet to be any specific biological investigations regarding the use of carpanone, a number of preliminary studies suggest that other lignans may be helpful for the prevention of breast and colon cancer. In addition, preliminary research suggests that flaxseeds which contain an abundance of naturally occurring lignans may help normalize cholesterol levels, decrease menopausal symptoms, and improve kidney function in kidney disease (specifically lupus nephritis and polycystic kidney disease).[234,235]

Carpanone is a lignan obtained from the carpano tree that initially appears as a deceptively complex structure.[236–238] While carpanone has been made using traditional chemistry, it has also recently been synthesized using immobilized reagents and was prepared by a relatively simple series of reactions from commercially available sesamol (**Scheme 40**).[239] Sesamol was *O*-alkylated with allyl bromide in the presence of a base (PS-BEMP). The subsequent

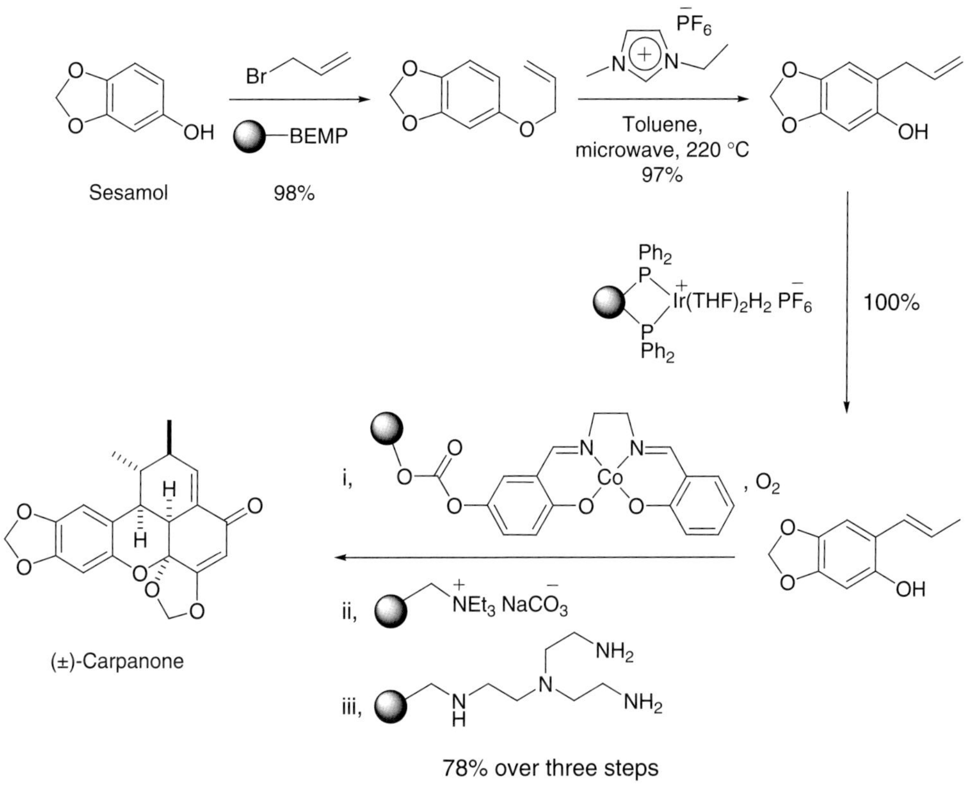

Scheme 40 Synthesis of carpanone.

Claisen rearrangement used ionic liquids, toluene and focused microwave heating. The combination of ionic liquids to absorb the microwave energy and an organic solvent, toluene, had been previously developed and provides a convenient binary heating mixture.[122,240] The product-containing toluene solution was separated using a Gilson liquid handling robot and was used directly in the next reaction step. Following a screening investigation, the double bond isomerization was achieved using an immobilized iridium catalyst specifically developed for this synthesis.[126,227] Both the yield and the selectivity in this process were extremely good, providing the phenolic styrene with a *trans* to *cis* ratio of 11:1. The final conversion of the phenolic styrene to carpanone was achieved using a supported cobalt catalyst in the presence of molecular oxygen,[237,241–243] which promoted an oxidative coupling, and via subsequent intramolecular Diels–Alder reaction, gave the natural product carpanone in excellent 78% yield. Prior to isolation, the spent catalyst was filtered off and any unreacted phenol and aldehyde by-products were scavenged from the reaction mixture with a supported trisamine and carbonate resin. The efficiency of this sequence allows gram quantities to be prepared, which could be useful for more extensive structural modification and biological evaluation. In addition, the cobalt catalyst used in this work can also be easily recycled and reused in other syntheses.

3.35.3.4.4 Plicamine and the related amaryllidaceae alkaloids plicane and obliquine

A recent botanical investigation of the Turkish snowdrop plant, *Galanthus plicatus* ssp. *byzantinus*, produced two new alkaloids, (+)-plicamine (**17**) and (−)-secoplicamine (**18**).[244] These compounds are the first reported structures in a new subgroup of Amaryllidaceae alkaloids possessing a dinitrogenous skeletal arrangement. While biological activity for either of these new alkaloids is yet to be reported, related structures (such as tazettine, pretazettine, criwelline, precriwelline, and 6a-epipretazettine) have already demonstrated an interesting array of medicinal properties, including anticholinergic, antitumor, immunosuppressive, and analgesic activity.[245–247]

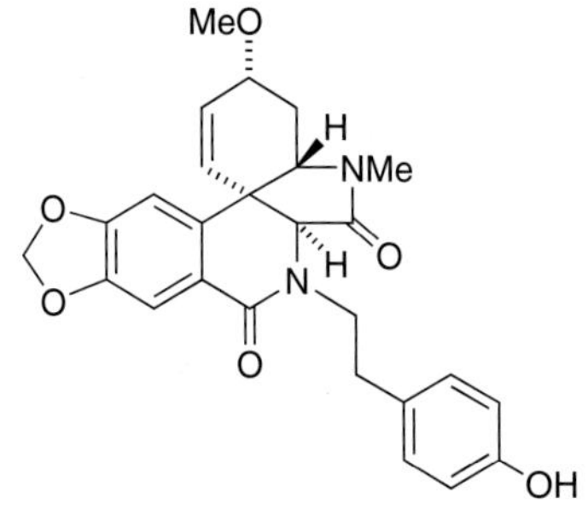

17
(+)-Plicamine

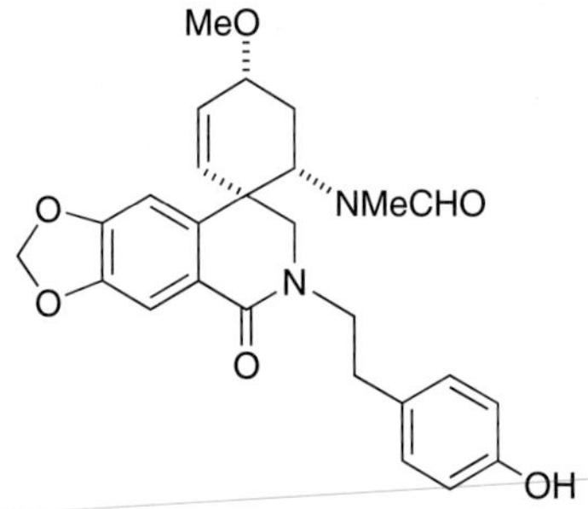

18
(−)-Secoplicamine

In terms of synthesis, the polycyclic alkaloid (+)-plicamine presents a technically more challenging target than carpanone (**Scheme 41**).[248] In this first synthesis of the molecule several general points are worthy of mention. First, the work made extensive use of parallel reaction equipment to optimize processes rapidly and progress the synthesis. Focused microwaves were also used in many of the reactions. Carrying out reactions on a gram scale also meant intermediates could be made which proved useful for other combinatorial programs.[249]

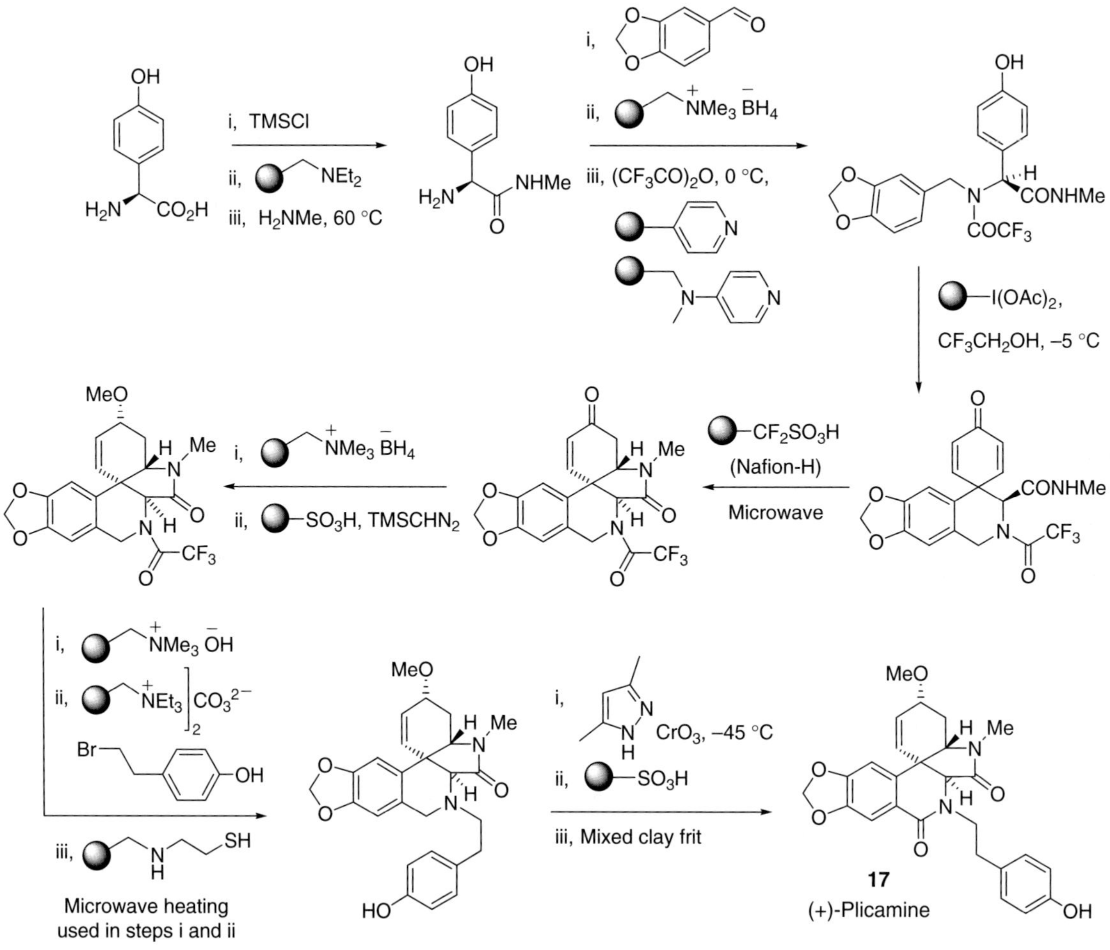

Scheme 41 The first synthesis of the Amaryllidaceae alkaloid plicamine.

The use of the supported hypervalent iodine reagent[212] in trifluoroethanol served well in the preparation of the initial dienone. The use of Nafion-H resin (a perfluoroalkanesulfonic acid) to bring about the cyclization of the dienone to the amide is also interesting. The use of Amberlyst 15 ion exchange resin (a strongly acidic resin with sulfonic acid functionality) in combination with trimethylsilyl diazomethane gave the methyl ether derivative. This is a reagent system recommended for methylation of hindered alcohols that should find wider application in other synthesis programs.

By appropriate selection of the absolute chirality of the initial starting amino acid the unnatural antipode of plicamine was also synthesized. Using batch-splitting methods intermediates were diverted for the preparation of other related alkaloids (plicane and obliquine (**Scheme 42**)).[250]

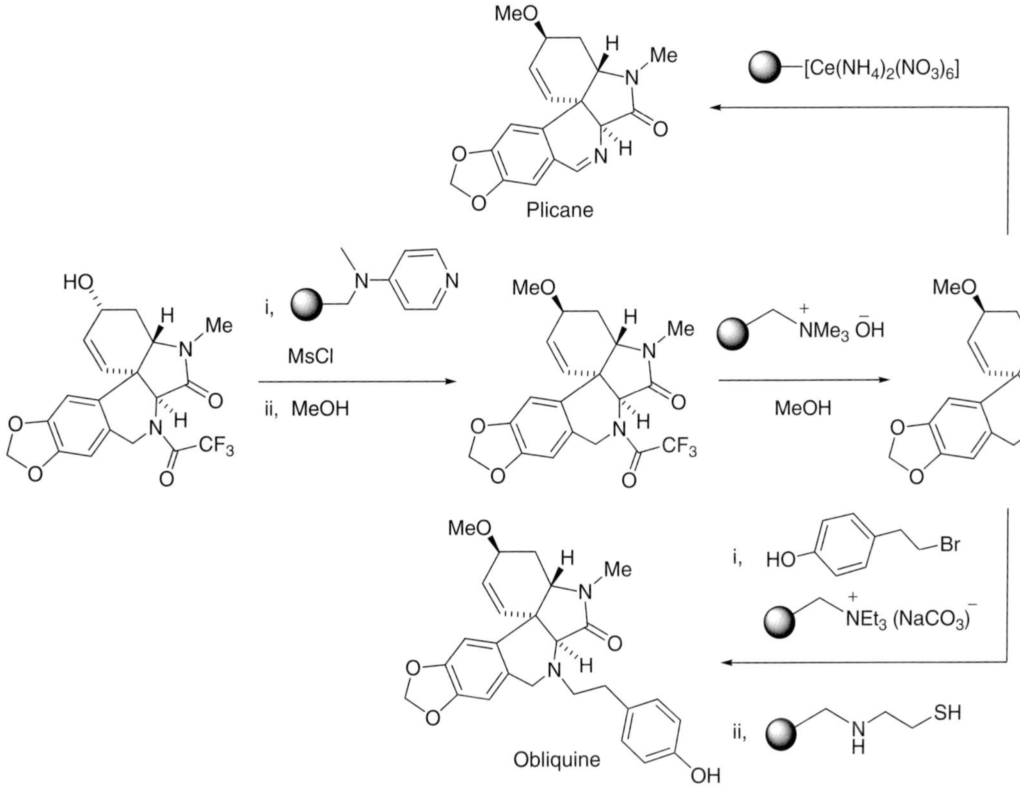

Scheme 42 Synthesis of related alkaloids plicane and obliquine.

3.35.3.4.5 The epothilones

The search for improved cytotoxic agents that can be directed toward ubiquitous cellular targets such as DNA or microtubules is of major importance in developing improved anticancer therapies. Among the most important clinical agents at present is the natural product paclitaxel (Taxol, **19**). The drug operates by stabilizing cellular microtubules and interfering with microtubule dynamics, thereby inhibiting human cancer cell growth. For a long time natural products of the taxane family were the only known inhibitors of microtubule depolymerization, until 1995, when the bacteria-derived epothilones (**20–22**) were discovered.[251–255]

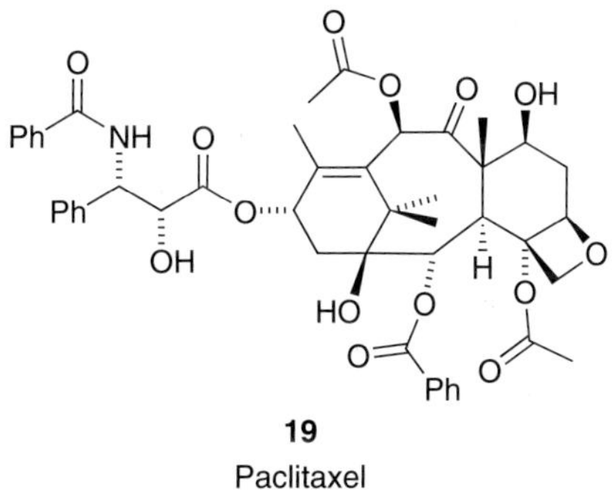

19
Paclitaxel

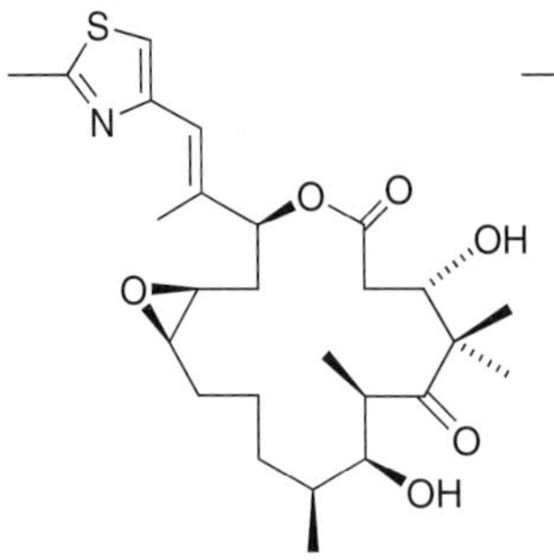

20
Epothilone A

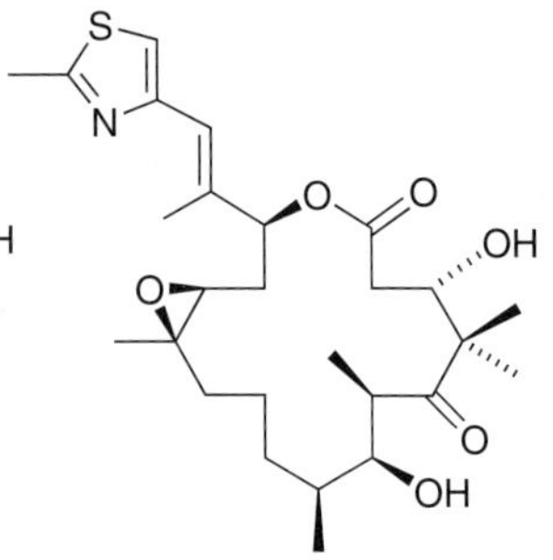

21
Epothilone B

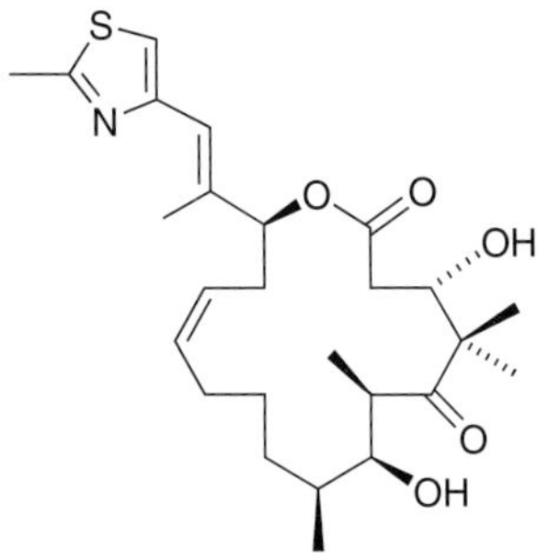

22
Epothilone C

The epothilones are naturally occurring microtubule stabilizers which inhibit the growth of human cancer cells in vitro at nanomolar and subnanomolar concentrations. In contrast to paclitaxel they are also active against different types of multidrug-resistant cancer cell lines in vitro. With their attractive preclinical profile, epothilones have become prominent lead structures in the search for improved cytotoxic anticancer drugs: epothilone B analogs are already undergoing clinical trials.

Epothilone structures lend themselves well to a convergent total synthesis approach by bringing together modular building blocks or fragments. This approach can be adapted readily to provide epothilone analogs for the probing of structure activity profiles. While total syntheses of these molecules are known,[256–258] it is the application of immobilized techniques and primarily supported reagents and scavengers that provides the final example in this section. The synthesis illustrates the current state of the art of these methods, leading, as it does, to the total synthesis of epothilone C in 29 steps; with the longest linear sequence of only 17 steps from readily available starting materials. In terms of overall yield (56%) and selectivity, the route compares with the best of the previously published conventional syntheses.

The route evolved considerably and several alternative strategies for the synthesis of the various fragments were also devised which are discussed in greater depth in the literature;[259,260] here only the shortest sequencc is discussed. The general strategy leads to the requirement for three fragments A, B, and C; these would ultimately be joined by appropriate methods to form the complete natural product (**Scheme 43**).

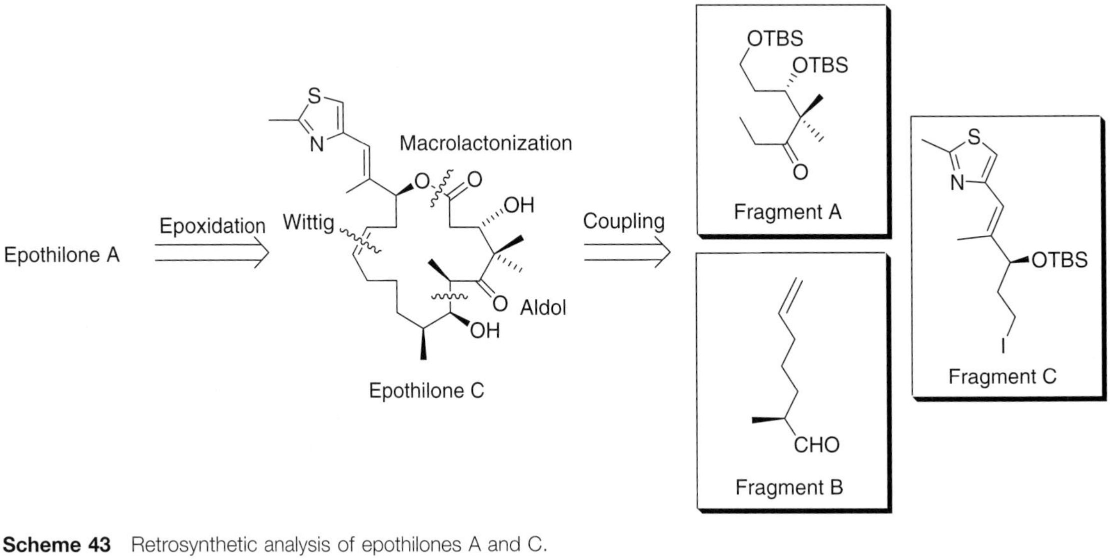

Scheme 43 Retrosynthetic analysis of epothilones A and C.

Of the three routes to fragment A that were investigated, the one shown in **Scheme 44** was shortest, most efficient, and proved to be the most easily scaled. A key feature of the route was the formation of the C2–C3 bond with the concomitant introduction of the C3 stereocenter. This was done via an asymmetric Mukaiyama aldol reaction using a borane complex with *N*-tosylphenylalanine as the chiral ligand. This reaction was followed by a work-up with a small

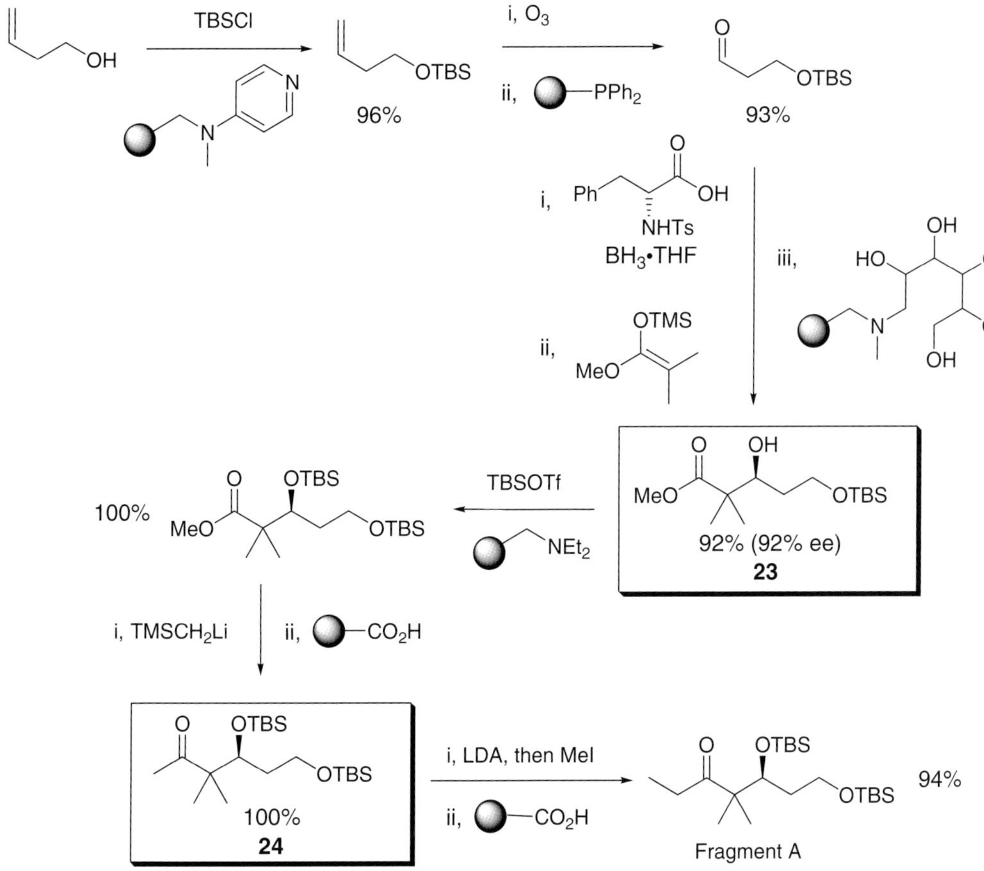

LDA, lithium diisopropylamide

Scheme 44 Synthesis of fragment A in the total synthesis of epothilone C.

amount of water and a boron selective scavenger, Amberlite IRA 743, to quench and remove the contaminating boric acid. This procedure led to the aldol product (**23**) in excellent yield and an enantiomeric excess of 92%. The ketone (**24**) obtained after *t*-butyldimethylsilyl (TBS) protection and reaction with (trismethylsilylmethyl)lithium using a scavenger quench was readily α-methylated via its lithium enolate. After a work-up with a supported carboxylic acid, fragment A was obtained in just six steps from the commercially available starting material.

Fragment B was obtained from the hydroxybromide available from the corresponding Roche ester (**Scheme 45**). Here, homologation of the intermediate iodide was accomplished with a Grignard-derived cuprate reagent (from 3-butenyl magnesium bromide). Work-up consisted of quenching with a mixture of a polymeric carboxylic acid and an amine resin to scavenge magnesium and copper residues. Straightforward tetrahydropyranyl deprotection and subsequent oxidation with immobilized pyridinium chlorochromate reagent furnished fragment B.

Fragment C was constructed in a convergent fashion from two readily available starting materials: (*S*)-α-hydroxyl-γ-lactone and chloromethyl thiazole hydrochloride (**Scheme 46**). The starting lactone was elaborated to the ketone, which was then coupled using a highly stereoselective Horner–Wadsworth–Emmons reaction, to the phosphonate derived from starting chloromethyl thiazole. A supported aldehyde was used to scavenge excess phosphonate from the reaction yielding the *bis*-TBS protected adduct (**23**). This compound was then taken through to fragment C. Ultimately, the iodide was captured onto a polymer-supported triphenylphosphine equivalent to produce the Wittig salt (**24**) as the coupling precursor of fragment C.

The final coupling of the three fragments was carried out in a sequential sequence of steps, the majority of which employed immobilized systems (**Scheme 47**). Of special note was the coupling of the polymer-bound Wittig species (the supported fragment C precursor, **24**) with the aldehyde to give the resultant alkene with complete stereoselectivity. The by-product of this Wittig coupling remains attached to the polymer and is simply removed by

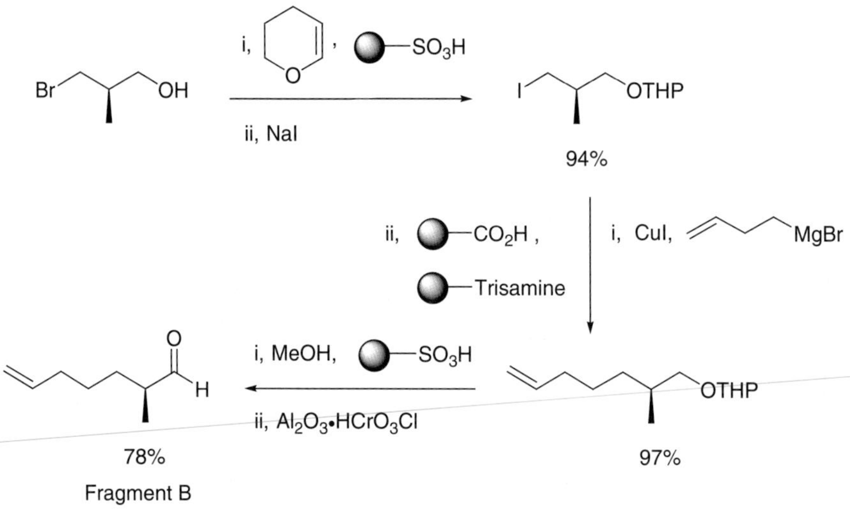

Scheme 45 Synthesis of fragment B of epothilone C.

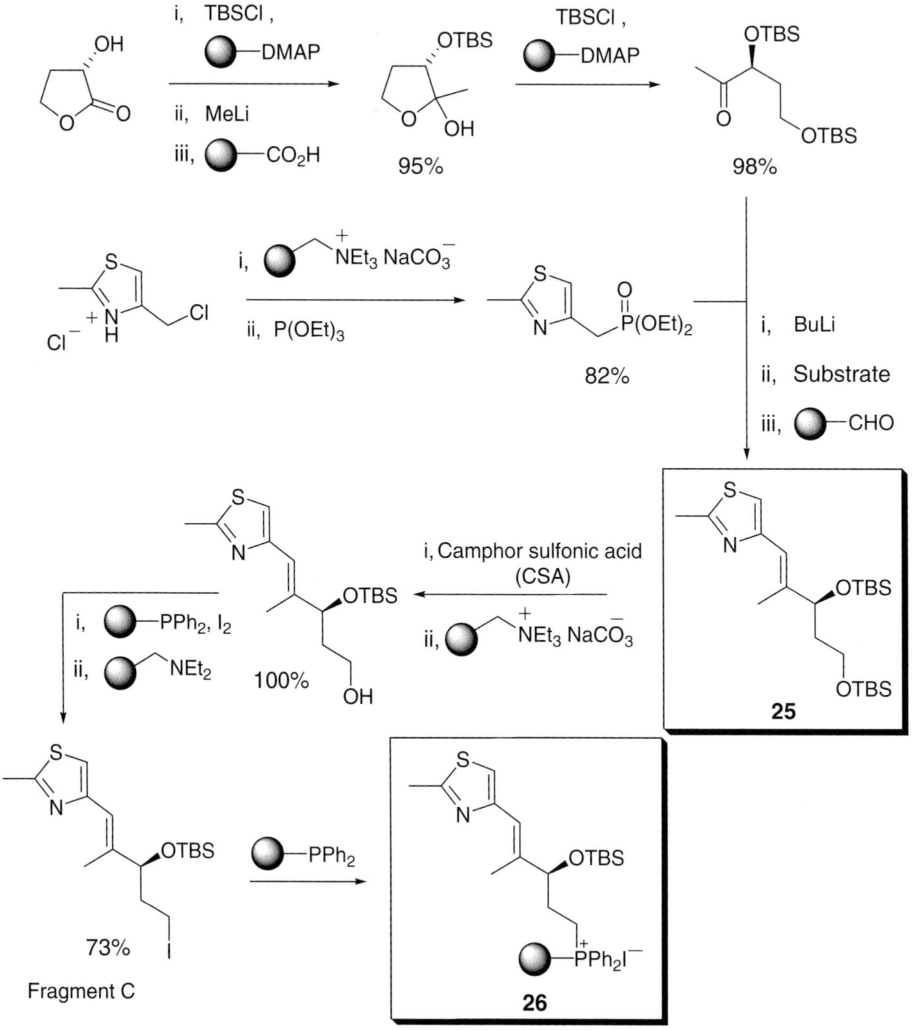

Scheme 46 Synthesis of fragment C of epothilone C.

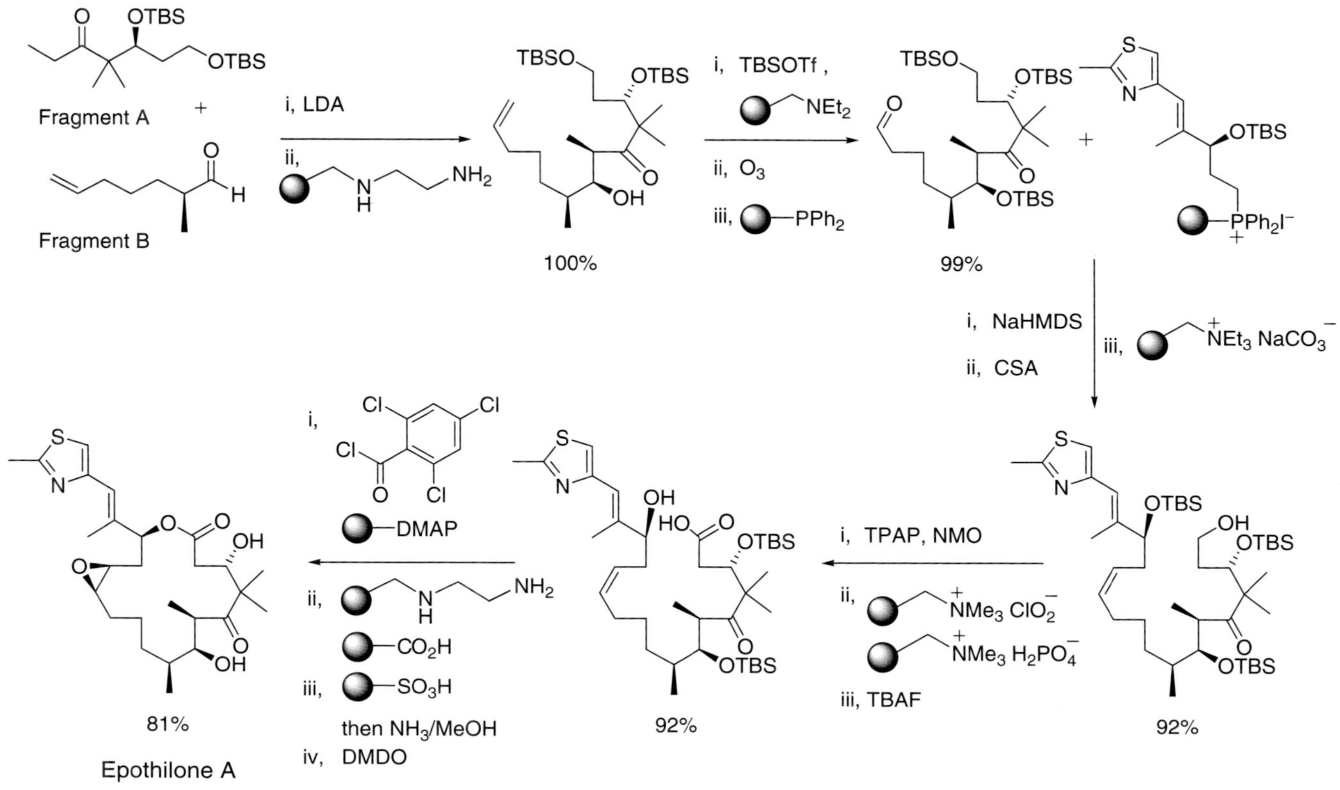

DMDO, dimethyldioxirane; NMO, *N*-methylmorpholine-*N*-oxide; TBAF, tetrabutylammonium fluoride; TPAP, tetra-*n*-propylammonium perruthenate, CSA, camphorsulfonic acid;

Scheme 47 Bringing together of the fragments in the total synthesis of epothilone C and A.

filtration, therefore avoiding product contamination common to the conventional Wittig process. Overall, this synthesis stands as a testament to these new immobilized methods and should provide considerable confidence for how they can be used effectively when incorporated into complex natural product synthesis programs. Obviously, given these effective methods, one can imagine that with greater use of automation and improved reaction optimization, these techniques will assist the assembly of other complex molecules and become an integrated part of future synthesis planning.

3.35.4 Flow Chemistry in Organic Synthesis

Flow processes, both on a microfluidic and mesofluidic scale, are receiving increased attention and offer a number of potential advantages over existing batch processes.[17–33] For example, reaction conditions (flow rate, stoichiometry, temperature, pressure) can be independently varied and precisely controlled. This leads to high reproducibility, and greatly facilitates the process of reaction optimization whereby a range of different conditions can be rapidly investigated. In addition, the incorporation of real-time in-line monitoring in combination with intelligent feedback loops offers considerable scope for fully automated reaction optimization, and, in a similar way, automated fast serial processing may be used to screen new transformations or generate compound libraries. Furthermore, the incorporation of in-line modules into flow devices presents opportunities to effect chromatographic separation, by-product scavenging, or the isolation of specific products by catch-and-release protocols. Flow processes are readily scaleable either by running the flow reactor for an extended time, or by employing multi-channel parallel reactors. However, there is a need to considerably expand the repertoire of reactions and processes that are possible using these techniques, in particular to enable multistep flow-through compound synthesis.

As traditional chemical reactors are being miniaturized we are witnessing micro- (nanoliter) and meso- (microliter) reactors becoming available with increasing sophistication and functionality. It is now possible to purchase fully automated flow synthesizers comprising high-precision dispensing pumps attached to chip and column-based mixing and reacting chambers that also have integrated in-line analytical ability. Such devices are now well within reach for the modern research laboratory. Indeed, the real future of supported reagents and related technologies will undoubtedly be how they are best integrated into flow processes.

The footprint of a typical meso-reactor setup allows it to sit comfortably in a standard laboratory fume cupboard. Once reagents and the appropriate solutions have been added to the equipment the reactor (operated for the most part by a software interface) is ready to be used. The flow-through processes can use cartridge or column-based reactors packed with supported reagents or catalysts.[261] The substrate carried in a solution is then pumped through the reactive cartridges carried in lines linking, in principle, one reactive cartridge to another. This principle is shown, in a simplistic way, in **Figure 7**.

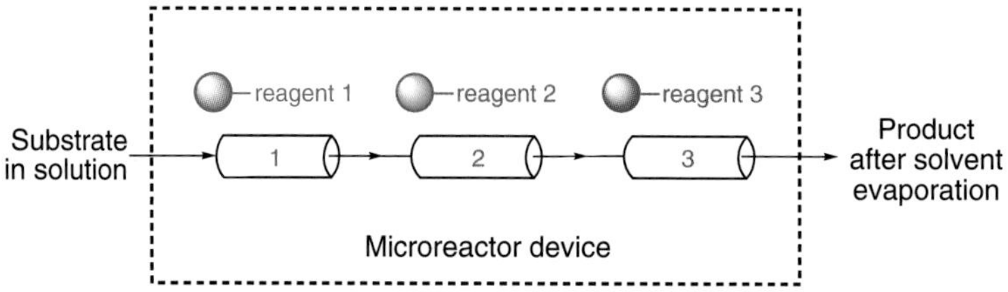

Figure 7 The principle behind a microreactor continuous-flow device.

3.35.4.1 Microreactors

A factor influencing the development of microreactors,[24] other than the trend in the miniaturization of technology in general, has been to do with safety. Chemical processing on a big scale typically involves using equipment that holds large amounts of potentially hazardous materials in the reactor vessels. Under many circumstances, exothermic reactions are especially difficult to control. If, however, the surface area is dramatically increased in relation to the amount of material reacted then temperatures can be controlled much more effectively. Within these miniaturized chemical reactors, reaction channels for transporting the dissolved substrates have proportionally very small apertures and therefore, there is a much greater surface area compared to flowing component volume.

3.35.4.2 Advantages of Flow Techniques in Practice

A significant advantage of working with a miniaturized flow domain is that chemicals that are hazardous to prepare, handle, or store can be produced on-site and on demand. A recent example of flow hydrogenation highlights this advantage in handling hydrogen gas in flow tubes (**Scheme 48**).[262]

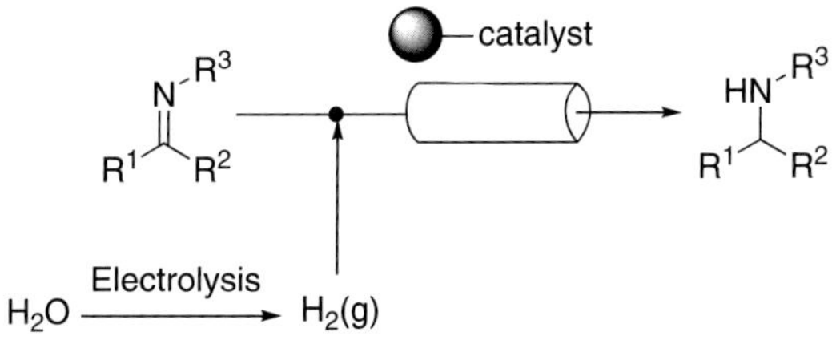

Scheme 48 Flow hydrogenation.

A flow-through strategy was used to effect the catalytic hydrogenation of imines with high chemoselectivity (**Scheme 48**). Reductive amination is a key transformation in the synthesis of many drug substances; however, problems arise due to reversibility, compound/functional group incompatibility, and over-reduction. In particular, the reduction of aryl imines often gives rise to secondary amines contaminated with the corresponding primary amine derived from over-reduction and debenzylation of the desired product.

A continuous flow-through hydrogenation was used to perform the reduction. The hydrogen required was delivered using a device that mixes hydrogen gas with the flowing stream in a T-piece mixer containing a titanium frit to ensure even gas dispersion (**Figure 8**). The hydrogen is generated internally on demand by the electrolysis of water, and the gas–liquid mixture is pumped through a suitable catalyst contained in an interchangeable metal cartridge. An integral heater enabled the cartridge to be heated at up to 100 °C, and a responsive back-pressure regulator allowed flow hydrogenation to be performed at a pressure up to 100 bar.

An important prerequisite in flow chemistry is the choice of an effective solvent which will prevent precipitation of either starting materials or products that may lead to blockages in the flow stream. In this case, tetrahydrofuran was found to be an effective solvent up to 0.5 M, and gave the best results under representative reaction conditions (10% Pd/C, 20 bar, 25 °C, 0.05 M, flow rate 1.0 mL min^{-1}), affording the amine in quantitative yield and >95% purity.

Optimization of the flow conditions for the conversion was systematically investigated by varying the concentration, flow rate, pressure, and temperature. This process was greatly facilitated under continuous-flow conditions and a number of runs were conducted rapidly in a sequential manner, without having to change or renew the catalyst cartridge. Increasing the flow rate resulted in a corresponding decrease in conversion, consistent with a reduction in residence time in the flow reactor. Notably, optimal conditions at a flow rate of 1 mL min^{-1} were obtained with a

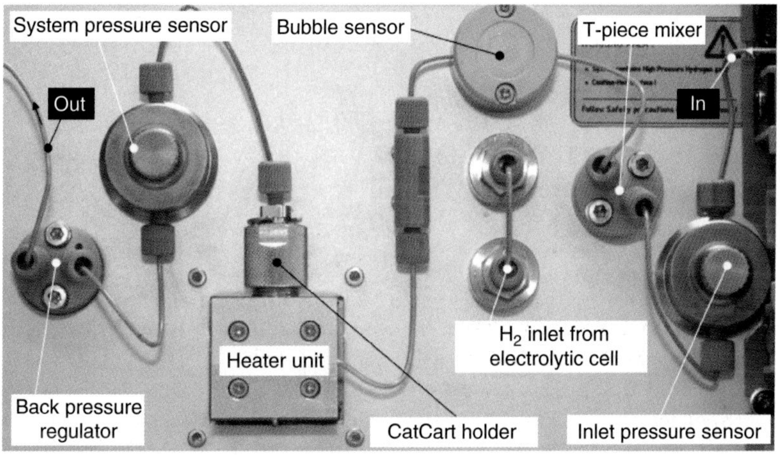

Figure 8 Front panel of the flow hydrogenator equipment.

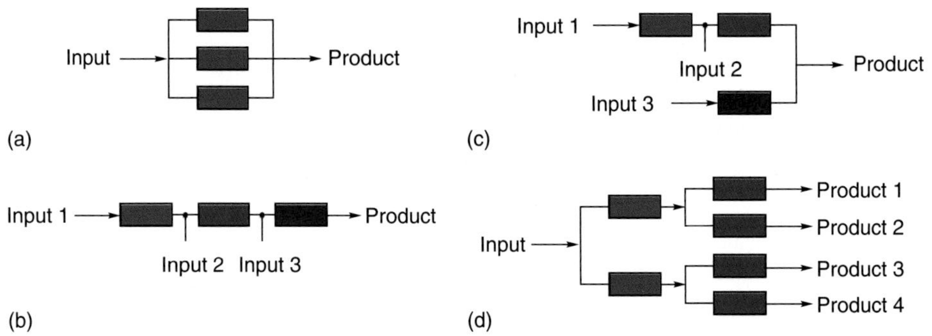

Figure 9 Possible configurations for flow-through reactors, creating opportunities for parallel synthesis as well as multistep synthesis: (a) scale out; (b) linear setup of flow-through reactors; (c) convergent setup of flow-through reactors; and (d) divergent setup of flow-through reactors.

substrate concentration of 0.05 M. Having established optimal conditions for the steady state production of pure amine, it was then demonstrated that the process could be scaled to provide a preparative quantity of material: a continuous-flow hydrogenation afforded 1 g of amine in quantitative yield and excellent purity (>95%). Again, a single 10% Pd/C catalyst cartridge was used throughout. The generality of flow-through hydrogenation for the selective reduction of a variety of imines derived from aromatic and aliphatic aldehydes and ketones were also investigated. We can anticipate many further uses of mixed-phase flow reactors in the future.

The reduced apertures of mixing chambers in the microreactor leads to more turbulent dispersion and therefore better amalgamation of the reaction species. Such a controlled and effective merging of reactive partners permits better reaction kinetic data to be collected which leads to faster optimizations. Microreactors are also versatile and inexpensive to use; their small physical size and minimal operation requirements mean only small amounts of reactants are needed; this is ideal for capitalizing on the use of supported reagents and catalysts, which can easily be incorporated as rapid-insert in-line plugs and cartridges. Indeed, the relatively large surface areas in the microreactors enables a wide range of controls to be incorporated in the devices, such as pressure, temperature, heating elements, and pH sensors. The immense utility of microreactors makes them highly suited to discovery chemistry bench-top processes. However, with the capacity of moving more swiftly into volume production (i.e., scale out rather than scale up), it becomes easy to imagine a scenario in which many parallel microreactors are operating together. By assembling different combinations of reactors linear, divergent, or convergent syntheses become possible (**Figure 9**).

3.35.4.3 Use of Flow Techniques in Organic Synthesis

An early example of a supported reagent being used in a flow mode is the oxidation of penicillin G to its corresponding sulfoxide (**Scheme 49**).[263] The column was regenerated in situ by treating it with hydrogen peroxide and

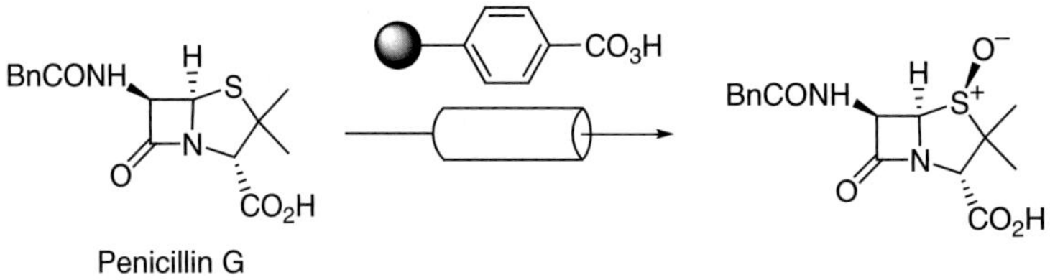

Scheme 49 Oxidation of Penicillin G.

methanesulfonic acid. Being able to regenerate expensive supported reagents is desirable and is something that can be assisted greatly by using flow processes in this way.

A further example is the derivatization of the *O*-silylated steroid to its amino derivative using three different supported reagents (**Scheme 50**).[264] The [2,2,6,6-tetramethyl-1-piperidinyloxy] TEMPO-catalyzed oxidation of alcohol via the bound bromate(I) anion in the first column gave the ketone. This was desilylated as it was pumped through a second immobilized tetraalkylammonium fluoride-packed reactor. Reductive amination was performed by

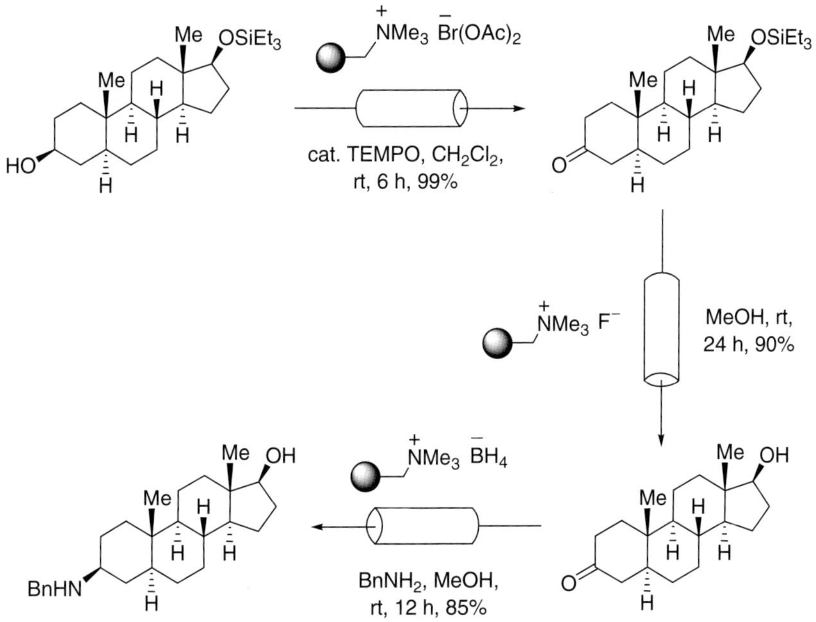

Scheme 50 Derivatization of an *O*-silylated steroid using three PASSflow reactor cartridges containing different supported anions.

in situ generation of the imine using benzylamine. The passing of this solution through a reactor containing a supported borohydride furnished the final amine.

Acyl-substituted β-lactams are inhibitors of many biological targets, including prostate-specific antigen[265] and cytomegalovirus protease.[266] In work directed toward developing a general method for synthesizing β-lactams by flow processes, a catalytic asymmetric reaction using a series of reaction columns, each packed with supported reagents or catalysts has been developed (**Scheme 51**).[183] The β-lactams were obtained pure with excellent enantio- and

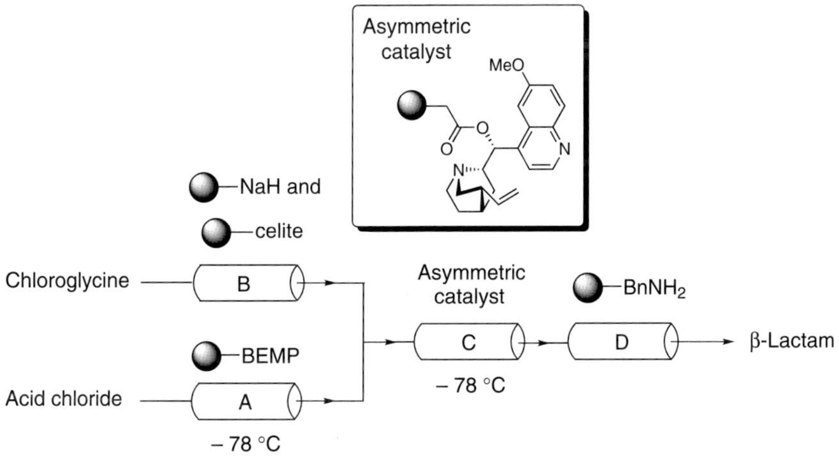

Scheme 51 Development of flow methods toward a general synthesis of β-lactams.

diastereoselectivity. The four steps involve formation of reactive ketenes in column A; formation of imines in situ in column B; catalysis of the condensation of the ketene and imine to form a β-lactam product in column C; and, finally, removal of unwanted by-products from the reaction stream using a scavenger resin in column D. Reaction solutions were allowed to percolate through the columns in a sequential fashion. After passing through the scavenger column the eluted reaction mixture was concentrated to afford the β-lactams.

A fully automated, sequential flow-through synthesis using a resin capture-and-release reactor column has been used in the preparation of a 44-member array of thioethers.[23] The acidic thiourea heterocycles were deprotonated using a strong polymer-supported base, such as PS-TBD, to generate an immobilized ionic complex on the column (**Scheme 52**).

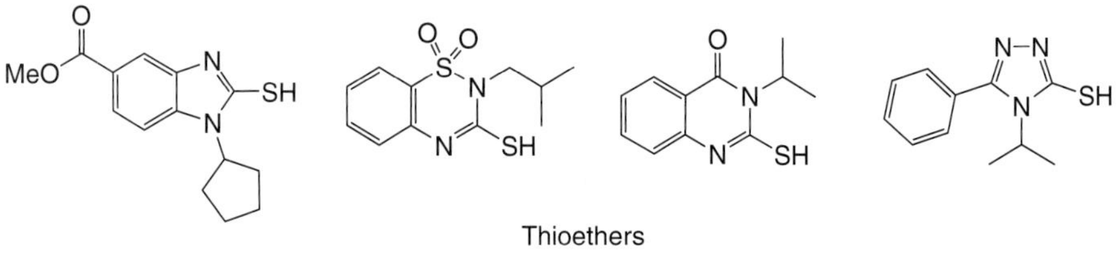

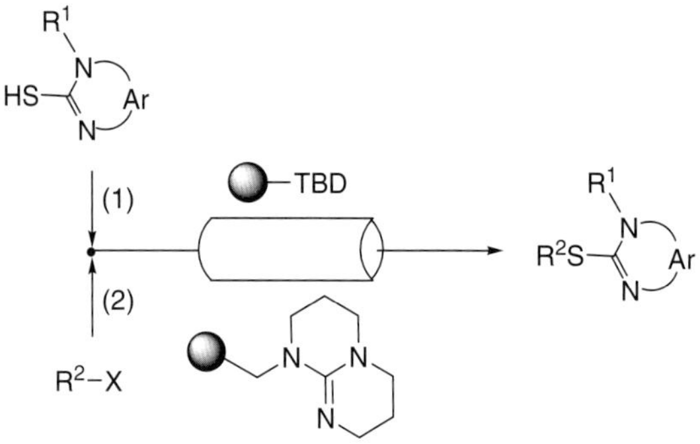

Scheme 52 Fully automated, sequential flow-through synthesis of thioethers.

Introduction of a substoichiometric amount of alkylating agent promoted an alkylative release of the corresponding *S*-thioether from the column. This process was repeated until the column was depleted. The PS-TBD column could then be regenerated by eluting with a solution of a stronger base, such as BEMP.

In an extension to this work, a 576-member combinatorial array was prepared in a sequential manner from a set of 24 heterocyclic thiourea and 24 alkylating agent building blocks, reusing the same reactor column over 40 times. Less than 2% of the unpurified library members fell below the 80% purity threshold, even when the yield was less impressive. This provides a clear demonstration of the power of flow processes to deliver high-purity compound libraries, particularly for chemistries that may be implemented as a catch-and-release strategy.

3.35.5 Conclusions

This chapter has summarized some of the developments and opportunities afforded by the application of immobilized systems for multistep synthesis programs, with particular emphasis on pharmaceutically relevant problems. Given that many of the reagents, scavengers, and quenching agents are now commercially available, we anticipate their continued use and integration into conventional synthesis strategies. However, there is still a need to improve their general recyclability in an effort to reduce costs and wastage. Many more reagents, and especially catalysts, need to be invented to drive this innovation. Nevertheless, we are beginning to see their impact across wider synthesis programs. Given that we will see many advances in microfluidics and flow processing, and that our ability to fabricate smaller and smaller devices is increasing, we can expect significant further advances. The effects of automation, informatics, data mining, and knowledge capture are acting as further stimuli for development. Linking product formation, and eventually product function and prospects, with real-time feedback of information should lead to self-optimizing systems. New reactor packs, reagents chips, and plug-and-play cartridges are also becoming available. This is an exciting time for chemistry, as we are continually adding to our toolkit to create new opportunities for molecular assembly.

The ease of immobilization of enzymes, and other species obtained by directed evolution, can be readily accommodated in a future vision of how synthesis may progress.[269] Moreover, the future ability to make and screen species on single workbenches will lead to a de-emphasis of the molecular structure at the expense of their function.[267]

Currently, all organic synthesis is conducted with a very limited range of less than 2000 commercially available reagents. This situation must change to embrace higher selectivity, by finding new catalysts and discovering processes not yet known. However, there is still enormous untapped potential for the future progress of molecular assembly methods; no one technique can be expected to be universally applicable.

References

1. Ley, S. V.; Baxendale, I. R. *Chem. Rec.* **2002**, *2*, 377–388.
2. Baxendale, I. R.; Storer, R. I.; Ley, S. V. In *Polymeric Materials in Organic Synthesis*; Buchmeiser, M. R., Ed.; Wiley-VCH: Berlin, Germany, 2003, pp 53–136.
3. Ley, S. V.; Baxendale, I. R.; Brusotti, G.; Caldarelli, M.; Massi, A.; Nesi, M. *Farmaco* **2002**, *57*, 321–330.
4. Baxendale, I. R.; Ley, S. V. *Nat. Rev. Drug Disc.* **2002**, *2*, 573–586.
5. Clapham, B.; Reger, T. S.; Janda, K. D. *Tetrahedron* **2001**, *57*, 4637–4662.
6. Kirschning, A.; Monenschein, H.; Wittenberg, R. *Angew. Chem. Int. Ed.* **2001**, *40*, 650–679.
7. Sherrington, D. C. *J. Poly. Sci. Poly. Chem.* **2001**, *39*, 2364–2377.
8. Ley, S. V.; Baxendale, I. R.; Bream, R. N.; Jackson, P. S.; Leach, A. G.; Longbottom, D. A.; Nesi, M.; Scott, J. S.; Storer, R. I.; Taylor, S. J. *J. Chem. Soc. Perkin Trans.* **2000**, *1*, 3815–4195.
9. Thompson, L. A. *Curr. Opin. Chem. Biol.* **2000**, *4*, 324–337.
10. Ley, S. V.; Baxendale, I. R. In *Supported Catalysts and Their Applications*; Sherrington, D. C., Kybett, A. P., Eds.; Royal Society of Chemistry: Cambridge, 2001, p 9.
11. Hall, B.; Haunert, F.; Scott, J.; Bolli, M.; Habermann, J.; Hinzen, B.; Gervois, A.-G.; Ley, S. V.; Patent WO 9958475, 1999.
12. Jenck, J. F.; Agterberg, F.; Droescher, M. J. *Green Chem.* **2004**, *6*, 544–556.
13. Haswell, S. J.; Watts, P. *Green Chem.* **2003**, *5*, 240.
14. Potoski, J. *Drug Disc. Today* **2005**, *10*, 115–120.
15. Glen, R. *Chem. Commun.* **2002**, *23*, 2745–2747.
16. Jamieson, C.; Congreve, M. S.; Emiabata-Smith, D. F.; Ley, S. V.; Scicinski, J. J. *Org. Process Res. Dev.* **2002**, *6*, 823–825.
17. Luckarift, H. R.; Nadeau, L. J.; Spain, J. C. *Chem. Commun.* 2005, 383–384.
18. Doku, G. N.; Verboom, W.; Reinhoudt, D. N.; van den Berg, A. *Tetrahedron* **2005**, *61*, 2733–2742.
19. Kautz, R. A.; Goetzinger, W. K.; Karger, B. L. *J. Comb. Chem.* **2005**, 7, 14–20.
20. Ratner, D. M.; Murphy, E. R.; Jhunjhunwala, M.; Snyder, D. A.; Jensen, K. F.; Seeberger, P. H. *Chem. Commun.* **2005**, *5*, 578–580.
21. Watts, P.; Haswell, S. J. *Drug Disc. Today* **2003**, *8*, 586–593.
22. Kirschning, A.; Jas, G. *Topics Curr. Chem.* **2004**, 209–239.
23. Jönsson, D.; Warrington, B. H.; Ladlow, M. *J. Comb. Chem.* **2004**, *6*, 584–595.
24. Jas, G.; Kirschning, A. *Chem. Eur. J.* **2003**, *9*, 5708–5723.
25. Watts, P.; Haswell, S. J. *Curr. Opin. Chem. Biol.* **2003**, 7, 380–387.
26. Fletcher, P. D. I.; Haswell, S. J.; Pombo-Villar, E.; Warrington, B. H.; Watts, P.; Wong, S. Y. F.; Zhang, X. *Tetrahedron* **2002**, *58*, 4735–4757.
27. Hafez, A. M.; Taggi, A. E.; Lectka, T. *Chem. Eur. J.* **2002**, *8*, 4115–4119.
28. Hafez, A. M.; Taggi, A. E.; Dudding, T.; Lectka, T. *J. Am. Chem. Soc.* **2001**, *123*, 10853–10859.
29. Anderson, N. G. *Org. Process Res. Dev.* **2001**, *5*, 613–621.
30. Sands, M.; Haswell, S. J.; Kelly, S. M.; Skelton, V.; Morgan, D. O.; Styring, P.; Warrington, B. H. *Lab Chip* **2001**, *1*, 64–65.
31. Haswell, S. J.; Middleton, R. J.; O'Sullivan, B.; Skelton, V.; Watts, P.; Styring, P. *Chem. Commun.* **2001**, *5*, 391–398.
32. Ehrfeld, W.; Hessel, V.; Löwe, H. *Microreactors: New Technology for Modern Chemistry*; Wiley-VCH: Weinheim, Germany, 2000.
33. Hafez, A. M.; Taggi, A. E.; Wack, H.; Drury, W. J., III.; Lectka, T. *Org. Lett.* **2000**, *2*, 3963–3965.
34. Baxendale, I. R.; Lee, A.-L.; Ley, S. V. Integrated Microwave Assisted Synthesis, Solid-Supported Reagents. In *Microwave-Assisted Organic Synthesis*; Tierney, J. P., Lidstrom, P., Eds.; Blackwell: Oxford, UK, 2005.
35. Kirschning, A.; Monenschein, H.; Wittenburg, R. *Chem. Eur. J.* **2000**, *6*, 4445–4450.
36. Keating, T. A.; Armstrong, R. W. *J. Am. Chem. Soc.* **1996**, *118*, 2574–2583.
37. Frechet, J. M. J.; Nuyens, L. J. *Can. J. Chem.* **1976**, *54*, 926–934.
38. Wong, J. Y.; Leznoff, C. C. *Can. J. Chem.* **1973**, *51*, 2452–2456.
39. Leznoff, C. C.; Wong, J. Y. *Can. J. Chem.* **1972**, *50*, 2892–2893.
40. Harrison, I. T.; Harrison, S. *J. Am. Chem. Soc.* **1967**, *89*, 5723–5724.
41. Letsinger, R. L.; Kornet, M. J. *J. Am. Chem. Soc.* **1963**, *85*, 3045–3046.
42. Merrifield, R. B. *J. Am. Chem. Soc.* **1963**, *85*, 2149–2154.
43. Sussman, S. *Ind. Eng. Chem.* **1946**, *38*, 1228.
44. Helferrich, F. *Ion Exchange*; McGraw-Hill: New York, 1962; Chapter 11, p 519.
45. Astle, M. J. In *Ion Exchangers in Organic and Biochemistry*; Calman, C., Kressman, T. R. E., Eds.; Interscience: New York, 1957; Chapter 36, p 658.
46. Bleicher, K. H.; Bohm, H. J.; Muller, K.; Alanine, A. I. *Nat. Rev. Drug Disc.* **2003**, *2*, 369–378.
47. Entzeroth, M. *Curr. Opin. Pharmacol.* **2003**, *3*, 522–529.
48. Battersby, B. J.; Trau, M. *Trends Biotech.* **2002**, *20*, 167–173.
49. Khandurina, J.; Guttman, A. *Curr. Opin. Chem. Biol.* **2002**, *6*, 359–366.
50. Hertzberg, R. P.; Pope, A. J. *Curr. Opin. Chem. Biol.* **2000**, *4*, 445–451.
51. Silverman, L.; Campbell, R.; Broach, J. R. *Curr. Opin. Chem. Biol.* **1998**, *2*, 397–403.
52. Fernandes, P. B. *Curr. Opin. Chem. Biol.* **1998**, *2*, 597–603.
53. Harding, D.; Banks, M.; Fogarty, S.; Binnie, A. *Drug Disc. Today* **1997**, *2*, 385–390.
54. Ley, S. V.; Ladlow, M.; Vickerstaffe, E. The Use of Polymer Assisted Solution Phase Synthesis, Automation for High Throughput Preparation of Biologically Active Materials. In *New Approaches to the Generation, Evaluation of Chemical Diversity*; Bartlett, P. A.; Entzeroth, M., Eds.; Royal Society of Chemistry: Cambridge UK, in press.

55. Ley, S. V.; Baxendale, I. R.; Myers, R. M. The Use of Polymer Supported Reagents, Scavengers in the Synthesis of Natural Products. In *Combinatorial Synthesis of Natural Products*; Boldi, A. M., Ed.; CRC Press: London, UK, in press.
56. Ley, S. V.; Baxendale, I. R.; Longbottom, D. A.; Myers, R. M. Natural Products as an Inspiration for the Discovery of New High Throughput Chemical Synthesis Tools. In *Drug Discovery & Development*; Chorghade, M. S., Ed.; Wiley: Chichester, UK, in press.
57. Baxendale, I. R.; Ley, S. V. *Mini Rev. Org. Chem.*, in press.
58. Ley, S. V. *Pure. Appl. Chem.* **2005**, 77, 1115–1130.
59. Toy, P. H.; Tanda, K. D. *Acc. Chem. Res.* **2000**, *3*, 546–554.
60. Ito, Y.; Manabe, S. *Curr. Opin. Chem. Biol.* **1998**, *2*, 701–708.
61. de la Hoz, A.; Diaz-Ortiz, A.; Moreno, A. *Chem. Soc. Rev.* **2005**, *34*, 164–178.
62. Kappe, C. O. *Angew. Chem. Int. Ed.* **2004**, *43*, 6250–6284.
63. Santagada, V.; Frecentese, F.; Perissutti, E.; Favretto, L.; Caliendo, G. *QSAR Combin. Sci.* **2004**, *23*, 919–944.
64. Wiesbrock, F.; Hoogenboom, R.; Schubert, U. S. *Macromol. Rapid Commun.* **2004**, *25*, 1739–1764.
65. Rajak, H.; Mishra, P. *J. Sci. Ind. Res.* **2004**, *63*, 641–654.
66. Nuchter, M.; Ondruschka, B.; Bonrath, W.; Gum, A. *Green Chem.* **2004**, *6*, 128–141.
67. Nuchter, M.; Muller, U.; Ondruschka, B.; Tied, A.; Lautenschlager, W. *Chem. Eng. Tech.* **2003**, *26*, 1207–1216.
68. Brase, S.; Lauterwasser, F.; Ziegert, R. E. *Adv. Syn. Catal.* **2003**, *345*, 869–929.
69. Benaglia, M.; Puglisi, A.; Cozzi, F. *Chem. Rev.* **2003**, *103*, 3401–3429.
70. Benaglia, M.; Cinquini, M.; Cozzi, F.; Puglisi, A.; Celentano, G. *Adv. Synth. Catal.* **2002**, *344*, 533–542.
71. Lee, C. K. Y.; Holmes, A. B.; Ley, S. V.; McConvey, I. F.; Al-Duri, B.; Leeke, G. A.; Santos, R. C. D.; Seville, J. P. K. *J. Chem. Soc. Chem. Commum.* **2005**, 2175–2177.
72. Okamoto, K.; Akiyama, R.; Kobayashi, S. *J. Org. Chem.* **2004**, *69*, 2871–2873.
73. Okamoto, K.; Akiyama, R.; Kobayashi, S. *Org. Lett.* **2004**, *6*, 1987–1990.
74. Mitchell, C.; Pears, D.; Ley, S. V.; Yu, J.-Q.; Zhou, W. *Org. Lett.* **2003**, *5*, 4665.
75. Ley, S. V.; Ramarao, C.; Lee, A.-L.; Østergaard, N.; Smith, S. C.; Shirley, I. M. *Org. Lett.* **2003**, *5*, 185–187.
76. Yu, J -Q.; Wu, H.-C.; Ramarao, C.; Spencer, J. B.; Ley, S. V. *J. Chem. Soc. Chem. Commun.* **2003**, 678–679.
77. Akiyama, R.; Kobayashi, S. *J. Am. Chem. Soc.* **2003**, *125*, 3412–3413.
78. Bremeyer, N.; Ley, S.V.; Ramarao, C.; Shirley, I. M.; Smith, S. C. *Synlett* **2002**, 1843–1844.
79. Ley, S. V.; Ramarao, C.; Gordon, R. S.; Holmes, A. B.; Morrison, A. J.; McConvey, I. F.; Shirley, I. M.; Smith, S. C.; Smith, M. D. *J. Chem. Soc. Chem. Commun.* **2002**, 1134–1135.
80. Ramarao, C.; Ley, S. V.; Smith, S. C.; Shirley, I. M.; DeAlmeida, N. *Chem. Commun.* **2002**, 1132.
81. Jansson, A. M.; Grøtli, M.; Halkes, K. M.; Meldal, M. *Org. Lett.* **2002**, *4*, 27–30.
82. Shimofure, S.; Koizumi, S.; Ichikawa, K.; Ichikawa, H.; Dobashi, T. *J. Microencapsulation* **2001**, *18*, 13–17.
83. Yeung, L. K.; Crooks, R. M. *Nano Lett.* **2001**, *1*, 14–17.
84. Tsuji, K. *J. Microencapsulation* **2001**, *18*, 137–147.
85. Uludag, H.; De Vos, P.; Tresco, P. A. *Adv. Drug Deliv. Rev.* **2000**, *42*, 29–64.
86. Jain, R. A. *Biomaterials* **2000**, *21*, 2475–2490.
87. Gelman, F.; Avnir, D.; Schumann, H.; Blum, J. *J. Mol. Catal. A* **1999**, *146*, 123–128.
88. Reetz, M. T.; Dugal, M. *Catal. Lett.* **1999**, *58*, 207–212.
89. Blum, J.; Avnir, D.; Schumann, H. *Chemtech* **1999**, *29*, 32–38.
90. Rosenfeld, A.; Avnir, D.; Blum, J. *J. Chem. Soc. Chem. Commun.* **1993**, 583–584.
91. Curran, D. P. *Angew. Chem. Int. Ed.* **1998**, *37*, 1174–1196.
92. Clarke, D.; Ali, M. A.; Clifford, A. A.; Parratt, A.; Rose, P.; Schwinn, D.; Bannwarth, W.; Rayner, C. M. *Curr. Topics Med. Chem.* **2004**, *4*, 729–771.
93. Ortholand, J. Y.; Ganesan, A. *Curr. Opin. Chem. Biol.* **2004**, *8*, 271–280.
94. Watts, P. *Curr. Opin. Drug Disc. Dev.* **2004**, 7, 807–812.
95. Solodenko, W.; Wen, H. L.; Leue, S.; Stuhlmann, F.; Sourkouni-Argirusi, G.; Jas, G.; Schonfeld, H.; Kunz, U.; Kirschning, A. *Eur. J. Org. Chem.* **2004**, *17*, 3601–3610.
96. Toy, P. H.; Janda, K. D. *Tetrahedron Lett.* **1999**, *40*, 6322–6329.
97. Pears, D. A.; Smith, S. C. *Aldrichimica Acta* **2005**, *38*, 24.
98. Peterson, D. S. *Lab Chip.* **2005**, *5*, 132–139.
99. Roy, S.; Bauer, T.; Al-Dahhan, M.; Lehner, P.; Turek, T. *Aiche J.* **2004**, *50*, 2918–2938.
100. Tripp, J. A.; Svec, F.; Frechet, J. M. J. *J. Comb. Chem.* **2001**, *3*, 604–611.
101. Tripp, J. A.; Stein, J. A.; Svec, F.; Frechet, J. M. J. *Org. Lett.* **2000**, *2*, 195–198.
102. Hird, N.; Hughes, I.; Hunter, D.; Morrison, M. G. J. T.; Sherrington, D. C.; Stevenson, L. *Tetrahedron* **1999**, *55*, 9575–9584.
103. Viklund, C.; Svec, F.; Frechet, J. M. J.; Irgum, K. *Chem. Mat.* **1996**, *8*, 744–750.
104. Svec, F.; Frechet, J. M. J. *Anal. Chem.* **1992**, *64*, 820–822.
105. Porcheddu, A.; Giacomelli, G.; Chighine, A.; Masala, S. *Org. Lett.* **2004**, *6*, 4925–4927.
106. Akelah, A.; Sherrington, D. C. *Eur. Polym. J.* **1982**, *18*, 301–305.
107. Davis, M. E. *Top. Catal.* **2003**, *25*, 3–7.
108. Davis, M. E. *Nature* **2002**, *417*, 813–821.
109. Davis, M. E. *Micropor. Mesop. Mat.* **1998**, *21*, 173–182.
110. Nikbin, N.; Watts, P. *Org. Proc. Res. Dev.* **2004**, *8*, 942–944.
111. Ciriminna, R.; Bolm, C.; Fey, T.; Pagliaro, M. *Adv. Syn. Catal.* **2002**, *344*, 159–163.
112. Bolm, C.; Fey, T. *Chem. Commun.* **1999**, 1795–1796.
113. Staudinger, H.; Meyer, J. *Helv. Chim. Acta* **1919**, *2*, 635.
114. Mitsunobu, O. *Bull. Chem. Soc. Jpn.* **1967**, *40*, 4235–4238.
115. Mitsunobu, O. *Tetrahedron* **1970**, *26*, 5731.
116. Wittig, G.; Schöllkopf, U. *Chem. Ber.* **1954**, *87*, 1318.
117. Schobert, R.; Jagusch, C.; Melanophy, C.; Mullen, G. *Org. Biomol. Chem.* **2004**, *2*, 3524–3529.
118. Bolli, M. H.; Ley, S. V. *J. Chem. Soc. Perkin Trans.* **1998**, *1*, 2243–2246.
119. McKinley, S. V.; Rakshys, J. W. *J. Chem. Soc. Chem. Commun.* **1972**, 134.

120. Camps, F.; Castells, J.; Fernando, M. J.; Font, J. *Tetrahedron Lett.* **1971**, 1713.
121. Pedersen, B. S.; Lawesson, S. O. *Bull. Soc. Chem. Belg.* **1977**, *86*, 693–697.
122. Ley, S. V.; Leach, A. G.; Storer, R. I. *J. Chem. Soc. Perkin Trans.* **2001**, *1*, 358–361.
123. Brose, A. K.; Manhas, M. S.; Ganguly, S. N.; Sharma, A. H.; Banik, B. K. *Synthesis* **2002**, 1578–1591.
124. Lew, A.; Krutzik, P. O.; Hart, M. E.; Chamberlin, A. R. *J. Comb. Chem.* **2002**, *4*, 95–105.
125. Lindstrom, P.; Tierney, J.; Wathey, B.; Westman, J. *Tetrahedron* **2001**, *57*, 9225–9283.
126. Baxendale, I. R.; Lee, A.-L.; Ley, S. V. *Synlett* **2001**, 1482–1484.
127. Ellis, W.; Keim, P.; Wasserscheid, P. *J. Chem. Soc. Chem Commun.* **1999**, 337–338.
128. Waffenschmidt, H.; Wasserscheid, P.; Vogt, D.; Keim, W. *J. Catal.* **1999**, *186*, 481–484.
129. Waffenschmidt, H.; Wasserscheid, P.; Keim, W. German Patent 19901524.4, 1994.
130. Welton, T. *Chem. Rev.* **1999**, *99*, 2071–2083.
131. Freemantle, M. *Chem. Eng. News* **1998**, *76*, 12.
132. Seddon, K. R. *J. Chem. Technol. Biotechnol.* **1997**, *68*, 351–356.
133. Seddon, K. R. *Kinet. Catal.* **1996**, *37*, 693–697.
134. Volkov, S. V. *Chem. Soc. Rev.* **1990**, *19*, 21–28.
135. Sundermeyer, W. *Chem. Unserer Zeit* **1967**, *1*, 150.
136. Sundermeyer, W. *Angew. Chem.* **1965**, 77, 241.
137. Sundermeyer, W. *Angew. Chem. Int. Ed. Engl.* **1965**, *4*, *222*.
138. Hodge, P. *Chem. Soc. Rev.* **1997**, *26*, 417–424.
139. Akelah, A. *J. Mat. Sci.* **1986**, *21*, 2977–3001.
140. Kraus, M. A.; Patchornik, A. *Chemtech* **1979**, 119.
141. Patchornik, A.; Kraus, M. A. *Pure Appl. Chem.* **1976**, *46*, 183–186.
142. Patchornik, A. *Chemtech* **1987**, 58.
143. Overberg, C. G.; Sannes, K. N. *Angew. Chem Int. Ed. Engl.* **1974**, *13*, 99–104.
144. Trost, B. M.; Warner, R. W. *J. Am. Chem. Soc.* **1983**, *105*, 5940–5942.
145. Hodge, P.; Waterhouse, J. *J. Chem. Soc. Perkin Trans.* **1983**, *1*, 2319–2323.
146. Leznoff, C. C. *Acc. Chem. Res.* **1978**, *11*, 327–333.
147. Goldwasser, J. M.; Leznoff, C. C. *Can. J. Chem.* **1978**, *56*, 1562–1568.
148. Fyles, T. M.; Leznoff, C. C. *Can. J. Chem.* **1976**, *54*, 935–942.
149. Fyles, T. M.; Leznoff, C. C.; Weatherston, J. *Can. J. Chem.* **1978**, *56*, 1031–1041.
150. Wong, J. Y.; Manning, C.; Leznoff, C. C. *Angew. Chem. Int. Ed. Engl.* **1974**, *13*, 666–667.
151. Leznoff, C. C.; Wong, J. Y. *Can. J. Chem.* **1973**, *51*, 3756–3764.
152. Kusama, T.; Hayatsu, H. *Chem. Pharm. Bull. Jpn.* **1970**, *18*, 319.
153. Rebek, J.; Gavina, F. *J. Am. Chem. Soc.* **1974**, *96*, 7112–7114.
154. Rebek, J., Jr.; Brown, D.; Zimmerman, S. *J. Am. Chem. Soc.* **1975**, *97*, 454–455.
155. Warshawsky, A.; Kalir, R.; Patchnornik, A. *J. Am. Chem. Soc.* **1978**, *100*, 4544–4550.
156. Pittman, C. U.; Smith, L. R. *J. Am. Chem. Soc.* **1975**, *97*, 1749–1754.
157. Cohen, B. J.; Kraus, M. A.; Patchornik, A. *J. Am. Chem. Soc.* **1981**, *103*, 7620–7629.
158. Cainelli, G.; Contento, M.; Manescalchi, F.; Regnoli, R. *J. Chem. Soc. Perkin Trans.* **1980**, *1*, 2516–2519.
159. Bessodes, M.; Antonakis, K. *Tetrahedron Lett.* **1985**, *26*, 1305–1306.
160. Hebert, N.; Beck, A.; Lennox, R. B.; Just, G. *J. Org. Chem.* **1992**, *57*, 1777–1783.
161. Cardillo, G.; Orena, M.; Porzi, G.; Sandri, S. *J. Chem. Soc. Chem. Commun.* **1982**, 1309–1311.
162. Parlow, J. J. *Tetrahedron Lett.* **1995**, *36*, 1395–1396.
163. Kaldor, S. W.; Siegel, M. G.; Fritz, J. E.; Dressman, B. A.; Hahn, P. J. *Tetrahedron Lett.* **1996**, *37*, 7193–7196.
164. Kaldor, S. W.; Fritz, J. E.; Tang, J.; McKinney, E. R. *Bioorg. Med. Chem. Lett.* **1996**, *6*, 3041–3044.
165. Starkey, G. W.; Parlow, J. J.; Flynn, D. L. *Bioorg. Med. Chem. Lett.* **1998**, *8*, 2385–2389.
166. Parlow, J. J.; Naing, W.; South, M. S.; Flynn, D. L. *Tetrahedron Lett.* **1997**, *38*, 7959–7962.
167. Siegel, M. G.; Hahn, P. J.; Dressman, B. A.; Fritz, J. E.; Grunwell, J. R.; Kaldor, S. W. *Tetrahedron Lett.* **1997**, *38*, 3357–3360.
168. Brown, S. D.; Armstrong, R. W. *J. Am. Chem. Soc.* **1996**, *118*, 6331–6332.
169. Ley, S. V.; Massi, A.; Rodríguez, F.; Horwell, D. C.; Lewthwaite, R. A.; Pritchard, M. C.; Reid, A. M. *Angew. Chem. Int. Ed.* **2001**, *40*, 1053–1056.
170. Porath, J.; Carlsson, J.; Olsson, I.; Belfrage, G. *Nature* **1975**, *278*, 598–599.
171. Hobbs DeWitt, S.; Kiely, J. S.; Stankovic, C. J.; Schroeder, M. C.; Reynolds, D. M.; Pavia, M. R. *Proc. Natl. Acad. Sci. USA* **1993**, *90*, 6909–6913.
172. Siu, J.; Baxendale, I. R.; Lewthwaite, R. A.; Ley, S. V. *Org. Biol. Chem.* **2005**, *3*, 3140–3160.
173. Schudt, C.; Dent, G.; Rabe, K. F. *Phosphodiesterase Inhibitors* **1996**, 228.
174. Baxendale, I. R.; Ley, S. V. *Bioorg. Med. Chem. Lett.* **2000**, *10*, 1983–1986.
175. Cantello, B. C. C.; Cawthorne, M. A.; Haigh, D.; Hindley, R. M.; Smith, S. A.; Thurlby, P. *Biorg. Med. Chem. Lett.* **1994**, *4*, 1181–1184.
176. Li, X.; Abell, C.; Warrington, B. H.; Ladlow, M. *Org. Biomol. Chem.* **2003**, *1*, 4392–4395.
177. Cantello, B. C. C.; Cawthorne, M. A.; Cottam, G. P.; Duff, P. T.; Haigh, D.; Hindley, R. M.; Lister, C. A.; Smith, S. A.; Thurlby, P. L. *J. Med. Chem.* **1994**, *37*, 3977–3985.
178. Bream, R. N.; Ley, S. V.; Procopiou, P. A. *Org. Lett.* **2002**, *4*, 3793–3796.
179. Bream, R. N.; Ley, S. V.; McDermott, B.; Procopiou, P. A. *J. Chem. Soc. Perkin Trans.* **2002**, *1*, 2237–2242.
180. Shih, S. R.; Chen, S. J.; Hakimelahi, G. H.; Liu, H. J.; Tseng, C. T.; Shia, K. S. *Med. Res. Rev.* **2004**, *24*, 449–474.
181. Graybill, T. L.; Thomas, S.; Wang, M. A. *Tetrahedron Lett.* **2002**, *43*, 5305–5309.
182. Boiani, M.; Gonzalez, M. *Rev. Med. Chem.* **2005**, *5*, 409–424.
183. Andrews, S. P.; Jönsson, D.; Warrington, B. H.; Ladlow, M. *Comb. Chem. HTS* **2004**, 7, 163–178.
184. Vickerstaffe, E.; Warrington, B. H.; Ladlow, M.; Ley, S. V. *J. Comb. Chem.* **2004**, *6*, 332–339.
185. Yun, Y. K.; Leung, S. S. W.; Porco, J. A., Jr. *Biotechnol. Bioeng. (Comb. Chem.)* **2000**, *71*, 9–18.
186. Hinzen, B.; Ley, S. V. *J. Chem. Soc. Perkin Trans.* **1997**, *1*, 1907–1908.
187. Hinzen, B.; Ley, S. V. *J. Chem. Soc. Perkin Trans.* **1998**, *1*, 1–2.
188. Hinzen, B.; Lenz, R.; Ley, S. V. *Synthesis* **1998**, 977–979.

189. Ley, S. V.; Bolli, M. H.; Hinzen, B.; Gervois, A. G.; Hall, B. J. *J. Chem. Soc. Perkin Trans.* **1998**, *1*, 2239–2242.
190. Bhattacharyya, S.; Fan, L.; Vo, L.; Labadie, J. *Comb. Chem. High Through. Screen.* **2000**, *3*, 117–124.
191. Ault-Justus, S. E.; Hodges, J. C.; Wilson, M. W. *Biotech. Bioeng. (Comb. Chem.)* **1998**, *61*, 17–22.
192. Bauer, U.; Egner, B. J.; Nilsson, I.; Berghult, M. *Tetrahedron Lett.* **2000**, *41*, 2713–2717.
193. Argentieri, D. C.; Ritchie, D. M.; Ferro, M. P.; Kirchner, T. *J. Pharm. Exp. Therap.* **1994**, *271*, 1399–1408.
194. Vickerstaffe, E.; Warrington, B. H.; Ladlow, M.; Ley, S. V. *J. Comb. Chem.* **2004**, *6*, 332–339.
195. Chen, J. S.; Dixon, B. R.; Dumas, J.; Brittelli, D. *Tetrahedron Lett.* **1999**, *40*, 9195–9199.
196. Dumas, J.; Brittelli, D.; Chen, J. S.; Dixon, B.; Hatoum-Mokdad, H.; Konig, G.; Sibley, R.; Witowsky, J.; Wong, S. *Bioorg. Med. Chem Lett.* **1999**, *9*, 2531–2536.
197. Ryder, T. R.; Hu, L. Y.; Rafferty, M. F.; Millerman, E.; Szoke, B. G.; Tarczy-Hornoch, K. *Biorg. Med. Chem. Lett.* **1999**, *9*, 1813–1818.
198. Parlow, J. J.; Dice, T. A.; Lachance, R. A.; Girard, T. J.; Stevens, A. M.; Stegeman, R. A.; Stallings, W. C.; Kurumbail, R. G.; South, M. S. *J. Med. Chem.* **2003**, *46*, 4043–4049.
199. South, M. S.; Dice, T. A.; Girard, T. J.; Lachance, R. M.; Stevens, A. M.; Stegeman, R. A.; Stallings, W. C.; Kurumbail, R. G.; Parlow, J. J. *Biorg. Med. Chem. Lett.* **2003**, *13*, 2363–2367.
200. Caldarelli, M.; Habermann, J.; Ley, S. V. *Biorg. Med. Chem. Lett.* **1999**, *9*, 2049–2052.
201. Vickerstaffe, E.; Warrington, B. H.; Ladlow, M.; Ley, S. V. *Org. Biomol. Chem.* **2003**, *1*, 2419–2422.
202. Habermann, J.; Ley, S. V.; Scott, J. S. *J. Chem. Soc. Perkin Trans.* **1998**, *1*, 3127–3130.
203. Mars, G. J.; Scher, H. B. In *Controlled Delivery of Crop Protecting Agents*; Wilkens, R. M., Ed.; Taylor and Francis: London, UK, 1990, p 65.
204. Wang, Y.; Sauer, D. R. *Org. Lett.* **2004**, *6*, 2793–2796.
205. Scher, H. B. US Patent 4,285,720, 1980.
206. Jessop, P. G.; Leitner, W. *Chemical Synthesis Using Supercritical Fluids*; Wiley-VCH: Weinheim, Germany, 1999.
207. Poliakoff, M.; Meehan, N. J.; Ross, S. K. *Chem. Ind.* **1999**, 750–752.
208. Gordon, R. S.; Holmes, A. B. *Chem. Commun.* **2002**, 640–641.
209. Early, T. R.; Gordon, R. S.; Carroll, M. A.; Holmes, A. B.; Shute, R. E.; McConvey, I. F. *Chem. Commun.* **2001**, 1966–1967.
210. Jin, Z.; Li, Z.; Huang, R. *Nat. Prod. Rep.* **2002**, *19*, 454.
211. Ley, S. V.; Schucht, O.; Thomas, A. W.; Murray, P. J. *J. Chem. Soc. Perkin Trans.* **1999**, *1*, 1251–1252.
212. Ley, S. V.; Thomas, A. W.; Finch, H. *J. Chem. Soc. Perkin Trans.* **1999**, *1*, 669–671.
213. Spande, T. F.; Garraffo, H. M.; Edwards, M. W.; Yeh, H. J. C.; Pannell, L.; Daly, J. W. *J. Am. Chem. Soc.* **1992**, *114*, 3475–3478.
214. Daly, J. W.; Garraffo, H. M.; Spande, T. F.; Decker, M. W.; Sullivan, J. P.; Williams, M. *Nat. Prod. Rep.* **2000**, *17*, 131–135.
215. Olivo, H. F.; Hemenway, M. S. *Org. Prep. Proc. Intl.* **2002**, *34*, 1.
216. Baraznenok, I. L.; Jonsson, E.; Claesson, A. *Bioorg. Med. Chem Lett.* **2005**, *15*, 1637–1640.
217. Fisher, M.; Huangfu, D.; Shen, T. Y.; Guyenet, P. G. *J. Pharmacol. Exp. Ther.* **1994**, *270*, 702–707.
218. Malpass, J. R.; White, R. *J. Org. Chem.* **2004**, *69*, 5328–5334.
219. Carroll, F. I.; Lee, J. R.; Navarro, H. A.; Ma, W.; Brieaddy, L. E.; Abraham, P.; Damaj, M. I.; Martin, B. R. *J. Med. Chem.* **2002**, *45*, 4755–4761.
220. Wei, Z.-L.; George, C.; Kozikowski, A. P. *Tetrahedron Lett.* **2003**, *44*, 3847–3850.
221. Baxendale, I. R.; Brusotti, G.; Matsuoka, M.; Ley, S. V. *J. Chem. Soc. Perkin Trans.* **2002**, *1*, 143–154.
222. Arison, B. H.; Hwang, S. B. *Acta Pharmacol. Sin.* **1991**, *5*, 345.
223. Zhang, H. J.; Tamez, P. A.; Hoang, V. D.; Tan, G. T.; Van Hung, N.; Xuan, L. T.; Huong, L. M.; Cuong, N. M.; Thao, D. T.; Soejarto, D. D.; Fong, H. H. S.; Pezzuto, J. M. *J. Nat. Prod.* **2001**, *64*, 772–777.
224. Ridley, R. G. *Science* **1999**, *285*, 1502–1503.
225. Greenwood, B.; Mutabingwa, T. *Nature* **2002**, *415*, 670–672.
226. Barata, L. E. S.; Santos, L. S.; Ferri, P. H.; Phillipson, J. D.; Paine, A.; Croft, S. L. *Phytochemistry* **2000**, *55*, 589–595.
227. Lee, A.-L.; Ley, S. V. *Org. Biomol. Chem.* **2003**, *1*, 3957–3966.
228. Baxendale, I. R.; Ernst, M.; Krahnert, W.-R.; Ley, S. V. *Synlett* **2002**, 1641–1644.
229. Bloedon, L. T.; Szapary, P. O. *Nutrition Rev.* **2004**, *62*, 18–27.
230. Adlercreutz, H.; Mazur, W. *Ann. Med.* **1997**, *29*, 95–120.
231. Prasad, K. *Circulation* **1999**, *99*, 1355–1362.
232. Clark, W. F.; Parbtani, A.; Huff, M. W.; Spanner, E.; Desalis, H.; Chinyee, I.; Philbrick, D. J.; Holub, B. J. *Kidney Int.* **1995**, *48*, 475–480.
233. Franco, O. H.; Burger, H.; Lebrun, C. E. I.; Peeters, P. H. M.; Larnberts, S. W. J.; Grobbee, D. E.; Van Der Schouw, Y. T. *J. Nutrition* **2005**, *135*, 1190–1195.
234. Lai, P. K.; Roy, J. *Curr. Med. Chem.* **2004**, *11*, 1451–1460.
235. Wilson, P. D. *New Engl. J. Med.* **2004**, *350*, 151–164.
236. For isolation see: Brophy, G. C.; Mohandas, J.; Slaytor, M.; Sternhell, S.; Watson, T. R.; Wilson, L. A. *Tetrahedron Lett.* **1969**, 5159.
237. Chapman, O. L.; Engel, M. R.; Springer, J. P.; Clardy, J. C. *J. Am. Chem. Soc.* **1971**, 6696.
238. Lindsley, C. W.; Chan, L. K.; Goess, B. C.; Joseph, R.; Shair, M. D. *J. Am. Chem. Soc.* **2000**, *122*, 422–423.
239. Baxendale, I. R.; Lee, A.-L.; Ley, S. V. *J. Chem. Soc. Perkin Trans.* **2002**, *1*, 1850–1857.
240. Westman, J. Patent WO 00/72956 A1, Personal Chemistry I Uppsala, 2000.
241. Matsumoto, M.; Kuroda, K. *Tetrahedron Lett.* **1981**, *22*, 4437–4440.
242. Nishiyama, A.; Eto, H.; Tevada, Y.; Iguchi, M.; Yamamura, S. *Chem. Pharm. Bull.* **1983**, *31*, 2834–2844.
243. Iyer, M. R.; Trivedi, G. K. *Bull. Chem. Soc. Jpn.* **1992**, *65*, 1662–1664.
244. Ünver, N.; Gözler, T.; Walch, N.; Gözler, B.; Hesse, M. *Phytochemistry* **1999**, *50*, 1255.
245. Missoum, A.; Sinibaldi, M.-E.; Vallee-Goyet, D.; Gramain, J.-C. *Synth. Commun.* **1997**, *27*, 453–466.
246. Antoun, M. D.; Mendoza, N. T.; Rios, Y. R.; Proctor, G. R.; Wickramaratne, D. B. M.; Pezzuto, J. M.; Kinghorn, A. D. *J. Nat. Prod.* **1993**, *56*, 1423–1425.
247. Furusawa, E.; Furusawa, S. *Onocology* **1988**, *45*, 180–186.
248. Baxendale, I. R.; Ley, S. V.; Nessi, M.; Piutti, C. *Angew. Chem. Int. Ed.* **2002**, *41*, 2194–2197.
249. Baxendale, I. R.; Ley, S. V.; Nessi, M.; Piutti, C. *Tetrahedron* **2002**, *58*, 6285–6304.
250. Baxendale, I. R.; Ley, S. V. *Ind. Eng. Chem. Res.* **2005**, *44*, 8588–8592.
251. Höfle, G.; Bedorf, N.; Gerth, K.; Reichenbach, H. German patent DE 4138042 (*Chem. Abstr.* 1993, *120*, 52841).
252. Gerth, K.; Bedorf, N.; Höfle, G.; Irschik, H.; Reichenbach, H. *J. Antibiot.* **1996**, *49*, 560–564.

253. Altmann, K. H. *Curr. Pharm. Des.* **2005**, *11*, 1595–1613.
254. Watkins, E. B.; Chittiboyina, A. G.; Jung, J. C.; Avery, M. A. *Curr. Pharm. Des.* **2005**, *11*, 1615–1653.
255. Altmann, K. H. *Org. Biomol. Chem.* **2004**, *2*, 2137–2152.
256. Rivkin, A.; Cho, Y. S.; Gabarda, A. E.; Fumihiko, Y. *J. Nat. Prod.* **2004**, *67*, 139–143.
257. Harris, C. R.; Kuduk, S. D.; Danishefsky, S. J. *J. Chem. 21st Century* 2001, 8.
258. Wessjohann, L. A. *Curr. Opin. Chem. Biol.* **2000**, *4*, 303–309.
259. Storer, R. I.; Takemoto, T.; Jackson, P. S.; Ley, S. V. *Angew. Chem. Int. Ed.* **2003**, *42*, 2521–2525.
260. Storer, R. I.; Takemoto, T.; Jackson, P. S.; Brown, D. S.; Baxendale, I. R.; Ley, S. V. *Chem. Eur. J.* **2004**, *10*, 2529–2547.
261. Fitch, W. L. *Mol. Div.* **1999**, *4*, 39–45.
262. Saaby, S.; Knudsen, K. R.; Ladlow, M.; Ley, S. V. *J. Chem. Soc. Chem. Commun.* **2005**, 2909–2911.
263. Harrison, C. R.; Hodge, P. *J. Chem. Soc. Perkin Trans.* **1976**, *1*, 2252–2254.
264. Kirschning, A.; Altwicker, C.; Drager, G.; Harders, J.; Hoffmann, N.; Hoffmann, U.; Schonfeld, H.; Solodenko, W.; Kunz, U. *Angew. Chem. Int. Ed.* **2001**, *40*, 3995–3998.
265. Adlington, R. M.; Baldwin, J. E.; Becker, G. W.; Chen, B.; Cheng, L.; Cooper, S. L.; Hermann, R. B.; Howe, T. J.; McCoull, W.; McNulty, A. M. et al. *J. Med. Chem.* **2001**, *44*, 1491–1508.
266. Bonneau, P. R.; Hasani, F.; Plouffe, C.; Malenfant, E.; LaPlante, S. R.; Guse, I.; Ogilvie, W. W.; Plante, R.; Davidson, W. C.; Hopkins, J. L. et al. *J. Am. Chem. Soc.* **1999**, *121*, 2965–2973.
267. Whitesides, G. M. *Angew. Chem. Int. Ed.* **2004**, *43*, 3632–3641.
268. Bapna, A.; Ladlow, M.; Vickerstaff, E.; Warrington, B. H.; Fan, T.-P.; Ley, S. V. *Org. Biomol. Chem.* **2004**, *2*, 611–620.
269. Baxendale, I. R.; Griffiths-Jones, C. M.; Ley, S. V.; Tranmer, G. K. *Synlett* **2006**, 427–430.
270. Andrews, S. P.; Stepan, A. F.; Tanaka, H.; Kaneko, K.; Ley, S. V.; Smith, M. D. *Adv. Synth. Catal.* **2005**, *347*, 647–654.

Biographies

Steven V Ley was born in Lincolnshire, England in 1945. He received his PhD from Loughborough University in 1972. After 2 years at Ohio State University with Leo Paquette, he returned to England to work with Derek Barton at Imperial College, London. In 1975 he joined the department as a lecturer and later became Head of Department in 1989. In 1992 he moved to take up the 1702 BP Chair of Organic Chemistry at the University of Cambridge and became a Fellow of Trinity College. He has been the President of the Royal Society of Chemistry (RSC) between 2000 and 2002 and was made a Commander of the British Empire in January 2002. He is also a Fellow of the Royal Society (London) and the Academy of Medical Sciences. He is distinguished for his creative work and innovative solutions in the art of organic synthesis. Professor Ley's research involves the discovery and development of new synthetic methods and their application to biologically active systems; the use of iron carbonyl complexes, organoselenium chemistry, the use of microwaves in organic chemistry, biotransformations for the synthesis of natural products, and strategies for oligosaccharide assembly. He is also developing new methods with the use solid-supported reagents in a sequential and multistep fashion in combination with advances in flow methods. He has made significant advances in the development of new organic catalysts for asymmetric synthesis. He is also the inventor of TPAP, a catalytic oxidant. He has received several inaugural prizes during his career and has received 12 awards from the Royal Society of Chemistry in the UK. His work has also been recognized by The Royal Society (London) with the Bakerian Lecture (1997) and the Davy Medal (2000). Others prizes include the ACS Ernest Guenther Award in the Chemistry of Natural Products (2003); the Adolf Windaus Medal from the Gesellschaft Deutscher Chemiker (1994); the August-Wilhelm-von Hofmann Medal (2001); and the Dr Paul Janssen Prize for Creativity in Organic Synthesis from the Janssen Research Foundation, Belgium (2004). He is also the first UK scientist to receive the Yamada-Koga Prize (2005). His most recent accolade has been the Award for Creative Work in Synthetic Organic Chemistry (2007).

Ian R Baxendale received his undergraduate degree at the University of Leicester where he remained for his PhD, which he did with Prof Pavel Kocovsky. The focus of his research was the synthesis and application of new catalyst systems for allyic substitution. After a brief spell in industry he returned to academia in 1999 to work with Steve Ley. He is currently senior researcher and director of the new Innovation Technology Centre at the Department of Chemistry at Cambridge.

Before starting her academic career **Rebecca M Myers** worked in the advertising industry in London. In 1994 she went to Imperial College, London and obtained a degree in Chemistry and in 1997 she moved to Cambridge with her family and began her PhD with Prof Chris Abell. The focus of her research was in the area of solid-phase chemistry and the development of new linker strategies. In 2001 she began post-doctoral work with Steve Ley, and apart from a year in the US, has remained in his group since.

Comprehensive Medicinal Chemistry II
ISBN (set): 0-08-044513-6

ISBN (Volume 3) 0-08-044516-0; pp. 791–836

3.36 Microwave-Assisted Chemistry

C O Kappe, Karl-Franzens-University Graz, Graz, Austria

3.36.1	**Introduction**	**837**
3.36.2	**Microwave Theory**	**838**
3.36.2.1	The Electromagnetic Spectrum	838
3.36.2.2	Microwave Dielectric Heating	839
3.36.2.3	Dielectric Properties	839
3.36.2.4	Microwave versus Conventional Thermal Heating	840
3.36.2.5	Microwave Effects	840
3.36.3	**Equipment Review**	**842**
3.36.3.1	Single-Mode Instruments	844
3.36.3.1.1	Biotage AB	844
3.36.3.1.2	CEM Discover platform	845
3.36.3.2	Multimode Instruments	848
3.36.3.2.1	Milestone s.r.l	848
3.36.3.2.2	CEM Corporation	850
3.36.3.2.3	Biotage AB	851
3.36.3.2.4	Anton Paar GmbH	851
3.36.4	**Microwave-Processing Techniques**	**852**
3.36.4.1	Solvent-Free Reactions	852
3.36.4.2	Phase-Transfer Catalysis (PTC)	853
3.36.4.3	Reactions Using Solvents	853
3.36.4.3.1	Open- versus closed-vessel conditions	853
3.36.4.3.2	Pre-pressurized reaction vessels	854
3.36.4.3.3	Nonclassical solvents	854
3.36.4.4	Parallel Processing	854
3.36.4.5	Scale-up in Batch and Continuous Flow	856
3.36.5	**Conclusion**	**857**
References		**857**

3.36.1 Introduction

High-speed microwave synthesis has attracted a considerable amount of attention in recent years. Since the first reports on the use of microwave heating to accelerate organic chemical transformations by the groups of Gedye and Giguere/Majetich in 1986,[1,2] more than 3000 articles have been published in the area of microwave-assisted organic synthesis (MAOS). The initial slow uptake of the technology in the late 1980s and early 1990s has been attributed to its lack of controllability and reproducibility, coupled with a general lack of understanding of the basics of microwave dielectric heating. The risks associated with the flammability of organic solvents in a microwave field and the lack of available systems for adequate temperature and pressure controls were major concerns. Although most of the early pioneering experiments in MAOS were performed in domestic, sometimes modified, kitchen microwave ovens, the current trend clearly is to use dedicated instruments for chemical synthesis, which have become available only in the last few years. Since 2000, the number of publications related to MAOS has therefore increased dramatically to a point where it might be assumed that, in a few years, most chemists will probably use microwave energy to heat chemical reactions on a laboratory scale. Not only is direct microwave heating able to reduce chemical reaction times from hours to minutes, but it is also known to reduce side reactions, increase yields, and improve reproducibility. Therefore, many medicinal chemists in academic and industrial research groups are already using MAOS as a forefront technology for rapid reaction optimization, for the efficient synthesis of new chemical entities, or for discovering and probing new chemical reactivity. A large number of review articles[3–33] and several books[34–37] provide extensive coverage of the subject.

Medicinal chemistry in general has benefited tremendously from the technological advances in the field of combinatorial chemistry and high-throughput synthesis (*see* 3.33 Combinatorial Chemistry; 3.34 Solution Phase Parallel Chemistry; 3.35 Polymer-Supported Reagents and Scavengers in Synthesis). This discipline has been the innovative machine for the development of methods and technologies that accelerate the design, synthesis, purification, and analysis of compound libraries. The bottleneck of conventional parallel/combinatorial synthesis is typically the optimization of reaction conditions to afford the desired products in suitable yields and purities. Since many reaction sequences require at least one or more heating steps for extended time periods, these optimizations are often difficult and time-consuming. Microwave-assisted heating under controlled conditions has been shown to be an invaluable technology for medicinal chemistry and drug discovery applications, since it often dramatically reduces reaction times, typically from days or hours to minutes or even seconds. Many reaction parameters can be evaluated in a few hours to optimize the desired chemistry. Compound libraries can then be rapidly synthesized either in a parallel or (automated) sequential format using this new, enabling technology. In addition, microwave synthesis allows for the discovery of novel reaction pathways, which serve to expand chemical space in general, and biologically relevant, medicinal chemistry space in particular.

Specifically, microwave synthesis has the potential to impact medicinal chemistry efforts in at least three major phases of the drug discovery process: lead generation, hit-to-lead efforts, and lead optimization. A common theme throughout the drug discovery and development process is speed. Speed equals competitive advantage, more efficient use of expensive and limited resources, faster exploration of structure activity relationships (SARs), enhanced delineation of intellectual property, and more timely delivery of critically needed medicines, and can ultimately determine positioning in the marketplace. To the pharmaceutical industry and the medicinal chemist, time truly does equal money, and microwave chemistry has become a central tool in this fast-paced, time-sensitive field.

Medicinal chemistry, like all sciences, consists of never-ending iterations of hypotheses and experiments, with results guiding the progress and development of projects. The short reaction times provided by microwave synthesis make it ideal for rapid reaction scouting and optimization, allowing proceeding very rapidly through the hypotheses–experiment–results iterations, resulting in more decision points per time unit. Not surprisingly, therefore, most pharmaceutical and biotechnology companies are already heavily using microwave synthesis as a frontline methodology in their medicinal chemistry programs, both for library synthesis and lead optimization, as they realize the ability of this enabling technology to speed chemical reactions and therefore the drug discovery process.

This chapter will discuss the basic principles and tools of this technology.

3.36.2 Microwave Theory

The physical principles behind and the factors determining the successful application of microwaves in organic synthesis are not widely familiar to medicinal chemists, possibly because electric field theory is generally taught to those studying engineering or physics rather than in chemistry or biology courses. Nevertheless, it is essential for the synthetic medicinal chemist involved in microwave-assisted synthesis to have at least a basic knowledge of the underlying principles of microwave–matter interactions and microwave effects. The basic understanding of macroscopic microwave interactions with matter was formulated by von Hippel in the mid-1950s.[38] In this chapter a brief summary on the current understanding of microwaves and their interactions with matter will be given. For more in-depth discussions on this quite complex field, the reader is referred to recent review articles.[39–43]

3.36.2.1 The Electromagnetic Spectrum

Microwave irradiation is electromagnetic irradiation in the frequency range 0.3–300 GHz, corresponding to wavelengths of 1 mm to 1 m. The microwave region of the electromagnetic spectrum therefore lies between the infrared and radio frequencies. The wavelengths between 1 and 25 cm are extensively used for radar transmissions, and the remaining wavelength range is used for telecommunications. All domestic kitchen microwave ovens and all dedicated microwave reactors for chemical synthesis that are commercially available today operate at a frequency of 2.45 GHz (corresponding to a wavelength of 12.25 cm) in order to avoid interference with telecommunication and cellular phone frequencies. Other frequency allocations for microwave heating applications exist (ISM frequencies),[39] but are not generally employed in reactors dedicated to synthetic chemistry. It is obvious that the energy of the microwave photon at a frequency of 2.45 GHz (0.0016 eV) is too low to cleave molecular bonds and is also lower than Brownian motion. It is therefore clear that microwaves cannot induce chemical reactions by direct absorption of electromagnetic energy, as opposed to ultraviolet and visible radiation (photochemistry).

3.36.2.2 Microwave Dielectric Heating

Microwave-enhanced chemistry is based on the efficient heating of materials by microwave dielectric heating effects.[41,42] Microwave dielectric heating is dependent on the ability of a specific material (e.g., a solvent or reagent) to absorb microwave energy and convert it into heat. Microwaves are electromagnetic waves that consist of an electric and a magnetic field component. For most practical purposes related to microwave synthesis, it is the electric component of the electromagnetic field that is of importance for wave–material interactions.[43]

The electric component of an electromagnetic field causes heating by two main mechanisms: dipolar polarization and ionic conduction. The interaction of the electric field component with the matrix is called the dipolar polarization mechanism.[41,42] For a substance to be able to generate heat when irradiated with microwaves it must possess a dipole moment. When exposed to microwave frequencies, the dipoles of the sample align in the applied electric field. As the applied field oscillates, the dipole field attempts to realign itself with the alternating electric field and, in the process, energy is lost in the form of heat through molecular friction and dielectric loss. The amount of heat generated by this process is directly related to the ability of the matrix to align itself with the frequency of the applied field. If the dipole does not have enough time to realign (high-frequency irradiation), or reorients too quickly (low-frequency irradiation) with the applied field, no heating occurs. The allocated frequency of 2.45 GHz used in all commercial systems lies between these two extremes, and gives the molecular dipole time to align in the field, but not to follow the alternating field precisely. Therefore, as the dipole reorients to align itself with the electric field, the field is already changing, and generates a phase difference between the orientation of the field and that of the dipole. This phase difference causes energy to be lost from the dipole by molecular friction and collisions, giving rise to dielectric heating. In summary, field energy is transferred to the medium, and electrical energy is converted into kinetic or thermal energy, and ultimately into heat. It should be emphasized that the interaction between microwave radiation and the polar solvent molecules that occurs when the frequency of the radiation approximately matches the frequency of the rotational relaxation process is not a quantum mechanical resonance phenomenon. Transitions between quantized rotational bands are not involved, and the energy transfer is not a property of a specific molecule but the result of a collective phenomenon involving the bulk.[41,42] The heat is generated by frictional forces occurring between the polar molecules, whose rotational velocity has been increased by the coupling with the microwave irradiation.

The second major heating mechanism is the ionic conduction mechanism.[41,42] During ionic conduction, as the dissolved charged particles in a sample (usually ions) oscillate back and forth under the influence of the microwave field, they collide with their neighboring molecules or atoms. This collision causes agitation or motion, creating heat. Thus, if two samples containing equal amounts of distilled water and tap water, respectively, are heated by microwave irradiation at a fixed radiation power, more rapid heating will occur for the tap water sample due to its ionic content. Such ionic conduction effects are particularly important when considering the heating behavior of ionic liquids in a microwave field. The conductivity principle is a much stronger effect than the dipolar rotation mechanism with regard to the heat-generating capacity.

3.36.2.3 Dielectric Properties

The heating characteristics of a particular material (e.g., a solvent) under microwave irradiation conditions are dependent on the dielectric properties of the material. The ability of a specific substance to convert electromagnetic energy into heat at a given frequency and temperature is determined by the so-called loss tangent, $\tan\delta$. The loss factor is expressed as the quotient, $\tan\delta = \varepsilon''/\varepsilon'$, where ε'' is the dielectric loss, indicative of the efficiency with which electromagnetic radiation is converted into heat, and ε' is the dielectric constant describing the ability of molecules to be polarized by the electric field. A reaction medium with a high $\tan\delta$ is required for efficient absorption, and, consequently, for rapid heating. Materials with a high dielectric constant such as water (ε' at 25 °C = 80.4) may not necessarily also have a high $\tan\delta$ value. In fact, ethanol has a significantly lower dielectric constant (ε' at 25 °C = 24.3), but heats much more rapidly than water in a microwave field due to its higher loss tangent ($\tan\delta$ for ethanol = 0.941, $\tan\delta$ for water = 0.123). The loss tangents for some common organic solvents are summarized in **Table 1**.[35] In general, solvents can be classified as high ($\tan\delta > 0.5$), medium ($\tan\delta$ 0.1–0.5), and low microwave-absorbing ($\tan\delta < 0.1$). Other common solvents without a permanent dipole moment such as carbon tetrachloride, benzene, and dioxane are more or less microwave-transparent. It has to be emphasized that a low $\tan\delta$ value does not preclude a particular solvent from being used in a microwave-heated reaction. Since either the substrates or some of the reagents/catalysts are likely to be polar, the overall dielectric properties of the reaction medium will in most cases allow sufficient heating by microwaves. Furthermore, polar additives such as alcohols or ionic liquids can be added to otherwise low-absorbing reaction mixtures in order to increase the absorbance level of the medium. It has to be noted that the loss tangent values are both frequency- and temperature-dependent.

Table 1 Loss tangents (tan δ) of different solvents (2.45 GHz, 20 °C)[a]

Solvent	*tan δ*	*Solvent*	*tan δ*
Ethylene glycol	1.350	*N*,*N*-Dimethylformamide	0.161
Ethanol	0.941	1,2-Dichloroethane	0.127
Dimethyl sulfoxide	0.825	Water	0.123
2-Propanol	0.799	Chlorobenzene	0.101
Formic acid	0.722	Chloroform	0.091
Methanol	0.659	Acetonitrile	0.062
Nitrobenzene	0.589	Ethyl acetate	0.059
1-Butanol	0.571	Acetone	0.054
2-Butanol	0.447	Tetrahydrofuran	0.047
1,2-Dichlorobenzene	0.280	Dichloromethane	0.042
1-Methyl-2-pyrrolidone	0.275	Toluene	0.040
Acetic acid	0.174	Hexane	0.020

[a] Data from [35].

The interaction of microwave irradiation with matter is characterized by three different processes: absorption, transmission, and reflection. Highly dielectric materials such as polar organic solvents lead to a strong absorption of microwaves, and consequently to a rapid heating of the medium. Nonpolar materials exhibit only small interactions with penetrating microwaves, and can thus be used as construction materials for reactors. If microwave radiation is reflected by the material surface, there is no or only negligible coupling of energy in the system. The temperature increases in the material only marginally. This holds true especially for metals with high conductivity.

3.36.2.4 Microwave versus Conventional Thermal Heating

Traditionally, organic synthesis is carried out by conductive heating with an external heat source (e.g., an oil bath or heating mantle). This is a comparatively slow and inefficient method for transferring energy into the system since it depends on convection currents and on the thermal conductivity of the various materials that must be penetrated, and results in the temperature of the reaction vessel being higher than that of the reaction mixture. In addition, a temperature gradient can develop within the sample, and local overheating can lead to product, substrate, or reagent decomposition.

In contrast, microwave irradiation produces efficient internal heating (in-core volumetric heating) by direct coupling of microwave energy with the molecules (solvents, reagents, catalysts) that are present in the reaction mixture. Since the reaction vessels employed are typically made out of (nearly) microwave-transparent materials such as borosilicate glass, quartz, or Teflon, the radiation passes through the walls of the vessel, and an inverted temperature gradient as compared with conventional thermal heating results (**Figure 1**). If the microwave cavity is well designed, the temperature increase will be uniform throughout the sample. The very efficient internal heat transfer results in minimized wall effects (no hot vessel surface), which may lead to the observation of so-called specific microwave effects, for example in the context of diminished catalyst deactivation. It should be emphasized that microwave dielectric heating and thermal heating by convection are totally different processes, and that any comparison between the two is inherently difficult.

3.36.2.5 Microwave Effects

Since the early days of microwave synthesis, the observed rate accelerations and sometimes altered product distributions compared with oil bath experiments have led to speculation on the existence of so-called specific or nonthermal microwave effects.[44–49] Historically, such effects were claimed when the outcome of a synthesis performed under microwave conditions was different from the conventionally heated counterpart at the same apparent

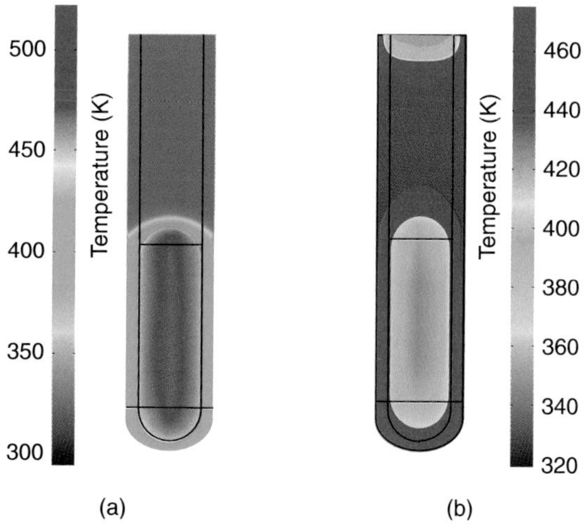

Figure 1 Inverted temperature gradients in microwave versus oil bath heating. Temperature profiles (finite-element modeling) after 1 min as affected by (a) microwave irradiation compared with (b) treatment in an oil bath. Microwave irradiation raises the temperature of the whole volume simultaneously (bulk heating), whereas in the oil-heated tube, the reaction mixture in contact with the vessel wall is heated first. (Reproduced from Kappe, C. O. *Angew. Chem. Int. Ed.* **2004**, *43*, 6250–6284, with permission from Wiley-VCH.)

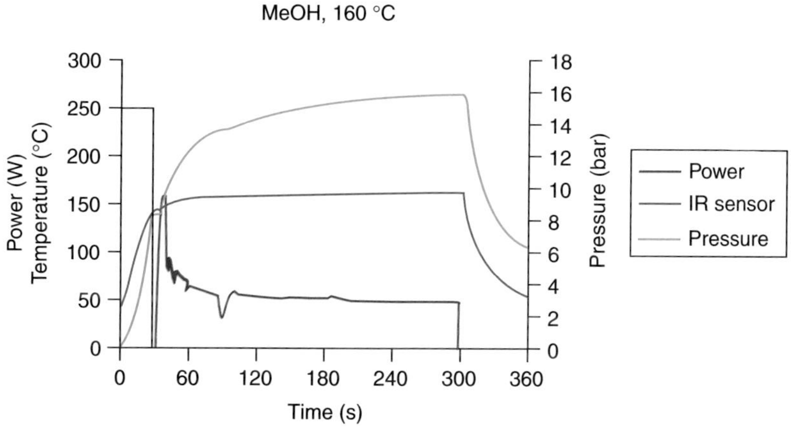

Figure 2 Temperature (*T*), pressure (*p*), and power (*P*) profile for a 3 mL sample of methanol heated under sealed-vessel microwave irradiation conditions. Single-mode microwave heating (250 W, 0–30 s), temperature control using the feedback from IR thermography (40–300 s), and active gas-jet cooling (300–360 s). The maximum pressure in the reaction vessel was *c.* 16 bar. After the set temperature of 160 °C is reached, the power regulates itself down to *c.* 50 W. (Reproduced from Kappe, C. O. *Angew. Chem. Int. Ed.* **2004**, *43*, 6250–6284, with permission from Wiley-VCH.)

temperature. Reviewing the present literature it appears that today most scientists agree that in the majority of cases the reason for the observed rate enhancements is a purely thermal/kinetic effect, that is, a consequence of the high reaction temperatures that can rapidly be attained when irradiating polar materials in a microwave field. As shown in **Figure 2**, a high-microwave-absorbing solvent such as methanol ($\tan\delta = 0.659$) can be rapidly superheated to temperatures >100 °C in excess of its boiling point when irradiated under sealed-vessel microwave conditions. The rapid increase in temperature can be even more pronounced for media with extreme loss tangents such as ionic liquids, where temperature jumps of 200 °C within a few seconds are not uncommon. Naturally, such temperature profiles are very difficult if not impossible to reproduce by standard thermal heating. Therefore, comparisons with conventionally heated processes are inherently troublesome.

Dramatic rate enhancements when comparing reactions that are performed at room temperature or under standard oil bath conditions (heating under reflux) with high-temperature microwave-heated processes have frequently been observed. As Mingos and Baghurst have pointed out based on simply applying the Arrhenius law ($k = A\exp(-E_a/RT)$), a transformation that requires 68 days to reach 90% conversion at 27 °C, will show the same degree of conversion within

Table 2 Relationship between temperature and time for a typical first-order reaction ($A=4\times10^{10}\,mol^{-1}\,s^{-1}$, $E_a=100\,kJ\,mol^{-1}$)[a]

Temperature (°C)	*Rate constant (k/s)*	*Time (90% conversion)*
27	1.55×10^{-7}	68 days
77	4.76×10^{-5}	13.4 h
127	3.49×10^{-3}	11.4 min
177	9.86×10^{-2}	23.4 s
227	1.43	1.61 s

[a] Data from [41].

1.61 s when performed at 227 °C (**Table 2**).[41] Due to the very rapid heating and extreme temperatures observable in microwave chemistry, it appears obvious that many of the reported rate enhancements can be rationalized by simple thermal/kinetic effects.

In addition to the above-mentioned thermal/kinetic effects, microwave effects that are caused by the uniqueness of the microwave dielectric heating mechanisms must also be considered. These effects should be termed specific microwave effects, and defined as accelerations that cannot be achieved or duplicated by conventional heating, but essentially are still thermal effects. In this category fall, for example (1) the superheating effect of solvents at atmospheric pressure,[50–52] (2) the selective heating of strongly microwave-absorbing heterogeneous catalysts or reagents in a less polar reaction medium,[53–58] (3) the formation of molecular radiators by direct coupling of microwave energy to specific reagents in a homogeneous solution (microscopic hot spots),[55] and (4) the elimination of wall effects caused by inverted temperature gradients (see **Figure 1**).[55] It should be emphasized that rate enhancements falling under this category are essentially still a result of a thermal effect (i.e., a change in temperature compared with heating by standard convection methods), although it may be difficult to experimentally determine the exact reaction temperature.

In contrast, some authors have suggested the possibility of nonthermal microwave effects (also referred to as athermal effects).[44–49] These should be classified as accelerations that cannot be rationalized by either purely thermal/kinetic or specific microwave effects. Essentially, nonthermal effects result from a proposed direct interaction of the electric field with specific molecules in the reaction medium. It has been argued that the presence of an electric field leads to orientation effects of dipolar molecules, and hence changes the pre-exponential factor A or the activation energy (entropy term) in the Arrhenius equation.[47] Furthermore, a similar effect should be observed for polar reaction mechanisms, where the polarity is increased going from the ground state to the transition state, resulting in an enhancement of reactivity by lowering of the activation energy.[47] Microwave effects are the subject of considerable current debate and controversy,[44–49] and it is evident that extensive research efforts will be necessary in order to truly understand these and related phenomena, such as the heating-while-cooling debate.[59] Since the issue of microwave effects is not the primary focus of this chapter, the interested reader is referred to more detailed surveys and essays covering this topic.[44–49]

3.36.3 Equipment Review

Although many of the early pioneering experiments in microwave-assisted organic synthesis have been carried out in domestic microwave ovens, the current trend undoubtedly is to use dedicated instruments for chemical synthesis. In a domestic microwave oven the irradiation power is generally controlled by on–off cycles of the magnetron (pulsed irradiation), and it is typically not possible to monitor the reaction temperature in a reliable way. Combined with the inhomogeneous field produced by the low-cost multimode designs and the lack of safety controls, the use of such equipment cannot be recommended. In contrast, all of today's commercially available dedicated microwave reactors for synthesis feature built-in magnetic stirrers, direct temperature control of the reaction mixture with the aid of fiber optic probes or infrared (IR) sensors, and software that enables on-line temperature/pressure control by regulation of microwave power output. Currently, two different philosophies with respect to microwave reactor design are emerging: multimode and monomode (also referred to as single-mode) reactors. In the so-called multimode instruments (conceptually similar to a domestic oven), the microwaves that enter the cavity are reflected by the walls and the load over the typically large cavity. In most instruments, a mode stirrer ensures that the field distribution is as homogeneous as possible. In the much smaller monomode cavities, only one mode is present, and the electromagnetic irradiation is directed through a precision-designed and built rectangular or circular waveguide onto the reaction vessel mounted at a fixed distance from the radiation source, creating a standing wave. The key difference between the two types of reactor systems is that whereas in multimode cavities several reaction vessels can be irradiated simultaneously in multi-vessel rotors (parallel synthesis), in monomode systems typically only one vessel can be irradiated at a time. In the latter case, high throughput can be achieved by integrated robotics that move individual reaction vessels in and out of the microwave cavity.

Most instrument companies offer a variety of diverse reactor platforms with different degrees of sophistication with respect to automation, database capabilities, safety features, temperature and pressure monitoring, and vessel design. Importantly, single-mode reactors processing comparatively small volumes also have a built-in cooling feature that allows for rapid cooling of the reaction mixture by compressed air after completion of the irradiation period (**Figure 2**). The dedicated single-mode instruments available today can process volumes ranging from 0.2 mL to *c.* 50 mL under sealed-vessel conditions (250 °C, *c.* 20 bar), and somewhat higher volumes (*c.* 150 mL) under open-vessel reflux conditions. In the much larger multi-mode instruments, several liters can be processed under both open- and closed-vessel conditions. For both single- and multimode cavities, continuous-flow reactors are nowadays available that already allow the preparation of kilograms of materials using microwave technology. This section provides a detailed description of the various commercially available microwave reactors that are dedicated to microwave-assisted organic synthesis.

With the growing interest in MAOS during the mid-1990s, the demand for more sophisticated microwave instrumentation, offering, for example, stirring of the reaction mixture, temperature measurement, and power control features, increased. For scientifically valuable, safe, and reproducible work, the utilized microwave instruments have to offer the following features: built-in magnetic or mechanical stirring, accurate temperature measurement, pressure control, continuous power regulation, efficient postreaction cooling, and computer-aided method programming.

A particularly difficult problem in microwave processing is the correct measurement of the reaction temperature during the irradiation phase. Classical temperature sensors (thermometers and thermocouples) will fail since they will couple with the electromagnetic field. Temperature measurement can be achieved either by an immersed temperature probe (fiber optic or gas balloon thermometer) or on the outer surface of the reaction vessels by a remote IR sensor. Due to the volumetric character of microwave heating, the surface temperature of the reaction vessel will not always reflect the actual temperature inside the reaction vessel.

Since the early applications in microwave-assisted synthesis were based on the use of domestic multimode microwave ovens, it was obvious that the primary focus in the development of dedicated microwave instruments was the improvement of multimode reactors. In general, one or two magnetrons create the microwave irradiation, which is typically directed into the cavity through a waveguide and distributed by a mode stirrer (**Figure 3**). The microwaves are reflected from the walls of the cavity, thus interacting with the sample in a chaotic manner. Multimode cavities may therefore show multiple energy pockets with different levels of energy intensity, thus resulting in hot and cold spots. To provide an equal energy distribution, the samples are continuously rotated within the cavity. Consequently, multimode instruments offer convenient applications for the increase of reaction throughput by the use of multivessel rotors for parallel synthesis or scale-up. A general problem for multimode instruments is the weak performance for small-scale experiments (<3 mL). While the generated microwave power is high (1000–1400 W), the power density of the field is generally rather low, making heating of small individual samples rather difficult, a major drawback especially for research and development purposes. Therefore, the general use of multimode instruments for small-scale synthetic organic research applications is not so extensive in comparison with the much more popular single-mode cavities.

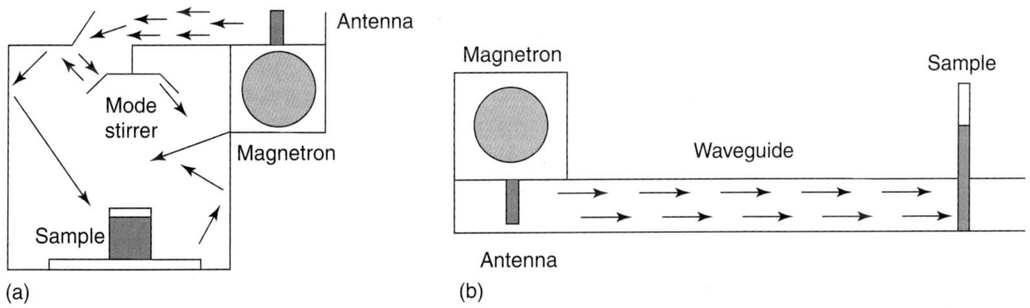

Figure 3 (a) Multimode and (b) single-mode cavities.

Monomode (also known as single-mode) instruments generate a single, highly homogeneous energy field of high power intensity. Thus, these systems couple efficiently with small samples, and the maximum output power is typically limited to 300 W. The microwave energy is created by a single magnetron, and directed through a rectangular waveguide to the sample, which is positioned at a maximized energy point (**Figure 3**). To create optimum conditions for different samples, a tuning device provides for the adjustment of the microwave field for variations in the samples (0.2–20 mL). Thus, the highly homogeneous field generally enables excellent reproducibility.

Recent advances and further improvements have led to a broad variety of applications for single-mode microwave instruments, offering flow-through systems as well as special features such as solid phase peptide synthesis or in situ online analytics. Thus, the use of single-mode reactors has tremendously increased since the year 2000, and these types of reactors have become very popular in many medicinal chemistry laboratories, both in industry and academia. However, it should be pointed out that it is rather a matter of the desired application and scale, than of the chemistry to be performed, which type of instrumentation is to be used. Both multimode and single-mode reactors are able to carry out chemical reactions efficiently and to improve classical heating protocols.

The following sections give a description of commercially available multimode and single-mode microwave reactors, including various accessories and special applications. Essentially, there are currently four instrument manufacturers that produce microwave reactors for laboratory scale organic synthesis. These are Anton Paar GmbH (Graz, Austria), Biotage AB (Uppsala, Sweden), CEM Corporation (Matthews, NC, USA), and Milestone s.r.l. (Sorisole, Italy).

3.36.3.1 Single-Mode Instruments

The first microwave instrument company offering single-mode cavities was the French company Prolabo.[60] In the early 1990s the Synthewave 402 was released, later followed by the Synthewave 1000. The instruments were designed with a rectangular waveguide, providing focused microwaves with a maximum output power of 300 W. The cavity was designed for the use of cylindrical glass or quartz tubes of several diameters for reactions at atmospheric pressure only. Temperature measurement was performed by an IR sensor at the bottom of the vessels that required calibration by a fiber optic probe. In 1999, all patents and microwave-based product lines were acquired by the CEM Corporation. However, a number of instruments are still in use, mainly in the French scientific community, and several publications per year describing the use of this equipment appear in the literature.

3.36.3.1.1 Biotage AB

Up to 2004, Biotage (formerly Personal Chemistry) offered the Emrys monomode reactor series of instruments (Emrys Creator, Optimizer, and Liberator). Although not commercially available any more, many instruments are currently still in use. The Emrys microwave unit is constituted of a closed rectangular waveguide tube, combined with a deflector device that, via a power sensor, physically maximizes the energy absorption by the reaction mixture (Dynamic Field Tuning). For that reason, the operation volume is limited (0.2–20 mL), but the uniform and high-density heating process results in fast and highly reproducible synthesis results. The output power is maximized to continuously delivered 300 W, sufficient for rapid heating of most reaction mixtures. The Emrys series is equipped with built-in magnetic stirring at a fixed level. Temperature measurement is achieved by an IR sensor perpendicular to the position of the vial in the waveguide, working in a measuring range of 60–250 °C. This arrangement requires a minimum filling height in each vessel type to receive accurate temperature values. Thus, the temperature is measured on the outer surface of the reaction vessels; no inside temperature measurement is available. The pressure limit for the Emrys instruments is 20 bar, due to the sealing mechanism utilizing Teflon-coated silicon seals in aluminum crimp tops. Pressure control is achieved by a sensor integrated in the closing lid of the cavity, sensing the deformation of the seal due to pressure built up. At 20 bar the instrument switches off, and the cooling mechanism is activated as the reaction is aborted. Efficient cooling is accomplished by a pressurized air supply with a rate of approximately 60 L min^{-1}, enabling cooling from 250 to 40 °C within approximately 1 min, depending on the heat capacity of the solvent used.

A very useful tool is the Emrys PathFinder Database, representing a collection of detailed protocols for reactions performed with Biotage (Personal Chemistry) instruments. To date, *c.* 4000 entries are available in this web-based tool.

The currently available instrumentation from Biotage is the Initiator reactor for small-scale reactions in a single-mode cavity (**Figure 4**). This instrument is closely related to the former Creator, but now equipped with a touch-screen for on-the-fly control or changes of parameters, no external PC is needed. The extended version, Initiator Sixty, in succession to the Optimizer EXP, and equipped with a robotic gripper, allows small-range scale-up using different vessels from 0.2 to 20 mL in operating volume (**Figure 5**). Between 24 to 60 reactions can be carried out sequentially and unattended with this automated upgrade.

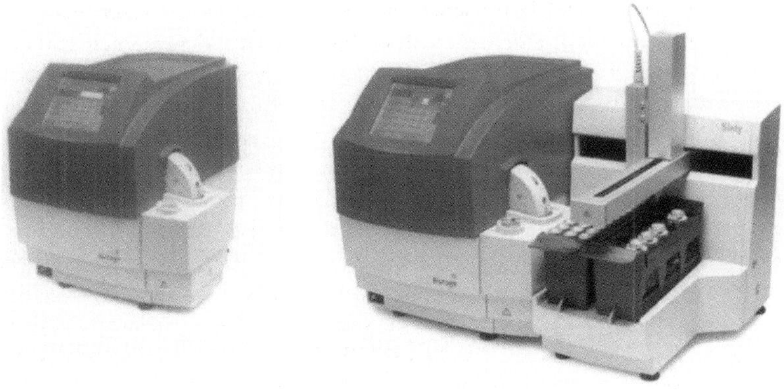

(a) (b)

Figure 4 (a) The Biotage Initiator. (b) The Biotage Initiator Sixty. (Courtesy of Biotage AB.)

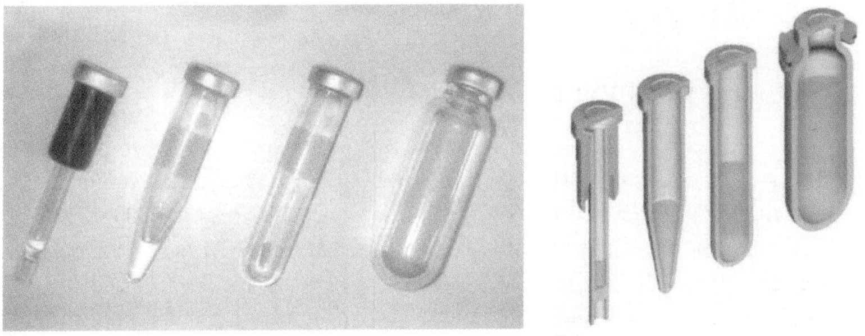

(a) (b)

Figure 5 (a) Biotage Initiator vessel types (0.2–0.5 mL, 0.5–2.0 mL, 2.0–5.0 mL, 5.0–20.0 mL (left to right)) and (b) schematic description of maximum volume. (Courtesy of Biotage AB.)

Similar to its predecessors of the Emrys series, the operation limits for the Initiator system are 60–250 °C at a maximum pressure of 20 bar. Temperature control is achieved in the same way by an IR sensor rectangular to the sample position. Thus, the temperature is measured on the outer surface of the reaction vessels, and no inside temperature measurement is available. Pressure measurement is conducted by a noninvasive sensor integrated into the cavity lid, measuring the deformation of the Teflon seal of the vessels. Efficient cooling is accomplished by a pressurized air supply with a rate of approximately $60\,L\,min^{-1}$, enabling cooling from 250 to 40 °C within 1 min.

As no external PC is needed, the characteristics of the software package have changed compared with the Emrys series. Via the touch-screen the user is now able to change the performed reaction protocols on-the-fly without aborting the experiment. Temperature and time changes can be done immediately and furthermore the power can be adjusted to a defined value over the process time.

3.36.3.1.2 CEM Discover platform

The CEM Discover platform, introduced in 2001, offers a single-mode instrument based on the self-tuning circular waveguide technique. This concept allows for the derivatization of a reactor to accommodate additional application-specific modules to address other common laboratory objectives. While the Discover reactor (**Figure 6**) covers a variety of reaction conditions in open- (up to 125 mL) and closed-vessel systems (maximum of 7 or 50 mL working volumes), this reactor can be easily upgraded for automation ($Explorer_{PLS}$, with a robotic gripper and up to 24-position autosampler racks), and for scale-up via a flow-through approach (Voyager). Offering a very economical choice in terms of cost of ownership and in footprint, the Discover BenchMate provides a fully featured entry-level system with reaction temperature and pressure management. Chemists have the ability to instantaneously change any reaction parameter during the reaction process, leading to improved instrument control and better reaction optimization, another valuable feature introduced by this line of reactors. The Investigator module brings real-time, in situ analysis of the reaction using an integrated Raman spectroscopy system to the platform. As with the automation and scale-up modules, the reactor core includes all of the necessary capabilities to accommodate the Raman module with minimal

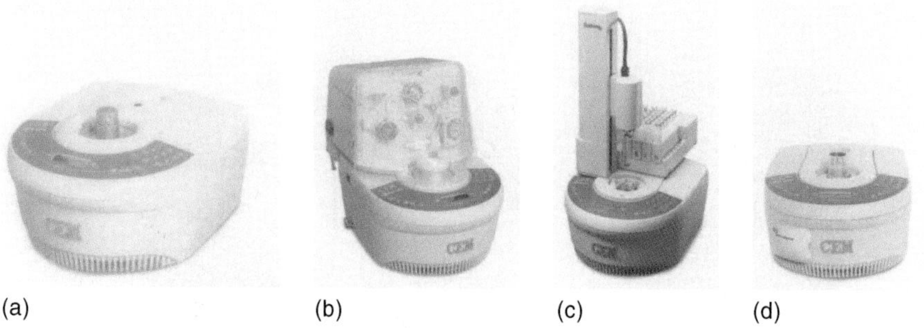

Figure 6 Available variations of the CEM Discover platform: (a) BenchMate, (b) Voyager, (c) Explorer$_{PLS}$, and (d) Investigator. (Courtesy of CEM Corp.)

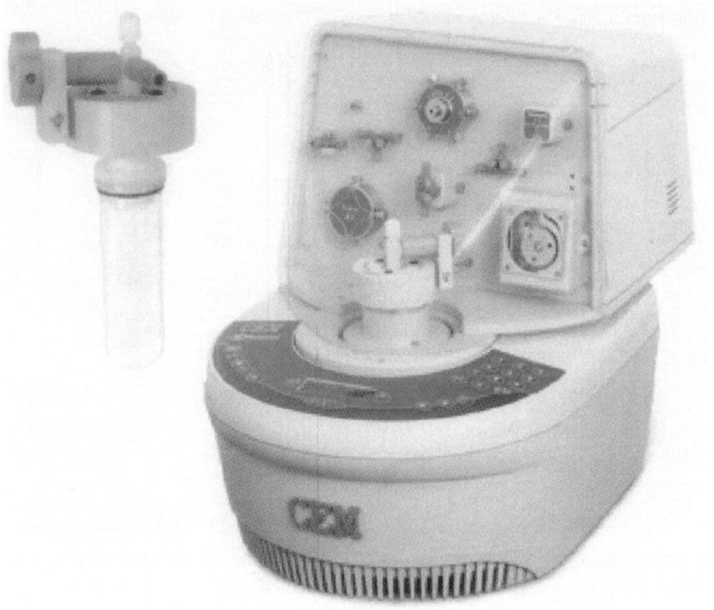

Figure 7 The CEM Voyager and its 80 mL reaction vessel. (Courtesy of CEM Corp.)

modification to the system as a whole. The analytical technique includes a proprietary fluorescence correction that overcomes the typical limitations that accompany Raman spectroscopy. The cavity design allows for the easy insertion of a quartz light pipe with focusing optics to enhance the Raman signal. The system software interrogates the Raman spectra, and uses this information as a feedback mechanism to assist in the optimization of the reaction conditions.

The Voyager system converts the standard reactor into a flow-through system, designed to allow the scale-up of reactions while still maintaining the advantages of single-mode energy transfer. While the technology accommodates both continuous and stop–flow formats, the stop–flow technique better accommodates the majority of scale-up applications encountered in today's synthesis laboratory. The Voyager system in stop–flow mode is operated with a special 80 mL vessel (**Figure 7**), utilizing reaction limits of 225 °C or 15 bar, and is applicable even for heterogeneous mixtures, slurries, and solid phase reactions.

For advanced high-throughput synthesis, the fully automated Navigator Microwave Compound Factory is available (**Figure 8**). This equipment offers dedicated XYZ robotics, including but not limited to the following options: a capping/decapping station, automated weighing, liquid handling (up to eight substrates), and solid reagent addition as well as solid/liquid sample withdrawal. A number of integrated analytics (solid phase extraction, high-performance liquid chromatography (HPLC), flash chromatography, Raman spectroscopy) can be delivered; the system is built according to customer requirements.

In addition, for solid phase peptide synthesis the Odyssey reaction system is available (**Figure 8**). This instrument, based on the Discover reactor core, enables the synthesis of up to 12 peptides unattended using a fluidics module to

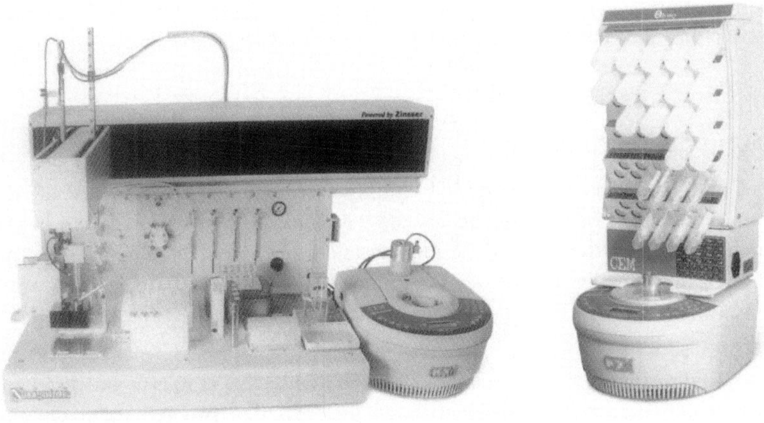

(a) (b)

Figure 8 (a) The CEM Navigator. (b) The CEM Odyssey. (Courtesy of CEM Corp.)

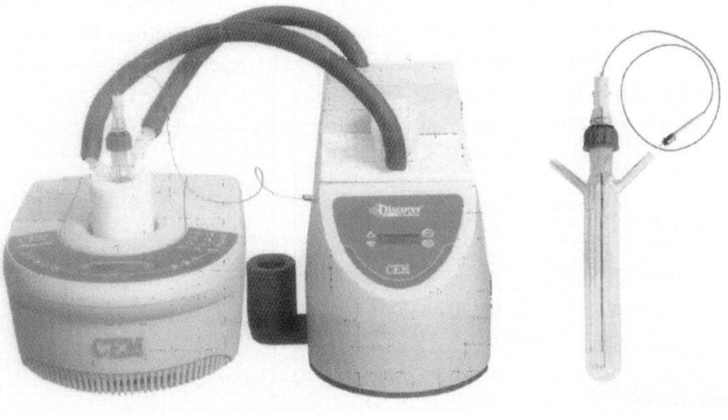

Figure 9 The CEM Discover CoolMate: a low-temperature microwave system. (Courtesy of CEM Corp.)

enable the controlled addition of resins, amino acids, coupling, deprotection, and washing reagents, as well as cleavage cocktails. Vessels with 10–100 mL volumes allow the synthesis of peptides at a 0.1–5 mmol scale, with up to 25 amino acids, accommodating both naturally occurring and non-natural amino acids. Typical cycle times with this system are 15 min, including all washings, for each residue addition. The Odyssey also allows programmable cleavage from the resin, either immediately after synthesis or at a later programmed time. Potential issues with racemization have been addressed through the use of precise temperature control.

Routine temperature measurement within the Discover series is achieved by an IR sensor positioned under the bottom of the cavity below the vessel. This allows accurate temperature control of the reaction while using minimum amounts of materials (0.2 mL). The platform also accepts an optional fiber optic temperature sensor system that addresses needs for temperature measurement where IR technology is not suitable, such as with subzero temperature reactions or with specialized reaction vessels. Pressure regulation is achieved by the IntelliVent pressure management technology. If the pressure in the vial exceeds 20 bar, the IntelliVent sensor allows for a controlled venting of the pressure, and reseals to maintain optimum safety and extend application scope. All Discover instruments are equipped with a built-in keypad for programming the reaction procedures and allowing for on-the-fly changes. All vessel types are equipped with corresponding stirring bars, ensuring optimum admixing of reagents. Flow cells afford mixing as the reagents travel along a circuitous path. The patented enhanced cooling system (PowerMAX) can be used during irradiation, therefore achieving simultaneous cooling for enhanced microwave synthesis, which results in a more efficient energy transfer into the sample and leads to improved results.

The latest extension in this context is the Discover CoolMate (**Figure 9**), a microwave system to perform subambient temperature chemistry. The reactor is equipped with a jacketed low-temperature vessel, and the microwave-transparent cooling media and chilling technology of the system keep the bulk temperature low (− 80 to + 35 °C). Thus, thermal degradation of the compounds is prevented while microwave energy is introduced to the reaction mixture.

3.36.3.2 Multimode Instruments

The development of multimode reactors for organic synthesis occurred mainly from already available microwave acid digestion/solvent extraction systems. Instruments for this purpose were first designed in the 1980s, and with the growing demand for synthesis systems, these reactors were subsequently adapted for organic synthesis applications. Therefore, there is still a close relationship between multimode microwave digestion systems and synthesis reactors.

3.36.3.2.1 Milestone s.r.l

Based on their microwave digestion system, Milestone offers the MicroSYNTH Labstation (also known as the ETHOS series) multimode instrument (**Figure 10**), available with various accessories. Two magnetrons deliver 1000 W microwave output power, and a patented pyramid-shaped microwave diffuser ensures homogeneous microwave distribution within the cavity.

The modular MicroSYNTH platform offers the diversity of different rotor and vessel systems, enabling reactions from 3 to 500 mL under open- and sealed-vessel conditions in batch and parallel manner up to 50 bar. The START package offers simple laboratory glassware for reactions at atmospheric pressure under reflux conditions. This system can be upgraded by a teaching kit (16 vessel rotor for reactions < 1.5 bar, **Figure 11**) or a research laboratory kit, equipped with the so-called MonoPREP module for single small-scale experiments (3–30 mL) in the multimode cavity. In addition, equipment for combinatorial chemistry approaches (CombiCHEM kit, microtiter well plates up to 96 × 1 mL), extraction, UV photoexcitation, and flow-through techniques is available.

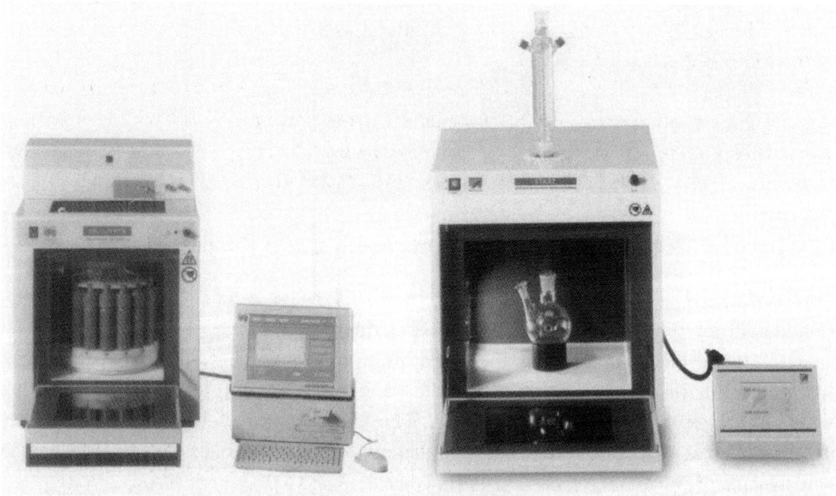

Figure 10 The Milestone MicroSYNTH/START Labstation. (Courtesy of Milestone s.r.l.)

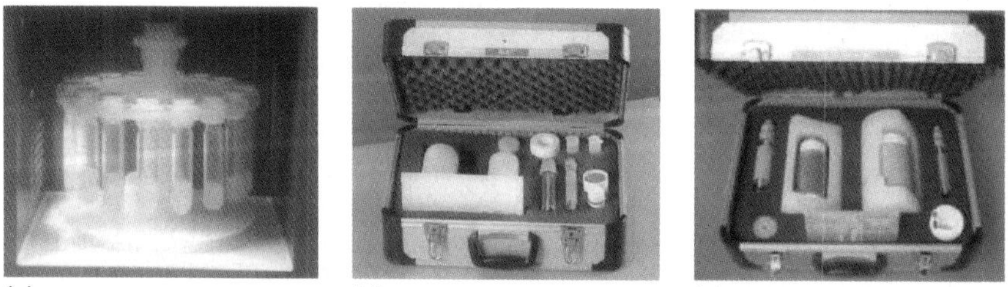

(a) (b) (c)

Figure 11 The Milestone (a) Teaching Kit, (b) Research Kit, and (c) MedChem Kit. (Courtesy of Milestone s.r.l.)

Post-reaction cooling of the reaction mixture is achieved by a constant airflow through the cavity. Due to the design of the thick plastic segments, the cooling of the high-pressure rotors is not very efficient, and external cooling by immersing the rotor in a water bath is recommended. Thus, vessels under pressure have to be handled outside the cavity, but a special cooling rack is available for this purpose.

Temperature measurement is achieved by the use of a fiber optic probe, immersed in one single reference vessel. Optionally available is an IR sensor for monitoring the outside surface temperature of each vessel, mounted in the side wall of the cavity, about 5 cm above the bottom. The reaction pressure is measured by a pneumatic sensor connected to one reference vessel. Therefore, the parallel rotors should be filled with identical reaction mixtures to ensure homogeneity.

For all MicroSYNTH systems, reaction monitoring is achieved via an external control terminal utilizing the EasyWAVE software packages. The runs can be controlled by temperature, pressure, or microwave power output in an up to 10-step defined program. The software enables on-line modifications of any method parameter, and the reaction process is monitored by an appropriate graphical interface. An included solvent library and electronic laboratory journal feature simplifies the experimental documentation.

As mentioned above, numerous rotors and vessel types for individual applications are available for the MicroSYNTH/ETHOS platform. Most of them are derived from the original digestion system and have been adapted for synthesis purposes.

For reactions at atmospheric pressure, standard laboratory glassware such as round-bottom flasks or simple beakers from 0.25 mL to 2 L can be used. A protective mount in the ceiling of the cavity enables the connection of reflux condensers or distillation equipment. An additional mount in the side wall allows for sample withdrawal, flushing of gas to create inert atmospheres, or live monitoring of the reaction with video cameras. Most of the published results in controlled MAOS are performed in sealed vessels, thus the following description mostly presents accessories for sealed-vessel reaction conditions:

- *MedChem Kit* (**Figure 11**). An accessory especially designed for medicinal chemistry laboratories covering working volumes from 2 to 140 mL. The package contains a 12 mL glass vial, a 50 mL quartz vessel, and single pressure reactor segments (100 and 270 mL, respectively).
- *MonoPREP module* (**Figure 12**). This tool is designed for single optimization runs at elevated pressure at comparatively small scale in multimode ovens. It comes with two glass vessels suitable for volumes from 3 mL to 30 mL at operation limits of 200 °C and 15 bar. A cooling mechanism achieves cooling down of the reaction mixture to 30 °C after the irradiation. For enhanced optimization the PRO-6 rotor, a parallel version of the MonoPREP module with six identical vessels, can be used.

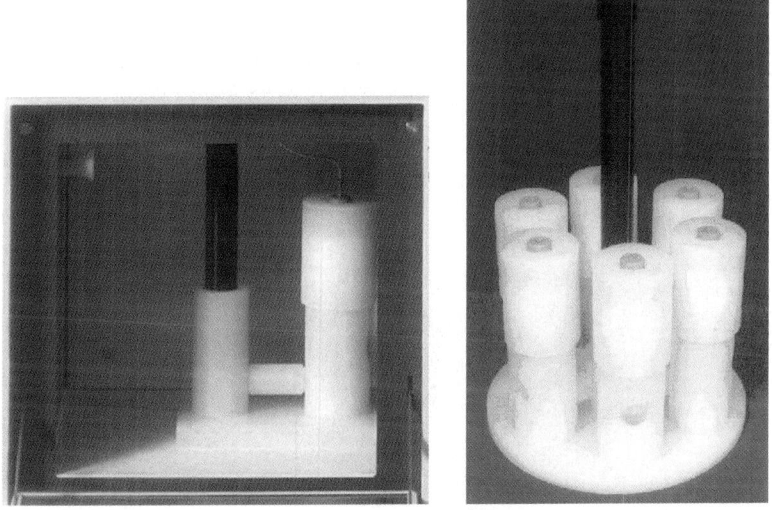

(a) (b)

Figure 12 (a) The Milestone MonoPREP module and (b) its parallel setup rotor PRO-6 (right). (Courtesy of Milestone s.r.l.)

- *CombiCHEM System* (**Figure 13**). For small-scale combinatorial chemistry applications this barrel-type rotor is available. It can hold two 24–96-well microtiter plates utilizing glass vials (0.5–4 mL) up to 4 bar at 150 °C. The plates are made of Weflon (graphite-doped Teflon) to ensure uniform heating, and are sealed by an inert membrane sheet. Axial rotation of the rotor tumbles the microwell plates to admix the individual samples. Temperature measurement is achieved by a fiber optic probe immersed in the center of the rotor.

3.36.3.2.2 CEM Corporation

The MARS Microwave Synthesis System (**Figure 14**) is based on the related MARS 5 digestion instrument, and offers different sets of rotor systems with several vessel designs and sizes for various synthesis applications.

For reactions at atmospheric pressure, standard laboratory glassware such as round-bottom flasks from 0.5 mL to 3 L can be used. A protective mount in the ceiling of the cavity allows for connection of a reflux condenser or distillation

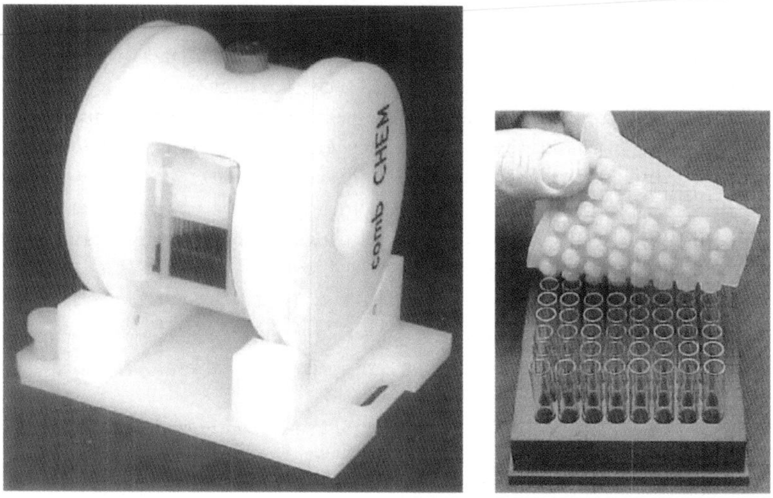

(a) (b)

Figure 13 The Milestone CombiCHEM system: (a) the rotor with (b) Weflon-made well plates. (Courtesy of Milestone s.r.l.)

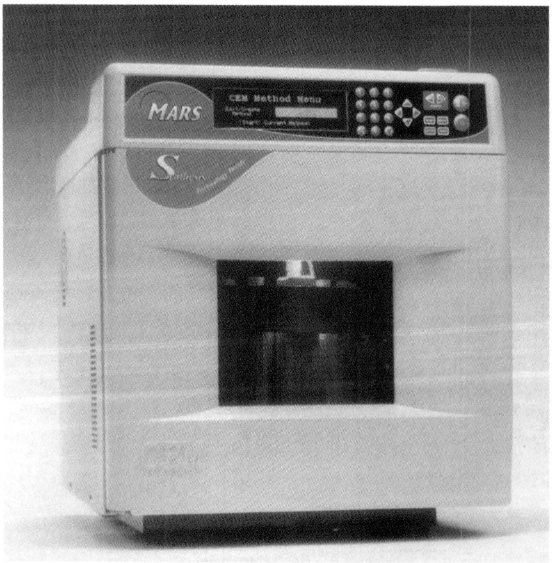

Figure 14 The CEM MARS Microwave Synthesis System. (Courtesy of CEM Corp.)

equipment as well as for addition of reagents and sample withdrawal. For parallel reactions at atmospheric pressure, turntables for up to 120 vials are available. Furthermore, a turntable for conventional 96-well titer plates is offered, allowing reactions at the minimum volume of 0.1 mL per vessel. Reactions under pressure can be carried out in various sealed-vessel rotor systems.

The general maximum output power of the instrument is 1200 W, but the MARS control panel offers two low-energy levels with unpulsed microwave output power of 300 and 600 W, respectively. This feature avoids overheating of the reaction mixture and unit, if just small amounts of reagents are used.

The MARS unit comes with a software package, operated by an integrated spill-proof keypad. The instrument can be connected to an external PC, but this is not required for most common operations. Methods and reaction protocols can be designed as temperature/time profiles or with precise control of constant power during the reaction.

3.36.3.2.3 Biotage AB

For scale-up applications, Biotage offers the Emrys Advancer batch reactor (**Figure 15**), serving a multimode cavity for operations with one 350–850 mL Teflon reaction vessel under high-pressure conditions. An operating volume of *c.* 50–500 mL at a maximum of 20 bar enables the production of 10–100 g product within one run. Homogeneous heating is ensured by a precise field-tuning mechanism and vigorous magnetic and/or overhead stirring of the reaction mixture. Direct scalability allows translating the optimized reaction conditions from the Emrys/Initiator system to larger scale.

The maximum output power of the Emrys Advancer is 1100 W, generating a heating rate of 0.5–4 °C s^{-1} to reach the maximum temperature of 250 °C for a 300 mL reaction volume in comparable times to the monomode experiments. Several connection ports in the chamber head (**Figure 15**) enable the addition of reagents during irradiation, sample removal for analysis, in situ monitoring by real-time spectroscopy, or creation of an inert/reactant gas atmosphere. Cooling is achieved by an effective definite gas expansion mechanism, to ensure drastically shortened cooling periods (from 180 to 40 °C for 200 mL of ethanol within 1 min). Due to the dimensions of the instrument (160 cm × 85 cm × 182 cm), extra laboratory space is required to make operations comfortable. This instrument is a custom-built, user-specified product, manufactured on request.

3.36.3.2.4 Anton Paar GmbH

The Anton Paar Synthos 3000 (**Figure 16**) is the most recent multimode instrument on the market. It is a microwave reactor dedicated to scale-up synthesis in quantities of up to approximately 250 g per run and designed for chemistry under high-pressure and high-temperature conditions. The instrument enables direct scalability of already elaborated and optimized reaction protocols from single-mode cavities without changing the reaction parameters.

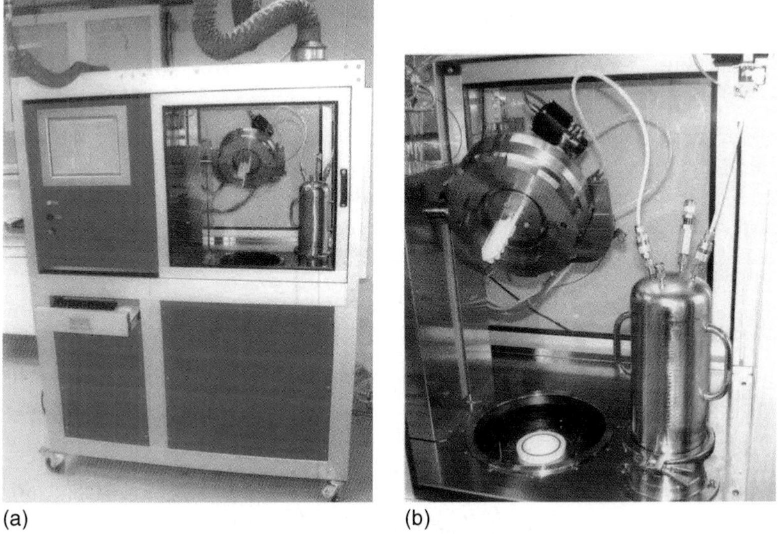

(a) (b)

Figure 15 (a) The Biotage Emrys Advancer scale-up instrument with (b) the multifunctional chamber head. (Courtesy of Biotage AB.)

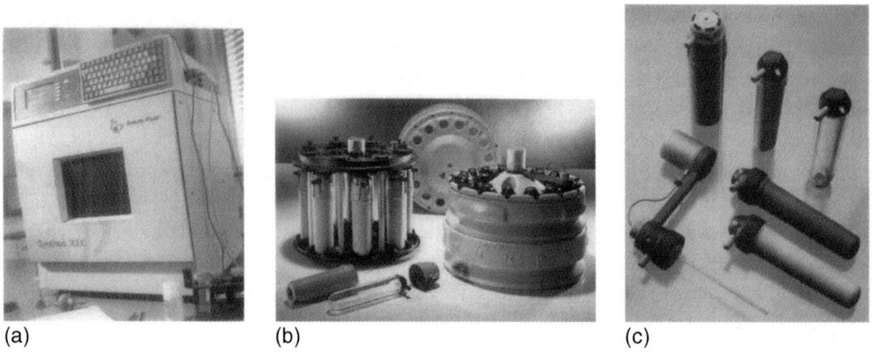

Figure 16 (a) The Anton Paar Synthos 3000 with (b) rotors and (c) vessel types. (Courtesy of Anton Paar GmbH.)

The use of two magnetrons (1400 W continuously delivered output power) allows upscaling of small-scale runs to produce large amounts of the desired compounds within a similar time frame. The homogeneous microwave field guarantees identical conditions at every position of the rotors (eight or 16 vessels), resulting in good reproducibility of experiments. Offering high operation limits (80 bar at 300 °C) the instrument facilitates the investigation of new reaction methods. The instrument can be operated with either an eight- or 16-position rotor, equipped with various vessel types (**Figure 16**) for different pressure and temperature conditions. Various accessories allow for special applications such as creation of inert/reactive gas atmosphere, reactions in pre-pressurized vessels, and chemistry in near-critical water as well as solid phase synthesis or photochemistry.

Temperature measurement is achieved by a remote IR sensor from the bottom on the outer surface of the vessels. The operation limit of the IR sensor is 400 °C, but regulated by the software safety features to 280 °C, as operation limits of the used materials are also at a maximum of 300 °C. For additional control, temperature measurement in a reference vessel by an immersed gas balloon thermometer is available. The operation limit of this temperature probe is 310 °C, suitable for reactions under extreme temperature and pressure conditions.

The pressure is measured by a hydraulic system, either in one reference vessel of the 16-vessel rotor or simultaneously for all vessels of the eight-vessel rotor. The operation limit is 86 bar, sufficient for synthetic applications. In addition, a pressure rate limit is set to $3.0\,\text{bar}\,\text{s}^{-1}$ by the included control software. Protection against sudden pressure peaks is achieved by metal safety disks, included in the vessel caps (safety limit of 70 or 120 bar, respectively) and by software regulations, dependent on the used rotor and vessel type.

All parameters are transmitted wireless by IR data transfer from the rotors to the system control computer of the instrument, to eliminate troublesome cables and hoses from inside the cavity. For the individual rotor types, different vessels made out of quartz or PTFE–TFM liners are available.

3.36.4 Microwave-Processing Techniques

In modern microwave synthesis a variety of different processing techniques can be utilized, aided by the availability of diverse types of dedicated microwave reactors. While in the past much interest has focused on, for example, solvent-free reactions under open-vessel conditions, it appears that nowadays most of the published examples in the area of controlled MAOS involve the use of organic solvents under sealed-vessel conditions. Despite this fact, a brief summary of alternative processing techniques is presented in the following sections.

3.36.4.1 Solvent-Free Reactions

A frequently used processing technique employed in microwave-assisted organic synthesis since the early 1990s involves solvent-less (dry-media) procedures,[61–66] where the reagents are preadsorbed onto either a more or less microwave-transparent (silica, alumina, or clay) or a strongly absorbing (graphite) inorganic support, which additionally can be doped with a catalyst or reagent. Particularly in the early days of MAOS, the solvent-free approach was very popular since it allowed the safe use of domestic household microwave ovens and standard open-vessel technology. While a large number of interesting transformations using dry-media reactions have been published in the literature,[61–66] technical difficulties relating to nonuniform heating, mixing, and the precise determination of the reaction temperature remain unsolved, in particular when scale-up issues need to be addressed.

One of the simplest methods involves the mixing of the neat reagents and subsequent irradiation by microwaves. In general, pure, dry solid organic substances do not absorb microwave energy, therefore almost no heating will occur. Thus, often small amounts of a polar solvent (e.g., *N,N*-dimethylformamide or water) need to be added to the reaction mixture in order to allow for dielectric heating by microwave irradiation.

An alternative technique utilizes microwave-transparent or only weakly absorbing inorganic supports such as silica, alumina, or clay materials.[61–66] These reactions are effected by the reagents/substrates immobilized on the porous solid supports, and have advantages over the conventional solution phase reactions because of their good dispersion of active reagent sites, associated selectivity, and easier work-up. The recyclability of some of these solid supports and freedom from the problems associated with the waste disposal of solvents means that these processes are more environment-friendly compared with solvent-based systems. In general, the substrates are preadsorbed onto the surface of the solid support and then exposed to microwave irradiation.

Apart from examples where the inorganic support merely acts as a catalyst, there are many instances where a solid-supported reagent can be used very effectively in the process.

In addition to cases where the inorganic support itself acts as a catalyst, or where a reagent has been impregnated on the solid support, it is also possible to dope the support material with, for example, metal catalysts.

In contrast to solvent-free (dry media) microwave processing involving (when properly dried) weak microwave-absorbing supports such as silica, alumina, clays, and zeolites, an alternative is to use strongly microwave-absorbing supports such as graphite. For reactions that require high temperatures, the idea of using a reaction support that takes advantage both of strong microwave coupling and strong adsorption of organic molecules has received increasing attention in recent years.[67] Since many organic compounds do not interact appreciably with microwave irradiation, such a support could be an ideal sensitizer, able to absorb, convert, and transfer energy provided by a microwave source to the chemical reagents. Most forms of carbon interact strongly with microwaves. Amorphous carbon and graphite, in their powdered form, irradiated at 2.45 GHz, rapidly reach *c.* 1000 °C within 1 min of irradiation. In addition to graphite being used as a 'sensitizer' (energy converter), there are also several examples in the literature where the catalytic activity of metal inclusions in graphite has been exploited (graphimets).[67]

3.36.4.2 Phase-Transfer Catalysis (PTC)

In addition to solvent-free processing, PTC conditions have also been widely employed as a processing technique in MAOS.[68,69] In PTC the reactants are situated in two separate phases, for example liquid–liquid or solid–liquid. Because the phases are mutually insoluble, ionic reagents are typically dissolved in the aqueous phase, while the substrate remains in the organic phase (liquid–liquid PTC). In solid–liquid PTC, on the other hand, ionic reagents can be used in their solid state as a suspension in the organic medium. Transport of the anions from the aqueous or solid phase to the organic phase is facilitated by phase transfer catalysts, typically quaternary onium salts or cation-complexing agents. PTC reactions are perfectly tailored for microwave activation, and the combination of solid–liquid PTC and microwave irradiation typically gives the best results in this area.[68,69] Numerous transformations in organic synthesis can be achieved under solid–liquid PTC and microwave irradiation in the absence of solvent, generally under atmospheric pressure in open vessels.[68,69]

3.36.4.3 Reactions Using Solvents

3.36.4.3.1 Open- versus closed-vessel conditions

Microwave-assisted syntheses can be carried out using standard organic solvents either under open- or sealed-vessel conditions. If solvents are heated by microwave irradiation at atmospheric pressure in an open vessel, the boiling point of the solvent (as in an oil bath experiment) typically limits the reaction temperature that can be achieved. In the absence of any specific or nonthermal microwave effects, the expected rate enhancements would be comparatively small. In order to nonetheless achieve high reaction rates, high-boiling microwave-absorbing solvents such as dimethyl sulfoxide, 1-methyl-2-pyrrolidone, 1,2-dichlorobenzene, or ethylene glycol have been frequently used in open-vessel microwave synthesis.[11,12] However, the use of these solvents presents serious challenges during product isolation. The approach has been adapted for lower-boiling solvents, such as toluene, by periodic interruption of heating. However, this modification not only precludes the high temperatures which are advantageous in microwave synthesis but also generates a potentially serious fire hazard when unmodified domestic microwave instruments are used. Because of the recent availability of modern microwave reactors with on-line monitoring of both temperature and pressure (see above), MAOS in sealed vessels has been celebrating a comeback in recent years. This is clearly evident surveying the recently published literature in the area of MAOS,[37] and it appears that the combination of rapid dielectric heating by microwaves with sealed-vessel technology (autoclaves) will most likely be the method of choice for performing MAOS in the future.

3.36.4.3.2 Pre-pressurized reaction vessels

Relatively little work has been performed with gaseous reagents in sealed-vessel microwave experiments. Although several publications describe this technique in the context of heterogeneous gas phase catalytic reactions important for industrial processes,[70–72] the use of pre-pressurized reaction vessels in conventional MAOS involving solvents is rare. Due to the design of modern single-mode microwave reactors and their reaction vessels, pre-pressurization is not possible. Several authors have, however, described the use of reactive gases in such experiments, and experimental techniques how to apply over-pressure to 10 bar.[73–76]

3.36.4.3.3 Nonclassical solvents

Apart from using standard organic solvents in conjunction with microwave synthesis, the use of either water or so-called ionic liquids as alternative reaction media has been increasingly popular in recent years.

Interest in the use of aqueous reaction media for organic reactions has been developing over recent decades.[77–81] Temperatures of 100 °C and below have been employed for synthesis, often to exploit the so-called hydrophobic effect. Conversely, at temperatures approaching the supercritical region ($T_c = 374$ °C), water has found degradative applications in addition to being a solvent still useful for synthetic purposes.[82–86] For microwave synthesis, the subcritical region (also termed near critical) at temperatures between 150 and 300 °C is particularly attractive.[87–89] Water has a dielectric constant ε'' that decreases from 78 at 25 °C to 20 at 300 °C, this latter value being comparable with that of typical organic solvents such as acetone at ambient temperature. Therefore, water behaves as a pseudo-organic solvent at elevated temperatures, allowing for the dissolution of many organic substrates. In addition to the environmental advantages of using water in place of organic solvents, isolation of products is normally facilitated by the decrease in the solubility of the organic reaction products upon postreaction cooling. Furthermore, the ionic product (dissociation constant, pK_W) of water is greatly influenced by temperature, increasing by three orders of magnitude between 25 and 250 °C.[90] Water is therefore a significantly stronger acid and base at subcritical conditions than at ambient conditions, which can be exploited for organic synthesis. Numerous organic transformations have been carried out in superheated water under microwave conditions, providing sometimes unexpected results.[91–101]

Room temperature ionic liquids (RTILs) are another class of solvents of importance in microwave synthesis. Ionic liquids made of organic cations and appropriate anions have attracted much recent attention as environmentally benign solvents for chemistry because they have melting points close or near to room temperature.[102] In some instances they have also been used as reagents. RTILs have negligible vapor pressure, and are immiscible with a range of organic solvents, meaning that organic products can be easily removed, and the ionic liquid can be recycled. In addition, they have a wide accessible temperature range (typically >300 °C), are nonflammable, and are relatively easy to use and to recycle. From the perspective of microwave chemistry, one of the points of key importance is their high polarity, and that this is variable depending on the cation and anion, and so can effectively be tuned to a particular application.

Ionic liquids interact very efficiently with microwaves through the ionic conduction mechanism, and are rapidly heated at rates easily exceeding 10 °C per second without any significant pressure build-up. Therefore, safety problems arising from over-pressurization of heated sealed reaction vessels can be minimized. The concept of performing microwave synthesis in RTILs as reaction media has been applied to several different organic transformations.[103]

As an alternative to the use of rather expensive ionic liquids as solvents, several research groups have used ionic liquids as doping agents for microwave heating of otherwise nonpolar solvents, such as hexane, toluene, tetrahydrofuran, or dioxane. This technique is becoming increasingly popular, as demonstrated by the many recently published examples.[103] Systematic studies on temperature profiles and the thermal stability of ionic liquids under microwave irradiation have shown that the addition of a small amount of an ionic liquid (0.1 mmol per milliliter of solvent) suffices to obtain dramatic changes in the heating profiles by changing the overall dielectric properties ($\tan\delta$ value) of the reaction medium.

It is worth noting that ionic liquids can also be prepared very rapidly and efficiently under microwave conditions.[104,105]

3.36.4.4 Parallel Processing

Parallel processing of synthetic operations has been one of the cornerstones of medicinal and high-throughput synthesis for years. In the parallel synthesis of compound libraries, compounds are synthesized using ordered arrays of spatially separated reaction vessels adhering to the traditional one vessel–one compound philosophy. The defined location of the compound in the array provides the structure of the compound. A commonly used format for parallel synthesis is the 96-well microtiter plate, and, today, combinatorial libraries comprising hundreds to thousands of compounds can be synthesized by parallel synthesis, often in an automated fashion.

As pointed out in other chapters, performing organic chemical transformations under microwave conditions often allows a reduction in reaction times from many hours or days to minutes or even seconds. For many organic or medicinal chemists in both academia and industry there is therefore little need to perform parallel synthesis any more, as carrying out rapid microwave chemistry in a sequential processing fashion almost becomes as efficient as performing traditional parallel synthesis. This is particularly true for the generation of small, focused libraries (50–100 compounds) utilizing fully automated sequential microwave synthesis platforms. Despite the growing trend in the pharmaceutical industry today to synthesize smaller, focused libraries, medicinal chemists in a high-throughput synthesis environment often still need to generate large compound libraries using a parallel synthesis approach.

Microwave-assisted reactions allow rapid product generation in high yield under uniform conditions. Therefore, they are ideally suited for parallel synthesis applications.

In a 1998 publication, the concept of microwave-assisted parallel synthesis in plate format was introduced for the first time. Using the three-component Hantzsch pyridine synthesis as a model reaction, libraries of substituted pyridines were prepared in a high-throughput parallel fashion.[106] Microwave irradiation was carried out in 96-well filter-bottom polypropylene plates, in which the corresponding eight 1,3-dicarbonyl compounds and 12 aldehyde building blocks were dispensed using a robotic liquid handler. Microwave irradiation of the 96-well plate containing aldehydes, 1,3-dicarbonyl compounds, and ammonium nitrate/clay in a domestic microwave oven for 5 min produced the expected pyridine library directly, after the desired products were extracted from the solid support by organic solvent and collected in a receiving plate. HPLC–mass spectrometry analysis showed that the reactions were uniformly successful across the 96-well reactor plate, without any starting material being present.[106]

In the majority of cases described before 2002, domestic multimode microwave ovens were used as heating devices, without utilizing specialized reactor equipment. Since reactions in household multimode ovens are notoriously difficult to reproduce due to the lack of temperature and pressure control, pulsed irradiation, uneven electromagnetic field distribution, and the unpredictable formation of hot spots, today in most of the published methods dedicated commercially available multimode reactor systems for parallel processing are used. These multivessel rotor systems have been described in detail above.

An important issue in parallel microwave processing is the homogeneity of the electromagnetic field in the microwave cavity. Inhomogeneities in the field distribution may lead to the formation of so-called hot and cold spots, resulting in different reaction temperatures in individual vessels or wells. Published examples of parallel microwave processing in dedicated multivessel rotor systems have included the generation of a 21-member library by parallel solid phase Knoevenagel condensations under open-vessel conditions.[107] Here the temperature was monitored with the aid of a shielded thermocouple inserted into one of the reaction containers. It has been confirmed by standard temperature measurements performed immediately after the irradiation period that the resulting end temperature in each vessel was the same within $\pm 2\,^{\circ}C$.[107] In a different study involving a 36-sealed-vessel rotor system, the uniformity of the reaction conditions in such a parallel setup was investigated.[108–110] All 36 individual vessels provided identical yields of product (65–70%) within experimental error.

Several articles in the area of microwave-assisted parallel synthesis have described the irradiation of 96-well filter-bottom polypropylene plates in conventional household microwave ovens for high-throughput synthesis. While some authors did not report any difficulties associated with the use of such equipment,[106] others have experienced problems in connection with the thermal instability of the polypropylene material itself, and with respect to temperature gradients developing between individual wells upon microwave heating.[111]

To overcome the problems associated with using conventional polypropylene deep-well plates in a microwave reactor, specifically designed well plates in 12-, 24-, 48-, and 96-reaction-well formats have been developed (see **Figure 13**).[110] These plates consist of a base of carbon-doped Teflon (Weflon) for better heat distribution, with glass inserts as reaction vessels. The vessels are sealed with Teflon-laminated silicon mats. For such a large number of reaction vessels, mixing with magnetic stirrers is impracticable, and therefore a new mixing method was developed using overhead shaking. Here, temperature measurement is performed with a fiber optic sensor inside the Weflon block, just underneath the reaction vessels.[110] Since several such devices can be mounted on top of each other, several hundred reactions may potentially be performed in one irradiation cycle. It is important to note that with this system the material used for the preparation of the plates (Weflon) absorbs microwave energy, which means that the sealed glass vials will be heated by microwave irradiation regardless of the dielectric properties of the reactants/solvents. The system is designed to interface with conventional liquid handler/dispensers to achieve a high degree of automation in the whole process.

The reaction homogeneity in a 24-well plate was investigated by monitoring the esterification of hexanoic acid with 1-hexanol at 120 °C for 30 min. The difference in conversion between the individual vessels was 4%, with a standard deviation of 2.4%.[110] It appears, therefore, that all individual reactions were irradiated homogeneously in the applied microwave field.

A different strategy to achieve high throughput in microwave-assisted reactions can be realized by performing automated sequential microwave synthesis in monomode microwave reactors. Since it is currently not feasible to have more than one reaction vessel in a monomode microwave cavity, a robotic system has been integrated into a platform that moves individual reaction vessels in and out of a specifically designed cavity. With some instruments, a liquid handler additionally allows dispensing of reagents into sealed reaction vials, while a gripper moves each sealed vial in and out of the microwave cavity after irradiation. Some instruments can process up to 120 reactions per run, with a typical throughput of 12–15 reactions/hour in an unattended fashion. In contrast to the parallel synthesis application in multimode cavities, this approach allows the user to perform a series of optimization or library production reactions with each reaction separately programmed.[112]

The issue of parallel versus sequential synthesis using multimode or monomode cavities, respectively, deserves special comment. While the parallel setup allows for a considerably higher throughput than can be achieved in the relatively short time frame of a microwave-enhanced chemical reaction, the individual control over each reaction vessel in terms of reaction temperature/pressure is limited. In the parallel mode, all reaction vessels are exposed to the same irradiation conditions. In order to ensure similar temperatures in each vessel, the same amount of identical solvent should be used in each reaction vessel because of the dielectric properties involved. As an alternative to parallel processing, the automated sequential synthesis of libraries can be a viable strategy if small focused libraries (20–200 compounds) need to be prepared. Irradiating each individual reaction vessel separately gives better control over the reaction parameters, and allows for the rapid optimization of reaction conditions. For the preparation of relatively small libraries, where delicate chemistries are to be performed, the sequential format may be preferable.

3.36.4.5 Scale-up in Batch and Continuous Flow

Most examples of microwave-assisted chemistry published to date have been performed on a less than 1 g scale (typically 1–5 mL reaction volume). This is in part a consequence of the recent availability of single-mode microwave reactors that allow the safe processing of small reaction volumes under sealed-vessel conditions by microwave irradiation. While these instruments have been very successful for small-scale organic synthesis, it is clear that once a compound in a medicinal chemistry program progresses from a lead structure to a development compound, much larger-scale MAOS techniques that can ultimately routinely provide products on a multi-kilogram (or even higher) scale need to be employed.

Keeping in mind some of the physical limitations of microwave heating technology such as magnetron power or penetration depth, two different approaches for microwave synthesis on a larger scale (>100 mL volume) have emerged. While some groups have employed larger batch-type multimode or monomode reactors (<1000 mL processing volume), others have used continuous-flow techniques (multi- and monomode) to overcome the inherent problems associated with MAOS scale-up. An additional key point in processing comparatively large volumes under pressure in a microwave field is the safety aspect, as any malfunction or rupture of a large pressurized reaction vessel may have significant consequences. In general, one should note that published examples of MAOS scale-up experiments are rare, in particular those involving complex organic reactions.

Modern single-mode microwave technology allows the performance of MAOS in very small reaction volumes (<0.2 mL). With today's commercially available single-mode cavities, the largest volumes that can be processed under sealed-vessel conditions are *c.* 50 mL, with different vessel types being available to upscale in a linear fashion from 0.05 to 50 mL (see **Figures 5** and **7**). Under open-vessel conditions, higher volumes (>1000 mL) have been processed under microwave irradiation conditions, not presenting any technical difficulties.

An important issue for the medicinal chemist is the potential of the direct scalability of microwave reactions, allowing rapid translation of previously optimized small-scale conditions to a larger scale. Several authors have reported independently the feasibility of directly scaling reaction conditions from small-scale single-mode (typically 0.5–5 mL) to larger-scale multimode batch microwave reactors (20–500 mL) without reoptimization of the reaction conditions.[113–116]

Mainly because of safety concerns and issues related to the penetration depth of microwaves into absorbing materials such as organic solvents, the preferable option for processing volumes of >1000 mL under sealed-vessel microwave conditions is a continuous-flow technique, although here the number of published examples using dedicated microwave reactors is limited.[117–121] In such a system the reaction mixture is passed through a microwave-transparent coil that is held in the cavity of a single- or multimode microwave reactor. The previously optimized reaction time under batch microwave conditions now needs to be related to a residence time (the time the sample stays in the microwave-heated coil) at a specific flow rate.

Current single-mode continuous flow microwave reactors allow processing of a comparatively small volume. Much larger volumes can be processed in continuous-flow reactors that are housed inside a multimode microwave system. One problem with continuous-flow reactors, however, is the clogging of lines and the difficulties in processing heterogeneous mixtures. Since many organic transformations involve some form of insoluble reagent or catalyst, single-mode so-called stop–flow microwave reactors have recently been developed, where peristaltic pumps – capable of pumping slurries and even solid reagents – are used to fill a batch reaction vessel (80 mL) with the reaction mixture (see **Figure 7**). After microwave processing in batch, the product mixture is pumped out of the system, which is then ready to receive the next batch of the reaction mixture.

Critically evaluating the currently available instrumentation for microwave scale-up in batch and continuous flow, one may argue that for processing volumes of <1000 mL a batch process may be preferable. By carrying out sequential runs in batch mode, kilogram quantities of product can easily be obtained. When larger quantities of a specific product need to be prepared on a regular basis, it may be worthwhile evaluating a continuous-flow protocol. Large-scale continuous-flow microwave reactors (flow rate of $20\,L\,h^{-1}$) are currently under development.[122] However, at the present time there are no documented published examples of the use of microwave technology for organic synthesis on a production scale level (>1000 kg), which is a clear limitation of this otherwise successful technology.[55]

3.36.5 Conclusion

The *c*. 3000 examples of microwave-assisted organic synthesis published since 1986 in the literature[3–37] should make it obvious that almost all types of chemical transformations can be carried out successfully under microwave conditions. A comprehensive summary of the most relevant reactions for the medicinal chemist is provided in a recent book.[37] Carrying out reactions by microwave heating does not necessarily imply that dramatic rate enhancements compared with a classical, thermal process will be observed in all cases, but the simple convenience of using microwave technology will make this nonclassical heating method a standard tool in every medicinal chemistry laboratory within a few years. In the past, microwaves were often used only when all other options to perform a particular reaction had failed, or when exceedingly long reaction times or high temperatures were required to complete a reaction. This practice is now slowly changing, and, due to the growing availability of microwave reactors in many laboratories, routine synthetic transformations are now also being carried out by microwave heating.

The benefits of controlled microwave heating, in particular in conjunction with using sealed-vessel systems, are manifold:

- Most importantly, microwave processing frequently leads to dramatically reduced reaction times, higher yields, and cleaner reaction profiles. In many cases the observed rate enhancements may simply be a consequence of the high reaction temperatures that can rapidly be obtained using this nonclassical heating method, or may result from the involvement of so-called specific or nonthermal microwave effects.
- The choice of solvent for a given reaction is not governed by the boiling point (as in a conventional reflux setup) but rather by the dielectric properties of the reaction medium, which can be easily tuned by, for example, addition of highly polar materials such as ionic liquids.
- The temperature/pressure monitoring mechanisms of modern microwave reactors allow for an excellent control of reaction parameters, which generally leads to more reproducible reaction conditions.
- Because direct in-core heating of the medium occurs, the overall process is more energy efficient than classical oil bath heating.
- Microwave heating can rapidly be adapted to a parallel or automatic sequential processing format. In particular, the latter technique allows for the rapid testing of new ideas and high-speed optimization of reaction conditions. The fact that a 'yes' or 'no' answer for a particular chemical transformation can often be obtained within 5–10 min (as opposed to several hours using a conventional protocol) has contributed significantly to the acceptance of microwave chemistry, in both industry and academia.

References

1. Gedye, R.; Smith, F.; Westaway, K.; Ali, H.; Baldisera, L.; Laberge, L.; Rousell, J. *Tetrahedron Lett.* **1986**, *27*, 279–282.
2. Giguere, R. J.; Bray, T. L.; Duncan, S. M.; Majetich, G. *Tetrahedron Lett.* **1986**, *27*, 4945–4958.
3. Abramovitch, R. A. *Org. Prep. Proced. Int.* **1991**, *23*, 685–711.

4. Caddick, S. *Tetrahedron* **1995**, *51*, 10403–10432.
5. Lidström, P.; Tierney, J.; Wathey, B.; Westman, J. *Tetrahedron* **2001**, *57*, 9225–9283.
6. Nüchter, M.; Ondruschka, B.; Bonrath, W.; Gum, A. *Green Chem.* **2004**, *6*, 128–141.
7. Nüchter, M.; Müller, U.; Ondruschka, B.; Tied, A.; Lautenschläger, W. *Chem. Eng. Technol.* **2003**, *26*, 1207–1216.
8. Kappe, C. O. *Angew. Chem. Int. Ed.* **2004**, *43*, 6250–6284.
9. Strauss, C. R.; Trainor, R. W. *Aust. J. Chem.* **1995**, *48*, 1665–1692.
10. Strauss, C. R. *Aust. J. Chem.* **1999**, *52*, 83–96.
11. Bose, A. K.; Banik, B. K.; Lavlinskaia, N.; Jayaraman, M.; Manhas, M. S. *Chemtech* **1997**, *27*, 18–24.
12. Bose, A. K.; Manhas, M. S.; Ganguly, S. N.; Sharma, A. H.; Banik, B. K. *Synthesis* **2002**, 1578–1591.
13. De la Hoz, A.; Díaz-Ortis, A.; Moreno, A.; Langa, F. *Eur. J. Org. Chem.* **2000**, 3659–3673.
14. Hamelin, J.; Bazureau, J.-P.; Texier-Boullet, F. Microwaves in Heterocyclic Chemistry. In *Microwaves in Organic Synthesis*; Loupy, A. Ed.; Wiley-VCH: Weinheim, Germanỳ, 2002; Chapter 8, pp 253–294.
15. Besson, T.; Brain, C. T. Heterocyclic Chemistry Using Microwave-Assisted Approaches. In *Microwave-Assisted Organic Synthesis*; Lidström, P.; Tierney, J. P., Eds.; Blackwell: Oxford, UK, 2005; Chapter 3.
16. Xu, Y.; Guo, Q.-X. *Heterocycles* **2004**, *63*, 903–974.
17. Elander, N.; Jones, J. R.; Lu, S.-Y., Stone-Elander, S. *Chem. Soc. Rev.* **2000**, 239–250.
18. Stone-Elander, S.; Elander, N. *J. Label. Compd. Radiopharm.* **2002**, *45*, 715–746.
19. Larhed, M.; Moberg, C.; Hallberg, A. *Acc. Chem. Res.* **2002**, *35*, 717–727.
20. Olofsson, K.; Larhed, M. Microwave Accelerated Metal Catalysis. Organic Transformations at Warp Speed. In *Microwave-Assisted Organic Synthesis*; Lidström, P.; Tierney, J. P., Eds.; Blackwell: Oxford, UK, 2005; Chapter 5.
21. Krstenansky, J. L.; Cotterill, I. *Curr. Opin. Drug Disc. Dev.* **2000**, *4*, 454–461.
22. Larhed, M.; Hallberg, A. *Drug Disc. Today* **2001**, *6*, 406–416.
23. Wathey, B.; Tierney, J. P.; Lidström, P.; Westman, J. *Drug Disc. Today* **2002**, *7*, 373–380.
24. Wilson, N. S.; Roth, G. P. *Curr. Opin. Drug Disc. Dev.* **2002**, *5*, 620–629.
25. Dzierba, C. D.; Combs, A. P. *Ann. Rep. Med. Chem.* **2002**, *37*, 247–256.
26. Lew, A.; Krutznik, P. O.; Hart, M. E.; Chamberlin, A. R. *J. Comb. Chem.* **2002**, *4*, 95–105.
27. Kappe, C. O. *Curr. Opin. Chem. Biol.* **2002**, *6*, 314–320.
28. Lidström, P.; Westman, J.; Lewis, A. *Comb. Chem. High Throughput Screen.* **2002**, *5*, 441–458.
29. Blackwell, H. E. *Org. Biomol. Chem.* **2003**, *1*, 1251–1255.
30. Al-Obeidi, F.; Austin, R. E.; Okonya, J. F.; Bond, D. R. S. *Mini-Rev. Med. Chem.* **2003**, *3*, 449–460.
31. Swamy, K. M. K.; Yeh, W.-B.; Lin, M. J.; Sun, C.-M. *Curr. Med. Chem.* **2003**, *10*, 2403–2423.
32. Stadler, A.; Kappe, C. O. Microwave-Assisted Combinatorial Chemistry. In *Microwave-Assisted Organic Synthesis*; Lidström, P.; Tierney, J. P., Eds.; Blackwell: Oxford, UK, 2005; Chapter 7, pp 405–434.
33. For online resources on MAOS, see: http://www.maos.net (accessed May 2006).
34. Loupy, A. *Microwaves in Organic Synthesis*; Weinheim: Wiley-VCH, 2002.
35. Hayes, B. L. *Microwave Synthesis: Chemistry at the Speed of Light*; CEM: Matthews, NC, 2002.
36. Lidström, P.; Tierney, J. P. *Microwave-Assisted Organic Synthesis*; Blackwell: Oxford, UK, 2005.
37. Kappe, C. O.; Stadler, A. *Microwaves in Organic and Medicinal Chemistry*; Wiley-VCH: Weinheim, Germany, 2005.
38. Von Hippel, A. R. *Dielectric Materials and Applications*; MIT Press: Cambridge, MA, 1954.
39. Stuerga, D.; Delmotte, M. Wave–Material Interactions, Microwave Technology and Equipment. In *Microwaves in Organic Synthesis*; Loupy, A., Ed.; Wiley-VCH: Weinheim, Germany, 2002; Chapter 1, pp 1–34.
40. Mingos, D. M. P. Theoretical Aspects of Microwave Dielectric Heating. In *Microwave-Assisted Organic Synthesis*; Lidström, P.; Tierney, J. P., Eds.; Blackwell: Oxford, UK, 2004, Chapter 1.
41. Baghurst, D. R.; Mingos, D. M. P. *Chem. Soc. Rev.* **1991**, *20*, 1–47.
42. Gabriel, C.; Gabriel, S.; Grant, E. H.; Halstead, B. S.; Mingos, D. M. P. *Chem. Soc. Rev.* **1998**, *27*, 213–223.
43. Neas, E.; Collins, M. Introduction to Microwave Sample. In *Preparation Theory and Practice*; Kingston, H. M.; Jassie, L. B., Eds.; American Chemical Society: Washington, DC, 1988.
44. Westaway, K. C.; Gedye, R. *J. Microwave Power Electromagn. Energy* **1995**, *30*, 219–230.
45. Langa, F.; de la Cruz, P.; de la Hoz, A.; Díaz-Ortiz, A.; Díez-Barra, E. *Contemp. Org. Synth.* **1997**, *4*, 373–386.
46. De la Hoz, A.; Díaz-Ortiz, A.; Moreno, A. *Chem. Soc. Rev.* **2005**, *34*, 164–178.
47. Perreux, L.; Loupy, A. *Tetrahedron* **2001**, *57*, 9199–9223.
48. Kuhnert, N. *Angew. Chem. Int. Ed.* **2002**, *41*, 1863–1866.
49. Strauss, C. R. *Angew. Chem. Int. Ed.* **2002**, *41*, 3589–3590.
50. Baghurst, D. R.; Mingos, D. M. P. *J. Chem. Soc. Chem. Commun.* **1992**, 674–677.
51. Saillard, R.; Poux, M.; Berlan, J.; Audhuy-Peaudecerf, M. *Tetrahedron* **1995**, *51*, 4033–4042.
52. Chemat, F.; Esveld, E. *Chem. Eng. Technol.* **2001**, *24*, 735–744.
53. Bogdal, D.; Lukasiewicz, M.; Pielichowski, J.; Miciak, A.; Bednarz, S. *Tetrahedron* **2003**, *59*, 649–653.
54. Lukasiewicz, M.; Bogdal, D.; Pielichowski, J. *Adv. Synth. Catal.* **2003**, *345*, 1269–1272.
55. Hajek, M. Microwave Catalysis in Organic Synthesis. In *Microwaves in Organic Synthesis*; Loupy, A., Ed.; Wiley-VCH: Weinheim, Germany, 2002; Chapter 10, pp 345–378.
56. Will, H.; Scholz, P.; Ondruschka, B. *Chem. Ing. Tech.* **2002**, *74*, 1057–1067.
57. Zhang, X.; Lee, C. S.-M.; Mingos, D. M. P.; Hayward, D. O. *Catal. Lett.* **2003**, *88*, 129–139.
58. Zhang, X.; Hayward, D. O.; Mingos, D. M. P. *Catal. Lett.* **2003**, *88*, 33–38.
59. Hayes, B. L. *Aldrichimica Acta* **2004**, *37*, 66–77.
60. Commarmot, R.; Didenot, R.; Gardais, J. F. (Rhone-Poulenc/Prolabo) Apparatus for wet chemical reaction on samples for analysis. French Patent 2560529, 1985. *Chem. Abstr.* **1986**, *105*, 17442.
61. Loupy, A.; Petit, A.; Hamelin, J.; Texier-Boullet, F.; Jacquault, P.; Mathé, D. *Synthesis* **1998**, 1213–1234.
62. Varma, R. S. *Green Chem.* **1999**, 43–55.
63. Kidawi, M. *Pure Appl. Chem.* **2001**, *73*, 147–151.
64. Varma, R. S. *Pure Appl. Chem.* **2001**, *73*, 193–198.

65. Varma, R. S. *Tetrahedron* **2002**, *58*, 1235–1255.
66. Varma, R. S. *Advances in Green Chemistry: Chemical Syntheses Using Microwave Irradiation*; Kavitha: Bangalore, India, 2002.
67. Laporterie, A.; Marquié, J.; Dubac, J. Microwave-Assisted Reactions on Graphite. In *Microwaves in Organic Synthesis*; Loupy, A., Ed.; Wiley-VCH: Weinheim, Germany, 2002; Chapter 7, pp 219–252.
68. Deshayes, S.; Liagre, M.; Loupy, A.; Luche, J.-L.; Petit, A. *Tetrahedron* **1999**, *55*, 10851–10870.
69. Loupy, A.; Petit, A.; Bogdal, D. Microwave and Phase-Transfer Catalysis. In *Microwaves in Organic Synthesis*; Loupy, A., Ed.; Wiley-VCH: Weinheim, Germany, 2002; Chapter 5, pp 147–180.
70. Will, H.; Scholz, P.; Ondruschka, B. *Chem. Ing. Tech.* **2002**, *74*, 1057–1067.
71. Zhang, X.; Lee, C. S.-M.; Mingos, D. M. P.; Hayward, D. O. *Catal. Lett.* **2003**, *88*, 129–139.
72. Zhang, X.; Hayward, D. O.; Mingos, D. M. P. *Catal. Lett.* **2003**, *88*, 33–38.
73. Van der Eycken, E.; Appukkuttan, P.; De Borggraeve, W.; Dehaen, W.; Dallinger, D.; Kappe, C. O. *J. Org. Chem.* **2002**, *67*, 7904–7909.
74. Kaval, N.; Dehaen, W.; Kappe, C. O.; Van der Eycken, E. *Org. Biomol. Chem.* **2004**, *2*, 154–156.
75. Miljanić, O. S.; Vollhardt, K. P. C.; Whitener, G. D. *Synlett* **2003**, 29–34.
76. Andappan, M. M. S.; Nilsson, P.; von Schenck, H.; Larhed, M. *J. Org. Chem.* **2004**, *69*, 5212–5218.
77. Breslow, R. *Acc. Chem. Res.* **1991**, *24*, 159–164.
78. Grieco, P. A. *Aldrichimica Acta* **1991**, *24*, 59–66.
79. Li, C.-J. *Chem. Rev.* **1993**, *93*, 2023–2035.
80. Li, C.-J.; Chan, T.-H. *Organic Reactions in Aqueous Media*; Wiley: New York, NY, 1997.
81. Lidström, U. M. *Chem. Rev.* **2002**, *102*, 2751–2772.
82. Bröll, D.; Kaul, C.; Krämer, A.; Krammer, P.; Richter, T.; Jung, M.; Vogel, H.; Zehner, P. *Angew. Chem.* **1999**, *111*, 3180–3196.
83. Savage, P. E. *Chem. Rev.* **1999**, *99*, 603–621.
84. Siskin, M.; Katritzky, A. R. *Chem. Rev.* **2001**, *101*, 825–836.
85. Katritzky, A. R.; Nichols, D. A.; Siskin, M.; Murugan, R.; Balasubramanian, M. *Chem. Rev.* **2001**, *101*, 837–892.
86. Akiya, N.; Savage, P. E. *Chem. Rev.* **2002**, *102*, 2725–2750.
87. Strauss, C. R.; Trainor, R. W. *Aust. J. Chem.* **1995**, *48*, 1665–1692.
88. Strauss, C. R. *Aust. J. Chem.* **1999**, *52*, 83–96.
89. Nolen, S. A.; Liotta, C. L.; Eckert, C. E.; Gläser, R. *Green Chem.* **2003**, *5*, 663–669.
90. Krammer, P.; Vogel, H. *J. Supercrit. Fluids* **2000**, *16*, 189–206.
91. Molteni, V.; Hamilton, M. M.; Mao, L.; Crane, C. M.; Termin, A. P.; Wilson, D. M. *Synthesis* **2002**, 1669–1674.
92. Vasudevan, A.; Verzal, M. K. *Synlett* **2004**, 631–634.
93. Baran, P. S.; O'Malley, D. P.; Zografos, A. L. *Angew. Chem. Int. Ed.* **2004**, *43*, 2674–2677.
94. Leadbeater, N. E.; Marco, M. *J. Org. Chem.* **2003**, *68*, 888–892.
95. Leadbeater, N. E.; Marco, M.; Tominack, B. J. *Org. Lett.* **2003**, *5*, 3919–3922.
96. Appukkuttan, P.; Dehaen, W.; Van der Eycken, E. *Eur. J. Org. Chem.* **2003**, 4713–4716.
97. Kaval, N.; Bisztray, K.; Dehaen, W.; Kappe, C. O.; Van der Eycken, E. *Mol. Divers.* **2003**, 7, 125–133.
98. Arvela, R. K.; Leadbeater, N. E.; Torenius, H. M.; Tye, H. *Org. Biomol. Chem.* **2003**, *1*, 1119–1121.
99. Bryson, T. A.; Stewart, J. J.; Gibson, J. M.; Thomas, P. S.; Berch, J. K. *Green Chem.* **2003**, *5*, 174–176.
100. Bryson, T. A.; Gibson, J. M.; Stewart, J. J.; Voegtle, H.; Tiwari, A.; Dawson, J. H.; Marley, W.; Harmon, B. *Green Chem.* **2003**, *5*, 177–180.
101. Westman, J.; Lundin, R. *Synthesis* **2003**, 1025–1030.
102. Wasserscheid, P.; Welton, T., Eds. *Ionic Liquids in Synthesis*; Wiley-VCH: Weinheim, Germany, 2003.
103. Leadbeater, N. E.; Torenius, E. M.; Tye, H. *Comb. Chem. High Throughput Screen.* **2004**, 7, 511–528.
104. Deetlefs, M.; Seddon, K. R. *Green Chem.* **2003**, *5*, 181–186.
105. Vo Thanh G.; Pegot, B.; Loupy, A. *Eur. J. Org. Chem.* **2004**, 1112–1116.
106. Cotterill, I. C.; Usyatinsky, A. Ya.; Arnold, J. M.; Clark, D. S.; Dordick, J. S.; Michels, P. C.; Khmelnitsky, Y. L. *Tetrahedron Lett.* **1998**, *39*, 1117–1120.
107. Strohmeier, G. A.; Kappe, C. O. *J. Comb. Chem.* **2002**, *4*, 154–161.
108. Nüchter, M.; Lautenschläger, W.; Ondruschka, B.; Tied, A. *LaborPraxis* **2001**, *25* (1), 28–31.
109. Nüchter, M.; Ondruschka, B.; Tied, A.; Lautenschläger, W.; Borowski, K. J. *Am. Gen./Proteom. Techn.* **2001**, *1*, 34–39.
110. Nüchter, M.; Ondruschka, B. *Mol. Divers.* **2003**, 7, 253–264.
111. Glass, B. M.; Combs, A. P. Rapid Parallel Synthesis Utilizing Microwave Irradiation. In *High-Throughput Synthesis. Principles and Practices*; Sucholeiki, I., Ed.; Marcel Dekker: New York, 2001; Chapter 4.6, pp 123–128.
112. Stadler, A.; Kappe, C. O. *J. Comb. Chem.* **2001**, *3*, 624–630.
113. Stadler, A.; Yousefi, B. H.; Dallinger, D.; Walla, P.; Van der Eycken, E.; Kaval, N.; Kappe, C. O. *Org. Process Res. Dev.* **2003**, 7, 707–716.
114. Iqbal, M.; Vyse, N.; Dauvergne, J.; Evans, P. *Tetrahedron Lett.* **2002**, *43*, 7859–7862.
115. Lehmann, F.; Pilotti, Å.; Luthman, K. *Mol. Divers.* **2003**, 7, 145–152.
116. Shackelford, S. A.; Anderson, M. B.; Christie, L. C.; Goetzen, T.; Guzman, M. C.; Hananel, M. A.; Kornreich, W. D.; Li, H.; Pathak, V. P.; Rabinovich, A. K. et al. *J. Org. Chem.* **2003**, *68*, 267–275.
117. Savin, K. A.; Robertson, M.; Gernert, D.; Green, S.; Hembre, E. J.; Bishop, J. *Mol. Divers.* **2003**, 7, 171–174.
118. Wilson, N. S.; Sarko, C. R.; Roth, G. *Org. Process Res. Dev.* **2004**, *8*, 535–538.
119. Shieh, W.-C.; Dell, S.; Repiè, O. *Org. Lett.* **2001**, *3*, 4279–4281.
120. Shieh, W.-C.; Lozanov, M.; Repiè, O. *Tetrahedron Lett.* **2003**, *44*, 6943–6945.
121. Pillai, U. R.; Sahle-Demessie, E.; Varma, R. S. *Green Chem.* **2004**, *6*, 295–298.
122. Bierbaum, R.; Nüchter, M.; Ondruschka, B. *Chem. Ing. Techn.* **2004**, *76*, 961–965.

Biography

Christian Oliver Kappe received his diploma degree (1989) and doctoral degree (1992) in organic chemistry from the Karl-Franzens-University in Graz, where he worked with Prof Gert Kollenz on cycloaddition and rearrangement reactions of acylketenes. After periods of postdoctoral research work on reactive intermediates and matrix isolation spectroscopy with Prof Curt Wentrup at the University of Queensland in Brisbane, Australia (1993–94), and on synthetic methodology–alkaloid synthesis with Prof Albert Padwa at Emory University in Atlanta, USA (1994–96) he moved back to the University of Graz in 1996 to start his independent academic career. He obtained his 'habilitation' in 1998 in organic chemistry, and currently holds the position of Associate Professor of Chemistry at the Karl-Franzens-University in Graz. In 2003, he spent a sabbatical at the Scripps Research Institute (La Jolla, USA) in the group headed by Prof K Barry Sharpless. He is currently a member of the Executive Board of the International Society of Heterocyclic Chemistry (ISHC), and a board member of the European Society of Combinatorial Sciences (ESCS). In addition he is the Editor of the journal *QSAR and Combinatorial Sciences*, and serves on the editorial/advisory boards of the *Journal of Combinatorial Chemistry*, *Molecular Diversity*, and a number of other journals. He is the co-author of more than 150 research publications in synthetic and mechanistic organic chemistry.

Comprehensive Medicinal Chemistry II
ISBN (set): 0-08-044513-6

ISBN (Volume 3) 0-08-044516-0; pp. 837–860

3.37 High-Throughput Purification

D B Kassel, Takeda San Diego, Inc., San Diego, CA, USA

3.37.1	**Introduction**	**861**
3.37.2	**Supporting Hit/Lead Generation**	**861**
3.37.3	**High-Performance Liquid Chromatography/Mass Spectrometry Analysis to Support Compound Characterization**	**862**
3.37.4	**Purity Assessment of Compound Libraries**	**863**
3.37.5	**Purification Technologies for Drug Discovery**	**864**
3.37.5.1	Ultraviolet-Directed Purification	864
3.37.5.2	Mass-Directed Preparative Purification	865
3.37.6	**Recent Innovations in High-Throughput Preparative High-Performance Liquid Chromatography/Mass Spectrometry**	**867**
3.37.6.1	Intelligent Method Development	867
3.37.6.2	Use of Monoliths and Turbulent Flow Media to Achieve Higher-Speed Serial Separations	867
3.37.6.3	Fluorous Split–Mix Library Synthesis and Preparative Liquid Chromatography/Mass Spectrometry Demixing	868
3.37.6.4	Parallel Analysis and Parallel Purification	868
3.37.7	**Streamlining the Purification Process**	**871**
3.37.8	**Conclusions**	**872**
References		**872**

3.37.1 Introduction

Focus within the pharmaceutical industry has been to increase the likelihood of successfully developing clinical candidates by optimizing each of the components of the discovery process (i.e., spanning target identification → chemical design → synthesis → compound analysis and purification → registration → biological and absorption, distribution, metabolism, and excretion screening. By optimizing each step in the iterative discovery process, it is expected that the compound attrition rate will be reduced dramatically as compounds advance from hit to lead to preclinical development. Although high-performance liquid chromatography/mass spectrometry (HPLC/MS) plays important roles throughout the discovery process, its role during the hit-to-lead stage of the discovery process has been perhaps the most significant. HPLC/MS is a key technology that has facilitated rapid, sensitive, selective, and high-resolution analysis and purification of compound libraries, and has become an indispensable tool for the medicinal chemist. Undeniably, advances in purification technology have helped to streamline the discovery phase of pharmaceutical drug discovery.

The field of combinatorial chemistry (encompassing parallel solution phase and solid phase synthesis) revolutionized the way in which chemists approach lead generation and lead optimization. Traditional tools to support the characterization and purification of these high-throughput synthesis techniques were simply not sufficient. Parallel solution phase and solid phase synthesis served as the catalyst for the analytical community to develop and implement new, faster, automated tools to support compound analysis and purification. Great advances have been made by the analytical community in the development of robust, automated purification technologies. A number of approaches have been employed successfully to purify compounds generated by high-throughput organic synthesis (HTOS) or parallel synthesis, and are presented in this chapter.

3.37.2 Supporting Hit/Lead Generation

The first step in the hit/lead generation process is to initiate high-throughput screening of a compound collection. Generally, pharmaceutical and biotechnology organizations initiate high-throughput screening against their corporate collections (compound archives). These corporate collections are generally diverse, but may be biased toward the specific therapeutic focus(es) of the organization. Consequently, the screening libraries are often augmented by the addition of commercially available screening libraries that are either gene family focused (G protein-coupled receptor-targeted libraries, kinase-targeted libraries, etc.) or are considered general diversity screening libraries. Further augmentation of the initial screening activities is to include custom synthesis compound libraries (typically produced by automated HTOS methodologies, such as those described by Nicolaou *et al.*[1]

One of the challenges with compound collections is that they are historical by nature. It is not uncommon for corporate collections of large pharmaceutical companies to include compounds that were synthesized more than 25 years ago. At the time of synthesis, it can be presumed that the compounds met the purity criteria for compound registration. However, it can also be presumed that a high likelihood exists that the compounds have degraded over the extended storage time. Another reason for the poor quality of compound collections is attributable to the fact that most compounds are stored as dimethyl sulfoxide (DMSO) stock solutions as opposed to storage as solid materials. Storage of compounds in DMSO is done primarily for the reasons that (1) DMSO is considered a 'universal' solvent and (2) solutions are much easier handle in plate-based high-throughput biological screening systems. However, the drawback to DMSO is that it is a very hygroscopic solvent, and unless the compounds are stored under inert conditions, they are prone to hydrolysis. Morand and co-workers evaluated the effect of freeze–thaw cycles on the stability of compounds stored in DMSO.[2]

Until very recently, with the introduction of high-throughput analytical technology, these compound sources were far too large to merit re-analysis and/or re-purification, and hence were screened 'as is.' The result was (and has been observed frequently) that hits could not be reconfirmed during follow-on bioassay screening, and subsequent evaluation of the compounds by techniques such as HPLC/MS and nuclear magnetic resonance (NMR) showed that the expected compound was not pure or, in some cases, completely absent! The adage 'garbage in, garbage out' became a mantra of many high-throughput screening laboratories, and forced companies to take a much more serious look at the quality of their compound collections. Morand and co-workers from Proctor and Gamble set out to fully assess the quality of their >500 000 compound corporate collection. They achieved this goal through the incorporation of a massive parallel flow injection/MS system, capable of analyzing a plate of samples in less than 2 min.[3] The throughput of their technique was one to two orders of magnitude faster than typical flow injection/MS systems used for reaction monitoring.[4]

In addition to quality control over compound collections, the issue of the purity of synthetic libraries derived using combinatorial chemistry quickly came under the microscope. In the early to mid-1990s, 'combi-chem' became a household word throughout the pharmaceutical industry, and was believed to be a key technology that would revolutionize drug discovery. The basis of combinatorial chemistry was the ability to perform split–mix synthesis on a solid support, and to take advantage of the combinatorial nature of the process to generate vast arrays of compounds. Combinatorial libraries were purported to be pure, owing to the fact that they were synthesized on a solid support and amenable to extensive washing to remove excess reagents, and therefore, directly amenable to high-throughput screening. However, these combinatorial libraries synthesized on a solid support suffered from the same problems that have long plagued solution phase synthesis, that is, the generation of unexpected and unwanted by-products. Due to the shear size of these compound libraries and the relatively small amounts available following resin cleavage, it was not possible to characterize, let alone purify, the expected products. Conventional split–mix combinatorial methods, though still popular with some bench chemists, have been replaced largely by the technique of directed parallel solution and parallel solid phase organic synthesis.

3.37.3 High-Performance Liquid Chromatography/Mass Spectrometry Analysis to Support Compound Characterization

Combinatorial chemistry paved the way for high-throughput, parallel organic synthesis techniques, now mainstream in the pharmaceutical and biotechnology industries for lead generation activities. The ability to synthesize compound libraries rapidly using automated solution phase and solid phase parallel synthesis has led to a dramatic increase in the number of compounds now available for high-throughput screening. The unprecedented rate by which compound libraries are now being generated has forced the analytical community to implement high-throughput methods for their analysis and characterization.

In the early 1990s, groups adopted high-speed, spatially addressable automated parallel solid phase and solution phase synthesis of discretes.[5–7] Both solution phase and solid phase parallel synthesis permit the production of large numbers as well as large quantities of these discrete compounds, eliminating the need for extensive de-coding of mixtures and re-synthesis following identification of 'active' compounds in high-throughput screening of combinatorial libraries. Importantly, parallel synthesis is performed readily in the microtiter plate format, amenable to direct biological screening, as was touched upon earlier. The relative ease of automation of parallel synthesis led to a tremendous increase in the number of compounds for lead discovery and lead optimization.

Almost all of the analytical characterization tools (e.g., HPLC, NMR, Fourier transform infrared (FTIR) spectroscopy, and LC/MS) are serial-based techniques, and parallel synthesis is inherently parallel. Consequently, this led rapidly to a new bottleneck in the discovery process (i.e., the analysis and purification of compound libraries). Parallel synthesis suffers from some of the same shortcomings of split and mix synthesis (e.g., the expected compound may not be pure, or even synthesized in sufficient quantities). The analytical community was faced with the decision of how to analyze these parallel synthesis libraries.

The traditional method for assessing compound purity has been to perform the following: purify the desired product to homogeneity by crystallization or column chromatography (e.g., reversed phase or normal phase HPLC); acquire a one- or two-dimensional NMR spectrum on the isolated product; obtain confirmatory molecular weight information by MS; perform a C, H, and N combustion analysis; generate an exact mass measurement (to within 5 ppm of the expected mass) by high-resolution MS; and determine the amount of isolated product by weight – all prior to compound submission and biological screening. In the era of high-throughput compound library synthesis, however, this extensive characterization is simply not possible. Therefore, groups have focused principally on a limited number of analytical measurements for compound identity and purity; in particular, LC/MS analysis incorporating orthogonal detection methods, such as ultraviolet (UV) and evaporative light-scattering detection (ELSD) and, in some instances, NMR.[8] The most commonly employed technique for characterizing compound libraries is to incorporate LC/MS with electrospray or atmospheric pressure chemical ionization with UV and ELSD, and more recently photoionization.[9]

LC/MS emerged as the method of choice for the quality control assessment to support parallel synthesis because the technique, unlike flow injection MS, provides the added measure of purity (and quantity) of the compound under investigation. In addition, 'universal-like' HPLC gradients (e.g., 10–90% acetonitrile in water in 5 min) have been found to satisfy the separation requirements for the vast majority of combinatorial and parallel synthesis libraries. Fast HPLC/MS has been found to serve as a good surrogate to conventional HPLC for assessing library quantity and purity.[10–13] Fast HPLC/MS is simple in concept. It involves the use of short columns (typically 4.6 mm i.d. × 30 mm in length) operated at elevated flow rates (typically 3–5 $mL\,min^{-1}$).

Typically, short columns are used for compound analysis because they allow for fast separations to be carried out at ultra-high flow rates. Also, these columns tend to be more robust than narrow-bore columns (1 mm and 2 mm i.d.) (i.e., less clogging is experienced and longer lifetimes are observed when these columns are subjected to unfiltered chemical libraries). A typical LC/MS analysis consists of the injection a small aliquot (10–30 μL) of the reaction mixture (total concentration of 0.1–1.0 $mg\,mL^{-1}$) and performing the separation using a 'universal' gradient of 10–90% buffer B in 2–5 min. Buffer A is typically H_2O containing 0.05% trifluoroacetic acid (or formic acid), and buffer B is typically acetonitrile containing 0.035% trifluoroacetic (or formic acid). HPLC columns are operated typically at flow rates of 3–5 $mL\,min^{-1}$ (depending on their dimensions), and the cycle time between injections is 3–5 min. Independent reports by Kyranos and Tiller[14,15] suggest that ballistic gradients and 'pseudo-chromatography' (in essence, step elution chromatography) provide a very rapid and reliable assessment of the quality of library synthesis.

3.37.4 Purity Assessment of Compound Libraries

The issue of compound purity has received a great deal of attention over the last several years as more and more chemists have adopted HTOS protocols but are unwilling to compromise the quality of the molecules submitted for biological evaluation. The general consensus target purity of a compound library compound before it is to be archived or screened for biological activity is between 90% and 95% pure. This purity criterion is more stringent than in the past, where 85–90% (based on UV detection) was considered acceptable. This may be attributed primarily to a shift toward smaller, focused (or biased) libraries than larger, diverse collections of compounds. The majority of mass spectrometer manufacturers now offer software packages that aid in the automatic determination of purity.

UV chromatograms are typically used, rather than the total ion current chromatogram, to assess purity. This is because the total ion current chromatogram is a measure of the 'ionizability' of a compound, which is well known to vary dramatically from one compound to the next. Orthogonal detection methods, such as chemical luminescence nitrogen detection (CLND)[16] and ELSD,[17,18] have been proposed to be more universal detection methods, and hence are also used to assess reaction yields and estimate purity.

CLND has been demonstrated to be a valuable tool for quantifying low quantities of material, and has been shown to be particularly well suited to normal phase HPLC and supercritical fluid chromatography/MS, for the principal reason that separations are carried out using solvents that do not contain nitrogen (i.e., CO_2 and CH_3OH). ELSD

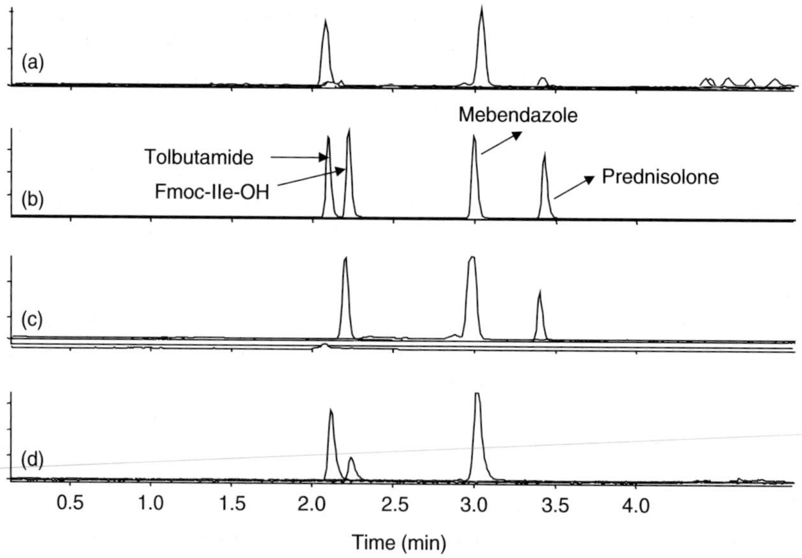

Figure 1 Separation of a four-component library. (a) Total ion current chromatogram showing that two of the four components ionize efficiently under electrospray ionization conditions. (b) ELSD chromatogram of the four components, all showing a comparable response. (c) The UV chromatogram (254 nm) shows some selectivity in detection, as does (d) CLND detection. The column flow rate was 5 mL min^{-1}. A portion of the column effluent was split to each of the three detectors (CLND, 200 μL min^{-1}; ELSD, 200 μL min^{-1}; MS, 100 μL min^{-1}). A make-up flow of 50/50 MeOH/H_2O (300 μL min^{-1}) was added to the flow stream diverted to the mass spectrometer ion source. Mass spectra were acquired using electrospray ionization with no special modifications to the ion source.

measures the mass (quantity) of the material directly, is often presented as being a molecular weight-independent detector, and is a tool that has gained wide-scale acceptance for the on-line quantification of compound libraries. An example of a separation of a four-component library incorporating UV, ELSD, CLND ,and MS detection is shown in **Figure 1**. Using these various detectors, chemists are able to obtain measures of purity of their libraries with greater confidence than when relying solely upon HPLC/UV/MS data. That being said, HPLC/MS (coupled with UV and ELSD detection) has been adopted as the method of choice for assessing the quality and quantity of material prepared by parallel synthesis techniques, and compound purity is most often reported by averaging the purities determined from the UV and ELSD responses.

3.37.5 Purification Technologies for Drug Discovery

Historically, it was believed that solid phase synthesis protocols would eliminate the need for purification because excess reagents could be removed readily by extensive washing. Unfortunately, even for solid phase peptide synthesis, the final products, acid cleaved from the resin, are found to be far from pure. Furthermore, parallel solution phase synthesis has found greater popularity, because it is readily automated and extends the 'portfolio' of reactions available to the chemist for high-throughput parallel synthesis. The limitations with solid phase synthesis and the movement toward parallel solution phase synthesis forced numerous groups to evaluate and implement a variety of purification strategies.

Numerous techniques are available to the organic chemist to support library purification. Techniques including liquid–liquid extractions,[19,20] liquid–solid extractions,[21–26] fluorous extractions,[27] and scavenger 'capture and release' resins[28–30] are fast and readily automated; however, they do not consistently provide satisfactory final product purity to support biological screening.[26,31] Because HPLC is a high-resolution separation method and is capable of being fully automated, it has become the method of choice for purifying compound libraries. Yang and co-workers published an extensive review touching on all of the above methods for compound library purification strategies.[32] This review focuses exclusively on HPLC- and HPLC/MS-based purification methods.

3.37.5.1 Ultraviolet-Directed Purification

Both activity and inactivity data are being used increasingly to generate structure–activity relationships and direct synthetic efforts. Consequently, organizations have recognized the importance of confirming the purity of compounds prior to biological screening, not only confirming the purity of those compounds for which activity is observed. In order to minimize false positives and false negatives, it is advantageous to assay only high-quality compounds. Therefore, great effort has been devoted to the development of automated purification technology designed to keep pace with the output of high-throughput combinatorial/parallel synthesis.

Automated methods are now available to the chemist to perform high-throughput purification. Although HPLC has long been a method available to the chemist for product purification, only recently were these systems developed for high-throughput operation. Weller and co-workers were one of the first groups to demonstrate 'walk-up' high-throughput purification of parallel synthesis libraries based on HPLC and UV detection.[33] An open-architecture software interface enabled chemists to select the appropriate separation method from a pull-down menu and initiate an unattended automated reversed phase UV-based fraction collection. Fractionation was achieved using a predetermined UV threshold. Multiple fraction collectors were coupled to one another so as to provide a sufficient footprint for fraction collection. Since the early work of Weller *et al.*, a number of commercial systems have been introduced for walk-up preparative LC/UV purification (including Gilson, Hitachi, and Shimadzu, to name but a few).

One of the challenges associated with UV-based purification systems is that multiple fractions are collected for every sample injected. Although user-defined adjustable triggering parameters (e.g., UV thresholds for initiating and terminating fraction collection) can be used to reduce the total number of fractions, all, to some extent, will contain impurities. The exact number of chromatographic peaks for a given sample will be hard to predict, and, therefore, the footprint for fraction collection will also be difficult to predict. Personal experience purifying compound libraries has shown that it is not uncommon for 5–10 fractions to be collected per injection. When purifying only a small number of samples (<10), it is neither too cumbersome to collect the fractions nor too time-intensive to perform postpurification analysis so as to identify the fraction(s) containing the desired product. However, when attempting to purify compound libraries (e.g., 96-well plates of samples), the number of fractions and the time it takes to identify the relevant fraction(s) fast becomes a bottleneck. Schultz and co-workers addressed the fraction collection issue, and streamlined post-fraction collection processing (including evaporation, reconstitution, and postpurification analysis) by collecting fractions directly into 48-well microtiter plates.[34] Their method was particularly well suited to semipreparative purifications (using smaller inner diameter columns to support low milligram quantities).

In order to gain further efficiencies into UV-based purification of compound libraries, numerous groups have developed automated high-throughput UV-based purification systems coupled with on-line MS detection. Kibbey was one of the first scientists to implement a fully automated preparative LC/MS system for combinatorial library purification.[35] His approach was to perform a scouting analytical run prior to purification so as to optimize the chromatographic method and fraction collection parameters. Fraction location and molecular weight information were captured through a custom laboratory information management system. The added MS information greatly facilitated deconvolution of collected fractions, and streamlined their purification process. Hochlowski described a service-based purification factory incorporating UV and ELSD detection coupled with MS that supports purification of over 200 compounds per day.[36]

More recently, intelligent UV-based systems for preparative scale purification of combinatorial libraries have been introduced, utilizing knowledge of the retention time of the expected product based on a pre-analytical evaluation followed by preparative HPLC with UV-based fractionation using a narrow time collection window. Yang and co-workers developed the 'accelerated retention window' method as a tool for improving high-throughput purification efficiency.[37] In this method, a high-throughput parallel LC/MS analysis is performed prior to preparative purification to confirm that the expected product is indeed contained within the well and to identify the approximate retention time of the expected synthetic product. Only those compounds found to be $>10\%$ pure based on the analytical run are candidates for final product purification. Further, the information from the high-throughput parallel analysis is uploaded to a stand-alone preparative LC system for final product purification. Fraction collection is initiated only during the window of time corresponding to the retention time of the compound (identified from the analytical LC/MS analysis), so as to reduce the number of fractions collected during the preparative HPLC analysis. Additional refinements of UV-based purification strategies have been made recently, allowing for further simplification of the fraction collection and post-purification analysis step. In one embodiment, Karancsi and co-workers implemented a 'main component' fraction collection method based on UV triggering that supports the ideal of high-throughput purification – the one compound/one fraction concept.[38]

3.37.5.2 Mass-Directed Preparative Purification

The technique of preparative LC/MS, introduced in the 1990s,[39–43] was the first technique to greatly simplify the purification process. Preparative LC/MS methods for the first time allowed the concept of one compound/one fraction to be realized.[40] In the preparative LC/MS mode, the mass spectrometer is used as a highly selective detector for mass-directed fractionation and isolation. This technique provides a means for reducing dramatically the number of HPLC fractions collected per sample and virtually eliminates the need for postpurification analysis to determine the mass of the UV-fractionated compound. Preparative LC/MS is now widely used in the pharmaceutical industry. Systems for preparative LC/MS are configured and operated in numerous ways, including an expert user mode, walk-up or open access mode, or in a project team setting, supporting small teams of chemists working on similar chemistries. All components of the system are under computer control, and are hence truly automated. Components of these systems are nearly identical to stand-alone HPLC systems, with the addition of a flow splitter device to divert a small portion of the column flow to the mass spectrometer for on-line detection and fraction collector triggering. Typical systems are configured in an automated analytical/preparative mode of operation. In this configuration, the chemist is able to select between a variety of column sizes for either analytical, semipreparative, or preparative separations. The HPLC instrument, switching valves, mass spectrometer, and fraction collector are under complete computer control, as shown in **Figure 2**. In some instances, a solvent pump is added to deliver a methanol make-up flow to the mass spectrometer. The flow splitter and extra solvent pump serve the primary purpose of reducing the potential for overloading of sample into the ion source. An advantage of the flow splitter and make-up pump is that it reduces the concentration of trifluoroacetic acid (ion pairing) in the ion source, which can affect the sensitivity of detection for acidic library components.

The very first published example of a mass-guided fractionation of a combinatorial library is shown in **Figure 3**. In this example, the crude reaction product is only about 30% pure. The component of interest shows a prominent single chromatographic peak when monitoring specifically for its corresponding mass. Postpurification analysis of this singly isolated fraction (based on mass-directed fractionation) demonstrates that the compound of interest was purified to greater than 90%. Had a UV-based fractionation system been used in this particular example, at least five individual fractions would have been isolated. Extending this to a 96-component library synthesized in the microtiter plate format (and assuming this compound was representative of the quality of the members of the library), a UV-based approach would have led to approximately 400–500 fractions, requiring reanalysis to pinpoint the desired product. This would not only be a time-consuming re-analysis process but would require significant time to transfer the appropriate fractions to a screening plate for biological assessment.

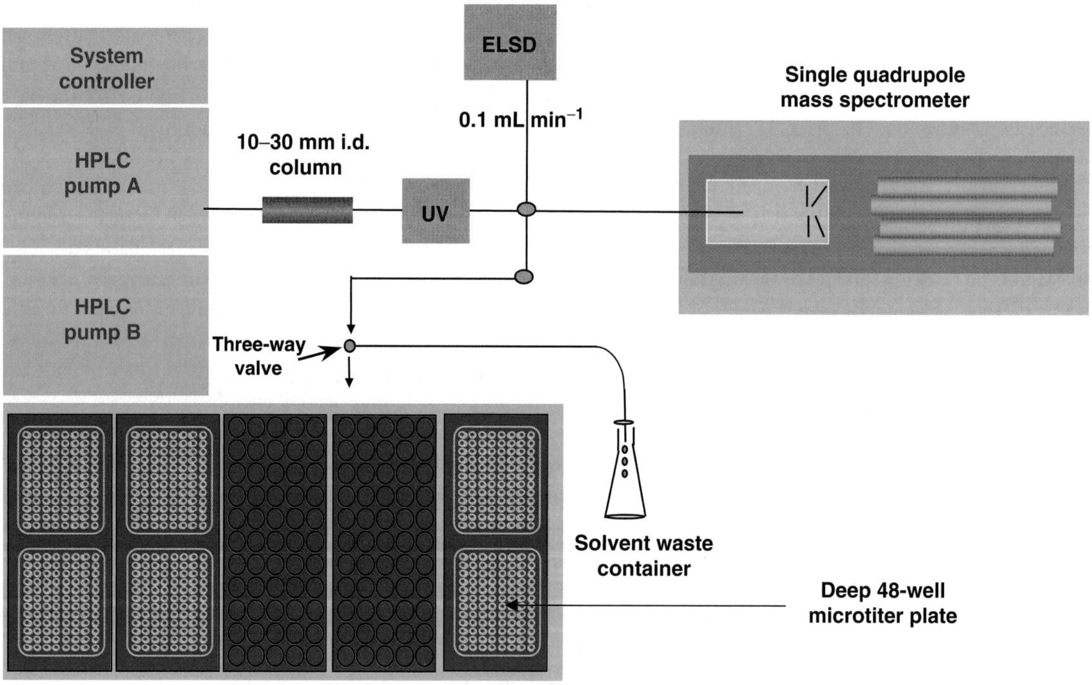

Figure 2 Preparative LC/MS systems on the market consist of a binary HPLC system, a combined autosampler/fraction collector (footprint of a Gilson 215 inject/collect liquid handler shown here), a switching valve to allow for analytical or preparative column selection, and a single quadrupole mass spectrometer.

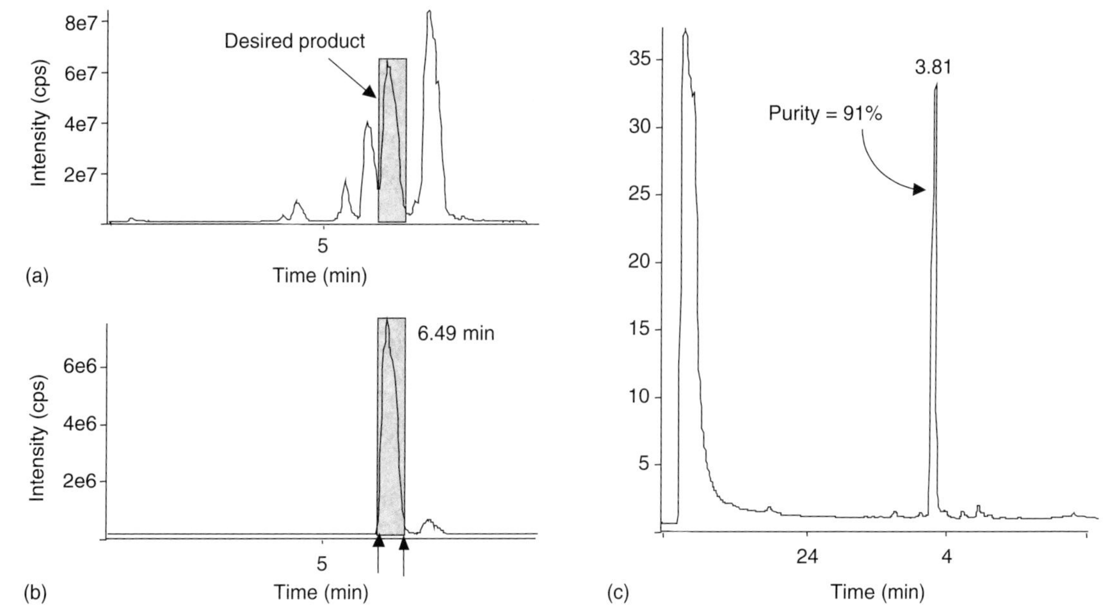

Figure 3 Mass-guided fractionation of a combinatorial library. Crude reaction product (50 mg) was solubilized in 1 mL of 50/50 MeOH/DMSO, and injected onto a 20 mm × 50 mm i.d. reversed phase column. Separation was achieved using a gradient of 10–90% acetonitrile in 7 min. (a) The total ion chromatogram shows five components well separated. (b) The extracted ion chromatogram (XIC) for the expected product shows a single, prominent peak at 6.49 min. Fraction collection was initiated and terminated, as indicated by the arrows directly below the XIC peak. (c) Post-purification analysis of the isolated component shows the compound was purified to approximately the 90% level.

Today, just about every major mass spectrometer instrument vendor offers mass-directed purification capabilities, and, through competition, and following years of development and improvements to hardware and software, these systems are now robust, affordable, and prevalent throughout the pharmaceutical industry.

3.37.6 Recent Innovations in High-Throughput Preparative High-Performance Liquid Chromatography/Mass Spectrometry

3.37.6.1 Intelligent Method Development

Because of the large number of samples generated by parallel synthesis there is little time available to optimize the chromatographic method for each compound, and, consequently, a single generic chromatographic method is selected for purifying the library. Although this approach works very well, in many instances a small percentage of compounds (typically 5–15%) fail to achieve a satisfactory level of purity. To address this, vendor software packages have been introduced that allow for intelligent selection of a preparative purification method based on the retention time of the target compound identified from a prepurification analysis. These software packages require the analyst to perform a pre-purification analysis of each of the compounds contained within the parallel synthesis library using a generic gradient (e.g., 10–90% acetonitrile in 3 min). Automated postdata acquisition processing identifies (1) the retention time of the expected product and (2) the percentage of organic modifier required to elute the desired product from the reverse phase column, and, using these parameters, automatically identifies (from an established list of methods) the most suitable preparative chromatographic method to support purification. The overhead associated with this approach is the necessity to perform a prepurification analysis. The benefit is that by automated selection of a purification method based on the observed retention time, the likelihood of isolating the compound with acceptable purity increases substantially.

3.37.6.2 Use of Monoliths and Turbulent Flow Media to Achieve Higher-Speed Serial Separations

The type of chromatography column most commonly used for the high-throughput purification of compound libraries has been 20 mm × 50 mm or 30 mm × 75 mm C18 reverse phase columns containing 5 μm particles. The columns are very robust (i.e., with appropriate care and use of guard columns), the C18 packing material works very well for the vast majority of drug-like molecules, the size of the columns permit multimilligram loadings, and the 5 μm particles allow for moderate- to high-resolution separations. One of the major challenges, though, is maintaining chromatographic integrity as a function of column loading. Very recently, monoliths and turbulent flow media have been evaluated for the high-throughput purification of compound libraries. Goetzinger and co-workers recently presented purification results for separations carried out on semipreparative monolith columns (10 mm × 100 mm length columns).[44] Owing to their mechanical stability and low operating backpressures, it was possible to reduce run times by an average of 25% while achieving superior separations to conventional C18 semipreparative columns. At the same meeting, Quinn and co-workers[45] showed purification results obtained using turbulent flow chromatography columns. Impressively, they were able to demonstrate very high loading capacity on 4.6 mm × 100 mm turbulent flow columns (200 mg loadings) while maintaining baseline separation for a mixture of drug substances.

3.37.6.3 Fluorous Split–Mix Library Synthesis and Preparative Liquid Chromatography/Mass Spectrometry Demixing

The interest in combinatorial chemistry (split–mix technology) as a means for generating large compound collections has plummeted over the past several years, due primarily for the reasons of poor quality control over synthesis and complex decoding strategies. Conventional split–mix technology was replaced by a number of parallel synthesis strategies that did not require the demixing or decoding step. Recently, a relatively new technique, so-called fluorous synthesis, developed by Curran[46] offers an interesting twist on the conventional split–mix combinatorial chemistry approach. In fluorous synthesis, a mixture of substrates is paired with a series of perfluoroalkyl phase tags is taken through multiple synthetic reaction steps, culminating in a mixture of tagged products. Demixing is achieved by tag-controlled fluorous HPLC preparative separation, the fractions stripped of solvent and the tag removed to yield individually purified and structurally distinct products. A recent publication by Zhang and co-workers.[47] describes both the synthetic and analytical HPLC strategy for a 420-component fluorous library. Kassel and Zeng, in collaboration with Zhang and co-workers, recently evaluated a higher-throughput mass-directed purification method to support the purification of fluorous tagged libraries.[48] **Figure 4** shows the UV and total ion current chromatogram for a pool of five compounds from the 420-component library. Separation of the five-component mixture was achieved in less than 6 min. Fraction collection was initiated when the expected $[M+H]^+$ ion for each of the tagged products exceeded the preset ion intensity threshold, and fraction collection was terminated when the ion intensity for the expected product(s) dropped below a second preset ion intensity threshold value. Combining split–mix fluorous synthesis with high-speed chromatography provides a means for rapidly generating large numbers and large quantities of highly purified drugable molecules.

3.37.6.4 Parallel Analysis and Parallel Purification

The synthetic throughput achievable by the medicinal chemist (having adopted parallel synthesis strategies) has rendered analysis and purification a rate-limiting step in the discovery process. Although advances in sample analysis throughput have been clearly demonstrated, there is a limit as to how fast a separation and analysis can be achieved while maintaining good separation efficiency and quality analysis. A number of techniques have been developed to increase throughput without compromising column chromatography, such as rapid column switching and regeneration systems for enhanced throughput serial based analysis and parallel chromatography methods.

Rapid column switching is achieved through a simple modification to the LC/MS instrumentation. Essentially, all that is required to achieve rapid column switching is to incorporate a set of switching valves and a third pump to reduce cycle time between injections, as shown in **Figure 5**. While one column is being used to perform the LC/MS analysis, the other column is being regenerated. An alternative use of 10-port switching valves is to allow for rapid serial sampling between columns. This technique works well for samples that are amenable to either isocratic or step elution. While one sample is being loaded onto one column, the contents of the other column are eluted into the ion source.

In order to increase sample throughput while maintaining high-quality analytical data, groups have begun to perform separations in parallel. Independently, Di Biasi *et al.*[49] and Kassel and co-workers[50] developed parallel MS sample introduction techniques. By performing analyses in parallel, chromatographic integrity can be maintained while effectively addressing sample throughput. These two groups presented novel ion source interfaces enabling 4–8 samples to be processed in parallel, thereby increasing the sample analysis throughput dramatically over conventional, serial-based LC/MS analyses. Commercially available parallel spray interfaces consist of a multiple spray head assembly and a blocking device (e.g., a rotating plate), enabling individual sprayers to be sampled at specific and defined time

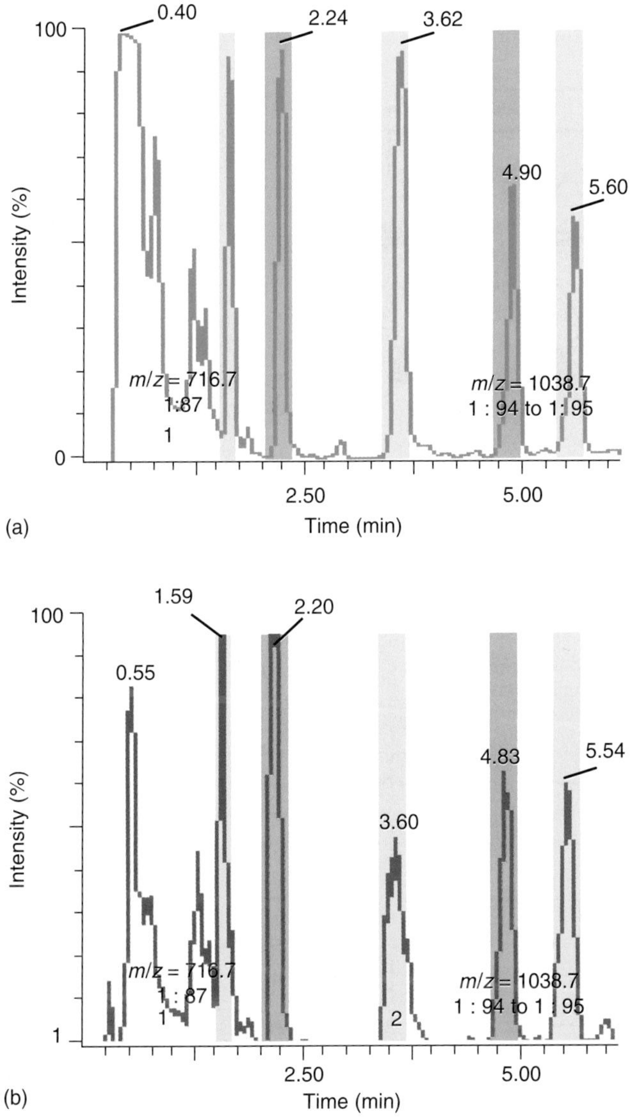

Figure 4 A five-component fluororus split–mix crude reaction mixture was injected into a 20 mm × 50 mm i.d. reversed phase column. (a) UV chromatogram and (b) Total ion current chromatogram. Compounds were purified using mass-directed fraction collection (peaks highlighted).

intervals. Although the multiple sprays are delivered to the mass spectrometer simultaneously, they are sampled in a time-dependent manner.

Performing parallel analysis of compound libraries offers many potential advantages over serial-based LC/MS analytical methods, the most obvious of which is dramatically increased compound analysis throughput. Using single-channel HPLC-based purification systems, routine sample throughput of up to 192 reaction mixtures per day was reported.,[51] With parallel HPLC systems, it has been reported that the theoretical throughput increases to 384 samples per day for a two-channel system, and to 768 samples per day for a four-channel system.

Parallel LC/MS analysis is achieved incorporating one or more of the following: a parallel autosampling device, a set of HPLC pumps configured to divert flow through an array of HPLC columns, a parallel UV detector, an array of analog detectors (e.g., ELSD), and a mass spectrometer configured either with or without an indexing device (i.e., a multiplexed (MUX) ion source) to facilitate independent sampling of individual sprayers. Some groups have found it possible to use a binary HPLC pumping system and split the flow equally between the array of columns using a simple Valco tee.

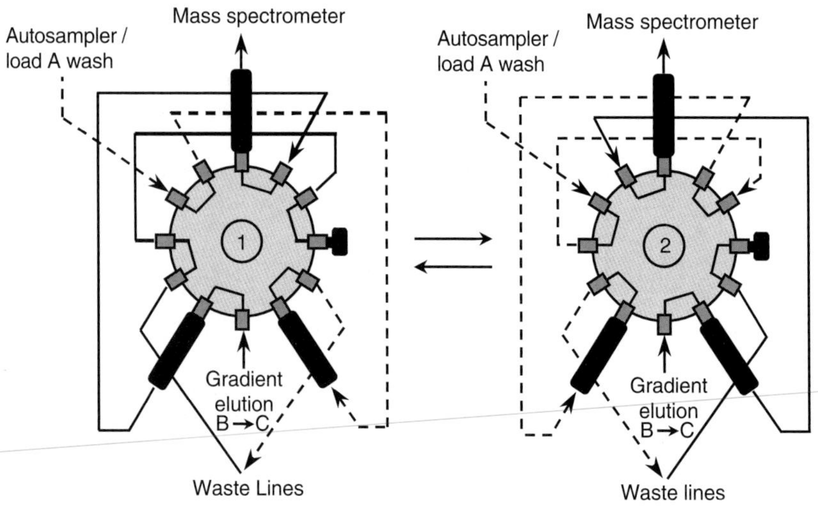

Figure 5 Schematic representation of a column switching configuration to support analysis from one column while the second column is equilibrating.

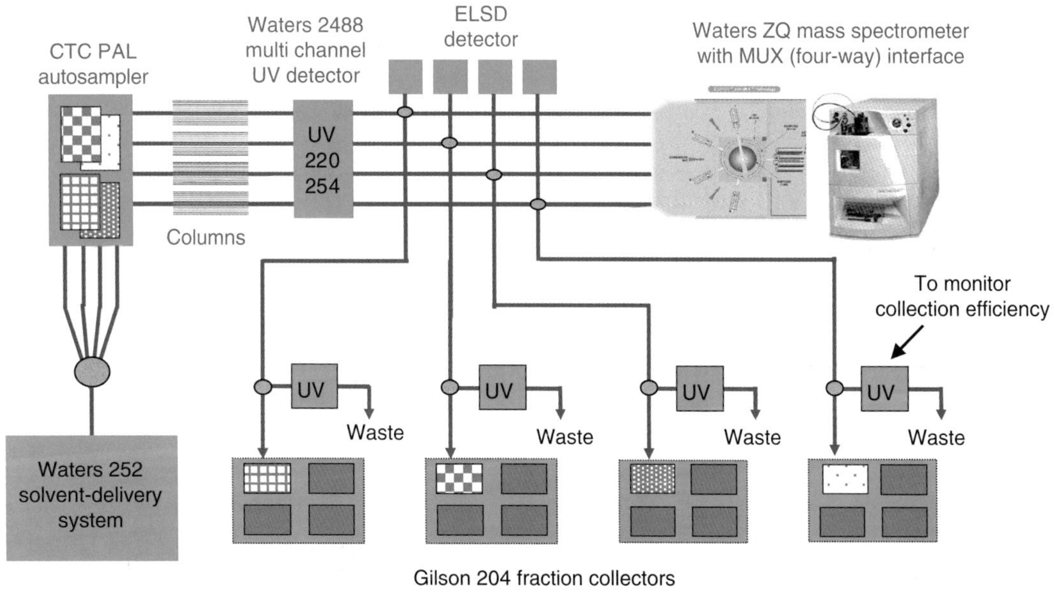

Figure 6 A four-channel LC/MS system supporting higher-throughput analysis and purification. A Waters 2525 solvent-delivery system is used to deliver flow to an array of C18 analytical or preparative columns.

A commercially available system configured for parallel analysis and purification is shown schematically in **Figure 6**. The four-channel parallel LC/MS purification system consists of a binary HPLC system, an autosampler configured with four injectors, a multichannel UV detector, a quadrupole mass spectrometer equipped with an MUX ion source that monitors four flow streams simultaneously, and four independently controlled fraction collectors. The binary HPLC pump is operated at a total flow rate either at 40 mL min^{-1} or 80 mL min^{-1} and the flow is split equivalently into four 10 mL min^{-1} or 20 mL min^{-1} flow streams, respectively, with the aid of a Valco manifold tee.

Each of the four streams passes through its own injection valve on the parallel autosampler and into the HPLC column array. The effluent from each column passes through the multichannel UV detector, where the separate chromatographic flow streams are monitored at two UV wavelengths (λ_{220} and λ_{254}). The column effluent is diverted through a parallel flow splitter unit at the outlet of the parallel UV detector. This parallel splitter diverts the majority of flow (99.5%) from each of the column effluents toward dedicated Gilson 204 fraction collectors. The remainder of the flow (i.e., 50 μL min^{-1} or 0.5% of each of the total column flows) is merged with one of the four 0.18 mL min^{-1} methanol streams, and detected in the MUX ion source. Xu *et al.* recently described advances in this laboratory for MUX-based parallel preparative LC/MS purification of compound libraries.[52] In this work, each of the flow streams is sampled in a time-dependent manner, and each of the individual sprayers and their associated ion signals are linked to unique fraction collectors, to facilitate independent parallel purification.

A limitation to this approach is duty cycle. The duty cycle for mechanical blocking devices is approximately 0.6 s (50 ms rotation time between each position of the spray assembly and 100 ms dwell/acquisition time at each of the spray positions). For preparative LC/MS analysis, peak widths of 5–10 s for analytical runs and 10–30 s for preparative analyses are routinely observed. Thus, this duty cycle should have only limited impact on compound analysis at present. Samples are sequentially transferred through the various valve positions, providing temporally spaced flow streams into the ion source. The duty cycle and cross-contamination between channels were not discussed in this preliminary report.

In order for parallel MS to become more mainstream, the chromatographic inlet side will require further development. Many groups have achieved flow splitting by incorporating a simple manifold splitting tee. This approach requires each column to be maintained at nearly identical backpressures in order to maintain constant and identical flow through each of the columns in the array. To achieve constant and identical flow through the array of parallel columns not only requires excellent quality control over column selection but careful plumbing as well. Further, the variation in backpressure becomes even more predominant in a gradient run.

Concomitant with the advances being made in the MS detection of parallel flow streams are the advances being made in parallel sampling and parallel chromatographic separations. Coffey and co-workers pioneered the Biotage Parallex HPLC system, a fully automated, high-throughput organic chemistry parallel purification system with parallel UV detection. In collaboration with scientists from GlaxoSmithKline, Coffey described an elegant parallel chromatography system coupled with on-line MS characterization, to support combinatorial library purification.[53] The Biotage Parallex HPLC system catalyzed the development of a number of parallel chromatography workstations now available for both intermediate and final product purification. Parallel HPLC systems now available commercially consist of multiple HPLC pumps, multiprobe autosamplers, parallel UV detection, and parallel mass spectrometer ion source interfaces. In addition to the Biotage system, there are three other commercially available systems available to support parallel analysis and purification, from Sepiatec, Inc. (Germany), Nanostream, Inc. (USA), and Eksigent (USA). The instrument vendors are continually being challenged by their customers to introduce more cost-effective and more compact systems.

3.37.7 Streamlining the Purification Process

The analytical community has focused primarily on developing higher-throughput analysis and purification tools that keep pace with parallel synthesis. The problem, however, is not just to provide fast methods for characterizing and purifying compound libraries: true high-throughput purification is only possible if the entire purification process is managed effectively. Between compound synthesis and compound registration are several important steps in addition to purification, including automated analysis, postdata acquisition processing, purification, postpurification analysis and reformatting for biological screening, and compound archiving. Following synthesis, a decision needs to be made as to which compounds should be (1) submitted directly for registration (2) purified prior to submission, or (3) discarded outright. This is accomplished by performing a high-throughput analysis of the reaction mixture, most typically by LC/MS with UV and/or evaporative light-scattering detection, as a method for assessing compound purity. Automated data processing facilitates the selection of compounds requiring purification prior to compound submission. Following library purification, a decision needs to be made as to which fractions should be submitted for postpurification LC/MS analysis, to verify purity and assess sample quantity. Again, sophisticated sample tracking and automated sample list generation greatly facilitates the next step in the process. Ultimately, the postpurification LC/MS data are collated (electronically), and those compounds satisfying both the purity and quantity criteria set by the project team (or organization) are registered and plated for biological screening.

Isbell and co-workers described both the steps and time involved from the receipt of crude samples for purification, through purification, postpurification analysis, reformatting, dissolution, and registration of the 'acceptably pure' fractions.[54] In their work, a custom parallel LC/MS purification system was used for all analytical and preparative analyses. The robustness of the parallel LC/MS system was assessed, purifying over 7800 reaction mixtures from several libraries over a period of 7 months. Two significant benefits were realized from this technology: the ability to collect only fractions likely to contain the product of interest and the ability to more easily track fractions due to the optional plate mapping feature of the system. When using technologies designed to facilitate fraction tracking and automated liquid handling, the time required to completely process samples, from receipt of sample plates for purification to submitting properly formatted plates of pure compounds for registration, required significantly more time than the LC/MS purification step itself. Overlooking these other important steps in the overall purification process may seriously underestimate the amount of time needed to purify a library, and may have serious implications on plans focused on a large-scale purification effort.

3.37.8 Conclusions

Undeniably, one of the greatest challenges of analytical chemistry in an era of high-throughput drug discovery has been to balance the need for high throughput while maintaining an analytical standard of high quality. Significant advances in sample analysis and purification throughput have been achieved, incorporating fast chromatography coupled with MS, and by performing these analyses both in series and in parallel. Throughput has also been impacted dramatically by the ability to seamlessly link all of the processes downstream of library synthesis, including automating the uploading of sample lists for automated data acquisition, automating the assessment of compound purity, and automating postpurification analysis and associated sample handling.

References

1. Nicolaou, K. C.; Pfefferkorn, J. A.; Mitchell, H. J.; Roecker, A. J.; Barluenga, S.; Cao, G. -Q.; Allfeck, R. L.; Lillig, J. E. *J. Am. Chem. Soc.* **2000**, *122*, 9954–9967.
2. Kozikowski, B. A.; Burt, T. M.; Tirey, D. A.; Williams, L. E.; Kuzmak, B. R.; Stanton, D. T.; Morand, K. L.; Nelson, S. L. *J. Biomol. Screen.* **2003**, *8*, 210–215.
3. Morand, K. L.; Burt, T. M.; Regg, B. T.; Chester, T. L. *Anal. Chem.* **2001**, *73*, 247–252.
4. Greaves, J. *J. Mass Spectrom.* **2002**, *37*, 777–785.
5. Hogan, J. *Nature* **1996**, *384*, 17–19.
6. Parlow, J. J.; Normansell, J. E. *Mol. Divers.* **1996**, *1*, 266–269.
7. Siegel, M. G.; Shuker, A. J.; Droste, C. A.; Hahn, P. J.; Jesudsan, D. C.; McDonald, J. H. I.; Matthews, D. P.; Rito, C. J.; Thorpe, A. J. *Mol. Divers.* **1997**, *3*, 113–116.
8. Lindon, J. C.; Nicholson, J. K.; Wilson, I. D. *J. Chromatogr. B.* **2000**, *748*, 233–258.
9. Syage, J. A.; Nies, B. J.; Evans, M. D.; Hanold, K. A. *J. Am. Soc. Mass Spectrom.* **2001**, *12*, 648–655.
10. Enjabal, C.; Martinez, J.; Aubagnac, J. L. *Mass Spectrom. Rev.* **2000**, *19*, 139–161.
11. Hughes, I.; Hunter, D. *Curr. Opin. Chem. Biol.* **2001**, *5*, 243–247.
12. Kyranos, J. N.; Cai, H.; Wei, D.; Goetzinger, W. K. *Anal. Biotechnol.* **2001**, 105–111.
13. Kyranos, J. N.; Lee, H.; Goetzinger, W. K.; Li, L. Y. *J. Comb. Chem.* **2004**, *6*, 796–804.
14. Goetzinger, W. K.; Kyranos, J. N. *Am. Lab.* **1998**, *30*, 27–37.
15. Romanyshyn, L. A.; Tiller, P. R. *J. Chromatogr. A* **2001**, *928*, 41–51.
16. Taylor, E. W.; Jia, W.; Bush, M.; Dollinger, G. D. *Anal. Chem.* **2002**, *74*, 3232–3238.
17. Kibbey, C. E. *Mol. Divers.* **1996**, *4*, 247–258.
18. Hsu, B. E.; Orton, E.; Tang, S. Y.; Carlton, R. A. *J. Chromatogr. B* **1999**, *725*, 103–112.
19. Peng, S. X.; Henson, C.; Strojnowski, M. J.; Golebiowski, A.; Klopfenstein, S. R. *Anal. Chem.* **2000**, *72*, 261–266.
20. Cheng, S.; Tarby, C. M.; Comer, D. D.; Williams, J. P.; Caporale, L. H.; Myers, P. L.; Boger, D. L. *Bioorg Med Chem.* **1996**, *4*, 727–737.
21. Parlow, J. J.; Vazquez, M. L.; Flynn, D. L. *Bioorg. Med. Chem. Lett.* **1998**, *8*, 2391–2394.
22. Moore, J. D.; Harned, A. M.; Henle, J.; Flynn, D. L.; Hanson, P. R. *Org. Lett.* **2002**, *4*, 1847–1849.
23. Hodges, J. C.; Harikrishnan, L. S.; Ault-Justus, S. *J. Comb. Chem.* **2000**, *2*, 80–88.
24. Parlow, J. J.; Flynn, D. L. *Tetrahedron* **1998**, *54*, 4013–4031.
25. Rabinowitz, M.; Seneci, P.; Rossi, T.; Dal Cin, M.; Deal, M.; Terstappen, G. *Bioorg. Med. Chem. Lett.* **2000**, *10*, 1007–1010.
26. Boger, D. L.; Goldberg, J.; Jiang, W.; Chai, W.; Ducray, P.; Lee, J. K.; Ozer, R. S.; Andersson, C. M. *Bioorg. Med. Chem.* **1998**, *6*, 1347–1378.
27. Luo, Z.; Zhang, Q.; Oderaotoshi, Y.; Curran, D. P. *Science* **2001**, *291*, 1766–1769.
28. Baxendale, I. R.; Ley, S. V.; Lumeras, W.; Nesi, M. *Comb. Chem. High Throughput Screen.* **2002**, *5*, 197–199.
29. Bookser, B. C.; Zhu, S. *J. Comb. Chem.* **2001**, *3*, 205–215.
30. Bhat, A. S.; Whetstone, J. L.; Brueggemeier, R. W. *J. Comb. Chem.* **2000**, *2*, 597–599.
31. Breitenbucher, J. G.; Arienti, K. L.; McClure, K. J. *J. Comb. Chem.* **2001**, *3*, 528–533.
32. Zhao, J.; Zhang, L.; Yan, B. Strategies and Methods for Purifying Organic Compounds and Combinatorial Libraries. In *Analysis and Purification Methods in Combinatorial Chemistry*; Yan, B., Ed.; John Wiley: Chichester, UK, 2004, pp 255–280.
33. Weller, H. N.; Young, M. G.; Michalczyk, S. J.; Reitnauer, G. H.; Cooley, R. S.; Rahn, P. C.; Loyd, D. J.; Fiore, D.; Fischman, S. *J. Mol. Diver.* **1997**, *3*, 61–70.
34. Schultz, L.; Garr, C. D.; Cameron, L. M.; Bukowski, J. *Bioorg. Med. Chem. Lett.* **1998**, *8*, 2409.
35. Kibbey, C. *Lab. Robotics Automation* **1998**, *9*, 309–321.
36. Hochlowski, J. High Throughput Purification: Triage and Optimization. In *Analysis and Purification Methods in Combinatorial Chemistry*; Yan, B., Ed.; John Wiley: Chichester, UK, 2004, pp 281–306.
37. Yan, B.; Collins, N.; Wheatley, J.; Irving, M.; Leopold, K.; Chan, C.; Shornikov, A.; Fang, L.; Lee, A.; Stock, M. et al. *J. Comb. Chem.* **2004**, *6*, 255–261.
38. Karancsi, T.; Godorhazt, L.; Szalay, D.; Darvas, F. *J. Comb. Chem.* **2005**, 7, 58–62.

39. Zeng, L.; Burton, L.; Yung, K.; Shushan, B.; Kassel, D. B. *J. Chromatogr. A* **1998**, *794*, 3–13.
40. Zeng, L.; Wang, X.; Wang, T.; Kassel, D. B. *Comb. Chem. High Throughput Screening* **1998**, *1*, 101–111.
41. Kiplinger, J. P.; Cole, R. O.; Robinson, S.; Roskanp, E. J.; O'Connell, H. J.; Ware, R. S.; Brailsford, A.; Batt, J. *Rapid Commun. Mass Spectrom.* **1998**, *12*, 658–664.
42. Nemeth, G.; Kassel, D. Existing and Emerging Strategies for the Analytical Characterization and Profiling of Compound Libraries. In *Annual Reports in Medicinal Chemistry*; Trainor, G., Ed.; Academic Press: Amsterdam, the Netherlands, 2001, pp 277–292.
43. Popa-Burke, I. G.; Issakova, O.; Arroway, J. D.; Bernasconi, P.; Chen, M.; Coudurier, L.; Galasinski, S.; Jadhav, A. P.; Janzen, W. P.; Lagasca, D. et al. *Anal. Chem.* **2004**, *76*, 7278–7287.
44. Goetzinger, W. K. High throughput HPLC-MS based purification – challenges in lead generation versus lead optimization. *Boston Society of Applied Therapeutics Meeting*, Boston, MA, 2005.
45. Quinn, H. M. High throughput liquid separations in porous media – a different view. *Boston Society of Applied Therapeutics Meeting*, Boston, MA, 2005.
46. Luo, Z.; Zhang, Q.; Oderaotoshi, Y.; Curran, D. P. *Science* **2001**, *291*, 1766–1769.
47. Zhang, W.; Lu, Y.; Chen, C. H.-T.; Curran, D. P.; Geib, S. *Eur. J. Org. Chem.* **2006**, 2055–2059.
48. Zeng, L.; Kassel, D. B.; Lu, Y.; Zhang, W. *J. Comb. Chem.* **2006**, *8*.
49. Di Biasi, V. D.; Haskins, N.; Organ, A.; Bateman, R.; Giles, K.; Jarvis, S. *Rapid Commun. Mass Spectrom.* **1999**, *13*, 1165–1168.
50. Wang, T.; Zeng, L.; Cohen, J.; Kassel, D. B. *Comb. Chem. High Throughput Screening* **1999**, *2*, 327–334.
51. Zeng, L.; Kassel, D. B. *Anal. Chem.* **1998**, *70*, 4380–4388.
52. Xu, R.; Wang, T.; Isbell, J.; Cai, Z.; Sykes, C.; Brailsford, A.; Kassel, D. B. *Anal. Chem.* **2002**, *74*, 3055–3062.
53. Edwards, C.; Liu, J.; Smith, T. J.; Brooke, D.; Hunter, D. J.; Organ, A.; Coffey, P. *Rapid Commun. Mass Spectrom.* **2003**, *17*, 2027–2033.
54. Isbell, J.; Xu, R.; Cai, Z.; Kassel, D. B. *J. Comb. Chem.* **2002**, *4*, 600–611.

Biography

Daniel B Kassel currently serves as Senior Director, Analytical and Discovery Technologies, Takeda San Diego, Inc. Dr Kassel's principal responsibilities include the development and implementation of state-of-the-art analytical and drug discovery technologies. His team focuses mainly in the areas of 'just-in-time' analysis, purification, in vitro ADME and in vivo pharmacokinetic profiling of chemical leads. Dr Kassel is also responsible for cardiovascular (hERG) safety assessment and serves as a project leader in a cardiovascular disease therapeutic area.

Prior to Takeda San Diego, Dr Kassel served as Senior Director of DuPont Pharmaceuticals Research Labs and previous to that as Director of Analytical Chemistry at CombiChem, Inc (acquired by DuPont). There, he and his research team pioneered the design, development and implementation of automated preparative HPLC/MS. In addition, his team developed parallel HPLC/MS technologies, culminating in the issuance of two key patents in the area of parallel (indexed) spray mass spectrometry. Prior to CombiChem, Dr Kassel held the positions of Senior Scientist, Research Investigator I and Research Investigator II at Glaxo Research Institute. There, he published one of the first papers coupling affinity column chromatography/MS for the identification of protein–ligand interactions. In addition, Dr Kassel pioneered the technique of perfusion chromatography/mass spectrometry and applied the technology to elucidate the role of phosphorylation in protein signaling pathways. Dr Kassel received his PhD in Analytical Chemistry at Michigan State University in 1988. Directly following, he held a post-doctoral fellowship at The Massachusetts Institute of Technology studying under the tutelage of Prof Klaus Biemann. Following his tenure there, Dr Kassel was awarded an NIH National Research Service Award fellowship at Harvard University. Dr Kassel has co-authored 60 peer-reviewed manuscripts and more than 100 abstracts, principally in the area of LC/MS, serves as a reviewer for many journals, and is editorial board member of Combinatorial Chemistry & High Throughput Screening and Journal of Combinatorial Chemistry.

Dr Kassel has been an invited speaker at numerous conferences both nationally and internationally, has served as chairperson for several conferences and organized and taught numerous short courses and workshops, including the ASMS short courses on 'Mass Spectrometry of Peptides and Proteins,' 'Mass Spectrometry and Combinatorial Chemistry' and 'Mass Spectrometry in Drug Discovery: From Target Identification to IND Enabling Studies.' Since 2005, Dr Kassel has served on the ASMS Board of Directors as Vice President for Arrangements.

Comprehensive Medicinal Chemistry II
ISBN (set): 0-08-044513-6

ISBN (Volume 3) 0-08-044516-0; pp. 861–874

3.38 Protein Crystallography in Drug Discovery

T Hogg and R Hilgenfeld, University of Lübeck, Lübeck, Germany

3.38.1	**Introduction: The Cost and Value of Macromolecular Crystallography in Drug Discovery**	**875**
3.38.2	**Macromolecular Crystallography and Rational Drug Design: A Historical Perspective**	**876**
3.38.3	**Crystallographic Milestones in Rational Drug Design**	**877**
3.38.3.1	Milestone 1: Gene to Protein	877
3.38.3.1.1	Domain mapping and rational deoxyribonucleic acid construct design	879
3.38.3.1.2	Affinity chromatography: finding the right tag	880
3.38.3.1.3	Choosing a suitable expression system	880
3.38.3.2	Milestone 2: From Protein to Crystal	880
3.38.3.2.1	Solubility optimization: the first hurdle between purification and crystallization	881
3.38.3.2.2	Protein characterization: fit for crystallization?	881
3.38.3.2.3	Dealing with heterogeneity	881
3.38.3.2.4	The protein as a variable	884
3.38.3.3	Milestone 3: From Crystal to Structure	888
3.38.3.3.1	Postcrystallization treatments to improve diffraction quality	888
3.38.3.3.2	Cryocrystallography in practice	889
3.38.3.3.3	Phasing methods for drug discovery	889
3.38.3.4	Milestone 4: Structure to Lead	890
3.38.3.4.1	Crystallography-driven lead characterization: the traditional view	890
3.38.3.4.2	Crystallography and 'fragonomics': the emerging area of fragment-based drug design	892
3.38.4	**The Era of High-Throughput Crystallography in Lead Discovery**	**894**
References		**894**

3.38.1 Introduction: The Cost and Value of Macromolecular Crystallography in Drug Discovery

Drug discovery has traditionally relied either on serendipitous observations (a classic example would be Sir Alexander Fleming's discovery of the antibacterial properties of penicillin[1]), or on screening of natural and synthetic compounds in combination with medicinal chemistry (*see* 1.07 Overview of Sources of New Drugs; 3.32 High-Throughput and High-Content Screening). While remaining indispensable to the drug discovery process, the conventional methods are beginning to give way to more rational approaches that utilize available knowledge on the three-dimensional (3D) structure of a drug target (*see* 4.24 Structure-Based Drug Design – The Use of Protein Structure in Drug Discovery).[2] This information can be generated experimentally by employing biophysical methods such as x-ray crystallography or nuclear magnetic resonance (NMR) spectroscopy, or theoretically by utilizing experimentally derived 3D structures of related macromolecules to generate 3D homology models of the drug target (*see* 3.21 Protein Crystallography; 3.22 Bio-Nuclear Magnetic Resonance; 4.10 Comparative Modeling of Drug Target Proteins).[3]

The method of x-ray crystallography has been the most productive generator of experimentally derived 3D structures of biological macromolecules. Of the approximately 14 000 nonredundant macromolecular structures (<95% sequence identity) deposited to the Protein Data Bank (PDB) at the time of this writing, more than 70% have been determined by x-ray crystallography, eclipsing all other biophysical methods combined (e.g., NMR spectroscopy, cryo-electron microscopy, neutron diffraction, and electron diffraction) (*see* 3.17 The Research Collaboratory for Structural Bioinformatics Protein Data Bank). Owing to a great number of technological advances in the field of x-ray crystallography in recent years (many of which will be reviewed in this chapter), the method is certain to maintain its dominant position in structure-based drug design (SBDD) (*see* 4.24 Structure-Based Drug Design – The Use of Protein

Structure in Drug Discovery). Recent estimates have underscored the value of an x-ray crystal structure in the drug discovery process: the average cost of $15–20 million required for a successful round of lead identification to investigational new drug filing can be reduced by an estimated 50% if 3D structure is centrally utilized in the discovery process.[4] The largest cost reductions stem from the increased quality of lead candidates and number of different pharmacophore series that can be designed with the aid of structural information. Considering the great value of a 3D structure, the relative cost of an x-ray structure determination is nearly negligible, with the elucidation of novel soluble targets averaging between $140 000 (bacterial) and $450 000 (human). On the other hand, the successful x-ray crystal structure determination of integral membrane proteins, which represent the largest class of human drug targets and are particularly recalcitrant to crystallization, can cost millions of dollars (*see* 2.19 Diversity versus Focus in Choosing Targets and Therapeutic Areas).[4] Ongoing developments in the field of soluble and membrane protein crystallization are poised to reduce further the attrition rates and average cost of x-ray structure determination in the coming years.

3.38.2 Macromolecular Crystallography and Rational Drug Design: A Historical Perspective

The utility of structural models in the rational design of therapeutically useful molecules had its humble beginnings in the mid-1970s, when two researchers at The Squibb Institute of Medical Research, David Cushman and Miguel Ondetti, attempted to design small molecule inhibitors of angiotensin-converting enzyme (ACE), a key enzyme involved in the regulation of blood pressure. ACE, a dipeptidyl carboxypeptidase, cleaves the C-terminal His-Leu dipeptide from the prohormone angiotensin I (Asp-Arg-Val-Tyr-Ile-His-Pro-Phe-His-Leu) to produce angiotensin II, a potent vasopressor octapeptide. Back then, only a handful of macromolecular x-ray crystal structures had been elucidated, one being carboxypeptidase A, a Zn^{2+}-containing exopeptidase, which had been a research focus of William Lipscomb's group at Harvard.[5–7] Although the structure of ACE was not known at the time, biochemical evidence suggested the two enzymes might be similar: both could be inhibited by metal-chelating agents such as EDTA, both could be reactivated by Zn^{2+}, Mn^{2+}, and Co^{2+}, and both were inactive against substrate analogs containing terminal carboxamide or dicarboxylic amino acids. The working hypothesis that Cushman and Ondetti had formulated was that ACE was a Zn^{2+}-metallopeptidase bearing an active site similar to carboxypeptidase A, with ACE cleaving dipeptides rather than single amino acids from the C-terminus of peptide substrates.

Guided by Lipscomb's structures[5–7] and published biochemical data reporting inhibition of carboxypeptidase A by D-benzylsuccinic acid,[8] Cushman and Ondetti sketched out a model of substrate and inhibitor binding to the ACE active site (**Figure 1**). In creating their model, they postulated that the potent inhibitory properties of D-benzylsuccinic acid against carboxypeptidase A ($K_i = 4.5 \times 10^{-7}$ M) originated from a specific binding mode that optimized interactions between the inhibitor and the important substrate-binding groups at the enzyme's active site. Several rationalizations were included in the design process:

1. ACE, like carboxypeptidase A, probably carries a positively charged group, which interacts with the carboxy-terminal carboxylate moiety of the ligand.
2. The active-site Zn^{2+} of ACE and carboxypeptidase A is suitably located to polarize the carbonyl group of the substrate's scissile peptide bond. The metal is likely coordinated by the anionic succinate moiety of D-benzylsuccinic acid in the case of carboxypeptidase A.
3. By virtue of ACE's dipeptidyl carboxypeptidase activity, the distance between its positively charged carboxyl-binding site and its active-site Zn^{2+} should be greater than the corresponding distance in carboxypeptidase A by the length of approximately one amino acid. An additional 'spacer' would therefore need to be incorporated into an ACE inhibitor.
4. All naturally occurring peptidic ACE inhibitors, such as teprotide (isolated from the venom of the Brazilian pit viper), carry a proline at the C-terminus, suggesting that this feature is critical and should be preserved in nonpeptidic inhibitors as well.

Armed with their simple pencil-sketch, Cushman and Ondetti set out to create potent nonpeptidic ACE inhibitors. Their first compound, succinyl-L-proline, indeed proved to be a specific inhibitor of ACE; however, it was only slightly active. By exploring different structural modifications, a simple methylation of the succinyl moiety was found to increase inhibitory potency by more than one order of magnitude. Further attempts to improve the binding affinity of this compound, 2-D-methylsuccinyl-L-proline, steered their decision to explore the role of the inhibitor's probable Zn^{2+}-binding moiety. Replacement of the succinyl carboxyl group with a sulfhydryl function (**Figure 1**) reduced the 50% inhibition constant (IC_{50}) by an additional three orders of magnitude.[9] The resulting

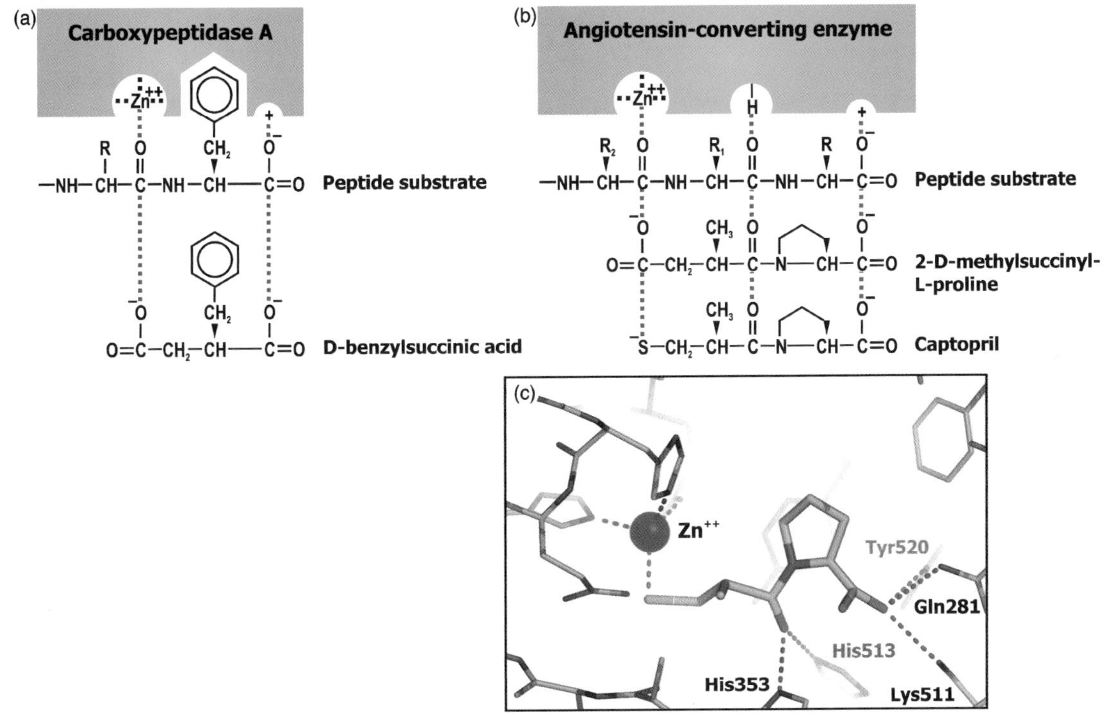

Figure 1 Structure-based design of captopril. (a) Schematic representation of substrate and inhibitor binding to the active site of pancreatic carboxypeptidase A derived from crystallographic data. (b) Cushman and Ondetti's hypothetical model[9] of the active site of ACE and proposed modes of substrate and inhibitor binding, leading to the design of captopril in 1977. Hydrogen bonds and ionic interactions are represented by dashed lines. (c) Confirmation of Cushman and Ondetti's model arising from the crystal structure elucidation of human testes ACE in complex with captopril in 2004.[11] Captopril is depicted in green. (Panels (a) and (b) are reprinted with permission from Ondetti, M. A.; Rubin, B.; Cushman, D. W. *Science* **1977**, *196*, 441–444. Copyright 1977 AAAS.)

compound, 2-D-methyl-3-mercaptopropanoyl-L-proline ('captopril', marketed under the tradename Capoten), turned out to be a potent oral ACE-specific antihypertensive, received FDA approval in the early 1980s, and became Squibb's first billion-dollar drug. Nearly three decades later, the recent crystal structure determinations of ACE:inhibitor complexes[10,11] finally confirm that – despite ACE bearing little overall structural similarity to carboxypeptidase A – the original pencil-and-paper models of ACE-inhibitor binding were remarkably accurate (**Figure 1**). Cushman and Ondetti's seminal work represents a landmark in the field of SBDD.

3.38.3 Crystallographic Milestones in Rational Drug Design

The principle stages of an x-ray crystallography project within an SBDD program are typically represented by the following milestones: (1) satisfactory overexpression, purification, and solubilization of the target macromolecule; (2) reproducible crystallization of the target in a form suitable for high-resolution x-ray data analysis; (3) collection of high-quality x-ray diffraction data (unliganded target and/or target:ligand complexes), successful phase determination, and production of a high-quality 3D structural model; and (4) crystallography-driven lead discovery and/or lead optimization (*see* 3.19 Protein Production for Three-Dimensional Structural Analysis; 3.20 Protein Crystallization; 3.21 Protein Crystallography; 3.23 Protein Three-Dimensional Structure Validation; 3.24 Problems of Protein Three-Dimensional Structures).

3.38.3.1 Milestone 1: Gene to Protein

One of the most challenging aspects of any crystallography project is the design of initial gene constructs coding for the protein of interest. Subsequent gene expression as well as purification and characterization of the recombinant gene product can also represent one of the most time-consuming stages in an SBDD program (**Figure 2**). Milligram quantities

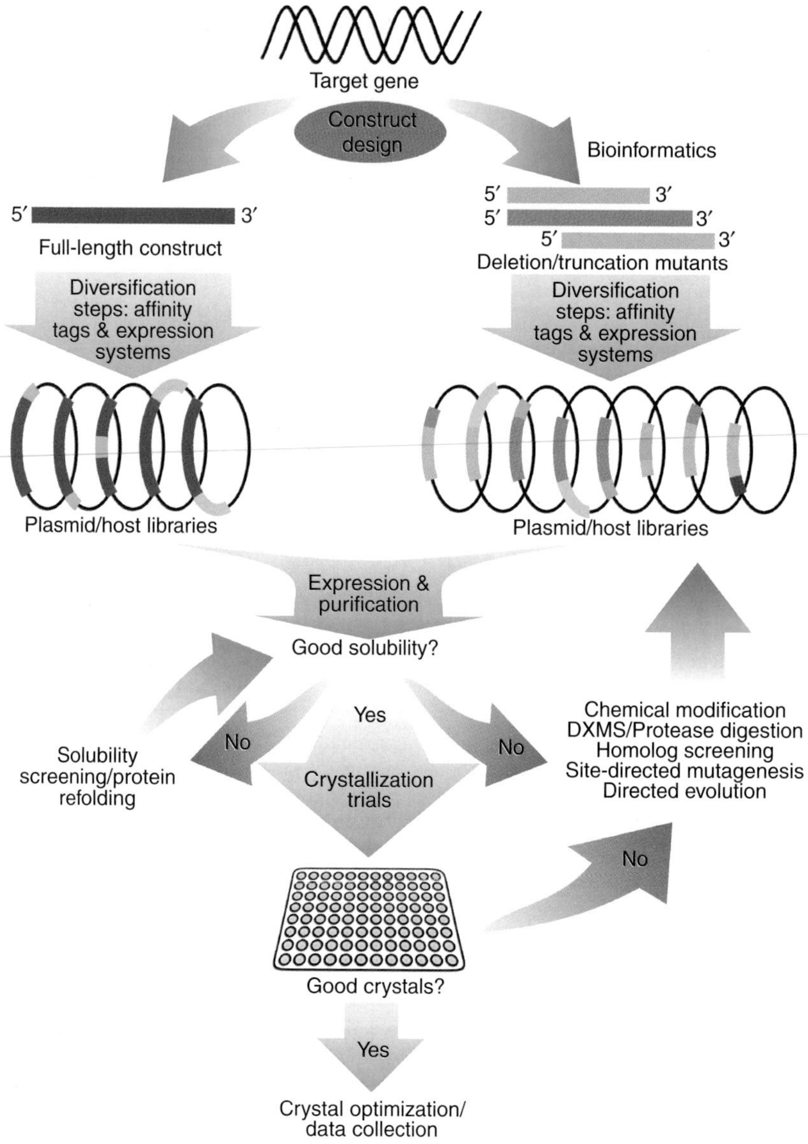

Figure 2 Flow chart illustrating the typical gene-to-crystal bottlenecks and milestones of an SBDD program. Because the probability of success is exponentially related to the number of independent protein-target variants screened, diversification should be emphasized at the stages of gene construct design, choice of expression system, and selection of affinity tags. Available biochemical, biophysical, and bioinformatical data should be exploited during the design of deletion/truncation mutants. Achievement of the first major milestone is commonly regarded as the obtainment of pure, homogeneous, monodisperse, and highly soluble samples of one or more recombinant proteins. Samples that fail to meet quality criteria can often be rescued by implementing feedback loops that incorporate different screening and optimization procedures such as high-throughput solubility screening. If extensive crystallization screening fails to produce initial crystal 'hits,' contingency pathways should be followed that involve modification protocols such as site-directed mutagenesis or directed evolution.

of pure protein (typically 0.5–5.0 mg) are routinely needed to screen a sufficient number of crystallization conditions in order to generate initial crystal 'hits.' In the past, a small number of deoxyribonucleic acid (DNA) constructs would be generated, subsequently tested for expression in the host cell, and the levels of soluble recombinant protein produced would be assayed by polyacrylamide gel electrophoresis (PAGE). This process was conducted sequentially for each individual gene construct, and the best were used to produce protein for downstream crystallization trials. Recent technological advances have now made it possible to rapidly generate and test multiple gene constructs in parallel.[12–14]

3.38.3.1.1 Domain mapping and rational deoxyribonucleic acid construct design

Because of the vast amount of uncertainties that go into designing initial gene constructs, structural information should be used whenever possible in the design process. Available 3D structures of homologs, orthologs, or paralogs are a rich source of information regarding domain boundaries and unstructured loop regions, which need to be considered when designing crystallizable targets (*see* 3.17 The Research Collaboratory for Structural Bioinformatics Protein Data Bank). Because multidomain proteins can exhibit a high degree of conformational flexibility, which might preclude crystallization, trimming down the target to a more compact form consisting of the catalytic core represents a logical strategy, particularly for the purposes of SBDD. Some recent examples pertaining to this approach include studies on the kinase domain of c-Abl,[15] and our own work on the bifunctional catalytic domain of the bacterial stringent response factor, RelA/SpoT.[16] Proteolytic mapping of the full-length protein with subsequent analysis by SDS-PAGE can be a useful technique for identifying stable domains. Typically, the full-length protein is digested with a panel of proteases, and samples are analyzed either at fixed time points with different protease concentrations, or at variable time points with fixed protease concentrations. The protein can also be digested in the presence of known ligands or inhibitors, since ligand-induced ordering of unstructured regions may lead to unique digestion maps and suggest alternative possibilities for DNA construct design. Characterization of stable fragments by N-terminal sequencing or mass spectrometry is subsequently carried out to provide a framework for gene construct optimization.[17]

Recent years have seen an explosion of useful bioinformatics tools, which can aid in gene construct design by predicting disordered regions in a given protein sequence (*see* 3.15 Bioinformatics; 3.16 Gene and Protein Sequence Databases). These include FoldIndex,[18] DisEMBL,[19] DISOPRED2,[20] DRIP-PRED (R. M. MacCallum, unpublished data), GlobPlot 2,[21] IUPred,[22] PONDR,[23] Prelink,[24] RONN,[25] and the VL2/VL3/VL3H/VL3E suite,[26,27] all of which utilize different methods (e.g., neural networks, support vector machines) for disorder prediction. The bi-annual CASP (Critical Assessment of Techniques for Protein Structure Prediction) challenge, which constantly evaluates and ranks the accuracy of the various structure prediction programs, has noted very significant progress in the field.[28] A curated database containing information on proteins with partial or complete disorder, DisProt, is under development.[29]

Perhaps one of the most exciting recent developments that promises to greatly assist the process of gene construct design is deuterium-exchange mass spectrometry (DXMS) (**Figure 3**).[30,31] This method, which produces high-resolution maps of ordered and disordered regions along the protein sequence, requires only micrograms of soluble protein, can detect disordered segments as little as four residues, can be used in combination with ligands and inhibitors, and has demonstrated success in producing crystallizable fragments.[31]

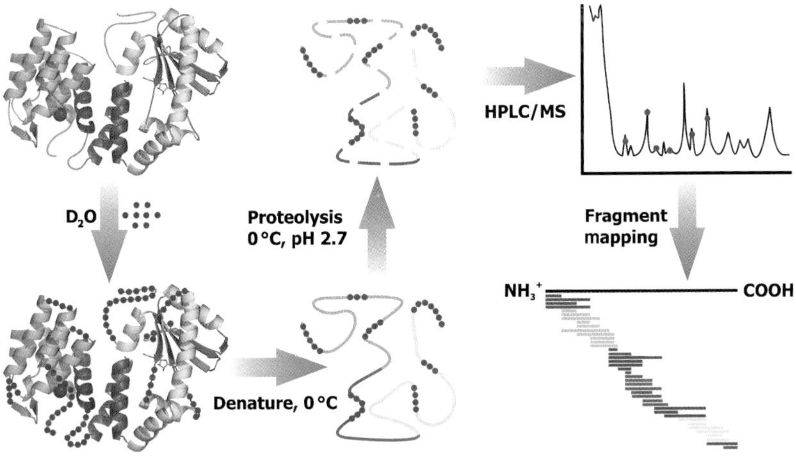

Figure 3 A schematic depiction of the hydrogen/deuterium-exchange MS (DXMS) procedure to aid the design of DNA constructs by identifying unstructured regions in the protein target. After establishment of initial protein fragmentation maps, the protein of interest is incubated in deuterium oxide (D_2O) for 10s at 0 °C, leading to rapid exchange at solvent exposed amide nitrogens. After rapid denaturation in an acidic quench solution, deuterated samples are proteolyzed by brief exposure to immobilized pepsin. Proteolytic fragments are separated by HPLC, analyzed by MS, and disordered regions are localized by interpretation of amide hydrogen/deuterium exchange maps.[30,31]

3.38.3.1.2 Affinity chromatography: finding the right tag

To aid in purification, recombinant protein targets are usually fused to an affinity tag, which can be added to the N- or C-terminus of the target sequence, or, if necessary, inserted within the sequence. To maximize the crystallization potential of the recombinant target, emphasis should be placed on diversification when selecting affinity tags. A vast number of affinity tags are available, and an ideal choice will depend on numerous factors including preferred tag location, tag size (ranging from a few residues to >100 kDa), and the desired balance of auxiliary tag characteristics such as gene expression and solubility of the tagged protein, ease of purification, overall purity of the eluted sample, or secreted expression.[12] Oligohistidine tags (His-tags) are often a first choice because of their relatively low cost, small size, and ease of use.[32] The incorporation of a His-tag permits simple one-step purification using an immobilized metal-affinity chromatography (IMAC) resin such as Ni^{2+}-nitrilotriacetate, but a second 'polishing step' using size-exclusion or ion-exchange chromatography is often needed to eliminate contaminating host proteins that have a natural affinity for IMAC resins. If subsequent tag removal is desired (e.g., if it is suspected that the tag might be interfering with crystallization or protein activity), vectors such as Qiagen's pQE-30 Xa are available that allow incorporation of a cleavable tag containing a flanking protease recognition site. A recent comparative study of different affinity tags has touted the use of the *StrepII* tag (Trp-Ser-His-Pro-Gln-Phe-Glu-Lys), which binds a modified streptavidin-coupled matrix (*Strep*-Tactin).[33] According to this study, the *StrepII* tag affords much higher purity over the His tag at a comparable cost. Large affinity tags containing fusion proteins, such as the maltose-binding protein,[34] thioredoxin,[35] or glutathione-*S*-transferase[36] can improve folding and enhance solubility of the coupled target. Moreover, the growing number of chimeric target-fusion crystal structures that are being reported suggests that fusion proteins can be useful tools for crystallization.[37]

3.38.3.1.3 Choosing a suitable expression system

Once the gene constructs have been designed, choosing the best expression system can be an equally troublesome procedure. The utilization of *Escherichia coli* as an expression host has long been the method of choice because it provides a fairly robust and easy means for producing recombinant proteins at a minimal expense. Despite these advantages, it should be pointed out that a large percentage (in some cases more than 50%) of cytosolic proteins encoded by bacterial and archaeal genomes appear to be insoluble when expressed in *E. coli*.[38] Production of eukaryotic proteins can be particularly problematic if posttranslational modifications such as phosphorylation or glycosylation are required for proper folding and activation of the recombinant target.[39] Eukaryotic expression systems offer the advantage of having the capability to phosphorylate and glycosylate recombinant proteins, but to various degrees.[40] Of these, the baculovirus system is particularly suitable to the high-throughput needs of an SBDD program, as insect cells are typically more durable than mammalian cells.[13] Recently, the yeast *Pichia pastoris* has been genetically engineered into a 'humanized' form with a full complement of *N*-glycosylation machinery,[41] a development which paves the way for large-scale production of recombinant human proteins with immediate applications ranging from protein therapeutics to SBDD.

To generate expression vector clones, newer cloning systems including the Novagen pTriEx, Invitrogen Echo, or Invitrogen Gateway kits are advantageous in that they allow the testing of multiple affinity tags in both *E. coli* and insect cells with minimal subcloning.[42] The production of misfolded protein can often be alleviated by utilizing chaperones and foldases[43] or factorial screening methods.[44] Coexpression of the target with stabilizing binding partners is now possible with newer expression systems such as Novagen's Duet coexpression vectors, which allow the simultaneous expression of up to eight proteins in the same cell. There are multiple cases illustrating the benefit of binding-partner coexpression if one protein component is insoluble or unstable when expressed on its own. For example, a complex of the polyomavirus internal protein VP2/VP3 with the pentameric major capsid protein VP1 was successfully prepared and crystallized only after coexpression of the components in *E. coli*. Coexpression was essential to obtain the complex since VP2 and VP3 fragments were found to be insoluble when expressed independently.[45]

3.38.3.2 Milestone 2: From Protein to Crystal

The protein-to-crystal milestone can represent the most formidable challenge in SBDD. Biological macromolecules are complex and dynamic entities, with their own unique set of physicochemical properties, and no universal recipe exists for their crystallization. A purified sample may seem at first to be 'intrinsically noncrystallizable,' and a great deal of time and resources may be required to find an optimal set of crystallization conditions or to construct a suitable variant that produces high-quality crystals. Several strategies for accelerating this phase of the SBDD pipeline are outlined in the following subsections.

3.38.3.2.1 Solubility optimization: the first hurdle between purification and crystallization

Once a successful purification protocol has been established, it is frequently the case that the final elution buffer is inappropriate for initiating crystallization trials. Moreover, it is not unusual to observe that the solubility of the recombinant target in standard elution buffers is much lower than the protein concentration normally required for crystallization (usually 5–20 $mg\,mL^{-1}$). Even though Franz Hofmeister observed the differential solubility of proteins in various salts more than 125 years ago[46,47] and a large amount of subsequent research has been conducted on this phenomenon,[48] protein solubility optimization can still be regarded somewhat as a 'trial-and-error' science because proteins are complex biological polyelectrolytes with fairly unpredictable solubility characteristics. The ordering of anions and cations according to their general ability to stabilize the structure of proteins has been referred to as the Hofmeister Series. Ammonium sulfate, which is a commonly used precipitant in protein crystallography, tends to both stabilize proteins in the folded state and drive them out of solution, while guanidinium chloride has the opposite tendencies. The differential effect of salts is one of the most important variables to screen when searching for conditions that enhance solubility of an overexpressed protein. Cosolvent additives, which can modulate protein solubility through direct interactions with the protein or by modifying the surrounding water structure,[49] are also worthwhile to explore for particularly recalcitrant proteins. Glycerol can enhance solubility without denaturing proteins.[48,50] Detergents can improve solubility by binding to hydrophobic surface patches on proteins.[51] Additional cosolvents such as small organic molecules, polyvalent ions, sugars, and polyhydric alcohols can also help to improve protein solubility and stability through different effects.[52–56] Screening for an ideal solubility buffer composition is therefore a complex multiparameter puzzle. Lindwall and co-workers[57] were one of the first groups to approach this problem by developing a sparse-matrix solubility screen that could be used on crude cell extracts carrying the recombinant protein. The screen is empirically designed based on known solubility enhancers, and has produced favorable results in our laboratory for proteins that appeared otherwise insoluble in standard lysis and purification buffers. Alternative solubility screening methods employing dynamic light scattering or photometric analysis have been developed specifically for crystallography applications.[58,59]

3.38.3.2.2 Protein characterization: fit for crystallization?

Protein purity is generally regarded as a critical determinant for successful crystallization. Of course there are always exceptions to this rule, as some proteins exhibit the propensity to crystallize even in the presence of high levels of impurities. The earliest account of successful protein crystallization by Friedrich Hünefeld in 1840 reported on the growth of 'blood crystals' from earthworm (earthworm hemoglobin) upon slow dehydration of raw blood samples.[60] Indeed, many of the crystal specimens that supplied the seminal macromolecular x-ray studies of the early 1930s were coaxed from crude biological mixtures.[61] As an example from our own laboratory, we have witnessed the remarkable propensity of one glycoprotein isozyme to crystallize out of an equal mixture of two different forms.[62]

As a rule-of-thumb, however, protein purity should be regarded with utmost stringency when preparing samples for crystallization trials. Purified protein samples should not only be as free as possible from macromolecular contaminants (proteins, DNA/RNA, complex carbohydrates, etc.), but should also be chemically and conformationally homogeneous, exhibit a monodisperse size distribution, and be free of any denatured species or other microheterogeneities that may preclude or adversely affect crystallization. Automated assessment of purity can be carried out with high-throughput SDS-PAGE[63] and matrix-assisted laser desorption ionization (MALDI) mass spectrometry.[64,65] Dynamic light scattering (DLS) measurements can be used to verify sample dispersity and this method has proven effective as a predictive tool for assessing the crystallizability of macromolecules.[66–68] It was estimated in one study that as much as 70% of proteins exhibiting a monodisperse size distribution in a light scattering experiment will crystallize using a standard 48–96 condition sparse-matrix crystallization screen.[69]

3.38.3.2.3 Dealing with heterogeneity

Posttranslational modifications (PTMs) are a major source of heterogeneity for proteins, and at least 150 different kinds of PTMs (e.g., glycosylation, methylation, phosphorylation, *S*-thiolation, etc.) are known,[70] with each affecting characteristics such as molecular weight, charge, and solubility of the modified protein. Although a PTM may be small, the modification may serve a physiological role and have a drastic impact on the overall conformation of the protein. From a crystallization standpoint, these different modified forms can be perceived as entirely foreign particles, and thus have similar adverse effects on the crystallization process. Chromatographic methods, enzymatic treatment, or mutagenesis can help to reduce or eliminate heterogeneity.

3.38.3.2.3.1 Glycosylation and phosphorylation

Glycosylation can be a particularly problematic source of heterogeneity when eukaryotic expression systems are used. The negative influence of attached oligosaccharides on the crystallizability of glycoproteins presumably stems from elevated surface entropy due to conformational flexibility of the oligosaccharide chains, as well as differences in glycan chain length, branching, and sugar composition. Removal of the attached oligosaccharides by enzymatic deglycosylation has been suggested as a way to overcome this problem,[71,72] and indeed several reports have emphasized the importance of complete deglycosylation for obtaining suitably diffracting crystals.[73–75] For complete removal of asparagine-linked (N-linked) oligosaccharides, peptide-*N*-glycosidases such as PNGase F or PNGase A are often used, which also cause deamidation of the asparagine residue to aspartic acid.[76] In our experience, however, leaving the innermost sugar residue attached to the protein may offer additional possibilities for sugar-mediated crystal-contact formation,[62] and different deglycosylation strategies (selective versus complete deglycosylation) should therefore be devised (**Figure 4**). As an example for selective *N*-deglycosylation, partial digestion might be attempted with selective endoglycosidases (such as endo F_1–F_3 or endo H), which leave the innermost *N*-acetyl glucosamine (GlcNAc) of the *N*-linked diacetylchitobiose glycan core attached to the modified Asn. In cases where the GlcNAc is fucosylated, a combined treatment with α-fucosidase might be warranted. Unfortunately, there is no enzyme comparable to PNGase F or PNGase A for removing *O*-linked glycans from Ser/Thr residues. Monosaccharides must be sequentially trimmed off by a series of exoglycosidases until the Gal-(1,3)-GalNAc core remains, at which point *O*-glycosidase can be used to remove the core structure without chemically modifying the Ser/Thr residue.[77] Modifications to the core structure can often block the action of *O*-glycosidase and may require the use of additional glycosidases such as α-2-(3,6,8,9)-neuraminidase.[78] It should be noted that a nonspecific galactosidase or β-(1,3)-galactosidase can be used to hydrolyze the galactose from the core glycan, leaving the O-linked GalNAc attached to the Ser/Thr and thereby providing an alternative to complete sugar removal.

Mutagenesis can be a suitable alternative to enzymatic or chromatographic approaches for generating homogeneous samples for crystallographic studies. For example, substitution of Asn or Ser/Thr in Asn–Xaa–Ser/Thr sequons can prevent *N*-glycosylation. Heterogeneous phosphorylation can be a problem, for instance, when expressing kinases in insect cells, and enzymatic treatment with alkaline phosphatase or phage λ protein phosphatase has been used successfully for obtaining completely dephosphorylated protein for crystallographic studies.[79–81] Alternatively, mutation of Ser/Thr- or Tyr-phosphorylation sites will eliminate heterogeneous phosphorylation altogether. When phosphorylation plays a critical modulatory role in the target protein's activity, a phosphorylated conformation might be of interest for SBDD. In such cases, generating a 'phosphorylated' conformation can often be accomplished by replacing the phosphorylated Ser/Thr residue with Asp or Glu, with the electronegative side chain carboxylate moiety acting as a phosphate mimic.[80,81]

3.38.3.2.3.2 Isoelectric heterogeneity

Microheterogeneities such as isoelectric heterogeneity can be overcome by additional purification steps such as ion-exchange chromatography or preparative isoelectric focusing techniques. The potential impact of microheterogeneity on protein crystallizability is exemplified by the work of Prongay and colleagues,[82] who crystallized a complex of human immunodeficiency virus capsid protein p24 with its antigen-binding fragment (Fab). In this work, recombinant p24 was purified to homogeneity and chromatographically separated into distinct isoelectric species, only one of which was able to crystallize in complex with the Fab. Moreover, different crystal forms of the p24:Fab complex could be obtained when different isoelectric species of the Fab were used in combination with the crystallizable p24 fraction.

3.38.3.2.3.3 Utilizing combined approaches to eliminating heterogeneity

A recent success story employing several of the methods described in this and preceding sections is the expression and structure determination of a truncated and deglycosylated form of human ACE from testes (tACE).[10,11,83] A truncated form of tACE, lacking the N-terminal 36 residues and the C-terminal transmembrane domain was expressed in the presence of *N*-butyl-deoxynojirimycin, an α-glucosidase I inhibitor. After expression, the mutant tACE (tACEΔ36NJ) was treated with endoglycosidase H to selectively remove all but the innermost GlcNAc residue at each *N*-glycosylation site. Although the expression was low, tACEΔ36NJ was shown to be homogeneous on SDS-PAGE, fully active in enzymatic assays, and produced crystals diffracting to 2.0 Å resolution. This led to the long-awaited structural elucidation of a human ACE in complex with captopril and related antihypertensives (as described previously in this chapter), and paved the way for design of next-generation inhibitors (**Figure 1**). The glycosylation sites (Asn–Xaa–Ser/Thr) on tACEΔ36 were mutated systematically by replacing Asn with Gln.[84] Two of the Asn→Gln tACEΔ36 mutants exhibited

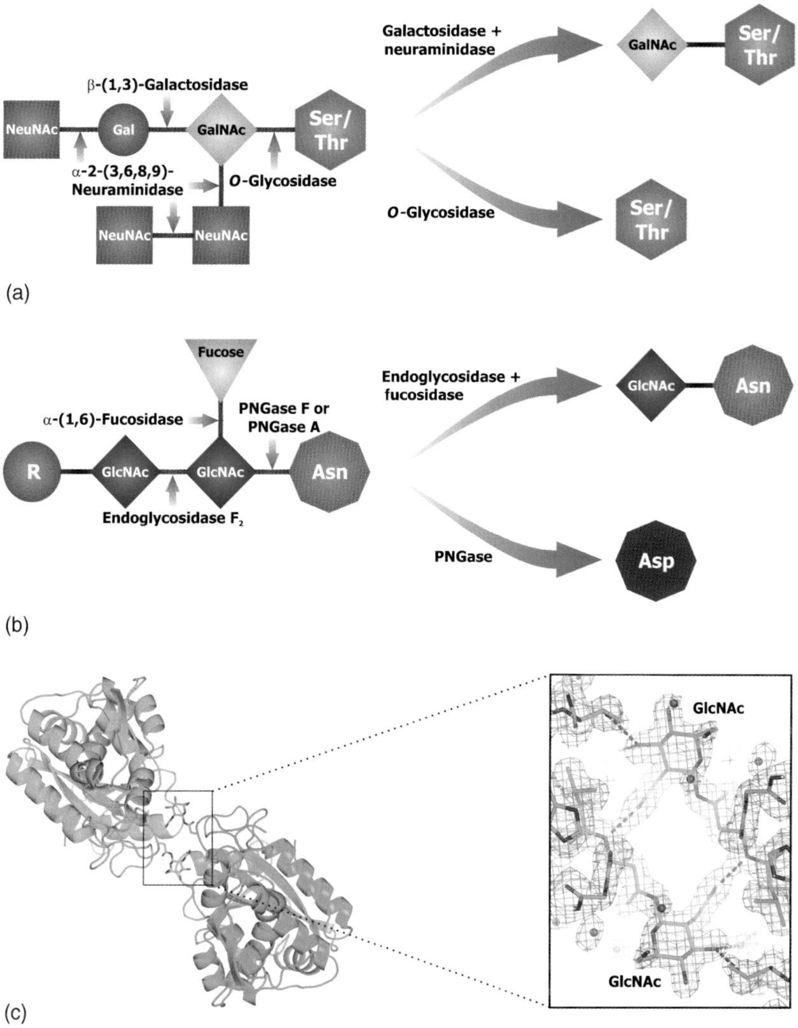

Figure 4 Examples of alternative deglycosylation strategies for *O*- and *N*-linked glycans. Cleavage sites of various glycosidases are indicated with red arrows. Sugar abbreviations used: NeuNAc, *N*-acetylneuraminic acid; Gal, galactose; GalNAc, *N*-acetylgalactosamine; GlcNAc, *N*-acetylglucosamine. (a) A trisialated *O*-linked glycan is digested with a combination of neuraminidase and galactosidase to yield a glycoprotein variant containing a residual *O*-linked GalNAc. A second protein variant is produced by complete glycan removal with *O*-glycosidase. (b) An N-linked biantennary or high-mannose glycan is trimmed with endoglycosidase and fucosidase to generate a glycoprotein with one remaining GlnNAc residue attached. Alternatively, treatment with peptide-*N*-glycosidase (PNGase) facilitates complete glycan removal with concomitant deamidation of the Asn to produce a variant bearing an Asp residue. (c) The crystal structure of pokeweed antiviral protein (PAP) reveals the potential benefit of *O*- and *N*-linked monosaccharides in promoting crystallization of glycoproteins.[62] Depicted are two crystallographically related PAP monomers engaged in a crystal contact via *N*-linked GlcNAc monosaccharides. The isoform lacking this modification failed to crystallize.

higher expression levels and produced crystals isomorphous with the tACEΔ36NJ construct, although the resolution was not as good (∼3.0 Å).

Another relevant success story involves the rational re-engineering of the anticancer target, human urokinase, for crystallographic studies.[85] Preliminary crystals of urokinase in complex with an inhibitor diffracted to 2.5 Å resolution, yet when the structure was solved, it was discovered that the active site was effectively shielded by tight intermolecular packing in the crystal. This unfortunate packing arrangement precluded the diffusion of small molecule inhibitors into the urokinase active site by crystal soaking methods. Using the initial crystal structure as a guide, all disordered or highly flexible polypeptide regions (as judged by missing electron density or high atomic temperature factors) were deleted from the gene construct. Moreover, a free Cys residue, which was exposed by the truncation, was mutated to Ala, and an additional Asn→Gln point mutation was introduced to remove an *N*-glycosylation site. The resulting modified urokinase, termed micro-urokinase, was subsequently crystallized in a better-diffracting crystal form, which was suitable for high-throughput soaking experiments and rational drug design.

3.38.3.2.4 The protein as a variable

The preceding section dealt with different strategies for improving crystallization probability by overcoming heterogeneity. Unfortunately, reality can (and often will) painfully dictate that some proteins, even in their purest, most monodisperse and homogeneous form(s), simply won't crystallize. For these recalcitrant types, fruitless screening of thousands upon thousands of crystallization conditions covering wide expanses of crystallization space will only galvanize the truth. The results of many structural genomics initiatives around the world have confirmed that the majority of proteins that have the propensity to crystallize will do so over an astonishingly narrow range of crystallization conditions.[86–88] Consequently, there is a growing emphasis on screening the protein target as a variable in protein crystallization, for example, either by homolog/ortholog screening, site-directed mutagenesis, directed evolution, or by chemical means.

The overall probability of crystallization of a target is exponentially related to the number of protein variants utilized. Assuming complete independence of variants, the overall crystallization probability, p_T, would be $1-(1-p_{ave})^n$, where p_{ave} is the average probability of crystallizing one of the n variants. As an example, if each variant of a target only had a 20% chance of crystallizing, but 20 variants were constructed, the overall probability of successful crystallization would be $1-(1-0.20)^{20}$, or nearly 99%. Wayne Hendrickson and colleagues were the first to systematically analyze the crystallization probability enhancement of different protein modifications and multiple combinations thereof on a specific target, gp120 (the envelope glycoprotein of type 1 human immunodeficiency virus), which had previously resisted crystallization.[89] By exploring different deglycosylation protocols, substituting different surface loops with tripeptide linkers of Gly-Ala-Gly, producing N- and C-terminal deletion mutants, and reducing conformational heterogeneity by using different protein ligands such as its receptor CD4 and Fabs from conformationally sensitive monoclonal antibodies, they were successful in growing six different types of gp120 crystals from 18 different variants. One of these, the ternary complex between gp120, CD4, and the Fab of the human neutralizing monoclonal antibody 17b, was determined at 2.5 Å resolution.[90]

3.38.3.2.4.1 Homolog screening and DNA shuffling

The value of using homologs for crystallization screening purposes was pioneered decades ago by John Kendrew, whose initially unsuccessful attempts at crystallizing horse heart myoglobin were rescued by choosing sperm whale as an alternative source for material.[91,92] In another piece of seminal work, Herman Watson and colleagues used protein from different sources to crystallize enzymes from the glycolytic pathway.[93] Systematic variation in the species of origin was also instrumental in the crystallization of the transcription initiation TATA-binding protein.[94] Recent structural proteomics data have corroborated this notion by showing that inclusion of homologous proteins of a target drastically increases crystallization success rates.[95] In a classic set of papers demonstrating the effectiveness of iterative structure-based drug discovery using a target homolog, the Agouron group designed potent antitumor agents targeting human thymidylate synthase by using the crystallizable *E. coli* enzyme as a surrogate.[96–98] An extension of this approach is exemplified by the SBDD of selective cyclin-dependent kinase (CDK) inhibitors. Because efforts to obtain crystals of CDK-4 for design of CDK-4-selective inhibitors were unsuccessful, Merck scientists mutated the ATP-binding pocket of the crystallizable CDK-2 homolog to create a CDK-4 active-site mimic. This approach resulted in the successful design of a CDK-4-specific inhibitor.[99] During the recent severe acute respiratory syndrome (SARS) outbreak, our group was able to integrate the structural data which we obtained on the main proteinases of human coronavirus 229E and transmissible gastroenteritis (corona)virus, with molecular modeling and biochemical analyses, to propose that derivatives of AG7088 (an inhibitor of the distantly related human rhinovirus 3C proteinase) would be promising lead candidates for the design of anti-SARS drugs targeting the SARS coronavirus main proteinase.[100]

When crystallization screening with homologs fails to generate hits, the directed evolution approach of 'DNA shuffling' may provide the necessary breakthrough (**Figure 5**). The DNA shuffling method relies on homologous recombination during the PCR reassembly of gene fragments from multiple parents to generate crossovers at points of high sequence identity. The procedure can generate a vast library of chimeras, which can be rapidly screened in parallel for expression, solubility, and activity. Sequence variants that meet or exceed solubility and activity criteria can then undergo crystallization trials. This strategy has recently paid off for Keenan and colleagues who implemented a directed evolution program for engineering a glyphosate *N*-acetyltransferase (GAT) superenzyme to confer glyphosate resistance

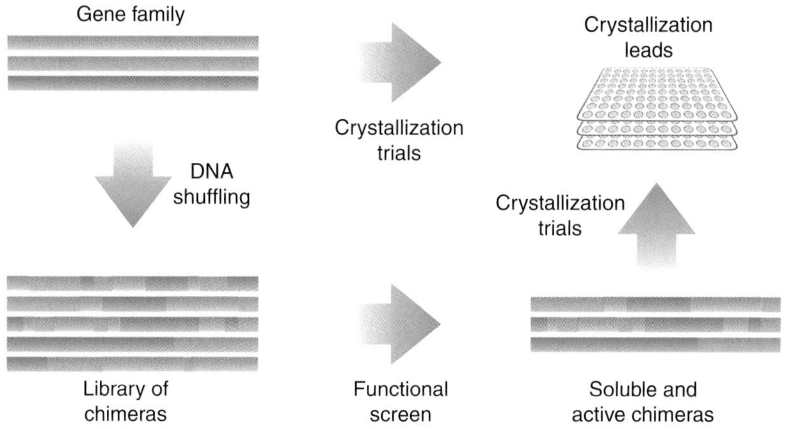

Figure 5 A diversified crystallization strategy employing a combination of homolog screening with DNA shuffling to improve success probability.

in transgenic plants.[101] Initial attempts to crystallize three wild-type GAT enzymes, both in the presence and absence of ligands, failed to produce crystals suitable for structure determination. To overcome this bottleneck, multiple rounds of DNA shuffling were carried out between three wild-type GAT genes in conjunction with high-throughput screening to identify GAT chimeras with decent expression, strong enzymatic activity, and high solubility. Eight randomly chosen GAT variants that met the above criteria were selected from the library, with each containing 5–25% exchanged amino acid content relative to the wild-type enzymes. Utilizing the same crystallization procedure as for the wild-type enzymes (96-condition sparse matrix screen conducted both in the presence and absence of ligands), two of the eight GAT chimeras produced suitable crystals, and the crystal structure of one of them was solved. Not surprisingly, the resulting 1.6 Å crystal structure showed that some of the exchanged regions in the shuffled GAT variant were involved in crystal contact formation.[101]

3.38.3.2.4.2 Metal-mediated crystallization

While orthologs and DNA-shuffled chimeras may exhibit multiple regions of sequence variability on their protein surfaces, which could provide unique areas for crystal packing, a single point mutation is often sufficient to drastically enhance the crystallizability of a target. Perhaps the earliest example demonstrating the effectiveness of this method was the rational genetic engineering of crystal contacts in human H ferritin.[102] The Czech physiologist Vilem Laufberger noted in 1937 that ferritin could be crystallized in situ by adding droplets of concentrated cadmium sulfate to fresh slices of horse spleen.[103] The crystal structures of horse spleen ferritin, rat liver ferritin, and recombinant rat L-chain ferritin – all crystallized in the presence of $CdSO_4$ – revealed intermolecular crystal packing via double Cd^{2+} bridges, with pairs of aspartate and glutamine side chains from neighboring ferritin molecules forming ligands to the metal ions.[104–106] Human and other H-rich ferritins, on the other hand, immediately form amorphous precipitate upon the addition of Cd^{2+}, even at low concentrations.[107] Sequence comparisons revealed that human H ferritin contained a lysine in place of the Cd^{2+}-coordinating glutamine residue. By constructing a Lys→Gln mutant of recombinant human ferritin, large well-diffracting crystals isomorphous to those of the horse and rat ferritins were obtained. Surprisingly, these crystals could only be grown in the presence of Ca^{2+}, although subsequent structure determination revealed that the two Ca^{2+} ions formed an identical intermolecular coordination arrangement as Cd^{2+} in the horse and rat ferritins.[102]

The general utility of metal ions and organometallic compounds to induce macromolecular crystal growth by forming metal-mediated crystal contacts cannot be overstated, and a growing number of success stories documenting this fact continue to fill the literature. The practice of using metal ions in protein crystallization dates as far back as 1925 when John J. Abel – perhaps fortuitously at first – used Zn^{2+} to obtain rhombohedral crystals of insulin.[108] In our laboratory, Zn^{2+} remains the divalent cation of choice for crystallization additive screening, with examples including the Zn^{2+}-mediated crystallization of the macrophage infectivity potentiator protein from *Legionella pneumophila*,[109] and a human carbonic anhydrase (TH, unpublished) (**Figure 6**). It should be emphasized, however, that Zn^{2+} will not be the best crystallization additive for every protein, and provided that enough protein sample is available, parallel screening of different metal salts is recommended. To illustrate this, a study examining the influence of various divalent cations on the crystallization of isoleucine/valine-binding protein and leucine-specific binding protein led to

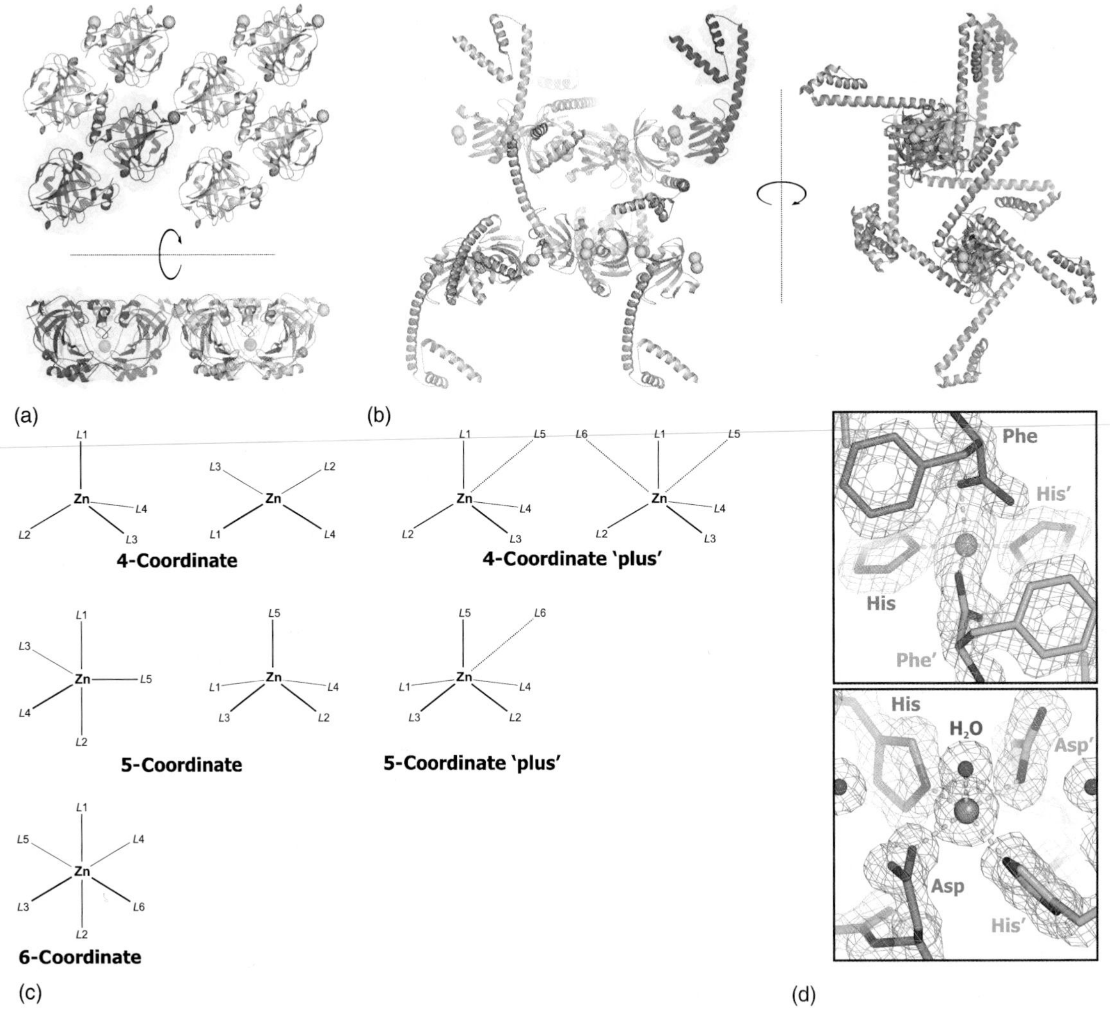

Figure 6 Zn^{2+}-mediated crystallization of drug targets, exemplifying the general utility of metals in crystallization additive screening. (a) Orthogonal views of the crystal lattice in P1 crystals of human carbonic anhydrase (TH, unpublished data). The crystallographic asymmetric unit contains two monomers (highlighted in blue), and the two Zn^{2+} ions involved in crystal contact formation are depicted as orange spheres. The tight crystal packing promoted by the intermolecular metal-bridging is visibly apparent. (b) Orthogonal views of the crystal lattice in tetragonal crystals of the macrophage infectivity potentiator protein (Mip) from *Legionella pneumophila*.[109] In this case, a dinuclear Zn^{2+} crystal-contact bridges the catalytic domains of symmetry-related monomers in an otherwise loosely packed crystal environment. In both of the above cases, the presence of Zn^{2+} in the crystallization buffer was mandatory for crystallization. (c) The various coordination states and ligation geometries observed for Zn^{2+} in protein crystals. 4-Coordinate and 5-coordinate states can be additionally decorated with weaker long-range interactions (denoted by 'plus'). (d) Two examples of Zn^{2+}-mediated crystal contacts observed at high resolution (TH, unpublished data). In the top example, the Zn^{2+} establishes a 4-coordinate tetrahedral crystal contact involving the C-terminal carboxylate moieties (Phe) and His side chains of symmetry-related monomers. The bottom example illustrates a 5-coordinate square-pyramidal Zn^{2+} crystal-contact ligated by the identical His/Asp residues in adjacent monomers. A water molecule occupies the apical position.

the observation that different metals promoted different crystal forms, with Cd^{2+} producing the best results.[110] This and a subsequent study with histidine-binding protein from *Salmonella typhimurium* also found that the optimal metal concentration for crystallization fell within a very narrow range.[111] It is as yet impossible to predict a priori which type of metal or organometallic complex, nor which concentration range (μM–M), will be most suitable for crystallizing a given protein; however, the remarkable propensity of Zn^{2+} to exhibit a wide array of coordination geometries (4-, 5-, and 6-coordinate geometry both with and without additional weak interactions) and a broad range of potential protein ligand functions (Asp, Glu, His, Cys, main-chain carbonyl, and C-terminal carboxylate),[112,113] together with its utility for multiple-wavelength anomalous dispersion phasing methods,[114] make it a logical first choice when metal-additive crystallization screening is considered (**Figure 6**). The Pfizer group has recently proposed engineering carboxylic protein surface residues (Asp and Glu) into proteins as a crystallization strategy to enable Zn^{2+}-mediated crystal contacts.[115]

3.38.3.2.4.3 Mutational strategies I: rational surface engineering

Since the original work by Lawson and co-workers on human H-chain ferritin,[102] a rapidly growing number of examples are appearing in which mutated proteins have crystallized when the wild-type counterparts have not.[116,117] In a classic example of 'serendipity dictating crystallizability,' the bacterial chaperonin GroEL was crystallized only after two mutations were accidentally incorporated by PCR.[118,119] In an elegant study by Ernest Villafranca's group at Agouron, nonconserved surface residues of human thymidylate synthase were rationally mutated to residues with altered polarity or charge.[120] All resulting point mutants had altered crystallization behavior, were generally more crystallizable than the wild-type protein, crystallized under unique conditions, and produced novel crystal forms that were of higher quality and more useful for inhibitor soaking studies. An analogous study by the Roche group with point mutants of the *E. coli* DNA gyrase B subunit produced similar improvements in crystallizability and crystal quality.[121]

The mind-boggling number of possibilities that present themselves when considering site-directed mutagenesis as a tool for crystallizing a new target raises a few fundamental questions: which sites in the protein sequence should be chosen for mutagenesis, and what kind of mutation will produce the best results? For the most part, the strategy has until now remained trial-and-error in nature, and the relative paucity of empirical data on the subject still leaves much to be answered. For example, solubility and crystallization screening of about 30 different hydrophobic→Lys point mutants of the catalytic domain of HIV integrase led to crystallization of the Phe185Lys mutant.[122] In another case, obtainment of diffraction-quality crystals of the mycobacterial outer membrane channel was possible only after preparing about a dozen point mutants – either within zones of interspecies sequence variation or at predicted surface regions – of which an Ala96Arg mutant produced crystals diffracting to 2.5 Å resolution.[123] A Trp100Glu point mutation in wild-type human leptin produced a soluble protein that crystallized and allowed structure determination to 2.4 Å.[124] A Lys133Ile mutant of cyclophilin D gave crystals that diffracted to 1.7 Å resolution.[125] What becomes clear at this point is there has been no real consensus on what works best; however, recent work by Zygmunt Derewenda's group has illustrated the consistent effectiveness of incorporating mutations that reduce conformational surface entropy.[117,126] The underlying concept stems from the idea that surface residues with high conformational entropy create an 'entropic shield' that prevents the formation of stable intermolecular interactions required for crystal contact formation. Therefore, the selective replacement of conformationally labile surface residues with small amino acids, such as Ala, could reduce the entropic penalty of crystal contact formation in these regions. Lysine and glutamate residues have been identified as good candidates for mutagenic replacement because they are usually located on the protein surface,[127] their side chains are typically characterized by high conformational entropy,[128] and they are disfavored at protein – protein interfaces.[129] The hypothesis was emphatically validated by mutational work on RhoGDI, the human Rho-specific GDP dissociation inhibitor, of which several mutants were constructed (each containing between one and four Lys→Ala and/or Glu→Ala exchanges) and subjected to crystallization trials.[130,131] Almost all mutants exhibited altered crystallization properties and several new crystal forms were obtained, with one Glu154Ala/Glu155Ala double-mutant producing crystals that diffracted to 1.25 Å resolution.[131] Crystal structure determinations of the different mutants revealed that the mutated epitopes participated directly in crystal contact formation. This surface engineering strategy has since produced successful results for a number of proteins, and the current trend involves making multiple mutations within short Lys/Glu-rich clusters.[117,126] A recent example of this approach toward SBDD is provided by work from Merck scientists who investigated the tyrosine kinase domain of the insulin-like growth factor-1 receptor (IGF-1R), an anticancer target. Crystals of the unphosphorylated apo form of the IGF-1R kinase domain diffracted to only 2.7 Å resolution, a resolution which is too low for high-throughput structure-based lead optimization. To overcome this hurdle, three different multiple mutants were prepared, Lys1025Ala/Lys1026Ala, Glu1067Ala/Glu1069Ala, and Lys1237Ala/Glu1238Ala/Glu1239Ala, with the second double-mutant producing high-quality crystals diffracting to 1.5 Å resolution.[132] In the event that the solubility of the mutant protein is critically compromised, alternative mutations to polar residues with less conformational entropy and/or smaller size (e.g., Lys→Arg; Lys→Asp) can also help to promote crystallization.[133,134] The concept of rationally designing proteins to crystallize has been extended by Wingren and colleagues who have shown it is possible to replace short stretches of residues in β-strand-containing proteins with so-called packing 'cassettes' – crystal packing motifs that promote crystallization by generating specific crystal packing interactions.[135] As with any mutational work, the functional integrity of the mutant proteins needs to be evaluated prior to SBDD.

3.38.3.2.4.4 Mutational strategies II: directed evolution

Directed evolution approaches to producing soluble and crystallizable targets have recently emerged as powerful alternatives to site-directed mutagenesis.[136] These methods allow large mutant libraries to be generated through error-prone PCR and DNA shuffling. Mutants can be expressed as green fluorescent protein (GFP) fusions and visibly assayed for solubility and proper folding. The great advantage of the GFP-based directed evolution approach compared to rational surface-engineering is that it eliminates all of the guesswork related to mutant design and samples mutational space in a much more efficient way. Yang and co-workers demonstrated the superior utility of GFP-based directed evolution for obtaining soluble and crystallizable mutants of the protein RV2002 from *Mycobacterium tuberculosis*.[137] Mutants of RV2002 generated by GFP-based directed evolution had solubilities of at least $\sim 15\,mg\,mL^{-1}$, compared to the wild-type protein, which was expressed in *E. coli* only as insoluble inclusion bodies. An important observation raised by the authors in their crystallographic study was that the underlying contributions of the individual mutations toward the enhanced solubility of the Ile6Thr/Val47Met/Thr69Lys triple mutant appeared to involve a combination of altered intrinsic solubility and folding kinetics.[137] It is very likely that it would have been much more difficult, if not impossible, to rationally engineer soluble mutants as quickly or effectively.

3.38.3.2.4.5 Chemical modifications

An alternative strategy that deserves consideration before embarking on the more time-consuming approaches of ortholog screening, rational surface engineering, or directed evolution (or perhaps even as a final recourse in the event that the above methods fail) is to chemically modify the target protein prior to crystallization. Reductive alkylation of the protein sample with formaldehyde and dimethylamine-borane complex (DMAB) has proven to be the most successful chemical treatment strategy established thus far for crystallization of recalcitrant samples. Although the technique was originally used to improve the quality of poorly diffracting crystals, its use as a tool for de novo crystallization was pioneered by Ivan Rayment for the structure determination of myosin subfragment 1,[138] and a methylation protocol of general utility has been subsequently published.[139] The net consequence of this treatment is the dimethylation of all solvent-accessible lysine side chains as well as the N-terminal amino group. The resulting charge of the modified protein is not changed, however, its isoelectric point may be shifted slightly. As with surface-entropy-reducing mutagenesis, it is believed that the crystallizability of the chemically modified target may be enhanced as a result of reduced side chain entropy of dimethylated lysines. Additional studies have demonstrated that methylated proteins can produce novel crystal forms differing from unmodified samples,[140,141] and that selenomethionine-substituted specimens as well as multimeric complexes are amenable to this form of chemical treatment.[142] Other chemical approaches for the purpose of crystallization remain less explored, although a recent study has demonstrated the utility of deliberate oxidation in this regard. Working on a bacterial Ppx/GppA phosphatase homolog, Kristensen and co-workers were able to obtain two unique crystal forms after incubating their protein samples with 0.1% H_2O_2 for 1 h at room temperature prior to crystallization. The oxidized crystal forms were unique and were found to exhibit increased physical stability and better diffraction compared to the crystal form obtained with untreated samples.[143]

3.38.3.3 Milestone 3: From Crystal to Structure

3.38.3.3.1 Postcrystallization treatments to improve diffraction quality

One of the major disappointments regularly encountered in an SBDD program employing crystallography is the realization that crystals of the protein target are of poor quality and unsuitable for diffraction studies. The phenomena of low-resolution or poor-quality diffraction (e.g., characterized by severe mosaicity or anisotropy) usually stems from loose molecular packing within the crystal and/or a high internal solvent content. These symptoms often arise from crystals that otherwise appear to be physically perfect when viewed under the light microscope, adding to a researcher's frustration and invariably leading to serious doubts about proceeding with the project. Before capitulating, however, there are several postcrystallization treatments at hand that can improve the diffraction quality of protein crystals, sometimes drastically.[144] The most successful methods currently employed for improving diffraction quality at this stage are crystal dehydration and 'crystal annealing.' The latter method arose as a way to repair crystal lattice damage inflicted by flash-cooling techniques (rapid cooling of crystals to cryogenic temperatures),[145,146] which are routinely used to limit crystal radiation damage induced by exposure to high-intensity x-ray sources.[147–155] Crystal annealing involves warming a flash-cooled crystal to room temperature followed by repeated flash cooling, and has been shown to improve crystal mosaicity and diffraction resolution.[156,157] The method of crystal dehydration effectively reduces crystal solvent content, enforcing tighter crystal packing and frequently promoting remarkable improvements in diffraction quality. Several dehydration methods have emerged, ranging from slow controlled dehydration of the crystallization droplet over a period of days or weeks, to quick crystal soaks in a dehydrating solution for a few minutes.

A novel device has been described that allows accurate control of crystal water content by regulating the relative humidity of a gas stream enveloping the crystal.[158] This device, which is now available commercially, allows the diffraction properties of the crystal to be monitored in real time while the crystal is dehydrated until diffraction is improved. A recent study has demonstrated that the method of fast crystal dehydration coupled with crystal annealing can lead to astonishing improvements in diffraction quality, sometimes extending the diffraction limit of some apparently worthless crystal specimens by nearly tenfold.[159] Other postcrystallization treatments for improving diffraction resolution, including crystal soaking and crystal cross-linking, have been reviewed recently.[144]

3.38.3.3.2 Cryocrystallography in practice

As mentioned, cryocrystallography is the method of choice when collecting diffraction data at high-intensity X-ray sources due to its dampening effect on the diffusion of harmful free radicals through the crystal, allowing most crystals to survive long enough in the x-ray beam for collection of a complete single-crystal high-resolution data set. A critical objective of cryocrystallography is the formation of amorphous ice within and around the crystal upon cryocooling, as crystalline ice formation yields spurious diffraction that obscures the useful protein diffraction. Amorphous ice formation can be facilitated by the addition of chemical cryoprotectants, such as glycerol or polyethylene glycol, to the crystallization mother liquor.[154,160] Sometimes the crystallization mother liquor will already contain a cryoprotective formulation of reagents, thereby facilitating the cryocooling of crystals directly from their growth solutions. Several commercial crystallization screening kits containing cryo-ready reagents have been developed based on this strategy.[161] One possible unwanted side effect with regard to SBDD, however, is the proclivity of cryoprotectant molecules to appear in the active sites of enzymes, in which case the screening of alternative cryoprotectants might be needed. A plausible workaround, which can eliminate the need for conventional cryoprotectants, is the use of cryoprotective oils such as Paratone-N or highly liquid paraffin oil, which work by forming a protective shield around the crystal.[148,162,163] Another cooling method that circumvents the need for penetrative cryoprotectants is high-pressure crystal cooling, which was first explored as early as 1973[164] and recently revisited.[165,166] A prototype pressure device has been constructed at the Cornell High Energy Synchrotron Source (CHESS) that allows for cryo-loop crystal mounting and high-pressure (200 MPa) cryocooling. Preliminary results with the device, which have shown significant improvement of diffraction quality for all protein crystals studied, are very encouraging.[166]

3.38.3.3.3 Phasing methods for drug discovery

Once protein crystals can be grown reproducibly in a high-quality crystal form suitable for SBDD, the next major bottleneck is finding the solution to the phase problem. If coordinates of a related homolog or domain are available, model-based phasing using the molecular replacement method is the preferred strategy because of its relative speed.[167] Modern molecular replacement software implementing improved search features such as likelihood-enhanced rotation and translation functions,[168,169] improved procedures for constructing molecular replacement search models,[170] and a burgeoning PDB databank of 3D macromolecular structure data,[171,172] are literally pushing the boundaries of what can be accomplished with the method.

When preexisting structural data is absent or insufficient for molecular replacement phasing, other methods must be employed. Historically, the most commonly used structure-solving technique was multiple isomorphous replacement (MIR), which relies on the preparation of derivative crystals in which one or more types of heavy atoms are bound specifically and uniformly to the macromolecules within the crystal. The outcome of the MIR method is additionally contingent on the heavy-atom derivatives being truly isomorphous, i.e., bearing no (or at most, extremely limited) alterations in molecular structure and unit-cell dimensions. Successful phase determination by MIR is consequently rate-limited by the speed with which suitable heavy-atom derivatives can be obtained. The advancement of multiple-wavelength anomalous dispersion phasing[114,173] and the construction of tunable synchrotron x-ray beamlines worldwide have made multiple-wavelength anomalous dispersion the method of choice for experimental phase determination, particularly for high-throughput (HT) SBDD applications. One of the big advantages multiple-wavelength anomalous dispersion phasing offers over MIR and related methods such as SIRAS and MIRAS (single or multiple-isomorphous replacement with anomalous scattering) is the reduction of systematic error since all data are measured on a single sample. The resulting phase angles are therefore more accurately determined and the resulting electron density maps are typically of higher quality than maps calculated with phases derived from isomorphous methods. Multiple-wavelength anomalous dispersion phasing requires the presence of a suitable number of atoms having an x-ray absorption edge in the energy range easily accessible by tunable synchrotron sources (typically $\lambda = 0.8–1.3\,Å$). Most multiple-wavelength anomalous dispersion phasing experiments are carried out using selenomethionine-substituted (SeMet) proteins,[174,175] although sometimes SeMet multiple-wavelength anomalous dispersion phasing can be an inappropriate method for phase determination.[176] In such situations, multiple-wavelength anomalous dispersion phasing can be carried out with bound elements of atomic numbers ~20–40 (or above ~60), as they have synchrotron-accessible absorption edges. These elements can often be found as natural metal cofactors in proteins (such as transition metals) or can be diffused into protein crystals using conventional heavy-atom soaking methods.[177–180] Prescreening for the degree of SeMet or heavy atom incorporation into the target protein can be rapidly carried out by mass spectrometry.[181] Alternatively, native polyacrylamide gel electrophoresis experiments allow for a quick and simple assessment of heavy atom binding, optimal binding concentrations, and impact on protein stability.[182]

Explorations of novel approaches for introducing anomalous scattering elements into protein crystals for the purpose of multiple-wavelength anomalous dispersion phasing has led to a robust collection of new phasing alternatives. For example, chemically modified ligands can be used as rational phasing tools, such as halogenated nucleotides for multiple-wavelength anomalous dispersion phasing of nucleotide-binding proteins[183] or selenium-substituted saccharides for carbohydrate-binding proteins.[184] Multiple-wavelength anomalous dispersion phasing with krypton has been demonstrated as a feasible technique,[185] as has cryosoaking with halides[186,187] or mono- and polyvalent cations.[188,189] Several other unique procedures for producing derivatives have appeared over the years and continued exploration in this field will surely generate additional phasing alternatives. Moreover, modern developments in the area of utilizing soft x-rays ($\lambda = 1.5$–3.0 Å) for sulfur-based anomalous phasing – a technique pioneered more than 20 years ago by Hendrickson & Teeter for the structure determination of crambin[190] – may eventually circumvent the need for derivatization in most cases altogether.[191–194]

3.38.3.4 Milestone 4: Structure to Lead

Once the crystal structure of a particular target has been solved, and before SBDD can begin, the crystal form should be thoroughly scrutinized for the presence of characteristic pathologies that might prolong or even prohibit the design process. Is the diffraction resolution high enough (i.e., a resolution of <2.5 Å, or preferably <2.2 Å) to obtain reliable protein:ligand bond distances, to unambiguously assign rotameric conformations of side chains and ligand functions, and to visualize water molecules? Is the active site unoccupied, or are there naturally occurring ligands or other nonwater molecules residing within it that may hamper ligand-soaking experiments? Is the active site open to solvent channels in the crystal or is it obstructed by crystal packing? Is the active site conformationally restricted or otherwise rendered incapable of undergoing structural rearrangements necessary for ligand binding? In situations where such pathologies do exist, the only solution may be to engineer a new variant and screen for a new crystal form, as was demonstrated for urokinase inhibitor design discussed earlier in this chapter (Section 3.38.3.2.3.3). Barring any of the aforementioned problems, the crystallographer can pursue a number of integrated strategies to generate and harvest the highly valued structural data. At this stage, an integrated effort between the crystallographers, the modelers, and the medicinal chemists cannot be overemphasized, as an early collaboration will undoubtedly speed the progression from hits to leads, and from leads to compounds with therapeutic potential (**Figure 7**).

3.38.3.4.1 Crystallography-driven lead characterization: the traditional view

The conventional approach to lead discovery involves screening the target against a high-throughput screening (HTS) library (*see* 3.32 High-Throughput and High-Content Screening). The HTS library can contain upwards of 2 million compounds and therefore implicitly requires high-throughput methods to make the time scale of the process manageable. A biochemical assay amenable to high throughput is normally devised for the screening, and compounds with an IC_{50} of $\leq 20\,\mu M$ are typically selected for further testing (*see* 3.27 Optical Assays in Drug Discovery; 3.28 Fluorescence Screening Assays; 3.29 Cell-Based Screening Assays). These 'hits' may be handed over to medicinal chemists for optimization of potency or other druglike properties, or delivered to the crystallographers for structural characterization of the target:ligand complex. Given the resources, both investigatory avenues might be pursued concurrently, and it is therefore vital for the crystallographers to keep apace with the medicinal chemists. The iterative nature of SBDD, in that each cycle of design, synthesis, and bioassay is followed by the crystal structure determination of the target:ligand complex, requires that the individual disciplines work effectively with one another (*see* 4.24 Structure-Based Drug Design – The Use of Protein Structure in Drug Discovery). This iterative approach can be applied to SBDD in several ways. A powerful technique is de novo drug design – the design of novel chemical structures from scratch – as guided by the crystal structure of an empty active site (*see* 4.13 De Novo Design). Another strategy is lead optimization, which begins with the crystal structure of a target:lead complex. The lead may have been discovered through the chemical or patent literature, through HTS of compound libraries, or sourced from de novo design. Once the complex of the lead and target protein has been elucidated crystallographically, the lead can be modified in order to optimize its chemical or biological properties (*see* 5.01 The Why and How of Absorption, Distribution, Metabolism, Excretion, and Toxicity Research).

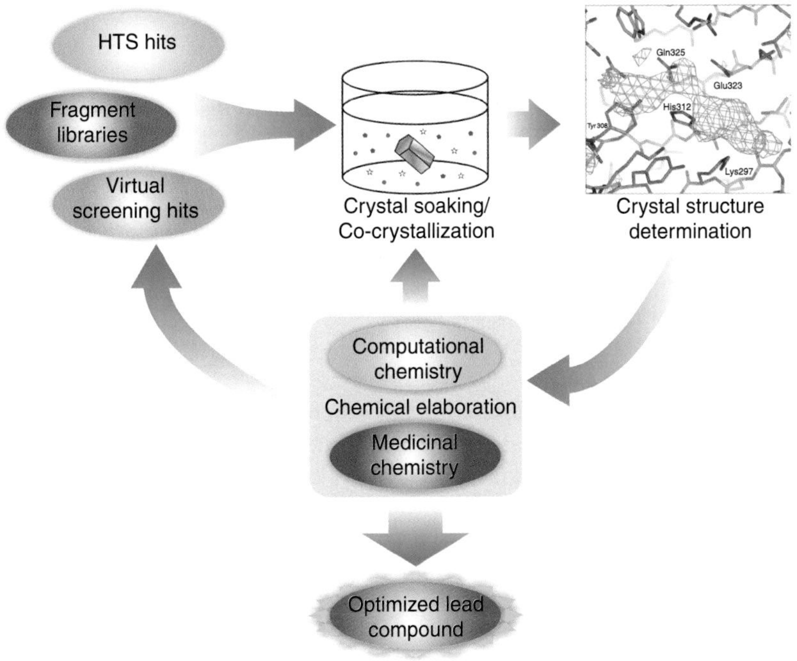

Figure 7 Flowpath illustrating the integrated approach to crystallography-driven SBDD. Small molecule ('fragment') libraries or promising compounds identified from conventional HTS or virtual screening are subjected to co-crystallization or crystal-soaking trials with the target macromolecule (*see* 3.32 High-Throughput and High-Content Screening; 3.41 Fragment-Based Approaches; 4.19 Virtual Screening). Crystal structure determination of the target:ligand complex provides critical data for computational exploration and medicinal chemistry decisions. Characterization of the target:ligand binding interactions may also guide selective screening of focused libraries, which can produce new hits for the SBDD cycle. Several iterative cycles of this process may lead to an optimized lead compound.

For the crystallographer, a fundamental component of the design cycle is successful structure determination of the complex to high resolution. Complexes are obtained by target:ligand co-crystallization or by diffusing the ligand into preformed crystals (soaking). Both methods have their particular strengths and weaknesses. Soaking takes advantage of the fact that macromolecular crystals are solvent rich[195] and contain integral networks of solvent channels; however, the size and configuration of the channels within the crystal lattice will place an upper limit on the size of ligands that may be diffused in (this is usually not an issue for small druglike molecules). As mentioned, crystal lattice restraints can sometimes hinder conformational rearrangements of the target, which might necessarily coincide with ligand binding or, on the other hand, any rearrangements that do take place may disrupt the crystal lattice, causing irreversible damage. With the co-crystallization approach the target:ligand complex is formed before crystallization is carried out, but any alterations in the physicochemical properties of the complex relative to the native protein may preclude crystallization of the complex under native conditions. Herein lie some of the perceived advantages of the soaking method: native crystals can be grown reproducibly, and the conservation of isomorphism between native and soaked crystals allows for rapid detection of the bound ligand by difference Fourier methods. Does this mean that co-crystallization methods are less amenable to high-throughput SBDD? On the contrary, the Plexxikon group has recently demonstrated the effectiveness of a co-crystallization approach in their SBDD program targeting human phosphodiesterases.[196]

3.38.3.4.1.1 Ligand soaking and co-crystallization: practical considerations

In order for a compound to be of value to SBDD, it must be present at a high enough concentration in the crystallization drop to promote a high occupancy of binding to the target and therefore produce interpretable electron density. A source of difficulty can be limited ligand solubility, which can impose a constraint on binding occupancy and limit the usefulness of the crystallographic data. If something is known about the target:ligand dissociation constant, K_D, the problem can be approached using the formula [ligand] = [protein] + nK_D, where [ligand] and [protein] are the concentrations in the crystallization drop, and n is an 'excess factor.' When $K_D \ll$ [protein], for example, when a ligand with a K_D in the low micromolar range is added to a drop containing a millimolar concentration of protein, then [ligand] ≅ [protein]. In this case, maximum binding occupancy should be achieved by adding only a slight molar excess of ligand. On the other hand, when the protein is crystallized at low concentration and/or a low affinity ligand is used (i.e., $K_D \cong$ [protein]), then the nK_D term becomes appreciable. A general guideline is to use an n value of at least 5–10, although trial-and-error comes into play here. An added complication is that the above calculation relies on the assumption that the measured K_D under assay conditions is similar to the K_D in the crystallization drop, which is very likely untrue. For a polar ligand, high salt concentrations will normally raise the K_D in the crystallization drop, while the opposite will be true for nonpolar ligands. Potentially adding to the puzzle is the effect of cryoprotectants and cosolvents, such as polyethylene glycol, which have been known to sometimes interfere with ligand binding by competing for the ligand-binding site. All considered, a 'more is better' philosophy for ligand soaking is often adhered to, however poorly soluble ligands may require the use of a suitable organic solvent such as dimethyl sulfoxide (DMSO), ethanol, methanol, or hexafluoropropanol.[180]

3.38.3.4.2 Crystallography and 'fragonomics': the emerging area of fragment-based drug design

Novel approaches to drug discovery have been developed recently that utilize biophysical methods to screen collections of basic chemical building blocks, termed 'fragments' (*see* 3.41 Fragment-Based Approaches).[197,198] Fragment libraries tend to contain ~100–1000 compounds and are therefore much smaller than conventional HTS libraries. The compounds themselves are also smaller and simpler (average molecular weight 100–250 Da; 8–18 nonhydrogen atoms) than typical HTS compounds and, consequently, fragments bind to target proteins weakly with an affinity that is below the detection limit of conventional HTS methods. Biophysical techniques such as NMR[199] (*see* 3.39 Nuclear Magnetic Resonance in Drug Discovery) or x-ray crystallography,[200–203] however, are powerful screening methods for fragment-based searches. It is now generally accepted that target:fragment interactions as low as millimolar K_D can be reliably detected using NMR or x-ray crystallography. X-ray-based screening has the added advantage of inherently producing high-resolution structural data that can provide a basis for computational design and medicinal chemistry decisions. But what constitutes a good fragment library? An analysis of fragment hits against a variety of targets has led the Astex group to formulate a 'rule of three'[204] for describing a typical fragment 'hit': molecular weight ≤300 Da; hydrogen bond donors ≤3; hydrogen bond acceptors ≤3; octanol/water partition coefficient (log P) ≤3; and number of rotatable bonds ≤3.

The first well-publicized crystal screening method, termed CrystalLEAD, was developed in the Abbott labs, and the practical application of the method was initially demonstrated on the anticancer target urokinase.[200,205] In this study, shape-diverse compound mixtures were soaked into crystals of an engineered form of the protein, termed micro-urokinase (Section 3.38.3.2.3.3). Four mixtures yielded evidence of binding by five novel ligands, with one mixture containing two binders where the less potent ligand was only discovered after rescreening with the most potent inhibitor removed. First-round optimization of one of the leads, 8-hydroxy-2-aminoquinoline, to 8-aminopyrimidyl-2-aminoquinoline, resulted in a 100-fold increase in inhibitor potency (K_i from 56 μM to 370 nM), and provided a favorable starting point for further lead development.

3.38.3.4.2.1 'Scaffold-based' discovery and tethering: variations on a theme

The demonstrated success of fragment-based methods has bolstered further exploration and development of the approach. The Plexxikon group has recently introduced a crystallography-driven screening technique called 'scaffold-based' drug design, which they implemented in their program to develop novel inhibitors of human phosphodiesterases (PDEs).[196] According to the Plexxikon definition, the main difference between libraries containing molecular scaffolds vs. molecular fragments is that the former comprises slightly larger molecules (average molecular weight ~250 Da) containing additional functional groups that enable them to bind to the target with an affinity detectable by conventional HTS. With this approach, a preliminary HTS assay is used at a compound concentration of ~200 μM for the initial selection of candidates. The target protein is then crystallized in the presence of 1 mM of the scaffold molecule, with ranges of pH, precipitants, and additives screened to help increase the chances of co-crystallization. The authors reported an overall co-crystallization success rate of 85% for the 316 chemical scaffolds selected for structural studies with different PDEs.[196]

Scientists at Sunesis have developed a proprietary fragment-based method, called Tethering, which offers a site-directed means to probe ligand binding to a target. Tethering takes advantage of native or engineered surface-exposed cysteines on the target to capture thiol-containing ligands by formation of a mixed disulfide.[206–208] Crystallographic analysis of the covalently linked complex provides a powerful means for rapid identification of fragments that can be converted into reversible inhibitors, as exemplified by the group's development of novel allosteric and active-site caspase inhibitors.[209,210]

3.38.3.4.2.2 Lead identification

Once the interactions between a fragment and its target have been characterized by crystallographic methods, iterative structure-driven chemistry can be applied to generate a drug-size lead compound (*see* 3.41 Fragment-Based Approaches). These subsequent steps can be subdivided into four distinct categories: fragment optimization; fragment linking; fragment evolution; and fragment self-assembly[198] (**Figure 8**). In fragment evolution, a lead molecule is iteratively generated, or 'evolved', by building away from the starting fragment into other regions of the binding pocket. Fragment linking defines the process of chemically joining two fragments that bind adjacent sites to generate a larger compound with greater potency. Fragment optimization involves altering properties of a lead molecule including selectivity or efficacy, such as a 'SAR by x-ray' approach.[211] The method of fragment self-assembly utilizes mixtures of fragments adorned with reactive groups (so-called 'dynamic combinatorial libraries'[212]), which link together when situated in close proximity (e.g., within a target's active site) (*see* 3.42 Dynamic Ligand Assembly). In this way, the target catalyzes the synthesis of its own – and presumably highest affinity – lead from a library of fragments. An x-ray-guided extension to fragment self-assembly, termed 'dynamic combinatorial x-ray crystallography' (DCX), allows self-assembled leads to be generated from dynamic combinatorial libraries in the presence of protein crystals, permitting direct observation of the binding interaction with the target.[213]

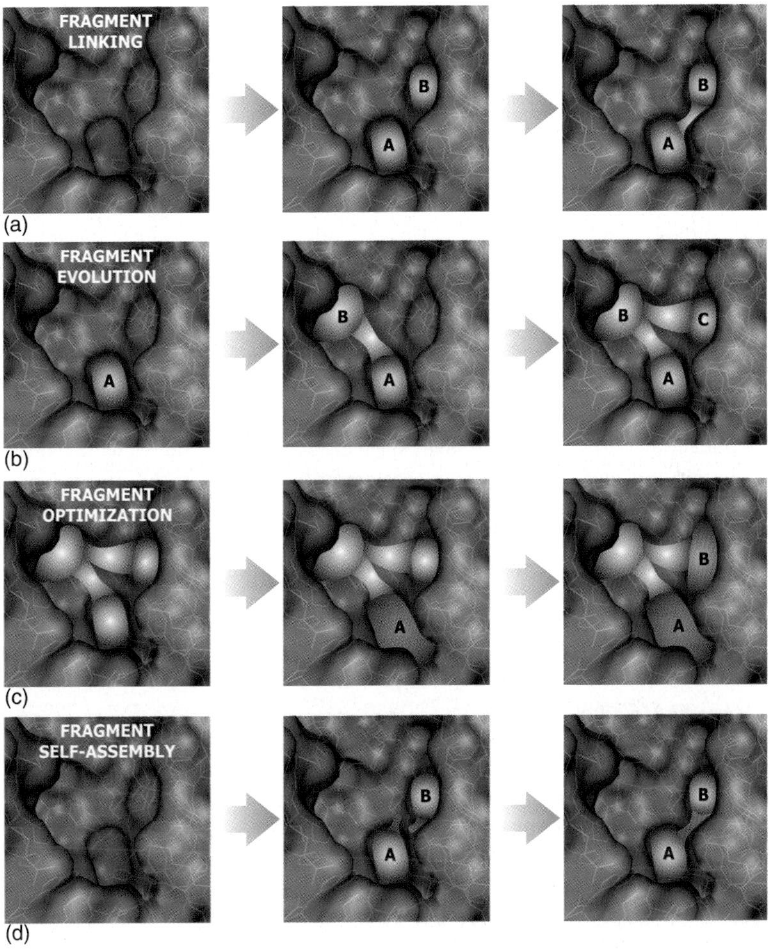

Figure 8 Fragment-based lead generation and optimization approaches. (a) Fragment linking involves the joining of fragments identified in vicinal binding sites with a chemical linker, leading to a larger molecule with higher binding affinity. (b) Fragment evolution proceeds by iteratively building out from a single starting fragment to create a larger, more complex molecule that interacts with neighboring pockets. (c) Fragment optimization entails the rational re-engineering of a lead molecule in order to improve or alter specific properties of the molecule (e.g., affinity, selectivity, oral bioavailability). (d) Fragment self-assembly exploits the ability of reactive fragments with complementary functional groups to assemble into a larger, more potent molecule when situated in proximal binding pockets on a target template.

3.38.4 The Era of High-Throughput Crystallography in Lead Discovery

In the past, the determination of a protein x-ray crystal structure typically required several years and, in difficult cases, even decades. The main bottleneck of obtaining suitable amounts of pure protein samples was drastically reduced with the advent of molecular biology tools developed in the 1980s and 1990s (*see* 3.19 Protein Production for Three-Dimensional Structural Analysis). The structure determination process, which has traditionally been a lengthy and laborious undertaking, has greatly benefited from sweeping technological and methodological breakthroughs. Advances in task automation, miniaturization, and parallelization have ushered in a new era of 'high-throughput crystallography.' The original notion of high-throughput crystallography was not conceived by the pharmaceutical sector (which had diminishing interests in x-ray crystallography and was instead focusing on other HT technologies), but by the structural genomics initiatives that were established with the goal of solving crystal structures of representatives from all known protein families (*see* 3.25 Structural Genomics).[214,215] Such ambitious goals bore the immediate need to establish new methodologies for producing hundreds of proteins in parallel, for creating HT systems to dispense and monitor thousands of crystallization experiments per day, and for determining the resultant x-ray structures in a fully automated fashion. The technological advances that have emerged from the various structural genomics initiatives have not only helped to improve our understanding of protein fold space, but has also substantially benefited the pharmaceutical industry by adding x-ray crystallography to a preexisting armamentarium of HT tools for drug discovery.

For the design of DNA constructs, a genomics approach is usually taken at the outset: bioinformatics tools are used to mine available databases for homologous sequences from various organisms, and multiple-sequence alignments are examined to predict a possible domain structure for the protein (*see* 3.15 Bioinformatics). Several orthologs are generated with different start/stop points delineating the domain(s) of interest, and appropriate oligonucleotide primers are either synthesized or ordered en masse. Microtiter plates (typically 96-well or higher) are used for automated HT-PCR, cloning, and testing of expression.[216–220] The attainability of large-scale automated production and purification of proteins has been demonstrated in the construction of an elaborate robotics system capable of handling 96 parallel 65–70 ml bacterial cultures with average yields of 10 mg of purified protein per sample.[14] Bacterial cell-free systems, such as those based on the *E. coli* S30 extract, offer opportunities for automation of the production of milligram quantities of protein in a test tube.[221,222]

Robotic systems for automated crystallization with smaller volumes and higher density plating configurations (96-, 384-, and 1536-well formats), along with imaging systems for automated crystallization drop monitoring and scoring, have enabled HT crystallization on a miniaturized scale and resulted in drastic reductions in space, material, and cost.[223–225] Hardware and software solutions for automated crystal mounting, alignment, and data collection can allow rapid, unmanned data collection for synchrotron[226–229] and laboratory[230,231] sources. Software developments for the automation of macromolecular structure determination are enabling a rapid and seamless transfer of diffraction data through the processing, phasing, and model-building pipeline.[232–237] Adding to this are new programs that provide solutions for automatic generation of ligand coordinates and topologies[238] and automated ligand building into electron-density maps.[239–243] Continued development of HT crystallography hardware and software applications and a global integration of the individual processes will further hasten the lead identification and optimization cycle.

References

1. Fleming, A. *B. J. Exp. Pathol.* **1929**, *10*, 226–236.
2. Hillisch, A.; Hilgenfeld, R. *Modern Methods of Drug Discovery*; Birkhäuser Verlag: Basel, Switzerland, 2003.
3. Hillisch, A.; Pineda, L. F.; Hilgenfeld, R. *Drug Disc. Today* **2004**, *9*, 659–669.
4. Stevens, R. C. *Drug Disc. World* **2003**, *4*, 35–48.
5. Steitz, T. A.; Ludwig, M. L.; Quiocho, F. A.; Lipscomb, W. N. *J. Biol. Chem.* **1967**, *242*, 4662–4668.
6. Lipscomb, W. N.; Hartsuck, J. A.; Quiocho, F. A.; Reeke, G. N., Jr. *Proc. Natl. Acad. Sci. USA* **1969**, *64*, 28–35.
7. Quiocho, F. A.; Lipscomb, W. N. *Adv. Prot. Chem.* **1971**, *25*, 1–78.
8. Byers, L. D.; Wolfenden, R. *Biochemistry* **1973**, *12*, 2070–2078.
9. Ondetti, M. A.; Rubin, B.; Cushman, D. W. *Science* **1977**, *196*, 441–444.
10. Natesh, R.; Schwager, S. L. U.; Sturrock, E. D.; Acharya, K. R. *Nature* **2003**, *421*, 551–554.
11. Natesh, R.; Schwager, S. L. U.; Evans, H. R.; Sturrock, E. D.; Acharya, K. R. *Biochemistry* **2004**, *43*, 8718–8724.
12. Stevens, R. C. *Structure* **2000**, *8*, R177–R185.
13. Gilbert, M.; Albala, J. S. *Curr. Opin. Chem. Biol.* **2001**, *6*, 102–105.
14. Lesley, S. A. *Protein Exp. Purif.* **2001**, *22*, 159–164.
15. Nagar, B.; Hantschel, O.; Young, M. A.; Scheffzek, K.; Veach, D.; Bornmann, W.; Clarkson, B.; Superti-Furga, G.; Kuriyan, J. *Cell* **2003**, *112*, 859–871.
16. Hogg, T.; Mechold, M.; Malke, H.; Cashel, M.; Hilgenfeld, R. *Cell* **2004**, *117*, 57–68.
17. Gao, X.; Bain, K.; Bonanno, J. B.; Buchanan, M.; Henderson, D.; Lorimer, D.; Marsh, C.; Reynes, J. A.; Sauder, J. M.; Schwinn, K.; Thai, C.; Burley, S. K. *J. Struct. Funct. Genomics* **2005**, *6*, 129–134.

18. Prilusky, J.; Felder, C. E.; Zeev-Ben-Mordehai, T.; Rydberg, E. H.; Man, O.; Beckmann, J. S.; Silman, I.; Sussman, J. L. *Bioinformatics* **2005**, *21*, 3435–3438.
19. Linding, R.; Jensen, L. J.; Diella, F.; Bork, P.; Gibson, T. J.; Russell, R. B. *Structure* **2003a**, *11*, 1453–1459.
20. Ward, J. J.; Sodhi, J. S.; McGuffin, L. J.; Buxton, B. F.; Jones, D. T. *J. Mol. Biol.* **2004**, *337*, 635–645.
21. Linding, R.; Russell, R. B.; Neduva, V.; Gibson, T. J. *Nucleic Acids Res.* **2003b**, *31*, 3701–3708.
22. Dosztanyi, Z.; Csizmok, V.; Tompa, P.; Simon, I. *J. Mol. Biol.* **2005**, *347*, 827–839.
23. Romero, P.; Obradovic, Z.; Li, X.; Garner, E. C.; Brown, C. J.; Dunker, A. K. *Proteins* **2001**, *42*, 38–48.
24. Coeytaux, K.; Poupon, A. *Bioinformatics* **2005**, *21*, 1891–1900.
25. Yang, Z. R.; Thomson, R.; McNeil, P.; Esnouf, R. M. *Bioinformatics* **2005**, *21*, 3369–3376.
26. Vucetic, S.; Brown, C. J.; Dunker, A. K.; Obradovic, Z. *Proteins* **2003**, *52*, 573–584.
27. Peng, K.; Vucetic, S.; Radivojac, P.; Brown, C. J.; Dunker, A. K.; Obradovic, Z. *J. Bioinform. Comput. Biol.* **2005**, *3*, 35–60.
28. Moult, J. *Curr. Opin. Struct. Biol.* **2005**, *15*, 285–289.
29. Vucetic, S.; Obradovic, Z.; Vacic, V.; Radivojac, P.; Peng, K.; Iakoucheva, L. M.; Cortese, M. S.; Lawson, J. D.; Brown, C. J.; Sikes, J. G.; Newton, C. D.; Dunker, A. K. *Bioinformatics* **2005**, *21*, 137–140.
30. Englander, J. J.; Del Mar, C.; Li, W.; Englander, S. W.; Kim, J. S.; Stranz, D. D.; Hamuro, Y.; Woods, V. L., Jr. *Proc. Natl. Acad. Sci. USA* **2003**, *100*, 7057–7062.
31. Pantazatos, D.; Kim, J. S.; Klock, H. E.; Stevens, R. C.; Wilson, I. A.; Lesley, S. A.; Woods, V. L., Jr. *Proc. Natl. Acad. Sci. USA* **2004**, *101*, 751–756.
32. Bornhorst, J. A.; Falke, J. J. *Methods Enzymol.* **2000**, *326*, 245–254.
33. Lichty, J. J.; Malecki, J. L.; Agnew, H. D.; Michelson-Horowitz, D. J.; Tan, S. *Protein Expr. Purif.* **2005**, *41*, 98–105.
34. Sachdev, D.; Chirgwin, J. M. *Methods Enzymol.* **2000**, *326*, 312–321.
35. LaVallie, E. R.; Lu, Z.; Diblasio-Smith, E. A.; Collins-Racie, L. A.; McCoy, J. M. *Methods Enzymol.* **2000**, *326*, 322–340.
36. Smith, D. B. *Methods Enzymol.* **2000**, *326*, 254–270.
37. Smyth, D. R.; Mrozkiewicz, M. K.; McGrath, W. J.; Listwan, P.; Kobe, B. *Protein Sci.* **2003**, *12*, 1313–1322.
38. Yee, A.; Chang, X.; Pineda-Lucena, A.; Wu, B.; Semesi, A.; Le, B.; Ramelot, T.; Lee, G. M.; Bhattacharyya, S.; Gutierrez, P. et al. *Proc. Natl. Acad. Sci. USA* **2002**, *99*, 1825–1830.
39. Morton, C. L.; Potter, P. M. *Mol. Biotechnol.* **2000**, *16*, 193–202.
40. Li, P.; Gao, X. G.; Arellano, R. O.; Renugopalakrishnan, V. *Protein Expr. Purif.* **2001**, *22*, 369–380.
41. Hamilton, S. R.; Bobrowicz, P.; Bobrowicz, B.; Davidson, R. C.; Li, H.; Mitchell, T.; Nett, J. H.; Rausch, S.; Stadheim, T. A.; Wischnewski, H. et al. *Science* **2003**, *301*, 1244–1246.
42. Sulzenbacher, G.; Gruez, A.; Roig-Zamboni, V.; Spinelli, S.; Valencia, C.; Pagot, F.; Vincentelli, R.; Bignon, C.; Salomoni, A.; Grisel, S. et al. *Acta Crystallogr.* **2002**, *D58*, 2109–2115.
43. Clark, E. D. B. *Curr. Opin. Biotechnol.* **1998**, *9*, 157–163.
44. Armstrong, N.; de Lencastre, A.; Gouaux, E. *Protein Sci.* **1999**, *8*, 1475–1483.
45. Chen, X. S.; Stehle, T.; Harrison, S. C. *EMBO J.* **1998**, *17*, 3233–3240.
46. Hofmeister, F. *Arch. Exp. Pathol. Pharmakol. (Leipzig)* **1888**, *24*, 247–260.
47. Kunz, W.; Henle, J.; Ninham, B. W. *Curr. Opin. Coll. Interface Sci.* **2004**, *9*, 19–37.
48. Arakawa, T.; Timasheff, S. N. *Methods Enzymol.* **1985**, *114*, 49–77.
49. Arakawa, T.; Bhat, R.; Timasheff, S. N. *Biochemistry* **1990**, *29*, 1924–1931.
50. Gekko, K.; Timasheff, S. N. *Biochemistry* **1981**, *20*, 4677–4686.
51. Womack, M. D.; Kendall, D. A.; MacDonald, R. C. *Biochim. Biophys. Acta* **1983**, *733*, 210–215.
52. Asakura, T.; Adachi, K.; Schwartz, E. *J. Biol. Chem.* **1978**, *253*, 6423–6425.
53. Hedman, P. O.; Gustafsson, J. G. *Anal. Biochem.* **1984**, *138*, 411–415.
54. Gerlsma, S. Y. *J. Biol. Chem.* **1968**, *243*, 957–961.
55. Lee, J. C.; Timasheff, S. N. *J. Biol. Chem.* **1981**, *256*, 7193–7201.
56. Arakawa, T.; Timasheff, S. N. *Biochemistry* **1982**, *21*, 6536–6544.
57. Lindwall, G.; Chau, M.-F.; Gardner, S. R.; Kohlstaedt, L. A. *Protein Eng.* **2000**, *13*, 67–71.
58. Jancarik, J.; Pufan, R.; Hong, C.; Kim, S. H.; Kim, R. *Acta Crystallogr.* **2004**, *D60*, 1670–1673.
59. Collins, B. K.; Tomanicek, S. J.; Lyamicheva, N.; Kaiser, M. W.; Mueser, T. C. *Acta Crystallogr.* **2004**, *D60*, 1674–1678.
60. Hünefeld, F. L.; *Der Chemismus in der thierischen Organisation*; F.A. Brockhaus: Leipzig, Germany, 1840, p 160.
61. McPherson, A. *Crystallization of Biological Macromolecules*; CSHL Press: New York, 1999, pp 1–10.
62. Hogg, T.; Smatanova, I. K.; Bezouska, K.; Ulbrich, N.; Hilgenfeld, R. *Acta Crystallogr.* **2002**, *D58*, 1734–1739.
63. Gevaert, K.; Vandekerckhove, J. *Electrophoresis* **2000**, *21*, 1145–1154.
64. Yates, J. R., III. *Trends Genet.* **2000**, *16*, 5–8.
65. Yates, J. R., III. *Annu. Rev. Biophys. Biomol. Struct.* **2004**, *33*, 297–316.
66. Ferre-D'Amare, A. R.; Burley, S. K. *Structure* **1994**, *2*, 357–359.
67. D'Arcy, A. *Acta Crystallogr.* **1994**, *D50*, 469–471.
68. Ferre-D'Amare, A. R.; Burley, S. K. *Methods Enzymol.* **1997**, *276*, 157–166.
69. Zulauf, M.; D'Arcy, A. *J. Cryst. Growth* **1992**, *122*, 102–106.
70. Han, K. K.; Martinage, A. *Int. J. Biochem.* **1992**, *24*, 19–28.
71. Baker, H. M.; Day, C. L.; Norris, G. E.; Baker, E. N. *Acta Crystallogr.* **1994**, *D50*, 380–384.
72. Grueninger-Leitch, F.; D'Arcy, A.; D'Arcy, B.; Chène, C. *Protein Sci.* **1996**, *5*, 2617–2622.
73. Hoover, D. M.; Schalk-Hihi, C.; Chou, C.; Menon, S.; Wlodawer, A.; Zdanov, A. *Eur. J. Biochem.* **1999**, *262*, 134–141.
74. Dale, G. E.; D'Arcy, B.; Yuvaniyama, C.; Wipf, B.; Oefner, C.; D'Arcy, A. *Acta Crystallogr.* **2000**, *D56*, 894–897.
75. Davis, S. J.; Ikemizu, S.; Collins, A. V.; Fennelly, J. A.; Harlos, K.; Jones, E. Y.; Stuart, D. I. *Acta Crystallogr.* **2001**, *D57*, 605–608.
76. Tarentino, A. L.; Plummer, T. H., Jr. *Methods Enzymol.* **1994**, *230*, 44–57.
77. Iwase, H.; Hotta, K. *Methods Mol. Biol.* **1993**, *14*, 151–159.
78. Uchida, Y.; Tsukada, Y.; Sugimori, T. *J. Biochem. (Tokyo)* **1979**, *86*, 1573–1585.
79. Binns, K. L.; Taylor, P. P.; Sicheri, F.; Pawson, T.; Holland, S. J. *Mol. Cell. Biol.* **2000**, *20*, 4791–4805.
80. Yang, J.; Cron, P.; Good, V. M.; Thompson, V.; Hemmings, B. A.; Barford, D. *Nat. Struct. Biol.* **2002**, *9*, 940–944.

81. Yang, J.; Cron, P.; Thompson, V.; Good, V. M.; Hess, D.; Hemmings, B. A.; Barford, D. *Mol. Cell.* **2002**, *9*, 1227–1240.
82. Prongay, A. J.; Smith, T. J.; Rossmann, M. G.; Ehrlich, L. S.; Carter, C. A.; McClure, J. *Proc. Natl. Acad. Sci. USA* **1990**, *87*, 9980–9984.
83. Yu, X. C.; Sturrock, E. D.; Wu, Z.; Biemann, K.; Ehlers, M. R.; Riordan, J. F. *J. Biol. Chem.* **1997**, *272*, 3511–3519.
84. Gordon, K.; Redelinghuys, P.; Schwager, S. L.; Ehlers, M. R.; Papageorgiou, A. C.; Natesh, R.; Acharya, K. R.; Sturrock, E. D. *Biochem. J.* **2003**, *371*, 437–442.
85. Nienaber, V.; Wang, J.; Davidson, D.; Henkin, J. *J. Biol. Chem.* **2000**, *275*, 7239–7248.
86. Kimber, M. S.; Vallee, F.; Houston, S.; Nečakov, A.; Skarina, T.; Evdokimova, E.; Beasley, S.; Christendat, D.; Savchenko, A.; Arrowsmith, C. H. et al. *Proteins* **2003**, *51*, 562–568.
87. Page, R.; Grzechnik, S. K.; Canaves, J. M.; Spraggon, G.; Kreusch, A.; Kuhn, P.; Stevens, R. C.; Lesley, S. A. *Acta Crystallogr.* **2003**, *D59*, 1028–1037.
88. Rupp, B. *J. Struct. Biol.* **2003**, *142*, 162–169.
89. Kwong, P. D.; Wyatt, R.; Desjardins, E.; Robinson, J.; Culp, J. S.; Hellmig, B. D.; Sweet, R. W.; Sodroski, J.; Hendrickson, W. A. *J. Biol. Chem.* **1999**, *274*, 4115–4123.
90. Kwong, P. D.; Wyatt, R.; Robinson, J.; Sweet, R. W.; Sodroski, J.; Hendrickson, W. A. *Nature* **1998**, *393*, 648–659.
91. Kendrew, J. C.; Parrish, R. G.; Marrack, J. R.; Orlans, E. S. *Nature* **1954**, *174*, 946–949.
92. Kendrew, J. C.; Parrish, R. G. *Proc. R. Soc. Lond. A* **1956**, *238*, 305–324.
93. Campbell, J. W.; Duee, E.; Hodgson, G.; Mercer, W. D.; Stammers, D. K.; Wendell, P. L.; Muirhead, H.; Watson, H. C. *Cold Spring Harbor Symp. Quant. Biol.* **1972**, *36*, 165–170.
94. Nikolov, D. B.; Hu, S. H.; Lin, J.; Gasch, A.; Hoffmann, A.; Horikoshi, M.; Chua, N. H.; Roeder, R. G.; Burley, S. K. *Nature* **1992**, *360*, 40–46.
95. Savchenko, A.; Yee, A.; Khachatryan, A.; Skarina, T.; Evdokimova, E.; Pavlova, M.; Semesi, A.; Northey, J.; Beasley, S.; Lan, N. et al. *Proteins* **2003**, *50*, 392–399.
96. Appelt, K.; Bacquet, R. J.; Bartlett, C. A.; Boothe, C. L. J.; Freer, S. T.; Fuhry, M. M.; Gehring, M. R.; Hermann, S. M.; Howland, E. F.; Janson, C. A. et al. *J. Med. Chem.* **1991**, *34*, 1925–1934.
97. Varney, M. D.; Marzoni, G. P.; Palmer, C. L.; Deal, J. G.; Webber, S.; Welsh, K. M.; Bacquet, R. J.; Bartlett, C. A.; Morse, C. A.; Booth, C. L. J. et al. *J. Med. Chem.* **1992**, *35*, 663–676.
98. Webber, S. E.; Bleckman, T. M.; Attard, J.; Deal, J. G.; Kathardekar, V.; Welsh, K. M.; Webber, S.; Janson, C. A.; Matthews, D. A.; Smith, W. W. et al. *J. Med. Chem.* **1993**, *36*, 733–746.
99. Ikuta, M.; Kamata, K.; Fukasawa, K.; Honma, T.; Machida, T.; Hirai, H.; Suzuki-Takahashi, I.; Hayama, T.; Nishimura, S. *J. Biol. Chem.* **2001**, *276*, 27548–27554.
100. Anand, K.; Ziebuhr, J.; Wadhwani, P.; Mesters, J. R.; Hilgenfeld, R. *Science* **2003**, *300*, 1763–1767.
101. Keenan, R. J.; Siehl, D. L.; Gorton, R.; Castle, L. A. *Proc. Natl. Acad. Sci. USA* **2005**, *102*, 8887–8892.
102. Lawson, D. M.; Artymiuk, P. J.; Yewdall, S. J.; Smith, J. M. A.; Livingstone, J. C.; Treffry, A.; Luzzago, A.; Levi, S.; Arosio, P.; Cesareni, G. et al. *Nature* **1991**, *349*, 541–544.
103. Laufberger, V. *Bull. Soc. Chim. Biol.* **1937**, *19*, 1575–1582.
104. Banyard, S. H.; Stammers, D. K.; Harrison, P. M. *Nature* **1978**, *271*, 282–284.
105. Ford, G. C.; Harrison, P. M.; Rice, D. W.; Smith, J. M.; Treffry, A.; White, J. L.; Yariv, J. *Philos. Trans. R. Soc. Lond. Ser. B* **1984**, *304*, 551–556.
106. Thomas, C. D.; Shaw, W. V.; Lawson, D. M.; Treffry, A.; Artymiuk, P. J.; Harrison, P. M. *Biochem. Soc. Trans.* **1988**, *16*, 838–839.
107. Arosio, P.; Gatti, G.; Bolognesi, M. *Biochim. Biophys. Acta* **1983**, *744*, 230–232.
108. Abel, J. J. *Proc. Natl. Acad. Sci. USA* **1926**, *12*, 132–136.
109. Riboldi-Tunnicliffe, A.; Konig, B.; Jessen, S.; Weiss, M. S.; Rahfeld, J.; Hacker, J.; Fischer, G.; Hilgenfeld, R. *Nat. Struct. Biol.* **2001**, *8*, 779–783.
110. Trakhanov, S.; Quiocho, F. A. *Protein Sci.* **1995**, *4*, 1914–1919.
111. Trakhanov, S.; Kreimer, D. I.; Parkin, S.; Ames, G. F.; Rupp, B. *Protein Sci.* **1998**, 7, 600–604.
112. Harding, M. M. *Acta Crystallogr.* **2000**, *D56*, 857–867.
113. Harding, M. M. *Acta Crystallogr.* **2001**, *D57*, 401–411.
114. Hendrickson, W. A. *Science* **1991**, *254*, 51–58.
115. Qiu, X.; Janson, C. A. *Acta Crystallogr.* **2004**, *D60*, 1545–1554.
116. Dale, G. E.; Oefner, C.; D'Arcy, A. *J. Struct. Biol.* **2003**, *142*, 88–97.
117. Derewenda, Z. S. *Structure* **2004**, *12*, 529–535.
118. Braig, K.; Otwinowski, Z.; Hegde, R.; Boisvert, D. C.; Joachimiak, A.; Horwich, A. L.; Sigler, P. B. *Nature* **1994**, *371*, 578–586.
119. Horwich, A. *Nat. Struct. Biol.* **2000**, 7, 269–270.
120. McElroy, H. E.; Sisson, G. W.; Schoettlin, W. E.; Aust, R. M.; Villafranca, J. E. *J. Cryst. Growth* **1992**, *122*, 265–272.
121. D'Arcy, A.; Stihle, M.; Kostrewa, D. A.; Dale, G. *Acta Crystallogr.* **1999**, *D55*, 1623–1625.
122. Dyda, F.; Hickman, A. B.; Jenkins, T. M.; Engelman, A.; Craigie, R.; Davies, D. R. *Science* **1994**, *266*, 1981–1986.
123. Faller, M.; Niederweis, M.; Schulz, G. E. *Science* **2004**, *303*, 1189–1192.
124. Zhang, F.; Basinski, M. B.; Beals, J. M.; Briggs, S. L.; Churgay, L. M.; Clawson, D. K.; DiMarchi, R. D.; Furman, T. C.; Hale, J. E.; Hsiung, H. M. et al. *Nature* **1997**, *387*, 206–209.
125. Schlatter, D.; Thoma, R.; Küng, E.; Stihle, M.; Müller, F.; Borroni, E.; Cesura, A.; Hennig, M. *Acta Crystallogr.* **2005**, *D61*, 513–519.
126. Derewenda, Z. S. *Methods* **2004**, *34*, 354–363.
127. Baud, F.; Karlin, S. *Proc. Natl. Acad. Sci. USA* **1999**, *96*, 12494–12499.
128. Avbelj, F.; Fele, L. *J. Mol. Biol.* **1998**, *279*, 665–684.
129. Lo Conte, L.; Chothia, C.; Janin, J. *J. Mol. Biol.* **1999**, *285*, 2177–2198.
130. Longenecker, K. L.; Garrard, S. M.; Sheffield, P. J.; Derewenda, Z. S. *Acta Crystallogr.* **2001**, *D57*, 679–688.
131. Mateja, A.; Devedjiev, Y.; Krowarsch, D.; Longenecker, K.; Dauter, Z.; Otlewski, J.; Derewenda, Z. S. *Acta Crystallogr.* **2002**, *D58*, 1983–1991.
132. Munshi, S.; Hall, D. L.; Kornienko, M.; Darke, P. L.; Kuo, L. C. *Acta Crystallogr.* **2003**, *D59*, 1725–1730.
133. Dasgupta, S.; Iyer, G. H.; Bryant, S. H.; Lawrence, C. E.; Bell, J. A. *Proteins* **1997**, *28*, 494–514.
134. Czepas, J.; Devedjiev, Y.; Krowarsch, D.; Derewenda, U.; Otlewski, J.; Derewenda, Z. S. *Acta Crystallogr.* **2004**, *D60*, 275–280.
135. Wingren, C.; Edmundson, A. B.; Borrebaeck, C. A. K. *Protein Eng.* **2003**, *16*, 255–264.
136. Waldo, G. S.; Standish, B. M.; Berendzen, J.; Terwilliger, T. C. *Nat. Biotechnol.* **1999**, *17*, 691–695.
137. Yang, J. K.; Park, M. S.; Waldo, G. S.; Suh, S. W. *Proc. Natl. Acad. Sci. USA* **2003**, *100*, 455–460.

138. Rayment, I.; Rypniewski, W. R.; Schmidt-Base, K.; Smith, R.; Tomchick, D. R.; Benning, M. M.; Winkelmann, D. A.; Wesenberg, G.; Holden, H. M. *Science* **1993**, *261*, 50–58.
139. Rayment, I. *Methods Enzymol.* **1997**, *276*, 171–179.
140. Rypniewski, W. R.; Holden, H. M.; Rayment, I. *Biochemistry* **1993**, *32*, 9851–9858.
141. Kurinov, I. V.; Mao, C.; Irvin, J. D.; Uckun, F. M. *Biochem. Biophys. Res. Commun.* **2000**, *275*, 549–552.
142. Schubot, F. D.; Waugh, D. S. *Acta Crystallogr.* **2004**, *D60*, 1981–1986.
143. Kristensen, O.; Laurberg, M.; Gajhede, M. *Acta Crystallogr.* **2002**, *D58*, 1198–1200.
144. Heras, B.; Martin, J. L. *Acta Crystallogr.* **2005**, *D61*, 1173–1180.
145. Kriminski, S.; Caylor, C. L.; Nonato, M. C.; Finkelstein, K. D.; Thorne, R. E. *Acta Crystallogr.* **2002**, *D58*, 459–471.
146. Juers, D. H.; Matthews, B. W. *Q. Rev. Biophys.* **2004**, *37*, 105–119.
147. Hope, H. *Acta Crystallogr.* **1988**, *B44*, 22–26.
148. Hope, H. *Annu. Rev. Biophys. Biophys. Chem.* **1990**, *19*, 107–126.
149. Young, A. C. M.; Dewan, J. C.; Nave, C.; Tilton, R. F. *J. Appl. Crystallogr.* **1993**, *26*, 309–319.
150. Rodgers, D. W. *Structure* **1994**, *2*, 1135–1140.
151. Rodgers, D. W. *Methods Enzymol.* **1997**, *276*, 183–203.
152. Watowich, S. J.; Skehel, J. J.; Wiley, D. C. *Acta Crystallogr.* **1995**, *D51*, 7–12.
153. Chayen, N. E.; Boggon, T. J.; Cassetta, A.; Deacon, A.; Gleichmann, T.; Habash, J.; Harrop, S. J.; Helliwell, J. R.; Nieh, Y. P.; Peterson, M. R. et al. *Q. Rev. Biophys.* **1996**, *29*, 227–278.
154. Garman, E. F.; Schneider, T. R. *J. Appl. Crystallogr.* **1997**, *30*, 211–237.
155. Garman, E. F.; Garman, E. F. *Acta Crystallogr.* **1999**, *D55*, 1641–1653.
156. Harp, J. M.; Timm, D. E.; Bunick, G. J. *Acta Crystallogr.* **1998**, *D54*, 622–628.
157. Harp, J. M.; Hanson, B. L.; Timm, D. E.; Bunick, G. J. *Acta Crystallogr.* **1999**, *D55*, 1329–1334.
158. Kiefersauer, R.; Than, M. E.; Dobbek, H.; Gremer, L.; Melero, M.; Strobl, S.; Dias, J. M.; Soulimane, T.; Huber, R. *J. Appl. Crystallogr.* **2000**, *33*, 1223–1230.
159. Abergel, C. *Acta Crystallogr.* **2004**, *D60*, 1413–1416.
160. Pflugrath, J. W. *Methods* **2004**, *34*, 415–423.
161. Garman, E. F.; Mitchell, E. P. *J. Appl. Crystallogr.* **1996**, *29*, 584–587.
162. Parkin, S.; Hope, H. *J. Appl. Crystallogr.* **1998**, *31*, 945–953.
163. Riboldi-Tunnicliffe, A.; Hilgenfeld, R. *J. Appl. Crystallogr.* **1999**, *32*, 1003–1005.
164. Thomanek, U. F.; Parak, F.; Mössbauer, R. L.; Formanek, H.; Schwager, P.; Hoppe, W. *Acta Crystallogr.* **1973**, *A29*, 263–265.
165. Urayama, P.; Phillips, G. N., Jr.; Gruner, S. M. *Structure* **2002**, *10*, 51–60.
166. Kim, C. U.; Kapfer, R.; Gruner, S. M. *Acta Crystallogr.* **2005**, *D61*, 881–890.
167. Rossmann, M. G. *Acta Crystallogr.* **1990**, *A46*, 73–82.
168. Storoni, L. C.; McCoy, A. J.; Read, R. J. *Acta Crystallogr.* **2004**, *D60*, 432–438.
169. McCoy, A. J.; Grosse-Kunstleve, R. W.; Storoni, L. C.; Read, R. J. *Acta Crystallogr.* **2005**, *D61*, 458–464.
170. Schwarzenbacher, R.; Godzik, A.; Grzechnik, S. K.; Jaroszewski, L. *Acta Crystallogr.* **2004**, *D60*, 1229–1236.
171. Berman, H. M.; Westbrook, J.; Feng, Z.; Gilliland, G.; Bhat, T. N.; Weissig, H.; Shindyalov, I. N.; Bourne, P. E. *Nucleic Acids Res.* **2000**, *28*, 235–242.
172. Berman, H. M.; Battistuz, T.; Bhat, T. N.; Bluhm, W. F.; Bourne, P. E.; Burkhardt, K.; Feng, Z.; Gilliland, G. L.; Iype, L.; Jain, S. et al. *Acta Crystallogr.* **2002**, *D58*, 899–907.
173. Hendrickson, W. A.; Ogata, C. M. *Methods Enzymol.* **1997**, *276*, 494–523.
174. Hendrickson, W. A.; Horton, J. R.; LeMaster, D. M. *EMBO J.* **1990**, *9*, 1665–1672.
175. Doublié, S. *Methods Enzymol.* **1997**, *276*, 523–530.
176. Smith, J. L.; Thompson, A. *Structure* **1998**, *6*, 815–819.
177. Blundell, T. L.; Johnson, L. N. In *Protein Crystallography*; Academic Press: London, UK, 1976, pp 183–239.
178. Petsko, G. A. *Methods. Enyzmol.* **1985**, *114*, 147–156.
179. Rould, M. A. *Methods Enzymol.* **1997**, *276*, 461–472.
180. Stura, E. A.; Gleichmann, T. In *Crystallization of Nucleic Acids and Proteins*; Ducruix, A., Giegé, R., Eds.; Oxford University Press: Oxford, UK, 1999, pp 365–390.
181. Cohen, S. L.; Chait, B. T. *Annu. Rev. Biophys. Biomol. Struct.* **2001**, *30*, 67–85.
182. Boggon, T. J.; Shapiro, L. *Structure* **2000**, *8*, R143–R149.
183. Gruen, M.; Becker, C.; Beste, A.; Reinstein, J.; Scheidig, A. J.; Goody, R. S. *Protein Sci.* **1999**, *8*, 2524–2528.
184. Buts, L.; Loris, R.; De Genst, E.; Oscarson, S.; Lahmann, M.; Messens, J.; Brosens, E.; Wyns, L.; De Greve, H.; Bouckaert, J. *Acta Crystallogr.* **2003**, *D59*, 1012–1015.
185. Cohen, A.; Ellis, P.; Kresge, N.; Soltis, S. M. *Acta Crystallogr.* **2001**, *D57*, 233–238.
186. Dauter, Z.; Dauter, M.; Rajashankar, K. R. *Acta Crystallogr.* **2000**, *D56*, 232–237.
187. Dauter, Z.; Dauter, M. *Structure* **2001**, *9*, R21–R26.
188. Nagem, R . A. P.; Dauter, Z.; Polikarpov, I. *Acta Crystallogr.* **2001**, *D57*, 996–1002.
189. Nagem, R. A. P.; Polikarpov, I.; Dauter, Z. *Methods. Enzymol.* **2003**, *374*, 120–137.
190. Hendrickson, W. A.; Teeter, M. M. *Nature* **1981**, *290*, 107–113.
191. Weiss, M. S.; Sicker, T.; Djinovic-Carugo, K.; Hilgenfeld, R. *Acta Crystallogr.* **2001**, *D57*, 689–695.
192. Weiss, M. S.; Sicker, T.; Hilgenfeld, R. *Structure* **2001**, *9*, 771–777.
193. Mueller-Dieckmann, C.; Polentarutti, M.; Djinovic-Carugo, K.; Panjikar, S.; Tucker, P. A.; Weiss, M. S. *Acta Crystallogr.* **2004**, *D60*, 23–38.
194. Mueller-Dieckmann, C.; Panjikar, S.; Tucker, P. A.; Weiss, M. S. *Acta Crystallogr.* **2005**, *D61*, 1263–1272.
195. Matthews, B. W. *J. Mol. Biol.* **1968**, *33*, 491–497.
196. Card, G. L.; Blasdel, L.; England, B. P.; Zhang, C.; Suzuki, Y.; Gillette, S.; Fong, D.; Ibrahim, P. N.; Artis, D. R.; Bollag, G. et al. *Nat. Biotechnol.* **2005**, *23*, 201–207.
197. Erlanson, D. A.; McDowell, R. S.; O'Brien, T. *J. Med. Chem.* **2004**, *47*, 3463–3482.
198. Rees, D. C.; Congreve, M.; Murray, C. W.; Carr, R. *Nat. Rev. Drug. Disc.* **2004**, *3*, 660–672.
199. Shuker, S. B.; Hajduk, P. J.; Meadows, R. P.; Fesik, S. W. *Science* **1996**, *274*, 1531–1534.

200. Nienaber, V. L.; Richardson, P. L.; Klighofer, V.; Bouska, J. J.; Giranda, V. L.; Greer, J. *Nat. Biotechnol.* **2000**, *18*, 1105–1108.
201. Blundell, T. L.; Jhoti, H.; Abell, C. *Nat. Rev. Drug Disc.* **2002**, *1*, 45–54.
202. Blundell, T. L.; Patel, S. *Curr. Opin. Pharmacol.* **2004**, *4*, 490–496.
203. Hartshorn, M. J.; Murray, C. W.; Cleasby, A.; Fredrickson, M.; Tickle, I. J.; Jhoti, H. *J. Med. Chem.* **2005**, *48*, 403–413.
204. Congreve, M.; Carr, R.; Murray, C.; Jhoti, H. *Drug. Disc. Today* **2003**, *8*, 876–877.
205. Nienaber, V. L.; Davidson, D.; Edalji, R.; Giranda, V. L.; Klinghofer, V.; Henkin, J.; Magdalinos, P.; Mantei, R.; Merrick, S.; Severin, J. M. et al. *Structure* **2000**, *8*, 553–563.
206. Hardy, J. A.; Wells, J. A. *Curr. Opin. Struct. Biol.* **2004**, *14*, 706–715.
207. Erlanson, D. A.; Hansen, S. K. *Curr. Opin. Chem. Biol.* **2004**, *8*, 399–406.
208. Erlanson, D. A.; Wells, J. A.; Braisted, A. C. *Annu. Rev. Biophys. Biomol. Struct.* **2004**, *33*, 199–223.
209. Hardy, J. A.; Lam, J.; Nguyen, J. T.; O'Brien, T.; Wells, J. A. *Proc. Natl. Acad. Sci. USA* **2004**, *101*, 12461–12466.
210. O'Brien, T.; Fahr, B. T.; Sopko, M. M.; Lam, J. W.; Waal, N. D.; Raimundo, B. C.; Purkey, H. E.; Pham, P.; Romanowski, M. J. *Acta Crystallogr.* **2005**, *F61*, 451–458.
211. Lesuisse, D.; Lange, G.; Deprez, P.; Bénard, D.; Schoot, B.; Delettre, G.; Marquette, J. P.; Broto, P.; Jean-Baptiste, V.; Bichet, P. et al. *J. Med. Chem.* **2002**, *45*, 2379–2387.
212. Ramstrom, O.; Lehn, J. M. *Nat. Rev. Drug Disc.* **2002**, *1*, 26–36.
213. Congreve, M. S.; Davis, D. J.; Devine, L.; Granata, C.; O'Reilly, M.; Wyatt, P. G.; Jhoti, H. *Angew. Chem., Int. Ed. Engl.* **2003**, *42*, 4479–4482.
214. Burley, S. K. *Nat. Struct. Biol.* **2000**, 7, 932–934.
215. Stevens, R. C.; Yokoyama, S.; Wilson, I. A. *Science* **2001**, *294*, 89–92.
216. Bussow, K.; Nordhoff, E.; Lubbert, C.; Lehrach, H.; Walter, G. *Genomics* **2000**, *65*, 1–8.
217. Christendat, D.; Yee, A.; Dharamsi, A.; Kluger, Y.; Savchenko, A.; Cort, J. R.; Booth, V.; Mackereth, C. D.; Saridakis, V.; Ekiel, I. et al. *Nat. Struct. Biol.* **2000**, 7, 903–909.
218. Nasoff, M.; Bergseid, M.; Hoeffler, J. P.; Heyman, J. A. *Methods Enzymol.* **2000**, *328*, 515–529.
219. Knaust, R. K.; Nordlund, P. *Anal. Biochem.* **2001**, *297*, 79–85.
220. Dobrovetsky, E.; Lu, M. L.; Andorn-Broza, R.; Khutoreskaya, G.; Bray, J. E.; Savchenko, A.; Arrowsmith, C. H.; Edwards, A. M.; Koth, C. M. *J. Struct. Funct. Genomics* **2005**, *6*, 33–50.
221. Kigawa, T.; Yabuki, T.; Yoshida, Y.; Tsutsui, M.; Ito, Y.; Shibata, T.; Yokoyama, S. *FEBS Lett.* **1999**, *442*, 15–19.
222. Kigawa, T.; Yabuki, T.; Matsuda, N.; Matsuda, T.; Nakajima, R.; Tanaka, A.; Yokoyama, S. *J. Struct. Funct. Genomics* **2004**, *5*, 6368.
223. Kuhn, P.; Wilson, K.; Patch, M. G.; Stevens, R. C. *Curr. Opin. Chem. Biol.* **2002**, *6*, 704–710.
224. Hosfield, D.; Palan, J.; Hilgers, M.; Scheibe, D.; McRee, D. E.; Stevens, R. C. *J. Struct. Biol.* **2003**, *142*, 207–217.
225. Weselak, M.; Patch, M. G.; Selby, T. L.; Knebel, G.; Stevens, R. C. *Methods Enzymol.* **2003**, *368*, 45–76.
226. Cohen, A. E.; Ellis, P. J.; Miller, M. D.; Deacon, A. M.; Phizackerley, R. P. *J. Appl. Crystallogr.* **2002**, *35*, 720–726.
227. Karain, W. I.; Bourenkov, G. P.; Blume, H.; Bartunik, H. D. *Acta Crystallogr.* **2002**, *D58*, 1519–1522.
228. Pohl, E.; Ristau, U.; Gehrmann, T.; Jahn, D.; Robrahn, B.; Malthan, D.; Dobler, H.; Hermes, C. *J. Synchrotron Radiat.* **2004**, *11*, 372–377.
229. Snell, G.; Cork, C.; Nordmeyer, R.; Cornell, E.; Meigs, G.; Yegian, D.; Jaklevic, J.; Jin, J.; Stevens, R. C.; Earnest, T. *Structure* **2004**, *12*, 537–545.
230. Muchmore, S. W.; Olson, J.; Jones, R.; Pan, J.; Blum, M.; Greer, J.; Merrick, S. M.; Magdalinos, P.; Nienaber, V. L. *Structure* **2000**, *8*, R243–R246.
231. Abad-Zapatero, C. *Acta Crystallogr.* **2005**, *D61*, 1432–1435.
232. Turk, D. Towards Automatic Macromolecular Crystal Structure Determination. In *Methods in Macromolecular Crystallography*; Turk, D., Johnson, L., Eds.; NATO Science Series I, Vol. 325; IOS Press: Amsterdam, Netherlands, 2001, pp 148–155.
233. Adams, P. D.; Gopal, K.; Grosse-Kunstleve, R. W.; Hung, L.; Ioerger, T. R.; McCoy, A. J.; Moriarty, N. W.; Pai, R. K.; Read, R. J.; Romo, T. D. et al. *J. Synchrotron Radiat.* **2004**, *11*, 53–55.
234. Kroemer, M.; Dreyer, M. K.; Wendt, K. U. *Acta Crystallogr.* **2004**, *D60*, 1679–1682.
235. Ness, S. R.; de Graaff, R. A.; Abrahams, J. P.; Pannu, N. S. *Structure* **2004**, *12*, 1753–1761.
236. Holton, J.; Alber, T. *Proc. Natl. Acad. Sci. USA* **2004**, *101*, 1537–1542.
237. Panjikar, S.; Parthasarathy, V.; Lamzin, V. S.; Weiss, M. S.; Tucker, P. A. *Acta Crystallogr.* **2005**, *D61*, 449–457.
238. Schüttelkopf, A. W.; van Aalten, D. M. F. *Acta Crystallogr.* **2004**, *D60*, 1355–1363.
239. Diller, D. J.; Pohl, E.; Redinbo, M. R.; Hovey, B. T.; Hol, W. G. *Proteins* **1999**, *36*, 512–525.
240. Oldfield, T. J. *Acta Crystallogr.* **2001**, *D57*, 696–705.
241. Oldfield, T. *Methods Enzymol.* **2003**, *374*, 271–300.
242. Zwart, P. H.; Langer, G. G.; Lamzin, V. S. *Acta Crystallogr.* **2004**, *D60*, 2230–2239.
243. Aishima, J.; Russel, D. S.; Guibas, L. J.; Adams, P. D.; Brunger, A. T. *Acta Crystallogr.* **2005**, *D61*, 1354–1363.

Biographies

Tanis Hogg was born in Saskatoon, Canada, and spent his formative years in the town of Kamloops, where he developed an early interest in the fields of biology and chemistry. He studied general sciences at Thompson Rivers University (Kamloops) and later transferred to the University of British Columbia (Vancouver), where he earned his BSc in biochemistry in 1993. After graduation, he took a temporary position with Baxter Pharmaceuticals in Burnaby, which stimulated his interest to pursue graduate studies in the field of protein crystallography and structure-based drug design. He moved to Jena, Germany, at the end of 1996 and studied under the supervision of Prof Rolf Hilgenfeld at the Institute of Molecular Biotechnology. There, he carried out crystallographic studies on ribosome-associated GTP-binding proteins including the bacterial drug target elongation factor Tu, and was awarded his PhD in 2001. After receiving his PhD, he became group leader in crystallography and structure-based drug design at JenaDrugDiscovery, GmbH. In 2004, he accepted a position at the Institute of Biochemistry, University of Lübeck, Germany, where his group is focused on macromolecular crystallography and rational design of novel antibiotics targeting pathogenic bacteria.

Rolf Hilgenfeld studied chemistry at the universities of Göttingen and Freiburg, Germany. He did his PhD in protein crystallography at the Free University of Berlin and after a postdoctoral stay at the Biocenter of the University of Basel, Switzerland, he joined Hoechst AG, the pharmaceutical company in Frankfurt, to build a macromolecular crystallography laboratory. During his 9 years in the company, he and his colleagues worked on the design of new insulin variants with the goal of improving the pharmacokinetics of the hormone. A major achievement of these efforts was the creation of a long-acting insulin, which has been introduced into the market as 'Lantus'. Rolf Hilgenfeld was also among the first scientists to determine the structure of the HIV-1 protease and to design inhibitors against this target. He also elucidated the structure of elongation factor Tu and studied its interaction with antibiotics. In 1995, he moved to the University of Jena to take over the Chair of Structural Biochemistry, in combination with the position of Head of the Crystallography Department at the newly founded Institute of Molecular Biotechnology. Since 2003, Rolf Hilgenfeld has been Full Professor of Biochemistry at the University of Lübeck, Germany. Today, his research focuses on the molecular basis of infectious diseases by bacteria such as *Legionella pneumophila* and *Chlamydia* and by RNA viruses. During the global SARS epidemic of 2003, he published the crystal structure of the coronavirus main proteinase and proposed a first inhibitor against the disease. His Lübeck laboratory follows an integrated approach to drug discovery against infectious agents, which includes comparative proteomics, molecular biology, x-ray crystallography, drug design, and chemical synthesis of inhibitors.

Comprehensive Medicinal Chemistry II
ISBN (set): 0-08-044513-6

ISBN (Volume 3) 0-08-044516-0; pp. 875–900

3.39 Nuclear Magnetic Resonance in Drug Discovery

J Klages and H Kessler, TU München, Garching, Germany

3.39.1	**Introduction**	**901**
3.39.2	**Binding Equilibria and their Relation to Nuclear Magnetic Resonance Observables**	**902**
3.39.2.1	Basic Binding Equilibria	902
3.39.2.2	Competition Binding Equilibria	903
3.39.2.3	Influence of Chemical Exchange on Nuclear Magnetic Resonance Observables	904
3.39.3	**Ligand versus Target-Based Techniques**	**907**
3.39.3.1	Ligand-Based Methods	907
3.39.3.1.1	Diffusion	908
3.39.3.1.2	Relaxation	909
3.39.3.1.3	Paramagnetic spin labels	909
3.39.3.1.4	^{19}F ligands	911
3.39.3.1.5	Intramolecular nuclear overhauser effect	911
3.39.3.1.6	Intermolecular nuclear overhauser effect	912
3.39.3.1.7	Competition-based screening	915
3.39.3.2	Target-Based Methods	915
3.39.3.2.1	Chemical shift perturbations	916
References		**917**

3.39.1 Introduction

It is becoming increasingly important for the pharmaceutical industry to shorten and optimize the drug discovery process. For efficient cost reduction the sensible use of diverse techniques is essential; beneficial synergetic effects may be achieved by combining several techniques, leading to improved drug development, and hence a decrease in costs.

NMR spectroscopy has been used more frequently since the 1990s as tool in the drug discovery process.[1–14] This change is based on instrumental as well as experimental improvements. Cryogenic probes, auto-samplers, and higher magnetic fields have reduced the time required for data acquisition. Also, developments in methodology have facilitated the search for potential drugs. In this chapter the focus is on the latter aspect, and current techniques, their requirements, and their limitations are discussed. For instrumental aspects, we refer the reader to other reviews.[15,16]

The goal of the drug discovery process can be reduced to the following: a molecule needs to be identified that binds specifically to the selected target molecule. This binder has to meet different criteria, such as a particular hydrophobicity or a molecular mass within certain limits.[17,18] Usually the drug discovery process is divided into different phases, to allow a more systematic approach. The incorporation of NMR spectroscopy into all of these phases is nowadays possible and common: NMR can have a profound impact on lead search and lead optimization; also, as proven by the work of Nicholsen and co-workers, NMR can assist the drug discovery process during the preclinical and clinical phase.[19–21] However, only techniques applicable to lead search and optimization will be discussed here.

The increased use of NMR spectroscopy in biomedical research results from its unique ability to detect weak binding, which is of major importance to the fragment-based approach.[22–26,99] During fragment-based screening initially only small molecules of low complexity are tested for binding. Most hits of this class of ligands show only a low affinity for the drug target, but if several of these hits in adjacent binding sites are combined, they can create a high-affinity binder. This approach drastically reduces the number of ligands that need to be screened, and is therefore highly favorable from a cost efficiency perspective. A detailed discussion of the fragment-based approach is given elsewhere (*see* 3.41 Fragment-Based Approaches).

Beside the ability to screen for weakly binding molecules, NMR spectroscopy allows structural information to be obtained at the atomic level, which is particularly relevant in lead optimization. The determination of the three-dimensional (3D) structure of biomacromolecules is presented elsewhere (*see* 3.22 Bio-Nuclear Magnetic Resonance).

Finally, we now turn to the advantages and disadvantages of NMR compared with other tools used in the process of drug development. For hit finding and lead generation, functional and fluorescence assays are frequently used. These techniques allow a large number of compounds to be screened within a short amount of time. Therefore, these techniques are often referred to high-throughput screening (HTS). However, binding assays do not allow a fragment-based approach because only high-affinity binders are detectable and give a positive response. Moreover, the development of a new assay is required for each new target, which can be a time-consuming task. It has been shown that the combination of NMR and assay techniques results in an improved screening process. NMR is especially helpful in the validation of HTS screening hits. Combinations with other techniques such as mass spectroscopy are also promising approaches.[27,28]

During lead optimization, where structural aspects come to the fore, only x-ray crystallography also provides structural information at the atomic level. Depending on the problem to be addressed, x-ray crystallography and NMR have different merits. While the structure determination by NMR can become tedious for high molecular weight targets (>30 kDa) x-ray crystallography is commonly not subject to size limitations. As x-ray analysis of drug–target complexes are usually obtained by immersing the ligand in the apo form, it is important that the crystal is stable under these conditions.[29] And, of course, either the target or the complex has to form crystals. On the other hand, information about the binding site is quickly obtained by NMR spectroscopy when the assignment of the NMR signals is available. If in addition the structure of the target molecule is known, new ligands can easily be docked into the binding pocket using NMR-derived data.[30,31] In conclusion, NMR spectroscopy is most usefully integrated into the drug discovery process as a complementary approach, rather than being used as 'stand-alone' technique.

In the next sections we will describe the kinetic principles underlying NMR screening experiments and then give an overview of commonly used screening techniques.

3.39.2 Binding Equilibria and their Relation to Nuclear Magnetic Resonance Observables

3.39.2.1 Basic Binding Equilibria

Most NMR screening experiments can be described by the following simple binding equilibrium:

$$\mathrm{L} + \mathrm{T} \underset{k_{\mathrm{off}}}{\overset{k_{\mathrm{on}}}{\rightleftharpoons}} \mathrm{LT} \qquad [\mathrm{I}]$$

This equation describes the dynamic equilibrium between the target T and the free ligand L on the one hand, and the ligand–target complex LT on the other. The bimolecular rate constant k_{on} is a measure for the frequency of productive collisions between the free ligand molecule and the target molecule. This rate is assumed to be diffusion limited, and varies between $1\times10^{8}\,\mathrm{M}^{-1}\,\mathrm{s}^{-1}$ to $1\times10^{9}\,\mathrm{M}^{-1}\,\mathrm{s}^{-1}$. As a result of a productive collision, the ligand docks into the binding pocket of the target, which leads to the formation of the ligand–target complex. This complex has a mean lifetime τ_{res}, the reciprocal value of which is the unimolecular rate constant k_{off}. The equilibrium can be universally described by the law of mass action and its corresponding dissociation constant K_{D}. The dissociation constant is also related to the rate constants k_{on} and k_{off}:

$$K_{\mathrm{D}} = \frac{[\mathrm{L}][\mathrm{T}]}{[\mathrm{LT}]} = \frac{k_{\mathrm{off}}}{k_{\mathrm{on}}} \qquad [1]$$

In this equation, the square brackets denote the concentration of the enclosed species; this convention will be used throughout this chapter.

It is instructive to consider the bound target fraction $P_{\mathrm{T,b}}$, which is defined as

$$P_{\mathrm{T,b}} = \frac{[\mathrm{LT}]}{[\mathrm{LT}] + [\mathrm{T}]} \qquad [2]$$

If the law of mass action is included in this definition, an equation is yielded that shows the hyperbolic dependence of $P_{T,b}$ on the amount of ligand added to the target:

$$P_{T,b} = \frac{[L]}{[L] + K_D} \quad [3]$$

Equation [3] describes how the target is saturated by an increasing amount of ligand, and basically represents the Langmuir binding isotherm. In this dose–response curve, three different ranges have to be distinguished, which can be easily identified in **Figure 1**. As long as $[L] \ll K_D$, the bound target fraction $P_{T,b}$ is proportional to the ligand concentration [L]. This changes when [L] approaches concentrations in the range of K_D, and the target becomes half-saturated. Then, 50% of the target molecules exist in a one-to-one complex with the ligand. This point is indicated by the horizontal dashed line in **Figure 1**. When the concentration of the ligand is increased further and $[L] \gg K_D$, nearly all the target molecules are complexed, and the bound target fraction adopts values close to 1. It is apparent from **Figure 1** that ligands with larger K_D are weaker binders and require a higher ligand concentration to saturate the target.

It is often favorable to express the concentrations [L] and [T] by the total ligand and target concentrations, $L_{tot} = [L] + [LT]$ and $T_{tot} = [T] + [LT]$, respectively, because these are the only variables which are in the control of the scientist. A subsequent substitution into the law of mass action (eqn [1]) yields

$$[LT] = \frac{1}{2}(T_{tot} + L_{tot} + K_D) - \frac{1}{2}\sqrt{(T_{tot} + L_{tot} + K_D)^2 - 4T_{tot}L_{tot}} \quad [4]$$

This equation can be used to theoretically predict the dose–response curves for given dissociation constants K_D and fixed target concentrations T_{tot}. Plots for three different dissociation constants have been calculated with this equation are given in **Figure 1**.

3.39.2.2 Competition Binding Equilibria

The theoretical background discussed above is sufficient for most screening experiments performed with NMR. Only in a few cases, where a bound ligand C is substituted by the ligand L from the binding site, is it necessary to expand the binding equilibrium of eqn [I] and to take another species into account.

$$CT \underset{k_{C,off}}{\overset{k_{C,on}}{\rightleftharpoons}} C + T + L \underset{k_{L,off}}{\overset{k_{L,on}}{\rightleftharpoons}} LT \quad [II]$$

The ligand C also binds to the target and forms the complex CTwith the target. It has to be pointed out that in the case considered the two different ligands L and C both compete for the same binding site of the target T. Hence, the ligand C is often called *competitor*. The rate constants of the two reactions are defined analogously to k_{on} and k_{off} in eqn [I], and

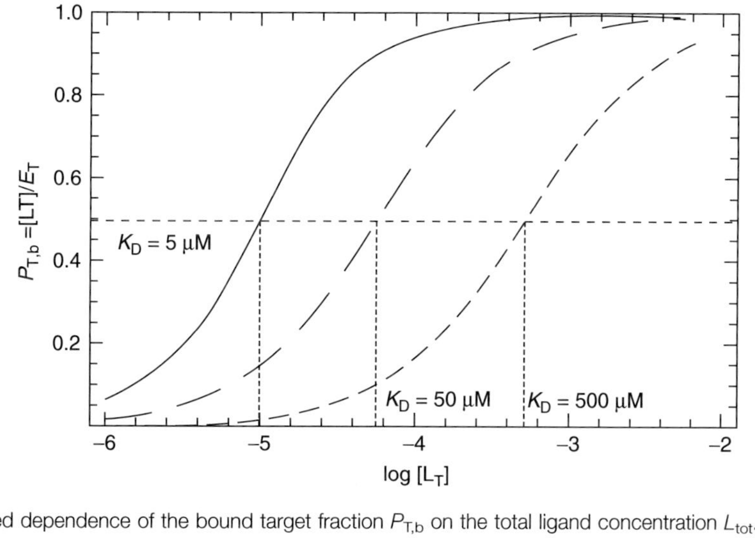

Figure 1 Calculated dependence of the bound target fraction $P_{T,b}$ on the total ligand concentration L_{tot}.[14]

consequently two separate laws of mass action are obtained:

$$K_{\mathrm{L}} = \frac{[\mathrm{L}][\mathrm{T}]}{[\mathrm{LT}]} = \frac{k_{\mathrm{L,off}}}{k_{\mathrm{L,off}}} \qquad [5a]$$

$$K_{\mathrm{C}} = \frac{[\mathrm{C}][\mathrm{T}]}{[\mathrm{CT}]} = \frac{k_{\mathrm{C,off}}}{k_{\mathrm{C,off}}} \qquad [5b]$$

The bound target fraction in the presence of the competitor $P_{\mathrm{T,b,C}}$ can now be examined. Note that only the target fraction bound to the ligand L is considered and not the target bound to the competitor. Thus, the definition for the bound target fraction (eqn [2]) does not change if a competitor is introduced:

$$P_{\mathrm{T,b,C}} = \frac{[\mathrm{LT}]}{[\mathrm{LT}] + [\mathrm{T}]} \qquad [6]$$

Substitution of eqns [5a] and [5b] into the definition of the bound target fraction (eqn [6]) leads to expressions comparable to eqn [3]:

$$P_{\mathrm{T,b,C}} = \frac{[\mathrm{L}]}{[\mathrm{L}] + K_{\mathrm{D,app}}} \qquad [7]$$

$$K_{\mathrm{D,app}} = K_{\mathrm{L}}\left(1 + \frac{[C]}{K_{\mathrm{C}}}\right) \qquad [8]$$

Because the factor $(1 + [\mathrm{C}]/K_{\mathrm{C}})$ is always $\geqslant 1$ and therefore $K_{\mathrm{D,app}} \geqslant K_{\mathrm{D}}$, the bound target fraction in the presence of the competitor $P_{\mathrm{T,b,C}}$ must be smaller than in the absence of the competitor ($P_{\mathrm{T,b,C}} \leqslant P_{\mathrm{T,b}}$). This finding emphasizes the fact that with the addition of a competitor the number of available binding sites is reduced, and accordingly the amount of target complexed with the ligand L is also reduced.

Again, it is desirable to replace the concentrations within the equilibrium by the corresponding total concentrations $\mathrm{T_{tot}}$, $\mathrm{L_{tot}}$, and $\mathrm{C_{tot}}$. Adequate substitutions and mathematical transformations yield an equation for the bound target fraction $P_{\mathrm{T,b,C}}$. As this expression is quite complex it is not given here.[14]

If the value for $P_{\mathrm{T,b,C}}$ is divided by $P_{\mathrm{T,b}}$ and plotted against the total competitor concentration C_{tot}, **Figure 2** is obtained. This diagram visualizes the displacement of the ligand L in the ligand–target complex LT when the concentration of the competitor C is increased. The corresponding curves belong to different values of K_{C}, where smaller values of K_{C} indicate stronger competitors. Therefore, strong competitors ($K_{\mathrm{C}} \gg K_{\mathrm{L}}$) will displace all ligand molecules L from the binding site of the target already at low concentrations ($\mathrm{C_{tot}} \ll \mathrm{L_{tot}}$). This effect can be used to screen for strong binders by adding the unknown ligand C to a solution of the target T and the known strong binder L. Only ligands with a higher affinity than L will displace it from the binding site and lead to an increase of the concentration [L].

3.39.2.3 Influence of Chemical Exchange on Nuclear Magnetic Resonance Observables

The chemical exchange between two species, such as the ligand in its free and bound forms, strongly influences the observable NMR parameters Q_{obs}. Depending on the timescale of the chemical exchange, these are modulated differently. The timescale can be divided into three regimes: fast, intermediate, and slow exchange. If Q_{f} and Q_{b} are values of the NMR parameter of interest in the free and bound states, respectively, then the chemical exchange is referred to be slow if $|Q_{\mathrm{f}} - Q_{\mathrm{b}}| \gg k_{\mathrm{ex}}$, intermediate if $|Q_{\mathrm{f}} - Q_{\mathrm{b}}| \approx k_{\mathrm{ex}}$, and fast if $|Q_{\mathrm{f}} - Q_{\mathrm{b}}| \ll k_{\mathrm{ex}}$. k_{ex} is the net rate constant of the reaction, and describes the relaxation of the system back to equilibrium if the concentrations are perturbed:

$$k_{\mathrm{ex}} = k_{\mathrm{on}}[\mathrm{T}] + k_{\mathrm{off}} \qquad [9]$$

As an example, the effects of chemical exchange are demonstrated on the chemical shifts of a ligand molecule. Assume that a signal within the free form has a chemical shift Ω_{f} and in the bound form Ω_{b}. Then these two signals are separated

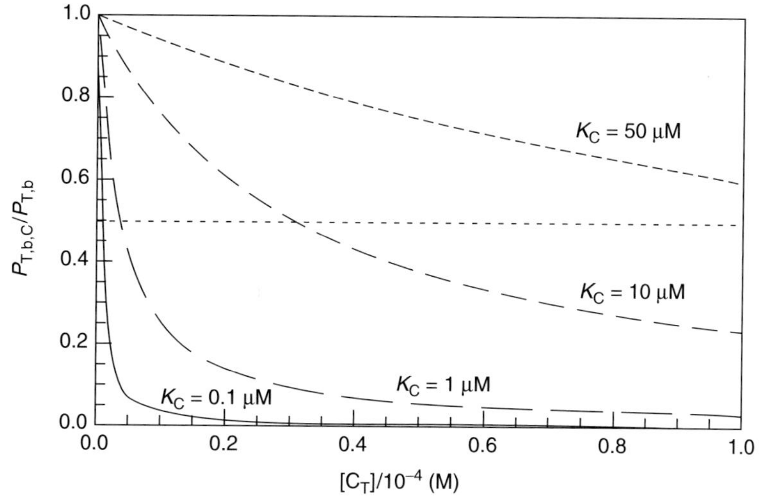

Figure 2 Displacement of the ligand L by the competitor C when the concentration of the competitor C_{tot} is increased.[14] The ratio between the bound target fraction in the presence $P_{T,b,C}$ and the absence of a competitor $P_{T,b}$ is shown. Four plots are displayed with different K_C values: solid line, $K_C = 0.1$ μM; long dashed line, $K_C = 1$ μM, medium dashed line, $K_C = 10$ μM; short dashed line, $K_C = 50$ μM. A dissociation constant $K_L = 50$ μM is assumed. The total concentration of the ligand L_{tot} is fixed at 100 μM, while the total concentration of the target T_{tot} is fixed at 1 μM. The dotted line indicates displacement of 50% by the competitor.

by $|\Omega_f - \Omega_b|$. In addition, we assume equal relaxation rates for both species for simplicity. Commonly, NMR screening experiments are carried out with an excess of ligand molecules compared with the target molecules. Here, we simply define the fraction of the free ligand as $P_{L,f} = 0.75$, and consequently the fraction of the bound ligand as $P_{L,b} = 0.25$. Now, when the net rate constant k_{ex} changes, the chemical shifts observable in the spectrum will also change. Their positions and the lineshape will be affected more strongly as the exchange becomes faster (**Figure 3**). In the slow exchange limit, two separated signals are observed, while in the fast exchange limit the two resonances fuse and form one sharp line. The point where the two resonances merge and show only one maximum ($k_{ex} \approx 2(P_{L,f} P_{L,b})^{1/2} |\Omega_f - \Omega_b|$) is called coalescence. It is important to note that in the fast exchange limit the NMR signal appears at the population-weighted average of the contributing resonances. The formal treatment of this phenomenon is considerably more complex, but is not the focus of this chapter (the interested reader is referred to standard textbooks[32–35]).

In most NMR screening experiments, ligands show only a low affinity for the drug target. This results in a short lifetime τ of the ligand–target complex, and consequently in a large dissociation rate constant k_{off}. It is apparent from eqn [9] that an increase in k_{off} will lead to an increase in the net rate constant k_{ex}. This increase in k_{ex} implies that the experimental conditions are forced toward the fast exchange limit. As shown above for the chemical shift under fast exchange, the observable NMR parameters are the population-weighted averages of the intrinsic values of the free and bound forms. This observation is generally valid for all NMR parameters (e.g., the chemical shift, the translational diffusion rate, the longitudinal and transverse relaxation rates, or the longitudinal (nuclear Overhauser effect, NOE) and transverse (rotating-frame Overhauser effect, ROE) cross-relaxation rates):

$$\Omega_{obs} = P_{L,b}\Omega_b + (1 - P_{L,b})\Omega_f \quad [10]$$

$$D_{obs} = P_{L,b}D_b + (1 - P_{L,b})D_f \quad [11]$$

$$R_{1,obs} = P_{L,b}R_{1,b} + (1 - P_{L,b})R_{1,f} \quad [12]$$

$$R_{2,obs} = P_{L,b}R_{2,b} + (1 - P_{L,b})R_{2,f} + R_{ex} \quad [13]$$

$$\sigma_{NOE,obs} = P_{L,b}\sigma_{NOE,b} + (1 - P_{L,b})\sigma_{NOE,f} \quad [14]$$

$$\sigma_{ROE,obs} = P_{L,b}\sigma_{ROE,b} + (1 - P_{L,b})\sigma_{ROE,f} \quad [15]$$

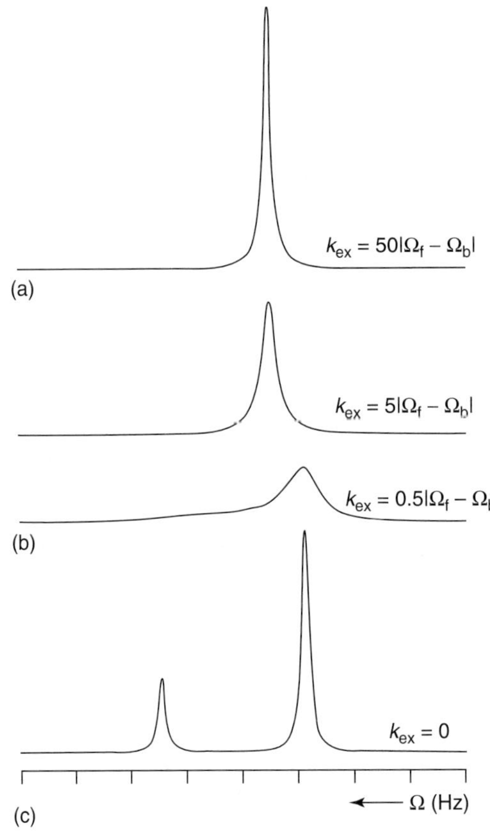

Figure 3 Illustration of the chemical exchange for a two-site system. Different populations of the two species (bound and free form) of the ligand are assumed ($P_{L,F} = 0.75$, $P_{L,b} = 0.25$) but equal transverse relaxation rates ($R_{2,f} = R_{2,b}$). (a) fast exchange, (b) intermediate exchange, and (c) slow exchange. The different panels belong to different exchange regimes (lowest panel, two middle panels, top panel).

From these equations two important facts are evident. First, the transverse relaxation rate may contain an additional relaxation term R_{ex}. This term arises from the formal derivation of these equations, and is usually negligible if standard screening conditions are considered. Then, k_{ex} is large, and the bound ligand fraction is small, leading to a very small contribution of R_{ex} to the observable relaxation rate R_{obs}.

$$R_{ex} = P_{L,b}\left(1 - P_{L,b}\right)\frac{4\pi^2(\Omega_f - \Omega_b)^2}{k_{ex}} \qquad [16]$$

Second, chemical exchange might be fast for the parameter Q_1 but slow for the parameter Q_2. If we consider the definitions of the terms 'fast' and 'slow,' this becomes plausible when the difference between the free and bound state values is considered. This difference varies for different NMR parameters. As a consequence, the same k_{ex} and therefore the same exchange correspond to different exchange regimes for different NMR parameters.

However, most NMR screening experiments are carried out under fast exchange conditions, which allows binding and nonbinding molecules to be distinguished. Molecules binding to the drug target will reveal exchange-averaged observables. If these differ significantly from the values of the free form, binding becomes detectable. As is evident from eqns [10–15], the exchange-averaged NMR parameter will only differ from the free state value if two requirements are satisfied: the bound ligand fraction must not be too small; and the difference between the free and the bound state values needs to be large. The latter condition is the reason why parameters relying on the difference in molecular mass are often chosen in NMR screening experiments. Usually the drug target has a much higher molecular mass than the ligand, which translates to a much smaller translational and rotational diffusion rate. **Figure 4** illustrates the impact on the NMR parameters of a ligand. In particular, the change in the rotational diffusion rate, which is inversely proportional to the correlation time τ_c, permits a broad range of applications. As shown in the next section, many screening techniques make use of the marked change in correlation time.

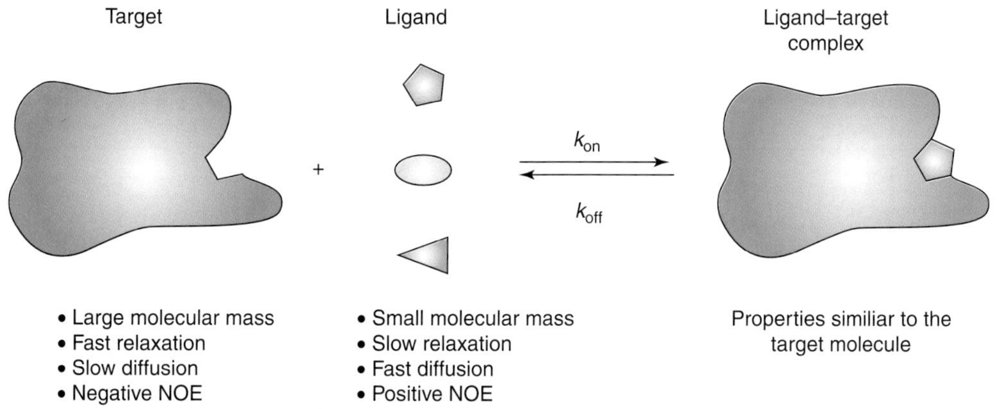

Figure 4 Alteration of the physicochemical properties of a ligand by complexation with a large target molecule.[7] The ligand appears to have a higher molecular weight while bound to the target molecule, and therefore the diffusion rate, relaxation rates, and cross-relaxation rates are changed.

Figure 4 also clarifies the reason for NMR screening experiments usually being ligand based and not target based. If the target molecule is considered, the NMR parameters relying on the properties shown in **Figure 4** do not change significantly upon binding. When a ligand molecule binds to a target molecule, the molecular mass of the target only undergoes a minor change from the perspective of the target molecule. Chemical shifts within the target are the only parameter showing a distinct deviation when the ligand binds to the target molecule. Therefore, chemical shift changes are the only meaningful measurements in target-based NMR screening.

Recapitulating, fast chemical exchange is essential for the successful application of ligand-based screening techniques. Moreover, particular attention is required in the choice of the observed NMR parameter. In the next section, ligand and target detection methods are discussed in detail.

3.39.3 Ligand versus Target-Based Techniques

NMR screening can be divided into two classes: ligand and target detection methods. Both techniques offer distinct advantages, and are to a certain degree complementary to each other. Ligand-based methods predominantly include simple one-dimensional (1D) spectra, and are therefore fairly fast, which leads to a higher throughput in screening. The detection of the ligand signals offers the opportunity to screen larger mixtures of ligands without the need for deconvolution, as long as signal overlap is avoided. Moreover, the amount of target in the screening process is relatively small, and the properties of the target are less restricted. Usually the target is not labeled, and upper size limitations are negligible; even immobilized targets can be investigated. Some of these techniques also provide crude information about the binding epitope and the binding mode. However, these methods only cover a small range of affinities, and sometimes give misleading results due to false positives originating from nonspecific binding.

Target-based methods can easily distinguish between specific and nonspecific binding; moreover, aggregation and pH effects can be excluded. Ligands with a broad range of affinities can be detected. However, their most important advantage is the ability to extract structural information, such as binding site or even complex structures. For target detection experiments, relatively large amounts of the target protein are needed. Usually these samples have to be labeled (^{15}N and/or ^{13}C), which makes these techniques quite expensive. In contrast to ligand detection methods, it is necessary to deconvolute the mixture when mixtures of potential ligands are used. Moreover, experimental time is significantly increased because of the need to include two-dimensional (2D) technologies to achieve a suitable resolution. The major drawback, though, is the size limitation of the target molecule due to the increasing transverse relaxation rate R_2.

A summary of the techniques discussed in this chapter is given in **Table 1**. For each technique limitations and the extractable information is listed.

3.39.3.1 Ligand-Based Methods

Detection of signals only from the ligands allows the direct identification of binders from a mixture without deconvolution. Common ligand-based techniques fundamentally rely on fast exchange, as only exchange-modulated

Table 1 Overview to common screening techniques[12]

Method	*Limits and requirements*			*Identification of*		
	Target MW limit	*Affinity limits*	*Labelled target req.*	*Binding site on target*	*Binding epitope on ligand*	*Binding comp. in mixtures*
Diffusion filtering	Lower	U/L[a]	No	No	No	Yes
Relaxation filtering	Lower	U/L	No	No	No	Yes
TrNOE	Lower	U/L	No	No	No	Yes
NOE pumping	Lower	U/L	No	No	No	Yes
Rev.[b] NOE pumping	Lower	U/L	No	No	Yes	Yes
WaterLOGSY	Lower	U/L	No	No	Yes	Yes
STD	Lower	U/L	No	No	Yes	Yes
^{19}F-Screening	None	None	^{19}F ligand	No	No	Yes
CSP	Upper	None	^{15}N or ^{13}C	Yes	No	No
Competition screening	None	None	No	Yes	No	No
C.-based ^{19}F screening	None	None	^{19}F ligand	Yes	No	No

[a] U/L = upper/lower.
[b] Rev. = reverse.

signals can give information on binding. Upon binding, the ligand adopts the physicochemical properties of the large target molecule for the duration of $1/k_{\text{off}}$. If k_{off} decreases, and consequently the residence time τ_{res} is prolonged, the exchange is shifted toward the intermediate or slow regime. Then, the ligand becomes indistinguishable from the target molecule, and binding will be unobservable. Moreover, ligands binding too weakly to the target will only result in a small bound ligand fraction $P_{\text{L,b}}$, and the exchange-modulated signal will not differ significantly from the free state value of the ligand. Therefore, the dissociation constant is subject to strict limitations unless special techniques (e.g., competition-based experiments) are used.

As many ligand detected techniques are based on the significant change in molecular mass when binding to a target molecule, difficulties arise if the size difference between the ligand and the target is small. Hence, targets with high molecular weights are favorable, while ligands preferably have low molecular masses.

In the following sections the most common experiments are briefly reviewed. Their basic concept will be outlined, and applications, requirements, and limitations will be discussed.

3.39.3.1.1 Diffusion

The translational diffusion rate is a classical parameter for monitoring the complex formation of the ligand and the target.[36–39] For this purpose, diffusion-edited spectra can be recorded, which differentiate the ligands within a mixture according to their diffusion coefficient. Complexed ligands have a smaller diffusion rate than noncomplexed ones. If refocusing pulsed-field gradients are applied (**Figure 5**), the signals of fast diffusing species are more attenuated than those of slow diffusing compounds, leaving only the signals of the complexed ligands observable. This procedure can be combined with standard 2D techniques resulting i.e., in diffusion-edited TOCSY (DECODES)[40] or COSY[41] sequences. The apparent diffusion coefficient can be extracted from a series of diffusion-edited spectra with changing gradient strength (diffusion-ordered spectroscopy; DOSY). If the diffusion coefficients of the free ligands and the target molecules are known, estimations about the dissociation constant can be made using via eqn [11].[42,43] Other applications, such as the NOE pumping experiment, use a diffusion filter before the actual screening experiment, to eliminate signals of species with low molecular mass (fast diffusing) (*see* Section 3.39.3.1.6).

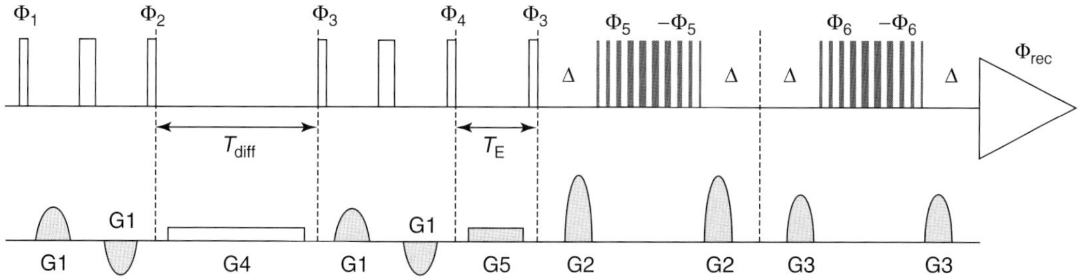

Figure 5 Pulse sequences for diffusion-edited screening experiments. The longitudinal eddy current delay (LED) sequence[105] suppresses signals of fast-diffusing components. Narrow (wide) empty bars denote 90° (180°) pulses and light gray bars, pulsed field gradients. Bipolar pulse pairs (G1) and a settling delay T_E (≈20 ms) are included to diminish eddy currents.[106,107] Water suppression (dark gray bars) is achieved via excitation sculpting[108] using WATERGATE W5[109] as the filter element. Gradients G4 and G5 are weak, and are used to eliminate unwanted transverse magnetization and reduce radiation damping. The diffusion time T_{diff} is in the range of about 100 ms.

Diffusion-based techniques suffer from their intrinsic lack of sensitivity, which results in an insufficient signal-to-noise ratio, or requires long, repetitive measurements. Moreover, the difference of diffusion coefficients between a large target and a small ligand is not that large, leaving weakly binding molecules undetectable due to the small fraction of bound ligand $P_{L,b}$ (see eqn [11]).

3.39.3.1.2 Relaxation

Measurements of relaxation rates (R_2, $R_{1\rho}$, and R_1) compare rather well with those of the diffusion coefficient.[7] While in diffusion-filtered experiments, translational diffusion rates are used to distinguish binding from nonbinding components, the alteration of the rotational diffusion rate is used in relaxation-filtered experiments.[36] A decrease in the rotational diffusion rate upon binding will result in a increased correlation time τ_c, which generally translates to an amplification of relaxation. Depending on the relaxation parameter of interest, different filter elements are applied, such as Carr–Purcell–Meiboom–Gill (CPMG),[44,45] spin lock,[46] or inversion recovery elements. Difference spectroscopy is usually used for the measurement of altered relaxation rates. Therefore, a reference spectrum of the ligand mixture without the target has to be acquired; hereof a spectrum of the mixture including the target is subtracted. This might entail subtraction artifacts, leading to false positives. These artifacts are due to chemical shift changes, pH effects, chemical exchange, or lineshape distortions. The latter is sometimes also a wanted feature, for instance in T_2-resolved spectra. As the linewidth is proportional to the transverse relaxation rate R_2, a significant line-broadening occurs upon binding. In T_2 experiments, a CPMG sequence[44,45] (**Figure 6a**) is usually included for filtering. The duration of the delay 2δ between the 180° refocusing pulses of the spin echo can be used to fine tune the experiment to strong binders (short delay) or weak binders (long delay). This type of experiment is very useful for targets of high molecular weight. When the molecular weight of the target decreases, a $T_{1\rho}$ experiment (**Figure 6b**) should be considered. This technique is based on alterations of the relaxation rate along a fixed axis defined by a spin lock field. Spin lock fields can also be used to suppress background signals of compounds with high molecular mass.

The longitudinal relaxation rate ($1/T_1$) may also serve as indicator for the binding process (**Figure 6c**). Essentially, experiments can be performed in two different ways. Either all signals of the ligands are inverted (nonselective) or just one single resonance is inverted (selective). This results in the measurement of $T_{1,ns}$ and $T_{1,s}$, respectively. It can be shown that $T_{1,ns}$ is not sensitive to the correlation time τ_c, and therefore cannot be recommended for screening experiments, whereas $T_{1,s}$ is a good measure for binding. The problem is to achieve selective inversion for a large amount of compounds within a screening mixture. For this reason, experiments measuring T_1 are mostly competition based (*see* Section 3.39.3.1.7).

3.39.3.1.3 Paramagnetic spin labels

Closely related to T_2 relaxation experiments are experiments with paramagnetic spin labels[47–50] and paramagnetic metal ions.[51] In both cases the transverse relaxation rate of binding ligands is increased, leading to a rapid decay of their signals. The difference to "classical" relaxation experiments (*see* Section 3.39.3.1.2), however, is the origin of the modified time constant T_2. The main relaxation mechanism with paramagnetic spin species is the electron–proton dipole–dipole interaction. Because of the large gyromagnetic ratio of the electron, the observed relaxation rate $R_{2,obs}$ is

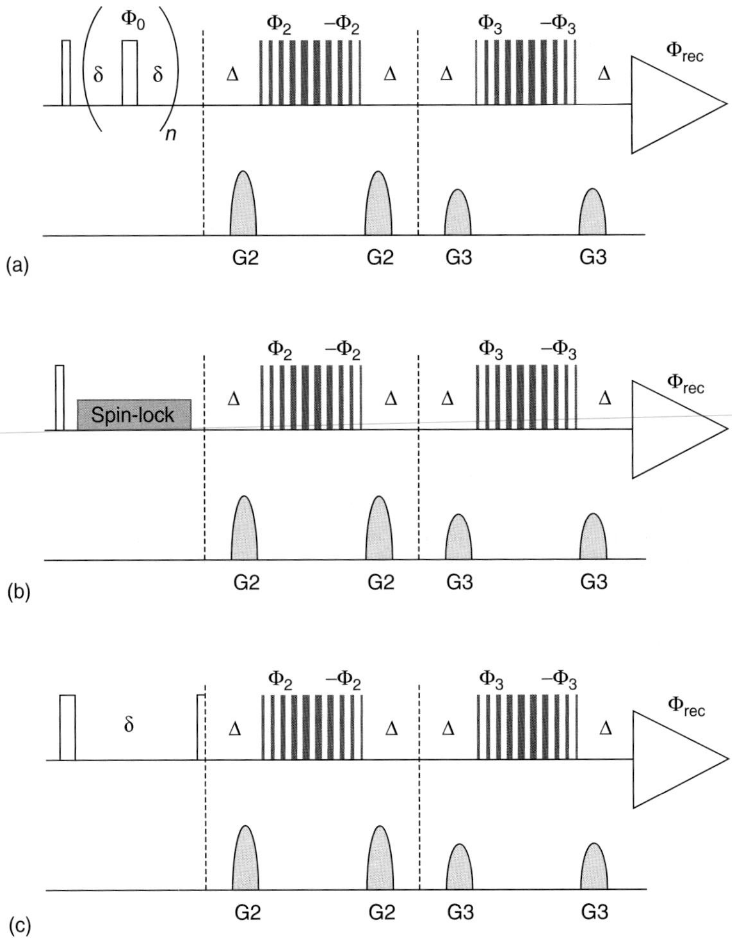

Figure 6 Pulse sequences for relaxation-edited screening experiments. Different filter elements are used prior to the water suppression element. (a) A CPMG pulse train[44,45] is used to suppress signals of fast relaxing species showing a large transverse relaxation rate $R_{2,obs}$. (b) The CPMG pulse train is replaced by a spin-lock filter. Spin-locking is easily achieved by continuous wave irradiation or adequate pulse sequences.[46] (c) A inversion recovery element is used for the measurement of longitudinal relaxation rates. In all three pulse sequences water suppression is achieved by excitation sculpting[108] and a WATERGATE W5 filter[109] (see **Figure 5**).

dominated by the term $R_{2,\mathrm{para}}$ (see eqn [17]).[49,52] Therefore, the experiment does not rely on the size difference between the target and the ligand, and thus larger ligands may be screened:

$$R_{2,\mathrm{obs}} = P_{\mathrm{L,b}}R_{2,\mathrm{b}} + \left(1 - P_{\mathrm{L,b}}\right)R_{2,\mathrm{f}} + R_{\mathrm{ex}} + P_{\mathrm{L,b}}R_{2,\mathrm{para}} \qquad [17]$$

As a consequence of the strong relaxation effect, smaller amounts of the target and ligand are needed, and also weaker binders (shorter contact times) become observable. The paramagnetic spin species has to be within a distance of about 9 Å[51] of the ligand binding site to enhance the relaxation. This can be realized in a number of ways: a paramagnetic spin label may be covalently attached to the target near the binding site, or a spin-labeled ligand introduced at a binding site remote from the binding site of interest. The latter is referred to as second-site screening[47,53] (**Figure 7**). An example of the first technique is the SLAPSTIC method (spin labels attached to protein side chains to identify interacting compounds). Both approaches can be performed by the use of $T_{1\rho}$ experiments as described in **Figure 6b**. For the SLAPSTIC technique, precise information about the 3D structure of the target is indispensable for the introduction of a spin label without modification of the binding site. Although established techniques for protein modification[54] can be used to attach spin labels, such as the paramagnetic TEMPO group (2,2,6,6-tetramethyl-piperidine-1-oxyl)[48] or metal ions such as lanthanides,[55] it can be a time-consuming task. An interesting alternative is the substitution of zinc ions by cobalt(II) ions in many metalloproteins, as the catalytic activity is maintained.[56]

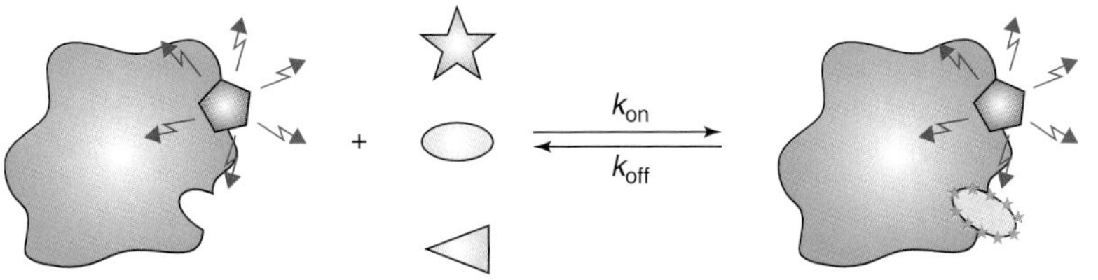

Figure 7 Influence of a paramagnetic spin label on the relaxation properties of a binding ligand. Red arrows indicate the strong magnetic dipole of the unpaired electron, which causes the nuclei in its proximity to relax faster. The red stars represent an enhanced relaxation rate R_2 on the bound ligand.

3.39.3.1.4 ^{19}F ligands

The screening of ligands containing ^{19}F labels is a comparably new and promising technique.[57–59] In contrast to 'classical' approaches, ^{19}F nuclei, not protons, are detected. The fluorine nucleus has some unique features, making it an attractive probe for screening. It has a high gyromagnetic ratio ($\gamma_F \approx 0.94\gamma_H$), and occurs at high natural abundance (100%), making it very sensitive for NMR experiments. Biological molecules and organisms lack ^{19}F, which enables the acquisition of simple 1D spectra without the need to suppress disturbing background resonances. In addition, the broad chemical shift dispersion allows the composition of large mixtures without signal overlap, thus reducing experimental time.

The transverse relaxation rate $R_{2,\mathrm{obs}}$ of the fluorine signals has an additional term originating from the large chemical shift anisotropy $\Delta\sigma$ of the ^{19}F nucleus.[60] This contribution is directly proportional to the rotational correlation time τ_c, and results in a significant line broadening when ligands bind to a large target.

Similarly to experiments with paramagnetic spin labels, fluorine-based experiments are particularly suitable for low-affinity binders. It is usually sufficient to acquire a 1D spectrum with and without the target molecule to screen for binding, but CPMG pulse sequences (**Figure 6a**) can also be employed. It is worth noting that high-affinity ligands can also be investigated, and fast exchange is not essential for fluorine screening.

The ^{19}F nucleus can be inserted into the ligand in the form of, for example, fluorinated aromatic or trifluoromethyl groups.[57,61,62] At first sight this might seem unfavorable, but fluorine is now often used to modulate the ADMET (absorption, distribution, metabolism, excretion, and toxicity) profile[63]; for example, it is common to introduce fluorine in an aromatic residue to prevent oxidation by CYP450 enzymes. Roughly 10% of all drugs on the market are fluorinated.[12] Moreover, fluorine often can be replaced by hydrogen, or vice versa, without loss of activity. Hence, ligands identified by screening fluorinated libraries can be converted into nonfluorinated binders. However, the best fluorine-containing binders can also be used as reporter ligands in competition-based screening to seek for higher-affinity binders (*see* Section 3.39.3.1.7).

3.39.3.1.5 Intramolecular nuclear overhauser effect

In transferred NOE spectra[64–68] the interaction between different protons within the ligand is detected (intramolecular NOE). The sign and the size of the NOE strongly depends on the molecular tumbling rate, which is equivalent to the rotational correlation time τ_c. Molecules with low molecular weights have small and positive cross-relaxation rates, resulting in weakly negative 2D cross-peak intensities. In contrast, compounds with high molecular mass have large, negative cross-relaxation rates, leading to intense positive cross-peaks in the 2D spectrum. When the ligands dissociate, magnetization, which is rapidly built up in the bound state by the NOE, is conserved in the free state. Here, we observe the NOE (caused by the nonequilibrium spin population created in the bound ligand) in the spectrum of the free molecule. Therefore, only binding ligands show a change in the sign and intensity of the cross-peaks.

These NOE measurements are commonly performed as 2D NOESY spectra, employing a relaxation filter to suppress residual signals from the large target molecule.[65,66] These spectra not only permit the detection of binding but also allow the investigation of the conformation of the ligand within the bound state.[68,69] This provides important structural information, and is worth noting, especially in the context of structure-based screening. With this technique, weak to medium affinity binders can be observed.

These experiments have several drawbacks. First, the molecular weight of the ligand is limited; ligands of higher molecular mass will exhibit positive cross-peaks in the spectrum even if they are not bound to the target. Also, 2D spectra are necessary to achieve sufficient resolution, which entails longer acquisition times. Moreover, strong diagonal peaks may lead to signal overlap and T_1 noise, reducing spectral quality and making analysis of the spectrum impracticable. Sometimes signal overlap can be overcome by introducing a third dimension, as in 3D TOCSYtrNOESY (transferred NOE spectroscopy).[70] Experiments monitoring the intramolecular NOE require comparatively large amounts of ligand, making them impractical for poorly soluble ligands.

3.39.3.1.6 Intermolecular nuclear overhauser effect

Some of the most powerful NMR screening experiments exploit the intermolecular NOE to seek for binding molecules. When high-molecular-weight molecules are considered, the NOE is strongly negative, whereas it is positive for low-molecular-weight molecules. Therefore, strong effects ($\sigma_{NOE,b} \gg \sigma_{NOE,f}$, see eqn [14]) are obtained at low concentrations and for very weakly binding ligands. This makes these experiments rather robust, and explains their widespread use.

In all these experiments, a nonequilibrium magnetization is created either on the target or on the ligand molecule. The nonequilibrium magnetization is subsequently transferred to the binding partner via spin diffusion. As the sign and size of the intermolecular NOE is different for binding and nonbinding ligands, their signals are perturbed differently. Hence, binding and nonbinding components can be discriminated within the NMR spectra. Different strategies can be used to create the nonequilibrium magnetization on the target or the ligand: either the target and the ligand are perturbed equally, and a filter element selectively removes the magnetization of one of the two components, or only the ligand/target is perturbed, and the nonequilibrium magnetization is created directly.

In the frequently used saturation transfer difference (STD) experiment,[11,71–73] the nonequilibrium magnetization is created by selective saturation of target resonances (on-resonance). Selective saturation is achieved via a radio-frequency pulse train (**Figure 8**) that is applied in a spectral window where only target resonances appear. The saturation rapidly spreads out across the target molecule due to spin diffusion, and is finally transferred to binding ligands via the intermolecular NOE (**Figure 9**). When these ligands dissociate back into solution, they retain their artificial spin state, and consequently will have a smaller z magnetization. By rotating the z magnetization into the transverse plane, binding ligands will exhibit lower signal intensities in the on-resonance experiment compared with nonbinding ligands. A relaxation filter is used to suppress signals from the target molecule. To visualize the loss of signal intensity, the on-resonance experiment is subtracted from a reference spectrum. In the reference experiment the radio-frequency pulse train is applied far off-resonance, where no resonances are saturated, leaving the ligand signals unperturbed. Hence, the difference spectrum will only contain the resonances of the binding components (**Figure 10**).

STD experiments are among the most commonly used experiments because of their diverse applications. Target molecules are not subject to any upper-size limitations, and even immobilized targets can be investigated.[74] For the analysis of complex ligand mixtures the STD experiment can be combined with any 2D or 3D technique,[75] making it a useful tool for screening experiments. Moreover, the dissociation constant K_D can be determined[11] by relating the ligand concentration [L] to the STD amplification in a competitive titration (*see* Section 3.39.3.1.7). Even crude structural information about the binding mode can be extracted from the experiments if the observed STD amplification is mapped to the structure of the ligand.[76,77] Recently, the STD experiment has been applied to an RGD

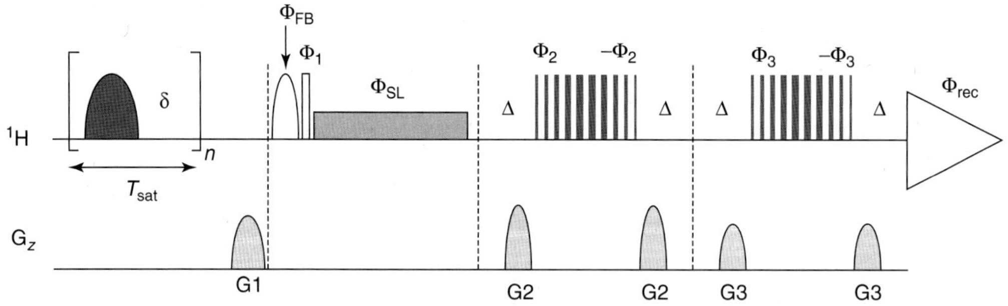

Figure 8 Typical pulse sequence for the measurement of saturation transfer difference.[76] Saturation is achieved via a repetitive ($n \approx 40$) selective 90° radio-frequency pulse (dark gray pulse, e.g., a 50 ms Gaussian pulse) followed by an interpulse delay δ ($\approx$1 ms). After the total saturation time T_{sat}, which is in the range of about 1–3 s, a gradient (G1) ensures that only z magnetization is present before the actual experiment starts. A water flip back pulse and excitation sculpting including WATERGATE W5 prior and after the spin lock element (see **Figure 6b**) are used to suppress the water signal. The irradiation frequency of the saturation pulse train is changed after every scan (on-resonance ≈ -0.4 ppm, off-resonance $\approx$30 ppm), and the subsequent subtraction of the two experiments is achieved by an appropriate phase cycle.

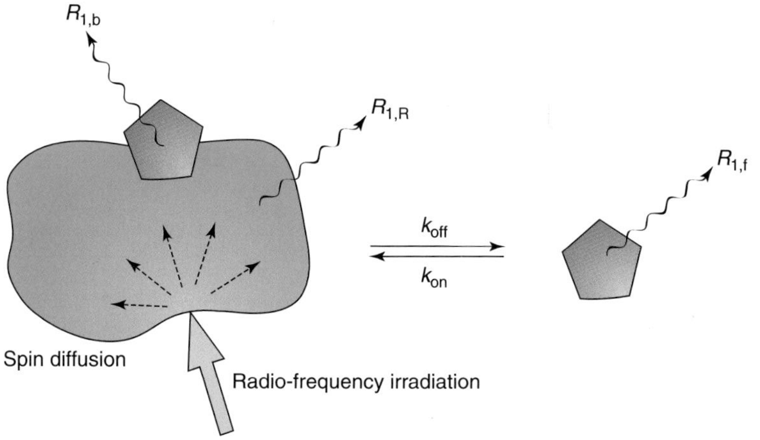

Figure 9 Illustration of the STD effect.[14] Selective irradiation of receptor resonances (yellow arrow) leads to the saturation of the whole target molecule via spin diffusion. The saturation is also transferred to binding ligands, which dissociate back into solution after the residence time τ_{res}. While the ligand is bound to the target, the ligand signals decay with a relaxation rate $R_{1,b}$, whereas in the free state they decay with a rate $R_{1,f}$. The relaxation rate $R_{1,b}$ can be much larger than $R_{1,f}$, and is therefore the main limiting factor for the transfer. The irradiation frequency of the saturation pulse train is changed after every scan (on-resonance ≈ -0.4 ppm, off-resonance ≈ 30 ppm) and the subsequent subtraction of the two experiments is achieved by an appropriate phase cycle.

peptide binding to membrane protein at an intact cell.[78] In this experiment the signals of the cell have to be suppressed as well as the misleading background binding events within the cell. Therefore, the technique is referred to as saturation transfer double difference (STDD), and might enable screening experiments in vivo.

Closely related to the STD approach is the waterLOGSY (water–ligand observed via gradient spectroscopy) experiment,[79,80] where the large bulk water magnetization is used to create a nonequilibrium spin state. Initially the water magnetization is either selectively inverted (**Figure 11**) or saturated; both strategies are feasible, although selective inversion seems to be favored. Subsequently, the magnetization is transferred from the water molecules to the binding ligands by one of three pathways.[14] First, the magnetization can be transferred directly by the proton–proton cross-relaxation of the water molecules within the binding pocket with the binding ligand. Secondly, exchangeable protons within the binding pocket of the receptor molecule can be substituted by the protons of the water molecules. Hence, the magnetization of these protons is also inverted. Their magnetization finally propagates to the binding ligand via the intermolecular NOE. The third pathway is the chemical exchange of the protons of the water molecules with exchangeable protons from the protein surface. The magnetization spreads across the protein by spin diffusion, and is transferred to the ligand. As a result of all of these pathways the magnetization of binding ligands is inverted while the magnetization of nonbinding ligands is not. Consequently, binding ligands appear negative within the spectrum, and nonbinding components appear positive. This introduces misleading results if signal overlap is not prevented.

The waterLOGSY experiment is especially useful if the target molecule lacks a high proton density. In this situation the mechanism of spin diffusion is insufficient to transfer the magnetization to the ligand, as is sometimes the case for DNA and RNA fragments.[14]

Intermolecular magnetization transfer from the target to the ligand is observed in NOE pumping experiments.[81] After unselective excitation, a diffusion filter prior to the NOE mixing time eliminates all ligand magnetization. Ligands bound to the target will regain some magnetization, whereas unbound ligands remain undisturbed. Finally, the magnetization of the ligands is detected. Bound compounds will exhibit a signal in the proton spectrum. No reference spectrum is needed for this experiment, and it is more sensitive to weakly binding ligands than standard diffusion experiments. However, the diffusion filter in NOE pumping causes a loss of signal intensity, and needs to be adjusted individually to every sample, making it a comparably insensitive and not universally applicable experiment.

If the diffusion filter is replaced by a relaxation filter,[82] the target magnetization will be removed instead of the ligand magnetization. In this case the magnetization passes from the ligand to the target, reducing the signal intensity of binding ligands. A reference spectrum, where the relaxation filter is swapped with the mixing period, is recorded and subtracted to visualize the changes induced by binding. The difference spectrum will yield signals of interacting components only. This technique is referred to as reverse NOE pumping, and does not suffer from the low sensitivity of NOE pumping. However, for both variants a large ligand excess is needed to discriminate ligand signals from target signals.

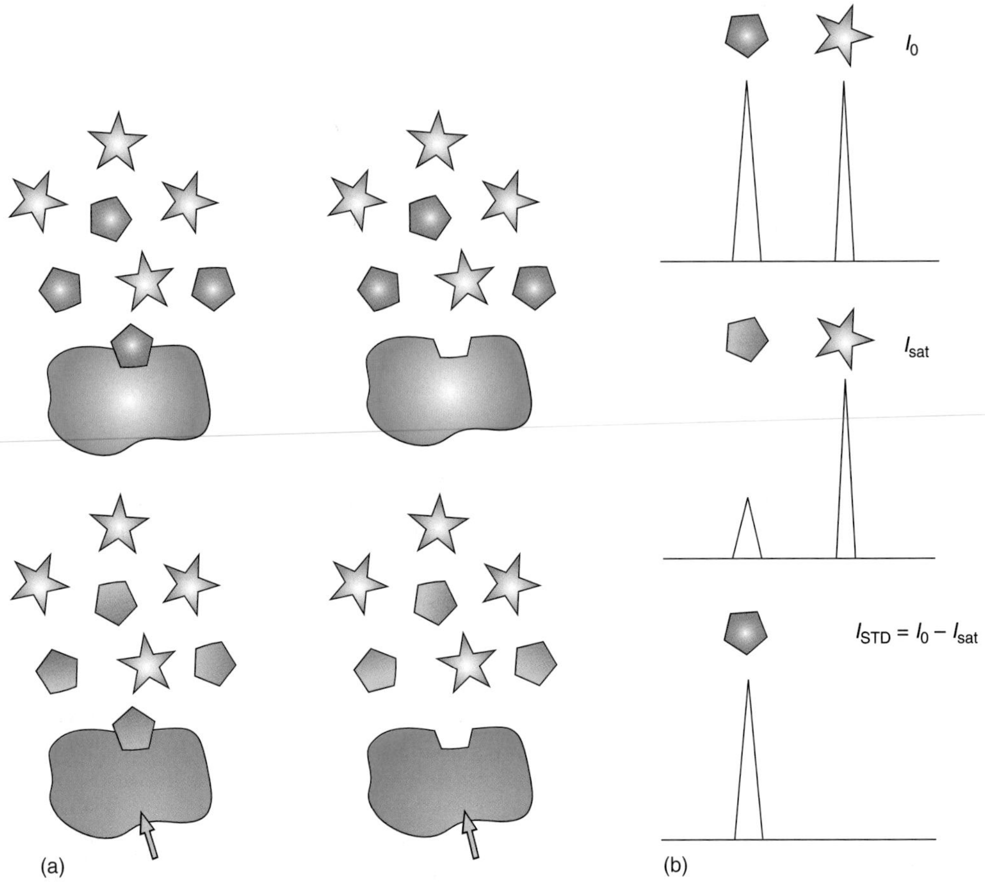

Figure 10 Schematic representation of the STD experiment.[14] (a) Off-resonance irradiation does not disturb any ligand or target signals. Therefore, a regular spectrum of the all ligands is obtained, while target resonances are suppressed by the spin lock filter. On-resonance irradiation leads to a decrease in the signal intensity of binding ligands. (b) Subtraction of the experiments. The subtraction yields signals from the binding components only.

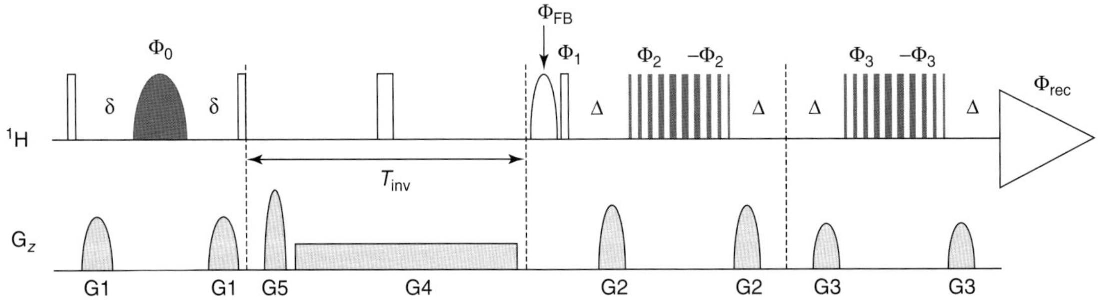

Figure 11 WaterLOGSY pulse sequence[80,110,111] for the observation of binding ligands. In the first part of the experiment the water resonance is selectively inverted while all other magnetization is dephased by the two gradients G1. Inversion of the water resonance is achieved by a selective 180° refocusing pulse (dark gray pulse, e.g., REBURP[112]). The inversion pulse may influence protein signals having the same chemical shift as the water resonance. Therefore, the delay δ can be extended to diminish residual protein signals via their large transverse relaxation rate. During the inversion time T_{inv}, magnetization is transferred from the bulk water to the ligands. The pulsed field gradients G5 and G4 are used to ensure that only z magnetization is present during the inversion time, and radiation damping effects are minimized. An optional nonselective 180° pulse during the inversion time can be included to reduce relaxation effects. To suppress the water signal prior to acquisition, a water flip back pulse and excitation sculpting are used.

3.39.3.1.7 Competition-based screening

One of the major drawbacks of ligand-based experiments is the inability to screen ligands with higher affinities. As fast exchange is essential for these techniques to visualize ligands binding to a target, high-affinity binders are unobservable and lead to false positives. Target-based techniques could be used instead, as they are not subject to limitations concerning the dissociation constant. Unfortunately, these techniques have other disadvantages (*see* Section 3.39.3.2), making them unsuitable for high molecular weight targets, for example. As an alternative, competition-based screening technologies can be used to screen for binders. These techniques combine the major advantages of ligand and target detection methods. With competition-based experiments, ligands with any affinity can be screened, and the target molecule is not subject to size limitations. In addition, only specific binders will be detected; nonspecific binders and binders of different binding sites than the reporter ligand (see below) remain invisible.

For these experiments a known medium-affinity binder (reporter ligand) is added to the solution of the target, and, afterwards, the so-called screening ligand (competitor) (**Figure 12**). Depending on their binding affinity and concentration, the screening ligand will displace the reporter ligand from its binding site (see **Figure 2**). This substitution will induce changes in the reporter ligand signal, which is monitored in competition-based screening experiments. It is important to note that only the reporter ligand needs to fulfill the condition of fast exchange. Essentially, there is no limit to the affinity of the screening ligand if an appropriate reporter ligand is known. If the binding constant K_L of the reporter ligand has been determined, competition screening experiments can be used to measure the binding constant K_C of the new ligand via a titration.[83] Of course, these techniques also have their intrinsic drawbacks: mixtures of screening ligands have to be deconvoluted to extract the binding ligand; moreover, competition-based screening negates the useful ability to detect subsites of the target molecule.

In principle, all ligand-based experiments can be replaced by competition-based experiments.[84,85] Nevertheless, it is reasonable to use techniques that are fast, effect major changes to the ligand signals upon binding, and are very sensitive even at low concentrations. WaterLOGSY,[86] STD,[87,88] and ^{19}F screening belong to this class of experiments, and are therefore especially powerful when competition-based techniques are considered. There are many examples that have proved applicability to lead search and lead optimization. Here, we focus on ^{19}F competition-based screening. This technique is of particular interest because no filter element is needed to suppress the signals of nonbinding components or the target molecule. In addition, it can be used on nonfluorinated libraries, as only the reporter ligand needs to be fluorinated. Dalvit *et al.*[89] published this approach under the acronym of FAXS (fluorine chemical shift anisotropy and exchange for screening). As seen in **Figure 13**, the displacement of a fluorinated ligand bound to riboflavine synthase by the stronger binder lumacine results in the reappearance of the ^{19}F signal in the spectrum. The spectra do not contain any distracting background signals, and are obtained in a comparably short amount of time.

3.39.3.2 Target-Based Methods

Unlike ligand-based techniques, target-based methods monitor the changes of the receptor signals. While ligand detection techniques are mainly based on changes in molecular weight, target-based experiments make use of the influence of binding ligands on the chemical shifts of the target (*see* Section 3.39.2.3). This is due to the fact that the molecular mass of the target does not change significantly upon binding, and consequently all NMR parameters relying on this alteration are not suitable for target detection methods. Nevertheless, this approach is very powerful, as

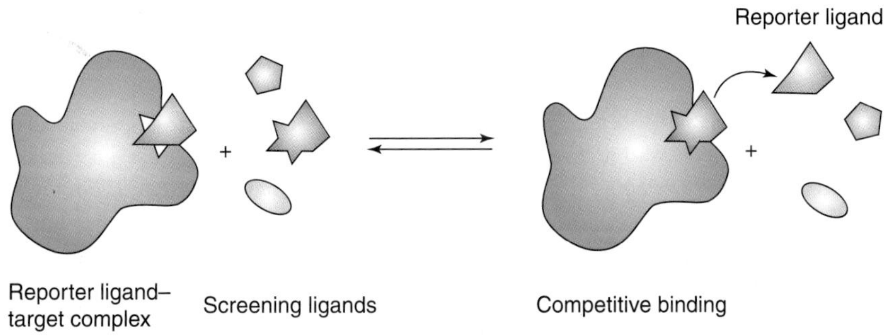

Figure 12 Illustration of a competition binding experiment.[12] The orange reporter ligand (medium to low affinity) is displaced by the magenta screening ligand (higher affinity). The other screening ligands do not displace the reporter ligand due to their low affinity for the target molecule.

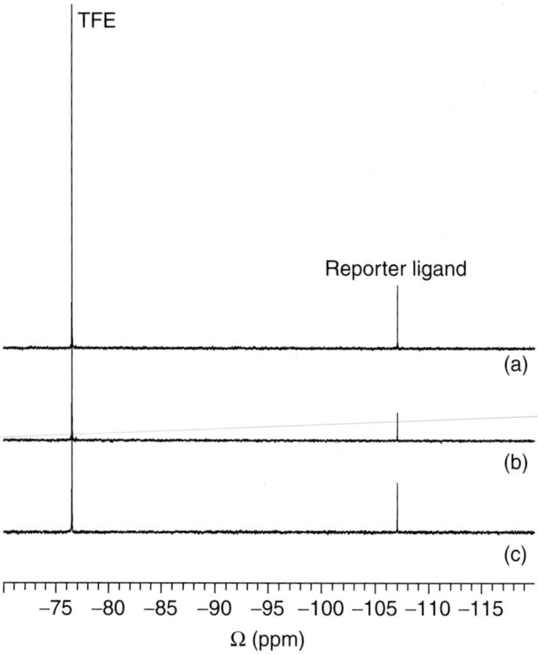

Figure 13 Displacement of a fluorinated reporter ligand by a screening ligand. (a) The ^{19}F spectrum of the reporter ligand with the internal standard TFE is shown. (b) Spectrum after addition of the target protein, showing the decrease in intensity of the reporter ligand signal due to line broadening. (c) Addition of the screening ligand causes the reporter ligand to be displaced from the binding pocket, and its signal intensity is restored.

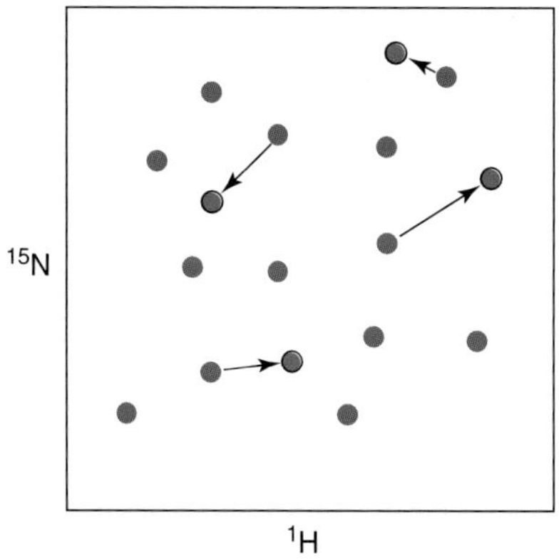

Figure 14 Depiction of a $^1H,^{15}N$ HSQC spectrum to trace chemical shift changes upon binding.[5] Only resonances belonging to residues within or close to the binding pocket are displaced as long as allosteric effects can be neglected. The amount of displacement may be different for all resonances that are shift.

discussed in the next section. Target-based techniques lack the ability to investigate high-molecular-weight targets because they are subject to fast relaxation. Even if techniques such as TROSY[90] (transverse relaxation optimized spectroscopy) or CRIPT[91] (cross-correlated relaxation-induced polarization transfer) are utilized, the size limit cannot be extended beyond ~100 kDa.

3.39.3.2.1 Chemical shift perturbations

Chemical shift changes are usually observed in 2D spectra,[1,23,92] as signal overlap obscures the observation in 1D spectra. Chemical shift perturbations are most easily monitored in 2D HSQC spectra (**Figure 14**), although HNCO[25] and homonuclear 2D NOESY spectra have also been used. Reference spectra are essential to detect displacements of resonances upon binding. If the displacement is correlated to the concentration of the added ligand, estimations of the dissociation constant K_D can be made.[93] In contrast to ligand-based methods high-affinity binders can also be investigated by target-based methods.

Fast exchange is not essential for the observation of binding molecules with target-based methods. As seen in **Figure 3**, in the slow exchange limit two separate resonances are observable. If the ligand is added to the target molecule in excess, only the signal set of the ligand–target complex will be visible, as the population of the free target is negligible. For the fast exchange limit the same rules hold as for ligand-based techniques – the observed chemical shift is the population-weighted average of the free and bound states of the target. Only the intermediate exchange regime will lead to problems in an analysis, as is apparent from **Figure 3**. When the resonances of the free and the bound forms coalesce, the signal intensity decreases, and the resonance becomes unobservable. To overcome this problem, the concentration ratio of the target and ligand has to be adjusted carefully to force the exchange either into the fast or slow exchange regime.

Chemical shift perturbations are usually observed in ^{15}N or ^{13}C HSQC spectra.[1,92] As spectra at natural abundance are not sufficient to achieve a reasonable sensitivity, the protein target needs to be enriched with ^{15}N and/or ^{13}C, respectively.[94–96] Once an appropriate overexpression system is established, ^{15}N labeling is comparably cheap, whereas ^{13}C labeling is costly. Recently developed labeling schemes[96–98] now allow the specific insertion of ^{13}C methyl groups into the side chains of valine, leucine, and isoleucine.[23,95] Owing to the threefold degeneracy of the methyl moiety, the signal-to-noise ratio is increased by a factor of three compared with amide protons. However, the chemical shift dispersion in ^{13}C HSQC using only the resonances of the above-mentioned methyl groups is smaller than in a ^{15}N HSQC.

The type of information that ^{13}C- and ^{15}N-detection methods provide is complementary. While amide protons are hydrophilic and interact significantly with negatively charged moieties of the binding ligand, methyl groups are hydrophobic and interact accordingly.[7] It is worth noting that ligands are often bound to hydrophobic pockets with a high probability of encountering methyl groups. Nevertheless, both techniques are suitable for seeking binding ligands, and are especially useful in a fragment-based approach.[1,99]

The location of the binding ligand can be determined if a sequential assignment is available. By optimizing and linking several ligands that bind to different binding sites, high-affinity ligands can be rationally designed. This technique is referred to as SAR-by-NMR,[1] and is discussed in detail elsewhere (*see* 3.41 Fragment-Based Approaches). Moreover, not only the binding site but also the binding mode can be investigated by target-based methods.[100–102] The theoretical prediction of target chemical shifts and the subsequent comparison with the experimentally obtained spectrum allows characterization of the orientation of the ligand within the binding pocket. Recently, the use of induced shift changes to determine ligand–target complex structures has been proposed, in a similar fashion to the docking of larger biomolecules using the HADDOCK program.[30,31]

When high-molecular-weight targets are investigated, signal overlap makes the analysis of the spectra difficult. The use of residue-selective labeling[97,103] may prevent this problem. Here, selectively labeled amino acids are introduced, leaving the spectra less crowded. Care must be taken in the analysis of these spectra, as the labeled amino acid might not be influenced by the binding ligand. If the binding site is already known, site-selective labeling[98,99,104] is applicable where only the amino acids of the binding pocket are labeled.

New and more complex labeling schemes have allowed the investigation of higher-molecular-weight targets and more complex circumstances. Therefore, the above-mentioned upper size limit of 100 kDa will certainly be exceeded in the near future.

References

1. Shuker, S. B.; Hajduk, P. J.; Meadows, R. P.; Fesik, S. W. *Science* **1996**, *274*, 1531–1534.
2. Keifer, P. A. *Curr. Opin. Biotechnol.* **1999**, *10*, 34–41.
3. Stockman, B. J. *Prog. Nucl. Magn. Reson. Spectrosc.* **1998**, *33*, 109–151.
4. Roberts, C. *Curr. Opin. Biotechnol.* **1999**, *10*, 42–47.
5. Roberts, G. *Drug Disc. Today* **2000**, *5*, 230–240.
6. Ross, A.; Schlotterbeck, G.; Klaus, W.; Senn, H. *J. Biomol. NMR* **2000**, *16*, 139–146.
7. Diercks, T.; Coles, M.; Kessler, H. *Curr. Opin. Chem. Biol.* **2001**, *5*, 285–291.
8. Heller, M.; Kessler, H. *Pure Appl. Chem.* **2001**, *73*, 1429–1436.
9. Stockman, B. J.; Dalvit, C. *Prog. Nucl. Magn. Reson. Spectrosc.* **2002**, *41*, 187–231.
10. Pellecchia, M.; Sem, D. S.; Wüthrich, K. *Nature Rev. Drug Disc.* **2002**, *1*, 211–219.
11. Meyer, B.; Peters, T. *Angew. Chem. Int. Ed.* **2003**, *42*, 864–890.
12. Coles, M.; Heller, M.; Kessler, H. *Drug Disc. Today* **2003**, *8*, 803–810.
13. Jahnke, W.; Widmer, H. *Cell. Mol. Life Sci.* **2004**, *61*, 580–599.
14. Peng, J. W.; Moore, J.; Abdul-Manan, N. *Prog. Nucl. Magn. Reson. Spectrosc.* **2004**, *44*, 225–256.
15. Kovacs, H.; Moskau, D.; Spraul, M. *Prog. Nucl. Magn. Reson. Spectrosc.* **2005**, *46*, 131–155.

16. Street, A. *Am. Lab.* **2003**, *35*, 28–30.
17. Lipinski, C. A.; Lombardo, F.; Dominy, B. W.; Feeney, P. J. *Adv. Drug Delivery Rev.* **1997**, *23*, 3–25.
18. Veber, D. F.; Johnson, S. R.; Cheng, H. Y.; Smith, B. R.; Ward, K. W.; Kopple, K. D. *J. Med. Chem.* **2002**, *45*, 2615–2623.
19. Nicholson, J. K.; Connelly, J.; Lindon, J. C.; Holmes, E. *Nature Rev. Drug Disc.* **2002**, *1*, 153–161.
20. Lindon, J. C.; Holmes, E.; Nicholson, J. K. *Prog. Nucl. Magn. Reson. Spectrosc.* **2004**, *45*, 109–143.
21. Lindon, J. C.; Holmes, E.; Nicholson, J. K. *Expert Rev. Mol. Diagn.* **2004**, *4*, 189–199.
22. Fejzo, J.; Lepre, C. A.; Peng, J. W.; Bemis, G. W.; Ajay; Murcko, M. A.; Moore, J. M. *Chem. Biol.* **1999**, *6*, 755–769.
23. Hajduk, P. J.; Gomtsyan, A.; Didomenico, S.; Cowart, M.; E. Bayburt, K.; Solomon, L.; Severin, J.; Smith, R.; Walter, K.; Holzman, T. F. et al. *J. Med. Chem.* **2000**, *43*, 4781–4786.
24. Peng, J. W.; Lepre, C. A.; Fejzo, J.; Abdul-Manan, N.; Moore, J. M. *Nucl. Mag. Reson. Biol. Macromol. A* **2001**, *338*, 202–230.
25. van Dongen, M.; Weigelt, J.; Uppenberg, J.; Schultz, J.; Wikstrom, M. *Drug Disc. Today* **2002**, *7*, 471–478.
26. Rees, D. C.; Congreve, M.; Murray, C. W.; Carr, R. *Nat. Rev. Drug Disc.* **2004**, *3*, 660–672.
27. Moy, F. J.; Haraki, K.; Mobilio, D.; Walker, G.; Powers, R.; Tabei, K.; Tong, H.; Siegel, M. M. *Anal. Chem.* **2001**, *73*, 571–581.
28. Corcoran, O.; Spraul, M. *Drug Disc. Today* **2003**, *8*, 624–631.
29. Blundell, T. L.; Jhoti, H.; Abell, C. *Nat. Rev. Drug Disc.* **2002**, *1*, 45–54.
30. Dominguez, C.; Boelens, R.; Bonvin, A. *J. Am. Chem. Soc.* **2003**, *125*, 1731–1737.
31. Schieborr, U.; Vogtherr, M.; Elshorst, B.; Betz, M.; Grimme, S.; Pescatore, B.; Langer, T.; Saxena, K.; Schwalbe, H. *Chembiochem* **2005**, *6*, 1891–1898.
32. Kessler, H. *Angew. Chem. Int. Ed.* **1970**, *9*, 219–235.
33. Binsch, G.; Kessler, H. *Angew. Chem.-Int. Ed.* **1980**, *19*, 411–428.
34. Cavanagh, J.; Fairbrother, W. J.; Palmer, A. G.; Skelton, N. J. *Protein NMR Spectroscopy – Principles and Practice,* 1st ed.; Academic Press: San Diego, CA, 1996.
35. Palmer, A. G.; Kroenke, C. D.; Loria, J. P. *Nucl. Mag. Reson. Biol. Macromol. B* **2001**, *339*, 204–238.
36. Hajduk, P. J.; Olejniczak, E. T.; Fesik, S. W. *J. Am. Chem. Soc.* **1997**, *119*, 12257–12261.
37. Bleicher, K.; Lin, M. F.; Shapiro, M. J.; Wareing, J. R. *J. Org. Chem.* **1998**, *63*, 8486–8490.
38. Anderson, R. C.; Lin, M. F.; Shapiro, M. J. *J. Comb. Chem.* **1999**, *1*, 69–72.
39. Lucas, L. H.; Price, K. E.; Larive, C. K. *J. Am. Chem. Soc.* **2004**, *126*, 14258–14266.
40. Lin, M. F.; Shapiro, M. J. *J. Org. Chem.* **1996**, *61*, 7617–7619.
41. Wu, D. H.; Chen, A. D.; Johnson, C. S. *J. Magn. Reson. Ser. A* **1996**, *121*, 88–91.
42. Lennon, A. J.; Scott, N. R.; Chapman, B. E.; Kuchel, P. W. *Biophys. J.* **1994**, *67*, 2096–2109.
43. Nesmelova, I. V.; Fedotov, V. D. *Molec. Biol.* **1998**, *32*, 549–552.
44. Carr, H. Y.; Purcell, E. M. *Phys. Rev.* **1954**, *94*, 630–638.
45. Meiboom, S.; Gill, D. *Rev. Sci. Instrum.* **1958**, *29*, 688–691.
46. Griesinger, C.; Ernst, R. R. *J. Magn. Reson.* **1987**, *75*, 261–271.
47. Jahnke, W.; Perez, L. B.; Paris, C. G.; Strauss, A.; Fendrich, G.; Nalin, C. M. *J. Am. Chem. Soc.* **2000**, *122*, 7394–7395.
48. Jahnke, W.; Rüdisser, S.; Zurini, M. *J. Am. Chem. Soc.* **2001**, *123*, 3149–3150.
49. Jahnke, W. *Chembiochem* **2002**, *3*, 167–173.
50. Jahnke, W. The Use of Spin Labels in NMR-Supported Lead Finding and Optimization. In *BioNMR in Drug Research*; Zerbe, O., Ed.; Wiley-VCH: Weinheim, Germany, 2003; Vol. 16, pp 341–354.
51. Bertini, I.; Fragai, M.; Lee, Y. M.; Luchinat, C.; Terni, B. *Angew. Chem. Int. Ed.* **2004**, *43*, 2254–2256.
52. Bertini, I.; Luchinat, C.; Parigi, G.;*Solution NMR of Paramagnetic Molecules*, 1st ed.; Elsevier: Amsterdam, the Netherlands, 2001; Vol. 2.
53. Jahnke, W.; Florsheimer, A.; Blommers, M. J. J.; Paris, C. G.; Heim, J.; Nalin, C. M.; Perez, L. B. *Curr. Topics Med. Chem.* **2003**, *3*, 69–80.
54. Kosen, P. A. *Methods Enzymol.* **1989**, *177*, 86–121.
55. Wohnert, J.; Franz, K. J.; Nitz, M.; Imperiali, B.; Schwalbe, H. *J. Am. Chem. Soc.* **2003**, *125*, 13338–13339.
56. Salowe, S. P.; Marcy, A. I.; Cuca, G. C.; Smith, C. K.; Kopka, I. E.; Hagmann, W. K.; Hermes, J. D. *Biochemistry* **1992**, *31*, 4535–4540.
57. Dalvit, C.; Fagerness, P. E.; Hadden, D. T. A.; Sarver, R. W.; Stockman, B. J. *J. Am. Chem. Soc.* **2003**, *125*, 7696–7703.
58. Dalvit, C.; Ardini, E.; Flocco, M.; Fogliatto, G. P.; Mongelli, N.; Veronesi, M. *J. Am. Chem. Soc.* **2003**, *125*, 14620–14625.
59. Tengel, T.; Fex, T.; Emtenas, H.; Almqvist, F.; Sethson, I.; Kihlberg, J. *Org. Biomolec. Chem.* **2004**, *2*, 725–731.
60. Peng, J. W. *J. Magn. Reson.* **2001**, *153*, 32–47.
61. Leone, M.; Rodriguez-Mias, R. A.; Pellecchia, M. *Chembiochem* **2003**, *4*, 649–650.
62. Dalvit, C.; Ardini, E.; Fogliatto, G. P.; Mongelli, N.; Veronesi, M. *Drug Disc. Today* **2004**, *9*, 595–602.
63. Böhm, H. J.; Banner, D.; Bendels, S.; Kansy, M.; Kuhn, B.; Müller, K.; Obst-Sander, U.; Stahl, M. *Chembiochem.* **2004**, *5*, 637–643.
64. Ni, F. *Prog. Nucl. Magn. Reson. Spectrosc.* **1994**, *26*, 517–606.
65. Meyer, B.; Weimar, T.; Peters, T. *Eur. J. Biochem.* **1997**, *246*, 705–709.
66. Henrichsen, D.; Ernst, B.; Magnani, J. L.; Wang, W. T.; Meyer, B.; Peters, T. *Angew. Chem. Int. Ed.* **1999**, *38*, 98–102.
67. Mayer, M.; Meyer, B. *J. Med. Chem.* **2000**, *43*, 2093–2099.
68. Post, C. B. *Curr. Opin. Struct. Biol.* **2003**, *13*, 581–588.
69. Yannopoulos, C. G.; Xu, P.; Ni, F.; Chan, L.; Pereira, O. Z.; Reddy, T. J.; Das, S. K.; Poisson, C.; Nguyen-Ba, N.; Turcotte, N. et al. Dionne. *Bioorg. Med. Chem. Lett.* **2004**, *14*, 5333–5337.
70. Herfurth, L.; Weimar, T.; Peters, T. *Angew. Chem. Int. Ed.* **2000**, *39*, 2097–2099.
71. Mayer, M.; Meyer, B. *Angew. Chem. Int. Ed.* **1999**, *38*, 1784–1788.
72. Meinecke, R.; Meyer, B. *J. Med. Chem.* **2001**, *44*, 3059–3065.
73. Mayer, M.; James, T. L. *J. Am. Chem. Soc.* **2002**, *124*, 13376–13377.
74. Klein, J.; Meinecke, R.; Mayer, M.; Meyer, B. *J. Am. Chem. Soc.* **1999**, *121*, 5336–5337.
75. Vogtherr, M.; Peters, T. *J. Am. Chem. Soc.* **2000**, *122*, 6093–6099.
76. Mayer, M.; Meyer, B. *J. Am. Chem. Soc.* **2001**, *123*, 6108–6117.
77. Haselhorst, T.; Weimar, T.; Peters, T. *J. Am. Chem. Soc.* **2001**, *123*, 10705–10714.
78. Claasen, B.; Axmann, M.; Meinecke, R.; Meyer, B. *J. Am. Chem. Soc.* **2005**, *127*, 916–919.
79. Dalvit, C.; Pevarello, P.; Tato, M.; Veronesi, M.; Vulpetti, A.; Sundstrom, M. *J. Biomol. NMR* **2000**, *18*, 65–68.
80. Dalvit, C.; Fogliatto, G.; Stewart, A.; Veronesi, M.; Stockman, B. *J. Biomol. NMR* **2001**, *21*, 349–359.

81. Chen, A.; Shapiro, M. J. *J. Am. Chem. Soc.* **1998**, *120*, 10258–10259.
82. Chen, A. D.; Shapiro, M. J. *J. Am. Chem. Soc.* **2000**, *122*, 414–415.
83. Jahnke, W.; Floersheim, P.; Ostermeier, C.; Zhang, X. L.; Hemmig, R.; Hurth, K.; Uzunov, D. P. *Angew. Chem. Int. Ed.* **2002**, *41*, 3420–3423.
84. Dalvit, C.; Flocco, M.; Stockman, B. J.; Veronesi, M. *Comb. Chem. High Throughput Screen.* **2002**, *5*, 645–650.
85. Dalvit, C.; Flocco, M.; Knapp, S.; Mostardini, M.; Perego, R.; Stockman, B. J.; Veronesi, M.; Varasi, M. *J. Am. Chem. Soc.* **2002**, *124*, 7702–7709.
86. Dalvit, C.; Fasolini, M.; Flocco, M.; Knapp, S.; Pevarello, P.; Veronesi, M. *J. Med. Chem.* **2002**, *45*, 2610–2614.
87. Wang, Y. S.; Liu, D. J.; Wyss, D. F. *Magn. Reson. Chem.* **2004**, *42*, 485–489.
88. McCoy, M. A.; Senior, M. M.; Wyss, D. F. *J. Am. Chem. Soc.* **2005**, *127*, 7978–7979.
89. Dalvit, C.; Flocco, M.; Veronesi, M.; Stockman, B. J. *Comb. Chem. High Throughput Screen.* **2002**, *5*, 605–611.
90. Pervushin, K.; Riek, R.; Wider, G.; Wüthrich, K. *Proc. Natl. Acad. Sci. USA* **1997**, *94*, 12366–12371.
91. Riek, R.; Wider, G.; Pervushin, K.; Wüthrich, K. *Proc. Natl. Acad. Sci. USA* **1999**, *96*, 4918–4923.
92. Hajduk, P. J.; Augeri, D. J.; Mack, J.; Mendoza, R.; Yang, J. G.; Betz, S. F.; Fesik, S. W. *J. Am. Chem. Soc.* **2000**, *122*, 7898–7904.
93. Fielding, L.; Rutherford, S.; Fletcher, D. *Magn. Reson. Chem.* **2005**, *43*, 463–470.
94. Kainosho, M. *Nat. Struct. Biol.* **1997**, *4*, 858–861.
95. Gardner, K. H.; Kay, L. E. *J. Am. Chem. Soc.* **1997**, *119*, 7599–7600.
96. Gardner, K. H.; Kay, L. E. *Annu. Rev. Biophys. Biomolec. Struct.* **1998**, *27*, 357–406.
97. Kigawa, T.; Yabuki, T.; Yoshida, Y.; Tsutsui, M.; Ito, Y.; Shibata, T.; Yokoyama, S. *FEBS Lett.* **1999**, *442*, 15–19.
98. Weigelt, J.; Wilkstrom, M.; Schultz, J.; van Dongen, M. J. P. *Comb. Chem. High Throughput Screen.* **2002**, *5*, 623–630.
99. Szczepankiewicz, B. G.; Liu, G.; Hajduk, P. J.; Abad-Zapatero, C.; Pei, Z. H.; Xin, Z. L.; Lubben, T. H.; Trevillyan, J. M.; Stashko, M. A.; Ballaron, S. J. et al. *J. Am. Chem. Soc.* **2003**, *125*, 4087–4096.
100. McCoy, M. A.; Wyss, D. F. *J. Am. Chem. Soc.* **2002**, *124*, 11758–11763.
101. Wang, B.; Raha, K.; Merz, K. M. *J. Am. Chem. Soc.* **2004**, *126*, 11430–11431.
102. Sanchez-Pedregal, V. M.; Reese, M.; Meiler, J.; Blommers, M. J. J.; Griesinger, C.; Carlomagno, T. *Angew. Chem. Int. Ed.* **2005**, *44*, 4172–4175.
103. Edwards, A. M.; Arrowsmith, C. H.; Christendat, D.; Dharamsi, A.; Friesen, J. D.; Greenblatt, J. F.; Vedadi, M. *Nat. Struct. Biol.* **2000**, *7*, 970–972.
104. Weigelt, J.; van Dongen, M.; Uppenberg, J.; Schultz, J.; Wikstrom, M. *J. Am. Chem. Soc.* **2002**, *124*, 2446–2447.
105. Wu, D. H.; Chen, A. D.; Johnson, C. S. *J. Magn. Reson. Ser. A* **1995**, *115*, 260–264.
106. Johnson, C. S. *Prog. Nucl. Magn. Reson. Spectrosc.* **1999**, *34*, 203–256.
107. Cohen, Y.; Avram, L.; Frish, L. *Angew. Chem. Int. Ed.* **2005**, *44*, 520–554.
108. Hwang, T. L.; Shaka, A. J. *J. Magn. Reson. Ser. A* **1995**, *112*, 275–279.
109. Liu, M. L.; Mao, X. A.; Ye, C. H.; Huang, H.; Nicholson, J. K.; Lindon, J. C. *J. Magn. Reson.* **1998**, *132*, 125–129.
110. Dalvit, C. *J. Magn. Reson. Ser. B* **1996**, *112*, 282–288.
111. Dalvit, C. *J. Biomol. NMR* **1998**, *11*, 437–444.
112. Geen, H.; Freeman, R. *J. Magn. Reson.* **1991**, *93*, 93–141.

Biographies

Jochen Klages was born in Braunschweig, Germany in 1978. He studied chemistry at the Technical University in Braunschweig, where he graduated 2003 in physical chemistry with Klaus-Dieter Becker. In 2003, he was an intern in the group of Prof Yoo at the Department of materials science at the Seoul National University. Currently he is working toward his PhD thesis in the group of Prof Kessler at the Technical University in Munich. His work is focused on the structural investigation of proteins and small molecules and their interaction.

Horst Kessler was born in Suhl (Thuringia, Germany) in 1940. He studied chemistry in Leipzig and Tübingen, where he received his PhD degree with Eugen Müller in 1966. He was appointed full professor for organic chemistry at the J. W. Goethe University in Frankfurt in 1971. In 1989 he moved to the Technische Universität München. Prof Kessler is the recipient of the Otto Bayer Award (1986), the Max Bergmann Medal for peptide chemistry (1988), the Emil Fischer Medal (1997), the Max-Planck-Forschungspreis (2001), the Vincent Du Vigneaud Award of the American Peptide Society (2002), the Hans Herloff Inhoffen Medal (2002), the Burkhardt Helferich Award (2005). In 2002 he received the honorary degree of the University of Leipzig. He is a member of the "Bayerische Akademie der Wissenschaft" and the "Deutsche Akademie der Naturforscher Leopoldina", Halle. Guest professorships lead him to Halifax, Tokyo, Madison, Haifa, Austin, and Jerusalem. His current interests are in the area of the development and application of new NMR techniques to peptides, proteins and nucleic acids as well as their complexes. Another field of interest is bioorganic and medicinal chemistry, with specific focus on the study of biological recognition phenomena and on conformationally oriented design of biologically active molecules, such as carbohydrates, peptides and peptidomimetics.

Comprehensive Medicinal Chemistry II
ISBN (set): 0-08-044513-6

ISBN (Volume 3) 0-08-044516-0; pp. 901–920

3.40 Chemogenomics

H Kubinyi, University of Heidelberg, Heidelberg, Germany

3.40.1	**Introduction**	**921**
3.40.2	**Chemical Biology**	**921**
3.40.3	**Chemical Genetics**	**923**
3.40.4	**Chemogenomics Strategies**	**923**
3.40.4.1	Agonists and Antagonists	923
3.40.4.2	Privileged Structures	925
3.40.4.3	Drugs from Side Effects – The Selective Optimization of Side Activities (SOSA) Approach	926
3.40.4.4	From Target Family-Directed Master Keys to Selective Drugs	927
3.40.5	**Chemogenomics Applications**	**929**
3.40.5.1	Protease Inhibitors	929
3.40.5.2	Kinase Inhibitors	930
3.40.5.3	Phosphodiesterase (PDE) Inhibitors	930
3.40.5.4	G Protein-Coupled Receptor Ligands	932
3.40.5.5	Nuclear Receptor Ligands	932
3.40.5.6	Integrin Ligands	933
3.40.5.7	Transporter Ligands	935
3.40.6	**Summary and Conclusions**	**935**
References		**935**

3.40.1 Introduction

Chemical biology, chemical genetics, and chemogenomics are recent strategies in drug discovery. Although their definitions in the literature are somewhat diffuse and inconsistent, a differentiation of the terms will be attempted here:

- Chemical biology can be defined as the study of biological systems (e.g., whole cells) under the influence of chemical libraries. If a new phenotype is discovered by the action of a certain substance, the next step is the identification of the responsible target.
- Chemical genetics is the dedicated study of protein function (e.g., signaling chains) under the influence of ligands that bind to certain proteins or interfere with protein–protein interaction; sometimes, orthogonal ligand–protein pairs are generated to achieve selectivity for a certain protein.
- Chemogenomics is defined, in principle, as the screening of the chemical universe (i.e., all possible chemical compounds) against the target universe (i.e., all proteins and other potential drug targets). Although this task can never be achieved, due to the almost infinite size of the chemical universe, the systematic screening of libraries of congeneric compounds against members of a target family offers unlimited possibilities in the search for compounds with significant target or subtype specificity.

3.40.2 Chemical Biology

In classical drug discovery, research was often based on vague hypotheses on structure–activity relationships. Compounds were synthesized and tested in whole animals. If a biological effect was observed, a medicinal chemistry project was started to optimize chemical structures with respect to activity, pharmacokinetic properties, and lack of toxic side effects. Later on, this approach was replaced by in vitro screening of defined targets, most often human proteins. Only in recent years has a more systematic investigation of druglike compounds in biological systems been developed – 'chemical biology.'

One illustrative example of the chemical biology approach is the discovery of monastrol (**1**) (**Figure 1**), a molecule that prevents spindle formation in mitotic cells by inhibiting the kinesin Eg5, a motor protein required for spindle bipolarity.[1] Acting in this way, monastrol stops cell division by mitotic arrest.

Another example of the concept of chemical biology is the discovery of synthetic small molecules that influence the fate of embryonic stem cells. A high-throughput phenotypic cell-based screen identified the 4,6-disubstitued pyrrolo-pyrimidine **2** (**Figure 1**), which induces the differentiation of embryonic stem cells to neurons.[2] Glycogen synthase kinase-3β has been identified as the target of this compound. Cardiogenol C (**3**), from a 100 000-member heterocycle library, induces the formation of cardiac muscle cells from embryonic stem cells.[3]

Reversine **4** (**Figure 1**) was discovered by screening kinase inhibitor libraries. It dedifferentiates adult murine myotube cells to mesenchymal progenitor cells, which can then be induced to form adipocytes (fat cells) or osteoblasts (bone cells).[4,5] However, the significance of these results has been criticized.[6]

On the other hand, general screening of compounds may not result in the desired results. The production of a 2.18 million compound natural product library[7,8] generated much hype on 'diversity-oriented organic synthesis' (DOS) but, so far, not the anticipated results with respect to biological activities. The author has since admitted that the chemical diversity of his library was too narrow[9]

> "the field of DOS has not yet come close to reaching its goals ... even a qualitative analysis of the members ... reveals that they are disappointingly similar. Of even greater concern is that the selection of compounds has so far been guided only by the organic chemist's knowledge of candidate reactions, creativity in planning DOS pathways, and intuition about the properties likely to yield effective modulators. Retrospective analyses of these compounds show that they tend to cluster in discrete regions of multidimensional descriptor space. Although algorithms exist to identify subsets of actual or virtual compounds that best distribute in chemical space in a defined way ... these are of little value to the planning of DOS."

This goes hand in hand with another problem: biologically active compounds seem to be distributed only in certain areas of chemical space, according to their physicochemical properties and structural features.[10] If we consider the chemical universe as a huge ocean, with small islands or groups of islands of biologically active compounds (e.g., the so-called 'privileged' compounds, *see* Section 3.40.4.2), we have to understand and accept that most chemistry-driven approaches will end up in the water, instead of discovering new islands. For the broad exploration of biology with small organic molecules,[11] the National Institute of Health has started an initiative to provide a repository of chemically diverse molecules for the public and private sector.[12]

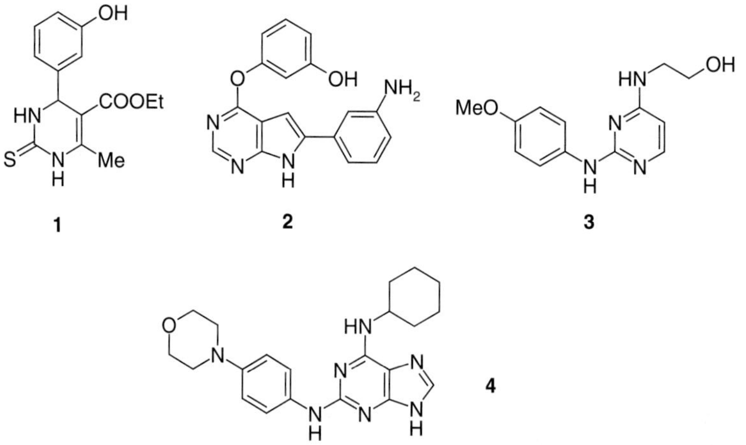

1 2 3

4

Figure 1 Compounds **1–4** were discovered by chemical biology approaches. Monastrol (**1**) inhibits cell division by mitotic arrest. The pyrrolo-pyrimidine **2** and compound **3** (cardiogenol C) induce differentiation of embryonic stem cells to neurons and cardiac muscle cells, respectively. Reversine (**4**) dedifferentiates adult murine myotube cells.

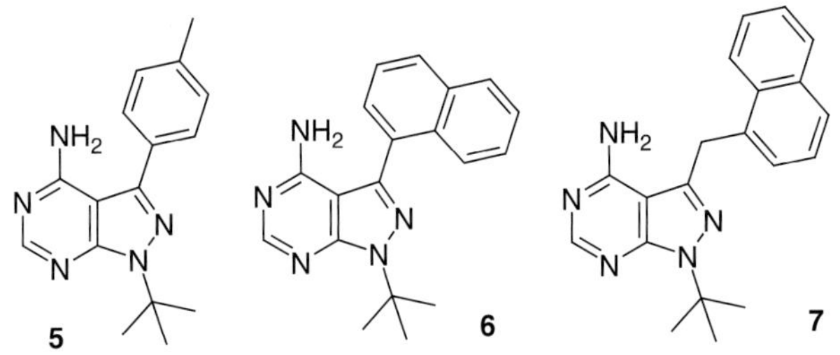

Figure 2 Compound **5** is a nonspecific kinase inhibitor. Analogs **6** and **7**, with sterically demanding side chains, have lower affinities for the wild-type kinases but bind with high affinity to several mutant kinases with larger binding pockets (**Table 1**).

Table 1 IC_{50} values (in μM) of compounds **5–7** for several wild-type and engineered kinases[14]

Wild-type kinase				*Engineered kinase*		
Kinase	*Compound* 5	*Compound* 6	*Compound* 7	*Kinase*	*Compound* 6	*Compound* 7
v-Src	2.2	1.0	28	v-Src I338G	0.0015	0.043
c-Fyn	0.050	0.60	1.0	c-Fyn T339G	0.0065	0.032
c-Abl	0.30	0.60	3.4	c-Abl T315A	0.0070	0.12
CDK-2	22	18	29	CDK-2 F80G	0.015	0.0050
CAMK II	17	22	24	CAMK II F89G	0.097	0.0080

3.40.3 Chemical Genetics

Classical genetics induces a (random) mutation (e.g., by irradiation), and attempts to determine the genotype from the appearance of new phenotypes. 'Chemical genetics' is another new term for a strategy that has also long been used in a less systematic manner; it describes the investigation of proteins by small molecules or libraries, for target identification (forward chemical genetics) or target validation (reverse chemical genetics).[13] If selective ligands are not available, sometimes orthogonal ligand–receptor pairs are constructed. Selective kinase inhibition has been achieved by specifically converting nonspecific low-affinity inhibitors (e.g., **5**) into larger analogs (e.g., **6** and **7**) (**Figure 2**) and constructing certain kinase mutants that specifically accommodate these originally less well-fitting ligands by their larger binding pocket (**Table 1**).[14] The specific inhibition of a given kinase can be studied in this manner, without having developed an inhibitor of comparable specificity against the wild-type kinase.

3.40.4 Chemogenomics Strategies

'Chemogenomics' also defines an approach that was used less systematically than at present. Since screening of the chemical universe against the target universe is impossible, due to the almost infinite number of potential druglike compounds, the more manageable screening of congeneric chemical libraries against certain target families (e.g., the G protein-coupled receptors (GPCRs), nuclear receptors, different protease families, kinases, phosphodiesterases, ion channels, and transporters) is instead undertaken[15–19]; this systematic strategy aims to discover highly potent selective ligands for functionally and evolutionary related targets, with the least effort.

3.40.4.1 Agonists and Antagonists

Several GPCR antagonists have been derived from the corresponding agonists by introducing (large) lipophilic groups (e.g., the first β-adrenergic antagonist dichloroisoproterenol (**8**) from isoproterenol (**9**) and the acetylcholine antagonist drofenine (**10**) from acetylcholine (**11**); ligands of the opiate receptors are generally antagonists, if they bear an *N*-allyl group (or an cyclopropyl- or cyclobutylmethylene group), such as **12**, instead of the *N*-methyl group of the agonist morphine (**13**)) (**Figure 3**).

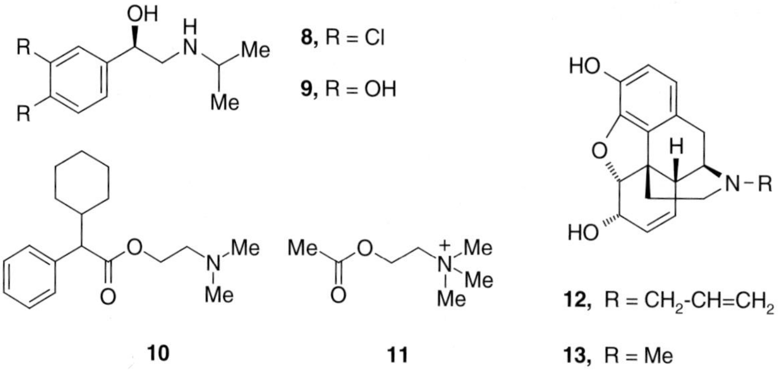

Figure 3 The first beta-blocker dichloroisoproterenol (**8**) is a lipophilic analog of the beta-adrenergic agonist isoproterenol (**9**). The anticholinergic drug drofenine (**10**) is, like many other neurotransmitter antagonists, a lipophilic analog of the corresponding agonist, in this case acetylcholine (**11**). The opiate antagonist nalorphine (**12**) and the agonist morphine (**13**) differ only in the size of the *N*-substituent.

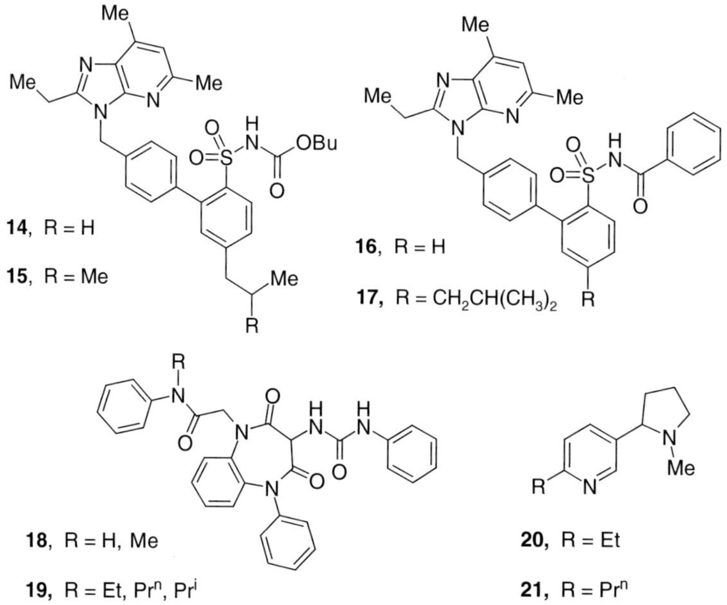

Figure 4 Whereas the 4-propyl-sulphonamide **14** is an AT_1 antagonist, its isobutyl analog **15** is an AT_1 agonist. The exchange of an aromatic hydrogen atom for an isobutyl group converts the highly selective AT_1 antagonist **16** into the nanomolar, nonselective AT_1/AT_2 receptor agonist **17**. An increase in size of the *N*-methyl group of the CCK_A antagonists **18** produces the CCK_A receptor antagonists **19**. A similar increase in size of an aromatic ethyl group of the nicotinic acetylcholine agonist **20** leads to the analog **21**, which discriminates between different in vivo activities of (–)-nicotine.

Only recently have agonists been derived from antagonists by the introduction of relatively small lipophilic groups. For example, the introduction of a methyl group into the angiotensin-1 receptor antagonist **14** (IC_{50} $AT_1 = 4.0$ nM) produces the agonist **15** (IC_{50} $AT_1 = 25$ nM); introduction of an isobutyl group into the highly specific angiotensin-1 receptor antagonist **16** (IC_{50} $AT_1 = 0.3$ nM, IC_{50} $AT_2 = 4500$ nM) yields the nonspecific agonist **17** (IC_{50} $AT_1 = 13$ nM, IC_{50} $AT_2 = 10$ nM); and the CCK_A antagonist **18** is converted into the agonist **19**, if the *N*-alkyl group is larger than the methyl group (**Figure 4**).[20–23] Some 6-alkyl derivatives of (–)-nicotine discriminate between the different biological activities of nicotine: (–)-6-ethylnicotine (**20**) retains a high affinity for nicotinic acetylcholine receptors ($K_i = 5.6$ nM), and produces the same in vivo actions as (–)-nicotine ($K_i = 1.2$ nM), whereas the 6-*n*-propyl analog **21** ($K_i = 22$ nM) antagonizes the antinociceptive effect in the mice tail-flick assay but not the spontaneous activity or discriminative stimulus effects of (–)-nicotine (**Figure 4**).[24]

3.40.4.2 Privileged Structures

Many drugs have been derived from certain chemotypes (e.g., phenethylamines (**Table 2**), tricyclics (**22–24**, **Figure 5**), steroids (**Table 3**), or benzodiazepines), whereas others have certain structural features in common (e.g., diphenylmethane, diphenylamine, or arylpiperazine groups).

The systematic chemical variation of benzodiazepines (e.g., the γ-amino-butyric acid (GABA)-agonist diazepam (**25**)) produced not only tranquilizers but also the GABA antagonist flumazenil (**26**), the inverse agonist Ro 15-3505 (**27**), and the strong κ-opiate receptor agonist tifluadom (**28**)[25]; devazepide (**29**) is an orally active cholecystokinin-B (CCK_B) antagonist (**Figure 6**).[26] In addition to these compounds, the benzodiazepine ring system is also the scaffold of some muscle relaxants, hypnotics, narcotics, NK-1 and vasopressin receptor antagonists, farnesyl transferase inhibitors, potassium channel modulators, and several tricyclic neuroleptics and antidepressants.

Since tifluadom (**28**) also has nanomolar affinity for the cholecystokinin receptor, Evans *et al.* concluded that[26] "these structures appear to contain common features which facilitate binding to various … receptor surfaces, perhaps through binding elements different from those employed for binding of the natural ligands," and that "what is clear is that certain 'privileged structures' are capable of providing useful ligands for more than one receptor and that judicious modification of such structures could be a viable alternative in the search for new receptor agonists and antagonists."[26]

Minor chemical modifications of such privileged structures (**Figure 7**)[27] may result in highly selective ligands or drugs (e.g., the estrogenic, gestagenic, androgenic, glucocorticoid, and mineralocorticoid steroids, or the α-adrenergic, β-adrenergic, and β-antiadrenergic phenethylamines). Others lack such target selectivity: the atypical neuroleptic

Table 2 Biological activities of phenethylamines

Compound (prototype)	*Biological activity*
Amphetamine	Stimulant
MDMA (ecstasy)	Hallucinogen
Dopamine	Dopaminergic agonist
Norepinephrine	α-Adrenergic agonist
Epinephrine	Mixed α,β-adrenergic agonist
Isoproterenol	β-Adrenergic agonist
Dobutamine	β_1-Specific adrenergic agonist
Salbutamol	β_2-Specific adrenergic agonist
Dichloroisoproterenol	β-Adrenergic antagonist
Metoprolol	β_1-Specific antagonist
Xamoterol	β_2-Specific partial agonist
Ephedrine	Indirect sympathomimetic agent
Norpseudoephedrine	Appetite suppressant

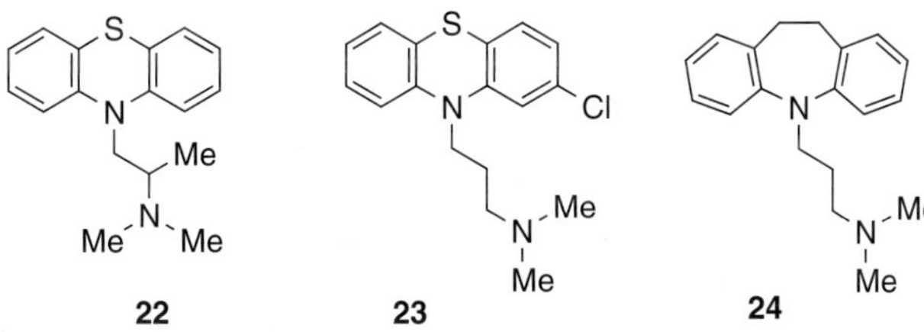

Figure 5 Promethazine (**22**), chlorpromazine (**23**), and imipramine (**24**) are chemically closely related but differ in their biological activities: **22** is an H_1 antihistaminic drug for the treatment of allergic inflammation, **23** is a dopamine antagonist, and **24** (as well as its *N*-desmethyl metabolite desipramine) is a neurotransmitter uptake blocker for the treatment of depression.

Table 3 Biological activities of steroids at nuclear receptors and other targets

Compound (prototype)	*Biological activity*
Estradiol	Estrogen
Clometherone	Antiestrogen
Danazole	Gonadotropin-releasing hormone (GRH) antagonist (indirect estrogen antagonist)
Formestane	Aromatase inhibitor
Progesterone	Progestogen
Mifepristone	Antiprogestogen
Testosterone	Androgen
Cyproterone acetate	Antiandrogen
Nandrolone decanoate	Anabolic
Finasteride	5α-Reductase inhibitor (testosterone biosynthesis inhibitor)
Cortisol	Nonselective gluco-/mineralocorticoid
Dexamethasone	Glucocorticoid
Aldosterone	Mineralocorticoid
Spironolactone	Aldosterone antagonist
Cholesterol	Membrane-modulating agent
Cholic acid	Choleretic
Chenodiol	Anticholelithogenic agent
Digoxin	H^+/K^+-ATPase inhibitor (cardiac glycoside)
Cholecalciferol (vitamin D_3)	Antirachitic vitamin
Pancuronium bromide	Muscle relaxant
Hydroxydione	Anesthetic
Allo-pregnanolone	Neuromodulator
Tirilazad	Lipid peroxidation inhibitor
Edifolone	Cardiac depressant/antiarrhythmic
Conessin	Antiamebic
α-Ecdysone	Insect juvenile hormone

olanzapine (**30**) (**Table 4**) is a highly promiscuous tricyclic ligand, with nanomolar affinities for various GPCRs, including $5HT_{2A}$, $5HT_{2B}$, $5HT_{2C}$, dopaminergic D_1, D_2, D_4, muscarinic M_1, M_2, M_3, M_4, M_5, adrenergic α_1, and histaminic H_1 receptors, as well as the $5HT_3$ ion channel.[28,29]

Privileged structures, even if they are promiscuous ligands, should not be confused with certain structural classes that seemingly bind with micromolar affinity to various enzymes. This unspecific binding behavior is caused by an aggregation of the ligands and clumping of these aggregates to the protein.[30–33]

3.40.4.3 Drugs from Side Effects – The Selective Optimization of Side Activities (SOSA) Approach

Many drugs of the past resulted from the experimental or clinical observation of side effects. Diuretic, antihypertonic, antiglaucoma, and antidiabetic drugs were derived from the bacteriostatic sulfonamides; the mood-improving effect of iproniazid was discovered when it was tested as an antituberculous drug; and the antidepressant inhibitors of neurotransmitter uptake, such as imipramine and desipramine, stem from the antipsychotic dopamine antagonist

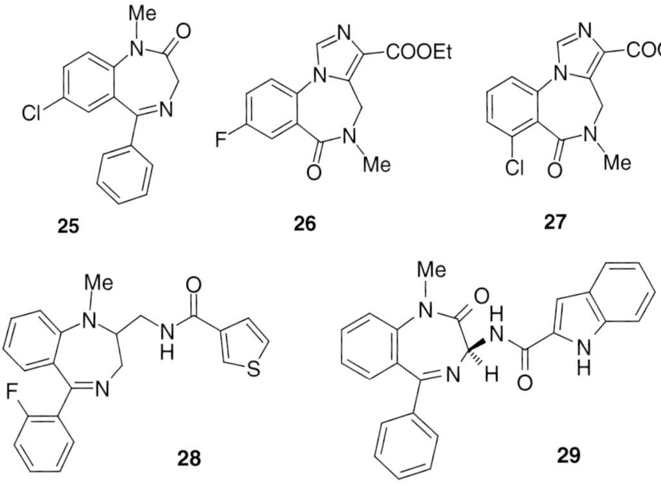

Figure 6 Diazepam (Valium (**25**)) was one of the first tranquilizers and the prototype of a series of other GABA receptor agonists, the antagonist **26**, and the inverse agonist **27**. The chemically related benzodiazepine tifluadom (**28**) is a κ-opiate receptor agonist and a nanomolar CCK receptor ligand. Devazepide (**29**) is a selective CCK_B receptor antagonist (K_i CCK_A rat = 1480 nM, K_i CCK_B human = 0.15 nM).

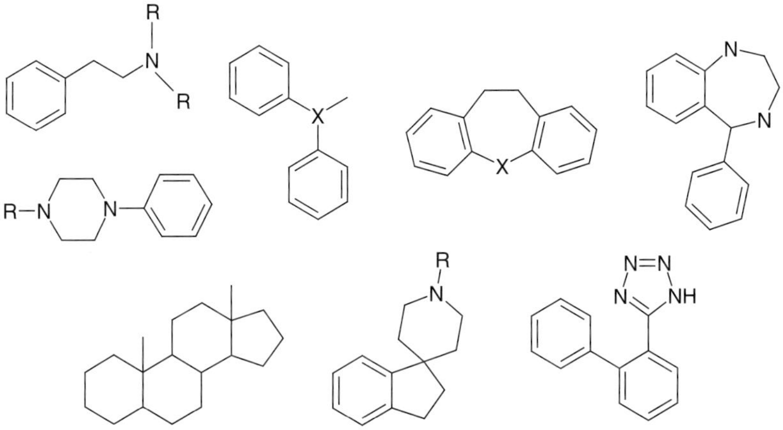

Figure 7 Privileged structures are scaffolds or substituents that frequently produce biologically active compounds, such as phenethylamines, diphenylmethyl, and diphenylamine compounds (X = C or N, respectively), tricyclic compounds (X = C or N), benzodiazepines, arylpiperidines, steroids, spiropiperidines, and tetrazolobiphenyls.

chlorpromazin, which itself was derived from H_1 antihistaminics (see **Figure 5**); there are many other similar examples.[34,35] Recently, Wermuth proposed a more systematic investigation of the side effects of drugs, through his SOSA approach.[36,37] When the side effect of a drug is observed, it might be possible to optimize the candidate to a selective drug with this observed biological activity, following a statement by Sir James Black that "the most fruitful basis for discovery of a new drug is to start with an old drug."[37] Among several other examples, Wermuth demonstrated the optimization of different weak side effects of the antidepressant minaprine (**31**) to the nanomolar muscarinic M_1 receptor ligand **32** and the reversible acetylcholinesterase inhibitor **33**[36,37]; and a closely related analog of minaprine was optimized to the nanomolar $5HT_3$ antagonist **34** (**Figure 8**).[38] Further examples are discussed in the literature.[35–37]

3.40.4.4 From Target Family-Directed Master Keys to Selective Drugs

Chemogenomics is mainly based on the master key concept of tailor-made privileged structures.[39,40] Starting from such master keys, selective ligands can be derived, either by classical medicinal chemistry or by systematic structural variation in combinatorial libraries. The master key concept will be illustrated by just one example: selective β_1 and β_2 agonists, as well as β antagonists (beta-blockers), were derived from the mixed α/β agonist epinephrine. Further chemical variation

Table 4 GPCR and $5HT_3$ binding affinities for olanzapine (**30**) in different in vitro models[28,29]

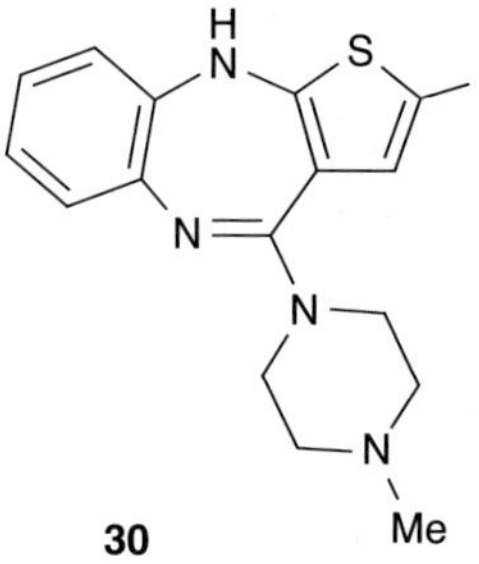

30

Receptor	K_i *(nM)*
$5HT_{2A}$	2.5, 4
$5HT_{2B}$	12
$5HT_{2C}$	11, 29
$5HT_3$	57
Dopaminergic D_1	31, 119
Dopaminergic D_2	11
Dopaminergic D_4	27
Muscarinic M_1	1.9, 2.5
Muscarinic M_2	18
Muscarinic M_3	13, 25
Muscarinic M_4	10, 13
Muscarinic M_5	6
Adrenergic α_1	19
Adrenergic α_2	230
Histaminic H_1	7

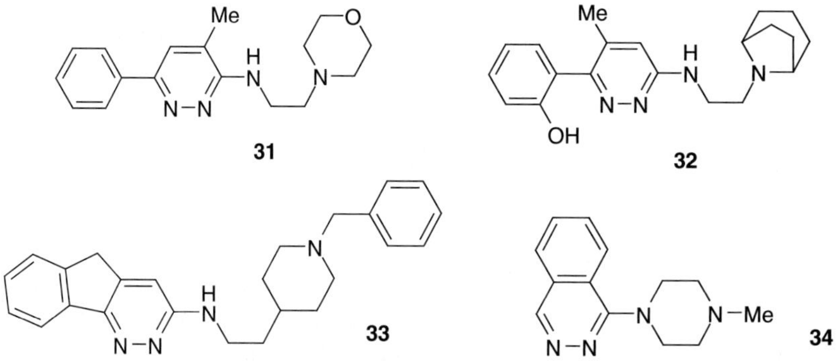

Figure 8 The antidepressant minaprine **31** is also a weak muscarinic M_1 receptor antagonist ($K_i = 17\,\mu M$) and an acetylcholinesterase inhibitor ($K_i = 600\,\mu M$). These activities could be enhanced by systematic structural variation to the nanomolar M_1 receptor antagonist **32** ($K_i = 3\,nM$) and the acetylcholinesterase inhibitor **33** ($K_i = 10\,nM$). A closely related analog of minaprine was optimized to the nanomolar $5HT_3$ receptor antagonist **34** ($IC_{50} = 10\,nM$).

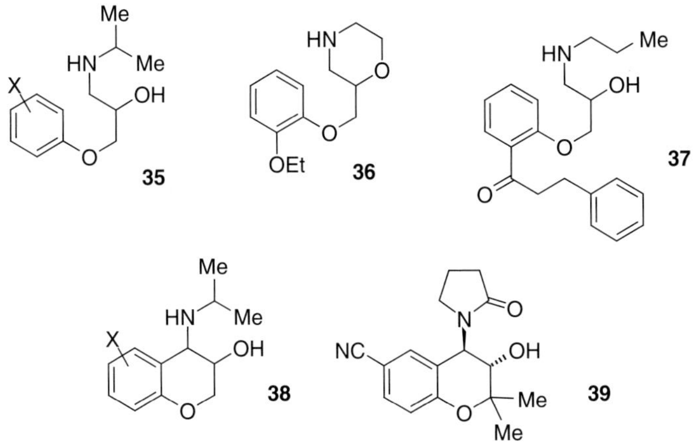

Figure 9 The beta-blocker prototype structure (**35**), phenyl-O-CH_2-CH(OR^1)-CH_2NHR^2, is also the key structural element of the antidepressant viloxacine (**36**) and the class Ic antiarrhythmic propafenone (**37**). Structural variation of a cyclic beta-blocker analog (**38**) yielded the potassium channel opener levcromakalim (**39**).

of the typical beta-blocker phenoxypropanolamine structure (**35**) yielded the antidepressant viloxacine (**36**) and the class Ic antiarrhythmic propafenone (**37**) (**Figure 9**). The optimization of a cyclic beta-blocker prototype (**38**) indeed produced an antihypertensive drug; however, levcromokalim (**39**) is no longer a beta-blocker but a vasodilatory potassium channel opener.[36,37] More examples are discussed in the following sections and in the literature.[19,35–37,39,40]

3.40.5 Chemogenomics Applications

3.40.5.1 Protease Inhibitors

Protease inhibitors are most often derived from the sequence of the amino acids in the positions next to the bond that is cleaved by the enzyme. A simple strategy for a first inhibitor is the conversion of the amide bond of the cleavage site into a noncleavable analog or a group that reacts or coordinates with the catalytic center of the enzyme; the P1, P2, ... and/or P1′, P2′ ... amino acids are kept constant.

The structural requirements of the individual protease classes are different:

- for aspartyl protease inhibitors it is necessary to attach some amino- and carboxy-terminal amino acid side chains to a group that mimics the transition state of the enzymatic cleavage;
- for metalloprotease inhibitors, a metal-coordinating group is introduced on the amino-terminal side of the peptide;
- for serine and cysteine protease inhibitors, the groups that interact with the catalytic center are not necessarily (but most often) at the carboxy-terminal end of the peptide.

The chemogenomics strategy in the design of protease inhibitors will be illustrated by four examples: the design of HIV protease inhibitors, thrombin and factor Xa inhibitors, selective angiotensin-converting enzyme (ACE) and dual zinc protease inhibitors, and 'dual warhead' metalloprotease (MMP)/cathepsin inhibitors.

Renin is an aspartyl protease that is involved in blood pressure regulation, by converting angiotensinogen into angiotensin I, the substrate of ACE; many years of research were invested to synthesize orally active peptidomimetics, without much success. When it became known that HIV protease is also an aspartyl protease, the accumulated experience on the design of transition state inhibitors was transferred to this new project.

The same situation applies to inhibitors of the serine protease thrombin; here, despite all the effort expended, there was only limited success in achieving orally active analogs. However, structural elements from inhibitors of another serine protease, elastase (i.e., the pyrimidone ring system as a substitute for a flexible amino acid), could also be applied to thrombin inhibitors. Later on, the search for inhibitors shifted from thrombin to factor Xa, a serine protease with similar specificity to thrombin.

Captopril **40** was the very first ACE inhibitor introduced into human therapy (**Figure 10**). A multitude of ACE-inhibiting analogs resulted from this drug (e.g., the ACE-specific inhibitor **41** and the dual ACE/NEP24.11 inhibitors **42** and**43**).[41]

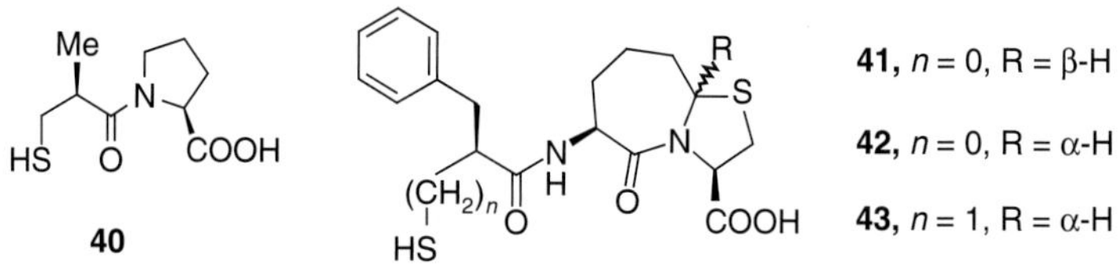

Figure 10 Captopril (**40**) was the very first marketed ACE inhibitor. The specific ACE inhibitor **41** ($n=0$, R=β-H; K_i ACE=11.5 nM, K_i NEP24.11=2820 nM) resulted from structural variation, as well as the dual zinc protease inhibitors **42** ($n=0$, R=α-H; K_i ACE=16 nM, K_i NEP24.11=11.5 nM) and **43** ($n=1$, R=α-H; K_i ACE=5.5 nM, K_i NEP24.11=1.1 nM).

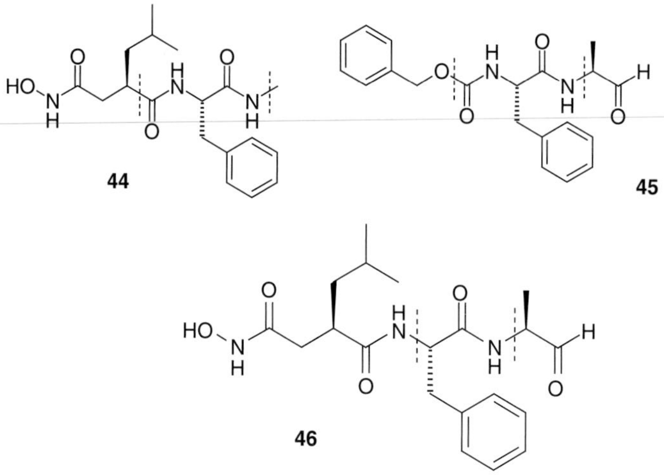

Figure 11 Compound **44** is a nanomolar metalloprotease inhibitor (IC_{50} MMP-1=3 nM; IC_{50} Cat L>1000 nM), whereas compound **45** is a nanomolar cysteine protease inhibitor (IC_{50} MMP-1 >1000 nM; IC_{50} Cat L=3 nM). 'Crossover' of the two structures produces the dual inhibitor **46** (IC_{50} MMP-1=25 nM; IC_{50} Cat L=15 nM); the dashed lines indicate the common core of all three molecules.

A 'dual warhead' inhibitor resulted from merging the structures of a selective matrix MMP inhibitor (**44**) with a cathepsin L inhibitor (**45**) (**Figure 11**). Despite the fact that MMP-1 is a zinc protease and cathepsin L is a cysteine protease, the resulting inhibitor (**46**), which bears both 'warheads,' inhibits both enzymes with nanomolar activity.[42]

3.40.5.2 Kinase Inhibitors

Kinases play a most important role in cell signaling. More than 500 different kinases are coded by the human genome; after activation, they phosphorylate either a tyrosine hydroxyl group (tyrosine kinases) or a serine or threonine hydroxyl group (serine/threonine kinases). Some kinase mutants are constitutionally active: they activate a signaling cascade without any external stimulus. Chronic myelogenous leukemia is caused by such a constitutionally active kinase. The coding regions of an *abl* tyrosine kinase at chromosome 9 and a *bcr* serine/threonine kinase at chromosome 22 form after reciprocal translocation a *bcr-abl* coding region in the new, shorter version of the chromosome 9, the so-called Philadelphia chromosome; the resulting *bcr-abl* tyrosine kinase is constitutionally active. At Novartis, a class of protein kinase C (PKC) inhibitors was optimized to the PKC inhibitor **47** (**Figure 12**). Amide analogs (**48**) of this compound showed activity against PKC and *bcr-abl* kinase; surprisingly, the methyl analog (**49**) inhibited only *bcr-abl* kinase; finally, an *N*-methyl-piperazine residue was added to increase solubility. Imatinib (Gleevec or Glivec, **50**) was clinically developed, and is successfully used for the treatment of chronic myelogenous leukemia.[43]

3.40.5.3 Phosphodiesterase (PDE) Inhibitors

In the early 1970 s, May & Baker worked on antiallergic xanthine derivatives.[44] Their first leads (**51** and **52**) (**Figure 13**) were about 40 times[44,45] and 1000 times[46] more active than the standard drug cromoglycate. The drug candidate zaprinast (M&B 22,948, **51**) was orally active, and showed, in addition to its 'mast cell stabilizing' activity against

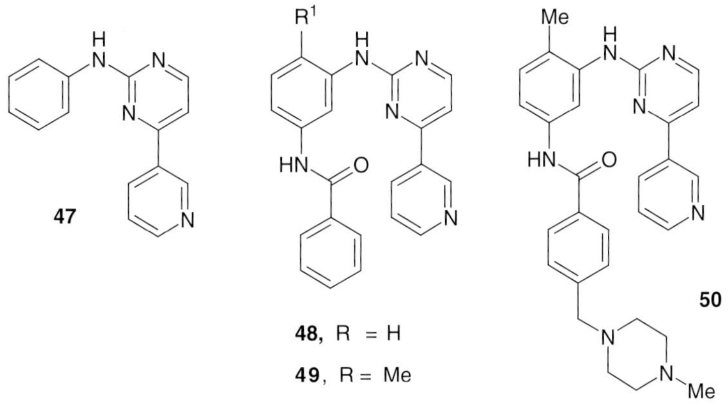

Figure 12 Structural variation of the PKC inhibitor **47** produced the dual PKC/*bcr-abl* inhibitor **48** (R = H). A minor structural modification to **49** (R = Me) abolished the undesired PKC activity. After the introduction of a methylpiperazine residue, to enhance aqueous solubility, the *bcr-abl* inhibitor imatinib (Glivec or Gleevec, **50**) resulted.

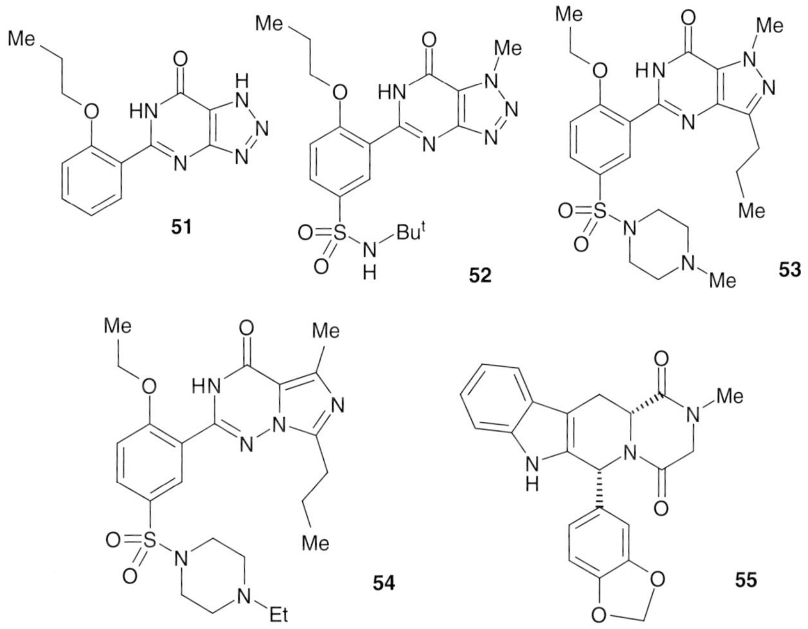

Figure 13 Zaprinast (**51**) and the sulfonamide **52** are early precursors of the PDE5 inhibitors sildenafil (Viagra; **53**) and vardenafil (**54**). Tadalafil (**55**) has a higher PDE5 selectivity than **53** and **54**.

histamine- and exercise-induced asthma, vasodilatory and antihypertensive activities. Later, Pfizer developed a working hypothesis that a new antihypertensive principle might result by enhancing the biological activity of the atrial natriuretic peptide (ANP). They attempted to prolong the action of cGMP, the second messenger of the corresponding receptor response.[47] Zaprinast was selected as the lead structure for further optimization, because it was one of the very few cGMP PDE inhibitors known at that time. The result of this optimization was the PDE5-selective analog sildenafil (**53**). Clinical tests showed that the drug was safe, but its clinical activity was disappointing. However, in 1992 a 10 day toleration study in healthy volunteers led to the serendipitous observation of penile erections. After further clinical investigation, sildenafil (Viagra; **53**) was approved for the treatment of male erectile dysfunction in March 1998.[47] Sildenafil and the close analog vardenafil (**54**) also inhibit PDE6 in the double-digit nanomolar range, which causes some undesirable side effects; the indoline derivative tadalafil (**55**) inhibits PDE6 only at considerably higher concentrations.[48]

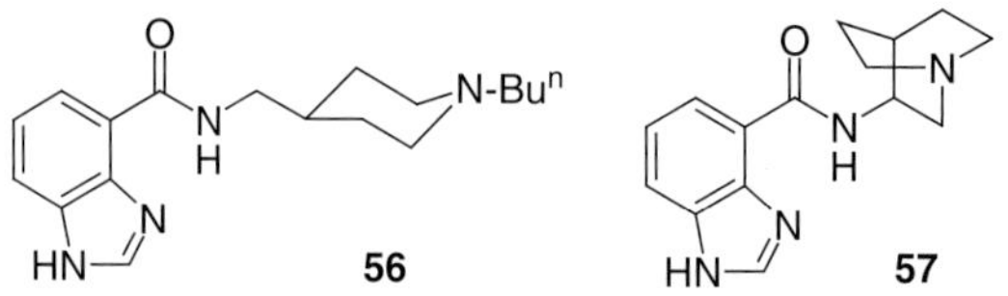

Figure 14 Compound **56** is a highly selective 5HT$_4$ antagonist (K_i 5HT$_3$ > 10 000 nM, K_i 5HT$_4$ = 13.7 nM), whereas the chemically closely related compound **57** is a selective 5HT$_3$ antagonist (K_i 5HT$_3$ = 3.7 nM, K_i 5HT$_4$ > 1000 nM).

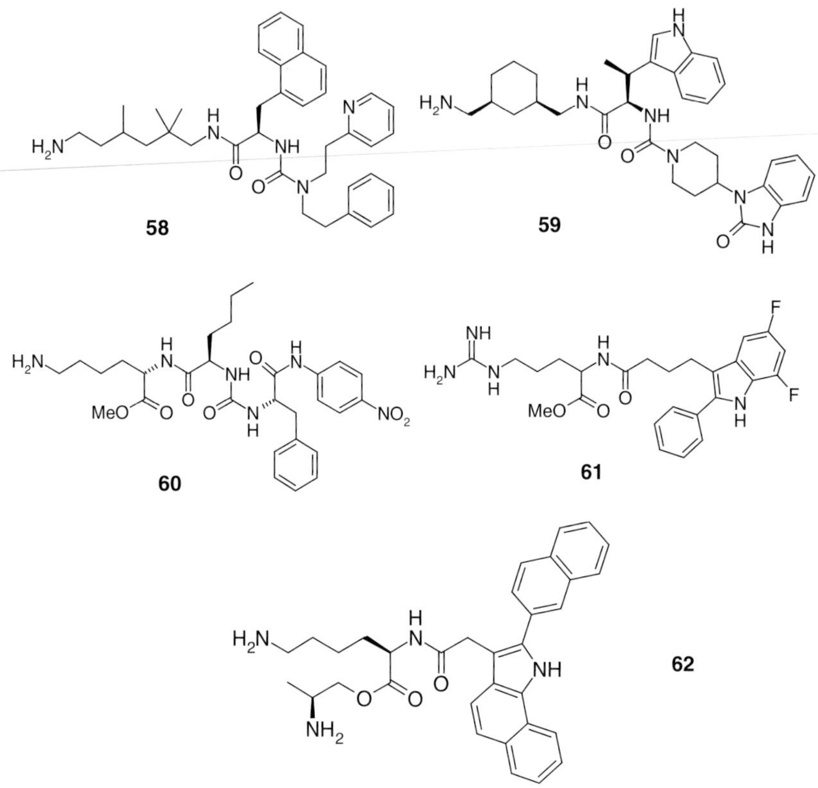

Figure 15 The β turn peptidomimetics **58–62** are highly selective ligands of the somatostatin receptor subtypes sst1 to sst5 (**Table 5**).

3.40.5.4 G Protein-Coupled Receptor Ligands

GPCRs are a large group of evolutionary related seven-transmembrane proteins. They are activated by different agents, such as light, ions, odorants, neurotransmitters, peptides, and proteins; the stimulus is transferred by inducing the dissociation of a G protein complex. Serotonin receptors are made up of 14 subtypes, of which 13 are GPCRs, whereas the 5HT$_3$ subtype is a ligand-controlled ion channel. From pharmacophore models, Lopez-Rodriguez *et al.* designed the structure of a highly selective 5HT$_4$ receptor ligand (**56**), which shows a selectivity difference of more than five orders of magnitude from its closely related 5HT$_3$-selective analog **57** (**Figure 14**).[49]

Somatostatin receptors are made up of five subtypes (sst1 to sst5). In their attempt to obtain selective peptidomimetic ligands for each subtype, Rohrer *et al.* synthesized four β-turn-mimicking combinatorial libraries, with up to 350 000 compounds per library. Highly specific ligands (**58–62**) resulted for all five receptor subtypes (**Figure 15** and **Table 5**).[50]

3.40.5.5 Nuclear Receptor Ligands

Nuclear receptors are another important receptor family. They comprise a ligand-binding domain and a DNA-binding domain. After activation by their specific ligands (e.g., the steroid hormones, the thyroid hormones, or retinoic acid), receptor dimers bind to DNA, and activate the expression of certain proteins.

Table 5 Somatostatin receptor subtype affinities of the tetradecapeptide somatostatin (K_i values in nM) and compounds **58–62** (see **Figure 18**; K_d values in nM)[50]

Compound	*Receptor subtype*				
	*sst*1	*sst*2	*sst*3	*sst*4	*sst*5
Somatostatin	0.4	0.04	0.7	1.7	2.3
58	1.4	1 875	2 240	170	3 600
59	2 760	0.05	729	310	4 260
60	1 255	> 10 000	24	8 650	1 200
61	199	4 720	1 280	0.7	3 880
62	3.3	52	64	82	0.4

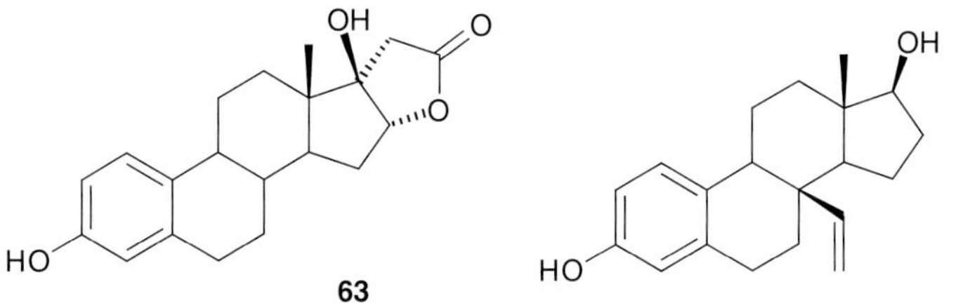

63 **64**

Figure 16 The estradiol analogs **63** (40% of the estradiol activity for the ERα receptor) and **64** (50% of the estradiol activity for the ERβ receptor) were designed as selective ERα and ERβ receptor ligands. They are slightly less active than estradiol, but they show 300-fold and 190-fold selectivity for the different receptor subtypes.

Estrogen receptors exist as two distinct subtypes (ERα and ERβ) that are relatively abundant in a number of tissues. As their function in all these organs and their potential interactions, forming ERα/ERβ heterodimers, have not so far been completely elucidated, it is most important to find selective ligands for both receptors. By homology modeling of the ligand-binding domain of ERβ, based on the corresponding three-dimensional structure of ERα, Hillisch *et al.* inspected the minor differences in the estradiol-binding site: in human ERβ, the leucine residue of ERα at the 'top' of the binding site ('top' refers to the β side of the steroid ring) is replaced by a flexible, sterically less demanding methionine residue, whereas at the 'bottom' of the binding site, close to ring D, a methionine residue in ERα is replaced by an isoleucine residue in ERβ. By applying this information to the narrower binding pockets above and below the estradiol-binding sites of ERα and ERβ, respectively, the selective ligands **63** and **64** could be designed (**Figure 16**).[51–53] Whereas **63** has only about 40% of the activity of estradiol for ERα, it shows a 300-fold increase in selectivity compared with ERβ; on the other hand, compound **64** has only 50% of the activity of estradiol for ERβ but a 190-fold increase in selectivity compared with ERα.

The thyroid hormone T3 (**65**) and its less active storage form T4 (**66**) are iodinated phenoxy-phenylalanines, which bind to two nuclear receptor subtypes TRα and TRβ (**Figure 17**). Unfortunately, the affinity of T3 is higher for TRα than for TRβ, which causes cardiac side effects, if hypothyroid patients are treated with T3. The alkyl analogs **67** and **68** are less active for TRα than for TRβ.[54] Compound **69** binds to both receptor subtypes, but has no agonistic activity for TRα, and is only a weak partial agonist for TRβ; consequently, this compound might be of use to treat hyperthyroid patients.[55] Other patients suffer from a R320C mutant of TRβ; due to the exchange of the strongly basic arginine side chain for a neutral cysteine residue, T3 binds with much lower affinity to this receptor, causing a hypothyroid condition. Treatment with T3 or compound **68** is impossible, due to its low affinity for the TRβ mutant and the high affinity of these compounds for TRα. Conversion of the acidic side chain of the ligands into the neutral alcohol **70** solved the problem: compound **70** has a higher affinity for the TRβ mutant than for TRα.[56]

3.40.5.6 Integrin Ligands

Integrins form another group of receptors. They are expressed at cell surfaces, and their endogenous ligands, such as fibrinogen for the glycoprotein IIb/IIIa integrin (also called the fibrinogen receptor) or vitronectin for the $\alpha_v\beta_3$ integrin

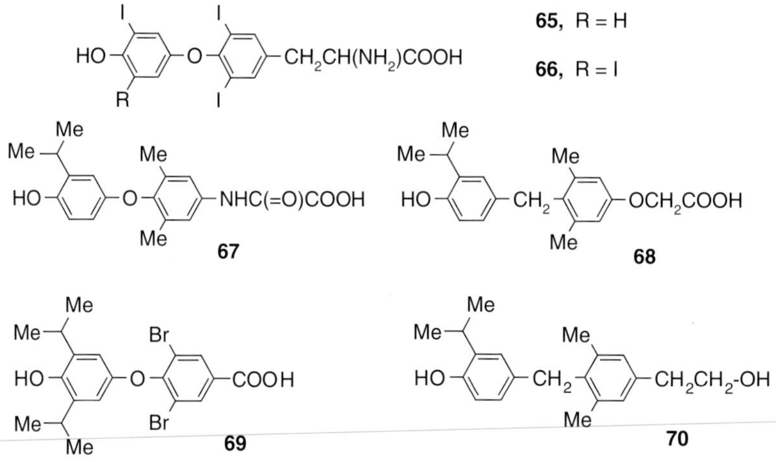

Figure 17 Compounds **67** (CGS-23425) and **68** (GC-1, UCSF) are alkyl analogs of the thyroid hormones T3 (**65**) and T4 (**66**); in contrast to T3, which has a higher activity at TRα, these analogs have higher activity at TRβ. Compound **69** is a thyroid hormone antagonist for TRα, and a weak partial agonist for TRβ. Neither T3 (EC_{50} hTRα = 0.14 nM, EC_{50} hTRβ = 0.66 nM, EC_{50} hTRβ R320C mutant = 4.3 nM) nor compound **68** (EC_{50} hTRα = 6.6 nM, EC_{50} hTRβ = 3.7 nM, EC_{50} hTRβ R320C mutant = 38 nM) have sufficient activity at a hTRβ R320C mutant. Compound **70** is a neutral, weakly active, but TRβ R320C mutant-selective thyromimetic (EC_{50} hTRα = 38 nM, EC_{50} hTRβ = 32 nM, EC_{50} hTRβ R320C mutant = 7.0 nM).

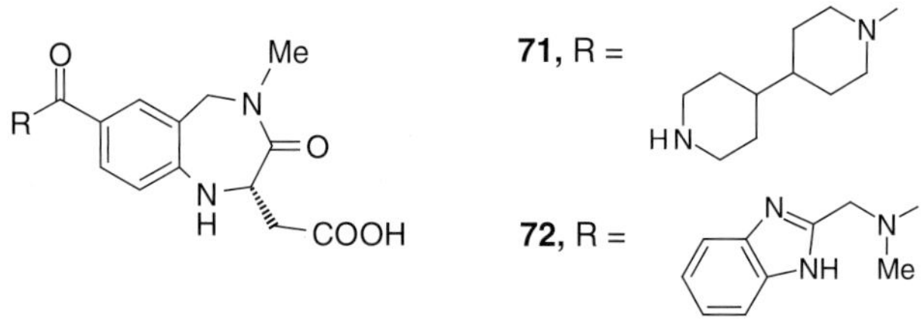

Figure 18 Compound **71** (lotrafiban, K_i GP IIb/IIIa = 2.5 nM, K_i $a_v\beta_3$ = 10 340 nM; failed in Phase III clinical trials) is a specific fibrinogen receptor antagonist, whereas compound **72** (K_i GP IIb/IIIa = 30 000 nM, K_i $a_v\beta_3$ = 2 nM) is a specific vitronectin receptor antagonist.

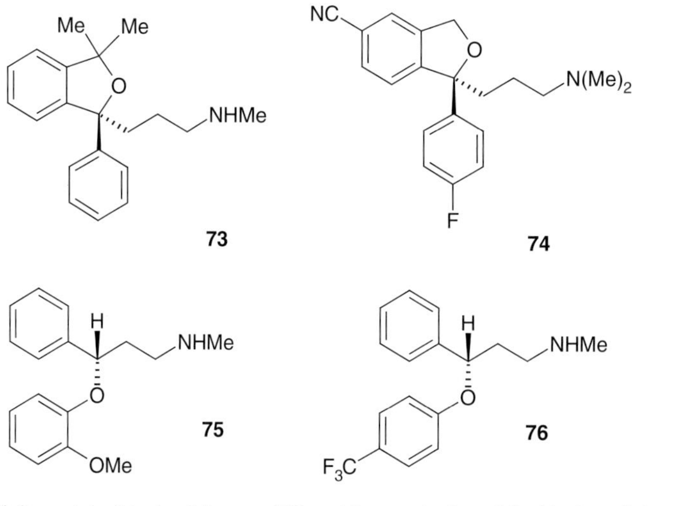

Figure 19 The norepinephrine uptake blocker talopram (**73**) and the serotonin uptake blocker citalopram (**74**) are chemically closely related but show a selectivity difference of several orders of magnitude. A similar selectivity difference is observed for the even more closely related analogs nisoxetine (**75**) and fluoxetine (**76**).

(also called the vitronectin receptor), mediate cell–cell contacts. The recognition motif of these two receptors is the Arg-Gly-Asp (RGD) sequence of the ligands, obviously in different conformations. Research at SmithKlineBeecham led to the discovery of ligands that showed, after minor chemical modification of a basic side chain, some selectivity for each of these two receptors.[57] After extensive structural modification, the highly selective ligands **71** (SB 214 857, lotrafiban) and **72** (SB 223 245) resulted (**Figure 18**)[57–59]; their selectivity profiles differ by more than seven orders of magnitude.

3.40.5.7 Transporter Ligands

Uptake inhibitors of neurotransmitters play an important role in the treatment of depression. Talopram (**73**) is a norepinephrine uptake blocker with a selectivity factor of about 550 compared with serotonin uptake inhibition, whereas the closely related analog citalopram (**74**) is a serotonin uptake blocker, with a selectivity of 3400 compared with norepinephrine uptake inhibition (**Figure 19**). A similar selectivity difference is observed for the even more closely related analogs nisoxetine (**75**) and fluoxetine (**76**), with a norepinephrine uptake selectivity of about 180 and a serotonin uptake selectivity of 54, respectively.[60]

3.40.6 Summary and Conclusions

Chemical similarity principles and bioisosterism are the guidelines for structural modification in classical medicinal chemistry. However, sometimes chemically similar compounds show very different biological activities and/or selectivities.[61] In the early years of combinatorial chemistry, its potential output was significantly overestimated. An unprecedented number of new drugs was expected from chemistry-driven combinatorial syntheses. However, the output was zero: sheer numbers did not contribute to drug discovery. The technology was able to deliver active analogs and to speed-up drug discovery only after a significant evolution. Instead of a combinatorial production of thousands of useless compounds, often in undefined mixtures, parallel syntheses of smaller libraries of single purified compounds are now performed, driven by medicinal chemistry. Used in this way, combinatorial chemistry is especially valuable in the very first steps of screening hit exploitation and lead structure optimization, in order to derive first structure–activity relationships and then to improve affinity, selectivity, and absorption, distribution, metabolism, and elimination properties up to a certain point.

Chemogenomics is a complementary strategy for the investigation of chemically related compounds and libraries against various members of a target family. It is largely based on the intelligent application of automated parallel synthesis. The advantages of such a systematic approach are manifold:

- specific analogs within a target family are discovered more easily;
- results from one target may be used to explore a related target;
- different subtype selectivities may be observed;
- structure–activity relationships are obtained earlier;
- coverage of chemical space and therefore patent coverage is more complete.

Of course, other rational approaches, such as molecular modeling, pharmacophore searches, virtual screening, and structure-based ligand design, support this new strategy. The final steps of drug optimization will always need dedicated structural modifications, using the accumulated knowledge base of classical medicinal chemistry.

Only a few examples of chemogenomics applications could be discussed in this review. More illustrative applications are presented in a recent monograph on chemogenomics in drug discovery.[19]

Acknowledgment

This chapter is an extended version of a chapter in *Chemical Genomics. Small Molecule Probes to Study Cellular Function* (ESRF Workshop 58),[62] and is reproduced with the kind permission of the Ernst Schering Research Foundation, Berlin, Germany.

References

1. Mayer, T. U.; Kapoor, T. M.; Haggarty, S. J.; King, R. W.; Schreiber, S. L.; Mitchison, T. J. *Science* **1999**, *286*, 971–974.
2. Ding, S.; Wu, T. Y.; Brinker, A.; Peters, E. C.; Hur, W.; Gray, N. S.; Schultz, P. G. *Proc. Natl. Acad. Sci. USA* **2003**, *100*, 7632–7637.
3. Wu, X.; Ding, S.; Ding, Q.; Gray, N. S.; Schultz, P. G. *J. Am. Chem. Soc.* **2004**, *126*, 1590–1591.

4. Ding, S.; Schultz, P. G. *Nat. Biotechnol.* **2004**, *22*, 833–840.
5. Chen, S.; Zhang, Q.; Wu, X.; Schultz, P. G.; Ding, S. *J. Am. Chem. Soc.* **2004**, *126*, 410–411.
6. Kim, S.; Rosania, G. R.; Chang, Y. T. *Mol. Intervent.* **2004**, *4*, 83–85.
7. Tan, D. S.; Foley, M. A.; Shair, M. D.; Schreiber, S. L. *J. Am. Chem. Soc.* **1998**, *120*, 8565–8566.
8. Schreiber, S. L. *Science* **2000**, *287*, 1964–1969.
9. Schreiber, S. L. *C&EN* **2003**, *81*, 51–61 [cf. Rouhi, A. M. *C&EN* **2003**, *81*, 104–107].
10. Lipinski, C.; Hopkins, A. *Nature* **2004**, *432*, 855–861.
11. Stockwell, B. R. *Nature* **2004**, *432*, 846–854.
12. Austin, C. P.; Brady, L. S.; Insel, T. R.; Collins, F. S. *Science* **2004**, *306*, 1138–1139.
13. Russell, K.; Michne, W. F. The value of chemical genetics in drug discovery. In *Chemogenomics in Drug Discovery. A Medicinal Chemistry Perspective*; Kubinyi, H., Müller, G., Eds. (Volume 22 of *Methods and Principles in Medicinal Chemistry*; Mannhold, R., Kubinyi, H., Folkers, G., Eds.); Wiley-VCH: Weinheim, Germany, 2004, pp 69–96.
14. Bishop, A. C.; Ubersax, J. A.; Petsch, D. T.; Matheos, D. P.; Gray, N. S.; Blethrow, J.; Shimizu, E.; Tsien, J. Z.; Schultz, P. G.; Rose, M. D. et al. *Nature* **2000**, *407*, 395–401.
15. Caron, P. R.; Mullican, M. D.; Mashal, R. D.; Wilson, K. P.; Su, M. S.; Murcko, M. A. *Curr. Opin. Chem. Biol.* **2001**, *5*, 464–470.
16. Bleicher, K. H. *Curr. Med. Chem.* **2002**, *9*, 2077–2084.
17. Jacoby, E.; Schuffenhauer, A.; Floersheim, P. *Drug News Perspect.* **2003**, *16*, 93–102.
18. Bredel, M.; Jacoby, E. *Nature Rev. Genet.* **2004**, *5*, 262–275.
19. Mannhold, R., Kubinyi, H., Folkers, G., Eds., *Methods and Principles in Medicinal Chemistry*; Wiley-VCH: Weinheim, Germany, 2004; Vol. 22.
20. Underwood, D. J.; Strader, C. D.; Rivero, R.; Patchett, A. A.; Greenlee, W.; Prendergast, K. *Chem. Biol.* **1994**, *1*, 211–221.
21. Perlman, S.; Costa-Neto, C. M.; Miyakawa, A. A.; Schambye, H. T.; Hjorth, S. A.; Paiva, A. C. M.; Rivero, R. A.; Greenlee, W. J.; Schwartz, T. W. *Mol. Pharmacol.* **1997**, *51*, 301–311.
22. Beeley, N. R. A. *Drug Discov. Today* **2000**, *5*, 354–363.
23. Ooms, F. *Curr. Med. Chem.* **2000**, 7, 141–158.
24. Dukat, M.; El-Zahabi, M.; Ferretti, G.; Damaj, M. I.; Martin, B. R.; Young, R.; Glennon, R. A. *Bioorg. Med. Chem. Lett.* **2002**, *12*, 3005–3007.
25. Römer, D.; Büscher, H. H.; Hill, R. C.; Maurer, R.; Petcher, T. J.; Zeugner, H.; Benson, W.; Finner, E.; Milkowski, W.; Thies, P. W. *Nature* **1982**, *298*, 759–760.
26. Evans, B. E.; Rittle, K. E.; Bock, M. G.; DiPardo, R. M.; Freidinger, R. M.; Whitter, W. L.; Lundell, G. F.; Veber, D. F.; Anderson, P. S.; Chang, R. S. L. et al. *J. Med. Chem.* **1988**, *31*, 2235–2246.
27. Patchett, A. A.; Nargund, R. P. *Annu. Rep. Med. Chem.* **2000**, *35*, 289–298.
28. Bymaster, F. P.; Calligaro, D. O.; Falcone, J. F.; Marsh, R. D.; Moore, N. A.; Tye, N. C.; Seeman, P.; Wong, D. T. *Neuropsychopharmacology* **1996**, *14*, 87–96.
29. Bymaster, F. P.; Nelson, D. L.; DeLapp, N. W.; Falcone, J. F.; Eckols, K.; Truex, L. L.; Foreman, M. M.; Lucaites, V. L.; Calligaro, D. O. *Schizophr. Res.* **1999**, *37*, 107–122.
30. McGovern, S. L.; Caselli, E.; Grigorieff, N.; Shoichet, B. K. *J. Med. Chem.* **2002**, *45*, 1712–1722.
31. McGovern, S. L.; Shoichet, B. K. *J. Med. Chem.* **2003**, *46*, 1478–1483.
32. McGovern, S. L.; Helfand, B. T.; Feng, B.; Shoichet, B. K. *J. Med. Chem.* **2003**, *46*, 4265–4272.
33. Seidler, J.; McGovern, S. L.; Doman, T. N.; Shoichet, B. K. *J. Med. Chem.* **2003**, *46*, 4477–4486.
34. Sneader, W. *Drug Prototypes and Their Exploitation*; John Wiley: Chichester, UK, 1996.
35. Kubinyi, H. Drug discovery from side effects. In *Chemogenomics in Drug Discovery. A Medicinal Chemistry Perspective*; Kubinyi, H., Müller, G., Eds. (Volume 22 of *Methods and Principles in Medicinal Chemistry*; Mannhold, R., Kubinyi, H., Folkers, G., Eds.); Wiley-VCH: Weinheim, Germany, 2004, pp 43–67.
36. Wermuth, C. G. *Med. Chem. Res.* **2001**, *10*, 431–439.
37. Wermuth, C. G. *J. Med. Chem.* **2004**, *47*, 1303–1314.
38. Rival, Y.; Hoffmann, R.; Didier, B.; Rybaltchenko, V.; Bourguignon, J. -J.; Wermuth, C. G. *J. Med. Chem.* **1998**, *41*, 311–317.
39. Müller, G. *Drug Discov. Today* **2003**, *6*, 681–691.
40. Müller, G. Target family-directed masterkeys in chemogenomics. In *Chemogenomics in Drug Discovery. A Medicinal Chemistry Perspective*; Kubinyi, H., Müller, G., Eds. (Volume 22 of *Methods and Principles in Medicinal Chemistry*; Mannhold, R., Kubinyi, H., Folkers, G., Eds.); Wiley-VCH: Weinheim, Germany, 2004, pp 7–41.
41. Slusarchyk, W. A.; Robl, J. A.; Taunk, P. C.; Asaad, M. M.; Bird, J. E.; DiMarco, J.; Pan, Y. *Bioorg. Med. Chem. Lett.* **1995**, *5*, 753–758.
42. Yamamoto, M.; Ikeda, S.; Kondo, H.; Inoue, S. *Bioorg. Med. Chem. Lett.* **2002**, *12*, 375–378.
43. Capdeville, R.; Buchdunger, E.; Zimmermann, J.; Matter, A. *Nature Rev. Drug Discov.* **2002**, *1*, 493–502.
44. Broughton, B. J.; Chaplen, P.; Knowles, P.; Lunt, E.; Marshall, S. M.; Pain, D. L.; Wooldridge, K. R. H. *J. Med. Chem.* **1975**, *18*, 1117–1122.
45. Fujita, T. The extrathermodynamic approach to drug design. In *Quantitative Drug Design*; Ramsden, C. A., Ed. (Volume 4 of *Comprehensive Medicinal Chemistry. The Rational Design, Mechanistic Study & Therapeutic Application of Chemical Compounds*; Hansch, C., Sammes, P. G., Taylor, J. B., Eds.); Pergamon Press: Oxford, UK, 1990, pp 497–560.
46. Wooldridge, K. R. H. Personal communication, 1976.
47. Kling, J. *Modern Drug Discov.*, **1998**, Nov./Dec., 31–38.
48. Hendrix, M.; Kallus, C. Phosphodiesterase inhibitors: a chemogenomic view. In *Chemogenomics in Drug Discovery. A Medicinal Chemistry Perspective*; Kubinyi, H., Müller, G., Eds. (Volume 22 of *Methods and Principles in Medicinal Chemistry*; Mannhold, R., Kubinyi, H., Folkers, G., Eds.); Wiley-VCH: Weinheim, Germany, 2004, pp 243–288.
49. Lopez-Rodriguez, M. L.; Morcillo, M. J.; Benhamu, B.; Rosado, M. L. *J. Comput.-Aided Mol. Design* **1997**, *11*, 589–599.
50. Rohrer, S. P.; Birzin, E. T.; Mosley, R. T.; Berk, S. C.; Hutchins, S. M.; Shen, D. M.; Xiong, Y.; Hayes, E. C.; Parmar, R. M.; Foor, F. et al. *Science* **1998**, *282*, 737–740 (erratum *Science* **1998**, *282*, 1646).
51. Hillisch, A.; Peters, O.; Kosemund, D.; Müller, G.; Walter, A.; Schneider, B.; Reddersen, G.; Elger, W.; Fritzemeier, K. H. *Mol. Endocrinol.* **2004**, *18*, 1599–1609.
52. Hillisch, A.; Peters, O.; Kosemund, D.; Müller, G.; Walter, A.; Elger, W.; Fritzemeier, K. H. Protein structure-based design, synthesis strategy and in vitro pharmacological characterization of estrogen receptor α and β selective compounds. In *New Molecular Mechanisms of Estrogen Action and Their Impact on Future Perspectives in Estrogen Therapy. ESRF Workshop 46*; Korach, K. S., Hillisch, A., Fritzemeier, K. H., Eds.; Springer: Berlin, Germany, 2004, pp 47–62.

53. Hillisch, A.; Pineda, L. F.; Hilgenfeld, R. *Drug Discov. Today* **2004**, *9*, 659–669.
54. Scanlan, T. S.; Yoshihara, H. A.; Nguyen, N.-H.; Chiellini, G. *Curr. Opin. Drug Disc. Dev.* **2001**, *4*, 614–622.
55. Baxter, J. D.; Goede, P.; Apriletti, J. W.; West, B. L.; Feng, W.; Mellstrom, K.; Fletterick, R. J.; Wagner, R. L.; Kushner, P. J.; Ribeiro, R. C. et al. *Endocrinology* **2002**, *143*, 517–524.
56. Ye, H. F.; O'Reilly, K. E.; Koh, J. T. *J. Am. Chem. Soc.* **2001**, *123*, 1521–1522.
57. Samanen, J. M.; Ali, F. E.; Barton, L. S.; Bondinell, W. E.; Burgess, J. L.; Callahan, J. F.; Calvo, R. R.; Chen, W.; Chen, L.; Erhard, K. et al. *J. Med. Chem.* **1996**, *39*, 4867–4870.
58. Keenan, R. M.; Miller, W. H.; Kwon, C.; Ali, F. E.; Callahan, J. F.; Calvo, R. R.; Hwang, S. M.; Kopple, K. D.; Peishoff, C. E.; Samanen, J. M. et al. *J. Med. Chem.* **1997**, *40*, 2289–2292.
59. Miller, W. H.; Keenan, R. M.; Willette, R. N.; Lark, M. W. *Drug Discov. Today* **2000**, *5*, 397–408.
60. Gundertofte, K.; Bogeso K. P.; Liljefors, T. A stereoselective pharmacophoric model of the serotonin re-uptake site. In *Computer-Assisted Lead Finding and Optimization. Proceedings of the 11th European Symposium on Quantitative Structure–Activity Relationships, Lausanne, 1996*; van de Waterbeemd, H., Testa, B., Folkers, G., Eds.; Verlag Helvetica Chimica Acta: Basel, Switzerland; VCH: Weinheim, Germany, 1997; pp 445–459.
61. Kubinyi, H. *Persp. Drug Design Discov.* **1998**, *9–11*, 225–252.
62. Kubinyi, H. Chemogenomics in drug discovery. In *Chemical Genomics. Small Molecule Probes to Study Cellular Function. ESRF Workshop 58*; Jaroch, S.; Weinmann, H., Eds.; Springer: Berlin, Germany, 2006.

Biography

Hugo Kubinyi is a medicinal chemist with 35 years of industrial experience, at KNOLL AG and BASF AG, Ludwigshafen, Germany. Since 1987, until his retirement in summer 2001, he was responsible for the Molecular Modelling, X-ray Crystallography and Drug Design group of BASF, since early 1998 also for Combinatorial Chemistry in the Life Sciences. He is Professor of Pharmaceutical Chemistry at the University of Heidelberg, former Chair of The QSAR and Modelling Society (1995–2000; from 2000–2010 Advisor to the Chair), and IUPAC Fellow. In 2006 he got the Herman Skolnik Award (CINF, ACS). From his scientific work resulted more than 100 publications and seven books on QSAR, 3D-QSAR, Drug Design, Chemogenomics, and Drug Discovery Technologies (Volume 3 of Comprehensive Medicinal Chemistry II). He is a member of several Scientific Advisory Boards, coeditor of the Wiley-VCH book series "Methods and Principles in Medicinal Chemistry", and member of the Editorial Boards of several scientific journals.

Comprehensive Medicinal Chemistry II
ISBN (set): 0-08-044513-6

ISBN (Volume 3) 0-08-044516-0; pp. 921–937

3.41 Fragment-Based Approaches

W Jahnke, Novartis Pharma AG, Basel, Switzerland

3.41.1	**Introduction**	**939**
3.41.1.1	What is Fragment-Based Ligand Design?	940
3.41.2	**Five Good Reasons for Fragment-Based Ligand Design**	**941**
3.41.2.1	Weak Hits Are Better Starting Points Than No Hits	941
3.41.2.2	High Hit Rate with Small Libraries	942
3.41.2.3	Fragment-Based Screening Covers Large Chemical Space	943
3.41.2.4	Favorable Lead Properties with High Ligand Efficiency	943
3.41.2.5	Fragment-Based Ligand Design Makes 'Undrugable' Targets Drugable	944
3.41.3	**Technologies for Fragment-Based Screening**	**944**
3.41.3.1	Nuclear Magnetic Resonance Spectroscopy	944
3.41.3.1.1	Protein observation	945
3.41.3.1.2	Ligand observation	946
3.41.3.1.3	Reporter ligand observation	946
3.41.3.1.4	Nuclear magnetic resonance detection of enzyme function	946
3.41.3.2	X-ray Crystallography	946
3.41.3.3	Mass Spectrometry	947
3.41.3.4	Tethering	947
3.41.3.5	Surface Plasmon Resonance	947
3.41.4	**Technologies for Structure Determination**	**948**
3.41.4.1	X-ray Crystallography	948
3.41.4.2	Nuclear Magnetic Resonance Spectroscopy	948
3.41.4.2.1	Inter-ligand nuclear Overhauser effects (ILOEs)	948
3.41.4.2.2	INPHARMA	948
3.41.5	**Fragment-Based Screening: Case Studies**	**949**
3.41.5.1	p38α Mitogen Activated Protein (MAP) Kinase, Case Study 1: X-ray and Fragment Elaboration	949
3.41.5.2	p38α Mitogen Activated Protein Kinase, Case Study 2: X-ray and Fragment Merging	950
3.41.5.3	Bcl-xL: Nuclear Magnetic Resonance and Fragment Linking	950
3.41.5.4	Caspase-3: Tethering and Linking	950
3.41.5.5	Leukocyte Function-Associated Antigen-1 (LFA-1): NMR and Reverse Screening	952
3.41.5.6	c-Src: Enzyme Assay and Fragment Linking	953
3.41.6	**Conclusions**	**954**
References		**955**

3.41.1 Introduction

A successful drug discovery program has two key molecules: a target, often a protein that is misregulated and is responsible for a disease phenotype, and a ligand molecule, the drug candidate, that binds to this target and modulates its activity to control or alleviate the disease or its symptoms. While suitable targets are often identified by basic research in academia, ligands to this target are generally discovered in the pharmaceutical or biotechnology industry.

The quest for a drug candidate is often divided into two steps. During the 'lead-finding stage,' a lead compound is sought. Lead compounds are typically small organic molecules that bind to and inhibit the target with medium to high affinity, and which possess molecular properties that indicate their suitability for a drug. During the subsequent 'lead optimization stage,' the affinity of this lead compound to its target, and its absorption, distribution, metabolism, excretion, and toxicity (ADMET) properties are optimized to match the specific requirements to treat a given disease.

For lead finding, high-throughput screening (HTS) of all available compounds in a corporate inventory is often performed. HTS produces 'hits,' which are validated, characterized, and optimized to become a 'lead.' HTS has been successful for the generation of lead compounds in some cases, but by far not in all cases.[1,2] Indeed, only a small percentage of leads stem from HTS. There is an urgent need for alternative lead-finding strategies. Fragment-based ligand design (FBLD) or fragment-based screening (FBS) is such an emerging strategy which holds high potential to generate lead compounds in a more rational and reliable manner.[3–7]

3.41.1.1 What is Fragment-Based Ligand Design?

Fragment-based screening is the initial step in FBLD, a technique to identify high-affinity ligands for a target protein. Unlike HTS, which screens fully assembled, lead-size or even drug-size molecules for their ability to bind to and inhibit the target protein, FBLD identifies these inhibitors in a modular way by assembling them piece by piece as the project progresses. In an initial screening step, small inhibitor fragments with a molecular weight around 250 Da or 300 Da are identified that bind to the target with weak but validated affinity. Due to the weak binding affinity, novel screening technologies must be employed, such as nuclear magnetic resonance (NMR) spectroscopy,[3,8–11] x-ray crystallography,[7,12–14] or mass spectroscopy.[15] Structural information on the protein–fragment complex is then gathered, and plays a crucial role in the evolution of this fragment into a lead and its concomitant increase in binding affinity. If the fragment library has been carefully designed and the fragment optimization is thoroughly carried out, the resulting lead truly possesses lead-like properties, and has a higher chance of being converted into a drug.[16,17]

Fragment-based ligand design is not a narrowly defined technique, and has been described with a variety of facets. However, in general, FBLD employs a technique for screening and a technique for structure determination. Screening techniques comprise NMR spectroscopy, x-ray crystallography, surface plasmon resonance, mass spectroscopy, or biochemical assays at high concentrations. For structure determination, x-ray crystallography or NMR spectroscopy is most commonly employed. All published examples of FBLD employ modular combinations of a screening technique and a structural technique (**Figure 1**).

Fragment-based ligand design is an iterative process with the goal of improving compound affinity in each iteration (**Figure 2**). After the identification of molecular starting points by fragment screening, structural information is

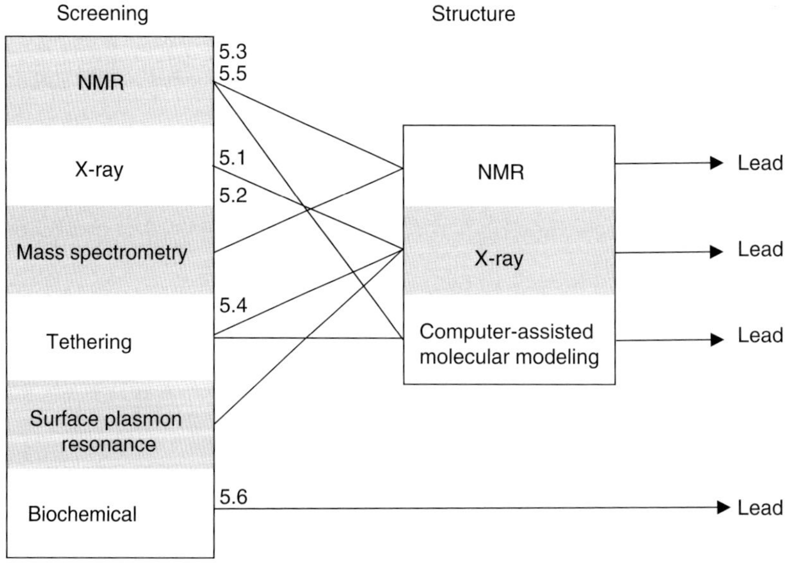

Figure 1 FBLD is the marriage between screening and structure-based design. Almost all published FBLD campaigns combine a screening module and a structural module, and almost all combinations are conceivable. The numbers refer to the examples in Section 3.41.5.

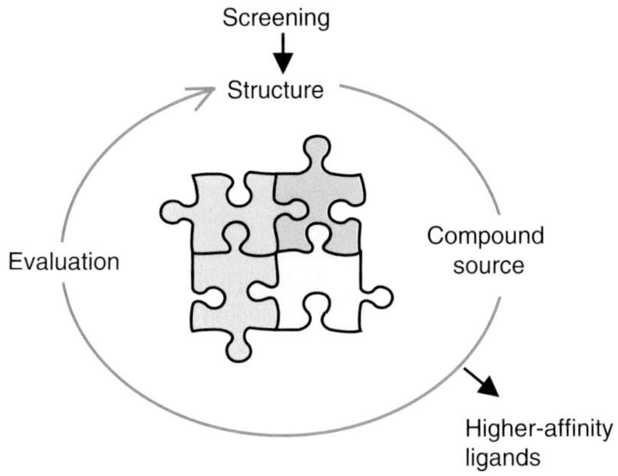

Figure 2 The FBLD cycle. Screening and structure techniques are detailed in **Figure 1** and Sections 3.41.3 and 3.41.4. 'Compound source' refers to cheminformatics and medicinal chemistry, and 'evaluation' refers to biochemical or biophysical assays to measure (relative) activity or binding affinity.

gathered to characterize the structure of the protein–ligand complex. Using this information, analogs of the initial fragment hits are identified by substructure or similarity searches, by other forms of data mining, or by chemical synthesis. These analogs are then tested in a functional assay or with a biophysical technique, and interesting compounds are again structurally characterized and pursued further.

The FBLD cycle shown in **Figure 2** describes the elaboration strategy in FBLD. It refers to the modification or decoration of a fragment to utilize and optimize additional binding interactions between the protein and ligand. Since elaboration of a fragment is the most straightforward strategy, it is most widely employed, and often yields compounds with the desired activity. Other strategies are available as well. The linked-fragment strategy is applicable to proteins with two or more distinct binding pockets, or with a single, large binding pocket. It seeks to identify and optimize a fragment for a first binding pocket, and then another fragment for a second binding pocket. Both fragments must bind simultaneously and in the same vicinity, and are subsequently linked to form a single high-affinity ligand. Linking two ligands should not only add the individual binding energies and thus multiply binding affinities,[3,18,19] but should also provide significant entropic advantages by reducing the translational entropy required for the formation of a ternary complex.[20] If fragments from two series have been identified that do not bind at distinct but at overlapping binding sites, they cannot be directly linked. Instead, the structure–activity relationships (SARs) of both series can be merged to combine the favorable interactions at one site with those of the neighboring site, in a process called the merged-fragment approach. Lastly, if a second and adjacent binding pocket exists but a suitable ligand for it has not yet been identified, a combinatorial library can be constructed by parallel synthesis, to probe the kind of ligand that best fits into this pocket (**Figure 3**).

Fragment-based strategies are not new to drug discovery. Jencks provided a theoretical framework for the linked-fragment approach more than 20 years ago,[18] and a few years later, Nakamura and Abeles provided experimental support with their work on HMG-CoA reductase.[21] Due to experimental limitations, however, fragment-based ideas were initially mostly realized computationally. In the early 1990s, docking programs such as DOCK[22] or LUDI[23] provided computational tools to identify and subsequently combine fragments on protein surfaces. Other computational approaches include the linked-fragment method proposed by Verlinde *et al.*,[24] fragment-based automated site-directed drug design,[25] dynamic ligand design,[26] a technique for docking both polar and nonpolar fragments,[27] and the 'core template' algorithm that connects two or more fragments.[28] The first rigorous experimental implementation of FBLD on an industrial scale was provided by Fesik's group at Abbott Laboratories. 'SAR-by-NMR'[3] was a pioneering strategy for experimental realization of FBLD using NMR spectroscopy. It paved the way for other applications and new developments of FBLD, as described below.

3.41.2 Five Good Reasons for Fragment-Based Ligand Design

3.41.2.1 Weak Hits Are Better Starting Points Than No Hits

High-throughput screening tends to give an 'all-or-nothing' result: either a suitable compound is present in the screening library, in which case a low micromolar or sub-micromolar hit will be identified which may be developed into a lead compound, or such a compound is not present in the library, in which case the HTS campaign does not provide any

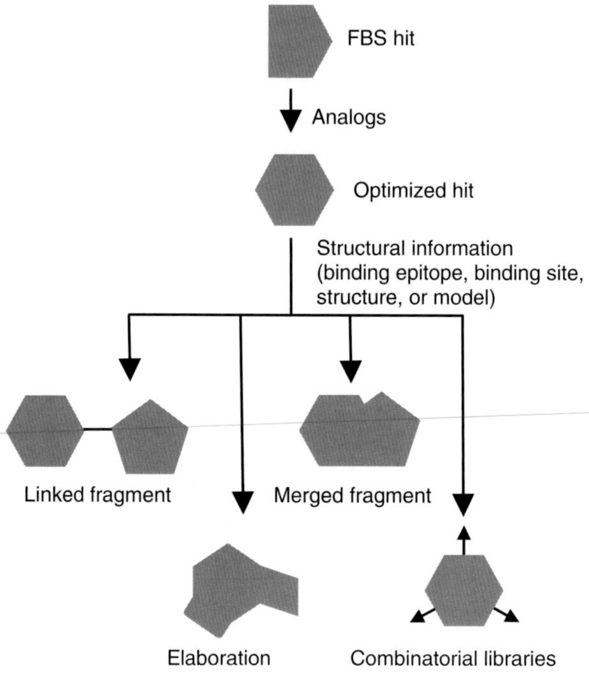

Figure 3 Follow-up strategies for FBLD.

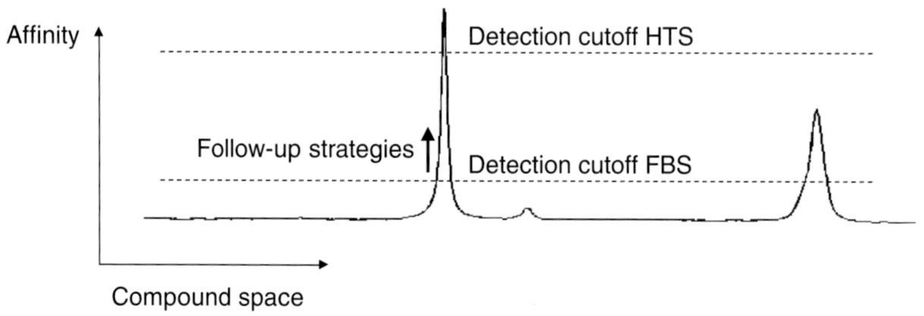

Figure 4 A hypothetical view of compound space in one dimension, with binding affinity to a given target indicated on the vertical axis. HTS has a high detection cut-off, and may miss interesting regions if the respective compounds are not present in the library. FBS has a lower detection cut-off, and will detect compounds on the edge of interesting regions. Follow-up strategies then provide ways to identify the most potent compounds in a directed search.

hits. The 'all-or-nothing' result in HTS is due to the low cut-off for hit detection in HTS, typically around 10 μM. Any compound with weaker but genuine affinity is not detected in HTS. FBS can detect much weaker ligands with up to millimolar affinity (**Figure 4**). Since such compounds are much more abundant than low micromolar ligands, FBS generally does yield fragment hits even for difficult targets, and is therefore not an 'all-or-nothing' method. Using an analogy in which chemical space corresponds to the surface topology on the earth and high-affinity ligands are represented by high mountains, HTS would probe the altitude of the underlying terrain for elevations higher than a cut-off, for example 4000 m. This would result in hits if any chemotypes corresponding to the Himalaya were probed, but would most likely miss all of the Alps except in lucky cases if Mont Blanc or the Matterhorn was present in the compound collection. FBS would probe the topology with many fewer probes, but would precisely detect elevations as low as a few hundred meters. This would detect not only the Himalaya or the Alps but also the highlands of Tibet as well as the foothills of the Alps. Any of those hills, or low-affinity ligands, are valuable starting points for a directed search for the highest mountains or highest-affinity ligands.

3.41.2.2 High Hit Rate with Small Libraries

Unlike HTS, which typically shows low hit rates between 0 and 0.2%, FBS often enjoys very high hit rates, typically between 3% and 10%.[29] As a practical consequence, this means that FBS libraries do not need to be large for the

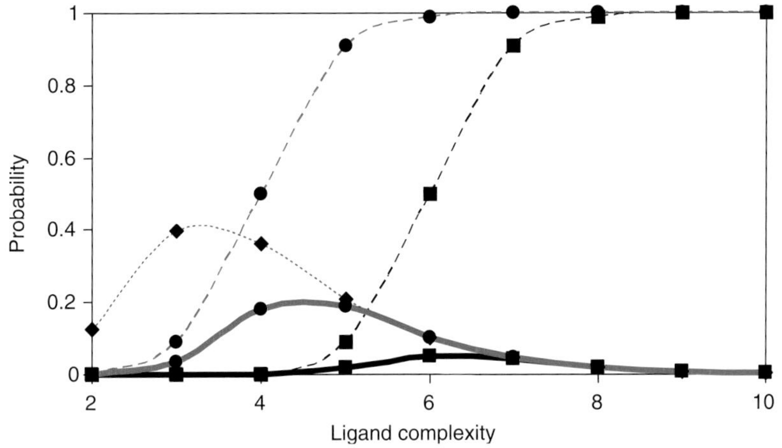

Figure 5 FBS yields higher hit rates than HTS. In a simple model proposed by Hann,[30] the probability for unique match of a fragment (blue dotted curve, diamonds), multiplied by the probability of detection (dashed curve), yields the hit rate (solid thick curve). Since FBS has higher sensitivity for detection and since the probability for unique match is higher with low molecular complexity, the FBS hit rate (red, circles) is higher than the HTS hit rate (black, squares).

identification of suitable starting points for FBLD. There are two reasons for such high hit rate. First, as stated above, the detection limit is higher than for HTS, which automatically translates into a larger number of hits. Most FBS hits are weakly binding ligands in the mid- to high-micromolar or even the low millimolar binding range. Secondly, FBS screens fragments that are of low molecular weight but also of low complexity. Small compounds with low complexity have a significantly higher probability of matching a given binding pocket. In a simple model proposed by Hann *et al.*,[30] ligands and binding pockets are described by a sequence of positive or negative bits. The complexity of a ligand corresponds to the length of this sequence. As the ligand becomes more complex, the probability of a unique match (**Figure 5**, dotted curve) decreases rapidly after an initial peak. This is because, with larger molecules, the chance of obtaining a mismatch (which sometimes completely abrogates binding) increases more rapidly than the chance of obtaining a beneficial interaction. The black solid curve with squares indicates the probability of detecting such a unique interaction using common HTS technologies. The product of both curves is the HTS hit rate in this simple model, which is relatively low and has its maximum with high-complexity ligands. The red solid curve with squares indicates the probability of detecting a unique interaction using common FBS techniques. The product of this curve and the dotted curve is the solid red curve with circles. It describes the FBS hit rate in this simple model, which is significantly higher than for FBS, and has its maximum at lower ligand complexity.

3.41.2.3 Fragment-Based Screening Covers Large Chemical Space

Although the libraries employed for FBS are generally small (of the order of 10^3 molecules), the combinatorial nature of FBLD entails that one is actually dealing with a large, albeit virtual, chemical library. This is particularly obvious with the linked-fragment approach: if a protein target has three subpockets in its binding site, and each fragment samples all three subpockets independently, a library of 1000 fragments is equivalent to a virtual library of $1000 \times 1000 \times 1000 = 1$ billion assembled compounds, if the fragments can subsequently be linked to form a single ligand. Rather than screening 1 billion compounds, however, only 1000 compounds are screened, and out of the virtual library, only those compounds that contain fragments with binding affinity are actually synthesized. FBS effectively accesses all combinations of fragments while avoiding the combinatorial explosion.

3.41.2.4 Favorable Lead Properties with High Ligand Efficiency

Many ligands fail to become a lead candidate, and many lead candidates fail to become a drug. There is much more to a drug than mere binding affinity for the target. A drug needs to be well absorbed, not prematurely metabolized (or metabolized in the desired manner in the case of a prodrug), distributed to the desired site of action, and excreted in a desired manner and with acceptable kinetics. Most importantly, the toxicity of the drug must be low enough to result in an acceptable therapeutic window. Problems with ADMET, together with lack of efficiency, are the major reasons for failure of a drug in development. Significant advances have been made in understanding causes of common drug failures. For example, rapid metabolism or drug–drug interactions are often the result of the compounds being good substrates or inhibitors of cytochrome P450 enzymes, and adverse affects on heart function such as QT prolongation have been found to be caused by hERG inhibition. Analysis of the reasons for poor oral drug absorption gave multiple parameters: compounds of high molecular weight (>500 Da), high lipophilicity (ClogP>5), or high numbers of hydrogen bond donors (>5) or acceptors (>10) tend to be poorly absorbed. It should be borne in mind that this 'rule of five'[31] is indeed only a rule, and not a stringent requirement, and should therefore not be overemphasized. The opportunity provided by this 'rule of five' is that the molecular parameters that influence oral absorption can be deliberately controlled by adequate library design. The 'rule of five' describes properties of 'drug-like' compounds. It was soon realized that slightly different properties need to be specified for 'lead-like' compounds, since properties of lead compounds are generally modified in the same directions (higher molecular weight, higher lipophilicity) in the lead optimization process.[32,33] Similarly, fragments again need a different specification since fragment hits experience even more drastic changes in properties in the fragment hit-to-lead process. A 'rule of three' has been proposed for the design of FBS libraries: fragments should have a molecular weight of <300 Da, ClogP<3, and the number of hydrogen bond donors and acceptors should each be $\leqslant 3$.[34] This 'rule of three,' or variants of it, is generally adhered to in the design of fragment libraries. As a consequence, fragment hits generally have high ligand efficiency[35] (binding energy per heavy atom) and good physicochemical properties. If care is taken to retain these favorable properties during FBLD, a lead compound from FBLD can indeed be superior to a lead compound from HTS.

3.41.2.5 Fragment-Based Ligand Design Makes 'Undrugable' Targets Drugable

High-throughput screening is often successful for highly drugable targets from a target class that has been worked on in the company's history (such as kinases or G protein-coupled receptors (GPCRs)), so that the corporate compound archive is enriched in structures belonging to chemotypes that are potential ligands for this target class. At the other extreme, poorly drugable targets without a working history within the company are not likely to produce any viable hits in HTS, which often means termination of the research project. It is clear that FBLD is particularly promising for targets that are drugable, that is, which have a binding pocket with affinity for small molecules, but for which the compound archive is not enriched with privileged scaffolds. For such targets, FBS generally delivers suitable starting points that are further optimized by FBLD. It is encouraging to see, however, that even poorly drugable targets have been successfully worked on using FBLD. These targets include protein–protein interactions such as IL2[36] and Bcl-xL.[37] These case studies suggest that FBLD can be successful in identifying high-affinity ligands even for difficult targets for which HTS has failed, be it because the corporate archive did not contain the proper chemotypes, or that those targets are indeed poorly drugable.

3.41.3 Technologies for Fragment-Based Screening

As outlined in **Figure 1**, a suitable screening technique is at the beginning of each FBLD project. This screening technique must be able to detect even weak binding interactions. It must be robust and reliable, so as to not yield false-positive hits: given the weak initial binding affinity of fragment hits used as starting points for FBLD, a fundamental prerequisite for FBLD is that these fragment hits are true and validated ligands for the target. Any false positives would lead the project in the wrong direction, and therefore must be avoided. False-negative detection should also be avoided, since this reduces the hit rate, and may lead to wrong conclusions regarding SARs. A number of technologies are now available that meet these requirements, and the most important ones are discussed below.

3.41.3.1 Nuclear Magnetic Resonance Spectroscopy

Nuclear magnetic resonance was the first experimental technique described for FBS,[3] and is still one of the most widely used techniques. Traditionally viewed as a method for structural elucidation of proteins and small molecules, NMR is also a robust tool to sensitively detect and characterize even weak protein–ligand interactions.[9–11,38] Much of the importance of NMR for FBLD stems from this latter asset.

Nuclear magnetic resonance detection of protein–ligand binding interactions can be achieved via three different approaches: protein observation, ligand observation, or reporter ligand observation (**Figure 6**). In addition, enzymatic protein function can be observed by NMR.[39]

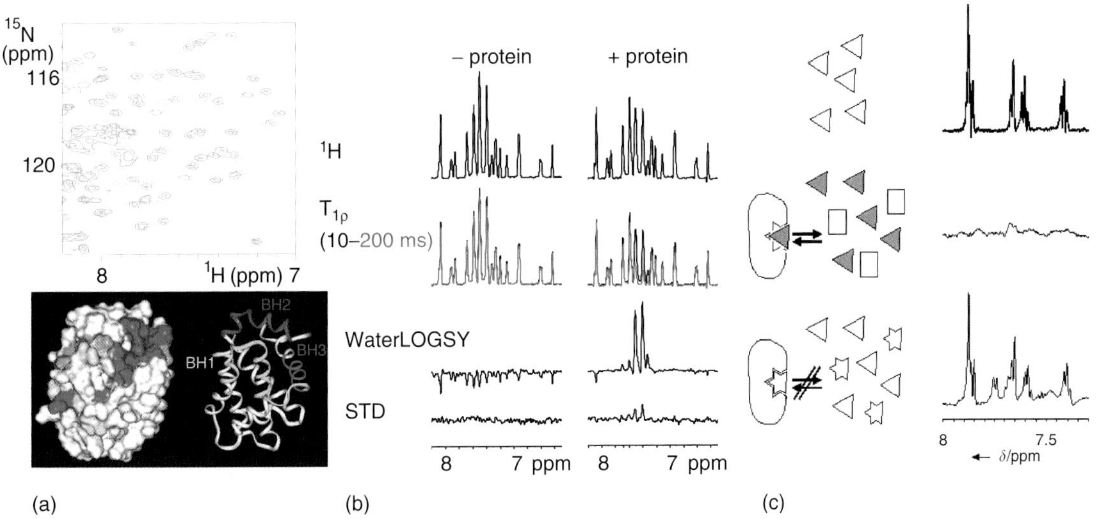

Figure 6 NMR techniques for the detection and characterization of protein–ligand interactions: (a) protein observation, (b) ligand observation, and (c) reporter screening.

The use of NMR spectroscopy as a screening tool allows a high degree of flexibility, since first-site screening as well as second-site screening can be performed.[3,40] Furthermore, NMR is a technique that is also useful for obtaining structural information on protein–ligand complexes (*see* Section 3.41.4.2).

3.41.3.1.1 **Protein observation**

The chemical shift of an NMR resonance is one of the most important and easy-to-measure NMR parameters. Chemical shifts depend sensitively on the local environment of the respective nucleus: it is dominated by the chemical nature of the directly bonded atoms, but also influenced by through-space interactions with other moieties that may be far in sequence but close in 3D space. Groups that strongly influence the chemical shift through space encompass aromatic ring systems or hydrogen bond partners. Also, dynamic processes have an influence on chemical shift. In principle, the full structure of a protein is encoded in its chemical shifts, although the theoretical understanding of the chemical shift is not yet advanced enough to calculate 3D protein structures solely on the basis of chemical shifts.[41,42] Since the structure and dynamics of a protein are encoded in its chemical shifts, changes in structure and dynamics, or more generally changes in local environment, are also sensitively detected by chemical shift changes. Binding of a ligand is a significant change in local environment, and can thus be detected by monitoring chemical shifts.

Since one-dimensional NMR spectra of proteins are very crowded, generally 2D NMR is used to resolve the spectrum. Two-dimensional ^{15}N,^{1}H-heteronuclear single-quantum coherence (HSQC) spectra are sensitive, and can often resolve most resonances of smaller to medium-size proteins. The ^{15}N,^{1}H-HSQC spectrum can therefore be used as a fingerprint spectrum of the protein and its ligation state. Since the natural abundance of ^{15}N is very low, recombinant ^{15}N-enriched proteins are required. For bigger proteins (>40 kDa), deuteration of all but the amide protons is required to achieve high sensitivity, and selective labeling is recommended to reduce signal overlap. Both ^{15}N labeling and ^{13}C-methyl labeling can be employed.[43,44]

Mere detection of ligand binding does not require resonance assignment to the individual protein protons. However, if resonances are assigned, the chemical shift changes can be mapped to the protein surface to reveal the ligand-binding site (**Figure 6a**). This assumes that changes in the local environment are mainly caused by direct contact with the ligand. While this is often true, some proteins exhibit greater conformational changes upon ligand binding, which leads to chemical shift changes at sites remote from the ligand-binding site. Qualitative information about the ligand-binding site can also be gained from comparing the chemical shift changes of a ligand with those of a reference ligand with a known binding site.

The advantages of the protein observation method are its high information content and its robustness. Its main disadvantage is the need for large amounts of isotopically labeled protein, which limits the application of this method in practice to small or medium-sized proteins that express well in *Escherichia coli*.

3.41.3.1.2 Ligand observation

Since most pharmaceutically relevant proteins are not small and well-expressed in *E. coli*, ligand observation methods have been developed. These method observe only the unbound ligand, but are able to detect the binding history of the ligand on the resonances of the unbound ligand. NMR parameters are observed that depend strongly on the correlation time, such as relaxation rates or nuclear Overhauser effects (NOEs). In its unbound state, a compound behaves like a small molecule since it is a small molecule: it relaxes slowly and shows small and positive NOEs. If the compound is bound to the protein, however, it temporarily behaves like a large molecule (like a protein), and relaxes fast and shows strong and negative NOEs. If the dissociation rate is fast compared with the NMR time-scale (milliseconds to seconds), properties of the unbound and bound ligand superimpose, and an averaged value is transferred to and detected on the resonances of the free ligand. The most popular methods are measurement of $T_{1\rho}$ relaxation rates,[45] NOEs between bound water and bound ligand (as in the waterLOGSY experiment),[46] and steady state NOEs transferred from the protein to the ligand (as in the saturation transfer difference experiment (**Figure 6b**)).[47] Other methods include detection of transferred NOEs or detection of protein-ligand NOEs edited by diffusion constants.

In order to confirm the specificity of binding, competition experiments with a known active-site ligand can be performed. If the test compound is displaced by the competitor, specific binding can be assumed.[48]

The advantages of ligand observation methods are reduced protein consumption, and the fact that these methods do not require isotopically labeled proteins, and work with proteins regardless of size – actually, the bigger the better. Their main disadvantage is the need for a fast dissociation rate for binding to be detected: ligands with a slow dissociation rate (half-life of seconds or longer) give false negatives with this detection method. NMR reporter screening is an important remedy for this shortcoming (see below). In practice, binding by ligand observation methods can generally be detected down to high-nanomolar ranges or weaker, although it must be kept in mind that it is not the dissociation constant but the dissociation rate that is the relevant parameter.

3.41.3.1.3 Reporter ligand observation

The false-negative detection of tightly binding ligands is a severe shortcoming of ligand observation methods. Competition-based experiments such as NMR reporter screening overcome this shortcoming.[49–51] In this format, a known, weakly binding 'reporter' ligand (or 'spy' molecule or probe) that binds to the desired site is added to all the test mixtures containing the protein and test compounds. It is not the binding of the test compounds that is monitored, but their ability to displace the reporter ligand from the binding site (**Figure 6c**). This format allows the detection of high-affinity ligands, and detects only active-site ligands while not detecting ligands binding to different (possibly irrelevant) binding sites. An elegant extension of this format is to choose reporter ligands with fluorine atoms, whose resonances can be observed without overlap from test compounds.[52] Spin labels can be introduced to further reduce protein consumption.[53]

3.41.3.1.4 Nuclear magnetic resonance detection of enzyme function

Most technologies described in this section concern binding assays, whereas virtually all 'traditional' methods for the detection of protein–ligand interactions use functional assays to detect enzyme inhibition. Often, these functional assays require chemical labels, or indirect detection using other components that can influence the read-out. Enzyme activity can also be measured by NMR by quantifying substrate and product concentrations.[54] Again, the addition of a fluorine atom or a trifluoromethyl group can help to increase sensitivity and resolve overlap.[39]

3.41.3.2 X-ray Crystallography

X-ray diffraction is an established and widely used technique to determine the crystal structures of proteins and protein–ligand complexes. Crystals of protein–ligand complexes can be obtained by either co-crystallizing the protein and the ligand, or by soaking the ligand in existing crystals. Soaking is often preferred, since it is faster. Due to the low binding affinities, high ligand concentrations need to be used, with final concentrations in solution of up to 25–100 mM. This requires fragment libraries with very high compound solubilities. It also requires sturdy protein crystals that are robust even in the presence of high ligand and dimethyl sulfoxide (or other solvent) concentrations. Protein crystals are sometimes made more robust by chemical cross-linking. Compound mixtures containing up to eight fragments are generally used. Upon data collection, the structure of a complex can generally be solved by molecular replacement, if the protein structure is known.

Although protein crystallography is an established tool, it was applied to the screening of compound libraries only a few years ago. Progress in the automation of data acquisition and analysis, including the interpretation of electron density maps, dramatically reduced the time in which multiple protein–ligand complexes could be solved, and concomitantly increased the throughput of protein crystallography.[7,12–14] Furthermore, parallelization in protein production increased the number of protein constructs and concomitantly crystallizable proteins, an obvious prerequisite for this technology.

Protein crystallography as primary screening technology has the great advantage that it is at the same time both a screening and a structural technology: identification of a fragment hit generally entails the precise 3D structure of the complex, which is of great help for the optimization of this hit. On the other hand, no information about the binding affinity (neither relative nor absolute) results from crystallographic screening. Protein crystallography can only prove but never disprove ligand binding, since an empty active site can have a variety of reasons besides lack of binding affinity.

3.41.3.3 Mass Spectrometry

Identification of weakly binding ligands can also be performed by mass spectrometry, if suitable ionization methods such as electrospray ionization are employed. In this case, complexes stay intact during the ionization, desolvation, and detection processes, so that noncovalent ligand binding can be detected directly by its molecular mass. By analyzing the observed mass and abundancy of the complexes, conclusions about the binding affinity and stoichiometry can be drawn. Also, competition experiments with known ligands can be performed. This appears to work particularly well for RNA,[55–57] but protein targets have also been successfully screened by mass spectrometry.[58,59]

The advantages of mass spectrometry as a screening tool are low protein consumption and a high degree of automation. Its disadvantages are lack of structural information and uncertainty to the degree to which interactions in the gas phase reflect interactions in solution.

3.41.3.4 Tethering

Tethering is an elegant combination of protein engineering, mass spectrometry, and protein crystallography.[60–62] First, a cysteine residue is introduced into the protein target near the active site. Second, a library of disulfide-containing fragments is exposed to the engineered protein under partially reducing conditions. This leads to disulfide formation between fragments and the cysteine residue. Importantly, disulfide formation is favored by the binding affinity of the fragment to a binding pocket, if this interaction is sterically possible. The reason for using partially reducing conditions is to govern the disulfide formation process solely by binding affinity rather than by thiol reactivity. In a third step, fragments with binding affinity for the target that form disulfide bonds are detected by mass spectrometry and investigated by protein crystallography. Both mass spectroscopy and crystallography are facilitated by disulfide bond formation.

Tethering is a site-directed screening method since the cysteine mutation can be introduced at the desired position. It is also very versatile – for example, the 'extender strategy' allows screening for ligands that bind adjacent to a known ligand.[63] In this approach, the known ligand in introduced with a masked thiol group. Upon deprotection, the protein/extender complex is screened against the library of thiol-containing fragments. In essence, this is second-site screening using the known ligand as a first-site ligand. After completion of screening and analysis of the x-ray structure, the disulfide linkage between both ligands can be replaced by another linker, and possibly rigidified.[63]

The site-directed aspect of tethering can also be used to probe binding sites of known ligands in the absence of high-resolution structural data. This is particularly relevant for GPCRs, for which available structural data are very limited. Experimental data on ligand-binding sites can be very valuable for modeling this interaction. Using the tethering strategy, the binding site of a ligand can be approximated if various cysteine mutants around the presumed binding site are produced and the reactivity of the ligand under investigation with each mutant is probed. Following the general concept of tethering, reaction should predominantly occur with mutants in which the mutated cysteine residue is in the correct position with respect to the ligand-binding site. This concept has been nicely demonstrated using the GPCR C5a receptor and a peptide trimer from C5a.[64,65]

3.41.3.5 Surface Plasmon Resonance

SPR provides a direct and label-free binding assay based on a different biophysical principle. The target protein has to be immobilized on an optical sensor surface, typically a glass support with a gold layer, coated by a dextran layer to facilitate protein immobilization. Compounds can then be added in solution, and their binding detected by SPR. Detection of ligand binding by SPR is based on changes in mass concentration and concomitant changes in refractive index. At an interface between two transparent media of different refractive indices (in this case, glass and an aqueous buffer), light coming from the side of higher refractive index (the glass) is partly reflected and partly refracted. Above a certain critical angle of incidence, no light is refracted across the interface, and total internal reflection is observed. Although incident light is totally reflected, the electromagnetic field component does penetrate a short (tens of nanometers) distance into the medium of lower refractive index, and creates an exponentially attenuated 'evanescent wave.' Due to resonance energy transfer between the evanescent wave and surface plasmons, the intensity of the reflected light is reduced at a specific incident angle, under conditions of surface plasmon resonance (SPR). The measurement of the SPR is often done in Biacore instruments using monochromatic and p-polarized light under conditions of total internal reflection. The resonance angle at which the intensity minimum occurs is a function of the refractive index of the solution close to the gold layer. Changes in mass, for example caused by ligand binding, cause a change in refractive index, and are thus detected in Biacore instruments.[66–70]

The advantages of SPR are low protein consumption and access to kinetic information about the protein–ligand interaction: the shape of the SPR response as a function of time allows measurement of the association and dissociation rates. In some cases, biological activities or pharmacological effects correlate better with one of the individual kinetic constants (on/off rates) than with the thermodynamic constant of binding affinity.[71]

In a different application, the identification of thrombin inhibitor fragments by SPR was recently described.[72]

3.41.4 Technologies for Structure Determination

3.41.4.1 X-ray Crystallography

Protein crystallography is an established and mature tool for the structural determination of proteins and protein–ligand complexes,[73,74] and will not be discussed in detail in this chapter. X-ray crystallography gives the most precise and comprehensive structural information, and a series of ligands can be investigated very fast. Protein crystallography can also be used as a screening tool, and is an integral part of the tethering strategy.

3.41.4.2 Nuclear Magnetic Resonance Spectroscopy

Proteins do not always crystallize, nor does every ligand lead to interpretable electron density in a co-crystallization or soaking experiment. If protein crystallography cannot be applied, structural information can also be obtained by NMR spectroscopy.[75,76] This can be the full 3D structural determination by NMR of a binary or ternary protein–ligand complex,[77,78] or it can be selected structural information aimed at answering very specific questions. Two methods are particularly useful since they can rapidly provide selected crucial information.

3.41.4.2.1 Inter-ligand nuclear Overhauser effects (ILOEs)

For the linked-fragment strategy, two ligands need to be identified that bind to the protein target at the same time and at neighboring binding sites. When this is achieved, those ligands should be linked together to form a high-affinity ligand. Structural information about the ternary complex is crucial to guide attachment points and linker length. Since x-ray structures of a protein complexed with two weak ligands are rare and hard to get, suitable NMR techniques are valuable in obtaining this information. An obvious method is a full 3D structural determination of the ternary complex,[3,37] but this is time consuming and yields much nonessential information. Alternatively, the transfer NOE experiment can be applied to the ternary complex. If both ligands bind in the same vicinity and have suitable binding kinetics, NOEs not only within each ligand but also between the two ligands can be detected. This identifies pairs of protons, one on each ligand, that are close in space in the ternary complex. Exactly this information is needed to guide linkage of the ligand. Since the discovery and validation of the method,[79] several applications have shown its usefulness for drug design.[40,80]

3.41.4.2.2 INPHARMA

While the measurement of ILOEs is used in the situation where two ligands bind simultaneously at different binding sites, the INPHARMA (inter-ligand NOEs for pharmacophore mapping) method is applied when two ligands bind competitively at the same binding site. In this case, cross-peaks between protons on different ligands can occur by spin diffusion via the protein. An NOE can originate from the first ligand, and be transferred to the protein. If the first ligand then dissociates and the second ligand binds, the NOE can return to the second ligand and be detected there. In essence, a cross-peak is detected that connects the two ligands that bind consecutively at similar positions relative to the protein. Pharmacophores for binding elements can be superimposed by this method.[81] Unfortunately, spin diffusion can complicate the interpretation of the experiment.

The experimental setup for ILOEs and INPHARMA measurements can be very similar, so that the competition behavior of both ligands must be carefully investigated to ensure correct interpretation of the results.

3.41.5 Fragment-Based Screening: Case Studies

The following case studies have been selected in order to cover the most common screening and optimization techniques (NMR, x-ray, mass spectrometry, and biochemical assay) and fragment optimization strategies (fragment elaboration, fragment merging, and fragment linking).

3.41.5.1 p38α Mitogen Activated Protein (MAP) Kinase, Case Study 1: X-ray and Fragment Elaboration

Apo crystals of p38α MAP kinase were prepared in which the kinase was nonphosphorylated.[82] The structure was solved, and it was ensured that the active site was not blocked by crystal contacts, and that solvent channels to the active site existed to make the crystals soakable. The crystals were then soaked with compound mixtures containing between two and eight fragments per mixture. Among the fragment hits was compound **1** (**Figure 7a**), which binds to the hinge region of the kinase and forms the well-known hydrogen bond with the backbone amide nitrogen of Met109. The in vitro potency of the compound was only 1.3 mM – however, it was chosen for follow-up due to its synthetic tractability and interesting binding mode: the phenyl side chain binds in a hydrophobic region which is not occupied by ATP and is known to be a key specificity pocket for p38α.

Optimization of this fragment hit was consequently guided by structural information, and took into account available SARs from different compound series. First, interaction of the ligand inside the hydrophobic specificity pocket was optimized. Two possibilities were discovered: 2,6-dichloro substitition of the phenyl ring, or replacement of the phenyl ring by naphthyl. Although the latter was somewhat more potent ($IC_{50} = 44\,\mu M$) than the former ($IC_{50} = 109\,\mu M$), the 2,6-dichloro substitution (compound **2**) was carried further due to its better balance of potency

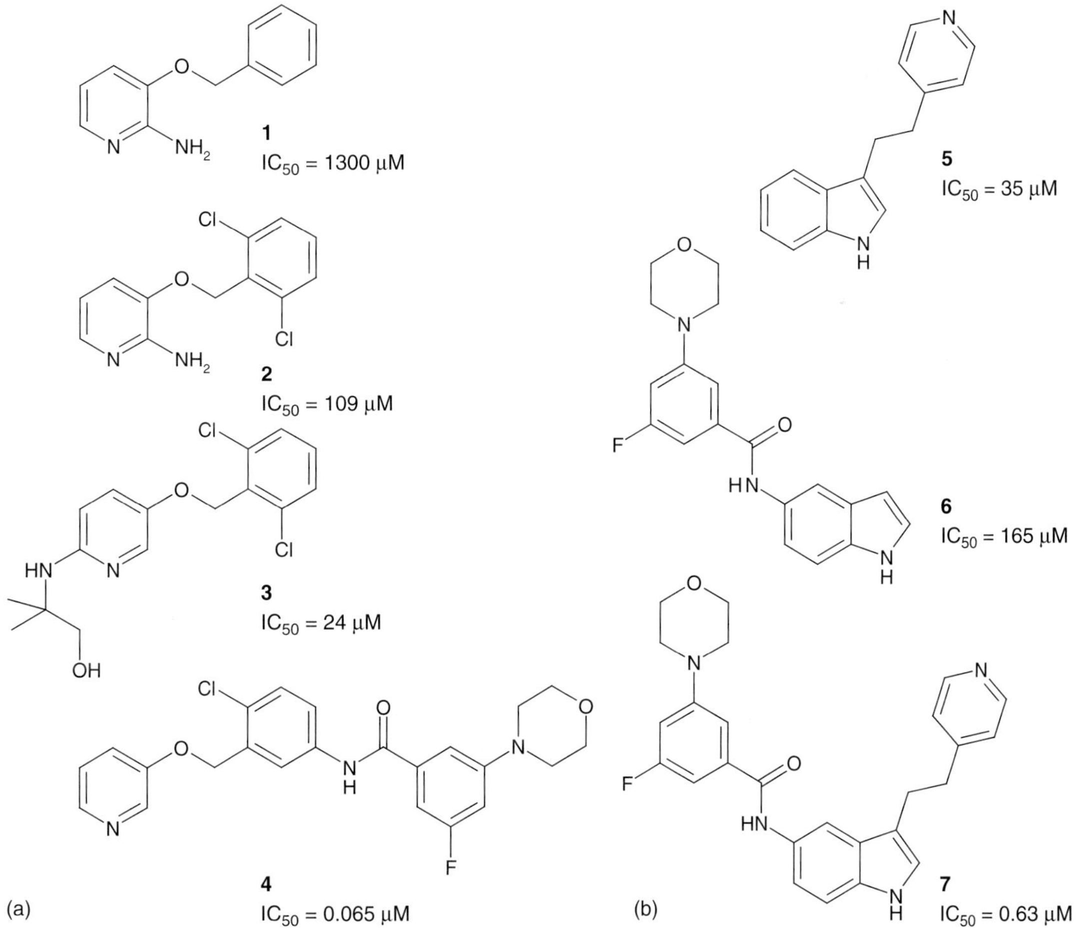

Figure 7 FBLD for p38α MAP kinase.

and physicochemical properties. Focusing on the pyridine part of the molecules, the amino group on compound **1** was surprisingly found unnecessary for potency and could be omitted. Based on the x-ray structure of **1**, substitution in the pyridine ring was attempted, and compound **3** was found to be slightly more active (24 μM).

Subsequently, structural information from known active p38 inhibitors was gathered and analyzed in the light of optimization of compound **3**. In particular, these studies encouraged exploitation of the conformational flexibility of the DFG motif, which is known to significantly rearrange in some complexes with more potent inhibitors. Structure-based design suggested an extension on the 2,6-dichloro-substituted phenyl ring. Indeed, compound **4** was significantly more potent ($IC_{50} = 0.065$ μM), and was found to induce the well-known 'DFG-out' conformation of the activation loop. The compound appears to be reasonably selective against other protein kinases, and demonstrated cellular activity in THP-1 cells.[82]

3.41.5.2 p38α Mitogen Activated Protein Kinase, Case Study 2: X-ray and Fragment Merging

Among the fragment hit list described in the previous section, indole derivative **5** (**Figure 7b**) was also considered promising because of its chemical tractability and binding mode: the indole ring is deeply buried in the hydrophobic specificity pocket, while the pyrimidine ring forms a hydrogen bond with the backbone amide nitrogen of Met109. The optimization of compound **5** took advantage of existing knowledge of p38 inhibitors in two respects. First, by adding an aromatic amide substituent to the 5-position of the indole while omitting the pyridyl substituent in the 3-position, a weaker inhibitor (compound **6**, $IC_{50} = 165$ μM) was created that induces the 'DFG-out' conformation described previously. This inhibitor is probably so weak since it does not contain chemical groups that form hydrogen bonds to the backbone amide nitrogen of Met109. Comparing the x-ray structures of compounds **5** and **6**, it seemed possible to attach such a moiety, and reintroduction of the pyridine group (compound **7**) indeed yielded sub-micromolar potency with the expected binding mode. Indole substitution in the 1,6-position instead of the 3,5-position yielded similar potencies.[82]

3.41.5.3 Bcl-xL: Nuclear Magnetic Resonance and Fragment Linking

A 10 000-compound fragment library with an average compound molecular weight of 210 Da was screened, and yielded the fluoro biaryl acid **8** as a ligand for Bcl-xL (**Figure 8**).[83] NMR-based titration yielded a K_D of about 300 μM for this compound. NMR-based structure determination indicated that **8** bound to Bcl-xL within the Bak peptide-binding site, at the position of its critical leucine residue. The compound could not be significantly optimized by investigating close analogs. Therefore, second-site screening was performed by NMR using a library of 3500 compounds (average molecular weight 125 Da), and identified naphthol analogs auch as compound **9**. This second-site screening was performed with saturating amounts of the first ligand **8**, so that the first binding site was not accessible. The K_D of the second-site ligand was 2 mM. In order to guide fragment linking, the 3D structure of the ternary complex between Bcl-xL, **8** and **9** was solved by NMR.[83] Based on this structure, a linked compound **10** using a *trans*-olefinic linker was designed and synthesized that inhibited Bcl-xL with $K_i = 1.4$ μM. Structural analysis indicated, however, that the *trans*-olefinic linker may not be optimal. Instead, an acylsulfonamide as an isoster for the *p*-carboxylate was tested, since this would allow linkage of the two fragments via the *para* position, while maintaining the key negative charge on fragment **8**. Using parallel synthesis, 120 acyl sulfonamides were produced, and submicromolar inhibition constants were obtained (**11**, $K_i = 0.25$ μM). Encouraged by this result, another round of parallel synthesis around the acylsulfonamide core was performed, which resulted in compound **12**, with a K_i of 36 nM. NMR structural analysis indicated that favorable π stacking involving the terminal phenyl ring and the nitrophenyl ring of the ligand as well as Y194 of Bcl-xL was responsible for the increased binding affinity of **12**. Compound **12** represented a lead compound in the search for Bcl-xL inhibitors, but its high affinity for human serum albumin precluded further development. The affinity for albumin was subsequently designed out by using structural information, and the resulting compound, ABT-737, had subnanomolar affinity ($K_i < 1$ nM) for Bcl-xL and was active in cellular and in vivo experiments.[37]

This Bcl-xL example is striking not only because it led to the most potent (in this case subnanomolar) compound from FBS reported so far in the literature, but also because the molecular target was involved in protein–protein interactions, which are notoriously difficult to tackle, rather than being a highly drugable enzyme such as a kinase.

3.41.5.4 Caspase-3: Tethering and Linking

Caspase-3, a cysteine-aspartyl protease that is one of the central 'executioners' of apoptosis, is an ideal target for tethering since the active-site cysteine residue can be readily covalently modified. The 'extender' strategy was chosen

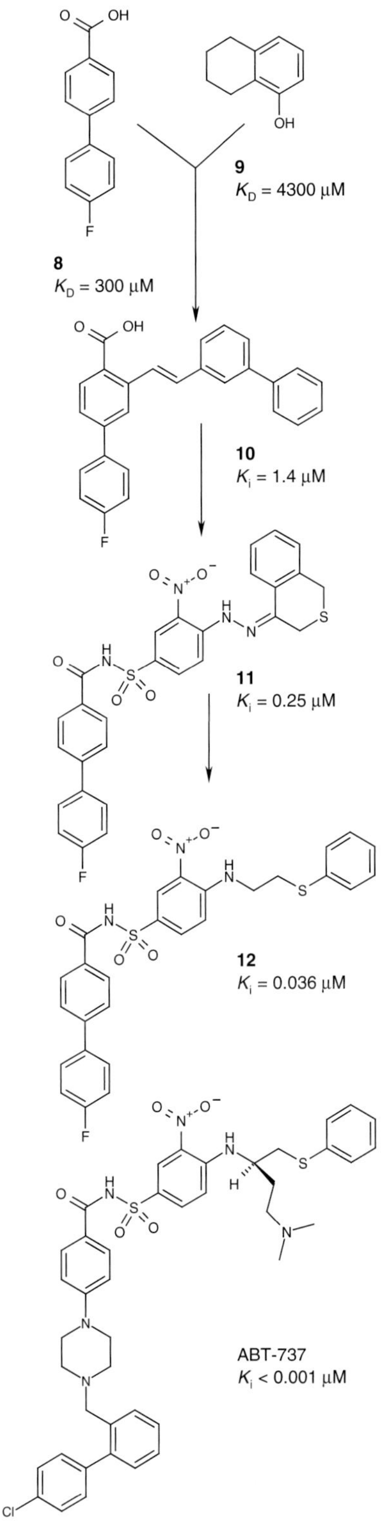

Figure 8 'SAR by NMR' used in the discovery of Bcl-xL antagonists.

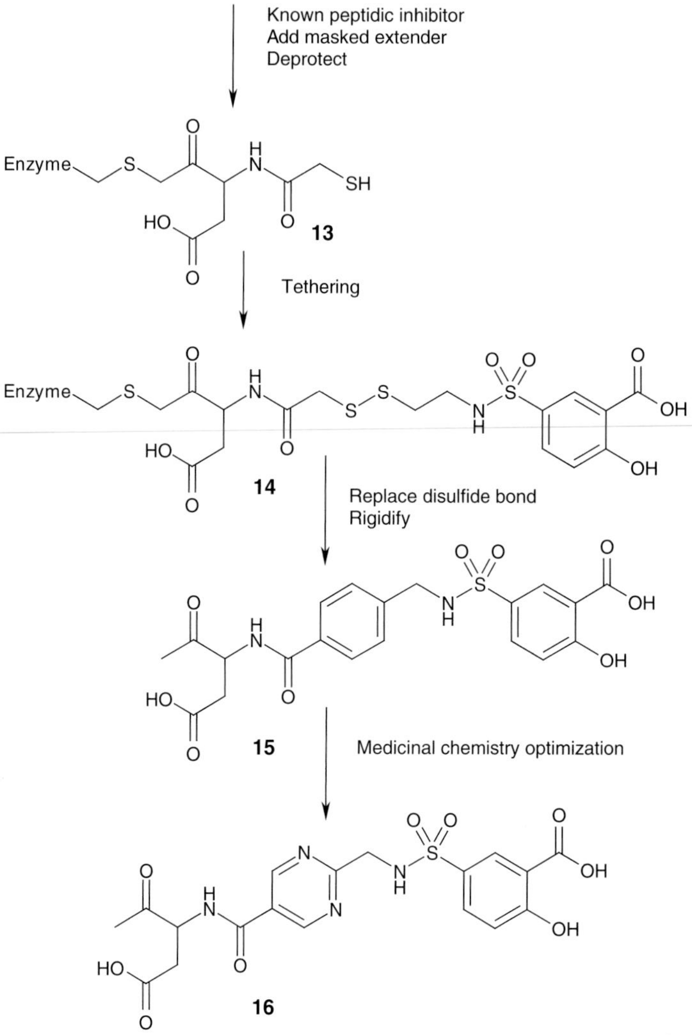

Figure 9 The discovery of caspase-3 inhibitors by tethering.

for the discovery of highly potent inhibitors (**Figure 9**). The 'extender', related to the 'first ligand' in a fragment-linking approach, was constructed from a known tetrapeptide-based inhibitor. It was converted into a smaller, protected thioester, which was then covalently reacted with caspase-3. Subsequently, the thioester was deprotected, to reveal a free thiol (adduct **13**), which was screened against a library of about 7000 disulfide-containing fragments. This 'second-site'-screening using mass spectrometric detection yielded one strong hit, a salicylic acid sulfonamide (adduct **14**). By replacing the disulfide bond with two methylene groups and replacing the irreversible 'warhead' with a reversible aldehyde, an inhibitor with $K_i = 2.8\,\mu M$ was created. By rigidifying the linker, the affinity was improved to 200 nM (compound **15**), and further medicinal chemistry yielded a 20 nM inhibitor (compound **16**).[84]

3.41.5.5 Leukocyte Function-Associated Antigen-1 (LFA-1): NMR and Reverse Screening

Fragment-based screening can also be used to substitute undesired parts in a lead molecule with a more suitable component. To this end, the undesired part is cleaved off the lead molecule, and a fragment screen is carried out in the presence of saturating concentrations of the truncated lead molecule. As a consequence, the fragment screen is a 'second-site' screen, since the 'first' binding site is occupied by the truncated lead molecule. This strategy has been successfully applied to inhibitors of adenosine kinase[85] and leukocyte function-associated antigen-1 (LFA-1).[86] In the latter case, the known diaryl sulfide lead molecule **16** was potent but poorly soluble, and showed no oral bioavailability

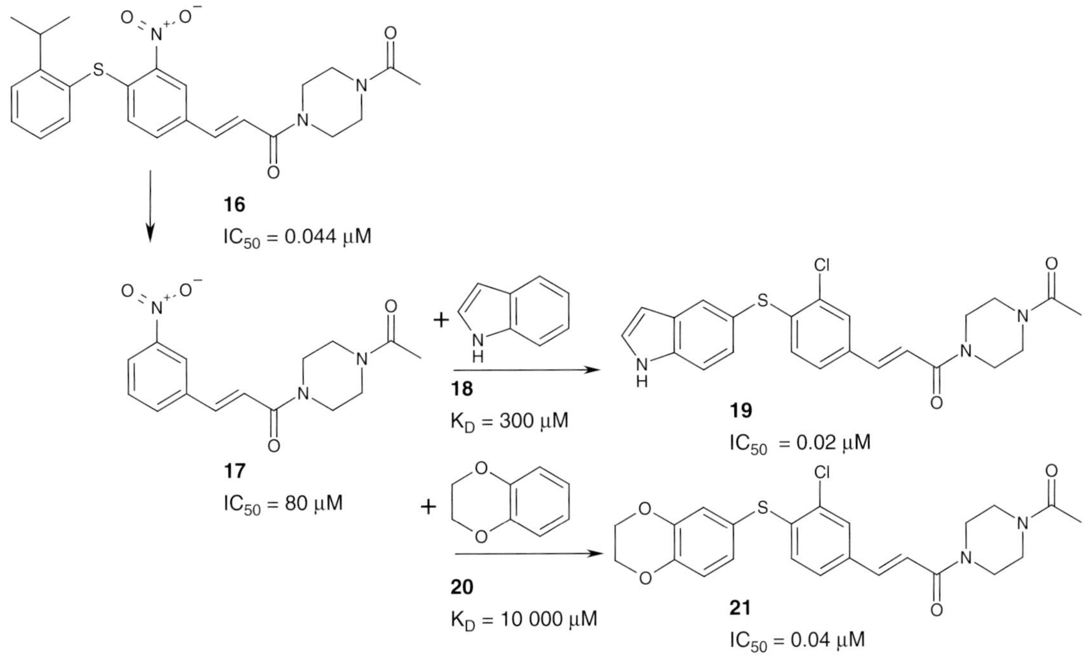

Figure 10 Reverse screening for LFA-1 antagonists.

(**Figure 10**). Based on an NMR-derived model of **16** in complex with LFA-1, the isopropyl group of **16** bound in the vicinity of two lysine residues, suggesting that it could be replaced by more hydrophilic groups. A truncated lead molecule (**17**) without the arylsulfide group was prepared, and a library of 2500 small fragments, each with a molecular weight <150, was screened by NMR in the presence of saturating amounts of **17**. Among the identified fragments were **18** and **20**. When these were linked with the truncated lead molecule **17**, the initial potency of **16** was recovered. Interestingly, both new compounds were equally potent, although the second-site fragment **18** was 30 times more potent than **20**. The solubility of **19** was about equal to **16**, but that of **21** was fourfold better. More significantly, oral delivery of **21** was possibly due to its better solubility.

3.41.5.6 c-Src: Enzyme Assay and Fragment Linking

Fragment-based screening has been introduced in this chapter emphasizing the necessity of a robust biophysical assay and structural information for fragment optimization. This is indeed how FBS is generally performed, with very few exceptions. However, for c-Src kinase, the robust biophysical assay was replaced by a biochemical assay running at high (1 mM) concentrations, and structural information was not used for optimization of the fragment hits. Instead, chemistry-focused approaches were used to rapidly create combinatorial libraries around the fragment hits. To this end, all fragments tested was equipped with an oxime as a common chemical linkage group. Out of 305 tested oximes, 37 were found active when screened at a concentration of 1 mM. A library was then constructed in which all pairwise combinations of these fragments were linked, using five different linker lengths. From this library, linked compound **24** was found to be a 64 nM inhibitor, whereas its unlinked fragments **22** and **23** had only 40 μM affinity (**Figure 11**).[87]

Another recent example of FBS using biochemical assays is provided by substrate activity screening.[88] This strategy is applicable to proteases. In a first step, substrate activity screening identifies weak protease substrates, rather than inhibitors. Based on the weak substrate hits, chemical libraries are rapidly produced and screened for improved substrates. Once the optimal substrate is found, it is readily converted to an inhibitor by replacement of the scissile bond by a known mechanism-based pharmacophore. Substrate activity screening has been applied to the cysteine protease cathepsin S.[88] Here, a library of 105 *N*-acylaminocoumarin candidate substrates was prepared with diverse, low molecular weight *N*-acyl fragments, and tested in a fluorescence-based assay. The fluorogenic properties of this substrate combined with the straightforward solid phase synthesis enable the rapid production of large chemical libraries and their subsequent evaluation in a high-throughput fluorescence assay. Weak substrates from the diverse library were identified, such as compounds **25** and **26** from the 1,4-disubstituted-1,2,3-triazole and the phenoxyacetyl

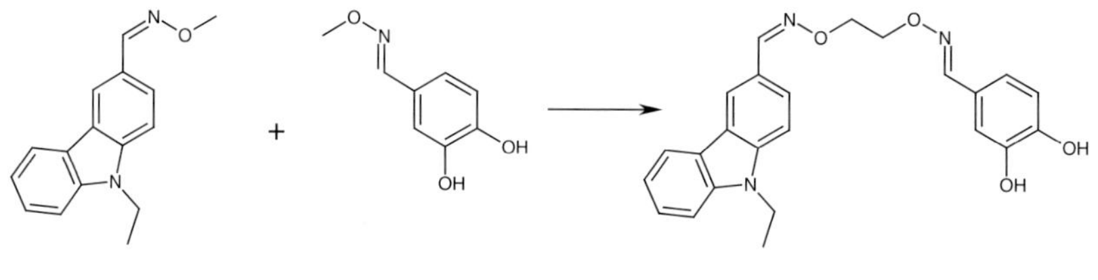

22
IC_{50} = 40 μM

23
IC_{50} = 41 μM

24
IC_{50} = 0.064 μM

Figure 11 The discovery of c-Src inhibitors.

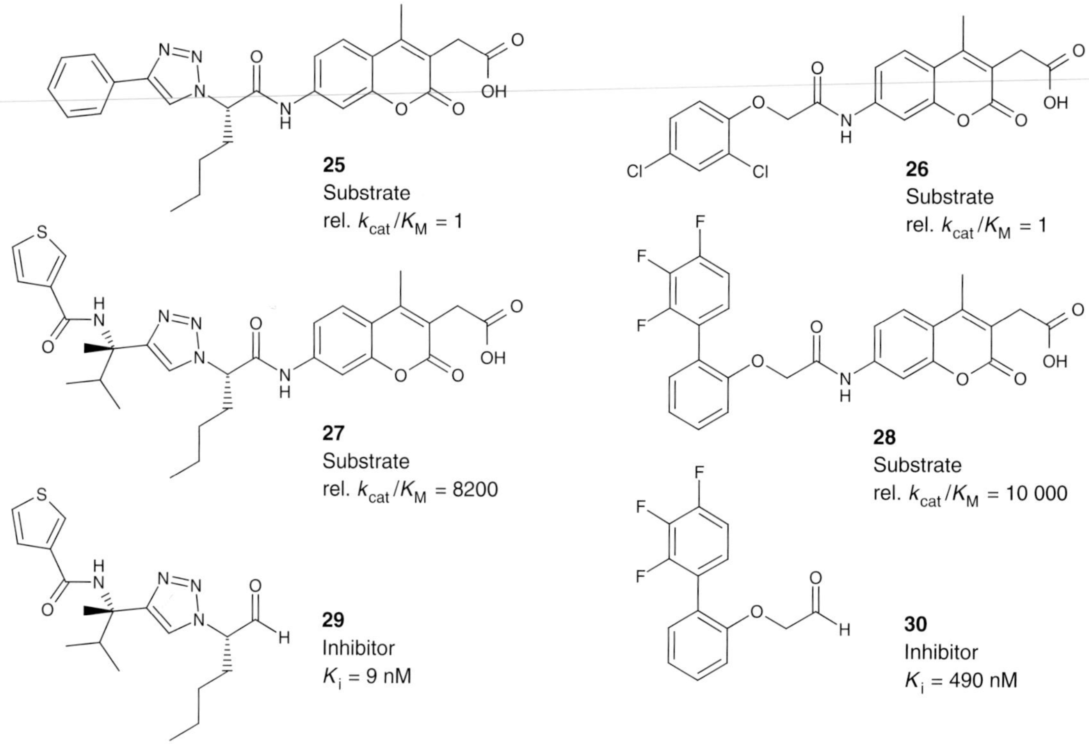

Figure 12 Substrate activity screening used in the discovery of covalent cathepsin S inhibitors.

series (**Figure 12**). Subsequently, focused libraries were prepared and tested. This resulted in 8000- and 10 000-fold increased cleavage efficiency for compounds **27** and **28**, respectively. Conversion from the substrate to the inhibitor was achieved by replacing the aminocoumarin group by a hydrogen atom. The resulting aldehyde then reacts covalently with the active-site cysteine residue, thus covalently inhibiting the enzyme. As in tethering,[62] the reaction between the aldehyde and the cysteine residue is expected to occur the more efficiently the higher the binding affinity of the remaining molecule is. With cathepsin S, the final inhibitors **29** and **30** had K_i values of 9 and 490 nM, respectively, which are remarkable activities given that the inhibitors possess novel scaffolds.[88] The strategy to initially screen for substrates rather than inhibitors alleviates one of the major drawbacks of biochemical screens at high concentrations, namely the occurrence of false positives. When screening for improved substrates, false-positive detection due to aggregation, protein precipitation, or nonspecific binding should not be observed, since both active enzyme and active-site binding are required for fluorescence detection.

3.41.6 Conclusions

Fragment-based screening has gained remarkable attention in the past decade, and now has its place in the lead-finding strategies of most pharmaceutical and several biotechnology companies. FBS is intuitively and conceptually attractive. Neither fully 'irrational' lead finding by HTS nor fully 'rational' lead finding by structure-based drug design has solved all problems in lead finding. FBS contains elements from screening and elements from structure-based design, and can be viewed as the marriage of both techniques.

While FBS generally combines a screening technique and a structural technique for hit identification and optimization, there are several alternatives for screening and structural techniques, and none has proven so superior as to make the others obsolete. As described in this review, NMR spectroscopy, x-ray crystallography, mass spectrometry, SPR, and other technique can be used for screening, while x-ray crystallography and NMR spectroscopy are most suited for obtaining structural information. The choice of techniques depends on the target and its suitability for the various biophysical techniques, and of course also on the availability of the different techniques in the company or laboratory. There is no standard operational procedure for FBS, and the field is still wide open to the development and application of novel methods and techniques.

In the author's judgment, FBS and HTS will be used more synergistically in the future, and both methods will approach each other. An example of the synergistic use of HTS and FBS is reverse screening, in which an 'assembled' hit from HTS is fragmented, and the undesired part is replaced by a fragment identified by FBS, as illustrated in Section 3.41.5.5. Additionally, hit lists from HTS and from FBS will be more synergistically analyzed and exploited in the future. Independently, HTS will pick up and integrate some of the basic ideas from FBS: it will put more emphasis on hit quality than on library size, and the compounds that are screened will tend to become smaller again. The c-Src example in Section 3.41.5.6 is a practical example of how the FBS concept and the HTS concept can be integrated. FBS will play a major role in making lead finding and the entire drug discovery process more efficient and reliable in the future.[89]

References

1. Aherne, G. W.; McDonald, E.; Workman, P. *Breast Cancer Res.* **2002**, *4*, 148–154.
2. Lahana, R. *Drug Disc. Today* **1999**, *4*, 447–448.
3. Shuker, S. B.; Hajduk, P. J.; Meadows, R. P.; Fesik, S. W. *Science* **1996**, *274*, 1531–1534.
4. Erlanson, D. A.; McDowell, R. S.; O'Brien, T. *J. Med. Chem.* **2004**, *47*, 3463–3482.
5. Rees, D. C.; Congreve, M.; Murray, C. W.; Carr, R. *Nat. Rev. Drug Disc.* **2004**, *3*, 660–672.
6. Carr, R. A.; Congreve, M.; Murray, C. W.; Rees, D. C. *Drug Disc. Today* **2005**, *10*, 987–992.
7. Blundell, T. L.; Jhoti, H.; Abell, C. *Nat. Rev. Drug Disc.* **2002**, *1*, 45–54.
8. Stockman, B. J.; Dalvit, C. *Prog. Nucl. Magn. Reson. Spectrosc.* **2002**, *41*, 187–231.
9. Meyer, B.; Peters, T. *Angew. Chem. Int. Ed. Engl.* **2003**, *42*, 864–890.
10. Pellecchia, M.; Sem, D. S.; Wuthrich, K. *Nat. Rev. Drug Disc.* **2002**, *1*, 211–219.
11. Jahnke, W.; Widmer, H. *Cell. Mol. Life Sci.* **2004**, *61*, 580–599.
12. Gill, A.; Cleasby, A.; Jhoti, H. *ChemBioChem* **2005**, *6*, 506–512.
13. Hartshorn, M. J.; Murray, C. W.; Cleasby, A.; Frederickson, M.; Tickle, I. J.; Jhoti, H. *J. Med. Chem.* **2005**, *48*, 403–413.
14. Lesuisse, D.; Lange, G.; Deprez, P.; Benard, D.; Schoot, B.; Delettre, G.; Marquette, J. P.; Broto, P.; Jean-Baptiste, V.; Bichet, P. et al. *J. Med. Chem.* **2002**, *45*, 2379–2387.
15. Swayze, E. E.; Jefferson, E. A.; Sannes-Lowery, K. A.; Blyn, L. B.; Risen, L. M.; Arakawa, S.; Osgood, S. A.; Hofstadler, S. A.; Griffey, R. H. *J. Med. Chem.* **2002**, *45*, 3816–3819.
16. Vieth, M.; Siegel, M. G.; Higgs, R. E.; Watson, I. A.; Robertson, D. H.; Savin, K. A.; Durst, G. L.; Hipskind, P. A. *J. Med. Chem.* **2004**, *47*, 224–232.
17. Wenlock, M. C.; Austin, R. P.; Barton, P.; Davis, A. M.; Leeson, P. D. *J. Med. Chem.* **2003**, *46*, 1250–1256.
18. Jencks, W. P. *Proc. Natl. Acad. Sci. USA* **1981**, *78*, 4046–4050.
19. Olejniczak, E. T.; Hajduk, P. J.; Marcotte, P. A.; Nettesheim, D. G.; Meadows, R. P.; Edalji, R.; Holzman, T. F.; Fesik, S. W. *J. Am. Chem. Soc.* **1997**, *119*, 5828–5832.
20. Murray, C. W.; Verdonk, M. L. *J. Comput.-Aided Mol. Des.* **2002**, *16*, 741–753.
21. Nakamura, C. E.; Abeles, R. H. *Biochemistry* **1985**, *24*, 1364–1376.
22. Shoichet, B. K.; Stroud, R. M.; Santi, D. V.; Kuntz, I. D.; Perry, K. M. *Science* **1993**, *259*, 1445–1450.
23. Bohm, H. J. *J. Mol. Recognit.* **1993**, *6*, 131–137.
24. Verlinde, C. L. M. J.; Rudenko, G.; Hol, W. G. J. *J. Comput.-Aided Mol. Des.* **1992**, *6*, 131–147.
25. Chau, P. L.; Dean, P. M. *J. Comput.-Aided Mol. Des.* **1992**, *6*, 385–396.
26. Stultz, C. M.; Karplus, M. *Proteins* **2000**, *40*, 258–289.
27. Majeux, N.; Scarsi, M.; Tenette-Souaille, C.; Caflisch, A. *Perspect. Drug Disc. Des.* **2000**, *20*, 145–169.
28. Leach, A. R.; Bryce, R. A.; Robinson, A. J. *J. Mol. Graph. Model.* **2000**, *18*, 358–367, 526.
29. Schuffenhauer, A.; Ruedisser, S.; Marzinzik, A. L.; Jahnke, W.; Blommers, M.; Selzer, P.; Jacoby, E. *Curr. Top. Med. Chem.* **2005**, *5*, 751–762.
30. Hann, M. M.; Leach, A. R.; Harper, G. *J. Chem. Inf. Comput. Sci.* **2001**, *41*, 856–864.
31. Lipinski, C. A.; Lombardo, F.; Dominy, B. W.; Feeney, P. J. *Adv. Drug Deliv. Rev.* **2001**, *46*, 3–26.
32. Teague, S. J.; Davis, A. M.; Leeson, P. D.; Oprea, T. *Angew. Chem. Int. Ed. Engl.* **1999**, *38*, 3743–3748.
33. Oprea, T. I.; Davis, A. M.; Teague, S. J.; Leeson, P. D. *J. Chem. Inf. Comput. Sci.* **2001**, *41*, 1308–1315.
34. Congreve, M.; Carr, R.; Murray, C.; Jhoti, H. *Drug Disc. Today* **2003**, *8*, 876–877.
35. Hopkins, A. L.; Groom, C. R.; Alex, A. *Drug Disc. Today* **2004**, *9*, 430–431.
36. Braisted, A. C.; Oslob, J. D.; Delano, W. L.; Hyde, J.; McDowell, R. S.; Waal, N.; Yu, C.; Arkin, M. R.; Raimundo, B. C. *J. Am. Chem. Soc.* **2003**, *125*, 3714–3715.

37. Oltersdorf, T.; Elmore, S. W.; Shoemaker, A. R.; Armstrong, R. C.; Augeri, D. J.; Belli, B. A.; Bruncko, M.; Deckwerth, T. L.; Dinges, J.; Hajduk, P. J. et al. *Nature* **2005**, *435*, 677–681.
38. Zartler, E. R.; Shapiro, M. J. *Curr. Opin. Chem. Biol.* **2005**, *9*, 366–370.
39. Dalvit, C.; Ardini, E.; Flocco, M.; Fogliatto, G. P.; Mongelli, N.; Veronesi, M. *J. Am. Chem. Soc.* **2003**, *125*, 14620–14625.
40. Jahnke, W.; Florsheimer, A.; Blommers, M. J.; Paris, C. G.; Heim, J.; Nalin, C. M.; Perez, L. B. *Curr. Top. Med. Chem.* **2003**, *3*, 69–80.
41. Wishart, D. S.; Case, D. A. *Methods Enzymol.* **2001**, *338*, 3–34.
42. Case, D. A. *Curr. Opin. Struct. Biol.* **1998**, *8*, 624–630.
43. Lian, L.-Y.; Middleton, D. A. *Prog. Nucl. Magn. Reson. Spectrosc.* **2001**, *39*, 171–190.
44. Goto, N. K.; Kay, L. E. *Curr. Opin. Struct. Biol.* **2000**, *10*, 585–592.
45. Hajduk, P. J.; Olejniczak, E. T.; Fesik, S. W. *J. Am. Chem. Soc.* **1997**, *119*, 12257–12261.
46. Dalvit, C.; Fogliatto, G.; Stewart, A.; Veronesi, M.; Stockman, B. *J. Biomol. NMR* **2001**, *21*, 349–359.
47. Mayer, M.; Meyer, B. *Angew. Chem. Int. Ed. Engl.* **1999**, *38*, 1784–1788.
48. Meinecke, R.; Meyer, B. *J. Med. Chem.* **2001**, *44*, 3059–3065.
49. Dalvit, C.; Flocco, M.; Knapp, S.; Mostardini, M.; Perego, R.; Stockman, B. J.; Veronesi, M.; Varasi, M. *J. Am. Chem. Soc.* **2002**, *124*, 7702–7709.
50. Siriwardena, A. H.; Tian, F.; Noble, S.; Prestegard, J. H. *Angew. Chem. Int. Ed. Engl.* **2002**, *41*, 3454–3457.
51. Jahnke, W.; Floersheim, P.; Ostermeier, C.; Zhang, X.; Hemmig, R.; Hurth, K.; Uzunov, D. P. *Angew. Chem. Int. Ed. Engl.* **2002**, *41*, 3420–3423.
52. Dalvit, C.; Fagerness, P. E.; Hadden, D. T.; Sarver, R. W.; Stockman, B. J. *J. Am. Chem. Soc.* **2003**, *125*, 7696–7703.
53. Jahnke, W. *ChemBioChem* **2002**, *3*, 167–173.
54. Brindle, K. M.; Campbell, I. D. *Q. Rev. Biophys.* **1987**, *19*, 159–182.
55. Hofstadler, S. A.; Griffey, R. H. *Curr. Opin. Drug Disc. Dev.* **2000**, *3*, 423–431.
56. Hofstadler, S. A.; Griffey, R. H. *Chem. Rev.* **2001**, *101*, 377–390.
57. Swayze, E. E.; Jefferson, E. A.; Sannes-Lowery, K. A.; Blyn, L. B.; Risen, L. M.; Arakawa, S.; Osgood, S. A.; Hofstadler, S. A.; Griffey, R. H. *J. Med. Chem.* **2002**, *45*, 3816–3819.
58. Ockey, D. A.; Dotson, J. L.; Struble, M. E.; Stults, J. T.; Bourell, J. H.; Clark, K. R.; Gadek, T. R. *Bioorg. Med. Chem.* **2004**, *12*, 37–44.
59. Breuker, K. *Angew. Chem. Int. Ed. Engl.* **2004**, *43*, 22–25.
60. Erlanson, D. A.; Braisted, A. C.; Raphael, D. R.; Randal, M.; Stroud, R. M.; Gordon, E. M.; Wells, J. A. *Proc. Natl. Acad. Sci. USA* **2000**, *97*, 9367–9372.
61. Erlanson, D. A.; Hansen, S. K. *Curr. Opin. Chem. Biol.* **2004**, *8*, 399–406.
62. Erlanson, D. A.; Wells, J. A.; Braisted, A. C. *Annu. Rev. Biophys. Biomol. Struct.* **2004**, *33*, 199–223.
63. Erlanson, D. A.; Lam, J. W.; Wiesmann, C.; Luong, T. N.; Simmons, R. L.; DeLano, W. L.; Choong, I. C.; Burdett, M. T.; Flanagan, W. M.; Lee, D. et al. *Nat. Biotechnol.* **2003**, *21*, 308–314.
64. Buck, E.; Wells, J. A. *Proc. Natl. Acad. Sci. USA* **2005**, *102*, 2719–2724.
65. Buck, E.; Bourne, H.; Wells, J. A. *J. Biol. Chem.* **2005**, *280*, 4009–4012.
66. Markgren, P.-O.; Hamalainen, M.; Danielson, U. H. *Anal. Biochem.* **1998**, *265*, 340–350.
67. Markgren, P.-O.; Hamalainen, M.; Danielson, U. H. *Anal. Biochem.* **2000**, *279*, 71–78.
68. Markgren, P.-O.; Lindgren, M. T.; Gertow, K.; Karlsson, R.; Hamalainen, M.; Danielson, U. H. *Anal. Biochem.* **2001**, *291*, 207–218.
69. Dickopf, S.; Frank, M.; Junker, H. D.; Maier, S.; Metz, G.; Ottleben, H.; Rau, H.; Schellhaas, N.; Schmidt, K.; Sekul, R. et al. *Anal. Biochem.* **2004**, *335*, 50–57.
70. Huber, W. *J. Mol. Recognit.* **2005**, *18*, 273–281.
71. Elg, M.; Gustafsson, D.; Deinum, J. *Thromb. Haemost.* **1997**, *78*, 1286–1292.
72. Neumann, T.; Junker, H.-D.; Keil, O.; Burkert, K.; Ottleben, H.; Gamer, J.; Sekul, R.; Deppe, H.; Feurer, A.; Tomandl, D. et al. *Lett. Drug Des. Disc.* **2005**, *2*, 590–594.
73. McPherson, A. *Introduction to Macromolecular Crystallography*; Wiley-VCH: Chichester, UK, 2002.
74. Drenth, J. *Principles of Protein X-Ray Crystallography*, 2nd ed.; Springer-Verlag: New York, 1999.
75. Pellecchia, M.; Meininger, D.; Dong, Q.; Chang, E.; Jack, R.; Sem, D. S. *J. Biomol. NMR* **2002**, *22*, 165–173.
76. Homans, S. W. *Angew. Chem. Int. Ed. Engl.* **2004**, *43*, 290–300.
77. Clore, G. M.; Gronenborn, A. M. *Curr. Opin. Chem. Biol.* **1998**, *2*, 564–570.
78. Montelione, G. T.; Zheng, D.; Huang, Y. J.; Gunsalus, K. C.; Szyperski, T. *Nat. Struct. Biol.* **2000**, 7, 982–985.
79. Kline, A.; Bravi, G.; Wikel, J. *NMR Newslett.* **1997**, *472*, 13.
80. Becattini, B.; Pellecchia, M. *Chemistry* **2006**, *12*, 2658–2662.
81. Sanchez-Pedregal, V. M.; Reese, M.; Meiler, J.; Blommers, M. J.; Griesinger, C.; Carlomagno, T. *Angew. Chem. Int. Ed. Engl.* **2005**, *44*, 4172–4175.
82. Gill, A. L.; Frederickson, M.; Cleasby, A.; Woodhead, S. J.; Carr, M. G.; Woodhead, A. J.; Walker, M. T.; Congreve, M. S.; Devine, L. A.; Tisi, D. et al. *J. Med. Chem.* **2005**, *48*, 414–426.
83. Petros, A. M.; Dinges, J.; Augeri, D. J.; Baumeister, S. A.; Betebenner, D. A.; Bures, M. G.; Elmore, S. W.; Hajduk, P. J.; Joseph, M. K.; Landis, S. K. et al. *J. Med. Chem.* **2006**, *49*, 656–663.
84. Choong, I. C.; Lew, W.; Lee, D.; Pham, P.; Burdett, M. T.; Lam, J. W.; Wiesmann, C.; Luong, T. N.; Fahr, B.; DeLano, W. L. et al. *J. Med. Chem.* **2002**, *45*, 5005–5022.
85. Hajduk, P. J.; Gomtsyan, A.; Didomenico, S.; Cowart, M.; Bayburt, E. K.; Solomon, L.; Severin, J.; Smith, R.; Walter, K.; Holxman, T. F. et al. *J. Med. Chem.* **2000**, *43*, 4781–4786.
86. Liu, G.; Huth, J. R.; Olejniczak, E. T.; Mendoza, R.; DeVries, P.; Leitza, S.; Reilly, E. B.; Okasinski, G. F.; Fesik, S. W.; von Geldern, T. W. *J. Med. Chem.* **2001**, *44*, 1202–1210.
87. Maly, D. J.; Choong, I. C.; Ellman, J. A. *Proc. Natl. Acad. Sci. USA* **2000**, *97*, 2419–2424.
88. Wood, W. J.; Patterson, A. W.; Tsuruoka, H.; Jain, R. K.; Ellman, J. A. *J. Am. Chem. Soc.* **2005**, *127*, 15521–15527.
89. Jahnke, W.; Erlanson, D. A., Eds. *Fragment-Based Approaches in Drug Discovery*; Wiley-VCH: Weinheim, Germany, 2006; Vol. 34.

Biography

Wolfgang Jahnke is a senior research investigator at the Novartis Institutes for Biomedical Research in Basel, Switzerland. Jahnke leads the fragment-based screening technology platform within the Protein Structure Unit at Novartis. During his 10 years at Novartis, he has contributed to shaping the role and impact of biomolecular nuclear magnetic resonance spectroscopy in drug discovery. He has received several honors, among them the appointment as Scientific Expert and the Leading Scientist Award. Jahnke received his PhD from TU München, working with Horst Kessler on the development and application of novel NMR methods. He subsequently carried out research with Peter Wright at the Scripps Research Institute in La Jolla, USA. He is the co-author of more than 50 scientific publications, review articles, book chapters, and patents.

Comprehensive Medicinal Chemistry II
ISBN (set): 0-08-044513-6

ISBN (Volume 3) 0-08-044516-0; pp. 939–957

3.42 Dynamic Ligand Assembly

O Ramström, Royal Institute of Technology, Stockholm, Sweden
J-M Lehn, ISIS-Université Louis Pasteur, Strasbourg, France

3.42.1	**Introduction**	**959**
3.42.2	**The Dynamic Combinatorial Chemistry Principle**	**960**
3.42.3	**Selection of Building Blocks**	**961**
3.42.4	**Dynamic Diversity Generation**	**961**
3.42.5	**Types of Dynamic Combinatorial Chemistry Systems**	**963**
3.42.5.1	Adaptive Dynamic Combinatorial Libraries	963
3.42.5.2	Pre-Equilibrated and Iterated Dynamic Combinatorial Libraries: Dynamic Deconvolution and Panning	963
3.42.5.3	Catalytic Approach to Dynamic Combinatorial Library Screening	964
3.42.5.4	Dynamic Combinatorial X-ray Crystallography (DCX)	965
3.42.5.5	Pseudodynamic (Deletion) Approach to Dynamic Combinatorial Library Screening	965
3.42.5.6	Tethering Approach to Fragment Screening	965
3.42.6	**Development of the Dynamic Combinatorial Chemistry Concept**	**966**
3.42.7	**Applications in Biological Systems**	**966**
3.42.7.1	Early Examples	966
3.42.7.2	Amino Acid-Based Dynamic Combinatorial Libraries	966
3.42.7.3	Nucleic Acid-Based Dynamic Combinatorial Libraries	968
3.42.7.4	Carbohydrate-Based Dynamic Combinatorial Libraries	969
3.42.7.5	Nonnatural Component Dynamic Combinatorial Libraries	969
3.42.8	**Conclusion and Future Prospects**	**974**
References		**974**

3.42.1 Introduction

Since its formulation in the early 1980s,[1] combinatorial chemistry has matured to become an important instrument in chemistry and biology in general, and in drug discovery processes in particular.[2,3] Combinatorial approaches have in a short time become a potent technology for creating large numbers of structurally related compounds and testing them for desirable properties. Initially developed to produce peptide libraries for screening against antibodies or receptors as candidates of optimal binding properties,[1,4–10] the technology has evolved rapidly to become a powerful technique primarily in drug discovery processes.[2] Numerous methodologies have been designed to produce chemical combinatorial libraries,[3,11–13] and library generation has become a valuable tool in many steps of modern drug development, notably in initial lead generation and refinement.[2,14] Combinatorial approaches are also gaining importance in the development of synthetic processes, in catalyst discovery, and in materials science.[15–17]

A highly schematic presentation of different approaches to compound library generation, inspired by Emil Fischer's lock-and-key metaphor,[18] is shown in **Figure 1**. Over time, chemical combinatorial libraries have been developed primarily using different formats of parallel syntheses. In particular, split–mix methodologies in which large libraries are formed progressively over several synthetic steps (**Figure 1a**) have been widely adopted, and discrete compounds are also frequently prepared in separate compartments, where all species are handled individually (**Figure 1b**). Many protocols have implemented resin-based chemistry, whereas the use of pools of discrete soluble substances in the same compartment has not made such rapid progress. The compounds are usually prepared as discrete, stable entities,

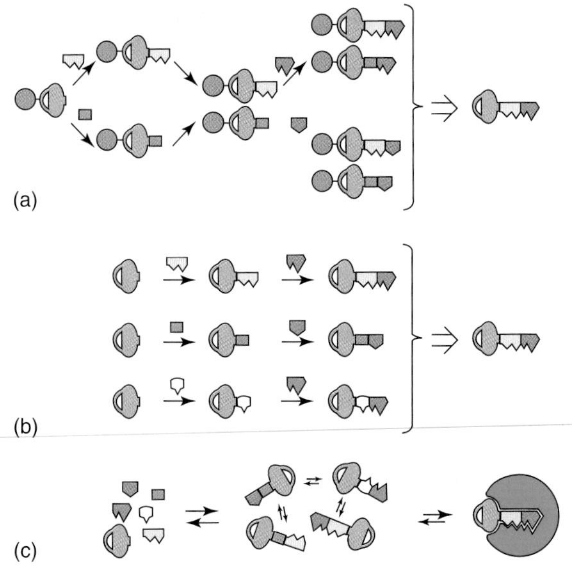

Figure 1 Schematic representation of different approaches to combinatorial chemistry. (a) Split–mix strategy: each reaction carried out in cycles of mixing and separation while compounds are immobilized on resin (gray circle); (b) parallel synthesis: each reaction step is carried out in separate vessels in parallel arrays; (c) dynamic combinatorial chemistry: reversible reactions allow for the generation of a dynamic pool of continuously exchanging constituents. Upon addition of a target species, the best binders may be selected.

virtually insensitive to changes in the environment, and all compounds are subsequently screened one at a time or in small sets using automated systems toward a chosen biological target.

Although these modern combinatorial techniques have allowed the synthesis of very large arrays of compounds in a short time, each individual compound needs nevertheless to be prepared, often over several synthetic steps. In addition, each compound needs to be sufficiently characterized in order to avoid ambiguities in activity screening. Automation techniques in combination with solid-phase synthesis and high-throughput analytical methods have enabled the development of such processes.

If, on the other hand, the target entity per se could be used to select an active ligand directly from a library pool, then the screening process would be greatly simplified and more efficient. In addition, if the library pool itself were able to undergo constitutional changes during this process, adapting itself to the system constraints, then the screening signal would be enhanced, facilitating detection and characterization. Furthermore, if the active species could be analyzed directly while bound to the receptor site, several synthetic steps could be avoided.

Constitutional dynamic chemistry (CDC),[19] and, in particular, its reversible covalent domain, designated as dynamic combinatorial chemistry (DCC),[20–26] have emerged as efficient supramolecular approaches to produce dynamic chemical diversity on both the molecular and supramolecular levels. DCC is a new paradigm in drug discovery that aims at producing flexible and adaptive libraries (**Figure 1c**).[20–24,27,28] In contrast to the regular static approaches to combinatorial chemistry, the library can instead be produced from a set of reversibly interchanging components that maintain dynamics into the system. In this case, a dynamic combinatorial library (DCL) can be generated,[20,21,23,24,26,28,29] where each library member affects, and is affected by, all other surrounding constituents and components. Such an approach is of supramolecular nature, being driven by the interactions of the library constituents with the target site, and relies on reversible reactions or interactions between sets of basic components to generate continually interchanging adducts. This gives access to virtual combinatorial libraries whose potentially accessible (virtual) constituents are all possible latent combinations of the components available.

3.42.2 The Dynamic Combinatorial Chemistry Principle

The DCC process can be divided in three simple steps (see **Figure 1c**):

1. Selection of initial building blocks, capable of connecting reversibly with one another. All building blocks contain functional groups capable of forming reversible bonds to one or several other components in the combinatorial

ensemble, thus allowing the formation of more or less complex molecular entities. This gives rise to true dynamic combinatorial libraries, since each library constituent is a combination of building components, and each member of the library interconverts with every other possible counterpart over time.

2. Establishment of dynamic library generation conditions, where the building blocks are allowed to form interchanging, individual molecular entities. This can be achieved by control of pH, temperature, solvent composition, auxiliary reagents, etc.
3. Subjection of the dynamic library to selection, for example binding affinity to a target protein. This is generally achieved by exposure of the library to the selector at the appropriate conditions. Coupled processes can here be associated to the selection procedure.

The first two steps, and especially the second generative reaction, are essential for the entire process to work, whereas the third selection step is a key corollary of the first two. In addition to the situation where ligands are screened toward a given receptor, (substrate casting), the corresponding situation where a synthetic receptor is selected by addition of a certain ligand can also be envisaged (receptor molding). The dynamic concept also lends itself to the production of virtual combinatorial libraries (VCLs),[30] inasmuch as conditions can be chosen so that the library constituents are only detectable in the presence of the selector.

3.42.3 **Selection of Building Blocks**

For a DCL to be efficiently produced, the building blocks need to fulfil several important characteristics. First and foremost, each element in the library must possess functional groups capable of undergoing reversible exchange to other elements. This function can be either symmetric, so that each interacting partner carries the same functionality (A–A), or orthogonal, where pools of elements carry two different, interacting functionalities (A–B) (**Figure 2**).

The DCL building blocks may further be divided into interactional and organizational units, respectively, in which each element may carry one or both characteristics. The former of these are building up the recognition pattern to the target entity, and the latter are creating the molecular scaffold. The library constituents must cover as completely as possible the geometrical and functional space of potential target sites, in particular by means of recognition groups that are potentially able to interact with the molecular features of the binding site. In common with traditional (static) library design, the choice of interacting groups may rely on experience or on careful study of the crystal structure of the target entity.

The recognition groups need to be organized geometrically for optimal binding to occur, and for this reason organizational units must be incorporated in the libraries. These structural elements may either be part of the recognition groups, or preferably represent separate components based on various types of molecular scaffolds undergoing dynamic decoration by reversible reaction with the recognition components. The organizational components may serve for establishing both the core geometry and the topicity (the number of reversible connections) of the DCL constituents. The use of separate components for organization and interaction allow the implementation of the same set of recognition elements for a variety of targets, while the geometry and topicity can be easily varied, and more pronounced amplification effects may be produced. The easier external control of the library is also an advantage, since tags/reporter groups or handles can be attached to the scaffold.

3.42.4 **Dynamic Diversity Generation**

Dynamic combinatorial libraries can be generated using essentially any type of reversible chemical mechanism, provided the interconverting states can be properly controlled and the final products identified. The most important processes involve molecular/supramolecular interchanges, where chemical bonds are continuously formed and broken. They can make use of a number of reversible connections, of either reversible covalent or noncovalent character.

Functional groups enabling reversible covalent bonds are of special importance in dynamic library generation, and a number of them are presented in **Figure 3**. Reactions at carbonyl groups or derived groups thereof are by far the most

$$A_i + A_j \rightleftharpoons A_i - A_j$$

$$A_i + B_j \rightleftharpoons A_i - B_j$$

Figure 2 Symmetric and orthogonal ditopic dynamic combinational libraries (DCLs).

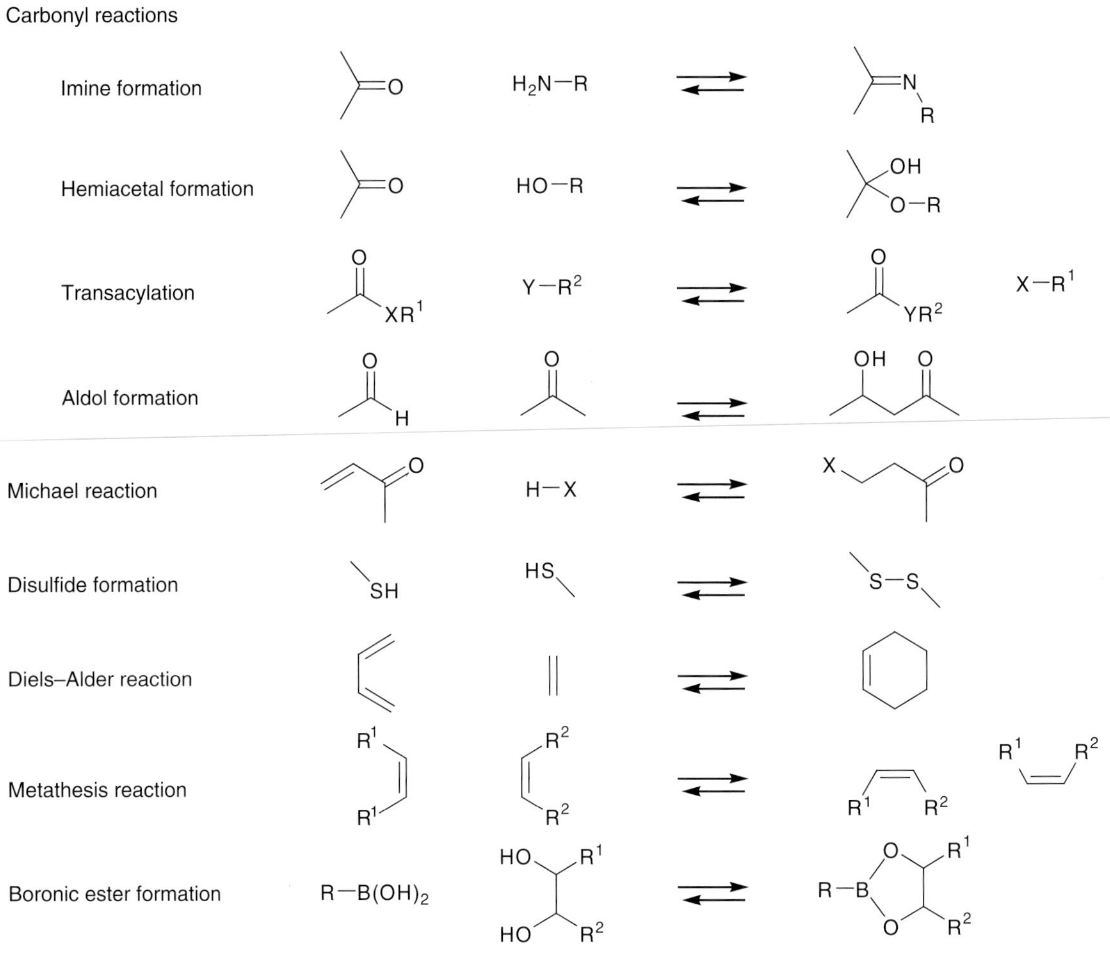

Figure 3 Selection of reversible covalent bond types for potential use in DCC systems.

important class of processes, where especially imine exchange and transacylation reactions have been used. Fine-tuning of the kinetics of both the formation and exchange can be achieved by changing the electronic properties of either the carbonyl compound and/or the nucleophile. For example, primary amines undergo rapid imine formation and exchange with aldehydes, but the equilibrium lies in this case normally toward the starting materials in aqueous medium. With hydroxylamines and acyl hydrazides, on the other hand, the situation is the opposite: the stability of the imines is high whereas the kinetics is slower. Common imines can, however, still be successfully used when coupled to an in situ reduction reaction, converting the imines to stable amines. Oximes and acyl hydrazones are more stable in aqueous solution and can be isolated without reduction and have proven particularly attractive, since they are reversible by mild acid catalysis, and sufficiently stable at neutral to alkaline pH. Conjugate (Michael) addition to carbonyl compounds may also be considered.

Cycloaddition reactions represent another class of reversible processes that may be employed to generate dynamic libraries. In the case of alkene metathesis, advances in catalyst development have enabled the use of this reaction in dynamic systems even in aqueous solutions, and Diels–Alder reactions have recently been exemplified.[31] Exchange reactions at noncarbon centers can also be envisaged, such as alcohol–bor(on)ate, and especially thiol–disulfide interconversion.

Metal coordination represents also a very versatile, often easily controlled class of reversible connections. The stability and the geometry of the complexes can be modulated by choice of metal ion and ligand type. The complexes can be rapidly disassembled by addition of efficient competing binding agents, and can potentially be switched by oxidation/reduction processes. A slight drawback lies in the size of the coordination spheres which may require a rather large binding site. In addition, since some biological target species are sensitive to coordinating ligands, capable of extracting essential metal ion cofactors, special caution in the choice of templating ligands has in this case to be observed.

In many of these reactions, however, the reactivities of the starting materials vary substantially with the substitution pattern of the reactants. This can be brought to an advantage in controlling the reactivities of the exchange reactions, but may also to some degree inhibit the generation of near isoenergetic systems.

A highly desirable feature of molecular/supramolecular systems is their susceptibility to potential fixation of the resulting DCLs, i.e., the freezing of the exchange process. This can be achieved either by changing the surrounding conditions (e.g., pH, temperature, solvent composition) or by adding quenching reagents (e.g., oxidation/reduction). In this way, such libraries can be more easily subjected to various analytical schemes, and the best binders more rapidly identified.

The generation of dynamic combinatorial libraries is not restricted to involving only one type of connection chemistry, but may also rely on two or more reactions or interactions, thus vastly extending the diversity of the library.[32,33] Dynamics can here operate in several dimensions, one for every reversible reaction type used, giving rise to highly complex multidimensional DCLs. The pharmacochemical space may in this case be covered much more efficiently, given a similar number of original building blocks. The chemistries used should preferably be chosen in an orthogonal manner so that they may be controlled separately at will, as for example in combinations of metal coordination and imine exchange. Also, the rapid increase in diversity will allow such libraries to be made from fewer building blocks, resulting in yet more compact, easily controlled sets of compounds.

3.42.5 Types of Dynamic Combinatorial Chemistry Systems

To date, different approaches to DCL generation and screening have been developed, all with a common first reversible generation step but differing in the screening/selection phase. These include: the adaptive approach, the pre-equilibrated approach, the iterative approach, the dynamic deconvolution approach, the x-ray crystallography approach, and the catalytic approach, all of which address different specific challenges of DCL screening in particular. A related, pseudodynamic method is the deletion approach.

3.42.5.1 Adaptive Dynamic Combinatorial Libraries

The original DCC concept addressed the potential of adaptive DCLs, where generation of the library constituents is performed in presence of the selection target in the same compartment (**Figure 4**).[30,34–37] All dynamic characteristics of the system are then utilized, and library adaptation and potential amplification may be obtained. The dynamic characteristics of the DCC system make it sensitive to disturbances and force it to adapt to internal changes or external triggers. These changes may be either physical or chemical in nature, for example addition or removal of components and changes in pH and temperature, potentially causing the system to adjust to the new prerequisites installed by the change. The adaptability of the system, enabled by the reversibility of the processes, also gives it the potential to be amplified according to Le Châtelier's principle. If one constituent in the DCL interacts better than all others with a certain target entity, this constituent will be withdrawn from the equilibrating pool and all components making up this constituent will also be masked by the interaction. The entire system has then to rearrange so as to produce more of this constituent at the expense of the other species in the library. Upon re-equilibration, the most active constituent (the best binder) will thus experience a certain degree of amplification in comparison to the situation where no target molecule was added.

3.42.5.2 Pre-Equilibrated and Iterated Dynamic Combinatorial Libraries: Dynamic Deconvolution and Panning

A second approach, denoted pre-equilibrated DCLs, divides the generation step and the screening step from each other.[38,39] The dynamic libraries are generated under reversible conditions, and the identification/screening is

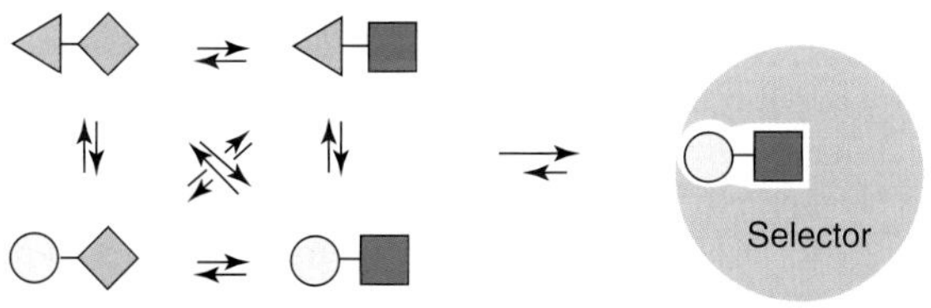

Figure 4 Adaptive DCL format.

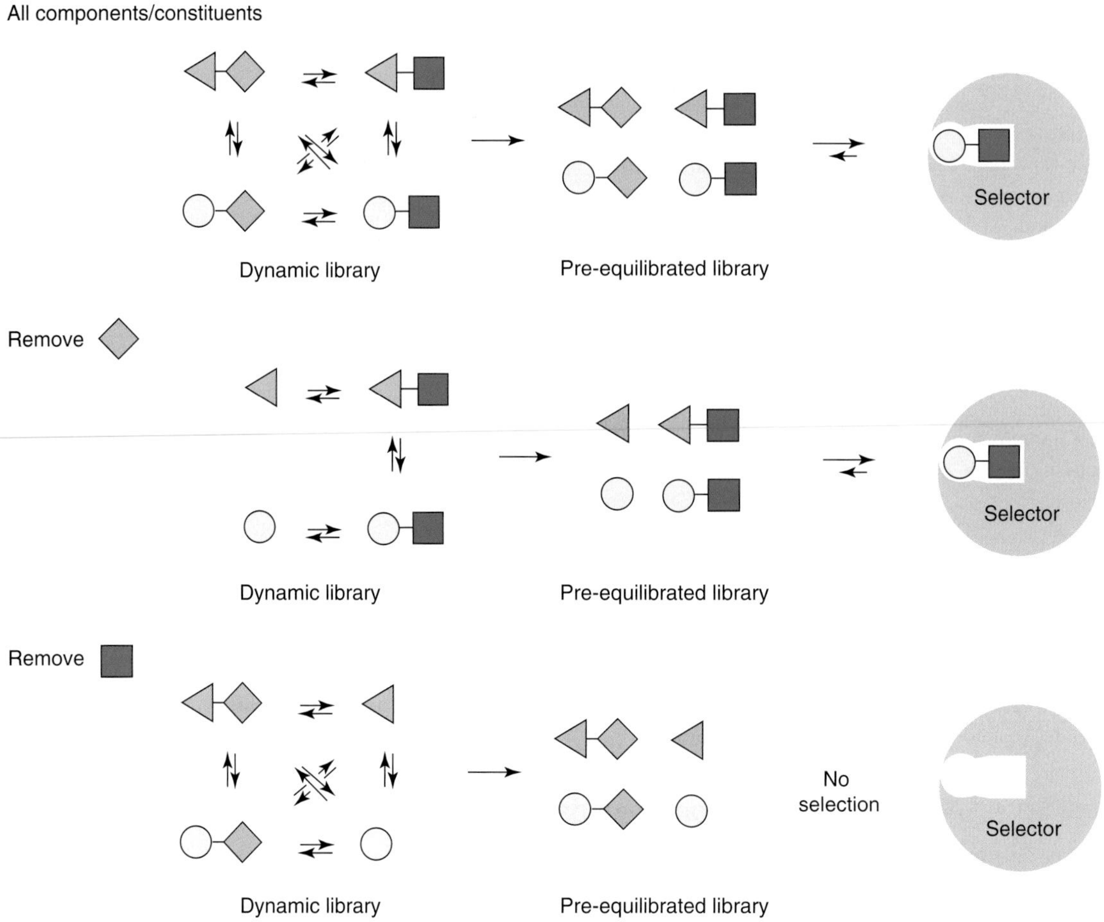

Figure 5 Pre-equilibrated DCL generation and dynamic deconvolution.

subsequently performed under static conditions. Stoichiometric amounts of selector is in this case not required, and this type of protocol is particularly useful when working with sensitive and delicate biological target species, unavailable in large amounts. Screening may be accomplished using common assay methods and identification of active components made from dynamic deconvolution protocols. Because of the dynamic nature of the library, such deconvolutions may be significantly simplified (**Figure 5**). The pre-equilibration procedure is also highly useful when the connecting reaction requires equilibration conditions that are incompatible with the biological target.

The pre-equilibration protocol is however also amenable to constituent amplification when run repeatedly (**Figure 6**). The DCLs are thus generated in one compartment under appropriate dynamic conditions, then in a subsequent step allowed to interact with the target species either in the same reaction chamber or separately. The bound constituents need to be separated from unbound ones using immobilized or entrapped target entities. The unbound species are then retransferred to the reaction chamber, rescrambled, and again allowed to interact with binding site. After several rounds of such a 'dynamic panning protocol,' the accumulated active species may be easily identified.

3.42.5.3 Catalytic Approach to Dynamic Combinatorial Library Screening

Dynamic combinatorial chemistry can also be coupled to catalysis and efficiently used to identify catalyst substrates (**Figure 7**).[40] By direct coupling of the dynamic library to the catalytic action of an enzyme, the best substrate candidates can be identified. The approach is obviously especially applicable to biocatalysis, but is in principle not restrained to enzyme catalysis and may rather be extended to any catalytic system. In particular, catalysts of unknown specificity may be easily mapped with such systems for example in proteomics, and catalyst development campaigns. This also enables screening of complex DCLs without the necessity of using equimolar amounts of targets.

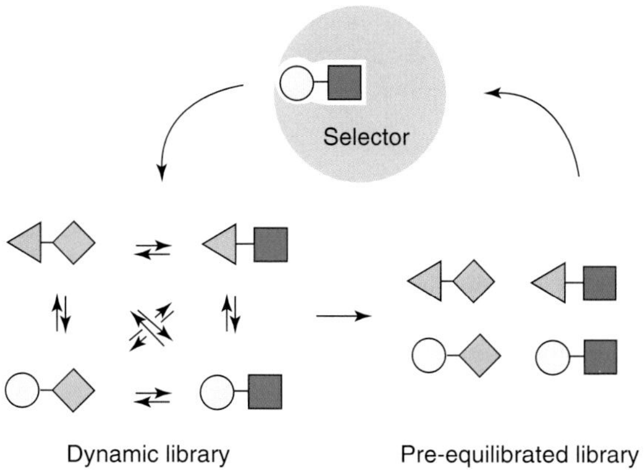

Figure 6 Pre-equilibrated DCL generation and iterative screening.

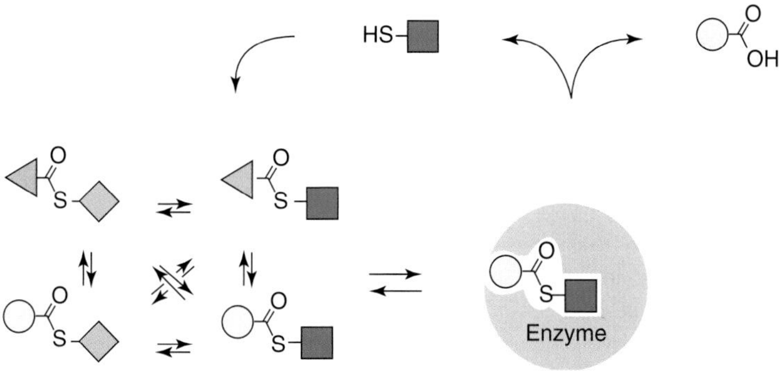

Figure 7 Schematic representation of the catalytic DCL screening process with thiolesters.

3.42.5.4 Dynamic Combinatorial X-ray Crystallography (DCX)

A highly interesting means of directly screening the DCLs in presence of the target was recently presented by Congreve and co-workers.[41] Protein crystals were exposed to combinatorial libraries, and the best constituents were directly bound to the active site in situ. The DCL constituents could be observed directly by x-ray crystallography, and interpretation of electron-density maps yielded information of the most potent inhibitor.

3.42.5.5 Pseudodynamic (Deletion) Approach to Dynamic Combinatorial Library Screening

Rather than selecting the best binder as proposed by the adaptive and pre-equilibrated approaches, Gleason, Kazlauskas and co-workers introduced an interesting alternative method, where the unbound constituents instead are deleted from the libraries.[42] With this concept, the formation and destruction of the libraries are separate irreversible reactions and target identification is enabled by the kinetic deletion of poor binders. Binding to the target entity shields the best binders from the deletion process and the relative ratio of good and poor ligands is increased. As the name implies, the libraries are in this case not truly dynamic in nature, inasmuch as the constituents are not spontaneously reformed in the process. Nevertheless, an improvement of the concept, in which the constituents were continuously resynthesized during the deletion process has also been demonstrated.[43]

3.42.5.6 Tethering Approach to Fragment Screening

A highly interesting related technique is the so-called 'tethering' approach proposed by Erlanson, Wells, and co-workers.[44,45] In order to screen for fragments of ligands to a specific binding site, the fragments are tethered to the vicinity of the binding site through a disulfide bond and the inhibitory activity measured. In cases where a suitable thiol functionality is absent from protein surface, the target is engineered to possess a cysteine residue in a position that otherwise does not interfere with the binding. Due to the reversible nature of the thiol–disulfide interchange, the process can also be performed under dynamic conditions, and in this case the biological target per se is made part of the library. The technique is furthermore especially useful for identifying weakly binding fragments of a potential ligand.

3.42.6 Development of the Dynamic Combinatorial Chemistry Concept

Dynamic combinatorial chemistry has mainly evolved from supramolecular recognition systems, held together by weak, reversible interactions. Many of these systems are composed of mixtures of various interactional partners in solution, and the control and understanding of the outcome of such interactions have been and still are fundamental issues in the field. In many cases, such supramolecular mixtures are reasonably well defined, and all discrete assemblies can be fully identified. Since a combinatorial library can be regarded as a controlled mixture, in which each independent constituent is known and can be addressed individually, many such supramolecular mixtures may actually be considered dynamic library precursors. Early examples of such systems, which may be regarded as DCC predecessors include for example, metal ion assisted imine macrocycle synthesis,[46,47] later nicely extended to a complete DCL study,[48] the reversible assembly of short nucleotide stretches by templating with a complementary oligonucleotide strand,[49] and episelection of boronic ester-based trypsin inhibitors.[50] The current DCC concept, however, arose in part from studies in metal coordination systems,[51] which was later presented in its generality,[21,37,52] and in part from studies involving protease-catalyzed transamidation,[53] as well as macrolactonization studies by transesterification.[54]

3.42.7 Applications in Biological Systems

Biological molecules are at the same time the most interesting and the most challenging target molecules for dynamic combinatorial systems. Often, they are only available in low amounts and are unstable to harsh treatment for longer periods of time. All studies have to be made in well-defined buffer systems, significantly limiting the choice of dynamic chemistry used. Thus, a number of reversible chemistries that are very efficient in organic phase, catalyzed by acids or bases, cannot be used. Instead, reversible reactions occurring under mild conditions have to be employed that do not interfere with the sensitive target molecules. To date, mainly reversible covalent connections such as imines, transacylations, disulfides, and to some extent metathesis as well as metal coordination interactions have been successfully used in these systems.

Even though dynamic systems in combinatorial chemistry have only been developed during a rather short period of time, their potential in biological systems has been demonstrated in several examples (**Table 1**). Several different classes of biomolecules have been targeted, including lectins, enzymes, oligonucleotides, etc., and libraries have been constructed using a range of different building blocks. Most of these have been based on nonnatural construction elements, but attempts have also been made with amino acids, nucleotides, and carbohydrates.

3.42.7.1 Early Examples

As mentioned, predecessors to the DCC concept can be found in biological systems. For example, DNA-templated oligonucleotide synthesis using reversible imine formation and subsequent reduction can be regarded as an early example of dynamic ligand assembly in presence of its target. In this case, however, no library was actually generated.[49] A system perhaps bearing more resemblance to the formulated concept is the so-called episelection of trypsin inhibitors.[50] In this case, a prototype boronic ester inhibitor was crystallized in the presence of trypsin and a few simple alcohols. The enzyme could preferentially select one of the boronic esters formed, to some degree demonstrating the ligand assembly during the process.

3.42.7.2 Amino Acid-Based Dynamic Combinatorial Libraries

The combinatorial chemistry field essentially emanated from libraries based on amino acids, and various oligopeptide library techniques have been thoroughly developed in classical combinatorial chemistry over the years. Peptides are also well suited for automated approaches, inasmuch as the chemistry can be generalized and easily controlled. In dynamic systems, attempts with reversibly interchanging peptide libraries have also been made. Although transamidations normally require rather severe conditions to proceed at any useful rate, the process can sometimes be catalyzed by enzymes.

Table 1 Applications of DCC with biological target species or self-recognition of biological constituents

Target	*Reversible chemistry*	*Library size*[a]	*Hit(s)*	*Reference*
Trypsin	Alcohol–boronate exchange	n.d.	Tripeptidyl-boronate	50
Anti-β-endorphin	Transamidation	n.d.	Peptide (YGG-FL)	53
Fibrinogen	Transamidation	n.d.	Peptides	53
Carbonic anhydrase	Transimination	12	Sulfamoylbenzaldimine	30
GalNAc-specific lectins	Metal coordination	4	Tris-GalNAc	55,56
DNA	Transimination	36	bis-Salicylaldimine–Zn(II) complex	57
DNA	Transimination + metal coordination	n.d.	Salicylaldimine–Zn(II) complex	33
Concanavalin A	Disulfide exchange	21	bis-Mannoside	39
Acetylcholinesterase	Acyl hydrazone exchange	66	bis-Pyridinium	38
RNA	Metal coordination	>27	Salicylamide–Cu(II) complex	58
Ac_2-L-Lys-D-Ala-D-Ala	Metathesis/disulfide exchange	36	bis-Vancomycin	59
Staphylococcus aureus	Disulfide exchange	3828	Psammaplin A analogs	60
Concanavalin A	Acyl hydrazone exchange	484	tris-Mannoside	61
HPr kinase	Acyl hydrazone exchange	440	bis-Benzimidazole	62
β-Galactosidase	Transimination	8	*N*-alkyl-piperidine	63,64
Neuraminidase	Transimination	>40 000	Tamiflu analogs	34
Gramicidin A	Disulfide exchange	3	Disulfide phospholipids	65
Carbonic anhydrase	–	–	Dipeptide	42
Carbonic anhydrase	'Transamidation'	8	Dipeptide	43
Wheat germ agglutinin	Aldol/retro-aldol	3	Sialic acid	66
CDK2	Hydrazone exchange	30	Oxindoles	41
Self-recognition	Metal coordination	11	Three-helix bundles	67
Self-recognition	Metal coordination	11	Three-helix bundles	68
Neuraminidase	Transimination	n.d.	Tamiflu analogs	69
Self-recognition	Disulfide exchange	3	β-Hairpin peptides	70
DNA/RNA	Transimination	n.d.	Oligonucleotide derivatives	71
DNA/RNA	Transimination	15	Oligonucleotide derivatives	72
DNA	Disulfide exchange	6	Peptide derivatives	73
Tripeptide	Disulfide exchange	3	Synthetic receptor	74
Acetylcholinesterase	Transthiolesterification	10	Substrates	40
Lysozyme	Transimination	n.d.	*N*-acetylglucosamine derivative	75

[a] n.d. denotes not determined.

The first true example of DCC in a biological system involved the generation of β-endorphin ligands by protease-catalyzed transamidation (**Figure 8**).[53] A soluble protease exhibiting broad specificity (thermolysin) was used in a mixed water–organic solvent system so as to establish both synthesis and hydrolysis, thus generating libraries of short peptides. When this process was performed in the presence of a target receptor, a monoclonal antibody specific for the *N*-terminus of β-endorphin, binding peptides could be selected and amplification detected. The target antibody was in

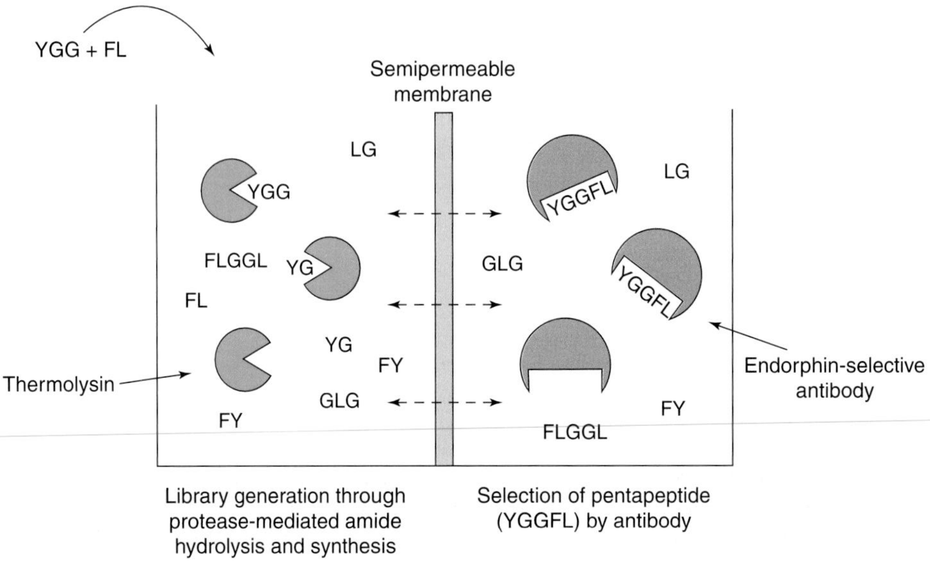

Figure 8 Generation and screening of β-endorphin peptides generated by protease-catalyzed transamidation.

this case protected from digestion by the protease using a semipermeable membrane, demonstrating the advantages of compartmentalization. Another biological entity targeted using the same principle was fibrinogen, also able to select discrete peptides from the transamidation pool.[53]

In addition to common imines, related structures based on the C=N motif have been studied that are more stable in aqueous solution and can be isolated without reduction. Acyl hydrazones have in this case proven particularly attractive, since they are reversible by mild acid catalysis, and sufficiently stable at neutral to alkaline pH. Amino acid containing libraries have thus been assayed by introducing acyl hydrazones as a reversible exchange reaction between amino acid derivatives and other building blocks.[76] In this case, the experience in peptide interaction mode with the target species could be used as a starting point in the library design, where the diversity of the amino acids residues could be combined in new arrangements with other building blocks. Thus, library components containing positively charged groups (e.g., Lys and Arg residues) and aromatic lipophilic groups were preferably chosen. Upon screening of these libraries against the enzyme thrombin, several active combinations were found.

Amino acids as building blocks have also successfully been applied in pseudodynamic systems.[42,43] In these cases, sulfonamide-containing dipeptides were synthesized in one compartment using common (nondynamic) synthetic protocols, and the libraries subsequently transferred to another compartment containing the target entity (carbonic anhydrase). A hydrolyzing enzyme (pronase) was also added, deleting all peptides that were poorly interacting with the carbonic anhydrase. In this way, the enzyme selected the best inhibitor and shielded it from deletion by the pronase, thus increasing the ratio relative to all other peptides.

3.42.7.3 Nucleic Acid-Based Dynamic Combinatorial Libraries

Similar to peptides, the use of nucleotides have experienced tremendous development in combinatorial chemistry, mainly in biological systems but also in various chemical approaches. Available tools and techniques, particularly automated synthetic strategies and means of amplification using the polymerase chain reaction (PCR), have allowed the development of fast and efficient library generation and screening protocols.

Despite this thorough exploitation of nucleic acid building blocks in combinatorial chemistry, and the fact that early examples indicated the potential of using nucleic acids as building blocks, these have only more recently been applied in DCC systems. Besides the fierce competition with a multitude of highly efficient static methods, a major reason for this is the high sensitivity in oligonucleotide recognition. Small changes in backbone geometry can radically reduce duplex formation, and introduction of reversible exchange reactions may severely hamper the interaction with natural DNA and RNA. Nevertheless, nucleic acid-based DCLs have been successfully exemplified. For example, dynamic hydrogels based on guanine-quartet formation were generated from a guanosine hydrazide and a series of aldehydes in presence of metal ions.[77] The hydrogels were truly dynamic and the gelation process could discriminate the best combination of the library.

In another example it was shown that nonnucleic acid residues appended to an oligonucleotide ligand stabilize the complex formed with its nucleic acid target. *N*-terminated oligonucleotides were used as building blocks together with a series of aldehydes, and the resulting imine DCL subjected to reduction in generating the amines. Nucleotide selection of the stabilizing aldehydes was demonstrated, using both DNA and RNA.[71]

These examples have shown that nucleic acid-based DCLs indeed have the expected potential, and may well inspire stronger development in this area in the nearer future.

3.42.7.4 Carbohydrate-Based Dynamic Combinatorial Libraries

Carbohydrate recognition plays an important role in many biological processes, particularly in cell–cell interactions and cell communication, and a multitude of regulatory processes are mediated by carbohydrates as ligands for endogenous lectins.[78,79] Carbohydrates are thus highly attractive tools for targeting biological entities, and many attempts have been made to evaluate the possibility of designing new lectin ligands and enzyme inhibitors.[80–82]

In contrast to other compound groups, for example amino acids and nucleic acids, the preparation of carbohydrate libraries by classical methods has not, however, seen a similarly rapid development.[83–86] There are numerous obstacles in carbohydrate synthesis, and simple automation protocols have been difficult to design. Although dynamic combinatorial chemistry indeed suffers from similar problems, it still offers a complementary route to carbohydrate libraries, perhaps especially in forming dynamic multivalent expression patterns.

In DCC, carbohydrates have been employed as building blocks in several examples. For example, a prototype DCL was constructed using metal coordination in aqueous media (**Figure 9a**).[55,56] *N*-acetyl-D-galactopyranoside (GalNAc) head groups were attached to a bipyridine unit, and upon addition of Fe(II), four different trivalent tris-bipyridine complexes were formed, expressing reversible exchange in solution. When screened against a series of GalNAc selective lectins, including the *Vicia villosa* B_4 and *Glycine max* lectins, one of the complexes was selected and amplified at the expense of the others.

In another example, disulfide interchange was used to generate a bivalent carbohydrate library (**Figure 9b**).[39] As mentioned, disulfides undergo an exchange reaction that is compatible with aqueous media, and that can be scrambled by mild control of the redox properties of the system. This was conveniently shown in the generation of a disulfide-based carbohydrate library, when screened against the common jack bean lectin concanavalin A. A bis-mannoside was selected in the process, as predicted from the lectin specificity, leaving all other structures essentially untouched. Generation of the library was performed at neutral to slightly basic pH, where the redox exchange is rapid, and the exchange could efficiently be stopped by lowering the pH. This bond type is especially advantageous for proteins devoid of any disulfide linkages, such as concanavalin A, and internal disulfide bridges in the proteins studied may present a problem. However, in an example where such a protein was used (acetylcholinesterase) no disturbances were seen.[87] Most likely, internal disulfide bridges are protected in the interior of the protein, and are not easily accessed by thiols at low concentration in the surrounding pool of solvent.

More recently, acyl hydrazone exchange was probed in the generation and screening of an oligotopic carbohydrate library against concanavalin A. A library composed of 474 different mono-, bi-, and tritopic species was easily formed, and upon screening against the lectin a trivalent tris-mannoside could be retrieved from the process.[61]

Lyzosyme was also recently targeted with a carbohydrate-based DCL, based on dynamic reduction imination (**Figure 10c**).[75] 4-Methylumbelliferyl-labeled derivatives of glucose (Glc) and *N*-acetylglucosamine (GlcNAc) were used as scaffolds, decorated with a series of aromatic aldehydes. A GlcNAc-derivative could be selected in the process.

3.42.7.5 Nonnatural Component Dynamic Combinatorial Libraries

Components based on natural building blocks such as amino acids, nucleosides, and carbohydrates have thus been successfully used in a variety of DCC protocols, but the majority of biological applications of the technique has rather made use of nonnatural entities. There are many reasons for this, but a major explanation is that many DCL studies to date are inspired by static systems, libraries or not, initially being proofs of the DCC principle. These systems have also been applied to a number of different biological target entities, ranging from enzymes and receptors, to DNA and bacteria.

As mentioned above, transimination (imine formation and exchange) is an especially attractive reaction that is compatible with water, characterized by a very rapid formation equilibrium and a fast exchange rate. Thus, in a seminal study, an imine system based on three different aldehydes and four different primary amines was used with the enzyme carbonic anhydrase, yielding a library of 12 different constituents (**Figure 10a**).[30] As for the situation with

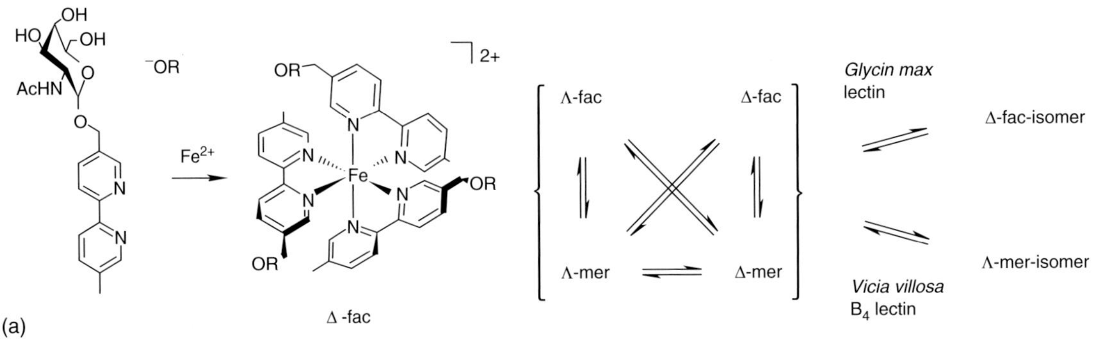

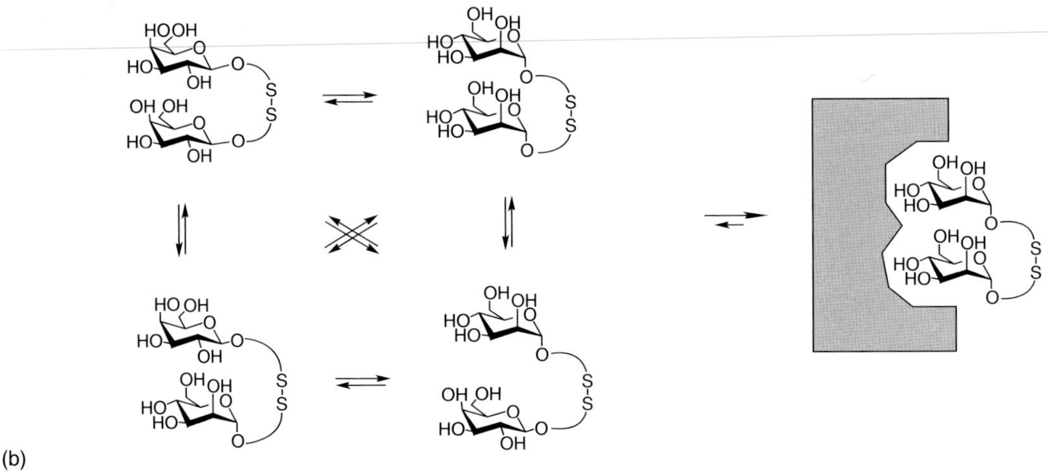

Figure 9 Carbohydrate libraries. (a) Dynamic generation around a metal coordination center. GalNAc-derivatized bipyridine units were coordinated by Fe^{2+}, resulting in four different isomers in dynamic equilibrium; upon addition of GalNAc-specific lectins, certain isomers could be selected. (b) Dynamic library of disulfide-containing carbohydrate structures. A bis-mannoside was selected in the presence of the jack bean lectin concanavalin A.

pseudodynamic DCLs mentioned above, the components of the library were in this case chosen to resemble known inhibitors of the enzyme, in part based of sulfonamide groups, in part based on lipophilic (aromatic) structures. Sulfonamides can in this case interact strongly the Zn(II)-binding site, and the lipophilic moieties can potentially form interactions with the neighboring hydrophobic site. A slight excess of amine was here applied in order to overcome the effect from amino groups at the surface of the enzyme, and a reducing agent (cyanoborohydride) was added to freeze out the formed imines to the corresponding secondary amines. One of the amine–aldehyde combinations, resulting from a sulfonamide aldehyde and a benzylamine, was found to bind preferentially to the enzyme, and its formation was furthermore markedly amplified with respect to the situation in absence of enzyme.

A similar approach of using dynamic reductive amination has also been applied in systems against the enzyme neuraminidase one of the key enzymes in influenza virus activity (**Figure 10b**).[34,69] In the first example, neuraminidase was targeted with a very large imine library (theoretically >40 000 constituents) based on components that were in part inspired by previous structure–activity relationship studies of the enzyme, in part based on a known drug against the enzyme (tamiflu). Potent inhibitors of the enzyme were identified from the DCC campaign, showing resemblance to the tamiflu structure, clearly demonstrating the potential of DCC in drug discovery. The virtual character of the library was in this case very impressive, and the final hit could only be detected upon reduction in the presence of enzyme, thus demonstrating a large amplification effect (>100). In a later study, ketones were used as components rather than aldehydes; these too were efficient in yielding neuraminidase inhibitors.

Enzymes have also been targeted using the acyl hydrazone system in another example (**Figure 11**).[38] A DCL composed of interconverting acyl hydrazones was generated and screened toward inhibition of acetylcholinesterase from the electric ray *Torpedo marmorata*. This enzyme has two binding sites, both of which selective for positively charged functionalities, such as quaternary ammonium groups. One active site is located at the bottom of a deep gorge, and a so-called peripheral site is situated near the rim of this gorge. By bridging the two sites, very efficient inhibitors

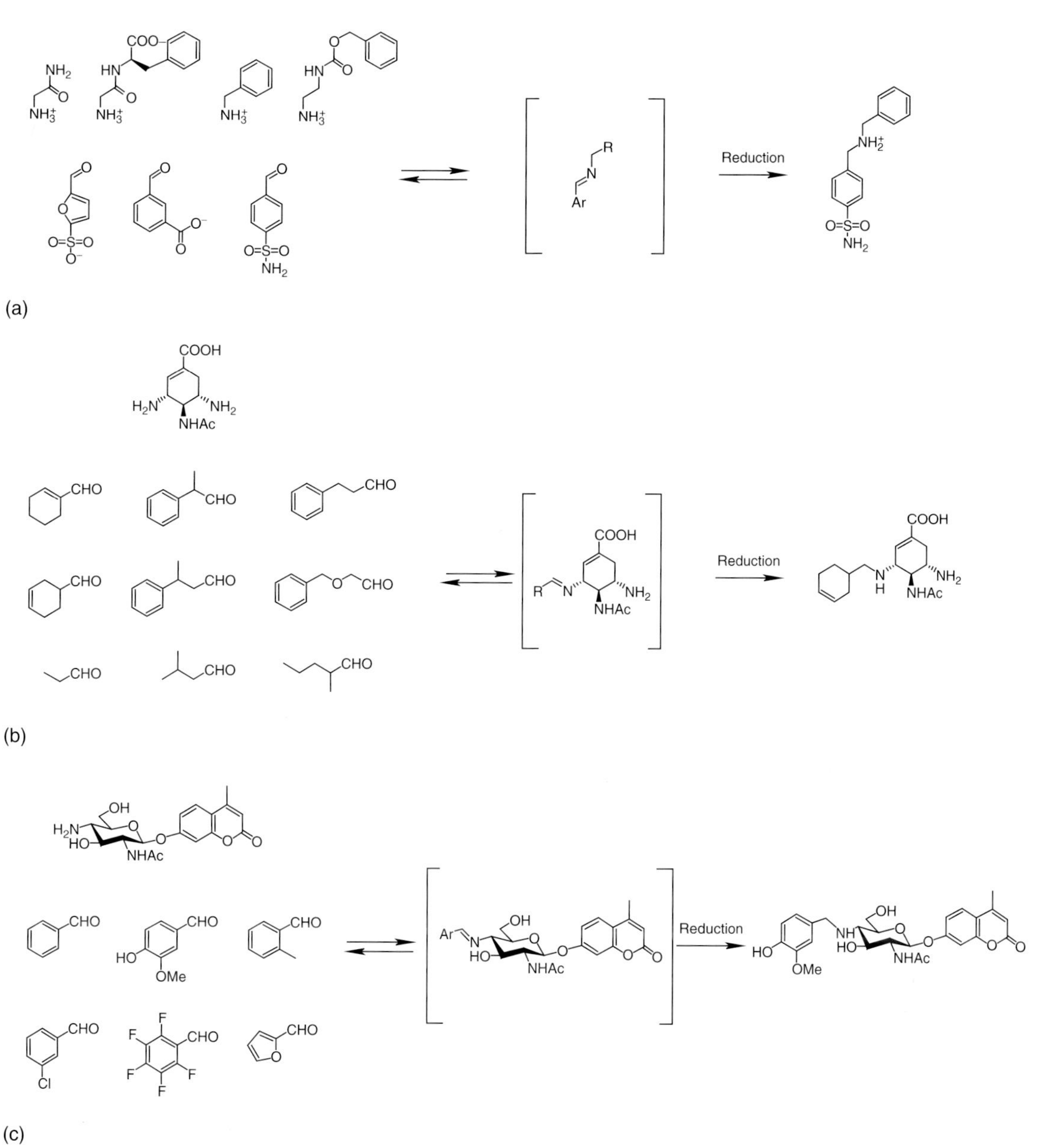

Figure 10 Examples of dynamic reductive amination in DCC. Systems designed against (a) carbonic anhydrase, (b) neuraminidase, and (c) lysozyme.

can be found. Thus, starting from 13 different hydrazide and aldehyde building blocks, some of which containing quaternary ammonium groups, a library of 66 different species could be obtained in a single operation. Of all possible acyl hydrazones formed, active compounds containing two terminal cationic recognition groups separated by a spacer of appropriate length could be rapidly identified using a dynamic deconvolution procedure based on the sequential removal of starting building blocks. A very potent bis-pyridinium inhibitor with a K_i-value in the nanomolar range was selected from the process ($K_i = 1.09\,\text{nM}$, $\alpha K_i = 2.80\,\text{nM}$) and the contribution of various structural features could be evaluated.

Acyl hydrazone exchange was chosen for ditopic heterocycle dynamic libraries constructed against HPr kinase, a serine kinase active in bacterial carbohydrate metabolism.[62] The libraries were in this case composed of all combinations resulting from the dynamic interconversion of 21 hydrazide and aldehyde building blocks, resulting in libraries containing up to 440 different constituents. Of the acyl hydrazones formed in this case, active lead compounds containing two terminal cationic heterocyclic recognition groups separated by a spacer of appropriate structure could be rapidly identified using a dynamic deconvolution procedure.

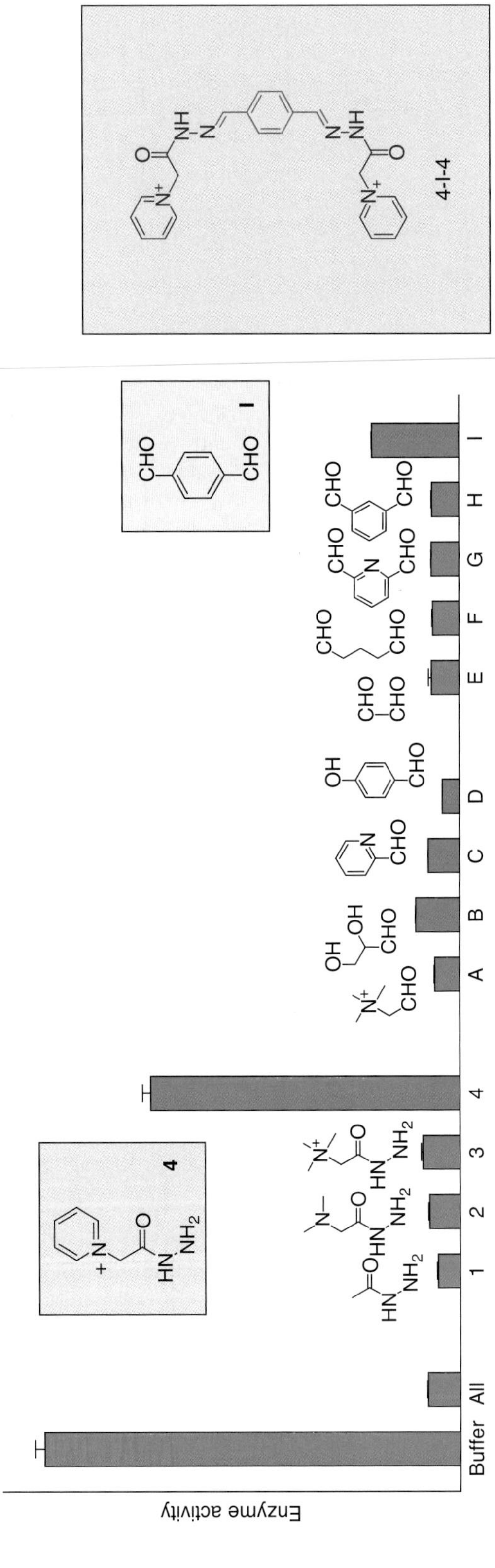

Figure 11 Dynamic deconvolution of acyl hydrazone libraries of potential inhibitors of acetylcholinesterase. Sequential removal of each building block results in identification of individual activities. The combination of **4** and **I** yielded the best inhibition.

Using the catalytic approach to DCL screening, the enzyme acetylcholinesterase was again targeted.[40] Libraries were generated from a pool of thiols and acyl functionalities through reversible transthiolesterification in aqueous media at neutral pH, resulting in the identification of two efficient substrates for the enzyme. The libraries were furthermore screened against a series of hydrolases for rapid substrate identification, clearly demonstrating the differences in selectivity.[88] The results show that transthiolesterification is a useful method to generate dynamic libraries, and that the catalytic approach is highly valuable for substrate screening.

By using the dynamic combinatorial x-ray crystallography approach, the cyclin-dependent kinase 2 (CDK-2) was targeted.[41] The DCL was based on phenyl hydrazone interconversion between a series of phenylhydrazines and a corresponding series of isatins. The library could in this study be generated in DMSO-rich solvent mixture, a prerequisite for not damaging the protein crystals, and the crystals were directly soaked with the DCL solution. By following the electron-density maps in the crystallography setup, the best binders could be identified, and several oxindole structures with inhibitory activities in the nanomolar range were found.

In addition to proteins, oligonucleotides have been targeted with dynamic libraries composed of nonnucleotide building blocks. Selective and high-affinity recognition of RNA and DNA by small molecules is not a trivial task, and combinatorial approaches have been tried to find new ligands.[89,90] In the DCL studies, metal coordination was adopted as a useful means to generate diversity under controlled conditions (**Figure 12**). Metal coordination interactions cover a very wide range of stabilities and a number of them are sufficiently prone to scrambling, in aqueous media. In this situation, Zn(II) was used in conjunction with a library of salicylaldimines and probed against binding to DNA.[57] Combinations of Zn complexes could be easily generated, and one of the species showed indeed a higher binding to the double-stranded oligo(A-T) nucleotide than all other library constituents. A binding constant in the lower micromolar range could also be recorded.

Not only biological macromolecules, but also cell surface ligands or other small biogenic species may be targeted with receptor DCLs. In an example, the bacterial cell wall peptide D-Ala-D-Ala was probed with a dynamic library of vancomycin-derived components.[59] The rationale behind this strategy was that the vancomycin dimer is known to bind to its ligand much more efficiently than the corresponding monomer, and DCLs were made by linking two peptide-binding units by a linker chain. Alkene metathesis and disulfide interchange were used to introduce reversibility in the system, and libraries of up to 36 members were efficiently generated. The final library constituents were subsequently tested for antibacterial activity against a series of vancomycin-resistant bacterial strains, in which several of the library components were found active.

DCC is however not restricted to discrete (macro)molecules, but can also be used to target whole cells. This was demonstrated in another type of large ditopic combinatorial library, composed of more than 3000 constituents, emanating from studies of psammaplin A, an antibacterial agent potent against methicillin-resistant *Staphylococcus aureus* (MRSA).[60] This marine natural product is composed of two identical subunits, linked together by a disulfide bridge, and with this structure as a starting point, a library could be generated from a range of thiol-containing building blocks, and several lead combinations were found active against MRSA.

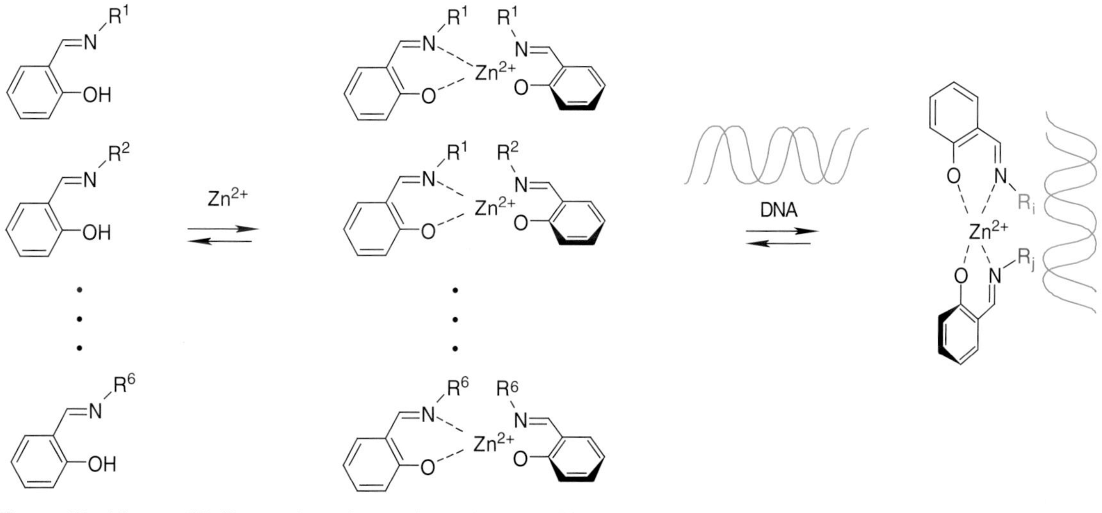

Figure 12 Library of Zn(II) complexes interacting with duplex DNA.

3.42.8 Conclusion and Future Prospects

In conclusion, dynamic combinatorial chemistry is a new tool that can be used to generate libraries of chemical compounds, identified as an efficient means of simultaneously producing molecular diversity, directly addressing the target, and self-screening the library. Not only can such libraries be produced in synthetic organic systems, in themselves highly interesting, but they are also applicable to biological entities such as enzymes and receptors.

However, some challenges still persist, perhaps especially for applications in biological systems and drug discovery. Although DCC protocols have now been demonstrated in many examples, some of which are mentioned above, a wider and improved variety of chemoselective reversible reactions is needed. New, efficient, and controllable chemistries that are entirely compatible with biological conditions, i.e., aqueous buffers, a specific range of temperature and pH, etc., and that show no interference with proteins and other essential components in the system, would add to the DCC toolbox and greatly enhance the chances of success. The constituents of most DCLs that have been developed to date are characterized by having rather extended structures, being composed of reversible connectors of some size, and for this reason more compact libraries are also of interest. Many biological receptor sites are quite limited in size, only capable of accommodating rather small ligands. On the other hand, extended binding sites are highly appropriate targets for dynamic combinatorial libraries, where larger ligands can be used. One such example is protein–protein interactions, of very high interest in biological communication and regulation mechanisms.

The dynamic ligand assembly concept has indeed opened new perspectives in drug discovery processes, offering a versatile and rapid targeting method endowed with self-screening capability. It emphasises the generation of informed diversity pointing toward the development of instructed, 'smart,' combinatorial libraries where the desired target species drives the assembling. The concept can thus be made compatible with drug discovery processes, and may as well be extended to materials science as for instance in supramolecular polymer chemistry.[91] It thus plays a major role in the emergence of adaptive chemistry.[92]

Acknowledgments

Financial support from the Swedish Research Council and the Carl Trygger Foundation is gratefully acknowledged. This study forms part of the EC Marie Curie Research Training Network 'DYNAMIC' (contract number MRTN-CT-2005-019561). The authors gratefully acknowledge the significant financial support of the European Commission.

References

1. Furka, A. Study on the possibilities of searching for pharmaceutically useful peptides. *Notarized report*, Budapest, Hungary, June 15, 1982.
2. Terrett, N. K. *Combinatorial Chemistry*; Oxford University Press: Oxford, UK, 1998.
3. Fenniri, H. *Combinatorial Chemistry*; Oxford University Press: Oxford, UK, 2000.
4. Geysen, H. M.; Meloen, R. H.; Barteling, S. J. *Proc. Natl. Acad. Sci. USA* **1984**, *81*, 3998–4002.
5. Houghten, R. A. *Proc. Natl. Acad. Sci. USA* **1985**, *82*, 5131–5135.
6. Furka, A.; Sebestyen, F.; Asgedom, M.; Dibo, G. *Abstr. 14th Int. Congr. Biochem. Prague, Czechoslovakia* **1988**, *5*, 47.
7. Scott, J. K.; Smith, G. P. *Science* **1990**, *249*, 386–390.
8. Devlin, J. J.; Panganiban, L. C.; Devlin, P. E. *Science* **1990**, *249*, 404–406.
9. McCafferty, J.; Griffiths, A. D.; Winter, G.; Chiswell, D. J. *Nature* **1990**, *348*, 552–554.
10. Lam, K. S.; Salmon, S. E.; Hersch, E. M.; Hruby, V. J.; Kazimierski, W. M.; Knapp, R. J. *Nature* **1991**, *354*, 82–84.
11. Abelson, J. N. *Methods in Enzymology: Combinatorial Chemistry*; Academic Press: London, 1996; Vol. 267.
12. Bunin, B. *The Combinatorial Index*; Academic Press: San Diego, CA, 1998.
13. Jung, G. *Combinatorial Chemistry*; VCH Verlagsgesellschaft: Weinheim, Germany, 1999.
14. Golebowski, A.; Klopfenstein, S. R.; Portlock, D. E. *Curr. Opin. Chem. Biol.* **2001**, *5*, 273–284.
15. Danielson, E.; Devenney, M.; Giaquinta, D. M.; Golden, J. H.; Haushalter, R. C.; McFarland, E. W.; Poojary, D. M.; Reaves, C. M.; Weinberg, H.; Wu, X. D. *Science* **1998**, *279*, 837–839.
16. Xiang, X.-D. *Annu. Rev. Mater. Sci.* **1999**, *29*, 149.
17. Brocchini, S. *Adv. Drug Deliv. Rev.* **2001**, *53*, 123–130.
18. Fischer, E. *Chem. Ber.* **1894**, *27*, 2985–2993.
19. Lehn, J. M. *Proc. Natl. Acad. Sci. USA* **2002**, *99*, 4763–4768.
20. Ganesan, A. *Angew. Chem., Int. Ed. Engl.* **1998**, *37*, 2828–2831; *Angew. Chem.* **1998**, *2110*, 2989–2992.
21. Lehn, J.-M. *Chem. Eur. J.* **1999**, *5*, 2455–2463.
22. Timmerman, P.; Reinhoudt, D. N. *Adv. Mater.* **1999**, *11*, 71–74.
23. Cousins, G. R. L.; Poulsen, S. A.; Sanders, J. K. M. *Curr. Opin. Chem. Biol.* **2000**, *4*, 270–279.
24. Lehn, J.-M.; Eliseev, A. V. *Science* **2001**, *291*, 2331–2332.
25. Cheeseman, J. D.; Corbett, A. D.; Gleason, J. L.; Kazlauskas, R. J. *Chem. Eur. J.* **2005**, *11*, 1708–1716.
26. Ramström, O.; Lehn, J.-M. *Nat. Rev. Drug Disc.* **2002**, *1*, 26–36.
27. Lehn, J.-M. In *Essays in Contemporary Chemistry: From Molecular Structure towards Biology*; Quinckert, G., Kisakürek, M. V., Eds.; Verlag Helvetica Chimica Acta: Zürich, 2001, pp 307–326.

28. Klekota, B.; Miller, B. L. *Trends Biotechnol.* **1999**, *17*, 205–209.
29. Huc, I.; Nguyen, R. *Comb. Chem. High Throughput Screen.* **2001**, *4*, 53–74.
30. Huc, I.; Lehn, J.-M. *Proc. Natl. Acad. Sci. USA* **1997**, *94*, 2106–2110.
31. Boul, P. J.; Reutenauer, P.; Lehn, J.-M. *Org. Lett.* **2005**, 7, 15–18.
32. Goral, V.; Nelen, M. I.; Eliseev, A. V.; Lehn, J.-M. *Proc. Natl. Acad. Sci. USA* **2001**, *98*, 1347–1352.
33. Klekota, B.; Miller, B. L. *Tetrahedron* **1999**, *55*, 11687–11697.
34. Hochgürtel, M.; Kroth, H.; Piecha, D.; Hofmann, M. W.; Nicolau, C.; Krause, S.; Schaaf, O.; Sonnenmoser, G.; Eliseev, A. V. *Proc. Natl. Acad. Sci. USA* **2002**, *99*, 3382–3387.
35. Otto, S.; Furlan, R. L.; Sanders, J. K. *Science* **2002**, *297*, 590–593.
36. Otto, S.; Furlan, R. L. E.; Sanders, J. K. M. *J. Am. Chem. Soc.* **2000**, *122*, 12063–12064.
37. Hasenknopf, B.; Lehn, J.-M.; Kneisel, B. O.; Baum, G.; Fenske, D. *Angew. Chem. Int. Ed. Engl.* **1996**, *35*, 1838–1840; *Angew. Chem.* **1996**, *1108*, 1987–1990.
38. Bunyapaiboonsri, T.; Ramström, O.; Lohmann, S.; Lehn, J.-M.; Peng, L.; Goeldner, M. *ChemBioChem* **2001**, *2*, 438–444.
39. Ramström, O.; Lehn, J.-M. *ChemBioChem* **2000**, *1*, 41–47.
40. Larsson, R.; Pei, Z.; Ramström, O. *Angew. Chem. Int. Ed. Engl.* **2004**, *43*, 3716–3718; *Angew. Chem.* **2004**, *3116*, 3802–3804.
41. Congreve, M. S.; Davis, D. J.; Devine, L.; Granata, C.; O'Reilly, M.; Wyatt, P. G.; Jhoti, H. *Angew. Chem. Int. Ed. Engl.* **2003**, *42*, 4479–4482; *Angew. Chem.* **2003**, *4115*, 4617–4620.
42. Cheeseman, J. D.; Corbett, A. D.; Shu, R.; Croteau, J.; Gleason, J. L.; Kazlauskas, R. J. *J. Am. Chem. Soc.* **2002**, *124*, 5692–5701.
43. Corbett, A. D.; Cheeseman, J. D.; Kazlauskas, R. J.; Gleason, J. L. *Angew. Chem. Int. Ed. Engl.* **2004**, *43*, 2432–2436; *Angew. Chem.* **2004**, *2116*, 2486–2490.
44. Erlanson, D. A.; Braisted, A. C.; Raphael, D. R.; Randal, M.; Stroud, R. M.; Gordon, E. M.; Wells, J. A. *Proc. Natl. Acad. Sci. USA* **2000**, *97*, 9367–9372.
45. Erlanson, D. A.; Lam, J. W.; Wiesmann, C.; Luong, T. N.; Simmons, R. L.; DeLano, W. L.; Choong, I. C.; Burdett, M. T.; Flanagan, W. M.; Lee, D.; Gordon, E. M.; O'Brien, T. *Nat. Biotechnol.* **2003**, *21*, 308–314.
46. Nelson, S. M.; Knox, C. V.; McCann, M.; Drew, M. G. B. *J. Chem. Soc., Dalton Trans.* **1981**, 1669–1677.
47. Nelson, S. M. *Inorg. Chim. Acta* **1982**, *62*, 39–50.
48. Storm, O.; Lüning, U. *Chem. Eur. J.* **2002**, *8*, 793–798.
49. Goodwin, J. T.; Lynn, D. G. *J. Am. Chem. Soc.* **1992**, *114*, 9197–9198.
50. Katz, B. A.; Finer-Moore, J.; Mortezaei, R.; Rich, D. H.; Stroud, R. M. *Biochemistry* **1995**, *34*, 8264–8280.
51. Krämer, R.; Lehn, J.-M.; Marquis-Rigault, A. *Proc. Natl. Acad. Sci. USA* **1993**, *90*, 5394–5398.
52. Hasenknopf, B.; Lehn, J.-M.; Boumediene, N.; Dupont-Gervais, A.; Van Dorsselaer, A. *J. Am. Chem. Soc.* **1997**, *119*, 10956–10962.
53. Swann, P. G.; Casanova, R. A.; Desai, A.; Freuenhoff, M. M.; Urbancic, M.; Slomczynska, U.; Hopfinger, A. J.; Le Breton, G. C.; Venton, D. L. *Biopolymers* **1996**, *40*, 617–625.
54. Brady, P. A.; Bonar-Law, R. P.; Rowan, S. J.; Suckling, C. J.; Sanders, J. K. M. *Chem. Commun.* **1996**, 319–320.
55. Sakai, S.; Shigemasa, Y.; Sasaki, T. *Tetrahedron Lett.* **1997**, *38*, 8145–8148.
56. Sakai, S.; Shigemasa, Y.; Sasaki, T. *Bull. Chem. Soc. Jpn.* **1999**, *72*, 1313–1319.
57. Klekota, B.; Hammond, M. H.; Miller, B. L. *Tetrahedron Lett.* **1997**, *38*, 8639–8642.
58. Karan, C.; Miller, B. L. *J. Am. Chem. Soc.* **2001**, *123*, 7455–7456.
59. Nicolaou, K. C.; Hughes, R.; Cho, S. Y.; Winssinger, N.; Smethurst, C.; Labischinski, H.; Endermann, R. *Angew. Chem., Int. Ed.* **2000**, *39*, 3823–3828; *Angew. Chem.* **2000**, *3112*, 3981–3986.
60. Nicolaou, K. C.; Hughes, R.; Pfefferkorn, J. A.; Barluenga, S.; Roecker, A. J. *Chem. Eur. J.* **2001**, 7, 4280–4295.
61. Ramström, O.; Lohmann, S.; Bunyapaiboonsri, T.; Lehn, J.-M. *Chem. Eur. J.* **2004**, *10*, 1711–1715.
62. Bunyapaiboonsri, T.; Ramström, H.; Ramström, O.; Haiech, J.; Lehn, J.-M. *J. Med. Chem.* **2003**, *46*, 5803–5811.
63. Borman, S. *Chem. Eng. News* **2001**, *79*, 49–63.
64. Thomas, N. R.; Quéléver, G.; Walsh, A. **2001**, unpublished.
65. Shibakami, M.; Inagaki, M.; Regen, S. *J. Am. Chem. Soc.* **1998**, *120*, 3758–3761.
66. Lins, R. J.; Flitsch, S. L.; Turner, N. J.; Irving, E.; Brown, S. A. *Angew. Chem. Int. Ed. Engl.* **2002**, *41*, 3405–3407; *Angew. Chem.* **2002**, *3114*, 3555–3557.
67. Case, M. A.; McLendon, G. L. *J. Am. Chem. Soc.* **2000**, *122*, 8089–8090.
68. Cooper, H. J.; Case, M. A.; McLendon, G. L.; Marshall, A. G. *J. Am. Chem. Soc.* **2003**, *125*, 5331–5339.
69. Hochgurtel, M.; Biesinger, R.; Kroth, H.; Piecha, D.; Hofmann, M. W.; Krause, S.; Schaaf, O.; Nicolau, C.; Eliseev, A. V. *J. Med. Chem.* **2003**, *46*, 356–358.
70. Krishnan-Ghosh, Y.; Balasubramanian, S. *Angew. Chem. Int. Ed. Engl.* **2003**, *42*, 2171–2173; *Angew. Chem.* **2003**, *2115*, 2221–2223.
71. Bugaut, A.; Toulme, J. J.; Rayner, B. *Angew. Chem. Int. Ed. Engl.* **2004**, *43*, 3144–3147; *Angew. Chem.* **2004**, *3116*, 3206–3209.
72. Bugaut, A.; Bathany, K.; Schmitter, J.-M.; Rayner, B. *Tetrahedron Lett.* **2005**, *46*, 687–690.
73. Whitney, A. M.; Ladame, S.; Balasubramanian, S. *Angew. Chem. Int. Ed. Engl.* **2004**, *43*, 1143–1146; *Angew. Chem.* **2004**, *1116*, 1163–1166.
74. Hioki, H.; Clark Still, W. *J. Org. Chem.* **1998**, *63*, 904–905.
75. Zameo, S.; Vauzeilles, B.; Beau, J. M. *Angew. Chem. Int. Ed. Engl.* **2005**, *44*, 965–969; *Angew. Chem.* **2005**, *2117*, 2987–2991.
76. Lohmann, S. Mise au point de bibliothèques combinatoires dynamiques pour l'application à des cibles biologiques. Doctoral Thesis, Université Louis Pasteur, Strasbourg, 2003.
77. Sreenivasachary, N.; Lehn, J. M. *Proc. Natl. Acad. Sci. USA* **2005**, *102*, 5938–5943.
78. Gabius, H. J. *Naturwissenschaften* **2000**, *87*, 108–121.
79. Gabius, H.-J.; Andre, S.; Kaltner, H.; Siebert, H. C. *Biochim. Biophys. Acta* **2002**, *1572*, 163–177.
80. Yarema, K. J.; Bertozzi, C. R. *Curr. Opin. Chem. Biol.* **1998**, *2*, 49–61.
81. Wong, C.-H. *Acc. Chem. Res.* **1999**, *32*, 376–385.
82. Chapleur, Y. *Carbohydrate Mimics: Concepts and Methods*; Wiley-VCH: Weinheim,Germany, 1998.
83. Schweizer, F.; Hindsgaul, O. *Curr. Opin. Chem. Biol.* **1999**, *3*, 291–298.
84. St. Hilaire, P. M.; Meldal, M. *Angew. Chem. Int. Ed.* **2000**, *39*, 1162–1179; *Angew. Chem.* **2000**, *1112*, 1210–1228.
85. Nishimura, S.-I. *Curr. Opin. Chem. Biol.* **2001**, *5*, 325–335.
86. Barkley, A.; Arya, P. *Chem. Eur. J.* **2001**, 7, 555–563.

87. Bunyapaiboonsri, T. Chimie combinatoire dynamique: exploration par des récepteurs biologiques. Doctoral Thesis, Université Louis Pasteur, Strasbourg, 2003.
88. Larsson, R.; Ramström, O. *Eur. J. Org. Chem.* **2006**, 285–291.
89. Alam, M. R.; Maeda, M.; Sasaki, S. *Bioorg. Med. Chem.* **2000**, *8*, 465–473.
90. Guelev, V. M.; Harting, M. T.; Lokey, R. S.; Iverson, B. L. *Chem. Biol.* **2000**, 7, 1–8.
91. Lehn, J.-M. In *Supramolecular Polymers*; Ciferri, A., Ed.; Marcel Dekker: New York, 2000, pp 615–641.
92. Lehn, J.-M. In *Supramolecular Science: Where It Is and Where It Is Going*; Ungaro, R., Dalcanale, E. Eds.; Kluwer Academic Publishers: Dordrecht, the Netherlands, 1999; Vol. 527, pp 273–286.

Biographies

Olof Ramström received his PhD degree in bioorganic chemistry/applied biochemistry from Lund Institute of Technology/Lund University, Sweden, under the supervision of Klaus Mosbach. After a period with Jean-Marie Lehn at the Université Louis Pasteur, Strasbourg, France, he joined the Royal Institute of Technology, Stockholm, Sweden, where he is now associate professor leading a group specializing in supramolecular chemistry and molecular recognition. In 2003, he was awarded a position as Senior Scientist of the Swedish Research Council.

Jean-Marie Lehn received his PhD in 1963 from the University of Strasbourg, France, under the supervision of Guy Ourisson. After a year as a postdoctoral associate with Robert Burns Woodward at Harvard University, he joined the Université Louis Pasteur in Strasbourg, where he became professor of chemistry in 1970. In 1979, he was elected to the Chair of Chemistry of Molecular Interactions at the Collège de France in Paris, and has since then been conducting research at both institutions. In 1987, he was awarded the Nobel Prize in Chemistry.

Comprehensive Medicinal Chemistry II
ISBN (set): 0-08-044513-6
ISBN (Volume 3) 0-08-044516-0; pp. 959–976

INDEX FOR VOLUME 3

Notes

Abbreviations

ADME – absorption, distribution, metabolism, excretion
CSD – Cambridge Structural Database
HCS – high-content screening
HPLC – high performance liquid chromatography
HTS – high throughput screening
MS – mass spectrometry
NMR – nuclear magnetic resonance

Cross-reference terms in italics are general cross-references, or refer to subentry terms within the main entry (the main entry is not repeated to save space). Readers are also advised to refer to the end of each article for additional cross-references – not all of these cross-references have been included in the index cross-references.

The index is arranged in set-out style with a maximum of three levels of heading. Major discussion of a subject is indicated by bold page numbers. Page numbers suffixed by T and F refer to Tables and Figures respectively. *vs.* indicates a comparison.

Names of scientists included in subentries refer to their development role, unless otherwise specified.

Radioactive isotopes are listed under the chemical symbol e.g. ^{131}I

This index is in letter-by-letter order, whereby hyphens and spaces within index headings are ignored in the alphabetization. Prefixes and terms in parentheses are excluded from the initial alphabetization.

Any method, model or other subject, associated with the name of the developer (e.g. name's model) does NOT imply that Elsevier, nor the indexers, have assumed the right to name models/methods after the authors of the papers in which they are described. This is merely a succinct phrase to refer to a model/method developed/described by the relevant author, so that the subentry could be alphabetized under the most pertinent name.

A

A_2, biological activity databases 302
Abel, John J, protein crystallization 885–887
Absorption
  cell-based assays 628
  dielectric heating 840
Absorption/emission- based optical assays 578
Abundant protein depletion, proteomics 31–32
Accelrys database 307
  BIOSTER 307
  Biotransformations 307
  Catalyst 295
  Kinase ChemBioBase 307
  Metabolism 307
Acceptor–donor distances, fluorescence resonance energy transfer 605
Accession number, UniProtKB flatfile 360
Accuracy
  comparative computational modeling 557
  fluorescence screening assays 600–601
  homology-based protein structure prediction 326
  NMR structures, problems 542
ACE inhibitors, discovery/development, protein crystallography, rational drug design 876, 877*F*
^{11}C Acetate, as metabolic probe 665
Acetic acid, microwave heating, loss tangent 840*T*
Acetone, microwave heating 840*T*
Acetonitrile, microwave heating 840*T*
Acetylcholine receptors, molecular probes 664, 664*T*
Acetylcholinesterase inhibitors, dynamic combinatorial libraries 970–971, 972*F*, 973
N-Acetylmuramic acid analogs, solid-phase pool libraries 705*F*
Acetylsalicylic acid (aspirin), molecular representation 236, 237*F*
Acid hydrazone exchange, dynamic combinatorial libraries 970–971
Acrylamides, PASP 792–793
Act 496 automatic synthesizer, PASP 806
Activating protein 1 (AP-1)
  cDNA functional screening 15–16
  node network theory 4

Active transport, cell-based assays 627
Activity-based protein profiling, proteomics 44–45
Acumen Explorer 686*T*, 687
Acute lymphoblastic leukemia (ALL), receptor tyrosine kinase FLT3 5
Acyl hydrazones
dynamic combinatorial libraries 968
exchange, dynamic combinatorial libraries (DCLs) 969
Acyl resorcinol carbamates, solid-phase discrete libraries 709*T*
Adaptive dynamic combinatorial libraries *see* Dynamic combinatorial libraries (DCLs)
Adenine phosphoribosyltransferase (Aprt), knockout animal models 161
mutagenicity testing 161
Adenosine, analogs, solid-phase discrete libraries 709*T*
Adenosine A_{2A} receptor
radioactive assays 586
scintillation proximity assay 586
Adenoviruses, late promoter, mammalian protein expression systems 117
Adenovirus vectors
RNA interference 17
siRNAs 177
Adis R&D Insight 278*T*
ADME
cell-based screening assays *see* Cell-based assays
drug toxicity markers 78
surface plasmon resonance 594
Administrative errors, protein 3D structure validation *see* Protein 3D structure validation
Adrenergic receptors, molecular probes 664, 664*T*
β_2-Adrenergic receptors, pharmacogenetics 58
Adverse drug reactions (ADRs), 'classical' pharmacogenetics 60
Adverse effects, pharmacogenetic effects *see* Pharmacogenetics
Aerosols, siRNAs 182
Affibodies
biomarker identification 73
derivation 73
Affinity, selection, target-oriented screening 141
Affinity chromatography
phenotype-oriented screening 138
protein purification 422
proteomics 31
Affinity tags
gene construct design 880
protein purification 421, 555
Affymetrix, microarrays 88, 89, 93, 97
Age-related macular degeneration (AMD), vascular endothelial growth factor 202
Agglomerative testing, microarrays 93
Agilent, microarrays 89
Aging, *Caenorhabditis elegans* model system 8
Agonist/antagonist screening, GPCRs 637
D-Alanyl-D-peptidase (*Streptomyces R61*) 532–533
structure 534*F*
Albumin, depletion, proteomics 31–32
Alcohol dehydrogenases (ALDs), yeast selection markers 117
Alcohol oxidase 1 (AOX1), yeast selection markers 117
Aldol formation, dynamic combinatorial library generation 962*F*
Algorithms, sequence alignment, bioinformatics 319–320
Alignment methods, regulatory regions, bioinformatics 322
Alignment scores, homology-based protein structure prediction 326
Alkane functionalization, PASP 796, 796*F*
Alkoxyprolines, solid-phase discrete libraries 709*T*
Alkylaminobenzanilides, solid-phase pool libraries 705*F*
Alkylating compounds, pharmacogenetics 61
Alkyl-tethered diisopropylarylsilane linker, solid-phase libraries 738*F*, 740
2-Alkylthiobenzimidazoles, PASP 805
Allergic inflammation, STAT (signal transducer and activator of transcription) 208
Alliance for Cellular Signaling, cell-based assays 643–644
Alteration of function (neomorphic), chemical mutagenesis 13–14
Alternative RNA splicing, STAT (signal transducer and activator of transcription) 204
Alumina, microwave-assisted chemistry 853
Alzheimer's disease, knockout animal models 160
Amaryllidaceae alkaloids, PASP 815, 816*F*, 819
Amberlyst 15 ion exchange resin, PASP 821
American National Standards Institute (ANSI), databases 294
American Type Culture Collection (ATCC) 622
Amide library construction, PASP 798, 799*F*
Amides, PASP 806, 807*F*
Aminamides, solid-phase discrete libraries 709*T*
Amine derivatization, PASP 806, 806*F*
Amino acid-based dynamic combinatorial libraries *see* Dynamic combinatorial libraries (DCLs)
Amino acid probes, small animal imaging 665
Amino acids
metabolism, small animal imaging 672
NMR structure determination 476, 478*F*
Amino alcohol synthesis, PASP 797, 797*F*
Aminoaryls, solid-phase pool libraries 705*F*
2-Aminopyrimidines, solid-phase pool libraries 705*F*
Aminoquinazolinones, solid-phase discrete libraries 709*T*
Aminothiazoles, solid-phase discrete libraries 709*T*
2-Aminothiazole scaffolds, PASP 805–806, 805*F*
Ammonium sulfate, protein crystallization 438, 881
Amplification, pre-equilibrated dynamic combinatorial libraries 964, 965*F*
Amyotrophic lateral sclerosis, treatment, vascular endothelial growth factor (VEGF) 203
Analogs, Ras/Raf protein interaction inhibitor 1 library construction 766
Analog sensitive kinase alleles (ASKA), orthogonal ligand–receptor pairs 218
Analysis software
see specific programs
high-content screening *see* High-content screening
Analytical Abstracts 271*T*
Analytical characterization, solution phase library synthesis 778
Analytical methods, solution phase library synthesis 771
Analytical ultracentrifugation, protein analysis 426
Anandamide, solid-phase libraries 703
Anatomical imaging, small animal *see* Small animal imaging
Androgen ablation imaging 669
[^{18}F]FDG 669
[^{3}H]2-deoxy-D-glucose 670
[^{99m}Tc]hexylmethylpropylene aminexone 670
[^{131}I]IUDR 670
Anecortave acetate 202
Anemia, *Danio rerio* model system 10
Anesthetics, target identification, *Caenorhabditis elegans* model system 8
Angiogenesis
cDNA functional screening 15–16
definition 197

inhibition, cancer treatment 202
probes, small animal imaging 665
vascular endothelial growth factor 198F, 200
Angiotensin-converting enzyme (ACE), crystallization 876, 882–883
Animal models
disadvantages 617–618
microarrays 93
proteomics 43
toxicogenomics 56
Anisotropic diffraction patterns, protein crystallography 441
Annotation
Genome Reviews 356
PDB 376
UniProtKB/Swiss-Prot 359–360
Anomalous dispersion *see* X-ray crystallography, phase problem solutions
Anomalous scattering, protein crystallography, rational drug design 890
ANOVA analysis, microarrays 101
Anti-apoptotic pathways, NF-κB pathway 196
Antibiotic resistance, gene expression vectors 107
Antibiotics, cell culture 622
Antibodies
affinity chromatography 422
development, biomarker identification 72
protein kinase measurement 72
specificity, ELISA 580
Antibody arrays, proteomics 40
Antibody-based purification, proteomics 32
Antibody capture assays, ELISA 579–580, 579F
Antibody staining, HCS labeling techniques 689
Antiemetic drugs, 5-HT_3 ligand gated ion channel blockage 633–634
Antigen capture assays, ELISA 579–580, 579F
Antigen processing, ubiquitin-proteasome pathway 190
Anti-phosphoserine antibodies, biomarker detection 72
Anti-phosphotyrosine antibodies
biomarker detection 72
protein kinase substrate purification 72
Antirhinoviral drugs
see also specific drugs
PASP 804, 804F
Antisense-based therapeutics, vascular endothelial growth factor 203
Antisense oligonucleotides, transgenic animal generation 152
Antisense phosphorodiamidate morpholino oligonucleotides, *Danio rerio* model system 10
Antitrypsin depletion, proteomics 31–32
Anton Paar GmbH, microwave-assisted chemistry 851
pressure measurement 852
Synthos 3000 851, 852F
temperature measurement 852
Anxiety disorders, corticotrophin-releasing factor receptor 93–94
Apopain *see* Caspase-3
Apoptosis
Caenorhabditis elegans model system 8
cell-based assays 629
cellular characteristics 195–196
extrinsic pathway (death receptors) 195–196
apoptosis-inducing factor (AIF) 196
Bak 196
Bax 196
cytochrome *c* 196
imaging, [^{111}In]-DTPA-PEG annexin 668
induction, tumor necrosis factor-α (TNF-α) 196–197
intrinsic pathway 195–196
NF-κB pathway 192, 195
small animal imaging 668
Apoptosis-inducing factor (AIF) 196
Applied Biosystems, microarrays 89
Aptamers
biomarker identification 73
conjugation, 5-bromo-deoxyuridine 73
AQUA 547T
Aqueous solutions, microwave-assisted chemistry 854
Arabidopsis thaliana, genome sequencing 317T
Archive of Obsolete PDB Entries 535–536
Aromatic polyamine amides, solid-phase discrete libraries 709T
Arrayed DNA-based siRNA libraries, RNA interference 18
ArrayExpress, cDNA microarrays 331
Array formats, cDNA functional screening 15–16, 16F
ArrayScan VTi 686T, 687
Artemin, knock-in animal models 160
Artifact correction, fluorescence screening assays 613
Aryl 1,5-enediols, solid-phase discrete libraries 709T
Arylamines, solid-phase discrete libraries 709T
2-Arylaminothiazoles, solid-phase discrete libraries 707F
Aryl boronic acid synthesis, PASP 813
Aryl bromide synthesis, PASP 813
Aryl diamidines, solid-phase discrete libraries 709T
5-Arylidine 4-thiazolidinones, PASP 806–807, 807F
Arylindoles, solid-phase pool libraries 705F
Arylmethyl amidines, solid-phase discrete libraries 709T
Arylmethyl cinnamic acids, solid-phase discrete libraries 709T
N-Aryloxazolidin-2-ones, solid-phase discrete libraries 709T
Aryloxoisobutyric acid, solid-phase discrete libraries 709T
Aryloxyprolines, solid-phase pool libraries 705F
Arylpiperazine derivatives, solid-phase discrete libraries 709T
N-Arylpiperazines, solid-phase discrete libraries 709T
Arylpyrimidine derivatives, solid-phase discrete libraries 709T
Arylsulfonamide hydroxamates
solid-phase discrete libraries 709T
solid-phase libraries 725–726, 726F
Arylsulfonamides, solid-phase discrete libraries 707F
Aryl tertiary amides, solid-phase discrete libraries 709T
Arylthioisobutyric acid, solid-phase discrete libraries 709T
Aryl vinyl sulfones, solid-phase discrete libraries 709T
Asparagine deaminations, heterogeneous protein crystallization 436
Aspartate aminotransferase, structure, monomer oddity 540
Assays
HTS automation 682
quality 91
quality evaluation, HTS *see* High-throughput screening (HTS)
Assembly algorithms, genome sequencing 317
Association of Laboratory Automation 680
Astbury, W T, protein crystallography 450
Asthma, disease basis
NF-κB pathway 197
STATs 208
Asthma treatment
see also specific drugs
imatinib mesylate 3
pharmacogenetics *see* Pharmacogenetics
Asymmetric units
crystals 451
x-ray structures, problems 540
Atom-by-atom substructures, CSD 390–391
Atomic *B*-factors, protein 3D structure 546

Atomic coordinate reliability *see* X-ray crystallography, problems
Atomic occupancies, protein 3D structure validation 512
Atomic order, protein crystallography *see* X-ray crystallography
ATOM record, protein 3D structure validation 513, 514*F*
ATP, measurement, bioluminescent assays 590, 591*F*
ATPase, assays, optical assays 580–581, 582
Audit trails, compound storage/management 561
Augmented Connectivity Molecular Formula, structure/substructure searching 238
Aureus Pharma, commercially available databases 308
Aurilide analogs, solid-phase discrete libraries 709*T*
Aurora-B kinase, RNAi 181
AurSCOPE ADME/Drug–Drug Interaction database 308
AurSCOPE database 308
  GPCR database 308
  hERG Channel database 308
  Ion Channel database 308
Author names, databases 301
Autocalibration, fluorescence resonance energy transfer 605
Autofluorescence
  fluorescence polarization 603
  fluorescence resonance energy transfer 605
  fluorescence screening assays 601, 602*F*, 611–612, 614
Autofocusing, high-content screening 687
Autoimmune diseases, STAT (signal transducer and activator of transcription) 208
Automation
  cell-based electrophysiology assays 624–625
  compound storage/management 566
    Pfizer 565
  high throughput purification 865
  HTS *see* High-throughput screening (HTS)
  LC-MS (peptide-based) proteomics 37, 39
  protein crystallization 440
  protein crystallography model generation 462–463
  protein expression, structural genomics 554–555
  protein purification 555
  protein 3D structure, protein production 415
  sequential synthesis, microwave-assisted chemistry 856
  upgrades, CEM Discover platform microwave-assisted chemistry 845–846
  x-ray crystallography 556
Auto-publishing, structural genomics 558
Avastin *see* Bevacizumab
Azarene pyrrolidin-2-ones, solid-phase discrete libraries 709*T*

B

BAC-based genome sequencing 318
Bacillus subtilis, protein expression systems 116
Bacteria, photoreaction center, crystallization 434
Bacteria-based screening assays, genotoxicity 160–161
Bacterial expression systems 415, 416, 417*F*
  cell lysis 420
  inclusion body purification 415
  NMR protein labeling 428
  protein solubilization 415
  purification 420
Bacterial Targets at IGS-CNRS (BIGS) 552*T*
Baculovirus expression systems 119, 415, 416, 880
  advantages 416–417
  cell lines 119, 417, 417*T*
  media 417*T*
  membrane proteins 115*F*, 119
  NMR protein labeling 429
  structural genomics 554
BAFF/BAFF-R complex, structure 543
Bak, apoptosis 196
Base composition, siRNAs 176
Base length, siRNAs 176
Base-paring patterns, RNA structure prediction, bioinformatics 327
BasePlate, compound storage/management (Pfizer) 569, 571
Basic binding equilibria, NMR *see* Nuclear magnetic resonance (NMR), drug discovery
Batch mode reactions, cell-free expression systems 121
Batch scale-up, microwave-assisted chemistry 856
Bax protein
  apoptosis 196
  ubiquitin-proteasome pathway inhibition 194
Bax repressor identification, *Saccharomyces cerevisiae* model system 8
Bcl-2 protein family, regulation, NF-κB pathway 196–197
BCL-2 transgenes, *Caenorhabditis elegans* model system 652
Bcl-xL, fragment-based drug discovery 950, 951*F*
BCUT (Burden, CAS and University of Texas), chemoinformatics, diversity analysis 255
Beilstein database 269–270, 277, 278*T*
Bengamides, target identification, proteomics 42–43
Benzamide, derivatives, solid-phase discrete libraries 709*T*
Benzazepine-3-one synthesis, solution phase library synthesis 782–783, 784*F*
Benzimidazole compounds
  PASP 804
  solution phase library synthesis 780, 782*F*
Benzimidazolin-2-ones, PASP 805, 805*F*
Benzimidazolones, solid-phase libraries 705*F*, 735, 736*F*
Benzo*[a]*[1,2,3]triazinones, solid-phase discrete libraries 709*T*
Benzodiazepine receptors, molecular probes 664, 664*T*
Benzodiazepines
  chemical variation 925
  PASP 799–800, 800*F*
  solid-phase libraries 703, 706*F*, 736–737, 736*F*
    discrete libraries 709*T*
Benzodiazepinones, solid-phase pool libraries 705*F*
Benzofuran derivatives, solid-phase discrete libraries 709*T*
Benzofurans, 2,3-substituted, solid-phase pool libraries 705*F*
Benzopyrans
  solid-phase discrete libraries 709*T*
  solid-phase pool libraries 707*F*
Benzylguanidine derivatives, solid-phase discrete libraries 709*T*
Berkeley DB 294
Berkeley Drosophila Genome Project, *Drosophila melanogaster* model system 9
Berkeley Structural Genomics Center (BSGC) 552*T*
Bernal, J D, protein crystallography 433–434, 450
Berzelius, Jons Jakob, proteomics 28
Beta-endorphin ligand generation, dynamic combinatorial libraries 967–968, 968*F*
Beta-secretase, knockout animal models 160
Bevacizumab
  cancer treatment 202
  colorectal cancer treatment 202
BHK cells, protein expression systems 117
BIAcore *see* Surface plasmon resonance
1,5-Biaryl pyrazoles, PASP 808, 809*F*

Biaryls, solid-phase pool libraries 705*F*
Bibliographic databases *see* Literature databases
Bibliographic references, databases *see* Databases
Bicistronic approach, *Escherichia coli* protein expression systems 113
Bicyclic amines, solid-phase libraries 706–708, 708*F*
Bicyclic guanidines, solid-phase pool libraries 705*F*
Bicyclic pyrimidines, solid-phase pool libraries 705*F*
BIDD Databases 305
  Drug ADME Associated Protein Database 305
  Drug Adverse Reaction Target database 305
  Kinetic Data of Biomolecular Interaction database 305
  National University of Singapore Bioinformatics Drug Design Group 305
  Therapeutically Relevant Multiple Pathways Database 305
  Therapeutic Target Database 305
Big Blue Mouse 161
Big Blue Rat 161
Biheterocyclic imidazoles, solid-phase pool libraries 705*F*
Biheterocyclic imidazolidinones, solid-phase pool libraries 705*F*
Biheterocyclic imidazolines, solid-phase pool libraries 705*F*
Binary formats, two-dimensional representations 298
Binase, NMR structure determination 498, 499*F*
BIND (Biomolecular Interaction Network Database)
  blueprint databases 305–306
  metalloproteomics 556–557
  optical biosensors 595
  protein function prediction, bioinformatics 337
  regulatory networks 330
Binding ligand location, NMR, drug discovery 917
Bioactivity databases **293–313**
  bibliographic references 296, 300, 302*F*
    author names 301
    BIOSIS 300
    Chemical Abstracts Service 300
    design 300
    Digital Object Identifier (DOI) 300–301
    entrance point 300–301
    hyperlinks 300
    internet references 301
    journal descriptors 301
    MEDLINE 300
    patent descriptors 301
  biological activity 301
    A_2 302
    EC_{50} 302
    ED_{50} 302
    IC_{50} 301
    K_D 302
    K_d 302
    multiple targets 302
    pharmacokinetic data 302
    standard error of mean 302
  biological activity databases 301*F*, 304
    BIDD Databases 305
    Blueprint databases 305
    Chembank 306
    *Ki*Bank 306
    Ligand Info 306
    PDSP K_i 306
    PubChem 307
    ZINC 307
  biological information databases 304
    Enzyme Structures Database (EC-PDB) 304–305
    European Bioinformatics Institute (EBI) 304–305
    GPCRDB 305
    MEROPS 305
    NucleaRDB 305
    Representivity of Target Families in the Protein data Bank (PDBRTF) 304–305
    TCDB Database 305
    TransportDB 305
    Transporter Classification (TC) System 305
    UniProtKB/Swiss-Prot Protein Knowledge database (Swiss-Prot) 304–305
    UniProtKB/TrEMBL 304–305
  chemical information 296, 297
    InChI 298
    molecular fingerprinting 298
    molecular similarity 298
    one-dimensional representations 298
    SMILES 298
    three-dimensional representations 298
    two-dimensional representations 298
  commercially available databases 307
    Accelrys 307
    Aureus Pharma 308
    CambridgeSoft 308
    Eidogen-Sertanty 308
    Elsevier MDL 308
    GVK Biosciences 309
    Jubilant Biosys 309
    Sunset Molecular Discovery Databases 309
  definition 294
  design guidelines 295
    see also specific topics
    biological *vs.* chemical data 296
    searchability 296
  examples 303
    see also specific examples
  external resource linking 301*F*
  hierarchical *vs.* relational 295, 297*F*
  homology models 544
  integration 302
    CABINET 303
    CORBA 303
    external identifiers 303
    heterogeneity 303
    HTML 303
    Java 303
    supplementary linking tables 303, 304*F*
    XML 303
  management systems 295
    Catalyst 295
    ChemFinder 295
    ChemoSoft 295
    definition 294
    JChem Base 295
    MDL Isentris framework 295
    MDL ISIS/Base 295
    THOR 295
    UNITY 295
  open-source DBMS 294
  options 294
    desktop databases 294
    server databases 294
    web-enabled databases 294
  protocol information 296, 299
    variation in 300
  specialized fields 303–304
  structured query language 294
    American National Standards Institute (ANSI) 294
    Data Definition Language (DDL) 294
    Data Manipulation Language (DML) 294
    relational database management system (RDBMS) 294
  target information 296, 299
    basic data 299
    full-text searching 300
    keywords 299
    synonyms 299–300
    Uniform Resource Locators 299
Bioequivalence, pharmacophore identification 242
Biofluids, biomarkers 45
Bioinformatics **315–347**
  definition 316
  domain mapping 879
  drug toxicity markers 78
  expression data 316, 330
    mRNA transcript 330–331
  future work 340
  genetic variation analysis *see* Genetic variation analysis

Bioinformatics (*continued*)
  genome sequencing 316, 317
    see also Genome sequencing
  molecular networks *see* Molecular networks, bioinformatics
  molecular sequence analysis *see* Molecular sequence analysis, bioinformatics
  molecular structure prediction 324
    see also Protein structure prediction; RNA structure prediction, bioinformatics
  protein function prediction *see* Protein function prediction
Biological Abstracts 274
Biological activity
  databases *see* Databases
  determination, cell-based assays 617
Biological connectivity, target validation 4, 4*F*
Biological information databases *see* Databases
Biological membranes
  protein analysis, 1D-SDS PAGE proteomics 39
  proteins
    baculovirus expression vector systems 115*F*, 119
    protein analysis 429
    Semliki Forest Virus expression systems 115*F*, 119
  targets, cell-based assays 618
Biological model, cell-based assays 623
Biological Process, PDB 377
Bioluminescent assays 578, 590
  ATP measurement 590, 591*F*
  hit validation 591, 592*F*
  microfluidic format 590, 591*T*
  small animal imaging 660*T*, 661, 668
BioMarker (database) 309
Biomarkers **69–85**
  advantages 82
  definition 70
    Biomarkers and Surrogate Endpoint Working Group 70
  downstream biological effects 73
    definition 73
    pathway characterization 73
    protein amounts 73
  drug toxicity markers 78
    ADME-related tissue 78
    bioinformatics 78
    genomics 78
    imaging technology 78
    luminescent imaging 78
    metabonomics 78
    oligonucleotide arrays 78
    proteomics 78
  examples 70
  fluorescent imaging 77
    fluorescent resonance energy transfer (FRET) 77
    green fluorescent protein (GFP) 77
    limitations 77
  functional imaging 76
    magnetic resonance imaging 77
    NMR 76
    positron emission tomography 77
  gene expression profiling 74
    experimental *vs.* clinical 75
    gene copy changes 75
    genomic alterations 74
    genomic amplification 75
    genomic deletion 75
    individual transcripts 74–75
    luminescent imaging 77–78
    quantitative reverse transcription polymerase chain reaction 75
  genetically engineered animals 154
  history 70
  identification/validation
    microarrays 104
    proteomics 43
    RNAi (RNA interference) 172
  lead identification 69
  luminescent imaging 77–78
    drug toxicity markers 78
    gene expression regulation 77–78
    luciferase 77–78
  metabonomics 76
    drug toxicity markers 78
    low weight components 76
    NMR 76
    thin layer chromatography 76
    toxicology 76
  preclinical models 76
  protein detection methods 72
    affibodies 73
    antibody development 72
    aptamers 73
    protein kinases 72
    surface plasmon resonance 73
  proteomics *see* Proteomics
  signal transduction 73
  target identification 69, 71
  target modulation 73
    definition 73
    flow cytometry 74
    immunoassays 73–74
    immunohistochemistry 74
  target validation 70, 71, 71*F*
    controls 70–71
    identified target 70–71
    novel target 70–71
  tissue microarrays 74
    biomarker identification 75
    immunohistochemistry 75
    pharmacodynamics 74–75
    reverse phase 75
    in situ hybridization 75
    structure 75
  2D-PAGE 45–46
  utility of 71, 72*T*
    reverse engineering 71–72
Biomarkers, clinical trials 79
  assays 79–80
    ELISA 79–80
    immunostaining 79–80
  endpoints 79
  pharmacodynamics 74–75, 79, 80, 81*F*
    definition 80
    dosage 80–81
  pharmacokinetics 80, 81*F*
    definition 80
  regulatory implications 81
    surrogate endpoints 81–82
  surrogate endpoints 80
    clinical validity 81–82
    definition 80
Biomarkers and Surrogate Endpoint Working Group 70
BioMart, proteome analysis 367, 368
Biomolecular probes
  ^{123}I 666
  ^{124}I 666
  ^{125}I 666
  ^{131}I 666
  ^{111}In 666
Bioreactors
  cell culture 622–623
  genetically engineered animals 151
BIOSIS (Biological Abstracts) 271*T*
  databases 300
BIOSTER software, Accelrys databases 307
Biotage AB, microwave-assisted chemistry 844, 851
  Emrys Advancer batch reactor 851, 851*F*
  Emrys Knowledge database 844
  Initiator reactor 844, 845*F*
Biotechnology and Bioengineering Abstracts 271*T*
Biotech Validation Suite 547*T*
Biotransformations, Accelrys databases 307
Biphenylalanines, solid-phase discrete libraries 709*T*
Biphenylamines, solid-phase discrete libraries 709*T*
Biplots, microarrays 92, 93–94, 94*F*
Birefringent crystals, protein crystallography 441
Bisarylthio *N*-hydroxypropionamides, solid-phase discrete libraries 709*T*
BL21 Origami strains, disulfide bond expression 116
BLAST database
  sequence alignment, bioinformatics 320
  siRNA design 176
Blindness, *Danio rerio* model system 10
Blood, cell-based assays 620

Blood–brain barrier (BBB), permeation, chemoinformatics, ADME property prediction 259
BLOSUM matrices, sequence alignment, bioinformatics 319–320
Blueprint databases 305
  Molecular Modeling Database (MMDB) 305–306
  SeqHound 305–306
  SMID (Small Interaction Database) 305–306
Bond angles
  protein 3D structure validation 511, 515
  x-ray structures, problems 537
Bond-by-bond substructures, CSD 390–391
Bond lengths
  CSD applications 400
  errors, ligand structure 538
  protein 3D structure validation 511, 515
  x-ray structures, problems 537
Bone morphogenic protein-9 (BMP-9), cDNA functional screening 15–16
Boronic ester-based trypsin inhibitors, dynamic combinatorial chemistry 966
Boronic ester formation, dynamic combinatorial library generation 962*F*
Bortezomib 194
  proteasome inhibition 194
  Ras/Raf/MAP kinase pathway activation 195
Botulinum A toxin, light chain structure, data overinterpretation 535, 536*F*
Boundary location, protein crystallography model generation 462, 463*F*
Bound ligand identification, surface plasmon resonance 593
Boyden chamber assays, cell-based assays 627–628
^{75}Br, PET 664*T*
^{76}Br
  PET 664*T*
  proliferation probes 665
BRAF studies, zebrafish (*Danio rerio*) 652–653
Bragg, Sir William Henry, protein crystallography 449–450
Bragg, William Lawrence, protein crystallography 449–450
BrdU incorporation, HCS cell cycle 691–692
Breast cancer
  disease basis, STATs 207–208
  pharmacogenetics 61
BRENDA database, metabolic networks 328–329
Brenner, Sydney, *Caenorhabditis elegans* as model system 8
Bright-field microscopy, HCS labeling techniques 686*T*, 688
5-Bromo-deoxyuridine, aptamer conjugation 73
Brookhaven National Laboratory, PDB 373
Browser, Integr8 368
Browsing capabilities, Protein Data Bank (PDB) *see* Protein Data Bank (PDB)
Building block selection, dynamic combinatorial chemistry (DCC) *see* Dynamic combinatorial chemistry (DCC)
Bump detection, protein 3D structure validation 517
1-Butanol, microwave heating 840*T*
2-Butanol, microwave heating 840*T*
Butenolides, solid-phase discrete libraries 709*T*
Byproduct scavenging, PASP 826

C

^{11}C
  PET 664*T*
  pharmacokinetic measurements 672–673
CAAT box, mammalian protein expression systems 117
CABINET, database integration 303
Caco-2 cell models, cell-based assays 628
Cadmium, protein crystallization 885
Caenorhabditis elegans model system 8, 652
  advantages 8
  aging 8
  anesthetic target identification 8
  apoptosis 8
  BCL-2 transgenes 652
  cell-cell signaling 8
  genome sequencing 647–649, 648*T*, 652
  history 647, 652
  muscular dystrophy model 8
  neurogenesis 8
  RNA interference technology *see* RNAi (RNA interference)
  small-molecule screening 652
Calcium-based screening assays 625–626, 626*F*
Calmodulin–C20W peptide complex, NMR structure determination 492, 493*F*, 497–498, 498*F*, 500*F*, 501
Cambridge Crystallographic data Centre (CCDC) *see* Cambridge Structural Database (CSD)
CambridgeSoft Corporation
  ChemFinder 295
  commercially available databases 308
Cambridge Structural Database (CSD) **389–410**
  advantages 399
  applications 398
    bond length values 400
    conformational analysis 400
    conformational preferences 400
    conformational variation 400*F*, 401
    crystal structure relevance 399
    intermolecular interactions *see below*
    intramolecular geometries 400
    metal coordination 401
    MIMUMBA 401
    valence angle values 400
  chemoinformatics, three-dimensional structure 240
  ConQuest Search 393, 394*F*
    conformational analysis 394*F*, 400, 400*F*
    query combinations 393–395
    3D geometrical parameters 393
    3D substructures 393
    2D substructures 393
  data acquisition 391
    Crystallographic Information File (CIF) 391–392
    personal communications 391–392
  drug design/development 399
    conformational information 399
    intermolecular interactions 399–400
    structure correlation principles 399
  information content 390, 390*T*, 391*F*
    atom-by-atom substructures 390–391
    bond-by-bond substructures 390–391
    PreQuest 391, 393
    2D structural representations 390–391
    validation 391
  intermolecular interactions 399–400, 401, 406
    computational paradigms 402
    halogens 404
    hydrogen bonds 401–402
    nonbonded interactions 402
    non-hydrogen bonds 404
    prediction 406
    strong hydrogen bonds 402, 403*F*, 404*F*
    SuperStar program 406, 407*F*
    weak hydrogen bonds 404
    X-SITE program 406
  IsoStar (intermolecular interactions) 395, 398, 398*F*
    applications 405
    contoured surface generation 405
    scatterplots 405
    'unlikelihood' 403*F*, 405, 405*F*

Cambridge Structural Database (CSD) (*continued*)
knowledge-based libraries 395
Mercury Visualizer 395, 395*T*, 396*F*
hydrogen bonds 395
nonbonded interactions 402, 402*F*
Mogul (intramolecular geometry) 395, 396
geometric features 396, 396*F*
hyperlinking 396–398
search tree generation 396
protein–ligand docking programs 406
FlexX 406
GOLD 406
LUDI 407
rational drug design 399
software components 393
statistical overview 392, 392*T*
growth rate 393*F*
number of entries 392
Vista data analysis 394*F*, 395
conformational analysis 394*F*, 400, 400*F*
WebCite 398
cAMP assays, EIA, G protein-coupled receptors (GPCRs) 634–636
cAMP response element-binding protein (CREB), identification, cDNA functional screening 15–16
Camptothecin (CPT), structure 700*F*
Cancer
see also Tumors
characterization, proteomics 43
genetically engineered animals *see* Genetically engineered animals
treatment
angiogenesis inhibition 202
bevacizumab 202
vascular endothelial growth factor (VEGF) *see* Vascular endothelial growth factor (VEGF)
Cancer cell targeting, RNAi (RNA interference) 184
Cancerlit 271*T*
Canonical activation pathway, NF-κB pathway 192, 193*F*
Canonical representations, structure/substructure searching 238
CAOLD, database interfaces 283
CAplus 271*T*, 273
database interfaces 283
Capoten, design/development 876–877, 877*F*, 929, 930*F*
Captopril, design/development 876–877, 877*F*, 929, 930*F*
Carbohydrate-based dynamic combinatorial libraries *see* Dynamic combinatorial libraries (DCLs)
Carbonic anhydrase, structure, coordinate errors 532
Carbonic anhydrase inhibitors, dynamic combinatorial libraries 969–970
Carbonyl groups, dynamic combinatorial library generation 961–962, 962*F*
Cardiovascular disease (CVD)
biomarkers 70
characterization, proteomics 43
Cardiovascular disease treatment, vascular endothelial growth factor 203
Carotenoid-processing enzyme
monotropic membrane protein crystallization 435, 435*F*
protein crystallization 435
Carpanone, PASP 818, 819*F*
CASP (Critical Assessment of Techniques for Protein Structure Prediction)
homology models reliability 544
protein structure prediction 324–325
Caspase-3
fragment-based drug discovery 950
optical assays 581, 581*F*
CASREACT 281–282, 282*T*
CAS registry, structure databases 280
Catalog databases 279
CATALYST program
Accelrys 295
database management systems 295
Catalytic screening dynamic combinatorial libraries *see* Dynamic combinatorial libraries (DCLs)
Catch-and-release principle, PASP 795, 796*F*, 799, 800*F*, 826
CATH (Class, Architecture, Topology and Homologous superfamily) 374
protein function prediction, bioinformatics 336–337
search capabilities 376–377
Cathepsin D, disease role 808
Cathepsin D inhibitors, PASP 808
Causative-acting drugs, pharmacogenetics 54
CBNB (Chemical Business News base) 271*T*, 275
CC (comments), UniProtKB flatfile 360, 361*F*
CCR5, pharmacogenetics 61
CCR5 antagonists, solid-phase libraries 727–735, 734*F*
cDNA functional screening 14
angiogenesis 15–16
array formats 15–16, 16*F*
'gain-of-function' screens 15, 15*F*
JAK gene isolation 14
limitations 15*F*
Mammalian Gene Collection 15–16
metastasis 15–16
NF-κB signaling pathways 14, 15–16
osteogenesis 15–16
overexpression pools 15
p53 15–16
signal transduction pathways 15–16
STAT gene isolation 14
TORC1 identification 15–16
Wnt signaling pathway 15–16
cDNA microarrays 88–89, 331
advantages 90
definition 331
disadvantages 90
experiment configuration 331
ArrayExpress 331
Gene Expression Omnibus (GEO) 331
MIAME standard 331
National Center for Computational Biology (NCBI) 331
normalization 331
quality control 331
low-level analysis 331
probe classification 332
gene co-expression 332
metabolic pathways 332
regulatory pathways 332
probe numbers 331
sample classification 332
decomposition 332
methods 332
single-channel *vs.* multi-channel 89
cDNA precursors, siRNAs 172
CE, protein function prediction, bioinformatics 336–337
CEABA-VTB 271*T*
Celecoxib, structure 808
Cell(s)
differentiation, phenotype-oriented screening 136
extraction/disruption, proteomics 30, 31
fusion, cell-based assays 618–619
HCS labeling techniques 688
immortalization, Epstein–Barr virus (EBV) 620
lysis, bacterial expression systems 420
proliferation
imaging 671
small animal imaging 671
siRNA targeting 182, 183
survival, HCS 691
Cell-based assays **617–646**
active transport 627
ADME 623, 628
absorption 628
excretion 629
metabolism 628, 629
advantages 618, 623
agonists *vs.* antagonists 618
biological model 623
endpoint measurements 623
membrane-bound targets 618
real-time measurements 618
signal pathways 618
test protocols 623
biological activity determination 617
calcium-based 625–626, 626*F*

cell culture 622
antibiotics 622
bioreactors 622–623
extracellular matrices 622–623
media 622
organ system mimics 623
supplements 622
cell migration/chemotaxis 627, 628
Boyden chamber assays 627–628
membrane inserts 627–628
cell sources/selection 619, 620*T*
blood 620
cell fusion 618–619
collections 622
history 618–619
immortalization techniques 620
insect cells 619
naturally proliferating cells 620–621
primary cells 619–620
transgenic lines 621–622
tumors 619
validation 620–621
variation of responses 620–621, 621*F*
Xenopus oocytes 619
cell surface transmitters 624*T*
co-culture, primary cells 619–620
early drug development 617–618
EIA 631
competitive assays 631
ELISPOT 632
enzymes used 632
fluorescence resonance energy transfer (FRET) 632
methodology determinants 632
radioimmunoassays *vs.* 632
sandwich assays 631
transcription factor production 632
electrophysiology 624
rubidium-86 625
throughput constraints 624–625
voltage clamp 624–625
examples 633
see also specific assays
flow cytometry 629, 630*F*
intracellular protein staining 630
fluorescent dyes 625
agonists *vs.* antagonists 627
drug interference 625
false positives 625
fluorescence resonance energy transfer (FRET) 618, 625
GPCRs 627
hERG potassium channel 625–626, 626*F*
future work 641
Alliance for Cellular Signaling 643–644
microenvironment changes 642
precursor cells 643
signaling proteins 643–644
stem cells 641–642, 643
technology changes 641
GPCRs 624*T*
high-throughput screens 618
ion fluxes 624, 624*T*
example *see* 5-HT_3 receptor
limitations
ADME 623
whole-organism 623
multiple outputs (multiplexing) 630
antibody-based 631
disadvantages 639
DNA-binding proteins 631
electrophoretic mobility shift assay (EMSA) 631
enzyme immunoassays 631
example 638
liquid assays 631
mRNA profiles 630–631
nucleic acid quantification 631
positional (flat) assays 631
protein expression 630–631
RNA-binding proteins 631
transcription factor activity 631
nuclear hormone receptors 624*T*
phenotype-oriented screening 136
proliferation assays 628
cell migration 628
contact inhibition 628
differentiation 628
nucleic acid quantification 628
protein kinases 624*T*
readouts 623, 624*T*
see also specific readouts
receptor binding 627
radioligands 627
sodium concentrations 627
regulatory systems 617–618
reporter genes 633
examples 633
response element sequence 633
technology improvements 618
toxicity 629
apoptosis 629
necrosis 629
Cell-based models, pharmacogenomics *vs.* 56
Cell-cell signaling, *Caenorhabditis elegans* model system 8
Cell culture, cell-based screening assays *see* Cell-based assays
Cell cycle
control, *Saccharomyces cerevisiae* model system 5–6
high-content screening *see* High-content screening
large-scale virus vector libraries 18–19
proteomics 30–31
ubiquitin-proteasome pathway 190
Cell-free expression systems 120, 120*F*, 415
advantages 416
components 121
design 121
batch mode reactions 121
continuous-exchange 121
continuous-flow 121
transcription–translation systems 121
disadvantages 416
disulfide-bridged proteins 123
extract preparation 122
bacterial sources 122
commercially available systems 122
protein synthesis using recombinant elements (PURE) system 122
reticulocytes 122
RNAse-deficient *Escherichia coli* 122
wheat germs 122
isotope labeling 120–121
membrane proteins 115*F*, 122
GPCRs 122–123
multidrug resistance transporters 122–123
NMR protein labeling 428
other systems *vs.* 113*T*
reaction parameters 120
Cell lab IC 100 686*T*
Cell lines
baculovirus expression systems 417, 417*T*
Drosophila melanogaster 651
Cell lysis
mammalian cell expression systems 420
protein purification 420
yeast expression systems 420
Cell migration/chemotaxis, cell-based screening assays *see* Cell-based assays
Cell proliferation assays, cell-based screening assays *see* Cell-based assays
Cell sources, cell-based screening assays *see* Cell-based assays
Cell surface receptors
dynamic combinatorial libraries 973
siRNAs 183
small animal imaging *see* Small animal imaging
Cell surface transmitters, cell-based assays 624*T*
Cellular metabolism, small animal imaging *see* Small animal imaging
Cellular retinoic-acid-binding protein type II (CRABP), structure, deliberately incorrect 535
Cellulose, PASP 793
CEM Corporation, microwave-assisted chemistry 850
MARS Microwave Synthesis System 850, 850*F*
power outputs 851
CEM Discover platform, microwave-assisted chemistry 845, 846*F*

automation upgrades 845–846
flow-through system 846, 846*F*
high-throughput synthesis 846, 847*F*
pressure regulation 847
Raman spectroscopy 845–846
solid-phase peptide synthesis 846–847, 847*F*
sub-ambient chemistry 847–848
temperature measurements 847
Center for Eukaryotic Structural Genomics 552*T*
Centralization, compound storage/management *see* Compound storage/management
Central nervous system (CNS)
imaging, [^{18}F]fallypride 667
small animal imaging 664*T*, 667
Ceramides, solid-phase libraries 746*F*, 751–752
discrete libraries 707*F*
cFLIP, NF-κB pathway regulation 196–197
Chain identifiers, protein 3D structure validation 513, 514*F*
Chaperone co-expression, *Escherichia coli* protein expression systems *see* *Escherichia coli* protein expression systems
Chaperonins, structure, problems with 543
Characterization studies, protein purification 555
Charge-coupled devices (CCDs)
Flash Plate assays 589
HCS image detection 685, 687
scintillation proximity assay 584
x-ray detection 468
Charge distribution, x-ray–crystal interaction 452–453
Checklist exchange, DDBJ/EMBL/GenBank 351–352
ChemAxon Ltd, JChem Base 295
Chembank 306
ChemFinder
CambridgeSoft Corporation 295
database management systems 295
Chemical Abstracts 271*T*, 273
databases 300
searching 274
Chemical biology **129–149**, 921
compound collections (libraries) 132
commercially available 132
federal sources 132
Molecular Libraries screening Center Network 132
National Cancer Institute (NCI) 132
National Institute of Neurological Disorders and Stroke (NINDS) 132
natural products 133
NIH Chemical Genomics Center (NCGC) 132
definition 129, 130, 921
drug discovery 146
academia/industry collaboration 146
human genome project 146
forward chemical genetics 130, 131*F*
HTS 130
historical perspective 130
Molecular Initiatives Roadmap 129
phenotype-oriented screens *see* Phenotype-oriented screening
profiling/computational analysis 144
cytologic profiling 144, 145*F*
genetic profiling 144, 144*F*
reverse chemical genetics 130, 131*F*
target-oriented screening *see* Target-oriented screening
Chemical Diversity Labs Inc, ChemoSoft 295
Chemical exchange effects *see* Nuclear magnetic resonance (NMR), drug discovery
Chemical genetics 923
definition 921, 923
drug discovery 923, 923*F*, 923*T*
target identification 700, 701, 701*F*
Chemical–genomic profiling, genomics 5
Chemical information databases *see* Databases
Chemical labeling, LC-MS (peptide-based) proteomics 38
Chemical Markup Language (CML), two-dimensional representations 298–299
Chemical modification, protein crystallization 888
Chemical mutagenesis, mouse models *see* Mouse models
Chemical proteomics *see* Proteomics
Chemical representation, pharmacophore identification 242
Chemical shifts
NMR spectroscopy 945
perturbations *see* Nuclear magnetic resonance (NMR), drug discovery
Chemical space, fragment-based drug discovery 943
Chemiluminescence nitrogen detection (CLND), high throughput purification 863
Chemiluminescent assays 578, 590
electrochemiluminescence 592
ELISA 592
ChemInformRX 281–282, 282*T*
Chemistry limitations, solution phase parallel chemistry 767
CHEMLIST 277
Chemogenomics **921–937**
agonists/antagonists 923
GPCRs 923, 924*F*
definition 921
GPCR ligands 932
integrin ligands 933, 933*F*
kinase inhibitors 930, 931*F*
nuclear receptor ligands 932, 933*F*, 934*F*
phosphodiesterase inhibitors 930, 931*F*
privileged structures 925, 925*F*, 926*T*
chemical modifications 925–926, 927*F*
protease inhibitors 929
SOSA approach 925*F*, 926
target family-directed master keys 927, 929*F*
transporter ligands 934*F*, 935
Chemoinformatics **235–264**
combinatorial library design 256
definition 256, 256*F*
optimization techniques 257
product-based selection 256–257
reactant- *vs.* product-based selection 256
definition 235
historical perspective 235
Chemoinformatics, ADME property prediction 257
computational filters 258
drug-like classifications 258
genetic algorithms 258
lead-likeness prediction 258
'rule-of-five' 258
physicochemical properties 259
blood–brain barrier penetration prediction 259
hydrophobicity prediction 259
polar surface area prediction 259
solubility prediction 259
throughput 257–258
Chemoinformatics, molecule representation 236
example 236, 237*F*
SMILES (simplified molecular input line entry specification) 236
structure/substructure searching 236, 239*F*
Augmented Connectivity Molecular Formula 238
canonical representations 238
connection tables 238
fragment bitstrings 238, 238*F*
graph theory 236–237
hashed fingerprints 239
molecular graphs 236–237, 237*F*
subgraph isomorphism algorithms 239
Ullmann algorithm 239–240
Chemoinformatics, three-dimensional structure 240
Cambridge Structural Database 240
compound numbers 240
Concord program 240
Corina program 240
PDB 240

searching 241
connection tables 241
rigid searching 242
subgraph isomorphs 242
substructure searching 241–242
target binding 241, 241*F*
3D graphs 241
structure generation programs 240
Chemoinformatics, virtual screening 244
computational techniques 245
definition 244–245
machine learning methods 249
classification techniques 250
recursive partitioning 250
support vector machine (SVM) 250
3D similarity searching 247
alignment-based 248
3D-vector 248
2D similarity searching 246
Extended Connectivity Fingerprints (ECFPs) 246
fragment methods 246
Functional Connectivity Fingerprints (FCFPs) 246
graph-based similarity 246
reduced graph similarity 247
ChemoSoft
Chemical Diversity Labs Inc 295
database management systems 295
χ-angle, protein 3D structure validation 517
Chinese hamster ovary cells *see* CHO cells
Chiral 1,3-oxazolidines, solid-phase libraries 722–724, 723*F*
Chlorobenzene, microwave heating 840*T*
Chloroform, microwave heating 840*T*
Chloromuconate cycloisomerase structure
incorrect structure 533–534
Ramachandran plots 533–534
R-factor 533–534
Chlorpromazine, discovery/development 925*F*, 926–927
CHO cells
cell-based assays 620–621
mammalian cell expression systems 418
protein expression systems 117
Cholesterol
as biomarker, cardiovascular disease and 70
tagging, siRNAs 183
Chromatography, PASP 826
Chromogenic substrates, target-oriented screening 142
Chymotrypsin, protein crystallization 434
CIN (Chemical Industry Notes) 275
Circular dichroism spectroscopy, protein 3D structure, protein analysis 427
Cis-peptide bonds, protein 3D structure validation 517
Citalopram, chemogenomics 934*F*, 935
c-kit, inhibition, imatinib mesylate 3
'Classical' pharmacogenetics *see* Pharmacogenetics
Classification techniques, machine learning methods 250
Clavulone analogs, solid-phase libraries 703
Clay, solvent-free reactions, microwave-assisted chemistry 853
Clinical trial(s)
drugs, RNAi (RNA interference) 172
pharmacogenetics 64
proteomics 42, 42*F*
Clone-by-clone approach, genome sequencing 318
Clozapine receptors, imaging 667
Cluster analysis, microarrays 93
Cluster-based compound selection *see* Chemoinformatics
Clustering 253, 254*F*
intermolecular similarities 255
Jarvis–Patrick clustering 254
microarrays 93
Ward's clustering 254–255
Cluster of orthologous genes (COGs), protein function prediction, bioinformatics 334
CluSTr database, Integr8 367
proteome analysis 367–368
c-myc, RNAi 180–181
^{55}Co, PET 664*T*
Co-crystallization 891
rational drug design 891
Codelink microarray 89
Codon usage, gene expression vectors *see* Gene expression vectors
CoE model, compound storage/management (Pfizer) 569, 570*F*
Cofactors, heterogeneous protein crystallization 437
Co-immunoprecipitation, proteomics 41–42
Collaborations, industry/academia, drug discovery 146
Collections, cell-based assays 622
Colony-formation assays, HCS 692
Colorectal cancer, bevacizumab 202
Colorimetric assays, protein concentration determination 426
CombiCHEM system, Milestone s.r.l. microwave-assisted chemistry 849, 850*F*
Combinatorial chemistry **697–760**
development 762, 763
discrete compound synthesis 763
compound design 763
compound quality 763
high throughput purification 861
HTS automation 681
natural products *vs.* 699, 700*F*
as models 699
reaction types 762
target identification 700
chemical genetics 700, 701, 701*F*
fast reading methods 702
terminology 763
Combinatorial libraries
chemical biology 133
diversity-oriented 133, 134*F*
privileged structure 133
design, chemoinformatics *see* Chemoinformatics
Combined fractional diagonal chromatography (COFRADIC), LC-MS (peptide-based) proteomics 38
Combretastatin A-4, [^{18}F]FDG imaging 668–669
Commercially available compound collections 132
Commercially available databases *see* Databases
Commercially-available drugs, PASP 801
Commercially available systems, cell-free expression systems 122
Common complex diseases, pharmacogenetics *see* Pharmacogenetics
Comparative computational modeling, structural genomics *see* Structural genomics
Comparative databases 363
see also specific databases
Comparative experiments, microarrays 90
Comparative genomics
hybridization, microarrays 90
repetitive elements, bioinformatics 324
Comparative sequence analysis, RNA structure prediction, bioinformatics 327
Comparison methods, 2D-PAGE protein quantitation 34*F*, 35–36, 36*F*
Compartmentalization, dynamic combinatorial libraries 967–968
Compatible dual-vector approach, *Escherichia coli* protein expression systems 113
Competition based ^{9}F screening, NMR, drug discovery 908*T*, 915, 915*F*
Competition binding equilibria, NMR, drug discovery 903
Competition experiments, NMR spectroscopy 946

Competitive enzyme immunoassay, cell-based assays 631
Competitive screening, NMR, drug discovery 908*T*
Complexes, protein crystallography, rational drug design 891
Compounds
numbers
chemoinformatics, three-dimensional structure 240
solution phase library synthesis 769–770
pharmacophore identification 244
quality, combinatorial chemistry 763
screening, mouse models 653
selection, chemoinformatics, diversity analysis 253
stability, HTS library composition 681
Compound storage/management **561–576**
acquisition/processing 561, 563
external suppliers 563–564
in-house synthesis 563–564
outsourcing 564
quality control 563–564
registration 564
audit trails 561
centralization 562
local sample handing 562
size of collection 562
dispensing 561
distribution 561
example *see* Compound storage/management (Pfizer)
future developments 573
liquid sample collections
media 567
storage conditions 567
temperature 567
processing for use 569
HTS 569
Compound storage/management (Pfizer) 562
CoE model 569, 570*F*
compressed sets 570, 572*F*
hardware selection 572
flexibility 573
human *vs.* automated 573
radio frequency identification tags 573
liquid sample collection 563, 566
automation 566
current practice 566
customer interaction 568
dispensing configuration 566, 566*F*
multiple replicates 566–567
quality control 568
primary screen supply 569, 570*F*
registration 564
Global electronic Compound Registration system 564
in-house synthesis 564
solid samples 565, 565*F*
automation 565
capacity 565
specific activity compounds 571
subsets 570
Global Diverse Representative Subset (GDRS) 570–571
random sets 570–571
Comprehensive Medicinal Chemistry (CMC), information on 308–309
Compressed sets, compound storage/management (Pfizer) 570, 572*F*
Computational filters *see* Chemoinformatics, ADME property prediction
Computed Ligand Binding Energy Database, BIDD Databases 305
Computed tomography (CT)
applications 659
functional small animal imaging 661
small animal imaging 660*T*, 661, 662
Computer-based technologies
chemoinformatics, virtual screening 245
solution phase library synthesis 770
Computing LLC, molecular fingerprinting 298
Concanavalin A library, dynamic combinatorial libraries 969
CONCORD program, chemoinformatics, three-dimensional structure 240
Conditional knockouts, knockout mouse model 13
'Conditional transgenic animals,' transgenic animal generation 152
Condition optimization, PASP 827–828
Confocal FLA, fluorescence lifetime assays 610, 611*F*
Confocal microscopy, fluorescence fluctuation techniques 607
Confocal systems, HCS *see* High-content screening (HCS)
Confocal TLA, fluorescence lifetime assays 611, 612*F*
Conformational analysis/searching
CSD 394*F*, 400, 400*F*
Vista data analysis 394*F*, 400, 400*F*
Conformational changes, heterogeneous protein crystallization 436
Conformation numbers, pharmacophore identification 243
Conformation-on-the-fly, pharmacophore identification 244
Conformations
CSD 399
flexibility, pharmacophore identification 242
preferences, CSD applications 400
variation, CSD applications 400*F*, 401
Confounding factors, genetically engineered animals 155*T*, 156
Connection tables
chemoinformatics, three-dimensional structure 241
structure/substructure searching 238
ConQuest Search *see* Cambridge Structural Database (CSD)
Consensus-based methods, regulatory regions, bioinformatics 322
Contact analysis *see* Protein 3D structure validation
Contact inhibition, cell-based assays 628
Contact number, protein 3D structure validation 525*F*, 526, 526*F*
Contents sensors, gene detection, bioinformatics 321–322
Continuous-exchange, cell-free expression systems 121
Continuous-flow
cell-free expression systems 121
scale up, microwave-assisted chemistry *see* Microwave-assisted chemistry
Contoured surface generation, IsoStar (intermolecular interactions) 405
Contrast drugs, small animal imaging 660
Control groups, microarrays 100–101
Controlled pore glass, PASP 793
Control probes, microarrays 90
Conventional thermal heating, microwaves *vs. see* Microwaves
Coomassie stains, 2D-PAGE protein quantitation 35, 35*T*
Coordinate errors
carbonic anhydrase structure 532
cycloamylase structure 532
estimate, protein 3D structure 546
rubredoxin structure 532
streptavidin structure 532
Copy numbers
gene expression vectors 107
mammalian protein expression systems 118
CORBA, database integration 303
CORINA program, chemoinformatics, three-dimensional structure 240
Corticotrophin-releasing factor overexpression, microarrays *see* Microarrays
Corticotrophin-releasing factor receptor, microarrays *see* Microarrays
Corynebacterium glutamicum, protein expression systems 116
CPP-32 *see* Caspase-3

CRF antagonist treatment, corticotrophin-releasing factor overexpression 98*F*
Crick, Francis, protein crystallography 450
CRIPT, NMR, drug discovery 915–916
Crohn's disease, NF-κB pathway 197
CrossFire Beilstein 281
CrossFire Gmelin 281
Cross-linking, PASP 792–793
Cross-relaxation rate, NMR structure determination 481
Crowfoot, Dorothy, protein crystallization 433–434
Cryocrystallography *see* Protein crystallization
CRYST1 record, protein 3D structure validation 511
Crystal engineering, protein crystallization *see* Protein crystallization
Crystal handling *see* X-ray crystallography
CrystalLEAD 892
Crystallographic Information File (CIF), CSD 391–392
Crystal packing
  insulin 442
  NADH peroxidase 442, 444*F*
Crystals *see* X-ray crystallography
Crystal size, protein crystallography *see* X-ray crystallography
CSF-1, STAT3 signaling 206
CSP, NMR, drug discovery 908*T*
^{13}C spectra, NMR, drug discovery 917
c-Src, fragment-based drug discovery *see* Fragment-based drug discovery
C-terminal domain, STAT (signal transducer and activator of transcription) 203
^{62}Cu, PET 664*T*
^{64}Cu
  biomolecular probes 666, 666*F*
  PET 664*T*
Curation
  RefSeq (NCBI Reference Sequences) 355–356
  secondary databases 350
  UniProtKB/Swiss-Prot 359–360
Current Contents 274
Cushman, David, protein crystallography, rational drug design 876
Customer interaction, compound storage/management 568
Cutoff values, protein 3D structure errors 509
Cyanoborohydride, PASP 806, 806*F*
Cyclic thioether peptidomimetics, solid-phase discrete libraries 709*T*
Cyclin-dependent kinase-1 (CDK1) substrates, orthogonal ligand–receptor pairs 218
Cyclin-dependent kinase 2 inhibitors, dynamic combinatorial libraries 973
Cyclin-dependent kinase-2 (CDK2) inhibitors, solution phase library synthesis 785, 785*F*
Cyclin-dependent kinases (CDKs)
  crystallography 884
  inhibitor identification
    crystallography 884
    Saccharomyces cerevisiae model system 8
Cycloaddition reactions, dynamic combinatorial library generation 962
Cycloamylase, structure 532
Cyclooxygenase-2 (COX-2), NF-κB pathway 197
Cyclooxygenase-5 (COX-5), expression, NF-κB pathway 197
Cyclooxygenase inhibitors, side effects 4
Cyclopent-2-enones, solid-phase libraries 708, 708*F*
Cyclopentanol derivatives, solid-phase discrete libraries 709*T*
Cyclophilin–cyclosporine A (CsA), orthogonal ligand–receptor pairs 217, 219*F*
Cyclosporine A (CsA), mechanism of action 2
Cystic fibrosis, as monogenetic disease 338
Cytochrome *c*, apoptosis 196
Cytochrome P-450 family
  radioactive assays 589
  scintillation proximity assay 589
Cytokines
  STAT signal transduction/activation 205–206, 206*T*
  vascular endothelial growth factor expression 200
Cytokinesis, RNAi (RNA interference) 179, 180
Cytokinesis inhibitor identification, *Drosophila melanogaster* model system 9
Cytologic profiling, chemical biology 144, 145*F*
Cytomegalovirus (CMV), microRNA coding 175
Cytometric beads, enzyme-linked immunosorbent assays (ELISAs) 73–74
Cytoplasmic fractions, proteomics 31
Cytoplasm targeting, *Escherichia coli* protein expression systems 114
Cytotoxic drugs
  [^{18}F]FDG 668
  small animal imaging 668
Cytotoxicity, Semliki Forest Virus expression systems 119

D

DALI, structural genomics 557
DALI/FSSP, protein function prediction, bioinformatics 336–337
Danio rerio model system *see* Zebrafish (*Danio rerio*)
Dansyl peptides, solid-phase pool libraries 705*F*
Data
  hierarchical databases 267
  interpretation 104
  microarrays, re-expression 91
  overinterpretation
    botulinum neurotoxin light chain structure 535, 536*F*
    x-ray structures, problems 535, 536*F*
  redundancy
    DDBJ/EMBL/GenBank 352
    protein 3D structure validation 521
  storage, HCS 692
Data access, pharmacogenetics 66–67
Data acquisition
  CSD *see* Cambridge Structural Database (CSD)
  DDBJ/EMBL/GenBank 354
  fluorescence fluctuation techniques *see* Fluorescence fluctuation techniques
  HCS 685
Data analysis
  HCS 692
  microarrays *see* Microarrays
Data application, pharmacogenetics 67
DataBase Management system (DBMS), definition 266
Databases **349–372**, 266*F*, 267
  see also Literature; *specific databases*
  access to 283
    historical perspective 283
  bioactivity *see* Bioactivity databases
  classification 269, 270*F*
  definition 266
  factual 269, 276, 278*T*
    catalog 279
    metadata 277
    numerical 277
    spectroscopic 279, 280*T*
    text 277
  generation, protein 3D structure validation *see* Protein 3D structure validation
  gene sequence *see* Nucleotide sequence databases
  hierarchical 267, 267*F*
    advantages 267
    data sequence 267
    disadvantages 267

Databases (*continued*)
history 349
hosts 284
historical perspective 284
licenses 284
information sources 285
in-house 284
integrated/comparative 363
integration 284
interfaces 283
internet-based 285
metabolic networks 328–329
network models 268, 268*F*
object-based models 268
relational database management system (RDBMS) 268
organization 266–267, 266*F*
outlook/future work 286
primary 350
experimental results 350
protein sequence *see* Protein sequences
reference linking 284
searching
mass spectrometry proteomics 33
protein 3D structure, protein production 413
secondary 350
curated reviews 350
sequence databases 281
structure databases 269, 279, 280, 282*T*
patents, with 281
three-dimensional 283
Database system (DBS), definition 266
Data Definition Language (DDL), databases 294
Data Manipulation Language (DML), databases 294
Data preprocessing, microarrays *see* Microarrays
Data protection, pharmacogenetics *see* Pharmacogenetics
Data quality, HCS *see* High-content screening (HCS)
Data visual inspection, microarrays *see* Microarrays
Daylight CIS
THOR 295
Traditional Chinese Medicine database 308
DcuS, NMR structure determination 490, 490*F*
DDBJ/EMBL/GenBank 351
checklist exchange 351–352
data acquisition 354
data redundancy 352
dynamics 351, 351*F*
editorial control 352
sequence identity 353–354
updating responsibility 354
format 352
feature table 352
flatfiles 352, 353*F*
header 352
sequence 352, 353*F*
history 350
structure 351
Third Party Annotation (TPA) 354
data contained 354–355
exchange of 355
flatfile 355, 355*F*
UniParc checking 354
UniProt checking 354
Deafness, *Danio rerio* model system 10
Death receptors, apoptosis *see* Apoptosis
Decomposition, cDNA microarrays 332
Defined patient populations, microarrays 93
Deglycobleomycin analogs, solid-phase discrete libraries 709*T*
Deglycosylation
protein crystallization 882, 883*F*
protein 3D structure, protein production 428
Deisenhofer, Johann, protein crystallography 451
DE (description) lines, UniProtKB flatfile 360
Delivery/administration route, genetically engineered animals 156
Denaturing conditions, 2D-PAGE 34
Density modification, protein crystallography model generation 462, 463*F*
Dephosphorylation
protein crystallization 882
ubiquitin protein targeting 192
Depletion, proteomics 31
Depression, corticotrophin-releasing factor receptor 93–94
Derewenda, Zygmunt, protein crystallization 887
Derwent Biotechnology Resource 271*T*
Derwent Drug File 271*T*
Derwent World Patent Index 275
Desferrioxamine analogs, solid-phase discrete libraries 709*T*
Desktop databases 294
FileMaker Pro 294
MS (Microsoft)-Access 294
Detergents
monotropic membrane protein crystallization 435
protein crystallization 438, 881
2D-PAGE 34
DETHERM 278*T*
Deuterium-exchange mass spectrometry (DXMS), domain mapping 879, 879*F*
Devazepide, discovery/development 925, 927*F*
Developmental biology, zebrafish (*Danio rerio*) 652
Diabetes mellitus
biomarkers
glucose 70
hemoglobin A1c 70
protein structures on PSD 382
RNAi 180–181
Diabetes mellitus management, imatinib mesylate 3
Diabetic retinopathy, vascular endothelial growth factor 202
DIALIGN, sequence alignment, bioinformatics 320
Dialkoxyindoles, solid-phase pool libraries 705*F*
N,N-Dialkyldipeptidylamines, PASP 808–809, 810*F*
Dialysis, buffer exchange 423
Diamines, solid-phase pool libraries 705*F*
Diaminopyrimidines, solid-phase pool libraries 705*F*
Diaminoquinazolinones, solid-phase discrete libraries 709*T*
Diaryl diamides, solid-phase pool libraries 705*F*
Diaryl ethers, solid-phase discrete libraries 709*T*
1,2-Diazepane-2-ones, solid-phase discrete libraries 709*T*
Diazepine derivatives, solid-phase discrete libraries 709*T*
Dibenz*[b,g]*1,5-oxazocines, solid-phase libraries 706, 707*F*
DICER *see* RNAi (RNA interference)
1,2-Dichlorobenzene, microwave heating 840*T*
Dichloromethane, microwave heating 840*T*
Dictyostelium discoidae, as model system 647
Dielectric heating, microwaves *see* Microwaves
Diels–Alder reaction, dynamic combinatorial library generation 962, 962*F*
Differential gene expression, microarrays *see* Microarrays
Differentiation, cell-based assays 628
Diffraction data resolution, x-ray–crystal interaction 456
Diffraction patterns
indexing 469
x-ray–crystal interaction 452
Diffraction symmetry, protein crystallography 441–442
Diffraction theory, protein crystallography 451
Diffuse gel electrophoresis (DIGE), 2D-PAGE protein quantitation 35, 35*T*

Diffusion filtering, NMR, drug discovery 908*T*, 913
Difunctional molecule synthesis, PASP 795
Digital Object Identifier (DOI), databases 300–301
Dihydrofolate reductase (DHFR), selection markers, mammalian protein expression systems 117, 118
Dihydroimidazolium salts, solid-phase pool libraries 705*F*
Dihydromotuporamine C, mechanism of action 7
Dihydropyrancarboxamides, solid-phase pool libraries 707*F*
Dihydrostilbene, solid-phase pool libraries 705*F*
4-4′-Dihydroxybenzil (DHB), orthogonal ligand–receptor pairs 227
Dimers, x-ray structures, problems 540
Dimethylbenzopyrans, solid-phase discrete libraries 707*F*
2,2-Dimethyl benzopyran scaffolds, solid-phase libraries 745*F*, 746
N,N-Dimethylformamide, microwave heating 840*T*
Dimethyl sulfoxide (DMSO)
  compound storage 566, 568
    measured concentrations 568
    stock solutions 862
  microwave heating 840*T*
2′-*O*-2,4-Dinitrophenyl derivatives, siRNAs modification 182–183
DIP database, regulatory networks 330
Dipeptide analogs, solid-phase pool libraries 705*F*
Dipolar polarization, dielectric heating 839
Dipolar relaxation, liquid-state NMR structure determination 481
DIPPR 278*T*
Directed evolution, protein crystallization 888
Directional atomic contact analysis, protein 3D structure validation 517–518
Directly bonded nuclei, NMR structure determination 485
Disallowed Ramachandran plot regions, x-ray structures, problems 537
Discovery-1 686*T*
Discrete compound synthesis, combinatorial chemistry *see* Combinatorial chemistry
Disease
  heterogeneity, pharmacogenetics 61
  mechanisms, proteomics 42–43
  models, zebrafish (*Danio rerio*) 652
  molecular mechanisms, genetically engineered animals 159–160
Disease program, PDB 377
DisEMBL, domain mapping 879
DISOPRED2, domain mapping 879
Dispensing, compound storage/management 561
Displacement assays, radioactive assays 585–586
DisProt, domain mapping 879
Dissimilarity-based compound selection (DCBS) 253, 254*F*
  sphere exclusion algorithms 253, 254*F*
Dissociation constant (K_D)
  biological activity databases 302
  determination, fluorescence screening assays 601
Distance-based methods, phylogeny construction, bioinformatics 321
Distribution, compound storage/management 561
3,4-Disubstituted isoxazoles, solid-phase discrete libraries 709*T*
3,5-Disubstituted isoxazoles, solid-phase discrete libraries 709*T*
Disulfide bridges
  cell-free expression systems 123
  dynamic combinatorial libraries 969, 970*F*
  dynamic combinatorial library generation 962*F*
  Escherichia coli protein expression systems *see Escherichia coli* protein expression systems
  protein crystallization 436, 437, 437*F*
Diversity, libraries 697
Diversity analysis 252
  compound selection methods 253
  optimization techniques 256
  partitioning 254*F*, 255
    BCUT (Burden, CAS and University of Texas) 255
    pharmacophore fingerprints 255
  screening test design 252–253
Diversity-oriented combinatorial libraries 133, 134*F*
Diversity-oriented libraries, solution phase parallel chemistry 767
Divinylbenzene, polystyrene resin, PASP 792
DNA
  HCS cell cycle 691–692
  repair, *Saccharomyces cerevisiae* model system 6
  shuffling
    glyphosate *N*-acetyltransferase crystallography 884–885
    protein crystallization 884–885, 885*F*
DNA-binding domain, STAT (signal transducer and activator of transcription) 203
DNA-binding proteins, cell-based assays 631
DNA Databank of Japan (DDBJ) *see also* DDBJ/EMBL/GenBank nucleotide sequence databases 350
DNA dyes, HCS labeling techniques 689
DNA helicase
  Flash Plate assays 589
  radioactive assays 588
DnaK/DnaJ/GrpE system, *Escherichia coli* protein expression systems 112–113
DNA polymerases
  Flash Plate assays 588
  radioactive assays 588
DNA primase
  Flash Plate assays 588
  radioactive assays 588
DNA-protein kinase (DNA-PK), ubiquitin-protein targeting 192
DNA-templated oligonucleotide synthesis, dynamic combinatorial libraries 966
DNA–transcription factor binding, ELISA 580
Docking
  program evaluation, structure-based virtual screening 251–252
  structure-based virtual screening 250–251
Docking, scoring functions, regression-based, structure-based virtual screening 251
DOCK program, structure-based virtual screening 251
Domains
  fusion, genomic context methods, bioinformatics 336
  mapping, gene construct design 879, 879*F*
Domestic microwaves, microwave-assisted chemistry 842–843
Dopamine receptor(s)
  imaging, [^{11}C]-*N*-methylspiperone 667
  molecular probes 664, 664*T*
DOQCS database, regulatory networks 330
Downstream biological effects, biomarkers *see* Biomarkers
DR (database crossreference), UniProtKB flatfile 360, 361*F*
DRESS (Database of Refined Solution NMR Structures), NMR structures, problems 542
DRIP_PRED, domain mapping 879
Drosophila melanogaster 9, 649
  Berkeley Drosophila Genome Project 9
  cell lines 651
  cytokinesis inhibitor identification 9
  dTSC1 gene 650–651, 651*F*

Drosophila melanogaster (*continued*)
ERK pathway 650, 650*F*
forward genetics 9
gene conservation 649–650
genome sequencing 317*T*, 647–649, 648*T*, 650
growth factor receptors 650
history 647
Huntington's disease 9
insulin-like growth factor receptor 650
large-scale genomic screens 9
oncogenesis 9
proteome complexity 29*T*
RAS proteins 9, 650
reverse genetic screens 651
RNA interference 9, 651
spinocerebellar ataxia 1 9
tumor suppresser genes 650–651
Wnt signaling 9
Drug(s)
action validation, RNAi (RNA interference) 181
efficacy, pharmacogenetics *see* Pharmacogenetics
ideal 2
interference, fluorescent cell-based assays 625
specificity 3
Drug ADME Associated Protein Database, BIDD Databases 305
Drug Adverse Reaction Target database, BIDD Databases 305
Drug Database 309
Drug design, combinatorial chemistry 763
Drug development
early, cell-based assays 617–618
pharmacogenetics 62
proteomics 42, 42*F*
Drug discovery
animal models, pharmacogenomics *vs.* 56
chemical genetics 923, 923*F*, 923*T*
definition 316
genetic variation analysis 339
genomics *see* Genomics
hit-to-lead, HTS 684
knockout mouse model 13
protein crystallography *see* X-ray crystallography, rational drug design
'Druggability'
definition 2
drug specificity 3
HTS target selection 681
'rule-of-five' 2
target assessment 3
Drug-induced haploinsufficiency, phenotype-oriented screening 139, 140*F*
'Druglikeness,' HTS library composition 680
Drug toxicity markers, biomarkers *see* Biomarkers
DsbA, gene expression vectors 110*T*
DsbC, gene expression vectors 110*T*
DsRNA dependent protein kinase, RNAi (RNA interference) 173
DSSP, protein structure prediction 325
DT (date) lines, UniProtKB flatfile 360
DTSC1 gene, *Drosophila melanogaster* 650–651, 651*F*
Dual-labeling, enzyme-linked immunosorbent assays (ELISAs) 73–74
Dual promoters, gene expression vectors 108
Dynamic combinatorial chemistry (DCC) **959–976**
see also Dynamic combinatorial libraries (DCLs)
applications 966, 967*T*
building block selection 960, 961
functional groups 961, 961*F*
interactional units 961
organizational units 961
recognition groups 961
definition 960
development 966
boronic ester-based trypsin inhibitors 966
metal ion assisted imine macrocycle synthesis 966
solution phase library synthesis *see* Solution phase synthesis, library preparation
types of 963
Dynamic combinatorial libraries (DCLs)
see also Dynamic combinatorial chemistry (DCC)
adaptive 963, 963*F*
selection target presence 963
system adaptability 963
amino acid-based 966
acyl hydrazones 968
beta-endorphin ligand generation 967–968, 968*F*
compartmentalization 967–968
pseudodynamic systems 968
carbohydrate-based 969
acyl hydrazone exchange 969
concanavalin A library 969
difficulties 969
disulfide interchange 969, 970*F*
lysozyme inhibition library 969, 971*F*
metal coordination 969, 970*F*
catalytic screening 963, 964, 965*F*
acetylcholinesterase inhibitors 973
definition 960
dynamic deconvolution 963
fixation 963
generation 960, 961
aldol formation 962*F*
boronic ester formation 962*F*
carbonyl groups 961–962, 962*F*
cycloaddition reactions 962
Diels–Alder reaction 962, 962*F*
disulfide formation 962*F*
hemiacetal formation 962*F*
imine formation 961–962, 962*F*
metal coordination 962
Metathesis reaction 962*F*
Michael reaction 961–962, 962*F*
multiple reactions 963
reversible covalent bonds 961, 962*F*
see also specific reactions
transacylation 962*F*
iterated 963
nonnatural components 969
acetylcholinesterase inhibitors 970–971, 972*F*
acid hydrazone exchange 970–971
carbonic anhydrase inhibitors 969–970
cell surface ligands 973
cyclin-dependent kinase 2 inhibitors 973
metal coordination 973, 973*F*
neuraminidase inhibitors 970, 971*F*
nonnucleotide building blocks 973
psammaplin A derivatives 973
transamination 969–970, 971*F*
whole cells 973
nucleic acid-based 968
DNA-templated oligonucleotide synthesis 966
dynamic hydrogels 968–969
nonnucleic acid residues 969
nonnucleotide building blocks 973
pre-equilibrated 963
amplification 964, 965*F*
biological targets 963–964
definition 963–964
screening 963–964, 964*F*
separation 964
pseudodynamic (deletion) screening 965, 968
selection conditions 960
see also specific methods
tethering screening 965
virtual 961
x-ray crystallography 963, 965
cyclin-dependent kinase 2 inhibitors 973
Dynamic contrast enhanced magnetic resonance imaging (DCE-MRI) 77
Dynamic deconvolution, dynamic combinatorial libraries 963

Dynamic hydrogels, dynamic combinatorial libraries 968–969
Dynamic ligand assembly **959–976**
see also Dynamic combinatorial chemistry (DCC); Dynamic combinatorial libraries (DCLs)
Dynamic light scattering (DLS)
protein analysis 426
protein purity determination 881
Dynamic programming paradigm, sequence alignment, bioinformatics 320
Dysidiolide, analogs, solid-phase discrete libraries 707*F*

E

EC_{50}, biological activity databases 302
Ecdysteroid–ecdysone receptor, orthogonal ligand–receptor pairs 229
E-Cell project 330
EcoCys database, metabolic networks 328–329
ED_{50}, biological activity databases 302
EDS 547*T*
EGFP, HCS cell cycle 691–692
Eidogen-Sertanty, commercially available databases 308
Electric field presence, microwave-assisted chemistry 842
Electrochemiluminescence, optical assays 592
Electromagnetic field homogeneity, microwave-assisted chemistry 855
Electron density, x-ray structures, problems 537
Electron density maps
autocorrelation 457
isomorphous replacement 457, 458*F*
protein crystallography model generation 464, 464*F*
protein 3D structure 519, 547
scaling 464, 464*F*
Electron Density Server 510
Electron microscopy, protein structure, problems *see* Protein 3D structure, problems
Electron oscillation, x-ray–crystal interaction 452, 455*F*
Electrophoretic mobility shift assay (EMSA) 631
Electrophysiology
cell-based screening assays *see* Cell-based assays
genetically engineered animals 155–156
Electroporation, siRNAs delivery 177
Electrospray, mass spectrometry proteomics 32
ELISPOT, cell-based assays 632
Elsevier-MDL
commercially available databases 308
MDL ISIS/Base 295
Embase 271*T*, 274
EMBL
see also DDBJ/EMBL/GenBank
nucleotide sequence databases 350
Embryogenesis, vascular endothelial growth factor 200–202
Embryonic lethality, knockout mouse model 13
EMOTIF database, protein function prediction, bioinformatics 334–335
Emrys Advancer batch reactor 851, 851*F*
Emrys Knowledge database 844
Endocytosis, RNA interference 17–18
Endometrial hyperplasia, vascular endothelial growth factor 202
Endometriosis, vascular endothelial growth factor 202
Endoxan, [^{18}F]FDG imaging 669
Endpoints, cell-based assays 623
Energetic criteria, comparative computational modeling 557
Enhancer regions, regulatory regions, bioinformatics 322
Ensembl
IPI (International Protein Index) 365–366
UniParc 358–359
Entrez Protein 358
Entropy, protein crystallization
change 438
minimization 434
Environmental effects
common complex diseases, pharmacogenetics *vs.* 54
PASP 793
pharmacogenetics *see* Pharmacogenetics
Enzyme(s)
function, NMR spectroscopy 946
target-oriented screening *see* Target-oriented screening
Enzyme Commission classification, protein function prediction 333–334
ENZYME database, metabolic networks 328–329
Enzyme I (*Escherichia coli*), structure 543
Enzyme immunoassays, cell-based assays 631
Enzyme immunosorbent assay (EIA), cell-based screening assays *see* Cell-based assays
Enzyme-linked immunosorbent assays (ELISAs) 578–579
advantages 578–579, 580
antibody capture assays 579–580, 579*F*
antibody specificity 580
antigen capture assays 579–580, 579*F*
biomarkers 79–80
chemiluminescent substrates 592
classification 579–580
competition 579–580
cytometric beads 73–74
design importance 580
disadvantages 580
washing steps 580
DNA–transcription factor binding 580
dual-labeling 73–74
genetically engineered animal characterization 153–154
high-throughput 580
disadvantages 580
indirect competition 579–580
PARP-1 inhibitors 589
Plasmodium falciparum drug sensitivity 580
protein quantitation 73–74
multiplexing 73–74
sandwich 579–580, 579*F*
protein quantitation 73–74
specificity 73–74
sulfated polysaccharide–protein interactions 580
technique 580
Enzyme Structures Database (EC-PDB) 304–305
Epibatidine, PASP 816, 817*F*, 818*F*
Epidermal growth factor (EGF)
STAT1 signaling 205–206
STAT3 signaling 206
STAT5 signaling 206
Epigenetics, genetically engineered animals 156
Epimaritidine, PASP 815, 816*F*
Epothilone A, structure 700*F*
Epothilones
NMR structure determination 494, 496*F*
PASP 821, 822*F*, 823*F*, 824*F*
Epstein–Barr virus (EBV)
cell immortalization 620
microRNA coding 175
ERK pathway, *Drosophila melanogaster* 650, 650*F*
ERRAT software, protein 3D structure validation 517–518
Erythromycin, structure 700*F*
Erythropoietin, STAT5 signaling 206
Escherichia coli
genome sequencing 317*T*
proteome complexity 29*T*
Escherichia coli protein expression systems 111, 880
chaperone co-expression 112
bicistronic approach 113
compatible dual-vector approach 113
DnaK/DnaJ/GrpE system 112–113
GroEL/GroES system 112–113

Escherichia coli protein expression systems (*continued*)
cost 111–112
disulfide bond expression 116
BL21 Origami strains 116
fermentation temperatures 116
expression stability 113
temperature 114
inclusion body formation 112
insolubility problems 880
membrane protein expression 114, 115*F*
strains 114
other systems *vs.* 113*T*
posttranslational modification 880
protein degradation 114
protease-free systems 114–116
protein precipitation 112
protein targeting 114
cytoplasm 114
periplasm 114
secretion 114
signal/leader sequences 114
requirements 112*T*
solubility tags 113
glutathione-*S*-transferase 113–114
mannose-binding protein 113–114
protein D 113–114
thioredoxin 113–114
strains 111–112
structural genomics 553*T*, 554
E-selectin, NF-κB pathway expression 197
Esp@cenet 276*T*
Estradiol
derivatives, solid-phase discrete libraries 709*T*
orthogonal ligand–receptor pairs 224*F*, 225–227, 225*F*, 226*F*
solid-phase libraries 708, 718*F*
Estrogen receptors (ERs), orthogonal ligand–receptor pairs 224, 224*F*, 225*F*, 226*F*
Estrogen receptor ligands
see also specific drugs
chemogenomics 933, 933*F*
Ethanol, microwave heating 840*T*
Ethical issues, pharmacogenetics *see* Pharmacogenetics
Ethyl acetate, microwave heating 840*T*
Ethylenediamino piperidines, solid-phase pool libraries 705*F*
Ethylene glycol, microwave heating 840*T*
Ethylmethane sulfonate (EMS), chemical mutagenesis 13–14
N-Ethyl-*N*-nitrosourea (ENU), chemical mutagenesis 13–14
European Bioinformatics Institute (EBI) 304–305
European Collection of Cell Cultures 622
Evaporative light-scattering detection (ELSD), high throughput purification 863–864, 864*F*
Evidence tags, Genome Reviews 356, 357*F*
Exchange rates, NMR, drug discovery 917
Excretion, cell-based assays 629
Exocyclic amino nucleosides, solid-phase discrete libraries 709*T*
Experimental groups, microarrays 100–101
Expressed proteins, posttranslational modifications 880
Expression libraries, RNAi (RNA interference) 180
Expression profiling, proteomics 42–43
Expression systems, gene construct design 880
Expression vectors
Semliki Forest Virus expression systems 119
yeast protein expression systems 117
Extended Connectivity Fingerprints (ECFPs), 2D similarity searching 246
External cross-references, UniProtKB 359
External identifiers, database integration 303
External resource linking, databases 301*F*
External suppliers, compound acquisition/processing 563–564
Extracellular matrix (ECM), cell culture 622–623
Extrapolation of data, microarrays 103
Extrinsic pathway (death receptors), apoptosis *see* Apoptosis

F

^{18}F
PET 661–662, 664*T*
pharmacokinetic measurements 672
proliferation probes 665
Factor VIIa inhibitors, PASP 809
Factual databases *see* Databases
[^{18}F]Fallypride, central nervous system imaging 667
False discovery rate (FDR), microarrays 98–99, 101–103
False-negatives
HTS 684
protein interaction networks, bioinformatics 335
False-positives
fluorescent cell-based assays 625
HTS 684
protein interaction networks, bioinformatics 335
proteomics 41–42
Fast acquisition schemes, NMR structure determination 491
FATCAT, protein function prediction, bioinformatics 336–337
^{123}I Fatty acids, as metabolic probe 665
[^{18}F]FdUrd, radiotreatment imaging 670
Feature identifiers, UniProtKB 359
Feature table, DDBJ/EMBL/GenBank 352
Federal sources, compound collections (libraries) 132
Female reproductive tract disorders, vascular endothelial growth factor (VEGF) *see* Vascular endothelial growth factor (VEGF)
Fermenters/fermentation 419
see also Protein production
disulfide bond expression 116
Ferroportin 1 identification, *Danio rerio* model system 10
[^{18}F]FET, infection imaging 670
[^{18}F]FDG imaging
cellular metabolism 668
androgen ablation 669
cytotoxic agents 668
infection 670
photodynamic treatment 669
viability 668
cellular proliferation imaging 671
5-fluorouracil 669
infections 670
as metabolic probe 665
photodynamic treatment 669
prostate cancer model 670
radiotreatment imaging 670
steroid hormone imaging 667
toremifene 669
1,3-[^{18}F]fluoro-αmethyl-tyrosine ([^{18}F]FMT) 665
O-(2-[^{18}F]fluoroethyl)-l-tyrosine ([^{18}F]FET) 665
16-[^{18}F]fluoro-l-DOPA ([^{18}F]DOPA) 665
2-[^{18}F]fluoro-l-tyrosine ([^{18}F]Tyr) 665
Fibroblast growth factor (FGF), receptor interaction 583
Figure of merit (FoM), isomorphous replacement 459
FileMaker Pro, desktop databases 294
Film, x-ray detection 468
Filter elements, NMR 909
Filtration assay *see* Radioactive assays
Final compound specification, solution phase library synthesis 771
Finances, genetically engineered animals 159
Fingerprint Models 298
Firebird SQL 294
Fischer, Emil, 'lock-and-key' metaphor 959–960, 960*F*
Fixation, dynamic combinatorial libraries 963

Flash Plate assays 585, 588
charge coupled devices 589
DNA helicase 589
DNA polymerase 588
DNA primase 588
GPCRs 585–586
HTS 587
kinases
LANCE technology *vs.* 587
serine/threonine kinases 587
small molecule inhibitors 587
tyrosine kinases 587
PARP-1 inhibitors 589
peptide–protein binding assays 589
surface coatings 585
washing steps 585
Flatfiles
DDBJ/EMBL/GenBank 352, 353*F*
Third Party Annotation (TPA) 355, 355*F*
FlexPharm program, structure-based virtual screening 252
FlexX program, protein–ligand docking programs 406
Flow cytometry
biomarkers 74
cell-based screening assays *see* Cell-based assays
RNAi detection 178
Flow-through system, CEM Discover platform microwave-assisted chemistry 846, 846*F*
[18F]FLT
cellular proliferation imaging 671
infection imaging 670
Fluorescence
HCS labeling techniques 688
small animal imaging 660*T*, 661
Fluorescence anisotropy, fluorescence polarization 603
Fluorescence correlation spectroscopy (FCS), fluorescence fluctuation techniques 608, 608*F*
Fluorescence fluctuation techniques 606
data acquisition 608
multichannel scaler (MCS) trace 608, 608*F*
fluorescence correlation spectroscopy 608, 608*F*
fluorescence intensity distribution analysis 609, 610
fluorescence intensity and lifetime distribution analysis (FILDA) 610
fluorescence intensity multiple distribution analysis (FIMDA) 610
quenching 609–610, 610*F*
technique 609
instrumentation 607
confocal optics 607
laser excitation 607–608
signal 607
two-dimensional fluorescence intensity distribution analysis (2D-FIDA) 610
technique 610
Fluorescence Imaging Plate Reader (FLIPR) 683
Fluorescence intensity and lifetime distribution analysis (FILDA) 610
Fluorescence intensity distribution analysis *see* Fluorescence fluctuation techniques
Fluorescence intensity multiple distribution analysis (FIMDA) 610
Fluorescence lifetime assays 605, 611*F*
frequency domain acquisition 605–606
future developments 610
confocal FLA 610, 611*F*
confocal TLA 611, 612*F*
time domain acquisition 605–606
Fluorescence polarization (FP) 602
autofluorescence 603
fluorescence anisotropy 603
molecular mass range 603, 604*F*
polarization definition 602
quenching 603, 604*F*
scintillation proximity assay *vs.* 590
target-oriented screening 142–143
Fluorescence resonance energy transfer (FRET) 604
acceptor–donor distances 605
autocalibration 605
autofluorescence 605
biomarker imaging 77
cell-based assays 632
fluorescent cell-based assays 618, 625
technique 604–605
Fluorescence screening assays **599–616**, 601
artifacts 601
autofluorescence 601, 602*F*
inner-filter effects 602
instrument 601
optics 602
quenching 601
bulk fluorescence techniques 602
fluorescence lifetime *see* Fluorescence lifetime assays
fluorescence polarization *see* Fluorescence polarization
fluorescence resonance energy transfer (FRET) *see* Fluorescence resonance energy transfer (FRET)
homogenous time-resolved fluorescence *see* Homogenous time-resolved fluorescence energy transfer (HTR-FRET)
time-resolved anisotropy analysis *see* Time-resolved anisotropy analysis
total fluorescence intensity (FLINT) 602
compound artifact correction 611, 612*F*
artifact correction 613
autofluorescence 611–612, 614
hit identification 613
identification step 612–613, 613*F*
quenching 611–612
read out parameters 614
statistical accuracy 612–613
fluctuation techniques *see* Fluorescence fluctuation techniques
single molecule detection 600
statistics 600
accuracy 600–601
compound activity rating 601
IC_{50} determination 601, 609*F*, 610*F*, 611*F*, 612*F*
K_D determination 601
suitability 600
Fluorescent dyes
cell-based screening assays *see* Cell-based assays
2D-PAGE protein quantitation 35, 35*T*
Fluorescent imaging, biomarkers *see* Biomarkers
Fluorescent labeled samples, 2D-PAGE 30–31
Fluorescent proteins, HCS labeling techniques 690
2-[18F]Fluoro-2-deoxy-D-glucose (FDG), positron emission tomography 77
16 alpha-[18F]Fluoroestradiol, steroid hormone imaging 667
Fluorogenic substrates, target-oriented screening 142
5-Fluorouracil
[18F]FDG imaging 669
mechanism of action 7
Fluorous split–mix library synthesis, high throughput purification 868
Fluoxetine, chemogenomics 934*F*, 935
Flux balance analysis, metabolic networks 329
Focused libraries, solution phase library synthesis 769
Focusing effects, protein crystallography 440
FoldIndex, domain mapping 879
Force-field molecular modeling, structure-based virtual screening 251
Form factor (*f*), anomalous dispersion 455*F*, 459
Formic acid, microwave heating 840*T*
N-Formyl-N-Met-L-Leu-L-Phe-OH, NMR structure determination 490

Förster resonance energy transfer (FRET) *see* Fluorescence resonance energy transfer (FRET)
Forward genetics
chemical biology 130, 131*F*
Danio rerio model system *see* Zebrafish (*Danio rerio*)
Drosophila melanogaster model system 9
human cells 14
Saccharomyces cerevisiae model system 5–6
Fractionated precipitation, protein purification 421
Fragmentation, genome sequencing 317
Fragment-based drug discovery **939–957**
Bcl-xL, fragment linking 950, 951*F*
caspase-3 950
chemical space coverage 943
c-Src 953, 954*F*
fragment linking 953, 954*F*
substrate activity screening (SAS) 953–954
definition 940, 940*F*
examples 949
follow-up strategies 941, 942*F*
fragment elaboration 941
historical perspective 941
HTS *vs.* 941
iterations 940–941, 941*F*
LFA-1, reverse screening 952, 953*F*
ligand efficiency 943
'rule of five' 943–944
'rule-of-three' 943–944
mass spectrometry 947
advantages 947
disadvantages 947
NMR spectroscopy 944
advantages 946
chemical shifts 945
competition experiments 946
enzyme function 946
example 952, 953*F*
fragment linking 950, 951*F*
ligand observation 945, 945*F*, 946
nuclear Overhauser effects (NOEs) 946
protein observation 945, 945*F*
relaxation rates 946
reporter ligand observation 945*F*, 946
two-dimensional 945
p38α MAP kinase
fragment elaboration 949, 949*F*
fragment mapping 949*F*, 950
SAR 941
screening technologies 940, 944
small libraries 942, 943*F*
structure determination 940
surface plasmon resonance 947
advantages 948
technique 947–948
technologies 944
see also specific technologies
tethering 947
definition 947
example 950
site-directed screening 947
technique 947
2D similarity searching 246
'undruggable' targets 944
weak hits 941
x-ray crystallography 946
advantages 947
example 949, 949*F*, 950
history 946–947
protein–ligand complexes 946
Fragment bitstrings
similarity searching 246
structure/substructure searching 238, 238*F*
Fragment elaboration, p38α MAP kinase 949, 949*F*
Fragment evolution, protein crystallography, rational drug design 893, 893*F*
Fragment linking
fragment-based drug discovery 953, 954*F*
NMR spectroscopy 950, 951*F*
protein crystallography, rational drug design 893, 893*F*
Fragment mapping, p38α MAP kinase 949*F*, 950
Fragment optimization, protein crystallography, rational drug design 893, 893*F*
Fragment self-assembly, protein crystallography, rational drug design 893, 893*F*
'Fragnomics,' protein crystallography, rational drug design 892
Franklin, Rosalind, protein crystallography 450
Free R factor monitoring, protein crystallography model generation 464
Frequency domain acquisition, fluorescence lifetime assays 605–606
Friedel's law, isomorphous replacement 459
^{19}F screening, NMR, drug discovery 908*T*, 910*F*, 911
F test, microarrays 92, 101–103
FT (feature) lines, UniProtKB flatfile 362, 362*F*
Full-text databases *see* Literature databases
Full-text searching, target databases 300
Functional analysis, protein sequence databases 357
Functional Connectivity Fingerprints (FCFPs), 2D similarity searching 246
Functional endpoints, genetically engineered animals 155–156
Functional groups, dynamic combinatorial chemistry 961, 961*F*
Functional imaging, biomarkers *see* Biomarkers
Functionalized pyrazole synthesis, PASP 797, 798*F*
Function-based target validation 3–4
Furan-4-ones
solid-phase discrete libraries 709*T*
solid-phase libraries 720–721, 720*F*
Furazano[3,4-*b*]pyrazines, solid-phase discrete libraries 709*T*
Fused bicyclic lactams, solid-phase discrete libraries 709*T*
Fusion proteins, protein crystallization 436–437

G

^{67}Ga
biomolecular probes 666
SPECT 663*T*
^{68}Ga
biomolecular probes 666
PET 664*T*
'Gain of function' mutants, chemical mutagenesis 13–14
'Gain-of-function' screens, cDNA functional screening 15, 15*F*
β-Galactosidase, genetically engineered animals 153–154
Galantamine (galanthamine), solid-phase pool libraries 707*F*
Galanthamine analogs, solid-phase libraries 745, 745*F*
Ganciclovir, [^{18}F]FDG imaging 669
Gap interpretation, homology-based protein structure prediction 326
Garrod, Archibald, pharmacogenetics 58
Gastroesophageal reflux disease (GERD), knockout mouse model 13
Gateway System, protein 3D structure, protein production 414
Gefitinib, biomarkers 79
GE Healthcare, microarrays 88, 89
Gel diffusion, protein crystallization 439
Gel stains, 2D-PAGE protein quantitation 35
GenBank
see also DDBJ/EMBL/GenBank
history 350
Gene(s)
activity, genetically engineered animal characterization 153
characterization, genetically engineered animals 152

co-expression, cDNA microarrays 332
conservation, *Drosophila melanogaster* 649–650
copy changes, biomarkers 75
duplication, protein function prediction 334
genetically engineered animal characterization, presence 153
genetically engineered animals, role definition 159
identification, genetic variation analysis 338–339
name standardization, UniProtKB/Swiss-Prot 359–360
neighborhood, genomic context methods, bioinformatics 336
order 336
RNA interference *vs.* 17
Saccharomyces cerevisiae model system
deletion combinations 6–7
function assessment 6
structure *322F*
Gene3D
InterPro, member of 363
signature method 363
Gene expression
biomarkers *see* Biomarkers
genetically engineered animal characterization 153
Gene expression microarrays, genomics 5
Gene Expression Omnibus (GEO), cDNA microarrays 331
Gene expression vectors 107
codon usage 109
host cell preference 109
strain specificity 109
copy number 107
definition 107
dual promoters 108
multiple cloning sites 108
multisystem expression vectors 108
promoters 108
inducible 108
isopropyl-β-D-1-thiogalactopyranoside (IPTG) 108
lac derived 108
leaky expression 108
P_{BAD}/AraC combination 108
pET expression 108
protein folding 108
T7-promoter 108
regulatory DNA sequences 109
ribosome binding site (RBS; Shine–Dalgarno sequence) 109
translation initiation signals 109
selection markers 107
antibiotic resistance genes 107
transcription termination 108
translational fusion constructs 109, 110*T*
benefits 109
DsbA 110*T*
DsbC 110*T*
glutathione-*S*-transferase 110, 110*T*
maltose binding protein (MBP) 110, 110*T*
NusA 110*T*
peptide tags 110
poly(His)$_x$ tag 110, 110*T*
protein D 110*T*, 111
protein G 110*T*
protein purification 110
protein tags 110
protein yield effects 110
restriction proteases 111, 111*T*
streptavidin 110, 110*T*
thioredoxin 110, 110*T*
ubiquitin 110
Gene knockout systems, *Danio rerio* model system 10
Gene Ontology Consortium
PDB 376–377
protein function prediction, bioinformatics 333–334
Gene silencing, RNAi *see* RNAi (RNA interference)
Genetic algorithms (GA), computational filters 258
Genetically engineered animals **151–170**
see also Knock-in animal models; Knockout animal models; Transgenic animals; *specific strains*
applications 158
financial constraints 159
gene role definition 159
HTS 159
applied research 160
genotoxicity 160
basic research 159
animal genetic backgrounds 160
disease molecular mechanisms 159–160
lead validation 159
signal transduction effects 160
target characterization 159
target validation 159
treatment effects 160
as bioreactors 151
carcinogenicity research 161
false positives 161–162
Ha-*ras* overexpression 161
p53 deletion 161, 163*F*
pathway correlation 161–162, 163*F*
tumor suppresser gene removal 161, 163*F*
generation 152
gene characterization 152
improvements 153
multimutants 153
speed phenotyping 153
genotypic analysis 153
ELISA 153–154
gene activity 153
gene expression 153
gene presence 153
histochemistry 154
Northern blot analysis 153–154
protein quantification 153–154
qualitative localization 154
reverse transcription polymerase chain reaction (RT-PCR) 153–154
ribonuclease protection assay 153–154
vector-specific markers 153–154
Western blots 153–154
limitations 163
extrapolation limitations 164–165
gene expression variation 163–164
spontaneous conditions 164
wild-type *vs.* 164
negative phenotype interpretation 157
assay quality 158
explanation 157–158
inter-laboratory variation 157–158
'no observable effect level' (NOEL) 157
phenotypic analysis 154, 155*T*
biomarker 154
electrophysiology 155–156
functional endpoints 155–156
high-throughput 154
lethal genes 155
metabolomics 155–156
screening plans 155
sexual dimorphism 154
structural abnormalities 155
surrogate endpoints 154
wild-type animal *vs.* 156
positive phenotype interpretation 156
analysis quality 156
confounding factors 155*T*, 156
delivery system toxicity 156
epigenetic factors 156
insertional mutagenesis 156
regulatory elements 156
severity estimate 156
target-condition correlation 156
species choice 151
'Genetic isolation' studies, common complex diseases 55
Genetic polymorphisms, siRNAs 176
Genetic profiling, chemical biology 144, 144*F*
Genetic redundancy, siRNAs 177
Genetic test definitions, pharmacogenetics 65

Genetic variation analysis 338
  association analysis 339
  pathogens 339
    highly active antiretroviral therapy (HAART) 340
  predispositions 338
    association analysis 339
    drug identification 339
    gene identification 338–339
    haplotype blocks 338
    Iceland Genotyping Program 339
    International HapMap Project 338
    linkage analysis 339
    molecular markers 338–339
    multiple genes 339
    pharmacogenomics 339
    single-nucleotide polymorphisms 338
    statistical genetics 338–339
    target identification 339
Genome functionalization through arrayed cDNA transduction (GFAcT) 3–4
Genome Reviews 356
  data enhancements 356
    additional data 356
    annotation 356
    mapping on protein sequences 356
  evidence tags 356, 357*F*
  further developments 357
  Integr8 367
  statistics 357
Genomes
  assembly checking 324
  IPI (International Protein Index) 365–366
  proteins per gene 29
  rearrangements, bioinformatics *see* Molecular sequence analysis, bioinformatics
  Saccharomyces cerevisiae model system 6, 6*T*
Genome sequencing 316
  Arabidopsis thaliana 317*T*
  assembly algorithms 317
  BAC-based 318
  bioinformatics 316, 317
  Caenorhabditis elegans model system 647–649, 648*T*, 652
  clone-by-clone approach 318
  definition 317
  Drosophila melanogaster 317*T*, 647–649, 648*T*, 650
  Escherichia coli 317*T*
  fragmentation 317
  genomes sequenced 316, 317*T*
  Haemophilus influenzae 317*T*
  hierarchical shotgun sequencing 318
  Homo sapiens 317*T*
  mouse models 317*T*, 648*T*
  repeats 317
  Saccharomyces cerevisiae 317*T*, 647–649
  zebrafish (*Danio rerio*) 10, 647–649, 648*T*, 652–653
Genomic amplification
  biomarkers 75
  mammalian protein expression systems 118
Genomic context methods, bioinformatics *see* Protein function prediction
Genomic deletion, biomarkers 75
Genomic Object Net 330
Genomics
  cDNA functional screening *see* cDNA functional screening
  computation chemical libraries 5
  drug discovery 5
  drug toxicity markers 78
  forward genetics *see* Mouse models, chemical mutagenesis
  gain-of-function analysis *see* Transgenic animals
  gene expression microarray 5
  HTS 5
  human cells 14
    see also specific techniques
    forward genetics 14
    reverse genetics 14
    techniques 14
  large-scale virus vector libraries 18
    cell cycle 18–19
    MISSION 19
    oncogenesis 18–19
    p53-dependent pathways 18–19
    PTEN identification 18–19
    REST/RSF identification 18–19
    retroviruses 18–19
    RNAi consortium (TRC) 19
    signal transduction 18–19
    stress signaling 18–19
    techniques 18–19
    TGFBR2 identification 18–19
    transcription regulation 18–19
  loss-of-function analysis *see* Knockout animal models
  models 5
    see also Caenorhabditis elegans model system; *Drosophila melanogaster*; *Saccharomyces cerevisiae* model system; Zebrafish (*Danio rerio*)
    Mus musculus see Mouse models
  oncology 5
  QTL analysis 14
    obese mouse strain 14
    single nucleotide polymorphisms 14
  receptor tyrosine kinase FLT3 identification 5
  reverse genetics *see* Knockout animal models
  RNA expression 5
  RNA interference *see* RNAi (RNA interference)
  target identification 5
  target validation 2
Genotoxicity
  bacteria-based screening assays 160–161
  genetically engineered animals 160
Genotypic analysis, genetically engineered animals *see* Genetically engineered animals
GenPept 357
Geometrical parameters, protein 3D structure validation *see* Protein 3D structure validation
Geometric features
  CSD 396, 396*F*
  PDB errors 522*F*, 523
GEPASI, metabolic networks 329
Gi assays, GPCRs 636, 638*F*
Glial cell-derived neurotrophic factor (GDNF), knockout animal models 156
Global Diverse Representative Subset (GDRS), compound storage/management (Pfizer) 570–571
Global Electronic Compound Registration system, compound storage/management (Pfizer) 564
GlobPlot 2, domain mapping 879
$\sim\alpha_{2U}$-Globulin, structure 535
Glucocorticoids, signaling pathway, corticotrophin-releasing factor overexpression 99
Glucose, as biomarker, diabetes mellitus 70
[^{3}H]2-deoxy-D-Glucose, androgen ablation imaging 670
2-deoxy-D-[1-^{14}C]-Glucose, radiotreatment imaging 671
Glutamine deaminations, protein crystallization 436
Glutathione-*S*-transferase (GST)
  as affinity tag 421
    Escherichia coli protein expression systems 113–114
    problems in crystallization 436–437
  gene expression vectors 110, 110*T*
Glycerol, protein crystallization 881
Glycine betaine analogs, solid-phase discrete libraries 709*T*
Glycosylation
  heterogeneous protein crystallization 882
  proteomics 41
Glyoxalase direct assay format, optical assays 581
Glyphosate *N*-acetyltransferase, crystallography 884–885
Gmelin database 277, 278*T*

GN (gene name) line, UniProtKB flatfile 360
GOLD, protein–ligand docking programs 406
GPCR Annotator 309
GPCRDB, biological information databases 305
GPCR Ligand Database 309
G protein-coupled receptors (GPCRs)
biological functions 229–231
cell-based assays *see* μ-Opiate receptor
cell-free expression systems 122–123
chemogenomics 923, 924*F*
ligands
chemogenomics 932
NMR structure determination 492, 494*F*
members 932
orthogonal ligand–receptor pairs *see* Orthogonal ligand–receptor pairs
PDB 374*F*, 376
radioactive assays *see* Radioactive assays
Saccharomyces cerevisiae expression system 8
scintillation proximity assay (SPA) *see* Scintillation proximity assay (SPA)
signal transduction, HCS receptor internalization 690–691
structure 634, 636*F*
surface plasmon resonance 593
Xenopus laevis 649
G protein-coupled receptors (GPCRs), assays
cell-based assays 624*T*, 634
agonist/antagonist screening 637
cAMP EIA 634–636
Gi assays 636, 638*F*
Gq assays 636, 637*F*
Gs response 634–636
drug design 3
see also specific drugs
filtration assay 585–586
Flash Plate assays 585–586
fluorescent cell-based assays 627
Gq assays, GPCRs 636, 637*F*
Granisetron, 5-HT_3 receptor binding 633–634
Granulocyte colony-stimulating factor (G-CSF), STAT3 signaling 206
Granulocyte-macrophage colony-stimulating factor (GM-CSF), STAT5 signaling 206
Graphite, microwave-assisted chemistry 853
Graphs
structure/substructure searching 236–237
2D similarity searching 246
Green fluorescent protein (GFP)
biomarker imaging 77
genetically engineered animals 153–154
HCS 690
history 690
protein crystallization 888
variants 690
GroEL, protein crystallization 887
GroEL/GroES system, *Escherichia coli* protein expression systems 112–113
Growth factor receptors
Drosophila melanogaster 650
STAT signal transduction/activation 204
Growth factors, vascular endothelial growth factor expression 200
Growth hormone (GH)
STAT3 signaling 206
STAT5 signaling 206
Gs subfamily, GPCRs 634–636
GTPases
biological functions 216–217
orthogonal ligand–receptor pairs 216, 219*F*
Guanidine derivatives, solid-phase discrete libraries 709*T*
Guanidinium chloride, protein crystallization 881
GVK Biosciences, commercially available databases 309

H

Haemophilus influenzae, genome sequencing 317*T*
Halogens, intermolecular interactions 404
Haloperidol receptors, imaging 667
Hanging drop method, protein crystallization 438–439
Haplotype blocks, genetic variation analysis 338
Haptoglobin depletion, proteomics 31–32
Ha-*ras* overexpression, genetically engineered animals 161
Harker construction, isomorphous replacement 457–459, 458*F*
Harvesting buffer, crystal handling 465–467
'Hashed fingerprints,' structure/substructure searching 239
Header, DDBJ/EMBL/GenBank 352
Head/neck cancer, STATs 207–208
Heart disease, *Danio rerio* model system *see* Zebrafish (*Danio rerio*)
Heat dissipation, x-ray generation 465, 466*F*
Heatmaps, microarrays 91
Heat shock protein-90 identification, *Saccharomyces cerevisiae* model system 5–6
Heat shock proteins, RNAi 181
Heck reaction, PASP 813
HEK-293 cells
cell-based assays 620–621
mammalian cell expression systems 418
protein expression systems 117
Hemiacetal formation, dynamic combinatorial library generation 962*F*
Hemoglobin A1c, diabetes mellitus biomarker 70
Hemophilia, as monogenetic disease 338
Hepatocytes, cell-based assays 629
hERG blocker screening, fluorescent cell-based assays 625–626, 626*F*
Herpes simplex viruses (HSV), transgenic mouse model promoters 11
Heterocompounds, protein 3D structure validation *see* Protein 3D structure validation
Heterocyclic libraries
solid-phase libraries 719–720
solution phase library synthesis *see* Solution phase synthesis, library preparation
Heterocyclic scaffolds, solid-phase libraries 703–706, 707*F*
Heterocyclic thiourea library preparation, PASP 830
Heterogeneity, database integration 303
Heterogeneous proteins
crystallization *see* Protein crystallization
protein crystallization *see* Protein crystallization
Hexane, microwave heating 840*T*
[^{99m}Tc]Hexylmethylpropylene amineoxone, androgen ablation imaging 670
HIC-Up database, protein 3D structure validation 514
Hidden Markov model (HMM)
protein function prediction, bioinformatics 335
sequence alignment, bioinformatics 320
Hierarchical databases *see* Databases
Hierarchical shotgun sequencing
genome sequencing 318
Human Genome Project 318
High concentrations, PASP 795
High-content screening (HCS) 685
advantages 693
applications 693
assay parameters 685
autofocusing 687
cell cycle 691
BrdU incorporation 691–692
DNA content information 691–692
EGFP 691–692
mitotic index 691–692

High-content screening (HCS) (*continued*)
cellular toxicity 691
colony-formation assays 692
confocal systems 687
Nipkow disc 687, 688*F*
resolution 687
signal-to-noise ratio 687
data acquisition 685
data analysis 687, 689*F*, 692
lack of standardization 687
data storage 692
hardware 685
history 685
image detection 686*T*, 687
CCD 685, 687
laser scanning cytometer 687
photomultiplier tubes 687
labeling techniques 688
antibody staining 689
bright-field microscopy 686*T*, 688
cellular stains 689
DNA dyes 689
fluorescence 688
fluorescent proteins 690
quantum dots 690
subcellular stains 689
micronucleus formation 692
multidimensional cytological profiling 693
neurite overgrowth 692
nuclear translocation assays 691
NF-κB 691
phenotype-oriented screening 138
receptor internalization 690–691
GPCR 690–691
translocation events 690
High-density lipoproteins (HDLs), as biomarker, cardiovascular disease 70
High gravity methods, protein crystallization 439
Highly active antiretroviral therapy (HAART), genetic variation analysis 340
High performance liquid chromatography (HPLC), high throughput purification *see* High throughput purification
High-performance liquid chromatography-mass spectrometry (HPLC-MS) 862
advantages 863
detection methods 863
historical perspective 862–863
parallel synthesis 863
proteomics 37
technology 863
High throughput purification **861–874**
combinatorial chemistry 861
fluorous split–mix library synthesis 868
hit generation support 861
intelligent method development 867
lead generation support 861
library purity assessment 863
chemiluminescene nitrogen detection (CLND) 863
evaporative light-scattering detection (ELSD) 863–864, 864*F*
target purity 863
UV chromatograms 863
mass-directed 865
preparative LC/MS 865–866, 866*F*, 867*F*
monoliths 867
parallel analysis/purification 868
advances 871
advantages 869
commercial systems 870
limitations 871
rapid column switching 868, 870*F*
technique 869
preparative liquid chromatography/MS demixing 868
streamlining 871
efficiency 871
robustness 871–872
techniques 864
see also specific techniques
turbulent flow media 867
UV-directed 864
automation 865
intelligent systems 865
MS coupling 865
multiple fractions 865
High-throughput screening (HTS) **679–696**
assay prerequisites 682
assay quality evaluation 683
false negatives 684
false positives 684
hit confirmation 684
nonspecific plate blank (NSB) 683–684
parallel screens 684
screening window coefficient 684
signal-to-background 684
signal-to-noise 684
target-unrelated effects 684
automation 681
advantages 682
assay dependency 682
combinatorial synthesis 681
large robotic system 682
operations needed 681
parallel synthesis 681
pipetting station/stacker dispensing 681
quality standards 682
small robotic system 682
workstations 681–682
cell-based assays 618
characteristics 599
chemical biology *see* Chemical biology
compound numbers 680
compound storage/management 569
Danio rerio model system 10
definition 679
development 130
Flash Plate assays 587
fragment-based drug discovery *vs.* 941
genetically engineered animals 159
genomics 5
historical perspective 680
Association of Laboratory Automation 680
Society for Bimolecular Screening 680
hit characterization 684
see also High-content screening
hit–to–lead process 684
see also High-content screening
hit validation 902
homogeneous soluble biological assays 600
importance 577
knockout animal models 159
lead identification 940
library composition 680
compound stability 681
'drug-like' properties 680
natural lead analogs 680
metalloproteomics 556–557
microplate formats 682–683
definition 682–683
high well densities 682–683
volumes 682–683
optical biosensors 594
phosphatase assays 582
problems 1
protein crystallography, rational drug design 890
purification *see* High throughput purification
scintillation proximity assay 587
single-parametric biochemical assays 683
homogeneous *vs.* heterogeneous assays 683
protein purity 683
simplification 683
single-parametric cellular assays 683
Fluorescence Imaging Plate Reader (FLIPR) 683
reporter gene assays 683
surface plasmon resonance 594
target selection 681
'druggability' 681
turbidimetric assays 595
High-throughput synthesis, CEM Discover platform microwave-assisted chemistry 846, 847*F*
High well densities, HTS 682–683
HinCyc database, metabolic networks 328–329

Hippocampus, corticotrophin-releasing factor overexpression 99
His_6 tag
affinity chromatography 436–437
problems in crystallization 436–437
His/Asn/Gln side chain flips, x-ray structures, problems 537
Histamine receptors
radioactive assays 586
scintillation proximity assay 586
Histochemistry, genetically engineered animal characterization 154
Histogram matching, protein crystallography model generation 462
Histone deacetylase (HDAC) inhibitors
PASP 811, 812*F*
solid-phase libraries 741
Histones, modifications, proteomics 41
Hit(s)
confirmation 684
generation support 861
identification
fluorescence screening assays 613
HTS 684
validation 591, 592*F*
HIV, protein structures on PSD 382
HIV infection, pharmacogenetics 61
HIV protease
Rous sarcoma virus *vs.* 508–509
structure 508–509
HIV protease inhibitors, discovery/development, surface plasmon resonance 593–594
Hodgkin, Dorothy Crowfoot, protein crystallography 450
Hofmeister Series, protein crystallization 881
Homodimers, STAT (signal transducer and activator of transcription) association 204
Homogenous samples, proteomics 30–31
Homogenous time-resolved fluorescence energy transfer (HTR-FRET) 605
scintillation proximity assay *vs.* 590
Homologous genes, protein function prediction, bioinformatics 334
Homologous recombination, knockout mouse model generation 11–13
Homolog screening
protein crystallization 884
protein structure prediction 325
Homology-based protein structure prediction *see* Protein structure prediction
Homology-based techniques, RNA structure prediction, bioinformatics 327
Homology models 544
databases 544
Live Bench website 545
model databases
ModBase 545
molecular dynamics (MD) 544–545
problems 544
reliability 544
CASP (Critical Assessment of tp Structural Prediction) 544
initial sequence alignment 544
SWISS-MODEL 544
Homo sapiens
genome sequencing 317*T*
proteome complexity 29*T*
Host cell preference, gene expression vectors 109
5-HT receptor(s), molecular probes 664*T*
5-HT receptor antagonists
see also specific drugs
discovery/development 932, 932*F*
$5\text{-}HT_{1A}$ receptor, knockout animal models 156
$5\text{-}HT_3$ receptor 624*T*, 633, 635*F*
antiemetic treatment 633–634
cell choice 633–634
receptor affinity 635*T*
distribution 633–634
$5\text{-}HT_3$ receptor antagonists
see also specific drugs
discovery/development 932, 932*F*
$5\text{-}HT_4$ receptor antagonists
see also specific drugs
discovery/development 932, 932*F*
HTML, database integration 303
Huber, Robert, protein crystallography 451
Human embryonic kidney cells *see* HEK-293 cells
Human genome, mouse genome *vs.* 10–11
Human Genome Project
drug discovery 146
hierarchical shotgun sequencing 318
Hünefeld, Friedrich, protein crystallization 881
Hunter-Bolton reagent, direct iodination 666
Huntington's disease
Drosophila melanogaster model system 9
knockout mouse model 11–13
Hybridization probe uniqueness, repetitive elements, bioinformatics 324
Hybridoma technology 118
Hydantoin derivatives, solid-phase discrete libraries 709*T*
Hydantoin synthesis
PASP 799–800, 800*F*
solution phase library synthesis 786, 786*F*
Hydrogen atoms, protein 3D structure validation 513
Hydrogen bonds
intermolecular interactions, CSD 401–402
Mercury Visualizer 395
protein 3D structure validation *see* Protein 3D structure validation
strong, CSD 402, 403*F*, 404*F*
weak, CSD 404
Hydrogen gas, PASP 800*F*, 827, 827*F*
Hydrogen/potassium ATPase, knockout mouse model 13
Hydrophobic interaction chromatography, protein purification 423
Hydrophobicity, prediction, chemoinformatics, ADME property prediction 259
Hydroxamic acids, derivatives, PASP 811, 811*F*
3,4-Hydroxymethyl isoxazoles, solid-phase discrete libraries 709*T*
Hydroxyproline derivatives, solid-phase discrete libraries 709*T*
5-Hydroxytryptamine receptors *see entries beginning 5-HT*
Hyperlinks
CSD 396–398
databases 300
Hypoxanthine phosphoribosyltransferase (Hprt), knockout animal models 161
Hypoxia
RNA interference 17–18
vascular endothelial growth factor expression 200
Hypoxia-inducible factor-1 (HIF-1), vascular endothelial growth factor expression 200

I

^{123}I
biomolecular probes 666
proliferation probes 665
SPECT 663*T*
^{124}I
biomolecular probes 666
PET 664*T*
^{125}I
biomolecular probes 666
proliferation probes 665
scintillation proximity assay 584
SPECT 663*T*
^{131}I
biomolecular probes 666
SPECT 663*T*
IBM DB2, server databases 294
IC_{50}
biological activity databases 301
determination, fluorescence screening assays 601, 609*F*, 610*F*, 611*F*, 612*F*

Iceland Genotyping Program 339
iCyte 686*T*, 687
ID lines, UniProtKB flatfile 360
IFIPAT 276*T*
IκB kinases (IKKs)
NF-κB pathway activation 195
structure 195
Image detection, HCS *see* High-content screening (HCS)
Image express 686*T*
Image formation, x-ray–crystal interaction 454
Image plates, x-ray detection 466*F*, 468
Imaging
drug toxicity markers 78
small animals *see* Small animal imaging
Imaging mass spectrometry, proteomics *see* Proteomics
Imatinib mesylate
c-kit inhibition 3
cross-reactivity 3
diabetes mellitus treatment 3
discovery/development 930, 931*F*
PDGFR inhibition 3
IMC-1C11 202–203
Imidazole derivatives, solid-phase discrete libraries 709*T*
Imidazolidinedione derivatives, solid-phase discrete libraries 709*T*
4-Imidazolidinones, solid-phase pool libraries 705*F*
Imidazolyl glycopyranosides, solid-phase pool libraries 705*F*
Imines
formation, dynamic combinatorial library generation 961–962, 962*F*
hydrogenation, PASP 827, 827*F*
Immobilized scavengers, PASP 795, 795*F*
Immobilized species *see* Polymer-assisted solution phase synthesis (PASP)
Immortalized cell lines, techniques 620
Immunoassays 578
see also specific types
biomarkers 73–74
Immunoblotting, protein quantitation 73–74
Immunofluorescence, RNAi detection 178
Immunoglobulin A (IgA), proteomics 31–32
Immunoglobulin G (IgG), proteomics 31–32
Immunohistochemistry, biomarkers 74, 75
Immunostaining, biomarkers 79–80
IMS New Product Focus 278*T*
IMS World Drug Monographs 278*T*
^{111}In
biomolecular probes 666
SPECT 663*T*
IN Cell 1000 686*T*
IN Cell 3000 686*T*, 687
InChI, chemical information databases 298
Inclusion bodies
formation, *Escherichia coli* protein expression systems 112
purification, bacterial expression systems 415
Indels, sequence alignment, bioinformatics 318–319
Indenopyrrolocarbazoles, solid-phase libraries 703, 705*F*
Indirect competition enzyme-linked immunosorbent assay 579–580
Indole derivatives, solid-phase discrete libraries 709*T*
Indolizino[8,7-b]-5-carboxylates, solid-phase discrete libraries 709*T*
Indomethacin analogs, solid-phase discrete libraries 707*F*
[^{111}In]-DTPA-PEG annexin, apoptosis imaging 668
Inducible nitric oxide synthase (iNOS), expression, NF-κB pathway 197
Inducible promoters
gene expression vectors 108
transgenic mouse model 11
Infection imaging
[^{18}F]FDG 670
[^{18}F]FET 670
[^{18}F]FLT 670
Infections
[^{18}F]FDG 670
small animal imaging 670
Inflammation, NF-κB pathway 192, 197
Inflammatory bowel disease (IBD), NF-κB pathway 197
Inflammatory mediators, cell-based assays, Toll receptors 639
Information extraction, protein function prediction, bioinformatics 337
Information integration, protein function prediction, bioinformatics 338
Information retrieval, protein function prediction, bioinformatics 337
Information sources, databases 285
Information systems **265–291**
see also Databases
definition 265
Informed consent, pharmacogenetics 65–66
Ingres 294
Inhibitor of NF-κB (IκB), ubiquitin-protein targeting 192
Inhibitors of apoptosis (IAPs), regulation, NF-κB pathway 196–197
In-house databases 284
In-house synthesis, compound acquisition/processing 563–564
Initial sequence alignment, homology models reliability 544
Initiation factors (IFs), scintillation proximity assay 588
Initiator reactor, Biotage AB microwave-assisted chemistry 844, 845*F*
Inner-filter effects, fluorescence screening assays 602
INPADOC (International Patent Documentation Center Database) 275, 276*T*
INPHARMA, NMR 948
Insect cells, cell-based assays 619
Insertional mutagenesis, genetically engineered animals 156
in situ hybridization, biomarkers 75
Insolubility problems, *Escherichia coli* as expression system 880
Instruments, fluorescence screening assays 601
Insulin
crystal packing 442
structure 382*F*
Insulin-like growth factor-1 receptor (IGF-1R)
Drosophila melanogaster 650
protein crystallization, rational surface engineering 887
Insurance risk, pharmacogenetics 67
Integr8 366
BioMart 367, 368
browser 368
CluSTr database 367
Genome Reviews 367
IPI 367
manual curation *vs.* 366
proteome analysis 367, 367*F*
classification 368
CluStr database 367–368
structural information links 368
structure 366
UniParc 367
Integral membrane proteins, protein crystallization *see* Protein crystallization
Integrated databases 363
see also specific databases
Integrating detectors, x-ray detection 468
Integration
databases 284
protein crystallography 468–469
Integrins, ligands, chemogenomics 933, 933*F*
Intelligent systems
development 867
high throughput purification 865
IntEnz database, metabolic networks 328–329
Interactional units, dynamic combinatorial chemistry 961
Intercellular adhesion molecule-1 (ICAM-1), expression 197

Interfaces, databases 283
Interfering RNA *see* RNAi (RNA interference)
Interferons
RNAi (RNA interference) 172, 176
STAT1 signaling 205–206
STAT2 signaling 206
Interleukin-1 (IL-1), NF-κB pathway activation 197
Interleukin-1 receptor activating kinase (IRAK), node network theory 4
Interleukin-2 (IL-2)
STAT3 signaling 206
STAT5 signaling 206
Interleukin-3 (IL-3), STAT5 signaling 206
Interleukin-4 (IL-4)
STAT5 signaling 206
STAT6 signaling 207
Interleukin-5 (IL-5)
STAT1 signaling 205–206
STAT3 signaling 206
STAT5 signaling 206
Interleukin-7 (IL-7)
STAT3 signaling 206
STAT5 signaling 206
Interleukin-9 (IL-9)
STAT3 signaling 206
STAT5 signaling 206
Interleukin-10 (IL-10)
STAT1 signaling 205–206
STAT3 signaling 206
Interleukin-11 (IL-11)
STAT1 signaling 205–206
STAT3 signaling 206
Interleukin-12 (IL-12)
STAT3 signaling 206
STAT4 signaling 206
Interleukin-13 (IL-13), STAT6 signaling 207
Interleukin-15 (IL-15)
STAT3 signaling 206
STAT5 signaling 206
Interleukin-21 (IL-21)
STAT1 signaling 205–206
STAT3 signaling 206
STAT5 signaling 206
Inter-ligand NOEs, NMR 948
Intermediate purification
solution phase library synthesis 776
solution phase parallel chemistry 767
Intermolecular interactions, CSD *see* Cambridge Structural Database (CSD)
Intermolecular NOE, NMR, drug discovery 908*T*, 912, 913*F*, 914*F*
Intermolecular similarities, cluster-based compound selection 255
Internal cell processes, small animal imaging *see* Small animal imaging
Internal heating, microwaves 840, 841*F*
Internal thermodynamic stability, siRNAs 176
International HapMap Project, genetic variation analysis 338
International Patent Classification (IPC) 275
International Tables for Crystallography Volume A 452
Internet
databases 285
references 301
PDB 374
Internuclear vectors, NMR structure determination 484, 485*F*
InterPro 363
InterProScan 364, 365*F*
member databases 363
members
Gene3D 363
PANTHER 363
PIRSuperFamily 363
PRINTS 363
ProDom 363
PROSITE 363
SMART 363
SUPERFAMILY 363
TIGRFAMS 363
signature integration 363
manual methods 363–364
Markov models 363
InterProScan 364, 365*F*
Intervector angle, NMR structure determination 486, 488*F*
Intracellular protein staining, flow cytometry 630
Intramolecular geometries, CSD applications 400
Intramolecular NOE, NMR, drug discovery 908*T*, 911
Intraocular neovascular syndromes, vascular endothelial growth factor (VEGF) *see* Vascular endothelial growth factor (VEGF)
Intrinsic pathway (mitochondrial), apoptosis 195–196
Intron splicing, mammalian protein expression systems 118
Invitrogen Echo 880
Invitrogen Gateway 880
in vivo gene silencing
RNAi (RNA interference) 182
siRNAs 182
in vivo mutagenicity testing, transgenic animals 161
Ion channels
surface plasmon resonance 593
Xenopus laevis 649
Ion exchange chromatography, protein purification 422
Ion fluxes, cell-based assays 624, 624*T*
Ionic conduction mechanism, dielectric heating 839
Ionic liquids, solvent reactions, microwave-assisted chemistry 854
IPA (International Pharmaceutical Abstracts) 271*T*
IPI (International Protein Index) 364
building of 365
Ensembl 365–366
genomes 365–366
Integr8 367
nonredundant complete UniProt proteome sets 365
predicted coding sequences 366
Irritable bowel syndrome, microarrays *see* Microarrays
Ischemia reperfusion induced lung apoptosis, RNAi (RNA interference) 182
ISI 282
Isoelectric focusing (IEF)
protein analysis 425, 425*F*
2D-PAGE 34
Isoelectric heterogeneity, heterogeneous protein crystallization 882
Isoenzymes, 'classical' pharmacogenetics 58
Isomorphous replacement *see* X-ray crystallography, phase problem solutions
Isopropyl-β-D-1-thiogalactopyranoside (IPTG), gene expression vectors 108
IsoStar (intermolecular interactions) *see* Cambridge Structural Database (CSD)
Isotope-coded affinity tags (ICAT), LC-MS (peptide-based) proteomics 37–38
Isotope labeling
cell-free expression systems 120–121
LC-MS (peptide-based) proteomics 38
Isotopic analog peak ratios, LC-MS (peptide-based) proteomics 38
Isotopic labels, NMR, structure determination 475, 477*F*
Isotropic *B*-factors, x-ray structures, problems 538
Iterated dynamic combinatorial libraries 963
[^{131}I]IUDR, androgen ablation imaging 670
IUPred, domain mapping 879

J

Jandagel, PASP 792–793
ϕ-angle, protein 3D structure validation 516–517
Janus kinase (JAK)
gene isolation 14

Janus kinase (JAK) (*continued*)
STAT signal transduction/activation 204
ubiquitin-protein targeting 192
JAPIO 276*T*
Jarvis–Patrick clustering, cluster-based compound selection 254
Java, database integration 303
JChem Base
ChemAxon Ltd 295
database management systems 295
JCSG Validation Tools 547*T*
JF959602, discovery/development 137
Joint Center for Structural Genomics (JCSG) 551–552, 552*T*
progress 553*T*
Journal descriptors, databases 301
Jubilant Biosys, commercially available databases 309
c-Jun-associated protein degradation, ubiquitin-proteasome pathway 192, 194*F*
Jun NH_2-kinase (JNK)
c-Jun phosphorylation 192–193
ubiquitin-protein targeting 192

K

Kalitoxin, NMR structure determination 493, 495*F*
Kallikrein, corticotrophin-releasing factor overexpression 97–98
KEGG database 328–329
Kendrew, John C 450, 884
Ketoamides
solid-phase libraries 741, 741*F*
solid-phase pool libraries 707*F*
Ketoimidazole derivatives, solid-phase discrete libraries 709*T*
α-Ketothiazole peptidyl protease inhibitors 809, 810*F*
Keywords
PDB searches 376
target databases 299
*Ki*Bank 306
Kinase ChemBioBase 309
Accelrys databases 307
Kinase Knowledgebase 308
Kinesin–myosin, orthogonal ligand–receptor pairs 220
Kinetic Data of Biomolecular Interaction database 305
Kinetics, optical biosensors 578
KineticScan 686*T*
Kleywegt, Gerard 535
K-means, microarrays 93
Knock-in animal models
artemin 160
generation 152
neuropathy treatment 160
Knockout animal models 11
adenine phosphoribosyltransferase (Aprt) 161
Alzheimer's disease 160
beta-secretase 160
conditional knockouts 13
drug discovery 13
embryonic lethality 13
gastroesophageal reflux disease 13
generation 11–13, 12*F*, 152
homologous recombination 11–13
glial cell-derived neurotrophic factor (GDNF) 156
HTS 159
Huntington's disease 11–13
hydrogen/potassium ATPase 13
hypoxanthine phosphoribosyltransferase (Hprt) 161
mouse models 654
nitric oxide synthase (NOS) 157–158
OPG 156, 157*F*
osteopetrosis 156, 157*F*
osteoporosis 156, 157*F*
peroxisome proliferator-activated receptor gamma (PPAR-γ) agonists 13
RANK 156
RANKL (*RANK* ligand) 156
serotonin 1A receptor (5-HT1A) 156
small animal test systems 649
target validation 13
time-specific gene deletions 152
tissue-specific gene deletions 152
KnowItAll program 279, 280*T*
Knowledge-based libraries, CSD 395
Knowledge-based scoring, structure-based virtual screening 251
Kojic acid analogs, solid-phase discrete libraries 709*T*
Kozak sequence, mammalian protein expression systems 118
KW (keyword), UniProtKB flatfile 361*F*, 362

L

Lac derived promoters, gene expression vectors 108
β-Lactamase, optical assays 580–581
β-Lactam synthesis, PASP 829, 829*F*
Lactic acid bacteria, protein expression systems 116
Lactobacillus lactis, protein expression systems 116
S-D-Lactoylglutathione hydrolysis, glyoxalase assays 581
LANCE technology, Flash Plate assays *vs.* 587
Langmuir binding isotherm 903
Language problems, protein function prediction, bioinformatics 337
Large, diverse libraries, solution phase library synthesis 769
Large ribosomal subunit (*Deinococcus radiodurans*) 535
Large robotic system, HTS automation 682
Large-scale genomic screens, *Drosophila melanogaster* model system 9
Large-scale virus vector libraries *see* Genomics
Laser capture microdissection (LCM), proteomics 30–31
Laser excitation, fluorescence fluctuation techniques 607–608
Laser scanning cytometer 687
Laue symmetry, molecular replacement 461–462
Laufburger, Vilem, protein crystallization 885
Lead(s)
identification/characterization
biomarkers 69
definition 939
fragment-based drug discovery *see* Fragment-based drug discovery
high throughput purification 861
HTS 940
NMR, drug discovery 901
protein crystallography *see* X-ray crystallography, rational drug design
surface plasmon resonance 593–594
likeness prediction, computational filters 258
optimization
genetically engineered animals 159
NMR, drug discovery 901, 902
protein crystallography, rational drug design 890
Lead discovery, solid-phase libraries 741, 742*T*–744
Leaky expression, gene expression vectors 108
Lentivirus vectors
RNA interference 17
siRNAs 177
Lethal genes
genetically engineered animals 155
identification, RNAi (RNA interference) 184
Leukemia
acute lymphoblastic (ALL), receptor tyrosine kinase FLT3 5
Danio rerio model system 10
Leukotrienes, pharmacogenetics 60
Levcromokalim, discovery/development 927–929, 929*F*
LFA-1, fragment-based drug discovery 952, 953*F*
Libraries
availability 130
composition, HTS *see* High-throughput screening (HTS)
construction
microwave-assisted chemistry 855

polymer-assisted solution phase synthesis (PASP) *see* Polymer-assisted solution phase synthesis (PASP)
Ras/Raf protein interaction inhibitor 1 *see* Ras/Raf protein interaction inhibitor 1
design, solution phase library synthesis *see* Solution phase synthesis, library preparation
dimethyl sulfoxide (DMSO) stock solutions 862
historical perspective 697, 862
diversity 697
numbers 697
quality 697
large, requirement for 763
purification, solution phase library synthesis 771, 772*T*–774
purity assessment, high throughput purification *see* High throughput purification
purity problems 862
re-analysis/purification 862
see also High throughput purification
size, solution phase library synthesis 769
Lidocaine analogs, solid-phase libraries 725, 725*F*
Life Sciences Collections 271*T*
Ligand(s)
design/docking, structural errors 538–539
errors, X-ray structure *see* X-ray crystallography
identification, *Saccharomyces cerevisiae* model system 8
NMR spectroscopy 945, 945*F*, 946
nomenclature standardization 539
pose prediction 252
Protein Data Bank (PDB) *see* Protein Data Bank (PDB)
structure
bond length errors 538
torsion angles 538
structure–function relationship, structural genomics 557–558
Ligand Depot, PDB 376
Ligand Info 306
Ligand soaking, protein crystallization 891
LigPro Ligand Explorer 377, 378*F*
Limited proteolysis combined with mass spectrometry (LP/MS) 555
Limited proteolysis studies, protein 3D structure, protein production 413
Line clogging, microwave-assisted chemistry 857
Linkage analysis, genetic variation analysis 339
Lipid kinases, radioactive assays 587
Liposomal transfection, siRNAs delivery 177
Lipoxygenase expression, NF-κB pathway 197
Lipscomb, William, protein crystallography, rational drug design 876
Liquid assays, cell-based assays 631
Liquid chromatography, proteomics *see* Proteomics, LC-MS (peptide-based)
Liquid–liquid extraction, solution phase library synthesis 776
Liquid–liquid phases, microwave-assisted chemistry 853
Liquid sample collection, Pfizer *see* Compound storage/management (Pfizer)
Literature databases 269, 270, 271*T*, 273
bibliographic 273
specialist 274
full-text 275
searching 275
patent 275, 276*T*
primary 270
searching 270
secondary 270
specific 275
tertiary 273
Live Bench website, homology models 545
Local correlation tests, protein 3D structure validation 510
Local sample handing, compound storage/management 562
'Lock-and-key' metaphor, Fischer, Emil 959–960, 960*F*
Locked nucleic acids, siRNAs modification 183
Long-chain fatty acyl coenzyme A synthase
radioactive assays 588
scintillation proximity assay 588
Longitudinal relaxation rate, NMR, drug discovery 909, 910*F*
Loss-of-function, chemical mutagenesis 13–14
Loss tangents of substance, dielectric heating 839, 840*T*
Lovastatin, structure 700*F*
Low copy repeats, repetitive elements, bioinformatics 324
Low-density lipoproteins (LDLs), as biomarker, cardiovascular disease 70
Low gravity methods, protein crystallization 439
Low-level analysis, cDNA microarrays 331
Low-pressure column chromatography 422
Low resolution, EM protein structure 542–543
Low weight components, biomarkers 76
Luciferase
biomarker luminescent imaging 77–78
LUDI, protein–ligand docking programs 407
Luminescence-based optical assays 578
Luminescent imaging
biomarkers *see* Biomarkers
drug toxicity markers 78
Lupeol derivatives, solid-phase discrete libraries 709*T*
Lymphoma, models, ^{99m}Tc imaging 668
Lysozyme
NMR structure determination 476, 478*F*
protein crystallization 434
Lysozyme inhibition library, dynamic combinatorial libraries 969, 971*F*

M

Machine learning methods *see* Chemoinformatics, virtual screening
MacKinnon, Roderick, protein crystallography 451
Macrocyclic peptidomimetics, solid-phase pool libraries 705*F*
Macrocyclization, PASP 795
Macrolactones
solid-phase discrete libraries 707*F*
solid-phase pool libraries 705*F*
Macrolides
solid-phase libraries 726, 727*F*
solid-phase pool libraries 705*F*
Macromolecular Crystallographic Information File (mmCIF), Research Collaboratory for Structural Bioinformatics (RCSB) 543
Macugen 202
Magic angle spinning (MAS), NMR, structure determination 475
Magnetic resonance imaging (MRI)
applications 659
small animal imaging 660*T*, 661
biomarkers 77
dynamic contrast enhanced magnetic resonance imaging (DCE-MRI) 77
Magnetic resonance spectrography (MRS), small animal imaging 660*T*
MALDI-TOFTOF analysis, 2D-PAGE 36
Malonyl-coenzyme A inhibitor identification, filtration assay 584

Maltose-binding protein (MBP)
as affinity tag 421
gene expression vectors 110, 110*T*
as sequence tag, problems in crystallization 436–437
Mammalian Gene Collection, cDNA functional screening 15–16
Mammalian protein expression systems 117, 418
cell lines 117
CHO cells 418
HEK-293 cells 418
cell lysis 420
modifications 118
copy numbers 118
genomic amplification 118
viral transcription/translation mechanisms 118
other systems *vs.* 113*T*
stable transfection 418
transient transfection 418
vectors 117
adenovirus late promoter 117
CAAT box 117
intron splicing 118
Kozak sequence 118
promoters 117
Rous sarcoma virus promoter 117
selection markers 117
SV40 promoter 117
target validation 3–4
TATA box 117
translation initiation 118
Mannose-binding protein
Escherichia coli protein expression systems 113–114
structure 540, 541*F*
Manual curation, Integr8 *vs.* 366
Manual methods, InterPro signature integration 363–364
MAP kinase (MAPK)
Saccharomyces cerevisiae model system 5–6
ubiquitin-protein targeting 192
Map skeletonization, protein crystallography model generation 462
Marker classes, pharmacogenetics *see* Pharmacogenetics
Markov model-based systems, InterPro signature integration 363
MARPAT 282*T*
Marseilles Structural Genomics Programme (MSGP) 552*T*
MARS Microwave Synthesis System, CEM Corporation microwave-assisted chemistry 850, 850*F*
Mass spectrometry (MS)
fragment-based drug discovery *see* Fragment-based drug discovery
high throughput purification *see* High throughput purification
protein analysis 427
proteomics *see* Proteomics
Matrix-assisted laser desorption ionization (MALDI)
protein purity determination, crystallization 881
proteomics 32
2D-PAGE 36
Matrix-attached regions, regulatory regions, bioinformatics 322
Matrix metalloproteinase(s) (MMPs), inhibitors, PASP 811
Matrix metalloproteinase-2 (MMP-2), NF-κB pathway 197
Matrix metalloproteinase-9 (MMP-9), NF-κB pathway expression 197
Matthews coefficient, molecular replacement 460
Maximum common subgraphs (MCS), pharmacophore identification 243
Maximum-likelihood approach, phylogeny construction, bioinformatics 321
Maximum-parsimony approach, phylogeny construction, bioinformatics 321
MDCK cell model, screening assays 628
MDL Drug Data Report (MDDR) 308–309
MDL Isentris framework, database management systems 295
MDL ISIS/Base
database management systems 295
Elsevier-MDL 295
MDL Patent Chemistry Database 308–309
MedChem kit, Milestone s.r.l. microwave-assisted chemistry 848*F*, 849
Media
cell culture 622
compound storage/management 567
insect cell expression systems (baculovirus) 417*T*
MediChem Database 308, 309
MEDLINE 271*T*, 274
databases 300
PDB searches 377
Melanoma, *Danio rerio* model system 10
MEL cells, protein expression systems 117
Member numbers, PDB 507–508, 508*F*
Membrane inserts, cell-based assays 627–628
Membrane proteins
cell-free expression systems *see* Cell-free expression systems
NMR structure determination 484
Mercury Visualizer *see* Cambridge Structural Database (CSD)
MEROPS
biological information databases 305
Universal Protein Resource 362–363
Mesa Analytics, molecular fingerprinting 298
Messenger RNA (mRNA)
amounts, RNAi detection 178
bioinformatics 330–331
cell-based assays 630–631
microarrays 90
Metabolic labeling, LC-MS (peptide-based) proteomics 38
Metabolic networks, bioinformatics *see* Molecular networks, bioinformatics
Metabolic pathways, cDNA microarrays 332
Metabolic probes, small animal imaging 665, 666*F*
Metabolism
Accelrys databases 307
cell-based assays 628, 629
Metabolomics, genetically engineered animals 155–156
Metabonomics
biomarkers *see* Biomarkers
drug toxicity markers 78
genetically engineered animals 155–156
models, mouse models 654
Metabotropic glutamate receptor(s) (mGluRs), orthogonal ligand–receptor pairs 231–232, 231*F*
MetaCyc database, metabolic networks 328–329
Metadata, factual databases 277
Metal coordination
CSD applications 401
dynamic combinatorial libraries 969, 970*F*, 973, 973*F*
generation 962
Metal ion assisted imine macrocycle synthesis, dynamic combinatorial chemistry 966
Metalloproteomics
BIND (Biomolecular Interaction Network Database) 556–557
MODBAS 556–557
New York Structural GenomiX Research (NYSGRC) 556–557
Southeast Collaboratory for Structural Genomics (SECSG) 556–557
Stanford Synchrotron Radiation Laboratory 556–557
structural genomics *see* Structural genomics
Metal-mediated crystallization, protein crystallization 885, 886*F*
Metastasis, cDNA functional screening 15–16

Metathesis reaction, dynamic combinatorial library generation 962*F*
METATOOL, metabolic networks 329
Metformin, mechanism of action 2
Methanol, microwave heating 840*T*
[^{14}C]Methionine, radiotreatment imaging 670
Methionine aminopeptidase, bengamide target identification 42–43
Methotrexate, target identification, *Saccharomyces cerevisiae* model system 8
2-D-Methyl-3-mercaptopropanoyl-L-proline, design/development 876–877, 877*F*, 929, 930*F*
1-Methyl-2-pyrrolidine, microwave heating 840*T*
[^{11}C]-*N*-Methylspiperone, dopamine receptor imaging 667
MIAME (Minimum Information About a Microarray Experiment) 89–90
  cDNA microarrays 331
MIAS-2 686*T*
[^{99m}Tc]MIBI, radiotreatment imaging 670
Michael reaction, dynamic combinatorial library generation 961–962, 962*F*
Michel, Hartmut, protein crystallography 451
Microarray Gene Expression Data (MGED) guidelines, microarrays 89–90
Microarrays **87–106**, 333
  applications 90
    comparative genome hybridization 90
    single nucleotide polymorphisms 90
  case studies 93
    see also specific case studies
    animal models 93
    defined patient populations 93
  cDNA microarray *see* cDNA microarrays
  commercially available
    advantages 90
    Affymetrix 88, 89, 93, 97
    Agilint 89
    Applied Biosystems 89
    Codelink microarray 89
    disadvantages 90
    GE Healthcare 88, 89
    Nimble Gen 89
  corticotrophin-releasing factor overexpression 97, 98*F*
    Affymetrix 97
    animal behavior 97
    CRF antagonist treatment 98*F*
    false discovery rate (FDR) 98–99
    glucocorticoid signaling pathway 99
    hippocampus 99
    kallikrein expression 97–98
    neurotensin receptor downregulation 100
    nucleus accumbens 99
    pituitary 99
    p-value 98–99
    quantitative PCR confirmation 99, 100
    q-value 98–99
    SAM algorithm 98–99
    sample preparation 97
    visual inspection 97–98, 99*F*
  corticotrophin-releasing factor receptor 93, 94*F*
    Affymetrix 93
    biplot 93–94, 94*F*
    confirmation 94
    quantitative PCR 94
  data analysis 91, 101, 102*F*
    cluster analysis 93
    preprocessing *see* Microarrays, data preprocessing
    sample classification 93
    visual inspection *see* Microarrays, data visual inspection
  data preprocessing 91
    assay quality 91
    data re-expression 91
    normalization 91
    quality checks 91
  data visual inspection 91
    biplots 92, 93–94, 94*F*
    heatmaps 91
    PCA 91
    spectral maps 92
  design 88, 89*F*, 90
    biological *vs.* technical replication 90
    comparative experiments 90
    control probes 90
    importance 88
    mRNA extraction/purification 90
    quality control 90
    statistics 90
  differential gene expression 92
    agglomerative testing 93
    clustering 93
    F test 92, 101–103
    k-means 93
    t test 92
  future work 104
    biomarker identification 104
    data interpretation 104
    microRNAs 104
  history 87–88
  irritable bowel syndrome 100, 101*F*
    ANOVA analysis 101
    control groups 100–101
    data analysis 101, 102*F*
    drug targets 100
    experimental groups 100–101
    false discovery rate 101–103
    F test 101–103
    neural specificity 100
    spectral map analysis 101, 102*F*
  limitations 103
    design importance 88
    extrapolation of data 103
    overcomplication of experiment 103
    timing 103
  oligonucleotide arrays 88, 89
    advantages 90
    disadvantages 90
    drug toxicity markers 78
    mismatch probes 89
    oligonucleotide length 89
    perfect match probes 89
  probes
    numbers 88
    oligonucleotide probes *see* Microarrays, oligonucleotide arrays
    PCR generated probes *see* cDNA microarrays
  sensitivity 88
  stringent *vs.* non-stringent buffers 88
  target identification 87, 88
  techniques 88
    advantages 90
    disadvantages 90
    Microarray Gene Expression Data (MGED) guidelines 89–90
  transgenic animals 97
Microencapsulation, polymer-assisted solution phase synthesis (PASP) *see* Polymer-assisted solution phase synthesis (PASP)
Microenvironment effects, cell-based assays 642
Microfluidic devices, bioluminescent assays 590, 591*T*
Micronuclei, formation, HCS 692
Microplates
  formats, HTS *see* High-throughput screening (HTS)
  replication 569, 570*F*
Micro-reactors, PASP 826, 826*F*
Microreticular resins, PASP 792–793
MicroRNA precursors *see* RNAi (RNA interference)
MicroRNAs, microarrays 104
Microscopic hot spot formation, microwave-assisted chemistry 842
Microscopy
  history 685
  small animal imaging 660*T*
Microtiter plates, microwave-assisted chemistry 850*F*, 854–855
Microwave-assisted chemistry **837–860**
  batch scale-up 856
  continuous flow scale up 856
    line clogging 857

Microwave-assisted chemistry (*continued*)
effects 840
electric field presence 842
microscopic hot spot formation 842
rapid heating 840–841
rate enhancements 841
selective heating 842
solvent properties 840–841
superheating effects 842
wall effect elimination 842
equipment 842
see also specific suppliers
domestic microwaves 842–843
monomode instruments 841*F*, 842–843
automated sequential synthesis 856
solid-phase peptide synthesis 843*F*
temperature measurement 843
multimode instruments 842–843, 843*F*, 848
see also individual manufacturers
optimization of conditions 838
parallel processing 854, 857
electromagnetic field homogeneity 855
library generation 855
microtiter plates 850*F*, 854–855
reaction homogeneity 856
sequential *vs.* 856
phase-transfer catalysis 853
liquid–liquid phases 853
solid–liquid phases 853
pressure monitoring 857
reaction times 857
safety
organic solvent flammability 837
pressure control 837
scale-up 856
reaction volumes 856
safety aspects 856–857
sequential processing 857
parallel processing *vs.* 856
single-mode instruments 844
Biotage AB *see* Biotage AB
CEM Discover platform *see* CEM Corporation
solution phase parallel chemistry 767
solvent choice 857
solvent-free reactions 852
alumina 853
clay 853
graphite 853
polar solvent addition 853
silica 853
solvent reactions 840–841, 853, 857
aqueous reaction media 854
ionic liquids 854
nonclassical solvents 854
open *vs.* closed vessel 853
pre-pressurized reaction vessels 854
room temperature ionic liquids 854
speed 838
techniques 852
temperature monitoring 857
yields 857
Microwaves 838
conventional thermal heating *vs.* 840
internal heating 840, 841*F*
dielectric heating 839
absorption 840
dipolar polarization 839
ionic conduction mechanism 839
loss tangents of substance 839, 840*T*
reflection 840
transmission 840
frequencies used 838
Midwest Center for Structural Genomics (MCSG) 551–552, 552*T*
progress 553*T*
Milestone s.r.l. 848, 848*F*
CombiCHEM system 849, 850*F*
MedChem kit 848*F*, 849
MonoPREP module 849, 849*F*
post-reaction cooling 849
reaction monitoring 849
temperature measurements 849
vessel range 848
MIMUMBA, CSD applications 401
Minaprine, discovery/development 926–927
Miniaturization
LC-MS (peptide-based) proteomics 39
PASP 826
Mini-batch procedure, protein crystallization 439
MIPS Functional Catalog, protein function prediction, bioinformatics 333–334
Mismatches, siRNAs 175
Mismatch probes, oligonucleotide arrays 89
Missing atoms, protein 3D structure validation 513
MISSION, large-scale virus vector libraries 19
Mitochondria
apoptosis 195–196
Saccharomyces cerevisiae model system 6
Mitotic index, HCS cell cycle 691–692
Mitotic spindle defects, RNA interference 18
Mitsunobu reaction, PASP 793, 798–799
Mobile regions, protein crystallization *see* Protein crystallization
MODBAS, metalloproteomics 556–557
ModBase, homology models 545
Model organisms
see specific models
phenotype-oriented screening *see* Phenotype-oriented screening
Models/modeling, bias, protein crystallography model generation 463–464
Mogul (intramolecular geometry) *see* Cambridge Structural Database (CSD)
Molecular diagnostics, proteomics 45
Molecular dynamics (MD)
homology models 544–545
protein crystallography model generation 463–464
Molecular fingerprints
chemical information databases 298
Fingerprint Models 298
Mesa Analytics 298
Molecular Function, PDB 377
Molecular graphs/graphics, structure/substructure searching 236–237, 237*F*
Molecular Initiatives Roadmap, chemical biology 129
Molecular Libraries Initiative (MLI), PubChem 307
Molecular Libraries screening Center Network, compound collections (libraries) 132
Molecular markers, genetic variation analysis 338–339
Molecular mass based methods, NMR, drug discovery 908
Molecular mass range, fluorescence polarization 603, 604*F*
Molecular Modeling Database (MMDB), blueprint databases 305–306
Molecular networks, bioinformatics 327
metabolic networks 328, 328*F*, 329*F*
BRENDA database 328–329
databases 328–329
definition 328
EcoCyc database 328–329
ENZYME database 328–329
flux balance analysis 329
GEPASI 329
HinCyc database 328–329
IntEnz database 328–329
KEGG database 328–329
MetaCyc database 328–329
METATOOL 329
MPW database 328–329
PseudoCyc database 328–329
regulatory networks 330
BIND database 330
DIP database 330
DOQCS database 330

nonstationary analysis 330
protein interaction networks 330
signal transduction pathways 330
TRANSPATH database 330
virtual cells 330
E-Cell project 330
Genomic Object Net 330
Molecular pathology, pharmacogenetics–environmental effects 57
Molecular probes, opiate receptors 664, 664*T*
Molecular replacement *see* X-ray crystallography, phase problem solutions
Molecular sequence analysis, bioinformatics 318
gene detection 321
contents sensors 321–322
signal sensors 321–322
genome rearrangements 324
fission/fusion 324
reversal 324
translocation 324
phylogeny construction 320, 321*F*
definition 320
distance-based approach 321
maximum-likelihood approach 321
maximum-parsimony approach 321
split diagrams 321, 321*F*
regulatory regions 322
alignment-based methods 322
consensus-based methods 322
enhancer regions 322
matrix-attached regions 322
promoter regions 322, 323*F*
silencer regions 322
repetitive elements 322
comparative genomics 324
genome assembly checking 324
hybridization probe uniqueness 324
low copy repeats 324
MUMmer 323
RepeatMasker 322–323
REPuter 323
sequence alignment 318, 319*F*
algorithms 319–320
BLAST 320
BLOSUM matrices 319–320
definition 318–319
DIALIGN 320
dynamic programming paradigm 320
hidden Markov model (HMM) 320
indels 318–319
multiple sequences 320
MUSCLE 320
PAM matrices 319–320
probabilistic models 319–320
T-COFFEE 320
Staden Package 318
Wisconsin Package (GCG) 318
Molecular similarity, chemical information databases 298
Molecular subtype identification, pharmacogenetics 62
Molecular transform generation, molecular replacement 460–461
Molecular variant screening, pharmacogenetics 63
Molecule numbers, molecular replacement 460
MOLPROBITY 547*T*
protein 3D structure validation 516–517
Monastrol, discovery/development 136, 136*F*, 693, 922, 922*F*
Monochromatic radiation, x-ray generation 465
Monoclonal antibodies (mAbs) 118
Monoliths
high throughput purification 867
PASP 793
Monomode instruments, microwave-assisted chemistry *see* Microwave-assisted chemistry
MonoPREP module, Milestone s.r.l. microwave-assisted chemistry 849, 849*F*
Monotropic membrane proteins, protein crystallization *see* Protein crystallization
Montreal-Kingson Bacterial Structural Genomics Initiative (BSGI) 552*T*
Morgan, Thomas Hunt, *Drosophila melanogaster* as model system 9
Morphinan analogs, solid-phase pool libraries 705*F*
Mosaic, compound storage/management (Pfizer) 571
Motifs
derivation 337
protein structure prediction 325
Mouse, genome sequencing 317*T*
Mouse mammary tumor virus (MMTV), transgenic mouse model promoters 11
Mouse models 10, 653
see also Knockout animal models; Transgenic animals
chemical mutagenesis 12*F*, 13
alteration of function (neomorphic) 13–14
ethylmethane sulfonate (EMS) 13–14
N-ethyl-*N*-nitrosourea (ENU) 13–14
gain-of-function (hypermorphic) 13–14
loss-of-function 13–14
reduction-of-function (hypomorphic) 13–14
compound screening 653
genome sequencing 648*T*
history 647
human genome *vs.* 10–11
knockout models 654
metabonomics 654
pharmacodynamics 653
pharmacokinetics 653
pharmacology 653
toxicogenomics 654
transgenic models 654
MPW database, metabolic networks 328–329
MS (Microsoft)-Access, desktop databases 294
MSSQL Server, server databases 294
MTRIX record, protein 3D structure validation 511, 512*F*
Mulder, Gerardus Johannes, proteomics 28
Multicellular organisms, phenotype-oriented screening 137
Multicenter Evaluation of in Vitro Cytotoxicity, cell-based assays, neurotoxicity 641
Multichannel scaler (MCS) trace, fluorescence fluctuation techniques 608, 608*F*
Multidimensional cytological profiling, HCS 693
Multidimensional protein identification technique (MudPIT)
biomarker identification 75–76
LC-MS (peptide-based) proteomics 37
Multidrug resistance (MDR), transporters, cell-free expression systems 122–123
Multienzyme complexes, structure 543
Multimode instruments, microwave-assisted chemistry 842–843, 843*F*, 848
Multimutants, genetically engineered animals 153
Multiple cloning sites, gene expression vectors 108
Multiple construct screening, gene construct design 877–878
Multiple genes, genetic variation analysis 339
Multiple isomorphous replacement (MIR)
isomorphous replacement 459
protein crystallography, rational drug design 889–890
Multiple-isomorphous replacement with anomalous scattering (MIRAS), protein crystallography, rational drug design 889–890
Multiple myeloma, STATs 207–208
Multiple outputs (multiplexing), cell-based screening assays *see* Cell-based assays

Multiple reactions, dynamic combinatorial library generation 963
Multiple replicates, compound storage/management 566–567
Multiple sequences, sequence alignment, bioinformatics 320
Multiple targets, biological activity databases 302
Multiple wavelength anomalous dispersion (MAD)
  anomalous dispersion 459–460
  protein crystallography, rational drug design 889–890
Multiplexing, enzyme-linked immunosorbent assays (ELISAs) 73–74
Multistep synthesis, polymer-assisted solution phase synthesis (PASP) *see* Polymer-assisted solution phase synthesis (PASP)
Multisystem expression vectors, gene expression vectors 108
MUMmer, repetitive elements, bioinformatics 323
Muramyl dipeptide analogs, solid-phase discrete libraries 709*T*
MUSCLE, sequence alignment, bioinformatics 320
Muscular dystrophy model, *Caenorhabditis elegans* model system 8
Mutagenesis, protein crystallization 882
Mutagenicity testing
  adenine phosphoribosyltransferase (Aprt) knockouts 161
  hypoxanthine phosphoribosyltransferase (Hprt) knockouts 161
Mutations
  PDB 381
  protein 3D structure validation 521
Mycobacterium tuberculosis, Structural Proteomic Project (XMTB) 552*T*
Myoglobin, protein crystallization 434
Myosin–actin complex, structure 543
MySQL 294
  server databases 294

N

^{13}N, PET 661–662, 664*T*
NADH peroxidase, crystal packing 442, 444*F*
NADH:ubiquinone oxidoreductase inhibitors, solid-phase libraries 751*F*, 752–753
$NADP^+$ reduction, oxidoreductase assays 581
NAD^+ reduction, oxidoreductase assays 581
Nafion-H resin, PASP 821
Naltrindole derivatives, solid-phase discrete libraries 709*T*
National Cancer Institute (NCI) 282*T*
  compound collections (libraries) 132
  database 308–309
  structure databases 281
National Center for Biotechnology Information (NCBI), nucleotide sequence databases 350
National Center for Computational Biology, cDNA microarrays 331
National Institute of General Medical Science (NIGMS) 551
National Institute of Neurological Disorders and Stroke (NINDS) 132
National University of Singapore Bioinformatics Drug Design Group, BIDD Databases 305
Natural lead analogs, HTS library composition 680
Naturally proliferating cells, cell-based assays 620–621
Natural products
  compound collections (libraries) 133
  disadvantages 133
  purification 133
  solid-phase libraries 718, 726, 728*T*, 740, 740*F*
  synthesis, PASP 814
  target identification 133
NCBI Reference Sequences *see* RefSeq (NCBI Reference Sequences)
Necrosis, cell-based assays 629
Negamycin analogs, solid-phase discrete libraries 709*T*
Negative phenotype interpretation, genetically engineered animals *see* Genetically engineered animals
Neolignans, PASP 818
Neonatal respiratory distress syndrome, vascular endothelial growth factor treatment 203
Neovastat 202–203
  psoriasis treatment 202
Network models, databases 268, 268*F*
Neural networks, protein function prediction, bioinformatics 335
Neural specificity, irritable bowel syndrome 100
Neuraminidase inhibitors, dynamic combinatorial libraries 970, 971*F*
Neurite overgrowth, HCS 692
Neurodegenerative diseases, animal models, *Danio rerio* model system 10
Neurogenesis, *Caenorhabditis elegans* model system 8
Neuropathy, knock-in animal models 160
Neurospora crassa, as model system 647
Neurotensin receptor downregulation, corticotrophin-releasing factor overexpression 100
Neurotoxicity, cell-based assays 639, 642*F*, 643*F*
  measures used 641
  Multicenter Evaluation of in Vitro Cytotoxicity 641
Neutron diffraction studies, crystal size 440
New York Structural GenomiX Research (NYSGRC) 551–552, 552*T*
  metalloproteomics 556–557
  progress 553*T*
NF-κB *see* Nuclear factor kappa B (NF-κB)
Nicotine, PASP 817
Nicotinic acid analogs, solid-phase discrete libraries 709*T*
NIH 3T3 cells, protein expression systems 117
NIH Chemical Genomics Center (NCGC), compound collections (libraries) 132
Nimble Gen, microarrays 89
Nipkow disc, HCS confocal systems 687, 688*F*
Nisoxetine, chemogenomics 934*F*, 935
NIST Databases 278*T*
NIST webBook 278*T*
Nitric oxide synthase (NOS), knockout animal models 157–158
Nitrilase and Nitrile Hydratase Knowledgebase 309
Nitrobenzene, microwave heating 840*T*
Node network theory
  nuclear factor kappa B (NF-κB) 4, 4*F*
  p53 4
  target validation 4, 4*F*
NOE assignments, NMR structures, problems 542
NOE pumping, NMR, drug discovery 908*T*, 913
NOESY, NMR structure determination 474, 482, 483*F*
Noise, protein crystallography model generation 462
Nomenclature
  errors, x-ray structures 539
  problems
    protein function prediction, bioinformatics 337
    protein 3D structure, problems 543
  protein 3D structure validation *see* Protein 3D structure validation
  validation 518
Nonapeptides, protein structure prediction 327
Nonbonded interactions
  CSD 402
  Mercury Visualizer 402, 402*F*

Nonclassical solvents, microwave-assisted chemistry 854
Noncrystallographic symmetry, protein crystallography 443
Non-Hodgkin's lymphoma, STATs 207–208
Non-hydrogen bonds, intermolecular interactions 404
Nonnatural components, dynamic combinatorial libraries (DCLs) *see* Dynamic combinatorial libraries (DCLs)
Nonnucleic acid residues, dynamic combinatorial libraries 969
Non-nucleotide building blocks, dynamic combinatorial libraries 973
Nonresponders, pharmacogenetics 64
Nonspecific immune response, RNAi (RNA interference) 182
Nonspecific plate blank (NSB), HTS 683–684
Nonstationary analysis, regulatory networks 330
Nonunique chain identifiers, PDB errors 521, 522*F*
'No observable effect level' (NOEL), genetically engineered animals 157
Norbinaltorphimine analogs, solid-phase pool libraries 705*F*
Normalization
  cDNA microarrays 331
  microarrays 91
Nornicotine, PASP 817
Northeast Structural Genomic Consortium (NESG) 551–552, 552*T*
  progress 553*T*
Northern blots
  genetically engineered animal characterization 153–154
  RNAi detection 178
Northwest Structural Genomics Centre (NMSGC) 552*T*
Novagen pTriEx 880
^{15}N spectra, NMR, drug discovery 917
N-terminal domain, STAT (signal transducer and activator of transcription) 203
NucleaRDB, biological information databases 305
Nuclear factor kappa B (NF-κB) 195
  alternative activation pathway 192
  anti-apoptotic pathways 196
  apoptosis 192, 195
  Bcl-2 protein family regulation 196–197
  canonical activation pathway 192, 193*F*
  cFLIP regulation 196–197
  in disease 197
    asthma 197
    Crohn's disease 197
    IBD 197
    rheumatoid arthritis 197
    ulcerative colitis 197
  gene isolation, cDNA functional screening 14
  HCS nuclear translocation assays 691
  induction of 196*F*
    cyclooxygenase-2 197
    cyclooxygenase-5 197
    E-selectin 197
    IκB kinases (IKKs) 195
    IL-1 197
    inducible nitric oxide synthase 197
    intercellular adhesion molecule-1 (ICAM-1) 197
    lipoxygenase 197
    matrix metalloproteinase-2 197
    matrix metalloproteinase-9 197
    TNF-α 197
    vascular cell adhesion molecule-1 (VCAM-1) 197
    vascular endothelial growth factor 197
  inflammation 192, 197
  inhibition 195
  inhibitors of apoptosis (IAPs) regulation 196–197
  node network theory 4, 4*F*
  Rel homology domain (RHD) 195
  signaling pathways, cDNA functional screening 15–16
  tumorigenesis 192
  viral replication 192
Nuclear fractions, proteomics 31
Nuclear hormone receptors
  biological functions 221
  cell-based assays 624*T*
  orthogonal ligand–receptor pairs *see* Orthogonal ligand–receptor pairs
Nuclear hormones, as drug targets 3
Nuclear magnetic resonance (NMR)
  biomarkers 76
  fragment-based drug discovery *see* Fragment-based drug discovery
  'fragnomics' 892
  INPHARMA 948
  inter-ligand NOEs 948
  protein labeling 428
    bacterial expression 428
    cell-free system 428
    insect cell production 429
  protein 3D structure validation 511
  small animal imaging 662
  tumors 77
Nuclear magnetic resonance (NMR), drug discovery **901–920**
  advantages 902
  basic binding equilibria 902, 903*F*, 905*F*
    Langmuir binding isotherm 903
  chemical exchange effects 904, 906*F*
    ligand effects 906, 907*F*
  chemical shift perturbations 916
    binding ligand location 917
    ^{13}C spectra 917
    exchange rates 917
    ^{15}N spectra 917
  competition based ^{9}F screening 908*T*, 915, 915*F*
  competition binding equilibria 903
  competitive screening 908*T*
  CSP 908*T*
  diffusion filtering 908*T*, 913
  diffusion rate measurement 908, 909*F*
    relaxation rate measurement *vs.* 909
    sensitivity 909
  disadvantages 902
  ^{19}F screening 908*T*, 910*F*, 911
  HTS hit validation 902
  intermolecular NOE 908*T*, 912, 913*F*, 914*F*
  intramolecular NOE 908*T*, 911
  lead optimization 901, 902
  lead search 901
  ligand-based methods
    labeling 907
    molecular mass based methods 908
    speed 907
    target-based techniques *vs.* 907
  NOE pumping 908*T*, 913
  paramagnetic spin labels 909
    second-site screening 910, 911*F*
    SLAPSTIC 910
  relaxation filtering 908*T*, 913
  relaxation rate measurement 909, 910*F*
    diffusion rate *vs.* 909
    filter elements 909
    longitudinal relaxation rate 909, 910*F*
    subtraction artifacts 909
  saturation transfer difference (STD) experiments 908*T*, 912, 912*F*, 913*F*, 914*F*
  target-based techniques 915
    CRIPT 915–916
    ligand-based methods 907
    pH effects 907
    receptor signal changes 915–916
    specific *vs.* non-specific binding 907
    structural information 907
    TROSY 915–916
  techniques 908*T*
  technology improvements 901
  trNOESY 908*T*
  WaterLOGSY 908*T*, 913, 914*F*
  weak binding detection 901–902
Nuclear magnetic resonance (NMR), structure determination **473–505**, 540, 902, 948

assignment 476
amino acid side chains 476, 478*F*
dipolar couplings 476, 478*F*
liquid *vs.* solid-state 476
scalar couplings 476, 478*F*
chemical shifts 486, 489*F*
solid-state 487
COSY 474
dipolar coupling 480*T*, 481, 483, 489
alignment 484–485
cross-relaxation rate 481
directly bonded nuclei 485
internuclear vectors 484, 485*F*
membrane proteins 484
NOESY 482, 483*F*
orientation information 484
recoupling 482
as restraints 484
dipolar relaxation, liquid-state 481
dynamics 495, 496*F*
μs to ms dynamics 497, 498*F*
ns to μs dynamics 499, 500*F*
relaxation dispersion 498
solid-state NMR 500*F*, 501, 501*F*
sub-correlation dynamics 496
examples 490
fast acquisition schemes 491
global quality measures 546
restraints per residue 546
interaction cross-correlation 486
intervector angle 486, 488*F*
side chain angles 486
torsional angles 486, 488*F*
isotopic labels 475, 477*F*
liquid-state 474, 474*F*
CSA 479–481
dipolar relaxation 481
examples 490
proton detection 479–481
relaxation optimized sequences 479–481
samples 476*F*
scalar coupling 486
solid-state *vs.* 476
magic angle spinning (MAS) 475
model ensembles 541
Olderado (On Line Database of Ensemble Representatives and Domains) 541–542
representative member 541
molecular weights 479
interactions 479, 480*F*, 480*T*
liquid *vs.* solid 476–477
solid-state 479
multidomain proteins 492
binding affinities 492
domain motion 500
preparation 492
tight binding 492
weak binding 493
multiple entries 543
NOE assignments 542
NOESY 474
numbers of 540
paramagnetic relaxation 482
parameters 481
protein–ligand complexes 492
binding affinities 492
preparation 492
tight binding 492
weak binding 493
protein–protein interactions 492
binding affinities 492
preparation 492
tight binding 492
weak binding 493
quality problems 542
accuracy 542
precision 542
x-ray crystallography *vs.* 542
re-refined structures 542
DRESS (Database of Refined Solution NMR Structures) 542
RECOORD 542
samples 475
concentrations 475
solid vs, liquid state 475, 476*F*
scalar coupling 485, 489
assignment 485–486
bonds 485–486, 487*F*
liquid-state 486
TOBSY spectra 486
solid-state
calculations 489
chemical shift anisotropy 479
dynamics 500*F*, 501, 501*F*
examples 490
INEPT 478
liquid *vs.* 476
magic angle spinning 473–474, 482
molecular tumbling 479
molecular weights 479
restrained molecular dynamics 489
samples 476*F*, 477
simulated annealing protocol 489
TOCSY 478
transfer schemes 477, 479*F*, 480*F*
structural genomics 555
TALOS routine 489
x-ray structure *vs.* 540
Nuclear Overhauser effects (NOEs), NMR spectroscopy 946
Nuclear receptor ligands, chemogenomics 932, 933*F*, 934*F*
Nuclear translocation assays, HCS *see* High-content screening (HCS)
Nucleic acid-based dynamic combinatorial libraries *see* Dynamic combinatorial libraries (DCLs)
Nucleic acids, cell-based assays 628, 631
Nucleoside analogs, solid-phase pool libraries 705*F*
Nucleotide-like libraries, solid-phase libraries 751–752
Nucleotides, solid-phase libraries 698
Nucleotide sequence databases 351
see also specific databases
history 350
RefSeq (NCBI Reference Sequences) 355
Nucleus accumbens, corticotrophin-releasing factor overexpression 99
Numerical databases 277
Numerical descriptors, similarity searching 245
NusA, gene expression vectors 110*T*

O

^{15}O, PET 661–662, 664*T*
O6-methylguanine DNA methylase methylation, pharmacogenetics 61
Obese mouse strain, QTL analysis 14
Obesity, *Caenorhabditis elegans* model system 9
Object-based models, databases *see* Databases
Obliquine, PASP 819
OC (organism classification) lines, UniProtKB flatfile 360
Ocular neovascularization, RNAi (RNA interference) 181–182
Off-target activity, RNAi (RNA interference) 178
OG (organelle) line, UniProtKB flatfile 360
Olderado (On Line Database of Ensemble Representatives and Domains), NMR structures 541–542
2′-5′ Oligoadenylate synthetase, RNAi (RNA interference) 173
Oligonucleotide arrays *see* Microarrays
Oligonucleotides, length, arrays 89
Oligosaccharide-binding fold, x-ray structures, problems 537
Omega angles, protein 3D structure validation 517
Omeprazole, mechanism of action, mouse knockout models 13
OMIM numbers, PDB 380
Oncogenes, activation, vascular endothelial growth factor (VEGF) 202
Oncogenesis
Drosophila melanogaster model system 9
large-scale virus vector libraries 18–19
RNAi 180–181
Oncology, target search genomics 5
Ondetti, Miguel, protein crystallography, rational drug design 876

1D-nuclear magnetic resonance, protein analysis 427
One-dimensional representations, chemical information databases 298
100A *271T*
One-pot reactions, polymer-assisted solution phase synthesis (PASP) *see* Polymer-assisted solution phase synthesis (PASP)
On-line design tools, siRNAs 176
Open-source DBMS, databases 294
Opera 686*T*, 687
OPG
  knockout animal models 156, 157*F*
  overexpression, transgenic animals 156, 157*F*
Opiate receptors, molecular probes 664, 664*T*
μ-Opiate receptor, assays 638*F*
  agonist response 637*F*
  cell-based assays 634
κ Opiate receptors (KOR), orthogonal ligand–receptor pairs 229–231, 229*F*
Optical assays **577–598**
  see also Optical biosensors; *specific assays*
  absorption/emission- based assays 578
  classification 578
  colored end products 580
    ATPase assays 580–581, 582
    caspase-3 direct assay format 581, 581*F*
    glyoxalase direct assay format 581
    β-lactamase assays 580–581
    oxidoreductase direct assay format 581
    phosphatase assays 580–581, 582, 582*F*, 582*T*, 583*T*
    phosphodiesterases 582
    protein–protein interactions (indirect format) 582
    pyrophosphatases 582
  immunoassays *see* Immunoassays
  luminescence-based 578
  photoluminescence 578, 590
  radioactive assays
    Flash Plate technology *see* Flash Plate assays
    scintillation proximity assay (SPA) technology *see* Scintillation proximity assay (SPA)
    targets
      see specific targets
  readouts 578
Optical biosensors 578, 592
  BIND (biomolecular interaction detection system) 595
  high-throughput applications 594
  kinetic measurements 578
  reflectometric interference spectroscopy (RIfS) 594–595
  surface plasmon resonance *see* Surface plasmon resonance
  technology 592
    label-free 592–593
Optics, fluorescence screening assays 602
Optimization of conditions, microwave-assisted chemistry 838
Optimization techniques, combinatorial library design 257
Oracle, server databases 294
Organic solvent flammability, microwave-assisted chemistry 837
Organic synthesis, PASP 828
Organizational units, dynamic combinatorial chemistry 961
Organ system mimics, cell culture 623
Orthogonal ligand–receptor pairs **215–234**
  analog sensitive kinase alleles (ASKA) 218
  applications 216
  biological function probing 216
  cyclophilin–cyclosporine A (CsA) 217, 219*F*
  definition 215
  GPCRs 229
    crystal structure importance 231–232
    metabotropic glutamate receptors (mGluRs) 231–232, 231*F*
    κ opiate receptors (KOR) 229–231, 229*F*
    rhodopsin 230*F*, 231–232
  GTPase family 216, 219*F*
  kinesin–myosin 220
  noncovalent interactions 215, 217*F*, 218*F*
  nuclear hormone receptors 221
    4-4′-dihydroxybenzil (DHB) 227
    ecdysteroid–ecdysone receptor 229
    estradiol 224*F*, 225–227, 225*F*, 226*F*
    estrogen receptor 224, 224*F*, 225*F*, 226*F*
    retinoid X receptors 221–222, 222*F*, 223*F*
    thyroid hormone receptors 227*F*, 228
    vitamin D receptor 228–229, 228*F*
  protein kinases 218
    CDK1 substrates 218
    inhibitors 218–220, 221*F*
    (4-amino-3-phenyl-1-*t*-butyl)pyrazolo[3,4-*d*]pyrimidine (PPI) derivatives 218–220, 221*F*
    v-Src 218, 220*F*
  technique 215
Orthostrapper, protein function prediction, bioinformatics 334
OS (organism species) line, UniProtKB flatfile 360
OSM, STAT3 signaling 206
Osteogenesis, cDNA functional screening 15–16
Osteopetrosis
  knockout animal models 156, 157*F*
  transgenic animals 156, 157*F*
Osteoporosis, knockout animal models 156, 157*F*
Outsourcing, compound acquisition/processing 564
Overexpression models, small animal test systems 649
Overexpression pools, cDNA functional screening 15
Oxalic acid arylamides, solid-phase discrete libraries 709*T*
Oxazolidines, solid-phase discrete libraries 709*T*
Oxazolines, solid-phase discrete libraries 709*T*
Oxford Protein Production Facility (OPPF) 552*T*
Oxidoreductases
  assays
    $NADP^+$ reduction 581
    NAD^+ reduction 581
  direct assay format, optical assays 581
OX (organism taxonomy) line, UniProtKB flatfile 360
Oxomaritidine, PASP 815, 816*F*
Oxytetracycline, discovery, HTS 680

P

P13 kinases, radioactive assays 587
p38α MAP kinase
  see specific drugs
  fragment-based drug discovery *see* Fragment-based drug discovery
p38 kinase inhibitors, chemical proteomics 44
p53
  as biomarker 72
  deletion, genetically engineered animals 161, 163*F*
  large-scale virus vector libraries 18–19
  node network theory 4
  protein degradation 193, 194*F*
  regulator identification 15–16
  RNAi 181
  ubiquitin-proteasome pathway inhibition 194
p561ck inhibitors, solution phase library synthesis 779–780, 781*F*
Packing arrangements, protein crystallography *see* X-ray crystallography
Packing densities, protein crystallography 442

Paclitaxel
analogs, solid-phase pool libraries 705*F*
proteomics, stathmin/OP-18 41
structure 700*F*
PagP, NMR structure determination 490, 491*F*
Palliative drugs, pharmacogenetics 54, 58
^{11}C Palmitate, as metabolic probe 665
PAM matrices, sequence alignment, bioinformatics 319–320
PANTHER
InterPro, member of 363
signature method 363
Parallel analysis/purification, high throughput purification *see* High throughput purification
Parallel chemistry
solid phase *see* Solid phase parallel synthesis
solution phase *see* Solution phase parallel chemistry
technology selection 768
factors 768
Parallel processing, microwave-assisted chemistry *see* Microwave-assisted chemistry
Parallel screens, HTS 684
Parallel synthesis, HTS automation 681
Paramagnetic relaxation, NMR structure determination 482
Paramagnetic spin labels *see* Nuclear magnetic resonance (NMR), drug discovery
PARP-1 inhibitors
ELISA 589
Flash Plate assays 589
radioactive assays 589
Particle size/aggregation studies, protein analysis *see* Protein 3D structure, analysis
Partitioning methods *see* Chemoinformatics
PATDPAFULL 276*T*
Patents
databases 275, 276*T*
descriptors 301
PathArt 309
Pathway characterization, biomarkers 73
Pathway HT 686*T*, 687
Patterson map
isomorphous replacement 457
molecular replacement 460–461, 461*F*
Patterson Search model, molecular replacement 460
Pauling, Linus, protein crystallography 450
PAZ domain, DICER 173
P_{BAD}/AraC combination, gene expression vectors 108
PCI (Patents Citation Index) 276*T*
PDBREPORT 547–548, 547*T*
protein 3D structure 546
PDBsum 547*T*
protein 3D structure 546
PDSP K_i 306
Pegaptanib 202
Pegylated polystyrenes, PASP 792–793
Penicillin sulfoxide synthesis, PASP 828
Pepticinnamin analogs, solid-phase discrete libraries 707*F*
Peptide mass fingerprinting, mass spectrometry proteomics 32, 33
Peptide nucleic acids (PNAs), solid-phase libraries 740–741
Peptide–protein binding assays
Flash Plate assays 589
radioactive assays 589
Peptides
proteomics *see* Proteomics, LC-MS (peptide-based)
solid-phase libraries 698
Peptide tags, gene expression vectors 110
Peptidomimetics
design 373
solid-phase discrete libraries 709*T*
solid-phase pool libraries 705*F*
Peptidomimetic scaffolds, solid-phase libraries 719–720, 720*F*
Peptidotriazoles, solid-phase pool libraries 705*F*
Peptoid drugs, solid-phase pool libraries 705*F*
Perfect match probes, oligonucleotide arrays 89
Peripheral blood mononuclear cells, Toll receptors 639
Periplasm targeting, *Escherichia coli* protein expression systems 114
Peroxisome proliferator-activated receptor gamma (PPAR-γ) agonists, knockout mouse model 13
Perruthenate, PASP 806, 806*F*
Personal communications, CSD 391–392
Personal disadvantage, pharmacogenetics 66
Perutz, Max F, protein crystallography 450
pET expression, gene expression vectors 108
Pfam5000 strategy
InterPro, member of 363
target selection, structural genomics 554
Pfizer, Warner Lambert, merger with 562
pH
NMR, drug discovery 907
protein crystallization 438
Pharmacodynamics
biomarkers *see* Biomarkers, clinical trials
'classical' pharmacogenetics 58, 60*F*
mouse models 653
Pharmacogenetics **51–68**
adverse effect avoidance 64
drug withdrawals 64–65
poor administration 64–65
applications 52
asthma treatment 58
β_2-adrenoreceptor variants 58
leukotriene synthesis inhibition 60
causative-acting drugs 54
'classical' 57
adverse effects 60
isoenzymes 58
pharmacodynamics 58, 60*F*
pharmacokinetic effects 58, 59*T*
pharmacokinetics 60*F*
common complex diseases 54
environmental factors *vs.* 54
'genetic isolation' studies 55
public education 55
complexity 61
data protection 65
explicit permission 65–66
genetic test definitions 65
informed consent 65–66
personal disadvantage 66
definition 53, 53*T*, 339
in diagnosis 62
molecular subtype identification 62
disease probability 62, 63*F*
molecular variant screening 63
polygenic disease 63
'responder' *vs.* 'nonresponder' 62–63
drug applicability 55
Herceptin 55
drug development 62
drug efficacy 64
business cases 64
environmental effects 56
interaction observation 57
molecular pathology 57
ethical issues 65, 66
data access 66–67
data application 67
insurance risk 67
predictive genotyping *vs.* 66
historical perspective 59*T*
Garrod, Archibald 58
marker classes 61
alkylating agent efficacy 61
O6-methylguanine DNA methylase methylation 61
molecular differential diagnosis 61
breast cancer 61
CCR5 cell surface receptors 61

disease heterogeneity 61
Herceptin 61
HIV infection 61
palliative drugs 54, 58
pharmacogenomics *vs.* 54
public education/information 66
regulatory aspects 63
clinical trials 64
nonresponders 64
safety trials 64
systemic classification 57, 57*T*
disease-related biological variation 57
Pharmacogenomics 56
definition 53, 53*T*, 339
drug discovery/development
animal models *vs.* 56
biological interactions 56
cell-based models *vs.* 56
disadvantages 56
efficacy effects 56
specificity of action 56
speed 56
genetic variation analysis 339
pharmacogenetics *vs.* 54
Pharmacokinetics
biological activity databases 302
biomarkers *see* Biomarkers, clinical trials
'classical' pharmacogenetics 58, 59*T*, 60*F*
mouse models 653
small animal imaging *see* Small animal imaging
Pharmacology, models, mouse models 653
Pharmacophores
fingerprints, chemoinformatics, diversity analysis 255
identification *see* Chemoinformatics, three-dimensional structure
Pharmacophores, detection 241*F*, 242, 243*F*
bioequivalence 242
chemical representation 242
compound choice 244
conformational flexibility 242
conformation numbers 243
conformation-on-the-fly 244
maximum common subgraphs (MCS) 243
programs 242
QSAR 242
rigid body alignments 243, 244*F*
Pharmaprojects 278*T*
Pharmline 271*T*
Phase ambiguity, isomorphous replacement 459
'Phase problem,' x-ray crystallography 556
Phase quality, protein crystallography 470
Phase shifts
anomalous dispersion 459
x-ray–crystal interaction 452
Phase switch concept, PASP 799–800
Phase-transfer catalysis, microwave-assisted chemistry *see* Microwave-assisted chemistry
Phasing information, x-ray structures, problems 533–534
Phasing methods, crystallography *see* X-ray crystallography, rational drug design
Phenotype-oriented screening 135
detection methods 138
high-content screening 138
uniform readout 138
goals 135
biological HTS 135
target identification 135
model organisms 135
cell-based assays 136
cell differentiation 136
human material 136
multicellular organisms 137
specificity 140
targetless mutant methods 140–141, 141*F*
techniques 140
target identification 138
affinity-based approaches 138
affinity chromatography 138
drug-induced haploinsufficiency 139, 140*F*
genetic approaches 139
protein microarrays 139
tagged libraries 139
three-hybrid reporter gene systems 140
Phenotypic analysis, genetically engineered animals *see* Genetically engineered animals
Phenoxypropanolamines, solid-phase pool libraries 705*F*
Phenylalkylamides
solid-phase libraries 737–738, 737*F*
solid-phase pool libraries 705*F*
Phenylethylamines, biological activities 925*T*
pH gradient, 2D-PAGE *see* 2D-PAGE
Philantotoxins
solid-phase discrete libraries 709*T*
solid-phase libraries 718–719, 719*F*
Phlorizin analogs, solid-phase pool libraries 705*F*
Phosphatase assays
high-throughput screens (HTS) 582
optical assays 580–581, 582, 582*F*, 582*T*, 583*T*
Phosphatidylinositol-3-kinase (PI3K)
activation, vascular endothelial growth factor receptor-2 (VEGF-R2) 200
ubiquitin-protein targeting 192
Phosphodiesterase(s)
as drug targets 3
inhibitors, chemogenomics 930, 931*F*
optical assays 582
Phospholipid-binding proteins, radioactive assays 587
Phosphorylation
heterogeneous protein crystallization 882
proteomics 41
Photoactive yellow protein (*Ectothiorhodospira halophila*) structure
correct 532, 533*F*
incorrect 532, 533*F*
Photodynamic treatment (PDT)
[^{18}F]FDG imaging 669
small animal imaging 669
Photoluminescence, optical assays 578, 590
Photomultiplier tubes
HCS image detection 687
scintillation proximity assay 584
Photon counters, x-ray detection 468
Phthalides
solid-phase discrete libraries 709*T*
solid-phase libraries 720*F*, 721
Phthalimido amides, solid-phase discrete libraries 709*T*
Phylogenetic profiles, genomic context methods, bioinformatics 336
Phylogeny construction, bioinformatics *see* Molecular sequence analysis, bioinformatics
Physical maps, *Danio rerio* model system 10
Physicochemistry, protein 3D structure validation 523, 525*T*
Pichia pastoris
eukaryotic protein expression system 880
protein expression 554
PINT, structure–function relationship, structural genomics 557
Piperazine derivatives
solid-phase discrete libraries 709*T*
solid-phase pool libraries 705*F*
N-substituted Piperidinone synthesis, solution phase library synthesis 780, 783*F*
Pipetting station/stacker dispensing, HTS automation 681
PIRSE, signature method 363
PIRSuperFamily, InterPro, member of 363
Pituitary gland, corticotrophin-releasing factor overexpression 99
Placental growth factor (PIGF) 198
see also Vascular endothelial growth factor (VEGF)
Plasmid construction, Semliki Forest Virus expression systems 119

Plasmodium falciparum, drug sensitivity, ELISA 580
Platelet-derived growth factor receptors (PDGFRs)
inhibition, imatinib mesylate 3
STAT3 signaling 206
STAT5 signaling 206
PLEX database, protein interaction networks, bioinformatics 336
Plexxikon, 'scaffold-based' drug design 892
Plicamine, PASP 819, 820*F*
Plicane, PASP 819
PNA-encoded tetrapeptides, solid-phase pool libraries 707*F*
Polar solvent, microwave-assisted chemistry 853
Polar surface area (PSA), prediction 259
Polyamines, solid-phase pool libraries 705*F*
Polycyclic compounds
solid-phase discrete libraries 707*F*
solid-phase pool libraries 707*F*
Polycystic kidney disease, *Danio rerio* model system 10
Polycystic syndrome, vascular endothelial growth factor 202
Polyethylene glycol (PEG), protein concentration 424
Polyfunctional 1,3-dioxanes, solid-phase pool libraries 707*F*
Polygenic disease, pharmacogenetics 63
Polyglutamine disease, *Saccharomyces cerevisiae* model system 6
Poly-His tag, as affinity tag 421
Polymerase chain reaction (PCR)
protein cloning 554
protein production 414
Polymer-assisted solution phase synthesis (PASP) **791–836**, 798–799
advantages 793
environmental advantages 793
Mitsunobu reaction 793
safety 793
separation simplicity 793
spent reagent removal 792, 793
Staudinger reaction 793
time saving 792
Wittig reaction 793, 793*F*
catch-and-release strategies 795, 796*F*
chemical library synthesis 804
2-alkylthiobenzimidazoles 805
amides 806, 807*F*
amine derivatization 806, 806*F*
2-aminothiazole scaffolds 805–806, 805*F*
antirhinoviral candidates 804, 804*F*
5-arylidine 4-thiazolidinones 806–807, 807*F*
benzimidazole derivatives 804
benzimidazolin-2-ones 805, 805*F*
1,5-biaryl pyrazoles 808, 809*F*
cathepsin D inhibitors 808
cyanoborohydride 806, 806*F*
N,N-dialkyldipeptidylamines 808–809, 810*F*
factor VIIa inhibitors 809
histone deacetylase inhibitors 811, 812*F*
hydroxamic acid derivatives 811, 811*F*
α-ketothiazole peptidyl protease inhibitors 809, 810*F*
matrix metalloproteinase inhibitors 811
perruthenate 806, 806*F*
scaffold decoration 811
substituted pyrazoles 807, 808*F*
3-thioalkyl-1,2,4-triazoles 804, 804*F*
thiomorphine analogs 811, 813*F*, 814*F*
triazole analogs 804, 804*F*
disadvantages, reaction kinetics 793–794
flow processes 826
advantages 826, 827–828
byproduct scavenging 826
catch-and-release protocols 826
chromatographic separation 826
condition optimization 827–828
heterocyclic thiourea library preparation 830
hydrogen gas 800*F*, 827, 827*F*
imine hydrogenation 827, 827*F*
β-lactam synthesis 829, 829*F*
micro-reactors 826, 826*F*
miniaturization 826
organic synthesis 828
penicillin sulfoxide synthesis 828
O-silylated steroid derivatization 828–829, 829*F*
thioester library 830, 830*F*
historical perspective 792
compound library construction 792
ligand bound synthesis *vs.* 792
ocadec-9-enoic butyl ester synthesis 792
immobilized scavengers 795, 795*F*
library generation 797–798, 799*F*
microencapsulation 812
aryl boronic acid synthesis 813
aryl bromide synthesis 813
definition 812
Heck reaction 813
polyurea microcapsules 813
preparation 812
Stille couplings 813
supercritical carbon dioxide ($seCO_2$) 813
Suzuki couplings 813
Suzuki–Miyaura coupling procedure 812
multistep synthesis 792, 800
Amaryllidaceae alkaloids 815, 816*F*, 819
carpanone 818, 819*F*
commercially-available drugs 801
see also specific drugs
epibatidine 816, 817*F*, 818*F*
epimaritidine 815, 816*F*
epothilones 821, 822*F*, 823*F*, 824*F*
natural product synthesis 814
see also specific compounds
neolignans 818
nicotine 817
nornicotine 817
obliquine 819
oxomaritidine 815, 816*F*
plicamine 819, 820*F*
plicane 819
polysyphorin 818
rhaphidecursinol B 818
rosiglitazone 802, 802*F*
salmeterol 802–804, 803*F*
sildenafil 801, 801*F*
one-pot reactions 794*F*, 795, 796
alkane functionalization 796, 796*F*
amino alcohol synthesis 797, 797*F*
functionalized pyrazole synthesis 797, 798*F*
propranolol synthesis 797, 798*F*
trihydroxy nucleoside synthesis 797
unsaturated nitrile synthesis 796–797
Wadsworth–Horner–Emmons alkenylation 796–797
'wolf-and-lamb' approach 796
reaction quenching 797
scavenging techniques 797
amide library construction 798, 799*F*
benzodiazepine synthesis 799–800, 800*F*
catch-and-release principle 799, 800*F*
chemical library generation 797–798, 799*F*
hydantoin synthesis 799–800, 800*F*
Mitsunobu reagents 798–799
phase switch concept 799–800
solution phase library construction 798–799
sulfonamide library construction 798, 799*F*
thiourea library construction 798, 799*F*
urea library construction 798, 799–800, 799*F*, 800*F*
site isolation effects 794, 794*F*
difunctional molecule synthesis 795

high concentrations 795
macrocyclization 795
one-pot transformations 794*F*, 795
three-dimensional environment 794–795
supports 792
acrylamides 792–793
Amberlyst 15 ion exchange resin 821
cellulose 793
controlled pore glass 793
cross-linking 792–793
divinyl benzene polystyrene resin 792
Jandagel 792–793
microreticular resins 792–793
monoliths 793
Nafion-H resin 821
PEGylated polystyrenes 792–793
polyureas 792–793
silica 793
zeolites 793
Polypeptides, as drug candidates 374–375
Polysyphorin, PASP 818
Polyureas, PASP 792–793
microcapsules 813
Poly(His)$_x$ tag, gene expression vectors 110, 110*T*
PONDR, domain mapping 879
Porin-like proteins, protein crystallization 434
Positional (flat) assays, cell-based assays 631
Positional cloning, *Danio rerio* model system 10
Positive phenotype interpretation *see* Genetically engineered animals
Positron emission tomography (PET)
applications 659
cancer research 77
functional small animal imaging 661
small animal imaging 660*T*, 661, 663–664
biomarkers 77
radionuclides 663*T*
see also specific nuclides
^{75}Br 664*T*
^{76}Br 664*T*
^{11}C 664*T*
^{55}Co 664*T*
^{62}Cu 664*T*
^{64}Cu 664*T*
^{18}F 661–662, 664*T*
2-[^{18}F]fluoro-2-deoxy-D-glucose (FDG) 77
^{68}Ga 664*T*
^{124}I 664*T*
^{13}N 661–662, 664*T*
^{15}O 661–662, 664*T*
^{82}Rb 661–662
^{94m}Tc 664*T*
^{86}Y 664*T*
Postcrystallization treatments *see* Protein crystallization
PostgreSQL 294
server databases 294
Post-reaction cooling, Milestone s.r.l. microwave-assisted chemistry 849
Post-source decay (PSD), mass spectrometry proteomics 33–34
Posttranscriptional gene silencing, RNAi (RNA interference) 173
Posttranslational modifications
Escherichia coli as expression system 880
expressed proteins 880
heterogeneous protein crystallization 881
proteome 29
proteomics 41
2D-PAGE 41
UniProtKB 359
Power outputs, CEM Corporation microwave-assisted chemistry 851
Precipitants, protein crystallization 438
Precision, NMR structures, problems 542
Preclinical trials, proteomics 42, 42*F*
Precursor cells, cell-based assays 643
Predicted coding sequences, IPI (International Protein Index) 366
Predictive genotyping, pharmacogenetics *vs.* 66
Pre-equilibrated dynamic combinatorial libraries *see* Dynamic combinatorial libraries (DCLs)
Prelink, domain mapping 879
Preparation, PASP 812
Preparative liquid chromatography/mass spectroscopy (LC/MS), high throughput purification 865–866, 866*F*, 867*F*
demixing 868
Pre-pressurized reaction vessels, microwave-assisted chemistry 854
PreQuest, CSD 391, 393
Pressure control, microwave-assisted chemistry 837
Pressure measurement, Anton Paar GmbH microwave-assisted chemistry 852
Pressure monitoring, microwave-assisted chemistry 857
Pressure regulation, CEM Discover platform microwave-assisted chemistry 847
Prevacid, mechanism of action, mouse knockout models 13
Primary cell lines, cell-based assays 619–620
Primary databases *see* Databases
Primary screen supply, compound storage/management (Pfizer) 569, 570*F*
Primary sequences, UniParc 358–359
Principal component analysis (PCA), microarrays 91
PRINTS
InterPro, member of 363
signature method 363
Privileged structures
chemogenomics 925, 925*F*, 926*T*
combinatorial libraries 133
definition 133–135
Probabilistic models, sequence alignment, bioinformatics 319–320
Probe classification, cDNA microarrays *see* cDNA microarrays
Procainamide, analogs, solid-phase libraries 725, 725*F*
PROCHECK 547–548
protein 3D structure 546
protein 3D structure validation 516–517
ProDom
InterPro, member of 363
signature method 363
PRODRG server, protein 3D structure validation 514–515
Product-based selection, combinatorial library design 256–257
Production levels, yeast protein expression systems 117
ProFun method, protein function prediction, bioinformatics 335
Proinflammatory cell-signaling pathways 195
see also specific pathways
regulation 196*F*
Prolactin, STAT5 signaling 206
Proliferation probes
^{123}I 665
^{125}I 665
small animal imaging 665
[^{3}H]thymidine 665
Proline puckering, protein 3D structure validation 516
Prolines, solid-phase discrete libraries 709*T*
Promoters
dependence, transgenic mouse model 11
gene expression vectors *see* Gene expression vectors
mammalian protein expression systems 117
regulatory regions, bioinformatics 322, 323*F*
RNA interference 17

Propafenone, discovery/development 927–929, 929*F*
2-Propanol, microwave heating 840*T*
Propionic acid derivatives, solid-phase discrete libraries 709*T*
Propranolol, synthesis, PASP 797, 798*F*
ProsaII, x-ray structures, problems 537
PROSA-II software, protein 3D structure validation 517–518
PROSITE database
  InterPro, member of 363
  protein function prediction, bioinformatics 334–335
  signature method 363
Prostate cancer, model, [^{18}F]FDG imaging 670
Prostate-specific antigen (PSA), as surrogate marker 80
Protease-free systems, *Escherichia coli* protein expression systems 114–116
Protease inhibitors
  see also specific drugs
  chemogenomics 929
Proteases
  radioactive assays 583–584
  structure 929
Proteasome inhibitors, bortezomib 194
Proteasomes, structure 190–191, 191*F*
Protein(s)
  amounts, biomarkers 73
  assessment
    biomarker identification 75–76
    NMR spectroscopy 945, 945*F*
    single-parametric biochemical assays 683
  characterization, proteomics *see* Proteomics
  concentrations, protein crystallization 438
  conservation, small animal test systems 647–649
  dynamics, x-ray structures, problems 538
  extraction/prefractionation, proteomics *see* Proteomics
  families, structural genomics 553–554
  flexibility, structure-based virtual screening 251
  folding
    gene expression vectors 108
    yeast protein expression systems 116
  identification, 2D-PAGE *see* 2D-PAGE
  interaction networks, regulatory networks 330
  precipitation, *Escherichia coli* protein expression systems 112
  quality effects, protein crystallography 440–441
  quantification
    enzyme-linked immunosorbent assays (ELISAs) 73–74
    genetically engineered animal characterization 153–154
    immunoblotting 73–74
    sandwich ELISA 73–74
    2D-PAGE *see* 2D-PAGE
  small molecule interactions, orthogonal ligand–receptor pairs *see* Orthogonal ligand–receptor pairs
  turnover
    siRNAs delivery 177
    ubiquitin-proteasome pathway 191–192
Protein A, Z-domain 73
Protein arrays, proteomics *see* Proteomics
Protein crystallization **433–447**
  see also X-ray crystallography
  apparatus 439
    drop size 439
    gel diffusion 439
    hanging drop method 438–439
    high gravity methods 439
    low gravity methods 439
    mini-batch procedure 439
    screening methods 439
    sitting drop method 439
  associations 436
  automation 440
  co-crystallization 891
  conditions 438
    ammonium sulfate 438
    detergents 438
    pH 438
    protein concentrations 438
  cryocrystallography 889
    enzymes 889
  crystal annealing 888–889
  crystal dehydration 888–889
  crystal engineering 438
    resolution 438
    surface residue mutation 437
  heterogeneous proteins 436, 437*F*, 881
    asparagine deaminations 436
    cofactors 437
    conformational changes 436
    disulfide bridges 436, 437, 437*F*
    fusion proteins 436–437
    glutamine deaminations 436
    glycosylation 882
    isoelectric heterogeneity 882
    phosphorylation 882
    posttranslational modifications 881
    recombinant proteins 436
  history 433
    Abel, John J 885–887
    Hünefeld, Friedrich 881
    Kendrew, John 884
    Laufburger, Vilem 885
    Watson, Herman 884
  improvements 555
  integral membrane proteins 434
    bacterial photoreaction center 434
    porin-like proteins 434
  ligand soaking 891
  mobile regions 436
    division of 436
    proton NMR 436
  modified proteins 436
  monotropic membrane proteins 435
    carotenoid-processing enzyme 435, 435*F*
    detergents 435
  postcrystallization treatments 888
    crystal annealing 888–889
    crystal dehydration 888–889
  protein modification 884
    chemical modification 888
    deglycosylation 882, 883*F*
    dephosphorylation 882
    directed evolution 888
    DNA shuffling 884–885, 885*F*
    homolog screening 884
    metal-mediated crystallization 885, 886*F*
    mutagenesis 882
    rational surface engineering 887
  solubility optimization 881
    ammonium sulfate 881
    detergents 881
    glycerol 881
    guanidinium chloride 881
    Hofmeister Series 881
  soluble proteins 434
    chymotrypsin 434
    entropy minimization 434
    lysozyme 434
    myoglobin 434
    ribonuclease 434
  success of 875–876
  theory 438
    entropy change 438
    precipitants 438
    water removal 438
Protein D
  Escherichia coli protein expression systems 113–114
  gene expression vectors 110*T*, 111
Protein Data Bank (PDB) **373–388**
  annotation 376
  browsing capabilities 377
    Biological Process 377
    Disease 377
    Molecular Function 377
  CATH (Class, Architecture, Topology and Homologous superfamily) *see* CATH (Class, Architecture, Topology and Homologous superfamily)
  chemoinformatics, three-dimensional structure 240

data deposition 376
information type 376
PDB Exchange Dictionary 376
PDB ID 376
Validation Server 376
disease implications 381
disease-related structures 381
OMIM numbers 380
structure summary 380, 380*F*
errors 521
see also Nuclear magnetic resonance (NMR), structure determination; Protein 3D structure, problems; Protein 3D structure validation; X-ray crystallography
geometrical parameters 522*F*, 523
nonunique chain identifiers 521, 522*F*
side-chain rotamer libraries 523
Z-scores 523, 524*F*
Exchange Dictionary 376
future work 382, 383*F*
history 373
Brookhaven National Laboratory 373
Internet effects 374
Research Collaboratory for Structural Bioinformatics (RCSB) 373
structure number 373, 374*F*
usage 374
worldwide PDB members 374–375
homology-based protein structure prediction 325
ID 376
large structure, PDB format problems 543
Ligand Depot 376
ligands 378*F*, 380
LigPro Ligand Explorer 377, 378*F*
major therapeutic target classes 376
GPCRs 374*F*, 376
member numbers 507–508, 508*F*
mutations 381
protein crystallography 450
protein–ligand interactions 380
related structures, comparative computational modeling 557
SCOP (Structural Classification of Proteins) *see* SCOP (Structural Classification of Proteins)
searching capabilities 377
Gene Ontology terms 376–377
keyword 376
MEDLINE 377
PubMed abstract 377
Query-by-Example 376
Search Ligands 377
Search Sequence 377
Search Unreleased 377
secondary databases 374
standards 376
protein 3D structure validation 511
structure error *see* Protein 3D structure, problems
Target DB 551
Universal Protein Resource 362–363
vintage model structure 536–537
Protein expression **107–128**, 112*T*, 113*T*
bacterial systems 116
Bacillus subtilis 116
Corynebacterium glutamicum 116
Escherichia coli see Escherichia coli protein expression systems
lactic acid bacteria 116
Lactobacillus lactis 116
cell-based assays 630–631
cell-free *see* Cell-free expression systems
Escherichia coli see Escherichia coli protein expression systems
mammalian *see* Mammalian protein expression systems
purification, gene expression vectors 110
solubilization, bacterial expression systems 415
structural genomics *see* Structural genomics
system comparison 113*T*
vector design/generation *see* Gene expression vectors
viral 118
see also Baculovirus expression systems; Semliki Forest Virus expression systems
yeast 116
advantages 116
expression vectors 117
production levels 117
protein folding 116
selection markers 117
species 116–117
yield effects, gene expression vectors 110
Protein function prediction 333
definition 333
Enzyme Commission classification 333–334
Gene Ontology Consortium 333–334
genomic context methods 335
domain fusion 336
gene neighborhood 336
gene order 336
phylogenetic profiles 336
information integration 338
methods 338
MIPS Functional Catalog 333–334
protein interaction networks 335
false negatives/positives 335
PLEX database 336
STRING database 335–336
yeast-two-hybrid method 335
sequence 334
cluster of orthologous genes (COGs) 334
EMOTIF database 334–335
gene duplication 334
hidden Markov models 335
homologous genes 334
neural networks 335
Orthostrapper 334
ProFun method 335
PROSITE database 334–335
PSSMs 335
sequence motifs 334–335
SMART 334
whole-genome sequence data 334
structure 336
CATH database 336–337
CE 336–337
comparison of 336
DALI/FSSP 336–337
FATCAT 336–337
motif derivation 337
SCOP 336–337
SSAP 336–337
surface analysis 337
text mining 337
BIND database 337
information extraction 337
information retrieval 337
language problems 337
nomenclature problems 337
Protein G
gene expression vectors 110*T*
ubiquitin complex, NMR structure determination 499, 500*F*
Protein Information Resource (PIR), history 349
Protein interaction networks, bioinformatics *see* Protein function prediction
Protein InTerfaces and assemblies (PITA) program, x-ray structures, problems 540
Protein kinase(s)
assays
cell-based assays 624*T*
filtration assay 586–587
biological functions 218
biomarker identification 72
Flash Plate assays *see* Flash Plate assays
measurement
antibodies 72
biomarker identification/ measurement 72
orthogonal ligand–receptor pairs *see* Orthogonal ligand–receptor pairs
radioactive assays *see* Radioactive assays
scintillation proximity assay (SPA) *see* Scintillation proximity assay (SPA)
small molecule inhibitors 587
solid-phase discrete libraries 707*F*

Protein kinase A (PKA), scintillation proximity assay 587
Protein kinase C (PKC)
 inhibitor development 930, 931*F*
 ubiquitin-protein targeting 192
Protein kinase inhibitors
 see also specific drugs
 chemical proteomics 44
 chemogenomics 930, 931*F*
 cross-reactivity 3
 RNA interference 17–18
Protein–ligand binding
 fragment-based drug discovery 946
 x-ray crystallography 946
Protein–ligand docking programs, CSD *see* Cambridge Structural Database (CSD)
Protein production **411–432**
 choice of 413
  database sequence search 413
  limited proteolysis studies 413
  sequence alignment/analysis 413
 cloning 413
  Gateway System 414
  PCR 414
 concentration 424
  polyethylene glycol 424
  ultrafiltration 424
 deglycosylation 428
 expression systems 415
  see also specific types
  automation 415
  choice of 414
  parallelization 415
 fermentation 418
  fermenters 419
  roller bottles 418
  shaking flasks 418
  spinner flasks 418
  wave bioreactors 419, 419*F*
 natural sources, production from 412
Protein production, purification 419
 affinity chromatography 422
 affinity tags 421
 buffer exchange 423
  dialysis 423
  size exclusion chromatography 424
  ultrafiltration 424
 cell lysis 420
 fractionated precipitation 421
 hydrophobic interaction chromatography 423
 ion exchange chromatography 422
 low-pressure column chromatography 422
 refolding 420, 421
 size exclusion chromatography 423, 423*F*
Protein–protein interactions
 optical assays (indirect format) 582
 proteomics *see* Proteomics
Protein Quaternary Structure (PQS) server, x-ray structures, problems 540
Protein sequences
 databases 357
  see also specific databases
  functional analysis 357
  history 349, 350
 RefSeq (NCBI Reference Sequences) 358
Protein Structural Initiative (PSI)
 see also Structural genomics
 center selection 551–552, 552*T*
 methodology 552, 553*F*
 second phase 551–552
 target selection 551
Protein Structure Factory (PSF) 552*T*
Protein structure prediction 324
 CASP (Critical Assessment of Structure Prediction) 324–325
 de novo 326
  nonapeptides 327
  ROSETTA 327
  thermodynamics 326
 homology-based 325
  accuracy 326
  alignment scores 326
  gap interpretation 326
  Protein Data Base 325
  protein threading 325, 326
  target sequence alignment 325
  template availability 325–326
  unknown structure 326
  validation 326
 secondary structure 325
  DSSP 325
  homolog searching 325
  motifs 325
  STRIDE 325
Protein synthesis using recombinant elements (PURE) system, cell-free expression systems 122
Protein tags, gene expression vectors 110
'Protein threading,' homology-based protein structure predication 325, 326
Protein 3D structure, analysis 424
 activity assays 427
 circular dichroism spectroscopy 427
 concentration determination 425
  colorimetric methods 426
  UV absorption 426
 electrophoretic methods 424
  isoelectric focusing 425, 425*F*
  SDS-PAGE 424, 425*F*
  Western blotting 425
 mass spectrometry 427
 membrane proteins 429
 1D-NMR 427
 particle size/aggregation studies 426
  analytical ultracentrifugation 426
  dynamic light scattering 426
  size exclusion chromatography 423*F*, 426
 sequence analysis 428
Protein 3D structure, problems **531–550**
 EM 542
  low resolution 542–543
  multiple entries 543
 global quality measures 545
  PDBREPORT 546
  PDBsum 546
  PROCHECK 546
  Ramachandran plots 546
  resolution 545
  R-factor 545
  stereochemical parameters 546
  websites 547*T*
  WHAT_CHECK 546
 homology models *see* Homology models
 local quality measures 546
  atomic *B*-factors 546
  coordinate error estimate 546
  electron density maps 547
  real-space *R*-factor plots 547
  stereochemical checks 547
  Uppsala Electron Density Server (EDS) 547
  websites 547*T*
 NMR *see* Nuclear magnetic resonance (NMR), structure determination
 predicted structures 545
  see also Homology models
 variation of 532
 very large structures 543
  chaperonins 543
  multienzyme complexes 543
  multiple entries 543
  nomenclature problems 543
  PDB format 543
  ribosomes 543
  virus structures 544
 X-ray structures *see* X-ray crystallography
Protein 3D structure errors
 cutoff values 509
 false positives *vs.* false negatives 509, 509*T*
 impact of 509
 quality effects 509
 sources 509
  method-specific 509
 standard conformation libraries 509
Protein 3D structure validation **507–530**, 532
 administrative errors 511
  atomic occupancies 512
  CRYST1 record 511
  example 512, 512*F*, 513*F*
  missing atoms 513

MTRIX record 511, 512*F*
PDB standards 511
SCALE record 511
water molecules 512
aim of 508
bond angles 511
bond length 511
classification 520
live *vs.* dead 520
contact analysis 517
bump detection 517
directional atomic contact analysis 517–518
ERRAT software 517–518
PROSA-II software 517–518
database generation 520
data redundancy 521
mutants 521
structure models 520
data problems 511
unsubmitted data 511
error types 511
see also specific errors
experimental data comparison 510
Electron Density Server 510
local correlation tests 510
NMR 511
QUEEN program 511
Real Space R-factor (RSR) 510
R-factor 510
experimental x-ray data 519
electron-density maps 519
example 519, 519*F*
real-space correlation coefficient 520
R-value 519
SFCHECK 519
future work 523
accuracy increase 523
contact number 525*F*, 526, 526*F*
new checks 523, 525*F*
physicochemical properties 523, 525*T*
geometry 511, 515
bond angles 515
bond lengths 515
χ-angle 517
cis-peptide bonds 517
examples 515, 516, 516*F*
ϕ-angle 516–517
MOLPROBITY 516–517
omega angles 517
PROCHECK 516–517
proline puckering 516
Ramachandran plots 516–517
side chain planarity 516
torsion angles 516
WHAT_CHECK 516–517
Ψ-angle 516–517
heterocompounds 514
example 514–515, 514*F*, 515*F*
HIC-Up database 514
PRODRG server 514–515
hydrogen bonds 518
nomenclature validation 518
positions 518
WHAT_CHECK 518
x-ray crystallography *vs.* NMR 518
nomenclature 513
ATOM record 513, 514*F*
chain identifiers 513, 514*F*
hydrogen atoms 513
SEQRES record 513
older structures 520, 521*T*
tools 507–508
torsion angles 511
WHAT_CHECK 510
Proteolytic mapping, domain mapping 879
Proteome 29
analysis, Integr8 *see* Integr8
complexity 29, 29*T*
definition 28
mapping 46
posttranslational modifications 29
proteins per gene 29, 29*T*
scoring 46
single nucleotide polymorphisms 29
splicing 29
Proteomics **27–50**
biomarkers 45, 75
abundance 45
biofluids 45
diagnostic implications 45–46
drug toxicity markers 78
immunoassays 45–46
multidimensional protein identification technique (MudPIT) 75–76
protein measurement 75–76
clinical testing 42, 42*F*
compound development 42, 42*F*
compound profiling (chemical proteomics) 42–43, 44
activity-based protein profiling 44–45
compound immobilization 44
kinase inhibitors 44
primary target abundance 44
SAR 44
definition 28, 330–331
drug toxicity markers 78
genomics *vs.* 5
history of 28
increase in use 28, 28*F*
imaging mass spectrometry 39
resolution 39–40
mass spectrometric identification 32
database searches 33
electrospray 32
matrix-assisted laser desorption ionization (MALDI) 32
peptide mass fingerprinting 32, 33
post-source decay (PSD) 33–34
tandem sequencing 33
tandem sequencing (MSMS) 33
time-of-flight (TOF) detectors 32
molecular diagnostics 45
1D-SDS PAGE 39
membrane protein analysis 39
peptide-based *see* Proteomics, LC-MS (peptide-based)
preclinical testing 42, 42*F*
protein arrays 40
antibody array 40
phenotype-oriented screening 139
reverse-phase arrays 40
protein characterization 41
glycosylation 41
histone modification 41
phosphorylation 41
posttranslational modifications 41
protein extraction/prefractionation 30
abundant protein depletion 31–32
affinity chromatography 31
antibody-based purification 32
cell cycle changes 30–31
cell disruption 31
cell extraction 30
depletion 31
homogenous samples 30–31
laser capture microdissection (LCM) 30–31
LC-MS 30–31
subcellular fractionation 30, 31
tissue extraction 30
2D-PAGE 30–31
protein–protein interactions 41
co-immunoprecipitation 41–42
'false positives' 41–42
systems biology 46
integration 46
prediction 46
proteome mapping 46
proteome scoring 46
target identification 42, 42*F*
animal models 43
biomarker validation 43
compound mechanism of action 42–43
disease mechanisms 42–43
expression profiling 42–43
novel biology 43
sample choice 43
target validation 42
technologies 30*F*
2D-PAGE *see* 2D-PAGE
Proteomics, LC-MS (peptide-based) 29–30, 30*F*, 37
2D-PAGE *vs.* 39
automation 37, 39
combined fractional diagonal chromatography (COFRADIC) 38
differential analysis 38
HPLC separation 37

Proteomics, LC-MS (peptide-based) (*continued*)
isotope-coded affinity tags (ICAT) 37–38
miniaturization 39
multidimensional protein identification technology (MudPIT) 37
peptide complexity reduction 37
prefractionation 37–38
protein extraction/ prefractionation 30–31
quantification 38
chemical labeling 38
isotope labeling 38
isotopic analog peak ratios 38
metabolic labeling 38
stable isotope labeling by amino acids in cell culture (SILAC) 38
trypsin-mediated ^{18}O incorporation 38–39
'shotgun proteomics' 37
'survey scan' 38
Proton detection, liquid-state NMR structure determination 479–481
Proton nuclear magnetic resonance, mobile protein regions 436
Prous Drug Data Report 278*T*
Psammaplin, derivatives, dynamic combinatorial libraries 973
PseudoCyc database, metabolic networks 328–329
Pseudodynamic (deletion) screening, dynamic combinatorial libraries 965, 968
Pseudodynamic systems, dynamic combinatorial libraries 968
Pseudoenergy, protein crystallography model generation 463–464
Psoriasis
pathogenesis, vascular endothelial growth factor 202
treatment, Neovastat 202
PSSMs, protein function prediction, bioinformatics 335
Psychiatric diseases/disorders
corticotrophin-releasing factor receptor 93–94
proteomics 43
PTEN, identification, large-scale virus vector libraries 18–19
PubChem 307
BioAssay database 307
Compounds database 307
Molecular Libraries Initiative (MLI) 307
Substances database 307
Public education
common complex diseases 55
pharmacogenetics 66
PubMed abstract, PDB searches 377
Purification
high throughput *see* High throughput purification
natural products 133
Ras/Raf protein interaction inhibitor 1 library construction 766
solid phase parallel synthesis 765
p-value, microarrays 98–99
Pyrazoles, solid-phase discrete libraries 709*T*
(4-amino-3-phenyl-1-*t*-butyl) Pyrazolo[3,4-*d*]pyrimidine (PPI) derivatives, orthogonal ligand–receptor pairs 218–220, 221*F*
Pyridines
solid-phase libraries 735, 735*F*
solid-phase pool libraries 705*F*
2-Pyridones, solid-phase discrete libraries 709*T*
Pyrimidines
solid-phase discrete libraries 709*T*
solution phase library synthesis 780, 782*F*, 786, 786*F*
Pyrophosphatases, optical assays 582
Pyrroles
solid-phase discrete libraries 709*T*
synthesis, solution phase library synthesis 780, 783*F*
Pyrrolidin-2,5-diones, solid-phase discrete libraries 709*T*
Pyrrolidines, solid-phase pool libraries 705*F*

Q

QTL analysis, target search genomics *see* Genomics
Quality, libraries 697
Quality checks, microarrays 91
Quality control
cDNA microarrays 331
compound acquisition/ processing 563–564
compound storage/management 568
microarrays 90
protein crystallography *see* X-ray crystallography
Quality standards, HTS automation 682
Quantitative PCR, microarrays 94, 99, 100
Quantitative reverse transcription polymerase chain reaction, biomarkers 75
Quantum dots, HCS labeling techniques 690
QUEEN program, protein 3D structure validation 511
Quenching
fluorescence intensity distribution analysis 609–610, 610*F*
fluorescence polarization 603, 604*F*
fluorescence screening assays 601, 611–612
Query-by-Example, PDB searches 376
Query combinations, CSD 393–395
Quest 205 Organic synthesizer, PASP 805–806
Quinazolinones
solid-phase discrete libraries 709*T*
solid-phase libraries 722, 722*F*
solid-phase pool libraries 705*F*
Quinoline-4-carboxamides, solid-phase discrete libraries 709*T*
Quinolines, synthesis, solution phase library synthesis 784, 784*F*
Quorum sensing effectors, solid-phase discrete libraries 709*T*
q-value, microarrays 98–99

R

RA (reference author), UniProtKB flatfile 360
Rac activation, vascular endothelial growth factor receptor-2 (VEGF-R2) 200
Radiation-induced sarcoma (RIF-1) cells, radiolabeled annexin imaging 667–668
Radioactive assays 578, 583
disadvantages 583–584
DNA helicase 588
DNA polymerase 588
DNA primase 588
filtration assay 584
GPCRs 585–586
kinases 586–587
malonyl-coenzyme A inhibitor identification 584
technique 583–584, 586–587
GPCRs 583–584, 585, 586*F*
adenosine 2a receptor 586
displacement assays 585–586
histamine receptors 586
human cytochrome P-450 s 589
kinases 583–584, 586
lipid kinases 587
P13 kinases 587
phospholipid-binding proteins 587
long-chain fatty acyl coenzyme A synthase 588
other formats *vs.* 590
PARP-1 inhibitors 589
peptide–protein binding assays 589
proteases 583–584
signal measurement 578
targets 585
translation factors 588
transporters 583–584
Radioimmunoassays (RIAs) 578–579
Radiolabeled annexin, cell surface receptors 667
Radiolabeled compounds, small animal imaging 666
Radioligands, cell-based assays 627

Radionuclides
PET 663*T*
SPECT 662*T*
Radiotreatment imaging
[^{18}F]FDG 670
[^{18}F]FdUrd 670
2-deoxy-D-[1-^{14}C]-glucose 671
[^{14}C]methionine 670
[^{99m}Tc]MIBI 670
small animal imaging 670
[^{3}H]thymidine 670
Ramachandran plots
chloromuconate cycloisomerase structure 533–534
protein 3D structure 516–517, 546
Raman spectroscopy, CEM Discover platform microwave-assisted chemistry 845–846
Random mutation screening, *Saccharomyces cerevisiae* model system 5–6
Random sets, compound storage/management (Pfizer) 570–571
Ranibizumab 202
RANK, knockout animal models 156
RANKL (*RANK* ligand), knockout animal models 156
Rapid column switching, parallel analysis/purification 868, 870*F*
Rapid heating, microwave-assisted chemistry 840–841
Ras, node network theory 4
RAS proteins, *Drosophila melanogaster* 650
signaling 9
Ras/Raf/MAP kinase pathway, activation, bortezomib 195
Ras/Raf protein interaction inhibitor 1, library construction 765, 765*F*
analog synthesis 766
purification 766
structural modifications 766, 766*F*
Rate enhancements, microwave-assisted chemistry 841
Rational drug design, protein crystallography *see* X-ray crystallography, rational drug design
Rational drug design (RDD), CSD 399
Rational surface engineering, protein crystallization 887
Rayment, Ivan, protein crystallization 888
^{82}Rb, PET 661–662
RC (reference comment), UniProtKB flatfile 360
^{186}Re, biomolecular probes 666
^{188}Re, biomolecular probes 666
Reaction conditions, solution phase library synthesis 770
Reaction databases 279, 281, 309
definition 279–280
Reaction homogeneity, microwave-assisted chemistry 856
Reaction kinetics, PASP 793–794
Reaction monitoring
Milestone s.r.l. microwave-assisted chemistry 849
solution phase library synthesis 775
Reaction quenching, PASP 797
Reaction times, microwave-assisted chemistry 857
Reaction types, combinatorial chemistry 762
Reaction volumes, microwave-assisted chemistry scale-up 856
Read out parameters, fluorescence screening assays 614
Real-space correlation coefficient, protein 3D structure validation 520
Real Space R-factor (RSR), protein 3D structure 510, 547
Real-time measurements, cell-based assays 618
Receptors
affinity, 5-HT$_3$ ligand gated ion channel assays 635*T*
cell-based screening assays *see* Cell-based assays
internalization, HCS *see* High-content screening (HCS)
signal changes, NMR, drug discovery 915–916
Receptor tyrosine kinase FLT3 identification, target search genomics 5
Reciprocal lattice, x-ray–crystal interaction 456–457
Recognition, dynamic combinatorial chemistry 961
Recombinant proteins, protein crystallization 436
RECOORD, NMR structures, problems 542
Recoupling, NMR structure determination 482
Recursive partitioning (RP), machine learning methods 250
Reduced graph similarity, 2D similarity searching 247
Reduction-of-function (hypomorphic), chemical mutagenesis 13–14
Reference linking, databases 284
Refinement, protein crystallography model generation 463
Reflection
dielectric heating 840
measurement, x-ray–crystal interaction 454
Reflection intensity strength, protein crystallography 470
Reflectometric interference spectroscopy (RIfS), optical biosensors 594–595
Refolding, protein purification 420, 421
RefSeq (NCBI Reference Sequences)
curation 355–356
data acquisition, DDBJ/EMBL/GenBank 354
features 355
nucleotide sequences 355
protein sequences 358
Registration, compound acquisition/processing 564
REGISTRY 282*T*
Regulatory aspects, pharmacogenetics *see* Pharmacogenetics
Regulatory DNA sequences, gene expression vectors *see* Gene expression vectors
Regulatory elements, genetically engineered animals 156
Regulatory networks, bioinformatics *see* Molecular networks, bioinformatics
Regulatory pathways, cDNA microarrays 332
Regulatory regions, bioinformatics *see* Molecular sequence analysis, bioinformatics
Regulatory systems, cell-based assays 617–618
Relational database management system (RDBMS)
databases 294
object-based databases 268
Relational databases 268, 269*F*
advantages 268
SQL (structured query language) 268
Relaxation dispersion, NMR structure determination 498
Relaxation filtering, NMR, drug discovery 908*T*, 913
Relaxation optimized sequences, liquid-state NMR structure determination 479–481
Relaxation rates, NMR spectroscopy 946
Rel homology domain (RHD), NF-κB 195
Reliability factor *see* *R*-factor
Rel protein family 195
Renin, biological functions 929
RepeatMasker, repetitive elements, bioinformatics 322–323
Repeats, genome sequencing 317
Repetitive elements, bioinformatics *see* Molecular sequence analysis, bioinformatics
Reporter genes/assays
cell-based screening assays *see* Cell-based assays
single-parametric cellular assays 683
small animal imaging 665
Reporter ligand observation, NMR spectroscopy 945*F*, 946

Representivity of Target Families in the Protein data Bank (PDBRTF) 304–305
REPuter, repetitive elements, bioinformatics 323
Research Collaboratory for Structural Bioinformatics (RCSB)
  data nomenclature 543
  ligand nomenclature standardization 539
  macromolecular Crystallographic Information File (mmCIF) 543
  PDB 373
Resin capture/release, solution phase library purification 777, 777*F*
Resin scavenging, solution phase library purification 776, 777*F*
Resolution
  HCS confocal systems 687
  imaging mass spectrometry 39–40
  protein crystallization 438
  protein crystallography model generation 462–463, 464*F*
  protein 3D structure 545
  2D-PAGE 34–35
Response element sequence, cell-based assays 633
Restraints per residue, NMR structures 546
Restriction proteases, gene expression vectors 111, 111*T*
REST/RSF, identification, large-scale virus vector libraries 18–19
Retaane 202
Reticulocytes, cell-free expression systems 122
Retinoblastoma protein (Rb), as biomarker 72
Retinoid X receptors (RXRs), orthogonal ligand–receptor pairs 221–222, 222*F*, 223*F*
Retinopathy of prematurity, vascular endothelial growth factor 202
Retroviruses
  large-scale virus vector libraries 18–19
  vectors
    RNA interference 17
    siRNAs 177
Reversal, genome rearrangements, bioinformatics 324
Reverse engineering, biomarkers 71–72
Reverse genetics
  chemical biology 130, 131*F*
  Drosophila melanogaster see Drosophila melanogaster
  human cells 14
  Saccharomyces cerevisiae model system *see Saccharomyces cerevisiae* model system
Reverse-phase arrays, proteomics 40
Reverse screening, fragment-based drug discovery 952, 953*F*
Reverse transcriptase polymerase chain reaction (RT-PCR)
  genetically engineered animal characterization 153–154
  RNAi detection 178
Reversible covalent bonds, dynamic combinatorial library generation 961, 962*F*
Reversine, discovery/development 922, 923*F*
R-factor
  chloromuconate cycloisomerase structure 533–534
  definition 546
  protein 3D structure 510, 545
  testing, x-ray structures, problems 535
R_{free}, testing of 535
RG (reference group), UniProtKB flatfile 360
Rhaphidecursinol B, PASP 818
Rheumatoid arthritis (RA)
  NF-κB pathway 197
  STATs 208
Rhinoviruses
  protein structures on PSD 384
  structures 382, 383*F*
Rhodopsin, orthogonal ligand–receptor pairs 230*F*, 231–232
RhoGDI, protein crystallization 887
Ribonuclease, protein crystallization 434
Ribonuclease protection assay, genetically engineered animal characterization 153–154
Ribosome binding site (RBS; Shine–Dalgarno sequence), gene expression vectors 109
Ribosomes, structure, problems with 543
Ribozymes, transgenic animal generation 152
Rigid body alignments, pharmacophore identification 243, 244*F*
Rigid searching, chemoinformatics, three-dimensional structure 242
RIKEN Structural Genomics Initiative (RSGI) 552*T*
RISC (RNA-induced silencing complex) *see* RNAi (RNA interference)
Risperidone receptors, imaging 667
RL (reference location), UniProtKB flatfile 360
RN (reference number), UniProtKB flatfile 360
RNA-binding proteins, cell-based assays 631
RNAi (RNA interference) 17
  see also Short interfering RNAs (siRNAs)
  adenovirus delivery 17
  applications 178
    biomarker identification/validation 172
    disease links 180
    disease therapy *see below*
    drug action validation 181
    ischemia reperfusion induced lung apoptosis 182
    kinase inhibition 17–18
    mitotic spindle defects 18
    ocular neovascularization 181–182
    stress responses 181
    target identification 172
    target validation 172, 181
  arrayed DNA-based siRNA libraries 18
  biological function 171
    viral infections 173
  Caenorhabditis elegans model system 8, 179
    advantages 179
    delivery 8, 179
    growth 179
    history 172
    obesity research 9
  DICER 173, 174*F*
    discovery 172
    PAZ domain 173
    species homologs 173
    structure 173
  Drosophila melanogaster 9, 179, 651
    cytokinesis 179
    delivery 179
  endocytosis 17–18
  experimental controls 177
  future work 183
    cancer cell targeting 184
    lethal gene identification 184
  gene product inhibition *vs.* 17
  gene silencing detection 178
    flow cytometry 178
    immunofluorescence 178
    mRNA 178
    Northern blot 178
    protein 178
    reverse transcriptase polymerase chain reaction (RT-PCR) 178
    Western blot analysis 178
  genome-wide screens 179
    see also specific models
    cytokinesis 180
    expression libraries 180
    humans 180
  history 172
    clinical trials 172
    technology development 172
  hypoxia response 17–18
  interferon response 172, 176
  in vivo gene silencing 182
  lentivirus delivery 17
  limitations 176, 182
    cellular uptake 182
    nonspecific immune response 182

off-target effects 178
stability 182–183
tissue distribution 182, 183
molecular mechanisms 173, 174*F*
dsRNA dependent protein kinase 173
message amplification 174
2′-5′ oligoadenylate synthetase 173
posttranscriptional gene silencing 173
promoters 17
reagent synthesis 18
retroviral delivery 17
RISC (RNA-induced silencing complex) 173, 174*F*
activated 173
composition 174
discovery 172
short hairpin RNA (shRNA) 17
siRNA 17
target search genomics 17
TRAIL inhibition 17–18
transgenic animal generation 152
RNAi consortium (TRC), large-scale virus vector libraries 19
RNA polymerase II promoters, siRNAs 177
RNA polymerase III promoters, siRNAs 177
RNAse-deficient *Escherichia coli*, cell-free expression systems 122
RNA structure prediction, bioinformatics 323*F*, 327
base-paring patterns 327
comparative sequence analysis 327
homology-based techniques 327
secondary structure motifs 327
thermodynamics 327
Roller bottles, protein fermentation 418
RONN, domain mapping 879
Röntgen, Wilhelm Conrad, protein crystallography 449–450
Room temperature
compound storage 567
crystal handling 467
Room temperature ionic liquids, microwave-assisted chemistry 854
Root mean square distance (RMSD), x-ray structures, problems 538
ROSETTA, protein structure prediction, bioinformatics 327
Rosiglitazone maleate, PASP 802, 802*F*
Rotation component, molecular replacement 462
Rous sarcoma virus promoter, mammalian protein expression systems 117
Rous sarcoma virus protease, HIV-1 protease *vs.* 508–509
Route validation, solution phase library synthesis *see* Solution phase synthesis, library preparation
RP (reference position), UniProtKB flatfile 360
RPMI cells, protein expression systems 117
RSCB PDB 547*T*
RT (reference title), UniProtKB flatfile 360
Rubidium-86, cell-based electrophysiology assays 625
Rubredoxin, structure, coordinate errors 532
'Rule-of-five'
computational filters 258
drug discovery, fragment-based 943–944
druggability 2
'Rule-of-three,' fragment-based drug discovery 943–944
RV2002, protein crystallization 888
R-value, protein 3D structure validation 519
RX (reference cross-reference), UniProtKB flatfile 360

S

Saccharomyces cerevisiae
genome sequencing 317*T*, 647–649
proteome complexity 29*T*
Saccharomyces cerevisiae model system 5, 647
cell cycle control 5–6
forward genetics 5–6
gene deletion combinations 6–7
gene function assessment 6
genome resources 6, 6*T*
hsp90 identification 5–6
human genome functional study model 8
Bax repressor identification 8
GPCR expression 8
ligand identification 8
induced haploinsufficiency (chemical genomics) 7
limitations 7
target identification 7
target validation 7
limitations 8
mitogen activated protein kinase (MAPK) signaling 5–6
random mutation screening 5–6
reverse genetics 6, 7*F*
DNA repair mechanisms 6
mitochondrial function 6
polyglutamine disease 6
viral replication 6
'synthetic lethality' 6–7
'three-hybrid system' 8
cyclin-dependent kinase inhibitor identification 8
methotrexate target identification 8
transcriptional profiling studies 6
two-hybrid assay 6
Saccharomyces Genome Deletion Project 139, 140*F*
Safety
microwave-assisted chemistry scale-up 856–857
PASP 793
pharmacogenetics 64
Salmeterol, PASP 802–804, 803*F*
SAM algorithm, microarrays 98–99
Sample analysis, preparation, microarrays 97
Sample classification, microarrays 93
Sandwich enzyme immunoassay, cell-based assays 631
Sandwich enzyme-linked immunosorbent assay 579–580, 579*F*
Saphenamycin analogs, solid-phase discrete libraries 709*T*
Saturation transfer difference (STD) experiments, NMR, drug discovery 908*T*, 912, 912*F*, 913*F*, 914*F*
SB-203580, chemical proteomics 44
Scaffold-based design
protein crystallography, rational drug design 892
solution phase library synthesis 779, 779*F*, 780*F*
Scaffold decoration, PASP 811
SCALE record, protein 3D structure validation 511
Scale-up, microwave-assisted chemistry *see* Microwave-assisted chemistry
Scaling, protein crystallography 468–469, 470
Scattering factor, x-ray–crystal interaction 452–453, 455*F*
Scatterplots, IsoStar (intermolecular interactions) 405
Scavenging techniques, polymer-assisted solution phase synthesis (PASP) *see* Polymer-assisted solution phase synthesis (PASP)
Schizophrenia, linkage studies, proteomics 43
SCI (Science Citation Index) 271*T*, 274
searching 274
SciFinder Scholar, database interfaces 283
Scintillation proximity assay (SPA) 584
bead type 584
fluorescence polarization (FP) *vs.* 590
GPCRs 585–586
adenosine 2a receptor 586
histamine receptors 586

Scintillation proximity assay (SPA) (*continued*)
- homogenous time-resolved fluorescence energy transfer (HTR-FRET) *vs.* 590
- human cytochrome P-450 s 589
- iodine-125 584
- kinases
  - HTS 587
  - protein kinase A 587
- long-chain fatty acyl coenzyme A synthase 588
- radioisotope characteristics 584
- signal measurement 584
  - charge-coupled device (CCD) 584
  - photomultiplier tube 584
- target-oriented screening 142–143
- technique 584, 585*F*
- translation factors 588
  - initiation factors (IFs) 588
- tritium 584

SCOP (Structural Classification of Proteins) 374
- protein function prediction, bioinformatics 336–337
- search capabilities 376–377

Scopus 271*T*, 274

Scoring functions, structure-based virtual screening 251

Screening
- cell-based *see* Cell-based assays
- fluorescence screening assays *see* Fluorescence screening assays
- pre-equilibrated dynamic combinatorial libraries 963–964, 964*F*
- protein crystallization 439
- small animal test systems *see* Small animal test systems

Screening test design, chemoinformatics, diversity analysis 252–253

Screening window coefficient, HTS 684

SDBS 280*T*

SDS-PAGE
- protein analysis 424, 425*F*
- protein purity determination 881
- proteomics *see* Proteomics

Search Ligands, PDB searches 377

Search Sequence, PDB searches 377

Search tree generation, CSD 396

Search Unreleased, PDB searches 377

Secondary databases *see* Databases

Secondary literature databases 270

Secondary structure, protein structure prediction *see* Protein structure prediction

Secondary structure motifs, RNA structure prediction, bioinformatics 327

β-Secretase, knockout animal models 160

Seed, Brian, cDNA functional screening 14

Selection conditions, dynamic combinatorial libraries 960

Selection markers
- gene expression vectors *see* Gene expression vectors
- mammalian protein expression systems 117
- yeast protein expression systems 117

Selection target presence, dynamic combinatorial libraries 963

Selective heating, microwave-assisted chemistry 842

Semliki Forest Virus expression systems 119
- cytotoxicity 119
- expression vectors 119
- membrane proteins 115*F*, 119
- plasmid construction 119
- safety 119
- two-vector systems 119

Sensitivity, microarrays 88

Separation
- PASP 793
- pre-equilibrated dynamic combinatorial libraries 964

SeqHound, blueprint databases 305–306

SEQRES record, protein 3D structure validation 513

Sequence
- DDBJ/EMBL/GenBank 352, 353*F*
- siRNAs 176

Sequence alignment
- bioinformatics *see* Molecular sequence analysis, bioinformatics
- protein 3D structure, protein production 413

Sequence databases 281

Sequence identity, DDBJ/EMBL/GenBank 353–354

Sequence motifs, protein function prediction, bioinformatics 334–335

Sequence–register shift errors
- large ribosomal subunit (*Deinococcus radiodurans*) structure 535
- X-ray structure *see* X-ray crystallography

Sequential processing, microwave-assisted chemistry 857

Serine proteases
- as drug targets 3

Serine–threonine protein kinases
- as drug targets 3
- Flash Plate assays 587

Server databases 294

Sexual dimorphism, genetically engineered animals 154

SFCHECK, protein 3D structure validation 519

SGD, Universal Protein Resource 362–363

SH2 domain, STAT (signal transducer and activator of transcription) 203

Short hairpin RNA (shRNA), RNA interference 17

Short interfering RNAs (siRNAs) 3–4, 17, **171–187**
- *see also* RNAi
- cDNA precursors 172
- delivery 177
  - aerosol 182
  - electroporation 177
  - liposomal transfection 177
  - protein turnover rate 177
  - stability problems 181–182
- design criteria 176
  - base composition 176
  - base length 176
  - genetic redundancy 177
  - internal thermodynamic stability 176
  - on-line design tools 176
  - polymorphisms 176
  - sequence 176
- generation 172
- history 172
- microRNA precursors 175, 175*F*
  - mismatches 175
  - transcription 175
  - translational repression 175
  - in viruses 175
- modification 182–183
  - 2′-*O*-2,4-dinitrophenyl derivatives 182–183
  - locked nucleic acids 183
- shRNA precursors (vectors) 172, 177
  - adenoviral vectors 177
  - in vivo gene silencing 182
  - lentivirus vectors 177
  - retroviral vectors 177
  - RNA polymerase III promoters 177
  - RNA polymerase II promoters 177
- stability problems 182
- synthetic 176
  - history 172
- systemic administration 181
- target site choice 176
  - specificity 176
- technical considerations 176
- tissue distribution 183
  - cell-surface receptors 183
  - cell targeting 183
  - cholesterol tagging 183
- topical administration 181

'Shotgun proteomics,' LC-MS (peptide-based) proteomics 37

ShRNA precursors (vectors) *see* Short interfering RNAs (siRNAs)

Sickle cell anemia, as monogenetic disease 338

Side chain angles, NMR structure determination 486

Side chain planarity, protein 3D structure validation 516
Side-chain rotamer libraries, PDB errors 523
Signaling pathways **189–214**
see also specific pathways
activation 190
cell-based assays 618
definition 190
Signaling proteins, cell-based assays 643–644
Signal/leader sequences, *Escherichia coli* protein expression systems 114
Signal measurement, radioactive assays 578
Signal sensors, gene detection, bioinformatics 321–322
Signal-to-background ratio, HTS 684
Signal-to-noise ratio
HCS confocal systems 687
HTS 684
Signal transducer and activator of transcription *see* STAT (signal transducer and activator of transcription)
Signal transduction
biomarkers 73
cDNA functional screening 15–16
genetically engineered animals 160
large-scale virus vector libraries 18–19
regulatory networks 330
ubiquitin-proteasome pathway 190
Signature method 363
Sildenafil citrate
discovery/development 930–931, 931*F*
PASP 801, 801*F*
Silencer regions, regulatory regions, bioinformatics 322
Silica
microwave-assisted chemistry 853
PASP 793
Silver stain, two-dimensional gel electrophoresis 35, 35*T*
O-Silylated steroid derivatization, PASP 828–829, 829*F*
Simian monkey fibroblasts(COS), cell-based assays 620–621
Simian virus 40 (SV40)
microRNA coding 175
promoter, mammalian protein expression systems 117
transgenic mouse model promoters 11
Similarity searching 245
coefficients 245
evaluation 248
fragment bitstrings 246
numerical descriptors 245
rationale 245
similarity coefficients 245
Tanimoto coefficient 246
Simplification, single-parametric biochemical assays 683
Simulated annealing, protein crystallography model generation 463
Single-domain prokaryotic proteins, structural genomics 554
Single-isomorphous replacement with anomalous scattering (SIRAS), protein crystallography, rational drug design 889–890
Single-mode instruments, microwave-assisted chemistry 844
Single molecule detection, fluorescence screening assays 600
Single nucleotide polymorphisms (SNPs)
genetic variation 338
microarrays 90
proteome 29
QTL analysis 14
Single-parametric biochemical assays, HTS *see* High-throughput screening (HTS)
Single-parametric cellular assays, HTS *see* High-throughput screening (HTS)
Single photon emission computed tomography (SPECT)
applications 659
functional small animal imaging 660*T*, 661, 663–664
radionuclides 662*T*, 663*T*
see also specific nuclides
small animal imaging 660*T*, 661, 663–664
^{99m}Tc 661–662, 663*T*
Single wavelength anomalous dispersion (SAD), anomalous dispersion 460
Site-directed screening, fragment-based drug discovery 947
Site isolation effects, polymer-assisted solution phase synthesis (PASP) *see* Polymer-assisted solution phase synthesis (PASP)
Sitting drop method, protein crystallization 439
Size exclusion chromatography
see also specific methods
buffer exchange 424
protein analysis 423*F*, 426
protein purification 423, 423*F*
SLAPSTIC, NMR, drug discovery 910
Small animal imaging **659–677**
amino acid metabolism 672
anatomical 661
CT 661
MRI 661
ultrasound 661
cell surface receptors 667
radiolabeled annexin 667
cellular metabolism 668
androgen ablation 669
cytotoxic agents 668
infection 670
photodynamic treatment 669
radiotreatment 670
viability 668
cellular proliferation 671
contrast agents 660
functional 661
CT 661
MRI 661
PET 661
SPECT 661
internal cell processes 665
amino acid probes 665
metabolic probes 665, 666*F*
proliferation probes 665
radiolabeled biomolecular probes 666
reporter genes 665
methods 660*T*
see also specific methods
microscopy 660*T*
molecular 661
angiogenesis probes 665
bioluminescence 661, 668
cell receptor probes 664
CT 662
fluorescence 661
NMR 662
PET 661, 663–664
see also Positron emission tomography (PET)
SPECT 661, 663–664
ultrasound 662
MRS 660*T*
pharmacokinetics 660, 672
direct measurements 672
indirect measurements 672
receptor occupancy 667
central nervous system 664*T*, 667
steroid hormones 667
SPECT 660*T*
transgenic animals 660
ultrasound 660*T*
Small animal test systems **647–657**
see also specific species
history 647
knockout models 649
overexpression models 649
protein conservation 647–649
transgenic models 649
Small libraries
fragment-based drug discovery 942, 943*F*
solution phase parallel chemistry 765
Small-molecule microarrays
surface plasmon resonance 594
target-oriented screening 142, 143*F*
Small organic molecules, solid-phase libraries 698

Small robotic system, HTS automation 682
SMART
InterPro, member of 363
protein function prediction, bioinformatics 334
signature method 363
SMID (Small Interaction Database), blueprint databases 305–306
SMILES (simplified molecular input line entry specification)
chemical information databases 298
chemoinformatics, molecule representation 236
THOR 295
two-dimensional representations 298
Society of Biomolecular Screening
HTS history 680
microplate standardization 682–683
Sodium concentrations, cell-based assays 627
Software
CSD 393
pharmacophore identification 242
Solid–liquid phases, microwave-assisted chemistry 853
Solid-phase libraries 702
chemical assessment 703, 704*T*–705, 738, 739*T*
alkyl-tethered diisopropylarylsilane linker 738*F*, 740
anandamide 703
benzodiazepine analogs 703, 706*F*
bicyclic amines 706–708, 708*F*
clavulone analogs 703
cyclopent-2-enones 708, 708*F*
heterocyclic scaffolds 703–706, 707*F*
indenopyrrolocarbazoles 703, 705*F*
natural products 740, 740*F*
dibenz[*b,g*]1,5-oxazocines 706, 707*F*
peptide nucleic acids (PNAs) 740–741
6,6-spiroketals 740*F*, 741
discretes 708, 709*T*, 746, 746*F*, 747*T*
arylsulfonamide hydroxamates 725–726, 726*F*
ceramides 746*F*, 751–752
chiral 1,3-oxazolidines 722–724, 723*F*
estradiols 708, 718*F*
furan-4-ones 720–721, 720*F*
heterocyclic libraries 719–720
lidocaine analogs 725, 725*F*
NADH:ubiquinone oxidoreductase inhibitors 751*F*, 752–753
natural products 718
nucleotide-like libraries 751–752
peptidomimetic scaffolds 719–720, 720*F*
philantotoxins 718–719, 719*F*
phthalides 720*F*, 721
procainamide analogs 725, 725*F*
quinazolinones 722, 722*F*
recurrent motifs 721–722, 721*F*
spiroimidazolidinones 722, 723*F*
tellimagrandin analogs 751*F*, 752
1,3,5-trisubstituted Triazines 724–725, 724*F*
Wnt antagonists 751–752
examples 738
nucleotides 698
peptides 698
pool libraries 726, 741
benzimidazolones 735, 736*F*
benzodiazepines 736–737, 736*F*
CCR5 antagonists 727–735, 734*F*
2,2-dimethyl benzopyran scaffolds 745*F*, 746
galanthamine analogs 745, 745*F*
histone deacetylase inhibitors 741
ketoamides 741, 741*F*
lead discovery 741, 742*T*–744
macrolides 726, 727*F*
natural products 726, 728*T*
phenylalkylamides 737–738, 737*F*
pyridines 735, 735*F*
tetrapeptide libraries 746
tyrphostins 727, 734*F*
ureido acids 737, 737*F*
size 698
small organic molecules 698
split-and-mix libraries 698
Solid phase parallel synthesis
applicability 764
intermediates 765
purification concerns 765
solution phase parallel chemistry *vs.* 763, 764*T*
Solid-phase peptide synthesis
CEM Discover platform microwave-assisted chemistry 846–847, 847*F*
microwave-assisted chemistry 843*F*
Solid samples, compound storage *see* Compound storage/management (Pfizer)
Solubility, prediction, chemoinformatics, ADME property prediction 259
Solubility tags, *Escherichia coli* protein expression systems *see Escherichia coli* protein expression systems
Soluble proteins, protein crystallization *see* Protein crystallization
Solution phase parallel chemistry **761–790**
see also specific drugs
applicability 765
advantages 767
chemistry limitations 767
diversity-oriented libraries 767
intermediate purification 767
microwave-assisted chemistry 767
small/focused libraries 765
target-focused libraries 767
validation times 767
future trends 786
solid phase *vs.* 763, 764*T*
Solution phase synthesis, library preparation 768
analytical characterization 778
dynamic combinatorial chemistry 785
CDK2 inhibitor synthesis 785, 785*F*
examples 778
fluorous chemistry 785, 785*F*
hydantoin synthesis 786, 786*F*
pyrimidine synthesis 786, 786*F*
heterocyclic libraries 779
benzazepine-3-one synthesis 782–783, 784*F*
benzimidazole synthesis 780, 782*F*
p561ck inhibitors 779–780, 781*F*
N-substituted piperidinone synthesis 780, 783*F*
pyrimidine synthesis 780, 782*F*
pyrrole synthesis 780, 783*F*
1,2,4-triazole synthesis 780, 783*F*
intermediate synthesis/ purification 771
novel intermediates 771, 775*F*
library design 768
compound numbers 769–770
computational chemistry 770
focused libraries 769
large, diverse libraries 769
library size 769
SAR 769
targeted libraries 769
multicomponent reactions 784
quinoline synthesis 784, 784*F*
parallel work-up 776
intermediate purification 776
liquid–liquid extraction 776
preproduction 771
purification 776
resin capture/release 777, 777*F*
resin scavenging 776, 777*F*
reaction monitoring 775
route validation 770
analytical techniques 771
final compound specification 771
library purification 771, 772*T*–774
reaction conditions 770
step assessment 770
scaffold derivatization 779, 779*F*, 780*F*
Solvent-free reactions, microwave-assisted chemistry *see* Microwave-assisted chemistry
Solvent reactions, microwave-assisted chemistry *see* Microwave-assisted chemistry
Southeast Collaboratory for Structural Genomics (SECSG) 552*T*
metalloproteomics 556–557

Space groups, protein crystallography *see* X-ray crystallography
Specificity, enzyme-linked immunosorbent assays (ELISAs) 73–74
SpecInfo 279, 280*T*
SPECT *see* Single photon emission computed tomography (SPECT)
Spectral maps, microarrays 92, 101, 102*F*
Spectroscopic databases 279, 280*T*
Speed
  microwave-assisted chemistry 838
  NMR, drug discovery 907
Speed phenotyping, genetically engineered animals 153
Spent reagent removal, PASP 792, 793
Sphere exclusion algorithms 253, 254*F*
Spinner flasks, protein fermentation 418
Spinocerebellar ataxia, *Drosophila melanogaster* model system 9
Spirohydantoins, solid-phase discrete libraries 709*T*
Spiroimidazolidinones
  solid-phase discrete libraries 709*T*
  solid-phase libraries 722, 723*F*
6,6-Spiroketals, solid-phase libraries 740*F*, 741
Spirooxindoles, solid-phase pool libraries 707*F*
Split-and-mix libraries, solid-phase libraries 698
Split diagrams, phylogeny construction, bioinformatics 321, 321*F*
SPRESI 281–282, 282*T*
SQL (structured query language), relational databases 268
SQ (sequence header) lines, UniProtKB flatfile 362, 362*F*
Src kinases, vascular endothelial growth factor receptor-2 (VEGF-R2) 200
SSAP, protein function prediction, bioinformatics 336–337
Stability, RNAi (RNA interference) 182–183
Stable isotope labeling by amino acids in cell culture (SILAC), LC-MS (peptide-based) proteomics 38
Stable transfection, mammalian cell expression systems 418
Staden Package, molecular sequence analysis, bioinformatics 318
STAN 547*T*
Standard conformation libraries, protein 3D structure errors 509
Standard error of mean, biological activity databases 302
Stanford Synchrotron Radiation Laboratory, metalloproteomics 556–557
STAT (signal transducer and activator of transcription) 203
  see also specific types
  activation pathway 205*F*
    homodimer association 204
  alternative RNA splicing 204
  chromosomal localization 203, 204*T*
  in disease 207
    allergic inflammation 208
    autoimmune disease 208
    tumorigenesis 207
  gene isolation, cDNA functional screening 14
  signal transduction/activation 203
    cytokines 205–206, 206*T*
    growth factor receptors 204
    Janus tyrosine kinases 204
  structure 203, 204*F*
    C-terminal transcriptional activation domain 203
    DNA-binding domain 203
    N-terminal domain (oligomerization domain) 203
    SH2 domain 203
  truncated isoforms 204
STAT1 203
  alternative RNA splicing 204
  asthma 208
  chromosomal localization 204*T*
  chromosomal location 203
  heterodimer formation 204
  rheumatoid arthritis 208
  signaling 205
    cytokines 205–206, 206*T*
  truncated isoforms 204
STAT2 203
  chromosomal location 203, 204*T*
  heterodimer formation 204
  signaling 206
    cytokines 206*T*
    interferons 206
STAT3 203
  alternative RNA splicing 204
  chromosomal location 203, 204*T*
  heterodimer formation 204
  knockout animals, embryonic lethals 206
  oncogenesis 207
  rheumatoid arthritis 208
  signaling 206
    cytokines 206, 206*T*
  truncated isoforms 204
STAT4 203
  chromosomal localization 204*T*
  chromosomal location 203
  signaling 207
    cytokines 206*T*
    IL-12 206
STAT5 203
  signaling 207
    cytokines 206
  truncated isoforms 204
STAT5a 203
  chromosomal location 203, 204*T*
  heterodimer formation 204
  signaling, cytokines 206*T*
STAT5b 203
  chromosomal location 203, 204*T*
  heterodimer formation 204
  natural killer cell development 207
  signaling, cytokines 206*T*
STAT6 203
  asthma 208
  chromosomal location 203, 204*T*
  signaling 207
    cytokines 206*T*, 207
Statistical genetics, genetic variation analysis 338–339
Statistics
  accuracy, fluorescence screening assays 612–613
  fluorescence screening assays *see* Fluorescence screening assays
  Genome Reviews 357
  microarrays 90
Staudinger reaction, PASP 793
Stem cells, cell-based assays 641–642, 643
Step assessment, solution phase library synthesis 770
Stereochemistry
  checks, protein 3D structure 547
  protein 3D structure 546
Steroid hormone imaging
  [^{18}F]FDG 667
  16 alpha-[^{18}F]fluoroestradiol 667
  small animal imaging 667
Steroid hormone receptors, molecular probes 664
Steroids, biological activities 926*T*
Stille couplings, PASP 813
Storage, compound storage/management 567
Strain specificity, gene expression vectors 109
Strategy, protein crystallography 468–469
Streisinger, George, *Danio rerio* as model system 10
StrepII tag, affinity chromatography 436–437
Strep-tag, as affinity tag 421
Streptavidin
  gene expression vectors 110, 110*T*
  structure, coordinate errors 532
Stress responses, RNAi (RNA interference) 181
Stress-responsive kinases, ubiquitin protein targeting 192
Stress signaling, large-scale virus vector libraries 18–19
STRIDE, protein structure prediction, bioinformatics 325

STRING database, protein interaction networks, bioinformatics 335–336
Strong hydrogen bonds
  CSD 402, 404*F*
  intermolecular interactions, CSD 403*F*
Structural abnormalities, genetically engineered animals 155
Structural criteria, comparative computational modeling 557
Structural errors, x-ray structures, problems 538
Structural genomics **551–560**
  see also Protein Structural Initiative (PSI)
  auto-publishing 558
  comparative computational modeling 557
    accuracy 557
    energetic criteria 557
    PDB related structures 557
    structural criteria 557
  future work 558
    technology improvement 558
  metalloproteomics 556
    HTS 556–557
    x-ray absorption spectroscopy (XAS) 556
  protein cloning 554
    PCR 554
  protein expression 554
    automation 554–555
    baculovirus 554
    Escherichia coli 553*T*, 554
    Pichia pastoris 554
    SUMO-based expression 555
  protein purification 555
    affinity tags 555
    automation 555
    characterization studies 555
  structure determination 555
    NMR 555
    x-ray crystallography *see* Structural genomics, x-ray crystallography
  structure–function relationship 557
    DALI 557
    ligand presence 557–558
    PINT 557
    VAST 557
  target selection 553
    biomedical interest 554
    Pfam5000 strategy 554
    protein families 553–554
    protein orthologs 554
    single-domain prokaryotic proteins 554
    structure determination methods 554
  x-ray crystallography 555
    automation 556
    crystallization improvements 555
    limited proteolysis combined with mass spectrometry (LP/MS) 555
    methodology determination 556
    'phase problem' solution 556
    synchrotron radiation 556
Structural Genomics for Pathogenic protozoa Consortium (SGPP) 552*T*
Structural information, Integr8 368
Structural similarity, sequence similarity *vs.* 462
Structure–activity relationships(SARs)
  chemical proteomics 44
  drug discovery, fragment-based 941
  solution phase library synthesis 769
Structure-based virtual screening 250–251
  aim 251
  docking program evaluation 251–252
  DOCK program 251
  FlexPharm program 252
  force-field scoring 251
  knowledge-based scoring 251
  ligand pose prediction 252
  protein flexibility 251
  regression-based scoring 251
  scoring functions 251
  ZINC database 252
Structure correlation principles, CSD 399
Structure databases *see* Databases
Structure determination
  NMR 948
  structural genomics 554
  x-ray crystallography 948
Structured query language (SQL), databases *see* Databases
Structure–function relationship, structural genomics *see* Structural genomics
Structure generation programs 240
Structure prediction, bioinformatics 324
Structure/substructure searching *see* Chemoinformatics, molecule representation
Structure to Function (S2F) 551, 552*T*
  Target DB 551
Sub-ambient chemistry, CEM Discover platform microwave-assisted chemistry 847–848
Subcellular fractions, proteomics 30, 31
Subcellular stains, HCS labeling techniques 689
Subgraph isomorphism algorithms
  chemoinformatics, three-dimensional structure 242
  structure/substructure searching 239
Substituted pyrazoles, PASP 807, 808*F*
Substrate activity screening (SAS), fragment-based drug discovery 953–954
Substructure searching, chemoinformatics, three-dimensional structure 241–242
Subtraction artifacts, NMR, drug discovery 909
Sulfahydantoins, solid-phase discrete libraries 709*T*
Sulfated polysaccharide–protein interactions, ELISA 580
Sulfonamides, library construction, PASP 798, 799*F*
N-Sulfonylpropyl amides, solid-phase discrete libraries 709*T*
Sulindac, analogs, solid-phase discrete libraries 707*F*
SUMO-based expression 555
Sunset Molecular Discovery Databases, commercially available databases 309
Supercritical carbon dioxide (scCO_2), PASP 813
SUPERFAMILY
  InterPro, member of 363
  signature method 363
Superheating effects, microwave-assisted chemistry 842
SuperStar program, intermolecular interactions 406, 407*F*
Supplementary linking tables, database integration 303, 304*F*
Supplements, cell culture 622
Supports, polymer-assisted solution phase synthesis (PASP) *see* Polymer-assisted solution phase synthesis (PASP)
Support vector machines (SVMs), machine learning methods 250
Surface analysis, protein function prediction, bioinformatics 337
Surface-enhanced laser desorption ionization (SELDI), proteomics 40
  limitations 40
Surface plasmon resonance (SPR) 593
  applications
    ADME 594
    biomarker identification 73
    bound ligand identification 593
    GPCRs 593
    high-content screening 684
    HIV protease inhibitor identification 593–594
    HTS 594
    ion channels 593
    lead characterization 593–594
    small molecule microarrays 594
    tyrosine kinase receptor ligands 593
  disadvantages 594
  evanescent wave 593
  fragment-based drug discovery *see* Fragment-based drug discovery

Surface residue mutation, protein crystallization 437
Surrogate endpoints
see also Biomarkers, clinical trials
clinical trials 81–82
trastuzumab 81–82
genetically engineered animals 154
Surrogate markers, prostate-specific antigen 80
'Survey scan,' LC-MS (peptide-based) proteomics 38
Suzuki couplings, PASP 813
Suzuki–Miyaura coupling procedure, PASP 812
SWISS-MODEL, homology models 544
Swiss-Prot, history 350
Symmetry elements, crystals 451, 453*F*
Synchrotron radiation
anomalous dispersion 460
x-ray crystallography 556
Synonyms, target databases 299–300
'Synthetic lethality,' *Saccharomyces cerevisiae* model system 6–7
Synthos 3000, Anton Paar GmbH microwave-assisted chemistry 851, 852*F*
α-Synuclein, NMR structure determination 490, 491*F*
Sypro ruby, 2D-PAGE protein quantitation 35*T*
System adaptability, dynamic combinatorial libraries 963
Systemic administration, siRNAs 181
Systemic gene delivery, transgenic animal generation 152
Systems biology, proteomics *see* Proteomics

T

T7-promoter, gene expression vectors 108
Tacrolimus (rapamycin)
discovery/development 140–141, 141*F*
structure 700*F*
Tadalafil, discovery/development 930–931, 931*F*
Tagged libraries, phenotype-oriented screening 139
Talopram, chemogenomics 934*F*, 935
TALOS routine, NMR structure determination 489
Tandem mass spectrometry (MSD/MS), proteomics 33
Tanimoto coefficient, similarity searching 246
Target(s)
assessment
genetically engineered animals 159
pharmacogenetics 52, 55
binding, chemoinformatics 241, 241*F*
definition 2, 87
G protein-coupled receptors (GPCRs) *see* G protein-coupled receptors (GPCRs)
microarrays 100
phosphodiesterases 3
pre-equilibrated dynamic combinatorial libraries 963–964
proteomics *see* Proteomics
purity, high throughput purification 863
sequence alignment, homology-based protein structure prediction 325
serine/threonine kinases 3
tyrosine kinases 3
validation 3
biological connectivity 4, 4*F*
biomarkers *see* Biomarkers
function-based 3–4
genetically engineered animals 159
genomics 2
knockout mouse model 13
mammalian expression vectors 3–4
node network theory 4, 4*F*
orthogonal ligand–receptor pairs *see* Orthogonal ligand–receptor pairs
proteomics 42
RNAi (RNA interference) 172, 181
Saccharomyces cerevisiae model system 7
signaling pathways *see* Signaling pathways
x-ray generation 465
zinc metallopeptidases 3
Target DB
PDB 551
Structure to Function (S2F) 551
Targeted libraries
solution phase library synthesis 769
solution phase parallel chemistry 767
Target family-directed master keys, chemogenomics 927, 929*F*
Target identification
biomarkers 69, 71
chemical biology *see* Chemical biology
Danio rerio model system 10
definition 316
HTS *see* High-throughput screening (HTS)
microarrays 87, 88
natural products 133
phenotype-oriented screening *see* Phenotype-oriented screening
protein expression *see* Protein expression
requirements 316
structural genomics *see* Structural genomics
Target identification, genomics 5
genetic variation analysis 339
pharmacogenetics/pharmacogenomics 52, 55
RNAi (RNA interference) 172
Targetless mutant methods, phenotype-oriented screening 140–141, 141*F*
Target-oriented screening 141
direct ligand binding 141
affinity selection 141
small-molecule microarrays 142, 143*F*
enzyme assays 142
chromogenic substrates 142
fluorescence polarization 142–143
fluorogenic substrates 142
scintillation proximity 142–143
goals 141
protein–protein interactions 142
fluorescence polarization 142
Genomics **1–25**
Target value errors, x-ray structures, problems 538
TATA-binding protein, crystallography 884
TATA box, mammalian protein expression systems 117
TB Structural Genomics consortium (TB) 552*T*
^{94m}Tc, PET 664*T*
^{99m}Tc
annexin, lymphoma models 668
biomolecular probes 666
SPECT 661–662, 663*T*
TCDB Database, biological information databases 305
T-COFFEE, sequence alignment, bioinformatics 320
TDT (THOR data tree), two-dimensional representations 298
Teleocidin analogs, solid-phase discrete libraries 707*F*
Tellimagrandin analogs, solid-phase libraries 751*F*, 752
Temperature, compound storage/management 567
Temperature measurement
Anton Paar GmbH microwave-assisted chemistry 852
CEM Discover platform microwave-assisted chemistry 847
microwave-assisted chemistry 843, 857
Milestone s.r.l. microwave-assisted chemistry 849
Template availability, homology-based protein structure prediction 325–326
Tepoxaline, structure 808
Tertiary literature databases 273
Test protocols, cell-based assays 623

Genomics (*continued*)
Tethering
'scaffold-based' drug design 892
screening, dynamic combinatorial libraries (DCLs) 965
Tetrahydrofuranones, solid-phase pool libraries 705*F*
Tetrahydrofurans
microwave heating 840*T*
solid-phase pool libraries 707*F*
Tetrahydrooxazepines, solid-phase discrete libraries 709*T*
Tetrahydroquinolines, solid-phase pool libraries 705*F*
Tetrakistriphenylphosphine palladium reagent, PASP 795
Tetramers, x-ray structures, problems 540
Tetrapeptide libraries, solid-phase libraries 746
Text databases 277
Text formats, two-dimensional representations 298
Text mining, protein function prediction *see* Protein function prediction
Therapeutically Relevant Multiple Pathways Database, BIDD Databases 305
Therapeutic Target Database, BIDD Databases 305
Thermodynamics
protein structure prediction, bioinformatics 326
RNA structure prediction, bioinformatics 327
Thiadiazole ethers, solid-phase pool libraries 705*F*
Thiazole peptide analogs, solid-phase discrete libraries 709*T*
Thin-layer chromatography (TLC), biomarkers 76
3-Thioalkyl-1,2,4-triazoles, PASP 804, 804*F*
Thioester library, PASP 830, 830*F*
Thiomorphine analogs, PASP 811, 813*F*, 814*F*
Thiophene aspartyl ketones, solid-phase discrete libraries 709*T*
Thioredoxin
Escherichia coli protein expression systems 113–114
gene expression vectors 110, 110*T*
Thiourea, PASP 798, 799*F*
Third Party Annotation (TPA), DDBJ/EMBL/GenBank *see* DDBJ/EMBL/GenBank
THOR
database management systems 295
Daylight CIS 295
SMILES 295
3D geometrical parameters, CSD 393
3D graphs, chemoinformatics 241
3D representations, chemical information databases 298
3D similarity searching *see* Chemoinformatics, virtual screening
3D-structure, databases 283
3D substructures, CSD 393
Three-dimensional environment, PASP 794–795
'Three-hybrid system'
see also Saccharomyces cerevisiae model system
phenotype-oriented screening 140
Thrombin inhibitors 929
[^{3}H]Thymidine
proliferation probes 665
radiotreatment imaging 670
Thyroid hormone receptor (TR), orthogonal ligand–receptor pairs 227*F*, 228
Thyroid hormone receptor inhibitors, chemogenomics 933, 934*F*
TIGRFAMS
InterPro, member of 363
signature method 363
TILLING, *Danio rerio* model system 10
Time domain acquisition, fluorescence lifetime assays 605–606
Time-of-flight (ToF) detectors, mass spectrometry proteomics 32
Time-resolved anisotropy analysis 606
mechanism of action 606
Time saving, PASP 792
Time-specific gene deletions, knockout animal models 152
Timing, microarrays 103
Tissue distribution, RNAi (RNA interference) 182, 183
Tissue extraction, proteomics 30
Tissue microarrays, biomarkers *see* Biomarkers
Tissue-specific gene deletions, knockout animal models 152
Tissue-specific promoters, transgenic animal generation 152
Titin mutation, *Danio rerio* model system 10
Toll-like receptors, cell-based assays 638, 640*F*
cytokine production 639
inflammatory mediators 639
peripheral blood mononuclear cells 639
Toluene, microwave heating 840*T*
Topology, x-ray structures, problems 532, 534*F*
TORC1, identification, cDNA functional screening 15–16
Toremifene, [^{18}F]FDG imaging 669
Torsion angles
ligand structure 538
NMR structure determination 486, 488*F*
protein 3D structure validation 511, 516
Total fluorescence intensity (FLINT), fluorescence screening assays 602
TOXCENTER 271*T*
Toxicity, cell-based screening assays *see* Cell-based assays
Toxicity Databases 309
Toxicogenomics 56
animal experimentation 56
mouse models 654
TOXNET 275, 278*T*
Traditional Chinese medicine (TCM), database 308
Daylight CIS 308
TRAIL, inhibition, RNA interference 17–18
Transacylation, dynamic combinatorial library generation 962*F*
Transamination, dynamic combinatorial libraries 969–970, 971*F*
Transcription
profiling, *Danio rerio* model system 10
regulation
large-scale virus vector libraries 18–19
ubiquitin-proteasome pathway 190
siRNAs 175
termination, gene expression vectors 108
Transcriptional profiling studies, *Saccharomyces cerevisiae* model system 6
Transcription factors, cell-based assays 631
production 632
Transcription–translation systems, cell-free expression systems 121
Transcriptomics, definition 330–331
Transferrin, depletion 31–32
Transforming growth factor-β receptor-2, large-scale virus vector libraries 18–19
Transgenic animals 11
see also Knock-in animal models; Knockout animal models
cell-based assays 621–622
gene expression 11
promoter dependence 11
generation 11, 12*F*, 152
antisense oligonucleotides 152
'conditional transgenic animals' 152

ribozymes 152
RNAi 152
systemic gene delivery 152
tissue-specific promoters 152
zygote microinjection 152
human genes 152–153
bioassays 152–153
microarrays 97
mouse models 654
OPG overexpression 156, 157*F*
osteopetrosis 156, 157*F*
small animal test systems 649
imaging 660
Transient transfection, mammalian cell expression systems 418
Translational fusion constructs, gene expression vectors *see* Gene expression vectors
Translational repression, siRNAs 175
Translation component, molecular replacement 462
Translation factors, radioactive assays 588
Translation function, molecular replacement 461–462, 461*F*
Translation initiation, mammalian protein expression systems 118
Translation initiation signals, gene expression vectors 109
Translocation
genome rearrangements, bioinformatics 324
HCS 690
Transmission, dielectric heating 840
TRANSPATH database, regulatory networks 330
TransportDB, biological information databases 305
Transporter Classification (TC) System, biological information databases 305
Transporters
chemogenomics 934*F*, 935
radioactive assays 583–584
Trastuzumab
clinical trials 81–82
biomarkers 79
pharmacogenetics 55, 61
pharmacokinetic measurements 673–674, 673*F*, 674*F*
TrEMBL, history 350
1,3,5-Triaminotriazines, solid-phase discrete libraries 709*T*
1,3,5-trisubstituted Triazines, solid-phase libraries 724–725, 724*F*
Triazole analogs, PASP 804, 804*F*
1,2,4-Triazole synthesis, solution phase library synthesis 780, 783*F*
Triazole-tethered pyrrolidones, solid-phase discrete libraries 709*T*
Tricyclic hydroxyindolines, solid-phase pool libraries 705*F*
Trident automated platforms, PASP 806
Trihydroxy nucleoside synthesis, PASP 797
Trimers, x-ray structures, problems 540
Tripos, UNITY 295
2,3,5-Trisubstituted indoles, solid-phase discrete libraries 707*F*
1,3,5-Trisubstituted triazines, solid-phase libraries 724–725, 724*F*
Tritium, scintillation proximity assay 584
TrNOESY, NMR, drug discovery 908*T*
TROSY, NMR, drug discovery 915–916
Truncated isoforms, STAT (signal transducer and activator of transcription) 204
Trypsin-mediated ^{18}O incorporation, LC-MS (peptide-based) proteomics 38–39
t test, microarrays 92
Tumorigenesis
NF-κB pathway 192
STAT (signal transducer and activator of transcription) 207
Tumor mass, as surrogate marker 80
Tumor necrosis factor-α (TNF-α?
apoptosis induction, NF-κB pathway 196–197
NF-κB pathway activation 197
Tumors
see also Cancer
cell-based assays 619
NMR 77
positron emission tomography 77
Tumor suppresser genes
Drosophila melanogaster 650–651
removal 161, 163*F*
TUniRef, UniProtKB 362
Turbidimetric assays 595
HTS 595
technique 595
Turbulent flow media, high throughput purification 867
Twinning, protein crystallography 470
2D-PAGE 29–30, 34
advantages/disadvantages 36
protein class 37
protein properties 39
biomarkers, diagnostic implications 45–46
denaturing conditions 34
detergents 34
fluorescent labeled samples 30–31
isoelectric focusing (IEF) (first dimension) 34
LC-MS (peptide-based) proteomics *vs.* 39
pH gradient 34–35, 34*F*
narrow range 34–35
protein characterization, posttranslational modifications 41
protein extraction/prefractionation 30–31
protein identification 36
MALDI analysis 36
MALDI-TOFTOF analysis 36
protein quantification 29–30, 35
comparison methods 34*F*, 35–36, 36*F*
Coomassie stain 35, 35*T*
diffuse gel electrophoresis (DIGE) 35, 35*T*
fluorescent dyes 35, 35*T*
gel stains 35
pre-separation labeling 35
replicates 35–36
silver stain 35, 35*T*
Sypro ruby 35*T*
resolution 34–35
2D similarity searching *see* Chemoinformatics, virtual screening
2D structural representations, CSD 390–391
2D substructures, CSD 393
Two-dimensional fluorescence intensity distribution analysis (2D-FIDA) *see* Fluorescence fluctuation techniques
Two-dimensional NMR spectroscopy, fragment-based drug discovery 945
Two-dimensional representations
chemical information databases 298
SMILES 298
TDT (THOR data tree) 298
text formats 298
Two-hybrid assay, *Saccharomyces cerevisiae* model system 6
Two-vector systems, Semliki Forest Virus expression systems 119
TWS119, target identification 138–139
Tyrosine kinase receptor ligands, surface plasmon resonance 593
Tyrosine kinases
as drug targets 3
Flash Plate assays 587
Tyrphostin analogs, solid-phase pool libraries 705*F*
Tyrphostins, solid-phase libraries 727, 734*F*

U

Ubiquitin
gene expression vectors 110

Genomics (*continued*)
NMR structure
determination 476, 478*F*, 490–491, 491*F*
structure 190–191
Ubiquitin-activating enzyme (E1) 190
Ubiquitin-conjugating enzyme (E2) 190
Ubiquitin-ligase enzyme (E3) 190
Ubiquitin-proteasome pathway 190, 191*F*
definition 190
functions 190
antigen processing 190
cell cycle 190
protein turnover regulation 191–192
signal transduction 190
transcription regulation 190
inhibition 194
see also Bortezomib
Bax protein 194
p53 levels 194
JNK-activated pathway
c-Jun-associated protein degradation 192, 194*F*
p53-associated protein degradation 193, 194*F*
NF-κB activation *see* Nuclear factor kappa B (NF-κB)
protein targeting 191–192
dephosphorylation 192
stress-responsive kinases 192
ubiquitin-activating enzyme (E1) 190
ubiquitin-conjugating enzyme (E2) 190
ubiquitin-ligase enzyme (E3) 190
Ulcerative colitis (UC), NF-κB pathway 197
Ullmann algorithm, structure/substructure searching 239–240
Ultrafiltration
buffer exchange 424
protein concentration 424
Ultrasound, small animal imaging 660*T*, 661, 662
Ultraviolet (UV) radiation
chromatograms, high throughput purification 863
protein concentration determination 426
'Undruggable' targets, fragment-based drug discovery 944
Uniform Resource Locators (URLs) 299
UniParc
see also Universal Protein Resource
DDBJ/EMBL/GenBank checking 354
Integr8 367
UniProtKB *see* Universal Protein Resource
UniProtKB flatfile *see* Universal Protein Resource
UniProtKB/Swiss-Prot *see* Universal Protein Resource
Unit cells 451, 452*F*
axis restrictions 451–452
UNITY software
database management systems 295
Tripos 295
Universal Protein Resource (Uni-Prot) 358
data acquisition 354
DDBJ/EMBL/GenBank checking 354
history 358
specialized protein databases 362
MEROPS 362–363
PDB 362–363
SGD 362–363
UniParc 358
design 358–359
Ensembl 358–359
primary sequences 358–359
sources 358–359
structure 358–359
UniProtKB 358, 359
design 359
external cross-references 359
feature identifiers 359
posttranslational modifications 359
TUniRef 362
UniProtKB flatfile 360, 361*F*
accession number 360
CC (comments) 360, 361*F*
DE (description) lines 360
DR (database crossreference) 360, 361*F*
DT (date) lines 360
FT (feature) lines 362, 362*F*
GN (gene name) line 360
ID lines 360
KW (keyword) 361*F*, 362
OC (organism classification) lines 360
OG (organelle) line 360
OS (organism species) line 360
OX (organism taxonomy) line 360
RA (reference author) 360
RC (reference comment) 360
RG (reference group) 360
RL (reference location) 360
RN (reference number) 360
RP (reference position) 360
RT (reference title) 360
RX (reference cross-reference) 360
SQ (sequence header) lines 362, 362*F*
UniProtKB/Swiss-Prot 359
annotation 359–360
curation 359–360
gene name standardization 359–360
UniProtKB/TremBL 359, 360
UniRef 358
'Unlikelihood' plotting 403*F*, 405, 405*F*
Unsaturated nitrile synthesis, PASP 796–797
Uppsala Electron Density Server (EDS), protein 3D structure 547
Urea, solid-phase discrete libraries 709*T*
Urea library construction, PASP 798, 799 800, 799*F*, 800*F*
Ureido acids
solid-phase libraries 737, 737*F*
solid-phase pool libraries 705*F*
Ureido hydantoins, solid-phase pool libraries 705*F*
Urokinase, crystallization 883–884
USPTO 276*T*
Uterine bleeding, vascular endothelial growth factor 202

V

VADAR 547*T*
Valence angle values, CSD applications 400
Validation
cell-based assays 620–621
criteria testing, x-ray structures, problems 535
CSD 391
homology-based protein structure prediction 326
Validation Server, PDB 376
Vancomycin, analogs, solid-phase discrete libraries 707*F*
Variation of responses, cell-based assays 620–621, 621*F*
Vascular cell adhesion molecule-1 (VCAM-1), NF-κB pathway 197
Vascular endothelial growth factor (VEGF) 198
see also Placental growth factor (PlGF)
actions 200
angiogenesis 198*F*, 200
embryogenesis 200–202
NF-κB pathway 197
cancer 202
antisense treatment 203
oncogene activation 202
cell-signaling pathways 197
NF-κB pathway 197
female reproductive tract disorders 202
dysfunctional uterine bleeding 202

endometrial hyperplasia 202
endometriosis 202
polycystic syndrome 202
gene expression regulation 200
cytokines 200
growth factors 200
hypoxia 200
hypoxia-inducible factor-1 (HIF-1) 200
gene structure 198
inflammatory disorders 202
psoriasis 202
intraocular neovascular syndromes 202
age-related macular degeneration 202
diabetic retinopathy 202
retinopathy of prematurity 202
overexpression 196*F*, 200–202
structure 198
subtypes 198
therapeutic uses 203
amyotrophic lateral syndrome 203
cardiovascular disease 203
neonatal respiratory distress syndrome 203
underexpression 196*F*, 200–202
Vascular endothelial growth factor receptor-1 (VEGF-R1) 198
Vascular endothelial growth factor receptor-2 (VEGF-R2) 198, 200
signaling pathways 201*F*
PI3 K activation 200
Rac activation 200
Src 200
VEGFR-associated protein (VRAP) 200
Vascular endothelial growth factor receptor-associated protein (VRAP), vascular endothelial growth factor receptor-2 (VEGF-R2) 200
Vascular endothelial growth factor receptors (VEGFR) 198
receptor dimer formation 198–199
structure 198, 199*F*
VAST 557
Vector-specific markers, genetically engineered animal characterization 153–154
Velcade *see* Bortezomib
Verify3D 547*T*
Viability
[^{18}F]FDG 668
small animal imaging 668
Villafranca, Ernest, protein crystallization 887
Viloxazine, discovery/development 927–929, 929*F*
VIPER (Virus Particle Explorer), virus structure 544
Virtual cells *see* Molecular networks, bioinformatics
Virtual dynamic combinatorial libraries 961
Viruses
coat proteins, x-ray structures 540
microRNA precursors 175
replication
NF-κB pathway 192
Saccharomyces cerevisiae model system 6
structure
problems with 544
VIPER (Virus Particle Explorer) 544
transcription/translation mechanisms 118
Virus infections, RNAi (RNA interference) 173
Vista data analysis *see* Cambridge Structural Database (CSD)
Visual inspection, microarrays 97–98, 99*F*
Vitamin D_3 analogs, solid-phase pool libraries 705*F*
Vitamin D receptor (VDR), orthogonal ligand–receptor pairs 228–229, 228*F*
Voltage clamp, cell-based electrophysiology assays 624–625
von Laue, Max 449–450
v-Src, orthogonal ligand–receptor pairs 218, 220*F*

W

Wadsworth–Horner–Emmons alkenylation, PASP 796–797
Walker, John E 451
Wall effect elimination, microwave-assisted chemistry 842
Ward's clustering, cluster-based compound selection 254–255
Water, microwave heating, loss tangent 840*T*
WaterLOGSY, NMR, drug discovery 908*T*, 913, 914*F*
Water molecules
protein 3D structure validation 512
x-ray structures, problems 539
Water removal, protein crystallization 438
Watson, Herman, protein crystallization 884
Watson, James D, protein crystallography 450
Wave bioreactors, protein fermentation 419, 419*F*
Weak binding detection, NMR, drug discovery 901–902
Weak hits, fragment-based drug discovery 941
Weak hydrogen bonds, CSD 404
WebCite, CSD 398
Web-enabled databases 294
Western blotting
genetically engineered animal characterization 153–154
protein analysis 425
RNAi detection 178
WHAT_CHECK
protein 3D structure 546
validation 510, 516–517, 518
x-ray structures, problems 537
Wheat germ, cell-free expression systems 122
Whole cells, dynamic combinatorial libraries 973
Whole-genome sequence data, protein function prediction, bioinformatics 334
Wilkins, Maurice, protein crystallography 450
Wisconsin Package (GCG), molecular sequence analysis, bioinformatics 318
Wittig reaction, PASP 793, 793*F*
Wnt antagonists, solid-phase libraries 751–752
Wnt signaling
cDNA functional screening 15–16
Drosophila melanogaster model system 9
'Wolf-and-lamb' approach, PASP 796
WOMBAT-PK database 309
Workstations, HTS automation 681–682
WPI (World Patent Index) 276*T*

X

Xanthines, solid-phase discrete libraries 709*T*
Xenopus 649
GPCR studies 649
history 647
ion channel studies 649
oocytes, cell-based assays 619
XML, database integration 303
X-ray absorption spectroscopy (XAS), metalloproteomics 556
X-ray–crystal interaction *see* X-ray crystallography
X-ray crystallography **449–472**
see also Protein crystallization
atomic order 441
anisotropic diffraction patterns 441
birefringent crystals 441
crystal agglomerates 441
crystalline defects 441
focusing effects 440
protein quality effects 440–441

Genomics (*continued*)
cryocrystallography 466*F*, 468
crystal handling 465
cryocrystallography 466*F*, 468
equipment 467
harvesting buffer 465–467
room temperature 467
crystals 451
asymmetric unit 451
International Tables for Crystallography Volume A 452
symmetry elements 451, 453*F*
unit cell 451, 452*F*
unit cell axis restrictions 451–452
crystal shape 439*F*, 440
crystal size 440
accuracy 440
intensity effects 440
neutron diffraction studies 440
data collection 468
crystal characterization 468–469
diffraction pattern indexing 469
integration 468–469
scaling 468–469, 470
space group assignment 469
strategy 468–469
twinning 470
diffraction theory 451
dynamic combinatorial libraries (DCLs) *see* Dynamic combinatorial libraries (DCLs)
fragment-based drug discovery *see* Fragment-based drug discovery
'fragnomics' 892
future work 470
history 433–434, 449
Astbury, W T 450
Bernal, J D 450
Bragg, Sir William Henry 449–450
Bragg, William Lawrence 449–450
Crick, Francis 450
Deisenhofer, Johann 451
Franklin, Rosalind 450
Hodgkin, Dorothy Crowfoot 450
Huber, Robert 451
Kendrew, John C 450
MacKinnon, Roderick 451
Michel, Hartmut 451
Pauling, Linus 450
PDB 450
Perutz, Max F 450
Röntgen, Wilhelm Conrad 449–450
von Laue, Max 449–450
Walker, John E 451
Watson James D 450
Wilkins, Maurice 450
model generation 462
automation 462–463
boundary location 462, 463*F*
density modification 462, 463*F*
density modification procedures 462
electron map scaling 464, 464*F*
free R factor monitoring 464
histogram matching 462
map skeletonization 462
model bias 463–464
molecular dynamics 463–464
noise 462
pseudoenergy 463–464
refinement 463
resolution 462–463, 464*F*
simulated annealing 463
NMR *vs.* 540
packing arrangements 443
contacts *vs.* interfaces 442
noncrystallographic symmetry 443
packing densities 442
quality control 470
NMR structures *vs.* 542
phase quality 470
reflection intensity strength 470
space groups 441*F*, 442
diffraction symmetry 441–442
structural genomics *see* Structural genomics
structure determination 948
x-ray–crystal interaction 452, 454*F*
charge distribution 452–453
diffraction data resolution 456
diffraction pattern 452
electron oscillation 452, 455*F*
image formation 454
phase shifts 452
reciprocal lattice 456–457
reflection measurement 454
scattering factor 452–453, 455*F*
x-ray detection 468
charge coupled devices (CCDs) 468
film 468
image plates 466*F*, 468
integrating detectors 468
photon counters 468
x-ray generation 465, 466*F*
heat dissipation 465, 466*F*
monochromatic radiation 465
target 465
X-ray crystallography, phase problem solutions 457
anomalous dispersion 459
form factor (*f*) 455*F*, 459
multiple wavelength anomalous dispersion (MAD) 459–460
phase shift 459
single wavelength anomalous dispersion (SAD) 460
synchrotron radiation 460
isomorphous replacement 457
electron density map 457, 458*F*
electron density map autocorrelation 457
figure of merit (FoM) 459
Friedel's law 459
Harker construction 457–459, 458*F*
multiple isomorphous replacement (MIR) 459
Patterson map 457
phase ambiguity 459
molecular replacement 460
Laue symmetry 461–462
Matthews coefficient 460
molecular transform generation 460–461
molecule numbers 460
Patterson map 460–461, 461*F*
Patterson Search model 460
rotation component 462
structural *vs.* sequence similarity 462
translation component 462
translation function 461–462, 461*F*
X-ray crystallography, problems 532
atomic coordinate reliability 538
isotropic *B*-factors 538
protein dynamics 538
RMSDs 538
current structures 536
data overinterpretation 535, 536*F*
deliberate 535
R-factor testing 535
validation criteria testing 535
improvement reliability 537
incomplete quaternary structure 540, 541*F*
asymmetric unit 540
dimers 540
Protein InTerfaces and assemblies (PITA) program 540
Protein Quaternary Structure (PQS) server 540
tetramers 540
trimers 540
virus coat proteins 540
ligand errors 538
nomenclature errors 539
structural errors 538
target value errors 538
obsolete structures 535
remaining errors 537
bond angles 537
bond lengths 537
disallowed Ramachandran plot regions 537
electron density 537
His/Asn/Gln side chain flips 537
WHAT_CHECK 537

sequence–register shift errors 534, 537
oligosaccharide-binding fold 537
ProsaII 537
severe errors 532
space group errors 533
phasing information 533–534
topology 532, 534*F*
totally incorrect structures 532, 533*F*
water molecules 539
X-ray crystallography, rational drug design **875–900**, 879*F*
cost 875
gene construct design 877, 878*F*
affinity tag 880
domain mapping 879, 879*F*
expression system 880
multiple construct screening 877–878
history 876
ACE inhibitor design 876, 877*F*
Cushman, David 876
Lipscomb, William 876
Ondetti, Miguel 876
lead identification/characterization 890, 891*F*
co-crystallization 891
complexes 891
fragment evolution 893, 893*F*
fragment linking 893, 893*F*
fragment optimization 893, 893*F*
fragment self-assembly 893, 893*F*
'fragnomics' 892
HTS 890
lead optimization 890
ligand soaking 891
'scaffold-based' discovery 892
phasing methods 889
anomalous scattering 890
multiple isomorphous replacement (MIR) 889–890
multiple-isomorphous replacement with anomalous scattering (MIRAS) 889–890
multiple-wavelength anomalous dispersion (MAD) 889–890
single-isomorphous replacement with anomalous scattering (SIRAS) 889–890
protein characterization 881
purity measurement 881
X-ray detection *see* X-ray crystallography
X-ray generation *see* X-ray crystallography
X-SITE program, intermolecular interactions 406

Y

^{86}Y, PET 664*T*
Yama *see* Caspase-3
Ψ-angle, protein 3D structure validation 516–517
Yeast, protein expression *see* Protein expression
Yeast expression systems 415, 416
cell lysis 420
Yeast Structural Genomics (YSG) 552*T*
Yeast-two-hybrid method, protein interaction networks, bioinformatics 335
Yields, microwave-assisted chemistry 857

Z

Zebrafish (*Danio rerio*) 10, 137–138, 652
advantages 10
antisense phosphorodiamidate morpholino oligonucleotides 10
blindness 10
BRAF studies 652–653
cDNA 10
deafness 10
developmental biology 652
disease biology 652
forward genetics 10
anemia 10
ferroportin 1 identification 10
target identification 10
gene knockdown 10
genome sequencing 10, 647–649, 648*T*, 652–653
heart disease 10
drug identification 10
titin mutation 10
history 647
HTS 10
leukemia 10
melanoma 10
neurodegeneration 10
physical maps 10
polycystic kidney disease 10
positional cloning 10
small-molecule screening 653
TILLING 10
transcription profiling 10
Zeolites, PASP 793
Zinc, protein crystallization 885–887, 886*F*
ZINC database 307
structure-based virtual screening 252
Zinc metalloproteinases as drug targets 3
Zinsser Sophas robotic synthesizer, PASP 805
Z-scores, PDB errors 523, 524*F*
Zygote microinjection, transgenic animal generation 152